The Blue Compendium

Jane Lubchenco • Peter M. Haugan
Editors

The Blue Compendium

From Knowledge to Action for a Sustainable Ocean Economy

Springer

Editors
Jane Lubchenco
Department of Integrative Biology
Oregon State University
Corvallis, OR, USA

Peter M. Haugan
Institute of Marine Research
Bergen, Norway

ISBN 978-3-031-16279-4 ISBN 978-3-031-16277-0 (eBook)
https://doi.org/10.1007/978-3-031-16277-0

World Resources Institute

This Springer imprint is published by the registered company Springer Nature Switzerland AG
The registered company address is: Gewerbestrasse 11, 6330 Cham, Switzerland

Preface

Home to over 80% of all life on Earth, the ocean is the world's largest carbon sink and a key source of food and economic security for billions of people. The relevance of the ocean for humanity's future is undisputed—though not usually fully appreciated. The ocean has much greater potential to drive economic growth and equitable job creation, sustain healthy ecosystems, and mitigate climate change than is realised today. Lack of awareness of the potential as well as management and governance challenges pose impediments. Until these impediments are removed, ocean ecosystems will continue to be degraded and opportunities for people lost. A transition and a clear path to a thriving and vibrant relationship between humans and the ocean are urgently needed.

This collection identifies a path that is inspired by science, energised by engaged people, and emboldened by visionary leaders. The papers and reports in this compendium are assessments of knowledge commissioned by a unique collaboration among global leaders who asked the question, 'How might we use the ocean wisely without using it up?' These leaders established the High Level Panel for a Sustainable Ocean Economy (Ocean Panel) in September 2018 as a unique initiative led by heads of state and government from around the world who are committed to building a sustainable ocean economy in which effective protection, sustainable production and equitable prosperity go hand in hand. Collectively, these 17 nations represent nearly 46% of the world's coastlines and at least 44% of the world's exclusive economic zones. The Ocean Panel's shared vision is to sustainably manage 100% of ocean areas under their national jurisdiction, guided by Sustainable Ocean Plans. In the *Transformations*[1] document, which was developed in a process informed by the knowledge in this collection, the Ocean Panel also set out a new ocean action agenda for the decade. This far-reaching political document is the result of broad and diverse engagement, collaboration and consultation, and an unprecedented scientific knowledge base coming together to result in actions that move 'from the purpose to the impact'. It identifies 15 outcomes and 74 bold yet pragmatic actions to be taken across five critical areas—ocean health, ocean wealth, ocean equity, ocean knowledge and ocean finance—to transform humanity's relationship with and impacts upon the ocean, and to ensure that the myriad benefits and opportunities that the ocean provides can be sustainably enjoyed by all.

Early on in their deliberations, and before considering action, the Ocean Panel intentionally set out to 'Start with science, with knowledge'. They identified a series of topics for which they would commission syntheses of knowledge that would inform their policy and action agenda. To ensure the high quality and intellectual integrity of the Ocean Panel's commissioned research, they established an Expert Group consisting of a global group of over 70 experts renowned for their exemplary contributions to the full range of ocean-related disciplines considered in the Ocean Panel's work. Together, more than 250 experts and authors, with 44% being women, representing 48 countries have contributed to Ocean Panel-commissioned research to date. The Secretariat of the Ocean Panel provided additional substantial contribu-

[1]Ocean Panel (High Level Panel for a Sustainable Ocean Economy). 2020. Transformations for a Sustainable Ocean Economy: A Vision for Protection, Production and Prosperity. High Level Panel for a Sustainable Ocean Economy. https://www.oceanpanel.org/ocean-action/files/transformationssustainable-ocean-economy-eng.pdf.

tions and coordination. The resulting series of 16 Blue Papers and 4 Special Reports responded to the request from the Ocean Panel and provided timely analyses of pressing challenges at the nexus of the ocean and the economy.

The Blue Papers and Special Reports, included in this Blue Compendium, showcase the latest leading-edge science, knowledge and state-of-the-art thinking. They offer innovative ocean solutions in technology, policy, governance and finance realms that could help accelerate a transition to a more sustainable and prosperous relationship with the ocean. The comprehensive assessments have already informed policy making at the highest levels of government and motivated an impressive array of responsive and ambitious action across a growing network of leaders in business, finance and civil society.

The 16 Blue Papers ranged from food, energy and mineral production, genetic resources and conservation, to climate change, plastic pollution, technology, equity, illegal fishing, organised crime in fisheries and ocean accounting. **'The Future of Food from the Sea'** considers the status and future trends of food production through fisheries and aquaculture at regional and global scales, identifies opportunities of ocean-based food in achieving SDG 2: Zero Hunger, and provides recommendations for how current barriers might be overcome to transition to more sustainable and abundant food production from the ocean. **'The Expected Impacts of Climate Change on the Ocean Economy'** addresses how the compounding hazards of climate change will impact the ocean economy, specifically marine fisheries, aquaculture and tourism; highlights opportunities for effective institutions and markets to reduce these impacts; and provides recommendations for how countries can achieve blue economic growth by implementing policies and infrastructure that reduce risks and build resilience to climate change. **'What Role for Ocean-Based Renewable Energy and Deep-Seabed Minerals in a Sustainable Future?'** examines how and to what degree energy from the ocean, current developments in green technology and the potential for deep-seabed minerals can help meet rising technological demand and contribute to the climate agenda and achievement of SDG 7: Affordable and Clean Energy. It identifies solutions and future policy options and their potential impact, as well as addressing related safety and environmental concerns. **'The Ocean Genome: Conservation and the Fair, Equitable and Sustainable Use of Marine Genetic Resources'** considers the existing and potential benefits associated with the ocean genome and the threats it is facing, and explores how efforts to promote inclusive innovation and governance can contribute to more equitable sharing of benefits derived from the use of marine genetic resources. **'Leveraging Multi-Target Strategies to Address Plastic Pollution in the Context of an Already Stressed Ocean'** examines the leakage of plastics and other pollutants into the ocean and the resulting impacts on marine ecosystems, human health and the economy. The paper explores the kind of regenerative global industry that needs to be built, as well as integrated solutions to reduce all pollutants of the ocean, and highlights the role of science-based targets in measuring progress on ocean pollution. **'Technology, Data and New Models for Sustainably Managing Ocean Resources'** explores existing and breakthrough technologies, such as drones, artificial intelligence and blockchains, and the associated challenges and possibilities they pose for ocean management and improving understanding of ecosystems and human interactions with the ocean. **'Coastal Development: Resilience, Restoration and Infrastructure Requirements'** examines trends in coastal behaviour, explores trade-offs between restoration and infrastructure development and makes an economic and security case for resilient coastlines providing much-needed recommendations for new models for shipping and tourism. **'National Accounting for the Ocean and Ocean Economy'** highlights the critical role of national accounting as a tool in achieving a sustainable ocean economy, identifies major gaps in how the ocean, ocean services and ocean assets are currently treated in national accounts, and offers the methods and a roadmap for measuring and valuing ocean assets. **'Ocean Finance: Financing the Transition to a Sustainable Ocean Economy'** explores the next generation of financing mechanisms and the role insurance can play in supporting the ocean transition in an inclusive manner and recommends approaches to be phased out, as well as new solutions that incentivise sustainable management. **'Critical Habitats and Biodiversity:**

Inventory, Thresholds and Governance' provides an inventory of the distribution of species and critical marine habitats exploring trends in drivers, pressures, impacts and responses; establishes thresholds for protecting biodiversity hotspots, as well as indicators to monitor change; and assesses the current legal framework, the gaps in ocean governance and management and the implications for achieving a sustainable ocean economy. **'The Human Relationship with our Ocean Planet'** illustrates the differing economic, legal, institutional, social and cultural relationships that people of varying cultures have with the ocean, through a historical lens, and charts a path towards inclusive ocean governance. **'The Ocean Transition: What to Learn from System Transitions'** examines past successes and failures and current dynamics of transitions, and explores alternative future transition pathways and policy responses that can drive to a more sustainable ocean. **'Towards Ocean Equity'** explores the distribution of the goods and services provided by the ocean; existing inequities and the resulting impacts spanning environmental, social and economic dimensions; and provides recommendations for addressing some of the underlying and systemic features of ocean inequities, as well as opportunities for policy to support a sustainable and just ocean economy. **'Integrated Ocean Management'** makes the case for integrated ecosystem-based management, one that combines value creation and the safeguarding of ecosystem health, identifying existing impediments in policy and practice and outlining steps and principles towards a successful integrated ocean management. **'Illegal, Unreported and Unregulated Fishing and Associated Drivers'** explores the drivers and consequences of illegal, unreported and unregulated (IUU) fishing, and provides a range of solutions to prevent and combat IUU fishing, from implementing international agreements to promoting new technologies and strengthening regional and international partnerships. **'Organised Crime in the Fisheries Sector'** presents the current state of knowledge on organised crime in fisheries and provides recommendations and best practices that promote an intelligence-led, skills-based cooperative law enforcement at a global level, facilitated by enabling legislative frameworks and increased transparency.

The four Special Reports illustrate how a sustainable ocean economy can create a healthy ocean, and vice versa, that provides solutions to global challenges. Collectively, they set out a new evidence-based narrative, in which the ocean is critical to achieving global targets to limit climate change and its detrimental effects to everyone's present and future, offers solutions for a sustainable and equitable recovery to current and future crises, and provides unparalleled opportunities to build a fair and just sustainable ocean economy. **'The Ocean as a Solution to Climate Change: Five Opportunities for Action'** evaluates the mitigation potential of a suite of ocean-based actions—renewable energy, transport, food production, ecosystems and carbon storage in the seabed—in 2030 and 2050 relative to a 1.5 °C and 2 °C pathway, explores their wider benefits to societies and economics, and highlights the enabling policy measures and research required for success. Building on this Special Report, **'A Sustainable Ocean Economy for 2050: Approximating Its Benefits and Costs'** examines the global net benefit and the benefit-cost ratio of implementing those sustainable ocean-based interventions, including conserving and restoring mangrove habitats, scaling up offshore wind production, decarbonising the international shipping sector and increasing the production of sustainably sourced ocean-based proteins, over a 30-year time horizon up to 2050. **'A Sustainable and Equitable Blue Recovery to the COVID-19 Crisis'** examines the impacts of the COVID-19 pandemic on the ocean economy and the role of ocean-based solutions in supporting sustainable and equitable recovery and enhancing resilience to future crises. Drawing on the latest scientific research and insights from the Blue Papers and the other Special Reports, **'Ocean Solutions That Benefit People, Nature and the Economy'** details a framework and a feasible action plan and practical solutions that when implemented could help achieve a sustainable ocean economy where people have more opportunities and better health, nature thrives and resources are distributed more equitably.

The impact of this collection of assessments of knowledge can be clearly seen in the *Transformations* announced by the Ocean Panel in December 2020. Many political leaders give lip service to grounding policy and action in science, evidence and knowledge. In this case, the connections are clear. Moreover, the benefit of basing commitments on expert knowledge continues in the Ocean Panel, as additional special reports are developed to inform subsequent action.

No transformative change could be possibly realised by just one actor, entity or sector. The governments of the Ocean Panel are leading by example on this transformative agenda, but are also working collaboratively with the public, private, financial, research and civil society sectors, to raise the profile of the ocean in international arenas, to develop a sustainable ocean economy and to successfully implement sustainable and equitable ocean management. The work of the Ocean Panel has triggered the formation of several coalitions and partnerships intended to promote and facilitate the Ocean Panel's ocean action agenda. Currently, there are ten multi-stakeholder initiatives, also called 'Action Groups', that collaborate to implement one or more of the priority actions in the *Transformations*, and whose strategies to tackle ocean issues have been informed by many of the Blue Papers and Special Reports.

Together, the 17 countries of the Ocean Panel recognise and promote the ocean as a smart investment with tremendous social, economic and environmental benefits. The ocean provides many of the urgent solutions humanity and the planet need, and it thus must be considered as our critical ally for global economic growth, climate resilience, social equity and future security and prosperity.

This Blue Compendium—representing one of the most comprehensive assessments in the ocean realm and already influencing policy and action—is the product of devoted efforts by numerous people. We are deeply grateful to the over 250 authors and reviewers who led, contributed to, and improved these knowledge assets. We also offer deep gratitude to the Secretariat of the Ocean Panel and colleagues at the World Resources Institute for their skilled guidance, editorial support, graphics, messaging and outreach. And we thank the active engagement of and trust placed in us by the Ocean Panel Leaders, their Sherpas, and teams. The partnerships, respect and new awareness that have emerged from the development of the Blue Compendium and Ocean Panel work are valued and they set an example for how governmental leaders and knowledge experts can engage productively to the benefit of society.

We close with the belief that the ocean is central to our collective future, that knowledge should inform action, and that partnerships will enable us to chart a course to use the ocean wisely without using it up.

Corvallis, OR, USA — Jane Lubchenco
Bergen, Norway — Peter M. Haugan

Contents

About the Editors

Jane Lubchenco University Distinguished Professor at Oregon State University, is a marine ecologist with expertise in the ocean, climate change and interactions between the environment and human well-being, received a B.A. in biology from Colorado College, a M.S. in zoology from the University of Washington and a Ph.D. in ecology from Harvard University.

Her academic career as a professor began at Harvard University (1975–1977) and continued at Oregon State University (1977–2009) until her appointment as NOAA Administrator (2009–2013). Thereafter, she was the 2013 Haas Distinguished Visitor in Public Service at Stanford University, then Oregon State University's University Distinguished Professor.

In recognition of her scientific contributions, Jane Lubchenco is an elected member of the National Academy of Sciences and many other distinguished academies. She has received numerous awards including 24 honorary doctorates, most recently from the University of Oxford.

She co-founded three organisations that train scientists to be better communicators and engage more effectively with the public, policy makers, media and industry. She also co-founded PISCO (an integrated research/monitoring/outreach programme), the National Ocean Protection Coalition and the MPA Project that seeks to advance smart use of effective Marine Protected Areas.

She has also served in multiple governmental capacities. She was the U.S. Under Secretary of Commerce for Oceans and Atmosphere and Administrator of the National Oceanic and Atmospheric Administration (NOAA) and an inaugural member of President Barack Obama's Science Team from 2009 to 2013. From 2014 to 2016, she was the first U.S. State Department Science Envoy for the Ocean, serving as a science diplomat to China, Indonesia, South Africa, Mauritius and the Seychelles. And beginning in 2021, she leads the Climate and Environment team at the White House Office of Science and Technology Policy. Dr. Lubchenco's contribution to this work was completed solely in her capacity as a Professor at Oregon State University. The views expressed are those of the authors, and do not necessarily represent the views of the U.S. Government.

She has served as the President of numerous professional scientific societies including the Ecological Society of America, the American Association for the Advancement of Science (AAAS) and the International Council for Science (ICSU). She has served on multiple national commissions including the Pew Oceans Commission, the Joint Ocean Commission Initiative and the Aspen Institute Arctic Commission. She has led or contributed to multiple regional, national and international scientific assessments on climate change, biodiversity, Marine Protected Areas, the ocean, and the intersection of science and society. Most recently, she co-chaired the Expert Group for the High Level Panel for a Sustainable Ocean Economy, a pioneering partnership across over a dozen serving heads of state or government to harness science and action to protect the ocean effectively, produce from it sustainably, and prosper equitably.

Peter M. Haugan is Policy Director at Institute of Marine Research, Norway, and professor of oceanography at the Geophysical Institute, University of Bergen. He started his career as a Research Engineer in the oil and gas industry in 1982 doing reservoir model development with a degree in applied mathematics. He turned to oceanography and climate research in 1987 joining the Nansen Environmental and Remote Sensing Center where he became its Deputy Director in 1994. He joined the Geophysical Institute in 1996 but also spent 2 years at the University Centre in Svalbard and was Deputy Director of the Bjerknes Centre for Climate Research from its establishment in 2000. From 2003 to 2011 he was Director of the Geophysical Institute widening its scope to lead university-wide efforts in renewable energy, notably offshore wind. From 2019, he has led international work on global development at the Institute of Marine Research (IMR). During 2021-2022 he was part time on loan from IMR to the Norwegian Ministry of Foreign Affairs before taking up his present position at IMR.

The Future of Food from the Sea

1

Global food demand is rising, and serious questions remain about whether supply can increase sustainably (FAO 2018). Land-based expansion is possible but may exacerbate climate change and biodiversity loss, and compromise the delivery of other ecosystem services (Olsen 2011; Foley et al. 2005, 2011; Mbow et al. 2019; Amundson et al. 2015). As food from the sea represents only 17% of the current production of edible meat, we ask how much food we can expect the ocean to sustainably produce by 2050. Here we examine the main food-producing sectors in the ocean—wild fisheries, finfish mariculture and bivalve mariculture—to estimate 'sustainable supply curves' that account for ecological, economic, regulatory and technological constraints. We overlay these supply curves with demand scenarios to estimate future seafood production. We find that under our estimated demand shifts and supply scenarios (which account for policy reform and technology improvements), edible food from the sea could increase by 21–44 million tonnes by 2050, a 36–74% increase compared to current yields. This represents 12–25% of the estimated increase in all meat needed to feed 9.8 billion people by 2050. Increases in all three sectors are likely, but are most pronounced for mariculture. Whether these production potentials are realized sustainably will depend on factors such as policy reforms, technological innovation and the extent of future shifts in demand.

Human population growth, rising incomes and preference shifts will considerably increase global demand for nutritious food in the coming decades. Malnutrition and hunger still plague many countries (FAO 2018; UNDP 2020), and projections of population and income by 2050 suggest a future need for more than 500 megatonnes (Mt) of meat per year for human consumption (Supplementary Information section 1.1.6). Scaling up the production of land-derived food crops is challenging, because of declining yield rates and competition for scarce land and water resources (Olsen 2011). Land-derived seafood (freshwater aquaculture and inland capture fisheries; we use seafood to denote any aquatic food resource, and food from the sea for marine resources specifically) has an important role in food security and global supply, but its expansion is also constrained. Similar to other land-based production, the expansion of land-based aquaculture has resulted in substantial environmental externalities that affect water, soil, biodiversity and climate, and which compromise the ability of the environment to produce food (Foley et al. 2005, 2011; Mbow et al. 2019; Amundson et al. 2015). Despite the importance of terrestrial aquaculture in seafood production (Supplementary Fig. 3), many countries—notably China, the largest inland-aquaculture producer—have restricted the use of land and public waters for this purpose, which constrains expansion (De Silva and Davy 2010). Although inland capture fisheries are important for food security, their contribution to total global seafood production is limited (Supplementary Table 1) and expansion is hampered by ecosystem constraints. Thus, to meet future needs (and recognizing that land based sources of fish and other foods are also part of the solution), we ask whether the sustainable production of food from the sea has an important role in future supply.

Food from the sea is produced from wild fisheries and species farmed in the ocean (mariculture), and currently accounts for 17% of the global production of edible meat (FAO Fisheries and Aquaculture Department 2019; Edwards et al. 2019; FAO 2020; Nijdam et al. 2012) (Supplementary Information section 1.1, Supplementary Tables 1–3). In addition to protein, food from the sea contains bioavailable micronutrients and essential fatty acids that are not easily found in land-based foods, and is thus uniquely poised to contribute to global food and nutrition security (Kawarazuka and Béné 2010; Allison 2011; Golden et al. 2016; Hicks et al. 2019).

Widely publicized reports about climate change, overfishing, pollution and unsustainable mariculture give the impres-

Originally published in:
Costello, C., Cao, L., Gelcich, S. et al. The future of food from the sea. Nature 588, 95–100 (2020). https://doi.org/10.1038/s41586-020-2616-y
Reprint by Springer International Publishing (2023) with kind permission
The license number is 5200691161865, in case this needs to appear.

J. Lubchenco, P. M. Haugan (eds.), *The Blue Compendium*, https://doi.org/10.1007/978-3-031-16277-0_1

sion that sustainably increasing the supply of food from the sea is impossible. On the other hand, unsustainable practices, regulatory barriers, perverse incentives and other constraints may be limiting seafood production, and shifts in policies and practices could support both food provisioning and conservation goals (Costello et al. 2016; Ye and Gutierrez 2017). In this study, we investigate the potential of expanding the economically and environmentally sustainable production of food from the sea for meeting global food demand in 2050. We do so by estimating the extent to which food from the sea could plausibly increase under a range of scenarios, including demand scenarios under which land-based fish act as market substitutes.

The future contribution of food from the sea to global food supply will depend on a range of ecological, economic, policy and technological factors. Estimates based solely on ecological capacity are useful, but do not capture the responses of producers to incentives and do not account for changes in demand, input costs or technology (Gentry et al. 2017; Troell et al. 2017). To account for these realities, we construct global supply curves of food from the sea that explicitly account for economic feasibility and feed constraints. We first derive the conceptual pathways through which food could be increased in wild fisheries and in mariculture sectors. We then empirically derive the magnitudes of these pathways to estimate the sustainable supply of food from each seafood sector at any given price (Costello et al. 2019). Finally, we match these supply curves with future demand scenarios to estimate the likely future production of sustainable seafood at the global level.

1 Sustainably Increasing Food from the Sea

We describe four main pathways by which food supply from the ocean could increase: (1) improving the management of wild fisheries; (2) implementing policy reforms of mariculture; (3) advancing feed technologies for fed mariculture; and (4) shifting demand, which affects the quantity supplied from all three production sectors.

Although mariculture production has grown steadily over the past 60 years (Fig. 1.1) and provides an important contribution to food security (Belton et al. 2018), the vast majority (over 80%) of edible meat from the sea comes from wild fisheries (FAO Fisheries and Aquaculture Department 2019) (Fig. 1.1b). Over the past 30 years, supply from this wild food source has stabilized globally despite growing demand worldwide, which has raised concerns about our ability to sustainably increase production. Of nearly 400 fish stocks around the world that have been monitored since the 1970s by the UN Food and Agriculture Organization (FAO), approximately one third are currently not fished within sustainable limits (FAO 2018). Indeed, overfishing occurs often in poorly managed ('open access') fisheries. This is disproportionately true in regions with food and nutrition security concerns (FAO 2018). In open-access fisheries, fishing pressure increases as the price rises: this can result in a 'backward-bending' supply curve (Copes 1970; Nielsen 2006) (the OA curve in Fig. 1.2a), in which higher prices result in the depletion of fish stocks and reduced productivity—and thus reduced equilibrium food provision.

Fishery management allows overexploited stocks to rebuild, which can increase long-term food production from wild fisheries (Hilborn and Costello 2018; Hilborn et al. 2020). We present two hypothetical pathways by which wild fisheries could adopt improved management (Fig. 1.2a). First, independent of economic conditions, governments can impose reforms in fishery management. The resulting production in 2050 from this pathway—assuming that fisheries are managed for maximum sustainable yield (MSY)—is represented by the MSY curve in Fig. 1.2a, and is independent of price. The second pathway explicitly recognizes that wild fisheries are expensive to monitor (for example, via stock assessments) and manage (for example, via quotas)—management reforms are adopted only by fisheries for which

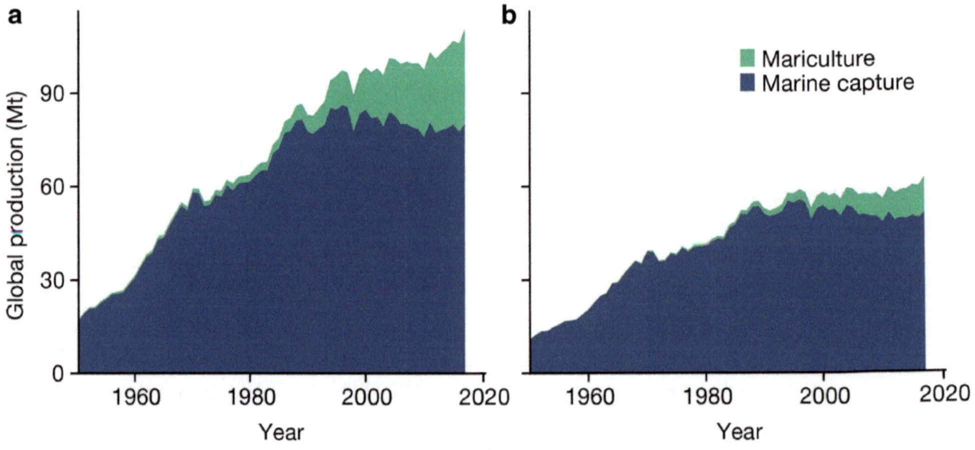

Fig. 1.1 Marine harvest and food from the sea over time (excluding aquatic plants). Data are from FAO Fisheries and Aquaculture Department (2019). (**a**, **b**) Harvests (live-weight production) (**a**) are converted to food equivalents (edible production) (Edwards et al. 2019) (**b**). In (**b**), there is also an assumption that 18% of the annual landings of marine wild fisheries are directed towards non-food purposes (Cashion et al. 2017)

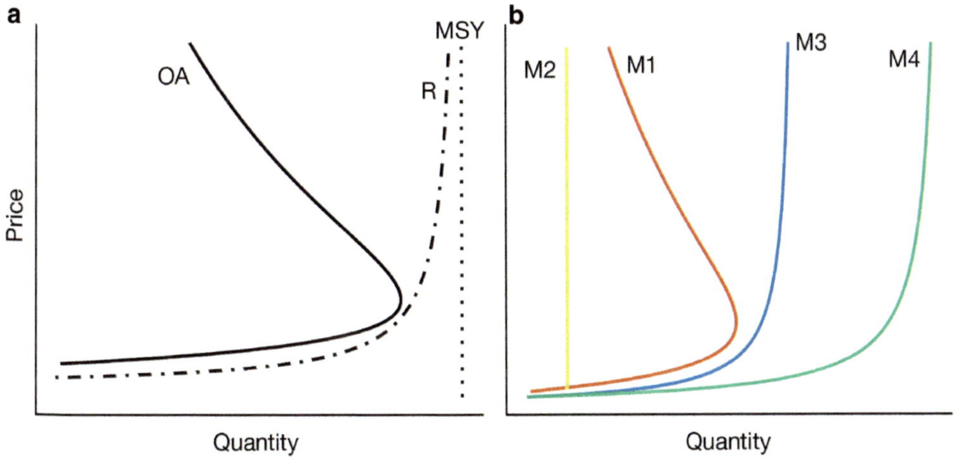

Fig. 1.2 Hypothetical supply curves for wild fisheries and mariculture, showing the influence of price on production quantity. (**a**) Wild fisheries. Curves represent poorly managed (open access) fisheries (OA); management reform for all fisheries (MSY); and economically rational management reform (R). (**b**) Mariculture. Curves represent weak regulations that allow for ecologically unsustainable production (M1); overly restrictive policies (M2); policies that allow for sustainable expansion (M3); and a reduced dependence on limited feed ingredients for fed-mariculture production (M4)

future profits outweigh the associated costs of improved management. When management entities respond to economic incentives, the number of fisheries for which the benefits of improved management outweigh the costs increases as demand (and thus price) increases. This economically rational management endogenously determines which fisheries are well-managed, and thus how much food production they deliver, resulting in supply curve designated R in Fig. 1.2a.

Although the production of wild fisheries is approaching its ecological limits, current mariculture production is far below its ecological limits and could be increased through policy reforms, technological advancements and increased demand (Gentry et al. 2017; Joffre et al. 2017). We present explanations for why food production from mariculture is currently limited, and describe how the relaxation of these constraints gives rise to distinct pathways for expansion (Fig. 1.2b). The first pathway recognizes that ineffective policies have limited the supply (Abate et al. 2016; Gentry et al. 2019). Lax regulations in some regions have resulted in poor environmental stewardship, disease and even collapse, which have compromised the viability of food production in the long run (curve M1 in Fig. 1.2b). In other regions, regulations are overly restrictive, convoluted and poorly defined (The Sea Grant Law Center 2019; Davies et al. 2019), and thus limit production (curve M2 in Fig. 1.2b). In both cases, improved policies and implementation can increase food production by preventing and ending environmentally damaging mariculture practices (the shift from M1 to M3 in Fig. 1.2b) and allowing for environmentally sustainable expansion (the shift from M2 to M3 in Fig. 1.2b).

The second pathway to sustainably increase mariculture production is through further technological advances in finfish feeds. Currently, most mariculture production (75%) requires some feed input (such as fishmeal and fish oil) that is largely derived from wild forage fisheries (FAO 2018).

If fed mariculture continues using fishmeal and fish oil at the current rate, its growth will be constrained by the ecological limits of these wild fisheries (Froehlich et al. 2018a). Alternative feed ingredients—including terrestrial plant- or animal-based proteins, seafood processing waste, microbial ingredients, insects, algae and genetically modified plants—are rapidly being developed and are increasingly used in mariculture feeds (Klinger and Naylor 2012; Cao et al. 2015; Little et al. 2016; Shah et al. 2018). These innovations could decouple fed mariculture from wild fisheries (but may refocus pressure on terrestrial ecosystems) and could catalyse considerable expansion in some regions (Troell et al. 2014; Froehlich et al. 2018b). This has already begun for many fed species, such as Atlantic salmon—for which fish-based ingredient use has been reduced from 90% in the 1990s to just 25% at present (Aas et al. 2019). A reduced reliance on fishmeal and fish oil is expected to shift the supply curve of fed mariculture to the right (curve M4 in Fig. 1.2b). The final pathway is a shift in demand (aggregated across all global fish consumers), which affects all three production sectors. When the sustainable supply curve is upward-sloping, an increase in demand (rightward shift; for example, from rising population, income or preferences) increases food production.

2 Estimated Sustainable Supply Curves

We estimate supply curves of food from the sea in 2050 for the three largest food sectors in the ocean: wild fisheries, finfish mariculture and bivalve mariculture. We construct global supply curves for marine wild fisheries using projected future production for 4702 fisheries under alternative management scenarios (Fig. 1.3a). We model future production with a bioeconomic model based on Costello et al. (2016), which tracks annual biomass, harvest and profit, and accounts for costs associated with extraction and management (see Methods and Supplementary Information for details). Managing all fisheries to maximize food production (MSY) would result in 57.4 Mt of food in 2050 (derived from 89.3 Mt of total harvest, hereafter noted as live-weight equivalent), representing a 16% increase compared to the current food production (Fig. 1.3a). Under a scenario of economically rational reform (in which the management approach and exploitation rate of fisheries depend on profitability), the price influences production (Fig. 1.3a). At current mean global prices, this scenario would result in 51.3 Mt of food (77.4 Mt live-weight equivalent)—a 4% increase compared to current food production. These management-induced shifts in supply are ultimately limited by the carrying capacity of the ecosystem. If current fishing pressure is maintained for each fish stock when profitable (*F* current, referring to the current fishing mortality rate), food production from wild fisheries is lower for most prices than under the two reform scenarios (owing to fishing too intensively on some stocks, and too conservatively on others) (Hilborn and Costello 2018): this supply curve is not backward-bending, as it reflects constant fishing pressures.

We estimate the production potential of mariculture at a resolution of 0.217° around the world for finfish and bivalves. Ecological conditions—sea surface temperature, dissolved oxygen and primary productivity—determine the suitability

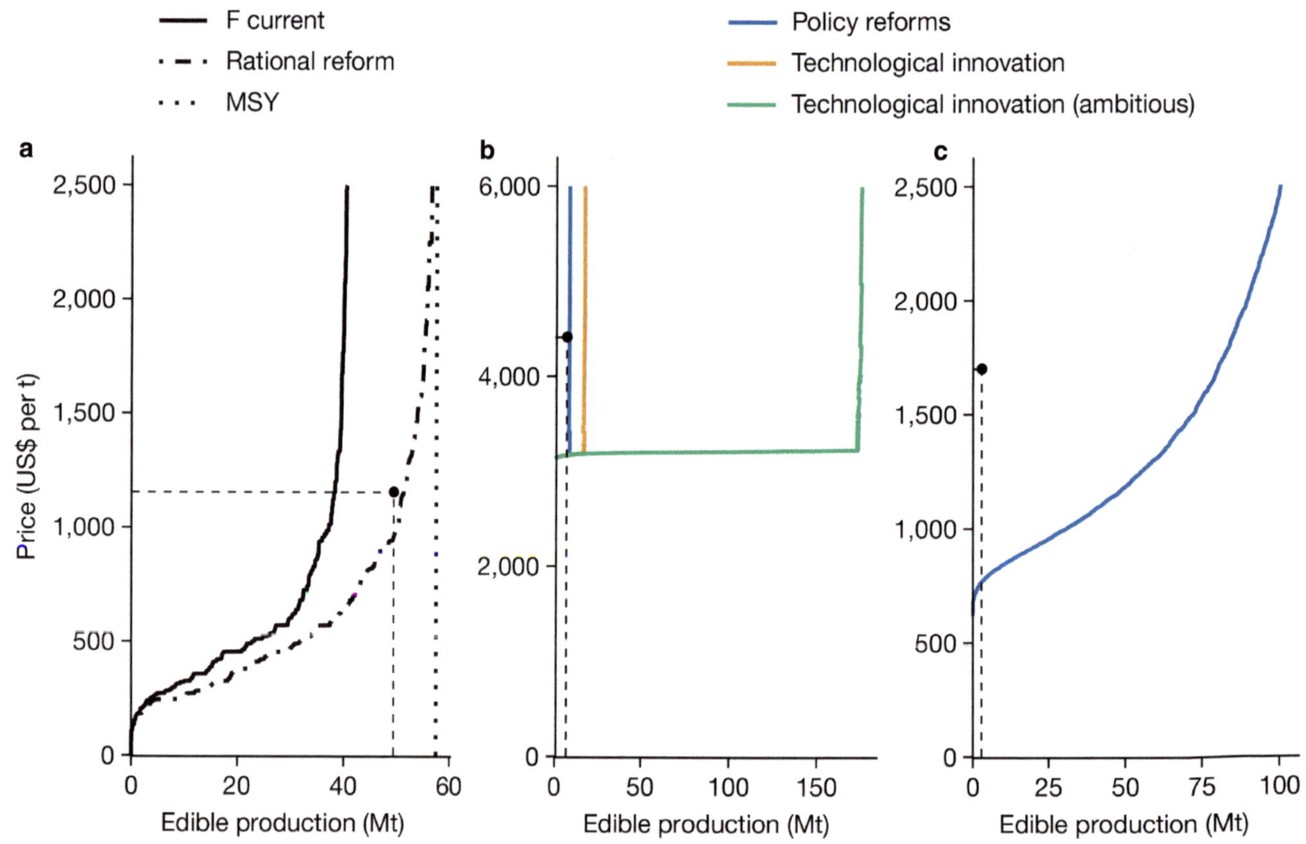

Fig. 1.3 Estimated sustainable supply curves for wild fisheries, finfish mariculture and bivalve mariculture. (**a**–**c**) Points represent current production and average price in each sector: marine wild fisheries (**a**), finfish mariculture (**b**) and bivalve mariculture (**c**). In (**a**), supply curves for annual steady-state edible production from wild fisheries are shown under three different management scenarios: production in 2050 under current fishing effort assuming that fishing only occurs in fisheries that are profitable (*F* current); the economically rational supply curve aimed at maximizing profitability (rational reform); and a reform policy aimed at maximizing food production, regardless of the economic considerations (MSY). In (**b**), supply curves for finfish (fed) mariculture show: future steady-state production under current feed assumptions and policy reform (policy reform); sustainable production assuming policy reform and a 50% reduction in fishmeal and fish oil feed requirements (technological innovation); and sustainable production assuming policy reform and a 95% reduction in fishmeal and fish oil feed requirements (technological innovation (ambitious)). In all cases, feed ingredients are from the economically rational reform of wild fisheries

of each pixel for mariculture production. We build on previous models (Gentry et al. 2017) by including economic considerations (including the capital costs of vessels and equipment, and the operating costs of wages, fuel, feed, insurance and maintenance; Supplementary Tables 5–7) to determine whether farming an ecologically suitable area is economically profitable at any given price. Summing economically viable production for each sector at the global level for different prices produces two mariculture supply curves. This approach assumes that the most profitable sites will be developed first, but does not explicitly include challenges such as the cost of public regulation and the delineation of property rights. Farm design is based on best practice for sustainable production, and we therefore interpret the results as an environmentally sustainable supply. We examine a range of assumptions regarding production costs, and explore different technological assumptions with respect to the species type farmed for finfish mariculture (Methods, Supplementary Information section 1.3, Supplementary Table 9). The supply curve for finfish mariculture differs substantially among future feed-technology scenarios, although all of these scenarios foretell a substantial increase in annual food supply in the future compared to the current production of the sector (6.8 Mt of food) (Fig. 1.3b). However, the policy reform scenario—which assumes mariculture policies are neither too restrictive nor lax (curve M3 in Fig. 1.2b), but that fishmeal and fish oil requirements match present-day conditions—produces a modest additional 1.4 Mt of food at current prices. In this scenario, marine-based feed inputs limit mariculture expansion even as the price increases considerably.

Two feed-innovation scenarios—representing policy reform plus a 50% or 95% reduction in fishmeal and fish oil requirements, which we refer to as 'technological innovation' and 'technological innovation (ambitious)', respectively—can substantially shift the supply curve.

At current prices, future supply under these scenarios is predicted to increase substantially to 17.2 Mt and 174.5 Mt of food for technological innovation and technological innovation (ambitious) scenarios, respectively (Fig. 1.3b). Bivalve mariculture is constrained by current policy but not by feed limitations, and is poised to expand substantially under policy reform scenarios. At current prices, economically rational production could lead to an increase from 2.9 Mt to 80.5 Mt of food (Fig. 1.3c). Even if our model underestimates costs by 50%, policy reforms would increase the production potential of both fed and unfed mariculture at current prices. For fed mariculture, this remains true even when evaluating mariculture species with different feed demands (Atlantic salmon, milkfish and barramundi).

3 Estimates of Future Food from the Sea

Our supply curves suggest that all three sectors of ocean food production are capable of sustainably producing much more food than they do at present. The quantity of seafood demanded will also respond to price. We present three demand-curve estimates, shown in Fig. 1.4 (Methods, Supplementary Information). The intersections of future demand and sustainable supply curves provide an estimate of future food production from the sea. Because it is a substantial contributor to fish supply and—in some instances—acts as a market substitute for seafood, we also account for land-based aquatic food production (from freshwater aquaculture and inland capture fisheries; Supplementary Information section 1.4, Supplementary Tables 10–12). Estimates of future production from this fourth sector ('inland fisheries') are shown side-by-side in Supplementary Fig. 3 and Supplementary Tables 13, 14 (for quantities of food) and in Supplementary Tables 15, 16 (for live-weight equivalents), and are discussed with the results on food from the sea.

Even under current demand curves (green curves in Fig. 1.4), the economically rational reform of marine wild fisheries and sustainable mariculture policies (stocking densities consistent with European organic standards (European Union 2008)) under the technological innovation (ambitious) scenario could result in a combined total of 62 Mt of food from the sea per year, 5% more than the current levels (59 Mt). But we know that demand will increase as incomes rise and populations expand. Under the 'future demand' scenario (purple curves in Fig. 1.4), total food from the sea is projected to increase to 80 Mt. If demand shifts even more (as represented by our 'extreme demand' scenario; red curves in Fig. 1.4), the intersection of supply and demand is expected to increase to 103 Mt of food. Using the approach used by the FAO to estimate future needs, the world will require an additional 177 Mt of meat by 2050 (Supplementary Information section 1.1.6)—our results suggest that additional food from the sea alone could plausibly contribute 12–25% of this need. Another possibility we consider is that future consumers will not distinguish between fish-producing sectors, such that all sources of fish (including land-based) would be substitutes for each other. Adopting that assumption alters the supply-and-demand equilibrium, and implies that the increase among all sources of fish (sea and land) relative to the present could be between 90–212 Mt of food; under this scenario, expansion of aquatic foods alone could possibly exceed the 177-Mt benchmark.

Our results also suggest that the future composition of food from the sea will differ substantially from the present (Fig. 1.5). Although wild fisheries dominate edible marine

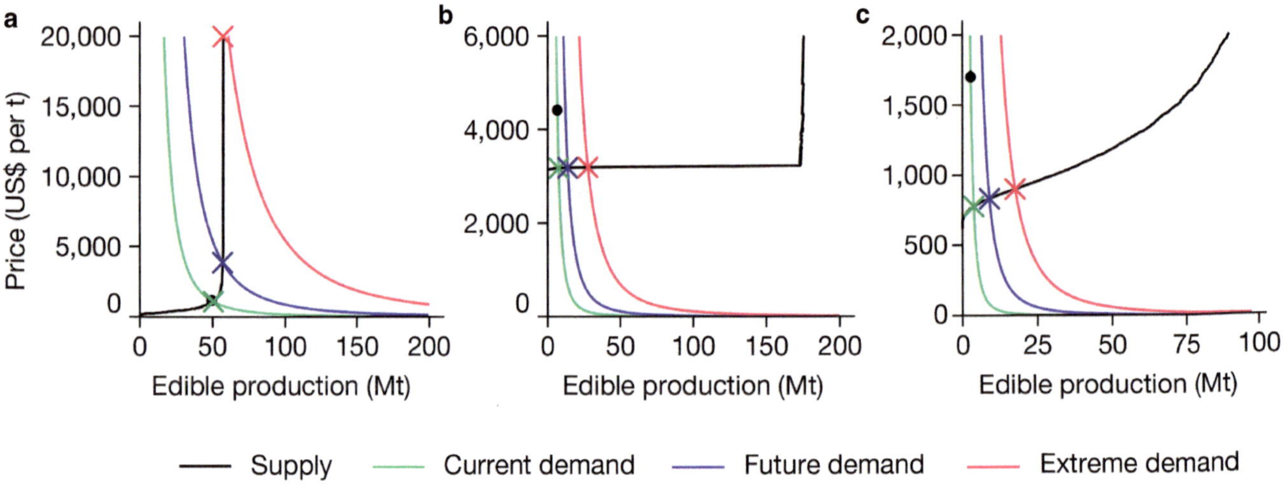

Fig. 1.4 Supply and demand curves of food from the sea for the three sectors. (**a–c**) Supply and demand curves for marine wild fisheries (**a**), finfish mariculture (**b**) and bivalve mariculture (**c**). In each panel, the solid black line is the supply curve from Fig. 1.3: for wild fisheries, the rational reform scenario is shown, and for finfish mariculture the technological innovation (ambitious) scenario is shown. Future demand refers to estimated demand in 2050; extreme demand represents a doubling of the estimated demand in 2050. The intersections of demand and sustainable supply curve (indicated with crosses) provide an estimate of the future food from the sea. Points represent current production and average price in each sector

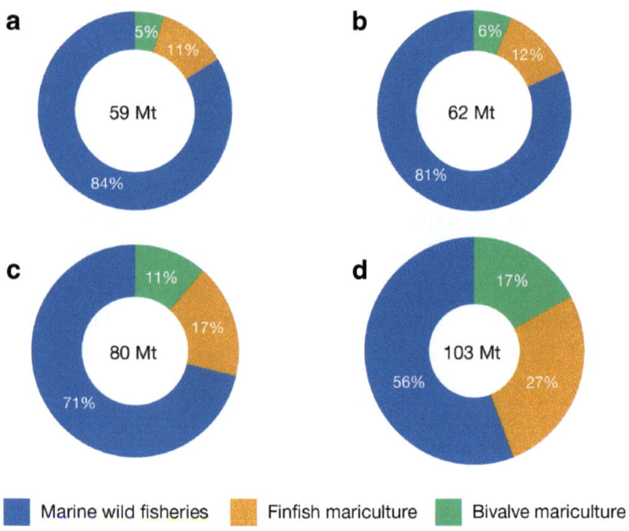

Fig. 1.5 Composition of current and future food from the sea under three alternative demand scenarios. (**a**) Composition of current (initial production) food from the sea. (**b–d**), Composition of future (2050) food from the sea under scenarios of current (**b**), future (**c**) and extreme (**d**) demand. The sustainable supply curves assumed for these predictions are: rational reform for wild fisheries; technological innovation (ambitious) for finfish mariculture; and policy reform for bivalve mariculture, as shown in Fig. 1.3. The total production of food from the sea per year is shown in the centre in each panel

production at present, we project that by 2050 up to 44% of edible marine production could come from mariculture (rising to 76% when all fish are substitutes and land-based fish are included under extreme demand scenarios (Supplementary Fig. 3, Supplementary Table 14)), although all sectors could increase production. Although even more substantial increases are technically possible (for example, fed mariculture alone is capable of generating at least the benchmark 177 Mt of additional meat), actually realizing these gains would require enormous shifts in demand.

Our models rely on a number of assumptions and parameters that are uncertain, and which may interact in nonlinear ways. To test the robustness of our main conclusions, we examine a range of scenarios and run an extensive sensitivity analysis (Supplementary Information). Across a wide range of cost, technology and demand scenarios, we find that sustainably harvested food from the sea: (1) has the potential to increase considerably in the coming decades; (2) will change in composition, with a greater future share coming from mariculture; and (3), in aggregate, could have an outsized role in meeting future meat demands around the world (Supplementary Figs. 1–4, Supplementary Tables 13–17).

4 Conclusions

Global food demand is rising, and expanding land-based production is fraught with environmental and health concerns. Because seafood is nutritionally diverse and avoids or lessens many of the environmental burdens of terrestrial food production, it is uniquely positioned to contribute to both food provision and future global food and nutrition security. Our estimated sustainable supply curves of food from the sea suggest substantial possibilities for future expansion in both wild fisheries and mariculture. The potential for increased global production from wild fisheries hinges on maintaining fish populations near their most-productive levels. For underutilized stocks, this will require expanding existing markets. For overfished stocks, this will require adopting or improving management practices that prevent overfishing and allow depleted stocks to rebuild. Effective management practices commonly involve setting and enforcing science-based limits on catch or fishing effort, but appropriate interventions will depend on the biological, socioeconomic, cultural and governance contexts of individual fisheries. Effective management will be further challenged by climate change, species composition changes in marine ecosystems and illegal fishing. Directing resources away from subsidies that enhance fishing capacity towards building institutional and technical capacity for fisheries research, management and enforcement will help to meet these challenges. Increased mariculture production will require management practices and policies that allow for environmentally sustainable expansion, while balancing the associated trade-offs to the greatest extent possible; this principle underpins the entire analysis. We find that substantial expansion is realistic, given the costs of production and the likely future increase in demand.

We have identified a variety of ways that sustainable supply curves can shift outward. These shifts interact with future demand to determine the plausible future equilibrium quantity of food produced from the sea. We find that although supply could increase to more than six times the current level (primarily via expanded mariculture), the demand shift required to engage this level of supply is unlikely. Under more realistic demand scenarios and appropriate reforms of the supply, we find that food from the sea could increase in all three sectors (wild fisheries, finfish mariculture and bivalve mariculture) to a total of 80–103 Mt of food in 2050 versus 59 Mt at present (in live-weight equivalents, 159–227 Mt compared to 102 Mt at present). When combined with projected inland production, this represents an 18–44% per decade increase in live-weight production, which is somewhat higher than the 14% increase that the Organisation for Economic Co-operation and Development (OECD) and the FAO project for total fish production during the next decade (OECD and Food and Agriculture Organization of the United Nations 2019). Under some scenarios, future production could represent a disproportionate fraction of the estimated total increase in global food production that will be required to feed 9.8 billion people by 2050. Substantial growth in mariculture will rely partly on public perceptions. Although there is some evidence of a negative public perception of aquaculture, it is highly variable by region and by context (Froehlich et al. 2017; Bacher 2015), and certifications and the provision of other information can help to alleviate concerns and expand demand (Bronnmann and Asche 2017).

These global projections will not have uniform implications around the world. For example, improved policies that shift the supply curve outward will decrease prices, but income-induced demand shifts will increase prices. Both effects increase production, but have vastly different consequences for low-income consumers. Bivalves may contribute substantially to food security by providing relatively low-cost and thus accessible food, because they have a high production potential at low costs compared to finfish production (Fig. 1.3). If all seafood is perfectly substitutable, bivalves could contribute 43% and 34% of future aquatic food under future and extreme demand scenarios, respectively (Supplementary Fig. 3)—which suggests potential large increases in production, provided demand is high enough. Trade also has an important role in distributing seafood from high-production to low-production regions, and in overcoming regional mismatches in price. The rate of international trade of seafood products has increased over past decades, and 27% of seafood products were traded in 2016 (FAO 2018), although major economic disruptions—such as the COVID-19 pandemic—can jointly reduce both supply and demand of traded seafood. On the other hand, trade may become increasingly relied upon as climate change alters regional productivity.

Substantially expanding the production of food from the sea will bring co-benefits and trade-offs, and will require national and inter-regional governance, as well as local capacity to ensure equity and sustainability. The improved management of wild fisheries can not only increase fish biomass, but also brings the co-benefit of improved livelihoods of fishers. However, there will be some short-term costs as overfished stocks rebuild to levels that support greater food provision. As mariculture expands, interactions with wild fisheries and other ecosystem services (via spatial overlaps, pollution and so on) must be constantly addressed. Ambitious technical innovation (that is, the substitution of marine ingredients with terrestrial-sourced proteins) can help to decouple fed mariculture from wild fisheries, but will probably refocus some pressure on terrestrial ecosystems. Climate change will further challenge food security. Estimates suggest that active adaptation to climate induced changes will be crucial in both wild fisheries (Gaines et al. 2018) and mariculture (Froehlich et al. 2018c). Climate-adaptive management of wild fisheries and decisions regarding mariculture production (for example, the type of feed used, species produced and farm siting) could improve food provision from the sea under conditions of climate change.

We have shown that the sea can be a much larger contributor to sustainable food production than is currently the case, and that this comes about by implementing a range of plausible and actionable mechanisms. The price mechanism—when it motivates improved fishery management and the sustainable expansion of mariculture into new areas—arises from change in demand, and acts on its own without any explicit intervention. The feed technology mechanism is driven by incentives to innovate, and thus acquire intellectual property rights to new technologies. When intellectual property is not ensured, or to achieve other social goals, there may be a role for public subsidies or other investments in these technologies. The policy mechanism pervades all three production sectors, and could make—or break—the ability of food from the sea to sustainably, equitably and efficiently expand in the future.

5 Methods

Sample size was a census of all available fisheries data. No experiments were conducted.

Here we describe our methods in brief: detailed methods, sensitivity analyses and robustness checks are provided in the Supplementary Information.

5.1 Sustainable Supply Curves

The supply of food from marine wild fisheries is jointly determined by ecosystem constraints, fishery policy and prevailing economic conditions. Estimated supply curves show the projected 2050 production quantity at a given price, incorporating harvesting costs, management costs and fishery-specific engagement decisions for individual fisheries. Current management of the 4702 marine fisheries included in our study range from open access to strong target-based management (Costello et al. 2016). Using data from the RAM Legacy Stock Assessment Database (Ricard et al. 2012), the FAO (FAO Fisheries and Aquaculture Department 2019; Costello et al. 2016; Melnychuk et al. 2017; Mangin et al. 2018), we calculate three supply curves that represent summed global production from established wild fisheries for a range of prices (Fig. 1.3). The first (*F* current) assumes that all fisheries in the world maintain their current fishing mortality rate if profitable (that is, fisheries for which current fishing pressure would result in steady-state profit < 0 are not fished). The second (rational reform) assumes that fisheries are reformed to maximize long-term food production (that is, adopt F_{MSY}, the fishing mortality rate that results in maximum sustain- able yield (MSY)), but only at prices for which reform results in greater future profit than that of current management. Importantly, adopting reform is associated with greater management costs for fisheries that are currently weakly managed. If a fishery is managed, its production changes, which alters the supply curve. Production occurs in a given fishery only if future profit > 0. The third supply curve (MSY) assumes that all fisheries are managed to maximize sustainable yield, regardless of the cost or benefit of doing so (Fig. 1.3). Supply curves under alternative cost assumptions yield results similar to those presented in Fig. 1.3 (Supplementary Fig. 1).

To construct supply curves for finfish and bivalve mariculture (which account for 83% of current production of edible animal products from mariculture (FAO 2020)), we use a previously published (Gentry et al. 2017) global suitability dataset at a resolution of 0.217°. Ecological conditions (that is, surface temperature, dissolved oxygen and primary productivity (bivalves only)) determine the suitability of different areas for production. We build on Gentry et al. (2017) by including economic considerations (for example, the capital costs of vessels and equipment and operating costs of wages, fuel, feed, insurance and maintenance; see Supplementary Information section 1.3, Supplementary Tables 5–7 for more details) to determine whether an ecologically suitable area is also economically profitable to farm at a given price. For any given price, we estimate the potential production and profitability of each pixel, and determine the global set of economically viable pixels for mariculture production of finfish and bivalves; we allow for production of both kinds of mariculture in the same pixel, provided the pixel is economically suitable for both. Summing production in this manner at the global level provides a point on the supply curve, at which farm design (Supplementary Table 4) is based on best practices for sustainable production (that is, stocking densities consistent with European organic standards (European Union 2008)). We then derive supply curves under different assumptions regarding mariculture policy and technological innovation, which affect the parameters of the supply model.

We estimate supply curves for finfish mariculture under three scenarios, all of which assume that wild fisheries are rationally managed; this pins down the potential supply of wild fish that can be used as feed in mariculture (Supplementary Table 8). We display three supply curves for fed mariculture (Fig. 1.3). The policy reforms scenario represents a future in which regulatory barriers are removed, unsustainable production is prevented and mariculture continues to use feed ingredients from wild fisheries at the current rate (that is, feed conversion ratios remain static, fishmeal and fish oil inclusion rates in feed remain the same, and feed availability depends on production from wild fisheries). This scenario represents the economically rational sustainable production given the current feed context. Two technological innovation scenarios represent policy reform plus a 50% and (a more ambitious) 95% reduction in fishmeal and fish oil requirements for fed mariculture production. The supply curve for bivalve (unfed) mariculture (Fig. 1.3) reflects production in the set of pixels for which unfed mariculture can be profitably produced at any given price.

5.2 Supply Meets Demand

To estimate how food from the sea might help to meet future increases in demand at the global level, we require estimates of the current and future demand curves of food from the sea. The intersection of future demand curves and our estimated sustainable supply curves provides an estimate of food from the sea in 2050. As a benchmark, we assume that the three sectors are independent, but that increases in demand are parametric, so each of the three sectors experiences a proportional increase in future demand—for example, as global population and per capita incomes rise (see Supplementary Information for detailed results, assuming all aquatic foods are perfect substitutes). We assume a straightforward structure in which each sector faces an isoelastic demand (for example, see Cai and Leung (2017), with own price elasticity = −0.382 (Muhammad et al. 2011); and sector-specific income elasticities estimated from Cai and Leung (2017)). Using these elasticities, the coefficient on current-demand curve in each sector (current, in Fig. 1.4) is tuned so the demand curve passes through the current price of seafood in that sector (averaged across fish from that sector) given the current global gross domestic product and population. Effectively, this approach assumes that all fish within a sector are substitutes. We do not explicitly estimate a current supply curve because it is not required to perform our calculations and—for reasons stated in the Article—we do not necessarily regard the current supply as sustainable. To project future demand at the global level, we develop two scenarios that we term future and extreme (Fig. 1.4). The future demand represents the demand curve for food from the sea in each sector given exogenous estimates of future population size and global income in 2050 (PwC 2017; United Nations 2017), which are entered as parameters in the demand curve (Supplementary Information). The extreme scenario doubles the quantity demanded at any given price in 2050, relative to the future scenario; we regard demand shifts larger than this amount as unlikely.

The Supplementary Information contains an extensive set of robustness checks and sensitivity analyses. One important alternative to the model in the Article is to allow all fish to be perfect substitutes in the future. Under that model, land-based fish production (aquaculture and capture) must be accounted for because those fish act as substitutes for food from the sea. Although this tends to increase the final estimates of food production from the sea, our qualitative findings are robust to this assumption and the Supplementary Information reports how this changes the model results described in the Article.

5.3 Reporting Summary

Further information on research design is available in the Nature Research Reporting Summary linked to this paper.

5.4 Data Availability

All datasets analysed during the current study are available in a Dryad repository at https://datadryad.org/stash/dataset/doi/10.25349/D96G6H.

5.5 Code Availability

All code used to conduct the study are available in a GitHub repository: https://github.com/emlab-ucsb/future_food_from_sea.

Acknowledgements This research is adapted from a Blue Paper commissioned by the High Level Panel for a Sustainable Ocean Economy entitled 'The Future of Food from the Sea'. We thank the high-level panel for a sustainable ocean economy, N. Frost, K. Teleki, T. Clavelle and A. Merkl for inspiration and comments. We thank SYSTEMIQ (C.C., C.M.F., T.M., E.O'R. and A.J.P.), World Resources Institute (C.C., C.M.F., T.M., E.O'R. and A.J.P.), the David and Lucile Packard Foundation (L.C. and S.G.), the European Research Council (679812) (E.O.), ANID PIA/BASAL 0002 (S.G.) and GAIN-Xunta de Galicia (E.O.) for financial support.

Author Contributions C.C., L.C., S.G. and A.J.P. conceived the study. C.C., L.C., C.M.F., H.E.F., S.G., T.M. and A.J.P. contributed to the study design. C.C., L.C., C.M.F., J.M., T.M., R.N. and A.J.P. contributed to the acquisition and analysis of data. C.C., L.C., M.Á.C.-M., C.M.F., H.E.F., S.G., T.M., R.N., A.J.P. and S.H.T. contributed to the interpretation of results. C.C., L.C., M.A.C., H.E.F., S.G., C.D.G., G.I., I.M.-S., J.M., T.M., M.C.M., M.M., C.L.d.M., R.N., L.N., E.O., E.O'R., A.M.P, A.J.P., J.L. and S.H.T. wrote and edited the manuscript.

Competing Interests C.C. serves as trustee for Environmental Defense Fund and Global Fishing Watch. H.E.F. serves as a scientific advisor on the Technical Advisory Group for the Aquaculture Stewardship Council. R.N. serves on the scientific advisory board for Oceana and *Nature Food*. C.L.d.M. has undertaken work funded by government agencies, fishery industry organizations and regional fisheries management organizations. C.D.G. serves on the scientific advisory board for Oceana.

Additional Information

Supplementary information is available for this paper at https://doi.org/10.1038/s41586-020-2616-y.

Correspondence and requests for materials should be addressed to C.C., L.C. or S.G.

Peer review information *Nature* thanks Dale Squires and the other, anonymous, reviewer(s) for their contribution to the peer review of this work.

nature research

Corresponding author(s): Christopher Costello

Last updated by author(s): 05/29/2020

Reporting Summary

Nature Research wishes to improve the reproducibility of the work that we publish. This form provides structure for consistency and transparency in reporting. For further information on Nature Research policies, see our Editorial Policies and the Editorial Policy Checklist.

Statistics

For all statistical analyses, confirm that the following items are present in the figure legend, table legend, main text, or Methods section.

n/a	Confirmed	
☒	☐	The exact sample size (n) for each experimental group/condition, given as a discrete number and unit of measurement
☒	☐	A statement on whether measurements were taken from distinct samples or whether the same sample was measured repeatedly
☒	☐	The statistical test(s) used AND whether they are one- or two-sided *Only common tests should be described solely by name; describe more complex techniques in the Methods section.*
☒	☐	A description of all covariates tested
☒	☐	A description of any assumptions or corrections, such as tests of normality and adjustment for multiple comparisons
☒	☐	A full description of the statistical parameters including central tendency (e.g. means) or other basic estimates (e.g. regression coefficient) AND variation (e.g. standard deviation) or associated estimates of uncertainty (e.g. confidence intervals)
☒	☐	For null hypothesis testing, the test statistic (e.g. F, t, r) with confidence intervals, effect sizes, degrees of freedom and P value noted *Give P values as exact values whenever suitable.*
☒	☐	For Bayesian analysis, information on the choice of priors and Markov chain Monte Carlo settings
☒	☐	For hierarchical and complex designs, identification of the appropriate level for tests and full reporting of outcomes
☒	☐	Estimates of effect sizes (e.g. Cohen's d, Pearson's r), indicating how they were calculated

Our web collection on statistics for biologists contains articles on many of the points above.

Software and code

Policy information about availability of computer code

Data collection	All data are from publicly available sources identified in the manuscript.
Data analysis	All data analysis was performed by the authors using our own code.

For manuscripts utilizing custom algorithms or software that are central to the research but not yet described in published literature, software must be made available to editors and reviewers. We strongly encourage code deposition in a community repository (e.g. GitHub). See the Nature Research guidelines for submitting code & software for further information.

Data

Policy information about availability of data

All manuscripts must include a data availability statement. This statement should provide the following information, where applicable:

- Accession codes, unique identifiers, or web links for publicly available datasets
- A list of figures that have associated raw data
- A description of any restrictions on data availability

All datasets analyzed during the current study are available in a Dryad repository: https://datadryad.org/stash/share/1hAjS-Q3nwsUAgrShYfVm6yNZSTF9oJpGWrT1_J0NyU [Note: this is currently a private repo but we will provide a public link prior to publication]. All code used to conduct the study are available in a GitHub repository: https://github.com/emlab-ucsb/future_food_from_sea.

Field-specific reporting

Please select the one below that is the best fit for your research. If you are not sure, read the appropriate sections before making your selection.

☐ Life sciences ☐ Behavioural & social sciences ☒ Ecological, evolutionary & environmental sciences

For a reference copy of the document with all sections, see nature.com/documents/nr-reporting-summary-flat.pdf

Ecological, evolutionary & environmental sciences study design

All studies must disclose on these points even when the disclosure is negative.

Study description	Analysis of existing data to derive supply of seafood.
Research sample	Publicly available data sources used: FAO Fishery and Aquaculture Statistics (FishstatJ); FAOSTAT; Costello, C., Ovando, D., Clavelle, T., Strauss, C.K., Hilborn, R., Melnychuk, M.C., Branch, T.A., Gaines, S.D., Szuwalski, C.S., Cabral, R.B. and Rader, D.N., 2016. Global fishery prospects under contrasting management regimes. Proceedings of the National Academy of Sciences, 113(18), pp.5125-5129; Mangin, T., Costello, C., Anderson, J., Arnason, R., Elliott, M., Gaines, S.D., Hilborn, R., Peterson, E. and Sumaila, R., 2018. Are fishery management upgrades worth the cost?. PLOS One, 13(9).; Cai, J. & Leung, P. Short-term projection of global fish demand and supply gaps. Food and Agriculture Organization of the United Nations, 2017; Gentry, R.R., Froehlich, H.E., Grimm, D., Kareiva, P., Parke, M., Rust, M., Gaines, S.D., Halpern, B.S. (2017) Mapping the global potential for marine aquaculture. Nature Ecology & Evolution 1(9) 1317-1324.
Sampling strategy	Sample is a census of all available fisheries data.
Data collection	Only pre-existing data was used in the analysis.
Timing and spatial scale	FAO Fishstat-J: 1950-2017; global scale Costello et al. 2016: historical 1950-2012; projections 2013-2050; global scale Mangin et al. 2018: 2012; global scale Cai et al. 2017: mean values based on mid-2010s to early 2020s; global scale Gentry et al. 2017: no temporal aspect; global scale
Data exclusions	No data were excluded from the analysis.
Reproducibility	Experimental replication was not attempted, as no experiment was performed.
Randomization	The study did not involve group allocation.
Blinding	Experiment was not performed, so blinding is not relevant to the study.

Did the study involve field work? ☐ Yes ☒ No

Reporting for specific materials, systems and methods

We require information from authors about some types of materials, experimental systems and methods used in many studies. Here, indicate whether each material, system or method listed is relevant to your study. If you are not sure if a list item applies to your research, read the appropriate section before selecting a response.

Materials & experimental systems

n/a	Involved in the study	
☒	☐	Antibodies
☒	☐	Eukaryotic cell lines
☒	☐	Palaeontology and archaeology
☒	☐	Animals and other organisms
☒	☐	Human research participants
☒	☐	Clinical data
☒	☐	Dual use research of concern

Methods

n/a	Involved in the study	
☒	☐	ChIP-seq
☒	☐	Flow cytometry
☒	☐	MRI-based neuroimaging

References

Aas TS, Ytrestøyl T, Åsgård T (2019) Utilization of feed resources in the production of Atlantic salmon (Salmo salar) in Norway: an update for 2016. Aquacult. Rep. 15:100216

Abate TG, Nielsen R, Tveterås R (2016) Stringency of environmental regulation and aquaculture growth: a cross-country analysis. Aquac. Econ. Manag. 20:201–221

Allison EH (2011) Aquaculture, fisheries, poverty and food security (Working Paper 2011–65). WorldFish Center, Penang

Amundson R et al (2015) Soil and human security in the 21st century. Science 348:1261071

Bacher K (2015) Perceptions and misconceptions of aquaculture: a global overview. FAO, Rome

Belton B, Bush SR, Little DC (2018) Not just for the wealthy: rethinking farmed fish consumption in the Global South. Glob. Food Secur. 16:85–92

Bronnmann J, Asche F (2017) Sustainable seafood from aquaculture and wild fisheries: insights from a discrete choice experiment in Germany. Ecol. Econ. 142:113–119

Cai J, Leung P (2017) Short-term projection of global fish demand and supply gaps. FAO, Rome

Cao L et al (2015) China's aquaculture and the world's wild fisheries. Science 347:133–135

Cashion T, Tyedmers P, Parker RWR (2017) Global reduction fisheries and their products in the context of sustainable limits. Fish Fish. 18:1026–1037

Copes P (1970) The backward-bending supply curve of the fishing industry. Scott. J. Polit. Econ. 17:69–77

Costello C et al (2016) Global fishery prospects under contrasting management regimes. Proc. Natl Acad. Sci. USA 113:5125–5129

Costello C et al (2019) The future of food from the sea. World Resources Institute, Washington. http://oceanpanel.org/future-food-sea

Davies IP et al (2019) Governance of marine aquaculture: pitfalls, potential, and pathways forward. Mar. Policy 104:29–36

De Silva S, Davy F (2010) Success stories in Asian aquaculture. Springer, Cham

Edwards P, Zhang W, Belton B, Little DC (2019) Misunderstandings, myths and mantras in aquaculture: its contribution to world food supplies has been systematically over reported. Mar. Policy 106:103547

European Union (2008) Commission Regulation (EC) No 889/2008 of 5 September 2008 laying down detailed rules for the implementation of Council Regulation (EC) No 834/2007 on organic production and labelling of organic products with regard to organic production, labelling and control. Off J Eur 250:1–84

FAO (2018) The state of world fisheries and aquaculture. FAO, Rome

FAO (2020) FAOSTAT. http://www.fao.org/faostat/en/#home

FAO Fisheries and Aquaculture Department (2019) FishStatJ – software for fishery and aquaculture statistical time series. http://www.fao.org/fishery/statistics/software/fishstatj/en

Foley JA et al (2005) Global consequences of land use. Science 309:570–574

Foley JA et al (2011) Solutions for a cultivated planet. Nature 478:337–342

Froehlich HE, Gentry RR, Rust MB, Grimm D, Halpern BS (2017) Public perceptions of aquaculture: evaluating spatiotemporal patterns of sentiment around the world. PLoS ONE 12: e0169281

Froehlich HE, Jacobsen NS, Essington TE, Clavelle T, Halpern BS (2018a) Avoiding the ecological limits of forage fish for fed aquaculture. Nat. Sustain. 1:298–303

Froehlich HE, Runge CA, Gentry RR, Gaines SD, Halpern BS (2018b) Comparative terrestrial feed and land use of an aquaculture-dominant world. Proc. Natl Acad. Sci. USA 115:5295–5300

Froehlich HE, Gentry RR, Halpern BS (2018c) Global change in marine aquaculture production potential under climate change. Nat. Ecol. Evol. 2:1745–1750

Gaines SD et al (2018) Improved fisheries management could offset many negative effects of climate change. Sci. Adv. 4:1378

Gentry RR et al (2017) Mapping the global potential for marine aquaculture. Nat. Ecol. Evol. 1:1317–1324

Gentry RR, Ruff EO, Lester SE (2019) Temporal patterns of adoption of mariculture innovation globally. Nat Sustain 2:949–956

Golden CD et al (2016) Fall in fish catch threatens human health. Nature 534:317–320

Hicks CC et al (2019) Harnessing global fisheries to tackle micronutrient deficiencies. Nature 574:95–98

Hilborn R, Costello C (2018) The potential for blue growth in marine fish yield, profit and abundance of fish in the ocean. Mar. Policy 87:350–355

Hilborn R et al (2020) Effective fisheries management instrumental in improving fish stock status. Proc. Natl Acad. Sci. USA 117:2218–2224

Joffre OM, Klerkx L, Dickson M, Verdegem M (2017) How is innovation in aquaculture conceptualized and managed? A systematic literature review and reflection framework to inform analysis and action. Aquaculture 470:129–148

Kawarazuka N, Béné C (2010) Linking small-scale fisheries and aquaculture to household nutritional security: an overview. Food Secur 2:343–357

Klinger D, Naylor R (2012) Searching for solutions in aquaculture: charting a sustainable course. Annu. Rev. Environ. Resour. 37:247–276

Little DC, Newton RW, Beveridge MCM (2016) Aquaculture: a rapidly growing and significant source of sustainable food? Status, transitions and potential. Proc. Nutr. Soc. 75:274–286

Mangin T et al (2018) Are fishery management upgrades worth the cost? PLoS ONE 13:e0204258

Mbow C et al (2019) Climate change and land (IPCC special report). IPCC, Geneva

Melnychuk MC, Clavelle T, Owashi B, Strauss K (2017) Reconstruction of global ex-vessel prices of fished species. ICES J. Mar. Sci. 74:121–133

Muhammad A, Seale JL Jr, Meade B, Regmi A (2011) International evidence on food consumption patterns: an update using 2005 International Comparison Program Data. Technical bulletin No. TB-1929. United States Department of Agriculture, Washington

Nielsen M (2006) Trade liberalisation, resource sustainability and welfare: the case of East Baltic cod. Ecol. Econ. 58:650–664

Nijdam D, Rood T, Westhoek H (2012) The price of protein: review of land use and carbon footprints from life cycle assessments of animal food products and their substitutes. Food Policy 37:760–770

OECD & Food and Agriculture Organization of the United Nations (2019) OECD-FAO agricultural outlook 2019–2028. OECD, Paris

Olsen Y (2011) Resources for fish feed in future mariculture. Aquacult. Environ. Interact. 1:187–200

PwC (2017) The Long view: how will the global economic order change by 2050? https://www.pwc.com/gx/en/world-2050/assets/pwc-the-world-in-2050-full-report-feb-2017.pdf

Ricard D, Minto C, Jensen OP, Baum JK (2012) Examining the knowledge base and status of commercially exploited marine species with the RAM Legacy Stock Assessment Database. Fish 13:380–398

Shah MR et al (2018) Microalgae in aquafeeds for a sustainable aquaculture industry. J. Appl. Phycol. 30:197–213

The Sea Grant Law Center (2019) Overcoming impediments to shellfish aquaculture through legal research and outreach: case studies. NOAA, Washington

Troell M et al (2014) Does aquaculture add resilience to the global food system? Proc. Natl Acad. Sci. USA 111:13257–13263

Troell M, Jonell M, Henriksson PJG (2017) Ocean space for seafood. Nat. Ecol. Evol. 1:1224–1225

UNDP (2020) Sustainable development goal 2, sustainable development goals. https://sustainabledevelopment.un.org/sdg2. Accessed 27 July 2020

United Nations (2017) World population projected to reach 9.8 billion in 2050, and 11.2 billion in 2100. World population projected to reach 9.8 billion in 2050, and 11.2 billion in 2100. https://www.un.org/development/desa/en/news/population/world-population-prospects-2017.html

Ye Y, Gutierrez NL (2017) Ending fishery overexploitation by expanding from local successes to globalized solutions. Nat. Ecol. Evol. 1:179

2 The Expected Impacts of Climate Change on the Ocean Economy

Highlights

- The ocean is critically important to our global economy. Collectively, it is estimated that ocean-based industries and activities contribute hundreds of millions of jobs and approximately US $2.5 trillion to the global economy each year, making it the world's seventh-largest economy when compared with national gross domestic products. In addition, the nonmarket services and benefits provided by the global ocean are significant and may in fact far exceed the value added by market-based goods and services.
- Climate change is altering ocean climate, chemistry, circulation, sea level and ice distribution. Collectively, these system changes have critical impacts on the habitats, biological productivities and species assemblages that underpin many of the economic benefits of the sea.
- Swift efforts to reduce anthropogenic greenhouse gas emissions are needed to maintain a robust ocean economy. The recent Intergovernmental Panel on Climate Change report estimates that climate-induced declines in ocean health will cost the global economy $428 billion/year by 2050 and $1.98 trillion/year by 2100.
- Climate change is reducing the productivities and changing the spatial distributions of economically important marine species and their habitats. All countries stand to gain significant benefits relative to a business-as-usual trajectory by implementing climate-adaptive fisheries management reforms that address both changes in species' distributions and productivities due to climate change. Many countries could maintain or improve profits and catches into the future with effective adaptation.
- The potential of marine aquaculture (mariculture) is likely to remain high under climate change and, with careful planning, mariculture could offset losses in food and income from capture fisheries in those countries that will experience losses in that sector. Expanding the potential for marine aquaculture will require enhancing technical capacities, defining best practices, easing undue regulatory burdens, increasing access to credit and insurance, breeding stocks for faster growth and improving feed technology.
- The combined effects of ocean warming and acidification result in predictions of negative impacts on coral reef cover and tourism values for all countries, with magnitudes dependent on the strength of climate change. For a high emissions scenario (Representative Concentration Pathway 8.5), coral cover is expected to decline by 72–87%, causing on-reef tourism values to decrease by over 90% in 2100.
- Climate change impacts will differ by country and sector and solutions must be context-specific. By exploring climate change impacts at the country level for fisheries, aquaculture and reef tourism, countries can assess what they stand to gain or lose due to climate change and understand how they might capitalise on these predictions to inform their investments and actions.
- Implementing certain key strategies will help build socioecological resilience to climate change and ensure the continued, or improved, provision of functions and services from the ocean, especially for the most vulnerable coastal nations. These strategies include the following:
  - **A focus on equity.** Climate change is likely to cause and exacerbate global inequities, reducing resilience and thereby likely worsening outcomes under all climate change scenarios. It will thus be profoundly important to examine the equity implications of all new and existing management decisions across all three sectors.

Originally published in:
Gaines, S., R. Cabral, C. Free, Y. Golbuu, et al. 2019. *The Expected Impacts of Climate Change on the Ocean Economy*. Washington, DC: World Resources Institute. Available online at www.oceanpanel.org/expected-impacts-climate-change-ocean-economy.
Reprint by Springer International Publishing (2023) with kind permission.
Published under license from the "World Research Institute".

J. Lubchenco, P. M. Haugan (eds.), *The Blue Compendium*, https://doi.org/10.1007/978-3-031-16277-0_2

- **Looking forward.** The future of the ocean economy is expected to drastically change given climate change, and the nature and magnitude of these changes can be highly variable. Each of these three sectors will need to work to understand risks and anticipate changes, and build precautionary and adaptive strategies into their management decisions.
- **Cooperating across boundaries.** As suitable habitats shift and change, marine species will move across jurisdictional boundaries and regional, national and international cooperative agreements will be necessary to ensure that these species are well-managed, and that the benefits are fairly distributed during and after the transitions.

1 Introduction

1.1 Overview

The ocean is critically important to our global economy. Collectively, it is estimated that ocean-based industries and activities contribute hundreds of millions of jobs and approximately US $2.5 trillion to the global economy each year, making it the world's seventh-largest economy when compared with national gross domestic products (GDPs) (Hoegh-Guldberg 2015; IPCC 2019). In addition, the nonmarket services and benefits provided by the ocean are significant and may in fact far exceed the value added by market-based goods and services (Costanza et al. 2014).

Anthropogenic climate change, driven by the exponential increase in emissions of greenhouse gasses (GHGs) since the industrial revolution, will continue to impact the ocean through a variety of channels. The severity of effects will depend greatly on the extent of warming reached through GHG emissions (IPCC 2018, 2019). The resulting changes to ocean processes and functioning have broad implications for our global economy that must be taken into account, both to inform adaptation efforts and motivate urgent mitigation strategies.

In this paper, we focus on those sectors of the ocean economy that are most in need of adaptation to ensure they can continue to provide valued functions as the climate changes: capture fisheries, marine aquaculture, and marine and coastal tourism. We also briefly discuss other marine-based sectors, some of which generate higher monetary value at a global scale, but either face less significant existential risks from the changing climate (e.g. shipping), or must be drastically transitioned to avoid worsening the climate crisis (e.g. oil and gas extraction). However, we leave deeper discussion of these important industries and the issues surrounding them to other Blue Papers (*Ocean Energy and Mineral Sources and Coastal Development*).

1.2 The Ocean Economy: Essentials

The ocean economy consists broadly of all ocean-based human activities that generate revenue, employment and other monetary and nonmonetary benefits (OECD 2016). Some of the ocean benefits, and the resources needed to generate them, are market-based in that they are traded on global markets and have market prices. Examples of market-based ocean benefits include the following: wild capture fisheries and marine aquaculture (also known as mariculture); pharmaceuticals; fossil fuel energy resources such as oil and gas; renewable energy resources such as wave, wind or thermal energy; the use of the ocean surface for transportation (shipping); ocean-based tourism; and emerging blue carbon markets. Following the Intergovernmental Science-Policy Platform on Biodiversity and Ecosystem Services (IPBES) framework, most of the marketable benefits mentioned are material contributions (e.g. food, energy sources, genetic resources), but other important marketable contributions to people are regulating services (e.g. carbon sequestration) and nonmaterial (e.g. tourism).

Many other ocean benefits are not traded on markets, and their values are thus far more difficult to assess. The set of nonmarket ocean benefits is very large (Polasky and Segerson 2009; Costanza et al. 2014) and ranges from different ecosystem services to the broader category of nonmaterial contributions to people. In terms of ecosystem services, nonmarket benefits include most of the ocean's cultural services (e.g. swimming, recreational fishing, observing sea life, the existence value of the ocean's diverse biota). In addition, ecosystem services include regulating services—the ocean's contribution to the global water, energy and chemical circulation systems, as well as the ocean's role in climate regulation, carbon dioxide (CO_2) uptake and coastal protection—which are typically not accounted for in existing markets. The IPBES framework further adds to the ocean's nonmaterial contributions by including learning and inspiration (i.e. education, scientific information), psychological experiences (i.e. relaxation, healing, aesthetic enjoyment), supporting identities (i.e. the basis for spiritual and social cohesion experiences, myths and traditional knowledge) and maintenance of options for future generations and innovations and needs (Díaz et al. 2015, 2018).

1.2.1 The Market-Based Ocean Economy

The Organisation for Economic Co-operation and Development (OECD) projects that market-based ocean industries will expand at least as fast as the global economy as a whole over the next decade. The OECD (2016) outlines the ocean industries that contribute the most in terms of production value and employment (see Table 2.1).

The rankings of ocean industries are quite different for these two economic outputs. Energy production, shipping

Table 2.1 Ocean industries contributing most to the ocean economy

	% of production value	% of employment
1. Offshore oil and gas	34	6
2. Marine and coastal tourism	26	22
3. Port activities	13	5
4. Maritime equipment	11	7
5. Fisheries, marine aquaculture and fish processing	6	49
6. Ocean transportation	5	4
7. Shipbuilding and repair	4	6
8. Offshore wind	1	1

Source: OECD (2016)
Note: Data are from 2010

and tourism dominate production values, while nearly half of all ocean employment arises from food production. Therefore, the impacts of climate disruptions on these industries can have quite disparate social and economic consequences.

1.2.2 The Nonmarket Ocean Economy

Despite the complexities and theoretical challenges, a number of researchers have attempted to calculate the value of the diverse ecosystem services provided by the ocean. Although there is much debate, these assessments generally conclude that nonmarket services from the ocean are nearly comparable in value to the entire market-based gross global product (i.e. from the entire global economy). For example, a prominent evaluation by Costanza et al. (2014) assessed the value of global ocean ecosystem services to be almost $50 trillion in 2011. This translates to more than 80% of the gross global product in that year, or 30 times more than the ocean-based gross value added. Recent initiatives, such as IPBES, broaden the concept of valuation of nonmarket goods and ecosystem services even further to the more inclusive Nature's Contributions to People (NCP). The ocean provides a number of these important contributions, which arise from a diversity of human-ocean relationships, including those of indigenous people and local communities (Díaz et al. 2015; Pascual et al. 2017). Although we focus on measuring the impacts of climate change on the market ocean economy in this assessment, it is clear that solutions to those challenges could generate far larger returns from the added benefits they provide to these nonmarket components of the ocean economy.

2 How Rising Greenhouse Gasses Alter the Ocean

Climate change is altering ocean climate, chemistry, circulation, sea level and ice distribution (Brander 2010; García Molinos et al. 2016; IPCC 2019). Collectively, these system changes have critical impacts on the habitats, biotic productivities and species assemblages (Doney et al. 2012; Poloczanska et al. 2013; Pinsky et al. 2013; Visser 2016; Bryndum-Buchholz et al. 2019; Lotze et al. 2019) that underpin many of the economic benefits of the sea (Barange et al. 2018; Cheung et al. 2010; Free et al. 2019a; Lam et al. 2016; Sumaila et al. 2011). They also affect the risks of various human activities and developments (Gattuso et al. 2015; de Suarez et al. 2014; Barange et al. 2014). Unprecedented ocean changes are already occurring across all latitudes (Barange et al. 2018; Friedrich et al. 2012; Holbrook et al. 1997; IPCC 2019; Kleisner et al. 2017; Walther et al. 2002), with a high risk of negative impacts to many ocean organisms, ecosystems and services (Gattuso et al. 2015; IPCC 2019; Lotze et al. 2019). These impacts are likely to increase dramatically toward the end of this century, depending on the extent of future GHG emissions, with potentially direct consequences for ecosystem services, the ocean economy and human welfare (IPCC 2019; Pecl et al. 2017). Below, we describe these effects individually, but many of these influences may synergistically or antagonistically interact, potentially with additional consequences (see, for example, Rosa and Seibel 2008).

Throughout this paper, we rely on the Representative Concentration Pathways (RCPs) (van Vuuren et al. 2011) adopted by the Intergovernmental Panel on Climate Change (IPCC) in its Fifth Assessment Report to describe potential GHG emission trajectories and associated climate futures. The RCP scenarios are named according to the projected radiative forcing experienced in 2100 (2.6, 4.5, 6.0 and 8.5 Watts per square metre [W/m^2], respectively). They roughly correspond to projected increases in planetary surface temperatures relative to 1850–1900 of 1.6, 2.5, 2.9 and 4.3 °C, respectively, by the end of this century (IPCC 2019).

2.1 Altered Ocean Temperatures and Disturbances

Climate change has already contributed to substantial warming of the ocean over most of the globe. The ocean has absorbed ~93% of additional heat, leading to significant warming of the upper ocean (above 700 metres [m]) and warming of deeper waters (700–2000 m), increasing in strength since the 1980s (Cheng et al. 2017). Sea surface temperatures have increased by an average of 0.7 °C globally since 1900 (Barange et al. 2018; Jewett and Romanou 2017). RCP scenarios suggest that these trends, which already exceed the range in natural seasonal variability in subtropical areas and the Arctic, will continue (IPCC 2014, 2019). Future upper ocean warming is expected to be most pronounced in tropical and Northern Hemisphere subtropical regions, while deep water warming is expected to be more pronounced in the Southern Ocean (Barange et al. 2018;

IPCC 2019; Gattuso et al. 2015). By 2100, the ocean as a whole is likely to have warmed by two to four times (RCP 2.6) to five to seven times (RCP 8.5) as much as the warming observed since 1970 (IPCC 2019).

As these warming trends continue, the suitable distribution ranges of many marine species are expected to shift poleward. In general, species that are able to move to cooler waters, and have suitable habitats to move to, will do so (Barange et al. 2018; Cheung et al. 2010; IPCC 2019; Pinsky et al. 2013). Organisms and habitats that cannot move will either adapt to the new conditions caused by climate change or become extirpated, unless extensive transplantation or other initiatives are mounted to prevent this. Significant habitat losses are predicted in many areas, especially in the Arctic and coral reef ecosystems, resulting in altered community assemblages, predator-prey mismatches and local extinctions (Doney et al. 2014; Free et al. 2019a; Gattuso et al. 2015; Holbrook et al. 1997; IPCC 2019).

Warming waters, along with an increase in episodic 'marine heat waves', ocean acidification (discussed below) and the spread of diseases, will lead to mass coral bleaching and mortality throughout the ranges of most coral species (Donner et al. 2005; FAO 2018; Gattuso et al. 2015; Hoegh-Guldberg 1999; IPCC 2019; Kubicek et al. 2019; McClanahan et al. 2002). Intense reshufflings of current biodiversity patterns are also anticipated in biogeographical transition zones, where local populations of multiple species are at or close to their thermal tolerance limits. As a result of these movements, studies have predicted 30–70% average increases in potential fish production at high latitudes, and decreases of up to 40% in the tropics (Barange et al. 2018; Cheung et al. 2010). Indeed, ongoing rapid replacement of cold-affinity species by warm-affinity species has been recently documented in tropical-to-temperate (Kumagai et al. 2018; Verges et al. 2014) and boreal-to-Arctic (Fossheim et al. 2015) regions.

Furthermore, tropical cyclones, extreme sea level events including storm surges and flooding and precipitation over the ocean are predicted to increase in intensity and frequency through the first half of this century due to ocean circulation changes (discussed below) (Barange et al. 2018; Hartmann et al. 2013; IPCC 2014, 2019; Kirtman et al. 2013; Kopp et al. 2014; Ren et al. 2013). In addition, recent models and observational data indicate that recurring climate patterns such as the El Niño-Southern Oscillation are likely to increase in frequency and intensity as the ocean warms (Barange et al. 2018; Cai et al. 2014, 2015; IPCC 2019; Wang et al. 2017), with potentially important impacts on fishing, aquaculture and tourism operations. River flows and flooding may also change with increased snowmelt and more variable land-based precipitation, reducing salinity, increasing sedimentation and impacting productivity in nearshore waters (IPCC 2019; Jha et al. 2006; Pervez and Henebry 2015; Siderius et al. 2013; Loo et al. 2015). Finally, ocean warming leads to increased stratification of the water column and reduced water circulation and mixing (Barange et al. 2018; FAO 2018; IPCC 2019; Jacox and Edwards 2011; Oschlies et al. 2018).

2.2 Sea Level Rise and an Altered Distribution of Ice

Polar areas have seen drastic changes including shifts in the timing of the annual melt seasons, changes in snow cover and changes in ice sheet and glacier mass, which have resulted in sea level rise. Globally, mean sea level rose on average by 0.16 m from 1902 to 2015, and estimates indicate that by 2100, the global mean sea level will rise between 0.29 m and 0.59 m under RCP 2.6, and between 0.61 m and 1.1 m under RCP 8.5 (Barange et al. 2018; IPCC 2019; Kopp et al. 2014). The rate of increase varies across regions—in the western Pacific, sea level is increasing at three times the global average, while the rate of increase in the eastern Pacific is null or negative (Barange et al. 2018; Dangendorf et al. 2017). The economic consequences of global sea level rise will therefore also be highly heterogeneous across regions, as well as across sectors, with likely significant impacts stemming from the modification of coastlines, reduced coastal productivity as reefs and seagrasses are submerged and increased flooding (Barange et al. 2018; IPCC 2019).

In the Arctic, annual sea ice extent has decreased at a rate of 3.5–4.1% per decade, plummeting to a rate of −13% in September, the month marking the end of the melt season. This strong downward trend in extent is accompanied by a progressive loss of multiyear sea ice with over 50% of its extent lost during the period 1999–2017 (Kwok 2018; IPCC 2019). Meanwhile, mass lost from the Antarctic ice sheet tripled between 2007 and 2016 relative to the previous decade, leading to the lowest average monthly and yearly Antarctic sea ice extents on record in 2017 (IPCC 2019; Parkinson 2019). The Greenland Ice Sheet's mass loss doubled over this same period, and the rates of mass loss for both Greenland and Antarctic sea ice are expected to increase throughout the twenty-first century and beyond (IPCC 2019). Together, these two ice sheets are projected to contribute 0.11 m to global mean sea level rise under RCP 2.6, and 0.27 m under RCP 8.5 (IPCC 2019). While reductions in sea ice have opened new routes for international shipping, potentially reducing costs to this sector, these changes have also resulted in losses to sea ice–based travel and tourism, and pose risks to cultural livelihoods such as subsistence fishing and hunting for polar species (IPCC 2019). Glaciers and land-based ice sheets across the world have also shrunk (Barange et al. 2018; IPCC 2019) and their combined influence was the dominant source of sea level rise between 2006 and 2015 (IPCC 2019).

Sea level rise, combined with increased storm frequency and intensity, is expected to have significant negative impacts on ocean and coastal economy infrastructure, including damage to ports, aquaculture operations and offshore energy structures, and added risks and constraints on shipping (IPCC 2019). These impacts are likely to be among the costliest and potentially most disruptive of all the climate-driven ocean changes. For example, global annual flood costs from sea level rise under RCP 8.5 are estimated at $14 trillion/year (Jevrejeva et al. 2018). Furthermore, although there is uncertainty around exact numbers, sea level rise and other climate-related ocean changes will likely lead to the displacement of millions of people worldwide, with the poorest households facing the greatest risk (IPCC 2019). Low-lying island nations, such as Maldives, Marshall Islands, Tuvalu and Nauru, are especially vulnerable, with sea level rise threatening their entire economies and populations.

2.3 Altered Ocean Chemistry

Ocean acidity has increased by 26% since the industrial revolution, with regional variability in severity and rate of change (Barange et al. 2018; Gattuso et al. 2015; IPCC 2014, 2019; Jewett and Romanou 2017).

This increase has been driven primarily by the oceanic absorption of CO_2, which lowers ocean pH (by increasing bicarbonate and hydrogen ion concentrations) and carbonate ion concentrations, and increases the partial pressure of CO_2 and dissolved inorganic carbon. These changes can impact many marine organisms, particularly in early life stages, but are especially detrimental to corals and organisms that form carbonate shells (Barange et al. 2018; FAO 2018; Pörtner et al. 2014), and perhaps beneficial for some photosynthetic, non-calcifying taxa (Kroeker et al. 2013). Observed trends of declining ocean pH already exceed the natural seasonal variability throughout most of the open ocean, and they are expected to continue throughout this century (Barange et al. 2018; Gattuso et al. 2015; Henson et al. 2017; IPCC 2019).

By 2100, surface ocean pH is projected to decline by 0.036–0.042 pH units under RCP 2.6, or 0.287–0.29 pH units under RCP 8.5. High-latitude waters, deep waters and upwelling regions will be the first to see carbonate ion concentrations drop below the 'saturation point' (meaning below the point at which shell and reef formation is possible; the Arctic Ocean, the northeastern Pacific and the California upwelling system already experience seasonally undersaturated conditions), while the tropical ocean (where current carbonate ion concentrations are higher) will experience the largest absolute decreases in carbonate ion concentration and pH (Barange et al. 2018; Harris et al. 2013). Warm water corals will be impacted by decreased carbonate ion saturation levels even where waters do not become undersaturated (Hoegh-Guldberg et al. 2017).

Even if global warming is limited to 1.5 °C, warm water corals are likely to suffer significant negative impacts, including changes to community composition and diversity, local extinctions and reductions in range and extent (IPCC 2019). Coastal seawater acidification can be intensified by additional carbon from riverine input or through coastal productivity stimulated from land-based nutrient inputs, or nutrients released from sediments, aquaculture, sewage discharges and other point sources (Gattuso et al. 2015). These impacts will have significant negative effects on coral reef–related tourism and fishery operations as well as on shellfish aquaculture operations (although see below for a discussion of the potential for aquaculture adaptation and expansion).

Climate change is also impacting the dissolved oxygen content in ocean systems across the globe. Warming-driven stratification of the water column, exacerbated by other physical and biogeochemical processes, reduces the dissolved oxygen content in ocean water (Barange et al. 2018; Breitburg et al. 2018; Gattuso et al. 2015; IPCC 2019; Jacox and Edwards 2011; Oschlies et al. 2018). In recent decades, oxygen concentration in coastal waters and the open ocean has decreased, while the prevalence and size of 'oxygen minimum zones' (OMZs), areas where oxygen consumption by sediment bacteria exceeds the availability of oxygen, have increased, especially in the tropics, although it is difficult to conclusively attribute these shifts to human activity in these regions (Barange et al. 2018; Breitburg et al. 2018; IPCC 2019; Oschlies et al. 2018; Stramma et al. 2010; Levin 2002). These trends are expected to continue, with the whole-ocean oxygen inventory expected to decrease by 1.6–2% (RCP 2.6) to 3.2–3.7% (RCP 8.5), and the global volume of OMZs expected to increase by $7.0 \pm 5.6\%$ by 2100 under RCP 8.5 (Barange et al. 2018; Fu et al. 2018; Gattuso et al. 2015; IPCC 2019). Increased deoxygenation will likely lead to habitat compression, shifts in distribution and losses in species abundance and biodiversity (Breitburg et al. 2018; Stramma et al. 2010; Levin 2002). Furthermore, observed deoxygenation is generally worse than modelled results, which emphasises the need to improve our understanding of the processes driving deoxygenation to reduce the model uncertainty in our projections (Bopp et al. 2013; Oschlies et al. 2018).

Deoxygenation and OMZs affect species in different ways and to different degrees depending on varying oxygen tolerances. While some hypoxia-adapted species may benefit, impacts on most fish and invertebrates will be negative, and may include restricted vertical and horizontal migration, compressed habitats, alterations to predator-prey interactions and increased competition, impairment of reproductive capacity, reduced growth, vision impairments, increased disease incidence, epigenetic changes and death from asphyxiation (Barange et al. 2018; Breitburg et al. 2018; Eby and Crowder 2002; Gattuso et al. 2015; IPCC 2019; Oschlies et al. 2018). The combination of ocean warming, increased acidity and decreased oxygen availability is predicted to result in significant decreases in both the average size and abundance of many important fishery species (Breitburg et al. 2018).

2.4 Altered Circulation Patterns

Water circulation in the ocean, known as the 'global conveyor belt', is responsible for the redistribution of heat and freshwater, influencing local climates, productivity levels and ocean chemistry. A warming climate increases inflows of warm freshwater (from increased precipitation and melting glaciers and sea ice), which can reduce the formation of sea ice and sinking of cold salt water. This influx slows parts of global conveyor belt circulation (Barange et al. 2018; IPCC 2019; Liu et al. 2017). The Atlantic Meridional Overturning Circulation and Gulf Stream, which are responsible for a significant portion of the redistribution of heat from the tropics to the middle and high latitudes as well as of the ocean's capacity to sequester carbon, are showing signs of weakening (Caesar et al. 2018; IPCC 2019; Thornalley et al. 2018; Barange et al. 2018) and may continue to do so under all RCP scenarios (IPCC 2019). In the Atlantic, this weakening is driving lower sea surface temperatures in the subpolar Atlantic Ocean and a warming and northward shift of the Gulf Stream, which is expected to further weaken in the coming decades (Caesar et al. 2018; Thornalley et al. 2018; Barange et al. 2018; Liu et al. 2017). These changes could lead to dramatic shifts in weather and local and regional climate patterns (IPCC 2019), which would have significant impacts on the ocean economy (e.g. through damage to infrastructure) and society as a whole.

All western boundary currents other than the Gulf Stream are expected to intensify in response to tropical atmospheric changes and shifts in wind patterns resulting from climate change and GHG concentrations, likely strengthening coastal storm systems (Barange et al. 2018; Yang et al. 2016). The intensity of the eastern boundary currents, responsible for the major coastal upwelling zones and thus for some of the most productive waters in the world, will also likely change, although there is more uncertainty around the severity and direction of these changes, as well as around the resulting impacts (Bakun et al. 2015; Barange et al. 2018; Brady et al. 2017). As the land and ocean warm at different rates, stronger upwelling-favourable winds may strengthen these patterns; however, increased thermal stratification may restrict the depth of upwelling waters, and thus limit the amount of nutrients brought with them (Bakun 1990; Barange et al. 2018; Jacox and Edwards 2011; Rykaczewski et al. 2015; Sydeman et al. 2014; Wang et al. 2015). The impacts of intensified upwelling may result in a net increase in nutrient inputs and primary productivity or, alternatively, increase the presence of low oxygen and more acidic waters along the continental shelf (Bakun et al. 2015; Barange et al. 2018). Changes in either direction will have critical impacts for the many valuable marine capture fisheries located in and around upwelling zones. The most recent estimations at a global scale show a decrease in primary productivity of 7–16% by 2100 for RCP 8.5, largely driven by changes to circulatory and upwelling patterns as well as thermal stratification (IPCC 2019). However, the interaction and relative importance of these forces, as well as of regional processes and seasonal variability, will vary across geographies (Barange et al. 2018; IPCC 2019), and thus local data collection and modelling will be necessary to inform management.

3 Connecting the Links Between Climate Change and the Ocean Economy

3.1 Capture Fisheries

3.1.1 Importance of Capture Fisheries to the Ocean Economy

In 2016, the United Nations Food and Agriculture Organization (FAO) estimated that marine capture fisheries produced 79.3 million metric tonnes (mmt) of landings, representing 46.4% of global seafood production (170.9 mmt) and $130 billion in first sale value (FAO 2018). It also estimated that approximately 30.6 million people participated—either full time, part time, or occasionally—in capture fisheries, operating approximately 4.6 million fishing vessels. Small-scale fisheries are the backbone of socioeconomic well-being in many coastal communities (Bene 2004; Béné et al. 2007, 2010), especially in the developing tropics where the majority of fish-dependent countries are located (Golden et al. 2016). Fish and fish products are also among the most traded food commodities in the world. In 2016, approximately 35% of production entered international trade for either human consumption or nonfood uses (FAO 2018). The 60 mmt ($143 billion) of fish products exported in 2016 constituted a 245% increase relative to 1976 exports ($8 billion). Over this time period, the rate of growth of exports from developing countries surpassed that from developed countries (FAO 2018). Finally, the average annual increase in fish consumption (3.2%) has outpaced the average annual increase in human population growth (1.6%), and demand for fish is projected to increase as the human population continues to grow and become increasingly wealthy (FAO 2018).

3.1.2 Impacts of Climate Change on Capture Fisheries

Climate change is significantly altering the ability for marine fisheries to provide food and income for people around the world (IPCC 2019). These changes are commonly viewed as occurring through impacts on either the distribution of fish stocks (i.e. where fish can be caught and by whom) or the productivity of fish stocks (i.e. how much fish can be caught). In general, productivity is predicted to decrease in tropical and temperate regions and increase toward the poles (Lotze et al. 2019) as marine organisms shift their distributions to maintain their preferred temperatures (Pinsky et al. 2013; Poloczanska et al. 2013; Poloczanska et al. 2016). These regional shifts in productivity, range and fishing opportunity are likely to result in regional discrepancies in food and profits from fisheries (Lam et al. 2016), with tropical developing countries and small island developing states exhibiting the greatest vulnerability to the climate change (Allison et al. 2009; Blasiak et al. 2017; Guillotreau et al. 2012).

In the remainder of this Sect. 3.1.2, we detail how both retrospective and forward-looking studies have revealed the impact of climate change on the distributions and productivities of marine fisheries and the implications of these observations and predictions for adapting fisheries management to climate change. In Sect. 3.1.3, we present results from a new study (Free et al. 2019b) that demonstrate the country-level economic and food provisioning benefits of reforming fisheries management to account for shifting distributions and productivities. Finally, in Sect. 3.1.4, we outline how fisheries could implement climate-adaptive reforms along a gradient of scientific, management and enforcement capacities.

Marine fish and invertebrates are shifting distributions to track their preferred temperatures. Adaptive international agreements that prioritise equitable outcomes will be necessary to ensure that management remains sustainable and just as species shift in and out of management jurisdictions.

Observed changes: As the ocean has warmed, marine fish and invertebrates have shifted their distributions to track their preferred temperatures (Perry et al. 2005; Dulvy et al. 2008; Poloczanska et al. 2013; Pinsky et al. 2013). In general, this has resulted in shifts poleward and into deeper waters. At a mean rate of 72 kilometres (km) per decade, marine species have been moving an order of magnitude faster than terrestrial species (Poloczanska et al. 2013). These distribution shifts are already generating management challenges (Pinsky et al. 2018). For example, a 'mackerel war' erupted in 2007 when the northeast Atlantic mackerel stock shifted from waters managed by the European Union, Norway and Faroe Islands into Icelandic and Greenland waters. Disagreements over the drivers of the shift, the expected duration of the shift, and appropriate catch reallocations resulted in the stock becoming increasingly overfished (Spijkers and Boonstra 2017).

Forecasted changes: The rate of distribution shifts and associated management conflicts are anticipated to increase under climate change. All studies forecast generally poleward shifts in species distribution and productivity under continued warming (Lotze et al. 2019), often with a decrease in species diversity in equatorial regions, an increase in diversity in poleward regions and the subsequent formation of novel marine communities (García Molinos et al. 2016; Cheung et al. 2016). These shifts are likely to increase the risk of management conflicts over transboundary stocks. For example, 23–35% of exclusive economic zones (EEZs) are expected to receive a new stock by 2100 under strong greenhouse gas mitigation (RCP 2.6) to business-as-usual mitigation (RCP 8.5) scenarios, respectively (Pinsky et al. 2018).

Implications for adaptation: Establishing and strengthening international institutions and agreements to better manage stocks shifting in and out of jurisdictions will be important. These agreements will need to be both adaptive, to ensure that management remains effective under continued uncertainty, and inclusive of all impacted groups, to ensure that outcomes are equitable. As with management decisions made at the fishery and community scales, these international agreements must engender procedural, distributional and recognitional equity if they are to be truly resilient (Matin et al. 2018; Meerow et al. 2019). See Opportunity for Action #3 in Sect. 3.1.4 for more detail.

Climate change is reducing the productivity of marine fisheries globally. Regional impacts are especially pronounced, with some regions experiencing large gains in productivity while others experiencing large losses. Resilience to climate change can be enhanced by implementing adaptive, inclusive and transparent 'primary fisheries management', by accounting for shifting productivity in assessment and management and by rebuilding overfished stocks. Solutions should be developed through processes that ensure procedural, distributional and recognitional equity at all stages.

Observed changes: Free et al. (2019a) estimate that ocean warming has already driven a 4.1% decline in the maximum sustainable yield (MSY), the maximum amount of catch that can be harvested for perpetuity, of 235 of the largest industrial fisheries over the past 80 years. The North Sea, which supports large commercial fisheries, and four East Asian marine ecoregions, which support some of the fastest-growing human populations, have experienced losses in MSY of 15–35%. Meanwhile, the Baltic Sea and other regions have seen increases in MSY of up to 15%. Changes in productivity are driven by changes in growth, mortality or recruitment rates resulting from changing environmental conditions, phenologies (i.e. mismatches in the timing of juvenile recruitment and food availability), disease or food web structures, as well as changes in carrying capacities resulting from distribution shifts or habitat degradation (Hollowed et al. 2013). In general, well-managed fisheries have been the most resilient to these changes while overexploited fisheries have been the most vulnerable (Britten et al. 2016; Free et al. 2019a).

Forecasted changes: An ensemble of six marine ecosystem models (Bryndum-Buchholz et al. 2019; Lotze et al. 2019) forecasts decreases in marine animal biomass of 4.8, 8.6, 10.4 and 17.2% by 2100 under RCPs 2.6, 4.5, 6.0 and 8.5, which represent increasingly severe greenhouse gas emissions scenarios. The ensemble model and its constituent models consistently predict reduced productivity in tropical to temperate regions and increased productivity at the poles. For example, marine animal biomass is forecast to decline by 15–30% in the North/South Atlantic, North/South Pacific and Indian Ocean basins by 2100 while increasing by 20–80% in the polar Arctic and Southern Ocean basins (Bryndum-Buchholz et al. 2019). Regional disparities in marine animal biomass become increasingly pronounced under increasingly severe emissions scenarios. The redistribution of catch potential will drive a concomitant redistribution of revenues (Lam et al. 2016) and nutrition (Golden et al. 2016; Hicks et al. 2019).

Implications for adaptation: First and foremost, in both low- and high-capacity fisheries systems, implementing general fisheries reforms will enhance resilience to climate change as well-managed fisheries are the most ecologically (Free et al. 2019a) and socioeconomically resilient to climate change. In low-capacity fisheries systems, this can be achieved through 'primary fisheries management' (Cochrane et al. 2011), which uses the best available science to inform precautionary management while building institutional capacity for adaptive and participatory co-management. To do so, adaptation policy should target the most vulnerable communities, which in fisheries are typically women and migrant fishers; those with highly fisheries-dependent livelihoods in terms of nutrition and income; and the agency of these individuals to adapt (Cinner et al. 2018). In high-capacity fisheries systems, this will involve accounting for shifting productivity in fisheries stock assessments and management procedures. See Opportunities for Action #1–2 and #4–5 in Sect. 3.1.4 for more detail.

3.1.3 Ability for management to mitigate the impacts of climate change

Most forecasts of the impacts of climate change on fisheries compare the maximum biological potential for food production today with that in the future (Cheung et al. 2010; Lam et al. 2016). While this is useful for understanding the biological limits of the ocean under climate change, it fails to consider the effects of alternative human responses (Barange 2019), which could either limit or exacerbate the impacts of climate change on society. The actions of fishers, management institutions and markets all influence the benefits derived from fisheries (Costello et al. 2016) and could mitigate many of the negative impacts of climate change (Gaines et al. 2018). *Thus, we present a recent analysis* (Free et al. 2019b)[1] *that documents the benefits countries stand to gain by implementing climate-adaptive fisheries management reforms that address both changes in species distribution and productivity due to climate change.*

Methods: Free et al. (2019b) forecasted the distributions and productivities of 779 harvested marine species out to 2100 under three greenhouse gas emissions scenarios (RCPs 4.5, 6.0 and 8.5), and compared the status of these fisheries and the amount of catch and profits derived from them under both climate-adaptive management and business-as-usual management. Under climate-adaptive management, fisheries management dynamically updates economically optimum harvest rates to match shifts in productivity, and transboundary institutions maintain management performance as shifts in distribution move stocks into new management jurisdictions. Under business-as-usual management, current (rather than economically optimal) harvest rates are initially applied and are gradually transitioned to open access as stocks shift into new management jurisdictions (see Free et al. 2019b for details on the management scenarios). Free et al. (2019b) then measured the extent to which climate-adaptive management could maintain catch and profits into the future and generate catch and profits relative to business-as-usual management.

Results: Even countries experiencing declines in fisheries productivity and catch potential would derive more catch and profits through climate-adaptive management than through business-as-usual management (Fig. 2.1). Furthermore, in many countries, adaptive management would not only reduce the impacts of climate change, but actually increase catch and profits relative to today (Fig. 2.1). Climate-adaptive fisheries management results in greater cumulative profits than business-as-usual management for 99% of countries under RCPs 6.0 and 8.5. It results in greater cumulative catches than business-as-usual management in 98% and 67% of countries in RCPs 6.0 and 8.5, respectively. Furthermore, under adaptive management, 71% and 45% of countries derive more catch and profits from fisheries in 2100 relative to today under RCPs 6.0 and 8.5, respectively. The impacts of climate change on fisheries and the opportunities and benefits of climate-adaptive fisheries management reforms can be explored for specific countries in an interactive web application created by the Sustainable Fisheries Group at the University of California, Santa Barbara (UCSB 2019).

[1]This paper is currently under peer review but a pre-print is publicly available on BioRxiv here: https://www.biorxiv.org/content/10.1101/804831v1.

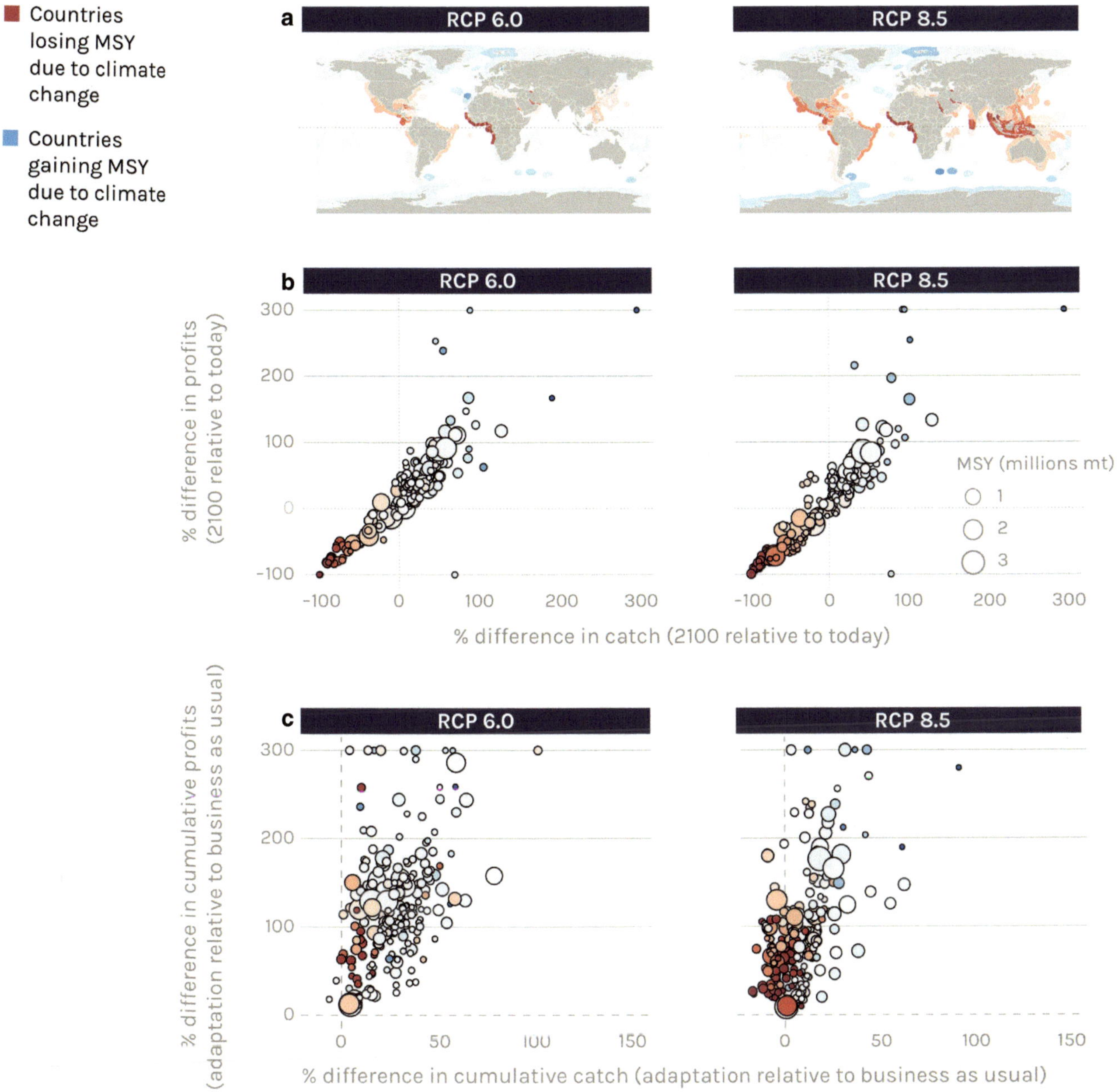

Fig. 2.1 Ability for adaptive fisheries management to mitigate impacts of climate change. *Notes:* (**a**) *shows that maximum sustainable yield (MSY) is forecast to decrease in equatorial exclusive economic zones (EEZs) and increase in poleward EEZs through 2100.* (**b**) *shows that adaptive management results in higher catch and profits in 2100 relative to today for many, but not all, EEZs despite climate change.* (**c**) *shows that adaptive management nearly always yields more cumulative profits than business-as-usual management and frequently yields more cumulative catches than business-as-usual management. In all panels, deeper reds show countries losing MSY and deeper blues show countries gaining MSY under climate change.* (Source: Adapted from Free et al. 2019b)

Implications for adaptation: Fisheries management that accounts for shifts in species distributions and productivities due to climate change will generate better outcomes than business-as-usual management in all countries, even those hardest hit by climate change.

Challenges for improving management include the lack of financial and technical capacity for monitoring and evaluating fisheries in many regions of the world, both for small-scale and industrial fisheries, and the conflicts emerging in fisheries due to climate change and other drivers (Spijkers et al. 2019). In the next section, we detail five key opportunities for action for implementing such reforms.

3.1.4 Opportunities for action and key conclusions

Building a socioecological system that is resilient to climate change is key to ensuring healthy, productive fisheries in the future. Below are five overarching, high-priority opportunities for designing fisheries management approaches in the context of a changing climate along a gradient of scientific, management and enforcement capacities:

1. **Implement best practices in fisheries management.** Historically, well-managed fisheries have been among the most resilient to climate change (Free et al. 2019a), and our results predict that well-intended, albeit imperfect, management will continue to confer climate resilience. Together, these results indicate that the wider implementation of best practices in fisheries management will mitigate many of the negative impacts of climate change.

   In higher-capacity systems, best practices include scientifically informed catch limits, accountability measures, regional flexibility in policy practices and the protection of essential fish habitats (Miller et al. 2018b). In the United States, such measures have contributed to dramatic declines in overfishing, increases in biomass and maintenance of catch and profits (NOAA 2018).

   In lower-capacity systems, best practices include implementing 'primary fisheries management' (Cochrane et al. 2011)—which uses the best available science and precautionary principles to manage data-poor and capacity-limited fisheries—and establishing local, rights-based management (Ojea et al. 2017) to incentivise sustainable stewardship.

   Rights-based management systems include catch share programmes, such as Individual Transferable Quotas (ITQs) and Territorial Use Rights in Fisheries (TURFs), which define property rights over catch and space, respectively (Costello et al. 2010). By giving users ownership of the resource, well-designed, rights-based management systems incentivise long term stewardship and have been shown to promote compliance, prevent overfishing and increase profits (Costello et al. 2016; Costello et al. 2008; Melnychuk et al. 2011). Enforcement and the strength of fishing pressure limits are also key for successful fisheries management (Melnychuk et al. 2017) and contribute to a precautionary approach in the face of climate change. Overall, fisheries best practices confer ecological resilience by maintaining healthy stock sizes, age structures, and genetic diversity and building socio-economic resilience by providing a portfolio of options to fishers and a buffer against climate-driven losses in any one target stock.
2. **Be dynamic, flexible and forward-looking.** Adapting to climate change will require dynamic, flexible and forward-looking management. This can be achieved by aligning management policies with the spatiotemporal scales of climate change, ecosystem change and socio-economic responses (Holsman et al. 2019). In higher-capacity systems, this could involve four broad strategies. First, managers can envision and prepare for alternative futures using tools such as forecasts (Hobday et al. 2016), structured scenario planning (Moore et al. 2013), holistic ecosystem models (Gaichas et al. 2016), risk assessments (Holsman et al. 2017) and climate vulnerability analyses (Hare et al. 2016). Second, the proliferation of near real-time biological, oceanographic, social and/or economic data can be harnessed for proactive and dynamic adjustments in spatial and temporal management actions (Hazen et al. 2018).

   Third, developing harvest control rules that account for or are robust to changing environmental conditions affecting productivity can increase catch while reducing the probability of overfishing (Tommasi et al. 2017). Finally, all of these management procedures should be simulation tested through management strategy evaluations (Punt et al. 2016) to measure the efficacy of alternative strategies and their robustness under different climate scenarios (Punt et al. 2014).

   In lower-capacity systems, forward-looking fisheries management could include precautionary management to buffer against uncertainty (Richards and Maguire 1998) as well as management strategies that preserve a population's resilience, age structure and genetic diversity. For example, size limits, seasonal closures and protected areas can be used to protect the big, old, fecund females (BOFFs) that disproportionately contribute to reproductive output (Hixon et al. 2014) and to maintain the genetic diversity required to promote evolutionary adaptations to climate change.
3. **Establish and strengthen international institutions and agreements to better manage stocks shifting in and out of jurisdictions**. **Shifting** distributions are already generating management challenges and the rates of these shifts and associated conflicts are expected to increase with climate change (Pinsky et al. 2018; Spijkers and Boonstra 2017; Spijkers et al. 2019). New or strengthened international institutions and agreements will be necessary to ensure that management remains sustainable as stocks shift between jurisdictions.

   First, this will require sharing data between regional fisheries management organisations or countries to identify, describe and forecast shifting stocks. Second, it will require a commitment to using these shared data to inform collaborative management. For example, these data could be used to regularly and objectively update national allocations of catch or effort based on changes in distribution rather than historical allocations (e.g. Havice 2013; Aqorau et al. 2018). An alternative approach could be to develop fisheries permits that are tradeable across political boundaries, which would provide future resource users access to fisheries not yet in their waters and incentivise good management (Serdy 2016). Finally, incentivising the cooperation necessary to establish data sharing and collaborative management will require overcoming prevailing management mentalities that one party 'wins' while the other 'loses' when stocks shift across boundaries. This could involve broadening negotiations to allow for alternative avenues of compensation or 'side payments' (Miller and Munro 2004). In cases where establishing international cooperation proves difficult, marine protected areas (MPAs) placed along country borders could buy time for negotiations by protecting stocks as they shift across borders (Roberts et al. 2017). A more precautionary approach would be to put new fishing areas on hold until adaptive management can be put in place, as illustrated by the Central Arctic Ocean Fisheries Agreement (Schatz et al. 2019).

4. **Build socioeconomic resilience.** The impact of climate change on fishing communities can be reduced through measures that increase socioeconomic resilience and adaptive capacity to environmental variability and changing fisheries (Cinner et al. 2018; Charles 2012; Fedele et al. 2019). Across low- to high-capacity systems, these measures include policies that do the following:
   (a) facilitate flexibility, such as by supporting access to multiple fisheries and alternative livelihoods
   (b) provide better assets, such as the enhancement of fisheries technology and capacity
   (c) provide better organisation in the system, including through multilevel governance, community-based management and other governance structures (Holsman et al. 2019; Ojea et al. 2017)
   (d) promote agency and learning (Cinner et al. 2018)

   For example, policies that promote access to multiple fisheries provide fishers with a portfolio of fishing opportunities that can buffer against variability (Kasperski and Holland 2013; Cline et al. 2017), while policies that help diversify livelihoods reduce reliance on fisheries (Cinner et al. 2009; Daw et al. 2012). Increased mobility through technological enhancements can increase social resilience by allowing fishers to follow shifting stocks (Cinner et al. 2018), but can also result in the migration of fishers. Multilevel governance promotes flexibility in resource governance by matching ecological resilience and management across scales (Hughes et al. 2005).

   Community-based management can increase adaptive capacity by incorporating local knowledge and can improve sustainability by fostering a sense of stewardship (Gutiérrez et al. 2011). Spatial rights-based approaches such as TURFs may confer social resilience insofar as they are often community managed and allow fishers to generate revenues through other compatible activities such as tourism, recreation and aquaculture (Moreno and Revenga 2014). On the other hand, ITQs may confer a different kind of resilience because rights are defined over fish catches, not spatial areas, so they may be more resilient to range shifts arising from climate change. Furthermore, all of these measures can be designed to reduce fishing pressure and promote ecological resilience to climate change.

5. **Use principles of fairness and equity to drive policy decisions.** The challenges of maintaining fairness and equity, such as adequately including the representation and needs of vulnerable marine livelihoods (i.e. those of women, migrants, indigenous peoples), are likely to be created or amplified by climate change. For example, on a regional level, we expect to see greater impacts in the equatorial region, which could exacerbate existing patterns of food insecurity and poverty. In the case of more informal or unregulated economies and fishing activities (e.g. shellfish gathering, fish processing), which are most times performed by women (Harper et al. 2017) and marginalised groups (Barange et al. 2018), there is a risk to being left out from regulations, leading to maladaptation.

At a more local level, climate change can shift the distribution of resources, thereby changing the impact on human populations from past patterns. Without an adequate response, these impacts could lead to inequalities, unrest and severe social disruption, thus likely worsening outcomes in the face of climate change. Addressing the inequities created by climate change is valuable in its own right to stem these potential negative consequences and deliver increased social resilience and stability. At the same time, using fairness and equity to guide policies can also help foster important buy-in to policies necessary for addressing climate change effects so that adoption is swifter and more complete. Finally, developing equitable solutions can help uncover and target the underlying drivers of both existing inequities and climate change itself, thereby allowing for wholesale system transformation when it is necessary to create equitable resilience (Cohen et al. 2019; Matin et al. 2018; Meerow et al. 2019; Mikulewicz 2019). Thus, equity is not just a valuable goal of management and policy reform; it is also a critical input into these decisions as it serves as a functional driver of climate resilience.

3.2 Marine Aquaculture

3.2.1 Importance of Mariculture to the Ocean Economy

Aquaculture, the cultivation of aquatic animals and plants, is one of the fastest-growing industries in the world and now produces more seafood than wild capture fisheries (FAO 2018). Although marine aquaculture, hereafter called 'mariculture', currently represents only one-third of total aquaculture production (freshwater/inland aquaculture represents the remainder), this proportion is increasing. In 2016, mariculture produced 38.6 mmt of seafood worth $67.4 billion at first sale. Over half of this production was shelled molluscs (58.8%), while finfish and crustaceans represented 23% and 17% of production, respectively (FAO 2018). When converted to edible food equivalents, finfish mariculture provides the most food by volume (Edwards et al. 2019). Additionally, fed aquaculture (including finfish and crustaceans), which requires feed inputs, is growing faster than unfed bivalve aquaculture due to increasing demand for these commodities (Tacon et al. 2011; Hasan 2017).

3.2.2 Impacts of climate change on mariculture

Mariculture production is vulnerable to climate change through impacts both on the cultivated organisms as well as on the cost and infrastructure of conducting mariculture operations. Like wild marine species, cultivated marine species are impacted by changing environmental conditions (Weatherdon et al. 2016), but unlike wild species, humans can induce accelerated adaptation in cultivated species through selective breeding (Sae-Lim et al. 2017). Unlike most wild capture fisheries, mariculture operations require a significant amount of shore- and ocean-based infrastructure for cultivating marine species through multiple life stages. Both shore- and ocean based infrastructure are vulnerable to storms, which are expected to increase in frequency and intensity under climate change (IPCC 2019), and ocean-based infrastructure such as lines, cages and pens must be actively moved in response to poor environmental conditions such as harmful algal blooms, hypoxia, or changing salinity or temperature, which increases costs and disproportionately impacts farmers unable to relocate (Dabbadie et al. 2018). As with capture fisheries, the impacts of climate change on aquaculture are expected to vary by location, species and method of production (Soto et al. 2018). The primary threats to unfed bivalve aquaculture and fed finfish and crustacean aquaculture are the following:

1. **Ocean warming** is expected to raise mortality rates and lower productivity for higher-trophic-level species (bivalves, finfish, crustaceans) (Rosa et al. 2014).
2. **Sea level rise** will increase the intrusion of saline water into deltas and estuaries compromising brackish-water aquaculture (De Silva 2012; Garai 2014), and shifting shoreline morphology could reduce habitat availability (bivalves, finfish, crustaceans).
3. **Increasing storm strength and frequency** pose risks to infrastructure (De Silva 2012), and increased weather variability has been associated with lower profits (bivalves, finfish, crustaceans) (Li et al. 2014).
4. **Ocean acidification** impedes the calcification of mollusc shells (Gazeau et al. 2013) resulting in reduced recruitment, higher mortality (Barton et al. 2012; Green et al. 2013) and increased vulnerability to disease and parasites (bivalves).
5. **Increasing rainfall** will raise the turbidity and nutrient loading of rivers, potentially causing more harmful algal blooms (HABs) that reduce production and threaten human health (bivalves, finfish, crustaceans) (Himes-Cornell et al. 2013; Rosa et al. 2014).
6. **The emergence, translocation and virulence of disease, pathogens and parasites** are impacted by climate change. For example, warming can increase susceptibility to disease, promote the influx of new pathogens (Rowley et al. 2014) and increase the toxicity of common pollutants (bivalves, finfish, crustaceans) (Fabbri and Dinelli 2014).
7. **Reduced feed availability** resulting from climate change and/or overfishing could challenge the growth potential for fed aquaculture (finfish, crustaceans) (Froehlich et al. 2018a).

3.2.3 Potential for mariculture production to grow under climate change

While marine capture fisheries production has stagnated over the past three decades, mariculture production has expanded rapidly, and is likely to become the source of new seafood production as the human population and demand for seafood grow (FAO 2018). However, the extent to which climate change could impede the ability for sustainable mariculture to meet growing food demand is unknown (IPCC 2019). **Although there are no global-scale estimates of how climate change is likely to impact mariculture profitability and productivity, four recent studies collectively suggest that the potential for sustainable and profitable mariculture is likely to remain high under climate change.**

First, Gentry et al. (2017) mapped the biological potential for mariculture and estimated that bivalve and finfish mariculture could respectively generate 767.7 mmt and 15.6 billion mt of production per year (>700 times more production than today). Second, the Blue Paper *The Future of Food from the Sea* (Costello et al. 2019) refined this analysis to account for economic feasibility and the limited availability of feed for fed finfish aquaculture, and estimated that bivalve and finfish mariculture could respectively generate 483.0 mmt and 10.5 mmt of production per year under current prices and feed compositions (~21 times more production than today). Third, Froehlich et al. (2018b) forecasted mariculture production potential under a high emissions scenario (RCP 8.5) and found only slight declines in suitable habitat and production potential across continents.

Finally, Klinger et al. (2017) suggest that breeding a larger proportion of mariculture stocks for fast growth could, on its own, more than offset the forecasted declines in productivity. In the remainder of this Sect. 3.2.3, we provide a brief overview of this chain of evidence.

1. **Enormous areas of the ocean are suitable for bivalve and finfish mariculture and the vast majority of countries would need to farm less than 1% of their exclusive economic zones to match current levels of seafood consumption.** Gentry et al. (2017) mapped the biological production potential for finfish and bivalve mariculture based on the growth potential of 180 mariculture species (120 finfish, 60 bivalves) constrained by their temperatures, dissolved oxygen levels, primary production tolerances and existing human uses (i.e. protected areas, shipping lanes and oil rigs). Overall, they estimated an enormous untapped potential for mariculture: bivalve and finfish mariculture could generate 767.7 mmt (over 2.5 million square kilometres [km^2] of suitable habitat) and 15.6 billion mt per year (over 11.4 million km^2), respectively. By comparison, bivalve and finfish mariculture currently produce only 15.3 and 7.7 mmt per year, respectively (FAO 2018). However, their analysis did not consider the economic feasibility of this production or the limited availability of feed for fed mariculture.
2. **Current mariculture production is far under capacity even after accounting for economic feasibility and limited feed availability. Advancements in feed technology would dramatically expand the production potential of finfish mariculture.** In their Blue Paper, Costello et al. (2019) refined the Gentry et al. (2017) analysis by calculating the cost and feed demand of their production estimates and assuming that mariculture production will occur only in profitable areas and that finfish mariculture production is capped by feed availability. They show that global- and country-level mariculture production is significantly under capacity. Bivalve production of 483.0 mmt should be possible at today's prices for maricultured bivalves ($1400 per mt of blue mussels). This is 467.7 mmt (>3000%) more than the current production of 15.3 mmt. Additionally, 10.5 mmt of finfish production should be possible at today's prices for maricultured finfish ($7000 per mt of Atlantic salmon) and today's feed composition. This is 2.8 mmt (36%) more than the current production of 7.7 mmt. However, technological advances resulting in a 95% reduction in the reliance of feed on fish ingredients (Oliva-Teles et al. 2015) would unlock a 209.6 mmt (>2700%) increase in finfish production to 217.3 mmt. The majority of these underages in mariculture production occur in equatorial countries (Fig. 2.2 on the following page), suggesting that mariculture expansion could mitigate the losses in capture fisheries productivity expected for these regions, potentially offsetting some of the inequities associated with these climate change impacts. Furthermore, mariculture operations can provide a critical source of jobs and income to local communities, especially to vulnerable groups such as unskilled workers (Irz et al. 2007) who might otherwise be made significantly worse off by climate change.
3. **Although climate change is expected to reduce mariculture production potential, the magnitude of this reduction is small relative to the sheer potential for production.** Froehlich et al. (2018b) extended the work of Gentry et al. (2017) to predict how finfish and bivalve mariculture will change from now to 2090 under the warming, acidification and primary productivity shifts associated with a high emissions scenario (RCP 8.5). They forecast a global increase in the suitable habitat available for finfish mariculture, particularly in polar and subpolar regions. Conversely, they forecast a global decrease in the suitable habitat available for bivalve mariculture due to the negative impact of ocean acidification. In both sectors, the growth and production potential of the suitable habitat decreases over time. As a result, global mariculture production is likely to decline by mid-century, with the greatest certainty around bivalve declines. However, the relevance of these declines is unclear, because Froehlich et al. (2018b) do not publish the nominal production potential (i.e. metric tonnes of food) for 2090. Even if climate change reduced the 495.5 mmt of mariculture production estimated to be economically feasible with today's feed technology (Costello et al. 2019) by 90%, mariculture would still be 28% more productive than it is today (49.4 mmt versus 38.6 mmt).
4. **Breeding a larger proportion of mariculture stocks for fast growth could more than offset the negative impacts of climate change on mariculture production potential.** Klinger et al. (2017) mapped the production potential of three important finfish mariculture species—Atlantic salmon (*Salmo salar*), gilthead seabream (*Sparus aurata*) and cobia (*Rachycentron canadum*)—under a

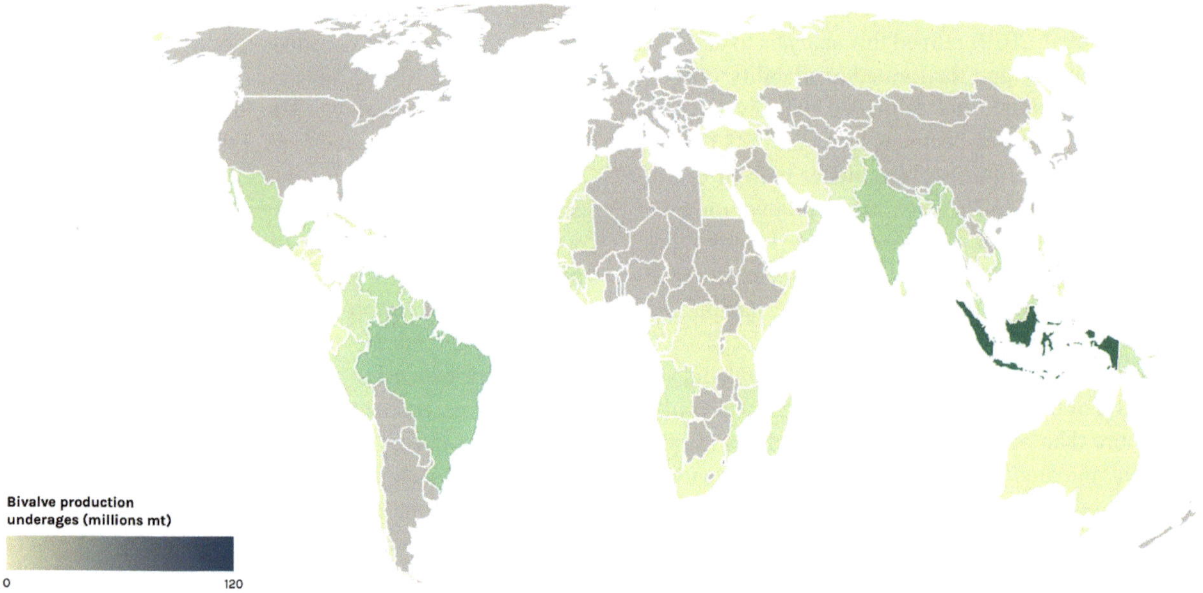

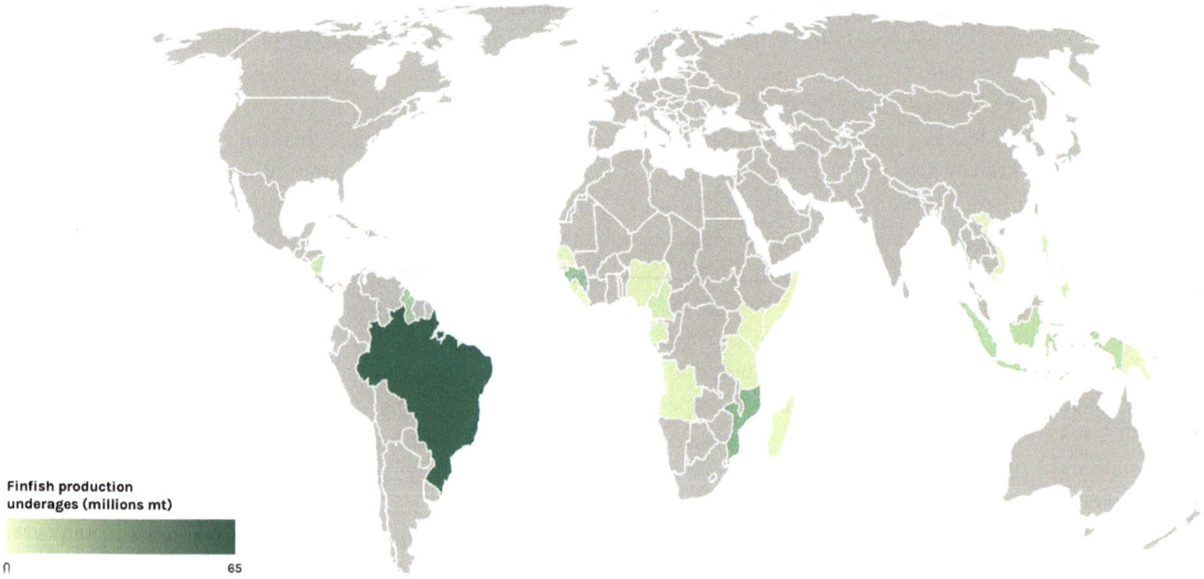

Fig. 2.2 Mariculture production underages for bivalves and finfish. *Notes: Mariculture production underages for bivalves at current prices ($1700/mt for blue mussels) (top map) and finfish at current prices ($7000/mt for Atlantic salmon) with a 95% reduction in the reliance of feed on fish ingredients (bottom map).* (Source: Adapted from Costello et al. 2019)

high emissions scenario (Geophysical Fluid Dynamics Laboratory Climate Model version 2.5 and estimated that increases in annual growth rates of 25–41% would be required to offset warming-induced declines in annual growth rates. They found that selective breeding programmes for faster growth in these species would increase growth rates by 10–15% per generation or by 100–200% over multiple generations—more than enough to offset the negative impacts of climate change. Given that only 10% of global mariculture production is currently derived from selectively bred stocks (Gjedrem et al. 2012), breeding a larger proportion of stocks for fast growth could, on its own, offset the negative impacts of climate change predicted by Froehlich et al. (2018b).

Although these four studies collectively present a chain of evidence to suggest that mariculture potential will remain high under climate change, they do not consider the social

(Froehlich et al. 2017), regulatory (Abate et al. 2016) or capacity barriers to mariculture development (Gentry et al. 2019); the challenges posed by climate-driven increases in HABs, diseases and storm frequency (IPCC 2019); or the environmental impacts of mariculture (Clavelle et al. 2019). In the next Sect. 3.2.4, we detail these challenges.

3.2.4 Barriers and Trade-Offs in the Expansion of Mariculture

If the potential for mariculture production is so large, why is current production so low? This gap is likely driven by two factors: a lack of expertise and capacity for conducting mariculture operations in many developing countries; and challenging regulatory barriers for developing mariculture operations in many developed countries. First, countries with low or crashed mariculture production exhibit lower GDPs and business friendliness scores than countries with stable or increasing mariculture production (Gentry et al. 2019). In Palau, for example, many mariculture operations have been initiated with outside funding but failed once the initial funding period ended. The longest-running mariculture operation in Palau is a government subsidised clam hatchery that would be unprofitable without government support (Y. Golbuu, personal communication). Second, countries with stricter environmental regulations have exhibited lower production and production growth than countries with more lenient regulations (Abate et al. 2016). For example, despite having one of the largest EEZs and longest coastlines, the United States produces only 1% of global mariculture (FAO 2018) due to precautionary regulations on mariculture zoning (Wardle 2017; Sea Grant California et al. 2019).

Mariculture operations can also pose a risk to marine ecosystems and the wild capture fisheries supported by these ecosystems (Clavelle et al. 2019). They can degrade habitats (Richards and Friess 2016), reduce water quality (Price et al. 2015), spread disease (Lafferty et al. 2015), hybridise with wild species (Lind et al. 2012) and introduce invasive species (Diana 2009). The expansion of mariculture should depend on adopting best practices for preventing or reducing these impacts (Klinger and Naylor 2012) including by doing the following:

1. using marine spatial planning to site mariculture in productive and profitable areas that minimise impacts on ecosystems
2. conducting offshore or integrated multitrophic mariculture to reduce eutrophication risk
3. expanding unfed bivalve mariculture, which has lower environmental impacts compared with fed finfish mariculture

See the Blue Paper *The Future of Food from the Sea* (Costello et al. 2019) for more details regarding the ecosystem impacts of mariculture and the opportunities for adaptation to reduce these impacts.

3.2.5 Adapting marine aquaculture to climate change

Selective Breeding for Fast Growth

Although selective breeding—the breeding of cultivated plants and animals to inherit specific traits—has historically been implemented less in aquaculture than in terrestrial farming (Gjedrem et al. 2012), aquaculture species are increasingly being bred to increase productivity and disease resistance (Gjedrem and Baranski 2009). The majority of breeding programmes have focused on increasing growth rates and maximising productivity and have been met with success. For example, Atlantic salmon breeding programmes have increased harvest weight by 12% per generation with cumulative genetic gains of ~200% over multiple generations (Janssen et al. 2016). Similarly, seabream breeding programmes have increased harvest weight by 10–15% per generation with cumulative genetic gains of ~100% over multiple generations (Janssen et al. 2016). These cumulative gains exceed the 25–41% total increase in annual growth rate thought to be necessary to offset the most extreme climate-induced decreases in mariculture productivity (Klinger et al. 2017); thus, selective breeding for fast growth rates alone could be sufficient to offset many of the negative impacts of climate change on mariculture.

Selective Breeding for Temperature Tolerance

Selective breeding for fast growth rates at elevated temperatures could further offset the impacts of climate change on mariculture but has yet to be widely implemented (Gjedrem et al. 2012) and has been met with mixed success (Gjedrem and Baranski 2009; Sae-Lim et al. 2015). Some selective breeding programmes have successfully resulted in increased temperature tolerances (Sae-Lim et al. 2017), but these breeding programmes can be costly (Ponzoni et al. 2008; Gjedrem et al. 2012). Furthermore, the use of selectively bred fish can pose risks to wild populations and ecosystems (Lind et al. 2012). Cultured fish frequently escape from aquaculture facilities (Jensen et al. 2010) and can interbreed with wild fish, leading to reduced genetic variability and a reduction in fitness in wild populations (Hutchings and Fraser 2008). However, in tropical countries where wild populations are projected to diminish (Lotze et al. 2019), this risk may be inherently reduced or deemed acceptable under climate change.

Risk-Based Planning and Environmental Monitoring Systems

The siting of mariculture farms based on risk-based zoning coupled with the active monitoring and responsive relocation of pens, cages and lines could help to minimise the impacts of both climate change and climate variability on mariculture production potential (Soto et al. 2018). To date, most mariculture site selections have been ad hoc, but a growing number of national and regional authorities are beginning to plan mariculture zoning using risk analysis (Aguilar-Manjarrez et al. 2017; Xinhua et al. 2017; Lester et al. 2018; Sainz et al. 2019). After siting mariculture farms in locations forecast to experience low climate risk, environmental monitoring systems could be used to track changes in environmental conditions, provide early warnings about oncoming environmental risks (e.g. HABs) and give farmers the opportunity to prepare for adverse conditions or relocate cages, pens and lines if logistically feasible (Soto et al. 2018).

Access to Affordable Credit And Insurance

Policies that increase mariculture farmers' access to credit and insurance options will also help promote the development and expansion of mariculture in the face of climate change (Soto et al. 2018). Access to affordable credit is necessary for funding both the upfront capital costs of establishing a mariculture farm as well as the annual operating costs required to adapt to or recover from climate-induced stressors (Karim et al. 2014). Increased access could be promoted through microfinance schemes or loan guarantee funds (Soto et al. 2018). Similarly, increasing storm frequency and intensity will necessitate providing more insurance options for mariculture farmers. Pilot programmes in China and Vietnam indicate that insuring small-scale farms, which are particularly vulnerable and also major contributors to food security, is a profitable investment (Nguyen and Pongthanapanich 2016; Xinhua et al. 2017).

The expansion of mariculture depends on it becoming a more efficient and lower-risk business endeavour and the insurance-pooled model used in these pilot programmes has helped raise production efficiencies while reducing production and market risks.

Reducing Feed Limitations for Fed Mariculture

Innovations in feed technology could greatly enhance the potential for fed mariculture (Costello et al. 2019; Froehlich et al. 2018a) and increase the opportunities for production under climate change. The amount of feed available for mariculture can be increased through a variety of mechanisms including the following:

1. ending over- and underfishing of the forage fish fisheries targeted for the production of fish meal (FM) and fish oil (FO) from whole fish (Froehlich et al. 2018a)
2. processing a larger proportion of landings for trimmings and diverting these by-products to the production of FM and FO (Jackson and Newton 2016)
3. reducing the amount of FM and FO used in the diets of non-carnivorous aquaculture species such as carp and other freshwater fishes, and terrestrially farmed species such as pigs and chickens (Froehlich et al. 2018a)
4. replacing fish ingredients with alternative sources of protein
5. increasing feed conversion rates

3.2.6 Opportunities for action and key conclusions

1. **Mariculture can provide food and income in countries losing access to capture fisheries.** Current mariculture production is far below potential production in many countries and the continued development of mariculture could provide food and employment in countries with climate-driven declines in capture fisheries.
2. **Expanding mariculture will require preventing, reducing and accepting the environmental trade-offs of mariculture.** Mariculture poses risks to marine ecosystems and capture fisheries and its expansion has frequently been impeded by these concerns. Expanding mariculture will depend on preventing and reducing these risks and establishing clear best practices that will help ease the regulatory burden.
3. **Finfish mariculture could generate more food and income through advancements in feed technology.** The production potential of finfish mariculture is challenged by the availability of fishmeal and fish oil from capture fisheries. Optimally managing forage fisheries, processing by-products for FM and FO, removing FM and FO from the diets of non-carnivorous fish and terrestrially farmed animals and replacing fish ingredients with alternative sources of protein would increase the viability of finfish mariculture.
4. **Mariculture species should be selectively bred for fast growth and robustness to climate change.** Despite the advantages of selective breeding, only 10% of global mariculture production is currently derived from selectively bred stocks (Gjedrem et al. 2012). Breeding a larger proportion of aquaculture stocks for fast growth could, on its own, offset the negative impacts of climate change on mariculture (Klinger et al. 2017). However, this will also necessitate increased efforts to reduce escapement, minimise pollution and mitigate other potential negative environmental impacts of mariculture.
5. **Increase access to financial services such as credit and insurance.** Mariculture is expected to become more expensive and riskier under climate change; increased access to credit and insurance for mariculture farmers will be necessary to assist with these costs and risks.
6. **Siting mariculture farms in low-risk areas and actively monitoring and responding to changing environmental conditions can enhance resilience to climate change.**

3.3 Marine and Coastal Tourism

3.3.1 Importance of marine tourism to the ocean economy

Marine and coastal tourism, referred to collectively as ocean tourism in this report, was the second-largest ocean-related economic sector in 2010, next to offshore oil and gas (OECD 2016). Ocean tourism is projected to be the top contributor of ocean industries by 2030 in terms of production value, when it will account for 26% of the ocean-based economy, compared with 21% for oil and gas (OECD 2016). Ocean tourism dwarfs the contribution of industrial capture fisheries, which constitute only 1% of ocean-based industries' production value (not accounting for artisanal fisheries, which are a critical component of the economies of Asia and Africa). The range of ocean tourism activities include beach tourism, recreational fishing, swimming, snorkelling, diving, whale watching, and taking cruises, among others. Ocean tourism's global direct value added was estimated at $390 billion in 2010, directly providing seven million full-time jobs. In addition, the ocean is a source of recreation for millions of people in the developed and developing worlds (Ghermandi and Nunes 2013; Arlinghaus et al. 2019). For comparison, the global value added of industrial capture fisheries was $21 billion in 2010 (OECD 2016), providing 11 million full-time jobs (artisanal fisheries not included).

Ocean tourism directly supports the livelihoods of millions of people and the economies of the developing tropics and many small island developing states. For example, coral reef tourism alone contributes over 40% of the gross domestic products of Maldives, Palau and St. Barthélemy (Spalding et al. 2017; Siegel et al. 2019). Despite the importance of ocean tourism in the economy, data and research on the impacts of climate change in the tourism sector are limited (Scott et al. 2012). Because coral reef tourism is one of the best-studied sectors (Scott et al. 2012), and potentially one of the most valuable ocean tourism options for many coastal nations, we focus our analysis on this sector.

Coral reef tourism is worth $35.8 billion globally every year (Spalding et al. 2017). We present a first- of-its-kind analysis of how climate change will affect coral reef tourism values at a country/territory level and explore options for nations and local communities to best prepare for the impacts of climate change.

3.3.2 Impacts of climate change on marine tourism

Weather conditions and attractiveness/uniqueness of the environment are key factors drawing people to ocean tourism (Moreno and Amelung 2009), and climate change impacts both. Understanding the potential impacts of climate change on tourism requires understanding how climate change will impact the physical and ecological resources on which tourism depends.

Marine heatwaves, or periods of extremely high ocean temperatures, have affected marine organisms and ecosystems (e.g. fisheries, coral reefs) in the last two decades and are expected to increase in frequency, intensity, duration and spatial extent (IPCC 2019). Marine heatwaves have critical impacts over habitat formation species (e.g. seagrasses, corals, kelps) that can disrupt the provision of ecosystem services (Smale et al. 2019). Future ocean warming will increase the frequency, intensity and spatial extent of bleaching events (Donner et al. 2005; IPCC 2019) that cause coral reef mortality (e.g. Arceo et al. 2001) and a subsequent reduction in reef fish diversity and numbers (e.g. Graham et al. 2007) that on-reef tourism depends on. Storms and storm surges are also expected to increase in intensity and become more frequent (IPCC 2019), causing a reduction in the desirability of a place for tourism, disrupting transportation (flights and ferries), and potentially destroying the coastal infrastructure that supports tourism. Sea level rise impacts coastal integrity and coastal assets and, together with extreme events, causes coastal erosion that, if constrained by urbanisation, can lead to coastal squeeze (Toimil et al. 2018; Scott et al. 2012). This has a known negative impact on visitors' perceptions and associated economic impacts (Scott et al. 2012). Ocean warming also affects fisheries productivity (Free et al. 2019a) and the migration patterns of species that are major draws for tourism (e.g. whales, sharks, turtles) (e.g. Lambert et al. 2010).

Climate change interacts with coral reef tourism through its direct impact on the following:

1. coral reefs and associated species on which some reef tourism directly depends (e.g. snorkelling, diving, recreational fishing)
2. weather conditions that drive a user's preference for the place
3. coastal infrastructure that supports tourism

For ocean tourism that directly depends on healthy coral reef ecosystems, such as diving and snorkelling (on-reef tourism), changes in reef conditions are expected to impact tourists' preferences and coral reef tourism's economic values. While activities that do not directly depend on reefs (i.e. reef-adjacent activities such as white sand beaches and sunbathing) are also expected to be affected by climate change (directly and indirectly through processes such as the wave attenuation role of reefs and coral reefs as a source of white sand), the magnitude of the impact is hard to measure.

3.3.3 Economic Impacts

Economic Impacts on Coral Reef Tourism

We use the coral reef tourism values per country and territory reported by Spalding et al. (2017) to represent current coral reef tourism values. These values are composed of on-reef and reef-adjacent tourism values.

Chen et al. (2015) performed a meta-analysis of how climate change impacts, in the form of changes in sea surface temperature (SST) and ocean acidification (using atmospheric CO_2 levels as a proxy), have affected and will continue to affect coral reef health and coral reef tourism values at the regional and global levels. We used their model to project how changes in SST and ocean acidification will change coral cover at the country level and how these changes in reef conditions would translate to changes in tourism values.

We project per-country future tourism value changes (with 2019 as a baseline) using the SST and CO_2 projections for RCP 2.6, 4.5, 6.0 and 8.5 climate scenarios from the CMIP5 Coupled Model Intercomparison Project (Taylor et al. 2012). For this report, we present the results for 2100 only to be consistent with the fisheries and aquaculture projections. These are the model's assumptions about how ocean warming and acidification affect coral reef cover and tourism values:

SST effect

- When the annual mean SST is less than 22.37°, a 1% increase in SST leads to a 0.67% increase in live coral coverage (relative to the percent of live corals available prior to changes in temperature).
- When SST is between 22.37° and 26.85°, a 1% increase in SST leads to a 1.59% increase in live coral coverage.
- When SST is greater than 26.85°, a 1% increase in SST leads to a 2.26% decrease in live coral coverage.

Ocean acidification effect

- Using atmospheric CO_2 as a proxy (Table 2.2), a 1% increase in CO_2 decreases live coral coverage by 0.61%.

Effect of changes in coral cover to coral reef tourism values

- A 1% decline in coral cover decreases coral reef value by 3.81%. We limit the effect of climate change to on-reef tourism values only.

Other factors not accounted for in the model above are the effects of climate change–associated increases in ocean disturbances such as storms, mass bleaching events that cause extensive reef mortality (Donner 2009; Frieler et al. 2013; Hughes et al. 2017, 2018), heat waves (Smale et al. 2019), sea level rise (Gattuso et al. 2018), algal blooms, jellyfish blooms, climate change–related diseases (Sokolow 2009) and water and electricity supply disruptions (Weatherdon et al. 2016). Also important and not included is the confounding effect of local stressors such as nutrient pollution and illegal and destructive fishing, which negatively impact tourism values.

Table 2.2 Global atmospheric CO_2 concentrations (ppm) for different RCPs using CMIP5

Year\RCP	2.6	4.5	6.0	8.5
2019	409.80	408.88	407.40	412.82
2030	430.78	435.05	428.88	448.83
2050	442.70	486.54	477.67	540.54
2100	420.90	538.36	669.72	935.87

Source: Royal Netherlands Meteorological Institute. "Time Series, Annual RCP45 CO2." KNMI Climate Explorer. http://climexp.knmi.nl/getindices.cgi?WMO=CDIACData/RCP45_CO2&STATION=RCP45_CO2&TYPE=i&id=someone@somewhere&NPERYEAR=1

Notes: PPM stands for parts per million, RCP for Representative Concentration Pathway and CMIP5 for Coupled Model Intercomparison Project 5

Nutrient enrichment has been shown to increase the susceptibility of coral reefs to bleaching (Wiedenmann et al. 2013), increase the severity of coral diseases (Bruno et al. 2003) and increase the vulnerability of coral reefs to ocean acidification (Silbiger et al. 2018). Furthermore, the poleward movement (Price et al. 2019), potential thermal evolution/adaptation (Speers et al. 2016; Donner 2009) and species-specific responses of corals (Fabricius et al. 2011) are not accounted for in our projections. All these additional climate change–induced stressors and the confounding effect of local stressors impact local and national economies (Hoegh-Guldberg et al. 2018).

The combined effect of warming (SST) and ocean acidification as factors affecting coral reef cover and tourism values results in predictions of negative effects for all countries, with magnitudes dependent on the climate pathways (Fig. 2.3, Table 2.3).

For the high-emissions scenario of RCP 8.5, which is characterised by considerable increases in greenhouse gas emissions, coral cover is expected to be reduced by 72–87% (relative to the present coral cover) and on-reef tourism values by over 90% from 2019 to 2100 due to combined ocean warming and acidification. The reduction will be less severe under a stabilisation scenario of RCP 4.5 with an expected reduction of 12–28% and 36–66% in coral cover and on-reef tourism values, respectively. Note that the reduction in coral cover is still conservative as other factors such as bleaching events, storms and other climate stressors, which are expected to intensify and become more frequent, are not included in the model.

Brander et al. (2012) projects that ocean acidification will cause a 27.5% reduction in global coral cover by 2100 under RCP 8.5 (with 2000 as the baseline year). This value is in line with Chen et al. (2015), which our projections are based on,

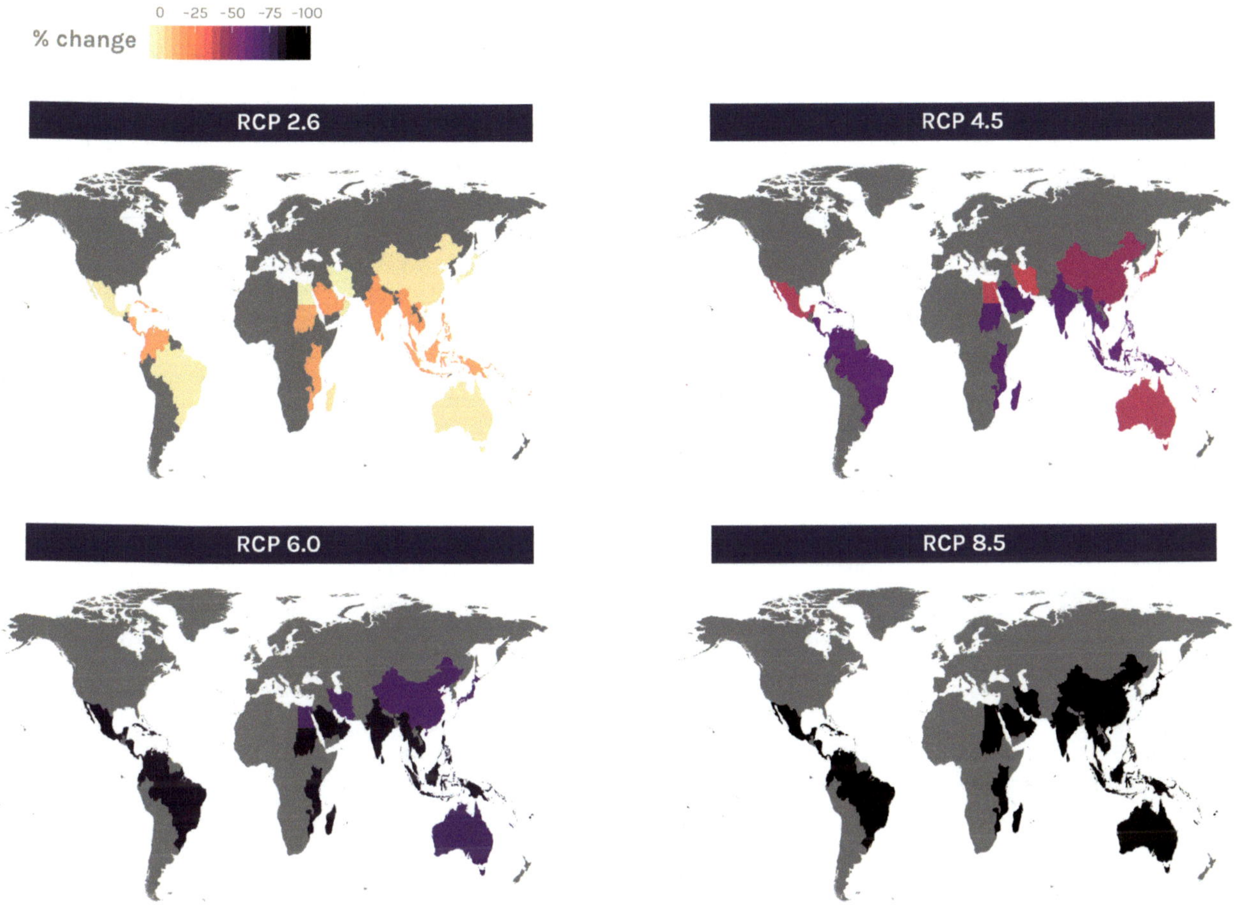

Fig. 2.3 Percent change in coral reef tourism values in 2100 for different climate projections. *Notes: Values in 2100 are relative to those in 2019. See* Table 2.3 *for country values*. (Source: Model adapted from Chen et al. 2015)

Table 2.3 Percent change in coral reef tourism values in 2100 for different climate projections, by country

Country	Total coral reef tourism value (US $1000 per year)	% on-reef	% change in coral cover (RCP 4.5)	% change in tourism values (RCP 4.5)	% change in coral cover (RCP 8.5)	% change in tourism values (RCP 8.5)
Egypt	6,917,028	86.3	−12.9	−39.4	−72.4	94.0
Indonesia	3,097,453	64.3	−25.2	−62.4	−81.7	−95.8
Mexico	2,999,883	44.8	−14.2	−42.4	−82.6	−96.0
Thailand	2,410,154	44.8	−25.4	−62.7	−81.8	−95.8
Australia	2,176,084	78.3	−14.1	−42.2	−73.1	−94.2
China	1,871,814	15.3	−16.2	−46.6	−71.8	−93.9
Philippines	1,385,144	67.4	−25.2	−62.4	−81.8	−95.8
Hawaii	1,230,891	44.8	−13.6	−41.1	−73.0	−94.1
Japan	1,177,549	53.9	−13.2	−40.2	−72.7	−94.1
Malaysia	1,148,955	64.3	−25.2	−62.4	−81.8	−95.8
Maldives	1,085,273	84.4	−25.9	−63.5	−82.2	−95.9
Puerto Rico	648,867	21.3	−26.4	−64.2	−82.1	−95.9
Brazil	612,864	8.3	−25.9	−63.4	−82.3	−95.9
Bahamas	526,058	60.5	−26.2	−63.9	−82.2	−95.9
Dominican Republic	511,669	26.5	−26.3	−64.0	−82.0	−95.9
India	464,082	15.3	−26.4	−64.1	−82.4	−95.9
Honduras	446,628	85.8	−26.0	−63.6	−82.1	−95.9
United Arab Emirates	445,654	15.3	−26.4	−64.2	−83.1	−96.0
Jamaica	333,386	35.1	−26.1	−63.8	−82.1	−95.9
Taiwan	323,440	15.3	−25.9	−63.5	−82.2	−95.9

(continued)

Table 2.3 (continued)

Country	Total coral reef tourism value (US $1000 per year)	% on-reef	% change in coral cover (RCP 4.5)	% change in tourism values (RCP 4.5)	% change in coral cover (RCP 8.5)	% change in tourism values (RCP 8.5)
Guam	323,244	64.3	−25.9	−63.4	−82.2	−95.9
Mauritius	312,389	47.4	−25.5	−62.8	−82.3	−95.9
Cayman Islands	292,794	83.2	−25.8	−63.4	−82.1	−95.9
Cuba	283,290	35.1	−25.8	−63.3	−82.0	−95.9
Venezuela	281,865	35.1	−26.2	−63.8	−82.0	−95.9
Virgin Islands of the United States	276,056	53.9	−26.3	−64.0	−82.0	−95.9
Saudi Arabia	268,681	49.7	−27.6	−65.8	−83.5	−96.1
Fiji	234,676	65.4	−24.9	−62.0	−81.9	−95.9
Bermuda	223,639	69.2	−13.3	−40.3	−73.3	−94.2
Oman	221,164	35.1	−27.2	−65.2	−83.1	−96.0
Aruba	218,226	35.1	−26.1	−63.7	−82.0	−95.9
Barbados	180,082	38.7	−26.0	−63.5	−81.9	−95.9
Costa Rica	169,518	35.1	−26.0	−63.6	−82.3	−95.9
Panama	154,178	38.7	−26.2	−63.9	−82.1	−95.9
Colombia	147,202	35.1	−26.3	−64.0	−82.3	−95.9
Vietnam	137,445	15.3	−25.6	−63.0	−81.9	−95.9
Tanzania	131,076	49.7	−26.3	−64.0	−82.5	−95.9
Kuwait	117,236	35.1	−12.4	−38.2	−71.7	−93.8
Bahrain	115,837	21.3	−11.5	−36.1	−86.6	−96.5
French Polynesia	113,657	63.1	−24.6	−61.6	−81.6	−95.8
Qatar	108,066	8.3	−25.8	−63.3	−83.2	−96.1
Turks and Caicos Islands	97,587	69.2	−26.4	−64.2	−82.3	−95.9
Palau	92,503	86.3	−25.1	−62.2	−81.9	−95.8
Guadeloupe	90,463	38.7	−26.3	−64.0	−81.9	−95.9
Martinique	89,337	35.1	−26.0	−63.6	−81.9	−95.9
Kenya	84,152	31.0	−26.4	−64.2	−82.6	−96.0
Sri Lanka	82,371	8.3	−26.0	−63.5	−82.2	−95.9
Belize	80,611	70.8	−25.7	−63.2	−81.3	−95.7
Seychelles	73,141	47.4	−26.1	−63.7	−82.5	−95.9
Mozambique	68,356	80.9	−26.6	−64.4	−82.5	−95.9
Northern Mariana Islands	61,302	73.0	−25.9	−63.4	−82.2	−95.9
Ecuador	58,883	60.5	−26.8	−64.7	−83.0	−96.0
Saint Lucia	56,574	41.9	−25.9	−63.4	−81.9	−95.8
Madagascar	50,496	47.4	−26.1	−63.8	−82.5	−95.9
Vanuatu	49,991	59.0	−24.7	−61.6	−82.0	−95.9
Papua New Guinea	32,024	73.0	−25.2	−62.4	−81.8	−95.8
Sudan	28,480	85.8	−27.1	−65.1	−83.0	−96.0
New Caledonia	28,465	57.4	−15.0	−44.2	−82.6	−96.0
Brunei	28,239	26.5	−24.9	−62.0	−81.8	−95.8
Grenada	23,150	53.9	−25.8	−63.4	−81.9	−95.8
Solomon Islands	21,984	79.5	−25.0	−62.1	−81.6	−95.8
Anguilla	19,685	41.9	−26.6	−64.4	−82.1	−95.9
Cook Islands	19,106	41.9	−25.0	−62.1	−81.6	−95.8
Cambodia	18,285	15.3	−25.6	−63.0	−81.9	−95.8
Micronesia	18,108	86.3	−25.3	−62.5	−81.9	−95.8
Haiti	15,206	31.0	−26.5	−64.3	−82.2	−95.9
Iran	13,345	0.0	−12.4	−38.2	−84.3	−96.2
Tonga	13,291	71.6	−15.2	−44.5	−82.1	−95.9
Samoa	12,490	31.0	−24.9	−62.0	−81.5	−95.8
Myanmar	11,581	51.9	−26.0	−63.6	−81.9	−95.8
Nicaragua	10,975	41.9	−25.9	−63.4	−81.9	−95.8

Source: Country-level tourism values data provided by M. Spalding. Model for change in coral cover adapted from Chen et al. (2015)

Notes: Climate change effect. Summary table for all countries and territories with over 50 square kilometres of reef, and total reef-related expenditures of more than $10 million per year. On-reef tourism value pertains to in-water activities such as diving, snorkelling and glass-bottom boats. Adjacent-reef tourism value captures a range of indirect benefits from coral reefs, including the provision of sandy beaches, sheltered water, seafood and attractive views which indicates a coral cover reduction of ~31% due to ocean acidification and ~28% due to ocean warming (for RCP 8.5). Our projections have not incorporated the effects of bleaching, which is expected to be more frequent in the future and can be a greater driver of coral mortality under climate change. Speers et al. (2016) modelled the effects of combined ocean warming, acidification and intensifying bleaching on changes in coral cover and projects that current global coral cover will be reduced by 92% by 2100 under RCP 8.5.

The top five countries with the highest coral reef tourism values are Egypt (~$7 billion/year), Indonesia (~$3.1 billion/year), Mexico (~$3 billion/year), Thailand (~$2.4 billion/year) and Australia (~$2.2 billion/year). These five countries have 45–86% of their coral reef tourism values based on on-reef activities (e.g. snorkelling and diving), and climate change impacts (ocean warming plus acidification) will reduce on-reef tourism values by over 90% in 2100 for RCP 8.5 (39–63% for RCP 4.5).

The projections above should be interpreted as the effect of climate change on future potential tourism values, holding all other factors equal. Our projections indicate that the degree of climate change impacts depends on the emissions pathways taken in the future, although any of the emissions scenarios would still negatively impact reef tourism values.

When most of a country's coral reef tourism value comes from reef-adjacent activities, climate change may not severely affect that country. The reef-adjacent values, however, will be affected by increased extreme weather events in the area, algal blooms and coastal erosion, which we have not yet incorporated into the current calculations.

We reported here how climate change impacts coral cover and the corresponding on-reef tourism values of several national economies. While the coral reef tourism values of all nations are projected to be negatively affected by climate change, nations can still incur positive tourism values in the future as our estimate has not accounted for increases in tourism demand and arrivals in the future—international arrivals are expected to increase 3–5% per year (UNWTO 2016; Lenzen et al. 2018). In accounting for the improvements in tourism values due to an increase in tourist arrivals, it should be noted that the tourism value is a hump shape, or concave function, of tourism arrivals. Additional arrivals increase tourism values up to some point after which the desirability of a place for tourism decreases as tourist numbers further increase. Future research can incorporate the Shared Socio-economic Pathways (SSPs) to future projections of tourism under climate change to account for not only ecosystem changes, but also changes in the demand for tourism.

Economic Impacts in Other Systems

Coral reef tourism is not the only tourism sector that will be impacted by climate change. Other non-reef coastal attractions such as the coastal glaciers in Ilulissat Icefjord, Denmark, a UNESCO World Heritage site, and places such as coastal cities like Venice, Italy (Moreno and Amelung 2009) or Alexandria, Egypt (Scott et al. 2012) will also be heavily affected by climate change. Beach tourism in tropical and temperate areas is expected to be significantly affected by climate change, especially due to the effect of sea level rise and storms on shoreline erosion (Scott et al. 2012). For example, in the Mexican Caribbean, the estimated total beach replenishment cost for the main five ocean tourism cities under a future 1 m sea level rise scenario is $330 million (Ruiz-Ramírez et al. 2019), and in the United States, the total beach nourishment cost for 2060 based a 0.32 m scenario amounts to $20.40 billion (Scott et al. 2012). The breaking of ice in the polar region also poses potential danger to cruise ships and navigation. For all these systems, 'last chance tourism' is emerging, attracting people to the most vulnerable areas (IPCC 2019).

Consequences need to be further explored to understand the implications and dimensions of this trend.

Quantifying the impacts of climate change on other ocean tourism activities and beyond will provide a more complete picture of the impacts of climate change on local and national economies, which could potentially motivate local, national and global actions.

Ocean Tourism and Equity

Ocean tourism has the potential to alleviate poverty, especially in coastal fishing and farming communities where poverty incidences are high. It can boost local and national economic development and improve local welfare. However, unregulated ocean tourism development can bring in several unwanted consequences, such as the degradation of the environmental resource base that the tourism industry depends on, destruction of local cultures and traditional livelihoods and inequitable distribution of economic benefits (Cabral and Aliño 2011). Actions that ensure an equitable and sustainable tourism industry include proper planning of tourism developments, promotion of ecotourism activities that respect local cultures and traditions (including indigenous peoples' rights over ancestral domains) and implementation of policies that ensure that economic benefits from tourism activities accrue locally (i.e. provide local opportunities).

3.3.4 Opportunities for action and key conclusions

1. **Enhance coral reef resilience to climate change.** Reducing the negative impacts of climate change and associated ocean disturbances to coastal economies requires improving the resilience of marine and coastal ecosystems to climate change (Gattuso et al. 2018; James et al. 2019; Weatherdon et al. 2016). Establishing marine protected areas and MPA networks can help improve the

ecological resilience of coral reefs. MPAs protect marine ecosystems and their services from environmental uncertainties (Roberts et al. 2017), help minimise the footprints of human activities such as fishing (Lester et al. 2009), secure the continuous supply of genetic materials and serve as climate refugia when sited in cooler, less-impacted areas (Roberts et al. 2017; Mcleod et al. 2009). Furthermore, MPAs help ensure that coral reefs and associated species that are important draws for tourism are protected.

However, conventional management approaches that include MPAs may be insufficient to protect global coral reefs under warming and acidifying ocean conditions (Anthony et al. 2017). Assisted relocation and evolution (van Oppen et al. 2017) together with new biotechnology practices can enhance the resilience of coral reefs, but with associated costs. Protection should prioritise ecosystem connectivity—while there are preferences for some physical attributes of coastal tourism, like white sand, and there is a tendency to alter the ecosystem to favour some components (e.g. removing mangroves to access sandy beaches) (e.g. Cabral and Aliño 2011), it is important to recognise the huge role these ecosystems play in maintaining coastal integrity. For example, protecting mangroves and seagrass beds—which serve as nursery areas for a number of coral reef fish species and protect coral reefs by trapping sediments—enhances reef health and productivity.

2. **Protect and regenerate natural habitats.** Preserving and restoring natural coastal habitats such as coral reefs, beaches and mangroves increases the resilience of coastal areas to climate change (James et al. 2019), providing an alternative to hard infrastructure that allows for wave attenuation and shoreline stabilisation (James et al. 2019; Gattuso et al. 2018), as well as providing additional protections from storm surges and excess flooding (Ruiz-Ramírez et al. 2019). Traditional infrastructures for tourism such as urbanised beach fronts are expected to suffer shoreline erosion (coastal squeeze) due to climate change (Toimil et al. 2018; Scott et al. 2012). In these cases, coastal natural habitats can allow for landward retreat; otherwise, beach nourishment will be required to maintain tourism in heavily urbanised areas at very high costs (Scott et al. 2012). The quality of nearby sand habitats can be important to reduce those costs (Ruiz-Ramírez et al. 2019).
3. **Diversify development portfolios.** Diversifying tourism activities and investments to include linked ecosystems will help maintain diverse ecosystem functions, while simultaneously capturing the tourism potential of various ecosystems. Ecotourism, or tourism activities that support nature conservation and education, should be prioritised. Pressures on and drivers of reef health are often associated with governance and the socioeconomic needs of the people dependent on reefs. Linking fisheries, aquaculture and tourism to local food and livelihood security will improve the portfolio of policies that can be applied to reduce climate change's impacts on local and national economies. Marine spatial planning will play a key role in maintaining healthy reefs by strategically siting activities in the ocean so that negative interactions can be reduced. Actions include properly siting tourism infrastructure and making investments that account for potential future coastal and ocean changes. Management plans should explicitly address the role of natural habitats functioning as buffers to climate change on tourism (Ruiz-Ramírez et al. 2019). Communities that directly depend on coral reef tourism for their livelihoods need to increase their adaptive capacities, as this sector is expected to be negatively impacted by the changing climate in all countries. Local governments, private investors and development agencies can help by improving and developing social and institutional arrangements that allow for learning (i.e. technical education and skills development) and diversifying livelihoods and income sources (Cinner et al. 2018) while incorporating local and indigenous knowledge into the planning and decision process.
4. **Ensure that waste is properly disposed of and that waste treatment facilities are included in coastal tourism infrastructure.** As described above, nutrient enrichment exacerbates ocean acidification.

   Controlling nutrient input from coastal and terrestrial activities will help reduce the impact of climate change on coral reefs and reef tourism. Strategies can include ensuring that waste management, such as waste treatment facilities/recycling, is included in tourism development plans. Pollution combined with overfishing that degrades coral reefs caused the Caribbean to lose \$95–140 million/year in net revenue from coral reef–associated fisheries, \$100–300 million/year in reduced tourism revenue and \$140–420 million/year in reduced coastal protection (Burke et al. 2011).
5. **Reduce the environmental footprint of tourism through ecotourism and clean energy investments.** While climate change will inevitably affect tourism, tourism is also a major contributor of greenhouse gas emissions (Scott et al. 2012). It is estimated that tourism contributes 8% of global GHG emissions, with transport, shopping and food as major contributors (Lenzen et al. 2018). With tourism expected to grow 3–5% per year, it is important to ensure that the environmental footprint of tourism is minimised. Future increases in international arrivals do not necessarily translate to economic benefits for countries; hence, policies that ensure optimal benefits

for national economies while reducing tourism's footprint, such as those that promote ecotourism activities, should be prioritised. Furthermore, investments in clean and efficient energy in the tourism sector help reduce tourism's environmental footprint.

3.4 Improving the Energy Efficiency of the Ocean Economy

Improving the energy efficiency of ocean-related industries, especially shipping/transportation, would generate climate change benefits as well as benefits to the industries themselves. While significant improvements to the offshore oil and gas industry would require extensive transitioning of investments away from exploration and extraction of fossil fuels and into renewable energy (Allison and Bassett 2015), the shipping industry can make relatively large energy efficiency gains using existing technologies (Allison and Bassett 2015; Ash and Scarbrough 2019; Hoegh-Guldberg et al. 2019). For example, switching international shipping to solar-generated, ammonia-based fuel would allow for significant reductions in greenhouse gas emissions (Ash and Scarbrough 2019). Related topics are discussed in more depth in the Blue Papers *Ocean Energy and Mineral Sources* and *Coastal Development*. Fisheries and aquaculture are already relatively energy efficient, especially when compared with the terrestrial production of animal protein (Allison and Bassett 2015; Hoegh-Guldberg et al. 2019), but there is great potential in the expansion of carbon- and energy-efficient shellfish aquaculture as well as in the reduction of overcapacity in fisheries (Allison and Bassett 2015). Finally, the tourism sector involves a diverse array of opportunities for improving energy efficiency—from increasing fuel efficiency and using carbon offsets for various modes of travel to improving the energy efficiency of hotels and other tourism destinations around the world (Allison and Bassett 2015).

4 Impacts of Climate Change Mitigation in the Sea

Global efforts to mitigate climate change include a variety of approaches that may themselves have impacts on ocean ecosystems, species assemblages and the ocean economy. Here, we discuss the potential marine impacts and opportunities of four major categories of climate change mitigation methods that directly affect the ocean: efforts to conserve and increase 'blue carbon' storage; expansion of ocean-based renewable energy generation; deep-sea mining to meet demand for rare earth elements; and geoengineering techniques. We limit our discussion of the three latter topics to their direct impact on the ocean.

4.1 Conserving and Expanding Blue Carbon

The term 'blue carbon' refers to the capacity of marine ecosystems to store organic carbon over centuries or millennia (Serrano et al. 2019). The ocean is the largest carbon sink on Earth; it has already absorbed more than 90% of Earth's additional heat and captured nearly one-third of all atmospheric CO_2 emissions since the 1700s (Gattuso et al. 2015). Through a process known as the 'biological pump', marine organisms convert CO_2 into biomass (referred to as carbon 'fixation') through photosynthesis. A portion of this carbon is deposited and buried on the seafloor, thus removing it from the atmospheric carbon cycle on a long enough time scale to constitute a carbon sink (at which point this carbon is referred to as having been 'sequestered') (Barange et al. 2017; Duarte et al. 2013; Mcleod et al. 2011; Serrano et al. 2019; Vaughan and Lenton 2011). Marine carbon sequestration occurs both in the open ocean and along the coast, and there are opportunities to increase the sequestration capacity and contribute to climate change mitigation in both areas. These opportunities are becoming an important sector of the ocean economy as efforts mature to quantify and monetise (e.g. with carbon pricing) marine ecosystem restoration and management for carbon sequestration (Alongi et al. 2016; Lavery et al. 2013; Lovelock et al. 2017; Mcleod et al. 2011; Pendleton et al. 2012). As this sector develops, it is critical to consider the implications for vulnerable and marginalised groups, including small-scale fishers, who may be overlooked in blue carbon decision-making (Cohen et al. 2019).

Vegetated coastal ecosystems—primarily seagrasses, mangrove forests and tidal marshes—occupy only 0.2% of the global ocean surface, but have an outsize capacity for carbon sequestration, contributing up to 50% of carbon burial in marine sediments (Duarte 2017; Duarte et al. 2013; Hoegh Guldberg et al. 2019; Mcleod et al. 2011; Serrano et al. 2019), far outpacing the capacity per unit area of terrestrial habitats (Hoegh-Guldberg et al. 2019; Serrano et al. 2019). Kelp and other macroalgal beds have also recently been identified as contributors to global blue carbon storage (Serrano et al. 2019), and although there is significant debate around whether coral reefs act as carbon sources or sinks, the presence of coral reefs adjacent to seagrass beds and mangrove forests may improve the blue carbon efficacy of the system as a whole (Watanabe and Nakamura 2019).

While the capacity to expand the existing inventories of fixed and sequestered carbon in vegetated coastal ecosystems is limited, there is a critical need to protect them from degradation and conversion to alternative land uses (Allison and Bassett 2015; Hoegh-Guldberg et al. 2019). These ecosystems are among the most threatened habitats on Earth, and their current and projected loss not only reduces global CO_2 uptake, but also releases large amounts of carbon currently stored in their biomass and soils (Allison and Bassett 2015; Duarte 2017; Gattuso et al. 2015; Hoegh-Guldberg et al. 2019; Serrano et al. 2019). There may be sizable blue carbon potential in the restoration of marine vegetation where large portions of the coastline have been lost to development, as well as in the expansion of macroalgae aquaculture (Duarte 2017). In addition to their carbon sequestration capacity, vegetated marine ecosystems provide coastal protection and sea level rise mitigation services, regulate water quality, provide critical habitat for many marine species including commercially important fishery targets and enhance system biodiversity and resilience (Serrano et al. 2019). Thus, their protection and restoration would have multiple synergistic benefits (Allison and Bassett 2015).

There are also potential opportunities to increase the open ocean's capacity to sequester carbon where the biological pump moves biogenic carbon to depths of 1000 m or more, capturing it for centuries or longer (Burd et al. 2016). The main sources of this biogenic carbon are faeces, mucus and dead organisms.

Researchers have recently suggested that fisheries could be managed to have higher standing stock biomass, even in the face of climate change (Gaines et al. 2018; Hilborn and Costello 2018), which could theoretically increase the input of organic matter (including carbon) to the biological pump, especially when cascading ecosystem impacts of increasing standing stock biomass are considered (Roman and McCarthy 2010). Fostering the recovery of larger, deeper-diving fish and marine mammals could also increase upward fluxes of fixed nitrogen and other limiting nutrients from the deep ocean, thereby spurring additional primary productivity and subsequent CO_2 fixation (Aumont et al. 2018). These potential deep-sea carbon sequestration opportunities have thus far been inadequately studied, and would benefit from further exploration.

4.2 Expanding Ocean Renewables

Marine renewable energy sources have significant potential for reducing human demand for fossil fuels and reducing climate-changing GHGs (Boehlert and Gill 2010; Hoegh-Guldberg et al. 2019). Technologies capable of producing energy from the ocean are vast and expanding, with most taking advantage of wind, waves, currents, tides or thermal gradients, collectively referred to as offshore renewable energy developments, or ORED (Boehlert and Gill 2010). As these technologies expand, they will impact the ocean both above and below the water's surface through the following six channels, discussed in depth in Boehlert and Gill (2010):

1. **Physical presence**: Stationary structures such as support pillars and cables will alter pelagic habitats and bottom communities. Structures not treated with anti-fouling chemicals will create new settlement habitats, essentially forming artificial reefs and de facto 'fish aggregation devices'. ORED structures may also create barriers to species migration above and below the water.
2. **Dynamic effects**: Structures with moving parts (e.g. wind energy devices and below-water turbines) may be especially hazardous to migratory birds, cetaceans and fish. Oscillating structures, such as buoys and rotors, will modify water movement, turbulence and stratification, potentially altering the associated movements of marine species.
3. **Chemical effects**: Anti-fouling and other chemicals used on ORED technologies can leach into the surrounding water. Constructing, servicing and decommissioning structures brings additional risk of chemical spills. Furthermore, the movement of deep water to the surface during ocean thermal energy conversion can change chemical conditions through the increased input of nutrients, heavy metals and carbon dioxide, which can also outgas to the atmosphere.
4. **Acoustic effects**: Acoustic ORED impacts will be most severe during survey and construction phases, but noise from moving ORED structures may impact marine species during the operational phase as well.
5. **Electromagnetic field effects**: The transmission of electricity from ORED structures to shore generates low-frequency electromagnetic fields in the surrounding water, which may change the behaviours of marine species that use natural electric and/or magnetic fields for a variety of behaviours. Electricity-transmitting cables may also increase the temperature of the surrounding water and sediment, but the effects of this are still unknown.
6. **Effects of the energy removal itself**: Removing energy from the water can change local water movement (e.g. seasonal or tidal opening and closing of estuary systems), more distant current patterns, tidal ranges and thermal regimes. All of these changes may impact productivity patterns and species movement.

Each of these impacts must be evaluated throughout the stages of development, and across spatial and temporal scales (i.e. local versus far-reaching, and short- versus long-term impacts). The cumulative impacts of multiple adjacent developments must also be understood (Boehlert and Gill 2010). In addition, both the feasibility and the potential impacts of marine renewable energy technologies may be altered by the effects of climate change, including sea level rise, increased storms and extreme events, and changes to wave and circulatory energy patterns. These eventualities will need to be considered, and operations will need to be designed for climate resilience if they are to be successful and sustainable.

4.3 Expanding Deep-Sea Mining to Meet Demand for Rare Earth Elements

Rare earth elements (a group of 17 elements comprised of 15 lanthanides, plus yttrium and scandium) are critical to the development and operation of a variety of renewable energy technologies, including solar cells, wind turbines and electric vehicles (Dutta et al. 2016), but current land-based supply streams may not meet growing demand (Dutta et al. 2016; Miller et al. 2018a). The deep-sea floor, especially areas around hydrothermal vents, contains relatively vast quantities of rare earths that could help to meet this demand, and mining contracts for deep-sea resources including rare earths have been awarded to a number of countries and companies (Kato et al. 2011; Miller et al. 2018a). However, the costs associated with extracting rare earth elements are thus far prohibitive, and no commercial-scale mines are as yet operational (Miller et al. 2018a).

In addition to the usual risks associated with mining and other extractive industries in the ocean (including the potential for the release of toxic elements, contamination from dredge spoils, increased noise, heat and light pollution, and loss of biodiversity), these deep-sea mining operations carry risks related to impacts to the fragile marine ecosystems and unique and endemic species communities found on the deep-ocean floor, many of which have been recognised as vulnerable (Miller et al. 2018a; Van Dover et al. 2017). Furthermore, impacts may extend many kilometres away from mining sites and the long-term impacts will be much more significant than in shallow water because deep-sea habitats can take decades to millennia to recover (Miller et al. 2018a). Finally, deep-sea mining carries additional challenges, such as the potential for conflict with other marine uses and the legal and political complexities of operating under international waters in the open ocean (Miller et al. 2018a).

4.4 Geoengineering Solutions

A variety of ocean-based geoengineering concepts have been suggested to help mitigate climate change including 'cloud brightening', by mechanical or biological means, to increase atmospheric albedo; fertilising patches of the ocean with limiting nutrients (iron, nitrogen or phosphorus) to enhance primary productivity and sequestration of carbon (see blue carbon discussion above); inducing upwelling to do the same; inducing downwelling to increase the sinking of CO_2-rich waters; and 'enhanced weathering', wherein materials such as carbonate or silicate are added to the water to increase alkalinity, thereby stimulating removal of CO_2 from the atmosphere (Allison and Bassett 2015; Vaughan and Lenton 2011). Together, these efforts could theoretically reduce global radiative forcing by an estimated ~4.2 W/m^2, with cloud brightening contributing the bulk of that reduction (Vaughan and Lenton 2011).

While the costs of implementing any of these techniques are currently prohibitive, and the carbon-balance effects are highly uncertain (Allison and Bassett 2015; Vaughan and Lenton 2011), even if they prove cost-effective and sequester substantial amounts of carbon they may result in unwanted ocean impacts. For example, ocean fertilisation could lead to increased deoxygenation and eutrophication, and making adjustments to natural upwelling and downwelling patterns could alter primary productivity and change community structures and functions (Vaughan and Lenton 2011). Increasing cloud cover could generate unwanted weather patterns (Irvine et al. 2010; Jones et al. 2009) and address only global temperature changes without reducing other impacts, such as ocean acidification (Gattuso et al. 2015; Vaughan and Lenton 2011; Williamson and Turley 2012). Each of these impacts could have significant consequences for other sectors of the ocean economy, as discussed above. Finally, there may be important ethical implications associated with many of these geoengineering options related to the uneven distribution of impacts (Allison and Bassett 2015; Jones et al. 2009; Vaughan and Lenton 2011). Thus, near-term efforts should be focused on drastically reducing CO_2 emissions while research into the risks and benefits of these geoengineering technologies continues.

5 Conclusions and Opportunities for Action

The ocean is critically important to the global economy. Collectively, it is estimated that ocean-based industries and activities contribute hundreds of millions of jobs and approximately \$2.5 trillion to the global economy each year, making it the world's seventh-largest economy when compared with national GDPs (Hoegh-Guldberg 2015; IPCC 2019). In this paper, we reviewed the impact of climate change on the three key components of the ocean ecosystem economy—fisheries, marine aquaculture and coral reef tourism—and the opportunities for effective institutions and markets to reduce these impacts.

Building on existing work, we developed three models to forecast the economic impacts of climate change and potential benefits of adaptation in each sector for every coastal country under diverse climate scenarios. For capture fisheries, we find that all countries would benefit from implementing climate-adaptive reforms and that many countries could maintain current profits and catches into the future with adaptation. For aquaculture, we show that production is under capacity in many countries and the negative effects of climate change could be more than offset by developing and expanding sustainable mariculture. For ocean tourism, we find that all countries will be negatively impacted, and both local and global actions that reduce the magnitude of climate change effects would help lessen the economic impacts.

Maintaining a robust ocean economy will depend on swift efforts to reduce greenhouse gas emissions. The recent IPCC (2019) report estimates that climate-induced declines in ocean health will cost the global economy $428 billion/year by 2050 and $1.98 trillion/year by 2100. The magnitude and inequity of these losses is highly sensitive to future greenhouse gas emissions across sectors of the ocean economy. The ability for climate-adaptive fisheries management to mitigate losses under climate change deteriorates under increasingly severe emissions scenarios. The ability for mariculture to be a viable substitute for declining capture fisheries is also diminished under increasingly severe climate futures. Finally, the magnitude of losses in marine and coastal tourism increases dramatically under increasingly severe emissions scenarios. In all cases, these impacts are especially pronounced in the tropical developing countries, which have contributed the least to growing greenhouse gas emissions. Thus, it will be the responsibility of the industrial nations to take a leadership role in curbing emissions and reducing the impacts of climate change on the ocean economy.

Since climate change impacts differ by country and sector, possible solutions will be context-specific. By exploring the climate change impacts at the country level for fisheries, aquaculture and reef tourism as described in this report, countries will be able to assess what they stand to gain or lose due to climate change. Below, we outline solutions for each sector based on whether a country will experience gains, no change or losses.

5.1 Capture Fisheries

An interactive web interface developed by the Sustainable Fisheries Group at the University of California, Santa Barbara, summarises the impact of climate change on marine fisheries around the world and the opportunities for countries to mitigate these impacts through climate-adaptive fisheries management reforms (UCSB 2019). It illustrates how the health of fisheries and the catches and profits provided by them will change under four increasingly severe climate change scenarios (+0.3 °C, +0.9 °C, +1.2 °C and +2.3 °C increases in sea surface temperature by 2100) with and without climate-adaptive fisheries reform. This tool can be used to determine whether a country is likely to experience negative, positive or neutral impacts of climate change.

1. **Lower-capacity countries** (often tropical, developing countries experiencing negative impacts of climate change) should implement or strengthen their fisheries management (see Cochrane et al. 2011) to enhance resilience to the negative effects of climate change.
2. **Higher-capacity countries** (often temperate, developed countries experiencing mixed impacts of climate change) should account for shifting productivity in fisheries stock assessments and management procedures (see Pinsky and Mantua 2014) to capitalise on the positive effects of climate change and mitigate the negative effects.
3. **All countries** will derive benefits from international cooperation that both ensures that management does not degrade as stocks shift distributions and results in fairness and equity in fisheries outcomes under climate change.

5.2 Aquaculture

1. **In countries with underdeveloped mariculture potential** (Fig. 2.2), the negative effects of climate change can be offset by both sustainably expanding current mariculture operations and investing in science and technologies that enhance mariculture efficiency and productivity amidst a changing climate.
2. **In countries with fully developed mariculture potential** (Fig. 2.2), mariculture production can be maintained by selectively breeding for fast growth or heat tolerance or by shifting portfolios of mariculture species to match the new thermal regime.
3. **In all countries**, studying the impact of large-scale mariculture on marine ecosystems will be essential to identifying and promoting best practices in sustainable mariculture. Making strategic investments and expanding mariculture operations can boost local food supply without interacting negatively with other ecosystem services.

5.3 Ocean Tourism

Climate change will reduce the potential of ocean tourism to boost the local economies of countries with coral reefs. The magnitude of the impact will depend on the realised global emissions pathways, confounding effects of local stressors, dependency of the local economy to ocean tourism and type of ocean tourism. While on-reef tourism (e.g. snorkelling and diving) will be more vulnerable than reef-adjacent tourism (e.g. sunbathing, white sand), the latter will also likely be affected, although the magnitude of the impact is uncertain. Table 2.3 summarises the predicted changes in coral cover and reef tourism values given climate change as well as the current on-reef and reef-adjacent tourism values of each coastal country with coral reefs.

1. **In countries with a high proportion of their local economy dependent on tourism**, such as Maldives, Palau and St. Barthélemy (i.e. over 40%of their GDPs are from reef tourism), options include slowly diversifying to other

industries, such as mariculture, and creating opportunities for alternative forms of tourism, such as wreck diving and other novel activities, while at the same time increasing investments in the management of and improvements to reef ecosystems, fisheries and ocean tourism.

2. **In countries with high reef-adjacent values and where ocean tourism is important**, it is still **imperative to improve and maintain coral reef health to secure the continuous provision of many of the ecological processes and services that support reef-adjacent activities (e.g. white sand from corals, wave attenuation function of coral reefs).**
3. **For countries with disproportionately high on-reef tourism values**, investments in reef-adjacent tourism activities and ecotourism activities can both enhance the economic potential of coral reefs and motivate more investments in protecting reef health.
4. **Coral reef tourism can be a viable industry in countries that are expected to experience losses in aquaculture and capture fisheries**. Although climate change will hinder countries' abilities to tap into the full potential of ocean tourism, that does not mean that coastal tourism cannot improve the local economy.
5. **Given that current ocean tourism activities impact future ocean tourism economic output and ecosystem health** (feedback loops), all countries must aim to efficiently enhance ocean tourism gains by prioritising high-economic-gain activities while reducing the ecological footprints of ocean tourism activities (i.e. by investing in ecotourism and clean and efficient energy).

Across each of the above sectors of the ocean economy, the recommendations to build socioecological resilience to climate change and ensure the continued, or improved, provision of valued functions and services can be captured in three high-level mandates:

(a) **Be forward looking:** The future of the ocean economy is expected to drastically change given climate change, and the nature and magnitude of these changes can be highly variable. It will no longer be appropriate (or possible) to make predictions based on historical benchmarks or to assume that our usual metrics for measuring outcomes will remain stable. As the climate changes, each of the above-discussed ocean sectors will need to work to understand risks, anticipate changes and make decisions aimed at improving ecosystem health. In many cases, the risks and changes will become increasingly uncertain, which means that all management decisions need to factor in the likelihood of increasing surprises by being a bit more precautionary. **For wild-capture fisheries**, looking forward will entail things like scenario planning and management strategy evaluation, while stock assessments, harvest controls, allocation systems and even marine protected areas will all need to be more flexible, adaptive and precautionary. **Mariculture operations** will need to invest in things like selective breeding, improvements to feed conversion ratios, and technologies that continue to reduce risks from increasingly frequent and stronger storms. **Ocean tourism operations** may need to engage in practices aimed at building ecosystem resilience and health and be efficient by catering to tourism activities that provide high economic returns and have smaller ecological footprints. The designs of spatial management systems should account for future shifts in species ranges and productivities to both facilitate the successful movement of species to other areas and enhance marine population resilience to environmental and social changes.

(b) **Cooperate across boundaries**: It will also be critical to expand the current boundaries of our management decisions to allow for effective systems-level problem identification and solution development. As suitable habitats shift and change, marine species will move across jurisdictional boundaries and regional, national and international cooperative agreements will be necessary to ensure that these species are well-managed, and that the benefits are fairly distributed during and after the transitions. **For mariculture**, it will be critical to incorporate other marine uses and sectors in the planning and implementation of operations. Whole-systems thinking would also benefit tourism by ensuring the durability of this sector into the future as well as taking advantage of tourism opportunities that emerge in new areas (i.e. for the case where new coral reefs may establish in subtropical areas). In addition, it will be critical to share lessons learned and tools applied across and between sectors and jurisdictions to ensure lower-capacity regions will not fall behind in the implementation of solutions.

(c) **Focus on equity**: Finally, it will be profoundly important to examine the equity implications of all new and existing management decisions across these sectors, as climate change is likely to cause and exacerbate global inequities. Inequity reduces resilience, thereby likely worsening outcomes under all climate change scenarios.

Furthermore, equity considerations should be an input to decision-making in terms of both the design and implementation of management reforms and the creation and execution of new international agreements. Equitable solutions are more likely to garner buy-in from impacted groups and will thus be more likely to be effectively implemented. Focusing on equity can also lead to the development of more effective solutions that target the underlying system dynamics and power differentials that are, in fact, the root drivers of climate change. These solutions should consider equity issues in access and participation in the blue economy, including through the provision of no-cost skills development opportunities, and they must involve different world views and knowledge systems, integrating local and indigenous knowledge and avoiding poverty traps and the marginalisation of already vulnerable groups.

Truly inclusive, representative, participatory decision-making processes are needed in all sectors to ensure procedural equity in all policy and management decisions. In addition, new solutions and interventions must seek to ensure distributional equity (i.e. equitable access to benefits and exposure to risks stemming from decisions) and to engender recognitional equity (i.e. recognition of and respect for differences within and between groups, and understanding of how these differences alter the perception and experience of impacts) if systems are to become equitably resilient to climate change.

It is imperative that countries explore the synergistic impacts of climate change across all three economic sectors (fisheries, mariculture and ocean tourism) and identify whether they are vulnerable to universally negative impacts, have options to offset negative impacts in some sectors through adaptation or could benefit from potentially positive impacts in other sectors. Countries should also note the magnitude of climate change impacts to the three major components of their ocean ecosystem economies to best plan their investments for climate change adaptation and mitigation strategies. While the solutions we put forward above are targeted to individual economic sectors, the three marine ecosystem economies are connected ecologically and socioeconomically, and positive actions to one sector often act synergistically with other sectors, especially when the actions are aimed at maintaining and enhancing ecosystem health.

Unregulated economic developments in fisheries, aquaculture and tourism have brought many unintended environmental and social consequences, including the degradation of non-use values and the provision of many other ecosystem services, both in developing and developed nations. While investments in these three sectors could improve national and local food and livelihood security amidst the challenges brought by anthropogenic climate change, sustaining the development and benefits they bring requires a development path that promotes and maintains a healthy ocean ecosystem. After all, the productivity and resilience of aquaculture, tourism and fisheries depend on clean water, intact habitats (e.g. mangroves and seagrass beds that serve as nursery grounds for commercial marine species) and diverse marine organisms, among others. Since this paper primarily focuses on ocean ecosystem sectors, the majority of the outlined recommendations and actions drive sustainable improvements in the ocean economy and, therefore, can provide positive synergistic effects for the underlying natural resource and its nonmarket values. Faster development and greater economic values in these three sectors can be realised if trade-offs between use and non-use values, which vulnerable communities often directly depend on, are avoided.

We expect that the variable directions of impacts of climate change across the three economic sectors for each country will draw new investments in some sectors while other sectors are expected to continually suffer.

It is imperative that developments are well-planned and properly regulated to avoid unwanted environmental impacts, degradation of local cultures and livelihoods, and the inequitable distribution of benefits. For instance, including access to technical education and skills development will ensure that resources are available for people to transition from one form of livelihood to another, hence ensuring that the economic benefits of local developments accrue locally. There is also huge potential for local investments in renewable energy and energy-efficient technologies that can improve local livelihoods, enhance local economic benefits and reduce the carbon footprints of human activities. Finally, we envision that our results will ultimately help guide new ocean investments and positive conservation actions by governments, nongovernmental organisations, development agencies, philanthropies and international communities.

Acknowledgements The authors thank Jane Lubchenco, Merrick Burden and Kate Bonzon for helpful comments on earlier drafts of this Blue Paper and Mark Spalding for reef tourism data. The paper's technical reviewers, Manuel Barange, Sarah Cooley, Salif Diop, Silvia Patricia González Díaz and Boris Worm, as well as its arbiter, Andreas Merkl, all provided helpful technical comments. The authors also thank World Resources Institute for providing support as the Ocean Panel Secretariat.

While our colleagues were very generous with their time and input, this report reflects the views of the authors alone. The authors thank Sarah DeLucia for copyediting and Jen Lockard for design.

About the Authors

Co-authors

Steven Gaines is the dean and a distinguished professor at the Bren School of Environmental Science & Management at the University of California, Santa Barbara (USA).

Reniel Cabral is an assistant researcher at the Bren School of Environmental Science & Management at the University of California, Santa Barbara (USA).

Christopher M. Free is a postdoctoral scholar at the University of California, Santa Barbara (USA).

Yimnang Golbuu is the CEO of the Palau International Coral Reef Center (Palau).

Contributing Authors

Ragnar Arnason is a professor of economics at the University of Iceland (Iceland).

Willow Battista is the manager of research, design, and engagement in the Fishery Solutions Center at the Environmental Defense Fund (USA).

Darcy Bradley is a lead scientist at the Environmental Market Solutions Lab at the University of California, Santa Barbara (USA).

William Cheung is a professor at the Institute for the Oceans and Fisheries at the University of British Columbia and the director of science for the Nippon Foundation-UBC Nereus Program (Canada).

Katharina Fabricius is a senior principal research scientist at the Australian Institute of Marine Science (Australia).

Ove Hoegh-Guldberg is a professor of marine science and director of the Global Change Institute at the University of Queensland (Australia).

Marie Antonette Juinio-Meñez is a professor at the Marine Science Institute at the University of the Philippines (Philippines).

Jorge García Molinos is an assistant professor at the Arctic Research Center at Hokkaido University (Japan).

Elena Ojea is a senior researcher at the Future Oceans Lab, CIM- UVigo, at the University of Vigo (Spain).

Erin O'Reilly is the projects and operations coordinator at the Environmental Market Solutions Lab at the University of California, Santa Barbara (USA).

Carol Turley is a senior merit scientist at the Plymouth Marine Laboratory (UK).

References

Abate TG, Nielsen R, Tveterås R (2016) Stringency of environmental regulation and aquaculture growth: a cross- country analysis. Aquac Econ Manag 20(2):201–221. https://doi.org/10.1080/13657305.2016.1156191

Aguilar-Manjarrez J, Soto D, Brummett RE (2017) Aquaculture zoning, site selection and area management under the ecosystem approach to aquaculture: a handbook. Food and Agriculture Organization of the United Nations and World Bank, Rome

Allison EH, Bassett HR (2015) Climate change in the oceans: human impacts and responses. Science 350(6262):778–782. https://doi.org/10.1126/science.aac8721

Allison EH, Perry AL, Badjeck MC, Adger WN, Brown K, Conway D, Halls AS et al (2009) Vulnerability of national economies to the impacts of climate change on fisheries. Fish Fish 10(2):173–196. https://doi.org/10.1111/j.1467-2979.2008.00310.x

Alongi DM, Murdiyarso D, Fourqurean JW, Kauffman JB, Hutahaean A, Crooks S, Lovelock CE et al (2016) Indonesia's blue carbon: a globally significant and vulnerable sink for seagrass and mangrove carbon. Wetl Ecol Manag 24(1):3–13

Anthony K, Bay LK, Costanza R, Firn J, Gunn J, Harrison P, Heyward A et al (2017) New interventions are needed to save coral reefs. Nat Ecol Evol 1(10):1420

Aqorau T, Bell J, Kittinger JN (2018) Good governance for migratory species. Science 361(6408):1208–1209. https://doi.org/10.1126/science.aav2051

Arceo HO, Quibilan MC, Aliño PM, Lim G, Licuanan WY (2001) Coral bleaching in philippine reefs: coincident evidences with mesoscale thermal anomalies. Bull Mar Sci 69(2):16

Arlinghaus R, Abbott JK, Fenichel EP, Carpenter SR, Hunt LM, Alós J, Klefoth T et al (2019) Opinion: governing the recreational dimension of global fisheries. Proc Natl Acad Sci 116(12):5209–5213. https://doi.org/10.1073/pnas.1902796116

Ash N, Scarbrough T (2019) Sailing on solar: could green ammonia decarbonise international shipping? Environmental Defense Fund, London

Aumont O, Maury O, Lefort S, Bopp L (2018) Evaluating the potential impacts of the diurnal vertical migration by marine organisms on marine biogeochemistry. Global Biogeochem Cycles 32(11):1622–1643. https://doi.org/10.1029/2018GB005886

Bakun A (1990) Global climate change and intensification of coastal ocean upwelling. Science 247(4939):198–201. https://doi.org/10.1126/science.247.4939.198

Bakun A, Black BA, Bograd SJ, Garcia-Reyes M, Miller AJ, Rykaczewski RR, Sydeman WJ (2015) Anticipated effects of climate change on coastal upwelling ecosystems. Curr Clim Change Rep 1(2):85–93. https://doi.org/10.1007/s40641-015-0008-4

Barange M (2019) Avoiding misinterpretation of climate change projections of fish catches. ICES J Mar Sci 76(6):1390–2. https://doi.org/10.1093/icesjms/fsz061

Barange M, Merino G, Blanchard JL, Scholtens J, Harle J, Allison EH, Allen JI et al (2014) Impacts of climate change on marine ecosystem production in societies dependent on fisheries. Nat Clim Change 4(3):211

Barange M, Butenschön M, Yool A, Beaumont N, Fernandes JA, Martin AP, Allen J (2017) The cost of reducing the North Atlantic Ocean biological carbon pump. Front Mar Sci 3:290. https://doi.org/10.3389/fmars.2016.00290

Barange M, Bahri T, Beveridge MCM, Cochrane KL, Funge-Smith S, Poulain F (eds) (2018) Impacts of climate change on fisheries and aquaculture: synthesis of current knowledge, adaptation and mitigation options. Food and Agriculture Organization of the United Nations, Rome

Barton A, Hales B, Waldbusser GG, Langdon C, Feely RA (2012) The Pacific Oyster, *Crassostrea Gigas*, shows negative correlation to naturally elevated carbon dioxide levels: implications for near-term ocean acidification effects. Limnol Oceanogr 57(3):698–710. https://doi.org/10.4319/lo.2012.57.3.0698

Béné C (2004) Poverty in small-scale fisheries: a review and some further thoughts. In: Neiland AE, Béné C (eds) Poverty and small-scale fisheries in West Africa. Springer, Dordrecht

Béné C, Macfadyen G, Allison EH (2007) Increasing the contribution of small-scale fisheries to poverty alleviation and food security. No. 481. Food and Agriculture Organization of the United Nations, Rome

Béné C, Hersoug B, Allison EH (2010) Not by rent alone: analysing the pro-poor functions of small-scale fisheries in developing countries. Dev Policy Rev 28(3):325–358. https://doi.org/10.1111/j.1467--7679.2010.00486.x

Blasiak R, Spijkers J, Tokunaga K, Pittman J, Yagi N, Österblom H (2017) Climate change and marine fisheries: least developed countries top global index of vulnerability. PLoS One 12(6):e0179632. https://doi.org/10.1371/journal.pone.0179632

Boehlert G, Gill A (2010) Environmental and ecological effects of ocean renewable energy development—a current synthesis. Oceanography 23(2):68–81. https://doi.org/10.5670/oceanog.2010.46

Bopp L, Resplandy L, Orr JC, Doney SC, Dunne JP, Gehlen M, Halloran P et al (2013) Multiple stressors of ocean ecosystems in the

21st century: projections with CMIP5 models. Biogeosciences 10:6225–6245. https://doi.org/10.5194/bg-10-6225-2013

Brady RX, Alexander MA, Lovenduski NS, Rykaczewski RR (2017) Emergent anthropogenic trends in california current upwelling. Geophys Res Lett 44(10):5044–5052. https://doi.org/10.1002/2017GL072945

Brander K (2010) Impacts of climate change on fisheries. J Mar Syst 79(3–4):389–402. https://doi.org/10.1016/j.jmarsys.2008.12.015

Brander LM, Rehdanz K, Tol RSJ, Van Beukering PJH (2012) The economic impact of ocean acidification on coral reefs. Clim Change Econ 3(1):1250002

Breitburg D, Levin LA, Oschlies A, Grégoire M, Chavez FP, Conley DJ, Garçon V et al (2018) Declining oxygen in the global ocean and coastal waters. Science 359(6371):eaam7240. https://doi.org/10.1126/science.aam7240

Britten GL, Dowd M, Worm B (2016) Changing recruitment capacity in global fish stocks. Proc Natl Acad Sci 113(1):134–139. https://doi.org/10.1073/pnas.1504709112

Bruno JF, Petes LE, Harvell CD, Hettinger A (2003) Nutrient enrichment can increase the severity of coral diseases. Ecol Lett 6(12):1056–1061

Bryndum-Buchholz A, Tittensor DP, Blanchard JL, Cheung WWL, Coll M, Galbraith ED, Jennings S et al (2019) Twenty-first-century climate change impacts on marine animal biomass and ecosystem structure across ocean basins. Glob Chang Biol 25(2):459–472. https://doi.org/10.1111/gcb.14512

Burd A, Buchan A, Church MJ, Landry MR, McDonnell AMP, Passow U, Steinberg DK et al (2016) Towards a transformative understanding of the ocean's biological pump: priorities for future research—report on the NSF Biology of the Biological Pump Workshop. Ocean Carbon and Biogeochemistry Program, Woods Hole. https://doi.org/10.1575/1912/8263

Burke L, Reytar K, Spalding M, Perry A (2011) Reefs at risk revisited. World Resources Institute, Washington, DC. https://www.wri.org/publication/reefs-risk-revisited

Cabral RB, Aliño PM (2011) Transition from common to private coasts: consequences of privatization of the coastal commons. Ocean Coast Manag 54(1):66–74. https://doi.org/10.1016/j.ocecoaman.2010.10.023

Caesar L, Rahmstorf S, Robinson A, Feulner G, Saba V (2018) Observed fingerprint of a weakening atlantic ocean overturning circulation. Nature 556(7700):191. https://doi.org/10.1038/s41586-018-0006-5

Cai W, Borlace S, Lengaigne M, Van Rensch P, Collins M, Vecchi G, Timmermann A et al (2014) Increasing frequency of extreme El Niño events due to greenhouse warming. Nat Clim Change 4(2):111. https://doi.org/10.1038/nclimate2100

Cai W, Santoso A, Wang G, Yeh S-W, An S-I, Cobb KM, Collins M et al (2015) ENSO and greenhouse warming. Nat Clim Change 5(9):849 https://doi.org/10.1038/nclimate2743

Charles A (2012) People, oceans and scale: governance, livelihoods and climate change adaptation in marine social–ecological systems. Curr Opin Environ Sustain 4(3):351–357. https://doi.org/10.1016/j.cosust.2012.05.011

Chen P-Y, Chen C-C, Chu LF, McCarl B (2015) Evaluating the economic damage of climate change on global coral reefs. Glob Environ Chang 30(January):12–20. https://doi.org/10.1016/j.gloenvcha.2014.10.011

Cheng L, Trenberth KE, Fasullo J, Boyer T, Abraham J, Zhu J (2017) Improved estimates of ocean heat content from 1960 to 2015. Sci Adv 3(3):e1601545. https://doi.org/10.1126/sciadv.1601545

Cheung WWL, Lam V, Sarmiento JL, Kearney K, Watson R, Zeller D, Pauly D (2010) Large-scale redistribution of maximum fisheries catch potential in the global ocean under climate change: climate change impacts on catch potential. Glob Chang Biol 16(1):24–35. https://doi.org/10.1111/j.1365-2486.2009.01995.x

Cheung WWL, Reygondeau G, Frölicher TL (2016) Large benefits to marine fisheries of meeting the 1.5°C global warming target. Science 354(6319):1591–1594. https://doi.org/10.1126/science.aag2331

Cinner JE, Daw T, McClanahan TR (2009) Socioeconomic factors that affect artisanal fishers' readiness to exit a declining fishery. Conserv Biol 23(1):124–130. https://doi.org/10.1111/j.1523--1739.2008.01041.x

Cinner JE, Adger WN, Allison EH, Barnes ML, Brown K, Cohen PJ, Gelcich S et al (2018) Building adaptive capacity to climate change in tropical coastal communities. Nat Clim Change 8(2):117–123

Clavelle T, Lester SE, Gentry R, Froehlich HE (2019) Interactions and management for the future of marine aquaculture and capture fisheries. Fish Fish 20(2):368–388. https://doi.org/10.1111/faf.12351

Cline TJ, Schindler DE, Hilborn R (2017) Fisheries portfolio diversification and turnover buffer Alaskan fishing communities from abrupt resource and market changes. Nat Commun 8(January):14042. https://doi.org/10.1038/ncomms14042

Cochrane KL, Andrew NL, Parma AM (2011) Primary fisheries management: a minimum requirement for provision of sustainable human benefits in small-scale fisheries: primary management of small-scale fisheries. Fish Fish 12(3):275–288. https://doi.org/10.1111/j.1467-2979.2010.00392.x

Cohen P, Allison EH, Andrew NL, Cinner JE, Evans LS, Fabinyi M, Garces LR et al (2019) Securing a just space for small-scale fisheries in the blue economy. Front Mar Sci 6:171

Costanza R, de Groot R, Sutton P, van der Ploeg S, Anderson SJ, Kubiszewski I, Farber S et al (2014) Changes in the global value of ecosystem services. Glob Environ Chang 26(May):152–158. https://doi.org/10.1016/j.gloenvcha.2014.04.002

Costello C, Gaines SD, Lynham J (2008) Can catch shares prevent fisheries collapse? Science 321(5896):1678–1681. https://doi.org/10.1126/science.1159478

Costello C, Lynham J, Lester SE, Gaines SD (2010) Economic incentives and global fisheries sustainability. Annu Rev Resour Econ 2(1):299–318. https://doi.org/10.1146/annurev.resource.012809.103923

Costello C, Ovando D, Clavelle T, Strauss CK, Hilborn R, Melnychuk MC, Branch TA et al (2016) Global fishery prospects under contrasting management regimes. Proc Natl Acad Sci 113(18):5125–5129. https://doi.org/10.1073/pnas.1520420113

Costello C, Cao L, Gelcich S et al (2019) The future of food from the sea. World Resources Institute, Washington, DC

Dabbadie L, Aguilar-Manjarrez J, Beveridge MC, Bueno PB, Ross LG, Soto D (2018) Effects of climate change on aquaculture: drivers, impacts and policies. In: Barange M, Bahri T, Beveridge MCM, Cochrane KL, Funge-Smith S, Poulain F (eds) Impacts of climate change on fisheries and aquaculture. Food and Agriculture Organization of the United Nations, Rome

Dangendorf S, Marcos M, Wöppelmann G, Conrad CP, Frederikse T, Riva R (2017) Reassessment of 20th century global mean sea level rise. Proc Natl Acad Sci 114(23):5946–5951. https://doi.org/10.1073/pnas.1616007114

Daw TM, Cinner JE, McClanahan TR, Brown K, Stead SM, Graham NAJ, Maina J (2012) To fish or not to fish: factors at multiple scales affecting artisanal fishers' readiness to exit a declining fishery. PLoS One 7(2):e31460. https://doi.org/10.1371/journal.pone.0031460

De Silva SS (2012) Aquaculture: a newly emergent food production sector—and perspectives of its impacts on biodiversity and conservation. Biodivers Conserv 21(12):3187–3220. https://doi.org/10.1007/s10531-012-0360-9

Diana JS (2009) Aquaculture production and biodiversity conservation. Bioscience 59(1):27–38

Díaz S, Demissew S, Carabias J, Joly C, Lonsdale M, Ash N, Larigauderie A et al (2015) The IPBES conceptual framework—connecting nature and people. Curr Opin Environ Sustain 14:1–16

Díaz S, Pascual U, Stenseke M, Martín-López B, Watson RT, Molnár Z, Hill R et al (2018) Assessing nature's contributions to people. Science 359(6373):270–272

Doney SC, Ruckelshaus M, Duffy JE, Barry JP, Chan F, English CA, Galindo HM et al (2012) Climate change impacts on marine ecosystems. Annu Rev 4:11–37. https://doi.org/10.1146/annurev-marine-041911-111611

Doney S, Rosenberg AA, Alexander M, Chavez F, Harvell CD, Hofmann G, Orbach M et al (2014) Chap. 24: Oceans and marine resources. In: Melillo JM, Richmond TC, Yohe GW (eds) Climate change impacts in the United States: the third national climate assessment. U.S. Global Change Research Program, Washington, DC, pp 557–578

Donner SD (2009) Coping with commitment: projected thermal stress on coral reefs under different future scenarios. PLoS One 4(6):e5712

Donner SD, Skirving WJ, Little CM, Oppenheimer M, Hoegh-Guldberg O (2005) Global assessment of coral bleaching and required rates of adaptation under climate change. Glob Chang Biol 11(2):2251–2265. https://doi.org/10.1111/j.1365-2486.2005.01073.x

Duarte CM (2017) Reviews and syntheses: hidden forests, the role of vegetated coastal habitats in the ocean carbon budget. Biogeosciences 14(2):301–310. https://doi.org/10.5194/bg-14-301-2017

Duarte CM, Losada IJ, Hendriks IE, Mazarrasa I, Marbà N (2013) The role of coastal plant communities for climate change mitigation and adaptation. Nat Clim Change 3(11):961–968. https://doi.org/10.1038/nclimate1970

Dulvy NK, Rogers SI, Jennings S, Stelzenmüller V, Dye SR, Skjoldal HR (2008) Climate change and deepening of the north sea fish assemblage: a biotic indicator of warming seas. J Appl Ecol 45:1029–1039. https://doi.org/10.1111/j.1365-2664.2008.01488.x

Dutta T, Kim K-H, Uchimiya M, Kwon EE, Jeon B-H, Deep A, Yun S-T (2016) Global demand for rare earth resources and strategies for green mining. Environ Res 150(October):182–190. https://doi.org/10.1016/j.envres.2016.05.052

Eby LA, Crowder LB (2002) Hypoxia-based habitat compression in the neuse river estuary: context-dependent shifts in behavioral avoidance thresholds. Can J Fish Aquat Sci 59(6):952–965. https://doi.org/10.1139/f02-067

Edwards P, Zhang W, Belton B, Little DC (2019) Misunderstandings, myths and mantras in aquaculture: its contribution to world food supplies has been systematically over reported. Mar Policy 106(August):103547. https://doi.org/10.1016/j.marpol.2019.103547

Fabbri E, Dinelli E (2014) Physiological responses of marine animals towards adaptation to climate changes. In: Goffredo S, Dubinsky Z (eds) The Mediterranean Sea: its history and present challenges. Springer, Dordrecht, pp 401–417. https://doi.org/10.1007/978-94-007-6704-1_23

Fabricius KE, Langdon C, Uthicke S, Humphrey C, Noonan S, De'ath G, Okazaki R et al (2011) Losers and winners in coral reefs acclimatized to elevated carbon dioxide concentrations. Nat Clim Change 1(3):165

FAO (Food and Agriculture Organization of the United Nations) (2018) The state of world fisheries and aquaculture: meeting the sustainable development goals. CC BY-NC-SA 3.0 IGO. FAO, Rome. http://www.fao.org/fishery/sofia/en

Fedele G, Donatti CI, Harvey CA, Hannah L, Hole DG (2019) Transformative adaptation to climate change for sustainable social-ecological systems. Environ Sci Policy 101(November):116–125. https://doi.org/10.1016/j.envsci.2019.07.001

Fossheim M, Primicerio R, Johannesen E, Ingvaldsen RB, Aschan MM, Dolgov AV (2015) Recent warming leads to a rapid borealization of fish communities in the Arctic. Nat Clim Change 5(7):673–677. https://doi.org/10.1038/nclimate2647

Free CM, Thorson JT, Pinsky ML, Oken KL, Wiedenmann J, Jensen OP (2019a) Impacts of historical warming on marine fisheries production. Science 363(6430):979–983. https://doi.org/10.1126/science.aau1758

Free CM, Mangin T, García Molinos J, Ojea E, Costello C, Gaines SD (2019b) Realistic fisheries management reforms could mitigate the impacts of climate change in most countries. BioRxiv. https://doi.org/10.1101/804831

Friedrich T, Timmermann A, Abe-Ouchi A, Bates NR, Chikamoto MO, Church MJ, Dore JE et al (2012) Detecting regional anthropogenic trends in ocean acidification against natural variability. Nat Clim Change 2(3):167. https://doi.org/10.1038/nclimate1372

Frieler K, Meinshausen M, Golly A, Mengel M, Lebek K, Donner SD, Hoegh-Guldberg O (2013) Limiting global warming to 2 C is unlikely to save most coral reefs. Nat Clim Change 3(2):165

Froehlich HE, Gentry RR, Rust MB, Grimm D, Halpern BS (2017) Public perceptions of aquaculture: evaluating spatiotemporal patterns of sentiment around the world. PLoS One 12(1):e0169281. https://doi.org/10.1371/journal.pone.0169281

Froehlich HE, Jacobsen NS, Essington TE, Clavelle T, Halpern BS (2018a) Avoiding the ecological limits of forage fish for fed aquaculture. Nat Sustain (June):1–6. https://doi.org/10.1038/s41893-018-0077-1

Froehlich HE, Gentry RR, Halpern BS (2018b) Global change in marine aquaculture production potential under climate change. Nat Ecol Evol 2(11):1745. https://doi.org/10.1038/s41559-018-0669-1

Fu W, Primeau F, Moore JK, Lindsay K, Randerson JT (2018) Reversal of increasing tropical ocean hypoxia trends with sustained climate warming. Global Biogeochem Cycles 32(4):551–564. https://doi.org/10.1002/2017GB005788

Gaichas SK, Seagraves RJ, Coakley JM, DePiper GS, Guida VG, Hare JA, Rago PJ et al (2016) A framework for incorporating species, fleet, habitat, and climate interactions into fishery management. Front Mar Sci 3:105. https://doi.org/10.3389/fmars.2016.00105

Gaines SD, Costello C, Owashi B, Mangin T, Bone J, García Molinos J, Burden M et al (2018) Improved fisheries management could offset many negative effects of climate change. Sci Adv 4(8):eaao1378. https://doi.org/10.1126/sciadv.aao1378

Garai J (2014) The impacts of climate change on the livelihoods of coastal people in Bangladesh: a sociological study. In: Filho WL, Alves F, Caeiro S, Azeiteiro UM (eds) International perspectives on climate change: Latin America and beyond. Springer, Cham, pp 151–163. https://doi.org/10.1007/978-3-319-04489-7_11

García Molinos J, Halpern BS, Schoeman DS, Brown CJ, Kiessling W, Moore PJ, Pandolfi JM et al (2016) Climate velocity and the future global redistribution of marine biodiversity. Nat Clim Change 6(1):83–88. https://doi.org/10.1038/nclimate2769

Gattuso J-P, Magnan A, Billé R, Cheung WWL, Howes EL, Joos F, Allemand D et al (2015) Contrasting futures for ocean and society from different anthropogenic CO_2 emissions scenarios. Science 349(6243):aac4722. https://doi.org/10.1126/science.aac4722

Gattuso J-P, Magnan AK, Bopp L, Cheung WWL, Duarte CM, Hinkel J, Mcleod E et al (2018) Ocean solutions to address climate change and its effects on marine ecosystems. Front Mar Sci 5:337. https://doi.org/10.3389/fmars.2018.00337

Gazeau F, Parker LM, Comeau S, Gattuso J-P, O'Connor WA, Martin S, Pörtner H-O et al (2013) Impacts of ocean acidification on marine shelled molluscs. Mar Biol 160(8):2207–2245. https://doi.org/10.1007/s00227-013-2219-3

Gentry RR, Froehlich HE, Grimm D, Kareiva P, Parke M, Rust M, Gaines SD et al (2017) Mapping the global potential for marine aquaculture. Nat Ecol Evol 1(9):1317. https://doi.org/10.1038/s41559-017-0257-9

Gentry RR, Ruff EO, Lester SE (2019) Temporal patterns of adoption of mariculture innovation globally. Nat Sustain 2(10):949–956. https://doi.org/10.1038/s41893-019-0395-y

Ghermandi A, Nunes PALD (2013) A global map of coastal recreation values: results from a spatially explicit meta-analysis. Ecol Econ 86:1–15

Gjedrem T, Baranski M (2009) Selective breeding in aquaculture: an introduction. Springer, Dordrecht

Gjedrem T, Robinson N, Rye M (2012) The importance of selective breeding in aquaculture to meet future demands for animal protein: a review. Aquaculture 350–53(June):117–129. https://doi.org/10.1016/j.aquaculture.2012.04.008

Golden CD, Allison EH, Cheung WWL, Dey MM, Halpern BS, McCauley DJ, Smith M et al (2016) Nutrition: fall in fish catch threatens human health. Nature 534(7607):317–320. https://doi.org/10.1038/534317a

Graham NAJ, Wilson SK, Jennings S, Polunin NVC, Robinson JAN, Bijoux JP, Daw TM (2007) Lag effects in the impacts of mass coral bleaching on coral reef fish, fisheries, and ecosystems. Conserv Biol 21(5):1291–1300

Green MA, Waldbusser GG, Hubazc L, Cathcart E, Hall J (2013) Carbonate mineral saturation state as the recruitment cue for settling bivalves in marine muds. Estuaries Coast 36(1):18–27. https://doi.org/10.1007/s12237-012-9549-0

Guillotreau P, Campling L, Robinson J (2012) Vulnerability of small island fishery economies to climate and institutional changes. Curr Opin Environ Sustain 4(3):287–291. https://doi.org/10.1016/j.cosust.2012.06.003

Gutiérrez NL, Hilborn R, Defeo O (2011) Leadership, social capital and incentives promote successful fisheries. Nature 470(7334):386–389. https://doi.org/10.1038/nature09689

Hare JA, Morrison WE, Nelson MW, Stachura MM, Teeters EJ, Griffis RB, Alexander MA et al (2016) A vulnerability assessment of fish and invertebrates to climate change on the Northeast U.S. continental shelf. PLoS One 11(2):e0146756. https://doi.org/10.1371/journal.pone.0146756

Harper S, Grubb C, Stiles M, Sumaila UR (2017) Contributions by women to fisheries economies: insights from five maritime countries. Coast Manag 45(2):91–106. https://doi.org/10.1080/08920753.2017.1278143

Harris KE, DeGrandpre MD, Hales B (2013) Aragonite saturation state dynamics in a coastal upwelling zone. Geophys Res Lett 40(11):2720–2725. https://doi.org/10.1002/grl.50460

Hartmann DL, Klein Tank AMG, Rusticucci M, Alexander LV, Brönnimann S, Abdul Rahman Charabi Y, Dentener FJ et al (2013) Observations: atmosphere and surface. In: Climate change 2013 the physical science basis: working group I contribution to the fifth assessment report of the Intergovernmental Panel on Climate Change. Cambridge University Press, Cambridge, pp 159–254

Hasan M (2017) Feeding global aquaculture growth. FAO aquaculture newsletter no. 56. Food and Agriculture Organization of the United Nations, Rome

Havice E (2013) Rights-based management in the Western and Central Pacific Ocean tuna fishery: economic and environmental change under the vessel day scheme. Mar Policy 42(November):259–267. https://doi.org/10.1016/j.marpol.2013.03.003

Hazen EL, Scales KL, Maxwell SM, Briscoe DK, Welch H, Bograd SJ, Bailey H et al (2018) A dynamic ocean management tool to reduce bycatch and support sustainable fisheries. Sci Adv 4(5):eaar3001. https://doi.org/10.1126/sciadv.aar3001

Henson SA, Beaulieu C, Ilyina T, John JG, Long M, Séférian R, Tjiputra J et al (2017) Rapid emergence of climate change in environmental drivers of marine ecosystems. Nat Commun 8:14682. https://doi.org/10.1038/ncomms14682

Hicks CC, Cohen PJ, Graham NAJ, Nash KL, Allison EH, D'Lima C, Mills DJ et al (2019) Harnessing global fisheries to tackle micronutrient deficiencies. Nature 574:1–4. https://doi.org/10.1038/s41586-019-1592-6

Hilborn R, Costello C (2018) The potential for blue growth in marine fish yield, profit and abundance of fish in the ocean. Mar Policy 87(January):350–355. https://doi.org/10.1016/j.marpol.2017.02.003

Himes-Cornell A, Allen S, Auad G, Boatman M, Clay PM, Herrick S, Kotowicz D et al (2013) Impacts of climate change on human uses of the ocean and ocean services. In: Griffis R, Howard J (eds) Oceans and marine resources in a changing climate: a technical input to the 2013 national climate assessment. NCA regional input reports. Island Press/Center for Resource Economics, Washington, DC, pp 64–118. https://doi.org/10.5822/978-1-61091-480-2_4

Hixon MA, Johnson DW, Sogard SM (2014) BOFFFFs: on the importance of conserving old-growth age structure in fishery populations. ICES J Mar Sci 71(8):2171–2185. https://doi.org/10.1093/icesjms/fst200

Hobday AJ, Spillman CM, Eveson JP, Hartog JR (2016) Seasonal forecasting for decision support in marine fisheries and aquaculture. Fish Oceanogr 25(S1):45–56. https://doi.org/10.1111/fog.12083

Hoegh-Guldberg O (1999) Climate change, coral bleaching and the future of the world's coral reefs. Mar Freshw Res 50(8):839–866. https://doi.org/10.1071/mf99078

Hoegh-Guldberg O (2015) Reviving the ocean economy: the case for action. WWF International, Gland

Hoegh-Guldberg O, Poloczanska ES, Skirving W, Dove S (2017) Coral reef ecosystems under climate change and ocean acidification. Front Mar Sci 4:158. https://doi.org/10.3389/fmars.2017.00158

Hoegh-Guldberg O, Jacob D, Taylor M, Bindi M, Brown S, Camilloni I, Diedhiou A et al (2018) Chapter 3: Impacts of 1.5°C global warming on natural and human systems. In: Global warming of 1.5°C: an IPCC special report on the impacts of global warming of 1.5°C above pre-industrial levels and related global greenhouse gas emission pathways, in the context of strengthening the global response to the threat of climate change. Intergovernmental Panel on Climate Change, Geneva. http://pure.iiasa.ac.at/id/eprint/15518/

Hoegh-Guldberg O, Caldeira K, Chopin T, Gaines S, Haugan P, Hemer M, Howard J et al (2019) The ocean as a solution to climate change: five opportunities for action. World Resources Institute, Washington, DC. http://www.oceanpanel.org/climate

Holbrook SJ, Schmitt RJ, Stephens JS (1997) Changes in an assemblage of temperate reef fishes associated with a climate shift. Ecol Appl 7(4):1299–1310. https://doi.org/10.1890/1051-0761(1997)007[1299:CIAAOT]2.0.CO;2

Hollowed AB, Barange M, Beamish RJ, Brander K, Cochrane K, Drinkwater K, Foreman MGG et al (2013) Projected impacts of climate change on marine fish and fisheries. ICES J Mar Sci 70(5):1023–1037. https://doi.org/10.1093/icesjms/fst081

Holsman K, Samhouri J, Cook G, Hazen E, Olsen E, Dillard M, Kasperski S et al (2017) An ecosystem-based approach to marine risk assessment. Ecosyst Health Sustain 3(1):e01256

Holsman KK, Hazen EL, Haynie A, Gourguet S, Hollowed A, Bograd SJ, Samhouri JF et al (2019) Towards climate resiliency in fisheries management. ICES J Mar Sci 76:fsz031. https://doi.org/10.1093/icesjms/fsz031

Hughes TP, Bellwood DR, Folke C, Steneck RS, Wilson J (2005) New paradigms for supporting the resilience of marine ecosystems. Trends Ecol Evol 20(7):380–386. https://doi.org/10.1016/j.tree.2005.03.022

Hughes TP, Kerry JT, Álvarez-Noriega M, Álvarez-Romero JG, Anderson KD, Baird AH, Babcock RC et al (2017) Global warming and recurrent mass bleaching of corals. Nature 543(7645):373–377. https://doi.org/10.1038/nature21707

Hughes TP, Anderson KD, Connolly SR, Heron SF, Kerry JT, Lough JM, Baird AH et al (2018) Spatial and temporal patterns of mass bleaching of corals in the anthropocene. Science 359(6371):80–83

Hutchings JA, Fraser DJ (2008) The nature of fisheries- and farming-induced evolution. Mol Ecol 17(1):294–313. https://doi.org/10.1111/j.1365-294X.2007.03485.x

IPCC (2018) In: Masson- Delmotte V, Zhai P, Pörtner H-O, Roberts D, Skea J, Shukla PR, Pirani A et al (eds) Global warming of 1.5°C. Intergovernmental Panel on Climate Change, Geneva

IPCC (2019) In: Pörtner HO, Roberts DC, Masson-Delmotte V, Zhai P, Tignor M, Poloczanska E, Mintenbeck K et al (eds) IPCC Special Report on the Ocean and Cryosphere in a Changing Climate (SROCC). Intergovernmental Panel on Climate Change, Geneva

IPCC (Intergovernmental Panel on Climate Change) (2014) Climate change 2014: synthesis report. Contribution of Working Groups I, II and III to the fifth assessment report on the Intergovernmental Panel on Climate Change. Intergovernmental Panel on Climate Change, Geneva. https://www.ipcc.ch/report/ar5/syr/

Irvine PJ, Ridgwell A, Lunt DJ (2010) Assessing the regional disparities in geoengineering impacts. Geophys Res Lett 37(18). https://doi.org/10.1029/2010GL044447

Irz X, Stevenson JR, Tanoy A, Villarante P, Morissens P (2007) The equity and poverty impacts of aquaculture: insights from the Philippines. Dev Policy Rev 25:495–516. https://doi.org/10.1111/j.1467--7679.2007.00382.x

Jackson A, Newton RW (2016) Project to model the use of fisheries by-products in the production of marine ingredients with special reference to omega-3 fatty acids EPA and DHA. University of Stirling and The Marine Ingredients Organization

Jacox MG, Edwards CA (2011) Effects of stratification and shelf slope on nutrient supply in coastal upwelling regions. J Geophys Res Oceans 116(C3):1–17. https://doi.org/10.1029/2010JC006547

James RK, Silva R, van Tussenbroek BI, Escudero-Castillo M, Mariño-Tapia I, Dijkstra HA, van Westen RM et al (2019) Maintaining tropical beaches with seagrass and algae: a promising alternative to engineering solutions. BioScience 69:136–142. https://doi.org/10.1093/biosci/biy154

Janssen K, Chavanne H, Berentsen P, Komen H (2016) Impact of selective breeding on European aquaculture. In: Aquaculture, international symposium on genetics in Aquaculture XII (ISGA XII), vol 472, pp 8–16. https://doi.org/10.1016/j.aquaculture.2016.03.012.

Jensen Ø, Dempster T, Thorstad EB, Uglem I, Fredheim A (2010) Escapes of fishes from Norwegian Sea-Cage Aquaculture: causes, consequences and prevention. Aquac Environ Interact 1(1):71–83. https://doi.org/10.3354/aei00008

Jevrejeva S, Jackson LP, Grinsted A, Lincke D, Marzeion B (2018) Flood damage costs under the sea level rise with warming of 1.5°C and 2°C. Environ Res Lett 13(7). https://iopscience.iop.org/article/10.1088/1748-9326/aacc76

Jewett L, Romanou A (2017) Ocean acidification and other ocean changes. In: Wuebbles DJ, Fahey DW, Hibbard KA, Dokken DJ, Stewart BC, Maycock TK (eds) Climate science special report: fourth national climate assessment, vol I. U.S. Global Change Research Program, Washington, DC, pp 364–392. https://science2017.globalchange.gov/chapter/13/

Jha M, Arnold JG, Gassman PW, Giorgi F, Gu RR (2006) Climate change sensitivity assessment on upper Mississippi River Basin stream flows using SWAT. J Am Water Resour Assoc 42(4):997–1015. https://doi.org/10.1111/j.1752-1688.2006.tb04510.x

Jones A, Haywood J, Boucher O (2009) Climate impacts of geoengineering marine stratocumulus clouds. J Geophys Res Atmos 114(D10). https://doi.org/10.1029/2008JD011450

Karim M, Castine S, Brooks A, Beare D, Beveridge M, Phillips M (2014) Asset or liability? Aquaculture in a natural disaster prone area. Ocean Coast Manag 96:188–197

Kasperski S, Holland DS (2013) Income diversification and risk for fishermen. Proc Natl Acad Sci 110(6):2076–2081. https://doi.org/10.1073/pnas.1212278110

Kato Y, Fujinaga K, Nakamura K, Takaya Y, Kitamura K, Ohta J, Toda R et al (2011) Deep-sea mud in the Pacific Ocean as a potential resource for rare-earth elements. Nat Geosci 4:535–539. https://doi.org/10.1038/ngeo1185

Kirtman B, Power SB, Adedoyin AJ, Boer GJ, Bojariu R, Camilloni I, Doblas-Reyes F et al (2013) Near-term climate change: projections and predictability. In: Climate change 2013: the physical science basis. Contribution of Working Group I to the fifth assessment report of the Intergovernmental Panel on Climate Change. Cambridge University Press, Cambridge, pp 953–1028

Kleisner KM, Fogarty MJ, McGee S, Hare JA, Moret S, Perretti CT, Saba VS (2017) Marine species distribution shifts on the US Northeast Continental Shelf under continued ocean warming. Prog Oceanogr 153:24–36. https://doi.org/10.1016/j.pocean.2017.04.001

de Suarez JM, Cicin-Sain B, Wowk K, Payet R, Hoegh-Guldberg O (2014) Ensuring survival: oceans, climate and security. Ocean Coast Manag 90:27–37. https://doi.org/10.1016/j.ocecoaman.2013.08.007

Klinger DH, Levin SA, Watson JR (2017) The growth of finfish in global open-ocean aquaculture under climate change. Proc R Soc B Biol Sci 284(1864):20170834. https://doi.org/10.1098/rspb.2017.0834

Klinger D, Naylor R (2012) Searching for solutions in aquaculture: charting a sustainable course. Annu Rev Env Resour 37(1):247–276. https://doi.org/10.1146/annurev-environ-021111-161531

Kopp RE, Horton RM, Little CM, Mitrovica JX, Oppen-Heimer M, Rasmussen DJ, Strauss BH, Tebaldi C (2014) Probabilistic 21st and 22nd century sea-level projections at a global network of tide-gauge sites. Earths Future 2(8):383–406. https://doi.org/10.1002/2014EF000239

Kroeker KJ, Kordas RL, Crim R, Hendriks IE, Ramajo L, Singh GS, Duarte CM, Gattuso J-P (2013) Impacts of ocean acidification on marine organisms: quantifying sensitivities and interaction with warming. Glob Chang Biol 19(6):1884–1896

Kubicek A, Breckling B, Hoegh-Guldberg O, Reuter H (2019) Climate change drives trait-shifts in coral reef communities. Sci Rep 9(1):3721. https://doi.org/10.1038/s41598-019-38962-4

Kumagai NH, García Molinos J, Yamano H, Takao S, Fujii M, Yamanaka Y (2018) Ocean currents and herbivory drive macroalgae-to-coral community shift under climate warming. Proc Natl Acad Sci 115(36):8990–8995. https://doi.org/10.1073/pnas.1716826115

Kwok R (2018) Arctic sea ice thickness, volume, and multiyear ice coverage: losses and coupled variability (1958–2018). Environ Res Lett 13(10):105005. https://doi.org/10.1088/1748-9326/aae3ec

Lafferty KD, Harvell CD, Conrad JM, Friedman CS, Kent ML, Kuris AM, Powell EN et al (2015) Infectious diseases affect marine fisheries and aquaculture economics. Ann Rev Mar Sci 7(1):471–496. https://doi.org/10.1146/annurev-marine-010814-015646

Lam VWY, Cheung WWL, Reygondeau G, Sumaila UR (2016) Projected change in global fisheries revenues under climate change. Sci Rep 6:32607. https://doi.org/10.1038/srep32607

Lambert E, Hunter C, Pierce GJ, MacLeod CD (2010) Sustainable whale-watching tourism and climate change: towards a framework of resilience. J Sustain Tour 18(3):409–427

Lavery PS, Mateo M-Á, Serrano O, Rozaimi M (2013) Variability in the carbon storage of seagrass habitats and its implications for global estimates of blue carbon ecosystem service. PLoS One 8(9):e73748

Lenzen M, Sun Y-Y, Faturay F, Ting Y-P, Geschke A, Malik A (2018) The carbon footprint of global tourism. Nat Clim Change 8(6):522. https://doi.org/10.1038/s41558-018-0141-x

Lester SE, Halpern BS, Grorud-Colvert K, Lubchenco J, Ruttenberg BI, Gaines SD, Airamé S et al (2009) Biological effects within no-take marine reserves: a global synthesis. Mar Ecol Prog Ser 384:33–46

Lester SE, Stevens JM, Gentry RR, Kappel CV, Bell TW, Costello CJ, Gaines SD et al (2018) Marine spatial planning makes room for offshore aquaculture in crowded coastal waters. Nat Commun 9(1):1–13. https://doi.org/10.1038/s41467-018-03249-1

Levin LA (2002) Deep-ocean life where oxygen is scarce: oxygen-deprived zones are common and might become more so with climate change. here life hangs on, with some unusual adaptations. Am Sci 90(5):436–444

Li S, Yang Z, Nadolnyak D, Zhang Y, Luo Y (2014) Economic impacts of climate change: profitability of freshwater aquaculture in China. Aquacult Res 47(5):1537–1548. https://doi.org/10.1111/are.12614

Lind CE, Ponzoni RW, Nguyen NH, Khaw HL (2012) Selective breeding in fish and conservation of genetic resources for aquaculture. Reprod Domest Anim 47(s4):255–263. https://doi.org/10.1111/j.1439-0531.2012.02084.x

Liu W, Xie S-P, Liu Z, Zhu J (2017) Overlooked possibility of a collapsed Atlantic Meridional Overturning Circulation in warming climate. Sci Adv 3(1):e1601666. https://doi.org/10.1126/sciadv.1601666

Loo YY, Billa L, Singh A (2015) Effect of climate change on seasonal monsoon in Asia and its impact on the variability of monsoon rainfall in Southeast Asia. Geosci Front 6(6):817–823. https://doi.org/10.1016/j.gsf.2014.02.009

Lotze HK, Tittensor DP, Bryndum-Buchholz A, Eddy TD, Cheung WWL, Galbraith ED, Barange M et al (2019) Global ensemble projections reveal trophic amplification of ocean biomass declines with climate change. Proc Natl Acad Sci 116(26):12907–12912. https://doi.org/10.1073/pnas.1900194116

Lovelock CE, Atwood T, Baldock J, Duarte CM, Hickey S, Lavery PS, Masque P et al (2017) Assessing the risk of carbon dioxide emissions from blue carbon ecosystems. Front Ecol Environ 15(5):257–265

Matin N, Forrester J, Ensor J (2018) What is equitable resilience? World Dev 109:197–205. https://doi.org/10.1016/j.worlddev.2018.04.020

McClanahan T, Polunin N, Done T (2002) Ecological states and the resilience of coral reefs. Conserv Ecol 6(2):18. https://doi.org/10.5751/ES-00461-060218

Mcleod E, Chmura GL, Bouillon S, Salm R, Björk M, Duarte CM, Lovelock CE et al (2011) A blueprint for blue carbon: toward an improved understanding of the role of vegetated coastal habitats in sequestering CO2. Front Ecol Environ 9(10):552–560

Mcleod E, Salm R, Green A, Almany J (2009) Designing marine protected area networks to address the impacts of climate change. Front Ecol Environ 7(7):362–370

Meerow S, Pajouhesh P, Miller TR (2019) Social equity in urban resilience planning. Local Environ 24(9):793–808. https://doi.org/10.1080/13549839.2019.1645103

Melnychuk MC, Peterson E, Elliott M, Hilborn R (2017) Fisheries management impacts on target species status. Proc Natl Acad Sci U S A 114(1):178–183. https://doi.org/10.1073/pnas.1609915114

Melnychuk MC, Essington TE, Branch TA, Heppell SS, Jensen OP, Link JS, Martell SJD et al (2011) Can catch share fisheries better track management targets? Fish Fish 13(3):267–290. https://doi.org/10.1111/j.1467-2979.2011.00429.x

Mikulewicz M (2019) Thwarting adaptation's potential? A critique of resilience and climate-resilient development. Geoforum 104:267–282. https://doi.org/10.1016/j.geoforum.2019.05.010

Miller KA, Munro GR (2004) Climate and cooperation: a new perspective on the management of shared fish stocks. Mar Resour Econ 19(3):367–393

Miller KA, Thompson KF, Johnston P, Santillo D (2018a) An overview of seabed mining including the current state of development, environmental impacts, and knowledge gaps. Front Mar Sci 4:418. https://doi.org/10.3389/fmars.2017.00418

Miller TJ, Jones CM, Hanson C, Heppell S, Jensen OP, Livingston P, Lorenzen K et al (2018b) Scientific considerations informing Magnuson-Stevens Fishery Conservation and Management Act Reauthorization. Fisheries 43(11):533–541. https://doi.org/10.1002/fsh.10179

Moore SS, Seavy NE, Gerhart M (2013) Scenario planning for climate change adaptation: a guidance for resource managers. Point Blue Conservation Science and the California Coastal Conservancy. http://www.prbo.org/refs/files/12263_Moore2013.pdf

Moreno A, Amelung B (2009) Climate change and coastal & marine tourism: review and analysis. J Coast Res II:1140–1144

Moreno A, Revenga C (2014) The system of territorial use rights in fisheries in Chile. The Nature Conservancy, Arlington

Nguyen KAT, Pongthanapanich T (2016) Aquaculture insurance in Viet Nam: experiences from the pilot programme. FAO fisheries and aquaculture circular no. 1133. Food and Agriculture Organization of the United Nations, Rome

NOAA (U.S. National Oceanic and Atmospheric Administration) (2018) Status of stocks 2017: annual report to congress on the status of U.S. Fisheries. NOAA, Silver Spring, pp 1–7

OECD (Organisation for Economic Co-operation and Development) (2016) The ocean economy in 2030. OECD Publishing, Paris

Ojea E, Pearlman I, Gaines SD, Lester SE (2017) Fisheries regulatory regimes and resilience to climate change. Ambio 46(4):399–412. https://doi.org/10.1007/s13280-016-0850-1

Oliva-Teles A, Enes P, Peres H (2015) Replacing fishmeal and fish oil in industrial aquafeeds for carnivorous fish. In: Feed and feeding practices in aquaculture. Porto University, Porto, pp 203–233. https://doi.org/10.1016/B978-0-08-100506-4.00008-8

Oschlies A, Brandt P, Stramma L, Schmidtko S (2018) Drivers and mechanisms of ocean deoxygenation. Nat Geosci 11(7):467. https://doi.org/10.1038/s41561-018-0152-2

Parkinson CL (2019) A 40-Y record reveals gradual Antarctic Sea ice increases followed by decreases at rates far exceeding the rates seen in the Arctic. Proc Natl Acad Sci 116:14414. https://doi.org/10.1073/pnas.1906556116

Pascual U, Balvanera P, Díaz S, Pataki G, Roth E, Stenseke M, Watson RT et al (2017) Valuing nature's contributions to people: the IPBES approach. Curr Opin Environ Sustain 26:7–16

Pecl GT, Araújo MB, Bell JD, Blanchard J, Bonebrake TC, Chen I-C, Clark TD et al (2017) Biodiversity redistribution under climate change: impacts on ecosystems and human well-being. Science 355(6332):eaai9214. https://doi.org/10.1126/science.aai9214

Pendleton L, Donato DC, Murray BC, Crooks S, Jenkins WA, Sifleet S, Craft C et al (2012) Estimating global 'blue carbon' emissions from conversion and degradation of vegetated coastal ecosystems. PLoS One 7(9):e43542

Perry AL, Low PJ, Ellis JR, Reynolds JD (2005) Climate change and distribution shifts in marine fishes. Science 308(5730):1912–1915. https://doi.org/10.1126/science.1111322

Pervez MS, Henebry GM (2015) Assessing the impacts of climate and land use and land cover change on the freshwater availability in the Brahmaputra River Basin. J Hydrol Reg Stud 3:285–311. https://doi.org/10.1016/j.ejrh.2014.09.003

Pinsky ML, Worm B, Fogarty MJ, Sarmiento JL, Levin SA (2013) Marine taxa track local climate velocities. Science 341(6151):1239–1242. https://doi.org/10.1126/science.1239352

Pinsky ML, Reygondeau G, Caddell R, Palacios-Abrantes J, Spijkers J, Cheung WWL (2018) Preparing ocean governance for species on the move. Science 360(6394):1189–1191. https://doi.org/10.1126/science.aat2360

Pinsky M, Mantua N (2014) Emerging adaptation approaches for climate-ready fisheries management. Oceanography 27(4):146–159. https://doi.org/10.5670/oceanog.2014.93

Polasky S, Segerson K (2009) Integrating ecology and economics in the study of ecosystem services: some lessons learned. Annu Rev Resour Econ 1(1):409–434. https://doi.org/10.1146/annurev.resource.050708.144110

Poloczanska ES, Brown CJ, Sydeman WJ, Kiessling W, Schoeman DS, Moore PJ, Brander K et al (2013) Global imprint of climate change on marine life. Nat Clim Change 3(10):919–925. https://doi.org/10.1038/nclimate1958

Poloczanska ES, Burrows MT, Brown CJ, García Molinos J, Halpern BS, Hoegh-Guldberg O, Kappel CV et al (2016) Responses of marine organisms to climate change across oceans. Front Mar Sci 3:62

Ponzoni RW, Nguyen NH, Khaw HL, Ninh NH (2008) Accounting for genotype by environment interaction in economic appraisal of genetic improvement programs in common carp Cyprinus carpio. Aquaculture 285(1):47–55. https://doi.org/10.1016/j.aquaculture.2008.08.012

Pörtner H-O, Karl DM, Boyd PW, Cheung W, Lluch-Cota SE, Nojiri Y, Schmidt DN et al (2014) Ocean Systems. In: Climate change 2014: impacts, adaptation, and vulnerability. Part A: global and sectoral aspects. contribution of Working Group II to the fifth assessment report of the Intergovernmental Panel on Climate Change. Cambridge University Press, Cambridge, pp 411–484

Price C, Black KD, Hargrave BT, Morris JA Jr (2015) Marine cage culture and the environment: effects on water quality and primary production. Aquac Environ Interact 6(2):151–174. https://doi.org/10.3354/aei00122

Price N, Muko S, Legendre L, Steneck R, van Oppen M, Albright R, Ang P Jr et al (2019) Global biogeography of coral recruitment: tropical decline and subtropical increase. Mar Ecol Prog Ser 621(July):1–17. https://doi.org/10.3354/meps12980

Punt AE, Butterworth DS, De Oliveira JAA, Haddon M (2016) Management strategy evaluation: best practices. Fish Fish 17(2):303–334. https://doi.org/10.1111/faf.12104

Punt AE, A'mar T, Bond NA, Butterworth DS, de Moor CL, De Oliveira JAA, Haltuch MA et al (2014) Fisheries management under climate and environmental uncertainty: control rules and performance simulation. ICES J Mar Sci 71(8):2208–2220. https://doi.org/10.1093/icesjms/fst057

Ren L, Arkin P, Smith TM, Shen SSP (2013) Global precipitation trends in 1900–2005 from a reconstruction and coupled model simulations. J Geophys Res Atmos 118(4):1679–1689. https://doi.org/10.1002/jgrd.50212

Richards DR, Friess DA (2016) Rates and drivers of mangrove deforestation in Southeast Asia, 2000–2012. Proc Natl Acad Sci 113(2):344–349

Richards LJ, Maguire J-J (1998) Recent international agreements and the precautionary approach: new directions for fisheries management science. Can J Fish Aquat Sci 55(6):1545–1552. https://doi.org/10.1139/f98-043

Roberts CM, O'Leary BC, McCauley DJ, Cury PM, Duarte CM, Lubchenco J, Pauly D et al (2017) Marine reserves can mitigate and promote adaptation to climate change. Proc Natl Acad Sci 114(24):6167–6175

Roman J, McCarthy JJ (2010) The whale pump: marine mammals enhance primary productivity in a coastal basin. PLoS One 5(10):e13255. https://doi.org/10.1371/journal.pone.0013255

Rosa R, Marques A, Nunes ML (2014) Mediterranean aquaculture in a changing climate. In: The Mediterranean Sea, pp 605–616. https://doi.org/10.1007/978-94-007-6704-1_37

Rosa R, Seibel BA (2008) Synergistic effects of climate-related variables suggest future physiological impairment in a top oceanic predator. Proc Natl Acad Sci 105(52):20776–20780. https://doi.org/10.1073/pnas.0806886105

Rowley AF, Cross ME, Culloty SC, Lynch SA, Mackenzie CL, Morgan E, O'Riordan RM et al (2014) The potential impact of climate change on the infectious diseases of commercially important shellfish populations in the Irish Sea—a review. ICES J Mar Sci 71(4):741–759. https://doi.org/10.1093/icesjms/fst234

Ruiz-Ramírez JD, Euán-Ávila JI, Rivera-Monroy VH (2019) Vulnerability of coastal resort cities to mean sea level rise in the Mexican Caribbean. Coast Manag 47(1):23–43. https://doi.org/10.1080/08920753.2019.1525260

Rykaczewski RR, Dunne JP, Sydeman WJ, García-Reyes M, Black BA, Bograd SJ (2015) Poleward displacement of coastal upwelling-favorable winds in the ocean's eastern boundary currents through the 21st century. Geophys Res Lett 42(15):6424–6431. https://doi.org/10.1002/2015GL064694

Sae-Lim P, Kause A, Mulder HA, Olesen I (2017) Breeding and genetics symposium: climate change and selective breeding in aquaculture. J Anim Sci 95(4):1801–1812. https://doi.org/10.2527/jas.2016.1066

Sae-Lim P, Mulder H, Gjerde B, Koskinen H, Lillehammer M, Kause A (2015) Genetics of growth reaction norms in farmed rainbow trout. PLoS One 10(8):e0135133

Sainz J, Di Lorenzo E, Bell TW, Gaines S, Miller R, Lenihan H (2019) Spatial planning of marine aquaculture under climate decadal variability: a case study for mussel farms in California. Front Mar Sci 6:253

Schatz VJ, Proelss A, Liu N (2019) The 2018 agreement to prevent unregulated high seas fisheries in the Central Arctic Ocean: a critical analysis. Int J Mar Coast Law 34(2):195–244. https://doi.org/10.1163/15718085-23342015

Scott D, Gössling S, Hall CM (2012) International tourism and climate change. Wiley Interdiscip Rev Clim Change 3(3):213–232. https://doi.org/10.1002/wcc.165

Sea Grant California; Sea Grant, University of Georgia; Virginia Coastal Policy Center, William & Mary Law School; Sea Grant Law Center; The University of Mississippi School of Law; Marine Affairs Institute; Roger Williams University School of Law; et al (2019) Overcoming impediments to shellfish aquaculture through legal research and outreach: case studies. Funded by the National Oceanic and Atmospheric Administration, U.S. Department of Commerce, p 118

Serdy A (2016) The new entrants problem in International Fisheries Law. Cambridge University Press, Cambridge. https://doi.org/10.1017/CBO9780511736148

Serrano O, Kelleway JJ, Lovelock C, Lavery PS (2019) Chapter 28: Conservation of blue carbon ecosystems for climate change mitigation and adaptation. In: Perillo GME, Wolanski E, Cahoon DR, Hopkinson CS (eds) Coastal wetlands. Elsevier, Amsterdam, pp 965–996. https://doi.org/10.1016/B978-0-444-63893-9.00028-9

Siderius C, Biemans H, Wiltshire A, Rao S, Franssen WHP, Kumar P, Gosain AK et al (2013) Snowmelt contributions to discharge of the Ganges. Sci Total Environ 468:S93–S101. https://doi.org/10.1016/j.scitotenv.2013.05.084

Siegel KJ, Cabral Reniel B, McHenry J, Ojea E, Owashi B, Lester SE (2019) Sovereign states in the Caribbean have lower social-ecological vulnerability to coral bleaching than overseas territories. Proc R Soc B Biol Sci 286(1897):20182365. https://doi.org/10.1098/rspb.2018.2365

Silbiger NJ, Nelson CE, Remple K, Sevilla JK, Quinlan ZA, Putnam HM, Fox MD et al (2018) Nutrient pollution disrupts key ecosystem functions on coral reefs. Proc R Soc B Biol Sci 285(1880):20172718

Smale DA, Wernberg T, Oliver ECJ, Thomsen M, Harvey BP, Straub SC, Burrows MT et al (2019) Marine heatwaves threaten global biodiversity and the provision of ecosystem services. Nat Clim Change 9(4):306. https://doi.org/10.1038/s41558-019-0412-1

Sokolow S (2009) Effects of a changing climate on the dynamics of coral infectious disease: a review of the evidence. Dis Aquat Organ 87(1–2):5–18

Soto D, Ross LG, Handisyde N, Bueno PB, Beveridge MC, Dabbadie L, Aguilar-Manjarrez J et al (2018) Climate change and aquaculture: vulnerability and adaptation options. In: Barange M, Bahri T, Beveridge MCM, Cochrane KL, Funge-Smith S, Poulain F (eds) Impacts of climate change on fisheries and aquaculture. Food and Agriculture Organization of the United Nations, Rome

Spalding M, Burke L, Wood SA, Ashpole J, Hutchison J, zu Ermgassen P (2017) Mapping the global value and distribution of coral reef tourism. Mar Policy 82(August):104–113. https://doi.org/10.1016/j.marpol.2017.05.014

Speers AE, Besedin EY, Palardy JE, Moore C (2016) Impacts of climate change and ocean acidification on coral reef fisheries: an integrated ecological–economic model. Ecol Econ 128:33–43

Spijkers J, Singh G, Blasiak R, Morrison TH, Le Billon P, Österblom H (2019) Global patterns of fisheries conflict: forty years of data. Glob Environ Chang 57:101921

Spijkers J, Boonstra WJ (2017) Environmental change and social conflict: the Northeast Atlantic Mackerel sispute. Reg Environ Change 17(6):1835–1851. https://doi.org/10.1007/s10113-017-1150-4

Stramma L, Schmidtko S, Levin LA, Johnson GC (2010) Ocean oxygen minima expansions and their biological impacts. Deep-Sea Res I Oceanogr Res Pap 57(4):587–595. https://doi.org/10.1016/j.dsr.2010.01.005

Sumaila UR, Cheung WWL, Lam VWY, Pauly D, Herrick S (2011) Climate change impacts on the biophysics and economics of world fisheries. Nat Clim Change 1(9):449

Sydeman WJ, García-Reyes M, Schoeman DS, Rykaczewski RR, Thompson SA, Black BA, Bograd SJ (2014) Climate change and wind intensification in coastal upwelling ecosystems. Science 345(6192):77–80. https://doi.org/10.1126/science.1251635

Tacon A, Hasan M, Metian M (2011) Demand and supply of feed ingredients for farmed fish and crustaceans: trends and prospects. FAO Fisheries and Aquaculture technical paper. Food and Agriculture Organization of the United Nations, Rome

Taylor KE, Stouffer RJ, Meehl GA (2012) An overview of CMIP5 and the experiment design. Bull Am Meteorol Soc 93(4):485–498. https://doi.org/10.1175/BAMS-D-11-00094.1

Thornalley DJR, Oppo DW, Ortega P, Robson JI, Brierley CM, Davis R, Hall IR et al (2018) Anomalously weak labrador sea convection and Atlantic overturning during the past 150 years. Nature 556(7700):227. https://doi.org/10.1038/s41586-018-0007-4

Toimil A, Díaz-Simal P, Losada IJ, Camus P (2018) Estimating the risk of loss of beach recreation value under climate change. Tour Manag 68:387–400. https://doi.org/10.1016/j.tourman.2018.03.024

Tommasi D, Stock CA, Pegion K, Vecchi GA, Methot RD, Alexander MA, Checkley DM (2017) Improved management of small pelagic fisheries through seasonal climate prediction. Ecol Appl 27(2):378–388. https://doi.org/10.1002/eap.1458

UCSB (University of California, Santa Barbara) (2019) Interactive web interface developed by the Sustainable Fisheries Group. https://sfg--ucsb.shinyapps.io/fishcast2/

UNWTO (United Nations World Tourism Organization) (2016) UNWTO tourism highlights, 2016 edition. World Tourism Organization, Madrid. https://doi.org/10.18111/9789284418145

Van Dover CL, Ardron JA, Escobar E, Gianni M, Gjerde KM, Jaeckel A, Jones DOB et al (2017) Biodiversity loss from sea mining. Nat Geosci 10:464–465. https://doi.org/10.1038/ngeo2983

van Oppen MJH, Gates RD, Blackall LL, Cantin N, Chakravarti LJ, Chan WY, Cormick C et al (2017) Shifting paradigms in restoration of the world's coral reefs. Glob Chang Biol 23(9):3437–3448

van Vuuren DP, Edmonds J, Kainuma M, Riahi K, Thomson A, Hibbard K, Hurtt GC et al (2011) The representative concentration pathways: an overview. Clim Change 109(1):5. https://doi.org/10.1007/s10584-011-0148-z

Vaughan NE, Lenton TM (2011) A review of climate geoengineering proposals. Clim Change 109(3):745–790. https://doi.org/10.1007/s10584-011-0027-7

Verges A, Steinberg PD, Hay ME, Poore AGB, Campbell AH, Ballesteros E, Heck KL et al (2014) The tropicalization of temperate marine ecosystems: climate-mediated changes in herbivory and community phase shifts. Proc R Soc B Biol Sci 281(1789):20140846. https://doi.org/10.1098/rspb.2014.0846

Visser ME (2016) Phenology: interactions of climate change and species. Nature 535(7611):236. https://doi.org/10.1038/nature18905

Walther G-R, Post E, Convey P, Menzel A, Parmesan C, Beebee TJC, Fromentin J-M et al (2002) Ecological responses to recent climate change. Nature 416(6879):389

Wang D, Gouhier TC, Menge BA, Ganguly AR (2015) Intensification and spatial homogenization of coastal upwelling under climate change. Nature 518(7539):390. https://doi.org/10.1038/nature14235

Wang G, Cai W, Gan B, Wu L, Santoso A, Lin X, Chen Z et al (2017) Continued increase of extreme El Niño frequency long after 1.5°C warming stabilization. Nat Clim Change 7(8):568. https://doi.org/10.1038/NCLIMATE3351

Wardle AR (2017) Farming the oceans: opportunities and regulatory challenges for U.S. Marine Aquaculture development, 39. Reason Foundation, Los Angeles

Watanabe A, Nakamura T (2019) Carbon dynamics in coral reefs. In: Kuwae T, Hori M (eds) Blue carbon in shallow coastal ecosystems: carbon dynamics, policy, and implementation. Springer, Singapore, pp 273–293. https://doi.org/10.1007/978-981-13-1295-3_10

Weatherdon LV, Magnan AK, Rogers AD, Sumaila UR, Cheung WWL (2016) Observed and projected impacts of climate change on marine fisheries, aquaculture, coastal tourism, and human health: an update. Front Mar Sci 3. https://doi.org/10.3389/fmars.2016.00048

Wiedenmann J, D'Angelo C, Smith EG, Hunt AN, Legiret F-E, Postle AD, Achterberg EP (2013) Nutrient enrichment can increase the susceptibility of reef corals to bleaching. Nat Clim Change 3(2):160

Williamson P, Turley C (2012) Ocean acidification in a geoengineering context. Philos Trans R Soc A Math Phys Eng Sci 370(1974):4317–4342

Xinhua Y, Pongthanapanich T, Zongli Z, Xiaojun J, Junchao M (2017) Fishery and aquaculture insurance in China. FAO Fisheries and Aquaculture circular no. 1139. Food and Agriculture Organization of the United Nations, Rome

Yang H, Lohmann G, Wei W, Dima M, Ionita M, Liu J (2016) Intensification and poleward shift of subtropical western boundary currents in a warming climate. J Geophys Res Oceans 121(7):4928–4945. https://doi.org/10.1002/2015JC011513

3 What Role for Ocean-Based Renewable Energy and Deep-Seabed Minerals in a Sustainable Future?

Abbreviations

APEI	Area of Particular Environmental Interest
BBNJ	biodiversity beyond national jurisdiction
BECCS	bioenergy with carbon capture and storage
CAGR	compound annual growth rate
CCS	carbon capture and storage
CCZ	Clarion-Clipperton Zone
CDR	carbon dioxide removal
EEZ	exclusive economic zone
FAIR	findable, accessible, interoperable, reusable
GW	gigawatt (10^9 watt)
GWh	gigawatt-hours
IEA	International Energy Agency
IPCC	Intergovernmental Panel on Climate Change
IRENA	International Renewable Energy Agency
ISA	International Seabed Authority
ITLOS	International Tribunal for the Law of the Sea
LCOE	levelised cost of electricity
LDAC	Long Distance Fleet Advisory Council
LED	low energy demand
LTC	Legal and Technical Commission, International Seabed Authority
MPA	Marine Protected Area
Mtoe	million tonnes of oil equivalent
MW	megawatt (10^6 watt)
MWh	megawatt-hours
OTEC	ocean thermal energy conversion
PV	photovoltaic
REE	rare earth element
REMP	Regional Environmental Management Plan
REY	rare earths and yttrium
SDG	Sustainable Development Goals (United Nations)
SDLO	Sustainable Development Licence to Operate
TIMES	The Integrated MARKAL-EFOM System model generator
TW	terawatt (10^{12} watt)
TWh	terawatt-hours
UNCLOS	United Nations Convention on the Law of the Sea
UNFCCC	United Nations Framework Convention on Climate Change
VME	vulnerable marine ecosystem

Originally published in:
Haugan, P.M., L.A. Levin, D. Amon, M. Hemer, H. Lily and F.G. Nielsen. 2020. What Role for Ocean-Based Renewable Energy and Deep Seabed Minerals in a Sustainable Future? Washington, DC: World Resources Institute. Available online at www.oceanpanel.org/blue-papers/ocean-energy-and-mineral-sources
Reprint by Springer International Publishing (2023) with kind permission.

Highlights

- This paper analyses the underlying tension between the need for rapid decarbonisation, including that required for scaling up ocean-based renewable energy, and the resource and environmental implications related to that metal demand, with particular attention on current proposals to mine the deep seabed.
- Building a sustainable global energy system is intimately linked to both scaling up renewable energy and finding a way to source and use rare minerals in a more sustainable way. Questions remain as to whether deep-seabed mining should be heralded as the key to a transition to a sustainable energy sector, based on whether it can be accomplished in a way that appropriately ensures a healthy and resilient ocean.
- Rapid transformation of our energy systems is required if we are to achieve the goals of the Paris Agreement and limit the global average temperature rise to 1.5 °C, or even 2 °C, above pre-industrial levels. In addition to expanding land-based renewable energy, the ocean offers significant potential for supporting this transition. However, new technologies must be implemented in a

J. Lubchenco, P. M. Haugan (eds.), *The Blue Compendium*, https://doi.org/10.1007/978-3-031-16277-0_3

sustainable way in order to avoid unintended consequences that could undermine other aspects of ocean health.

- Ocean-based renewable energy sources include offshore wind (near-surface as well as high-altitude), floating solar, marine biomass and ocean energy, which encompasses tidal range, tidal stream, wave, ocean thermal energy conversion (OTEC), current and salinity gradient.
- Offshore wind (near-surface, i.e. based on bottom- fixed or floating support structures) is presently more developed than other marine renewable energy and has reached cost parity with fossil sources of electricity.
- The trend for newer multi-megawatt wind turbine generators is to use direct-drive systems with permanent magnet generators. Since most other ocean-based renewable energy technologies are still in early phases of development with little deployment, few studies have been completed on what materials will be needed to scale up the use of these technologies. If these technologies have similar metal requirements to modern wind turbines, which is likely, implementation will rapidly increase the demand for many metals, such as lithium, cobalt, copper, silver, zinc, nickel and manganese, and rare earth elements (REEs).
- The demand for specific metals to serve the global energy transition is highly dependent on their cost. Often, alternatives to specific metals can be found. The industry is continually developing solutions that can use cheaper and more abundant resources avoiding specific costly metals.
- Selected metals and minerals are increasingly difficult to find in large quantities or high grades on land, but are present in higher concentrations in some parts of the deep seabed. As such, the deep seabed resource potential has attracted interest in mining for copper, cobalt, nickel, zinc, silver, gold, lithium, REEs and phosphorites.
- The potential to mine the deep seabed raises various environmental, legal and governance challenges, as well as possible conflicts with the United Nations Sustainable Development Goals.
- Greater knowledge of the potential environmental impacts and measures to mitigate them to levels acceptable to the global community will be crucial.
- Full analysis of the perceived positive and negative impacts is required before there can be confidence that engaging in industrial-scale deep-seabed mining would achieve a global net benefit.

1 Introduction

Scenarios for sustainable transformation of the global economy to near zero greenhouse gas emissions in 2050 in line with the Paris Agreement and the UN 2030 Agenda for Sustainable Development rely strongly on renewable energy. Offshore wind shows potential to become a globally significant supplier of electricity in these scenarios. Floating solar energy and direct ocean energy sources, such as wave, tidal and ocean thermal energy, may also contribute significantly in a range of locations, but require more policy support and understanding of potential environmental impacts in order to become significant in the transition to a sustainable global energy system.

The expanding use of batteries to electrify the transport sector is leading to increasing demand for a range of rare minerals. Renewable energy technologies, such as solar panels and wind turbines, along with electronic products and cell phones, also use these various minerals. One potential new source of minerals is the deep seabed. But the mining of these minerals raises potentially serious environmental, legal, social and rights-based challenges, as well as potential conflicts with UN Sustainable Development Goals 12, 13 and 14.

This Blue Paper focuses on the extent to which a selected subset of ocean resources, ocean-based renewable energy and deep-seabed minerals can contribute to sustainable development. Options for harvesting ocean-based renewable energy and the needs for ocean-based minerals are reviewed with a focus on scenarios where anthropogenic global warming in the twenty-first century is limited to 1.5–2 °C – in other words, where decarbonisation of the global economy has to happen fast. The deep-seabed minerals case is discussed in some detail in order to spell out the steps that would be required if deep-seabed mining were to be developed, and to weigh up the benefits, risks and alternatives.

The introductory section briefly explains the basic characteristics of ocean-based renewable energy, discusses the expected demands for minerals from ocean-based renewable energy and global energy system transformation, and ends with an introduction to deep-seabed mining. In Sect. 2, 1.5 °C scenarios, both with and without carbon capture and storage (CCS) and negative emissions in the later part of the century, are described. In Sect. 3, ocean-based renewable energy options, their technological and cost status, and projections for future development are reviewed. In Sect. 4, deep-seabed minerals and the motivations for mining them are addressed. Section 5 focuses on sustainability, including the environmental impacts of ocean-based renewable energy and deep-seabed mining. Section 6 deals with governance issues, before moving into the opportunities for action in Sect. 7.

1.1 What is Ocean-Based Renewable Energy?

Ocean-based renewable energy sources (often called marine renewable energy) include offshore wind (near-surface as well as high-altitude), floating solar, marine biomass and ocean energy, which encompasses tidal range, tidal stream, wave, ocean thermal energy conversion (OTEC), current and salinity gradient. All of these are considered in this paper except marine biomass. Harvesting of naturally growing marine biomass, as well as industrial production, is ongoing in several locations, mostly motivated by demands for food, feed or pharmaceuticals. The by-products of such production may be combusted for energy purposes, and thereby reduce the need for other energy sources. However, based on current knowledge, the global long-term significance as an energy source is believed to be limited.

Offshore wind (near-surface, i.e. based on bottom-fixed or floating support structures) is presently much more developed than the others and has reached cost parity with fossil sources of electricity in recent contracts. Offshore wind is therefore dealt with separately in Sect. 3.1. Of the others, technology for exploiting tidal range is well developed in some locations, tidal stream is developing rapidly now, and wave energy has a long history of research but no clear technology winner. OTEC, which has potential in the tropics, requires significant investment in order to capitalise on the economy of scale. Salinity gradient, which has potential where fresh water meets saline seawater, has only seen experimental-scale testing. Ocean currents, exploiting the energy contained in large-scale thermohaline ocean circulation, has considerable potential, but has challenges relating to proximity to demand, in combination with the early stage of technology. Floating solar has so far been mostly developed in fresh water for reservoirs and dams but has clear potential for ocean scale-up. High-altitude wind can be scaled up offshore once key technology has been validated, presumably first onshore. These energy sources are further described in Sect. 3.2.

1.2 Renewable Energy and the Demand for Metals

Key elements of a low-carbon emissions future are the accelerated use of wind power, solar energy and the electrification of the energy sector, including use of electric vehicles.

Construction of offshore wind turbines requires significant amounts of conventional materials, in particular steel. However, rare earth elements (REEs) are also needed, in particular in the construction of the direct-drive permanent magnet generators that are currently preferred. For offshore wind, it is the use of REEs in the generators that appears to be the biggest potential challenge when it comes to supply of minerals. Wilburn (2011) states that each megawatt (MW) of installed capacity needs 42 kilograms (kg) of neodymium and 3000 kg of copper.

Stegen (2015) provides an overview of REEs and permanent magnets in connection with renewable energies. Stegen notes that present wind turbines using direct-drive permanent magnet generators are favoured over conventional heavy gearboxes since the latter require more steel and concrete. The reduced weight of permanent magnet generators and increased reliability and efficiency is particularly attractive offshore. Permanent magnets typically use neodymium, dysprosium, praseodymium and terbium. For turbines above 10 MW, which are now beginning to be applied offshore, superconducting generators may be preferred over permanent magnet generators, again because of costs and weight. However, greater deployment of superconductors will increase demand for yttrium, another element typically considered together with REEs (included in REEs or expressed as REY, rare earths and yttrium).

Pavel et al. (2017) discuss substitution strategies for REEs in wind turbines, noting the variety of designs that are being considered and the potential for material efficiency. They do not consider the deep seabed as a source, but still conclude that the wind industry is well prepared for potential shortages in REEs in both the short and medium term. For the longer term, superconductors are being considered. A considerable amount of REEs, including yttrium at high concentration in seafloor mud, was recently documented in the Japanese exclusive economic zone (EEZ) (Takaya et al. 2018).

Goodenough et al. (2018) note that very little mineral-processing research on REEs took place outside of China during the 1980s, 1990s and 2000s, but this research has been accelerating in recent years after China introduced export restrictions. It remains a challenge to develop the value chain from mining through processing and separation to end-uses. Goodenough et al. (2018) also note that, within 10 years, new technological developments are likely to drive substantial changes in both processing of, and demand for, REEs.

Moving to the further requirements from the energy sector as a whole, a recent IPCC report indicates that 70–85 percent of all electricity must be from renewable sources by 2050 to limit global warming to 1.5 °C (IPCC 2018). Implementation of these renewable technologies will rapidly increase demand for many metals, including lithium, cobalt, copper, silver, zinc, nickel and manganese, and REEs and others (Arrobas et al. 2017; Sovacool et al. 2020). The projected metal demand varies greatly for the different energy sources under scenarios involving different amounts of renewable energy at different rates over the next 30 years (Arrobas et al. 2017; Dominish et al. 2019). For example, the demand for metals, such as aluminium, cobalt, nickel, lithium, iron and lead, coming from solar and wind will be twice as high under a 2 °C warming scenario than under a 4 °C scenario, but the demand from batteries would be more than 10 times higher. Offshore wind energy generation requires more metals than onshore wind due to the use of magnets; differing solar technologies use different amounts of silver, zinc and indium; and for cars, fully electric, hybrid and hydrogen fuel cells differ in their demands for lithium, lead and platinum (Arrobas et al. 2017). There is general agreement that electric car batteries will be the greatest source of increased metal demand.

Deetman et al. (2018) study scenarios for copper, tantalum, neodymium, cobalt and lithium demand up to 2050 and find that in a stringent climate policy scenario (1.5–2 °C), the demand from cars rises more rapidly than that from appliances and energy technologies. In particular, this applies to cobalt and lithium. Boubault and Maizi (2019) extend the well-known TIMES energy system model tool for electricity generation to metal need is expected to diminish as the industry transitions to even larger turbines with superconductors. The energy sector as a whole has a wider set of mineral needs but also larger flexibility to switch between alternative technological solutions. Trends and demands for the coming decade can be estimated, but it is very difficult to deduce a minimal set of required metals to enable energy system transition to a 1.5–2 °C global temperature rise requirements for the power sector using a life cycle approach. Cost-optimal deployments of different electricity generation sources in a 2 °C scenario to 2100 provide corresponding metal needs. In comparison with the baseline scenario, cobalt and aluminium are among those that increase the most.

Limiting the global average temperature rise to 1.5 °C using 100 percent renewable energy is projected to increase demand in 2050 to more than four times the existing reserves for cobalt, almost three times the reserves for lithium, and slightly more than the existing reserves for nickel (Dominish et al. 2019). Cobalt and nickel, whose demand could exceed current production rates by 2030, are driving the rapidly rising interest in mineral mining on the deep seafloor. Cobalt in particular has highly concentrated production and reserves (especially in the Democratic Republic of the Congo) and thus poses the greatest supply risk; cobalt contamination also causes severe health impacts for miners and surrounding communities (Dominish et al. 2019).

Attempts to compare various modelling studies of energy systems and metal needs (Boubault and Maizi 2019) are complicated by the different choices made in terms of scenarios, assumptions and the degree of resolution in the metals covered by each model. In conclusion, there are large uncertainties about metal needs over time horizons of longer than a decade. A hot topic for offshore wind is REEs for permanent magnets. However, this across the timeframe of 2050 to 2100. Integrated energy system models that include metal needs in a life cycle approach (Hertwich et al. 2015) are useful tools but rely on bottom-up estimates of costs of energy sources and energy conversion processes. The search for alternative technologies is intense, driven by actual costs as well as projections of future costs.

The increase in metal mining needed to address climate change (and the transition to renewable energy) is drawing increasing attention (Arrobas et al. 2017) and has led to a proposal that nationally determined contributions under the Paris Agreement identify critical minerals for energy security options and identify sourcing challenges (Sovacool et al. 2020). Population growth and rising consumption associated with an increased standard of living globally creates additional increased demand for metals, independent of climate change (Graedel et al. 2015; Ali et al. 2017).

1.3 Minerals on the Deep Seafloor

Metals and minerals of interest on the deep seafloor include primarily copper, cobalt, nickel, zinc, silver, gold, lithium, REEs and phosphorites (see Sect. 4). Many of the metals are found in polymetallic nodules on abyssal plains (covering 38 million square kilometres (km^2) at water depths of 3000–6500 metres (m)), on cobalt-rich crusts which occur on seamounts (covering over 1.7 million km^2 at 800–2500 m), and in polymetallic sulphides near mid-ocean ridges and in back-arc basins (covering 3.2 million km^2) (Fig. 3.1) (Levin et al. 2016; Miller et al. 2018; Hein and Koschinsky 2014; Petersen et al. 2016). Phosphorites, of interest for fertiliser, occur as modern deposits or fossil beds along productive continental margins (slopes) (Baturin 1982). These resources occur both within and beyond national jurisdictions (Fig. 3.1), with the exception of phosphorites, which are targeted only within EEZs. However, while 42 percent of areas with massive sulphides and 54 percent of areas with cobalt-rich crusts fall within EEZs, only 19 percent of known polymetallic nodules are within EEZs. More information on their formation and distribution is provided in Fig. 3.1 and by Petersen et al. (2016) and Jones et al. (2017).

Mining of the deep seabed (below 200 m) has not yet taken place. Extraction of minerals from the seafloor is planned to involve either modified dredging (for nodules) or cutting (for massive sulphides and crusts), and transport of the material as a slurry in a riser or basket system to a surface support vessel (Fig. 3.2). The mineral-bearing material will be processed on board a ship (cleaning and dewatering—with the waste water and sediment being returned to the ocean) and transferred to a barge for transport to shore where it will be further processed to extract the target metals (Collins et al. 2013; Brown 2018) (Fig. 3.2). Relative to mining on land, there is less overburden to remove and no permanent mining infrastructure required for deep-seabed mining (Lodge and Verlaan 2018).

However, there is likely to be solid waste material left after metal extraction, and disposal mechanisms for this

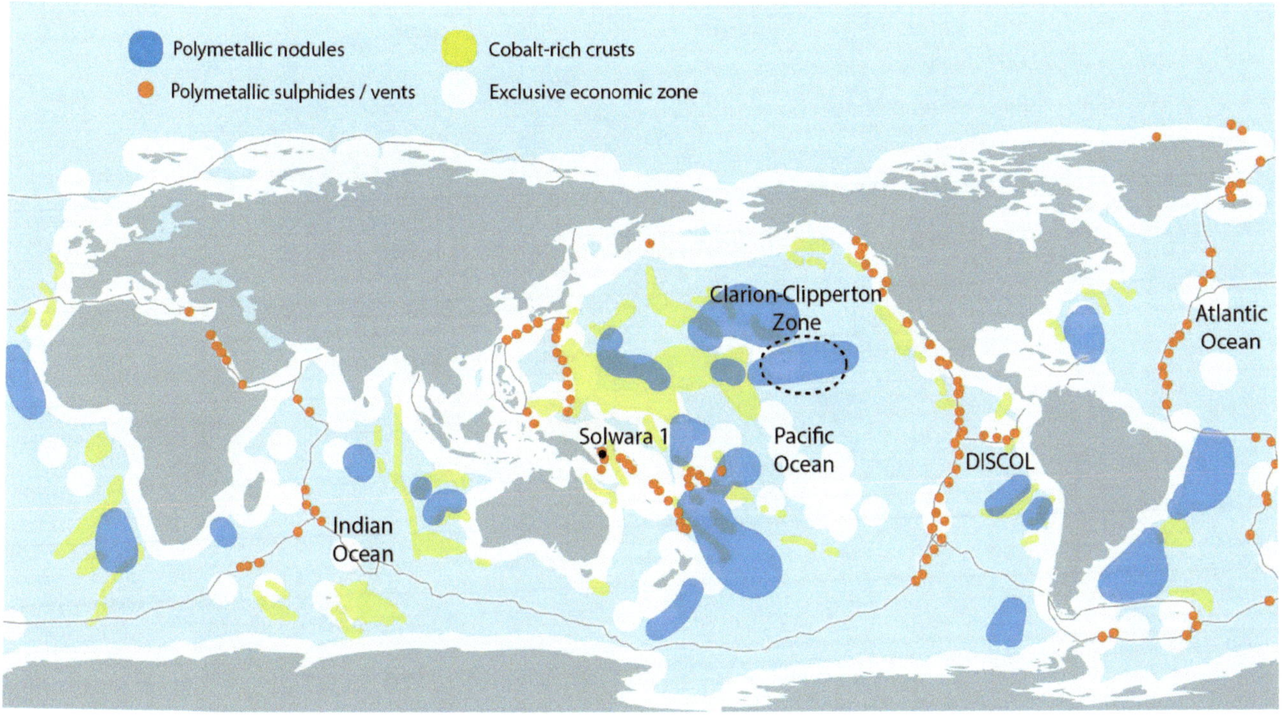

Fig. 3.1 Distribution of Polymetallic Nodules, Polymetallic Sulphides and Cobalt-Rich Crust Resources in the Deep Sea. (Note: The white area around Antarctica is not an exclusive economic zone but rather governed by an international commission. Source: Miller et al. 2018; Hein et al. 2013)

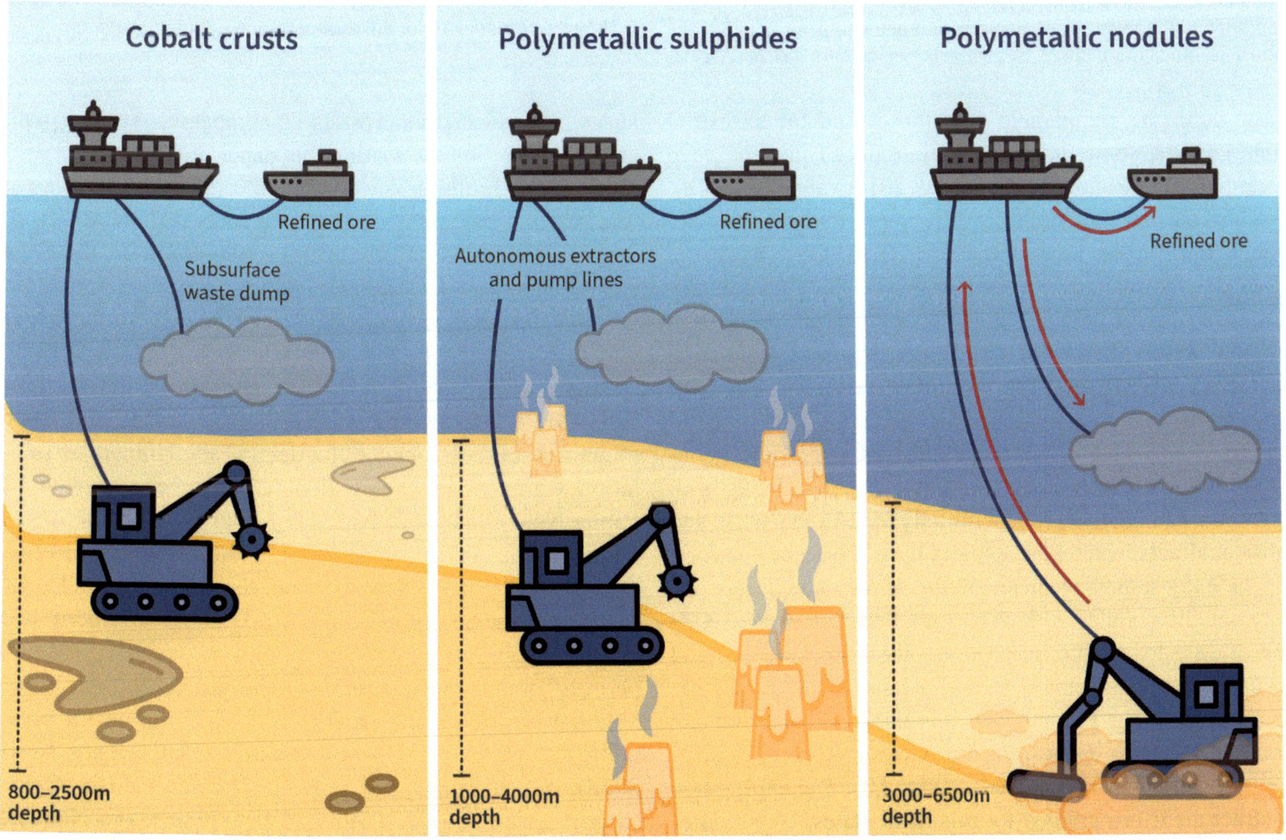

Fig. 3.2 Schematic Illustrating Deep-Seabed Mining for the Three Resources. (Source: Modified from Fleming et al. 2019)

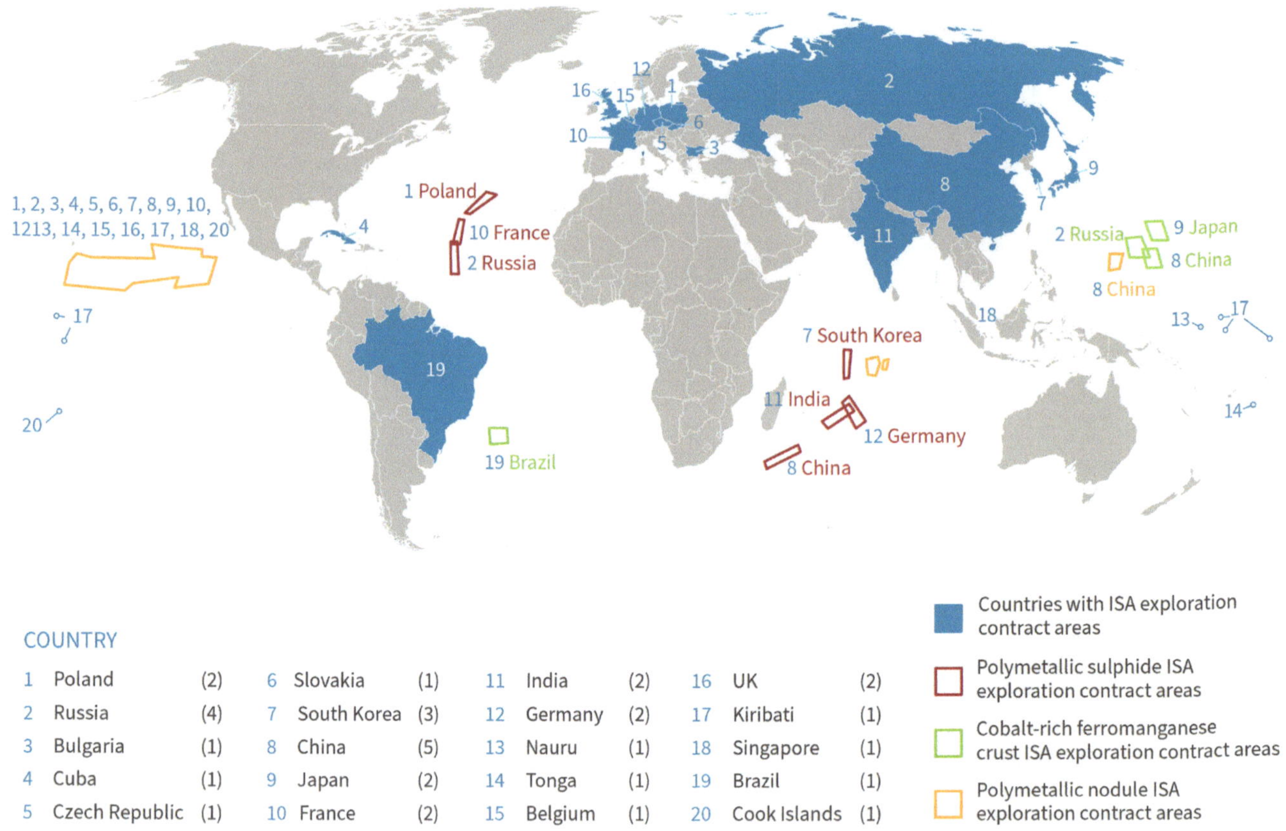

Fig. 3.3 International Exploration Contracts from the ISA. (Note: Countries with international exploration contracts from the ISA are shown in blue, the number of contracts per country (as of 2019 is depicted in the legend), and the general location of contracts in the Area is shown schematically for different resources. Source: Authors)

waste could be comparable with those used for terrestrial mine tailings, some of which are introduced into the deep ocean via pipe (Ramirez-Llodra et al. 2015; Vare et al. 2018).

The current governance structure under the UN Convention on the Law of the Sea (UNCLOS; UN 1982) gives the International Seabed Authority (ISA) regulatory responsibility for both the minerals on the seafloor in international waters (the Area) *and* the protection of the marine environment from the effects of mining in the Area. The minerals of the Area are designated as "the common heritage of [hu]mankind" (UN 1982). Since 2001, 30 *exploration* contracts for deep-seabed minerals in the Area have been approved. These were granted initially for 15 years each, and those contracts which have expired have been renewed for a 5-year extension. Seventeen of the ISA contracts are for polymetallic nodules in the Clarion-Clipperton Zone (CCZ) and two are for nodules elsewhere; others are for crusts and seafloor massive sulphides, and occur on West Pacific Seamounts (in the Prime Crust Zone), the Mid-Atlantic and Southwest Indian Ridges, the Rio Grande Rise off Brazil,[1] and in the Central Indian Ocean (Fig. 3.3). The exploration contract areas are granted to individual states, consortia of states, state-owned enterprises or companies working with states. At the time of writing this paper, the contracts cover more than 1.3 million km^2 (or 500,000 sq. miles), equivalent to about 0.3 percent of the abyssal seabed (Petersen et al. 2016). No contracts for mineral *exploitation* in the Area exist. Regulations for the exploitation of seabed minerals and for associated environmental management are currently under development by the ISA.

Roughly 70 percent of the 154 coastal states have significant deep ocean within their EEZs; many of these contain mineral resources. Licences for deep-seabed mineral exploitation within national jurisdictions have been granted by Papua New Guinea (to Nautilus Minerals) and by Sudan/Saudi Arabia (Diamond Fields International) (Miller et al. 2018). Additionally, New Zealand, the Kingdom of Tonga, Japan, Fiji, the Solomon Islands and Vanuatu have permitted research to assess the mining viability or issued exploration permits for national seafloor polymetallic sulphides, although some of them have lapsed. Exploration for polymetallic nodules in the Cook Islands (Cook Islands News 2018), cobalt crusts and polymetallic nodules in Brazil (Marques and Araújo 2019), and phosphorites in Namibia and South Africa (NMP n.d.; Levin et al. 2016) are also under consideration.

Sand is another resource mined in the ocean. Demand for sand, used in building and transportation, has increased

[1]Brazil has more recently indicated that the site in question falls within national jurisdiction (not the ISA's jurisdiction), according to an extended continental shelf claim, lodged by Brazil subsequent to the award of their ISA contract.

23-fold from 1900 to 2010, and is now seen as a scarce resource, the extraction of which can cause environmental degradation, health risks and social disruption (Torres et al. 2017). Sand occurs in shallow marine waters, is not closely tied to energy industries and is not a mineral per se, so will not be considered further here.

2 Transition to a Sustainable Global Energy System—1.5 °C Scenarios

2.1 Characteristics of 1.5 °C Scenarios

A recent special report from the Intergovernmental Panel on Climate Change (IPCC 2018) describes two main pathways to a 1.5 °C global average temperature rise by 2100. In the first pathway, global warming stabilises and stays at or below 1.5 °C. The second pathway sees some overshoot around mid-century before returning to a 1.5 °C rise. Scenarios with long and large overshoot typically rely heavily on technologies for removing CO_2 from the atmosphere. Such negative emission scenarios are treated in Sect. 2.2, but it should be noted that related technologies have not yet been deployed at scale and it remains to be seen if they will be applicable and cost-competitive. For example, Reid et al. (2019) raise a series of issues with bioenergy and argue against a path dependency and lock-in that would be implicated by substituting bioenergy for fossil fuel in scenarios involving bioenergy with carbon capture and storage (BECCS). Scenarios that do stay continuously below a rise of 1.5 °C typically require more rapid and larger deployment of renewable energy, as well as stronger energy efficiency and demand-side measures. Such scenarios are characterised by electrification of the global energy system and the stabilisation in or even reduction of global final energy use, despite delivering modern and sufficient energy to a growing world population (IPCC 2018). They are therefore low energy demand (LED) scenarios compared with fossil-based business-as-usual scenarios even if they deliver the same energy services.

IPCC LED scenarios (IPCC 2018) typically see a reduction in final energy use of 15 percent in 2030 and 30 percent in 2050, compared with 2010. Renewables deliver approximately 60 percent of electricity in 2030 and 80 percent in 2050. This translates to an increase of more than 400 percent in non-biomass renewables from 2010 to 2030 and more than 800 percent from 2010 to 2050 (IPCC 2018). IPCC LED scenarios (IPCC 2018) with no overshoot show 10–15 percent reduction in the global use of biomass renewables for energy, and employ a limited amount of afforestation but use no other carbon dioxide removal (CDR) technologies.

Jacobson and colleagues in a series of publications (most recently Jacobson et al. 2017, 2018, 2019) construct scenarios requiring 100 percent of global energy to come from wind, water (including ocean energy, hydropower and geothermal) and solar energy by 2050 (Fig. 3.4).

Jacobson et al. (2017) provide detailed specifications of their modelled contributions from different energy sources and grid components, such as batteries, heat and cold storage and heat pumps. Jacobson et al. (2018) confirm that the energy systems modelled provide stable energy services, despite relying heavily on variable wind and solar. While the scenarios by Jacobson et al. (2017, 2018) have previously

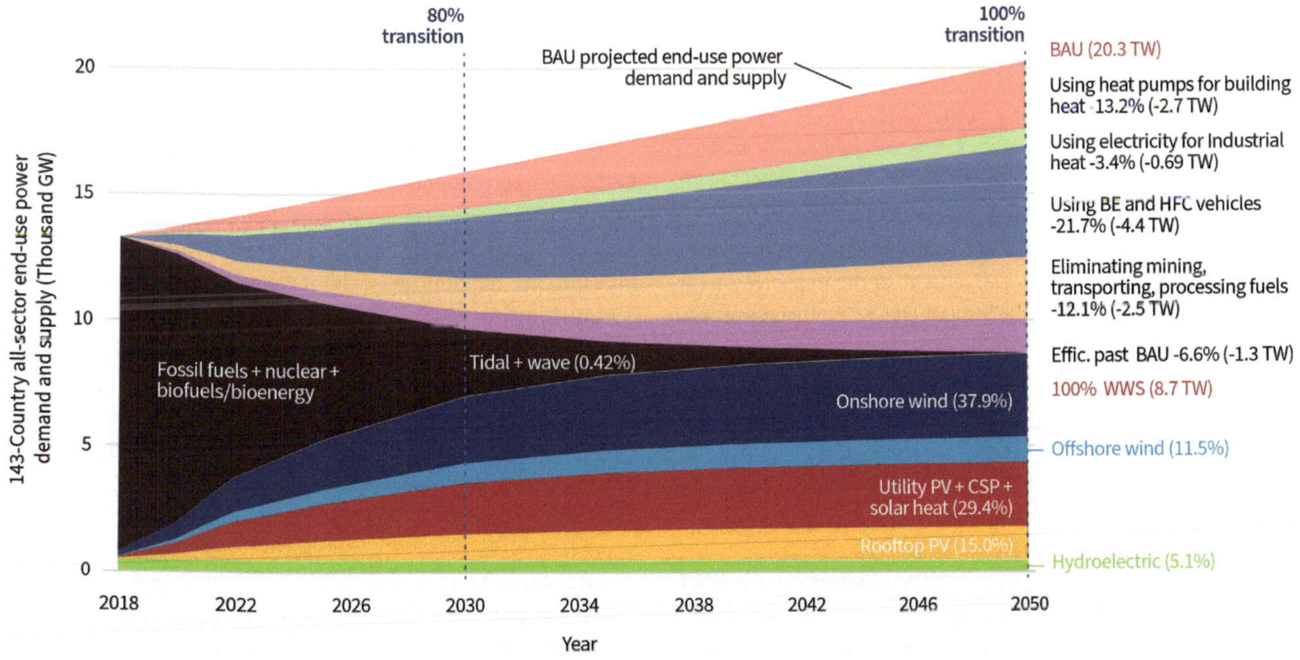

Fig. 3.4 Development of Wind, Solar and Other Energy Sources in a Low Energy Demand Transition to 100 Percent Wind, Water and Solar. 143-Country all-sector end-use power demand and supply (Thousand GW). (Note: An earlier study (Jacobson et al. 2017) gave less drastic reductions in final energy use to 11.8 TW in 2050, of which 13.6% or 1.6 TW was offshore wind. Source: Jacobson et al. 2019)

been considered extreme and have been criticised (Clack et al. 2017), other recent studies, notably Grubler et al. (2018) with a different modelling approach, achieve even larger reduction in global final energy demand in 2050, based on improved service efficiencies and demand-side transformation. Beneficial effects on other UN Sustainable Development Goals (SDGs) include better health via reduced pollution (SDG 3), reduced bioenergy and larger forest areas (SDG 15) and reduced ocean acidification (SDG 14). Environmental impacts are discussed in Sect. 5. Grubler et al. (2018) allow for some bioenergy, fossil fuel and nuclear energy. Their requirements for solar and wind energy are therefore lower than those of Jacobson et al. (2017, 2018, 2019), even though they deal with all countries and regions of the world.

Solar photovoltaic (PV) and wind energy are particularly implicated in use of certain minerals (Sect. 1.2). The installed capacities (i.e. nameplate capacities or full-load outputs) of solar PV and wind in 2050 from Jacobson et al. (2018) of approximately 30 and 17 terawatts (TW), respectively, are assumed to be upper bounds on the possible demands for installed solar PV and wind in a sustainable energy future. This includes onshore and offshore installations. The installed offshore wind capacity is estimated at about 4 TW. Note that these installed capacities are 2.5 to 6 times larger than the average utilised capacities in Fig. 3.4, reflecting a varying capacity factor (ratio between the energy delivered over a time period and the energy that would have been delivered if the turbine was running at maximum, i.e. installed capacity) due to variable winds and sun. In comparison, Teske et al. (2015), in their Advanced Energy [R] evolution scenario (ADV ER) arrive at approximately 9 TW installed capacity for solar PV and 8 TW installed capacity for wind in 2050. Teske et al. (2016) claim that this scenario is ambitious and may not guarantee to keep the global temperature rise below 1.5 °C, but may be the maximum transformation that is realistically achievable.

IEA (2019a) presents two scenarios, a stated policy scenario (SPS) and a sustainable development scenario (SDS). In the two cases, the global installed capacity of offshore wind in 2040 is estimated to be 340 and 560 gigawatts (GW), respectively. With a significant improvement in the capacity factors over the coming 20 years, the annual energy contribution from offshore wind in 2040 is estimated to be 1400 and 2350 terawatt-hours (TWh) per year for the two scenarios respectively.

2.2 Negative Emissions and Carbon Capture and Storage

As mentioned in Sect. 2.1, many of the scenarios in the IPCC report (IPCC 2018) rely on negative emissions in the later part of the present century in order to repair the overshoot and get back to a global temperature rise of less than 1.5 °C. Overshoot would imply potentially damaging impacts on the ocean and its ecosystems.

Geoengineering through solar radiation management would, if successful, limit global warming, but to avoid ocean acidification atmospheric CO_2 needs to be limited too. Several CDR technologies which would capture CO_2 from the air have been proposed. However, IPCC (2018) states, with high confidence, that: "CDR deployment of several hundreds of $GtCO_2$ is subject to multiple feasibility and sustainability constraints." Afforestation and BECCS are the options most widely studied.

BECCS consists of harvesting biological material, burning it for energy purposes in an energy plant (power or combined heat and power) and adding facilities for CCS. A few BECCS pilot plants exist (IPCC 2018). More research experience is available on CCS from fossil fuel power plants and some from transport and storage of CO_2 for other purposes or from other sources (IPCC 2005). Storage of CO_2 is taking place also in the subseabed, notably for more than two decades on the Norwegian continental shelf (Furre et al. 2017). While CCS research and application has been promoted in several countries over the past decades, questions still remain on the practicality and cost-competitiveness. In Europe, developments in new renewable energy, notably wind for production of electricity, mean that it is steadily becoming cheaper and is already cost- competitive with fossil fuel without CCS.

With CCS, there is the added investment in capture facility, transport and storage, and the related energy penalty (increase in energy and fuel use for running the CCS process) which tends to sit around 20–25 percent (IPCC 2005). Research and development continues, however. Active projects in Norway are directed at CO_2 from other industries like cement and incineration of waste. There are also studies on the separation of CO_2 from natural gas and on delivering hydrogen for energy purposes. Related efforts may lead to an increase in the interest in storing CO_2 offshore in the subseabed and development of technology that could be transferred to BECCS. However, the energy penalty (use of more biological material to provide energy to run the process) and investments in facilities cannot be avoided. In view of the diminishing costs of electricity based on renewables, competition on cost appears to be difficult. Furthermore, the carbon capture process is never 100 percent effective so some CO_2 release has to be accepted. In a sustainable energy future with very tight restrictions on CO_2 emissions, it appears that non-biomass renewables—wind, water, solar and in some locations geothermal—have to replace the lion's share of the energy services presently served by fossil fuel.

Overshoot in itself may lead to irreversible damage to the climate system. No CDR technologies have yet been scaled up. Costs and environmental implications are uncertain. The modelling approaches used in scenario calculations assume learning curves and discount rates that tend to favour shifting of costs to the distant future. Ethical and hard science aspects of these questions are interlinked and hotly debated in the popular media as well as scientific forums (Anderson and Peters 2016). It appears that relying on negative emission technologies in the future is optimistic and could be deemed irresponsible. In the context of this report, the 1.5 °C scenarios discussed in Sect. 2.1. are those considered to be representative of a sustainable future.

3 Ocean-Based Renewable Energy

The status and costs of the various technologies—in other words, their technical and economic potential—are addressed in this section, while the environmental impacts and wider sustainability issues are discussed in Sect. 5. Since offshore wind is considerably further advanced in its implementation than the other technologies, offshore wind is treated separately.

3.1 Offshore Wind

3.1.1 Technical Potential

When considering the available wind energy resources across the global ocean, a geophysical potential may be estimated from knowledge of the global wind field. This global potential remains theoretical, however, and of little practical interest. For example, it is considered unrealistic to deploy wind turbines in the Southern Ocean, not only because of the difficult operating conditions, but also because of the distance to users of the electricity. The cost and even the energy expenditure associated with the manufacturing and laying of electric cables, the deployment of floating turbines at great ocean depths and the loss in transmission would prohibit any such project. A more interesting consideration is the technical potential (Fig. 3.5). The technical potential takes into account technical limitations and excludes inaccessible resources. What these technical limitations are will depend on technology developments and trends. Assessments therefore vary depending on the assumptions made.

Bosch et al. (2018) estimate the global and regional offshore wind power potential. They consider three different water depth ranges (0–40 m, 40–60 m and 60–1000 m) within the EEZ of each country. Various exclusion zones are accounted for. They find that the worldwide technical potential for power production from offshore wind amounts to about 330,000 TWh/year as compared with the world's electric energy production in 2018 of about 26,700 TWh/year (IEA 2018) and the modelled offshore wind contribution in 2050 in Fig. 3.4 which corresponds to 9000 TWh/year. Bosch et al. (2018) also review resource estimates made by others. The global total estimates range from 157,000 TWh/year to 631,000 TWh/year, depending upon the assumptions made.

A similar study performed by Eurek et al. (2017) estimated the global potential for offshore wind deployment while including various exclusion zones related to water depth, distance to shore, protected areas and sea ice. They ended up with an estimated potential of 315,000 TWh/year using a capacity factor of 0.285.

IEA (2019b) has also made estimates on the technical potential for offshore wind, using somewhat different criteria for exclusion zones. The results are summarised in Table 3.1. The total global technical resources are found to be about 420,000 TWh/year.

The above estimates for the global potential for offshore wind are 6 to 23 times the present global electricity consumption. Most of the estimates also exceed the pres-

Table 3.1 Offshore Wind Potential (TWh/year)

	Shallow water (depth < 60 M)		Deeper water (depth 60 M–2000 M)		
	Near shore	Far shore	Near shore	Far shore	Total Potential
North America	9907	13,238	22,819	58,937	104,901
Central and South America	3847	4438	6439	37,144	51,869
Europe	2629	2390	14,817	52,009	71,845
Africa	1123	572	7699	17,107	26,502
Middle East	478	673	600	1791	3543
Eurasia	9382	17,402	9943	48,735	85,462
Asia Pacific	8508	12,451	14,440	41,357	76,757
WORLD	**35,875**	**51,166**	**76,757**	**257,081**	**420,878**

"Near shore" denotes sites less than 60 km from the shore and "far shore" denotes sites at a distance of 60–300 km from the shore. Source: IEA 2019b

Fig. 3.5 Geophysical, Technical, Economic and Social/Political Potential of Wind or other Energy Resources across the Global Ocean. (Source: Adapted from Hoegh-Guldberg et al. 2019)

ent global total primary energy consumption (14,314 million tonnes of oil equivalent (Mtoe) = 166,470 TWh in 2018; IEA (2019b)). The above estimates do not consider limitations due to costs and make some assumptions on technological elements. The economic potential depends on the costs (see next section) of these technologies in relation to competing technologies. The economic potential will be smaller than the technical potential. Figure 3.4 shows one example of an estimate of economic potential given certain assumptions. Social/ political and environmental considerations discussed in Sect. 5 may limit the potential further (Fig. 3.5).

3.1.2 Status of Technology and Costs

While there is abundant technical potential for offshore wind energy generation, the economics of deploying energy offshore limit the capacity that might be installed. In future low-carbon scenarios, technologies with similar GHG mitigation potential compete. A conservative approach based on a range of earlier published scenarios was chosen by Hoegh-Guldberg et al. (2019), resulting in an estimate of up to 3500 TWh/year in 2050 from offshore wind. The 1.6 TW yearly average offshore wind power from Jacobson et al. (2017), corresponding to approximately 14,000 TWh/year, and the 1.0 TW figure (Jacobson et al. 2019) corresponding to approximately 9000 TWh/year (Fig. 3.4), are estimates of the economic potential for offshore wind in a future low-emission scenario. While the numbers cited in Sect. 3.1 do take into account areas that would be unavailable for offshore wind, they are still theoretical and not likely to ever be achieved. However, theoretical estimates are at least an order of magnitude larger than those of Jacobson et al. (2017), indicating that there are no resource constraints on offshore wind installations.

By the end of 2018, the total worldwide installed capacity of wind energy amounted to 564 GW, of which only 23 GW were offshore (IRENA 2019c). The yearly electrical power production from offshore wind amounted to about 77 TWh (IEA 2018). For offshore wind turbines, bottom-fixed turbines in shallow water depth (< 40 m water depth) dominate. Deep-water, floating support structures are used in one wind farm only, a 0.03 GW wind farm on the east coast of Scotland. This wind farm was installed in 2017. Europe presently has the majority of the offshore wind installations, with an installed capacity of 18.5 GW, while Asia has 4.6 GW. It has been anticipated that China will have more installed capacity than Europe by 2021 (Backwell 2019). However, according to IEA (2019a) estimates, China will overtake Europe in the early 2030s. It is expected that North America will be number three after Asia and Europe.

As the wind conditions in general are better offshore – the wind is more stable—the utilisation of the installed generator capacity is generally higher than onshore.

In Europe, the capacity factor for offshore wind farms commissioned in 2018 was 43 percent, increasing from 38 percent in 2010. Onshore, the comparable global averages are 34 percent and 27 percent, respectively (IRENA 2018a, b). IEA (2019b) expects that, by 2040, the capacity factors for good offshore sites will move towards 60 percent, while the worldwide average will be close to 50 percent. Over the last decade, the cost per MW of installed power has been reduced and the capacity factor for new installations has increased. The operation and maintenance costs per produced megawatt-hour (MWh) are also expected to decline as the turbines are designed to be more robust and fit for the offshore environment. All three factors contribute to a reduced levelised cost of electricity (LCOE; the ratio between the discounted costs over the lifetime of an electricity-generating plant and the sum of actual energy amounts delivered). However, the single most important factor to reduce LCOE is the cost of capital or the discount rate. Reduced project uncertainties and favourable financing terms will contribute to a reduced LCOE. IEA (2019a) shows that using an average discount rate of 4 percent rather than 8 percent may reduce the LCOE for offshore wind projects from US $140/MWh to $100/MWh. As the number of shallow-water, bottom-fixed support structures, mainly monopiles, has increased, the cost reduction due to mass production has been significant. Bottom-fixed offshore wind turbines are thus considered mature and have reached commercial scale. The costs have reached parity with fossil sources of electricity in recent contracts, down towards $50/MWh, without transmission costs. The Dogger Bank Creyke Beck A wind farm won a UK government auction for renewable power in September 2019, with a strike price of $51/MWh (IEA 2019a; Dogger Bank Wind Farms 2019).

IRENA (2019b) shows similar LCOE figures. Stehly et al. (2018) found that the 2017 average in the United States was $124/MWh for bottom-fixed and $146/MWh for floating. In Europe, the LCOE for projects commissioned in the period 2010–2015 shows very moderate decline. However, after that, a significant drop in LCOE for new projects is observed. In 2012, the European Union set an ambitious aim of LCOE of $110/MWh in 2020. This aim has already been achieved for several projects. For projects commissioned in 2018, the European average was $134/MWh and for projects in China $105/MWh (IRENA 2019b). However, contracts with record low costs have been signed in the Netherlands ($55/MWh to $73/ MWh) and Denmark ($65/MWh) for a near-shore project, excluding grid connection costs. No data are available for floating systems as only one small wind farm has been realised. Ørsted (2019) indicates a cost reduction in offshore wind of 18 percent per doubling of capacity.

Bottom-fixed support structures are designed for site-specific conditions. Worldwide, there are limited large, shallow-water areas suitable for wind-power development. Bosch et al. (2018) estimate the potential wind power production from shallow-water areas (<40 m) to be less than one-third of the potential production from the deeper areas (60–1000 m water depth). IEA (2019a) estimates the shallow-water areas (<60 m water depths) to be about 20 percent of the total areas available (see Table 3.1). Deep water requires floating support structures. Such solutions are less mature than the bottom-fixed solutions and are presently more expensive than the shallow-water bottom-fixed support structures. Floating support structures are well suited for standardisation and mass production as they do not depend upon site-specific conditions at sea bottom. In a scenario with large-scale deployment of floating offshore wind turbines, it is thus expected that the LCOE will be comparable with that of bottom-fixed support structures.

The increased size of turbines and wind farms, as well as the learning rate of the offshore wind industry, have all contributed to reduced LCOE. However, moving into deeper water and farther from shore has partly outweighed the cost reductions. In Fig. 3.6, an approximate split of the capital costs of offshore wind turbines completed in 2018 is given.

3.1.3 Future Development Scenarios

According to IRENA (2019c), the rate of wind energy deployment (2017–18) is 54 GW/year globally. To achieve the required energy transformation (increased electrification, reduced emissions) a significant speed-up in wind energy installations is required. IRENA (2019c) indicate 200 GW/year in 2030, increasing to 240 GW/year in 2050 worldwide. How much of this growth can be taken offshore is uncertain. However, the resources are not a limitation.

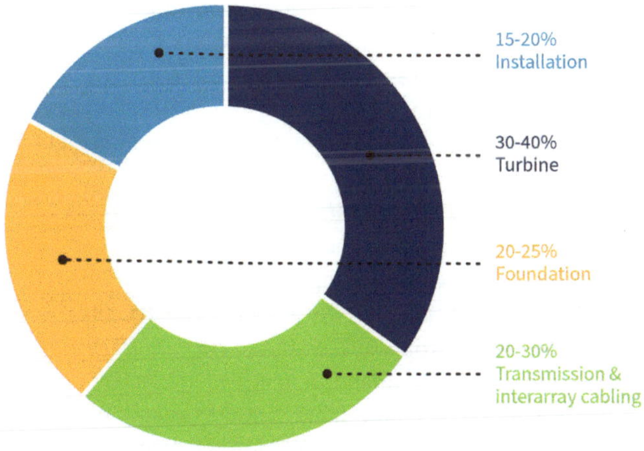

Fig. 3.6 Approximate Split of the Capital Costs of Offshore Wind Turbines Completed in 2018. (Source: IEA 2019b; IEA analysis based on IRENA 2019a, IJGlobal 2019 and BNEF 2019)

According to IEA (2017), offshore wind generation has grown five-fold over the period 2010–2015 and is expected to double over the period 2015–2020. IEA (2019b) in their SDS has a compound annual growth rate (CAGR) of 16.9 percent for the global offshore wind market in the period 2018–2040, while Bloomberg (2019) forecasts growth at a CAGR of over 18 percent over the period 2019–2023. Between 2020 and 2025, offshore wind generation needs to triple to be fully on track with the 2 °C target. By 2025 about 2785 TWh/year of electricity should be produced from offshore wind to be in line with the SDS. The corresponding figure for 2040 is 6950 TWh/year. It is indicated that in 2040 the electric energy produced from other ocean-based renewable energy sources could contribute more than 1200 TWh/year.

Assuming that it comes from offshore wind alone, this requires an installed capacity of about 326 GW of offshore wind in 2040. To achieve this, 15 GW of offshore wind has to be installed every year for 20 years. Using the Jacobsen et al. (2017) figures, the contribution from offshore wind is larger. To achieve 3800 GW of installed offshore wind capacity (corresponding to 1600 GW average power) in 2050, an installation rate of 127 GW/year is required over 30 years. In other words, the 2 scenarios require substantially different installation rates, almost 10 times greater for the Jacobsen et al. (2017) scenario. This difference is mainly due to differences in the assumptions regarding the future contribution of offshore wind to the electricity supply. Both scenarios require an accelerated development of new ocean areas for offshore wind. Development of deep-water areas with floating wind turbines can make a significant contribution to achieve this goal. Even further acceleration would be needed to ensure only a global temperature rise of 1.5 °C. It is to be noted that the European Commission (2018) presents a strategic roadmap which would lead to an even larger contribution of offshore wind in their region.

3.2 Other Ocean-Based Renewable Energy

3.2.1 Technical Potential

There are several other renewable energy technologies which exploit the available resources of the offshore environment. The technologies that harness energy directly from the ocean itself (i.e. water-based technologies) have particular advantages, such as the power density of moving water (much larger than that of air), the predictability and consistency of the resource (notably tides), and the fact that the resource can typically deliver at times when other renewable energy resources do not. Floating solar photovoltaics and high-altitude wind have different characteristics. The range of other ocean-based renewable energy technologies, summarised in Table 3.2, include:

Table 3.2 Geophysical and Technical Potential Estimates for Ocean-Based Renewable Energy Technologies, with Technology Readiness Levels Estimated

Technology	Geophysical energy potential (TWh/year)	Technical potential (TWh/year)	Technology readiness level
Tidal range	26,000	6000	9
Tidal stream	–	150	8
Wave	32,000	1750–5550	7
OTEC	38,000	–	4
Salinity gradient	–	1650	3
Ocean current	–	–	3
Floating PV	700,000,000	–	7*
High-altitude wind	–	–	6

Note: The technology readiness level (TRL) scale used here is based on the guidance principles for TRLs for ocean-based energy technologies, as defined by the European Commission (Appendix A in Magagna et al. 2018*), ranging from TRL 1 (Basic principles observed), to TRL 9 (Actual system proven in operational environment). The actual assessment of TRLs for each technology is our own. Offshore wind would appear with TRL 9*

** Very recent developments (Oceans of Energy* 2020*) could justify lifting the TRL of floating PV to 8*

- **Tidal range energy:** Tidal range energy technologies include tidal barge energy systems and tidal lagoon energy systems. Tidal range systems represent the bulk of existing installed ocean-based renewable energy, having been in operation for decades.

  Tidal range technologies act effectively as low-head 0.22 percent of the world ocean. Taking into account the impracticality of ice-covered regions, the global annual potential energy from tidal range technologies is approximately 6000 TWh, with 90 percent distributed across five countries (O'Neill et al. 2018).
- **Tidal stream energy:** With the rise and fall of tidal water elevation that occurs twice a day, tidal currents are generated. Tidal stream energy converters harvest the energy of these currents and convert it to electrical energy. Many technologies are in development, but convergence towards horizontal axis turbines has occurred. These tidal energy converters are intended to be modular, to be deployed in subsurface arrays. No reasonable estimate for the total global geophysical tidal stream potential is known, but best estimates of the total global technical tidal stream energy potential is approximately 150 TWh/year (with high uncertainty; Yan 2015).
- **Wave energy:** Wave power converts the kinetic and potential energy of the surface wind-waves of the ocean into electrical energy (or some usable commodity, such as desalinated water). Wave energy converters are designed to be deployed in arrays, similar to wind farms. Many concepts are in development, with little to no convergence in technologies. The total geophysical wave energy potential is estimated to be 32,000 TWh/year (Mørk et al. 2010), with estimates of the global technical potential ranging from 1750 (Sims et al. 2007) tohydropower systems—in their simplest form, water is constrained on the high tide (by barrage or lagoon) and powers a water turbine on release. The estimated global annual geophysical tidal range potential is around 25,880 TWh (constrained to regions with water depth < 30 m, and a reasonable threshold for energy output). The distribution of this resource, however, is confined to just 5550 (Krewitt et al. 2009) TWh/year.
- **Ocean thermal energy conversion (OTEC)**: OTEC exploits the temperature gradient between the cold deep ocean and the warmer surface waters and converts it into electricity or other commodities, such as desalinated water, heating and cooling, or nutrient supply for other marine applications. A temperature gradient in excess of 20 degrees is required, which constrains interest in OTEC to the tropics (+/−20 degrees latitude). An upper limit of the long-term steady-state global resource has been estimated to about 38,000 TWh/year (Nihous 2018). This is from a theoretical study assuming all OTEC facilities have optimal discharge depth and efficient generators. The technical potential is very uncertain.
- **Salinity gradient energy**: This technology converts energy produced from the chemical pressure that results from the difference in salt concentration between freshwater and saltwater. It can be exploited at river mouths where freshwater and saltwater meet. The technical potential for power generation has been estimated at 1650 TWh/year (Lewis et al. 2011).
- **Ocean current energy**: These technologies operate along a similar concept to tidal stream energy, harvesting the flow of water in motion. However, the targeted ocean currents for these technologies are the deep-water currents of the thermohaline circulation (e.g., the western boundary currents such as the Kuroshio, the Gulf Stream and the East Australian current). These currents have less variability than the tidal currents but are less accessible. No estimate of technical potential has been made, but it is an area of interest for innovators.
- **Floating solar photovoltaics (PV)**: Over the past three years, the installed capacity of floating solar (e.g., PV panels deployed on floating platforms) has increased at a CAGR of 168 percent, to a total capacity of 1.3 GW (World Bank Group, ESMAP and SERIS 2019). This is predominantly on inland waterways (reservoirs, canals, etc.), with the offshore market still nascent. While there are unique challenges for offshore, the available resource presents an opportunity for a growing market. No global estimate of the offshore solar resource is available. However, with the ocean representing 70 percent of the

earth's surface, a very rough estimate of the geophysical potential is 70 percent of the almost 1 billion TWh (WEC 2013) of solar radiation reaching the earth's surface each year, which represents an abundant resource.

- **High-altitude wind:** Technologies exploiting wind at high altitudes are under development, notably using kites (Lunney et al. 2017). One advantage of kites compared with conventional turbines is their low demand for materials. Testing in offshore environments has recently begun from a floating platform (Norwegian Offshore Wind Cluster n.d.). Since most ocean-based renewable energy technologies are still in early phases without much deployment, few studies have been done on life cycle and material needs assessment (Uihlein 2016). It seems that the major metal requirements of these technologies would be similar to those of offshore wind. Specific requirements for floating solar PV would be similar to those for land-based PV (Arrobas et al. 2017).

3.2.2 Status of Technology and Costs

At the end of 2018, the total installed capacity of ocean-based energy technologies was 532.1 MW (IRENA 2019a), consisting mainly of tidal barrage technology at two sites. Installed capacity in 2016 was 523.3 MW, which generated 1023.3 GWh electricity (IRENA 2019a), implying a mean capacity factor of 0.23 across the sector. Estimates of the LCOE are subject to a range of parameters, including the local conditions which increase costs. The estimated LCOE for wave energy is in the range of $360–690/MWh (IRENA 2014). Tidal stream energy LCOE is presently in the range of $275–520/MWh (IRENA 2014b), contingent upon sufficient current speeds. LCOE of OTEC is in the range of $600–940/MWh (IRENA 2014c). Learning rates for ocean-based technologies are typically assumed at around 15 percent (OES 2015), with average LCOEs for wave energy and tidal energy of $165–220/MWh by 2030 (Cascajo et al. 2019; SI Ocean 2013). Due to the capital intensity of OTEC, interest and discount rates have a high impact on LCOE estimates. Economies of scale are anticipated to bring LCOE into a range of $70–190/MWh for installed capacities exceeding 100 MW (IRENA 2014c; OES 2015).

3.2.3 Future Development Scenarios

Electricity generation from marine technologies increased an estimated 3 percent in 2018 (IEA 2019c). This rate of growth is not on track to meet the IEA SDS target for ocean-based technologies of 15 TWh/year in 2030 (IEA 2019c), which would require an annual growth rate of 24 percent to meet. The IEA SDS corresponds with an emissions target of approximately 25 GT CO_2e by 2030. By 2050, the range of projected power generation from ocean-based technologies for various scenarios (reference technology scenario/two degrees scenario/beyond two degrees scenario) is 108/536/637 TWh/year (2050 emissions 40/13/4.7 GT CO2e), corresponding to annual growth rates from present of 15/21/22 percent (IEA 2019c). The full range of projections currently being put forward for other ocean-based technologies extends up to a max of 1943 TWh/year (Teske et al. 2010).

4 Motivations for Deep-Seabed Mining

As mentioned above (Sect. 1.2), there is increased global demand for metals and REEs from emerging technology industries (Table 3.3). For example, renewable energy production requires significant amounts of a range of metals, generally more than required for production of energy from fossil fuels (IRP 2019; Giurco et al. 2019). Many of the required metals and elements occur together—not only in large amounts but also at higher concentrations than on land—in minerals precipitated in the deep ocean. The higher concentration makes them attractive for mining operations and contributes to their resource potential (Petersen et al. 2016).

Mining on land has significant environmental and social impacts (IRP 2019). Among these, displacement of communities, contamination of rivers and groundwater from tailings, damage to communities from tailings slides, violation of land rights, mining community repression and unfavourable child labour/slavery practices (Church and Crawford 2018; Sovacool et al. 2020) have all provided the incentive to look to the ocean as a source of minerals (Batker and Schmidt 2015; IRP 2019). A large fraction of the minerals required for renewable energy technologies are produced in states with corrupt or fragile governance (Church and Crawford 2018). The social impacts of deep-seabed mining is a topic less considered, although concerns have been expressed in the Pacific region about the potential for deep-seabed mining to interfere with local traditional practices, local communities' property, food sources and lifestyle, and that deep-seabed mining could exacerbate social tensions and even lead to political instability (SPC 2012; Aguon and Hunter 2018). Also, the extraction of deep-seabed minerals from offshore sites should not be considered in isolation from the infrastructure development, and the transfer and processing of ore, which would occur on land and could also have impacts similar to mining on land (SPC 2013).

Table 3.3 Major Uses, Production and Potential Supply in Selected Seabed Deposits Relative to Land-Based Reserves for Metals Targeted for Deep-Seabed Mining

Metal	Uses	Deep-sea sources	Annual production in 2017 in thousands of metric tonnes (top 3 land producers)	Annual projected demand in 2050 in Thousands of metric tonnes from low-carbon energy technology	Metal supply in the clarion-clipperton zone in thousands of metric tonnes (% of land-based reserves)#	Metal supply in the prime crust zone in thousands of metric tonnes** (% of land-based reserves)#	Inferred metal supply in seafloor massive sulphides in thousands of metric tonnes*** (% of land-based reserves)[#]
Copper (Cu)	Used in electricity production and distribution – wires, telecommunication cables, circuit boards. Non- corrosive Cu-Ni alloys are used as ship hulls	Polymetallic sulphides at hydrothermal vents, polymetallic nodules on abyssal plains	19,700 (Chile, Peru, USA)	1378	226,000* (23–30% of land-based reserves)	7400 (0.7% of land-based reserves)	21,600 (2% of land-based reserves)
Cobalt (Co)	Used to produce high-temperature super alloys (for aircraft gas turbo-engines, rechargeable lithium-ion batteries)	Cobalt-rich crusts on seamounts, polymetallic nodules on abyssal plains	110 (Democratic Republic of Congo, Australia, China)	644	44,000 (340–600% of land-based reserves)	50,000 (380% of land-based reserves)	N/A
Zinc (Zn)	Used to galvanise steel or iron to prevent rusting, in the production of brass and bronze, paint, dietary supplements	Polymetallic sulphides at hydrothermal vents	12,800 (China, Peru, Australia)	N/A	N/A	N/A	47,400 (21% of land-based reserves)
Manganese (Mn)	Used in construction for sulphur fixing, deoxidizing, alloying properties	Cobalt-rich crusts on seamounts, polymetallic nodules on abyssal plains	16,000 (China, Australia, South Africa)	694	5,922,000 (114% of land-based reserves)	1,714,000 (33% of land-based reserves)	N/A
Silver (Ag)	Used in mobile phones, personal computers, batteries. Also in mirrors, jewellery, cutlery and for antibiotic properties	Polymetallic sulphides at hydrothermal vents	25 (Peru, China, Mexico)	15	N/A	N/A	69 (4.3% of land-based reserves)
Gold (Au)	Used in jewellery, electrical products (metal-gold alloys)	Polymetallic sulphides at hydrothermal vents	2.5–3 (China, Australia, USA)	N/A	N/A	N/A	1.02 (0.002%
Lithium (Li)	High-performance alloys for aircraft; electrical, optical, magnetic and catalytic applications for hybrid and electric cars	Cobalt-rich crusts on seamounts, marine sediments	43 (Chile, Australia, China)	415	2800 (25% of land-based reserves)	20	N/A
Nickel (Ni)	Stainless steel (automobiles, construction), weapons, armour	Cobalt-rich crusts on seamount, polymetallic nodules on abyssal plains	2100 (Russia, Indonesia, Canada)	2,268	274,000* (180–340% of land-based reserves)	32,000 (21% of land-based reserves)	N/A

Note: The land-based reserves are known with enough certainty that they can be mined economically whereas the seafloor estimates are far from this level of certainty

* *India's 75,000 km2 nodule claim in the Indian Ocean contains another 7000 thousand metric tonnes of Cu and Ni*

** *Based on 7,533,000 thousand metric tonnes in the Prime Crust Zone*

*** *Based on 600,000 thousand metric tonnes in the neovolcanic zone with grades determined as averages of analysis of surface samples. Source: Compiled from Hein et al.* 2013; *Petersen et al.* 2016; *Miller et al.* 2018; *Hannington et al.* 2010; *Fleming et al.* 2019; *Sovacool et al.* 2020

4.1 Will Deep-Seabed Mining Help Address Climate Change?

Deep-seabed mining could lead to an increased global supply of cobalt, copper, nickel, silver, lithium and REEs (Hein et al. 2013), which could make solar energy, wind turbines and electric cars more affordable and/or prevalent, potentially aiding the transition to renewable energy (Dominish et al. 2019). Mining deep-sea polymetallic nodules is calculated to release less CO_2 per kg than mining on land (Van der Voet et al. 2019). A recent report commissioned by a deep-seabed mining company involved with three exploration tenements in the CCZ suggests that extracting half of the CCZ nodules would provide the manganese, nickel, cobalt and copper needed to electrify 1 billion cars, while releasing only 30 percent of the greenhouse gases of land mining (Paulikas et al. 2020).

This conclusion has been questioned under various future global energy scenarios. Teske et al. (2016) conclude that an energy revolution, required to combat climate change, could take place without deep-seabed mining. Increasing mineral production rates in combination with more recycling (e.g., of lithium-ion batteries) and research into alternative technologies that reduce or completely eliminate the use of lithium, silver, neodymium and dysprosium – the critical elements under the greatest resource pressure – would advance this option. Recycling costs and thus incidence is a function of energy and raw material costs, which are affected by collection and transportation efficiency; in many cases, where the mass of the desired mineral is small in the waste stream, product redesign would be required for recycling to become effective.

4.2 Can Metal Demand Be Reduced to Avoid Deep-Seabed Mining?

Key to reducing metal demand is the concept of a circular economy, which acts through improved product design, reduced demand, reuse, recycling, reclassification of materials and use of renewable energy for production (Ghisellini et al. 2016). With REEs and metals, it is particularly hard to achieve economies of scale in recycling and reuse, because of the limited quantity of elements contained, the long lifespan of some products using these elements, and metal separation issues requiring complex and energy- intensive processes (Schüler et al. 2011). The materials added to improve product quality and durability can make metal recovery from electronic products even more difficult (Tansel 2017).

Models for increasing metal demand often assume growth in demand based on recent rates of increase, or based on current technology status, which may in fact become obsolete quickly. Commodity price forecasts are notoriously inaccurate. As an example: it is reported that Massachusetts Institute of Technology, commissioned by the ISA to undertake financial modelling for nodule mining in the CCZ, in the space of several months revised its estimate of the likely value of a metric tonne of one target metal in nodules (electrolytic manganese metal) from \$3500 to \$1561 (Africa Group 2019). Future demand for resources could be lower than expected, including through saturation in material use as countries move through stages of development (Bleischwitz et al. 2018). This has been documented for copper in the United States, United Kingdom, Japan and Germany, and may be especially relevant for emerging economies, such as China, that are undergoing changing growth patterns that could stabilise demand in the future (Bleischwitz et al. 2018).

5 Sustainability Challenges and Enabling Conditions

5.1 Environment, Vulnerabilities and Costs

5.1.1 Environmental effects of ocean-based renewable energy deployment

The potential benefits of ocean-based renewable energy to contribute to future low-carbon energy generation have been specified in the sections above. However, given the early stages of development of these technologies, there remain environmental risks to the marine environment from their deployment, particularly when considered at the scale required to make a decisive contribution to the future energy system.

As offshore wind is a more mature technology, with greater installed capacity, the risks it poses are slightly better known than for the less mature ocean-based technologies. However, there are still large knowledge gaps in the field of environmental impacts of offshore wind. Considerable lack of baseline data may be a key limitation when evaluating impacts, depending on location and whether there have been any prior studies in the area for other purposes such as oil and gas or fisheries. Baseline data provide information on the state of the marine environment prior to construction, and are used as a basis for comparison over time during the construction and operational phases. Such data may include information on distribution of important and vulnerable species and habitats, and migration routes for marine mammals, fish and birds. Baseline research on species abundance and distribution over annual cycles, population structures and status, and assessment of ecosystem dynamics are necessary.

The literature on the environmental impacts of ocean-based renewable energy was very limited before 2000, but it has increased considerably in the last 20 years (Mendoza et al. 2019; Zydlewski et al. 2015). Boehlert and Gill (2010) provide an early overview of literature with recommendations for needed environmental research on ocean-based renewable energy developments including offshore wind. Impacts range from effects on bird migration, physical habitat change on the seafloor, chemical spills and sound (in air and water) to electromagnetic disturbance from submarine power cables.

The primary environmental concerns of ocean-based renewable energy deployment are typically common to both offshore wind and most of the other ocean- based technologies. Key concerns relate to possible interactions between aspects of the energy conversion systems (turbines, anchors, foundations, mooring lines, etc.) and marine ecosystems. As the installed capacity of these offshore energy systems increases, additional concerns relating to ecosystem processes may arise, such as concerns over changes to atmospheric mixing and climate implications from offshore wind (Wang and Prinn 2011) or concerns around changes in sediment transport and coastal stability implications from other ocean-based technologies (Contardo et al. 2018). OTEC is rather different from the other ocean-based energy technologies. Water discharged in the upper part of the water column would cool and change the environment and may cause concern well before reaching a new steady state consistent with the maximum geophysical potential of OTEC (Nihous 2018).

Regulators and other stakeholders for ocean-based energy projects have identified several possible interactions and potential effects of ocean-based energy devices. These include some that have been evaluated and deemed less critical, such as release of chemicals from coatings or oil spills from devices. As the evidence base grows, there has also been progress towards "retiring" some of the environmental concerns that have been assigned to ocean-based energy developments, such as the effects of electromagnetic fields on marine organisms (Copping et al. 2019). Noise and fauna–device interactions, however, remain key environmental concerns.

Environmental impacts will vary among the technologies. For bottom-fixed offshore wind, noise from piling during construction is of particular concern. Noise associated with pile-driving of foundations can lead to changes in the behaviour of a range of sea animals. For example, porpoise populations have been found to temporarily migrate during construction of offshore wind farms, with population density returning to normal following construction (Carstensen et al. 2006). Based on measurements from wind farms in the German Bight, Brandt et al. (2018) find that harbour porpoises avoid the construction site for up to two days after piling activities, and observable declines in porpoise detections are found up to 17 km away during actual piling activities.

Noise mitigation systems have reduced the impacts and such systems are being further developed.

Noise also affects fish. Hammar et al. (2014) studied impacts on cod in an area between Sweden and Denmark addressing impacts from pile-driving, working vessels and cable-trenching during construction, as well as from turbine noise, turbine lubricants and cable electric fields during operation. They found that noise from pile-driving was the most significant stressor and that ecological risks can be significantly reduced by avoiding particular construction events during the cod recruitment period.

While sound intensities of noise from shipping and installation of wind turbines, notably pile- driving, will be considerably higher than during operation, noise from operation of wind turbines is also of concern.

During the operation phase, the noise from the wind turbines varies with the strength of the wind. Noise arises in the turbine gearbox and generator, and is transmitted through the structure to the water and to the ground. Clearly the noise will depend on the type of gearbox and on the fundament or anchoring.

During the operational phase, the noise from ocean- based energy technologies might be considered comparable with other offshore industries; however, the characteristics of the sound will differ from the sound from other industries (e.g., slower rotational speeds). For subsurface technologies (e.g., wave and tidal devices), marine mammals may be disturbed by certain frequencies of noise and potentially avoid the area.

Wahlberg and Westerberg (2005) review pertinent aspects of underwater sound and hearing abilities of fish, noting that despite decades of increasing anthropogenic noise in the ocean due to maritime traffic and other human activities, the knowledge about fish response to noise is very limited. They conclude that fish can detect offshore wind turbines and that the noise may have a significant impact on the maximum acoustic signalling distances by fish within a range of a few tens of kilometres. The noise level and characteristics are expected to vary between types of wind turbine and fundament, and the hearing abilities at different sound frequencies vary among fish species.

Despite considerable efforts on understanding the impacts of noise from seismic investigations for offshore oil and gas, marine noise management in general is still in its infancy. De Jong et al. (2018) provide experimental evidence for the negative effects of noise on acoustic communication and spawning success for fish. But it remains to be investigated in the field and with noise characteristics from offshore wind activities. Electrification of the service vessels in the wind farm will reduce the noise level and other ship traffic will be minimised in the wind farm area. A risk-based approach integrating noise from different human activities (Faulkner et al. 2018) is proposed as a component of marine spatial planning. Offshore infrastructure may also create habitats acting as artificial reefs that enhance biodiversity and protect the area against heavy fishing including bottom trawling. Current regulations for the North Sea oil and gas installations require decommissioning at end of life, but complete decommissioning is not favoured by most experts (Fowler et al. 2018), neither for oil and gas nor for offshore wind installations. Both with respect to reef effects and noise, it is worth remembering that the distance between individual turbines in an offshore wind farm is in the order of 6–10 rotor diameters. For state-of-the-art turbines with a rotor diameter of 160–220 m, the distance between the turbines will be in the range of 1–2 km.

Collisions with offshore wind turbines are a notable risk for some seabird species, if turbines are placed such that they disconnect important roosting and feeding sites, or in migratory routes. However, recent research, spanning a two-year monitoring period at the Vattenfall Thanet offshore wind farm (one of the United Kingdom's largest offshore wind farms) has shown the risk of seabirds colliding with offshore wind turbines is lower than previously predicted (Skov et al. 2018), with six strikes recorded during the two-year monitoring period. For other ocean-based energy technologies, collision between devices and marine mammals is a key concern (Copping et al. 2016).

With increasing deployment of offshore wind, the potential environmental risks associated with offshore wind are much more clearly understood and there is growing consensus towards the position that offshore wind farms can be constructed without significantly damaging the environment. However, to achieve this requires proper planning and putting in place mitigation measures (WWF 2014). Other ocean-based technologies, being less mature, with fewer deployments from which to monitor potential risks, have much greater scientific uncertainty surrounding the probability of occurrence, and/or the severity of consequences, specified as the potential risk.

The combination of collecting proper baseline data, careful monitoring of interactions, effective device design and proper marine spatial planning for projects will be required to ensure that potential risks are mitigated. Ecosystem modelling is being used to determine impacts on ecosystem indicators (Raoux et al. 2018). Various approaches and methods for marine spatial planning with specific focus on offshore wind have been proposed (Pinarbasi et al. 2019).

No wind energy projects in the high seas have been proposed up to now. However, the resource is considerable and may be of interest in the future. Elsner and Suarez (2019) make the point that important justice questions remain concerning access and benefits. Even if the UN Convention for the Law of the Sea (UNCLOS; UN 1982) is recognised as the legal basis for any offshore wind deployments in the high seas, there is a danger that flag states may undercut environmental and safety standards for offshore wind energy installations (Elsner and Suarez 2019). Marine spatial planning approaches and the establishment of cooperative mechanisms are needed to safeguard against such developments.

5.1.2 Environmental Effects of Deep-Seabed Mining

Environmental unknowns, vulnerabilities and costs are some of the most challenging aspects of deep-seabed mining (Thompson et al. 2018). The remoteness of most of the deep ocean combined with the harsh operating conditions (high pressure, low temperatures and darkness), requiring expensive and highly technical equipment, have resulted in limited exploration and scientific research. These constraints, and the vastness of the area in question, mean that the majority of the deep ocean, both within and beyond national jurisdictions, are poorly characterised and understood, or still completely unexplored.

Of the three habitat types vulnerable to mining – abyssal plains with polymetallic nodules, hydrothermal vents with massive sulphides and seamounts with cobalt-rich ferromanganese crusts, the last – especially in the Prime Crust Zone (an area in the West Pacific identified as of the greatest economic interest for mining cobalt-rich crusts)— are the least explored, hence their biodiversity has not yet been characterised (Morgan et al. 2015). Even in polymetallic nodule zones, thought to be bereft of life only 40–50 years ago when UNCLOS Part XI was crafted, four decades of research by contractors and scientific organisations in the nodule-rich CCZ show that environments and associated biodiversity remain largely undiscovered or unidentified. For example, in the eastern CCZ, over 50 percent of species over two centimetres (cm) in size collected by Amon et al. (2016) in 2013, and 34 of the 36 species of xenophyophores (large single-celled organisms) collected by Gooday et al. (2017) in 2015, were new to science. And while hydrothermal vents are the most characterised and understood of the three habitats, many species at vents appear to be rare (comprising <5 percent of the total abundance in samples), and poorly known (Van Dover et al. 2018). Finally, the connections of these habitats to the wider global functioning is poorly understood, although new studies have begun to shed some light on this (Sweetman et al. 2019; Ardyna et al. 2019)

5.1.3 The impacts of deep-seabed mining remain unknown

Deep-seabed mining is expected to create environmental impacts that involve the following (Van Dover 2014; Levin et al. 2016; Vanreusel et al. 2016; Gollner et al. 2017; Boetius and Haeckel 2018):

- Direct removal of the resources which act as a substrate for specialized faunal communities, including at least half of the species larger than 0.5 millimetres (mm) in size inhabiting these ecosystems – as a result, the animals will be killed or crushed.
- Changes to the geochemical and physical properties of the seafloor.
- Sediment plumes created from the disturbance on the seafloor as well as from the return waterdeposited in the water column that may smother or clog feeding apparatus and limit visibility.
- Contaminant release and changes to water properties.
- Increases in sound, vibration and light.

Several large programmes (such as MIDAS and JPI Oceans Mining Impact) have addressed likely mining impacts, but in the absence of disturbance studies on appropriately large scales (across space and time), the intensity, duration and consequences of the impacts of commercial mining remain speculative. Regulators can set rules designed to minimise environmental impacts, such as requiring processed water and sediment to be returned to the ocean at certain depths in order to minimise the creation of a sediment plume in the water column. However, deep-seabed mining poses a risk for biodiversity loss, forced species migrations and loss of connectivity, potentially leading to species extinctions in the deep ocean (Van Dover et al. 2017; Niner et al. 2018). This is of particular concern as many deep-sea species may have genetic compounds that could have biotechnical or pharmaceutical use in the future. There could also be impacts to ecosystem services, such as to fisheries, climate regulation, detoxification and nutrient cycling, but the potential risks have not yet been quantified (Le et al. 2017).

Another poorly understood issue is the length of time that biological communities affected by deep-seabed mining will take to recover. There have been no tests undertaken on a scale that would replicate commercial mining in any of the three habitats, and it is likely that recovery times will differ among ecosystems. However, information gleaned from small-scale experiments, as well as from other industries such as deep-sea trawling, point to lengthy recovery times in each system. Bluhm (2001), Vanreusel et al. (2016), Jones et al. (2017) and Miljutin et al. (2011) have shown that, while there is always some recovery in faunal density and diversity, communities have still not returned to baseline conditions two decades after tests in nodule areas. Simon-Lledó et al. (2019) echoed these findings, showing that, in disturbed areas of the Peru Basin, both the presence of suspension feeders (corals, sponges, etc.) and diversity generally remained significantly reduced after 26 years. Instead, the community was dominated by deposit feeders and detritivores. They concluded that, if the results of the DISCOL experiment in the Peru Basin could be extrapolated to the CCZ, the impacts of nodule mining (taking into account the area directly impacted, as well as the plume deposition area) may be greater than expected, and could lead to an irreversible loss of some ecosystem functions. As nodule mining will remove the nodules, which take millions of years to form, full-scale recovery will likely take a period of time on that scale. Sites identified as being the most favourable for nodule mining are estimated to span 38 million km^2 (Petersen et al. 2016); individual nodule exploration contracts, of which there are 19 in international waters, each cover 75,000 km^2.

On seamounts, where cobalt-rich ferromanganese crusts are located, cold-water corals and other sessile suspension feeders are extremely susceptible to physical disturbances, such as those already caused by bottom- trawling fisheries (Kaiser et al. 2006; Clark and Tittensor 2010; Williams et al. 2010), because they grow extremely slowly (a few mm to ~1 mm per year) and are long-lived (decades to thousands of years) (Roark et al. 2006; Clark et al. 2016). Most seamounts with high trawling impact have coral cover reduced to below 30–50 percent of the coral cover estimated as necessary to maintain habitat viability (Clark and Tittensor 2010). Impact by trawling fisheries is likely to differ from mining, where the entire substrate will be removed. For organisms dependent on cobalt-rich ferromanganese crusts on seamounts, recovery from substrate removal could require thousands to millions of years, given the rate of formation of crusts (Gollner et al. 2017). Sites identified as being the most favourable for crust mining are estimated to cover 1.7 million km^2 (Petersen et al. 2016); each contractor (there are presently five) may have contracts that cover up to 3000 km^2 consisting of 150 blocks, each no greater than 20 km^2. Polymetallic crusts on seamounts may be the most technically difficult resource to mine and the one most likely to support active fisheries.

At hydrothermal vents, distinct global faunal patterns, vent site distances and natural background disturbance regimes make it currently impossible to predict recovery rates using volcanic eruptions in other regions as an analogy for deep-seabed mining (Gollner et al. 2017). Recent observations of decadal stability and longevity at vents in the Pacific back-arc basins indicate recovery periods may be longer than initially thought (Du Preez and Fisher 2018). Active hydrothermal vents have been proposed by scientists to be set off limits to mining (Van Dover et al. 2018), but no regulations currently limit mining at active hydrothermal vents and many active sites inside exploration contract areas (both within and beyond national jurisdiction) are vulnerable to impacts from mining at nearby inactive vent sites. There is currently little baseline information and no data available for recovery times at inactive vent sites, making predictions there difficult (Gollner et al. 2017). Sites identified as being the most favourable for seafloor massive sulphide mining are estimated to cover 3.2 million km^2 (Petersen et al. 2016); individual contracts may cover up to 10,000 km^2, with up to 100 blocks of 100 km^2.

5.1.4 Deep-Seabed Mining Could Result in Loss of Species and Functions Before They Are Understood

The danger of biodiversity loss is of particular concern given the lack of baseline knowledge of the communities in habitats vulnerable to deep-seabed mining (Van Dover et al. 2017; Van Dover 2019; Niner et al. 2018). It is expected that there will be local extinctions, because many of the fauna inhabiting vents, nodule-rich abyssal plains and encrusted seamounts rely on the resources to be extracted as substrate (Vanreusel et al. 2016). For example, Amon et al. (2016) observed that half of the species over 1 cm in size in the eastern CCZ relied on the nodules as an attachment surface. Strong environmental control and prevalence of rare species makes the smallest invertebrates (meiofauna) in the CCZ vulnerable to the risk of extinction from nodule extraction (Macheriotou et al. 2020).

If mining was to go ahead with the current state of knowledge, species and functions could be lost before they are known and understood. A consideration of scale, placement and connectivity is key to prevention of biodiversity loss. In vast, contiguous systems such as the CCZ, cumulative impacts from more than one mining operation may threaten species persistence, depending on their location or timing. The same may be true for vents along a mid-ocean ridge or for seamounts in a chain. For this reason, the series of Regional Environmental Management Plans (REMPs), which the ISA has commenced developing as strategic environmental management tools (ISA 2019b), will need clear environmental objectives (Tunnicliffe et al. 2018). The purpose of REMPs, broadly, is to provide region-specific information, measures and procedures in order to ensure the effective protection of the marine environment in accordance with Article 145 of UNCLOS (UN 1982). To this end, REMPs should establish environmental management measures, including the designation of protected areas (in ISA nomenclature, Areas of Particular Environmental Interest or APEIs) prior to or independent of contract placement and periodic reassessment (Wedding et al. 2013; Mengerink et al. 2014; Dunn et al. 2018), and should be used as management tools which feed into regulatory decisions and actions. REMPs should take into account cumulative effects from multiple mine sites, or synergistic effects from different marine uses or stressors, and seek to manage potential conflicts occurring in the same region. Consideration of climate change in REMP development will help to inform spatial management and environmental impact assessment, and ensure that monitoring programmes can differentiate climate from mining impacts (Levin et al. 2020).

5.1.5 The Challenges of Mitigation and Restoration of Ecosystems

It is difficult to anticipate how best to mitigate the potential impacts of deep-seabed mining because there have been so few studies investigating mining impacts that resemble those actually caused by mining activity, as well as none on the scale on which deep-seabed mining would take place (Jones et al. 2017; Cuvelier et al. 2018). It is likely that the mitigation hierarchy (avoid, minimise, remediate and offset) used in terrestrial and shallow-water extractive activities is not applicable in the deep ocean (Van Dover et al. 2017). Challenges associated with restoration and recovery include the slow recruitment and growth of deep-sea species, the potentially vast scale of mining impacts, and the limited understanding of the requirements for proper ecosystem functions (Gollner et al. 2017). Additionally, the likely high cost of deploying assisted regeneration techniques, such as the use of artificial substrates, the transplantation or seeding of larvae and the artificial eutrophication of the ocean surface, may also be insurmountable (Van Dover 2014; Niner et al. 2018). Furthermore, no restoration strategies proposed have been tested, and even if benthic remediation were technically feasible, the financial commitment required may be extensive (Niner et al. 2018).

Offsetting is the last stage in the mitigation hierarchy and includes the protection of a similar type and equivalent amount of habitat under threat from other existing or planned activities (e.g. preventing trawling in cobalt- rich ferromanganese crust communities), and the creation or restoration of biodiversity of a similar type in a different location to that lost to ensure no net loss. It also includes compensatory mechanisms – for example, the creation of biodiversity of a different type and/or in a different location, such as in shallow or coastal environments – or additional actions that do not provide biodiversity gains ecologically linked to biodiversity losses, such as capacity-building. All of these options are currently unable to replicate biodiversity and ecosystem services lost through deep-seabed mining, so cannot be considered true offsets (Niner et al. 2018). This is, in part, due to gaps in current ecological knowledge and restoration abilities in the deep sea (Niner et al. 2018).

If deep-seabed mining moves forward, it must be approached in a precautionary and adaptive manner, so as to integrate new knowledge and avoid and minimise harm to habitats, communities and functioning (Jaeckel 2017; Niner et al. 2018). There are a number of ways in which this can be done, with each option informed by goals, standards, indicators and monitoring protocols. Avoiding harm altogether is unlikely to be achievable, given the destructive nature of deep-seabed mining, which will heavily impact the immediate mining sites. The size of mined sites would vary by deposit type, but a single operation might mine polymetallic nodules over about 8500 km^2 of seafloor over several decades (Ellis 2001; Van Dover 2014; Petersen et al. 2016; Jones et al. 2017; Van Dover et al. 2017; Niner et al. 2018). Some impacts may be avoided at a project level by reducing the footprint of mining within a contracted area and/or by leaving some minerals with associated fauna in place and undisturbed (protected areas or refugia). However, given that the effects of mining will be three-dimensional and diffuse, are poorly understood, and will involve impacts from sediment plumes as well as toxicity and noise, the identification of refugia that are free from impacts will not be straightforward, and biodiversity loss will likely still occur (Ellis 2001; Thiel et al. 2001; Van Dover 2014; Niner et al. 2018).

Minimising losses of biodiversity and other ecosystem damage to the greatest extent possible includes technologies and practices that may be developed and applied to reduce these risks. There is currently limited technological capacity to minimise harm but possible adaptations include instrument optimisation to limit sediment-plume dispersal, longevity and toxicity, to avoid seabed compaction, and to reduce light and noise pollution (Niner et al. 2018). The effectiveness of such measures at reducing biodiversity losses requires testing and will rely upon a strong regulatory framework, with monitoring and enforcement capabilities. Adaptive management has been identified as a useful regulatory approach that could be applied to deep-seabed mining operations once other challenges are addressed (Jaeckel 2016).

5.2 Economic, Societal and Cultural Costs and Benefits

5.2.1 Benefits of Ocean-Based Renewable Energy

Ocean-based renewable energy provides several benefits in comparison with other sources of energy. It has very low CO_2 emissions over the life cycle of deployment. Decarbonising the transport and construction in the sector will further reduce its CO_2 footprint. It also has negligible emissions of mercury, SO_2 and NO_2, and no waste generation. Estimating the total social cost of carbon emissions is a widely discussed topic in the literature, and is beyond the scope of the present paper. But it is clear that, if substituting ocean-based renewable energy for coal-fired power, the direct and indirect benefits for human health and well-being would be considerable.

In terms of employment opportunities, offshore wind provides more jobs than fossil fuel electricity. IRENA (2018b) estimates that a total of 2.1 million person- days is needed to develop an offshore wind farm of 500 MW capacity. The largest part of this effort is in manufacturing and procurement (59 percent), but even for countries that do not aim to stimulate production locally, operation and maintenance (24 percent) and installation and grid connection (11 percent) offer considerable local job opportunities. Gender balance is generally better in renewable energy jobs than in fossil fuel. Training and re-skilling of the oil and gas workforce is an attractive opportunity given the relevance of many skills. Much less information is available for other ocean- based renewable energy but a study of tidal stream and wave energy in the United Kingdom suggests that these can deliver similar employment opportunities to offshore wind when being scaled up (Offshore Renewable Energy Catapult 2018). Jobs will mostly be in coastal areas, some of which may currently suffer from a lack of employment opportunities.

Since, in contrast to thermal power plants, there is no water usage associated with ocean-based renewable energy, there will also be significant water savings. This can be important in many areas where water resources are scarce and costly. A brief overview of the impacts of accelerated deployment of ocean-based renewable energy on all the 17 SDGs was given in Hoegh-Guldberg et al. (2019), showing positive impacts on all. Only for the ocean goal SDG 14 was there a potential red flag, associated with negative impacts on marine life and biodiversity.

The deployment of power plants offshore can create conflicts about the use of ocean space for other human activities, such as maritime transport, offshore oil and gas, fisheries and potentially also offshore fish farming, as well as for marine protection (with Marine Protected Areas or MPAs). Baseline ecosystem mapping and marine spatial planning that takes the various interests into account will be required. In some cases, combinations could be fruitful; for example, wind farms with traffic and fisheries restrictions could usefully delineate MPAs. Such considerations will vary from place to place based on local conditions, and ultimately decisions will be based on what is socially and politically acceptable (the social/ political potential – see Fig. 3.5).

5.2.2 Benefits of Deep-Seabed Mining

Deep-seabed mining will bring increased metal supply to consumers globally and is likely to benefit the exploitation company, shareholders and members of the supply chain through financial profits (Kirchain and Roth 2019). Deep-seabed mining within a state's national jurisdiction or in the Area under a state's sponsorship has the potential to benefit that state by contributing to government revenues (through taxes and/or royalties).

The quantum may be significant. The UK Prime Minister's assessment of the UK sponsorship of an ISA contract as "a 40 billion £ opportunity" has been called "an overly cautious estimate" by one fellow Parliamentarian (House of Commons 2019). The Cook Islands government has valued their national seabed mineral resources at "hundreds of billions of dollars": a significant sum for any nation, let alone one with a population of fewer than 18,000 people (*Cook Islands News* 2018).

Further benefits may include creating jobs and training opportunities, strengthening the domestic private sector, encouraging foreign investment, funding public- service or infrastructure improvements, introducing a new supply of metals, and supporting other economic sectors (SPC 2012; World Bank 2017). Those benefits may not be large, but may nonetheless be significant, as, for example, in the case of small island developing states with limited land resources and economic options (Wakefield and Myers 2018).

Deep-seabed mining in the Area will bring revenue to humankind, collected and managed on humankind's behalf by the ISA. The quantum and form of that revenue will depend on the system of payments for contractors that is currently under negotiation in the ISA. An initial royalty of 2 percent (rising later to 6 percent) has been proposed for the ISA under an economic model based on contractor profits and contractor data. This could lead to the mining company receiving around 70 percent of the total project profits, and the ISA around 6 percent (with the remainder going to the sponsoring state or whichever state is receiving profit taxes from the mining company) (Africa Group 2018b). Some stakeholders have expressed concern with the principles used in that economic model, and the low royalty rate and return to the ISA. Opponents include all of the 47 African countries who are members of the ISA, and who calculated that the proposed payment regime would lead to a return to humankind of less than $100,000 per annum per country, which they did not deem to be fair compensation (African Group 2019). The international seabed regime established by UNCLOS (UN 1982) is predicated on the basis that mining be carried out (only) in such a manner as to "foster healthy development of the world economy and balanced growth of international trade, and to promote international co-operation for the overall development of all countries, especially developing States" (Article 150). So a regime that would see benefits from mining in the Area flow principally to developed states, or to wealthy shareholders of the companies that are conducting the mining should not be permitted (African Group 2018b).

Other benefits may involve technological innovation and the advancement of deep-sea science. Exploration and impact monitoring may expand scientific knowledge that is currently lacking (if levels of data quality and public-sharing are improved) (Pew Charitable Trusts 2017). Similarly, research associated with deep-seabed mining could also increase our understanding of genetic resources, with the potential for use in pharmaceuticals, industrial agents, biomedical products or bioinspired materials (Le et al. 2017).

Economic development is a key driver for most states, but many resource-rich developing states exhibit slow economic growth. The type of windfall income streams that may be generated if successful deep-seabed mining occurs in significant quantities, if not handled carefully, could have negative effects on a state's economic status (Taguchi and Khinsamone 2018). Commentators observe that the risk of this "resource curse" may be combated by sound revenue management, and an integrated resource management approach, grounded in transparent and non-discretionary policy and law, with funds that are generated by deep-seabed mining being used both for long-term investments in infrastructure or socio-economic projects, and also safeguarded for future generations ("intergenerational equity") (SPC 2016). Some Pacific Island countries (Cook Islands, Tonga, Kiribati, Tuvalu) have addressed this challenge by requiring by law the establishment of a ring-fenced sovereign wealth fund in which any proceeds from seabed mining within national waters must be invested.

The ISA has a different revenue management challenge: how to distribute the proceeds from mining in the Area equitably, and for the benefit of all of humankind (Feichtner 2019). This potentially complex aspect of the ISA's regime has received little attention to date, while the more immediately urgent operational rules for mining and the specific payment rate for contractors are under focus. Different models may include: direct distribution of a share of proceeds to individual member states, or some kind of ISA-managed fund to which states can apply for grants (ISA 2013). However, the proceeds available for distribution may not be large amounts, and may also be depleted by the need to cover operational costs of running the ISA (Thiele et al. 2019).

5.2.3 Costs

The economic costs of ocean-based renewable energy has been treated in Sect. 3 and costs to the environment in Sect. 5.1. It is clear that, provided that the environmental impact assessment is performed with an integrated ecosystem approach to avoid areas of particular value, ocean-based renewable energy, in particular offshore wind will increasingly become cost- competitive with other sources of electricity. This will be a driving force for expansion in more and more areas around the world, even if none of the indirect benefits to climate, human health and other aspects of sustainable development discussed in Sect. 5.2 are used as a rationale for policies or incentives for transitioning from fossil fuels to renewables. A cost–benefit analysis incorporating all of these aspects is beyond the scope of this paper but a preliminary assessment was included in the Annex to Hoegh-

Guldberg et al. (2019) and further analysis is available in Konar and Ding (n.d.).

Little cost–benefit analysis has been done for deep- seabed mining projects (SPC and Cardno 2016), although there have been recent calls for such analyses. For example, the UK government has recently committed to analyse the potential economic value to the United Kingdom of the two ISA contracts granted to UK Seabed Resources Ltd. under its sponsorship in the CCZ (House of Commons 2019). Pacific Island finance ministers and civil society organisations also agreed at a meeting in May 2019 that an independent regional study on deep- seabed mining and its implication for Pacific economies, the environment and ocean biodiversity, and people's livelihoods would provide a helpful evidence base to inform countries' policy decisions on seabed mining (Pacific Islands Forum Secretariat 2019).

The primary benefits of seabed mining are presumed to be economic, and the primary costs ecological. There may, however, also be economic costs to a state engaging in a deep-seabed mining operation (e.g., DSM Observer 2018), and in regulating it (SPC and Cardno 2016; World Bank 2017). In the Area, if third-party harm or unforeseen damages occur, then either the mining company or the sponsoring state will be liable to cover the costs of compensation or remediation (Craik et al. 2018). UNCLOS (UN 1982) specifically provides that mining in the Area must not adversely affect the economies of developing countries derived from terrestrial mining, or must compensate them (sections 1(5)I and 7(1) of the Annex to the 1994 Agreement). This may mean that proceeds flowing to the ISA from royalties for mining cobalt in the Area, for example, will be used to compensate countries such as the Democratic Republic of Congo (where most cobalt on land originates) for any losses caused by the ISA contract. This has the potential to limit financial benefits flowing to other parties (apart from the contractor and the compensated country). Alternatively, mining in the Area may occur in addition to terrestrial mining for the same metals (which obfuscates arguments about advantages of offshore mining relative to land mining, and adds to overall adverse impacts rather than replacing the existing ones). An increased supply of minerals could drive metal prices down, which again may require ISA proceeds to be used to compensate developing countries whose economies suffer as a result.

Although the mining activities will largely occur at sea, transporting and processing of minerals is likely to occur on land. There are concerns that associated land- based activities will adversely affect local communities' property, food sources and lifestyle (Aguon and Hunter 2018). Equally, local communities may seek to host industrial facilities or support services in the interests of attaining employment or building infrastructure and so on. There may also be concern that coastal communities in countries who permit deep-seabed mining within national waters, or whose national waters lie adjacent to deep-seabed mining sites under international or another state's jurisdiction, and who rely heavily on the sea for their food and income will be affected by deep-seabed mining through the disruption of fragile and diverse ecosystems, through displacement of fisheries, or through failure to respect the rights of indigenous peoples (SPC 2012; Aguon and Hunter 2018). In extreme cases, and particularly in the absence of strong governance systems, other extractive industry activity has been seen to worsen social tensions and even lead to political instability, such as the Bougainville Civil War in Papua New Guinea, which cost thousands of lives. It has also been noted that deep-seabed mining may cause a loss of cultural or spiritual value associated with a pristine ocean, or traditional sense of ownership of or identification with the ocean and its resources (World Bank 2017). Given strong ecological connectivity between waters in areas beyond national jurisdiction (ABNJ) and coastal zones (Popova et al. 2019), concerns have been expressed about transboundary impacts, whereby a mining operation within one jurisdiction causes deleterious effects to the marine environment or coastal communities of a neighbouring country (Singh and Pouponneau 2018). The international legal framework currently contains lacunae with regards to identifying and enforcing liability for compensation, clean-up or remediation (Craik et al. 2018; ITLOS 2011).

5.2.4 Environmental Costs, Ecosystem Services Valuation, Tradeoffs and Intergenerational Equity

When the value of the seafloor environment – for example, in terms of ecosystem services – is weighed against the value of the minerals residing on the seafloor, this comparison is almost always made in terms of monetary value. The value of minerals can be estimated based on past, current and predicted future market prices. The living environment can be valued for the services it provides to humans in the form of food, although other provisioning services (such as pharmaceuticals, industrial agents, biomaterials) may be discovered. Regulating services in the form of carbon sequestration or nutrient recycling are modelled rather than measured (e.g., Burdige 2007) and new elements such as dark carbon fixation are being uncovered. In 2015, the Nautilus Solwara 1 project at hydrothermal vents in Papua New Guinea – a very small pilot mine site – was estimated to have \$245 million worth of gold and \$397 million worth of copper which could be mined over 2.5 years. Earth Economics conducted a social benchmarking study to monetise the impacts of mining at

Solwara 1 on ecosystem services and determined that the dollar value of natural capital assets impacted was far lower for Solwara 1 than for a comparable terrestrial mine (Batker and Schmidt 2015). They used the UN Environment Program The Economics of Ecosystems and Biodiversity and Millennium Ecosystem Assessment in utilising a landscape and seascape approach to natural capital valuation based on the land cover type and area disrupted with conservative overestimates. It was determined that the value of ecosystem services of an acre of Solwara 1 hydrothermal vent (valued at $24,724) was either 80 or 1733 times less than two comparison terrestrial mines. The Earth Economics valuation method has been employed subsequently in social cost–benefit analyses of mining in Papua New Guinea, Cook Islands and the Republic of the Marshall Islands (Wakefield and Myers 2018), while also being criticised for its methodology, given the difficulty of quantifying impacts which have not been or are only just beginning to be studied (World Bank 2017).

The use of terrestrial metrics to quantify deep-sea services overlooks functions not found on land and fails to recognise that deep-sea vent ecosystems are small and among the rarest on the planet (active vents are estimated to cover a total area of less than 50 km^2 globally). Each hydrothermal vent is different from others, and the biodiversity and vent functions remain poorly known (Van Dover et al. 2018). For example, it was only recently found that skates lay their egg cases at vents (Salinas-de-León et al. 2018). In general, the ecosystem services of the deep sea are poorly known (Armstrong et al. 2012; Thurber et al. 2014), so are underconsidered in cost–benefit analyses and are rarely addressed in the development of mining regulations (Le et al. 2017). Particularly under-represented are the non- monetary social, cultural and livelihood values of seabed ecosystems or possible downsteam impacts where the minerals are landed. A greater focus is needed on how to value non-monetary assets linked to the existence and aesthetic and educational uses of biodiversity, as well as the functions and services not yet discovered.

Disruptions caused by deep-seabed mining at the seafloor, in the overlying water column and where ore is brought to land can cause conflict with other economic sectors and threaten loss of non-market ecosystem services (Thompson et al. 2018). Noise, light, sediment plumes with contaminants, and oil leakages can threaten both commercial and subsistence fisheries (Miller et al. 2018). In the case of phosphorites, there is often direct spatial overlap between fisheries and the mineral resource, as well as potential disruption in the overlying waters caused by extraction (Levin et al. 2016). It is also possible that mining activity could prevent future use of the mining site for other purposes. Seafloor substrates targeted for mining may hold genetic resources that could be lost (Le et al. 2017; Van Dover et al. 2018). These are subject to the Nagoya Protocol within national waters, and are the subject of negotiations in international waters (UN General Assembly 2018), but are not currently regulated in the Area (Vierros et al. 2016). Deep-seabed mining could disrupt carbon cycling linked to iron flux from hydrothermal vents, which plays a role in stimulating primary production and carbon drawdown from the atmosphere (German et al. 2015; Ardyna et al. 2019), and by removal of autotrophic microbes that fix carbon, and fauna that bury carbon in sediments (Sweetman et al. 2019). Loss of tourism from the threat of mining is feared in diverse settings such as Papua New Guinea, Fiji, Portugal and Spain (Thompson et al. 2018). Since no full-scale mining impacts have occurred, the nature and extent of these tradeoffs cannot be studied and thus remain speculative.

The value of lost ecosystem services due to mining impacts could appear in the financial code as a form of monetary compensation (e.g., to the common heritage of humankind) or be factored into the amount of the royalty payable by the miner. Built into the concept of the common heritage of humankind is the principle of intergenerational equity, in which, in addition to sharing the benefits of resources, the resources in the natural environment are preserved for generations to come (Jaeckel et al. 2017). The idea of partitioning resources among current and future generations is an important component of sustainability (and intergenerational equity) for non-renewable resources.

5.2.5 Decisions to mine

Most discussions of deep-seabed mining address where, when and how to conduct deep- seabed mining, as well as what the impacts might be, but not.

whether to mine (Kim 2017). The distribution of metal resources and their production creates geopolitical uncertainties that were to be solved by designating minerals in the Area as the "Common Heritage of Mankind" (UN 1982). In 2012, most cobalt (68 percent) was mined in the Democratic Republic of the Congo, Chile produced 32 percent of copper, China 90 percent of REEs and almost 78 percent of terrestrial manganese resources were found in South Africa (Brown et al. 2014). Extraction of these minerals creates social and environmental problems on land (Kim 2017; IRP 2019), which proponents of deep-seabed mining have argued are mostly absent in the ocean (Lodge and Verlaan 2018).

Currently, opinions on whether deep-seabed mining should proceed span a broad spectrum (Box 3.1). At one end, there is the adamant opposition to any deep- seabed mining, with the claim that adverse effects on the environment will outweigh the benefit of additional metals (Kim 2017). This perspective argues that seabed minerals are not needed (Teske et al. 2016) and suggests that "we should do more with less" via a circular economy that advances recycling, reuse and extended product lifetimes.

Box 3.1 Scenarios for Deep- Seabed Mining

Scenario 1. Full steam ahead on current knowledge

Accept environmental and economic risks, social and equitability concerns and proceed with mining the seabed within and beyond national jurisdiction as soon as legally possible. Biodiversity and its ecological functions in the areas impacted could be lost, possibly irreparably. The scale and ramifications of those impacts – as well as the extent to which the mining will lead to overall benefit for humankind – are hard to predict on current knowledge.

Scenario 2. Slow the transition from exploration to exploitation – precautionary pause

Allow more time to fully assess and understand the environmental risks, including through additional scientific study prior to issuing exploitation contracts; design spatial protections carefully (including the identification of regions to set aside from mining) and develop additional methods to promote resilience; further clarify the need for deep-seabed metals; develop mining regulations in a careful, thorough and transparent manner, with independent expert input and engagement of all stakeholders. Stop issuing new exploration contracts, and do not grant any mining contracts, unless and until the above has been undertaken.

Scenario 3. Indefinite moratorium on deep- seabed mining

Deep-seabed mining does not move forward. Refocus on initiatives that enable transition to circular economies with emphasis on metal demand reduction through reuse, recycling, alternative materials, extended product lifetimes and behavioural change.

In the middle, there are calls for pilot testing and further scrutiny of the issue, as well as for a moratorium or precautionary pause to allow more scientific study and to see the highest environmental standards and the precautionary approach embodied in deep-seabed mining regulations and guidelines. A pause may in practice prevent the issue of additional mining contracts unless and until there is scientifically supported evidence – not currently available – that the impacts will be outweighed by the benefits. Concerns have been voiced that the process has gone too fast relative to the state of knowledge and the ISA's capacity for environmental management. Various bodies have proposed different forms of such a precautionary pause or moratorium on deep-seabed mining in international waters, namely the European Parliament, the UK House of Commons Environment Audit Committee, the Long Distance Fleet Advisory Council (LDAC) of the European Union (LDAC 2019), and the UN Secretary General's Special Envoy for the Ocean. A major aim of such a pause is to allow scientific research to advance, possibly in conjunction with the Decade of Ocean Science for Sustainable Development (Johnson 2019). Additionally, Fiji has proposed, and Papua New Guinea and Vanuatu may be considering, a similar moratorium within their national jurisdictions.

At the other end, there is the stance that deep-seabed mineral projects should be facilitated and incentivised (essentially the current position of the ISA). To date, no requests for exploration contracts have been denied by the ISA and there is a current push to develop the Mining Code (regulations, guidelines and procedures) by 2020 so that exploitation may commence (ISA 2017, n.d.).

The complexity of the stakeholder input to decisions about deep-seabed mining cannot be underestimated (Box 3.2). The most vocal are states with exploitation contracts (Fig. 3.3), and the mining companies that partner with them. Those states with mines on land, and those with a history of ocean conservation have also weighed in, while civil society and the public in general have had a limited voice to date (Fleming et al. 2019).

Box 3.2 Stakeholders for Deep-Seabed Mining

At this time, those expressing the greatest interest in deep-seabed mining, both actively and passively (and not necessarily always to propel the industry forward) include the following groups:

- Nations that have ISA exploration contracts (e.g., China, India, Japan, Russia, South Korea and various EU countries).
- Countries that have deep-sea mineral deposits of commercial interest within national jurisdictions (e.g., Papua New Guinea, Tonga, Cook Islands, Namibia, Japan, Kiribati).
- Countries that actively mine the same minerals on land (e.g., Democratic Republic of Congo, Chile, South Africa).
- Mining companies that have claims within EEZs or have partnered with states on international exploration claims (e.g. Nautilus Minerals, UK Seabed Resources Ltd, Global Sea Mineral Resources, Deep Green).

- Research institutions and scientific networks (e.g. JPI Oceans, the Deep-Ocean Stewardship Initiative, InterRidge, and the Deep Ocean Observing Strategy) interested in bringing science to decision-making and the development of regulations, and in providing sustained observations that can help to address outstanding scientific questions.
- States, environmental advocacy groups, intergovernmental organisations (IGOs) and non-governmental organisations (NGOs) focused on conservation and biodiversity maintenance (e.g. International Union for Conservation of Nature, Deep-Sea Conservation Coalition, Greenpeace, WWF, The Pew Charitable Trusts).
- Other components of the blue economy, such as the deep-sea fishing industry and underwater cabling companies, with potential conflict or spatial overlap.
- Civil society and religious groups that are largely active within EEZs and wary of exploitation of local and indigenous peoples and threats to their local environment and culture (e.g. the Holy See, Deep Sea Mining Campaign, the Pacific Conference of Churches, Alliance of Solwara Warriors, Fair Ocean, Misereor, Brot für die Welt).

It is possible, as has been shown with other habitats, that with time and education, civil society may be willing to pay to forgo blue industrial growth for conservation of the deep sea in order to preserve ecosystem services (Aanesen and Armstrong 2019). Deep-sea scientists are a growing constituency that is increasingly engaged as part of baseline surveys for contractors or discussions with the ISA via organisations such as the Deep-Ocean Stewardship Initiative and InterRidge.

Regulatory sectors overlap in areas targeted for or potentially impacted by deep-seabed mining. Water-column impacts in international waters falling under ISA jurisdiction will intersect with management by regional fisheries management organisations under the Food and Agriculture Organization of the United Nations or the regime overseen by the International Maritime Organization which regulates contaminants and dumping (and which is implemented via individual “flag states” to whom vessels are registered). Current negotiations on the conservation and sustainable use of marine biological diversity of areas beyond national jurisdiction (BBNJ), with its focus on spatial protections, environmental impact assessment, marine genetic resources and technology transfer and capacity-building, has large potential overlap with the ISA for the Area (which includes the entire seafloor). The designation and nature of spatial protections, applications of ecosystem- based management, the disposition and accessibility of data, and the development of shared goals and objectives to achieve sustainability are all areas where the ISA will need to work across sectors both within the United Nations, and with industry, academia and civil society.

6 Governance and Regulatory Framework for Deep-Seabed Mining

6.1 State Level

A state should adopt appropriate measures to exercise control over any seabed mineral activities under its jurisdiction and to secure compliance with international standards. State laws relating to the management of seabed mineral activities must be “no less effective than international rules, regulations and procedures” (UNCLOS; UN 1982)—such as the Mining Code of the ISA, currently under negotiation (ISA n.d., 2019d).

Direct obligations under international law in respect of seabed mining include: applying the precautionary approach, employing best environmental practice, and conducting prior environmental impact assessment (ITLOS 2011). These obligations apply to states regardless of their individual wealth or capacity (ITLOS 2011). A number of states, particularly in the Pacific region, have implemented national legislation to govern seabed mineral activities (both within national and international jurisdiction) (e.g. Fiji, Tonga, Tuvalu, Kiribati, Cook Islands, Federated States of Micronesia, Nauru, United Kingdom, Belgium, United States, Japan, Germany, China) (Lily 2018; World Bank 2017). It is notable, however, that several states actively engaged in exploration activities as yet have no detailed legal regime in place (e.g. India, France, South Korea, Brazil, Russia, Poland) (Lily 2018; ISA 2019a).

The creation of adequate legislative frameworks by states, while essential, is not sufficient in itself: implementation and enforcement of the rules created are also crucial (ITLOS 2011). This point is supported by international law (e.g., UNCLOS, Articles 214 and 215; UN 1982), which requires appropriate environmental standards not only to be governed by domestic legislation, but also to be implemented through monitoring and enforcement. Strong institutions are particularly important to the oversight of seabed mining; legal, fiscal and environmental matters will all require dedicated public administration capacity. This may be particularly challenging for small developing states with limited administrative and technical capabilities. Provision should also be made for independent oversight and public notification of, and participation in, decision-making (SPC 2012; United Nations Conference on Environment and Development 1992).

To date, little scrutiny has been applied to the states who sponsor ISA activities, including the extent to which relevant measures are in place to ensure ISA contractor compliance (and compensation for third- party damages) via domestic regulation (Lily 2018), and the nature of the arrangements between the state and the ISA contract-holder (Rojas and Phillips 2019). There is little information in the public domain as to the extent to which the sponsoring state, or another state, stands to benefit financially from the contract – which may be deemed of particular importance where the sponsoring state is a developing state.

6.2 International Level

The ISA is tasked to "organise and control" contractors to "secure compliance" with ISA rules, including those rules designed to deliver on the ISA's mandate to "protect and preserve the marine environment" (UNCLOS; UN 1982). Much of the oversight authority within the ISA rests with the Council and the Legal and Technical Commission (LTC) which provides initial recommendations regarding rules, regulations and procedures, as well as recommendations on applications for mining contracts (Box 3.3). In some instances, it is difficult for the Council to take a decision contrary to an LTC recommendation. For example, in order to decide not to approve an application for a mining contract where the LTC recommends approval, a two-thirds majority of the 36 Council member states would be required. Even then, any one of four chambers within the Council could veto that disapproval decision (UN 1994, Annex, section 3, para. 11(a)). For this reason, the potential for a mining "approval bias" at the ISA has been noted (Greenpeace 2019; Pew Charitable Trusts 2019), and the composition, election, expertise and capacity of the LTC are often under scrutiny. The fact that only 3 of the 30 commissioners currently in post appear to have ecological science backgrounds has been remarked upon as a particular challenge, given the ISA's environmental protection mandate, and the LTC's immediate task to review environmental impact assessment reports, to develop environmental management plans, and to draft regulations, standards and guidelines pertaining to environmental management and thresholds. Criticisms of the LTC have also extended to a lack of transparency and potential conflict of interests (Greenpeace 2019; Ardron et al. 2018; Seascape 2016).

Box 3.3 The International Seabed Authority (ISA)

The ISA is an intergovernmental agency created by the UN Convention on the Law of the Sea (UNCLOS; UN 1982), with a structure that includes the following organs: the Assembly, the Council, the Legal and Technical Commission, the Finance Committee, the Economic Planning Commission, The Enterprise and the Secretariat (Fig. 3.7).

The executive body of the ISA is its "Council", comprising 36 member states. These states are elected in a number of different groups, designed to ensure a diversity of nations, representing different interests. These groups include major consumers or importers of the relevant metals, the largest investors in deep-seabed mining in the Area, major exporters of the relevant metals from land-based sources, developing countries with special interests (e.g. land-locked, geographically disadvantaged, islands), and five regional geographic groupings (Africa, Asia-Pacific, Eastern Europe, Latin America and Caribbean, and Western Europe and Others). The groups are then organised into four chambers, for decision-making purposes (UN 1994, Annex, section 3, para. 15).

The Council reports to the Assembly, which comprises all 168 ISA member states. Both organs meet at least annually at the ISA's headquarters in Kingston, Jamaica.

The ISA is supported by a Secretariat, also based in Jamaica, headed by a Secretary-General who is the chief administrative officer of the ISA, and required to support all ISA meetings and to perform such other administrative functions as may be instructed (UNCLOS, Article 166).

Another key organ within the ISA is the Legal and Technical Commission (LTC). This is a group of, currently, 30 experts, serving in their individual capacities, who meet bi-annually with responsibility to prepare recommendations and advisory inputs to the Council. The LTC's mandate includes the provision of recommendations on applications for ISA contracts, and preparing drafts of rules, regulations and procedures of the ISA, for Council consideration or adoption (UNCLOS, Article 165).

The Finance Committee oversees the ISA's administrative budget. The Economic Planning Commission is tasked with examining the impacts of mining in the Area on land-based mining economies; its function is currently being covered by the LTC. The Enterprise is envisaged to be an in-house mining arm of the ISA, who will commence operations via joint ventures with other contractors. The Enterprise has not yet been operationalised.

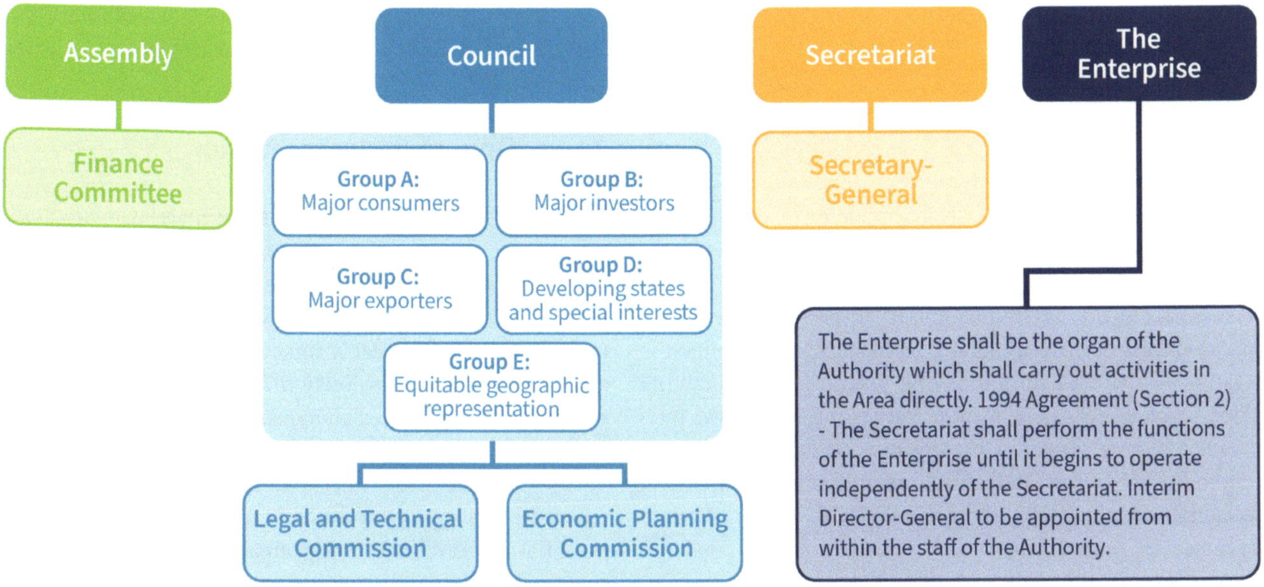

Fig. 3.7 The International Seabed Authority Governance Structure. Source: Adapted from Grid Arendal (https://www.grida.no/resources/6311)

There is no other precedent of an international intergovernmental treaty body (with 168 members, each with their own political priorities and interests) attempting to act as a minerals licensing, monitoring and enforcement, and revenue collection agency – as is required of the ISA (French and Collins 2019). UNCLOS even envisages an in-house mining wing of the ISA called "The Enterprise" (Article 170). When The Enterprise comes into existence, the ISA will be required to issue exploration or mining contracts to, and regulate, itself. These are functions that within national jurisdictions are usually performed by a raft of different government agencies operating under separate mandates. The ISA also faces constraints from the infrequency of meeting, a lack of funding and the fact that the same governments may be represented simultaneously in the ISA's advisory body, decision-making organ and as mining contractors. The challenges of conflict management and capacity constraint will be exacerbated if and when the ISA operates as a mining company itself ("The Enterprise"), as envisioned by UNCLOS (African Group 2018a). Different stakeholders have previously raised concerns with regards to the ISA due process and governance practice (Seascape 2016; Ardron et al. 2018; Belgium Government 2018; German Government 2018). Noting the capacity limitations and other constraints of the existing ISA structures, several parties have called for better incorporation of science and external, independent expertise in the ISA's development of regulations, rules and procedures, and in its regulatory oversight of contracts (Pew Charitable Trusts 2019).

The regulations for mining within the Area are under negotiation at the ISA currently. While there is a political push for these to be finalised by 2020 (ISA 2017), there appears to be a large amount of work still required to reach agreement on all necessary elements of the regime (ISA 2019c; Pew Charitable Trusts 2019), and at the ISA Annual Sessions in July 2019 and February 2020, several member states called for "quality over haste".

6.3 Mining in the Context of the UN Sustainable Development Goals

Several SDGs that affect materials use and natural resources (presumably designed for land use) are relevant to the ocean, including SDG 8.4, which addresses decoupling of materials use and environmental degradation, and SDG 12.2, which considers efficient use and sustainable management of natural resources (OECD 2018). Deep-seabed mining could contribute positively to several SDGs. Financial and economic benefits could help to relieve poverty (SDG 1), in the least developed countries such as Kiribati as an ISA-sponsoring state, or Solomon Islands as a state with sovereign rights over minerals within national jurisdiction, for example. But the benefit- sharing mechanisms have yet to be determined and are likely to be modest for non-mining countries (Kim 2017). Benefits from a greater availability of metals will almost certainly accrue to the most industrialised (or industrialising) nations, but could contribute to clean energy (SDG 7), which would counter climate change (SDG 13). These benefits all come with tradeoffs for the ocean environment under SDG 14, "Conserve and sustainably use the oceans, seas and marine resources for sustainable development", along with questions over the extent to which seabed mining can meet SDG 12.2, the target to "achieve the sustainable management and efficient use of natural resources" by 2030.

A recent report on mining governance on land introduces the concept of a Sustainable Development Licence to Operate

(SDLO). The SDLO adopts principles, standards of behaviour and best practices compatible with SDGs and their targets (IRP 2019). However, many struggle to understand what sustainability looks like in the context of deep-seabed mining. Obtaining maximum economic benefit in return for the extraction of minerals, and applying this to the long-term development goals of the poorest populations would seem to be a prerequisite.

Within SDG 14 targets, sustainability seems to encompass protecting ecosystems, conservation, economic benefits, scientific knowledge, and governance. When defined in relation to the extraction of living resources, sustainability often involves eco- certification, harvesting low on the food chain, avoiding government subsidies, technology innovation to avoid bycatch, and management to achieve maximum sustainable yield (Carr 2019). Could there be parallels for deep-seabed minerals? Could mining practices undergo review for a certification of limited damage to the environment? This is already effectively the mandate of the ISA. Could metal-containing end-products, such as mobile phones, come with source information about the metals in order to enable consumers to base their purchase choices on informed and ethical grounds? Could miners select mineral substrates of lesser value to biota or leave a significant fraction of the hard substrate on the seabed? What technologies can minimise the intensity, area or duration of impact on the environment? Given that most of the targeted minerals precipitate very slowly (e.g., 1–10 mm/million years for polymetallic crusts and nodules (Hein et al. 2013)), would there be an equivalent of maximum sustainable yield?

The drafters of UNCLOS appeared to pre-empt some of these issues, by stipulating that the ISA's production policy should be based on the principle that "there shall be no discrimination between minerals derived from the Area and from other sources. There shall be no preferential access to markets for such minerals or for imports of commodities produced from such minerals", while also requiring that state subsidies be avoided (UN 1994, Annex, section 6). These principles may be difficult to implement and police in practice.

Just as vulnerable marine ecosystems (VMEs) (e.g., dense corals and sponges) are protected from bottom fishing in international waters (UN General Assembly 2006), it has been proposed that active hydrothermal vents, which function as VMEs, should be protected from deep-seabed mining (Van Dover et al. 2018). There is a wholesale ban on bottom trawling in deep water (>800 m) in the European Union and elsewhere to prevent major habitat destruction. Although there are mineral resources of value in the EEZs of many countries, especially island nations in the West Pacific, none have permanently banned seabed mining. Notably, a precautionary pause/moratorium has recently been proposed by Fiji (Fiji Sun 2019) and mining licences for phosphorites have been denied in Mexico and New Zealand (Miller et al. 2018).

7 Opportunities for Action

Considering the above analyses, some high-level opportunities for action regarding ocean-based renewable energy (Sect. 7.1) and deep-seabed mining (Sect. 7.2) are presented here. The development of the global energy system referred to in Sect. 7.1 is intimately linked to both renewable energy and the use of minerals, Section 7.2. The Appendix provides further elaboration of challenges, detailed opportunities for action and associated benefits and some alternative or additional options. These are designed to ensure that ocean-based renewable energy is harvested in a manner that exploits its potential to contribute to sustainable development, and to ensure that the ocean, particularly in the context of deep-seabed mining, remains healthy and resilient for future generations.

7.1 Ocean-Based Renewable Energy and the Global Energy System

As discussed in Sect. 2.1, ocean-based renewable energy plays a significant role in cost-optimised models for transitioning the global energy system to a global temperature increase of 1.5 °C, in line with the Paris Agreement. In particular, offshore wind has the potential for further cost reductions and for the upscaling of implementation over the coming decade (see Sect. 3.1). The actual development path will depend upon several factors, including access to areas, grid connections, financing models, ownership and, in some cases, regulation of cross-border electric cables and legal conditions. While the expansion of offshore wind is well under way, the speed of development and implementation of further cost-reducing technologies, such as floating large turbines in deeper waters with bigger wind resources, depends on government incentives.

WindEurope has listed several challenges to be addressed to scale up offshore wind (WindEurope 2019). They also state six policy recommendations for Europe (WindEurope 2019, 66–67):

- Governments should set ambitious maritime spatial planning policies to deliver 450 GW by 2050.
- Governments should ensure that permitting and other relevant authorities have the necessary expertise and resources to consent enough sites.

- Governments should accelerate the expansion of the necessary on- and offshore grid infrastructure.
- The EU should elaborate a regulatory framework for offshore hybrid projects (e.g., hydrogen).
- Governments should accelerate the electrification of transport, heating and industrial processes.
- Governments should ensure visibility and confidence in volumes and revenue schemes.

ETIP Ocean (2019) describes a set of challenges and actions that would help lift ocean-based renewable energy in Europe towards delivering 100 GW by 2050.

Offshore wind and other ocean-based renewable energy does use metals, including some REEs, but it is not a major driver for deep-seabed mining exploration and exploitation. New technological solutions for components of offshore wind installations change the specific demands from one resource to another, showing the adaptability of the industry. The environmental impact of ocean-based renewable energy can slow down or limit its expansion. Baseline surveys and marine spatial planning exercises involving all stakeholders are required. Noise remains a concern but floating structures are expected to be more environmentally benign.

Other ocean-based renewable energy technologies should be developed to provide a wider range of energy sources in the future, particularly in areas with more limited wind resources. The low-carbon future energy system, as well as human activities in general, have to be developed and operated within given resource limitations. Minerals must be used in a way that is compatible with sustainable development in all its dimensions. In particular, the use of deep- seabed mining implies many potentially negative side effects and uncertainties and should be avoided at least until more knowledge has been gathered.

Opportunities for action are:

- **Strengthen research, development and demonstration programmes** to scale up offshore wind, in particular to make floating offshore wind cost-competitive more quickly.
- **Strengthen research and development and economic incentives to favour a less mineral-intensive global energy system**, including ocean-based renewable energy.
- **Support marine spatial planning and sustainable ocean economy plans** with taxation schemes and regulations that stimulate investments in variable renewable energy supply from the ocean to the shore.
- **Strengthen research and development for other ocean-based renewable energy technologies** to make them more mature and available to contribute significantly in later decades.

7.2 Deep-Seabed Mining

Deep-seabed mining represents a sustainability conundrum. The significant environmental and social impacts of mining on land (IRP 2019; Church and Crawford 2018) could be improved with focused effort, yet presently provide incentives to look to the ocean as a source of minerals (Batker and Schmidt 2015; IRP 2019). But extreme knowledge gaps remain, particularly in understanding how deep-ocean ecosystems will respond to industrial-scale mining disturbance. There is an inherent conflict between a duty to protect the marine environment, and a call to mine the deep sea for metals. The remote nature of the deep ocean and its unfamiliarity to most people raise the challenge of ensuring the participation of all relevant stakeholders to inform decisions taken at the international and state level that relate to areas out of sight. How society moves past these crossroads, and the decisions taken on behalf of humankind by governments at the ISA, will likely have a lasting impact on our ocean.

Because no deep-seabed mining has occurred yet, but substantial policy-making is in progress, four opportunities for additional action are proposed:

- **Develop and execute a road map to build the regulatory capacity of the isa to ensure effective protection for the marine environment** from harmful effects of mining in a transparent and inclusive manner. This would include the creation of environmental consents, evidence, inspectorate and enforcement functions, and would involve a slower process of transitioning from exploration to exploitation.
- **Establish an international research agenda and timeline, in conjunction with the un decade of ocean science for sustainable development**, to collect and synthesise high- quality deep-sea scientific data to fill identified gaps in knowledge required for decision-making and environmental management, before any deep-seabed mining takes place.
- **Promote the identification, declaration and enforcement of spatial protections** (including large, biologically representative, fully protected no-mining zones established in perpetuity prior to any award of exploitation contracts), **Across all ocean regions under isa jurisdiction**.

This would enable states to demonstrate efforts towards their international duties to ensure effective protection for the marine environment from mining's harmful effects (UNCLOS; UN 1982), to achieve in-situ conservation (Convention on Biological Diversity; UN 1992) and to conserve a percentage of marine areas (SDG 14.5 and Aichi Biodiversity Target 11). More time could also allow new opportunities to emerge for industry and scientists to partner on testing technological and conceptual innovations for mineral recovery that minimise harm to the marine environment.

- **Create incentives and remove barriers to implement a circular economy**, which acts through improved product design, reduced demand, reuse, recycling, reclassification of materials and use of renewable energy for production (Ghisellini et al. 2016). For metals targeted by deep-seabed mining, this would require independent research and long-term planning with attention focused on Life Cycle Sustainability Analysis (Van der Voet et al. 2019). Alternative energy technologies are already under investigation which reduce the use of lithium, silver, neodymium and dysprosium. New solid-state battery designs avoid the use of cobalt and nickel and have great durability and longevity. Redesign of existing batteries is required to avoid additives that improve product quality and durability but make metal recovery from electronic products even more difficult (Tansel 2017). More government policy focus, consumer awareness and behaviour change to favour a less mineral-intensive renewable energy system will also be crucial.

Acknowledgements The authors thank the paper's technical reviewers, Saleem Ali, Matthias Haeckel, Siegfried Husak and Andrew Thaler, its arbiters, Mari Elka Pangestu and Kristian Teleki, and Jane Lubchenco, who all provided helpful technical comments.

The authors also thank World Resources Institute for providing support as the Ocean Panel Secretariat.

Support for authors contributing to the seabed mining discussion was provided by The PEW Charitable Trusts (HL), World Resources Institute (LL), JM Kaplan Fund (HL), Marc and Lynne Benioff (DA), and the European Union's Horizon 2020 research and innovation program under a Marie Sklodowska-Curie grant (DA).

While our colleagues were very generous with their time and input, this report reflects the views of the authors alone.

The authors thank Sarah Chatwin for copyediting and Shannon Collins for design.

Appendix: Detailed Opportunities for Action

The *Opportunities for Action* in the main document are expanded upon below.

- To ensure that ocean-based renewable energy is harvested in a manner that exploits its potential to contribute to sustainable development.
- To ensure that the ocean, particularly in the way it is considered for deep-seabed mining, remains healthy and resilient for future generations.

  Each recommendation is prefaced by a specific challenge to be addressed, the recommendation itself, what following the recommendation would imply and the benefits to be achieved. In some cases, alternative or additional options are also described.

Detailed Opportunities for Action for Ocean-Based Renewable Energy

Detailed Opportunity for Action 1

Challenge 1: *Ocean-based renewable energy could contribute significantly to sustainable global energy supply, but the development is too slow for timely phase- out of fossil fuel.*

Detailed opportunity for action 1: Strengthen research, development and demonstration programmes and financing, taxation and legal regimes to scale up ocean-based renewable energy, in particular market incentives to make floating offshore wind cost- competitive faster, but also research and development to make other ocean-based renewable energy technologies more mature.

By doing this: The harmful effects of CO_2 emissions on the climate and on the ocean will be reduced. The urgent transformation of the global energy system will accelerate.

Associated Benefits: A new sector of the ocean economy will develop including new jobs. Less mineral- intensive options will reduce pressure on mining.

Alternative (or Additional) Option: Create regional national or international programmes focusing on different energy technologies, recognising that the various ocean-based renewable energy sources are unequally distributed because of varying wind resources, wave climate, tidal range and so on. Create floating or remote plants converting renewable electricity to hydrogen to supply fuel for shipping and transport to shore.

Detailed Opportunity for Action 2

Challenge 2: *Rapid transformation of the energy system helps to save the climate, but contributes to the demand for rare minerals and the pressure to accelerate deep-seabed mining during the decarbonisation phase, before re-use and the circular economy can be deployed.*

Detailed opportunity for action 2: Strengthen research and development and economic incentives to favour a less mineral-intensive renewable energy system.

By Doing This: The transformation of the global energy system can take place without risking harmful effects on the deep-sea environment, its ecosystem services or other potential resources.

Associated Benefits: The global energy system will develop in a more sustainable way with less material use overall. Note that the recommendation has system-wide implications beyond the choice of renewable energy source. For example, the need for batteries as well as the curtailment of electricity from renewable energy can be reduced in an energy system with well-developed demand management and other types of energy storage than batteries.

Detailed Opportunities for Action Specifically for Deep- Seabed Mining

A rapidly growing literature on deep-seabed mining has generated many ideas for the development of protection and management schemes for the marine environment, on governance options, ways to approach initial mining operations, and on alternatives to deep-seabed mining. The series of recommendations below emerge from the synthesis presented here, informed in part by the information gaps and scientific uncertainties associated with deep-seabed mining.

Detailed Opportunity for Action 3A and 3B

Challenge 3: *Unclos was written in a time of limited knowledge about deep- sea ecosystems, their vulnerabilities and the services they provide (e.g., hydrothermal vents had not yet been discovered). extreme knowledge gaps remain, particularly in understanding how deep-ocean ecosystems will respond to industrial-scale mining disturbance.*

Detailed opportunity for action 3A: Slow the process of transitioning from exploration to exploitation, and take the time necessary to fully develop – in a transparent and inclusive manner – the ISA rules, regulations and procedures for mining (possibly extending the ISA-imposed deadline from 2020 to 2030), and institute a precautionary pause in the issuance of new contracts by the ISA during this period.

By Doing This: The (legally required) precautionary approach is applied. More time is allowed for scientific study, and appropriate scientific input into regulations and decision-making, including the development of environmental goals and objectives, and identification of science-based indicators and thresholds.

Associated Benefits: More time would also allow for broader stakeholder input, and building of ISA capacity, to include data access and management mechanisms, access to relevant independent expertise, and regulatory capacity.

Alternative (Or Additional) option: Develop rules, regulations and procedures at the ISA that set highly stringent and prescriptive environmental standards, and give ISA decision-makers appropriate agility and powers to reject applications to mine, and to amend required conduct from contractors where there is a threat of serious, irreversible or otherwise unacceptable harm to the marine environment, including through cumulative impact. This should include the adoption by the ISA of a conscious policy of a controlled, staged development approach to exploitation: initially cautious about the number and size of sites licensed for mining activities – with new projects not authorised until existing ones are completed and the impacts measured.

Detailed opportunity for Action 3B: Create as soon as possible an international research agenda to collect and synthesise high-quality scientific data (during the Decade of Ocean Science for Sustainable Development, 2021–2030), which answers strategic questions about deep-sea ecosystems required for decision-making and environmental management related to deep-seabed mining.

By doing this: The deep-sea environment can be better understood before taking decisions that could irreparably affect it. Current knowledge of species distributions, connectivity, habitat requirements, ecological functions and ecosystem services, vulnerability to mining impacts (including cumulative impacts, sediment plumes, noise and light), resilience, recovery and mitigation potential, and the influence of cumulative impacts from climate stressors can be expanded. The agenda should also support and engage existing sustained observing programmes to enhance relevant deep-sea data acquisition, improve understanding of natural variability, and develop standards around the acceptable level of statistical power for monitoring impacts of seabed mineral activities. New opportunities can emerge for industry and scientists to partner on testing technological and conceptual innovations for mineral extraction techniques that minimise harm to the marine environment.

The agenda should promote FAIR data principles (findable, accessible, interoperable, reusable) and facilitate data portals that create compatibility across networks and agencies (e.g., the ISA, Intergovernmental Oceanographic Commission clearinghouse, regional fisheries management organisations).

Associated benefits: Greater understanding of deep-sea environments, and improved data quality and sharing will also assist with governance decisions beyond those relevant to seabed mining, including the negotiations at the Intergovernmental Conference on Marine Biodiversity of Areas Beyond National.

Jurisdiction (BBNJ), climate change talks at the United Nations Framework Convention on Climate Change (UNFCCC) and conservation initiatives, and will constitute important work towards SDG 14, including Target 14.A ("increase scientific knowledge, develop research capacity and transfer marine technology").

Alternative (or Additional) option: ISA marine scientific research, data management and strategic environmental assessment functions should be strengthened, and additional requirements or incentives be exerted by the ISA upon contractors and member states, encouraging multilateral cooperation, so that more science across wider biogeographic areas is collected, analysed, published and used to inform ISA policy and regulation.

Detailed Opportunities for Action 4A and 4B

Challenge 4: *There is an inherent conflict between a duty to protect the marine environment, and a call to mine the deep seabed for metals.*

Detailed opportunity for action 4A: Enable as soon as possible an expert and independent environmental and scientific committee to handle ISA environmental regulations and decision-making, to assess monitoring and impact assessment and to identify triggers for regulatory action or cessation of mining.

By doing this: The ISA can bolster its capacity and expertise to manage its mandated environmental stewardship function. This should be run separately from other ISA functions (such as contract award and management, revenue collection and distribution, and direct engagement in mining, through The Enterprise).

Associated benefits: A science-driven, expert-led, transparent, independent, consistent and consultative regulatory agency will garner greater public and investor trust and confidence, which should enhance the ISA's ability to meet its bifurcated duty both to develop the mineral resources of the Area and to protect and preserve the marine environment. Any steps taken to strengthen the ISA's regulatory capacity will contribute towards the goal of preventing serious harm to the marine environment, and minimising other harmful effects from mining.

Detailed opportunity for action 4B: Ensure the declaration (by 2022) and enforcement of a network of large, biologically representative, fully protected no- mining zones established in perpetuity prior to any award of exploitation contracts, across all ocean regions under ISA jurisdiction. These should be designed according to scientific principles, and placed on the basis of physical, geochemical, ecological and social analyses. Ideally, they should cover at least 30 percent of the Area, ensure connectivity, be representative of habitats that will be lost to mining and protect particularly vulnerable habitats.

By doing this: The precautionary approach to environmental management of deep-seabed mining is enacted by ensuring that representative benthic habitats and associated ecosystems are protected from harm on regional scales. This is particularly important given uncertainties regarding the severity, frequency and spatial extent of mining impacts.

Associated benefits: Protected areas can serve as refugia for marine species, offer climate resilience and preserve ecosystem functions. Their declaration would enable states to demonstrate efforts towards their international duties to ensure that the marine environment is effectively protected from mining's harmful effects (UNCLOS; UN 1982), to achieve in-situ conservation (Convention on Biological Diversity; UN 1992) and to conserve a percentage of marine areas (SDG 14.5 and Aichi Biodiversity Target 11).

Detailed Opportunity for Action 5

Challenge 5: *The remote nature of the deep ocean and its unfamiliarity to most people raise the challenge of ensuring the participation of all relevant stakeholders to inform decisions taken at the international and state level that relate to areas out of sight.*

Detailed opportunity for action 5: *The ISA, member governments and non-governmental bodies should cooperate immediately to enhance societal awareness of the choices associated with deep-seabed mining (through social media, traditional media, formal educational programmes and other forms of outreach) and diverse and inclusive opportunities for interested parties to have their views heard and considered in deep-seabed mining decision-making processes.*

The ISA regime, and states with mining interests, should maximise opportunities for public and expert consultation, including during the contract application, approval and review process. Non-governmental observers should be facilitated to attend ISA and state meetings. Such meetings should be supported by technical advisory inputs that are comprehensive and fully explained (with dissenting views noted) and produced in a timely fashion. Meeting documents, contracts, financial information, compliance information and environmental data should all be made publicly available immediately.

By doing this: Better and more durable decisions will be taken. It will enable the collection of comprehensive relevant information by decision-makers and will enhance public understanding, consent and commitment to implementation. Trust and confidence in the ISA's decisions will be improved.

Associated benefits: Consultation with as wide a group of experts and stakeholders as possible will assist national and international policy-makers to take the complex and momentous judgement calls that are inherent in deciding what degree of environmental harm is deemed acceptable in order to facilitate access to metals. It will also be a means for governments to operationalise commitments made at the international level (various regional environmental treaties, the Rio Declaration (UN Conference on Environment and Development 1992) and Rio + 20), as well as the ISA's duty to act on behalf of all of humankind (UNCLOS; UN 1982).

Alternative (or additional) option: Member governments should:

- attend ISA meetings at which crucial decisions are taken (for example, approval of the ISA's Exploitation Regulations currently under negotiation, or a decision whether or not (and on what terms) to approve or disapprove the first mining application made tothe ISA);
- hold meaningful prior national consultations on the relevant issues, before attending; and
- reflect the results of those national consultations in their positions at the ISA.

Detailed Opportunity for Action 6

Challenge 6: *The growing global demand for metals is threatening to push extractive practices beyond planetary boundaries.*

Detailed opportunity for action 6: Engage urgently in independent research and long-term planning to facilitate a circular economy for targeted rare metals and rare earth elements, initially with a five-year programme (2020–25). Focus attention on Life Cycle Sustainability Analysis (Van der Voet et al. 2019) and developing alternative methods to address the metal demand. Create incentives and reduce barriers to promote the following:

- Recycle, reduce, re-use opportunities
- Product redesign that enables improved metal recycling or extended product lifetime
- Demand reduction via use of alternatives and consumer behaviour change
- Improved sustainability of on-land mining practices
- More sustainable metal waste disposal practices and less resulting pollution

By doing this: The negative environmental impacts of land-based mining can be minimised and the need for deep-seabed mining is reduced, while human development is supported, in line with the SDGs. Social, economic, behavioural and technical issues can be addressed together.

Associated benefits: More government policy focus, and consumer awareness and demand about metal sourcing and use, should stimulate innovation and lead to better environmental and human rights practices by extractive industries. The circular economy and enhanced secondary production of metals will reduce energy use and carbon emissions. It should also enhance competitiveness and economic growth, and new employment opportunities. This can also identify possible less harmful alternatives to deep- seabed mining.

About the Authors

Lead Authors

Peter M. Haugan is programme director at the Institute of Marine Research, Norway, and professor at the Geophysical Institute, University of Bergen, Norway.

Lisa A. Levin is a Distinguished Professor at the Scripps Institution of Oceanography, University of California, San Diego, USA

Contributing Authors

Diva Amon is a scientific associate at the Natural History Museum, London, UK.

Mark Hemer is a principal research scientist at CSIRO, Australia.

Hannah Lily is an independent consultant and lawyer based in the UK.

Finn Gunnar Nielsen is a professor and head of Bergen Offshore Wind Centre at the University of Bergen, Norway.

References

Aanesen M, Armstrong CW (2019) Trading off co-produced marine ecosystem services: natural resource industries versus other use and non-use ecosystem service values. Frontiers in Marine Science 6:102. https://doi.org/10.3389/fmars.2019.00102

African Group. 2018a. "Statement by Algeria on Behalf of the African Group to the International Seabed Authority 6 July 2018." https://www.isa.org.jm/document/statement-algeria-obo-african-group-1

African Group. 2018b. "Statement by Algeria on Behalf of the African Group to the International Seabed Authority 9 July 2018." https://www.isa.org.jm/document/statement-algeria-obo-african-group-2

African Group. 2019. "Statement by Algeria on Behalf of the African Group to the International Seabed Authority 25 February 2019." https://www.isa.org.jm/document/algeria-obo-african-group

Aguon J, Hunter J (2018) Second Wave Due Diligence: The Case for Incorporating Free, Prior, and Informed Consent into the Deep Sea Mining Regulatory Regime. Stanf Environ Law J 38:3–56

Ali SH, Giurco D, Arndt N, Nickless E, Brown G, Demetriades A, Durrheim R et al (2017) Mineral Supply for Sustainable Development Requires Resource Governance. Nature 543(7645):367–372

Amon DJ, Ziegler AF, Dahlgren TG, Glover AG, Goineau A, Gooday AJ, Wiklund H, Smith CR (2016) Insights into the Abundance and Diversity of Abyssal Megafauna in a Polymetallic-Nodule Region in the Eastern Clarion-Clipperton Zone. Sci Rep 6:30492. https://doi.org/10.1038/srep30492

Anderson K, Peters G (2016) The trouble with negative emissions. Science 354(6309):182–183. https://doi.org/10.1126/science.aah4567

Ardron JA, Ruhl HA, Jones DOB (2018) Incorporating Transparency into the Governance of Deep-Seabed Mining in the Area beyond National Jurisdiction. Marine Policy 89:58–66. https://doi.org/10.1016/j.marpol.2017.11.021

Ardyna M, Lacour L, Sergi S, d'Ovidio F, Sallée JB, Rembauville M, Blain S, Tagliabue A, Schlitzer R, Jeandel C, Arrigo KR, Claustre H (2019) Hydrothermal Vents Trigger Massive Phytoplankton Blooms in the Southern Ocean. Nat Commun 10(1):2451. https://doi.org/10.1038/s41467-019-09973-6

Armstrong CW, Foley NS, Tinch R, van den Hove S (2012) Services from the Deep: Steps towards Valuation of Deep Sea Goods and Services. Ecosystem Services 2:2–13. https://doi.org/10.1016/j.ecoser.2012.07.001

Arrobas DLP, Hund KL, Mccormick MS, Ningthoujam J, Drexhage JR (2017) The Growing Role of Minerals and Metals for a Low Carbon Future (English). World Bank Group, Washington, DC. http://documents.worldbank.org/curated/en/207371500386458722/The-Growing-Role-of-Minerals-and-Metals-for-a-Low-Carbon-Future

Backwell B (2019) Globalization of Offshore Wind. In: Paper presented at GWEC, Tokyo, 8 March 2019. Global Wind Energy Council, Brussels

Batker, David and Rowan Schmidt. 2015. "Environmental and Social Benchmarking Analysis of Nautilus Minerals Inc: Solwara 1 Project." Earth Economics. https://mining.com/wp-content/uploads/2015/06/Earth-Economics-Environmental-Social-Benchmarking-Solwara-1-2015.pdf

Baturin, Gleb Nikolaevich. 1982. Phosphorites on the Sea Floor Origin, Composition and Distribution. Vol. 33 of Developments in Sedimentology. Elsevier Scientific. https://www.sciencedirect.com/bookseries/developments-in-sedimentology/vol/33/suppl/C

Belgian Government. 2018. "Strengthening Environmental Scientific Capacity of the ISA." Submission by Belgium to the International Seabed Authority, 22 June 2018. https://ran-s3.s3.amazonaws.com/isa.org.jm/s3fs-public/files/documents/belgium-capacity.pdf

Bleischwitz R, Nechifor V, Winning M, Huang B, Geng Y (2018) Extrapolation or Saturation – Revisiting Growth Patterns, Development Stages and Decoupling. Global Environ Change 48:86–96. https://doi.org/10.1016/j.gloenv-cha.2017.11.008

Bloomberg. 2019. "Global Offshore Wind Power Market Will Grow at a CAGR of 18% During 2019-2023|Technavio." Press release, 28 May. https://www.bloomberg.com/press-releases/2019-05-28/global-off-shore-wind-power-market-will-grow-at-a-cagr-of-18-during-2019-2023-technavio

Bluhm H (2001) Re-Establishment of an Abyssal Megabenthic Community after Experimental Physical Disturbance of the Seafloor. Deep Sea Res 2 Top Stud Oceanogr 48(17):3841–3868. https://doi.org/10.1016/S0967-0645(01)00070-4

BNEF (Bloomberg New Energy Finance) (2019) Offshore Wind Projects June 2019. BNEF, London

Boehlert G, Gill A (2010) Environmental and Ecological Effects of Ocean Renewable Energy Development: A Current Synthesis. Oceanography 23(2):68–81

Boetius A, Haeckel M (2018) Mind the Seafloor. Science 359(6371): 34–36

Bosch J, Stafell I, Hawkes A (2018) Temporally Explicit and Spatial- ly Resolved Global Offshore Wind Energy Potentials. Energy 163:766–781

Boubault A, Maizi N (2019) Devising Mineral Resource Supply Pathways to a Low-Carbon Electricity Generation by 2100. Deep Sea Res 2 Top Stud Oceanogr 8(1):33. https://doi.org/10.3390/resources8010033

Brandt M, Dragon A, Diederichs A, Bellmann M, Wahl V, Piper W, Nabe-Nielsen J, Nehls G (2018) Disturbance of Harbour Porpoises During Construction of the First Seven Offshore Wind Farms in Germany. Mar Ecol Prog Ser 596:213–232. https://doi.org/10.3354/meps12560

Brown CG (2018) Mining at 2,500 Fathoms under the Sea: Thoughts on an Emerging Regulatory Framework. Ocean Sci J 53(2):287–300. https://doi.org/10.1007/s12601-018-0033-z

Brown TJ, Idoine NE, Raycraft ER, Shaw RA, Deady EA, Rippingale J, Bide T, Wrighton CE, Rodley J (2014) World Mineral Production 2008–2012. Keyworth: British Geological Survey, Nottingham

Burdige D (2007) Preservation of Organic Matter in Marine Sediments: Controls, Mechanisms, and an Imbalance in Sediment Organic Carbon Budgets? Chem Rev 107(2):467–485

Carr, Sarah. 2019. "All Things to All People: What Different Groups Mean When They Talk about the Blue Economy." The Skimmer on Marine Ecosystems and Management. 18 March. https://meam.openchannels.org/news/skimmer-marine-ecosystems-and-management/all-things-all-people-what-different-groups-mean-when

Carstensen J, Henriksen OD, Teilmann J (2006) Impacts of Off- shore Wind Farm Construction on Harbour Porpoises: Acoustic Monitoring of Echolocation Activity Using Porpoise Detectors (T-PODs). Marine Ecol Prog Series 321:295–308. https://doi.org/10.3354/meps321295

Cascajo R, García E, Quiles E, Correcher A, Morant F (2019) Integration of marine wave energy converters into seaports: a case study in the port of valencia. Energies 12(5):787. https://doi.org/10.3390/en12050787

Church C, Crawford A (2018) Green Conflict Minerals. International Institute for Sustainable Development, Winnipeg

Clack CTM, Qvist SA, Apt J, Bazilian M, Brandt AR, Caldeira K, Davis SJ et al (2017) Evaluation of a proposal for reliable low-cost grid power with 100% wind, water, and solar. Proc Natl Acad Sci U S A 114(26):6722–6727. https://doi.org/10.1073/pnas.1610381114

Clark MR, Tittensor DP (2010) An index to assess the risk to stony corals from bottom trawling on seamounts. Mar Ecol 31(S1):200–211. https://doi.org/10.1111/j.1439-0485.2010.00392.x

Clark MR, Althaus F, Schlacher TA, Williams A, Bowden DA, Rowden AA (2016) The impacts of deep-sea fisheries on benthic communities: a review. ICES J Mar Sci 73(S1):i51–i69. https://doi.org/10.1093/icesjms/fsv123

Collins PC, Croot P, Carlsson J, Colaço A, Grehan A, Hyeong K, Kennedy R et al (2013) A primer for the environmental impact assessment of mining at seafloor massive sulfide deposits. Marine Policy 42:198–209. https://doi.org/10.1016/j.marpol.2013.01.020

Contardo S, Hoeke R, Hemer M, Symonds G, McInnes K, O'Grady J (2018) In situ observations and simulations of coastal wave field transformation by wave energy converters. Coast Eng. 140:175–188

Cook Islands News. 2018. Ocean floor holds rich future. http://www.cookislandsnews.com/item/71099-ocean-floor-holds-rich-future

Copping A, Sather N, Hanna L, Whiting J, Zydlewski G, Staines G, Gill A, Hutchison I et al. 2016. Annex IV 2016 State of the Science Report: Environmental Effects of Marine Renewable Energy Development Around the World. Pacific Northwest National Laboratory. https://tethys.pnnl.gov/publications/state-of-the-science-2016

Copping A et al., 2019. European Wave and Tidal Energy Conference Proceedings. https://ewtec.org/proceedings/

Craik A. Neil, Alfonso Ascencio-Herrera, Andres Rojas, Christopher Brown, Dire Tladi, Eden Charles, Elie Jarmache et al. 2018. Legal Liability for Environmental Harm: Synthesis and Overview. Liability Issues for Deep Seabed Mining Series, Paper No. 1, July. Centre for International Governance Innovation. https://www.cigionline.org/publications/legal-liability-environmental-harm-synthesis-and-overview

Cuvelier D, Gollner S, Jones DOB, Kaiser S, Arbizu PM, Menzel L, Mestre NC et al (2018) Potential mitigation and restoration actions in ecosystems impacted by seabed mining. Front Marine Sci 5:22

De Jong K, Amorim MCP, Fonseca PJ, Fox CJ, Heubel KU (2018) Noise Can Affect Acoustic Communication and Subsequent Spawning Success in Fish. Environ Pollut 237:814–823

Deetman S, Pauliuk S, van Vuuren DP, van der Voet E, Tukker A (2018) Scenarios for demand growth of metals in electricity generation technologies, cars, and electronic appliances. Environ Sci Technol 52:4950–4959. https://doi.org/10.1021/acs.est.7b05549

Dogger Bank Wind Farms. 2019. "World's Largest Wind Farm Successful in CfD Auction." Press release, 20 September. https://doggerbank.com/downloads/CfD-results-press-release-FINAL.pdf

Dominish E, Florin N, Teske S (2019) Responsible Minerals Sourcing for Renewable Energy. Report prepared for Earthworks by the Institute for Sustainable Futures. University of Technology Sydney, Australia

DSM Observer. 2018. "The Writing is on the Wall for Solwara 1—PNG Should Withdraw." January. http://dsmobserver.com/2018/01/solwara-1-png-withdraw

Du Preez C, Fisher CR (2018) Long-term stability of back-arc basin hydrothermal vents. Front Mar Sci 5:54. https://doi.org/10.3389/fmars.2018.00054

Dunn DC, Van Dover CL, Etter RJ, Smith CR, Levin LA, Morato T, Colaço A et al (2018) A strategy for the conservation of biodiversity on mid-ocean ridges from deep-sea mining. Sci Adv 4(7):eaar4313. https://doi.org/10.1126/sciadv.aar4313

Ellis DV (2001) A review of some environmental issues affecting marine mining. Marine Georesour Geotechnol 19(1):51–63. https://doi.org/10.1080/10641190109353804

Elsner P, Suarez S (2019) Renewable Energy from the High Seas: Geo-Spatial Modelling of Resource Potential and Legal Implications for Developing Offshore Wind Projects Beyond the National Jurisdiction of Coastal States. Energy Policy 128:918–929. https://doi.org/10.1016/j.enpol.2019.01.064

ETIP Ocean. 2019. Powering Homes Today, Powering Nations Tomorrow. https://www.etipocean.eu/resources/powering-homes-today-powering-nations-tomorrow

Eurek K, Sullivan P, Gleason M, Hettinger D, Heimiller D, Lopez A (2017) An Improved Global Wind Resource Estimate for Integrated Assessment Models. Preprint. National Renewable Energy Laboratory, Golden, CO

European Commission (2018) A Clean Planet for All. A European strategic long-term vision for a prosperous, modern, competitive and climate neutral economy. European Commission, Brussels

Faulkner R, Merchant N, Martinez R (2018) Marine Noise Budgets in Practice. Conserv Lett 11(3):1–8. https://doi.org/10.1111/conl.12420

Feichtner I (2019) Mining for humanity in the deep sea and outer space: the role of small states and international law in the extraterritorial expansion of extraction. Leiden J Inter Law 32(2):255–274. https://doi.org/10.1017/S0922156519000013

Fiji Sun. 2019. "Fiji Budget 2019: A Budget That Is Securing Our Nation A Sustainable Future." 10 June. https://fijisun.com.fj/2019/06/10/fiji-budget-2019-a-budget-that-is-securing-our-nation-a-sustainable-future

Fleming J, Ford L, Hornsby E (2019) Facing the Abyss: The Future of Deep Sea Mining. Environmental Change Institute, Oxford University, Oxford. https://doi.org/10.6084/m9.figshare.7564976.v1

Fowler A, Jørgensen A, Svendsen J, Macreadie P, Jones D, Boon A, Booth D et al (2018) Environmental benefits of leaving offshore infrastructure in the ocean. Front Ecol Environ 16(10):571–578. https://doi.org/10.1002/fee.1827

French D, Collins R. 2019. A Guardian of Universal Interest or Increasingly Out of Its Depth? The International Seabed Authority Turns 25. International Organizations Law Review, April. http://eprints.lincoln.ac.uk/34948/

Furre A-K, Eiken O, Alnæs H, Vevatne JN, Kiær AF (2017) 20 years of monitoring CO_2-injection at sleipner. Energy Procedia 114:3916–3926. https://doi.org/10.1016/j.egypro.2017.03.1523

German CR, Legendre LL, Sander SG, Niquil N, Luther GW, Bharati L, Han X, Le Bris N (2015) Hydrothermal Fe Cycling and Deep Ocean Organic Carbon Scavenging: Model-Based Evidence for Significant POC Supply to Seafloor Sediments. Earth Planetary Sci Lett C 419:143–153. https://doi.org/10.1016/j.epsl.2015.03.012

German Government. 2018. "Statement by Germany to the International Seabed Authority 27 June 2018." https://www.isa.org.jm/document/isba24c18

Ghisellini P, Cialani C, Ulgiati S (2016) A review on circular economy: the expected transition to a balanced interplay of environmental and economic systems. J Clean Prod 114:11–32. https://doi.org/10.1016/j.jclepro.2015.09.007

Giurco D, Dominish E, Florin N, Watari T, McLellan B (2019) Requirements for Minerals and Metals for 100% Renewable Scenarios. In: Teske S (ed) Achieving the Paris Climate Agreement Goals. Springer, New York, pp 437–457. https://doi.org/10.1007/978-3-030-05843-2_11

Gollner S, Kaiser S, Menzel L, Jones DOB, Brown A, Mestre NC, van Oevelen D et al (2017) Resilience of benthic deep-sea fauna to mining activities. Mar Environ Res 129:76–101. https://doi.org/10.1016/j.marenvres.2017.04.010

Gooday AJ, Holzmann M, Caulle C, Goineau A, Kamenskaya O, Weber AA-T, Pawlowski J (2017) Giant protists (Xenophyophores, Foraminifera) are exceptionally diverse in parts of the abyssal eastern pacific licensed for polymetallic nodule exploration. Biol Conser 207:106–116. https://doi.org/10.1016/j.biocon.2017.01.006

Goodenough KM, Wall F, Merriman D (2018) The rare earth elements: demand, global resources, and challenges for resourcing future generations. Nat Resour Res 27(2):201–216. https://doi.org/10.1007/s11053-017-9336-5

Graedel TE, Harper EM, Nassar NT, Reck BK (2015) On the Ma- terials Basis of Modern Society. Proc Natl Acad Sci U S A 112:6295. https://doi.org/10.1073/pnas.1312752110

Greenpeace. 2019. In Deep Water: The Emerging Threat Of Deep Sea Mining. https://storage.googleapis.com/planet4-international-stateless/2019/06/f223a588-in-deep-water-greenpeace-deep-sea-mining-2019.pdf

Grubler A, Wilson C, Bento N, Boza-Kiss B, Krey V, McCollum DL, Rao ND et al (2018) A Low Energy Demand Scenario for Meeting the 1.5 Degrees C Target and Sustainable Development Goals without Negative Emission Technologies. Nat Energy 3:515–527. https://doi.org/10.1038/s41560-018-0172-6

Hammar L, Wikström A, Molander S (2014) Assessing ecological risks of offshore wind power on kattegat cod. Renew Energy 66:414–424

Hannington M, Jamieson J, Monecke T, Petersen S (2010) Modern Sea-Floor Massive Sulfides and Base Metal Resources: Toward an Estimate of Global Sea-Floor Massive Sulfide Potential, vol 15. Society of Economic Geologists, Inc. Special Publication, Littleton, Colorado, pp 317–338

Hein JR, Koschinsky A (2014) Deep-ocean Ferromanganese Crusts and Nodules. In: Holland HD, Turekian KK (eds) Treatise on geochemistry, 2nd edn. Elsevier, Amsterdam, pp 273–291

Hein JR, Mizell K, Koschinsky A, Conrad TA (2013) Deep-Ocean Mineral Deposits as a Source of Critical Metals for High- and Green-Technology Applications: Comparison with Land-Based Resources. Ore Geol Rev 51:1–14

Hertwich EG, Gibon T, Bouman EA, Arvesen A, Suh S, Heath GA et al (2015) Integrated Life-Cycle Assessment of Electricity-Supply Scenarios Confirms Global Environmental Benefit of Low-Carbon Technologies. Proc Natl Acad Sci U S A 112(20):6277e82

Hoegh-Guldberg O et al (2019) The Ocean as a Solution to Climate Change: Five Opportunities for Action. World Resources Institute, Washington, DC. http://www.oceanpanel.org/climate

House of Commons. 2019. Hansard, 20 February, vol. 654, col. 572WH. https://hansard.parliament.uk/Commons/2019-02-20/debates/19022027000002/UKDeepSeaMiningIndustry

IEA (International Energy Agency) (2017) Energy Technology Perspectives 2017: Catalysing Energy Technology Transformations. International Energy Agency, Paris. www.iea.org/etp2017

IEA. 2018. Offshore Energy Outlook. World Energy Outlook Series. Paris: International Energy Agency

IEA. 2019a. World Energy Outlook 2019. www.iea.org/reports/world-energy-outlook-2019

IEA. 2019b. Offshore Wind Outlook 2019. World Energy Outlook Special Report. https://www.iea.org/reports/offshore-wind-outlook-2019

IEA. 2019c. Global Energy & CO_2 Status Report (GECO): The Latest Trends in Energy and Emissions in 2018. https://www.iea.org/reports/global-ener gy-co2-status-report-2019

IJGlobal. 2019. Asset Data. http://ijglobal.com

IPCC (Intergovernmental Panel on Climate Change) (2005) In: Metz B, Davidson O, de Coninck HC, Loos M, Meyer LA (eds) IPCC Special Report on Carbon Dioxide Capture and Storage. Prepared by Working Group III of the Intergovernmental Panel on Climate Change. Cambridge University Press, Cambridge, UK and New York

IPCC (2018) In: Masson-Delmotte PZ, Pörtner H-O, Roberts D, Skea J, Shukla PR, Pirani A, Moufouma-Okia W, Péan C, Pidcock R, Connors S, Matthews JBR, Chen Y, Zhou X, Gomis MI, Lonnoy E, Maycock T, Tignor M, Waterfield T (eds) Global Warming of 1.5°C. An IPCC Special Report on the impacts of global warming of 1.5°C above pre-industrial levels and related global greenhouse gas emission pathways, in the context of strengthening the global response to the threat of climate change, sustainable development, and efforts to eradicate poverty. Cambridge University Press, Cambridge, UK and New York

IRENA (International Renewable Energy Agency) (2014) Ocean Thermal Energy Conversion: Technology Brief, Abu Dhabi, IRENA. www.irena.org/DocumentDownloads/Publications/Ocean_Thermal_Energy_V4_web.pdf

IRENA (2014b) Tidal Energy: Technology Brief. IRENA, Abu Dhabi. www.irena.org/documentdownloads/publications/tidal_energy_v4_web.pdf

IRENA (2014c) Wave Energy: Technology Brief. IRENA, Abu Dhabi. www.irena.org/documentdownloads/publications/wave-energy_v4_web.pdf

IRENA (2018a) Offshore Wind Investment, Policies and Job Creation: Review of Key Findings for G7 Ministerial Meetings, Halifax, Canada. IRENA, Abu Dhabi. www.irena.org/-/media/Files/IRENA/Agency/Publication/2018/Sep/IRENA_offshore_wind_note_G7_2018.pdf?la=en&hash=B186614D923AB1F0A07D7285612C4B037057A0C0

IRENA (2018b) Renewable Energy Benefits Leveraging Local Capacity for Offshore Wind. IRENA, Abu Dhabi. www.irena.org/-/media/Files/IRENA/Agency/Publication/2018/May/IRENA_Leveraging_for_Offshore_Wind_2018.pdf

IRENA (2019a) Global Energy Transformation. A Roadmap to 2050. 2019 Edition. IRENA, Abu Dhabi. www.irena.org/publications/2019/Apr/Global-energy-transformation-A-roadmap-to-2050-2019Edition

IRENA (2019b) Renewable Energy: A Gender Perspective. IRENA, Abu Dhabi. https://irena.org/-/media/Files/IRENA/Agency/Publication/2019/Jan/IRENA_Gender_perspective_2019.pdf

IRENA (2019c) Renewable Capacity Statistics 2019. IRENA, Abu Dhabi. https://www.irena.org/-/media/Files/IRENA/Agency/Publication/2019/Mar/IRENA_RE_Capacity_Statistics_2019.pdf

IRP (International Resource Panel) (2019) Mineral Resource Governance in the 21st Century: Gearing Extractive Industries Towards Sustainable Development. In: Ayuk ET, Pedro AM, Ekins P, Oberle B, Gatune J, Milligan B et al (eds) . United Nations Environment Programme, Nairobi. https://www.resourcepanel.org/reports/mineral-resource-governance-21st-century

ISA (International Seabed Authority) (2013) Implementation of Article 82 of the United Nations Convention on the Law of the Sea. ISA, Kingston, Jamaica

ISA (2017) Report of the Chair of the Legal and Technical Commission on the Work of the Commission at Its Session in 2017. Document IS-BA/23/C/13. ISA, Kingston, Jamaica. https://www.isa.org.jm/document/isba23c13

ISA (2019a) Comparative study of the existing national legislation on deep seabed mining. ISA, Kingston, Jamaica. https://www.isa.org.jm/document/comparative-study-existing-national-legislation-deep-sea- bed-mining-0

ISA (2019b) Implementation of the Authority's Strategy for the Development of Regional Environmental Management Plans for the Area. Document ISBA/25/C/13. ISA, Kingston, Jamaica. https://www.isa.org.jm/document/isba25c13

ISA (2019c) Draft Regulations on Exploitation of Mineral Resources in the Area. Document ISBA/25/C/18. ISA, Kingston, Jamaica. https://www.isa.org.jm/document/isba25c18

ISA (2019d) Draft Regulations on Exploitation of Mineral Resources in the Area. ISBA/25/C/WP.1. ISA, Kingston, Jamaica. https://www.isa.org.jm/document/isba25cwp1-0

ISA (n.d.) The Mining Code. ISA, Kingston, Jamaica. https://www.isa.org.jm/mining-code

ITLOS (International Tribunal for the Law of the Sea). 2011. "Responsibilities and Obligations of States Sponsoring Persons and Entities with Respect to Activities in the Area—List of Cases: No. 17." Advisory Opinion, 1 February. International Tribunal for the Law of the Sea. https://www.itlos.org/en/cases/list-of-cases/case-no-17/#c587

Jacobson MZ, Delucchi MA, Bauer ZAF, Goodman SC, Chapman WE, Cameron MA et al (2017) 100% Clean and Renewable Wind, Water, and Sunlight All-Sector Energy Roadmaps for 139 Countries of the World. Joule 1:108–121. https://doi.org/10.1016/j.joule.2017.07.005

Jacobson MZ, Delucchi MA, Cameron MA, Mathiesen BV (2018) Matching Demand with Supply at Low Cost in 139 Countries Among 20 World Regions with 100% Intermittent Wind, Water, and Sunlight (WWS) for all Purposes. Renewable Energy 123:236–248. https://doi.org/10.1016/j.renene.2018.02.009

Jacobson MZ, Delucchi MA, Cameron MA, Manogaran IP, Shu Y, von Krauland A-K (2019) Impacts of green new deal energy plans on grid stability, costs, jobs, health, and climate in 143 countries. One Earth 1(4):449–463

Jaeckel A (2016) Deep Seabed Mining and Adaptive Management: The Procedural Challenges for the International Seabed Authority. Mar Policy 70:205–211

Jaeckel AL (2017) The International Seabed Authority and the Precautionary Principle: Balancing Deep Seabed Mineral Mining and Marine Environmental Protection. Publications on Ocean Development, vol 83. Brill, Nijhoff. https://brill.com/view/title/33967

Jaeckel A, Gjerde KM, Ardron JA (2017) Conserving the Common Heritage of Humankind—Options for the Deep-Seabed Mining Regime. Mar Policy 78:150–157. https://doi.org/10.1016/j.marpol.2017.01.019

Johnson DE (2019) Protecting the Lost City Hydrothermal Vent System: All Is Not Lost, or Is It? Mar Policy 107:103593. https://doi.org/10.1016/j.marpol.2019.103593

Jones DOB, Kaiser S, Sweetman AK, Smith CR, Menot L, Vink A, Trueblood D et al (2017) Biological Responses to Disturbance from Simulated Deep-Sea Polymetallic Nodule Mining. PLoS One 12(2):e0171750. https://doi.org/10.1371/journal.pone.0171750

Kaiser M, Clarke K, Hinz H, Austen M, Somerfield P, Karakassis I (2006) Global Analysis and Recovery of Benthic Biota to Fishing. Mar Ecol Prog Ser 311:1–14. https://doi.org/10.3354/meps311001

Kim R (2017) Should Deep Seabed Mining Be Allowed? Mar Policy 82:134–137. https://doi.org/10.1016/j.marpol.2017.05.010

Kirchain, Randolph and Richard Roth. 2019. "MIT Presentation: Decision Analysis Framework & Review of Cash Flow Approach." Paper presented at the Financial Payment System Working Group Meeting, February. https://www.isa.org.jm/document/mit-presentation-decision-analy-sis-framework-review-cash-flow-approach

Konar M, Ding H (n.d.) A Sustainable Ocean Economy for 2050: Approximating Its Benefits and Costs. World Resources Institute, Washington, DC

Krewitt W, Nienhaus K, Kleßmann C, Capone C, Stricker E, Graus W, Hoogwijk M, Supersberger N, von Winterfeld U, Samadi S (2009) Role and potential of renewable energy and energy efficiency for global energy supply. Climate Change 18:336

LDAC (Long Distance Advisory Council) (2019) LDAC Advice on Deepsea Mining R.04.19.WG5. Long Distance Advisory Council, Madrid. https://ldac.eu/en/publications/706-ldac-advice-on-deepsea-mining-r-04-19-wg5

Le JT, Levin LA, Carson RT (2017) Incorporating Ecosystem Services into Environmental Management of Deep-Seabed Mining. Deep Sea Res Part II Top Stud Oceanogr 137:486–503. https://doi.org/10.1016/j.dsr2.2016.08.007

Levin LA, Mengerink K, Gjerde KM, Rowden AA, Van Dover CL, Clark MR, Ramirez-Llodra E et al (2016) Defining 'Serious Harm' to the Marine Environment in the Context of Deep-Seabed Mining. Marine Policy 74:245–259. https://doi.org/10.1016/j.marpol.2016.09.032

Levin LA, Wei C-L, Dunn DC, Amon DJ, Ashford OS, Cheung W, Colaço A et al (2020) Climate Change Considerations are Fundamental to Management of Deep-Sea Resource Extraction. Glob Chang Biol 26(9):4664–4678

Lewis A, Estefen S, Huckerby J, Musial W, Pontes T, Torres-Martinez J (2011) Ocean Energy. In: Edenhofer O, Pichs-Madruga R, Sokona Y, Seyboth K, Matschoss P, Kadner S, Zwickel T et al (eds) IPCC Special Report on Renewable Energy Sources and Climate Change Mitigation. Cambridge University Press, Cambridge, UK, and New York

Lily H. 2018. Sponsoring State Approaches to Liability Regimes for Environmental Damage Caused by Seabed Mining. Liability Issues for Deep Seabed Mining Series, Paper No. 3, December. Centre for Interna- tional Governance Innovation. https://www.cigionline.org/publications/sponsoring-state-approaches-liability-regimes-environmental-dam-age-caused-seabed

Lodge MW, Verlaan PA (2018) Deep-Sea Mining: International Regulatory Challenges and Responses. Elements 14(5):331–336. https://doi.org/10.2138/gselements.14.5.331

Lunney E, Ban M, Duic N, Foley A (2017) A State-of-the-Art Review and Feasibility Analysis of High Altitude Wind Power in Northern Ireland. Renew Sustain Energy Rev 68:899–911. https://doi.org/10.1016/j.rser.2016.08.014

Macheriotou L, Rigaux A, Derycke S, Vanreusel A (2020) Phylogenetic clustering and rarity imply risk of local species extinction in prospective deep-sea mining areas of the clarion–clipperton fracture zone. Proc Biol Sci 287:20192666. https://doi.org/10.1098/rspb.2019.2666

Magagna D, Margheritini L, Alessi A, Bannon E, Boelman E, Bould D et al (2018) Workshop on identification of future emerging technologies in the ocean energy sector. Luxembourg: European Union, Ispra, Italy. https://doi.org/10.2760/23207

Marques S, de Araújo TCM (2019) Survey and Assessment of Seabed Resources from the Brazilian Continental Shelf by the Law of the Sea: From National to International Jurisdictions. Ocean Coastal Manag 178:104858

Mendoza E, Lithgow D, Flores P, Felix A, Simas T, Silva R (2019) A Framework to Evaluate the Environmental Impact of OCEAN Energy Devices. Renewable Sustainable Energy Rev 112:440–449

Mengerink KJ, Van Dover CL, Ardron J, Baker M, Escobar-Briones E, Kristina G, Anthony Koslow J et al (2014) A call for Deep-Ocean Stewardship. Science 344:696–698

Miljutin DM, Miljutina MA, Arbizu PM, Galéron J (2011) Deep-Sea Nematode Assemblage Has Not Recovered 26 Years after Experimental Mining of Polymetallic Nodules (Clarion-Clip- perton Fracture Zone, Tropical Eastern Pacific). Deep-Sea Res I Oceanogr Res Pap 58(8):885–897. https://doi.org/10.1016/j.dsr.2011.06.003

Miller KA, Thompson KF, Johnston P, Santillo D (2018) An overview of seabed mining including the current state of development, environmental impacts, and knowledge gaps. Front Mar Sci 4:418. https://doi.org/10.3389/fmars.2017.00418

Morgan NB, Cairns S, Reiswig H, Baco AR (2015) Benthic Megafaunal Community Structure of Cobalt-Rich Manganese Crusts on Necker Ridge. Deep-Sea Res I Oceanogr Res Pap 104:92–105. https://doi.org/10.1016/j.dsr.2015.07.003

Mørk G, Barstow S, Kabuthand A, Pontes MT. 2010. Assessing the Global Wave Energy Potential. Paper presented at OMAE2010 29th International Conference on Ocean, Offshore Mechanics and Arctic Engineering, Shanghai, China, 6–11 June

Nihous GC (2018) A Preliminary Investigation of the Effect of Ocean Thermal Energy Conversion (OTEC) Effluent Discharge Options on Global OTEC Resources. J Marine Sci Eng 6(1):25. https://doi.org/10.3390/jmse6010025

Niner HJ, Ardron JA, Escobar EG, Gianni M, Jaeckel A et al (2018) Deep-Sea Mining With No Net Loss of Biodiversity—An Impossible Aim. Front Mar Sci 5:53. https://doi.org/10.3389/fmars.2018.00053

NMP (Namibian Marine Phosphate). n.d. "About NMP." http://www.nam- phos.com. Accessed 13 May 2020)

Norwegian Offshore Wind Cluster. n.d. "Makani's First Offshore Flight Last Week." https://offshore-wind.no/news/makanis-first-offshore-flight-last-week. Accessed 14 May 2020

Oceans of Energy. 2020. "Offshore Floating Solar Farm Installed and Operational at the Dutch North Sea." February. https://oceansofenergy.blue/north-sea-1-offshore-solar-project

OECD (Organisation for Economic Co-operation and Development) (2018) Global Material Resources Outlook to 2060 – Economic drivers and environ- mental consequences. OECD Publishing. oe.cd/materials-outlook, Paris

OES (Ocean Energy Systems). 2015. International Levelised Cost of Energy for Ocean Energy Technologies. Ocean Energy Systems. https://www.ocean-energy-systems.org/oes-projects/levelised-cost-of-energy-assessment-for-wave-tidal-and-otec-at-an-international-level/

Offshore Renewable Energy Catapult. 2018. Tidal Stream and Wave Energy Cost Reduction and Industrial Benefit. https://ore.catapult.org.uk/?orecatapultreports=tidal-stream-and-wave-energy-cost-reduction-and-industrial-benefit

O'Neill SP, Angeloudis A, Robins PE, Walkington I, Wardalan SL, Masters I, Lewis MJ et al (2018) Tidal range energy resource and optimization: past perspectives and future challenges. Renew Energy 127:763–778. https://doi.org/10.1016/j.renene.2018.05.007

Ørsted. 2019. Making Green Energy Affordable. Whitepaper. Gentofte, Denmark: Ørsted. https://orsted.com/en/Explore/Making-green-energy-affordable

Pacific Islands Forum Secretariat. 2019. "2019 Femm Action Plan." 2019 Forum Economic Ministers Meeting, Pacific Islands Forum Secretariat Conference Centre, Suva, Fiji, 8–9 May

Paulikas, D., S. Katona, E. Ilves, G. Stone and A. O'Sullivan. 2020. Where Should Metals for the Green Transition Come From? Comparing Environmental, Social, and Economic Impacts of Supplying Base Metals from Land Ores and Seafloor Polymetallic Nodules. https://www.autoevolution.com/pdf/news_attachements/ocean-floor-could-be-the-next-el-dorado-for-ev-battery-raw-materials-142915.pdf

Pavel CC, Lacal-Arántegui R, Marmier A, Schüler D, Tzimas E, Buchert M, Jenseit W, Blagoeva D (2017) Substitution strategies for reducing the use of rare earths in wind turbines. Resources Policy 52:349–357. https://doi.org/10.1016/j.resourpol.2017.04.010

Petersen S, Krätschell A, Augustin N, Jamieson J, Hein JR, Hannington MD (2016) News from the Seabed – Geological Characteristics and Resource Potential of Deep-Sea Mineral Resources. Mar Policy 70(August):175–187. https://doi.org/10.1016/j.marpol.2016.03.012

Pew Charitable Trusts (2017) First Report of the Code Project: Developing International Seabed Authority Environmental Regulations. Pew Charitable Trusts, Philadelphia. http://pew.org/2wiPnNk

Pew Charitable Trusts (2019) Fourth Report of the Code Project: Summary of Stakeholder Comments on the 2018 ISA Draft Regulations. Pew Charitable Trusts, Philadelphia. https://pew.org/2GqmMyN

Pinarbaşi K, Galparsoro I, Depellegrin D, Bald J, Pérez-Morán G, Borja A (2019) A Modelling Approach for Offshore Wind Farm Feasibility with Respect to Ecosystem-Based Marine Spatial Planning. Sci Total Environ 667:306–317. https://doi.org/10.1016/j.scitotenv.2019.02.268

Popova E, Vousden D, Sauer WHH, Mohammed EY, Allain V, Downey-Breedt N, Fletcher R et al (2019) Ecological Connectivity between the areas beyond national jurisdiction and coastal waters: safeguarding interests of coastal communities in developing countries. Mar Policy 104:90–102. https://doi.org/10.1016/j.marpol.2019.02.050

Ramirez-Llodra E, Trannum HC, Evenset A, Levin LA, Andersson M, Finne TE, Hilario A et al (2015) Submarine and Deep-Sea Mine

Tailing Placements: A Review of Current Practices, Environmental Issues, Natural Analogs and Knowledge Gaps in Norway and Internationally. Marine Pollut Bullet 97:13–35. https://doi.org/10.1016/j.marpolbul.2015.05.062
Raoux A, Lassalle G, Pezy J, Tecchio S, Safi G, Ernande B et al (2018) Measuring Sensitivity of Two OSPAR Indicators for a Coastal Food Web Model under Offshore Wind Farm Construction. Ecol Indic 96(1):728–738. https://doi.org/10.1016/j.ecolind.2018.07.014
Reid WV, Ali MK, Field CB (2019) The future of bioenergy. Global Change Biol 26:1–13. https://doi.org/10.1111/gcb.14883
Roark EB, Guilderson TP, Dunbar RB, Lynn Ingram B (2006) Radiocarbon-Based Ages and Growth Rates of Hawaiian Deep-Sea Corals. Mar Ecol Prog Ser 327:1–14. https://doi.org/10.3354/meps327001
Rojas, Andres and Freedom-Kai Phillips. 2019. Effective Control and Deep Seabed Mining: Toward a Definition. Liability Issues for Deep Seabed Mining Series, Paper No. 7, February. Centre for International Governance Innovation. https://www.cigionline.org/publications/effective-control-and-deep-seabed-mining-toward-definition-1
Salinas-de-León P, Phillips B, Ebert D, Shivji M, Cerutti-Pereyra F, Ruck C, Fisher CR, Marsh L (2018) Deep-Sea Hydrothermal Vents as Natural Egg-Case Incubators at the Galapagos Rift. Sci Rep 8(1):1788. https://doi.org/10.1038/s41598-018-20046-4
Schüler D, Buchert M, Liu R, Dittrich S, Merz C (2011) Study on Rare Earths and Their Recycling. Öko-Institut, Freiburg, Germany. http://www.ressourcenfieber.eu/publications/reports/Rare%20earths%20study_Oe-ko-Institut_Jan%202011.pdf
Seascape Consultants. 2016. Periodic Review of the International Sea- bed Authority pursuant to UNCLOS Article 154. ISBA/22/A/CRP.3 (1)
Simon-Lledó E, Bett BJ, Huvenne VAI, Köser K, Schoening T, Greinert J, Jones DOB (2019) Biological Effects 26 Years after Simulated Deep-Sea Mining. Sci Rep 9(1):8040. https://doi.org/10.1038/s41598-019-44492-w
Sims REH, Schock RN, Adegbululgbe A, Fenhann J, Konstantinaviciute I, Moomaw W, Nimir HB et al (2007) Energy Supply. In: Metz B, Davidson OR, Bosch PR, Dave R, Meyer LA (eds) Climate Change 2007: Mitigation of Climate Change. Contribution of Working Group III to the Fourth Assessment Report of the Intergovernmental Panel on Climate Change. Cambridge University Press, Cambridge, UK, pp 251–322
Singh P, Pouponneau A (2018) Comments to the Draft Regulations on Exploitation of Mineral Resources in the Area: Transboundary harm and the rights of Coastal States adjacent to the Area. International Seabed Authority, Kingston, Jamaica. https://webcache.google-usercontent.com/search?q=cache:PDmK51pCqlwJ:https://ran-s3.s3.amazonaws.com/isa.org.jm/s3fs-public/files/documents/comments_0.pdf+&cd=1&hl=en&ct=clnk&gl=uk
SI Ocean. 2013. Ocean Energy: Cost of Energy and Cost Reduction Opportunities. SI Ocean Project
Skov H, Heinanen S, Norman T, Ward R, Mendez-Roldan S, Ellis I. 2018. ORJIP Bird Collision and Avoidance Study—Final Report April 2018. The Carbon Trust
Sovacool BK, Ali SH, Bazilian M, Radley B, Nemery B, Okatz J, Mulvaney D (2020) Sustainable minerals and metals for a low-carbon future. Science 367(6473):30–33. https://doi.org/10.1126/science.aaz6003
SPC (Secretariat of the Pacific Community). 2012. "Pacific-ACP states regional legislative and regulatory framework for deep sea minerals exploration and exploitation." Prepared under the SPC-EU EDF10 Deep Sea Minerals Project
SPC. 2013. Deep Sea Minerals: Deep Sea Minerals and the Green Economy. Edited by E. Baker and Y. Beaudoin. Vol. 2. https://nicholasinstitute.duke.edu/sites/default/files/publications/chapter_4_-_sustainable_economic_development_and_deep_sea_mining.pdf
SPC. 2016. Pacific-ACP states regional financial framework for deep sea minerals exploration and exploitation. Prepared under the SPC-EU EDF10 Deep Sea Minerals Project
SPC and Cardno. 2016. An Assessment of the Costs and Benefits of Mining Deep-sea Minerals in the Pacific Island Region: Deep-sea Mining Cost-Benefit Analysis. https://www.sprep.org/attachments/VirLib/Regional/deep-sea-mining-cba-PICs-2016.pdf
Stegen KS (2015) Heavy rare earths, permanent magnets, and renewable energies: an imminent crisis. Energy Policy 79:1–8. https://doi.org/10.1016/j.enpol.2014.12.015
Stehly T, Beiter P, Heimiller D, Scott G (2018) 2017 Cost of Wind Energy Review. National Renewable Energy Laboratory, Golden, CO
Sweetman AK, Smith CR, Shulse CN, Maillot B, Lindh M, Church MJ, Meyer KS, van Oevelen D, Stratmann T, Gooday AJ (2019) Key Role of Bacteria in the Short-Term Cycling of Carbon at the Abyssal Seafloor in a Low Particulate Organic Carbon Flux Region of the Eastern Pacific Ocean. Limnol Oceanogr 64(2):694–713. https://doi.org/10.1002/lno.11069
Taguchi H, Khinsamone S (2018) Analysis of the 'Dutch Disease' Effect on the Selected Resource-Rich ASEAN Economies. Asia Pacific Policy Studies 5(2):249–263. https://doi.org/10.1002/app5.233
Takaya Y, Yasukawa K, Kawasaki T et al (2018) The Tremendous Potential of Deep-Sea Mud as a Source of Rare-Earth Elements. Sci Rep 8:5763. https://doi.org/10.1038/s41598-018-23948-5
Tansel B (2017) From Electronic Consumer Products to E-Wastes: Global Outlook, Waste Quantities, Recycling Challenges. Environ Int 98:35–45. https://doi.org/10.1016/j.envint.2016.10.002
Teske S, Pregger T, Simon S, Naegler T, Graus W, Lins C (2010) Energy [R]evolution: A sustainable world energy outlook. Energ Effic 4:409. https://doi.org/10.1007/s12053-010-9098-y
Teske S, Sawyer S, Schafer O. 2015. Energy [r]evolution: A sustain- able world energy outlook 2015. Greenpeace International, Global Wind Energy Council, Solar Power Europe. 5th Edition 2015 World Energy Scenario. https://www.greenpeace.org/archive--international/Global/international/publications/climate/2015/Energy-Revolution-2015-Full.pdf
Teske, Sven, Nick Florin, Elsa Dominish and Damien Giurco. 2016. Renewable Energy and Deep Sea Mining: Supply, Demand and Scenarios. Report Prepared by ISF for J. M. Kaplan Fund, Oceans 5 and Synchronicity Earth. Sydney: Institute for Sustainable Futures. https://opus.lib.uts.edu.au/bitstream/10453/67336/1/DSM%20-%20RE%20resource%20Report_9_FINAL%20DRAFT-NEWTITLE-ANDNAME.pdf
Thiel H, Schriever G, Ahnert A, Bluhm H, Borowski C, Vopel K (2001) The Large-Scale Environmental Impact Experiment DISCOL—Reflection and Foresight. Deep Sea Res Part II 48(17):3869–3882. https://doi.org/10.1016/S0967-0645(01)00071-6
Thiele T, Ginzjy H, Christiansen S, Damian HP. 2019. A Benefit Sharing Mechanism Appropriate for the Common Heritage of Mankind, UBA/IASS Workshop Summary, Potsdam. Project No. (FKZ) 3717 25 227 0. Institute for Advanced Sustainability Studies e.V
Thompson KF, Miller KA, Currie D, Johnston P, Santillo D (2018) Seabed mining and approaches to governance of the deep seabed. Front Marine Sci 5. https://doi.org/10.3389/fmars.2018.00480
Thurber AR, Sweetman AK, Narayanaswamy BE, Jones DOB, Ingels J, Hansman RL (2014) Ecosystem function and services provided by the deep sea. Biogeosciences 11(14):3941–3963. https://doi.org/10.5194/bg-11-3941-2014
Torres A, Brandt J, Lear K, Liu J (2017) A looming tragedy of the sand commons. Science 357:970–971. https://doi.org/10.1126/science.aao0503
Tunnicliffe V, Metaxas A, Le J, Ramirez-Llodra E, Levin LA (2018) Strategic environmental goals and objectives: setting the basis for environmental regulation of deep seabed mining. Marine Policy 114:103347. https://doi.org/10.1016/j.marpol.2018.11.010
Uihlein A (2016) Life Cycle Assessment of Ocean Energy Technologies. Int J Life Cycle Assess 21:1425–1437
United Nations. 1982. "Convention on the Law of the Sea." https://www.un.org/depts/los/convention_agreements/texts/unclos/unclos_e.pdf

United Nations. 1992. "Convention on Biological Diversity." https://www.cbd.int/doc/legal/cbd-en.pdf

United Nations. 1994. Agreement Relating to the Implementation of Part XI of the Convention. https://www.un.org/Depts/los/convention_agree-ments/texts/unclos/closindxAgree.htm

United Nations Conference on Environment and Development. 1992. Rio Declaration on Environment and Development. https://www.cbd.int/doc/ref/rio-declaration.shtml

United Nations General Assembly. 2006. Resolution 61/105, 8 December. Sustainable fisheries, including through the 1995 Agreement for the Implementation of the Provisions of the United Nations Convention on the Law of the Sea of 10 December 1982 relating to the Conservation and Management of Straddling Fish Stocks and Highly Migratory Fish Stocks, and related instruments. United Nations document A/RES/61/105

United Nations General Assembly. 2018. Resolution 72/279, 31 May. Re- positioning of the United Nations development system in the context of the quadrennial comprehensive policy review of operational activities for development of the United Nations system. United Nations document A/RES/72/279

der Voet V, Ester L v O, Verboon M, Kuipers K (2019) Environmental Implications of Future Demand Scenarios for Metals: Methodology and Application to the Case of Seven Major Metals. J Ind Ecol 23(1):141–155. https://doi.org/10.1111/jiec.12722

Van Dover CL (2014) Impacts of Anthropogenic Disturbances at Deep-Sea Hydrothermal Vent Ecosystems: A Review. Mar Environ Res 102:59–72. https://doi.org/10.1016/j.marenvres.2014.03.008

Van Dover CL (2019) Inactive sulfide ecosystems in the deep sea: a review. Front Marine Sci 6:461. https://doi.org/10.3389/fmars.2019.00461

Van Dover CL, Ardron JA, Escobar E, Gianni M, Gjerde KM, Jaeckel A, Jones DOB et al (2017) Biodiversity loss from deep-sea mining. Nat Geosci 10(7):464–465. https://doi.org/10.1038/ngeo2983

Van Dover CL, Arnaud-Haond S, Gianni M, Helmreich S, Huber JA, Jaeckel AL, Metaxas A et al (2018) Scientific Rationale and International Obligations for Protection of Active Hydrothermal Vent Ecosystems from Deep-Sea Mining. Marine Policy 90:20–28. https://doi.org/10.1016/j.marpol.2018.01.020

Vanreusel A, Hilario A, Ribeiro PA, Menot L, Arbizu PM (2016) Threatened by Mining, Polymetallic Nodules Are Required to Preserve Abyssal Epifauna. Sci Rep 6:26808. https://doi.org/10.1038/srep26808

Vare LL, Baker MC, Howe JA, Levin LA, Neira C, Ramirez-Llodra EZ, Reichelt-Brushett A et al (2018) Scientific Considerations for the Assessment and Management of Mine Tailings Disposal in the Deep Sea. Front Mar Sci 5:17. https://doi.org/10.3389/fmars.2018.00017

Vierros M, Suttle CA, Harden-Davies H, Burton G (2016) Who Owns the Ocean? Policy Issues Surrounding Marine Genetic Resources. Limnol Oceanogr Bull 25(2):29–35. https://doi.org/10.1002/lob.10108

Wahlberg M, Westerberg H (2005) Hearing in Fish and Their Reactions to Sounds from Offshore Wind Farms. Mar Ecol Prog Ser 288:295–309

Wakefield JR, Myers K (2018) Social Cost Benefit Analysis for Deep Sea Minerals Mining. Mar Policy 95:346–355. https://doi.org/10.1016/j.marpol.2016.06.018

Wang C, Prinn RG (2011) Potential Climatic Impacts and Reliability of Large-Scale Offshore Wind Farms. Environ Res Lett 6:2. https://doi.org/10.1088/1748-9326/6/2/025101

WEC (World Energy Council). 2013. Chapter 8: Solar. In World Energy Re- sources: 2013 Survey. London: World Energy Council. https://www.world-energy.org/assets/images/imported/2013/09/Complete_WER_2013_Sur-vey.pdf

Wedding LM, Friedlander AM, Kittinger JN, Watling L, Gaines SD, Bennett M, Hardy SM, Smith CR (2013) From Principles to Practice: A Spatial Approach to Systematic Conservation Planning in the Deep Sea. Proc R Soc B Biol Sci 280(1773):20131684. https://doi.org/10.1098/rspb.2013.1684

Wilburn DR 2011. Wind Energy in the United States and Materials Re- quired for the Land-Based Wind Turbine Industry From 2010 Through 2030. Scientific Investigations Report 2011–5036. US Department of the Interior and US Geological Survey

Williams A, Schlacher TA, Rowden AA, Althaus F, Clark MR, Bowden DA, Stewart R, Bax NJ, Consalvey M, Kloser RJ (2010) Seamount Megaben- thic Assemblages Fail to Recover from Trawling Impacts. Mar Ecol 31(S1):183–199. https://doi.org/10.1111/j.1439--0485.2010.00385.x

WindEurope. 2019. Our Energy, Our Future: How Offshore Wind Will Help Europe Go Carbon-Neutral. WindEurope and BVG Associates. https://windeurope.org/wp-content/uploads/files/about-wind/reports/WindEurope-Our-Energy-Our-Future.pdf

World Bank (2017) Precautionary Management of Deep Sea Minerals. World Bank, Washington, DC. http://documents.worldbank.org/curated/en/349631503675168052/Precautionary-management-of-deep-sea-minerals

World Bank Group, ESMAP and SERIS (2019) Where Sun Meets Water: Floating Solar Market Report. World Bank, Washington, DC. http://www.seris.sg/doc/publications/ESMAP_FloatingSolar_TEXT-A4-WEB.pdf

WWF (2014) Environmental Impacts of Offshore Wind in the North Sea – A Literature Review. WWF, Oslo. https://www.wwf.no/assets/attach-ments/84-wwf_a4_reporthavvindrapport.pdf

Yan J (2015) Handbook of Clean Energy Systems, vol 5. Wiley, New York

Zydlewski G, Copping A, Redden A (2015) Special Issue: Renewable Ocean Energy Development and the Environment. Estuar Coasts 38(1):156–158. https://doi.org/10.1007/s12237-014-9922-2

The Ocean Genome: Conservation and the Fair, Equitable and Sustainable Use of Marine Genetic Resources

4

Abbreviations

ABNJ	Areas beyond national jurisdiction
ABS	Access and benefit sharing
antiSMASH	Antibiotics and Secondary Metabolite Analysis Shell
BBNJ	Biodiversity in areas beyond national jurisdiction
CBA	Critical biodiversity area
CBD	Convention on biological diversity
cDNA	Complementary DNA
CRISPR	Clustered Regularly Interspaced Short Palindromic Repeats
DHA	Docosahexaenoic acid
DNA	Deoxyribonucleic acid
eDNA	Environmental DNA
EEZ	Exclusive economic zone
EIA	Environmental impact assessment
EPA	Eicosapentaenoic acid
EPS	Extracellular polymeric substances
ESA	Ecological support area
GFP	Green fluorescent protein
IPBES	Intergovernmental Science-Policy Platform on Biodiversity and Ecosystem Services
IUCN	International Union for the Conservation of Nature
LBSAP	Local Biodiversity Strategy and Action Plan
MGR	Marine genetic resources
MPA	Marine protected area
NBA	National Biodiversity Assessment
NBSAP	National Biodiversity Strategy and Action Plan
NGS	Next-generation sequencing
OECM	Other effective area-based conservation measure
RNA	Ribonucleic acid
RRI	Responsible research and innovation
SDG	Sustainable development goal
SEA	Strategic environmental assessment
TALENs	Transcription activator-like effector nucleases
TRIPS	The Agreement on Trade-Related Aspects of Intellectual Property Rights
UN	United Nations
UNCLOS	United Nations Convention on the Law of the Sea
WIPO	World Intellectual Property Organization

Highlights

- A sustainable ocean economy prioritises the conservation and sustainable use of the 'ocean genome' and leads to equitable outcomes for all.
- Marine life is incredibly diverse—having existed in the ocean for three times as long as life has existed on land—and comprises a minimum of 2.2 million existing eukaryotic marine species, of which some 91% remain undescribed.
- The ocean genome is the genetic material present in all marine biodiversity, including both the physical genes and the information they encode. It determines the abundance and resilience of biological resources, including fisheries and aquaculture, which collectively form a pillar of global food security and human well-being. It is the foundation upon which all marine ecosystems, including their functionality and resilience, rest.

Originally published in:
Blasiak, R., R. Wynberg, K. Grorud-Colvert, S. Thambisetty, et al. 2020. *The Ocean Genome: Conservation and the Fair, Equitable and Sustainable Use of Marine Genetic Resources*. Washington, DC: World Resources Institute. Available online at www.oceanpanel.org/blue-papers/ocean-genome-conservation-andfair-equitable-and-sustainable-use-marine-genetic.
Reprint by Springer International Publishing (2023) with kind permission.
Published under license from the World Research Institute.

J. Lubchenco, P. M. Haugan (eds.), *The Blue Compendium*, https://doi.org/10.1007/978-3-031-16277-0_4

- The ocean genome is threatened by overexploitation, habitat loss and degradation, pollution, impacts from a changing climate, invasive species and other pressures, as well as their cumulative and interacting effects.
- Fully and highly protected marine protected areas (MPAs) are proven tools for safeguarding genetic diversity at the ecosystem level, along with other effective area-based conservation measures (OECMs)—if they are effectively designed and managed. Yet, only 2.5% of the ocean is in MPAs considered fully or highly protected. Urgent action is required to apply measures based on scientific evidence and meet internationally agreed targets along with growing calls to fully or highly protect at least 30% of the ocean to support ocean health, productivity and resilience.
- In parallel, significant efforts are needed to ensure that genetic diversity in areas outside of MPAs and OECMs is conserved. These include effectively managing the sustainable use of resources; preventing habitat degradation; cautiously using previously unexploited places; enforcing and complying with regulations; and protecting rare, threatened and endangered species and populations.
- Rapid advances in sequencing technologies and bioinformatics have enabled exploration of the ocean genome. These new findings are informing innovative approaches to conservation and a growing number of commercial biotechnology applications, from anticancer treatments to cosmetics and industrial enzymes.
- At the same time, the environmental, social and ethical risks arising from using existing and new biotechnologies such as CRISPR (Clustered Regularly Interspaced Short Palindromic Repeats) remain under-investigated and poorly known, especially in marine environments.
- The capacity to undertake genomic research and to access, download and analyse massive amounts of sequence data relating to marine genetic resources is inequitably distributed among countries. There is an urgent need to build capacity; increase access to affordable innovation and technologies; and ensure that research and innovation is ethically and socially acceptable, environmentally sustainable, and delivering solutions to the problems of the poorest and most marginalised communities and income groups.
- Scientific and commercial benefits arising from using the ocean genome must be fairly and equitably shared. Reforms to intellectual property rights should support this shift.
- International legal measures governing aspects of the conservation and use of the ocean genome must comprehensively, actively and persistently engage with scientists and other actors from both commercial and noncommercial sectors. This will ensure that regulations reflect up-to-date scientific knowledge and understanding, are needs based and enable a shared sense of responsibility to conserve and protect the ocean genome.
- This paper takes a holistic approach to evaluating the prospects for conservation and sustainable use of the ocean genome. It does this by analysing our understanding of the genetic diversity of life within the ocean, the threats posed to such diversity, the benefits provided by genetic diversity and the ecosystems it supports in the context of a changing world, as well as tools and approaches for ensuring fair and equitable sharing of these benefits.
- The paper concludes with opportunities for action that, if followed, would improve our understanding of the ocean genome and support its conservation as well as its sustainable and equitable use.

1 Introduction

1.1 Overview

The 'ocean genome' is the foundation upon which all marine ecosystems rest and is defined here as the ensemble of genetic material present in all marine biodiversity, including both the physical genes and the information they encode. The dynamics of the ocean genome enable organisms to adapt to diverse ecological niches and changing environmental conditions. The ocean genome also determines the productivity and resilience of biological resources, including fisheries and aquaculture, which collectively support global food security, human well-being and a sustainable ocean economy.

A deeper understanding of the ocean genome has contributed to an increased awareness of the pressures facing marine biodiversity, including those from habitat loss and degradation; overfishing and other extractive activities such as mining; climate change and the spread of invasive species. Rapid advances in sequencing technologies and bioinformatics have enabled exploration of the ocean genome, which is informing the designation of marine protected areas as well as innovative approaches to conservation such as the establishment and incorporation of temporal genetic monitoring datasets into conservation planning and management as well as the sustainable use of resources. Exploring the ocean genome has also enabled a growing number of commercial biotechnology applications, extending from multiple anticancer treatments to cosmetics and industrial enzymes. At the same time, the environmental, social and ethical risks arising from the use of existing and new biotechnologies such as CRISPR (Clustered Regularly Interspaced Short Palindromic Repeats) remain under-investigated and poorly known, especially in marine environments.

As awareness of the unique nature and consequent value of the ocean genome grows and the importance of ensuring its conservation and sustainable use becomes more pressing, so too has the complexity of the national and transnational legal, institutional and ethical contexts that govern it. Within national jurisdictions, the Convention on Biological Diversity (CBD) and its Nagoya Protocol comprise key governance mechanisms for the conservation and sustainable use of marine biodiversity. For biodiversity in areas beyond national jurisdiction (BBNJ), a new United Nations (UN) agreement is under negotiation focusing on issues of crucial importance to the ocean genome, including area-based management tools, access to and intellectual property protection of marine genetic resources and their commercial exploitation, as well as capacity building.

Sharing benefits arising from the use of the ocean genome is a central issue. There is an urgent need to promote inclusive and responsible research and innovation that addresses equity differentials and fosters enhanced capacity and access to technology while facilitating the realisation of commitments to conserve and sustainably use the ocean's genetic diversity.

1.2 Scope and Ambition

This Blue Paper takes a holistic approach to the issue of the ocean genome and addresses

- our understanding of the genetic diversity of life within the ocean;
- the threats posed to these building blocks of life;
- the many benefits this diversity provides for functional ocean ecosystems, humanity and the biosphere in the context of a changing world;
- the tools and approaches that have been demonstrated to protect and restore genetic, species and ecosystem diversity; and
- stumbling blocks and opportunities for achieving sustainable and equitable use.

After introducing the ocean genome and the ecological benefits it provides, we present an overview of the expanding range of commercial activities it enables. This is followed by a description of the challenges facing the conservation and sustainable use of the ocean genome, including the primary anthropogenic threats to marine biodiversity. We then discuss the pathways to solutions, spanning novel conservation approaches, efforts to promote inclusive and responsible research and innovation and equitable access and benefit sharing from the use of marine genetic resources.

Ultimately, individuals, communities, companies and states have all contributed to different degrees to the degraded state of marine ecosystems. Their reliance on and stewardship of these resources varies, as do the benefits they derive from the ocean genome. Equity and sustainability are therefore crosscutting themes, and attention is given not only to evidence of inequitable and unsustainable practices, but also to the institutional and informal approaches and tools available to address these challenges.

The paper concludes with a number of opportunities for action that, if adopted, would improve our understanding of the ocean genome, and contribute to ensuring its conservation as well as its sustainable and equitable use.

1.3 What Is the Ocean Genome and Why Is It Uniquely Important?

The ocean covers 70% of the Earth's surface and represents 99% of the habitable space on the planet by volume (Costanza 1999). Life has existed in the ocean for at least 3.7 billion years, over three times as long as on land (Pearce et al. 2018; Strother et al. 2011). This long evolutionary history has resulted in some 2.2 million existing eukaryotic marine species (estimates range from 0.3 to 10 million species), of which 230,000 are confirmed (Mora et al. 2011; Louca et al. 2019). Marine species have been discovered at a higher rate than terrestrial species since the 1950s (Costello et al. 2012); indeed, the ocean harbours unique biodiversity that dwarfs the biodiversity found on land. For example, of the 34 major known animal phyla, 33 are found in the ocean while only 12 are found on land (Jaume and Duarte 2006). On land, a single phylum accounts for 90% of all terrestrial animal species (Arthropoda—including insects and arachnids), but in the ocean 90% of the animal species are distributed across 8 phyla (Mollusca, Arthropoda, Chordata, Annelida, Nematoda, Cnidaria, Bryozoa and Porifera), showing a remarkable range of biodiversity at higher taxonomic levels (Jaume and Duarte 2006). Depending on the taxon group, some 24–98% of eukaryotic marine species remain undescribed. Even less is known about prokaryotic marine life (bacteria and archaea) and viruses, which form the majority of life in the ocean by weight—some 1.2×10^{29} prokaryote cells (Bar-On et al. 2018) and 1.3×10^{30} virus particles (Cobián Güemes et al. 2016). The estimated number of microbial species (operational taxonomic units of bacteria, archaea and microscopic fungi) in the ocean ranges widely, due to extrapolation based on scaling laws, from 1.0×10^{6} to 3.0×10^{27} (Locey and Lennon 2016; Louca et al. 2019; Mora et al. 2011).

Genetic diversity—the total number of genetic characters in the genetic makeup of a species—is a foundational component of biodiversity, strongly determining the biogeography of species distribution and allowing us to indirectly retrace the history of life and its evolution on Earth. Its conservation is necessary for evolution and, through genetic variability, for greater population fitness and potential to adapt and recover (Reed and Frankham 2003). Such attributes are especially critical in the context of rapid environmental change (Hilborn et al. 2003; Ellegren and Galtier 2016). Genetically diverse fish stocks, for instance, may be able to exploit a range of environments and have a better ability to withstand anomalous conditions, and are therefore of key interest to fishery managers (Schindler et al. 2010; Ruzzante et al. 2006). Genetic diversity is also important for understanding long-term climate resilience, such as the ability of some corals to be heat resistant in the face of mass bleaching events (Norström et al. 2016; Cornwall 2019; Morikawa and Palumbi 2019).

The 'ocean genome' is defined here as the ensemble of genetic material present in all marine biodiversity, including both the physical genes and the information they encode. While discussions of genetic resources typically centre on physical resources, the informational component of genes has become increasingly important. This is due to the possibility of storing the nucleotide sequences of DNA (deoxyribonucleic acid) and RNA (ribonucleic acid) as digital information and using this information to create proteins, molecular processes, innovation and even organisms (Gibson et al. 2010; Hutchison et al. 2016). Patent and ownership claims now often centre on using genetic sequence data in addition to the physical genetic material from which they were extracted. Patent applications require sequences to be disclosed, depending on what is being patented, and many scientific journals also require sequences to be deposited and an accession number to be supplied prior to publishing associated research (Giles 2011; Blasiak et al. 2019). Limiting genetic resources to their material representation does not encompass the diverse ways in which these resources are used and commercially exploited; therefore, for the purposes of this paper, we conceptualise genetic material, and by extension the ocean genome, to include both physical molecules and their genetic sequence information (Elkin-Koren and Netanel 2002). Figure 4.1, developed by Broggiato et al. (2014), provides an illustration of the pathways that can lead to using marine genetic resources (MGR) after sampling and identifying interesting applications.

Box 4.1 A Note on Scientific Terminology and Legal Scope

Following the negotiations and signing of the Convention on Biological Diversity (CBD), one observer claimed that 'biodiversity is dead' on the basis that its definition was simply too inclusive and non-specific.[a] Over two decades later, the term 'genetic resources' is causing similar disquiet due to its scope,[b] and an expanding library of sometimes overlapping terminology is complicating the global task of governing the access, use and circulation of genetic resources. The following is a brief guide to the current or emerging legal terminology relevant to this paper:

Biodiversity (from CBD): The variability among living organisms from all sources including, among others, terrestrial, marine and other aquatic ecosystems and the ecological complexes of which they are part; this includes diversity within species, between species and of ecosystems.

Genetic resources (from CBD): Genetic material of actual or potential value.

Genetic material (from CBD): Any material of plant, animal, microbial or other origin containing functional units of heredity, such as individual genes or genetic sequences.

Digital sequence information (or data): Used in association with research and development, and the use of genetic resources, this is a placeholder term in international discussions under the CBD. As used, it includes various types of information including nucleic acid sequences; information on sequence assembly that may describe whole genomes, individual genes or fragments; single nucleotide polymorphisms; information on gene expression structures including morphological data and phenotype; data on macromolecules and cellular metabolites; information on ecological relationships and abiotic factors of the environment; behavioural data; information related to taxonomy; and modalities of use. The term is typically used in negotiating processes linked to international agreements such as the CBD, United Nations Convention on the Law of the Sea (UNCLOS) and the International Treaty on Plant Genetic Resources for Food and Agriculture.

Genetic sequence data: The order of nucleotides found in nucleic acid molecules—DNA (deoxyribonucleic acid) or RNA (ribonucleic acid)—which contain the genetic information that determines the biological characteristics of an organism or a virus. The term is widely used in the scientific community, and is preferred by some parties to the CBD.

Nucleotide sequence data: The arrangement of nucleotides on strands of naturally occurring DNA or RNA. Information about the genetic resources arises through analysis of these data.

Marine genetic resources: The genetic material of marine plant, marine animal, microbial or other origin containing functional units of heredity, which have an actual or potential value. The scope of this term is subject to negotiations related to biodiversity in areas beyond national jurisdiction, but, as such, it is not defined or used in UNCLOS.

In many ways, the battle over terminology is central to the effective and equitable governance of genetic resources. The terms above straddle environmental and biotechnological norms, and are therefore defined both by tangible parameters such as location and place, as well as intangible parameters such as information and function.

Notes:

[a] *Lautenschlager* (1997)

[b] *Thambisetty* (2020)

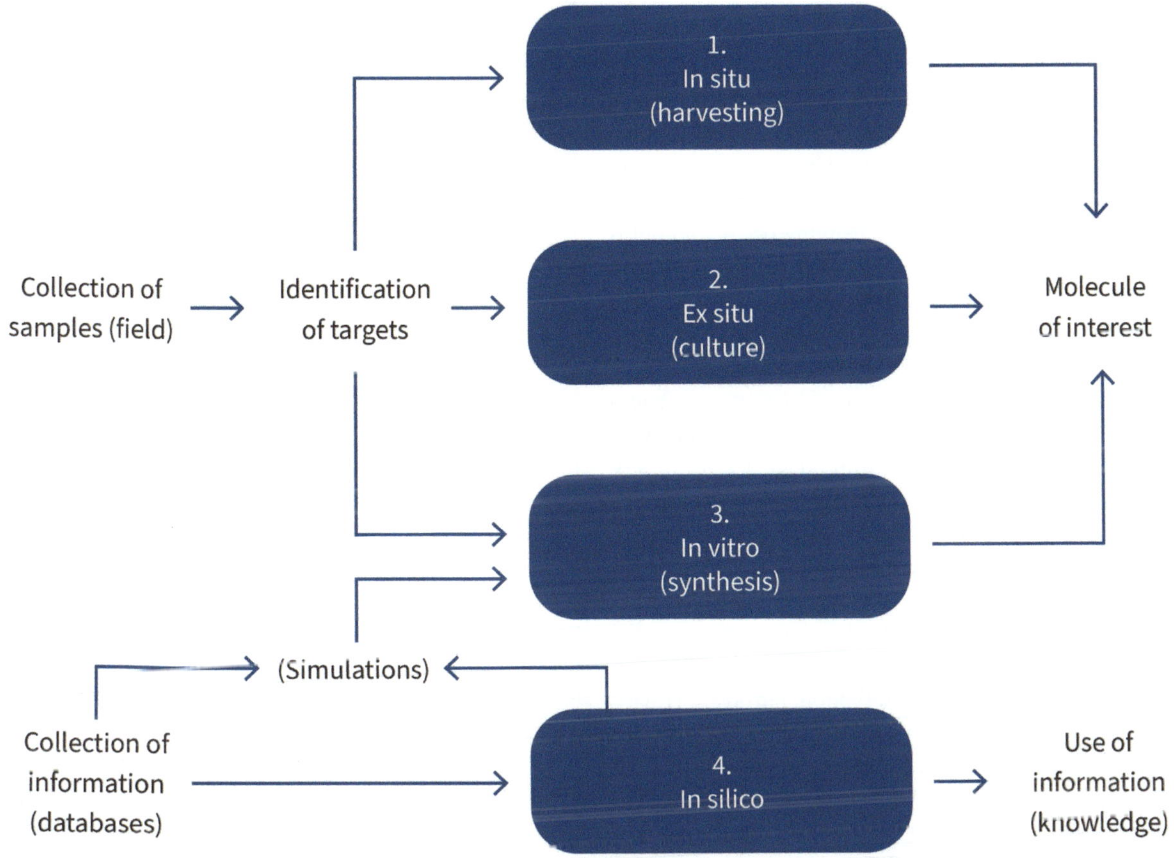

Fig. 4.1 Pathways for using marine genetic resources. *Notes:* (1) *Harvest of in-situ biological material. (2) Ex situ culture of biological material. (3) In vitro laboratory synthesis of interesting molecules. (4) Use of information in databases (in silico), sometimes also leading to the use of this information for in-vitro synthesis. (Source: Broggiato et al.* 2014)

1.4 How Do We Benefit from the Ocean Genome?

The ocean genome is the foundation upon which all marine ecosystems rest and is therefore integrally linked to the existence of all life on Earth, including humanity. Throughout human history, diverse cultures, societies and knowledges have evolved that are integrally linked to marine and coastal biodiversity, leading over time to the emergence of a rich diversity of social-ecological systems and worldviews in coastal regions around the world. As the custodians of many coastal areas, and the repositories of associated traditional knowledge, local and traditional communities have played a critical role in contributing such knowledge toward our food, medicines, cosmetics and emotional connections to the ocean.

Maintaining the health of ocean ecosystems is critical; these ecosystems provide over 50% of the oxygen on the planet, sustain vast fisheries generating 17% of the animal protein we consume and shape and regulate global climate patterns (FAO 2018; IPBES 2019). The genetic diversity within the ocean genome contributes to the capacity of species and populations to adapt to a changing ocean and helps mitigate the impacts associated with realised and projected climate change (Reed and Frankham 2003).

Marine plants, animals, fungi and microorganisms have evolved to occupy a variety of niches, being able to thrive in the extremes of heat, cold, water chemistry and darkness found in the ocean. The resulting adaptations are recorded in their genetic codes, enabling them to produce a wide variety of primary and secondary metabolites with significant biological activities that have attracted growing commercial interest from a range of industries (Blasiak et al. 2018; Arnaud-Haond et al. 2011; Arrieta et al. 2010). Applications include the development of industrial enzymes, pharmaceuticals, cosmeceuticals, nutraceuticals, antifoulants, adhesives and tools for research and conservation purposes (Leary et al. 2009). Over 34,000 marine natural products—naturally occurring molecules produced by marine organisms—have been discovered (MarinLit 2020), many with remarkable levels of bioactivity, resulting in rates of drug discovery from marine organisms that are up to 2.5 times the industry average (Carroll et al. 2019; Gerwick and Moore 2012; Arrieta et al. 2010).

In addition to these commercial uses, a range of noncommercial applications based on the ocean genome has also emerged. Through the use of genetic sequence data, a substantial and growing body of work has been done in the fields of evolution and ecology to inform our knowledge on taxonomy, connectivity, demography and evolution, while new techniques, such as the sampling of environmental DNA (eDNA), are enhancing our understanding of marine taxonomy and enabling noninvasive study methods (Hansen et al. 2018). DNA barcodes have also been used to help identify mislabeled seafood and fight wildlife trafficking (Di Muri et al. 2018). Finally, the potential of gene editing tools like CRISPR (Clustered Regularly Interspaced Short Palindromic Repeats) as novel conservation techniques is now being explored (Phelps et al. 2019), although its application remains theoretical. Moreover, environmental, social and ethical risks remain under-investigated and poorly known, especially for the marine environment (CSS et al. 2019; Jasanoff et al. 2015).

1.5 How Is the Ocean Genome at Risk?

Multiple threats face the ocean genome, largely through overexploitation, habitat destruction, pollution, invasive species and, increasingly, the degradation of marine ecosystems, all of which are additionally impacted by a changing climate (Aburto-Oropeza et al. 2020). Land-based activities such as high-input industrialised agriculture are leading to pollution and eutrophication from excessive nutrient runoff and expanding low-oxygen dead zones around river deltas. Coastal aquaculture (mariculture) can also have significant environmental impacts due to high nutrient inputs, chemical pollution, the removal of large amounts of fry from the wild and the large-scale destruction of coastal habitats such as mangroves, among other issues (Hamilton 2013; Ahmed and Glaser 2016). Mariculture has also created an immediate threat to the genetic diversity of native fish populations, most prominently perhaps in the southern hemisphere where salmonids, absent from the native fauna (Arismendi et al. 2009), have been introduced as aquaculture escapees and in areas supporting wild capture salmon fisheries (McGinnity et al. 2003, 2009). Shipping activities and the flow of ballast water and waste into the ocean have contributed to the spread of invasive species and pathogens, and to the creation of anoxic, no-oxygen zones and toxic red tide algal blooms (Pitcher and Probyn 2016).

Ocean-based activities like trawl fisheries, mining, dredging and the construction of artificial islands are drastically reducing biodiversity and completely reshaping some marine environments (Halpern et al. 2008; Du Preez et al. 2020). Overfishing has led to the collapse of major fisheries like the Newfoundland cod fishery, where a regime shift has subsequently resulted in a restructuring of regional food webs (Pedersen et al. 2017). Overfishing is also damaging the genetic diversity of fish and bycatch species, with one study suggesting that overfished species carry about 18% fewer unique genetic variations than their lightly fished relatives (Pinsky and Palumbi 2014). In some cases, multiple threats combine synergistically; for instance, aquaculture salmon broodstock escapes and fishing of wild populations have led to the loss of genetic diversity (Waples et al. 2012).

At a global scale, the burning of fossil fuels has generated greenhouse gas emissions and led to climate change—and the ocean absorbs 93% of the increased heat associated with these greenhouse gas emissions (Resplandy et al. 2018). The majority of marine species have narrower windows of thermal tolerance compared with those of terrestrial species, and local extinctions of marine species have been twice as common as those of species on land based on a global dataset of the range-edge positions of species on land and in the sea (Pinsky et al. 2019). Increased heating of the ocean remains the biggest climate impact to date, with the absorption of excess carbon also resulting in ocean acidification, which has negatively impacted marine ecosystems as it interacts synergistically with other drivers of loss such as direct exploitation and pollution (IPBES 2019).

About 20 marine species are known to have gone extinct over the past 500 years (McCauley et al. 2015), yet this is likely an underestimate given that little is known about how many species inhabit the marine environment. Some marine species have not been observed for decades and could already be extinct, while others, including 25% of marine mammals, sharks and rays, are at risk of extinction or are globally threatened (IUCN 2019; Dulvy et al. 2014). Although advances in working with ancient DNA may still allow species' genomes to be recovered from remains held in museums (McCormack et al. 2017), these would be devoid of the variability present in viable existing populations.

In the face of these threats to marine biodiversity and the ocean genome, adequate protection lags significantly. For most of human history the ocean was largely a de facto fully protected area that was too remote, too distant and too deep to exploit based on technological, economic and social limitations to access (Lubchenco and Gaines 2019). The creation of marine protected areas (MPAs) is in recognition of the need to reestablish places that are protected from exploitation. International targets including the Convention on Biological Diversity's Aichi Target 11 and the United Nations' Sustainable Development Goal 14 call for protecting 10% of the ocean by 2020 in MPAs and other effective area-based conservation measures (OECMs). Yet only 8% of the ocean is in any kind of designated MPA, including 5% in implemented MPAs, and only 2.5% is in fully or highly protected, implemented MPAs (Sala et al. 2018, updated via Marine Conservation Institute 2020). Further, there are growing calls from the scientific community for at least 30% of the ocean to be fully to highly protected to maintain a healthy, productive and resilient ocean (O'Leary et al. 2016; Gaines et al. 2010). In parallel, significant efforts are also needed to ensure that genetic diversity in areas outside MPAs and OECMs is conserved. This would include ensuring the sustainable use of resources; preventing habitat degradation; cautiously using previously unexploited places; and protecting rare, threatened and endangered species and populations.

Marine science has contributed significantly to revolutionary scientific and technological transformations in the life sciences and microbiology over the past two decades. Advances in genomic technologies, with sequencing costs declining 4000-fold over the past decade (Green et al. 2017), mean that millions of DNA fragments can be sequenced simultaneously and inexpensively, creating an intensely data-rich field (see Fig. 4.2) (Pevsner 2015). While such innovations have rapidly expanded the boundaries of our knowledge, vast knowledge gaps remain (Wetterstrand 2019). For instance, a large fraction of predicted genes from marine prokaryotes cannot be assigned functions (Sunagawa et al. 2015), and the functions of some 90% of genetic sequences collected from viruses remain unknown (Hurwitz and Sullivan 2013).

The rapidly growing field of synthetic biology now allows genes from different organisms, from different parts of the world, and from the ocean, soil and rivers to be combined into new patented organisms, including some synthesised components. Although the full contribution of MGR remains unknown, the Synthetic Biology Project reports at least 116 synthetic biology products and applications to be near to or on the market. The pace has rapidly increased over the past 5 years due to the introduction of fast, reliable and low-cost genome-editing techniques such as CRISPR, gene drives, TALENs (transcription activator-like effector nucleases) and oligonucleotide-directed mutagenesis techniques (Doudna and Charpentier 2014).

While such developments could yield important benefits, they also carry significant and often unknown risks. It is especially hard to predict the ecological consequences of introducing transformed organisms into marine environments. Containing introduced organisms is likely impossible and escaped transgenic fish or bacteria may establish viable populations in the wild, leading to altered natural ecosystems (Li et al. 2015). For instance, simple, commercially available kits (GeneArt® *Synechococcus* Engineering Kits) allow the photosynthetic cyanobacteria *Synechoccocus*—responsible for up to 80% of the photosynthetic production in the oligotrophic ocean (Campbell et al. 1994)—to be genetically manipulated. Accidental release of genetically modified strains in the ocean could, if viable, generate significant risks to the entire biosphere. The introduction of genome-editing techniques heightens such concerns, raising a suite of important questions about the governance and regulation of such technologies, about how problems are framed and solved, about how decisions get made about the release of modified organisms, and about the ethical considerations of international, intergenerational and interspecies justice (CSS et al. 2019).

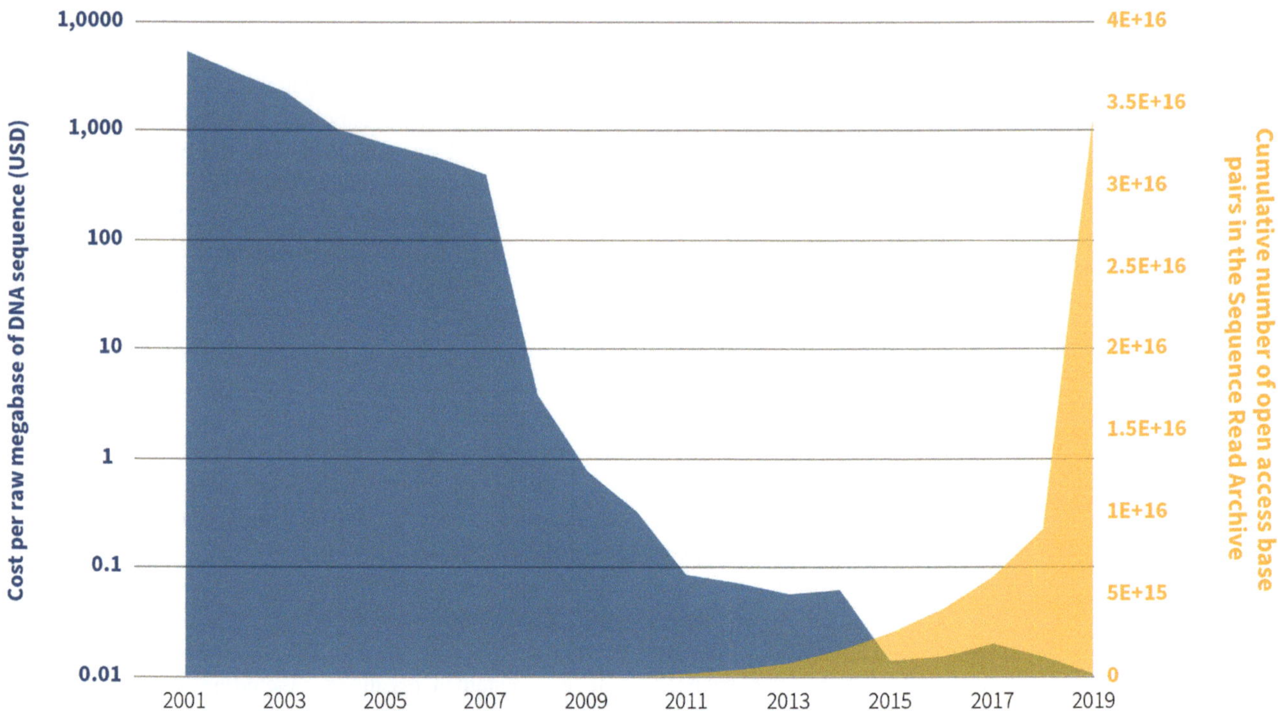

Fig. 4.2 Growth in GenBank Sequence Read Archive Records, and Trend in Average Cost of Sequencing. *Notes: GenBank is the genetic sequence database of the United States National Institutes of Health. It maintains the Sequence Read Archive, a bioinformatics database of sequencing data, particularly the short reads of fewer than 1000 base pairs typical of high-throughput sequencing methods. Also note that a logarithmic scale is used on the left axis. DNA stands for deoxyribonucleic acid. (Source: Data from National Center for Biotechnology Information. 2018. "Sequence Read Archive." Last updated August 2, 2018.* https://trace.ncbi. nlm.nih.gov/Traces/sra/sra.cgi; *Wetterstrand, K.A. 2019. "DNA Sequencing Costs: Data." National Human Genome Research Institute, Genome Sequencing Program. Last updated October 30, 2019.* https://www.genome.gov/about-genomics/fact-sheets/DNA-Sequencing-Costs-Data)

1.6 How Is the Ocean Genome Governed and Regulated?

Governance of the ocean genome is complex. This is due, not least, to its conceptual broadness, the lack of boundaries for the spread of species in the ocean, the diversity of threats it faces and the mix of its commercial and noncommercial dimensions. The United Nations Convention on the Law of the Sea (UNCLOS), concluded in 1982, acts as a sort of 'constitution' for the ocean, specifying provisions for the protection of the marine environment. The convention itself does not refer to either biodiversity or genetic resources, but it does refer to the 'conservation and utilization of living resources' (Articles 61 and 62), including on the high seas (Articles 116 and 117). UNCLOS combines elements of different conceptions of property, with a governance approach to achieving social objectives (Allott 1992).

UNCLOS also defined a series of maritime zones and jurisdictional claims (Fig. 4.3). Of particular relevance is the distinction between exclusive economic zones (EEZs) and areas beyond national jurisdiction (ABNJ), which comprise roughly 36% and 64% of the ocean's area, respectively (Smith and Jabour 2018). According to UNCLOS, each coastal state has 'sovereign rights [within its EEZ] for the purpose of exploring and exploiting, conserving and managing the natural resources'. In the case of straddling and highly migratory fish stocks, many of which are found across multiple EEZs as well as ABNJ, the UN Fish Stocks Agreement applies (UN 1995). According to that agreement, states 'shall apply the precautionary approach widely to conservation, management and exploitation [. . .] to protect the living marine resources and preserve the marine environment'.

Two international agreements under the CBD are of particular relevance: the Cartagena Protocol on Biosafety (2000), which aims to protect biological diversity from risks associated with biotechnology innovations, including genetically modified organisms; and the Nagoya Protocol on Access and Benefit-Sharing (2010), which aims to operationalise the CBD's third objective. The 196 parties to the CBD have agreed to a wide range of obligations and relevant global targets, including safeguarding genetic diversity, operationalising the Nagoya Protocol and ensuring the integrity of ecosystems. Meeting these targets entails committing to protect 10% of coastal and marine areas by 2020 (Aichi Target 11) and maintaining the genetic diversity of wild ani-

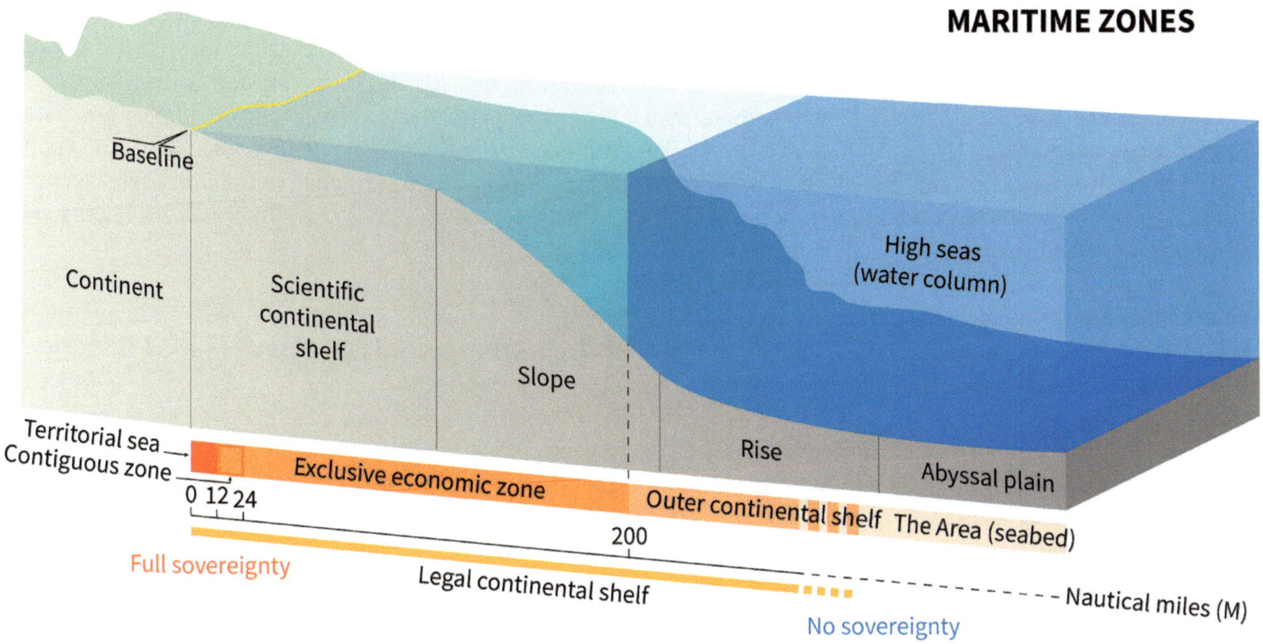

Fig. 4.3 Maritime zones and levels of sovereignty. (*Source: Courtesy of Riccardo Pravettoni, cartographer. Pravettoni, R. 2010. "Maritime Zones." UN Environment Programme.* http://www.grida.no/resources/7923)

mals, in addition to domesticated species, by using strategies to minimise genetic erosion (Aichi Target 13).

As mandated by UN General Assembly Resolution 72/249, an intergovernmental conference began in 2018 with the aim of negotiating a new legally binding international treaty on biodiversity in ABNJ (BBNJ). The negotiations cover four elements of a package: marine genetic resources including issues related to access and benefit sharing; measures such as area-based management tools, including MPAs; environmental impact assessments (EIAs); and capacity building and technology transfer.

These negotiations have been complicated by a focus on marine scientific research and the noncommercial aspects of research and development of MGR as well as the intersection with intellectual property issues. These controversies arise from the potential commercial exploitation of these resources, along with the desire to 'not undermine' mandates of the World Intellectual Property Organization (WIPO) and the World Trade Organization's Agreement on Trade-Related Aspects of Intellectual Property Rights (TRIPS) (see Sect. 4.2). Additionally, definitions from the CBD and Nagoya Protocol have not been embraced by state parties in the negotiations. Strong views on the scopes and definitions of terms such as 'genetic resources', 'access', 'digital sequence information', 'derivatives' and even the need for, and scope of, a definition of 'utilisation' of genetic resources threaten the possibility of multilateral consensus positions (see Box 4.1).

2 Existing and Potential Benefits

2.1 Ecological Benefits Associated with Marine Genetic Diversity

The genomes of organisms encode the biological, morphological, behavioural and physiological attributes that define their structures and roles within ecosystems. In the ocean, functioning ecosystems supported by this genetic diversity contribute essential services, including producing and recycling organic matter, channelling energy across food webs, providing food, maintaining water quality, regulating climate, establishing cultural values and providing recreational opportunities and other ecosystem services that benefit humanity (Worm et al. 2006). The ecological benefits of the ocean genome are vast (see Sect. 1.2), with their scope broadly organised into two equally important and interrelated themes.

First, genetic diversity in the ocean is critical because it stabilises ecosystems, as well as the species and ecological processes they encompass and the ecosystem services they provide. Genetic variability, including single nucleotide base pair substitution, insertion-deletion and structural variability, can result in the presence of species with redundant functions (Li et al. 2015) as well as genotypes within species that encode variable responses to environmental pressures. These support ecosystem stability and ensure that ecosystems remain functional even if unpredictable changes lead to the loss of some species, or the loss of within-species genetic diversity at the population level (Webster et al. 2017). For example, genetic variability across more than 100 discrete populations of Bristol Bay salmon in Alaska entails greater heterogeneity and resilience to anomalous conditions, resulting in lower variability in fisheries production and greater stability than that of a homogeneous population. This has also led to fewer fisheries closures for the fishing community (Schindler et al. 2010). At the ecosystem level, nutrients from the salmon spread throughout the system as predators feed on the population during spawning season.

Genetic diversity can also stabilise populations during restoration efforts. Seagrass restoration experiments in North America and Indonesia showed plots with higher genetic diversity had increased survival, density and/or growth (Reynolds et al. 2014). In the Chesapeake Bay, restoration efforts were linked to ecosystem services, including increased primary production and nutrient retention (Reynolds et al. 2012).

Second, genetic diversity enables biological variability and drives genetic potential, which allow species to persist in changing environmental conditions and to evolve as environments change over time. Overexploitation of and declines in marine populations can lead to dwindling population sizes and greater potential for lost genes compared with the greater standing genetic variation in larger populations. This variation helps species persist and adapt to perturbations (Thornburg et al. 2018), including those associated with anthropogenic changes. For example, an experiment on marine phytoplankton showed that cultures with higher genetic diversity were better able to withstand low salinities. High diversity in the simulated populations corresponded to the highest primary production and greatest nitrogen uptake under salinity stress (Sjöqvist and Kremp 2016). This is particularly important alongside increasing evidence that adaptation in some species can take place faster than previously thought; adaptation has been shown to occur in only 200 generations of short-lived species such as tropical diatoms (Jin and Agustí 2018). Corals also provide context for this adaptive capacity with their ability to respond relatively quickly via symbiont and microbiome shuffling, phenotypic plasticity, acclimatisation and adaptation. Some corals may have already adapted to ocean warming since the Industrial Revolution (Webster et al. 2017).

The ecosystem stability and adaptive potential afforded by genetic diversity are already vital to species, populations and communities as we know them. Yet their future values may go beyond these, as systems change at rates that are unprecedented and in ways that are unexpected, involving additive and synergistic effects. This underscores the benefits of conserving the ocean genome (see Sect. 4.1), particularly in areas that are minimally explored but may harbour high genetic diversity and isolated populations, including the fragile communities on seamounts and in the deep sea (Taylor and Roterman 2017; Zeng et al. 2017).

2.2 Commercial Benefits of Marine Genetic Resources

An intact and healthy ocean genome provides not only ecological benefits but also the foundation that has enabled and supported a growing range of commercial applications. Although the monetary benefits associated with these innovations are notoriously difficult to quantify (see, for example, Fig. 4.4), it is important to emphasise how these innovations contribute to human well-being. For instance, bioactive compounds from marine microorganisms associated with sea sponges are considered promising candidates for the development of novel antibiotics, which are relevant in the context of increasing antimicrobial resistance (El Samak et al. 2018). Likewise, the venoms of species such as cone snails are of interest for the development of new drugs (see, for example, Table 4.1) that could replace opioids and consequently lower instances of misuse (Zachos 2017).

2.2.1 Marine Drug Discovery

The targeted search for compounds with biological activity against human diseases began in the late 1960s, but structures of compounds with high potency and selectivity were not defined until the 1980s. Extensive funding by the United States' National Cancer Institute along with its commitment to collect MGR globally meant that the focus was on the treatment of cancer, using compounds mostly collected from shallow tropical reefs and derived from marine invertebrates (Thornburg et al. 2018). As a result, five out of the eight clinically approved drugs derived from MGR are treatments for cancer; the remaining three are treatments for neuropathic pain, *Herpes simplex* virus and hypertriglyceridemia (Table 4.1). Out of these, seven are derived from marine invertebrates and one is derived from an oily fish. Development of and approval for all of these took many years.

As is the case for most drugs derived from MGR, the issue of a sustainable supply of the raw material/compound needs to be addressed. Attempts to solve this have involved

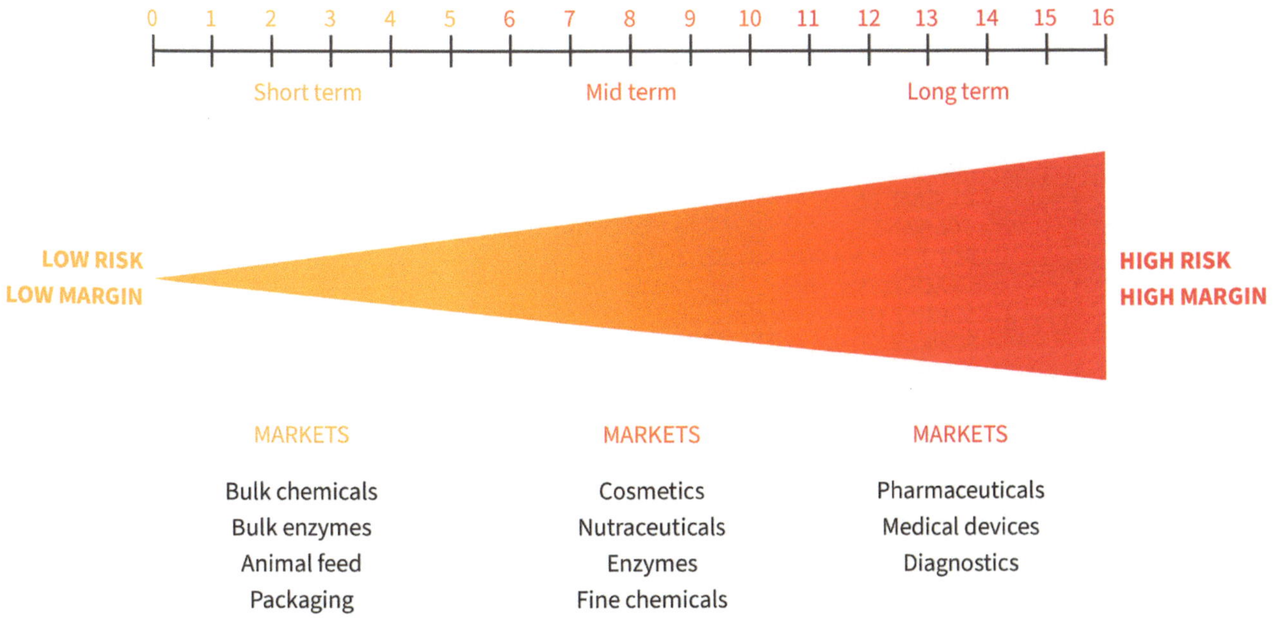

Fig. 4.4 Risk, profit margins and timelines for commercial activities associated with marine genetic resources. (*Source: Authors*)

several approaches, the most common being total chemical synthesis. Biotechnological approaches have also been employed, including hybrid synthetic/biotechnological approaches (Table 4.1). As an example, the case of Yondelis is described in Box 4.2. Two of the compounds in Table 4.1, Vidarabine and Cytarabine, now have second-generation analogues, Fludarabine (Fludara) and Nelarabine (Arranon), respectively (Alves et al. 2018). The European Medicines Agency has approved some over-the-counter medications based on MGR, such as Carragelose, a broadly effective antiviral drug that can be used to treat respiratory viruses such as the common cold (Alves et al. 2018). Currently, 28 marine-derived products are in clinical trials with a further 250 in preclinical investigation, all from around 33,000 reported marine natural products (MarinLit 2020). This is an astounding success rate when compared with terrestrial natural products. This success may be due in part to the vast taxonomic diversity in marine environments. For sessile marine invertebrates, the lack of an evolved immune system, combined with the pressures of preventing predation and competing for space and resources, may have led to the evolution of a chemical arsenal for survival.

Because of the supply issue associated with marine invertebrate-derived pharmaceutical candidates, the marine natural product research community has also focused on investigating marine microorganisms as sources of bioactive compounds. The long time lag between discovery and development (see Fig. 4.4) means that most of these are still under preclinical investigation, with a smattering of microbial compounds in human clinical trials (Mayer et al. 2017) and many more at preclinical stages of development. The ability to sequence genomes quickly and cheaply coupled with bioinformatics tools—such as antiSMASH, which enables the rapid identification of secondary metabolite gene clusters in bacteria and fungi—often renders the inherent capacity for microorganisms to produce chemicals predictable even before testing begins (Medema et al. 2011). Challenges remain when genes are of completely unknown function, as is the case for many marine viruses. Advances in chemoinformatics, such as Global Natural Products Social Molecular Networking, allow scientists to verify this latent talent, massively speeding up the biodiscovery process. Finally, advances in assay technology mean we use less material in bioassays while obtaining better quality data with higher information content (e.g. Caicedo et al. 2017). Compound isolation and structure determination, the final stages of the biodiscovery process that were previously a bottleneck, have also improved over the last decade (Chhetri et al. 2018). Much development is also focused on finding secure methods other than chemical synthesis to reliably and sustainably generate and modify bioactive compounds, using, for instance, synthetic biology (e.g. for the plant-derived natural product artemisinin, see Paddon and Keasling 2014) and enzymes in synthesis (e.g. for chemoenzymatic synthesis of cyanobactins, see Houssen et al. 2014).

Table 4.1 Marine-derived compounds currently in clinical use

Plitidepsin	TRADEMARK YEAR	Aplidin® (2018)
	MARINE ORGANISM	Tunicate
	SOURCE	Mediterranean
	CHEMICAL CLASS	Depsipeptide
	MOLECULAR TARGET	eEF1A2
	DISEASE AREA	Cancer: multiple myeloma, leukaemia, lymphoma
	COMPANY	Pharmamar
	ROUTE OF MANUFACTURE	Synthesis
Trabectedin (ET-743)	TRADEMARK YEAR	Yondelis® (2015)
	MARINE ORGANISM	Tunicate
	SOURCE	Caribbean
	CHEMICAL CLASS	Alkaloid
	MOLECULAR TARGET	Minor groove of DNA
	DISEASE AREA	Cancer: soft tissue sarcoma, ovarian
	COMPANY	Pharmamar
	ROUTE OF MANUFACTURE	Semi-synthesis
Brentuximab vedotin (SGN-35)	TRADEMARK YEAR	Adcetris® (2011)
	MARINE ORGANISM	Mollusk/cyanobacterium
	SOURCE	Mauritius
	CHEMICAL CLASS	ADC (MMAE)
	MOLECULAR TARGET	CD30 and microtubules
	DISEASE AREA	Cancer: anaplastic large T-cell systemic malignant lymphoma, Hodgkin's disease
	COMPANY	Seattle Genetics
	ROUTE OF MANUFACTURE	Synthesis/biotechnology
Eribulin Mesylate (E7389)	TRADEMARK YEAR	Halaven® (2010)
	MARINE ORGANISM	Sponge
	SOURCE	Japan
	CHEMICAL CLASS	Macrolide
	MOLECULAR TARGET	Microtubules
	DISEASE AREA	Cancer: metastatic breast cancer
	COMPANY	Eisai Inc.
	ROUTE OF MANUFACTURE	Synthesis
Omega-3-acid ethyl esters	TRADEMARK YEAR	Lovaza® (2004)
	MARINE ORGANISM	Fish
	SOURCE	Undisclosed but manufactured in the United States
	CHEMICAL CLASS	Omega-3 fatty acids
	MOLECULAR TARGET	Triglyceride-synthesising enzymes
	DISEASE AREA	Hypertriglyceridemia
	COMPANY	GlaxoSmithKline
	ROUTE OF MANUFACTURE	Refined from fish oils
Ziconotide	TRADEMARK YEAR	Prialt® (2004)
	MARINE ORGANISM	Cone snail
	SOURCE	Philippines
	CHEMICAL CLASS	Peptide
	MOLECULAR TARGET	N-Type calcium channel
Ziconotide	DISEASE AREA	Pain: severe chronic pain
	COMPANY	Jazz Pharmaceuticals
	ROUTE OF MANUFACTURE	Biotechnology
Vidarabine (Ara-A)	TRADEMARK YEAR	Vira-A® (1976)
	MARINE ORGANISM	Sponge
	SOURCE	United States
	CHEMICAL CLASS	Nucleoside
	MOLECULAR TARGET	Viral DNA polymerase
	DISEASE AREA	Antiviral: herpes simplex virus
	COMPANY	Mochida Pharmaceutical Co.
	ROUTE OF MANUFACTURE	Synthesis

Table 4.1 (continued)

Cytarabine (Ara-C)	TRADEMARK YEAR	Cytosar-U® (1969)
	MARINE ORGANISM	Sponge
	SOURCE	United States
	CHEMICAL CLASS	Nucleoside
	MOLECULAR TARGET	DNA polymerase
	DISEASE AREA	Cancer: leukaemia
	COMPANY	Pfizer
	ROUTE OF MANUFACTURE	Synthesis

Source: Adapted from Midwestern University. n.d. "Approved Marine Drugs." https://www.midwestern.edu/departments/marinepharmacology/clinical-pipeline.xml

Despite these developments, there is a lagging interest from major pharmaceutical companies to explore marine and terrestrial natural products as potential sources of new leads. Most large pharmaceutical companies have closed their natural product discovery sections, while small and medium-sized companies are filling this gap and leading the way in the development of innovative new treatments using MGR. Large pharmaceutical companies will often buy small companies that have developed potential treatments to a certain stage of development, thus reducing their own risk while gaining access to the most recent innovations. The redefinition of the industrial landscape and the development of new tools and processes to investigate and develop MGR-derived bioactive compounds is thus critical for realising the overall potential of MGR for pharmaceutical discovery.

However, the benefits of marine biodiscovery extend far beyond the successful development of a product. Acknowledging the potential commercial value of biodiversity may lead to better funding for biodiversity surveys that access a broad range of marine life and assess these for bioactivity, which may lead to improved biodiversity conservation measures (Van Soest et al. 2012, see Sect. 4.1.3). A study carried out by the UN on the collaboration between Griffith University in Queensland, Australia, and the large pharmaceutical company AstraZeneca clearly articulates the regional benefits of engaging in biodiscovery research (Laird et al. 2008). These benefits include the availability of biorepositories of local species for further investigation; access to sophisticated bioassay and analytical equipment; the availability of highly skilled researchers and expertise; an improved publication profile and an enhanced research reputation. All of these together can boost the capacity of a region to thrive via multiple medical and biotechnological industries while contributing to the protection and sustainable use of biodiversity itself.

Box 4.2 Development of the Anticancer Agent Yondelis (Trabectedin)

The discovery of the active pharmaceutical ingredient in Yondelis, Ecteinascidin-743, from the Caribbean ascidian (seasquirt) *Ecteinascidia turbinata*, was first reported by two research groups in 1990. It was shown to have antineoplastic activity in cell-based and animal models, being particularly effective against soft tissue sarcoma, for which no good treatment options existed at that time. It was shown to have a unique mechanism of action, interfering with DNA (deoxyribonucleic acid) transcription by binding to the minor groove of DNA, which together with the new structure offered a strong commercial outlook. It was licensed to the Spanish company PharmaMar, which started the development process in the early 1990s. Initially, material was produced by aquaculture (Fig. 4.5, photo a), but this avenue was abandoned due to variability in production coupled with low yields, contamination issues and the high cost of infrastructure, among other reasons. Nevertheless, much of the clinical data were obtained using this aquaculture-derived material. To ensure a continuity of supply as well as quality of material, a semi-synthetic process was developed, modifying the fermentation product cyanosafracin-B to produce Yondelis economically. In 2007, the European Medicines Agency approved the use of Yondelis for advanced soft tissue sarcoma, but it took a further 8 years for the U.S. Food and Drug Administration to follow suit (Fig. 4.5, photo b). A combination treatment of Yondelis/Doxil is also being investigated as a second- and third-line treatment for ovarian cancer.

Fig. 4.5 Successful marine drug development. (**a**) The product packaging for Yondelis (PharmaMar). (**b**) The Caribbean ascidian (seasquirt) *Ecteinascidia turbinata* in aquaculture. (*Source: Text and photo* ***b****: used with permission from PharmaMar; photo* ***a****: S. Nash, Flickr*)

2.2.2 Nutraceuticals

The original definition of nutraceuticals, or functional foods, was given as 'food, or parts of a food, that provide medical or health benefits, including the prevention and treatment of disease' (Mannion 1998). Regulation for nutraceuticals is currently changing, with stricter rules being developed in many jurisdictions to prevent unrealistic claims of possible benefits. Marine resources have a huge nutraceutical potential (Bonfanti et al. 2018; Hill and Fenical 2010; Suleria et al. 2015). Indeed, due to their genomic diversity, they comprise a very wide range of enzymes and, as a consequence, of metabolic pathways. These in turn yield an extreme diversity of bioactive compounds with possible positive effects on health and well-being. These compounds encompass specific oligo- and polysaccharides; fatty acids and more complex lipids; proteins (including enzymes); and peptides, vitamins, minerals, phenolic substances, carotenoids, halogenated compounds and many others (Suleria et al. 2015).

Omega-3 fatty acids—EPA (eicosapentaenoic acid) and DHA (docosahexaenoic acid)—are of particular importance to the overlapping areas of nutrition and highly bioactive substances. Microalgae (as well as fish and some crustaceans, due to their consumption of microalgae) are a rich source of these polyunsaturated fatty acids (Rincón-Cervera et al. 2019; Ryckebosch et al. 2012), which are known for their positive health effects on inflammatory conditions (more importantly, EPA), cardiovascular disease (see Table 4.1 for examples of highly purified fish oils approved for clinical use, though the health benefits need confirmation; Manson et al. 2019) and neurocognitive development and health (DHA) (Echeverría et al. 2017). As a result, fish, crustacean and algal oils are deemed the best sources of EPA and DHA. Algal oils are a more recent development and a response to the overexploitation of fish resources. Microalgae may be produced under controlled conditions and in large quantities, and are rich in lipids (Rodolfi et al. 2009). Thraustochytrids, large-celled marine heterokonts classified as oleaginous microorganisms, are an important example of algal sources for their lipid productivity and particular richness in EPA and DHA (Gupta et al. 2012). The technological advantages of phototrophic microalgae have fostered research that has led to a growing number of applications. While the global supply of fish oil has stabilised at around one million metric tonnes every year and is constrained by overexploitation, phototrophic microalgae can be cultivated using renewable resources such as sunlight, carbon dioxide and cheap and plentiful nutrient sources (Chauton et al. 2015).

The potential for lipid production and the ability to produce EPA and DHA differ across microalgal species (Chauton et al. 2015). Strains with a desirable fatty acid profile can be obtained by selection (Rodolfi et al. 2009). For more ambitious targets, microalgae strains can be modified by inserting genes that enhance EPA and DHA synthesis or, alternatively, by silencing gene expression in competing metabolic routes (Mühlroth et al. 2013). Provided that environmental and social risks as well as ethical and safety concerns are fully addressed, such transgenic approaches may become solutions for the production of these invaluable marine nutrients, especially if new genes and corresponding enzymes from not yet prospected resources are drawn into a rigorous research and development effort.

2.2.3 Cosmetics

Cosmetics are big business—worth US $532 billion worldwide in 2017, growing at around 7.14% annually and expected to reach above $800 billion by 2023 (Orbis 2018). There is growing interest in naturally sourced products, including those derived from marine biodiversity. Additionally, products that show verifiable effects such as reducing wrinkles and protecting skin from the damaging effects of ultraviolet or infrared radiation attract a price premium and are placed at the high end of the market; these are often referred to as cosmeceuticals. The first true marine cosmeceutical, formulated in Estée Lauder's Resilience line, is a mixture of pseudopterosins derived from the Caribbean gorgonian (seawhip) *Pseudopterogorgia elisabethae*. These compounds were originally discovered as potent anti-inflammatory agents, but their physical properties meant they were unsuitable for systemic administration and they were therefore used topically. The material for the Resilience products is derived from environmentally managed seawhip farms, with population-level effects of the harvest studied in detail (Lasker 2013).

Two cosmeceuticals derived from vent bacteria have been commercialised, Abyssine 657 (Meyer/L'Oreal) and Venuceane (Sederma/Croda). The active product in Abyssine, Deepsane, is an anti-inflammatory polysaccharide obtained from a deep-sea bacterium *Alteromonas macleodi*, which is isolated from an annelid worm collected from a hydrothermal vent in the East Pacific Rise at 2625 metres (m) depth (in ABNJ) (Rogers et al. 2015). Venuceane, a product marketed as anti-ageing, detoxifying and moisturising, screens damaging infrared radiation and is derived from another hyperthermophile bacterium, *Thermus thermophilus*, obtained at 2000 m depth in the Guaymas Basin in the Gulf of California (Marteinsson et al. 1999). It has also been shown to screen ultraviolet radiation to prevent radical damage of DNA, thus protecting skin.

2.2.4 Aquaculture and New Food Products

Whereas marine aquaculture, developed originally in Egypt, spans 4000 years (Duarte et al. 2007), industrial aquaculture was initiated 40 years ago with the development of mussel raft aquaculture and fish aquaculture, along with the closing of the life cycle of salmon in captivity. Controlled food production from land organisms predates aquaculture by about 10,000 years, yet the number of marine species that have already been domesticated (about 270) matches that on land (about 294) (Duarte et al. 2007). Moreover, the domestication of new land species for food has remained nearly stagnant for the past two centuries, while about one-third of new marine species were domesticated in the past decade. The number of domesticated marine species continues to grow at a pace of about ten new species introduced to marine aquaculture every year (Duarte et al. 2007). The spider crab (*Maja brachydactyla*) (Pazos et al. 2018) and the common octopus (*Octopus vulgaris*) (Cerezo Valverde et al. 2019) are examples of two species domesticated in the past 2 years. Indeed, there is significant potential to domesticate all 3000 species harvested from the ocean as human food (Duarte et al. 2007).

Land species typically require a long selection process to achieve suitability for farming. On the other hand, the existing genetic diversity of marine species means that many mariculture-suitable species already exist (though selection often occurs when those growing these marine species select for specific traits—e.g. faster growth or better color). While natural and cultured populations of South African abalone (*Haliotis midae*) register similar levels of genetic diversity, cultured populations are genetically distinct from wild abalone, potentially as a result of selective pressures particular to each mariculture facility (Rhode et al. 2012). These findings highlight the need to maintain genetically diverse natural populations to support the mariculture industry, and to make provisions to ensure that commercially grown abalone are not released, accidentally or otherwise, into natural systems, as the latter poses a serious risk to the genetic integrity of an already vulnerable stock (Rhode et al. 2012; Bester-van der Merwe et al. 2011). Moreover, ongoing genetic monitoring is required for these species to maintain the genetic integrity of wild populations and to prevent genetic erosion, especially with the ongoing and largely uncontrolled release of cultivated organisms to the wild (da Silva and van Vuuren 2019). Thus far, only one aquaculture species, salmon, has been genetically modified for production (Waltz 2017), while on land genetically modified crops that are commercialised include maize, soya, cotton and canola, among others (Abberton et al. 2016).

Aquaculture now supplies almost half of the fish consumed worldwide (Troell et al. 2014), releasing some pressure on wild stocks. Yet sustainability within the sector and issues of genetic diversity within the industry will need to be addressed more comprehensively given projected expansions, linked to increased demand (Oyinlola et al. 2018).

The advantages of MGR also extend to new food products. Indeed, poorly known resources may yet be evaluated for their nutritional value and become subject to exploitation, which may then lead to cultivation. Some MGR may also provide novel functional food ingredients (Shahidi and Ambigaipalan 2015). These may encompass chitosans, specific carbohydrates, enzymes and protein hydrolysates given their ability to confer new properties to foods (e.g. altering their texture) or extend their shelf life (e.g. protein hydrolysates) (Shahidi and Ambigaipalan 2015). Recent developments in nanotechnology also bring new possible applications, such as the preparation of biogenic nanoparticles of marine algae for antioxidant and stabilisation effects on food matrices through active packaging (Gu et al. 2018; He et al. 2019). In addition, marine microbial enzymes encompassing agarases, cellulases, collagenases, lipases and proteases display valuable properties and offer various applications. The biochemical diversity of marine microorganisms makes these enzymes possible tools for food processing (Beygmoradi and Homaei 2017).

Considerable research is still needed to explore the use of MGR for engineering new foods. Indeed, only a little more than 40 fish species and even fewer in other marine taxonomic groups have had their genomes fully sequenced (Zhu and Ge 2018). These sequenced genomes and all genetic engineering tools, including recently available gene editing techniques such as CRISPR, may pave the way for a new generation of cultured seafood products (Zhu and Ge 2018), although questions of consumer acceptability, environmental risk and social desirability remain paramount. At present, genetic transformation of fish is mainly directed toward individual growth enhancement to increase the economic advantages of aquaculture. Atlantic salmon (*Salmo salar*) is the species that has been most targeted by genetic engineering efforts (Hafsa et al. 2016). For example, a transgenic Atlantic salmon (AquAdvantage)—recently approved by the U.S. Food and Drug Administration (Ledford 2015) after a 20-year review process—has a gene construct consisting of growth hormone cDNA (complementary DNA) from Chinook salmon (*Oncorhynchus tshawytscha*) that is regulated with anti-freeze protein gene sequences obtained from an ocean pout (*Zoarces americanus*), leading to growth rates that are much higher than those of non-transgenic salmon, with fish reaching market size in 16–18 months instead of 3 years (Smith et al. 2010; Waltz 2016). Questions remain, however, about the overall impacts of such enterprises, given that carnivorous fish such as salmonids and Asian bass still require significant quantities of fishmeal and fish oil in their pelleted diets. Limited attention has been given in aquaculture to seaweeds and lower-trophic-level organisms such as bivalves, which might offer more sustainable targets for aquaculture and might also bring additional benefits such as removing nutrients that cause eutrophication or particulates in seawater (Aburto-Oropeza et al. 2020).

2.2.5 Bulk Chemicals

MGR-derived products and processes could make a big impact in the bulk market, which includes bulk chemicals, enzymes for industrial processes and laundry detergents, probiotics in animal feed and packaging and further applications being researched to replace plasticisers in plastics with renewable resources. One of the largest markets is in alginates obtained from brown algae by wild harvest and aquaculture, which are used extensively as stabilisers and emulsifiers in food production as well as in specialty bandages for burns. Alginates are now being used to generate biodegradable drinks and food packaging, such as the Oohos produced by the Skipping Rocks Lab (Ooho Water n.d.). Its model is based around the product (e.g. ketchup) being put in the packaging at the retail outlet and being produced for that day's needs, as the material degrades in less than 6 weeks. Seaweed polymers are gaining attention as a source of sustainable bioplastics (Guedes et al. 2019) across a range of commercial applications, ranging from seaweed-based straws (Beygmoradi and Homaei 2017) to flip-flops (Algenesis Materials n.d.). The use of seaweed products as probiotics extends beyond human consumption, and with the 2006 banning of in-feed antibiotics given to animals in the European Union, using probiotics to prevent bacterial infections in livestock has been proposed as a sustainable solution. Sulfated polysaccharides prevent bacterial infections in pigs and other animals, thus reducing animal suffering and economic damage. Recent evidence also shows that the addition of ~1% red seaweed to the feed of ruminants reduces methane emissions by over 50% (Roque et al. 2019), thereby offering an opportunity to mitigate this significant component of global greenhouse gas emissions. However, concerns exist about the ozone-depleting properties of bromoform, a secondary metabolite produced by these seaweeds, if industrial-scale production for animal feed is pursued (Carpenter and Liss 2000).

The marine environment offers important opportunities for cold- and heat-adapted enzymes. The former is of utility in low temperature laundry detergents to reduce electrical costs during washing. One example of using heat-adapted enzymes in the bulk market is a thermostable enzyme from a hydrothermal vent organism that can be used in the production of bioethanol. Dubbed 'Fuelzyme' and licensed to the German chemical company BASF by Verenium, it is a genetically modified version of the original enzyme that is able to function over a wide temperature and pH range, thus improving the efficiency and economics of bioethanol production (Synthetic Biology Project n.d.).

2.2.6 Other Applications

Additional commercial applications of MGR relate to the capacity of certain marine microorganisms to produce extra cellular polymeric substances (EPSs), which are naturally occurring polymers. EPSs can be used as vehicles of bioremediation due to their capacity to detoxify heavy metals and other pollutants (Pal and Paul 2008). ArcticZymes, a developer and marketer of enzymes for highly specialised research applications, has developed a family of isothermal polymerases—enzymes of marine origin that can be used to synthesise DNA and RNA molecules under high salinity conditions and across a flexible temperature range (Ward 2018).

Fouling of ship hulls by marine plants and animals slows vessels and increases costs, while fouling of nuclear power plant cooling water intake by mussels and other species can compromise operations (Rittschof 2017). Antifoulants such as organotin have been banned by the International Maritime

Organization due to their broad toxicity and environmental impacts. There is therefore considerable interest in understanding the complexity of active and passive biological fouling processes and developing nontoxic, environmentally benign marine antifoulants. Research has focused on marine bacteria and antifoulant biomolecules, including at least 198 marine invertebrates such as gorgonians and soft corals, and over a dozen synthetic analogues with the capacity to adhere to a variety of substrates (Wang et al. 2017; Leary et al. 2009; Qi and Ma 2017).

The bioluminescence in a jellyfish discovered in the North Atlantic (*Aequorea victoria*) was found to be due to interactions between two proteins, namely aequorin and green fluorescent protein (GFP). A broad range of applications of GFP have emerged over the years, including as a reporter of gene expression, in the tagging (and subsequent photomicrography) of proteins and as a biosensor indicating levels of environmental toxicity. The scientists responsible for discovering GFP and developing its initial applications were recognised with the 2008 Nobel Prize in Chemistry, 1 of more than 20 Nobel Prizes linked to the ocean and ocean biology (Rogers 2019).

3 Challenges

3.1 Threats to Conserving the Ocean Genome

Human activities have been intensifying globally, threatening marine species' survival, contributing to the rapid loss of genetic diversity and weakening species' adaptive capacities (Laikre and Ryman 1996; Law 2007; Allendorf et al. 2008; Palkovacs et al. 2011; Jouffray et al. 2020). Fishing has significant negative impacts on marine ecosystems and biodiversity, which has implications for species extinction (Dulvy et al. 2014; de Mitcheson et al. 2013) and the reduction of genetic diversity or selection at specific *loci* (Pinsky and Palumbi 2014; Czorlich et al. 2018; Madduppa et al. 2018). Unsustainable coastal development, land- and sea-based pollution and growing interest in deep-sea exploration and mining constitute additional significant threats to biodiversity that often compound those from overharvesting (Devine et al. 2006; Prouty et al. 2011; Nielsen et al. 2016).

Climate change is leading to a warmer, more acidic and less oxygenated ocean, directly affecting all stages of marine life (Pörtner and Peck 2010) across all latitudes (Doney et al. 2012; Barton et al. 2016; Scheffers et al. 2016; Pratchett et al. 2018). Specific responses have included geographic distribution shifts to higher latitudes and deeper water, advances in spring phenology and increases in the abundance of warm-water species (Poloczanska et al. 2016). Climate change affects biodiversity through changes in the distribution of genetic variants in space and time, changes to the degree of phenotypic plasticity (the individual characteristics of organisms that result from interacting with the environment) as well as changes in the ability of organisms to adapt over time to changing environmental conditions (Hoffmann and Sgrò 2011). Realised climate change has already had substantial deleterious impacts across a range of biological processes and taxa, including critical habitat-forming species such as corals (Carpenter et al. 2008; Davidson et al. 2012; Spalding and Brown 2015; Ainsworth et al. 2016). Arguably, one of the most documented impacts of climate change has been a redistribution of species as they track their preferred environmental niches (Perry et al. 2005; Pinsky et al. 2013; Pecl et al. 2017; Morley et al. 2018). Such shifts are likely to be associated with differences in genetic variability between the historical and range extension zones as well as within the extended ranges themselves (Ramos et al. 2018). Critically, patterns in genetic diversity, connectivity and population size associated with species shifts are important determinants of whether species will be able to continue shifting, adapting, establishing and persisting in their new ranges (Ramos et al. 2018). Knowledge of how genetic variation is distributed across a species' range is of particular significance, as historic refuges often harbour a large proportion of total diversity (Hampe and Petit 2005), yet are also often threatened by climate change (Provan and Maggs 2012).

While the loss of certain marine species due to human impacts has been documented (see below), this is likely an underestimate as humans have been responsible for ecological, commercial and local extinctions (McCauley et al. 2015), and the substantial decline of genetic diversity within species and across populations (Chapin et al. 2000; Pinsky and Palumbi 2014). Loss of both types of variation in genetic diversity has pervasive impacts on ecosystem processes as well as on species' capacities to respond and adapt to change (McNaughton 1977; Allendorf et al. 2008; Grorud-Colvert et al. 2014). Continued loss of genetic diversity contributes to reduced population viability and ultimately can lead to extinction (Dawson et al. 2011).

Our activities have altered life in the ocean substantially, impacting the ability of ocean systems to provide ecological, socioeconomic and cultural benefits (Worm et al. 2006; Halpern et al. 2008). Such impacts have eroded the genetic base of biological diversity, and may make it more difficult to sustainably harvest and manage marine species (Walsh et al. 2006).

3.1.1 Species Extinctions

While extinction rates in the ocean currently appear far lower than species loss in the terrestrial realm (McCauley et al. 2015), species extirpations due to climate change are likely to be twice as common in the ocean as on land due to the narrow thermal range tolerated by marine species (Pinsky et al. 2019). Estimates of marine extinctions are likely to be conservative—little is known about how many species inhabit the marine environment and there is a lack of monitoring or specific assessments of extinction risk under the IUCN Red List. Heavy use of the maritime space by humans has led to dramatic declines in the abundance of the baiji river dolphin (*Lipotes vexillifer*) and the vaquita (*Phocoena sinus*), leading the former to be declared functionally extinct (Smith and Jabour 2018) and the latter to become the most endangered cetacean in the world as declared by the International Union for Conservation of Nature (IUCN) (Roche et al. 2016). IUCN has recorded 15 marine species extinctions, including the Caribbean monk seal (*Monachus tropicalis*), the Japanese sea lion (*Zalophus japonicus*) and the sea mink (*Neovison macrodon*) (IUCN 2019). Some species have not been observed for several decades and may be extinct. Based on available data, IUCN considers 25% of marine mammals at risk of extinction (Davidson et al. 2012). Eight percent of marine bony fishes from the Arabian/Persian Gulf are also considered regionally threatened due to fishing and loss of habitat—an estimate twice that of other regions where such assessments have been undertaken (Buchanan et al. 2019). In addition, 25% of sharks, rays and chimaeras are globally threatened (Dulvy et al. 2014). Smaller-size organisms may have a similar risk of extinction due to habitat destruction, introduction of invasive species, exploitation and the effects of climate change (Cowie et al. 2017). Yet census and extinction inventories are largely lacking for smaller marine species.

Many parts of the ocean remain unexplored (Van Dover 2014). For instance, scientific expeditions to the deep sea regularly encounter new species—a 3-week expedition off the coast of Costa Rica in early 2019 led to the discovery of at least 4 new species of deep-sea corals and 6 other animals (Schmidt Ocean Institute 2019). Commercial deep-sea mining activities may result in the loss of habitat, leading to potentially irreversible negative impacts on the biodiversity of vulnerable deep-sea communities (Van Dover et al. 2017). In June 2019, the scaly-foot snail (*Chrysomallon squamiferum*) became the first species at risk of extinction in the event of future deep-sea mining (two of the three hydrothermal vent systems where it is found are within areas under exploratory mining licenses), and it is expected to soon be joined by at least a dozen more hydrothermal vent species on the IUCN Red List (Sigwart et al. 2019).

3.1.2 Loss of Populations

Population extirpations and declines in abundance due to unsustainable fishing practices, habitat destruction and pollution have led to contractions in the ranges of many fish species—including large pelagics—and invertebrates (Musick et al. 2000; Hutchings and Reynolds 2004; Worm and Tittensor 2011). Salmon have suffered significant declines in numbers and now occur over a much smaller range than historically documented (Levin and Schiewe 2001). Several sockeye salmon (*Oncorhynchus nerka*) subpopulations are classified as extinct as a result of the construction of impassible dams throughout the Columbia River basin. Columbia River chinook salmon (*Oncorhynchus tshawytscha*) have been documented to have lost up to two-thirds of their genetic diversity (Johnson et al. 2018). Declines in population diversity have been shown to increase the variability in salmon returns (Hilborn et al. 2003; Schindler et al. 2010). Declines in the size or density of individual populations also result in greater fluctuations in the frequency of certain genotypes due to the loss of certain genes over time. This process, known as genetic drift, is magnified in smaller populations (Palstra and Ruzzante 2008)—or in larger populations with a reduced number of adults who can reproduce (Hauser et al. 2002; Hare et al. 2011).

In addition, as ecosystem connectivity decreases among marine populations due to habitat fragmentation as well as lower dispersal via ocean currents, which are projected to shift with climate change, a potential loss of populations is predicted, which could result in decreased genetic connectivity (i.e. through genetic drift leading to increased isolation by distance) (Hastings and Botsford 2006; Hellberg 2009; Gerber et al. 2014; Carr et al. 2017). Genetic drift and subpopulation losses both lead to declines in genetic diversity, in turn undermining a species' ability to recover, adapt and survive in changing conditions (Walsh et al. 2006; Hare et al. 2011). This is particularly critical as species face increasingly variable environmental conditions as a result of climate change; climate change itself is projected to have impacts on populations' and species' genetic diversities, further lowering their stress resistance and adaptive potentials (Frankham 2005).

3.1.3 Invasive Species

Aquaculture and shipping are two important means by which species are being translocated around the world, leading to a rise in invasive species. While the introduced species often do not survive, when they do, they may outcompete native species or prey on them, leading to cascading changes in native communities (Sorte et al. 2010; Green et al. 2012). While aquaculture is rapidly becoming a critical component to ensuring food security (Béné et al. 2016; Thilsted et al. 2016), the sector presents important concerns with regard to genetic diversity (Weir and Grant 2005). Aquaculture often breeds species (which are often introduced) by favouring certain traits that give them an advantage over native species in the wild (Fleming et al. 2002). While the environment in which cultured species are grown tends to be carefully contained and monitored, escape events do happen. Such events can lead to farmed species interbreeding with native species (genetic introgression) and rapid genetic homogenisation (Fleming et al. 2000), resulting in the irreversible reduction in genetic diversity and fitness of wild fish (McGinnity et al. 2003; Weir and Grant 2005; Waples et al. 2012; Glover et al. 2017)—and hence lowering their capacities to adapt to environmental change. Farming can also facilitate the spread of pathogens (Naylor et al. 2005), placing further pressure on stocks and posing a serious challenge to the management of farmed and wild populations (Karlsson et al. 2016).

3.1.4 Cumulative Effects

It is important to recognise that many marine species and communities are now under pressure from more than one direct or indirect human impact (Jouffray et al. 2020; Halpern et al. 2019). While species can be resilient to a single impact or even several, the additive or synergistic effects of multiple pressures or interactions between them can drive decreases in populations; affect spatial genetic structures and gene flow, including impacts on connectivity; and drive large-scale regime changes at the community level. Fishing, for instance, has led to rapid changes in growth and reproduction schedules (e.g. earlier maturation at a smaller size, smaller adult body size). And climate change—particularly changes in temperature and dissolved oxygen—is expected to have evolutionary consequences that are qualitatively similar to those observed from exploitation (Hutchings and Fraser 2008; Waples and Audzijonyte 2016; Czorlich et al. 2018; Duncan et al. 2019). One example of synergistic effects is the interactions between eutrophication, overfishing and invasive species in the Black Sea (Oguz and Velikova 2010). Another is between aquaculture—through the release of fingerlings and escapes of broodstock—and fishing of salmon, where both activities have reduced genetic variability of wild populations (Waples et al. 2012). Such effects can prove difficult to reverse (an ecological state called hysteresis) and can lead to the occurrence of new/alternative stable states in marine ecosystems (Fauchald 2010; Fung et al. 2011).

3.2 Impediments to the Equitable Use of the Ocean Genome

3.2.1 Impediments to Innovation, Equity and Benefit Sharing

Investments in marine biodiscovery are typically costly and risky due in part to the extreme expense of sampling in areas like the deep sea, the low chances of success and the significant regulatory hurdles for product approval (Broggiato et al. 2014; Morgera 2018). Moreover, each stage of the research, development and commercialisation process requires high levels of technical, financial and scientific investment, with costs depending on the form and ease of access, the type of technology required to collect the material and undertake the research, and the sector or envisaged product involved (Laird and Wynberg 2012). Equipment costs remain high, although the costs of molecular technologies have decreased considerably in recent decades, alongside an increase in speed, efficiency and capacity. Marine biotechnology remains a rapidly developing and fast-moving sector (Leary et al. 2009; Broggiato et al. 2014). The nature of the research enterprise is also changing, as research shifts toward bioinformatics—the collection, classification, storage and analysis of complex biological data—and the mining and exploration of these vast and growing datasets of genetic information, which requires advanced computational resources that are not broadly available (Muir et al. 2016).

The considerable costs involved in marine bioprospecting research, alongside the advanced technologies and expertise required, have meant that most exploration has been undertaken by high-income countries. Notably, these are the United States, United Kingdom, Australia, Canada, Japan, Germany and Russia—but, as Fig. 4.6 indicates, with the sampling often conducted in low- or middle-income tropical countries, and Australasia in particular (Greiber 2012; Leal et al. 2012; Oldham et al. 2013).

As an indicator, Fig. 4.7 illustrates the global distribution of research efforts focused on marine genetic resources, using scientific publications as a proxy. Similarly, studies of patents associated with marine genes demonstrate disparities in capacity to engage in commercial activities associated with these resources, although such studies do not distinguish whether the genes are for reference or are claimed in the filings. Arnaud-Haond et al. (2011) found that patents citing marine genes originated from only 31 of the 194 countries in the world, with 10 countries responsible for 90% of them. By 2017, this imbalance had grown, with the share of the top 10 countries increasing to 98%, and 70% filed by researchers or companies in the United States, Germany and Japan (Blasiak et al. 2018). Approximately 1600 patent sequences were derived from species associated with the deep sea and hydrothermal vent systems, commonly found in ABNJ, and are of particular relevance in the context of the ongoing BBNJ negotiations. Greater specification is hampered by the lack of a legal obligation to disclose sample origin or source in patent filings, and the tendency for applicants to not volunteer such information (Blasiak et al. 2019).

Other researchers emphasise that studies of patent filings actually highlight how limited commercial interest in MGR has been (Leary 2018). Blasiak et al. (2018), for instance, analysed 7.3 million sequences and identified only 12,998 of marine origin from 862 species. A text-mining analysis employing more liberal definitions of what constitutes a 'marine' species identified only 1464 marine species in the patent system (Oldham et al. 2013). However, while patents are an indication, they should not be taken as a proxy for the

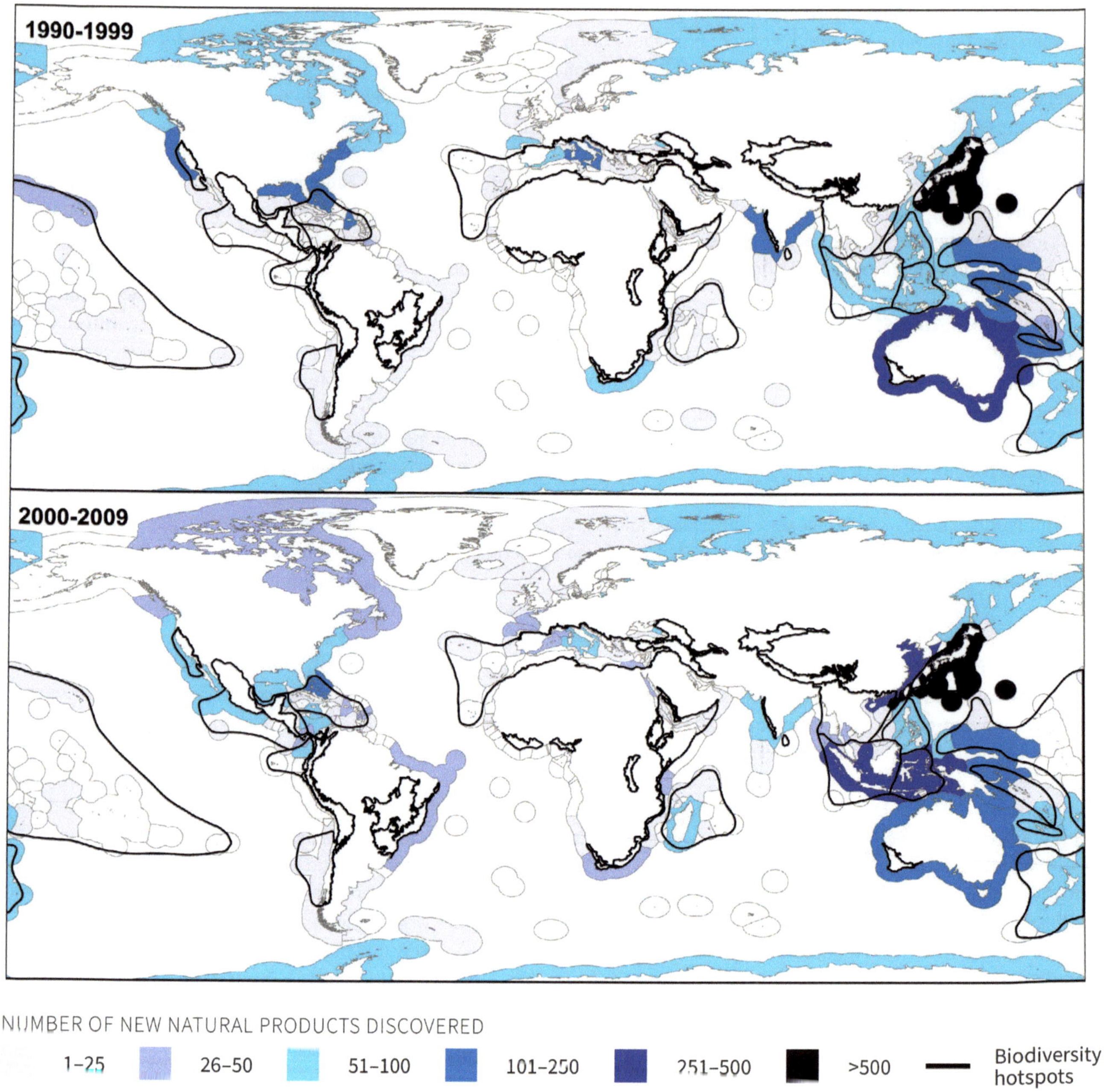

Fig. 4.6 Sources of natural products from marine invertebrates. *Note: This figure shows the number of new natural products from marine invertebrates found in exclusive economic zones during the 1990s and 2000s, as well as boundaries of biodiversity hotspots.* (*Source:* Leal et al. 2012)

full scale of mature commercial interest, or research and innovation that might be pre-competitive. Not all inventions are patentable, many that are patented will never be commercialised and there are strategies other than intellectual property to protect competitive or commercial advantages, including publication (Merges 2004; Thambisetty 2007; Herrera and Schroth 2000; Quah 2002).

Disparities in research capacity, technology and finances represent major constraints that prevent the inclusion of low- and middle-income countries in marine biotechnology efforts. Biodiversity and molecular expertise is unevenly spread (Hendriks and Duarte 2008); research vessels or submersibles are typically owned by only a few high-income nations and entail substantial operational costs (Stokstad 2018); and while there are growing numbers of collaborations between high-income and lower-income countries (Kyeremeh et al. 2020), the model of international collaboration is still characterised by a pharmaceutical or biotech company working with established centres of excellence located in high-income countries. As an example, despite active marine biodiscovery programmes in the Western Indian Ocean, with the exception of South Africa and, to a

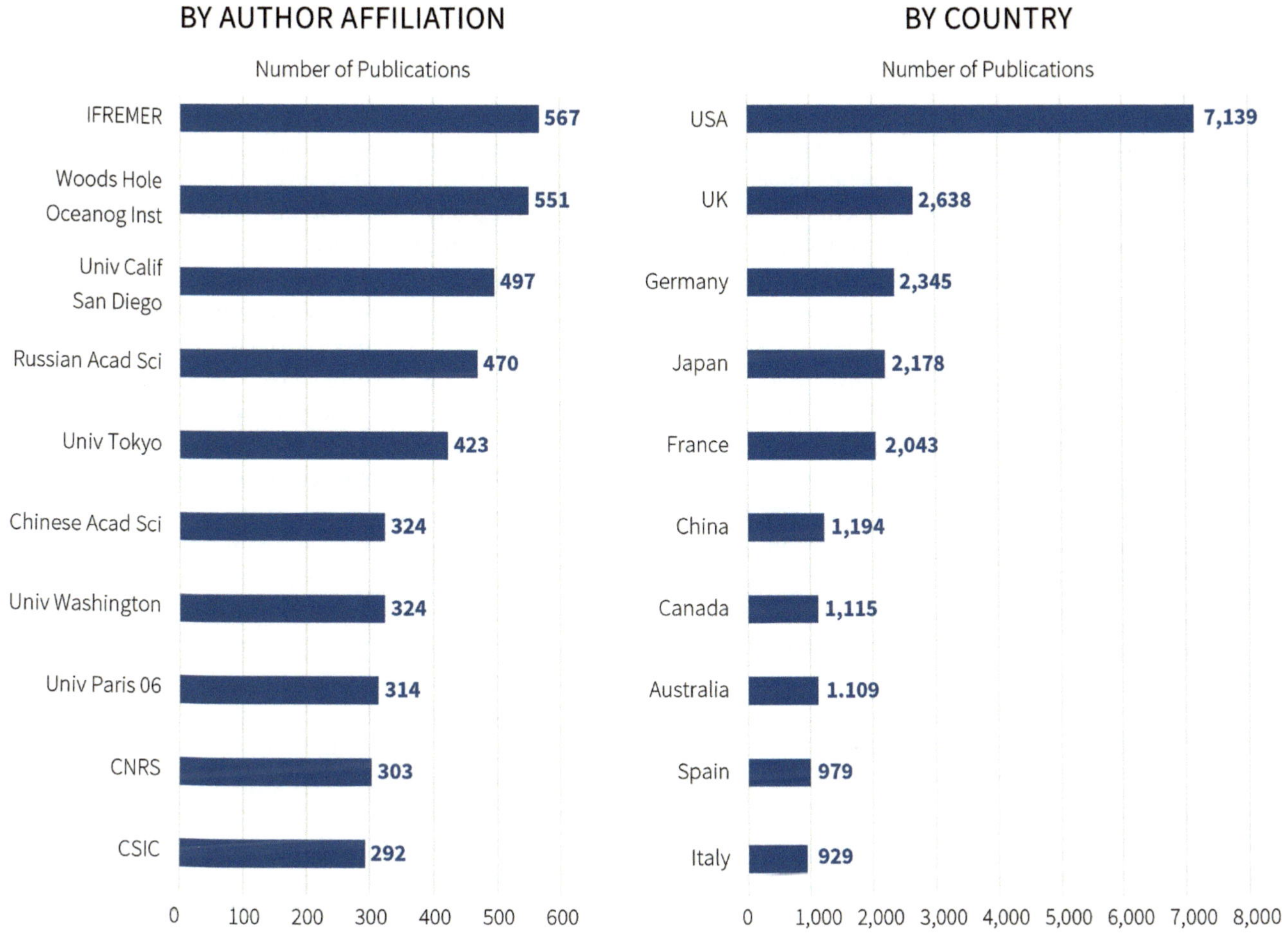

Fig. 4.7 Author and country affiliations for scientific literature focused on marine genetic resources. *Notes: The full names and locations of the authors' affiliate institutions include, from top to bottom, French Research Institute for Exploitation of the Sea, France; Woods Hole Oceanographic Institution, United States; University of California, San Diego, United States; Russian Academy of Sciences, Russia; University of Tokyo, Japan; Chinese Academy of Sciences, China; University of Washington, United States; University of Paris 06, France; French National Center for Scientific Research, France; Spanish National Research Council, Spain.* (*Source:* Oldham et al. 2014)

lesser extent, Kenya, few African countries have engaged actively as research collaborators in international endeavours (Wynberg 2016). A particular concern across countries is the gender imbalance in marine biotechnology (and science in general) and the attrition of women in this male-dominated field (Ceci and Williams 2011; Kitada et al. 2015).

3.2.2 Regulating Fair and Equitable Access and Benefit Sharing

The CBD, Nagoya Protocol and International Treaty on Plant Genetic Resources for Food and Agriculture together provide an important platform around which new models of equitable research partnerships can evolve, on the basis of the presence or absence of national sovereign rights over biological resources. As described earlier, marine biodiscovery depends in part upon access to marine organisms, which in turn is governed by multiple legal regimes and national and international laws (Fig. 4.3). Under UNCLOS, coastal states have the exclusive right to regulate, authorise and conduct marine scientific research in their territorial sea (Article 245). MGR found within the EEZ are subject to domestic measures implemented under the Nagoya Protocol or directly under the CBD. This means that coastal states that choose to regulate marine bioprospecting in their EEZ can specify conditions of access to this material, including mutually agreed terms on access and benefit sharing (ABS). As noted by Oldham et al. (2013), natural product research has historically concentrated on marine invertebrates *inside* national jurisdictions, with most marketed products derived from organisms found there—with limited exceptions for enzymes from extremophiles and Antarctic krill (*Euphausia superba*) as a source of nutraceuticals.

In practice, the CBD has spawned a number of approaches to regulating genetic resources, but a common element across these approaches is the requirement that researchers abide by local conditions of access to and use of genetic resources. The evolving nature of ABS governance—and negotiated compliance in different contexts and gaps in workable policies in many countries—makes this difficult terrain to navigate (Morgera 2018). From a legal perspective, perhaps the greatest current challenge is determining the full scope of the term 'genetic resources' and discussing whether this includes digital sequence information/genetic sequence data. For some countries, not incorporating genetic sequence data within the scope of ABS approaches undermines sovereign control over genetic resources. Other countries insist that the publication of sequence information in open-access databases can be seen as a globalised and important form of benefit sharing (Laird and Wynberg 2018).

Monitoring is an important concern given that informational resources are highly mobile and malleable and are more difficult to track than physical genetic resources, with most data held in databases typically lacking identification and origin information (Garrity et al. 2009). Several groups are working to improve monitoring by attaching information on origin to sequences, and by including stronger links between physical samples and sequences. But monitoring has grown increasingly difficult over time as sequences pass through multiple hands, are modified or have their identities eroded (Garrity et al. 2009; Slobodian et al. 2015). The technological gap described above is exacerbated by a failure to capture traceability in legal frameworks to support appropriate law and policy.

Specialised ABS rules for MGR from ABNJ have not yet been developed and it is one of the four main issues within the BBNJ negotiations, which are ongoing under the parent treaty, UNCLOS (Leary 2018; Thambisetty 2019, 2020). The negotiations should clarify the status of MGR found beyond national jurisdiction, including whether they are to be regarded as the common heritage of humankind and what implications that would have for the private appropriation of the ocean genome in tangible and informational forms.

Such discussions also extend to the scope of regulation of marine scientific research in ABNJ. Due to the open nature of the ocean, biogeographical ranges of marine species are typically large (including those of prokaryotes, which contribute the bulk of marine genes) with connectivity driven by very large population sizes and ocean transport systems resulting in large distributions (Villarino et al. 2018). Therefore, MGR are often shared among the EEZs of multiple nations and ABNJ, which renders delineation over different ownership and governance regimes cumbersome. Central questions include whether the benefits arising from the commercial use of these resources should be shared by the entire international community; the scope of the obligation on states and corporations with the technological capacities to exploit these resources to share benefits; and whether those who first locate and describe MGR should be given certain rights of priority.

A central issue—and one that is not confined to MGR—is the blurring between noncommercial and commercial research as the academic community and governments increasingly partner with industry, and patent laws change patterns of appropriability. Most sequences move fluidly between commercial and noncommercial institutions, and if uploaded to public databases might be available for all to use without the original providers aware of or involved in this process. Most benefit sharing under the Nagoya Protocol occurs through bilateral arrangements between users and providers who are obligated by local and international laws to enter into mutually agreed terms on benefit sharing, often when research moves from an academic to a commercial phase. The performance of contracts cannot easily be monitored by provider countries (Young and Tvedt 2017). Additionally, if scientific data and information were treated solely in a bilateral, benefit-sharing manner, countries would not benefit from information generated from non-endemic species, or from *ex situ* collections. Environmental management in particular benefits from increasing the quantity of available data.

Bilateralism presents other problems in the marine context given the challenges of delineating ownership. A multilateral mechanism such as that found in Article 10 of the Nagoya Protocol may become salient in the context of the BBNJ negotiations. The complexity of the regulatory environment demands fresh approaches that can help shape and negotiate ethical and responsible conduct on the part of marine scientists. Initiatives such as voluntary codes of conduct, good practices, training for younger scientists, funding incentives by research councils and grantmaking bodies, mentoring and other initiatives can speed up the process to internalise new behaviours and norms of research.

An important concern stems from the overregulation or poorly implemented application of ABS laws, especially given the blur between commercial and noncommercial use. Although the CBD and Nagoya Protocol explicitly support research for biodiversity conservation and enhanced scientific knowledge, national ABS legislation has often had unintended negative impacts on basic biodiversity research (Bockmann et al. 2018; Prathapan et al. 2018). It is important that new laws to regulate the use of MGR learn from these experiences to ensure that basic biodiversity research to support conservation efforts, the advancement of knowledge and equitable benefit sharing is promoted, rather than hindered.

4 Pursuing Solutions

4.1 Conservation

4.1.1 Managing Competing Interests in the Ocean to Conserve Biodiversity

Despite recognition by the CBD, genetic diversity is still largely neglected in policies and management and conservation plans (Laikre 2010). Much greater attention is needed to embed genetic diversity in policies, plans and programmes and to ensure that holistic strategies are developed to use the ocean sustainably and maintain the genetic diversity that underpins biodiversity and the benefits it provides (Karlsson et al. 2016). The distribution of those uses and benefits is of particular importance when considering how to manage the many interests and stakeholders at the table. In marine systems, there are opportunities for change via key tools, among them ecosystem-based approaches to fisheries management, spatial planning, effective quotas, MPAs, protecting and managing key marine biodiversity areas, reducing runoff pollution into the ocean and working closely with producers and consumers (IPBES 2019). The conservation of genetic diversity is embedded in all of the above.

The goal of 'conserving' genetic diversity can differ depending on the perspective of each stakeholder. What is more, what constitutes high biodiversity in an area may mean different things to different people, especially as baselines shift and successive generations consider increasingly degraded systems to be the norm. Different stakeholders will also have inherently different interests, yet may benefit from using the same approach to conservation. A representative from a biotech firm may be primarily interested in protecting the highest diversity of marine genes possible to discover and develop new products. An ocean manager may desire the same outcome, with an interest in conserving a diversity of species in the ecosystem to provide resilience and adaptive capacity to environmental change. Many conservation goals exist, encompassing different species with distinct distributions of genetic diversity and patterns of connectivity, yet there are also multiple management strategies that balance trade-offs with positive outcomes (Ingeman et al. 2019). If these strategies move forward from an agreed upon set of minimum conservation imperatives, multiple interests can be supported as long as incentives are in place to support participation (Lubchenco et al. 2016).

For example, North Atlantic right whales (*Eubalaena glacialis*) were historically decimated by whaling and have consistently declined after a brief increase in population size that peaked in 2010 (Corkeron et al. 2018). Anthropogenic impacts including historical whaling have reduced their abundance and the genetic variability within the small breeding population (Kraus et al. 2005). The current primary causes of mortality are ship strikes and entanglement in fishing gear—every individual North Atlantic right whale shows evidence of entanglement in fishing gear at some point in their life (Corkeron et al. 2018). If the minimum conservation imperative is to recover this species to a viable population level that prevents a further genetic bottleneck, emerging management strategies—such as real-time whale position data to prevent ship strikes and ropeless fishing gear that remains free of vertical lines until the time of retrieval—can still support multiple activities (Ingeman et al. 2019). Using real-time data also allows flexibility as management needs evolve.

This management flexibility can be critical as ecosystems change, due to both natural cycles and increasing anthropogenic impacts. At its core, adaptive management assumes that activities and regulations need to be recalibrated as changes in the system occur. Yet tighter feedback loops will be required to keep pace with the changing ocean and to acknowledge the impacts if genetic diversity is diminished (Ingeman et al. 2019). As opposed to predict-and-prescribe approaches—which require a thorough scientific understanding of the dynamics within a system to predict how that system will change—scenario planning can help identify a number of alternative strategies that could potentially arise within a system (Schindler and Hilborn 2015). Management appropriate to one scenario may shift to another, making it necessary to combine the range of conditions encompassed by these alternate scenarios with decision-making structures that are streamlined for faster responses (Ingeman et al. 2019). Governance structures must match the flexibility required for this approach with the use of impact assessments that account for biodiversity and at the appropriate scales needed to conserve genetic diversity, whether at the scale of ecosystems, species, populations or individual genes. As technologies such as eDNA evolve, and the understanding of genomics increases, it will become increasingly feasible to implement such requirements for genetic diversity.

Prioritising interventions to conserve biodiversity, and the underlying genetic diversity, requires taking a robust approach based on sound science and available data. Yet ocean genome data over space and time are largely lacking, even though this scientific information is critical for evaluating the status and future outlook for genetic diversity, such as for fisheries encompassing multiple populations or when protecting areas of particularly high biodiversity. In the absence of data, reasonable surrogates may serve as a proxy for genetic diversity (e.g. guild-level diversity or representation of species within given taxonomic families), yet these should be coupled with the incorporation of genetic monitoring into preexisting programmes, and the creation of targeted genetic monitoring programmes for species and areas of particular interest. Such activities must go beyond simply documenting what genetic material is where, and how it is being extracted and used, to also encompass the changes in this genetic diversity and the trends in those changes over time. This requires having a baseline understanding of the genetic variability of each species. Coupled modelling and empirical approaches will also be increasingly important.

However, waiting until comprehensive datasets are available before making interventions also runs the risk of losing rapidly deteriorating storehouses of genetic information due to the overharvesting of species and habitat degradation. Interventions are already proceeding and a precautionary approach is needed to stem the loss of marine genetic resources, including those that are not well protected by area-based management, such as pelagic species with large home ranges. Ecosystem-based fisheries management; reduction of gear impacts and bycatch; and consideration of species' life histories, population genetics and historical exploitation are all important aspects of sustainable fisheries management. As technological advancements enable the exploration and exploitation of new areas in the ocean, including the deep sea, permitting and extraction limits will need to ensure both sustainability of the resource as well as conservation of its ecosystem. Caution is needed when approving new or expanded uses of managed areas for extractive activities such as mining, particularly in areas where biodiversity is not well characterised or is potentially vulnerable. The potential loss of rare, threatened and endangered species and populations poses a serious risk of contributing to an overall loss in genetic diversity, and such populations require continual monitoring and conservation efforts to ensure their persistence. In addition, safeguarding areas of high biodiversity or those of particular importance to exploited species in fully or highly protected areas is a key strategy for protecting genetic diversity, both in the short and long terms, while scientific monitoring and evaluation keep pace with the rapidly changing ocean.

To meet the needs and uses of multiple actors, protected areas should be balanced with those set aside to support sustainable use for key services such as harvesting genes for product development by industry, or wilderness areas to protect pristine habitat that provides key ecosystem services for those actors (Schleicher et al. 2019; see also Österblom et al. 2020). Although conflicting uses can be balanced across ocean spaces in particular contexts, this is not always possible. Commercial activities are being carried out across the majority of the ocean, yet only 8% is set aside for biodiversity conservation, of which only 2.5% is fully or highly protected and implemented (Sala et al. 2018, updated via Marine Conservation Institute 2020). This falls significantly short of the targets to effectively protect 10% of the ocean by 2020 (see Box 4.3) while also leaving open the conversation around sustainability and conservation of marine genetic resources in the other 90% of the ocean, of which two-thirds is in ABNJ. This points to the urgent need to prioritise decisions toward biodiversity conservation, given the foundational role this plays both for ecosystem health and for the well-being of human and nonhuman species.

Many countries fail to explicitly address the genetic level of biodiversity in fisheries policy and legislation (Dulvy and Reynolds 2009). Therefore, in enacting strategies for conservation and sustainable use, genetic biodiversity should be integrated or mainstreamed into the planning and decision-making of multiple sectors that may impact and benefit from the ocean genome, including from species that are new to science (Manuel et al. 2016). This includes fisheries, mariculture, mining, shipping and marine biodiscovery. New approaches should help integrate genetic biodiversity into ocean use planning; environmental authorisations such as licenses, permits and registrations; and environmental management. Maintenance of genetic diversity needs more explicit consideration and planning in food systems policies and management, including for wild capture fisheries and mariculture. In addition to legal and policy instruments, industry collaboration is also needed to prevent genetic erosion, prevent and manage marine invasive species and increase the benefits from genetic diversity through inclusive and responsible research and innovation. Mainstreaming may also include strategies through which activities in production sectors may actually benefit biodiversity. For example, mariculture could relieve pressure on commonly harvested wild species if undertaken in a sustainable and responsible manner (FAO 2016).

4.1.2 Protecting Storehouses of Genetic Diversity

Genetic diversity in the ocean is important and needs to be conserved and managed to protect the resources it provides and the people it sustains. Many have embraced this imperative at the local, national, regional and international levels, as reflected in various commitments, goals and targets for biodiversity conservation (Grorud-Colvert et al. 2019). For example, the CBD's Aichi Target 11 and the United Nations' Sustainable Development Goal 14 call to conserve ocean areas 'through effectively and equitably managed, ecologically representative and well-connected systems of protected areas and other effective area-based conservation measures.' These protected areas—MPAs and OECMs—are central tools for protecting marine genetic diversity (Fig. 4.8). They have been rapidly growing in number and extent over the last few decades (Lubchenco and Grorud-Colvert 2015; Sala et al. 2018), but many are poorly enforced and the total area remains below global targets, far below what scientists have recommended, and is not representative of the full range of habitats and ecosystems.

Decades of data from scientific research conducted in hundreds of fully and highly protected MPAs around the globe show clear ecological trends (Sala and Giakoumi 2018). MPAs tend to lead to positive ecological outcomes and often result in social and cultural benefits if they are properly designed, managed and sustained to ensure that full protection is real and lasting (Gill et al. 2017; Giakoumi et al. 2018). Key to achieving these benefits is an open and transparent planning process that engages stakeholders representing diverse perspectives and that integrates science-based solutions (Ruiz-Frau et al. 2015; Twichell et al. 2018).

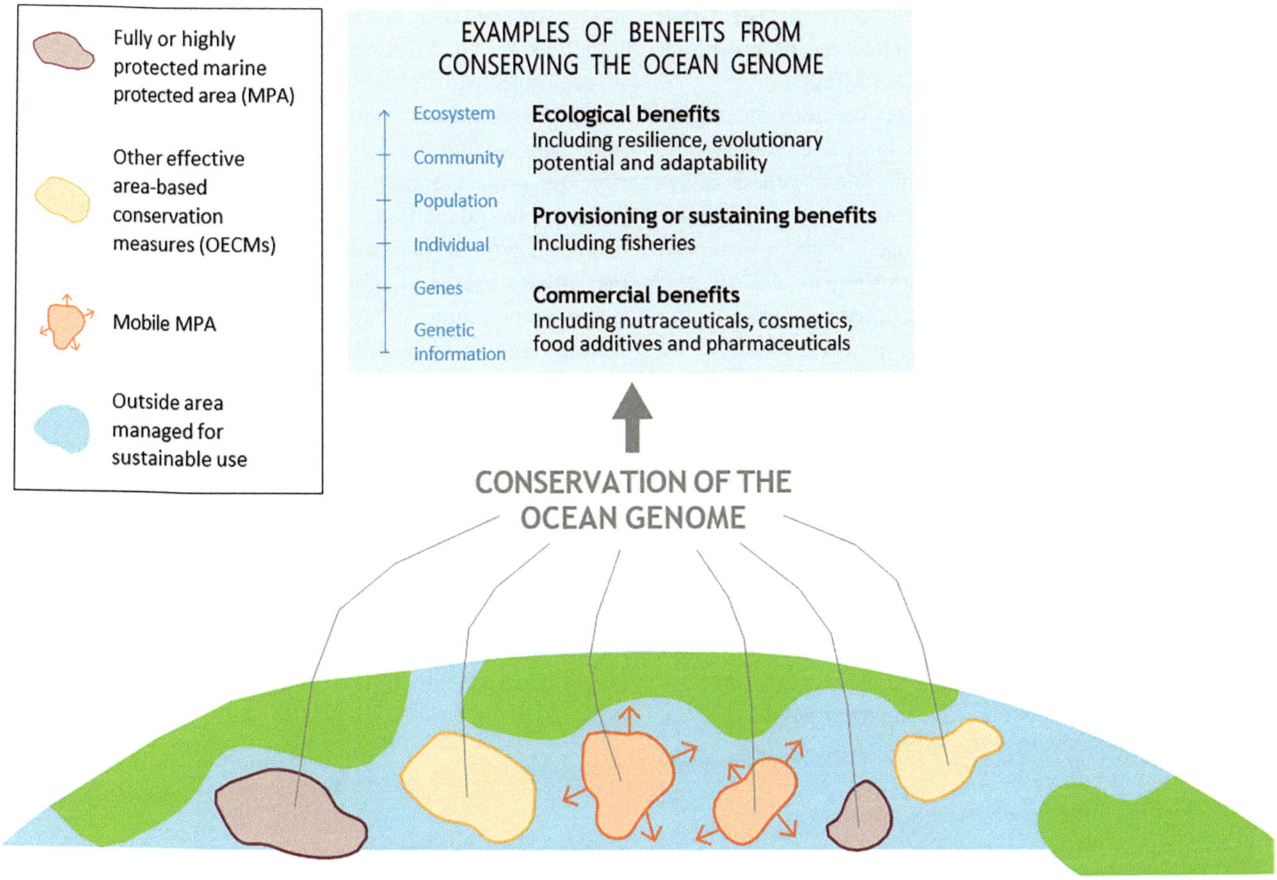

Fig. 4.8 How area-based conservation measures and sustainable use help conserve the ocean genome and its associated benefits. *Note: This figure depicts a portfolio approach for conserving the ocean genome and its associated benefits. Effective conservation hinges on using multiple tools, including area-based conservation measures such as fully and highly protected MPAs, that provide the greatest protection from the impacts of extractive and destructive activities. Coupling these with effective management of sustainable use can ensure wide-ranging benefits that are ecological, sustaining, provisioning and commercial. (Source: Developed by the authors. Designed by J. Lokrantz/Azote)*

When users are involved in MPA planning, compliance with regulations tends to be higher, boosting the ecological and social benefits (e.g. Viteri and Chávez 2007; Weeks and Jupiter 2013; Giakoumi et al. 2018).

When ecosystems, habitats and species are fully protected from all extractive and destructive activities within their borders, ecological communities tend to be more diverse, and formerly targeted individual species tend to be more numerous and larger in size, and have greater reproductive capacities and higher potential to move outside the MPA borders into areas beyond (Claudet et al. 2008; Halpern et al. 2009; Lester et al. 2009; PISCO and UNS 2016). When well-designed and managed, fully and highly protected MPAs result in greater abundance and size of previously exploited species, restoration of ecological interactions, habitat recovery, enhanced reproductive output due to larger body size of previously exploited species, greater resilience inside the MPA and higher potential for adaptation to climate and other environmental changes (e.g. Roberts et al. 2017; Hastings et al. 2017; Magris et al. 2018; Sala and Giakoumi 2018; Cheng et al. 2019). These ecological outcomes are also integrally tied to outcomes for human well-being. These can include income generated from tourism to fully and highly protected areas that preserve higher biodiversity and spectacular seascapes (e.g. Sala et al. 2013) as well as spillover from the MPA to augment catches in fished areas outside (e.g. Vandeperre et al. 2011). Fully protected areas tend to have more positive human well-being outcomes than MPAs with lower protection levels (Ban et al. 2019), provided key enabling conditions are met to ensure good governance, sound ecological and social design, and ongoing management. Fully and highly protected areas also provide reference areas for evaluating the impacts of extraction outside them, a buffer against accidental mismanagement or environmental changes, and often some enhancement of fisheries outside the MPA (e.g. Allison et al. 2003; McCook et al. 2009; De Leo and Micheli 2015; Di Lorenzo et al. 2016). Although the impacts of MPA networks on genetic diversity are implied and theoretically supported (e.g. Costello 2014; National Academies of Sciences, Engineering, and Medicine

2019; see also McInerney et al. 2012), there is an urgent need for research to test the potential outcomes of protecting genetic diversity in multiple, connected MPAs.

To effectively protect the ocean genome, fully and highly protected MPAs must be sufficiently large to encompass the relevant ecosystem and the full distribution of genetic diversity within it. Yet, in many contexts this is not possible, such as in the Mediterranean Sea, where use is high, coastal populations are dense and many countries share the sea's waters (Giakoumi et al. 2017). To support effective conservation while working within these realities, networks of MPAs are frequently used to protect multiple sites that are connected through the movement of adult or young marine organisms (Allison et al. 2003; Roberts et al. 2003). MPAs in a network can collectively encompass a large area and protect genetic diversity represented by different species while still allowing for sustainable use outside. A network also provides redundancy in the event that one MPA is impacted by a disturbance that reduces its ability to sufficiently protect the genetic diversity of the species inside. Networks of MPAs can have synergistic effects that lead to even greater ecological benefits than separate, unconnected MPAs that are not networked (Grorud-Colvert et al. 2014). When fully and highly protected, MPA networks provide a unique opportunity to protect storehouses of genetic diversity in a changing ocean. As organisms adapt to these changing conditions (see Sect. 2.1), adaptation networks can be established to identify and protect areas where genetic diversity and/or the potential for adaptation is high (Webster et al. 2017). For example, in coral reef systems, adaptation networks may be particularly useful as corals are increasingly threatened by rising temperatures, ocean acidification, pollution and overfishing (Hughes et al. 2018) while simultaneously showing quantifiably high rates of adaptation (Munday et al. 2013). A single species of coral can have a wide geographic range and inhabit different reef environments where genetic diversity is high across scales as small as less than 100 m (Barshis et al. 2013; Webster et al. 2017). These species may benefit from networks of protected areas that span different depths and allow for redistribution across latitudes. Future research should test the rate and limit of different adaptive responses for coral species across latitudes to better understand the ranges these adaptation networks need to encompass (Logan et al. 2014).

Box 4.3 Marine Protected Areas

One of the most effective tools to protect marine genetic diversity at an ecosystem scale is through implemented and fully or highly protected marine protected areas (MPAs). Because MPAs provide place-based protection, they can conserve not only target species and genetic material, but also all associated biodiversity within that habitat. International targets such as the Convention on Biological Diversity's Aichi Target 11 and the United Nations' Sustainable Development Goal 14.5 recognise the importance of using MPAs and other effective area-based conservation measures (OECMs) to protect biodiversity in 10% of the ocean by 2020. However, there are growing calls from the scientific community to fully or highly protect at least 30% of the ocean to achieve conservation goals, and for a corresponding post-2020 target to be formulated (see Sect. 1.3). But what exactly is an MPA or OECM? And which types are most effective for protection?

An MPA is a clearly defined geographical space, recognised, dedicated and managed through legal or other effective means to achieve the long-term conservation of nature with associated ecosystem services and cultural values.[a]

OECMs also provide conservation benefits, but biodiversity conservation is not their primary goal.[b] They are sites that are not by definition protected areas but are governed and managed in ways that achieve positive and sustained long-term outcomes for the in situ conservation of biodiversity, with associated ecosystem functions and services and, where applicable, cultural, spiritual, socioeconomic and other locally relevant values.[c]

MPA and OECM are broad terms that encompass many types of areas. MPAs are a focal tool for protecting genetic diversity because by definition these areas have biodiversity conservation as their primary goal. Yet MPAs can have different levels of protection and may be at different stages of establishment. *The MPA Guide*[d] provides a common language for describing the types of MPAs and the outcomes arising from areas with different protection levels.

Based on these definitions, 'fully and highly protected areas' are the only protection levels that are expected to deliver sufficient biodiversity conservation to protect genetic diversity.

In fully protected areas, no extractive or destructive activities are allowed, and all impacts are minimised.[e] In highly protected areas, only light extractive activities are allowed, and other impacts are minimised to the extent possible.[f] These may be stand-alone MPAs or fully and/or highly protected zones within multi-use MPAs.

Lightly or minimally protected areas allow for multiple uses and activities that have moderate to high impacts on species and habitats. Thus, these are not recommended for the goal of preserving genetic diversity within a system.

Further, for biodiversity conservation to occur, an MPA cannot be merely proposed or committed through an informal announcement; an MPA must be more than designated by law or other authoritative rule on paper. An MPA must be implemented—with regulations in force on the water such that users know to comply.[g] It is critical for public consultations and appropriate notification and transparency measures, as well as up-to-date scientific information, to become part of the designation and management of MPAs. Ideally, such areas should be actively managed with monitoring, enforcement and frequent review of management goals and outcomes.

Notes:
[a] *IUCN and WCPA* (2018)
[b] *CBD* (2018)
[c] *CBD* (2018)
[d] *Oregon State University et al.* (2019), *Grorud-Colvert et al.* (2019)
[e] *Oregon State University et al.* (2019)
[f] *Oregon State University et al.* (2019)
[g] *Oregon State University et al.* (2019)

By networking fully and highly protected areas of high diversity and ensuring connectivity among sites as well as sufficient spatial scale and ecosystem representation, it is possible to mitigate the risk of species moving outside the protected areas as their ranges shift in response to changing environmental conditions. Adaptation networks can also provide an insurance policy against ecosystem and species loss if they are sufficiently replicated within the system (Allison et al. 2003). MPAs in any effective network, including an adaptation network, should encompass a range of environmental conditions and habitat types—including both disturbed and pristine areas—that are sufficiently replicated. Networking can also accommodate different species distributions, as well as their underlying genetic diversity, by supporting species ranges and patterns of connectivity with multiple MPAs of varying sizes and distances from each other (Pujolar et al. 2013; Jonsson et al. 2016). Connectivity is particularly vital for ensuring adaptation pathways in a network (Almany et al. 2009; Blowes and Connolly 2012). Sites should be at appropriate sizes and distances from each other to promote the exchange of genes as young organisms disperse in the plankton or adults migrate out of the protected areas. Connected areas also provide sources of population replenishment within the network if one or more sites are compromised by local disturbances or become insufficient for protection due to shifting species ranges.

The existing coverage of MPAs should be continuously evaluated, especially in the case of MPAs functioning as a network, to identify areas where urgent protection of genetic diversity is needed. MPA planning processes should identify gaps, including areas of high genetic diversity that are currently unprotected and areas where highly variable systems have led to higher adaptation rates and possibly greater capacity for adaptation in the future.

The BBNJ process is now debating the declaration and functioning of MPAs in ABNJ as a tool for area-based management. There are divergent views on whether MPAs could be used to achieve long-term biodiversity conservation and sustainable use, and whether decision-making related to MPAs should be informed by strategic environmental assessments (SEAs) (High Seas Alliance 2019). This would include broader factors relating to social and economic considerations, traditional knowledge and cultural values. The management of ABNJ is not currently designed to protect genetic diversity, and MPAs could provide a mechanism to do so (Protected Planet 2020). These should be coupled with other facets of ecosystem-based management such as sustainable fisheries, habitat restoration efforts, pollution reduction and climate mitigation. Agreements on area-based management tools would in turn need to align with EIA and SEA processes under existing national, regional and international regimes.

4.1.3 Leveraging Biotechnology for Conservation and Biodiversity Management

Starting in the late 1970s, Sanger sequencing became the primary genetic technology employed to generate organisms' genetic information. Though it produces only a single DNA sequence for a given gene region (Sanger et al. 1977), it is still considered a highly valuable tool and is often used in wildlife biology, conservation and management. For example, it remains the gold standard in seafood surveillance and for identifying biological invasion pathways and sources of introductions (e.g. Roman and Darling 2007; Dlugosch and Parker 2008; Barbuto et al. 2010; Cawthorn et al. 2012; Di Pinto et al. 2013; Xiong et al. 2016; Tinacci et al. 2018). However, over the past two decades, key advances in molecular markers, new sequencing technologies and new statistical methods have enabled researchers to tackle a wider range of questions and issues to better inform species conservation and management.

Next-generation sequencing (NGS) offers improved resolution relative to early molecular markers and Sanger sequencing, provides high throughput and better enables large-scale spatial and temporal syntheses for single species as well as entire community assemblages through DNA metabarcoding (e.g. Taberlet et al. 2012; Lindeque et al. 2013; Aylagas et al. 2016; Pitz et al. 2017; Djurhuus et al. 2018). Moreover, because multiple regions across the genome can be sequenced using NGS, fewer samples are needed to acquire a wide breadth of the genetic diversity available within populations or species—a key benefit for studying marine taxa, which often occur in small numbers or are difficult to access (Xiong et al. 2016; Arulandhu et al. 2017).

With respect to seafood surveillance, NGS has proven effective at identifying multispecies seafood products (Giusti et al. 2017), and may even prove instrumental in identifying whether certified stocks have been swapped with uncertified stocks of the same species (Barendse et al. 2019). For marine invasions, using NGS with transcriptomic and epigenetic markers provides an unprecedented opportunity for identifying adaptive variation within and among native and nonindigenous populations, uncovering candidate genes responsible for certain adaptive traits and understanding the mechanism of epigenetic variation in plastic responses to novel environments (Sherman et al. 2016; Chan et al. 2017). Moreover, NGS coupled with environmental DNA can be used for early detection and monitoring of marine invasive species (e.g. Ardura et al. 2015; Carugati et al. 2015; Simmons et al. 2015; Zaiko et al. 2018), as well as the monitoring of rare, threatened and difficult-to-study or detect species (e.g. Bakker et al. 2017; Weltz et al. 2017; Boussarie et al. 2018; Pikitch 2018; Parsons et al. 2019).

Environmental DNA is a molecular approach that uses a passive sampling technique to acquire DNA from specific species or entire community assemblages. As species interact with their environments, their DNA is continuously being shed into their surroundings—be it soil, sediment or water—via their faeces, saliva, urine and skin cells (Baird and Hajibabaei 2012; Rees et al. 2014; Thomsen and Willerslev 2015; Deiner et al. 2016). As such, it is not necessary to have visual signs of the species under investigation, a requirement of more traditional sampling methods. The primary focus of eDNA has been to acquire species' presence and absence data to quantify their distributions, extents and connectivities (e.g. Weltz et al. 2017; Jeunen et al. 2019). Furthermore, given that several tens of species (from microbes to vertebrates) can be identified in a single sample, this technique can help identify areas of high species richness, which could prove instrumental in informing MPA design and ecosystem-level monitoring (e.g. Andruszkiewicz et al. 2017; Deiner et al. 2017; Pitz et al. 2017; Djurhuus et al. 2018; Stefanni et al. 2018). Moreover, eDNA has a very short life-span of hours or days in seawater, so analysis provides near real-time insight into the presence of species. The ability of eDNA to detect multiple species also holds great promise for rapid biodiscovery (Heidelberg et al. 2010; Chang and Brady 2012). However, the effectiveness of eDNA is fundamentally dependent on the availability of reference collections (e.g. in museums and aquaria) and a genetic reference library, which may not exist and may be difficult to create for elusive marine species. Recently, the focus of eDNA studies has evolved beyond simple presence/absence to studies quantifying the abundance of species (Stewart 2019), which holds great value for threatened and invasive species monitoring and response planning. Moreover, there is a growing body of research focused on quantifying population genetic structures from eDNA in marine species (e.g. Jeunen et al. 2019; Parsons et al. 2019).

The latest molecular technology with a potential conservation application is CRISPR. Considering the discovery of CRISPR as a genome-editing technology was only first reported in 2012 (Jinek et al. 2012), it is still very much in its infancy and its application in threatened species conservation has yet to be tested (Johnson et al. 2016; Piaggio et al. 2017; Phelps et al. 2019). Moreover, beyond unease about the manipulation of human germline cells, significant ethical and governance concerns remain about the use of the technology. Gaps in knowledge with regard to the environmental, social and economic impacts heighten such concerns, alongside fears about the stability of modified genomes (Caplan et al. 2015; Jasanoff et al. 2015; CSS et al. 2019). The interconnectivity of marine environments in particular underpins the importance of having full and adequate knowledge before moving forward with any applications.

Despite coral reefs being among the oldest ecosystems on Earth (Roark et al. 2009), many have suffered unprecedented losses. Although their decline is partly attributed to human-mediated disturbances such as land-based pollution, introductions of invasive species and overexploitation of coral reef ecosystems (e.g. Johannes 1975; Grigg and Dollar 1990; Wilkinson and Buddemeier 1994; Roberts 1995), the rapid decline is also likely linked to the rapid changes in the Earth's climate over the past century (National Academies of Sciences, Engineering, and Medicine 2019). Many coral populations may not have the capacity to adapt to these altered conditions. The plethora of benefits that coral populations provide (see Moberg and Folke 1999)—including as sources of medicine to treat various infections and diseases (e.g. Bruckner 2002)—underscores the importance of supporting their persistence and resilience. Gene editing could theoretically provide an opportunity to increase genetic diversity within populations to allow them to adapt to a changing environment, or permit selection of traits that may improve the resilience of coral populations and species (van Oppen et al. 2015; National Academies of Sciences, Engineering, and Medicine 2019).

The jury is still out as to the costs and benefits of CRISPR, and discussions of its usage are highly controversial. For instance, one proposition by Phelps et al. (2019) is to apply genome-editing in a manner that mirrors the threat level classifications of the IUCN, whereby CRISPR is used primarily as a means of slowing the rate of decline without altering the underlying genetic diversity of species with 'near threatened' or 'vulnerable' statuses (e.g. via genetic barcoding for enhanced monitoring of populations). For more threatened species where genetic erosion is evident (e.g. those that are categorised as 'critically endangered'), this would imply a focus on enhancing the adaptive capabilities of the species within its environment. In such cases, Phelps et al. (2019) propose making genetic modifications in the form of targeted beneficial mutations and gene replacements as potential tools for species survival. However, understanding the genetic underpinnings of these adaptations (e.g. via transcriptomics and epigenetics) is critical before any such steps are explored. For example, while CRISPR may be technically feasible to apply to corals, little knowledge exists regarding candidate genes on which it could operate to increase resilience, and whether it may translate to phenotypic changes (National Academies of Sciences, Engineering, and Medicine 2019). For corals, such work has already begun with differences in genome expression found between corals that were sensitive or resilient to thermal stress (Barshis et al. 2013; Palumbi 2014). From a broader perspective, arguments have also been made about addressing the root cause of the problem rather than relying on technological 'fixes' that might well go awry (CSS et al. 2019) (Box 4.4).

Box 4.4 South Africa Case Study: A Lack of Knowledge and Techniques Limits Our Ability to Assess the Risks to the Genetic Component of Marine Biodiversity

South Africa has an established community of biodiversity assessment and planning practitioners whose collective experience led them to establish spatial plans for ecological sustainability. A series of spatial biodiversity layers have been used to support the Marine Protected Area Expansion program, and the National Biodiversity Assessment (NBA), which is used to inform policies and management decisions. This allowed for systematic assessments of the state of biodiversity in 2004 and 2011. In addition to the statuses and trends of ecosystems and species, the 2018 NBA reports on the state of genetic diversity.[a]

From a genetics perspective, the general outcome was a clear lack of temporal genetic diversity datasets and indicators—a finding mirrored throughout the globe.

Although genetic studies have been conducted on several species, these data typically represent a snapshot of a species' genetic diversity and are applicable to only a limited portion of the species' range. Although still highly informative, the lack of a temporal component prevents the tracking of genetic changes and limits the assessment of genetic risks to marine biodiversity; however, efforts are underway to rectify this.

Within the past two decades, a strong baseline understanding of the spatial genetic patterns in various coastal species and offshore commercially exploited fish stocks has been established.[b] This work is primarily based on mitochondrial DNA and, to a lesser extent, microsatellite markers.[c] More recently, with the advent of NGS, research is being directed toward epigenetics[d] and genome-wide scanning of various coastal species to identify intraspecific variability and structure. Given the heterogeneous marine environment of South Africa, which spans a variety of ecological gradients (e.g. temperature, primary productivity, oxygen, salinity), such work is likely to provide insights into signals of local adaptation and population connectivity. In doing so, areas of evolutionary importance, persistence and resilience may be identified, which could inform marine spatial planning. Moreover, environmental DNA coupled with metabarcoding is assisting with large-scale foundational surveys to quantify the vast and mostly unexplored portions of the marine environment. These data can act as a baseline for more targeted monitoring and assist in amassing phylogenies on specific taxonomic groups for national-level monitoring.

Although single species are typically the focus of genetic monitoring studies, the ability to track genetic diversity across species for a given taxonomic group at a seascape or ecosystem level could greatly inform biodiversity planning at a national scale. South Africa is developing a Critical Biodiversity Area (CBA) map as a spatial plan for ecological sustainability, including identification of CBAs and Ecological Support Areas (ESAs), which together with protected areas are important for landscape and seascape functioning. To bring in the genetic diversity component to this planning process, work has already begun on intertidal chitons using phylogenetic diversity to help prioritise areas of high genetic diversity for marine spatial planning.[e] However, additional metrics should also be considered, such as phylogenetic endemism, evolutionary distinctiveness, and evolutionarily distinct and globally endangered. Each of these metrics can be useful for evaluating biodiversity under different scenarios, and the choice of metric depends on the conservation objectives.

A recent study used all four metrics to examine patterns of genetic diversity across South Africa for terrestrial reptiles.[f] Similar studies focusing on marine taxa would be of great value.

To help guide genetic monitoring research, South Africa is developing a National Genetic Diversity Monitoring Framework to ensure that comparable long-term datasets can be established and used to better inform biodiversity management. This framework will outline how to strategically prioritise taxa; identify the most appropriate genetic markers and metrics to use for national-, ecosystem- and species-level monitoring; and provide advice on the frequency of monitoring. It will also inform the spatial plan currently in revision to refine the boundaries of existing CBAs, ESAs and MPAs.

Notes:

[a] *da Silva and van Vuuren* (2019), *Skowno et al.* (2019)

[b] *E.g. von der Heyden et al.* (2007, 2010), *Henriques et al.* (2017)

[c] *von der Heyden* (2009), *Teske et al.* (2011), *Wright et al.* (2015)

[d] *Baldanzi et al.* (2017)

[e] *Volkmann et al.* (2014)

[f] *Tolley and Šmíd* (2019)

4.2 Toward Responsible and Inclusive Research and Innovation

Important conceptual approaches to take forward these ideas are responsible research and innovation (RRI) and inclusive innovation. RRI envisages a transparent, interactive process by which societal actors and innovators become mutually responsive to each other with a view to the ethical acceptability, sustainability and societal desirability of the innovation process and its marketable products (Von Schomberg 2013). As observed by Laird and Wynberg (2018), the CBD and ABS provisions of the Nagoya Protocol already encapsulate the principled basis of RRI, although by default rather than by design. Inclusive innovation is an alternative, and perhaps a more contextually appropriate, framing of RRI. It explicitly includes those who have been excluded from the development mainstream (Foster and Heeks 2013), and refers to the production and delivery of innovative solutions to the problems of the poorest and most marginalised communities (Heeks et al. 2013).

For example, the extent to which MGR are used to treat neglected diseases has not been as prominent as the search for treatments for cancer where the direction of research has been influenced by major funders such as the U.S. National Cancer Institute (Mayer et al. 2017). However, the funding landscape seems to be changing due to, among other things, the growing prominence of philanthropic organisations. The potential for philanthropy to help fill gaps left by a lack of focus from national science programmes or demand from the market is one of several positive contributions to ocean science: A growing fleet of research vessels are operated with philanthropic support, and some are offering access to scientists from developing countries. Yet a lack of coordination as well as a tendency for philanthropies to have narrow missions suggests the potential for more added value if efforts were aligned with global agendas such as the UN Decade of Ocean Science or the Sustainable Development Goals (SDGs). Where appropriate, these efforts could also be aligned with national initiatives such as the United Kingdom's Global Challenges Research Fund, which aims at a more inclusive approach to meeting the needs of developing countries in a range of areas, including through efforts to discover novel pharmaceuticals for neglected diseases. There is a clear need to forge more equitable research partnerships between industrialised and developing countries—and between users and providers of MGR—centred on scientific capacity, technology transfer and adequate finance. But it is also important to look at new models of partnerships driven by scientific advances that are changing the way researchers work. These are enabling the creation of dynamic knowledge hubs, and diffuse scientific collaborations, with increasing reliance on data and information (Broggiato et al. 2014). As marine genomics increasingly enters the big data realm, the challenges in equitable access are increasingly loaded toward computational and bioinformatics capacity, a trend that will continue in the future. This trend also underscores the need to resolve what some have termed the 'definitional mistake' of the CBD and Nagoya Protocol, which is the challenge of moving beyond the physical dimension of genetic resources (Ruiz Muller 2015).

The use of genetic sequence data presents both opportunities and challenges for benefit sharing, and is an increasingly central issue within several multilateral fora and organisations, including UNCLOS, the CBD, the World Health Organization and the Food and Agriculture Organization of the United Nations (Blasiak 2019; Laird et al. 2020). Dramatic changes in science and technology have also shifted the nature of benefits (Wynberg and Laird 2018). An important benefit has emerged in the form of publicly available databases, but it has also raised questions about the monetary and nonmonetary benefits that accrue to hosting countries (typically those that can provide funds, expertise and technological capacity) and the lack of access to such databases by countries that lack sufficient molecular research capacity or biotechnology infrastructure (Rabone et al. 2019). Concerns have also been expressed about the loss of control and benefits over national patrimony when DNA is sent overseas for more affordable sequencing and loaded onto public or open-access databases (Elbe and Buckland-Merrett 2017). Progress toward creating standards for the publication of genomic data and making scientific data open access, which is now mandatory for projects funded by public funds in many places (e.g. European Union, United States, Australia), has led to massive growth in publicly available data on the ocean genome. This has become big data (Stephens et al. 2015), with several petabytes of sequence data available, including hundreds of millions of predicted genes (e.g. Sunagawa et al. 2015; Carradec et al. 2018; Gregory et al. 2019). This development is leading toward the consideration of the global ocean genome sequence catalogue as a universal resource, although this risks exacerbating inequity due to widely differing technological capacities to benefit from such shared access. At the same time, it may well be that enabling virtual access to data and the ability to use it might prove an easier task than equalising physical access to marine genetic resources.

Industry sequencing efforts are generally excluded from benefit-sharing obligations unless supported with public funds and/or published in peer-reviewed scientific literature. This provides industry with the advantage of accessing publicly funded sequence data for the global ocean genome without any corresponding obligations to share the data they generate. This raises serious questions about equitable use and distributional justice. In addition, this development is also redefining the challenge of access—from advanced ocean sampling and sequencing technologies, to advanced computational resources and enhanced predictive modelling capacities. These modelling capacities require bandwidth to access and download massive amounts of sequence data, which in turn requires high-speed broadband connections, supercomputers to mine and analyse the sequence data and scientists with advanced bioinformatics skills to query the datasets (Quince et al. 2017).

4.3 Equitable Governance and Benefit Sharing

Capacity building, access to and the transfer of marine technology, and information exchange are critical components of responsible and inclusive research and innovation and benefit sharing (Broggiato et al. 2018; Morgera 2018; Collins et al. 2019). The low chance of commercial success from biodiscovery, combined with the long time frame for potential financial returns, means that some of the most significant benefits are nonmonetary, emerging from the research process itself rather than from commercial products. These might include scientific training; access to research infrastructure; and increased collaboration and cooperation in marine science through data collection, technical exchange and the development of joint scientific research projects and programmes. The complexities of MGR governance mean that in addition to the scientific, institutional and legal capacities necessary to develop and administer international and national regulatory frameworks, capacity is also needed to negotiate equitable agreements, resolve disputes and untangle the knotty problems of ownership and access. A deepened social and ethical understanding (Morgera 2018), focused on the role of marine scientists, is also required to manage the use of commonly shared MGR in a sustainable and equitable manner.

Independent of the legal status of MGR, a more principled approach toward benefit sharing should be adopted, in turn fostering 'deeper and cosmopolitan cooperation' via existing UNCLOS obligations on scientific research, capacity building, technology transfer and environmental protection. Such a principled approach would see equitable benefit sharing as an emerging principle of international law of which the human right to science is a part (Morgera 2018).

Current frameworks, including the intersection between environment and intellectual property norms, are extrapolated from constructs that apply on land, where boundaries are more tangible and organisms tend to have restricted ranges. These frameworks neglect the open nature of the ocean, where flows transport organisms across vast distances, including microbes aerosolised from the sea surface to be deposited back in the ocean thousands of kilometres away (Mayol et al. 2017; Ramesh et al. 2019). The 200-nautical-mile legal boundary that separates most national exclusive economic zones from areas beyond national jurisdiction lacks a biological rationale or scientific basis, and a successful mechanism regulating access and benefit sharing with regard to marine genetic resources will need to address this, possibly through collaborative mechanisms between the CBD and UNCLOS.

It is important that the BBNJ process does not replicate the implementation challenges that follow from the wide disparities in domestic measures under the Nagoya Protocol. One way to avoid the pitfalls of disparate implementation would be to agree on what equitable benefit sharing means as a principle of international law, rather than as a mere modality that has polarised the ABS debate. With benefit sharing as a freestanding principle of international law, the links between other global mandates would become clearer, including as an aspect of the human right to science (Article 15 of the International Covenant on Economic, Social and Cultural Rights), contribution to other human rights such as those to food and health, and therefore significant for the realisation of SDGs 2 (hunger) and 3 (health and well-being). It could also be linked to UNCLOS's preambular language—'just and equitable international economic order which takes into account the interests and needs of [hu]mankind as a whole'—as this was the basis for UNCLOS benefit-sharing provisions in relation to outer continental shelf resources and deep-seabed mineral resources. These are issues that require international political will (Morgera 2018), and are subject to negotiations in the upcoming intergovernmental conference.

A key question that threatens swift progress in these negotiations is the issue of intellectual property rights over marine genetic resources and their commercialisation, as well as in relation to capacity building and technology transfer. It is important to note that given existing disparities in technical capabilities to engage in marine scientific research in ABNJ, leaving intellectual property regimes unchanged would likely lead to an exacerbation of technology gaps and inequity due to differential access to MGR and technologies arising from marine scientific research. It is in this context that negotiations related to intellectual property rights and marine genetic resources in the BBNJ process are particularly significant (Thambisetty 2020) for progress toward conservation and sustainable use goals.

One of the main pillars of disagreement and a significant challenge for research on MGR is the inability of the CBD and other international processes to agree on the use of disclosure requirements in the international patent system. The patent specification is a technical and legal document that contains clear and specific information about the invention seeking to be patented. Often these specifications will include information about the source or origin of biological material. As a mandatory measure, such disclosure could facilitate bilateral, global and multilateral benefit sharing. It could also help resolve the artificial distinction between physical and informational genetic resources, inhibit the possibility of public domain or open-access information ending up in private patents, improve trust and ease the global compliance burden of marine scientists.

In the context of the BBNJ negotiations, a global multilateral benefit-sharing mechanism would go some way toward ensuring that the commercial exploitation and use of MGR from ABNJ, whether in physical or intangible form, are subject to benefit-sharing obligations. A multilateral mechanism is particularly important as some countries are advocating for MGR of unknown provenance to be deemed to be from ABNJ. Unless benefit-sharing obligations in the new instrument match or go beyond those in the CBD, this assumption is likely to lead to a race to the bottom of lax benefit-sharing regimes. Some scientists are also urging a rethink of existing rules on disclosing the origin of genetic resources (Blasiak et al. 2019; Chiarolla 2019) while ensuring that intellectual property rights including patents, copyright trade secrets and database rights do not impede capacity building around valuable information.

One of the lessons of the CBD and Nagoya Protocol is the inadequacy of international legal measures to actively engage with scientists and researchers. This in turn negatively impacts confidence in domestic regulatory authorities and the ability to develop laws based on up-to-date scientific understanding. Such concerns highlight the need for scientists to take a more active role in self-regulation, and to instigate training, particularly for younger researchers. Global engagement by scientists and other researchers across jurisdictional boundaries is potentially a powerful dynamic that can, with the right kinds of support and incentives, catalyse effective and equitable governance, and strengthen a shared sense of responsibility to conserve and protect the ocean genome.

5 Conclusion and Opportunities for Action

The ocean genome is the genetic material present in all marine biodiversity, determining the abundance and resilience of biological resources—such as fisheries and aquaculture—that collectively form a pillar of global food security and human well-being. It is the foundation upon which all marine ecosystems, including their functionality and their resilience, rest. Thus, protecting and conserving the ocean genome is crucially important not only for the functioning, stability and integrity of ocean ecosystems and the life within these systems, but for the biosphere and humanity. Yet the ocean genome is also being degraded and eroded through overexploitation, habitat loss and degradation, pollution, impacts from a changing climate such as ocean acidification, invasive species and other pressures, as well as their cumulative and interacting effects.

Simultaneously, exploration of the ocean at a genetic level has resulted in new insights into taxonomy and adaptive capacity that can help optimise conservation efforts, while also spawning a growing number of marine biotechnology applications of commercial importance, from anticancer treatments to cosmetics and industrial enzymes. Initiatives to commercialise the ocean genome should be coupled with considerations regarding conservation, with attention given to both monetary and nonmonetary benefits, and associated environmental, social and ethical risks.

Ensuring that the ocean genome is both conserved and used in a sustainable, fair and equitable manner is urgently important, particularly through the implementation of fully and highly protected areas in the ocean. The sustainable ocean economy is underpinned by the conservation and sustainable use of the ocean genome and a focus on equitable outcomes for all. Yet effective conservation, sustainable use and economic benefits from the ocean genome are challenged by a fragmented ocean governance landscape, gaps in scientific understanding and a world in which the capacity to access and share in the benefits of utilisation of marine genetic resources and associated information varies widely across states. Addressing these issues requires the adoption of effective national and transnational legal measures that ensure both incentives for research and development as well as equitable technology diffusion. Better coordination is needed to ensure that the resources available for promoting conservation, capacity development and other activities associated with the ocean genome are effectively used and equitably shared.

Following from these conclusions, we have identified the following eight opportunities for action to address these issues:

5.1 Opportunities for Action

5.1.1 Protect Marine Genetic Diversity as Part of Conservation Measures and Monitor Outcomes

- **Protect at least 30 % of the ocean in implemented, fully or highly protected MPAs** to effectively conserve genetic diversity and ensure ocean health, productivity and resilience. Support this progress by connecting with existing international commitments in the post-2020 framework such as those in the CBD and UN SDGs, and through new voluntary commitments, as well as with support from philanthropies.
- **Ensure the conservation of genetic diversity beyond the boundaries of MPAs and other area-based management** by supporting the sustainable use of resources; avoiding habitat and ecosystem degradation; affording special protections for rare, vulnerable, threatened or endangered genotypes and species; and using precautionary approaches when initiating exploitation of species or places.
- **Incorporate considerations for marine genetic diversity directly into the management plans of industry/production sectors and conservation,** and support monitoring under existing and new international mechanisms. **Form a joint working group of scientists, legal experts and practitioners** with expertise spanning geography, ecoregions and sectoral international institutions (CBD, UNCLOS, World Trade Organization, WIPO) to advise on best practices in genetic monitoring, planning and management.
- **Use strategic environmental assessments** to manage conflicting uses, address the cumulative effects of multiple human activities and guide marine spatial planning and EIAs.
- **Report on the conservation and use of marine genetic diversity in national and local biodiversity strategies and action plans (NBSAPs/LBSAPs).**

5.1.2 Support Greater Equity in Genomics Research and Commercialisation

- **Ensure that marine science capacity building, information exchange, collaboration and appropriate technology transfer are given adequate attention,** including through their integration into access and benefit sharing (ABS) approaches, research agreements and funder policies. Ensure that new and additional funding streams are employed beyond repackaging existing funds.
- **Facilitate the implementation of domestic legal measures** to ensure that intellectual property norms support an equitable ocean economy. Mechanisms include limitations to the exercise of intellectual property rights through fair, nonexclusive licensing terms, and in ways that do not hinder capacity building, technology transfer or affordable access to technologies.
- **Build the above components into national research policies, plans and programmes and innovation strategies. Increase efforts to ensure that biodiscovery programmes are aware of capacity-building priorities,** and that users and providers of marine genetic resources and associated information are brought into discussions about how best to implement these actions. Make analytical platforms freely available to anyone able to access an internet connection.

5.1.3 Promote Inclusive and Responsible Research and Innovation in Marine Genomics Research

- **Support a transparent, interactive process by which societal actors, innovators and scientists become mutually responsive** to each other with a view to the ethical acceptability, environmental sustainability and societal desirability of the innovation process and its marketable products.
- **Provide incentives for research that are targeted toward important, underfunded objectives,** for example, diseases afflicting the global South. Ensure a focus on lower-income countries, the most marginalised and vulnerable communities, women and environmental concerns.
- **Support scientists to enable their engagement in socially responsive processes, including through the development of new communication tools, to determine key needs and priorities and feed these into national research agendas.**

5.1.4 Embed Conservation of the Ocean Genome Within Research and Commercialisation, Including Benefit-Sharing Approaches and Agreements

- **Develop a global, multilateral benefit-sharing mechanism** for the fair and equitable use of marine genetic resources beyond national jurisdiction. This could include a review of international voluntary codes of conduct, and the cataloguing of examples where conservation outcomes have been achieved through such efforts.
- **Enhance the legal capacity of developing countries** to domestically address issues emerging from multilateral processes, including those related to intellectual property, benefit sharing, capacity building and technology transfer.
- Parties to the Convention on Biological Diversity should **develop benefit-sharing agreements with mutually agreed terms** focused on conservation and sustainable and equitable use outcomes when granting access to marine genetic resources within national jurisdictions, and support countries in monitoring the performance of such contracts.
- **Funders of research related to the ocean genome should require applicants to explain the potential conservation, sustainability and equity applications and benefits of their research.**

5.1.5 Disclose the Biological and Geographical Origins of Genetic Material as a Norm Across All Associated Commercial and Noncommercial Activities

- **Modify procedural aspects of international patent law to require disclosure of the origin of genetic material in patent filings.**
- **Encourage and incentivise the disclosure of the origin of genetic material among marine scientists and private institutions as an aspect of responsible research and innovation.**
- Regardless of legal obligations, funding bodies, genetic sequence database administrators and journal editors should **require disclosure of the origin of genetic material.**

5.1.6 Increase Financial and Political Support to Improve Knowledge of the Ocean Genome

- **Build support for integrative taxonomic research** aimed at understanding the ocean genome by making this a key element of the UN Decade of Ocean Science for Sustainable Development.
- **Support the research needed for genetic monitoring as part of existing environmental assessments.** Research and share results on the links between genetic diversity and adaptive capacity in the context of global change.
- **Support research on the functional biology of the ocean**, including the systematic unveiling of gene function, gene networks and species interactions.
- **Prioritise the allocation of resources to build scientific capacity** using approaches such as environmental DNA, DNA metabarcoding and other emerging genetic monitoring techniques, as well as to develop more cost-efficient methods.

5.1.7 Comprehensively Assess the Risks and Benefits of Transgenic Marine Organisms as well as the Use of New Molecular Engineering Technologies: Such as CRISPR-Cas (Gene Editing) and Gene Drives—In the Marine Environment

- **Initiate a deliberative process, beginning with a working group,** to gather scientists, ethicists, environmentalists, policymakers and other actors to develop principles and debate approaches for whether and how genetic technologies should be used in the marine environment. Address the limits and directions of current research and development activities, assess risks and wider impacts and engage in dialogue about associated ethical considerations.

5.1.8 Strengthen the Role of Philanthropy in Providing Infrastructure and Funding for Marine Science

- **Establish a network to better coordinate privately funded initiatives, align their priorities with those of states** that are acquiring knowledge for societal needs and improve the transparency of philanthropic funding.
- **Encourage financial supporters of ocean science, including philanthropies, to publish and comply with an ethical code of conduct, and sign a 'Declaration for Coordinated Ocean Action'** based on the principles set forth in the Paris Declaration on Aid Effectiveness and the Accra Agenda for Action to ensure that support is aligned and coordinated with the objectives of the UN Decade of Ocean Science for Sustainable Development, the SDGs and priorities identified by developing countries.

Acknowledgements The authors thank the paper's technical reviewers, Sophie Arnaud-Haond, Elisa Morgera and Jorge Torre, its arbiter, Kristian Teleki, and Ocean Panel Expert Group co-chairs Jane Lubchenco and Peter Haugan, who all provided helpful technical comments.

The authors also thank World Resources Institute for providing support as the Ocean Panel Secretariat, and particularly Katie Flanagan for guiding the paper through to its publication. A. Rogers wishes to thank REV Ocean for its financial support. R. Wynberg's contribution is supported by the South African Research Chairs Initiative of the Department of Science and Innovation and National Research Foundation of South Africa.

While our colleagues were very generous with their time and input, this report reflects the views of the authors alone.

The authors thank Sarah DeLucia for copyediting and Billie Kanfer for design.

About the Authors

Lead Authors

Robert Blasiak is a researcher at the Stockholm Resilience Centre at Stockholm University in Sweden and a visiting researcher at the Graduate School of Agricultural and Life Sciences at the University of Tokyo in Japan.

Rachel Wynberg is a professor and bio-economy SARChI research chair in the Department of Environmental & Geographical Science at the University of Cape Town in South Africa.

Kirsten Grorud-Colvert is an associate professor in the Department of Integrative Biology at Oregon State University in the United States.

Siva Thambisetty is an associate professor in the Department of Law at the London School of Economics and Political Science in the United Kingdom.

Contributing Authors

Narcisa M. Bandarra is a researcher at and the head of the Aquaculture, Upgrading and Bioprospecting Division of the Portuguese Institute for Sea and Atmosphere in Portugal.

Adelino V.M. Canário is professor at and president of CCMAR, the Centre for Marine Sciences, at the University of Algarve in Portugal.

Jessica da Silva is a researcher at the South African National Biodiversity Institute and the Centre for Ecological Genomics and Wildlife Conservation at the University of Johannesburg in South Africa.

Carlos M. Duarte is a professor at the Red Sea Research Center and Computational Bioscience Research Center at King Abdullah University of Science and Technology in Saudi Arabia, and a member of the Expert Group for the High Level Panel for a Sustainable Ocean Economy.

Marcel Jaspars is a professor at and founder of the Marine Biodiscovery Centre in the Department of Chemistry at the University of Aberdeen in the United Kingdom.

Alex D. Rogers is science director of REV Ocean and a senior research fellow at Somerville College, University of Oxford in the United Kingdom.

Kerry Sink is a scientist at the South African National Biodiversity Institute and a professor at the Institute for Coastal and Marine Research at Nelson Mandela University in South Africa.

Colette C.C. Wabnitz is a senior research scientist at the Institute for the Oceans and Fisheries at the University of British Columbia in Canada and at the Stockholm Resilience Centre at Stockholm University in Sweden.

Appendix

To support success, we include in this table potential barriers to implementation and strategies to overcome them (Table 4.2).

Table 4.2 Opportunities for action to conserve and use marine genetic resources fairly, equitably and sustainably

	Theme	Opportunities for action	Barriers to implementation	Overcoming barriers
1.	Protect marine genetic diversity as part of conservation measures and monitor their outcomes.	International level	Securing funding to establishing a joint working group.	Connect with existing commitments (e.g. under UNCLOS and within the Sustainable Development Goals [SDGs]), voluntary commitments (e.g. from UN Ocean Conference) and philanthropy (see Opportunity for Action 8).
		Post-2020 Convention on Biological Diversity (CBD) targets on marine protected areas (MPAs) should follow the scientific evidence showing that protecting at least 30% of the ocean in fully to highly protected, implemented MPAs is needed to conserve biodiversity and genetic diversity and to sustain ocean health, productivity and resilience.	Lack of capacity at national, regional and local levels to engage in genetic monitoring activities.	On lack of capacity, see Opportunity for Action 4; on gaps in taxonomic knowledge, see Opportunity for Action 1.
		Form a joint working group of scientists, legal experts and practitioners with expertise spanning geography, ecoregions, and sectoral international institutions (CBD, United Nations Convention on the Law of the Sea [UNCLOS], World Trade Organization, World Intellectual Property Organization) to mainstream genetic monitoring into existing international mechanisms (e.g. International Seabed Authority [ISA] mining code for prospecting and exploration) and new international mechanisms (e.g. Biodiversity in Areas Beyond National Jurisdiction treaty, ISA mining code for exploitation).	Gaps in taxonomic knowledge and datasets to enable genetic monitoring activities.	
		With respect to identifying priorities for conservation in areas beyond national jurisdiction, strategic environmental assessments, comprehensively understood, can help avoid conflicting uses, address cumulative effects of multiple human activities, and guide environmental impact assessments for specific current and proposed activities.		
		The CBD should issue guidance on how to incorporate aspects of genetic diversity into National Biodiversity Strategies and Action Plans.		
		National, regional and local levels		
		Marine genetic diversity should be explicitly incorporated into the design and management of conservation measures, including by establishing fully and highly protected MPAs, as well as subsequently monitoring their outcomes.		

Table 4.2 (continued)

Theme	Opportunities for action	Barriers to implementation	Overcoming barriers
2. Support greater equity in genomics research and commercialisation.	International level	Ensure that prioritisation of allocating resources for researching the ocean genome results in new funding streams rather than a simple repackaging of existing funds.	See Opportunity for Action 8 on developing a 'Declaration for Coordinated Ocean Action'.
	Ensure that marine science capacity building, information exchange, collaboration and appropriate technology transfer are given adequate attention in international research programmes, and that priorities are well articulated in CBD and UNCLOS decisions.		
	Articulate and facilitate internationally the implementation of hard-edged domestic legal measures such as limitations to the exercise of intellectual property rights through fair, nonexclusive licensing terms; market authorisations that take note of compliance with benefit-sharing mechanisms; and the application of international legal norms that facilitate technology transfer and affordable access to technologies.		
	National level		
	Build these components into national research policies, plans and programmes and innovation strategies. **Ensure that biodiscovery programmes are aware of capacity-building priorities**, and that users and providers of marine genetic resources and associated information are brought into discussions about how best to implement these actions. Make analytical platforms available to anyone able to access an internet connection.		
	Explore the full range of limitations and exceptions to intellectual property rights so that capacity building and technology transfer are not precluded by exclusive intellectual property rights.		
3. Promote inclusive innovation in marine genomics research.	International level	Funding for research and development programmes is often driven by commercial entities, with products geared toward affluent markets rather than to either broader societal needs or diseases afflicting the global South.	See Opportunity for Action 8 on developing a 'Declaration for Coordinated Ocean Action'.
	Support a transparent, interactive process by which societal actors and innovators become mutually responsive to each other with a view to the ethical acceptability, sustainability and societal desirability of the innovation process and its marketable products.		
	Provide incentives for research that are targeted toward societally important yet underfunded objectives. Ensure a focus on lower-income countries, the most marginalised and vulnerable communities, women and environmental concerns.		
	National level		
	Support scientists to enable their engagement in socially responsive processes that determine key needs and priorities and feed these into national research agendas. Ensure a focus on the most marginalised and vulnerable communities, women and on key environmental concerns. **Develop communication tools to improve linkages between societal actors.**		

(continued)

Table 4.2 (continued)

Theme	Opportunities for action	Barriers to implementation	Overcoming barriers
4. Embed conservation of the ocean genome within research and commercialisation, including through benefit-sharing approaches and agreements.	International level	No legal obligation exists to undertake such actions, so states and funding bodies would be acting in a voluntary manner. Resistance to depart from the status quo.	Develop international voluntary codes of conduct, and catalogue case studies and best practices when conservation outcomes are achieved through such efforts.
	Facilitate a fair, equitable, global, multilateral benefit-sharing mechanism for the use and exploitation of marine genetic resources beyond national jurisdiction.		Develop opportunities for legal pluralism for specific problems through training and the exchange of legal expertise.
	Enhance the legal capacity of developing countries to address domestically issues emerging from multilateral processes including those related to intellectual property, benefit sharing, capacity building and technology transfer.		
	National level		
	Parties to the Convention on Biological Diversity should **include benefit-sharing agreements with mutually agreed terms focused on conservation and sustainable and equitable use outcomes** when granting access to marine genetic resources.		
	When allocating funding for **research associated with marine genetic resources**, grant-making bodies and research councils **should require applicants to explain the potential conservation, sustainability and equity applications and benefits of their research.**		
5. Disclose the origins (species and geographical area where organisms were extracted) of genetic material as a norm across all associated commercial and noncommercial activities.	International level	Slow pace of consensus building within relevant international forums.	Reputational benefits accrued by voluntary disclosure of origin by scientists, with potential to shape norms of best practice.
	Modify procedural aspects of international patent law to require the disclosure of the origins (species and geographical area where organisms were extracted) of genetic material in patent filings. This could be achieved through either in-application disclosure or the development of new categories in the international patent classification system. Such measures could help identify cases of noncompliance with the Nagoya Protocol and ensure compliance with existing and emerging access and benefit-sharing obligations.		
	National, regional and local levels		
	Regardless of legal obligations, funding bodies, genetic sequence database administrators and journal editors should **require disclosure of origin.**		

Table 4.2 (continued)

	Theme	Opportunities for action	Barriers to implementation	Overcoming barriers
6.	Increase financial and political support to improve knowledge of the ocean genome.	International level	Convincing policymakers and funding bodies to prioritise taxonomic research and genetic monitoring approaches.	Communicate the range of benefits associated with improved knowledge of the ocean genome (for both conservation and commercial purposes).
		Build support for taxonomic research aimed at understanding the ocean genome by making this a key element of the UN Decade of Ocean Science.		See Opportunity for Action 8.
		National, regional and local levels		
		Responsible ministries, departments, research councils and other relevant actors should **support research needed for basic taxonomic knowledge, genetic monitoring** as part of existing environmental assessments, and **research** on the links between genetic diversity and adaptive capacity in the context of global change.		
		All levels		
		Funding agencies should **prioritise the allocation of resources to support the building of scientific capacity** to enhance understanding using the range of available resources, including environmental DNA, DNA metabarcoding and other emerging techniques to enable genetic monitoring.		
7.	Comprehensively assess the risks and benefits of transgenic marine organisms as well as the use of new technologies— such as CRISPR-Cas (gene editing) and gene drives—in the marine environment.	International level	Different worldviews and knowledge systems are difficult to bring together.	Ensure that scientific information is effectively translated into accessible language; improve interdisciplinary understandings; build awareness among policymakers.
		Initiate a deliberative process or 'observatory' think tank to bring together scientists, ethicists, environmentalists, policymakers and other actors to develop principles and debate approaches to using genetic technologies in the marine environment, and to engender robust conversations about the limits and directions of research and development, risk assessments, and wider impacts as well as ethical considerations.	Rigid positions may be adopted by different actors.	
			Communication between actors remains a major challenge.	

(continued)

Table 4.2 (continued)

	Theme	Opportunities for action	Barriers to implementation	Overcoming barriers
8.	Increase the role of philanthropy in providing infrastructure and funding for marine science.	International level	Hesitance by national research councils, philanthropies or others to commit to a coordinated and aligned approach in their financial support.	Take a stepwise approach, first asking signatories to recommit to existing development frameworks, and then seeking more ambitious commitments to align and coordinate support over time.
		Establish a network to better coordinate privately funded initiatives with those of states that are acquiring knowledge for societal needs, as outlined by global agendas such as the SDGs and the UN Decade of Ocean Science.		
		Financial supporters of ocean science, including philanthropies, sign a 'Declaration for Coordinated Ocean Action' that is based on the principles set forth in the Paris Declaration on Aid Effectiveness and the Accra Agenda for Action to ensure that support is aligned and coordinated with the objectives of the UN Decade of Ocean Science and the SDGs.		
		Communicate with the Organisation for Economic Co-operation and Development's Development Assistance Committee for data illustrating the impact of the Paris Declaration on Aid Effectiveness and Accra Agenda for Action, and use these experiences to communicate the added value of a coordinated approach.		

References

Abberton M, Batley J, Bentley A, Bryant J, Cai H, Cockram J, Costa de Oliveira A et al (2016) Global agricultural intensification during climate change: a role for genomics. Plant Biotechnol J 14(4):1095–1098. https://doi.org/10.1111/pbi.12467

Aburto-Oropeza O, Rogers A et al (2020) Critical habitats and biodiversity: inventory, thresholds and governance. World Resources Institute, Washington, DC

Ahmed N, Glaser M (2016) Coastal aquaculture, mangrove deforestation and blue carbon emissions: is REDD+ a solution? Mar Policy 66:58–66

Ainsworth TD, Heron SF, Ortiz JC, Mumby PJ, Grech A, Ogawa D, Eakin CM et al (2016) Climate change disables coral bleaching protection on the Great Barrier Reef. Science 352(6283):338–342. https://doi.org/10.1126/science.aac7125

Algenesis Materials (n.d.) Our technology. https://www.algenesismaterials.com/technology

Allendorf FW, England PR, Luikart G, Ritchie PA, Ryman N (2008) Genetic effects of harvest on wild animal populations. Trends Ecol Evol 23(6):327–337. https://doi.org/10.1016/j.tree.2008.02.008

Allison GW, Gaines SD, Lubchenco J, Possingham HP (2003) Ensuring persistence of marine reserves: catastrophes require adopting an insurance factor. Ecol Appl 13(1):S8–S24

Allott P (1992) Mare Nostrum: a new international law of the sea. Am J Int Law 86(4):764–787

Almany GR, Connolly SR, Heath DD, Hogan JD, Jones GP, McCook LJ, Mills M et al (2009) Connectivity, biodiversity conservation and the design of marine reserve networks for coral reefs. Coral Reefs 28(2):339–351

Alves C, Silva J, Pinteus S, Gaspar H, Alpoim MC, Botana LM, Pedrosa R (2018) From marine origin to therapeutics: the antitumor potential of marine algae-derived compounds. Front Pharmacol 9:777. https://doi.org/10.3389/fphar.2018.00777

Andruszkiewicz EA, Starks HA, Chavez FP, Sassoubre M, Block BA, Boehm AB (2017) Biomonitoring of marine vertebrates in Monterey bay using eDNA metabarcoding. PLoS One 12(4):e0176343. https://doi.org/10.1371/journal

Ardura A, Zaiko A, Martinez JL, Samulioviene A, Semenova A, Garcia-Vazquez E (2015) eDNA and specific primers for early detection of invasive species—a case study on the Bivalve Rangia Cuneata, currently spreading in Europe. Mar Environ Res 112:48–55. https://doi.org/10.1016/j.marenvres.2015.09.013

Arismendi I, Soto D, Penaluna B, Jara C, Leal C, León-Muñoz J (2009) Aquaculture, non-native salmonid invasions and associated declines of native fishes in northern Patagonian Lakes. Freshw Biol 54(5):1135–1147

Arnaud-Haond S, Arrieta JM, Duarte CM (2011) Marine biodiversity and gene patents. Science 331(6024):1521–1522

Arrieta JM, Arnaud-Haond S, Duarte CM (2010) What lies underneath: conserving the oceans' genetic resources. Proc Natl Acad Sci 107(43):18318–18324

Arulandhu AJ, Staats M, Hagelaar R, Voorhuijzen MM, Prins TW, Scholtens I, Costessi A et al (2017) Development and validation of a multi-locus DNA metabarcoding method to identify endangered species in complex samples. Gigascience 6(10):1–18. https://doi.org/10.1093/gigascience/gix080

Aylagas E, Borja Á, Irigoien X, Rodríguez-Ezpeleta N (2016) Benchmarking DNA metabarcoding for biodiversity-based monitoring and assessment. Front Mar Sci 3. https://doi.org/10.3389/fmars.2016.00096

Baird DJ, Hajibabaei M (2012) Biomonitoring 2.0: a new paradigm in ecosystem assessment made possible by next-generation DNA sequencing. Mol Ecol 21:2039–2044

Bakker J, Wangensteen OS, Chapman DD, Boussarie G, Buddo D, Guttridge TL, Hertler H et al (2017) Environmental DNA reveals tropical shark diversity in contrasting levels of anthropogenic impact. Sci Rep 7:1–11. https://doi.org/10.1038/s41598-017-17150-2

Baldanzi S, Watson R, McQuaid CD, Gouws G, Porri F (2017) Epigenetic variation among natural populations of the South African Sandhopper Talorchestia Capensis. Evol Ecol 31(1):77–91. https://doi.org/10.1007/s10682-016-9877-9

Ban NC, Gurney GG, Marshall NA, Whitney CK, Mills M, Gelcich S, Bennett NJ et al (2019) Well-being outcomes of marine protected areas. Nat Sustain 2(6):524

Barbuto M, Galimberti A, Ferri E, Labra M, Malandra R, Galli P, Casiraghi M (2010) DNA barcoding reveals fraudulent substitutions in shark seafood products: the Italian case of 'Palombo' (*Mustelus* Spp.). Food Res Int 43(1):376–381. https://doi.org/10.1016/j.foodres.2009.10.009

Barendse J, Roel A, Longo C, Andriessen L, Webster LMI, Ogden R, Neat F (2019) DNA barcoding validates species labelling of certified seafood. Curr Biol 29(6):R198–R199. https://doi.org/10.1016/j.cub.2019.02.014

Bar-On YM, Phillips R, Milo R (2018) The biomass distribution on earth. Proc Natl Acad Sci 115(25):6506–6511

Barshis DJ, Ladner JT, Oliver TAA, Seneca FO, Traylor-Knowles N, Palumbi SR (2013) Genomic basis for coral resilience to climate change. Proc Natl Acad Sci 110(4):1387–1392. https://doi.org/10.1073/pnas.1210224110

Barton AD, Irwin AJ, Finkel ZV, Stock CA (2016) Anthropogenic climate change drives shift and shuffle in North Atlantic phytoplankton communities. Proc Natl Acad Sci 113(11):2964–2969

Béné C, Arthur R, Norbury H, Allison EH, Beveridge M, Bush S, Campling L et al (2016) Contribution of fisheries and aquaculture to food security and poverty reduction: assessing the current evidence. World Dev 79:177–196

Bester-van der Merwe AE, Roodt-Wilding R, Volckaert FAM, D'Amato ME (2011) Historical isolation and hydrodynamically constrained gene flow in declining populations of the South-African Abalone, Haliotis Midae. Conserv Genet 12(2):543–555. https://doi.org/10.1007/s10592-010-0162-0

Beygmoradi A, Homaei A (2017) Marine microbes as a valuable resource for brand new industrial biocatalysts. Biocatal Agric Biotechnol 11:131–152

Blasiak R (2019) International regulatory changes poised to reshape access to marine genes. Nat Biotechnol 37(4):357

Blasiak R, Jouffray J-B, Wabnitz CCC, Sundström E, Österblom H (2018) Corporate control and global governance of marine genetic resources. Sci Adv 4(6):eaar5237

Blasiak R, Jouffray J-B, Wabnitz CCC, Österblom H (2019) Scientists should disclose origin in marine gene patents. Trends Ecol Evol 34(5):392–395

Blowes SA, Connolly SR (2012) Risk spreading, connectivity, and optimal reserve spacing. Ecol Appl 22(1):311–321

Bockmann FA, Rodrigues MT, Kohsldorf T, Cobra Straker L, Grant T, Cardoso de Pinna MC, Medina Mantelatto FL et al (2018) Brazil's government attacks biodiversity. Science 360(6391):865

Bonfanti C, Cardoso C, Afonso C, Matos J, Garcia T, Tanni S, Bandarra NM (2018) Potential of microalga Isochrysis galbana: bioactivity and bioaccessibility. Algal Res 29:242–248

Boussarie G, Bakker J, Wangensteen OS, Mariani S, Bonnin L, Juhel J-B, Kiszka JJ et al (2018) Environmental DNA illuminates the dark diversity of sharks. Sci Adv 4(5):eaap9661. https://doi.org/10.1126/sciadv.aap9661

Broggiato A, Arnaud-Haond S, Chiarolla C, Greiber T (2014) Fair and equitable sharing of benefits from the utilization of marine genetic resources in areas beyond national jurisdiction: bridging the gaps between science and policy. Mar Policy 49:176–185. https://doi.org/10.1016/j.marpol.2014.02.012

Broggiato A, Vanagt T, Lallier LE, Jaspars M, Burton G, Muyldermans D (2018) Mare geneticum: balancing governance of marine genetic resources in international waters. Int J Mar Coastal Law 33(1):3–33. https://doi.org/10.1163/15718085-13310030

Bruckner AW (2002) Life-saving products from coral reefs. Issues Sci Technol 18(3):39–44. https://issues.org/p_bruckner/.

Buchanan JR, Ralph GM, Krupp F, Harwell H, Abdallah M, Abdulqader E, Al-Husaini M et al (2019) Regional extinction risks for marine bony fishes occurring in the Persian/Arabian Gulf. Biol Conserv 230:10–19

Caicedo JC, Cooper S, Heigwer F, Warchal S, Qiu P, Molnar C, Vasilevich AS et al (2017) Data-analysis strategies for image-based cell profiling. Nat Methods 14(9):849

Campbell L, Nolla HA, Vaulot D (1994) The importance of Prochlorococcus to community structure in the Central North Pacific Ocean. Limnol Oceanogr 39(4):954–961

Caplan AL, Parent B, Shen M, Plunkett C (2015) No time to waste—the ethical challenges created by CRISPR. EMBO Rep 16(11):1421–1426

Carpenter LJ, Liss PS (2000) On temperate sources of bromoform and other reactive bromine gases. J Geophys Res Atmos 105:20539–20547

Carpenter KE, Abrar M, Aeby G, Aronson RB, Banks S, Bruckner A, Chiriboga A et al (2008) One-third of reef-building corals face elevated extinction risk from climate change and local impacts. Science 321(5888):560–563

Carr MH, Robinson SP, Wahle C, Davis G, Kroll S, Murray S, Schumacker EJ et al (2017) The central importance of ecological spatial connectivity to effective coastal marine protected areas and to meeting the challenges of climate change in the marine environment. Aquat Conserv Mar Freshwater Ecosyst 27:6–29

Carradec Q, Pelletier E, Da Silva C, Alberti A, Seeleuthner Y, Blanc-Mathieu R, Lima-Mendez G et al (2018) A global ocean atlas of eukaryotic genes. Nat Commun 9(1):1–13. https://doi.org/10.1038/s41467-017-02342-1

Carroll AR, Copp BR, Davis RA, Keyzers RA, Prinsep MR (2019) Marine natural products. Nat Prod Rep 36:122–173

Carugati L, Corinaldesi C, Dell'Anno A, Danovaro R (2015) Metagenetic tools for the census of marine meiofaunal biodiversity: an overview. Mar Genomics 24:11–20. https://doi.org/10.1016/j.margen.2015.04.010

Cawthorn D-M, Steinman HA, Witthuhn RC (2012) DNA barcoding reveals a high incidence of fish species misrepresentation and substitution on the South African Market. Food Res Int 46(1):30–40

CBD (Convention on Biological Diversity) (2018) Protected areas and other effective area-based conservation measures. Recommendation adopted by the subsidiary body on scientific, technical, and technological advice. CBD/SBSTTA/REC/22/5. Convention on biological diversity

Ceci SJ, Williams WM (2011) Understanding current causes of women's underrepresentation in science. Proc Natl Acad Sci 108(8):3157–3162

Cerezo Valverde J, Rodríguez-González T, Granero-Fernández MD, Aguado-Giménez F, García García B (2019) Successful rearing of common octopus (*Octopus Vulgaris*) fed a formulated feed in an offshore cage. Aquacult Res 50(3):968–972

Chan FT, Briski E, Chan F (2017) An overview of recent research in marine biological invasions. Mar Biol 164(6):1–10. https://doi.org/10.1007/s00227-017-3155-4

Chang F-Y, Brady SF (2012) Cloning and characterization of an eDNA derived gene cluster that encodes the biosynthesis of the antitumor substance BE-54017. J Am Chem Soc 133(26):9996–9999. https://doi.org/10.1021/ja2022653.Cloning

Chapin FS III, Zavaleta ES, Eviner VT, Naylor RL, Vitousek PM, Reynolds HL, Hooper DU et al (2000) Consequences of changing biodiversity. Nature 405(6783):234

Chauton MS, Inge Reitan K, Norsker NH, Tveterås R, Kleivdal HT (2015) A techno-economic analysis of industrial production of marine microalgae as a source of EPA and DHA-rich raw material for aquafeed: research challenges and possibilities. Aquaculture 436:95–103

Cheng BS, Altieri AH, Torchin ME, Ruiz GM (2019) Can marine reserves restore lost ecosystem functioning? A global synthesis. Ecology 100(4):e02617. https://doi.org/10.1002/ecy.2617

Chhetri BK, Lavoie S, Sweeney-Jones AM, Kubanek J (2018) Recent trends in the structural revision of natural products. Nat Prod Rep 35(6):514–531

Chiarolla C (2019) Intellectual property from a global environmental law perspective: lessons from patent disclosure requirements for genetic resources and traditional knowledge. Transnatl Environ Law 8(3):503–521

Claudet J, Osenberg CW, Benedetti-Cecchi L, Domenici P, Garcia-Charton JA, Perez-Ruzafa A, Badalamenti F et al (2008) Marine reserves: size and age do matter. Ecol Lett 11(5):481–489

Cobián Güemes AG, Youle M, Cantú VA, Felts B, Nulton J, Rohwer F (2016) Viruses as winners in the game of life. Annu Rev Virol 3(1):197–214. https://doi.org/10.1146/annurev-virology-100114-054952

Collins JE, Harden-Davies H, Jaspars M, Thiele T, Vanagt T, Huys I (2019) Inclusive innovation: enhancing global participation in and benefit sharing linked to the utilization of marine genetic resources from areas beyond national jurisdiction. Mar Policy 109:103696. https://doi.org/10.1016/j.marpol.2019.103696

Corkeron P, Hamilton P, Bannister J, Best P, Charlton C, Groch KR, Findlay K et al (2018) The recovery of north Atlantic right whales, *Eubalaena Glacialis*, has been constrained by human-caused mortality. R Soc Open Sci 5(11):180892. https://doi.org/10.1098/rsos.180892

Cornwall W (2019) The reef builders. Science 363(6433):1264–1269. https://doi.org/10.1126/science.363.6433.1264

Costanza R (1999) The ecological, economic, and social importance of the oceans. Ecol Econ 31(2):199–213

Costello MJ (2014) Long live marine reserves: a review of experiences and benefits. Biol Conserv 176:289–296. https://doi.org/10.1016/j.biocon.2014.04.023

Costello MJ, Wilson S, Houlding B (2012) Predicting total global species richness using rates of species description and estimates of taxonomic effort. Syst Biol 61(5):871

Cowie RH, Regnier C, Fontaine B, Bouchet P (2017) Measuring the sixth extinction: what do mollusks tell us. Nautilus 131(1):3–41

CSS, ENSSER and VDW (Critical Scientists Switzerland, European Network of Scientists for Social and Environmental Responsibility and Vereinigung Deutscher Wissenschaftler) (2019) Gene drives: a report on their science, applications, social aspects, ethics and regulation

Czorlich Y, Aykanat T, Erkinaro J, Orell P, Primmer CR (2018) Rapid sex-specific evolution of age at maturity is shaped by genetic architecture in Atlantic Salmon. Nat Ecol Evol 2(11):1800–1807. https://doi.org/10.1038/s41559-018-0681-5

da Silva J, van Vuuren B (2019) South African National Biodiversity Assessment 2018 technical report volume 4: marine realm. South African National Biodiversity Institute, Pretoria

Davidson AD, Boyer AG, Kim H, Pompa-Mansilla S, Hamilton MJ, Costa DP, Ceballos G et al (2012) Drivers and hotspots of extinction risk in marine mammals. Proc Natl Acad Sci 109(9):3395–3400. https://doi.org/10.1073/pnas.1121469109

Dawson TP, Jackson ST, House JI, Prentice IC, Mace GM (2011) Beyond predictions: biodiversity conservation in a changing climate. Science 332(6025):53–58. https://doi.org/10.1126/science.1200303

De Leo GA, Micheli F (2015) The good, the bad and the ugly of marine reserves for fishery yields. Philos Trans R Soc B Biol Sci 370(1681):20140276. https://doi.org/10.1098/rstb.2014.0276

de Mitcheson YS, Craig MT, Bertoncini AA, Carpenter KE, Cheung WWL, Choat JH, Cornish AS et al (2013) Fishing groupers towards extinction: a global assessment of threats and extinction risks in a billion dollar fishery. Fish Fish 14(2):119–136. https://doi.org/10.1111/j.1467-2979.2011.00455.x

Deiner K, Fronhofer EA, Mächler E, Walser J-C, Altermatt F (2016) Environmental DNA reveals that rivers are conveyer belts of biodiversity information. Nat Commun 7:12544. https://doi.org/10.1038/ncomms12544

Deiner K, Bik HM, Mächler E, Seymour M, Lacoursière-Roussel A, Altermatt F, Creer S et al (2017) Environmental DNA metabarcoding: transforming how we survey animal and plant communities. Mol Ecol 26(21):5872–5895. https://doi.org/10.1111/mec.14350

Devine JA, Baker KD, Haedrich RL (2006) Deep-sea fishes qualify as endangered. Nature 439(7072):29. https://doi.org/10.1038/439029a

Di Lorenzo M, Claudet J, Guidetti P (2016) Spillover from marine protected areas to adjacent fisheries has an ecological and a fishery component. J Nat Conserv 32:62–66. https://doi.org/10.1016/j.jnc.2016.04.004

Di Muri C, Vandamme SG, Peace C, Barnes W, Mariani S (2018) Biodiversity defrosted: unveiling non-compliant fish trade in ethnic food stores. Biol Conserv 217:419–427. https://doi.org/10.1016/j.biocon.2017.11.028

Di Pinto A, Di Pinto P, Terio V, Bozzo G, Bonerba E, Ceci E, Tantillo G (2013) DNA barcoding for detecting market substitution in salted cod fillets and battered cod chunks. Food Chem 141(3):1757–1762. https://doi.org/10.1016/j.foodchem.2013.05.093

Djurhuus A, Pitz K, Sawaya NA, Rojas-Márquez J, Michaud B, Montes E, Muller-Karger F et al (2018) Evaluation of marine zooplankton community structure through environmental DNA metabarcoding. Limnol Oceanogr Methods 16(4):209–221. https://doi.org/10.1002/lom3.10237

Dlugosch KM, Parker IM (2008) Founding events in species invasions: genetic variation, adaptive evolution, and the role of multiple introductions. Mol Ecol 17:431–449. https://doi.org/10.1111/j.1365-294X.2007.03538.x

Doney SC, Ruckelshaus M, Duffy JE, Barry JP, Chan F, English CA, Galindo HM et al (2012) Climate change impacts on marine ecosystems. Annu Rev Mar Sci 4(1):11–37. https://doi.org/10.1146/annurev-marine-041911-111611

Doudna JA, Charpentier E (2014) Genome editing. The new frontier of genome engineering with CRISPR Cas9 Science 346(6213):1258096. https://doi.org/10.1126/science.1258096

Du Preez C, Swan KD, Curtis JMR (2020) Cold-water corals and other vulnerable biological structures on a North Pacific Seamount after half a century of fishing. Front Mar Sci 7:17

Duarte CM, Marbá N, Holmer M (2007) Rapid domestication of marine species. Science 316(5823):382–383. https://doi.org/10.1126/science.1138042

Dulvy NK, Reynolds JD (2009) Skates on thin ice. Nature 462(7272):417. https://doi.org/10.1038/462417a

Dulvy NK, Fowler SL, Musick JA, Cavanagh RD, Kyne PM, Harrison LR, Carlson JK et al (2014) Extinction risk and conservation of the world's sharks and rays. ELife 3:e00590. https://doi.org/10.7554/eLife.00590

Duncan MI, Bates AE, James NC, Potts WM (2019) Exploitation may influence the climate resilience of fish populations through removing high performance metabolic phenotypes. Sci Rep 9(1):1–10

Echeverría F, Valenzuela R, Hernandez-Rodas MC, Valenzuela A (2017) Docosahexaenoic acid (DHA), a fundamental fatty acid for the brain: new dietary sources. Prostaglandins Leukot Essent Fatty Acids 124:1–10. https://doi.org/10.1016/j.plefa.2017.08.001

El Samak M, Solyman SM, Hanora A (2018) Antimicrobial activity of bacteria isolated from red sea marine invertebrates. Biotechnol Rep 19:e00275. https://doi.org/10.1016/j.btre.2018.e00275

Elbe S, Buckland-Merrett G (2017) Data, disease and diplomacy: GISAID's innovative contribution to global health. Glob Challenges 1:33–46. https://doi.org/10.1002/gch2.1018

Elkin-Koren N, Netanel NW (eds) (2002) The commodification of information, 1st edn. Kluwer Law International, The Hague

Ellegren H, Galtier N (2016) Determinants of genetic diversity. Nat Rev Genet 17(7):422–433. https://doi.org/10.1038/nrg.2016.58

FAO (2018) The state of world fisheries and aquaculture: meeting the sustainable development goals. FAO, Rome

FAO (Food and Agriculture Organization of the United Nations) (2016) Voluntary guidelines for mainstreaming biodiversity into policies, programmes and national and regional plans of action on nutrition. Commission on Genetic Resources for Food and Agriculture, Rome. http://www.fao.org/family-farming/detail/en/c/416118/

Fauchald P (2010) Predator–prey reversal: a possible mechanism for ecosystem hysteresis in the north sea? Ecology 91(8):2191–2197. https://doi.org/10.1890/09-1500.1

Fleming IA, Hindar K, Mjølnerød IB, Jonsson B, Balstad T, Lamberg A (2000) Lifetime success and interactions of farm salmon invading a native population. Proc R Soc Ser B Biol Sci 267(1452):1517–1523. https://doi.org/10.1098/rspb.2000.1173

Fleming IA, Agustsson T, Finstad B, Johnsson JI, Thrandur Björnsson B (2002) Effects of domestication on growth physiology and endocrinology of Atlantic Salmon (*Salmo Salar*). Can J Fish Aquat Sci 59(8):1323–1330

Foster C, Heeks R (2013) Conceptualising inclusive innovation: modifying systems of innovation frameworks to understand diffusion of new technology to low-income consumers. Eur J Dev Res 25(3):333–355. https://doi.org/10.1057/ejdr.2013.7

Frankham R (2005) Stress and adaptation in conservation genetics. J Evol Biol 18(4):750–755. https://doi.org/10.1111/j.1420-9101.2005.00885.x

Fung T, Seymour RM, Johnson CR (2011) Alternative stable states and phase shifts in coral reefs under anthropogenic stress. Ecology 92(4):967–982. https://doi.org/10.1890/10-0378.1

Gaines SD, White C, Carr MH, Palumbi SR (2010) Designing marine reserve networks for both conservation and fisheries management. Proc Natl Acad Sci 107(43):18286–18293

Garrity GM, Thompson LM, Ussery DW, Paskin N, Baker D, Desmeth P, Schindel DE et al (2009) Studies on monitoring and tracking genetic resources: an executive summary. Stand Genomic Sci 1(1):78–86. https://doi.org/10.4056/sigs.1491

Gerber LR, Del Mar Mancha-Cisneros M, O'Connor MI, Selig ER (2014) Climate change impacts on connectivity in the ocean: implications for conservation. Ecosphere 5(3):1–18

Gerwick WH, Moore BS (2012) Lessons from the past and charting the future of marine natural products drug discovery and chemical biology. Chem Biol 19(1):85–98. https://doi.org/10.1016/j.chembiol.2011.12.014

Giakoumi S, Scianna C, Plass-Johnson J, Micheli F, Grorud-Colvert K, Thiriet P, Claudet J et al (2017) Ecological effects of full and partial protection in the crowded Mediterranean sea: a regional meta-analysis. Sci Rep 7(1):8940. https://doi.org/10.1038/s41598-017-08850-w

Giakoumi S, McGowan J, Mills M, Beger M, Bustamante RH, Charles A, Christie P et al (2018) Revisiting 'success' and 'failure' of marine protected areas: a conservation scientist perspective. Front Mar Sci 5:223. https://doi.org/10.3389/fmars.2018.00223

Gibson DG, Glass JI, Lartigue C, Noskov VN, Chuang R-Y, Algire MA, Benders GA et al (2010) Creation of a bacterial cell controlled by a chemically synthesized genome. Science 329(5987):52–56. https://doi.org/10.1126/science.1190719

Giles PB (2011) How to claim a gene: application of the patent disclosure requirements to genetic sequences. Georgia State Univ Law Rev 27:35

Gill DA, Mascia MB, Ahmadia GN, Glew L, Lester SE, Barnes M, Craigie I et al (2017) Capacity shortfalls hinder the performance of marine protected areas globally. Nature 543(7647):665–669. https://doi.org/10.1038/nature21708

Giusti A, Tinacci L, Sotelo CG, Marchetti M, Guidi A, Zheng W, Armani A (2017) Seafood identification in multispecies products: assessment of 16SrRNA, Cytb, and COI universal primers' efficiency as a preliminary analytical step for setting up metabarcoding next-generation sequencing techniques. J Agric Food Chem 65(13):2902–2912. https://doi.org/10.1021/acs.jafc.6b05802

Glover KA, Solberg MF, McGinnity P, Hindar K, Verspoor E, Coulson MW, Hansen MM et al (2017) Half a century of genetic interaction between farmed and wild Atlantic salmon: status of knowledge and unanswered questions. Fish Fish 18(5):890–927. https://doi.org/10.1111/faf.12214

Green SJ, Akins JL, Maljković A, Côté IM (2012) Invasive lionfish drive Atlantic coral reef fish declines. PLoS One 7(3):e32596. https://doi.org/10.1371/journal.pone.0032596

Green ED, Rubin EM, Olson MV (2017) The future of DNA sequencing. Nat News 550(7675):179

Gregory AC, Zayed AA, Conceição-Neto N, Temperton B, Bolduc B, Alberti A, Ardyna M et al (2019) Marine DNA viral macro- and microdiversity from pole to pole. Cell 177(5):1109–23.e14. https://doi.org/10.1016/j.cell.2019.03.040

Greiber T (2012) An explanatory guide to the Nagoya protocol on access and benefit-sharing, vol 83. International Union for Conservation of Nature, Gland

Grigg RW, Dollar SJ (1990) Natural and anthropogenic disturbance on coral reefs. In: Dubinsky Z (ed) Ecosystem of the world, vol 25. Elsevier, New York, pp 439–452

Grorud-Colvert K, Claudet J, Tissot BN, Caselle JE, Carr MH, Day JC, Friedlander AM et al (2014) Marine protected area networks: assessing whether the whole is greater than the sum of its parts. PLoS One 9(8):e102298. https://doi.org/10.1371/journal.pone.0102298

Grorud-Colvert K, Constant V, Sullivan-Stack J, Dziedzic K, Hamilton SL, Randell Z, Fulton-Bennett H et al (2019) High-profile international commitments for ocean protection: empty promises or meaningful progress? Mar Policy 105:52–66. https://doi.org/10.1016/j.marpol.2019.04.003

Gu H, Chen X, Chen F, Zhou X, Parsaee Z (2018) Ultrasound-assisted biosynthesis of CuO-NPs using brown alga Cystoseira Trinodis: characterization, photocatalytic AOP, DPPH scavenging and antibacterial investigations. Ultrason Sonochem 41:109–119. https://doi.org/10.1016/j.ultsonch.2017.09.006

Guedes AC, Amaro HM, Sousa-Pinto I, Malcata FX (2019) Algal spent biomass—a pool of applications. In: Pandey A et al (eds) Biofuels from algae, 2nd edn. Elsevier, Amsterdam, pp 397–433

Gupta A, Barrow CJ, Puri M (2012) Omega-3 biotechnology: thraustochytrids as a novel source of omega-3 oils. Biotechnol Adv 30(6):1733–1745. https://doi.org/10.1016/j.biotechadv.2012.02.014

Hafsa AB, Nabi N, Salem Zellama M, Said K, Chaouachi M (2016) A new specific reference gene based on growth hormone gene (GH1) used for detection and relative quantification of Aquadvantage® GM Salmon (*Salmo Salar L.*) in food products. Food Chem 190:1040–1045

Halpern BS, Walbridge S, Selkoe KA, Kappel CV, Micheli F, D'Agrosa C, Bruno JF et al (2008) A global map of human impact on marine ecosystems. Science 319(5865):948–952. https://doi.org/10.1126/science.1149345

Halpern BS, Lester SE, Kellner JB (2009) Spillover from marine reserves and the replenishment of fished stocks. Environ Conserv 36(4):268–276

Halpern BS, Frazier M, Afflerbach J, Lowndes JS, Micheli F, O'Hara C, Scarborough C et al (2019) Recent pace of change in human impact on the world's ocean. Sci Rep 9(1):1–8. https://doi.org/10.1038/s41598-019-47201-9

Hamilton S (2013) Assessing the role of commercial aquaculture in displacing mangrove forest. Bull Mar Sci 89(2):585–601

Hampe A, Petit RJ (2005) Conserving biodiversity under climate change: the rear edge matters. Ecol Lett 8(5):461–467. https://doi.org/10.1111/j.1461-0248.2005.00739.x

Hansen BK, Bekkevold D, Worsøe Clausen L, Nielsen EE (2018) The sceptical optimist: challenges and perspectives for the application of environmental DNA in marine fisheries. Fish Fish 19(5):751–768. https://doi.org/10.1111/faf.12286

Hare MP, Nunney L, Schwartz MK, Ruzzante DE, Burford M, Waples RS, Ruegg K et al (2011) Understanding and estimating effective population size for practical application in marine species management. Conserv Biol 25(3):438–449. https://doi.org/10.1111/j.1523-1739.2010.01637.x

Hastings A, Botsford LW (2006) Persistence of spatial populations depends on returning home. Proc Natl Acad Sci 103(15):6067–6072

Hastings A, Gaines SD, Costello C (2017) Marine reserves solve an important bycatch problem in fisheries. Proc Natl Acad Sci 114(34):8927–8934. https://doi.org/10.1073/pnas.1705169114

Hauser L, Adcock GJ, Smith PJ, Bernal Ramírez JH, Carvalho GR (2002) Loss of microsatellite diversity and low effective population size in an overexploited population of New Zealand snapper (*Pagrus Auratus*). Proc Natl Acad Sci 99(18):11742–11747. https://doi.org/10.1073/pnas.172242899

He Y, Wu T, Sun H, Sun P, Liu B, Luo M, Chen F (2019) Comparison of fatty acid composition and positional distribution of microalgae triacylglycerols for human milk fat substitutes. Algal Res 37:40–50. https://doi.org/10.1016/j.algal.2018.11.004

Heeks R, Amalia M, Kintu R, Shah N (2013) Inclusive innovation: definition, conceptualisation and future research priorities. Working paper no. 53. Centre for Development Informatics, Manchester

Heidelberg KB, Gilbert JA, Joint I (2010) Minireview marine genomics: at the interface of marine microbial ecology and biodiscovery. Microb Biotechnol 3(5):531–543. https://doi.org/10.1111/j.1751-7915.2010.00193.x

Hellberg ME (2009) Gene flow and isolation among populations of marine animals. Annu Rev Ecol Evol Syst 40:291–310

Hendriks IE, Duarte CM (2008) Allocation of effort and imbalances in biodiversity research. J Exp Mar Biol Ecol 360(1):15–20. https://doi.org/10.1016/j.jembe.2008.03.004

Henriques R, Nielsen ES, Durholtz D, Japp D, von der Heyden S (2017) Genetic population sub-structuring of kingklip (*Genypterus Capensis*–Ophidiidae), a commercially exploited demersal fish off South Africa. Fish Res 187:86–95

Herrera H, Schroth E (2000) Profitable innovation without patent protection: the case of credit derivatives. Working paper 01-02. Center for Entrepreneurial Studies, Berkeley

High Seas Alliance (2019) Strategic environmental assessments (SEAs) under the new BBNJ instrument. HSA Briefing. http://highseasalliance.org/sites/highseasalliance.org/files/SEA%20brief%20final2.clean.pdf

Hilborn R, Quinn TP, Schindler DE, Rogers DE (2003) Biocomplexity and fisheries sustainability. Proc Natl Acad Sci 100(11):6564–6568. https://doi.org/10.1073/pnas.1037274100

Hill RT, Fenical W (2010) Pharmaceuticals from marine natural products: surge or Ebb? Curr Opin Biotechnol 21(6):777–779. https://doi.org/10.1016/j.copbio.2010.10.007

Hoffmann AA, Sgrò CM (2011) Climate change and evolutionary adaptation. Nature 470(7335):479–485. https://doi.org/10.1038/nature09670

Houssen WE, Bent AF, McEwan AR, Pieiller N, Tabudravu J, Koehnke J, Mann G et al (2014) An efficient method for the in vitro production of azol(in)e-based cyclic peptides. Angew Chem Int Ed 53(51):14171–14174. https://doi.org/10.1002/anie.201408082

https://www.orbisresearch.com/reports/index/global-cosmetic-products-market-research-report-and-forecast-to-2019-2023

https://schmidtocean.org/new-deep-sea-animal-discoveries-protections-costa-rica/

Hughes TP, Kerry JT, Baird AH, Connolly SR, Dietzel A, Eakin CM, Heron SF et al (2018) Global warming transforms coral reef assemblages. Nature 556(7702):492. https://doi.org/10.1038/s41586-018-0041-2

Hurwitz BL, Sullivan MB (2013) The Pacific Ocean Virome (POV): a marine viral metagenomic dataset and associated protein clusters for quantitative viral ecology. PLoS One 8(2):e57355. https://doi.org/10.1371/journal.pone.0057355

Hutchings JA, Fraser DJ (2008) The nature of fisheries- and farming-induced evolution. Mol Ecol 17(1):294–313. https://doi.org/10.1111/j.1365-294X.2007.03485.x

Hutchings JA, Reynolds JD (2004) Marine fish population collapses: consequences for recovery and extinction risk. BioScience 54(4):297–309. https://doi.org/10.1641/0006-3568(2004)054[0297:MFPCCF]2.0.CO;2

Hutchison CA, Chuang R-Y, Noskov VN, Assad-Garcia N, Deerinck TJ, Ellisman MH, Gill J et al (2016) Design and synthesis of a minimal bacterial genome. Science 351(6280):aad6253. https://doi.org/10.1126/science.aad6253

Ingeman KE, Samhouri JF, Stier AC (2019) Ocean recoveries for tomorrow's earth: hitting a moving target. Science 363(6425):eaav1004. https://doi.org/10.1126/science.aav1004

IPBES (Intergovernmental Science-Policy Platform on Biodiversity and Ecosystem Services) (2019) Summary for policymakers of the global assessment report on biodiversity and ecosystem services—unedited advance version. IPBES, Bonn

IUCN (International Union for Conservation of Nature) (2019) The IUCN red list of threatened species. IUCN, Gland

IUCN and WCPA (International Union for Conservation of Nature and World Commission on Protected Areas) (2018) Applying IUCN's global conservation standards to marine protected areas (MPAs). Delivering effective conservation action through MPAs, to secure ocean health & sustainable development. Version 1.0. IUCN and WCPA, Gland

Jasanoff S, Hurlbut JB, Saha K (2015) CRISPR democracy: gene editing and the need for inclusive deliberation. Issues Sci Technol 32(1). https://issues.org/crispr-democracy-gene-editing-and-the-need-for-inclusive-deliberation/

Jaume D, Duarte CM (2006) General aspects concerning marine and terrestrial biodiversity. In: Duarte CM (ed) The exploration of marine biodiversity—scientific and technological challenges. Fundación BBVA, Bilbao, pp 17–30. http://imedea.uib-csic.es/damiajaume/DamiaJaumewebpage_archivos/PDFs/BBVA-inglés.pdf

Jeunen G-J, Knapp M, Spencer HG, Lamare MD, Taylor HR, Stat M, Bunce M et al (2019) Environmental DNA (eDNA) metabarcoding reveals strong discrimination among diverse marine habitats connected by water movement. Mol Ecol Resour 19(2):426–438. https://doi.org/10.1111/1755-0998.12982

Jin P, Agustí S (2018) Fast adaptation of tropical diatoms to increased warming with trade-offs. Sci Rep 8(1):17771. https://doi.org/10.1038/s41598-018-36091-y

Jinek M, Chylinski K, Fonfara I, Hauer M, Doudna JA, Charpentier E (2012) A programmable dual-RNA-guided DNA endonuclease in adaptive bacterial immunity. Science 337(6096):816–821. https://doi.org/10.1126/science.1225829

Johannes RE (1975) Pollution and degradation of coral reef communities. In: Ferguson Wood EJ, Johannes RE (eds) Tropical marine pollution. Elsevier, Amsterdam, pp 13–50

Johnson JA, Altwegg R, Evans DM, Ewen JG, Gordon IJ, Pettorelli N, Young JK (2016) Is there a future for genome-editing technologies in conservation? Anim Conserv 19:97–101. https://doi.org/10.1111/acv.12273

Johnson BM, Kemp BM, Thorgaard GH (2018) Increased mitochondrial DNA diversity in ancient Columbia River Basin Chinook Salmon *Oncorhynchus Tshawytscha*. PLoS One 13(1):e0190059. https://doi.org/10.1371/journal.pone.0190059

Jonsson PR, Jacobi MN, Moksnes P-O (2016) How to select networks of marine protected areas for multiple species with different dispersal strategies. Divers Distrib 22(2):161–173

Jouffray J-B, Blasiak R, Norström AV, Österblom H, Nyström M (2020) The blue acceleration—the trajectory of human expansion into the ocean. One Earth 2(1):43–54. https://doi.org/10.1016/j.oneear.2019.12.016

Karlsson S, Diserud OH, Fiske P, Hindar K (2016) Widespread genetic introgression of escaped farmed Atlantic salmon in wild salmon populations. ICES J Mar Sci 73(10):2488–2498. https://doi.org/10.1093/icesjms/fsw121

Kitada M, Williams E, Froholdt LL (eds) (2015) Maritime women: global leadership. Springer, Berlin

Kraus SD, Brown MW, Caswell H, Clark CW, Fujiwara M, Hamilton PK, Kenney RD et al (2005) North Atlantic right whales in crisis. Science 309(5734):561–562. https://doi.org/10.1126/science.1111200

Kyeremeh K, Flint A, Jaspars M, Sunassee SN, Amewu R, Dziwornu GA, Acquah KS et al (2020) Making north–south collaborations work: facilitating natural product drug discovery in Africa. In: Ramutsindela M, Mickler D (eds) Africa and the sustainable development goals, Sustainable development goals series. Springer International Publishing, Cham, pp 257–266. https://doi.org/10.1007/978-3-030-14857-7_24

Laikre L (2010) Genetic diversity is overlooked in international conservation policy implementation. Conserv Genet 11(2):349–354

Laikre L, Ryman N (1996) Effects on intraspecific biodiversity from harvesting and enhancing natural populations. Ambio 25(8):504–509

Laird S, Wynberg R (2012) Bioscience at a crossroads: implementing the Nagoya Protocol on access and benefit sharing in a time of scientific, technological and industry change. Secretariat of the Convention on Biological Diversity, Montreal

Laird S, Wynberg R (2018) A fact-finding and scoping study on digital sequence information on genetic resources in the context of the convention on biological diversity and the Nagoya Protocol. Secretariat of the Convention on Biological Diversity, Montreal, pp 2–79

Laird S, Monagle C, Johnston S, Queensland Biodiscovery Collaboration (2008) The Griffith University AstraZeneca partnership for natural product discovery. An access and benefit sharing case study. United Nations University, Institute for the Advanced Study of Sustainability, Tokyo

Laird S, Wynberg R, Rourke M, Humphries F, Muller MR, Lawson C (2020) Rethink the expansion of access and benefit sharing. Science 367(6483):1200–1202

Lasker HR (2013) Recruitment and resilience of a harvested Caribbean octocoral. PLoS One 8(9):e74587. https://doi.org/10.1371/journal.pone.0074587

Lautenschlager RA (1997) Biodiversity is dead. Wildlife Soc Bull (1973-2006) 25(3):679–685

Law R (2007) Fisheries-induced evolution: present status and future directions. Mar Ecol Prog Ser 335:271–278

Leal MC, Puga J, Serôdio J, Gomes NCM, Calado R (2012) Trends in the discovery of new marine natural products from invertebrates over the last two decades—where and what are we bioprospecting? PLoS One 7(1):e30580. https://doi.org/10.1371/journal.pone.0030580

Leary D (2018) Marine genetic resources in areas beyond national jurisdiction: do we need to regulate them in a new agreement? Maritime Saf Sec Law J 19(5):22–47

Leary D, Vierros M, Hamon G, Arico S, Monagle C (2009) Marine genetic resources: a review of scientific and commercial interest. Mar Policy 33(2):183–194. https://doi.org/10.1016/j.marpol.2008.05.010

Ledford H (2015) Salmon approval heralds rethink of transgenic animals. Nat News 527(7579):417

Lester SE, Halpern BS, Grorud-Colvert K, Lubchenco J, Ruttenberg BI, Gaines SD, Airame S et al (2009) Biological effects within no-take marine reserves: a global synthesis. Mar Ecol Prog Ser 384:33–46

Levin PS, Schiewe MH (2001) Preserving salmon biodiversity: the number of Pacific salmon has declined dramatically. But the loss of genetic diversity may be a bigger problem. Am Sci 89(3):220–227

Li L, Pitcher TJ, Devlin RH (2015) Potential risks of trophic impacts by escaped transgenic salmon in marine environments. Environ Conserv 42(2):152–161. https://doi.org/10.1017/S0376892914000319

Lindeque PK, Parry HE, Harmer RA, Somerfield PJ, Atkinson A (2013) Next generation sequencing reveals the hidden diversity of zooplankton assemblages. PLoS One 8(11):1–14. https://doi.org/10.1371/journal.pone.0081327

Locey KJ, Lennon JT (2016) Scaling laws predict global microbial diversity. Proc Natl Acad Sci 113(21):5970–5975

Logan CA, Dunne JP, Eakin CM, Donner SD (2014) Incorporating adaptive responses into future projections of coral bleaching. Global Change Biol 20(1):125–139. https://doi.org/10.1111/gcb.12390

Louca S, Mazel F, Doebeli M, Parfrey L (2019) A census-based estimate of earth's bacterial and archaeal diversity. PLoS Biol 17(2):e3000106

Lubchenco J, Gaines SD (2019) A new narrative for the ocean. Science 364(6444):911. https://doi.org/10.1126/science.aay2241

Lubchenco J, Grorud-Colvert K (2015) Making waves: the science and politics of ocean protection. Science 350(6259):382–383. https://doi.org/10.1126/science.aad5443

Lubchenco J, Cerny-Chipman EB, Reimer JN, Levin SA (2016) The right incentives enable ocean sustainability successes and provide hope for the future. Proc Natl Acad Sci U S A 113(51):14507–14514. https://doi.org/10.1073/pnas.1604982113

Madduppa HH, Timm J, Kochzius M (2018) Reduced genetic diversity in the clown Anemonefish *Amphiprion Ocellaris* in exploited reefs of Spermonde Archipelago, Indonesia. Front Mar Sci 5:80. https://doi.org/10.3389/fmars.2018.00080

Magris RA, Andrello M, Pressey RL, Mouillot D, Dalongeville A, Jacobi MN, Manel S (2018) Biologically representative and well-connected marine reserves enhance biodiversity persistence in conservation planning. Conserv Lett 11(4):e12439. https://doi.org/10.1111/conl.12439

Mannion M (1998) Nutraceutical revolution continues at foundation for innovation in medicine conference. Am J Nat Med 5:30–33

Manson JE, Cook NR, Lee I-M, Christen W, Bassuk SS, Mora S, Gibson H et al (2019) Marine N−3 fatty acids and prevention of cardiovascular disease and cancer. N Engl J Med 380(1):23–32. https://doi.org/10.1056/NEJMoa1811403

Manuel J, Maze K, Driver M, Stephens A, Botts E, Parker A, Tau M, et al (2016) Key ingredients, challenges and lessons from biodiversity mainstreaming in South Africa: people, products, processes. Environment working paper no. 107. Organisation for Economic Co-operation and Development, Paris. https://doi.org/10.1787/5jlzgj1s4h5h-en

Marine Conservation Institute (2020) Atlas of marine protection. http://mpatlas.org

MarinLit (2020) A database of marine natural products literature. http://pubs.rsc.org/marinlit/

Marteinsson VT, Birrien JL, Raguénès G, da Costa MS, Prieur D (1999) Isolation and characterization of thermus thermophilus Gy1211 from a deep-sea hydrothermal vent. Extremophiles 3(4): 247–251

Mayer AMS, Rodríguez AD, Taglialatela-Scafati O, Fusetani N (2017) Marine pharmacology in 2012–2013: marine compounds with antibacterial, antidiabetic, antifungal, anti-inflammatory, antiprotozoal, antituberculosis, and antiviral activities; affecting the immune and nervous systems, and other miscellaneous mechanisms of action. Mar Drugs 15(9):273. https://doi.org/10.3390/md15090273

Mayol E, Arrieta JM, Jiménez MA, Martínez-Asensio A, Garcias-Bonet N, Dachs J, González-Gaya B et al (2017) Long-range transport of airborne microbes over the global tropical and subtropical ocean. Nat Commun 8(1):1–9. https://doi.org/10.1038/s41467-017-00110-9

McCauley DJ, Pinsky ML, Palumbi SR, Estes JA, Joyce FH, Warner RR (2015) Marine defaunation: animal loss in the global ocean. Science 347(6219):1255641. https://doi.org/10.1126/science.1255641

McCook LJ, Almany GR, Berumen ML, Day JC, Green AL, Jones GP, Leis JM et al (2009) Management under uncertainty: guidelines for incorporating connectivity into the protection of coral reefs. Coral Reefs 28(2):353–366

McCormack JE, Rodríguez-Gómez F, Tsai WLE, Faircloth BC (2017) Transforming museum specimens into genomic resources. In: Webster MS (ed) The extended specimen: emerging frontiers in collections-based ornithological research. CRC Press, Boca Raton, pp 143–156

McGinnity P, Prodöhl P, Ferguson A, Hynes R, Maoiléidigh NÓ, Baker N, Cotter D et al (2003) Fitness reduction and potential extinction of wild populations of Atlantic Salmon, *Salmo Salar*, as a result of interactions with escaped farm Salmon. Proc R Soc Ser B Biol Sci 270(1532):2443–2450. https://doi.org/10.1098/rspb.2003.2520

McGinnity P, Jennings E, deEyto E, Allott N, Samuelsson P, Rogan G, Whelan K et al (2009) Impact of naturally spawning captive-bred Atlantic Salmon on wild populations: depressed recruitment and increased risk of climate-mediated extinction. Proc R Soc B Biol Sci 276(1673):3601–3610. https://doi.org/10.1098/rspb.2009.0799

McInerney CE, Allcock AL, Johnson MP, Prodohl PA (2012) Ecological coherence in marine reserve network design: an empirical evaluation of sequential site selection using genetic structure. Biol Conserv 152:262–270. https://doi.org/10.1016/j.biocon.2012.03.009

McNaughton SJ (1977) Diversity and stability of ecological communities: a comment on the role of empiricism in ecology. Am Nat 111(979):515–525. https://doi.org/10.1086/283181

Medema MH, Blin K, Cimermancic P, de Jager V, Zakrzewski P, Fischbach MA, Weber T et al (2011) antiSMASH: rapid identification, annotation and analysis of secondary metabolite biosynthesis gene clusters in bacterial and fungal genome sequences. Nucleic Acids Res 39(Suppl 2):W339–W346

Merges RP (2004) A new dynamism in the public domain. Univ Chicago Law Rev 71:183–203

Moberg F, Folke C (1999) Ecological goods and services of coral reef ecosystems. Ecol Econ 29(2):215–233. https://doi.org/10.1016/S0921-8009(99)00009-9

Mora C, Tittensor DP, Adl S, Simpson AGB, Worm B (2011) How many species are there on earth and in the ocean? PLoS Biol 9(8):e1001127. https://doi.org/10.1371/journal.pbio.1001127

Morgera E (2018) Fair and equitable benefit-sharing in a new treaty on marine biodiversity: a principled approach towards partnership building? Maritime Saf Secur Law J. 5/2018-19 – Special Issue on Ocean Commons. pp 48–77. http://dx.doi.org/10.2139/ssrn.3234346

Morikawa MK, Palumbi SR (2019) Using naturally occurring climate resilient corals to construct bleaching-resistant nurseries. Proc Natl Acad Sci 116(21):10586–10591

Morley JW, Selden RL, Latour RJ, Frölicher TL, Seagraves RJ, Pinsky ML (2018) Projecting shifts in thermal habitat for 686 species on the North American continental shelf. PLoS One 13(5):e0196127. https://doi.org/10.1371/journal.pone.0196127

Mühlroth A, Li K, Røkke G, Winge P, Olsen Y, Hohmann-Marriott MF, Vadstein O et al (2013) Pathways of lipid metabolism in marine algae, co-expression network, bottlenecks and candidate genes for enhanced production of EPA and DHA in species of chromista. Mar Drugs 11(11):4662–4697. https://doi.org/10.3390/md11114662

Muir P, Li S, Lou S, Wang D, Spakowicz DJ, Salichos L, Zhang J et al (2016) The real cost of sequencing: scaling computation to keep pace with data generation. Genome Biol 17(1):53. https://doi.org/10.1186/s13059-016-0917-0

Munday PL, Warner RR, Monro K, Pandolfi JM, Marshall DJ (2013) Predicting evolutionary responses to climate change in the sea. Ecol Lett 16(12):1488–1500. https://doi.org/10.1111/ele.12185

Musick JA, Harbin MM, Berkeley SA, Burgess GH, Eklund AM, Findley L, Gilmore RG et al (2000) Marine, estuarine, and diadromous fish stocks at risk of extinction in North America (exclusive of Pacific Salmonids). Fisheries 25(11):6–30. https://doi.org/10.1577/1548-8446(2000)025<0006:MEADFS>2.0.CO;2

National Academies of Sciences, Engineering, and Medicine (2019) A research review of interventions to increase the persistence and resilience of coral reefs. The National Academies Press, Washington, DC. https://doi.org/10.17226/25279

Naylor R, Hindar K, Fleming IA, Goldburg R, Williams S, Volpe J, Whoriskey F et al (2005) Fugitive salmon: assessing the risks of escaped fish from net-pen aquaculture. BioScience 55(5):427–437. https://doi.org/10.1641/0006-3568(2005)055[0427:FSATRO]2.0.CO;2

Nielsen J, Hedeholm RB, Heinemeier J, Bushnell PG, Christiansen JS, Olsen J, Ramsey CB et al (2016) Eye lens radiocarbon reveals centuries of longevity in the Greenland Shark (*Somniosus Microcephalus*). Science 353(6300):702–704. https://doi.org/10.1126/science.aaf1703

Norström AV, Nyström M, Jouffray J-B, Folke C, Graham NAJ, Moberg F, Olsson P et al (2016) Guiding coral reef futures in the anthropocene. Front Ecol Environ 14(9):490–498. https://doi.org/10.1002/fee.1427

O'Leary BC, Winther-Janson M, Bainbridge JM, Aitken J, Hawkins JP, Roberts CM (2016) Effective coverage targets for ocean protection. Conserv Lett 9(6):398–404

Oguz T, Velikova V (2010) Abrupt transition of the northwestern Black Sea Shelf ecosystem from a eutrophic to an alternative pristine state. Mar Ecol Prog Ser 405:231–242. https://doi.org/10.3354/meps08538

Oldham P, Hall S, Forero O (2013) Biological diversity in the patent system. PLoS One 8(11):e78737. https://doi.org/10.1371/journal.pone.0078737

Oldham P, Hall S, Barnes C, Oldham C, Cutter M, Burns N, Kindness L (2014) Valuing the deep: marine genetic resources in areas beyond national jurisdiction. Department for Environment, Food and Rural Affairs, London. http://randd.defra.gov.uk/Document.aspx?Document=12289_ValuingTheDeepDEFRAFinalVersion-One28092014.pdf

Ooho Water (n.d.) Home. http://www.oohowater.com.

Oregon State University, World Database on Protected Areas, IUCN-Marine, National Geographic Pristine Seas, and Marine Conservation Institute's MPAtlas (2019) The MPA Guide. Wcmc.io/8408

Österblom H, Wabnitz CCC, Tladi D et al (2020) Towards ocean equity. World Resources Institute, Washington, DC. www.oceanpanel.org/blue-papers/how-distribute-benefits-ocean-equitably

Oyinlola MA, Reygondeau G, Wabnitz CCC, Troell M, Cheung WWL (2018) Global estimation of areas with suitable environmental conditions for mariculture species. PLoS One 13(1):e0191086. https://doi.org/10.1371/journal.pone.0191086

Paddon CJ, Keasling JD (2014) Semi-synthetic artemisinin: a model for the use of synthetic biology in pharmaceutical development. Nat Rev Microbiol 12(5):355–367. https://doi.org/10.1038/nrmicro3240

Pal A, Paul AK (2008) Microbial extracellular polymeric substances: central elements in heavy metal bioremediation. Indian J Microbiol 48(1):49–64. https://doi.org/10.1007/s12088-008-0006-5

Palkovacs EP, Wasserman BA, Kinnison MT (2011) Eco-evolutionary trophic dynamics: loss of top predators drives trophic evolution and ecology of prey. PLoS One 6(4):e18879. https://doi.org/10.1371/journal.pone.0018879

Palstra FP, Ruzzante DE (2008) Genetic estimates of contemporary effective population size: what can they tell us about the importance of genetic stochasticity for wild population persistence? Mol Ecol 17(15):3428–3447. https://doi.org/10.1111/j.1365-294x.2008.03842.x

Palumbi SR (2014) Mechanisms of reef coral resistance to future climate change. Science 344(6186):895–898. https://doi.org/10.1126/science.1251336

Parsons KM, Meredith E, Marilyn D, Linda P (2019) Water, water everywhere: environmental DNA can unlock population structure in elusive marine species. R Soc Open Sci 5(8):180537. https://doi.org/10.1098/rsos.180537

Pazos G, Fernández J, Linares F, Sánchez J, Otero JJ, Iglesias J, Domingues P (2018) The complete life cycle in captivity of the spider crab, *Maja Brachydactyla*, Herbst 1788. Aquacult Res 49(7):2440–2445. https://doi.org/10.1111/are.13703

Pearce BKD, Tupper AS, Pudritz RE, Higgs PG (2018) Constraining the time interval for the origin of life on earth. Astrobiology 18(3):343–364. https://doi.org/10.1089/ast.2017.1674

Pecl GT, Araújo MB, Bell JD, Blanchard J, Bonebrake TC, Chen I-C, Clark TD et al (2017) Biodiversity redistribution under climate change: impacts on ecosystems and human well-being. Science 355(6332):eaai9214. https://doi.org/10.1126/science.aai9214

Pedersen EJ, Thompson PL, Ball RA, Fortin M-J, Gouhier TC, Link H, Moritz C et al (2017) Signatures of the collapse and incipient recovery of an overexploited marine ecosystem. R Soc Open Sci 4(7):170215. https://doi.org/10.1098/rsos.170215

Perry AL, Low PJ, Ellis JR, Reynolds JD (2005) Climate change and distribution shifts in marine fishes. Science 308(5730):1912–1915. https://doi.org/10.1126/science.1111322

Pevsner J (2015) Bioinformatics and functional genomics. John Wiley & Sons, Oxford

Phelps MP, Seeb LW, Seeb JE (2019) Transforming ecology and conservation biology through genome editing. Conserv Biol 34(1):54–65. https://doi.org/10.1111/cobi.13292

Piaggio AJ, Segelbacher G, Seddon PJ, Alphey L, Bennett EL, Carlson RH, Friedman RM et al (2017) Is it time for synthetic biodiversity conservation? Trends Ecol Evol 32(2):97–107. https://doi.org/10.1016/j.tree.2016.10.016

Pikitch EK (2018) A tool for finding rare marine species: environmental DNA analysis shows promise for studying rare and elusive marine species. Science 360(6394):1180–1182

Pinsky ML, Palumbi SR (2014) Meta-analysis reveals lower genetic diversity in overfished populations. Mol Ecol 23(1):29–39. https://doi.org/10.1111/mec.12509

Pinsky ML, Worm B, Fogarty MJ, Sarmiento JL, Levin SA (2013) Marine taxa track local climate velocities. Science 341(6151):1239–1242. https://doi.org/10.1126/science.1239352

Pinsky ML, Eikeset AM, McCauley DJ, Payne JL, Sunday JM (2019) Greater vulnerability to warming of marine versus terrestrial ectotherms. Nature 569(7754):108

PISCO and UNS (Partnership for Interdisciplinary Studies of Coastal Oceans and University of Nice Sophia Antipolis) (2016) The science of marine reserves. 3rd ed. Mediterranean Version

Pitcher GC, Probyn TA (2016) Suffocating phytoplankton, suffocating waters—red tides and anoxia. Front Mar Sci 3. https://doi.org/10.3389/fmars.2016.00186

Pitz KJ, Closek CJ, Djurhuus A, Michisaki RP, Walz K, Boehm AB, Breitbart M et al (2017) Rewards and challenges of eDNA sequencing with multiple genetic markers for marine observation programs. Biodivers Inf Sci Standards 1:e20548

Poloczanska ES, Burrows MT, Brown CJ, García Molinos J, Halpern BS, Hoegh-Guldberg O, Kappel CV et al (2016) Responses of marine organisms to climate change across oceans. Front Mar Sci 3. https://doi.org/10.3389/fmars.2016.00062.

Pörtner HO, Peck MA (2010) Climate change effects on fishes and fisheries: towards a cause-and-effect understanding. J Fish Biol 77(8):1745–1779. https://doi.org/10.1111/j.1095-8649.2010.02783.x

Pratchett MS, Thompson CA, Hoey AS, Cowman PF, Wilson SK (2018) Effects of coral bleaching and coral loss on the structure and function of reef fish assemblages. In: van Oppen MJH, Lough JM (eds) Coral bleaching: patterns, processes, causes and consequences. Springer, Cham, pp 265–293. https://doi.org/10.1007/978-3-319-75393-5_11

Prathapan KD, Pethiyagoda R, Bawa KS, Raven PH, Rajan PD (2018) When the cure kills—CBD limits biodiversity research. Science 360(6396):1405–1406

Protected Planet (2020) Explore the world's marine protected areas. https://www.protectedplanet.net/marine

Prouty NG, Roark EB, Buster NA, Ross SW (2011) Growth rate and age distribution of deep-sea black corals in the Gulf of Mexico. Mar Ecol Prog Ser 423:101115. https://doi.org/10.3354/meps08953

Provan J, Maggs CA (2012) Unique genetic variation at a species' rear edge is under threat from global climate change. Proc Biol Sci 279(1726):39–47. https://doi.org/10.1098/rspb.2011.0536

Pujolar JM, Schiavina M, Di Franco A, Melià P, Guidetti P, Gatto M, De Leo GA et al (2013) Understanding the effectiveness of marine protected areas using genetic connectivity patterns and Lagrangian simulations. Divers Distrib 19(12):1531–1542

Qi S-H, Ma X (2017) Antifouling compounds from marine invertebrates. Mar Drugs 15(9):263. https://doi.org/10.3390/md15090263

Quah D (2002) Technology dissemination and economic growth: some lessons for the new economy. Technol New Econ 3:95–156

Quince C, Walker AW, Simpson JT, Loman NJ, Segata N (2017) Shotgun metagenomics, from sampling to analysis. Nat Biotechnol 35(9):833–844. https://doi.org/10.1038/nbt.3935

Rabone M, Harden-Davies H, Collins JE, Zajderman S, Appeltans W, Droege G, Brandt A et al (2019) Access to marine genetic resources (MGR): raising awareness of best-practice through a new agreement for biodiversity beyond national jurisdiction (BBNJ). Front Mar Sci 6. https://doi.org/10.3389/fmars.2019.00520

Ramesh N, Rising JA, Oremus KL (2019) The small world of global marine fisheries: the cross-boundary consequences of larval dispersal. Science 364(6446):1192–1196. https://doi.org/10.1126/science.aav3409

Ramos JE, Pecl GT, Moltschaniwskyj NA, Semmens JM, Souza CA, Strugnell JM (2018) Population genetic signatures of a climate change driven marine range extension. Sci Rep 8(1):1–12. https://doi.org/10.1038/s41598-018-27351-y

Reed DH, Frankham R (2003) Correlation between fitness and genetic diversity. Conserv Biol 17(1):230–237. https://doi.org/10.1046/j.1523-1739.2003.01236.x

Rees HC, Maddison BC, Middleditch DJ, Patmore JRM, Gough KC (2014) The detection of aquatic animal species using environmental DNA: a review of eDNA as a survey tool in ecology. J Appl Ecol 51(5):1450–1459. https://doi.org/10.1111/1365-2664.12306

Resplandy L, Keeling RF, Eddebbar Y, Brooks MK, Wang R, Bopp L, Long MC et al (2018) Quantification of ocean heat uptake from changes in atmospheric O2 and CO2 composition. Nature 563(7729):105–108. https://doi.org/10.1038/s41586-018-0651-8

Reynolds LK, McGlathery KJ, Waycott M (2012) Genetic diversity enhances restoration success by augmenting ecosystem services. PLoS One 7(6):e38397. https://doi.org/10.1371/journal.pone.0038397

Reynolds PL, Richardson JP, Duffy JE (2014) Field experimental evidence that grazers mediate transition between microalgal and seagrass dominance. Limnol Oceanogr 59(3):1053–1064. https://doi.org/10.4319/lo.2014.59.3.1053

Rhode C, Hepple J-A, Jansen S, Davis T, Vervalle J, Bester-van der Merwe AE, Roodt-Wilding R (2012) A population genetic analysis of abalone domestication events in South Africa: implications for the management of the abalone resource. Aquaculture 356–357:235–242. https://doi.org/10.1016/j.aquaculture.2012.05.012

Rincón-Cervera MÁ, González-Barriga V, Valenzuela R, López-Arana S, Romero J, Valenzuela A (2019) Profile and distribution of fatty acids in edible parts of commonly consumed marine fishes in Chile. Food Chem 274:123–129. https://doi.org/10.1016/j.foodchem.2018.08.113

Rittschof D (2017) Off the shelf fouling management. Mar Drugs 15(6):176. https://doi.org/10.3390/md15060176

Roark EB, Guilderson TP, Dunbar RB, Fallon SJ, Mucciarone DA (2009) Extreme longevity in proteinaceous deep-sea corals. Proc Natl Acad Sci 106(13):5204–5208. https://doi.org/10.1073/pnas.0810875106

Roberts CM (1995) Effects of fishing on the ecosystem structure of coral reefs. Conserv Biol 9(5):988–995

Roberts CM, Branch G, Bustamante RH, Castilla JC, Dugan J, Halpern BS, Lafferty KD et al (2003) Application of ecological criteria in selecting marine reserves and developing reserve networks. Ecol Appl 13(1):S215–S228

Roberts CM, O'Leary BC, McCauley DJ, Cury PM, Duarte CM, Lubchenco J, Pauly D et al (2017) Marine reserves can mitigate and promote adaptation to climate change. Proc Natl Acad Sci 114(24):6167–6175. https://doi.org/10.1073/pnas.1701262114

Roche RC, Harvey CV, Harvey JJ, Kavanagh AP, McDonald M, Stein-Rostaing VR, Turner JR (2016) Recreational diving impacts on coral reefs and the adoption of environmentally responsible practices within the Scuba Diving Industry. Environ Manag 58(1):107–116. https://doi.org/10.1007/s00267-016-0696-0

Rodolfi L, Chini Zittelli G, Bassi N, Padovani G, Biondi N, Bonini G, Tredici MR (2009) Microalgae for oil: strain selection, induction of lipid synthesis and outdoor mass cultivation in a low-cost photobioreactor. Biotechnol Bioeng 102(1):100–112. https://doi.org/10.1002/bit.22033

Rogers AD (2019) The deep: the hidden wonders of our oceans and how we can protect them. Wildfire, London

Rogers AD, Brierley A, Croot P, Cunha MR, Danovaro R, Devey C, Hoel AH, Ruhl H, Sarradin P-M, Trevisanut S, van den Hove S, Vieira H, Visbeck M (2015) Delving deeper: critical challenges for 21st century deep-sea research. In: Larkin KE, Donaldson K, McDonough N (eds) Position paper 22 of the European Marine Board. Ostend, Belgium, 224 pp. ISBN 978-94-920431-1-5

Roman J, Darling JA (2007) Paradox lost: genetic diversity and the success of aquatic invasions. Trends Ecol Evol 22(9):454–464. https://doi.org/10.1016/j.tree.2007.07.002

Roque BM, Salwen JK, Kinley R, Kebreab E (2019) Inclusion of Asparagopsis Armata in lactating dairy cows' diet reduces enteric methane emission by over 50 percent. J Clean Prod 234:132–138. https://doi.org/10.1016/j.jclepro.2019.06.193

Ruiz Muller M (2015) Genetic resources as natural information. Implications for the convention on biological diversity and Nagoya protocol, 1st edn. Routledge, London

Ruiz-Frau A, Possingham HP, Edwards-Jones G, Klein CJ, Segan D, Kaiser MJ (2015) A multidisciplinary approach in the design of marine protected areas: integration of science and stakeholder based methods. Ocean Coast Manag 103:86–93

Ruzzante DE, Mariani S, Bekkevold D, André C, Mosegaard H, Clausen LAW, Dahlgren TG et al (2006) Biocomplexity in a highly migratory pelagic marine fish, Atlantic herring. Proc R Soc B Biol Sci 273(1593):1459–1464. https://doi.org/10.1098/rspb.2005.3463

Ryckebosch E, Bruneel C, Muylaert K, Foubert I (2012) Microalgae as an alternative source of omega-3 long chain polyunsaturated fatty acids. Lipid Technol 24(6):128–130. https://doi.org/10.1002/lite.201200197

Sala E, Giakoumi S (2018) No-take marine reserves are the most effective protected areas in the ocean. ICES J Mar Sci 75(3):1166–1168. https://doi.org/10.1093/icesjms/fsx059

Sala E, Costello C, Dougherty D, Heal G, Kelleher K, Murray JH, Rosenberg AA et al (2013) A general business model for marine reserves. PLoS One 8(4):e58799

Sala E, Lubchenco J, Grorud-Colvert K, Novelli C, Roberts C, Sumaila UR (2018) Assessing real progress towards effective ocean protection. Mar Policy 91:11–13. https://doi.org/10.1016/j.marpol.2018.02.004

Sanger F, Nicklen S, Coulson AR (1977) DNA sequencing with chain-terminating inhibitors. Proc Natl Acad Sci 74(12):5463–5467

Scheffers BR, De Meester L, Bridge TCL, Hoffmann AA, Pandolfi JM, Corlett RT, Butchart SHM et al (2016) The broad footprint of climate change from genes to biomes to people. Science 354(6313):aaf7671. https://doi.org/10.1126/science.aaf7671

Schindler DE, Hilborn R (2015) Prediction, precaution, and policy under global change. Science 347(6225):953–954. https://doi.org/10.1126/science.1261824

Schindler DE, Hilborn R, Chasco B, Boatright CP, Quinn TP, Rogers LA, Webster MS (2010) Population diversity and the portfolio effect in an exploited species. Nature 465(7298):609–612. https://doi.org/10.1038/nature09060

Schleicher J, Zaehringer JG, Fastré C, Vira B, Visconti P, Sandbrook C (2019) Protecting half of the planet could directly affect over one billion people. Nat Sustain 2(12):1094–1096

Shahidi F, Ambigaipalan P (2015) Novel functional food ingredients from marine sources. Curr Opin Food Sci 2:123–129. https://doi.org/10.1016/j.cofs.2014.12.009

Sherman CDH, Lotterhos KE, Richardson MF, Tepolt CK, Rollins LA, Palumbi SR, Miller AD (2016) what are we missing about marine invasions? Filling in the gaps with evolutionary genomics. Mar Biol 163(10):198. https://doi.org/10.1007/s00227-016-2961-4

Sigwart JD, Chen C, Thomas EA, Allcock AL, Böhm M, Seddon M (2019) Red listing can protect deep-sea biodiversity. Nat Ecol Evol 3(8):1134. https://doi.org/10.1038/s41559-019-0930-2

Simmons M, Tucker A, Chadderton WL, Jerde CL, Mahon AR (2015) Active and passive environmental DNA surveillance of aquatic invasive species. Can J Fish Aquat Sci 73(1):76–83. https://doi.org/10.1139/cjfas-2015-0262

Sjöqvist CO, Kremp A (2016) Genetic diversity affects ecological performance and stress response of marine diatom populations. ISME J 10(11):2755–2766. https://doi.org/10.1038/ismej.2016.44

Skowno AL, Poole CJ, Raimondo DC, Sink KJ, Van Deventer H, Van Niekerk L, Harris L, et al (2019) National biodiversity assessment 2018: the status of South Africa's ecosystems and biodiversity. Synthesis report. South African National Biodiversity Institute, an entity of the Department of Environment, Forestry and Fisheries, Pretoria

Slobodian L, Kinna R, Kambu A, Ognibene L (2015) Bioprospecting in the global commons: legal issues brief. United Nations Environment Programme. Division of Environmental Law and Conventions. Environmental Law and Governance Branch, Nairobi

Smith D, Jabour J (2018) MPAs in ABNJ: lessons from two high seas regimes. ICES J Mar Sci 75(1):417–425. https://doi.org/10.1093/icesjms/fsx189

Smith MD, Asche F, Guttormsen AG, Wiener JB (2010) Genetically modified salmon and full impact assessment. Science 330(6007):1052–1053

Sorte CJB, Williams SL, Carlton JT (2010) Marine Range shifts and species introductions: comparative spread rates and community impacts. Glob Ecol Biogeogr 19(3):303–316. https://doi.org/10.1111/j.1466-8238.2009.00519.x

Spalding MD, Brown BE (2015) Warm-water coral reefs and climate change. Science 350(6262):769–771. https://doi.org/10.1126/science.aad0349

Stefanni S, Stanko D, Borme D, De Olazabal A, Juretić T, Pallavicini A, Tirelli V (2018) Multi-marker metabarcoding approach to study mesozooplankton at basin scale. Sci Rep 8:1–13. https://doi.org/10.1038/s41598-018-30157-7

Stephens ZD, Lee SY, Faghri F, Campbell RH, Zhai C, Efron MJ, Iyer R et al (2015) Big data: astronomical or genomical? PLoS Biol 13(7):e1002195. https://doi.org/10.1371/journal.pbio.1002195

Stewart KA (2019) Understanding the effects of biotic and abiotic factors on sources of aquatic environmental DNA. Biodivers Conserv 28(5):983–1001. https://doi.org/10.1007/s10531-019-01709-8

Stokstad E (2018) Norwegian billionaire funds deluxe deep ocean research ship. Science. https://www.sciencemag.org/news/2018/11/norwegian-billionaire-funds-deluxe-deep-ocean-research-ship

Strother PK, Battison L, Brasier MD, Wellman CH (2011) Earth's earliest non-marine eukaryotes. Nature 473(7348):505–509. https://doi.org/10.1038/nature09943

Suleria HAR, Osborne S, Masci P, Gobe G (2015) Marine-based nutraceuticals: an innovative trend in the food and supplement industries. Mar Drugs 13(10):6336–6351. https://doi.org/10.3390/md13106336

Sunagawa S, Coelho LP, Chaffron S, Kultima JR, Labadie K, Salazar G, Djahanschiri B et al (2015) Structure and function of the global ocean microbiome. Science 348(6237):1261359. https://doi.org/10.1126/science.1261359

Taberlet P, Coissac E, Pompanon F, Brochmann C, Willerslev E (2012) Towards next-generation biodiversity assessment using DNA metabarcoding. Mol Ecol 21(8):2045–2050. https://doi.org/10.1111/j.1365-294X.2012.05470.x

Taylor ML, Roterman CN (2017) Invertebrate population genetics across earth's largest habitat: the deep-sea floor. Mol Ecol 26(19):4872–4896. https://doi.org/10.1111/mec.14237

Teske PR, von der Heyden S, McQuaid CD, Barker NP (2011) A review of marine phylogeography in Southern Africa coastal phylogeography. South Afr J Sci 107:43–53. https://doi.org/10.4102/sajs.v107i5/6.514

Thambisetty S (2007) Patents as credence goods. Oxford J Legal Stud 27(4):707–740

Thambisetty S (2019) Due diligence and ABS compliance under EUR 511/2014. In: SSRN scholarly paper ID 3323389. Social Science Research Network, Rochester. https://papers.ssrn.com/abstract=3323389

Thambisetty S (2020) Biodiversity beyond national jurisdiction: (intellectual) property heuristics. In: Biodiversity beyond national jurisdiction: intractable challenges & potential solutions. Brill Nijhoff, Leiden

Thilsted SH, Thorne-Lyman A, Webb P, Bogard JR, Subasinghe R, Phillips MJ, Allison EH (2016) Sustaining healthy diets: the role of capture fisheries and aquaculture for improving nutrition in the post-2015 era. Food Policy 61:126–131. https://doi.org/10.1016/j.foodpol.2016.02.005

Thomsen PF, Willerslev E (2015) Environmental DNA: an emerging tool in conservation for monitoring past and present biodiversity. Biol Conserv 183:4–18. https://doi.org/10.1016/j.biocon.2014.11.019

Thornburg CC, Britt JR, Evans JR, Akee RK, Whitt JA, Trinh SK, Harris MJ et al (2018) NCI program for natural product discovery: a publicly-accessible library of natural product fractions for high-throughput screening. ACS Chem Biol 13(9):2484–2497. https://doi.org/10.1021/acschembio.8b00389

Tinacci L, Guidi A, Toto A, Guardone L, Giusti A, D'Amico P, Armani A (2018) DNA barcoding for the verification of supplier's compliance in the seafood chain: how the lab can support companies in ensuring traceability. Ital J Food Saf 7(2):6894. https://doi.org/10.4081/ijfs.2018.6894

Tolley KA, Šmíd J (2019) Chapter 4: Evaluating the status and trends of landscape level genetic diversity: a case study using South African Reptiles. In: Tolley KA, da Silva JM, Jansen van Vurren B (eds) National biodiversity assessment 2018 technical report volume 7: genetic diversity. South African National Biodiversity Institute, Pretoria

Troell M, Naylor RL, Metian M, Beveridge M, Tyedmers PH, Folke C, Arrow KJ et al (2014) Does aquaculture add resilience to the global food system? Proc Natl Acad Sci 111(37):13257–13263. https://doi.org/10.1073/pnas.1404067111

Twichell J, Pollnac R, Christie P (2018) Lessons from Philippines MPA management: social ecological interactions, participation, and MPA performance. Environ Manag 61(6):916–927

UN (United Nations) (1995) Agreement for the implementation of the provisions of the United Nations Convention on the Law of the Sea of 10 December 1982 relating to the conservation and management of straddling fish stocks and highly migratory fish stocks. United Nations, Division for Ocean Affairs and the Law of the Sea, New York. https://documents-dds-ny.un.org/doc/UNDOC/GEN/N95/274/67/PDF/N9527467.pdf

Van Dover CL (2014) Impacts of anthropogenic disturbances at deep-sea hydrothermal vent ecosystems: a review. Mar Environ Res 102:59–72. https://doi.org/10.1016/j.marenvres.2014.03.008

Van Dover CL, Ardron JA, Escobar E, Gianni M, Gjerde KM, Jaeckel A, Jones DOB et al (2017) Biodiversity loss from deep-sea mining. Nat Geosci 10(7):464–465. https://doi.org/10.1038/ngeo2983

van Oppen MJH, Oliver JK, Putnam HM, Gates RD (2015) Building coral reef resilience through assisted evolution. Proc Natl Acad Sci 112(8):2307–2313. https://doi.org/10.1073/pnas.1422301112

Van Soest RWM, Boury-Esnault N, Vacelet J, Dohrmann M, Erpenbeck D, De Voogd NJ, Santodomingo N et al (2012) Global diversity of sponges (porifera). PLoS One 7(4):e35105. https://doi.org/10.1371/journal.pone.0035105

Vandeperre F, Higgins RM, Sánchez-Meca J, Maynou F, Goñi R, Martín-Sosa P, Pérez-Ruzafa A et al (2011) Effects of no-take area size and age of marine protected areas on fisheries yields: a meta-analytical approach. Fish Fish 12(4):412–426

Villarino E, Watson JR, Jönsson B, Gasol JM, Salazar G, Acinas SG, Estrada M et al (2018) Large-scale ocean connectivity and planktonic body size. Nat Commun 9(1):1–13. https://doi.org/10.1038/s41467-017-02535-8

Viteri C, Chávez C (2007) Legitimacy, local participation, and compliance in the Galápagos Marine Reserve. Ocean Coast Manag 50(3–4):253–274

Volkmann L, Martyn I, Moulton V, Spillner A, Mooers AO (2014) Prioritizing populations for conservation using phylogenetic networks. PLoS One 9(2):e88945

von der Heyden S (2009) Why do we need to integrate population genetics into South African marine protected area planning? Afr J Mar Sci 31(2):263–269. https://doi.org/10.2989/AJMS.2009.31.2.14.886

von der Heyden S, Lipinski MR, Matthee CA (2007) Mitochondrial DNA analyses of the cape hakes reveal an expanding, panmictic population for Merluccius capensis and population structuring for mature fish in Merluccius paradoxus. Mol Phylogenet Evol 42(2):517–527. https://doi.org/10.1016/j.ympev.2006.08.004

von der Heyden S, Lipinski MR, Matthee CA (2010) Remarkably low MtDNA control region diversity in an abundant demersal fish. Mol Phylogenet Evol 55:1183–1188. https://doi.org/10.1016/j.ympev.2009.09.018

Von Schomberg R (2013) A vision of responsible research and innovation. In: Owen R, Bessant J, Heintz M (eds) Responsible innovation. Wiley, Hoboken, pp 51–74. https://doi.org/10.1002/9781118551424.ch3
Walsh MR, Munch SB, Chiba S, Conover DO (2006) Maladaptive changes in multiple traits caused by fishing: impediments to population recovery. Ecol Lett 9(2):142–148. https://doi.org/10.1111/j.1461-0248.2005.00858.x
Waltz E (2016) GM salmon declared fit for dinner plates. Nature Publishing Group, San Francisco
Waltz E (2017) First genetically engineered salmon sold in Canada. Nat News 548(7666):148. https://doi.org/10.1038/nature.2017.22116
Wang K-L, Wu Z-H, Wang Y, Wang C-Y, Xu Y (2017) Mini-review: antifouling natural products from marine microorganisms and their synthetic analogs. Mar Drugs 15(9). https://doi.org/10.3390/md15090266
Waples RS, Audzijonyte A (2016) Fishery-induced evolution provides insights into adaptive responses of marine species to climate change. Front Ecol Environ 14(4):217–224. https://doi.org/10.1002/fee.1264
Waples RS, Hindar K, Hard JJ (2012) Genetic risks associated with marine aquaculture. U.S. Department of Commerce. NOAA Technical Memo NMFS-NWFSC-119
Ward A (2018) Exploring a new enzymatic tool for AAV production: balancing optimal nuclease activity with effective DNA digestion. Genet Eng Biotechnol News 38(3):22
Webster MS, Colton MA, Darling ES, Armstrong J, Pinsky ML, Knowlton N, Schindler DE (2017) Who should pick the winners of climate change? Trends Ecol Evol 32(3):167–173. https://doi.org/10.1016/j.tree.2016.12.007
Weeks R, Jupiter SD (2013) Adaptive comanagement of a marine protected area network in Fiji. Conserv Biol 27(6):1234–1244
Weir LK, Grant JWA (2005) Effects of aquaculture on wild fish populations: a synthesis of data. Environ Rev 13:145–168
Weltz K, Lyle JM, Ovenden J, Morgan JAT, Moreno A, Semmens JM (2017) Application of environmental DNA to detect an endangered marine skate species in the wild. PLoS One 12(6): e0178124
Wetterstrand KA (2019) DNA sequencing costs: data. National Human Genome Research Institute. https://www.genome.gov/about-genomics/fact-sheets/DNA-Sequencing-Costs-Data
Wilkinson CR, Buddemeier RW (1994) Global climate change and coral reefs: implications for people and reefs. Report of the UNEP-IOC-ASPEI-IUCN Global Task Team on the Implications of Climate Change on Coral Reefs. IUCN, Gland
Worm B, Tittensor DP (2011) Range contraction in large pelagic predators. Proc Natl Acad Sci 108(29):11942–11947
Worm B, Barbier EB, Beaumont N, Duffy JE, Folke C, Halpern BS, Jackson JBC et al (2006) Impacts of biodiversity loss on ocean ecosystem services. Science 314(5800):787–790. https://doi.org/10.1126/science.1132294
Wright D, Bishop JM, Matthee CA, Von Der Heyden S (2015) Genetic isolation by distance reveals restricted dispersal across a range of life histories: implications for biodiversity conservation planning across highly variable marine environments. Divers Distrib 21:698–710. https://doi.org/10.1111/ddi.12302
Wynberg R (2016) Marine genetic resources and bioprospecting in the Western Indian Ocean. In: Regional state of the coast report. United Nations Environment Programme, Nairobi Convention. https://doi.org/10.18356/7a82e6cd-en
Wynberg R, Laird SA (2018) Fast science and sluggish policy: the herculean task of regulating biodiscovery. Trends Biotechnol 36(1):1–3. https://doi.org/10.1016/j.tibtech.2017.09.002
Xiong X, Guardone L, Giusti A, Castigliego L, Gianfaldoni D, Guidi A, Andrea A (2016) DNA barcoding reveals chaotic labeling and misrepresentation of cod (鳕, Xue) products sold on the Chinese market. Food Control 60:519–532. https://doi.org/10.1016/j.foodcont.2015.08.028
Young TR, Walløe Tvedt M (2017) Drafting successful access and benefit-sharing contracts. Brill, Leiden
Zachos E (2017) To fight opioid addiction, scientists see promise in sea snail venom. National Geographic. https://www.nationalgeographic.com/news/2017/12/fight-opioid-addiction-scientists-see-promise-in-sea-snail-venom-spd/
Zaiko A, Pochon X, Garcia-Vazquez E, Olenin S, Wood SA (2018) Advantages and limitations of environmental DNA/RNA tools for marine biosecurity: management and surveillance of non-indigenous species. Front Mar Sci 5:322
Zeng C, Rowden AA, Clark MR, Gardner JPA (2017) Population genetic structure and connectivity of deep-sea stony corals (order Scleractinia) in the New Zealand region: implications for the conservation and management of vulnerable marine ecosystems. Evol Appl 10(10):1040–1054. https://doi.org/10.1111/eva.12509
Zhu B, Ge W (2018) Genome editing in fishes and their applications. Gen Comp Endocrinol 257:3–12. https://doi.org/10.1016/j.ygcen.2017.09.011

Leveraging Multi-target Strategies to Address Plastic Pollution in the Context of an Already Stressed Ocean

5

Highlights

- Plastic is the newest pollutant to be entering the ocean in significant quantities. It joins nonplastic solid waste; nutrients; antibiotics, parasiticides and other pharmaceuticals; heavy metals; industrial chemicals including persistent organic pollutants; pesticides; and oil and gas, each of which has a longer history of scholarship and greater body of existing research as an ocean pollutant than does plastic.
- There are four major sources that discharge pollutants into the ocean: municipal, agricultural (including aquaculture), industrial and maritime. These pollutants have damaging impacts on ecosystems and marine life, human health and the economy.
- The presence of plastic in the ocean in growing quantities is symptomatic of many societal challenges that are relevant to the other pollutants and pollution pathways: the lack of access to sanitation and wastewater and stormwater processing for millions of people around the world, the need for safe use and disposal of chemicals, the development and degradation of coastal zones, the need for an efficient use of natural resources, and the need for improved access to safe food and water.
- This paper proposes seven holistic approaches for the reduction of pollutants in the ocean: improve wastewater management; improve stormwater management; adopt green chemistry practices and new materials; implement coastal zone improvements; practice radical resource efficiency; recover and recycle the materials we use; and build local systems for safe food and water.
- These seven approaches address the major sources of pollution entering the ocean and contribute to multiple United Nations Sustainable Development Goals.
- Each of the approaches identified are cross-sectoral and system-level in nature, making them perfect candidates for delivery through public-private partnerships, innovative financing arrangements and leveraging capital from a range of sources.
- To solve the pollution challenge we need to start with the premise that there is no such thing as waste. The Earth is a closed system and there is nowhere for damaging pollution to go that won't harm ecosystems, plant and animal life and, ultimately, human life.
- Once we adopt a no-waste approach, our economy will be very effective at finding the most efficient ways to stop the problem of pollution

1 Introduction

1.1 Overview

The ocean is the ultimate sink for anthropogenic pollution. According to the HydroSHED model, over 80% of the land mass on Earth is in a watershed that drains directly to the ocean (Lehner and Grill 2013). Until recently, the ocean seemed to be endlessly able to absorb all the waste that human activity has discharged into it. The Ocean Health Index (OHI) scores the health of the ocean on a range of criteria, from how clean the water is to the ability of the ocean to continue providing services such as food provision, carbon storage, tourism and recreation, and biodiversity (Halpern et al. 2012). The 2019 combined global ocean score was 71 out of 100 (as it has been for the last five years), showing that significant impairment has occurred, but that many of the functions and services of the ocean remain and must be better managed (OHI 2019). The Clean

Originally published in:
Jambeck, J., E. Moss, B. Dubey et al. 2020. *Leveraging Multi-Target Strategies to Address Plastic Pollution in the Context of an Already Stressed Ocean.* Washington DC: World Resources Institute. Available online at: https://oceanpanel.org/blue-papers/pollution-and-regenerative-economy-municipal-industrial-agricultural-and-maritime-waste.
Reprint by Springer International Publishing (2023) with kind permission.
Published under license from the World Research Institute.

J. Lubchenco, P. M. Haugan (eds.), *The Blue Compendium*, https://doi.org/10.1007/978-3-031-16277-0_5

Water section of the OHI includes details on the statuses and pressures of chemical, nutrient, pathogen and trash pollution. It also includes social pressure as a further pressure. Indicators of resilience were based upon the Convention on Biological Diversity (in particular for marine ecosystems) and quality of governance (using Worldwide Governance Indicators). The score for Clean Water has tracked closely to the overall score, remaining at 70 for the past five years (OHI 2019). With an estimated 91% of all temperate and tropical coasts predicted to be heavily developed by 2050 (Nellemann et al. 2008), this is a critical time to significantly reduce and prevent anthropogenic pollution to the ocean.

Pollutants enter the ocean in four ways: They may be discharged directly into the ocean, discharged into rivers which flow to the ocean, washed from land by stormwater into rivers or directly into the ocean or deposited from the air onto land to be washed into waterways or directly into the ocean.

There are many anthropogenic sources of pollution, and this paper focuses on pollution inputs to the ocean from four sectors: municipal, agricultural, industrial and maritime. This paper focuses first on plastic, as the newest and least well understood pollutant, and puts plastic pollution in the context of an ocean already receiving significant pollution from nutrients, heavy metals, persistent organic pollutants (POPs), pesticides and oil.

While successful implementation of all the United Nations (UN) Sustainable Development Goals (SDGs) would help protect the ocean, SDG 14: Life Below Water is the primary SDG directly related to the ocean. But there are several other SDGs that are very relevant to pollution reaching the ocean: SDG 2: Zero Hunger, SDG 3: Good Health and Well-Being, SDG 6: Clean Water and Sanitation, SDG 8: Decent Work and Economic Growth, SDG 9: Industrial Innovation and Infrastructure, SDG 11: Sustainable Cities and Communities and SDG 12: Responsible Consumption and Production

1.2 Context

Plastic is the newest pollutant to be entering the ocean in significant quantities. It joins nonplastic solid waste; nutrients (nitrogen, phosphorous); antibiotics, parasiticides and other pharmaceuticals; heavy metals; industrial chemicals including persistent organic pollutants; pesticides; and oil and gas, each of which has a longer history of scholarship and greater body of existing research as an ocean pollutant than does plastic. This paper seeks to put ocean pollution from plastic into the context of total pollutant inputs to the ocean and identify the interventions that can have the greatest total impact on all pollution to the ocean, capitalising on the current global attention on plastic pollution.

In this Blue Paper, four major sectors that create pollutants are explored—municipal, agricultural (including aquaculture), industrial and maritime—and three types of impacts are characterised—ecosystems and marine life, human health and economic. The impacts on ecosystems include harm to marine life from ingestion of and entanglement from plastic, eutrophication and hypoxia, and biomagnification of chemicals. The human health impacts from direct or indirect exposure to these pollutants include reproductive, developmental, behavioural, neurologic, endocrine and immunologic adverse health effects; acute or chronic toxicity; cancer; increased exposure to pathogens and mosquito-borne diseases; and risk of entanglement or entrapment. The economic impacts come from impaired productivity of fisheries, loss of seafood supply resulting from toxicity and reduced tourism and recreation in coastal areas.

The presence of plastic in the ocean in growing quantities is one symptom of a set of societal challenges that are also relevant to the other pollutants and pollution pathways: the lack of access to sanitation and wastewater and stormwater processing for millions of people around the world; the need for safe use and disposal of chemicals; the development and degradation of coastal zones; the need for an efficient use of natural resources; and the need for improved access to safe food and water.

At the heart of these challenges is recognising that the notion that things can be thrown away is a myth—there is no 'away' where pollutants can safely go.

This paper proposes seven intervention approaches that lead with reducing plastic inputs to the ocean but also seek to maximise the reduction of other pollutants as co-benefits. Four types of actions were considered: innovation, infrastructure, policy and mindset. Specific actions of each type were identified across the sectors and pollutants described in the report. These actions were then bundled into the following seven holistic opportunities for action (not in ranked order):

1. Improve wastewater management
2. Improve stormwater management
3. Adopt green chemistry practices and new materials
4. Implement coastal zone improvements
5. Practice radical resource efficiency
6. Recover and recycle the materials we use
7. Build local systems for safe food and water

These seven opportunities for action address the major sources of pollution entering the ocean, and contribute to achieving the United Nations Sustainable Development Goals (SDGs). They would directly influence SDG targets 2.1, 2.3, 3.9, 6.1, 6.2, 6.3, 6.B, 8.3, 11.6, 12.2, 12.4, 12.5 and 14.1 and indirectly influence a number of others, such as through expanded economic opportunities, benefits to people's livelihoods and increased well-being. The cross-sector, system-level nature of these challenges makes them perfect candidates for public-private partnerships, innovative financing arrangements and leveraging capital from a range of sources.

Finally, while the body of research on plastic is growing rapidly, there remain significant data gaps both on inputs and impacts. More research is needed to better understand and document the scope and scale of plastic pollution, as well as its impacts on ecosystem and human health. Given the global nature of the problem, open data protocols that can facilitate the aggregation and sharing of compatible data are critical.

2 Sources of Ocean Pollution

This paper includes pollution inputs from land and sea, grouped into four sectors: municipal, agricultural, industrial and maritime.

Municipal sources are residential and commercial solid waste and wastewater as well as runoff from roads and landscaping activities. Additionally, debris entering the ocean as a result of natural disasters is included here.

Land-based agricultural activities impacting the ocean include plastic, pesticide and nutrient use as well as waste management for animal agriculture. **Ocean-based aquaculture**'s pollution impacts include the use of antibiotics and parasiticides, antifoulants containing heavy metals, loss of equipment and management of fish waste.

The **industrial** sector includes manufacturing, mining and energy production. Pollutants coming from this sector include plastic pellets and waste, other solid waste, dredge spoils, industrial chemicals including POPs, heavy metals, pharmaceuticals and pharmaceutical waste products, and oil and gas.

Maritime pollution comes from the shipping, cruise and fishing industries and from recreational boating. Pollution from these sources includes litter, food waste, sewage and accident debris.

Figure 5.1 shows the primary sources of pollution in the marine environment from these sectors. Table 5.1 summarises the types of pollution entering the ocean and the ways that each sector contributes to ocean pollution.

Other than specific pollutants regulated by international treaties in certain situations—e.g. International Convention for the Prevention of Pollution from Ships (MARPOL) for plastic discharges; Stockholm Convention for specific chemicals; Basel Convention for waste exportation; London Convention and Protocol for ocean dumping—and acts that have regulated discharge nationally and locally—e.g. total maximum daily loads under the Clean Water Act in the United States—pollutants continue to enter the ocean without consistent and global limits or regulation.

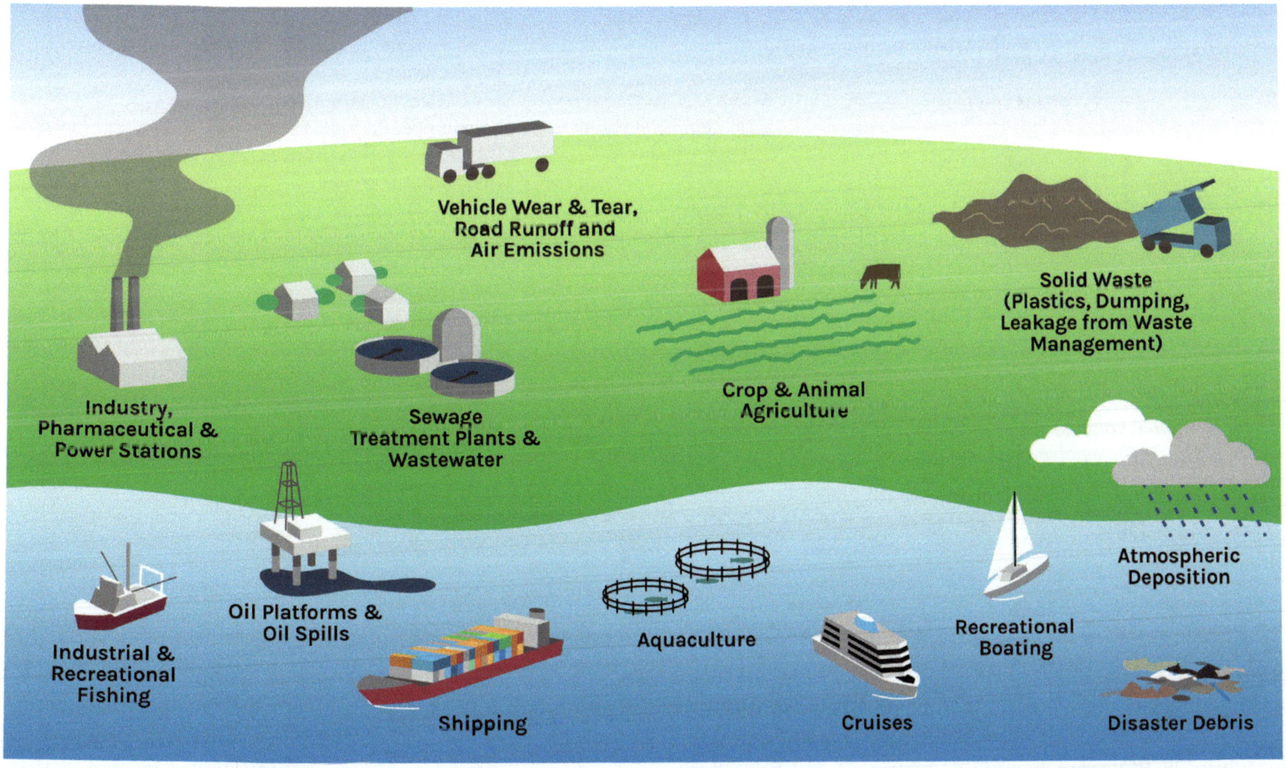

Fig. 5.1 Sources of ocean pollution. (Source: Graphic developed by K. Youngblood)

Table 5.1 Sources of pollutant discharges into the ocean

Subcategories	Municipal (coastal or near rivers) Residential, commercial	Agricultural and aquacultural Crops, animal land, aquaculture	Industrial Manufacturing, energy	Maritime Fishing, cruise, shipping, recreation
Microplastics (<5 millimetres [mm])[a]	Microbeads, microfibres, tire dust, fragments in runoff from land	Slow release fertiliser pellets, plastic mulch fragments	Industrial pellets	Pellets lost at sea in shipping accidents, dredged materials and breakdown of other wastes dumped at sea[b]
Macroplastics (>5 mm)[a]	Unmanaged plastic waste within 50 kilometres (km) of river or ocean[1]	Aquaculture infrastructure and equipment, greenhouses, plastic sheeting and associated equipment	Unknown	Fishing gear, lines and lures; litter from ships and boats; debris from shipping accidents
Other solid waste	Unmanaged solid waste within 50 km of river or ocean, disaster debris, wood, food waste dumping[c]	Lost/unmanaged aquaculture infrastructure and equipment, manure and biosolids land application	Dredge spoils	Fishing gear, litter from ships and boats, debris from shipping accidents, food waste discharge from ships
Pesticides[2]	Residential and commercial landscaping and gardening	Crop-based agriculture	Minimal	Minimal
Nutrients (N, P)	Untreated municipal wastewater, residential and commercial landscaping and gardening, airborne nitrogen from vehicle exhaust deposition into ocean	Crop-based agriculture, lagoon leakage, aquaculture fish waste	Airborne nitrogen from energy production deposition into ocean	Sewage discharges into ocean
Antibiotics, parasiticides, other pharmaceuticals	Treated and untreated wastewater	Aquaculture/mariculture, land-based animal agricultural runoff	Pharmaceutical production waste-water	Treated and untreated wastewater from ships
Heavy metals	Urban runoff: copper, chromium, nickel; mismanaged electronic waste	Aquaculture/mariculture: arsenic, mercury, cadmium, lead	Mining manufacturing: copper, zinc, lead, cadmium, chromium, nickel, arsenic, mercury	Paints and pigments: zinc, tributyltin, lead, cadmium
Industrial chemicals and persistent organic pollutants[c]	Treated and untreated wastewater, urban runoff	Use of organochlorine pesticides	Regulated and unregulated discharge from manufacturing	Treated and untreated wastewater from ships
Oil and gas	Urban runoff	Accidental discharge from agricultural equipment use and maintenance	Spills, water contamination, and improper disposal from oil refineries and logistics (pipelines, rail, trucks)	Drilling rigs, bilge water and fuel release, tanker spills, shipping

Table includes both point source (e.g. specific discharge points) and nonpoint source (e.g. stormwater runoff) forms of pollution

[a]*Macroplastics are any plastics larger than 5 mm. Microplastics are small pieces or fragments of plastic smaller than 5 mm* (Galgani et al. 2010; SAPEA 2019)

[b]*Wastes allowed to be dumped at sea according to the London Convention and Protocol include dredged materials; sewage sludge; fish waste, or material resulting from industrial fish processing operations; vessels and platforms or other man-made structures at sea; inert or inorganic geological material; organic matter of natural origin; bulky items comprising primarily iron, steel, concrete or non-harmful materials; and carbon dioxide streams from carbon dioxide capture processes for sequestration*

[c]*Jambeck et al. (2015)*

[d]*Persistent organic pollutants are organic compounds that are resistant to environmental degradation through chemical, biological and photolytic processes. They include polybrominated diphenyl ethers (PBDEs), per- and polyfluoroalkyl substances (PFASs), polychlorinated biphenyls (PCBs) and organochlorine (OC) pesticides.*

Sources:

1. *Jambeck et al. (2015)*
2. *Weibel et al.* 1966

Past emissions of ocean pollution remain relevant today, especially in the case of persistent pollutants such as plastics, heavy metals and POPs, as they remain in the ocean interacting with each other and the marine environment. For example, while 28 POPs are banned or restricted and have been for a number of years (12 since 2004, 16 since 2010), they are readily absorbed by plastic in the ocean, which creates a new mechanism for them to interact with the marine ecosystem (Rochman et al. 2013, 2014b; Rochman 2015). Heavy metals have also been found to adhere to plastic in the ocean as biofilms accumulate on its surface (Rochman et al. 2014a; Richard et al. 2019).

It should be noted that the ocean is also subject to other forms of pollution, including acidification (see Blue Paper 2, Gaines et al. 2019) and other nonphysical forms like thermal, noise and biological pollution. Thermal pollution is a change in temperature in the ocean water from discharges, often warmer water from powerplant cooling, that can change both physical and chemical properties of the ocean, impacting, for example, bivalves since they are stationary (Dong et al. 2018). Noise pollution in the ocean from shipping, oil and gas exploration and military activities can also impact marine life (Francis and Barber 2013). The International Whaling Commission and Convention on Biological Diversity have groups working on noise pollution. Biological pollution is the transfer of, for example, invasive species, which has been exacerbated by evolving habitats due to climate change and ocean acidification (Miranda et al. 2019), topics covered in Blue Paper 2 (Gaines et al. 2019). The transport of invasive species by plastic is covered in this paper. While these other pollution sources are out of scope for this paper, it is worth noting them here as they underscore the high number of stressors that ocean ecosystems are facing.

2.1 Plastic Pollution

Plastic is a material that has permanently changed our world since its introduction into mainstream society (in some countries) after World War II; global annual plastic production has increased from 1.7 million metric tonnes per year (mmt/yr) in 1950 to 422 mmt/yr. in 2018 (Geyer et al. 2017; PlasticsEurope 2019). Along with a steep increase in production, we have seen a resulting increase in plastic in the waste stream from 0.4% in 1960 to 13.2% in 2017 (by mass) in the United States (EPA 2014, 2019). In 1966, two U.S. Fish and Wildlife Service employees, Karl W. Kenyon and Eugene Kridler, were among the first scientists to document plastic and wildlife interactions when they discovered plastic had been consumed by seabird (albatross) chicks that died in the Hawaiian Islands National Wildlife Refuge (Kenyon and Kridler 1969). Since then, analysing plastic material flows (especially the waste streams), contamination in our environment and the economics of the material has become a recognised scientific discipline, with rapid increases in the science, especially in the last five years (Beaumont et al. 2019). But as a relevantly young scientific discipline, there are still many gaps in knowledge and a lack of information for solutions to plastic pollution (Bucci et al. 2019; Forrest et al. 2019). Even with knowledge gaps, plastic pollution has quickly become one of the most salient topics of late—people around the world passionately care about and want to address this issue.

2.1.1 Municipal Plastic Pollution

Plastic pollution is often subdivided into macroplastics and microplastics (e.g. the U.S. National Oceanic and Atmospheric Administration uses this division), and although there is much discussion internationally about terminology (GESAMP 2015), these size categories are currently used extensively around the world. Macroplastics are any plastics larger than 5 millimetres (mm) and can include both short-use items (e.g. food packaging and foodservice disposables) and longer-use items (e.g. flip flops, printer cartridges, synthetic textiles). Microplastics are small pieces or fragments of plastic less than 5 mm (Galgani et al. 2010; SAPEA 2019) that enter the environment as a consequence of either the direct release of small particles such as microbeads from cosmetic products; the fragmentation of larger items of litter in the environment; or the wear or abrasion of products during use, such as the release of fibres from textiles or particles from car tires (Law and Thompson 2014). The term microplastic was first used in this context in 2004 (Thompson et al. 2004) and the identification of microplastics is a relatively new field (Shim et al. 2017), with nanoscale plastics (not even yet formally defined) especially challenging to identify because of limits to the capabilities of the current instrumentation used for environmental samples. As a consequence, quantifying inputs has been challenging (Koelmans et al. 2015; Rist and Hartmann 2017; SAPEA 2019).

As of 2017, 8 billion metric tonnes of plastic had been produced for human use. Because a large quantity was used for packaging (about 40%) and single-use items, 6.4 billion metric tonnes had already become waste by 2015 (Geyer et al. 2017). Many packaging and single-use materials are composed of polyethylene (high and low density, HDPE and LDPE), polypropylene and polyethylene terephthalate (PET). These polymers are often the materials used in the most common items found littering the environment, especially on coastlines: cigarette butts, plastic bottles, plastic food wrappers, straws, plastic bags and bottle caps (Ocean Conservancy 2018).

The total quantity of plastic entering the ocean every year is still unknown. While there have been estimates of some sources (e.g. municipal waste), there are more sources that do not have current estimates. While many scientists would agree that a large portion of mismanaged plastic comes from land, even the 80% from land is a questionable statistic since the true total from all sources remains unknown. Some of the sources have been quantified. Jambeck et al. (2015) found that the annual input from mismanaged solid waste on land (one of the major sources) in 2010 was between 4.8 and 12.7 mmt/yr. Other estimates have come from riverine input and other geographic information system (GIS) analyses, which have found that from 0.41 to 4 mmt of plastic is entering the ocean every year from rivers (a subset of the total quantity entering the ocean) (Lebreton et al. 2017; Schmidt et al. 2017). Up to 99 mmt of mismanaged plastic waste has been estimated to be available to enter waterways around the world (Lebreton and Andrady 2019). The estimate of 8 mmt as a middle estimate for input to the ocean (Jambeck et al. 2015) remains the most widely used value for land-based input of plastic waste into the ocean, although this is likely conservative. Forrest et al. (2019) built on the existing research by incorporating additional estimates of plastic waste flows to the ocean arising from imported waste by developing countries from wealthier consumer economies. This export/import imbalance was initially outlined in (Brooks et al. 2018), which describes the plastic import ban, more commonly known as the National Sword policy, imposed by China and its impacts on global plastic scrap trade. Forrest et al. (2019) estimated current plastic flows to the ocean from all sources to be at least 15 mmt/yr.

There are at least two more global baseline estimates in the process of being calculated for plastic, one by a working group through the National Socio-Environmental Synthesis Center funded by the U.S. National Science Foundation and one through The Pew Charitable Trusts and SYSTEMIQ, which, while not available before publishing this document, will make it possible to measure the impacts of interventions at the global and country levels, similar to the wedges approach developed for climate change (Pacala and Socolow 2004). Clearly topography and proximity to the ocean are relevant for land-based or riverine plastic, but some of the biggest data gaps in modelling and measuring quantities entering the ocean exist for these pathways. The most credible current estimates nonetheless indicate that the quantities of plastic entering the ocean are significant. The only regulatory limits on plastic concentrations in the ocean are the total maximum daily load limits in aquatic systems in the United States (Smith 2000; MDOE and DCDOE 2010), the Convention on the Prevention of Marine Pollution by Dumping of Wastes and Other Matter 1972 (the 'London Convention' for short, and then later the 'London Protocol' upon its revision in 1996) (International Maritime Organization n.d.), and MARPOL Annex V, all of which have zero tolerance for plastic pollution.

2.1.2 Agricultural Plastic Pollution

Land-based agricultural plastic use typically includes greenhouse or hoop house sheeting, netting, plastic mulch (film), irrigation tape and piping, agrochemical containers, silage, fertiliser bags and slow release fertiliser pellets. The best current estimate of agricultural plastic usage extrapolates from the European Union's (EU's) demand for agricultural plastics of 1.6 million tonnes annually to place world demand at approximately eight to ten million tonnes in 2015 (Cassou et al. 2018). A separate calculation projected that the global agricultural film market would reach 7.4 million tonnes in 2019 (Sintim and Flury 2017). At the end of the growing season, plastic mulch should be recovered from fields but this is difficult because it shreds easily, so it is common practice to till plastic mulch into the soil (Steinmetz et al. 2016). Depending on the proximity to the ocean or ocean-bound waterways, this improper end-of-life management of the mulch could contribute to inputs of plastic, especially microplastic, into the ocean.

Aquaculture also contributes significantly to marine plastic pollution. Several studies have reported abandoned, lost and discarded aquaculture gear in coastal waters or on shores (Heo et al. 2013; Liu et al. 2013; Hong et al. 2014). Near aquaculture centres, beaches often contain large amounts of lost or discarded plastic materials (Fujieda and Sasaki 2005; Hinojosa and Thiel 2009; Andréfouët et al. 2014; Jang et al. 2014b; Bendell 2015). Lost aquaculture gear that is floating at the sea surface can also be transported over long distances, potentially bringing non-native species to other ecoregions (Astudillo et al. 2009). One of the few studies that has estimated the losses from aquaculture activities and their contribution to marine plastic debris has been conducted in South Korea (Jang et al. 2014b). The authors showed that lost aquaculture gear contributes a significant amount of plastic litter (mostly expanded polystyrene, or EPS) in the coastal waters of South Korea.

2.1.3 Industrial Plastic Pollution

Plastic resin pellets, the raw material from which plastic items are made, continue to leak into the ocean despite voluntary industry campaigns like Operation Clean Sweep that encourage secure handling of the pellets. Pellet pollution in the ocean has been further documented because they are used to study POPs and bacteria as well (Heskett et al. 2012; Rodrigues et al. 2019). While quantities of inputs have not been published on a global scale, one case study quantified inputs from a facility along the west coast of Sweden (Karlsson et al. 2018). While most of the pellet pollution was reported to be localised, 3 to 36 million pellets (above 300 μm) were estimated to enter the waterways surrounding the production facility annually. Karlsson et al. (2018) also stated that while there are regulatory frameworks that can be applied to reduce this pollution, they are not being effectively applied or enforced. Lechner and Ramler (2015) found that the regulations in Austria still allowed a production facility to legally discharge 200 g of pellets per day and up to 200 kilograms (kg) during a high rainfall event. An important legal precedent was just set in the United States with Formosa Plastics agreeing to pay a US $50 million settlement for a lawsuit against them for discharging resin pellets into Lavaca Bay and other nearby waterways (Collier 2019). Besides paying the settlement, it has to adhere to a 'zero discharge' policy moving forward with fines that increase over time for any future discharges (Collier 2019).

2.1.4 Maritime Plastic Pollution

Fisheries activities contribute to pollution through the accidental or intentional discarding of nets, ropes, buoys, lines and other equipment, also known as 'abandoned, lost or otherwise discarded fishing gear' (ALDFG) (see Box 5.3 for a discussion of aquaculture). Historic fishing nets were made from biodegradable, locally sourced natural materials like cotton, flax or hemp, but as materials like nylon and other polymers were introduced, fishing practices (and efficiencies) were increased, as early as 1951 in the United States and Canada (Pycha 1962). United Nations General Assembly and United Nations Environment Assembly resolutions have addressed ALDFG (UNEP, 2017), encouraging the reduction of impacts from this marine debris that is designed to capture and kill marine animals (Gilman 2015; Gilman et al. 2016). The Food and Agriculture Organization of the United Nations' (FAO's) Committee on Fisheries, the FAO Code of Conduct for Responsible Fisheries and the FAO's Voluntary Guidelines for the Marking of Fishing Gear have also presented on marking fishing gear and ALDFG reporting and recovery (Gilman et al. 2016). Richardson et al. (2019) reviewed 68 publications from 1975 to 2017 that contain quantitative information about fishing gear losses and found that at an annual rate, all net studies reported gear loss rates from 0% to 79.8%, all trap studies reported loss rates from 0% to 88%, and all line studies reported loss rates from 0.1% to 79.2%. Based upon this review, Richardson et al. (2019) performed a meta-analysis estimating global fishing gear losses for major gear types, finding that 5.7% of all fishing nets, 8.6% of all traps, and 29% of all lines are lost around the world each year. Abandoned, lost or discarded fishing gear can ensnare or entangle marine wildlife, have economic consequences due to losses of commercially important food fish and can smother sensitive coral reef ecosystems (Macfadyen et al. 2009; Gunn et al. 2010; Wilcox et al. 2013; Richardson et al. 2018). Commercial shipping and discharge from ocean-going vessels result in plastic inputs through accidental releases of cargo during ocean transit, which may occur during rough weather or when containers are insufficiently secured during transport (World Shipping Council 2017).

Box 5.1 Spotlight on Africa's Current and Future Rapid Growth

Africa's contribution to waste generation is currently low by global standards.[a] However, the continent is set to undergo a major social and economic transformation over the coming century as its population explodes, cities urbanise and consumer purchasing habits change.[b] These changes will lead to significant growth in waste and wastewater generation, including nutrient exports to coastal waters,[c] with sub-Saharan Africa forecast to become the dominant region globally in terms of municipal solid waste generation.[d] This will put significant strain on already constrained public and private sector services and infrastructure.[e]

As noted by Yasin et al. (2010) and UNEP (2018a), there are limited reliable, geographically comprehensive waste and water quality data for Africa. This makes it extremely difficult to assess the potential impacts of waste and wastewater systems locally and regionally. However, anthropogenic sources of nutrients in rivers, including agricultural sources and human sewage (often untreated) from urban centres, will become more important than natural sources in large parts of Africa.[f] Furthermore, with growing population comes increased waste generation and changing waste types.[g] As such, in the absence of reliable waste and water quality data, population growth and economic development can provide signals of potential 'geographic areas of concern' with regard to plastic, industrial, agricultural and municipal wastes. According to the United Nations' Department of Economic and Social Affairs, more than half of the world's projected population growth between 2017 and 2050 is expected to come from only 10 countries, with 6 of these in Africa—Nigeria, the Democratic Republic of the Congo, Ethiopia, Tanzania, Uganda and Egypt (ordered by their expected contribution to global growth).[h]

Where the impacts of plastic and nutrients on coastal systems in Africa have been modelled, the models have forecasted significant growth in waste generation and potential impact.[i] Tonnages of mismanaged plastic waste is expected to increase significantly between 2010 and 2025, particularly in coastal countries such as Nigeria, Egypt, Algeria, South Africa, Morocco and Senegal (ordered by their forecasted 2025 mismanaged plastic).[j] The nutrient risk for large marine ecosystems forecast for 2050 shows very high coastal eutrophication risk off the coast of West Africa around the Gulf of Guinea.[k]

While waste volumes produced in Africa are currently low, waste is impacting the environment due to a number of factors, including limited environmental regulation and often weak enforcement, inadequate waste and wastewater systems and the transport of waste into Africa, often from developed countries.[l] With an average municipal solid waste collection rate of only 55% for Africa,[m] the potential for plastic to leak into the environment is high. There is growing citizen and government concern around the leakage of plastic waste into the environment, resulting in many African countries moving to ban single-use plastics as a way of limiting their negative impacts. According to UNEP (2018b), 29 countries in Africa, predominantly coastal countries, have already implemented some sort of regulation against plastics. Currently, these regulations vary from a ban on single-use (thin) plastic bags, with associated requirements for bag thickness, to a complete ban on all plastic carrier bags. However, the growing concern around plastic waste is sparking discussions in many African countries on possible further bans on other single-use plastic products, such as PET beverage bottles and food service industry products such as straws, cups, containers and utensils.

There is, however, a growing response from a number of brand owners, retailers and convertors to address the current waste problems in Africa. South Africa, for example, has had voluntary industry initiatives in place for over a decade aimed at growing the local plastic recycling industry. Initiatives such as the South African PET Recycling Company, which has achieved a 65% post-consumer PET bottle recycling rate in South Africa,[n] are now being rolled out in Kenya, with plans to launch in Ethiopia and Uganda.[o] There are also a number of social innovations emerging in Africa to deal with the plastic waste problem. These often focus on innovative community-driven collection systems and associated financial rewards for recyclables, such as Wecyclers in Nigeria and Packa-ching in South Africa.

Notes:

[a] *Kaza et al. (2018)*

[b] *African Development Bank (2012), UNDESA (2015a, b)*

[c] *Yasin et al. (2010), UNEP (2015)*

[d] *Hoornweg et al. (2015)*

[e] *UNEP (2015)*

[f] *Yasin et al. (2010)*

[g] *UNEP (2015)*

[h] *UN (2017)*

[i] Jambeck et al. (2017), UNEP (2018a)

[j] *Jambeck et al. (2015)*

[k] *Seitzinger and Mayorga (2016)*

[l] *Brooks et al. (2018), UNEP (2018a)*

[m] *UNEP (2018a)*

[n] *PETCO (2018)*

[o] *Coca Cola (2019)*

2.2 Other Pollutants Compounding Ocean Stress

Pollution in this category stems from anthropogenic development (including in rural and urban areas). Municipal sources of pollution can be especially high where population densities are high. Lack of infrastructure that can handle sanitation and waste management in rapidly growing cities, especially near the coasts, is a large source of ocean pollution. Sources in this sector include residential and commercial solid waste and wastewater as well as runoff from roads and landscaping activities. Additionally, debris entering the ocean as a result of natural disasters is included here.

2.2.1 Other Municipal Solid Waste Pollution

The World Bank estimates that 2 billion metric tonnes of municipal waste are generated globally with 33% (663.3 mmt) being managed by 'open dumping' (Kaza et al. 2018). Approximately 50% or more of this waste is organic waste (e.g. food waste) in many places except for Europe and North America, which generate around 30% organic waste. In high-income countries (as ranked by the World Bank), 51% of the waste stream is plastic, paper, cardboard, metal and glass, while in low-income countries, only 16% of the waste stream is estimated to be dry waste and able to be recycled (Kaza et al. 2018). These statistics do not even include special waste materials like medical and electronic waste (e-waste), which pose even further management challenges beyond municipal waste. While regulated by the Basel Convention in international trade, e-waste continues to be processed in areas without adequate infrastructure or protection for workers; to access the metal, the plastic housing and coatings on wires are often burned, releasing toxic emissions impacting ecosystems and human health (Asante et al. 2019).

Box 5.2 Waste Management in Indonesia

The Indonesian government, through President Act No. 83 in 2018 regarding marine debris management, has committed to reducing plastic waste up to 70% by 2025.[a] To support this effort, the Coordinating Ministry of Maritime and Investment Affairs plans to build a protocol to collect marine debris data from several big cities in Indonesia, including Banjarmasin, Balikpapan, Bogor and Denpasar, and has taken action through the Mayor Act (Peraturan Wali Kota) and Governor Act (Peraturan Gubernur) to regulate the reduction of single-use plastic. While some regulations regarding waste reduction, segregation, collection and transport already existed, the lack of enforcement has caused them to be poorly implemented. To amplify efforts to reduce plastic waste, the national government has also constructed a cross-government collaboration approach through a National Plan of Action (Rencana Aksi Nasional) on marine plastic debris for 2018–2025, which includes five main actions: change behaviour, reduce land-based leakage, reduce sea-based leakage, enhance law enforcement and financial support, and increase research and development.[b]

In addition to regulatory solutions, some villages are setting up their own waste management facilities. In 2018, Muncar, a small village in East Java, worked with a private organisation named SYSTEMIQ on a pilot project called Project STOP, which, if successful, can be implemented in other villages throughout Indonesia. For this project, they built a waste management system in the area that focuses on waste segregation in households and capacity building through a sorting centre. The plan has five strategies, including optimised waste collection, behaviour change, regulation setting, village waste management, institutional capacity building and optimised waste processing for both inorganic and organic waste. In December 2019, 47,500 people received waste collection, mostly for the first time, from two facilities established by the project. These facilities have collected 3000 tonnes of waste so far and employ 80 local people.[c]

Indonesia is also looking for alternatives to landfills for plastic waste that cannot be recycled. One option being investigated is a plastic road tar that uses plastic waste, mainly LDPE and HDPE. The plastics are shredded, melted and added into road-tar mix. In 2017, this method was piloted at Udayana University, Bali, where they laid a 700-metre-long plastic road. However, an evaluation hasn't yet been done assessing the potential for contamination into the environment.

Notes:

[a] *Purba et al.* (2019)

[b] *Coordinating Ministry for Maritime Affairs* (2018)

[c] *National Geographic* (2020)

Another contribution beyond municipal waste is disaster debris. With climate change increasing both the intensity and frequency of storms around the globe, this pollution input may increase in the future. One quantified example of disaster debris originated from the 2011 Japanese tsunami that washed out 3.6 mmt of debris, with 0.91 mmt floating across the Pacific Ocean and portions of it reaching the western shores of North America (NOAA 2013).

2.2.2 Pesticide Pollution

Municipal pesticide pollution has been recognised in non-point source stormwater runoff since the 1960s (Weibel et al. 1966). It is sourced from use in commercial and residential landscaping and wastewater (Sutton et al. 2019). Pesticide use and pollution can be significant in densely populated areas where use is common, but it is often on a smaller scale compared with agriculture use. One study of the Marne River in France determined that urban uses of pesticides were considerably lower (47 tonnes/yr) than agricultural ones (4300 tonnes/yr) (Blanchoud et al. 2007), with similar trends observed in eight urban streams in the United States (Hoffman et al. 2000).

Agricultural pesticides represent a category of human-made or human-appropriated chemicals that are used to prevent, destroy, repel or mitigate any pest, or as a plant regulator, defoliant or desiccant (U.S. Code 1947). Pesticides are categorised based on the target class of organisms they are designed to impact. The most common categories include herbicides, insecticides, fungicides, rodenticides, algaecides and antimicrobials.

The ocean is exposed to pesticides through air, water, soil and biota. The air transports pesticides globally, documented as early as 1968 (Risebrough et al. 1968; Seba and Prospero 1971), and has resulted in detectable levels of pesticides in every part of the biosphere, including in arctic ice (Pućko et al. 2017; Rimondino et al. 2018). Pesticide transport through surface runoff occurs in both the liquid phase, where the pesticide is solubilised in the runoff water, and the solid phase, where the pesticide is bound with soil particles that erode with surface runoff. Both mechanisms transport pesticides from their application sites to the ocean. More areas are likely to face high pesticide pollution risk as global population grows and the climate warms, likely requiring even higher rates of pesticide use for increased agricultural activity and crop pests (Ippolito et al. 2015).

2.2.3 Nutrient Pollution

Untreated sewage carries a large volume of pollutants to the ocean (Islam and Tanaka 2004) and wastewater itself contains a number of pollutants: nutrients, pathogens, plastics, chemicals, pharmaceuticals and other suspended solids. On a volume basis, raw sewage discharge is of most concern where sanitation infrastructure is still developing. For example, in Southeast Asia, more than 600,000 tonnes of nitrogen are discharged annually from the major rivers. These numbers may become further exacerbated as coastal population densities are projected to increase from 77 people per square kilometre (people/km^2) to 115 people/km^2 in 2025 (Nellemann et al. 2008). The global anthropogenic nitrogen (N) load to fresh water systems from both diffuse and point sources in the period 2002–2010 was 32.6 mmt/yr (Mekonnen and Hoekstra 2015), though only a portion of this might reach the ocean.

The accumulated anthropogenic N loads related to gray water footprints in the period 2002–2010 was 13×10^{12} cubic metres per year, with China contributing about 45% to the global total. Twenty-three percent came from domestic point sources and 2% from industrial point sources (Nellemann et al. 2008). From 2002 to 2010, the global total phosphorous (P) load to freshwater systems from the sum of anthropogenic diffuse and point sources was estimated to be 1.47 mmt/yr, though only a portion of this might reach the ocean. About 62% of this total load was from point sources (domestic, industrial) while diffuse sources (agriculture) contributed the remainder. China contributed most to the total global anthropogenic P load, about 30%, followed by India (8%), the United States (7%), and Spain and Brazil (6% each) (Bouwman et al. 2011).

A global indicator of wastewater treatment to inform the SDGs has been recently created: Wastewater treatment was normalised by connections to wastewater systems around the world. The regions with the greatest average scores (i.e. the most comprehensive wastewater treatment) are Europe (66.14 ± 4.97) and North America (50.32 ± 17.42). The Middle East and North Africa (36.45 ± 6.33), East Asia and the Pacific (27.06 ± 6.91), Eastern Europe and Central Asia (18.34 ± 5.40), and Latin America and the Caribbean (11.37 ± 2.51) had scores falling in the middle, with some infrastructure lacking. Sub-Saharan Africa (3.96 ± 1.50) and South Asia (2.33 ± 1.34) have the lowest scores with extensive needs for wastewater treatment improvements (Malik et al. 2015). Even where treatment facilities exist, they may sometimes discharge untreated sewage into waterways and the ocean due to decayed infrastructure, facility malfunctions or heavy rainfall events that overwhelm systems using combined sewers and stormwater drains (known as combined sewer overflows).

Nutrient pollution from agricultural sources comes from using synthetic nitrogen and phosphorus fertilisers and from discharging animal waste into the ocean, either via direct runoff, rivers or disaster events (e.g. hurricanes). Globally, humans increased the application of synthetic nitrogen fertilisers by ninefold and phosphorous fertilisers by threefold between the 1960s and the 2000s (Sutton et al. 2013). The global agricultural system fixed 50–70 Teragrams (Tg) of N biologically, while nearly double that, 120 Tg per year of N, was added as synthetic fertilisers to support the production of crops and grasses as well as feedstock for industrial animal agriculture (Galloway et al. 2008; Herridge et al. 2008). A large share of the human-applied N is lost, including some 40–66 Tg N/yr exported from rivers to the ocean from 2000 to 2010 (Seitzinger et al. 2005, 2010; Voss et al. 2011, 2013). Estimates show an increase in the total N and P exports to coastal waters by almost 20% and over 10%, respectively, from 1970 to 2000 (Seitzinger et al. 2010). Diffuse sources, including agriculture, contributed about 28% of the global total P load to freshwater systems, which eventually lead to the ocean.

Global crop production is often seen as the primary accelerator of N and P cycles. However, the demand for animal feed produced from different crops and by-products of the food industry has rapidly increased in the past century. At present, about 30% of global arable land is used for producing animal feed, probably also involving a similar fraction of fertiliser use to produce crops for human consumption (Steinfeld et al. 2006). In addition, total N and P in animal manure generated by livestock production exceed the global N and P fertiliser use (Mekonnen and Hoekstra 2018). Livestock production has increased rapidly in the past century, with a gradual intensification that has influenced the composition of livestock diets. In general, intensification is accompanied by decreasing dependence on open range feeding in ruminant systems and increasing use of concentrate feeds, mainly feed grains grown with fertiliser and fed to animals at feedlots with concentrated manure to manage.

2.2.4 Antibiotics and Other Pharmaceuticals

Antibiotics and other pharmaceuticals are present in most wastewater both from improper disposal (flushing down sinks or toilets) and from human waste. Where wastewater treatment facilities exist, treatment primarily removes solids and pathogens, but is not typically able to remove pharmaceuticals without advanced treatment (Keen et al. 2014). A rapid increase (up 65% in defined daily doses) of antibiotic use between 2000 and 2015 was seen globally, with the largest increases in lower-middle-income countries where wastewater treatment may be less available (Klein et al. 2018).

Box 5.3 The Impacts of Aquaculture

The four primary discharges to the ocean from ocean-based aquaculture, as identified and quantified by the Global Aquaculture Performance Index, are antibiotics, antifoulants (primarily copper), parasiticides and uneaten feed and faeces, the last of which impacts the biochemical oxygen demand of the water.[a] There are two additional biological impacts—escaped fish and pathogens—that are considered out of scope for this paper. Plastics discharged by aquaculture are presented at the beginning of this section. The relative volume and impacts of these four discharges vary by species, geography and type of aquaculture, with impacts ranging from relatively benign to quite damaging for the marine environment and marine life. The index identified the worst-performing sector as marine finfish in tropical and subtropical water, such as groupers, red drum and cobia, and the worst geography as Asia, with Asian countries holding the lowest 15 spots in the species-country ranking. These countries tended to score particularly poorly on biochemical oxygen demand and use of antibiotics and parasiticides.[b]

Notes:

[a] *Volpe et al.* (2013)

[b] *Volpe et al.* (2013)

2.2.5 Heavy Metals, Persistent Organic Pollutants and Oil and Gas

Urban runoff, especially roadway runoff, is the primary source of heavy metals, POPs and oil and other chemicals from municipal sources, although some of these can also be contained in wastewater. One recent example from China shows road runoff contains significant cadmium, chromium, copper, manganese, nickel, lead and zinc when classifying with a pollution load index, and that roadways have two to six times greater metal concentrations than rooftop runoff (Shajib et al. 2019).

Pollution from industry refers to any discharges of hazardous substances, which may be a result of effluent discharges from manufacturing operations and cleaning equipment and any accidental spills. Industrial activities may generate waste that contains heavy metals, carcinogenic hydrocarbons, dioxins, pesticides, and noxious organic and inorganic substances. Hazardous substances are used to produce electrical equipment, oil and petrochemicals, organic and inorganic chemicals, pesticides and heavy metals (mercury, arsenic, lead, cadmium), and are used by the wood/pulp processing and electroplating industries. Additionally, by-products of industrial processes include toxic dioxins (e.g. $C_4H_4O_2$) produced in the manufacture of certain herbicides and chlorine from paper pulp bleaching. Hazardous materials can be explosive, toxic or carcinogenic, and must be treated and managed appropriately. Like other pollutant pathways already discussed, industrial pollutants can enter the ocean directly through point discharges or by flowing in rivers (water or sediment transport) to the ocean, but may also come from atmospheric deposition as illustrated in a river and estuary source and transport case study of organochlorine compounds by Wu et al. (2016).

Industrial water consumption comprises 22% of global water use (UN-Water 2018). In 2009, industrial water use in Europe and North America was 50% of total water use compared with 4–12% in developing countries, but it is expected to increase by a factor of five in the next 10–20 years in rapidly industrialising countries (UN-Water 2018). As far back as 2002, 160,000 factories were estimated to discharge between 41,000 and 57,000 tonnes of toxic organic chemicals and 68,000 tonnes of toxic metals into coastal waters (UNDP 2002). Globally, 80% of wastewater, including some industrial wastewater, is discharged into the environment without treatment (UN-Water 2018). In the United States, around 60% of coastal rivers and bays had already been degraded by 2006 (UNEP/GPA 2006). The Mediterranean coastline has faced major environmental pressures from industrial development, with wastewater flows from the mineral, chemical and energy sectors (GRID Arendal 2013). Meanwhile, China has discharged approximately 20–25 billion tonnes per year of industrial wastewater since 2000 (Jiang et al. 2014). The real number may be even higher, due to underreporting and a mismatch in both water quality standards and wastewater standards. In 2018, only about 71% of the industrial wastewater was treated in Vietnam—craft villages near Hanoi, for example, were discharging 156,000 cubic metres of water a day into the Red River Delta near the coast (World Bank 2019). The World Bank (2019) also states that treating 22 million cubic metres of wastewater from industrial clusters along the Nhue-Day River could considerably improve coastal water quality. The Ganga River, despite being a sacred river, is heavily polluted by untreated industrial activities. Seven hundred sixty four units of industry generate 501 million litres of wastewater from tanneries, textile mills, paper, pulp and other sources (India Ministry of Water Resources 2017). The Tiram River in Malaysia had high levels of toxics due to the improper treatment of industrial effluent in 2015 (Asri 2015). Only one-third of Philippine river systems are considered suitable for public water supply due to untreated domestic and industrial wastewater (Asian Development Bank 2009). These polluted rivers stream to the ocean and threaten the coastal resources in the Philippines. Monitoring of fish and macroinvertebrates in Manila Bay, Philippines, showed the content of cadmium, lead and chromium were considerable (Sia Su et al. 2009). Heavy metal pollution for lead and hexavalent chromium had accounted for 99.2% of disease burden from toxic exposure among those in India, Indonesia and the Philippines (Chatham-Stephens et al. 2013). Seawater along the coast of the Korean Peninsula was analysed for heavy metal concentrations three times from 2009 to 2013 and copper and zinc concentrations were found to exceed acceptable standards all three times (Lee et al. 2017). Untreated industrial discharges threaten not only ecosystem services, but potentially billions of people.

Box 5.4 Jakarta Bay Struggles with Industrial Pollution

Jakarta Bay is on the northern coast of Jakarta Metropolitan City, Indonesia. Three large rivers, the Citarum, Ciliwung and Cisadane, flow into Jakarta Bay. These rivers are used by inhabitants as well as industry in the Jakarta, West Java and Banten Provinces. There has been significant anthropogenic impact on the Citarum River dating back to the increase in use of the area for industrial activities in the early 1980s.[a] Septiono et al. (2016) discovered heavy metals—namely cadmium, chromium hexavalent, zinc, mercury, lead and copper—exceeding the national concentration standards in the river. The concentrations of lead and copper in the sediment of Jakarta Bay increased five and nine times, respectively, between 1982 and 2002 (Arifin 2004). In 2006–2007, sampling found that sediment distribution in the estuary of Jakarta Bay consisted mostly of black clay, which is indicative of anthropogenic influences from the Jakarta River Basin.[b] Sampling done from June 2015 to June 2016 showed that around 97,000 debris items entered the bay daily through nine rivers, and about 59% of it was macroplastic,[c] a further stressor on Jakarta Bay.

Thousands of people, such as fishers in North Jakarta and those along the Thousand Islands, depend on the ecosystem goods and services provided by the river. However, the extreme pollution of toxic chemicals, eutrophication and sediment load in the area, as well as overexploitation of marine resources, are threatening coastal communities. Production of the capture fishery sector decreased in the last five years. Fish production continuously declined from about 35,000 tonnes in 1999 to almost 18,000 tonnes in 2002.[d] Jakarta Bay is under stress from both intensive fishing and degraded water quality due to pollution from both land and marine sources. Mercury content in green mussels and arsenic concentrations in green mussels and tuna samples in Jakarta Bay are above the national standard concentrations (1.0 milligram per kilogram),[e] yet the polluted green mussels can be found in local markets. Despite being highly used for food and to support livelihoods, Jakarta Bay is a sea of wastewater and solid waste.

Notes:

[a] *Bukit (1995), Parikesit et al. (2005), Dsikowitzky et al. (2017)*

[b] *Tejakusuma et al. (2009)*

[c] *Cordova and Nurhati (2019)*

[d] *Arifin (2004)*

[e] *Koesmawati and Arifin (2015)*

2.2.6 Maritime Pollution

Pollution into the ocean does not arise only from land; the ocean is also impacted by ocean-sourced pollution. Pollution other than plastic (see Sect. 2.1.4 for a discussion of plastic pollution), results from fishing, shipping and transportation, cruises, recreational boating, ocean exploration and other maritime activities. Similar to land-based sources, wastewater and grey water contribute to nutrient and chemical loading in the ocean, and unique to ocean-going vessels, improper management of bilge water can also cause pollution. Sewage and grey water are regulated under MARPOL Annex IV and bilge water under Annex I. Beyond that, oil spills are one of the most evident forms of ocean pollution due to large areas that may be impacted and the visible consequences for seabirds and other marine wildlife (Palinkas et al. 1993). Most maritime oil spills occur due to transportation mishaps or accidents on oil rigs. Less frequently, a sunken vessel or discharge of oil-containing bilge or ballast water may be released. Because of policies by the International Maritime Organization (IMO) and goals to improve safety and reduce environmental risk, the overall trend of oil spills from tankers (not including rigs and platforms) has decreased over time (Kontovas et al. 2010). However, in 2010 BP's Deepwater Horizon oil spill resulted in 4.9 million barrels of oil entering the ocean, the largest oil spill in the history of the petroleum industry; thousands of scientific papers have assessed the impacts of this oil spill since it occurred.

2.3 Compounding Effects of Multiple Pollutants

More than one source and pollutant can cause a complex mix of stressors on the ecosystem and marine life, with sometimes synergistic effects (the impact of the two together is greater than the sum of their individual impacts).

Distinct pollutants may also enter the ocean through similar pathways. The municipal sector, for example, is a source of both plastic waste and wastewater. In general, wastewater also carries all the contaminants of urban stormwater runoff in addition to pollutants in sewage. When municipal infrastructure for handling solid and liquid wastes is lacking, rapid economic development exacerbates pollution. In some cases, open sewage canals are sometimes used to 'manage' wastewater in urban systems, yet solid waste on land washes into these canals and drains into waterways that can lead to the ocean. In other cases, aging infrastructure incapable of handling stormwater leaks both sewage and plastic into waterways from combined sewer overflow events. Just as negative synergies exist, so do positive ones: Waste management of the residual solids from wastewater treatment are often managed within the solid waste management sector, and development of infrastructure to manage biosolids can help properly manage other solid waste, including plastic waste.

The agriculture sector has the highest input of nutrients to the ocean. In one of the largest river basins, the Mississippi River, fertiliser use delivered 64% and 41% of the N and P, respectively, to the Gulf of Mexico. Pasture use delivered another 5% and 38% of the contribution, for a total N and P from agriculture of 70 to 80% of the total (by comparison, urban use is 9 to 11%) (Alexander et al. 2008). The research also found that source reductions on land near large rivers (nearly 1:1) or quickly flowing streams (2:1) had the greatest reduction of overall nutrient loading to the Gulf (Alexander et al. 2008). This means that in large river basins, it is possible to get a nearly kg per kg reduction to the ocean by decreasing fertiliser use and adjusting management of grazelands. Figure 5.2 shows use of N and P on land, as well as all the major watersheds that drain to the ocean.

Box 5.5 Spotlight on Vietnam

Vietnam has a coastline of 3260 km with over 3000 islands and 114 river mouths and estuaries. Due to the rapid rate of population increase, urbanisation and industrialisation, a large amount of pollution has been introduced into the coastal zone in recent decades. The major sources of pollution discharges into the ocean include untreated or incompletely treated effluents from the municipal and industrial sectors, as well as waste from agriculture activities and seaport and tourism activities.

The total amount of domestic wastewater in both urban and rural areas in Vietnam is estimated to be 8.7 million cubic metres per day (million m^3/day).[a] Major pollutants are nutrients, organic matter, suspended solids and nitrogen-containing organic substances. According to the Vietnam Ministry of Construction, the total designed capacity of 39 domestic wastewater treatment plants over the country is approximately 907,950 m^3/day, which covers only 11% of domestic wastewater.[b] In Ha Noi capital and Ho Chi Minh City, the two largest cities in the country, the percentage of all domestic wastewater processed by centralised wastewater treatment plants is 20.6% and 13% of the total wastewater, respectively.[c] By the end of 2016, 344 industrial zones had been established with the amount of industrial wastewater varying in the regions, and 220 industrial zones were in operation of which 86% had a centralised wastewater treatment plant. Only 98 of 620 industrial clusters, or 16%, were designed with a wastewater treatment system—and those treatment systems have been shown to have a number of limitations. In addition, wastewater from handicraft villages also contributes to marine pollution.[d]

Waste from agricultural activities also contributes to marine pollution, especially from the livestock, aquaculture and crop sectors. The estimated livestock solid waste— including nutrients, suspended solids, organic matter, pathogens and pharmaceuticals—was reported to be 47 million tonnes in 2016, of which 40–70% was treated and the rest discharged into lakes, streams and rivers.[e] For instance, 70–90% of the wastewater from one pig farm, comprised of nutrients (nitrogen), minerals, heavy metals and pharmaceuticals, was reported to be excreted into the environment. Aquaculture activities also release a large amount of untreated waste directly into the ocean with high levels of nitrogen and phosphorus. In 2014, more than 10 billion cubic metres of wastewater containing 51,336 metric tonnes of nitrogen and 16,070 metric tonnes of phosphorus in a pangasius fish farm were estimated to be discharged to local canals to eventually end up in the Mekong Delta River.[f]

The use of pesticides and chemical fertilisers in agricultural production is another major source of surface water pollution. Fertilizer use is increasing in Vietnam. From 1983 to 2013, fertiliser consumption increased nearly sevenfold to 26 mmt in 2013, and about 80,000–100,000 tonnes of pesticides, herbicides and fungicides were used from 2012 to 2014.[g] On average, 20–30% of pesticides and chemical fertilisers applied will not be retained by plants and will be washed by rainwater and irrigation water into surface water resources as well as accumulate in the soil and groundwater in the form of residues. In summary, the pollutants released by these activities include, among others, nutrients, organic chemicals, sediments and pesticides, which ultimately end up in the sea of Vietnam. In addition, wastewater is also discharged from ocean-going ships, other maritime facilities, ship building and repair plants, seaports and freight yards and stores.

The two major river basins in Vietnam, the Mekong and the Red River, annually discharge approximately 500 million and 137 billion cubic metres of water into the ocean, respectively.[h] Sediment is discharged from the Mekong alone at a rate of 36 mmt/yr, although this is a decrease from previous estimates since dams are now reducing that transport.[i] However, both of these water and sediment flows can transport pollutants from the anthropogenic activities in the river catchment and coastal areas to the ocean.[j] About 13 mmt of solid waste is mismanaged in Vietnam each year, with 1.8 mmt of that plastic, and an estimated 0.28–0.73 mmt entering the ocean from Vietnam each year.[k]

In Vietnam, not many studies on plastics, including microplastics, have been conducted, although Vietnam is one of the top countries in the world in terms of plastic waste.[k] The plastic industry during 2010–2015 was the third-largest industry in terms of growth, with an annual increase of 16–18% (following the telecommunications and textile industries). The amount of plastic used per capita increased from 3.8 kg/year in 1990 to over 41 kg/year in 2015.[l] Although there are no official statistics on the amount and varieties of plastic in the Vietnamese sea, plastic waste, originating from wastewater and solid waste from the mainland, can enter the ocean through 114 river mouths and estuaries.

Fishing, aquaculture and on-sea activities are also major sources of plastic in the Vietnamese sea. Every day, about 80 tonnes of plastic waste and bags are thrown away in Ho Chi Minh and Ha Noi combined.[m] In Ho Chi Minh, microplastics were found in urban canals with 172,000 to 519,000 items/m^3,[n] and in the surface water in Can Gio Sea at a rate of 0.176 ± 0.0 items/m^3.[o]

Vietnam is addressing the plastic issue on both the national and regional scales. The government has released a national action plan for marine litter (Government of Vietnam 2020). Regionally, the Lower Mekong Initiative, a multinational partnership among Cambodia, Laos, Myanmar, Thailand, Vietnam and the United States to create integrated subregional cooperation among the five Lower Mekong countries, launched in 2009, is now also working to address plastic contamination upstream before it gets to the ocean.

To more effectively address plastic waste, more research is needed. In particular, research that provides a more complete characterisation of macro and microplastics at sea is needed, as well as further study on effective strategies for managing plastic waste—particularly microplastics (including microbeads).

Notes:

[a] *MONRE (*2016*)*

[b] *Nam (*2016*)*

[c] *MONRE (*2017*)*

[d] *MONRE (*2017*)*

[e] *MONRE (*2016*), World Bank Group (*2017*)*

[f] *World Bank Group (*2017*)*

[g] *MONRE (*2014*), World Bank Group (*2017*)*

[h] *World Bank (*2019*)*

[i] *Thi Ha et al. (*2018*)*

[j] *World Bank (*2019*)*

[k] *Jambeck et al. (*2015*)*

[l] *VPAS (*2019*)*

[m] *Viet Nam News (*2019*)*

[n] *Lahens et al. (*2018*)*

[o] *Hien et al. (*2019*)*

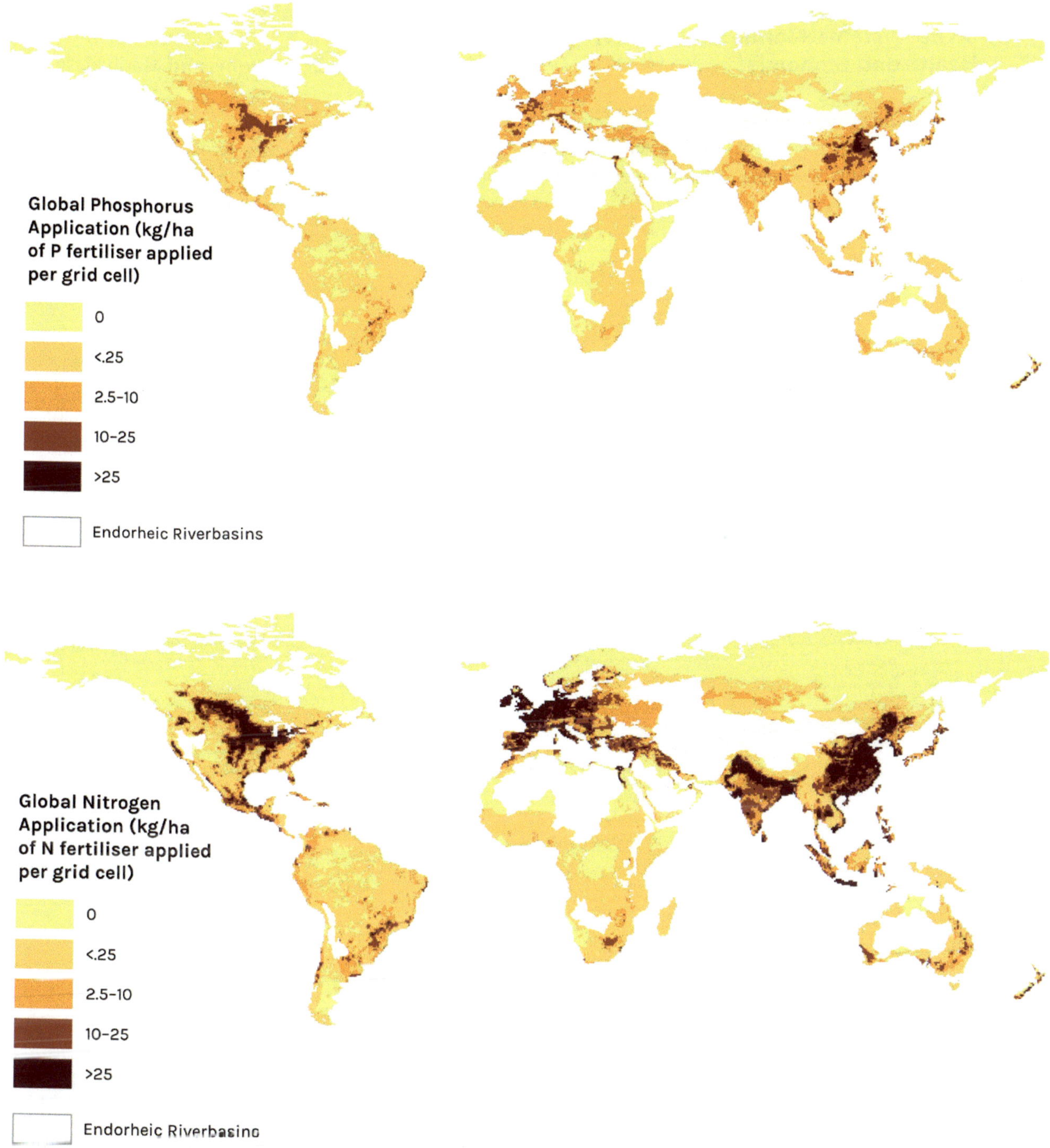

Fig. 5.2 Global Nitrogen and Phosphorous Applications (minus endorheic basins). Notes: These applications could impact the ocean based upon runoff and drainage. Kg/ha stands for kilogram per hectare. N stands for nitrogen, and P for phosphorous. As used here, an endorheic basin is a body of water that has no outflow to other bodies of water, such as rivers or the ocean. [Sources: Potter et al. (2010), Potter et al. (2011a), and Potter et al. (2011b). Map created by A. Brooks]

3 Impacts of Ocean Pollution on Ecosystems, Marine Life, Human Health and Economies

There are a multitude of potential impacts pollutants can have on the ocean, which we have categorised into four types: ecosystem, marine life, human health and economic. See Table 5.2 for a brief outline of these impacts.

3.1 Impacts of Plastic

3.1.1 Impacts on Ecosystems and Marine Life

Microplastics

Microplastics have accumulated across a wide range of environmental compartments including marine, terrestrial and freshwater habitats as well as in the air (SAPEA 2019; Eerkes-Medrano et al. 2015). These areas also include remote

Table 5.2 Potential ecosystem, marine life, human health and economic impacts from ocean pollution

Pollutant	Ecosystem impacts	Marine life impacts	Human health impacts	Economic impacts
Microplastics	• Potential to alter the distribution of sediment dwelling organisms in assemblages • Can provide surface vectors that facilitate the transport of potentially harmful microorganisms	• Negative effects on food consumption, growth, reproduction and survival across a wide range of organisms at the individual level • Starvation (due to ingestion) • Potential of exposure to toxic substances (in or absorbed by plastics) • Trophic transfer	• Unknown impact of ingestion through consumption of marine animals with microplastics in their tissues • Unknown exposure to toxic chemicals due to ingestion • Unknown exposure to pathogens	• Reduction in global marine ecosystem services has been estimated at US $0.5–2.5 trillion[1]
Macroplastics	• Smothering and impact on coral reefs • Transport of invasive species	At the individual level: • Starvation (due to ingestion) • Entanglement • Chemical exposure	• Increase in mosquito-borne diseases • Potential for exposure to pathogens	• Estimated $40 billion in negative externalities annually[2] • Global damage to marine environments from plastic pollution estimated at a minimum $13 billion per year[3] • Aggregated estimates across the plastics life cycle concluded that annual damages from plastic production and the current stock of plastic waste in the ocean amount to $2.2 trillion[4] • Fishermen lose time and efficiency from catching trash in nets • Damage to maritime industries in the APECa region was estimated at $1.26 billion per year[5] • Loss of revenue from tourism, e.g. reducing marine debris by 100% was estimated to improve the savings and welfare of local residents by $148 million over the three-month summer period[6]
Other solid waste	• Additive nutrients to the ocean as source of hypoxia (from organic waste) • Source of heavy metals (e-waste)	• Ingestion, entrapment or entanglement causing impairment or death • Transport of invasive species	• 15 million people worldwide work informally in waste management in poor, unhealthy conditions[7] • Risk of entrapment/bodily injury • Heavy metal contamination and exposure (e-waste) • Pathogen exposure (medical waste)	• Fishermen lose time and efficiency from catching trash in nets • Debris in water can damage fishing gear and nets • Loss of revenue from tourism

Table 5.2 (continued)

Pollutant	Ecosystem impacts	Marine life impacts	Human health impacts	Economic impacts
Pesticides	• Reduced photosynthetic efficiency of sea grass, corals and algae (herbicides), resulting in chronic stress • Measurable impacts on seagrass productivity, especially when combined with light attenuation from high sediment loads from agricultural runoff • Can restrict or fully inhibit coral settlement and metamorphosis at concentrations as low as one part per billion, and at higher concentrations can cause coral branch death	• Death, cancers tumours and lesions on fish and animals, reproductive inhibition or failure, suppression of immune system, disruption of endocrine system, cellular and molecular damage, teratogenic effects, poor fish health marked by low red to white blood cell ratio, excessive slime on fish scales and gills, intergenerational effects, and other physiological effects such as egg shell thinning[b]	• Toxicity via consumption of marine sumption of marine bioaccumulated or biomagnified pesticides in their tissue. Most at risk are vulnerable populations (children, elderly) in communities with high levels of seafood consumption	• Loss of productivity and resiliency of seagrass beds and coral reefs due to pesticide pollution impacts global economic security by reducing provision of ecosystem services that are essential for human society. While exact level of damage is not known, if we assume a reduction in productivity of these ecosystems by 25%, the annual economic impact of those pesticides in the ocean would be $200 billion per year[8]
Nutrients (N, P)	• Eutrophication and hypoxia • Biodiversity losses • Ecosystem losses	• Fish kills, red tides • Decreases in population and species diversity with benthic and fish communities • Release of ammonia and hydrogen sulfide, which can be toxic to marine life	• Respiratory irritation from harmful algal blooms (HABs), e.g. red tides • Illness from consuming seafood exposed to HABs	• Black Sea fishery value was reduced by 90% (from roughly $2 billion). Other economic impacts included an estimated loss of $500 million in tourism revenue[9] • A major and extensive red tide outbreak occurred along the coast of Hong Kong and south China, covering an area of more than 100 km^2. Over 80% (3400 tonnes) of mariculture fish were killed, and the total loss was over $40 million[10] • Major economic impacts on fisheries, aquaculture and tourism
Antibiotics, parasiticides, other pharmaceuticals	• The occurrence of subtherapeutic doses of antibiotics on bacteria over a prolonged period leads to resistance, which is a threat to the environment	• The occurrence of subtherapeutic doses of antibiotics on bacteria over a prolonged period leads to resistance, which is a threat to the environment	• Carcinogenic, mutagenic and re-productive toxicity potential • Endocrine system	• Unknown

(continued)

Table 5.2 (continued)

Pollutant	Ecosystem impacts	Marine life impacts	Human health impacts	Economic impacts
Heavy metals	• Toxicity to some micro-organisms and animals, cancer in animals, uptake by plants	• Increase in the permeability of the cell membrane in phytoplankton and other marine algae, leading to the loss of intracellular constituents and, therefore, cellular integrity • These include lymphocytic infiltration, lesions and fatty degeneration • In addition, cadmium, lead and mercury are potential immunosuppressants; of concern is the buildup of mercury, which marine mammals tend to accumulate in the liver	• Acute toxicity at high doses, chronic toxicity, cancer, impacts to the nervous system and behaviour (especially lead)	• Unknown
Industrial chemicals and persistent organic pollutants[c]	• Food chain interactions, birth defects, cancer, accumulation and transformations in the environment	• Abnormal behaviour, birth defects in fish, birds, mammals • Biomagnification in the food chain	• Reproductive, developmental, behavioural, neurologic, endocrine, and immunologic adverse health effects	• Unknown
Oil and gas	• Coat and smother benthic areas	• Death • Negative impacts on reproductive health • Carcinomas and papillomas on the lips of bottom-feeding fish, as well as changes in the cell membrane • Severe eye irritation with subsequent blindness in seals • Individual birds become unable to swim or fly and nervous system abnormalities can occur • Population-level effects of oil toxicity on aquatic birds occur through the loss of egg viability	• Localised health impacts from immediate exposure, potential for longer-term impacts from exposure, e.g. cancer, mental health issues if fisheries and livelihoods are impacted	• BP's Deepwater Horizon spill in the Gulf of Mexico is estimated to have cost the company $61.6 billion in penalties and fines; cleanup and remediation; and payments to affected companies, communities and individuals[11] • The 'true' cost of the 2010 Deepwater Horizon oil spill including loss of tourism, cost of cleanup, and loss of fisheries is estimated to be $144.89 billion[12]

Notes: Pathogens present in human and animal waste discharged to the ocean can infect marine animals, but this is considered out of scope for this analysis

[a]*APEC stands for Asia-Pacific Economic Cooperation*

[b]*These effects are not necessarily caused solely by exposure to pesticides or other organic contaminants, but may be associated with a combination of environmental stresses such as eutrophication and pathogens.*

Sources:

1. *Beaumont et al.* (2019)
2. *Ellen MacArthur Foundation et al. (2016)*
3. *UNEP* (2014)
4. *Forrest et al.* (2019)
5. *McIlgorm et al.* (2008)
6. *Leggett et al.* (2014)
7. *Medina* (2008)
8. *Cesar et al.* (2003)
9. *World Bank* (2009)
10. *Yang and Hodgkiss* (2004)
11. *Mufson* (2016)
12. *Islam and Tanaka* (2004)

locations far from population centres such as in the deep sea (Woodall et al. 2015) and the Arctic (Obbard et al. 2014). There is clear evidence that microplastics are ingested by a wide range of species including marine mammals, birds, fish and small invertebrates at the base of the food chain (Law and Thompson 2014; Lusher 2015). While it has been shown that particles can pass through the digestive system and be excreted, it has also been established that some particles can be retained in the body for several weeks (Browne et al. 2008; Ory et al. 2018). Microplastics can also transfer between prey and predator species within food webs (Watts et al. 2015; Chagnon et al. 2018). Many of the species that have been shown to be contaminated with microplastics are commercially important for human consumption (Lusher et al. 2013).

Laboratory experiments indicate that at high doses ingesting microplastics can induce physical and chemical toxicity (SAPEA 2019). The physical presence of microplastic particles has been shown to have negative effects on food consumption, growth, reproduction and survival across a wide range of organisms, and there is evidence that zooplankton, non-mollusc invertebrates and juvenile fish are particularly sensitive (Cole et al. 2015). For example, a reduction in feeding efficiency has been demonstrated for zooplankton, lugworms and fish. In addition, when ingested, microplastics can transfer potentially harmful chemicals to biota; this can occur as a consequence of the transfer of hydrophobic chemicals from the surrounding water or the release of additive chemicals incorporated at the time of manufacture (Teuten et al. 2007; Tanaka et al. 2013). While the transfer of chemicals by plastics to biota has been demonstrated, it is the dose that determines the poison. In a recent bird feeding experiment, Roman et al. (2019) found that plastic ingestion caused higher frequencies of male reproductive cysts and minor delays in chick growth and sexual maturity, but did not affect ultimate survival or reproductive output. With regard to the transfer of chemicals by plastics from seawater, recent work has shown that other pathways including direct uptake from water and natural foods are likely to be more important pathways than microplastics (Bakir et al. 2016; Koelmans et al. 2016). Less is known about the risks associated with the release of additive chemicals from plastic. Determining the release of additives is particularly challenging since chemical formulations are not typically in the public domain (SAPEA 2019).

Most experimental work on effects has focused on those on individuals, but there is some evidence of wider ecological effects including the potential to alter the distribution of sediment-dwelling organisms in assemblages (Green 2016) and to influence the sinking rates of faecal material to the seabed (Cole et al. 2013). Microplastics also provide a surface that can readily become colonised by microorganisms including pathogens and there are concerns that microplastic particles may therefore provide vectors facilitating the transport of potentially harmful microorganisms (Zettler et al. 2013; Kirstein et al. 2016).

Plastic is rapidly colonised by microorganisms in a marine environment (Harrison et al. 2014). Plastic surface habitat has even been defined as the 'plastisphere' in recognition of the unique communities it harbours (Zettler et al. 2013). In fact, litter items made with many materials appear to have unique biofilm communities (Woodall et al. 2018). These communities include potentially harmful pathogens such as *Vibrio* spp. (Kirstein et al. 2016) and *E. coli* (Rodrigues et al. 2019) and are known to colonise the surfaces of submerged plastic surfaces, similar to how they colonise other hard submerged surfaces (Shikuma and Hadfield 2010). A submerged plastic cup laid on a seagrass meadow can serve as a home for more than 500 individual meiofauna, which potentially affects meiofauna community structure (Susetiono 2019). These communities might also impact biogeochemical cycles (Cornejo-D'Ottone et al. 2020).

It is important to recognise that most studies of physical and particle toxicity have been conducted using concentrations and/or particle sizes that are not typical of those currently recorded in the environment (Lenz et al. 2016). There are challenges since environmental concentrations are not known with confidence, especially for particles smaller than 300 μm, which are less likely to be collected from water using conventional net sampling. Plastics can fragment because of environmental exposure and so the abundance of very small particles in the nano-size range could be considerable. These particles are currently too small to detect in environmental samples, but laboratory studies show the potential for these particles to transfer from the gut to the circulatory system with the potential to rapidly become widely distributed in organisms (Brandelli 2020). More work is needed to understand the potential toxicological impacts of this. Despite the uncertainties about environmental concentrations in relation to evidence of harm, there is some consensus based on risk assessment approaches that if microplastic emissions to the environment remain the same or increase the ecological risk may become widespread within a century (SAPEA 2019).

Macroplastics

To date, around 700 species of marine life have been demonstrated to interact with plastic (Gall and Thompson 2015), with the main impacts occurring through entanglement, ingestion and chemical contamination (Wilcox et al. 2015). Far more is known about harm to individuals through interaction with plastic than is known about harm to populations, species and ecosystems within the marine environment (Rochman 2015).

Entanglement in Plastic Debris

Impacts on marine systems from entanglement are most commonly associated with abandoned, lost, or derelict fishing gear. Called 'ghost fishing', derelict fishing nets can continue to indiscriminately catch fish (and other marine organisms) for weeks, months or decades, which, in addition to impacting ecosystems and marine life, results in food security issues through lost resources to feed the world's population (and the associated economic consequences of lost revenue). With an estimated 640,000 tonnes of gear lost to the ocean each year per a census taken a decade ago (Macfadyen et al. 2009), some areas have reported up to three tonnes of derelict nets per kilometre of coastline in a given year (Wilcox et al. 2013). Derelict nets have been reported to ensnare or entangle invertebrates, crabs, fish, sharks, rays, sawfish, turtles, seabirds, crocodiles, dugongs, whales, dolphins and numerous other marine taxa. Ghost nets can also damage fragile habitats (such as by smothering or breaking coral reefs (Sheavly and Register 2007), entangle propellers, cause navigation hazards to other vessels (Gunn et al. 2010; Hong et al. 2017) and transport invasive species (Macfadyen et al. 2009). Impacts can be substantial—it has been estimated that in the Gulf of Carpentaria in northern Australia alone, derelict nets have likely entangled more than 10,000 sea turtles (Wilcox et al. 2013).

Ingestion of Plastic Debris

There are numerous demonstrated effects of plastic ingestion by marine fauna. These may include not only death (van Franeker 1985; Schuyler et al. 2012; Wilcox et al. 2015) but also reduction in body mass (Schuyler et al. 2012), starvation that may result from the physical blockage of the gut (van Franeker 1985; Laist 1987; Acampora et al. 2014; Hardesty et al. 2015), ulceration or perforation of the digestive tract (van Franeker 1985; Laist 1987; Schuyler et al. 2012) and potential toxicity due to sorption of chemicals contained within and sorbed to the plastic (Teuten et al. 2009). In some studies, incidence of plastic ingestion was as high as 60–80% or more of individuals sampled [crustaceans as reported by Murray and Cowie (2011); green turtles in Brazil as reported by Bugoni et al. (2001) and deep sea species as reported by Jamieson et al. (2019)].

Chemical Contamination from Plastic Debris

At present, far less is known and understood about the effects of chemical contamination (which takes place through ingesting plastic) than impacts from entanglement. In laboratory experiments, it has been demonstrated that ingested plastic can induce hepatic stress in fish (Rochman et al. 2013). Plasticisers (softening and other chemical agents such as dibutyl phthalate and diethylhexyl phthalate that are often added to plastics) have been detected in the preen gland oil of wild-caught seabirds, with higher levels of plasticisers found in birds that had eaten more plastic pieces (Hardesty et al. 2015). Polystyrene, heavily used in fisheries and aquaculture, is also of particular concern, as styrenes have been shown to leach into marine systems (Kwon et al. 2015). Jamieson et al. (2019) found plastics in animals in some of the deepest parts of the ocean. Endocrine-disrupting compounds leaching into tissues from plastics are of increasing concern, not only for wildlife (Olivares-Rubio et al. 2015), but also for humans (Meeker et al. 2009; Halden 2010).

3.1.2 Human Health Impacts

The risk of marine plastic debris to human health can be measured by the likely exposure of humans to marine plastic multiplied by the potential for harm by the plastic. This is not a simple equation, as plastics comprise many and diverse chemical additives in addition to their primary polymer component. The limitless combinations of polymers and additives mean that each plastic product has a different combination of chemicals, uses and disposal pathways with varying levels of risk to humans. As a result, plastics should not be treated as a single product, and need to be addressed separately (Lithner et al. 2011). To understand the risk, potential exposure should be identified and quantified, and the potential for harm, including from factors such as the concentration of chemical additives, size fraction (Smith et al. 2018) and ageing (Kedzierski et al. 2018), should also be determined. Because there are so many confounding variables and ethical issues, and a lack of a control group, studying human exposure to various plastic materials and forms is challenging. This section outlines exposure pathways, but without reliable measures for all exposure pathways (pre and post waste) it is not possible to calculate the relative risk of plastic waste on human health.

Humans have been exposed to plastics and their constituent components since they were first mass produced in the 1940s and 1950s. The growing use of plastics in primary food packaging has resulted in increased exposure to them over recent years, and the increased waste has resulted in more plastic entering the environment (Jambeck et al. 2015). Consequently, a host of recent studies have reported microplastics found in nonmarine foodstuffs—e.g. honey (Liebezeit and Liebezeit 2013), beer (Liebezeit and Liebezeit 2014) and seafood (Rochman et al. 2015)—and the air (Dris et al. 2016). However, realistic measures of humans' exposure to plastics have neither been taken nor modelled (Koelmans et al. 2017).

Potential Pathways of Harm

Ingestion

A recent review (Wright and Kelly 2017) concluded that toxicity from chemical constituents could occur via leaching from plastics ingested by eating seafood, and this also could result in the chronic exposure of some chemicals due to the bioaccumulation of toxins in tissues. It is known that additives such as plasticisers (e.g. phthalates) and bisphenol (BPA) can cause harm directly or from their breakdown products. For example, BPA, which has received the most interest to date, can migrate out of polycarbonate to contaminate food and drink products (Guart et al. 2013). Once internalised, this chemical interacts with hormone receptors, resulting in a complex bodily response (Koch and Calafat 2009). Plastics are also known to adsorb persistent organic pollutants and heavy metals once they have become waste in the natural environment. With a larger surface area–to–volume ratio, microplastics can act as a conduit and/or sink for these chemicals, and hence can transport them into humans through ingestion. Physical interactions between internal tissues and microplastics may also be problematic. Smaller particles have been flagged as the most concerning (reviewed in Galloway 2015), but, again, knowledge gaps mean the potential for harm is unknown.

Inhalation

Inhaling fibrous material is known to be hazardous to human health at high concentrations; consequently, this type of exposure has been monitored by industry for many years. These studies have shown that fibres (natural and synthetic), once inhaled, can cause chronic irritation and inflammation (reviewed by Prata 2018). The harm caused at the exposure level generally found in the environment is unknown.

Littering and Human Health

The connection between human well-being and ocean proximity has only recently been investigated, and studies have revealed that coastal proximity and blue spaces positively affect well-being (Wheeler et al. 2012; White et al. 2010). However, beach litter and microplastics are considered a risk to well-being (Gollan et al. 2019), and are one of the biggest threats to the benefits local communities receive from the marine estate. Litter can undermine the positive effects of a coastal estate and inhibit beach use (Wyles et al. 2016; Rangel-Buitrago et al. 2018), potentially reducing enjoyment outdoors and exercise, both of which are known to positively affect mental and physical health (Gladwell et al. 2013). As society begins to better recognise mental health challenges, this is an area that requires more research as it could be the most important influence marine plastic has on human health.

3.1.3 Economic Impacts

Plastic pollution in the ocean also has broad economic consequences. All sectors of the economy use plastic, and, across sectors, plastic waste is generated in near proportion to the level of use (Lin and Nakamura 2019). The full life cycle cost of plastics is not reflected in the pricing of plastic products (Oosterhuis et al. 2014). Plastic production is therefore not a fully costed system. Instead, the economic costs of plastic pollution are predominantly borne by the environment and by society (United Nations Environment Assembly of the United Nations Environment Programme 2017; Forrest et al. 2019).

The costs of plastic pollution can be broadly divided into two categories: direct and indirect. The direct costs of plastic pollution include prevention (e.g. environmentally sound waste management, awareness-raising, behaviour change campaigns), remediation (e.g. beach grading, fishing-for-litter programmes) and direct damage (e.g. lost productivity from fish mortality or reduced ecosystem services, repairs to equipment). The indirect costs of plastic pollution have proven difficult to quantify, partly due to differences in the values held by individuals (such as the importance of a clean beach), but also due to the challenges in placing an economic value on a healthy environment. Irrespective of the categorisation and estimation methodologies, the above direct and indirect costs are 'avoidable costs' (McIlgorm et al. 2008).

Direct and Indirect Costs of Plastic Pollution

The impact of plastic not being a fully costed system is highlighted by the particularly problematic plastic packaging sector. It produces a conservatively estimated $40 billion annually in negative externalities, such as degradation of natural systems and greenhouse gas emissions, outstripping the profits of the sector (Ellen MacArthur Foundation et al. 2016). Including plastic products, the total environmental cost in 2015 to society from plastics was estimated to be over $139 billion, which equated to nearly 20% of revenues in the plastic manufacturing sector (Lord et al. 2016).

The UN Environment Programme has estimated the global damage to marine environments from plastic pollution to be a minimum of $13 billion per year (UNEP 2014). Moving beyond damage costs to the environment, the reduction in global marine ecosystem services has been estimated at $0.5–2.5 trillion, based on 2011 stocks of marine plastic pollution (Beaumont et al. 2019). Forrest et al. (2019) aggregated estimates across the plastics life cycle to conclude that annual damages from plastic production and the current stock of plastic waste in the ocean amount to $2.2 trillion. The European Parliament's new measures to regulate single-use plastics cite benefits including avoiding the emission of 3.4 million tonnes of carbon dioxide equivalent and environmental damages equivalent to €22 billion by 2030, as well as an estimated savings to consumers of €6.5 billion (European Commission 2019).

Marine litter and plastics in particular both originate mainly from sea-based and coastal activities (fishing, aquaculture, tourism, shipping) and can, in turn, significantly impact these economic sectors (Newman et al. 2015; Krelling et al. 2017). For example, fishermen report nets fouled with plastic litter (Wiber et al. 2012; Brennan and Portman 2017) sometimes even reaching levels that cause them to move to areas less polluted with plastic litter (Nash 1992). Litter accumulating in the net may also affect the efficiency of the nets (Eryaşar et al. 2014). Fishermen lose time cleaning litter out of nets but surprisingly then dump the same litter overboard (Neves et al. 2015). Similar to cultured species, commercially caught fish may have ingested microplastics (see, for example, Rochman et al. 2015), which could affect the health of the fisheries and, eventually, the economic value of the catches.

Aquaculture may suffer from marine litter through fouled holding cages and health risks to the cultured species, which may ingest small microplastics. There is special concern regarding cultured bivalves, which have been shown to contain microplastics in their tissues in several independent studies (De Witte et al. 2014; Van Cauwenberghe and Janssen 2014; Davidson and Dudas 2016; Li et al. 2016, 2018a,b; Naji et al. 2018; Phuong et al. 2018; Cho et al. 2019; Teng et al. 2019). While this is of concern for consumers, there are other sources of microplastic ingestion (e.g. from air on food) that might far exceed those taken up by bivalves (Catarino et al. 2018). Interestingly, ingesting small microplastics (between 1 and 10 μm) by oyster larvae had no effect on the survival or growth of those larvae (Cole et al. 2015), but a similar study on mussel larvae showed detrimental effects of microplastic ingestion (Rist et al. 2018).

There is also concern of trophic transfer of microplastics (Nelms et al. 2018), but a recent study suggested that large predators rapidly egest microplastics taken up with their small prey organisms (Chagnon et al. 2018). Commercially important crustaceans can contain large numbers of microplastics, but it is suggested that they significantly reduce their accumulated microplastic load during moulting (Welden and Cowie 2016). In addition, the risk of ingesting microplastics is reduced when the gut is removed (such as those of fish, crustaceans and most other species) prior to consumption by humans (Lusher et al. 2017).

Shipping can be severely impacted as vessels can get entangled with marine litter, causing high risk of damage to the ships and injury to mariners and travellers (Newman et al. 2015; Hong et al. 2017). These risks might be exacerbated in harbour waters where the same structures that protect the harbour from wave exposure accumulate large quantities of marine litter (Aguilera et al. 2016), including fishing lines (Farias et al. 2018), which ships can become entangled in.

McIlgorm et al. (2008) estimated damage to maritime industries in the Asia-Pacific Economic Cooperation (APEC) region to be $1.26 billion per year in 2008 terms. For comparison, the gross domestic product for this same region of 21 member countries was $29 billion in 2008 (McIlgorm et al. 2008). McIlgorm et al. (2020) have updated these numbers, now estimating $10.8 billion in damage per year to industries in the marine economy attributable to marine debris. This is eight times greater than the previous estimate due to improved data, growth in the marine economy and an increase in the amount of plastic in the ocean over that time. By 2050, this damage is projected to be $216 billion (McIlgorm et al. 2020).

Beach litter may cause annoyance among beach visitors (Schuhmann et al. 2016; Brouwer et al. 2017; Shen et al. 2019) or even induce people to abandon a heavily littered beach (Krelling et al. 2017) and travel to more distant, cleaner beaches (Leggett et al. 2014). A study in South Korea showed that following a litter event (rains flushing inland litter onto coastal beaches) visitor numbers decreased dramatically; the authors estimated income losses of millions of dollars (Jang et al. 2014a). On tourist beaches, large amounts of litter are removed daily (Williams et al. 2016), incurring substantial costs for local municipalities (de Araújo and Costa 2006). Interestingly, several studies show that people would be willing to pay to visit beaches if they were cleaned (Brouwer et al. 2017; Shen et al. 2019). Besides the impact on the aesthetic value of beaches (Rangel-Buitrago et al. 2018), litter can also pose a health risk to visitors (Campbell et al. 2016), especially to young children (Campbell et al. 2019).

In California, modelling indicated that a 25% reduction in marine debris on all 31 of its beaches would improve the welfare of local residents by $32 million over three summer months by improving the welfare value of beach visits by residents and increasing the number of visits made. Improving marine debris reduction to 100% raised the savings to $148 million for the same period (Leggett et al. 2014). The chemical burden and disease cost of endocrine-disrupting chemicals within the European Union has been estimated at €119 billion (Trasande et al. 2015), of which some daily contact is likely via plastics (Feldman 1997; Magliano and Lyons 2013). The environmental costs of marine plastic pollution are not fully understood. The concern, however, is of such gravity that the issue is now being considered within the realm of a planetary boundary threat (Villarrubia-Gómez et al. 2018).

3.2 Impacts of Other Solid Waste

Inadequate waste collection and uncontrolled dumping or burning of solid waste still occurs around the world, but primarily where waste infrastructure is lacking, often in low- and middle-income countries (Kaza et al. 2018). This other waste includes all other municipal waste, medical waste, e-waste and disaster debris, and mismanagement of it has a range of impacts. Inadequate sanitation and mismanagement of organic waste and medical waste can cause exposure to pathogens and disease, and e-waste mismanagement results in the release of heavy metals into the environment. For example, the plastic used to house wires and cases is often open burned where informal processing takes place, releasing dioxin, particulate matter and heavy metals into the air (Asante et al. 2019).

3.2.1 Impacts on Ecosystems and Marine Life

Leachate (liquid that accumulates from waste containing organic compounds as well as heavy metals and POPs) can drain directly into the ocean (depending on the proximity of the waste) or into rivers, groundwater and the soil (Yadav et al. 2019), further contributing to ocean pollution. Organic waste from garbage can also contribute to nutrient loading in waterways and the ocean, and the open burning of solid waste gives off particulate matter and emissions (Wiedinmyer et al. 2014) that can contribute to atmospheric deposition into the marine ecosystem. According to the International Solid Waste Association, greenhouse gas emissions across the economy—which indirectly impact the ocean through climate change effects—can be reduced by 15–20% with improved global waste management (UNEP 2015).

3.2.2 Human Health Impacts

Inadequate waste management, especially open burning and dumping, around the world produces pollution (Vasanthi et al. 2008; Wiedinmyer et al. 2014) that can impact people living near management facilities and those working directly with solid waste. About 15 million people globally, often called waste pickers (who include men, women, children, migrants and the underemployed), work informally in the waste sector (Medina 2008). In China alone, it is estimated that 3.3 to 5.6 million people work informally in the recycling of solid waste (Linzner and Salhofer 2014), and Forrest et al. (2019) acknowledge the millions of people working in poor conditions for little money in jobs that would not qualify as decent work by the International Labour Organization. While these issues must be addressed, it is also important to recognise that waste and plastic management constitute the livelihoods of millions of people. Any interventions used to address plastic and other waste must incorporate the views and participation of informal workers, and especially waste pickers, so that millions of people aren't negatively impacted through the unintended consequences of 'traditional' infrastructure, such as eliminating a crucial source of income (Dias 2016). Women can be disproportionately harmed by the formalisation of waste management, as they are typically excluded from formal employment in the formalized sector. But they can be helped through inclusive improved recycling operations, capacity building, provision of equipment, formal training and awareness building, financial assistance and health insurance since they have high levels of participation in the informal sector but often have less access to these kinds of benefits (Krishnan and Backer 2019).

3.2.3 Economic Impacts

While there are global data on the cost of plastic pollution (see Sect. 3.1, Impacts of Plastic), there is not a global number for the cost of mismanaged waste. The World Bank estimates that proper waste management infrastructure would cost $50–100 per metric tonne (Kaza et al. 2018), which is in the same range as tipping fees charged for municipal solid waste disposal in the United States. In Palau, where the ocean is extremely important to the economy, the cost of waste-related pollution, or mismanaged waste (not the cost of waste management which is estimated at $87 per tonne), was estimated at be $1.9 million per year, which is 1.6% of the country's gross domestic product and equates to an annual cost of $510 per household (Hajkowicz et al. 2005).

3.3 Impacts of Pesticides

Pesticide mixtures include active and inert ingredients; both are important, as the active ingredient is the toxicant for the target organism, and the inert ingredient often amplifies the exposure mechanism. For example, an herbicide with an active ingredient might be mixed with an inert ingredient that is water soluble to more effectively penetrate soil, while the same active ingredient can be mixed with a non-water-soluble oil to more effectively penetrate the leaf. The same active ingredient can have different toxic effects on the target organisms, and potential environmental effects, based on the carrier or inert ingredient. In the United States, only the active ingredients must be disclosed in pesticide labelling, making impact assessments very difficult to conduct.

Pesticides are very effective at improving the efficiency of agricultural production by reducing crop and animal losses. However, there are risks associated with pesticide applications to nontarget organisms. Nontarget organisms include the people who apply the pesticides, process the products and consume the products. There are also risks to nontarget organisms in the fields and paddocks where these pesticides are applied. Broad spectrum insecticides kill desirable insects such as pollinators and the biological predators of undesirable insects. Some pesticides persist in the environment and move through the food chain, resulting in toxic impacts on nontarget organisms including song birds, raptors, rodents, reptiles and fish (UNEP 2019).

3.3.1 Impacts on Ecosystems and Marine Life

Pesticides that reach the ocean can impact nontarget organisms in several ways, depending on the active ingredient pesticide category, inert ingredient mediator, transport mechanism and depositional environment. The toxic impact of pesticides is generally proportional to the concentration, so very low concentrations often have very low impacts. However, pesticides can be bioconcentrated and biomagnified through the food chain to result in cumulatively higher impacts on predators and scavengers (including humans). Bioconcentration is the process of uptake of a chemical by an organism from the abiotic environment, resulting in higher concentrations in that organism than in the environment (LeBlanc 1995). Bioconcentration of pesticides occurs when the active ingredient persists in the environment long enough to be ingested by an organism such as krill, where it is either metabolised, excreted or stored in fatty tissues (Cincinelli et al. 2009). The pesticides that are stored in fatty tissues can persist through many cycles of ingestion, and thus accumulate in the organism. Biomagnification is the process whereby the amount of the pesticide is amplified up the food chain, and the active ingredient can be concentrated in the fatty tissues of top predators such as swordfish, sharks and tuna. These concentrations can be amplified over 1000-fold through this process. Most modern pesticides have been designed to not persist in the environment, and thus are less prone to bioconcentration. However, early twentieth-century pesticides, which are banned in Europe and the United States but are still manufactured and used in many countries, can last over 100 years in the environment and are very prone to bioconcentration and biomagnification (Dromard et al. 2018). In general, organochlorine pesticides (OCPs), which were developed in the early-to-mid twentieth century, are the world's most persistent legacy pollutants in the ocean. These include dichlorodiphenyltrichloroethanes (DDTs), hexachlorocyclohexanes, heptachlor, aldrin, alpha and beta-endosulfans, dieldrin, endrin, endrin aldehyde, endrin ketone, methoxychlor, endosulfan sulfate and heptachlor epoxide (Guo et al. 2007).

Pesticides have been documented to reduce the photosynthetic efficiency of sea grass, corals and algae (herbicides), resulting in chronic stress (Brodie et al. 2017). Certain herbicides in common use, including Diuron, Atrazine, Hexazinone and Tebuthiuron, have been shown to have measurable impacts on seagrass productivity, especially when combined with light attenuation from high sediment loads from agricultural runoff (Flores et al. 2013). Seagrass beds are critical habitats for many marine species and support global fisheries. Insecticides, including organophosphates, organochlorines, carbamates and pyrethroids, as well as fungicides, have been shown to restrict to fully inhibit coral settlement and metamorphosis at concentrations as low as one part per billion (Markey et al. 2007). Concentrations just ten times that amount have caused coral branch death. These concentrations are at or below detection levels for conventional laboratory analyses, rendering these pesticides virtually invisible to investigators

3.3.2 Human Health Impacts

The primary exposure mechanism to humans from ocean-borne pesticides is through ingestion of species that biomagnify those pollutants. The most common pesticides found in seafood at concentrations above background levels are OCPs. Communities whose diets are seafood-based are most at risk given their higher rates of fish consumption. Consuming fatty piscivores such as hairtail, mackerel and tuna in South Korea was shown to increase exposure of vulnerable populations (children and elderly) to increased OCPs (Moon et al. 2009). In general, these pesticide concentrations are below chronic toxicity levels for most people (Smith and Gangolli 2002). Toxicants of concern in fish from biomagnification include heavy metals (mercury, cadmium and lead—see Sect. 3.7 on heavy metals), and legacy organochlorines from industry such as polychlorinated biphenyls (PCBs) (Storelli 2008).

3.3.3 Economic Impacts

The economic impacts of pesticides in the ocean are largely through decreased productivity rather than human toxicity. The loss of productivity and resiliency of seagrass beds and coral reefs is having a significant impact on global economic security. These critical ecosystems provide a portfolio of ecosystem services that are essential for human society, including the provision of food, water, energy and other resources, and tourism. The estimated net present value for 2050 of Earth's coral reefs was almost $800 billion (Cesar et al. 2003). If pesticides are reducing the productivity of these ecosystems by only 25%, the annual economic impact of those pesticides in the ocean would still be $200 billion per year. These critical ecosystems are also stressed by other pollutants, sediment and climate change. Some estimates suggest that under a high greenhouse gas emissions scenario, more than 90% of coral reef communities would be lost by 2100 (Speers et al. 2016). Cumulatively, these pose imminent threats to Earth's ocean ecosystems.

3.4 Impacts of Nutrient Pollution

3.4.1 Impacts on Ecosystems and Marine Life

Nutrient pollution, which occurs when anthropogenic sources of primarily N and P are discharged into marine systems, leads to eutrophication, algal blooms, dead zones and fish kills in freshwater and coastal waters. Scientists have estimated that about 80% of large marine ecosystems in the world already suffer from serious eutrophication, hypoxia and anoxia in coastal waters (Selman et al. 2008; Diaz et al. 2011; STAP 2011). In addition, related incidences of toxic algal blooms such as 'red tides' have become more frequent (Rabalais 2002). Eutrophication also leads to habitat changes and the loss of species of high value (Heisler et al. 2008).

Many species can be impacted directly or indirectly by nutrients in marine ecosystems as nutrient inputs have altered the abundances and distributions of marine species (e.g. through algal blooms). Eutrophication and oxygen depletion (often referred to as 'dead zones' when affecting a large area) have direct adverse effects on coral reefs, seagrass beds, fish and shellfish (Bouwman et al. 2011). Diaz et al. (2011) identified more than 770 eutrophic and hypoxic coastal systems worldwide, where 70% of the areas had documented hypoxia and almost 30% were developing hypoxia. The dead zone in the Gulf of Mexico resulting from agricultural runoff into the Mississippi River has been studied extensively, but there is less data on these zones in developing countries, so these estimates are likely conservative.

One example of the direct impact of increased nutrients in the ocean is the world's largest macroalgal bloom, which was recorded from 2011 to 2018 (the most recent data available). Using satellite images, (Wang et al. 2019) showed that since 2011, the free-floating mats of brown macroalgae called *Sargassum* spp. have increased both in density and size, generating a long belt of 8880 km extending from West Africa to the Caribbean Sea and Gulf of Mexico. *Sargassum* is a naturally occurring seaweed that provides a critical habitat to a diverse array of species in this ecosystem. However, when the *Sargassum* mats overcrowd the coasts, it can impact the movement of some marine species. When the excess *Sargassum* dies and sinks to the ocean bottom in large quantities, corals and seagrasses can be smothered. On the beach, rotten *Sargassum* releases a strong smell, potentially imposing health challenges for people who have asthma. *Sargassum* blooms and their adverse effects could reduce the number of tourists during a bloom. For example, in 2018, Barbados had to declare a national emergency because of a bloom.

The ocean is becoming more stratified, and while there is still some discussion over coastal marine ecosystems being N- or P-limited, (Elser et al. 2007) found that, for coastal systems, N and P limitations play a similar role, implying that reducing the discharges of both N and P is important for alleviating pollution in coastal areas. This is exactly what was shown by (Beman et al. 2005), who found that areas that are nitrogen deficient were especially vulnerable to nitrogen pollution. They also found that agricultural runoff had a strong and consistent influence on biological processes, stimulating algal blooms 80% of the time within days of fertilisation in the Gulf of California. They then projected that by 2050, 27–59% of all nitrogen fertiliser would be applied in developing regions upstream of nitrogen-deficient marine ecosystems. These ecosystems are especially vulnerable to agricultural runoff and nitrogen pollution impacts (Beman et al. 2005).

3.4.2 Human Health Impacts

Some important drinking water sources (e.g. Lake Erie) cannot be used during algal blooms, as the toxins either increase the cost of treatment or make it impossible to treat. Other human health impacts come from direct or indirect exposure to toxins resulting from algal blooms—for example, a red tide can cause ciguatera poisoning, paralytic shellfish poisoning, neurotoxic shellfish poisoning (NSP), amnesic shellfish poisoning and diarrhetic shellfish poisoning, which are the five most commonly recognised illnesses related to harmful algal blooms (HABs). Exposure to the toxins from HABs is mediated through the consumption of contaminated fish and shellfish, or through exposure to aerosolised NSP toxins near water bodies where a bloom is occurring (Grattan et al. 2016).

3.4.3 Economic Impacts

Attempts to evaluate the monetary impacts of eutrophication have been made over the last two decades. Studies indicate a variety of impacts and costs that are quantifiable fairly directly, for instance, when cities of hundreds of thousands of people are deprived of drinking water for several days. One example is the toxic algal bloom that occurred in the western Lake Erie basin in 2011, which led to a disruption of water supplies for 400,000 people (Watson et al. 2016). In another example, a major and extensive red tide outbreak occurred along the coast of Hong Kong and south China, covering an area of more than 100 km^2. Over 80% (3400 tonnes) of mariculture fish were killed, and the total loss was over $40 million (Yang and Hodgkiss 2004). On the other hand, integrating all the environmental, health and socioeconomic impacts in the calculations of indirect effects poses more of a challenge.

3.5 Impacts of Antibiotics, Parasiticides and Other Pharmaceuticals

3.5.1 Impacts on Ecosystems and Marine Life

Our understanding of the impacts of emerging contaminants is limited to what has been learned by studying specific instances where they have been found and identified; impacts on the overall marine environment are not well-understood. Pharmaceutical substances have been examined worldwide in surface water, groundwater, tap/drinking water, manure, soil and other environmental matrices. (aus der Beek et al. 2016) reviewed 1016 articles and found that pharmaceuticals or their transformation products have been detected in the environment of 71 countries covering all continents. Six hundred thirty-one pharmaceutical substances were found at levels above the detection limit of the respective analytical methods employed. Residues of 16 pharmaceutical substances were detected in each of the five UN regions, and the antibiotic tetracycline was detected in wastewater treatment plant effluents in all UN regions. Regional patterns of pharmaceutical leakage to the environment emerged as well: Antibiotics were most prevalent in Asia, analgesics edged out other pharmaceuticals as most prevalent in Eastern Europe, lipid-lowering drugs were highest in Europe and Latin America, oestrogens were found most in Africa, and the 'other pharmaceutical' category was predominant in Western Europe (aus der Beek et al. 2016).

Research presented at the 2019 annual meeting of the Society of Environmental Toxicology and Chemistry found that of the sites monitored, 65% of them contained at least one of the 14 most commonly used antibiotics. These sites were located in rivers in 72 countries across six continents. The concentration of one antibiotic, metronidazole, found by the researchers at a site in Bangladesh was 300 times the 'safe' level. (The AMR Industry Alliance recently established 'safe' levels of antibiotics in the environment, ranging from 20 to 32,000 nanograms per litre depending on the antibiotic.) Ciprofloxacin, a general antibiotic used to treat various bacterial infections, most frequently exceeded safe levels, surpassing the safety threshold in 51 places.

Geographically, 'safe' limits were most frequently exceeded in Asia and Africa (to the greatest degree in Bangladesh, Kenya, Ghana, Pakistan and Nigeria), but sites in Europe, North America and South America also had levels of concern, showing that antibiotic contamination is a global problem. Sites with the highest risk of contamination were typically adjacent to wastewater treatment systems and waste or sewage dumps and in areas of political turmoil, including the Israeli and Palestinian border (University of York 2019).

Benzophenone-2 (BP-2) is an additive to personal-care products and commercial solutions that protects against the damaging effects of ultraviolet (UV) light. BP-2 is an 'emerging contaminant of concern' that is also often released as a pollutant through municipal and boat/ ship wastewater discharges and landfill leachates, as well as through residential septic fields and unmanaged cesspits. Although BP-2 may be a contaminant on coral reefs, its environmental toxicity to reefs is unknown. This poses a potential management issue, since BP-2 is a known endocrine disruptor as well as a weak genotoxicant (Downs et al. 2014).

There is concern over the impacts of commonly used organic UV filters, including oxybenzone (benzophenone-3), 4-methylbenzylidene camphor, octocrylene and octinoxate (ethylhexyl methoxycinnamate), on the marine environment. Oxybenzone, octocrylene, octinoxate and ethylhexyl salicylate have been identified in water sources around the world, and are not easily removed by wastewater treatment plant techniques (Schneider and Lim 2019). Oxybenzone has been specifically linked to coral reef bleaching. In addition, 4-methylbenzylidene camphor, oxybenzone, octocrylene and octinoxate have been identified in various species of fish worldwide, which has possible consequences for the food chain (Schneider and Lim 2019). Danovaro et al. (2008) found that even low concentrations of sunscreens caused bleaching of corals. The organic UV filter induces the lytic viral cycle in symbiotic zooxanthellae with latent infections. Therefore, sunscreens may be playing an important role in coral bleaching by promoting viral infections in areas with high recreational use by humans (Danovaro et al. 2008).

3.5.2 Human Health Impacts

Wastewater treatment plants are a main source of antibiotics released into the environment. An overabundance of antibiotics in wastewater may generate antibiotic resistance genes and antibiotic resistant bacteria. Some scientists are concerned that wastewater treatment plants are becoming hot spots for resistant genes and bacteria, which has implications for human health should people get infections that are then resistant to typical antibiotics (Rizzo et al. 2013).

3.6 Impacts of Industrial Chemicals Including Persistent Organic Pollutants

3.6.1 Impacts on Ecosystems and Marine Life

Polybrominated diphenyl ethers (PBDEs) and POPs are toxic, are not easily degradable in the environment, bioaccumulate in the food chain and undergo long-range transport (European Environment Agency 2019). Many industrial chemicals and POPs are known to be poisonous and to damage the environment and the organisms living in the affected ecosystems. These pollutants have become distributed throughout the ocean and have been found in seemingly pristine environments. These pollutants also bioaccumulate in marine organisms such as fish and invertebrates such as corals, which can lead to various physiological impairments, varying from subcellular changes such as direct effects on DNA (deoxyribonucleic acid) to metabolic stress (Logan 2007; van Dam et al. 2011).

3.6.2 Human Health Impacts

POPs and PBDEs can cause cancer and toxicity in the liver, kidneys and reproductive system (Qing Li et al. 2006). The main impacts of industrial pollution to human health are derived from making direct contact with contaminated water. The direct contact with polluted water puts people at risk when the toxins are heavy metals. The chemical content in the water, whether carcinogenic or not, may nevertheless play a role in contributing to cancer mortality risk (Hendryx et al. 2012). Bathing in contaminated water increases the risk of respiratory disease and skin problems.

Human consumption of marine organisms that have been contaminated with polluted water is one major impact of industrial pollution on humans. Many of the fish that are a primary food source for the indigenous people in the Canadian Arctic are heavily contaminated by POPs (Dewailly 2006). While some persistent organic pollutants have started to decrease in humans and food in monitored Arctic locations because of international restrictions, levels of oxychlordane, hexachlorobenzene, polybrominated diphenyl ether and perfluorinated compounds are not decreasing (Abass et al. 2018).

Greenland has some of the highest concentrations of POPs in humans in the Arctic—with the exception of PBDEs, Greenland populations had the highest measured levels of POPs than any other Arctic country (Gibson et al. 2016).

3.6.3 Economic Impacts

Studies indicate a variety of economic impacts from industrial pollution. The tangible economic impacts include those that occur during pollution incidents as well as from activities undertaken to prevent, mitigate, manage, clean up or remedy pollution incidents. The global economic cost related to the pollution of coastal waters is $16 billion annually, largely due to human health impacts (UNEP 2006). An additional source of cost is the loss of earnings caused by damage to natural resources. The intangible costs are the loss of marine biodiversity and the provision of other environmental services caused by industrial pollution.

3.7 Impacts of Heavy Metals

3.7.1 Impacts on Ecosystems and Marine Life

Exposure to heavy metals can increase the permeability of the cell membrane in phytoplankton and other marine algae, leading to the loss of intracellular constituents and cellular integrity, and inhibiting metabolism (Sunda 1989; González-Dávila 1995; Hindarti and Larasati 2019). High trace element burdens in marine mammals have been associated with lymphocytic infiltration, lesions and fatty degeneration in bottlenose dolphins, and decreasing nutritional states and lung pathologies in other marine mammals (Siebert et al. 1999). In addition, cadmium, lead and mercury are potential immunosuppressants; of particular concern is the buildup of mercury, which marine mammals tend to accumulate in the liver.

3.7.2 Human Health Impacts

Mercury and Arsenic Methylmercury is a neurotoxic compound responsible for microtubule destruction, mitochondrial damage, lipid peroxidation and accumulation of neurotoxic molecules such as serotonin, aspartate and glutamate (Patrick 2002). Consumption of contaminated aquatic animals is the major route of human exposure to methylmercury (Trasande et al. 2015). Seafood contaminated by heavy metals or metalloids such as mercury and arsenic can contribute to human health risk (Harris et al. 2014; Gao et al. 2018). One unforgettable case was the mass poisoning of people in Minamata, Japan, in the 1950s, when 2252 people were impacted by the contamination and 1043 died (Harada 1995).

Cadmium and Lead Consuming fish containing to cadmium and lead can cause major diseases in humans such as renal failure, liver damage and symptoms of chronic toxicity in the kidney (Bosch et al. 2016; Gao et al. 2016).

Chromium Because of its mutagenic properties, hexavalent chromium is a carcinogen that humans can get exposed to through soils, sediment and surface waters, as well as some fish (Copat et al. 2018; Tseng et al. 2019).

3.7.3 Economic Impacts

Heavy metal pollution results in substantial economic impacts to the fishing sector. Bioaccumulation of metals in fish limits the species that can be safely eaten and the frequency that those fish can be eaten, and as a result can limit imports and exports. For example, in 2006, European Commission Regulation 1881/2006 established the maximum levels for cadmium, lead and mercury in food products. High quantities of heavy metals in fish are one of the principal reasons why fish are detained at EU borders and the main problem that importers from non-EU countries must address. The economic losses deriving from EU border detentions amount to hundreds of millions of euros each year (FAO n.d.).

3.8 Impacts of Oil and Gas

3.8.1 Impacts on Ecosystems and Marine Life

Oil spills tend to disproportionately impact sea birds, which can be harmed and killed by exposure to oil. Individual birds become unable to swim or fly and nervous system abnormalities can occur. Population-level effects of oil toxicity on aquatic birds occur through the loss of egg viability. Because it is inherently poisonous, oil in the marine environment has the potential to harm any creature that comes in contact with it. This includes larger animals such as sea turtles, which are sensitive to chemical exposure at all stages of life and lack an avoidance behaviour, and seals, which can become blind, as well as smaller organisms, such as zooplankton and larval fish. Oil spills, and their associated responses, can be particularly damaging to fragile but vital marine ecosystems such as coral reefs and mangroves, but are believed to damage life throughout the water column. Heavier oils settle and can coat and smother benthic areas. In areas impacted by oil spills, bottom-feeding fish have developed carcinomas and papillomas on their lips, as well as changes in their cell membranes. Spilled oil can persist in the environment, continuing to injure and kill marine life. More research is needed to fully understand the less obvious impacts of oil spills on the marine environment (NOAA OR&R 2019).

3.8.2 Human Health Impacts

A 2016 review article on the human health impacts of oil spills looked at mental health effects; physical and physiological effects; and genotoxicity, immunotoxicity and endocrine toxicity. While there exist a number of obstacles to calculating human health impacts—such as challenges to determining exposure levels and the level of effectiveness of personal protective gear as well as a reliance on self-reported health symptoms and variations in genetic sensitivities to chemical exposure— the authors concluded that there is sufficient evidence to establish a relationship between exposure to oil spills and the development of adverse health effects in exposed individuals (Laffon et al. 2016).

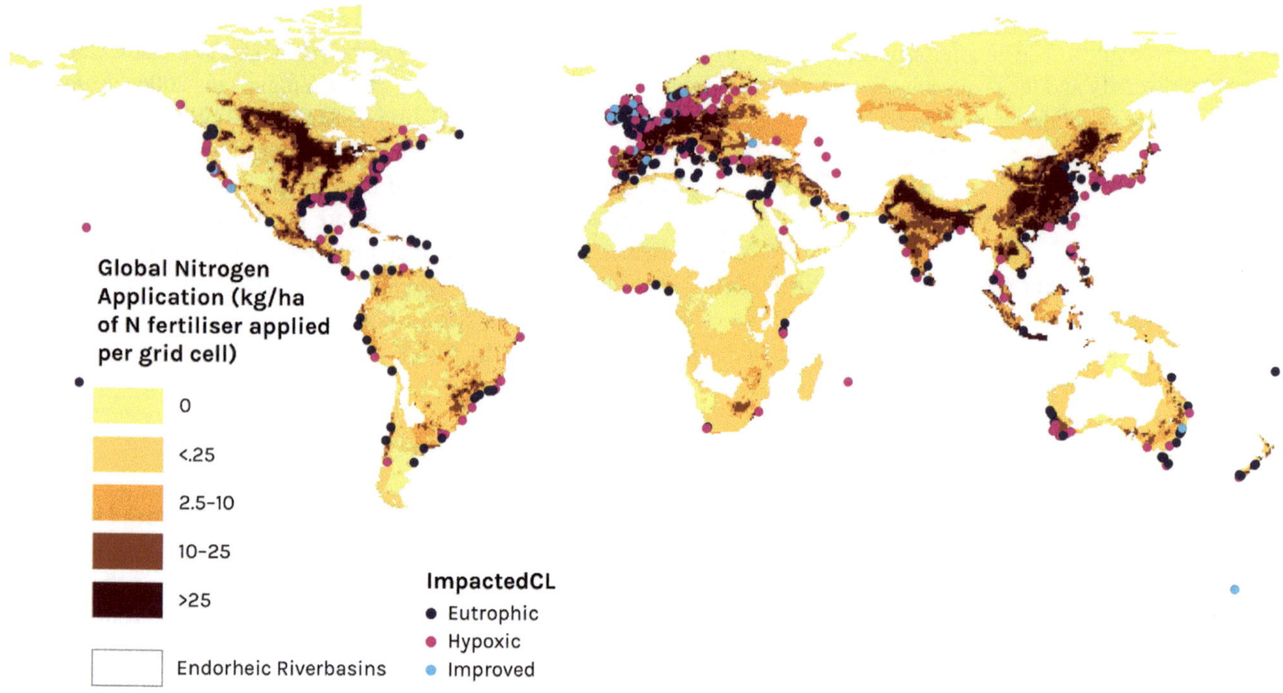

Fig. 5.3 Global nitrogen use and hypoxic areas in the ocean. Note: Mismanaged nitrogen use on land and incidences of eutrophication and hypoxia. [Sources: Data compiled from Potter et al. (2011a) and World Resources Institute (2013). Map created by A. Brooks]

3.8.3 Economic Impacts

Oil spills can be very costly to the responsible companies as well as to the fishing and tourism industries affected by the spill. For example, BP's Deepwater Horizon spill in the Gulf of Mexico is estimated to have cost the company $61.6 billion in penalties and fines; cleanup and remediation; and payments to affected companies, communities and individuals (Mufson 2016). The sinking of the Prestige oil tanker in November 2002 off the coast of Galicia, Spain, resulted in estimated losses to the Galician fishing sector of €76 million by December 2003 (Surís-Regueiro et al. 2007). Kontovas et al. (2010) calculated a per metric tonne cost for oil spills based upon a regression of 38 years of oil spill and cost data—the average value being $4118 per metric tonne in 2009.

3.9 Impacts Summary

3.9.1 Inputs Lead to Impacts

Based upon the literature reviewed for this Blue Paper, it is evident that all pollutants discussed in this report are concerning to our ocean, though some may be more urgent or easier to address than others. In addition, multiple pollutants can act synergistically, creating a greater effect on the ocean than the sum of their individual impacts. Exploring the present and future impacts, as was done in this section, is one way to start to prioritise which pollutants to tackle first. At this moment, the plastic pollution crisis is very salient—the issue is tangible and understandable, and countries around the world are working on solutions because of marine ecosystem and economic impacts. It is also evident from this work that nutrient pollution is of great concern to the ocean. Nutrients contribute to harmful algal blooms and create low-oxygen hypoxic zones and stratification, ultimately impacting the health of marine life and humans. Without changes to either of these two pollutant input systems in a business-as-usual trajectory, the impacts from them will get only worse as populations grow and economies continue to develop. Figure 5.3 provides a global map showing nitrogen use, along with the drainage basins and the impact of this drainage by showing hypoxic areas in the ocean. Urgent action is needed to protect the ocean from further impacts from pollution.

4 Human Dimensions

The issue of pollutants leaking into the ocean is entirely a consequence of human decisions and behaviours. It is determined by individuals, communities, companies and politicians, to name but a few of the actors within the social-environmental system (Pahl and Wyles 2017; SAPEA 2019). These actors have varying perceptions, goals and values that motivate existing practices (and can also be harnessed for change). For example, a farmer might decide to employ a pesticide to increase yield and be willing to accept adverse effects on wildlife. A cosmetics company might decide to replace natural ingredients with plastic microbeads to save money and reduce allergens in their products.

Within the social and behavioural sciences' research on environmental pollution, the focus has been on principles of risk perception and determinants of behaviour (Pahl and Wyles 2017). In other words, how does a person or community decide that a pollutant poses a risk, and what factors motivate behaviour change (including not just individual actions but also demand for legislation and policy change)?

Researchers have found that how experts assess risk is different from how non-experts assess risk (see (Böhm and Tanner 2013). Experts apply scientific methods of risk assessment that focus on specific thresholds or outcomes such as fatalities or concentrations, whereas non-experts judge risk levels using a wide range of factors such as moral evaluation of the issue, perceived fairness, perceived control and positive and negative emotions such as dread and pride. These discrepancies can contribute to conflict between stakeholders. Mental model approaches are useful in this context because they can illustrate different expectations about the sources, pathways and impacts of pollution, which can provide triggers for change. However, it has also been noted that perception of risk in itself is not strongly linked to action, and if too strong, could even undermine action (Peters et al. 2013). However, when the risk is associated with an emergency event such as a natural disaster, this may encourage people to take action, depending on personal agency, community capacity and resilience (Brown and Westaway 2011). It is important to understand risk perception differences among stakeholder groups because they can influence how media reporting is interpreted, and should be taken into consideration when policies and interventions are developed.

Behavioural practices can contribute to pollution but are rarely quantified. For example, the dosage of fertilisers and timing of applications might vary according to practices and knowledge available to farmers, and fine-tuning practices could greatly reduce environmental (and health) impacts. Behaviour is determined by a range of factors beyond mere knowledge. To illustrate, most people understand healthy lifestyles but few eat very healthily and regularly exercise. This is similar in the environmental domain, where knowledge is one factor that can motivate behaviour change, but other factors are more powerful, including perceived control, social approval and moral norms, among others. In addition, contextual factors, such as the accessibility and design of the waste disposal system and availability of materials, are important. For example, if there is no recycling bin nearby, a person needs to have a strong motivation to recycle to put in the extra effort to find one (Pahl and Wyles 2017; SAPEA 2019).

To change perceptions and behaviours, a multipronged approach can target actors individually. Laws, bans and restrictions are powerful tools that can signal a social norm of undesirable behaviour. While outlawing a particular substance can be the most powerful tool, some materials, such as plastics, are so widespread that a simple ban would fall short or could be applied only to certain products. Education and public outreach campaigns are necessary to accompany policy change and are powerful instruments in their own right. Good campaigns build on behavioural science insights and integrate key elements that have been shown to work, e.g. empowering individuals, making specific suggestions for behavioural solutions that are effective and socially acceptable. It is important not to crowd out intrinsic motivation but rather to build on personal norms and values and develop a pro-environmental identity as this could spill over into other domains and behaviours. Effective interventions link to the target group's understanding of the issue and to their motivations and concerns, and build on existing social networks and channels. Often, there is initial reluctance to change (e.g. introduction of seat belts, smoking bans), but early adopters may forge the path. Trusted members of a community can trigger wider change and could be empowered as change agents. Change can happen top down and bottom up; to target plastic pollution, for example, there are many examples of community-led actions, voluntary efforts in the retail sector (e.g. bans on plastic bags) and nonprofit initiatives.

5 Opportunities for Action

Over the last several years, marine plastic pollution has captured the world's attention and inspired hundreds of commitments from governments, businesses and nongovernmental organisations (NGOs); dozens of innovation challenges; hundreds of start-up companies seeking to create solutions; and millions of citizens taking action, whether as citizen scientists, as part of a beach clean-up or by changing their own consumption choices.

It is extremely challenging, at least with available data, to weigh the damage done by marine plastic pollution against the harmful impacts of nonplastic pollution from municipal, agricultural, industrial and maritime sources, though the latter group has been more exhaustively studied. A more helpful question to ask, however, might be this: How can action to address plastic pollution be leveraged to maximise the benefits across as many other ocean pollutants as possible? If plastic pollution is uniquely able to catalyse action on solutions, how can we prioritise and design solutions to also stop the flow of other pollutants into the ocean?

The seven approaches developed from this research and presented below begin to address these questions. Each approach includes recommendations for interventions and actions to address ocean pollution through four levers: infrastructure, policy, mindset and innovation. These levers consider actions that may be taken by companies large and small, by elected officials and policymaking staff, by citizens and by innovators. There is likely a role for some form of voluntary collective action from the biggest producers and users of plastics. In fact, hundreds of companies have signed on to frameworks such as the New Plastics Economy, facilitated by the Ellen MacArthur Foundation, and/or have set goals regarding how they will address the problem of plastic pollution. This paper does not speculate about the precise paths companies will take, but rather focuses on the specific actions most likely to move the needle on plastic and other types of pollution reaching our ocean. After the details of the approaches are introduced, they are then summarised and compared based on their breadth of mitigation across pollutants and sectors.

In this section, the list of key interventions and actions are mapped to the following:

- Sectors: Municipal (M), agricultural (A), industrial (I), maritime (Mar)
- Types: Infrastructure, Policy, Mindset and Innovation
- Pollutants: Sourced from Table 5.1. Given below each corresponding intervention table
- Relevant UN Sustainable Development Goals (SDGs)

Improve wastewater management			
Infrastructure	Policy	Mindset	Innovation
1. Create or expand wastewater treatment capacity (M) 2. Add tertiary treatment for nutrients and microplastics (M) 3. Install toilets (wet or dry) where needed to prevent open defecation (M) 4. Install septic tanks where access to municipal wastewater systems is limited (M) 5. Ensure industrial wastewater is appropriately treated, whether through municipal or other infrastructure (I)	1. Ensure supporting policies for wastewater improvements and sustainability of infrastructure over time are in place (M)	1. See wastewater as a natural resource, especially in water-constrained regions (M)	1. Develop washing machine filters for microplastic fibres (M) 2. Innovate ways to remove pharmaceuticals and antibiotics from wastewater effectively and affordably (M)

Sectors: *Municipal (M), industrial (I)*
Pollutants: *Macroplastics; microplastics; other solid waste; nutrients; antibiotics, parasiticides and other pharmaceuticals; heavy metals; and industrial chemicals and POPs*
SDGs: *6.2, 6.3*

Improve stormwater management			
Infrastructure	Policy	Mindset	Innovation
1. Use natural filters such as berms and clay to minimise runoff into the ocean (A, M) 2. Implement stormwater and storm drain filtration and river mouth trash collection (M)	1. Set total maximum daily loads (TMDLs) for trash (M) 2. Impose regulatory limits, TMDLs for discharge (I) 3. Employ stormwater permitting (M) 4. Regulate animal waste lagoons that have the potential to discharge into the ocean (A) 5. Regulate use of pesticides, herbicides and nutrients for residential and commercial use (M) 6. Require nutrient management plans and pesticide management plans (A) 7. Require reporting of and/or limit usage of nutrients and pesticides (A)	1. Change cultural norms around having manicured lawns to reduce the use of pesticides, herbicides and fertilisers used for residential and commercial landscaping (M) 2. Create a culture of responsibility regarding picking up dog feces (M) 3. Change habit of washing with excessive soap, shampoo and products that contain high levels of nitrogen and phosphorus (M)	1. Conduct research and development in stormwater and other treatment systems (M, A, I) 2. Change crops, seeds and farming practices to minimise nutrient application prone to leakage (A)

Sectors: *Municipal (M), agricultural (A), industrial (I)*
Pollutants: *Macroplastics; microplastics; other solid waste; pesticides; nutrients; antibiotics, parasiticides and other pharmaceuticals; heavy metals; industrial chemicals and POPs; oil and gas*
SDGs: *None*

Adopt green chemistry practices and new materials			
Infrastructure	Policy	Mindset	Innovation
1. Construct treatment facilities with 'green engineering' principles (M) 2. Develop infrastructure for the production of new or alternative materials	1. Ban or limit the use of chemicals of concern and hazardous materials (I) 2. Ban hard-to-manage materials (M) 3. Require tracking/manifest of chemicals of concern (I)	1. Adopt green chemistry principles as a practice for companies (I) 2. Change cultural norms around having manicured lawns to reduce the use of pesticides, herbicides and fertilisers used for residential and commercial landscaping (M)	1. Develop new materials that maintain the desirable performance characteristics of plastics but not the problematic ones, e.g. true biodegradables (M, A) 2. Develop alternative cleaning products, e.g. phosphate-free soap and detergents (M) 3. Use fish waste or seaweed to make biopolymers for fishing gear (A) 4. Support research and development in green chemistry and alternative chemicals (I) 5. Reduce and prevent tire wear and tire dust by using new materials or other mechanisms 6. Use new materials for fishing gear, e.g. biodegradable components (Mar) 7. Support the development of products and services that do not use any chemicals of concern (I)

Sectors: *Municipal (M), agricultural (A), industrial (I), maritime (Mar)*
Pollutants: *Macroplastics; microplastics; other solid waste; pesticides; heavy metals; industrial chemicals and POPs*
SDGs: *3.9, 12.4*

Practice radical resource efficiency			
Infrastructure	Policy	Mindset	Innovation
1. Enable the development of circular business models through shared infrastructure, for example, reverse logistics or commercial washing services for reusable foodservice items (M)	1. Impose fees on single-use or other high leakage items (M) 2. Encourage industry voluntary contributions to reduce fossil-fuel-based plastics (M, A, I, Mar) 3. Support policies that allow personal container use in shopping and dining (M) 4. Enable treatment and use of food and human waste in appropriate applications (M, A)	1. Change cultural norms around waste generation/consumption and reuse, in particular to reduce the use of single-use plastic items (M)	1. Design zero-packaging grocery stores or include 'packaging free' or 'plastic free' aisles in regular grocery stores (M) 2. Develop new purchasing models that end reliance on single-use plastics (e.g. packaging as a service, reuse models) (M) 3. Pricing structure/business model for nutrients and pesticides to optimise outcomes and minimise waste (A) 4. Require fishing gear tracking (Mar)

***Sectors:** Municipal (M), agricultural (A), industrial (I), maritime (Mar)*
***Pollutants:** Macroplastics; microplastics; other solid waste; pesticides; nutrients*
***SDGs:** 8.4, 12.2, 12.5*

Recover and recycle the materials we use (formal and informal sectors)			
Infrastructure	Policy	Mindset	Innovation
1. Implement systems for compliance with bale contamination standards in exported/imported waste (M) 2. Deploy technology for advanced waste drop-off facilities (M) 3. Use materials that are recyclable and retain value (M) 4. Improve technology used at recycling facilities (M) 5. Use equipment and processes to recover and recycle chemicals and materials (I)	1. Implement extended producer responsibility laws (M) 2. Provide incentives for waste segregation and recycling (M) 3. Strengthen markets for recycled plastics (e.g. mandate use, secure demand, create price premiums) (M) 4. Implement Fishing for Litter programmes (Mar)	1. Change cultural norms around proper sorting and recycling (M) 2. Expand home composting (M) 3. Promote and expand commercial composting infrastructure (M)	1. Invest in tracking technology to combat illegal dumping (M) 2. Develop and scale on-demand waste collection (M)

***Sectors:** Municipal (M), agricultural (A), industrial (I)*
***Pollutants:** Macroplastics; microplastics; other solid waste; nutrients; industrial chemicals and POPs*
***SDGs:** 8.3, 8.8, 11.6, 12.2, 12.5*

Implement coastal zone improvements			
Infrastructure	Policy	Mindset	Innovation
1. Provide for sediment/dredge material removal and treatment (I, Mar) 2. Conduct sediment remediation with in situ mats (Mar) 3. Improve wastewater and solid waste management on ships (Mar) 4. Build ships and rigs to prevent and minimise oil spills (Mar) 5. Improve infrastructure at ports to manage waste generated from ships, including making waste management affordable (M, I, Mar) 6. Land solid waste where infrastructure is available (Mar)	1. Enforce international dumping agreements (M, Mar) 2. Strengthen oil spill prevention policies (M) 3. Restrict locations and types of coastal and open-ocean aquaculture (A)	1. Engage people to adhere to MARPOL to reduce illegal discharge (Mar) 2. Ensure that shipping/maritime developments prioritise marine protection (M, Mar) 3. Operate and manage oil rigs and ships to minimise oil spills (I) 4. Encourage participation in beach cleanups, Adopt-a-Beach programmes and clean beach certifications such as Blue Flag and Project Aware (M) 5. Use citizen science apps such as the Debris Tracker to engage citizens on pollution issues (M)	1. Innovate equipment and methods for managing wastewater and solid waste on ships (Mar) 2. Develop new oil spill prevention technology (Mar) 3. Conduct research and development in individual pollutant cleanup systems (I, Mar) 4. Shift to land-based aquaculture systems (A)

***Sectors:** Municipal (M), agricultural (A), industrial (I), maritime (Mar)*
***Pollutants:** Macroplastics; microplastics; other solid waste; pesticides; nutrients; antibiotics, parasiticides and other pharmaceuticals; heavy metals; industrial chemicals and POPs; oil and gas*
***SDGs:** None*

Table 5.3 Summary of interventions and pollutants addressed across sectors and SDGs

	(1) Improve wastewater management	(2) Improve stormwater management	(3) Adopt green chemistry practices and new materials	(4) Practice radical resource efficiency	(5) Recover and recycle	(6) Improve coastal zones	(7) Build local systems for safe food and water
SDGS	6.2, 6.3	NONE	3.9, 12.4	8.3, 8.8, 11.6, 12.2, 12.5	8.3, 8.8, 11.6, 12.2, 12.5	NONE	6.1, 6.B, 2.1, 2.3
Microplastics	**M**	**M**	**M, A**	**M, A, I, Mar**	**M, A, I, Mar**	M, Mar	M, A
Macroplastics	**M**	**M**	**M, A, Mar**	**M, A, I, Mar**	M, A, Mar	M, Mar	**M, A**
Other solid waste	**M**	**M**		**M**	**M, A, Mar**	M, Mar	M, A
Pesticides		**A**	**M, A**				**A**
Nutrients (N, P)	M, A	**A**		M, A	M, A	**A**	**M, A**
Antibiotics, parasiticides, other pharmaceuticals	M, I	A				**A**	**A**
Heavy metals	M, I	M, A, I	M, A, I, Mar			**A, I**, Mar	**A**
Industrial chemicals and POPs	M, I	**M, A**	M, **A, I**, Mar		I	**I**	
Oil and gas		M, A, I		I, Mar	I	**M, I, Mar**	

Notes: Sectors are municipal (M), agricultural (A), industrial (I), maritime (Mar)
Bold sectors are the primary scope of influence, non-bold are secondary; cells are shaded progressively darker as more sectors are impacted.
Source: Authors

Build local systems for safe food and water			
Infrastructure	Policy	Mindset	Innovation
1. Expand drinking water infrastructure (M) 2. Develop municipal composting systems to support local food production (M, A)	1. Ensure adequate drinking water standards (M)	1. Use technology to raise awareness and provide practical solutions, e.g. Fill it Forward and apps to locate water fountains 2. Encourage local sourcing of food (e.g. people, restaurants, government) (M) 3. Encourage people to bring their own pack-aging to purchase local food (M) 4. Use sustainable methods of food production (both on land and aquaculture) and minimise pesticide and nutrient use (A)	1. Use multitrophic aquaculture production—'waste' from one aquatic species becomes food for another (A) 2. Farm mussels, sea grass or other nutrient-ab-sorbing species for nutrient equilibrium (A)

Sectors: *Municipal (M), agricultural (A)*
Pollutants: *Macroplastics; microplastics; other solid waste; pesticides; nutrients; antibiotics, parasiticides and other pharmaceuticals; heavy metals; industrial chemicals and POPs; oil and gas*
SDGs: *6.1, 6.B. 2.1, 2.3*

For comparison purposes, the scope of each intervention approach is presented in Table 5.3. As the data do not exist today to quantitatively compare the value of one approach versus another, this table focuses on showing the reach of each intervention by sector for each pollutant and those directly related SDGs.

Figure 5.4 presents spider graphs of each intervention to visually compare their effects on each class of pollutants across the sectors. These graphs do not illustrate a score for each intervention, but show the extent to which they impact pollutants across single or multiple sectors (depicted by how far the shape spreads outward). In general, the overall impact

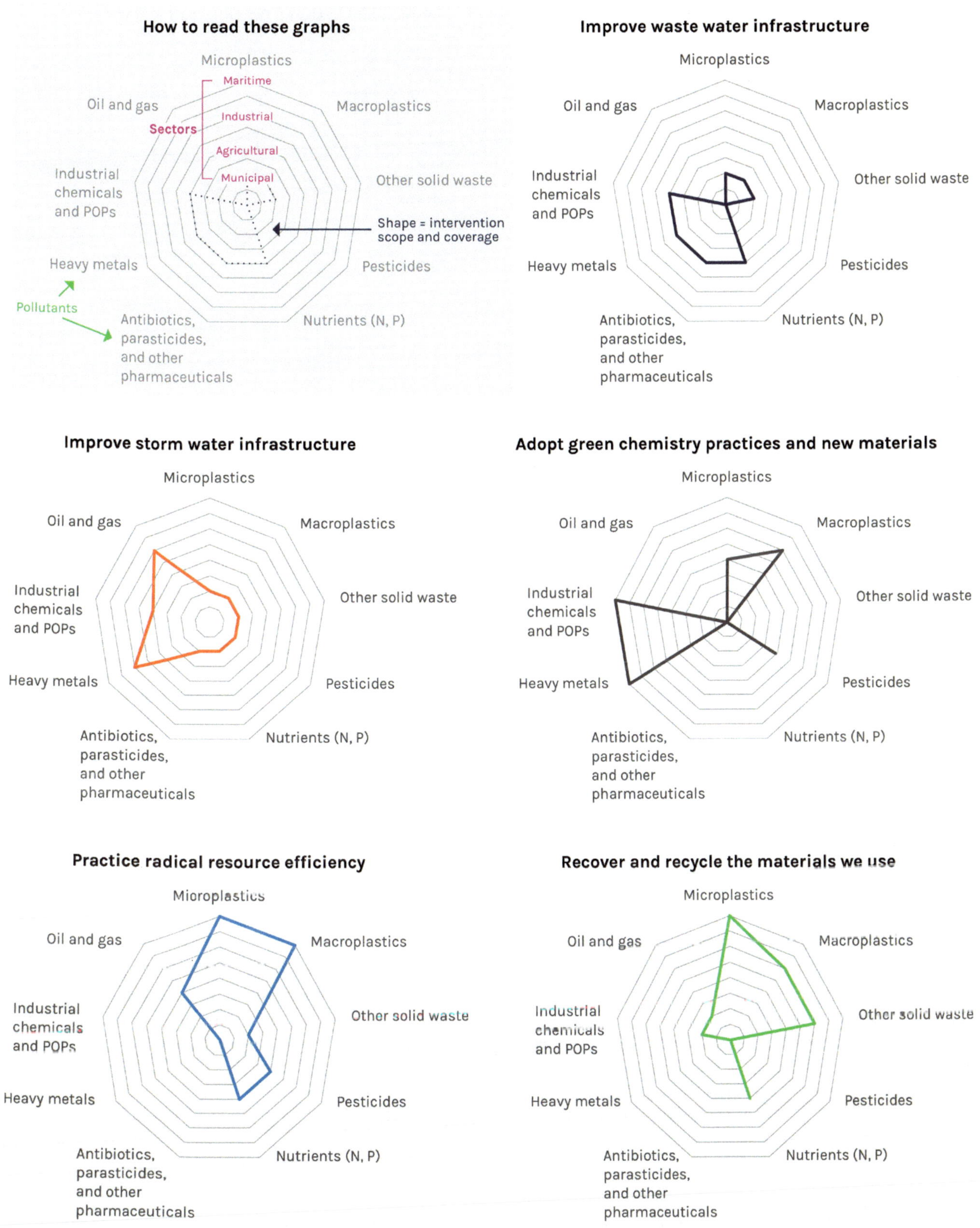

Fig. 5.4 Spider graph illustrations of approaches 1–7 by pollutant and sector. (Source: Authors)

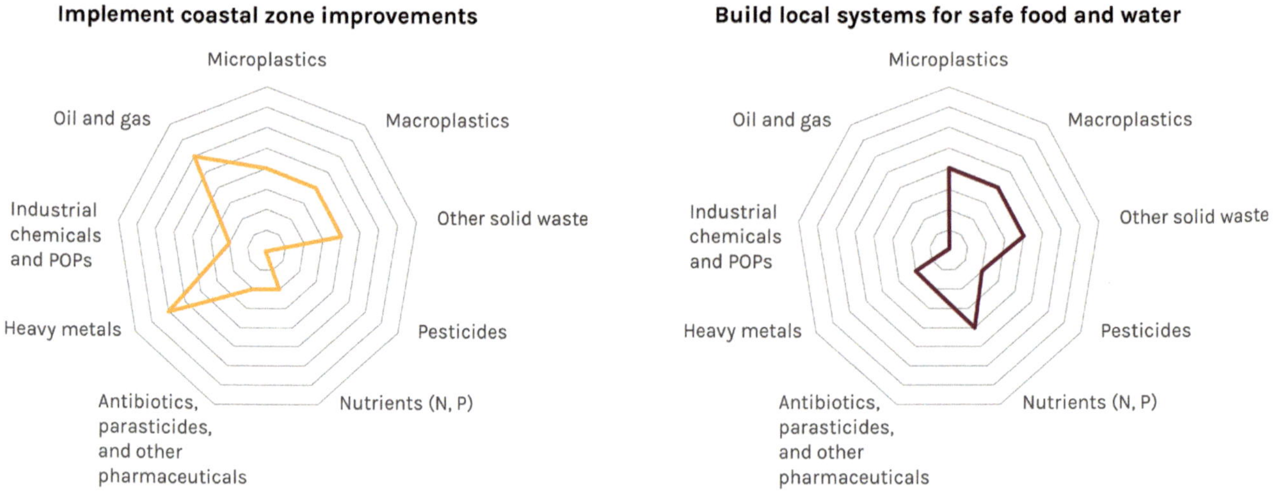

Fig. 5.4 (continued)

increases as more pollutants and sectors are impacted, but the metrics of mass quantities, discrete counts and values, as well as risk and impact, are not able to be taken into account in these illustrations. However, synergies in addressing other pollutants while addressing plastic pollution are illustrated.

Governments, together with businesses, investors, individuals, communities and NGOs, can have a major impact on changing the trajectory of pollution discharges into the ocean—with the opportunity to address other intersectional social and environmental challenges in the process. Solutions will come from innovative policies, support for research and innovation, investment in wastewater and solid waste infrastructure and shifting mindsets and behavioural practices. Many companies that are facing increased costs—or are taking responsibility for costs that they have historically imposed on others—will inevitably claim that these actions will only result in a loss of jobs, profits and economic prosperity.

It is important that we don't confuse the minimisation of harmful pollution with a reduction in quality of life, livelihood opportunities or economic success. In fact, the reality can be quite the opposite. Pollution in the ocean is already negatively impacting human health, economic prosperity for ocean-based businesses and marine ecosystems on which humans depend for essential ecosystem services. Solutions to ocean pollution can create jobs, reduce costs to many businesses and governments and improve the health and prosperity of millions of people.

Pollution is an externality of a linear economy. In creating an economic system where product costs nearly always exclude the environmental impacts for those products (whether during their creation, useful life or end of life), we have effectively designed our economies to maximise pollution, in service of maximising profits. We have invented the idea of 'throwing things away'—and the vastness of the ocean has enabled this fiction to persist for a very long time.

Alternative economic systems, such as the circular economy or regenerative economy, begin with the premise that there is no such thing as waste; that in a closed system like that of Earth, there is nowhere for damaging pollution to go that won't end up harming ecosystems, plant and animal life and, ultimately, human life. The branding of an economic model is less important than this fundamental premise: There is no 'away,' so we must design our economic system to recognise complete life cycle costs. Once the boundaries of the economic system are fixed, the machinery of the economy itself will be very effective at finding the most efficient ways to stop the problem of pollution.

How One Place Can Make a Difference

While no single community or country can solve the problem of ocean pollution alone, a single country can be a first mover in adopting innovative policies and solutions that show the way for others to follow. One barrier innovators face is helping to bridge the imagination gap between today's and tomorrow's realities. A community, country or region can bring the vision of a pollution-free future to life and make it easier for others to begin to adopt the same solutions.

Regional Strategies

Smaller communities and countries can consider adopting regional strategies to help achieve critical mass for certain types of innovations, investments and infrastructure. For example, regions that align their requirements for companies to innovate around packaging, end-of-life responsibility and other issues can make it more compelling and less complex for multinational companies to comply.

Global Collaboration
The ocean is a global resource impacted by all actions everywhere. Given this, it would be appropriate and effective to organise a global compact or commitment to improving the health of the ocean so the ocean can better support all life. International treaties have had success in the past at reducing some impacts on the ocean (e.g. Montreal Protocol, Stockholm Convention). As communications and technologies make the world feel like a smaller place and emphasise the interconnectedness of humanity and our environment, there may be openings to build global support for such an agreement. At a minimum, current declarations from the G7 and G20, as well as United Nations Environment Assembly of the United Nations Environment Programme and other UN initiatives, can be built upon.

Further Research
While much has been learned about the scope, scale and impacts of marine plastic pollution in recent years, there remain significant gaps that could help inform and prioritise solutions. There are multiple significant research efforts underway that were not published in time to be referenced in this paper. It is the authors' hope that these studies will be completed and released as soon as possible as they are expected to contribute significantly to the state of knowledge on this topic. Ongoing research on ocean plastic is also needed, and would be greatly facilitated by the creation of open data protocols to aggregate and share data globally for scientific scholarship.

Finally, just as we see synergies in the solutions to ocean plastic and other pollutants in the ocean, more research is needed to understand their other interactions in the ocean as well as their implications.

Acknowledgements The paper's technical reviewers, Grant Collins, Rob Kaplan, Chris Reddy and David Tickler, as well as its arbiter, Martin Stuchtey, all provided helpful technical comments.

The authors thank World Resources Institute for providing support as the Ocean Panel Secretariat.

The authors would also like to acknowledge Amy Brooks for GIS analysis and map images, Kathryn Youngblood for technical graphic assistance, and Connor Keisling for referencing assistance.

While our colleagues were very generous with their time and input, this report reflects the views of the authors alone. The authors thank Sarah DeLucia for copyediting and Jen Lockard for design.

About the Authors

Co-authors

Jenna Jambeck is a professor at the University of Georgia in the United States.

Ellie Moss is an independent consultant and researcher in the United States.

Brajesh Dubey is a professor at the Indian Institute of Technology Kharagpur in India.

Contributing Authors

Zainal Arifin is deputy chairman for Earth science at the Indonesian.

Institute of Sciences in Indonesia and at the Research Center for Oceanography.

Linda Godfrey is a principal researcher at the Council for Scientific and Industrial Research in South Africa.

Britta Denise Hardesty is a principal research scientist at the Commonwealth Scientific and Industrial Research Organization,

Oceans and Atmosphere, in Australia.

I. Gede Hendrawan is a researcher and lecturer in the Department of Marine Science, Faculty of Marine Science and Fisheries, at Udayana University in Indonesia.

To Thi Hien is dean at VNUHCM-University of Science in Vietnam.

Liu Junguo is professor at the Southern University of Science and Technology in China.

Marty Matlock is the executive director of the University of Arkansas Resiliency Center at the University of Arkansas in the United States.

Sabine Pahl is a professor at the University of Plymouth in the United Kingdom.

Karen Raubenheimer is a lecturer at the University of Wollongong in Australia.

Martin Thiel is a professor at the Catholic University of the North in Chile.

Richard Thompson is a professor at the University of Plymouth in the United Kingdom.

Lucy Woodall is a senior research fellow at the University of Oxford and principal scientist at Nekton in the United Kingdom.

References

Abass K, Emelyanova A, Rautio A (2018) Temporal trends of contaminants in arctic human populations. Environ Sci Pollut Res 25(29):28834–28850

Acampora H, Schuyler QA, Townsend KA, Hardesty BD (2014) Comparing plastic ingestion in juvenile and adult stranded short-tailed shearwaters (*Puffinus tenuirostris*) in Eastern Australia. Mar Pollut Bull 78(1–2):63–68

AfDB (African Development Bank) (2012) Urbanization in Africa

Aguilera MA, Broitman BR, Thiel M (2016) Artificial breakwaters as garbage bins: structural complexity enhances anthropogenic litter accumulation in marine intertidal habitats. Environ Pollut 214:737–747

Alexander RB, Smith RA, Schwarz GE, Boyer EW, Nolan JV, Brakebill JW (2008) Differences in phosphorus and nitrogen delivery to the gulf of Mexico from the Mississippi river basin. Environ Sci Technol 42(3):822–830

Andréfouët S, Thomas Y, Lo C (2014) Amount and type of derelict gear from the declining black pearl oyster aquaculture in Ahe atoll lagoon, French Polynesia. Mar Pollut Bull 83(1):224–230

Arifin Z (2004) Local millennium assessment: condition and trend of the greater Jakarta bay ecosystem. Report submitted to Assistant Deputy for Coastal and Marine Ecosystem, Ministry of Environment, Republic of Indonesia

Asante KA, Amoyaw-Osei Y, Agusa T (2019) E-waste recycling in Africa: risks and opportunities. Curr Opin Green Sustain Chem 18:109–117

Asian Development Bank (2009) Country environmental analysis 2008—Philippines. Asian Development Bank, Mandaluyong City

Asri NHBM (2015) Impact of industrial effluents on water quality at the Tiram River, Kuantan Malaysia. Universiti Malaysia Pahang, Gambang

Astudillo JC, Bravo M, Dumont CP, Thiel M (2009) Detached aquaculture buoys in the SE pacific: potential dispersal vehicles for associated organisms. Aquat Biol 5(3):219–231

aus der Beek T, Weber F-A, Bergmann A, Hickmann S, Ebert I, Hein A, Küster A (2016) Pharmaceuticals in the environment—global occurrences and perspectives. Environ Toxicol Chem 35(4): 823–835

Bakir A, O'Connor IA, Rowland SJ, Hendriks AJ, Thompson RC (2016) Relative importance of microplastics as a pathway for the transfer of hydrophobic organic chemicals to marine life. Environ Pollut 219:56–65

Beaumont NJ, Aanesen M, Austen MC, Börger T, Clark JR, Cole M, Hooper T et al (2019) Global ecological, social and economic impacts of marine plastic. Mar Pollut Bull 142:189–195

Beman MJ, Arrigo KR, Matson PA (2005) Agricultural runoff fuels large phytoplankton blooms in vulnerable areas of the ocean. Nature 434(7030):211–214

Bendell L (2015) Favored use of anti-predator netting (APN) applied for the farming of clams leads to little benefits to industry while increasing nearshore impacts and plastics pollution. Mar Pollut Bull 91(1):22–28

Blanchoud H, Moreau-Guigon E, Farrugia F, Chevreuil M, Mouchel J (2007) Contribution by urban and agricultural pesticide uses to water contamination at the scale of the Marne watershed. Sci Total Environ 375(1–3):168–179

Böhm G, Tanner C (2013) Environmental risk perception. In: Steg L, de Groot JIM (eds) Environmental psychology: an introduction. BPS Blackwell, pp 15–25

Bosch AC, O'Neill B, Sigge GO, Kerwath SE, Hoffman LC (2016) Heavy metals in marine fish meat and consumer health: a review. J Sci Food Agric 96(1):32–48

Bouwman AF, Pawłowski M, Liu C, Beusen AH, Shumway SE, Glibert P, Overbeek C (2011) Global hindcasts and future projections of coastal nitrogen and phosphorus loads due to shellfish and seaweed aquaculture. Rev Fish Sci 19(4):331–357

Brandelli A (2020) The interaction of nanostructured antimicrobials with biological systems: cellular uptake, trafficking and potential toxicity. Food Sci Human Wellness 9(1):8–20

Brennan RE, Portman ME (2017) Situating Arab-Israeli artisanal fishermen's perceptions of marine litter in a socio-institutional and socio-cultural context. Mar Pollut Bull 115(1):240–251

Brodie JE, Lewis SE, Collier CJ, Wooldridge S, Bainbridge ZT, Waterhouse J, Rasheed MA et al (2017) Setting ecologically relevant targets for river pollutant loads to meet marine water quality requirements for the great barrier reef, Australia: a preliminary methodology and analysis. Ocean Coast Manag 143:136–147

Brooks AL, Wang S, Jambeck JR (2018) The Chinese import ban and its impact on global plastic waste trade. Sci Adv 4(6)

Brouwer R, Hadzhiyska D, Ioakeimidis C, Ouderdorp H (2017) The social costs of marine litter along European coasts. Ocean Coast Manag 138:38–49

Brown K, Westaway E (2011) Agency, capacity, and resilience to environmental change: lessons from human development, well-being, and disasters. Annu Rev Environ Resour 36(1):321–342

Browne MA, Dissanayake A, Galloway TS, Lowe DM, Thompson RC (2008) Ingested microscopic plastic translocates to the circulatory system of the mussel, *Mytilus edulis* (L). Environ Sci Technol 42:5026–5031

Bucci K, Tulio M, Rochman CM (2019) What is known and unknown about the effects of plastic pollution: a meta-analysis and systematic review. Ecol Appl 30(2):e02044

Bugoni L, Krause LG, Petry MVN (2001) Marine debris and human impacts on sea turtles in Southern Brazil. Mar Pollut Bull 42(12):1330–1334

Bukit NT (1995) Water quality conservation for the Citarum River in West Java. Water Sci Technol 31(9):1–10

Campbell ML, Slavin C, Grage A, Kinslow A (2016) Human health impacts from litter on beaches and associated perceptions: a case study of 'clean' Tasmanian beaches. Ocean Coast Manag 126:22–30

Campbell ML, Peters L, McMains C, de Campos MCR, Sargisson RJ, Blackwell B, Hewitt CL (2019) Are our beaches safe? Quantifying the human health impact of anthropogenic beach litter on people in New Zealand. Sci Total Environ 651:2400–2409

Cassou E, Jaffee SM, Ru J (2018) The challenge of agricultural pollution : evidence from China, Vietnam, and the Philippines : Main report. Directions in development; environment and sustainable development. World Bank Group. Washington, DC

Catarino AI, Macchia V, Sanderson WG, Thompson RC, Henry TB (2018) Low levels of microplastics (MP) in wild mussels indicate that MP ingestion by humans is minimal compared to exposure via household fibres fallout during a meal. Environ Pollut 237:675–684

Cesar H, Burke L, Pet-Soede L (2003) The economics of worldwide coral reef degradation. World Wildlife Fund, Arnhem

Chagnon C, Thiel M, Antunes J, Ferreira JL, Sobral P, Ory NC (2018) Plastic ingestion and trophic transfer between easter island flying fish (*Cheilopogon rapanouiensis*) and yellowfin tuna (*Thunnus albacares*) from Rapa Nui (Easter Island). Environ Pollut 243:127–133

Chatham-Stephens K, Caravanos J, Ericson B, Sunga-Amparo J, Susilorini B, Sharma P, Landrigan PJ et al (2013) Burden of disease from toxic waste sites in India, Indonesia, and the Philippines in 2010. Environ Health Perspect 121(7):791–796

Cho Y, Shim WJ, Jang M, Han GM, Hong SH (2019) Abundance and characteristics of microplastics in market bivalves from South Korea. Environ Pollut 245:1107–1116

Cincinelli A, Martellini T, Del Bubba M, Lepri L, Corsolini S, Borghesi N, King MD et al (2009) Organochlorine pesticide air– water exchange and bioconcentration in krill in the Ross Sea. Environ Pollut 157(7):2153–2158

Coca Cola (2019) Coca-Cola fast-tracks collection and recycling of PET plastic bottles across Africa. Coca Cola Journey. https://www.coca-colaafrica.com/stories/coca-cola-fast-tracks-collection-and-recycling-of-pet-plastic-bottles#

Cole M, Lindeque P, Fileman E, Halsband C, Galloway TS (2015) The impact of polystyrene microplastics on feeding, function and fecundity in the marine copepod *Calanus helgolandicus*. Environ Sci Technol 49(2):1130–1137

Cole M, Lindeque P, Fileman E, Halsband C, Goodhead R, Moger J (2013) Microplastic ingestion by zooplankton. Environ Sci Technol 47(12)

Collier K (2019) Retired Texas shrimper wins record-breaking $50 million settlement from plastics manufacturing giant. The Texas Tribune, Dec 3. https://www.texastribune.org/2019/12/03/texas-judge-approves-settlement-agreement-water-pollution-formosa/

Coordinating Ministry for Maritime Affairs (2018) National plan of action on marine plastic debris. President Act No 83. Indonesia

Copat C, Grasso A, Fiore M, Cristaldi A, Zuccarello P, Signorelli SS, Conti GO et al (2018) Trace elements in seafood from the mediterranean sea: an exposure risk assessment. Food Chem Toxicol 115:13–19

Cordova MR, Nurhati IS (2019) Major sources and monthly variations in the release of land-derived marine debris from the Greater Jakarta area, Indonesia. Sci Rep 9(1):18730

Cornejo-D'Ottone M, Molina V, Pavez J, Silva N (2020) Greenhouse gas cycling by the plastisphere: the sleeper issue of plastic pollution. Chemosphere 246:125709

Danovaro R, Bongiorni L, Corinaldesi C, Giovannelli D, Damiani E, Astolfi P, Greci L et al (2008) Sunscreens cause coral bleaching by promoting viral infections. Environ Health Perspect 116(4): 441–447

Davidson K, Dudas SE (2016) Microplastic ingestion by wild and cultured Manila clams (Venerupis philippinarum) from Baynes sound, British Columbia. Arch Environ Contam Toxicol 71(2):147–156

de Araújo MCB, Costa MF (2006) Municipal services on tourist beaches: costs and benefits of solid waste collection. J Coast Res 22(5):1070–1075

De Witte B, Devriese L, Bekaert K, Hoffman S, Vandermeersch G, Cooreman K, Robbens J (2014) Quality assessment of the blue mussel (Mytilus edulis): comparison between commercial and wild types. Mar Pollut Bull 85(1):146–155

Dewailly É (2006) Canadian Inuit and the arctic dilemma. Oceanography 19(2):88–89

Dias SM (2016) Waste pickers and cities. Environ Urban 28(2):375–390

Diaz R, Selman M, Chique C (2011) Global eutrophic and hypoxic coastal systems. Eutrophication and hypoxia: nutrient pollution in coastal waters. Map Dataset. World Resources Institute, Washington, DC. http://www.wri.org/project/eutrophication

Dong Z, Chen Y, Ge H, Li X, Wu H, Wang C, Hu Z et al (2018) Response of growth and development of the pacific oyster (*Crassostrea gigas*) to thermal discharge from a nuclear power plant. BMC Ecol 18(1):31

Downs CA, Kramarsky-Winter E, Fauth JE, Segal R, Bronstein O, Jeger R, Lichtenfeld Y et al (2014) Toxicological effects of the sunscreen UV filter, benzophenone-2, on planulae and in vitro cells of the coral, *Stylophora pistillata*. Ecotoxicology 23(2):175–191

Dris R, Gasperi J, Saad M, Mirande C, Tassin B (2016) Synthetic fibers in atmospheric fallout: a source of microplastics in the environment? Mar Pollut Bull 104(1–2):290–293

Dromard CR, Guéné M, Bouchon Navaro Y, Lemoine S, Cordonnier S, Bouchon C (2018) Contamination of marine fauna by chlordecone in guadeloupe: evidence of a seaward decreasing gradient. Environ Sci Pollut Res 25(15):14294–14301

Dsikowitzky L, Hagemann L, Ariyani F, Irianto HE, Schwarzbauer J (2017) Complex organic pollutant mixtures originating from industrial and municipal emissions in surface waters of the megacity Jakarta—an example of a water pollution problem in emerging economics. Environ Sci Pollut Res 24(35):27539–27552

Eerkes-Medrano D, Thompson RC, Aldridge DC (2015) Microplastics in freshwater systems: a review of the emerging threats, identification of knowledge gaps and prioritisation of research needs. Water Res 75:63–82

Ellen MacArthur Foundation, World Economic Forum and McKinsey Center for Business and Environment (2016) The new plastics economy: rethinking the future of plastics. https://www.ellenmacarthurfoundation.org/publications/the-new-plastics-economy-rethinking-the-future-of-plastics

Elser JJ, Bracken ME, Cleland EE, Gruner DS, Harpole WS, Hillebrand H, Ngai JT et al (2007) Global analysis of nitrogen and phosphorus limitation of primary producers in freshwater, marine and terrestrial ecosystems. Ecol Lett 10(12):1135–1142

EPA (Environmental Protection Agency, United States) (2014) Municipal solid waste generation, recycling, and disposal in the United States: tables and figures for 2012. Office of Resource Conservation and Recovery. EPA, Washington, DC

EPA (2019) Advancing sustainable materials management: 2017 fact sheet assessing trends in material generation, recycling, composting, combustion with energy recovery and landfilling in the United States. EPA, Washington, DC

Eryaşar AR, Özbilgin H, Gücü AC, Sakınan S (2014) Marine debris in bottom trawl catches and their effects on the selectivity grids in the north eastern Mediterranean. Mar Pollut Bull 81(1):80–84

European Commission (2019) Circular economy: commission welcomes european parliament adoption of new rules on single–use plastics to reduce marine litter. Mar 27. European Commission

European Environment Agency (2019) Contaminants in Europe's seas: moving towards a clean, non-toxic marine environment. European Environment Agency, Copenhagen

FAO (Food and Agriculture Organization of the United Nations) (n.d.) EU protection of public health: metal contamination detention issues at borders. GLOBEFISH - Information and Analysis on World Fish Trade. http://www.fao.org/in-action/globefish/fishery-information/resource-detail/en/c/456789/. Accessed 25 July 2019

Farias EGG, Preichardt PR, Dantas DV (2018) Influence of fishing activity over the marine debris composition close to coastal jetty. Environ Sci Pollut Res 25(16):16246–16253

Feldman D (1997) Editorial: estrogens from plastic—are we being exposed? Endocrinology 138(5):1777–1779

Flores F, Collier CJ, Mercurio P, Negri AP (2013) Phytotoxicity of four photosystem ii herbicides to tropical seagrasses. PLoS One 8(9):e75798

Forrest A, Giacovazzi L, Dunlop S, Reisser J, Tickler D, Jamieson A, Meeuwig JJ (2019) Eliminating plastic pollution: how a voluntary contribution from industry will drive the circular plastics economy. Front Mar Sci 6:627

Francis CD, Barber JR (2013) A framework for understanding noise impacts on wildlife: an urgent conservation priority. Front Ecol Environ 11(6):305–313

Fujieda S, Sasaki K (2005) Stranded debris of foamed plastic on the coast of ETA Island and Kurahashi Island in Hiroshima Bay. Nippon Suisan Gakkaishi 71(5):755–761

Gaines S, Cabral R, Golbuu Y et al (2019) The expected impacts of climate change on the ocean economy. World Resources Institute, Washington, DC

Galgani F, Fleet D, Van Franeker J, Katsanevakis S, Maes T, Mouat J, Oosterbaan L, et al (2010) Marine strategy framework directive-task group 10 report: marine litter. Report on the identification of descriptors for the good environmental status of European seas regarding marine litter under the marine strategy framework directive. Office for Official Publications of the European Communities, Luxembourg

Gall SC, Thompson RC (2015) The impact of debris on marine life. Mar Pollut Bull 92(1–2):170–179

Galloway JN, Townsend AR, Erisman JW, Bekunda M, Cai Z, Freney JR, Martinelli LA et al (2008) Transformation of the nitrogen cycle: recent trends, questions, and potential solutions. Science 320(5878):889–892

Galloway TS (2015) Micro- and nano-plastics and human health. In: Bergmann M, Gutow L, Klages M (eds) Marine anthropogenic litter. Springer, Chamsss, pp 343–366

Gao M, Klerks PL, Wu X, Chen H, Xie L (2016) Metal concentrations in sediment and biota of the Huludao coast in Liaodong bay and associated human and ecological health risks. Arch Environ Contam Toxicol 71(1):87–96

Gao Y, Baisch P, Mirlean N, Rodrigues da Silva Júnior FM, Van Larebeke N, Baeyens W, Leermakers M (2018) Arsenic speciation in fish and shellfish from the North Sea (Southern bight) and Açu Port area (Brazil) and health risks related to seafood consumption. Chemosphere 191:89–96

GESAMP (Joint Group of Experts on the Scientific Aspects of Marine Environmental Protection) (2015) Sources, fate and effects of microplastics in the marine environment: a global assessment. Reports and Studies 90. IMO/FAO/UNESCO-IOC/UNIDO/WMO/IAEA/UN/UNEP/UNDP Joint Group of Experts on the Scientific Aspects of Marine Environmental Protection, London

Geyer R, Jambeck JR, Law KL (2017) Production, use, and fate of all plastics ever made. Sci Adv 3(7):e1700782

Gibson J, Adlard B, Olafsdottir K, Sandanger TM, Odland JØ (2016) Levels and trends of contaminants in humans of the arctic. Int J Circumpolar Health 75:33804

Gilman E (2015) Status of international monitoring and management of abandoned, lost and discarded fishing gear and ghost fishing. Mar Policy 60:225–239

Gilman E, Chopin F, Suuronen P, Kuemlangan B (2016) Abandoned, lost and discarded gillnets and trammel nets: methods to estimate ghost fishing mortality, and the status of regional monitoring and management. FAO Fisheries and Aquaculture Technical Paper. Food and Agriculture Organization of the United Nations, Rome

Gladwell VF, Brown DK, Wood C, Sandercock GR, Barton JL (2013) The great outdoors: how a green exercise environment can benefit all. Extrem Physiol Med 2(1):3

Gollan N, Voyer M, Jordan A, Barclay K (2019) Maximising community wellbeing: assessing the threats to the benefits communities derive from the marine estate. Ocean Coast Manag 168:12–21

González-Dávila M (1995) The role of phytoplankton cells on the control of heavy metal concentration in seawater. Mar Chem 48(3):215–236

Government of Vietnam (2020) National action plan for management of marine plastic litter by 2030. via United Nations Development Programme. https://www.vn.undp.org/content/vietnam/en/home/library/environment_climate/national-action-plan-for-management-of-marine-plastic-litter-by-.html

Grattan LM, Holobaugh S, Morris JG (2016) Harmful algal blooms and public health. Harmful Algae 57:2–8

Green DS (2016) Effects of microplastics on European flat oysters, ostrea edulis and their associated benthic communities. Environ Pollut 216:95–103

GRID Arendal (2013) Industrial hazardous waste in the Mediterranean countries. https://www.grida.no/resources/5891

Guart A, Wagner M, Mezquida A, Lacorte S, Oehlmann J, Borrell A (2013) Migration of plasticisers from Tritan™ and polycarbonate bottles and toxicological evaluation. Food Chem 141(1):373–380

Gunn R, Hardesty BD, Butler J (2010) Tackling 'Ghost Nets': local solutions to a global issue in Northern Australia. Ecol Manage Restor 11(2):88–98

Guo L, Kelley K, Goh KS (2007) Evaluation of sources and loading of pesticides to the Sacramento River, California, USA, during a storm event of winter 2005. Environ Toxicol Chem 26(11):2274–2281

Hajkowicz SA, Tellames K, Aitaro J (2005) The costs of solid waste pollution to Palau. A consultancy report to the office of environmental response and coordination of the Republic of Palau. Draft Report. CSIRO Sustainable Ecosystems, Brisbane

Halden RU (2010) Plastics and health risks. Annu Rev Public Health 31(1):179–194

Halpern BS, Longo C, Hardy D, McLeod KL, Samhouri JF, Katona SK, Kleisner K et al (2012) An index to assess the health and benefits of the global ocean. Nature 488(7413):615

Harada M (1995) Minamata disease: methylmercury poisoning in Japan caused by environmental pollution. Crit Rev Toxicol 25(1):1–24

Hardesty BD, Good TP, Wilcox C (2015) Novel methods, new results and science-based solutions to tackle marine debris impacts on wildlife. Ocean Coast Manag 115:4–9

Harris CA, Scott AP, Johnson AC, Panter GH, Sheahan D, Roberts M (2014) Principles of sound ecotoxicology. Environ Sci Technol 48(6):3100–3111

Harrison JP, Schratzberger M, Sapp M, Osborn AM (2014) Rapid bacterial colonization of low-density polyethylene microplastics in coastal sediment microcosms. BMC Microbiol 14(1):232

Heisler J, Glibert PM, Burkholder JM, Anderson DM, Cochlan W, Dennison WC, Dortch Q et al (2008) Eutrophication and harmful algal blooms: a scientific consensus. Harmful Algae 8(1):3–13

Hendryx M, Conley J, Fedorko E, Luo J, Armistead M (2012) Permitted water pollution discharges and population cancer and non-cancer mortality: toxicity weights and upstream discharge effects in US rural-urban areas. Int J Health Geogr 11(1):9

Heo NW, Hong SH, Han GM, Hong S, Lee J, Song YK, Jang M et al (2013) Distribution of small plastic debris in cross-section and high strandline on Heungnam Beach, South Korea. Ocean Sci J 48(2):225–233

Herridge DF, Peoples MB, Boddey RM (2008) Global inputs of biological nitrogen fixation in agricultural systems. Plant Soil 311(1–2):1–18

Heskett M, Takada H, Yamashita R, Yuyama M, Ito M, Geok YB, Ogata Y et al (2012) Measurement of persistent organic pollutants (POPs) in plastic resin pellets from remote islands: toward establishment of background concentrations for international pellet watch. Mar Pollut Bull 64(2):445–448

Hien TT, Nguyen TTN, Nguyen TN (2019) Preliminary investigation of microplastics in the surface seawater in Can Gio, Viet Nam. In: Analytica conference

Hindarti D, Larasati AW (2019) Copper (Cu) and cadmium (Cd) toxicity on growth, chlorophyll-a and carotenoid content of phytoplankton Nitzschia Sp. IOP Conf Ser Earth Environ Sci 236:012053

Hinojosa IA, Thiel M (2009) Floating marine debris in fjords, gulfs and channels of southern Chile. Mar Pollut Bull 58(3):341–350

Hoffman RS, Capel PD, Larson SJ (2000) Comparison of pesticides in eight U.S. urban streams. Environ Toxicol Chem 19(9):2249–2258

Hong S, Lee J, Lim S (2017) Navigational threats by derelict fishing gear to navy ships in the Korean seas. Mar Pollut Bull 119(2):100–105

Hong S, Lee J, Kang D, Choi H-W, Ko S-H (2014) Quantities, composition, and sources of beach debris in Korea from the results of nationwide monitoring. Mar Pollut Bull 84(1–2):27–34

Hoornweg D, Bhada-Tata P, Kennedy C (2015) Peak waste: when is it likely to occur? J Ind Ecol 19(1):117–128

India Ministry of Water Resources (India Ministry of Water Resources, River Development and Ganga Rejuvenation) (2017) Effluents polluting ganga river. https://pib.gov.in/PressReleasePage.aspx?PRID=1483959

International Maritime Organization (n.d.) Convention on the prevention of marine pollution by dumping of wastes and other matter. http://www.imo.org/en/OurWork/Environment/LCLP/Pages/default.aspx. Accessed 24 Feb 2020

Ippolito A, Kattwinkel M, Rasmussen JJ, Schäfer RB, Fornaroli R, Liess M (2015) Modeling global distribution of agricultural insecticides in surface waters. Environ Pollut 198:54–60

Islam MS, Tanaka M (2004) Impacts of pollution on coastal and marine ecosystems including coastal and marine fisheries and approach for management: a review and synthesis. Mar Pollut Bull 48(7–8):624–649

Jambeck J, Hardesty BD, Brooks AL, Friend T, Teleki K, Fabres J, Beaudoin Y et al (2017) Challenges and emerging solutions to the land-based plastic waste issue in Africa. Mar Policy 96:256–263

Jambeck JR, Geyer R, Wilcox C, Siegler TR, Perryman M, Andrady A, Narayan R et al (2015) Plastic waste inputs from land into the ocean. Science 347(6223):768–771

Jamieson AJ, Brooks LSR, Reid WDK, Piertney SB, Narayanaswamy BE, Linley TD (2019) Microplastics and synthetic particles ingested by deep-sea amphipods in six of the deepest marine ecosystems on Earth. R Soc Open Sci 6(2):180667

Jang YC, Lee J, Hong S, Lee JS, Shim WJ, Song YK (2014a) Sources of plastic marine debris on beaches of Korea: more from the ocean than the land. Ocean Sci J 49(2):151–162

Jang YC, Hong S, Lee J, Lee MJ, Shim WJ (2014b) Estimation of lost tourism revenue in Geoje island from the 2011 marine debris pollution event in South Korea. Mar Pollut Bull 81(1):49–54

Jiang L, Lin C, Lin P (2014) The determinants of pollution levels: firm-level evidence from Chinese manufacturing. J Comp Econ 42(1):118–142

Karlsson TM, Arneborg L, Broström G, Almroth BC, Gipperth L, Hassellöv M (2018) The unaccountability case of plastic pellet pollution. Mar Pollut Bull 129(1):52–60

Kaza S, Yao L, Bhada-Tata P, Van Woerden F (2018) What a waste 2.0: a global snapshot of solid waste management to 2050. Urban Development Series. World Bank, Washington, DC

Kedzierski M, d'Almeida M, Magueresse A, Le Grand A, Duval H, César G, Sire O et al (2018) Threat of plastic ageing in marine environment. adsorption/desorption of micropollutants. Mar Pollut Bull 127:684–694

Keen OS, Bell KY, Cherchi C, Finnegan BJ, Mauter MS, Parker AM, Rosenblum JS et al (2014) Emerging pollutants - part II: treatment. Water Environ Res 86(10):2036–2096

Kenyon KW, Kridler E (1969) Laysan albatrosses swallow indigestible matter. Auk 86(2):339–343

Kirstein IV, Kirmizi S, Wichels A, Garin-Fernandez A, Erler R, Martin L, Gerdts G (2016) Dangerous hitchhikers? Evidence for potentially pathogenic Vibrio spp. on microplastic particles. Mar Environ Res 120:1–8

Klein EY, Van Boeckel TP, Martinez EM, Pant S, Gandra S, Levin SA, Goossens H et al (2018) Global increase and geographic convergence in antibiotic consumption between 2000 and 2015. Proc Natl Acad Sci 115(15):E3463–E3470

Koch HM, Calafat AM (2009) Human body burdens of chemicals used in plastic manufacture. Philos Trans R Soc B Biol Sci 364(1526):2063–2078

Koelmans AA, Bakir A, Burton GA, Janssen CR (2016) Microplastic as a vector for chemicals in the aquatic environment: critical review and model-supported reinterpretation of empirical studies. Environ Sci Technol 50(7):3315–3326

Koelmans AA, Besseling E, Foekema E, Kooi M, Mintenig S, Ossendorp BC, Redondo-Hasselerharm PE et al (2017) Risks of plastic debris: unravelling fact, opinion, perception, and belief. Environ Sci Technol 51(20):11513–11519

Koelmans A, Besseling E, Shim W (2015) Nanoplastics in the aquatic environment: critical review. In: Bergmann M, Gutow L, Klages M (eds) Marine anthropogenic litter. Springer, Cham

Koesmawati TA, Arifin Z (2015) Mercury and arsenic content in seafood samples from the Jakarta Fishing Port, Indonesia. Mar Res Indones 40(1):9–16

Kontovas CA, Psaraftis HN, Ventikos NP (2010) An empirical analysis of IOPCF oil spill cost data. Mar Pollut Bull 60(9):1455–1466

Krelling AP, Souza MM, Williams AT, Turra A (2017) Transboundary movement of marine litter in an estuarine gradient: evaluating sources and sinks using hydrodynamic modelling and ground truthing estimates. Mar Pollut Bull 119(1):48–63

Krishnan S, Backer A (2019) The role of gender in waste management: gender perspectives on waste in India, Inodonesia, the Philippines and Vietnam. GA Circular. Commissioned by Ocean Conservancy, Singapore

Kwon BG, Koizumi K, Chung S-Y, Kodera Y, Kim J-O, Saido K (2015) Global styrene oligomers monitoring as new chemical contamination from polystyrene plastic marine pollution. J Hazard Mater 300:359–367

Laffon B, Pásaro E, Valdiglesias V (2016) Effects of exposure to oil spills on human health: updated review. J Toxicol Environ Health B 19(3–4):105–128

Lahens L, Strady E, Kieu-Le T-C, Dris R, Boukerma K, Rinnert E, Gasperi J et al (2018) Macroplastic and microplastic contamination assessment of a tropical river (Saigon River, Vietnam) transversed by a developing megacity. Environ Pollut 236: 661–671

Laist DW (1987) Overview of the biological effects of lost and discarded plastic debris in the marine environment. Mar Pollut Bull 18(6):319–326

Law KL, Thompson RC (2014) Microplastics in the seas. Science 345(6193):144–145

LeBlanc GA (1995) Are environmental sentinels signaling? Environ Health Perspect 103(10):888–890

Lebreton L, Andrady A (2019) Future Scenarios of Global Plastic Waste Generation and Disposal. Palgrave Commun 5(1)

Lebreton LCM, van der Zwet J, Damsteeg JW, Slat B, Andrady A, Reisser J (2017) River plastic emissions to the world's oceans. Nat Commun 8:15611

Lechner A, Ramler D (2015) The discharge of certain amounts of industrial microplastic from a production plant into the River danube is permitted by the Austrian legislation. Environ Pollut 200: 159–160

Lee S, Chung J, Kim C, Lee Y (2017) Distribution of heavy metals in Korean coastal areas correlated to emission from industrial operations. Environ Eng Sci 34(8):569–576

Leggett C, Scherer N, Curry M, Bailey R (2014) Assessing the economic benefits of reductions in marine debris: a pilot study of beach recreation in orange county, California. NOAA Marine Debris Program and Industrial Economics, Inc

Lehner B, Grill G (2013) Global river hydrography and network routing: baseline data and new approaches to study the world's large river systems. Hydrol Process 27(15):2171–2186

Lenz R, Enders K, Nielsen TG (2016) Microplastic exposure studies should be environmentally realistic. Proc Natl Acad Sci 113(29):E4121–E4122

Li H-X, Ma L-S, Lin L, Ni Z-X, Xu X-R, Shi H-H, Yan Y et al (2018a) Microplastics in oysters Saccostrea cucullata along the Pearl River Estuary, China. Environ Pollut 236:619–625

Li J, Green C, Reynolds A, Shi H, Rotchell JM (2018b) Microplastics in mussels sampled from coastal waters and supermarkets in the United Kingdom. Environ Pollut 241:35–44

Li J, Qu X, Su L, Zhang W, Yang D, Kolandhasamy P, Li D et al (2016) Microplastics in mussels along the coastal waters of China. Environ Pollut 214:177–184

Liebezeit G, Liebezeit E (2013) Non-pollen particulates in honey and sugar. Food Addit Contam A 30(12):2136–2140

Liebezeit G, Liebezeit E (2014) Synthetic particles as contaminants in German beers. Food Addit Contam A 31(9): 1574–1578

Lin C, Nakamura S (2019) Approaches to solving China's marine plastic pollution and CO2 emission problems. Econ Syst Res 31(2):143–157

Linzner R, Salhofer S (2014) Municipal solid waste recycling and the significance of informal sector in urban China. Waste Manag Res 32(9):896–907

Lithner D, Larsson Å, Dave G (2011) Environmental and health hazard ranking and assessment of plastic polymers based on chemical composition. Sci Total Environ 409(18):3309–3324

Liu T-K, Wang M-W, Chen P (2013) Influence of waste management policy on the characteristics of beach litter in Kaohsiung, Taiwan. Mar Pollut Bull 72(1):99–106

Logan DT (2007) Perspective on ecotoxicology of PAHs to fish. Hum Ecol Risk Assess Int J 13(2):302–316

Lord R, et al (2016) Plastics and sustainability: a valuation of environmental benefits, costs and opportunities for continuous improvement. Prepared by TruCost for the American Chemical Council

Lusher AL (2015) Microplastics in the marine environment: distribution, interactions and effects. In: Bergmann M, Gutow L, Klages M (eds) Marine anthropogenic litter. Springer, Cham

Lusher AL, McHugh M, Thompson RC (2013) Occurrence of microplastics in the gastrointestinal tract of pelagic and demersal fish from the English channel. Mar Pollut Bull 67(1–2):94–99

Lusher AL, Welden NA, Sobral P, Cole M (2017) Sampling, isolating and identifying microplastics ingested by fish and invertebrates. Anal Methods 9:1346–1360

Macfadyen G, Huntington T, Cappell R (2009) Abandoned, lost or otherwise discarded fishing gear. FAO Fisheries and Aquaculture Technical Paper No 523/UNEP Regional Seas Reports and Studies No 185

Magliano DJ, Lyons JG (2013) Bisphenol A and diabetes, insulin resistance, cardiovascular disease and obesity: controversy in a (plastic) cup? J Clin Endocrinol Metabol 98(2):502–504

Malik OA, Hsu A, Johnson LA, de Sherbinin A (2015) A global indicator of wastewater treatment to inform the sustainable development goals (SDGs). Environ Sci Pol 48:172–185

Markey KL, Baird AH, Humphrey C, Negri AP (2007) Insecticides and a fungicide affect multiple coral life stages. Mar Ecol Prog Ser 330:127–137

McIlgorm A, Campbell H, Rule M (2008) Understanding the economic benefits and costs of controlling marine debris in the APEC region (MRC 02/2007). A report to the Asia-Pacific Economic Cooperation Marine Resource Conservation Working Group. National Marine Science Centre (University of New England and Southern Cross University), Coffs Harbour

McIlgorm A, Raubenheimer K, McIlgorm DE (2020) Update of 2009 APEC report on economic costs of marine debris to APEC economies. A report to the APEC Ocean and Fisheries Working Group by the Australian National Centre for Ocean Resources and Security (ANCORS). University of Wollongong, Wollongong

MDOE, DCDOE (Maryland Department of the Environment and District of Columbia Department of the Environment) (2010) Total maximum daily loads of trash for the Anacostia River Watershed, Montgomery and prince George's counties, Maryland and the District of Columbia. Submitted to the U.S. Environmental Protection Agency, Region 3, Water Protection Division

Medina M (2008) The informal recycling sector in developing countries: organizing waste pickers to enhance their impact. Gridlines No 44. World Bank, Washington, DC

Meeker JD, Sathyanarayana S, Swan Shanna H (2009) Phthalates and other additives in plastics: human exposure and associated health outcomes. Philos Trans R Soc B Biol Sci 364(1526):2097–2113

Mekonnen MM, Hoekstra AY (2015) Global gray water footprint and water pollution levels related to anthropogenic nitrogen loads to fresh water. Environ Sci Technol 49(21):12860–12868

Mekonnen MM, Hoekstra AY (2018) Global anthropogenic phosphorus loads to freshwater and associated grey water footprints and water pollution levels: a high-resolution global study. Water Resour Res 54(1):345–358

Miranda RJ, Coleman MA, Tagliafico A, Rangel MS, Mamo LT, Barros F, Kelaher BP (2019) Invasion-mediated effects on marine trophic interactions in a changing climate: positive feedbacks favour kelp persistence. Proc R Soc B Biol Sci 286(1899):20182866

MONRE (Ministry of Natural Resources and Environment, Vietnam) (2014) Vietnam: rural environment. National State of Environment 2014

MONRE (2016) Vietnam. National State of Environment 2016

MONRE (2017) Vietnam: waste management. National State of Environment 2017

Moon H-B, Kim H-S, Choi M, Yu J, Choi H-G (2009) Human health risk of polychlorinated biphenyls and organochlorine pesticides resulting from seafood consumption in South Korea, 2005–2007. Food Chem Toxicol 47(8):1819–1825

Mufson S (2016) BP's big bill for the world's largest oil spill reaches $61.6 billion. Washington Post

Murray F, Cowie PR (2011) Plastic contamination in the decapod crustacean Nephrops norvegicus (Linnaeus, 1758). Mar Pollut Bull 62(6):1207–1217

Naji A, Nuri M, Vethaak AD (2018) Microplastics contamination in molluscs from the northern part of the Persian Gulf. Environ Pollut 235:113–120

Nam BH (2016) Current situation and policies on wastewater management in Vietnam. Vietnam Environment Administration, Ministry of Natural Resources and Environment

Nash AD (1992) Impacts of marine debris on subsistence fishermen: an exploratory study. Mar Pollut Bull 24(3):150–156

National Geographic (2020) The place where waste management is a form of worship. Partner content for the alliance to end plastic waste. https://www.nationalgeographic.com/science/2020/01/partner-content-place-where-waste-management-is-form-of-worship/

Nellemann C, Hain S, Alder J (2008) In dead water: merging of climate change with pollution, over-harvest, and infestations in the world's fishing grounds. UNEP/Earthprint, Nairobi

Nelms SE, Galloway TS, Godley BJ, Jarvis DS, Lindeque PK (2018) Investigating microplastic trophic transfer in marine top predators. Environ Pollut 238:999–1007

Neves D, Sobral P, Pereira T (2015) Marine litter in bottom trawls off the Portuguese coast. Mar Pollut Bull 99(1):301–304

Newman S, Watkins E, Farmer A, ten Brink P, Schweitzer J-P (2015) The economics of marine litter. In: Bergmann M, Gutow L, Klages M (eds) Marine anthropogenic litter. Springer, Cham

NOAA (National Oceanic and Atmospheric Administration, United States) (2013) Severe marine debris event report: Japan tsunami marine debris. Overview and update to congress

NOAA OR&R (National Oceanic and Atmospheric Administration, Office of Response and Restoration) (2019) How oil harms animals and plants in marine environments. https://response.restoration.noaa.gov/oil-and-chemical-spills/oil-spills/how-oil-harms-animals-and-plants-marine-environments.html

Obbard RW, Sadri S, Wong YQ, Khitun AA, Baker I, Thompson RC (2014) Global warming releases microplastic legacy frozen in arctic sea ice. Earth's Future 2(6):315–320

Ocean Conservancy (2018) Top 10 items collected. 2018 International coastal cleanup by the numbers. Infographic. https://oceanconservancy.org/trash-free-seas/international-coastal-cleanup/annual-data-release/

OHI (Ocean Health Index) (2019) Ocean health index 2018 global assessment. Ocean health index. University of California, Santa Barbara. National Center for Ecological Analysis and Synthesis, Santa Barbara, CA

Olivares-Rubio HF, Dzul-Caamal R, Gallegos-Rangel ME, Madera-Sandoval RL, Domínguez-López ML, García-Latorre E, Vega-López A (2015) Relationship between biomarkers and endocrine-disrupting compounds in wild *Girardnichthys viviparus* from two lakes with different degrees of pollution. Ecotoxicology 24(3):664–685

Oosterhuis F, Papyrakis E, Boteler B (2014) Economic instruments and marine litter control. Ocean Coast Manag 102:47–54

Ory NC, Gallardo C, Lenz M, Thiel M (2018) Capture, swallowing, and egestion of microplastics by a planktivorous juvenile fish. Environ Pollut 240:566–573

Pacala S, Socolow R (2004) Stabilization wedges solving the climate problem for the next 50 years with current technologies. Science 305(5686):968–972

Pahl S, Wyles KJ (2017) The human dimension: how social and behavioural research methods can help address microplastics in the environment. Anal Methods 9:1404–1411

Palinkas LA, Downs MA, Petterson JS, Russell J (1993) Social, cultural, and psychological impacts of the 'Exxon Valdez' oil spill. Hum Organ 52(1):1–13

Parikesit SH, Triharyanto E, Gunawan B, Sunardi, Abdoellah OS, Ohtsuka R (2005) Multi-source water pollution in the Upper Citarum watershed, Indonesia, with special reference to its spatiotemporal variation. Environ Sci 12(3):121–131

Patrick L (2002) Mercury toxicity and antioxidants: part I: role of glutathione and alpha-lipoic acid in the treatment of mercury toxicity. (Mercury toxicity). Altern Med Rev 7(6):456–471

PETCO (2018) A new way of thinking. Review of PETCO activities 2017

Peters G-JY, Ruiter RAC, Kok G (2013) Threatening communication: a critical re-analysis and a revised meta-analytic test of fear appeal theory. Health Psychol Rev 7(Suppl 1):S8–S31

Phuong NN, Poirier L, Pham QT, Lagarde F, Zalouk-Vergnoux A (2018) Factors influencing the microplastic contamination of bivalves from the French Atlantic coast: location, season and/or mode of life? Mar Pollut Bull 129(2):664–674

PlasticsEurope (2019) Plastics the facts—2019

Potter P, Ramankutty N, Bennet E, Donner S (2011a) Global fertilizer and manure, version 1: nitrogen fertilizer application. NASA Socioeconomic Data and Applications Center, Pallisades, NY

Potter P, Ramankutty N, Bennett E, Donner S (2011b) Global fertilizer and manure, version 1: phosphorus fertilizer application. NASA Socioeconomic Data and Applications Center (SEDAC), Palisades, NY

Potter P, Ramankutty N, Bennett EM, Donner SD (2010) Characterizing the spatial patterns of global fertilizer application and manure production. Earth Interact 14(2):1–22

Prata JC (2018) Airborne microplastics: consequences to human health? Environ Pollut 234:115–126

Pućko M, Stern GA, Burt AE, Jantunen LM, Bidleman TF, Macdonald RW, Barber DG et al (2017) Current use pesticide and legacy organochlorine pesticide dynamics at the ocean-sea ice-atmosphere interface in resolute passage, Canadian arctic, during winter-summer transition. Sci Total Environ 580:1460–1469

Purba NP, Handyman DIW, Pribadi TD, Syakti AD, Pranowo WS, Harvey A, Ihsan YN (2019) Marine debris in Indonesia: a review of research and status. Mar Pollut Bull 146:134–144

Pycha RL (1962) The relative efficiency of nylon and cotton gill nets for taking lake trout in Lake Superior. J Fish Res Board Can 19(6):1085–1094

Qing Li Q, Loganath A, Seng Chong Y, Tan J, Philip Obbard J (2006) Persistent organic pollutants and adverse health effects in humans. J Toxic Environ Health A 69(21):1987–2005

Rabalais NN (2002) Nitrogen in aquatic ecosystems. AMBIO J Hum Environ 31(2):102–113

Rangel-Buitrago N, Williams A, Anfuso G (2018) Killing the goose with the golden eggs: litter effects on scenic quality of the Caribbean coast of Colombia. Mar Pollut Bull 127:22–38

Richard H, Carpenter EJ, Komada T, Palmer PT, Rochman CM (2019) Biofilm facilitates metal accumulation onto microplastics in estuarine waters. Sci Total Environ 683:600–608

Richardson K, Hardesty BD, Wilcox C (2019) Estimates of fishing gear loss rates at a global scale: a literature review and meta-analysis. Fish Fish 20(6):1218–1231

Richardson K, Gunn R, Wilcox C, Hardesty BD (2018) Understanding causes of gear loss provides a sound basis for fisheries management. Mar Policy 96:278–284

Rimondino GN, Pepino AJ, Manetti MD (2018) Latitudinal distribution of OCPs in the open ocean atmosphere between the Argentinian coast and Antarctic Peninsula. Environ Sci Pollut Res 25(13):13004–13013

Risebrough RW, Huggett RJ, Griffin JJ, Goldberg ED (1968) Pesticides: transatlantic movements in the northeast trades. Science 159(3820):1233

Rist S, Almroth BC, Hartmann NB, Karlsson TM (2018) A critical perspective on early communications concerning human health aspects of microplastics. Sci Total Environ 626:720–726

Rist S, Hartmann N (2017) Aquatic ecotoxicity of microplastics and nanoplastics: lessons learned from engineered nanomaterials. In: Wagner M, Lambert S (eds) Freshwater microplastics. The handbook of environmental chemistry, vol 58. Springer, Cham

Rizzo L, Manaia C, Merlin C, Schwartz T, Dagot C, Ploy MC, Michael I, Fatta-Kassinos D (2013) Urban wastewater treatment plants as hotspots for antibiotic resistant bacteria and genes spread into the environment: a review. Sci Total Environ 447:345–360

Rochman C (2015) The complex mixture, fate and toxicity of chemicals associated with plastic debris in the marine environment. In: Bergmann M, Gutow L, Klages M (eds) Marine anthropogenic litter. Springer, Cham

Rochman CM, Tahir A, Williams SL, Baxa DV, Lam R, Miller JT, Teh F-C et al (2015) Anthropogenic debris in seafood: plastic debris and fibers from textiles in fish and bivalves sold for human consumption. Sci Rep 5:14340

Rochman CM, Hentschel BT, Teh SJ (2014a) Long-term sorption of metals is similar among plastic types: implications for plastic debris in aquatic environments. PLoS One 9(1):e85433

Rochman CM, Hoh E, Kurobe T, Teh SJ (2013) ingested plastic transfers hazardous chemicals to fish and induces hepatic stress. Sci Rep 3(3263)

Rochman CM, Kurobe T, Flores I, Teh SJ (2014b) Early warning signs of endocrine disruption in adult fish from the ingestion of polyethylene with and without sorbed chemical pollutants from the marine environment. Sci Total Environ 493:656–661

Rodrigues A, Oliver DM, McCarron A, Quilliam RS (2019) Colonisation of plastic pellets (nurdles) by *E. coli* at public bathing beaches. Mar Pollut Bull 139:376–380

Roman L, Lowenstine L, Parsley LM, Wilcox C, Hardesty BD, Gilardi K, Hindell M (2019) Is plastic ingestion in birds as toxic as we think? Insights from a plastic feeding experiment. Sci Total Environ 665:660–667

SAPEA (Science Advice for Policy by European Academies) (2019) A Scientific perspective on microplastics in nature and society. SAPEA, Berlin. https://doi.org/10.26356/microplastics

Sato T, Qadir M, Yamamoto S, Endo T, Zahoor A (2013) Global, regional, and country level need for data on wastewater generation, treatment, and use. Agric Water Manag 130:1–13

Schmidt C, Krauth T, Wagner S (2017) Export of plastic debris by rivers into the sea. Environ Sci Technol 51(21):12246–12253

Schneider SL, Lim HW (2019) Review of environmental effects of oxybenzone and other sunscreen active ingredients. J Am Acad Dermatol 80(1):266–271

Schuhmann PW, Bass BE, Casey JF, Gill DA (2016) Visitor preferences and willingness to pay for coastal attributes in Barbados. Ocean Coast Manag 134:240–250

Schuyler Q, Hardesty BD, Wilcox C, Townsend K (2012) To eat or not to eat? Debris selectivity by marine turtles. PLoS One 7(7): e40884

Seba DB, Prospero JM (1971) Pesticides in the lower atmosphere of the northern equatorial Atlantic Ocean. Atmos Environ 5(12):1043–1050

Seitzinger S, Mayorga E (2016) Chapter 7.3: nutrients inputs from river systems to coastal waters. In: IOC-UNESCO and UNEP. Large marine ecosystems: status and trends. United Nations Environment Programme, Nairobi, pp 179–195. http://onesharedocean.org/public_store/publications/lmes_report_chapters/TWAP_VOLUME_4_TECHNICAL_REPORT_Chapter7.3_1DEC2016.pdf

Seitzinger S, Mayorga E, Bouwman A, Kroeze C, Beusen A, Billen G, Van Drecht G et al (2010) Global river nutrient export: a scenario analysis of past and future trends. Glob Biogeochem Cycles 24(4)

Seitzinger S, Harrison J, Dumont E, Beusen AH, Bouwman A (2005) Sources and delivery of carbon, nitrogen, and phosphorus to the coastal zone: an overview of Global Nutrient Export from Watersheds (NEWS) models and their application. Glob Biogeochem Cycles 19(4)

Selman M, Greenhalgh S, Diaz R, Sugg Z (2008) Eutrophication and hypoxia in coastal areas: a global assessment of the state of knowledge. World Resources Institute, Washington, DC. https://www.wri.org/publication/eutrophication-and-hypoxia-coastal-areas

Septiono M, Roosmini D, Salami I, Ariesyady H, Lufiandi L (2016) Industrial activities and its effects to river water quality (case study citarum, bengawan solo and brantas), an Evaluation for java island as an economic corridor in master plan of acceleration and expansion of indonesia economic development (MP3EI) 2011–2025

Shajib MTI, Hansen HCB, Liang T, Holm PE (2019) Metals in surface specific urban runoff in Beijing. Environ Pollut 248:584–598

Sheavly SB, Register KM (2007) Marine debris & plastics: environmental concerns, sources, impacts and solutions. J Polym Environ 15(4):301–305

Shen M, Mao D, Xie H, Li C (2019) The social costs of marine litter along the east China sea: evidence from ten coastal scenic spots of Zhejiang Province, China. Sustainability 11(6)

Shikuma NJ, Hadfield MG (2010) Marine biofilms on submerged surfaces are a reservoir for *Escherichia coli* and *Vibrio cholerae*. Biofouling 26(1):39–46

Shim WJ, Hong SH, Eo SE (2017) Identification methods in microplastic analysis: a review. Anal Methods 9:1384–1391

Sia Su G, Martillano KJ, Alcantara TP, Ragragio E, De Jesus J, Hallare A, Ramos G (2009) Assessing heavy metals in the water, fish and macro invertebrates in Manila Bay, Philippines. J Appl Sci Environ Anital 4(3):187–195

Siebert U, Joiris C, Holsbeek L, Benke H, Failing K, Frese K, Petzinger E (1999) Potential relation between mercury concentrations and necropsy findings in cetaceans from German waters of the North and Baltic seas. Mar Pollut Bull 38(4):285–295

Sintim HY, Flury M (2017) Is biodegradable plastic mulch the solution to agriculture's plastic problem? Environ Sci Technol 51(3):1068–1069

Smith AG, Gangolli SD (2002) Organochlorine chemicals in seafood: occurrence and health concerns. Food Chem Toxicol 40(6):767–779

Smith M, Love DC, Rochman CM, Neff RA (2018) Microplastics in seafood and the implications for human health. Curr Environ Health Rep 5(3):375–386

Smith RF (2000) Total maximum daily load (TMDL) for debris in the waters of duck creek in Mendenhall Valley, Alaska. United States Environmental Protection Agency, Seattle

Speers AE, Besedin EY, Palardy JE, Moore C (2016) Impacts of climate change and ocean acidification on coral reef fisheries: an integrated ecological–economic model. Ecol Econ 128:33–43

STAP (Scientific and Technical Advisory Panel) (2011) Hypoxia and nutrient reduction in the coastal zone: advice for prevention, remediation and research. An advisory document of the scientific and technical advisory panel. Global Environment Facility, Washington, DC

Steinfeld H, Gerber P, Wassenaar T, Castel V, Rosales M, de Haan C (2006) Livestock's long shadow. Environmental issues and options. Methane to markets partnership expo. FAO, Rome

Steinmetz Z, Wollmann C, Schaefer M, Buchmann C, David J, Troeger J, Munoz K et al (2016) Plastic mulching in agriculture. trading short-term agronomic benefits for long-term soil degradation? Sci Total Environ 550:690–705

Storelli MM (2008) Potential human health risks from metals (Hg, Cd, and Pb) and polychlorinated biphenyls (PCBs) via seafood consumption: estimation of target hazard quotients (THQs) and toxic equivalents (TEQs). Food Chem Toxicol 46(8):2782–2788

Sunda WG (1989) Trace metal interactions with marine phytoplankton. Biol Oceanogr 6(5–6):411–442

Surís-Regueiro JC, Garza-Gil MD, Varela-Lafuente MM (2007) The prestige oil spill and its economic impact on the Galician fishing sector. Disasters 31(2):201–215

Susetiono (2019) Personal Communication between author Z. Arifin and Susetiono, Indonesia Institute of Sciences

Sutton MA, Bleeker A, Howard C, Erisman J, Abrol Y, Bekunda M, Datta A et al (2013) Our nutrient world. The challenge to produce more food & energy with less pollution. Centre for Ecology & Hydrology, Edinburgh

Sutton R, Xie Y, Moran KD, Teerlink J (2019) Occurrence and sources of pesticides to urban wastewater and the environment. In: Goh KS, Gan J, Young DF, Luo Y (eds) Pesticides in surface water: monitoring, modeling, risk assessment, and management, vol 1308. American Chemical Society, Washington, DC, pp 63–88

Tanaka K, Takada H, Yamashita R, Mizukawa K, Fukuwaka M-A, Watanuki Y (2013) Accumulation of plastic-derived chemicals in tissues of seabirds ingesting marine plastics. Mar Pollut Bull 69(1–2):219–222

Tejakusuma IG, Adi S, Nugrahadi MS, Rahmania D, Yanagi T (2009) Sediment and suspended particulate matter of Jakarta Bay, Indonesia. Indonesia Water J 5(2)

Teng J, Wang Q, Ran W, Wu D, Liu Y, Sun S, Liu H et al (2019) Microplastic in cultured oysters from different coastal areas of China. Sci Total Environ 653:1282–1292

Teuten EL, Saquing JM, Knappe DR, Barlaz MA, Jonsson S, Björn A (2009) Transport and release of chemicals from plastics to the environment and to wildlife. Philos Trans R Soc Lond Ser B 364:2027–2045

Teuten EL, Rowland SJ, Galloway TS, Thompson RC (2007) Potential for plastics to transport hydrophobic contaminants. Environ Sci Technol 41(22):7759–7764

Thi Ha D, Ouillon S, Van Vinh G (2018) Water and suspended sediment budgets in the lower Mekong from high frequency measurements (2009–2016). Water 10(7)

Thompson RC, Olsen Y, Mitchell RP, Davis A, Rowland SJ, John AW (2004) Lost at sea: where is all the plastic? Science 304

Trasande L, Zoeller RT, Hass U, Kortenkamp A, Grandjean P, Myers JP, DiGangi J et al (2015) Estimating burden and disease costs of exposure to endocrine-disrupting chemicals in the European union. J Clin Endocrinol Metabol 100(4):1245–1255

Tseng C-H, Lee IH, Chen Y-C (2019) Evaluation of hexavalent chromium concentration in water and its health risk with a system dynamics model. Sci Total Environ 669:103–111

UN (United Nations) (2017) World population prospects: the 2017 revision, key findings and advance tables. Population Division, Department of Economic and Social Affairs

UNDESA (United Nations Department of Economic and Social Affairs) (2015a) world urbanization prospects: the 2014 revision. ST/ESA/SER.A/366

UNDESA (2015b) Urban and rural areas 2014. Wall Chart

UNDP (United Nations Development Programme) (2002) Conserving biodiversity sustaining livelihoods: experiences from GEF-UNDP biological diversity projects. E.S.D.G. Bureau for Development Policy, United Nations Development Programme, Global Environment Facility, Nairobi

UNEP (United Nations Environment Programme) (2006) Marine and Coastal ecosystems and human well-being: a synthesis report based on the findings of the millennium ecosystem assessment. United Nations Environment Programme, Nairobi

UNEP (2014) Valuing plastics: the business case for measuring, managing and disclosing plastic use in the consumer goods industry. United Nations Environment Programme, Nairobi

UNEP (2015) Global waste management outlook. United Nations Environment Program, Nairobi

UNEP (2017) Draft resolution on marine litter and microplastics, 3rd Session. United Nations Environment Assembly of the United Nations Environment Programme, Nairobi. https://papersmart.unon.org/resolution/uploads/k1709154.docx

UNEP (2018a) Africa waste management outlook. U.N.E. Programme. United Nations Environment Programme, Nairobi

UNEP (2018b) Single-use plastics: a roadmap for sustainability. United Nations Environment Programme, Nairobi

UNEP (2019) Highly hazardous pesticides. https://www.unenvironment.org/explore-topics/chemicals-waste/what-we-do/emerging-issues/pesticides

UNEP/GPA (2006) The state of the marine environment: trends and processes. United Nations Environment Program/Global Programme of Action for the Protection of the Marine Environment from Land-Based Activities, The Hague

University of York (2019) Antibiotics found in some of the world's rivers exceed 'safe' levels, global study finds. ScienceDaily, May 27. https://www.sciencedaily.com/releases/2019/05/190527094120.htm

UN-Water (2018) Factsheet on water quality and wastewater. Water Program, United Nations, Geneva

U.S. Code (1947) Title 7. Agriculture, chapter 6. Insecticides and environmental pesticide control, subchapter II. Environmental pesticide control, section 136. Definitions. U.S. Government

Van Cauwenberghe L, Janssen CR (2014) Microplastics in bivalves cultured for human consumption. Environ Pollut 193:65–70

van Dam JW, Negri AP, Uthicke S, Muller JF (2011) Chemical pollution on coral reefs: exposure and ecological effects. In: Sanchez-Bayo F, van den Brink PJ, Mann RM (eds) Ecological impacts of toxic chemicals. Bentham Science Publishers, Sharjah, pp 187–211

van Franeker JA (1985) Plastic ingestion in the North Atlantic fulmar. Mar Pollut Bull 16(9):367–369

Vasanthi P, Kaliappan S, Srinivasaraghavan R (2008) Impact of poor solid waste management on ground water. Environ Monit Assess 143(1–3):227–238

Viet Nam News (2019) Vietnam takes action to reduce plastic waste. Feb 12. https://vietnamnews.vn/environment/505164/viet-nam-takes-action-to-reduce-plastic-waste.html#oGB1RwvF1pP3khp0.97

Villarrubia-Gómez P, Cornell SE, Fabres J (2018) Marine plastic pollution as a planetary boundary threat: the drifting piece in the sustainability puzzle. Mar Policy 96:213–220

Volpe J, Gee J, Ethier V, Beck M, Wilson A, Stoner J (2013) Global aquaculture performance index (GAPI): the first global environmental assessment of marine fish farming. Sustainability 5(9): 3976–3991

Voss M, Baker A, Bange HW, Conley D, Deutsch B, Engel A, Heiskanen A-S et al (2011) Nitrogen processes in coastal and marine ecosystems. In: Sutton M, Howard C, Erisman J, Billen G, Bleeker A, Grennfelt P et al (eds) The European nitrogen assessment: sources, effects and policy perspectives. Cambridge University Press, Cambridge, pp 147–176

Voss M, Bange HW, Dippner JW, Middelburg JJ, Montoya JP, Ward B (2013) The marine nitrogen cycle: recent discoveries, uncertainties and the potential relevance of climate change. Philos Trans R Soc B Biol Sci 368(1621):20130121

VPAS (Vietnam Plastics Association) (2019) Overview of Vietnam plastics industry. Vietnam Plastics Association, Ho Chi Minh

Wang M, Hu C, Barnes BB, Mitchum G, Lapointe B, Montoya JP (2019) The great atlantic sargassum belt. Science 365(6448):83–87

Watson SB, Miller C, Arhonditsis G, Boyer GL, Carmichael W, Charlton MN, Confesor R et al (2016) The re-eutrophication of Lake Erie: harmful algal blooms and hypoxia. Harmful Algae 56:44–66

Watts AJ, Urbina MA, Corr S, Lewis C, Galloway TS (2015) Ingestion of plastic microfibers by the crab *Carcinus maenas* and its effect on food consumption and energy balance. Environ Sci Technol 49(24):14597–14604

Weibel SR, Weidner RB, Cohen JM, Christianson AG (1966) Pesticides and other contaminants in rainfall and runoff. J Am Water Works Assoc 58(8):1075–1084

Welden NAC, Cowie PR (2016) Environment and gut morphology influence microplastic retention in langoustine, Nephrops norvegicus. Environ Pollut 214:859–865

Wheeler BW, White M, Stahl-Timmins W, Depledge MH (2012) Does living by the coast improve health and wellbeing? Health Place 18(5):1198–1201

White M, Smith A, Humphryes K, Pahl S, Snelling D, Depledge M (2010) Blue space: the importance of water for preference, affect, and restorativeness ratings of natural and built scenes. J Environ Psychol 30(4):482–493

Wiber MG, Young S, Wilson L (2012) Impact of aquaculture on commercial fisheries: fishermen's local ecological knowledge. Hum Ecol 40(1):29–40

Wiedinmyer C, Yokelson RJ, Gullett BK (2014) Global emissions of trace gases, particulate matter, and hazardous air pollutants from open burning of domestic waste. Environ Sci Technol 48(16): 9523–9530

Wilcox C, Hardesty B, Sharples R, Griffin D, Lawson T, Gunn R (2013) Ghostnet impacts on globally threatened turtles, a spatial risk analysis for Northern Australia. Conserv Lett 6(4):247–254

Wilcox C, Van Sebille E, Hardesty BD (2015) Threat of plastic pollution to seabirds is global, pervasive, and increasing. Proc Natl Acad Sci 112(38):11899–11904

Williams AT, Randerson P, Di Giacomo C, Anfuso G, Macias A, Perales JA (2016) Distribution of beach litter along the coastline of Cádiz, Spain. Mar Pollut Bull 107(1):77–87

Woodall LC, Jungblut AD, Hopkins K, Hall A, Robinson LF, Gwinnett C, Paterson GL (2018) Deep-sea anthropogenic macrodebris harbours rich and diverse communities of bacteria and archaea. PLoS One 13(11):e0206220

Woodall LC, Gwinnett C, Packer M, Thompson RC, Robinson LF, Paterson GL (2015) Using a forensic science approach to minimize environmental contamination and to identify microfibres in marine sediments. Mar Pollut Bull 95(1):40–46

World Bank (2009) Environment matters at the world bank: valuing coastal and marine ecosystem services. World Bank, Washington, DC, pp 32–35

World Bank (2019) Vietnam: toward a safe, clean, and resilient water system. World Bank, Washington, DC

World Bank Group (2017) An overview of agricultural pollution in Vietnam: summary report 2017. World Bank, Washington, DC

World Shipping Council (2017) Containers lost at sea 2017 update. http://www.worldshipping.org/industry-issues/safety/Containers_Lost_at_Sea_-_2017_Update_FINAL_July_10.pdf

WRI (World Resources Institute) (2013) Interactive map of eutrophication & hypoxia

Wright SL, Kelly FJ (2017) Plastic and human health: a micro issue? Environ Sci Technol 51(12):6634–6647

Wu Y, Wang X, Ya M, Li Y, Hong H (2016) Distributions of organochlorine compounds in sediments from Jiulong River Estuary and adjacent Western Taiwan Strait: implications of transport, sources and inventories. Environ Pollut 219:519–527

Wyles KJ, Pahl S, Thomas K, Thompson RC (2016) Factors that can undermine the psychological benefits of coastal environments: exploring the effect of tidal state, presence, and type of litter. Environ Behav 48(9):1095–1126

Yadav H, Kumar P, Singh VP (2019) Hazards from the municipal solid waste dumpsites: a review. In: Singh H, Garg P, Kaur I (eds) Proceedings of the 1st international conference on sustainable waste management through design. Springer, Cham

Yang Z, Hodgkiss I (2004) Hong Kong's worst "red tide"— causative factors reflected in a phytoplankton study at port shelter station in 1998. Harmful Algae 3(2):149–161

Yasin JA, Kroeze C, Mayorga E (2010) Nutrients export by rivers to the coastal waters of africa: past and future trends. Glob Biogeochem Cycles 24(4)

Zettler ER, Mincer TJ, Amaral-Zettler LA (2013) Life in the 'plastisphere': microbial communities on plastic marine debris. Environ Sci Technol 47(13):7137–7146

Technology, Data and New Models for Sustainably Managing Ocean Resources

6

Abbreviations

AI	Artificial intelligence
AIS	Automated identification system
DARPA	US Defense Advanced Research Projects Agency
EBM	Ecosystem-based management
FAO	Food and Agriculture Organization of the United Nations
HF	High-frequency
HLP	High Level Panel for a Sustainable Ocean Economy
IOC	Intergovernmental Oceanographic Commission
IOM	Integrated ocean management
IoT	Internet of things
IUU fishing	Illegal, unreported and unregulated fishing
JAMSTEC	Japan Agency for Marine-Earth Science and Technology
ML	Machine learning
MPA	Marine protected area
MOU	Memorandum of understanding
NOAA	United States National Oceanic and Atmospheric Administration
OECD	Organisation for Economic Co-operation and Development
PSMA	Port State Measures Agreement
R&D	Research and development
RBM	Rights-based management
SeaBOS	Seafood Business for Ocean Stewardship
TCP/IP	Transmission Control Protocol/Internet Protocol
TED	Turtle exclusion devices
USGS	United States Geological Survey
VM	Virtual machine

Originally published in:
Leape, J., M. Abbott, H. Sakaguchi et al. 2020. *Technology, Data and New Models for Sustainably Managing Ocean Resources*. Washington, DC: World Resources Institute. Available online at www.oceanpanel.org/Technology-data-and-new-models-for-sustainably-managing-ocean-resources
Reprint by Springer International Publishing (2023) with kind permission.
Published under license from the World Research Institute.

Highlights

- Effective management of resources has been hindered by a lack of information about how humans are impacting the ocean.
- There is an explosion in new data and technology for the ocean at the moment, and with it enormous potential for advances in the understanding and stewardship of ocean resources.
- Coordinated efforts by industry, researchers and governments can create advanced sensor networks that provide high-resolution, real-time information about the ocean to anyone who needs it, an "Internet of Things" for the ocean.
- However, significant technical and non-technical barriers exist to creating an equitable, open and accessible digital ecosystem for the ocean. To capitalise on the revolution in data and technology, breakthroughs are needed on several fronts.
- Vast stores of ocean data are in the hands of governments, researchers and industry but are unstructured, inaccessible and unusable. These data should by default be made open and available through data tagging, federated networks and, where possible, data lakes.
- Technology can leverage vital innovations in management. Real-time information and automation can allow robust and nimble adaptation to changing conditions and create new accountabilities in government and in business. An urgent priority is to ensure that these new capabilities are available to all ocean stakeholders.
- Overcoming market barriers is critical to fostering successful innovation that supports science and management

J. Lubchenco, P. M. Haugan (eds.), *The Blue Compendium*, https://doi.org/10.1007/978-3-031-16277-0_6

in the future. Capturing the extraordinary potential of technology will require action by governments and others to foster the needed innovations for all those who have a role in ocean stewardship, by creating new market incentives for innovation, new public–private instruments for investment and new business models.

1 Introduction

We are in the middle of an explosion in new data on the ocean, creating enormous potential for advances in our understanding and stewardship of ocean resources. An exponential increase in the number and variety of ocean observing systems and other new data sources has created the prospect of a digital ocean ecosystem. Advances in processing techniques and visualisation are rapidly expanding our ability to extract information from those data, and are enabling a wide array of tools to provide real-time information in actionable form to decision-makers, such as policymakers, resource managers, resource users, consumers and citizens.

To capitalise on this revolution in data and technology, we will need breakthroughs on several fronts. A first imperative is to end the balkanisation of data to create a new era of open and automated data access—so that the data now locked in the servers of government agencies, businesses or researchers are much more broadly available—and to enable the flowering of an ocean Internet of Things (IoT). A second priority is to harness this revolution to support vital innovations in management. Real-time information and automation can allow robust and nimble adaptation to changing conditions and create new accountabilities in government and in business. A third priority is to create the incentives, investments and business models that will support the innovations that are needed not just by wealthy governments and resource users but by all who depend on the ocean and have a role in sustaining the ocean's future. In this paper, we outline the most promising avenues to create this open, actionable and equitable digital ecosystem for the ocean.

2 The Data Explosion

2.1 Fostering New Scientific Understanding of the Ocean

Walter Munk once said that the twentieth century would be known as "the century of undersampling" (Munk 2012). The ocean is 10 trillion times more opaque to light than the atmosphere. This means that we cannot observe the ocean system by looking at it, as we can with terrestrial ecosystems. Instead, we must place our devices inside the ocean itself. The ocean and its ecosystems change on both small and large scales in time and space. A typical phytoplankton growth rate is to double every 1–10 days, and while the average ocean depth is about 3700 m, most of its photosynthesis occurs in the upper 100 m. At the same time, ocean currents move slowly both horizontally and vertically, causing the ocean to act as the "memory" of the Earth system. Organic carbon that is created in the upper ocean may be buried in deep ocean sediments for millennia. Changes in our land and atmosphere will have an ocean signature for decades or centuries. To end "the century of undersampling" will require a fundamental transformation of our observing systems. We need to sample the ocean on its own intrinsic scales, not on the scales that are dictated by our current technical capabilities.

Over the last three decades, there has been an exponential increase in the number and variety of ocean observing systems. From profiling floats such as Argo (e.g. Freeland and Cummins 2005) to cabled observatories (e.g. Kelly 2014), our understanding of ocean dynamics has been transformed through these new tools. And these observing systems are not just in the ocean, but they are also in space. Beginning with the launch of SeaSat and the launch of the Coastal Zone Color Scanner on NIMBUS-7 in 1978, ocean remote sensing has moved from experimental missions in support of the research community to continuously operating systems that support a wide range of management and application needs.

New communication pathways are opening up a vision of a connected ocean, although the fundamental physical properties of seawater will never enable the same level of ubiquitous communications that we have with land and atmospheric observing systems. Cabled observatories, such as the US Ocean Observatories Initiative (Smith et al. 2018), now bring data ashore directly to the Internet. Acoustic modems, although limited in data throughput, can provide a level of connectivity that may eventually enable heterogeneous "swarms" of platforms to behave as a coordinated network. Hybrid systems of both underwater and ocean surface vehicles are now being tested, with the surface vehicles acting as data "mules," receiving low-bandwidth acoustic data streams from the underwater vehicles and converting them into high-bandwidth radio data streams for transmission to aircraft or satellites. With the emergence of high-bandwidth communications based on networks of hundreds to thousands of small satellites, there is promise of gigabit/second networks everywhere over the surface of the world ocean.

With advances in microelectronics and mechanical design, there has been a rapid increase in the type of measurements that can now be made in the undersea environment. Beginning with measurements of physical properties (temperature, conductivity, velocity, etc.), we can now measure a wide variety of chemical and biological properties in the ocean environment. For example, flow cytometry, which was originally designed as a tool for human blood cell analysis, is now being used in situ to identify a wide variety of microorganisms in the ocean (e.g. Lambert et al. 2016). These instruments are being used to identify harmful algal blooms (HAB), as well as in a wide range of ecological studies (Seltenrich 2014). Environmental DNA analysis is becoming a powerful tool for understanding ecosystem composition, and such analyses can now be made in situ, not just through laboratory analysis of water samples (Kelly et al. 2017).

These examples can be viewed as adapting traditional lab- based techniques to the ocean environment through processes such as miniaturisation, lowering power requirements and automation. However, there are also sensing tools that are fundamentally new. For example, new methods of manufacturing fibre-optic cables are enabling sensors to be embedded within the fibre (Rein et al. 2018). Undersea fibre-optic cables are critical conduits of global information flows, carrying over 95 percent of international data, and more are rapidly being added as bandwidth demands increase, creating huge opportunities to expand ocean sensing (Wrathall 2010). Designers are exploring the possibility of embedding both processing and communication semiconductors within these fibre-optic fabrics, thus creating a dense network of smart sensors and allowing fibre-optic cables to act as both sensors and platforms. Fibre-optic sensors in sea-floor cables are also being used for a wide range of environmental sensing, including seismic activity (Joe et al. 2018).

The variety and capability of these new sensing systems are continuing to increase, and they are now being deployed on a broader range of platforms. For decades, sensors were mounted on fixed buoys or attached on ships. With miniaturisation and power reduction, sensors are now being deployed on underwater passive platforms, such as Lagrangian drifters or buoyancy- driven gliders, or on self-propelled devices, such as the REMUS (Stokey et al. 2005). The same holds for platforms on the sea surface. The Wave Glider (Thomson and Girton 2017) can traverse entire ocean basins, and also remain in areas that are simply too hostile for conventional ships. Saildrone (Cokelet et al. 2015) is pursuing a different model for ocean data acquisition. Rather than sell individual vehicles that are managed by the end user, Saildrone provides "mission as a service," where the user defines the mission plan (types of data, location, etc.) and then Saildrone designs and manages the mission.

These new platforms have greatly expanded our sampling "footprint" in both time and space. We can sample over longer time periods and greater spatial distances than with fixed buoys and a few ships.

The "always on, always connected" ocean (Abbott and Sears 2006) could soon be a reality, with the decreasing costs, improved performance and increasing availability of data. Munk's "century of undersampling" could be drawing to a close. However, there remain both technical obstacles and opportunities.

On the technology side, power availability continues to be challenging. Slow-moving or passive devices, such as floats and gliders, can sample the ocean for many months but they can only cover a small area. Therefore, their ability to observe rapidly changing processes or to map large areas is severely limited. Self-propelled systems require significant power to move through the ocean, as power requirements increase non-linearly with speed. Such systems simply run out of battery power.

Power-harvesting systems are being developed for platforms that operate on the ocean surface, such as the Wave Glider or Saildrone. These platforms can harvest wind and solar energy as well, thus enabling them to remain working for months to years. Bottom- mounted systems that rely on microbial fuel cells are being deployed as well. These fuel cells harvest energy by taking advantage of the natural oxidation of organic material at the sea floor (Reimers and Wolf 2018). New approaches in battery technology, such as aluminium- based systems that use seawater, show promise for greatly increasing battery capacity.

Along with power, the undersea environment is challenging for communication and navigation. Unlike the terrestrial environment where radio frequencies can support WiFi and cellular networks as well positioning systems such as GPS, the ocean lacks such fundamental infrastructure. The ocean is nearly opaque to electromagnetic radiation, and therefore we must rely on acoustic signals and other approaches to provide the basics of communication and navigation.

Acoustic modems are increasing their capability to transmit data, but the amount of data that can be transmitted remains substantially smaller than what we can achieve on land. However, as microprocessors continue to decrease in size and power requirements, and increase in computational performance, we are beginning to develop on-board systems that process and analyse the data on the platform and transmit only the results rather than the entire observed data stream. For example, a resource manager may only need to know if a harmful algae species is present or not, rather than detailed information on every species of microbe in the water. Long fibre-optic cables may string together swarms of platforms that can then communicate with a single data "mule," which can carry the data to the surface. Next-generation Internet-capable microsats are capable of delivering high bandwidth anywhere over the world ocean. While the ocean will always be a difficult environment for high-bandwidth communication systems, distributed intelligence in undersea networks shows promise in overcoming this basic physical obstacle.

Navigation systems are showing similar signs of improvement. A small number of precisely located undersea beacons could serve as anchor points for platform swarms and networks that rely on relative distances from each other to create a precise "cooperative" map. Following Metcalfe's Law of networks, the value of the network increases non- linearly with the number of nodes in the network. Thus, such smart swarms show promise in delivering increasing value with regard to navigation and operational efficiency.

In addition to the technical obstacles to our vision of an always on, always connected ocean, there are many non-technical barriers as well. Our ability to sustain long-term ocean observing systems is always under threat. A recent report by the US National Academies of Sciences, Engineering, and Medicine (NAS 2017) documents both the importance of long-term ocean observing systems and the inability of governments to sustain these systems. Numerous reports on the global ocean observing system also highlight these issues. Long time series have enabled significant growth in our understanding of ocean processes, but every year is a struggle to sustain costly and often remote infrastructure. Even the Argo system, with roughly 3800 floats, must expend significant political and financial resources to try to make modest increases in the number and capabilities of these profiling floats (Roemmich et al. 2009).

The majority of ocean instrumentation is developed primarily to meet the requirements of the science community, and therefore the requirements of cost and schedule are often restricted with respect to the science needs. Most ocean instruments and platforms are expensive and often crafted by hand. There is only a small commercial market to counteract the pressures from the science community to build state- of-the-art, one-of-a-kind instruments. Even systems that are "transitioned" from the science lab to the commercial sector often remain focused on the small market of ocean science. There is little incentive (or pressure) for the funding agencies to engage in any sort of sustained design effort that would encourage an extensible architecture that supports the development of multipurpose instrument systems. Instead, solutions are generally monolithic, with their design focused on meeting the specific needs of a specific science question. Thus, technology lock-in and a relatively slow pace of instrument system evolution are characteristics of scientific ocean observing tools and the generally undercapitalised commercial instrument developers in the field.

Buck et al. (2019) describe a parallel environment in the world of data systems that are built around "portal and download," with little regard to how data will be used within a framework of user-driven services. They propose a fundamental rethinking of data systems architecture, where data are democratised, enabling users to build their own knowledge systems. In a sense, rather than a pre-defined data organisation structure, tagged data would reside in unstructured data lakes where the schema are written as the data are accessed. Much as data lakes are transforming machine learning and analytics, a similar development environment needs to be created for ocean observing systems that would enable knowledge services to be driven by the user.

There is considerable work to do to define and realise such a vision, but if we are to develop adaptive and flexible management approaches to our changing ocean, we will need to rethink how we both collect and deliver data. Much like natural ecosystems, these knowledge ecosystems will deliver critical services.

2.2 Monitoring Human Activity

Technology is changing our ability to understand ocean ecosystems, and how humans are using (and abusing) them. Effective management of resources has been stymied by a dearth of information about how humans are impacting the ocean. The big advances that are generating new opportunities for scientific data collection present parallel opportunities to improve oversight of human activity at global and local scales.

At the global level, increasing access to satellite technologies has enabled real-time, precise vessel tracking. Where once ships operated largely out of sight of regulators, the ubiquity of GPS has allowed governments to mandate that most commercial vessels carry Automated Identification System (AIS) devices, which automatically track and transmit their location. Knowledge products, such as Deep Sea Mining Watch and Global Fishing Watch, publish this information online, allowing anyone to look at what vessels are doing on the world ocean.

The proliferation of increasingly powerful imaging satellites has also been an important development in understanding global impacts on the ocean. Imaging satellites can track changes to coastal and ocean ecosystems, and can be used to understand coastal development patterns, monitor nutrient run off and track pollution from ships.

Drones offer similar imaging at a more granular level. Drones are a cost-effective way of reaching offshore areas, allowing managers to see what is happening at a distance through real-time video streaming.

Drones can also be equipped with chemical sensors, supporting a wide variety of management uses. In Denmark, drones are being flown over the exhaust of shipping vessels, for example, allowing enforcement agencies to determine whether ships are using legally mandated low-sulphur fuels.

Drones are also being used in the water. Autonomous underwater vehicles and swarms of sensors can gather visual and chemical information on vessels. Drones and buoys equipped with acoustic sensors are particularly powerful in understanding human activity. Sound travels great distances in the ocean and different types of vessels have different acoustic signatures. Acoustic sensors can allow managers both to identify when vessels are operating in areas where no vessels are allowed, such as marine protected areas (MPAs), and to identify specific malefactor vessels.

Sensors on vessels provide another level of detail. Video cameras on fishing vessels and even on fishing nets can be used to monitor fish catch and potentially to identify labour abuses (Michelin et al. 2018). These cameras can be coupled with gear sensors that activate when fishing gear is deployed, giving regulators robust insight into where fishing is actually taking place.

Chemical sensors on smokestacks and in the water are being used to monitor water and air pollution to determine compliance with environmental regulations. These sensors also contribute important scientific data to world meteorological organisations, which use sensors on ships for critical in situ data from remote areas to support weather forecasting.

Connected sensors are also a building block for efforts to create traceability in supply chains. The IoT opens the door to robust tracking of all types of maritime goods from the moment they are harvested or produced through ports to their destinations throughout the value chain. Digital tracking will introduce critical efficiency and transparency in global supply chains.

Lastly, social media and the increasing connectivity between people give new insights into human actions. Mining social media data and the dark web can illuminate labour abuses and other illegal activity that historically has been nearly impossible to penetrate (Greenemeier 2015). Online forums can illuminate how and why resource users are flouting regulations, information not generally communicated accurately to regulators but critical for developing effective management (Shiffman et al. 2017). Social media is also providing new sources of data for scientists. Citizen science apps allow members of the public to submit photos for species identification, leading to updated species distribution maps as well as the discovery of new species (Silverman 2016). Photo submission can also help regulators target problem areas: in Los Angeles, citizen tracking of plastic pollution along the Los Angeles River identified the most important spots for intervention (Thompson 2019). Scientists are using Twitter reports of flooding to generate high-resolution urban flooding maps to improve model accuracy and forecasting (Wang et al. 2018).

2.3 A Vision of an "IoT" for the Ocean

The dramatic increase in intelligent, connected devices is enabling a vast array of new services on land. The IoT phenomenon is in its infancy, but the prospect of trillions of connected devices is driving technologies in both network communications (e.g. 5G) and microprocessors. This is not just a simple scaling up of the Internet; it will require a fundamental shift in our software design and network architectures. Developers will no longer think solely of "dumb" sensors feeding high-speed data ingestion systems. Instead, computational power will be pushed out to these "edge" sensors. Workflows will be intelligent, driven by the services being provided. The pressures of near real-time data flows and derived services will require that "time to insight" becomes a fundamental metric. While the traditional historical analyses (and associated data ingestion engines) will continue to be important, these new real-time flows will grow hugely in significance.

Thinking about an IoT for the ocean will still require new approaches to data communications and sensor location. Terrestrial systems can rely on satellite-based positioning systems and radio networks, whereas ocean systems cannot. But, over the next decade, we can expect that an IoT model will begin to become a reality (Fig. 6.1). The availability of powerful microprocessors that consume small amounts of energy will enable networks that transmit small, but information-rich messages (e.g. sensors that identify harmful algal bloom species on board and then transmit a simple presence/ absence message). And as the number of these sensing platforms increases, and they communicate with each other, Metcalfe's Law of networks, where the value of every node in the network increases with each new node added, will come into play in the ocean.

The vision of an IoT for the ocean will only be realised if the private sector, governments and researchers ensure that ocean sensors are interoperable and network architectures support connected, smart sensors (Cater and O'Reilly 2009). Without concerted efforts to achieve these goals, business as usual could lead to a plethora of disconnected sensors all generating proprietary data types that do little to achieve the potential of a connected IoT for the ocean. It is also essential that smart sensor networks are compatible with different types of data access regimes, including open access. New platform and sensor types may minimise the need for researchers and managers to gather their own data, but these platforms are often costly. Effort must be made to ensure that, where possible, the data generated by these platforms are available to relevant researchers and managers and not locked in high-cost proprietary systems.

IoT sensors are also vulnerable to attack. While the security and privacy concerns that are relevant for smart sensors located in the home are less pressing in the ocean, the vulnerability of sensor networks could make large-scale manipulation of data inputs relatively easy (Li et al. 2015). Governments, industry and researchers must work together to develop network architectures that overcome these concerns.

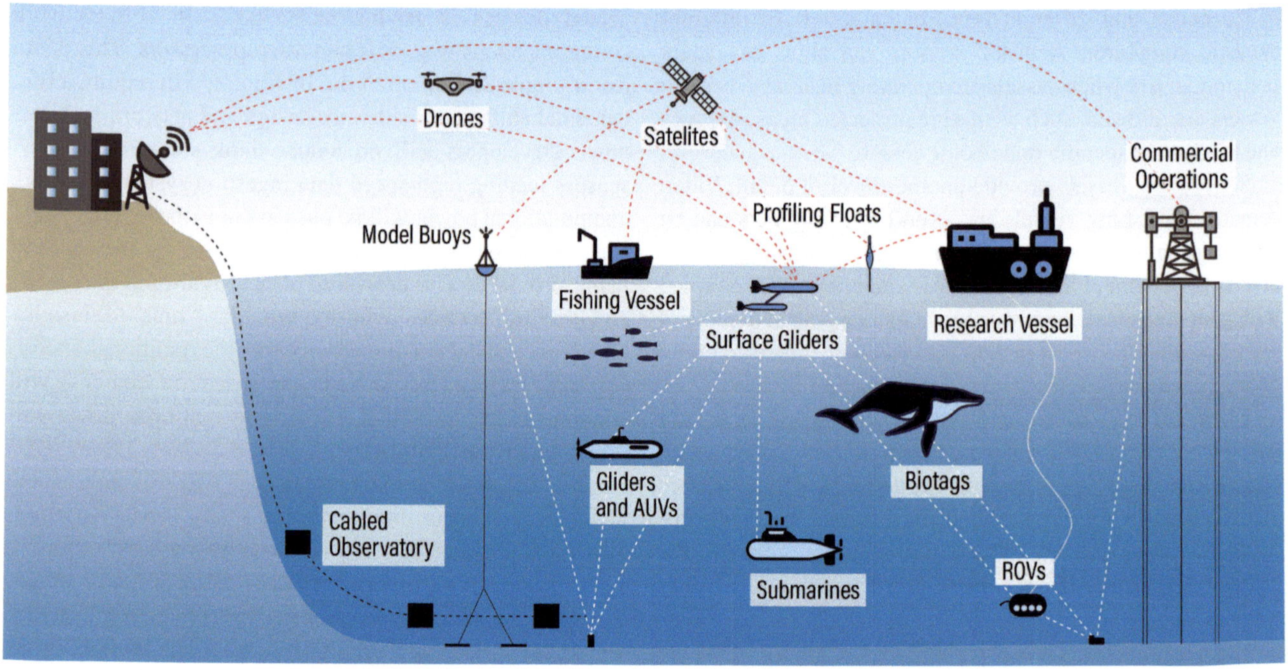

Fig. 6.1 An ocean internet of things. *Source: Authors*

3 Tapping into the Explosion in Data Sources

The explosion in new data about the ocean has the potential to reshape how we understand and manage the ocean. Ocean management has long been impeded, and often defeated, by a lack of timely, accurate and relevant information on the condition of ocean resources (Cvitanovic et al. 2015) and on human activities and their impacts. New technologies are vastly increasing the collection of data, and the urgent challenge is to ensure that these data are available and useful to ocean management.

Data alone are not inherently useful (Kelly 2014). Relevant information must be extracted, combined with information from other sources, and translated into a form that is easily understandable, timely, actionable and accessible for decision-makers (Bradley et al. 2019). The importance of effective knowledge translation cannot be overstated amid the rise of "big data" in the ocean, but historically it has been a weakness in the science–policy interface (ELI 2014). The key challenge ahead is to create a "digital ecosystem for the ocean," which makes diverse ocean datasets available and translates that data into actionable information for decision-makers.

3.1 Making Data Available

"Water, water everywhere, nor any drop to drink." Although Coleridge was referring to the ocean, the same could be said about ocean data. We may be drowning in a sea of data but cannot find the information we need to increase our knowledge or to make science-informed decisions. Quantitatively, the amount of unstructured data gathered and managed annually by organisations within the government, research and business sectors is growing exponentially. Qualitatively, this shift is even more radical, as the conceptual framework for data management moves from a historic, disaggregated and static model to one that is based on dynamic, unstructured and collaborative use. Knowledge extraction will require new tools to enable new levels of collaboration, visualisation and synthesis—this is not just scaling up traditional workflows to accommodate greater volumes. Data will be broadly dispersed, as will the teams that come together to work on specific economic and science issues, and these many-to-many networks will constantly be changing as the needs for collaboration change. As a result, new frameworks are required that provide a systematic basis for data management, analysis and collaboration, rather than ad hoc aggregations of independent components (Buck et al. 2019.)

In the next 10 years, frontier efforts are aiming to create a "digital ecosystem for the environment" (Jensen and Campbell 2018), which aggregates many sources of data to provide timely and high-quality information to decision-makers. There are numerous initiatives that have set out to create this digital ecosystem, from the Global Ocean Observing System hosted by the Intergovernmental Oceanographic Commission of UNESCO (IOC-UNESCO) ("a truly integrated global ocean observing system that delivers the essential information needed for our sustainable development, safety, wellbeing and prosperity," GOOS

2019) to private sector efforts like REV Ocean's Ocean Data Platform ("a global, unifying ocean data platform [that] will enable unbiased research and facilitate a data-driven debate, leading to better decision-making and enable more successful conservation and utilization of ocean resources," REV Ocean 2019). Most current efforts focus on combining datasets into one centralised database, which is a more powerful version of the traditional portal–download data model (Buck et al. 2019).

Efforts to create unified data platforms have faced daunting challenges, however. Datasets are often not consistent or interoperable. Data holders are often reluctant to share data because, once data are combined, they lose control over how their data are accessed and used (Piwowar et al. 2007). Lastly, there are few incentives (either financial or professional) to expend the considerable effort necessary to make datasets available on a sustained basis.

Outside of the ocean, Google and other technology companies have created various tools, such as Google's BigQuery, that crawl the web combing and combining diverse datasets to mine insights. These tools provide new ways to access datasets that previously would not have been interoperable, but they face many of the same challenges as ocean-focused solutions. Researchers and governments do not share their data in ways that allow these tools to access the information, and the incentives needed to tailor these tools to ocean problems do not exist.

We must now rethink our fundamental strategy (and culture) and move decisively towards a data architecture that allows diverse datasets to be accessed automatically by researchers and managers. Universal data tagging standards are the essential foundation for this new wave of ocean data infrastructure, allowing data to be combined in federated data networks and data lakes that support verified and automated global access. Federated data networks offer the potential to liberate ocean data that are currently locked in private sector and government databases, while data lakes create new opportunities to combine data in ways that support real-time management needs and enable the development of new (and sometimes unanticipated) data-driven services.

3.1.1 Tagging Standards

Standardised data tagging and metadata protocols are the first step in making ocean data globally accessible. Standardised metadata include normal indicators, such as where and when data were collected, and how. Tags build on this, indicating whether and how data can be stored, transmitted and used, and its suitability for management and enforcement decision-making. Data tagged appropriately can be made automatically available to users that meet the criteria specified in the tags. Data owners can update data tags at any time, ensuring that access restrictions can be changed as needed. Some have raised concerns about reliance on federated networks for scientific purposes, namely that federating the data removes the connection between the data provider and user and may make it difficult to convey the nuances of how the data were collected (Buck et al. 2019). Tagging can overcome these concerns (Bar-Sinai et al. 2016; Crosas et al. 2015; Sweeney and Crosas 2015).

Creating data networks based on tagging may also allow new types of knowledge to be included more comprehensively in management decisions. Traditional knowledge that does not meet standardised scientific requirements, but which is increasingly recognised as an important part of management decisions, can be included with the appropriate tags (Berkes et al. 2010). Historical data from diverse sources, such as ships' logs, newspapers and menus, can be included to bolster understanding of historical baselines (Thurstan et al. 2015).

3.1.2 Federated Data Networks

Tagged data can be stored and connected through federated data networks, allowing researchers and managers access to diverse ocean data. Global standards allow disparate datasets to be queried and relevant information extracted (WEF 2019). A trusted broker creates and maintains the system, including access verification and other trust-promoting tools (Buck et al. 2019).

Federated data networks can be used to overcome commercial and other confidentiality concerns. They are currently being used successfully in several contexts. They have been particularly attractive to those in healthcare as they provide a way to access data without violating the many privacy laws that govern how health data are shared. Creating systems where the actual data are not shared, but instead external queries can gather the needed information from the data, allows researchers critical access to healthcare data while protecting the privacy of patients.

3.1.3 Data Lakes

Where users are willing to relinquish some control over their data storage, data lakes can be included as nodes within larger federated data networks. Data lakes move data onto cloud architecture, which is designed to scale and bring data closer to the processing pipelines. This type of computing architecture and the workflow pipelines running on top of these cloud solutions is not new. From early mainframes to the virtual machine operating system released by IBM in the 1970s, the concept of shared access to services has emerged and evolved because of the commodification of the entire Internet ecosystem (from microprocessors to services).

Data lakes rely on service-driven data schema rather than pre-defined schema used in "data warehouses" and are particularly promising for scientific data where compute needs are intensive and concerns over data privacy are low (Stein and Morrison 2014). This presents a significant change in the way data users access and use data by implementing, at scale, tightly coupled compute and storage, as well as services. This pipeline creates more efficient access to data and the ability to produce insights at scale.

Anticipated growth in observing technologies driven by advancements in radio telecommunications (5G, satellite and other radio technologies) pose significant challenges for data ingest and archive volumes that are growing exponentially. Distribution for the science community has become a logistics problem of moving assets in order to produce useable products. Efficient utilisation of a data lakes architecture places data close to compute and provides access to countless building block services that enable and expedite science discovery for data users.

Data lakes can enable new workflows that will change the way science is done across multiple domains. These new workflows will create new modelling approaches that help address algorithmic and analytical variability, which has led to reproducibility errors in the present system of science workflows. Adopting a cloud services approach through data lakes eliminates downloading and data transfers, thus allowing researchers and the public to interact and work with data directly, and move only the finished derived products or user experiences to achieve scale.

Data lakes present a path forward for the scientific community, and when built on universal tagging standards can be integrated into ocean data networks that allow automated data access and use for a diverse set of stakeholders. The United States Geological Survey (USGS) and National Oceanic and Atmospheric Administration (NOAA) have successfully transitioned some of their satellite remote sensing data into cloud-based data lakes, and have seen their user base rise exponentially as a result (NCE 2018). Data lakes can unlock new value by allowing users to analyse many data types and opening ocean data to a broader range of users.

Together, tagging, federated data networks and data lakes offer the promise of vastly expanding the ocean data available, and broadening access:

- **Access to more data:** Data tagging coupled with federated data networks enables the liberation of data that are currently locked away because of security, commercial or privacy concerns. The most notable of these data are those collected by defence departments and private sector companies, many of which have collected robust, long-term datasets on ocean conditions for decades. These data are sometimes classified (in the case of defence departments) or confidential (in the case of industry), even when much of the data are on oceanographic conditions with no associated security risk. New standards for data tagging could allow data collected by industry and militaries to automatically be available to researchers, for instance, after any security or time embargos have been met.
- **Accessible to more users:** Tagging allows automation of data access and thus makes it both simpler and more efficient (Sweeney and Crosas 2015). Currently, researchers and managers rely on one-off agreements between parties to allow access to needed data. In robust tagged systems, these agreements can be built into the data from the beginning. If parties are verified research institutions, for example, data tagged with "academic research" as an allowable use will automatically be available to these institutions on specified terms.
- This type of automated access also creates avenues for more equitable access to data. Currently, many marine datasets are in principle available to other researchers. In practice though, these datasets are often only shared with known research partners or top academic institutions. Executing complex Memoranda of Understanding (MOUs), which often take months to be agreed on, is an insurmountable barrier to entry for smaller institutions and resource- constrained managers.
- When combined with the reach provided by federated data networks, automating data use can provide managers with access to actionable information as they need it. Specialised apps can be built on top of data networks that are tailor-made to address common management questions and provide robust knowledge solutions.
- **Access globally:** Federated data networks and data lakes can enable global data access for scientists, managers, communities, consumers and others, but it is essential that they are built with these goals in mind. Without coordinated efforts by governments, research institutions and technology service providers, there is the danger of these solutions becoming additional siloed pieces in an already fragmented ocean data landscape.

As these solutions come online, governments and others must also ensure that data networks and lakes are accessible to everyone. Federated networks and data lakes are promising in part because of the business models they enable, which allow data to be stored for free while the knowledge services built on top of the data, or the increased speed generated by storing data closer to computations, generate revenue. These models, discussed further in Sect. 4, can support widespread, free access to data. Governments must work with web service providers to ensure that these systems are fulfilling this promise and not just providing data access to those that can pay (Borowitz 2019). The data-scarce areas where additional data are most needed to guide marine management are also the ones that are least likely to be able to pay for data access.

Beyond ensuring equitable access to data, governments also need to address the important privacy and security concerns raised by open data. Network architectures must ensure that data integrity is protected throughout the data lifecycle, including quality assurance mechanisms that prevent false data from being added to data networks (Buck et al. 2019). As personal devices, such as mobile phones, and video monitoring tools are increasingly sources of data for ocean management, it is essential that the privacy of users is built into management systems. Additionally, as governments open up access to ocean data, they need to be mindful of potential social and economic costs—open access may provide a de facto subsidy to some private sector actors, for example, or provide avenues for policy influence to those that are best equipped to make use of the data (Johnson et al. 2017).

Opening up access to data will require new incentives for governments, companies and researchers to make their data available. Government can lead the way directly—by taking bold steps to help create and contribute to federated data networks. Governments can also require that a condition of access to public resources—whether the resources are fish stocks and mineral deposits or funds for coastal management or for research—is a commitment to sharing the data produced.

International cooperation around the UN Decade of Ocean Science for Sustainable Development (2021–2030) provides a unique opportunity for concerted action to overcome existing barriers and make real progress towards integrated ocean data (Ryabinin et al. 2019; UNESCO and IOC 2019). It is essential that this opportunity is not wasted.

3.2 Extraction of Information and Translation

Recent innovations are improving our capacity to translate data into useful information. Advanced processing techniques coupled with new visualisation portals enable a wide array of digital decision support tools aimed at providing actionable information to decision-makers (Lathrop et al. 2017).

Rapid advances in artificial intelligence and machine learning (AI/ML), including the emergence of deep learning methods such as neural networks and machine vision, have great promise for ocean data (LeCun et al. 2015). As the variety and volume of ocean data increase, there are similar efforts to use AI/ML tools to derive insights and, more importantly, predictions regarding complex processes, such as large-scale rainfall patterns or severe storms, and eventually even more complex systems that involve ecosystem resilience and human activities. For these complex systems, where deriving mathematical formulations and collecting reproducible data are extremely difficult, big data and AI/ML have become especially appealing.

Within the physical domain, AI/ML have shown potential as a means to substantially improve traditional methods for systems predictions. For example, the US Bureau of Reclamation recently sponsored a contest on sub- seasonal climate forecasting for rainfall patterns in the western United States. The best-performing team relied on AI/ML methods to outperform the benchmark forecast model (Soeth 2019). NOAA is developing a comprehensive strategy to integrate its enormous volume of data with its numerical models using AI/ML approaches to tackle long-standing challenges in Earth system forecasting, such as hurricane tracks and intensity (Bayler 2019).

Much of the appeal of these new methods rests on the fundamental difficulty of developing a mathematical framework for complex, multiscale processes. For example, the microphysics of clouds cannot be resolved at the scales possible in global climate models. Moreover, the processes are difficult to measure as well. However, these processes cannot be ignored and therefore must be parameterised. New methods rely on stochastic formulations of these processes, which are then coupled with the deterministic models of larger- scale processes (e.g. Palmer and Williams 2008). With the advent of AI/ML techniques, it is a fairly straightforward intellectual leap to move from stochastic/deterministic models to AI/ML models.

Phenomenal improvements in AI/ML have enabled better understanding of complex processes, such as language, than is possible with traditional approaches. This has led some scientists to claim that "big data" represents a new scientific paradigm (e.g. Hey et al. 2009). In complex, multiscale processes, AI/ML appears to overcome the challenges in understanding the linkages between these processes, where traditional scientific approaches have been unable to provide any conceptual foundation or mathematical framework. In fact, some have asserted that this means the end of the scientific method, which is based on the connection between reason-driven experiment (or data collection) and analysis based on mathematics and modelling.

Coveney et al. (2016) and Succi and Coveney (2019) provide an extensive review of the interplay between big data and the scientific method. These authors argue that "big data" must work in partnership with "big theory," even when the work of mathematical formulations is difficult and slow. AI-based models are extremely fragile, rarely working outside of the specific data domain in which they are developed. Succi and Coveney (2019) note four key points:

1. Complex systems are rarely based on Gaussian distributions.
2. Complex systems are highly sensitive to small errors, so datasets are never "big enough."
3. Correlation does not imply causation, especially as the links become more remote as the size of the datasets increase.
4. Too many data are as bad as no data.

While we can expect AI/ML to help guide our observing systems and our analyses, we must continue with the fundamental science and mathematics of complex systems.

Beyond better forecasting and analysis of scientific datasets, AI/ML have also unlocked new potential for management. Advancements in computer vision, for instance, allow marine species to be automatically identified from video footage. This opens the door to a new era of electronic management in fisheries, replacing human observers—who are often harassed and in some cases even killed—with video cameras. ML algorithms can automatically review the video footage captured by these cameras to determine what species are being caught and whether vessels are operating legally, at a much lower human and monetary cost than taking observers on board.

More powerful AI/ML analysis techniques also support the creation of advanced knowledge products to support key ocean management needs. Global Fishing Watch, for example, provides a global window on fishing, by providing visualisation of fishing activity through the GPS devices (AIS) required on large vessels. Using ML algorithms to analyse the large amount of data coming from these vessels, Global Fishing Watch can identify when and where a vessel is engaged in fishing activity, classify the type of fishing, and detect other behaviours such as trans-shipments and potentially illegal incursions into protected areas. Similar techniques are being employed by a large new class of enforcement tools that use ML to identify illegal behaviour on the ocean. AI/ML capabilities are foundational to analysing the volumes of data pro vided by emerging technologies and newly networked data, supporting a new generation of knowledge products for managers.

AI has enormous potential to translate the growing flood of ocean data into information that is relevant—and vital—for research, and for the use and management of ocean resources. To realise the potential will require better access to data, through the federated networks and data lakes described above. It will also require innovations in ML. While current methods to train neural networks require vast labelled datasets, emergent methods are able to learn from relatively few labelled points (Reichstein et al. 2019). These methods provide a path forward for many ocean problem sets where the quantity of labelled data are extremely low. As these new methods come online, predictive modelling for ocean management will become exponentially more powerful.

Beyond issues of data availability, current ML suffers from intensive computational requirements. The future will see exponential increases in available compute power, enabling more powerful understanding of our ocean. However, increases in compute are fuelled by significant energy expenditures. The future of ML compute must come from renewable sources.

AI/ML solutions are currently highly tailored to specific ocean problems. For instance, image recognition algorithms are trained to identify individual fish species and may be very difficult to adapt to recognise other fish species. Computational and methodological improvements unlock new possibilities to move beyond hyper-specific ML prediction to generate new cross- cutting understanding of ocean conditions. Advances in modelling that combine ML techniques with physical modelling can combine both data-driven and theoretical insights to generate robust, interpretable results that are testable against physical realities (Reichstein et al. 2019). Applying these methods to broad datasets can move beyond single-problem insights to demonstrate new relationships between diverse ocean conditions.

While ML shows promise, there are significant issues of bias that also need to be addressed before it is widely adopted in management. ML outcomes are only as good as the data they learn from. Existing inequity can be exacerbated in cases where complex machine learning algorithms are being used to identify illegal behaviour (such as in the case of many advanced tools for monitoring illegal fishing) (Sajin 2018). If, for instance, an algorithm looks at past enforcement actions to build a model that predicts the likelihood of future illegal activity, this algorithm will solidify any historical bias in which types or flags of vessels have been most often targeted for enforcement. AI/ML algorithms can also be susceptible to false or "spoofed" data. Small pieces of inaccurate or manufactured data can lead to erroneous results and inferences from these complex, but fragile, algorithms (Amodei et al. 2016). Emerging work in AI interpretability may help to overcome these issues by allowing managers to see into the black box of AI to identify systemic biases and to elucidate the basis for management outcomes so that they can be legally enforceable.

4 Harnessing the Technology Revolution to Transform Ocean Management

In recent decades, there have been important innovations in ocean management and in using markets to incentivise more sustainable use of ocean resources. Technological advances

Table 6.1 Technology enables innovations in management

		Management Innovations			
		Dynamic and automated management	Integrated ocean management	Rights based management	Harnessing the market
Enabling technologies	Sensors	In-situ, remote and vessel-based sensors enable highly granular observations of current ocean conditions	Autonomous vehicles, profiling floats and other new sensor platforms allow previously unreachable areas to be studied	Low cost sensors support community management of marine resources	DNA barcoding and other biotechnology tools can verify product identity throughout the supply chain
	Communication networks	5G networks and satellites enable real-time transmission of ocean data to managers and resource users	Acoustic networks, cabled observatories and satellite transmission can link distant sensors to shore	5G and cellular networks allow fishers and other resource user to access resource user to access management	Apps that use blockchain can create an immutable record of product movement
	Data systems	Data lakes and federated networks provide access to the data from different sources needed to support dynamic management	Data lakes can give scientists access to unstructured data that supports many different kinds of analysis	Local data networks allow resource users to share and access relevant data on resource use and conditions	Federated data networks allow industry to share relevant data while respecting privacy and ownership concerns
	Data processing	Advanced modelling analytics support near real-time data processing and analysis	Machine learning enables new analysis of large and previously disparate datasets	Modelling can better predict resource use and allocations	Machine learning can be used to analyze large volumes of industry information for compliance
	Knowledge tools	Blockchain combined with near real-time sensor data can be used to create smart contracts that automate management decisions	Near real-time vizualizations of ocean conditions provide critical information for managers	Daily maps based on new data and modeling are being used in fisheries to maximize catch and reduce protected bycatch	Apps and other tools illuminate the supply chain for consumers at the point of sale

offer the opportunity to leverage those innovations, creating new capabilities, new incentives and new accountabilities (Table 6.1).

4.1 Public Management

Historically, the ocean has been managed as a public good. Public management has had limited tools, and has been constrained by politics, practicalities and a profound lack of information. The result of these limitations has been a reliance on regimes that are static and often crude, and that sometimes create perverse incentives.

4.1.1 Innovations in Management

In recent years, there has been increasing emphasis on ecosystem-based management (EBM) for managing marine systems. EBM shifts away from traditional, siloed management of individual resources or uses to consider the ecosystem as a whole and the full range of human activities within it (Long et al. 2015). Successful EBM regimes require a wealth of scientific data to understand and predict the complex relationships and dynamics in marine systems. EBM must also be nimble in responding to changing ecosystems and stakeholder needs and interests, requiring an integrated approach to ocean management.

Two innovations in governance—dynamic management and rights-based management—have shown particular promise in aligning capabilities and incentives with sustainability. Emerging technologies can leverage these policy tools to increase the effectiveness of marine management.

Dynamic Management Ocean management has always been challenged by the fact that resources and conditions are constantly changing. With the increase in climate and other stressors, that challenge will only grow. Yet ocean management has typically been static—relying on fixed areas, seasons and catch limits. Dynamic management strategies allow managers to make near real-time adjustments as conditions change (Maxwell et al. 2015). In fisheries, this has meant a transition from, for example, static spatial limits on fishing that are set at the beginning of a season, to dynamic closures where the allowed fishing area can be adjusted based on the status of stocks, the presence of bycatch species and other key indicators. Dynamic management is the essential underpinning of a new generation of responsive, ecosystem-based marine spatial planning.

Technological innovations have made dynamic management possible. New tools for monitoring ocean conditions and for communicating with geographically dispersed resource users allow managers to make rapid decisions and disseminate them widely. In one striking example of dynamic management in action, a series of hydrophones were attached to buoys in the busy shipping lane approach to Boston Harbor. When the hydrophones detect the song of endangered right whales, this information is automatically transmitted to ships approaching the harbour and reduced speed limits are imposed (Laist et al. 2014). By allowing vessels to maintain high speeds when whales are not in the area, this approach reduces ship strikes on whales while maximising shipping efficiency. Other examples of dynamic management include the dissemination of real-time information on high-risk areas for turtle bycatch to fishers in Hawaii. A recent study found that in the California drift gillnet fishery, a highly dynamic fishery that is difficult to manage, implementing dynamic spatial closures could significantly reduce the percentage of total area closed to fishing to achieve the same conservation goals (Hazen et al. 2018).

Rights-Based Management Policies that focus on shifting incentives to achieve management goals represent another important frontier in marine policy (Lubchenco et al. 2016). For fisheries, many jurisdictions have taken steps to better align the incentives of resource users with long-term sustainability by instituting rights-based management (RBM). RBM regimes seek to eliminate the traditional problems associated with common pool resources by assigning property rights in the resource to the resource users (Nyborg et al. 2016), either through quota systems that assign a percentage of fish catch to each user (Individual Transferable Quotas) or through territorial rights that give stakeholder groups exclusive rights to fish in a specific area (Territorial Use Rights for Fishing (TURFs)).

When designed correctly, RBM has proven to be a highly effective management solution (Lubchenco et al. 2016). To succeed, leaders must build consensus among stakeholders before policies are implemented. They should develop a regime that combines strong property rights with reputational and behavioural incentives and ensure that rights are protected with enforceable sanctions (Crona et al. 2017).

RBM is not a silver bullet to solve fisheries management, however. Some note that giving fishers a quota of fish stocks is not the same as a true property right, and may lead to continuing management issues in the future as incentives for fishers are not fully aligned with the long-term viability of the fishery (Bromley 2016). Others note that inequity may be reinforced by the distributional choices made in allocating quotas, which are often based on historical catches, rewarding those with the most economic clout (Guyader and Thébaud 2001).

Some systems have found creative solutions to these challenges. In some industrial fisheries in the Bering Sea, for example, a percentage of the fish catch is allocated to coastal communities as Community Development Quotas (Haynie 2014). Coastal communities are able to fish or lease their quotas to fishing companies and invest the revenues. These programmes have been successful in helping to alleviate some of the largest equity concerns around the privatisation of fisheries (Carothers 2015).

These new models of governance—ecosystem-based, rights-based and dynamic—are helping managers meet the challenges of managing the many pressures on ocean resources. New technologies—from more powerful sensors to smart contracts—offer opportunities to build on these policy innovations, creating a new era in ocean management that transforms both capabilities and incentives.

4.1.2 Making Management Robust and Nimble

The years ahead will see significant advances in our ability to collect data on resource conditions and uses with high spatial and temporal resolution, and to translate those data into actionable information for users and managers. The continued proliferation of satellites and ocean-going drones will expand capability to monitor activities on and in the water. Video cameras on fishing boats and on nets will allow fishers to more precisely control their catch and will enable increasingly granular management and accountability. Flocks of communicating sensors in the water will be able to identify emergent problems and swarm to investigate (Jaffe et al. 2017).

These capabilities will become increasingly vital to effective, ecosystem-based ocean management as climate change and other stressors disrupt ocean systems. It will be essential to have real-time information on ocean conditions to be able to manage heatwaves, shifting fish stocks, harmful algal blooms and other upheavals.

New technologies enable a better understanding of how humans are using marine ecosystems. Monitoring data on human use can guide enforcement efforts, allowing more targeted deployment of enforcement solutions focused on providing data in near real time that meets legal evidentiary requirements. New options, from drones that allow visual monitoring of distant water areas (e.g. ATLAN Space) to mandatory tamper-proof GPS-enabled devices on fishing vessels, provide this information to enforcement officials.

Real-time data supports integrated approaches to ocean management. Integrated ocean management (IOM) creates comprehensive management plans to reconcile competing uses of the ocean and ensure ecosystem health (See Blue Paper 14 on "Integrated Ocean Management"). IOM tools, such as marine spatial planning, are important pieces of the ocean management landscape but require extensive data on both ecosystem baselines and human uses of the ocean.

Technological advances could have profound value for helping fishing communities manage their resources. In small-scale fisheries, for instance, small GPS trackers enable fishers to accurately track where they fish each day. Apps like mFish allow fishers to use their smartphones to receive critical data on weather, market prices and other conditions, while at the same time using their phones to collect key data about what they catch and where. Fishcoin allows buyers to compensate small-scale fishers for collecting data they need, paying them in mobile-phone minutes through a blockchain. Blockchain technology can also help small producers connect to global supply chains.

A future of robust management based on better information is not assured. Even when relevant data are available, managers often do not get the information they need because data are not available to them, or because they do not have scientists working with the data to address the most policy-relevant questions (McConney et al. 2016). Even decision-support tools designed explicitly for marine managers are often so technical that only programmers are able to use them (Stelzenmüller et al. 2013). Non-governmental organisations (NGOs) and interdisciplinary research organisations have been important players in bridging the science–policy divide, allowing research priorities to be developed collaboratively with scientists and managers (Sutherland et al. 2011).

4.1.3 Automating Management Through Smart Contracts

In the next decade, technology will not only expand the potential for dynamic management regimes, but also open new frontiers for completely automated management. Dynamic management still typically relies on the human process of translating data into management decisions. Coupling dynamic management with the possibilities opened up by smart contracts, among other technologies, creates the opportunity to automate some areas of marine management.

In other industries, smart contracts are the cutting edge of regulatory compliance efforts. Smart contracts rely on verification—once the agreed conditions have been met, smart contracts execute automatically (Le Seve et al. 2018). For instance, smart contracts for travel insurance can automatically send compensation to passengers when online flight trackers report that their flights have been delayed by a pre-agreed amount. These smart contracts are generally based on distributed ledger technologies, so that they are immutable and tamper- proof. Automatic execution reduces opportunities for corruption and fosters transparency.

When these contracts are connected to environmental sensors, there is the potential to automate aspects of environmental management (Jensen and Campbell 2018). Smart contracts have already been used to facilitate peer-to-peer water management in Australia (Le Seve et al. 2018). Water rights are notoriously complex to manage and transfer. Smart contracts allow for easy transfer of water quotas between users depending on agreed upon conditions (for example, if a user uses less than their monthly allotment, a sensor can automatically detect this and transfer the remainder immediately to another party at an agreed rate). In the case of ocean pollution control, for example, sensors placed on ship exhaust could automatically fine companies when the concentration is above allowable levels.

Combining the technological innovation of smart contracts with the policy innovation of dynamic management has the potential to reshape how marine management functions. Replacing tasks that currently require human verification with smart contracts and other tools can free up management resources to be spent in more critical oversight functions that require human attention.

In fisheries, governments and industry working could create near-automated port entry systems based on increasingly powerful monitoring capabilities. This "global entry" system could provide expedited entry into port for fishing vessels that meet predetermined transparency requirements, such as sharing of AIS data, electronic monitoring on board the vessel, and release of information on permits and ownership. Fisheries agencies can use these data to ensure that the vessels are at low risk of illegal, unregulated and unreported (IUU) fishing, and in turn provide preferential clearance and processing while in ports. This type of system can incentivise good behaviour by fishers, while at the same time reducing the impact of corruption by port officials.

Box 6.1: Case Study: Preventing Bycatch

Japan Agency for Marine-Earth Science and Technology (JAMSTEC) uses high-frequency (HF) radar data to understand the relationship between the sea state and the small Pacific bluefin tuna (<30 kg) catch by the setnet. The observations are acquired in quasi real time, every 30 min, and are posted immediately (usually within 1 h) as a surface current map on JAMSTEC website. These setnets are able to register when current patterns are likely to lead to mass bycatch of restricted tuna and alert the local fishers of the potential risk of young tuna entering their setnets in large numbers. The setnet fishers can, therefore, prepare themselves for releasing the young tunas based on the alert. See Appendix 1 for more detail.

Automated systems can also be used to strengthen mitigation measures. Timely detection and forecasting of environmental threats, such as storms, heatwaves and harmful algal blooms, can be directly linked to automated systems that preemptively protect ecosystems. These systems are already beginning to be used in storm water management: predictions of impending storms or detection of water quality parameters outside the normal range automatically trigger additional treatment measures to prevent nutrient loading (Klenzendorf et al. 2015; Lenhart et al. 2018), as well as in the power sector where heatwave predictions trigger cooling curtailments. This should be expanded to link real-time threat detection and forecasting to automatic mitigation action in other contexts. Threat forecasting for harmful algal blooms, for instance, has become highly advanced in order to prevent human health impacts. Linking bloom forecasts with automatic reduction in fertiliser application in neighbouring areas or increased water treatment could help to not just predict but mitigate these and other types of environmental threats.

Automated systems have the potential to make RBM an even more powerful, and more equitable, management tool by facilitating effective enforcement and efficient exchange of fisheries rights. Additionally, new tools like blockchain provide for new, more transparent and reliable ways to transfer quotas quickly without many of the transaction costs that have plagued these systems in the past.

Blockchain advocates go further, pointing towards a future of decentralised management and the complete disappearance of the state (Atzori 2015). With governance based entirely on smart contracts, they argue, managers are no longer needed to create regulations and ensure compliance. In the marine governance system a future where managers are completely removed from the picture is unlikely. Creating regulations is a complex process that involves negotiations among many stakeholders, coupled with an understanding of ecosystem dynamics, which requires human decision-making. Knowledgeable managers thus remain important. One can envision a future, however, in which much of the burden of implementation and enforcement is alleviated by automation.

There are hazards. Although blockchain-based options create immutable records, these records are only as good as the information put into them. Smart contract solutions thus must include robust measures for assuring the accuracy of the data upon which they depend. Stakeholder participation can be part of a data verification system (Jensen and Campbell 2018). If industrial permits, for instance, rely on the clean-up of certain environmental conditions, local stakeholders can verify that conditions have been met by submitting evidence such as photos.

Automated management also raises the spectre of a dystopian future where decisions are made based on complex and opaque algorithms with no human judgement. Governments should only adopt automated management when they have robust processes in place for dispute and review of automated decisions. Automated management should also only be applied to management problems where metrics are quantitatively verifiable (e.g. changes in ocean temperature) and results do not compromise fundamental civil liberties. These criteria need to be evaluated for each proposed application on a case-by-case basis. In the case of alterations to fishing areas or allowable gear types, for example, automated management can allow rapid, real-time changes as oceanographic conditions change without compromising protected legal rights. On the other hand, while AI algorithms can be used to identify probable illegal fishing vessels based on their behaviour, they cannot be a sufficient basis for automated enforcement action because the basis of the determination is unspecified and the consequences could be criminal liability.

Automated management can shift human management resources from routine, numerical determinations to more complex ecosystem-level analysis and decision- making. When coupled with stakeholder engagement, incentive-shifting and improved baseline data, automated and dynamic management will help to support successful ocean governance and integrated ecosystem-based management.

4.2 Harnessing the Market

In the private sector, the transparency and traceability enabled by technological advances can create new incentives for more sustainable practices.

Over the past 20 years, the Sustainable Seafood Movement has demonstrated the potential for market actors—including consumers, retailers, processors, fishers—to incentivise better management of fisheries. Independent certification of fisheries and chain of custody through supply chains, such as through the Marine Stewardship Council, and ratings systems, such as Seafood Watch, help buyers to identify seafood from well-managed fisheries. A growing number of multinational companies have taken increasingly active roles in promoting sustainable seafood, including: retailers such as Walmart and Tesco; the leading tuna processors, through the International Seafood Sustainability Foundation; and 10 of the largest seafood companies, through Seafood Business for Ocean Stewardship (SeaBOS).

In recent years, growing consumer concern over fish provenance, coupled with corporate interest in supply chain control, have sparked significant momentum towards supply chain traceability. In 2017, 66 companies signed the Tuna 2020 Traceability Declaration, pledging that all tuna they buy will be completely traceable by 2020. More than 30 major companies, including SeaBOS, have signed up to the Global Dialogue on Seafood Traceability, specifying the key data elements to be collected in their supply chains and creating standards for IT platforms to ensure interoperability.

Many are looking to blockchain and other distributed ledger technologies to support supply chain traceability. As noted above, however, these systems depend on the reliability of the data on the provenance of goods entering the system, and therefore depend on the market creating strong incentives for driving transparency through far-flung supply chains (Hardt et al. 2017). Emerging technology offers the prospect of increasingly robust transparency—providing near real-time information on where boats are fishing and what they are catching—and traceability from the moment of catch to the supermarket shelf.

Publicly available vessel tracking data are now being used to track larger vessels (Kroodsma et al. 2018). As more countries share the more granular data they already collect, and as satellite surveillance capabilities expand, a much larger proportion of the global fishing fleet will be tracked. Global Fishing Watch, for example, aims to expand its coverage from 60,000 vessels today to 300,000 by 2029. Continued progress in developing AI and ML tools to process data from video monitors and satellites will also expand the ability to monitor fishing activity. This growing transparency will be matched by continued improvements in traceability, through genetic tools, sensors and electronic tags or QR codes that can be used to track fish through supply chains and verify source and species.

These data systems have the potential to enable buyers, such as processors and retailers, to ensure that the fish they buy is legal and meets their environmental and social standards. Providing actionable information at the moment of the decision may also spur sustainable choices on the part of consumers. Apps at the point of sale can display these data for consumers, showing them where fish is caught and how it has been processed and shipped. Allowing consumers access to data on whether fish have been illegally caught or are contaminated with mercury or microplastics, for instance, could inspire more informed decisions.

Historically, fishers have closely guarded information about where they are fishing. High-level information on the most productive areas has been available for decades, however, leading to the globalisation of effort by the major fishing nations (McCauley et al. 2018). Global Fishing Watch and other platforms protect more granular information on where vessels are moving in response to daily fluctuations in fish stocks by placing a 72-h delay on the release of vessel location.

Stringent transparency and traceability requirements can make it harder for small-scale fishers to sell into global supply chains. The cost of vessel tracking systems is already out of reach for most small fisheries. Low- cost traceability apps built on smartphones provide a promising option for these small-scale fishers, but companies will need to accommodate these types of solutions in their traceability systems. Agreement on global standards, like the Global Dialogue on Seafood Traceability, can also facilitate the development of tools.

As technology continues to improve and leaders in the seafood industry act on their commitments, there is the clear prospect that full transparency and traceability will become the expectation of the marketplace and the cost of doing business, and usher in a new era of accountability.

4.3 Ensuring That Technology Promotes Sustainability

Over the course of history, advances in technology have generally led to increased exploitation of ocean resources—more powerful boats and fishing gear have transformed fishing from a coastal activity to a global industry and driven many fish stocks into decline; deepwater platforms and drilling innovations have enabled massive extraction of oil resources and soon, possibly, minerals on the seafloor. The rapidly expanding capabilities in information technology described above could similarly accelerate exploitation—helping fishers track down every last fish, for example. These new capabilities thus come with two imperatives. The first is management—as the ability to exploit resources expands, effective management of those resources will be ever more vital. The second is accountability—information on resource conditions and use must be public, so that users of public resources are accountable to governments, to markets and to the public.

To realise the potential of new technology to support sustainability, it will be essential that these new capabilities are available not only to well-funded governments, companies and institutions, but also to governments and communities with more limited means. This requires both that ocean data are widely accessible and that the hardware and software to access those data are available and affordable. Low-cost technologies based on smartphone capabilities are one promising avenue, taking advantage of the increasing ubiquity of smartphones to allow both access to global information and the generation of locally relevant data. This can enable better management and increased accountability, and facilitate access to global markets. However, capacity-building is needed to ensure that the physical and intellectual infrastructure exists to support these advances in all areas of the globe.

In this report, we have focused principally on the explosion in new data on ocean health, resources and resource use—from new sensors and other sources—and the increasingly powerful technologies for extractin action. Advances in genetics and biotechnology mean that those fields also have great potential to play a central role in sustaining ocean resources. Research on the genetics of coral, for example, is helping scientists identify species that are more resilient to heatwaves, and thus better equipped to thrive in a warming ocean. Researchers have developed new microbes that can break down plastics in the ocean or oil from oil spills.

Biotech may also have a role in mitigating the environmental impacts of aquaculture, including: the destruction of coastal habitats to build fish farms; pollution from the use of pesticides and antibiotics; and a massive increase in demand for fishmeal and fish oil, harvested from wild stocks, to use in feed. New strains of fish, bred to be resistant to disease can reduce the need for antibiotics. New plant-based feeds are reducing the need for fishmeal and fish oil.

Gene drives can eliminate invasive species and restore ecosystems by introducing altered genes that promote the inheritance of a certain genetic variant (in the case of invasive species, often a variant that makes organisms infertile) (Esvelt and Gemmell 2017). These solutions have the potential to eliminate invasive species populations that have wreaked havoc on ecosystems and been nearly impossible to control using conventional methods. However, introducing altered genes is akin to introducing another invasive species into an ecosystem—one that can invade any viable population with consequences beyond what we are capable of predicting.

Some innovators are now aiming to reduce overfishing by producing seafood without relying on fish. Companies, such as Finless Foods, Wild Type and BlueNalu, are cultivating tuna, shrimp and other seafood in laboratories. Cultured seafood has the potential to protect wild fish stocks while having a significantly lower overall environmental footprint and a reduced risk of contamination (a major problem in high trophic-level fish species due to the bioaccumulation of mercury and other heavy metals in wild populations) (Stephens et al. 2018).

5 Fostering Technological Innovations for the Ocean

Sustainable use of the ocean will require new technologies for researchers, managers, resource users, coastal communities, companies, consumers and others who have a stake and a role in ocean stewardship. Technologies that are important for ocean stewardship typically face significant barriers, however—debilitating start-up capital costs, regulatory constraints and lack of clear revenue streams (OECD 2019). Technological innovation in the ocean has therefore been largely driven by government and large-scale commercial interests. For some other needs, such as scientific instrumentation, small markets have often led to hyper-specific solutions that lack commercial applicability, creating an environment of technology lock-in. Many needs are simply unserved.

Overcoming these market barriers is critical to fostering successful innovation that supports science and management in the future. The landscape of innovation is complex. To capture the extraordinary potential of technology to enable ocean stewardship will require action by governments and others to create market incentives for innovation, as well as new public–private instruments for investment and new business models.

5.1 Creating Market Incentives for Innovation and Diffusion

Both governments and private actors have critical roles to play in incentivising the technological innovations that will be needed to safeguard the health and sustainable use of the ocean.

5.1.1 Governments

The history of environmental policy has shown that strong, technology-forcing regulations drive innovation. Regulations that place limits on pollution, such as automobile or power-plant emissions, for example, have repeatedly spurred technological innovation by industry to lower the cost of reducing emissions. In the same way, the International Convention for the Prevention of Pollution from Ships (MARPOL) has incentivised innovation across that sector. In addition, the recent International Maritime Organization mandate requiring the global shipping fleet to halve its greenhouse gas emissions by 2050 has already spurred major technological advances in vessel propulsion, creating the prospect that zero-emission vessels may enter into service by 2030. Similarly, government requirements for monitoring and safety provisions on vessels have created markets for technologies that enable companies to achieve and demonstrate compliance.

Government regulation can also be vital in driving the diffusion of new technologies into large-scale application. In recent years, for example, there have been many innovations that could significantly reduce bycatch in fisheries, but many have not been widely implemented. Stronger government restrictions on bycatch could quickly drive the widespread adoption of those solutions.

The ocean is a patchwork of regulatory jurisdictions, but experience in other sectors demonstrates that actions by individual authorities can nonetheless drive progress. Measures to promote the use of solar energy in Germany and a few other jurisdictions spurred massive innovation in that sector globally, for example. The US mandate that required shrimp catchers to use turtle exclusion devices (TEDs) led to global adoption and innovation in TEDs (Yaninek 1995). Individual governments can incentivise innovation in the ocean by adopting forward-looking technology-forcing regulations, without waiting for international action.

Specifically, governments should prioritise forward- looking technology-forcing regulations that target real- time monitoring of fishing, shipping emissions, mineral development, coastal development and pollution, and that create public accountability. Some technology solutions already exist in these areas. Government could radically increase innovation by building on these tools. In the case of fisheries, mandates by major seafood-catching countries (such as European Union countries, the United States and Japan) that all vessels use electronic monitoring, for example, could spur a wave of innovation, speeding up the translation of existing AI expertise from the technology sector to ocean management.

Governments can also drive innovation in less direct ways. The barriers to innovation are often information gaps: the technology community is unaware of the specific problems that managers need to solve, while managers do not have the technical expertise to know what solutions are possible. By bringing together managers and technology companies, governments can catalyse the development of innovative management tools that use readily available resources. For example, in the Caribbean, MPA managers and technology experts worked together to develop low-cost acoustic sensors that are being used, together with smartphones, to detect vessel activity in areas that are off limits to boats. When the sensors detect an acoustic signature, the mobile phones are programmed to send a text to local enforcement agencies, allowing effective, low-cost enforcement of MPAs.

Creating a national account for the ocean can make the economic benefits of innovation in the ocean clear. Current GDP-based models of national accounting do not effectively capture these benefits, and as a result ocean innovation is often undervalued. Using a suite of indicators to understand ocean production, income and sustainability can spur economic investment, innovation and stewardship (see Blue Paper 8, "National Accounting for the Ocean & Ocean Economy").

Trade and import controls extend a government's influence beyond its own territory. Requirements to ensure that imported products were legally produced or comply with labour or environmental standards spur innovations to create transparency and traceability in supply chains. The US Lacey Act, for example, has required importers to demonstrate compliance with the laws of producing countries. Under the EU 2008 IUU fishing regulation, the European Commission has blocked imports from countries with inadequate controls on illegal seafood products, and has issued "yellow cards" to others as a warning that imports will be blocked unless stronger measures are put in place.

5.1.2 Private Sector

Crucially, private sector action can often play a similar role in creating market incentives for innovation. Over the past two decades, many global companies have begun to address issues of environmental impacts and labour conditions in their businesses and in the far reaches of their supply chains. The Sustainable Seafood Movement, described above, is a leading example. The Global Plastic Action Partnership is another. As companies drive changes in their own operations and raise standards for their suppliers, they create opportunities for innovators to develop technologies that can improve environmental performance or provide greater accountability and sustainability across supply chains.

Commitments by companies to transparency and traceability in their supply chains illustrate the potential. Companies are beginning to capitalise on the rapidly expanding capabilities for monitoring activities on the ocean— through remote sensing, for example, and video or other monitoring on board vessels and in the water—in order to gain greater visibility and stronger accountability across their businesses. In this way, they can drive both improvements in technology and reductions in cost, and these capabilities will then become increasingly available to less-developed markets. Similarly, growing corporate interest in traceability spawns new solutions, such as the recently launched blockchain platform OpenSC. Tech innovators partnering with NGOs and big seafood companies can extend that capability to small-scale fisheries, as Fishcoin is now pioneering, using blockchain to compensate fishers for the collection of key data on their fishing and enabling traceability.

5.1.3 International Standards

Finally, both governments and the private sector can play important roles in setting the standards for technology that enable a fertile ecosystem for innovation. There are many examples of past collaborative efforts between the private sector, governments and academia to create new standards, but the Internet is one of the most useful examples (Abbate 1999). In this case, a government agency (the US Defense Advanced Research Projects Agency—DARPA) worked with a small number of academic researchers to create the basic structures of the Transmission Control Protocol/ Internet Protocol (TCP/IP) to serve ARPAnet, the forerunner of the Internet. TCP/IP was then widely adopted by the Internet community as a result of a DARPA mandate to all of its contractors to use ARPAnet. The initial standard- setting by government, and the subsequent buy-in by the private sector, was successful in launching a standardised Internet platform and unleashing a wave of innovation.

International agreements can also play a role in creating global market demand for new technological innovation for the ocean. The Port State Measures Agreement (PSMA) by the Food and Agriculture Organization of the United Nations (FAO), for example, creates new requirements for port monitoring and control that are applied globally and that will require technological innovation in data collection and sharing to achieve. Agreements like PSMA also often include goals for technology transfer and capacity-building that commit governments to ensuring that developing countries have the same access to promising management solutions (Harden-Davies 2017).

5.2 Mobilising Investment

The current landscape of ocean innovation is centred in highly capitalised private sector industries, such as oil and gas, industrial fishing and shipping, and government-funded defence departments. This has been the case for the past century, and consequently many of the technologies now used by scientists and managers were developed under government defence contracts or for marine industrial use. Examples of this include many deep-sea submersibles and autonomous vehicles, with technological underpinnings pioneered by defence departments before being adopted by scientists. Similarly, innovations in the oil, gas and fishing industry that allow companies to work on submerged infrastructure or increase detection abilities of fish schools have been widely adopted beyond these industries. As with government defence efforts, these profitable industries are able to support significant research and development (R&D) expenses beyond what is generally feasible for marine researchers or managers.

This model has been successful in many ways. Capitalising on the market power of industry and government to develop technological solutions for the ocean has allowed scientists and managers to take advantage of innovation without high capital expenditures. The government model of investing in early-stage technologies has led to important advances. This happens both with investments through R&D programmes as well as through direct investment in the innovations needed for government purposes, particularly the defence industry. Both of these avenues have yielded critical marine innovations without which managers would have significantly less technological capacity than they do today.

Relying on the trickle-down of commercial and defence technologies is not sufficient to fill the needs of marine managers and other ocean stakeholders. For instance, gaps in information about marine ecosystems that are not commercially valuable may not be filled by technologies aimed at efficient oil extraction or target detection. The development of technologies to fill these gaps lags behind those incentivised by the strong market forces of industry.

Overall, environmental innovations have been notoriously underrepresented in the new wave of technological innovation. In the United States, for example, total federal expenditure on R&D is about US $125 billion. Of these expenditures, the amount spent on space flight and space research is about $10 billion; less than $2 billion is spent on the ocean sciences. Moreover, in the United States and elsewhere, government funding tends to go to early-stage research and dries up in later stages of development (OECD 2019).

In recent years, private investment has expanded beyond traditional marine industry R&D, with venture capital funding and start-up accelerators focused on ocean innovation. These avenues lag far behind the funding available in other industries, such as energy and healthcare, but provide potential avenues for scaling up technology solutions with strong business models.

Several specialised technology accelerators focused on the ocean are providing early-stage funding to innovative technologies that advance the sustainable use and management of marine resources (e.g. Katapult Ocean and the Sustainable Ocean Alliance). The start- ups funded are tackling issues ranging from seafood traceability to the development of bioplastics and wave energy. These solutions present important steps towards solving ocean issues in cases where innovation offers the potential for strong market returns.

Large prizes are also incentivising ocean technology innovation. These prizes are funded by a mix of individuals, companies and large foundations. XPRIZE, for instance, has been successful in incentivising the development of breakthrough technologies such as private spaceflight and autonomous ocean mapping robots. While these prizes have spurred important progress technologically, there are significant concerns about whether these developments will be able to scale given current market constraints (Kremer and Williams 2015).

Considerable academic research has been devoted to identifying the driving frameworks for innovation. These frameworks are complex, adaptive systems that rely on the participation of a wide range of actors, including public, private and research institutions. Other sectors provide a roadmap for what this ecosystem could look like. Agriculture has faced many of the same problems in technology innovation and adoption as the ocean has, including a fragmented producer landscape, lack of technology incubation support and high resource investment requirements. Partnerships that bring together a mix of institutions, from private sector investment to government incubators and philanthropic efforts, are able to overcome many of these barriers (WEF 2018).

For the ocean, the Organisation for Economic Co- operation and Development (OECD) has specifically recommended bringing together a diverse group of actors to spur innovation in "ocean economy innovation networks" (OECD 2019). These networks provide many potential benefits by leveraging complementary innovations at different points in the innovation stack and by providing technology transfer to developing countries. These multisector approaches are more likely to foster complementary innovation that increases the potential impact and uptake of new technologies. By combining multiple technologies in layered systems, the impact of technologies can be exponentially increased (OECD 2019). For example, innovation ecosystems that allow developments in sensor processing to happen in parallel with new communication and platform tools both unlocks unique collaboration but also ensures that emerging technologies are plugged into larger ecosystems of innovation.

Technology clusters such as those recommended by the OECD have already been successful in moving innovation in ocean industries from early, government- funded stages to thriving multi-commercial markets. The Norwegian Centres of Expertise Maritime CleanTech cluster, for example, has been pivotal in driving the adoption of clean energy innovations in cruise and ferry lines. By creating a platform for collaboration between emerging players innovating in the clean energy space, established industry, and government and academic researchers, this cluster drove the development of the first fully electric car ferry, among other innovations in zero-emission and hybrid vessels. Moving forward, similar blue technology innovation clusters should be created to help emerging technology solutions achieve adoption and market penetration. On the other hand, although there is a plethora of these clusters, many of them have struggled to achieve sufficient momentum to be self-sustaining. In these early days, it is essential that governments focus on enabling market demand as well as market supply. Too often these innovation clusters rely solely on a "build it and they will come" model. Creating partnerships between market pull and market push is a role that government should be encouraged to perform.

Box 6.2: Case Study: Creating New Market Opportunities

In Japan, the declining number of operational fishing boats together with the declining number of fishers—due to ageing and other factors—is emerging as an important issue, especially for sustaining the exploration-type fisheries on the coast and offshore. For this reason, it has become more difficult to search for fishing grounds, and the fishers are forced to continue with their inefficient fishing operations. One of the solutions for bringing back the efficiency in the operation is to deliver highly accurate information about fishing grounds to reduce fuel consumption. With that aim, JAMSTEC started research and development on the advancement of fishery forecasting technology for squid, which is one of the most important species for the fisheries of Aomori Prefecture. The outcome was a squid fishing ground forecasting system that provided fishing ground information in real time, and was so successful with fishers that it was transferred to the private sector for routine operational distribution of the information.

5.3 Creating New Business Models

Beyond investment and regulation, innovation in business models can also create new ways to make the economics work to support data access and collection by marine managers and other stakeholders. There are also opportunities to further exploit existing market opportunities that are currently underdeveloped. Research in energy and other markets has shown that the pace of innovation is highly related not only to public investment in R&D, but also to market growth (Bettencourt et al. 2013).

The provision of ocean data by governments is viewed as an important public good, but the costs associated with this can be significant. In addition to the direct economic costs, additional indirect costs of open data include the potential subsidy of private sector activities and the creation of inroads for corporate influence, and the need to be considered in relation to the purpose and potential benefits of open access data (Johnson et al. 2017). For ocean and environmental data, several models exist to help support research and management databases.

Most existing research databases rely on public funding, from governments, universities or other research institutions, with a minority also generating revenue through use and access fees (OECD 2017).

The cost of storing large quantities of data can be prohibitively high. Several creative solutions exist though. NOAA, for instance, reached an agreement with Amazon Web Services (AWS) for storage of key ocean data. Having NOAA data on AWS servers brought data significantly closer to the computation needed to support key knowledge services—for example, weather forecasting—and drove traffic to AWS (Barr 2015). In return, NOAA was able to store petabytes of data on the AWS servers at no cost to the taxpayer.

Innovation in business models can create solutions that are able to meet both management and industry needs. Several approaches are showing promise.

5.3.1 Segmentation

Existing commercial markets for satellite data, for instance, are strong. Many new companies, such as Planet Labs and others, provide slightly degraded data free of cost to researchers. The cost of collecting these data is borne by the commercial entities paying for the data, and the degraded data are of sufficiently high quality to support research use. These secondary markets are important opportunities for ocean management and other uses.

5.3.2 Data Services

Data networks can be supported by the knowledge products built using them. Already, ocean and climate data are being used as the basis for complex insurance decisions, targeted weather forecasts for precision agriculture, and other lucrative knowledge products. Companies like Descartes Labs and others have been successful in this model (Jensen and Campbell 2018). These "data as a service" models can also create opportunities to sustainably support research databases over time (OECD 2017).

Markets for data and knowledge services can also support new innovations for gathering data. Low-cost and distributed sensor systems that are able to gather data at very high resolutions, which directly support commercially valuable knowledge outcomes, for example, have clear market use.

5.3.3 Innovations in Payment

Innovations in payment can drive data collection and traceability throughout the supply chain. Fishcoin, described above, is one example—paying fishers for their data with mobile-phone minutes. Other blockchain- based solutions in agriculture show promise in linking consumers directly to small-scale producers, allowing consumers to directly pay small-scale farmers, for instance, that use desired production techniques. Coupling these payment innovations with new data services can allow citizens to participate more directly in environmental conservation. In China, a tree planting app that allows citizens to donate money to reforestation efforts and then track their growth over time using satellite imagery has already planted over 13 million trees (Thompson 2019).

6 Opportunities for Action

We are poised on the threshold of a digital ocean. To realise that vision, and to enable a flowering of new capabilities to understand and steward ocean resources, governments, companies, researchers and civil society must each do their part. There are six critical steps:

1. **Capitalise on the UN Decade of Ocean Science for Sustainable Development to create a global data network that provides broad and automated access to ocean data.**

   Vast stores of ocean data currently in the hands of governments, researchers, industry and others can be made available to all through data tagging, federated networks and, where possible, data lakes.

   (a) UNESCO should build on existing efforts to establish global standards for metadata, query and data tagging that allow existing datasets to be interconnected and automatically accessed.

   (b) Governments, industry and research institutions should use those standards to make their data broadly available in a global federated data network.

   (c) Data holders and cloud service providers should collaborate to create data lakes within that network to facilitate access to large scientific datasets and enable development of new data services.

   (d) Investment in capacity-building should ensure that these data are available, useful and affordable to all ocean users.

2. **Liberate ocean data.**

   Enabled by federated networks, data holders should establish a new default—that ocean data are broadly available to other users unless there are compelling security, proprietary or other interests.

   (a) Governments should:

   - provide public access to all data collected by defence and security agencies that can be shared without compromising security interests;
   - mandate use of AIS and share essential data on fisheries, including vessel ownership, licences and tracking for all fishing vessels; and
   - require that any user of ocean resources, such as fisheries, minerals or coastal land, is required to make their environmental data available to the public.

(b) Industry should make the environmental data they collect accessible to scientists, managers and the public.
(c) Scientific researchers should, by default, make their data available to all.

3. **Create an "Internet of Things" for the ocean.**

Coordinated efforts by industry, researchers and governments can create advanced sensor networks that provide high-resolution, real-time information about the ocean to anyone who needs it.

(a) Governments should develop new open standards for underwater communications and positioning.
(b) The private sector should work with governments and researchers to ensure that sensors are interoperable and data are generated in standardised formats.
(c) Security and privacy standards need to be developed for terrestrial IoT systems, and these should be adopted for marine IoT systems as well.

4. **Automate ocean management based on near real-time data on ocean conditions and resource use.**

(a) Governments should expand use of dynamic management and, where possible, automate management with smart contracts. These solutions are particularly promising in fisheries management, where stock limits, fishing areas and allowable gear types can be automatically updated based on changing conditions.
(b) Governments should automate mitigation measures to create immediate responses to acute environmental threats, from storms to heatwaves to nutrient fluxes. Forecasts that show impending harmful algal blooms or storms, for instance, could automatically trigger reductions in fertiliser application and increased storm water treatment to proactively protect ecosystems.
(c) Governments and companies should collaborate to create mechanisms for data-based proof of compliance. A voluntary "global entry" system for fishing vessels, for instance, could allow expedited access to ports for vessels that provide information on their ownership, permits and activities to managers—creating incentives for transparency and compliance.

5. **Create incentives for innovation.**

Existing markets do not incentivise many of the technological innovations that are needed for ocean stewardship and research. Governments and companies can change that.

(a) In regulating ocean activities, governments should design regulations to spur innovations that will enable more effective management, such as requiring real-time monitoring of fishing, shipping emissions, mineral development, coastal development and pollution.
(b) Companies should require full transparency and traceability in their operations and supply chains—to spur both better management of resources and innovation in technology, and enable consumers to hold producers accountable and reward better management.
(c) Governments should partner with the private sector to create innovation clusters in areas of market demand that support cross-sectoral collaboration and link emerging technology research and innovation with established industry players.
(d) Governments and companies should support innovative business models that combine commercial viability with support for management, such as governments and large companies who are buying data from, for example, private satellite and drone providers, making that data available in delayed or slightly degraded form for research and management uses.

6. **Mobilise capital for technologies for under-served markets.**

Many markets for ocean technologies do not offer commercial returns. We thus need innovative financial instruments that can leverage the different expectations and risk tolerances of different investors. Governments, philanthropies and private investors should join forces to:

(a) create blended finance facilities that combine risk reduction, impact capital and market capital; and
(b) invest in the development of low-capital technologies and training for developing countries, coastal communities, citizens and consumers to conserve, manage and sustainably use ocean resources.

Acknowledgements The authors thank Annie Brett for her extensive contributions to the research and writing of this report. The paper's technical reviewers, Zainal Arafin, Geir Huse and Sergio Lillo, as well as its arbiter, Peter Haugan, all provided helpful technical comments. The authors also thank the World Resources Institute for providing support as the Ocean Panel Secretariat.

While our colleagues were very generous with their time and input, this report reflects the views of the authors alone. The authors thank Sarah Chatwin for copyediting and Shannon Collins for design.

Appendix 1: Case Studies of Technology Deployment by JAMSTEC

The ocean, seas and coastal zones have diverse and vibrant ecosystems as well as other resources vital for the sustenance of human lives on Earth. In the spirit of sustainable management of these resources, scientists at the Japan Agency for Marine-Earth Science and Technology (JAMSTEC) have conducted a few pilot studies. In a specific case study, high-frequency (HF) radar data were applied to understand the relationship between the sea state and the small Pacific bluefin tuna (<30 kg) catch by the setnet. JAMSTEC has been observing the spatial distribution of surface current velocity in the eastern Tsugaru Strait and the surrounding area since 2014 with an HF ocean radar system (Fig. 6.2).

The observations are acquired in quasi real time, every 30 min, and are posted immediately (usually within 1 h) as a surface current map on the JAMSTEC website (http://www.godac.jamstec.go.jp/morsets/e/top/). The data are publicly available and the maps can be accessed by desktop or mobile/smartphone devices. An analysis of the website's access logs suggests that the fishers working in this area might be the main users of this website.

In the fall of 2017, mass bycatch of small tunas was registered by just two setnets near the HF radar measurement area. The surface current pattern observed by the HF radar at the time indicated a typical current pattern in this area. The catches of such small tunas are strictly restricted to maintain the stocks of the prized fish. Based on this pilot study, the local current data along the coast from the HF radar are now routinely used for safely releasing small tunas from the setnets. For example, when a surface current pattern similar to 2017 was observed in August 2018, a researcher at a local fisheries research institute, Hakodate Research Center for Fisheries and Oceans, immediately alerted the local fishers of the potential risk of young tuna entering their setnets in large numbers. The setnet fishers could, therefore, prepare themselves for releasing the young tunas based on the alert.

JAMSTEC researchers also try to apply the numerical simulation techniques to fisheries using a super- computer. The declining number of operational fishing boats together with the declining number of fishers—due to ageing and other factors—is emerging as an important issue, especially for sustaining the exploration-type fisheries on the coast and offshore. For this reason, it has become more difficult to search for fishing grounds, and the fishers are forced to continue with their inefficient fishing operations. One of the solutions for bringing back the efficiency into the operation is to deliver highly accurate information about fishing grounds to reduce fuel consumption. With that aim, in the financial year 2010, JAMSTEC started research and development on the advancement of fishery forecasting technology for squid, which is one of the most important species for the fisheries of Aomori Prefecture. In this research, JAMSTEC developed a squid fishing ground forecasting system and provided fishing ground information in real time. JAMSTEC conducted a demonstration experiment to deliver ocean forecasts to fishers through a web-based system. An ocean circulation forecast was conducted every week for two fishing seasons (June–August and January– March), a mathematical model was applied to estimate the fishing ground based on a statistical relationship between the ocean environment and the fishing ground and catches, and the results were provided to fishers through our website (Fig. 6.3). In addition, fishing ground positions and fish catches reported by fishers in real time every day were used to fine-tune the model to reproduce the information in our predictions. This demonstration experiment made us realise that there is a strong aspiration from fishers to continuously receive fishing ground forecast information in real time. In order to meet the operational demand in real time and to maintain sustainable

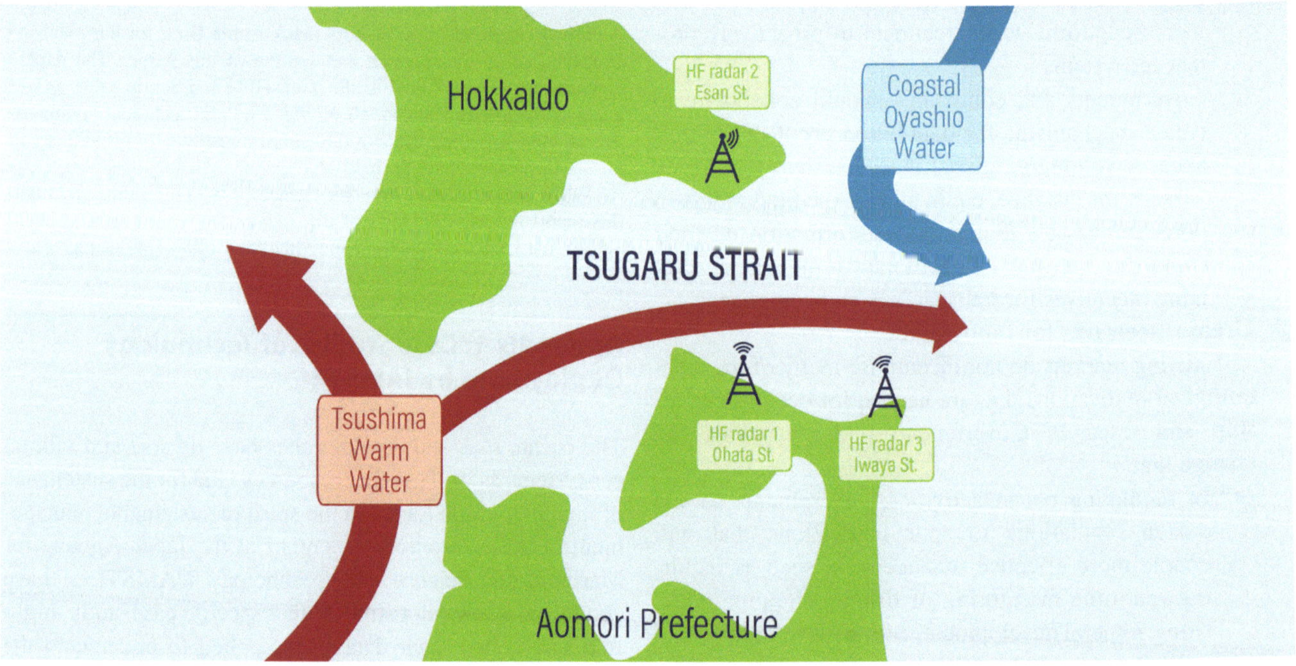

Fig. 6.2 Locations of the HF radars. *Source: Mutsu Institute for Oceanography (MIO)/RIGC/JAMSTEC 2019*

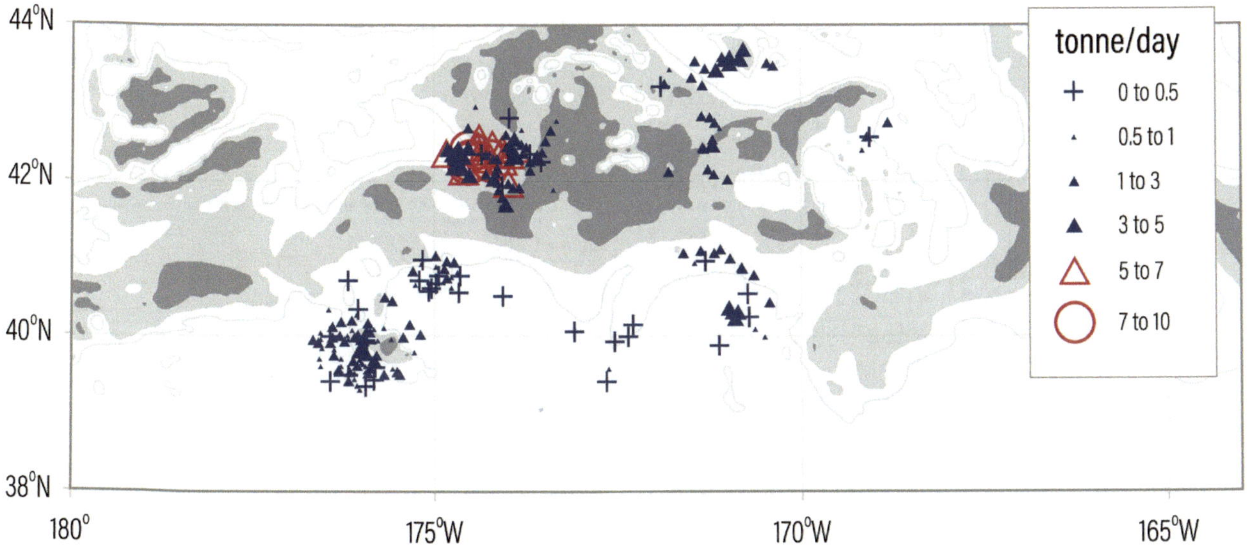

Fig. 6.3 Map of the potential fishing ground on 20 July 2012. *Notes: The potential fishing ground and fishing points in the central North Pacific from 38 degree North to 44 degree North in latitude, from 164 degree West to Dateline in longitude. The fishing point and the amount reported by fishing vessels are denoted by symbols (plus, triangle and circle). The potential fishing ground is shown as the Habitat Suitability Index (HSI, contours), which is normalised between 0 and 1. The contour interval is 0.2. Light grey shading indicates HSI values over 0.6 and dark grey shading indicates those over 0.8. (Source: Information Engineering Program (IEP)/VAiG/JAMSTEC*

fishing, the developed technologies were transferred to the private sector for routine operational distribution of the information.

JAMSTEC is also operating a set of ocean state forecasting models on a super-computer targeting a wide range of spatio-temporal scales from global/seasonal to nearshore/hourly for various other marine applications. The seasonal forecast aims at representing the effects of global climate modes, which are important for seasonal forecasts of basin-scale sea surface temperature variations, obtained from several atmosphere–ocean coupled model forecasts. Nowcast/forecast operations of the ocean currents and the mesoscale eddies are performed by high-resolution ocean circulation models driven by atmospheric weather forecasting products. A main target region of the ocean current forecast is the North Western Pacific around Japan. Detailed behaviours of the major ocean currents, including the Kuroshio/ Oyashio path variations, are predicted every day, and the resulting information is provided to shipping companies for planning optimal ship routes and safe navigation. In addition, currents in some of the targeted areas are highly resolved by utilising downscaling techniques. Figure 6.4 shows an example of downscaling applied to Sukumo Bay, which is located in the Shikoku region of the western part of Japan. The local ocean currents in the bay are forecast every day with a 200-m resolution, and the forecast information is directly provided to the local fishers for their use (In Japanese) (http://www.jamstec.go.jp/jcope/vwp/sukumo500/).

JAMSTEC has from time to time held meetings with the fishers of the area, to exchange views and to explain the coastal environment based on our research results. Based on the outcome of these discussions on such occasions, it became apparent that local fishers wish to stabilise their profit rather than maximise the catch; in other words, they wish to ensure production consistency. More specifically, some of their desires are to:

- reduce the number of days with no catch, which would prevent wasting fuel;
- avoid extreme over-catch to avoid the fall in prices; and
- avoid catching juveniles to increase cost- effectiveness.

All these desires are key to sustainable fishery and it is very impressive that fishers have already recognised them through personal experience. To achieve such a sustainable direction, fishers have requested that JAMSTEC provide the following information and data in real time for their operations:

- Three-dimensional distributions of temperature, currents and current-rip from a few hours ahead for coastal fishers to a few days ahead for offshore fishers, to avoid going to an unsuitable area for fishing
- Positions of "hot spots" of specific fish species to avoid over-fishing
- Details of spawning grounds and juvenile habitats to avoid fishing there

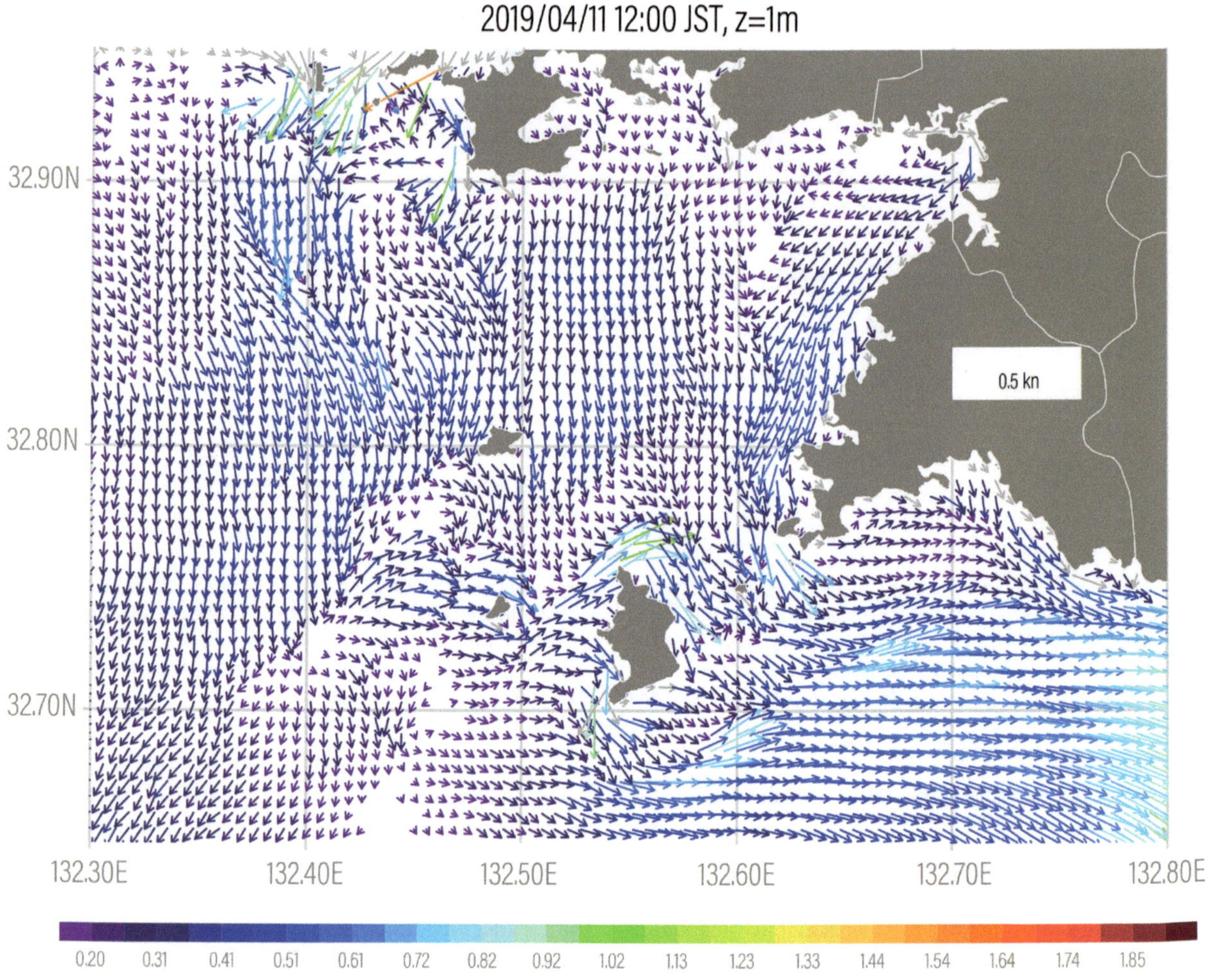

Fig. 6.4 Map of the potential fishing ground on 20 July 2012. *Notes: Arrows and colours indicate direction and magnitude (in knot) of surface ocean currents, respectively. The figure shows surface current. Source: APL/VAiG/JAMSTEC*

JAMSTEC is now at the initial stage of such R&D to meet these requirements and hopes to provide those data and information to the fishers in the near future. In spite of the enormous scientific and technical challenges, research towards such a sustainable goal should be one of the most important missions for science and societal well-being. Therefore, JAMSTEC researchers are now exploring the possibility of forecasting surface current velocity several hours ahead in the Tsugaru Strait by harmonic and pattern analyses as the first step to respond to the requests of local fishers.

A more comprehensive real-time data acquisition system from wider areas of the ocean, as well as advanced simulation models, is required to produce practically useful forecasts. In order to realise such a system, the development of lightweight automated observational instruments (sufficiently easy to use that they can be mounted on fishing boats) and the improvement of technology in data aggregation, processing, large-scale high-speed computation and information distribution services are indispensable. Furthermore, there is a scope to develop overseas non-commercial and commercial applications in the future after domestic operationalisation of the system and its nationwide adoption.

About the Authors

Lead Authors

Jim Leape is William and Eva Price Senior Fellow and Co-Director of the Center for Ocean Solutions at the Stanford Woods Institute for the Environment.

Mark R. Abbott is President and Director of Woods Hole Oceanographic Institution.

Hide Sakaguchi is Executive Director at Japan Agency for Marine-Earth Science and Technology (JAMSTEC).

Contributing Authors

Annie Brett is an André Hoffmann Fellow at Stanford Center for Ocean Solutions and the World Economic Forum Centre for the Fourth Industrial Revolution.

Ling Cao is Associate Professor in Marine Habitat Conservation and Fisheries at the Shanghai Jiao Tong University (China).

Kevin Chand is a Legal Fellow at the Stanford Center for Ocean Solutions and a Lecturer at Stanford University.

Yimnang Golbuu is CEO at the Palau International Coral Reef Center.

Tara Martin is a Research Group Leader in the Oceans and Atmosphere Business Unit of the Commonwealth Scientific and Industrial Research Organisation.

Juan Mayorga is a researcher with the Sustainable Fisheries Group at UC Santa Barbara, and with National Geographic Society's Pristine Seas project.

Mari S. Myksvoll is a senior researcher at the Institute of Marine Research (Norway).

References

Abbate J (1999) Inventing the internet (inside technology). MIT Press, Cambridge, MA

Abbott MR, Sears CE (2006) The always-connected world and its impacts on ocean research. Oceanography 19:14–21

Amodei D, Olah C, Steinhardt J, Christiano P, Schulman J, Mané D (2016) Concrete problems in AI safety. http://arxiv.org/abs/1606.06565

Atzori M (2015) Blockchain technology and decentralized governance: is the state still necessary? https://papers.ssrn.com/sol3/papers.cfm?ab-stract_id=2709713

Barr J (2015) Announcing the NOAA big data project. AWS Blog, 21 Apr. https://aws.amazon.com/blogs/aws/announcing-the-noaa-big-data-project

Bar-Sinai M, Sweeney L, Crosas M (2016) DataTags, data handling policy spaces and the tags language. In: 2016 IEEE Security and Privacy Workshops (SPW). IEEE, pp 1–8. https://doi.org/10.1109/SPW.2016.11

Bayler E (2019) NOAA research and development plan. Feb 27. https://sab.noaa.gov/LinkClick.aspx?fileticket=wcRxtbwANj0%3D&por-talid=10&timestamp=1550757834232

Berkes F, Colding J, Folke C (2010) Rediscovery of traditional ecological knowledge as adaptive management. Ecol Appl 10(5):1251 1262

Bettencourt LMA, Trancik JE, Kaur J (2013) Determinants of the pace of global innovation in energy technologies. PLoS One 8(10):e67864. https://doi.org/10.1371/journal.pone.0067864

Borowitz M (2019) Government data, commercial cloud: will public access suffer? Science 363(6427):588–589. https://doi.org/10.1126/science.aat5474

Bradley D, Merrifield M, Miller KM, Lomonico S, Wilson JR, Gleason MG (2019) Opportunities to improve fisheries management through innovative technology and advanced data systems. Fish Fish 20(3):564–583. https://doi.org/10.1111/faf.12361

Bromley DW (2016) Rights-based fisheries and contested claims of ownership: some necessary clarifications. Mar Policy 72:231–236. https://doi.org/10.1016/j.marpol.2016.07.002

Buck JJH, Bainbridge SJ, Burger EF, Kraberg AC, Casari M, Casey KS, Darroch L et al (2019) Ocean data product integration through innovation – the next level of data interoperability. Front Mar Sci 6:32. https://doi.org/10.3389/fmars.2019.00032

Carothers C (2015) Equity and access to fishing rights: exploring the community quota program in the Gulf of Alaska. Hum Organ 70(3):213–223. https://doi.org/10.17730/humo.70.3.d686u2r7j2267055

Cater NE, O'Reilly T (2009) Promoting interoperable ocean sensors the smart ocean sensors consortium. In: Oceans 2009, vols. 1–3. p. 214–19

Cokelet ED, Meinig C, Lawrence-Slavas N, Stabeno PJ, Mordy CW, Tabisola H, Jenkins R, Cross J (2015) The use of Saildrones to examine spring conditions in the Bering Sea: instrument comparisons. Sea Ice Meltwater and Yukon River Plume Studies

Coveney PV, Dougherty ER, Highfield RR (2016) Big data need big theory too. Philos Trans R Soc A Math Phys Eng Sci 374:20160153. https://doi.org/10.1098/rsta.2016.0153

Crona B, Gelcich S, Bodin Ö (2017) The importance of interplay between leadership and social capital in shaping outcomes of rights-based fisheries governance. World Dev 91:70–83. https://doi.org/10.1016/j.worlddev.2016.10.006

Crosas M, King G, Honaker J, Sweeney L (2015) Automating open science for big data. Ann Am Acad Pol Soc Sci 659(1):260–273. http://datatags.org/publications/automat-ing-open-science-big-data

Cvitanovic C, Hobday AJ, Van Kerkhoff L, Marshall NA (2015) Overcoming barriers to knowledge exchange for adaptive resource management; the perspectives of Australian marine scientists. Mar Policy 52:38–44

ELI (2014) Big data and environmental protection: an initial survey of public and private initiatives. Washington, DC

Esvelt KM, Gemmell NJ (2017) Conservation demands safe gene drive. PLoS Biol 15(11):e2003850. https://doi.org/10.1371/journal.pbio.2003850

Freeland HJ, Cummins PF (2005) Argo: a new tool for environ- mental monitoring and assessment of the world's oceans: an example from the NE pacific. Prog Oceanogr 64:31–44. https://doi.org/10.1016/j.pocean.2004.11.002

GOOS (Global Ocean Observing System) (2019) The global ocean observing system 2030 strategy. https://www.goosocean.org/index.php?option=com_content&view=article&id=235:the-global-ocean-observing-system-2030-strategy&catid=22&Itemid=207

Greenemeier L (2015) Human traffickers caught on hidden internet. Scientific American, 8 Feb. https://www.scientificamerican.com/article/human-traffickers-caught-on-hidden-internet

Guyader O, Thébaud O (2001) Distributional issues in the operation of rights-based fisheries management systems. Mar Policy 25(2):103 112. https://doi.org/10.1016/S0308-597X(00)00041-5

Harden-Davies H (2017) Capacity building and technology transfer for marine biodiversity in areas beyond national jurisdiction. Proceed ASIL Ann Meet 111:243–245. https://doi.org/10.1017/amp.2017.75

Hardt MJ, Flett K, Howell CJ (2017) Current barriers to large- scale interoperability of traceability technology in the seafood sector. J Food Sci 82:A3–A12. https://doi.org/10.1111/1750-3841.13796

Haynie AC (2014) Changing usage and value in the western alaska community development quota (CDQ) program. Fish Sci 80(2):181–191. https://doi.org/10.1007/s12562-014-0723-0

Hazen EL, Kohin S, Eguchi T, Scales KL, Welch H, Bograd SJ, Bailey H et al (2018) A dynamic ocean management tool to reduce bycatch and support sustainable fisheries. Sci Adv 4(5):eaar3001. https://doi.org/10.1126/sciadv.aar3001

Hey T, Tansley S, Tolle K (eds) (2009) The fourth paradigm: data-intensive scientific discovery. Microsoft Research, Redmond, WA

Jaffe JS, Franks PJS, Roberts PLD, Mirza D, Schurgers C, Kastner R, Boch A (2017) A swarm of autonomous miniature underwater robot drifters for exploring submesoscale ocean dynamics. Nat Commun 8(1):14189. https://doi.org/10.1038/ncomms14189

Jensen D, Campbell J (2018) The case for a digital ecosystem for the environment, UN Science Policy Business Forum Discussion Paper. Available at: https://un-spbf.org/wp-content/uploads/2019/03/Digital-Ecosystem-final.pdf

Joe H-E, Yun H, Jo S-H, Jun MBG, Min B-K (2018) A review on optical fiber sensors for environmental monitoring. Int J Precis Engineer Manufactur Green Technol 5:173–191

Johnson PA, Sieber R, Scassa T, Stephens M, Robinson P (2017) The cost(s) of geospatial open data. Trans GIS 21(3):434–445. https://doi.org/10.1111/tgis.12283

Kelly RP (2014) Will more, better, cheaper, and faster monitoring improve environmental management? Environ Law 44:1111

Kelly RP, Closek CJ, O'Donnell JL, Kralj JE, Shelton AO, Samhouri JF (2017) Genetic and manual survey methods yield different and complementary views of an ecosystem. Front Mar Sci 3:283

Klenzendorf B, Barrett M, Christman M, Quigley M (2015) Water quality and conservation benefits achieved via real time control retrofit of stormwater management facilities near Austin, Texas Brandon. StormCon 2015:1–10

Kremer M, Williams H (2015) Incentivizing innovation: adding to the tool kit. Innov Policy Econ 10(1):1–17. https://doi.org/10.1086/605851

Kroodsma DA, Mayorga J, Hochberg T, Miller NA, Boerder K, Ferretti F, Wilson A et al (2018) Tracking the global footprint of fisheries. Science 908:904–908. https://doi.org/10.1126/science.aao5646

Laist DW, Knowlton AR, Pendleton D (2014) Effectiveness of mandatory vessel speed limits for protecting North Atlantic right whales. Endanger Species Res 23(2):133–147

Lambert BS, Olson RJ, Sosik HM (2016) A fluorescence-activated cell sorting subsystem for the imaging flow cytobot. Limnol Oceanograp Methods 15:94–102

Lathrop RG, Odell J, MacDonald T, Vilacoba K, Bognar J, Trimble J, Bruce C et al (2017) The role of mid-Atlantic ocean data portal in supporting ocean planning. Front Mar Sci 4:1–8. https://doi.org/10.3389/fmars.2017.00256

Le Seve MD, Mason N, Nassiry D (2018) Delivering blockchain's potential for environmental sustainability. ODI Briefing Note. Overseas Development Institute, London. https://www.odi.org/sites/odi.org.uk/files/resource-documents/12439.pdf

LeCun Y, Bengio Y, Hinton G (2015) Deep learning. Nature 521: 436–444

Lenhart J, Biondi S, Arnoffi M, Sambo F, Favaro F (2018) Real time control of highway runoff for water quality and spill response. In: World environmental and water resources congress 2018, p. 121

Li H, He Y, Cheng X, Zhu H, Sun L (2015) Security and privacy in localization for underwater sensor networks. IEEE Commun Mag 53:56–62

Long RD, Charles A, Stephenson RL (2015) Key principles of marine ecosystem-based management. Mar Policy 57:53–60. https://doi.org/10.1016/j.marpol.2015.01.013

Lubchenco J, Cerny-Chipman EB, Reimer JN, Levin SA (2016) The right incentives enable ocean sustainability successes and provide hope for the future. Proc Natl Acad Sci 113(51):14507–14514. https://doi.org/10.1073/pnas.1604982113

Maxwell SM, Hazen EL, Lewison RL, Dunn DC, Bailey H, Bograd SJ, Briscoe D et al (2015) Dynamic ocean management: defining and conceptualizing real-time management of the ocean. Mar Policy 58:42–50. https://doi.org/10.1016/j.marpol.2015.03.014

McCauley DJ, Allison EH, Joyce FH, Mayorga J, Jablonicky C, Golden CD, Kroodsma D (2018) Wealthy countries dominate industrial fishing. Sci Adv 4(8):eaau2161. https://doi.org/10.1126/sciadv.aau2161

McConney P, Fanning L, Mahon R, Simmons B (2016) A first look at the science-policy interface for ocean governance in the wider Caribbean region. Front Mar Sci 2:1–16. https://doi.org/10.3389/fmars.2015.00119

Michelin M, Elliott M, Bucher M, Zimring M, Sweeney M (2018) Catalyzing the growth of electronic monitoring in fisheries. building greater transparency and accountability at sea. opportunities, barriers, and recommendations for scaling the technology. California Environmental Associates, San Francisco. https://www.ceaconsulting.com/wp-content/uploads/CEA-EM-Report-9-10-18-download.pdf

Munk W (2012) Oceanography before, and after, the advent of satellites. In: Halpern D (ed) Satellites, oceanography, and society. Elsevier, Amsterdam and New York, pp 1–5

NAS (National Academies of Sciences, Engineering, and Medicine) (2017) Sustaining ocean observations to understand future changes in Earth's climate. The National Academies Press, Washington, DC. https://doi.org/10.17226/24919

NCE (National Centers for Environmental Information) (2018) NOAA expands public access to big data, 10 Dec. https://www.ncei.noaa.gov/news/noaa-expands-big-data-access

Nyborg K, Anderies JM, Dannenberg A, Lindahl T, Schill C, Schluter M, Adger WN et al (2016) Social norms as solutions. Science 354(6308):42–43. https://doi.org/10.1126/science.aaf8317

OECD (2017) Business models for sustainable research data repositories. OECD Science, Technology and Innovation Policy Papers No. 47. OECD Publishing, Paris. http://www.oecd-ilibrary.org/docserver/download/302b12bb-en.pdf?expires=1513799615&id=id&accname=guest&checksum=6F7679BD4EA184E8A34D95C7C6DD6287

OECD (2019) Rethinking innovation for a sustainable ocean economy. OECD Publishing, Paris. https://doi.org/10.1787/9789264311053-en

Palmer TN, Williams PD (2008) Stochastic physics and climate modelling. Philos Trans R Soc A Mathem Phys Eng Sci 366(1875):2421–2427. https://doi.org/10.1098/rsta.2008.0059

Piwowar HA, Day RS, Fridsma DB (2007) Sharing detailed research data is associated with increased citation rate. PLoS One 2(3). https://doi.org/10.1371/journal.pone.0000308

Reichstein M, Camps-Valls G, Stevens B, Jung M, Denzler J, Carvalhais N, Prabhat (2019) Deep learning and process understanding for data-driven earth system science. Nature 566(7743):195–204. https://doi.org/10.1038/s41586-019-0912-1

Reimers CE, Wolf M (2018) Power from benthic microbial fuel cells drives autonomous sensors and acoustic modems. Oceanography 31(1):98–103. https://doi.org/10.5670/oceanog.2018.115

Rein M, Favrod VD, Hou C, Khudiyev T, Stolyarov A, Cox J, Chung CC et al (2018) Diode fibres for fabric-based optical communications. Nature 560(7717):214

REV Ocean (2019) Ocean data platform. https://www.revocean.org/platform/oceandata/. Accessed 26 Nov

Roemmich D, Johnson GC, Riser S, Davis R, Gilson J, Brechner Owens W, Garzoli SL et al (2009) The Argo program: observing the global ocean with profiling floats. Oceanography 22(S2):34–43

Ryabinin V, Barbière J, Haugan P, Kullenberg G, Smith N, McLean C, Troisi A et al (2019) The UN decade of ocean science for sustainable development. Front Mar Sci 6:470. https://doi.org/10.3389/fmars.2019.00470

Sajin S (2018) Preventing machine learning bias. Towards Data Science, 31 October. https://towardsdatascience.com/preventing-machine-learning-bias-d01adfe9f1fa

Seltenrich N (2014) Keeping tabs on HABs: new tools for detecting, monitoring, and preventing harmful algal blooms. Environ Health Perspect 122:207–213

Shiffman DS, Macdonald C, Ganz HY, Hammerschlag N (2017) Fishing practices and representations of shark conservation issues among users of a land-based shark angling online forum. Fish Res 196:13–26. https://doi.org/10.1016/J.FISHRES.2017.07.031

Silverman L (2016) The app that aims to gamify biology has amateurs discovering new species. All Tech Considered: NPR, 6 Aug. https://www.npr.org/sections/alltechconsidered/2016/08/06/488830352/the-app-that-aims-to-gamify-biology-has-amateurs-discovering-new-species
Smith LM, Barth JA, Kelley DS, Plueddemann A, Rodero I, Ulses GA, Vardaro MF, Weller R (2018) The ocean observatories initiative. Oceanography 31(1):16–35. https://doi.org/10.5670/oceanog.2018.105
Soeth P (2019) Teams complete bureau of reclamation's sub-seasonal climate forecast rodeo—outperforming the baseline forecasts. CISION, PR Web, 8 Mar. https://www.prweb.com/releases/teams_com-plete_bureau_of_reclamation_s_sub_seasonal_climate_forecast_rodeo_outperforming_the_baseline_forecasts/prweb16155779.htm
Stein B, Morrison A (2014) The enterprise data lake: better integration and deeper analytics. Technol Forec Rethink Integrat (1):1–9
Stelzenmüller V, Lee J, South A, Foden J, Rogers SI (2013) Practical tools to support marine spatial planning: a review and some prototype tools. Mar Policy 38:214–227. https://doi.org/10.1016/J.MARPOL.2012.05.038
Stephens N, Di Silvio L, Dunsford I, Ellis M, Glencross A, Sexton A (2018) Bringing cultured meat to market: technical, socio-political, and regulatory challenges in cellular agriculture. Trends Food Sci Technol 78:155–166. https://doi.org/10.1016/j.tifs.2018.04.010
Stokey R, Roup A, von Alt C, Allen B, Forrester N, Austin T, Goldsborough R, Purcell M, Jaffre F, Packard G, Kukulya A (2005) Development of the REMUS 600 autonomous underwater vehicle. Proceed OCEANS 2005 MTS/IEEE 2:1301–1304. https://doi.org/10.1109/OCEANS.2005.1639934
Succi S, Coveney PV (2019) Big data: the end of the scientific method? Philos Trans R Soc A Mathem Phys Eng Sci 377:20180145. https://doi.org/10.1098/rsta.2018.0145
Sutherland WJ, Fleishman E, Mascia MB, Pretty J, Rudd MA (2011) Methods for collaboratively identifying research priorities and emerging issues in science and policy. Methods Ecol Evol 2(3):238–247. https://doi.org/10.1111/j.2041-210X.2010.00083.x
Sweeney L, Crosas M (2015) An open science platform for the next generation of data. SSRN. https://doi.org/10.2139/ssrn.2302084
Thompson C (2019) We need a data-rich picture of what's killing the planet. Wired
Thomson J, Girton J (2017) Sustained measurements of southern ocean air-sea coupling from a wave glider autonomous surface vehicle. Oceanography 30(2):104–109. https://doi.org/10.5670/oceanog.2017.228
Thurstan RH, McClenachan L, Crowder LB, Drew JA, Kittinger JN, Levin PS, Roberts CM et al (2015) Filling historical data gaps to foster solutions in marine conservation. Ocean Coast Manag 115:31–40. https://doi.org/10.1016/j.ocecoaman.2015.04.019
UNESCO and IOC (Intergovernmental Oceanographic Commission) (2019) Accelerating ocean science for a better world: the UN decade of ocean science for sustainable development 2021–2030. Decade Paper No. 1. An Information paper developed by the Executive Planning Group, IOC
Wang RQ, Mao H, Wang Y, Rae C, Shaw W (2018) Hyper-resolution monitoring of urban flooding with social media and crowdsourcing data. Comput Geosci 111:139–147. https://doi.org/10.1016/j.cageo.2017.11.008
WEF (World Economic Forum) (2018) Innovation with a purpose: the role of technology innovation in accelerating food systems transformation. WEF, Cologny
WEF (World Economic Forum) (2019) Federated data systems: balancing innovation and trust in the use of sensitive data. White Paper. WEF, Cologny
Wrathall L (2010) Vulnerability of subsea infrastructure to underwater attack: legal shortcomings and the way forward. San Diego Int Law J 12:223. http://heinonlinebackup.com/hol-cgi-bin/get_pdf.cgi?handle=hein.journals/sdintl12&section=10
Yaninek KD (1995) Turtle excluder device regulations: laws sea turtles can live with. North Carol Centr Law J 21(2):256–299

Coastal Development: Resilience, Restoration and Infrastructure Requirements

7

Abbreviations

ADB	Asian Development Bank
CBD	Convention on Biological Diversity
CO_2	Carbon dioxide
EEZ	Exclusive economic zone
GSLR	Global mean sea level rise
ICM	Integrated coastal management
ICZM	Integrated coastal zone management
IPBES	Intergovernmental Science-Policy Platform on Biodiversity and Ecosystem Services
IPCC	Intergovernmental Panel on Climate Change
LMMA	Locally managed marine area
MPA	Marine protected area
NbS	Nature-based Solutions
NDC	Nationally determined contribution
NGO	Non-governmental organisation
OECM	Other effective conservation measure
PES	Payments for ecosystems services
RCP	Representative Concentration Pathway
SDG	Sustainable Development Goal
SEEA	System of Environmental Economic Accounting
TURF	Territorial use right for fisheries
SIDS	Small island developing states
UNFCCC	United Nations Framework Convention on Climate Change

Originally published in:
Steven, A.D.L., Appeaning Addo, K., Llewellyn, G., Vu, T.C. et al. 2020. *Coastal Development: Resilience, Restoration and Infrastructure Requirements*. Washington, DC: World Resources Institute. www.ocean-panel.org/blue-papers/coastal-development-resilience-restoration-and-infrastructure-requirements.
Reprint by Springer International Publishing (2023) with kind permission.
Published under license from the "World Research Institute".

Highlights

- Resilient coastal ecosystems are central to the realisation of a sustainable, inclusive, prosperous, and equitable ocean economy, as coastal areas are home to more than 40% of the world's population and host most of the transport, commercial, residential and national defence infrastructure of more than 200 nations and territories.
- Coastal ecosystems are undergoing profound changes, as they are challenged by climate change, threatened by urbanisation and poor upstream agriculture and extractive industry practices, increasing sprawl of coastal infrastructure, and over-exploitation of resources.
- Failure to properly manage our coastal ecosystems will result in continued environmental damage, compromised development of established and emerging ocean sectors, disadvantaged nations and peoples, as well as inadequate infrastructure to meet the demands of changing demographics and climate change impacts.
- To ensure the environmental, economic and social sustainability of our space-constrained coastal ecosystems, ongoing development of our coasts must be balanced across multiple competing uses.
- The full range of economic, social, cultural and environmental values of coastal ecosystems must be balanced through enduring partnerships and active stewardship from government, industry and communities, and supported through innovation and research.
- This Blue Paper focuses on how to enhance coastal ecosystem resilience and enable sustainable pathways for economic, infrastructure and social development, without compromising the integrity and benefits of coastal ecosystems, or disadvantaging the people who rely upon them.
- This paper identifies opportunities for nations to cooperate by building upon past success to realise a sustainable ocean economy through championing the following four coastal opportunities for action: build ecosystem resilience; mitigate impacts of terrestrial and extractive activi-

J. Lubchenco, P. M. Haugan (eds.), *The Blue Compendium*, https://doi.org/10.1007/978-3-031-16277-0_7

ties on coastal ecosystems; advance sustainable, future-proofed blue infrastructure; and enhance community resilience, equity and access.
- With COVID-19 creating an imperative for stimulating economic activity, there is a unique window of opportunity to ensure that relevant policy and investment decisions address the challenges faced by coastal ecosystems and communities, as well as foster sustainable economic pathways. This involves supporting the recovery and development of impacted communities, building the resilience of coastal ecosystems and safeguarding the services they provide and future-ready built infrastructure.

1 Introduction

More than 200 countries have a coastline, and this forms the basis for their claims to territorial waters and exclusive economic zones (EEZs). Globally, about 40% of the world's population live within the "near coastal zone"—the area below an elevation of 100 metres (m) and closer than 100 kilometres (km) from the coast (Kummu et al. 2016). The vast majority of resources for current and emerging sectors that comprise the "ocean— or blue—economy" are concentrated along coastal areas within these EEZs and must operate within a complex, multiple-use and often space-constrained context. The near coastal zone is also where the majority of many coastal nations' commercial, residential, transport and national defence infrastructure is situated, and it is the backbone to domestic and international supply chains that deliver the marine goods and services upon which we increasingly rely.

Coasts sustain livelihoods for hundreds of millions of people in work that ranges from artisanal small-scale fisheries and aquaculture to transnational fishing, shipping, energy and tourism industries. Our increasingly urbanised societies are highly dependent upon coastal resources for food, energy, minerals and pharmaceuticals. Consequently, the coastal economy—which is much broader in its accounting than the ocean economy because it includes not only the sum of outputs from ocean resources but also employment on or near the coast—makes a disproportionately high contribution to the economies of many countries, and to the global ocean economy (He et al. 2014; Mohanty et al. 2015; NOEP 2016; Voyer et al. 2018). A significant, but mostly unquantified, informal or grey economy also occurs within coastal settings and underpins the livelihoods of some of the most disadvantaged populations. In addition to providing these important provisioning goods, the biodiversity and natural functions of intact coastal ecosystems provide regulating, supporting and cultural services that also underpin the ocean economy. These services are recognised as nature's contributions to people (NCP), as they are central to links between nature and people and their culture knowledge systems (Pascual et al. 2017; Diaz et al. 2018; IPBES 2019).

Coastal environments occur where the land and the ocean meet, and they are the place where, historically, people have concentrated and prospered. These environments are intrinsically dynamic—shaped as they are by the interaction of marine, terrestrial and atmospheric processes. However, they are also profoundly changing across human timescales, as they are challenged by extreme climate events that are escalating in frequency and severity, and threatened by increasing population growth and urbanisation, poor upstream land practices, conversion of coastal habitats, and environmental impacts from industry, pollutants and over-exploitation of resources. These changes are direct and physical through the loss, fragmentation and alteration of many ecosystems, but also functional, through a loss of resilience that diminishes the capability of coastal environments to resist and recover from such perturbations. Poorly designed and operated infrastructure can also create harmful environmental and social impacts, increase vulnerability to natural disasters and can sometimes leave an unserviceable burden of debt.

Future projections over the coming decades of our accelerating use and dependence on the coastal zone for living space and resources highlight that, unless we change the way we manage and adapt our use of coastal environments, there will be profound consequences for the resilience of coastal environments and the communities that rely upon them. To avoid the realisation of these projections requires innovative approaches to increase the resilience of coastal environments, and to ensure that the services they provide are sustained. Nature-based solutions are increasingly being adopted as complementary approaches to bridging this adaptation gap, to make infrastructure more resilient to climate change effects and add longer-term value to infrastructure assets.

They are also critical to our aspirations for achieving a sustainable ocean economy and many of the Sustainable Development Goals (SDGs). To realise a sustainable ocean economy that protects, produces and prospers, fundamental issues of equity, inclusion and access must be addressed by developing better governance, participatory, finance and capability-enhancing mechanisms. While COVID-19 has had a profound impact on the economies and social fabric of many nations, under the banner of "build back better", there are significant opportunities to address many of the challenges confronting coastal environments, by adopting approaches that support both a sustainable ocean economy and associated livelihoods to create win-win outcomes for governments and coastal communities. For example, promoting natural infrastructure and grey-green infrastructure nation-building projects provides jobs and builds coastal resilience, while establishing local supply chains for fisheries supports community resilience in low-income countries.

This Blue Paper reviews the major human activities that have increased pressure on coastal ecosystems and reduced their resilience. Our focus is principally on reviewing and identifying practicable solutions that can be implemented to enhance coastal ecosystem resilience and enable sustainable pathways for economic and infrastructure development, without compromising the integrity and benefits of coastal ecosystems or disadvantaging the people who rely upon them. Thus, we use the Intergovernmental Science-Policy Platform on Biodiversity and Ecosystem Services (IPBES) conceptual framework, which rationalises inclusive valuation of nature's contributions to people in decision-making, and we consider resilience not only in physical and ecological contexts but also in terms of social, institutional and financial resilience (see Table 7.1).

We have drawn upon a number of recent intergovernmental reports, notably the Intergovernmental Panel on Climate Change (IPCC) *Special Report on the Ocean and Cryosphere* (IPCC 2019) and the IPBES *Global Assessment Report* (2019), which provide comprehensive global assessments of current and projected conditions of coastal environments.

Table 7.1 Coastal resilience definitions adopted for this report

Type of resilience	Characteristics
Physical resilience	Resilience of existing and planned infrastructure, including through risk-sensitive land-use planning, incorporation of structural resilient measures into infrastructure projects, investments in structural risk reduction measures, and improved operation and maintenance of infrastructure as pathways to building physical resilience
Financial resilience	Improved financial management and timely provision of adequate flows through contingency financing, increased availability and coverage of insurance and capital market risk transfer solutions. Comprehensive risk financing solutions and enhanced capabilities to use financing effectively
Social and institutional resilience	Governance and the promotion of poverty reduction and social protection programmes that build community resilience and channel support to affected poor households. In particular, building women's resilience through greater access to technologies and finance, diversification of livelihoods, and increased participation in women-led solutions
Ecological resilience	Natural ecosystems play multiple roles in conferring resilience. Examples of this type of resilience are enhancing support for nature-based climate and disaster solutions, including upper watershed restoration, wetlands restoration, mangrove rehabilitation, and installation of detention basins and retention ponds to reduce flooding, storm surges and coastal erosion

Source: Adapted from ADB (2019)

Likewise, there are a number of other excellent reports that develop solutions spanning a number of fields: innovation, finance, engineering and material science, and behavioural psychology. Many of the coastal issues and their potential solutions can only be briefly considered here, and several companion Blue Papers provide more detailed analysis.

2 Coastal Changes and Challenges

For millennia, coastal environments have been the location of many civilisations, providing resources and materials for local use, as well as trade along sea routes with other nations (Paine 2014). Today, however, the scale of coastal use and resource demand, driven by rapid population growth and increasing urbanisation, is unprecedented and has been referred to as the *blue acceleration*—a race among diverse and often competing interests for ocean food, material and space (Jouffray et al. 2020). Concurrently, coastal environments, which have always been shaped by climate processes, are now the frontline of anthropogenic climate change, with these environments and their dependent human communities already experiencing the impacts of both extreme climate events and slow-onset changes, such as sea level rise. Together, these climate-induced changes and the accelerating demand for coastal space and resources, as well as the forms of pollution that result (e.g. litter, wastewater), are threatening the extent, condition and biodiversity of many coastal ecosystems, and the goods and services we derive from them. Below, we briefly summarise global patterns of change in climate conditions and human demand for coastal resources and space over the last 50 or so years, and projections for the coming three decades that will profoundly shape and alter our coastal environments.

2.1 Climate Changes and Coasts

Rising carbon dioxide (CO_2) and greenhouse gas emissions have led to well-documented global increases in sea level and sea temperatures, which have resulted in stormier and more extreme sea conditions. The IPCC *Special Report on the Ocean and Cryosphere* (IPCC 2019; and key chapters: Bindoff et al. 2019; Magnan et al. 2019; Oppenheimer et al. 2019) provides the most current and authoritative analysis of recent (1950–present) observed changes in the climate system, and future projections (to 2100) based on low and high Representative Concentration Pathway (RCP) emission scenarios (RCP2.6 and RCP8.5, respectively). Table 7.2

Table 7.2 Summary of key observation and trends in climate change

Parameter	Observed trends	Near-term (2030–2080) and end-of-century projections	Physical effects on coastal ecosystems and settlements	Key references
Global mean sea level	Rate change from 1.38 mm/year during 1901–1990 to 3.16 mm/year during 1993–2015	Up to 2050, global mean sea level will rise between 0.24 m and 0.32 m In 2100, the numbers are 0.59 m and 1.10 m, respectively	Permanent submergence, flood damage, erosion, saltwater intrusion, rising water tables/impeded drainage, ecosystem loss (and change)	Storlazzi et al. (2018), Vitousek et al. (2017), Donnelson Wright et al. (2019a, b), Becker et al. (2020), Oppenheimer et al. (2019)
Regional mean sea level	Rising and accelerating	Increased regional relative sea level nearly everywhere (RCP8.5)	Coastal flooding, submergence, erosion, salinisation	Oppenheimer et al. (2019), Minderhoud et al. (2020)
Extreme sea levels	Increase due to increase in storm intensity	More frequent extreme sea level events as a consequence of sea level rise at many locations by the end of the century (RCP8.5)	Coastal flooding, erosion, saltwater intrusion	Mentaschi et al. (2018)
Waves	Small increases in significant wave height with larger increases in extreme conditions and largest increase in the Southern Ocean	Low confidence for projections overall but medium confidence for Southern Ocean increases in wave height	Coastal erosion, overtopping and coastal flooding	Young and Ribal (2019), Reguero et al. (2019a), Camus et al. (2017)
Winds	Small increases in wind velocity with larger increases in extreme conditions and largest increase in the Southern Ocean	General trend of reduction in wind velocity in summer, autumn and spring, but increase in winter in Northern and Central Europe. General increase in extreme conditions	Wind waves, storm surges, coastal currents, land coastal infrastructure damage	Young and Ribal (2019), Zheng et al. (2019)
Storms, tropical cyclones, extra-tropical cyclones	Regionally variable but increase in annual global proportion of tropical cyclones reaching Category 4 or 5 intensity	Decrease in global tropical cyclone frequenc but proportion of cyclones that reach Category 4 or 5 intensity will increase by 1–10% (RCP8.5)	Higher storm surge levels and storm waves, coastal flooding, erosion, saltwater intrusion, rising water tables/impeded drainage, wetland loss (and change). Coastal infrastructure damage and flood defence failure	Kossin et al. (2020)
Sea surface temperature	SST warming rates highest near the ocean surface (>0.1 °C per decade in the upper 75 m from 1971 to 2010) decreasing with depth	0–2000 m layer of the ocean projected to warm by 900 zettajoules (ZJ) (RCP2.6) and 2150 ZJ (RCP8.5)	Increase in number of coral bleaching events, number of coastal bottom dead zones due to density stratification, harmful algal bloom events, altered ecosystem structure, increased stress to coastal ecosystems	Bindoff et al. (2019)
Marine heatwaves	Doubled since 1980s	Projected to increase (high confidence)	Changes to stratification and circulation, reduced incidence of sea ice at higher latitudes, increased coral bleaching and mortality, increased poleward species migration, decrease in the abundance of kelp forests, massive sea bird die-off and harmful algal bloom	Bindoff et al. (2019), Oliver et al. (2019)
Freshwater input	Declining trend in annual volume of freshwater input	Increase in high latitude and wet tropics and decrease in other tropical regions	Altered flood risk in coastal lowlands, water quality, salinity, fluvial sediment supply, circulation and nutrient supply	Wang et al. (2019), Llovel et al. (2019)
Sea ice and perma-frost thaw	A loss of soil carbon of 5.4% per year across the site Arctic sea ice loss of over 40% over the last last 40 years	By 2100, thaw-affected carbon increase 3-fold (RCP4.5) to 12-fold (RCP8.5)	More storm surges, increasing ocean swells, coastal erosion and land loss in the Arctic and Antarctica regions	Nitzbon et al. (2020), Plaza et al. (2019), Rignot et al. 2019
Ocean acidification	Ocean surface water pH is declining by a very likely range of 0.017–0.027 pH units per decade, since 1980	pH drops of between 0.1 (RCP2.6) and 0.3 (RCP8.5) pH units by 2100, with regional and local variability, exacerbated in polar regions	Increased CO_2 fertilisation, decreased seawater pH and carbonate ion concentration. Enhancing coral reef dissolution and bioerosion, affecting coral species distribution and community	Bindoff et al. (2019), Agostini et al. (2018), Gao et al. (2019)

summarises the historical changes and future projections for climate drivers and ocean and coastal conditions, while below, we focus on the consequences and implications of rising sea levels, warmer and more acidic water, and a greater frequency of extreme climate events, for coastal environments.

Changes in the observation record—which extends back to the early 1900s for tide gauges and more recently for measurements from satellites—from the ocean around the world are clear: sea surface temperatures, wave energy, storminess and acidity have all risen, in many cases doubled, and have continued to accelerate, particularly in the last 30 years (Table 7.2). Near-term and end-of-century model projections all predict, with a high degree of certainty, that these trends will continue to increase and to accelerate. What is unclear is the magnitude, extent and timing of slow-onset climate drivers, such as global mean sea level rise (GSLR), and the frequency of occurrence and magnitude of extreme climate events, including inundation and marine heatwaves.

These changes in ocean state result from both changes occurring directly within the ocean, such as the changes in heat content, density stratification and circulation patterns, and cryosphere changes that include the melting of glaciers, particularly in Greenland and the Antarctic, and sea ice. Both of these factors can act to dilute the salinity of seawater, leading to changes in density and circulation patterns, but only glacial melt will increase the volume of the ocean.

Changes in ocean condition and state are magnified in shallow coastal environments, where tidal and wave energy have their greatest impact on shorelines, and extend across the regional tidal range, and can result in: increased frequency of inundation and subsidence, changes in wetlands, increased erosion of beaches and soft cliffs, and the salinisation of surface and groundwater. Here, the local or relative sea level is complicated and compounded by activities occurring within the coastal zone that affect land elevation, such as subsidence, as well as prevailing winds and water circulation.

While there are significant regional variations, GSLR over the coming century (to 2100) could increase by between 0.43 m (c.4 millimetres (mm)/year) under RCP2.6, and 0.84 m (c.15 mm/year) under RCP8.5. Locally high sea levels, which historically only occurred once per century (historical centennial events or HCE), are projected by 2050 to occur at least annually in many locations, inundating many low-lying areas, including deltaic regions (e.g. Bangladesh and the Mekong Delta), coastal megacities (e.g. Jakarta and Manila) and small islands (e.g. Oceania), impacting their coastal ecosystems, economic development and habitability (Vitousek et al. 2017; Storlazzi et al. 2018; Minderhoud et al. 2019; Oppenheimer et al. 2019; Donnelson Wright et al. 2019a, b; Becker et al. 2020).

In conjunction with sea level rise, greater wave action (wave height, period) and changes in direction and intensity, and more frequent and intense storm surges will affect many coastal areas that were previously never, or infrequently, exposed to such events. These changes can result in cascading impacts on coastal infrastructure and communities living in coastal areas, which are considered further in Sect. 3.3. Projected changes in sea level and wave action, and storm surges will be important considerations for how we build future climate-ready coastal infrastructure (Bhatia et al. 2018; Morss et al. 2018; Abram et al. 2019; Bindoff et al. 2019; Fernández-Montblanc et al. 2019; Kim et al. 2019; Marcos et al. 2019; Magnan et al. 2019; Morim et al. 2019; Oppenheimer et al. 2019; Reguero et al. 2019b).

Coastal shelf waters, from polar regions to the tropics, are also undergoing profound changes as a result of changes in patterns of water circulation and stratification, warmer sea surface temperatures, deoxygenation and more acidic conditions. Rising sea surface temperatures have led to well-documented and rapid changes in the distributions of many marine taxa, including fish, birds and mammals, while changes in circulation and upwelling events have affected the productivity of many eastern boundary systems of the Pacific and Atlantic (Bakun et al. 2015; Champion et al. 2018).

Prolonged extreme ocean warming events—also known as marine heatwaves—over the period 1982–2016 have doubled in frequency and have become longer lasting, more intense and more extensive. Climate models project further increases in the frequency of marine heatwaves, notably in the Arctic Ocean and tropical oceans. Marine heatwaves can severely impact marine ecosystems, resulting in losses of species and habitats from ecosystems as varied as coral reefs, kelp forests, seagrass meadows and mangrove forests, and indirect effects like disruption to sediment-nutrient dynamics and carbon storage (Hughes et al. 2017, 2018; Arias-Ortiz et al. 2018; Hoegh-Guldberg et al. 2018; Oliver et al. 2019; Smale et al. 2018, 2019; Babcock et al. 2019; Garcias-Bonet et al. 2019; Hebbeln et al. 2019; Holbrook et al. 2019, 2020; Sanford et al. 2019; Thomsen et al. 2019; Wernberg et al. 2019).

Deoxygenation in coastal regions results not only from rising sea temperatures but also over-fertilisation and associated runoff from agriculture and from sewage outputs into coastal waters, which leads to algal blooms that consume oxygen once they die and decay. Since the mid-twentieth century, over 700 coastal sites have reported new or worsening low-oxygen conditions. Such oxygen minimum zones can cause widespread changes to marine ecosystems, including loss of invertebrate and fish species and changes in biogeochemical cycling.

Climate models confirm this decline and predict continuing and accelerating ocean deoxygenation (Breitburg et al.

2018; Laffoley and Baxter 2018; Oschlies et al. 2018; Limburg et al. 2020; Rodríguez-Martínez et al. 2020).

Over the last 25 years, the pH in the surface waters of the ocean has reduced as they have absorbed more CO_2, and it is projected to decline further during this century, leading to under-saturation of the carbonate system in the Arctic Ocean, major parts of the Southern, North Pacific and Northwestern Atlantic Oceans (Orr et al. 2005; Hauri et al. 2015; Sasse et al. 2015; Bindoff et al. 2019). As a result, primary productivity of calcifying and non-calcifying plankton species are projected to decrease, while the calcification of corals and bivalves can be impeded, making them more brittle and susceptible to damage, which causes higher mortality, reduced recruitment, increased vulnerability to disease and increasing sensitivity to warming (Fabricius et al. 2011; Doropoulos et al. 2012; Nagelkerken and Connell 2015; Mollica et al. 2018; Gao et al. 2019; Hall-Spencer and Harvey 2019; Liao et al. 2019). In coastal waters, carbonate chemistry is also affected by freshwater runoff which lowers pH due to leachate from acid sulphate soils and humic acids from groundwaters. The extent of coastal acidification can be exacerbated by sea level rise, catchment driven flooding and land runoff, and has had significant impacts on the shellfish industry—a US $19 billion global industry—and can lead to intermittent fish-kills (Salisbury et al. 2008; Barton et al. 2015; Gledhill et al. 2015; Fitzer et al. 2018).

2.2 Changes to Coastal Environments and Ecosystems

Coastal ecosystems are diverse, forming a mosaic of interconnected seascapes, which vary latitudinally from the tropics to the poles, across intertidal and cross-shelf gradients from land to ocean, and in relation to the amount of tidal and wave energy. These coastal ecosystems are most often classified by their geomorphic landform (e.g. estuaries, sandy beaches and rocky shores) or by their foundation species, which can be wetland vegetation (e.g. saltmarshes, seagrass meadows, mangrove forests) or biogenic structures such as coral and shellfish reefs. Many of these ecosystems, particularly over the last 50 years, have undergone massive worldwide reductions in their extent and in their functional resilience, which are the combined consequence of various human activities (clearing and fragmentation of vegetation, hydrological alterations, decreased coastal sediment supply, pollution and emplacement of coastal infrastructure) as well as climate change. Combined with other coastal pressures, such as pollution, most countries are experiencing increased cumulative impacts in their coastal areas, with islands in the Caribbean and mid-latitudes of the Indian Ocean experiencing the greatest impacts (Halpern et al. 2015, 2019). In this section, we summarise observed global changes to these ecosystems, while Fig. 7.1 represents the global extent of these

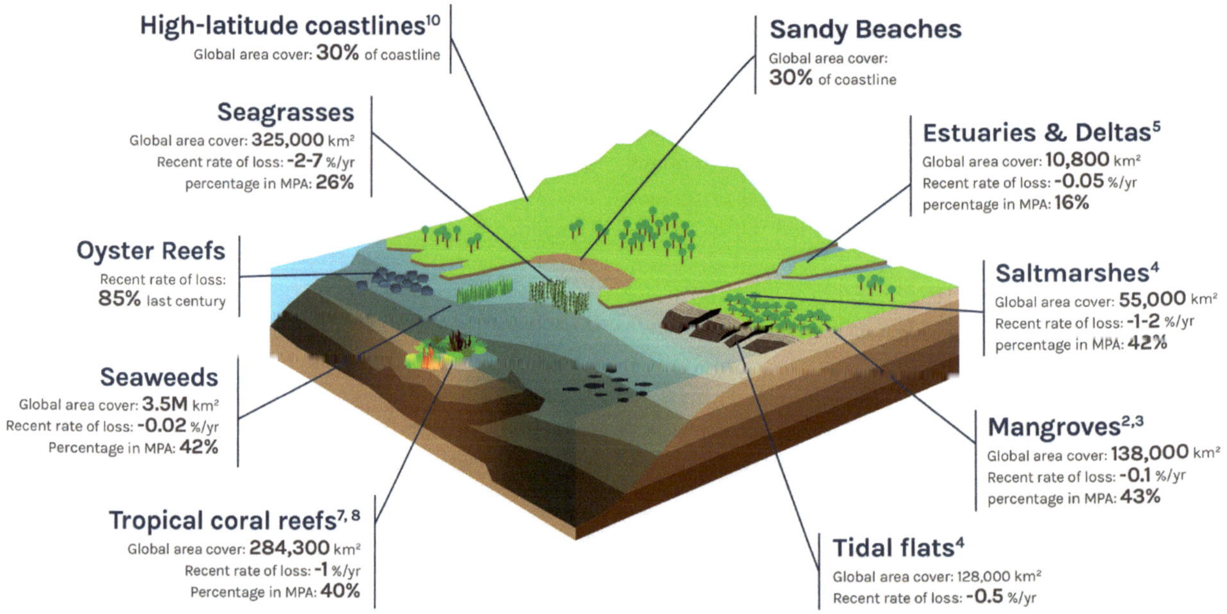

Fig. 7.1 Areal extent and historical and projected losses of major coastal ecosystems. (Source: CSIRO. (1) Beck et al. (2011); (2) Bunting et al. (2018); (3) Goldberg et al. (2020); (4) Mcowen et al. (2017); (5) Murray et al. (2018); (6) Nienhuis et al. (2020); (7) Rogers et al. (2020); (8) UNEP (2020); (9) Vousdoukas et al. (2020); (10) Wernberg et al. (2019)

changes. The companion Blue Paper *Critical Habitats and Biodiversity* (Rogers et al. 2020) provides comprehensive analyses of these changes in habitats and biodiversity.

2.2.1 Coastal Landforms

Coastlines comprise a variety of coastal landforms—beaches, dunes, cliffs, reefs, estuaries, rias, fjords, bays and headlands—that have developed at the dynamic interface of land and sea and have evolved over multiple timescales from quasi-cyclical patterns of erosion and accretion that occur under varying climatic, oceanographic and geological conditions.

Along exposed open coasts erosion is the dominant process weathering these environments. About 50% of the world's coastlines are rocky and sandy beaches. Rocky coasts form where harder more stable substrates predominate, forming reefs that are often covered subtidally by seaweeds and shellfish beds, which in turn support biodiverse communities. Intertidal areas are exposed to strong environmental gradients and these ecosystems are highly sensitive to ocean warming, acidification and extreme heat exposure during low tide emersion. While rocky coastlines form a physical barrier between the land and the sea, softer lithologies are more susceptible to both physical and biological erosion, with significant morphological changes often following extreme events such as storms or tropical cyclones (Hawkins et al. 2016; Ciavola and Coco 2017; Young and Carilli 2019).

Muddy depositional environments, such as estuaries, deltas and tidal flats, are highly dynamic as they are affected by natural and/or human-induced processes originating from both the land and the sea. In addition to GSLR, changes occurring in adjacent catchments that affect sediment supply can result in land subsidence or coastal erosion, or introduce pollution. This in turn can lead to flooding, land loss, salination of coastal aquifers and river reaches, with consequences for properties, agricultural production and food security, especially in agriculture-dependent coastal countries (Khanom 2016).

Shoreline erosion leads to loss of coastal habitats and can, together with sea level rise, contribute to "coastal squeeze" when the intertidal region is constrained by infrastructure built above high water. Shoreline erosion increases the risk of increased flooding and damage to coastal infrastructure and anthropogenic activities, such as upstream dam construction, and river and coastal sand mining, while coastal infrastructure development can significantly alter depositional processes that lead to increased erosion and subsequently diminish the resilience of coastal habitats and increase risks to infrastructure (Naylor et al. 2010; Brooks and Spencer 2012; Pontee 2014; Koehnken and Rintoul 2018).

Satellite-based observational records, from the 1980s to the present, demonstrate changes in the global extent of coastal landforms and show strong regional patterns—with some areas eroding and others accreting—that reflect a dynamic balance between prevailing sea conditions and the extent of catchment and hydrological modification. When globally aggregated, these patterns can be less discernible, which belies the significance of regional changes. Over the last 40 years:

- The loss of *permanent land* in coastal areas is almost 28,000 km^2, which is almost twice as large as land gained within the same period; more than 50% of this net loss of 14,000 km^2 occurred along Asian and Caspian coasts (Mentaschi et al. 2018).
- Twenty-four percent of the world's *sandy beaches* have eroded at rates exceeding 0.5 m/year, but other areas either accreted (28%) or were stable. It is projected that by 2050 13–15% of the world's sandy beaches could face severe erosion, but in low-elevation coastal zones the figure is more than 30%. A number of countries, including the Democratic Republic of the Congo, Gambia, Jersey, Suriname, Comoros, Guinea-Bissau, Pakistan, could face extensive sandy beach erosion issues by the end of the twenty-first century (Luijendijk et al. 2018; Vousdoukas et al. 2020). Worldwide, sandy beaches show vegetation transformations caused by erosion following locally severe wave events with the original dense vegetation being replaced by sparser vegetation and often resulting in a regime shift in the beach morphology and shifts in the associated fauna composition. Coastal dunes are extensive along the world's sandy shorelines and back the majority of beaches forming a linked system. Human disturbances, especially tourism and recreation that have increased foot and vehicular traffic, have increased erosion rates on sandy beaches and dunes, while coastal squeeze has constrained sediment supply and accretion capacity. Paradoxically, vegetation cover on sand-dunes has increased substantially on multiple, geographically dispersed, coastal dune fields on all continents in the period 1984–2017 and points to enhanced dune stability and storm buffering effects (Jackson et al. 2019; Nayak and Byrne 2019).
- *Tidal flats* are intertidal, muddy, sedimentary habitats, often flanking estuaries, and are widely distributed, with a present global extent of 128,000 km^2, of which 70% occur in three continents (Asia 44%; North America 15.5%; South America 11%). Since 1984, it is estimated that 16% of tidal flats have been lost, principally from coastal development and coastal erosion due to reduced sediment delivery from major rivers and sinking of riverine deltas. In China, massive losses of tidal flats have resulted from reclamation, or conversion to other activities, principally aquaculture (Murray et al. 2018).
- *Deltas* account for less than 1% of global land area, yet are home to more than half a billion people and some of

the largest cities. Deltas have a dynamic and evolving geomorphology, formed by the accumulation of unconsolidated river-borne fine sediments (mud, silt and clay) and so are particularly sensitive to anthropogenic activities which influence the mobilisation, delivery, deposition and erosion of sediment to and from a delta. Over the past 30 years, despite sea level rise, deltas globally have experienced a net land gain of 54 km²/year with deltas being responsible for 30% of all net land area gains that result principally from deforestation-induced increases in fluvial sediment supply. Yet, for nearly 1000 deltas, river damming has resulted in a severe (more than 50%) reduction in anthropogenic sediment flux, resulting in global deltaic land loss of 12 km²/year. In many of the major deltas (e.g. Mekong, Irrawaddy, Ganges-Brahmaputra), this decline of sediment supply due to upstream dam construction, combined with land-use changes, river sand mining and over-abstraction of groundwater, has led to deltaic subsidence rates at least twice the concurrent rate of GMSL rise (3 mm/year). As a consequence, flooding now routinely occurs in many deltas around the world, with an estimated 260,000 km² of delta temporarily submerged in the 1990s/2000s, and leads to saline or brackish water intrusion that increases residual salinity of potable and irrigated water. Intensive human activities around estuaries and river deltas have also substantially increased nutrient and organic matter inputs since the 1970s resulting in eutrophication (Ericson et al. 2006; Nicholls et al. 2020; Nienhuis et al. 2020).

- Some of the most significant effects of climate change are occurring along *high latitude* (polar) coastlines that occur to the north and south of 60° (IPCC 2019). Whereas Arctic coastlines represent about one-third of the world's coastlines and occur over a range of geological and oceanographic settings, Antarctic coastlines are often permanently covered in ice. Rapid and accelerated Arctic sea ice loss, which has averaged 10% per decade over the last 40 years, is attributed to the impacts of land–ocean warming and the northward heat advection into the Arctic Ocean. The possibility of a nearly ice-free Arctic summer within the next 15 years has led to speculation as to whether this will create new shipping channels between Asia and Europe. With longer open-water periods during summer, extra wave activity is expected to result in higher erosion rates along many high-latitude shorelines, while warmer temperatures and increased frequencies of extreme storms may trigger landscape instability, increase sediment and nutrient supply, change carbon fluxes, affect the structure and composition of pelagic communities and benthic habitats and the well-being of dependent human populations. Given the rapidity of these changes, adequate governance frameworks need to be urgently implemented (Moline et al. 2008; Krause-Jensen and Duarte 2014; Kroon et al. 2014; Bull et al. 2019; Gardner et al. 2018; Bendixen et al. 2019b; Oppenheimer et al. 2019; Rignot et al. 2019; Ouyang et al. 2020; Kumar et al. 2020; Hugelius et al. 2020; Peng et al. 2020).

2.2.2 Vegetated Coastal Ecosystems

Vegetated coastal ecosystems, including saltmarshes, mangroves, seagrasses, and kelp and other seaweed, are wetland systems that form important interconnected habitats which support high biodiversity and provide valuable ecosystem services, such as fisheries production, sediment and nutrient trapping, storm protections and carbon storage. Mangroves typically grow between the low and high tide, and reach their highest abundance and diversity in the tropics, predominantly in the Indo-Pacific region. Saltmarshes occur particularly in middle to high latitudes but often overlap with mangrove distributions, resulting in dynamic transitions between these two communities. Seagrasses rooted in unconsolidated sediments grow in shallow coastal waters to 60 m depth and have a global distribution. Seaweeds attach to solid reef substrates, with some species such as kelp—a brown algae—forming large canopies present in more than 40% of the world's marine ecoregions.

However, these ecosystems have been extensively modified by human activities and must also adapt to accelerating rates of climate change. For example, it is estimated that eustatic sea level rise could result in the loss of 22% of the world's coastal wetlands by 2080, and in the Indo-Pacific region, where sediment delivery has declined due to damming of rivers, existing mangrove forests at sites with low tidal range and low sediment supply could be submerged as early as 2070 (Waycott et al. 2009; Duarte et al. 2013; Blankespoor et al. 2014; Copertino et al. 2016; Lovelock et al. 2016; Kelleway et al. 2017; van Oosterzee and Duke 2017; Besset et al. 2019; Serrano et al. 2019a, b).

The current extent and historical loss of these ecosystems are summarised in Fig. 7.1, and below.

- Globally more than 6000 km² of mangroves were cleared between 1996 (142,795 km²) and the present (137,000 km²). Contemporary (2000–present) global losses (0.2–0.6%/year) of mangroves are an order of magnitude less than losses during the late twentieth century, and have resulted primarily from land-use change, usually through conversion but also fragmentation. In Southeast Asia, mangrove loss has been recorded at twice the global rate, where conversion of mangroves to shrimp aquaculture accounted for more than 50% of losses, while more recently oil palm plantations and coastal erosion are leading to further losses. In Brazil, Puerto Rico, Cameroon, China and Singapore, large areas of mangroves have been lost to urban development. Significant declines in the delivery of upstream sediment supply have further diminished the ability of mangroves to expand and to keep pace with rising sea levels (Richards and Friess 2016; Woodroffe et al. 2016; Hamilton and Casey 2016; Bunting et al. 2018; Romañach et al. 2018; Worthington and Spalding 2018; Agarwal et al. 2019; Friess et al. 2019; Goldberg et al. 2020; Richards et al. 2020; Turschwell et al. 2020a).

- Saltmarshes, with a present global extent of c.56,000 km^2, are declining around the world, having lost between 25 and 50% of their global historical coverage through conversion to agriculture, urban and industrial land uses. Many saltmarshes are also being squeezed between an eroding seaward edge and fixed flood defence walls, and agricultural grazing has a marked effect on the structure and composition of saltmarsh vegetation, reducing its height and the diversity of plant and invertebrate species (Bromberg Gedan et al. 2009; Crooks et al. 2011; Mcowen et al. 2017; Thomas et al. 2017).
- Seagrass meadows, with a present global distribution of about 300,000 km^2, are estimated to have been lost at rates of 110 km^2 per annum between 1980 and 2006. Current losses are particularly high in East and Southeast Asia, principally as the consequence of coastal development: poor water quality resulting from watershed siltation, physical disturbance such as dredging and coastal reclamation, and the degradation of food webs from aquaculture and fisheries (Waycott et al. 2009; Short et al. 2011; Erftemeijer and Shuail 2012; McKenzie et al. 2020).
- Loss of macroalgal forests over the last half-century has been significant, although spatially variable; kelps have declined by 38% in some ecoregions, but have either grown or remained stable in other regions such as southern South America. Temperature is a key determinant of the biogeographic distribution of many seaweeds, so increases in sea temperatures have led to changes in range and abundance. Kelp die-off from marine heatwaves has been reported along the coasts of Europe, South Africa and Australia, and the kelp is replaced by a less diverse turf-dominated ecosystem (Ling et al. 2015; Krumhansl et al. 2016; Vergés et al. 2016; Wernberg et al. 2016; Piñeiro-Corbeira et al. 2018; Filbee-Dexter and Wernberg 2018; Smale et al. 2019; Wernberg et al. 2019; Wernberg and Filbee-Dexter 2019; Friedlander et al. 2020).

2.2.3 Coral and Shellfish Reefs

Coral reefs occur throughout tropical latitudes and are one of the most diverse and productive ecosystems, providing services that support almost 30% of the world's marine fish species fisheries, and 500 million people who depend on them for work, food and coastal protection in more than 100 countries across Australasia, Southeast Asia, the Indo-Pacific, the Middle East, the Caribbean and the tropical Americas. Coral reefs throughout the world are today one of the most endangered habitats, threatened by a combination of climate change and human activities that weaken the natural resilience of coral reefs.

Activities such as over-exploitation and destructive fishing, watershed and marine-based pollution, and coastal infrastructure development have had an impact on reef population structure and biodiversity by reducing coral recruitment, survival and growth, and hindering community recovery (Fabricius 2005; Roff et al. 2012; Otaño-Cruz et al. 2017; Lam et al. 2018; MacNeil et al. 2019; Vo et al. 2019).

Since 1998, marine heatwaves have bleached, or killed, corals on many reefs across the Indo-Pacific, Atlantic and Caribbean. In 2016 and 2017, heat stress associated with consecutive El Niño events triggered the third major global coral bleaching event, resulting in severe coral bleaching of around 70% of the world's reefs throughout all three tropical ocean basins; in the Great Barrier Reef, the world's largest reef system, half of the corals died. Further projected increase in sea level, storm intensity, marine heatwaves, turbidity, nutrient concentration due to floods may contribute to the degradation trend of a majority of coral reefs worldwide and require comprehensive management and intervention responses (Hoegh-Guldberg and Bruno 2010; Hughes et al. 2017, 2018; Magel et al. 2019; Morrison et al. 2019, 2020).

Shellfish reef ecosystems have, until recent times, been overlooked as an important estuary habitat. Historically, dense aggregations of bivalves, their shells, associated species and accumulated sediments were a dominant habitat in temperate and subtropical estuaries around the world. Oyster reefs provide numerous ecosystem services, such as improvements to water quality through filtration, shoreline stabilisation and fisheries productivity. Dredging, habitat degradation, including poor water quality and altered species interactions, disease outbreaks and habitat loss, have contributed to the drastic decline in bivalve habitats with an estimated 85% of oyster reefs lost over the last century, as well as largely unquantified losses of other habitat-forming bivalves, such as the formerly widespread green-lipped mussel (*Perna canaliculus*) beds in New Zealand, which now occur at less than 1% of historical levels (Lenihan and Peterson 1998; Newell and Koch 2004; Piehler and Smyth 2011; Scyphers et al. 2011; Beck et al. 2011; Grabowski et al. 2012; Paul 2012).

2.3 Coastal Development Changes

The key global economic trends relevant to maritime sectors are increasing energy demand, increasing food and water demand, and increasing population growth and urbanisation, all of which depend on coastal infrastructure. Factors such as adaptation to climate change, developing economies seeking an increasing share of global growth, growing expectations around health and safety and human rights, and technological innovations are also relevant to maritime and coastal development trends and coastal infrastructure. Energy production, food production and water demand, as well as urbanisation and population growth, represent over a third of the global economy and provide up to two-thirds of jobs. While natural resources make human development possible and underpin economic growth, our accelerating demand for coastal space and resources, as well as the forms of pollution that result (e.g. litter, wastewater), threatens the extent, condition and biodiversity of many coastal ecosystems (IPBES 2019; WEF 2020a). Below, we summarise the major trends in coastal development and discuss the potential consequences.

2.3.1 Population Growth and Urbanisation

About 40% of the world's population lives within 100 km of the coast and 11% live in low-lying coastal areas that are less than 10 m above sea level. While the majority of these populations are based on continental coastal areas, small island developing states (SIDS) are home to 65 million people, while 4 million people live within the Arctic region. Coastal population growth has been increasing at around twice the rate of national growth and is the result of population and demographic changes, as well as migration from rural areas to cities, and displacement of some indigenous and other disaffected communities. Over the next decade, population growth will occur most significantly in Africa (380 million) and Asia (373 million), where the urban population is expected to grow by 2.5 billion over the next 30 years (Creel 2003; McGranahan et al. 2007; Ford et al. 2015; Neumann et al. 2015; Kummu et al. 2016; Jones and O'Neill 2016; Merkens et al. 2016).

Population growth has been accompanied by rapid urbanisation, and today 55% of the global urban population lives in coastal settlements, and 16 of the world's 31 megacities—those with over 10 million inhabitants—are coastal, including New York City, Tokyo, Jakarta, Mumbai, Shanghai and Lagos. Asia has the greatest intensification of coastal population, property and infrastructure, with 10 of the world's megacities, and 20 of the top 30 most populated coastal cities. Even in many SIDS, urbanisation is a growing concern, where 38 million (59%) already live in urban settlements. Globally, from 1985 to 2015, urbanisation expanded on average by 9687 km^2/year, with nearly 70% of this development occurring in Asia and North America (Small and Nicholls 2003; Jongman et al. 2012; UN-Habitat 2015; Liu et al. 2020).

2.3.2 Infrastructure Development

Coastal infrastructure systems form the backbone of every society, providing essential services that include coastal defence, trade, tourism, fisheries and aquaculture, energy, water, waste management, transport, telecommunications and other industries. Urbanisation is, however, not only a land-based problem, and coastal development has led to a proliferation of coastal infrastructure, commonly referred to as "ocean sprawl", that is occurring worldwide along coastlines and in near-shore waters, and is more recently expanding offshore as industries seek to utilise new resources and access space to operate. Along and adjacent to coastal foreshores, infrastructure for defence, residential and commercial developments, transport and tourism/ recreation are common, while moving further offshore infrastructure for aquaculture, oil and gas, offshore renewable energy, mineral extraction and desalination occur.

Although this proliferation of structures provides a suite of economic, social and even ecological benefits, it also replaces natural habitats and can modify environmental conditions critical to habitat persistence at regional scales. Catchment-based infrastructure, such as dams, that affect the natural patterns of hydrological discharge and sediment transport to the coast, can also affect downstream coastal ecosystems.

As of 2018, the physical footprint of built structures was at least 32,000 km^2 worldwide, and is expected to increase by at least 23% (7300 km^2) to cover 39,400 km^2 by 2028. The global area of seascape that is modified around these structures is estimated to be in the order of 1.0–3.4 million km^2 globally, an area comparable to the global extent of urban land (Bugnot et al. 2020). This concentration of structures close to the shore means that many coastal habitats are affected by multiple structures.

There are also substantial regional differences in the amount of different types of marine infrastructure. Proportionally, China, Democratic People's Republic of Korea and the Philippines have the largest marine infrastructure footprints; nearly half of all oil and gas rigs are located in the US Gulf of Mexico, while wind and tidal farms are spread along the coasts of North America, India, the United Kingdom, Germany and in the Asian North Pacific (Bugnot et al. 2020).

Table 7.3 represents the current extent and projected growth of various infrastructure and activities occurring in, and adjacent to, coastal environments. There is also a growing number of regional-scale transnational infrastructure projects under way that will fundamentally change the use of the coastal zone and marine water offshore and in areas beyond national jurisdictions and, unless carefully managed, these present serious threats to biodiversity (see Box 7.1).

Table 7.3 Crowded coasts: global growth of major coastal infrastructure

Category	Type	Footprint	Category	Type	Footprint
Reclamation	Coastal reclaimed land[2]	Area: 3370 km^2	Water Infrastructure	Large dams[8]	Number: 58,000
	Artificial islands[1]	Number: 480		Desalination plants[3]	Number: 16,000
		Area: 1267 km^2			Growth rate: 10.5%/year
	Artificial reefs[1]	Area: 36,000 km^2	Ports and Shipping	Commercial harbours	Number: 4700
	Cemented shorelines	Length: >14,000 km			Area: 4500 km^2
Coastal Defence	Cemented shorelines	Length: >14,000 km		Marinas[1]	Number: 9628
	Breakwaters[1]	Number: 268			Area: 776 km^2
		Area: 577 km^2		Commercial vessels[7]	Number: 95,402
	Coastal canals[4]	Area: 4000 km			Growth rate: 2.6%/year
Energy Infrastructure	Oil rigs[1]	Number: 5179		Cruise ships[5]	Number: 272
		Area: 89,964 km^2			Growth rate: 6%/year
		Growth rate: 1.2%/year		Fishing vessels[6]	Number: 4,600,000
	Oil pipelines[1]	Length: 136,000 km		Motor vessels[6]	Number: 67,800
		Growth rate: 1.2%/year	Miscellaneous	Coastal aquaculture and mariculture[1]	Number: 78,240
	Offshore wind energy[1]	Number: 6000			Area: 22,927 km^2
		Area: 30%/year			Growth rate: 3%/year
	Offshore wave and tidal energy[1]	Growth rate: 208%/year		Telecom cables[1]	Number: 428
					Length: 39,304 km
					Growth rate: 8.2%/year

Source: CSIRO. (1) Bugnot et al. (2020); (2) Donchyts et al. (2016); (3) Jones et al. (2019); (4) Waltham and Connolly (2011); (5) CLIA (2019); (6) FAO (2020a); (7) UNCTAD (2020b); (8) Mulligan et al. (2020)

Since investment in infrastructure is at an all-time high globally, an ever-increasing number of decisions are being made now that will lock in patterns of development for future generations (Bromberg Gedan et al. 2009; Aerts et al. 2011; Sekovski et al. 2012; Jennerjahn and Mitchell 2013). Such infrastructure, unless carefully planned to account for future climate conditions, constructed using environmentally sensitive methods, and operated with appropriate regulations, can pose significant environmental risks to coastal environments, including: changes in coastal morphology from disruption to natural sedimentary processes, destruction and fragmentation of coastal habitat, and impacts on resident and migratory wildlife through disruption to established connectivity pathways or from "accidents" with infrastructure (Dafforn et al. 2015; Firth et al. 2016; Hughes 2019; Hughes et al. 2020). Below, the major forms of infrastructure, their extent and projected growth, and known impacts on coastal ecosystems are summarised.

Coastal defence structures: With increased urbanisation, rising sea levels and stormier seas, shorelines worldwide have dramatically changed as they become increasingly "hardened" with a proliferation of coastal armouring infrastructure, constructed to protect coastal populations and their property, transport infrastructure, industry and commerce, and amenity and recreational areas. Seawalls, breakwaters, jetties, piers and related infrastructure have replaced once natural shorelines by more than 50% in some cities and countries; for example, wetlands along China's 34,000-km coastline have been replaced with 13,830 km of hard engineering structures (Luo et al. 2015). Such coastal defence structures can have a variety of negative effects on adjacent coastal ecosystems. These structures are typically designed to reflect waves and reduce coastal flooding and erosion; consequently, they can alter wave exposure, interfere with the spatial dynamics of sediment transport, and impede animal movement and connectivity between habitats. Over the longer term, this can cause changes in sediment, current and wave dynamics that accelerate erosion, leading to the loss of beaches and other coastal habitats they were intended to protect. Artificial structures may also produce larger-scale impacts through their alteration of ecological connectivity, which restricts the movement or dispersal of organisms, and which may in turn, influence the genetic structure and size of populations, the distribution of species, community structure and ecological functioning (Bulleri and Chapman 2010; Nordstrom 2014; Bishop et al. 2017; Leo et al. 2019).

Box 7.1. Regional Coastal Infrastructure Projects

Belt and Road Initiative and the Maritime Silk Road: The Belt and Road Initiative is a long-term Chinese government vision for improved global connectivity, expanded production and trade chains, and closer overall cooperation. Potentially spanning 72 countries, the Belt and Road Initiative is the largest infrastructure project of all time (valued at over $8 trillion by 2049) and seeks to create connections between core cities and key ports across Eurasia, Asia and parts of the African continent through infrastructure development in the transport, energy, mining, IT and communications sectors. First announced in 2013, the twenty-first Century Maritime Silk Road is the maritime/coastal component of the Belt and Road Initiative, and focuses on creating a network (string of pearls) of ports, through construction, expansion or operation, and the development of portside industrial parks and special economic zones that link China's coastal ports through the South China Sea to the Indian Ocean, extending to Africa and Europe; and potentially to the Pacific Ocean (Fig. 7.2). To date, deep-water ports projects have been initiated in Africa (Tunisia, Senegal, Tanzania, Djibouti, Gabon, Mozambique, Ghana), Asia (Pakistan, Sri Lanka, Myanmar, Indonesia) the Middle East (Oman, Israel) and Europe (Greece). Studies looking at the potential environmental impact of the Belt and Road Initiative have identified that over 400 threatened marine species, including mammals, could be affected by port infrastructure, while over 200 threatened species are at risk from an increase in shipping traffic and noise pollution (Huang 2016; Hughes 2019; Hughes et al. 2020; Narain et al. 2020; Turschwell et al. 2020b).

LAPSSET Corridor, Africa: The LAPSSET Corridor Program is a regional project intended to provide transport and logistics infrastructure aimed at creating seamless connectivity between the eastern African countries of Kenya, Ethiopia and South Sudan. The project connects a population of 160 million people in the three countries and is part of the larger land bridge that will connect the East African coast (at Lamu Port) to the West African coast (at Douala Port). The LAPSSET Corridor is intended to operate as an Economic Corridor with the objective of providing multiple eastern African nations access to a large-scale economic trade system, thereby promoting socioeconomic development in the region. The LAPSSET Corridor Program consists of several subsidiary projects, including the development of deep-water ports, railway lines and highways connecting cities in Kenya, South Sudan and Ethiopia, oil refineries and pipelines, and international airports and resort cities (LAPSSET Corridor Development Authority 2016; Okafor-Yarwood et al. 2020).

Bangladesh Delta Plan 2100: The Bangladesh Delta Plan 2100 is the combination of long-term strategies and subsequent interventions for ensuring long-term water and food security, economic growth and environmental sustainability. It aims to effectively reduce vulnerability to natural disasters and build resilience to climate change and other delta challenges through robust, adaptive and integrated strategies, and equitable water governance. Six hotspot areas were identified: coastal zone; Barind and drought-prone areas; Haor region (flash-flood areas); Chittagong hill tracts and coast; major rivers and estuaries, and urban areas (Bangladesh Planning Commission 2018).

The Red Sea Project: The Red Sea Project is a large-scale luxury tourism development that will extend over 28,000 km^2 along the shores of the Kingdom of Saudi Arabia. The area includes the Al-Wajh lagoon, a large lagoon area with 92 islands, valuable ecosystems and rich biodiversity, including species of global conservation importance. The Red Sea Development Company, responsible for the execution of the Red Sea Project, has committed to achieving a net-positive impact on biodiversity. To grow tourism, which currently represents only 3% of the economy, it will create a special economic zone that is expected to attract 1 million people every year, create 70,000 new jobs and add $5.9 billion to the Saudi GDP (Chalastani et al. 2020).

Ports and harbours: Seaports are nodal hubs in the maritime transportation network, enabling more than 90% of world trade. A growing reliance on marine transport for international trade has led to the construction of more ports and harbours, and the expansion and deepening of existing facilities to accommodate larger vessels. Today, there are more than 4700 commercially active ports worldwide, which are used by more than 50,000 international merchant ships, manned by over a million seafarers, and carry more than 90% (>10 billion tonnes in 2015) of global trade by weight. The development and operation of ports and harbours have been associated with a number of negative environmental and social impacts on coasts, including principally altering regional coastal processes which disturb the sediment balance and exposing down-drift areas to increased erosion. Oil, sewage and noise pollution can result from port operations and can seriously impact surrounding marine life and disrupt social amenity (Zanuttigh 2014; Lee et al. 2015; Johnston et al. 2015; IAPH 2016; IMO 2017; Camus et al. 2019; Leo et al. 2019; Santana-Ceballos et al. 2019; Vaughan 2019; Valdor et al. 2020).

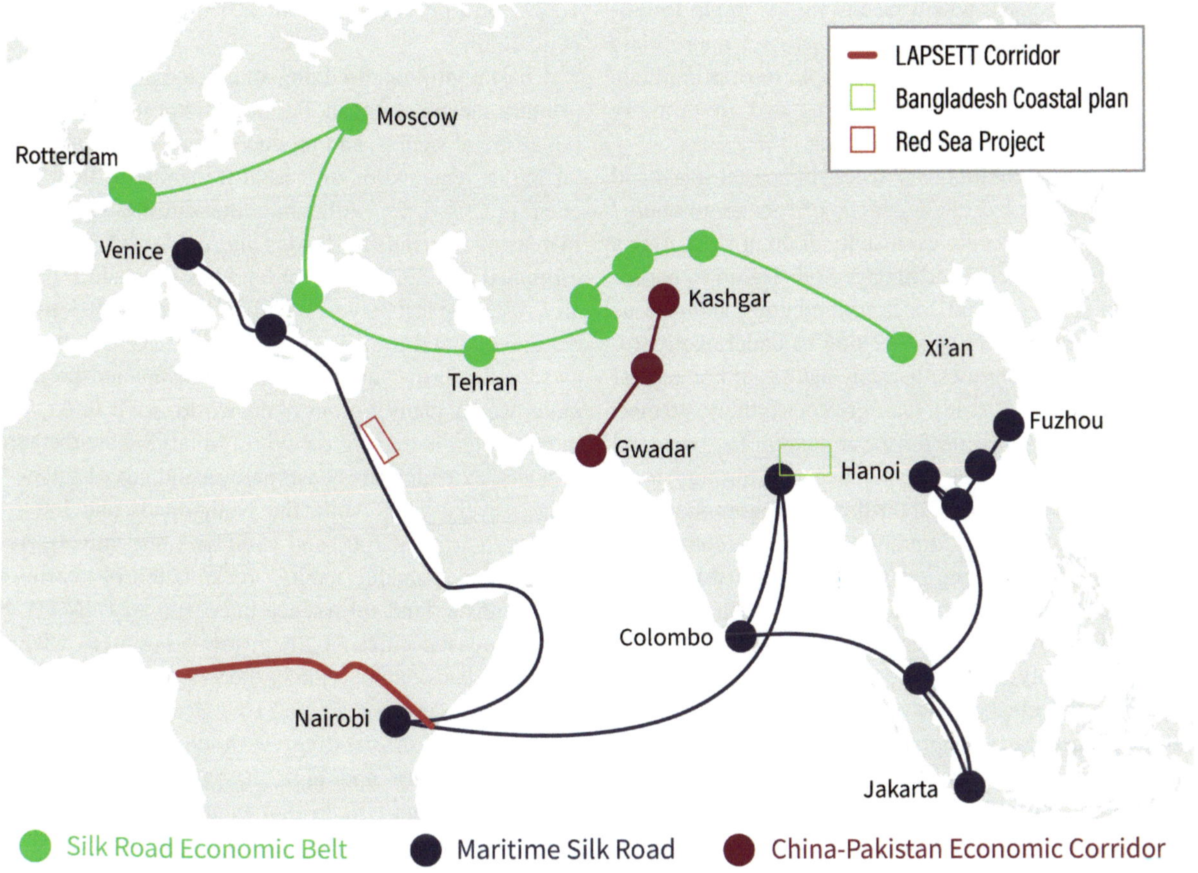

Fig. 7.2 Map of belt and road initiative and twenty-first-century maritime silk road

Energy infrastructure: Conventional oil and gas platforms and associated pipelines, and increasingly offshore renewable energy technologies, including wind farms and tidal power, are common infrastructure in coastal and offshore environments. Tidal farms are located closest to the shoreline, with 41% closer than 2 km in 2018, while nearly half (47%) of all wind farms were located within 10–50 km of the coast, and half of all oil and gas fields were located within 40 km of the shoreline.

Catchment infrastructure: Infrastructure such as dams and weirs for the impoundment of water, irrigation, hydroelectric power generation and flood protection results in hydrologic alteration of the quantity and timing of river flow. Decreased fluvial sediment transfer to coastal regions can lead to sand-starved beaches, and accelerated coastal erosion of deltas and loss of mangrove forests. Construction of embankments and navigation structures can result in rivers becoming disconnected from their floodplains, disrupting natural sediment fluxes, reducing marine and ecological connectivity. Coastal aquifers are more vulnerable to groundwater extraction than to predicted sea level rise under a wide range of hydrogeologic conditions, and over-pumping has led to saltwater intrusion, subsidence and loss of the water supply for future use. Lack of appropriate sewage processing facilities in coastal areas can increase the nutrient pollution and consequent degradation of coastal ecological systems (Poulos and Collins 2002; Giannico and Souder 2005; Al-Bahry et al. 2009; Dafforn et al. 2015; Reopanichkul et al. 2009; Ferguson and Gleeson 2012; Martínez et al. 2014; Rovira et al. 2014; Firth et al. 2016; Chee et al. 2017; Smith et al. 2017; Appeaning Addo et al. 2018; Tessler et al. 2018; Silva et al. 2019; Luijendijk et al. 2020).

2.3.3 Competition for Coastal Space

Today, coasts are an increasingly crowded space, where various sectors of the economy vie for access to areas within territorial and EEZ waters, not only for food and materials, but for a number of other activities, including tourism and leisure, transport and telecommunications. Other activities, such as aquaculture and renewable energies, seek to produce rather than extract seafood and energy, but require coastal space with environmental conditions conducive to their operation. The growth and success of these emerging industries are central to the predicted growth and significance of the ocean economy over the next few decades (see Box 7.2) and will place further demands on access to coastal space.

The allocation of space and the management of associated resources is the responsibility of various government entities, often with overlapping jurisdictions and sometime incompatible mandates, which results in a struggle to balance conservation and sustainable use, to set appropriate operational limits for individual sectors, to resolve conflicts between overlapping incompatible uses and to properly assess the cumulative impact of activities.

Competition and conflict over access to coastal space and resources have often led to allegations of "ocean grabbing" by powerful actors to secure exclusivity or dominance over a resource, and have often disadvantaged other groups, particularly livelihood-dependent local communities, indigenous and artisanal peoples, or those seeking to undertake recreational and cultural activities. In many nations today, coastal developments and industries face greater scrutiny of their environmental and social performance than in the past, and many businesses now seek "social licence" through local environmental and social responsibility programmes that create benefits for the local community, but this is sometimes viewed negatively as nothing more than "green-washing".

To accommodate increasing urban and industrial development, agriculture, silviculture (notably oil palm) and aquaculture (notably shrimp), coastal space has been created by both the clearing of existing coastal vegetation, including mangroves, saltmarshes, coastal dunes and tidal flats, and the reclamation of intertidal and subtidal areas. Urban land expansion rates are significantly higher in coastal areas than in adjacent hinterland, and it is estimated that by 2030, global urban land cover will increase to 1,527,000 km^2 (Brown et al. 2013; Seto et al. 2011; Liu et al. 2020).

Coastal infilling, dredging and reclamation have been the primary means of expanding the coastal foreshore or creating artificial islands, and produce direct loss of marine habitat. While reclamation of coastal wetlands and tidal flats has been practised for millennia, the current scale to meet increasing demands for land is unprecedented. Globally, it is estimated that 33,700 km^2 of land has been added to coastal areas over the last 30 years, with more than 1250 km^2 of land being reclaimed from 16 megacities between the mid-1980s and 2017. Further ambitious land reclamation projects are under way in many regions of the world (see Box 7.1). China, in particular, is leading the world in large-scale reclamation projects, extending its coastline by hundreds of square kilometres every year, while the Netherlands and India have reclaimed areas of 7000 and 1500 km^2, respectively. As well as removing valuable habitat, reclamation of coastal areas contributes to land subsidence in coastal areas. Many of the world's coastal cities—built in low-lying areas where soft sediments can compress under the weight of infrastructure as cities grow—are now sinking (see Box 7.5). This results in increased risk of flooding, with consequences including structural damage and high maintenance costs for urban infrastructure and risk to human livelihood (Waltham and Connolly 2011; Wang et al. 2014; Donchyts et al. 2016; Reyna et al. 2016; Tian et al. 2016; Sengupta et al. 2017).

Box 7.2. Key Coastal Growth Sectors of the Ocean Economy

Coastal (beach-based) and maritime (water-based) tourism is the second largest employer in the ocean economy, providing 1 in 11 jobs worldwide and generating more than $1.5 trillion in trade income or 9.2% of global GDP, and it is the dominant sector in an increasing number of national economies. For some island states, tourism can comprise 25% of national GDP. Globally, over 350 million people annually travel to the coral reef coast, and the coral reef tourism sector has an estimated annual value of $36 billion globally, with over 70 countries and territories having "million dollar reefs"—reefs that generate over $1 million in tourism spending. Cruise-ship tourism has been growing (at least until the COVID-19 pandemic, see Box 7.3) at 7.7%, and can account for more than 90% of international visitors to some destinations. By 2030, coastal and maritime tourism is expected to comprise 26% of the total ocean industry value-added, and will employ c.1.5 million more people than it does at present. Coastal and maritime tourism generates indirect land activities linked to infrastructure construction that are responsible for pollution and destruction of natural habitats, as well as for pressure on natural resources, such as water, but also sand, limestone and wood (OECD 2016; Spalding et al. 2017; Tonazzini et al. 2019; WTTC 2019; UNWTO 2020a, 2020b).

Aquaculture, the farming of aquatic animals (e.g. finfish, molluscs and crustaceans) and seaweeds, is the fastest-growing food production sector in the world, with an average annual growth rate of 5.8% during the period 2000–2016. In 2016, global aquaculture production reached 80 million tonnes of food fish, with coastal aquaculture and mariculture (i.e. aquaculture in a marine environment) producing 28.7 million tonnes or 36% of this production. Aquaculture is mainly practised in tropical and subtropical regions and globally more than 60,000 km^2 of coastal areas is used for aquaculture. Asia accounted for 89% of global aquaculture production in 2016, much of which is produced in areas of former tidal flats and near-shore areas; China ranked first, followed by India, Indonesia, Viet Nam, Bangladesh, Egypt, Norway, Chile and Myanmar. Comparatively, Africa contributes the least of any continent to total global aquaculture production, yet the continent's aquaculture sector is growing faster than anywhere in the world, and accounts for 8% of the 12.3 million Africans employed in the fisheries sector.

Globally, the potential for onshore and offshore mariculture is large, and seafood production is expected to be predominantly sourced through mariculture by 2050. Significant areas of coast have been identified as areas that are suitable for further aquaculture development, including environmentally sensitive areas such as southern Patagonian coastal waters. The environmental impacts of aquaculture are well recognised and include the clearance of mangroves for shrimp ponds, eutrophication leading to disruptions to the surrounding benthic communities and increased phytoplankton and harmful algal blooms and disease (Kapetsky et al. 2013; Waite et al. 2014; Tenório et al. 2015; FAO 2018, 2020a; Obiero et al. 2019; Agarwal et al. 2019; Ahmed and Thompson 2019).

Offshore renewable energy, particularly offshore wind, is projected to grow significantly over the next decades. Under the International Energy Agency's (IEA) Stated Policies Scenario, installed capacity of global offshore wind is set to expand by at least 13%/year, increasing more than 15-fold over the 2018 installed capacity of 23 GW by 2040 (IEA 2019). Further extension of policy targets and falling technology costs may drive even greater uptake with over 560 GW installed capacity, accounting for 5% of global electricity supply, by 2040 in the Sustainable Development Scenario (IEA 2019). The UK Government 2020 levelised costs update shows continual reduction of offshore wind costs, being now lower than new gas and other fossil fuel generation, and projected to be less than onshore wind by the mid-2030s, owing to the relative strength and consistency of resource, and the large-size turbines able to be deployed offshore (BEIS 2020). Technical resource potential in shallow (<60 m) water depths, accessible to current fixed-bottom foundation wind technologies, is more than 87,000 terawatt-hours (TWh)/year. The emergence of floating foundation wind technologies removes depth constraints, and could provide access to another 330,000 TWh/year; 70% of the most accessible wind resource (20–60 km from shore) is located in water depths greater than 60 m (IEA 2019), which reflects the size of the opportunity to floating technologies. Other ocean-based renewable energy sources, including tidal, wave, floating solar and others, are less developed than offshore wind but also have significant potential for many regions where other drivers or advantages occur (see Haugan et al. 2020).

Over the coming decades, a number of industries are seeking to move further from the coasts in search of space to operate, or additional or more stable energy resources. Aquaculture, common in many inshore areas, will be much more prevalent offshore where larger, more complex, infrastructure designed to withstand the rigours of these environments will be required. Likewise, marine renewable energy infrastructures to harness wind waves and tidal power will become much more common.

2.3.4 Resource Extraction

The demand for food and materials—some traditional and others novel—from coastal environments has expanded rapidly in the last 50 years and will continue to do so over the coming decades, as growing coastal populations and a rising middle class seek greater protein in their diets, increased fresh water and more materials to build infrastructure. Box 7.2 briefly summarises three sectors—aquaculture, tourism and offshore renewable energy—that will see significant growth in coastal regions throughout the world over the coming decade.

The growing demand for global seafood still relies predominantly on coastal fisheries—those that occur less than 50 km from inhabited coastlines, or in waters less than 200 m deep. Despite significant declines over the last 60 years in a large number of exploited fish and invertebrate populations, coastal fisheries (see Gaines et al. 2019) still accounted for 55% (50–60 million tonnes/year) of global marine fisheries in the period 2010–2014. About 36% of this catch is from small-scale fisheries, undertaken mainly in developing countries and engaging more than 47 million people, nearly 50% of whom are women. These statistics not only highlight the global importance of coastal fisheries, but also the prominent role of small-scale fisheries in supporting coastal livelihoods, food security, nutrition and human well-being (World Bank et al. 2012; Monfort 2015; FAO 2020a, b; Palomares and Pauly 2019; Palomares et al. 2020).

Demand for fresh water for human consumption and agricultural and industrial use has rapidly increased and led to greater impoundment and extraction from coastal rivers and aquifers in drier areas, or where there is no longer sufficient water, the use of desalination plants has become commonplace. In 2000, there were c.45,000 reservoirs installed, and, as of 2014, at least 3700 major dams, each with a capacity of more than 1 megawatt (MW), were either planned or under construction, primarily in countries with emerging economies in Southeast Asia, South America and Africa.

While this construction will increase the present global hydroelectricity capacity by 73% to about 1700 gigawatts (GW), these impoundments will reduce the number of free-flowing large rivers by about 21% and trap 25–30% of the total global sediment load—and all of the bed load—that might otherwise be delivered to the coasts. Desalination facilities worldwide include about 16,000 operational plants with a global capacity of more than 95 million m^3 per day and the majority of these are from seawater or brackish water (21%). New ocean-water desalination projects are on the rise, including floating desalination plants constructed on ships and offshore structures, which have the advantage of being mobile (Vörösmarty et al. 2003; Syvitski et al. 2009; Grill et al. 2019).

Over the next 30 years, greater areas of irrigated agricultural land will be required, which, unless carefully managed, will have negative consequences for downstream coastal ecosystems. While today c.70% of irrigated areas are in Asia, under business-as-usual scenarios, by 2030 the total harvested irrigated area is expected to increase by at least 12%, to 394 million hectares (ha) (and perhaps as high as 1.8 billion ha), with approximately 9% of this growth expected to be in developing countries, especially those in Sub-Saharan Africa, but also South Asia and Latin America and the Caribbean. Irrigation is responsible for significant groundwater depletion in many regions, with about 11% of this resource embedded in the international food trade (Dalin et al. 2017; Ringler 2017; Puy et al. 2020).

Aggregates, such as sand—a key ingredient of concrete, asphalt, glass and electronics—and gravel, are the largest proportion of primary material inputs used in building and transport infrastructure (79% or 28.6 gigatonnes/year in 2010) and are the most extracted group of materials worldwide, exceeding fossil fuels and biomass. In most regions, sand is a common-pool resource, and even when sand mining is regulated, it is often subject to illegal extraction and trade. As a result, sand scarcity is an emerging issue with major sociopolitical, economic and environmental implications. Continued urban expansion and large infrastructure projects, as well as increasing trade and consumption, are pressuring sand deposits, causing conflicts, and compromising environmental and human systems. Activities such as unregulated sand mining of riverbeds, particularly in developing countries, can accelerate erosion and destabilise riverbanks and shorelines, and can harm benthic habitats, either through direct removal during dredging operations or from sedimentation. Transport of sand may also lead to increased biosecurity risks (Torres et al. 2017; Schandl et al. 2018; Koehnken and Rintoul 2018; Bendixen et al. 2019a; UNEP 2019).

More than 30% of current global energy demands are met by marine oil and gas reserves, and collectively the oil and gas sector accounts for one-third of the total value of the ocean economy. There are currently more than 6000 offshore and a few coastal (<200) platforms in service worldwide. As shallow-water fields become depleted and novel technologies emerge, production is moving towards greater depths and new territories. Other unconventional forms of gas, such as shale and natural hydrates, are also being increasingly exploited, as the technology to cost-effectively extract these reserves develops (Arthur and Cole 2014; US Department of Energy 2015; OECD 2016).

Despite the rhetoric of a sustainable ocean economy, there is growing scepticism that a business-as-usual scenario, favouring industrial and economic expansion of established and emerging industries, is being progressed without adequately addressing the equity, inclusion, access and benefit-sharing rights of those who also hold rights to the same resource (Selig et al. 2019; Bennett et al. 2019; Cisneros-Montemayor et al. 2019; Cinner and Barnes 2019; Hodgson et al. 2019; Cohen et al. 2019; Lau et al. 2019).

2.4 Summary

Large-scale declines in the extent of coastal landforms, vegetated ecosystems and biogenic structures over the last 40 years have occurred in many regions, and these declines have diminished coastal ecosystems' natural resilience to recover from a range of climate and anthropogenic threats, and to the biodiversity and services they support. The primary agents occurring on local to regional scales are the direct consequences of land-clearing and fragmentation, the degradation of these ecosystems from pollution, and imbalance in natural sediment supplies. Most of the remaining regions with a natural coastline are found in Africa and Asia, and these regions are also projected to experience the highest coastal population and urbanisation growth in the decades to come. Coastal ecosystems have been affected to varying degrees by sea level rise, ocean warming and acidification, and extreme weather and these effects are projected to be more significant in the future. Over the coming decades, further urbanisation and adaptation to rising sea levels and intensified storms will require even more coastal infrastructure.

This will require more material extraction, such as for aggregates to build infrastructure, new methods of environmentally sensitive construction with designs capable of withstanding future climate conditions. New forms of coastal infrastructure will also be required over the coming years to meet increased demand and access to coastal space. In an era of scarcity and increasing demand for fresh water, desalinisation plants will become much more common.

3 Risks to Coastal Resilience

Globally, coastal systems are undergoing profound, rapid and undesirable environmental changes, driven by the combined consequences of climate change, coastal development pressures and pollution, which leads to habitat loss and fragmentation—subdivision of habitat into smaller and more isolated patches. This degrades the ability of these ecosystems to provide essential ecosystem services. Anthropogenic threats to coastal systems can be exacerbated due to connectivity between marine, freshwater and terrestrial ecosystems, complicating the task of governance across the land–sea interface. Likewise, coastal settlements, their people, infrastructure and economies are increasingly at risk, as they struggle to adapt to these changes. In this section, we summarise the potential impacts on coastal ecosystem and services that arise from activities related to coastal development and industries, and review the risk to human populations, settlements, infrastructure and economies.

3.1 Threats to Coastal Ecosystems

3.1.1 Habitat Clearing and Fragmentation

The globally significant net loss of coastal landforms and vegetated and biogenic habitats that has occurred over the last half-century was summarised in Sect. 2.2, and includes erosion of depositional coastlines, loss of coastal vegetated ecosystems (50% of saltmarshes, 35% of mangroves), and coral (30%) and shellfish reefs (85%). These losses vary between regions, with some of the greatest losses occurring in Asia and Africa. While some of these changes have been incremental—although cumulative over time—in other cases, rapid/abrupt and potentially irreversible changes have also occurred. In some cases, such as mangroves and saltmarshes, the rates of loss in recent decades has decreased relative to changes that occurred 40–50 years ago, but other habitats, notably coral reefs and kelp forests, faced with the likelihood of more frequent and severe marine heatwaves in the future, are likely to see further significant and widespread losses.

However, the primary factor responsible for losses to date has been the clearing of coastal vegetated habitats to make way for agricultural, urban and industrial uses, and the reclamation of intertidal and subtidal areas (see Sect. 2.2 for more details). Less obvious, but equally pervasive, are the consequences of incremental fragmentation of these habitats, which, as has been highlighted in a number of recent publications, accrues significant cumulative net losses, impairs a number of ecosystem functions and services, and leads to declines in biodiversity for a range of taxa that rely on large intact areas for their home range, or as wildlife corridors on migratory routes. Patterns of fragmentation do not necessarily correlate with deforestation, or clearing, and relate to differing land-use and extractive activities. For example, in Ca Mau province, Viet Nam, over a 24-year period, the number of mangrove patches increased by 58% but the mean patch size decreased by 52%, and fish diversity was 1.8 times lower than in less fragmented mangrove forests (Tran and Fischer 2017; Jacobson et al. 2019; Bryan-Brown et al. 2020).

Other human activities, most notably alterations to natural patterns of river and surface water discharge, and the sediment, nutrients and pollutant loads that these carry to the coast, can have detrimental impacts on adjacent coastal habitats. While ecosystems, such as seagrass, oysters and coral reefs, are particularly sensitive to too-much sediment, in depositional coastal areas an adequate supply of sediment from upstream will be required to ensure the stabilisation of shorelines and the ongoing accretion of mangrove and saltmarsh habitat.

3.1.2 Pollution

An estimated 80% of pollution load in coastal environments originates from industrial, agricultural, urban/rural and other land-based activities, and is a key threat to biodiversity (IPBES 2019). While sediment and nutrients (principally nitrogen and phosphorus) occur naturally in the environment, excessive levels released from point sources (wastewater effluent, storm-water outfalls and runoff from waste storage) and nonpoint sources (deforestation, land conversion and runoff from agriculture or ranching) into rivers and estuaries, or directly into coastal and marine ecosystems, are considered serious threats.

Among developed nations, it is estimated that more than 70% of wastewater is treated with discharges to sewered connections centralised in wastewater treatment plants, where remedial technologies improve the quality of the effluent to differing standards—tertiary treatment, which removes nutrients, being the best. The quality of the discharge is often regulated by the setting and reporting of established concentration or load-based criteria. Over recent decades, this has resulted in significant reductions in nutrient loads from major coastal cities discharged into rivers, estuaries and coastal waters. However, among developing nations only 8% of generated wastewater is treated and most people rely on some form of decentralised or self-provided services. With increasing urbanisation, especially in Africa and Asia where the urban population is expected to grow by 2.5 billion over the next 30 years, there is an urgent need to better treat urban waste (Sato et al. 2013; Gallego-Schmid and Tarpani 2019).

As detailed in the companion Blue Paper, *Leveraging Multi-Target Strategies to Address Plastic Pollution in the Context of an Already Stressed Ocean* (Jambeck et al. 2020), marine litter is a global environmental concern, entering the ocean largely through storm-water runoff, but is also dumped on shorelines or directly discharged at sea from ships. Between 61 and 87% of this litter is plastics, and since the 1950s this has increased dramatically, with current estimates of between 4.8 and 12.7 million tonnes of land-based plastic waste ending up in the ocean every year, while in the next two decades, the amount of plastics produced is expected to double (Jambeck et al. 2015, 2020; Geyer et al. 2017; Löhr et al. 2017; Barboza et al. 2019; Walker et al. 2019; Galgani et al. 2019).

3.1.3 Bio-invasions and Disease in Coastal Ecosystems

In an increasingly tele-coupled world, invasive alien species—most commonly introduced via shipping and associated coastal infrastructure—threaten global biodiversity, economies and human health. Shipping is the primary vector for 60–90% of marine invasions globally, transporting marine species, including plankton, crustaceans and molluscs, in ballast water or attached to ships' hulls. Terrestrial pest species can also be transported in carried goods and their packaging, and upon arrival at destination ports can be rapidly spread inland along transportation chains. Once introduced, alien species can rapidly establish, particularly in areas that have already been disturbed, displacing native species, altering ecosystem structure and functions such as nutrient cycling and carbon sequestration. Well-known examples include invasions of coastal wetlands, dunes and saltmarshes by vascular plant species, marine algae and plankton, which increasingly result in occurrences of harmful algal blooms, by molluscs such as the Asian green mussel (*Perna viridis*) and by echinoderms such as the Northern Pacific seastar (*Asterias amurensis*). Projected increases in global maritime traffic of 240–1200% by 2050 are forecast to lead to a 3- to 20-fold increase in global invasion risk and this will occur mainly in middle-income countries. Significantly, Northeast Asia will not only be disproportionally affected but will also be the primary vector source to other geo-regions (Pyšek et al. 2008; Early et al. 2016; Seebens et al. 2015; Carrasco et al. 2017; IPBES 2019; Sardain et al. 2019).

Marine organisms serve as hosts for a diversity of parasites and pathogens affecting not only the host population that can include vertebrates, invertebrates and plants, but can also cascade through ecosystems altering the structure and function of marine ecosystems. Marine diseases can become emergencies when they result in significant ecological, economic or social impacts, so understanding the factors responsible for the genesis and timing of diseases will be increasingly important as our use of coastal and marine resources accelerates. The billions of dollars lost in the early 1990s as a result of a global pandemic of white spot syndrome in penaeid shrimp aquaculture, and the environmental and economic impacts of coral diseases that led to widespread mass mortality in Caribbean reefs and reduced ecotourism, are salient examples.

Marine disease emergencies can also have significant social impacts, capable of disrupting public safety, threatening human health, or decreasing the resilience of local human communities. The probability of humans acquiring infections from marine mammals, avian influenza from marine birds, and cryptosporidiosis and vibriosis from consumption of shellfish is also expected to increase unless carefully managed, with better surveillance, impact mitigation, and adaptive and responsive strategies. It should be noted that COVID-19 is not considered infectious to marine organisms (Ward and Lafferty 2004; Groner et al. 2016; Mordecai and Hewson 2020).

3.2 Risks to Coastal Ecosystem Services

Coastal ecosystems, their biodiversity and functions provide important provisioning goods, as well as regulating, supporting and cultural services that underpin the ocean economy and that also have values that are not explicitly economic. Provisioning goods, such as the harvesting of fish or timber from coastal habitats, represent products that are consumed. Growing demand for these products is a key driver in the conversion of habitats for these provisioning goods. Regulating services represent intangible benefits provided when ecosystems are left intact, such as flood and erosion reduction, and underpin provisioning goods such as fisheries production. Coastal areas also provide for uses that are considered aesthetic, spiritual and cultural services, such as sacred sites or points of historic interest. Such services are not easily valued in economic terms and thus lead to questions as to whether the concept of ecosystem services is an overly transactional view of nature, and whether the benefits that people receive can be represented better by frameworks that are less anthropocentric.

A central dilemma facing coastal ecosystems, and achieving a sustainable ocean economy more generally, is reconciling the competing demands for provisioning goods and services with the need for regulating, maintenance and cultural services (HM Treasury 2020). Loss or impairment to coastal ecosystems can result in a concomitant, although often non-linear, loss of service(s). It is notable that the most recent IPBES Global Assessment Report and World Economic Forum Global Risks Report both ranked biodiversity loss and ecosystem collapse in the top five risks to the global economy (IPBES 2019; WEF 2020a).

While provisioning services can be readily measured and valued, regulatory, supporting and cultural services are much harder to quantify and only rarely are they directly accounted for in coastal management because their services are not quantified in terms familiar to decision-makers, such as (loss of) annual expected benefits (Beck et al. 2018a). There are several competing lines of thought about this conundrum. Some argue that we should accept that some values cannot, and perhaps should not, be measured and monetised, and that we need to invoke other frameworks to accommodate the different types of values (Sagoff 2008). Others argue that incorporation into systems derived from economic accounting is an efficient way to ensure that resources are devoted to conserving the ecosystems. From the latter are emerging global standards, such as the System of Environmental-Economic Accounting (SEEA), which uses natural capital accounting frameworks and associated methodologies to classify and place a monetary value on even intangible services. These frameworks can be integrated with traditional economic national accounts, allowing them to be explicitly considered in resource and environmental decision-making. They might also facilitate development of financial instruments, such as payments for ecosystem services (PES), to incentivise the conservation or repair of natural assets. More details are provided in the companion Blue Paper *National Accounting for the Ocean and Ocean Economy* (Fenichel et al. 2020), while below we briefly summarise the key coastal services, the value they provide and the risks if they are diminished or lost.

3.2.1 Coastal Protection

Coastal vegetation and reefs can contribute significantly to coastal protection by absorbing the energy of wind and waves and providing a buffer that helps to minimise erosion and limit the intrusion of storm surges and damaging floodwater. As such, they provide significant annual flood protection savings for people and property, particularly from the most frequent storms. Globally, and averaged across these ecosystems, it is estimated that they can together reduce wave heights between 35% and 71%, with mangroves and reefs providing annual storm and flood protection benefits exceeding $65 billion and $4 billion, respectively.

Along the Northeastern seaboard of the United States, saltmarshes avoided costs of $625 million in direct flood damages resulting from Hurricane Sandy in 2012. In the Philippines, it is estimated that annually mangroves reduce flood-risk for more than 613,500 people, 23% of which live below the poverty line, and avert damages of $1 billion to residential and industrial property. Coral reefs protect more than 18,000 people from flood damage and avoid costs of $272 billion.

Without mangroves, it is estimated that a further 15 million people would be potentially exposed to flooding annually across the world, while the absence of reefs would more than double the expected damage from flooding, and costs from frequent storms would triple. Many countries (notably Bangladesh, Cuba, China, India, Indonesia, Malaysia, Mexico, Philippines, USA and Viet Nam) are estimated to gain annual expected flood savings exceeding $400 million, while some small (20-km) coastal stretches, particularly those near cities, receive more than $250 million in flood protection benefits from nearby mangrove forests (Spalding et al. 2014; Narayan et al. 2016, 2017; Beck et al. 2018a, b; Reguero et al. 2019b; Storlazzi et al. 2019; Menendez et al. 2020).

3.2.2 Carbon Sequestration

Coastal vegetated wetlands, including saltmarshes, mangrove forests and seagrass meadows, are considered to be the main blue carbon habitats, due to their ability to sequester and store large amounts of carbon within their root systems and in the underlying soil in which they grow. Despite their relatively small global extent (equivalent to 0.2% of the ocean surface), these vegetated coastal ecosystems contribute approximately 50% of the carbon sequestered in marine sediments, absorb CO_2 up to 40 times faster than terrestrial forest and consequently are globally equivalent to c.10% of the entire net residual land sink. Consequently, it is now well recognised by many nations and organisations that the conservation and restoration of these blue carbon ecosystems is an effective climate solution that could deliver substantial mitigation of CO_2 through storage and sequestration, as well as delivering other important benefits, like enhancing livelihoods and reducing risks from storms. Other research commissioned by the Ocean Panel estimates that coastal blue ecosystems could, by 2030, contribute avoided emissions of 0.32–0.89 billion tonnes of CO_2e per annum and this would increase to 0.50–1.38 billion tonnes of CO_2e by 2050. However, impairment or loss of these blue carbon ecosystems can contribute significant emissions. For example, as a result of global net losses in mangrove ecosystems between 1996 and 2016, global mangrove carbon stocks have declined by 1.5% (0.16 billion tonnes) with greatest losses occurring in Indonesia, which has the largest areal extent of mangroves, but also in countries such as Myanmar, where mangrove clearing rates today still remain high. Internationally, many countries with large blue carbon stocks seek to recognise the mitigation potential as part of their national emission reduction commitments. For example, preventing mangrove deforestation in Indonesia could reduce emissions from land-use change by 30%. Efforts to halt and reverse this trend could be supported by the private sector, as many companies look to offset their carbon emissions though investments in blue carbon protection or restoration (McLeod et al. 2011; Atwood et al. 2017; Serrano et al. 2019b; Spivak et al. 2019; Lovelock and Reef 2020; Richards et al. 2020).

3.2.3 Fisheries Productivity

Another important service of mangroves, marshes, reefs and seagrass beds is that they provide breeding and nursery habitat for a number of commercially important inshore and offshore fisheries. The complex structure of these habitats provides juveniles with refuge from predators and access to a variety of food sources that sustain their growth into adulthood. The fisheries value can be highly site specific and often more than one of these habitats may contribute to the life cycle of the fishery, including larval dispersal and migration to offshore habitats. The economic values of these fisheries also vary according to specific costs associated with each fishery, proximity to economic markets and levels of utilisation. Inshore, small-scale fisheries target a variety of mollusc, crustacean and fish species, often for domestic use or in local markets, and are of relatively low economic value relative to offshore fisheries harvested mainly by industrial-scale operations. In many tropical coastal communities, gleaning—fishing for invertebrates such as sea cucumbers in water that is shallow enough to walk in—is often done by women and children, and provides a source of essential protein, micronutrients and income. Mangrove habitats adjacent to large river mouths, where freshwater and high nutrients enhance productivity, generate the highest numbers of juvenile fish (e.g. mangrove jack, *Lutjanus aregentimaculatus*) and commercially important species like shrimp. While many coastal small-scale fishers are fully aware of their reliance on mangroves, larger commercial fisheries, such as the shrimp industry, operating in offshore waters often overlook the mangroves on which they depend.

Box 7.3. Enabling Coastal Resilience in a COVID World

COVID-19 is having serious and significant impacts on national economic growth trajectories, including coastal economies. The hardening of borders, limited movement of people, shrinking income opportunities, disruption of globalised supply chains and rise in restrictive trade policies are emerging as early consequences of the global pandemic that are relevant to coastal economic sectors. Poor urban coastal communities are most vulnerable to the pandemic since they live in crowded areas in low sanitary conditions often at the water's edge. The reduction of income for coastal residents, social distancing and quarantine, and even the provision of basic food supplies to coastal communities, are proving difficult. The impacts are most profound for marginalised groups and increase the social and environmental stressors, as well as exacerbating the challenges of disaster response in coastal contexts (CIRAD 2020; UNCTAD 2020a; WEF 2020b).

Coastal and ocean economy sectors, such as fisheries, aquaculture and seafood processing, tourism and recreation, maritime transport and logistics, are most impacted by the pandemic. Restrictions to ship docking, limited road transport and access to ports, falling demand for fish products and for tourism and recreation all reduce the income of the coastal and ocean economy sectors and associated jobs, as well as impacting on those who work on board vessels, with accounts of crew being stranded at sea for months (Bennett et al. 2020; Gössling et al. 2020; OECD 2020a, b).

COVID-19 has exposed weaknesses in the complex global fisheries and seafood production system and supply chains. Impacts on the hospitality sector and live export markets has led to international demand for fresh fish dropping dramatically and prices dropping accordingly. At the same time, demand for canned tuna has been maintained as it is seen as desirable as a source of shelf-stable protein, and some markets have seen increased demand (FAO 2020c, d; OECD 2020a, b).

The small-scale fisheries sector has been particularly hard hit, especially where perishable product is dependent on being sold through wet markets and then processed locally. Small-scale fisheries and fish processing are high-employment, low-wage sectors, with a high proportion of women working in fish processing facilities, and where proximity puts workers at risk of COVID-19. Entrepreneurial vendors are using digital technology to connect directly to customers, but the closure of wet markets and the closure of processing facilities has meant that a large proportion of product has no pathway to market (Bennett et al. 2020; CIRAD 2020; Davey and Steer 2020; FAO 2020c, d; OECD 2020a, b).

Positive stories have emerged from several Pacific Island nations, where practices such as food sharing have restarted and local food networks have been revived, and where collective actions have worked to safeguard rights. There are also stories of increased pressure on natural resources, through more fishing effort, regulations being relaxed and areas being opened up to fishing, including as people move back to their home communities from major cities, because of the loss of jobs. Using coastal resources as a social safety net, and relaxing rules during times of economic crisis is a high-risk strategy and could lead to greater problems in the long term (Bennett et al. 2020; CIRAD 2020; Davey and Steer 2020).

Tourism is one of the economic sectors hardest hit by COVID-19. Economies and communities with a high dependence on international tourism receipts have been badly affected by travel bans and restrictions. Tourism is a high-employment, low-wage sector, often employing large numbers of young people, and is particularly important as a source of GDP for many SIDS economies. Many coastal hotels and recreation facilities are facing bankruptcy, and stimulus options are urgently needed for this sector to preserve the long-term potential and to engage the affected workforce. Reskilling in digital technologies, engagement in natural resource recovery programmes or mobilising the workforce into nation-building sustainable natural infrastructure programmes are all options that could be explored (Gössling et al. 2020; OECD 2020a, b; Vianna et al. 2020; WEF 2020b).

Seagrasses also support substantial fisheries, both from small-scale fisheries that target species that rely on seagrass for most of their life (e.g. rabbitfish) or species that rely on seagrass in early life stages before they move offshore (e.g. northern Atlantic cod). Seagrass meadows are also popular locations for small-scale mariculture, like sea cucumbers and seaweeds (Benzeev et al. 2017; Carrasquilla-Henao and Juanes 2017; Worthington and Spalding 2018; Waltham et al. 2019; Jinks et al. 2020; Vianna et al. 2020; Waltham et al. 2020).

3.3 Risks to Coastal Populations, Infrastructure and Economies

Coastal communities, built infrastructure, and established and emerging economic sectors are significantly affected through the disruption of coastal physical processes resulting from climate change and coastal and upstream development. Globally, around 10 million people experience coastal flooding due to storm surges, cyclones and heavy rainfall every year with impacts ranging from displacement of households and destruction of sources of livelihoods, to disruption of national economies. The World Economic Forum's two most recent Global Risks Reports ranked extreme weather, preparing cities for sea level rise biodiversity loss, and ecosystem collapse in the top five risks.

While the consequences of COVID-19 (see Box 7.3) on the resilience of coastal ecosystems will continue to unfold over many years, the immediate impacts on coastal-dependent industries, such as tourism, and the flow-on effects on the economies of nations and livelihoods of local communities, are profound (Vitousek et al. 2017; Bergillos et al. 2019; Hino et al. 2019; DasGupta and Shaw 2015; Betzold and Mohamed 2017; Kramer et al. 2017; Hagedoorn et al. 2019; WEF 2019, 2020a).

3.3.1 Populations

An estimated 310 million people, and $11 trillion in GDP, are exposed globally to the risk of a 100-year flood event. Risk is expected to increase, due to rising sea levels and other climate-related threats concurrent with population growth. If no mitigation measures are undertaken, by 2050, c.9 million of the world population, concentrated in more than 570 coastal cities, situated in low elevation areas, notably in China, Bangladesh and Indonesia, could suffer from enhanced inundation and increased coastal erosion. By 2060, up to 411 million people could be exposed to the risk of a 100-year flood event (Ericson et al. 2006; Hallegatte et al. 2013; Hinkel et al. 2014; Wong et al. 2014; Neumann et al. 2015; Reguero et al. 2015; Arnell and Gosling 2016; Lumbroso 2017; Brown et al. 2018; Barnard et al. 2019; Nicholls et al. 2020).

Both sea level rise and extreme coastal events cause massive and existential displacement of populations. Sea level rise has already affected many low-lying islands, such as Kiribati, and Isle de Jean Charles in Louisiana, USA, and resettlements of populations are either under way or planned. After the Indian Ocean tsunami, in the coastal areas of the provinces of Aceh and North Sumatra in Indonesia, over half a million people, including some 300,000 living in severely damaged areas, were displaced. The task of resettling these residents, while keeping their sense of community, serves as a test case for future events (UNDP 2005; McGranahan et al. 2007; Birkmann et al. 2013; Gray et al. 2014; Wilkerson et al. 2016; Oliver-Smith 2019; Visessri and Ekkawatpanit 2020).

The risk posed to coastal populations depends not only on the exposure to the hazard, but also on social conditions (susceptibility) and capacities to respond (resilience) and together describe the vulnerability of societies. As a result, countries have different risks, with tropical states and SIDS in the Caribbean and Oceania and coastal areas in Southeast Asia (Bangladesh, Myanmar, Papua New Guinea and Timor-Leste) being most at risk (Fig. 7.3). Countries in Africa have high overall risk, as vulnerability scores are high and exposure to coastal hazards and adaptation are generally low; in contrast, countries like the Netherlands and Japan have high exposure rates but are more resilient (Beck 2014).

Coastal indigenous peoples, particularly those inhabiting islands or archipelagos, are some of the most vulnerable populations to coastal hazards. Often their traditional and customary use areas are not recognised and their access to cultural and spiritual sites of importance is not upheld, including where national and multinational interests seek access to the coast (see Box 7.4).

To mitigate the impacts of the pandemic, government and industry need to address the immediate economic and social hardships caused by the pandemic and enable coastal communities to maintain their resilience and rapid after-pandemic recovery, while maintaining their long-term goals of protecting coastal natural resources, the coastal environment and ecosystems. This can be done by supporting the incomes of and providing healthcare to the most vulnerable groups and ensuring that evidence-based management remains in place and is enforced. It is estimated that $10–20 trillion of public funding will be mobilised into the world economy in the next 2–3 years to support and stimulate economic recovery, including the recovery of coastal economic sectors. Therefore, a unique window of opportunity exists to engage and influence relevant policy and investment decisions and ensure stimulus funds foster sustainable ocean economic pathways and support the recovery and development of impacted communities. For example, coastal restoration can be used to help economic recovery from the COVID-19 pandemic while providing co-benefits of ecosystem services, community cohesion and climate adaptation (ADB 2020; OECD 2020a).

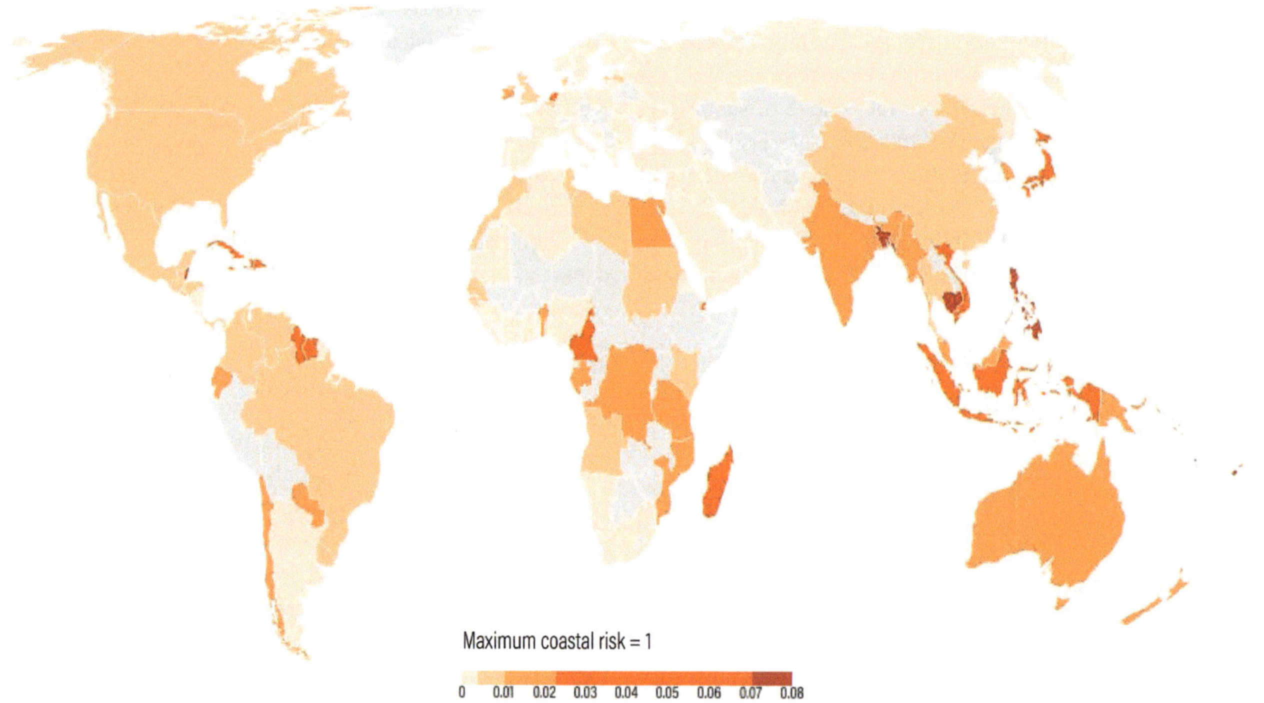

Fig. 7.3 Coastal risks to nations and geo-regions. (Source: Beck (2014). Data from The Atlas of Ocean Wealth (https://oceanwealth.org))

Box 7.4. Vulnerabilities of Coastal Indigenous Peoples and People from Traditional Communities

Coastal indigenous peoples, including those of SIDS, comprise some 370 million people, or 5% of the global population. As they rely on ocean resources and are highly vulnerable to ecosystem and economic change, the exploitation of fish resources and climate hazards pose distinct threats to these communities (Cisneros-Montemayor et al. 2016). Coastal indigenous peoples consume approximately 1.9 million tonnes of fish per year, approximately 2% of the global catch, and this seafood demand is concentrated around equatorial regions in Africa and Asia, and in the Arctic. In many of these areas' fisheries, stocks (e.g. of Pacific tuna; Bell et al. 2015) are changing migration and distribution patterns in response to global environmental changes, and traditional fisheries areas are under mounting pressure from foreign and domestic fishing fleets. Already, people in the 22 small island nations and territories of the southwest Pacific have increased their reliance on imported foods, including canned meats and packaged products, in part because of depleted fish stocks. Food imports to countries such as Samoa and Tonga now exceed total exports. These societal shifts have strong negative implications for the health and well-being of indigenous peoples. For instance, deaths in the Pacific from preventable non-communicable diseases, such as diabetes, cardiovascular disease and cancer, have risen, in part because of the dietary and lifestyle changes that have accompanied people's increased reliance on food imports (Morrison et al. 2019).

Coastal indigenous peoples are some of the most vulnerable populations to coastal hazards, such as storms, cyclones and tsunamis. While efforts to mitigate the impacts of these hazards mainly focus on defence infrastructure development, or early warning systems, the traditional and local knowledge of these communities has been found to increase their resilience and help them to manage crises—be it natural hazards, economic problems or political conflicts (Hiwasaki et al. 2014). Furthermore, many indigenous communities live in regions without strong governance, although a number of international agreements and bodies (e.g. United Nations Declaration on the Rights of Indigenous Peoples; Convention on Biological Diversity; Intergovernmental Panel on Climate Change) recognise preferential access rights for indigenous peoples, their vulnerability to climate and food security, and the value of traditional ecological knowledge (Cisneros-Montemayor et al. 2016; Vierros et al. 2020).

3.3.2 Infrastructure

Building resilient communities is a shared challenge for the world's population living along the coast now and in the future. To address this challenge, communities typically engineer barriers along the coast. However, there is growing understanding that traditional approaches to coastal protection (e.g. seawalls, bulkheads) are unsustainable. Hardened shorelines can be expensive to build and maintain, and can lead to unintended shoreline erosion, degradation or loss of habitat, impacting on communities that depend on healthy coastal ecosystems for protection, subsistence and livelihoods. However, decision-makers often lack basic information about where and under what conditions ecosystems reduce risk to coastal hazards and who would benefit from the protective function conferred by those ecosystems (Adger et al. 2005; McGranahan et al. 2007; Kron 2013).

The proportion of the world's gross domestic product (GDP) annually exposed to tropical cyclones has increased from 3.6% in the 1970s to 4.3% in the first decade of the 2000s. Flood assessment of 136 major coastal cities shows that average flood losses in 2005 were about $6 billion/year, and in the last 10 years insurers have paid out more than $300 billion for coastal storm damage. Considering the risks from sea level rise and sinking land, both the World Bank and the Organisation for Economic Co-operation and Development (OECD) estimate that, by 2050, global flood damage in large coastal cities could cost $1 trillion a year, while climate-induced declines in coastal and ocean health will cost the global economy $428 billion/year, and global infrastructure investment of more than $94 trillion will be required to reduce these risks (UNISDR 2011; Hallegatte et al. 2013; Diaz 2016; Oxford Economics 2017; IPCC 2019; ORRAA 2019; WEF 2019).

Box 7.5. Sinking Cities

Land subsidence is one of the world's under-rated problems, yet its impact on many coastal cities is increasingly apparent. Many of the world's sinking cities are built on low-lying marshes, flood plains or river deltas, where soft sediments compress under the weight of infrastructure, and this is exacerbated by groundwater or oil/gas extraction for human use, as well as reductions in sediment supply due to dams and impoundments. The increased frequency and magnitude of extreme weather events and changing sea levels further increase the risk of flooding, the consequences of which include structural damage to infrastructure, drains and sewage systems and high maintenance costs for roads and railways.

Cities that have grown rapidly, or have failed to curb groundwater usage, are particularly at risk, most notably in Asia (e.g. Jakarta, Guangzhou, Shanghai, Dhaka, Ho Chi Minh, Bangkok), but also on other continents, including the Americas (e.g. Mexico City, Houston, New Orleans), Africa (e.g. Lagos) and Europe (e.g. London, Rotterdam, Venice). Jakarta is the world's fastest-sinking city, at a rate of c.25.4 centimetres (cm)/year.

Around 40% of the city now lies below current sea levels and some coastal districts have sunk as much as 4.3 m in recent years. With further population growth, urbanisation, intensification of economic activities in deltas, and climate change, the problem is set to accelerate. Stopping the pumping of groundwater is one of the practical and local actions that can be readily implemented. A century ago, Tokyo was sinking at a greater rate than Jakarta is now. Following the Second World War, laws limiting pumping and a programme to re-inject water back into the ground has stabilised land subsidence such that, by the early 2000s, the city's subsidence slowed to 1 cm a year (Sato et al. 2006; Kramer 2018).

3.4 Summary

Coastal environments and dependent human communities are already experiencing the impacts of climate-related changes from extreme events and slow-onset changes, and the consequences of rapidly growing and urbanising populations that demand great access to greater resources, built infrastructure and services, and space. These climate and development changes can act synergistically and result in cascading and hard-to-predict impacts, as the world has seen with the global COVID-19 pandemic, in which core vulnerabilities have been exposed with devastating consequences.

While across the globe there is regional variability in how coastal environments will be affected, these changes will continue for many decades irrespective of actions taken, while no action will result in disproportionately higher risks, and a return to previous conditions should not be expected.

4 Building Coastal Resilience

To effectively manage the challenges wrought by coastal development and climate change, there are four main management strategies that can be used to secure the integrity and resilience of coastal ecosystems and their contributions to people:

1. *Protection strategies* use regulations and area-based management, to designate where and how much of specified activities can and cannot occur in coastal environments and in the adjacent catchment, and legislate areas for conservation such as marine protected areas (MPAs) or implement area, habitat and species-specific conservation plans.
2. *Mitigation strategies* aim to reduce local stressors caused by human action through the use of technology, regulation and the promotion of stewardship to minimise the introduction of pollutants, the over-exploitation of resources or activities that will otherwise harm coastal environments.
3. *Adaption strategies* explicitly consider the coastal social-ecological system and are implicitly related to resilience; adaptation leads to resilience and resilience is a property needed for having capacity to adapt (Nelson 2011). They use principles of ecosystem-based adaptation and ecological engineering to incorporate natural infrastructure into existing grey infrastructure, relocate at-risk activities and populations away from the coast, and also use incentives to change behaviours and practices.
4. *Repair strategies* seek to restore damaged ecosystems by restoring the composition and/or function of lost or fragmented habitats, restoring (reinstating) the natural hydrology, sediment and nutrient balance entering and cycling through coastal ecosystems, or by assisted evolution.

Figure 7.4 represents 17 actions that can be taken under these four strategies and highlights the enabling conditions needed to ensure their success.

All four strategies broadly fall under the umbrella framework of nature-based solutions (NbS), which are defined by the IUCN as "actions to protect, sustainably manage, and restore natural or modified ecosystems, that address societal challenges effectively and adaptively, simultaneously providing human well-being and biodiversity benefits" (Cohen-Shacham et al. 2016). NbS is an area that covers a range of ecosystem-related approaches to protect (i.e. area-based conservation), to manage holistically (e.g. integrated coastal management or ICM), to address specific issues and to repair and restore ecosystems.

NbS approaches are now being used to reframe policy debates on biodiversity conservation, climate change adaptation and mitigation, and sustainable use of natural resources, to address conflicts and trade-offs associated with use and management of ecosystem services, and to invest in blue infrastructure and ocean finance (World Bank 2008; United Nations 2013; Nesshover et al. 2017; Thiele et al. 2020; WEF 2020c). But not all strategies are applicable in a given situation, and evaluating a broad range of actions, and combinations of actions, can help decision-makers to estimate the trade-offs of different management approaches and to maximise the co-benefits. In fact, comprehensively tackling issues, such as reducing pollution or preventing clearing of mangroves and saltmarshes, will require a mix of all four strategies.

The success of any of these strategies is predicated on the presence of a number of enabling factors or conditions that encompass the dimensions of technical readiness, social equity, economic viability and environmental sustainability. Some of these are shown in the outside ring of Fig. 7.4 and are summarised in Sect. 4.6; they also form the basis for many of the opportunities for actions outlined in Sects. 5.2 and 5.3. In particular, integrated management is listed here as an enabling factor as it provides the framework with which these various strategies can be applied across the terrestrial-coast ocean continuum, between institutional lines of responsibility, as well as integrating with other relevant agendas, such as those for climate action and urban transitions. Approaches to coastal integrated management are discussed below but are considered in detail in the companion Blue Papers *Integrated Ocean Management* (Winther et al. 2020) and *The Ocean Transition* (Swilling et al. 2020).

As part of the four strategies outlined above, the approaches and activities most useful to ensuring coastal resilience are evaluated below and form the basis for the opportunities for actions presented in the following sections: area-based measures for protecting coastal ecosystems; mitigating terrestrial impacts on coastal environments; adapting infrastructure; and restoring coastal habitats.

Fig. 7.4 Four strategies and actions for building coastal resilience and the enabling conditions to achieve them. (Source: CSIRO)

4.1 Protecting Coastal Ecosystems

The first line of defence in ensuring coastal resilience is to provide adequate protection of coastal habitats from inappropriate forms, or unsustainable levels, of human use, and to secure the rights of peoples with recognised tenure and customary access rights. Protecting coastal habitats is more cost-effective and has better ecological outcomes than rehabilitating lost habitat. For example, protection of mangroves provides an immediate benefit–cost ratio of 88, compared with restoration activities which have a ratio of 2, because they require higher logistical costs and take decades to realise the benefits (Konar and Ding 2020).

Protection can only occur where there is a clear, effective and enforceable regulatory framework in place, with national and subnational policies and regulations that, among other things, forbid the clearance of natural coastal lands for other activities, designate appropriate human activities that are allowed to occur within defined coastal areas while minimising harm, and set limits to levels of resource extraction or activity. In all cases, such regulations are most effective when ownership is clearly established. However, as many of these regulations pertain to single areas, sectors or individual developments, a priority for further enhancing protection of coastal ecosystems is to improve legislation, policies and planning frameworks to better consider multiple pressures and cumulative impacts from marine and land-based activities. Ensuring there is a comprehensive monitoring program and research agenda in place to assess and predict potential impacts and develop effective management strategies is also required (Griffiths et al. 2019).

There are a number of international conventions and agreements that relate to various aspects of coastal management, including conservation of coastal environments and biodiversity (e.g. Convention on Biological Diversity (CBD), Ramsar Convention on Wetlands) and controlling pollution (MARPOL Convention, United Nations Framework Convention on Climate Change (UNFCCC)), which can be built upon to provide greater levels and greater breadth of protection for coastal ecosystems and their services. Both the SDGs (14.5) and the CBD Aichi Targets (11) commit nations to conserve at least 10% of their coastal and marine areas by 2020; and it is now being advocated that at least 30% will need to be protected by 2030, with the remaining areas also under environmental management (World Conservation Congress 2016; Laffoley et al. 2019; Roberts et al. 2020). The UNFCCC nationally determined contributions (NDCs) for greenhouse gas emission reductions under the Paris Agreement and the Sendai Agreement for Disaster Risk Reduction 2015–2030 may secure better protection of coastal ecosystems through recognition of their carbon sequestration and climate protection services. However, many of these agreements are usually voluntary and non-binding and, as Winther et al. (2020) note, "it is failure to implement these existing international instruments at national levels that is one of the most important weaknesses of ocean governance".

Over the last 30 years, a number of integrated planning frameworks, conservation and spatial management tools and processes have been developed and implemented to protect coastal ecosystems, and minimise multi-sector competition for resources or space. Best known is integrated coastal management (ICM), also known as integrated coastal zone management (ICZM), which aims to balance the complexities and potential conflict of growing uses of the coastal zone through the use of relevant legislation and policy and spatial and conservation management tools to integrate planning, decision-making and management across sectors and across land and sea estates, and aspires to consider cumulative effects and trade-offs (Álvarez-Romero et al. 2011; Bernal 2015; Cicin-Sain 1993; Katona et al. 2017; Stephenson et al. 2019).

ICM principles and frameworks have been implemented at global, regional and national scales. Many countries have sought to implement ICM in several forms and with various degrees of success. For example, many countries in East Asia, including Viet Nam, the Philippines, China, and the Republic of Korea, have institutionalised ICM in national legislation, and this has supported countries in the region to improve coastal management and to enhance the effectiveness of use and conservation of coastal natural resources and environment. Regionally, intergovernmental cooperation, such as the Partnerships in Environmental Management for the Seas of East Asia (PEMSEA), has for more than 25 years applied ICM solutions in dozens of sites across East Asia, covering around 38% of the region's coastline, across 12 countries (see Box 7.6).

Success in implementing ICM in some countries has been hindered by the absence, or limited presence, of key enabling factors, including inadequacies with legal frameworks, poor cooperation between different sectors or government departments, lack of personnel, capacity and access to knowledge (White et al. 2006; Shipman and Stojanovic 2007; Borja et al. 2008; Nguyen and Bui 2014; Elmgren et al. 2015; Candel 2017; Liu and Xing 2019; PEMSEA 2020).

ICM is a dynamic process and continues to evolve, with greater emphasis on better management across the catchment–coast–ocean continuum, coupling coastal, water and urban frameworks, integrating climate and disaster risk reduction and management. A terrestrial–ocean integrated climate policy is part of a larger changing narrative about the ocean and the recognition of its untapped potential for climate regulation, mitigation and adaptation, and our aspirations for a sustainable ocean economy. There are significant opportunities for alignment with Integrated Water Resources Management initiatives, including the UN's 2018–2028 Water Action Decade and the urban sustainability agenda (discussed further in Sect. 4.2).

Today, an integrated management framework, coupled with an ecosystem-based approach to management and supported by marine spatial planning, including the use of MPAs and other effective conservation measures (OECMs), is recognised as best practice. Ecosystem-based approaches and management are based on the application of scientific methodologies, focused on levels of biological organisation, which encompass the essential structure, processes, functions and interactions among organisms and their environment. These approaches have been most widely applied and institutionalised into fisheries management. For example, Indonesia and the Philippines have both recently adopted ecosystem-based fisheries management and spatial closures by designating a number of Fisheries Management Areas (Mokhtar and Aziz 2003; Levin et al. 2009; Saad et al. 2012; Ureta et al. 2016; Altenburg et al. 2017; Gelcich et al. 2018; Muawanah et al. 2018; Alexander et al. 2019; Alexander and Haward 2019; Kirkfeldt 2019).

MPAs are one of the most widely implemented area-based management tools used by countries to protect valuable or representative coastal and marine areas, and, increasingly, areas of the high seas. MPAs vary in levels of protection, from marine reserves and parks that provide full protection to multiple-use areas that restrict some activities in some areas, such as no-take areas. In most countries, multiple-use MPAs are the most common form (>75% in 2013). More than 40% of mangroves and warm water coral reefs are placed within gazetted MPAs, while seagrasses and estuaries are the habitats with the lowest proportion of area (<30%) contained within MPAs (Toonen et al. 2013; Costello and Ballantine 2015; Jacobsen 2019; Bryan-Brown et al. 2020; Rogers et al. 2020; UNEP 2020).

Box 7.6. Regional Coastal Management Strategies

Partnerships in Environmental Management for the Seas of East Asia (PEMSEA): PEMSEA is an intergovernmental organisation operating in East Asia to foster and sustain a healthy and resilient ocean, as well as coasts, communities and economies across the region. PEMSEA serves as the regional coordinating mechanism for the shared regional coastal and marine strategy, Sustainable Development Strategy for the Seas of East Asia (SDS-SEA), adopted by 14 countries (Brunei, Cambodia, China, DPR Korea, Indonesia, Japan, Lao PDR, Malaysia, the Philippines, RO Korea, Singapore, Thailand, Timor-Leste and Viet Nam) (Fig. 7.5). The strategy is a package of applicable principles, relevant existing regional and international action programmes, agreements and instruments, as well as implementation approaches, for achieving sustainable development of the Seas of East Asia. It offers a regional framework for the interested countries and other stakeholders to implement, in an integrated or holistic manner, the commitments they have already made, without assuming new legal obligations. It addresses linkages among social, cultural, economic and environmental issues and embodies the shared vision of the countries and other stakeholders for the Seas of East Asia, and the ways by which they will achieve that shared vision. PEMSEA has developed an ICM system that addresses complex coastal management concerns, covering governance and various sustainable development aspects. In November 2015, PEMSEA country partners committed to scale up the ICM to cover 25% of the region's coastline by 2021. To date, PEMSEA has exceeded that target and secured about 37.9% of the region's coastline, having a significant impact on 86,284 km of coastline and over 150 million people living in coastal and watershed areas. As part of ICM implementation towards achieving blue economies in the region, PEMSEA is committed to improving coastal and ocean governance, and implements programmes on climate change mitigation, disaster risk reduction, habitat protection restoration and management, water use and supply management, food security and livelihood management.

West Africa Coastal Areas Management Program (WACA): WACA was established by the World Bank in 2015 in response to demands from countries in the region to manage their growing coastal erosion and flooding problems. Countries already participating in the programme include Benin, Côte d'Ivoire, Ghana, Mauritania, Sao Tome and Principe, Senegal, and Togo, and discussions are under way with other countries. WACA is designed to improve the livelihoods of coastal communities in West Africa by reducing the vulnerability of its coastal areas and promoting climate-resilient integrated coastal management. The programme's mix of technical assistance and investments will seek to preserve and rehabilitate the natural coastal resources essential for livelihoods; spur economic development and increase social welfare; and support the sustainable development of key growth sectors, such as agro-industry, fisheries, offshore petroleum exploration and production, and tourism. WACA is also a convening platform to help countries obtain the finance and expertise they need to sustainably manage their coastal areas. It also serves as a forum within which countries and regions can share lessons learned.

Southeast Pacific Data and Information Network in Support to Integrated Coastal Area Management (SPINCAM): SPINCAM is an IOC-UNESCO/Flanders and Permanent Commission for the South Pacific (CPPS) initiative, created in 2008 to develop a framework of indicators in various pilot sites of the southeast Pacific (Chile, Colombia, Ecuador, Panama and Peru). SPINCAM supports the development of decision-making tools and implementation of ICM through regional and national investment for improved data and information management capacity, knowledge, communication and networking at national and regional level (COI-UNESCO and CPPS 2016). Main outputs so far have been the development of information systems, in the form of substantial ICM atlas and web-based portals for the associated metadata. The main outcomes expected from SPINCAM include: institutionalisation of coastal and marine governance at national and regional level; improved regional networks on coastal and marine issues; regional strategic recommendations on marine spatial planning, sustainable blue growth, monitoring systems and decision support tools; reduction of national technical disparities on capacity development; and improved communication and participatory processes.

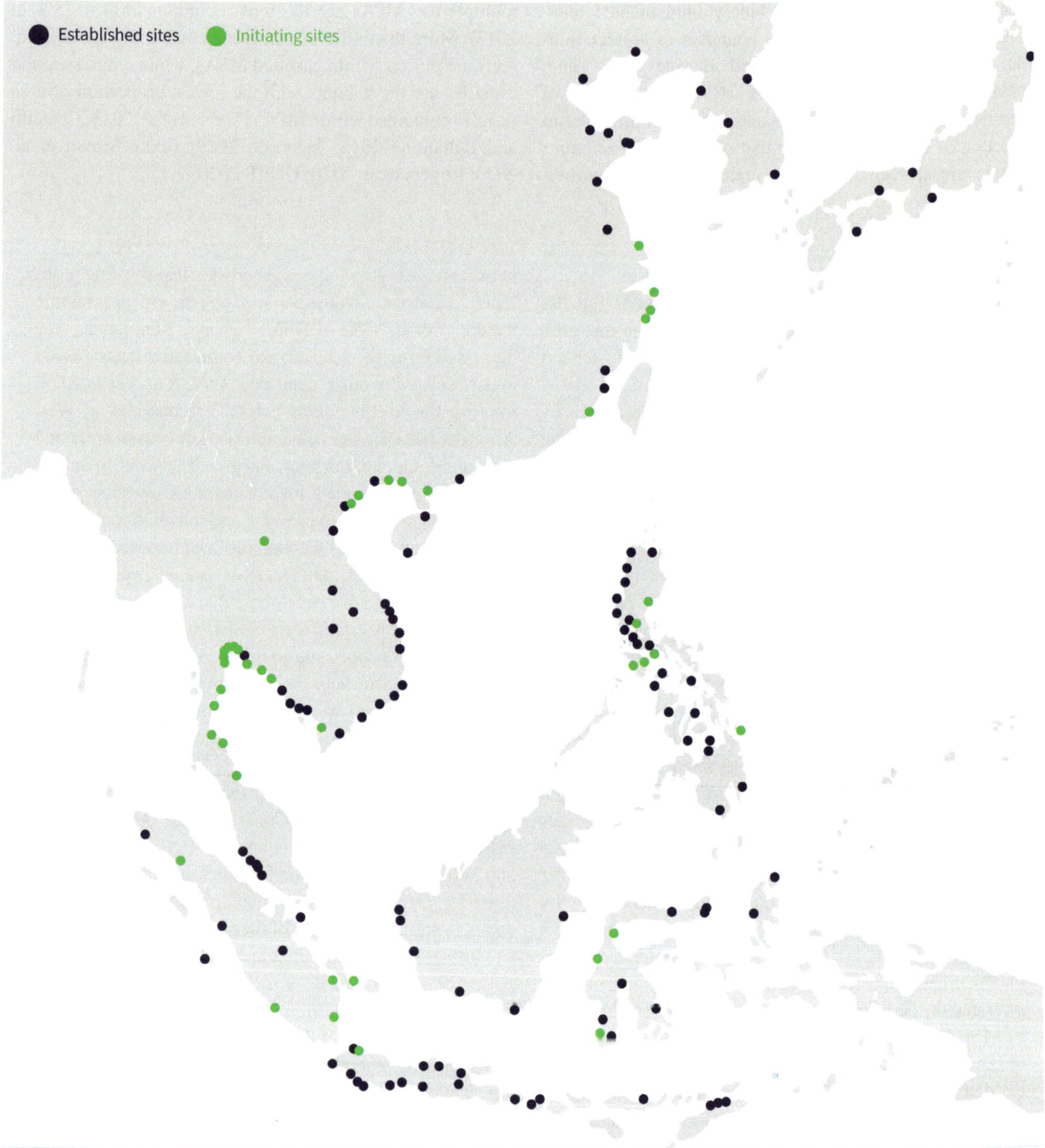

Fig. 7.5 PEMSEA ICM sites

There is now a recognition that conservation is enhanced when the people and communities dependent on resources take on some of the responsibility for managing (making decisions about) those resources. The most widely used OECM is locally managed marine areas (LMMAs), whereby coastal communities limit or prohibit extractive or destructive practices within a defined area. One example is the Territorial Use Right for Fisheries (TURFs), where local communities, or associations or cooperatives, of fishers have exclusive property rights to harvest resources within defined areas.

TURFs, in combination with no-take-areas, are now being implemented throughout the Americas, Oceania and Southeast Asia, and demonstrate a range of positive effects, including increased yields, ancillary biodiversity conservation, and social and ecological enabling conditions for local stewardship. For example, in Chile, the combination of TURFs and small-scale aquaculture are showing promising results for livelihood diversification, production and food security. LMMAs generally operate on more limited spatial (1–10 km^2) scales than contemporary MPAs, potentially reducing their conservation effectiveness. There are also, however, a number of less-encouraging aspects, including biases towards only reporting positive results and focusing on sedentary biota, lack of effective enforcement, misalignment between yields and sharing agreements, and operating as isolated silos that can't meet ecological and economic expectations (Christy 1982; Jupiter et al. 2014; Afflerbach et al. 2014; Albert et al. 2016; Viana et al. 2017; Andrachuk et al. 2019; Sepulveda et al. 2019; Villaseñor-Derbez et al. 2019; Aceves-Bueno et al. 2020; Halim et al. 2020; Nguyen et al. 2017).

Recognition of indigenous rights to, and ownership of, significant coastal estate in some countries (e.g. Australia, Canada, New Zealand) is having a growing role in marine governance and conservation and aspirations for blue economy livelihood opportunities. Across many cultures, traditional owner communities have often long practised conservation of coastal ecosystems and resources through application of traditional ecological knowledge, such as spatial and seasonal closures, and there is growing recognition of the need to incorporate such knowledge within modern conservation practices (Kerr et al. 2015; Charles 2017; Ban and Frid 2018; von der Porten et al. 2019a, 2019b; Rist et al. 2019; Reid and Rout 2020).

To empower and to incentivise local communities as custodians to protect and restore local coastal areas, the application of payments for ecosystems services (PES) is increasingly being adopted. With this approach, stewards (traditional owners or community groups generally) of a coastal area are incentivised (paid) to carry out activities that preserve or enhance the provision of ecosystem services. Those who pay for PES are motivated by direct benefits (e.g. environmental protection helps a business) or indirect benefits (e.g. offsetting carbon footprint), and PES transactions are generally regulated by independent organisations who certify that measurable units of benefit (e.g. carbon sequestered) have been created by the project's activities and allow the resulting credits to be sold or traded in relevant marketplaces (UNEP 2020).

Beyond designating areas with a level of protected status, many factors can define success or failure of individual MPAs. Without careful governance, planning and execution, MPA designations can amount to little more than "paper parks". Multi-stakeholder engagement is considered a critical factor affecting success, as is whether zoning and plans identify and resolve conflicts among users, and whether effective performance monitoring and evaluation occurs. Even when MPAs are effective, issues can arise with unmet expectations by communities, upfront costs from decreased fishing in new protected or regulated grounds, loss or change of cultural uses, and unequal distribution of resources (Cinner et al. 2012; Fox et al. 2012; Ehler 2018; Giakoumi et al. 2018).

Marine Protected Area design continues to evolve as it seeks to meet a range of emerging challenges. Irrespective of the level of protection afforded from human-use impacts, effective MPA management must now also consider the consequences of a changing climate (recurrent coral bleaching, for example) and the role of MPAs in addressing the impacts on biodiversity—for example, through creating refugia and connected networks of "bright spots" and incorporate projected future distributions of coastal ecosystems rather than focusing on past conditions.

4.2 Mitigating Catchment Impacts Through Terrestrial Reform

Achieving a sustainable ocean economy relies on the adequacy of upstream urban and hinterland infrastructure to provide the transport, energy and water services required to support ocean industries and their supply chains. Equally important, however, is addressing the downstream impacts of inappropriately designed and operated infrastructure and activities on coastal ecosystems. The activities of concern are those that clear, convert or modify coastal ecosystems to other land uses; extract resources such as surface water, groundwater and sand; and introduce land-based pollutants, such as excessive nutrients, sediments and manufactured chemicals (e.g. agricultural, industrial pharmaceuticals and personal care products), and litter.

In relation to the last of these, the companion Blue Paper *Leveraging Multi-Target Strategies to Address Plastic Pollution in the Context of an Already Stressed Ocean* (Jambeck et al. 2020) proposes several relevant interventions that would reduce pollutant inputs: improve wastewater and storm-water management, adopt green chemistry and new materials, recover and recycle materials, implement coastal zone improvements, and build local systems for safe food and water.

Given growing water scarcity worldwide, there is also opportunity for better reuse of wastewater to meet these demands as well as to reduce impacts on coastal environments. There are well-established technologies that can be deployed to increase the amount of wastewater that is recycled and reused. As a result, cities across the globe are establishing ambitious targets and developing policies to support zero discharge concepts (UN Water 2017).

As noted in Sect. 4.1, there are significant opportunities for closer integration with current global water, urban and climate agendas and initiatives.

The High Level Panel on Water articulated an agenda for water reform that encompassed: establishing a *foundation for action* based on better understanding, valuing and managing water; *leading an integrated agenda* to provide sustainable and universal access to safe water and sanitation, build more resilient societies and economies, invest more and more effectively in water-related infrastructure, and build sustainable cities and human settlements; and *catalysing change, building partnerships and international cooperation* to encourage innovation, promote partnerships and strengthen cooperation, and leverage finance and institutional support. The High Level Panel on Water highlighted the need to consider "urban deltas, coastal areas and other environmentally sensitive areas" and "to integrate appropriate measures into sustainable urban and territorial planning and development" (High Level Panel on Water 2017).

To support these efforts, the UN General Assembly proclaimed the period 2018–2028 as the Water Action Decade, which—given that it overlaps with the 2021–2030 UN Decade of Ocean Science for Sustainable Development—means that there are considerable opportunities to harmonise initiatives to develop a "source-to-sea" approach that explicitly considers downstream impacts of terrestrial infrastructure and activities on coastal ecosystems. The strength of source-to-sea management is that it considers the entire system, highlighting upstream and downstream environmental, social and economic linkages, and stimulating coordination across sectors and across different authority levels. Source-to-sea approaches have been implemented in many countries, often under different names (e.g. catchment-to-coast, ridge-to-reef), while globally this approach is recognised as essential to addressing SDG implementation by ensuring that linkages between the different goals and their targets are considered (Mathews et al. 2019; Singh et al. 2020).

Mitigating the impacts of diffuse land-based sources of pollution, including the application of nutrient fertilisers and agricultural chemicals (pesticides and herbicides), as well as the erosion of sediment on sensitive adjacent coastal ecosystems (seagrass beds and coral reefs), is now the principal concern among developed nations, and globally. For example, in Australia, management of activities to mitigate the loads on nutrients, sediment and pesticides discharged from catchments adjacent to the Great Barrier Reef have been in place since the 1990s and encompass regulations that range from setting end-of-catchment load reduction targets to regulating and incentivising land-practice activities, such as precision and regenerative farming that retain soils on-farm, and minimise the use of agrochemicals. Further downstream, the retention of vegetative buffer strips along the banks of rivers, estuaries and shoreline, and the use of natural and constructed wetlands to trap sediments and filter nutrients, are also effective ways of minimising discharges to coastal environments (Brodie et al. 2012; Day 2018; Adame et al. 2019; Saderne et al. 2020).

While ecosystems such as seagrass, oysters and coral reefs are particularly sensitive to too much sediment, in depositional coastal areas an adequate supply of sediment from upstream will be required to ensure the stabilisation of shorelines and the ongoing accretion of mangrove and saltmarsh habitat. Regulation of the amount of water used by upstream activities, removal of unnecessary impoundments and barriers, sustainable sediment management in reservoirs, and the setting of dedicated natural environmental hydrological flows that can reach the coast unimpeded are needed to ensure that deltas and estuaries can keep pace with sea level and erosion (Kondolf et al. 2014; Anthony and Goichot 2020).

The carbon sequestration and storage of areas of mangrove, saltmarsh and seagrass is now widely considered by many countries with large blue carbon stocks as part of their national emission reduction commitments, and they are now active in conserving and restoring these ecosystems (see Sect. 4.4).

Emerging initiatives are focusing attention on the importance of action to curb the over-extraction of sand from rivers and coastal areas and stop critical deltas from "sinking and shrinking". For example, WWF's Resilient Asian Deltas initiative (WWF 2019) focuses on Asia's six largest delta systems—Ganges–Brahmaputra–Meghna, Indus, Irrawaddy, Mekong, Pearl and Yangtze—with an emphasis on the importance of building with nature and the benefits nature provides as a key solution for delta and coastal resilience. From restoring fluvial and coastal sediment flows to creating more room for rivers, from reconnecting floodplains to restoring mudflats, mangroves and other wetlands, from minimising the impact of new infrastructure on river flows to creating ponds and sponge cities to compensate for expanding areas of impermeable concrete, building with nature across their river basins would transform the future of these deltas.

Likewise, over-abstraction of groundwater, leading to subsidence in low-lying areas and cities, requires a comprehensive approach to better manage these resources. Given that these abstraction and extraction activities can occur

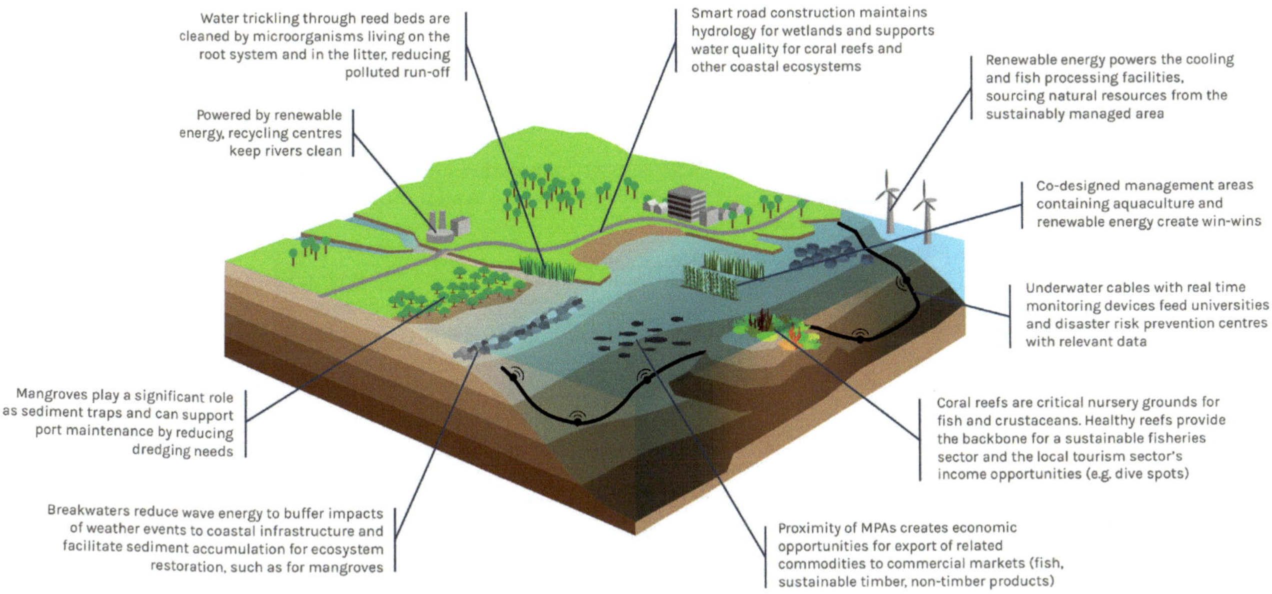

Fig. 7.6 Benefits of implementing blue infrastructure. (Source: Thiele et al. (2020). IUCN)

many hundreds of kilometres from the coast, and sometimes in adjacent countries, regional, or transnational source-to-sink approaches are required.

4.3 Adapting Coastal Infrastructure

The coastal adaptation strategies considered here are principally concerned with using nature-based approaches to adapting infrastructure to increase resilience to changing conditions, and to minimise the loss of ecosystem services. This also requires a change in behaviours or practices by which we make use of coastal environments and the resources they provide, and the way in which we value the direct and indirect benefits derived from these ecosystems.

Traditional coastal infrastructure is typically built with "hard" or "grey" engineering techniques and materials (e.g. concrete, steel) and designed to specifications for withstanding probabilistic exceedances that are based on the assumption that the past can reliably predict the future; as discussed in Sect. 2.1, this is no longer the case and puts many low-lying settlements at risk. These hard approaches have also left a legacy of environmental impacts affecting the structure, function and connectivity among adjacent coastal habitats and diminishing biodiversity. The next generation of coastal infrastructure will have a critical role in meeting these increased climate-driven challenges, as well as accommodating continued urbanisation and the needs of blue economy industries. To ensure that the right infrastructure is built, we must adopt resilient approaches, and policymakers will need to establish long-term visions for sustainable national infrastructure systems, informed by the SDGs (Thacker et al. 2019).

Softer, natural approaches—often labelled "green" for terrestrial or "blue" for marine—that apply ecological engineering principles are increasingly being used to build coastal defence structures that "mimic" natural coastal areas, including dynamic coastal landforms, such as beaches, barrier islands and dunes; coastal vegetation, such as mangroves, seagrasses, dune vegetation, saltmarshes and kelp forests; and reef systems, such as mussel beds, oyster reefs and coral reefs. Figure 7.6 illustrates the benefits of implementing blue infrastructure, and Table 7.4 summarises the advantages and disadvantages of each form of infrastructure.

The direct benefits of natural infrastructure are principally protection from flooding and from erosion. Consequently, such approaches are now recognised as a way of balancing continuing development with solutions that deliver climate change resilience and adaptation benefits, alongside multiple ecosystem benefits, including enhancing biodiversity and carbon sequestration and improving water quality by filtering storm water (Francis 2010; Lai et al. 2015; Perkins et al. 2015; Sutton-Grier et al. 2015; Firth et al. 2016; Vikolainen et al. 2017; Gracia et al. 2018; Burt et al. 2019; Browder et al. 2019; Conger and Chang 2019; Liu et al. 2019; Thacker et al. 2019).

Table 7.4 Summary of selected management approaches advantages and disadvantages

Advantages	Disadvantages
Grey infrastructure	
• Significant expertise exists on how to design and build such approaches at large scales	• Does not adapt with changing conditions such as sea level rise
• Decades of experience with implementation	• Weakens with time and has a built-in lifetime
• Excellent understanding of how these approaches function and what level of protection will be provided by different types of structures built to specific engineering standards	• Can disrupt longshore coastal sediment transport and cause downdrift coastal erosion
• Infrastructure is ready to withstand a storm event as soon as it is constructed	• Can cause coastal habitat loss and have negative impacts on the ecosystem services provided by nearby coastal ecosystems
	• May sustain more damage during small storm events
	• Only provides storm protection benefits when a storm is approaching; no co-benefits accrue in good weather
	• Needs continuous monitoring and regular maintenance
	• Barrier to dispersal and movement of fauna and flora, resulting in loss of ecosystem connections
Natural and hybrid infrastructure	
• Capitalises on best characteristics of built and natural	• Little data on how well these systems perform to date
• Allows for innovation in designing coastal protection systems	• Does not provide the same benefits that natural systems provide
• Provides some co-benefits besides coastal protection	• More research is needed to design the best hybrid systems
• Can provide a greater level of confidence than natural approaches alone	• Growing but limited expertise in the coastal planning and development community on which approaches to use
• Can be used in areas where there is little space to implement natural approaches alone	• Hybrid systems, due to the built part of them, can still have some negative impacts on species diversity
• Balances conservation with development	• Uncertainty in cost-effectiveness and long-term performance
	• Permitting for hybrid projects can be a more difficult process than for built projects
	• Response to native species colonisation is unpredictable
Ecosystem restoration	
• Provides many co-benefits in addition to coastal protection, including fishery habitat, water quality improvements, and carbon sequestration and storage, and can provide these benefits to coastal communities all the time, not just during storm events	• The development of best practices for how to restore ecosystems is needed, according to a set of starting criteria
• Ecosystem grows stronger with time as establishes	• Provides variable levels of coastal protection (non-linearity of the provisioning of coastal protection benefits), depending on the ecosystem, geography and also on the type and severity of storm events; more research is needed to better understand how to estimate or predict the coastal protection provided
• Has the potential to self-recover after a storm or other disturbance event	• In the case of restored ecosystems, it can take a long time for ecosystems to get established so that the natural systems can provide the necessary level of coastal protection
• Can keep pace with sea level rise	• Likely requires a substantial amount of space to implement natural approaches (such as ecosystem restoration or protection of existing ecosystems), which may not be possible in highly urban or industrial contexts
• Can be cheaper to construct	• Uncertainty in cost-effectiveness and long-term performance
• Increased CO_2 storage capacity in created, maintained or restored ecosystems; reduction of urban heatwave island effect; improvement in water quality	• Permitting for natural projects can be a more difficult process than for built projects
• Can enhance tourism, recreational and local employment opportunities included in establishment and maintenance	• Uncertainty over responsibility for ownership and maintenance
• Enhances the natural environment and implicit value • Saves raw materials and improves public health	• Uncertainty in assessing levels of risk for insurance cover and premiums for coastal assets

However, as the design and performance of this natural infrastructure is often influenced by local ecological, social and political conditions, increasingly *hybrid* approaches blending strategic use of natural assets and ecological principles with grey-engineered techniques and existing infrastructure are being adopted. Hybrid approaches provide cost-effective hazard protection solutions and are increasingly being adopted in urban areas where green approaches may be insufficient to meet the rising impacts of climate change, or where space is limited. There are now numerous examples and guidance on applying these approaches to applications that range from the landscape scale to individual breakwaters. Restoration of wetlands, sand-dunes and beaches can be integrated with supporting grey structures for flood or erosion management (e.g. levees, breakwaters and seawalls), providing a solution that is more comprehensive, robust and cost-effective than either solution alone. Small-scale engineering interventions to coastal defence structures can be implemented at relatively low cost, in intertidal and shallow subtidal zones to increase faunal and algal abundance and diversity. The modification of these structures, by adding grooves, pits, ledges and texture, can be incorporated into the design of coastal defence structures or retrofitted to existing structures. For example, the Living Seawalls project (see Box 7.7) is fitting seawalls with various shaped tiles—made with 3D printing technology—that enhance relief and facilitate settlement of a variety of benthic organisms, or create habitat for small cryptic fishes (Borsje et al. 2011; Depietri and McPhearson 2017; Strain et al. 2018a, b; Browder et al. 2019; Conger and Chang 2019).

Another area of adaptation is the development and use of building materials that are more environmentally benign. There are now a number of green concretes—made with waste material as a partial or complete replacement for cement or aggregate, including recycled demolition waste aggregate, blast furnace slag, manufactured sand, glass aggregate and fly ash. While green concrete requires less energy for its production and produces less CO_2, the higher cost of reinforcement, and the shorter life of buildings constructed with green concrete are limitations that are being addressed (Zhang et al. 2014; Khazaleh and Gopalan 2019; UNEP 2019).

Many cities around the world are now developing and implementing green urban infrastructure plans and demonstrating that urban transitions integrating green, blue and grey infrastructure are possible and affordable, and lead to more efficient, multipurpose infrastructure. Recognition that these solutions can be applied in other parts of the world has led to a number of international city networks, notably the C40 Cities Climate Leadership Group, the World Mayors Council on Climate Change and the Urban Climate Change Research Network, now actively collaborating and learning from each other to improve their adaptive capacity.

Similarly, many seaports around the world, facing growing environmental concerns about their construction and operation, have sought to align their performance with sustainability considerations, as well as planning protection from the impact of climate change. A shift to greener, integrated ports is now recognised as a long-term economic choice, and an increasing number of ports now implement a range of in-port operations, including energy conservation, environmental protection and development planning that considers the adjacent environment, other coastal operations and nearby cities. A range of incentives are also used to reduce emissions, such as using shore-based electricity for ships at berth, requiring slow vessel speeds, and incentivising rail and barge transport, rather than roads, from ports. Some ports also reduce fees based on indices that assess the environmental performance of individual vessels, such as the Environmental Ship Index. However, the voluntary nature of such schemes means that progress on significant emission reductions remains slow. Consequently, regulators and policymakers must be prepared not merely to nudge and incentivise but to take more concrete action (PIANC 2014; Gonzalez et al. 2018; Bergqvist and Monios 2019; Psaraftis 2019; de Boer et al. 2019; Dundas et al. 2020; WPSP 2020; UNCTAD 2020b).

With growing offshore sprawl, there are opportunities to find synergy in sharing infrastructure between industry sectors that might previously have been in conflict. For example, combining aquaculture with wind or solar operations, and even conventional oil and gas platforms, is now increasingly common. Such multifunctional use is, however, still very much in its infancy and requires technical and economic feasibility as a basic prerequisite, as well as alignment among sectors and national jurisdictions of environmental, safety and regulatory regimes and practices. Similarly, a growing legacy of ageing marine (e.g. oil and gas platforms and pipelines) and catchment (e.g. small dams) infrastructure that must be decommissioned in the near to medium future is driving the development of policy and science that seeks to minimise environmental harm while ensuring cost-effectiveness (Buck and Langan 2017; Buck et al. 2018).

Multilateral funding and investment agencies and the insurance industry now recognise that integrating blue and grey infrastructure can help to fill the need for the next generation of climate-resilient infrastructure solutions and allow for the devising of new risk financing for nature-based solutions, such as the recent insurance for the Mesoamerican Reef (Reguero et al. 2019b). There is a large and growing pool of funding for natural infrastructure—although the availability is geographically uneven—with the largest opportunities in the redirection of post-disaster recovery funds to pre-disaster investments in risk reduction. However, the largest barriers for securing adequate resources are identifying locations where natural infrastructure can play a significant role in flood risk reduction, developing the experience and standards to overcome institutional biases that favour grey infrastructure, and developing institutional arrangements capable of matching available funding with the needs of individual situations (Colgan et al. 2017).

Policy support for green/blue/hybrid infrastructure can make good politics and has an important social dimension, as adoption will be most successful when it meets the needs and interests of local stakeholders and communities. However, much clearer integrated policy pathways to promote adoption of blue and hybrid infrastructure are required, and while "green" and "blue" are often used to delineate terrestrial and marine approaches, in fact, "teal" approaches are what is required to effectively address coastal development (Browder et al. 2019; Dundas et al. 2020).

A body of policy and practitioner guidance has emerged in recent years that provides tools to enable integration and the use of natural infrastructure solutions, lessons from implementation, and policy recommendations to ensure that infrastructure meets sustainable development objectives (Browder et al. 2019; Conservation International 2019; Thiele et al. 2020). At a macro level, the 2019 G20 Principles for Quality Infrastructure Investment provide clear policy guidance for the consideration that needs to be taken around infrastructure planning, including that in the coastal zone. The principles include a focus on maximising the positive impact of infrastructure on achieving sustainable growth and development through the positive economic, environmental, social and development impact of infrastructure and encourages the use of a virtuous circle of economic activities, including the use of ecosystem-based adaptation where possible. They highlight the need for comprehensive disaster risk management planning in the design of infrastructure, including in terms of considering the re-establishment of essential services, as well as the need to ensure long-term adaptability and to build for infrastructure resilience against natural disasters and the slow onset of environmental changes. Finally, they highlight the importance of finance and insurance mechanisms, including well-designed disaster risk finance, to help incentivise resilient infrastructure through the financing of preventive measures, and the need to make transparent the additional benefits of sustainable infrastructure projects to enable the use of green finance instruments, including in the delivery of NDCs (G20 2019).

Box 7.7. Living Seawalls: A Green Engineering Solution with Global Significance

The Living Seawalls project enhances the ecological value of seawalls by using modular habitat panels, constructed using 3D design and printing technology, to mimic features of natural shorelines (SIMS 2020). Panels with crevices and ridges, in New South Wales, Australia, enhance native biodiversity and the survival of Sydney rock oysters, a native habitat-forming and economically important species (Strain et al. 2018a, b). Individual panels can be designed to mimic the natural habitat features of a locality, and panels of multiple designs can be configured in mosaic arrangements to provide a variety of habitats to maximise diversity.

To date, panels of multiple designs have been installed at six locations in Sydney Harbour to create "living seawalls". Within hours of installation, panels were inhabited by microbes and mobile macro-invertebrates, and within just a few months, the complex panels supported more diverse and abundant marine communities than flat surfaces. This project, a collaboration between marine biologists, designers and engineers, was made possible by a forward-thinking local council (North Sydney), which has long supported seawall greening and provided access to their seawalls (Fig. 7.7).

Urban stakeholders are supportive of green engineering initiatives and local stakeholders reported a greater sense of well-being associated with these initiatives. The enormous potential of the Living Seawalls habitat panels to transform seawalls around the world has captured the attention of local and state governments, consultants, marine managers and ecologists from around Australia and abroad. At present, the main barriers to implementation are the lack of clarity on seawall ownership due to jurisdictional boundaries within the intertidal and shallow subtidal environment, confusion around required documentation for the permitting process, and the slow rate at which these questions are resolved.

4.4 Repairing Coastal Ecosystems

It is now widely accepted that protection is not enough to reverse trends in coastal habitat loss and degradation, and efforts to repair these ecosystems, through at-scale habitat restoration efforts and by re-establishing natural coastal and hydrological processes, are required. It is also increasingly accepted that these efforts can deliver substantial environmental co-benefits (Sect. 4.5), including biodiversity protection, coastal protection, coastal carbon storage and fisheries production, as well as direct and indirect employment co-benefits related to installation, maintenance, recreation, tourism and education. Several studies have begun to quantify the singular and bundled value of the direct and indirect benefits that accrue from repairing coastal ecosystems, and demonstrate substantial economic gains and cost avoidance relative to business-as-usual scenarios.

Restoration science and practice is also fundamental to creating new nature-based infrastructure for coastal defence (Sect. 4.3). Recent analysis commissioned by the High Level Panel for a Sustainable Ocean Economy notes that restoration activities provided a benefit of four dollars for every one invested, but due to higher logistical costs and the longer

Fig. 7.7 Living seawalls. Living Seawalls panels can be affixed to existing seawalls that are generally flat and featureless, and otherwise provide little protection for marine life (**a**). Panels have been designed using 3D printing technology to incorporate complex texture and a variety of microhabitat features (**b**). Seawalls can be retrofitted in a variety of configurations to suit site conditions, ecological objectives or aesthetic preferences (**c**). Within months, the complex panels were colonised by a variety of invertebrate and macroalgae and fish (**d, e**). (Source: Maria Vozzo)

timescales taken to realise the benefits, this ratio is at least 20-fold less than implementation of protection measures (Konar and Ding 2020).

Globally, there are a number of initiatives actively seeking to scale up restoration. The UN General Assembly has declared 2021–2030 the UN Decade on Ecosystem Restoration, promoting global cooperation on the restoration of degraded ecosystems to combat climate change, protect biodiversity, assist with food security and deliver clean water for the planet. The Bonn Challenge seeks to restore 350 million ha of the world's degraded and deforested lands by 2030, while the Global Mangrove Alliance has set a target of increasing global mangrove extent by 20% within this time period.

While the reasons for restoration are varied, it should be understood that the aim of restoration activities is not to return degraded coastal ecosystems to any particular past reference point, but rather to focus on increasing the extent and abundance of key habitats and keystone species, and use metrics that include presence of structure, functions, resilience and ecosystem services to evaluate success (Bayraktarov et al. 2016, 2019, 2020a; Friess et al. 2019; Duarte et al. 2020).

Depending on the habitat to be restored, as well as local conditions, a variety of restoration methods have been used and there are now numerous examples of successful and unsuccessful projects that have allowed the development of extensive practical guidance on restoration (Gann et al. 2019). Box 7.8 summarises some of these principles and learnings, and Boxes 7.10–7.13 provide relevant examples of mangrove, seagrass, coral and shellfish reef restoration.

For coastal marine ecosystems, mangrove restoration is the most well established and is widely undertaken throughout the world. Mangrove restoration occurs predominantly by planting seedlings and saplings in projects that vary from small (<1 km^2) to large (1000 km^2) scale, and by 2010, nearly 4000 km^2 had been restored. Online tools, such as the Mangrove Restoration Potential Map (maps.oceanwealth.org/mangrove-restoration/), allow users to explore at global, regional and national levels the opportunities for mangrove restoration.

The map identifies c.8120 km^2, or 6%, of former mangrove area as restorable, with the greatest opportunities in Southeast Asia (Worthington and Spalding 2018). The Global Mangrove Alliance (http:// www.mangrovealliance.org/gma/) provides practical advice on mangrove restoration.

The reinstatement of natural hydrological conditions for rivers, as well as tidal areas that have been restricted, is an important pre-condition for restoration in coastal marine habitats. In many cases, the removal of bunds and other structures restricting natural tidal flows can be sufficient to assist revegetation of coastal areas that had previously been cleared for other land use, including agriculture and aquaculture (Kelleway et al. 2020).

Technically, to improve the success of restoration efforts, the rigorous application of science to design and select areas that are suitable for restoration is needed, and the use of "big data" can be utilised for such assessments. Continual evaluation of project progress with metrics that assess effectiveness rather than effort will help to ensure that lessons are learned from past failures and successes so that restoration practices are improved and resources can be maximised in the most cost-effective manner. Harnessing knowledge of the life histories of the habitat-forming organisms, using technologies such as drones to identify suitable areas for restoration and to disperse pods into ideal locations, or using commercial vessels equipped with oil booms to collect wild coral-spawn slicks for re-seeding target reefs (see Box 7.11) are just a few examples to help achieve the scale of restoration required (Fairhead et al. 2012; Baker and Eckerberg 2013; Doropoulos et al. 2019a, b; Vanderklift et al. 2020; Worthington et al. 2020; Waltham et al. 2020).

Apart from the technical challenges of undertaking restoration at ecologically meaningful scales, restoration must operate within a complex and dynamic interplay between technical decision-making, legal constraints, social licence to operate, ideologies and politics. As a result, many efforts are considered value-laden, context-driven and prone to disagreement and compromise. In developing countries, restoration projects must also operate and respect the cultural norms and traditional ownership/rights issues relevant to the project area, while at the same time addressing perceptions of "green grabbing". Governance and institutional issues can also hamper rehabilitation if there is poor coordination among agencies, many of whom often have conflicting production/development and environmental protection mandates.

In addition to the ecosystem services that restoration of coastal habitats can provide, there are also significant flow-on benefits through the creation of new jobs and supporting local economies. Marine habitat restoration is recognised as a "jobs-intensive" industry and strong driver of economic growth, creating immediate employment in transport, construction, marine engineering, project management, science and aquaculture. For example, the economic impact of 50 coastal habitat restoration projects funded through the American Recovery and Reinvestment Act (2009) created on average 17 jobs per million dollars spent, which was higher than traditional industries, including coal and gas, roads and energy generation. Many jobs are created in rural and regional coastal areas and offer a range of skilled and low-skilled positions, considerably enhancing economic opportunities in regional areas. Longer-term employment can be created through the flow-on benefits of these ecological improvements to new and increased opportunities for fishing, aquaculture and tourism and their service sectors (Edwards et al. 2013; Powell et al. 2018).

Marine habitat restoration is also almost unparalleled in its capacity to deliver collaborative, partnership-based approaches for restoration. Active involvement and meaningful consultation between practitioners, local communities and the science sector that leads to integration of best-practice science and local knowledge is essential for effective implementation. Factors for success include local government support, community involvement, property rights, education and preparation, and supplementary livelihoods. Citizen science activities are regularly incorporated into projects to reduce costs and expand community participation and education. Engagement with traditional landowners can result in shared learning, application of traditional ecological knowledge and improved coastal management and indigenous engagement (Diefenderfer and Adkins 2003; Stojanovic et al. 2004; Ismadi and Yamindago 2014; Dharmawan et al. 2016; McLeod et al. 2018; Powell et al. 2018; Waltham et al. 2020).

Box 7.8. Success Factors for Coastal Restoration

1. Planning: Careful planning is necessary and should include identifying the causes of degradation and conducting preliminary small-scale interventions to test effectiveness prior to applying any full-scale restoration activities.
2. Create the right preconditions: Removal or mitigation of stressors, such as poor water quality, and limiting conditions, such as lack of suitable substrate or inadequate supply of propagules, is necessary before natural recovery can occur. Stressors that enhance mortality, such as disease and predation, particularly during early stages of growth, also need to be minimised.
3. Consider the right scale and context: The need to scale up restoration activities means that the patch-based approaches must consider processes at the broader landscape and regional scales—for example, movement of water or dispersal of biota.
4. Location: Ensuring restoration takes place in the locations that maximise success for the system being

restored, in terms of considering scale, access, disturbance history and forecasting, and downstream benefits, is vital.

5. Focus on tangible outcomes, not targets: While ambitious area-based targets (e.g. size of area planted, number of seedlings planted) for restoration are being widely advocated, these should be reframed to focus on success criteria linked to environmental outputs (e.g. percentage survival, vegetation densities similar to natural forests) and incorporate social-ecological outcomes of restoration.
6. Engage partners and community: Active involvement and meaningful consultation between practitioners, local communities and the science sector—that leads to integration of best-practice science and local knowledge—is essential for successful implementation and longevity.
7. Harness technology: Technology must be developed and utilised to effectively scale up restoration efforts. Remote sensing technology opens new ways to monitor and inform conservation and restoration.
8. Long-term monitoring and adaptive management: It is important to plan for, commit to and invest in long-term monitoring, so that small issues can be quickly rectified.
9. Investment: Besides public investment, restoration efforts clearly need private investment, and this investment could be accessed via new financial instruments, including payment for ecosystem services, green bonds, biodiversity offsets, carbon credits, debt-for-nature swaps, and water quality credit markets.

Box 7.9. Mangrove Protection and Restoration: Nature-Based Solutions to Multiple Problems

Mangrove conservation—including actions that both protect and restore—is becoming a priority for international policy, in part because mangroves provide multiple benefits, including carbon sequestration, coastal protection and fish habitats. Currently, around 36% of the world's mangroves have some form of legal protection, and they are also implicitly or explicitly included under multiple international policy frameworks, including the Ramsar Convention on Wetlands of International Importance.

Many nations are developing policies and legislation that afford increased protection. For example, the island nation of Sri Lanka—one of the countries most affected by climate change—has implemented legal protection for all of its mangrove areas, as well as a policy to rehabilitate 10,000 ha of mangrove forest, while Indonesia aims to restore 50,000 ha of mangroves by 2024. However, policy frameworks still include incentives (such as expansion of aquaculture) that contribute to mangrove degradation and loss, and removing such perverse incentives is key to reversing decline.

Efforts to restore mangroves have taken many forms, from using seedlings grown in pots or directly inserting mangrove propagules into the soil, to simply allowing the tide to return and letting nature take its course. The approach has varied, depending on the purpose, such as whether the focus is on stabilising an eroding coast or generating carbon credits. Many of these initiatives often fail completely (for example, all the seedlings die), or they do not achieve the intended result. However, many successful initiatives exist, which shows the enormous potential of restoration. For example, in Bali, Indonesia, restoration of abandoned aquaculture ponds has yielded excellent results over more than a decade, including high rates of carbon sequestration. Breaching the barriers around the ponds (i.e. pond walls and gates) has allowed the tide to return, promoting rapid natural mangrove regeneration and accumulation of carbon-rich soil (Fig. 7.8).

In southwest Madagascar—a nation that lost 21% of its mangroves in the 20 years from 1990 to 2010 alone—coastal communities are almost entirely reliant on the resources they get from the sea. Blue Ventures has worked with these local communities using participatory approaches to develop a suite of activities designed to encourage sustainable use of mangroves, including development of sustainable alternative ways of generating income. Among the activities is the implementation of a locally managed marine area, alongside local regulations (Dina) to prevent overharvesting mangroves. The project also includes mangrove restoration by directly inserting into the soil the viviparous propagules of *Rhizophora mucronata, Ceriops tagal* and *Bruguiera gymnorhiza,* which are collected from parent trees nearby. The survival rate of planted mangroves is high, and measurements also include the carbon content of mangroves and the underlying soil, to develop carbon credits for sale in the voluntary carbon market, and so generate an additional source of income for local people.

Fig. 7.8 Mangrove restoration in Bali, Indonesia. (Source: Mangrove Nusantara)

Box 7.10. Seagrass Protection, Adaptation and Restoration

Seagrasses globally have been degraded over recent decades, and there is ample evidence from well-studied parts of Australia, North America and Europe showing that millions of hectares of seagrass meadows have died around the world (Waycott et al. 2009).

Can we begin to reverse this pattern through restoration? Advances in seagrass restoration techniques suggest that we can. Broadly, there are two main ways of restoring seagrass, which take advantage of the way that seagrasses (like grasses on land) can multiply both asexually and sexually. In asexual growth, seagrasses send out rhizomes (structures like horizontal stems) that colonise new areas; sometimes parts of an adult plant can break off and be transported to a distant area through sea currents, where it can then establish and grow. This characteristic of seagrass has been harnessed for decades in attempts at seagrass restoration, by methods which involve taking shoots from a healthy meadow, and planting them elsewhere. It can be laborious, and sometimes survival is low. But, if circumstances are right, it can be very successful. One example is Oyster Harbour, an enclosed embayment on the southern coast of Australia. After the original causes of seagrass death were ameliorated, efforts were made to transplant rhizomes of *Posidonia australis*, it with its characteristic large leaves attached. These were replanted in areas that once hosted seagrass, taking care to bury the rhizomes below the sediment surface, holding them in place with a wire hook. Survival was high, and the transplanted seagrass began to extend outwards. After 8 years, individual transplants could not be distinguished and meadows of transplanted *Posidonia* had begun to merge with existing natural meadows. When rates of carbon burial were measured 18 years after the original planting, rates inside meadows that grew from transplanted seagrass were similar to those in natural seagrass—further demonstration of the success of that project (Bastyan and Cambridge 2008; Marbà et al. 2015; Serrano et al. 2020).

Another restoration method yielding promising results harnesses the use of the seeds that seagrasses produce. In this method, seeds are dispersed into areas where seagrass once grew. Although only a small proportion survive and grow, many seeds can be dispersed, so that the overall chances of success are improved. In coastal bays of Virginia (USA), a project started in 1999, which involved scattering seeds of eelgrass (*Zostera marina*) from a boat across 125 ha over several years, had, by 2010, formed seagrass meadows covering greater than 1700 ha (Orth et al. 2012). Similar successes are now being reported at multiple locations around the world, highlighting that this method offers considerable promise (Fig. 7.9).

Fig. 7.9 Seagrass (Syringodium isoetifolium) in Mauritius. (Source: Mat Vanderklf)

Box 7.11. Restoration and Adaptation of Coral Reefs

With widespread and more frequent bleaching events, it is now widely held that conventional management approaches are not enough to protect coral reefs, and that restoration at ecologically meaningful scales is urgently needed to aid and accelerate recovery of damaged reefs.

Restoration methods developed over the last 40 years have traditionally involved transplanting coral fragments or adding artificial substrate, with other approaches such as larval addition, rubble stabilisation or algal removal infrequently applied (Boström-Einarsson et al. 2020). The coral gardening approach was pioneered in the 1990s and programmes using this approach now operate in more than 150 coral nurseries across 20 countries. Most interventions have traditionally been small, labour intensive and costly (replanting coral fragments grown in a nursery costs between $1 million and $4 million per ha) and have had mixed results (Rinkevich 1995; Edwards and Gomez 2007; Lirman and Schopmeyer 2016; Bayraktarov et al. 2016, 2019, 2020b; Anthony et al. 2017; van Oppen et al. 2017; Ladd et al. 2018; National Academies of Sciences, Engineering and Medicine 2019).

Some recent studies have begun to demonstrate longer-term and larger-scale (around 1–2 ha) successes (Fox et al. 2019; Williams et al. 2019; Bayraktarov et al. 2020a, b). One promising approach shown below is the harvesting, culturing and release of wild coral-spawn slicks to targeted reefs. Recent studies in Australia have demonstrated the feasibility of such large-scale restoration, and have been accomplished by incorporating technologies used in oil spill remediation, dredging operations and land-based aquaculture. Such an approach allows for long-distance translocation of corals and maintenance of coral diversity, and has virtually no impact at source (Doropoulos et al. 2019a, b).

Assisted evolution, such as selective breeding, assisted gene flow, conditioning or epigenetic programming, and manipulation of microbiome could also help coral reefs, which are particularly sensitive to warmer water temperatures (van Oppen et al. 2017). Moreover, including strategic decision science (Doropoulos and Babcock 2018) alongside novel interventions (Anthony et al. 2017) is necessary to maximise the long-term effectiveness of restoration activities (Fig. 7.10).

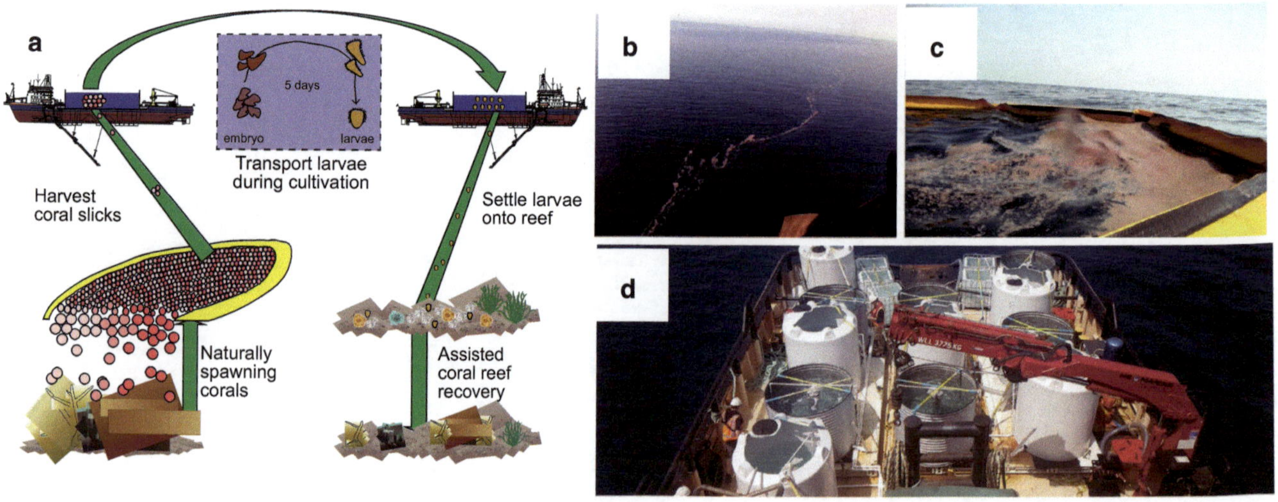

Fig. 7.10 Coral-Spawn Slicks. (**a**) Conceptual diagram for the harvesting of wild coral-spawn slicks following mass spawning events for transport during cultivation and release onto degraded reefs to assist in recovery. (**b**) Kilometre-long slick seen from the sky. (**c**) Slick contained in an oil boom. (**d**) Slick cultivated on a floating aquaculture system built on a commercial tug-boat in the first field trial. (Source: CSIRO)

Box 7.12. Shellfish Reef Restoration

Shellfish restoration has been successfully undertaken in several countries at scale and has employed approaches ranging from natural regeneration, assisted regeneration and reconstruction approaches. Shellfish reef restoration now frequently occurs at large scales (>10 ha), engages across government, non-governmental organisations (NGOs) and local communities and is innovative in addressing financing options. Examples of shellfish restoration around the globe include:

- In the United States, the Chesapeake Bay Executive Order requires oyster populations of 20 Chesapeake Bay tributaries to be restored by 2025. Three estuaries have been restored thus far, including 390 ha of restored reef at a projected total cost of $72 million. The economic investment was returned in less than 5 years through the increased catch of commercial fish and crab fisheries as a result of increased productivity from restored reefs (Knoche and Ihde 2019).
- In China in 2004, over 20 tonnes of hatchery-reared seed oysters were successfully transplanted onto two around 50-km concrete dykes previously constructed in the Yangtze River (Quan et al. 2009).
- In South Australia in 2018, a 20-ha native flat oyster reef was restored at a cost of c.$3 million, to support recreational fishing tourism and regional jobs at an employment ratio of 8.5 jobs per million invested. Key success factors included using case studies of the environmental and social benefits of reef restoration (particularly from the United States) to help educate the community and government stakeholders on the benefits of natural habitat restoration compared with artificial reefs; identifying a clear social beneficiary stakeholder (i.e. recreational fishers) and economic beneficiary stakeholder (i.e. local service businesses that financially benefit from the predicted increase in recreational fishers in the region); and successfully articulating marine ecosystems as natural infrastructure, which is synonymous with built infrastructure in terms of providing a beneficial service to communities and which can be quantified like other types of infrastructure (Econsearch 2016).

The benefits of restoring shellfish reefs to coastal communities and industries are well quantified, with the economic value of the full suite of ecosystem services derived from natural oyster reefs in North America estimated to be as high as $100,000 ha per annum (Grabowski et al. 2012) and including job creation and economic development, fish production, water filtration and dentification, coastal protection and providing habitat for many other marine species (Fig. 7.11).

Fig. 7.11 Shellfish Reef Restoration at Windara Reef in South Australia. (Source: Chris Gillies)

Box 7.13. Promoting Gender Equality for Coastal Resilience
Promoting gender equality is essential for ensuring coastal resilience, as women play key roles in many marine sectors and are important negotiators and decision-makers. Women can make up more than half of the workforce in some marine industries, especially small-scale fisheries, aquaculture and processing plants. However, women, particularly in developing countries are often disadvantaged through gender inequalities caused by unequal power relations and structures, lack of training, discriminatory laws and customs, and unequal access to and control of resources, and as a result, there are very few women in leadership positions. Women are also more vulnerable than men to climate change and natural disaster impacts.

Many examples from developing countries show that, where women are empowered and can contribute to decision-making processes, social well-being is enhanced and conflict reduced, the health and education of children is improved, and the environment is better protected. Thus, developing and implementing education programmes and capacity building, not only for women but men in the community as well, and establishing women's cooperatives and advocacy groups are needed (Tschakert and Machado 2012; Alston 2013; Monfort 2015; CARE 2016; Dah-gbeto and Villamor 2016; Smucker and Wangui 2016; Tran et al. 2016; de la Torre-Castro et al. 2017; MFF 2018; UNFCCC 2019; Stacey et al. 2019; Ravera et al. 2020).

Box 7.14. Incentivising Coastal Development and a Sustainable Ocean Economy

The Asian Development Bank (ADB) gives us an example of how a multilateral development bank is moving to incentivise coastal development and a sustainable ocean economy. In its operational plan, ADB highlights the importance of building resilience as part of its overarching vision for a "prosperous, inclusive, resilient, and sustainable Asia and the Pacific". Many Asia Pacific countries, particularly low-lying nations and SIDS, are highly exposed and vulnerable to natural hazards and the impacts of climate change. Disaster losses are already growing due to insufficient regard for climate and disaster risk in either the design or location of new infrastructure. A clear priority is planning and delivering infrastructure that builds resilience in a climate and disaster risk resilience context, with a number of different categories of resilience being identified (see Box 7.1).

ADB's Action Plan for Healthy Oceans and Sustainable Blue Economies, launched in 2019 (ADB 2020), is an example of an integrated approach to promoting ocean health and sustainable coastal development. It includes a commitment of $5 billion in investments and technical assistance in focus areas that include sustainable infrastructure, blue economy livelihoods, ecosystem management and pollution control management, supported by an ocean financing initiative.

4.5 Coastal Co-benefits and Trade-Offs

4.5.1 Co-benefits

All too often coastal management strategies are undertaken in order to meet a single objective, without recognising the multiple other benefits that can result from an action. Only by accounting for these can we truly place the full value of ecosystem services into a planning and management context. Value can be expressed in many ways other than direct monetary metrics, including food security, health and cultural values. The benefits that people receive from ecosystems may accrue far from where they are produced. In the last decade, there have been significant advances in developing methods to quantify non-market benefits of coastal ecosystems and to map additional benefits that cascade from them.

Today, ecosystem service valuation is increasingly being used as a tool to assist coastal planning and management to achieve better informed and more holistic decision-making about resource use, and identify opportunities for effective conservation. For example, ecosystem-service approaches can help to inform coastal and marine planning by modelling the likely outcomes of management strategies for objectives expressed in terms of value to people, whether monetary or otherwise.

A number of studies demonstrate that spatially explicit scenario modelling of ecosystem services allows stakeholders and policymakers to better refine zones of human use, identify how different regions may contribute to the flow of services on a larger scale, and test the efficacy of different management strategies. One such recent global analysis finds that, under business as usual, the biggest economic impacts that could result from the loss of nature would be increased coastal vulnerability, followed by loss of carbon sequestration potential, while a "global conservation" scenario would deliver economic gains that result principally from improved natural coastal defences (Johnson et al. 2020). These results suggest that one clear opportunity for action is to focus on protecting and rehabilitating natural infrastructure. These types of nature-based solutions are increasingly being viewed as critical actions to reduce societal risk regarding a number of complex problems, from coastal protection to food security (Whelchel et al. 2018).

4.5.2 Trade-Offs

A key challenge in coastal marine conservation and management is how to manage trade-offs between social and ecological goals, so that both benefits and costs can be distributed equitably across individuals or communities (Halpern et al. 2013). For example, the decision to protect a mangrove to avoid carbon emissions or to slow erosion may have an impact on current timber harvesting or future opportunities to develop the coast for aquaculture or urban expansion. For the people who rely on these for their livelihoods, there is no obvious benefit and therefore little incentive, unless alternative sources of income can be provided.

Globally, climate and coastal development projections over the coming decades mean that we are inevitably faced with compelling circumstances requiring trade-offs to maintain viable environmental conditions and standards of living (Whelchel et al. 2018). Navigating these trade-offs will require thoughtful consideration of the distribution of costs and benefits, and development of mechanisms that protect the livelihoods of those least able to bear the cost. For example, in southwestern Madagascar, efforts to reduce mangrove deforestation have involved developing partnerships with local communities that include finding alternative fuel sources, and alternative ways of generating food and income (Rakotomahazo et al. 2019).

Understanding trade-offs can be complex and cannot be limited to assessing only quantifiable costs and benefits, but needs to consider less obvious factors that can result from complex social-ecological interactions, or that arise because the trade-offs affect marginalised individuals. Concepts of social equity, justice and human rights need to be incorporated in assessing these trade-offs and co-benefits, especially within the wider global discourse on governing the blue economy. While efforts to meet SDG 14 will typically be compatible with those for other SDGs, protecting and conserving coasts to meet SDG 14 targets can also lead to social and economic trade-offs and the downstream effects of such trade-offs can be especially pronounced in low-income coastal communities (Allison et al. 2012; Daw et al. 2015; Galafassi et al. 2017; McClanahan et al. 2016; Nippon Foundation–Nereus Program 2017; Gattuso et al. 2018; Kittinger et al. 2017; Davies et al. 2018; Singh et al. 2018; Bennett 2019; Cohen et al. 2019; Lombard et al. 2019).

To avoid these effects, consideration of trade-offs requires a deliberative approach engaging stakeholders through participatory processes, and harnessing marine spatial planning and scenario modelling tools that allow multiple perspectives and objectives to be considered. As a result, final decisions may reflect open debates about trade-offs and can inform solutions that balance multiple objectives—and surprising synergies may occur (e.g. developing infrastructure to meet multiple uses) that transform a trade-off to a co-benefit (Arkema et al. 2015).

It is important to adopt a long-term perspective when considering trade-offs. For example, short-term losses of livelihoods or income resulting from the creation of MPAs can lead in the long term to ecological, socioeconomic and cultural benefits upon the recovery of fish populations and marine habitats. Intergenerational equity must also be an essential criterion when balancing short-term trade-offs and long-term benefits.

Unintended consequences can also arise. For example, a focus on gross area targets for MPAs may promote the creation of very large marine protected areas, which, by virtue of their size, are generally located in offshore areas, where space is available, tenure arrangements are less complicated and the numbers of stakeholders involved are lower. This may, however, discourage the further protection of mangroves, saltmarshes and seagrass, as their coastal location, often fringing and disjunct distributions, and their location along coasts with multiple land uses and stakeholders make it more complicated to create large protected areas (Friess et al. 2019).

4.6 Enabling Conditions

The coastal zone is crowded, jurisdictionally complex, contains an extremely diverse set of user and interest groups and is subject to multiple competing demands, particularly for space and access. It is a complex socioeconomic system, where achieving sustainable ocean economy outcomes that are resilient to current and future shocks will depend on strong institutions, clear and appropriate governance and finance, an inclusive and equitable approach, and a set of information and science needs. These enabling elements are by nature cross-cutting and are listed below.

- *Strengthening governance and recognising customary rights:* A key influence on the choice and likely success of management options is the existing regulatory framework, through which management authorities, such as permitting and other approvals, are distributed across local, regional, state and/ or federal entities. Most coastal landscapes in the tropics have complex and unclear land tenure and sea use arrangements, especially for indigenous peoples and traditional communities. Furthermore, in many countries indigenous peoples and traditional communities have traditional and customary tenure and rights to significant coastal assets, often defined by LMMAs. Ensuring that these rights are respected and indigenous peoples and traditional communities are engaged in the stewardship of these coastal assets and the creation of alternative livelihood opportunities will be essential.
- *Multilateral partnerships:* Any decision pathway necessarily involves multiple stakeholders who will be interested and involved in the decisions surrounding interventions that sustain or repair coastal ecosystems: practitioners, science and engineering, regulators, industry, investment community, traditional owners and local communities. Developing ecological engineering solutions will require much closer collaboration between scientists and engineers, plus the funding for and a commitment to scientifically test a range of bold innovations at sufficient scale. Where successful, this knowledge should be shared to understand how these innovations could be applied in other settings. Globally, the private sector is seen to have a major role in the implementation of the SDGs and in conserving coastal ecosystems. The International Chamber of Commerce has explicitly stated that sustainability is no longer a luxury business investment, but a core driver of business productivity and growth.
- *Valuing and accounting for coastal assets and ecosystem services:* Coasts and coastal natural infrastructure are essential to the economies of countries and the livelihood of their inhabitants. Impairment of coastal ecosystems that leads to a reduction in resilience or productivity can be a significant cost to the economy. Many ecosystem assets and services remain unquantified. Better methods for valuing non-market assets and services, and applying these consistently within national Systems of Environmental-Economic Accounting (SEEAs), will better inform choices relating to what areas or assets can be developed and what needs to be protected.
- *Quantify co-benefits and trade-offs:* As discussed in Sect. 4.5, analysing trade-offs requires a deliberative approach,

with stakeholder values at the centre. Obtaining full stakeholder involvement through participatory integrated and ecosystem-based marine planning is an important component of assessing trade-offs because it allows for the articulation of multiple perspectives that can inform solutions that balance multiple objectives (Galafassi et al. 2017; Gattuso et al. 2018; Lombard et al. 2019).

- *Science, technology and innovation:* Implementing these various strategies and actions must be underpinned by multidisciplinary science that informs wise decision-making. Although many of these issues encompass the complexity of human decision-making, institutional structures and governance arrangements, science is pivotal to developing more sophisticated and evidence-based policy and management. Integrated management must be underpinned by a deeper understanding of how biophysical and human systems are coupled and an understanding of singular and cumulative impacts. Technological innovation underpins the emerging "science of solutions" that will guide the choice of interventions chosen to safeguard and restore coastal ecosystem resilience. Novel approaches have originated from the growing understanding of biology and ecology, inspiring new theories (e.g. positive species interactions) on which new interventions can be built and tested at scale. There is an important role for the social sciences to be included in future intervention study design, implementation efforts and the collection of evaluation effectiveness metrics.
- *Monitoring and assessment:* Ongoing synoptic and finer-scale observations are required to assess changes in the coastal ecosystems and the surveillance of activities that are occurring within and adjacent to coastal zones. A new generation of in situ sensors, observational platforms, environmental satellite capabilities and informatics provide unprecedented capability and are increasingly accessible and affordable.
- *Capacity building and sharing knowledge:* Supporting the capacity and adaptability of nations—especially least developed and small island states—to successfully implement these strategies requires ongoing, not one-off, capability development that includes both training and access to best-practice information.
- *Financing the future:* Financing for green and grey-green infrastructure is in an exciting growth phase as private investors and development banks increasingly recognise the high potential of this type of infrastructure to tackle development challenges. Initiatives such as the WWF's Sustainable Blue Economy Finance Principles (WWF 2018) lay the groundwork for such investments and need to be broadly and full adopted by public and private sector finance organisations. Strong and effective national sustainable blue economy strategies or plans, based on a clear vision for and definition of a sustainable blue economy, help to foster an enabling environment that reduces risk and builds investor confidence. *The* creation of targeted finance and investment opportunities, such as blue bonds, blended finance, public–private partnerships, insurance payments for risk reduction and corporate stewardship, have emerged as novel ways to build resilience, restore natural capital, and reduce environmental, social and economic risks and investor risk (Herr et al. 2015; Colgan et al. 2017; Niehörster and Murnane 2018; Sumaila et al. 2020; Thiele et al. 2020).

5 Conclusions and Opportunities for Action

5.1 Conclusions

The coastal zone has the world's highest population density, is where the vast majority of resources that underpin the world's ocean and maritime economic sectors are located, and where the majority of many coastal nations' commercial, residential, transport and national defence infrastructure is situated. Coasts sustain livelihoods for hundreds of millions of people in endeavours that range from artisanal small-scale fisheries and aquaculture to transnational fishing, shipping, energy and tourism industries.

Over the last 30–50 years, there have been significant, and, in many cases, rapid/abrupt and irreversible, changes across all of the world's coastal ecosystems. These have included erosion of depositional coastlines, loss of coastal vegetated ecosystems (50% of saltmarshes, 35% of mangroves), and coral (30%) and shellfish reefs (85%), and significant reduction in system resilience.

Coastal ecosystems have evolved in dynamic spatial contexts and many are adapted to disturbance and perturbation or perform a stabilisation and energy dissipation function. Climate change impacts are increasing the physical stress and damage to coastal habitats from storms, flooding and inundation, and are also directly affecting ecosystems through warming and changing ocean chemistry on the abundance and distribution of species and ecosystems.

Humans are also directly affecting coastal ecosystems, with pressures from increasing population growth and urbanisation, poor upstream land practices, alteration of freshwater and sediment flows, habitat conversion, water quality degradation, litter, pollution and over-exploitation of resources. Agriculture operations in catchments can lead to alteration of flows, and increased sediment, nutrient and chemical loads, while coastal fisheries and aquaculture can have direct and indirect effects on coastal ecosystems and habitats. Energy production and resource extraction infrastructure have high freshwater requirements, while urban infrastructure growth leads to habitat conversion, hardening of coastlines, channelisation of flow, and sand-mining in upstream catchments altering sediment budgets at the coast. The result is direct physical loss, fragmentation and alteration of many ecosystems, as well as a functional loss of resilience—which diminishes their ability to resist and recover from such perturbations—that is unprecedented historically. The drivers of this change in coastal ecosystems are complex and interconnected and result from unsustainable levels of human modification to, and resource extraction from, coastal ecosystems.

A rapidly growing and urbanising coastal population, and expansion of existing and new industries, has generated additional demand for coastal space and resources, while incompatibility between different uses—and sometimes ideologies—has also led to conflict in coastal environments. Coastal population growth and increasing urbanisation, catchment degradation and mismanagement, loss of coastal foreshore amenities, environmental impacts from industry, incompatible or unsustainable resource use, and climate change are some of the major challenges that can result in conflict and require careful management.

The physical loss of ecosystems and habitats leads to the loss of their ecological function within the coastal zone socioeconomic system, including provisioning and protection functions. Coastal communities, especially those that are poor and vulnerable and that rely directly on coastal resources for food security, nutrition and livelihoods, are often those most at risk of climate impacts or disasters. COVID-19 has shown us the vulnerabilities that exist in many coastal economic sectors, and again it is those who are poor and vulnerable who appear to be most at risk.

If current trends continue unabated, or without significant interventions, projections over the next 10, 30 and 80 years comprehensively demonstrate the widespread and potentially catastrophic risks to coastal ecosystems, human populations, built infrastructure and economies that will result. The rapid population growth that will occur across Asia, and, even more significantly, Africa, will increase demand for coastal resources and services, and potentially expose coastal cities and settlements to increased impacts. It is here, as well as in the many small island nations spanning the Indo-Pacific and in the Caribbean, that the greatest risks occur.

Failure to properly manage our coastal resources will result in further significant environmental damage to coastal environments, loss of economic well-being for existing industries that operate in the coastal zone (and disincentive for new industries to invest), and inadequate infrastructure development to meet the demands of changing demographics and climate change impacts.

Actions that aim to deploy protection as the "first line of defence" are no longer enough; strategies and technology solutions that mitigate threats, assist in the adaptation of human activities, infrastructure and behaviours, or seek to repair coastal natural systems through restoration and facilitated adaptation will be required. Over the coming decade, implementing these actions *at scale* must be accelerated and assistance to less-developed countries will need to be stepped up.

There are, however, a range of positive policy, planning and coastal infrastructure developments that are cause for cautious optimism as we look towards 2030. Nature-based and hybrid approaches are increasingly being used to adapt existing, and design new, infrastructure to increase resilience to changing climate conditions, and to minimise the loss of ecosystem services. There is great interest, and a large pool of funds, from the investment, insurance and business sectors to implement natural and hybrid approaches for the next generation of climate-resilient infrastructure, and to empower nations and communities to protect coastal ecosystems through a range of financial mechanisms that remunerate for the protection and enhancement of ecosystem services. Intergovernmental bodies, funding agencies (the World Bank, Global Environment Facility, Green Climate Fund), the insurance industry and investment banks all recognise the need for investing in nature-based solutions. However, the availability of support is geographically uneven and there are many barriers to implementation of such approaches at scale.

Building the socioeconomic resilience of those who are most vulnerable, and empowering and engaging natural resource users and coastal communities, especially those who rely directly on coastal resources for food security, nutrition and livelihoods, are critical aspects of ensuring healthy coastal ecosystems and realising a sustainable ocean economy.

5.2 Opportunities for Action

To ensure environmental, economic and social sustainability of our space-constrained coastal systems, the great challenge will be to balance ongoing development and multiple competing uses. By realising the following opportunities for action, it will be possible to reverse the trend of degradation, including from terrestrial and extractive activities, and instead optimise the benefits of healthy ecosystems, natural infrastructure, and inclusive and equitable approaches, to build a coastal zone and coastal economy that is robust, sustainable and resilient.

5.2.1 Building Coastal Resilience

The resilience of coastal ecosystems, and the people who rely upon them, can be enhanced through actions that increase their ability to withstand pressures, and actions that help them to recover when damage occurs.

- Coastal ecosystems must be better protected by strengthening regulations and increasing area-based conservation to halt the net loss, increase the extent and improve the condition of critical coastal habitats, such as sand-dunes, saltmarshes, mangroves, seagrass, and coral and shellfish reefs.
- At-scale habitat restoration, and re-establishing natural coastal and hydrological processes, are required in order to repair many damaged coastal ecosystems and restore functional resilience.
- Restoration also delivers significant co-benefits that extend beyond ecosystem goods and services by creating jobs related to restoration activities, and once established, livelihood opportunities from tourism, enhanced fisheries and payment for ecosystem services, such as carbon sequestration and storage.

5.2.2 Creating Coastal Community Resilience, Equity and Access

Actions that build the socioeconomic resilience of communities, including gender equity and social inclusion, are important in mitigating and recovering from climate and disaster risks and shocks, such as COVID-19, where the impacts are greatest among the poor and vulnerable.

- The multiple benefits coastal communities derive for ocean and coastal services should be included in the valuing and accounting of the ocean economy.
- Communities and coastal fishers should be recognised as legitimate resource users and also stewards of marine ecosystems. This is particularly true for SIDS and remote coastal regions, where communities are often isolated from major governance centres and where marine tenure has remained or is being reinvigorated.
- Governance approaches must be inclusive, incorporating indigenous and local knowledge in planning and decision-making processes.
- It is vital to ensure that business processes are inclusive and that incentives exist to protect and restore coastal ecosystems and enhance local livelihood opportunities.
- Local supply chains should be prioritised, so that pregnant women and infants, and those at risk of malnutrition or hunger, gain access to the nutritional benefits from locally sourced sustainable fish.
- Governments should prioritise poverty reduction and social protection programmes that build community resilience, including to climate change and disasters, and channel post-disaster support to affected poor households. In particular, they should build the resilience of those who are most vulnerable, especially by promoting gender equality and empowering women.

While the consequences of COVID-19 for the resilience of coastal communities will continue to unfold over many years to come, as nations begin to rebuild their economies, there is a unique window of opportunity to ensure relevant policy and investment decisions also address these coastal challenges. In addition, they must foster sustainable economic pathways that support the recovery and development of impacted communities and build the resilience of coastal ecosystems, safeguarding the services they provide.

- Coastal fiscal and economic stimulus and recovery packages must be designed with a sustainable and equitable ocean and coastal economy outcome as a primary objective, and meet multiple SDGs.
- High-employment sectors should be prioritised if they are essential services, or support sustainable ocean economy opportunities. Options include micro-canneries for domestic consumption, mangrove restoration for disaster risk reduction, and investments in effective waste management systems that reduce disease prevalence.
- Vulnerabilities in coastal economic sectors and supply chains should be prioritised for investment and innovation. Examples include the development of product alternatives that have a longer shelf life, using digital means to connect to customers and local markets, and adopting electronic and digital verification systems in supply chains.
- Climate change projections and impacts should be incorporated into all aspects of COVID-19 recovery planning and sustainable infrastructure design. This includes the protection and restoration of coastal ecosystems and fisheries as part of building resilience.

5.2.3 Mitigation of Terrestrial and Extractive Activities on Coastal Ecosystems

The impacts of terrestrial and extractive activities on coastal ecosystems may be cumulative and may be amplified by climate change effects, while the downstream impacts of upstream activities can lead to conflicts among user groups.

- Integrated management underpinned by good spatial planning and coastal ecosystem planning must be fully integrated into urban, catchment and land-use planning frameworks.
- Urban and agriculture water use should be managed to ensure that coastal ecosystems receive healthy surface flows and that coastal groundwater reserves are maintained.
- Upstream catchment diversions and dams should be managed to ensure that adequate freshwater flow and adequate sediment supply is maintained to the coast. Promotion of

alternatives to mega dams, such as building small dams with sediment release facilities, is a priority.

- Regional multi-sector dialogues should be initiated to address upstream sand extraction and sand scarcity, particularly in relation to coastal city subsidence and stability of urbanised deltas.
- Closer integration should be pursued between the current global water, food and energy nexus, and the water, urban and climate agendas and initiatives, including the High Level Panel on Water, and the overlapping UN decade initiatives for Oceans, Water, and Ecosystem Restoration.

5.2.4 Sustainable, Future-Ready Blue Infrastructure

The following opportunities for action are designed as ones that industry, government, scientists and communities can take to promote the uptake, resourcing and deployment of natural infrastructure.

- Identify locations where natural or hybrid infrastructure can play a significant role in natural hazard risk reduction, and adapt and upgrade existing coastal infrastructure through the adoption of nature-based approaches for more sustainable designs, including retrofitting coastal defence structures.
- Develop and scale cost-effective, hybrid approaches that enhance resilience by integrating nature into mainstream infrastructure systems. Encourage closer collaboration between scientists and engineers, and dedicate funding to develop eco-engineering opportunities.
- Build the skills and capacity of government staff in the design of sustainable ocean economy recovery programmes and in the design and maintenance of sustainable green coastal infrastructure, such that there is a common understanding of the benefits and opportunities.
- Embed opportunities for future-ready blue infrastructure and nature-based solutions within existing planning and management approaches, including within spatial management tools such as marine spatial planning, ecosystem-based integrated ocean management, marine protected areas and community-based natural resource management tools and approaches.
- Support the restoration of coastal ecosystems, including mangrove forests, saltmarshes, seagrass meadows, kelp and other seaweed forests, and coral and shellfish reefs, to optimise their function for coastal defence, coastal stabilisation or as part of hybrid coastal defence structures. Recognise that coral reef and mangrove restoration in particular offer cost-effective options for risk reduction of climate hazards.
- Develop the experience and standards to overcome institutional biases that favour grey infrastructure, and develop institutional arrangements capable of matching available funding with the needs of individual situations.
- Design new and innovative financial instruments to provide the pathways for investors to direct private finance into nature-based solutions, including through public–private investments.
- Establish standards and principles for developing and financing blue infrastructure and appropriate blended finance instruments, a good example of which are the Sustainable Blue Economy Finance Principles.
- Enable the use of green finance instruments, including in the delivery of NDCs, and use them to promote the uptake of natural infrastructure and sustainable infrastructure projects, including in developing and low-income countries seeking financing from multilateral development banks.

5.3 Enabling Conditions to Support Coastal Resilience

For any of the above actions to be successful in delivering coastal resilience, a number of enabling conditions need to occur. These were summarised in Sect. 4.6, while enabling actions specific to the coastal context are given below.

Strengthening governance and recognising customary rights: The enabling conditions for inclusive and effective local governance must be put in place, so coastal communities can effectively advocate for their rights to access coastal resources. Power imbalances must be acknowledged and addressed, to allow coastal communities the necessary influence and impact in governance and policy fora. As suggested in SDG 14, Target 14.b, the will and needs of coastal communities should be recognised, respected and reflected in policymaking and decision-making, and in the implementation of the SDGs. Local and national policies recognise the role of communities in the management of coastal resources, incorporate advice from community members in decision-making, and facilitate more equitable and inclusive access of communities to natural resources and markets.

Science, technology and innovation: The cross-disciplinary nature of grey-green infrastructure and natural infrastructure brings together ecology and engineering in the emerging discipline of ecological engineering, in designing societal services such that they benefit society and nature. As the implementation of hybrid and grey-green infrastructure solutions grows, there needs to be a body of research on all aspects and at a range of scales, in order to optimise the design of individual projects and to facilitate scaling. It is an area that is ripe for the application of technological innovation, consistent with the use of intelligent and smart building design in green buildings. COVID-19 has stress-tested contemporary coastal economic and logistical systems, and identified weaknesses and vulnerabilities that need to be prioritised for research, innovation and technological solutions.

Multilateral partnerships: Ensuring traditional owners and local communities have a voice and are engaged in the co-design and development of strategies and plans will be essential for gaining social licence. Ensuring a role in the stewardship and day-to-day management of activities that use these coastal assets and creation of alternative livelihood opportunities must be a priority.

Capacity building and sharing knowledge: Translating coastal research to best practice and "how to" guidance on coastal issues, such as dredging, coastal modelling, water and sediment quality standards, restoration methodologies, coastal and eco-engineering, and emergency preparedness, is required. Making this information available through a clearinghouse of coastal information will encourage adoption by regulators, environmental consulting and analysis sectors and organisations, and communities seeking to undertake restoration activities.

Financing the future: Green infrastructure and hybrid infrastructure designed with co-benefits in mind opens up a range of possible finance options in addition to the standard government financing model. It allows projects to be promoted to governments, the private sector or development agencies as stand-alone investment opportunities, matched to particular motivations. The private sector has the ability to provide substantial investment to support nature-based solutions, including through bonds and other novel instruments. However, the amount they currently invest is small because of constraints such as limited evidence of the returns on investment and lack of appropriate financial instruments. Development agencies with core mandates of climate resilience, poverty reduction and environmental sustainability also have a strong motivation to invest in such projects. The next decade should see significant growth in green and grey infrastructure, as investment pipelines grow, the capacity for designing and managing such investments is increased in target countries, and as challenges to scaling are overcome.

Acknowledgements We wish to thank the many people who generously provided material, advice and review for the report: Russ Babcock, Elisa Bayraktarov, Melanie Bishop, Katherine Dafforn, Dorothee Herr, Alexander Goad, Mariana Mayer-Pinto, Peter Steinberg and Torsten Thiele. The paper's technical reviewers, Audrey Legat, Borja Reguero, Julien Rochette, Alan Simcock and Sharon Tatman, as well as its arbiter, Jane Lubchenco, all provided helpful technical comments.

While our colleagues were very generous with their time and input, this report reflects the views of the authors alone, and is not necessarily the views of their organisations. Thank you to Em Muirhead, Lauren Hardiman and Derek Fulton for providing administrative, editing and graphic design support.

The authors thank Sarah Chatwin for copyediting and Romain Warnault for design.

Glossary

Ecological engineering The design of sustainable ecosystems that integrate human society with its natural environment for the benefit of both. The approach has developed over the last 30 years, rapidly over the last 10 years, and its goals include the restoration of ecosystems that have been substantially disturbed by human activities and the development of new sustainable ecosystems that have both human and ecological values.

Green grabbing The appropriation of land and resources for environmental ends, where "green" credentials are called upon to justify appropriations of land, and the restructuring of rules and authority in the access, use and management of these resources may have profoundly alienating effects.

Green infrastructure Green infrastructure (also sometimes called natural infrastructure or engineering with nature) intentionally and strategically preserves, enhances or restores elements of a natural system, such as forests, agricultural land, floodplains, riparian areas, coastal forests (such as mangroves), among others, and combines them with grey infrastructure to produce more resilient and lower-cost services.

Grey infrastructure Traditionally used to manage coastal hazards, often constructed out of concrete with a uniform and smooth texture, often costly to install and maintain, usually has low flexibility, and when it fails can generate catastrophic impacts on social and ecological domains.

Nature-based solutions (NbS) Actions to protect, manage and restore natural or modified ecosystems, which address societal challenges, effectively and adaptively, providing human well-being and biodiversity benefits. IUCN defines nature-based solutions as actions to protect, sustainably manage and restore natural or modified ecosystems, that address societal challenges effectively and adaptively, simultaneously providing human well-being and biodiversity benefits.

Ocean economy Also known as the blue economy, encompasses a sustainable economy for the ocean-based marine environment, related biodiversity, ecosystems, species and genetic resources, including marine living organisms (from fish and algae to microorganisms) and natural resources in the seabed, while ensuring their sustainable use and hence conservation.

Rehabilitation The replacement of structural or functional characteristics of an ecosystem that have been diminished or lost.

Resilience The capacity of social, economic and environmental systems to cope with a hazardous event or trend or disturbance, responding or re-organising in ways that maintain their essential function, identity and structure, while also maintaining the capacity for adaptation, learning and transformation.

Restoration The process of assisting the recovery of an ecosystem that has been degraded, damaged or destroyed.

Social-ecological Refers to systems that emphasise humans as part of nature and stress that the delineation between social systems and ecological systems is artificial and arbitrary. While resilience has somewhat different meanings in social and ecological contexts, the social-ecological approach holds that social and ecological systems are linked through feedback mechanisms, and that both display resilience and complexity.

Source-to-sea A source-to-sea system is the land area that is drained by a river system, its lakes and tributaries (the river basin), connected aquifers and downstream recipients, including deltas and estuaries, coastlines and nearshore waters, as well as the adjoining sea and continental shelf and the open ocean. A source-to-sea system can also be defined at a larger scale to include a sea and its entire drainage area, which may include several river basins.

About the Authors

Co-authors

Andy Steven is the Research Director of Coasts at the Commonwealth Scientific and Industrial Research Organisation (CSIRO), Australia.

Kwasi Appeaning Addo is an associate professor in Coastal Processes and Delta Studies and the Director of the Institute for Environment and Sanitation Studies, University of Ghana.

Ghislaine Llewellyn is Deputy Practice Leader – Oceans, with WWF International, Australia.

Vu Thanh Ca is a principal lecturer at the Ha Noi University of Natural Resources and Environment.

Contributing Authors

Isaac Boateng is an associate professor of Coastal Engineering at the Department of Construction and Wood, University of Education, Winneba, Ghana.

Rodrigo Bustamante is a research scientist at CSIRO, Australia.

Christopher Doropoulos is a research scientist at CSIRO, Australia.

Chris Gillies is the Program Director, Oceans, at The Nature Conservancy, Australia Program Office, Australia.

Mark Hemer is a principal research scientist at CSIRO, Australia.

James Kairo is a principal scientist at the Kenya Marine and Fisheries Research Institute, Kenya.

Priscila Lopes is an associate professor at the Universidade Federal do Rio Grande do Norte, Brazil.

Munsur Rahman is a professor at the Institute of Flood Management and Water at the Bangladesh University of Engineering and Technology, Bangladesh.

Lalao Aigrette Ravaoarinorotsihoarana is a National Technical Advisor for Mangrove Programme, Blue Ventures, Madagascar.

Megan Saunders is a research scientist at CSIRO, Australia.

Frida Sidik is a senior researcher at the Institute for Marine Research and Observation, Indonesian Ministry of Marine Affairs and Fisheries, Indonesia.

U. Rashid Sumaila is Director of the Fisheries Economics Research Unit at the Institute for the Oceans and Fisheries, University of British Columbia, Canada.

Louise Teh is a research associate at the Fisheries Economics Research Unit at the Institute for the Oceans and Fisheries, University of British Columbia, Canada.

Mat Vanderklift is a research scientist at CSIRO, Australia.

Maria Vozzo is a research associate at the Sydney Institute of Marine Science, Australia.

References

Abram N, Gattuso J-P, Prakash A, Cheng L, Chidichimo MP, Crate S, Enomoto H, Garschagen M, Gruber N, Harper S, Holland E, Kudela RM, Rice J, Steffen K, von Schuckmann K (2019) Framing and context of the report. In: Pörtner H-O, Roberts DC, Masson-Delmotte V, Zhai P, Tignor M, Poloczanska E, Mintenbeck K et al (eds) IPCC special report on the ocean and cryosphere in a changing climate. Intergovernmental Panel on Climate Change, Geneva

Aceves-Bueno E, Miller SJ, Cornejo-Donoso J, Gaines SD (2020) Cooperation as a solution to shared resources in territorial use rights in fisheries. Ecol Appl 30(1):e02022

Adame MF, Roberts ME, Hamilton DP, Ndehedehe CE, Reis V, Lu J, Griffiths M, Curwen G, Ronan M (2019) Tropical coastal wetlands ameliorate nitrogen export during floods. Front Mar Sci 6:671

ADB (Asian Development Bank) (2019) Operational priority 3: tackling climate change, building climate and disaster resilience, and enhancing environmental sustainability. Asian Development Bank, Manila

ADB (Asian Development Bank) (2020) COVID-19 recovery: a pathway to a low-carbon and resilient future. Asian Development Bank, Manila. https://www.adb.org/publications/covid-19-recovery-low-carbon-resilient-future

Adger WN, Hughes TP, Folke C, Carpenter SR, Rockström J (2005) Social-ecological resilience to coastal disasters. Science 309(5737):1036–1039

Aerts JCJH, Botzen WJW, Bowman MJ, Ward PJ, Dircke P (2011) Climate adaptation and flood risk in coastal cities. Earthscan, London

Afflerbach JC, Lester SE, Dougherty DT, Poon SE (2014) A global survey of 'TURF-reserves', territorial use rights for fisheries coupled with marine reserves. Glob Ecol Conserv 2:97–106

Agarwal N, Bonino C, Deligny A, El Berr L, Festa C, Ghislain M, Homolova K (2019) Getting the Shrimp's share. Mangrove deforestation and shrimp consumption, assessment and alternatives. IDDRI and Sciences Po, Paris

Agostini S, Harvey BP, Wada S, Kon K, Milazzo M, Inaba K, Hall-Spencer JM (2018) Ocean acidification drives community shifts towards simplified non-calcified habitats in a subtropical–temperate transition zone. Sci Rep 8(1):1–11

Ahmed N, Thompson S (2019) The blue dimensions of aquaculture: a global synthesis. Sci Total Environ 652:851–861

Al-Bahry SN, Mahmoud IY, Al-Belushi KI, Elshafie AE, Al-Harthy A, Bakheit CK (2009) Coastal sewage discharge and its impact on fish with reference to antibiotic resistant enteric bacteria and enteric pathogens as bio-indicators of pollution. Chemosphere 77(11):1534–1539. https://doi.org/10.1016/j.chemosphere.2009.09.052

Albert S, Tawake A, Vave R, Fisher P, Grinham A (2016) Indicators of herbivorous fish biomass in community-based marine management areas in Fiji. Pac Conserv Biol 22:20–28

Alexander KA, Haward M (2019) The Human Side of Marine Ecosystem-Based Management (EBM): 'Sectoral interplay' as a challenge to implementing EBM. Mar Policy 101:33–38

Alexander KA, Hobday AJ, Cvitanovic C, Ogier E, Nash KL, Cottrell RS, Fleming A, Fudge M, Fulton EA, Frusher S, Kelly R (2019) Progress in integrating natural and social science in marine ecosystem-based management research. Mar Freshw Res 70(1):71–83

Allison EH, Ratner BD, Åsgård B, Willmann R, Pomeroy R, Kurien J (2012) Rights-based fisheries governance: from fishing rights to human rights. Fish Fish 13(1):14–29

Alston M (2013) Women and adaptation. Wiley Interdiscip Rev Clim Change 4(5):351–358

Altenburg T, Fischer C, Huck K, Kruip A, Müller S, Sörensen S (2017) Managing coastal ecosystems in the Philippines: what cash for work programmes can contribute. DIE Studies No. 94. German Development Institute/Deutsches Institut für Entwicklungspolitik (DIE), Bonn

Álvarez-Romero JG, Pressey RL, Ban NC, Vance-Borland K, Willer C, Klein CJ, Gaines SD (2011) Integrated land-sea conservation planning: the missing links. Annu Rev Ecol Evol Syst 42: 381–409

Andrachuk M, Armitage D, Hoang HD, Le NV (2019) A network perspective on spatially clustered Territorial Use Rights for Fishers (TURFs) in Vietnam. Coast Manag 47(3):292–311

Anthony E, Goichot M (2020) Sediment flow in the context of mangrove restoration and conservation. a rapid assessment guidance manual. BMZ (Federal Ministry for Economic Cooperation and Development), IUCN (International Union for Conservation of Nature) and WWF, Bonn

Anthony K, Bay LK, Costanza R, Firn J, Gunn J, Harrison P, Heyward A, Lundgren P, Mead D, Moore T, Mumby PJ, van Oppen MJH, Robertson J, Runge MC, Suggett DJ, Schaffelke B, Wachenfeld D, Walshe T (2017) New interventions are needed to save coral reefs. Nat Ecol Evol 1:1420–1422

Appeaning Addo K, Nicholls RJ, Codjoe SNA, Mumuni A (2018) A biophysical and socio-economic review of the Volta Delta, Ghana. J Coast Res 34(5):1216–1226. https://doi.org/10.2112/JCOASTRES-D-17-00129.1

Arias-Ortiz A, Serrano O, Masqué P, Lavery PS, Mueller U, Kendrick GA, Rozaimi M, Esteban A, Fourqurean JW, Marbà N, Mateo MA (2018) A marine heatwave drives massive losses from the world's largest seagrass carbon stocks. Nat Clim Change 8(4):338

Arkema KK, Verutes GM, Wood SA, Clarke-Samuels C, Rosado S, Canto M, Rosenthal A, Ruckelshaus M, Guannel G, Toft J, Faries J, Silver JM, Griffin R, Guerry AD (2015) Embedding ecosystem services in coastal planning leads to better outcomes for people and nature. Proc Natl Acad Sci 112:7390–7395

Arnell NW, Gosling SN (2016) The impacts of climate change on river flood risk at the global scale. Clim Change 134(3):387–401. https://doi.org/10.1007/s10584-014-1084-5

Arthur MA, Cole DR (2014) Unconventional hydrocarbon resources: prospects and problems. Elements 10:257–264

Atwood TB, Connolly RM, Almahasheer H, Carnell PE, Duarte CM, Ewers Lewis CJ, Irigoien X, Kelleway JJ, Lavery PS, Macreadie PI, Serrano O, Sanders CJ, Santos I, Steven ADL, Lovelock CE (2017) Global patterns in Mangrove soil carbon stocks and losses. Nat Clim Change 7:523–528. https://doi.org/10.1038/nclimate3326

Babcock RC, Bustamante RH, Fulton EA, Fulton DJ, Haywood MD, Hobday AJ, Kenyon R, Matear RJ, Plaganyi EE, Richardson AJ, Vanderklift MA (2019) Severe continental-scale impacts of climate change are happening now: extreme climate events impact marine habitat forming communities along 45% of Australia's Coast. Front Mar Sci 6:411

Baker S, Eckerberg K (2013) A policy analysis perspective on ecological restoration. Ecol Soc 18(2):17

Bakun A, Black BA, Bograd SJ, García-Reyes M, Miller AJ, Rykaczewski RR, Sydeman WJ (2015) Anticipated effects of climate change on coastal upwelling ecosystems. Curr Clim Change Rep 1:85–93

Ban NC, Frid A (2018) Indigenous peoples' rights and marine protected areas. Mar Policy 87:180–185. https://doi.org/10.1016/j.marpol.2017.10.020

Bangladesh Planning Commission (2018) Bangladesh Delta Plan 2100: Bangladesh in the 21st century, vol 1: Strategy. General Economics Division, Bangladesh Planning Commission

Barboza LGA, Cózar A, Gimenez BC, Barros TL, Kershaw PJ, Guilhermino L (2019) Macroplastics pollution in the marine environment. In: Shephard C (ed) World seas: an environmental evaluation. Academic Press, Cambridge, pp 305–328

Barnard PL, Erikson LH, Foxgrover AC, Finzi Hart JA, Limber J, O'Neill AC, van Ormondt M, Vitousek S, Wood N, Hayden MK, Jones JM (2019) Dynamic flood modeling essential to assess the coastal impacts of climate change. Sci Rep 9(1):1–13. https://doi.org/10.1038/s41598-019-40742-z

Barton A, Waldbusser G, Feely R, Weisberg S, Newton J, Hales B, Cudd S, Eudeline B, Langdon C, Jefferds I, King T, Suhrbier A, McLauglin K (2015) Impacts of coastal acidification on the Pacific Northwest Shellfish Industry and adaptation strategies implemented in response. Oceanography 25:146–159

Bastyan GR, Cambridge ML (2008) Transplantation as a method for restoring the seagrass Posidonia australis. Estuar Coast Shelf Sci 79:289–299

Bayraktarov E, Saunders MI, Abdullah S, Mills M, Beher J, Possingham HP, Mumby PJ, Lovelock CE (2016) The cost and feasibility of marine coastal restoration. Ecol Appl 26:1055–1074

Bayraktarov E, Stewart-Sinclair PJ, Brisbane S, Boström-Einarsson L, Saunders MI, Lovelock CE, Possingham HP, Mumby PJ, Wilson KA (2019) Motivations, success and cost of coral reef restoration. Restor Ecol 27:981. https://doi.org/10.1111/rec.12977

Bayraktarov E, Brisbane S, Hagger V, Smith CS, Wilson KA, Lovelock CE, Gillies C, Steven ADL, Saunders MI (2020a) Priorities and motivations of marine coastal restoration research. Front Mar Sci 7:484. https://doi.org/10.3389/fmars.2020.00484

Bayraktarov E, Banaszak AT, Montoya Maya P, Kleypas J, Arias-González JE, Blanco M, Calle-Triviño J, Charuvi N, Cortés-Useche C, Galván V, García Salgado MA, Gnecco M, Guendulain-García SD, Hernández Delgado EA, Marín Moraga JA, Maya MF, Mendoza Quiroz S, Mercado Cervantes S, Morikawa M, Nava G, Pizarro V, Sellares-Blasco RI, Suleimán Ramos SE, Villalobos Cubero T, Villalpando MF, Frías- Torres S (2020b) Coral reef restoration efforts in Latin American countries and territories. PloS One 15:e0228477

Beck MW (2014) Coasts at risk: an assessment of coastal risks and the role of environmental solutions. A joint publication of United Nations University–Institute for Environment and Human Security (UNU-EHS), The Nature Conservancy (TNC) and the Coastal Resources Center (CRC) at the University of Rhode Island Graduate School of Oceanography

Beck MW, Brumbaugh RD, Airoldi L, Carranza A, Coen LD, Crawford C, Defeo O, Edgar GJ, Hancock B, Kay MC, Lenihan HS, Luckenbach MW, Toropova CL, Zhang G, Guo X (2011) Oyster reefs at risk and recommendations for conservation, restoration, and management. Bioscience 61(2):107–116

Beck MW, Narayan S, Trespalacios D, Pfliegner K, Losada IJ, Menéndez P, Espejo A, Torres S, Díaz-Simal P, Fernandez F, Abad S, Mucke P, Kirch L (2018a) The global value of mangroves for risk reduction. Summary report. The Nature Conservancy, Berlin. https://doi.org/10.7291/V9930RBC

Beck MW, Losada IJ, Menéndez P, Reguero BJ, Díaz-Simal P, Fernández F (2018b) The global flood protection savings provided by coral reefs. Nat Commun 9:2186. https://doi.org/10.1038/s41467-018-04568-z

Becker M, Papa F, Karpytchev M, Delebecque C, Krien Y, Khan JU, Ballu V, Durand F, Le Cozannet G, Saiful Islam AKM, Calmant S, Shum CK (2020) Water level changes, subsidence, and sea level rise in the Ganges–Brahmaputra–Meghna Delta. Proc Natl Acad Sci 117(4):1867–1876. https://doi.org/10.1073/pnas.1912921117

BEIS (Department for Business, Energy and Industrial Strategy) (2020) Electricity generation costs 2020. BEIS, London. https://assets.publishing.service.gov.uk/government/uploads/system/uploads/attachment_data/file/911817/electricity-generation-cost-report-2020.pdf

Bell JD, Allain V, Allison EH, Andréfouët S, Andrew NL, Batty MJ, Blanc M, Dambacher JM, Hampton J, Hanich Q, Harley S (2015) Diversifying the use of tuna to improve food security and public health in Pacific Island countries and territories. Mar Policy 51: 584–591

Bendixen M, Best J, Hackney C, Iversen LL (2019a) Time is running out for sand. Nature 571(7763):29–31

Bendixen M, Overeem I, Rosing MT, Bjørk AA, Kjær KH, Kroon A, Zeitz G, Iversen LL (2019b) Promises and perils of sand exploitation in Greenland. Nat Sustain 2(2):98–104

Bennett G (2019) Dilemmas: coping with environmental problems. Routledge, London. https://doi.org/10.4324/9780429201752

Bennett NJ, Cisneros-Montemayor AM, Blythe J, Silver JJ, Singh G, Andrews N, Calò A, Christie P, Di Franco A, Finkbeiner EM, Gelcich S (2019) Towards a sustainable and equitable blue economy. Nat Sustain 2:991–993. https://doi.org/10.1038/s41893-019-0404-1

Bennett NJ, Finkbeiner EM, Ban NC, Belhabib D, Jupiter SD, Kittinger JN, Mangubhai S, Scholtens J, Gill D, Christie P (2020) The COVID-19 pandemic, small-scale fisheries and coastal fishing communities. Coast Manag 48(4):336–347. https://doi.org/10.1080/08920753.2020.1766937

Benzeev R, Hutchinson N, Friess DA (2017) Quantifying fisheries ecosystem services of mangroves and tropical artificial urban shorelines. Hydrobiologia 803(1):225–237

Bergillos RJ, Rodriguez-Delgado C, Iglesias G (2019) Wave farm impacts on coastal flooding under sea-level rise: a case study in Southern Spain. Sci Total Environ 653:1522–1531. https://doi.org/10.1016/j.scitotenv.2018.10.422

Bergqvist R, Monios J (2019) Green ports in theory and practice. In: Bergqvist R, Monios J (eds) Green ports. Elsevier, Cambridge, pp 1–17

Bernal PA (2015) State ocean strategies and policies for the open ocean. In: Smith HD, Suárez de Vivero JL, Agardy TS (eds) Routledge handbook of ocean resources and management. Routledge, London, pp 33–54

Besset M, Gratiot N, Anthony E, Bouchette F, Goichot M, Marchesiello P (2019) Mangroves and shoreline erosion in the Mekong River Delta, Viet Nam. Estuar Coast Shelf Sci 226:106263. https://doi.org/10.1016/j.ecss.2019.106263

Betzold C, Mohamed I (2017) Seawalls as a response to coastal erosion and flooding: a case study from Grande Comore, Comoros (West Indian Ocean). Reg Environ Change 17(4):1077–1087

Bhatia K, Vecchi G, Murakami H, Underwood S, Kossin J (2018) Projected response of tropical cyclone intensity and intensification in a global climate model. J Climate 31(20):8281–8303

Bindoff NL, Cheung WWL, Kairo JG, Arístegui J, Guinder VA, Hallberg R, Hilmi N, Jiao N, Karim MS, Levin L, O'Donoghue S, Purca Cuicapusa SR, Rinkevich B, Suga T, Tagliabue A, Williamson P (2019) Changing ocean, marine ecosystems, and dependent communities supplementary material. In: Pörtner H-O, Roberts DC, Masson-Delmotte V, Zhai P, Tignor M, Poloczanska E, Mintenbeck K et al (eds) IPCC special report on the ocean and cryosphere in a changing climate. Geneva, Intergovernmental Panel on Climate Change

Birkmann J, Garschagen M, Fernando N, Tuan V, Oliver-Smith A, Hettige S (2013) Dynamics of vulnerability: relocation in the context of natural hazards and disasters. In: Birkmann J (ed) Measuring vulnerability to natural hazards, 2nd edn. United Nations University Press, New York, pp 505–550

Bishop MJ, Mayer-Pinto M, Airoldi L, Firth LB, Morris RL, Loke LHL, Hawkins SJ, Naylor LA, Coleman RA, Chee SY, Dafforn KA (2017) Effects of ocean sprawl on ecological connectivity: impacts and solutions. J Exp Mar Biol Ecol 492:7–30. https://doi.org/10.1016/j.jembe.2017.01.021

Blankespoor B, Dasgupta S, Laplante B (2014) Sea-level rise and coastal wetlands. Ambio 43:996–1005. https://doi.org/10.1007/s13280-014-0500-4

de Boer WP, Slinger JH, A.K. Wa Kangeri, H.S.I. Vreugdenhil, P. Taneja, K.A. Addo and T. Vellinga. (2019) Identifying ecosystem-based alternatives for the design of a Seaport's marine infrastructure: the case of Tema Port expansion in Ghana. Sustainability 11(23):6633. https://doi.org/10.3390/su11236633

Borja A, Bricker SB, Dauer DM, Demetriades NT, Ferreira JG, Forbes AT, Hutchings P, Jia X, Kenchington R, Marques JC, Zhu C (2008) Overview of integrative tools and methods in assessing ecological integrity in estuarine and coastal systems worldwide. Mar Pollut Bull 56(9):1519–1537. https://doi.org/10.1016/j.marpolbul.2008.07.005

Borsje BW, van Wesenbeeck BK, Dekker F, Paalvast P, Tjeerd J, Marieke B, van Katwijk M, de Vries MB (2011) How ecological engineering can serve in coastal protection. Ecol Eng 37(2): 113–122

Boström-Einarsson L, Babcock RC, Bayraktarov E, Ceccarelli D, Cook N, Ferse SCA, Hancock B, Harrison P, Hein M, Shaver E, Smith A, Suggett D, Stewart-Sinclair PJ, Vardi T, McLeod IM (2020) Coral restoration – a systematic review of current methods, successes, failures and future directions. PloS One 15(1):e0226631

Breitburg D, Levin LA, Oschlies A, Grégoire M, Chavez FP, Conley DJ, Garçon V, Gilbert D, Gutiérrez D, Isensee K, Jacinto GS (2018) Declining oxygen in the global ocean and coastal waters. Science 359(6371):eaam7240

Brodie JE, Kroon FJ, Schaffelke B, Wolanski EC, Lewis SE, Devlin MJ, Bohnet IC, Bainbridge ZT, Waterhouse J, Davis AM (2012) Terrestrial pollutant runoff to the Great Barrier Reef: an update of issues, priorities and management responses. Mar Pollut Bull 65(4–9):81–100

Bromberg Gedan K, Silliman BR, Bertness MD (2009) Centuries of human-driven change in salt marsh ecosystems. Ann Rev Mar Sci 1:117–141. https://doi.org/10.1146/annurev.marine.010908.163930

Brooks SM, Spencer T (2012) Shoreline retreat and sediment release in response to accelerating sea level rise: measuring and modelling cliffline dynamics on the Suffolk Coast, UK. Global Planet Change 80–81:165–179. https://doi.org/10.1016/j.gloplacha.2011.10.008

Browder G, Ozment S, Bescos IR, Gartner T, Lange GM (2019) Integrating green and gray: creating next generation infrastructure.

World Bank and World Resources Institute, Washington, DC. https://openknowledge.worldbank.org/handle/10986/31430

Brown S, Nicholls RJ, Woodroffe CD, Hanson S, Hinkel J, Kebede AS, Neumann B, Vafeidis AT (2013) Sea-level rise impacts and responses: a global perspective. In: Finkl CW (ed) Coastal hazards. Springer, Dordrecht, pp 117–149. https://doi.org/10.1007/978-94-007-5234-4_5

Brown MI, Tristan P, Javier L, Sidle R, Wilson R (2018) Using remote sensing and traditional ecological knowledge (TEK) to understand mangrove change on the Maroochy River, Queensland, Australia. Appl Geogr 94:71–83. https://doi.org/10.1016/j.apgeog.2018.03.006

Bryan-Brown DN, Connolly RM, Richards DR, Adame F, Friess DA, Brown CJ (2020) Global trends in mangrove forest fragmentation. Sci Rep 10(1):1–8

Buck BH, Langan R (2017) Aquaculture perspective of multi-use sites in the open ocean: the untapped potential for marine resources in the Anthropocene. Springer, Basel

Buck BH, Troell MF, Krause G, Angel DL, Grote B, Chopin T (2018) State of the art and challenges for offshore integrated multi-trophic aquaculture (IMTA). Front Mar Sci 5:165

Bugnot AB, Mayer-Pinto M, Airoldi L, Heery EC, Johnston EL, Critchley LP, Strain EMA, Morris RL, Loke LHL, Bishop MJ, Sheehan EV, Coleman RA, Dafforn KA (2020) Current and projected global extent of marine built structures. Nat Sustain. https://doi.org/10.1038/s41893-020-00595-1

Bull DL, Frederick J, Mota A, Thomas MA, Jones BM, Jones CA, Flanary C, Kasper J, Choens R, Bristol E, McClelland JW (2019) Development of a tightly coupled multi-physics numerical model for an event-based understanding of Arctic coastal erosion. In: AGU Fall Meeting 2019, pp C12B–04

Bulleri F, Chapman MG (2010) The introduction of coastal infrastructure as a driver of change in marine environments. J Appl Ecol 47:26–35

Bunting P, Rosenqvist A, Lucas RM, Rebelo LM, Hilarides L, Thomas N, Hardy A, Itoh T, Shimada M, Finlayson CM (2018) The global mangrove watch—a new 2010 global baseline of mangrove extent. Remote Sens (Basel) 10(10):1669

Burt JA, Killilea ME, Ciprut S (2019) Coastal urbanization and environmental change: opportunities for collaborative education across a global network university. Reg Stud Mar Sci 26:100501

Camus P, Losada IJ, Izaguirre C, Espejo A, Menéndez M, Pérez J (2017) Statistical wave climate projections for coastal impact assessments. Earth's Futur 5(9):918–933

Camus P, Tomás A, Díaz-Hernández G, Rodríguez B, Izaguirre C, Losada IJ (2019) Probabilistic assessment of port operation downtimes under climate change. Coast Eng 147:12–24

Candel JJ (2017) Holy grail or inflated expectations? The success and failure of integrated policy strategies. Policy Stud 38(6):519–552

CARE (2016) Enhancing resilience through gender equality. Gender equality and women's voice in Asia-Pacific Resilience Programming. Research report. CARE, Canberra

Carrasco LR, Chan J, McGrath FL, Nghiem LTP (2017) Biodiversity conservation in a telecoupled world. Ecol Soc 22(3):24. https://doi.org/10.5751/ES-09448-220324

Carrasquilla-Henao M, Juanes F (2017) Mangroves enhance local fisheries catches: a global meta-analysis. Fish Fish 18(1):79–93

Chalastani VI, Manetos P, Al-Suwailem AM, Hale JA, Vijayan AP, Pagano J, Williamson I, Henshaw SD, Albaseet R, Butt F, Brainard RE, Coccossis H, Tsoukala VK, Duarte CM (2020) Reconciling tourism development and conservation outcomes through marine spatial planning for a Saudi Giga-Project in the Red Sea (The Red Sea Project, Vision 2030). Front Mar Sci 7:168. https://doi.org/10.3389/fmars.2020.00168

Champion C, Hobday AJ, Tracey SR, Pecl GT (2018) Rapid shifts in distribution and high-latitude persistence of oceanographic habitat revealed using citizen science data from a climate change hotspot. Glob Chang Biol 24(11):5440–5453

Charles A (2017) The big role of coastal communities and small-scale fishers in ocean conservation. In: Levin P, Poe M (eds) Conservation for the Anthropocene Ocean, interdisciplinary science in support of nature and people. Academic Press, Cambridge, pp 447–461

Chee SY, Othman AG, Sim YK, Adam ANM, Firth LB (2017) Land reclamation and artificial islands: walking the tightrope between development and conservation. Glob Ecol Conserv 12:80–95

Christy FT (1982) Territorial use rights in marine fisheries: definitions and conditions. FAO fisheries technical paper 227. Food and Agriculture Organization of the United Nations, Rome. http://www.fao.org/docrep/003/t0507e/T0507E00.htm

Ciavola P, Coco G (eds) (2017) Coastal storms: processes and impacts. Wiley, Hoboken

Cicin-Sain B (1993) Sustainable development and integrated coastal management. Ocean Coast Manag 21(1–3):11–43

Cinner JE, Barnes ML (2019) Social dimensions of resilience in social-ecological systems. One Earth 1:51–56

Cinner JE, McClanahan TR, MacNeil MA, Graham NAJ, Daw TM, Mukminin A, Feary DA, Kuange J (2012) Comanagement of coral reef social-ecological systems. Proc Natl Acad Sci 109(14):5219–5222. https://doi.org/10.1073/pnas.1121215109

CIRAD (Agricultural Research for Development) (2020) Covid-19 and food security | How African agriculture could hold its own in the face of the crisis. CIRAD, 2 June 2020. https://www.cirad.fr/en/news/all-news-items/articles/2020/science/covid-food-security-african-agriculture-resilient-in-face-of-crisis

Cisneros-Montemayor AM, Pauly D, Weatherdon LV, Ota Y (2016) A global estimate of seafood consumption by coastal indigenous peoples. PloS One 11(12):e0166681

Cisneros-Montemayor AM, Moreno-Báez M, Voyer M, Allison EH, Cheung WWL, Hessing-Lewis M, Oyinlola MA, Singh GG, Swartz W, Ota Y (2019) Social equity and benefits as the nexus of a transformative blue economy: a sectoral review of implications. Mar Policy 109:103702

CLIA (2019) 2020 State of the cruise industry outlook. Cruise Lines International Association, Washington, DC. https://cruising.org/-/media/research-updates/research/state-of-the-cruise-industry.ashx

Cohen PJ, Allison EH, Andrew NL, Cinner J, Evans LS, Fabinyi M, Garces LR, Hall SJ, Hicks CC, Hughes TP, Jentoft S, Mills DJ, Masu R, Mbaru EK, Ratner BD (2019) Securing a just space for small-scale fisheries in the blue economy. Front Mar Sci 6:1–8. https://doi.org/10.3389/fmars.2019.00171

Cohen-Shacham E, Walters G, Janzen C, Maginnis S (eds) (2016) Nature-based solutions to address global societal challenges. International Union for Conservation of Nature, Gland. https://doi.org/10.2305/IUCN.CH.2016.13.en

COI-UNESCO and CPPS (2016) Experiencias locales en el manejo costero integrado: Casos piloto SPINCAM en el Pacífico Sudeste. In: Serie Técnica 127 - Dossier ICAM 9. UNESCO, Paris. http://fust.iode.org/spincam

Colgan CS, Beck MW, Narayan S (2017) Financing natural infrastructure for coastal flood damage reduction. Lloyd's Tercentenary Research Foundation, London

Conger T, Chang SE (2019) Developing indicators to identify coastal green infrastructure potential: the case of the Salish Sea Region. Ocean Coast Manag 175:53–69. https://doi.org/10.1016/j.ocecoaman.2019.03.011

Conservation International (2019) A practical guide to implementing green-gray infrastructure. https://www.conservation.org/docs/default-source/publication-pdfs/a-practical-guide-to-implementing-green-gray-infrastructure_aug2019.pdf

Copertino MS, Creed JC, Lanari MO, Magalhães K, Barros K, Lana PC, Sordo L, Horta PA (2016) Seagrass and submerged aquatic vegetation (VAS) habitats off the Coast of Brazil: state of knowledge, conservation and main threats. Braz J Oceanogr 64(SPE2):53–80. https://doi.org/10.1590/S1679-875920161036064sp2

Costello MJ, Ballantine B (2015) Biodiversity conservation should focus on no-take marine reserves: 94% of marine protected areas allow fishing. Trends Ecol Evol 30(9):507–509. https://doi.org/10.1016/j.tree.2015.06.011

Creel L (2003) Ripple effects: population and coastal regions. Population Reference Bureau, Washington, DC. https://www.prb.org/wp-content/uploads/2003/09/RippleEffects_Eng.pdf

Crooks S, Herr D, Tamelander J, Laffoley D, Vandever J (2011) Mitigating climate change through restoration and management of coastal wetlands and near-shore marine ecosystems: challenges and opportunities. Environment Department Papers No. 121. Marine ecosystem series. World Bank, Washington, DC

Dafforn KA, Glasby TM, Airoldi L, Rivero NK, Mayer-Pinto M, Johnston EL (2015) Marine urbanisation: an ecological framework for designing multifunctional artificial structures. Front Ecol Environ 13:82–90

Dah-gbeto AP, Villamor GB (2016) Gender-specific responses to climate variability in a semi-arid ecosystem in Northern Benin. Ambio 45(3):297–308

Dalin C, Wada Y, Kastner T, Puma MJ (2017) Groundwater depletion embedded in international food trade. Nature 543:700–704

DasGupta R, Shaw R (2015) An indicator based approach to assess coastal communities' resilience against climate related disasters in Indian Sundarbans. J Coast Conserv 19(1):85–101

Davey E, Steer A (2020) After COVID-19: how we can improve the global food system. Blog 21, Nutrition Connect, 8 May. https://nutritionconnect.org/resource-center/blog-21-after-covid-19-how-we-can-improve-global-food-system

Davies TE, Epstein G, Aguilera SE, Brooks CM, Cox M, Evans LS, Maxwell SM, Nenadovic M, Ban NC (2018) Assessing trade-offs in large marine protected areas. PloS One 13(4):1–14. https://doi.org/10.1371/journal.pone.0195760

Daw TM, Coulthard S, Cheung WWL, Brown K, Abunge C, Galafassi G, Peterson GD, McClanahan TR, Omukoto JO, Munyi L (2015) Evaluating taboo trade-offs in ecosystems services and human well-being. Proc Natl Acad Sci 112(22):6949–6954. https://doi.org/10.1073/pnas.1414900112

Day JC (2018) How effective is the management of the great barrier reef? ICES J Mar Sci 75:1188–1190

Depietri Y, McPhearson T (2017) Integrating the grey, green and blue in cities: nature-based solutions for climate change adaptation and risk reduction. In: Kabisch N, Korn H, Stadler J, Bonn A (eds) Nature-based solutions to climate change adaptation in urban areas. Springer, Cham, pp 91–109. https://doi.org/10.1007/978-3-319-56091-5

Dharmawan B, Böcher M, Krott M (2016) The failure of the mangrove conservation plan in Indonesia: weak research and an ignorance of grassroots politics. Ocean Coast Manag 130:250–259. https://doi.org/10.1016/j.ocecoaman.2016.06.019

Diaz DB (2016) Estimating global damages from sea level rise with the coastal impact and adaptation model (CIAM). Clim Change 137(1–2):143–156. https://doi.org/10.1007/s10584-016-1675-4

Díaz S, Pascual U, Stenseke M, Martín-López B, Watson RT, Molnár Z, Hill R, Chan KM, Baste IA, Brauman KA, Polasky S (2018) Assessing nature's contributions to people. Science 359(6373):270–272

Diefenderfer HL, Adkins J (2003) Systematic approach to coastal ecosystem restoration. Report prepared for National Oceanic and Atmospheric Administration. Battelle Memorial Institute, Columbus. https://www.researchgate.net/publication/241472480

Donchyts G, Baart F, Winsemius H, Gorelick N, Kwadijk J, van de Giesen N (2016) Earth's surface water change over the past 30 years. Nat Clim Change 6(9):810–813

Donnelson Wright L, Wu W, Morris J (2019a) Coastal erosion and land loss: causes and impacts. In: Donelson Wright L, Reid Nichols C (eds) Tomorrow's coasts: complex and impermanent. Springer, Cham, pp 137–150. https://doi.org/10.1007/978-3-319-75453-6

Donnelson Wright L, Syvitski JPM, Reid Nichols C (2019b) Sea level rise: recent trends and future projections. In: Donelson Wright L, Reid Nichols C (eds) Tomorrow's coasts: complex and impermanent. Springer, Cham, pp 47–57

Doropoulos C, Babcock RC (2018) Harnessing connectivity to facilitate coral restoration. Front Ecol Environ 16:558–559

Doropoulos C, Ward S, Diaz-Pulido G, Hoegh-Guldberg O, Mumby PJ (2012) Ocean acidification reduces coral recruitment by disrupting intimate larval-algal settlement interactions. Ecol Lett 15:338–346

Doropoulos C, Elzinga J, ter Hofstede R, van Koningsveld M, Babcock RC (2019a) Optimizing industrial-scale coral reef restoration: comparing harvesting wild coral spawn slicks and transplanting gravid adult colonies. Restor Ecol 27:758–767

Doropoulos C, Vons F, Elzinga J, ter Hofstede R, Salee K, van Koningsveld M, Babcock RC (2019b) Testing industrial-scale coral restoration techniques: harvesting and culturing wild coral-spawn slicks. Front Mar Sci 6:658

Duarte CM, Losada IJ, Hendriks IE, Mazarrasa I, Marbà N (2013) The role of coastal plant communities for climate change mitigation and adaptation. Nat Clim Change 3(11):961–968

Duarte CM, Agusti S, Barbier E, Britten GL, Castilla JC, Gattuso J-P, Fulweiler RW, Hughes TP, Knowlton N, Lovelock CE, Lotze HK, Predragovic M, Poloczanska E, Roberts C, Worm B (2020) Rebuilding marine life. Nature 580:39–51

Dundas SJ, Levine AS, Lewison RL, Doerr AN, White C, Galloway AWE, Garza C, Hazen EL, Padilla-Gamiño J, Samhouri JF, Spalding A, Stier A, White JW (2020) Integrating oceans into climate policy: any green new deal needs a splash of blue. Conserv Lett 13:e12716

Early R, Bradley BA, Dukes JS, Lawler JJ, Olden JD, Blumenthal DM, Gonzalez P, Grosholz ED, Ibanez I, Miller LP, Sorte CJ, Tatem AJ (2016) Global threats from invasive alien species in the twenty-first century and national response capacities. Nat Commun 7:12485

Econsearch (2016) Economic analysis for the blue infrastructure initiative. A report to The Nature Conservancy, Primary Industries and Regions South Australia and Department of Environment, Water and Natural Resources. Econsearch, Marryatville

Edwards AJ, Gomez ED (2007) Reef restoration, concepts and guidelines: making sensible management choices in the face of uncertainty. The Coral Reef Targeted Research & Capacity Building for Management Program, St Lucia

Edwards PET, Sutton-Grier A, Coyle GE (2013) Investing in nature: restoring coastal habitat blue infrastructure and green job creation. Mar Policy 38:65–71. https://doi.org/10.1016/j.marpol.2012.05.020

Ehler CN (2018) Marine spatial planning: an idea whose time has come. In: Yates KL, Bradshaw CJA (eds) Offshore energy and marine spatial planning. Routledge, Abingdon, pp 6–17

Elmgren R, Blenckner T, Andersson A (2015) Baltic sea management: successes and failures. Ambio 44(3):335–344

Erftemeijer PLA, Shuail DA (2012) Seagrass habitats in the Arabian Gulf: distribution, tolerance thresholds and threats. Aquat Ecosyst Health Manage 15(supp 1):73–83. https://doi.org/10.1080/14634988.2012.668479

Ericson J, Vorosmarty C, Dingman S, Ward L, Meybeck M (2006) Effective sea-level rise and deltas: causes of change and human dimension implications. Global Planet Change 50(1–2):63–82. https://doi.org/10.1016/j.gloplacha.2005.07.004

Fabricius KE (2005) Effects of terrestrial runoff on the ecology of corals and coral reefs: review and synthesis. Mar Pollut Bull 50:125–146

Fabricius KE, Langdon C, Uthicke S, Humphrey C, Noonan S, De'ath G, Okazaki R, Muehllehner N, Glas MS, Lough JM (2011) Losers and winners in coral reefs acclimatized to elevated carbon dioxide concentrations. Nat Clim Change 1:165–169

Fairhead J, Leach M, Scoones I (2012) Green grabbing: a new appropriation of nature? J Peasant Stud 39(2):237–261. https://doi.org/10.1080/03066150.2012.671770

FAO (Food and Agriculture Organization of the United Nations) (2018) The state of world fisheries and aquaculture: meeting the sustainable development goals. FAO, Rome. http://www.fao.org/publications/card/en/c/I9540EN

FAO (Food and Agriculture Organization of the United Nations) (2020a) The state of world fisheries and aquaculture 2020. Sustainability in action. FAO, Rome. https://doi.org/10.4060/ca9229en

FAO (Food and Agriculture Organization of the United Nations) (2020b) Voluntary guidelines for securing sustainable small-scale fisheries in the context of food security and poverty eradication. FAO, Rome. http://www.fao.org/voluntary-guidelines-small-scale-fisheries/ihh/en

FAO (Food and Agriculture Organization of the United Nations) (2020c) Summary of the impacts of the COVID-19 pandemic on the fisheries and aquaculture sector: addendum to the state of world fisheries and aquaculture 2020. FAO, Rome. https://doi.org/10.4060/ca9349en

FAO (Food and Agriculture Organization of the United Nations) (2020d) How is COVID-19 affecting the fisheries and aquaculture food systems. FAO, Rome. https://doi.org/10.4060/ca8637en

Fenichel EP, Milligan B, Porras I et al (2020) National accounting for the ocean and ocean economy. World Resources Institute, Washington, DC. https://www.oceanpanel.org/blue-papers/national-accounting-ocean-ocean-economy

Ferguson G, Gleeson T (2012) Vulnerability of coastal aquifers to groundwater use and climate change. Nat Clim Change 2(5):342–345. https://doi.org/10.1038/nclimate1413

Fernández-Montblanc T, Vousdoukas MI, Ciavola P, Voukouvalas E, Mentaschi L, Breyiannis G, Feyen L, Salamon P (2019) Towards robust pan-european storm surge forecasting. Ocean Model 133:129–144. https://doi.org/10.1016/j.ocemod.2018.12.001

Filbee-Dexter K, Wernberg T (2018) Rise of turfs: a new battlefront for globally declining kelp forests. Bioscience 68(2):64–76

Firth L, Knights A, Bridger D, Evans A, Mieszkowska N, Moore P, O'Connor N, Sheehan EV, Thompson RC, Hawkins SJ (2016) Ocean sprawl: challenges and opportunities for biodiversity management in a changing world. Oceanogr Mar Biol 54:189–262

Fitzer SC, Torres Gabarda S, Daly L, Hughes B, Dove M, O'Connor W, Potts J, Scanes P, Byrne M (2018) Coastal acidification impacts on shell mineral structure of bivalve mollusks. Ecol Evol 8: 8973–8984

Ford JD, Berrang-Ford L, Bunce A, McKay C, Irwin M, Pearce T (2015) The status of climate change adaptation in Africa and Asia. Reg Environ Change 15(5):801–814

Fox HE, Mascia MB, Basurto X, Costa A, Glew L, Heinemann D, Karrer LB, Lester SE, Lombana AV, Pomeroy RS, Recchia CA, Roberts CM, Sanchirico JN, Pet-Soede L, White AT (2012) Reexamining the science of marine protected areas: linking knowledge to action. Conserv Lett 5(1):1–10. https://doi.org/10.1111/j.1755-263X.2011.00207.x

Fox HE, Harris JL, Darling ES, Ahmadia GN, Estradivari, Razak TB (2019) Rebuilding coral reefs: success (and failure) 16 years after low-cost, low-tech restoration. Restor Ecol 27(4):862–869

Francis RA (2010) Wall ecology: a frontier for urban biodiversity and ecological engineering. Prog Phys Geogr Earth Environ 35(1):43–63. https://doi.org/10.1177/0309133310385166

Friedlander AM, Ballesteros E, Bell TW, Caselle JE, Campagna C, Goodell W, Hune M, Munoz A, Salinas-de-Leon P, Sala E, Dayton PK (2020) Kelp forests at the end of the earth: 45 years later. PloS One 15(3):e0229259

Friess DA, Rogers K, Lovelock CE, Krauss KW, Hamilton SE, Lee SY, Lucas R, Primavera J, Rajkaran A, Shi S (2019) The state of the world's mangrove forests: past, present, and future. Annu Rev Env Resour 44:89–115

G20 Japan (2019) 2019 Principles for quality infrastructure. https://www.g20-insights.org/related_literature/g20-japan-principles-quality-infrastructure-investment/; https://ec.europa.eu/maritimeaffairs/befp_en#:~:text=The%20Sustainable%20Blue%20Economy%20Finance%20Principles&text=promote%20the%20implementation%20of%20the,existing%20frameworks%20for%20responsible%20investment

Gaines S, Cabral R, Free C, Golbuu Y et al (2019) The expected impacts of climate change on the ocean economy. World Resources Institute, Washington, DC. https://www.oceanpanel.org/blue-papers/expected-impacts-climate-change-ocean-economy

Galafassi D, Daw TM, Munyi L, Brown K, Barnaud C, Fazey I (2017) Learning about social-ecological trade-offs. Ecol Soc 22(1). https://doi.org/10.5751/ES-08920-220102

Galgani L, Beiras R, Galgani F, Panti C, Borja A (2019) Editorial: impacts of marine litter. Front Mar Sci 6:208

Gallego-Schmid A, Tarpani RRZ (2019) Life cycle assessment of wastewater treatment in developing countries: a review. Water Res 153:63–79

Gann GD, McDonald T, Walder B, Aronson J, Nelson CR, Jonson J, Hallett JG, Eisenberg C, Guariguata MR, Liu J, Hua F, Echeverría C, Gonzales E, Shaw N, Decleer K, Dixon KW (2019) International principles and standards for the practice of ecological restoration. second edition. Restor Ecol 27(S1):S1–S46

Gao K, Beardall J, Häder D-P, Hall-Spencer JM, Gao G, Hutchins DA (2019) Effects of ocean acidification on marine photosynthetic organisms under the concurrent influences of warming, UV radiation, and deoxygenation. Front Mar Sci 6:322. https://doi.org/10.3389/fmars.2019.00322

Garcias-Bonet N, Vaquer-Sunyer R, Duarte CM, Marbà N (2019) Warming effect on nitrogen fixation in Mediterranean macrophyte sediments. Biogeosciences 16:167–175. https://doi.org/10.5194/bg-16-167-2019

Gardner AS, Moholdt G, Scambos T, Fahnstock M, Ligtenberg S, van den Broeke M, Nilsson J (2018) Increased West Antarctic and unchanged East Antarctic ice discharge over the last 7 years. Cryosphere 12:521–547. https://doi.org/10.5194/tc-12-521-2018

Gattuso JP, Magnan AK, Bopp L, Cheung WWL, Duarte CM, Hinkel J, Mcleod E, Micheli F, Oschlies A, Williamson P, Bille R, Chalastani VI, Gates RD, Irisson JO, Middleburg JJ, Portner HO, Rau GH (2018) Ocean solutions to address climate change and its effects on marine ecosystems. Front Mar Sci 5:337. https://doi.org/10.3389/fmars.2018.00337

Gelcich S, Reyes-Mendy F, Arriagada R, Castillo B (2018) Assessing the implementation of marine ecosystem based management into national policies: insights from agenda setting and policy responses. Mar Policy 92:40–47

Geyer R, Jambeck JR, Law KL (2017) Production, use, and fate of all plastics ever made. Sci Adv 3(7):e1700782

Giakoumi S, McGowan J, Mills M, Beger M, Bustamante RH, Charles A, Christie P, Fox M, Garcia-Borboroglu P, Gelcich S, Guidetti P (2018) Revisiting success and failure of marine protected areas: a conservation scientist perspective. Front Mar Sci 5:223

Giannico G, Souder JA (2005) Tide gates in the Pacific Northwest. Oregon Sea Grant, Corvallis

Gledhill D, White M, Salisbury J, Thomas H, Misna I, Liebman M, Mook B, Grear J, Candelmo A, Chambers RC, Gobler C, Hunt C, King A, Price N, Signorini S, Stancioff E, Stymiest C, Wahle R, Waller J, Rebuck N, Wang Z, Capson T, Morrison JR, Cooley S, Doney S (2015) Ocean and coastal acidification off New England and Nova Scotia. Oceanography 25:182–197

Goldberg L, Lagomasino D, Thomas N, Fatoyinbo T (2020) Global declines in human-driven mangrove loss. Glob Chang Biol 00:1–12

Gonzalez AM, Bergqvist R, Monios J (2018) A global review of the hinterland dimension of green port strategies. Transport Res D Transp Environ 59:23–34

Gössling S, Scott D, Hall CM (2020) Pandemics, tourism and global change: a rapid assessment of COVID-19. J Sustain Tour. https://doi.org/10.1080/09669582.2020.1758708

Grabowski JH, Brumbaugh RD, Conrad RF, Keeler AG, Opaluch JJ, Peterson CH, Piehler MF, Powers SP, Smyth AR (2012) Economic valuation of ecosystem services provided by oyster reefs. Bioscience 62(10):900–909

Gracia A, Rangel-Buitrago N, Oakley JA, Williams AT (2018) Use of ecosystems in coastal erosion management. Ocean Coast Manag 156:277–289

Gray C, Frankenberg E, Gillespie T, Sumantri C, Thomas D (2014) Studying displacement after a disaster using large scale survey methods: Sumatra after the 2004 Tsunami. Ann Assoc Am Geogr 104(3):594–612

Griffiths LL, Connolly RM, Brown CJ (2019) Critical gaps in seagrass protection reveal the need to address multiple pressures and cumulative impacts. Ocean Coast Manag 183:104946

Grill G, Lehner B, Thieme M, Geenen B, Tickner D, Antonelli F, Babu S, Borrelli P, Cheng L, Crochetiere H, Ehalt Macedo H, Filgueiras R, Goichot M, Higgins J, Hogan Z, Lip B, McClain ME, Meng J, Mulligan M, Nilsson C, Olden JD, Opperman JJ, Petry P, Reidy Liermann C, Sáenz L, Salinas-Rodríguez S, Schelle P, Schmitt RJP, Snider J, Tan F, Tockner K, Valdujo PH, van Soesbergen A, Zarfl C (2019) Mapping the world's free-flowing rivers. Nature 569:215–221. https://doi.org/10.1038/s41586-019-1111-9

Groner ML, Maynard J, Breyta R, Carnegie RB, Dobson A, Friedman CS, Froelich B, Garren M, Gulland FM, Heron SF, Noble RT, Revie CW, Shields JD, Vanderstichel R, Weil E, Wyllie-Echeverria S, Harvell CD (2016) Managing marine disease emergencies in an era of rapid change. Philos Trans R Soc Lond B Biol Sci 371(1689):20150364. https://doi.org/10.1098/rstb.2015.0364

Hagedoorn LC, Brander LM, van Beukering PJH, Dijkstra HM, Franco C, Hughes L, Gilders I, Segal B (2019) Community-based adaptation to climate change in small island developing states: an analysis of the role of social capital. Clim Dev 11(8):723–734

Halim A, Loneragan NR, Wiryawan B, Fujita R, Adhuri DS, Hordyk AR, Sondita MFA (2020) Transforming traditional management into contemporary territorial-based fisheries management rights for small-scale fisheries in Indonesia. Mar Policy 116:103923

Hallegatte S, Green C, Nicholls RJ, Corfee-Morlot J (2013) Future flood losses in major coastal cities. Nat Clim Change 3(9):802–806. https://doi.org/10.1038/nclimate1979

Hall-Spencer JM, Harvey B (2019) Ocean acidification impacts on coastal ecosystem services due to habitat degradation. Emerg Topics Life Sci 3:197–206. https://doi.org/10.1042/ETLS20180117

Halpern BS, Klein CJ, Brown CJ, Beger M, Grantham HS, Mangubhai S, Ruckelshaus M, Tulloch VJ, Watts M, White C, Possingham HP (2013) Achieving the triple bottom line in the face of inherent trade-offs among social equity, economic return, and conservation. Proc Natl Acad Sci 110(15):6229–6234. https://doi.org/10.1073/pnas.1217689110

Halpern BS, Longo C, Lowndes JSS, Best BD, Frazier M, Katona SK, Kleisner KM, Rosenberg AA, Scarborough C, Selig ER (2015) Patterns and emerging trends in global ocean health. PloS One 10(3):e0117863

Halpern BS, Frazier M, Afflerbach J, Lowndes JS, Micheli F, O'Hara C, Scarborough C, Selkoe KA (2019) Recent pace of change in human impact on the world's ocean. Sci Rep 9(1):1–8

Hamilton SE, Casey D (2016) Creation of a high spatio-temporal resolution global database of continuous mangrove forest cover for the 21st century (CGMFC-21). Glob Ecol Biogeogr 25(6):729–738. https://doi.org/10.1111/geb.12449

Haugan PM, Levin LA, Amon D, Hemer M, Lily H, Nielsen FG (2020) What role for ocean-based renewable energy and deep seabed minerals in a sustainable future? World Resources Institute, Washington, DC. www.oceanpanel.org/blue-papers/ocean-energy-and-mineral-sources

Hauri C, Friedrich T, Timmermann A (2015) Abrupt onset and prolongation of aragonite undersaturation events in the southern ocean. Nat Clim Change 6:172. https://doi.org/10.1038/nclimate2844

Hawkins SJ, Evans AJ, Firth LB, Genner MJ, Herbert RJ, Adams LC, Moore PJ, Mieszkowska N, Thompson RC, Burrows MT, Fenburg PB (2016) Impacts and effects of ocean warming on intertidal rocky habitats. In: Laffoley D, Baxter JM (eds) Explaining ocean warming: causes, scale, effects and consequences. International Union for Conservation of Nature, Gland, pp 147–176. https://doi.org/10.2305/IUCN.CH.2016.08.en

He Q, Bertness MD, Bruno JF, Li B, Chen G, Coverdale TC, Altieri AH, Bai J, Sun T, Pennings SC, Liu J, Ehrlich PR, Cui B (2014) Economic development and coastal ecosystem change in China. Sci Rep 4(1):1–9. https://doi.org/10.1038/srep05995

Hebbeln D, da Costa Portilho-Ramos R, Wienberg C, Jürgen Titschack J (2019) The fate of cold-water corals in a changing world: a geological perspective. Front Mar Sci 6:119

Herr D, Agardy T, Benzaken D, Hicks F, Howard J, Landis E, Soles A, Vegh T (2015) Coastal "blue" carbon. A revised guide to supporting coastal wetland programs and projects using climate finance and other financial mechanisms. International Union for Conservation of Nature, Gland

High Level Panel on Water (2017) Effective models of resilient and integrated urban water management. https://sustainabledevelopment.un.org/content/documents/hlpwater/03-EffectModelsResilntegrUrbanW aterManag.pdf

Hinkel J, Lincke D, Vafeidis AT, Perrette M, Nicholls RJ, Tol RS, Marzeion B, Fettweis X, Ionescu C, Levermann A (2014) Coastal flood damage and adaptation costs under 21st century sea-level rise. Proc Natl Acad Sci 111(9):3292–3297

Hino M, Belanger ST, Field CB, Davies AR, Mach KJ (2019) High-tide flooding disrupts local economic activity. Sci Adv 5(2):eaau2736. https://doi.org/10.1126/sciadv.aau2736

Hiwasaki L, Syamsidik EL, Shaw R (2014) Process for integrating local and indigenous knowledge with science for hydro-meteorological disaster risk reduction and climate change adaptation in coastal and small island communities. Int J Disaster Risk Reduct 10(Part A):15–27. https://doi.org/10.1016/j.ijdrr.2014.07.007

HM Treasury (2020) The Dasgupta review – independent review on the economics of biodiversity. Interim Report

Hodgson EE, Essington TE, Samhouri JF, Allison EH, Bennett NJ, Bostrom A, Cullen AC, Kasperski S, Levin PS, Poe MR (2019) Integrated risk assessment for the blue economy. Front Mar Sci 6:609

Hoegh-Guldberg O, Bruno JF (2010) The impact of climate change on the world's marine ecosystems. Science 328:1523–1528

Hoegh-Guldberg O, Kennedy EV, Beyer HL, McClennen C, Possingham HP (2018) Securing a long-term future for coral reefs. Trends Ecol Evol 33(12):936–944. https://doi.org/10.1016/j.tree.2018.09.006

Holbrook NJ, Scannell HA, Gupta AS, Benthuysen JA, Feng M, Oliver EC, Alexander LV, Burrows MT, Donat MG, Hobday AJ, Moore PJ (2019) A global assessment of marine heatwaves and their drivers. Nat Commun 10(1):2624

Holbrook NJ, Sen Gupta A, Oliver ECJ, Hobday AJ, Benthuysen JA, Scannell HA, Smale DA, Wernberg T (2020) Keeping pace with marine heatwaves. Nat Rev Earth Environ 1:482–493

Huang Y (2016) Understanding China's belt & road initiative: motivation, framework and assessment. China Econ Rev 40:314–321. https://doi.org/10.1016/j.chieco.2016.07.007

Hugelius G, Loisel J, Chadburn S, Jackson RB, Jones M, MacDonald G, Marushchak M, Olefeldt D, Packalen M, Siewert MB, Treat C, Turetsky M, Voigt C, Yu Z (2020) Large stocks of peatland carbon and nitrogen are vulnerable to permafrost thaw. Proc Natl Acad Sci 117(34):20438–20446

Hughes AC (2019) Understanding and minimizing environmental impacts of the belt and road initiative. Conserv Biol 33(4):883–894. https://doi.org/10.1111/cobi.13317

Hughes TP, Kerry JT, Álvarez-Noriega M, Álvarez-Romero JG, Anderson KD, Baird AH, Babcock RC, Beger M, Bellwood DR, Berkelmans R, Bridge TC, Butler IR, Byrne M, Cantin NE, Comeau S, Connolly SR, Cumming GS, Dalton SJ, Diaz-Pulido G, Eakin CM, Figueira WF, Gilmour JP, Harrison HB, Heron SF, Hoey AS, Hobbs JA, Hoogenboom MO, Kennedy EV, Kuo CY, Lough JM, Lowe RJ, Liu G, McCulloch MT, Malcolm HA, McWilliam MJ, Pandolfi JM, Pears RJ, Pratchett MS, Schoepf V, Simpson T, Skirving WJ, Sommer B, Torda G, Wachenfeld DR, Willis BL, Wilson SK (2017) Global warming and recurrent mass bleaching of corals. Nature 543(7645):373–377. https://doi.org/10.1038/nature21707

Hughes TP, Anderson KD, Connolly SR, Heron SF, Kerry JT, Lough JM, Baird AH, Baum JK, Berumen ML, Bridge TC (2018) Spatial and temporal patterns of mass bleaching of corals in the Anthropocene. Science 359:80–83

Hughes AC, Lechner AM, Chitov A, Horstmann A, Hinsley A, Tritto A, Chariton A, Li BV, Ganapin D, Simonov E, Morton K, Toktomushev K, Foggin M, Tan-Mullins M, Orr MC, Griffiths R, Nash R, Perkin S, Glémet R, Kim M, Yu DW (2020) Horizon scan of the belt and road initiative. Trends Ecol Evol 35(7):583–593. https://doi.org/10.1016/j.tree.2020.02.005

IAPH (International Association of Ports and Harbors) (2016) IAPH annual report 2016–17. IAPH, Tokyo. https://www.iaphworldports.org/news/4226

IEA (International Energy Agency) (2019) Offshore wind outlook 2019. World energy outlook special report. IEA, Paris

IMO (International Maritime Organization) (2017) Consideration of how to progress the matter of reduction of GHG emissions from ships. IMO, London. http://www.imo.org/en/OurWork/Environment/PollutionPrevention/AirPollution/Documents/ISWG-GHG%201-2.pdf

IPBES (Intergovernmental Science-Policy Platform on Biodiversity and Ecosystem Services) (2019) The global assessment report on biodiversity and ecosystem services. IPBES, Bonn. https://ipbes.net/global-assessment

IPCC (Intergovernmental Panel on Climate Change) (2019) In: Pörtner H-O, Roberts DC, Masson-Delmotte V, Zhai P, Tignor M, Poloczanska E, Mintenbeck K et al (eds) IPCC special report on the ocean and cryosphere in a changing climate. IPCC, Geneva

Ismadi R, Yamindago A (2014) Restoring coastal ecosystems – a case study Malang and Gresik regency, Indonesia. J Coast Conserv 19:119–130. https://doi.org/10.1007/s11852-015-0373-0

Jackson DWT, Costas S, González-Villanueva R, Cooper A (2019) A global 'greening' of coastal dunes: an integrated consequence of climate change? Global Planet Change 182:103026

Jacobsen R (2019) Beyond seawalls. Sci Am 320(4):32–37. https://dialnet.unirioja.es/servlet/articulo?codigo=6939491

Jacobson AP, Riggio J, Tait AM, Baillie JEM (2019) Global areas of low human impact ('low impact areas') and fragmentation of the natural world. Sci Rep 9:14179

Jambeck JR, Geyer R, Wilcox C, Siegler TR, Perryman M, Andrady A, Narayan R, Law KL (2015) Plastic waste inputs from land into the ocean. Science 347:768–771

Jambeck J, Moss E, Dubey BK et al (2020) Leveraging multi-target strategies to address plastic pollution in the context of an already stressed ocean. World Resources Institute, Washington, DC. https://www.oceanpanel.org/blue-papers/leveraging-target-strategies-to-address-plastic-pollution-in-the-context

Jennerjahn T, Mitchell S (2013) Pressures, stresses, shocks and trends in Estuarine ecosystems – an introduction and synthesis. Estuar Coast Shelf Sci 130:1–8. https://doi.org/10.1016/j.ecss.2013.07.008

Jinks KI, Rasheed MA, Brown CJ, Olds AD, Schlacher TA, Sheaves M, York PH, Connolly RM (2020) Saltmarsh grass supports fishery food webs in subtropical Australian estuaries. Estuar Coast Shelf Sci:238. https://doi.org/10.1016/j.ecss.2020.106719

Johnson JA, Baldos U, Hertel T, Liu J, Nootenboom C, Polasky S, Roxburgh T (2020) Global futures: modelling the global economic impacts of environmental change to support policy-making. Technical report. WWF, Gland. https://www.wwf.org.uk/globalfutures

Johnston EL, Hedge LH, Mayer-Pinto M (2015) The urgent global need to understand port and harbour ecosystems. Mar Freshw Res 66(12):i–ii

Jones B, O'Neill BC (2016) Spatially explicit global population scenarios consistent with the shared socioeconomic pathways. Environ Res Lett 11(8):084003

Jones E, Qadir M, van Vliet MTH, Smakhtin V, Kang SM (2019) The state of desalination and brine production: a global outlook. Sci Total Environ 657:1343–1356

Jongman B, Ward PJ, Aerts JCJH (2012) Global exposure to river and coastal flooding: long-term trends and changes. Glob Environ Change Hum Policy Dimens 22(4):823–835. https://doi.org/10.1016/j.gloenvcha.2012.07.004

Jouffray J-B, Blasiak R, Norström AV, Österblom H, Nyström M (2020) The blue acceleration: the trajectory of human expansion into the ocean. One Earth 2:43–54

Jupiter SD, Cohen PJ, Weeks R, Tawake A, Govan H (2014) Locally-managed marine areas: multiple objectives and diverse strategies. Pac Conserv Biol 20:65–179

Kapetsky JM, Aguilar-Manjarrez J, Jenness J (2013) A global assessment of potential for offshore mariculture development from a spatial perspective. FAO fisheries and aquaculture technical paper no. 549. FAO, Rome

Katona S, Polsenberg J, Lowndes J, Halpern BS, Pacheco E, Mosher L, Kilponen A, Papacostas K, Guzmán-Mora AG, Farmer G, Mori L, Andrews O, Taei S, Carr S (2017) Navigating the seascape of ocean management: waypoints on the voyage toward sustainable use. https://www.researchgate.net/publication/315788211_Navigating_the_seascape_of_ocean_management_waypoints_on_the_voyage_toward_sustainable_use

Kelleway JJ, Cavanaugh K, Rogers K, Feller IC, Ens E, Doughty C, Saintilan N (2017) Review of the ecosystem service implications of mangrove encroachment into salt marshes. Glob Chang Biol 23(10):3967–3983

Kelleway JJ, Serrano O, Baldock JA, Burgess R, Cannard T, Lavery PS, Lovelock CE, Macreadie PI, Masque P, Newnham M, Saintilan N, Steven ADL (2020) A national approach to greenhouse gas abatement through blue carbon management. Glob Environ Chang 63:102083

Kerr S, Colton J, Johnson K, Wright G (2015) Rights and ownership in sea country: implications of marine renewable energy for indigenous and local communities. Mar Policy 52:108–115. https://doi.org/10.1016/j.marpol.2014.11.002

Khanom T (2016) Effect of salinity on food security in the context of interior coast of Bangladesh ocean coast. Ocean Coast Manag 130:205–212. https://doi.org/10.1016/j.ocecoaman.2016.06.013

Khazaleh M, Gopalan B (2019) Eco-friendly green concrete: a review. Paper prepared for international conference on innovative applied energy, Oxford

Kim I, Lee H-H, Lee D (2019) Development of a new tool for objective risk assessment and comparative analysis at coastal waters. J Int Marit Saf Environ Aff Shipp 2:58–66

Kirkfeldt TS (2019) An ocean of concepts: why choosing between ecosystem-based management, ecosystem-based approach and ecosystem approach makes a difference. Mar Policy 106:103541

Kittinger JN, Teh LCL, Allison EH, Bennett NJ, Crowder LB, Finkbeiner EM, Hicks C, Scarton CG, Nakamura K, Ota Y, Young J, Alifano A, Apel A, Arbib A, Bishop L, Boyle M, Cisneros-Montemayor AM, Hunter P, Le Cornu E, Levine M, Jones RS, Koehn JZ, Marschke M, Mason JG, Micheli F, McClenachan L, Opal C, Peacey J, Peckham SH, Schemmel E, Solis-Rivera V,

Swartz W, Wilhelm T (2017) Committing to socially responsible seafood. Science 356(6341):912–913. https://doi.org/10.1126/science.aam9969

Knoche S, Ihde TF (2019) Estimating ecological benefits and socio-economic impacts from oyster reef restoration in the Choptank River Complex, Chesapeake Bay. Final report to The National Fish and Wildlife Foundation & The NOAA Chesapeake Bay Office. Morgan State University, PEARL Report #11-05

Koehnken L, Rintoul M (2018) Impacts of sand mining on ecosystem structure, process and biodiversity in rivers. WWF, Gland

Konar M, Ding H (2020) A sustainable ocean economy for 2050. Approximating its benefits and costs. High Level Panel for a Sustainable Ocean Economy; World Resources Institute, Washington, DC. https://oceanpanel.org/sites/default/files/2020-07/Ocean%20Panel_Economic%20Analysis_FINAL.pdf

Kondolf GM, Gao Y, Annandale GW, Morris GL, Jiang E, Zhang J, Cao Y, Carling P, Fu K, Guo Q, Hotchkiss R, Peteuil C, Sumi T, Wang H-W, Wang Z, Wei Z, Wu B, Wu C, Yang CT (2014) Sustainable sediment management in reservoirs and regulated rivers: experiences from five continents. Earths Futur 2(5):256–280. https://doi.org/10.1002/2013EF000184

Kossin JP, Knapp KR, Olander TL, Velden CS (2020) Global increase in major tropical cyclone exceedance probability over the past four decades. Proc Natl Acad Sci 117(22):11975–11980

Kramer K (2018) Sinking cities, rising seas: a perfect storm of climate change and bad development choices. Christian Aid, London. https://www.christianaid.org.uk/sites/default/files/2018-10/Christian-Aid-Sinking-cities-rising-seas-report.pdf

Kramer DB, Stevens K, Williams NE, Sistla SA, Roddy AB, Urquhart GR (2017) Coastal livelihood transitions under globalization with implications for trans-ecosystem interactions. PloS One 12(10):e0186683

Krause-Jensen D, Duarte CM (2014) Expansion of vegetated coastal ecosystems in the future arctic. Front Mar Sci 1:77. https://doi.org/10.3389/fmars.2014.00077

Kron W (2013) Coasts: the high-risk areas of the world. Nat Hazards 66(3):1363–1382

Kroon FJ, Schaffelke B, Bartley R, R. (2014) Informing policy to protect coastal coral reefs: insight from a global review of reducing agricultural pollution to coastal ecosystems. Mar Pollut Bull 85(1):33–41. https://doi.org/10.1016/j.marpolbul.2014.06.003

Krumhansl KA, Okamoto DK, Rassweiler A, Novak M, Bolton JJ, Cavanaugh KC, Connell SD, Johnson CR, Konar B, Ling SD, Micheli F, Norderhaug KM, Perez-Matus A, Sousa-Pinto I, Reed DC, Salomon AK, Shears NT, Wernberg T, Anderson RJ, Barrett NS, Buschmann AH, Carr MH, Caselle JE, Derrien-Courtel S, Edgar GJ, Edwards M, Estes JA, Goodwin C, Kenner MC, Kushner DJ, Moy FE, Nunn J, Steneck RS, Vasquez J, Watson J, Witman JD, Byrnes JE (2016) Global patterns of kelp forest change over the past half-century. Proc Natl Acad Sci 113(48):13785–13790. https://doi.org/10.1073/pnas.1606102113

Kumar A, Yadav J, Mohan R (2020) Global warming leading to alarming recession of the arctic sea-ice cover: insights from remote sensing observations and model reanalysis. Heliyon 6(7):e04355

Kummu M, de Moel H, Salvucci G, Viviroli D, Ward PJ, Varis O (2016) Over the hills and further away from coast: global geospatial patterns of human and environment over the 20th–21st centuries. Environ Res Lett 11(3):034010

Ladd MC, Miller MW, Hunt JH, Sharp WC, Burkepile DE (2018) Harnessing ecological processes to facilitate coral restoration. Front Ecol Environ 16(4):239–247. https://doi.org/10.1002/fee.1792

Laffoley D, Baxter JM (eds) (2018) Ocean deoxygenation – everyone's problem: causes, impacts, consequences and solutions. International Union for Conservation of Nature and Natural Resources, Gland

Laffoley D, Baxter JM, Amon DJ, Currie DEJ, Downs CA, Hall- Spencer JM, Harden-Davies H, Page R, Reid CP, Roberts CM, Rogers A, Thiele T, Sheppard CRC, Sumaila RU, Woodall LC (2019) Eight urgent, fundamental and simultaneous steps needed to restore ocean health, and the consequences for humanity and the planet of inaction or delay. Aquat Conserv Mar Freshw Ecosyst 30(1):194–208

Lai S, Loke LHL, Hilton MJ, Bouma TJ, Todd PA (2015) The effects of urbanisation on coastal habitats and the potential for ecological engineering: a Singapore case study. Ocean Coast Manag 103:78–85. https://doi.org/10.1016/j.ocecoaman.2014.11.006

Lam VY, Chaloupka M, Thompson AA, Doropoulos C, Mumby PJ (2018) Acute drivers influence recent inshore great barrier reef dynamics. Proc R Soc B Biol Sci 285:20182063

LAPSSET Corridor Development Authority (2016) Brief on LAPSSET Corridor Project. Building transformative and game changing infrastructure for a seamless connected Africa. https://s3-eu-west-1.amazonaws.com/s3.sourceafrica.net/documents/118442/LAPSSET-Project-Report-July-2016-1.pdf

Lau JD, Hicks CC, Gurney GG, Cinner JE (2019) What matters to whom and why? Understanding the importance of coastal ecosystem services in developing coastal communities. Ecosyst Serv 35:219–230

Lee K, Boufadel M, Chen B, Foght J, Hodson P, Swanson S, Venosa A (2015) The behaviour and environmental impacts of crude oil released into aqueous environments. Royal Society of Canada, Ottawa

Lenihan HS, Peterson CH (1998) How habitat degradation through fishery disturbance enhances impacts of hypoxia on oyster reefs. Ecol Appl 8(1):128–140

Leo KL, Gillies CL, Fitzsimons JA, Hale LZ, Beck MW (2019) Coastal habitat squeeze: A review of adaptation solutions for saltmarsh, mangrove and beach habitats. Ocean Coast Manag 175:180–190

Levin PS, Fogarty MJ, Murawski SA, Fluharty D (2009) Integrated ecosystem assessments: developing the scientific basis for ecosystem-based management of the ocean. PLoS Biol 7(1):e1000014

Liao H, Yang Z, Dou Z, Sun F, Kou S, Zhang Z, Huang X, Bao Z (2019) Impact of ocean acidification on the energy metabolism and antioxidant responses of the Yesso Scallop (Patinopecten yessoensis). Front Physiol 9:1967. https://doi.org/10.3389/fphys.2018.01967

Limburg KE, Breitburg D, Swaney DP, Jacinto G (2020) Ocean deoxygenation: a primer. One Earth 2:24–29

Ling SD, Scheibling RE, Rassweiler A, Johnson CR, Shears N, Connell SD, Salomon AK, Norderhaug KM, Pérez-Matus A, Hernández JC, Clemente S, Blamey LK, Hereu B, Ballesteros E, Sala E, Garrabou J, Cebrian E, Zabala M, Fujita D, Johnson LE (2015) Global regime shift dynamics of catastrophic sea urchin overgrazing. Philos Trans R Soc B Biol Sci 370(1659):20130269

Lirman D, Schopmeyer S (2016) Ecological solutions to reef degradation: optimizing coral reef restoration in the Caribbean and Western Atlantic. PeerJ 4:e2597. https://doi.org/10.7717/peerj.2597

Liu D, Xing W (2019) Analysis of China's coastal zone management reform based on land-sea integration. Mar Econ Manag 2(1):39–49. https://doi.org/10.1108/MAEM-03-2019-0001

Liu X, Wang Y, Costanza R, Kubiszewski I, Xu N, Yuan M, Geng R (2019) The value of China's coastal wetlands and seawalls for storm protection. Ecosyst Serv 36:100905

Liu X, Huang Y, Xu X et al (2020) High-spatiotemporal-resolution mapping of global urban change from 1985 to 2015. Nat Sustain 3:564–570. https://doi.org/10.1038/s41893-020-0521-x

Llovel W, Purkey S, Meyssignac B, Blazquez A, Kolodziejczyk N, Bamber J (2019) Global ocean freshening, ocean mass increase and global mean sea level rise over 2005–2015. Sci Rep 9(1):1–10

Löhr A, Savelli H, Beunen R, Kalz M, Ragas A, van Belleghem F (2017) Solutions for global marine litter pollution. Curr Opin Environ Sustain 28:90–99

Lombard F, Boss ES, Waite AM, Vogt M, Uitz J, Stemmann L, Sosik HM, Schulz J, Romagnan JB, Picheral M, Pearlman J, Ohman MD, Niehoff B, Möller KO, Miloslavich P, Lara-Lpez A, Kudela RM, Lopes RM, Kiko R, Karp-Boss L, Jaffe JS, Iversen MH, Irisson JO, Fennel K, Hauss H, Guidi L, Gorsky G, Giering SLC, Gaube P, Gallager SM, Dubelaar G, Cowen RK, Carlotti F, Briseño-Avena C, Berline L, Benoit-Bird KJ, Bax N, Batten S, Ayata SD, Artigas LF, Appeltans W (2019) Globally consistent quantitative observations of planktonic ecosystems. Front Mar Sci 6:196. https://doi.org/10.3389/fmars.2019.00196

Lovelock CE, Reef R (2020) Variable impacts of climate change on blue carbon. One Earth 3(2):195–211

Lovelock CE, Krauss KW, Osland MJ, Reef R, Ball MC (2016) The physiology of mangrove trees with changing climate. In: Goldstein G, Santiago LS (eds) Tropical tree physiology. Springer, Cham, pp 149–179. https://doi.org/10.1007/978-3-319-27422-5_7

Luijendijk A, Hagenaars G, Ranasinghe R, Baart F, Donchyts G, Aarninkhof S, S. (2018) The state of the world's beaches. Sci Rep 8(1):1–11

Luijendijk E, Gleeson T, Moosdorf N (2020) Fresh groundwater discharge insignificant for the World's oceans but important for coastal ecosystems. Nat Commun 11:1260

Lumbroso D (2017) Coastal surges. In: Vinet F (ed) Floods. ISTE Press, London, pp 209–223. https://doi.org/10.1016/b978-1-78548-268-7.50012-2

Luo S, Cai F, Liu H, Lei G, Qi H, Su X (2015) Adaptive measures adopted for risk reduction of coastal erosion in the People's Republic of China. Ocean Coast Manag 103:134–145

MacNeil MA, Mellin C, Matthews S, Wolff NH, McClanahan TR, Devlin M, Drovandi C, Mengersen K, Graham NAJ (2019) Water quality mediates resilience on the Great Barrier Reef. Nat Ecol Evol 3(4):620–627

Magel JMT, Burns JHR, Gates RD, Baum JK (2019) Effects of bleaching-associated mass coral mortality on reef structural complexity across a gradient of local disturbance. Sci Rep 9(1):1–12. https://doi.org/10.1038/s41598-018-37713-1

Magnan AK, Garschagen M, Gattuso J-P, Hay JE, Hilmi N, Holland E, Isla F, Kofinas G, Losada IJ, Petzold J, Ratter B, Schuur T, Tabe T, van de Wal R (2019) Cross-Chapter Box 9: "Integrative cross-chapter box on low-lying islands and coasts". In: Pörtner H-O, Roberts DC, Masson-Delmotte V, Zhai P, Tignor M, Poloczanska E, Mintenbeck K et al (eds) IPCC special report on the ocean and cryosphere in a changing climate. Special report. Geneva, Intergovernmental Panel on Climate Change

Marbà N, Arias-Ortiz A, Masqué P, Kendrick GA, Mazarrasa I, Bastyan GR, Garcia-Orellana J, Duarte CM (2015) Impact of seagrass loss and subsequent revegetation on carbon sequestration and stocks. J Ecol 103:296–302

Marcos M, Rohmer J, Vousdoukas MI, Mentaschi L, Le Cozannet G, Amores A (2019) Increased extreme coastal water levels due to the combined action of storm surges and wind waves. Geophys Res Lett 46(8):4356–4364

Martínez ML, Mendoza-González G, Silva-Casarín R, Mendoza-Baldwin E (2014) Land use changes and sea level rise may induce a 'coastal squeeze' on the Coasts of Veracruz, Mexico. Glob Environ Chang 29:180–188. https://doi.org/10.1016/j.gloenvcha.2014.09.009

Mathews RE, Tengberg A, Sjödin J, Liss-Lymer B (2019) Implementing the source-to-sea approach: a guide for practitioners. Stockholm International Water Institute, Stockholm

McClanahan TR, Sebastián CR, Cinner JE (2016) Simulating the outcomes of resource user- and rule-based regulations in a coral reef fisheries-ecosystem model. Glob Environ Chang 38:58–69. https://doi.org/10.1016/j.gloenvcha.2016.02.010

McGranahan G, Balk D, Anderson B (2007) The rising tide: assessing the risks of climate change and human settlements in low elevation coastal zones. Environ Urban 19(1):17–37. https://doi.org/10.1177/0956247807076960

McKenzie LJ, Nordlund LM, Jones BL, Cullen-Unsworth LC, Roelfsema C, Unsworth RKF (2020) The global distribution of seagrass meadows. Environ Res Lett 15:074041

McLeod E, Chmura GL, Bouillon S, Salm R, Bjork M, Duarte CM, Lovelock CE, Schlesinger WH, Silliman BR (2011) A blueprint for blue carbon: toward an improved understanding of the role of vegetated coastal habitats in sequestering CO2. Front Ecol Environ 9:552–560

McLeod I, Schmider J, Creighton C, Gillies C (2018) Seven pearls of wisdom: advice from traditional owners to improve engagement of local indigenous people in shellfish ecosystem restoration. Ecol Manage Restor 19(2):98–101

Mcowen CJ, Weatherdon LV, Bochove JV, Sullivan E, Blyth S, Zockler C, Stanwell-Smith D, Kingston N, Martin CS, Spalding M, Fletcher S (2017) A global map of saltmarshes. Biodivers Data J 5:e11764. https://doi.org/10.3897/bdj.5.e11764

Menendez P, Losada IJ, Torres-Ortega S, Narayan S, Beck MW (2020) The global flood protection benefits of mangroves. Sci Rep 10:4404

Mentaschi L, Vousdoukas MI, Pekel JF, Voukouvalas E, Feyen L (2018) Global long-term observations of coastal erosion and accretion. Sci Rep 8(1):1–11. https://doi.org/10.1038/s41598-018-30904-w

Merkens JL, Reimann L, Hinkel J, Vafeidis AT (2016) Gridded population projections for the coastal zone under the shared socioeconomic pathways. Global Planet Change 145:57–66

MFF (Mangroves for the Future) (2018) Coastal resilience in action. IUCN Pakistan, Karachi. https://www.mangrovesforthefuture.org/assets/Repository/Documents/Coastal-Resilience-in-Action-Lessons-Learned-2014-2018-Final.pdf

Minderhoud PSJ, Coumou L, Erkens G, Middelkoop H, Stouthamer E (2019) Mekong delta much lower than previously assumed in sea-level rise impact assessments. Nat Commun 10:3847. https://doi.org/10.1038/s41467-019-11602-1

Minderhoud PSJ, Middelkoop H, Erkens G, Stouthamer E (2020) Groundwater extraction may drown mega-delta: projections of extraction-induced subsidence and elevation of the Mekong Delta for the 21st century. Environ Res Commun 2(1):011005. https://doi.org/10.1088/2515-7620/ab5e21

Mohanty SK, Dash P, Gupta A, Gaur P (2015) Prospects of blue economy in the Indian Ocean. Research and Information System for Developing Countries, New Delhi. http://ris.org.in/pdf/Final_Blue_Economy_Report_2015-Website.pdf

Mokhtar M, Aziz SAG (2003) Integrated coastal zone management using the ecosystems approach, some perspectives in Malaysia. Ocean Coast Manag 46(5):407–419. https://doi.org/10.1016/S0964-5691(03)00015-2

Moline MA, Karnovsky NJ, Brown Z, Divoky GJ, Frazer TK, Jacoby CA, Torres JJ, Fraser WR (2008) High latitude changes in ice dynamics and their impact on polar marine ecosystems. Ann N Y Acad Sci 1134:267–319

Mollica NR, Guo W, Cohen AL, Huang K-F, Foster GF, Donald HK, Solow AR (2018) Ocean acidification affects coral growth by reducing skeletal density. Proc Natl Acad Sci 115(8):1754–1759. https://doi.org/10.1073/pnas.1712806115

Monfort MC (2015) Globefish research programme: the role of women in the seafood industry. FAO Report No. 119. Food and Agriculture Organization of the United Nations, Rome

Mordecai GJ, Hewson I (2020) Coronaviruses in the sea. Front Microbiol 11:1795. https://doi.org/10.3389/fmicb.2020.01795

Morim J, Hemer M, Wang XL, Cartwright N, Trenham C, Semedo A, Young I, Bricheno L, Camus P, Casas-Prat M, Erikson L, Mentaschi L, Mori N, Shimura T, Timmermans B, Aarnes O, Breivik O, Behrens A, Dobrynin M, Menendez M, Staneva J, Wehner M, Wolf J, Kamranzad B, Webb A, Stopa J, Andutta F (2019) Robustness and uncertainties in global multivariate wind-wave climate projections.

Nat Clim Change 9(9):711–718. https://doi.org/10.1038/s41558-019-0542-5

Morrison TH, Hughes TP, Adger WN, Brown K, Barnett J, Lemos MC (2019) Save reefs to rescue all ecosystems. Nature 573(7774):333–336. https://doi.org/10.1038/d41586-019-02737-8

Morrison TH, Adger N, Barnett J, Brown K, Possingham H, Hughes T (2020) Advancing coral reef governance into the Anthropocene. One Earth 2:64–74

Morss RE, Cuite CL, Demuth JL, Hallman WK, Shwom RL (2018) Is storm surge scary? The influence of hazard, impact, and fear-based messages and individual differences on responses to Hurricane risks in the USA. Int J Disaster Risk Reducton 30:44–58. https://doi.org/10.1016/j.ijdrr.2018.01.023

Muawanah U, Yusuf G, Adrianto L, Kalther J, Pomeroy R, Abdullah H, Ruchimat T (2018) Review of national laws and regulation in Indonesia in relation to an ecosystem approach to fisheries management. Mar Policy 91:150–160

Mulligan M, van Soesbergen A, Sáenz L (2020) GOODD, a global dataset of more than 38,000 georeferenced dams. Sci Data 7:31. https://doi.org/10.6084/m9.figshare.10538486

Murray NJ, Phinn SR, DeWitt M, Ferrari R, Johnston R, Lyons MB, Clinton N, Thau D, Fuller RA (2018) The global distribution and trajectory of tidal flats. Nature 565(7738):222–225

Nagelkerken I, Connell SD (2015) Global alteration of ocean ecosystem functioning due to increasing human CO2 emissions. Proc Natl Acad Sci 112(43):13272–13277. https://doi.org/10.1073/pnas.1510856112

Narain D, Maron M, Teo HC, Hussey K, Lechner AM (2020) Best-practice biodiversity safeguards for belt and road initiative's financiers. Nat Sustain 3:650–657

Narayan S, Beck MW, Reguero BG, Losada IJ, van Wesenbeeck B, Pontee N, Sanchirico JN, Ingram JC, Lange GM, Burks-Copes KA (2016) The effectiveness, costs and coastal protection benefits of natural and nature-based defences. PloS One 11(5):e0154735

Narayan S, Beck MW, Wilson P, Thomas CJ, Guerrero A, Shepard CC, Reguero BG, Franco G, Ingram JC, Trespalacios D (2017) The value of coastal wetlands for flood damage reduction in the Northeastern USA. Sci Rep 7(1):1–12

National Academies of Sciences, Engineering, and Medicine (2019) A decision framework for interventions to increase the persistence and resilience of coral reefs. The National Academies Press, Washington, DC. https://doi.org/10.17226/25424

Nayak PM, Byrne ML (2019) Impact of land use land cover change on a sand dune ecosystem in Northwest Beach, Point Pelee National Park, Canada. J Great Lakes Res 45(6):1047–1054

Naylor LA, Stephenson WJ, Trenhaile AS (2010) Rock coast geomorphology: recent advances and future research directions. Geomorphology 114(1–2):3–11

Nelson DR (2011) Adaptation and resilience: responding to a changing climate. Wiley Interdiscip Rev Clim Change 2:113–120

Nesshover C, Assmuth T, Irvine KN, Rusch GM, Waylen KA, Delbaere B, Haase D, Jones-Walters L, Keune H, Kovacs E, Krauze K, Kulvik M, Rey F, van Dijk J, Vistad OI, Wilkinson ME, Wittmer H (2017) The science, policy and practice of nature-based solutions: an interdisciplinary perspective. Sci Total Environ 579:1215–1227

Neumann B, Vafeidis AT, Zimmermann J, Nicholls RJ (2015) Future coastal population growth and exposure to sea-level rise and coastal flooding - a global assessment. PloS One 10(3):1–34. https://doi.org/10.1371/journal.pone.0118571

Newell RIE, Koch EW (2004) Modelling seagrass density and distribution in response to changes in turbidity stemming from bivalve filtration and seagrass sediment stabilization. Estuaries 27:793–806

Nguyen CH, Bui TTH (2014) Integrated spatial planning and management for marine and coastal sustainability in Viet Nam. International Union for Conservation of Nature, Gland

Nguyen T, Quynh C, Schilizzi S, Hailu A, Iftekhar S (2017) Territorial use rights for fisheries (TURFs): state of the art and the road ahead. Mar Policy 75:41–52

Nicholls RJ, Adger WN, Hutton CW, Hanson SE (2020) Deltas in the Anthropocene. Springer, London

Niehörster F, Murnane R (2018) Ocean risk and the insurance industry. XL Catlin Services SE, London

Nienhuis JH, Ashton AD, Edmonds DA, Hoitink AJF, Kettner AJ, Rowland JC, Tornqvist TE (2020) Global-scale human impact on delta morphology has led to net land area gain. Nature 577(7791):514–518

Nippon Foundation–Nereus Program (2017) Restored ocean will alleviate poverty, provide jobs, and improve health, finds report. EurekAlert! Science News, AAAS. https://eurekalert.org/pub_releases/2017-05/nfp-row052917.php

Nitzbon J, Westermann S, Langer M, Martin LC, Strauss J, Laboor S, Boike J (2020) Fast response of cold ice-rich permafrost in Northeast Siberia to a warming climate. Nat Commun 11(1):1–11

NOEP (National Ocean Economic Program) (2016) State of the U.S. ocean and coastal economies. Middlebury Institute of International Studies at Monterey, Monterey

Nordstrom KF (2014) Living with shore protection structures: a review. Estuar Coast Shelf Sci 150:11–23. https://doi.org/10.1016/j.ecss.2013.11.003

Obiero K, Meulenbroek P, Drexler S, Dagne A, Akoll P, Odong R, Kaunda-Arara B, Waidbacher H (2019) The contribution of fish to food and nutrition security in Eastern Africa: emerging trends and future outlooks. Sustainability 11(6):1636. https://doi.org/10.3390/su11061636

OECD (Organisation for Economic Co-operation and Development) (2016) The ocean economy in 2030. OECD Publishing, Paris. https://doi.org/10.2166/9781780408927

OECD (Organisation for Economic Co-operation and Development) (2020a) Fisheries, aquaculture and COVID-19: issues and policy responses. OECD Publishing, Paris. http://www.oecd.org/coronavirus/policy-responses/fisheries-aquaculture-and-covid-19-issues-and-policy- responses-a2aa15de/

OECD (Organisation for Economic Co-operation and Development) (2020b) Economic outlook June 2020. OECD Publishing, Paris. https://www.oecd.org/economic-outlook/june-2020/

Okafor-Yarwood I, Kadagi NI, Miranda NAF, Uku J, Elegbede IO, Adewumi IJ (2020) The blue economy–cultural livelihood–ecosystem conservation triangle: the African experience. Front Mar Sci 7:586

Oliver ECJ, Burrows MT, Donat MG, Sen Gupta A, Alexander LV, Perkins-Kirkpatrick SE, Benthuysen JA, Hobday AJ, Holbrook NJ, Moore PJ, Thomsen MS, Wernberg T, Smale DA (2019) Projected marine heatwaves in the 21st century and the potential for ecological impact. Front Mar Sci 6:734

Oliver-Smith A (2019) Disasters and large-scale population dislocations: international and national responses. In: Oxford research encyclopedia of natural hazard science. https://doi.org/10.1093/acrefore/9780199389407.013.224

van Oosterzee P, Duke N (2017) Extreme Weather Likely Behind Worst Recorded Mangrove Dieback in Northern Australia. Conversation 14:1–6. http://theconversation.com/extreme-weather-likely-behind-worst-recorded-mangrove-dieback-in-northern-australia-71880

van Oppen MJH, Gates RD, Blackall LL, Cantin N, Chakravarti LJ, Chan WY, Cormick C, Crean A, Damjanovic K, Epstein H, Harrison PL, Jones TA, Miller M, Pears RJ, Peplow LM, Raftos DA, Schaffelke B, Stewart K, Torda G, Wachenfeld D, Weeks AR, Putnam HM (2017) Shifting paradigms in restoration of the world's coral reefs. Glob Chang Biol 23:3437–3448

Oppenheimer M, Glavovic BC, Hinkel J, van de Wal R, Magnan AK, Abd-Elgawad A, Cai R, Cifuentes-Jara M, DeConto RM, Ghosh T, Hay J, Isla F, Marzeion B, Meyssignac B, Sebesvari Z (2019) Sea level rise and implications for low-lying islands, coasts and communities. In: Pörtner H-O, Roberts DC, Masson-Delmotte V, Zhai P, Tignor M, Poloczanska E, Mintenbeck K et al (eds) IPCC special report on the ocean and cryosphere in a changing climate. Intergovernmental Panel on Climate Change, Geneva

Orr JC, Fabry VJ, Aumont O, Bopp L, Doney SC, Feely RA, Gnanadesikan A, Gruber N, Ishida A, Joos F, Key RM, Lindsay K, Maier-Reimer E, Matear R, Monfray P, Mouchet A, Najjar RG, Plattner G-K, Rodgers KB, Sabine CL, Sarmiento JL, Schlitzer R, Slater RD, Totterdell IJ, Weirig M-F, Yamanaka Y, Yool A (2005) Anthropogenic ocean acidification over the twentyfirst century and its impact on calcifying organisms. Nature 437(7059):681–686. https://doi.org/10.1038/nature04095

ORRAA (Ocean Risk and Resilience Action Alliance) (2019) Concept paper. https://wedocs.unep.org/bitstream/handle/20.500.11822/28829/InnoSolutions.pdf?sequence=1&isAllowed=y

Orth RJ, Moore KA, Marion SR, Wilcox DJ, Parrish DB (2012) Seed addition facilitates eelgrass recovery in a coastal bay system. Mar Ecol Prog Ser 448:177–195

Oschlies A, Brandt P, Stramma L, Schmidtko S (2018) Drivers and mechanisms of ocean deoxygenation. Nat Geosci 11:467–473

Otaño-Cruz A, Montañez-Acuña AA, Torres-López V, Hernández-Figueroa EM, Hernández-Delgado EA (2017) Effects of changing weather, oceanographic conditions, and land uses on spatio-temporal variation of sedimentation dynamics along near-shore coral reefs. Front Mar Sci 4:1–17. https://doi.org/10.3389/fmars.2017.00249

Ouyang Z, Qi D, Chen L, Takahashi T, Zhong W, DeGrandpre MD, Chen B, Gao Z, Nishino S, Murata A, Sun H, Robbins LL, Jin M, Cai W-J (2020) Sea-ice loss amplifies summertime decadal CO2 increase in the Western Arctic Ocean. Nat Clim Change 10:678–684

Oxford Economics (2017) Global infrastructure outlook. Global infrastructure hub: a G20 initiative. https://www.oxfordeconomics.com/publication/open/283970

Paine L (2014) The sea and civilization: a maritime history of the world. Atlantic Books, London

Palomares ML, Pauly D (2019) Coastal fisheries: the past, present, and possible futures. In: Wolanski E, Day JW, Elliott M, Ramachandran R (eds) Coasts and estuaries. Elsevier, Oxford, pp 569–576. https://doi.org/10.1016/B978-0-12-814003-1.00032-0

Palomares MLD, Froese R, Derrick B, Meeuwig JJ, Nöel S-L, Tsui G, Woroniak J, Zeller D, Pauly D (2020) Fishery biomass trends of exploited fish populations in marine ecoregions, climatic zones and ocean basins. Estuar Coast Shelf Sci 243:106896. https://doi.org/10.1016/j.ecss.2020.106896

Pascual U, Balvanera P, Díaz S, Pataki G, Roth E, Stenseke M, Watson RT, Başak Dessane E, Islar M, Kelemen E, Maris V, Quaas M, Subramanian SM, Wittmer H, Adlan A, Ahn S, Al-Hafedh YS, Amankwah E, Asah ST, Berry P, Bilgin A, Breslow SJ, Bullock C, Cáceres D, Daly- Hassen H, Figueroa E, Golden CD, Gómez-Baggethun E, González- Jiménez D, Houdet J, Keune H, Kumar R, Ma K, May PH, Mead A, O'Farrell P, Pandit R, Pengue W, Pichis-Madruga R, Popa F, Preston S, Pacheco-Balanza D, Saarikoski H, Strassburg BB, van den Belt M, Verma M, Wickson F, Yagi N (2017) Valuing nature's contributions to people: the IPBES approach. Curr Opin Environ Sustain 26–27:7–16

Paul LJ (2012) A history of the firth of Thames dredge fishery for mussels: use and abuse of a coastal resource. New Zealand aquatic environment and biodiversity report no. 94. Ministry of Agriculture and Forestry, Wellington. https://www.mpi.govt.nz/dmsdocument/4016/direct

PEMSEA (2020) Integrated coastal management. http://pemsea.org/our-work/integrated-coastal-management. Accessed 24 March 2020

Peng G, Matthews JL, Wang M, Vose R, Sun L (2020) What do global climate models tell us about future Arctic Sea Ice coverage changes? Climate 8(1):15

Perkins MJ, Ng TPT, Dudgeon D, Bonebrake TC, Leung KMY (2015) Conserving intertidal habitats: what is the potential of ecological engineering to mitigate impacts of coastal structures? Estuar Coast Shelf Sci 167:504–515

Pianc W (2014) Harbour approach channels – design guidelines. Report of Marcom Working Group, 49. https://www.pianc.org/publications/marcom/harbour-approach-channels-design-guidelines

Piehler MF, Smyth AR (2011) Habitat-specific distinctions in estuarine denitrification affect both ecosystem function and services. Ecosphere 2:1–17

Piñeiro-Corbeira C, Barreiro R, Cremades J, Arenas F (2018) Seaweed assemblages under a climate change scenario: functional responses to temperature of eight intertidal seaweeds match recent abundance shifts. Sci Rep 8(1):1–9

Plaza C, Pegoraro E, Bracho R, Celis G, Crummer KG, Hutchings JA, Pries CEH, Mauritz M, Natali SM, Salmon VG, Schädel C (2019) Direct observation of permafrost degradation and rapid soil carbon loss in Tundra. Nat Geosci 12(8):627–631

Pontee N (2014) Accounting for siltation in the design of intertidal creation schemes. Ocean Coast Manag 88:8–12

von der Porten S, Ota Y, Cisneros-Montemayor A, Pictou S (2019a) The role of indigenous resurgence in marine conservation. Coast Manage 47:527–547

von der Porten S, Corntassel J, Mucina D (2019b) Indigenous nationhood and herring governance: strategies for the reassertion of indigenous authority and inter-indigenous solidarity regarding marine resources. Alternative 15(1):62–74. https://doi.org/10.1177/2F1177180118823560

Poulos SE, Collins MB (2002) Fluviatile sediment fluxes to the Mediterranean Sea: a quantitative approach and the influence of dams. Geol Soc Lond Spec Publ 191(1):227–245. https://doi.org/10.1144/GSL.SP.2002.191.01.16

Powell EJ, Tyrrell MC, Milliken A, Tirpak JM, Staudinger MD (2018) A review of coastal management approaches to support the integration of ecological and human community planning for climate change. J Coast Conserv 23(1):1–18

Psaraftis HN (2019) Decarbonization of maritime transport: to be or not to be? Marit Econ Logist 21:353–371. https://link.springer.com/article/10.1057/s41278-018-0098-8

Puy A, Lo Piano S, Saltelli A (2020) Current models underestimate future irrigated areas. Geophys Res Lett 47(8):e2020GL087360. https://doi.org/10.1029/2020GL087360

Pyšek P, Richardson DM, Pergl J, Jarosík V, Sixtová Z, Weber E (2008) Geographical and taxonomic biases in invasion ecology. Trends Ecol Evol 23:237–244. https://doi.org/10.1016/j.tree.2008.02.002

Quan WM, Zhu JX, Ni Y, Shi L-Y, Chen Y-Q (2009) Faunal utilization of constructed intertidal oyster (Crassostrea rivularis) Reef in the Yangtze River Estuary, China. Ecol Eng 35:1466–1475

Rakotomahazo C, Ravaoarinorotsihoarana LA, Randrianandrasaziky D, Glass L, Gough C, Boleslas Todinanahary GG, Gardner CJ (2019) Participatory planning of a community-based payments for ecosystem services initiative in Madagascar's mangroves. Ocean Coast Manag 175:43–52

Ravera F, Tarrasón D, Hubacek K, Molowny-Horas R, Sendzimir J (2020) Participatory modelling in adaptive environmental management: a case study in Semi-Arid Northern Nicaragua. In: Ninan KN (ed) Environmental assessments. Edward Elgar Publishing, Cheltenham, pp 231–248

Reguero BG, Losada IJ, Diaz-Simal P, Mendez FJ, Beck MW (2015) Effects of climate change on exposure to coastal flooding in Latin America and the Caribbean. PloS One 10(7):e0133409

Reguero BG, Losada IJ, Mendez FJ (2019a) A recent increase in global wave power as a consequence of oceanic warming. Nat Commun 10:205

Reguero BG, Secaira F, Toimil A, Escudero M, Díaz-Simal P, Beck MW, Silva R, Storlazzi C, Losada IJ (2019b) The risk reduction benefits of the Mesoamerican Reef in Mexico. Front Earth Sci 7:125. https://doi.org/10.3389/feart.2019.00125

Reid J, Rout M (2020) The implementation of ecosystem-based management in New Zealand–a Māori perspective. Mar Policy 117:103889. https://doi.org/10.1016/j.marpol.2020.103889

Reopanichkul P, Carter RW, Worachananant S, Crossland CJ (2009) Wastewater discharge degrades coastal waters and reef communities in Southern Thailand. Mar Environ Res 69(5):287–296. https://doi.org/10.1016/j.marenvres.2009.11.011

Reyna J et al (2016) Land-sea physical interaction. In: Inniss L et al (eds) The first global integrated marine assessment. world ocean assessment I. United Nations, New York

Richards DR, Friess DA (2016) Rates and drivers of mangrove deforestation in Southeast Asia, 2000–2012. PNAS 113(2):344–349

Richards DR, Thompson BS, Wijedasa L (2020) Quantifying net loss of global mangrove carbon stocks from 20 years of land cover change. Nat Commun 11(1):1–7

Rignot E, Mouginot J, Scheuchl B, van den Broeke M, van Wessem MJ, Morlighem M (2019) Four decades of Antarctic ice sheet mass balance from 1979–2017. Proc Natl Acad Sci 116(4):1095–1103. https://doi.org/10.1073/pnas.1812883116

Ringler C (2017) Investments in irrigation for global food security. Policy note. International Food Policy Research Institute, Washington, DC. https://www.ifpri.org/publication/investments-irrigation-global-food-security

Rinkevich B (1995) Restoration strategies for coral reefs damaged by recreational activities: the use of sexual and asexual recruits. Restor Ecol 3:241–251

Rist P, Rassip W, Yunupingu D, Wearne J, Gould J, Dulfer-Hyams M, Bock E, Smyth D (2019) Indigenous protected areas in sea country: indigenous-driven collaborative marine protected areas in Australia. Aquat Conserv Mar Freshw Ecosyst 29:138–151. https://doi.org/10.1002/aqc.3052

Roberts CM, O'Leary BC, Hawkins JP (2020) Climate change mitigation and nature conservation both require higher protected area targets. Philos Trans R Soc Lond B Biol Sci 375:20190121

Rodríguez-Martínez RE, Roy PD, Torrescano-Valle N, Cabanillas-Terán N, Carrillo-Domínguez S, Collado-Vides L, van Tussenbroek BI (2020) Element concentrations in Pelagic Sargassum along the Mexican Caribbean Coast in 2018-2019. PeerJ 8:e8667

Roff G, Clark TR, Reymond CE, Zhao JX, Feng Y, McCook LJ, Done TJ, Pandolfi JM (2012) Palaeoecological evidence of a historical collapse of corals at Pelorus Island, Inshore Great Barrier Reef, following European settlement. Proc R Soc B Biol Sci 280(1750):20122100

Rogers A, Aburto-Oropeza O et al (2020) Critical habitats and biodiversity: inventory, thresholds and governance. World Resources Institute, Washington, DC. www.oceanpanel.org/blue-papers/critical-habitats-and-biodiversity-inventory-thresholds-and-governance

Romañach SS, DeAngelis DL, Koh HL, Li Y, Teh SY, Barizan RSR, Zhai L (2018) Conservation and restoration of mangroves: global status, perspectives, and prognosis. Ocean Coast Manag 154:72–82. https://doi.org/10.1016/j.ocecoaman.2018.01.009

Rovira A, Ballinger R, Ibáñez C, Parker P, Dominguez MD, Simon X, Lewandowski A, Hochfeld B, Tudor M, Vernaeve L (2014) Sediment imbalances and flooding risk in European Deltas and estuaries. J Soil Sediment 14(8):1493–1512. https://doi.org/10.1007/s11368-014-0914-4

Saad J, Hiew K, Gopinath N (2012) Review of Malaysian Laws and Policies in relation to the implementation of ecosystem approach to fisheries management in Malaysia. The USAID Coral Triangle Support Partnership, Print, Honolulu

Saderne V, Cusack M, Serrano O, Almahasheer H, Krishnakumar PK, Rabaoui L, Qurban MA, Duarte CM (2020) Role of vegetated coastal ecosystems as nitrogen and phosphorous filters and sinks in the coasts of Saudi Arabia. Environ Res Lett 15:034058

Sagoff M (2008) The economy of the Earth: philosophy, law, and the environment, 2nd edn. Cambridge University Press, Cambridge

Salisbury J, Green M, Hunt C, Campbell J (2008) Coastal acidification by rivers: a threat to shellfish? EOS 89(50):513

Sanford E, Sones JL, García-Reyes M, Goddard JH, Largier JL (2019) Widespread shifts in the coastal biota of Northern California during the 2014–2016 marine heatwaves. Sci Rep 9(1):4216

Santana-Ceballos J, Fortes CJ, Reis MT, Rodríguez G (2019) Wave overtopping and flood risk assessment in harbours: the Port of Las Nieves and its future expansion. Int J Environ Impacts Manage Mitig Recov 2(1):59–71. https://doi.org/10.2495/ei-v2-n1-59-71

Sardain A, Sardain E, Leung B (2019) Global forecasts of shipping traffic and biological invasions to 2050. Nat Sustain 2:274–282

Sasse TP, McNeil BI, Matear RJ, Lenton A (2015) Quantifying the influence of CO2 seasonality on future aragonite undersaturation onset. Biogeosciences 12(20):6017–6031. https://doi.org/10.5194/bg-12-6017-2015

Sato C, Michiko H, Nishinoc J (2006) Land subsidence and groundwater management in Tokyo. Int Rev Environ Strateg 6(2):403

Sato T, Qadir M, Yamamoto S, Endo T, Zahoor A (2013) Global, regional, and country level need for data on wastewater generation, treatment, and use. Agric Water Manag 130:1–13

Schandl H, Fischer-Kowalski M, West J, Giljum S, Dittrich M, Eisenmenger N, Geschke A, Lieber M, Wieland H, Schaffartzik A, Keausmann F, Gierlinger S, Hosking K, Lenzen K, Tanikawa H, Alessio M, Fishman T (2018) Global material flows and resource productivity: forty years of evidence. J Ind Ecol 22(4):827–838. https://doi.org/10.1111/jiec.12626

Scyphers SB, Powers SP, Heck KL Jr, Byron D (2011) Oyster reefs as natural breakwaters mitigate shoreline loss and facilitate fisheries. PloS One 6(8):e22396

Seebens H et al (2015) Global trade will accelerate plant invasions in emerging economies under climate change. Glob Chang Biol 21:4128–4140

Sekovski I, Newton A, Dennison WC (2012) Megacities in the coastal zone: using a driver-pressure-state-impact-response framework to address complex environmental problems. Estuar Coast Shelf Sci 96:48–59. https://doi.org/10.1016/j.ecss.2011.07.011

Selig ER, Hole DG, Allison EH, Arkema KK, McKinnon MC, Chu J et al (2019) Mapping global human dependence on marine ecosystems. Conserv Lett 12:e12617. https://doi.org/10.1111/conl.12617

Sengupta D, Chen R, Meadows M (2017) Building beyond land: an overview of coastal land reclamation in 16 global megacities. Appl Geogr 90:229–238

Sepulveda C, Rivera A, Gelcich S, Stotz WB (2019) Exploring determinants for the implementation of mixed TURF-aquaculture systems. Sci Total Environ 682:310–317

Serrano O, Kelleway J, Lovelock C, Lavery PS (2019a) Conservation of blue carbon ecosystems for climate change mitigation and adaptation. In: Perillo G, Wolanski E, Cahoon DR, Hopkinson CS (eds) Coastal wetlands. Elsevier, Cambridge, pp 965–996. https://doi.org/10.1016/b978-0-444-63893-9.00028-9

Serrano O, Lovelock CE, Atwood TB, Macreadie PI, Canto R, Phinn S, Arias-Ortiz A, Bai L, Baldock J, Bedulli C, Carnell P, Connolly RM, Donaldson P, Esteban A, Lewis CJE, Eyre BD, Hayes MA, Horwitz P, Hutley LB, Kavazos CRJ, Kelleway JJ, Kendrick GA, Kilminster K, Lafratta A, Lee S, Lavery PS, Maher DT, Marbà N, Masque P, Mateo MA, Mount R, Ralph PJ, Roelfsema C, Rozaimi M, Ruhon R, Salinas C, Samper-Villarreal J, Sanderman J, Sanders CJ, Santos I, Sharples C, Steven ADL, Cannard T, Trevathan-Tackett SM, Duarte CM (2019b) Australian vegetated coastal ecosystems as global hotspots for climate change mitigation. Nat Commun 10:4313. https://doi.org/10.1038/s41467-019-12176-8

Serrano O, Lavery PS, Bongiovanni J, Duarte CM (2020) Impact of seagrass establishment, industrialization and coastal infrastructure on seagrass biogeochemical sinks. Mar Environ Res 160:104990
Seto KC, Fragkias M, Güneralp B, Reilly MK (2011) A meta-analysis of global urban land expansion. PloS One 6(8):1–9. https://doi.org/10.1371/journal.pone.0023777
Shipman B, Stojanovic T (2007) Facts, fictions, and failures of integrated coastal zone management in Europe. Coast Manage 35: 375–398
Short FT, Polidoro B, Livingstone SR, Carpenter KE, Bandeira S, Bujang JS, Calumpong HP, Carruthers TJ, Coles RG, Dennison WC, Erftemeijer PL (2011) Extinction risk assessment of the world's seagrass species. Biol Conserv 144(7):1961–1971
Silva AN, Taborda R, Andrade C, Ribeiro M (2019) The future of insular beaches: insights from a past-to-future sediment budget approach. Sci Total Environ 676:692–705. https://doi.org/10.1016/j.scitotenv.2019.04.228
SIMS (Sydney Institute of Marine Science) (2020) Living aeawalls. https://www.sims.org.au/page/130/living-seawalls-landing
Singh G, Cisneros-Montemayor AM, Swartz W, Cheung WJ, Guy A, Kenny TA, McOwen CJ (2018) A rapid assessment of co-benefits and trade-offs among sustainable development goals. Mar Policy 93:223–231. https://doi.org/10.1016/j.marpol.2017.05.030
Singh GG, Eddy IMS, Halpern BS, Neslo R, Satterfield T, Chan KMA (2020) Mapping cumulative impacts to coastal ecosystem services in British Columbia. PloS One 15:e0220092
Smale DA, Moore PJ, Queirós AM, Higgs ND, Burrows MT (2018) Appreciating interconnectivity between habitats is key to blue carbon management. Front Ecol Environ 16(2):71–73. https://doi.org/10.1002/fee.1765
Smale DA, Wernberg TW, Eric CJO, Thomsen M, Harvey BP, Straub SP, Burrows MT, Alexander L, Benthuysen J, Donat MG, Feng M, Hobday A, Holbrook J, Perkins-Kirkpatrick N, Scannell H, Gupta AS, Payne BL, Moore PJ (2019) Marine heatwaves threaten global biodiversity and the provision of ecosystem services. Nat Clim Change 9(4):306–312. https://doi.org/10.1038/s41558-019-0412-1
Small C, Nicholls RJ (2003) A global analysis of human settlement in coastal zones. J Coast Res 19:584–599
Smith SL, Cunniff SE, Peyronnin NS, Kritzer JP (2017) Prioritizing coastal ecosystem stressors in the Northeast United States under increasing climate change. Environ Sci Policy 78:49–57. https://doi.org/10.1016/j.envsci.2017.09.009
Smucker TA, Wangui EE (2016) Gendered knowledge and adaptive practices: differentiation and change in Mwanga District, Tanzania. Ambio 45(3):276–286
Spalding MD, McIvor AL, Beck MW, Koch EW, Möller I, Reed DJ, Rubinoff P, Spencer T, Tolhurst TJ, Wamsley TV, van Wesenbeeck BK, Wolanski E, Woodroffe CD (2014) Coastal ecosystems: a critical element of risk reduction. Conserv Lett 7(3):293–301
Spalding M, Burke L, Wood SA, Ashpole J, Hutchisone J, Ermgassen P (2017) Mapping the global value and distribution of coral reef tourism. Mar Policy 82:104–113
Spivak AC, Sanderman J, Bowen JL, Canuel EA, Hopkinson CS (2019) Global-change controls on soil-carbon accumulation and loss in coastal vegetated ecosystems. Nat Geosci 12(9):685–692
Stacey N, Gibson E, Loneragan NR, Warren C, Wiryawan B, Adhuri D, Fitriana R (2019) Enhancing coastal livelihoods in Indonesia: an evaluation of recent initiatives on gender, women and sustainable livelihoods in small-scale fisheries. Marit Stud 18:359–371
Stephenson RL, Hobday AJ, Cvitanovic C, Alexander KA, Begg GA, Bustamante RH, Dunstan PK, Frusher S, Fudge M, Fulton EA, Haward M (2019) A practical framework for implementing and evaluating integrated management of marine activities. Ocean Coast Manag 177:127–138
Stojanovic T, Ballinger RC, Lalwani CS (2004) Successful integrated coastal management: measuring it with research and contributing to wise practice. Ocean Coast Manag 47(5–6):273–298. https://doi.org/10.1016/j.ocecoaman.2004.08.001
Storlazzi C, Gingerich S, Dongeren AP, Cheriton O, Swarzenski P, Quataert E, Voss C (2018) Most atolls will be uninhabitable by the mid-21st century because of sea-level rise exacerbating wave-driven flooding. Sci Adv 4:eaap9741. https://doi.org/10.1126/sciadv.aap9741
Storlazzi CD, Reguero BG, Cole AD, Lowe E, Shope JB, Gibbs AE, Nickel BA, McCall RT, van Dongeren AR, Beck MW (2019) Rigorously valuing the role of U.S. In: coral reefs in coastal hazard risk reduction. US geological survey open-file report 2019–1027. US Geological Survey, Santa Cruz. https://doi.org/10.3133/ofr20191027
Strain E, Olabarria C, Mayer-Pinto M, Cumbo V, Morris R, Bugnot A, Dafforn K, Heery E, Firth L, Brooks P, Bishop M (2018a) Eco-engineering built infrastructure for marine and coastal biodiversity: which interventions have the greatest ecological benefit? J Appl Ecol 55:426–441
Strain EMA, Morris RL, Coleman RA, Figueira WF, Steinberg PD, Johnston EL, Bishop MJ (2018b) Increasing microhabitat complexity on seawalls can reduce fish predation on native oysters. Ecol Eng 120:637–644
Sumaila UR, Walsh M, Hoareau K, Cox A et al (2020) Ocean finance: financing the transition to a sustainable ocean economy. World Resources Institute, Washington, DC. https://oceanpanel.org/blue-papers/ocean-finance
Sutton-Grier AE, Wowk K, Bamford H (2015) Future of our coasts: the potential for natural and hybrid infrastructure to enhance the resilience of our coastal communities, economies and ecosystems. Environ Sci Policy 51:137–148. https://doi.org/10.1016/j.envsci.2015.04.006
Swilling M, Ruckelshaus M, Brodie Rudolph T et al (2020) The ocean transition: what to learn from system transitions. World Resources Institute, Washington, DC. www.oceanpanel.org/blue-papers/ocean-transition-what-learn-system-transitions
Syvitski JPM, Kettner AJ, Overeem I, Hutton EW, Hannon MT, Brakenridge RG, Vörösmarty C, Saito Y, Giosan L, Nicholls RJ (2009) Sinking deltas due to human activities. Nat Geosci 2:681. https://doi.org/10.1038/ngeo629
Tenório G, Souza-Filho PW, Ramos EMLS, de Oliveira Alves PJ (2015) Mangrove shrimp farm mapping and productivity on the Brazilian Amazon Coast: environmental and economic reasons for coastal conservation. Ocean Coast Manag 104:65–77. https://doi.org/10.1016/j.ocecoaman.2014.12.006
Tessler ZD, Vörösmarty C, Overeem I, Syvitski JMP (2018) A model of water and sediment balance as determinants of relative sea level rise in contemporary and future deltas. Geomorphology 305:209–220. https://doi.org/10.1016/j.geomorph.2017.09.040
Thacker S, Adshead D, Fay M, Hallegatte S, Harvey M, Meller H, O'Regan N, Rozenberg J, Watkins G, Hall JW (2019) Infrastructure for sustainable development. Nat Sustain 2:324–331
Thiele T, Alleng G, Biermann A, Corwin E, Crooks S, Fieldhouse P, Herr D, Matthews N, Roth N, Shrivastava A, von Unger M, Zeitlberger J (2020) Blue infrastructure finance: a new approach, integrating nature-based solutions for coastal resilience. International Union for Conservation of Nature, Gland
Thomas N, Lucas R, Bunting P, Hardy A, Rosenqvist A, Simard M (2017) Distribution and drivers of global mangrove forest change, 1996-2010. PloS One 12(6):1–14. https://doi.org/10.1371/journal.pone.0179302
Thomsen MS, Mondardini L, Alestra T, Gerrity S, Tait L, South PM, Lilley SA, Schiel DR (2019) Local extinction of Bull Kelp (Durvillaea spp.) due to a marine heatwave. Front Mar Sci 6:84

Tian B, Wu W, Yang Z, Zhou Y (2016) Drivers, trends, and potential impacts of long-term coastal reclamation in China from 1985 to 2010. Estuar Coast Shelf Sci 170:83–90

Tonazzini D, Fosse J, Morales E, González A, Klarwein S, Moukaddem K, Louveau O (2019) Blue tourism. Towards a sustainable coastal and maritime tourism in world marine regions. Eco-Union, Barcelona

Toonen JT, Wilhelm A, Maxwell SM, Wagner D, Bowen BW, Sheppard RC, Taei SM (2013) One size does not fit all: the emerging frontier in large-scale marine conservation. Mar Pollut Bull 77(1–2):7–10. https://doi.org/10.1016/j.marpolbul.2013.10.039

de la Torre-Castro M, Fröcklin S, Börjesson S, Okupnik J, Jiddawi NS (2017) Gender analysis for better coastal management – increasing our understanding of social-ecological seascapes. Mar Policy 83:62–74

Torres A, Brandt J, Lear K, Liu J (2017) A looming tragedy of the sand commons. Science 357(6355):970–971. https://doi.org/10.1126/science.aao0503

Tran LX, Fischer A (2017) Spatiotemporal changes and fragmentation of mangroves and its effects on fish diversity in Ca Mau Province (Vietnam). J Coast Conserv 21(3):355–368

Tran TA, Tran VGP, Nguyen TKH, Dinh QC (2016) Gender analysis in building climate resilience in Da Nang challenges and solutions. Asian cities climate resilience working paper series 35. IIED, London

Tschakert P, Machado M (2012) Gender justice and rights in climate change adaptation: opportunities and pitfalls. Ethics Soc Welf 6(3):275–289. https://doi.org/10.1080/17496535.2012.704929

Turschwell MP, Tulloch VJD, Sievers M, Pearson RM, Andradi- Brown DA, Ahmadia GN, Connolly RM, Bryan-Brown D, Lopez-Marcano S, Adame MF, Brown CJ (2020a) Multi-scale estimation of the effects of pressures and drivers on mangrove forest loss globally. Biol Conserv 247:108637

Turschwell MP, Brown CJ, Pearson RM, Connolly RM (2020b) China's belt and road initiative: conservation opportunities for threatened marine species and habitats. Mar Policy 112:103791

UNCTAD (United Nations Conference on Trade and Development) (2020a) The COVID-19 pandemic and the blue economy: new challenges and prospects for recovery and resilience. UNCTAD/DITC/TED/INF/2020/2. UNCTAD, Geneva

UNCTAD (United Nations Conference on Trade and Development) (2020b) Review of maritime transport 2019. UNCTAD, Geneva

UNDP (United Nations Development Programme) (2005) Survivors of the tsunami: one year later-UNDP assisting communities to build back better. UNDP, New York

UNEP (United Nations Environment Programme) (2019) Sand and sustainability: finding new solutions for environmental governance of global sand resources. GRID-Geneva and UNEP, Geneva. https://wedocs.unep.org/bitstream/handle/20.500.11822/28163/SandSust.pdf?sequence=1&isAllowed=y

UNEP (United Nations Environment Programme) (2020) Out of the blue: the value of seagrasses to the environment and to people. UNEP, Nairobi

UNFCCC (United Nations Framework Convention on Climate Change) (2019) Differentiated impacts of climate change on women and men; the integration of gender considerations in climate policies, plans and actions; and progress in enhancing gender balance in national climate delegations. Synthesis report by the secretariat. FCCC/SBI/2019/INF.8. UNFCCC, Bonn

UN-Habitat (2015) Urbanization and climate change in small island developing states. UN-Habitat, Nairobi. http://unhabitat.org/cities-and-climate-change-initiative/

UNISDR (United Nations Office for Disaster Risk Reduction) (2011) In: UNISDR (ed) Global assessment report on disaster risk reduction: revealing risk, redefining development, Geneva. https://www.undrr.org/publication/global-assessment-report-disaster-risk-reduction2011#:~:text=Revealing%20risk%2C%20redefining%20development%3A,progress%20in%20disaster%20risk%20reduction

United Nations (2013) Integrating nature-based solutions into urban planning can help lead to better water future, Secretary-General says in message for day of biodiversity. Press release, 17 May, Department of UN Secretary General, New York. https://www.un.org/press/en/2013/sgsm15032.doc.htm

UNWTO (World Tourism Organization) (2020a) How are countries supporting tourism recovery? UNWTO briefing note – tourism and COVID-19, issue 1. UNWTO, Madrid. https://doi.org/10.18111/9789284421893

UNWTO (World Tourism Organization) (2020b) Tourism in SIDS – the challenge of sustaining livelihoods in times of COVID-19. UNWTO briefing note – tourism and COVID-19, issue 2. UNWTO, Madrid. https://doi.org/10.18111/9789284421916

Ureta JC, Lasco R, Sajise AJ, Calderon M (2016) A ridge-to-reef ecosystem-based valuation approach to biodiversity conservation in Layawan Watershed, Misamis Occidental, Philippines. J Environ Sci Manag 19(2):64–75

US Department of Energy (2015) Chapter 7: Advancing systems and technologies to produce cleaner fuels. In: 2015 quadrennial technology review. US Department of Energy, Washington, DC. https://www.energy.gov/downloads/chapter-7-advancing-systems-and--technologies-produce-cleaner-fuels

Valdor PF, Gómez AG, Steinberg P, Tanner E, Knights AM, Seitz RD, Airoldi L, Firth LB, Arvanitidis C, Ponti M, Chatzinikolaou E (2020) A global approach to mapping the environmental risk of harbours on aquatic systems. Mar Policy 119:104051

Vanderklift MA, Doropoulos C, Gorman D, Leal I, Minne AJP, Statton J, Steven ADL, Wernberg T (2020) Using propagules to restore coastal marine ecosystems. Front Mar Sci 7:724

Vaughan A (2019) Moving away from the coast. New Sci 243(3245):8. https://doi.org/10.1016/S0262-4079(19)31601-X

Vergés A, Doropoulos C, Malcolm HA, Skye M, Garcia-Pizá M, Marzinelli EM, Campbell AH, Ballesteros E, Hoey AS, Vila-Concejo A, Bozec Y-M, Steinberg PD (2016) Long-term empirical evidence of ocean warming leading to tropicalization of fish communities, increased herbivory, and loss of kelp. Proc Natl Acad Sci 113:13791–13796

Viana DF, Halpern BS, Gaines SD (2017) Accounting for tourism benefits in marine reserve design. PloS One 12(12):e0190187. https://doi.org/10.1371/journal.pone.0190187

Vianna GMS, Zeller D, Pauly D (2020) Fisheries and policy implications for human nutrition. Curr Environ Health Rep 7(3):161–169

Vierros MK, Harrison A-L, Sloat MR, Crespo GO, Moore JW, Dunn DC, Ota Y, Cisneros-Montemayor AM, Shillinger GL, Watson TK, Govan H (2020) Considering indigenous peoples and local communities in governance of the global ocean commons. Mar Policy 119:104039

Vikolainen V, Flikweert J, Bressers H, Lulofs K (2017) Governance context for coastal innovations in England: the case of sandscaping in North Norfolk. Ocean Coast Manag 145:82–93. https://doi.org/10.1016/j.ocecoaman.2017.05.012

Villasenor-Derbez JC, Aceves-Bueno E, Fulton S, Suarez A, Hernandez-Velasco A, Torre J, Micheli F (2019) An interdisciplinary evaluation of community-based TURF-reserves. PLoS One 14:e0221660

Visessri S, Ekkawatpanit C (2020) Flood management in the context of climate and land-use changes and adaptation within the Chao Phraya River Basin. J Disaster Res 15(5):579–587

Vitousek S, Barnard P, Fletcher CH, Frazer N, Erikson L, Storlazzi CD (2017) Doubling of coastal flooding frequency within decades due to sea-level rise. Nat Sci Rep 7(1):1–9, Article number: 1399

Vo ST, Hua TT, Phan KH (2019) A study of coral reef resilience and implications of adaptive management and rehabilitation in Khanh Hoa Province, Vietnam. Acta Oceanol Sin 38(1):112–117

Vörösmarty CJ, Meybeck M, Fekete B, Sharma K, Green P, Syvitski JPM (2003) Anthropogenic sediment retention: major global impact

from registered river impoundments. Global Planet Change 39(1–2):169–190. https://doi.org/10.1016/S0921-8181(03)00023-7
Vousdoukas MI, Ranasinghe R, Mentaschi L, Plomaritis TA, Athanasiou P, Luijendijk A, Feyen L (2020) Sandy coastlines under threat of erosion. Nat Clim Change 10(3):260–263
Voyer M, Schofield C, Azmi K, Warner R, McIlgorm A, Quirk A (2018) Maritime security and the blue economy: intersections and interdependencies in the Indian Ocean. J Indian Ocean Reg 14(1):28–48. https://doi.org/10.1080/19480881.2018.1418155
Waite R, Beveridge M, Brummett R, Castine S, Chaiyawannakarn N, Kaushik S, Mungkung R, Nawapakpilai S, Phillips M (2014) Improving productivity and environmental performance of aquaculture. World Resources Institute, Washington, DC. https://www.wri.org/publication/improving-aquaculture
Walker TR, Olubukola A, Aguila Feijoo MD, Elhaimer E, Tahazzud H, Edwards SH, Morrison CE (2019) Environmental effects of marine transportation. In: Shephard C (ed) World seas: an environmental evaluation. Academic Press, Cambridge, pp 505–530. https://doi.org/10.1016/b978-0-12-805052-1.00030-9
Waltham NJ, Connolly RM (2011) Global extent and distribution of artificial, residential waterways in estuaries. Estuar Coast Shelf Sci 94(2):192–197. https://doi.org/10.1016/j.ecss.2011.06.003
Waltham NJ, Burrows D, Wegscheidl C, Buelow C, Ronan M, Connolly N, Groves P, Marie-Audas D, Creighton C, Sheaves M (2019) Lost floodplain wetland environments and efforts to restore connectivity, habitat, and water quality settings on the great barrier reef. Front Mar Sci 6:71
Waltham NJ, Elliott M, Lee SY, Lovelock C, Duarte CM, Buelow C, Simenstad C, Nagelkerken I, Claassens L, Wen CK-C, Barletta M, Connolly RM, Gillies C, Mitsch WJ, Ogburn MB, Purandare J, Possingham H, Sheaves M (2020) UN decade on ecosystem restoration 2021–2030—what chance for success in restoring coastal ecosystems? Front Mar Sci 7:71. https://doi.org/10.3389/fmars.2020.00071
Wang W-X, Pan K, Tan Q, Guo L, Simpson SL (2014) Estuarine pollution of metals in China: science and mitigation. Environ Sci Tech 48(17):9975–9976. https://doi.org/10.1021/es503549b
Wang J, Beusen AH, Liu X, Bouwman AF (2019) Aquaculture production is a large, spatially concentrated source of nutrients in Chinese freshwater and coastal seas. Environ Sci Technol 54(3):1464–1474
Ward JR, Lafferty KD (2004) The elusive baseline of marine disease: are diseases in ocean ecosystems increasing? PLoS Biol 2:E120
Water UN (2017) Integrated monitoring and reporting processes for global SDG 6 indicators. UN Water, Geneva
Waycott M, Duarte CM, Carruthers TJB, Orth RJ, Dennison WC, Olyarnik S, Calladine A, Fourqurean JW, Heck KL Jr, Hughes AR, Kendrick GA, Kenworthy WJ, Short FT, Williams SL (2009) Accelerating loss of seagrass across the globe threatens coastal ecosystems. Proc Natl Acad Sci 106(30):12377–12381
WEF (World Economic Forum) (2019) The global risks report 2019, 14th edn
WEF (World Economic Forum) (2020a) The global risks report 2020, 15th edn
WEF (World Economic Forum) (2020b) COVID-19 risks outlook a preliminary mapping and its implications. In partnership with Marsh & McLennan and Zurich Insurance Group
WEF (World Economic Forum) (2020c) Nature risk rising: why the crisis engulfing nature matters for business and the economy. New Nature Economy series, Book I. World Economic Forum in Collaboration with PwC
Wernberg T, Filbee-Dexter K (2019) Missing the marine forest for the trees. Mar Ecol Prog Ser 612:209–215
Wernberg T, Bennett S, Babcock RC, De Bettignies T, Cure K, Depczynski M, Dufois F, Fromont J, Fulton CJ, Hovey RK, Harvey ES (2016) Climate-driven regime shift of a temperate marine ecosystem. Science 353(6295):169–172
Wernberg T, Krumhansl K, Filbee-Dexter K, Pedersen MF (2019) Status and trends for the world's kelp forests. In: Shephard C (ed) World seas: an environmental evaluation. Academic Press, Cambridge, pp 57–78
Whelchel AW, Reguero BG, van Wesenbeeck B, Renaud FG (2018) Advancing disaster risk reduction through the integration of science, design, and policy into eco-engineering and several global resource management processes. Int J Disaster Risk Reduct 32:29–41
White AT, Gomez E, Alcala AC, Russ G, Vincent A (2006) Evolution and lessons from fisheries and coastal management in the Philippines. In: McClanahan T, Castilla JC (eds) Fisheries management: progress towards sustainability. Blackwell Publishing, Oxford, pp 88–108
Wilkerson E, Kirbyshire A, Mayhew L, Batra P, Milan A (2016) Climate-induced migration and displacement: closing the policy gap. Briefing. Overseas Development Institute, London
Williams SL, Sur C, Janetski N, Hollarsmith JA, Rapi S, Barron L, Heatwole SJ, Yusuf AM, Yusuf S, Jompa J, Mars F (2019) Large-scale coral reef rehabilitation after blast fishing in Indonesia. Restor Ecol 27:447–456
Winther J-G, Dai M et al (2020) Integrated ocean management. World Resources Institute, Washington, DC. https://oceanpanel.org/blue-papers/integrated-ocean-management
Wong PP, Losada IJ, Gattuso JP, Hinkel J, Khattabi A, McInnes KL, Saito Y, Sallenger A (2014) Coastal systems and low-lying areas. In: White LL, Field CB, Barros VR, Dokken DJ, Mach KJ, Mastrandrea MD, Bilir TE, Chatterjee M, Ebi KL, Estrada YO, Genova RC, Girma B, Kissel ES, Levy AN, MacCracken S, Mastrandrea PR (eds) Climate change 2014: impacts, adaptation, and vulnerability. Part A: global and sectoral aspects. Contribution of Working Group II to the fifth assessment report of the Intergovernmental Panel on Climate Change. Cambridge University Press, Cambridge, pp 361–409
Woodroffe C, Rogers K, McKee K, Lovelock C, Mendelssohn I, Saintilan N (2016) Mangrove sedimentation and response to relative sea-level rise. Ann Rev Mar Sci 8:243–266
World Bank (2008) Biodiversity, climate change, and adaptation: nature- based solutions from the World Bank Portfolio. World Bank, Washington, DC. https://openknowledge.worldbank.org/handle/10986/6216
World Bank, Food and Agriculture Organization and WorldFish (2012) Hidden harvests: the global contribution of capture fisheries, economic and sector. World Bank, Washington, DC
World Bank Group (2015) WACA: West Africa Coastal Areas Management Program. A partnership for saving West Africa's Coastal Assets. http://pubdocs.worldbank.org/en/622041448394069174/1606426-WACA-Brochure.pdf
World Conservation Congress (2016) Resolution No. WCC-2016-Res-050-EN of the World Conservation Congress, at its session in Hawaii, 1–10 Sept 2016
Worthington T, Spalding M (2018). Mangrove restoration potential: a global map highlighting a critical opportunity. The Nature Conservancy and IUCN
Worthington TA, Andradi-Brown DA, Bhargava R, Buelow C, Bunting P, Duncan C, Fatoyinbo L, Friess DA, Goldberg L, Hilarides L, Lagomasino D, Landis E, Longley-Wood K, Lovelock CE, Murray NJ, Narayan S, Rosenqvist A, Sievers M, Simard M, Thomas N, van Eijk P, Zganjar C, Spalding M (2020) Harnessing big data to support the conservation and rehabilitation of mangrove forests globally. One Earth 2:429–443
WPSP (World Ports Sustainability Program) (2020) World ports sustainability report. https://sustainableworldports.org/wp-content/uploads/WORLD-PORTS-SUSTAINABILITY-REPORT-2020-FIN.pdf
WTTC (World Travel and Tourism Council) (2019) Economic impact. https://www.wttc.org/economic-impact

WWF (2018) Declaration of the sustainable blue economy finance principles. WWF, Gland. https://www.wwf.org.uk/sites/default/files/2018-03/Declaration%20of%20the%20Sustainable%20Blue%20Economy%20Finance%20Principles_Brochure%20Insert_2018.pdf
WWF (2019) Summary of the regional forum on building resilient Asian deltas. WWF, Gland. https://wwfeu.awsassets.panda.org/downloads/rad_conference_summary_brief_15nov2019_3.pdf
Young AP, Carilli JE (2019) Global distribution of coastal cliffs. Earth Surf Process Landf 44(6):1309–1316
Young IR, Ribal A (2019) Multiplatform evaluation of global trends in wind speed and wave height. Science 364:548–552
Zanuttigh B (2014) Editorial. Coast Eng 87:1–3. https://doi.org/10.1016/j.coastaleng.2014.01.003
Zhang Z, Provis JL, Reid A, Wang H (2014) Geopolymer foam concrete: an emerging material for sustainable construction. Construct Build Mater 56:113–127
Zheng M, Li XM, Sha J (2019) Comparison of sea surface wind field measured by HY-2A scatterometer and windsat in global oceans. J Oceanol Limnol 37(1):38–46

National Accounting for the Ocean and Ocean Economy

8

Highlights

- Organised information provides the power to make good decisions and justify them. National accounts contain and organise the information that describes our economies and helps decision-makers, and the public, understand near-term policy outcomes and long-term sustainability. However, currently only a small fraction of the information in national accounts is used because the focus has been overly narrow—producing a gross domestic product (GDP) metric.
- Planning for and managing a sustainable ocean economy requires tapping into the rich information that national accounts can provide. A system of national accounts can provide information in three critical areas to ocean economy decision-making: output or national means—a measure of production; outcomes or policy ends—a measure of real income and its distribution; and sustainability—indicated by changes in the national balance sheet.
- Many countries already produce an ocean GDP, but ocean GDP is usually the wrong metric for measuring the outcomes of ocean policy or the sustainability of the ocean economy. Efforts to calculate ocean GDP or measure the ocean economy with GDP will often be misleading because of fundamental features of GDP.
- This paper discusses a system of national accounts with multiple indicators and how they should be applied to the sustainable ocean economy. The paper emphasises the need to develop the underlying data structures to anticipate unintended consequences of decisions such as inequity and resource depletion.
- The paper proposes four principles of accounting for a sustainable ocean economy, including a set of Opportunities for Action for unlocking the information from national accounts needed to secure a sustainable ocean economy.
  - Assess policy options and decisions about the ocean and ocean economy in terms of their impacts on (1) real income and its distribution, (2) ocean production and (3) changes in ocean wealth, including ecosystems. Changes in ocean wealth are the most important indicator of sustainability.
  - Develop ocean accounts that build on the existing internationally agreed framework and standards for national accounting.
  - Avoid overreliance on GDP, which is not a sustainability indicator or measure of benefits to people from economic activity.
  - Lead or contribute to collaboration efforts to improve national ocean accounting systems, including global partnerships to share best practices and build capacity.
- The paper concludes that developing national accounts to guide economic development for the ocean is critical but not as daunting as it may seem. Many of the data already exist in national accounts, in government agencies or in scientific databases. The knowledge to build the connections also exists but is dispersed throughout government, academia, business and nongovernmental organisations (NGOs). Furthermore, most governments have already committed to many of these steps, with the gaps largely in implementation.

Originally published in:
Fenichel, E.P., B. Milligan, I. Porras et al. 2020. *National Accounting for the Ocean and Ocean Economy*. Washington, DC: World Resources Institute. Available online at https://www.oceanpanel.org/blue-papers/national-accounting-ocean-ocean-economy.
Reprint by Springer International Publishing (2023) with kind permission.
Published under license from the World Research Institute.

1 Introduction

Realising the goal of the High Level Panel for a Sustainable Ocean Economy (Ocean Panel) to catalyse the transition to a sustainable ocean economy depends on coordinating and

J. Lubchenco, P. M. Haugan (eds.), *The Blue Compendium*, https://doi.org/10.1007/978-3-031-16277-0_8

managing humanity's relationship with the ocean and the broader environment. This task requires organising information that currently is often disorganised, spread across multiple government agencies or in a few cases not yet available. National ocean accounts would provide countries with the information needed to guide ambitious and broad- based plans to develop ocean economies and to capitalise on marine opportunities (European Union Directorate-General of Maritime Affairs and Fisheries and Joint Research Centre 2018; Economist Intelligence Unit 2015), while protecting the ocean for generations to come in accordance with the Sustainable Development Goals, most notably SDG 14, 'Life below Water'. The 'blue-ing' of the ocean economy—or making the ocean economy sustainable—requires ensuring that the ocean continues to provide at least the current levels of opportunity; 'measuring the ocean economy gives a country a first-order understanding of the economic importance of the seas' (Economist Intelligence Unit 2015). The old adage goes that 'what gets measured, gets managed', or more accurately, that 'if you cannot measure it, you cannot improve it'. Sound decision- making requires organised information.

National ocean accounts provide a system to organise and process information to guide sustainable development. They can be characterised as a specific application and extension of the existing standardised System of National Accounts (European Commission et al. 2009) used by most countries, whose main objective is 'to provide a comprehensive conceptual and accounting framework that can be used to create a macroeconomic database suitable for analysing and evaluating the performance of an economy. The existence of such a database is a prerequisite for informed, rational policymaking and decision-taking.'

Box 8.1 Ocean or Marine to Blue

We use ocean, but marine could be used in its place throughout this report, to refer to large aquatic systems. This could include large lake systems. A sustainable ocean economy, or "blue economy" for short, accounts for biophysical processes and may include production outside of current national accounting boundary. A blue economy is one potential form of an ocean economy.

The ocean must be fully accounted for in this system to enable decision-makers around the world to balance between using the ocean today and conserving, restoring or enhancing it for the future to strengthen productivity, create jobs and reinforce food security and regional stability. Something as complex as the ocean economy cannot be managed by a single indicator. A complete set or 'sequence' of national ocean accounts provides three key high-level indicators: ocean product, changes in the ocean balance sheet and ocean income (Fig. 8.1):

1. **Ocean product** measures the 'outputs' of human efforts on the ocean to provide 'means' or 'inputs' into achieving other social and economic goals; monetary components of the ocean product account aggregate to ocean 'gross domestic product (GDP)' or 'net domestic product (NDP)'. In environmental accounting standards, physical accounts are also important.
2. **Change in the ocean balance sheet** provides a sustainability indicator. A stable or increasing balance sheet is necessary for sustainability. This is because the balance sheet reports current and future potential for the ocean to provide products and benefits. The ocean balance sheet includes 'natural capital' like live fish populations, coastal wetlands and seabed minerals, which fall under the heading of 'non- produced' assets, in addition to 'produced assets', such as port infrastructure. Changes in the balance sheet integrate physical and monetary changes.
3. **Ocean income** measures benefits to nationals (people of a nation) from the ocean, the 'ends' or 'outcomes' of policy; income accounts aggregate to net national income (NNI), though in practice national statistics offices usually produce gross national income (GNI). Importantly, income measures can be disaggregated to show the importance of the ocean for different segments of the population. Furthermore, income can include non-monetary types of income, though these are often expressed in monetary equivalents.

The most important thing world leaders can do is to request reports on all three indicators and discuss information on national income and changes to national balance sheets along with changes in GDP in public addresses and policy meetings. All three indicators are important for directing a sustainable economy, in the same way that altitude, airspeed and fuel in the tank are important for flying a plane (Matson et al. 2016). Certainly, the ocean economic system is at least as complex as an airplane. The primacy of GDP dates to the World War II crises and a need to measure means to carry out the war and rebuild after (Pilling 2018). If a plane is crashing, one would focus only on altitude for a short time, but flying from crisis to crisis is not the way to direct an economy. It is important to track relatively rapid changes in production. Changes in the balance sheet tell the story of sustainability, but balance sheets need to exist for a period of time before this information truly becomes useful. Few people would invest in a company without inspecting its balance sheet, yet countries' balance sheets are often an

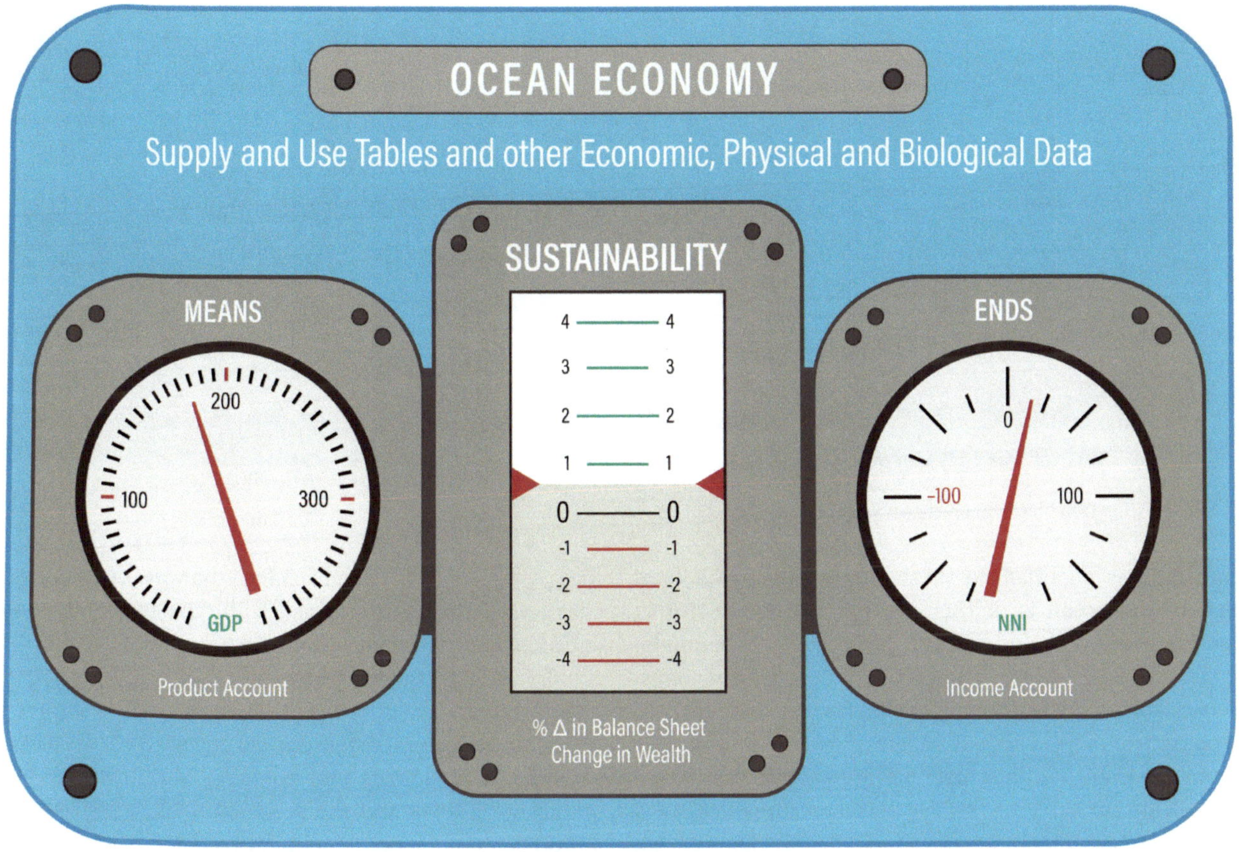

Fig. 8.1 National Accounts: a dashboard for assessing the ocean economy. (*Source: Jamie Ficker with input from the authors*)

afterthought, and few include ocean assets. This is despite agreement that national balance sheets should include produced and non-produced assets. Greater leadership in asking about the ocean in national accounts and on the national balance sheet can change this.

National accounts connect information about the processes of generating, producing, consuming, saving and building wealth within an information system. The strengths of the System of National Accounts lie in its data, as well as the data's organisation and consistency, which enable comparisons, especially through time within a country. While imperfect, national accounts are uniquely able to connect existing ocean-related data systems so they can provide information on economic activities and guide decisions. This is a logical place to use the information generated as part of the UN Decade of Ocean Science and similar initiatives. Ocean accounts can support coherent and holistic assessment and reporting on a wide range of social, economic and environmental conditions related to the ocean. National accounts for the ocean provide information in a form consistent with the needs of macroeconomic decision making to achieve sustainable development.

National ocean accounts provide three services. First, national accounts are a nation's information system. Aggregates such as GDP are representative of this information system, but GDP is just the tip of the iceberg. **It is important to avoid overreliance on GDP.** Second, national accounts provide a structured set of data about relationships among entities that provides the information needed to analyse policy, including ocean policy. For the ocean, many of these data exist, but they are currently distributed across different government agencies and international repositories. Third, the valuation component of national accounts facilitates analysis of policy trade-offs by organising ocean biological and physical data, many of which currently exist in disparate units, into a harmonised structure, evaluated in monetary terms with other economic data. Economic valuation helps answer important value related questions such as the following:

- How is the value stored in the ocean changing through time?
- What is the expected net present value associated with current and alternative management of the ocean?
- How is income generated in an ocean sector interconnected with other ocean and non-ocean income?
- How could changes in ocean policy impact tax revenue?

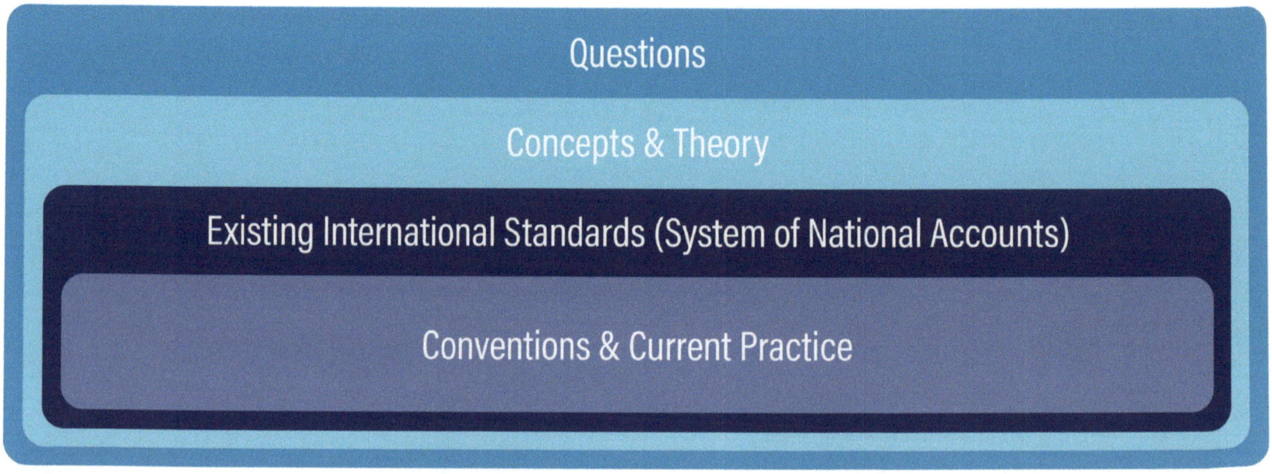

Fig. 8.2 Nesting from needs to practice. (*Source: Authors*)

However, it is important to stress that **the idea of a total value of the ocean is neither meaningful nor useful in practice**. Without the ocean, life on Earth would be fundamentally different. To paraphrase World Bank economist Michael Toman (1998), attempting the find the total value of the ocean would be 'a serious underestimate of infinity'.

Existing economic and national accounting theory and concepts inform many sustainable development policy questions. However, there are questions that economic and accounting theory do not answer. Moreover, the current existing international standards for national accounts—the 2008 System of National Accounts (SNA) and System of Environmental Economic Accounting (SEEA)—only partially address the concepts and theory. Some limitations result from economic questions that are not accounting questions, but others stem from design decisions in national accounts that merit revisiting. Furthermore, accounting practice often only partially implements the agreed international standards (Fig. 8.2). This Blue Paper identifies the gaps between these layers to provide Opportunities for Action.

Section 2 of the paper reviews questions about the ocean economy that national accounts can inform. It then turns to concepts and theory and how these are, or are not, addressed in the existing, internationally agreed System of National Accounts (European Commission et al. 2009). This helps identify formal changes needed to the SNA and SEEA to guide a sustainable ocean or blue economy. At the end of Sect. 2, we address some important ancillary issues such as the role of technology and concerns about equity. In Sect. 3, we examine the gaps between the formal SNA and current convention and practice in order to understand the need for leadership to modify norms and practices. Section 4 provides Opportunities for Action.

Now is the time to upgrade national accounting to provide information about the sustainability of economic activities. A focus on the ocean can lead the way. Economies are changing. Policy is concerned with outcomes and sustainability, not simply managing monetary inflation, and '21st century progress cannot be measured with 20th century statistics' (Agarwala 2019). On the one hand, bringing the environment, natural resources and ecosystems into national economic assessments and planning is critical for future human well-being and the persistence of natural systems, and all parts of the ocean are now impacted by human activities (Díaz et al. 2019). On the other hand, the SEEA is being revised, there is discussion of revising the internationally agreed system of national accounts to focus on sustainability (UN Stats 2019), the 'valuation of natural resources' is an active area of discussion within national accounting (UN Stats 2017), and the development and pilot testing of technical guidance for ocean accounting is underway (UN-ESCAP n.d.).

2 Questions, Concepts and Standards for Ocean Accounting

'What is the value of the ocean?' There are many reasons to ask this question, from concerns about specific ocean- related sectors to international commitments such as:

- The 2030 Agenda for Sustainable Development, including Sustainable Development Goal (SDG) 14 on 'Life below Water'.
- SDG Target 15.9, which calls for the integration of ecosystem and biodiversity values into national and local planning, development processes, poverty reduction strategies and accounts by 2020.
- SDG Target 17.19, which calls for efforts building on existing initiatives to develop measurements of progress

on sustainable development that *complement gross domestic product* and support statistical capacity building in developing countries by 2030.

However, this question it is far too imprecise for national accounts to answer. Notions of 'value' and 'ocean' are variable. National ocean accounting, in contrast, can answer specific questions like the following six, which we will return to throughout this paper:

1. How do industries, somehow connected to the ocean, create resources and products for use elsewhere? What jobs do these industries provide?
2. How do biological, chemical and physical ocean processes contribute to products for use elsewhere?
3. How does the ocean contribute to livelihoods and for whom?
4. How does the ocean provide welfare directly and for whom?
5. Is the ocean economy being developed sustainably?
6. How will a policy change affect aspects of the ocean economy? How will changes in the ocean affect the economy, or how will a use of the ocean in one location influence other industries and residents?

Questions 1–6 align with the dashboard in Fig. 8.1. Questions 1 and 2 relate to ocean production and GDP metrics, though they require some disaggregation, with the difference between these two questions reflecting the gap between the 'ocean economy' and the 'blue economy'. Questions 3 and 4 relate to national income and welfare metrics, which are closely connected, but they also require disaggregation in some cases. These are questions about 'development'. Question 5 relates to future opportunities—sustainable development, which is a question about the national ocean balance sheet. Addressing Question 6 requires understanding relationships within and between ocean processes and the economy, which depend on information in national supply-and-use tables and the broader national information system.

Fortunately, there is a full a set, or 'sequence', of national accounts rather than a single, all-encompassing account. There are sub-accounts and satellite accounts to help answer all of these questions. It is easiest to focus on the four pieces illustrated in Fig. 8.1: a product account, an income account, a balance sheet and an information structure derived from the economy itself. The actual system of national accounts has even more pieces that facilitate the reliable creation of these accounts. What is and what is not in measures differs somewhat from account type to account type. This, in theory, allows the different accounts to address different questions: questions about means, ends or outcomes, sustainability and forecasting the impact of changes. All but Question 6 are retrospective in nature. It is important to ask the questions of the right pieces of the system of national accounts, otherwise the answers can be misleading.

Focusing on questions such as the six above narrows the question 'What is the value of the ocean?' For example, one could mean, 'How important is the ocean to indigenous cultures?' or 'What opportunities are being lost by current ocean management?' The first question is beyond the scope of economic theory and national accounts. The second requires assumptions about alternative ocean management. National accounts will reflect changes in the economic sphere made to address non-economic concerns, such as cultural preservation. Furthermore, national accounting information is useful for assessing the forgone economic opportunities associated with policies that advance non-market policy priorities. The accounts will not tell decision-makers what the correct trade-off is, but national account data help leaders to identify trade-offs and make informed, purposeful, defensible decisions by holding a mirror to past decisions. They do not tell leaders what choice to make, any-more than an airplane dashboard tells a pilot what the destination should be.

Many of the issues of national ocean accounting bring broader national accounting and national sustainability assessment issues into focus. However, the fact the ocean often contains the physical boundaries between countries leads to a unique challenge for developing national ocean accounts. In the context of production, gross domestic product is different from gross national product (GNP). GDP uses the physical boundaries of a country, whereas GNP uses its people as the basis for calculations. This raises the question of how to account for activities on the high seas or other areas beyond national jurisdiction. In practice, this can lead to confusion about how to account for activities within countries' exclusive economic zones (EEZs). These conceptual and practical challenges extend across the sequence of accounts. Furthermore, areas beyond national jurisdiction influence ocean processes within national ocean waters (Popova et al. 2019). This makes it hard to know where national data begin and end, which suggests the need for international cooperation to develop data systems.

A seemingly natural place to start is by asking, 'What is the ocean economy?' and 'What needs to be included and what does not for a sustainable or blue ocean economy?' These questions, however, are rooted in an early twentieth century reporting paradigm in which computing sums and aggregates was a major bottleneck to statistical reporting. Modern technology is changing this. Computing power, algorithmic development and new data-management structures make it increasingly possible to align the exact sectorial boundary with the question being asked. Apportioning certain industries into and out of the ocean economy is challenging; for example, should a seaside coffee shop be included? This concern is secondary and addressed at the end of this section. A modern national accounting infrastructure makes it easy for decision-makers to ask if such an apportioning decision is material to a specific policy question. Developing the information system so that it is robust to these shifts addresses this challenge.

2.1 Production as Means

Simon Kuznets, Lillian Epstein and Elizabeth Jenks (1934) compiled one of the earliest modern national accounts in focusing on the 'industrial branches' of the U.S. economy during the Great Depression.[1] Given, the dominance of Keynesian monetary policy in the 1940s, the system focused on the balance between total supply and demand and investment in produced or man-made capital (Maler 1991). This effort evolved into modern GDP. Formally, the gross domestic product is the monetised value of new goods and services that could in principle be exchanged in a market—value added. In other words, it is the means a country has at its disposal at a point in time, but it says nothing about the ends. Product measures are used to understand a country's tax base, how sectors are changing in relation to other sectors in the economy, how productive certain sectors are and how demand for capital in various sectors might influence available money supply and inflation. These are all important questions, but aside from the ones about how the ocean sectors change in relationship to others, influencing potential tax bases, none of these questions are unique to the ocean or relevant to ocean policy.

Question 1 focuses on 'ocean-linked' market-like production, and Question 2 focuses on 'ocean-based' environmental contributions to market-like production. Neither is about outcomes for people or households. Colgan (2016) documents and discusses the confusion in various ocean accounting efforts between these two questions. Question 1 is the most common. It focuses on a group of firms or industries in an 'ocean' cluster and asks what means these firms generate. One can think of this as creating an ocean-affiliated industry class. The challenge is that often these firms are tangentially related to ocean processes (e.g. law firms provide maritime law services), and how much of any one industry to include is challenging. In practice, Question 1 is commonly used by industries for lobbying purposes—something along the lines of, 'We are an ocean industry, ocean industries are X percent of GDP, so the government should or should not enact a specific policy that will impact ocean industry.' The logic chain here is weak. First, means are not ends, and GDP measures the production or mobilisation of means. Public policy is concerned with means and ends. Second, ocean industries often are differentially influenced by policy. The purpose of lumping them together can, at times, be to inflate any one industry's perceived importance. It is often possible to analyse the affected industry directly and to examine changes to other influenced sectors. This errant lumping effect is even more perverse when nongovernmental organisations (NGOs) use aggregate ocean GDP to argue for the conservation of ocean biodiversity, which is not included in GDP at all. This use confuses the distinction between ocean based production that is dependent on environmental processes and mere ocean-linked production, which may or may not depend on environmental processes. Most ocean GDP calculations focus on the latter.

Question 2 is harder to answer and less commonly asked. It addresses how production based on the condition of the ocean ripples through the economy to create means during a certain period of time. Answering this question requires connecting the detailed information contained within national accounts with biophysical data.[2] Few national statistics offices or marine affairs offices have the capacity to do this on their own—they often must collaborate. Such collaboration requires removing barriers between agencies with ocean and biophysical data expertise and national statistics offices with expertise and access to often sensitive economic data. This leads to two interconnected challenges, beyond the principal challenge of increasing collaboration. Marine affairs agencies may have a regulatory role that access to private economic data (e.g. tax returns) could enhance, therefore there is a need to (1) develop confidentiality protocols and (2) establish clear institutional separation between measurement and reporting functions, on the one hand, and regulatory functions, on the other.

In practice, the product account records activities producing goods and services—this piece of the account provides information for GDP. The scope of product accounts is defined by a 'production boundary', which is 'understood to be a physical process, carried out under the responsibility, control and management of an institutional unit, in which labour and assets are used to transform inputs of goods and services into outputs of other goods and services' (European

[1] In 1947 the United Nations began chairing the design of a standardised system of national accounts to measure the total product and income of a nation over a specific period of time (the first release was in 1953). Subsequently, the system, adopted by virtually all UN member states, has been used (often erroneously) to make broader statements about social progress. The international system has been revised multiple times. The current version is the 2008 System of National Accounts (European Commission et al. 2009).

[2] These include biological, ecological, physical and chemical data.

Commission et al. 2009). This definition creates a challenge for ocean products. According to the SNA,

> *A necessary condition for an activity to be treated as productive is that it must be carried out under the instigation, control and responsibility of some institutional unit that exercises ownership rights over whatever is produced. For example, the natural growth of stocks of fish in the high seas not subject to international quotas is not counted as production: the process is not managed by any institutional unit and the fish do not belong to any institutional unit. On the other hand, the growth of fish in fish farms is treated as a process of production in much the same way that rearing livestock is a process of production.*

This illustrates the need for national accountants to pair with ocean specialists to understand relevant governance structures that often bring marine resources within the scope of the production boundary. Most national waters have an institutional unit that regulates and 'exercises ownership rights over' marine resources, putting these resources inside the production boundary (Obst et al. 2019). For areas beyond national jurisdiction, the activities of regional fisheries management organisations like the Inter-American Tropical Tuna Commission arguably move relevant marine resources within the production and asset boundary.

An example of how ocean product enters a national account is helpful. The value added of harvested ocean resources, like wild fish, is measured using resource rent (Table 8.1). The basic value of production is equivalent to total revenues generated by fishers. The intermediate uses are the values of goods and services consumed or used up as inputs in production, such as fuel costs. Taxes on products are regarded as a part of the value that is created by the industry when the resource is extracted, while a product-specific subsidy is considered part of the costs of extracting the resource. A product specific tax paid by the specific resource industry is added to the resource rent, while product-specific subsidies, including price supports, are subtracted. Industry-specific taxes and subsidies are not included in the calculation of the resource rent because they are a transfer of the resource rent between the government and the industry and do not affect the bottom-line value of the resource rent.

Table 8.1 Calculation of the realised resource rent

Sign	Term
+	Basic value of production
−	Intermediate uses
+	Taxes on products
−	Subsidies on products
=	Gross product
−	Non-industry-specific taxes
+	Non-industry-specific subsidies
−	Compensation of employees
−	Return on fixed capital
−	Capital consumption
=	Resource rent of the sector

Source: Authors

Singular focus on the production account can be misleading. Repeated illegal fishing is formally within the scope of production accounts (European Commission et al. 2009). This is because illegally caught fish provide additional means in the current period, and if the illegal fishing is ongoing national accountants understand this as if the government, acting as a trustee, were voluntarily (implicitly) giving up fish to the unlawful fishers. Irregular piracy is not included in the production boundary because piracy does not create new means but shifts them involuntarily. However, defensive government expenditures preventing piracy are production in the current period. If increases in piracy increase government expenditure, then piracy indirectly adds to the product account. Increasing piracy or illegal fishing are not policy goals. These are just a few cautionary examples of why measures of means are not equivalent to measures of ends or outcomes.

2.2 Income as Ends

Economists—such as Dasgupta (2001); Jorgenson (2018); Kuznets (1973); Nordhaus and Tobin (1972); Solow (1993); Stiglitz et al. (2010); and Weitzman (1976)—have long understood the shortcomings of GDP for measuring human welfare or the ends or outcomes of policy. GDP is merely production, a 'means'. It is not an 'end' (Nordhaus and Tobin 1972), such as consumption or benefits to people, or sustainability (Solow 1993). Income is often associated with livelihoods. Livelihoods support household consumption (see Question 3 above). This is closer to the outcome goals of modern policy.

The standard starting place for considering the role of national accounts in measuring well-being is Nobel laureates William Nordhaus and James Tobin's (1972) 'measure of economic welfare', which responded to Kuznet's earlier calls to complete the consumption or well-being portion of national accounts (Jorgenson 2018). The renewed efforts by Joseph Stiglitz, Amartya Sen and Jean-Pierre Fitoussi (2010) to expand national accounts to provide welfare measures, 'beyond GDP', in their report to former French president Nicholas Sarkozy, are summarised by Marc Fleurbaey and Didier Blanchet (2013), both members of the Stiglitz commission. However, to our knowledge none of these efforts explicitly focused on ocean well-being, income, consumption or expenditure. If country leadership wants to link the ocean to well-being, then it is important to (1) support beyond GDP efforts and (2) prioritise their construction in a way that enables a disaggregation focused on ocean-related activities. Doing so may require more individual and time-use surveys along with expansion of the income or expenditure boundary.

Most scholars (e.g., Heal 1998) and national accountants define income following John Hicks's (1939) income concept, which applied to the ocean would define 'blue income' as the maximum amount a society can take from the ocean 'and still be as well off at the end of the week as at the beginning'. This includes 'nonmonetary' benefits to being 'well off' (Krutilla 1967). Question 4 differs from Question 3 by acknowledging that services not acquired through market or market-like production matter. The ocean contributes many such services, such as leisure. Most economic theory related to national measures of income accommodates these services (Fleurbaey and Blanchet 2013).[3] This creates a challenge in accounting theory because it means that the 'boundary' of the income account is broader than the production account, yet the two are expected to balance. Nevertheless, it is possible to create balancing items to address this challenge. National accountants already face this challenge when calculating gross national income in comparison to gross domestic product.

Continuing with the wild harvest fishery example from the end of Sect. 2.1, when calculating compensation of employees, it is a goal to use wage rates that reflect the alternative value of the fishers. This analysis uses the likely wage of fishers if they had to find a job elsewhere at the start of their working career, such as the average wage rate on the mainland. Clearly, this creates a challenge in subsistence settings, and it imposes a set of highly restrictive assumptions about labour mobility. The compensation of employees is calculated as the number of hours worked times this wage rate. Vessel owner income is included as the number of hours worked multiplied by the employee wage rate. This illustrates the current shortcomings of national income accounts. Payroll taxes and other finely resolved data are used by countries that have those data.

To capture the contribution of the ocean to national welfare or income requires including market and non-market benefits to people. Yet the divide between the market and the non-market is often the boundary for national accounts, leaving out economic activities, such as home production and flows from environmental public goods, that are often thought of as services. Insofar as these activities represent substitutes for market activities, their inclusion is necessary. Nordhaus (2006) writes, 'Probably the most difficult issue in design of augmented accounts is, where to draw the border.' Expanding the income boundary is important in enabling national ocean accounts to capture many of the services that lead people to care about the ocean. If the boundary is adjusted, then various methods to estimate the implicit income from non-market ocean services exist (Freeman 2003; Phaneuf and Requate 2017). It is more complicated to apply these methods than to use market data. Furthermore, the data analysis is often highly localised, and transferring results from one region to another is challenging (Boyle et al. 2010). Finally, the current version of income accounts is not a true measure of social welfare or economic well-being because they do not address distributional concerns (Fleurbaey 2009). Nevertheless, completing these accounts, with a broader boundary, would represent a substantial advance, and new technologies are enabling disaggregation. Dale Jorgenson (2018) argues that much greater information on distribution is needed for income, consumption and expenditure accounts. This is true for non-market services like many important leisure opportunities provided by the ocean and in cases where the ocean provides substantial subsistence opportunities. What is in and out of the account imposes a binary equity weighting. The ability of dashboards to enable disaggregation goes a long way towards addressing, or at least enabling informed discussion of, distributional concerns of 'fair allocation' of benefits associated with the ocean.

Leaders interested in policy outcomes, or 'ends', should be more interested in net national income (NNI) than GDP. NNI calculations require attention to the valuation of often hard-to-value assets, and NNI over a period of time is expected to balance with changes in national wealth reflected on a balance sheet.

2.3 Sustainable Development and the Balance Sheet

Production provides means, income is ends, but a sequence of balance sheets provides information to assess whether development is sustainable (Arrow et al. 2004; Hamilton and Clemens 1999; Maler 1991) and whether ocean development is sustainable or 'blue'.[4]

The balance sheet shows a country's wealth—the present value of the country's current and future economic opportunities conditioned on the current or most likely future institutional arrangements. Changes in national balance sheets are

[3]The 2008 SNA admits to the arbitrary nature of including household produced goods but excluding household produced services. This is done to support traditional monetary and fiscal policy concerns.

[4]It is often suggested to reformulate non-declining wealth as non-declining per capita wealth. However, it is unclear that per capita is the "correct" normalisation (Jorgenson 2018), for two reasons. First, some ocean services are non-rival, and all individuals experience the same service level. Therefore, the more people, the more service, the more wealth, and in such a case we should not divide by the total population. Second, per capita normalisation carries a certain distributional element that implies that distribution of access to ocean capital takes a certain form, but it is possible to increase per capita measures while reducing the most common (median) experience.

expected to balance with net changes in net national income. Ocean balance sheets reflect current and future economic opportunities afforded by the ocean. **Changes in balance sheets provide the sustainable development report card**, that is, 'meeting the needs of current generations without compromising the ability of future generations to meet their needs' (World Commission on Environment and Development 1987).[5] Kirk Hamilton and Michael Clemens (1999) put it succinctly, 'Achieving sustainable development necessarily entails creating and maintaining wealth.' A physical account complements the balance sheet that shows the current stock of assets.

Infrastructure and environmental assets, including natural resources, belong on national balance sheets (European Commission et al. 2009; Hulten 2006). This includes marine capital. Port infrastructure falls under the heading of produced assets. Other ocean assets from live fish populations to coral reefs to deep-water oil reserves are non-produced assets. The inclusion of natural capital in national accounts is not a novel or controversial idea. The idea of natural capital was well established by the early 1900s, long before the term natural capital was used. Irving Fisher (1906) used an ocean asset, Newfoundland fish stocks, as the first example of capital in his seminal 1906 text. U.S. president Theodore Roosevelt (1910) spoke of natural resources as assets as early as 1910. The current system of national accounts makes frequent mention of natural resources as capital (European Commission et al. 2009). Many Nobel laureates in economics, including William Nordhaus, Joseph Stiglitz, Robert Solow, James Tobin, Amartya Sen and Kenneth Arrow, have advocated greater inclusion of the natural environment in national accounts. Comprehensively completing the balance sheet is currently being piloted as 'wealth accounting' by the World Bank and UN Environment (Lange et al. 2018; Managi and Kumar 2018). The key innovation in these comprehensive wealth measures is that human and natural assets are given equal footing with produced assets. Recent versions of these reports include some ocean assets. The indicators for a sustainable ocean future will be contained in an ocean account balance sheet. Canada, Australia and other countries are already producing wealth reports, but we are unaware of any that are well developed for ocean sectors.

The boundary of the balance sheet is one of the most challenging pieces of national accounts (Hulten 2006). The 2008 SNA (European Commission et al. 2009) states that 'natural resources such as land, mineral deposits, fuel reserves, uncultivated forests or other vegetation and wild animals [fish] are included in the balance sheets provided that institutional units are exercising effective ownership rights over them, that is, are actually in a position to be able to benefit from them.'

Most countries exercise effective ownership over their marine assets, by virtue of their assertion of national maritime zones, and related management activities. It is telling, however, that in the current system of national account documents (European Commission et al. 2009, §10.167), 'ocean' only appears in the mention of 'certain naturally occurring resources, however, maybe such that it is not feasible to establish ownership over them, for example air, or oceans.' This suggests that informal conventions within national accounting require amendment to improve their consistency with prevailing realities of ocean governance.

Producing comprehensive balance sheets, including non-produced assets, is a first step to verifiable sustainable development. In the ocean economy, non- produced assets are especially important given the role of non-produced ocean assets in growing food, storing minerals, sequestering carbon and generating many other services. It is also not possible to calculate net measures or income or production without measuring changes in stored wealth.

An important challenge to creating balance sheets is the valorisation of ocean assets. National accounts primarily focus on the consumption of fixed capital, which is the amount of an asset used to produce a good or service (Obst et al. 2019). For produced capital, consumption of fixed capital is often computed using market prices or the perpetual inventory model, and consumption of fixed capital does not include depletion or degradation of nonproduced capital (European Commission et al. 2009).

In the fishery example from Sect. 2.1, what capital makes it to the balance sheet? The focus is on port infrastructure, the fishing vessels and other 'fixed capital'—not the fish population. When there are no market prices, the perpetual inventory method works by adding capital each year based on the cost of new investments (e.g. spending on boats or port maintenance), and capital is subtracted based on an estimate of the lifetime and depreciation profile. In practice for fishery capital, the lifetime is set to 20 years, and the depreciation profile is geometric, with a 10% annual rate. This is assumed to reflect the wear of this kind of capital. The claim is that the analysis takes a long-term perspective and essentially asks what the return on the capital would have been if it were not invested in the fishery sector in the first place. However, this is inconsistent with assessing the current state of the world. For something like fishing capital, this clearly ignores complementarities with the non-produced capital, which is the

[5]Nobel laureate Kenneth Arrow et al. (2004) formalise the definition of sustainable development as requiring constant or increasing opportunities, where the concept of wealth has evolved to be a measure of future opportunities. Wealth itself is the net present value of income.

fish stock. Ignoring marine non-produced capital can lead to errors in valuing marine-produced capital, such as port infrastructure. New Zealand has introduced a novel solution by creating a market place for the rights to use non-produced capital, that is, fish stocks, known as individual tradable quotas or catchshares. These programs were developed to align fisher incentives with regulatory goal (Grafton et al. 2000), but they create the added benefits of enabling the living fish population to be tracked on the national balance sheet (Hammond 2005).

The 2008 SNA provides little guidance for valuing non-produced assets, but methods exist. Fenichel et al. (2018) and Fenichel and Obst (2019) provide guidance for valuing non-produced assets in the form of natural capital, which can be applied to ocean non-produced assets. Yun et al. (2017a) provide a software package, called capital asset pricing for nature (capn) to facilitate implementation of these techniques. These techniques use observed behaviours but do not assume a constant flow of services. The approach accounts for economic and ecological feedbacks in the valuation process. The core challenge is to group strongly interacting pieces of the ocean ecosystem and economy to capture the most important feedbacks. Yun et al. (2017b) apply these techniques to develop balance sheet components for the Baltic Sea cod-herring-sprat fishery based on Polish data. When all data are not available, simplifying assumptions may be used that are as reasonable as those used in the perpetual inventory model.

Charles Hulten (2006) makes the real problem clear: 'When it comes to capital, however, it is more a question of what to do than how to do it.'

2.4 Analysing Policy and Marine Planning

Headline indicators, the gauges in Fig. 8.1, are retrospective measures. They can provide lessons from the past, but 'past performance is no guarantee of future results'. National accounts organise data to enable analyses that can inform future decision-making. This is the main goal of national accounts (European Commission et al. 2009). Question 6 above is about the future. A national ocean account can provide information to develop economy-wide models, including economy- wide models with fine spatial resolution. This is because national accounts are the system for processing information to coordinate national activities, provide business forecasts and evaluate policy outcomes.

National accounts provide a commonly agreed set of facts for shared understanding and decision-making. These accounts are built on extensive data, with high resolution, potentially down to a beachside ice cream parlour's tax reports. New technologies and reporting paradigms are making data increasingly easy to access and disaggregate to answer questions about specific sectors of the economy—including the ocean economy.

The three gauges in Fig. 8.1 report the condition of the national (ocean) economy. The detailed data are stored in many structures, chief among them a set of supply- and-use tables. These tables provide the material to produce the aggregated, sector-level input-output tables commonly used in economic analysis and projects.

These are critical for understanding the interconnections within an economy and connecting the science of ocean processes with the traditional economy. Furthermore, supply-and-use tables are regularly produced at fine spatial scales. Indeed, in many countries the limits of publicly available spatial disaggregation are set by ethical and confidentiality concerns rather than data resolution.

The supply-and-use tables record the production and demand structure of an economy by describing the goods and services brought in through domestic production or through imports from outside the economy. The tables describe how those goods and services are used, such as through intermediate consumption, final consumption at the household or government level, gross capital formation or exports (Department of Economic and Social Affairs 2018; Kazemier et al. 2012). The tables provide the foundation for developing input-output (IO) tables. IO tables and supply-and-use tables may be in physical or monetary units. Supply-and-use tables allow analysts to verify that the underlying data used in national aggregate calculations are consistent, complete and balanced.

IO tables aggregate goods and services to industry or sector levels and track value flows between and within industries or sectors for intermediate consumption and final expenditure. Therefore, the IO tables are used in all sorts of economic analyses and forecasts. The supply-and-use tables can be expanded to include services and consumption currently outside of the income or production boundary to get a better handle on true national income and on ocean income and can ultimately be linked to similar structures for environ mental processes taking place in the ocean. Natural production from the ocean could be treated as an economic sector.

Analysis to support sustainable ocean economic policy requires reducing the barriers between experts and data generators from different agencies. **Connecting assets with supply-and-use tables will make it easy for analysts to analyse how economic activity changes the ocean and how changes in the ocean influence economic activity.** Scientists already build models of the marine environment, such as Atlantis and EcoPath/EcoSim (Audzijonyte et al. 2018; Collie et al. 2016; Steenbeek et al. 2016), that use structures similar to IO tables. Establishing a central accounts structure could enable macro-environmental-economic policy analysis based on an integrated platform that links data and models and brings the environment into standard macroeconomic modelling frameworks (Finnoff and Tschirhart 2003). Currently, the expertise for much of this work sits outside national statistics offices, while those with biophysical knowledge struggle to connect their data and understanding to macroeconomic models.

Connecting ocean and economic experts is imperative because the ocean economy is strongly influenced by the performance of non-produced assets. Many natural assets may directly interact and influence each other's value, just like firms interact in an economy where automobile manufacturers influence the market value of suppliers—predators affect the value of prey. The value of services or the value of assets depends on substitutes and complements to those services or those assets. The ocean generates many services, but these services are produced through interconnected processes, and some of these services are regenerating assets. These connections can enable substitutions or create complementarities. Interactions can be physical or biological (such as ecosystem interactions between species), technical or market-driven. In an era of globalisation, markets connect the incentives for using various components of the ocean worldwide (e.g. fisheries, tourism). For example, coastal resources enable swimming and recreational fishing, which may be complements in producing tourism services. Sites that enable both may be of greater value than the sum of sites that only enable swimming or recreational fishing. In another example, unharvested prey fish biomass may seem of little value but actually have great value in supporting a harvested predator fish (Yun et al. 2017b). At the same time, one species of prey fish may be a good substitute for another species of prey fish, so the value of that prey fish species in a system with many species may be lower than if that prey fish species were the only prey source. This means that changes in the value of ecosystems may not be the sum of the changes in the value of the parts if the parts are measured independently. Measuring the parts independently may lead to double counting or undercounting. It is important to account for interactions, which often depend on policy decisions and institutions as well as ecology and natural processes.

2.5 Satellite Accounts

The term *satellite account* is used for separate accounts of interest that are not part of the central structure of the System of National Accounts. Most satellite accounts are rearrangements of items already included in a central account. They do not influence national aggregates. However, some satellite accounts allow items to be treated differently, such as with a different boundary than the central accounts.

One important system of satellite accounting is the System of Environmental Economic Accounting (SEEA), which is coordinated by the UN Statistical Division. The SEEA Central Framework (SEEA-CF) is an internationally agreed standard for accounting for environmental assets and their supply to and use in the economy. It provides guidance for services from non-produced assets, such as fisheries, in greater detail than the System of National Accounts. The SEEA-CF provides the specific guidance on fisheries, forests and agriculture, which reflects the SNA guidance with additional details for natural resources. The SEEA also has a system of Experimental Ecosystem Accounting (SEEA-EEA) that is currently being revised, with the goal of establishing an international standard by 2021. The experimental ecosystem accounts focus on the biophysical condition of ecosystems and interactions among non-produced assets. The SEEA-EEA will likely also provide guidance on ecosystem services that can be counted as income beyond the current income boundary, though this guidance is still in development. The revisions working groups have produced working papers, which are available on the SEEA webpage, https://seea.un.org/.

A second important set of satellite accounts consists of satellite ocean accounts developed by individual countries with guidance from the Organisation for Economic Co-operation and Development (OECD), EuroStat or in coordination with the UN Economic and Social Commission for Asia and the Pacific (UN-ESCAP). The UN-ESCAP program is also associated with ongoing efforts, coordinated by the UN Statistics Division, to maintain and develop the SEEA. Some countries also produce satellite transportation and tourism accounts with ocean-related components or coverage.

2.6 Related Issues

Before assessing the current state of national ocean accounting, we should examine some common issues that we have not yet addressed. These include boundaries and existing discussions of the ocean economy, data and technology, and equity concerns.

2.6.1 Conceptual and Spatial Boundaries

Ocean accounts need to address three types of boundaries: (1) accounting boundaries, which determine what types of services to include and which we have already discussed, (2) the marine economy boundary and (3) spatial boundaries within the marine system. This section focuses on the second and third boundary types.

Of the six established and five emerging blue sectors outlined by the European Union Directorate-General of Maritime Affairs and Fisheries and Joint Research Centre (2018) and by the OECD (OECD 2016), three major sectors—extraction of marine living resources, coastal tourism and biotechnology—are likely impacted, in some cases severely, by changes in the ocean's biological condition. These sectors depend critically on the biological natural capital of marine ecosystems. All sectors may be impacted by physical changes that alter access to the ocean by changing the distribution of storms, waves, wind and so on. Therefore, all sectors depend on physical natural capital, but it is less clear that the physical capital is ocean capital as opposed to climate.[6] It is likely that all sectors influence changes in the biological and physical condition of the ocean, which ultimately influence the accounting price of critical forms of ocean capital. Finally, 'marine and coastal' protection is often included as a sector of the blue economy. But this sector would be better thought of as investments or maintenance of ocean natural capital, which is how the current System of Environmental Economic Accounting treats this sector.

From shipbuilding to biotechnology to clean energy, the ocean spurs innovation and encourages human capital formation. Of course, the ocean is one of many contributors. More work is needed to partition the incremental contributions of the ocean to knowledge generation. Ocean accounting initiatives should be integrated with accounts that cover broader sections of the economy. Experiences with individual tradable permits for fisheries suggest there are regulatory structures that increase the value of natural capital (Fenichel and Abbott 2014) while increasing the value of human capital through safety improvements (Birkenbach et al. 2017; Pfeiffer and Gratz 2016). Such property rights may be important in marine mining and other extractive industries as well (Libecap 1994).

The issue of national boundaries, made acute by the ocean, is a somewhat unique issue for national accounts. Currently there is no institution maintaining a balance sheet for ocean areas beyond national jurisdiction. Many countries do not even include assets in their own EEZs on their balance sheets. Another concern is vessels operating in the territorial waters of other countries. The production and income are usually attributed to the vessel's home country, while any impacts to the balance sheet would occur to the geographic location. This could lead to the changes in wealth not balancing with NNP.

A global emerging issue is marine spatial planning.[7] National account data are useful to marine spatial planning in ways that parallel regional development modelling—a common use of national account data. National account data enable input-output, integrated assessment and computable general equilibrium modelling. These sorts of models have a role in marine spatial planning.

2.6.2 Data and the Digital Revolution

The key strength of national accounts is their organisation of data. The digital revolution is changing the way people interact with data, and this is especially relevant for national ocean accounts.[8]

National accountants already use 'big data' and detailed business statistics from multiple sources, and they are experimenting with remote sensing. Aggregates are often built from very fine scale measurements, such as business receipts. This is important because environmental data will likely also not come from a single source. However, the national account reporting paradigm, with a heavy focus on headline GDP, is based on 1930s technology, now in the early stages of a radical update involving online digital dashboarding that makes headline numbers less essential (Fig. 8.3).

Marine conditions and activities are often 'far away' from observers, but remote sensing and in situ techniques are making it easier to observe the ocean. There is substantial untapped potential to monitor and measure the biophysical condition of the ocean through 'earth observation' (Ramirez-Reyes et al. 2019), and technology exists for these data to flow directly into national ocean accounts. Earth observation is defined as the union of diverse data sources, including from satellite, airborne, in situ platforms and citizen observatories (GEO 2015), for improved monitoring and forecasting of Earth's physical, chemical and biological conditions. The Group on Earth Observations (GEO) provides physical, chemical and biological information at increasingly fine scales, including at a few metres and hourly. Earth observation provides rapid, repeated and long-term synoptic observations that provide a platform for a nested ocean observing framework at global, basin, regional and local scales.

The Framework for Ocean Observing (Lindstrom et al. 2012), implemented under the auspices of the Intergovernmental Oceanographic Commission of the United Nations Educational, Scientific and Cultural Organization and coordinated by the Global Ocean Observing System, seeks to meet the need for ocean data that support governance, management, science and other ocean uses. It

[6]The Blue Paper on "The Expected Impacts of Climate Change on the Ocean Economy" discusses the strong link between the ocean and climate.

[7]The Blue Paper on "Integrated Ocean Management" focuses on marine spatial planning.

[8]The Blue Paper on "Technology, Data and New Models for Sustainably Managing Ocean Resources" focuses on data and emerging technologies.

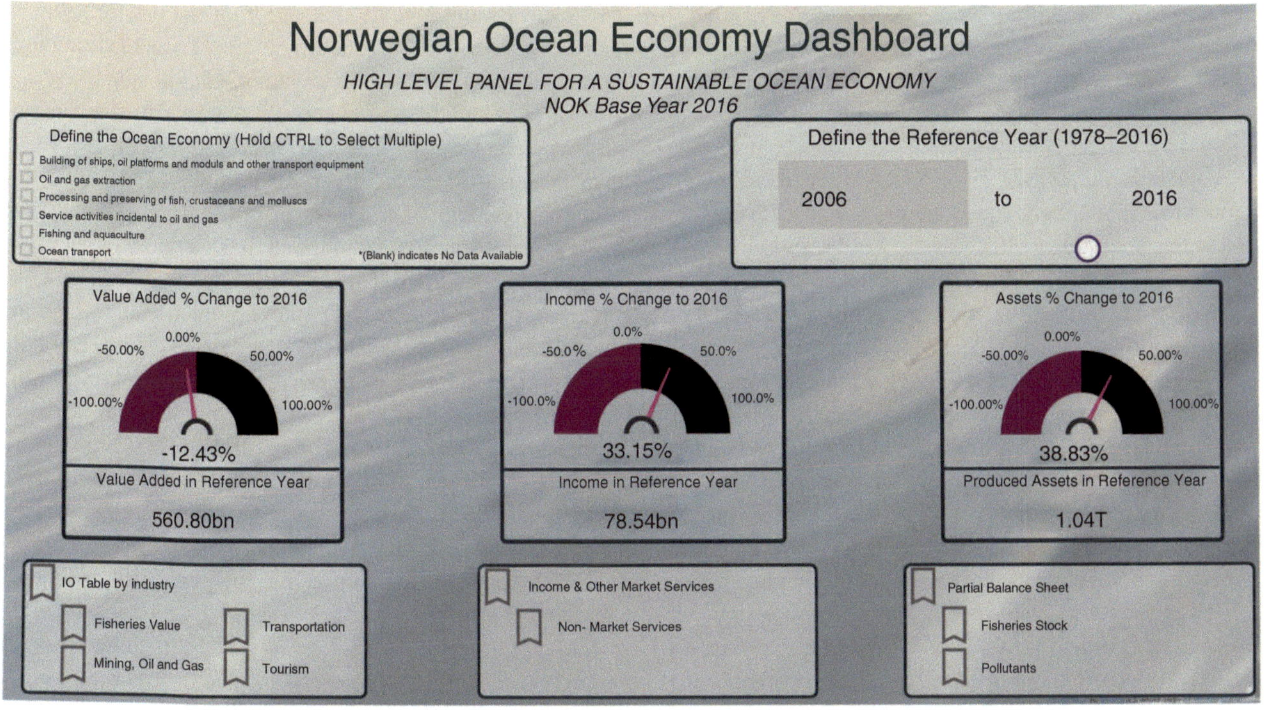

Fig. 8.3 Example of a Live Interactive Digital Dashboard for Norway. (*Source: Working version at* https://environment.yale.edu/data-science/norwegian-ocean-economy-dashboard/)

proposes the coordination and integration of routine and sustained observations of physical, biogeochemical, geological and biological essential ocean variables (Bojinski et al. 2014).

Through its marine and coastal (GEO Blue Planet) and biodiversity observatory network initiatives, the international Group on Earth Observations is working to improve the availability, access and use of ocean-related Earth observations. This includes work on a framework for a set of essential biodiversity variables for use in monitoring programs to understand patterns and changes in Earth's biodiversity (Navarro et al. 2017; Pereira et al. 2013) as well as on ecosystem essential ocean variables, a set of observable ecological quantities that contribute to the assessment of the ocean ecosystem (Miloslavich et al. 2018).

These efforts categorise specific ocean parameters that should be monitored continuously in order to identify key processes and determine the sustainability of the ecosystem as a whole, thereby addressing the challenge of evaluating the ocean's status in a synergistic way (Muller-Karger et al. 2018).

The biophysical data present two main challenges. First, expertise for working with Earth observations (which include ocean and coastal data) often resides outside national statistical offices, though some national statistics offices do possess this expertise (e.g. Canada's). It is imperative that national accountants collaborate with Earth observation experts to acquire physical data of ocean flows and measures of non-produced ocean assets. Coverage can vary, and some countries lack capacity for accessing these data altogether. Many habitats, including the deep sea, ocean trenches, ice- bound waters, methane seeps and even coral reefs, remain poorly studied at the global scale. Costello et al. (2010) shows that geographic gaps in biodiversity data are particularly acute for many parts of the global ocean, including coastal areas of the Indian Ocean, the southern and eastern Mediterranean Sea, polar seas and much of the South American coastal ocean.

There is a critical need to inventory data to quantify natural stocks, audit the data's usability for accounting for non-produced assets and identify priority data gaps. Data gaps need to be articulated with clear measurable and feasible observable units; such measures should be prioritised over derived measures like biodiversity. Data quality needs to be checked against academic data sets such as 'The Sea around Us' (http://www.seaaroundus.org/) (Pauly and Zeller 2017), and discrepancies should be documented, explained or remedied. Physical measures need to be linkable to human transactions and decisions for valuation. As part of the UN System of Environmental Economic Accounting, substantial progress has been made in defining the extent of ecosystem and other natural assets within basic spatial units (BSUs).[9] But data must go beyond spatial delineation to track condition through time. For example, the spatial extent of the Great Barrier Reef has not changed much over the last 30 years, but the reef's biological condition has.

Second, the amount of Earth observation data leads to computation challenges. The Copernicus Earth Observation Programme Sentinel satellites of the European Union (EU) produce approximately 20 terabytes of data per day (Esch et al. 2018). The geospatial community has developed solutions that 'bring the user to the data instead of the data to the user'. Technological advances in cloud technologies, the development of data cube technologies, the availability of analysis-ready datasets and the development of web- based platforms providing access to these services make this possible. These solutions may not work for national statistics offices that may need to match ocean data with confidential microeconomic data. National accountants and statisticians, led by economy and finance ministers in cooperation with transportation, marine affairs and fishery ministers, need to negotiate a platform that serves the needs of ocean accounts.

The digital revolution is aiding the understanding of human activity on the ocean and the implicit income that people gain or lose as the ocean changes. Many vessels are tracked by satellite. The International Maritime Organisation monitors maritime traffic through a regulation requiring the Automated Identification System (AIS) on all ships over 300 gross tonnes on international voyages, on cargo ships over 500 gross tonnes and all passenger ships irrespective of size. AIS reports the ship's identity, type, position, course, speed, navigational status and other safety-related information—automatically to appropriately equipped shore stations, other ships and aircraft. Vessels engaged in fisheries activities also report their locations. The vessel monitoring system (VMS) is a satellite-based monitoring system that provides data to the fisheries authorities on the location, course and speed of vessels (https://globalfishingwatch.org/). AIS and VMS data are key elements for measuring maritime transport.

Human transactions increasingly involve a digital footprint, and these data are increasingly useful for imputing the non-market income received from the environment, such as social media posts, administrative time-use surveys, voluntary recording on recreation, activity tracking and digital consumption of complement and substitute market goods. A number of difficult ethical concerns must be addressed with these data, but national accountants already grapple with these issues for business reporting data. All of these data could greatly improve determination of the precise value of non-market services provided by the ocean and nature more broadly. Digital transactions are already improving the precision of market data, and in some countries national accountants and economists are already working with these sorts of data for measuring the ocean economy. National statistics offices increasingly invest in the infrastructure and algorithms to support information from the digital world and lower barriers to bring in data from other data-collection agencies without loss of resolution. In the context of the ocean, this means that agencies must find ways to incorporate biophysical data and associate shore-based transactions with the marine physical environment. It is also important that national statistics agencies draw on the expertise of marine sector experts to understand the complex institutional arrangements and assignments of 'economic ownership', which often differs from 'physical ownership' in the marine context.

New technology makes national account data more accessible and more useful for policy analysis. For instance, an ocean proto-account for Norway can be displayed as an interactive dashboard (Fig. 8.3), and the United States hosts an interactive ocean proto-account (https://coast.noaa.gov/digitalcoast/tools/enow.html). Such dashboarding aligns with the recommendations of Stiglitz et al. (2010) for going beyond GDP. New interactive dashboarding technology makes decisionmakers less dependent on specific aggregates like gross or net ocean product and enables them to drill down quickly to indicators of interest.

2.6.3 Equity and National Accounts

Equity and inclusion are cornerstones of the sustainable development agenda, and distributional concerns are a limitation of only focusing on national income aggregates—though when used with care these can be an important piece of addressing equity (Fleurbaey 2009; Jorgenson 2018; Jorgenson and Slesnick 2014).[10] 'Equity' refers to the distribution of benefits and costs of resources (distributional justice). Conservation and changes in wealth are central to intergenerational equity (Dasgupta 2007; Solow 1974).[11] Intragenerational equity is also important (Adler 2013, Hart 1974; Sikor 2013; Stiglitz et al. 2009), and the ocean can contribute to poverty alleviation, especially for small island developing states and coastal least developed countries, providing food, jobs, livelihoods and cultural spaces (World Bank and UN-DESA 2017). It is a reasonable aspiration for ocean accounting to support 'equity measures', while being agnostic as to the 'correct' distribution or measure.

[9] Ideally, BSUs covering marine and coastal locations should designate a three-dimensional volume including the ocean, the seabed and subsoil, combined with a shoreline vector delineating the ocean from land. Conditions that can be assigned to and accounted for within BSUs include, for example, acidification (pH), eutrophication (BOD), temperature (°C), and plastics (T), and the abundance of various species.

[10] Fair sharing of the ocean is addressed in the Blue Paper on "How to Distribute the Benefits of the Ocean Equitably".

[11] Measuring sustainability with balance sheets requires considering access to assets and going beyond per capita measures (Jorgenson 2018).

National accounts are denominated in national monetary currencies and thus depend on the distribution of money income and wealth. While not reporting directly on equity, national accounts can provide some data to generate equity indicators and help countries meet international sustainable development reporting commitments. In order to do so, it is important to be able to disaggregate and apply politically chosen equity weights. Microeconomic and survey data are also important (Jorgenson 2018). The boundary of the account limits potential equity weights. Anything left out of the account is implicitly weighted at zero.

A clear limitation for policy analysis concerned with distributional outcomes is the ability to disaggregate reporting. Current national accounts vary from country to country in this respect, but technology is making it easier and easier to disaggregate data. The technical challenge is to build the data structure in a way that it can be disaggregated and recalled quickly. However, perhaps the greater change will ultimately be balancing ethical issues that emerge from the identifiability of fine-scale disaggregation (e.g. linked to data protection) with the ethical imperatives of using disaggregation to address equity concerns.

2.7 Aspirations for the System of Ocean Accounts

The ocean plays a major role in market and non-market services. The ocean unites and divides countries, and it links people through a common heritage and regulated climate. It also brings people together through trade and travel. A substantial number of services from the ocean rely on production underpinned by natural capital. In principle, much of this natural capital should already be on national balance sheets and within national accounts. The asset and production boundaries of national accounts may require adjustment to justify adding other stocks of natural capital to the balance sheet. **However, of first-order importance is generating balance sheets with the produced and non-produced ocean assets currently within the production and asset boundaries. This is not being done, but it would provide an immediate gauge of the ocean economy's sustainability.**

Furthermore, many ocean-provided services are not traded on the market. Therefore, they are missing from production and income accounts. Account boundary adjustments are required to provide clearer measures of the 'ends' in terms of economic well-being. It is **important to work towards a broader income boundary that includes broader ocean services**, such as household-produced services, leisure services and carbon sequestration and storage. The money boundary is a subset of this broader boundary.

Finally, the accounts must be more than summary statistics. Analysts must be able to get into the details of the data. Integrating the economic and biophysical data into a single platform will make it easy for economic analysts to consider the role of the ocean and perhaps help physical scientists better understand the economic trade-offs with biophysically based recommendations. Turning attention towards the data structures and away from the aggregates is imperative to address environmental concerns while answering forward looking policy and business questions.

3 Current State of Accounts for the Ocean

It is not enough to review concepts and investigate official guidance for national accounting for the ocean. We must also look at what countries are doing with respect to ocean accounting. This section surveys the current state of ocean accounting and relates practices to frameworks for ocean accounting and the suite of actors implementing frameworks. The goal is to identify gaps between the formal structure and practice. Assessing the current state of the accounts helps (a) show what is currently feasible, (b) identify important gaps where alternatives may exist or where resources are required and (c) identify areas where novel approaches to ocean accounting are needed.

There are three main components to the current practice of ocean accounting:

1. The set of internationally agreed frameworks for national accounting systems.
2. Countries that engage with these frameworks to provide national accounting information related to the ocean.
3. Programs and outside actors who link, filter or otherwise engage and support the set of existing frameworks and/or countries that are producing these national accounts.

3.1 Internationally Agreed Frameworks

A growing range of ocean accounting initiatives, frameworks and studies exists. It is useful to think about their articulation with the System of National Accounts (European Commission et al. 2009). Most countries' national accounts comply with this system. Relevant frameworks developed through intergovernmental systems include the following:

- The System of Environmental Economic Accounting (SEEA) Central Framework, developed through a UN Statistics Division process.
- The SEEA Agriculture, Forestry and Fisheries (AFF), which applies and expands on the SEEA–CF. This system is developed through the Food and Agriculture Organization of the United Nations (FAO).
- The SEEA Experimental Ecosystem Accounting (SEEA-EEA), which incorporates physical indicators of ecosystem conditions and services, as well as measures of ecosystem value. This system is developed through a UN Statistics Division process.
- The UN Technical Guidance on Ocean Accounting for Sustainable Development, which applies the SNA and SEEA, with additional guidance focusing on accounting for ocean governance and social circumstances within an integrated Ocean Accounts Framework. This guidance is developed through the UN Economic and Social Commission for Asia and the Pacific (UN-ESCAP), in collaboration with several governments and other actors.
- The Integrated Maritime Policy Database, a proposal/pilot refinement of ESA 2010 guidance that is a European-tailored version of the SNA.

The broadest of the international accounting frameworks is the System of National Accounts (European Commission et al. 2009), developed through the Inter-secretariat Working Group on National Accounts. The SNA aims to provide a framework for creating a sequence of national accounts that is 'comprehensive', 'consistent' and 'integrated'. Much of Sect. 2 focused on this system, and the SNA is clearly relevant for the ocean economy. Sections 6.136–42 address the output produced by sectors that operate in part within the marine economy. Many of the same challenges addressed by the SNA—for example, those having to do with home production—apply in the marine economy.

The SEEA-CF informs monetary measurement of economic activity related to the environment as well as physical measurement of environmental stocks and flows. The SEEA-CF complements and expands the SNA. Physical asset accounts are a key way in which the SEEA-CF expands the boundaries defined by the SNA. The SEEA-AFF provides more specific standards for physical and monetary accounting and measurement of fish and other aquatic products within the SNA. SEEA-CF adopts the notion of countries, firms or asset owners as economic units. The SEEA-EEA, in contrast, takes an ecosystem-centric perspective focusing on spatial units grounded in ecological rather than administrative boundaries (Chow 2016; FAO n.d.). The UN Technical Guidance on Ocean Accounting for Sustainable Development, developed through UN-ESCAP, focuses on the application of the SNA and SEEA in marine and coastal contexts, providing methods and approaches for developing satellite accounts for the ocean environment and economy that allow for spatial disaggregation. It also provides experimental guidance on accounting for contextual factors such as ocean-related social circumstances and current modes of governance. Ocean systems, given variation in depth, currents and boundary types, present specific challenges to the notion of an ecosystem-based spatial unit for a given terrestrial system. The UN-ESCAP guidance includes ecological and technological detail needed to define ocean spatial units and physical measurement standards tailored to measure marine assets. The community of practice surrounding UNESCAP ocean accounts includes Australia, Canada, China, Fiji, Indonesia, Malaysia, Portugal, Samoa, Thailand, Vanuatu and Vietnam. These accounts can be constructed at the national or subnational level. Together, the SNA, SEEA-CF and SEEA-AFF guidance constitute the internationally agreed framework applicable to ocean accounting. The SEEA-EEA and UN-ESCAP provide more detailed guidance produced through the same mechanism, but they have not yet been adopted as international standards.

3.2 Implementation of Ocean Accounting

The conceptual design of national accounts suggests that it should be possible to extract substantial information about the state of the ocean and the ocean economy. Greater detail requires more complex national accounts. It is important to develop a consistent framework for categorising economic activities to prevent double counting of flows from economic activities. Double counting and undercounting are surprisingly easy traps because of the many ways countries can group these activities. Increasing the level of detail in national accounts exposes important linkages across industries and early indicators of economic health.

The Ocean Panel member countries are diverse and clearly invested in the ocean, but not all have high-profile national ocean accounts. Therefore, they constitute a useful sample to examine the state of national ocean accounts. We focus on these 14 countries in this section and in the following section review other selected high profile efforts. A survey of the Ocean Panel countries' treatment of the ocean in their national accounts provides a representative, if optimistic, view of the state of national ocean accounting. We provide specific examples, so that practitioners can find examples of steps being implemented. We investigate four questions:

1. Do member countries explicitly account for the ocean economy? If so, to what extent?
2. Which accounting tools—production, income, balance sheets and supply-and-use tables—are produced? Are the

	Production Accounts Disaggregated to Ocean Economy	Physical Production Accounts Only at Level of Ocean Economy	Physical/Monetary Production Accounts Somewhat Disaggregated	Contribution of Ocean Economy to Production Unable to be Reported
Fisheries	13	1	0	0
Tourism	10	1	2	1
Transportation	3	4	1	6
Mining, Oil and Petroleum	2	1	0	11

Fig. 8.4 Ocean sector aggregates for ocean panel countries. (*Source: Authors' tally based on publicly available national count data*)

accounts usable to inform services, sustainability, and conduct economic analysis or are only production accounts produced?
3. How are non-produced ocean assets (ocean natural capital) treated in the accounts?
4. Is the current level of national account detail sufficient to produce a set of satellite accounts and aggregate statistics for the ocean economy?

We were able to find ocean-related data in national accounts for all 14 Ocean Panel member countries.[12] National ocean accounting is not starting from zero in any of these countries. Nevertheless, Ocean Panel member countries' national ocean accounting data vary greatly and are only comparable at a broad level. A variety of specialised reporting is already evident.

For example, Fiji's national accounts maintain detailed reporting on the *bêche-de-mer* (sea cucumber) industry and have an extensive structure for the harvest of biological organisms from the ocean. The same economic activity in other countries' national accounts in principle may only exist in an aggregate of the entire agriculture, fishing and forestry sector. Harmonising ocean accounting would facilitate inter-country comparison; more important, it would also facilitate capacity building and knowledge sharing.

3.2.1 Product and Income Accounts

Macroeconomic *production aggregates* exist across three of the four principal ocean-related sectors: (1) fisheries and (2) mining/oil and gas and (3) transportation or commerce (Fig. 8.4). Many, but not all countries report aggregates for tourism or hospitality, however, these estimates are often provided in a satellite account given the potential for double counting. Furthermore, the existing data are reorganisations of data from the countries' main sequence of national accounts and do not extend the income boundary. Therefore, accounting for the ocean does not change countries' headline GDP. Countries with explicit ocean accounts include additional sectors in their 'ocean accounts' beyond the four we focus on. These may include all coastal activities, maritime law, research on the ocean, restoration activities, ocean governance, bio-prospecting, and the list goes on. Such accounts aim to answer Question 1 in the initial set of 'value of the ocean' questions.

Some countries use spatial data on reporting location to partition marine-related coastal tourism and hospitality. Some countries, such as Portugal and Canada, go a step further and provide dedicated satellite accounts for the ocean (see INE n.d.; and Fisheries and Oceans Canada n.d.-a). Other countries, such as Fiji and Indonesia, have expressed interest in or are in the process of developing ocean satellite accounts. The production accounts are relatively complete, sufficient to provide marine GDP, if the boundary of the marine economy can be defined and data can be disaggregated. Marine GDP can be, and often is, created by reorganising items contained in standard national accounts, and many countries already produce a marine GDP.[13] The statistical offices for countries such as

[12] Our analysis is based on data we could locate online, so gaps in the analysis may reflect that the data are not easily located through the internet rather than that they are missing.

[13] See volume 2, issue 2, of the Journal of Ocean and Coastal Economics for country-specific experiences.

Norway have computed a statistic that is essentially marine GDP as a onetime exercise. In the marine affairs agencies of other countries, such as Canada, the Department of Fisheries and Oceans generates this sort of calculation. These marine GDP aggregates do not account for depreciation or degradation of marine produced assets (e.g. port infrastructure) or non-produced assets (e.g. fish stocks), because GDP calculations do not consider capital depreciation or degradation of any kind. Existing marine GDP statistics leave out changes in ocean capital because of the design of GDP and not necessarily because of a lack of information or an effort to conceal or ignore these changes. GDP is the wrong tool for assessing the sustainability of the ocean economy.

There are supply-and-use tables for the included sectors products, but these seldom connect to underlying ocean processes. It is not clear how ocean processes influence tourism, but ocean processes likely influence fisheries, and physical ocean processes may influence transportation.

With respect to the ocean, most national accounting effort goes into the national production account. Marine GDP does not provide insights into the well-being people derive from the ocean or ocean sustainability. This is insufficient for the accounts to inform how ocean policy is or is not contributing to well-being or whether or not ocean policy ensures a sustainable ocean future. Including the non-market contribution to welfare, which would generally sit in the income, consumption or expenditure account, is important for understanding well-being, even if it is not part of 'economic production'.

The Australian Bureau of Statistics (ABS) has presented headline indicators for the environment, characterising trends in terrestrial biodiversity, the atmosphere and land use for 2006. Yet, for the marine and coastal regions, the ABS (2007) simply states that 'these regions are also important to Australian society and the economy. Many of the ways in which we use the ocean, beaches and estuaries can affect the quality of the ocean's water and the diversity of life within it.' However, no headline indicator is reported for this sector through the program's reporting, which ends in 2012. There are national accounts efforts to track wages in industries that can be identified as ocean-related, such as in Portugal. Other sectors, such as tourism, can be difficult to disaggregate. Furthermore, it is often impossible to tell what fraction of the wage is attributable to various attributes of the ocean. For example, even if beachfront resources were their own category, it would be impossible to tell how much the ocean's biological capital was contributing to wages or revenue without greater survey data that exist in national accounts.

Furthermore, the few attempts to measure income beyond the current boundary (Jorgenson 2018) have not focused on the ocean or have taken place outside the purview of formal national accounts, and often not at a scale sufficient for national accounting.

3.2.2 Balance Sheets, Natural Capital and Supply-and-Use Tables

Balance sheets are essential for national ocean accounting. All 14 countries make an official national balance sheet available online. Six countries include non-produced assets, and Japan and Mexico include non-produced assets that are potentially non-produced ocean assets (Fig. 8.5) (OECD 2019). However, a number of other countries reference programs that might involve natural capital accounting or mea-

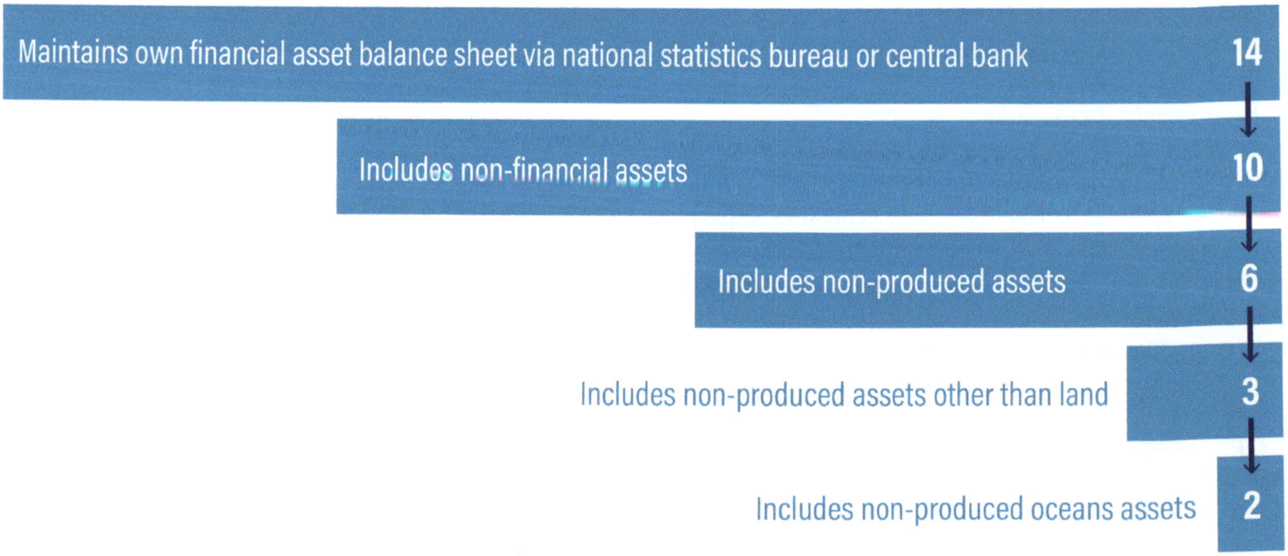

Fig. 8.5 Balance sheets among ocean panel countries. (*Note: Descriptions are not sufficiently precise to classify fully all non-produced assets. There may be some misclassification, but the pattern appears robust to misclassification. Source: Authors tally based on publicly available national count data*)

sure non-produced assets (e.g. Jamaica and Canada). However, these efforts do not appear to make it to the official statistics on the national balance sheet.

Half of the 14 Ocean Panel countries have some form of physical account paired with their monetary accounts—physical production sheets—for these aggregated sectors, but we found no complete balance sheets with ocean assets. For example, Kenyan fisheries accounts track physical and monetary flows that are disaggregated by freshwater fisheries, marine fisheries and aquaculture. Ideally, physical accounts would be paired with indicators of the quality or condition of the assets on balance sheets. These are typically not included in national accounts but are critical for natural capital accounts and are part of the UN System of Environmental Economic Accounting guidance.

Indicators, such as those for environmental quality, can provide a missing link between physical and monetary accounts. Ocean acidity, for example, can impact oyster biomineralisation, leading to smaller and therefore less valuable oysters (Fitzer et al. 2018). Indeed, these sorts of linkages are similar to those described above about how the condition of a prey fish stock may raise or lower the value of the predator stock. The 2008 System of National Accounts is clear that prices that come from markets include these 'general equilibrium' interactions (European Commission et al. 2009). Price- influencing interactions are also important for 'non- produced' natural asset valuation. Qualitative changes matter in physical accounts of resources and on the asset balance sheet.

Environmental processes, much like income distributions, are often not characterised adequately by a single number. Namibia is the only member country of the Ocean Panel to link environmental indicators such as sea surface temperature and plankton abundance explicitly with fisheries. The ocean transport sector influences regional ocean acidification (Hassellöv et al. 2013). However, for many countries the data for the transportation or commerce sector in supply-and-use (or input-output tables) are not disaggregated by transport mode. As a result, the share of commercial activity that occurs via maritime transportation is not available in the account. The link between ocean transport, regional ocean acidification and seafood product value (i.e. shellfish), is lost due to low data resolution. The greater the aggregation in the supply-and-use tables, the less useful they are for economic modelling and forecasting and the harder it is to link economic activity and biophysical changes in the ocean.

The ability to disaggregate monetary accounts, physical accounts and environmental indicators is important for characterising the overall state of the ocean economy. For some countries or sectors, maritime activity likely comprises such a large share of the sector that disaggregation is unnecessary. For example, mining and oil extraction in Norway is almost exclusively marine.

3.2.3 Satellite Accounts

Ten of the 14 countries have a tourism satellite account. In contrast, only two Ocean Panel countries, Portugal and Canada, have dedicated ocean satellite accounts. Only Portugal's is currently produced by the national statistics office. Portugal's efforts often are promoted as a national ocean accounting example, so it is useful to discuss them in a bit more detail.

Understanding what share of the tourism sector's contribution to the economy is due to ocean-related products and services depends, in part, on which ocean related product and service values are considered. Portugal's ocean account considers 65 different products and services across nine groups. The recreation, sports, culture and tourism group captures a range of activities including recreational and sport boating, cultural events related to the sea, coastal tourism (including state spending on advertising) and imputed rents from second homes on the coast.

Portugal creates a complete set of production, expenditure and income accounts and is able to produce a set of balanced national aggregates for the ocean economy. Portugal includes standard wage and employment data and household consumption information as part of the satellite account. Portugal includes non-produced assets on its national balance sheet, but these do not appear to include non-produced ocean assets.

Portugal's ocean account is one of the most advanced in the world, and is the model for many of the ocean accounting efforts underway. However, its headline numbers address Question 1, and it is less clear that additional effort has been made to address the other types of questions. Of the 65 sectors included, many are only tangentially related to the ocean. For example, shares of 'computer programming, consultancy and related services', 'legal and accounting services' and 'leather and related products' are included in the ocean account. These are industries that can be linked to the ocean but are hardly production from the ocean—Question 2. Using Portugal's 2013 numbers, we find that only 8% of the value added of 'sea products' seems to be clearly from the ocean, with another 34% possibly being from the ocean, as opposed to related industries. Linking industries to the ocean can mislead about the benefits from the ocean. For example, insurance is included in Portugal's ocean account. This is presumably insurance against ocean storms. It seems that, if anything, this is a cost of the ocean, not a benefit. This highlights the need to be clear about the question and enable disaggregation.

3.3 Country Implementation Globally

A growing number of countries are implementing national ocean accounting, with a focus on product accounts (Colgan 2016). Some individual countries or blocs of countries have further developed or expanded the frameworks for their own use. Eurostat's European System of Accounts 2010, for example, provided much of the methodological guidance for Portugal's Satellite Account of the Sea. However, for some countries interest or capability drives a wedge between framework and implementation. Other countries' capabilities and interest enable them to go beyond the international frameworks, providing experience, lessons-learned and guidance for future refinement of frameworks. These countries strive for backwards compatibility with internationally agreed frameworks, as in the case of China (Wang 2016). The guidance in international frameworks is seldom sufficient to address every scenario and provide complete production accounts for a nation's ocean economy, let alone asset balance sheets. In the case of asset balance sheets, it is likely that the lack of availability of guidance and data to create such balance sheets (which do not currently exist for any country's account of its marine economy) is a 'chicken or the egg' problem. Nevertheless, revision of internationally agreed frameworks is critical to avoid issues of interoperability of national accounts and the challenges of double counting (De Maio and Irwin 2016). Separation of physical and economic data also poses a challenge for balance sheets.

Consider the ocean accounting efforts of the United States, China, New Zealand, Portugal, the Netherlands and Australia. The National Oceanic and Atmospheric Administration's *Report on the U.S. Ocean and Great Lakes Economy* divides ocean productivity into six sectors: marine construction, offshore mineral extraction, tourism and recreation, living resources, ship and boat building, and marine transportation (NOAA 2019). Each of these sectors includes direct and indirect ocean production, where indirect contributions can be inferred using tools like input-output tables. Large gains are achievable using data already collected for national accounting or other national statistical purposes. The NOAA Economics: National Ocean Watch explorer represents a reorganisation of employment data from the U.S. Bureau of Labor Statistics that, coupled with imputed sectorial employment to GDP ratios, provides a first-order glimpse of contributions to the ocean economy by sector at the county (sub-state) level within the United States. Simultaneously, the U.S. Bureau of Labor Statistics reports wages in most marine sectors.

Canada divides ocean production into direct, indirect and induced ocean production (Fisheries and Oceans Canada n.d.-b). Direct and indirect production flows may depend on produced and/or non-produced assets.

China's Gross Ocean Product uses 12 major sectors to measure the gross value added of China's ocean economy.[14] Nearly half of this index is coastal tourism, just under 20% is transportation, and marine fisheries account for just under 15%.

New Zealand is a leader producing national balance sheets, but Stats NZ (2018) states, 'The SNA08 [2008 System of National Accounts] conceptually includes a wide range of natural resources beyond those included in New Zealand's accumulation accounts. The omitted natural resources need to be quantified and valued.' New Zealand focuses on land as a non-produced asset, like many other countries. However, New Zealand produces a satellite physical and monetary fish stock account, enabled by New Zealand's broad adoption of individual trade quotas (ITQs) for managing fisheries, which creates a market for the fish asset. Stats NZ claims this is an added benefit of ITQ management (Hammond 2005).

'Natural Capital Accounts for the North Sea: The Physical SEEA EEA Accounts', a pilot project in the Netherlands, represents an advancement towards paired physical and monetary asset accounts. Major headway in this project was achieved by defining boundaries with respect to economic and ecosystem activities and collating and repurposing existing data from Statistics Netherlands and external data sources. The conclusion of the pilot study was that it is feasible to pursue natural capital accounts for marine ecosystems and that it is possible to complete much of the work using extant data sources.

Determining which industries are and are not included in the ocean sector is not the challenge for the methodology employed in China, Canada, New Zealand and most all other countries that produce these aggregate measures, which define industries in a way that can be linked to the International Standard Industrial Classification of All Economic Activities (Wang 2016). However, some countries, such as the United States, attempt to partition at a scale of sub-classification schemes. It is clear that not all countries are making the same decisions, which is why within-country comparisons through time are more salient than cross-country comparisons. It is also clear that the aggregates do

[14] China's Gross Environmental Product index alters the production account boundaries, whereas the Gross Ocean Product is a conventional satellite account produced by China's Ministry of Natural Resources.

not provide sufficient information to evaluate questions of sustainability, but ideally measures of ocean production are comparable through time within a country.[15]

In practice, national ocean accounting remains a somewhat bespoke process. Reporting systems and frameworks have also been developed to either more easily use existing data sources at the country level, address areas of national interest or tackle unique country-specific accounting challenges.

3.4 Supporting Programs and Other Actors

Supporting programs and actors is a broad group of entities only connected by their interaction with at least two of the following: the ocean, countries and the formal internationally agreed frameworks for national accounting. Some groups exist to support their member countries and the suite of methodologies and other tools available to them (e.g. OECD, EU). Others have specialised agendas, such as the World Wildlife Foundation or the Great Barrier Reef Foundation. There are groups that aim to share information and expertise around national accounting among business and practitioners in a 'bottom-up' approach, such as the Capitals Coalition or the London Group on Environmental Accounting. There are finance organisations or country supporting partnerships aimed at developing technical capacity, such as the World Bank's Wealth Accounting and the Valuation of Ecosystem Services (WAVES), the UN Development Program's Biodiversity Finance Initiative (BIOFIN), the European Union's Mapping and Assessment of Ecosystems and Their Services and the partnership of the UN Development Programme, the Economics of Ecosystems and Biodiversity and the Convention on Biological Diversity (UNEP TEEB CBD).

These organisations can function as filters or mediators through which countries engage with central international accounting frameworks. Finance and capacity-building organisations like WAVES (https://www.wavespartnership.org/) facilitate development of institutional capacity. Regional supporting organisations like Partnerships in Environmental Management for the Seas of East Asia (PEMSEA) can provide management solutions and partnerships for promoting sustainability among groups of countries.

[15]Cross-country comparisons are challenging, as Jorgenson (2018) explains. Market exchange rates can be misleading, which leads the World Bank to produce purchasing power parity conversions. However, it is unclear how purchasing power parity can be developed when some goods or income-generating services are not market-based.

Many of these organisations solicit academics, other NGOs and outside consultants to produce fact-finding, momentum-building reports or both that address gaps or areas of interest in national accounting practices for the marine sector. Alongside these reports exist evaluations, methodologies and estimates produced in peer-reviewed academic literature. Below, we address a few of these reports and studies produced by noncountry organisations as they pertain to production accounts, balance sheets and income accounts. This is not an exhaustive survey, but it represents the use and misuse of national accounting for the ocean. We provide illustrations of three types of reports, though the categories are fluid: motivating reports, illustrative reports and policy reports. Decision-makers interested in the performance of the ocean or blue economy should

- make sure the results align with the question the decision-maker is asking;
- prefer a repeated series of reports or reports that document changes and enable disaggregation; and
- assess the agenda of the report's producer and if the claims align with the statistics and data used.

National statistics offices should pay attention to these reports to understand the information demands, especially demands that national statistics offices might be failing to satisfy. Jorgenson (2018) suggests this is a substantial problem once one moves beyond production questions.

3.4.1 Motivating Reports

Most people have seen at least one of the motivating reports. The thesis of these reports is that the ocean is important, the ocean provides opportunity or the ocean is valuable. These reports at times misuse national accounting because of the belief that GDP or some economic number implies importance. Others, like the UN working group, the World Bank and other stakeholders' high-resolution Blue Economy report, highlight the importance of ocean resources for least developed countries and small island developing states without promoting a single metric or calculating an aggregate value. The Blue Economy report characterises the 'blue economy' by assembling a diverse reference list of sectors and constructing a case for their importance. It advocates expanding the boundary of the ocean economy beyond fisheries to include the extraction of marine non-renewables, commerce and trade, and indirect contributions to economic activities. The report provides a framework for mapping *ocean-related* activities to sectors and then to major drivers of demand and growth in these sectors. The **headline policy recommendation of the Blue Economy report urges countries to accurately value the contribution of natural ocean capital to welfare to better guide policy decisions**

and trade-offs. This means focusing on net income and balance sheets, not GDP. Many countries have taken up this challenge.

Other motivating reports attempt to produce oneoff 'large number' monetary valuations of ocean environments to attract attention. For example, the 2015 World Wildlife Fund (WWF) report *Reviving the Ocean Economy* argues that 'the future of humanity depends on oceans' healthy living conditions', drawing attention to the crucial point that ocean biodiversity contributes to human well-being (WWF 2015). The report presents an indicator of the annual value generated from the global ocean economy or a 'Gross Marine Product' of ~US $2.5 trillion and a total ocean asset value of US $24 trillion.

Another example is the 2017 Deloitte report on the value of the *Great Barrier Reef, at What Price: The Economic, Social and Icon Value of the Great Barrier Reef*, which provides a headline indicator that the Great Barrier Reef contributes A$6.4 billion to Australian GDP.

These reports should be treated as little more than 'calls to action'. First, because they tend to be one-offs, they provide little useful information about trends, though some reports do look at changes over time. More critical is the risk of their undercutting the message that ocean products and assets are uniquely important, and sometimes irreplaceable. In *Reviving the Ocean Economy* and similar studies, one of the chief methods for computing the total value of gross marine product is rescaling the gross value added from the ocean economy available from G20 countries. Despite countries' use of different sectorial boundaries, it is clear that for most countries shipping, tourism and recreation, and other activities only tangentially related to the biological condition of the ocean contributed the most to the gross value added figures (NOAA 2019). Most of the asset value is transportation, coastal capital and other forms of produced capital. These are important, but do not speak directly to the importance of the biology or 'living conditions' of the ocean.

Assessing the sustainability of the ocean economy requires monitoring changes in the asset values or change in the balance sheet over time, coupled with assessment of the role of biodiversity in net national income, or development of a biological non-produced asset account tracked through time.

The numbers presented in the 2017 Deloitte report imply a gross value added of A$18,354 per square kilometre, given that the Great Barrier Reef covers approximately 350,000 square kilometres. This is almost equivalent to Australian agricultural GDP per square kilometre when dividing official statistics by the total agricultural land area (ABS 2018; Trading Economics n.d.). The report also provides an estimated asset value of A$56 billion, an asset value of A$1606/hectare, right around the median price of agriculture land in Australia (ABC 2018). This would make the value of the Great Barrier Reef equivalent to that of the tenth-largest Australian public company, right behind BHP Billiton. While not trivially small, these numbers are shockingly small in the context of something as unique and irreplaceable as the Great Barrier Reef. Of course, if asset management can be improved, then the value will be lower than expected under optimal or improved management (Fenichel and Abbott 2014).

3.4.2 Illustrative Reports

A second set of reports are illustrative reports. Good examples of these are the World Bank's *The Changing Wealth of Nations* and UN Environment's *Inclusive Wealth Reports* (Lange et al. 2018; Managi and Kumar 2018). These reports illustrate how comprehensive national balance sheets could be used to assess sustainability. However, they do not focus on ocean or blue assets. Moreover, the data used to produce these reports enter in a relatively aggregate form. National statistics offices should be able to do much better. Most academic studies fall into this category as well, where the goal is to illustrate methods rather than to provide regularly produced, definitive numbers.

3.4.3 Policy Reports

The third set are policy reports. These reports would ideally take ocean accounts as their starting point, but historically they have had to generate national-accounts-style data that were not readily available. A good example is *The Sunken Billions: The Economic Justification for Fisheries Reform*, jointly published in 2009 by the World Bank and FAO (Arnason et al. 2009). The report focuses on the contribution of wild marine fisheries to the global economy and the economic production lost due to overfishing and depleted stocks by comparing potential and realised economic benefits. The report models the world's fisheries as a single stock and uses global aggregate data to estimate the production deficit at around US $50 billion per year (in 2004$). This number is similar in magnitude to the 1992 FAO study *Marine Fisheries and the Law of the Sea: A Decade of Change*, which estimated the aggregate production deficit incurred by the world's fishing fleet at $22 billion (in 1989$). Adding in the capital cost, this early FAO report estimated the deficit at $54 billion per year. To put these numbers in context, the estimated gross revenues of the global marine fish harvest in 1989 were $70 billion. The methodology employed a single-stock model to estimate the deficit, leaving many questions open about the spatial heterogeneity of the operating deficit.

The follow-up, *Sunken Billions Revisited*, followed the same approach as its predecessor but delved further into the regional analysis to provide more disaggregated impacts and policy recommendations. This study found an $83 billion production deficit for the year 2012. **Sunken Billions is a benefit-cost analysis that highlights the potential gains of a policy change. This is not part of most national accounts.** However, robust national accounts would be a good starting point for this sort of analysis.

4 Guidance for a Path Forward

Developing national accounts to guide economic development for the ocean is critical but less daunting than it may seem. Many of the data already exist in national accounts, in government agencies or in scientific databases; the knowledge to build the connections exists, but it is dispersed throughout government, academia, business and NGOs. As we have seen, many countries already produce reports that are or are nearly marine GDP. These reports, however, make clear that GDP is about means, not about ends or sustainability. The ocean's biophysical assets are valuable. But marine GDP calculations do not and cannot measure this. Even as a measure of income, the dominance of shipping and coastal development in these sums could obscure the mostly unmeasured non-market income components. Academics and international organisations, such as the World Bank, do not have access to the fine-level data that most countries' statistics offices can access. Therefore, **country level statistics offices need to develop a sequence of accounts reflecting** Fig. 8.1, then partition out the ocean sections with reporting tools that enable adjustments to the ocean economy boundary. Changes in the country's ocean balance sheet are the country's 'ocean wealth index' for assessing the sustainability of blue development.

Experience implementing the System of Environmental Economic Accounting (United Nations et al. 2014) shows that even incomplete accounts can inform policy. For example, countries have started water accounts with available data on municipally supplied water. Subsequent revisions have added estimates of household, industrial and other use. Incomplete accounts highlight critical areas of data gaps and provide bounds useful for making policy decisions. It is likely that in the near future many more data sources will be available to populate ocean accounts. Indeed, this is a clear case of needing to plan for the data of the future rather than plan around existing data or the data of the past. With this in mind, we offer crosscutting Opportunities for Action for developing national ocean accounts.

4.1 Four Principles of Accounting for a Sustainable Ocean Economy

1. Ask multiple questions and expect multiple answers, especially questions about income and sustainability (balance sheets) in addition to production. This means that the impacts of policies and decisions about the ocean economy should be evaluated based on at least three indicators: income, production and ocean wealth.
2. Build on the existing structure of the System of National Accounts and System for Environmental-Economic Accounting so that ocean accounts are compatible with existing national accounts, and with international statistical standards.
3. Avoid an overreliance on GDP, which is not an indicator of either sustainability or the societal ends of economic activity. Do not use a hammer when you need a wrench.
4. Lead or contribute to collaboration efforts to improve national ocean accounting systems, including global partnerships to share best practices and build capacity. Such efforts will likely involve creating new integrated data management systems for ocean accounting and other purposes.

4.2 Crosscutting Opportunities for Action for Developing Credible Ocean Accounting

Eleven general crosscutting Opportunities for Action support the implementation of these principles, with additional detailed Opportunities for Action in the areas of physical measurement and valuation:

1. National statistical offices, in partnership with marine agencies, need to develop a complete sequence of national ocean accounts: product, income, balance sheets and supply-and-use tables. This should be achievable by 2025. It is important to aggregate these to a few key headline indicators (Fig. 8.1) and be able to disaggregate to examine specific sectors and constituencies nested within the ocean economy. The sequences of accounts provide a commonly agreed set of facts about the ocean and its relationship to human benefits. This is the starting point for ocean policy discussions.
2. Countries need to be able to track their own progress through time. Cross-country comparisons are of secondary importance and substantially more challenging to make.

3. Leadership needs to ask the right questions. National ocean accounts are only useful if national leaders use them to ask questions about the state of the ocean system and ocean economy. This needs to start now. Information on ocean income and changes in the ocean balance sheet, in addition to ocean GDP, needs to be considered in the decision-making process.[16] If asking for a hammer when you need a wrench is not helpful, worse is to then use the hammer to drive in the bolt. That aptly describes what is currently being done with GDP with respect to economic well-being and sustainability.
4. Avoid one-off accounts or reports. National ocean accounts increase in value the longer they are kept and the more frequently they are updated. The value of national ocean balance sheets may take years to fully materialise, but they would greatly enhance a country's ability to make decisions compatible with sustainable development.
5. The sequence of ocean accounts needs to be a structured compilation of consistent and comparable information concerning marine and coastal environments, social circumstances and economic activity. Standardisation enables a degree of third-party verification.
6. Ensure the compatibility of ocean accounting efforts with international statistical standards and approaches, mainly the System of National Accounts (SNA), the System of Environmental Economic Accounting (SEEA) and also other broadly accepted initiatives, such as those reviewed by Jorgenson (2018) (UN Stats n.d.-a; SEEA n.d.).
7. Ensure the compatibility of ocean accounting efforts with the 10 Fundamental Principles of Official Statistics endorsed by the UN General Assembly in January 2014. These principles were designed as a reference point to ensure that official statistics are fit- for-purpose given their critical role in policy decision- making in support of sustainable development and securing public trust in governance (UN Stats n.d.-b; UN-ESC 2013).
8. National governments should ensure that their national accountants, economic analysts and marine scientists participate in the workshops organised by the UN Statistical Division and associated organisations for developing ocean accounts. This helps to maintain standards and increase credibility. Furthermore, these international organisations need to evolve to provide a degree of third-party verification of accounts coupled with capacity-building assistance.
9. National statistics offices should use interactive dashboards (e.g. Fig. 8.3) for ocean account reporting. Ocean accounts need to address a variety of questions broader than Questions 1–6. Therefore, it is important that users have the ability to explore the data, aggregate and disaggregate sectors and groups of people, alter the account boundaries and access ethically acceptable disaggregation by digital means.
10. National leaders need to take the time to 'play with and explore' these dashboards to learn about the state of the ocean economy. This recommendation is intended to empower decision-makers. In the past, such dynamic structures were not feasible and would have required volumes of reports that no decision-maker had time to read. Old print media required statisticians to make decisions to generate 'hard copy'. This constraint is vanishing rapidly. New data-management and visualisation software is allowing these changes to be made through a user-friendly interface in real time, which allows the important political decisions to be shifted back to the policymakers and away from national statisticians and scientists. Corporations are already shifting to interactive dashboards for decision-making. National governments need to do so as well. This transition requires decisions by leaders to dedicate funds in national budgets for upgrades to national account reporting.
11. Governments need to invest in data architecture and engineering at levels surpassing global multinational companies. These investments are necessary to connect fine-scale data about the marine environment with detailed economic data into supply-and-use structures and other data structures for national accounting and forecasting the ocean economy. These investments should build on existing Earth observation programs when possible. Investment must also include investments in people. Hardware and software alone will not solve the problem.

4.2.1 Know the Condition of the Ocean

- National statistics offices need to work with marine scientists, agencies or organisations to identify marine data and audit their feasibility for use in national accounts.
- Direct digital pipelines need to be developed from marine agencies to national statistics offices without first aggregating. For example, fish stock assessment data should be matchable to valuation data. Surveys conducted by marine agencies, such as fishing log books, need to be accessible to national statistics offices. There are confidentiality issues with such data, but many national statistics offices already access micro-level tax data. Safeguards and appropriate regulatory frameworks for data privacy, anonymisation and use need to be put in place.
- National accountants and country scientists need to assemble physical account measurements to provide

[16] Some national statistics offices produced these or similar indicators in the past but stopped because they were not used.

easy-to-use data structures for prospective economic forecasts such as regional development analyses, general equilibrium models that include feedbacks with the environment (Kerry Smith 2012) and other forms of integrated assessment modelling (Kling et al. 2017). Decision-makers need to ask how non-market effects are treated in economic analyses.

- Not all data need to be produced locally. There is an increasing role for remotely sensed data. Various national governments are increasingly creating and using fine-scale global marine data sets derived from satellite-based sensors. Countries should consider using these data, but it would also be good for multiple nations or coalitions to produce and certify some of these products to reduce duplicate effort. This is not limited to geographic data but also includes physical, geo-chemical and biological data. Data should be assembled on a regular basis at reasonable time scales.

4.2.2 Use Valuation to Understand Economic Interconnections and Trade-Offs

Valuation is critical in order to enable analyses in comparable units and to analyse explicit or implicit trade-offs. Furthermore, valuation forces society to 'look in the mirror' and observe the trade-offs being made. Valuation is not without controversy. Part of the confusion is that valuation is often misused in an attempt to estimate a 'total value' where the thought experiment asks what society would be willing to pay to avoid losing the natural asset or ecosystem service completely. This is fundamentally different from the value added of a production process connected to the ocean, the additional economic well-being individuals experience with a change in the condition of the ocean or the expected change in net present value available given of an ocean natural asset under a specific management policy. These last three questions align with the types of questions one can query of national account data. National statistics offices should focus on these latter three questions, and decision-makers should interpret valuations as such.

- Heads of government need to start asking about changes in ocean balance sheets that contain produced and non-produced assets today. National statistics offices need to start producing these balance sheets. It is also important to accurately value produced marine capital. Some forms of produced marine capital, such as ships, are relatively easy to account for. There are market prices, but even some forms of marine produced capital are challenging, such as port infrastructure. For others it is important that national statistics agencies use methods to impute value (Hulten 2006) for produced and non-produced assets in general equilibrium systems (Carbone and Smith 2013; Fenichel et al. 2018). Including ocean non-produced natural assets on the balance sheet is important for two reasons:
  - The ocean's natural capital, non-produced assets, stores substantial wealth that is important for a sustainable ocean economy.
  - The valuation of produced ocean assets is influenced by the condition of ocean natural assets. Excluding natural assets runs the risk of mis-valuing produced ocean assets. For example, the value of fish-processing machinery may be influenced by a processor's ability to secure fish or the value of port infrastructure may depend on barrier islands and other natural protective features. Rouhi Rad et al. (2019) shows that the value of the locks in the Panama Canal, which transit close to 5% of global marine shipping, increases with the amount of water in the canal system during the dry season. Complementarity between natural and produced capital could be common.
- Heads of government and other policy leaders should encourage their national statistics offices (NSOs) to incorporate a broad definition of income to address ends because NSO heads are already engaging in these conversations. This should be in addition to a more restricted money income boundary to balance with produced means. The SNA's income boundary is governed by the production boundary (European Commission et al. 2009). This is a shortcoming, because 'measures of welfare are needed to appraise the outcomes of changes in economic policies and evaluate the results' (Jorgenson 2018). Irving Fisher (1906) defined income as services, and the ocean provides substantial services outside of the market economy. These services are also income.[17] Heads of state should start asking heads of NSOs about revisions to capture these sources of income.
- The international national accounting community should provide technical guidance for country-level statistics offices on welfare measures beyond the current income boundary. This guidance should adapt the vast literature on non-market valuation intended for benefit-cost analysis (e.g., Freeman 2003) and be developed to make use of available micro-data. An important issue related to research is that valuation for national accounting needs to focus on existing or agreed-to institutions, even when these are 'inefficient'. This means care must be taken not to use hypothetical changes in management to compute

[17]The SNA makes the argument for excluding non-market income because the information is not useful for monetary policy. However, national accounts are used for much more than monetary policy.

potential changes in value. It is unreasonable to assume 'optimal' management that is inconsistent with prevailing institutions (Fenichel and Hashida 2019). There is a need to map into actual situations existing valuation methods that focus on potential changes, and to develop benefits-transfer libraries (Boyle et al. 2010).

- The accounts should be used to track progress over time, provide data to evaluate past programs and provide the starting point to analysis of alternative ocean policies. Furthermore, aggregate income statistics need to be able to be disaggregated because income and consumption is where equity is reflected (Jorgenson 2018). There is a need to know what money and non-money income stems from the ocean and to whom.

5 Conclusion

When the Wright brothers took flight in 1903 there was no dashboard. As planes became more complex, gauges and indicators were added. Today, nobody would fly on a plane where the pilot only looked at the air speed. It is unfathomable that the joint economic- physical-biological system of the ocean economy is not at least as complex as an airplane. So how can we expect to develop a sustainable ocean economy, the 'blue economy', with a single indicator, ocean GDP?

The simple answer is we cannot. In this Blue Paper we have discussed **a system of national accounts with multiple indicators** and how they should be applied to the sustainable ocean economy. We have emphasised the need to develop the underlying data structures to anticipate unintended consequences of decisions that may increase production in the present at great cost to the opportunities afforded to future generations, or increase production to a select organised few at a cost to the great disorganised many. The opportunities for the ocean to spur production bring this challenge into focus. On the one hand, the OECD (2016) and others have raised the prospect of the ocean's spurring new means of production. On the other, there is great concern over the future of biophysical ocean processes, as highlighted in SDG 14 on 'Life under Water'. Without an accounting system capable of producing multiple, well-designed indicators it is unclear if these causes align, compete or simply coexist. Multiple indicators are needed, and the existing ***system*** of national accounts is a good place to start to look for them.

While terrestrial asset accounts capture the greatest fraction of the human population and manufactured capital, the sphere of influence that ocean assets have in governing global environmental systems (e.g. climate and weather) is unmatched. In addition to direct economic activities involving ocean resources, the ocean links the impacts of climate warming at the poles to critically important sources of food via ocean acidification, the resilience of infrastructure via sea level rise and many other facets of the global economy via interactions with atmospheric processes and weather events. Measuring the ocean economy in national accounts requires addressing the full suite of challenges of developing measures to determine if society is meeting the needs of current generations without compromising the ability of future generations to meet their needs. Using national accounts to measure 'ocean development' can be a model for using national accounts to measure 'sustainable development'.

Acknowledgements The authors are grateful for input from Jeffery Adkins and Christopher Lauer of the U.S. National Oceanic and Atmospheric Administration, Rebecca Blank of the University of Wisconsin, Ted Morgan of the U.S. Bureau of Economic Analysis, Jane Lubchenco of Oregon State University, the UN Statistics Environmental Economic Accounts Section, Alessandra Affieri and Bram Edens, Carl Obst, participants in the 2019 Glen Cove SEEA Conference and Charles Colgan. The paper's technical reviewers, Linwood Pendleton, Hugh Possingham, Michael Vardon and Rintaro Yamaguchi, as well as its arbiter, Adrien Vincent, all provided helpful technical comments. The authors also thank the World Resources Institute for providing support as the Ocean Panel Secretariat.

Fenichel and Addicott were additionally supported by the Knobloch Family Foundation. Figure 8.1 artwork done by Jamie Ficker and funded by the Yale School of Forestry and Environmental Studies.

The authors thank Alex Martin for copyediting and Shannon Collins for design.

About the Authors

Lead Authors

Eli P. Fenichel is Knobloch Family Chair of Natural Resource Economics at Yale University. His email address is eli.fenichel@yale.edu.

Ben Milligan is a Scientia Fellow based at the Environmental Law Cluster, Centre for Applied Economic Research, and Centre for Ecosystem Science at the University of New South Wales. His email address is b.milligan@unsw.edu.au.

Ina Porras is an international economic advisor on environment, climate and sustainability based in the United Kingdom. Her email is ina.porras@gmail.com.

Contributing Authors

Ethan T. Addicott is a Doctoral Student in Environmental and Natural Resource Economics at Yale University.

Ragnar Árnasson is a Professor, Department of Economics, at the University of Iceland.

Michael Bordt is Regional Adviser on Environment Statistics (Ret.) at the United Nations Economic and Social

Commission for Asia and the Pacific and Senior Economic Adviser for Ocean Accounts at Fisheries and Oceans Canada.

Samy Djavidnia is Senior Project Officer, Integrated Services Design at the European Maritime Safety Agency and Steering Committee Member at GEO Blue Planet.

Anthony Dvarskas is Chief Environmental Scientist at Office of the New York State Attorney General and Adjunct Professor at Stony Brook University's School of Marine and Atmospheric Sciences.

Erica Goldman is the Science Policy Director at the National Council for Science and the Environment in Washington, DC.

Kristin Grimsrud is a Researcher at Statistics Norway.

Glenn-Marie Lange is a Senior Environmental Economist at the World Bank.

John Matuszak is a Senior Fellow for International Strategy at the National Council for Science and the Environment in Washington, DC.

Umi Muawanah is a Researcher at the Indonesian Ministry of Marine Affairs.

Martin Quaas is Head of Biodiversity Economics Research Group at the German Centre for Integrative Biodiversity Research and Professor of Environmental, Resource, and Ecological Economics, Department of Economics, Christian-Albrechts- University of Kiel in Kiel, Germany.

François Soulard is Chief, Research and Development, Environment Accounts and Statistics Program, at Statistics Canada.

Niels Vestergaard is a Professor, Department of Sociology, Environmental and Business Economics, at the University of Southern Denmark.

Junjie Zhang is an Associate Professor and Director of the Environmental Research Center at Duke Kunshan University in Kunshan, China.

References

ABC (Australian Broadcasting Corporation) (2018) Median Australian farmland values grow 7pc nationally, prices highest on city fringes. https://www.abc.net.au/news/rural/2018-05-02/farmland-values-report-shows-median-price-rise/9713822. Accessed 1 May 2018

ABS (Australian Bureau of Statistics) (2007) Measures of Australia's progress: summary indicators, 2006. https://www.abs.gov.au/AUSSTATS/abs@.nsf/Previousproducts/53F94C7B3713502C-CA2571770016194C?opendocument#Oceans%20and%20estuaries_0. Accessed 24 Apr 2006

ABS (2018) Australian System of National Accounts, 2017–18. https://www.abs.gov.au/AUSSTATS/abs@.nsf/Detail-sPage/5204.02017-18?OpenDocument. Accessed 26 Oct

Adler MD (2013) The Pigou-Dalton principle and the structure of distributive justice. SSRN eLibrary, May 1

Agarwala M (2019) Natural capital accounting for sustainability measurement. PhD dissertation. London School of Economics and Political Science

Arnason R, Kelleher K, Willmann R (2009) The Sunken Billions: the economic justification for fisheries reform. World Bank, Washington, DC

Arrow K, Dasgupta P, Goulder L, Daily G, Ehrlich P, Heal G, Levin S et al (2004) Are we consuming too much? J Econ Perspect 18:147–172

Audzijonyte A, Gorton R, Kaplan I, Fulton EA (2018) Atlantis User's Guide Part I: general overview, physics & ecology. Commonwealth Scientific and Industrial Research Organisation, Canberra

Birkenbach AM, Kaczan DJ, Smith MD (2017) Catch shares slow the race to fish. Nature 544:233–226

Bojinski S, Verstraete M, Peterson TC, Richter C, Simmons A, Zemp M (2014) The concept of essential climate variables in support of climate research, applications, and policy. Bull Am Meteorol Soc 95:1431–1433

Boyle KJ, Kuminoff NV, Parmeter CF, Pope JC (2010) The benefit-transfer challenges. Ann Rev Resour Econ 2:161–182

Carbone JC, Kerry Smith V (2013) Valuing nature in a general equilibrium. J Environ Econ Manag 66:72–89

Chow J (2016) Introduction to core accounting principles on SEEA and SNA. System of Environmental Economic Accounting, UN Statistics Division. https://unstats.un.org/unsd/envaccounting/workshops/Malaysia_2016/2.%20Introduction%20to%20Core%20Accounting%20Principles%20on%20SEEA%20and%20SNA.pdf

Colgan CS (2016) Measurement of the ocean economy from national income accounts to the sustainable blue economy. J Ocean Coast Econ 2:12

Collie JS, Botsford LW, Hastings A, Kaplan IC, Largier JL, Livingston PA, Plaganyi E et al (2016) Ecosystem models for fisheries management: finding the sweet spot. Fish Fish 17:101–125

Costello MJ, Coll M, Danovaro R, Halpin P, Ojaveer H, Miloslavich P (2010) A census of marine biodiversity knowledge, resources, and future challenges. PloS One 5:e12110

Dasgupta P (2001) Human well-being and the natural environment. Oxford University Press, New York

Dasgupta P (2007) Human well-being and the natural environment. Oxford University Press, New York

De Maio A, Irwin C (2016) From the orderly world of frameworks to the Messy world of data: Canada's experience measuring the economic contribution of maritime industries. J Ocean Coast Econ 2:9

Department of Economic and Social Affairs, Statistics Division (2018) Handbook on supply, use and input-output tables with extensions and applications. Studies in methods, handbook of national accounting, vol 74. United Nations, New York

Díaz S, Settele J, Brondizio E, Ngo HT, Gueze M, Agard J, Arneth A et al (2019) Summary for policymakers. In: da Cunha MC, Mace G, Mooney H (eds) Global assessment report on biodiversity and ecosystem services of the Intergovernmental Science-Policy Platform on Biodiversity and Ecosystem Services. Intergovernmental Science-Policy Platform on Biodiversity and Ecosystem Services, Bonn

Economist Intelligence Unit (2015) The Blue Economy: growth, opportunity and a sustainable ocean economy. Economist Intelligence Unit, London

Esch T, Üreyen S, Zeidler J, Metz-Marconcini A, Hirner A, Asamer H, Tum M et al (2018) Exploiting big earth data from space-first experiences with the timescan processing chain. Big Earth Data 2:36–55

European Commission, International Monetary Fund, Organisation for Economic Co-operation and Development, United Nations and World Bank (2009) System of national accounts 2008. European Commision, International Monetary Fund, Organisation for Economic Co-operation and Development, United Nations, World Bank, New York

European Union Directorate-General for Maritime Affairs and Fisheries, and Joint Research Centre (2018) The 2018 annual economic report on the EU Blue Economy. Publications Office of the European Union, Luxembourg

FAO (Food and Agriculture Organization of the United Nations, Department of Economic and Social Affairs) (n.d.) System of environmental-economic accounting for agriculture, forestry and fisheries: SEEA AFF. FAO, Department of Economic and Social Affairs—Statistics Division United Nations, Rome

Fenichel EP, Abbott JK (2014) Natural capital from metaphor to measurement. J Assoc Environ Resour Econ 1:1–27

Fenichel EP, Hashida Y (2019) Choices and the value of natural capital. Oxf Rev Econ Policy 35:120–137

Fenichel EP, Obst C (2019) A framework for the valuation of ecosystem assets. In: Discussion paper 5.3. Presented at Expert Meeting on Advancing the Measurement of Ecosystem Services for Ecosystem Accounting, Glen Cove, NY

Fenichel EP, Abbott JK, Yun SD (2018) The nature of natural capital and ecosystem income. In: Dasgupta P, Pattanayak SK, Kerry Smith V (eds) Handbook of environmental economics. North-Holland, Amsterdam, pp 85–142

Finnoff D, Tschirhart J (2003) Harvesting in an eight-species ecosystem. J Environ Econ Manag 45:589–611

Fisher I (1906) The nature of capital and income. Norwood, Norwood, MA

Fisheries and Oceans Canada (n.d.-a) GDP contribution to provincial and territorial economies. http://www.dfo-mpo.gc.ca/stats/maritime/tab/mar-tab2-eng.htm

Fisheries and Oceans Canada (n.d.-b) Direct, indirect, and induced gross domestic product. https://www.dfo-mpo.gc.ca/stats/maritime/tab/mar-tab4-eng.htm

Fitzer SC, Torres Gabarda S, Daly L, Hughes B, Dove M, O'Connor W, Potts J et al (2018) Coastal acidification impacts on shell mineral structure of bivalve mollusks. Ecol Evol 8:8973–8984

Fleurbaey M (2009) Beyond GDP: the quest for a measure of social welfare. J Econ Lit 47:1029–1075

Fleurbaey M, Blanchet D (2013) Beyond GDP measuring welfare and assessing sustainability. Oxford University Press, New York

Freeman AMI (2003) The measurement of environmental and resource values: theory and methods, 2nd edn. Resources for the Future, Washington, DC

GEO (Group on Earth Observations) (2015) GEO Strategic Plan 2016–2025: implementing GEOSS. Document 11 (Rev1). November 11–12. https://www.earthobservations.org/documents/GEO_Strategic_Plan_2016_2025_Implementing_GEOSS_Reference_Document.pdf

Grafton RQ, Squires D, Fox KJ (2000) Efficiency: a study of a common-pool resource. J Law Econ 43:679–714

Hamilton K, Clemens M (1999) Genuine savings rates in developing countries. World Bank Econ Rev 13:333–356

Hammond K (2005) Statistical benefits of individual transferable quotas for valuing natural capital 2005. In: EASDI Conference 2005. Stats NZ, Prague

Hart DK (1974) Social equity, justice, and the equitable administrator. Public Adm Rev 34:3–11

Hassellöv IM, Turner DR, Lauer A, Corbett JJ (2013) Shipping contributes to ocean acidification. Geophys Res Lett 40:2731–2736

Heal G (1998) Valuing the future: economic theory and sustainability. Columbia University Press, New York

Hicks JR (1939) Value and capital: an inquiry into some fundamental principles of economic theory. Oxford University Press, New York

Hulten CR (2006) The 'Architecture' of capital accounting: basic design principles. In: Jorgenson DW, Landefeld JS, Nordhaus WD (eds) A new architecture for the U.S. National Accounts. University of Chicago Press, Chicago, pp 193–214

INE (Insitituto Nacional de Estatística) (n.d.) Statistical tables. https://www.ine.pt/xportal/xmain?xpid=INE&xpgid=cn_quadros&boui=261914910

Jorgenson DW (2018) Production and welfare: progress in economic measurement. J Econ Lit 56:867–919

Jorgenson DW, Slesnick DT (2014) Measuring social welfare in the US National Accounts. In: Measuring economic sustainability and progress. University of Chicago Press, Chicago, pp 43–88

Kazemier B, Driesen CH, Hoogbruin E (2012) From input-output tables to supply-and-use tables. Econ Syst Res 24:319–327

Kling CL, Arritt RW, Calhoun G, Keiser DA (2017) Integrated assessment models of the food, energy, and water nexus: a review and an outline of research needs. Ann Rev Resour Econ 9:143–163

Krutilla JV (1967) Conservation reconsidered. Am Econ Rev 57:777–786

Kuznets S (1973) Modern economic growth: findings and reflections. Am Econ Rev 63:247–258

Kuznets S, Epstein L, Jenks E (1934) National income, 1929–1932. National Bureau of Economic Research, Cambridge

Lange G-M, Wodon Q, Carey K (2018) The changing wealth of nations 2018: building a sustainable future. World Bank, Washington, DC

Libecap GD (1994) Contracting for property rights. Cambridge University Press, New York

Lindstrom E, Gunn J, Fischer A, McCurdy A, Glover LK (2012) A framework for ocean observing. Task Team for an Integrated Framework for Sustained Ocean Observing, Paris

Maler K-G (1991) National accounts and environmental resources. Environ Resource Econ 1:1–15

Managi S, Kumar P (2018) Inclusive wealth report 2018: measuring progress towards sustainability. Routledge, New York

Matson P, Clark WC, Andersson K (2016) Pursuing sustainability: a guide to the science and practice. Princeton University Press, Princeton, NJ

Miloslavich P, Bax N, Simmons S, Klein E, Appeltans W, Aburto-Oropeza O, Anderson-García M et al (2018) Essential ocean variables for sustained observations of marine biodiversity and ecosystems. Glob Chang Biol 24:2416–2433

Muller-Karger FE, Miloslavich P, Bax NJ, Simmons S, Costello MJ, Sousa Pinto I, Canonico G et al (2018) Advancing marine biological observations and data requirements of the complementary essential ocean variables (EOVs) and essential biodiversity variables (EBVs) frameworks. Front Mar Sci 5:211

Navarro L, Fernández N, Guerra C, Guralnick R, Kissling WD, Londoño MC, Muller-Karger F et al (2017) Monitoring biodiversity change through effective global coordination. Curr Opin Environ Sustain 29:158–169

NOAA (National Oceanic and Atmospheric Administration) (2019) Report on the U.S. Ocean and Great Lakes Economy: Regional and State Profiles. NOAA, Office for Coastal Management, Charleston, SC

Nordhaus WD (2006) Principles of national accounting for nonmarket accounts. In: Jorgenson DW, Landefeld JS, Nordhaus WD (eds) A new architecture for the U.S. National Accounts. University of Chicago Press, Chicago, pp 143–160

Nordhaus WD, Tobin J (1972) Is growth obsolete? In: Economic research: retrospect and prospect, vol 5. National Bureau of Economic Research, Cambridge, MA, pp 1–80

Obst C, van der Ven P, Tebrake J, Lawrence JS, Edens B (2019) Issues and options in accounting for ecosystem degradation and enhancement. In: Discussion paper 5.4. Presented at Expert Meeting on Advancing the Measurement of Ecosystem Services for Ecosystem Accounting, Glen Cove, NY

OECD (2019) 9B. Balance sheets for non-financial assets. https://stats.oecd.org/index.aspx?DataSetCode=SNA_TABLE9B

OECD (Organisation for Economic Co-operation and Development) (2016) The Ocean Economy in 2030. OECD, Paris

Pauly D, Zeller D (2017) Comments on FAO's State of World Fisheries and Aquaculture (SOFIA 2016). Mar Policy 77:176–181

Pereira HM, Ferrier S, Walters M, Geller GN, Jongman RHG, Scholes RJ, Bruford MW et al (2013) Essential biodiversity variables. Science 339:277–278

Pfeiffer L, Gratz T (2016) The effect of rights-based fisheries management on risk taking and fishing safety. Proc Natl Acad Sci 113:2615–2620

Phaneuf DJ, Requate T (2017) A course in environmental economics theory, policy, and practice. Cambridge University Press, New York

Pilling D (2018) The growth delusion: wealth, poverty, and the well-being of nations. Tim Duggan, New York

Popova E, Vousden D, Sauer WH, Mohammed EY, Allain V, Downey-Breedt N, Fletcher R et al (2019) Ecological connectivity between the areas beyond national jurisdiction and coastal waters: safeguarding interests of coastal communities in developing countries. Mar Policy 104:90–102

Ramirez-Reyes C, Brauman KA, Chaplin-Kramer R, Galford GL, Adamo SB, Anderson CB, Anderson C et al (2019) Reimaging the potential of earth observations for ecosystem services assessments. Sci Total Environ 665:1053–1063

Roosevelt T (1910) Conservation: speech at Denver before the Colorado Live Stock Association. In: Lewis AH (ed) Compilation of the messages and speeches of Theodore Roosevelt. Bureau of National Literature and Art, New York

Rouhi Rad M, Adamowicz WVL, Entem A, Fenichel EP, Lloyd-Smith P (2019) The empirical case for complementarities between natural and produced capital. Colorado State University, Yale University, University of Alberta and University of Saskatchewan

SEEA (System of Environmental Economic Accounting) (n.d.) What is the SEEA? United Nations. https://seea.un.org

Sikor T (2013) The justices and injustices of ecosystem services. Routledge, New York

Smith VK (2012) Reflections: in search of crosswalks between macroeconomics and environmental economics. Rev Environ Econ Policy 6:298–317

Solow RM (1974) Intergenerational equity and exhaustible resources. Rev Econ Stud 41:29–45

Solow R (1993) An almost practical step towards sustainability. Resour Policy 19:162–172

Stats NZ (2018) Sources and methods for the accumulation accounts (changes in assets and liabilities). Stats NZ, Wellington, New Zealand

Steenbeek J, Buszowski J, Christensen V, Akoglu E, Aydin K, Ellis N, Felinto D et al (2016) Ecopath with Ecosim as a model-building toolbox: source code capabilities, extensions, and variations. Ecol Model 319:178–189

Stiglitz JE, Sen A, Fitoussi J-P (2009) Report by the commission on the measurement of economic performance and social progress. Commission on the Measurement of Economic Performance and Social Progress, Paris

Stiglitz JE, Sen A, Fitoussi J-P (2010) Mismeasuring our lives: why GDP doesn't add up, the report by the commission on the measurement of economic performance and social progress. New Press, New York

Toman M (1998) Why not to calculate the value of the world's ecosystem services and natural capital. Ecol Econ 25:57–60

Trading Economics (n.d.) Australia: agricultural land (% of land area). https://tradingeconomics.com/australia/agricultural-land-percent-of-land-area-wb-data.html

UN Stats (2019) Report on the Fiftieth Session. Official records. United Nations Economic and Social Council, New York

UN Stats (United Nations Statistical Commission) (2017) Issues under discussion. Updated August. https://unstats.un.org/unsd/nationalac--count/UnderDiscussion.asp

UN-ESC (United Nations Economic and Social Council) (2013) Resolution adopted by the Economic and Social Council on 24 July 2013. October 28. https://unstats.UN-ESC.2013un.org/unsd/dnss/gp/FP-Rev2013-E.pdf

UN-ESCAP (United Nations Economic and Social Commission for Asia the Pacific) (n.d.) Regional ocean accounts platform. https://unstats.un.org/unsd/nationalaccount/UnderDiscussion.asp

United Nations, European Union, Food and Agriculture Organization of the United Nations, International Monetary Fund, Organisation for Economic Co-operation and Development, and World Bank (2014) System of Environmental Economic Accounting 2012: Central Framework. United Nations, New York

UNStats (n.d.-a) The System of National Accounts (SNA). https://unstats.un.org/unsd/nationalaccount/sna.asp

UNStats (n.d.-b) Fundamental principles of official statistics. https://unstats.un.org/unsd/dnss/gp/fp-english.pdf

Wang X (2016) The ocean economic statistical system of China and understanding of the Blue Economy. J Ocean Coast Econ 2:10

Weitzman ML (1976) On the welfare significance of national product in a dynamic economy. Q J Econ 90:156–162

World Bank and UN-DESA (United Nations Department of Economic and Social Affairs) (2017) The potential of the blue economy: increasing long-term benefits of the sustainable use of marine resources for small island developing states and coastal least developed countries. World Bank and UN-DESA, Washington, DC

World Commission on Environment and Development (1987) Our common future. Oxford University Press, New York

WWF (World Wildlife Fund) (2015) Reviving the oceans economy: the case for action—2015. April 22. https://www.worldwildlife.org/publications/reviving-the-oceans-economy-the-case-for-action-2015

Yun SD, Fenichel EP, Abbott JK (2017a) CAPN: Capital Asset Pricing for Nature. https://cran.r-project.org/web/packages/capn/index.html

Yun SD, Hutniczak B, Abbott JK, Fenichel EP (2017b) Ecosystem based management and the wealth of ecosystems. Proc Natl Acad Sci 114:6539–6544

Ocean Finance: Financing the Transition to a Sustainable Ocean Economy

9

Abbreviations

ABNJ	Areas beyond national jurisdiction
ADB	Asian Development Bank
CFF	Climate Finance Facility
CZMT	Coastal Zone Management Trust
ESG	Environmental, social and governance
EU	European Union
GDP	Gross domestic product
IIX	Impact investment exchange
ISB	IIX sustainability bond
IUU	Illegal, unreported and unregulated
MPA	Marine protected area
NGO	Non-governmental organisation
ODA	Official development assistance
OECD	Organisation for Economic Co-operation and Development
PSI	Principles for Sustainable Insurance
SDG	Sustainable development goal
SME	Small and medium enterprise
SOE	Sustainable ocean economy
TCFD	Task Force on Climate-related Financial Disclosures
TNC	The Nature Conservancy
UN	United Nations
UNEP	United Nations Environment Programme
WWF	World Wide Fund for Nature

Originally published in:
Sumaila, U.R., M. Walsh, K. Hoareau, A. Cox, et al. 2020. Ocean Finance: Financing the Transition to a Sustainable Ocean Economy. Washington, DC: World Resources Institute. www.oceanpanel.org/blue-papers/ocean-finance-financing-transition-sustainable-ocean-economy
Reprint by Springer International Publishing (2023) with kind permission
Published under license from the World Research Institute

Highlights

- The ocean economy is currently at risk from multiple stressors, ranging from overextraction, direct habitat damage, pollution and climate change. Continuing with this 'business-as-usual' trajectory poses great risks to the health and integrity of the ocean and therefore to the world's population.
- Ocean finance can play a critical role in changing this trajectory and helping to achieve a sustainable ocean economy (SOE).
- However, current investments fall well below what is needed to transition to a SOE. In the last 10 years, less than 1% (US $13 billion) of the total value of the ocean has been invested in sustainable projects through philanthropy and official development assistance.
- Of the public and private sector investments already committed, a significant proportion are targeted at larger-scale economic activities that are often unsustainable and counter to the delivery of Sustainable Development Goal 14.
- To achieve a SOE, this sustainable finance gap needs to be closed. To close the gap, improved SOE policies, incentives, tools and approaches will need to be designed and established, knowledge and innovations proactively shared and capacities built to address environmental, social and economic risks, mobilise new forms of finance and redirect mainstream finance towards a SOE that empowers local people and supports responsible business and long-term societal goals.
- Several barriers are preventing the growth in financing of the SOE. Capacity constraints, data challenges, regulatory gaps and a lack of transparency all create a riskier enabling environment and negatively affect large-scale private sector finance. Most notably, complicated tenure and ownership and a lack of monitoring and enforcement increase the risk profile.
- Projects lack the appropriate deal size and risk-return ratios to match capital, making scaling and replication more complex than in familiar terrestrial sectors. There is a lack of familiarity with ocean-based project develop-

J. Lubchenco, P. M. Haugan (eds.), *The Blue Compendium*, https://doi.org/10.1007/978-3-031-16277-0_9

ment and financing by both the business and finance sectors. Capacity gaps, particularly in developing countries, exist regarding how to access sustainable ocean finance.
- This paper finds that many potential actions can be taken by the private and public sectors to remove existing barriers and open the pipeline to investment in a SOE.
- The paper proposes priority opportunities for action to remove existing barriers and open the pipeline to investment into a sustainable ocean economy, including:
  - Adopt clear principles to redirect mainstream finance towards a SOE.
  - Create a supportive and inclusive enabling environment.
  - Proactively strengthen and scale up the pipeline of investible projects.
  - Explore new financing mechanisms and tools.
- Achieving a robust ocean finance supportive of a SOE requires that the public and private sectors create and better mobilise a full suite of financial tools and approaches, insurance, and fiscal and market incentives as well as strengthen key aspects of the enabling environment. These actions will support the transition to an ocean economy that is sustainable and inclusive by making the benefits it generates available to all, especially women, youth and marginalised communities.

1 Introduction

The ocean covers more than 70% of Earth's surface and plays a crucial role in providing ecosystem goods and services that sustain life and support the well-being of billions of people worldwide (Teh and Sumaila 2013; FAO 2018; Hoegh-Guldberg et al. 2019; IPBES (Intergovernmental Science-Policy Platform on Biodiversity and Ecosystem Services) 2019). The ocean holds many economic opportunities, and many experts are recognising the need for a coordinated and sustainable approach to its use. Ocean finance can play a vital role in supporting sustainable development of the ocean economy by directing investments to sustainable development pathways that minimise ocean risks and maximise social equity, human well-being and environmental health.

The objective of this Blue Paper is to (1) inform about the inadequacy of the current financing of the sustainable ocean economy (SOE), (2) synthesise the barriers and challenges in financing the SOE and (3) propose concrete solutions to overcome these barriers.

Although a globally accepted definition of a SOE is still not agreed upon—with many organisations and entities providing varying definitions (see Appendix A)—refer to the SOE here as: 'development of the ocean economy in a way that balances the needs of people, planet and prosperity'. This is amplified by Winther et al. (2020) as ensuring 'long-term, sustainable use of ocean resources in ways that preserve the health and resilience of marine ecosystems and improve livelihoods and jobs, balancing protection and prosperity'. The size of the ocean economy is valued at an estimated 2.5% of global gross value added and is growing rapidly. In 2010, prior to the COVID-19 pandemic, the Organisation for Economic Co-operation and Development (OECD) projected that the estimated US $1.5 trillion in global gross value added of ocean industries—including fishing, shipping, offshore wind, maritime and coastal tourism and marine biotechnology—would increase to $3.0 trillion by 2030 (OECD 2016). Hoegh-Guldberg et al. (2015) concluded that if the ocean were a country, it would rank seventh in the world in terms of gross domestic product (GDP). This translates into significant contributions to national economies, generating millions of jobs in many countries. Note, however, that many valuations do not account for the ocean's full range of ecosystem goods and services, such as cultural and social values. In order to avoid undervaluation, ocean ecosystems must be valued as critical natural capital that underpins the vast public goods and ecosystem services they provide (Fenichel et al. 2020).

It is highly likely that COVID-19 will negatively impact this estimate (OECD n.d.), especially for the shipbuilding and tourism sectors of the ocean economy. However, the net effect of COVID-19 is not a given since governments worldwide are spending billions on short-term measures to prop up their economies in the face of the pandemic (Vivid Economics 2020).

The ocean economy is currently at risk from multiple stressors, ranging from overextraction, direct habitat damage, pollution and climate change (Hernández-Delgado 2015; Gaines et al. 2019; IPBES (Intergovernmental Science-Policy Platform on Biodiversity and Ecosystem Services) 2019; IPCC (Intergovernmental Panel on Climate Change) 2019).

Continuing with this 'business-as-usual' trajectory poses great risks to the health and integrity of the ocean and therefore to the world's population, especially the future well-being of hundreds of millions of people in coastal and island communities. These risks undermine the 2030 Agenda for Sustainable Development and, in particular, the Sustainable Development Goals (SDGs) focused on 'life below water' (SDG 14) as well as others, including 'no poverty' (SDG 1), 'zero hunger' (SDG 2), 'decent work and economic growth' (SDG 8) and 'climate action' (SDG 13) (Wright et al. 2017). To change this trajectory, it is imperative to build ocean resilience and minimise ocean risks by restoring, protecting and effectively managing human use of and impacts to ocean ecosystems. Nevertheless, the bulk of investments in the ocean economy have been directed not at transitioning to a SOE but rather at large-scale infrastructure, energy, transport, commercial fisheries, aquaculture, biotechnology and tourism (Vivid Economics 2020).

The costs for not conserving and sustainably using the ocean are high. For instance, invasive species are estimated to cause $100 billion in economic damages to infrastructure, ecosystems and livelihoods each year (OECD 2017a). The total estimated cost of coastal protection, relocation of people and loss of land to sea level rise is projected to range from about $200 billion to $1 trillion annually by 2100, depending on the increase in sea level (0.5–1.0 m). Already, it appears that a 1-metre sea level rise is more than likely by the turn of the century (IPCC (Intergovernmental Panel on Climate Change) 2019). Noone et al. (2013) state that in the absence of proactive measures to mitigate climate change, the cost of damage to the ocean could be an additional $322 billion a year by 2050. The 2019 *IPCC Special Report on the Ocean and Cryosphere in a Changing Climate* (IPCC (Intergovernmental Panel on Climate Change) 2019) and Gaines et al. (2019) both report significant impacts of climate change on the ocean economy.

Despite the huge costs of inaction and the substantive investments in the ocean economy, current levels of investments in the 'sustainable' ocean economy are inadequate (Fig. 9.1). Although an estimated $8 billion from philanthropy and $5 billion from official development assistance (ODA) were invested during the last 10 years (de Vos and Hart 2020), this level is insufficient to drive the change needed to achieve a SOE. Most significantly, there is limited finance available to achieve the restoration, protection and sustainable management of the ocean—to ensure the building blocks for achieving a SOE are in place (Fig. 9.1). Evidence from the general conservation finance literature indicates that the ocean finance gap is most likely very high. Huwyler et al. (2014) estimated the gap in conservation finance across all major biomes at $300 billion globally, though the proportion of this relevant to the ocean has not been identified and the financing needs of the SOE extend significantly beyond conservation objectives. Furthermore, it has been estimated that to achieve the global need for conservation funding, investible cash flows from conservation projects need to be at least 20–30 times greater than they are today. Sumaila et al. (2017) report that currently about 0.002% of global GDP is invested in conserving and sustainably using biodiversity, and about 4 times the current level of investment is required to meet conservation needs. Although these estimates are for biodiversity in general, the available data suggest inadequate investments in ocean sustainability. We provide a detailed description of current funding gaps for marine protected areas (MPAs) in Appendix B.

Fig. 9.1 A major gap in ocean finance for supporting a sustainable ocean economy. Note: All figures are in US $. *ODA* official development assistance. Source: Authors. Designed by Patricia Tiffany Angkiriwang

The reasons for the low levels of sustainable financing and investment in SDG 14 and the SOE are manifold, and if addressed, they could result in real and sustained change in the way our ocean ecosystems are utilised and managed. These issues are discussed in this Blue Paper. Section 2 summarises the study method, Sect. 3 provides evidence of current challenges to financing and investing in a SOE, Sect. 4 discusses opportunities for actions that can be taken to overcome these challenges and Sect. 5 concludes.

2 Research Methodology

The methodology consists of reviewing the literature to identify what information is currently available on finance flows for the conservation and sustainable use of the ocean; understanding existing fiscal instruments and financing options, including insurance; and identifying opportunities for how these instruments for a SOE can be scaled up. Available data is compiled from various sources, and key data gaps are highlighted. Examples are used to illustrate good practice. We review and analyze the literature on the use of fiscal instruments to support ocean economic activities and governance. We specifically query the literature to help us understand the types of subsidies, fees and taxes currently applied to the ocean economic sector at different scales and ocean economic activity types, and we analyse how these can be re-designed and re-directed to make them support a SOE. We also review current practices of the insurance industry and ask pertinent questions, such as does insurance as currently practiced support the goal of promoting a SOE? What role could insurance have in accelerating the transition towards a new ocean economy? Finally, we review the literature on ocean risk and resilience in the context of the wider discussions about climate risks to identify how insurance products can best help to deliver ocean and coastal finance solutions. Since an important aim of this Blue Paper is to provide the most current information possible, we also rely on the expert knowledge of contributing authors about recent and ongoing initiatives to address current challenges and barriers to financing a SOE.

3 Current Barriers and Challenges to a SOE

Several root causes can explain the current low levels of sustainable financing and investment in a SOE (Fig. 9.2).

3.1 Inadequate Frameworks and Taxonomies

Current frameworks and taxonomies to guide which investments support a SOE—that is, 'blue' investments—do not adequately communicate with each other and are not yet being guided by universally adopted principles. This is necessary to establish a classification system of activities that are considered to comply with the principles of a SOE, thereby guiding investment decisions and development policy towards a SOE. Efforts towards common frameworks and taxonomies are under way (ADB (Asian Development Bank) 2020), with several notable examples outlined below.

3.1.1 The Sustainable Blue Economy Finance Principles

The European Commission, the World Wide Fund for Nature (WWF), the European Investment Bank and the Prince of Wales' International Sustainability Unit developed the Sustainable Blue Economy Finance Principles (WWF (World Wide Fund for Nature) 2018). Designed to build on and complement existing sets of principles for sustainable finance, such as the Equator Principles, these 14 principles aim to fill the current gaps associated with a SOE. If widely adopted, the principles could help to positively transform the way in which ocean ecosystems are used and managed.

3.1.2 The United Nations Environment Programme Finance Initiative

The United Nations Environment Programme (UNEP) recently launched the Sustainable Blue Economy Finance Initiative, a platform targeted at the finance, investment and insurance sectors (UNEP FI (United Nations Environment Programme Finance Initiative) n.d.). Adopting the Sustainable Blue Economy Finance Principles as a guiding framework provides practical information on SOE issues and will support the development of more granular guidance. UNEP works with financial institutions to incorporate environmental, social and governance (ESG) issues into their business principles and to integrate sustainability principles into financial market practices. This is done through frame-

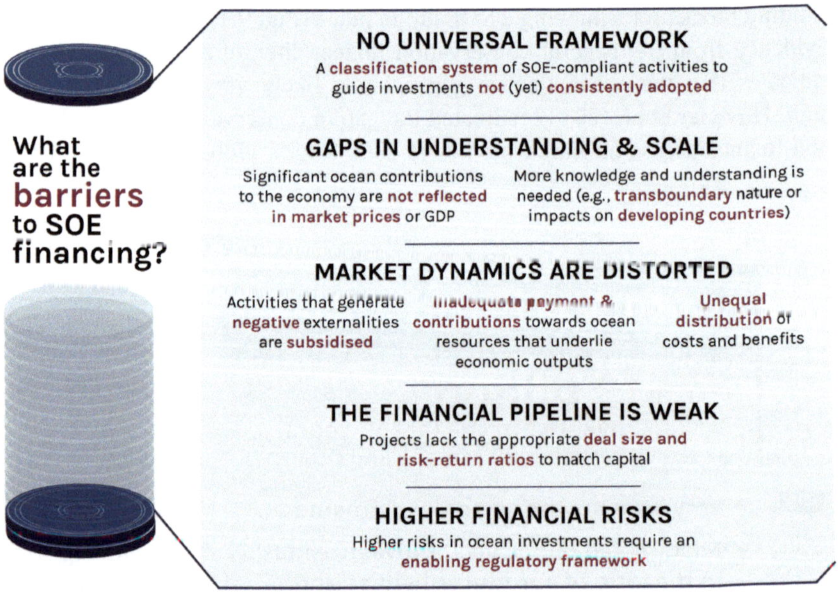

Fig. 9.2 Barriers to marshalling adequate funding for a SOE. *GDP* gross domestic product, *SOE* sustainable ocean economy. Source: Authors. Designed by Patricia Tiffany Angkiriwang

works such as the Principles for Sustainable Insurance (PSI), the Principles for Responsible Banking and the Principles for Responsible Investment. Other complementary sectoral principles include the Principles for Investment in Sustainable Wild-Caught Fisheries (EDF, Rare/Meloy Fund, and Encourage Capital 2018) and the Poseidon Principles (Poseidon Principles 2019) aimed at the shipping sector.

3.1.3 The European Union Taxonomy

The European Union (EU) taxonomy (European Commission 2020)—which includes a blue component—is the first to develop such a framework. It is likely that this taxonomy will significantly influence the creation of a common global taxonomy, which will be required to standardise decision-making across global markets and support the delivery of a SOE. Complementary frameworks are being developed by institutions such as multilateral development banks to guide the screening and selection of ocean investments by defining the sectors, segments and objectives that are allowable.

Notably, the Asian Development Bank (ADB) framework also stipulates rules to help reduce 'blue-washing' (ADB (Asian Development Bank) 2020).

To design guidance that is both ambitious and pragmatic, strong collaboration between a range of actors from the science, policy, non-governmental organisation (NGO) and finance communities will be critical.

3.2 Gaps and Mismatches in Information, Awareness, Capacity and Scale

3.2.1 Inadequate Information and Awareness

Information about the ocean and its economic, social and environmental value is missing or inadequate.

For appropriate and adequate finance to flow into the ocean economy, its overall contribution needs to be understood and measured more comprehensively than it is currently. Notably, humans derive multiple ecosystem services from the ocean whose values are generally not reflected in market prices and are therefore barely captured within the GDP. These include ecosystem services such as those linked to cultural values and benefits as well as natural hazard protection, carbon sequestration and climate mitigation and pollution buffering (OECD 2017a). The values of these services can be extremely high. For instance, Rogers et al. (2014) estimated the carbon sequestration value of marine life in the high seas to be 10 times the revenue generated by high seas fish catch.

3.2.2 Mismatched Capacity and Scale

Ocean finance systems do not yet have adequate capability to match the governance needs of a shared global ocean. Therefore, another prerequisite for finance to catalyse the transition to a SOE is the availability of comprehensive information about the shared and transboundary nature of the ocean. The ocean is ecologically and physically connected across the entire globe, and impacts to the ocean and marine ecosystem services in areas beyond national jurisdiction (ABNJ)—the high seas (Sala et al. 2013), for instance—can have large impacts on marine resources within exclusive economic zones, and vice versa (White and Costello 2014; Sumaila et al. 2015; Popova et al. 2019). Large pelagic stocks, for example, cross the boundaries of several countries, and financing schemes that take this into account are necessary for ensuring that the ocean economy is sustainable.

Given the large scales and complex connectivity of ocean systems, managing ocean resources requires regional and global initiatives. A cooperative approach for maritime security (including climate change and illegal activities), trade and investment, and transboundary ocean resource management (including consideration of ABNJ) are critical to achieving sustainable ocean/blue economy outcomes. While regional sustainable ocean strategies are being developed (Table 9.1), these initiatives need to cover ocean basins currently overlooked by working better together.

Ocean financing is needed that matches the scales of these large and complex ocean governance initiatives. The ongoing negotiations at the United Nations (UN) on the governance of ABNJ may result in new regimes for governing these areas and impact the kind of financing instruments that would be needed.

3.3 The Market Dynamics Are Distorted

3.3.1 Fiscal Policies Undermine a SOE

Ocean economic activities that generate negative externalities, such as fossil energy extraction, unsustainable fishing and aquaculture and non-green shipping (i.e., vessels that are not eco-friendly (Lee and Nam 2017)), receive subsidies.

Table 9.1 Examples of regional collaboration in sustainable ocean governance

Geographical region	Sustainable ocean initiative/strategy
Africa	African Union's Blue Economy Strategy United Nations Economic Commission for Africa's Blue Economy Regional Action Plan
Asia	Asia Development Bank's Action Plan for Healthy Oceans and Sustainable Blue Economies Indonesia's Sustainable Oceans Programme
Baltic Sea	Baltic Sea Initiative
Europe	European Union's Blue Growth Strategy
Indian Ocean	Indian Ocean Rim Association's Blue Economy Declaration
Pacific Ocean	Pacific Regional Oceanscape Program

Source: Authors

Fossil fuel support measures in the ocean economy are common and are in place in most countries (OECD 2019b). The International Monetary Fund estimates that 6.3% of global GDP ($4.7 trillion) was provided as fossil fuel subsidies in 2015 (Fig. 9.1), including uninternalized externalities (Coady et al. 2019). About $35 billion in subsidies is allocated to global marine fisheries alone each year, of which $22 billion is allotted to harmful subsidies (Sumaila et al. 2019). According to OECD estimates, governments spend on average an amount equal to 20% of the value of fisheries landings in support of the sector, amounting to $7 billion per year in the OECD region and reaching an estimated $35 billion worldwide. These harmful fisheries subsidies prop up fishing operations which would otherwise be unprofitable, thereby facilitating excessive fishing capacity which perpetuates the overexploitation of fisheries resources (Sumaila et al. 2019).

A large percentage of fisheries subsidies are currently allocated to large-scale industrial fishing fleets (Schuhbauer et al. 2017), which can make small-scale fishing fleets less economically viable (Schuhbauer and Sumaila 2016). Given the important food security, livelihoods and cultural roles that small-scale fisheries play worldwide (Österblom et al. 2020), public policies should not proactively disadvantage them if the aim is to meet the SDGs, especially SDGs 1–5 and 10 ('no poverty', 'zero hunger', 'good health and well-being', 'quality education', 'gender equality', and 'reduced inequality').

In particular, because relatively more women are small-scale than large-scale fishers (Harper et al. 2020), everything being equal, eliminating and/or redirecting harmful subsidies could improve gender equality by empowering female fishers. This is because most harmful fisheries subsidies go to large-scale fisheries, but women work proportionately more in small-scale fisheries (Harper et al. 2020). Certain subsidy policies, particularly those related to fuel use, also have the potential to disproportionately encourage large fishing operations to increase effort, thereby reducing the catches, food and livelihood opportunities available to small-scale fishers (Martini and Innes 2018).

Beneficiaries Do not Adequately Pay for Access or Management of Ocean Resources

Maritime countries are generating large economic outputs from the ocean economy, but the cost of ocean management is currently not being borne by those exploiting it, including direct harvesters and consumers. Consequently, there is underfunding of effective ocean governance and reviving and maintaining the health and integrity of ocean ecosystems required to sustain the ocean's economic outputs. Although comprehensive economic outputs are not always fully measured and accounted for, or may not be considered at all (e.g., support for emerging sectors), current figures available indicate that in Australia, the ocean economy is valued at 4.3% of the GDP (AIMS (Australian Institute of Marine Science) 2018). In Mauritius, the ocean economy accounts for over 10% of GDP.

In China, the ocean economy accounts for nearly 10% of GDP and is rising steadily (EDB (Economic Development Board Mauritius) 2020). The fishing sector alone is worth 10% of the GDP in the Pacific Islands region, and this does not include all of the other ocean sectors. In East Asian countries, the ocean economy accounts for 15–20% of GDP (PEMSEA (Partnerships in Environmental Management for the Seas of East Asia) 2009). In the United States, the ocean economy is growing twice as fast as the U.S. economy as a whole (NOAA (National Oceanic and Atmospheric Administration) 2019). Despite the significance of the ocean economy to maritime countries and to the global economy, public investments to ensure that the ocean economy is sustainable are inadequate. For instance, Binet et al. (2015) estimated that Mediterranean countries were facing an annual financing gap of $776.4 million for effective management of marine protected areas in the Mediterranean (Appendix B).

The private sector also benefits from, as well as impacts, the ocean, but it generally does not contribute sufficiently to investments or initiatives that could improve the sustainability of the ocean economy.

Recognising the need to diversify funding sources and increase blending, the 2015 Financing for Development conference in Addis Ababa encouraged the use of different streams of funding to meet global challenges and the SDGs. 'Turning billions into trillions' requires mobilizing private finance alongside public capital to achieve sustainable outcomes. However, the scale of current financial flows is insufficient mainly because private finance is constrained by risk-return requirements (Appendix C), and the volume of public sector and philanthropic finance is inherently limited.

3.3.2 An Unequal Distribution of Costs and Benefits

Access to ocean finance is limited and not well understood by potential recipients, especially in developing countries. Österblom et al. (2020) found that ocean resources and sectors are 'rarely equitably distributed', and many of their benefits are captured by a few. At the same time, most of the costs from ocean- based economic activities, such as the environmental impacts from pollution, are borne by women, youth and marginalised communities. Women are particularly disadvantaged because they face inequity worldwide (Österblom et al. 2020). Further, the lack of enabling conditions in many developing countries means that access to finance is more limited to begin with, and it is not always fully understood by potential recipients.

The inequity identified by Österblom et al. (2020) results from the provision of subsidies to the fossil fuel sector to the tune of $4.7 trillion globally, or 6.3% of global GDP, in 2015

(Coady et al. 2019). Such subsidies to big business only serve to increase inequality, which ultimately leads to the unfair distribution of ocean economic values and benefits to small-scale actors, women and minority groups (Österblom et al. 2020).

Clearly, existing inequalities need to be solved so that the ocean economy can help reduce these inequities around the world. Österblom et al. (2020) make the important argument that promoting equity, both within and between countries, is integral to a SOE. Equitably allocating ocean finance to all groups in society (including women and minority groups) is a necessary, if not sufficient, condition for tackling inequality in the ocean economy. New innovative and inclusive investments that are fair and accessible to all members of society are needed.

The potential impacts of expanding the SOE in developing countries are not well captured. Economic inequality between nations has resulted in divergent progress in ocean activities, and this has affected millions of people worldwide who depend on the ocean for their livelihoods and culture. For example, seabed mining and fishing activities financed with capital from a range of countries affect the well-being of people in the developing world when they target, for instance, migratory fish stocks as well as stocks shared with developing countries (Jouffray et al. 2020). It is important that the impacts of these activities at all scales are studied and understood in order to provide the information base to ensure that finance, subsidies and insurance are designed to avoid supporting activities with negative impacts on people and nature.

The available data are too aggregated in existing national accounts and need to be disaggregated more, but this is not an easy task (Fenichel et al. 2020). The current effort by the UN System for Environmental Economic Accounting, which provides a standardised framework to account for environmental protection and management expenditure (and taxes or subsidies) in a manner that is interoperable, is a good effort that needs to be expanded to include ocean finance information more comprehensively (Fenichel et al. 2020). Similarly, the recently established Global Ocean Accounts Partnership has yet to include ocean financial flows in its framework for ocean accounting and capacity-building activities (UN ESCAP (United Nations Economic and Social Commission for Asia and the Pacific) n.d.).

3.4 The Investment Pipeline Is Weak

3.4.1 Limited Availability of High-Quality, Investible Projects

Although there is no shortage of investment capital available globally, the immediate lack of high-quality, investible projects that would contribute to a SOE is a substantial challenge (Koh et al. 2012; UN-OHRLLS (United Nations Office of the High Representative for the Least Developed Countries) 2013; PEMSEA 2015; Fritsch 2020). The majority of sustainable ocean interventions currently require grant capital and do not generate sufficient, if any, financial returns. For the minority of projects that do benefit the ocean and generate a financial return, many are (1) too small to be financially viable once the costs of due diligence are considered and/or (2) too high in the risk-return profile (see more on ocean risk in Sect. 3.5). This is exacerbated by the fact that many sustainable ocean interventions have low potential economic returns (see Appendix C). The good news, however, is that there is a growing number of ocean-focused start-up companies.[1]

3.5 High Risks Without an Enabling Regulatory Environment

3.5.1 Environmental Complexities, Untested Interventions, and Patchy Regulatory and Governance Frameworks

Historically, ocean economic sectors have operated under relatively more unpredictable conditions than those based on land due to the ocean's vast size, physical environment, and comparative lack of ownership and responsibility in the ocean. For instance, the fluid and interconnected nature of the sea means that pollutants and alien species can be carried across much greater distances than on land, thus creating unanticipated impacts in far-off places. Likewise, because water is less transparent than air, remote sensing technology is unable to penetrate deep down to the sea's surface, thereby making it a lot more difficult and expensive to understand what is occurring in the seabed and water column. To overcome the higher risk profile associated with the ocean environment and attract investments and new forms of finance, a number of challenging enabling conditions will need to be addressed. These include capacity constraints, data challenges and higher-risk operating and governance environments. In addition, structural challenges related to investment in the ocean make scale and replication more complex than in more familiar terrestrial sectors (notably, related to tenure and ownership, monitoring and enforcement). To attract large-scale investments, it is critical to find ways to de- risk the enabling environment associated with ocean- based sustainable development projects and activities. De-risking would help catalyse and catapult hundreds of promising sustainable ocean ventures that are already in development globally. The majority of these projects are still in their very early days, are small and are hampered by high risk levels that conventional venture capital investors are unwilling to take.

[1] For example, see Katapult Ocean (https://katapultocean.com/).

4 Opportunities for Action

Here, we suggest a number of ways and means by which the challenges and barriers identified can be removed through actions by governments, private entities and individuals.

4.1 Set Up and Implement New (Shared) Rules, Guardrails and Guidelines

To guide investment decisions and develop SOE policies, it is critical that effective guardrails and guidelines are in place and are widely adopted. An essential element of this emerging SOE finance ecosystem will also be the creation of ocean-based finance taxonomies, which, in effect, will create classification systems of those activities considered to comply with strong principles for a SOE. The definition of a SOE as the 'development of the ocean economy in a way that balances the needs of people, planet and prosperity' is a good working definition that could be adopted more widely in developing such guidelines. Ultimately, the goal should be to ensure that existing frameworks and guidelines bridge and speak to each other and identify commonalities and differences that exist between them. Finally, it is very important to make sure that the frameworks developed are actually implemented.

New standards and metrics need to be developed to encourage and support stronger transparency and consistent reporting across the SOE finance community. Adequate governance, tracking and monitoring of flows, as well as principles and policy frameworks, are needed to ensure that innovative financial mechanisms support the scaling up of blended finance and private funds that are effectively aligned with inclusive and sustainable development.

The Sustainable Blue Economy Finance Principles are a strong set of scientifically and economically sound principles, and wider adoption by private and public entities should be encouraged. The principles are very high level and, therefore, are relevant across many contexts. However, due to the high-level nature of the principles, more detailed guidance and common blue taxonomies are still needed. The EU taxonomy can provide an important first step in creating a common global taxonomy.

The Task Force on Climate-related Financial Disclosures (TCFD) is a private sector–led task force that provides a standardised disclosure framework so that carbon- related assets and climate risks can be better assessed and decisions better informed. Within an ocean context, the disclosure framework can help minimise unanticipated impacts arising from climate change, such as financial losses to coastal real estate and infrastructure resulting from sea level rise. With strong endorsement and leadership, in particular from central banks, uptake has already been high, with 1,068 supporters as of February 2020 and a market capitalization of over $12 trillion (TCFD (Task Force on Climate-related Financial Disclosures) 2020). Building on the success of the TCFD, dialogue is now ongoing around the potential to extend this approach to the risks associated with the loss of nature through a new Task Force for Nature-related Disclosures. These tools will need to align with science and the post-2020 framework on biodiversity.

4.2 Strengthen Knowledge, Data and Capacity in Ocean Health and Finance

This will allow decision-making processes and activities to adapt to new knowledge of the potential risks, cumulative impacts and opportunities associated with business activities. Moreover, information on the status of the natural asset being invested in is required for meeting rigorous criteria during a project's due diligence phase and throughout its life cycle. Consequently, strengthening knowledge is especially pertinent in developing countries, where data and information gaps are key challenges to attracting finance for investments, such as for fisheries reform (Holmes et al. 2014).

At the national level, investing in a coordinated research and development framework is a way to leverage funds and expertise to grow knowledge and human capital and to advance 'blue' technology for ocean sustainability. For instance, Australia's Blue Economy Cooperative Research Centre brings together government, public and private enterprises in the aquaculture and renewable energy sectors to develop sustainable offshore technologies to increase Australia's food and renewable energy production. A central focus is investment in higher education and research to increase knowledge and human capital to enable further engagement in a SOE.

SDG 17 on 'means of implementation' identifies targets and indicators that can be used to track financial flows for sustainable development. These are broad but nonetheless applicable to a SOE: (1) increasing domestic resource mobil isation, including through international support to developing countries to increase capacity for tax and revenue raising (17.1); (2) mobilising additional financial resources for developing countries from multiple sources (17.3); (3) assisting developing countries to attain long-term debt sustainability (17.4); and (4) achieving the target of 0.7% of gross national income to developing countries as ODA.

A recent tracking tool launched in 2019 by the Our Ocean conference, which also records monetary commitments to ocean conservation and sustainability, is beginning to fill this need (Our Ocean 2019). The OECD is developing estimates of financial support provided to the SOE, especially with respect to the role of ODA and blended finance in supporting sustainable ocean activities, and the finance flowing from the use of economic instruments (such as fees, taxes and charges). Finally, the philanthropic- and grant-funding tracker FundingtheOcean.org is seeking to shed greater light on the size of this source of finance.

Efforts should be made to more consistently and comprehensively monitor and report on finance for the conservation and sustainable use of the ocean, across both the public and private sectors. These efforts should involve better tracking and monitoring of financial flows for biodiversity, which cover finance for both terrestrial and ocean ecosystems (OECD 2019a). Countries should—individually and regionally—invest in data and analysis more generally. Government budgets need to be able to track spending on ocean-based sustainable development. Developing ocean data architecture at sufficient granularity will support adaptive management to assist with ocean health and social equity and will help private investors have sufficient information to make key investment decisions. It will also help local entrepreneurs and support good business plans and practices.

National ocean accounts are a major sub-component of the data infrastructure required. The integration of environmental and economic information through a sequence of ocean accounts is one means of improving the data situation highlighted here. For example, the time series of financial flows can be correlated with ecosystem changes within an integrated national accounting framework (Fenichel et al. 2020).

4.3 Strengthen the Enabling Environment, Increase Inclusivity and Correct Market Distortions

4.3.1 Strengthen the Enabling Environment

Effective and stable regulatory and policy environments will do a better job of attracting investment. To maintain and potentially increase the flow of economic benefits from the ocean economy, governments need to continuously provide a supportive enabling environment. Governments and multilateral agencies have critical roles to play, therefore, in creating attractive financing conditions by reforming policies and creating regulations that strengthen the sustainable management and governance of natural capital and facilitate and incentivise social enterprise and new forms of capital (UNDP (United Nations Development Programme) and GEF (Global Environment Facility) 2012; Morgan and GIIN (Global Impact Investing Network) 2014; Whisnant and Reyes 2015). This might include national policies that secure tenure and establish robust enforcement mechanisms in the fishing sector (FAO (Food and Agriculture Organization of the United Nations) 2013) or that support technology transfer and incentivise renewable energy (Thiele and Gerber 2017; IRENA (International Renewable Energy Agency) 2018). Ocean policymakers and managers should provide greater clarity regarding their policy objectives and approaches and maintain a high level of transparency and consultation with stakeholders at all levels.

Investing in improving and streamlining policies—such as those related to (1) transitioning shipping to become more green; (2) building renewal energy infrastructure; (3) nature-based solutions, such as the effective protection of habitats and ecosystems (e.g., coral reefs and mangroves) that provide essential ecosystem services (including coastal protection and carbon sequestration); (4) supporting multi-sectoral collaboration; and (5) the implementation of marine spatial planning to reduce user conflicts and ensure that cumulative impacts of activities do not exceed the carrying capacity of the ecosystem—would be an excellent way to improve the enabling environment.

4.3.2 Increase Inclusivity

Given the importance of small and medium enterprises (SMEs) to portfolio development, governments should also create conditions that provide access to financing, savings, micro-insurance and other services (Grace and van Anrroy 2019). Sovereign insurance products can also substantially improve the risk profile of projects.

Improved disaster and shock-related insurance, such as the Caribbean Catastrophe Risk Insurance Facility, is also critical to strengthen private investor confidence.

Other factors limiting ocean finance include the lack of intermediation capacity and transition capital.

Capacity building, training and tertiary education needs to support leaders, managers and local entrepreneurs who can speak both the language of finance and the language of ocean science. National and international organisations can build the capacities in support of sustainable ocean finance through information provision, training and networking.

Building the kind of information needed to attract investments into the ocean economy requires a significant increase in human capacity for acquiring, investing and aligning ocean finance in many developing maritime countries. Effective capacity building in the areas of ocean finance, insurance and the application of fiscal instruments—especially from the international community, such as multilateral organisations or bilateral aid—is urgently needed to support investment for a SOE.

4.3.3 Correct Market Distortions

A resilient ocean economy requires rigorous and comprehensive ocean governance, which is not cheap and therefore needs continuous funding. A greater proportion of ocean economic output needs to be allocated to multi-sector and multinational ocean governance strategies. In addition, market distortions need to be corrected through taxation, the pricing of services and the re-purposing of harmful subsidies to more sustainable and equitable uses.

The mechanisms by which countries could capture such revenue will vary according to the country context and include a combination of domestic taxes, levies, fines, fees and other mechanisms that monetize ocean benefits and ocean impacts. Once collected, these funds could be allocated, in a transparent way to multi-sector ocean governance strategies and marine spatial plans, including management of all significant threats and impacts to ocean health (Appendix D) as determined by the country.

Fiscal (e.g., subsidies, fees and taxes) and non-fiscal (e.g., tradable permits and social norms) incentives should also be deployed to ensure that the effects of negative externalities are eliminated while those of positive externalities are promoted. Environmentally or socially harmful subsidies could be diverted to support the move to renewable energy or related sectors such as sustainable aquaculture (Fig. 9.3).

Market-based incentives, such as certification, can increase the investment potential for projects by providing some assurances on sustainability throughout the supply chain and implementing more transparent monitoring approaches (Lubchenco et al. 2016). Governments again have a role to play in creating stronger incentives for certification, whilst the conservation and development sector will need to provide technical assistance. A key challenge for these products is their accessibility to developing countries. The costs of certification schemes, for instance, are usually high enough to make them out of reach for many developing countries. Also, the reporting requirements for these products may be too demanding for many countries, thereby limiting the ability to scale them up globally.

Environmental fiscal reform (i.e., taxation and pricing measures that can raise revenues while furthering environmental goals (OECD (Organisation for Economic Co-operation and Development) 2006)) provides an opportu-

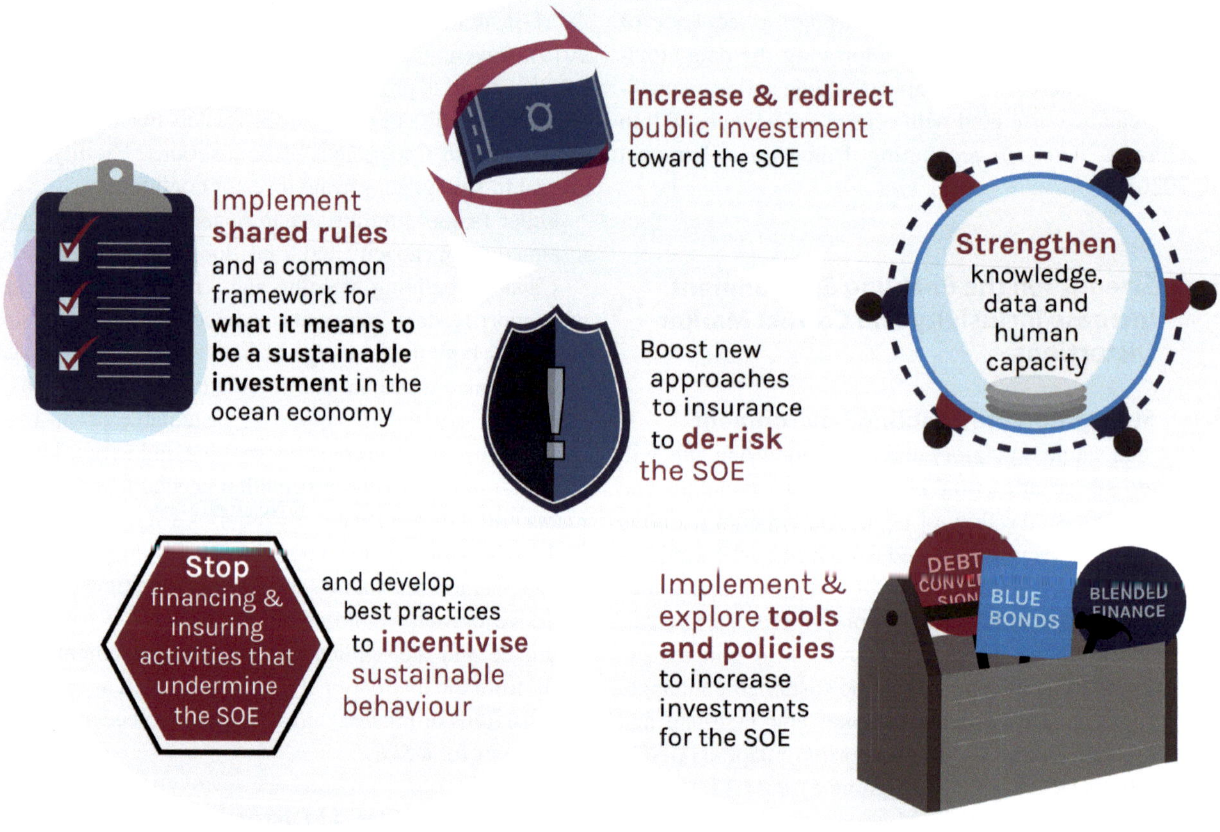

Fig. 9.3 Examples of opportunities for action by the private and public sectors in support of a SOE. *SOE* sustainable ocean economy. Source: Authors. Designed by Patricia Tiffany Angkiriwang

nity to align public and private incentives in the ocean economy. It is also a mechanism to share the wealth of ocean resources more broadly in society. The imposition of taxes, levies and fees on ocean economic activities, in combination with proper management measures—which may include assigning rights to or limiting access appropriately to these resources—can generate revenues to help bring about a SOE.

Auctions for access to ocean resources can help measure their value and generate funds to use for sustainable management and for the benefits of communities at large. The vessel day scheme of the Parties to the Nauru Agreement provides an example whereby vessel days are pooled and then auctioned to distant fishing nations. This ensures that the owners receive the full value of these ocean resources from users. Auctions need to consider community, customary and indigenous rights—for example, by reserving quotas for indigenous or local fishing communities or by establishing license banks and funding mechanisms for community fishing associations (Sumaila 2010).

Existing funds can also be used more wisely. Redirecting harmful subsidies to beneficial uses is an opportunity to catalyse a SOE and improve gender and other equalities (Österblom et al. 2020). For example, international negotiations and mandates, such as the Asia-Pacific Economic Cooperation, the Group of Twenty, SDG 14.6, the World Trade Organization negotiations and the Group of Seven, have called repeatedly to phase out inefficient fuel subsidies and distortive support measures (OECD 2018). This momentum for reform can be channeled into better policies for a SOE.

4.4 Stimulate the Pipeline of Investible Sustainable Projects

Some recent studies estimate the potential economic benefits of transition investments (e.g., WWF (World Wildlife Fund) 2019) and find that the return on investments can be high (Sumaila et al. 2012; Costello et al. 2016; World Bank 2017; Waldron et al. 2020). This opens the opportunity for finance to capture future economic gains in exchange for helping to pay for and smooth the transition. It is unlikely that a single financial transaction or institution will be responsible for bringing a green shipping business or fishery all the way through the policy reform process to environmental and economic sustainability. Yet a variety of mechanisms can blend early stage grant funding and concessional finance from philanthropic organisations and development finance institutions with later-stage capital from the private sector (EDF (Environmental Defense Fund) and Nicholas Institute for Environmental Policy Solutions 2018).

Investments into these kinds of SOE tools and approaches should be considered to be an essential part of any business or nation's risk-reduction and resilience-building plan. In particular, large-scale sustainable—and, more importantly, natural—ocean and coastal infrastructure spending must become a priority to address climate adaptation and build the resilience of the ocean to cumulative impacts. The Energy Sector Management Assistance Program, funded by the World Bank and the governments of World Bank Group members, helps low- and middle-income countries develop environmentally sustainable energy solutions, including offshore wind energy. Importantly, it does this by focusing on addressing poverty and knowledge-creation needs. The private sector can also play a key role in delivering sustainable coastal infrastructure at a local scale. For example, Swimsol, a company based in Europe, set up the first floating solar panels in the Maldives. It achieves a 3–8% rate of return from its investment by engaging in a long-term power purchasing agreement with its clients (either hotels or utilities). Both parties benefit from this agreement because Swimsol's solar power is 10–50% cheaper than its clients' current power generation costs, which are based on diesel generators.[2]

Efforts to deliver debt, equity and grants to key initiatives—including finance for the implementation of the high seas or ABNJ agreements currently being developed by the UN, the next phase of delivery on SDG 14 and the Decade of Ocean Science for Sustainable Development—would be achieved through the broad adoption of a common implementation framework and guidance that aligns with the Sustainable Blue Economy Finance Principles. In this regard, the Ocean Financing Initiative spearheaded by the ADB supports Asia-Pacific countries in developing bankable investments needed to meet SDG 14.

International institutions, such as the Intergovernmental Oceanographic Commission of the United Nations Educational, Scientific and Cultural Organization; the Food and Agriculture Organization of the United Nations; the United Nations Development Programme; UNEP; the OECD; and the United Nations Economic Commission for Africa, as well as international NGOs and others, also have a role in clearly communicating ocean challenges to their respective sectors (IOC-UNESCO (Intergovernmental Oceanographic Commission of the United Nations Educational, Scientific and Cultural Organization), IMO (International Maritime Organization), FAO (Food and Agriculture Organization of the United Nations), and UNDP (United Nations Development Programme) 2011).

By doing so, the impacts of a changing ocean on food, human health, development and the environment are highlighted and can be used to deliver funding strategies. Such dialogues will also present a good opportunity to develop an understanding with the finance community on the scale and

[2]Based on personal communication from D. Schmitz of Swimsol.

urgency of the issues that need to be tackled, of the importance of building in flexibility in investment time horizons, and of including a diverse stakeholder group (from decision-makers to the wider community) in governance and equity ownership.

In addition to addressing specific inequalities related to ocean governance and developed versus developing countries (e.g., capping carbon limits for developed countries; improving fisheries access regimes), developed countries should financially support developing maritime countries with sustainable ocean management. Individual projects and regulations that address specific components of ocean governance are important and necessary, but there is a larger picture of whole-domain ocean governance that requires significant financial capital. In addition, for SMEs beset with problems associated with economies of scale and high transaction costs, business technology and innovation incubators are needed.

4.5 Explore New Financing Mechanisms and Tools

New innovative financing vehicles can be created and launched by the private sector alone or in partnership with public entities in developed and developing countries alike. In fact, the latter can leapfrog to innovative financial instruments in support of a SOE, climate change mitigation and adaptation, and the environment in general. Two recent examples are described in Box 9.1.

Box 9.1. Innovative Financial Instruments

Example 1: Ghana launches funds to attain Sustainable Development Goals (SDGs). On 1 August 2019, Ghana introduced two funds with the objective of attracting finance to the country's effort to achieve the SDGs of the United Nations. The funds known as the SDG Green Fund and the SDG Delivery Fund will be mobilised and managed by the private sector, with government support—a kind of public-private partnership arrangement. The SDG Green Fund is geared towards the provision of clean and renewable energy ('Think Ocean') for use by industry, whereas the SDG Delivery Fund will draw finance to fund climate-smart activities. The funds are expected to raise billions of Ghanaian cedis (US $1 = 5.40 cedis) from voluntary contributions and corporate social responsibility initiatives from the private sector to support the country's efforts towards achieving the SDGs.

Example 2: The Development Bank of Southern Africa establishes a 2-billion-rand (US $142.86 million) Climate Finance Facility (CFF). The CFF will be available to infrastructure projects and businesses that mitigate or adapt to climate change. The finance facility raises capital from both private sector commercial banks and other development finance institutions. It co-funds projects by offering credit-enhancing products in the form of subordinated funding and/or tenor extension. The CFF provides risk mitigation where new technology is involved or the project and businesses are still in a developmental phase. This initiative applies the green bank concept. Green banks have been established in the developed world, but South Africa is probably the first to establish it in the developing world. The goal of these banks is to support the Paris Agreement and the SDGs. A similar example is the Seychelles blue bond. This blended finance combines a World Bank–guaranteed Global Environmental Facility concessionary loan and private investment to support a transition to sustainable fisheries and is implemented through the independent Seychelles Conservation and Climate Adaptation Trust and the national Development Bank of Seychelles.

New financing tools and access to capital markets are needed to act as a positive incentive for sustainable, inclusive and climate-resilient ocean activities.

Innovative mechanisms that bring new forms of finance into the system and are more accessible to communities in developing countries, particularly women, youth and marginalised communities, will need to be created while reducing the overexploitation of ocean resources. These tools can also facilitate effective management and governance while promoting the security of the ocean space in a context of increased access to new ocean resources. In the Caribbean, programs are currently being developed to provide small-scale fishers with micro-insurance policies to protect them from extreme weather risks and to provide governments with parametric policies that would help with the recovery of fisheries after an event with a severity that surpassed an agreed-upon threshold point (McConney et al. 2015).

In Madagascar, a partnership between a commercial seafood exporter and a local research institute produces juvenile sea cucumbers for locally managed sea cucumber aquaculture farms, thereby enabling locals to earn income while reducing exploitation of wild stocks. It is expected that these schemes will also indirectly strengthen conservation and fisheries management.

A noteworthy financial mechanism developed and implemented by Impact Investment Exchange (IIX) and the IIX Foundation USA, known as the IIX Sustainability Bonds (ISBs), explicitly targets the inclusion of women in economic activities. These bonds are debt securities that effectively mobilise large-scale private sector capital by pooling together a basket of high-impact entities (underlying borrowers) into a single structure. These instruments are then sold to impact investors and are paid back over time by the underlying borrowers, with a reasonable rate of interest. These ISBs can be listed on a social stock exchange to add an additional layer of transparency for both financial and impact performance.

Green/blue/climate bonds have to meet investment criteria and accountability requirements (e.g., Green Bond Principles; ESG criteria; and investment principles for sustainable fisheries) and certification to qualify for such labels and ensure the integrity of markets in the investment community. The Climate Bonds Initiative has put in place a number of sector criteria, including for marine energy and water utilities.[3] Other relevant initiatives include the Blue Natural Capital Positive Impacts Framework and the technical guideline for blue bonds (Roth et al. 2019).[4] Innovative financial instruments (e.g., green/blue bonds) are increasingly attractive and can generate new capital for sustainable ocean economic activities built on a healthy and well- managed ocean resource base (Hudson and Glemarec 2012; Miller et al. 2016; Thiele and Gerber 2017; Iyer et al. 2018; Walsh 2018; Jouffray et al. 2019).

Under a debt conversion program, also known as debt restructuring and formerly known as debt-for-nature swaps, negotiations take place whereby a portion of the debt owed to creditors is restructured and converted into agreed-upon initiatives that address, for instance, marine conservation and climate change. The debtors are then obligated to execute the initiatives. As an example, the government of Seychelles entered into a debt conversion program with the Paris Club, with the assistance of The Nature Conservancy (TNC). One of the conditions linked to the debt conversion was the development of the Seychelles Marine Spatial Plan. A new act was also passed to create the Seychelles Conservation and Climate Adaptation Trust, which provides a well-governed funding mechanism for the long-term financing of activities related to the stewardship of Seychelles' ocean resources and blue economy. We suggest that additional debt conversions be designed and implemented to support developing countries to implement ocean governance priorities. This will only be possible with full government commitment due to the comprehensive negotiations and related obligations.

Having the right kind of investment structure is critical to the success of innovative finance mechanisms. Trust funds and endowments, in particular, have a strong track record in assuring long-term funding sources for conservation and development projects (de Vos and Hart 2020). Ultimately, the most appropriate financing mechanism depends on many factors, including scale and debt/equity mix. It will be important to showcase the ability of these mechanisms in achieving beneficial financial, social and environmental returns (Bladon et al. 2014; Baumann et al. 2017; Fitzgerald et al. 2020).

Along with having the right model, a trusted project entity is needed to manage and distribute the funds across aggregated projects, reducing the overall project risk and transaction costs, especially when projects are small scale (Bladon et al. 2014). Local business communities can achieve this by acting collectively in networks and forming cooperatives (Lubchenco et al. 2016) that can substantially lower transaction costs, identified as a key priority for investments into fisheries (WWF (World Wildlife Fund) 2019). However, unless cooperatives have strong governance in place, they may not be suitable for large investment structures.

Although micro-finance continues to be important to many communities in the global south and east, innovative sustainable finance mechanisms should also play an important role in attracting and sustaining new forms of finance. These may include tried-and-tested models, such as payments for ecosystem services, debt- for-nature or adaptation swaps, new SOE investment funds or emerging MPA-financing models. Seychelles' innovative and blended financing mechanism has provided an important model at a national scale and has shown that developed countries have a strong role to play in supporting debt conversions that enable maritime developing countries to effectively invest in a SOE. The green finance space, which considers wider terrestrial sustainable finance challenges, has had a head start on sustainable ocean finance and may offer a wealth of experience, examples and best practices to adapt and apply to finance for the SOE.

For instance, blended finance can offer substantial opportunities to improve investor confidence by providing up-front low-interest or grant-based investments to strengthen the enabling environment—such as strengthening the governance and regulatory environment and restoring the resource base—towards reducing the risk profile and improving investor confidence. This might include investing in (1) improved fisheries policies as well as monitoring control and surveillance at sea, including ABNJ, to reduce illegal, unreported and unregulated (IUU) fishing and strengthen sustainable management of fisheries; (2) the effective protection of habitats and ecosystems, such as coral reefs, sea grass and man-

[3] For more information about the Climate Bonds Initiative, see https://www.climatebonds.net.

[4] To learn more about the Blue Natural Capital Positive Impacts Framework, see https://bluenaturalcapital.org.

groves that provide essential ecosystem services, including coastal protection and carbon sequestration; (3) technology and capacity for implementation of marine spatial planning to reduce user conflicts and ensure that the cumulative impacts of activities do not exceed the carrying capacity of the ecosystem; and (4) setting up investible entities that can substantially lower transaction costs and aggregate sustainable projects in a way that they become more investible. Larger projects can be structured as blue bonds and blended finance, whereas smaller projects may be more suitable for other forms of impact investing or other incremental investment approaches. The OECD's *Principles on Blended Finance* provides an important reference in this context (OECD 2017b).[5]

4.6 Stop Insuring Non-compliance and Develop Best Practices to Incentivise Sustainable Behaviour

The understanding and design of policies to deal with the negative effects of externalities is fundamental to achieving a SOE. This is because externalities underlie many aspects of our unsustainable behaviour. Whilst the SOE finance ecosystem develops, immediate action should be taken to avoid financing practices that support illegal and significantly harmful activities in the SOE, such as illegal fishing (Sumaila et al. 2020; Widjaja et al. 2020) and pollution (Jambeck et al. 2020), and work towards incentivising positive behaviour at both macro and micro levels.

Such activities pose risks to the ocean and have significant costs to people, the private sector (e.g., insurance companies) and governments. In 2017, an industry-wide statement against IUU fishing was launched, confirming the commitment of insurers, brokers and agents to not knowingly insure or facilitate the insuring of IUU fishing vessels (Miller et al. 2018). Economic instruments such as subsidies and taxes are cost-effective mechanisms that can help eliminate the effects of negative externalities while promoting positive externalities (Milazzo 1998; Akerlof and Kranton 2000; Ellickson 2001; Kübler 2001; Clark et al. 2005; Sumaila and Pauly 2007; Sumaila et al. 2010).

Institutional investors can use their influence to promote transparency and best practices in seafood and other ocean sectors, such as those implemented by the Fisheries Transparency Initiative (FiTi). Supply chain traceability is fundamental to the ability of any investor to exercise due diligence. The arguably low traceability in many marine seafood supply chains currently impedes this capacity of corporate investors to steer investments towards more sustainable practices. Yet catch documentation and certification schemes offer mechanisms to enhance traceability. To ensure scrutiny of corporate behaviour, investors should demand that companies show demonstrable efforts at achieving full-chain traceability, and that they fully declare their product mix and sourcing (including area and supplier). Furthermore, investors should insist on the systematic disclosure of metrics such as biomass produced, amount of antibiotics used and percentage of eco-certified products.

Corporate debt also provides a powerful source of influence for banks to promote sustainability in all ocean sectors. Although the literature on bank lending and environmental sustainability remains limited (Coulson and Monks 1999; Thompson and Cowton 2004; Jouffray et al. 2019), a recent review of the plausible power of banks in setting a sustainability agenda suggests that covenants are a key mechanism to examine further (Jouffray et al. 2019). By regulating the actions of the borrower, covenants can be important mechanisms through which banks can incentivise and steer companies towards implementing improved sustainability measures. Unlike blue or green loans, which are earmarked to finance a specific project, sustainability-linked covenants can be used for general corporate purposes. Providers of corporate debt should develop loan covenants based on the best available practices. Whereas such covenants will require tailoring to fit specific sectors, initiatives such as the Principles for Investment in Sustainable Wild-Caught Fisheries (EDF, Rare/Meloy Fund, and Encourage Capital 2018) and the Ocean Disclosure Project can serve as valuable baselines to be further developed.[6]

A good example of how to overcome these challenges is the introduction of special green investment funds in the Netherlands that are exempt from income tax, thus allowing investors in green projects to contract loans at reduced interest rates (usually around 2% less than commercial rates). Some of the approved funds include investments that aim to improve ocean sustainability by encouraging green shipping or incentivising cruise ships to use electric energy. These Dutch green funds have attracted more investment than can be utilised in the available schemes (Mountford and Keppler 1999), which is a very good sign for the future prospects of such initiatives.

4.7 Boost New Approaches to Insurance

The insurance industry has the potential to play three important roles: as risk managers, risk carriers, and investors. As risk managers and carriers who rely on research, modelling

[5]For an overview of some other instruments and how they enable investment, please consult *The Ocean Finance Handbook* (de Vos and Hart 2020).

[6]For more information about the Ocean Disclosure Project, see www.oceandisclosureproject.org.

and data analysis, insurers can recommend more sustainable practices to their clients and the communities they service. Insurers are also major institutional investors, and in this role, they can elect to support only those clients or projects that contribute to a SOE and to divest from those that do not. There is also an opportunity for all levels of government—local, national and international—to work with the insurance industry to promote the development of a SOE. At the local level, this could involve making improvements in risk modelling; at the national and international levels, policy and regulatory frameworks could be reshaped to incentivise responsible and sustainable maritime industry practices (Carr 2018; Laffoley and Baxter 2018; Niehörster and Murnane 2018).

Insurance can also be a source of finance that can be used both to leverage private investments such as blended finance and/or directly invest in the conservation and sustainable use of the ocean.

Mechanisms to address local externalities can be designed and implemented by a single country, but it is much more challenging to address global externalities.

In June 2019, the first-ever insurance policy on natural infrastructure was put in place on a portion of the Mesoamerican Reef in the Mexican Caribbean. Created through an initiative led by TNC, this policy secures funding to repair damages to the reef following a hurricane, preventing long-term damage and enhancing protection of the onshore community. Studies have been completed that estimate the monetary values of the protection offered by coastal habitats, and these findings justify the development of the Mesoamerican Reef policy and the future creation of insurance policies elsewhere, which TNC hopes to pursue.

Insurance companies could design novel products that proactively seek commercial opportunities. For example, TNC, Swiss Re and several partners have developed a coral reef insurance mechanism in Quintana Roo, Mexico, to provide finance to build the resilience of coral reefs to storm damage and to fund restoration activities in the event of a large storm event (Iyer et al. 2018) (Fig. 9.4).

There appears to be great potential for the creation of additional insurance products and services that can contribute to the creation of a SOE. The insurance industry could become a leader in tackling the issue of marine plastic pollution (Lau et al. 2020), for example. Insurers could provide insurance statistics to show the extent to which plastic debris enters the ocean from at-sea sources and its economic impact. Similar to what has been done on the topic of IUU fishing, insurance industry guidelines and other strategies for influencing reduced plastic marine pollution from at-sea sources, seabed mining and biotechnology investment could be developed.

The Pacific Ocean Finance Program is funding the analysis and development of novel ocean insurance products for the Pacific Islands region in partnership with Willis Towers Watson (Wharton and Young Forthcoming). Three draft concepts for the potential application of parametric insurance to support ocean health are in development, including (1) parametric cyclone insurance for a segment of the Great Sea Reef in Fiji to incentivise preparedness and finance rapid response and early recovery after major cyclone shock events, (2)

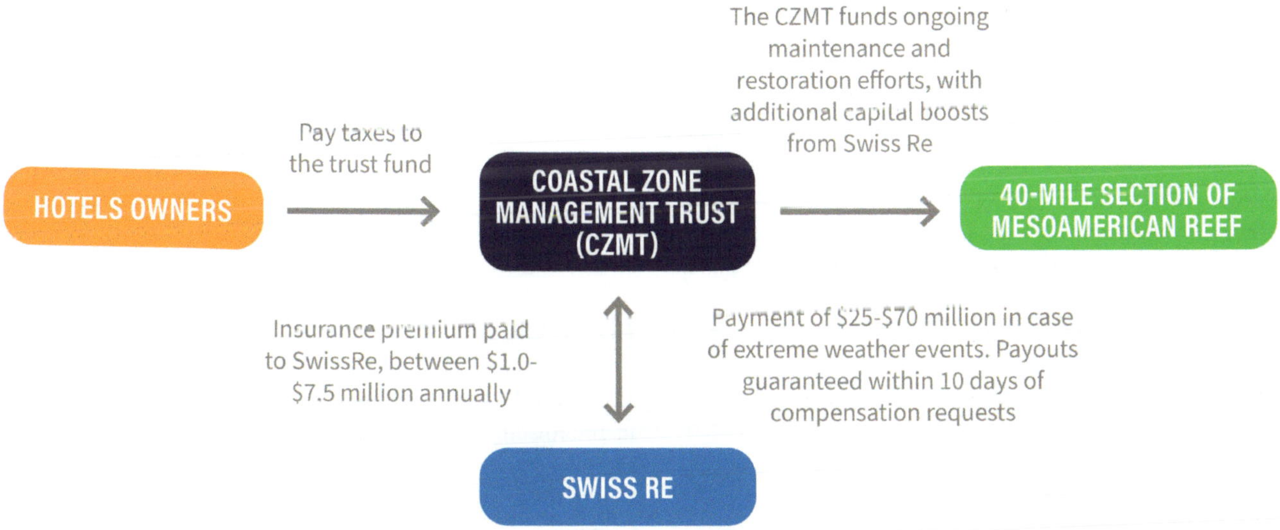

Fig. 9.4 Simplified conceptual diagram of coral reef insurance for the Mesoamerican reef. Note: All figures are in US $. Source: Iyer et al. (2018)

parametric insurance for marine thermal shock events using a sea surface temperature index to help mitigate the economic consequences of tourism revenue decline due to sudden natural asset degradation in Palau and (3) livelihood protection as a social benefit through parametric insurance to support fishers' resilience and incentivise improved fisheries management in Vanuatu.

In addition, the ADB, the Global Environment Facility, and TNC are collaborating on developing natural capital insurance products for coral reefs in Asia and the Pacific Islands. Another relevant initiative is the Ocean Risk and Resilience Action Alliance, launched at the UN Secretary-General's Climate Action Summit in 2019, which is designed to help drive the development of innovative finance products to regenerate coastal natural capital and build resilience in the world's most exposed and vulnerable regions and communities.

The marine insurance sector has already begun implementing strategies to manage and reduce ocean- related risks. The development of this statement was facilitated by the environmental non-profit organisation Oceana and UNEP's PSI initiative. To date, more than 30 insurers, insurance market bodies and key stakeholders, spread across five continents, have signed and supported this statement, including some of the world's largest companies. Working again with the PSI Secretariat and with contributions from industry stakeholders, Oceana developed risk management guidelines that were launched in February 2019 (Miller et al. 2018). These guidelines help insurers avoid contracts associated with IUU fishing and improve transparency and accountability within the global fishing sector. The FiTi and other transparency programs can help guide and support appropriate investment in the ocean economy.

Looking to the future, insurers can also follow guidance issued by the PSI to determine how they can best contribute to a SOE and manage ESG risks. Together with the support of industry contributors, the PSI Secretariat recently developed a draft guide for managing ESG risks in the non-life insurance business (UNEP FI and PSI (Principles for Sustainable Insurance) 2019). The guide contains heatmaps indicating areas of potential elevated risk within various economic sectors and lines of insurance, and it also provides a list of standards and technical guidelines that are available to help insurers identify, assess and mitigate risks.

5 Conclusions

A healthy ocean that supports a SOE requires a range of interventions to improve governance, science and management; finance is an important enabler of a SOE and the major driver behind all ocean-based commercial activities. The best ocean policies and practices can be undone by inadequate financing and by economic externalities that undermine conservation and sustainable use.

This Blue Paper provides an evaluation of how economic instruments and finance mechanisms can be applied to realise a SOE. To turn ocean sustainability challenges into opportunities, the public and private sectors need to create and better mobilise a full suite of financial tools and approaches, insurance, and fiscal and market incentives. Additionally, they need to strengthen key aspects of the enabling environment to support the transition to an ocean economy that is sustainable and inclusive; this can be accomplished by making the benefits the SOE generates available to all, especially women, youth and marginalised communities.

The most significant action will be to influence future mainstream finance. By providing clear principles, guiding frameworks and metrics, and by proactively avoiding the financing of known illegal and harmful activities, trillions of dollars of ocean finance could be redirected towards sustainable development pathways, creating long-term and positive systemic change.

If our suggestion to allocate a higher proportion of ocean GDP to attaining a SOE is followed by half of the world's maritime countries, that alone could generate the seed money needed to incentivise the kind of public and private investments needed to ensure a SOE. The big message from this contribution is that a significant increase in sustainable ocean finance will be required to ensure a SOE that benefits all, including a broad section of society and businesses.

Acknowledgements The authors thank the paper's technical reviewers, Chip Cunliffe, Klaas de Vos, Justin Mundy and Kelly Wachowicz; its arbiters, Mari Elka Pangestu and Adrien Vincent; as well as the Expert Group co-chairs, who all provided helpful technical comments. The authors also thank World Resources Institute for providing support as the Ocean Panel Secretariat.

Although numerous colleagues were very generous with their time and input, this report reflects the views of the authors alone and does not necessarily reflect the views of the organisations or their countries of employment. Finally, we thank Patricia Tiffany Angkiriwang for the design of Figs. 9.1, 9.2, and 9.3. The authors thank Lauri Scherer for copyediting and Romain Warnault for design.

About the Authors

U. Rashid Sumaila is Professor and Canada Research Chair in Interdisciplinary Oceans and Fisheries Economics at the Institute for the Oceans and Fisheries and the School of Public Policy and Global Affairs at the University of British Columbia.

Melissa Walsh is a Blue Finance Specialist with the Asian Development Bank and also a principal at Marine Conservation Finance Consulting.

Kelly Hoareau is the Director of the James Michel Blue Economy Research Institute at the University of Seychelles.

Anthony Cox is the Deputy Director of the Environment Directorate of the OECD.

Patrízia Abdallah is Professor and Acting Director of the Institute of Economy, Administration and Accounting at the Federal University of Rio Grande in Brazil.

Wisdom Akpalu is Associate Professor of Economics and the Dean of the School of Research and Graduate Studies at the Ghana Institute of Management and Public Administration.

Zuzy Anna is the Executive Director of the Center for Sustainable Development Goals Studies at Universitas Padjadjaran in Bandung, Indonesia.

Dominique Benzaken is a Senior Policy Expert and PhD candidate on the sustainable blue economy, ocean governance and islands with the Australian National Centre for Ocean Resources and Security.

Beatrice Crona is Executive Director of the Global Economic Dynamics and the Biosphere Programme at the Royal Swedish Academy of Sciences as well as Associate Professor and Deputy Science Director at the Stockholm Resilience Centre.

Timothy Fitzgerald is Senior Director of Oceans Climate Strategies at the Environmental Defense Fund.

Louise Heaps is the Global Lead on the sustainable blue economy for WWF and is based at WWF-UK.

Katia Karousakis is an environmental economist at the OECD.

Glenn-Marie Lange is a Senior Environmental Economist with the Environment and Natural Resources Global Practice and Chief Technical Adviser at the World Bank.

Amanda Leland is Executive Vice President of the Environmental Defense Fund and oversees programs including climate, energy, health, ecosystems, and oceans.

Dana Miller is a Senior Policy Advisor for Oceana in Europe.

Karen Sack is currently President and Chief Executive Officer of Ocean Unite. She was formerly the Senior Director for International Oceans at Pew Charitable Trusts, where she helped initiate the Global Ocean Commission.

Durreen Shahnaz is the founder and Chief Executive Officer of Impact Investment Exchange in the United States.

Louise Teh is a Research Associate in Interdisciplinary Oceans and Fisheries Economics at the Institute for the Oceans and Fisheries and the School of Public Policy and Global Affairs at the University of British Columbia.

Torsten Thiele is a Visiting Fellow at the Institute of Global Affairs at the London School of Economics and Political Science, where he leads on blue finance, and a Senior Research Associate at the Institute for Advanced Sustainability Studies in Potsdam.

Niels Vestergaard is Professor and Head of the Department of Sociology, Environmental and Business Economics at the University of Southern Denmark.

Nobuyuki Yagi is Professor at the University of Tokyo and Member of the Science Council of Japan.

Junjie Zhang is an Associate Professor of Environmental Economics at Duke Kunshan University and in the Nicholas School of the Environment at Duke University.

Appendix A: Definitions of Sustainable Ocean Economy and Blue Economy

Asian development bank	The environmental, social and economic sustainability of sectors that impact and/or derive economic activity from the ocean
Center for the Blue Economy	It uses three related but distinct meanings: (1) the overall contribution of the ocean to economies, (2) the need to address the environmental and ecological sustainability of the oceans, and (3) the ocean economy as a growth opportunity for both developed and developing countries
Economist Intelligence Unit (2015 adapted working definition)	A sustainable ocean economy emerges when economic activity is in balance with the long-term capacity of ocean ecosystems to support this activity and remain resilient and healthy
Organisation for Economic Co-operation and Development	The ocean economy can be defined as the sum of the economic activities of ocean-based industries as well as the assets, goods and services of marine ecosystems
United Nations Development Program	Protecting and restoring ocean resources while increasing the economic activity derived from the ocean
World Bank	The sustainable use of ocean resources for economic growth, improved livelihoods and jobs while preserving the health of the ocean ecosystem
World Wide Fund for Nature	A sustainable blue economy is one which provides social and economic benefits for current and future generations; restores, protects and maintains diverse, productive and resilient ecosystems; and is based on clean technologies, renewable energy and circular material flows

Appendix B: The MPA Financing Gap

Binet et al. (2015) assessed financing needs and gaps for MPAs in the Mediterranean and found that the 14 countries studied funded their MPA systems to the tune of nearly $60.5 million per year. The authors also found that the financing needs for effective management of these MPAs was much higher, resulting in a financing gap (available funds minus financial needs) of $776.4 million per year. Data reported in Sumaila et al. (2019) reveals that the total cost of establishing and maintaining MPAs in 2018 worldwide was $2.3 billion. It is known that not all of the currently declared MPA area of 7.3% of the ocean surface is adequately protected. In fact, only 2.3% is currently 'highly' or 'fully protected', and most of the remaining 5% is not protected at all (Sala et al. 2018). Hence, to get to 10% of highly or fully protected areas from the current 2.3%, $7.7 billion is needed globally. Clearly, and as suggested by Laffoley et al. (2020), adequate, comprehensive and effective funding mechanisms will need to be put in place to deliver the actions required for integrated ocean management in support of a SOE.

Although the costs of establishing and running MPAs (which should be more appropriately seen as investments) are high, there are numerous benefits for implementing MPAs. Effectively managed and located MPAs reduce fishing pressure and increase habitat protection and ecosystem resilience (Costello 2014). This can lead to ecological benefits such as an increase in the abundance, diversity, size and biomass of fish and invertebrate species (McClanahan et al. 2007; Russ et al. 2008; Lester et al. 2009). MPAs may also help marine organisms, ecosystems and societies adapt to climate change by protecting habitats from harm and degradation, thereby reducing the effects of climate change. For example, intact coastal ecosystems can reduce the risks arising from more frequent and severe storms and flooding (Roberts et al. 2017). The ecological benefits from MPAs can translate into economic benefits. For instance, fisheries benefits can arise from the spillover of fish biomass from inside the MPA to fished areas outside the MPA (Russ et al. 2004; Goñi et al. 2008). Well-implemented MPAs can also benefit the tourism and recreation sectors (Ballantine 2014) and provide ecosystem goods and services (e.g., coastal protection from coral reefs, mangroves and sea grass) (Davis et al. 2019). These studies show the insurance, market and non-market values of protecting a significant portion of the ocean portfolio and highlight the fact that establishing MPAs not only support social equity and ocean health but also make economic sense.

Appendix C: The Types and Sources of Capital for Financing a SOE

Several capital types are currently available that can be used to either finance a SOE or serve as the basis for developing new innovative ones that can better serve the ocean economy. A broad categorisation of capital types is provided by de Vos and Hart (2020). The deployment of these different types of capital by investors depends on the expected returns from the investment, which, in turn, depend on the risk-return equation faced by investors (Table 9.2).

Table 9.2 Capital types and their use depending on expected returns

Capital type	Description	Expected return (market or below market, including <0 return)
Impact only • Corporate social responsibility investment • Public grants • Philanthropic grants • Public financing • Official development assistance	This is usually long term but small-scale in comparison to larger types of commercial finance	Below-market return
Debt • Loans • Bonds	This is a low-risk, low-reward type of cap- ital. Debt providers do not have the same level of influence over an investment as equity investors	Market return
Equity • Public equity • Equity investment	Equity is based on taking an ownership stake in an investment; some types of equity (e.g., venture capital) are high risk, high reward	Greater-than-market return
Blended finance	This combines official development assistance with other private or public resources in order to 'leverage' additional funds from other actors	Below-market return

Note: See de Vos and Hart (2020) for more details and examples

Some of the needed investments in a SOE are likely to generate competitive expected **market returns** and should be better promoted/simplified for private investors. Other investments should expect positive but **below-market returns**. In this case, the blending of private and public capital can still deliver adequate returns to investors, such as impact funds. Other investments are 'pure costs' and need socially beneficial subsidies (i.e., those that help society eliminate negative externalities or reinforce positive ones), public investments and philanthropy (grants) to work (Table 9.2).

Appendix D: Threats to the Ocean Economy

The following environmental and social impacts have the potential to undermine the sustainability of the ocean economy.

Environmental Impacts

Fishing and Capture of Marine Life

Overfishing, whaling and shark finning; habitat destruction from fishing gear and practices; ghost fishing; discards, bycatch and entanglement

Marine Pollution

Plastics, mercury and other heavy metals; garbage; and land-based pollution, including nutrients and agri-chemicals, directly harm marine organisms and ecosystems (e.g., ingestion of plastics, algal blooms, eutrophication)

Climate Change

Ocean warming; ocean acidification, sea level rise, more frequent events, hypoxia and dead zones; melting ice caps

Marine Mining, Offshore Oil and Gas

Biodiversity and habitat loss; oil spills

Fish Farming

Escapees; overuse of antibiotics; excessive use of fishmeal and oil

Coastal and Marine Tourism

Habitat destruction and damage from the construction and operation of tourism infrastructure; impact of tours and activities on habitat and biodiversity; wastewater and garbage pollution from tourists

Coastal Development

Habitat destruction and damage from the construction and operation of coastal infrastructure; impact of coastal cities on habitat and biodiversity; wastewater and garbage pollution from coastal populations

Ports and Shipping

Habitat damage; ship groundings; anchor damage; the dumping of rubbish; invasive species from ballast water; oily waste

Social Impacts

Ocean Grabbing/Blue-Washing

Delineation of ocean space that marginalises certain groups, resulting in loss of livelihoods, and compromises food security; hasty planning and limited resources can impact integrated ocean management.

Perverse Economic Incentives

Fisheries and other sectoral subsidies that harm marine ecosystems while favouring industrial scale operators, compromise food security and put small-scale operators and women at a disadvantage.

Global Markets

The pursuit to service global markets can jeopardise locals' access to ocean resources, compromising food security and livelihoods; gains generated in distant markets also rarely trickle down to local producers.

References

ADB (Asian Development Bank) (2020) Explainer: the role of ocean finance in transitioning to a blue economy in asia and the pacific. Development Asia. https://development.asia/explainer/role-ocean-finance-transitioning-blue-economy-asia-and-pacific

AIMS (Australian Institute of Marine Science) (2018) The AIMS index of marine industry. AIMS, Townsville. https://www.aims.gov.au/sites/default/files/2018%20AIMS%20Marine%20Index.pdf

Akerlof GA, Kranton RE (2000) Economics and identity. Q J Econ 115(3):715–753. https://doi.org/10.1162/003355300554881

Ballantine B (2014) Fifty years on: lessons from marine reserves in New Zealand and principles for a worldwide network. Biol Conserv 176(August):297–307. https://doi.org/10.1016/j.biocon.2014.01.014

Baumann K, Havemann T, Werneck F, Negra C, Nair S (2017) Capitalising conservation: how conservation organisations can engage with investors to mobilise capital. Clarmondial AG, Zurich. http://cpicfinance.com/wp-content/uploads/2018/05/Capitalising_Conservation_Clarmondial_WWF.pdf

Binet T, Diazabakana A, Hernandez S (2015) Sustainable financing of marine protected areas in the Mediterranean: a financial analysis. MedPAN, Marseille. http://www.rac-spa.org/sites/default/files/doc_medmpanet/final_docs_regional/55_study_on_the_sustainable_financing_of_mediterranean_mpas.pdf

Bladon A, Mohammed EY, Milner-Gulland EJ (2014) A review of conservation trust funds for sustainable marine resources management: conditions for success. Working paper. International Institute for Environment and Development, London. https://pubs.iied.org/pdfs/16574IIED.pdf

Carr S (ed) (2018) Can we insure our way to healthier oceans and communities? Skimmer on Marine Ecosystems and Management (blog). https://meam.openchannels.org/news/skimmer-marine-ecosystems-and-management/can-we-insure-our-way-healthier-oceans-and-ocean

Clark CW, Munro GR, Sumaila UR (2005) Subsidies, buybacks, and sustainable fisheries. J Environ Econ Manag 50(1):47–58. https://doi.org/10.1016/j.jeem.2004.11.002

Coady D, Parry I, Le NP, Shang B (2019) Global fossil fuel subsidies remain large: an update based on country-level estimates. Working paper 19/89. International Monetary Fund, Washington. https://www.imf.org/en/Publications/WP/Issues/2019/05/02/Global-Fossil-Fuel-Subsidies-Remain-Large-An-Update-Based-on-Country-Level-Estimates-46509

Costello MJ (2014) Long live marine reserves: a review of experiences and benefits. Biol Conserv 176(August):289–296. https://doi.org/10.1016/j.biocon.2014.04.023

Costello C, Ovando D, Clavelle T, Strauss CK, Hilborn R, Melnychuk MC, Branch TA et al (2016) Global fishery prospects under contrasting management regimes. Proc Natl Acad Sci U S A 113(18):5125–5129. https://doi.org/10.1073/pnas.1520420113

Coulson AB, Monks V (1999) Corporate environmental performance considerations within bank lending decisions.

Eco-Manage Audit 6(1):1–10. https://doi.org/10.1002/(SICI)1099-0925(199903)6:1<1::AID-EMA93>3.0.CO;2-M

Davis KJ, Vianna GMS, Meeuwig JJ, Meekan MG, Pannell DJ (2019) Estimating the economic benefits and costs of highly-protected marine protected areas. Ecosphere 10(10):e02879. https://doi.org/10.1002/ecs2.2879

de Vos K, Hart B (2020) The ocean finance handbook: increasing finance for a healthy ocean. Friends of Ocean Action, Geneva. http://www3.weforum.org/docs/WEF_FOA_The_Ocean_Finance_Handbook_April_2020.pdf

EDB (Economic Development Board Mauritius) (2020) Blue economy: the development of a new pole of growth through the sustainable use of our maritime zone. https://www.edbmauritius.org/opportunities/ocean-economy/

EDF (Environmental Defense Fund) and Nicholas Institute for Environmental Policy Solutions (2018) Financing fisheries reform: blended capital approaches in support of sustainable wild-capture fisheries. EDF, New York. https://www.edf.org/sites/default/files/documents/financing-fisheries-reform.pdf

EDF, Rare/Meloy Fund, and Encourage Capital (2018) Principles for investment in sustainable wild-caught fisheries. EDF and Encourage Capital, New York. http://www.fisheriesprinciples.org/files/2019/05/updated-PrinciplesInvestmentWEB_final.pdf

Ellickson RC (2001) The market for social norms. Am Law Econ Rev 3(1):1–49. https://www.jstor.com/stable/42705380

European Commission (2020) Taxonomy: final report of the technical expert group on sustainable finance. Technical Expert Group on Sustainable Finance, European Commission, Brussels. https://ec.europa.eu/info/sites/info/files/business_economy_euro/banking_and_finance/documents/200309-sustainable-finance-teg-final-report-taxonomy_en.pdf

FAO (2018) The state of world fisheries and aquaculture 2018: meeting the sustainable development goals. FAO, Rome. http://www.fao.org/3/i9540en/i9540en.pdf

FAO (Food and Agriculture Organization of the United Nations) (2013) Implementing improved tenure governance in fisheries: a technical guide to support the implementation of the voluntary guidelines on the responsible governance of tenure of land, fisheries and forests in the context of national food security. FAO, Rome. http://www.fao.org/3/i3420e/i3420e.pdf

Fenichel EP, Milligan B, Porras I, Addicott ET, Árnasson R, Bordt M, Djavidnia S et al (2020) National accounting for the ocean and ocean economy. World Resources Institute, Washington. https://oceanpanel.org/sites/default/files/2020-08/National%20Accounting%20for%20the%20Ocean%20Final%20Full%20Paper.pdf

Fitzgerald TP, Higgins PR, Quilligan E, Sethi SA, Tobin-de la Puente J (2020) Catalyzing fisheries conservation investment. Front Ecol Environ 18(3):151–158. https://doi.org/10.1002/fee.2147

Fritsch D (2020) Investors and the blue economy. Credit Suisse, Zurich

Gaines S, Cabral R, Free C, Golbuu Y, Arnason R, Battista W, Bradley D et al (2019) The expected impacts of climate change on the ocean economy. World Resources Institute, Washington. https://www.oceanpanel.org/sites/default/files/2019-12/expected-impacts-climate-change-on-the-ocean-economy.pdf

Goñi R, Adlerstein S, Alvarez-Berastegui D, Forcada A, Reñones O, Criquet G, Polti S et al (2008) Spillover from six western Mediterranean marine protected areas: evidence from artisanal fisheries. Mar Ecol Prog Ser 366(August):159–174. https://doi.org/10.3354/meps07532

Grace L, van Anrroy R (2019) Guidelines for micro-finance and credit services in support of small-scale fisheries in Asia: a handbook for finance and fisheries stakeholders. FAO, Rome. http://www.fao.org/3/ca5128en/CA5128EN.pdf

Harper S, Adshade M, Lam VWY, Pauly D, Sumaila UR (2020) Valuing invisible catches: estimating the global contribution by women to small-scale marine capture fisheries production. PLoS ONE 15(3):e0228912. https://doi.org/10.1371/journal.pone.0228912

Hernández-Delgado EA (2015) The emerging threats of climate change on tropical coastal ecosystem services, public health, local economies and livelihood sustainability of small islands: cumulative impacts and synergies. Mar Pollut Bull 101(1):5–28. https://doi.org/10.1016/j.marpolbul.2015.09.018

Hoegh-Guldberg O et al (2015) Reviving the ocean economy: the case for action—2015. WWF International, Gland. https://c402277.ssl.cf1.rackcdn.com/publications/790/files/original/Reviving_Ocean_Economy_REPORT_low_res.pdf?1429717323

Hoegh-Guldberg O et al (2019) The ocean as a solution to climate change: five opportunities for action. World Resources Institute, Washington. https://oceanpanel.org/sites/default/files/2019-10/HLP_Report_Ocean_Solution_Climate_Change_final.pdf

Holmes L, Strauss CK, de Vos K, Bonzon K (2014) Towards investment in sustainable fisheries: a framework for financing the transition. Environmental Defense Fund, New York. https://www.edf.org/sites/default/files/content/towards-investment-in-sustainable-fisheries.pdf

Hudson A, Glemarec Y (2012) Catalyzing ocean finance: transforming markets to restore and protect the global ocean, vol 1. United Nations Development Programme, New York. https://www.undp.org/content/undp/en/home/librarypage/environment-energy/water_governance/ocean_and_coastalareagovernance/catalysing-ocean-finance.html

Huwyler F, Kappeli J, Serafimova K, Swanson E, Tobin J (2014) Conservation finance moving beyond donor funding toward an investor-driven approach. Credit Suisse, Zurich

IOC-UNESCO (Intergovernmental Oceanographic Commission of the United Nations Educational, Scientific and Cultural Organization), IMO (International Maritime Organization), FAO (Food and Agriculture Organization of the United Nations), and UNDP (United Nations Development Programme) (2011) A blueprint for ocean and coastal sustainability. IOC-UNESCO, Paris

IPBES (Intergovernmental Science-Policy Platform on Biodiversity and Ecosystem Services) (2019) Global assessment report on biodiversity and ecosystem services of the intergovernmental science-policy platform on biodiversity and ecosystem services. IPBES Secretariat, Bonn. https://ipbes.net/global-assessment

IPCC (Intergovernmental Panel on Climate Change) (2019) Summary for policymakers. In: Pörtner HO, Roberts DC, Masson-Delmotte V, Zhai P, Tignor M, Poloczanska E, Mintenbeck K et al (eds) IPCC special report on the ocean and cryosphere in a changing climate. IPCC, Geneva. https://www.ipcc.ch/site/assets/uploads/sites/3/2019/11/03_SROCC_SPM_FINAL.pdf

IRENA (International Renewable Energy Agency) (2018) Offshore innovation widens renewable energy options: opportunities, challenges and the vital role of international cooperation to spur the global energy transformation. IRENA, Abu Dhabi. https://www.irena.org/publications/2018/Sep/Offshore-innovation-widens-renewable-energy-options

Iyer V, Mathias K, Meyers D, Victurine R, Walsh M (2018) Finance tools for coral reef conservation: a guide. Wildlife Conservation Society, New York. https://static1.squarespace.com/static/57e1f17b37c58156a98f1ee4/t/5c7d85219b747a7942c16e01/1551730017189/50+Reefs+Finance+Guide+FINAL-sm.pdf

Jambeck J, Moss E, Kumar Dubey B et al (2020) Pollution and a regenerative economy: municipal, industrial, agricultural, and maritime waste, its impacts, and solutions. World Resources Institute, Washington

Jouffray J-B, Crona B, Wassénius E, Bebbington J, Scholtens B (2019) Leverage points in the financial sector for seafood sustainability. Sci Adv 5(10):eaax3324. https://doi.org/10.1126/sciadv.aax3324

Jouffray J-B, Blasiak R, Norström AV, Österblom H, Nyström M (2020) The blue acceleration: the trajectory of human expansion into the ocean. One Earth 2(1):43–54. https://doi.org/10.1016/j.oneear.2019.12.016

Koh H, Karamchandani A, Katz R (2012) From blueprint to scale: the case for philanthropy in impact investing. Monitor Group, New York. https://acumen.org/wp-content/uploads/2017/09/From-Blueprint-to-Scale-Case-for-Philanthropy-in-Impact-Investing_Full-report.pdf

Kübler D (2001) On the regulation of social norms. J Law Econ Org 17(2):449–476. https://www.jstor.org/stable/3554935

Laffoley D, Baxter JM (2018) Ocean connections: an introduction to rising risks from a warming, changing ocean. International Union for Conservation of Nature, Gland. https://portals.iucn.org/library/sites/library/files/documents/2018-021-En.pdf

Laffoley D, Baxter JM, Amon DJ, Currie DEJ, Downs CA, Hall- Spencer JM, Harden-Davies H et al (2020) Eight urgent, fundamental and simultaneous steps needed to restore ocean health, and the consequences for humanity and the planet of inaction or delay. Aquat Conserv Mar Freshwat Ecosyst 30(1):194–208. https://doi.org/10.1002/aqc.3182

Lau WWY, Shiran Y, Bailey RM, Cook E, Stuchtey MR, Koskella J, Velis CA et al (2020) Evaluating scenarios toward zero plastic pollution. Science July:eaba9475. https://doi.org/10.1126/science.aba9475

Lee T, Nam H (2017) A study on green shipping in major countries: in the view of shipyards, shipping companies, ports, and policies. Asian J Ship Logist 33(4):253–262. https://doi.org/10.1016/j.ajsl.2017.12.009

Lester SE, Halpern BS, Grorud-Colvert K, Lubchenco J, Ruttenberg BI, Gaines SD, Airamé S, Warner RR (2009) Biological effects within no-take marine reserves: a global synthesis. Mar Ecol Prog Ser 384(May):33–46. https://doi.org/10.3354/meps08029

Lubchenco J, Cerny-Chipman EB, Reimer JN, Levin SA (2016) The right incentives enable ocean sustainability successes and provide hope for the future. Proc Natl Acad Sci U S A 113(51):14507–14514. https://doi.org/10.1073/pnas.1604982113

Martini R, Innes J (2018) Relative effects of fisheries support policies. Food, agriculture and fisheries paper 115. Organisation for Economic Co-operation and Development, Paris. https://doi.org/10.1787/bd9b0dc3-en

McClanahan TR, Graham NAJ, Calnan JM, MacNeil MA (2007) Toward pristine biomass: reef fish recovery in coral reef marine protected areas in Kenya. Ecol Appl 17(4):1055–1067. https://doi.org/10.1890/06-1450

McConney P, Cox S-A, Parsram K (2015) Building food security and resilience into fisheries governance in the Eastern Caribbean. Reg Environ Chang 15(7):1355–1365. https://doi.org/10.1007/s10113-014-0703-z

Milazzo M (1998) Subsidies in world fisheries: a reexamination. Technical paper 406. World Bank, Washington. http://documents.worldbank.org/curated/en/133031468776403491/Subsidies-in-world-fisheries-a-re-examination

Miller DD, Sumaila UR, Copeland D, Zeller D, Soyer B, Nikaki T, Leloudas G, Fjellberg ST, Singleton R, Pauly D (2016) Cutting a lifeline to maritime crime: marine insurance and IUU fishing. Front Ecol Environ 14(7):357–362. https://doi.org/10.1002/fee.1293

Miller DD, Bacani B, Fabry O (2018) Risk assessment and control of IUU fishing for the marine insurance industry: guidelines to control or mitigate the risk of insuring vessels and companies associated with illegal, unreported and unregulated (IUU) fishing. Oceana, Madrid. https://eu.oceana.org/sites/default/files/oceana-psi_insurance_insdustry_guidelines_for_iuu_fishing_final_with_doi.pdf

Morgan JP, GIIN (Global Impact Investing Network) (2014) Spotlight on the market: the impact investor survey. J.P. Morgan and GIIN, New York. https://thegiin.org/assets/documents/pub/2014MarketSpotlight.PDF

Mountford H, Keppler JH (1999) Financing incentives for the protection of biodiversity. Sci Total Environ 240(1–3):133–144. https://doi.org/10.1016/S0048-9697(99)00312-5

Niehörster F, Murnane RJ (2018) Ocean risk and the insurance industry. XL Catlin Services, London. https://www.oceanrisksummit.com/Content/press-releases/FALK-MAIN-REPORT-FINAL-LOW-RES.pdf

NOAA (National Oceanic and Atmospheric Administration) (2019) NOAA report on the US ocean and great lakes economy. NOAA Office for Coastal Management, Charleston. https://coast.noaa.gov/data/digitalcoast/pdf/econ-report.pdf

Noone KJ, Sumaila UR, Diaz RJ (eds) (2013) Managing ocean environments in a changing climate: sustainability and economic perspectives. Elsevier, London

OECD (2016) The ocean economy in 2030. OECD, Paris. https://doi.org/10.1787/9789264251724-en

OECD (2017a) Marine protected areas: economic, management and policy mixes. OECD, Paris. https://doi.org/10.1787/9789264276208-en

OECD (2017b) Principles on blended finance. OECD, Paris. http://www.oecd.org/dac/financing-sustainable-development/blended-finance-principles/

OECD (2018) OECD companion to the inventory of support measures for fossil fuels 2018. OECD, Paris. https://doi.org/10.1787/9789264286061-en

OECD (2019a) Biodiversity: finance and the economic and business case for action. OECD, Paris. https://www.oecd.org/environment/resources/biodiversity/G7-report-Biodiversity-Finance-and-the-Economic-and-Business-Case-for-Action.pdf

OECD (2019b) Environment at a glance indicators. OECD, Paris. https://doi.org/10.1787/ac4b8b89-en

OECD (n.d.) OECD work in support of a sustainable ocean. https://www.oecd.org/ocean/#:~:text=The%20ocean%20covers%20two%2Dthirds,vital%20for%20human%20well%2Dbeing.&text=Working%20with%20both%20developed%20and,a%20sustainable%20and%20inclusive%20basis. Accessed 1 September 2020

OECD (Organisation for Economic Co-operation and Development) (2006) Development co-operation report 2005. Efforts and policies of the members of the development assistance committee. OECD, Paris. https://doi.org/10.1787/dcr-2005-en

Österblom H, Wabnitz C, Tladi D et al (2020) Towards ocean equity. World Resources Institute, Washington. https://www.oceanpanel.org/sites/default/files/2020-04/towards-ocean-equity.pdf

Our Ocean (2019) Commitments 2019. https://ourocean2019.no/commitments/

PEMSEA (2015) Integrated coastal management and sustainable development of coasts and oceans. PEMSEA, Quezon City

PEMSEA (Partnerships in Environmental Management for the Seas of East Asia) (2009) The marine economy in times of change. Trop Coasts 16(1):1. http://pemsea.org/sites/default/files/tc-v16n1.pdf

Popova E, Vousden D, Sauer WHH, Mohammed EY, Allain V, Downey-Breedt N, Fletcher R et al (2019) Ecological connectivity between the areas beyond national jurisdiction and coastal waters: safeguarding interests of coastal communities in developing countries. Mar Policy 104(June):90–102. https://doi.org/10.1016/j.marpol.2019.02.050

Poseidon Principles (2019) Poseidon principles: a global framework for responsible ship finance. Poseidon Principles, Copenhagen. https://www.poseidonprinciples.org/download/Poseidon_Principles.pdf

Roberts CM, O'Leary BC, McCauley DJ, Cury PM, Duarte CM, Lubchenco J, Pauly D et al (2017) Marine reserves can mitigate and promote adaptation to climate change. Proc Natl Acad Sci U S A 114(24):6167–6175. https://doi.org/10.1073/pnas.1701262114

Rogers AD, Sumaila UR, Hussain SS, Baulcomb C (2014) The high seas and us: understanding the value of high-seas ecosystems. Global Ocean Commission, Oxford

Roth N, Thiele T, von Unger M (2019) Blue bonds: financing resilience of coastal ecosystems: key points for enhancing finance action. International Union for Conservation of Nature, Gland. https://www.4climate.com/dev/wp-content/uploads/2019/04/Blue-Bonds_final.pdf

Russ GR, Alcala AC, Maypa AP, Calumpong HP, White AT (2004) Marine reserve benefits local fisheries. Ecol Appl 14(2):597–606. https://doi.org/10.1890/03-5076

Russ GR, Cheal AJ, Dolman AM, Emslie MJ, Evans RD, Miller I, Sweatman H, Williamson DH (2008) Rapid increase in fish numbers follows creation of world's largest marine reserve network. Curr Biol 18(12):514–515. https://doi.org/10.1016/j.cub.2008.04.016

Sala E, Costello C, Dougherty D, Heal G, Kelleher K, Murray JH, Rosenberg AA, Sumaila R (2013) A general business model for marine reserves. PLoS ONE 8(4):e58799. https://doi.org/10.1371/journal.pone.0058799

Sala E, Lubchenco J, Grorud-Colvert K, Novelli C, Roberts C, Sumaila UR (2018) Assessing real progress towards effective ocean protection. Mar Policy 91(May):11–13. https://doi.org/10.1016/j.marpol.2018.02.004

Schuhbauer A, Sumaila UR (2016) Economic viability and small-scale fisheries: a review. Ecol Econ 124(April):69–75. https://doi.org/10.1016/j.ecolecon.2016.01.018

Schuhbauer A, Chuenpagdee R, Cheung WWL, Greer K, Sumaila UR (2017) How subsidies affect the economic viability of small-scale fisheries. Mar Policy 82(August):114–121. https://doi.org/10.1016/j.marpol.2017.05.013

Sumaila UR (2010) A cautionary note on individual transferable quotas. Ecol Soc 15(3):36. https://www.ecologyandsociety.org/vol15/iss3/art36/

Sumaila UR, Pauly D (2007) All fishing nations must unite to cut subsidies. Nature 450(December):945. https://doi.org/10.1038/450945a

Sumaila UR, Khan AS, Dyck AJ, Watson R, Munro G, Tydemers P, Pauly D (2010) A bottom-up re-estimation of global fisheries subsidies. J Bioecon 12(October):201–225. https://doi.org/10.1007/s10818-010-9091-8

Sumaila UR, Cheung W, Dyck A, Gueye K, Huang L, Lam V, Pauly D et al (2012) Benefits of rebuilding global marine fisheries outweigh costs. PLoS ONE 7(7):e40542. https://doi.org/10.1371/journal.pone.0040542

Sumaila UR, Lam VWY, Miller DD, Teh L, Watson RA, Zeller D, Cheung WWL et al (2015) Winners and losers in a world where the high seas is closed to fishing. Sci Rep 5(February):8481. https://doi.org/10.1038/srep08481

Sumaila UR, Rodriguez CM, Schultz M, Sharma R, Tyrrell TD, Masundire H, Damodaran A et al (2017) Investments to reverse biodiversity loss are economically beneficial. Curr Opin Environ Sustain 29(December):82–88. https://doi.org/10.1016/j.cosust.2018.01.007

Sumaila UR, Ebrahim N, Schuhbauer A, Skerritt D, Li Y, Kim HS, Mallory TG, Lam VWL, Pauly D (2019) Updated estimates and analysis of global fisheries subsidies. Mar Policy 109(November):103695. https://doi.org/10.1016/j.marpol.2019.103695

Sumaila UR, Zeller D, Hood L, Palomares MLD, Li Y, Pauly D (2020) Illicit trade in marine fish catch and its effects on ecosystems and people worldwide. Science. Advances 6(9):eaaz3801. https://doi.org/10.1126/sciadv.aaz3801

TCFD (Task Force on Climate-related Financial Disclosures) (2020) Task force on climate-related financial disclosures overview. TCFD, New York. https://www.fsb-tcfd.org/wp-content/uploads/2020/03/TCFD_Booklet_FNL_Digital_March-2020.pdf

Teh LCL, Sumaila UR (2013) Contribution of marine fisheries to worldwide employment. Fish Fish 14(1):77–88. https://doi.org/10.1111/j.1467-2979.2011.00450.x

Thiele T, Gerber LR (2017) Innovative financing for the high seas. Aquat Conserv Mar Freshwat Ecosyst 27(S1):89–99. https://doi.org/10.1002/aqc.2794

Thompson P, Cowton CJ (2004) Bringing the environment into bank lending: implications for environmental reporting. Br Account Rev 36(2):197–218. https://doi.org/10.1016/j.bar.2003.11.005

UN ESCAP (United Nations Economic and Social Commission for Asia and the Pacific) (n.d.) Global dialogue on ocean accounting and first annual meeting of the global ocean accounts partnership. https://www.unescap.org/sites/default/files/Global_Dialogue_on_Ocean_Accounting-Draft_Workshop_Report_for_comment.pdf. Accessed 14 January 2020

UNDP (United Nations Development Programme) and GEF (Global Environment Facility) (2012) Transforming markets to restore and protect the global ocean. Volume 1 of catalysing ocean finance. UNDP, New York

UNEP FI (United Nations Environment Programme Finance Initiative) (n.d.) Sustainable blue economy. https://www.unepfi.org/ecosystems/sustainable-blue-economy-finance/. Accessed 1 September 2020

UNEP FI and PSI (Principles for Sustainable Insurance) (2019) Underwriting environmental, social and governance risks in non-life insurance business. PSI Working Paper. Nairobi, UNEP. https://www.unepfi.org/psi/wp-content/uploads/2019/02/PSI-Guidance-for-non-life-insurance-underwriting.pdf

UN-OHRLLS (United Nations Office of the High Representative for the Least Developed Countries) (2013) Small island developing states in numbers. https://wedocs.unep.org/bitstream/handle/20.500.11822/9271/-Small%20Island%20Developing%20States%20in%20numbers-2013SIDS_IN_NUMBERS_121813_FA_WEB.pdf?sequence=3&amp%3BisAllowed

Vivid Economics (2020) Integrating climate change and biodiversity into the response to COVID-19: bailout measures. https://www.vivideconomics.com/wp-content/uploads/2020/04/200427-greening-COVID-corporate-bailouts.pdf

Waldron A, Adams V, Allan J, Arnell A, Asner G, Atkinson S, Baccini A et al (2020) Protecting 30% of the planet for nature: costs, benefits and economic implications. Campaign for Nature, Washington. https://www.conservation.cam.ac.uk/files/waldron_report_30_by_30_publish.pdf

Walsh M (2018) Ocean finance: definition and actions. Pacific Ocean Finance Program, Pacific Islands Forum Fisheries Agency, Honiara. https://gallery.mailchimp.com/b37d1411f778c043250da5ab5/files/f1a910e2-32f9-4aed-ad35-e2bccab6cf12/Ocean_Finance_Definition_Paper_Walsh_June_2018.pdf

Wharton JMH, Young S (Forthcoming) Pacific ocean finance program—insurance. Report by Willis Towers Watson for the Pacific Ocean Finance Program. London, United Kingdom

Whisnant R, Reyes A (2015) Blue economy for business in East Asia: towards an integrated understanding of blue economy. Partnerships in Environmental Management for the Seas of East Asia, Quezon City. http://www.pemsea.org/sites/default/files/Blue_Economy_for_Business_in_East_Asia.pdf

White C, Costello C (2014) Close the high seas to fishing? PLoS Biol 12(3):e1001826. https://doi.org/10.1371/journal.pbio.1001826

Widjaja S, Wirajuda H, Long T et al (2020) IUU fishing and associated drivers. World Resources Institute, Washington. https://oceanpanel.org/sites/default/files/2020-02/HLP%20Blue%20Paper%20on%20IUU%20Fishing%20and%20Associated%20Drivers.pdf

Winther JG, Dai M, Douvere F, Fernandes L, Halpin P, Hoel AH, Juinio-Meñez MA et al (2020) Integrated ocean management. World Resources Institute, Washington. https://oceanpanel.org/sites/default/files/2020-05/BP14%20IOM%20Full%20Paper%20Final%20Web.pdf

World Bank (2017) The sunken billions revisited: progress and challenges in global marine fisheries. World Bank, Washington. https://openknowledge.worldbank.org/handle/10986/24056

Wright G, Schmidt S, Rochette J, Shackeroff J, Unger S, Waweru Y et al (2017) Partnering for a sustainable ocean: the role of regional ocean governance in implementing sustainable development goal 14. Institute for Advanced Sustainability Studies, Potsdam. https://www.iddri.org/sites/default/files/import/publications/online_iass_report_report_170524.pdf

WWF (World Wide Fund for Nature) (2018) The sustainable blue economy finance principles. https://www.wwf.org.uk/updates/sustainable-blue-economy-finance-principles

WWF (World Wildlife Fund) (2019) Risk and opportunity in the seafood sector: the business case for sustainability. WWF, Washington. https://seafoodsustainability.org/wp-content/uploads/2019/06/Business-Case-for-Sustainability-2019.pdf

10 Critical Habitats and Biodiversity: Inventory, Thresholds and Governance

Abbreviations

ABNJ	Areas beyond national jurisdiction
AUV	Autonomous underwater vehicle
eDNA	Environmental DNA (molecular tool for assessing biodiversity)
BBNJ	Biodiversity beyond national jurisdiction (refers to negotiations to establish an international legally binding instrument under UNCLOS on the conservation and sustainable use of marine biological diversity of areas beyond national jurisdiction)
BEF	Biodiversity and ecosystem function
B_{msy}	Biomass at maximum sustainable yield
CBD	Convention on Biological Diversity
CCAMLR	Convention for the Conservation of Antarctic Marine Living Resources
CCRF	Code of Conduct for Responsible Fishing
CoML	Census of Marine Life
COPEPOD	Coastal and Oceanic Plankton Ecology Production and Observation Database
CPUE	Catch per unit effort
DD	Data deficient (Red List category)
EBSA	Ecologically and biologically significant area
EEZ	Exclusive economic zone
EOV	Essential Ocean Variable
EuroGOOS	European Global Ocean Observing System
FAIR	Findable, accessible, interoperable and reusable (principles for data sharing)
FAO	Food and Agricultural Organization of the United Nations
GEO	Group on Earth Observations Biodiversity Observation Network
GOBI	Global Ocean Biodiversity Initiative
GOOS	Global Ocean Observing System
GOOS	BioEco Biology and Ecosystems Panel of the Global Ocean Observing System
GDP	Gross domestic product
IMOS	Integrated Marine Observing System (Australia)
IOC	Intergovernmental Oceanographic Commission
IODE	International Oceanographic Data and Information Exchange
IOOS	Integrated Ocean Observing System (United States)
IPBES	Intergovernmental Science-Policy Platform on Biodiversity and Ecosystem Services
IPCC	Intergovernmental Panel on Climate Change
ISA	International Seabed Authority (UN agency charged with managing mining in the area; seabed in ABNJ)
ITIS	Integrated Taxonomic Information System
IUCN	International Union for the Conservation of Nature
IUU	Illegal, unregulated and unreported
KBA	Key biodiversity area
MBON	Marine Biodiversity Observation Network (part of the GEO BON program)
MPA	Marine protected area
NAGISA	Natural Geography in Shore Areas (CoML project)
NCP	Nature's contribution to people

Originally published in:
Rogers, A., O. Aburto-Oropeza, et al. 2020. *Critical Habitats and Biodiversity: Inventory, Thresholds and Governance*. Washington, DC: World Resources Institute. Available online at www.oceanpanel.org/blue-papers/critical-habitats-and-biodiversity-inventory-thresholds-and-governance.
Reprint by Springer International Publishing (2023) with kind permission.
Published under license from the World Research Institute.

J. Lubchenco, P. M. Haugan (eds.), *The Blue Compendium*, https://doi.org/10.1007/978-3-031-16277-0_10

NEAFC	North-East Atlantic Fisheries Commission
NEOLI	No-take, enforced, old, large, isolated (refers to MPAs; Edgar et al. 2014)
NGO	Non-governmental organisation
OBIS	Ocean Biogeographic Information System
OECM	Other effective area-based marine conservation measure
OSPAR	Oslo Paris Commission
PR	Performance review (in the context of fisheries management organisations)
RFMO	Regional Fisheries Management Organisation
RLS	Reef Life Survey
ROV	Remotely operated vehicle
SDG	Sustainable Development Goal
TURF	Territorial use rights for fishing programs
UN	United Nations
UNCLOS	United Nations Convention on the Law of the Sea
UNEP	United Nations Environment Programme
UNESCO	UNESCO
VME	Vulnerable marine ecosystems
WCMC	World Conservation Monitoring Centre
WoRMS	World Register of Marine Species

Highlights

- Evidence suggests that ocean biodiversity at all levels is being lost as a result of the direct and indirect impacts of human pressures. The main drivers of biodiversity loss are overexploitation and human pressures in coastal environments (development, habitat loss, pollution, disturbance). Increasingly, climate change and ocean acidification are and will be drivers of biodiversity loss especially in sensitive coastal ecosystems.
- Despite advances in understanding the distribution of species and habitats in the ocean, many aspects of marine biodiversity remain poorly understood. As a result, changes in marine biodiversity are difficult to ascertain and there is a critical need to establish current baselines and trends through survey and monitoring activities.
- There needs to be a concerted effort to increase funding and capacity for marine biodiversity research, especially in developing countries which are rich in biodiversity. There also needs to be an increase in collaboration across scientific disciplines and other data users and measures to make data collection and analysis interoperable and repeatable to ensure that we can enjoy the benefits of ecosystem services which underpin the blue economy whilst ensuring that biodiversity is conserved. These efforts should be focused on the already established international networks for biodiversity monitoring that include the Biology and Ecosystems Panel of the Global Ocean Observing System (GOOS BioEco), the Group on Earth Observation Biodiversity Observation Network (GEO BON), the Marine Biodiversity Observation Network (MBON), and global data integrators such as the Ocean Biogeographic Information System (OBIS) of the International Oceanographic Data and Information Exchange (IODE) programme of the Intergovernmental Oceanographic Commission of the United Nations Educational, Scientific and Cultural Organization (UNESCO-IOC) and the Ocean Data Viewer of the United Nations Environment Programme's World Conservation Monitoring Centre (UNEP-WCMC).
- There has been a significant apparent increase in the coverage of marine protected areas (MPAs). However, most MPAs are only lightly to minimally protected, with many lacking even management plans and very few classified as fully protected. Maximum environmental and societal benefits accrue only when 30–40% of key marine ecosystems are represented in fully or highly protected and implemented MPAs. We estimate that only 3% of the key habitats explored in this study lie in fully protected MPAs, and for some habitats, no countries have placed them in fully protected MPAs. Hence, opportunities abound to strengthen protection in existing MPAs and create new highly to fully protected MPAs, paying close attention to positive enabling conditions, good design principles and adequate enforcement and funding.
- It is critical to establish a legal framework for the conservation of biodiversity in the whole ocean, including areas beyond national jurisdiction. For this reason, reaching a strong agreement for the new international legally binding instrument on the conservation and sustainable use of biological diversity of areas beyond national jurisdiction (BBNJ) is essential.
- The ability of wealthier countries to implement conservation measures within their exclusive economic zones (EEZs) is higher and might need to compensate for less wealthy countries with higher biodiversity and higher pressures. Achieving the 30–40% target in fully or highly protected areas, especially in developing countries, will be greatly enhanced by capacity building, financial support and development of alternate economically viable options for employment.
- Marine ecosystems often exhibit tipping points where pressures lead to a major regime shift that results in an alternative and less productive state. Recognising such tipping points and incorporating them as reference points in fisheries management can greatly improve marine species conservation as well as the functioning and resilience of marine ecosystems.
- Accelerated and expanded reform of fisheries management practices are required if the food and nutritional needs of a growing human population are to be met with-

out permanent and long-lasting biodiversity loss resulting in the erosion of ecosystem services. It is especially important that these reforms include greatly improved monitoring of catch and bycatch in fisheries; the elimination of illegal practices in industrial fisheries through improved enforcement; a reduction in the fishing capacity where it is contributing to overfishing and/or damage to biodiversity whilst ensuring that basic needs for food, nutrition and livelihoods are met in coastal communities; and better incorporation of biodiversity considerations into all levels of fisheries management and the fishing industry. There must be better collaboration with the environmental sector for government departments and also with intergovernmental and non-governmental organisations.

1 Overview

Marine habitats are extremely valuable in many ways (e.g., economically, culturally or for subsistence) and provide many necessary services for humans (Costanza et al. 1997, 2014). Despite their importance, coastal and oceanic habitats are increasingly threatened by fishing, climate change, oil and gas exploration, pollution and coastal development (Jackson et al. 2001; Halpern et al. 2008, 2019; Heery et al. 2017; Harris 2020). Habitat degradation and loss from these threats are not uniformly distributed and are cumulative with poorly understood interactions between pressures (Halpern et al. 2008). Despite the enormous impacts humans have had on marine ecosystems in the global ocean over the past 50 years, they tend to appear not as the complete extinction of individual species (Dulvy et al. 2003) but rather as changes in ecosystem composition and in the relative abundance and ecological status of individual species, along with more regional or local extirpations (Worm and Tittensor 2011). A species need not become globally extinct to radically alter the composition of the ecosystem ('ecological extinction'), disappear from the local environment ('local extinction') or become commercially non-viable ('commercial extinction'). Biodiversity loss is a globally significant symptom of unsustainable exploitation of Earth's natural environment and a major threat to the ecosystem services on which we, and future generations, depend.

The ocean's natural capacity to provide ecosystem services such as food, coastal protection and carbon sequestration are being eroded as a result of the above changes (Cheung et al. 2010, 2013; Barange et al. 2014; Spalding et al. 2014; Arias-Ortiz et al. 2018). Over 500 million people worldwide live in the coastal zone and are afforded protection by ecosystems such as coral reefs, mangroves forests, seagrass beds and kelp forests. In the case of coral reefs, the reduction in damage to terrestrial assets conferred through coastal protection is estimated at US $4 billion annually (Beck et al. 2018). For the top five countries that benefit from reef protection, this is the equivalent benefit of $400 million annually in mitigated damage to society (Beck et al. 2018). Without reefs, the economic impact of flooding would more than double, with the area of land affected increasing by 69% and people affected by 81% (Beck et al. 2018). The loss of this critical ecosystem, which is estimated to result in a 1–10% reduction of its former range under the most optimistic future scenarios (IPCC 2018), is a looming crisis of vast ecological and social dimensions.

In response to habitat degradation, losses in biodiversity and associated impacts, there has been an international effort towards conserving marine ecosystems. The Strategic Plan for Biodiversity 2011–2020 from the Convention on Biological Diversity (CBD) has resulted in an accelerated effort to increase the protection of marine areas. Specifically, Aichi Biodiversity Target 11 calls for the conservation by 2020 of 'at least 10% of coastal and marine areas, especially areas of particular importance for biodiversity and ecosystem services … through effectively and equitably managed, ecologically representative and well-connected systems of protected areas and other effective area-based conservation measures'. A body of scientific literature suggests that the Aichi Biodiversity Target should be a first step. More ambitious targets of ocean protection (e.g., 30%), have been proposed and discussed in the scientific literature for many years (Gell and Roberts 2003; Balmford et al. 2004). Recent meta-analyses indicate that maximum environmental and societal benefits do not accrue until 30–40% of representative marine ecosystems are protected (Gell and Roberts 2003; Gaines et al. 2010; O'Leary et al. 2016; Sala et al. 2018a). This call for an enhanced scope for protection was endorsed by Resolution 50 of the International Union for the Conservation of Nature (IUCN) at the World Conservation Congress in 2016 'to designate and implement at least 30% of each marine habitat in a network of highly protected MPAs and other effective area-based conservation measures, with the ultimate aim of creating a fully sustainable ocean'. This call included specific reference to implementing protected areas in the exclusive economic zone (EEZ) of countries and in areas beyond national jurisdiction (ABNJ) (IUCN 2016).

Spatial conservation measures such as marine protected areas (MPAs) are one way of addressing these problems and have become the most recognised area-based marine conservation measure worldwide. An abundance of evidence suggests that if they are well designed, enforced and financed, fully protected MPAs can provide an abundance of benefits, including increases in biodiversity, size and abundance of previously targeted species (Halpern 2003; Lester and Halpern 2008; Lester et al. 2009; Edgar et al. 2014; Sala and Giakoumi 2017); enhanced spillover of juveniles and adults to adjacent fished areas (Halpern et al. 2010; Di Lorenzo et al. 2016); and restoration of ecological interactions within the protected area (Micheli et al. 2004; Mumby et al. 2007). More recent studies report additional benefits, including enhanced resilience to environmental and climate changes (Mumby and Harborne 2010; Micheli et al. 2012; Roberts et al. 2017; Bates et al. 2019). It is important to note here that biodiversity may benefit even further if more than 30–40% of representative habitats are protected by networks of MPAs. However, because of trade-offs between ocean conservation and uses such as fisheries, placing 30–40% of habitats in highly or fully protected MPAs is viewed as the optimal balance between protection of biodiversity and ecosystem service provision (Gaines et al. 2010). Also, to attain a representative coverage of 30% of marine habitats in fully or highly protected MPAs, a larger area may be required than 30% of the ocean to attain representativeness (O'Leary et al. 2018; see Jones et al. 2020 for an assessment based on species ranges lying within MPAs). Other effective area-based marine conservation measures (OECMs), such as locally managed marine areas, territorial use rights for fishing programs (TURFs), fisheries restricted areas, particularly sensitive sea areas, and areas of particular environmental interest, among others, have proven successful in conserving important areas for biodiversity and ecosystem services that include food security and poverty alleviation, such as in Northern Mozambique (Diz et al. 2018). The IUCN has created guidelines to recognise and report OECMs (IUCN-WCPA 2019) to incentivise robust long-term conservation and management of biodiversity. OECMs are an important but complementary tool to supplement an existing MPA network; however, they are not necessarily (or generally) mandated with a biodiversity conservation objective (Tittensor et al. 2019).

Therefore, this Blue Paper focuses on MPAs because they are supported by an important body of peer-reviewed literature indicating their effectiveness as fisheries management and conservation tools. Furthermore, MPAs can protect biodiversity but can also restore ecosystem structure, function and potentially services (Cheng et al. 2019) that mitigate and promote adaptation to climate change (Mumby and Harborne 2010; Micheli et al. 2012; Roberts et al. 2017). Therefore, implementing MPAs preserves habitats and their biodiversity and allows the maintenance of valuable ecosystem services (Costanza et al. 2014). We can roughly divide MPAs into no-take areas (where no fishing is allowed) and multiuse areas. Although, in some cases, the latter category does generate some benefits, in others, MPAs fail to reach their conservation objectives completely (Agardy et al. 2011). Scientific evidence is now accumulating in favour of fully protected MPAs (also known as marine reserves), which are dubbed most effective in environmental management (McClanahan et al. 2008; Edgar et al. 2014; MacNeil et al. 2015; Sala and Giakoumi 2017). Fully protected marine reserves, besides prohibiting fishing activities, also remove or minimise other human pressures that enable species to maintain or recover their abundance, biomass and diversity (Lester et al. 2009). It is notable, however, that MPAs are often not well designed, enforced or financed (Gill et al. 2017; Dureuil et al. 2018), which impacts their effectiveness, and there is particular concern for regions of high marine biodiversity, such as the marine biodiversity hot spot in Southeast Asia, where many species are reduced and destructive exploitation is expanding largely unchecked even within MPAs.

The High Level Panel for a Sustainable Ocean Economy has a vision of a productive and protected ocean, which will play a major role in achieving the Sustainable Development Goals (SDGs). Continued loss of marine biodiversity will undermine our ability to achieve a number of the SDGs, especially SDG 14 (to conserve and sustainably use the ocean), but also other goals (e.g., SDG 2, hunger and food security; SDG 9, resilient infrastructure). This Blue Paper addresses the topic of critical habitats and marine biodiversity with the following specific aims:

- Synthesise knowledge presenting the most recent inventory of marine habitats and biodiversity in the global ocean.
- Provide a brief overview of the impacts of habitat degradation and biodiversity loss in reducing ecosystem services.
- Review evidence of how biodiversity relates to ecosystem function and exploitation/degradation tipping points.
- Identify the range of measures undertaken by governments and industrial sectors to monitor, protect and address loss of marine biodiversity and their effectiveness.
- Determine opportunities for action to improve the sustainability of blue economic activities with respect to maintaining, and, where possible, restoring, the ocean's habitats and biodiversity.

We use the Convention on Biological Diversity's definition of biodiversity as the variability among living organisms, including diversity within species, between species and of ecosystems. The topics of marine biodiversity and ecosystem integrity are complicated by a lack of data, which pervades almost all aspects of our understanding of its distribution and trends. By necessity, therefore, we have been driven to examine specific aspects of the topic, such as well-studied groups of organisms or habitats as well as particular case studies.

This underlines the need for more scientific work on many aspects of ocean biodiversity, from variation within species and connectivity of populations to processes at the level of habitats and entire ecosystems, the sum of which underpin the functioning of Earth.

2 An Inventory of Marine Habitats and Biodiversity

2.1 Species

Globally, it is estimated that only 10–25% of marine species have been described (Mora et al. 2011; Appeltans et al. 2012), and some of the least known groups are likely to have thousands to over a hundred thousand undescribed species (e.g., Isopoda, Gastropoda, Tanaidacea). The geographic distributions of even fewer species are known (Gagné et al. 2020). Genomic approaches, coupled with large-scale sampling of the upper layers of the ocean (e.g., the Tara expedition), have also revealed tens of thousands of uncharacterised microbes, including eukaryotes, prokaryotes and viruses (de Vargas et al. 2015; Sunagawa et al. 2015). However, it is estimated that about half of the major taxonomic groupings (e.g., Vertebrata) have identified more than 50% of their known species already, and with the current rate of description of new species (average of 2000 new species described per year), those groups might have all their species described by the end of the century (Appeltans et al. 2012).

Knowledge of marine biodiversity varies markedly across regional, national and, more importantly, trophic levels (Costello et al. 2010). Data from the Census of Marine Life (CoML) programme is available in the ever-growing Ocean Biogeographic Information System (OBIS)[1] of the Intergovernmental Oceanographic Commission (IOC) of the United Nations Educational, Scientific and Cultural Organization (UNESCO). The CoML data suggest that, in relative terms, China, Australia and Europe have the best knowledge base of marine species with the tropical western Atlantic, tropical eastern Pacific and Canadian Arctic regions being poorly studied (Costello et al. 2010). Ecosystems that are particularly poorly known include the deep sea, coral reefs, icecovered areas and chemosynthetic habitats (Costello et al. 2010). Knowledge of the identity and distribution of commercially exploited taxa is greater than that of non-extracted taxa, and larger organisms tend to be better known than smaller organisms (Fautin et al. 2010; Worm and Tittensor 2018). Currently, only a handful of species are considered to have enough independent records that describe their full geographic distribution (about 50,000 species; Gagné et al. 2020). Emblematic (mammals, corals or fish) and exploited species (fish and invertebrates) are among the most well-documented spatially. Other patterns of biodiversity, including intraspecific genetic variation and habitat diversity, are also not well described (Fautin et al. 2010; Blasiak et al. 2020), with some exceptions. The Global Ocean Biodiversity Initiative (GOBI), which uses CoML and OBIS as primary sources of data, has participated in the CBD effort to identify ecologically and biologically significant areas (EBSAs) in the ocean.[2] These areas can be characterised by high biological diversity, but they also include a number of other criteria, including unique or rare species or communities; importance for the life history stages of marine species; importance for threatened or endangered species or habitats; vulnerability, fragility or slow recovery; biological productivity; and naturalness (CBD 2009). Geographic areas with the best knowledge of marine biodiversity do not match well with areas of highest diversity, reflecting both historical and present-day scientific capacity for taxonomy. Historically, highly sampled regions are often located in the Northen Hemisphere in the coastal regions around developed countries. It is crucial to account for such sampling bias when examining the distribution of biodiversity (Tittensor et al. 2010; Gagné et al. 2020). The common approaches to provide an unbiased picture of marine biodiversity consist of (1) removing species with not enough records to describe their full distribution and (2) applying statistical methodologies on known species records to correct for bias. The main hot spots of marine biodiversity have been recognized in the Indo-Pacific Coral Triangle and a lower peak in the Caribbean (Briggs 2007; Worm and Tittensor 2018; see Box 10.1). A general decline in biodiversity from the tropics to the polar latitudes has also been hypothesised, although there is debate on whether some taxa show more bimodal patterns (Thorson 1952, 1957; Fischer 1960; Stehli et al. 1967, 1972; Clarke and Crame 1997; Williamson 1997; Roy et al. 1998; Tittensor et al. 2010; Edgar et al. 2017; Worm and Tittensor 2018; Box 10.1). Hypothesised explanations include speciation and extinction rates over geological timescales as correlated with latitude (Crame and Clarke 1997; Jablonski et al. 2006, 2013) and ecological drivers such as habitat area, land versus ocean area by latitude, sea surface temperature (Worm and Tittensor 2018), and intrinsic biological traits such as larval development mode and interspecies interactions (Roy et al. 1998; Pappalardo and Fernández 2014; Edgar et al. 2017).

[1] For more information, see the OBIS website, https://obis.org

[2] To learn more about GOBI, visit its website, http://gobi.org/

Box 10.1 Estimating Global Patterns of Biodiversity

Using the biodiversity data found in Reygondeau (2019) and Gagné et al. (2020), the authors developed a standardised database drawing on online websites with records of the global distribution of marine species with sufficient records to have a robust distribution. Specifically, the database was populated with species data for which at least 10 spatially informed occurrences were available. Occurrence data originated from the Ocean Biogeographic Information System (OBIS);[a] Intergovernmental Oceanographic Commission (IOC) of the United Nations Educational, Scientific and Cultural Organization (UNESCO);[b] the Global Biodiversity Information Facility (GBIF);[c] Fishbase;[d] the Coastal and Oceanic Plankton Ecology Production and Observation Database (COPEPOD);[e] the Jellyfish Database Initiative;[f] and the International Union for the Conservation of Nature (IUCN).[g] The full filtering methodology can be found in Gagné et al. (2020).

From the initial data set (more than one billion entries), we removed records (1) with spatial location as "not assigned" (NA) or null values, (2) not identified to species level and (3) replicated among databases (i.e., records with the same species name, coordinates, and sampling details). The remaining records (731,329,129 records; more than 101,000 species) were assigned full taxonomic information using the Taxize library4 in R Studio. We also used this procedure to update all species' synonyms to valid names, as officially recognised by the Integrated Taxonomic Information System (ITIS)[h] and the World Register of Marine Species (WoRMS).[i] Next, we explored the relationship between the number of independent records (independent in time and area of sampling) and latitudinal range and thermal range for species with well-known global coverage and ecology (number of observations greater than 2000; 1196 species). For each known species, we randomly selected n records (number of observations from 1 to 1000) within the global pool, and for each selected number of records ($n = 1$ to 1000 records), we computed the species' latitudinal range and thermal range. The procedure was replicated 1000 times. We then confronted the simulated latitudinal range and thermal range (1000 simulations) to values obtained using all the information gathered on the species. We computed an interval of confidence of known range by quantifying the difference between the 1st and the 99th percentile of observed latitude coordinates and thermal value, and we assumed that the acceptable number of records to capture the latitudinal and thermal range was obtained when more than 950 randomly selected records were included within the confidence interval determined from the global pool of records. The median number of points found to capture the latitudinal range was 33 ± 4 records and 41 ± 3 records for thermal range. All species with less than 41 independent records were removed from further analysis.

Thus, the final data set on which all analyses presented in this study are based comprises up-to-date taxonomic information and filtered occurrences for 41,625 species, for a total of 51,459,235 records representing 17% of all accepted marine and non-fossil species.

Notes:

[a] *OBIS, http://www.iobis.org*

[b] *UNESCO-IOC, http://ioc-unesco.org/*

[c] *GBIF, http://www.gbif.org*

[d] *FishBase, http://www.fishbase.org*

[e] *COPEPOD, http://www.st.nmfs.noaa.gov/plankton*

[f] *Jellyfish Database Initiative, http://people.uncw.edu/condonr/JeDI/JeDI.html*

[g] *IUCN, http://www.iucnredlist.org/technical-documents/spatial-data*

[h] *ITIS, http://www.itis.gov*

[i] *WORMS, http://www.marinespecies.org*

[j] *For more information see WORMS*

The distribution of biodiversity in the global ocean has been described for numerous taxa, particularly in recent years as more observations have been synthesised into large-scale patterns (Tittensor et al. 2010; Reygondeau 2019). While there is consistency across many groups, it is important to bear in mind that there remains a significant taxonomic bias in our understanding.

There are some groups that we know well (typically those species in which we have a keen commercial interest or which are charismatic, such as vertebrates, or those which form biogenic habitats such as corals and seagrasses), but there are many for which we have very limited information (numerous invertebrate groups, most deep-sea taxa, and much of the microbial biosphere). In Box 10.1 we present a new analysis of the global pattern of marine biodiversity which is aimed at reducing bias from the issue of uneven sampling of species from different parts of the ocean.

At a global scale, the biodiversity distribution estimated from our study appears to be relatively consistent with other studies, resolutions and analyses (Fig. 10.1; Tittensor et al. 2010; Asch et al. 2018; Reygondeau 2019). The pattern across multiple taxa is primarily tropical to subtropical peaks

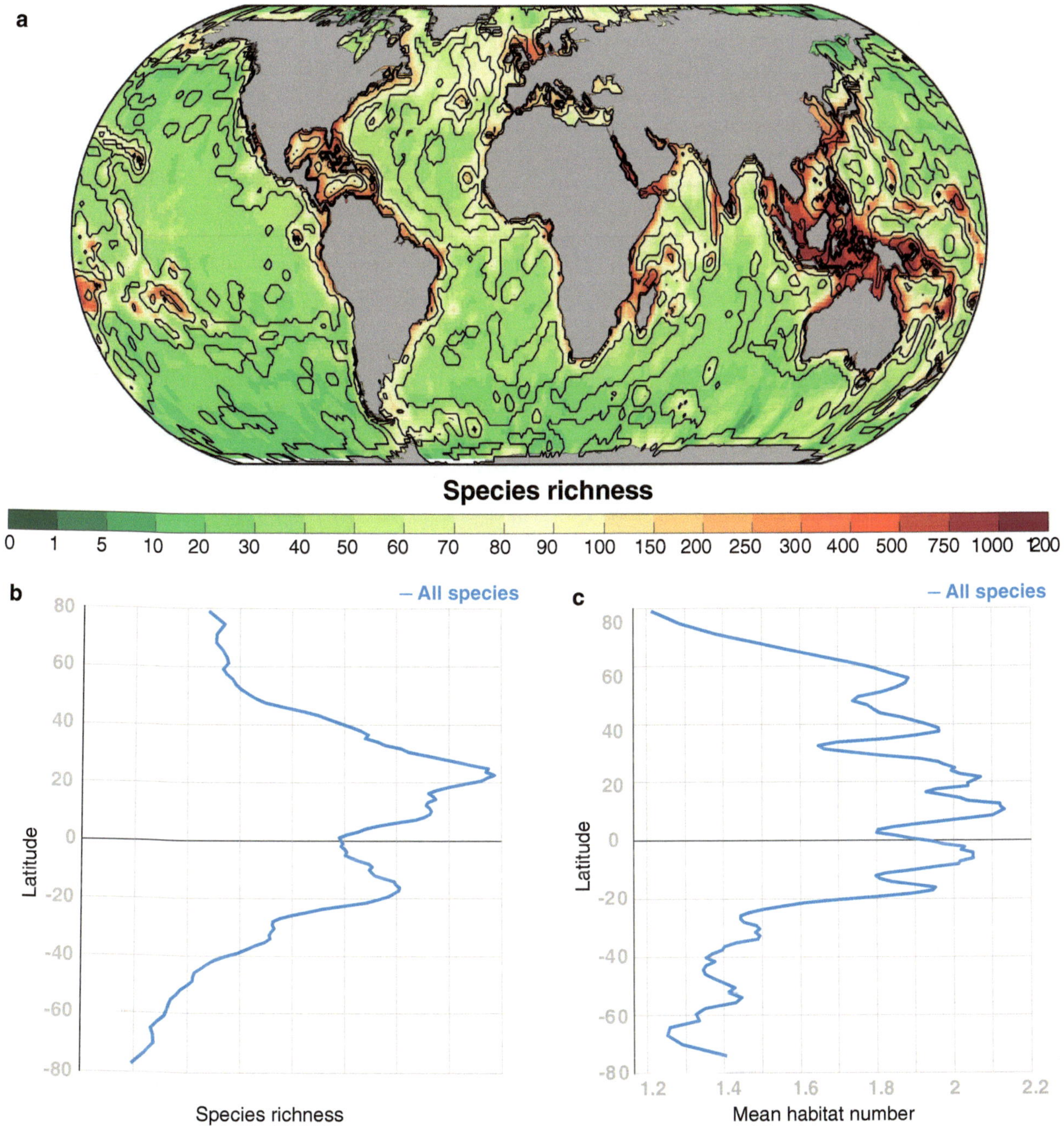

Fig. 10.1 Global patterns of biodiversity and habitat richness. *Notes: Map of species richness* (**a**) *is on a 100 × 100 km equal-area grid with a superimposed contour map of the number of habitats per geographical cell (see Sect. 2.2). Latitudinal gradient of species richness* (**b**) *is of all marine species.* (**c**) *Plot of the average number of habitats versus latitude). (Source: Authors)*

in species biodiversity, particularly for coastal species; but there are steep longitudinal gradients in diversity, with an increase from both east and west towards Southeast Asia, and from east to west in the tropical Atlantic. The Indo-Pacific Coral Triangle, central and western Indian Ocean, Red Sea, South West Pacific Islands (i.e., the Bismarck Archipelago, the Great Sea Reef of Fiji, New Caledonia, New Guinea, the Solomon Islands, Vanuatu) and Southeast Asia show the highest levels of species richness as indicated in previous studies (e.g., Selig et al. 2014). The Caribbean also has a relatively high species richness, but not as high as the aforementioned areas and parts of the northeast Atlantic, such as the North Sea, are as diverse. This latter result may reflect the high number of species records in the northeast

Atlantic, introducing some bias into the overall picture of the distribution of species richness given the exclusion of species with less than 41 samples. Also, small areas, such as tropical or subtropical islands, which are characterized by a high species diversity may be unresolved because of the spatial resolution of this analysis (as for Selig et al. 2014). Individual taxonomic groups and different parts of the ocean (coastal, pelagic, deep sea) can show differing distributions. Taxa that follow the general pattern, albeit with some variation in relative intensity of hot spots, include reef-building corals, coastal fishes, shallow-water ophiuroids (brittle stars), cone snails, mangroves, coastal cephalopods, lobsters and gastropods. Seagrasses have a more temperate-skewed distribution of richness, perhaps reflecting their improved ability to tolerate cold water, relative to reef-building corals and mangroves.

Macroalgae (seaweeds such as kelp) are less well-known in terms of distribution at the species level, but at the genus level again appear to peak at more temperate or subtropical latitudes (Gaines and Lubchenco 1982; Kerswell 2006; Short et al. 2007; Tittensor et al. 2010; Keith et al. 2014; Worm and Tittensor 2018). Coastal sharks show a similar pattern to other coastal fishes, but their distribution is more centered around temperate latitudes (Lucifora et al. 2011). Deviations from the general patterns described include coastal marine mammals, whose endothermy has enabled them to develop a metabolic advantage in colder waters (Pompa et al. 2011; Grady et al. 2019). Pinnipeds (seals, sea-lions and walrus) show an inverse pattern with peak species diversity in subpolar and polar environments (Tittensor et al. 2010; Pompa et al. 2011).

Biodiversity in the open ocean shows a generally bimodal pattern (Chaudhary et al. 2016), with pelagic zooplankton such as foraminifera, copepods and euphausiids, open ocean fishes such as tuna and billfishes, pelagic sharks, and cetaceans all showing a mid-latitudinal peak in species richness, generally between latitudes 30 and 40° (Tittensor et al. 2010). Some differences between these taxa are apparent, including cetaceans being widely distributed in terms of richness peaks across latitudinal bands, whereas pelagic shark hot spots tend to skew towards the coast. Marine bacteria and phytoplankton diversity patterns remain much less well-known at a global scale, though modelling has predicted an intermediate latitude peak in phytoplankton, and there may be a similar gradient in bacteria, though more data and analyses are needed to confirm this for both groups (Worm and Tittensor 2018). Pelagic cephalopods are undersampled, but they appear to show a similar intermediate latitudinal peak, albeit only in the Northern Hemisphere (Tittensor et al. 2010). Pelagic seabirds (such as albatross and petrels) show a mid-latitude peak, but only in the Southern Hemisphere (Davies et al. 2010).

Deep-sea biodiversity is far less known, and whilst regional patterns have been described for multiple groups (Rex and Etter 2010), global patterns are far less well understood at the species level (though model predictions of habitat suitability are available at higher taxonomic levels for other taxa, such as cold-water corals; Tittensor et al. 2009). A global pattern has been described only for the ophiuroids (brittle-stars), which, as mentioned above, show a relatively typical shallowwater pattern of a peak in diversity at low latitudes on the continental shelf and slope, but they have a markedly different distribution in deep waters (more than 2000 m; Woolley et al. 2016). Deep-water ophiuroids show maximum richness at temperate latitudes (between latitudes 30 and 50°), with diversity higher in regions closer to continental margins where particulate organic material export from the surface, used as a food source by most deepsea organisms, is higher. The deep sea is an extremely food-limited, lightless environment, with relatively shallow gradients of temperature over large distances horizontally, and these environmental factors may shape different patterns, though more information is needed to ascertain whether these patterns hold across multiple taxonomic groups.

Biodiversity metrics, other than species richness, that have been assessed at a global scale are few. The global distribution of functional richness in fishes appears similar to species richness, but evenness shows an opposite pattern (increasing with latitude), and functional diversity appears highest in the tropical eastern Pacific (Stuart-Smith et al. 2013). The fish food web is globally connected and suggests a higher vulnerability to species extinctions in the open ocean compared to coastal areas (Albouy et al. 2019).

In summary, known patterns (based on a biased sample of taxonomic groups) indicate that species biodiversity appears to peak in the tropical Indo-Pacific, with a secondary peak in the Caribbean, and a general tropical or subtropical peak in richness. Coastal species tend to match this pattern more closely than oceanic species, which tend to show bimodal peaks at intermediate latitudes; yet whilst deep-sea taxa remain poorly known, one group (brittle stars) shows a markedly different distribution with temperate peaks close to continental margins and in areas of high food export from the surface ocean.

2.2 Habitats

Using previously published spatial data sets (Table 10.1), we synthesised information at the global level to produce patterns of habitat diversity (see Fig. 10.2). Because of their ecological and socio-economic importance, and the relative availability of information, we focused on the following marine habitats ordered from their distance to the coast: estuaries, mangroves, saltmarshes, seagrasses, coral reefs, kelp

Table 10.1 Spatially referenced habitat data for coastal and oceanic ecosystems included in the habitat diversity analysis

Habitat	Time span	Data type	Source
Estuaries	2003	Polygon	Alder (2003)–updated by UNEP-WCMC
Mangroves	1997–2000	Polygon	Giri et al. (2011)–updated by UNEP-WCMC
Saltmarsh	1973–2015	Point	McOwen et al. (2017)–updated by UNEP-WCMC
Seagrasses	1934–2015	Polygon	UNEP-WCMC and Short (2018)
Coral reefs	1954–2018	Polygon	UNEP-WCMC et al. (2018)
Kelp forests	NA	Point	Jorge Assis, research in progress
Shelf valley and canyons	1950–2009	Polygon	Harris et al. (2014)
Cold coral reefs	1915–2014	Point	Freiwald et al. (2017)–updated by UNEP-WCMC
Seamounts and guyots	1950–2009	Polygon	Harris et al. (2014)
Trenches	1950–2009	Polygon	Harris et al. (2014)
Hydrothermal vents	1994–2019	Point	Beaulieu and Szafranski (2018) (InterRidge Vents Database)
Ridges	1950–2009	Polygon	Harris et al. (2014)

Source: Authors

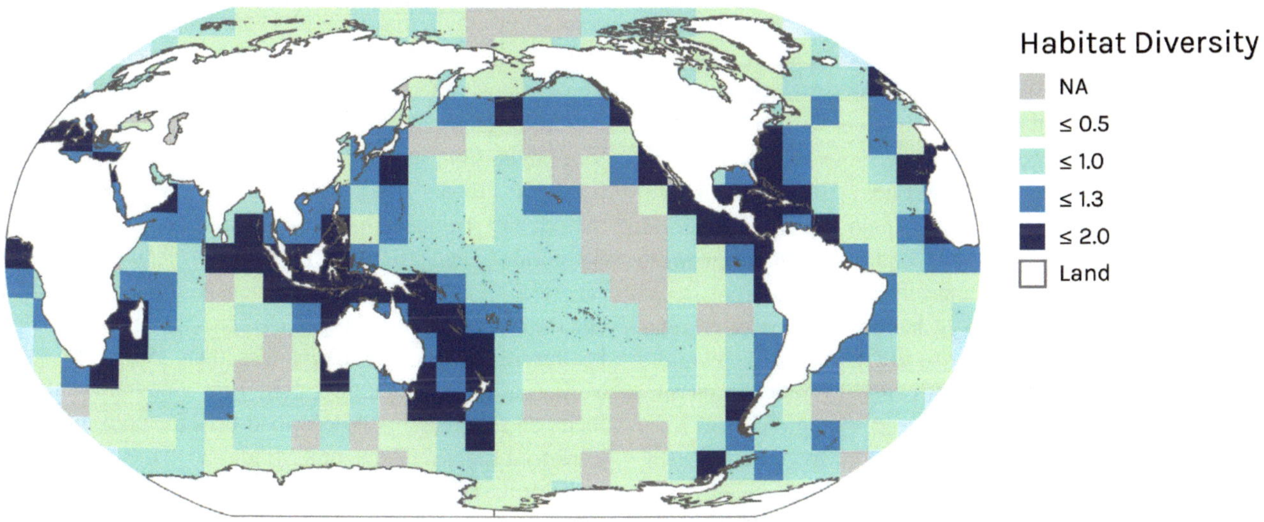

Fig. 10.2 Global habitat diversity. *Note: Habitat diversity calculated with Shannon-Wiener diversity index for habitats studied. Habitat diversity is displayed for 1000-km pixels. (Source: Authors)*

forests, shelf valley and canyons, cold-water corals (deep sea corals), seamounts and guyots, trenches, hydrothermal vents and ridges (Table 10.1).

The global habitat diversity index was based on the 12 habitats in Table 10.1. First, these habitats were converted into binary rasters at a 1-km resolution and projected into the World Robinson projection. A constant raster was created at a resolution of 1000 km by 1000 km. Next, these rasters were imported into R Studio. The packages 'raster', 'sp', 'rgdal', and 'tidyverse' were used to work with the data. Within each cell of the constant raster, the number of 1 km pixels that contained a habitat were summed. Each of the cells of the constant raster was then viewed as a community, and the Shannon Index of diversity was used to calculate a diversity value for each cell using the number of cells of each habitat as species counts. These values were then transformed into a raster and were uploaded into ArcGIS Pro 2.4 to create Fig. 10.2.

Coastal areas had a much higher diversity, because of the occurrence of 6 of the 12 habitats considered. The other 6 habitats occur in deeper waters, where many areas remain understudied. Although our technological capability is increasing through efforts like the global Seabed 2030 mapping project,[3] there are still large gaps in our understanding of deepwater habitat distribution (Rogers et al. 2015). Hence, although the data considered (Table 10.1) are the current best-available representation of the extent of global habitats, the progressive use of improved large-scale mapping technologies will improve our knowledge of global habitat diversity patterns.

Based on the habitat diversity analysis, the Indo-Pacific Coral Triangle, the eastern seaboard of Australia and the Caribbean are hot spots for habitat diversity (Fig. 10.2), a

[3]Information about the Seabed 2030 project can be found at https://seabed2030.gebco.net/

pattern which is similar to that for species diversity (Fig. 10.1). The distribution of these data skews to the right, with fewer areas with higher diversity. The United States, Australia and Indonesia have the highest area of analysed habitats with an average of 6.94%, 5.81% and 5.05% of the global total, respectively. Unsurprisingly, there is a strong and significant correlation with EEZ area, explaining 63% of the variation. Russia, which also has a very large EEZ, does not seem to follow this trend—probably because much of its coastline lies at polar latitudes.

3 Biodiversity Loss

3.1 Evaluating the Loss of Species

The dominant pressures on the ocean are direct exploitation by fisheries, followed by land and sea use change (Costello et al. 2010; IPBES 2019). These pressures were identified by the Global Assessment Report on Biodiversity and Ecosystem Services of the Intergovernmental Science-Policy Platform on Biodiversity and Ecosystem Services (IPBES) and by previous studies. Of the three other main drivers considered, invasive species, climate change and pollution are growing in importance. Climate change impacts arise from ocean warming, acidification, deoxygenation, changes in currents and circulation, and sea level rise (IPCC 2019). Temperature rise is correlated with global shifts in distribution, generally away from the tropics but influenced by regional and local oceanography (Cheung et al. 2009; Burrows et al. 2011, 2014; Poloczanska et al. 2013, 2016; Humphries et al. 2015; Molinos et al. 2016). This is driving the large-scale alteration of marine communities at middle to high latitudes (e.g., the Atlantification of the Barents Sea; Fossheim et al. 2015; Oziel et al. 2017; Vihtakari et al. 2018) and may be exacerbated by geographic patterns of thermal tolerance in marine species (Stuart-Smith et al. 2015). Deoxygenation of the ocean has already caused a shift in the vertical and horizontal distribution of pelagic species such as marlins and squid (Stramma et al. 2012; Stewart et al. 2013; reviewed in Breitburg et al. 2018). Climate change is also a significant driver of ecosystem damage, including on coral reefs (Hoegh-Guldberg et al. 2007; Gattuso et al. 2015; Hughes et al. 2018a) and seagrass beds (Thomson et al. 2015; Arias-Ortiz et al. 2018).

To evaluate such impacts on biodiversity, we analysed the IUCN Red List for 12 marine invertebrate and vertebrate taxa. This list comprises analyses of the current status of populations of species with respect to extinction risk, and it considers population decline, negative changes in range (e.g., range of occupancy and/or levels of fragmentation of populations), and whether populations of a species are very small (IUCN 2017). For marine invertebrates and vertebrates, data were extracted from the IUCN online Summary Statistics.[4]

To reduce bias, the assessment was restricted to taxa with more than 10 species assessed. Whilst these taxa represent a relatively small proportion of those living in marine environments, they are the best studied to date; therefore, they present a good (if taxonomically biased) data set on which to assess the threat of extinction and its causes across a range of marine ecosystems (Webb and Mindel 2015). Only around 3% of the roughly 240,000 described marine species have been assessed for the Red List (Sullivan et al. 2019).

3.2 Invertebrates

There are 3081 marine invertebrate species in seven classes across four phyla that have had some representative assessment on the IUCN Red List (see Fig. 10.3 and Table 10.2). The numbers reflect the extremely low level of assessment of marine invertebrates, a total of 2.6% of species across these four phyla, from as low as 0.5% for Arthropoda to 7.5% for Cnidaria (Table 10.2). Furthermore, these samples are biased: 839 species of hard corals (order Scleractinia) and 16 fire corals (genus *Millepora*) make up 97% of the cnidarians assessed, all from a single assessment (Carpenter et al. 2008), and the 686 Cephalopoda species represent 44% of all marine Mollusca assessed but likely less than 1% of all marine Mollusca. By their nature, Red List assessments tend to focus on relatively well-described taxa for both marine and terrestrial species (Webb and Mindel 2015).

With these caveats and the challenge of data deficiency, the proportion of species threatened ranges from a lower bound of 11% to an upper bound of 46%. The most speciose invertebrate classes (Anthozoa, Gastropoda, Malacostraca) as well as the Cephalopoda show the lowest levels of threat. The criteria used for assessment are indicative of marine species characteristics: of the 326 species listed in one of the three 'threatened' categories (vulnerable, endangered, and critically endangered), over 75% (254) are listed on the basis of estimated population decline (Criterion A, for the past, present and/or future), 14% were listed on the basis of small range and decline (Criterion B), and 7% were listed for their very small population size or range (Criterion D). Only 5 species were listed under more than one criterion.

3.3 Vertebrates

Compared to invertebrates, marine vertebrates are relatively well represented in the IUCN Red List (Fig. 10.3). Reptiles, birds and mammals have been fully assessed, and among

[4]See IUCN Red List, https://www.iucnredlist.org/search

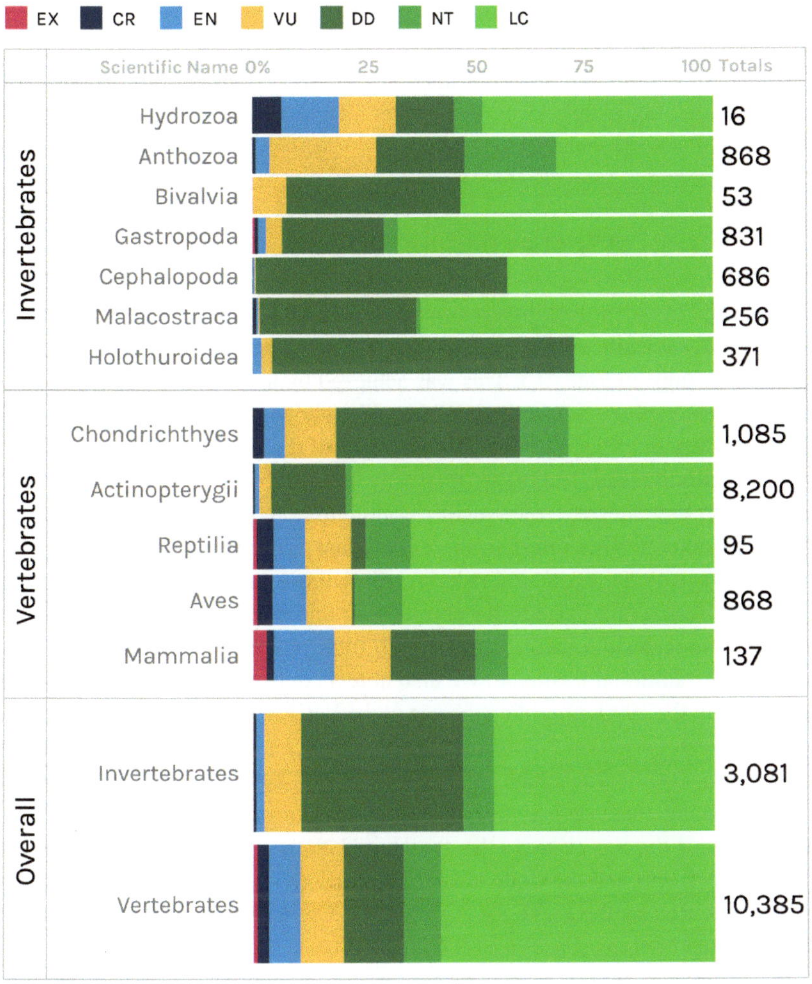

Fig. 10.3 IUCN Red List threat categories for marine species. *Note: These taxa have more than 10 species assessed. Data deficient (DD) species are depicted between the threatened categories (CR critically endangered, EN endangered, VU vulnerable) and non-threatened categories (NT not-threatened, LC least concern). EX extinct in the wild. Numbers on the right of the bars represent the total number of species assessed per taxon group. (Source: Authors)*

Table 10.2 Proportion of invertebrate species assessed on the IUCN Red List compared to the total number of species currently described on the World Register of Marine Species

Phylum	Number of species described	Number of species assessed	% Assessed
Arthropoda	56,479	266	0.5
Cnidaria	11,744	884	7.5
Echinodermata	4408	372	5.0
Mollusca	48,275	1570	3.3
TOTAL	120,906	3092	2.6

Source: WoRMS (n.d.)

marine fishes, of the approximately 18,000 described to date, just over 50% have been assessed (9285 species of sharks, rays and bony fish). Of these, there are 8200 marine actinopterygians, from 30 different orders, for which at least 10 species have been assessed. The two fish classes included in this analysis make up 79% of all assessed marine vertebrates and compose 70% of marine vertebrates listed as threatened. However, the actinopterygians have the lowest overall proportion of threatened species (4%) compared to other marine vertebrate taxa (20–30%). The chondrichthyan extinction risk at this taxonomic level of analysis is substantially higher than for most other vertebrates, and only about one-third of species are considered safe (Dulvy et al. 2014). We note that all species of marine turtles are currently threatened with extinction.

The actinopterygians are less well understood than marine reptiles, birds and mammals, and, as a result, have by far the highest proportion (and number) of species listed as 'data deficient' (DD; see Fig. 10.3); some of these DD species may also be threatened but the lack of data prohibits this assessment from being made.

This situation highlights the poor overall understanding we have of many fish species, even some that are heavily exploited, such as many deepwater and coral reef fishes; examples include the deepwater orange roughy (*Hoplostethus atlanticus*), coral reef groupers and snappers (Epinephelidae and Lutjanidae), coastal and estuarine groups such as croakers (Sciaenidae), and cold-water wolf-fishes (*Anarhichas*). The documentation of these species should be a priority from the perspective of population (status, distribution and trends) and use (i.e., fisheries catches). However, for all taxa there is also a need to collect data on less well-understood aspects of impacts on populations, such as from unintentional catch/

bycatch or through destruction of key life history areas such as spawning or nursery grounds. Such data are collected for some fisheries but by no means all, and data are often aggregated at higher taxonomic levels that render them useless for species-level assessments.

3.4 Drivers of Species Decline

We analysed the identified drivers of extinction risk for species listed as critically endangered, endangered or vulnerable for the 12 groups in Fig. 10.3. This was achieved by looking at each threatened species in the IUCN Red List and recording the drivers of extinction risk. Whilst many of the IUCN drivers of biodiversity decline are relatively straightforward to interpret, the category 'biological resource use' requires some explanation. This refers to the effects that harvesting activities have on the extinction risk, including those caused by targeted catch and bycatch for commercial and artisanal fisheries, the aquarium trade, marine curio trade, shell collecting and traditional medicine. We also note a controversy that began in the 1990s regarding the use of the IUCN extinction threat categories for commercially fished species (Rice and Legacè 2007). The main policy instruments used for fisheries management such as the United Nations Convention on the Law of the Sea (UNCLOS), the United Nations Fish Stocks Agreement and the Code of Conduct for Responsible Fishing (CCRF) by the Food and Agricultural Organization of the United Nations (FAO) all highlight biomass at maximum sustainable yield (B_{msy}) as a target for sustainable fisheries management. Under a sustainable management regime, it is possible to reduce a stock size to below levels which would trigger categorising a species or stock as threatened with extinction under the IUCN Red List criterion of decline in population size while other fisheries management reference points indicate the stock can still be exploited (Rice and Legacè 2007).

Whilst this has been a subject of debate (see Rice and Legacè 2007), more recent studies have demonstrated that conservation metrics as assessed by Red List criteria align well with fisheries assessments of stock status (e.g., Davies and Baum 2012; Fernandes et al. 2017). Thus, it can be concluded that threat categories identified through the Red List criteria do not exaggerate extinction or extirpation risk and occurrences of disagreement between the two approaches are rare (Davies and Baum 2012; Fernandes et al. 2017). The IUCN has specifically identified this issue in the guidelines for applying extinction risk criteria (IUCN 2017).

For invertebrates, the most significant threat for mobile taxa was biological resource use (Fig. 10.4), including overexploitation of populations through directed fishing (Holothuroidea), bycatch (Cephalopoda) or for shell collecting (Gastropoda). For sessile taxa, Anthozoa and Hydrozoa, drivers of extinction risk are evenly distributed amongst multiple drivers, reflecting a range of anthropogenic stressors in coastal ecosystems. The assessed Gastropoda are also predominantly coastal, and this is reflected in the broader range of drivers of extinction risk in this taxon. Other contributing factors to extinction risk included small geographic range (e.g., cone shells; Peters et al. 2013), life history factors (e.g., Cephalopoda, Holothuroidea; Bruckner et al. 2003; Collins and Villanueva 2006) and high commercial value (e.g., Holothuroidea; Purcell et al. 2014). We also note that the first assessment of threat from deep-sea mining has just occurred, with the first of 14 hydrothermal vent invertebrates (a snail) being listed as 'endangered' (Sigwart et al. 2019). This assessment was on the basis of the small geographic range and number of populations of this species, an attribute shared by other vent-endemic taxa. Deep-sea mining is currently controversial, and regulations for environmental management of this activity are still being formulated by the International Seabed Authority (ISA) of the United Nations. Whether these measures will be sufficient to protect vent-endemic species with small ranges from the effects of exploitation of seabed massive sulphides remains to be seen (Durden et al. 2018; Washburn et al. 2019).

Across the marine vertebrate taxa assessed (except birds), the major driver of extinction risk is resource use, including by both small- and large-scale fisheries and both targeted and incidental catch (Fig. 10.4). This is in general agreement with the key messages of the IPBES Global Assessment Report (2019). In particular, larger species at higher trophic levels have been heavily reduced by exploitation whether as high-value target species or because they are taken incidentally as bycatch, and many have shown a sharp decline (Christensen et al. 2014; Suazo et al. 2014; Fernandes et al. 2017). However, the full impacts of incidental catch are little understood for smaller fish species and many invertebrates, because catch data poorly documents them at the species level. Despite little evidence that overexploitation or bycatch have caused global extinctions, local extinctions and commercial extinctions (in which a species is reduced to a level at which it is no longer commercially viable) are much more common (Dulvy et al. 2003). In addition, overexploitation has dramatically reduced the abundance of numerous species worldwide, both large and small (McCauley et al. 2015), caused large range contractions (Worm and Tittensor 2011) and impacted body mass (Ward and Myers 2005). At the ecosystem level, overexploitation has triggered trophic cascades (Worm and Myers 2003; Frank et al. 2005; Daskalov et al. 2007), reduced total community biomass (Ward and Myers 2005) and degraded habitat structure (Thrush and Dayton 2002; Clark et al. 2016). Within species, it has also affected

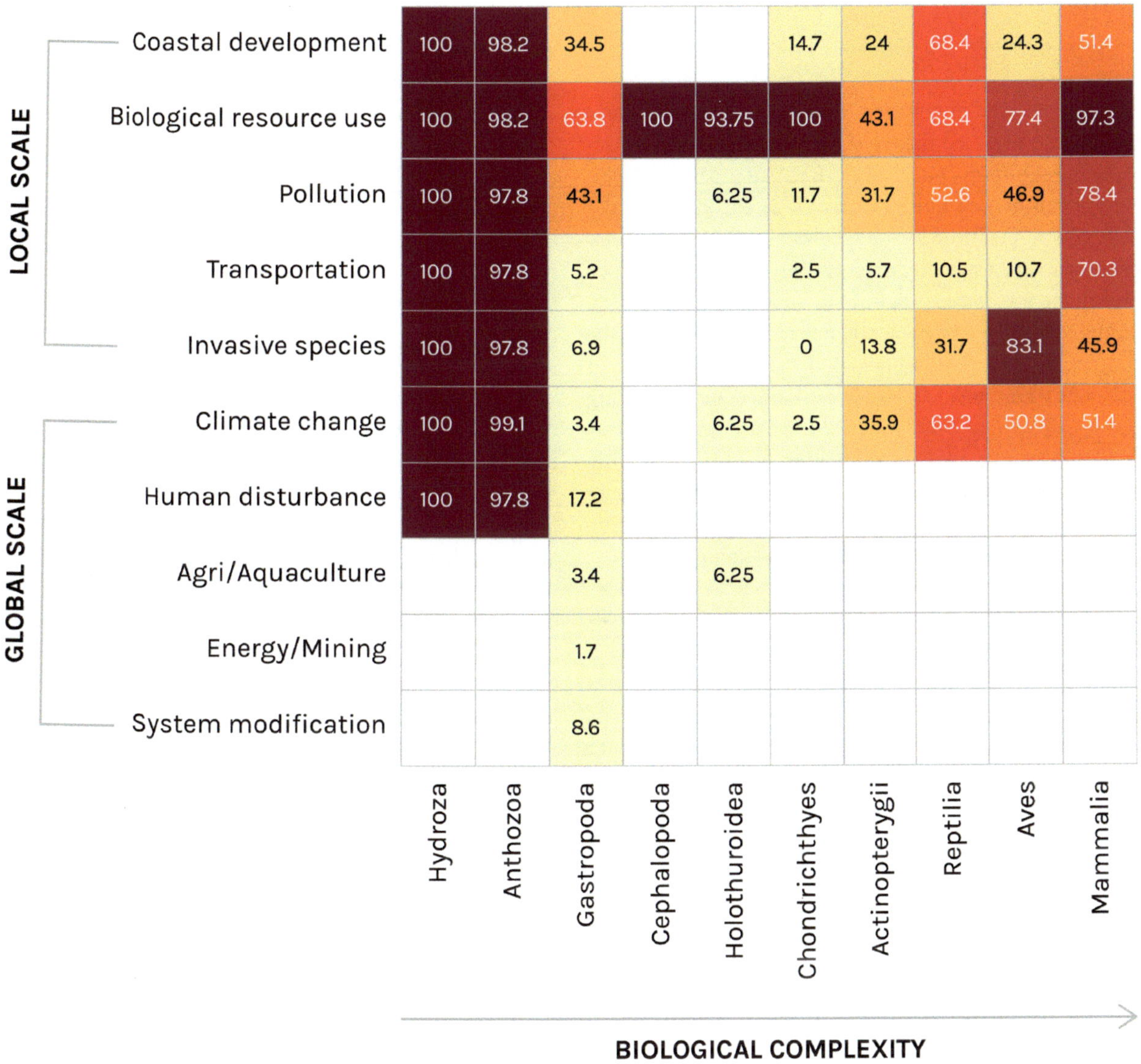

Fig. 10.4 The proportion of the threatened species of each taxon affected by different drivers of extinction risk. *Note: The percentage is reported within each cell. Threatened species out of those assessed for each taxon were: 5 out of 16 Hydrozoa; 226 out of 868 Anthozoa; 58 out of 831 Gastropoda; 5 out of 52 Cephalopoda; 16 out of 371 Holothuroidea; 197 out of 1085 Chondrichthyes; 334 out of 8200 Actinopterygii; 19 out of 95 Reptilia; 177 out of 868 Aves; 37 out of 137 Mammalia. Note that drivers are drawn from the IUCN (2019) Red List. Several drivers are often listed for an individual species. (Source: IUCN Red List)*

genetic diversity and induced evolutionary effects (Pinsky and Palumbi 2014; Heino et al. 2015; Kuparinen and Festa-Bianchet 2017), both of which can potentially reduce the capacity of populations to adapt to threats such as climate change (Blasiak et al. 2020).

A growing number of species are part of high-value consumer markets. As with the Holothuroidea (Purcell et al. 2014), greater rarity pushes their value even higher, which means that they continue to be sourced even if they become more difficult to procure (Courchamp et al. 2006; Sadovy de Mitcheson et al. 2018). Examples of this include shark fins and fish swim bladders, exotic pet species and a range of animals highly valued as luxury food, traditional medicines or ornamentals.

Box 10.2 Fish Spawning Aggregations as Key Biodiversity Areas

To illustrate the importance of key biodiversity areas (KBAs), we selected fish spawning aggregations to contextualise the term "site" in the KBAs,a a seascape unit that (1) can be delimited on maps, (2) encompasses the important habitat used by the species of conservation concern and (3) can actually or potentially be managed as a single unit for conservation. Fish spawning aggregation 'timing' is also part of the context of KBAs. Unlike the conspicuous and betterunderstood breeding colonies of birds and mammals, or the well-known turtle nesting beaches, spawning aggregations of fish are relatively less well-known. But like bird colonies and turtle nesting beaches, they can remain consistent from year to year in time and space and are often appealing targets for fishing because catchability can be particularly high.

Many medium- to large-sized demersal and benthopelagic species in the global ocean form temporary aggregations solely for the purpose of reproduction; these gatherings are the only occasions known for locating a mate and spawning. In the case of tropical groupers (Fig. 10.5a, b) and snappers, many aggregations are highly predictable both spatially and temporally; typically, they form for a week or two over several consecutive months each year. Among temperate species, of the top 25 fishes by weight supplying global fisheries,b many undergo regular spawning migrations, aggregate to spawn for short or extended periods in small or extensive areas, and are exploited at these times. Examples range from Alaska (walleye) pollock (*Theragra chalcogramma*), Atlantic cod (*Gadus morhua*) and Atlantic mackerel (*Scomber scombrus*) to largehead hairtail (*Trichiurus lepturus*) and European pilchard (*Sardina pilchardus*).

Overfishing of spawning aggregations, or of migrations towards these, was a major factor in several fishes declining to threatened status, including the Nassau grouper (*Epinephelus striatus*), the totoaba croaker (*Totoaba macdonaldi*) and the 74 sparid, *Polysteganus undulatus* and other species, none of which were effectively managed prior to declines. Aggregation fishing is likewise implicated for certain populations of orange roughy (*Hoplostethus atlanticus*) (Fig. 10.5c), barred sand bass (*Paralabrax nebulifer*) and large yellow croaker (*Larimichthyes crocea*).[c] Spatial concentration from spawning was also identified by fuzzy logic as an intrinsic extinction vulnerability factor in marine fishes.[d] A global assessment of the known status of 948 spawning aggregations (mainly reef fishes) shows that 26% are decreasing (as determined by reduced catches or underwater visual census counts), 13.5% are unchanged and 3–4%, each, are either increasing or have disappeared entirely; the remaining 53% are of unknown status (Fig. 10.6). These aggregations occur in the global ocean, in over 50 countries, in almost 50 families and in more than 300 fish species.

As productivity hot spots that support a massive proportion of fish biomass, spawning aggregations are key components of the marine ecosystem. Because they are particularly vulnerable to fishing—yet are important to fisheries—they need more conservation and management attention than they have attracted to date, especially from spatial and/or seasonal protective measures.[e] Although conventional management controls may be used for aggregating species—such as minimum sizes, fishing effort or gear controls—and assessments consider maximum sustainable yield or recruitment overfishing, the spawning aggregations themselves are not often explicitly the focus of management, partly because they are so appealing to target. Their management, for example, is not included as a criterion in the Marine Stewardship Council fishery assessment Principle 1, except in relation to habitat protection or access to spawning grounds. However, given issues such as hyperstability and possible depensatory effects at low population levels associated with assessing and managing exploited aggregating species, a specific focus on protecting spawning fish deserves higher priority and special management consideration, especially for species forming large aggregations.[f] On the other hand, well-managed spawning aggregations can support valuable fisheries and contribute to food security as well as conserving biodiversity.

Sources:

[a] Edgar et al. (2008)

[b] FAO (2018)

[c] Sadovy de Mitcheson (2016)

[d] Cheung et al. (2005)

[e] Erisman et al. (2015); Sadovy de Mitcheson (2016)

[f] van Overzee and Rijnsdorp (2015); Sadovy de Mitcheson (2016)

Fig. 10.5 (**a**) Spawning aggregation of the camouflage grouper, *Epinephelus polyphekadion* in French Polynesia. (Photo © Yvonne Sadovy-Micheson). (**b**) Gravid female camouflage grouper at spawning site. (Photo © Stan Shea). (**c**) Orange roughy, *Hoplostethus atlanticus*, a deep-sea species which aggregates around the summits and upper flanks of seamounts for spawning when it is targeted for fishing. (Photo © IUCN Seamounts Project, AD Rogers)

Loss or compromise of key biodiversity areas (such as key egg-laying, nesting, pupping or mating grounds) can quickly reduce populations (see Box 10.2). The finding that biological resource use is the number-one driver of species decline, both in this study and in the IPBES Global Assessment Report (2019), suggests that Aichi Biodiversity Target 6[5] of the Strategic Plan for Biodiversity 2011–2020 has not been attained across the fisheries sector. This is a surprise considering the reported stabilisation and rebuilding of many fish stocks resulting from improved management practices in recent decades (Fernandes and Cook 2013; Hilborn and Ovando 2014; Fernandes et al. 2017; Hilborn et al. 2020). Findings of stabilisation of fisheries are also in contrast to observations that the overall trend, globally, is one of increased overfishing (Pauly and Zeller 2016; FAO 2018). One explanation of the global trends of fisheries declines is the massive increase in the size of the global fishing fleet from 1950 to the present (2015 figures) from 1.7 to 3.7 million vessels (Rousseau et al. 2019). As a result of improving technology (e.g., vessel power) over this period, fishing effort has increased almost exponentially, and catch per unit effort (CPUE) has declined exponentially (Rousseau et al. 2019). The catches from artisanal fishing fleets are often not reported in official government figures, and yet globally the total power levels of these fishing fleets are comparable to those of industrial fishing fleets; they are also less well managed (see below; Rousseau et al. 2019). Asian fishing fleets, in particular, have increased dramatically in both numbers of vessels and fishing power (Rousseau et al. 2019).

Fishing fleets in Europe and North America were reduced in the 2010s, and evidence suggests that it is in these regions CPUEs have stabilised and the decline has also decreased in Oceania as a result of improved fisheries management (Rousseau et al. 2019). Despite a continued increase in overfishing and the decline in CPUEs, global fishing fleets have continued to increase in size and power (Rousseau et al. 2019). If past trends continue, a million more motorized vessels could appear in global marine fisheries in the coming decades.

Both small-scale fisheries and those undertaken by developing states are performing worse than those of developed

[5] By 2020 all fish and invertebrate stocks and aquatic plants are managed and harvested sustainably, legally and applying ecosystem-based approaches, so that (i) overfishing is avoided, (ii) recovery plans and measures are in place for all depleted species, (iii) fisheries have no significant adverse impact on threatened species and vulnerable ecosystems, and (iv) the impacts of fisheries on stocks, species and ecosystems are within safe biological limits.

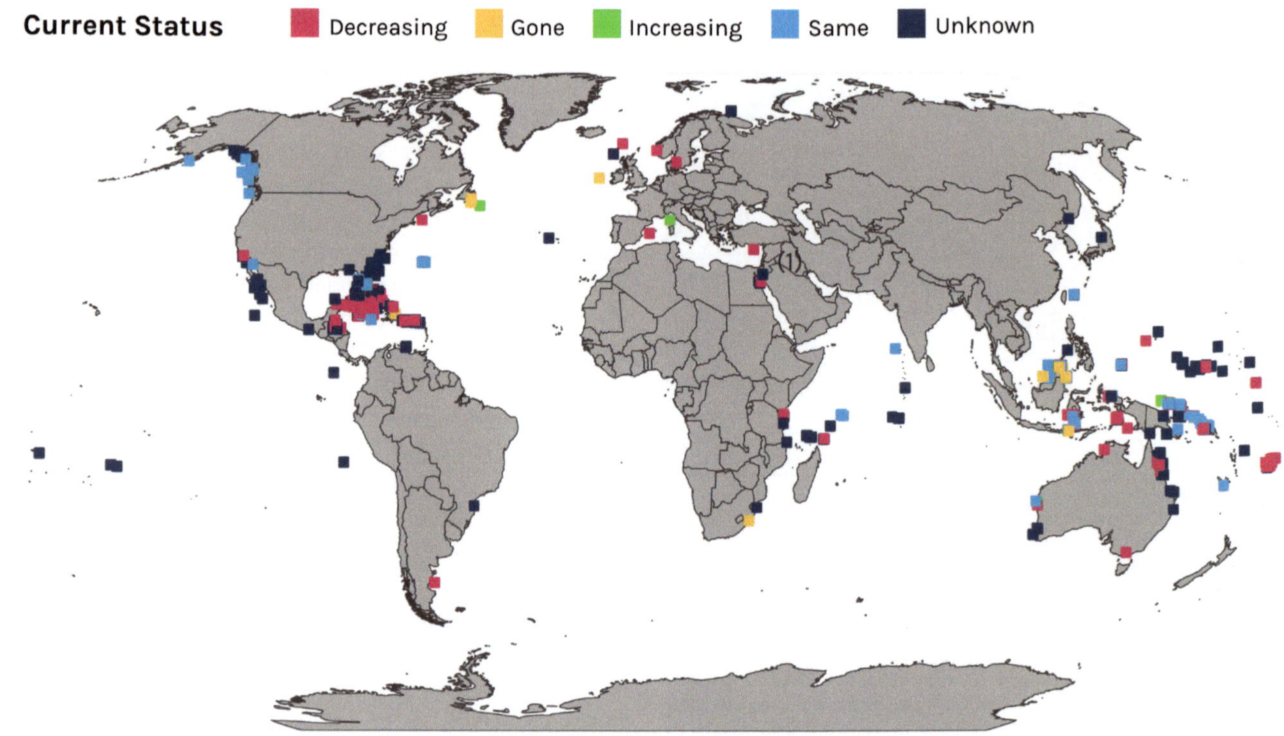

Fig. 10.6 Proportion of invertebrate species assessed on the IUCN Red List compared to the total number of species currently described on the World Register of Marine Species. *Note: A total of 948 documented spawning aggregations are shown. The database is weighted towards tropical reef fish species and underrepresents non-reef and temperate or polar regions. (Source: Science and Conservation of Fish Aggregations (database),* https://www.*SCRFA*.org. *Accessed 14 July 2019)*

states (Hilborn and Ovando 2014; Ye and Gutierrez 2017). A conservative estimate that 23% of global fish catch comes from unassessed fisheries indicates that the lack of data gathering is a significant barrier to sustainable management of target and non- target (bycatch) species (Costello et al. 2012; Gilman et al. 2014; Rousseau et al. 2019). Unassessed fisheries perform poorly in terms of sustainable management compared to those which are subject to scientific stock assessment (Hilborn and Ovando 2014). A large proportion (though not all) of the unassessed fisheries are small, mostly coastal and often artisanal, and many of them are located in the developing world. The costs of scientific fisheries assessments are high and therefore may be uneconomical for implementation in small fisheries, particularly for developing coastal states. In such cases, methods for data-poor fisheries assessment—which rely on broader life history characteristics and/or catch trends, including catch-per-unit-effort estimates—may be a more cost-effective and practical means of management (Hilborn and Ovando 2014), although less reliable (Edgar et al. 2019). Ecosystem-based fisheries management may also be appropriate for small-scale, multispecies fisheries but there is a challenge between the need for complex data with that of practical implementation (Hilborn and Ovando 2014).

Studies that have found standards of fisheries management to be generally poor amongst coastal states with many fisheries exhibiting overcapacity, capacity-enhancing subsidies, problems with foreign access agreements and issues around the transparency of management and decision-making, show that such problems are worse within developing states (Mora et al. 2009; Pitcher et al. 2009). This emphasises the lack of capacity to manage fisheries in these countries (Pitcher et al. 2009; Hilborn and Ovando 2014; Ye and Gutierrez 2017). This situation is magnified because developed countries either import fish from other regions of the world or establish fisheries partnership agreements, effectively externalising their costs for fisheries management (Ye and Gutierrez 2017). As with smallscale fisheries, investment in management methods that are appropriate for developing countries are needed to establish more even standards for global fisheries sustainability. However, this may need reciprocal arrangements between developed and developing countries, especially where the former benefit from the fisheries resources of the latter, to enhance fisheries management capacity through finance, training and technology transfer (Ye and Gutierrez 2017). Seafood trading mechanisms that promote sustainability may also be useful for addressing the management of fisheries in developing countries. Carrot-and-stick approaches may be useful as well, such as marketbased measures (e.g., certification or eco-labelling) which promote sustainable fishing or impose import restrictions on overfished stocks (Ye and Gutierrez 2017).

We also point out that overfishing is by no means restricted to developing states, and a cursory examination of the literature indicates that even in the waters of regions such as Europe, a significant number of stocks are in decline or are overfished, especially smaller stocks (Fernandes et al. 2017). Studies of fisheries sustainability also often neglect to acknowledge that even modern fish stock assessment methods have levels of uncertainty associated with them and relatively few use, or are validated by, fisheries independent data (Edgar et al. 2019). Improvements in catch efficiency in fleets may also be difficult to represent in stock assessments (Edgar et al. 2019). An increasing issue is also that stock assessments are often based on historical assessments when current climate change means that the environment is changing rapidly and such data may not reflect alterations in stock dynamics or distribution (Edgar et al. 2019). Stock assessments also concentrate on management of single species or stocks, ignoring interspecies interactions (e.g., with predators and prey) and other aspects of ecosystem structure, function and health (Edgar et al. 2019).

There have been increasing measures to incorporate biodiversity considerations into fisheries management (Garcia 2010; Rice and Ridgeway 2010; Friedman et al. 2018). These measures can be seen as part of a broader shift in societal views on the use of natural resources from one of straightforward economic exploitation to one of sustainable development whereby the use of ecosystem goods and services must be traded off against the resilience of the environment (Garcia 2010; Friedman et al. 2018). These concepts were introduced into the arena of resource management following World War II, but they were significantly strengthened through the adoption of the World Conservation Strategy in the 1980s, the outcomes of the United Nations Conference on Environment and Development and the Brundtland Commission (1983–87), culminating in the CBD which entered into force in 1993 (Friedman et al. 2018). UNCLOS and the subsequent 1995 Fish Stocks Agreement both included specific provisions with respect to sustainability of both target fish stocks and the wider ecosystem. At this point, states began to incorporate increasing measures to address sustainability and to decrease the environmental impact of fishing. These measures have been reviewed on a regular basis through the United Nations General Assembly, and biodiversity considerations have been gradually mainstreamed in fisheries management through a variety of voluntary agreements and measures by the FAO (e.g., the CCRF; international plans of action to reduce fishing impacts on sharks, seabirds and turtles; see Friedman et al. 2018 for a more comprehensive list). Likewise, the fisheries management and environmental sectors have increased their collaboration to improve the environmental performance of fisheries (Friedman et al. 2018). However, given the impact on extinction risk in marine species (this study and the IPBES Global Assessment Report 2019), there is clearly a long way to go in improving the environmental sustainability of marine capture fisheries. It is also notable that reducing overfishing would in itself reduce impacts on threatened species affected by bycatch (e.g., mammals, seabirds and turtles; Burgess et al. 2018).

Uneven implementation at the global level is also an issue with measures to conserve biodiversity from the destructive effects of fishing. For example, the Convention for the Conservation of Antarctic Marine Living Resources (CCAMLR), which manages fisheries in the Southern Ocean, has worked with the nongovernmental organisation (NGO) Birdlife International to massively reduce interactions (often fatal) of albatrosses and petrels with longline fishing in the region by 67,000 per annum (Friedman et al. 2018). However, at present it is estimated that seabird bycatch in longline fisheries globally range from an average of 160,000 to an upper range of 320,000 per annum and is a major driver of the decline of albatrosses and petrels (Anderson et al. 2011; Dias et al. 2019). Technical measures for longline fishing, including setting lines at night, are known to decrease bycatch and have been successful at reducing this source of mortality in albatrosses and petrels in areas of the Southern Ocean such as South Georgia (Anderson et al. 2011; Phillips et al. 2016). Yet recent analysis of the behaviour of pelagic longline fishing vessels in the southern Atlantic, Indian and Pacific Oceans indicate that less than 5% of vessels may be complying with requirements south of latitude 25° south by setting in the daytime (Winnard et al. 2018). We point out that obtaining data on fisheries bycatch is problematic for many fisheries, especially on the high seas and where observer coverage is low and reporting mechanisms are weak (Anderson et al. 2011; Gilman et al. 2014; Phillips et al. 2016), while the impact of purse-seine fisheries, such as for forage fish, have not been properly evaluated.

Another example of uneven implementation of actions to conserve biodiversity has been in the uptake of the FAO's International Guidelines for the Management of Deep-Sea Fisheries in the High Seas (FAO 2008). These guidelines were established to protect vulnerable marine ecosystems (VMEs), such as deep-sea cold-water coral reefs and seamounts, from the impacts of bottom trawling as well as to improve the management of low-productivity deepwater fisheries. The guidelines have resulted in significant actions to protect biodiversity by regional fisheries management organisations (RFMOs) or agreements through the use of spatial conservation measures, gear restrictions and encounter rules, which require a vessel to move away from an area where VMEs are encountered and to report the encounter (Rogers and Gianni 2010; Wright et al. 2015; Bell et al.

2019). There have also been efforts to implement biodiversity conservation measures using RFMOs and Regional Seas Agreements to implement biodiversity conservation measures (Friedman et al. 2018).

A good example is the action by the North-East Atlantic Fisheries Commission (NEAFC) and the Oslo Paris (OSPAR) Commission to initiate MPAs in areas beyond national jurisdiction, such as the Mid-Atlantic Ridge (Wright et al. 2015). The collaboration between the NEAFC and the Oslo Paris (OSPAR) Commission was formalised first through a memorandum of understanding (NEAFC/OSPAR 2008) and then through a collective arrangement (NEAFC/OSPAR 2014). However, implementation of the FAO guidelines has progressed much more slowly and unevenly with other RFMOs and agreements (Rogers and Gianni 2010; Wright et al. 2015) with some showing poor progress even to the present (Bell et al. 2019). In some cases, this seems to be linked to a lack of capacity and financial support to achieve a better performance of fisheries in areas beyond national jurisdiction in terms of sustainability of stocks and protection of biodiversity (Bell et al. 2019).

For birds, the major threats are invasive species for breeding colonies and unintentional bycatch at sea, with the latter being solely responsible for many species becoming threatened (Paleczny et al. 2015; Dias et al. 2019). For mammals, it is notable that transportation corridors are a major threat given the increasing impacts of ship strikes on cetacean populations (Ritter and Panigada 2019). Climate change and extreme weather are also significant threats for four of the five vertebrate groups assessed. Additional threats include coastal activities such as residential and commercial development and pollution.

Box 10.3 The Global Risk to Marine Biodiversity

In order to estimate the patterns of global risk to biodiversity, we overlaid spatial data on human impacts from Halpern et al. (2008) onto the data on species diversity used to generate Fig. 10.1. Human impact index data were regridded on a 110-by-110-km equal area grid and overlaid with the species richness data (Fig. 10.7a). The relationship between species richness and the corresponding human impact index was assessed by computing the centroid of the relationship in a log-log dimension (Fig. 10.7b). Based on the position of the geographical cell, we established four categories: high richness and high impact in red, low richness and high impact in violet, high richness and low impact in green and, finally, low richness and low impact in blue. Then the Euclidean distance among each geographical cell to the centroid of each category was computed, and the shades of colour in Fig. 10.7b represent these distance intervals.

The multitude of impacts from human society on the ocean have been summarised at a global level, showing alteration of all marine ecosystems.[a] The examination of the relationship between biodiversity[b] and anthropogenic pressures[c] (Fig. 10.7a, b), reveal four different scenarios:

1. Regions where the level of biodiversity and human impact are very high include the Indo-Pacific Coral Triangle; Southeast Asia, including the seas off Thailand, China and Korea; northern Australia; the western Indian Ocean; the Mediterranean; the coasts of northern Europe (North Sea); and some western Pacific Islands. Although this analysis specifically aimed to reduce sampling bias, the levels of sampling for species from different regions of the ocean vary dramatically. Therefore, it is likely that sampling bias has resulted in some areas being classified as having a high biodiversity as a result of intense sampling rather than in having a high inherent species richness; the North Sea is the most obvious example. Some areas have been identified as high impact with a high diversity in other studies (e.g., Indo-Pacific Coral Triangle, northern Australia, some of the western Pacific Islands, areas of the Indian Ocean).[d] In some cases, they are also in locations where there is a rapid increase in human pressures (e.g., Australia and parts of the Indo-Pacific Coral Triangle).[e] The explanation for some areas of the ocean having high levels of diversity and impact vary. In some cases, there is a high coastal population and/or high levels of direct (e.g., fisheries) and indirect (e.g., pollution) exploitation of coastal and offshore marine ecosystems. These waters include those of both developed and developing coastal states.
2. Areas where human pressures and biodiversity are moderately high include the central Indian Ocean and Caribbean, the eastern seaboard of the United States and Canada, and the western coast of the United States as well as northern Brazil. Some of these areas have been identified as high impact with a high diversity in other studies (i.e., the Caribbean and parts of the Indian Ocean).[f] The recent rapid increase in human pressures has also been observed for the coast of Brazil.[g]
3. Areas of high biodiversity and low human pressure include some of the islands in the western and central tropical Pacific, parts of Hawaii, the Galapagos Islands, the Seychelles and areas of the open ocean, Russian Arctic and Alaska. Some regions with a high diversity and a low level of human threat include those in which

significant fully or highly protected MPAs have been established and have reduced pressures as well as being relatively remote (e.g., Kiribati and the Galapagos Islands).

4. We note that there is also a lack of areas which have a lower diversity which are highly impacted (i.e., points in the lower right quadrant of Fig. 10.7b). This may be explained by relatively low observed impacts in polar and open ocean ecosystems, regions with a lower diversity than the tropics and coastal ecosystems. Lack of data on human impacts may be a factor here.

In conclusion, more than half of the ocean is considered to be heavily perturbed by human activities; this includes more than half of the hot spots of marine species richness.

Sources:

[a] Halpern et al. (2008, 2019)
[b] Reygondeau (2019)
[c] Halpern et al. (2019)
[d] Jenkins and Van Houten (2016)
[e] Halpern et al. (2019)
[f] Jenkins and Van Houten (2016)
[g] Halpern et al. (2019)

Climate change, especially increasing temperature and habitat impacts, is predicted to become an increasing threat to many invertebrate and vertebrate species (IPCC 2019), but there are uncertainties in terms of projections. The upper thermal tolerance limits for shallow tropical reef-building corals have been exceeded in multiple global stress events from 1998 to 2017 (Hughes et al. 2018a; Stuart-Smith et al. 2015), resulting in large-scale coral loss, local and regional scale shifts in species composition and ultimately reef function. This impacts ecosystem function and the provisioning of ecosystem services (Hughes et al. 2017), and as waters warm, such thermal limits will be more frequently exceeded. There is already evidence that reproductive synchrony in broadcast-spawning corals is breaking down (Shlesinger and Loya 2019), and in fish species, spawning times could be radically affected; these are often temperature-associated changes, and they may impact reproductive success (Asch and Erisman 2018). Some fish appear to go deeper, tracking cooler waters in warming seas, illustrating the rapid responses of marine life to ocean warming (Burrows et al. 2019). It is also stressed here again that the taxa assessed for the IUCN Red List are a biased sample often focusing on those which are heavily exploited (e.g., the Holothuroidea for the invertebrates). Many groups of organisms, especially poorly known invertebrates, are likely to be significantly impacted by climate change either directly as their environmental tolerances are exceeded or indirectly as their habitat is destroyed. The overall impacts of climate change on marine biodiversity is therefore likely to be currently underestimated.

Particularly in the coastal zone, development and pollution, which are often connected, have been the other major drivers of species declines. As with the lack of information on small-scale fisheries, it is notable that the monitoring of biodiversity within the waters of coastal states is weak despite it being a requirement in several of the conventions and agreements reviewed in this report. An indicator of this is the number of species categorised as DD in Red List assessments (see Fig. 10.3). There is overwhelming evidence from a broad range of taxa that loss of habitats formed by foundation species, including corals, mangrove forests, seagrass beds, saltmarshes and kelp forests, continues unabated in many regions of the world (see Sect. 3.5), despite specific agreements or conventions which are aimed at conserving such ecosystems (e.g., the Ramsar Convention on Wetlands).

In summary, biodiversity impacts in the ocean have generally manifested as population declines, habitat degradation and loss, and ecosystem-level changes rather than as global extinctions. Although overexploitation has been the primary driver of loss to date for many groups, it is notable that habitat destruction through coastal development and pollution are major contributors to species being added to the Red List's threatened categories. Climate change impacts are expected to grow in the future.

Although few marine extinctions have been observed (Dulvy et al. 2003), in the best-assessed groups of marine species around 11–46% are judged to be at risk of extinction, a range that spans the proportion of threatened terrestrial species in well assessed groups (20–25%; Webb and Mindel 2015) with individual groups falling above and below this range. Global extinctions in the marine environment are relatively rarely documented (Dulvy et al. 2003; McCauley et al. 2015), and trends in the species richness of local communities can be relatively flat, though with turnover in species composition (Dornelas et al. 2014). OBIS currently holds over 50 million occurrence records of 125,000 marine species; about half of the total number of marine species described to date according to the World Register of Marine Species (WoRMS). Given this, extinction rates in marine species may be higher than previously estimated because they have simply not been documented.

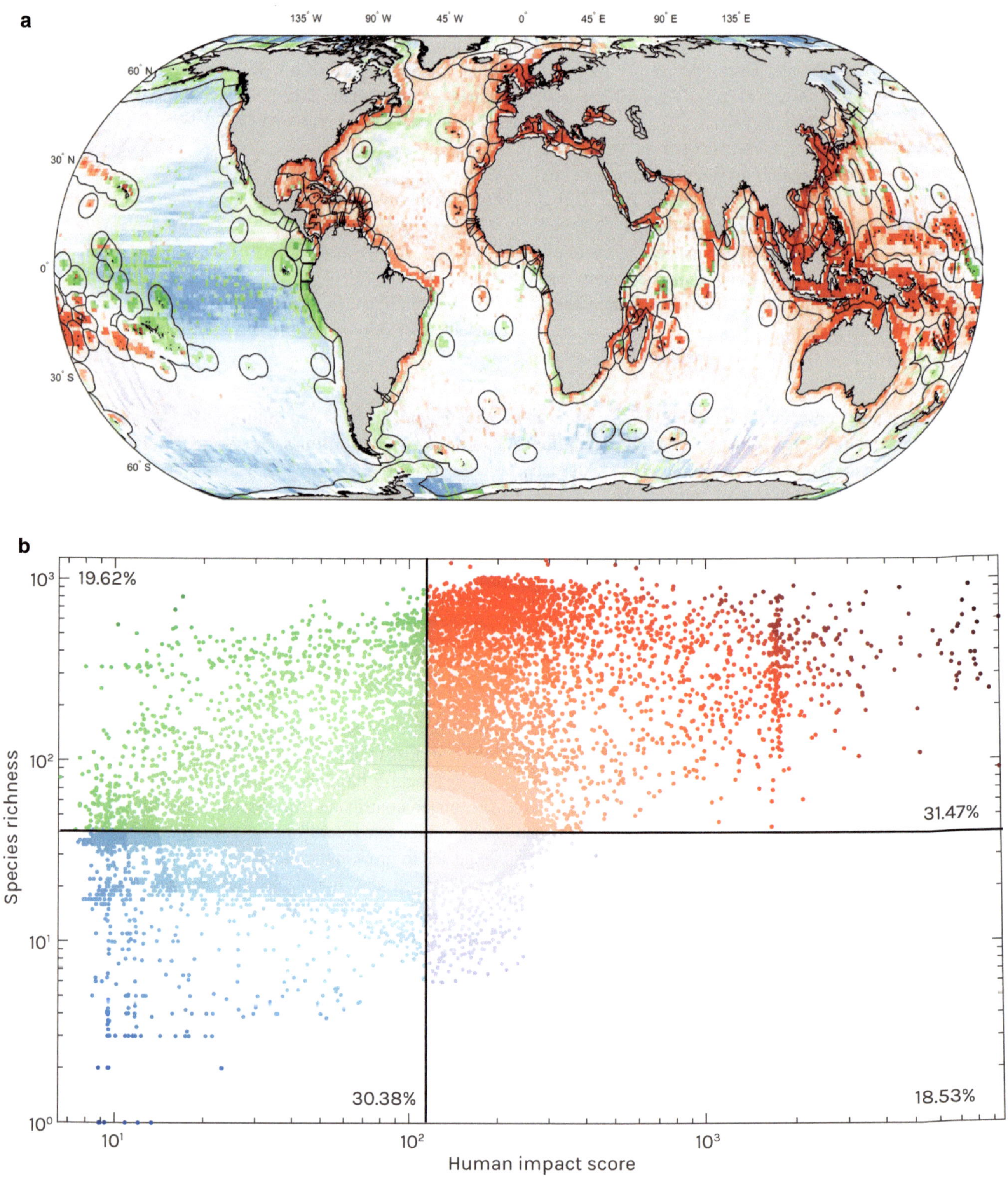

Fig. 10.7 Marine biodiversity in relation to human impacts. *Note: Map* (**a**) *and scatter plot* (**b**) *of the relationship between marine biodiversity and the human impact score. Each quadrant has been computed based on the centroid of the relation in a log-log dimension. Colour shades are computed as the Euclideian distance of the geographical pixel from the centroid of the relation. (Sources: Based on Halpern et al.* 2008 *and Reygondeau* 2019*)*

3.5 Habitat Degradation and Its Drivers

The IPBES Global Assessment Report (2019) summarised key threats to the ocean. Overall, 66% of the ocean is experiencing increasing cumulative impacts (Halpern et al. 2015). The area of ocean still classified as 'wilderness', characterised by having a low impact across a range of anthropogenic stressors, is as low as 13% (Jones et al. 2018), and the area with no discernible human footprint is down to 3% (Halpern et al. 2015). Seagrass meadows decreased in extent by over 10% per decade from 1970 to 2000, the global cover of mangroves has declined about 40% (Thomas et al. 2017) and that of saltmarshes by an estimated 60% (Gedan et al. 2009). Regionally, kelp forests have also shown significant declines in distribution, such as in the Great Southern Reef area of Australia, where they are associated with a high level of endemism (species restricted to a specific geographic location) (Bennett et al. 2016). However, kelp forests are highly dynamic ecosystems, and globally the picture is more complicated; whereas in some areas no trends are apparent, in others, kelp forests are extending their range (Krumhansl et al. 2016).

The role of coral reefs as a flagship ecosystem is characterised by their high biodiversity (Fisher et al. 2015) and their benefits to people (Wilkinson et al. 2016; IPBES 2019). Coral reefs have lost half of their live coral cover since the 1870s, and losses have accelerated over the last two to three decades as a result of the direct effects of climate change (Wilkinson et al. 2016; IPCC 2019) and the indirect effects of other drivers, such as predator outbreaks or disease epidemics, some of which are exacerbated by climate change (Wilkinson et al. 2016; IPBES 2019, BG 4, 5).

Projections for coral reef loss—even at the most optimistic climate change scenarios—are dire: corals could be reduced to 10–30% of their former abundance at warming of 1.5°C, and they could be reduced to only 1% at 2 °C (IPCC 2018). Estimates of coral loss generally conflate loss of cover with loss of reefs. Most reefs will endure, but coral cover on them will decline.

Marine habitats have experienced significant reductions in area in the past century. Coastal reclamation and land-use change, together with pollution and, more recently, climate change, have led to the vast loss of many valuable coastal habitats, estimated at an average of 30–50% (Pandolfi et al. 2003; Polidoro et al. 2010; Waycott et al. 2009; Barbier 2017; Duarte et al. 2020). The first large-scale loss of coastal habitats was documented in China more than a millennium ago and in Europe around the 14th century, when seawalls were built to prevent tidal inundations and to transform saltmarshes into agricultural land (Loke et al. 2019). Such coastal development sprawls throughout much of the world, leading to significant saltmarsh losses in Europe, the United States, Canada and Asia. In China, for instance, more than 60% of the coastline is now artificial (Liu et al. 2018). Land reclamation and conversion to aquaculture ponds and rice paddies has led to much of the observed mangrove loss (Richards and Friess 2016).

Eutrophication and physical impacts, such as dredging, are responsible for much of global seagrass losses (Waycott et al. 2009). It is important to note that as well as causing the loss of ecosystems such as mangroves and saltmarshes, coastal engineering can also prevent such ecosystems from adapting to climate change by preventing the landward migration of such habitats as sea level rises which is known as transgression (Hughes 2004; Alongi 2015; Lovelock et al. 2015).

The first losses of coral reefs were driven by siltation derived from the deforestation of tropical watersheds, overfishing and reduced water quality from sewage and excess nutrient inputs from agricultural land (Pandolfi et al. 2003; MacNeil et al. 2019; Williams et al. 2019). Recent global bleaching events, driven by El Niño warming events exacerbated by anthropogenic ocean warming (Hughes et al. 2017, 2018a, b; Claar et al. 2018; Lough et al. 2018), have now emerged as a major driver of present, and future, coral loss. Under such a multiplicity of detrimental anthropogenic stressors, coral reefs have a tendency to convert to alternative stable states, such as dominance by fleshy algae or cyanobacterial mats (Ford et al. 2018a). This can be associated not only with loss of positive ecosystem services, such as coastal protection or fisheries, but also the potential for negative impacts on coastal human communities (e.g., an elevated risk of ciguatera or ciguatera-like diseases; Ford et al. 2018a).

Upwelling regions, where most of the fisheries for forage fish are located, have also been degraded by overfishing. This results in food chain alterations and the risk of trophic structure breakdown, particularly when small pelagic fish—which are the link between primary producers and the secondary consumers in the typical wasp-waist trophic structure—are removed from the food web (Cury et al. 2000). Such examples have already been observed affecting top predators and lower trophic levels (Velarde et al. 2015a, b).

Overfishing of small pelagic or forage fish results in increased population fluctuations (Cisneros-Mata et al. 1996; Hsieh et al. 2006) and reduces their resilience to natural environmental periodic changes such as the El Niño Southern Oscillation and the Pacific Decadal Oscillation, rendering these forage fish populations more vulnerable to these natural variations, risking their final collapse. Furthermore, more than one million square kilometres (km^2) of the seabed are subject to bottom trawling each year (about 14% of the total trawlable area of 7.8 million km^2 which lies shallower than 1000 m depth; Amoroso et al. 2018). This degrades seabed communities through physical impact, affecting biodiversity and ecosystem function (Thrush and Dayton 2002; Pusceddu et al. 2014; Ashford et al. 2018) and significantly alters ecosystem processes such as sedimentation at large scales (Puig et al. 2012; Pusceddu et al. 2014). Deep-sea ecosystems can be especially vulnerable to the effects of fishing. Seafloor ecosystems are fragile and have low resilience (Clark et al. 2016; Rogers 2018) and the targeting of deep-sea fish species and the effects of bycatch have been observed to rapidly overexploit stocks (Norse et al. 2012; Victorero et al. 2018). The deep sea is increasingly contaminated with litter (Pham et al. 2014; Woodall et al. 2015) and in the near future, it will experience increased temperatures, stratification, decreased oxygen concentrations, and ocean acidification (Rogers 2015; Sweetman et al. 2017). The increasing demand for raw metals and minerals, coupled with the depletion of terrestrial resources, is making deep-sea mining more attractive economically (Petersen et al. 2016; Miller et al. 2018). The impacts of this industry are likely to be extremely severe (Niner et al. 2018).

3.6 Reducing the Provisioning of Ecosystem Services

Biodiversity plays a significant role in ecosystem functioning, which underpins nature's contribution to people (NCP). The concept of NCP is elaborated in the IPBES Global Assessment Report (2019), as the positive and negative contributions of living nature to people's lives. Here, we focus specifically on positive ecosystem services, 'the benefits people obtain from ecosystems' (MA 2005), a subset of NCP. This is because we focus on the potential negative consequences of biodiversity loss in the ocean, and the positive provisioning of ecosystem services has been widely discussed in the context of the marine environment. The benefits of ecosystem services include provisioning services; the production of goods and materials such as food, raw materials and pharmaceuticals; regulatory services; the control of climate, atmosphere and other aspects of the environment that maintain the Earth system; supporting services; those that enable the provision of direct and indirect ecosystem services to humankind; and cultural services, including recreation, tourism, inspiration for art, culture, spiritual experience and cognitive development (de Groot et al. 2012; Costanza et al. 2014; Barbier 2017).

There have been various attempts to estimate a monetary value for marine ecosystem services (Costanza et al. 1997, 2014; WWF 2015; Martin et al. 2016), demonstrating that conservation of species and ecosystems is economically advantageous (Costanza et al. 2014). Specific examples include the use of natural ecosystems for coastal defence (Narayan et al. 2017; Hooper et al. 2019) and for sustainable fisheries management (Costello et al. 2016, 2019; World Bank 2017). Valuations for ecosystem services have been developed for land-based systems where the 'value' of natural capital (abiotic and biotic elements of nature) can easily be estimated from the areas of different types of habitat. Such valuation methods run into difficulties when applied to marine ecosystems, which are three-dimensional; contain many mobile elements, both spatially and temporally; are highly connected and often data-poor (Hooper et al. 2019). Ecosystem services are also provided at different scales—from the individual to human society as a whole (Pendleton et al. 2016; Small et al. 2017)—and, as such, are often public goods or the product of common assets that lead to problems with simplistic systems of monetisation (Costanza et al. 2014). Also, whilst ecosystem services are generally positive, nature can also generate negative impacts on people depending on spatial, temporal, social and cultural contexts (IPBES 2019). This is particularly complicated by the fact that many ecosystem services are strongly linked; thus, enhancing provisioning services, for example, can have a negative impact on regulating services (Raudsepp-Hearne et al. 2010). This can be assessed through analysis of bundles of ecosystem services and the trade-offs between them (Raudsepp-Hearne et al. 2010). The cost-benefit analysis approach inherent in the monetary valuation of ecosystem services can be useful in some contexts, but a more comprehensive methodology is required to establish a value for ecosystem services that takes into account more than just instrumental values (Colyvan et al. 2010; Hooper et al. 2019). One way of counteracting some of the difficulties in valuing ecosystem services can be the development of a risk register, which identifies those ecosystems and their services in danger of loss (Mace et al. 2015).

The relationship between Biodiversity and Ecosystem Functioning (the BEF curve), and thus the provisioning of ecosystem services, is not well understood but is generally observed to be positive (Hector and Bagchi 2007; Harrison et al. 2014; Lefcheck et al. 2015), including in marine ecosystems (Stachowicz et al. 2007; Danovaro et al. 2008; Gamfeldt et al. 2014; Duffy et al. 2016). The shape of the BEF curve has major implications for the impacts of biodiversity loss on ecosystem functioning and service provision and can be saturating, linear or accelerating (Fig. 10.8). These studies provide some scientific understanding of the mechanisms that may underlie the degradation of ecosystem services when biodiversity is lost, including biomass production, resilience to disturbance and biological invasions (Stachowicz et al. 2007; Duffy et al. 2016).

The impacts of biodiversity loss on ecosystem services are multi-faceted. Regional changes in biodiversity have been shown to affect fisheries and other services and generate risks, including harmful algal blooms, oxygen depletion, coastal flooding, and species invasions (Worm et al. 2006).

High biodiversity may also result in greater resistance to climate change impacts, potentially mitigating the effects on fishery yields (Duffy et al. 2016). On coral reefs, ecosystem

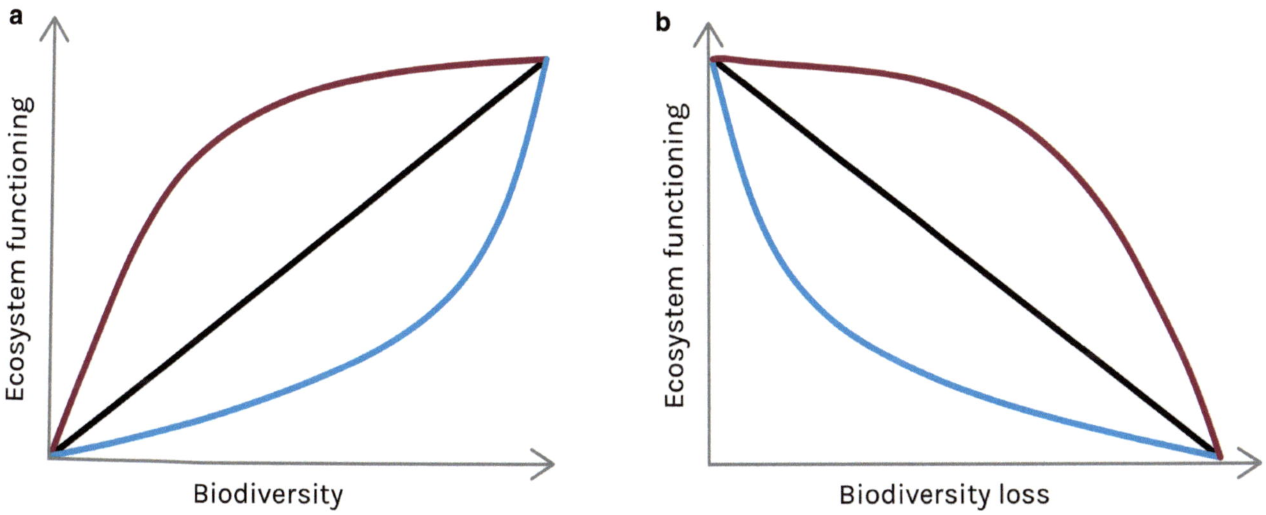

Fig. 10.8 Three types of positive biodiversity-ecosystem functioning relationships. *Notes:* (**a**) *Ecosystem functioning relationships: saturating (red), linear (black), and accelerating (blue).* (**b**) *Relationship between biodiversity loss and the three types of biodiversity-ecosystem functioning relationships. (Source: Modified from Naeem 2002; Strong et al. 2015)*

functioing has been suggested to scale with biodiversity, with human population density impacting both biodiversity and ecosystem functioning (Mora et al. 2018). The loss of coastal habitats renders coastlines more vulnerable to flood risks from sea level rise (Guannel et al. 2016) and cyclones (Barbier 2017; Hochard et al. 2019). In the case of coral reefs, the reduction in damage to terrestrial assets conferred through coastal protection is estimated at $4 billion annually (Beck et al. 2018). For the top five countries that benefit from reef protection (Indonesia, Philippines, Malaysia, Mexico, Cuba), this is the equivalent benefit of $400 million annually in mitigated damage (Beck et al. 2018). Annual expected damage from flooding would double, and costs from frequent storms would triple without coral reefs (Beck et al. 2018). The global loss of coral reefs has been estimated to have an economic impact of more than $10 trillion per annum (Costanza et al. 2014). Coastal habitats are important habitats and nursery sites for many species, so their losses result in reductions in fisheries and coastal food production (Aburto-Oropeza et al. 2008; Barbier 2017; Robinson et al. 2019; Unsworth et al. 2019), and they increase threats to species with a fragile conservation status.

Seagrasses, saltmarshes and mangroves are the three internationally recognised blue carbon habitats that actively sequester and store organic carbon from the environment (Nellemann et al. 2009; Duarte et al. 2013a, b). Mangroves are able to sequester more organic carbon on average than seagrasses and slightly more than saltmarshes (Mcleod et al. 2011). However, seagrasses have an area of around 180,000 km^2 globally, more than twice the area of mangroves, highlighting their importance as a significant carbon sink in comparison to mangroves. However, some of the carbon that is stored in these marine macrophytes has an allochthonous source from other habitats. Kelp beds and other macroalgae communities (Wernberg and Filbee-Dexter 2019) are only recently being considered important in blue carbon storage (Trevathan-Tackett et al. 2015; Krause-Jensen and Duarte 2016; Krause-Jensen et al. 2018). This may not only be through the existence of natural kelp and macroalgal communities but also through kelp aquaculture, where a significant amount of carbon is sequestered prior to harvesting (Duarte et al. 2017). Therefore, it is critical to focus on filling the gaps in knowledge of the extent, distribution and role of macroalgae in a global context, for both climate mitigation and adaptation, and as providers of crucial ecosystem goods and services.

Projected reductions in overall marine biomass associated with climate change may further impact ecosystem services such as fishery yields (Lotze et al. 2019). Any impact on fishery yields may have knock-on effects on food security. It is possible that some countries are likely to face a 'double jeopardy' of impacts on both agricultural and fisheries sectors as a result of climate change (Blanchard et al. 2017).

4 Thresholds and Tipping Points

There are ecological thresholds and other reference points that—if exceeded through the alteration of marine habitats, the exploitation of living marine resources or other human impacts on marine ecosystems—could result in negative and irreversible changes to ecosystems and the broader services they provide (Rockström et al. 2009; Lenton 2013).

The ecosystem approach to management of marine resources aims to preserve the integrity and resilience of marine ecosystems by reconciling conservation and exploitation (Pikitch et al. 2004). Under heavy fishing and climate pressures, many ecosystems are facing severe and abrupt regime shifts. This results in alternate ecosystem states that are most often less productive for fisheries, more prone to booms and busts, weakly reversible and thus less manageable (Pine et al. 2009; Estes et al. 2011, Travis et al. 2014). In this context, a major challenge for research and management is understanding evolving species interactions while identifying critical thresholds and tipping points involved in the disruption of marine ecosystems.

4.1 Changes in Marine Ecosystems

Climate patterns have long been recognised as responsible for regime shifts in both pelagic and benthic marine ecosystems. Empirical evidence has accumulated to indicate that shifts in species composition are initiated by large environmental changes, such as in the California Current (Hooff and Peterson 2006), the Gulf of Alaska (McGowan et al. 1998), the northern Pacific (Hare and Mantua 2000), the northern Atlantic (Aebischer et al. 1990) or the Humboldt Current (Chavez et al. 2003). Likewise, regime shifts between tropical coral reefs and algal-dominated reefs have been reported in response to thermal anomalies associated with El Niño events (Hughes et al. 2007; Diaz-Pulido et al. 2009), now compounded with anthropogenic ocean warming (Graham et al. 2015).

Long-term ocean warming and acidification—as well as extreme events that are becoming more frequent, more intense and longer lasting—alter the structure of ecosystems and cause mortality and community reconfiguration. This is particularly noticeable for sessile organisms that are impacted by discrete, prolonged, anomalous warm-water events known as marine heat waves (Hobday et al. 2018). The widespread bleaching and mortality of reef-building corals (e.g., in the Great Barrier Reef, the Caribbean Sea and the Gulf of Mexico), seagrass meadows and kelp forests have been strongly affected by localised, extreme warming of the ocean (Smale et al. 2019). The density and diversity of corals on reefs are declining, leading to vastly reduced habitat complexity, loss of biodiversity and domination by macroalgae that form stable communities relatively resistant to a return to coral domination (Wilson et al. 2006; Hoegh-Guldberg et al. 2007).

Climate change reinforces the frequency and strength of ecosystem shifts by affecting the distribution of marine life. Geographical shifts in marine species, from plankton and fishes to mammals and seabirds, occur as the result of ocean warming and have changed the distribution by hundreds of kilometres or more since the 1950s (Poloczanska et al. 2013, 2016; IPCC 2019). Ocean warming and heat waves also cause a poleward expansion of corals, leading to a phase shift from kelps to corals in South West Australia, facilitated by the poleward expansion of tropical herbivorous fish that prevent kelp from reestablishing (Wernberg et al. 2016). A poleward shift in species distributions is the most commonly observed pattern; it leads to changes in community structure, resulting in cascading impacts on ecosystem structure (IPCC 2019). The tropics may be particularly sensitive to this phenomenon as well as the transition zone between tropical and temperate communities, where the rate and magnitude of change will be highest. However, in the Humboldt upwelling system off the coast of Chile, most fish species do not show expansion of their southern endpoint because of a weak warming trend, reinforcing the hypothesis that temperature is a major determinant of species range dynamics (Rivadeneira and Fernandez 2005).

A global decrease in abundance and biodiversity of marine species driven by ocean warming is projected to diminish the catch potential for global fisheries in the 21st century (Britten et al. 2017; IPCC 2019). Global rates of biomass production as well as standing stocks are projected to decrease in ocean ecosystems at all depths, from the surface to the deep seafloor. The large-scale redistribution of global fish and invertebrate species biomass is expected to occur by 2055, with an average increase of 30–70% in high-latitude regions and a drop of up to 40% in the tropics under climate change scenarios (Cheung et al. 2010).

These changes in distribution are already affecting the species composition of catches. Fisheries are catching an ever-increasing percentage of warm-water marine species, a phenomenon identified as the 'tropicalisation' of the world catch (Cheung et al. 2013). Displacement of tropical herbivorous fish to temperate habitats also drives a similar tropicalisation of benthic habitats (Vergés et al. 2014; Wernberg et al. 2016). Using an ensemble of multiple climate and ecosystem models, it is projected that even without considering fishing impacts, mean global marine animal biomass will decrease by 5% (±4% standard deviation) under low emissions and 17% (±11% standard deviation) under high emissions by 2100, with an average 5% decline for every 1 °C of warming (Lotze et al. 2019).

In ecosystems stressed by overexploitation and climate change, cascading effects that have promoted regime shifts have been thoroughly documented in diverse marine ecosystems, ranging from upwelling systems to coral reefs. In the upwelling system of Namibia, following the collapse of the forage fish during the 1970s, namely sardines (*Sardinops sagax*) and anchovies (*Engraulis encrasicolus*), the ecosystem became dominated by two species of very low caloric value: the bearded goby (*Sufflogobius bibarbatus*) and a jellyfish (Cnidaria, Medusozoa). The latter reached a biomass estimated at more than 40 million tonnes during the 1980s and 12 million tonnes during the

2000s. As a consequence, the predators of these forage fish, the African penguin (*Spheniscus demersus*) and the Cape gannet (*Morus capensis*), suffered a lack of adequate prey and declined by 77% and 94%, respectively. Juvenile penguin survival was found to be approximately 50% lower than in proximate areas that were not food depleted, revealing the extent and effect of marine ecological traps. Cape hake (*Merluccius capensis*) and deepwater hake (*Merluccius paradoxus*) catches declined from 295,000 tonnes in 1972 to 150,000 tonnes since 1990, and the production of Cape fur seal (*Arctocephalus pusillus*) pups was strongly affected (Roux et al. 2013; Sherley et al. 2017).

In the Gulf of California, elegant terns (*Thalasseus elegans*) experience low or failed breeding and nesting distribution changes during years of positive sea surface temperature anomalies associated with increased sardine fishing effort by the local industrial fleet (Velarde et al. 2015b). In the Black Sea ecosystem, intense fishing of large predators and eutrophication of the ecosystem resulted in an outburst of an invasive comb jelly, *Mnemiopsis leidyi*, in a system-wide trophic cascade (Daskalov et al. 2007). Likewise, Wanless et al. (2005) observed that the major reproductive failure of birds in the North Sea during the 1990s was caused by a change in the dominant trophic pathway, which forced the birds to feed on sprats rather than sand eels, with the latter constituting higher-energy feed. A comprehensive fishery-independent data set of North Pacific seabird tissues was recently used to inform pelagic ecosystem trends over thirteen decades (from the 1890s to the 2010s), revealing a long-term shift from higher trophic level prey to lower trophic level prey, from fishes to squids (Gagné et al. 2018).

Most Caribbean reefs experienced a rapid shift from coral to algal dominance during the 1980s. The regime shift was initiated by a decline in the abundance of herbivorous fish caused by overexploitation. The role of herbivory was replaced by the urchin *Diadema antillarum*, but populations of this animal were severely depleted by a disease epidemic. Macroalgae proliferated over the reefs, thereby reducing reef coral recruitment.

Key interactions among four major tropical taxa—coral, macroalgae, fish and urchins—have created a self-perpetuating process that locked reef ecosystems into an alternative, nearly coral-free state (Travis et al. 2014), sometimes together with increased nutrients, to cause and perpetuate regime shifts cascading down to microbial components (Bozec et al. 2016; Haas et al. 2016; Zaneveld et al. 2016). Similarly, in the Humboldt upwelling system, the influence of overfishing of carnivores has favoured the increase in the biomass of herbivores, which subsequently changed the structure of kelp forests (Pérez-Matus et al. 2017).

4.2 Quantifying Tipping Points

The above examples illustrate the need to quantify connectivity in food webs, particularly the strength of predator-prey interactions in order to identify thresholds that push marine ecosystems past their tipping points.

Small pelagic fish exert a major control on the trophic dynamics of upwelling ecosystems and constitute mid-trophic level, 'wasp-waist' populations (Cury et al. 2000; Bakun 2006). These small- and medium-sized pelagic species are the primary food source of many marine mammals, larger fishes and seabirds, transferring energy from plankton to larger predators. They also are grazers/predators in marine ecosystems, feeding upon phytoplankton, zooplankton, and, in some cases, the early life stages of their predators. Using 72 ecosystem models, a global meta-analysis quantified the required forage fish biomass to sustain all fish predators in marine ecosystems, including marine mammals (Pikitch et al. 2012). A minimum precautionary biomass of 40% of forage fish is required to sustain predators.

The cascading effect of the overexploitation of forage fish is particularly detrimental to seabirds. The global and substantial overlap and competition between small pelagic fisheries and seabirds represents 48% of all marine areas, notably in the Southern Ocean, Asian shelves, Mediterranean Sea, Norwegian Sea, and California coast (Grémillet et al. 2018). Behind all of the diversity and complexity of the world's marine ecosystems and the multitude of adverse drivers in bird declines, a striking pattern relating seabird breeding success and their fish prey abundance was found for 14 bird species within the Atlantic, Pacific, and Southern Oceans (Cury et al. 2011). A threshold in prey (fish and krill, termed forage fish) abundance, equivalent to one-third of the maximum prey biomass, was found below which there is the occurrence of consistently reduced and more variable seabird breeding success. This threshold is also equivalent to the long-term average prey abundance and constitutes an evolutionary stable strategy for marine birds. This empirically derived guiding principle embraces the ecosystem approach to management aimed at sustaining the integrity of predator-prey interactions and marine food webs. In well-documented ecosystems, this universal threshold can be revisited and sometimes adapted according to specific ecological and environmental constraints, such as the quality of food or the existence of specific reproductive habitats that are accessible to birds (Guillemette et al. 2018).

Coral bleaching events resulting from global warming and ocean acidification will compromise carbonate accretion, with corals becoming increasingly rare on reef systems (Hoegh-Guldberg et al. 2007). Consequently, policies that result in atmospheric levels of carbon dioxide above 500 parts per million, appear extremely risky for the future of coral reefs and should be strongly avoided. Moreover, near-future increases in local sea temperature of as little as 0.5°C will result in the protective mechanism of coral reefs being lost, which may increase the rate of degradation of local coral reefs (Ainsworth et al. 2016). The loss of ecological resilience occurs because coral cover increases more slowly after disturbances but also when competitive interactions with macroalgae become more frequent and longer in duration. To reduce those interactions, coral reefs require higher levels of grazing to exhibit recovery trajectories (i.e., about 40% of the reef being grazed; Hoegh-Guldberg et al. 2007). Maintaining resilient coral reefs similarly requires harvest limitations and maintaining the minimum biomass of grazing fish species playing a key role, such as parrotfish (with a harvest limitation of less than 10% of virgin fishable biomass combined, with an enforceable size restriction of more than 30 cm) (Bozec et al. 2016).

4.3 Fisheries Management Perspective

With climate change and overexploitation, ecosystems are more vulnerable to changes that previously could be absorbed and may suddenly shift from desired to less desired states in their capacity to generate ecosystem services (Folke et al. 2004). Recovering ecosystems that have experienced regime shifts and have moved past their tipping points appears very difficult, to almost impossible (Haas et al. 2016), so that adaptive practices work only poorly or not at all (IPCC 2019).

For sustainable exploitation and conservation, it is crucial to fully appreciate the fact that ecosystems have tipping points, identify the potential thresholds, and implement them into management (Suding and Hobbs 2009; Travis et al. 2014). In a global change context, multiple and confounding factors influence the state of marine ecosystems. Reliable detection and attribution appear to be fundamental to our understanding of ecosystem changes (IPCC 2019), however, the confident attribution of tipping points in ecosystem dynamics remains challenging. Overexploitation and climate change can promote tipping points and can potentially act in synergy within ecosystems, increasing the risk of irreversible changes. Marine conservation and adaptive management approaches must consider long-term persistent warming and acidification as well as consequent discrete extreme events that are pivotal in shaping ecosystems. The limitation of CO_2 emissions appears to be a strong constraint in the preservation of marine ecosystems, despite the difficulty in reaching the Paris Agreement targets. However, the growing threat of abrupt and irreversible climate change must compel political and economic action on carbon emissions (Lenton et al. 2019).

Fisheries management will have to consider the structuring role of key species, such as small pelagics in upwelling systems or herbivorous fishes in coral reef ecosystems. To avoid regime shifts, the ecosystem approach would greatly benefit from the integration of readily available limit reference points, defined by predator-prey interactions between species, into fisheries management strategies. Examples of such ecosystem-based management approaches which go beyond the traditional single-species stock assessment are plentiful. For example, the CCAMLR has the principle embodied in its articles to ensure that target stocks and their dependent and related species are all maintained at productive levels (Constable 2011). This has steered the management of krill fisheries in the Southern Ocean to ensure that stocks are fished sustainably but also that the predators of this keystone species are supplied with ample prey (Constable 2011). Similar approaches are used to manage finfish in the Antarctic (Constable 2011). Other successes of the CCAMLR ecosystem approach include technical measures to prevent the mortality of albatrosses and petrels in longline fisheries for Patagonian and Antarctic toothfish (*Dissostichus eliginoides* and *Dissostichus mawsoni*; Friedman et al. 2018). Many fisheries, including those in the CCAMLR, employ observer programmes to estimate the bycatch of endangered species or non-target species which may be vulnerable to fishing mortality and to alter fishing practices to reduce such impacts should they be detected (see Gilman et al. 2017). Integration of such ecosystem-based indicators will help to sustain desired ecosystem states while protecting marine species.

5 Monitoring

Humans and climate change continue to impact the marine world and its resources. Thus, when evaluating policy and management approaches, it is vital to be guided by indicators that can capture the status, trends and drivers of ocean health (Block et al. 2011; Miloslavich et al. 2018b; Cubaynes et al. 2019). The main indicators used in marine conservation planning relate to habitat extent, species diversity and extinction risk. Nevertheless, quantifying habitat extent and its associated diversity is difficult because of the high technical and logistical requirements as well as funding constraints; therefore, results are limited in statistical power and often fail to provide the required spatial- temporal dimension (Palmer et al. 2002).

5.1 How Can We Effectively Monitor and Manage Biodiversity and Enjoy the Benefits of a Sustainable Blue Economy in a Changing World?

Ocean monitoring and surveillance have been identified as components of the blue economy needed to respond to ocean health challenges (EIU 2015). The Framework for Ocean Observing (Lindstrom et al. 2012; Tanhua et al. 2019) provided key concepts based on the delivery of a multidisciplinary system, focused on the use of Essential Ocean Variables (EOVs). EOVs act as the common focus for observations to generate data and information products based on the scientific and social requirements. Biological EOVs, which are highly focused on understanding biodiversity trends, were identified based on their relevance to address such social and scientific requirements and their feasibility for global measurement in terms of cost, available technologies and human capabilities (Miloslavich et al. 2018a). The sustained observation of these EOVs will serve as the foundation for implementing management and policy based on science to promote a healthy and sustainable ocean, from local to regional to global scales. These biological EOVs also support the global climate observing system as plankton communities and some coastal ecosystems (e.g., coral reefs, seagrass beds, and mangrove forests) are considered to be essential climate variables (WMO 2016). Planning is currently underway for the internationally coordinated and global networks that measure these biological EOVs. Such planning includes (1) identifying existing data sets for each EOV at all geographical scales; (2) reviewing technological monitoring approaches and standard operating procedures along with the capacity needed to use them; and (3) recommending approaches for data and metadata consolidation in findable, accessible, interoperable and reusable (FAIR) systems. Building the system required to achieve the sustainability of marine diversity and ecosystems, which is critical for the blue economy, will require governance, broad communication and establishing partnerships. It will also require the development of new technologies and of human capacity. Investing in people and their institutions, particularly for developing countries, is required to build infrastructure and long-term support networks with enhanced access to data, tools and technologies. Additionally, collaborations that combine multiple knowledges, including indigenous knowledge, can provide an important role in understanding species distribution (Skroblin et al. 2019) and may play an increasing role in enhancing our capacity to have a more holistic understanding of ecology (Ens et al. 2015).

This can be facilitated by international initiatives, but it will require the long-term engagement of national institutions and local communities as well as funding, including major contributions from philanthropists and the private sector if it is to be sustained (Bax et al. 2018; Miloslavich et al. 2018b).

5.2 What Are the Technological Tools for Biodiversity Monitoring?

The methods for monitoring marine biodiversity are quite extensive and specific to the taxonomic group, type of ecosystem and/or spatial scale of the monitoring effort. Some of the persistent technical challenges of marine biodiversity monitoring include the need for clearly defined and standardised best practices and interoperable observation technologies. Data are collected through a combination of remote sensing and in situ observations (see Canonico et al. 2019 for a recent review). Remote sensing allows for observations at broad, global scales repeatedly, with a resolution highly dependent on the sensor. It provides information on functional phytoplankton groups and on the cover and distribution of some coastal habitats, such as coral reefs, seagrass beds, mangroves and macroalgae, and some structured habitats such as floating macroalgae (e.g., *Sargassum*). In situ observations include a variety of methods, from simple visual survey and/or sample collection to the use of sensors, instruments, and platforms. At the most basic level these observations rely on survey and/or sampling either on shore or in shallow water using scuba divers. Large-scale application of such methods can be used to tackle global questions about spatial differences in coastal marine communities or for monitoring over time if protocols are standardised (e.g., the Natural Geography in Shore Areas, or NAGISA, sampling protocol used in the CoML; Iken and Konar 2003; Cruz-Motta et al. 2010). Some of the most-used newer technologies include acoustic monitoring, which supports biomass and abundance estimates among other parameters; animal telemetry for animal movement in combination with environmental descriptions; 'omic' approaches to report on biodiversity across scales and taxa; and video/photo imagery from automated underwater vehicles (AUVs), remotely operated vehicles (ROVs), submersibles and divers. These technologies are already generating big data, which will require the use of artificial intelligence and machine learning processes, improved (real-time) quality control and enhanced data capabilities (Edgar et al. 2016). In the next decade, it will be critical to develop technologies that enable increasingly automated real-time biological observations.

In this context, satellite based remote sensing is frequently proposed as a cost-effective tool to lower the costs of obtaining spatially and temporally relevant information and monitoring changes (Mumby et al. 1999, 2004; Green et al. 2000). As technology continues to advance, improving the resolution and accuracy of satellite imagery, our knowledge of the distribution of habitats is improving. Although there has been a progression in monitoring a number of coastal habitats (Mumby et al. 2004; Giri et al. 2011), remote sensing has certainly not reached its full potential (Andréfouët 2008) because of technical limitations and difficulties classifying habitats (Zoffoli et al. 2014). Often there is a need to supplement this with existing field data and/or expert knowledge to obtain a more complete picture (Andréfouët 2008). Moreover, only the shallower component of subtidal critical habitats, such as seagrass meadows and algal stands, can be resolved by even the most advanced remote sensing technologies (e.g., hyperspectral satellite imaging; Wicaksono et al. 2019). Likewise, important habitats, such as deep-sea corals, are beyond the reach of existing or future airborne remote sensing technologies. The mapping of seabed topography at a relatively coarse scale can be undertaken using satellite gravity mapping (e.g., for seamounts; Yesson et al. 2011).

Habitats at shelf depths and in the deep sea were traditionally mapped by using plumb lines which had a wad of tallow in a cavity at the bottom of the plummet (the weight at the end of the line). The tallow would pick up fragments of whatever was on the seabed and a notation of the seabed type was added to nautical charts, providing a navigational aid for mariners.

As modern oceanographic science developed in the 19th century, habitat mapping was undertaken by trawling, dredging or other forms of seabed sampling. A significant advancement in seabed mapping was the development of single-beam sonar. Using this technology, Bruce Heezen and Marie Tharp constructed the first global topography maps of the seafloor. In the present day, the main tool of habitat mapping in coastal and deep waters is multibeam acoustic survey (Harris and Baker 2012; Lamarche et al. 2016). These sophisticated sounders not only accurately measure the depth of the seafloor but also give information on the hardness of substrata through the strength of acoustic return as well as seafloor microtopography (roughness) and volume heterogeneity, which relates to sediment grain size and composition (Harris and Baker 2012; Lamarche et al. 2016). This information can be used to identify seafloor texture, whether it is made of rock or sediment, for example, and can be used to classify habitat (Lamarche et al. 2016). Coupled with the use of seabed sampling using surface deployed gear (e.g., trawls or cores) and/or image-based surveying using towed cameras, ROVs, AUVs or submersibles for groundtruthing, these methods can provide accurate maps of seabed habitats (Harris and Baker 2012; Lamarche et al. 2016). An issue with this approach is that it is time consuming and expensive, and coverage tends to be restricted to areas targeted for specific study for scientific or industrial purposes. The global Seabed 2030 mapping project is currently collecting multibeam data to produce a more comprehensive map of seafloor topography than previously available.

Although it will certainly allow the identification of larger-scale geomorphological structures such as seamounts, canyons and plains, the extent this will be used in mapping of finer-scale habitats is unclear. An alternative technology to multibeam bathymetry is side-scan sonar. This produces a photograph-like sonar image of the seabed and can be particularly useful in imaging small objects and finer-scale structures on the seabed (e.g., sand waves; Lamarche et al. 2016). This technology is cheaper than multibeam systems but has a poor georeferencing capability, and backscatter calibration is usually not possible (Lamarche et al. 2016). A relatively new technology now being carried by AUVs is synthetic aperture sonar which provides very high resolution imagery but at a longer range than side-scan sonar (Hansen 2011). AUVs with hyperspectral imaging capabilities are now being developed to extend remote sensing capabilities to deeper waters for high-resolution habitat identification (Bongiorno et al. 2018; Foglini et al. 2019).

Many marine habitats and areas of the world still remain under-studied at larger scales, such as rocky reefs, algae beds, and large areas of the deep ocean for which there are no publicly available global distribution maps at present (Rogers et al. 2015). For the habitats where spatially referenced and processed information are available, often data sets relate to one point in time with very little indication of changes through time (Halpern et al. 2015). This limits their utility in understanding how, where and when the natural world is changing. As new technology is made available, such as the Google Earth Engine platform (Gorelick et al. 2017; Traganos et al. 2018; Nijland et al. 2019),[6] and barriers for information sharing are removed, there is a great opportunity to increase our capacity to understand, monitor and develop evidence-based policies and management plans to protect marine ecosystems.

Satellite remote sensing has had a significant impact on assessing the levels of fishing effort in the global ocean. Access to fisheries data is often denied for reasons of commercial confidentiality, but in a world where fisheries are sustainably managed, it is not necessary to hide what is taken or conceal the location, whether in national waters or in ABNJ. Satellite surveillance is increasingly useful as a means of spotting problems such as illegal fishing and transhipments; it is also a useful way to assess patterns of fishing even in the remotest parts of the ocean (Eigaard et al. 2017; Amoroso et al. 2018; Boerder et al. 2018; Elvidge et al. 2018; Ford et al. 2018b, c; Kroodsma et al. 2018;

[6]For more information about the Google Earth engine, see https://earthengine.google.com/

Longépé et al. 2018; Rowlands et al. 2019). The development of online platforms such as the Global Fishing Watch has exposed the industry to societal oversight where previously it did not exist, especially in waters far from the coast.[7]

The new model of fisheries surveillance has been taken up by several coastal states, such as Chile, Indonesia and Panama. These countries have now committed to making the tracking data of vessels carrying their flags available to public scrutiny. Such data can only improve the sustainability of fishing; it will not only identify where and when fishing is taking place but also provide insight into the enforcement of MPAs (Rowlands et al. 2019) and destructive fishing practices (Winnard et al. 2018).

5.3 Overseeing the Monitoring of Biodiversity

At the intergovernmental level, two major organisations provide a governance framework for marine biodiversity observations.

The first, the IOC of UNESCO, through the Global Ocean Observing System (GOOS), has led the implementation of the Framework for Ocean Observing (Lindstrom et al. 2012) with the goal of serving users across climate, operational services and ocean health (Tanhua et al. 2019). GOOS is also co-sponsored by the World Meteorological Organization, the United Nations Environment Program, and the International Science Council. Within GOOS, marine biodiversity observations are coordinated by the Biology and Ecosystems Panel, or GOOS BioEco (Miloslavich et al. 2018a). GOOS also provides governance at the regional level through the GOOS Regional Alliances, examples of which are the Integrated Marine Observing System (IMOS) of Australia, the Integrated Ocean Observing System (IOOS) of the United States and the European Global Ocean Observing System (EuroGOOS) in Europe. Through expert panels, regional alliances, the Observations Coordination Group, and affiliated projects, GOOS supports a broad observing community, from individual scientists and research organisations to governments, UN agencies, and international programmes.

The second major organisation is the Marine Biodiversity Observation Network (MBON) framed in the Group on Earth Observations Biodiversity Observation Network (GEO BON), which facilitates the coordination between individual monitoring programmes and existing networks (Muller-Karger et al. 2018). Both MBON and GOOS BioEco share common goals and encourage the use of best practices for marine biodiversity monitoring, the contribution of data to open access data systems and provide a framework for data management, communication and applications (Canonico et al. 2019).

Based on these shared goals, these organisations have signed an agreement together with OBIS, which operates under the IOC's International Oceanographic Data and Information Exchange (IODE) programme, to work together to advance sustained, globally consistent observations of marine biodiversity with the commitment to open access and data sharing, implementing best practices and international standards and enhancing global capacity (Miloslavich et al. 2018a). Having this overarching governance in place is a major step; however, much work still needs to be done. To achieve the required level of coordination and communication across all networks, programmes and countries, the organisations need to ensure the interoperability of the data and that the data contributes to the development of indicators to address policy and management requirements. Specifically related to governance in coastal zones, an assessment carried out by the Economist Intelligence Unit across 20 countries found that the Coastal Governance Index is uneven, with developed countries doing relatively well but still requiring work. Other important factors that contribute to better coastal policies include participatory inclusion in decision-making and accountability, the level of economic development, having the capacity required for the implementation of policies, and having marine spatial planning policies (EIU 2015).

With the proper training and quality control, citizen science can be used both as a way of communication and as a way for data collection on a broad range of scales. An excellent success story of citizen science is the Reef Life Survey (RLS) programme.[8] The RLS was established in Australia in 2008 to collect data on the biodiversity of benthic and fish communities on rocky and coral reefs through trained volunteer scuba divers (Stuart-Smith et al. 2017). Since its establishment, it has expanded globally to more than 3000 sites in nearly 50 countries, providing invaluable data for ecosystem management and conservation (Stuart-Smith et al. 2018). Furthermore, the Biodiversity Indicators Partnership, which promotes the development and delivery of biodiversity indicators to measure progress on Aichi Biodiversity Targets and SDGs, has recently accepted two of the RLS indicators (the 'Large Reef Fish Indicator' and the 'Reef Fish Thermal Index') to inform Aichi Biodiversity Targets 6, 10 and 11 and also SDG 14.2 (RLS 2019).

[7]See the Global Fishing Watch, https://globalfishingwatch.org/

[8]More information about the Reef Life Survey can be found on its website, https://reeflifesurvey.com

6 Gaps and Challenges in Habitat Protection

6.1 How Much of Key Marine Habitats Are Protected?

To understand how MPAs are currently distributed across the key habitats considered (Table 10.1), the March 2020 version of the World Database of Protected Areas (UNEP-WCMC and IUCN 2020) was used to calculate the extension of all the coastal protected areas and MPAs (hereafter collectively referred to as MPAs), or the number of reported locations of each habitat, inside of an MPA within EEZs. We considered three scenarios for the analyses: (1) all areas designated as MPAs without distinction, (2) only MPAs reporting a management plan and (3) only fully protected MPAs (labelled in the database as 'no-take zones').

We estimate that 12% of the habitats considered in this study lie within an MPA. However, when we considered only the MPAs with management plans, only 6% of the habitats are included, and just 3% are in fully protected MPAs at a global level. An example of how these three scenarios overlap is provided by kelps, where more than 40% of the world extent of these habitats are recorded as protected within all forms of MPAs (Fig. 10.9a). However, kelp protection decreases to only 24% under MPAs with management plans and only 1% in fully protected MPAs (Fig. 10.9a).

The deeper habitats show a similar trend, with the habitat with most of its area protected being cold-water corals. They have 24% lying within MPAs, which drops when only managed and fully protected MPAs are considered to 14% and 4%, respectively.

It is important to consider that coastal habitats have arguably received historically higher levels of human pressures

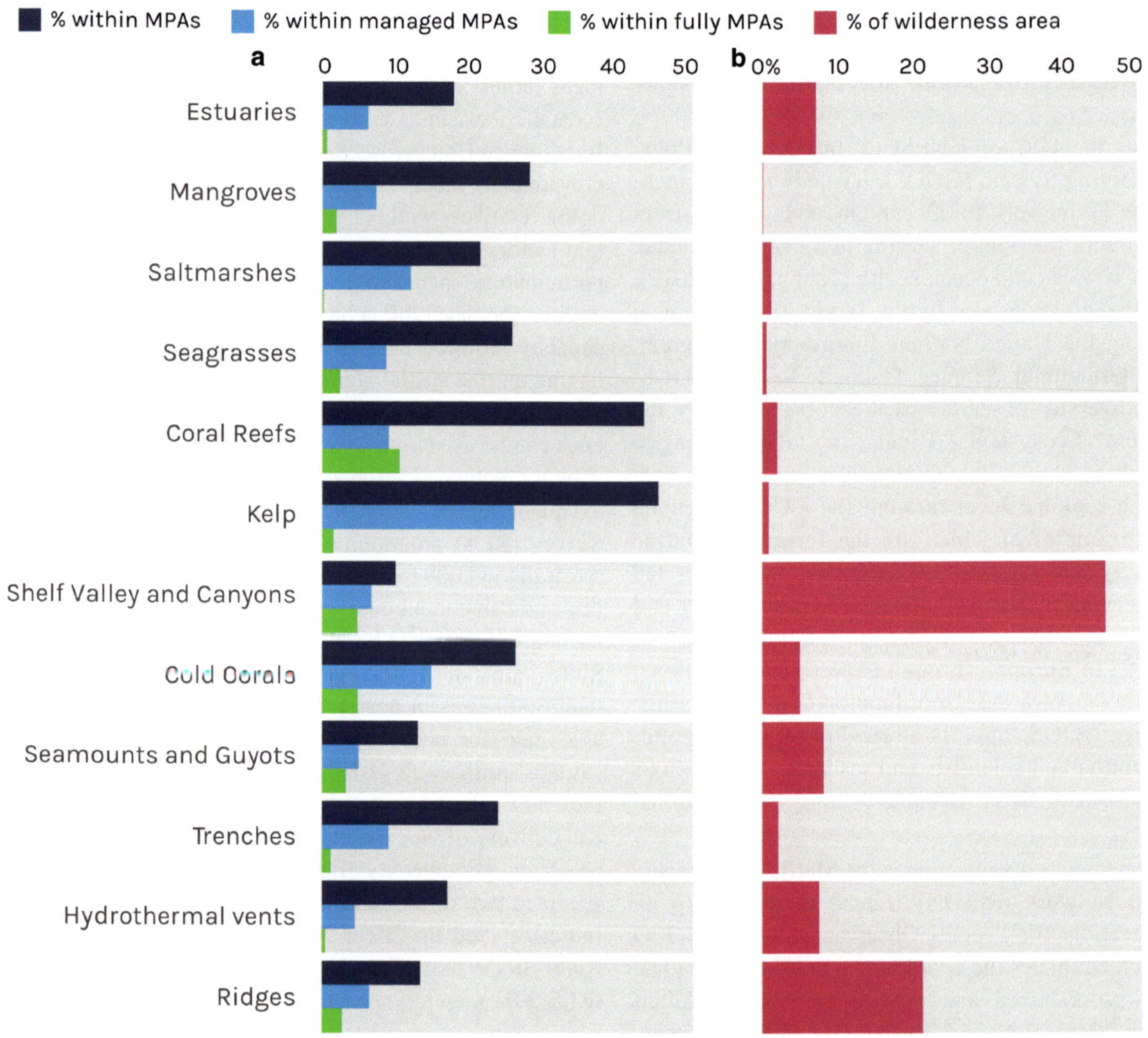

Fig. 10.9 Current conservation efforts for key selected habitats. *Notes: Habitats on the x-axis are ordered according to their distance to the coast, as a proxy for their average depth.* (**a**) *The bars represent the percentage of the habitat within MPAs, within MPAs with a management plan, and fully protected MPAs.* (**b**) *The percentage of wilderness inside the habitat area. (Source: Authors)*

compared to oceanic habitats. Evidence of the destruction of coastal habitats (see Sect. 3.5, Habitat Degradation and Its Drivers), which has already severely reduced their original distributional area, should be taken into account when considering the percentage of the current habitat extent in MPAs.

Estuaries and saltmarshes are the coastal habitats with the lowest proportion in fully protected MPAs (Fig. 10.9a) despite their importance in habitat provision for a wide range of species and ecosystem services (e.g., carbon sequestration, nutrient cycling, coastal protection; Barbier 2017).

The area of the selected habitats lying within designated MPAs declines moving from the coast to offshore (Fig. 10.9a). However, this pattern is much less obvious for MPAs with management plans and non-existent for fully protected MPAs (Fig. 10.9a). This suggests that both coastal and offshore habitats are equally poorly represented within fully protected MPAs. The offshore habitats had on average a higher proportion in marine wilderness (based on the area estimated by Jones et al. 2018); most likely the result of decreased accessibility (Fig. 10.9b). At present the global coverage of MPAs is 7.43%, with 17.22% of national waters designated as MPAs, but this figure falls to 1.18% in ABNJ (UNEP-WCMC and IUCN 2020; accessed on 30 March 2020). The discrepancy between the coverage of MPAs in EEZs and ABNJ results from the lack of a coherent international legal framework for the establishment of marine protected areas on the high seas, putting at risk largely unknown biodiversity (O'Leary et al. 2012; Rogers et al. 2015). International efforts towards protecting habitats such as seamounts in ABNJ have been made in regional or sub-regional organisations such as RFMOs (e.g., New England seamounts protected from bottom trawling by the Northwest Atlantic Fisheries Organization), and the ongoing negotiations to manage marine biological diversity in ABNJ, which are aimed at establishing a new legal framework for protection of biodiversity in international waters and on the seafloor.

Additionally, the existence of a habitat inside of an area designated as an MPA does not mean it is protected. As can be seen from the above analyses, many MPAs lack a management plan, and even where such plans exist, MPA objectives and management might not involve the habitat, and permitted activities may even be destructive and/or poorly enforced (e.g., trawling in MPAs; Dureuil et al. 2018). In many meta-analyses of MPA effectiveness, there are benefits to conservation even where protection is partial (i.e., MPAs where not all activities are banned; e.g., Lester and Halpern 2008; Sciberras et al. 2013; Gill et al. 2017; Sala and Giakoumi 2017). Our analyses suggest that despite the apparent progress reported in MPA designation (UNEP-WCMC and IUCN 2020), reaching the Aichi Biodiversity Target 11 of having 10% of representative habitats of our oceans being well protected is still a remote target, as has been found in other studies (Klein et al. 2015; Jenkins and Van Houten 2016; Sala et al. 2018a; Jones et al. 2020). Key shortfalls and key features that can hinder and enhance MPA effectiveness, respectively, have been recognised in current literature (Edgar et al. 2014; Gill et al. 2017). In particular, the NEOLI features identified the most important characteristics of an MPA: being No-take (i.e., fully protected), well Enforced, Old (more than 10 years), Large (more than 100 km^2) and Isolated. The main issue is that MPAs that fulfill some or all of these features, are not common globally (Edgar et al. 2014; Sala et al. 2018a). Although, most existing MPAs could improve in some of the NEOLI features by increasing the no-take area, fostering compliance and enforcement, and extending the boundaries to isolate key habitats to protect, these features are difficult to achieve. Our analyses indicate that to reach international goals and markedly increase the conservation benefits of the global MPA network, it is important to improve existing MPAs while also creating new ones.

6.2 Protection Gaps in EEZs

Humans are exerting pressures on marine habitats throughout the world, often leading to significant damage to them as well as loss of associated biodiversity (Halpern et al. 2015). To understand this on a global scale, we calculated the average biodiversity value for each EEZ, using biodiversity data from Reygondeau and Dunn (2018), and found the sum of ecological and social factors that decrease the health of the ocean. This analysis reveals that countries that have higher biodiversity also experience higher pressure (p-value <0.001, $R^2 = 0.165$; see Fig. 10.7). One might expect that countries with high gross domestic product (GDP) would be capable of protecting a larger fraction of their EEZ. Although we found a significant relationship, GDP explains very little of the variation in the area of MPAs that are implemented in the national waters of each country (p-value <0.001, $t = 0.11$; see Fig. 10.10a). We would expect that countries with higher investment capacities (i.e., GDP) would show a higher relative area of MPA coverage, especially because EEZs and GDP tend to be related. Furthermore, although there are considerable conservation efforts and investments—reflected in MPA coverage—biodiversity and the relative MPA area to each country's EEZ are not correlated (p-value >0.05; see Fig. 10.10b). These results indicate that areas with high biodiversity should be prioritised for protection not only for their biodiversity per se but also to create resilience from the high pressures they experience.

However, representative biodiversity from all regions must be included in a global network of fully or highly protected MPAs, and this must be complemented by sustainable management of all human activities in the ocean (see below; Margules and Pressey 2000). The lack of correlation between

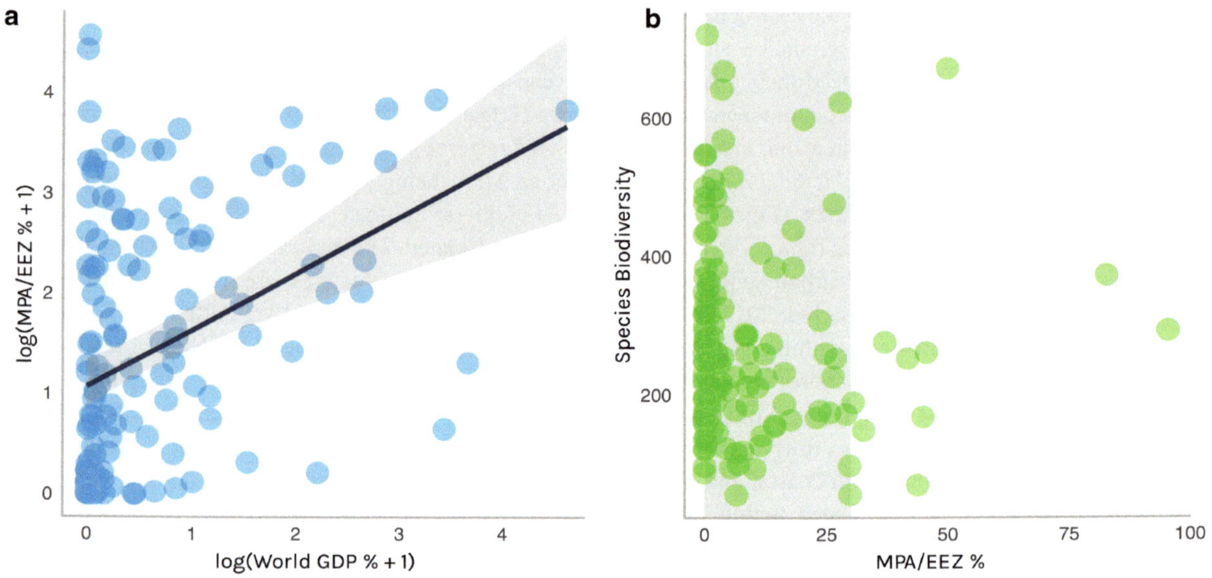

Fig. 10.10 Relationships between biodiversity, GDP and MPA extent. *Notes: Panel* (**a**) *shows the gross domestic product (GDP) that a country has relative to the world and the amount of their exclusive economic zone (EEZ) that is covered by marine protected areas (MPAs). Panel* (**b**) *reveals that the relative size of a country's MPAs are not correlated with their biodiversity. The grey region in Panel* (**b**) *represents the countries with less than 30% of their EEZ with MPA coverage. (Source: Authors)*

the biodiversity within an MPA and the amount of the EEZ that is protected by a coastal state suggests that biodiversity-rich countries do not develop more MPAs than biodiversity-poor countries.

Further, Kuempel et al. (2019) found that MPAs with the strictest protection were 6.3 times more likely to be found in low-threat ecoregions, indicating that countries focus conservation efforts in the least threatened areas as opposed to areas with high threats to biodiversity. Additionally, areas with lower biodiversity can still be highly productive and valuable in terms of ecosystem services provision to coastal states as well as in ABNJ.

Even when considering the best-case scenario, using all the MPAs reported and assuming that these have at least some benefit to protect habitats, it is possible to see that between 45% and 90% of countries are protecting less than 30% of habitat extent (Table 10.3). The numbers worsen, in area terms, when the two other scenarios are considered, with at best 23.3% of countries with 30% or more of a habitat lying within a managed MPA (saltmarsh) and 4.2% in fully protected MPAs (hydrothermal vents; Table 10.3).

For saltmarshes and estuaries, no countries include 30% of the area of habitat in fully protected MPAs (Table 10.3). Indeed, if we break down the conservation effort for each country for the category of all MPAs, there is a large gap where some countries are committing more effort whereas others are not performing as well. Here, we propose to measure the proportional conservation efforts amongst countries by using measures of central tendency, the mean and the median percentage of habitat protected globally, as an alternative to absolute measures of habitat area. The overall protection effort is 'fair' when the mean and median percentage of habitat protected globally coincide to form a normal distribution of the conservation efforts (Fig. 10.11). The mean and the median percentages are reported as blue and red circles, respectively, which show that for most habitats there is a wide gap between area present and area protected. This indicates that current global conservation efforts are inadequate. Most countries are protecting very little (less than 1%) of the habitats they could protect, and conservation efforts are unevenly distributed. If MPAs with management plans are considered, for some habitats the 'effort gap' metric is even worse (e.g., saltmarshes, kelps and coral reefs; Table 10.3). In other cases, the effort gap appears to decrease, but this is mainly because the amount of habitat in managed MPAs is so small compared to all MPAs. A very small amount of habitat lies within fully protected MPAs, rendering the effort gap metric very small as all states are equally performing badly. Through this effort gap metric, we see that for fair habitat conservation globally, countries need to cooperate to reach international goals, thereby compensating for the effort gap either by increasing their MPAs and/or aiding conservation programmes in less wealthy countries or regions. The effort gap highlights how even if some countries are contributing towards achieving a 'total conservation target', the majority of countries are under-performing.

This proportional conservation approach could also be applied to properly measure the effort each country should give to the protection of the high seas. This approach can be useful in a context where the use of ABNJ is emerging and

Table 10.3 Summary of the habitat protection target proposed

Habitat	Percentage of countries below 30%	Percentage of area below 30%	Mean percentage effort	Median percentage effort	Effort gap
Saltmarshes	51.2	87.9	41.0	28.1	12.9
	(76.7/100)	(92.0/100)	(21.1/0.5)	(1.8/0)	(19.3)
Kelps	45.3	37.2	36.3	37.7	−1.4
	(77.4/98.1)	(52.6/100)	(17.0/2.88)	(1.0/0)	(16)
Coral Reefs	61.6	44.5	30.7	17.5	13.2
	(86.6/97.3)	(91.7/95.8)	(22.3/1.1)	(0/0)	(22.3)
Hydrothermal vents	64.6	62.6	29.5	0.0	29.5
	(85.4/95.8)	(94.1/99.3)	(13.9/3.2)	(0/0)	(13.9)
Mangroves	59.1	59.2	29.3	19.9	9.4
	(86.0/98.9	(92.3/97.3)	(9.3/1.1)	(0/0)	(9.3)
Seagrasses	70.3	58.6	24.4	6.68	17.7
	(89.0/99.2)	(86.6/98.5)	(9.0/0.8)	(0/0)	(9)
Estuaries	72.8	76.2	20.2	5.7	14.5
	(88.8/100)	(94.8/100)	(8.5/0.1)	(0/0)	(8.5)
Cold Corals	76.6	77.5	18.7	0.0	18.7
	(91.2/98.5)	(87.3/99.8)	(7.98/1.5)	(0/0)	(7.98)
Trenches	80.4	74.8	18.3	0.0	18.3
	(93.5/97.8)	(91.4/100)	(6.59/2.34)	(0/0)	(6.59)
Ridges	82.0	73.1	16.4	0.0	16.
	(92.6/98.4)	(89.6/97.8)	(7.77/2.26)	(0/0)	(7.77)
Seamounts and guyots	81.4	59.1	14.4	0.0	14.4
	(92.0/96.5)	(84.7/86.6)	(6.3/2.5)	(0/0)	(6.3)
Shelf Valley and Canyons	90.6	97.1	11.1	0.1	11.0
	(95/98.9)	(98.0/99.9)	(5.3/1.0)	(0/0)	(5.3)

Notes: For each habitat, the percentage of countries that have granted less than 30% protection is shown ('Percentage of Countries below 30%') for all MPAs and then, in parentheses, the figure for managed MPAs/fully protected MPAs that is below 30% protection. The 'Mean' and 'Median Percentage Effort' refers to the percentage of habitat countries protect on average. The differences between these two values is reported as the 'Effort Gap', representing the percentage by which countries below the threshold should ideally increase their protection to make a fair contribution to conservation. We did not calculate this for fully protected MPAs as the amount of habitat lying within this category of protected area is so low that the effort for all countries is equally very poor
Source: Authors

presents serious governance challenges (Merrie et al. 2014). For example, each country should deploy a conservation effort relative to its use of ABNJ across all sectors (e.g., fishing, shipping). ABNJ are a special case of global commons management. In these areas, establishing and enforcing conservation measures will require new financing mechanisms, such as a levy on the use of the resources and/or by establishing an international trust fund under the new legally binding instrument for the conservation and sustainable use of biological diversity of ABNJ. It is important that ABNJ are managed fairly by a proportional conservation measure rather than international goals with total conservation targets, which might disproportionately favour some countries over others and imperil the health of the high seas.

Whilst we have emphasised the use of MPAs in biodiversity protection mainly because their implementation can be quantified and analysed spatially to some extent, MPAs are not the only management measure that can conserve biodiversity (Duarte et al. 2020). It has been argued that the ocean can be compared to a frontier system, both to within EEZs and in ABNJ, where there is open access to resources, larger and less differentiated jurisdictions than on land and fewer laws that constrain human activity (Norse 2005). This situation has led to a free, open access scramble for resources. This has resulted in increasingly unsustainable levels of exploitation of marine living and other resources and the impacts on biodiversity that have been documented here and in other studies (Norse 2005). Marine reserves by themselves do not necessarily reduce overfishing, competition amongst fishers or the growth of global fishing fleets, and they may even increase competition amongst fishers by reducing areas available to fish, possibly even displacing fishing effort to areas where levels of fishing have been low or nonexistent (Kaiser 2005; Norse 2005; Agardy et al. 2011; FAO 2011; Hilborn 2018).

Marine reserves also provide little protection from threats such as long-range pollutants (e.g., many persistent organic pollutants; Agardy et al. 2011) or invasive species (e.g., Burfeind et al. 2013). The connectivity of populations of marine species and between habitats also means that even if fully protected MPAs are designed to ensure maximum con-

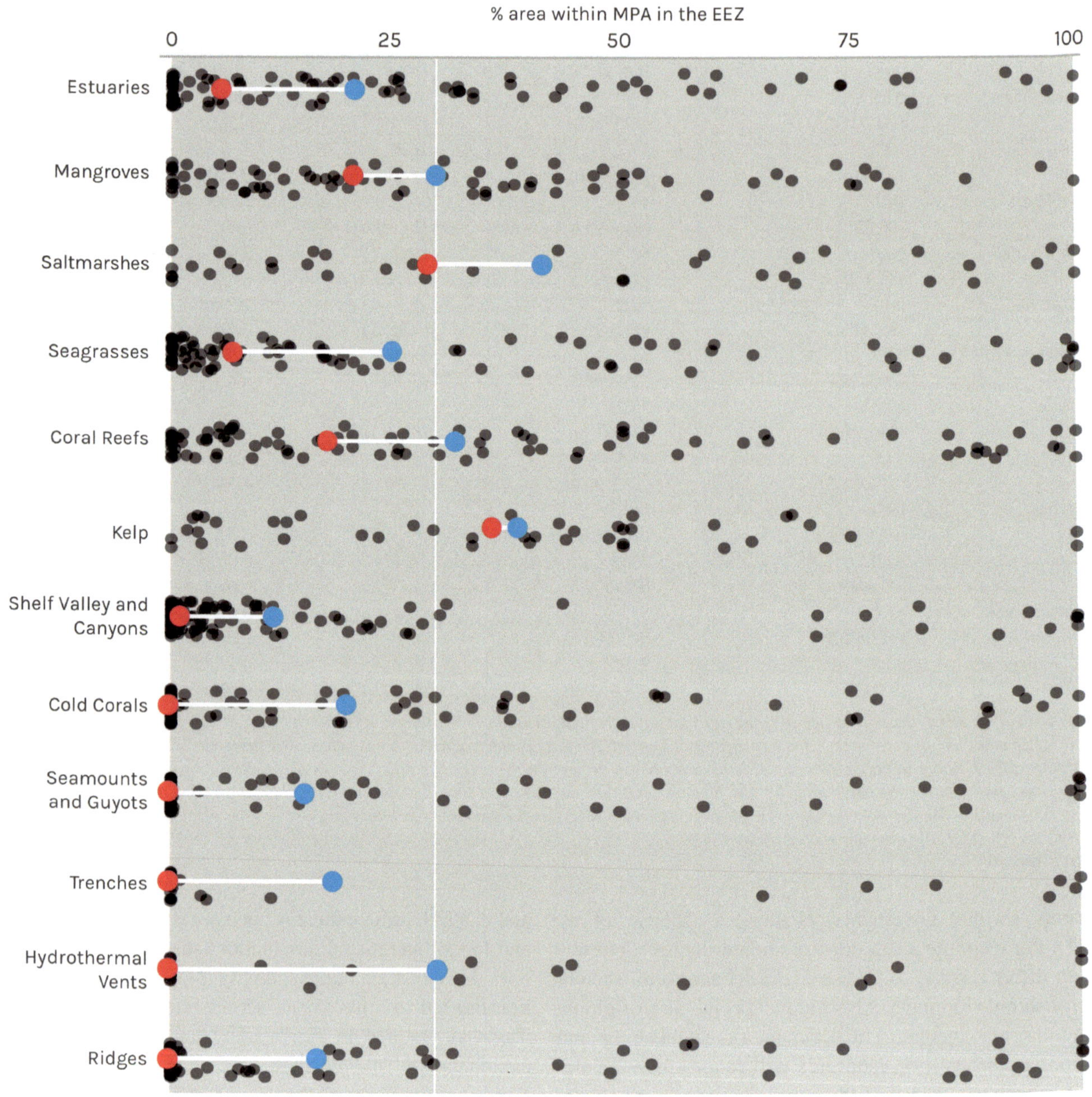

Fig. 10.11 Current conservation efforts for key selected habitats. *Notes: Best-case scenario, using all the MPAs reported. Habitats on the x-axis are ordered according to their distance to the coast, as a proxy for their average depth. Black circles represent countries hosting one of the key habitats. The y-axis represents the percentage of area that each country is protecting of that habitat within its exclusive economic zone (EEZ). Most of the countries are below the 30% target (white line), which has been identified as a threshold to ensure the maintenance of the ecosystem services of a habitat. The blue circles represent the mean percentage of all the countries' protection efforts for that habitat, whereas the red circles are the median percentage of all the countries' protection efforts. (Source: Authors)*

servation effectiveness, other measures are required outside of reserves to ensure success (e.g., Lipcius et al. 2005; Gaines et al. 2010). This concern applies also and increasingly to climate change and ocean acidification. It is therefore important that all areas of the ocean are managed, including global measures to improve the sustainability of fisheries and aquaculture (Costello et al. 2019; Widjaja et al. 2020; Duarte et al. 2020), as well as of industries extracting non-living resources. As such, it will be important to implement zoning or marine spatial planning to include all areas of EEZs and ABNJ to reduce competition between ocean uses (e.g., Norse 2005) and to reduce the occurrence of pollution from all sources (Duarte et al. 2020) as well as opportunities for alien species to invade non-native ecosystems (Molnar et al. 2008). Reducing and mitigating greenhouse gas emissions to hold global temperature increases to 1.5oC or below is also a priority (IPCC 2019; Duarte et al. 2020) in which the ocean has a role to play (Hoegh-Guldberg et al. 2019).

7 International Conventions and Agreements

We have identified 23 international conventions and agreements that relate to protection of the marine environment and biodiversity (Table 10.4). It is important to consider that these conventions and agreements are not exhaustive in terms of the binding obligations on states. Below the level of international conventions and agreements are regional and sub-regional conventions and agreements (e.g., for RFMOs) as well as voluntary actions such as the CCRF (for a list of examples, see Friedman et al. 2018). Also, decisions under the governance framework of such conventions and agreements, as well as by their implementing agencies, put further binding obligations on states. Added to this is national legislation which provides a complex and interacting web of marine legislation (for an example based on Europe, see Boyes and Elliott 2014). Therefore, the absence of a 'yes' in Table 10.4 does not necessarily mean that a signatory state is not obliged to conform to the activity in the column.

Table 10.4 Characteristics of the International conventions and agreements to protect marine biodiversity and environments

Convention/Agreement	A	B	C	D	E	F	G	H	I	J	K
1. IWC	Yes	Yes		Yes		Yes					
2.Convention on fishing	Yes									Yes	
3. Convention on high seas oil casualties								Yes		Yes	
4. Ramsar	Yes	Yes	Yes	Yes		Yes				Yes	Yes
5. Dumping convention						Yes		Yes		Yes	Yes
6.Heritage Convention			Yes	Yes						Yes	Yes
7. CITES	Yes	Yes				Yes				Yes	
8. Marine pollution (not oil)								Yes		Yes	
9. Marpol						Yes		Yes		Yes	
10. CMS	Yes	Yes	Yes	Yes		Yes				Yes	
11. UNCLOS	Yes	Yes	Yes			Yes	Yes	Yes	Yes	Yes	Yes
12. Basel						Yes		Yes		Yes	Yes
13. CBD	Yes	Yes	Yes	Yes		Yes	Yes		Yes	Yes	Yes
14. High seas fisheries compliance	Yes									Yes	Yes
15. Part XI UNCLOS							Yes			Yes	Yes
16. Straddling stocks agreement	Yes	Yes			Yes	Yes		Yes		Yes	Yes
17. Protocol marine pollution					Yes	Yes	Yes	Yes		Yes	Yes
18. Cartegena							Yes		Yes	Yes	Yes
19. Stockholm						Yes	Yes	Yes		Yes	Yes
20. Antifouling					Yes	Yes	Yes	Yes		Yes	
21. Ballast						Yes	Yes		Yes	Yes	Yes
22. Port state measures	Yes									Yes	Yes
23. Nagoya	Yes	Yes	Yes							Yes	Yes

A. Sustainable management of living resources; B. Sustainable management of unexploited species, C. Habitat management or protection; D. Implement protected areas; E. Precautionary principle; F. Monitoring of species, habitats or environment; G. Environmental impact assessment; H. Prevention of environmental pollution; I. Biosecurity; J. Encourage or impel international cooperation; K. Capacity building

Notes: a. Where trade in that species may impact on an endangered species. The conventions and agreements are as follows: (1) International Whaling Convention (1946); (2) Convention on Fishing and Conservation of the Living Resources of the High Seas (1958); (3) International Convention Relating to Intervention on the High Seas in Cases of Oil Pollution Casualties (1969); (4) Convention on Wetlands of International Importance Especially as Waterfowl Habitat (Ramsar; 1971); (5) Convention on the Prevention of Marine Pollution by Dumping of Wastes and Other Matter (1972); (6) Convention Concerning the Protection of the World Cultural and Natural Heritage (1972); (7) Convention on International Trade in Endangered Species of Wild Flora and Fauna (CITES; 1973); (8) Protocol Relating to Intervention on the High Seas in Cases of Marine Pollution by Substances Other than Oil (1973); (9) Protocol of 1978 Relating to the International Convention for the Prevention of Pollution from Ships, 1973, (Marpol); (10) Convention on the Conservation of Migratory Species of Wild Animals (CMS or Bonn Convention; 1979); (11) United Nations Convention on the Law of the Sea (UNCLOS; 1982); (12) Basel Convention on the Control of Transboundary Movements of Hazardous Wastes and Their Disposal (1989); (13) Convention on Biological Diversity (CBD; 1992); (14) Agreement to Promote Compliance with International Conservation and Management Measures by Fishing Vessels on the High Seas (1993); (15) Agreement Relating to the Implementation of Part XI of the United Nations Convention on the Law of the Sea of 10 December 1982 (1994); (16) Agreement for the Implementation of the Provisions of the United Nations Convention on the Law of the Sea of 10 December 1982 Relating to the Conservation and Management of Straddling Fish Stocks and Highly Migratory Fish Stocks (1995); (17) Protocol to the Convention on the Prevention of Marine Pollution by Dumping of Wastes and Other Matter, 1972 (1996); (18) Cartegena Protocol on Biosafety to the Convention on Biological Diversity (2000); (19) Stockholm Convention on Persistent Organic Pollutants (2001); (20) International Convention on the Control of Harmful Anti-Fouling Systems on Ships (2001); (21) International Convention for the Control and Management of Ships' Ballast Water and Sediments (2004); (22) Agreement on Port State Measures to Prevent, Deter and Eliminate Illegal, Unreported and Unregulated Fishing (2009); (23) Nagoya Protocol on Access to Genetic Resources and the Fair and Equitable Sharing of Benefits Arising from Their Utilization to the Convention on Biodiversity (2010)
Source: Authors

Notwithstanding this, Table 10.4 provides an overview at the highest level of what what ocean management measures states have enacted to protect marine biodiversity.

The 23 international treaties to protect the marine environment and conserve marine biodiversity were analysed using clustering and were found to fall into in three hierarchal groups (Fig. 10.12a): those that aim to protect biodiversity, those dedicated to fisheries and regulation of anthropogenic activities (navigation, ballast waters, etc.) and those regulating pollution.

Beginning more than 60 years ago, the International Whaling Convention (1946) was aimed at the sustainable management of whaling but also concerns protected areas specifically targeted at whale conservation. Almost all the international treaties since then have required cooperation between countries; capacity building; monitoring of species, habitat or the environment; and the management of living resources (Fig. 10.12b). In the last three decades, they have evolved to include a wider range of considerations, including prevention of pollution, conservation of non-commercial species and habitats and biosecurity (Fig. 10.12c). However, many of these treaties focused on specific sectors (e.g., pollution or fisheries management; see Fig. 10.12a) with some specifically dealing with a narrow range of issues (e.g., the Cartagena Protocol relating to biosecurity of organisms modified through biotechnology). Of the 23 conventions, 11 represent the sustainable management of living resources in the ocean and 10 pertain to preventing damage to the marine environment by pollution. It is notable that only 8 conventions and agreements deal with managing or conserving species which are not fished commercially, and only 6 protect marine habitats. Five of the conventions or agreements specifically require the implementation of MPAs.

7.1 Fisheries Governance, Sustainability and Impacts on Biodiversity

On the face of it, the range of international and sub-international conventions and agreements would appear to adequately manage the marine environment and biodiversity. However, as outlined in Sect. 3 of this report, marine species and habitats are in decline, and this amounts to a loss in the provisioning of ecosystem services. For fisheries, this has a significant impact in economic terms; for example, the Sunken Billions report suggests that lost revenue resulting from overfishing amounted to $83 billion in 2012 (World Bank 2017).

Improved management and judicious conservation of wild fisheries would lead to increased biomass in the ocean, higher profits for fishers and greater food provision (40% more production in the future than under business as usual and 20% more than now; Costello et al. 2019; see also World Bank 2017).

No fewer than 11 conventions and agreements deal with the sustainable management of living resources, and all but 3 of them also cover non-target species (Table 10.4). This does not include the large number of regional and sub-regional agreements and additional binding measures that states are committed to for fisheries (Friedman et al. 2018). As already indicated in Sect. 6 the problem in fisheries management is one of uneven implementation of measures to increase sustainability of catches of target species and to prevent harm to biodiversity. There are many aspects of fisheries management where this unevenness of implementation is apparent. For example, compliance to the FAO's CCRF, one of the primary pillars in placing biodiversity measures in fisheries management (Friedman et al. 2018), is better in developed countries than in developing ones, but for most it falls far short of 'good' (Pitcher et al. 2009). Likewise, RFMOs have been widely criticized for their performance both in terms of managing target fish stocks on the high seas and also bycatch (Cullis-Suzuki and Pauly 2010; Polacheck 2012; Gilman and Kingma 2013; Gjerde et al. 2013; Gilman et al. 2014; Clark et al. 2015; Leroy and Morin 2018; Pentz et al. 2018). Since 2006, the United Nations General Assembly has called for the development of performance reviews (PRs) for RFMOs (Haas et al. 2019). By 2016, all RFMOs which had entered into force by 2012 had undergone PRs, and some have been reviewed twice (Haas et al. 2019). There is evidence that these reviews have led to improvements, particularly in the areas of compliance and enforcement, conservation and management and international cooperation (Haas et al. 2019). Decision-making and dispute settlement and financial and administrative issues were areas where lower improvement scores were obtained (Haas et al. 2019). Other recent reviews of RFMO performance reveal a more mixed picture of improvement (Gjerde et al. 2013; Gilman et al. 2014; Pons et al. 2018).

An analysis of the drivers of management effectiveness in tuna RFMOs identified that those with a greater number of member countries, a greater economic dependency on the fisheries, a lower mean GDP, a greater number of fishing vessels and a higher proportion of small vessels had lower levels of research, management and enforcement (e.g., the Indian Ocean Tuna Commission; Pons et al. 2018). There are multiple issues within RFMOs, but those most pertinent to biodiversity conservation include the fact that fisheries management has paid insufficient attention to the environmental management of a broader range of natural assets (Gilman et al. 2014; Hooper et al. 2019). In the analysis on tuna RFMOs by Pons et al. (2018), it was noted that scores

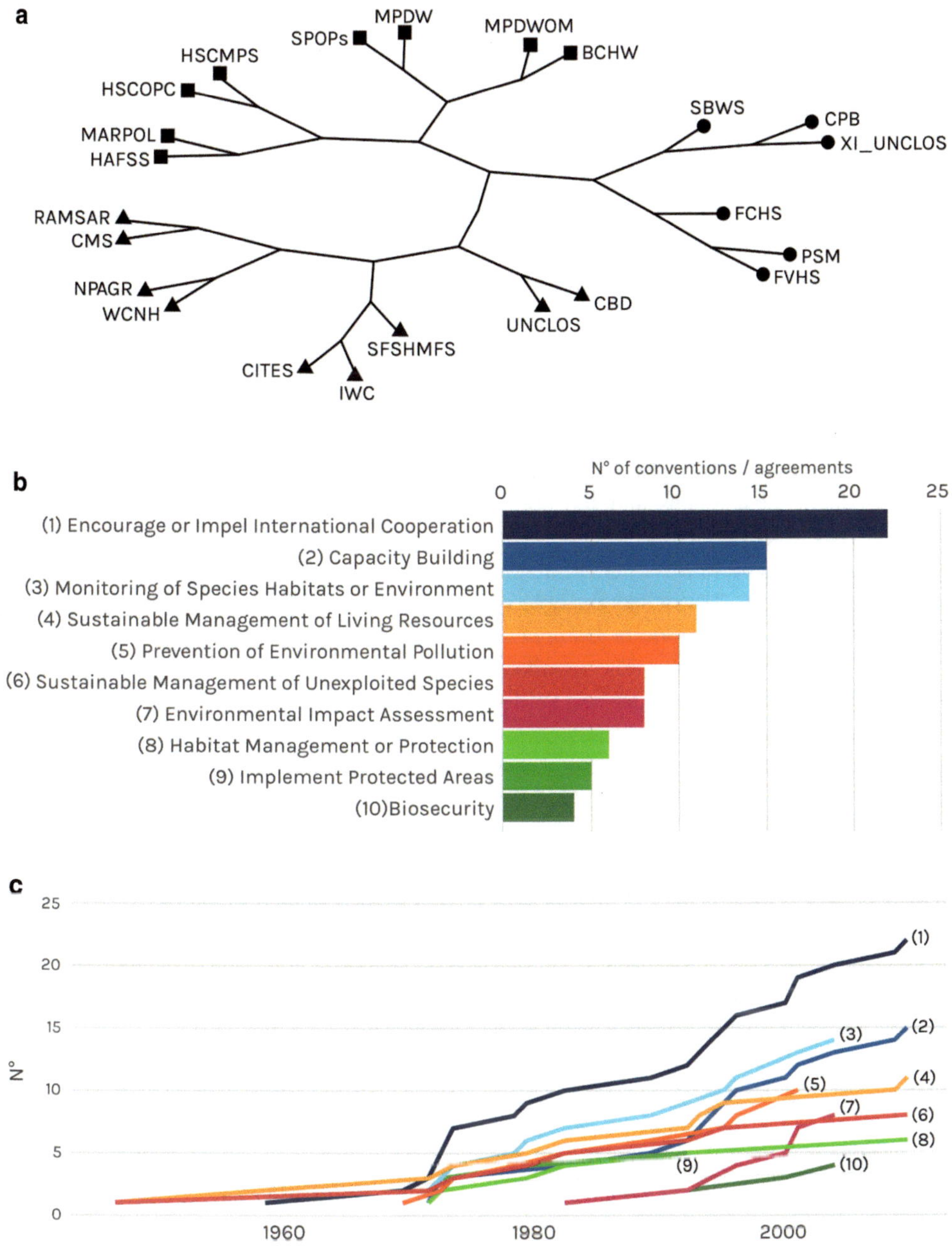

Fig. 10.12 Analysis of 23 International treaties to protect the marine environment and conserve marine biodiversity. *Notes: Panel* (**a**) *shows Ward's hierarchical clustering with Euclidean distance of international conventions/agreements according to their mission topics; the convention acronyms are as follows: BCHW basel convention on the control of transboundary movements of hazardous wastes and their disposal, CBD convention on biological diversity, CITES convention on International trade in endangered species of wild flora and fauna, CMS convention on the conservation of migratory species of wild animals (or Bonn), CPB cartegena protocol on biosafety to the convention on biological diversity, FCHS convention on fishing and conservation of the living resources of the high seas, FVHS agreement to promote compliance with international conservation and management measures by fishing vessels on the high seas, HAFSS International convention on the control of harmful anti-fouling systems on ships, HSCMPS protocol relating to intervention on the high seas in cases of marine pollution by substances other than oil, HSCOPC International convention relating to intervention on the high seas in cases of oil pollution casualties, IWC International Whaling Commission, Marpol protocol of 1978 relating to the International convention for the prevention of pollution from ships, MPDW protocol to the convention on the prevention of marine pollution by dumping of wastes and other matter, MPDWOM convention on the prevention of marine pollution by dumping of wastes and other matter, Ramsar convention on wetlands of international importance especially as waterfowl habitat, SFSHMFS agreement for the implementation of the provisions of the United Nations convention on the law of the sea of 10 December 1982 relating to the conservation and management of straddling fish stocks and highly migratory fish stocks, SPOPs Stockholm convention on persistent organic pollutants, UNCLOS United Nations Convention on the Law of the Sea, WCNH convention concerning the protection of the world cultural and natural heritage, XI_UNCLOS agreement relating to the implementation of Part XI of the United Nations Convention on the Law of the Sea of 10 December 1982. Panel* (**b**) *shows the number of conventions/agreements associated towards a main goal as listed in Table 10.4; Panel* (**c**) *shows how the number of each conventions/agreements changed over time for each main goal. (Source: Authors)*

for fisheries management in general were low and, in particular, for discarding and bycatch measures. This was attributed to a lack of severe consequences for exceeding bycatch quotas, with the result that non-target species such as marlins and sharks scored low for all management dimensions (Pons et al. 2018). Application of the precautionary principle can be useful in such cases, but this has been included in few international agreements or conventions (Table 10.4), although its use in RFMOs is spreading (de Bruyn et al. 2013).

Illegal, unregulated and unreported (IUU) fisheries contribute significantly to the overexploitation of fish stocks as well as impacts on biodiversity. They are a particular problem for commercial species, which acquire a high value because of their increasing scarcity. Examples of such species include several croakers, giant clams and red corals (Zhang and Wu 2017). These IUU vessels do not adopt fishing practices to avoid bycatch or other forms of environmental damage (Petrossian et al. 2018). A very sad example of this is the imminent extinction of the vaquita (*Phocoena sinus*), a porpoise found in the Sea of Cortez. The vaquita is suffering high mortality as bycatch in illegal gill nets set for the totoaba (*Totoaba macdonaldi*), a croaker whose swim bladder is prized in Chinese medicine and which is also endangered with extinction (Jaramillo-Legorreta et al. 2019).

What is less recognised is the role of state-corporate crime in marine fisheries (Standing 2015). This is an issue in developing coastal states where fisheries access agreements are used to allow foreign fishing vessels into their waters. There is ample evidence that the licensing coastal states and the vessels' flag states often ignore overfishing, corruption and the significant losses to the livelihoods and incomes of local small-scale fisher folk (e.g., Belhabib et al. 2015; Standing 2015; Zhang and Wu 2017). States can use their political and economic power to impose such agreements on countries, even where there is awareness of the likely outcome in terms of overfishing and negative societal impact (Standing 2015; Zhang and Wu 2017). There is also a significant role in such activities by business elites and global investment companies (Standing 2015). This is further exacerbated when political issues arise, such as in the disputed waters of the South China Sea (Zhang and Wu 2017).

Whilst fisheries impacts are not the only drivers of loss of species and habitats in the ocean, they illustrate the barriers to tackling the biodiversity crisis. Setting specific targets as policy objectives and then ensuring that their progress is monitored and reported on is crucial. Despite the objectives of increasing MPAs under the CBD (and other conventions and agreements), it was the adoption of Aichi Biodiversity Target 11 that has spurred the international community to reach a specific goal of 10% of coastal and marine areas, which are in ecologically representative and well-connected protected areas or other forms of spatial conservation management.

Likewise, SDG 14 has reinforced Aichi Biodiversity Target 11 by also calling for the protection of 10% of coastal and marine areas (SDG 14.5); the elimination of overfishing, IUU fishing, and destructive fishing practices (SDG 14.4); and the prohibition of fishing subsidies which enhance overcapacity and overfishing and which contribute to IUU fishing (SDG 14.6).[9] These targets also come with indicators against which progress can be monitored. By setting such clear goals and guidelines for reporting progress, coastal and flag states can better manage their ecosystems (Lidström and Johnson 2019).

Along with the clear setting of targets for achieving standards of fisheries sustainability, biodiversity and environmental protection, high seas fisheries management organisations should be operating to clear international standards and a system of monitoring progress to achieve such standards should also be put in place. Further improvement in the sustainability of fisheries can also be achieved by using innovative technologies to improve the monitoring of fishing activities and catches (Kroodsma et al. 2018; Bradley et al. 2019) as well reducing bycatch (Avery et al. 2017) and other environmental impacts of large- and small-scale fisheries. Implementing these measures will require adequate funding and increased capacity, especially amongst developing coastal states (Friedman et al. 2018).

A significant improvement in fisheries management would also be attained through the adoption of several voluntary codes and guidelines as clear international standards for management of fisheries (e.g., the FAO's CCRF, 1995, and Voluntary Guidelines for Flag State Performance, 2014), but again, without mechanisms for monitoring and reporting such standards will be slow in improving performance.

The implementation of new conventions and agreements should also be more rapid, and we note that the Agreement on Port State Measures to Prevent, Deter and Eliminate Illegal, Unreported and Unregulated Fishing 2009) to date only has 61 parties. A new implementing agreement for UNCLOS, known as the biodiversity beyond national jurisdiction (BBNJ) agreement, currently under negotiation, represents a step forward in putting in place a framework for spatial conservation and other measures to protect biodiversity in ABNJ. The text of this agreement contains strong provisions for monitoring and reporting on progress in implementation as well as the establishment for international standards through the operation of a Scientific and Technical Committee and a decision body (e.g., a Conference of Parties

[9]For more information on SDG 14, see https://sustainabledevelopment.un.org/sdg14

in collaboration with existing agreements and implementing agencies). It also includes the precautionary principle and significant improvements in transparency and the involvement of civil society in aspects of decision-making, particularly in processes related to environmental impact assessment. The inclusion of provisions for capacity building and technology transfer among states in the BBNJ agreement may also be extremely important not just for improving the capacity of developing states to monitor and manage biodiversity in ABNJ but also within their own coastal waters.

8 Opportunities for Action

The IPBES Global Assessment Report identifies that biodiversity is declining faster than at any other time in human history, and rates of species extinction are likely tens to hundreds of times higher than any time in the last ten million years (IPBES 2019). Despite the data limitations, we have presented evidence in this paper that marine ecosystems, like their terrestrial and freshwater counterparts, are suffering from severe habitat degradation, species population reductions and ecosystem impacts at multiple levels, with significant consequences to society through loss of ecosystem services provision which is the cause of direct economic losses, impacts on livelihoods and ultimately on human health and security.

Although these findings present a gloomy prospect for the future there are notable successes in reversing the decline of marine species through strong management and conservation measures (Duarte et al. 2020). The most notable of these is the recovery of populations of the great whales following the moratorium of whaling imposed by the IWC (Duarte et al. 2020). As related in the present report, reduction of fishing fleet capacity, coupled with modern fisheries management approaches and strong monitoring, control and enforcement has led to the stabilisation and recovery of fish stocks in the waters of Europe, the United States and elsewhere (Fernandes et al. 2013, 2017; Hilborn and Ovando 2014; Rousseau et al. 2019; Hilborn et al. 2020). Some habitats have also showed some recovery from past losses, an example being the recovery in seagrass beds in northern Europe (de los Santos et al. 2019).

This recovery was attributed to management actions including those reducing coastal pollution, measures to prevent anchoring and trawling in seagrass beds, as well as natural recovery (de los Santos et al. 2019). There are also examples of habitat restoration leading to local rehabilitation of habitats such as mangrove forests in the Mekong Delta (Duarte et al. 2020). Duarte et al. (2020) suggest that strong management action could lead to substantial recovery of abundance of species and structure, function of communities with increased provision of ecosystem services by 2050. Given the evidence for strong recovery of species and some recovery of specific habitats over decadal timescales we believe that such optimism is justified. However, recovery will only take place at large scales following strong and coordinated management action. Based on this evidence and our analysis of drivers of biodiversity loss, we find these opportunities for urgent action at local to international levels.

There are opportunities to improve monitoring, increase efficiency in MPAs, and achieve sustainable ecosystem-based fisheries management. Some specific actions/ deliverables for these high-level policy decisions include no net loss of habitat; establishing a blue bond market for investing in marine environmental sustainability; marine spatial planning to identify (on a regional basis) best options to increase no-take areas, including in the vicinity of offshore renewable energy projects; moving intensive aquaculture operations offshore, where feasible; and planning conservation responses to future coastline inundations (e.g., determining where the new sea grass meadows and mangroves will exist with sea level rise). Bringing the entire ocean under sustainable management is also a critical element in reducing open access and overexploitation of resources which has led to declines in marine species and ecosystems (Norse 2005).

8.1 Technology for Mapping

Technological advancements in remote sensing, including satellites, lidar, unmanned aerial vehicles, AUVs, and the computational ability to process such multidimensional big data in the past few decades has drastically expanded our capacity to understand the world. With increasing spatial and temporal resolution of the data captured, there is a large opportunity to further enhance our understanding of the status and trends in marine habitats and ecosystems, the drivers of change and the impacts of degradation on their contribution to people and, thus, improved visualisation and maps to support the decision-making process. The advancements in the field of artificial intelligence have also paved the way for the application of data mining and natural language processing into biodiversity and ecosystem studies. Therefore, marine scientists have the unique opportunity to extract knowledge from historical and unstructured sources (e.g., text, images, audio), store complex information in machine-readable formats and connect with expert systems to set up knowledge bases—all areas of marine science that have yet to be well explored. For effective management, governments need to know where, what, why, and how much of an activity is sustainable because anthropogenic impacts expand into deeper and deeper waters (Baker and Harris 2020).

However, there are challenges to overcome with regard to harnessing the above-mentioned technological advancements into global marine studies. Utilising the technological advancements into a thematic discipline requires multidisciplinary experts, dialogue and knowledge exchange across disciplines as well as basic scientific programming skills and knowledge of machine-readable data and metadata formats. The lack of interoperable web services and a catalogue for referencing remote sensing products and geospatial data sets limits the smooth communication of needs from a thematic discipline to the technology developers.

There is an opportunity for NGOs, industry, researchers, and government institutions to collaborate to increase the application of current advancements in technological capacity. To accomplish this cross-disciplinary discussion, there needs to be an exchange of knowledge, and scientists need to be trained to make their analysis and work interoperable. Streamlined services are also needed to support the production of standard essential variables and indicators in the field, including a catalogue of key data sets, which would integrate a wide variety of primary data, and standardised processing services (i.e., web rest services), which would improve access and maintain frequently used data resources.

We envision that by 2030 a catalogue of marine habitats, including those that we currently have limited information on, such as kelp forests and rocky reefs, will have their EOVs monitored spatially and temporally, and variation and distribution changes within them will be automatically generated over time and publicly accessible. We support the development of a comprehensive ocean observing system which has been identified as a priority for the United Nations Decade of Ocean Science for Sustainable Development and GOOS. With this information accessible, organisations can effectively monitor the global distributions of economically important marine habitats, such as coral reefs, mangroves and seagrasses. On a local level, governments should collaborate with industry and NGOs to effectively map drivers of habitat degradation and ground truth the data produced from the global habitat mapping efforts. Such mapping and monitoring of marine ecosystems has been among recommendations for improved management of marine biodiversity for almost 30 years (Norse 1993).

To be able to develop the collaborations and technological capacity to make this vision a reality, we suggest the following high-priority opportunities for action:

- The present intergovernmental organisations (e.g., UNESCO-IOC), biodiversity monitoring networks (GOOS BioEco, GEO BON/MBON), databases (e.g., OBIS) and philanthropic efforts involved in gathering and making ocean data available for management purposes (e.g., Google Earth Engine; Ocean Data Foundation)[10] require a coordinated approach to face the challenge of comprehensive and global monitoring of biodiversity. These organisations, under the leadership of UNESCO-IOC, in partnership with national ocean biodiversity monitoring networks (e.g., IMOS, IOOS, EuroGOOS) and the CBD, should— through workshops or other means—create maps of both habitat extent and environmental drivers to identify conflicts and gaps in knowledge, including in the distribution of marine habitats, technological limitations and solutions with explicit goals and institutions/organisations assigned to meeting the goals. These efforts should include multidisciplinary scientists, including, but not limited to, marine, artificial intelligence and data experts.
- The Decade of Ocean Science for Sustainable Development provides an ideal jump-off point for such a coordinated approach to ocean biodiversity monitoring, especially as it recognises the importance of producing actionable data but will also produce significant new data sets on species and habitat distribution in the ocean.
- By 2025 this should culminate in collaborative research platforms where global habitat maps and EOVs can be compiled based on interoperable data sources, be visualised and be made publicly available in a way that facilitates ecosystem-based management of human activities in the ocean whilst enabling biodiversity conservation.
- By 2028, integration of novel technological developments with quality-control standards increase temporal resolution of habitat maps and drivers so that quality annual maps of habitat extent and impacts are made available.
- Throughout 2020–30, knowledge bases and technology transfer between governments is promoted to equip all countries with the tools necessary to sustainably manage and map the ocean. Capacity-building efforts are targeted at providing all countries with the expertise to access and act upon biodiversity data for meeting international targets and ocean management needs.

- By accomplishing these goals, we believe there will be numerous additional benefits past increasing our understanding of the planet, including improved environmental and biodiversity monitoring plans, technological advancements, the training of new generations of scientists from diverse backgrounds and increased collaboration between stakeholders.

8.2 Addressing the Biodiversity Data Gap

There is a pressing need for a greater coordinated effort to gather information on marine biodiversity and extinction risk, from baselines of diversity and ecosystems to the long-term monitoring of population genetics, species, habitats and

[10] Information about the Ocean Data Foundation can be found on its website, https://www.oceandata.earth/

ecosystems. Again, despite recommendations to develop such coordinated knowledge gathering on marine biodiversity, as well as improving the capacity to do so by all nations nearly 30 years ago (Norse 1993), this has not happened to date. The IUCN Red List shows that although there are a good range of assessments for marine vertebrates (fish, seabirds, marine mammals), extinction risk assessments on marine invertebrates are restricted to a few scattered groups.

There is now an opportunity for states, intergovernmental organisations, foundations and other philanthropic organisations to invest in the infrastructure, including human resources, to meet their international commitments (e.g., under the CBD) to establish baselines of biodiversity and long-term monitoring of the status of species and habitats both within their EEZs and in ABNJ, especially where their flagged vessels are or will be undertaking activities such as fishing or other extractive activities. Such an effort should focus on the already established networks for biodiversity monitoring, including GOOS BioEco and the marine component of GEO BON, MBON.

The first has developed a framework and a globally coordinated strategy for monitoring biodiversity change using biological EOVs which are complemented by the EBVs coordinated by the latter. Data repositories already exist to receive such information (e.g., OBIS; Navarro et al. 2017). GOOS BioEco is facilitating the establishment of coordinated networks to implement monitoring of these essential variables. These will be established in collaboration with MBON and will include oceanographic research centres, government institutions and universities, and natural history museums. These networks should also build on existing efforts, such as the Global Coral Reef Monitoring Network.

By establishing such networks, states will be able to establish a baseline of marine biodiversity in their waters and in ABNJ, allowing the subsequent monitoring of changes in biodiversity through time. This will enable the continual assessment of the success of measures to reduce biodiversity loss by states and allow them to actively manage their activities to mitigate or reverse biodiversity loss. For developing states, assistance in capacity building will be required. Associated benefits from such an effort will include

- maintenance or enhancement of marine ecosystem services provision (e.g., fisheries, coastal protection, tourism);
- identification of marine genetic resources (Blasiak et al. 2020);
- the training of a new generation of marine scientists;
- increased opportunities for citizen science and education; and
- increased effectiveness of investment in biodiversity conservation through specific targeting of interventions.

At present, there are no alternative measures to achieve such a goal, and without it, undocumented biodiversity loss will continue in the face of pressures arising from poverty, the increasing human population and the drive for economic development. We envision a pathway to improved biodiversity monitoring to include the following milestones:

- The identification or establishment of national centres for marine biodiversity monitoring and developed capacity in taxonomy and field ecology, including training in new taxonomic tools such as environmental DNA (eDNA) and other emerging technologies, to undertake baseline assessments and long-term monitoring.
- A baseline biodiversity inventory and the establishment of key monitoring sites as part of the GOOS BioEco networks or of an existing MBON and expanding geographic coverage through the establishment of new MBON sites/regions (2023–25).
- The coordination of biodiversity monitoring activities at a regional basis implementing best practices to exchange knowledge, deliver FAIR and open-access data and share resources where appropriate (2020–25).
- The establishment of a marine biodiversity programme that feeds into national policies and management actions to mitigate biodiversity loss as well as into regional organisations, such as RFMOs, to manage activities in a way as to protect and conserve biodiversity. Biodiversity management becomes embedded into national institutions and legislation and into regional bodies (2025–30+).

There are a range of habitats formed by foundation species that are overwhelmingly important to biodiversity because they are connected to ecosystem functions over a wider geographic area than their immediate occurrence. These include, most notably, coral reefs, mangrove forests, seagrass beds, saltmarshes, kelp forests and other coastal ecosystems. In ABNJ, these are probably strongly represented within EBSAs and may include habitats such as seamounts.

We recommend that coastal states and regional ocean management organisations should adopt a policy of zero net loss for such ecosystems. Because the costs of habitat restoration are often much higher than conservation (Friess et al. 2019), such a policy should prioritise avoidance of activities which lead to significant damage in the first place.

We believe that by establishing or further developing a national MBON coordinated at a regional level, including ABNJ, it could—if used to support effective management and conservation—help to improve and secure economic and other societal gains from the provisioning of ecosystem services. Additional benefits from developing marine genetic resources (Blasiak et al. 2020) and improving environmental awareness and education within society are difficult to estimate but would certainly be positive.

8.3 Citizen Science and Education Programmes

Citizen science provides a great opportunity to increase public participation in science, overcome significant barriers to the scientific process and improve natural resource management (Theobald et al. 2015; McKinley et al. 2017). Citizen science and environmental education programmes are also scientific projects that can produce reliable information in which members of the public directly engage in research to answer particular questions (Parrish et al. 2018; McKinley et al. 2017). Biodiversity-related projects have been shown to span greater geographic and temporal ranges than conventional academic research, engaging millions of volunteers and generating up to $2.5 billion in kind annually (Theobald et al. 2015). There are many goals and benefits for citizen science, spanning publishing results in peer-reviewed journals, education, community empowerment and personal fulfilment (Parrish et al. 2018).

Despite many long-term citizen science projects creating robust data sets,[11] many academic researchers still show a bias against citizen science (Bonney et al. 2014). Theobald et al. (2015) found that only about 12% of projects out of 388 provide data to scientific publications. Therefore, methods of quality assurance (actions taken to ensure the quality of measurements taken) and quality control (post hoc actions to ensure the quality of results) are pivotal to many projects where the primary goal is science generation and should continue to be developed (Bonney et al. 2014; McKinley et al. 2017). A participant's time and success in mastering a task is a function of the complexity of the task (Sauermann and Franzoni 2015), which supports that projects should be simply designed at scale, and projects at smaller scales, with higher complexity, can be more involved (Parrish et al. 2018).

Citizen science programmes can also generate significant social outcomes, including increasing science education, engagement in policy and collaboration. As such, they represent the following opportunities for action:

- Governments increase general science education in line with SDG 4 to 'Ensure inclusive and equitable quality education and promote lifelong learning opportunities for all' (2020–25).
- Citizen science programmes coordinate and organise to ensure that the wealth of information gathered is accessible, usable, known to decision-makers and connected with networks of biodiversity monitoring, including GOOS BioEco and the marine component of GEO BON, MBON, starting in 2023.
- Industry and governments that benefit from this information provide increased funding for the development of community-based programmes in developing countries to increase exposure to science and raise a new generation of scientists by 2025.
- Academia generates best practices and resources to increase the amount citizen science can be used to generate robust data and science, thus removing the bias against this information by 2030.

By accomplishing the previous recommendations, we see a future defined by increased scientific literacy around the world, improved efficiency of moving conservation science into conservation action, and higher awareness and knowledge of the planet around us.

8.4 Well-Enforced, Green-Listed, Fully Protected Marine Reserves

There is strong evidence that the implementation of well-enforced, fully protected MPAs that include 30–40% of key marine habitats will conserve biodiversity, enhance biomass and abundance of marine life as well as improve the resilience of marine ecosystems (Roberts et al. 2001; Lester and Halpern 2008; Gaines et al. 2010; Sciberras et al. 2013; Edgar et al. 2014; Mellin et al. 2016; Sala and Giakoumi 2017). These MPAs can also benefit fisheries (Roberts et al. 2001; Gaines et al. 2010; Di Franco et al. 2016; Ban et al. 2017), provide coastal protection (Roberts et al. 2017) and improve the resilience of ecosystems against the impacts of climate change (Mellin et al. 2016; Roberts et al. 2017). However, poor capacity for the enforcement of MPAs (Gill et al. 2017) and poverty alleviation—specifically, the generation of jobs (Cinner et al. 2009; Gurney et al. 2014)—can undermine MPA objectives. Additionally, the social impacts of protected areas are poorly understood largely because MPA evaluations have tended to focus on one or very few outcomes, and few have had the requisite data to assess causal effects (Gurney et al. 2014). Opportunities over the next two years (e.g., the BBNJ agreement and the CBD Conference of Parties in 2021) offer the chance to adopt a new target beyond the 10% of marine protection and to accelerate the slow progress made to date. Whatever targets for biodiversity protection are set, they must represent the full range of marine ecosystems and species. The aims should include no net loss of important habitats which structure marine ecosystems, such as coral reefs, mangrove forests, seagrass beds, saltmarshes and others.

Experts, conservation practitioners, philanthropic organisations and representatives from government should come together convened by the IUCN, the United Nations Environment Programme (UNEP) and the CBD to establish

[11] See eBird (https://ebird.org/home), COASST (https://coasst.org/) and Zooniverse (https://www.zooniverse.org/)

the best strategy for increasing and improving existing MPAs on the basis of the approach we have outlined in this paper for coastal states.

Strategies tailored for each group of countries—and ultimately each individual country—can be developed, and international assistance, including economic, capacity building and technical advice, can be targeted to effectively achieve global, regional and national targets. For ABNJ, a different approach can target areas of conservation importance whilst balancing these with economic need. The framework developed by O'Leary et al. (2018), with input from the CBD EBSA process, offers a practical approach to achieve this. We envision the pathway as follows:

- The MPA targets are established internationally, at the CBD's Conference of Parties or (for the ocean) at the United Nations Ocean Conference in 2021.
- An implementation conference is initiated to identify specific targets at global, regional and national levels to protect representative marine ecosystems and the best strategic approaches and practical measures to achieve these targets. The conference should be convened by the IUCN, UNEP and the CBD, with attendance from experts and governmental, intergovernmental and non-governmental organisations as well as potential funders (Global Environment Facility, government-funding agencies, private philanthropists and foundations). The target year for the conference is 2022.
- By 2022, a large campaign and economic support should be in place to involve communities and stakeholders to implement community-based MPAs (Pollnac et al. 2001; Aburto-Oropeza et al. 2011). By 2023, a global map to implement community- based MPAs should be generated by states. In the Philippines, where government policy, international aid, universities and NGOs have invested a great effort to implement community-based MPAs, there are over 400 of these management areas. Although only 25% of them are effective in the protection of the resources, clear common factors have been described as the path to successful community-based MPAs: (1) relatively small communities, (2) community census statistics to prioritise targeted interventions, (3) overfishing challenges, (4) movement to alternative income projects, (5) increased level of community participation in decision-making, (6) strong local leadership, (7) receiving scientific and MPA-implementing advice and (8) closely working with local or municipal governments (Pollnac et al. 2001; Crawford et al. 2006; Rossiter and Levine 2014). These small but successful examples of community- based MPAs have proven that not only is it possible to recover marine biodiversity in a short time period (one decade), but they are also producing significant economic benefits for local communities. Cabo Pulmo National Park in Mexico is considered a success according to both biological and social measures: the MPA has seen significant recovery of biomass (Aburto-Oropeza et al. 2011) and demonstrable community engagement and participation, along with extensive socio-political support (and media attention) at the local, national and international levels. Cabo Pulmo has achieved a kind of symbolic power in the world of marine conservation (Anderson 2019), and it has influenced the transition of a governance system into a new, adaptive tourism model (Langle-Flores et al. 2017). There is a need for scaling up community-based MPAs to increase the social and ecological benefits for coastal areas. Evaluating approaches has demonstrated that 'opportunistic approaches' and 'donor-assisted approaches' do not create the necessary outcomes requested by global conservation targets. Rather, a systematic conservation planning approach of community-based MPAs can improve ecological and social outcomes, particularly if this planning incorporates equity for stakeholders (Kockel et al. 2019).
- The implementation conference should lay out a clear road to attaining established targets, with appropriate milestones (2023–30). We suggest that a single agency be tasked with measuring progress towards milestones and the final targets (e.g., UNEP- WCMC). Reports should be produced for the CBD's Conferences of Parties in 2024, 2026 and 2028 prior to 2030. Reporting should also extend to other relevant meetings (e.g., the Our Ocean and United Nations Ocean Conferences).

Balmford et al. (2004) estimated the costs of running a global MPA network covering 20–30% of the ocean at \$5–\$19 billion per annum. However, the potential gain in direct enhancement of fisheries and tourism and the avoided costs in environmental damage through reduction/mitigation of coastal inundation is likely to dwarf these costs. This is without accounting for other ecosystem services, such as CO_2 sequestration, nutrient cycling, waste remediation, protection of marine genetic resources and cultural services, which represent a value in the trillions of dollars overall (Costanza et al. 2014).

Furthermore, we point to the already estimated erosion in the value of marine ecosystem services as a result of the erosion of habitats which amount to a loss of more than \$10 trillion per annum in just over a decade between 1997 and 2011. Much of this loss was focused on coastal ecosystems, with coral reefs losing nearly half their value as a result of the loss of this habitat (Costanza et al. 2014).

8.5 Ecosystem-Based Fisheries Management

There is an extreme urgency to eliminate IUU fishing and accelerate the reform of fisheries management to reflect modern ecosystem-based concepts where biodiversity is managed sustainably alongside target stocks. Both the IPBES Global Assessment Report (IPBES 2019) and our own analyses indicate that overfishing, illegal fishing and destructive fishing practices are the prime drivers of biodiversity loss in the ocean. Whilst much progress has been made in sustainable ecosystem-based fisheries management (Hilborn and Ovando 2014; Friedman et al. 2018; Hilborn et al. 2020), progress remains fragmented. The fishing power of the global fishing fleet is continuing to grow and underlies overfishing in much of the global ocean (Rousseau et al. 2019). We have identified clear barriers to accelerating progress in fisheries sustainability and increasing consideration of biodiversity conservation in fisheries. These barriers include a lack of capacity and funding, whether being associated with institutions or developing states, and overwhelming pressure in some parts of the world to exploit living marine resources exacerbated by growing industrial and small-scale fishing fleets. There is also evidence that in some states, elements of the fishing industry and financial institutions are complicit in allowing overfishing and illegal fishing to continue (Standing 2015; Zhang and Wu 2017). This is not only immensely damaging to biodiversity but also leads to massive economic losses (Costello et al. 2016; World Bank 2017) and the loss of livelihoods and impacts food security (Sumaila et al. 2013; Standing 2015; Freduah et al. 2017). In the face of climate change impacts, overfishing will exacerbate these problems (Badjeck et al. 2010). If biodiversity loss in the ocean is to be halted or reversed, this elephant in the room cannot be ignored.

The reform of fisheries management practices and of the institutions charged with their management is already under way (Friedman et al. 2018). This reform process must be accelerated and driven through the adoption of appropriate targets by the competent authorities. The most important of these reforms include the following:

- Good data underlies all fisheries management both in the context of target species, bycatch species and the environmental impact of fishing. Given the development of modern technologies, from remote sensing to mobile computing and phones, there is an opportunity to greatly improve the monitoring of catches of target and bycatch species in all industrial fisheries. Given the importance of small-scale fisheries in terms of global fishing power, special measures to include these in fisheries catch statistics as well as fisheries management (including co- management/community management arrangements) is critical. Such measures will also allow an assessment of the nutritional and economic benefits of small-scale fisheries at the national level so they are accounted for in decisions on fisheries policy.
- Uniformly adopting modern principles of ecosystem-based fisheries management and the precautionary principle for all fisheries management as expressed in the UN conventions and agreements, the FAO's CCRF and other FAO guidelines and codes.
- Eliminating IUU fishing and other illegal practices in fishing through improved monitoring, control and enforcement. It is especially important that measures to eliminate IUU fishing are adopted rapidly by all fishing and port states.
- Stabilising, and then reducing, fishing pressure should be a priority in regions where growth in fishing capacity continues, undermining efforts to sustainably manage fisheries pressure and to conserve biodiversity. It is critical to ensure that measures to reduce fishing capacity protect the basic needs for food, nutrition and livelihoods in coastal communities, particularly in developing countries.

We also note the opportunities for other important reforms in fisheries management:

- Develop and fund infrastructure and human capacity to enable sustainable management of biodiversity as well as target fish stocks.
- Reform decision-making processes and adopt greater transparency by fisheries management organisations to speed up progress in eliminating overfishing.
- Make all fisheries data public, including data on vessel tracking, catch and bycatch within 12 months of collection.
- Specify measures to address issues of overfishing by developing states and in small-scale fisheries, including investment in data-poor stock assessment methods and the use of reciprocal mechanisms to enhance institutional, management and governance capacity in developing states through finance, training and technology transfer.
- Establish community-based fisheries management to assist in increasing the biological and socio-economic sustainability of fisheries.
- Continue efforts to merge and coordinate the objectives of the fisheries and environmental sectors at all levels of fisheries management (international to local).
- Develop a set of investment standards for the investment in fisheries, and especially infrastructure such as vessels, so only sustainable fisheries/fishing operations are financed.
- Initiate a formal regular review of RFMOs, ensuring they are meeting new standards of fisheries management; the

following areas specifically require attention: (1) updating conventions and agreements to implement modern standards of ecosystem-based fisheries management, including specific provisions for the conservation and protection of biodiversity; (2) further convergence between fisheries and environmental sector governance structures to integrate biodiversity considerations into fisheries management; (3) implementing mechanisms to ensure the rapid and accurate reporting of catches of target and bycatch species; (4) more rigorous target-based efforts to ensure rapid implementation of rules and recommendations; (5) a transformation of transparency for both fisheries-related data and decision-making processes; (6) reforming decision- making structures to prevent 'opt-out' or lowest- common-denominator regulations within fisheries management organisations; and (7) greater clarity on participatory rights, such as allocation of catch levels or fishing effort (Gjerde et al. 2013; Friedman et al. 2018).

- Develop a set of minimum standards for fisheries partnership agreements to ensure (1) sustainable fishing; (2) fair and equitable financial benefits for parties; (3) clear financial structures and reporting arrangements to ensure licence fees or other financial benefits flow to society; (4) adequate arrangements for monitoring, control, surveillance and enforcement of fisheries; and (5) formal structures for dispute resolution amongst partners with arbitration by an impartial third party.

Aichi Biodiversity Target 6 and SDG 14 embody specific targets for fisheries sustainability, and the measures above will clearly help to attain these goals. The SDGs are set for 2030 (with some interim targets due in 2020), but the CBD post-2020 biodiversity framework also provides a timetable for achievement of these goals and an opportunity to finally achieve the objectives of Aichi Biodiversity Target 6. We view the next decade, therefore, as critical in accelerating reforms of fisheries and biodiversity objectives to protect marine living resources.

By adopting these reforms, overfishing and IUU fishing will be eliminated, and fish stocks and associated ecosystems should be able to rebuild. The financial benefits of this just in fisheries revenue alone has been estimated at $83 billion per annum (World Bank 2017). Broader benefits will include increasing fish catches (Costello et al. 2016) and securing both livelihoods and food supplies as well as increasing their resilience to climate change impacts for the future. Given that destructive fishing impacts, such as bycatch, are the main drivers of biodiversity loss for a number of marine species, the benefits of reducing extinction risk and restoring ecosystem function and services provision will be enormous. This will also increase ecosystem resilience against climate change and other impacts.

9 Limitations of the Paper and Conclusions

As identified in several parts of this study, a lack of FAIR and open data on marine biodiversity is problematic when trying to identify patterns of species and habitat diversity as well as changes in these parameters over time. For example, in the IUCN Red List data, many species are classified as DD, and many groups of invertebrates have not been assessed at all. Without this information, it is very difficult to estimate the current state of, and trends in, marine biodiversity in the ocean.

There are significant gaps in our analyses because comparable global data sets were not available for many coastal habitats, including rocky reefs. Within the available data sets, there are many gaps and sampling biases, leading to higher diversity values in areas which likely do not correspond to species or habitat diversity. Likewise, particularly for deep-sea and offshore parts of the ocean, only large-scale oceanic habitats that can be identified through physical features (e.g., seamounts) could be identified, and the water column, the largest ecosystem on Earth, was largely neglected in this study. A trend analysis for the marine habitats examined here was not possible with the current publicly available data but should be pursued in future efforts as outlined in Sect. 8.

Despite these gaps, we have sufficient information to understand the broad state of marine species and habitat diversity to generate effective management responses. However, to reduce habitat loss and degradation, we need an increase in multi-decadal monitoring because it is essential to be able to understand, prevent future damage and monitor potential recoveries of marine ecosystems (Bayraktarov et al. 2016; Gangloff et al. 2016). Monitoring will establish baselines so that we can quantify changes in habitat extent and impacts from anthropogenic activities and use this information effectively to manage our natural resources.

A lack of adequate funding and capacity—particularly in developing countries but also in the organisations charged with sustainably managing economic activities in the ocean—is repeatedly highlighted in this study. Urgent measures are required to build capacity, transfer technology and build the global financial supporting structures so the blue economy can grow in a sustainable fashion that neither depletes marine species or habitats nor undermines the ecosystem services on which humankind relies. Current biodiversity loss in the ocean is at least partially due to a lack of equitability in states' ability to monitor biodiversity and manage activities within their EEZs and ABNJ.

The current crisis of biodiversity loss in the ocean may require developing and implementing further international agreements and national measures to protect habitats and species. A new legally binding instrument under UNCLOS to conserve and sustainably use marine biodiversity of areas beyond national jurisdiction (the BBNJ agreement) is currently being negotiated and should become an important legal framework for the conservation of 50% of Earth's surface area. In addition, new protocols could be developed as part of existing conventions, specifically the CBD, the World Heritage Convention and the Convention on Migratory Species, among others. Such protocols should include provisions that human activities should not result in the long-term or permanent loss of biodiversity in the ocean, with clear mandates for monitoring their effectiveness. They should also lay out renewed commitments for implementing biodiversity protection measures as well as monitoring and data-gathering activities which are already embodied in existing conventions and agreements. These new protocols should apply to all sectors operating in the ocean and should include the broad family of UN specialized agencies, including the FAO and associated RFMOs, the International Maritime Organization and the ISA.

The fisheries reforms described in this Blue Paper would likely cost millions to tens of millions of dollars on a state-by-state basis; yet in economic returns from fisheries alone, there is the potential for billions of dollars in return. Not undertaking these reforms will lead inevitably to commercial and/or local to wide-scale extinction of both exploited and non-target species, undermining ecosystem resilience and service provision. By extending this to the broader values to society and to the restoration of biodiversity and ecosystem services, reforms could be transformative.

The speed of the decline of marine species and habitats means that the opportunities for action we have identified should be taken up with urgency. Such an international effort, spanning all sectors involved in the blue economy as well as the implementing organisations involved in their management, may require a coordinated effort on the scale of that currently addressing climate change. A large-scale global plan of action for ocean biodiversity conservation may be required to expedite these opportunities with the speed required.

Acknowledgements The paper's technical reviewers, Miriam Fernandez Bergia, Stuart Kininmonth and Boris Worm, as well as its arbiter, Martin Stuchtey, all provided helpful technical comments.

The authors also thank the World Resources Institute for providing support as the Ocean Panel Secretariat.

While our colleagues were very generous with their time and input, this report reflects the views of the authors alone. Jorge Assis was supported by the Foundation for Science and Technology (Portugal) through the projects UIDB/04326/2020, BIODIVERSA/004/2015 and the transitional norm—DL57/2016/CP1361/CT0035.

The authors thank LSF Editorial for copyediting and Rosie Ettenheim for design.

About the Authors

Lead Authors

Alex D. Rogers is the Science Director of REV Ocean, a not-for-profit working for a healthier ocean. He is also a Visiting Professor of the Department of Zoology, University of Oxford, a Researcher of the Oxford Martin School and a Senior Research Fellow of Somerville College, University of Oxford. Prof. Rogers has a background in marine ecology, systematics and molecular ecology.

Octavio Aburto-Oropeza is an Associate Professor at Scripps Institution of Oceanography (SIO), a professional photographer associate with the International League of Conservation Photographers and a National Geographic Explorer. His research and photographs have focused on marine reserves and commercially exploited marine species and their fisheries in Latin America and the United States.

Contributors

Ward Appeltans is the marine biodiversity focal point of UNESCO- IOC and since 2012 is the head of the secretariat of OBIS. Before joining UNESCO-IOC he managed WoRMS and was a project leader at the data centre of the Flanders Marine Institute and was involved in several European marine biodiversity projects.

Jorge Assis is a research fellow at the Centre of Marine Sciences and lecturer at the University of Algarve. He has a background in marine ecology and data science and currently focuses his research in addressing the potential consequences of global climate changes (past, ongoing and future) to marine biodiversity.

Lisa T. Ballance is Director of Oregon State University's Marine Mammal Institute, Endowed Chair for Marine Mammal Research and a former marine ecologist with the National Oceanic and Atmospheric Administration (NOAA). Her research focuses on ecology and conservation biology of cetaceans and seabirds, largely in the Pacific and Southern Oceans far from shore.

Philippe Cury is senior scientist at IRD (Institut de Recherche pour le Développement; French National Research Institute for Sustainable Development) and the IRD representative for Europe in Brussels at CLORA (Club des Organismes de Recherche Associés). He is the President of the Scientific Council of the Monaco Institute of Oceanography and one of the two scientific Directors of Euromarine Consortium (www.euromarinenetwork.eu). His research focuses on sustainable management of marine resources in an ecosystem perspective.

Carlos M. Duarte is Professor of Marine Science and Tarek Ahmed Juffali Research Chair in Red Sea Ecology at King Abdullah University of Science and Technology (KAUST), and Professor at the Arctic Research Center, Aarhus University, Denmark. His research addresses marine ecology, conservation and the impacts and responses to human pressures for ecosystems ranging from polar to tropical and from coastal to the deep sea. He is a member of the Expert Group supporting the HLP commissioning this series of Blue Papers.

Fabio Favoretto is a Marine Ecologist at the Universidad Autónoma de Baja California Sur in La Paz, Mexico. He has a background in benthic ecology and currently works on remote sensing applications in coastal environments.

Joy Kumagai is a Research Associate at SIO. Her research focuses on the impacts of anthropogenic activities on ecosystem services and marine biodiversity. Her background is in marine biology and applying Geographical Information Systems to answer research questions.

Catherine Lovelock is a Professor at the School of Biological Sciences at the University of Queensland, Australia. She has investigated the physiology and ecology of mangroves and other coastal plant communities over their global distribution, focusing on climate change and its impacts as well as the role of coastal wetlands in climate change adaptation and mitigation.

Patricia Miloslavich is the Executive Director of the Scientific Committee on Oceanic Research (SCOR), an international NGO to help address interdisciplinary science questions related to the ocean. She is a retired professor from the Universidad Simón Bolívar in Venezuela. Her background is in marine biology, biodiversity and global ocean biological observations.

Aidi Niamir is a scientist at the Data and Modelling Centre of the Senckenberg Biodiversity and Climate Research Institute in Germany. His research and services mainly focus on spatiotemporal models, application of earth observation in biodiversity and ecosystem studies and open geo-information tools to support science-policy interactions. He is a lead author of the IPBES Global Assessment Report on Biodiversity and Ecosystem Services and the head of the IPBES technical support unit on data.

David Obura is a Director of CORDIO (Coastal Oceans Research and Development in the Indian Ocean) East Africa, a Kenyan research organisation focused on coral reefs of the Western Indian Ocean. From a foundation in coral reef resilience, he increasingly works to integrate conservation and development in ways that will be necessary for a sustainable future for coastal systems. He works with a number of global working groups, including the Earth Commission and GOOS BioEco, and chairs the IUCN Coral Specialist Group.

Bethan C. O'Leary is a Research Associate at the University of York and a Senior Research Fellow at the University of Salford. Her research focuses on sustainable management of marine resources, particularly through marine protected areas, and the integration of scientific research with applied environmental management and policy.

Gabriel Reygondeau is a research associate in the Changing Ocean Research Unit at the University of British Columbia and in the Department of Ecology and Evolution at Yale University. He is an oceanographer, ecologist and biogeographer who explores the relationship between climate, biodiversity and human activities.

Callum Roberts is Professor of Marine Conservation at the University of York. His research focuses on threats to marine ecosystems and species and on finding the means to protect them. His team helped design the world's first network of high seas protected areas and provided the scientific underpinning for a new ocean protection target—30% by 2030—which is gaining widespread international support.

Yvonne Sadovy was a Professor at the School of Biological Sciences of the University of Hong Kong for over 25 years after directing the government fishery lab in Puerto Rico. She founded and co-chairs the IUCN Groupers & Wrasses Specialist Group and co-founded and directs Science and Conservation of Fish Aggregations. Her specialty is reef fishes and fisheries, their conservation and management

Tracey T. Sutton is Professor in the Halmos College of Natural Sciences and Oceanography at Nova Southeastern University (Dania Beach, Florida) and Assistant Chair of the Department of Marine and Environmental Sciences. His research entails oceanic ecosystem structure, food web dynamics, organismal ecology, biogeography, fish taxonomy, and deep ocean stewardship, with emphasis on human impacts. He is a member of the Expert Panel for the first and second United Nations World Ocean Assessments.

Derek P. Tittensor is the Jarislowsky Chair in Marine Ecosystem Forecasting at the Department of Biology, Dalhousie University, Canada. He is a marine ecologist with a research focus on biodiversity and human impacts on marine ecosystems. He is also coordinator of the Fisheries and Marine Ecosystem Model Intercomparison Project (Fish-MIP).

Enriqueta Velarde is Senior Researcher/Professor at the Institute of Marine Sciences and Fisheries at the Universidad Veracruzana in Mexico and has 42 years of experience in research projects related to seabird ecology. Main areas of interest include generating a long-term time series of biological and ecological seabird parameters providing information to support demographic, trophic and fisheries predictive models; marine protected area planning as well as genetic and life history research of endangered species in natural protected areas. Other work areas focus on ethno-ornithology, indigenous knowledge and citizen science.

Interactive versions of the major figures in this paper with statistics broken down to coastal state EEZs are available at: https://octopus. zoo.ox.ac.uk/studies/critical-habitats-2020

References

Aburto-Oropeza O, Ezcurra E, Danemann G, Valdez V, Murray J, Sala E (2008) Mangroves in the Gulf of California increase fishery yields. Proc Natl Acad Sci U S A 105(30):10456–10459. https://doi.org/10.1073/pnas.0804601105

Aburto-Oropeza O, Erisman B, Galland GR, Mascareñas-Osorio I, Sala E, Ezcurra E (2011) Large recovery of fish biomass in a no-take marine reserve. PloS One 6(8):e23601. https://doi.org/10.1371/journal.pone.0023601

Aebischer NJ, Coulson JC, Colebrook JM (1990) Parallel long-term trends across four marine trophic levels and weather. Nature 347(October):753–755. https://doi.org/10.1038/347753a0

Agardy T, di Sciara GN, Christie P (2011) Mind the gap: addressing the shortcomings of marine protected areas through large scale marine spatial planning. Mar Policy 35:226–232. https://doi.org/10.1016/j.marpol.2010.10.006

Ainsworth TD, Heron SF, Ortiz JC, Mumby PJ, Grech A, Ogawa D, Eakin CM, Leggat W (2016) Climate change disables coral bleaching protection on the great barrier reef. Science 352(6283):338–342. https://doi.org/10.1126/science.aac7125

Albouy C, Archambault P, Appeltans W, Araujo MB, Beauchesne D, Cazelles K, Cirtwill AR et al (2019) The marine fish food web is globally connected. Nat Ecol Evol 3(August):1153–1161. https://doi.org/10.1038/s41559-019-0950-y

Alder J (2003) Putting the coast in the sea around us project. Sea Around Us Newsl 15:1–2. http://seaaroundus.org/newsletter/Issue15.pdf

Alongi DM (2015) The impact of climate change on mangrove forests. Curr Clim Change Rep 1(March):30–39. https://doi.org/10.1007/s40641-015-0002-x

Amoroso RO, Pitcher CR, Rijnsdorp AD, McConnaughey RA, Parma AM, Suuronen P, Eigaard OR et al (2018) Bottom trawl fishing footprints on the world's continental shelves. Proc Natl Acad Sci U S A 115(43):E10275–E10282. https://doi.org/10.1073/pnas.1802379115

Anderson RB (2019) Beyond 'Success': community, governance, and the future of Cabo Pulmo National Park. Hum Organ 78(2):147–157. https://doi.org/10.17730/0018-7259.78.2.147

Anderson ORJ, Small CJ, Croxall JP, Dunn EK, Sullivan BJ, Yates O, Black A (2011) Global seabird bycatch in longline fisheries. Endanger Species Res 14(2):91–106. https://doi.org/10.3354/esr00347

Andréfouët S (2008) Coral reef habitat mapping using remote sensing: a user vs producer perspective. implications for research, management and capacity building. J Spatial Sci 53(1):113–129. https://doi.org/10.1080/14498596.2008.9635140

Appeltans W, Ahyong ST, Anderson G, Angel MV, Artois T, Bailly N, Bamber R et al (2012) The magnitude of global marine species diversity. Curr Biol 22(23):2189–2202. https://doi.org/10.1016/j.cub.2012.09.036

Arias-Ortiz A, Serrano O, Masqué P, Lavery PS, Mueller U, Kendrick GA, Rozaimi M et al (2018) A marine heatwave drives massive losses from the world's largest seagrass carbon stocks. Nat Clim Change 8(April):338–344. https://doi.org/10.1038/s41558-018-0096-y

Asch RG, Erisman B (2018) Spawning aggregations act as a bottleneck influencing climate change impacts on a critically endangered reef fish. Divers Distrib 24(112):1712–1728. https://doi.org/10.1111/ddi.12809

Asch RG, Cheung W, Reygondeau G (2018) Future marine ecosystem drivers, biodiversity, and fisheries maximum catch potential in Pacific Island countries and territories under climate change. Mar Policy 88:285–294. https://doi.org/10.1016/j.marpol.2017.08.015

Ashford OS, Kenny AJ, Froján CRS, Bonsall MB, Horton T, Brandt A, Bird GJ, Gerken S, Rogers AD (2018) Phylogenetic and functional evidence suggests that deep-ocean ecosystems are highly sensitive to environmental change and direct human disturbance. Proc R Soc Lond B Biol Sci 285(1884):20180923. https://doi.org/10.1098/rspb.2018.0923

Avery JD, Aagaard K, Burkhalter JC, Robinson OJ (2017) Seabird longline bycatch reduction devices increase target catch while reducing bycatch: a meta-analysis. J Nat Conserv 38(July):37–45. https://doi.org/10.1016/j.jnc.2017.05.004

Badjeck M-C, Allison EH, Halls AS, Dulvy NK (2010) Impacts of climate variability and change on fishery-based livelihoods. Mar Policy 34(3):375–383. https://doi.org/10.1016/j.marpol.2009.08.007

Baker EK, Harris PT (2020) Habitat mapping and marine management. In: Harris PT, Baker EK (eds) Seafloor geomorphology as benthic habitat. Elsevier, Amsterdam, pp 17–33. https://doi.org/10.1016/B978-0-12-814960-7.00002-6

Bakun A (2006) Wasp-waist populations and marine ecosystem dynamics: navigating the 'predator pit' topographies. Prog Oceanogr 68(2–4):271–288. https://doi.org/10.1016/j.pocean.2006.02.004

Balmford A, Gravestock P, Hockley N, McClean CJ, Roberts CM (2004) The worldwide costs of marine protected areas. Proc Natl Acad Sci U S A 101(26):9694–9697. https://doi.org/10.1073/pnas.0403239101

Ban NC, Davies TE, Aguilera SE, Brooks C, Cox M, Epstein G, Evans LS, Maxwell SM, Nenadovic M (2017) Social and ecological effectiveness of large marine protected areas. Glob Environ Chang 43(March):82–91. https://doi.org/10.1016/j.gloenvcha.2017.01.003

Barange M, Merino G, Blanchard JL, Scholtens J, Harle J, Allison EH, Allen JI, Holt J, Jennings S (2014) Impacts of climate change on marine ecosystem production in societies dependent on fisheries. Nat Clim Change 4(March):211–216. https://doi.org/10.1038/nclimate2119

Barbier EB (2017) Marine ecosystem services. Curr Biol 27(11):R507–R510. https://doi.org/10.1016/j.cub.2017.03.020

Bates AE, Cooke RSC, Duncan MI, Edgar GJ, Brunoi JF, Benedetti-Cecchi L, Côté IM et al (2019) Climate resilience in marine protected areas and the 'Protection Paradox'. Biol Conserv 236(August):305–314. https://doi.org/10.1016/j.biocon.2019.05.005

Bax NJ, Appeltans W, Brainard R, Duffy JE, Dunstan P, Hanich Q, Harden Davies H et al (2018) Linking capacity development to GOOS monitoring networks to achieve sustained ocean observation. Front Mar Sci 5(September):346. https://doi.org/10.3389/fmars.2018.00346

Bayraktarov E, Saunders MI, Abdullah S, Mills M, Beher J, Possingham HP, Mumby PJ, Lovelock CE (2016) The cost and feasibility of marine coastal restoration. Ecol Appl 26(4):1055–1074. https://doi.org/10.1890/15-1077

Beaulieu SE, Szafranski K (2018) InterRidge global database of active submarine hydrothermal vent fields. Version 3.4. http://vents-data.interridge.org

Beck MW, Losada IJ, Menéndez P, Reguero BG, Díaz-Simal P, Fernández F (2018) The global flood protection savings provided by coral reefs. Nat Commun 9(June):2186. https://doi.org/10.1038/s41467-018-04568-z

Belhabib D, Sumaila UR, Lam VWY, Zeller D, Le Billon P, Abou Kane E, Pauly D (2015) Euros vs. Yuan: comparing European and Chinese fishing access in West Africa. PloS One 10(3):e0118351. https://doi.org/10.1371/journal.pone.0118351

Bell JB, Guijarro-Garcia E, Kenny A (2019) Demersal fishing in areas beyond national jurisdiction: a comparative analysis of regional fisheries management organisations. Front Mar Sci 6(October):596. https://doi.org/10.3389/fmars.2019.00596

Bennett S, Wernberg T, Connell SD, Hobday AJ, Johnson CR, Poloczanska ES (2016) The 'Great Southern Reef': social, ecological and economic value of Australia's neglected kelp forests. Mar Freshw Res 67(1):47–56. https://doi.org/10.1071/MF15232

Blanchard JL, Watson RA, Fulton EA, Cottrell RS, Nash KL, Bryndum-Buchholz A, Büchner M et al (2017) Linked sustainability challenges and trade- offs among fisheries, aquaculture and agriculture. Nat Ecol Evol 1(September):1240–1249. https://doi.org/10.1038/s41559-017-0258-8

Blasiak R, Wynberg R, Grorud-Colvert K, Thambisetty S, Bandarra N, Canário AVM, da Silva J et al (2020) The ocean genome: conservation and the fair, equitable and sustainable use of marine genetic resources. In: Blue Paper 4, High level panel for a sustainable ocean economy. World Resources Institute, Washington, DC

Block BA, Jonsen ID, Jorgensen S, Winship AJ, Shaffer SA, Bograd SJ, Hazen EL et al (2011) Tracking apex marine predator movements in a dynamic ocean. Nature 475(July):86–90. https://doi.org/10.1038/nature10082

Boerder K, Miller NA, Worm B (2018) Global hot spots of transhipment of fish catch at sea. Sci Adv 4(7):eaat7159. https://doi.org/10.1126/sciadv.aat7159

Bongiorno DL, Bryson M, Bridge TCL, Dansereau DG, Williams SB (2018) Coregistered hyperspectral and stereo image seafloor mapping from an autonomous underwater vehicle. J Field Robot 35(3):312–329. https://doi.org/10.1002/rob.21713

Bonney R, Shirk JL, Phillips TB, Wiggins A, Ballard HL, Miller-Rushing AJ, Parrish JK (2014) Next steps for citizen science. Science 343(6178):1436–1437. https://doi.org/10.1126/science.1251554

Boyes SJ, Elliott M (2014) Marine legislation: the ultimate 'Horrendogram': International Law, European Directives & National Implementation. Mar Pollut Bull 86(1–2):39–47. https://doi.org/10.1016/j.marpolbul.2014.06.055

Bozec YM, O'Farrell S, Bruggemann JH, Luckhurst BE, Mumby PJ (2016) Tradeoffs between fisheries harvest and the resilience of coral reefs. Proc Natl Acad Sci U S A 113(16):4536–4541. https://doi.org/10.1073/pnas.1601529113

Bradley D, Merrifield M, Miller KM, Lomonico S, Wilson JR, Gleason MG (2019) Opportunities to improve fisheries management through innovative technology and advanced data systems. Fish Fish 20(3):564–583. https://doi.org/10.1111/faf.12361

Breitburg D, Levin LA, Oschlies A, Grégoire M, Chavez FP, Conley DJ, Garçon V et al (2018) Declining oxygen in the global ocean and coastal waters. Science 359(6371):eaam7240. https://doi.org/10.1126/science.aam7240

Briggs JC (2007) Marine longitudinal biodiversity: causes and conservation. Divers Distrib 13(5):544–555. https://doi.org/10.1111/j.1472-4642.2007.00362.x

Britten GL, Dowd M, Kanary L, Worm B (2017) Extended fisheries recovery timelines in a changing environment. Nat Commun 8(May):15325. https://doi.org/10.1038/ncomms15325

Bruckner AW, Johnson KA, Field JD (2003) Conservation strategies for sea cucumbers: can a CITES Appendix II listing promote sustainable international trade? SPC Beche-de-mer Inform Bull 18(May):24–33

Burfeind DD, Pitt KA, Connolly RM, Byers JE (2013) Performance of non-native species within marine reserves. Biol Invasions 15:17–28. https://doi.org/10.1007/s10530-012-0265-2

Burgess MG, McDermott GR, Owashi B, Peavey Reeves LE, Clavelle T, Ovando D, Wallace BP et al (2018) Protecting marine mammals, turtles, and birds by rebuilding global fisheries. Science 359(6381):1255–1258. https://doi.org/10.1126/science.aao4248

Burrows MT, Schoeman DS, Buckley LB, Moore P, Poloczanska ES, Brander KM, Brown C et al (2011) The pace of shifting climate in marine and terrestrial ecosystems. Science 334(6056):652–655. https://doi.org/10.1126/science.1210288

Burrows MT, Schoeman DS, Richardson AJ, Molinos JG, Hoffmann A, Buckley LB, Moore PJ et al (2014) Geographical limits to species-range shifts are suggested by climate velocity. Nature 507(March):492–495. https://doi.org/10.1038/nature12976

Burrows MT, Bates AE, Costello MJ, Edwards M, Edgar GJ, Fox CJ, Halpern BS et al (2019) Ocean community warming responses explained by thermal affinities and temperature gradients. Nat Clim Change 9(December):959–963. https://doi.org/10.1038/s41558-019-0631-5

Canonico G, Buttigieg PL, Montes E, Muller-Karger FE, Stepien C, Wright D, Benson A et al (2019) Global observational needs and resources for marine biodiversity. Front Mar Sci 6(July):367. https://doi.org/10.3389/fmars.2019.00367

Carpenter KE, Abrar M, Aeby G, Aronson RB, Banks S, Bruckner A, Chiriboga A et al (2008) One-third of reef-building corals face elevated extinction risk from climate change and local impacts. Science 321(5888):560–563. https://doi.org/10.1126/science.1159196

CBD (Convention on Biological Diversity) (2009) Azores scientific criteria and guidance for identifying ecologically or biologically significant areas and designing representative networks of marine protected areas in open ocean waters and deep sea habitats. CBD Secretariat, Quebec. https://www.cbd.int/marine/doc/azores-brochure-en.pdf

Chaudhary C, Saeedi H, Costello MJ (2016) Bimodality of latitudinal gradients in marine species richness. Trends Ecol Evol 31(9):670–676. https://doi.org/10.1016/j.tree.2016.06.001

Chavez FP, Ryan J, Lluch-Cota SE, Niquen M (2003) From anchovies to sardines and back: multidecadal decadal change in the Pacific. Science 299(5604):217–221. https://doi.org/10.1126/science.1075880

Cheng BS, Altieri AH, Torchin ME, Ruiz GM (2019) Can marine reserves restore lost ecosystem functioning? a global synthesis. Ecology 100(4):e02617. https://doi.org/10.1002/ecy.2617

Cheung WWL, Pitcher TJ, Pauly D (2005) A fuzzy logic expert system to estimate intrinsic extinction vulnerabilities of marine fishes to fishing. Biol Conserv 124(1):97–111. https://doi.org/10.1016/j.biocon.2005.01.017

Cheung WWL, Lam VW, Sarmiento JI, Kearney K, Watson R, Pauly D (2009) Projecting global marine biodiversity impacts under climate change scenarios. Fish Fish 10(3):235–251. https://doi.org/10.1111/j.1467-2979.2008.00315.x

Cheung WWL, Lam VW, Sarmiento JI, Kearney K, Watson R, Zeller D, Pauly D (2010) Large-scale redistribution of maximum fisheries catch potential in the global ocean under climate change. Glob Chang Biol 16(1):24–35. https://doi.org/10.1111/j.1365-2486.2009.01995.x

Cheung WWL, Watson R, Pauly D (2013) Signature of ocean warming in global fisheries catch. Nature 497(May):365–369. https://doi.org/10.1038/nature12156

Christensen V, Coll M, Piroddi C, Steenbeek J, Buszowski J, Pauly D (2014) A century of fish biomass decline in the ocean. Mar Ecol Prog Ser 512(October):155 166. https://doi.org/10.3354/meps10946

Cinner J, McClanahan T, Daw TM, Graham NAJ, Maina J, Wilson SK, Hughes TP (2009) Linking social and ecological systems to sustain coral reef fisheries. Curr Biol 19(3):206–212. https://doi.org/10.1016/j.cub.2008.11.055

Cisneros-Mata MA, Montemayor-López G, Nevárez-Martínez MO (1996) Modeling deterministic effects of age structure, density dependence, environmental forcing, and fishing in the population dynamics of *Sardinops sagax* caeruleus in the Gulf of California. Calif Cooper Ocean Fish Investig Rep 37(October):201–208. https://www.calcofi.org/publications/calcofireports/v37/CalCOFI_Rpt_Vol_37_1996.pdf

Claar DC, Szostek L, McDevitt-Irwin JM, Schanze JJ, Baum JK (2018) Global patterns and impacts of El Niño events on coral reefs: a meta-analysis. PloS One 13(2):e0190957. https://doi.org/10.1371/journal.pone.0190957

Clark NA, Ardron JA, Pendleton LH (2015) Evaluating the basic elements of transparency of regional fisheries management organizations. Mar Policy 57(July):158–166. https://doi.org/10.1016/j.marpol.2015.03.003

Clark MR, Althaus F, Schlacher TA, Williams A, Bowden DA, Rowden AA (2016) The impacts of deep-sea fisheries on benthic communities: a review. ICES J Mar Sci 73(1):51–69. https://doi.org/10.1093/icesjms/fsv123

Clarke A, Crame JA (1997) Diversity, latitude and time: patterns in the Shallow sea. In: Ormond RFG, Gage JD, Angel MV (eds) Marine biodiversity: patterns and processes. Cambridge University Press, Cambridge, pp 122–147

Collins MA, Villanueva R (2006) Taxonomy, ecology and behaviour of the cirrate octopods. In: Gibson RN, Atkinson RJA, Gordon JDM (eds) Oceanography and marine biology: an annual review, vol 44. CRC, Boca Raton, FL, pp 277–222. https://doi.org/10.1201/9781420006391

Colyvan M, Justus J, Regan HM (2010) The natural environment is valuable but not infinitely valuable. Conserv Lett 3(4):224–228. https://doi.org/10.1111/j.1755-263X.2010.00118.x

Constable AJ (2011) Lessons from CCAMLR on the implementation of the ecosystem approach to managing fisheries. Fish Fish 12(2):138–151. https://doi.org/10.1111/j.1467-2979.2011.00410.x

Costanza R, D'Arge R, de Groot R, Farber S, Grasso M, Hannon B, Limburg K et al (1997) The value of the world's ecosystem services and natural capital. Nature 387(May):253–260. https://doi.org/10.1038/387253a0

Costanza R, de Groot R, Sutton P, van der Ploeg S, Anderson SJ, Kubiszewski I, Farber S, Turner RK (2014) Changes in the global value of ecosystem services. Glob Environ Chang 26(May):152–158. https://doi.org/10.1016/j.gloenvcha.2014.04.002

Costello MJ, Coll M, Danovaro R, Halpin P, Ojaveer H, Miloslavich P (2010) A census of marine biodiversity knowledge, resources, and future challenges. PloS One 5(8):e12110. https://doi.org/10.1371/journal.pone.0012110

Costello C, Ovando D, Hilborn R, Gaines SD, Deschenes O, Lester SE (2012) Status and solutions for the world's unassessed fisheries. Science 338(6106):517–520. https://doi.org/10.1126/science.1223389

Costello C, Ovando D, Clavelle T, Strauss CK, Hilborn R, Melnychuk MC, Branch TA et al (2016) Global fishery prospects under contrasting management regimes. Proc Natl Acad Sci U S A 113(18):5125–5129. https://doi.org/10.1073/pnas.1520420113

Costello C, Cao L, Gelcich S, Cisneros MA, Free CM, Froehlich HE, Galarza E et al (2019) The future of food from the sea. World Resources Institute, Washington, DC. www.oceanpanel.org/future-food-sea

Courchamp F, Angulo E, Rivalan P, Hall RJ, Signoret L, Bull L, Meinard Y (2006) Rarity value and species extinction: the anthropogenic allee effect. PLoS Biol 4(12):e415. https://doi.org/10.1371/journal.pbio.0040415

Crame JA, Clarke A (1997) The historical component of marine taxonomic diversity gradients. In: Ormond RFG, Gage JD, Angel MV (eds) Marine biodiversity: patterns and processes. Cambridge University Press, Cambridge, pp 258–273

Crawford B, Kasmidi M, Korompis F, Pollnac RB (2006) Factors influencing progress in establishing community-based marine protected areas in Indonesia. Coast Manag 34(1):39–64. https://doi.org/10.1080/08920750500379300

Cruz-Motta JJ, Miloslavich P, Palomo G, Iken K, Konar B, Pohle G, Trott T et al (2010) Patterns of spatial variation of assemblages associated with intertidal rocky shores: a global perspective. PloS One 5(12):e14354. https://doi.org/10.1371/journal.pone.0014354

Cubaynes HC, Fretwell PT, Bamford C, Gerrish L, Jackson JA (2019) Whales from space: four mysticete species described using new VHR satellite imagery. Mar Mamm Sci 35(2):466–491. https://doi.org/10.1111/mms.12544

Cullis-Suzuki S, Pauly D (2010) Failing the high seas: a global evaluation of regional fisheries management organizations. Mar Policy 34(5):1036–1042. https://doi.org/10.1016/j.marpol.2010.03.002

Cury P, Bakun A, Crawford RJM, Jarre-Teichmann A, Quiñones RA, Shannon LJ, Verheye HM (2000) Small pelagics in upwelling systems: patterns of interaction and structural changes in 'wasp-waist' ecosystems. ICES J Mar Sci 57(3):603–618. https://doi.org/10.1006/jmsc.2000.0712

Cury PM, Boyd IL, Bonhommeau S, Anker-Nilssen T, Crawford RJM, Furness RW, Mills JA et al (2011) Global seabird responses to forage fish depletion—one-third for the birds. Science 334(6063):1703–1706. https://doi.org/10.1126/science.1212928

Danovaro R, Gambi C, D'll'anno A, Corinaldesi C, Fraschetti S, Vanreusel A, Vincx M, Gooday AJ (2008) Exponential decline of deep-sea ecosystem functioning linked to benthic biodiversity loss. Curr Biol 18(1):1–8. https://doi.org/10.1016/j.cub.2007.11.056

Daskalov GM, Grishin AN, Rodionov S, Mihneva V (2007) Trophic cascades triggered by overfishing reveal possible mechanisms of ecosystem regime shifts. Proc Natl Acad Sci U S A 104(25):10518–10523. https://doi.org/10.1073/pnas.0701100104

Davies TD, Baum JK (2012) Extinction risk and overfishing: reconciling conservation and fisheries perspectives on the status of marine fishes. Sci Rep 2(August):561. https://doi.org/10.1038/srep00561

Davies RG, Irlich UM, Chown SL, Gaston KJ (2010) Ambient, productive and wind energy, and ocean extent predict global species richness of procellariiform seabirds. Glob Ecol Biogeogr 19(1):98–110. https://doi.org/10.1111/j.1466-8238.2009.00498.x

de Bruyn P, Murua H, Aranda M (2013) The precautionary approach to fisheries management: how this is taken into account by Tuna Regional Fisheries Management Organisations (RFMOs). Mar Policy 38(March):397–406. https://doi.org/10.1016/j.marpol.2012.06.019

de Groot R, Brander L, van der Ploeg S, Costanza R, Bernard F, Braat L, Christie M et al (2012) Global estimates of the value of ecosystems and their services in monetary units. Ecosyst Serv 1(1):50–61. https://doi.org/10.1016/j.ecoser.2012.07.005

de los Santos CB, Krause-Jensen D, Alcoverro T, Marbà N, Duarte CM, van Katwijk MM, Pérez M et al (2019) Recent trend reversal for declining european seagrass meadows. Nat Commun 10:3356. https://doi.org/10.1038/s41467-019-11340-4

de Vargas C, Audic S, Henry N, Decelle J, Mahé F, Logares R, Lara E et al (2015) Eukaryotic plankton diversity in the sunlit ocean. Science 348(6237):1261605. https://doi.org/10.1126/science.1261605

Di Franco A, Thiriet P, Di Carlo G, Dimitriadis C, Francour P, Gutiérrez NL, de Grissac AJ et al (2016) Five key attributes can increase marine protected areas performance for small-scale fisheries management. Sci Rep 6(December):38135. https://doi.org/10.1038/srep38135

Di Lorenzo M, Claudet J, Guidetti P (2016) Spillover from marine protected areas to adjacent fisheries has an ecological and a fishery component. J Nat Conserv 32(July):62–66. https://doi.org/10.1016/j.jnc.2016.04.004

Dias MP, Martin R, Pearmain EJ, Burfield IJ, Small C, Phillips RA, Yates O et al (2019) Threats to seabirds: a global assessment. Biol Conserv 237(September):525–537. https://doi.org/10.1016/j.biocon.2019.06.033

Diaz-Pulido G, McCook LJ, Dove S, Berkelmans R, Roff G, Kline DI, Weeks S et al (2009) Doom and boom on a resilient reef: climate change, algal overgrowth and coral recovery. PloS One 4(4):e5239. https://doi.org/10.1371/journal.pone.0005239

Diz D, Johnson D, Riddell M, Rees S, Battle J, Gjerde K, Hennige S, Roberts JM (2018) Mainstreaming marine biodiversity into the SDGs: the role of other effective area-based conservation measures (SDG 14.5). Mar Policy 93(July):251–261. https://doi.org/10.1016/j.marpol.2017.08.019

Dornelas M, Gotelli NJ, McGill B, Shimadzu H, Moyes F, Sievers C, Magurran AE (2014) Assemblage time series reveal biodiversity change but not systematic loss. Science 344(6181):296–299. https://doi.org/10.1126/science.1248484

Duarte CM, Losada IJ, Hendriks IE, Mazarrasa I, Marbà N (2013a) The role of coastal plant communities for climate change mitigation and adaptation. Nat Clim Change 3(November):961–968. https://doi.org/10.1038/nclimate1970

Duarte CM, Sintes T, Marbà N (2013b) Assessing the CO2 capture potential of seagrass restoration projects. J Appl Ecol 50(6):1341–1349. https://doi.org/10.1111/1365-2664.12155

Duarte CM, Wu J, Xiao X, Bruhn A, Krause-Jensen D (2017) Can seaweed farming play a role in climate change mitigation and adaptation? Front Mar Sci 4(April):100. https://doi.org/10.3389/fmars.2017.00100

Duarte CM, Agusti S, Barbier E, Britten GL, Castilla JC, Gattuso J-P, Fulweiler RW, Hughes TP et al (2020) Rebuilding marine life. Nature 580:39–51. https://doi.org/10.1038/s41586-020-2146-7

Duffy JE, Lefcheck JS, Stuart-Smith RD, Navarrete SA, Edgar GJ (2016) Biodiversity enhances reef fish biomass and resistance to climate change. Proc Natl Acad Sci U S A 113(22):6230–6235. https://doi.org/10.1073/pnas.1524465113

Dulvy NK, Sadovy Y, Reynolds JD (2003) Extinction vulnerability in marine populations. Fish Fish 4(1):25–64. https://doi.org/10.1046/j.1467-2979.2003.00105.x

Dulvy NK, Fowler SL, Musick JA, Cavanagh RD, Kyne PM, Harrison LR, Carlson JK et al (2014) Extinction risk and conservation of the world's sharks and rays. Elife 3:e00590. https://doi.org/10.7554/eLife.00590

Durden JM, Lallier LE, Murphy K, Jaeckel A, Gjerde K, Jones DOB (2018) Environmental impact assessment process for deep-sea mining in 'the area'. Mar Policy 87(January):194–202. https://doi.org/10.1016/j.marpol.2017.10.013

Dureuil M, Boerder K, Burnett KA, Froese R, Worm B (2018) Elevated trawling inside protected areas undermines conservation outcomes in a global fishing hot spot. Science 362(6421):1403–1407. https://doi.org/10.1126/science.aau0561

Edgar GJ, Langhammer PF, Allen G, Brooks TM, Brodie J, Crosse W, De Silva N et al (2008) Key biodiversity areas as globally significant target sites for the conservation of marine biological diversity. Aquat Conserv Mar Freshw Ecosyst 18(6):969–983. https://doi.org/10.1002/aqc.902

Edgar GJ, Stuart-Smith RD, Willis TJ, Kininmonth S, Baker SC, Banks S, Barrett NS et al (2014) Global conservation outcomes depend on marine protected areas with five key features. Nature 506(February):216–220. https://doi.org/10.1038/nature13022

Edgar GJ, Bates AE, Bird TJ, Jones AH, Kininmonth S, Stuart-Smith RD, Webb TJ (2016) New approaches to marine conservation through scaling up of ecological data. Ann Rev Mar Sci 8(January):435–461. https://doi.org/10.1146/annurev-marine-122414-033921

Edgar GJ, Alexander TJ, Lefcheck JS, Bates AE, Kininmonth SJ, Thomson RJ, Duffy JE, Costello MJ, Stuart Smith RD (2017) Abundance and local-scale processes contribute to multi-phyla gradients in global marine diversity. Sci Adv 3(10):e1700419. https://doi.org/10.1126/sciadv.1700419

Edgar GJ, Ward TJ, Stuart-Smith RD (2019) Weaknesses in stock assessment modelling and management practices affect fisheries sustainability. Aquat Conserv Mar Freshw Ecosyst 29:2010–2016. https://doi.org/10.1002/aqc.3161

Eigaard OR, Bastardie F, Hintzen NT, Buhl-Mortensen L, Buhl-Mortensen P, Catarino R, Dinesen GE et al (2017) The footprint of bottom trawling in european waters: distribution, intensity, and seabed integrity. ICES J Mar Sci 74(3):847–865. https://doi.org/10.1093/icesjms/fsw194

EIU (Economist Intelligence Unit) (2015) The blue economy: growth, opportunity and a sustainable ocean economy. EIU, New York

Elvidge CD, Ghosh T, Baugh K, Zhizhin M, Hsu FC, Katada NS, Penalosa W, Hung BQ (2018) Rating the effectiveness of fishery closures with visible infrared imaging radiometer suite boat detection data. Front Mar Sci 5(April):132. https://doi.org/10.3389/fmars.2018.00132

Ens EJ, Pert P, Clarke PA, Budden M, Clubb L, Doran B, Douras C et al (2015) Indigenous biocultural knowledge in ecosystem science and management: review and insight from Australia. Biol Conserv 181(January):133–149. https://doi.org/10.1016/j.biocon.2014.11.008

Erisman B, Heyman W, Kobara S, Ezer T, Pittman S, Aburto-Oropeza O, Nemeth RS (2015) Fish spawning aggregations: where well-placed management actions yield big benefits for fisheries and conservation. Fish Fish 18(1):121–144. https://doi.org/10.1111/faf.12132

Estes JA, Terborgh J, Brashares JS, Power ME, Berger J, Bond WJ, Carpenter SR et al (2011) Trophic downgrading of planet earth. Science 333(6040):301–306. https://doi.org/10.1126/science.1205106

FAO (2011) FAO technical guidelines for responsible fisheries: fisheries management—4. Marine protected areas and fisheries. No. 4, Suppl. 4. FAO, Rome

FAO (2018) The State of world fisheries and aquaculture 2018—meeting the sustainable development goals. FAO, Rome. http://www.fao.org/3/i9540en/i9540en.pdf

FAO (Food and Agriculture Organization of the United Nations) (2008) The FAO International guidelines for the management of deep-sea fisheries in the high seas. Fisheries and Aquaculture Department, FAO, Rome

Fautin D, Dalton P, Incze LS, Leong JAC, Pautzke C, Rosenberg A, Sandifer P et al (2010) An overview of marine biodiversity in United States waters. PloS One 5(8):e11914. https://doi.org/10.1371/journal.pone.0011914

Fernandes PG, Cook RM (2013) Reversal of fish stock decline in the Northeast Atlantic. Curr Biol 23(15):1432–1437. https://doi.org/10.1016/j.cub.2013.06.016

Fernandes PG, Ralph GM, Nieto A, Criado MG, Vasilakopoulos P, Maravelias CD, Cook RM et al (2017) Coherent assessments of Europe's marine fishes show regional divergence and megafauna loss. Nat Ecol Evol 1(May):0170. https://doi.org/10.1038/s41559-017-0170

Fischer AG (1960) Latitudinal variations in organic diversity. Evolution 14(1):64–81. https://doi.org/10.1111/j.1558-5646.1960.tb03057.x

Fisher R, O'Leary RA, Low-Choy S, Mengersen K, Knowlton N, Brainard RE, Caley MJ (2015) Species richness on coral reefs and the pursuit of convergent global estimates. Curr Biol 25(4):500–505. https://doi.org/10.1016/j.cub.2014.12.022

Foglini F, Grande V, Marchese F, Bracchi VA, Prampolini M, Angeletti L, Castellan G et al (2019) Application of hyperspectral imaging to underwater habitat mapping, Southern Adriatic Sea. Sensors 19(10):2261. https://doi.org/10.3390/s19102261

Folke C, Carpenter S, Walker B, Scheffer M, Elmqvist T, Gunderson L, Holling CS (2004) Regime shifts, resilience and biodiversity in ecosystem management. Annu Rev Ecol Evol Syst 35(December):557–581. https://doi.org/10.1146/annurev.ecolsys.35.021103.105711

Ford AK, Bejarano S, Nugues MM, Visser PM, Albert S, Ferse SCA (2018a) Reefs under siege—the rise, putative drivers, and consequences of benthic cyanobacterial mats. Front Mar Sci 5(February):18. https://doi.org/10.3389/fmars.2018.00018

Ford JH, Bergseth B, Wilcox C (2018b) Chasing the fish oil—do bunker vessels hold the key to fisheries crime networks? Front Mar Sci 5(August):267. https://doi.org/10.3389/fmars.2018.00267

Ford JH, Peel D, Hardesty BD, Rosebrock U, Wilcox C (2018c) Loitering with intent—catching the outlier vessels at sea. PloS One 13(7):e0200189. https://doi.org/10.1371/journal.pone.0200189

Fossheim M, Primicerio R, Johannesen E, Ingvaldsen RB, Aschan MM, Dolgov AV (2015) Recent warming leads to a rapid borealization of fish communities in the Arctic. Nat Clim Change 5(July):673–677. https://doi.org/10.1038/nclimate2647

Frank KT, Petrie B, Choi JS, Leggett WC (2005) Trophic cascades in a formerly cod-dominated ecosystem. Science 308(5728):1621–1623. https://doi.org/10.1126/science.1113075

Freduah G, Fidelman P, Smith TF (2017) The impacts of environmental and socio-economic stressors on small scale fisheries and livelihoods of fishers in Ghana. Appl Geogr 89(December):1–11. https://doi.org/10.1016/j.apgeog.2017.09.009

Freiwald A, Rogers A, Hall-Spencer J, Guinotte JM, Davies AJ, Yesson C, Martin CS, Weatherdon LV (2017) Global distribution of cold-water corals. Version 5.0. Fifth update to the data set in Freiwald et al. (2004). http://data.unep-wcmc.org/datasets/3

Friedman K, Garcia SM, Rice J (2018) Mainstreaming biodiversity in fisheries. Mar Policy 95(September):209–220. https://doi.org/10.1016/j.marpol.2018.03.001

Friess DA, Rogers K, Lovelock CE, Krauss KW, Hamilton SE, Lee SY, Lucas R, Primavera J et al (2019) The State of the World's mangrove forests: past, present, and future. Annu Rev Env Resour 44:89–115. https://doi.org/10.1146/annurev-environ-101718-033302

Gagné TO, Hyrenbach KD, Hagemann ME, Van Houtan KS (2018) Trophic signatures of seabirds suggest shifts in oceanic ecosystems. Sci Adv 4(2):eaao3946. https://doi.org/10.1126/sciadv.aao3946

Gagné TO, Reygondeau G, Jenkins CN, Sexton JO, Bograd SJ, Hazen EL et al (2020) Towards a global understanding of the drivers of marine and terrestrial biodiversity. PloS One 15(2):e0228065. https://doi.org/10.1371/journal.pone.0228065

Gaines SD, Lubchenco J (1982) A unified approach to marine plant-herbivore interactions II. Biogeographic patterns. Annu Rev Ecol Syst 13:111–138. https://www.jstor.org/stable/2097064

Gaines SD, White C, Carr MH, Palumbi SR (2010) Designing marine reserve networks for both conservation and fisheries management. Proc Natl Acad Sci U S A 107:18286–18293. https://doi.org/10.1073/pnas.0906473107

Gamfeldt L, Lefcheck JS, Byrnes JEK, Cardinale BJ, Duffy JE, Griffin JN (2014) Marine biodiversity and ecosystem functioning: what's known and what's next? Oikos 124(3):252–265. https://doi.org/10.1111/oik.01549

Gangloff MM, Edgar GJ, Wilson B (2016) Imperilled species in aquatic ecosystems: emerging threats, management and future prognoses. Aquat Conserv Mar Freshw Ecosyst 26(5):858–871. https://doi.org/10.1002/aqc.2707

Garcia SM (2010) Governance, science, and society: the ecosystem approach to fisheries. In: Grafton RQ, Hilborn R, Squires D, Tait M, Williams M (eds) Handbook of marine fisheries conservation and management. Oxford University Press, Oxford, pp 87–98

Gattuso JP, Magnan A, Billé R, Cheung W, Howes EL, Joos F, Allemand D et al (2015) Contrasting futures for ocean and society from different anthropogenic CO2 emissions scenarios. Science 349(6243):aac4722. https://doi.org/10.1126/science.aac4722

Gedan KB, Silliman BR, Bertness MD (2009) Centuries of human-driven change in salt marsh ecosystems. Ann Rev Mar Sci 1:117–141. https://doi.org/10.1146/annurev.marine.010908.163930

Gell FR, Roberts CM (2003) Benefits beyond boundaries: the fishery effects of marine reserves. Trends Ecol Evol 18(9):448–455. https://doi.org/10.1016/S0169-5347(03)00189-7

Gill DA, Mascia MB, Ahmadia GN, Glew L, Lester SE, Barnes M, Craigie I et al (2017) Capacity shortfalls hinder the performance of marine protected areas globally. Nature 543(March):665–669. https://doi.org/10.1038/nature21708

Gilman E, Kingma E (2013) Standard for assessing transparency in information on compliance with obligations of regional fisheries management organizations: validation through assessment of the Western and Central Pacific Fisheries Commission. Ocean Coast Manag 84(November):31–39. https://doi.org/10.1016/j.ocecoaman.2013.07.006

Gilman E, Passfield K, Nakamura K (2014) Performance of regional fisheries management organizations: ecosystem-based governance of bycatch and discards. Fish Fish 15(2):327–351. https://doi.org/10.1111/faf.12021

Gilman E, Weijerman M, Suuronen P (2017) Ecological data from observer programmes underpin ecosystem-based fisheries management. ICES J Mar Sci 74(6):1481–1495. https://doi.org/10.1093/icesjms/fsx032

Giri C, Ochieng E, Tieszen LL, Zhu Z, Singh A, Loveland T, Masek J, Duke N (2011) Status and distribution of mangrove forests of the world using earth observation satellite data: status and distributions of global mangroves. Glob Ecol Biogeogr 20(1):154–159. https://doi.org/10.1111/j.1466-8238.2010.00584.x

Gjerde KM, Currie D, Wowk K, Sack K (2013) Ocean in Peril: reforming the management of global ocean living resources in areas beyond national jurisdiction. Mar Pollut Bull 74(2):540–551. https://doi.org/10.1016/j.marpolbul.2013.07.037

Gorelick N, Hancher M, Dixon M, Ilyushchenko S, Thau D, Moore R (2017) Google Earth Engine: planetary-scale geospatial analysis for everyone. Remote Sens Environ 202(1):18–27. https://doi.org/10.1016/j.rse.2017.06.031

Grady JM, Maitner BS, Winter AS, Kaschner K, Tittensor DP, Record S, Smith FA et al (2019) Metabolic asymmetry and the global diversity of marine predators. Science 363(6425):eaat4220. https://doi.org/10.1126/science.aat4220

Graham NAJ, Jennings S, MacNeil MA, Mouillot D, Wilson SK (2015) Predicting climate-driven regime shifts versus rebound potential in coral reefs. Nature 518(February):94–97. https://doi.org/10.1038/nature14140

Green EP, Mumby PJ, Edwards AJ, Clark CD (2000) Remote Sensing Handbook Tropical Coastal Management. United Nations Educational, Scientific and Cultural Organization, Paris

Grémillet D, Ponchon A, Paleczny M, Palomares MLD, Karpouzi V, Pauly D (2018) Persisting worldwide seabird-fishery competition despite seabird community decline. Curr Biol 28(24):4009–4013. https://doi.org/10.1016/j.cub.2018.10.051

Guannel G, Arkema K, Ruggiero P, Verutes G (2016) The power of three: coral reefs, seagrasses and mangroves protect coastal regions and increase their resilience. PloS One 11(7):e0158094. https://doi.org/10.1371/journal.pone.0158094

Guillemette M, Grégoire F, Bouillet D, Rail JF, Bolduc F, Caron A, Pelletier D (2018) Breeding failure of seabirds in relation to fish depletion: is there one universal threshold of food abundance? Mar Ecol Prog Ser 587(January):235–245. https://www.int-res.com/articles/meps_oa/m587p235.pdf

Gurney GG, Cinner J, Ban N, Pressey RL, Campbell SJ, Tasidjawa S, Setiwan F (2014) Poverty and protected areas: an evaluation of a marine integrated conservation and development project in Indonesia. Glob Environ Chang 26(May):98–107. https://doi.org/10.1016/j.gloenvcha.2014.04.003

Haas AF, Fairoz MFM, Kelly LW, Nelson CE, Dinsdale EA, Edwards RA, Giles S et al (2016) Global microbialization of coral reefs. Nat Microbiol 1(April):16042. https://doi.org/10.1038/nmicrobiol.2016.42

Haas B, Haward M, McGee J, Fleming A (2019) The influence of performance reviews on regional fisheries management organizations. ICES J Mar Sci 76(7):2082–2089. https://doi.org/10.1093/icesjms/fsz088

Halpern BS (2003) The impact of marine reserves: do reserves work and does reserve size matter? Ecol Appl 13(Sp1):S117–S137. https://doi.org/10.1890/1051-0761(2003)013[0117:TIOMRD]2.0.CO;2

Halpern BS, Walbridge S, Selkoe KA, Kappel CV, Micheli F, D'Agrosa C, Bruno JF et al (2008) A global map of human impact on marine ecosystems. Science 319(5865):948–952. https://doi.org/10.1126/science.1149345

Halpern BS, Lester SE, Kellner JB (2010) Spillover from marine reserves and the replenishment of fished stocks. Environ Conserv 36(4):268–276. https://doi.org/10.1017/S0376892910000032

Halpern BS, Frazier M, Potapenko J, Casey KS, Koenig K, Longo C, Lowndes JS et al (2015) Spatial and temporal changes in cumulative human impacts on the world's ocean. Nat Commun 6(July):7615. https://doi.org/10.1038/ncomms8615

Halpern BS, Frazier M, Afflerbach J, Lowndes JS, Micheli F, O'Hara C, Scarborough C, Selkoe KA (2019) Recent pace of change in human impact on the world's ocean. Sci Rep 9(August):11609. https://doi.org/10.1038/s41598-019-47201-9

Hansen RE (2011) Introduction to synthetic aperture sonar. In: Kolev NZ (ed) Sonar systems. InTech, Rijeka, Croatia, pp 3–28. https://doi.org/10.5772/23122

Hare SR, Mantua NJ (2000) Empirical evidence for North Pacific Regime Shifts in 1977 and 1989. Prog Oceanogr 47(2–4):103–145. https://doi.org/10.1016/S0079-6611(00)00033-1

Harris PT (2020) Anthropogenic threats to benthic habitats. In: Harris PT, Baker EK (eds) Seafloor geomorphology as benthic habitat: Geohab atlas of seafloor geomorphic features and benthic habitats. Elsevier, London, pp 35–60. https://doi.org/10.1016/B978-0-12-385140-6.00003-7

Harris PT, Baker EK (2012) Why Map Benthic Habitats? In: Harris PT, Baker EK (eds) Seafloor geomorphology as benthic habitat: Geohab atlas of seafloor geomorphic features and benthic habitats. Elsevier, London, pp 3–22. https://doi.org/10.1016/B978-0-12-385140-6.00001-3

Harris PT, Macmillan-Lawler M, Rupp J, Baker EK (2014) Geomorphology of the oceans. Mar Geol 352(1):4–24. https://doi.org/10.1016/j.margeo.2014.01.011

Harrison PA, Berry PM, Simpson G, Haslett JR, Blicharska M, Bucur M, Dunford R et al (2014) Linkages between biodiversity attributes and ecosystem services: a systematic review. Ecosyst Serv 9(September):191–203. https://doi.org/10.1016/j.ecoser.2014.05.006

Hector A, Bagchi R (2007) Biodiversity and ecosystem multifunctionality. Nature 448(7150):188–190. https://doi.org/10.1038/nature05947

Heery EC, Bishop MJ, Critchley LP, Bugnot AB, Airoldi L, Mayer-Pinto M, Sheehan EV et al (2017) Identifying the consequences of ocean sprawl for sedimentary habitats. J Exp Mar Biol Ecol 492(July):31–48. https://doi.org/10.1016/j.jembe.2017.01.020

Heino M, Díaz Pauli B, Dieckmann U (2015) Fisheries-induced evolution. Annu Rev Ecol Evol Syst 46(December):461–480. https://doi.org/10.1146/annurev-ecolsys-112414-054339

Hilborn R (2018) Are MPAs effective? ICES J Mar Sci 75(3):1160–1162. https://doi.org/10.1093/icesjms/fsx068

Hilborn R, Ovando D (2014) Reflections on the success of traditional fisheries management. ICES J Mar Sci 71(5):1040–1046. https://doi.org/10.1093/icesjms/fsu034

Hilborn R, Amoroso RO, Anderson CM, Baum JK, Branch TA, Costello C, de Moor CL et al (2020) Effective fisheries management instrumental in improving fish stock status. Proc Natl Acad Sci U S A 117(4):2218–2224. https://doi.org/10.1073/pnas.1909726116

Hobday AJ, Oliver ECJ, Sen Gupta A, Benthuysen JA, Burrows MT, Donat MG, Holbrook NJ et al (2018) Categorizing and naming marine heatwaves. Oceanography 31(2):162–173. https://doi.org/10.5670/oceanog.2018.205

Hochard JP, Hamilton S, Barbier EB (2019) Mangroves shelter coastal economic activity from cyclones. Proc Natl Acad Sci U S A 116(25):12232–12237. https://doi.org/10.1073/pnas.1820067116

Hoegh-Guldberg O, Mumby PJ, Hooten AJ, Steneck RS, Greenfield P et al (2007) Coral reefs under rapid climate change and ocean acidification. Science 318(5857):1737–1742. https://doi.org/10.1126/science.1152509

Hoegh-Guldberg O, Caldeira K, Chopin T, Gaines S, Haugan P, Hemer M, Howard J et al (2019) The ocean as a solution to climate change: five opportunities for action. World Resources Institute, Washington, DC. http://www.oceanpanel.org/climate

Hooff RC, Peterson WT (2006) Copepod biodiversity as an indicator of changes in ocean and climate conditions of the Northern California current ecosystem. Limnol Oceanogr 51(6):2607–2620. https://doi.org/10.4319/lo.2006.51.6.2607

Hooper T, Börger T, Langmead O, Marcone O, Rees SE, Rendon O, Beaumont N, Attrill MJ, Austen M (2019) Applying the natural capital approach to decision making for the marine environment. Ecosyst Serv 38(August):100947. https://doi.org/10.1016/j.ecoser.2019.100947

Hsieh CH, Reiss CS, Hunter JR, Beddington JR, May RM, Sugihara G (2006) Fishing elevates variability in the abundance of exploited species. Nature 443(October):859–862. https://doi.org/10.1038/nature05232

Hughes RG (2004) Climate change and loss of saltmarshes: consequences for birds. Ibis 146(s1):21–28. https://doi.org/10.1111/j.1474-919X.2004.00324.x

Hughes TP, Rodrigues MJ, Bellwood DR, Ceccarelli D, Hoegh-Guldberg O, McCook L, Moltschaniwsky N et al (2007) Phase shifts, herbivory, and the resilience of coral reefs to climate change. Curr Biol 17(4):360–365. https://doi.org/10.1016/j.cub.2006.12.049

Hughes TP, Barnes ML, Bellwood DR, Cinner JE, Cumming GS, Jackson JBC, Kleypas J et al (2017) Coral reefs in the anthropocene. Nature 546(June):82–90. https://doi.org/10.1038/nature22901

Hughes TP, Anderson KD, Connolly SR, Heron SF, Kerry JT, Lough JM, Baird AH et al (2018a) Spatial and temporal patterns of mass bleaching of corals in the anthropocene. Science 359(6371):80–83. https://doi.org/10.1126/science.aan8048

Hughes TP, Kerry JT, Baird AH, Connolly SR, Dietzel A, Eakin CM, Heron SF et al (2018b) Global warming transforms coral reef assemblages. Nature 556(April):492–496. https://doi.org/10.1038/s41586-018-0041-2

Humphries GRW, Velarde E, Anderson DW, Haasey B, Sydeman WJ (2015) Seabirds as early warning indicators of climate events in the North Pacific. PICES Press 23(1):18–20. http://aquaticcommons.org/21569/1/PPJanuary2015.pdf

Iken K, Konar B (2003) Natural Geography in Nearshore Areas (NaGISA): the nearshore component of the census of marine life. Gayana 67(2):153–160. https://doi.org/10.4067/S0717-65382003000200004

IPBES (Intergovernmental Science-Policy Platform on Biodiversity and Ecosystem Services) (2019) In: Díaz S, Brondízio JSES, Ngo HT, Guèze M, Agard J, Arneth A et al (eds) Summary for policymakers of the global assessment report on biodiversity and ecosystem services of the Intergovernmental Science-Policy Platform on Biodiversity and Ecosystem Services. IPBES Secretariat, Bonn. https://ipbes.net/sites/default/files/2020-02/ipbes_global_assessment_report_summary_for_policymakers_en.pdf

IPCC (2019) In: Pörtner HO, Roberts DC, Masson-Delmotte V, Zhai P, Tignor M, Poloczanska E, Mintenbeck K et al (eds) IPCC special report on the ocean and cryosphere in a changing climate. IPCC, Geneva

IPCC (Intergovernmental Panel on Climate Change) (2018) Global warming of 1.5°C. In: Masson-Delmotte V, Zhai P, Pörtner HO, Roberts D, Skea J, Shukla PR, Pirani A et al (eds) An IPCC special report on the impacts of global warming of 1.5°C above pre-industrial levels and related global greenhouse gas emission pathways, in the context of strengthening the global response to the threat of climate change, sustainable development, and efforts to eradicate poverty. IPCC, Geneva

IUCN (International Union for Conservation of Nature) (2016) Increasing marine protected area coverage for effective marine biodiversity conservation. In: Prepared for the World Conservation Congress 2016, Resolution 050, Hawaii, September 1–10. https://portals.iucn.org/library/sites/library/files/resrecfiles/WCC_2016_RES_050_EN.pdf

IUCN (2017) Guidelines for using the IUCN Red List categories and criteria. Version 13. IUCN, Standards and Petitions Subcommittee, Gland

IUCN (2019) The IUCN Red List of threatened species. IUCN, Gland. https://www.iucnredlist.org

IUCN-WCPA (World Commission on Protected Areas) (2019) Recognising and reporting other effective area-based conservation measures. IUCN, Gland. https://portals.iucn.org/library/sites/library/files/documents/PATRS-003-En.pdf

Jablonski D, Roy K, Valentine JW (2006) Out of the tropics: evolutionary dynamics of the latitudinal diversity gradient. Science 314(5796):102–106. https://doi.org/10.1126/science.1130880

Jablonski D, Belanger CL, Berke SK, Huang S, Krug AZ, Roy K, Tomasovych A, Valentine JW (2013) Out of the tropics, but how? Fossils, bridge species, and thermal ranges in the dynamics of the marine latitudinal diversity gradient. Proc Natl Acad Sci U S A 110(26):10487–10494. https://doi.org/10.1073/pnas.1308997110

Jackson JBC, Kirby MX, Berger WH, Bjorndal KA, Botsford LW, Bourque BJ, Bradbury RH et al (2001) Historical overfishing and the recent collapse of coastal ecosystems. Science 293(5530):629–638. https://doi.org/10.1126/science.1059199

Jaramillo-Legorreta AM, Cardenas-Hinojosa G, Nieto-Garcia E, Rojas-Bracho L, Thomas L, Ver Hoef JM, Moore J et al (2019) Decline towards extinction of Mexico's Vaquita Porpoise (*Phocoena sinus*). R Soc Open Sci 6(7):190598. https://doi.org/10.1098/rsos.190598

Jenkins CN, Van Houten KS (2016) Global and regional priorities for marine biodiversity protection. Biol Conserv 204(part B (December)):333–339

Jones KR, Klein CJ, Halpern BS, Venter O, Grantham H, Kuempel CD, Shumway N et al (2018) The location and protection status of earth's diminishing marine wilderness. Curr Biol 28(15):2506–2512. https://doi.org/10.1016/j.cub.2018.06.010

Jones KR, Klein CJ, Grantham HS, Possingham HP, Halpern BS, Burgess ND, Butchart SHM, Robinson JG et al (2020) Area requirements to safeguard earth's marine species. One Earth 2(2):188–196

Kaiser MJ (2005) Are marine protected areas a red herring or fisheries panacea? Can J Fish Aquat Sci 62(5):1194–1199. https://doi.org/10.1139/f05-056

Keith SA, Kerswell AP, Connolly SR (2014) Global diversity of marine macroalgae: environmental conditions explain less variation in the tropics. Glob Ecol Biogeogr 23(5):517–529. https://doi.org/10.1111/geb.12132

Kerswell AP (2006) Global biodiversity patterns of benthic marine algae. Ecology 87(10):2479–2488. https://doi.org/10.1890/0012-9658(2006)87[2479:gbpobm]2.0.co;2

Klein CJ, Brown CJ, Halpern BS, Segan DB, McGowan J, Beger M, Watson JEM (2015) Shortfalls in the global protected area network at representing marine biodiversity. Sci Rep 5:17539. https://doi.org/10.1038/srep17539

Kockel A, Ban NC, Costa M, Dearden P (2019) Evaluating approaches for scaling up community-based marine protected areas into socially equitable and ecologically representative networks. Conserv Biol 34(1):137–147. https://doi.org/10.1111/cobi.13368

Krause-Jensen D, Duarte CM (2016) Substantial role of macroalgae in marine carbon sequestration. Nat Geosci 9(October):737–742. https://doi.org/10.1038/ngeo2790

Krause-Jensen D, Lavery P, Serrano O, Marbà N, Masque P, Duarte CM (2018) Sequestration of macroalgal carbon: the elephant in the blue carbon room. Biol Lett 14(6):20180236. https://doi.org/10.1098/rsbl.2018.0236

Kroodsma DA, Mayorga J, Hochberg T, Miller NA, Boerder K, Ferretti F, Wilson A et al (2018) Tracking the global footprint of fisheries. Science 359(6378):904–908. https://doi.org/10.1126/science.aao5646

Krumhansl KA, Okamoto DK, Rassweiler A, Novak M, Bolton JJ, Cavanaugh KC, Connell SD et al (2016) Global patterns of kelp forest change over the past half-century. Proc Natl Acad Sci U S A 113(48):13785–13790. https://doi.org/10.1073/pnas.1606102113

Kuempel CD, Jones KR, Watson JEM, Possingham HP (2019) Quantifying biases in marine-protected-area placement relative to abatable threats. Conserv Biol 33(6):1350–1359. https://doi.org/10.1111/cobi.13340

Kuparinen A, Festa-Bianchet M (2017) Harvest-induced evolution: insights from aquatic and terrestrial systems. Philos Trans R Soc B 372(1712):20160036. https://doi.org/10.1098/rstb.2016.0036

Lamarche G, Orpin AR, Mitchell JS, Pallentin A (2016) Benthic habitat mapping. In: Clark MR, Consalvey M, Rowden AA (eds) Biological sampling in the deep sea. Blackwell, Chichester, pp 80–102. https://doi.org/10.1002/9781118332535.ch5

Langle-Flores A, Ocelík P, Pérez-Maqueo O (2017) "The role of social networks in the sustainability transformation of cabo pulmo: a multiplex perspective," in "Coastal resilience: exploring the many challenges from different viewpoints," edited by M.L. Martinez, A. Taramelli, and R. Silva, special issue. J Coast Res 77(77):134–142. https://doi.org/10.2112/SI77-014.1

Lefcheck JS, Byrnes JEK, Isbell F, Gamfeldt L, Griffin JN, Eisenhauer N, Hensel MJS et al (2015) Biodiversity enhances ecosystem multifunctionality across trophic levels and habitats. Nat Commun 6(April):6936. https://doi.org/10.1038/ncomms7936

Lenton TM (2013) Environmental tipping points. Annu Rev Env Resour 38(October):1–29. https://doi.org/10.1146/annurev-environ-102511-084654

Lenton TM, Rockström J, Gaffney O, Rahmstorf S, Richardson K, Steffen W, Joachim Schellnhuber H (2019) Climate tipping points—too risky to bet against. Nature 575(7784):592–595. https://doi.org/10.1038/d41586-019-03595-0

Leroy A, Morin M (2018) Innovation in the decision-making process of the RFMOs. Mar Policy 97(November):156–162. https://doi.org/10.1016/j.marpol.2018.05.025

Lester SE, Halpern BS (2008) Biological responses in marine no-take reserves versus partially protected areas. Mar Ecol Prog Ser 367(September):49–56. https://doi.org/10.3354/meps07599

Lester SE, Halpern BS, Grorud-Colvert K, Lubchenco J, Ruttenberg BI, Gaines SD, Airamé S, Warner RR (2009) Biological effects within no-take marine reserves: a global synthesis. Mar Ecol Prog Ser 384(May):33–46. https://doi.org/10.3354/meps08029

Lidström S, Johnson AF (2019) Ecosystem-based fisheries management: a perspective on the critique and development of the concept. Fish Fish 21(1):216–222. https://doi.org/10.1111/faf.12418

Lindstrom E, Gunn J, Fischer A, McCurdy A, Glover LK (2012) A Framework for Ocean Observing. Intergovernmental Oceanographic Commission Information Document 1284. United Nations Educational, Scientific and Cultural Organization, Paris. https://unesdoc.unesco.org/ark:/48223/pf0000211260

Lipcius RN, Crowder LB, Morgan LE (2005) Metapopulation structure and marine reserves. In: Norse EA, Crowder LB (eds) Marine conservation biology: the science of maintaining the sea's biodiversity. Island, Washington, DC, pp 328–345

Liu L, Xu W, Yue Q, Teng X, Hu H (2018) Problems and countermeasures of coastline protection and utilization in China. Ocean Coast Manag 153(March):124–130. https://doi.org/10.1016/j.ocecoaman.2017.12.016

Loke LHL, Heery EC, Todd PA (2019) Shoreline defences. In: Sheppard C (ed) World seas: an environmental evaluation. Ecological issues and environmental impacts, vol 3, 2nd edn. Academic Press, London, pp 491–504

Longépé N, Hajduch G, Ardianto R, de Joux R, Nhunfat B, Marzuki MI, Fablet R et al (2018) Completing fishing monitoring with spaceborne vessel detection system (VDS) and automatic identification system (AIS) to assess illegal fishing in Indonesia. Mar Pollut Bull 131(B):33–39. https://doi.org/10.1016/j.marpolbul.2017.10.016

Lotze HK, Tittensor DP, Bryndum-Buchholz A, Eddy TD, Cheung WWL, Galbraith ED, Barange M et al (2019) Global ensemble projections reveal trophic amplification of ocean biomass declines with climate change. Proc Natl Acad Sci U S A 116(26):12907–12912. https://doi.org/10.1073/pnas.1900194116

Lough JM, Anderson KD, Hughes TP (2018) Increasing thermal stress for tropical coral reefs: 1871–2017. Sci Rep 8(April):6079. https://doi.org/10.1038/s41598-018-24530-9

Lovelock CE, Cahoon DR, Friess DA, Guntenspergen GR, Krauss KW, Reef R, Rogers K et al (2015) The vulnerability of Indo-Pacific mangrove forests to sea-level rise. Nature 526(October):559–563. https://doi.org/10.1038/nature15538

Lucifora LO, García VB, Worm B (2011) Global diversity hot spots and conservation priorities for sharks. PloS One 6(5):e19356. https://doi.org/10.1371/journal.pone.0019356

MA (Millennium Ecosystem Assessment) (2005) Ecosystems and human well-being: synthesis. Island, Washington, DC

Mace GM, Hails RS, Cryle P, Harlow J, Clarke SJ (2015) Towards a risk register for natural capital. J Appl Ecol 52(3):641–653. https://doi.org/10.1111/1365-2664.12431

MacNeil MA, Graham NAJ, Cinner JE, Wilson SK, Williams ID, Maina J, Newman S et al (2015) Recovery potential of the world's coral reef fishes. Nature 520(April):341–344. https://doi.org/10.1038/nature14358

MacNeil MA, Mellin C, Matthews S, Wolff NH, McClanahan TR, Devlin M, Drovandi C, Mengersen K, Graham NAJ (2019) Water quality mediates resilience on the great barrier reef. Nat Ecol Evol 3(April):620–627. https://doi.org/10.1038/s41559-019-0832-3

Margules CR, Pressey RL (2000) Systematic conservation planning. Nature 405:243–253. https://doi.org/10.1038/35012251

Martin SL, Ince L, Grooves T (2016) An ecosystem services perspective for the oceanic eastern tropical pacific: commercial fisheries, carbon storage, recreational fishing, and biodiversity. Front Mar Sci 3(April):50. https://doi.org/10.3389/fmars.2016.00050

McCauley DJ, Pinsky ML, Palumbi SR, Estes JA, Joyce FH, Warner RR (2015) Marine defaunation: animal loss in the global ocean. Science 347(6219):1255641. https://doi.org/10.1126/science.1255641

McClanahan TR, Cinner JE, Maina J, Graham NAJ, Daw TM, Stead SM, Wamukota A et al (2008) Conservation action in a changing climate. Conserv Lett 1(2):53–59. https://doi.org/10.1111/j.1755-263X.2008.00008_1.x

McGowan JA, Cayan DR, Dorman LM (1998) Climate-ocean variability and ecosystem response in the Northeast Pacific. Science 281(5374):210–217. https://doi.org/10.1126/science.281.5374.210

McKinley DC, Miller-Rushing AJ, Ballard HL, Bonney R, Brown H, Cook-Patton SC, Evans DM et al (2017) Citizen science can improve conservation science, natural resource management, and environmental protection. Biol Conserv 208(April):15–28. https://doi.org/10.1016/j.biocon.2016.05.015

Mcleod E, Chmura GL, Bouillon S, Salm R, Björk M, Duarte CM, Lovelock CE, Schlesinger WH, Silliman BR (2011) A blueprint for blue carbon: toward an improved understanding of the role of vegetated coastal habitats in sequestering CO2. Front Ecol Environ 9(10):552–560. https://doi.org/10.1890/110004

McOwen C, Weatherdon LV, Bochove J, Sullivan E, Blyth S, Zockler C, Stanwell-Smith D et al (2017) A global map of saltmarshes. Biodivers Data J 5(March):e11764. https://doi.org/10.3897/BDJ.5.e11764

Mellin C, MacNeil MA, Cheal AJ, Emslie MJ, Caley MJ (2016) Marine protected areas increase resilience among coral reef communities. Ecol Lett 19(6):629–637. https://doi.org/10.1111/ele.12598

Merrie A, Dunn DC, Metian M, Boustany AM, Takei Y, Elferink AO, Ota Y et al (2014) An ocean of surprises—trends in human use, unexpected dynamics and governance challenges in areas beyond national jurisdiction. Glob Environ Chang 27(July):19–31. https://doi.org/10.1016/j.gloenvcha.2014.04.012

Micheli F, Halpern BS, Botsford LW, Warner RR (2004) Trajectories and correlates of community change in no-take marine reserves. Ecol Appl 14(6):1709–1723. https://doi.org/10.1890/03-5260

Micheli F, Saenz-Arroyo A, Greenley A, Vazquez L, Espinoza Montes JA, Rossetto M, De Leo GA (2012) Evidence that marine reserves enhance resilience to climatic impacts. PloS One 7(7):e40832. https://doi.org/10.1371/journal.pone.0040832

Miller KA, Thompson KF, Johnston P, Santillo D (2018) An overview of seabed mining including the current state of development, environmental impacts, and knowledge gaps. Front Mar Sci 4(January):418. https://doi.org/10.3389/fmars.2017.00418

Miloslavich P, Bax NJ, Simmons SE, Klein E, Appeltans W, Aburto-Oropeza O, Shin YJ (2018a) Essential ocean variables for global sustained observations of biodiversity and ecosystem changes. Glob Chang Biol 24(6):2416–2433. https://doi.org/10.1111/gcb.14108

Miloslavich P, Seeyave S, Muller-Karger F, Bax N, Ali E, Delgado C, Evers-King H et al (2018b) Challenges for global ocean observation: the need for increased human capacity. J Oper Oceanogr 12(S2):137–156. https://doi.org/10.1080/1755876X.2018.1526463

Molinos JG, Halpern BS, Schoeman DS, Brown CJ, Kiessling W, Moore PJ, Pandolfi JM et al (2016) Climate velocity and the future global redistribution of marine biodiversity. Nat Clim Change 6(January):83–87. https://doi.org/10.1038/nclimate2769

Molnar JL, Gamboa RL, Revenga C, Spalding MD (2008) Assessing the global threat of invasive species to marine biodiversity. Front Ecol Environ 6(9):485–492

Mora C, Myers RA, Coll M, Libralato S, Pitcher TJ, Sumaila RU, Zeller D et al (2009) Management effectiveness of the world's marine fisheries. PLoS Biol 7(6):e1000131. https://doi.org/10.1371/journal.pbio.1000131

Mora C, Tittensor DP, Adl S, Simpson AGB, Worm B (2011) How many species are there on earth and in the ocean? PLoS Biol 9(8):e1001127. https://doi.org/10.1371/journal.pbio.1001127

Mora C, Spirandelli D, Franklin EC, Lynham J, Kantar MB, Miles W, Smith CZ et al (2018) Broad threat to humanity from cumulative climate hazards intensified by greenhouse gas emissions. Nat Clim Change 8(December):1062–1071. https://doi.org/10.1038/s41558-018-0315-6

Muller-Karger FE, Miloslavich P, Bax NJ, Simmons S, Costello MJ, Sousa Pinto I, Canonico G et al (2018) Advancing marine biological observations and data requirements of the complementary essential ocean variables (EOVs) and essential biodiversity variables (EBVs) frameworks. Front Mar Sci 5(June):211. https://doi.org/10.3389/fmars.2018.00211

Mumby PJ, Harborne AR (2010) Marine reserves enhance the recovery of corals on Caribbean reefs. PloS One 5(1):e8657. https://doi.org/10.1371/journal.pone.0008657

Mumby PJ, Green EP, Edwards AJ, Clark CD (1999) The cost-effectiveness of remote sensing for tropical coastal resources assessment and management. J Environ Manage 55(3):157–166. https://doi.org/10.1006/jema.1998.0255

Mumby PJ, Hedley JD, Chisholm JRM, Clark CD, Ripley H, Jaubert J (2004) The cover of living and dead corals from airborne remote sensing. Coral Reefs 23(July):171–183. https://doi.org/10.1007/s00338-004-0382-1

Mumby PJ, Harborne AR, Williams J, Kappel CV, Brumbaugh DR, Micheli F, Holmes KE et al (2007) Trophic cascade facilitates coral recruitment in a marine reserve. Proc Natl Acad Sci U S A 104(20):8362–8367. https://doi.org/10.1073/pnas.0702602104

Naeem S (2002) Ecosystem consequences of biodiversity loss: evolution of a paradigm. Ecology 83(6):1537–1552. https://doi.org/10.1890/0012-9658(2002)083[1537:ECOBLT]2.0.CO;2

Narayan S, Beck MW, Wilson P, Thomas CJ, Guerrero A, Shepard CC, Reguero BG et al (2017) The value of coastal wetlands for flood damage reduction in the Northeastern USA. Sci Rep 7(August):9463. https://doi.org/10.1038/s41598-017-09269-z

Navarro LM, Fernández N, Guerra C, Guralnick R, Kissling WD, Londoño MC, Muller-Karger F et al (2017) Monitoring biodiversity change through effective global coordination. Curr Opin Environ Sustain 29(December):158–169. https://doi.org/10.1016/j.cosust.2018.02.005

NEAFC/OSPAR (2014) Collective arrangement between competent International Organisations on Cooperation and Coordination regarding selected areas in areas beyond National Jurisdiction in the North-East Atlantic. NEAFC, London

NEAFC/OSPAR (North-East Atlantic Fisheries Commission/Oslo Paris Commission) (2008) Memorandum of Understanding between the North East Atlantic Fisheries Commission (NEAFC) and the OSPAR Commission. NEAFC, London. https://www.ospar.org/site/assets/files/1357/mou_neafc_ospar.pdf

Nellemann C, Corcoran E, Duarte CM, Valdés L, De Young C, Fonseca L, Grimsditch G (2009) Blue carbon: the role of healthy oceans in binding carbon. United Nations Environmental Programme, GRID-Arendal, Arendal, Norway. https://cld.bz/bookdata/WK8FNPt/basic-html/index.html#1

Nijland W, Reshitnyk L, Rubidge E (2019) Satellite remote sensing of canopy-forming kelp on a complex coastline: a novel procedure using the landsat image archive. Rem Sens Environ 220(January):41–50. https://doi.org/10.1016/j.rse.2018.10.032

Niner HJ, Ardron JA, Escobar EG, Gianni M, Jaeckel A, Jones DOB, Levin LA et al (2018) Deep-sea mining with no net loss of biodiversity—an impossible aim. Front Mar Sci 5(March):53. https://doi.org/10.3389/fmars.2018.00053

Norse EA (1993) Global marine biological diversity: a strategy for building conservation into decision making. Island, Washington, DC

Norse EA (2005) Ending the range wars on the last frontier: zoning the sea. In: Norse EA, Crowder LB (eds) Marine conservation biology: the science of maintaining the sea's biodiversity. Island, Washington, DC, pp 422–443

Norse EA, Brooke S, Cheung WWL, Clark MR, Ekeland I, Froese R, Gjerde KM et al (2012) Sustainability of deep-sea fisheries. Mar Policy 36(2):307–320. https://doi.org/10.1016/j.marpol.2011.06.008

O'Leary BC, Brown RL, Johnson DE, von Nordheim H, Ardron J, Packeiser T, Roberts CM (2012) The first network of marine protected areas (mpas) in the high seas: the process, the challenges and where next. Mar Policy 36(3):598–605. https://doi.org/10.1016/j.marpol.2011.11.003

O'Leary BC, Winther-Janson M, Bainbridge JM, Aitken J, Hawkins JP, Roberts CM (2016) Effective coverage targets for ocean protection. Conserv Lett 9(6):398–404. https://doi.org/10.1111/conl.12247

O'Leary BC, Allen HL, Yates KL, Page RW, Tudhope AW, McClean C, Rogers AD, Hawkins JP, Roberts CM (2018) 30 x 30: a blue print for ocean protection—how we can protect 30% of our oceans by 2030. Greenpeace, London. https://storage.googleapis.com/planet4-new-zealand-stateless/2019/04/f136cbfb-5db0f88b-greenpeace-30x30-blueprint-report1.pdf

Oziel L, Neukermans G, Ardyna M, Lancelot C, Tison JL, Wassmann P, Sirven J, Ruiz-Pino D, Gascard JC (2017) Role for Atlantic inflows and sea ice loss on shifting phytoplankton blooms in the Barents sea. J Geophys Res Oceans 122(6):5121–5139. https://doi.org/10.1002/2016JC012582

Paleczny M, Hammill E, Karpouzi V, Pauly D (2015) Population trend of the world's monitored seabirds, 1950-2010. PloS One 10(6):e0129342. https://doi.org/10.1371/journal.pone.0129342

Palmer MW, Earls PG, Hoagland BW, White PS, Wohlgemuth T (2002) Quantitative tools for perfecting species lists. Environmetrics 13(2):121–137. https://doi.org/10.1002/env.516

Pandolfi JM, Bradbury RH, Sala E, Hughes TP, Bjorndal KA, Cooke RG, McArdle D et al (2003) Global trajectories of the long-term decline of coral reef ecosystems. Science 301(5635):955–958. https://doi.org/10.1126/science.1085706

Pappalardo P, Fernández M (2014) Mode of larval development as a key factor to explain contrasting effects of temperature on species richness across oceans. Glob Ecol Biogeogr 23(1):12–23. https://doi.org/10.1111/geb.12115

Parrish JK, Burgess H, Weltzin JF, Fortson L, Wiggins A, Simmons B (2018) Exposing the science in citizen science: fitness to purpose and intentional design. Integr Comparat Biol 58(1):150–160. https://doi.org/10.1093/icb/icy032

Pauly D, Zeller D (2016) Catch reconstructions reveal that global marine fisheries catches are higher than reported and declining. Nat Commun 7(January):10244. https://doi.org/10.1038/ncomms10244

Pendleton LH, Thébaud O, Mongruel RC, Levrel H (2016) Has the value of global marine and coastal ecosystem services changed? Mar Policy 64(February):156–158. https://doi.org/10.1016/j.marpol.2015.11.018

Pentz B, Klenk N, Ogle S, Fisher JAD (2018) Can regional fisheries management organizations (RFMOs) manage resources effectively during climate change? Mar Policy 92(June):13–20. https://doi.org/10.1016/j.marpol.2018.01.011

Pérez-Matus A, Carrasco S, Gelcich S, Fernandez M, Wieters E (2017) Exploring the effects of fishing pressure and upwelling intensity over subtidal kelp forest communities in Central Chile. Ecosphere 8(5):e01808. https://doi.org/10.1002/ecs2.1808

Peters H, O'Leary BC, Hawkins JP, Carpenter KE, Roberts CM (2013) Conus: first comprehensive conservation red list assessment of a marine gastropod mollusc genus. PloS One 8(12):e83353. https://doi.org/10.1371/journal.pone.0083353

Petersen S, Krätschell A, Augustin N, Jamieson J, Hein JR, Hannington MD (2016) News from the seabed—geological characteristics and resource potential of deep-sea mineral resources. Mar Policy 70(August):175–187. https://doi.org/10.1016/j.marpol.2016.03.012

Petrossian GA, Deby RA, Clarke RV (2018) Illegal longline fishing and albatross extinction risk. Oryx 52(2):336–345. https://doi.org/10.1017/S0030605316000818

Pham CK, Ramirez-Llodra E, Alt CHS, Amaro T, Bergmann M, Canals M, Company JB et al (2014) Marine litter distribution and density in European seas, from the shelves to deep basins. PloS One 9(4):e95839. https://doi.org/10.1371/journal.pone.0095839

Phillips RA, Gales R, Baker GB, Double MC, Favero M, Quintana F, Tasker ML et al (2016) The conservation status and priorities for albatrosses and large petrels. Biol Conserv 201(September):169–183. https://doi.org/10.1016/j.biocon.2016.06.017

Pikitch EK, Santora C, Babcock EA, Bakun A, Bonfil R, Conover DO, Dayton P et al (2004) Ecosystem-based fishery management. Science 305(5682):346–347. https://doi.org/10.1126/science.1098222

Pikitch E, Boersma PD, Boyd IL, Conover DO, Cury P, Essington T, Heppell SS et al (2012) Little fish, big impact: managing a crucial link in ocean food webs. Lenfest Ocean Program, Washington, DC. https://www.oceanconservationscience.org/foragefish/files/Little%20Fish,%20Big%20Impact.pdf

Pine WE, Martell SJD, Walters CJ, Kitchell JF (2009) Counter intuitive responses of fish populations to management actions: some common causes and implications for predictions based on ecosystem modelling. Fisheries 34(4):165–180. https://doi.org/10.1577/1548-8446-34.4.165

Pinsky ML, Palumbi SR (2014) Meta-analysis reveals lower genetic diversity in overfished populations. Mol Ecol 23(1):29–39. https://doi.org/10.1111/mec.12509

Pitcher T, Kalikoski D, Pramod G, Short K (2009) Not honouring the code. Nature 457(February):658–659. https://doi.org/10.1038/457658a

Polacheck T (2012) Politics and independent scientific advice in RFMO processes: a case study of crossing boundaries. Mar Policy 36(1):132–141. https://doi.org/10.1016/j.marpol.2011.04.006

Polidoro BA, Carpenter KE, Collins L, Duke NC, Ellison AM, Ellison JC, Farnsworth EJ et al (2010) The loss of species: mangrove extinction risk and geographic areas of global concern. PloS One 5(4):e10095. https://doi.org/10.1371/journal.pone.0010095

Pollnac RB, Crawford BR, Gorospe MLG (2001) Discovering factors that influence the success of community-based marine protected areas in the Visayas, Philippines. Ocean Coast Manag 44(11–12):683–710. https://doi.org/10.1016/S0964-5691(01)00075-8

Poloczanska ES, Brown CJ, Sydeman WJ, Kiessling W, Schoeman DS, Moore PJ, Brander K et al (2013) Global imprint of climate change on marine life. Nat Clim Change 3(October):919–925. https://doi.org/10.1038/nclimate1958

Poloczanska ES, Burrows MT, Brown CJ, García Molinos J, Halpern BS, Hoegh-Guldberg O, Kappel CV et al (2016) Responses of marine organisms to climate change across oceans. Front Mar Sci 3(May):62. https://doi.org/10.3389/fmars.2016.00062

Pompa S, Ehrlich P, Ceballos G (2011) Global distribution and conservation of marine mammals. Proc Natl Acad Sci U S A 108(33):13600–13605

Pons M, Melnychuk MC, Hilborn R (2018) Management effectiveness of large pelagic fisheries in the high seas. Fish Fish 19(2):260–270. https://doi.org/10.1111/faf.12253

Puig P, Canals M, Company JB, Martín J, Amblas D, Lastras G, Palanques A, Calafat AM (2012) Ploughing the deep sea floor. Nature 489(September):286–289. https://doi.org/10.1038/nature11410

Purcell SW, Polidoro BA, Hamel JF, Gamboa RU, Mercier A (2014) The cost of being valuable: predictors of extinction risk in marine invertebrates exploited as luxury seafood. Proc R Soc B 281(1781):20133296. https://doi.org/10.1098/rspb.2013.3296

Pusceddu A, Gambi C, Corinaldesi C, Scopa M, Danovaro R (2014) Relationships between meiofaunal biodiversity and prokaryotic heterotrophic production in different tropical habitats and oceanic regions. PloS One 9(3):e91056. https://doi.org/10.1371/journal.pone.0091056

Raudsepp-Hearne C, Peterson GD, Bennett EM (2010) Ecosystem service bundles for analyzing tradeoffs in diverse landscapes. Proc Natl Acad Sci U S A 107(11):5242–5247. https://doi.org/10.1073/pnas.0907284107

Rex MA, Etter RJ (2010) Deep-sea biodiversity: pattern and scale. Harvard University Press, Cambridge, MA

Reygondeau G (2019) Current and future biogeography of exploited marine exploited groups under climate change. In: Cisneros-Montemayor AM, Cheung WWL, Ota Y (eds) Predicting future oceans: sustainability of ocean and human systems amidst global environmental change. Elsevier, Oxford, pp 87–101

Reygondeau G, Dunn D (2018) Data science and its relationship to big data and data-driven decision making. Big Data 1(1):51–59. https://doi.org/10.1089/big.2013.1508

Rice JC, Legacè È (2007) When control rules collide: a comparison of fisheries management reference points and IUCN criteria for assessing risk of extinction. ICES J Mar Sci 64(4):718–722. https://doi.org/10.1093/icesjms/fsm011

Rice J, Ridgeway L (2010) Conservation of biodiversity and fisheries management. In: Grafton RQ, Hilborn R, Squires D, Tait M, Williams M (eds) Handbook of marine fisheries conservation and management. Oxford University Press, Oxford, pp 139–149

Richards DR, Friess DA (2016) Rates and drivers of mangrove deforestation in Southeast Asia, 2000–2012. Proc Natl Acad Sci U S A 113(2):344–349. https://doi.org/10.1073/pnas.1510272113

Ritter F, Panigada S (2019) Collisions of vessels with cetaceans—the underestimated threat. In: Sheppard C (ed) World seas: an environmental evaluation. Ecological issues and environmental impacts, vol 3, 2nd edn. Academic Press, London, pp 531–547

Rivadeneira MM, Fernandez M (2005) Shifts in southern endpoints of distribution in rocky intertidal species along the South-Eastern Pacific Coast. J Biogeogr 32(2):203–209. https://doi.org/10.1111/j.1365-2699.2004.01133.x

RLS (Reef Life Survey) (2019) RLS data informing global biodiversity targets. https://reeflifesurvey.com/rls-data-informing-global-biodiversity-targets/. Accessed 25 Jun 2019

Roberts CM, Bohnsack JA, Gell F, Hawkins JP, Goodridge R (2001) Effects of marine reserves on adjacent fisheries. Science 294(5548):1920–1923. https://doi.org/10.1126/science.294.5548.1920

Roberts CM, O'Leary BC, McCauley DJ, Cury PM, Duarte CM, Lubchenco J, Pauly D et al (2017) Marine reserves can mitigate and promote adaptation to climate change. Proc Natl Acad Sci U S A 114(24):6167–6175. https://doi.org/10.1073/pnas.1701262114

Robinson JPW, Wilson SK, Robinson J, Gerry C, Lucas J, Assan C, Govinden R et al (2019) Productive instability of coral reef fisheries after climate-driven regime shifts. Nat Ecol Evol 3(February):183–190. https://doi.org/10.1038/s41559-018-0715-z

Rockström J, Steffen W, Noone K, Persson Å, Chapin FS, Lambin EF, Lenton TM et al (2009) A safe operating space for humanity. Nature 461(September):472–475. https://doi.org/10.1038/461472a

Rogers AD (2015) Environmental change in the deep ocean. Annu Rev Env Resour 40(November):1–38. https://doi.org/10.1146/annurev-environ-102014-021415

Rogers AD (2018) The biology of seamounts: 25 years on. Adv Mar Biol 79:137–224. https://doi.org/10.1016/bs.amb.2018.06.001

Rogers AD, Gianni M (2010) The implementation of UNGA resolutions 61/105 and 64/72 in the management of deep-sea fisheries on the high seas. Report prepared for the deep-sea conservation coalition. International Programme on State of the Ocean, London. http://www.savethehighseas.org/publicdocs/61105-Implemention-finalreport.pdf

Rogers AD, Brierley A, Croot P, Cunha MR, Danovaro R, Devey C, Hoel AH et al (2015) Delving deeper: critical challenges for 21st century deep-sea research. Position Paper 22. European Marine Board, Ostend, Belgium. https://www.marineboard.eu/sites/marineboard.eu/files/public/publication/EMB_PP22_Web_v4.pdf

Rossiter JS, Levine A (2014) What makes a 'successful' marine protected area? The unique context of Hawaii's fish replenishment areas. Mar Policy 44(February):196–203. https://doi.org/10.1016/j.marpol.2013.08.022

Rousseau Y, Watson RA, Blanchard JL, Fulton EA (2019) Evolution of global marine fishing fleets and the response of fished resources. Proc Natl Acad Sci U S A 116(25):12238–12243

Roux JP, van der Lingen C, Gibbons MJ, Moroff NE, Shannon LJ, Smith AD, Cury PM (2013) Jellification of marine ecosystems as a consequence of overfishing small pelagic fish: lessons from the Benguela. Bull Mar Sci 89(1):249–284. https://doi.org/10.5343/bms.2011.1145

Rowlands G, Brown J, Soule B, Boluda PT, Rogers AD (2019) Satellite surveillance of the Ascension Island exclusive economic zone and marine protected area. Mar Policy 101(March):39–50. https://doi.org/10.1016/j.marpol.2018.11.006

Roy K, Jablonski D, Valentine JW, Rosenberg G (1998) Marine latitudinal diversity gradients: tests of causal hypotheses. Proc Natl Acad Sci U S A 95(7):3699–3702. https://doi.org/10.1073/pnas.95.7.3699

Sadovy de Mitcheson Y (2016) Mainstreaming fish spawning aggregations into fishery management calls for truly precautionary approach. BioScience 66(4):295–306. https://doi.org/10.1093/biosci/biw013

Sadovy de Mitcheson Y, Andersson AA, Hofford A, Law CSW, Hau LCY, Pauly D (2018) Out of control means off the menu: the case for ceasing consumption of luxury products from highly vulnerable species when international trade cannot be adequately controlled; shark fin as a case study. Mar Policy 98(December):115–120. https://doi.org/10.1016/j.marpol.2018.08.012

Sala E, Giakoumi S (2017) No-take marine reserves are the most effective protected areas in the ocean. ICES J Mar Sci 75(3):1166–1168. https://doi.org/10.1093/icesjms/fsx059

Sala E, Lubchenco J, Grorud-Colvert K, Novelli C, Roberts C, Sumaila UR (2018a) Assessing real progress towards effective ocean protection. Mar Policy 91(May):11–13. https://doi.org/10.1016/j.marpol.2018.02.004

Sauermann H, Franzoni C (2015) Crowd science user contribution patterns and their implications. Proc Natl Acad Sci U S A 112(3):679–684. https://doi.org/10.1073/pnas.1408907112

Sciberras M, Jenkins SR, Kaiser MJ, Hawkins SJ, Pullin AS (2013) Evaluating the biological effectiveness of fully and partially protected marine areas. Environ Evid 2(February):4. https://doi.org/10.1186/2047-2382-2-4

Selig ER, Turner WR, Troëng S, Wallace BP, Halpern BS, Kaschner K, Lascelles BG, Carpenter KE, Mittermeier RA (2014) Global priorities for marine biodiversity conservation. PloS One 9(1):e82898

Sherley RB, Ludynia K, Dyer BM, Lamont T, Makhado AB, Roux JP, Scales KL, Underhill LG, Votier SC (2017) Metapopulation tracking juvenile penguins reveals an ecosystem-wide ecological trap. Curr Biol 27(4):563–568. https://doi.org/10.1016/j.cub.2016.12.054

Shlesinger T, Loya Y (2019) Breakdown in spawning synchrony: a silent threat to coral persistence. Science 365(6457):1002–1007. https://doi.org/10.1126/science.aax0110

Short F, Carruthers T, Dennison W, Waycott M (2007) "Global seagrass distribution and diversity: a bioregional model," in "The biology and ecology of seagrasses," edited by B.W. Touchette, special issue. J Exp Mar Biol Ecol 350(1–2):3–20. https://doi.org/10.1016/j.jembe.2007.06.012

Sigwart JD, Chen C, Thomas EA, Allcock AL, Böhm M, Seddon M (2019) Red listing can protect deep-sea biodiversity. Nat Ecol Evol 3(August):1134. https://doi.org/10.1038/s41559-019-0930-2

Skroblin A, Carboon T, Bidu G, Chapman N, Miller M, Taylor K, Taylor W, Game ET, Wintle BA (2019) Including indigenous knowledge in species distribution modelling for increased ecological insights. Conserv Biol 35:587. https://doi.org/10.1111/cobi.13373

Smale DA, Wernberg T, Oliver ECJ, Thomsen MS, Harvey BP, Starub SC, Burrows MT et al (2019) Marine heatwaves threaten global biodiversity and the provision of ecosystem services. Nat Clim Change 9(April):306–312. https://doi.org/10.1038/s41558-019-0412-1

Small N, Munday M, Durance I (2017) The challenge of valuing ecosystem services that have no material benefits. Glob Environ Chang 44(May):57–67. https://doi.org/10.1016/j.gloenvcha.2017.03.005

Spalding MD, Ruffo S, Lacambra C, Meliane I, Hale LZ, Shepard CC, Beck WW (2014) The role of ecosystems in coastal protection: adapting to climate change and coastal hazards. Ocean Coast Manag 90(March):50–57. https://doi.org/10.1016/j.ocecoaman.2013.09.007

Stachowicz JJ, Bruno JF, Duffy JE (2007) Understanding the effects of marine biodiversity on communities and ecosystems. Annu Rev Ecol Evol Syst 38(December):739–766. https://doi.org/10.1146/annurev.ecolsys.38.091206.095659

Standing A (2015) Corruption and state-corporate crime in fisheries. U4 Issue Paper 15. U4 Anti-Corruption Resource Centre, Chr. Michelsen Institute, Bergen, Norway. https://www.u4.no/publications/corruption-and-state-corporate-crime-in-fisheries.pdf

Stehli FG, McAlester AL, Helsey CE (1967) Taxonomic diversity of recent bivalves and some implications for geology. Geol Soc Am Bull 78(4):455–466. https://doi.org/10.1130/0016-7606(1967)78[455:TDORBA]2.0.CO;2

Stehli FG, Douglas RG, Kafescioglu A (1972) Models for the evolution of Planktonic Foraminifera. In: Schopf TJM (ed) Models in paleobiology. Freeman, Cooper, San Francisco, pp 116–128

Stewart JS, Field JC, Markaida U, Gilly WF (2013) Behavioral ecology of Jumbo Squid (*Dosidicus gigas*) in relation to oxygen minimum zones. DeepSea Res II Top Stud Oceanogr 95(15):197–208. https://doi.org/10.1016/j.dsr2.2012.06.005

Stramma L, Prince ED, Schmidtko S, Luo J, Hoolihan JP, Visbeck M, Wallace DWR, Brandt P, Körtzinger A (2012) Expansion of oxygen minimum zones may reduce available habitat for tropical pelagic fishes. Nat Clim Change 2(January):33–37. https://doi.org/10.1038/nclimate1304

Strong JA, Andonegi E, Bizsel KC, Danovaro R, Elliott M, Franco A, Garces E et al (2015) Marine biodiversity and ecosystem function relationships: the potential for practical monitoring applications. Estuar Coast Shelf Sci 161(August):46–64. https://doi.org/10.1016/j.ecss.2015.04.008

Stuart-Smith RD, Bates AE, Lefcheck JS, Duffy JE, Baker SC, Thomson RJ, Stuart-Smith JF et al (2013) Integrating abundance and functional traits reveals new global hot spots of fish diversity. Nature 501(September):539–542. https://doi.org/10.1038/nature12529

Stuart-Smith RD, Edgar GJ, Barrett NS, Kininmonth SJ, Bates AE (2015) Thermal biases and vulnerability to warming in the world's marine fauna. Nature 528(December):88–92. https://doi.org/10.1038/nature16144

Stuart-Smith RD, Edgar GJ, Barrett NS, Bates AE, Baker SC, Bax NJ, Becerro MA et al (2017) Assessing national biodiversity trends for rocky and coral reefs through the integration of citizen science and scientific monitoring programs. BioScience 67(2):134–146. https://doi.org/10.1093/biosci/biw180

Stuart-Smith RD, Brown CJ, Ceccarelli DM, Edgar GJ (2018) Ecosystem restructuring along the great barrier reef following mass coral bleaching. Nature 560(August):92–96. https://doi.org/10.1038/s41586-018-0359-9

Suazo CG, Cabezas LA, Moreno CA, Arata JA, Luna-Jorquera G, Simeone A, Adasme L et al (2014) Seabird bycatch in chile: a synthesis of its impacts, and a review of strategies to contribute to the reduction of a global phenomenon. Pac Seabirds 41(1–2):1–12. http://www.pacificseabirdgroup.org/publications/PacificSeabirds/VOL_41_1-2.pdf

Suding KN, Hobbs RJ (2009) Threshold models in restoration and conservation: a developing framework. Trends Ecol Evol 24(5):271–279. https://doi.org/10.1016/j.tree.2008.11.012

Sullivan JM, Constant V, Lubchenco J (2019) Extinction threats to life in the ocean and opportunities for their amelioration. In: Dasgupta P, Raven P, McIvor A (eds) Biological extinction: new perspectives. Cambridge University Press, Cambridge, pp 113–137. https://doi.org/10.1017/9781108668675

Sumaila UR, Dyck A, Cheung WWL (2013) Fisheries subsidies and potential catch loss in sids exclusive economic zones: food security implications. Environ Dev Econ 18(4):427–439. https://doi.org/10.1017/S1355770X13000156

Sunagawa S, Coelho LP, Chaffron S, Kultima JR, Labadie K, Salazar G, Djahanschiri B et al (2015) Structure and function of the global ocean microbiome. Science 348(6237):1261359. https://doi.org/10.1126/science.1261359

Sweetman AK, Thurber AR, Smith CR, Levin LA, Mora C, Wei CL, Gooday AJ et al (2017) Major impacts of climate change on deep-sea benthic ecosystems. Elementa Sci Anthropocene 5:4. https://doi.org/10.1525/elementa.203/

Tanhua T, McCurdy A, Fischer A, Appelatans W, Bax N, Currie K, DeYoung B et al (2019) What have we learned from the framework for ocean observing: evolution of the global ocean observing system. Front Mar Sci 6(August):471. https://doi.org/10.3389/fmars.2019.00471

Theobald EJ, Ettinger AK, Burgess HK, DeBey LB, Schmidt NR, Froehlich HE, Wagner C et al (2015) Global change and local solutions: tapping the unrealized potential of citizen science for biodiversity research. Biol Conserv 181(January):236–244. https://doi.org/10.1016/j.biocon.2014.10.021

Thomas N, Lucas R, Bunting P, Hardy A, Rosenqvist A, Simard M (2017) Distribution and drivers of global mangrove forest change, 1996–2010. PloS One 12(6):e0179302. https://doi.org/10.1371/journal.pone.0179302

Thomson JA, Burkholder DA, Heithaus MR, Fourqurean JW, Fraser MW, Stratton J, Kendrick GA (2015) Extreme temperatures, foundation species, and abrupt ecosystem change: an example from an iconic seagrass ecosystem. Glob Chang Biol 21(4):1463. https://doi.org/10.1111/gcb.1269474

Thorson G (1952) Zur jetzigen lage der marinen bodentier-okolgie. Zool Anz S16:276–327

Thorson G (1957) Bottom communities (sublittoral or shallow shelf). In: Hedgepeth JW (ed) Treatise on marine ecology and paleoecology. Geological Society of America, Ithaca, NY, pp 461–534. https://doi.org/10.1130/MEM67V1-p535

Thrush SE, Dayton PK (2002) Disturbance to marine benthic habitats by trawling and dredging: implications for marine biodiversity. Annu Rev Ecol Syst 33(November):449–473. https://doi.org/10.1146/annurev.ecolsys.33.010802.150515

Tittensor DP, Baco-Taylor AR, Brewin P, Clark MR, Consalvey M, Hall-Spencer J, Rowden AA et al (2009) Predicting global habitat suitability for stony corals on seamounts. J Biogeogr 36(6). https://doi.org/10.1111/j.1365-2699.2008.02062.x

Tittensor DP, Mora C, Jetz W, Lotze HK, Ricard D, Berghe EV, Worm B (2010) Global patterns and predictors of marine biodiversity across taxa. Nature 466(August):1098–1101. https://doi.org/10.1038/nature09329

Tittensor DP, Beger M, Boerder K, Boyce DG, Cavanagh RD, Cosandey-Godin A, Crespo GO et al (2019) Integrating climate adaptation and biodiversity conservation in the global ocean. Sci Adv 5(11):eaay9969. https://doi.org/10.1126/sciadv.aay9969

Traganos D, Aggarwal B, Poursanidis D, Topouzelis K, Chrysoulakis N, Reinartz P (2018) Towards global-scale seagrass mapping and monitoring using sentinel-2 on google earth engine: the case study of the Aegean and Ionian Seas. Remote Sens (Basel) 10(8):1227. https://doi.org/10.3390/rs10081227

Travis J, Coleman FC, Auster PJ, Cury PM, Estes JA, Orensanz J, Peterson CH et al (2014) Integrating the invisible fabric of nature into fisheries management. Proc Natl Acad Sci U S A 111(2):581–584. https://doi.org/10.1073/pnas.1305853111

Trevathan-Tackett SM, Kelleway J, Macreadie PI, Beardall J, Ralph P, Bellgrove A (2015) Comparison of marine macrophytes for their contributions to blue carbon sequestration. Ecology 96(11):3043–3057. https://doi.org/10.1890/15-0149.1

UNEP-WCMC (United Nations Environment Programme World Conservation Monitoring Centre) and IUCN (International Union for Conservation of Nature) (2020) Marine protected planet. UNEP WCMC and IUCN, Cambridge. www.protectedplanet.net

UNEP-WCMC, Short FT (2018) "Global Distribution of Seagrasses." Version 6.0. Sixth update to the data layer used in Green and Short (2003). UNEP-WCMC, Cambridge. http://data.unep-wcmc.org/datasets/7

UNEP-WCMC, WorldFish Centre, WRI, and TNC (2018) "Global Distribution of Warm-Water Coral Reefs, Compiled from Multiple Sources Including the Millennium Coral Reef Mapping Project." Version 4.0. Includes contributions from IMaRS-USF and IRD (2005), IMaRS-USF (2005), and Spalding et al. (2001). UNEP-WCMC, Cambridge. http://data.unep-wcmc.org/datasets/1

Unsworth RKF, Nordlund LM, Cullen-Unsworth LC (2019) Seagrass meadows support global fisheries production. Conserv Lett 12(1):e12566. https://doi.org/10.1111/conl.12566

van Overzee HMJ, Rijnsdorp AD (2015) Effects of fishing during the spawning period: implications for sustainable management. Rev Fish Biol Fish 25(March):65–83. https://doi.org/10.1007/s11160-014-9370-x

Velarde E, Ezcurra E, Anderson DW (2015a) "Seabird Diet Predicts Following-Season Commercial Catch of Gulf of California Pacific Sardine and Northern Anchovy," in "California Current System—Predators and the Preyscape," edited by D.G. Ainley, P.B. Adams, and J. Jahncke, special issue. J Mar Syst 146(June):82–88. https://doi.org/10.1016/j.jmarsys.2014.08.014

Velarde E, Ezcurra E, Horn MH, Patton RT (2015b) Warm oceanographic anomalies and fishing pressure drive seabirds nesting north. Sci Adv 1(5):e1400210. https://doi.org/10.1126/sciadv.1400210

Vergés A, Steinberg PD, Hay ME, Poore AG, Campbell AH, Ballesteros E, Heck KL Jr et al (2014) The tropicalization of temperate marine ecosystems: climate- mediated changes in herbivory and community phase shifts. Proc R Soc B Biol Sci 281(1789):20140846. https://doi.org/10.1098/rspb.2014.0846

Victorero L, Watling L, Deng Palomares ML, Nouvian C (2018) Out of sight, but within reach: a global history of bottom-trawled deep-sea fisheries from >400m depth. Front Mar Sci 5(April):98. https://doi.org/10.3389/fmars.2018.00098

Vihtakari M, Welcker J, Moe B, Chastel O, Tartu S, Hop H, Bech C, Descamps S, Gabrielsen GW (2018) Black-legged Kittiwakes as messengers of atlantification in the Arctic. Sci Rep 8(January):1178. https://doi.org/10.1038/s41598-017-19118-8

Wanless S, Harris MP, Redman P, Speakman JR (2005) Low energy values of fish as a probable cause of a major seabird breeding failure in the North Sea. Mar Ecol Prog Ser 294(June):1–8. https://doi.org/10.3354/meps294001

Ward P, Myers RA (2005) Shifts in open-ocean fish communities coinciding with the commencement of commercial fishing. Ecology 86(4):835–847. https://doi.org/10.1890/03-0746

Washburn TW, Turner PJ, Durden JM, Jones DOB, Weaver P, Van Dover CL (2019) Ecological risk assessment for deep-sea mining. Ocean Coast Manag 176(June):24–39. https://doi.org/10.1016/j.ocecoaman.2019.04.014

Waycott M, Duarte CM, Carruthers TJB, Orth RJ, Dennison WC, Olyarnik S, Calladine A et al (2009) Accelerating loss of seagrasses across the globe threatens coastal ecosystems. Proc Natl Acad Sci U S A 106(30):12377–12381. https://doi.org/10.1073/pnas.0905620106

Webb TJ, Mindel BL (2015) Global patterns of extinction risk in marine and non-marine systems. Curr Biol 25(4):506–511. https://doi.org/10.1016/j.cub.2014.12.023

Wernberg T, Filbee-Dexter K (2019) Missing the marine forest for the trees. Mar Ecol Prog Ser 612(March):209–215. https://doi.org/10.3354/meps12867

Wernberg T, Bennett S, Babcock RC, De Bettignies T, Cure K, Depczynski M, Dufois F et al (2016) Climate-driven regime shift of a temperate marine ecosystem. Science 353(6295):169–172. https://doi.org/10.1126/science.aad8745

Wicaksono P, Fauzan MA, Kumara ISW, Yogyantoro RN, Lazuardi W, Zhafarina Z (2019) Analysis of reflectance spectra of tropical seagrass species and their value for mapping using multispectral satellite images. Int J Rem Sens 40(23):8955–8978. https://doi.org/10.1080/01431161.2019.1624866

Widjaja S, Long T, Wirajuda H, Van As H, Bergh PE, Brett A, Copeland D, Fernandez M, Gusman A, Juwana S, Ruchimat T, Trent S, Wilcox C (2020) Illegal, unreported and unregulated fishing and associated drivers. World Resources Institute, Washington, DC. https://ocean-panel.org/blue-papers/iuu-fishing-and-associated-drivers

Wilkinson C, Salvat B, Eakin CM, Brathwaite A, Francini-Filho R, Webster N, Ferreira BP, Harris P (2016) Tropical and sub-tropical coral reefs. In: Iniss L, Simcock A (eds) The first global integrated marine assessment. World Ocean Assessment I, Division for Ocean Affairs and Law of the Sea, New York, p 42

Williams GJ, Graham NAJ, Jouffray JB, Norström AV, Nyström M, Gove JM, Heenan A, Wedding LM (2019) Coral reef ecology in the anthropocene. Funct Ecol 33(6):1014–1022. https://doi.org/10.1111/1365-2435.13290

Williamson M (1997) Marine biodiversity in its global context. In: Ormond RFG, Gage JD, Angel MV (eds) Marine biodiversity: patterns and processes. Cambridge University Press, Cambridge, pp 1–17

Wilson SK, Graham NAJ, Pratchett MS, Jones GP, Polunin NVC (2006) Multiple disturbances and the global degradation of coral reefs: are reef fishes at risk or resilient? Glob Chang Biol 12(11):2220–2234. https://doi.org/10.1111/j.1365-2486.2006.01252.x

Winnard S, Hochberg T, Miller N, Kroodsma D, Small C, Augustyn P (2018) A new method using AIS data to obtain independent compliance data to determine mitigation use at sea. In: Paper prepared for the Thirteenth Meeting of the CCSBT Compliance Committee, October 11–13, Noumea, New Caledonia. https://www.ccsbt.org/en/system/files/CC13_Info03_BLI_AIS_MitigationUse_Rev1.pdf

WMO (World Meteorological Organization) (2016) The global observing system for climate: implementation needs. WMO, Geneva. https://library.wmo.int/opac/doc_num.php?explnum_id=3417

Woodall LC, Robinson LF, Rogers AD, Narayanaswamy BE, Paterson GLJ (2015) Deep-sea litter: a comparison of seamounts, banks and a ridge in the Atlantic and Indian Oceans reveals both environmental and anthropogenic factors impact accumulation and composition. Front Mar Sci 2(February):3. https://doi.org/10.3389/fmars.2015.00003

Woolley SNC, Tittensor DP, Dunstan PK, Guillera-Arroita G, Lahoz-Monfort JJ, Wintle BA, Worm B, O'Hara TD (2016) Deep-sea diversity patterns are shaped by energy availability. Nature 533(7603):393–396. https://doi.org/10.1038/nature17937

World Bank (2017) The Sunken Billions revisited: progress and challenges in global marine fisheries. World Bank, Washington, DC. https://openknowledge.worldbank.org/handle/10986/24056

Worm B, Myers RA (2003) Meta-analysis of Cod–Shrimp interactions reveals top-down control in oceanic food webs. Ecology 84(1):162–173. https://doi.org/10.1890/0012-9658(2003)084[0162:MAOCSI]2.0.CO;2

Worm B, Tittensor DP (2011) Range contraction in large pelagic predators. Proc Natl Acad Sci U S A 108(29):11942–11947. https://doi.org/10.1073/pnas.1102353108

Worm B, Tittensor DP (2018) A theory of global biodiversity. In: Monographs in Population Biology 60. Princeton University Press, Princeton, NJ

Worm B, Barbier EB, Beaumont N, Duffy E, Folke C, Halpern BS, Jackson JBC et al (2006) Impacts of biodiversity loss on ocean ecosystem services. Science 314(5800):787–790. https://doi.org/10.1126/science.1132294

WoRMS (World Register of Marine Species) (n.d.) Statistics. http://www.marinespecies.org/aphia.php?p=stats. Accessed 22 Mar 2020

Wright G, Ardron J, Gjerde K, Currie D, Rochette J (2015) Advancing marine biodiversity protection through regional fisheries management: a review of bottom fisheries closures in areas beyond national jurisdiction. Mar Policy 61(November):134–148. https://doi.org/10.1016/j.marpol.2015.06.030

WWF (World Wildlife Fund) (2015) Reviving the ocean economy: the case for action—2015. WWF International, Gland. https://c402277.ssl.cf1.rackcdn.com/publications/790/files/original/Reviving_Ocean_Economy_REPORT_low_res.pdf?1429717323

Ye Y, Gutierrez NL (2017) Ending fishery overexploitation by expanding from local successes to globalized solutions. Nat Ecol Evol 1(June):0179. https://doi.org/10.1038/s41559-017-0179

Yesson C, Clark MR, Taylor M, Rogers AD (2011) The global distribution of seamounts based on 30-second bathymetry data. Deep-Sea Res I Oceanogr Res Pap 58(4):442–453. https://doi.org/10.1016/j.dsr.2011.02.004

Zaneveld JR, Burkepile DE, Shantz AA, Pritchard CE, McMinds R, Payet JP, Welsh R et al (2016) Overfishing and nutrient pollution interact with temperature to disrupt coral reefs down to microbial scales. Nat Commun 7(June):11833. https://doi.org/10.1038/ncomms11833

Zhang H, Wu F (2017) China's marine fishery and global ocean governance. Global Pol 8(2):216–226. https://doi.org/10.1111/1758-5899.12419

Zoffoli ML, Frouin R, Kampel M (2014) Water column correction for coral reef studies by remote sensing. Sensors 14(9):16881–16931. https://doi.org/10.3390/s140916881

11 The Human Relationship with Our Ocean Planet

1 Highlights

- The human relationship with the ocean is diverse and complex. It is built on values that are often non-monetary, and which contribute to non-material dimensions of well-being. These values are essential to broader human flourishing. They include contributions to cultural and social and legal identity; a sense of place; occupational pride and self-respect; spirituality; mental and bodily health; and human security.
- The plurality of these values and interests matters to individuals and societies and could be more strongly represented in high-level ocean policy discussions.
- A sustainable ocean economy must be built on these diverse relationships, in ways that encourage equity and inclusion and that recognise the non-material aspects of well-being.
- How we govern the ocean will determine who accesses and benefits from the ocean space. A heavily privatised, zoned and securitised ocean undermines the human-ocean relationship. Building upon existing institutional foundations, ones that provide livelihoods and well-being benefits to all citizens, will foster a more constructive long-term engagement with the ocean.
- There is no sole human relationship with the ocean with which all people will identify: each individual has different interests, experiences, economic stakes, emotional investments and cultural and social ties to different aspects of the ocean. To increase the ocean's contribution to both material and non-material well-being globally, we need to build a sustainable ocean economy based on this plurality of values.
- This paper identifies and focuses on the relationships with the ocean that contribute to human well-being. In doing so, it outlines these relationships in new ways and identifies the means to ensure that the plurality of 'ocean values' is represented in processes of planning and implementing a sustainable ocean economy.
- The paper suggests five key actions to assist states and international organisations in supporting and improving humanity's diverse relationships with the ocean by fostering participatory democratic governance: (1) humanise the new ocean narrative by focusing economic development on the objective of increasing human well-being; (2) foster diversity and inclusion in the sustainable ocean economy; (3) engage in partnerships with a broad constituency of ocean supporters, including small-scale fisherfolk, community elders and next-generation social and environmental activists, Indigenous Peoples, and women who work in the maritime economy and who steward marine environments; (4) build the capacity of meso-level institutions below the level of the national government and above the level of the individual citizen-consumer; and (5) ensure that responses to COVID-19 consider the well-being of ocean-dependent people and economic sectors.
- Governing the ocean is a 'collective responsibility of humanity' and can only be achieved by ensuring that those who have lived in, worked on and stewarded coastal and continental waters for centuries or millennia—prominent among them small-scale fisherfolk—are included in decisions on its future governance. These 'ocean citizens' and the institutions they have forged are pivotal to achieving a sustainable ocean economy. As such, maintaining ocean health and maintaining ocean access should be the dual aims of governing the future ocean.

> Just as the sea is an open and ever flowing reality, so should our oceanic identity transcend all forms of insularity, to become one that is openly searching, inventive, and welcoming. (Epeli Hau'ofa 2008)

Originally published in:
Allison EH, Kurien J, Ota Y, et al. 2020. *The Human Relationship with Our Ocean Planet*. Washington, DC: World Resources.
Reprint by Springer International Publishing (2023) with kind permission.
Published under license from the "World Research Institute".

J. Lubchenco, P. M. Haugan (eds.), *The Blue Compendium*, https://doi.org/10.1007/978-3-031-16277-0_11

2 Introduction

People across the world have diverse economic, sociolegal, institutional, social and cultural relationships with the ocean—both its littoral zones and the open sea spaces through which people have traditionally navigated, migrated, fished, traded, played and sought solace, spiritual enlightenment, adventure, material enrichment, social identity, cultural expression, artistic inspiration or good health. These relationships are reflected in formal and informal institutions (polices, laws, social norms) that regulate many of these activities, including those that regulate access to resources. These institutions represent a series of prior claims and rights to the use and enjoyment of the ocean by coastal and maritime societies.

By taking account of the range of ways coastal and maritime societies use, enjoy and govern coastal seas and ocean basins, we are better placed to design a sustainable ocean economy that is fair and equitable and that reflects 'the future we want' (UNCSD 2012). This paper argues that policymakers should consider the full range of human relationships with the ocean. The economic investment strategies and governance actions envisaged in contemporary ocean policy and planning can transform those relationships (Swilling et al. 2020) and will thus change the nature and distributions of the values that humanity derives from its interactions with the oceanic realm.

How can humanity's diverse relationships with the ocean be supported to flourish in the future, so that the ocean can make sustainable contributions to human well-being? This is the overarching policy question to which this paper responds. Policy research has made significant advances in assessing the ocean's ability to generate economic goods and services. The complementary perspectives presented here aim to draw attention to the wider role that the ocean has played—and will continue to play—in sustaining and reproducing other human values such as social and cultural identity, individual and collective well-being, sense of place and belonging, and human emotions such as curiosity, spirituality, awe and a sense of adventure.

From a brief survey of the past and current range of human relationships with the ocean and how they contribute to human well-being, and by examining the economic and policy implications of these relationships, we will argue that a sustainable ocean economy can contribute not only to the sustainable and equitable growth of economic goods and services but also to human well-being and flourishing more generally. Thus, the ocean can play a catalytic role in the next phase of human development, enhancing human capabilities and freedoms (Sen 1999, 2001), and thereby contribute to meeting the UN Sustainable Development Goals (SDGs) (Singh et al. 2018; Nash et al. 2020).

It is not our intent here to document every way that people and the ocean interact, for good or ill. Other papers in this series examine in detail how we might sustain and grow marine food production (Costello et al. 2019); how climate change has impacted the ocean and how humanity may respond (Gaines et al. 2019); how we might better deal with human rights violations and other criminal activities and inequities at sea (Widjaja et al. 2020; Österblom et al. 2020); how pollution threatens the ocean and how we might control it better (Jambeck et al. 2020); what opportunities exist to improve the financing and governance (Swilling et al. 2020; Winther et al. 2020) of the ocean economy, and so on. These issues and solution pathways all impact the plurality of people-ocean relationships and may undermine some and enhance others, in part depending on how they affect existing ocean-related economic inequalities (Österblom et al. 2020). Our point here is that the relational and subjective elements of people-ocean relationships have not yet been fully articulated in policy arenas and are therefore not yet fully considered in plans to respond to these ocean threats or to seize ocean economic and conservation opportunities.

Drawing on brief overviews of representative social and legal institutions that have developed in different maritime societies, we identify how different societies have governed oceanic spaces and volumes and how these governance mechanisms reflect the diversity of 'ocean values' held by different peoples. We use these overviews of the diversity of human relationships with the ocean, the examples of historically and culturally grounded sea tenure arrangements, and contemporary policy debates around the 'blue economy' (Voyer et al. 2018), 'blue justice' (Bennett et al. 2019) and 'blue degrowth' (Ertör and Hadjimichael 2020), to identify a series of opportunities for action to build a sustainable ocean economy and a future human relationship with the ocean that reflects the breadth and plurality of world views and values of current and future ocean citizens, and that acknowledges the diversity of social identities of the people for whom the ocean matters. At the time of this writing, the world was reeling from the impacts of the COVID-19 pandemic, which, by 16 August 2020, had infected around 21.3 million people and resulted in 761,779 deaths (WHO 2020a). We felt it necessary to consider how relationships between people and the ocean may be affected by the public health measures taken to slow the spread of the virus and the economic and social consequences of both the disease itself and measures taken to contain it. Accordingly, we briefly consider what is known about impacts on the maritime economy and on human-ocean relationships. It also cannot be overlooked that humanity is embarking on an ocean governance transformation at a time when action on climate change is critical. The ocean offers many opportunities to reduce greenhouse gas emissions and increase carbon capture and storage (Hoegh-Guldberg et al. 2019). Ocean-related climate change impacts are likely to exacerbate existing inequalities within coastal communities, with vulnerable populations being those living in low-lying areas of the tropics, on small oceanic islands and in the Arctic, as well as those whose livelihoods are tied to fisheries affected by global environmental change (Adger et al. 2005; Barbier 2015). Most sectors of the ocean economy will be negatively impacted by climate change, and tele-connected climate and economic processes mean that oceanic changes also have impacts inland (Allison and Bassett 2015). Investments in building adaptive capacity in ways that respond to different peoples' values will be required, and these should be kept in mind when considering how the human relationship with the ocean is understood, assessed and governed.

2.1 Conceptual Development

This paper draws on multiple disciplines, theories and conceptual frameworks, reflecting the wide scope of the paper's subject and the wide range of the authors' disciplinary backgrounds. Grounded largely in human, cultural and economic geography, economic history, economic and legal anthropology and political ecology, the paper also includes contributions from critical literary studies, rural sociology, psychology, Indigenous studies and development studies, as well as fisheries science and conservation biology.

The way the ocean is being studied and thought about is changing, with a 'new thalassology'[1] emergent that draws on cross-cultural world histories to examine ocean basins from a human historical perspective (e.g. Paine 2013). This has particularly enriched the study of the Mediterranean and Indian Ocean regions (Horden and Purcell 2006; Vink 2007). Earlier foundational work on oceanic trade, from the perspective of historical economic anthropology, concerns itself with studies of mercantilism and colonisation and its continuing societal impacts (Curtin 1984). This scholarship is expanding into a more socially differentiated research agenda, teaching us more about how gender, class, ethnicity, race and colonial history—and their intersections—have shaped the experiences and either constrained or enhanced the possibilities of different people's encounters with the ocean (e.g. Amrith 2013; Catterall and Campbell 2012; McKay 2007).

The recent 'blue turn' in human and cultural geography (Peters and Anderson 2016) has brought the analysis of human-nature spatial relations into the oceanic realm, informing the emergent interest in marine spatial planning (Fairbanks et al. 2019) but also explaining why the realities of ocean space—its fluidity, its fourth dimension (volume) and the challenges in identifying fixed boundaries—are important determinants of our maritime imaginings and our practical ability to govern the ocean (Steinberg and Peters 2015). The arts and humanities have gone blue, too, with an immersion into how the ocean has shaped our history, science, languages, aesthetics and sensibilities (Mentz 2009; Guo-jun 2013; Mack 2013; Alaimo 2019). More broadly, recent influential dialogues in the humanities and social sciences have also disrupted binary distinctions between the categories of nature and culture, generating new possibilities for living in the 'Anthropocene' (Tsing et al. 2017).

We also note a flourishing of scholarship on and policy attention to the question of Indigeneity and Indigenous knowledges in what heralds an 'indigenous resurgence' (Alfred 2009; Corntassel 2012). Some of this analysis and activism for 'decolonisation' is focused on the interactions of Indigenous and colonised peoples with the ocean (e.g. Hauʻofa 1994; von der Porten et al. 2019). Parallel to this we see a growing interest in the legal pluralism that affects coastal and oceanic regions (Bavinck and Gupta 2014).

We see an opportunity to bring all this vigorous and exciting intellectual and political endeavour to bear in discussing the economic development and governance of the future ocean in high-level and intergovernmental and governmental forums.

Drawing on a range of disciplines beyond economics to address the issue of ocean futures highlights a range of values beyond monetary or market ones. It allows the exploration of human relationships with the ocean in terms of what people value about it and their societies' interactions with it, and what motivates their actions with respect to future ocean governance. This shift from thinking about (economic) value to thinking about human values more broadly is in part informed by cross-cultural studies in psychology (e.g. Schwartz 2012) which examine individuals' motivations to act in ways that either oppose or embrace change, satisfy their own needs or the needs of others and of nature. These ideas have had few direct applications in ocean decision-making to date (e.g. Slimak and Dietz 2006; Bidwell 2017) but are likely to become more important as public attention turns towards the ocean and concepts such as the 'social license to operate' (Voyer and van Leeuwen 2019) inform decision-making on what kind of oceanic economic activities different societies will support or oppose.

This shift in thinking also responds to calls from the Intergovernmental Science-Policy Platform on Biodiversity and Ecosystem Services (IPBES) to develop further the concept of 'ecosystem services' so that plural world views, values and knowledges can be better recognised in assessments

[1] Before the ocean sciences became known as 'oceanography' (derived from Latin) they were sometimes referred to as thalassology (derived from Ancient Greek). This older term has been revived by scholars in the arts and humanities.

of nature's contributions to people (and indeed people's contributions to non-human nature), alongside scientific and economic assessments (Fig. 11.1a–c; Pascual et al. 2017).

Here we adopt a pluralistic valuation approach (Fig. 11.1a), drawing on multiple world views, knowledges and values. Our focus is on anthropocentric values, as we are concerned with the human relationship with the ocean, rather than intrinsic values (Fig. 11.1b). We include instrumental values (see Sect. 11.3.2, Table 11.1), but our focus is on relational and subjective values (Fig. 11.1b, orange shading), which we further unpack using a social well-being framework (see Sect. 11.2.2). Our purpose is to raise awareness of the diversity, range and nature of the ocean's contributions to people (Fig. 11.1c, step 1). Our analysis highlights differing world views and types of value (Fig. 11.1c, step 2), we draw on a range of disciplines, methods and knowledge systems and we highlight and discuss (but do not fully assess) key potential trade-offs among types of values and power relations among holders of values (Fig. 11.1c, step 3). We begin the process of integrating and bridging Indigenous and local knowledge, the arts, humanities, social sciences, policy and management sciences and natural sciences (Fig. 11.1c, step 4) to communicate the range and nature of ocean values that contribute to human well-being in all its dimensions

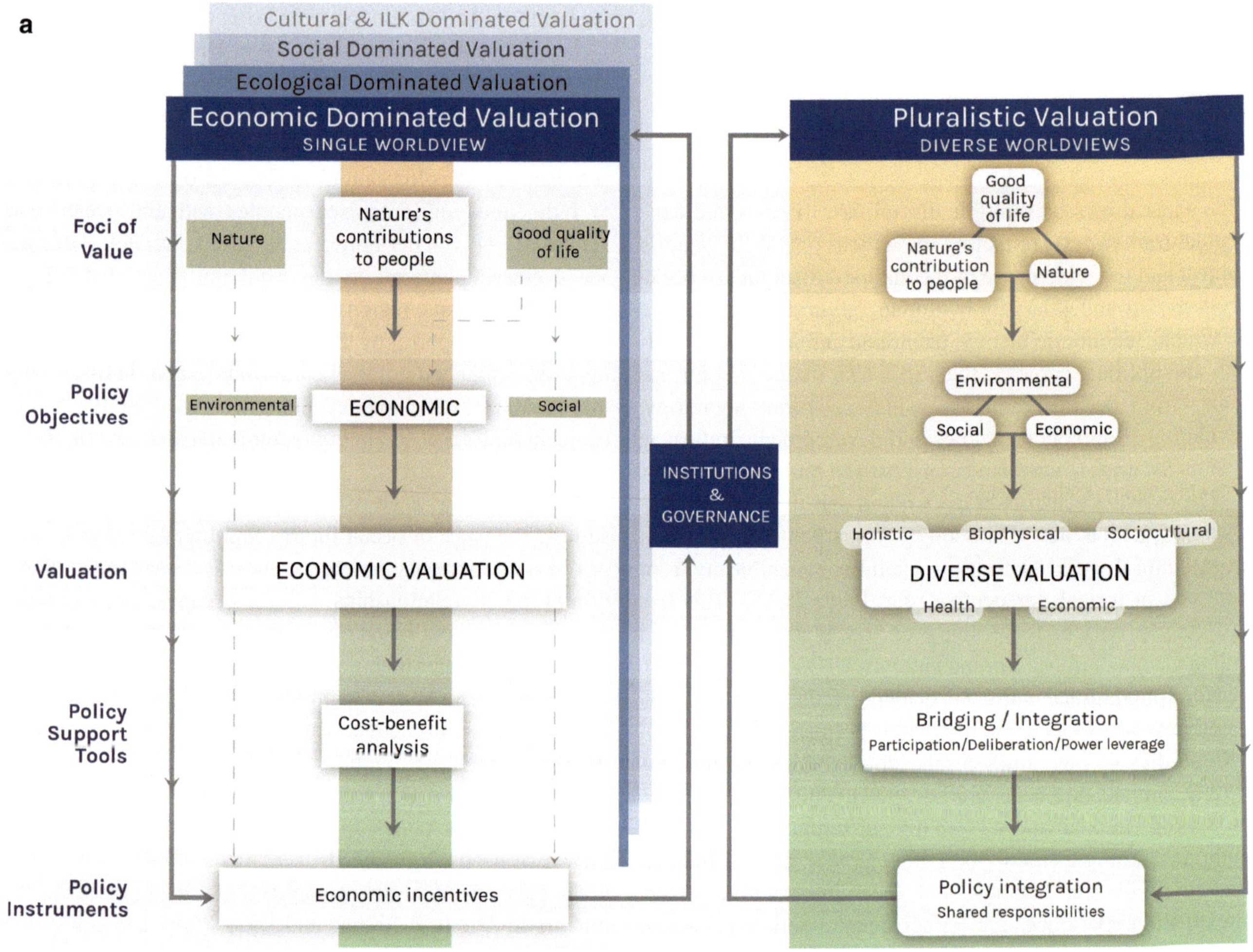

Notes: The panel on the right emphasises the importance of a pluralistic valuation approach, compared with value monism or unidimensional valuation approaches to human–nature relationships represented in the panel on the left. The pluralistic approach is adopted in this paper.

ILK = Indigenous and local knowledge.

Source: Redrawn from Pascual et al. 2017, for IBPES.

Fig. 11.1 (**a**) Contrasting approaches to the process of valuation. (**b**) Diverse values related to nature, nature's contributions to people and a good quality of life. (**c**) The IPBES approach for assessing values and conducting valuation

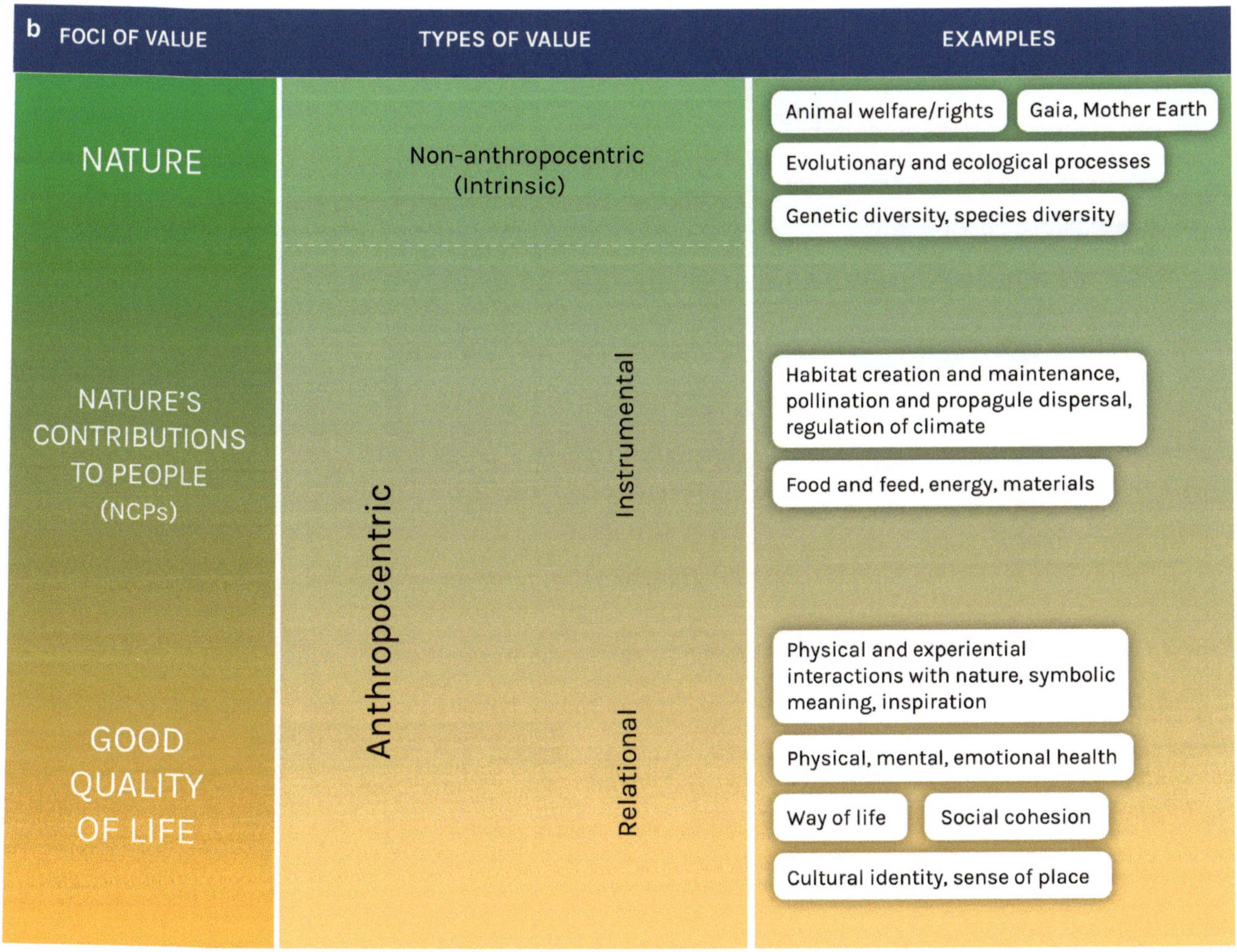

Note: The grading in the colours indicates that both instrumental and relational values can be ascribed to the value of NCPs, and to highlight that NCPs are intertwined with nature and a good quality of life. In this paper, we emphasise the anthropocentric relational values.

Source: Redrawn from Pascual et al. 2017, for IPBES.

Fig. 11.1 (continued)

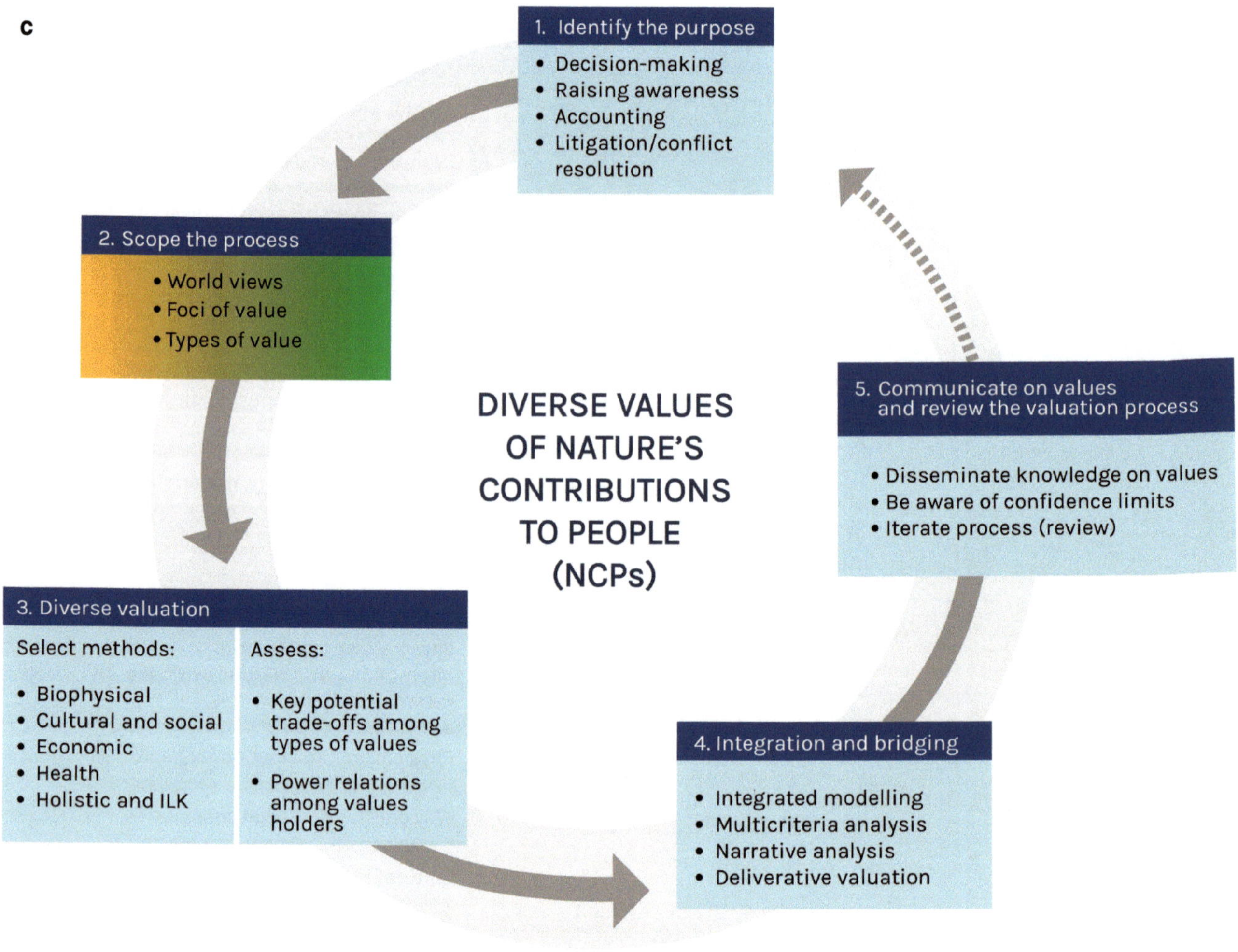

Notes: Orange and green colours in step 2 indicate that the scoping applies to methods for both valuation and integrating or bridging diverse values (steps 3 and 4).

ILK = Indigenous and local knowledge.

Source: Redrawn from Pascual et al. 2017, for IPBES.

Fig. 11.1 (continued)

Table 11.1 Ocean contributes to material dimensions of well-being

Contribution to human well-being	Mechanism or rationale	Examples of contributions
Food	Seafood is rich in essential micro-nutrients and a key component of the human diet, particularly for coastal and island societies. Healthy diets reduce risks of non-communicable diseases such as cancer, diabetes and poor heart health (Golden et al. 2016; Hicks et al. 2019).	Calls for a global transition to sustainable food systems include increasing seafood consumption and decreasing that of land- based meats. The ocean offers great potential for improved food productivity, particularly from aquaculture of low tropic-level species such as bivalve shellfish (Costello et al. 2019).
Water	Freshwater supplies are becoming limited in some parts of the world as groundwater is depleted and surface water intensively used or contaminated. Large-scale desalinisation is a potential solution.	Large-scale desalinisation provides an increasing contribution to freshwater needs of coastal and small island states in dry areas: 48% of the world's 95 million m^3 daily production for human use is in the Middle East and North Africa region. The large volume of brine produced as a result is an environmental concern (Jones et al. 2019).

Table 11.1 (continued)

Contribution to human well-being	Mechanism or rationale	Examples of contributions
Energy	Clean sustainable energy sources are needed to decarbonise the economy and, more generally, to help decouple economic growth from increasing demand for environmental goods and services (Sachs et al. 2019). The ocean provides a range of opportunities for generating energy from clean and renewable sources.	Wind energy has traditionally been used for voyaging at sea but is now increasingly used to generate electricity for land-based human activities through offshore wind farms (Esteban and Leary 2012). Where land for solar panels is limited, ocean-based solar power is an option (Sahu et al. 2016), and there is increasing interest in using tidal, current and wave energy (Khan et al. 2017; Weiss et al. 2018). Sub-sea fossil-fuel reserves provide about a third of the world's current oil and gas and remain a target for exploration and exploitation, with 42% of the estimated undiscovered oil and gas reserves being offshore (Narula 2019). Macroalgae are potential sources of marine bioenergy (Ghadiryanfar et al. 2016).
Materials and minerals	Human societies use a wide range of non-food materials to sustain and improve quality of life and for building homes and infrastructure. Minerals are used in a wide range of industries and commodities. These include rare earth minerals needed for a transition to a renewable energy economy (Takaya et al. 2018). The ocean is seen as an underutilised source of raw materials for contemporary societies.	The ocean's calcified organisms, such as corals and molluscs, have long been a source of building materials; lime and 'coral rag' have been used to construct Mayan and Swahili cities, for example (Russell and Dahlin 2007; Fleisher et al. 2015), and are still in use today in places such as East Africa (Dulvy et al. 1995) and Indonesia (Caras and Pasternak 2009). Deep-sea mining for rare earth minerals receives a lot of policy attention (Carver et al. 2020), but much more significant, in both economic and environmental terms at present, is coastal and shelf-sea gravel extraction to meet the construction demands of rapidly urbanising and industrialising nations that are investing heavily in infrastructure development (Peduzzi 2014).
Transportation of people and goods	The ocean provides a key means of transporting goods and people. Trade and human mobility are seen as necessary to sustain our current economic system—one that has lifted billions of people out of poverty, despite its shortcomings in addressing inequality. Containerisation of goods has greatly reduced the cost and improved the reliability of sea transport (Notteboom and Rodrigue 2008).	Maritime transport enabled the generation and accumulation of capital through state-sponsored trading firms like the British and Dutch East India Companies. Mercantilism provided one of the foundations of modern economies but also began the process of resource exploitation and colonisation (Wallerstein 2011). The sea has enabled the conversion of natural resources to wealth by bringing labour to exploit the resources: the migrations of Europeans to settler colonies in the Americas, Oceania and South Africa, and the involuntary migrations of enslaved Africans, are the best-known (Gillis 2012), but there are others: between 1840 and 1940, 25 million Indians migrated to Ceylon (Sri Lanka), Malaya and Burma, with others moving to Indonesia and Indochina (Amrith 2013, 62).
Income, business profits and resource rents	Shipping and fishing have long been contributors to generating wealth and jobs and supporting livelihoods in coastal and island economies. Licenses and taxes of these maritime activities have helped finance governments. The blue economy embodies a set of principles to guide the next phase of maritime economic development for human well-being.	Income, livelihoods, profits and resource rents (access and license fees, tax revenues) are generated by jobs and investments in established maritime industries such as fishing, shipping, oil and gas extraction, military and security forces, coastal and marine tourism, as well as emergent ones such as offshore renewable energy, mariculture, deep-sea mining and bioprospecting.
Physical health	A healthy population is the basis for a productive workforce and economy. The ocean contributes to health through three main pathways: (1) as source of healthy food (see above); (2) as a source of pharmaceutical compounds for the prevention and treatment of disease; and (3) as an arena to enjoy outdoor physical activity.	The ocean has provided anti-cancer drugs and other medically useful biocompounds that contribute to human health. Coordinated plans for bioprospecting and pharmaceuticals development are underway in India (Malve 2016) and Europe (PharmSea, a program discussed in Jaspars et al. 2016). People's interactions with a healthy ocean—as tourists, recreationalists, retirees and incomers to coastal communities—have measurable benefits to their health, with the seaside sometimes referred to as a 'therapeutic landscape' (Finlay et al. 2015) or a 'blue gym' (Depledge and Bird 2009).
A physical and biological environment conducive to human flourishing	Coastal ecosystems provide services to humanity which are not easily included in monetary- based decisions, such as coastal stabilisation, regulation of coastal water quality, biodiversity conservation, spawning habitats, carbon sinks, dilution of pollution and buffering of changes in biogeochemical cycles (Baker et al. 2019).	A healthy ocean is critical to stabilising the global climate. Ocean-based mitigation options could reduce global greenhouse gas emissions by nearly 4 billion tonnes of CO_2 equivalents per year in 2030 and by more than 11 billion tonnes per year in 2050, relative to projected business-as-usual emissions. The five main options for doing so are in ocean-based renewable energy, ocean-based transport, coastal and marine ecosystems, shifting human diets towards food from the ocean while improving fisheries and aquaculture, and carbon storage in the seabed (Hoegh-Guldberg et al. 2019).

(Fig. 11.1c, step 5). We recognise that much work remains to be done on enumerating and identifying these values and their ranges in specific local, national and regional contexts.

2.2 Methodology

We use a social well-being framework (Fig. 11.2) to weave together the multiple strands of inter- and trans-disciplinary scholarship outlined in Sect. 11.3.1. A key point of origin for the concept of well-being in an economic development context lies in the work of Amartya Sen. Well-being represents a broadening of welfare economics and a further development of Sen's 'capabilities approach' (Sen and Nussbaum 1993). The material dimensions of well-being (adequate food, health, shelter, income) have been the focus of much economic policy, but it is now well known from work on the 'economics of happiness' that material wealth alone does not deliver improvements in human well-being once basic material needs are satisfied (Kahneman and Krueger 2006; Kahneman et al. 2006). This implies a need to consider how to avoid an overly narrow focus on the generation of material wealth from the ocean and thereby risking trading off ocean contributions to the other dimensions of well-being in societies.

A sense of belonging, having social status, good social relations and a sense of personal fulfilment are key contributors to human well-being. To elucidate the concepts, the framework we have chosen here is the three-dimensional social well-being framework widely applied in the field of international development (e.g. White 2010) and in the study of small-scale fisheries (e.g. Weeratunge et al. 2014; Johnson et al. 2018). It has mostly been applied at the individual, household and community level, while here we extend the concepts to include higher levels of political, social and economic organisation (Fig. 11.2). In doing so, we build on the Millennium Ecosystem Assessment (MEA) and its coupling of 'ecosystem services' with 'human well-being' (Leemans and de Groot 2003).

The ecosystem services concept acknowledges and values non-monetary human uses of nature, including 'cultural ecosystem services'. We extend the MEA's conceptualisation of human well-being by further unpacking its multiple dimensions. Responding to perspectives from Indigenous Peoples, we also adopt the Intergovernmental Science-Policy Platform on Biodiversity and Ecosystem Services (IPBES) language of

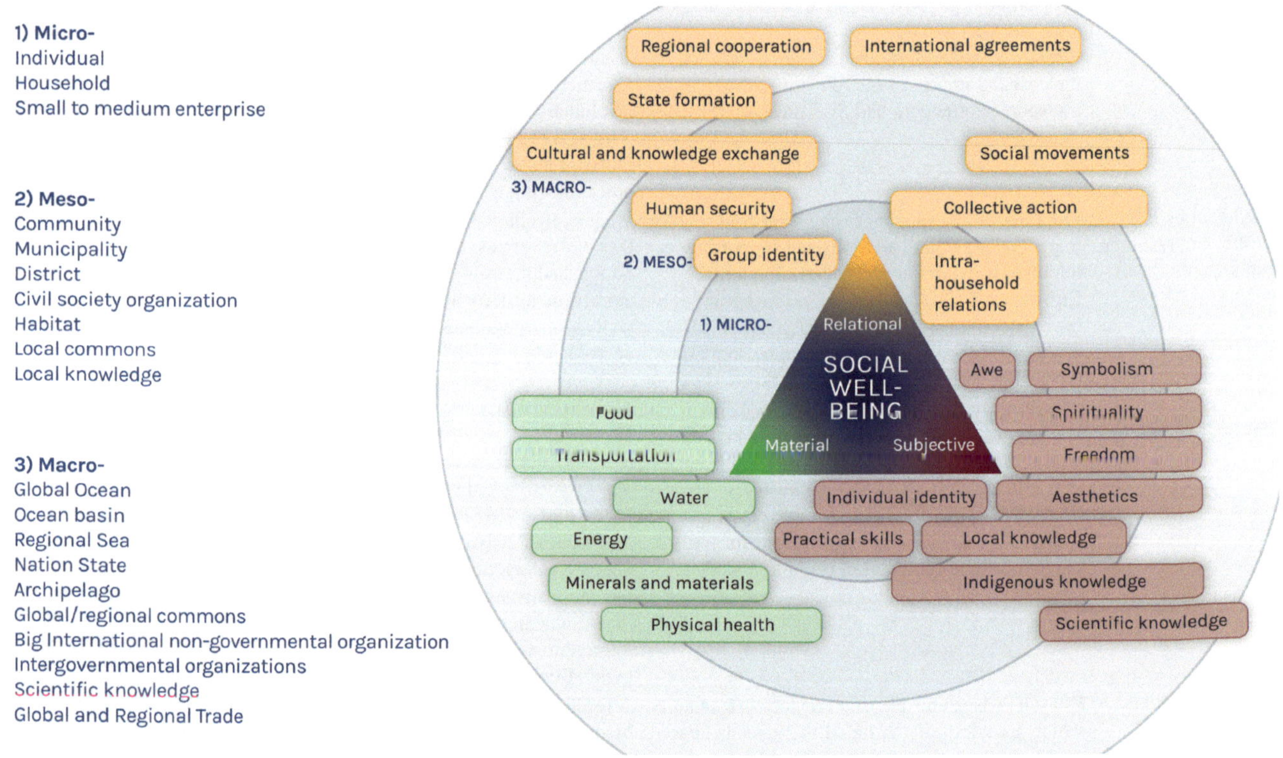

Fig. 11.2 Framework to identify human relationships with the ocean. *Note: Scales at which these relationships take place range from the micro (individual, community, local place or small enterprise) to the macro (nation, citizenry, region, ocean basin or whole ocean, large domestic or multinational firm). Source: Modified from Weeratunge et al.* (2014)

'nature's contribution to people' (Díaz et al. 2018) to reinforce the shift towards considering multiple value systems in human-relationships on an equal footing, rather than using economic values as the benchmark against which to compare all others (see Fig. 11.1 and Sect. 11.2.1). The intent here is not to replace economic valuation and ocean accounting as tools for ocean governance but to draw attention to other values, so that they may, in the future, also be fully accounted for. We therefore refer to a plurality of 'values' but we make no attempt to convert them to a universal monetary equivalent value.

The High Level Panel for a Sustainable Ocean Economy acknowledges that no one-size-fits-all solution is appropriate for the ocean. A similar case for a diversity of solutions to fisheries governance problems has also been made (Young et al. 2018). People with an interest in the ocean include those in traditional maritime occupations such as merchant seamen and fisherfolk, workers in newer offshore economies such as the energy sector, coastal Indigenous Peoples, seaside residents and tourists, the cosmopolitan populations of port cities, and seafood consumers everywhere. A common vision for the relationship between humanity and the ocean must be broad enough to accommodate this diversity of interests. It must also be democratic enough to include the interests of those who have little influence on global economic systems, state and intergovernmental policies or global ocean science. This paper brings some of those voices to the surface.

After considering the variety of relationships humanity has with the ocean from a well-being perspective (Sect. 11.3), we identify some examples from the huge variety of existing, remnant or (re)nascent regional and local governing systems (Sect. 11.4) that have evolved in response to this diversity of ocean values. We then turn to how a more beneficial and sustainable human relationship with the ocean can be pursued by building on these values and institutions, to develop a global commitment to a sustainable ocean economy and future (Sect. 11.5).

3 Human Relationships with the Ocean and Their Contributions to Well-being

3.1 The Ocean Economy and Its Acceleration

The 'blue economy'—a term that has emerged in the past decade—attempts to embrace the opportunities associated with the ocean, whilst recognising, accounting for and addressing the threats posed by such an economy. It is essentially the ocean equivalent of the 'green economy'—a vision for a decarbonised, regenerative and more equitable economic system.

The 'blue economy' gained prominence at the 2012 UN Convention on Sustainable Development (UNCSD), or Rio+20 conference, when small island developing states began emphasising the importance of the ocean and marine economy in response to land-focused calls for a 'green economy' (Silver et al. 2015; Dornan et al. 2018). Since then use of the term has become increasingly common (Mulazzani and Malorgio 2017), although the narratives surrounding the blue economy diverge considerably across different actors (Voyer et al. 2018). Strategies for implementing the blue economy vary enormously across jurisdictions and organisations.

Despite this, they commonly focus on encouraging private sector development of the ocean using innovation and investment strategies, supported by macro-level calculations and projections of the current and future 'worth' of maritime industries to global markets (Hadjimichael 2018).

It is clear that the ocean is entering a new phase of large-scale industrialisation (Box 11.1). This creates challenges for ocean governance in how to manage the increasingly private use of what has traditionally been considered a common pool resource (Arbo et al. 2018; Hadjimichael 2018). A key question for the blue economy is how to manage this acceleration without also accelerating unintended societal costs (Arbo et al. 2018; Kooiman et al. 2005; Jentoft et al. 2010; Alongi et al. 2015). This requires active consideration of questions of power and marginalisation, agency and values (Ratner 2004). Without adequate and integrated consideration of social and cultural objectives, the blue economy may become a tool for 'ocean grabbing' and marginalisation and dispossession of traditional cultural, recreational and small-scale commercial uses and users (Bennett et al. 2015; Hadjimichael 2018). Adopting a well-being perspective and emphasising the achievement of SDGs can help ensure that this does not happen.

The cumulative process of marginalisation can often be unintended, incremental and hidden. For example, as our coasts and shorelines are increasingly enclosed and developed for luxury residential and tourist uses, this 'coastal squeeze' (Cohen et al. 2019) reduces the operating space of those using the ocean and coastal commons. In particular, it marginalises small-scale fishers and lower-income recreational visitors to the coast, and confines them to less desirable, more polluted or industrialised and degraded sites. It thus risks reallocating the well-being benefits of the ocean—that common heritage of humankind—to the wealthy.

Whether we use the terms *blue economy*, *blue growth* or *sustainable ocean economy*, there is an urgent need to focus on equity—both currently and intergenerationally—as a crucial component of ocean development (Cisneros-Montemayor et al. 2019; Österblom et al. 2020). Mainstreaming and foregrounding equity within sustainability narratives will require active consideration of alternative and diverse visions for the future ocean economy. Some of these 'counter-narratives' are already being articulated and include steady-state economics (Daly 1991) and de-growth (Kallis 2011; Ertör and Hadjimichael 2020). These merit critical consideration alongside promising but less radical alternatives for future environmental sustainability, such as the circular economy, which calls for 'a new relationship with our goods and materials', emphasising a shift away from disposability and planned obsolescence towards durability and reuse (Stahel 2016). Other alternative visions, such as Indigenous ocean economies, likely exist, but uncovering and understanding them may require work with a diverse array of stakeholders.

Here we aim to ensure that any blue economy or future ocean economy vision considers the objectives of the ocean economy in social terms. An economy ultimately exists to support people's needs and aspirations.

People's needs and aspirations are more than monetary, so it follows that the economy should be built upon broader values too. Here we suggest that economic policy consider social well-being as a way to identify and categorise human relationships with the sea.

Box 11.1 'Blue Acceleration': The Urbanisation and Industrialisation of the Ocean

In the last 50 years, there has been rapid growth in new ocean industries such as mariculture, deep ocean drilling for hydrocarbons and minerals, desalinisation and offshore wind farms. Existing maritime communications, transport and tourism industries have expanded rapidly, as have the territorial claims and information needs around the ocean. Together, these have been described as a 'blue acceleration' (Jouffray et al. 2020). The impacts of this economic growth on ocean-dependent people and economies have not yet been assessed.

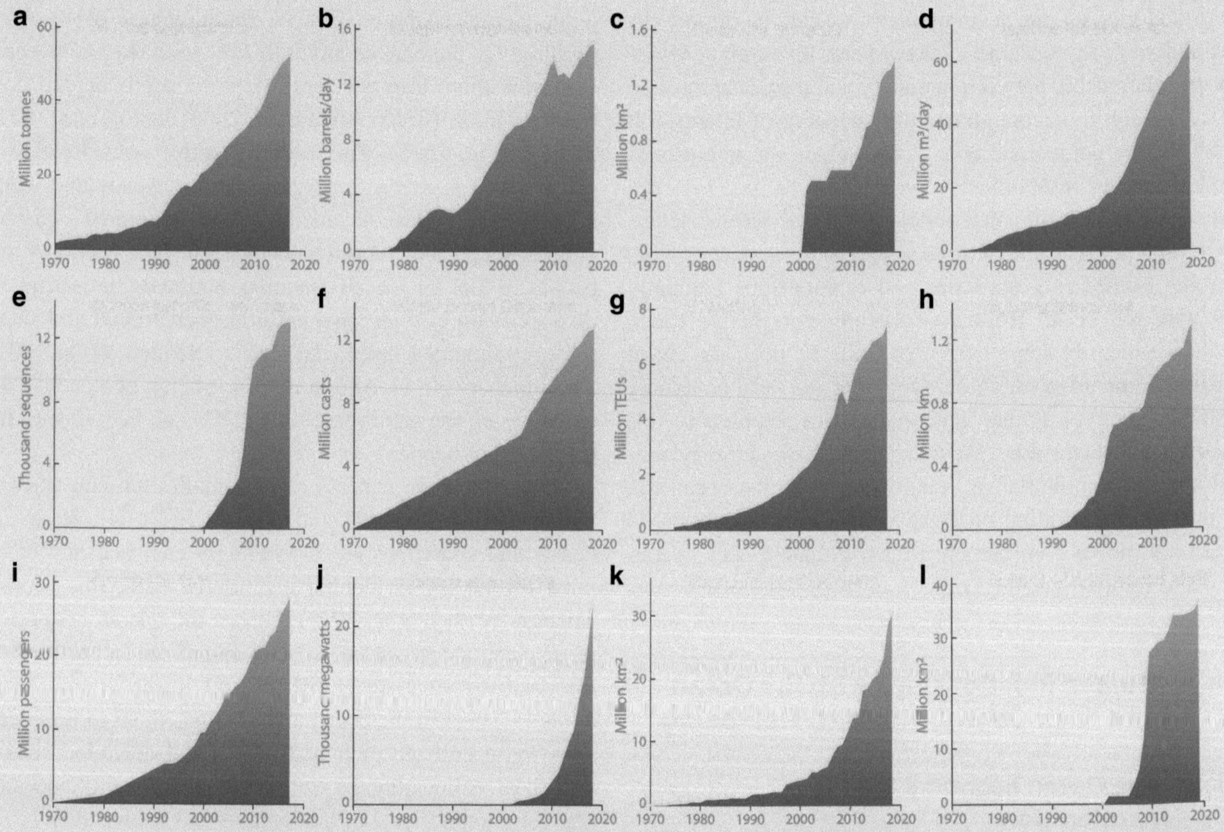

Note: Global trends in (a) marine aquaculture production; (b) deep offshore hydrocarbon production, including gas, crude oil and natural gas liquids below 125 m; (c) total area of seabed under mining contract in areas beyond national jurisdiction; (d) cumulative contracted seawater desalination capacity; (e) accumulated number of marine genetic sequences associated with a patent with international protection; (f) accumulated number of casts added to the World Ocean Database; (g) container port traffic measured in 20-foot equivalent units (TEUs); (h) total length of submarine fibre-optic cables; (i) number of cruise passengers; (j) cumulative offshore wind energy capacity installed; (k) total marine area protected; and (l) total area of claimed extended continental shelf.

Source: Redrawn from Jouffray et al. (2020), *where further information on data sources can be found.*

3.2 Social Well-being and How the Ocean Contributes to It

Economic activity in the ocean is growing rapidly (Box 11.1). If the upsurge in economic activity is to lead to an upsurge in human well-being, then its emergent and potential future impacts must be understood. In order to build that understanding, we must first 'unpack' the concept of well-being and identify the ways the ocean contributes to well-being in all its dimensions. It is these 'human relationships with the ocean' that we seek to characterise, drawing on the framework introduced in Sect. 11.2.2.

Ocean values and their contributions to human well-being at multiple levels are outlined in Tables 11.1–11.3, with the material (Table 11.1), relational (Table 11.2) and subjective (Table 11.3) dimensions outlined. These tables provide explanation and examples of the concepts outlined in the methodology (Fig. 11.2). Although we include an outline of ocean contributions to material well-being (Table 11.1), the focus of this paper is on the ocean's contribution to relational and subjective dimensions of well-being (Tables 11.2 and 11.3). Other Blue Papers focus mostly on the material dimensions of well-being when they connect to human values. However, human well-being is only achieved if attention is

Table 11.2 Ocean contribution to relational dimensions of well-being

Contribution to human well-being	Mechanism or rationale	Examples
Cultural and knowledge-exchange	The ocean has played a key role in sustaining the flow of knowledge and social and cultural exchange among societies and countries. Sea-voyaging, exploration and coastal trade and population interchange have provided key pathways for such exchanges. Before the rapid growth of air travel and the internet in the late twentieth century, ships were the most important tool of globalisation (Frykman et al. 2013). Port cities, receiving and sending ships to different destinations, were and are sites of cosmopolitanism and multi- or polyculturalism, of coastal and maritime trading societies that have been and could once again be a foundation for efforts at improved labour force diversity and inclusion, regional and global cooperation, the building of a shared ocean vision and a reinvigorated multilateralism.	Coastal trade gave rise to distinctive seaboard civilisations in the Red Sea, Arabian Sea and Bay of Bengal. These societies began to connect, some 5000 to 3000 years ago, when traders learned to use the monsoon to trade across the ocean rather than along coasts (Gillis 2012). These movements of people created relationships and exchanges of ideas and languages as well as goods and services. Religions spread along coasts more readily than between inland civilisations, and local and universal religions blended to create distinct cultures like the Swahili of coastal East Africa (Fleisher et al. 2015). Mediterranean port cities have frequently been described as 'cosmopolitan', with their merchants and populations being diverse, tolerant, multilingual and outward-looking, in contrast to land-based elites in the nation-states in which they were located. While such generalisations bear closer critical scrutiny, cities such as Izmir, Alexandria and Trieste flourished, to some degree, as centres of cultural and intellectual as well as material exchange—functions that port cities may still play, despite increasing physical separation of ports from downtown cities, under containerisation (Driessen 2002).
International agreements	Because of the fluid and interconnected nature of the ocean, and because areas beyond natural jurisdiction are both the 'common goods of humanity' and the 'collective responsibility of humanity', the ocean has played a prominent role in building the system of international law—particularly environmental law—needed to sustain humanity in the long term	Small island developing states have led advocacy for strong climate action, notably through the Alliance of Small Island States (AOSIS), and were a key group in negotiations leading up to the Paris Agreement on Climate Change. Despite their heterogeneity, they built a common diplomatic discourse and strategy for influencing policy, enabling them to mobilise political leaders, negotiators and advisers (Ourbak and Magnan 2018).
Regional cooperation	While there is 'one ocean', there are four ocean basins (Atlantic, Arctic, Indian, Pacific) and many seas. Regardless of the extent of globalisation, there are long-standing and continuing relationships between countries sharing an ocean basin or a semi-enclosed sea. Sharing these resources can help maintain regional political and economic stability, which contributes to well-being of populations around regional seas and ocean rims	The European Union binds together in economic and political union a region housing historically warring nation-states. By extending its collaborative governance arrangements into its shared seas, regional cooperation is further reinforced. The European Union has a legally binding framework, supported by EU financing and technical capacity, for establishing maritime spatial plans in the exclusive economic zones of its member states by 2021 (Friess and Grémaud-Colombier 2019). This provides the basis for creating lasting mechanisms for cross-border cooperation.

(continued)

Table 11.2 (continued)

Contribution to human well-being	Mechanism or rationale	Examples
State formation	The ocean has played an important role in the formation of modern states, with the extension of maritime territory into exclusive economic zones (EEZs), providing the basis for future national projects in the blue economy, future prosperity and therefore future well-being.	The sea has been particularly important to countries labelled as 'small island developing states'. The law of the sea has enabled them to claim large EEZs, which, together with their historical connections to the ocean, has encouraged them to position themselves in ocean policy dialogues as 'large ocean states' (e.g. Chan 2018). This relational repositioning has enabled oceanic island nations to shift from an emphasis on their vulnerability and small size to highlighting their stake in—and sovereignty over—30% of the ocean, thereby positioning them more strongly in ocean economy and governance policymaking.
Human security	Ocean peoples have long shared many social norms, many of which are now codified in modern law. These maritime codes of conduct were designed to improve safety and well-being at sea and facilitate travel and commerce. Global norms around neutrality, calling truces and the practice of rendering assistance to those in distress all have maritime origins and provide important contributions to human security and to our frameworks for moral conduct more generally.	The sea has been particularly important to countries labelled as 'small island developing states'. The law of the sea has enabled them to claim large EEZs, which, together with their historical connections to the ocean, has encouraged them to position themselves in ocean policy dialogues as 'large ocean states' (e.g. Chan 2018). This relational repositioning has enabled oceanic island nations to shift from an emphasis on their vulnerability and small size to highlighting their stake in—and sovereignty over—30% of the ocean, thereby positioning them more strongly in ocean economy and governance policymaking. The early twentieth-century Hague Conventions on Neutrality are based on the rules set out in the Consolato del Mare of 1494, which in turn is partly based on ancient Greek (Rhodian) sea law (Bauslaugh 1991, xiii). The dangers and solitudes of sea voyages in the ancient world led to a set of unwritten codes of maritime conduct, including those regarding the 'sacred duty of hospitality' included in Hugo Grotius's Mare Liberum. This, in turn, forms the basis for the 1979 Convention on Maritime Search and Rescue and the 1974 International Convention for the Safety of Life at Sea. There is much concern that these principles are being eroded with the rise of human trafficking, with coastal states failing in the legal duty to assist migrants and others lost at sea, and in their obligations to disembark rescued persons in a place of safety (e.g. Aalberts and Gammeltoft-Hansen 2014). These dialogues point to the continuing importance of maritime social norms in shaping the moral basis for governing human affairs.
Social movements	Shared experience of the ocean can help to build solidarity among boat crews, among port cities and between maritime nations that give rise to 'social movements'. Such social movements—and their protests—are a primary means by which social justice has been achieved historically. The equity dimension of the blue economy can be fulfilled by working together to overthrow tyranny; or to create safe working conditions, fair wages and equitable access to the benefits of the sustainable ocean economy, and to exert influence on democratic governance. Social movements may coalesce into representative organisations, such as the International Transport Workers Federation, the World Forum of Fisherpeople and the World Forum of Fishworkers. The Missions to Seamen, which provides spiritual, pastoral and practical care for seafarers, has become particularly important to crews stranded on ships amid the COVID-19 pandemic (Rynd 2020).	A series of connections, formed over centuries of trade among seafaring peoples of the Americas, Europe and Africa, led to loose coalitions of fugitives from state violence and exploitative work. These maritime societies, and the experiences embodied in their stories, existed over 'vast spaces and spans of time' (Linebaugh and Rediker 2000) and were characterised by their mobility and multi-ethnicity—giving rise to the expression 'a motley crew'. They nurtured an Atlantic 'maritime radicalism' characterised by collectivism, anti-authoritarianism and egalitarianism, that connected revolutionary impulses in the United States, France and Haiti (Frykman et al. 2013). These values and the transnational coalitions that spread them are models for today's globalised anti-racist and anti-capitalist protests. Social movements have long begun in seaports, including the Solidarity movement that eventually led to the election of one of its leaders, Lech Walesa, as Polish president. It began among workers in the Gdańsk shipyard in 1980 and, in 1989, succeeded in ending Soviet communism in Poland (Garton Ash 2002).

Table 11.2 (continued)

Contribution to human well-being	Mechanism or rationale	Examples
Group identity and 'belongingness'	Identifying with a nation, ethnic group, community or locality ('belongingness') is associated with well-being (Helne and Hirvilammi 2015). Some societies and cultures identify strongly with the ocean as their 'place' and with fellow ocean users as their people; the well-being of these maritime societies and sea people depends on their access to a healthy ocean, a maritime lifestyle and to relationships with each other.	Many of the Solomoni of Fiji are descended from nineteenth-century indentured labourers. They have acquired and negotiated their sense of belonging through active engagement with their oral history (tukuni), particularly around the notion of tauvu ('springing from the same ancestor') with taukei (Native/Indigenous Fijian). This has helped to break down group stereotypes, overcome initial disadvantage and enable coexistence and intermarriage (Mateiviti-Tulavu 2013). The construction of shared Pacific Islander identity continues at larger scales as the region faces common threats, such as climate change, and as Islanders meet in regional forums and diaspora populations extend across the world—particularly in Australia, New Zealand and the United States. A shared history with the ocean forges an evolving trans-Pacific identity (Hau'ofa 1994).
Cooperation and collective action	Because the ocean is largely governed as a commons, most ocean users cooperate to share access to ocean resources and spaces, and to reduce conflict. This is particularly the case with fishing communities. Building good relationships within and between communities contributes to well-being.	Many coastal fishing societies around the world have at some time developed cooperative ways to manage the marine resources upon which they depend. These social institutions do more than regulate access to resources; they provide the basis for harmonious community life; they perpetuate culture and they provide social security through risk-sharing and asset-sharing mechanisms (Ruddle 1988; McGoodwin 2001). Some of these traditional institutions have been eroded by the switch to state-based fisheries management, but there is revival of traditional management blended with formal government in a range of 'co-management' arrangements, including locally managed marine areas networks in the Pacific (Techera et al. 2009) and territorial use rights for shellfish in Chile (Fernández et al. 2011). Further examples of traditional institutions for marine resource management are given in Sect. 11.4 of this paper.
Intra-household and intra-community relations	In many contemporary cultures, beaches and the seaside have strong associations with family holidays and childhood memories (Marschall 2015), with romance and with togetherness in old age (Huntsman 2001). Within fishing communities, boats crews often have strong kinship and friendship ties. These are all forms of 'social capital' that contribute to relational well-being at more intimate levels, as well as ensuring the intergenerational exchange of knowledge and skills in maritime households.	Australians make or break romances at the beach, they marry and take honeymoons at the beach, they go on holiday with their children at the beach, and in vast numbers retire by the sea. —Robert Drewe, quoted in Huntsman (2001, 2) In the artisanal fishing community of Lobitos, Peru, 'fishermen spend a great deal of time socializing with each other at the harbor, during communal celebrations and activities, on boats and at the usual meeting places, talking about the state of the sea and fishing activity', but they only share fishing secrets among kin. The boats are often crewed by kin: the most common crew combinations are groups of brothers (30.6%) and parents and children (26.5%). Fishing tasks are taught and learned through these family and community relations, with children involved from their pre-teens (Maya-Jariego et al. 2017).

paid to all three dimensions. Different individuals and cultures ascribe different levels of priority to these values, but no society discounts them entirely.

Subjective well-being has entered national economic policy as measures of economic and social performance (Stiglitz et al. 2018), also popularly known as the 'economics of happiness' (Easterlin 2001; Clark 2018).

The material, relational and subjective dimensions of well-being are, however, interconnected or 'co-constitutive' (White 2017), and, like all such classifications, the boundaries between categories are porous. For example, seafood provides for material needs for nutrients, protein, energy, income and profit, particularly in maritime South and Southeast Asia, coastal West Africa and the Pacific islands. But it also contributes to relational and subjective well-being through association with religious observance (e.g. fish at Easter in Catholic societies in South America), status (e.g. seafood banquets in Chinese culture) and feelings of connection to place (e.g. the importance of wild salmon to both Native/First Nation and settler coastal populations in the U.S. and Canadian Pacific Northwest). Note also that most well-being classifications are made at a single scale (e.g. the

Table 11.3 Ocean contributions to subjective dimensions of well-being

Contribution to human well-being	Mechanisms or rationale	Examples
Scientific and scholarly knowledge	The scientific exploration of the ocean has provided the foundations to secure a sustainable and prosperous future for the ocean economy. It allowed the exploration of the ocean in search of the materials that support contemporary human societies. Ocean science has also identified the nature and scale of threats to human well- being posed by ocean ecosystem degradation and has illuminated a number of solution pathways in the form of conservation measures, pollution control and abatement technologies. While ocean sciences provide indirect support to the realisation of relational and subjective well-being, they are not primarily suited for this purpose; additional scholarly disciplines, such as the ones we draw on in this paper, are also required to identify ways to support ocean values and their contributions to well-being. Conceiving of the ocean as a peopled space has been an important first step towards attracting the social sciences and humanities to study the ocean (Steinberg 2001) and how it might be better governed to support human well-being.	Scientific knowledge of the ocean has helped find ways to support sustainable fisheries (Hilborn and Hilborn 2019), ocean-based climate change mitigation solutions (Hoegh-Guldberg et al. 2019), countering pollution threats and conserving biodiversity (Knowlton 2020). Science underpins the search for ways to ensure that material well- being is sustained. Beyond the ocean (natural) sciences, the social sciences, arts, humanities and professional disciplines (e.g. law, finance, management and policy studies) can provide insight into how to support all three dimensions of ocean- related human well-being. Inter- and trans-disciplinary research facilitates knowledge integration and the provision of useful policy advice (Claudet et al. 2020).
Phronetic (practical) wisdom and skills	Knowing how to row a boat, fix an engine, cross the surf or spear a fish are among the many practical skills that imbue a sense of pride in maritime occupations. Skills and practical sea knowledge are useful for sustaining material well-being from the ocean, for gaining admission to maritime society by earning the respect and recognition of other mariners, and as a source of personal pride and of belonging to the ocean. These apply whether you are a navigator, fisher, diver, surfer or marine biologist.	In the highly gendered world of commercial fishing, women in Alaska's Bristol Bay salmon fishery take pride in their seagoing and fishing skills as well as their role in safeguarding and deepening traditional knowledge in their fishing communities. They have earned the right, in their view, to be called 'fishermen' (Lavoie et al. 2019).
Freedom and adventure	A sense of freedom and a sense of adventure both contribute to autonomy, which is an important component of subjective well-being . Freedom evokes a sense of choice and possibility, as well as agency . Adventure involves challenging oneself mentally and physically, learning to calculate and confront risk, and experiencing contrast with everyday life . Exposure to nature is correlated with a sense of autonomy (Passmore and Howell 2014). The search for these subjective mental states has motivated individuals and societies to embark on ocean voyages, to explore the coast and undersea environments and to challenge themselves by big wave surfing, deep-sea exploration or blue ocean sailing.	The sense of freedom and adventure of being on or in the ocean is described by well-known and accomplished participants in marine adventure sports as restorative, connective to nature, and contributing to building their autonomy and resilience (MacIntyre et al. 2019). Fishers often emphasise their need for independence and choose to stay in the fisheries, even when more lucrative work is available, because they value the independence and freedom of working for themselves, or with friends and family, and are unwilling to submit to working indoors, being bound to a timetable and reporting to a boss (Pollnac and Poggie 2008). In contrast, 'adventure' is not one of the reasons listed for the revival of Pacific voyaging, though such voyages through the vast ocean in a small craft would be most people's idea of an adventure. Instead, the 'five values' motivating such voyages are a mixture of the relational and the subjective: knowledge, the pursuit of excellence, the exercise of rights and responsibilities, acting morally and selflessly, and nurturing relationships to the ocean and nature (Herman 2016).
Awe	To land-based people, the ocean inspires a range of emotions, from fear to curiosity. The human psyche flourishes when there is opportunity to confront our anxieties, reflect on our place in the cosmos and experience a feeling of something larger and more permanent than ourselves. The ocean seems to prompt such reflection, due to its vastness and relative unknowability.	Feelings of fear and awe have characterised some of our land-based cultures' views of the ocean, such as Judeo-Christian and Hindu ones (Connery 2006; Andaya 2017), while for those more at home on the waves, the ocean invites awe and respect but is also regarded as a source of comfort and familiarity (King and Robinson 2019). Fear gave way to fascination through the eighteenth- and nineteenth-century notion of the 'sublime' in European culture, where rough seas, ocean depths and rocky coasts are reinvented as sources of aesthetic and sensory pleasure (Gillis 2012).

Table 11.3 (continued)

Contribution to human well-being	Mechanisms or rationale	Examples
Aesthetics	The ocean and coasts have inspired the visual and creative arts, and humans have felt the need for aesthetic expression since at least the time the first cave paintings were created. A continuing cultural relationship with the ocean is important to sustaining this inspiration in future generations.	Marine animals appear in early rock paintings, even in continental interiors in Africa and Australia. Perhaps one of the most recognised marine images comes from Japan: The Great Wave off Kanagawa, an early nineteenth-century woodblock print by Hokusai. Paintings of ships and beach- side scenes are popular in European art. See Annex 2 for sources and further discussion.
Belongingness (subjective elements)	While belongingness has relational components (being part of a group, see Table 11.2), the sense of belonging with the ocean and being 'of the sea' also has subjective elements that contribute to self-actualisation and well-being.	It is often said that the Bajo-Laut people of maritime Southeast Asia feel sick if they spend too much time on land, or away from the ocean. They maintain a rich Indigenous marine cosmology and ritual practice, with belief in supernatural beings—ancestors of the ocean—that live in and control the universe of the ocean, and all the creatures in it (Stacey 2007). See Annex 1 for a more in-depth review. A sense of well-being that comes from feeling at one with the ocean may also be achieved by those who are not of the ocean: 'The term "blue mind" describes the mildly meditative state we fall into when near, in, on or under water. It's the antidote to what we refer to as "red mind," which is the anxious, over-connected and over-stimulated state that defines the new normal of modern life … spending time near the water is essential to achieving an elevated and sustained happiness' (Nichols 2015).
Ocean identity	While the ocean is the place where you go to do your job, it may also be where you feel most free, most in control of your own destiny, most competent and most valued by others. When jobs are evaluated and compared only in terms of returns on investment or labour productivity, they are seen as fully substitutable; they are not (Pollnac and Bavinck 2008). Occupational pride and place attachment are associated with well-being. To know yourself, to be known and respected by others, and to belong to a place are all important to people's subjective well-being. Maritime identities, whether as fisherfolk, fish traders, mariners, islanders or 'waterpersons' contribute to these feelings.	'Fish mammies' are wealthy and respected independent entrepreneurs who finance the fishing operations of men in coastal West Africa, as well as run fish processing and trading operations. It is a title and social position adopted with pride and not one that comes from working as an employee in an industrial fish processing plant. The identity of a person of the ocean or a 'waterman', first associated in wider culture with Hawaiian surfing pioneer Duke Kahanamoku (Davis 2015), has been applied both to those whose work involves physically entering the water (maritime rescuers, commercial divers) and to dedicated practitioners of marine sports, whether they are amateur or professional. It indicates both high levels of competence in and affinity with the ocean.
Symbolism	All human cultures create and value symbols, as material representations of abstract or subjective concepts that are important to us, like love, loyalty, faith, belonging, status and power (Callahan 2013). Many important cultural symbols relate to the ocean and ocean animals. The ocean's power to inspire symbols contributes to our well-being.	Sharks in the Pacific islands were imbued with spiritual powers, considered as ancestor guardians and/or gods who offered protection from the unpredictable forces of the ocean. As symbolic animals, sharks appear to signify both the fear of the unknown vastness and depths of the ocean as well as its bounty to those who respect its ways and powerful creatures. See Annex 3 for sources and further details. At independence, the Caribbean island nation of Barbados adopted an everyday food item of the poor, the flying fish, as symbolic of Bajan identity; it appears on flags, as the name of national sports teams and as the personification of national identity; a person is said to be 'as Bajan as flying fish' if he or she exhibits traditional mores and behaviours (Cumberbatch and Hinds 2013). See Annex 4 for sources and further details.
Spirituality	Spirituality and religiosity are positive predictors of subjective well-being (Villani et al. 2019). The ocean has played an important part in the development of human spirituality—for example, many cultures have 'sea gods', some of which are benign and others which warn of dangers (Andaya 2017). Both benign and danger-warning deities help make sense of fate and provide a psychological means to cope with mortality . Judeo-Christian religious texts are generally hostile to the ocean (Connery 2006), but many non-Western maritime cultures and religions have more complex and affirmative spiritual relationships with it (Andaya 2017; Hauʻofa 2008).	In Nordic mythology, the sea exceeds the land, in power and expanse, and is depicted as 'a treacherous surface to be traversed for fame and gain', its treachery personified by the female sea-deity Rán. A robber of life from men, she personifies death by drowning (Quinn 2014). The lives of peoples reliant on the ocean are fraught with unpredictability; belief in and appeal to sea gods and spirits are means of ensuring both good fortune and protection (Andaya 2017). The ocean people of Southeast Asia believe well-being is assured by seeking protection from benign spirits and gods, and offering propitiation to malign ones. With the advent of long-distance trading and voyaging into the oceanic realm, spirituality has been extended to include the gods of the major religions, notably Islam and Christianity, whose protection is sought in distant reaches of the ocean (Andaya 2017) .

individual, household or nation-state) while here we consider multiple scales. This classification is therefore schematic and each dimension and scale is illustrated by a small number of examples only, due to considerations of space.

The aim here is to establish a new conceptual framework that links ocean services or benefits to human well-being in ways that account for the plurality of human values.

Note that ocean services or benefits described here extend beyond ocean *ecosystem* services. An ecosystem services approach values only some of what the ocean contributes to human well-being. Many of the ocean's contributions are not directly related to its ecology but instead relate to the ocean as a space—both material (having area, volume and fluidity) and non-material (as a place of consciousness and imagination). These distinctions and their importance will become clear through examination of the processes and examples in Tables 11.1–11.3.

> The ocean does not divide us, it connects us.
> Pacific island aphorism that has become a contemporary cultural meme

> When anxious, uneasy and bad thoughts come, I go to the sea, and the sea drowns them out with its great wide sounds, cleanses me with its noise, and imposes a rhythm upon everything in me that is bewildered and confused. Rainer Maria Rilke (1969 [1903])

Tables 11.1–11.3 illustrate the many ways the ocean has and will continue to contribute to human well-being. While many of these contributions, across all three dimensions, rely on maintaining ocean ecosystem health, it is too simple to say that all of them do. Ensuring ocean health is, by itself, not enough to safeguard and improve human well-being. It is also important to continue to maintain and build the kinds of social and cultural connections to the ocean that have improved human knowledge, understanding, cooperation, security, meaning and happiness in the past. If the majority of those who would benefit from an ecologically healthy ocean are excluded from it, this will not lead to improved human well-being for all. Thus, maintaining ocean health *and* maintaining inclusive ocean access should be the dual aims of governing the future ocean.

Although we have separated out the dimensions of well-being in order to explain and explore them, it is important to reinforce that they are interrelated or 'co-constitutive' (i.e. each dimension builds on the others). As Sarah White (2017, 133) notes, 'Rather than dividing "subjective" from "objective", subjective, material and relational dimensions of well-being are revealed as co-constitutive. Wellbeing is emergent, the outcome of accommodation and interaction that happens in and over time through the dynamic interplay of personal, societal and environmental structures and processes, interacting at a range of scales, in ways that are both reinforcing and in tension' (White 2017, 133).

Such reinforcing feedbacks between dimensions of well-being can be found in the ways Pacific Islanders think about their relationship with the ocean. As Epeli Hau'ofa (1994, 153) wrote, the ocean provides material, relational and subjective 'goods' in inseparable and historically constructed ways: '"Oceania" denotes a sea of islands with their inhabitants. The world of our ancestors was a large sea full of places to explore, to make their homes in, to breed generations of seafarers like themselves. People in this environment were at home in the sea.'

Similarly, a sense of being part of a community and a sense of place contribute to the 'social embeddedness' of coastal communities engaged in small-scale fishing (Jentoft 2019). Coastal and sea-dwelling communities have strong social ties and distinct cultures from which they derive well-being. These identities and concepts of embeddedness straddle both the relational and subjective dimensions of well-being—the sense of belonging to a group, such as fishers or a coastal community, of being able to depend on your group during emergencies, times of loss and crisis due to the strength of social relations and networks, but also subjective feelings of pride and self-worth in one's occupation, community or ethnicity.

In Sect. 11.4 of this paper, we will outline some case-study examples of institutions—both contemporary and traditional—that illustrate how the different dimensions of well-being and the multiple spatial scales at which they accrue and are intertwined in the institutions that have evolved to govern human-ocean interactions in practice. First, we consider each set of relationships separately.

Table 11.1 summarises the many ways the ocean contributes to provisioning human needs and regulating the earth's environment to ensure human flourishing and biodiversity conservation. These contributions are linked to the ocean as both a place and as an ecological system.

While the ability to extract minerals or transport goods is largely independent of ocean ecosystem health, these activities certainly impact upon it. The challenge for the future ocean economy is to ensure that governance of provisioning and regulatory goods and services—such as food production and climate mitigation measures—do not threaten ocean health. Future ocean governance also has to ensure that human relational and subjective well-being are supported rather than undermined by the 'blue acceleration'. Most modern, state-based and global ocean governance institutions have formed with the goal of managing access to and use of ocean resources and ocean space for material use and, to a degree, to regulate relationships between private enterprises (e.g. property rights and trade and commercial laws) and between states (e.g. regional seas agreements, freedom of navigation agreements). They seldom consider subjective well-being.

Ocean governance institutions are also mostly designed to regulate commercial activity at sea. However, people do more than extract resources, trade goods or migrate across the ocean, they also interact with the marine environment and the marine species and ecosystems in a multitude of ways that may be rooted in material elements of the ocean, such as seafood, beaches, waves and reefs, but which are enjoyed, both consumptively and non-consumptively, for non-material purposes. These interactions—whether we experience them directly as beachcombers, rock-poolers, snorkelers, divers or recreational fishers, or vicariously through aquarium visits or viewing television series such as David Attenborough's *The Blue Planet*—create sets of relationships with ocean nature that respond to a range of human material, relational, spiritual and emotional needs. As people rise out of poverty globally, such interactions may engage an increasing number of us.

Prior to COVID-19, a burgeoning area of tourism research was devoted to understanding ways to cater to the preferences of China's growing number of newly middle-class beach tourists, both domestically and internationally (e.g. Liu et al. 2019; Jie Li and Carr 2004; Liao and Aguilera 2020). It is encouraging that over 80% of surveyed beach tourists in Qingdao, China, would be willing to pay a tourist tax in order to maintain beach and water quality at their destination (Liu et al. 2019). The global rise of beach and marine tourism, instead of being seen solely as a threat, might be considered an opportunity to bring the well-being benefits of the ocean to a growing proportion of the global population, and to engage ever more people in the cause of sustaining the global ocean.

The ways the ocean contributes to relational well-being (Table 11.2) are more concrete and better understood at smaller scales: the social cohesion of traditional fishing communities and how this contributes to economic, social and cultural life is well-studied, understood and increasingly legally mandated in the form of devolved management and community resource rights. At higher spatial scales, the relationships are more abstract but nevertheless important: for example, the need to share the oceanic realm has fostered certain moral norms that have spread onto land, such as the principles of neutrality, truce and rendering assistance to others in need. These were all codified at sea before they became part of the broader moral and legal framework for inter-state governance, and in some cases (e.g. rendering assistance to others in need) they remain more strongly upheld in oceanic than terrestrial contexts. This became very evident when fishers in the Indian state of Kerala took the lead to rescue thousands of inland folk in the 2018 floods because they felt it was the 'right thing to do' (*OnManorama* 2018).

Of the three main dimensions of well-being, the elements of 'subjective' well-being described in Table 11.3 are the most difficult to ascribe monetary value to and therefore to incorporate into traditional sectoral economic planning, though some of them have been considered in social accounts and happiness and well- being indices (e.g. Stiglitz et al. 2018). We know that these are some of the concepts and emotions that give life meaning, purpose and value beyond the meeting of basic physiological and economic needs and beyond the sociopolitical necessities of cooperating with others. For these reasons, they are worthy of policy consideration.

Because they are difficult to value—and even to articulate—the subjective elements of well-being may be dismissed as unimportant. Yet people have used symbols of belief or identity as a pretext to fight wars or have gone to war driven by socially constructed moral concepts such as honour (O'Neill 2001). Political and legal regimes are built around symbols such as flags (Posner 1998). As this paper is being written, statues that symbolise economic and social progress to some and colonial oppression and enslavement to others are being fought over as the U.S.-initiated Black Lives Matter movement ignites a worldwide reckoning on racism and colonial history (Grovier 2020). The symbolic value of the ocean and its organisms to coastal societies—and the extent to which people from these societies are willing to defend them—should therefore not be dismissed lightly, since it provides opportunity for both conservation and development.

Several policy implications arise from an understanding of the symbolic value of marine organisms. The first is that dominant global sensibilities and relationships to animals may be regarded as an imposition of cultural values if forced upon all people. There are lessons for wider global ocean governance from attempts to implement universal bans on the harvest of whales and other marine mammals, with nation-states and Indigenous Peoples who pursue traditional whaling activity resisting these bans in various ways and maintaining their cultural relationship to whales as food as well as cultural keystone species invested with complex symbolic meaning (Paul 2000). The principle of free prior informed consent (PIC) is relevant here. PIC is a negotiated or treaty-based procedural right for Indigenous Peoples in relation to development or natural resource exploitation proposals and their effect on Indigenous lands, culture and traditions (Rosenthal et al. 2006). PIC relates to the public trust doctrine—the main legal concept for governments' fiduciary obligation to protect and sustainably manage natural places held in common by the public citizenry. It is especially relevant as legal support for citizen participation in official decisions made about the marine space when government trustee obligations are breached.

Subjective well-being is also driven by anxieties, with psychologists identifying six existential ones: identity, happiness, isolation, meaning in life, freedom and death (Passmore and Howell 2014). All these anxieties can be either confronted or relieved (or both) in our relationships with the ocean—and with nature more generally—whether that relationship is professional, residential, consumptive or recreational (Tables 11.1–11.3). We observe that groups whose lives are closely entwined with marine resource use (fisherfolk, mariners, Indigenous Peoples, marine tourism and recreation professionals) have complex, multidimensional relationships with the ocean which are often deeply spiritual (in Southeast Asia; see, e.g., Andaya 2017) and strongly inform social and cultural identities (in the Pacific islands; see, e.g., Hau'ofa 2008).

While separating out the different dimensions of well-being enables them to be identified in any policy context, it is also useful to consider how they relate to and reinforce each other in a sectoral context. Consider, for example, the values embodied in small-scale fisheries (Box 11.2).

Human health also combines all three dimensions of well-being. The physical and mental aspects of people's health is affected positively by a clean ocean, which can be enjoyed by seafood consumption, trips to the seaside, swimming or pleasure cruises on the ocean. It is negatively affected by a polluted ocean (mercury and microplastics in seafood chains, oil spills, coastal industries, etc.). The relational aspect of health has to do with a sense of community, social cohesion, and so on, for example following the disintegration of coastal communities due to loss of fish stocks, as happened with the collapse of the Canadian cod in Newfoundland in the 1990s (Gien 2000). The subjective aspect of health has to do with the emotional state of being, in this case with the kind of feelings towards the ocean that are evoked by relationships to the environment and to marine species. By examining issues from both a sectoral perspective (in this case, health) and a well-being perspective, the ramifications of different policy choices can be examined, and synergies and trade-offs between dimensions of well-being can be identified.

Box 11.2 Ten New Ways to Valorise Small-Scale Fishers

- For their phenomenal vernacular ecological knowledge.
- For their innate contribution to biodiversity conservation through convivial technologies.
- For their largely owner-operated and collegial harvesting which fosters greater equity and comradery in work.
- For their cost effective and energy-efficient operations with lower carbon footprints.
- For their entrepreneurial prowess in providing high private and social returns despite limited means.
- For their greater contribution to food security and wholesome nutrition for local rural consumers at affordable prices.
- For their generation of inclusive livelihoods—particularly among women—along these short value chains.
- For their provision of localised physical protection and security to coastal and riparian territory.
- For their vital contributions to the economy of their countries with minimal subsidies.
- For their protection of balanced life both below and above water.

Source: Kurien (2019).

Finally, it is important to reinforce that cultures, along with their symbols, spirituality, aesthetics and ethics, are not static. Even the ways emotions are elicited and expressed—how we show anger, fear, hope or love—change over time (Zeldin 2012). Governing the ocean to maintain well-being is not, therefore, about preserving the status quo or returning to the past. It is about finding ways to maintain a diverse and inclusive set of relationships with the ocean and among ocean nations and peoples. It is these relationships that have generated—and will continue to generate—curiosity, awe, wonder, spirituality and aesthetic appreciation, as well as food, energy and wealth. Supporting these 'ocean contributions to people' means allowing people to (re)discover and interact with the ocean in ways that build on their own histories and their existing maritime relationships. Such relationships may be highlighted and promoted under existing slogans and campaigns such as those extolling 'ocean pride' (Indonesia Ocean Pride 2020), 'ocean optimism' (Knowlton 2020) and 'ocean literacy' (Schoedinger et al. 2010), though they may need extending to become more inclusive. The aim of such campaigns should be to reconnect people with the ocean, and raise awareness of its importance to our history, our present and our future.

3.3 Addressing Social Difference Within Maritime Societies: Towards a Diverse and Inclusive Future Ocean Economy

Moving towards a more diverse and inclusive future ocean economy, and understanding how plans to develop a sustainable ocean economy might affect people with different identities and circumstances differently, requires that we understand a diversity of people's experiences and relationships with the ocean. Such analyses of the contemporary ocean economy are few, and collating the scattered accounts of diverse people's lives at and with the ocean, upon which such an analysis could be based, is beyond the scope of this study.

Box 11.3 How Are Blue Economy Narratives Gendered?
The 2019 World Oceans Day (June 8) focused on promoting gender equality in all ocean-related activities, linking UN Sustainable Development Goals 5 and 14:

- An inclusive blue economy must, by definition, place fairness and equity at its core. It must consider the different needs and challenges faced by women and men. Research that includes gender data disaggregation or women specific studies, while valuable, is not enough.
- In research and policymaking, gender must be mainstreamed, from project inception and design to implementation, whether talking about fiscal policy and incentive-based management, or natural capital accounting, or impact investment or plastic pollution (Merayo 2019).

There is, for example, no substantive analysis of gender dimensions of the blue economy beyond calls for gender-inclusive development (Box 11.3), though there is a growing historical appreciation of the role of gender relations in shaping the maritime past (e.g. Creighton and Norling 1996) and an evolving historiographic gender research agenda (Stanley 2002). There is an extensive literature on gender and fisheries (e.g. Frangoudes et al. 2019) and an emergent one on gender, ports and shipping, much of it the work of the gender research group at the UN World Maritime University (Kitada et al. 2015). This work could be extended to understand the different relationships men and women (and other gender identities) have with the ocean, how they can inform and enrich governance, and the ways future economic growth and governance reforms could influence those relationships to improve women's social and relational well-being as well as their material economic circumstances.

The idea of maritime work as the domain of men is relatively recent and not present in every culture (Gillis 2012). Prior to the nineteenth century in Europe, women had been very much a part of the coastal world, involved in virtually every aspect of trading and fishing. This was forgotten as the ocean came to perform the metaphorical service of reinforcing and intensifying gender differences that were being eroded on land as women gained access to education, political power, rights to land and professional employment (Steinberg 2001, 191).

Gender inequality is a continuing constraint to improving well-being and meeting the UN Sustainable Development Goals, but it is not the only one. Wealth inequality, after falling in much of the world during the mid-twentieth century, has risen once again, particularly in the 'anglosphere' (the United States, United Kingdom and Australia), as growth has proved not to be the 'rising tide that lifts all boats' (Österblom et al. 2020). The ocean, central to capitalist wealth accumulation through mercantilism, colonialism and globalisation, has not historically offered equal opportunity to improve well-being.

Recent work on equity and inclusion in Washington State's maritime economy (Maritime Blue 2020; Arbow 2019) points to the need to diversify the state's maritime workforce—which is aging, white and male, and faces potential labour shortages that may limit its growth potential. Workplace barriers to greater inclusivity include a climate of discrimination against women and people of colour in the sector and maritime labour unions, with a history of protecting jobs by limiting access to certain ethnic or racial groups. This has led to a pattern familiar throughout the global economy: recent immigrants, people of colour and women are all overrepresented in seafood processing and other low-wage, low-status jobs but underrepresented in senior and middle management positions and skilled or better-remunerated jobs, such as longshoremen in ports. This is likely to be a wider issue in the maritime sector and planning, for a sustainable ocean economy will require specific attention to equity and inclusion in the expanding maritime workforce. The development of new offshore industries provides an opportunity for inclusive maritime vocational training to create a diverse workforce in this emergent maritime economy.

We have so far provided a few illustrative examples of social differences *within* coastal societies and maritime workforces. However, there are also inequalities in ocean-related well-being among geographies. In a sea of exclusive economic zones (EEZs) and nation-states, among the most vulnerable 'ocean citizens' are, ironically, those with the longest histories and closest ties to the ocean. These include Indigenous coastal people around the world (Fig. 11.3) and mobile populations who cross national frontiers or live upon the ocean, like the Sama-Bajau people in Southeast Asia (see Annex 1). Diverse relationships with the ocean in different cultural and geographical settings are explored further in Sect. 11.4.

While we have emphasised sense of freedom, adventure, autonomy and self-actualisation as part of the ocean's contribution to subjective well-being (Sect. 11.3.2, Table 11.3), it is important to recognise that inequalities—and the histories of how these have been produced and reproduced—shape this aspect of our relationship with the ocean. Western literature portrays the ocean as a source of freedom, but in doing

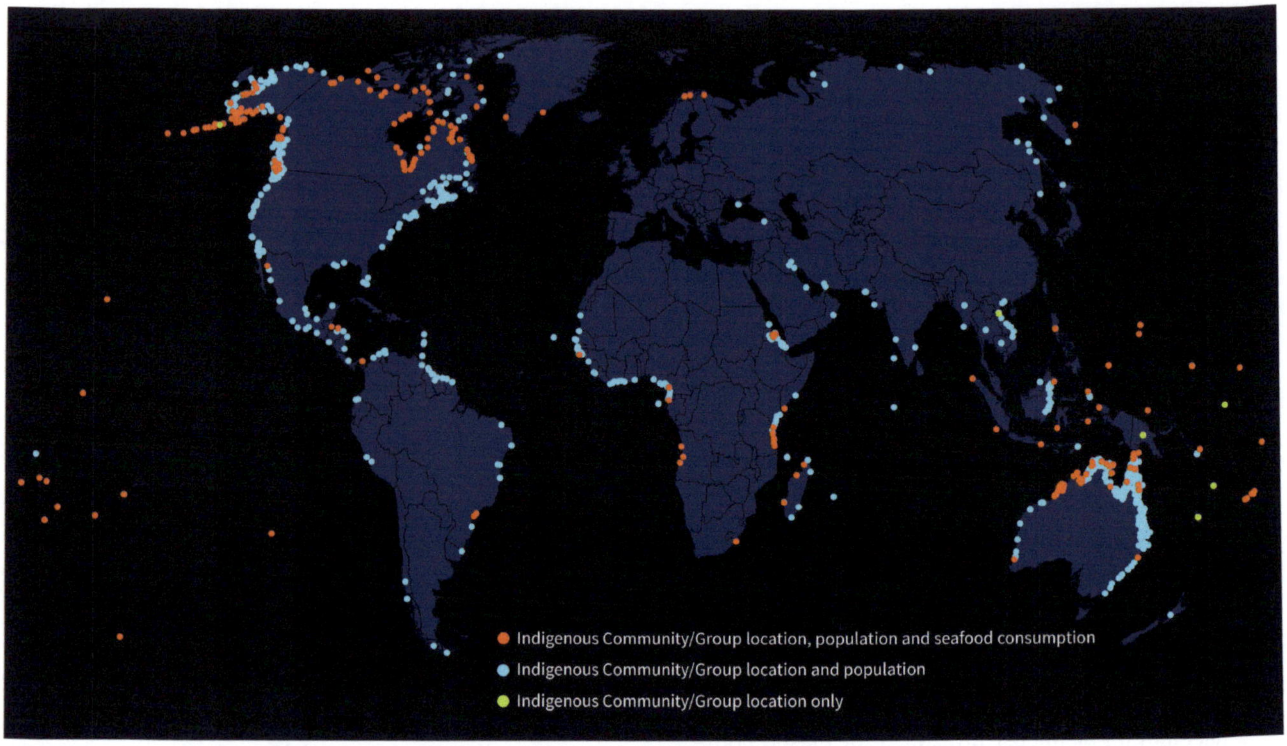

Fig. 11.3 Global map of the location of known indigenous coastal populations, prepared in order to estimate their seafood consumption. *Note: Groups for whom fish catch and population data are known are differentiated from those for whom only location is recorded. Source: Cisneros-Montemayor et al.* (2016)

so it does not represent the experience of the millions of men, women and children who crossed the ocean in shackles, as slaves. For them, and 'for the indentured labourers, convicts and refugees, oceans have never figured as spaces of freedom, exchange and connectedness, but of unfreedom, objectification and a separation from family and homelands' (Bartels et al. 2019, 81).

Discourses about shared ocean values and campaigns for greater ocean literacy should not neglect the legacies of past exploitation and the denial of others' values and knowledges. Nations that built their economies and societies through mercantilism and colonialism, and the nations that were exploited or colonised by them, will have differing perspectives and priorities in governing their ocean estate. Ocean governance futures are likely to be negotiated amid calls for continuing decolonisation (Vásquez-Fernández 2020), strengthening moral and legal arguments for reparations to the descendants of slaves (Araujo 2017) and upholding the rights of formerly colonised states to fully develop resources in their EEZs without hindrance from former colonising powers. Such a 'right to development' (Udombana 2000) is difficult to deny on moral grounds when these nations were denied the right to use their land-based resources for autonomous development, and when that land and those resources were taken by colonisers who also subjugated, displaced or killed the original inhabitants. The legacy of past injustice remains visible in the greater vulnerability of marginalised populations to COVID-19's health and economic impacts (see Sect. 11.3.4), and it drives current global anti-racism protests.

An ocean-based governance reform process that learns from these land-based upheavals will be more successful at addressing the linked problems of inequity and unsustainability than a process that ignores them.

3.4 The Emergent Impact of COVID-19 on Human-Ocean Relationships

On 16 August 2020, confirmed infections with SARS-CoV-2 approached 21.3 million people, resulting in 761,779 deaths (WHO 2020a). As of this date, the pandemic was still growing, in the United States, Latin America, the Middle East, India and parts of sub-Saharan Africa, while much of South and Southeast Asia, Oceania and Europe were seeing control measures decreasing the number of new cases, though with signs of a 'second wave' of infections leading to a return to more stringent social distancing. No vaccine has yet been developed.

Making a rapid assessment of its impacts is complicated by the 'infodemic'—the viral proliferation of information, some of it accurate, some not, that confronts the policy analyst.[2]

As well as the human and economic costs of the public health crisis, the measures to contain the spread of the virus—which have included the closure of businesses, workplaces and schools, and restrictions on travel—have led to mass unemployment, large-scale government expenditures on bailout and stimulus packages, a decrease in trade and economic activity, a fall in the value of financial markets and a global recession (Fernandes 2020). The International Monetary Fund estimates that the global GDP will shrink by 4.9% in 2020 and grow by 5.4% in 2021, to attain a value 6.1% below its pre-COVID 19 projected value (IMF 2020). Low-income households are particularly imperilled, reversing the significant global progress made in reducing extreme poverty since the 1990s (IMF 2020). This makes achieving the SDGs that much harder.

The crisis has also prompted much thought about what kind of society is to be rebuilt in future, with an exhortation to 'build forward better' and to invest in making our societies more equitable and sustainable and more resilient to future pandemics (van Barneveld et al. 2020).

Table 11.4 summarises documented impacts and responses to the COVID-19 pandemic by the maritime economy.

While impacts by country will vary depending on the role that the maritime economy plays, the three most prominent global impacts are on (1) global trade, where over 80% of the world's goods (by weight; 70% by value) are transported by sea (UNCTAD 2020); (2) the global food system, where 3.3 billion people depend on seafood for at least 20% of their animal protein intake (FAO 2020a, 67); and (3) the world's tourism and leisure industries, 50% of which are coastal and marine-associated. As a whole, worldwide tourism and leisure, pre COVID-19, generated nearly $1.5 trillion in receipts and $250 billion in transportation (totalling around 10% of global GDP in 2008) and employed 319 million people (UNCTAD 2020; TNC 2019). The impacts on each of these sectors, and the societies and economies that depend on them, are complex and take place through multiple pathways, as Fig. 11.4 illustrates for the seafood system.

What are impacts of this linked public health and economic crisis on the current and future ocean economy? Who in the ocean economy is most vulnerable? How have governments and ocean economy sector actors responded? How have these impacts and responses affected our future relationships with the ocean? These questions deserve more scrutiny than we can give them here (but see Table 11.4). We therefore recommend, as an opportunity for action (see Sect. 11.5), the application of a social well-being approach to understanding how to build a resilient and equitable set of relationships with the ocean, to complement the more ecological and economic focus of policy advice to date (McCauley et al. 2020; UNCTAD 2020). Bennett et al.'s (2020) proposals for small-scale fisheries address some of the relational and subjective elements of COVID-related impacts on well-being, as does the ILO's (2020) policy briefing for maritime industry employees. The ILO documents mariners' inability to go ashore for medical treatment, to receive medical and safety equipment and to return home. All of these have led to increased fatigue, stress, isolation and social pressures for seafarers and their families.

The COVID-19 pandemic has revealed many things about subjective and relational well-being, as social distancing and travel restrictions have reconfigured our human relationships, our relationships with nature and our sense of what is important for a good and meaningful life. One piece of evidence for the importance of the ocean to our quality of life is that, as COVID-19 lockdowns eased in May 2020, people in Europe, the United States and Australia rushed to beaches (e.g. Wood 2020). The sense of freedom, of tranquillity or sociability, the sense of renewal and new possibility that being by the ocean invokes are powerful and continue to have value. If anything, this highlights that such non-material values are more resilient than the material ones. Much has been made of the monetary value of ocean ecosystem services—estimated at $2.8 trillion a year (McCauley et al. 2020). And yet the initial economic stimulus package to address short-term economic losses from COVID-19 in the United States alone, was over $3 trillion (Delevingne and Schneider 2020)—more than the whole of the ocean is apparently worth, in monetary terms, to the whole of humanity. Given this, perhaps the call to calculate ocean ecosystem service values is not the best or only way to draw policy attention to the values of the ocean to humanity.

As thoughts turn to how to rebuild economies and restart social life, the coming months will provide opportunities to reinforce how important 'blue spaces' are to people and to ensure that people have access to them for their well-being.

[2]The World Health Organization held its first infoepidemiology conference in June–July 2020 (WHO 2020b).

Table 11.4 Some observed COVID-19 impacts, responses and proposed longer-term outlooks for maritime economic sectors and their governance

Sector or issue	Impacts	Responses	Possible futures
Fisheries and aquaculture	Demand impacts of COVID-19-related lockdowns first hit high-value seafood in the restaurant sector in China and then worldwide. Job losses and supply disruptions affected low-income households' ability to access fish and other nutrient-rich foods, particularly in developing countries. In wealthier countries, demand for tinned and frozen fish has risen. Industrial fishing activity fell by 50% in several European countries in February–April 2020; several epidemics broke out on fishing vessels and in fish-processing factories. (Summarised from Love et al. 2020.)	Governments (e.g. Russia, Canada, South Africa) designated fishers, processors and retailers as 'essential workers' during the pandemic lockdown, allowing many in the sector to keep working. Multilateral organisations (Food and Agriculture Organization of the United Nations, World Bank, etc.) are providing technical, policy and financial support to the sector for adaptation, with a focus on fish-dependent, low-income, food-deficit countries. In the private sector, companies are investing in personal protective equipment for their workers, redesigning workplaces for social distancing and quarantining fishers before they set out for distant waters. Small-scale fisher organisations, marketing cooperatives and local catch schemes have mobilised to support small producers reach local consumers with their produce, to support livelihoods (e.g. in the Indian state of Kerala and the United States). (Summarised from Love et al. 2020.)	Policy emphasis is on building resilience by continuing to address pressures on wild fish stocks (including IUU fishing, subsidies and habitat degradation), building resilience to climate change, investing in mariculture, focusing on small-scale producers most vulnerable to economic shocks, improving traceability of seafood and biosafety protocols, and maintaining diverse product forms in seafood (e.g. by canning, drying, smoking, freezing) as fresh value chains are more vulnerable (e.g. Love et al. 2020; FAO 2020a; Bennett et al. 2020; McCauley et al. 2020). Aquaculture futures depend on whether the post-COVID-19 world is oriented more towards growth or more towards sustainability, and whether trade is globalised or localised (Gephart et al. 2020).
Marine trade and transportation	13–32% downturns in container ship traffic in 2020, relative to 2019, due to slowdowns in both production of and demand for goods. Secondary effects include impacts on the marine insurance industry (Willis Towers Watson 2020). The International Labor Organization (ILO 2020) reports that seafarers face considerable problems joining and leaving their ships in port (each month around 100,000 seafarers are involved in crew changes), due to quarantine and movement restrictions.	Governments have supported continued marine transportation of food, medicines and other essential supplies to land- based populations. Measures have been taken by governments, the private sector and civil society organisations to protect the health and well-being of seafarers and port workers, classified as essential workers (UNCTAD 2020; ILO 2020).	Maritime trade is likely to continue to grow, but the rate at which it does so will depend on demand- recovery as economies reopen. Given its importance, the sector is likely to receive policy attention in areas such as opportunities for decarbonisation, improved crew working conditions and quarantine and sanitary measures on ships and in ports.
Marine tourism and leisure	Cruise tourism slowed dramatically in the first quarter of 2020 and has halted this $40-billion-a-year industry in the second quarter, as early shipboard epidemics drew attention to the vulnerability of cruise ships to epidemic disease outbreaks. Tourism revenue is forecast to decrease by between $300 billion and $2.1 trillion (UNCTAD 2020), with up to half of that value represented by coastal and marine tourism. With 80% of the tourism sector's 319 million employees being seasonal and/or in small and medium enterprises, the welfare impacts of this decline are substantial.	Where tourism is an important source of revenue, states have used stimulus payments to support tourist sector businesses. Airlines, hotel and restaurant groups have moved quickly to monopolise the bailout funds (Renaud 2020). As lockdowns eased in many countries in May and June 2020, businesses reopened while maintaining new distancing and hygiene regulations that restricted their profitability. Populations in lockdown have flocked to beaches when allowed to do so, causing concerns that a second wave of COVID-19 infections would manifest in the Northern Hemisphere summer. Meanwhile, island nations such as Fiji that were virus-free encouraged exclusive 'billionaire tourism' to restart their industries (Doherty 2020).	The need for 'blue space' for health and well-being has been reinforced by lockdowns. Crowding on beaches has exposed the effects of the creeping privatisation of coastal zones around the world, exemplified by the United States (Rao 2020). Investment in improving public access to the foreshore and coast could be part of a strategy to protect public access to natural landscapes for well-being. The COVID-19 crisis has prompted critical examination of 'overtourism', the environmental costs of air travel, and, in the maritime context, the social and ecological impacts of cruise tourism. In addition to overcoming its ships' reputation as 'floating petri dishes', the cruise ship industry will need to examine its impacts on the marine environment, destination ports and their resident populations, and its own labour force, as well as on the safety of passengers (Renaud 2020).

Table 11.4 (continued)

Sector or issue	Impacts	Responses	Possible futures
Marine biodiversity conservation	Reports of the return of marine wildlife to heavily used inshore and coastal habitats were widespread, with quieter areas of the ocean being good for whales, too (McVeigh 2020). A reduction in at-sea observers raises the possibility of increased IUU fishing, endangering seabirds and marine mammals. The COVID-19 pandemic has disrupted what was to have been 'a big year for ocean conservation' (Dineen 2020), leading to the postponement of the UN World Ocean Conference, the World Conservation Congress and the Glasgow meeting of the Intergovernmental Panel on Climate Change, at a critical time for the ocean and climate change. Cancellation of in-person meetings, including the 2020 UN Ocean Summit, may have slowed the building of ocean policy constituencies. The COVID-19 pandemic has diverted policy attention away from the ocean at a critical time	The probable origins of the SARS-CoV-2 virus in the wild animal meat trade have drawn global attention to wildlife conservation (Dineen 2020). While seafood is not implicated in SARS-CoV-2 transmission (Bondad-Reantaso et al. 2020), the wider focus on zoonotic disease risks from animal source foods is impacting confidence in seafood safety too, prompting greater investment in food safety procedures. The inability to directly monitor fisheries and conservation is leading to greater use of remote sensing technologies and calls for greater traceability in wildlife meat and seafood sectors. The rise of virtual conferences and meetings, online shared working platforms for routine meetings of regional fishery management agencies, transnational corporations, international agencies and nongovernmental organisations have shown that multilateralism can be pursued with less travel cost. The Virtual Ocean dialogues and related processes have kept the ocean visible in policies for sustainability beyond the current COVID-19 pandemic.	The COVID-19 pandemic seems likely to lead to renewed focus on environmental conservation and the linkages between environmental health and human health. If this is translated into effective policies, it could strengthen the case for investment in ocean health to support human well-being (Franke et al. 2020). It is too early to say how the current crisis will affect multilateral governance. The world's focus on our interconnectedness and the fragilities this brings could either lead to more investment in making multilateral governance and globalised economies work, or to a retreat from both of them. Some analysts have predicted the rise of China in global affairs and the decline of U.S. influence (Campbell and Doshi 2020), which would change the direction of multilateral ocean governance. In the arena of marine biodiversity conservation, 'the potential of multilateral spaces to change the world order for the better is the main reason why we should care about their future after the COVID-19 crisis has passed and find ways to strengthen their legitimacy' (Vadrot 2020).

3.5 Avoiding 'Taboo Trade-Offs' and the Need for Inclusive Ocean Policymaking to Improve Human Well-being

Once a largely coastal species (Gillis 2012), we are once again returning to the littoral. Coastal populations have been growing about twice as fast as national growth rates, and population densities there (ca. 80 persons km^2) are twice the world's average (Steven et al. 2020). Many of the world's megacities are coastal, and seafront land and properties with sea views attract price premiums in real estate markets all over the world (e.g. Jim and Chen 2009). While the majority of the world's people—even the coastal ones—experience the ocean as an alternative to a largely terrestrial existence, there are people who are 'at home on the waves' and for whom 'oceans persistently constitute the principal organizing spaces through which many communities dwell in the world' (King and Robinson 2019, 1). This range of 'ocean citizens' and the rest of humanity, all of us directly or indirectly connected to the ocean though our climate, trade, economic and food systems, derive well-being from very different material, relational and subjective relationships with the ocean.

The examples in Tables 11.1–11.3 represent some of the many ways people interact with and benefit from the ocean. All these interactions have economic dimensions and policy and legislative implications. Governing them needs to go beyond regulating the flow of material goods. Our ocean relationships engage with all the other things that make us human: our need for identity, a sense of social belonging, an attachment to place, our sense of being and doing good in our community or our world, of fulfilling our varying needs for adventure, inspiration, comfort, calm, satisfaction of curiosity and refuge from fear (Schwartz 2012).

Whether they were surfers thousands of years ago in the Hawaiian Islands (Finney and Houston 1996), fishers and coastal traders decorating their canoes in Ghana (Verrips 2002) or Inuit hunters in the Arctic making miniature carvings out of the bones of the marine mammals they caught (Laugrand and Oosten 2008), our forebears had relationships with the ocean that were complex, emotional, spiritual and artistic, as well as material and transactional. We, their descendants, will continue to forge these complex relationships with the ocean, provided we have continuing access to it.

A key planning challenge is to find ways to consider this plurality of ocean-experience and ocean-values in formulating economic development and marine conservation plans.

Fig. 11.4 COVID-19 disruptions and impacts on seafood supply chains. *Note: Disruptions to production, labour, distribution, supply and demand create a range of impacts. The colour gradient indicates the hypothesised relative impacts to different components of, or actors within, seafood supply chains. The ordering of groups is based on multiple data streams collected through May 2020 but is not intended to be a quantitative or absolute ranking. Source: Love et al.* (2020)

The perspectives of the people most familiar with, most socially and culturally attached to, and most dependent upon the ocean—namely, small-scale fisherfolk, coastal Indigenous Peoples, island peoples and sea-dwelling peoples—must surely feature prominently in marine spatial and economic planning and the formulation of maritime policy and law, both within nation-states and globally. In the interests of legitimacy and equity, this should be so even when utilitarian ethics are applied and the greatest good for the greatest number of people is sought. Recognition of this diversity of interest and incorporation of the knowledge systems and values of this broad 'ocean citizenry' is an important first step towards an equitable and sustainable ocean economy.

In addition to traditional ocean users, the contemporary industrial ocean has its temporary sea-dwellers such as cruise-ship passengers, oil rig workers, merchant and naval seafarers and deep-sea fishers, each of whom may be drawn from less maritime populations and may not share the same set of 'ocean values' as those with longer and more culturally embedded relations. The future ocean may also include blue carbon investors, offshore aquaculturalists, workers in the renewable energy sector and deep-sea miners, each of whom will bring new conceptions of the ocean, new values and new priorities. Ecomodernist visions of the future ocean often incorporate plans for floating cities in coastal and open waters (Riffat et al. 2016), or on insulated ice floes (Bolonkin 2011); these too, have governance implications. While considering how to add new ocean populations and their values to ocean accounting and management, including those who may live or work beyond national exclusive economic zones, there are opportunities to support and learn from the dwindling numbers of historically maritime populations, including those whose extraterritoriality has posed challenges for land-based governments and left them marginalised and sometimes without sovereignty (Stacey and Allison 2019).

The decisions societies make on how to govern the ocean will not be determined entirely by monetary cost-benefit analysis. Power struggles have shaped the ocean governance regime and will continue to do so. Relational and subjective considerations will play their part too: the role of emotion in policy decision- making is often overlooked, but, for example, much of our nation-building seeks to draw on emotional responses to symbols of nationhood (flags, anthems, commemorations and celebrations), and our global stock markets are governed by traders' moods ('behavioural finance'; Nofsinger 2005). Our emotional responses to the ocean are thus also likely to shape our decisions on how to govern it.

While expanding the ocean economy can create new opportunities to improve welfare, it can also lead to unequal endowments, reinforced discrimination, or inequality of status (Satz 2004). Trade-off decisions between use and users will need to be made in an increasingly busy 'blue future'. Ideally, we solve trade-off decisions with the constructive objective of equitable outcomes, where the allocation distribution is envy- free and where no individual would prefer having what another person has (Arrow 1951; Kolm 1972). But even when a socially inclusive and holistic view of a blue future is taken, it is likely to show that trade-offs are pervasive, some are hidden and some are 'taboo' (see Box 11.4). Trade-off analysis has to adequately consider ethical and moral values to prevent individuals and societies from having to make such 'taboo trade-offs'.

Box 11.4 Taboo Trade-Offs at Sea

Taboo trade-offs result when a sacred value is asked to be traded for values that are secular. Sacred values possess infinite or transcendental significance and are inviolable and absolute. Sacred values preclude comparisons with bounded or secular values (Tetlock et al. 2000). Taboo trade-offs are often ignored by managers because the sacred values that people ascribe (whether to possessions or natural assets) are likely to be very different. Not only do the sacred values differ in measure, but the actual values they consider sacred are likely to vary culturally, spatially, and demographically. Generalisation of sacred values is therefore near impossible, and this diversity can breed substantial conflict. Nevertheless, social values and the psychological context within which taboo trade-offs decisions are made must be considered to ensure an equitable, envy- and conflict-free blue future.

When people are asked to trade their sacred values for secular values, they often experience this as deeply offensive. People have an aversion to making taboo trade-offs (Stikvoort et al. 2016), and they are likely to display insensitivity to a strict cost-benefit analysis of the exchange. They are likely to exhibit moral outrage, express anger and disgust, and become increasingly inflexible in negotiations. Examples of taboo trade-offs include being asked to exchange locally held cultural values for something secular like the profit of a fishing or oil exploration company, or, for Indigenous communities, being offered money not to fish or have access to their traditional waters. Contrary to classic economic theory's assumption that financial incentives motivate behaviour, bringing economics into the equation and trading sacred values for money can make people recoil.

Ecosystem management that doesn't acknowledge uncomfortable truths and the taboo nature of some trade-offs is likely to fail. In order to deal better with trade-off decisions, we must be cognisant of how we present and frame decisions and aim to predict decision difficulty and better anticipate resultant behaviour (Daw et al. 2015).

How we feel about the ocean will influence what we choose to do with it, in it and on it. Within and between nation-states, there is heterogeneity in people's emotional response to the ocean, so it follows that there will not be complete agreement within or between countries about what kind of sustainable ocean economy we want. Emotive relationships with the coast and the ocean will influence the 'social license' granted to governments and private sector actors to develop new ocean uses. Understanding people's values (Schwartz 2012)—and the emotional responses elicited when those values are challenged—therefore become a key part of the ocean governance process.

4 Governing Humanity's Relationships with the Ocean: Some National and Regional Perspectives

Governance systems for coastal waters, territorial waters, regional seas and ocean basins have emerged from the turbulent mixing of historic, geographical, sociocultural, legal, political and economic relations. They are based on a rich foundation of traditional or local ecological knowledge and reflect the attempts of different societies to ensure that their relationships with the ocean support their well-being in all dimensions. These governance systems are under threat, but they are part of humanity's social and legal legacy, and they are now being overlaid by governmental and international law. A just and equitable, diverse and inclusive sustainable ocean economy will not allow this legacy to be swept aside but will nurture and support it.

Legal and customary systems in the ocean have also evolved through struggle and litigation. In some cases, modern states have neglected or actively undermined pre-contact and pre-colonial institutions and sovereign systems and attempted to replace them with newer forms of state-based governance. States' capacity to govern their 'marine estates' is limited, particularly in the large ocean states and in low-income, food-deficit countries where ocean governance competes with many other priorities for limited government spending. In these circumstances, instrumental as well as moral reasons may justify encouraging the revival and continued evolution of the traditional institutions of governance that predated modern state-formation.

There is increasing recognition that the knowledges, cosmologies and traditional institutions for environmental governance developed in Indigenous and traditional coastal societies are empirically valid, have contemporary relevance and can be (and have already) been mobilised for current contexts (Eckert et al. 2018; Jentoft et al. 2019). This is complicated by the history of active denigration and suppression of these practices (see, e.g., Maldonado 2014). The current wave of 'decolonisation' of thought systems (as well as territory) responds to this recognition that colonisation has marginalised local and Indigenous people and their knowledge systems and replaced them with human- environment relationships that are less functional and responsive. Other national and international environmental programs have begun to respond to these calls for decolonisation—notably the Intergovernmental Panel on Biodiversity and Ecosystem Services with its shift from talking about ecosystem services to 'nature's contributions to people' (Díaz et al. 2018).

Here we briefly introduce some of the different ways people around the world have built and institutionalised a 'human relationship with the ocean'. We do not present them as blueprints or exemplars of good practice. We merely offer them to illustrate a few of the diverse ways different (non-Western) societies with long histories of living with the ocean have developed ways of life and institutions that could be built upon in the places where they occur. They also illustrate that large economies like China and the African Union have their own plans for the ocean economy; these may differ from a vision that comes from a model of global collaboration and consensus. Finally, we also indicate how contemporary international instruments—in this case the World Heritage process of the UN Educational, Scientific and Cultural Organization (UNESCO)—can support relational and subjective elements of well-being that are embodied in traditional maritime cultures, societies and practices.

We have picked only a few examples to illustrate a diversity of approaches and highlight whether and how (or how not) the institutions of these groups and nations support the relational and subjective dimensions of well-being derived from human-ocean interactions. Space precludes a more comprehensive or globally representative treatment. Every coastal and maritime society will have its own historical and cultural foundations to build upon when it comes to developing an equitable, diverse and inclusive ocean economy that supports human well-being. Whatever those foundations, a guiding principle for 'blue justice' can be that of 'participatory democracy', which implies decision-making devolved to more local scales (communities, municipalities and districts), active roles for a civil society distinct from governmental politics, and a more active engagement of citizens in the political process than is the norm in 'representative democracy' (Barber 2014).

4.1 An Indigenous Perspective: Aboriginal and Torres Strait Island Peoples' Connection to 'Saltwater Country'

Identified as Aboriginal or Torres Strait Island people, Indigenous Peoples in Australia comprise approximately 3% (or about 650,000) of the total population. The majority of coastal Indigenous Peoples live in remote or very remote locations, most of which are part of the Indigenous estate, Aboriginal-owned and managed land and sea, held under some form of Indigenous tenure.

Indigenous Peoples view the ocean, islands and coastal environments as part of their 'sea country' or 'saltwater country' and often refer to themselves as 'saltwater people' (Smyth and Isherwood 2016).[3]

[3]Aboriginal interests also include inland waters and wetlands.

These features relate to ownership of traditional clan estates and marine resources under traditional law, and recognised in some state, territory and commonwealth legislation. The saltwater peoples of the Arnhem Land region continue to rely on coastal and marine environments and resources providing food, cultural identity, health and well-being, and as part of domestic and commercial trade economies. Opportunities for employment are very much determined by good land and sea management and the ability to harvest natural resources. The harvesting of plants and animals for food, ceremonial or celebratory purposes (community feasts), art production or the performance of sacred duties in natural and cultural management, including maintaining social and economic relationships within and between tribal groups, all play an important and central role in Aboriginal livelihoods, belief systems and well-being in Australia's Northern Territories.[4]

There is no overarching legal fisheries management framework for Indigenous customary fishing rights in Australia. Instead, fishing rights have been intricately tied to developments in land and native rights with different local arrangements across states and territories (Schnierer and Egan 2016).[5] This has meant that in most cases Indigenous customary fishing is exempt from fisheries management frameworks and laws. The effect is that Indigenous fishing has not been recognised and consequently engages low numbers of Indigenous people in fisheries and associated businesses (Fleming et al. 2015; Productivity Commission 2016). In most cases,[6] a definition of Indigenous customary fishing does not include fishing for commercial purposes, even though it is recognised that Aboriginal people have fished commercially (i.e. to sell, exchange or barter fish) according to traditional laws and customs. As such, land or Native title-holders who fish commercially have been subject to the same commercial fishing laws and regulations as the rest of the population. In this case, the Australian state's fisheries governance system fails to adequately understand, account for and support the relational and subjective benefits associated with Aboriginal marine resource use.

4.2 A 'Big State' Perspective: China's 'Ocean Dream'

While China's coastlines have historically hosted diverse peoples and cultures (e.g. Anderson 1970; He and Faure 2016), the Chinese state had until recently directed its attentions mainly to the land. In the last four decades, however, state policy has turned increasingly towards the ocean. This recalls

[4]Sacred sites are also registered in these clan estates.

[5]One exception is the Torres Strait Islands, where legal rights for customary and commercial fishing are recognised under Native Title legislation (Lalancette 2017).

[6]Torres Strait is again an exception (Lalancette 2017).

the early fifteenth-century voyages of the great Chinese navigator, diplomat and maritime leader Zheng He, whose seven voyages, with a flotilla of 'treasure ships' and a retinue of 20,000, sought to extend China's trading influence throughout the Indian Ocean and strike awe in all who saw them (Dreyer 2006). Almost a century before Columbus and Vasco da Gama, Zheng He's flotilla sailed throughout the Indo-Malayan archipelago, to India, to the mouth of the Persian Gulf and to the East African coast. As China turns to the ocean once more in the twenty-first century, Heng Ze's exploits are part of a state narrative of a tradition of peaceful trading and knowledge-sharing that China contrasts with European imperial sea voyaging (Holmes 2006). As Western disquiet grows over China's growing modern maritime presence—not least in the disputed territories of the South China Sea—so too does Western historical research that seeks to reinterpret Zheng He's voyages as more overtly imperialistic in intent (Wade 2005).

China's overall domestic strategy aims to mobilise its people to support the 'Chinese Dream': achieving 'national rejuvenation', after what the central authority views as a period of global domination and humiliation by the West, and building a 'moderately prosperous society'. Thus, the national development policy has both relational, subjective and material dimensions. China's 'ocean dream', therefore, is to achieve the 'Chinese Dream' through an ocean-based economy. As such, China's 'blue economy' is state-led, rather than private-sector-driven, and ultimately serves the purposes of the state and bolsters the legitimacy of the Chinese Communist Party.

The Chinese central authority referred to the 'blue economy' in the 13th five-year plan in 2016, but contemporary policy on China's ocean economy dates to the start of the 'opening and reform' period in the late 1970s. Blue economic development in China accelerated around the turn of the twenty-first century. Since China ratified the UN Convention of the Law of the Sea (UNCLOS) in 1996, the state has established numerous exclusive economic zones, called for 'implementing ocean development' and issued various five-year plans for ocean economic development. The focus has culminated with the explicit goal of becoming a 'maritime power' (海洋强国), possessing military defence capabilities, a strong ocean economy and advanced marine science and technology. In 2019, Premier Li Keqiang summarised the state vision of China's blue economy as to 'vigorously develop the blue economy, protect the ocean environment, and construct a great maritime nation' (大力发展蓝色经济, 保护海洋环境, 建设海洋强国) (Li 2019).

Development of China's 'blue economy' has indeed been vigorous: In 2018, it accounted for 9.3% of GDP.

China has focused on transforming its ocean economy from the primary production of raw commodities, such as capture fisheries, to secondary and tertiary production and service industries, such as processing and tourism. The country is also placing increasing emphasis on developing ocean-related technology and innovation; 'blue economic pilot zones', such as the Qingdao Blue Silicon Valley (青岛蓝色硅谷), are built so that demarcated parts of the coast can serve as (industrial or technological) parks for ocean-focused research and development.

The state-led nature of China's blue economy is also apparent in its more ideologically oriented policies.

To bolster blue economic development and China's identity as a maritime power, the state has endeavoured to increase 'ocean consciousness' among the citizenry.

Through outreach and education, the state is working to raise public knowledge about China's ocean nature, economy, culture and politics. Museums dedicated to cultural relics like Mazu, a goddess of the sea, have been newly created. Tourist destinations have likewise been established elsewhere in the country, such as fishing villages and festivals, in an effort to build interest in and understanding of China's maritime heritage. While this might at first sight appear to be directed at building non-material values and recognising tradition, these aims are martialled to support a top-down vision, and their instrumental nature is unlikely to confer the types of well-being that a bottom-up approach to culture and tradition might foster.

China also dreams of achieving marine environmental sustainability. Policy rhetoric certainly emphasises environmental sustainability; 'ecological civilisation' (生態文明), a framework to live in harmony with nature, is highly promoted, even constitutionalised. However, much of 'ecological civilisation' acts as a means to intensify production in some areas while restricting production activities in other, mostly rural, landscapes (e.g. Brown 2014; Hong 2018). This policy better protects some ecologically sensitive areas but also presents social and economic challenges, as it can displace livelihoods (Chen et al. 2017). Challenges in marine environmental governance also remain, including pollution, overfishing, subsidies, implementation and enforcement, and varied levels of government capacity (Zhang et al. 2016; Cao et al. 2017; Mallory 2016).

China's focus on its ocean economy also has international dimensions. The twenty-first-century Maritime Silk Road, a component of China's Belt and Road Initiative, aims to recreate historical international partnerships through investments, trade and aid. China's aspirations to become a maritime power have intensified a range of maritime disputes, such as in the South China Sea. China is also increasingly involved in ocean developments in the global commons, such as in the polar regions and on the high seas (e.g. seabed mining and distant water fishing), where it sees significant security and resource opportunities (Brady 2019; Mallory 2013).

In particular, the increasingly international scope of China's ocean economy means that its vision of becoming a strong maritime power is now interacting with alternative visions for a sustainable ocean economy currently being developed. While elements of relational and subjective well-being are addressed in China's 'Ocean Dream', they are invoked in the service of state power rather than individual or group agency and well-being.

4.3 Japan's Satoumi System: A Socially Negotiated Institution for Sub-national Governance

Satoumi is a Japanese concept describing a mosaic of interacting marine ecosystems and coastal human communities, where the livelihoods of people and the blessings of nature harmoniously co-exist. They are built on traditional ecological knowledge. It therefore addresses several facets of social well-being outlined in Tables 11.1–11.3. It emphasises ecosystem linkages and thus is somewhat incompatible with the species- or stock-based approach of more conventional fisheries management. The delineation of ocean space under *satoumi* is driven not by the need to differentiate ocean space based on user purposes but by the zoning promoted by fishers and other stakeholders to manage and conserve ocean resources and the ocean environment. Thus, there are fundamental differences in both objectives and approaches between spatial management and *satoumi*.

Attempts to characterise *satoumi* as a model of co-governance are inappropriate as they fail to fully appreciate the historic context and social dynamics encapsulated in the decision-making process.

Specifically, unlike the self-governance framework, which limits their scope to collective decision-making by a stakeholder 'community', the scope of *satoumi* extends beyond ocean policymaking, as a greater, regional governance framework. A key feature of *satoumi* is that it is explicitly based on information-sharing among social actors, both administrative and economic. Through a system of information-sharing—including among fishers—*satoumi* facilitates management consensus. It also builds on the belief that fishers must be the stewards and protectors of the ocean; it defines their role and thus the objectives of local coastal governance and marine conservation efforts (Takehiro 2018).

Administratively speaking, Japan's Fisheries Act was the institutionalisation of a historic system of localised fisheries governance and provided rights to local coastal communities. Collective fishing rights are allocated to fishers and fisheries cooperatives, which are responsible for managing their adjacent waters.

Satoumi can, therefore, be considered a social contract between fishers and local communities founded upon the awareness of complex interactions between ocean and humans. Nevertheless, *satoumi* is not a concept that is universally recognised or defined but rather one that is dynamically applied on a case-by-case basis. Ocean management that is described as *satoumi* does not implement a specific set of measures. While *satoumi*'s key functions are described as 'productivity enhancement', 'conservation of environment', 'promotion of communication' and 'cultural succession', it is not necessary for all these functions to be explicitly identified in a system described as *satoumi*. Under the *satoumi* model, conservation is a means to improve fisheries productivity and mobilise social networks for conservation effort, with the understanding that such effort supports fisheries. If *satoumi* is to be considered a form of social contract that transcends conservation and fisheries aspirations, it is critical that further discussions focus on the ocean as an integral part of the coastal community, rather than on the environment that the coastal community occupies and utilises.

4.4 Indonesia: Diverse Marine Ecosystems Support Diverse Maritime Cultures and Societies

Indonesia, a centre of global marine biodiversity and at the heart of the Coral Triangle, has five main maritime populations: Bajo, Bugis, Butonese, Makassarese and Madurese (from the Raas and Tondok islands in west Java), plus a number of smaller populations (Stacey 2007). Since before the earliest European presence in Indonesia, these populations have engaged in migratory behaviour as part of artisanal commercial fisheries for local (e.g. dried and fresh reef fish) and international trade (e.g. trepang, shark fin, live reef fish), the latter strongly driven by demand from China and other Southeast Asian countries for highly prized marine products. Since the seventeenth century, fishing, sailing, fishing and trading strategies have permeated as far south as northern Australia, throughout eastern Indonesia to the north to Malaysia and Singapore, and even as far east as the Island of Palau in the northwest Pacific (Stacey 2007). Over the last three centuries, these groups have adapted diverse sailing, fishing and trading livelihood strategies in response to the island environment, political processes and alliances, and commercial trading networks.

Mobility and adaptability underlie the social and economic life of these maritime populations and are key features of these communities. People move regularly and frequently between home villages and transient or semi-permanent settlements across the archipelago, staying for short or extended periods in settlements. Mobility is facilitated by kinship, economic ties (patron-client) and historical antecedents (Stacey 2007). The most mobile and specialised of all seafaring groups are the Sama-Bajau (commonly referred to in the academic and popular literature as 'sea nomads'). They are also the most vulnerable and marginalised due to lack of secure sea tenure, landlessness and their status as a minority Indigenous group (ILO Convention no. 169 [1989]; see Annex 1).

Mobility among Indonesian fishers does not mean an absence of resource governing institutions, however, and these provide the basis for both fisheries management and marine conservation and economic development, including the traditional community-based coastal resource management sasi system of the Maluku archipelago. Sasi, which means 'to prohibit', regulates the harvesting of certain biological resources in the estuarine and nearshore coastal areas, in an effort to protect their quality and population. Sasi also operates to maintain patterns of social life, through the equal distribution among all local citizens of the benefits from the surrounding natural resources (Kissya 1995). As an institution it has never been static, changing with the coming of state and church organisational structures into the islands and varying from village to village. The governing and enforcing authorities may be traditional, church, local government or private individuals holding the harvest rights to coastal land and aquatic resources. In certain areas sasi has evolved to accommodate significant commercial transactions involving the natural resources and a spectrum of claimants.

Consequently, the rules that define how the players in sasi work together are a mixture of tradition and modern innovations and demonstrates attention to relational and subjective issues that make it more than a technical means to manage fishing effort. This has been important to the resilience of the institution and its continued relevance to culture and well-being.

The modern state apparatus in Indonesia was keen to make marine fisheries an important source of foreign-exchange earnings. Extending state control over the coastal waters of the archipelago was a prerequisite for this. State patronage of modern fishing technologies (such as the bottom trawl), with investors from the Chinese communities taking the lead in the mid-1960s, resulted in the gradual and extensive spread of bitter conflict with coastal fishermen using small- scale, artisanal techniques. Institutions like sasi were initially deemed irrelevant to handling these new forms of conflict. State-supported legislation and zoning arrangements were introduced to contain the conflict, but these centrally administered regulatory regimes were costly to implement and largely ineffective in enforcement given the geographic spread of the islands of Indonesia. Moreover, they had no legitimacy in coastal communities that were marginalised from their traditional fishing grounds. This led to a revival of interest in the coastal villages for more community- oriented arrangements for protection and nurturing of the natural assets of the coastal waters. The sasi system takes on new meaning in this context.

Sasi does not cover the entire fishery. It is applied only in small inshore areas and to a few species. However, these areas and species can be considered to be keystones for the health of the ecosystem. This important ecological fact, together with the sociocultural foundations of sasi in Maluku, provides a robust rationale for supporting sasi where it continues to be vibrant, and for efforts to revive it where it faces extinction. Since collaboration, trust and legitimacy are the pillars of the sasi system, these are also crucial elements of any new institutional arrangements (Novaczek et al. 2001).

4.5 South India: Sea Courts and Legal Pluralism

Two case studies from South India illustrate the strong institutional 'interplay, fit and scale-relevance that characterise effective and resilient institutions' (Young and Gasser 2002). They also illustrate that the processes by which decisions on natural resource governance are made affect social well-being; having a voice, belonging to a group and exercising local autonomy are key values that support these institutions.

4.5.1 Kadakkodi: The Court of the Sea

Among the Hindu fishing communities scattered in the predominantly Muslim districts of Kozhikode, Kannur and Kasargode, in the Indian state of Kerala, the age-old traditional community institution called the *kadakkodi*, or 'sea court', is closely associated with the temples located on the beach. The fishery of the region exhibits great seasonality and is marked by bumper harvests. Confusion and conflicts among fishing units are inherent to the very nature of the fishery. The 'court' consists of the 'elders' and certain number of 'functionaries' who implement the decisions. The court meets openly. All fishers participate in the discussions on issues that relate to access, conservation and conflict resolution, thus nurturing senses of status and belonging and fairness.

The elders make the decisions, which are considered final. The implementation and the sanctions against offenders are the responsibility of the whole community. The *kadakkodi* has more recently been subjected to considerable pressure by rapid technological changes, new organisational forms promoted by the government, new political divisions among fishing communities and the greater involvement of formally educated youth in fishing operations. Yet its basic scaffolding is still in place. With the government emphasising decentralised resource management and governance, there is latent interest in reviving and strengthening the institution, albeit in a new form. In this context, communities with a history of traditional institutions will have an important edge in the negotiation of any new stewardship contract between state and community (Kurien 2001).

4.5.2 Uur Panchayats and Legal Pluralism

The Coromandel Coast of southeastern India has a long history of settlement and particularly strong expressions of customary law (Bavinck 2001a). The fishing communities belonging to the Pattinavar caste had established village councils (*uur panchayats*) for formulating and implementing a broad spectrum of customary laws. Along this coastline fishers law often prevails, with civil servants having to walk on tiptoe.

Uur panchayats' prime responsibility is social and territorial in nature and ultimately anchored in control over people as well as territory (Bavinck and Vivekanandan 2017). All adult men are considered members of the community and are considered equal—a fact mirrored not only in the structure of village meetings (members seated in a circle at the same level) but in the consensus approach to decision-making (Bavinck 2001b). All members contribute taxes to the village and not only follow the rules that are collectively adopted but also enforce them. The position of the elders used to be hereditary. But today most councils organise regular elections to fill these positions.

The *uur panchayat* controls adjacent waters, regulating the kinds of fishing practiced and also taxing outsiders who wish to land their catches locally. Residential lands and common lands are in principle still held collectively under the jurisdiction of the *uur panchayat*. Beaches adjoining fishing settlements are also controlled by the village councils. All tenure rights are held under community common property regime, with rights flowing to members of the village population as a whole. These rights are not transferable and so protect the group against unwanted encroachment and infiltration.

According to India's constitution, the use of marine waters is regulated by the state alone. In the case of fishing, the clash between the state and the customary legal systems such as the *uur panchayat* has been pronounced and sometimes violent. The introduction of semi-industrial fishing by the state caused uproar in the *uur panchayats*, since this new category of fishers was blatantly infringing upon village fishing territories, causing significant damage and indignation. Recently *co-management* has become a buzz word in government circles, but though the *uur panchayats* are not considered legal and held in suspicion as a relic of the past, on an informal basis, government representatives have been shown to negotiate regularly with *uur panchayats*, recognising the need to come to terms with the legal plurality for the sake of maintaining peace, equity and better management of fishery resources.

4.6 African Maritime Economies: Historical Foundations and Contemporary Visions

African nations, under the leadership of High Level Panel member states Kenya, Ghana and Namibia, together with South Africa, are leading the global charge to embrace the growth possibilities of the 'blue economy'. The Seychelles has pioneered the use of 'blue bonds': an innovative financial instrument to support the conservation of its ocean estate and its associated ecosystem goods and services, building on debt-for-nature swaps (Schutter and Hicks 2019). While conservation benefits are a primary motivation for some of their development partners (such as The Nature Conservancy), African nations are looking at the blue economy largely in the context of an imperative to industrialise, grow their economies and provide employment opportunities and food for their citizens.

African nations' success at lifting their citizens out of poverty is crucial to meeting SDGs. But the ocean around Africa is not empty, and it has a history. That history—and the human relationships with the ocean that have persisted through or been shaped by the mercantile and colonial periods—provide a basis to build on. The Swahili, for example, are regarded as a coastal or maritime culture (Fleisher et al. 2015). While their languages, religion, customs and sailing technologies have reached the heart of the African continent, they are originally a cosmopolitan maritime trading society, whose East African port cities of Lamu (Kenya) and Stone Town Zanzibar (Tanzania) are listed as UNESCO World Heritage sites (see Annex 5), attracting international tourism in the past 50 years. Plans to develop a new port near Lamu may make trans-oceanic trade once again its most important economic activity (Lesutis 2019). Whether Africa's future ocean economy is to be built on expanding ports and transportation infrastructure (much of it funded by China's Maritime Silk Road initiative; Lim 2015) or whether it will be built on 'blue bonds' and rationalised fisheries remains to be seen. What is clear, however, is that Africa's citizens will benefit most if any such development is also built on Africa's existing maritime cultures and institutions. Here we introduce the Yoruba littoral in Nigeria to illustrate that new ocean economy initiatives will be layered upon a rich tradition of maritime trade and interactions with and dependencies on the ocean, even in societies usually associated with land-based activities.

4.6.1 West Africa and the Atlantic: The Yoruba Littoral

The Yoruba are found mainly in the western part of Nigeria, but also with substantial numbers in the neighbouring republics of Benin and Togo. Descendants of Yoruba slaves exported to the Americas between the seventeenth and nineteenth centuries live in Brazil, Cuba and several Caribbean and Latin American nations. With an estimated population today of around 56 million in Nigeria, the Yoruba have lived in large settlements for centuries—they were the most urbanised African ethnic nationality before European colonisation in the nineteenth century. The bulk of the Yoruba population lives in the hinterland, while a much smaller percentage lives on the Atlantic coast.

The Yoruba littoral is not exclusively Yoruba. Other Nigerian and non-Nigerian coastal peoples also live, fish and trade in the zone. Among these are the Ijo (Izon) and Egun of Nigeria, the Ewe (Keta) of Togo and Ghana, and the Fante of western Ghana. It is claimed that migrant fishermen from Gold Coast/Ghana have been in the Lagos section of the Yoruba littoral since the late eighteenth century (Adewusi 2017, 165). The diversity of people speaks to the multicultural character of the Yoruba littoral and the range of criss-crossing migrations, over hundreds of kilometres across the Gulf of Guinea— westwards to Côte d'Ivoire and eastwards to Cameroon, Gabon and Angola. Fishing techniques and methods of fish processing have been borrowed or transferred by migrants as they fished and traded beyond their homelands. Migrations, economic enterprise, war and peace, and cultural exchanges have defined identity and inter-group relations in the Yoruba littoral. The zone was the site of power relations between proximate and distant neighbours, commercial enterprise, oracular and religious activities, all of which are critical to the daily lives of the people. There were hierarchies of power relationships, especially between political and economic nodes, on the one hand, and their satellites, on the other.

The sea (*okun*), lagoon (*osa*), lake (*adagun-odo*) and river (*odo*) are connected in Yoruba cosmology. The sea features in Yoruba legends of migrations and festivals. All over the coast, festivals dedicated to sea goddesses are commemorated either to celebrate fish harvests or to placate hostile natural forces so that the environment may be conducive to fishing. The association of the Yoruba with the ocean is embodied by the deity Olokun (Lord of the Sea) and Yemoja (the sea goddess). Rituals and festivals are performed by the Yoruba and non-Yoruba ethnic groups in Nigeria in honour of this deity. The iconic Ori Olokun mask—adopted as symbol of the global Festival of Black and African Arts and Culture in 1977—represents the Yoruba, though Olokun is not exclusive to the Yoruba. The deity is represented in the pantheon of the religious and cultural systems of the neighbouring non-Yoruba Edo, Itsekiri and Western Igbo communities in midwestern Nigeria. The multi-ethnic character of the Yoruba littoral has led to extensive mutual cultural borrowings and overlaps in language registers, broad similarities in songs and festivals, dress styles, family, personal and place names (though the same words might have different contextual meanings) and inter-ethnic marriages. Festivals for Olokun are celebrated across ethnic and social divides. Okosi, an annual festival of rowing and sailing, is a boat regatta to appease Olokun and thereby ensure a greater catch, as well as safety for fishermen on the lagoons and the ocean (Adeyeri 2012).

Ayelala, strongly entrenched among the Ilaje and their Ikale (fellow Yoruba) neighbours, is another sea-themed deity that cuts across the Yoruba littoral. Ayelala is 'not only worshipped with pomp and pageantry but also highly revered and respected by its worshippers [It] is linked to conflict resolution and retribution against societal impropriety' (Raheem and Famiyesin 2017, 233). It has a fearsome reputation among the Ikale, Ilaje and Ijaw-Apoi for dispensing swift justice in cases of sexual infidelity or murder. Its religious and oracular network extends far beyond the cultural and geographical boundaries of the Yoruba littoral The deity thus plays a role in fostering peaceful inter-group relations among communities across the zone. 'Seeing the efficacy of Ayelala's powers,' it has been noted, 'other neighboring and distant communities such as Ikaleland were quickly attracted to the goddess and infused it as part of their traditional deity' (Raheem and Famiyesin 2017, 244).

An overview of the economic activities of the Yoruba coast, with an emphasis on fishing, is given in Annex 6 fishing has been the driving force behind the lateral movements along the coast It remains the centre of littoral social and economic life Fishing is the impetus behind Ilaje, Izon, Ogu and Keta migrations and settlement across the Gulf of Guinea and beyond The fishing industry has witnessed significant innovations, while remaining largely artisanal, since the 1890s Both Indigenous religion and Christianity have contributed to the political, economic and social development of the Yoruba littoral.

The key lesson from the way the Yoruba have institutionalised their relationship to the ocean is that coastal trading societies need ways to foster cohesion, cooperation and mutual understanding among distinct but regularly interacting groups if they are to benefit from trade relationships Their cosmology is built on fostering and maintaining these interactions and attending to relational, spiritual and practical needs.

4.7 Island Communities' Leadership in Governing the Ocean Locally and Globally

Island nations in the Pacific, Caribbean, North Atlantic and Indian Ocean have featured prominently both in the redevelopment of local conservation and resource management institutions and in ensuring that the ocean is prominent in global governance mechanisms and policy discourses around 'greening' the economy and reaching agreement on global greenhouse gas emissions.

The nations of the Pacific Community have been at the forefront of campaigns for local autonomy in marine and coastal natural resource management and for local and Indigenous knowledges as the basis for systems of governance. The 'New Song', or Nouméa Strategy (SPC 2015), calls for an enhanced focus on coastal fisheries management and related development activities in the Pacific region. It is a community-driven approach built upon traditional Melanesian, Micronesian and Polynesian institutions, supported by national governments and all other stakeholders, to provide direction and encourage coordination, cooperation and an effective use of regional and other support services to develop management of coastal fisheries. At the regional level it coordinates initiatives and stakeholders with a shared vision of coastal fisheries management. At the national and subnational level, it seeks political recognition of the value of coastal fisheries to food security and rural development. It thus addresses what have been seen as key limitations of local governance initiatives: coordination and scalability.

Pacific island nations, such as Solomon Islands, have worked to include locally managed marine areas and other forms of customary marine tenure in their legal systems, which are based on state laws, themselves a mix of English law dating from the colonial era and post-independence legal development (Schwarz et al. 2020). This pluralistic legal system is itself also nested within regional treaties and agreements of the Pacific Community and international law, including the law of the sea. The 'New Song' supporting local coastal fisheries is mostly compatible with the FAO Voluntary Guidelines for Securing Sustainable Small-Scale Fisheries (SSF Guidelines) (FAO 2015), with areas of divergence occurring around gender relations and human rights (Song et al. 2019), illustrating both the strengths and weaknesses of global policy harmonisation: it fosters coordination but potentially undermines cultural autonomy. The governance regime for the Pacific is thus both polycentric and plural, with the complex legal and policy environment responding to evolving relationships between people and the ocean at multiple jurisdictional levels.

At the level of global environmental governance, the Alliance of Small Island States (AOSIS) has been prominent in four ocean-related arenas: global climate change agreements, evolving high seas fisheries governance, agreements governing deep-sea mining and the conceptual roots of the blue economy. In all these cases, the negotiating and diplomatic tactics employed by AOSIS have relied extensively on appeal to subjective and relational values, as we briefly discuss here with respect to climate change and the blue economy.

In the climate change arena, AOSIS, building on the work of many civil society advocacy groups and regional and international nongovernmental organisations (NGOs), built a narrative around islands as vulnerable to sea level rise and other ocean-related climate change but not responsible for climate change. This narrative also rejected victimhood and stressed resilience, equity, rights and justice as values embodied in their calls for industrialised nations to commit to curbing carbon emissions and funding adaptation. In doing so, they influenced the outcome of the UN Framework Convention on Climate Change's 21st Conference of the Parties, resulting in the Paris Agreement (McGregor and Yerbury 2019).

The blue economy was originally an appeal by the 'large ocean states' (as SIDS now prefer to be called) of the Pacific Community and the Caribbean for investment and policy attention around greening the economy to move beyond the land. AOSIS argued that most of the technologies, policy measures, investments and attention around greening the economy—through energy efficiency and decarbonisation, improved land-use and recycling materials—were focused on terrestrial technologies (Voyer et al. 2018). The blue economy narrative has since shifted and broadened to encompass a range of views (see Sect. 11.3.1), but it was originally conceived around ways small island states, with their small-scale community customary fisheries and large ocean territories, might participate actively and equitably in the greening of the global economy. Thus, the blue economy concept was originally built on island nations' concerns for equity and participation as well as the growth potential and environmental sustainability of the ocean.

4.8 Maritime Examples of UNESCO's 'Intangible Cultural Heritage' Designations

While many traditional systems of ocean governance and some more recent state-driven ones demonstrate how ocean resources and the ocean economy are governed with relational and subjective elements of well-being in mind, the contemporary global governance instrument that most explicitly seeks to consider these non-material dimensions of ocean-related contributions to people is UNESCO's 1972 Convention Concerning the Protection of the World's Cultural and Natural Heritage.

UNESCO (2020a) defines heritage as 'our legacy from the past, what we live with today, and what we pass on to future generations. Our cultural and natural heritage are both irreplaceable sources of life and inspiration'.

UNESCO evaluates, designates and maintains lists of natural and built-environmental sites as well as intangible (i.e. non-material) cultural heritage. While nation-states invest in protecting their own heritage, what makes the concept of World Heritage exceptional is its universal application. 'World Heritage sites belong to all the peoples of the world, irrespective of the territory on which they are located'

(UNESCO 2020a). The process of identifying and designating such sites draws attention and resources to their conservation and to support their continued contribution to humanity as 'irreplaceable sources of life and inspiration'. These listings are supported by legislative protection derived from the Convention Concerning the Protection of the World Cultural and Natural Heritage, including the 2003 UNESCO Convention on Intangible Cultural Heritage.

Sites and cultural systems may receive multiple UNESCO World Heritage designations due to their landscape or biodiversity significance, their historical and material significance or their contributions to less tangible elements of heritage such as ritual, belief, language, skills, identity and social organisation—many of which this paper has discussed.

The Marine program within the UNESCO World Heritage program (UNESCO 2020b), under which a number of marine and coastal sites have been designated, is helping to raise the profile of World Heritage designations in ocean and coastal governance. Although much of the program's work is focused on natural heritage, several designations of coastal and marine sites include either historical and archaeological heritage or living human cultural practices and meanings as criteria for listing. These include Papahānaumokuākea (United States), an isolated linear cluster of small, low-lying islands and atolls, with their surrounding ocean, roughly 250 km to the northwest of the main Hawaiian archipelago and extending over some 1931 km^2. The area has deep cosmological and traditional significance for living Native Hawaiian culture, as an ancestral environment, an embodiment of the Hawaiian concept of kinship between people and the natural world, and the place where it is believed that life originates and to which the spirits return after death. On two of the islands, Nihoa and Makumanamana, archaeological remains testify to pre-European settlement and use. Much of the monument is made up of pelagic and deep-water habitats, with notable features such as seamounts and submerged banks, extensive coral reefs and lagoons. It is one of the largest marine protected areas in the world.

An overview of the application of UNESCO World Heritage designations to oceanic and coastal sites is given in Annex 5, with examples of maritime- and ocean-associated inscriptions under the 2003 UNESCO Convention on Intangible Cultural Heritage.

5 Opportunities for Action to Support Pluralistic and Inclusive Ocean Values

There have been calls from scientific leaders for a new narrative about the ocean that replaces indifference or despair at the state of the ocean with optimism and innovation, to 'reset expectations and liberate ingenuity' (Lubchenco and Gaines 2019, 911). Calls for new narratives also come from the grass roots, with small-scale fishers' representatives calling for a more positive story about the role of small-scale fisheries in contemporary and future society (FAO 2020b). We suggest that both these narratives are calls for recognition of the many and diverse contributions the ocean makes to human well-being.

5.1 Synthesis and Conclusions

People have multiple economic, political, social, cultural, spiritual and emotional relationships with the ocean. These 'blue relationships' are a product of geography, history and human diversity—including diversity of economic circumstances. Because there is unprecedented attention to ocean policy at present, and because—amid a health and economic crisis—there is a wider global reflection on human values and trajectories and a rising tide of protest against economic and racial injustice, the present moment offers great opportunity to take the bold political actions needed to develop a sustainable ocean economy built on diverse relationships, in ways that encourage equity and inclusion and that recognise the non-material aspects of well-being.

The human relationship with the ocean tends to change with distance from land. More of humanity is acquainted with the coast than with the high seas. Our relationships with shorelines, estuaries and tidal waters are intimate, while our relationships with blue waters may be warier and more transactional. This gradation of relationships also requires a gradation in the scale of legal and policy frameworks. Local and traditional systems are prevalent in inshore waters and could provide a basis for building governing institutions with legitimacy and in ways that sustain both the material and non-material contributions the ocean makes to well-being. The legal regime governing EEZs and the high seas is largely in place and supported by UNCLOS, but it requires strengthening and dialogue to accommodate new ocean uses as countries make ocean economic plans and engage in marine spatial planning. It also requires recognition that all peoples and nations should have a say in how this 'common heritage of humanity' is *governed*. This principle—and the right to participate in decision-making about areas beyond national jurisdiction—is increasingly exercised by landlocked states, local coastal communities and Indigenous Peoples (Vierros et al. 2020).

A sustainable economy will be accessed through the land-water interface. How we govern coasts will determine who accesses and benefits from blue space. A heavily privatised, zoned and securitised coast will exacerbate the separation of humanity and the ocean and risk alienating and reducing access for lower-income visitors and residents, as well as long-term ocean resource users and stewards, such as fisherfolk and Indigenous Peoples. An accessible coastal commons, built upon existing institutional foundations and providing livelihood and well-being benefits to all citizens, is likely to foster a more constructive long-term engagement with the ocean.

Our overarching conclusions from our review of the myriad human relations with the ocean, across time and space, are threefold:

- First, there is no simple human relationship with the ocean with which all people will identify. Fishers inshore and offshore, refugees attempting to cross the Mediterranean, beach and adventure tourists, seaside condominium buyers, oil rig workers, Ghanaian fish mammies and the oceanographers all have different interests, experiences, economic stakes, emotional investments and cultural and social ties to different aspects of the coast and ocean. Building on this plurality of values to forge a diverse and inclusive sustainable ocean economy provides opportunities for increasing the ocean's contribution to both material and non-material well-being on a planetary scale. Doing this in practice is likely to require a participative style of democracy with the active engagement of ocean and coastal citizens.
- Second, ocean economic development plans that specifically address equity and inclusion will help reduce existing gender, class, ethnic, North-South and racial inequalities. While the ocean's regulatory and support services benefit all humanity, it is particularly important to address the concerns and interests of the majority of those working on the ocean or stewarding the ocean, such as Indigenous Peoples using diverse natural resources, municipal authorities maintaining clean water and beaches, and those working in and managing fishery resources and conservation areas. Collectively, small-scale fisherfolk, sea nomads and coastal Indigenous Peoples are 'too big to ignore' (Chuenpagdee 2011).
- Third, in the industrialising ocean, maintaining open and natural spaces such as coastal beaches and seas contributes to citizens' access to opportunities to gaze at, walk, swim or play near, in or on the ocean. These access rights and resources should be preserved as far as possible and extended where they have been eroded or unjustly encroached upon. Losing this access to 'blue nature' will result in an incalculable loss to human well-being.

This is therefore a once-in-a-generation opportunity to pause and carefully consider our complex relationship with the ocean, and to rethink it and remake it to meet the challenges future generations will face. The Sustainable Development Goals themselves represent more than a reduction in the incidence and prevalence of the material dimensions of poverty—important as these are. They represent a set of pathways to human well-being and a transition to planetary sustainability. Achieving them will require humanity to have a rich, diverse, engaged and evolving relationship with our ocean planet.

5.2 Opportunities for Action

We present action opportunities as linked steps, starting from a reframing of the human-ocean relationship, progressing through an engagement with a wider ocean constituency, and finally supporting that constituency to establish diversity and inclusion as core elements of a sustainable ocean economy.

5.2.1 Humanise the Ocean Narrative

Narratives motivate and inform political action (Lubchenco and Gaines 2019). Narratives that celebrate the rich diversity of human social, cultural, cognitive and emotional relationships with the ocean and emphasise the relationship between human well-being and ocean ecosystem flourishing could help broaden the political consensus around a sustainable ocean economy. We therefore suggest several ways that such narratives can be developed, progressing from short- to longer-term action.

Shifting Language and Frames of Reference

Three shifts in frames can help create a context for more inclusive actions in support of positive human-ocean relationships:

- *Reclaim the idea of the blue economy.* It has energised governments, NGOs and the private sector but alienated some in civil society. Originally developed as a means to bring the principles of equitable 'green growth' to small island developing states, it has been co-opted by many different interests. Its 'greenness' and its commitment to equity need to be reasserted, while the growth imperative itself requires critical scrutiny.
- *Reframe ocean ecosystem services as 'nature's contributions to people'* (NCPs), following the IPBES (Díaz et al. 2015, 2018). The emphasis on converting all of nature's goods, services and gifts to humans to their equivalent in monetary values (even if only conceptually) is both politically polarising and problematic for many outside of Judeo- Christian cultural influences, particularly Indigenous Peoples. NCPs constitute a more inclusive frame that recognises a diversity of knowledges and value systems (Pascual et al. 2017).
- *Emphasise the ocean's contributions to meeting all SDGs.* Here it is particularly useful to move beyond the achieve-

ment of SDG 14 to consider potential ocean economy contributions not only to food security and poverty reduction but also to gender equity, youth employment and development partnerships. Tools for mapping ocean economy plans against potential contributions to SDGs are being developed and could be applied to sustainable ocean economy development plans at the national and subnational levels (e.g. Obura 2020).

Broadening the Knowledge Base: Informing Blue Futures

Local and Indigenous knowledges and the environmental humanities can, along with natural and social sciences, inform the sustainable ocean economy and reflect the diversity of human-ocean relationships. Here are tangible ways to achieve this representation:

- Ensure that historians, anthropologists and local and Indigenous knowledge-holders are part of national ocean economy planning teams.
- Allocate marine research funding to the arts and humanities as well as the natural and social sciences.
- Encourage institutions such as maritime museums to consider broader representations of the ocean sector.[7]
- Work with communities, municipalities and private sector sponsors to support cultural festivals that celebrate coastal and sea life, or arts, theatre and film festivals[8] with ocean themes.
- Identify coastal and maritime candidates for UNESCO World Heritage listing. Such listings not only help preserve and reproduce cultural and social values but also attract investment in tourism and renewal of coastal towns' housing stock and regeneration of waterfronts.
- Document fishers' knowledge to broaden the knowledge base. *The IUCN Guidelines: Gathering of Fishers' Knowledge for Policy Development and Applied Use* (to be published shortly) sets out protocols for doing this.

[7]Stories of the past are integral to shaping values and world views—as the current challenges to ethnocentric histories and cultural symbols are showing. Exhibitions on slavery, immigration and maritime labour unions are increasingly found in the maritime museums of former colonial powers (e.g. in Liverpool and Amsterdam), to balance the more traditional tales of heroic exploration and naval victories. These attract more diverse visitors and, through a better understanding of the past, allow a reimagined future.

[8]Community theatre, dance, song and poetry are widely popular globally (e.g. Buck and Rowe 2017) and used to both celebrate and explore contentious issues. Pêcheurs du Monde (Fishers of the World) is a film festival held biannually in Lorient, Brittany, France. The 2019 British film *Bait*, about gentrification in a Cornish fishing village, has won a number of festival awards; see https://en.wikipedia.org/wiki/Bait_(2019_film).

All these activities draw attention to both positive and contentious human relationships with the ocean and complement the more widely known public presentations of the ocean that focus on ocean wildlife or present the ocean as despoiled wilderness.

5.2.2 Engage Key Constituencies in the Development of Future Ocean Visions and Planning Processes at National and International Levels

After some 40 years of global consensus on globalised and liberalised economies, vulnerabilities in that system revealed by the COVID-19 pandemic, concerns for rising inequalities and inadequate action on climate change are leading to radical calls for transformative economic and social policy. There are a diversity of visions and values for both the global economy and the global ocean being developed ('blue justice', 'blue degrowth') but these are largely outside the UN process. The proponents of these visions, which include small-scale fisher representative organisations, could be included in dialogues and planning for the sustainable ocean economy. Specific actions may include the following:

- *Set up intergenerational dialogues on ocean futures at the national level.* Youth are taking a lead on global climate action and could bring that leadership to the ocean policy arena, in dialogue with elders who, in many Indigenous and traditional societies, are keepers of stories, knowledge and authority.
- *Engage with organisations proposing alternative visions for human-ocean relationships.* Civil society organisations representing some 'ocean citizen' interests have expressed concerns about the process and vision of developing the sustainable ocean economy. They have watched what has happened to their fellow citizens in previous agrarian and industrial revolutions and fear what follows from the 'blue acceleration'. Their concerns are legitimate, and their institutions, knowledge and policy advice could help prevent such outcomes in the future ocean economy. Indigenous Peoples are key constituents here in many ocean states; so too are trade unions and NGOs in the human rights, community development and environment and conservation arenas.
- *Choose between commons or private ocean.* This should include holding national-level dialogues on the institutional and financial means to achieve a diverse, equitable and sustainable ocean economy that delivers well-being to the greatest number of people. The debate centres principally on the use of communitarian, public sector or privatisation approaches to governance. Balancing local and global responsibilities is the critical challenge.
- *Support small-scale fisherfolk.* Small-scale fishers are currently the largest population group directly economi-

cally dependent on the ocean and are part of the private sector. They feel squeezed out of coastal zones that they have occupied, used and stewarded, in some cases for centuries. They could be powerful allies for ocean stewardship.

- *Engage port cities and coastal local governments in inclusive ocean governance.* Port cities have played major roles in shaping the human relationship with the ocean, often both driving economic growth and leading social change movements. Leaders of major cities are already key actors in global policy arenas such as climate action, sustainable food systems and equity and justice (Pearson et al. 2014), so they are well positioned to play a greater role in ocean governance. Ports, waterfronts and ocean tourism are key elements of many coastal city economies, which increasingly look to a 'blue urbanism' to improve their citizens' quality of life (Beatley 2014). Coastal local governments may also be better than national marine spatial planning processes at accommodating diverse local interests.

The dialogues with small-scale fishers and with coastal Indigenous people could be brokered through FAO's civil society platforms, through which they have been able to develop the SSF Guidelines (FAO 2015)—a voluntary global governance instrument with significant buy-in from the world's fishing communities and, increasingly, governments, coastal Indigenous Peoples and environmental NGOs (Jentoft et al. 2017). This process of engagement could begin with the 2021 meeting of the FAO Committee on Fisheries. An equally important mission is to ensure that the democratic and human rights–based guiding principles and provisions of the SSF Guidelines are mainstreamed and embedded in fisheries large and small, and in the numerous blue economy programs at the national and international levels.

Table 11.5 summarises the rationale for developing the knowledge base, narratives and engagement strategies described above.

5.2.3 Create Policies and Mobilise Finance for Actions in Support of an Inclusive Ocean-Society Relationship

Inclusive governance is best supported by participatory democracy, which requires an active and capable civil society. We propose three actions to safeguard and facilitate engagement of, and action by, a wide spectrum of ocean interests, including those who have been historically or recently marginalised by the economic development process. This entails support for such communities and populations to enable them to participate fully in the sustainable ocean economy and to retain and expand their historical rights and responsibilities in the ocean. The effort will require the development of adaptive leadership in both the public and private sectors.

Table 11.5 Synthesis of means to foster narratives, dialogues and knowledge to mobilise key constituents to support inclusive, pluralistic ocean governance

Areas of benefit				
Goal for action	Discourse	Knowledge	Value	Alignment
Enrich the diversity of ocean policy actions by humanising ocean narratives.	Reclaim the original meanings of the 'blue economy', centred on equity and sustainability.	Broaden the knowledge base informing 'blue futures' to include indigenous and local knowledge, the arts and humanities.	Shift from valuing ocean ecosystem services to valuing the ocean's contributions to society.	Emphasise the contributions of the ocean to meeting all SDGs, not just SDG 14.
	Stress the relational and subjective elements of ocean contributions to societal well-being.	Ensure equity of access to information by marginalised and/or vulnerable groups.		
Ensure Ocean inclusivity by engaging key constituencies.	Set up intergenerational dialogues on ocean futures.	Engage overlooked or marginalised constituencies for the ocean.	Engage with religions and spiritual beliefs about the oceans	Develop dialogues with small-scale organisations representing fisheries.
			Recognise the sovereign claims of indigenous peoples.	Engage with leadership of the governance bodies of major coastal cities.
	Dialogue and facilitation		Mainstreaming and policy alignment	

1. **Create a publicly funded community-based knowledge depository for use in dialogue with new ocean economy interests**. In order to safeguard their own knowledge and perspectives that connect intergenerational interests and communities' needs for a sustainable ocean economy, we need to *build on traditional institutions, values and narratives of the ocean in different regions and nation-states and support communities* **to continue to evolve their ocean relationships**. Public funds to support these relationships could include support for scientific monitoring and data collection, maintenance of social institutional knowledge and practices, and support for civil society organisations and Indigenous governments as they chart their own courses in the future ocean.

   ACTION: Establish long-term public funding streams dedicated to the creation and maintenance of national-level knowledge repositories in which community-level experiences of disasters, toxic exposures and development in general can be archived. Build institutionalised connections between such repositories and communities or their legitimate representatives in order to increase community-level access to salient information and prior experience, and level the playing field for communities negotiating with external interests.[9]

2. **Provide statutory protection to prevent private equity from undermining community capacity and interests**. Powerful corporate interests have the capital and political influence to shape community-government relationships in their favour. Governments have a duty of care to ensure that their citizens' interests are fairly considered in future ocean economic development. Governments need to create legal and expert support aid for ocean-dependent communities and civil society organisations to facilitate their capacity (including their internal decision-making process and legal base) to negotiate with external interests in order to represent their cultural world views in formal negotiation processes. Current ocean-related funding reflects global conservation and development values but leaves little space for local voices and values. The funding will prevent powerful actors from buying their 'social licence to operate'.

   ACTION: In recognition of clear conflicts of interest, and in order to improve social license for development as well as development outcomes and impacts, disallow industry and development interests from funding the negotiators representing the communities and decision-makers whose interests they seek to influence. Instead, establish institutionalised, national-level sources of expert legal support that may be freely accessed by communities approached by development interests. Ensure that legal support is fluent in and supportive of the diverse priorities and cultural world views of the range of represented community groups.

3. Re-evaluate development funding structures and mechanisms to support meso-level institutions.

   In reformist agendas such as those explored by the Ocean Panel, there is a tendency to look to either high-level processes and powerful actors (technological change, global laws, standards and conventions, intergovernmental organisations, national governments, national policy instruments, market forces, large corporations or industry associations, international conservation NGOs) or shifts in individual behaviour (diversifying out of fishing, eliminating single-use plastics). These are the two extreme levels in multi-level governance. In between are devolved, municipal and local government, Indigenous sovereignty, community, civil society organisations, trade unions, pluralistic legal systems, public institutions, small and medium enterprises, and the trade and cooperative organisations that bind them together.

   ACTION: Invest in the capacity of socially and culturally embedded meso-level institutions to govern both traditional and emergent ocean industries, in partnership with government and intergovernmental and international NGO actors where appropriate. This includes support for institutions that enable legal protection of sea tenure, uphold existing human rights (gender, labour, rights to food, rights to livelihood, etc.) and facilitate access to new opportunities in the ocean economy by marginalised groups in society.

4. **Ensure that responses to COVID-19 are based on an understanding of well-being in all its dimensions:** COVID-19 impacts are linked to many other issues, including building resilient economies and food systems and adapting to and mitigating climate change. This nexus of responses needs to build on an understanding of what is required to support subjective and relational well-being as well as material needs.

   ACTION: Invest funding in social well-being assessments as part of COVID-19 recovery planning. Support inclusion of ocean-dependent communities in recovery planning in other sectors, principally transportation, tourism and fisheries.

Finally, the Ocean Panel has the opportunity to reassert a commitment to multilateralism as part of any attempt to bring greater order to the ocean and to use the members' diplomatic channels to engage other states in this endeavour: The 14 countries' call for harmonised ocean governance comes at a time of resurgent nationalism, when many of the world's largest economies are either turning away from multilateral treaties and institutions or seeking to control and influence them in their

[9]Namati, who 'put the power of law in people's hands' around the world, is an example of such a group: https://namati.org.

own interests. There is a delicate balance between recognising countries' rights to use their ocean assets for their national priorities, on the one hand, and representing the 'common good of humanity' and the rights of non-human nature, on the other. It will take bold political and social leadership to develop such a governance architecture.

6 Conclusion

The Yoruba littoral has experienced the profound impact of waves of human migrations over the centuries.

Migrations took place along two axes: north-south, dominated by fellow Yoruba-speaking peoples; and east- west, dominated by the Ilaje, Izon, Ogu and Keta (Ewe).

The littoral is, therefore, 'Yoruba' only in the sense of the overwhelming dominance of Yoruba-speaking people in the zone. Most Yoruba littoral communities share social and cultural institutions and practices. These include the insignia of chieftaincy and traditional institutions; divination by individuals and communities; the Olokun, Malokun and Okosi festivals; and the pervasive presence of Ayelala. Nevertheless, non-Yoruba elements have contributed substantially to the zone's socioeconomic development. Hence, the region is a linguistic and cultural continuum, a pan-Yoruba commonwealth, which includes Yoruboid (Itsekiri and Edo) and non-Yoruba peoples as indigenes at both ends of the zone and as residents all over it.

Fishing has been the driving force behind the lateral movements along the coast. It remains the centre of littoral social and economic life. Fishing is the impetus behind Ilaje, Izon, Ogu and Keta migrations and settlement across the Gulf of Guinea and beyond. The fishing industry has witnessed significant innovations, while remaining largely artisanal, since the 1890s. Both Indigenous religion and Christianity have contributed to the political, economic and social development of the Yoruba littoral.

The Yoruba, with the exception of the Ilaje and Awori subgroups, are largely riparian landlubbers, but the ocean has shaped their world view (Olukoju 2000, 2017). The Ilaje are the only truly maritime community among the Yoruba, followed in importance by the waterside Ijebu and Awori. Ilaje colonies all over the Yoruba littoral, and far beyond, testify to their unique status in the zone.

The littoral has not been a dominant force in the geopolitical and economic systems of the Yoruba of western Nigeria. It has been marginal in terms of political influence and state formation. In pre-modern times, Indigenous communities in littoral Yorubaland were subject to political and cultural domination from hinterland power centres, such as Ife, Ijebu and Benin. Such control or influence varied over time. It was minimal in post-1800 Lagos but pronounced in the case of Ejinrin and other lagoon communities vis-à-vis the Ijebu Kingdom. Badagry and Epe were havens or rear bases for political fugitives from Lagos before 1862.

Inter-group relations between indigenes and residents of the Yoruba littoral are complex and dynamic. In general, ethnic affinity between host communities (Awori and Ijebu) and fellow Yoruba immigrants (especially the Ilaje) has mitigated potential conflict and produced generally cordial relationships. In contrast, ethnic diversity has compounded ordinarily tense relations centred on competition over land, fishing and other economic rights. This is illustrated by the Ilaje-Izon conflict in southern Ondo State in the late 1990s, and clashes between the Yoruba and Ogu in the Ajah axis of metropolitan Lagos in 2016.

Multifaceted relationships characterise the Yoruba littoral. These have been underpinned and moderated by homogenising forces of religious affinity, cultural exchanges, commercial relations, intermarriage and the linguistic and cultural accommodation of immigrant elements in local communities.

Acknowledgements We thank the participants in a discussion seminar on 'The Blue Economy and the Future Human Relationship with the Sea', MARE conference, July 2019.

Name	Affiliation
Edward H. Allison[a] (seminar organiser)	Nippon Foundation Ocean Nexus and WorldFish, Penang, Malaysia
Fabio Boschetti	CSIRO, Australia
Ruth Brennan	Trinity College, Dublin, Ireland
Jens Christiansen	University of Lancaster, UK
Philippa Cohen	WorldFish, Penang, Malaysia
Anna Farmery	ANCORS, University of Wollongong, Australia
Madeleine Gustavsson	University of Exeter, UK Mariam de Haan
Derek Johnson	University of Manitoba, Canada
John Kurien[a]	ICSF and WorldFish, Penang, Malaysia
Cynthia McDougall	WorldFish, Penang, Malaysia Borya Nogue Alguero
Camilla Novaglio	CSIRO, Australia
Qurban Rochani	Rhodes University, Grahamstown, South Africa
Irna Sari	SARI, Indonesia
Marleen Schutter	Lancaster Environment Centre, University of Lancaster, UK
Michelle Scobie	University of the West Indies, Trinidad and Tobago Sarah Taylor
Michelle Voyer[a]	ANCORS, University of Wollongong, Australia
Stella Williams	Mundus Maris, Port Harcourt, Nigeria

[a] Lead author or contributing author of Blue Paper 11

Jonathan Barton, Janice Cumberbatch, Christina Hicks, Moeniba Isaacs and Kimberley Peters contributed ideas and support to early drafts of this paper. The work also benefited from having co-authors in common with other Blue Papers related to human relationships with the ocean. Allison and Fabinyi were also involved in BP13, on equity, and Allison was involved in BP12, on governance transformations. We thank the author teams on those papers for support and cross- fertilisation of ideas. The work was completed while Allison was a visiting fellow at the Australian National Centre for Ocean Resources and Security, University of Wollongong.

The authors thank the paper's technical reviewers, Olanike Adeyemo, Tekau Frere, Hugh Govan, Danika Kleiber, Ghislaine Llewelyn, Jorge Torre and Joeli Veitayaki, as well its arbiter, Jane Lubchenco, who all provided helpful technical comments. The authors also thank World Resources Institute for providing support as the Ocean Panel Secretariat, its insights, support and patience during the preparation of this paper.

While our colleagues were very generous with their time and input, this report reflects the views of the authors alone. The authors thank Alex Martin for copyediting and Carni Klirs for design.

Annex 1. Social Well-being and Values of Sama-Bajau

Scattered throughout island Southeast Asia are three groupings of specialist maritime populations commonly referred to in academic and popular literature as 'sea nomads' and 'sea gypsies', 'a designation at once romantic and derogatory' (Gaynor 2005, 90). These ethno-linguistic groups are the Moken, Orang-Laut and the Sama-Bajau. Each of these groups is geographically, linguistically and culturally distinct and has developed its own modes of adaptation and livelihoods on Southeast Asia's highly biodiverse island, coral reef and ocean environments to support its livelihoods. The Sama-Bajau are the most widely dispersed ethno-linguistic group indigenous to insular Southeast Asia, scattered over a maritime zone 3.25 million square kilometres in extent, stretching from the Philippines to Indonesia. Between 750,000 and 1.1 million Sama-Bajau speakers are estimated to live in Southeast Asia (Stacey and Allison 2019). Generally landless, the Sama-Bajau spend their entire lives in the vicinity of the ocean, in a marine environment that constitutes 'culturally defined living spaces' (Chou 1997). It is often said by Sama, and by other Indigenous groups with whom they reside, that Sama feel sick if they spend too much time on land, or away from the ocean. They maintain a rich Indigenous marine cosmology and ritual practice, with belief in supernatural beings—ancestors of the ocean—that live in and control the universe of the ocean and all the creatures in it for Sama-Bajau people (Stacey 2007).

Sama-Bajau culture is intimately connected to marine environments on which they depend for subsistence and cash income, as well as their cultural identity.

Culturally defined patterns of fishing activity (including migratory expeditions) unite all sectors of Sama-Bajau communities through catching, consuming, processing and trading of marine resources. Fishing and gathering of shellfish and other strand resources provide the focus for individual and communal relations within villages and across extensive kin and trading networks. The maintenance and transmission of Indigenous language and knowledge from one generation to the next occurs through socialisation into livelihoods and related social and cultural activities. As such, customary beliefs and practices in relation to boats and sea spirits endure among the Sama-Bajau, and are primarily oriented to ensuring return on fishing effort (Stacey et al. 2018).

The perception of the ocean as an open space of living and trading, as well as the Sama-Bajau's notion of freely moving through that space as performing their identity and generating their world (Pauwelussen 2015), stands in contrast to the politically fragmented seascapes we see on maps. As Cynthia Chou (2006, 1) notes, 'Whichever translation one is inclined toward, the heart of the matter is that the space which others have named "Southeast Asia", comprising a number of bordered nation-states, is, in contrast, a space of deep emotional and personal meaning for the sea nomads'. The islands and sea which they occupy constitute 'living spaces' (Chou 1997, 613) for the generally landless Sama-Bajau. The movement of Sama-Bajau is entirely over water, whether commuting in dugouts between neighbouring households, visiting the 'mainland' or going fishing. Extensive pressures undermine Sama-Bajau fishing practices and their way of life, which in turn erode values of social well-being. The majority of Sama-Bajau in Southeast Asia are marginalised Indigenous groups, and in many instances their contributions as fishing peoples are not recognised by the region's governments. In many cases, they remain on the outskirts of mainstream societies in the countries they inhabit and are often stigmatised as being unruly, lazy and backward aliens. It is unlikely, then, that mainstream societies will recognise the societal loss associated with the erosion of the Sama-Bajau access to fisheries and the consequent transformations in Indigenous knowledge, cultural practices and diversity. However, loss of the fundamental values that underpin Sama-Bajau culture and social well-being will have significant impact beyond obvious implications for the Sama-Bajau themselves. In a country like Indonesia, for example, which prides itself on the national motto *Bhinneka Tunggal Ika* ('Unity in Diversity') and the concept of *Nusantara* as an archipelago where seas connect society, the loss of a unique maritime way of life such as that of the Sama-Bajau is an erosion of the very principle that constitutes the nation's desired identity. Moreover, at a local level, many rural fish markets in areas where Sama-Bajau reside are largely stocked by fishing activities from these Sama-Bajau groups and provide land-based ethnic groups with important sources of nutrition. Furthermore, marine conservation and sustainable fisheries management initiatives across the region increasingly recognise that effective and locally relevant measures need to flow from increased ground-level co-production of knowledge and practices drawing from Western technical management principles and local traditional knowledge. The presence of the Sama-Bajau at sea and their knowledge of the ocean should be acknowledged as a significant asset not only for any marine resource management initiative but also for the national identities of the region's states.

Annex 2. The Arts and the Ocean

The sea, the ships that navigate it and the fish and mammals that inhabit the marine space are a limitless source of artistic inspiration. Artistic representations of the ocean can create similar emotions that we experience from seeing, hearing, smelling and interacting with the ocean. Art can simply be a record of a place, time or event, but its purpose is usually to create an emotional response. Contemplating marine art can 'enrich … us, help us see how precious all our lives are' (Krupinski 2019, 9).

Marine (or maritime) art has a prominent place in European history (and much has been written on this topic) but equally so in other parts of the globe. Perhaps one of the most recognised marine images comes from Japan: an early nineteenth-century woodblock print by the Japanese artist Hokusai, *The Great Wave off Kanagawa*. Marine art in Europe initially focused on harbours (e.g. *A Calm* [1654], by Dutch artist Jan van de Cappelle) and sea battles but gradually progressed to works where the ocean and the shores were depicted as playgrounds for recreation, bathing and sailing Later a fascination with heavy industry and shipbuilding became a focus. The last century saw more depiction of things below the surface, as these have become more known and accessible.

Marine art is not only produced near the ocean. Prehistoric rock paintings in the interior of Africa depict marine mammals and fish (van Riet Lowe 1947). In Australia, Indigenous art has historically been influenced by the *ocean, and it still is today. Indigenous Australian marine art has recorded many events in time*. For instance, in the Northern Territory rare Indigenous rock art may depict the first seafarers to reach Australian shores (Middleton 2013). Marine art can change history. It has played an important role in self-determination and gaining sea rights for a group of Australian Indigenous people. The success of the Yolŋu community's native sea title claim was underpinned by bark paintings that evidenced a long relationship with the ocean (https://hyperallergic.com/412659/sea-rights-bark-paintings-australia/).

Annex 3. Sharks as Symbolic Animals

Shark binders and charmers in the Gulf of Mannar linking South India and Sri Lanka are believed to have operated for six centuries to protect pearl divers from shark attacks. In 1885 the then colonial administration forced them to discontinue what it saw as a superstitious practice linked to payments that exacted an undue toll on the pearl industry (Cordiner 1807; Kunz and Stevenson 2001 [1908]).

Sharks in the Pacific islands were believed to have spiritual powers, as ancestor guardians and/or gods offering protection from the unpredictable forces of the ocean, or as malevolent spirits in the form of shape- shifting 'shark men', who needed to be appeased so they would not devour human beings venturing into their domain (Beckwith 1970; Grimble 1972; Hviding 1996; Barry 2002; Montgomery 2006; Hilmi et al. 2016).

As symbolic animals, sharks appear to signify both the fear of the ocean's vast unknown depths and the bounty available to those who respect the ocean's ways and its powerful creatures.

Annex 4. Flying Fish as Symbol of Barbadian Identity

As Bajan as flying fish. Local saying

To Barbados the flying fish (*Hirundichthys affinis* and other species) is a quintessential aspect of intangible heritage: a symbol of Bajan 'pride and industry' (the country's motto). It adorns the silver dollar coin and the logo of the Barbados Tourism Authority, as well as being the mascot for some national sports teams.

Bajan cuisine has its own unique ways of preparing and cooking the flying fish. It is also a tangible part of Bajan culture, as a source of livelihoods and nutrition in an important fishery. Long a mainstay of local fisheries and diets (its bones have been found in archaeological digs of Indigenous people's middens), the flying fish was officially recognised as one of Barbados's icons after the island gained independence in 1966 and the government began looking for symbols that generated a sense of pride among the populace, something of which islanders could say, 'This belongs to us and nobody else'.

But the availability of flying fish is at risk, as are its associated traditions, and there is no regulatory framework to protect this icon of national heritage. Threatened by a maritime boundary and fisheries dispute, it struggles under the legacy of price controls that hamper the profitability of the fishing industry. It is uncertain whether 50 years from now there will still be people to harvest the species, or people who know how to de-bone it in the traditional way. Significant variability in abundance could affect Barbadians' access to flying fish, and climate change could even lead to the species' local extinction.

A small survey (100 people) of Barbadians' cultural attachment to flying fish showed that as a symbol of national identity it lags behind the broken trident on the national flag, is a close second (34 vs. 35 respondents) to the much larger dolphinfish (*Coryphaena hippurus*) as the preferred fish to eat and as a staple food is challenged by the rising consumption of chicken throughout the Caribbean.

Source: Adapted from Cumberbatch and Hinds (2013).

Photo by Angie Torres from Flickr

Photo by Roshan Kamath from Pexels

Annex 5. The Use of UNESCO World Heritage Instruments to Support the Conservation of Plurality of Values Humans Derive from Interactions with the Ocean

Marine and coastal world heritage sites that are UNESCO-listed also include St. Kilda and surrounding islands (Scotland, UK), Ibiza (Spain), the Trang An Landscape complex in the Red River Delta (Vietnam) and the Rock Islands Southern Lagoon (Palau). Some sites that could perhaps be listed as mixed cultural and natural heritage, such as the Sundarbans in Bangladesh or the Great Barrier Reef in Australia, are currently designated as Natural Heritage only.

Africa is generally underrepresented in the designation of World Heritage sites (Breen 2007). Its coastal and marine representation includes the port cities and island trading centres of Lamu Old Town (Kenya); Stone Town Zanzibar (Tanzania); the Island of Mozambique; Robben Island (South Africa); and the islands of Gorée and Saint-Louis (Senegal). Many of these are listed partly or mostly because of their historical role in the Indian Ocean and transatlantic slave trades or, in the case of Robben Island, because it is where Nelson Mandela, was held for much of his 27 years as a political prisoner of South Africa's apartheid-era government. More positive representation of cultural and maritime heritage is an unrealised (and so far, little-discussed) opportunity within the 'blue economy' movement in Africa (Table 11.6).

Table 11.6 Examples of maritime and ocean-associated inscriptions under the 2003 UNESCO convention on intangible cultural heritage

Country, year listed, list category	Maritime, coastal or ocean-related cultural heritage
China, 2009	**Mazu belief and customs.** As the most influential goddess of the sea in China, Mazu is at the Centre of a host of beliefs and customs, including oral traditions, religious ceremonies and folk practices, throughout the country's coastal areas. Deeply integrated into the lives of coastal Chinese and their descendants, belief in and commemoration of Mazu is an important cultural bond that promotes family harmony, social concord and the social identity of these communities.
South Korea, 2016 (1)	**Culture of *Jeju haenyeo*** (women divers). In Jeju Island, there is a community of women, some in their 80 s, who go diving 10 meters under the surface to gather shellfish, such as abalone or sea urchins, for a living without the help of oxygen masks. With knowledge of the ocean and marine life, they harvest for up to 7 h a day, 90 days a year, holding their breath for just 1 min for every dive and making a unique verbal sound when resurfacing. Before a dive, prayers are said to the Jamsugut, goddess of the sea, to ask for protection and an abundant catch. Knowledge is passed down to younger generations in families, schools, local fishery cooperatives which have the area's fishing rights, *haenyeo* associations, a school and a museum. The culture of *Jeju haenyeo* has also contributed to the advancement of women's status in the community and promoted environmental sustainability with its eco-friendly methods and community involvement in management of fishing practices.
Indonesia, 2017	**Art of Pinisi boatbuilding, South Sulawesi.** *Pinisi* refers to the rig and sail of the famed 'Sulawesi schooner'. The construction and deployment of such vessels are part of the millennia-long tradition of Austronesian boatbuilding and navigation that has brought forth a broad variety of sophisticated watercraft. Shipbuilding and sailing are not only the communities' economic mainstay but also the focus of daily life and identity. The reciprocal cooperation between the communities of shipwrights and their relations with their customers strengthen mutual understanding among the parties involved.

(continued)

Table 11.6 (continued)

Country, year listed, list category	Maritime, coastal or ocean-related cultural heritage
Islamic Republic of Iran, 2011 (2)	**Traditional skills of building and sailing Lenj boats in the Gulf of Iran**. Lenj vessels are hand-built and used by inhabitants of the northern coast of the Persian Gulf for sea journeys, trading, fishing and pearl diving. The traditional knowledge surrounding Lenjes includes oral literature, performing arts and festivals, in addition to the navigation techniques and terminology and the weather forecasting closely associated with sailing, as well as the skills of wooden boat-building itself. The philosophy, ritualistic background, culture and traditional knowledge of sailing in the Persian Gulf are gradually fading.
Belgium, 2013	**Shrimp fishing on horseback in Oostduinkerke**. Brabant horses walk breast-deep in the surf, parallel to the coastline, pulling funnel-shaped nets held open by two wooden boards. A chain dragged over the sand creates vibrations, causing the shrimp to jump into the net. Shrimpers place the catch in baskets hanging at the horses' sides. A good knowledge of the ocean and the sand strip, coupled with a high level of trust and respect for one's horses, are the shrimpers' essential attributes. The tradition gives the community a strong sense of collective identity and plays a central role in social and cultural events. The shrimp fishers function on principles of shared cultural values and mutual dependence. Experienced shrimpers demonstrate techniques and share their knowledge of nets, tides and currents with beginners and the up to 10,000 visitors who attend the annual shrimp festival.
Cyprus, Croatia, Spain, Greece, Italy, Morocco and Portugal, 2013 (1)	**The Mediterranean diet** involves a set of skills, knowledge, rituals, symbols and traditions concerning crops, harvesting, fishing, animal husbandry, conservation, processing, cooking and particularly the sharing and consumption of food. Eating together is the foundation of the cultural identity and continuity of communities throughout the Mediterranean basin. It is a moment of social exchange and communication, an affirmation and renewal of family, group or community identity. The Mediterranean diet emphasises values of hospitality, neighbourliness, intercultural dialogue and creativity, and a way of life guided by respect for diversity. It plays a vital role in cultural spaces, festivals and celebrations, bringing together people of all ages, conditions and social classes. Women play an important role in transmitting knowledge of the Mediterranean diet: They safeguard its techniques, respect seasonal rhythms and festive events, and transmit the diet's values to new generations. Markets also play

Note: Three types of designation are included on these lists: (1) representative heritage sites and practices, (2) those in need of urgent safeguarding and (3) those representing good safeguarding practices
Source: UNESCO (2020b)

Annex 6. Economic Activities of the Yoruba Coast

Commercial Activities in Coastal Yorubaland

Trade across the Yoruba littoral was anchored in the lagoon ports of Badagry, Lagos (well before modern port development began in the late 1890s), Epe, Ejinrin and Atijere. While slave trafficking dominated the external trade with the Americas up to the mid-nineteenth century, forest produce, which accompanied the slave exports, became dominant in the second half of the century with British colonial rule. The lagoon ports were melting pots of culture, as epitomised by the history and peopling of Lagos, Epe and Ejinrin during the nineteenth and twentieth centuries. In its formative years, Epe drew migrant hunters, fishermen and political adventurers from the Yoruba towns of Ile-Ife, Ilara, Ibeju and other outlying settlements in the Epe region, and from Benin, a non-Yoruba kingdom. The Ilaje from the waterside of Okitipupa Division and the non-Yoruba Ijo [Izon] were also represented in the population before 1900. Izon men caught fish, while their wives produced a local staple, *garri*, from cassava flour, and manufactured local gin, *ogogoro*, from palm wine.

Though Epe people engaged in other economic activities—cassava, rice and maize farming; boat- building (for deep-sea fishing and water transportation); and commerce—fishing has been their defining economic activity. This earned them the nickname 'Epe Eleja', meaning, 'Epe, community of fishermen'.

Ejinrin, the port of the Ijebu (a Yoruba kingdom) was 'a strategic link between Lagos and the rest of the southeastern Yoruba hinterland' (Olubomehin 1990: 128). It was also a point of convergence for traders from other lagoon settlements, such as Epe and Atijere, and from as far in the hinterland as Ode Ondo. The hinterland of Ejinrin expanded after the defeat of the Ijebu by the British in 1892: it received an influx of traders from the Yoruba hinterland towns of Ife, Oyo, Ilesha, Gbongan, Ado-Ekiti, Owo and Ilorin (Olubomehin 1990, 132). This made it an important feeder for the port of Lagos, which, until 1914, relied on the lagoon network for produce for local consumption and export, in which Ejinrin was a lynchpin.

Fishing and Other Economic Activities in Littoral Yorubaland: Focus on the Ilaje

Coastal Yoruba exhibit occupational specialisation in accordance with variations in ecological setting and resource endowments. The people engage in seafaring, fishing, saltmaking and boat-making, though on an artisanal scale.

However, fishing is the dominant activity in the zone. It is characterised by large-scale migrations of fisherfolk across the Gulf of Guinea, which have facilitated exchange of technology of fishing and fish processing. These migrations have been classified into six types: internal, short-term, seasonal, long-term, permanent and contractual. These have been further grouped into two broad categories: internal (within a country) and trans-border or international (Adewusi 2017, 165). Human migrations, especially seasonal ones, tend to follow the migration of fish species.

Fishing and commercial activities across the Yoruba littoral have generally been peaceful. The ecological context has promoted economic symbiosis between the upland areas conducive for arable farming and the littoral that is dominated by fishing. Foodstuffs, especially yams and cassava tubers, and palm oil produced outside of the coastal mangrove swamp, are exchanged for fish and salt produced in the latter zone. The range of migrations and commercial exchanges is indicated as follows: the 'agriculturally poor but fish-rich eastern Delta [homeland of the Ilaje] exchanged fish for agricultural products from the adjoining communities, including the Ijebu in the west, the Ijaw in the north and the Ikale in the east, which supply most of the food items like cassava (gari), yam, plantain, and palm oil' (Onipede 2017, 181). These and other commodities of the lagoon trade, such as prawns, coconuts and cassava flour, are conveyed over a wide area beyond the coastal zone.

The spatial spread of these activities runs hundreds of kilometres, reaching places as distant as Port Harcourt, Onitsha and Enugu in eastern Nigeria, and Lagos, Ibadan, Ondo, Oyo, Ode-Aye (Ikale) and several Ijebu and Ekiti towns (Onipede 2017, 182). A major item of trade during the twentieth century, second only to fish in importance, is local gin (*ogogoro*). Once declared illicit by the British colonial government and the successor nationalist government, it is now widely consumed in the coastal and upland areas of southern Nigeria, for both social and religious purposes (Oluwapayimo 2017).

The positive impact of non-Indigenous migrants on the domestic economy is indicated by claims that the arrival of the Fante 'boosted fishing' in Ijebu Waterside, while the arrival of the Keta (Ewe) and Ilaje had the same effect on Badagry (Olukoju 2000, 72). The Orimedu community east of Lagos also benefited from the settlement of Gold Coast/Ghanaian fishermen, from whom they adopted the seine 'drag net' fishing technique the Ghanaians introduced on their arrival in the 1890s. It is claimed that while the Yoruba hosts have continued to practice subsistence fishing for family consumption, the Ghanaians run theirs as a business. The Yoruba continue to fish in small canoes, unlike the Ghanaians, who employ 'boats fitted with outboard engines, synthetic nettings such as beach seiners, set nets, gill nets and long lines' (Adewusi 2017, 166–67).

However, occasional conflict interrupts peaceful commercial and social exchanges on the Yoruba littoral. The most protracted and catastrophic was the Ilaje-Ijo war of 1998–1999, fought over claims to a territory rich in hydrocarbon deposits. The war proved ruinous to both parties, who suffered heavy human and material losses (Ehinmore 2014). Analysis has shown that conflicts are more intense in intergroup relations between Yoruba and non-Yoruba, and rarely between two Yoruba groups (Olukoju 2000).

The Ilaje are the most dominant fishing and migrant Yoruba subgroup in the zone. Outside of their homeland on the eastern edge of the Yoruba littoral, they have established colonies all over the Yoruba littoral and the entire Gulf of Guinea. Ilaje migrant fishermen and their families overlap with Indigenous communities as well as Izon and Ogu migrant fishing groups. In the port city of Lagos, the Ilaje occupy lagoon-side neighbourhoods, especially at Bariga, where a common Yoruba identity has ensured cordial relations with their Awori hosts. Ilaje have intermarried with the Awori and now bury their dead in Lagos, contrary to age-old practice.

Ilaje fishing enterprise and extensive migrations eastwards of their homeland towards the Congo, westwards to Senegal and southwards to Angola have been described as 'perhaps "the greatest inter-regional movement in the modern history of the Yorubas"'. It has been stated that they are 'always on the move … in search of fish' (Olukoju 2000, 70). Ilaje fishing enterprise is complex, comprising artisanal fishery, local aquaculture and industrial fishing (Ikuejube 2005). Artisanal fishing remains dominant, carried out in local canoes with traditional fishing instruments: *awo* (nets with hooks laced with bait of earthworms), *akase* or *eporo* (bamboo spears) and *iyanma* (baskets used in shallow streams and creeks) (Onipede 2017, 183). But Ilaje fishing has also incorporated innovations from external sources. The Ilaje learned the *dogbo* method of trawling from the Ijebu in the 1930s and borrowed the *itahun*, 'a new method of fishing net floatation,' from Gabon in the early 1970s (Ehinmore 1998, 31, 37). A turning point in Ilaje fishing enterprise was the 1960s expansion from freshwater to coastal fishing. This necessitated the use of larger watercraft, with carrying capacities varying with the distances covered and the quantity of merchandise carried. Engine-powered boats are now deployed in high-sea fishing, thus extending Ilaje fishing enterprise beyond the lagoon. In addition, the Ilaje in metropolitan Lagos have, since their initial settlement in the 1930s, expanded their fishing enterprise and diversified into other business ventures—quarrying sand from the lagoon, lumbering and water transportation (Olukoju 2000; Onipede 2017, 184).

About the Authors

Lead Authors

Edward H. Allison is Research Director of the Nippon Foundation Ocean Nexus Center, and Research Chair for Equity and Justice in the Blue Economy at WorldFish, Penang, Malaysia.

John Kurien is a founding member of the International Collective in Support of Fishworkers, a Visiting Professor at the Azim Premji University in Bengaluru, India and Honorary Fellow at WorldFish, Penang, Malaysia.

Yoshitaka Ota is Director of the Nippon Foundation Ocean Nexus Center and Research Assistant Professor at the School of Marine and Environmental Affairs, University of Washington, Seattle, USA.

Contributing Authors

Dedi S. Adhuri is coordinator of the Maritime Study Group at the Indonesian Institute of Sciences, Jakarta, Indonesia.

J. Maarten Bavinck is Professor of Maritime Studies at the University of Amsterdam, the Netherlands.

Andrés Cisneros-Montemayor is Deputy Director of the Nippon Foundation Ocean Nexus Program at the University of British Columbia, Vancouver, British Columbia, Canada.

Michael Fabinyi is Associate Professor of Anthropology in the School of Communications, University of Technology, Sydney, Australia.

Svein Jentoft is Professor of Sociology at the University of Tromsø, Norwegian College of Fisheries Science, Tromsø, Norway.

Sallie Lau is a recent graduate of the Masters in Marine Affairs program at the School of Marine and Environmental Affairs, University of Washington, Seattle, USA.

Tabatha Grace Mallory is founder and Chief Executive Officer of China Ocean Institute, Seattle, Washington, USA.

Ayodeji Olukoju, is Professor of History at the University of Lagos, Nigeria.

Natasha Stacey is Associate Professor at the Research Institute for the Environment and Livelihoods, College of Engineering, IT and Environment, Charles Darwin University, Darwin, Northern Territories, Australia.

Ingrid van Putten is an Environmental Economist with the Commonwealth Scientific and Industrial Research Organisation in Hobart, Tasmania, Australia.

Michelle Voyer is a Research Fellow at the Australian National Centre for Ocean Resources and Security, University of Wollongong, Australia.

Nireka Weeratunge is an anthropologist and independent scholar, affiliated as Research Fellow to the International Centre for Ethnic Studies in Colombo, Sri Lanka.

References

Aalberts TE, Gammeltoft-Hansen T (2014) Sovereignty at sea: the law and politics of saving lives in *Mare Liberum*. J Int Relat Dev 17(4):439–468

Adewusi A (2017) Ghanaian settlers in Orimedu: Oju Ota, gender, and christianity in a coastal fishing community. Yoruba Stud Rev 2(1):161–175

Adeyeri JO (2012) Socio-economic significance of cultural festivals in Epe division of Lagos State: an overview. Int J Res Arts Soc Sci 4:147–152

Adger WN, Hughes TP, Folke C, Carpenter SR, Rockström J (2005) Social-ecological resilience to coastal disasters. Science 309(5737):1036–1039

Alaimo S (2019) Introduction: science studies and the Blue Humanities. Configurations 27(4):429–432

Alfred T (2009) Peace, power, righteousness: an indigenous manifesto, 2nd edn. Oxford University Press, Don Mills, ON

Allison EH, Bassett HR (2015) Climate change in the oceans: human impacts and responses. Science 350(6262):778–782

Alongi DM, Murdiyarso D, Fourqurean JW, Kauffman JB, Hutahaean A, Crooks S, Lovelock CE et al (2015) Indonesia's blue carbon: a globally significant and vulnerable sink for seagrass and mangrove carbon. Wetl Ecol Manag 24(1):3–13

Amrith SS (2013) Crossing the Bay of Bengal: the furies of nature and the fortunes of migrants. Harvard University Press, Cambridge, MA

Andaya BW (2017) Seas, oceans and cosmologies in Southeast Asia. J Southeast Asian Stud 48(3):349–371

Anderson EN (1970) The boat people of South China. Anthropos H.1./2:248–256

Araujo AL (2017) Reparations for slavery and the slave trade: a transnational and comparative history. Bloomsbury Acad, New York

Arbo P, Knol M, Linke S, St K, Martin. (2018) The transformation of the oceans and the future of marine social science. Maritime Stud 17:295–304

Arbow T (2019) Washington maritime blue and the blue economy: using diversity and inclusion to advance social justice in the maritime industry. MA thesis, University of Washington, Seattle

Arrow KJ (1951) Social choice and individual values. Yale University Press, New Haven, CT

Baker E, Mensah A, Rice J, Grellier J, Alkhatlan AA (2019) Oceans and coasts global environment outlook (GEO-6): healthy planet, healthy people. In: Global environment outlook (GEO-6): healthy planet, healthy people, vol 7. United Nations Environment Programme and Cambridge University Press, Nairobi and Cambridge, pp 175–198

Barber BR (2014) Participatory democracy. In: The encyclopedia of political thought. Wiley, Hoboken, NJ, pp 2650–2654

Barbier EB (2015) Climate change impacts on rural poverty in low-elevation coastal zones. World Bank, Washington, DC

Barry R (2002) Voyaging and interaction in ancient East Polynesia. Asian Perspect 41(2):183–194

Bartels A, Eckstein L, Waller N, Wiemann D (2019) Postcolonial oceans. In: Postcolonial literatures in English: an introduction, 81-06. J.B. Metzler, Stuttgart

Bauslaugh RA (1991) The concept of neutrality in classical Greece. University of California Press, Berkeley

Bavinck M (2001a) Marine resource management: conflict and regulation in the fisheries of the Coromandel coast. Sage Publications, New Delhi

Bavinck M (2001b) Caste *panchayats* and the regulation of fisheries along Tamil Nadu's Coromandel Coast. Econ Polit Wkly 36:1088–1094

Bavinck M, Gupta J (2014) Legal pluralism in aquatic regimes: a challenge for governance. Curr Opin Environ Sustain 11:78–85

Bavinck M, Vivekanandan V (2017) Qualities of self-governance and wellbeing in the fishing communities of northern Tamil Nadu, India: the role of Pattinavar Uur panchayats. Maritime Stud 16(1):16

Beatley T (2014) Blue urbanism: exploring connections between cities and oceans. Island, Washington, DC

Beckwith M (1970) Hawaiian mythology. University of Hawai'i Press, Honolulu

Bennett NJ, Govan H, Satterfield T (2015) Ocean grabbing. Mar Policy 57:61–68

Bennett NJ, Cisneros-Montemayor AM, Blythe J, Silver JJ, Singh G, Andrews N, Calò A et al (2019) Towards a sustainable and equitable blue economy. Nat Sust 2:991–993

Bennett NJ, Finkbeiner EM, Ban NC, Belhabib D, Jupiter SD, Kittinger JN, Mangubhai S et al (2020) The COVID-19 pandemic, small-scale fisheries and coastal fishing communities. Coastal Manag J 48(4):336–347

Bidwell D (2017) Ocean beliefs and support for an offshore wind energy project. Ocean Coastal Manag 146:99–108

Bolonkin AA (2011) Floating cities. In: Brunn SD (ed) Engineering earth: the impacts of megaengineering projects. Springer, Dordrecht, pp 967–983

Bondad-Reantaso MG, Mackinnon B, Bin H, Jie H, Tang-Nelson K, Surachetpong W, Alday-Sanz V et al (2020) SARS-CoV-2 (the cause of COVID-19 in humans) is not known to infect aquatic food animals nor contaminate their products. Asian Fish Sci 33:1

Brady A-M (2019) Facing up to China's military interests in the Arctic. China Brief 19:21. Jamestown Foundation. https://jamestown.org/program/facing-up-to-chinas-military-interests-in-the-arctic/

Breen C (2007) Advocacy, international development and world heritage sites in sub-Saharan Africa. World Archaeol 39(3):355–370

Brown W (2014) Mapping China's latest supercity. Foreign Policy. https://foreignpolicy.com/2014/05/30/mapping-chinas-latest-supercity/

Buck R, Rowe N (eds) (2017) Moving oceans: celebrating dance in the South Pacific. Routledge, London

Callahan T (2013) Symbols and symbolic thought. In: Zelazo PD (ed) The oxford handbook of developmental psychology, vol. 1, body and mind. Oxford University Press, Oxford, pp 974–1005

Campbell KM, Doshi R (2020) The coronavirus could reshape global order: China is maneuvering for international leadership as the United States falters. Foreign Aff. 18 March. https://www.foreignaffairs.com/print/node/1125706

Cao L, Chen Y, Dong S, Hanson A, Huang B, Leadbitter D, Little DC et al (2017) Opportunity for marine fisheries reform in China. Proc Natl Acad Sci 114(3):435–442

Caras T, Pasternak Z (2009) Long-term environmental impact of coral mining at the Wakatobi Marine Park, Indonesia. Ocean Coastal Manag 52(10):539–544

Carver R, Childs J, Steinberg P, Mabon L, Matsuda H, Squire R, McLellan B, Esteban M (2020) A critical social perspective on deep sea mining: lessons from the emergent industry in Japan. Ocean Coastal Manag 193:105242

Catterall D, Campbell J (2012) Women in port: gendering communities, economies, and social networks in Atlantic port cities, 1500–1800. Brill, Leiden

Chan N (2018) 'Large Ocean States': sovereignty, small islands, and marine protected areas in global oceans governance. Global Governance 24(4):537–555

Chen J-C, Zinda JA, Yeh ET (2017) Recasting the rural: state, society and environment in contemporary China. Geoforum 78:83–88

Chou C (1997) Contesting the tenure of territoriality: the orang Suku Laut. Bijdragen tot de Taal-, Land- en Volkenkunde 153(4):605–629. https://doi.org/10.2307/27865391

Chou C (2006) Research trends on southeast Asian Sea Nomads. Kyoto Rev Southeast Asia 7:1–11

Chuenpagdee R (2011) Too big to ignore: global research network for the future of small-scale fisheries. In: Chuenpagdee R (ed) World small-scale fisheries contemporary visions. Eburon, Utrecht, pp 383–394

Cisneros-Montemayor AM, Pauly D, Weatherdon LV, Ota Y (2016) A global estimate of seafood consumption by coastal indigenous peoples. PLoS One 11(12):e0166681. https://doi.org/10.1371/journal.pone.0166681

Cisneros-Montemayor A, Moreno-Báez M, Voyer M, Allison EH, Cheung W, Hessing-Lewis M, Oyinlola M et al (2019) Social equity and benefits as the Nexus of a transformative blue economy: a sectoral review of implications. Mar Policy 109:103702

Clark AE (2018) Four decades of the economics of happiness: where next? Rev Income Wealth 64(2):245–269

Claudet J, Bopp L, Cheung WW, Devillers R, Escobar-Briones E, Haugan P, Heymans JJ et al (2020) A roadmap for using the UN decade of ocean science for sustainable development in support of science, policy, and action. One Earth 2(1):34–42

Cohen PJ, Allison EH, Andrew NL, Cinner J, Evans LS, Fabinyi M, Garces LR et al (2019) Securing a just space for small-scale fisheries in the blue economy. Front Mar Sci 6(March):171. https://doi.org/10.3389/fmars.2019.00171

Connery C (2006) There was no more sea: the suppression of the ocean, from the Bible to cyberspace. J Hist Geogr 32(3):494–511

Cordiner J (1807) A description of Ceylon: containing an account of the country, inhabitants and natural productions, vol 2. Longman, Hurst, Rees and Orme, London

Corntassel J (2012) Re-envisioning resurgence: indigenous pathways to decolonization and sustainable self-determination. Decolon Indig Edu Soc 1(1):86–101

Costello C, Cao L, Gelcich S et al (2019) The future of food from the sea. World Resources Institute, Washington, DC. www.oceanpanel.org/blue-papers/future-food-sea

Creighton MS, Norling L (eds) (1996) Iron men, wooden women: gender and seafaring in the Atlantic world, 1700–1920. Johns Hopkins University Press, Baltimore

Cumberbatch JA, Hinds CJ (2013) Barbadian bio-cultural heritage: an analysis of the flying fish. Int J Intang Heritage 8:118–134

Curtin PD (1984) Cross-cultural trade in world history. Cambridge University Press, Cambridge

Daly HE (1991) Steady-state economics: with new essays. Island, Washington, DC

Davis D (2015) Waterman: the life and times of Duke Kahanamoku. University of Nebraska Press, Lincoln

Daw TM, Coulthard S, Cheung WW, Brown K, Abunge C, Galafassi D, Peterson GD et al (2015) Evaluating taboo trade-offs in ecosystems services and human well-being. Proc Natl Acad Sci 112(22):6949–6954

Delevingne L, Schneider H (2020) U.S. stimulus package is biggest ever, but may not be big enough. Reuters, London. 30 March. https://www.reuters.com/article/us-health-coronavirus-fed-stimulus-analy/u-s-stimulus-package-is-biggest-ever-but-may-not-be-big-enough-idUSKBN21H0E7

Depledge MH, Bird WJ (2009) The blue gym: health and wellbeing from our coasts. Mar Pollut Bull 58(7):947

Díaz S, Demissew S, Carabias J, Joly C, Lonsdale M, Ash N, Larigauderie A et al (2015) The IPBES conceptual framework: connecting nature and people. Curr Opin Environ Sustain 14:1–16

Díaz S, Pascual U, Stenseke M, Martín-López B, Watson RT, Molnár Z, Hill R et al (2018) Assessing nature's contributions to people. Science 359(6373):270–272

Dineen J (2020) COVID-19 disrupts a major year for biodiversity policy and planning. Mongabay. 3 April. https://news.mongabay.com/2020/04/covid-19-disrupts-a-major-year-for-biodiversity-policy-and-planning/

Doherty B (2020) 'Escape the pandemic in paradise': Fiji opens its borders seeking billionaires. The Guardian. 27 June. https://www.theguardian.com/world/2020/jun/28/escape-the-pandemic-in-paradise-fiji-opens-its-borders-seeking-billionaires

Dornan M, Morgan W, Newton Cain T, Tarte S (2018) What's in a term? 'Green growth' and the 'Blue–green economy' in the Pacific Islands. Asia Pacific Policy Stud 5:408–425

Dreyer EL (2006) Zheng He: China and the oceans in the early ming dynasty, 1405–1433. Pearson Longman, New York

Driessen H (2002) Mediterranean port cities: cosmopolitanism reconsidered. Hist Anthropol 16(1):129–141

Dulvy NK, Stanwell-Smith D, Darwall WR, Horrill CJ (1995) Coral mining at Mafia Island, Tanzania: a management dilemma. Ambio 24(6):358–365

Easterlin RA (2001) Income and happiness: towards a unified theory. Econ J 111(473):465–484

Eckert LE, Ban NC, Frid A, McGreer M (2018) Diving back in time: extending historical baselines for yelloweye rockfish with indigenous knowledge. Aquat Conserv Mar Freshwat Ecosyst 28(1):58–66

Ehinmore OM (1998) A history of fishing in Ondo State, 1950–1997: a case study of the Ilaje coastal area. MA research project, University of Lagos

Ehinmore O (2014) Boundary conflict and security challenges in the Western Coast of The Niger Delta: the Ilaje-Ijo War Factor, 1998–99. Int J Humanit Soc Sci 4(8):276–283

Ertör I, Hadjimichael M (2020) Blue degrowth and the politics of the sea: rethinking the blue economy. Sustain Sci 15(1):1–10

Esteban M, Leary D (2012) Current developments and future prospects of offshore wind and ocean energy. Appl Energy 90(1):128–136

Fairbanks L, Boucquey N, Campbell LM, Wise S (2019) Remaking oceans governance: critical perspectives on marine spatial planning. Environ Soc 10(1):122–140

FAO (2020a) The state of world fisheries and aquaculture. FAO, Rome

FAO (2020b) Proceedings of the International symposium on fisheries sustainability: strengthening the science-policy Nexus. FAO headquarters, 18–21 November 2019. FAO fisheries and aquaculture proceedings no. 65. FAO, Rome

FAO (Food and Agriculture Organization of the United Nations) (2015) Voluntary guidelines for securing sustainable small-scale fisheries in the context of food security and poverty eradication. FAO, Rome

Fernandes N (2020) Economic effects of coronavirus outbreak (COVID-19) on the World Economy. Soc Sci Res Network. https://papers.ssrn.com/sol3/papers.cfm?abstract_id=3557504

Fernández GG, Stotz W, Aburto J, Mondaca C, Vera K (2011) Emerging commons within artisanal fisheries: the Chilean territorial use rights in fisheries (TURFs) within a broader coastal landscape. Int J Commons 5(2):459–484

Finlay J, Franke T, McKay H, Sims-Gould J (2015) Therapeutic landscapes and wellbeing in later life: impacts of blue and green spaces for older adults. Health Place 34:97–106

Finney BR, Houston JD (1996) Surfing: a history of the ancient Hawaiian sport. Pomegranate Artbooks, San Francisco

Fleisher J, Lane P, LaViolette A, Horton M, Pollard E, Morales EQ, Vernet T, Christie A, Wynne-Jones S (2015) When did the Swahili become maritime? Am Anthropol 117(1):100–115

Fleming AE, Petheram L, Stacey N (2015) Australian indigenous women's seafood harvesting practices and prospects for integrating aquaculture. J Enterp Commun 9(2):156–181

Frangoudes K, Gerrard S, Kleiber D (2019) Situated transformations of women and gender relations in small-scale fisheries and communities in a globalized world. Maritime Stud 18(3):241–248

Franke A, Blenckner T, Duarte CM, Ott K, Fleming LE, Antia A et al (2020) Operationalizing Ocean health: toward integrated research on ocean health and recovery to achieve ocean sustainability. One Earth 2(6):557–565

Friess B, Grémaud-Colombier M (2019) Policy outlook: recent evolutions of maritime spatial planning in the European Union. Mar Policy 132:103428. https://doi.org/10.1016/j.marpol.2019.01.017

Frykman N, Anderson C, van Voss LH, Rediker M (2013) Mutiny and maritime radicalism in the age of revolution: an introduction. Int Rev Soc Hist 58(S21):1–14

Gaines S, Cabral R, Free C, Golbuu Y et al (2019) The expected impacts of climate change on the ocean economy. World Resources Institute, Washington, DC. https://www.oceanpanel.org/expected-impacts-climate-change-ocean-economy

Garton Ash T (2002) Solidarity: the Polish revolution, 3rd edn. Yale University Press, New Haven, CT

Gaynor JL (2005) The decline of small-scale fishing and the reorganization of livelihood practices among Sama people in eastern Indonesia. Michigan Disc Anthropol 15:90–149

Gephart JA, Golden CD, Asche F, Belton B, Brugère C, Froehlich H, Fry J et al (2020) Scenarios for global aquaculture and its role in human nutrition. Rev Fish Sci Aquacult July:1–17. https://doi.org/10.1080/23308249.2020.1782342

Ghadiryanfar M, Rosentrater KA, Keyhani A, Omid M (2016) A review of macroalgae production, with potential applications in biofuels and bioenergy. Renew Sust Energ Rev 54:473–481

Gien LT (2000) Land and sea connection: the east coast fishery closure, unemployment and health. Can J Public Health 91(2):121–124

Gillis JR (2012) The human shore: seacoasts in history. University of Chicago Press, Chicago

Golden CD, Allison EH, Cheung WWL, Dey MM, Halpern BS, McCauley DJ, Smith M, Vaitla B (2016) Nutrition: fall in fish catch threatens human health. Nature 534:317–320

Grimble R (1972) Migrations, myth and magic from the Gilbert Islands: early writings of Sir Arthur Grimble. Routledge, London

Grovier K (2020) Black lives matter protests: why are statues so powerful? BBC Culture. 12 June. https://www.bbc.com/culture/article/20200612-black-lives-matter-protests-why-are-statues-so-powerful

Guo-jun YAO (2013) A touch of blue in Guangdong literature: on the sea charm and ocean color of Zhanjiang local literature. J Guangdong Ocean Univ 33(5):10

Hadjimichael M (2018) A call for a blue degrowth: unravelling the European Union's fisheries and maritime policies. Mar Policy 94:158–164

Hau'ofa E (1994) Our sea of islands. Contemp Pacific 6(1):148–161

Hau'ofa E (2008) We are the ocean. University of Hawai'i Press, Honolulu

He X, Faure D (eds) (2016) The fisher folk of late imperial and modern China: an historical anthropology of boat-and-shed living. Routledge, Abingdon

Helne T, Hirvilammi T (2015) Wellbeing and sustainability: a relational approach. Sustain Dev 23(3):167–175
Herman RDK (2016) Traditional knowledge in a time of crisis: climate change, culture and communication. Sustain Sci 11(1):163–176
Hicks CC, Cohen PJ, Graham NA, Nash KL, Allison EH, D'Lima C, Mills DJ et al (2019) Harnessing global fisheries to tackle micronutrient deficiencies. Nature 574(7776):95–98
Hilborn R, Hilborn U (2019) Ocean recovery: a sustainable future for global fisheries? Oxford University Press, Oxford
Hilmi N, Bambridge T, Safa A, Quinquis B, D'Arcy P (2016) Socioeconomic significance of fisheries in the small Island developing states: natural heritage or commodity? In: Fache E, Pauwels S (eds) Fisheries in the pacific: the challenges of governance and sustainability. Pacific-Credo, Marseille, pp 175–197
Hoegh-Guldberg O, Caldeira K, Chopin T, Gaines S, Haugan P, Hemer M, Howard J et al (2019) The ocean as a solution to climate change: five opportunities for action. World Resources Institute, Washington, DC. http://www.oceanpanel.org/climate
Holmes JR (2006) 'Soft power' at sea: Zheng He and China's maritime diplomacy. Southeast Rev Asian Stud 28:95
Hong G (2018) Islands of enclavisation: eco-cultural Island tourism and the relational geographies of near-shore islands. Area 52(1):47–55
Horden P, Purcell N (2006) The Mediterranean and 'the New Thalassology'. Am Hist Rev 111(3):722–740
Huntsman L (2001) Sand in our souls: the beach in Australian history. Melbourne University Press, Melbourne
Hviding E (1996) Guardians of Marovo Lagoon: practice, place, and politics in maritime Melanesia. University of Hawai'i Press, Honolulu
Ikuejube G (2005) Ilaje: the Yoruba fishing people of The Niger delta. Novec, Ondo
ILO (International Labor Organization) (2020) Impact of Covid-19 on maritime shipping and fishing. Policy brief. ILO, Geneva
IMF (International Monetary Fund) (2020) World economic outlook 2020. IMF, Washington, DC. 24 June update. https://www.imf.org/en/Publications/WEO
Indonesia Ocean Pride (2020) Join us and celebrate #PrideInTheOcean. https://sanctuaries.noaa.gov/heritage/pride-in-the-ocean.html
Jambeck J, Moss E, Dubey B et al (2020) Leveraging multi-target strategies to address plastic pollution in the context of an already stressed ocean. World Resources Institute, Washington, DC. https://oceanpanel.org/blue-papers/pollution-and-regenerative-economy-municipal-industrialagricultural-and-maritime-waste
Jaspars M, De Pascale D, Andersen JH, Reyes F, Crawford AD, Ianora A (2016) The marine biodiscovery pipeline and ocean medicines of tomorrow. J Mar Biol Assoc U K 96(1):151–158
Jentoft, S. 2019. Life above water: essays on human experiences of small-scale fisheries. TBTI global book series 1. St. John's, NL: Too Big to Ignore
Jentoft S, Onyango P, Islam MM (2010) Freedom and poverty in the fishery commons. Int J Commons 4:345–366
Jentoft S, Chuenpagdee R, Franz N, Barragán-Paladines MJ (2017) Implementing the voluntary guidelines for securing small-scale fisheries. In: Jentoft S, Chuenpagdee R, Barragán-Paladines MJ, Franz N (eds) The small-scale fisheries guidelines: global implementation. Springer, New York, pp 3–13
Jentoft S, Stacey N, Sunde J, González M (2019) The small-scale fisheries of indigenous peoples: a struggle for secure tenure rights. In: Chuenpagdee R, Jentoft S (eds) Transdisciplinarity for small-scale fisheries governance. Springer, Cham, pp 263–282. MARE Publication Series 21
Jie Li JW, Carr N (2004) Visitor satisfaction: an analysis of Mainland Chinese tourists on the Australian Gold Coast. Int J Hosp Tour Adm 5(3):31–48
Jim CY, Chen WY (2009) Value of scenic views: hedonic assessment of private housing in Hong Kong. Landsc Urban Plan 91(4): 226–234
Johnson D, Acott T, Stacey N, Urquhart J (eds) (2018) Social wellbeing and the values of small-scale fisheries. Springer, Cham
Jones E, Qadir M, van Vliet MT, Smakhtin V, Kang SM (2019) The state of desalination and Brine production: a global outlook. Sci Total Environ 657:1343–1356
Jouffray J-B, Blasiak R, Norström AV, Österblom H, Nyström M (2020) The blue acceleration: the trajectory of human expansion into the ocean. One Earth 2(1):43–54
Kahneman D, Krueger AB (2006) Developments in the measurement of subjective well-being. J Econ Perspect 20:3–24
Kahneman D, Krueger AB, Schkade D, Schwarz N, Stone AA (2006) Would you be happier if you were richer? A focusing illusion. Science 312:1908–1910
Kallis G (2011) In defence of degrowth. Ecol Econ 70(5):873–880
Khan N, Kalair A, Abas N, Haider A (2017) Review of ocean tidal, wave and thermal energy technologies. Renew Sust Energ Rev 72:590–604
King TJ, Robinson G (eds) (2019) At home on the waves: human habitation of the sea from the Mesolithic to today. Berghahn, New York
Kissya E (1995) Sasi Aman Haru-Ukui: traditional management of sustainable natural resources in Haruku. SEJATI Publications, Jakarta
Kitada M, Williams E, Froholdt LL (eds) (2015) Maritime women: global leadership. Springer, Berlin
Knowlton N (2020) Ocean optimism: moving beyond the obituaries in marine conservation. Annu Rev Mar Sci 13:479–499
Kolm SC (1972 [1998]) Justice et équité. Paris: Editions du Centre National de la Recherche Scientifique. English translation [Justice and equity]. MIT Press, Cambridge, MA
Kooiman J, Bavinck M, Jentoft S, Pullin R (eds) (2005) Fish for life: interactive governance for fisheries. MARE publication series no. 3. Amsterdam University Press, Amsterdam
Krupinski L (2019) Of fish and fisheries: an artistic reverence for a bountiful sea. Am Soc Marine Artists 41. https://americansocietyofmarineartists.com/resources/SiteGraphics/NewsnJournal/FullPDFs/2019FallJournal.pdf
Kunz GF, Stevenson CH (2001 [1908]) Book of the pearl: its history, art, science and industry. Dover, Mineola, NY
Kurien J (2001) The socio-cultural aspects of fisheries: implications for food and livelihood security—a case study of Kerala State, India. In: McGoodwin JR (ed) Understanding the cultures of fishing communities: a key to fisheries management and food security. FAO fisheries technical paper 401. FAO, Rome
Kurien J (2019) Fisheries sustainability leaving none behind: three key policy opportunities for the 21st century. Keynote address to the FAO/UN international symposium on fisheries sustainability. FAO, Rome. November (Summary published in FAO 2020)
Lalancette A (2017) Creeping in? Neoliberalism, indigenous realities and tropical rock lobster (Kaiar) management in Torres Strait, Australia. Mar Policy 80:47–59
Laugrand F, Oosten J (2008) When toys and ornaments come into play: the transformative power of miniatures in Canadian inuit cosmology. Mus Anthropol 31(2):69–84
Lavoie A, Lee J, Sparks K, Hoseth G, Wise S (2019) Engaging with women's knowledge in Bristol Bay fisheries through oral history and participatory ethnography. Fisheries 44(7):331–337
Leemans R, de Groot RS (2003) Millennium ecosystem assessment: ecosystems and human well-being: a framework for assessment. Island, Washington, DC
Lesutis G (2019) How to understand a development corridor? The case of Lamu port–South Sudan–Ethiopia–transport corridor in Kenya. Area 2(3):600–608. https://doi.org/10.1111/area.12601

Li K (2019) Annual policy blueprint, Government of China (Press release)

Liao X, Aguilera DS (2020) Chinese tourism in the Caribbean: destination image and future expectations. In: Luo Y, Jiang J, Bi D (eds) Tourism product development in China, Asian and European countries. Springer, Singapore, pp 121–135

Lim ACH (2015) Africa and China's 21st century maritime silk road. National University of Singapore, East Asian Institute

Linebaugh P, Rediker M (2000) The many-headed hydra: sailors, slaves, commoners, and the hidden history of the revolutionary Atlantic. Beacon, Boston

Liu J, Liu N, Zhang Y, Qu Z, Yu J (2019) Evaluation of the non-use value of beach tourism resources: a case study of Qingdao coastal scenic area, China. Ocean Coastal Manag 168:63–71

Love DC, Allison EH, Asche F, Belton B, Cottrell RS, Froehlich HE, Gephart JA et al (2020) Emerging COVID-19 impacts, responses, and lessons for building resilience in the seafood system. Global Food Security Preprint, July 6. https://osf.io/preprints/socarxiv/x8aew/

Lubchenco J, Gaines SD (2019) A new narrative for the ocean. Science 346:911

MacIntyre TE, Walkin A, Calogiuri G, Oliver G, Donnelly AA, Gritzka S, Warrington G (2019) Immersion, watersports and blueways and the blue mind. In: Donnelly AA, MacIntyre TE (eds) Physical activity in natural settings: green and blue exercise. Routledge, Abingdon, pp 190–203

Mack J (2013) The sea: a cultural history. Reaktion, London

Maldonado JK (2014) A multiple knowledge approach for adaptation to environmental change: lessons learned from coastal Louisiana's tribal communities. J Politc Ecol 21:61–82

Mallory TG (2013) China's distant water fishing industry: evolving policies and implications. Mar Policy 38:99–108

Mallory TG (2016) Fisheries subsidies in China: quantitative and qualitative assessment of policy coherence and effectiveness. Mar Policy 68:74–82

Malve H (2016) Exploring the ocean for new drug developments: marine pharmacology. J Pharm Bioallied Sci 8(2):83–91

Maritime Blue (2020) A strategic alliance for maritime innovation and sustainability. https://maritimeblue.org/alliance/

Marschall S (2015) Travelling down memory lane': personal memory as a generator of tourism. Tour Geogr 17(1):36–53

Mateiviti-Tulavu E (2013) Connecting identities and relationships through indigenous epistemology: the Solomoni of Fiji. PhD Diss, University of Auckland. https://pdfs.semanticscholar.org/56eb/a2ef0eef7a0626892ae66f7614658565b64e.pdf

Maya-Jariego I, Querevalú-Miñán JF, Varela LG, Ávila J (2017) Escape the lion cage: social networks by catch zones of small-scale fisheries in the oil settlement of Lobitos (Peru). Mar Policy 81:340–349

McCauley D, Teleki C, Fluxà Thienemann G (2020) 8 ways to rebuild a stronger ocean economy after COVID-19. World Economic Forum blog. https://www.weforum.org/agenda/2020/05/how-to-build-a-bluer-ocean-economy-after-covid-19/

McGoodwin JR (2001) Understanding the cultures of fishing communities: a key to fisheries management and food security. Fisheries technical paper 401. Food and Agriculture Organization of the United Nations, Rome

McGregor I, Yerbury H (2019) Politics of rising tides: governments and nongovernmental organizations in small-Island developing states. In: Harris PG (ed) Climate change and ocean governance: politics and policy for threatened seas. Cambridge University Press, Cambridge, pp 118–132

McKay SC (2007) Filipino Sea men: constructing masculinities in an ethnic labour niche. J Ethn Migr Stud 33(4):617–633

McVeigh K (2020) Silence is golden for whales as lockdown reduces ocean noise. The Guardian. 27 April. https://www.theguardian.com/environment/2020/apr/27/silence-is-golden-for-whales-as-lockdown-reduces-ocean-noise-coronavirus

Mentz S (2009) Toward a blue cultural studies: the sea, maritime culture, and early modern English literature. Literat Compass 6(5):997–1013

Merayo E (2019) Steering gender to the Centre of the blue economy. Blog post, 6 June. International Institute for Environment and Development, London. https://www.iied.org/steering-gender-centre-blue-economy

Middleton A (2013) Aboriginal rock art may depict first sea arrivals. Aust Geogr. https://www.australiangeographic.com.au/news/2013/08/aboriginal-rock-art-may-depict-first-sea-arrivals/

Montgomery C (2006) The shark god: encounters with ghosts and ancestors in the South Pacific. HarperCollins, Glasgow

Mulazzani L, Malorgio G (2017) Blue growth and ecosystem services. Mar Policy 85:17–24

Narula K (2019) Oceans as a source of hydrocarbon energy. In: Narula K (ed) The maritime dimension of sustainable energy security. Springer, Singapore, pp 145–162

Nash KL, Blythe JL, Cvitanovic C, Fulton EA, Halpern BS, Milner-Gulland EJ, Addison PF et al (2020) To achieve a sustainable blue future, progress assessments must include interdependencies between the sustainable development goals. OneEarth 2(2):161–173

Nichols WJ (2015) Blue mind. Little, Brown, Columbus, GA

Nofsinger JR (2005) Social mood and financial economics. J Behav Finance 6(3):144–160

Notteboom T, Rodrigue JP (2008) Containerisation, box logistics and global supply chains: the integration of ports and liner shipping networks. Maritime Econom Logist 10(1–2):152–174

Novaczek I, Harkes IHT, Sopacua J, Tatuhey MDD (2001) An institutional analysis of Sasi Laut in Maluku, Indonesia. ICLARM Publications, Penang

O'Neill B (2001) Honor, symbols, and war. University of Michigan Press, Ann Arbor

Obura DO (2020) Getting to 2030: scaling effort to ambition through a narrative model of the SDGs. Mar Policy 117:103973

Olubomehin OO (1990) The fluctuating fortunes of Ejinrin market in the era of Yoruba warfare, 1860–1892. Oye Ogun J Arts 3: 128–135

Olukoju A (2000) Fishing, migrations and inter-group relations in the Gulf of Guinea (Atlantic Coast of West Africa) in the nineteenth and twentieth centuries. Itinerario Eur J Overseas History 24(2): 69–85

Olukoju A (2017) Making sense of the Yoruba littoral. Yoruba Stud Rev 2(1):45–60

Oluwapayimo DZ (2017) The illicit production and consumption of Ogogoro in coastal Yorubaland and The Niger Delta. Yoruba Stud Rev 2(1):195–208

Onipede KJ (2017) By canoe and speedboat: the recent expansion of the Ilaje fishing trade. Yoruba Stud Rev 2(1):177–194

OnManorama (2018) Kerala fishermen rescued 65,000 people during floods. 27 August. https://www.onmanorama.com/news/kerala/2018/08/26/kerala-flood-fishermen-rescue.html

Österblom H, Wabnitz CCC, Tladi D et al (2020) Towards Ocean equity. World Resources Institute, Washington, DC. https://www.oceanpanel.org/sites/default/files/2020-04/towards-ocean-equity.pdf

Ourbak T, Magnan AK (2018) The Paris agreement and climate change negotiations: small islands, big players. Reg Environ Chang 18:2201–2207

Paine L (2013) The sea and civilisation: a maritime history of the world. Alfred A. Knopf, New York

Pascual U, Balvanera P, Díaz S, Pataki G, Roth E, Stenseke M, Watson RT et al (2017) Valuing nature's contributions to people: the IPBES approach. Curr Opin Environ Sustain 26:7–6

Passmore HA, Howell AJ (2014) Eco-existential positive psychology: experiences in nature, existential anxieties, and well-being. Humanist Psychol 42(4):370–388

Paul JR (2000) Cultural resistance to global governance. Michigan J Int Law 22(1):2–84

Pauwelussen AP (2015) The moves of a Bajau Middlewoman: understanding the disparity between trade networks and marine conservation. Anthropol Forum 25(4):329–349. https://doi.org/10.1080/00664677.2015.1054343

Pearson L, Newton P, Roberts P (eds) (2014) Resilient sustainable cities: a future. Routledge, Abingdon

Peduzzi P (2014) Sand, rarer than one thinks. Environ Dev 11: 208–218

Peters K, Anderson J (eds) (2016) Water worlds: human geographies of the ocean. Routledge, Abingdon

Pollnac RB, Bavinck M (2008) Alternative livelihoods and job satisfaction among fishermen: a cross-national study. In: EOCST: ecosystems, societies, consilience, precautionary principle: development of an assessment method of the societal cost for best fishing practices and efficient public policies. D4SA ecoregion report on sociological aspects: results of the job satisfaction surveys of all case-studies. Brussels: CORDIS, European Commission Report. 3–13

Pollnac RB, Poggie JJ (2008) Happiness, well-being and psychocultural adaptation to the stresses associated with marine fishing. Hum Ecol Rev 15(2):194–200

Posner EA (1998) Symbols, signals, and social norms in politics and the law. J Leg Stud 27(S2):765–797

Productivity Commission (2016) Marine fisheries and aquaculture. Draft report. Productivity Commission, Canberra

Quinn J (2014) Mythologizing the sea: the Nordic Sea-deity Rán. In: Tangherlini TR (ed) Nordic mythologies: interpretations, intersections, and institutions. North Pinehurst, Berkeley, CA, pp 71–99

Raheem O, Famiyesin M (2017) Controlling the boundaries of morality: the history and powers of Ayelala deity. Yoruba Stud Rev 2(1):231–247

Rao A (2020) Race, wealth and public spaces: US beaches are a new flashpoint of the lockdown. The Guardian (U.S. edition). 24 May. https://www.theguardian.com/us-news/2020/may/24/us-beaches-covid-19-florida-access-parks

Ratner BD (2004) "Sustainability" as a dialogue of values: challenges to the sociology of development. Sociol Inq 74:50–69

Renaud L (2020) Reconsidering global mobility: distancing from mass cruise tourism in the aftermath of COVID-19. Tour Geogr 22(3):679–689. https://doi.org/10.1080/14616688.2020.1762116

Riffat S, Powell R, Aydin D (2016) Future cities and environmental sustainability. Future Cities Environ 2:1. https://doi.org/10.1186/s40984-016-0014-2

Rilke RM (1969 [1903]) Letter to Clara Rilke, 27 March 1903. In Letters of Rainer Maria Rilke, 1892–1910, translated by J.B. Greene and M.D. Herter-Norton, 42. Norton Library, New York

Rosenthal J, Bannister K, Berlin B, Berlin E, Brush S, Field L, Greene S et al (2006) Politics, culture, and governance in the development of prior informed consent in indigenous communities. Curr Anthropol 47(1):119–142

Ruddle K (1988) Social principles underlying traditional inshore fishery management systems in the Pacific Basin. Mar Resour Econ 5(4):351–363

Russell BW, Dahlin BH (2007) Traditional burnt-lime production at Mayapán, Mexico. J Field Archaeol 32(4):407–423

Rynd C (2020) Seafarers: governments need to step up on repatriations. Maritime Execut. 4 May. https://www.maritime-executive.com/editorials/seafarers-governments-need-to-step-up-on-repatriation

Sachs JD, Schmidt-Traub G, Mazzucato M, Messner D, Nakicenovic N, Rockström J (2019) Six transformations to achieve the sustainable development goals. Nat Sust 2(9):805–814

Sahu A, Yadav N, Sudhakar K (2016) Floating photovoltaic power plant: a review. Renew Sust Energ Rev 66:815–824

Satz D (2004) Noxious markets: why some things should not be for sale. In: Cullenberg S, Pattanaik PK (eds) Globalization, culture and the limits of the market: essays in economics and philosophy. Oxford University Press, New Delhi, pp 11–38

Schnierer S, Egan H (2016) Composition of the aboriginal harvest of fisheries resources in Coastal New South Wales, Australia. Rev Fish Biol Fish 26(4):693–709

Schoedinger S, Tran LU, Whitley L (2010) From the principles to the scope and sequence: a brief history of the ocean literacy campaign. NMEA Special Rep 3:3–7

Schutter MS, Hicks CC (2019) Networking the blue economy in Seychelles: pioneers, resistance, and the power of influence. J Pol Ecol 26(1):425–447

Schwartz SH (2012) An overview of the Schwartz theory of basic values. Online Readings Psychol Cult 2(1):11. https://doi.org/10.9707/2307-0919.1116

Schwarz AM, Gordon J, Ramofafia C (2020) Nudging statutory law to make space for customary processes and community-based fisheries management in Solomon Islands. Maritime Stud 19:475–487. https://doi.org/10.1007/s40152-020-00176-0

Sen A (1999) Commodities and capabilities. Oxford University Press, Oxford

Sen A (2001) Development as freedom. Oxford University Press, Oxford

Sen A, Nussbaum M (eds) (1993) Capability and well-being: the quality of life. Clarendon, Oxford

Silver JJ, Gray NJ, Campbell LM, Fairbanks LW, Gruby RL (2015) Blue economy and competing discourses in international oceans governance. J Environ Dev 24:135–160

Singh GG, Cisneros-Montemayor AM, Swartz W, Cheung W, Guy JA, Kenny TA, McOwen CJ et al (2018) A rapid assessment of co-benefits and trade-offs among sustainable development goals. Mar Policy 93:223–231

Slimak MW, Dietz T (2006) Personal values, beliefs, and ecological risk perception. Risk Anal 26(6):1689–1705

Smyth D, Isherwood M (2016) Protecting Sea country: indigenous people and marine protected areas in Australia. In: Wescott G, Fitzsimons J (eds) Big, bold and blue: lessons from Australia's marine protected areas. CSIRO Publishing, Canberra, pp 307–325

Song AM, Cohen PJ, Hanich Q, Morrison TH, Andrew N (2019) Multi-scale policy diffusion and translation in Pacific Island coastal fisheries. Ocean Coast Manag 168:139–149

SPC (Secretariat of the Pacific Community) (2015) A new song for coastal fisheries—pathways to change: the Noumea strategy. SPC, Nouméa. https://spccfpstore1.blob.core.windows.net/digitallibrary-docs/files/fe/fedc2bcffdee2b46bbb2ef08caad7e54.pdf?sv=2015-12-11&sr=b&sig=nfxrlmoDCJmdhtsKY1gbw%2B%2BnKgeHXZd0utZXcclz9t0%3D&se=2021-02-17T20%3A12%3A26Z&sp=r&rscc=public%2C%20max-age%3D864000%2C%20max-stale%3D86400&rsct=application%2Fpdf&rscd=inline%3B%20filename%3D%22Anon_2015_New_song_for_coastal_fisheries.pdf%22

Stacey N (2007) Boats to burn: Bajo fishing activity in the Australian fishing zone. Australian National University Press, Canberra

Stacey N, Allison EH (2019) Sea nomads: Sama-Bajau mobility, livelihoods and marine conservation in Southeast Asia. In: King TJ, Robinson G (eds) At home on the waves: human habitation of the sea from the Mesolithic to today. Berghahn, New York, pp 309–331

Stacey N, Steenbergen DJ, Clifton J, Acciaioli G (2018) Understanding social wellbeing and values of small scale fisheries amongst the Sama-Bajau of Insular Southeast Asia. In: Johnson D, Acott T, Urquart J, Stacey N (eds) Social wellbeing and the values of small scale fisheries. MARE publication series 17. Springer, Cham, pp 97–123

Stahel WR (2016) The circular economy. Nature 531(7595):435–438

Stanley J (2002) And after the cross-dressed cabin boys and whaling wives? Possible futures for Women's maritime historiography. J Transp History 23(1):9–22

Steinberg PE (2001) The social construction of the ocean. Cambridge University Press, Cambridge

Steinberg P, Peters K (2015) Wet ontologies, fluid spaces: giving depth to volume through oceanic thinking. Environ Plann D Soc Space 33(2):247–264

Steven ADL et al (2020) Coastal development: resilience, restoration and infrastructure requirements. World Resources Institute, Washington, DC. https://oceanpanel.org/blue-papers/coastal--development-managing-resilience-restoration-and-infrastructure-coastlines

Stiglitz J, Fitoussi JP, Durand M (2018) Beyond GDP: measuring what counts for economic and social performance. OECD Publishing, Paris. https://doi.org/10.1787/9789264307292-en

Stikvoort B, Lindahl T, Daw TM (2016) Thou shalt not sell nature: how taboo trade-offs can make us act pro-environmentally, to clear our conscience. Ecol Econ 129:252–259

Swilling M, Ruckelshaus M, Rudolph TB, Allison EH, Gelcich S, Mbatha P, Österblom H (2020) The ocean transition: what to learn from system transitions. World Resources Institute, Washington, DC. www.oceanpanel.org/blue-papers/ocean-transition-what-learn-system-transitions

Takaya Y, Yasukawa K, Kawasaki T, Fujinaga K, Ohta J, Usui Y, Nakamura K et al (2018) The tremendous potential of deep-sea mud as a source of rare-earth elements. Sci Rep 8(5763):1–8

Takehiro T (2018) Satoumi and fisheries rights system, special feature: opinions and concerns about fishery policy reform plan. Aquanet 8:21–51. [In Japanese]

Techera E, Govan H, Tawake A, Tabunakawai K, Jenkins A, Lasgorceix A, Schwarz A et al (2009) Status and potential of locally-managed marine areas in the South Pacific: meeting nature conservation and sustainable livelihood targets through wide-spread implementation of LMMAs. SPREP/WWF/WorldFish-Reefbase/CRISP, Suva

Tetlock PE, Kristel O, Elson B, Green M, Lerner J (2000) The psychology of the unthinkable: taboo trade-offs, forbidden base rates, and heretical counterfactuals. J Pers Soc Psychol 78(5):853–870

TNC (The Nature Conservancy). (2019) Mapping ocean wealth: recreation and tourism. https://oceanwealth.org/ecosystem-services/recreation-tourism/

Tsing AL, Bubandt N, Gan E, Swanson HA (eds) (2017) Arts of living on a damaged planet: ghosts and monsters of the Anthropocene. University of Minnesota Press, Minneapolis

Udombana NJ (2000) The third world and the right to development: agenda for the next millennium. Hum Rights Q 22:753–787

UNCSD (United Nations Conference on Sustainable Development) (2012) The future we want: outcome document of the United Nations Conference on Sustainable Development. Rio de Janeiro, 20–22 June. https://sustainabledevelopment.un.org/content/documents/733FutureWeWant.pdf

UNCTAD (United Nations Conference on Trade and Development) (2020) The Covid-19 pandemic and the blue economy: new challenges and prospects for recovery and resilience. UNCTAD/DITC/TED/INF/2020/2. https://unctad.org/en/PublicationsLibrary/ditctedinf2020d2_en.pdf

UNESCO (UN Educational, Scientific and Cultural Organization) (2020a) UNESCO World Heritage Centre. UNESCO, Paris. https://whc.unesco.org/en/about

UNESCO (UN Educational, Scientific and Cultural Organization) (2020b) Intangible cultural heritage lists. UNESCO, Paris. https://ich.unesco.org/en/lists

Vadrot ABM (2020) Multilateralism as a 'Site' of struggle over environmental knowledge: the North-South divide. Crit Policy Stud 14(2):233–245

van Barneveld K, Quinlan M, Kriesler P, Junor A, Baum F, Chowdhury A, Junankar PN et al (2020) The COVID-19 pandemic: lessons on building more equal and sustainable societies. Econ Labour Relat Rev 31(2):133–157

van Riet Lowe C (1947) Rock paintings of marine animals in the interior of South Africa. South Afr Archaeol Bullet 2(6):41–45

Vásquez-Fernández AM (2020) Resurgence of relationality: reflections on decolonizing and indigenizing 'sustainable development'. Curr Opin Environ Sustain 43:65–70

Verrips J (2002) Ghanaian canoe decorations. MAST 1(1):43–66

Vierros MK, Harrison A-L, Sloat MR, Ortuño Crespo G, Moore JW, Dunn DC, Ota Y et al (2020) Considering indigenous peoples and local communities in governance of the global ocean commons. Mar Policy 119:104039

Villani D, Sorgente A, Iannello P, Antonietti A (2019) The role of spirituality and religiosity in subjective well-being of individuals with different religious status. Front Psychol 10:1525

Vink MP (2007) Indian ocean studies and the 'new Thalassology.'. J Glob Hist 2(1):41–62

von der Porten S, Ota Y, Cisneros-Montemayor A, Pictou S (2019) The role of indigenous resurgence in marine conservation. Coast Manag 47(6):527–547

Voyer M, van Leeuwen J (2019) Social license to operate in the blue economy. Resour Policy 62:102–113

Voyer M, Quirk G, McIlgorm A, Azmi K (2018) Shades of blue: what do competing interpretations of the blue economy mean for oceans governance? J Environ Policy Plan 20(5):595–616

Wade G (2005) The Zheng He Voyages: a reassessment. J Malays Branch Royal Asiatic Soc 78(1):37–58

Wallerstein I (2011) The modern world-system ii: mercantilism and the consolidation of the European World-Economy, 1600–1750, 2nd edn. University of California Press, Berkeley

Weeratunge N, Béné C, Siriwardane R, Charles A, Johnson D, Allison EH, Nayak PK, Badjeck MC (2014) Small-scale fisheries through the wellbeing lens. Fish Fish 15(2):255–279

Weiss CV, Guanche R, Ondiviela B, Castellanos OF, Juanes J (2018) Marine renewable energy potential: a global perspective for offshore wind and wave exploitation. Energy Convers Manag 177:43–54

White SC (2010) Analysing wellbeing: a framework for development practice. Dev Pract 20(2):158–172

White SC (2017) Relational wellbeing: re-centering the politics of happiness, policy and the self. Policy Polit 45(2):121–136

WHO (2020a) Coronavirus (Covid-19 disease) situation report 209, 16 August 2020. WHO, Geneva. https://www.who.int/docs/default-source/coronaviruse/situation-reports/20200816-covid-19-sitrep-209.pdf

WHO (2020b) 1st WHO international infodemiology conference. WHO, Geneva. https://www.who.int/teams/risk-communication/infodemic-management/1st-who-infodemiology-conference

Widjaja S, Long T, Wirajuda H, Van As H, Bergh PE, Brett A, Copeland D et al (2020) Illegal, unreported and unregulated fishing and associated drivers. World Resources Institute, Washington, DC. https://www.oceanpanel.org/blue-papers/illegal-unreported-and-unregulated-fishing-and-associated-drivers

Willis Towers Watson (2020) COVID-19: impact on the marine and energy industry. https://www.willistowerswatson.com/en-GB/Insights/2020/04/covid-19-willis-re-impact-report

Winther J-G, Dai M, Douvere F, Fernandes L, Halpin P, Hoel AH, Juinio-Meñez MA et al (2020) Integrated ocean management. World Resources Institute, Washington, DC. www.oceanpanel.org/blue-papers/integrated-ocean-management

Wood J (2020) Drones, disinfectant, distancing: Europe's beaches open up. World Economic Forum Coronavirus Briefing. 20 May. https://www.weforum.org/agenda/2020/05/coronavirus-europe-beach-reopen-restrictions-sun-pandemic-tourism/

Young OR, Gasser L (2002) The institutional dimensions of environmental change: fit, interplay, and scale. MIT Press, Cambridge, MA

Young OR, Webster DG, Cox ME, Raakjær J, Blaxekjær LØ, Einarsson N, Virginia RA et al (2018) Moving beyond panaceas in fisheries governance. Proc Natl Acad Sci 115(37):9065–9073

Zeldin T (2012) An intimate history of humanity. Random House, London

Zhang K, Dearing JA, Tong SL, Hughes TP (2016) China's degraded environment enters a new normal. Trends Ecol Evol 31(3):175–177

The Ocean Transition: What to Learn from System Transitions

12

Highlights

- The ocean is a global commons on which the prosperity and well-being of humanity rests. Business as usual will result in the collapse of key biophysical ocean functions, which will have significant implications for the global economy, societies and people.
- Science has demonstrated that ecosystems on land, rivers, deltas, estuaries and the ocean are intrinsically linked. Therefore, to transition to a sustainable ocean economy that protects the ocean and provides for humanity, a more holistic approach to ocean governance is needed. In short, humanity needs to redefine its relationship with the ocean.
- Ocean governance currently consists of different forms of plural and multilevel institutions responsible for designing solutions for common resources in the ocean. However, the nature of these institutions limits their ability to consider outcomes at different scales, and the ability of resource users to devise livelihood strategies within integrated systems. These weaknesses in the current system generate significant problems for the conservation, sustainable use and equitable sharing of the ocean.
- Major efforts to better manage the ocean as a common resource are needed. These efforts will require greater willingness and cooperation, from local communities to national and international action. Stronger accountability, transparency and participation mechanisms will be required to resolve conflicts and enable equitable sharing among different users, particularly in areas beyond national jurisdiction.
- This paper considers what governance configurations would facilitate the better management of the ocean as a global commons. To do so, it considers the conditions that have facilitated societal transitions in the past, from the information and communications technology revolution in the 1970s to the more recent Paris Agreement of the UN Framework Convention on Climate Change, and the governance arrangements that have enabled them. The paper's authors are optimistic about these shifts, pointing to system transitions that are already occurring at the local level through 'niche innovations' that communities, governments and business are implementing, to governments negotiating new agreements at the global level.
- This paper proposes four key opportunities for action to strengthen ocean governance: support current UN ocean processes (e.g. ratification of UNCLOS); reconfigure nation-state authority as it relates to the ocean (e.g., establish a global 'ocean agency' that supports polycentric, 'bottom up' governance innovations); support civil society's ability to play a more significant role (e.g. by recognising access to a healthy environment as a human right and establishing a new 'wiki-type' interactive ocean knowledge commons for co-creating solutions); and integrate property rights with stewardship responsibilities (e.g. establish local user rights programs).
- A balance of civil society rights and stronger government leadership from the state is crucial to avoid overburdening citizens with securing their future in the ocean system, or with inequitable access opportunities and benefit distribution resulting from policy interventions that fail to consider their implications.

1 Introduction

The ocean is the ultimate commons. Sustainability narratives now recognise what science continues to demonstrate—that ecosystems on land, rivers, deltas, estuaries and the ocean are intrinsically linked (Mathews et al. 2019). There is a growing consensus that the prosperity and well-being of

Originally published in:
Swilling, M., M. Ruckelshaus, T. Brodie Rudolph et al. 2020. *The Ocean Transition: What to Learn from System Transitions*. Washington, DC: World Resources Institute. www.oceanpanel.org/blue-papers/ocean-transition-what-learn-system-transitions
Reprint by Springer International Publishing (2023) with kind permission
Published under license from the World Research Institute

J. Lubchenco, P. M. Haugan (eds.), *The Blue Compendium*, https://doi.org/10.1007/978-3-031-16277-0_12

humanity depends on the health of the ocean environment, including the ocean-climate nexus (OECD 2016; IPCC 2019). Critical indicators reveal that business as usual is going to result in the collapse of key biophysical ocean functions, with major implications for the global economy and societies (IPCC 2019). Science has demonstrated that these close systemic interlinkages in and among ecological, economic and social systems require solutions which are responsive and flexible, robust yet elastic (SDG 2019).

It is also evident that time is of the essence (Steffen et al. 2018; IPCC 2019). A new relationship between humanity and the ocean is thus required. It follows that the transition to a new and effective governance system for the ocean should ensure that it 'does justice to humanity's obligations to itself, and to the planet which is its home' (International Court of Justice judge C.G. Weeramantry in Gabčíkovo-Nagymaros Project (Hungary v Slovakia) 1997).

This paper proceeds from the assumption that the ocean is a commons. The problem this paper seeks to address is the complex challenge of governing the ocean as a commons. Governance systems since the dawn of modernity have evolved to govern city-states, nation-states and international relations. But the transformations to sustainability require governing interlinkages and interactions that have not previously existed across sectors, and scales with multiple actors. Sustainable development is not only a laboratory for governance innovation (How will goals be achieved?) but also for policy innovation (Which concrete goals need to be set in a specific situation?) (Meuleman 2019). Governance configurations to craft the required policies that are appropriate for a global commons like the ocean are nascent at best. It follows, therefore, that this paper must address this key question: In response to the threats to the ocean's biophysical functions and life-support services, what transition dynamics are emerging at different levels (local, regional and global) that suggest appropriate governance configurations for the future?

What governance configurations could be established to govern the ocean as a commons?

The scale of the contemporary transition that is required now can be compared to that of the transition from hunter-gatherers to agrarian societies, and from agrarian societies to industrial societies (Haberl et al. 2011). Over the past 500 years, societal transitions of this scale have typically resulted in a change in governance with respect to socio-political arrangements, territorial authority, representation, rights, regulatory authority and accountability (Jessop 2016). In addition, since the beginning of human civilisation, people have collaborated to secure and protect common natural resources they have depended on for their survival (Ostrom 1990). During the modern industrial era, the commons has gradually been replaced by private ownership, on the one hand, and the public goods owned or controlled by states, on the other. The sustainability- oriented transitions that characterise the twenty-first century have once again brought into focus the crucial importance of the commons (Dasgupta et al. 2019). This has been made most clear with respect to the global scientific consensus about climate change, which is, arguably, about protecting the most important commons of all.

Societal transitions are not random events. Following the well-known 'multilevel perspective' on sustainability transitions, they emerge from a specific constellation of conditions that interact in complex ways (Geels and Schot 2007; Geels 2010; Grin 2010; Geels et al. 2017; Schot and Kanger 2018). Transitions come about when landscape pressures (e.g. population growth, technological change, climate change) result in a realisation that existing socio-technical regimes (e.g. fossil fuel-based energy systems, mobility systems based on the private car, industrial fishing) are inappropriate to address potentially destructive pressures or achieve a set of broader goals that previously did not exist (e.g. mitigating climate change, degradation of the ocean). For example, overfishing may be allowed by a legalised fishing regime, incentivised by an economic system and promoted by a political system. Yet in too many cases, the governance system as a whole does not resolve the problem of the ultimate collapse of the fish stocks as predicted by scientific research on wider landscape-level system dynamics (Cullinan 2014). At the same time, 'niche innovations' can open up as key networks of innovators respond to changing conditions by designing systems that aim to respond to the emerging landscape pressures (e.g. sustainable fishing regimes in the ocean context, or renewable energy in the context of climate change). Sometimes existing governance regimes engage with and absorb the niche innovations as their way of responding to the landscape pressures (e.g. the decision of large coal-based energy companies like Italy's Enel to become major renewable energy providers; Swilling 2019). At other times, regimes resist change, thus creating space for niches to coalesce and emerge as an alternative regime (e.g. the emergence of the organic food movement in response to the persistent dominance of the conventional global food system) (Smith et al. 2005). Sometimes niches are too weak and alternative regimes refuse to reform, resulting in landscape pressures causing social-ecological system breakdowns (as is emerging in water systems in many city-regions; Smith et al. 2005).

If, however, the dynamics of change are such that a societal transition becomes possible (e.g. the commitment by 57 countries to meet 100% of their energy needs with renewables, led by countries like Costa Rica and Uruguay; REN21 2018), much will depend on the emergence of governance configurations that are fit for purpose.

Incumbent governance arrangements can often prevent the implementation of known solutions (e.g. despite the rapid growth in renewables worldwide, there is no significant decline in the use of fossil fuels, largely because of path dependencies that existing tax and subsidy regimes reinforce; Geels 2014). Given the focus on a new set of global environmental goods such as the climate, the ocean and water, new governance arrangements appropriate for managing a commons will better enable the necessary transitions to occur in these sectors.

The approach to sustainability transitions used in this paper draws on complex adaptive systems theory (Levin 1998; Norberg et al. 2008; Preiser et al. 2018), many aspects of which are relevant to ocean systems (Lubchenco et al. 2016). The ocean system, from this perspective, is portrayed not as deterministic, predictable and mechanistic but as a diverse range of complex processes underway at multiple scales (global, regional, local). These processes catalyse self-organising dynamics that produce a new set of emergent and adaptive patterns. Ex post facto, data- intensive modelling may track past patterns, but future predictions cannot capture many critical drivers and responses. Scenario analysis and other forms of modelling may be useful for certain purposes, such as exploring global pathways to mitigate coastal risk through nature-based solutions (Chaplin-Kramer et al. 2019) or quantifying a subset of interacting ecosystem components for coordinated sectoral management, such as coastal habitat-fishery and disaster reduction linkages (Mumby et al. 2004; Arkema et al. 2015; Guannel et al. 2017; Rogers and Mumby 2019), but such approaches are unlikely to help us fully understand the highly complex dynamics of the ocean transition (Levin et al. 2012).

By identifying landscape, regime and niche dynamics, sustainability transition theory instead uses narrative analyses to help create useful categories for describing different kinds of transition dynamics, how they intersect and probable future pathways.

In order to describe what a transition to a more sustainable ocean system might look like, we first characterise the system through a brief outline of the current governance regime (Sect. 2) for the ocean. Establishing a governance baseline is required in order to build a narrative which can be used to understand ocean dynamics, and assist in identifying the nature of the changes that are needed for the transition to a sustainable ocean economy. Next, we define the stakeholders for several key sectors and describe the dynamics at play at multiple scales in the various sectors comprising the ocean economy (Sect. 3). We then demonstrate how regimes can respond to landscape pressures by diagnosing the principal drivers of change, including global and local scales (Sect. 4). Thereafter we demonstrate how the ocean system has evolved as a result of these drivers of change.

Hundreds of niche innovations have emerged, some of which we detail for illustrative purposes (Sect. 5). We propose theories of change which are suggestive of future trajectories, which in turn highlight the benefits of protecting and regenerating the ocean commons (Sect. 6). We offer paths forward, with examples of actions taken at local, national, regional and global levels which demonstrate successful transitions (Sect. 7). We conclude with opportunities for nation-states and other stakeholders in the ocean economy to contribute to a purposive transition to a thriving and vibrant relationship between humans and the ocean founded on a clear vision of the changes required, and an agreed future pathway for bringing about those changes (Sect. 8).

2 Current Governance Baseline

Governance is recognised as one of the key enablers of sustainable transformation (SDG 2019; TWI2050 2018; Pretlove and Blasiak 2018). The United Nations' 2030 Agenda represents a new mode of governance, one defined not through binding legal agreements but through the Sustainable Development Goals (SDG 2019). Current governance models and arrangements, whether global, regional, national or institutional, are ill-suited to develop, oversee or implement truly integrated, multidimensional sustainable development agendas such as the SDGs (Vidas 2011; Kotzé 2017, 2019; TWI2050 2018). Ocean governance is currently too fragmented across administrative boundaries and sectors to provide integrated responses (IPCC 2019; IPBES 2019).

An analysis of ocean governance includes institutions with mandates related to land use (urban, rural, coastal), freshwater management (surface and groundwater; quantity and quality), natural resource use (agriculture, horticulture, silviculture, mining, fisheries), environmental protection (including protected areas in terrestrial, freshwater and marine environments), development policies (e.g. economic, energy, transportation), human-environment interactions and the policies, procedures and regulations within and across segments of the source-to-sea ocean continuum (Mathews et al. 2019).

2.1 Ocean Governance

The current regime for ocean governance is complex (WCED 1987). As with international environmental law in general, governance internationally is comprised of two branches: customary law (judicial precedent, government policy and practice) and conventional law (international conventions).

The UN Convention on the Law of the Sea (UNCLOS) is at the centre of ocean governance. A key provision in UNCLOS is the distinction between various maritime zones of coastal states in contradistinction with the high seas (UNCLOS 1982). Maritime zones endow coastal states with either full sovereign jurisdiction or more limited sovereign rights (depending on the zone), in contradistinction to the principles of freedom of navigation and freedom of fishing which operate in the high seas, otherwise known as areas beyond national jurisdiction (ABNJs).

ABNJs generally are subject to weaker governance and poorer management. These waters are home to some of the rarest and most charismatic species on the planet—but all countries have the right to navigate, fly over, carry out scientific research and fish on the high seas with limited restriction (High Seas Alliance Treaty Tracker 2019). This, by definition, is why the high seas are a global commons. However, unlike the natural commons that is the focus of extensive work on commons governance today (influenced by Ostrom's research), the high seas as commons lack integrated commons governance.

In ABNJs, no nation-state is vested with jurisdiction, nor is any single international body vested with a strong mandate or effective means to secure a holistic, sustainable approach to managing the high seas (Pretlove and Blasiak 2018). In addition, human activities, water conditions and migratory species in nations' exclusive economic zones (EEZs) and territorial waters affect, and are affected by, activities in ABNJs. Ocean currents and species do not heed governance boundaries; and economic sectors commonly overlap in ocean spaces and share inputs to production (Klinger et al. 2018). Even the significant bodies such as the International Seabed Authority (ISA) or the various regional fisheries management organisations (RFMOs) are not mandated to implement the holistic comprehensive approach that is required to secure sustainable management of humankind's most important global commons. Peggy Kalas from the High Seas Alliance argues, 'The current high seas governance system is weak, fragmented and unfit to address the threats we now face in the twenty-first century from climate change, illegal and overfishing, plastic pollution and habitat loss' (High Seas Alliance Treaty Tracker 2019).

A promising development is the new agreement being negotiated under the provisions of UNCLOS, known as the internationally legally binding instrument for conservation and sustainable use of biological diversity in areas beyond national jurisdiction (the BBNJ).

A working draft of the BBNJ, released in September 2019, addresses a package of four topics, namely, the conservation and sustainable use of marine biological diversity of areas beyond national jurisdiction; marine genetic resources, including benefit-sharing; area- based management tools, including marine protected areas, environmental impact assessments and capacity building; and the transfer of marine technology (A/CONF.232/2019/6). During the first Marine Regions Forum, held in October 2019, participants emphasised the potential of universal participation in the BBNJ (Tsioumani et al. 2019). Proposals included listed mandatory environmental impact assessments; innovative options for the management of high seas biodiversity; improved coordination and cooperation among key stakeholders; and ongoing inter-regional exchange (Tsioumani et al. 2019). Concerns remain, however, as to whether the final agreement will be sufficiently robust to overcome the tension in a process characterised by individuated state and (institutional) interests. It will need to do so in order to solve a collective problem—the protection of ocean biodiversity in ABNJs. Several challenges may compromise a meaningful agreement. First, the agreement will require some inherent flexibility in its design in order to provide a useful framework for regulating decision- making in circumstances characterised by uncertainty; second, in an already crowded ocean governance space, existing institutions need to be rationalised in ways that increase coherence and effectiveness; and, finally, the new agreement needs to respond to the complex set of multiple, multilevel and systemic threats to marine biodiversity in areas beyond national jurisdiction (including, among other things, overfishing, plastic pollution and climate change) (De Santo et al. 2019) (see Appendix A1 for a more detailed discussion of the BBNJ).

2.2 Coastal Governance

Coastal regions naturally fall within the sovereign jurisdiction of coastal states. The interface between humans and the ocean is keenly experienced at the coast. More than 1.9 billion people lived in coastal areas in 2010, mostly in developing countries (Kummu et al. 2016). Coastal ecosystems provide services to humanity which are not easily included in monetary-based decisions, such as coastal stabilisation, regulation of coastal water quality, biodiversity, spawning habitats, carbon sinks, buffering and livelihood resources (Baker et al. 2019). The ocean is also an integral part of the global climate system (IPCC (Intergovernmental Panel on Climate Change) 2013; IPCC 2019).

From a governance perspective, integrated coastal management (ICM) is an approach developed to manage, in an integrated way, the complex and dynamic system encompassing the many interactions between people and ecosystems (Bremer and Glavovic 2013). The underlying key principle in ICM is the recognition that the traditional sectoral approach to managing human activities in the coastal zone, characterised by competing needs and overlapping mandates, has significant negative impacts on the environment. The aim of ICM is then to provide integrated governance for guiding coastal area development in an ecologically sustainable fashion (FAO 2019a). ICM has been applied in regional applications with marine ecosystem programs and regional seas programs, for example the EU Integrated Maritime Policy, and in a growing number of nations (see Winther et al. 2020).

However, regulatory challenges for effective ICM arise as a result of institutional and sectoral inertia, lack of flexible decision frameworks to manage the complexity, uncertainty and difficulties managing the trade-offs inherent in ICM (Vierros and Buonomo 2017).

Figure 12.1 illustrates the complexity of overlapping and competing interests and mandates which exist in the ocean economy.

Ngeta (2014, p. 28) confirms this by stating that 'the complexity of the actor constellation tends to increase as one moves up the governance ladder from the local to the global'. Complexity of this nature in coastal governance has implications for resource and livelihood sustainability (Agrawal and Perrin 2009). For this reason, implementing ICM has proven to be a challenge in different parts of the world (Ngoran et al. 2016; Cantasano et al. 2017; Warnken and Mosadeghi 2018).

Figure 12.1 demonstrates that, from a national perspective, overlapping mandates in the ocean economy arise primarily within coastal states' exclusive economic zones, but they can also occur on the high seas. An integrated ocean management (IOM) approach to policymaking is designed to address challenges for ocean management in EEZs that can include conflicts between sectors (e.g. tourism versus hydrocarbon extraction), across different scales of organisations (such as local, municipal, regional) and biophysical features (local water bodies, shared watercourses, regional seas, global ocean) and across time (current and future uses) (Klinger et al. 2018; Winther et al. 2020). Another obstacle to integrated management is a lack of information on how sectors interact, and how changes in one affect incentives and actions in others (Klinger et al. 2018). Anticipated economic growth from the blue economy is likely to lead to additional cross-sector conflicts and could bring environmental degradation, inefficient natural resource use and other socially undesirable outcomes (McCauley et al. 2015; Winther et al. 2020). A FAO study on the application of IOM in a number of countries including Norway, Indonesia and Angola found that multi-sectoral coordination, created through a grouping of marine ministries, was very effective in Indonesia (Torrie 2016). Information- gathering through a decentralised power structure and incorporating stakeholder engagement through a network approach (see Sect. 7.1.2)

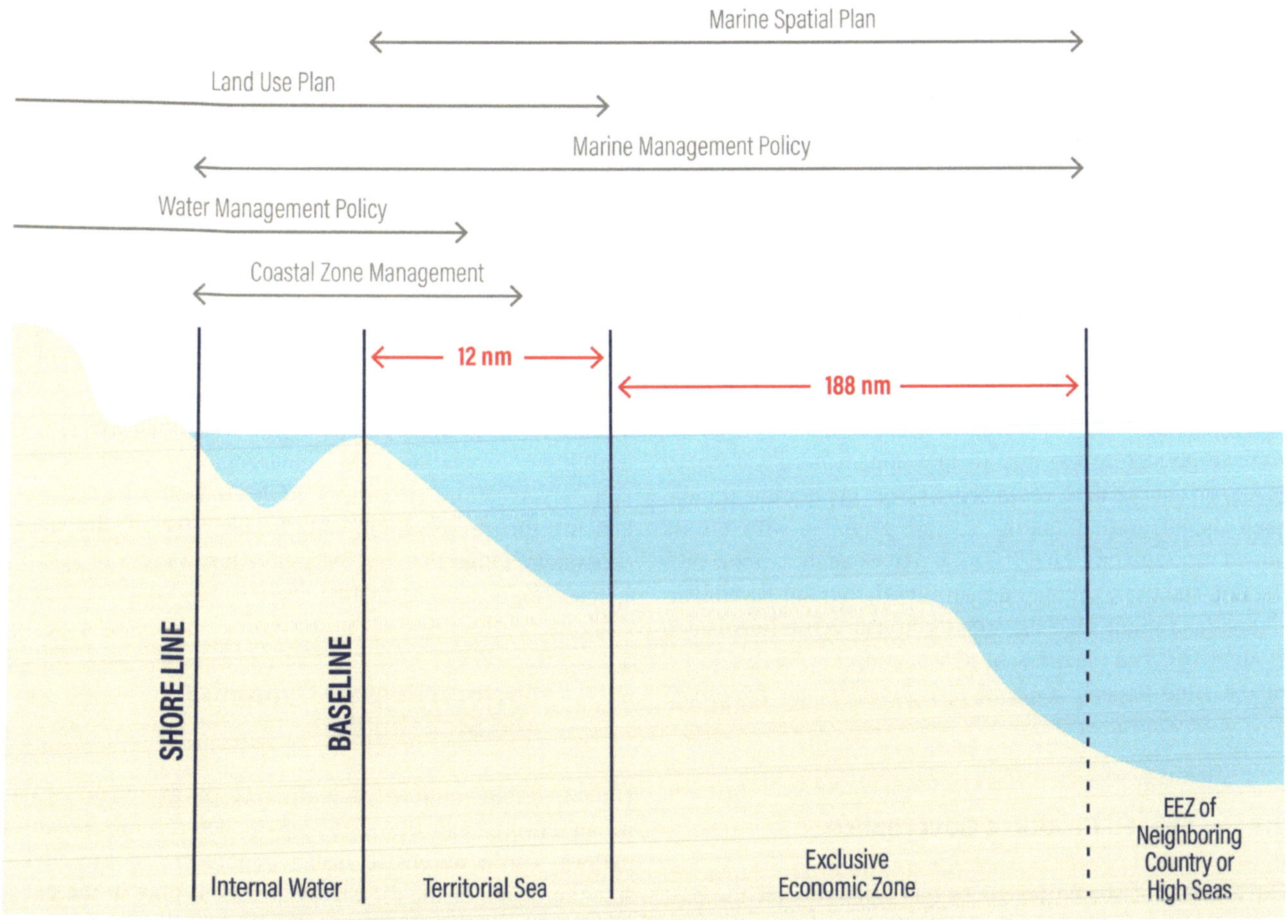

Fig. 12.1 Independent and overlapping management frameworks in the source-to-sea system in Sweden

provided highly valuable information (Torrie 2016). An advantage in Norway has been the relative speed and flexibility with which the government is able to draft, pass and enforce laws, which has contributed to effective ocean governance focused on the health of fish stocks and ecosystems (Torrie 2016).

Many of these principles will be echoed throughout this paper. In addition, the knowledge commons for sharing information and establishing transparency (discussed in Sect. 7.2.2) will facilitate policy development in the national context. Polycentric governance processes within the boundaries of national laws and policies will facilitate coordination, the sharing of knowledge and information and the identification of possible solutions for trade-offs.

Enhancing understanding about the governance interactions, as well as the manner in which governance outcomes may influence resource and livelihood sustainability, is needed in order to elucidate understanding about how different forms of commons governance produce different outcomes for livelihoods and well-being of marginalised communities (Brockington and Wilkie 2015). South Africa's first UN Educational, Scientific and Cultural Organization (UNESCO) World Heritage Site, Simangaliso, provides an example of the governance challenges which arise as a result of the co-existence of plural and multilevel governance systems and institutions governing the coast (see Appendix A2 for this case study).

Sectoral institutions still dominate in national governments and in the UN system. This gap presents one important opportunity for future policy development—a UN or another version of a supranational body could, through access to a knowledge commons platform (discussed later in this paper), analyse the functioning of existing laws and institutions, articulate a flexible framework (which can provide for diverse region-specific conditions) on agreed ICM principles, as well as provide monitoring and oversight to achieve greater coherence. To that end, sectorally focused management regimes would benefit from greater integration, such as integrating marine spatial planning with efforts aimed at regulating land-based activities such as food production and the resulting nutrient runoff, tourism-based livelihoods and small-scale fisheries (Sale et al. 2014; Lubchenco et al. 2016). The functions of ICM are more fully developed in the Blue Papers "*Coastal Development* and *Integrated Ocean Management*".

2.3 Shared Resource Governance

Governance of shared resources has matured over the past several decades. The lucrative global fisheries for migratory and straddling fish stocks in the high seas, such as tuna, are managed through several measures. Regional fisheries management organisations (RFMOs) were set up under the UN Fish Stocks Agreement (UNFSA), and sector-specific measures were taken by the UN Food and Agriculture Organization (FAO), including the FAO Code of Conduct for Responsible Fisheries, the International Plan of Action to combat illegal, unreported and unregulated (IUU) fishing, and the Agreement on Port State Measures to Prevent, Deter and Eliminate IUU Fishing. While some RFMOs have been effective, severe challenges result from a lack of cooperation among states, conflicting interests in resource utilisation and conservation, fragmented responsibilities, lack of political will, lack of effective monitoring and enforcement and perverse economic incentives for 'free riders' to cheat the system (Figueres et al. 2014).

River-basin management of shared transboundary water-basins provides useful lessons for joint stewardship of the ocean. In circumstances of shared international watercourses, states have been required to share sovereignty in a kind of 'joint ownership' of the water body (Rieu-Clarke and Spray 2013). For example, the Mekong River Basin, a 4909 km river system which flows through six countries, is collaboratively managed by the Mekong River Commission. The UN Watercourse Convention 1997 (UNWC), is widely recognised as a pivotal document in international water law (Litke and Rieu-Clarke 2014). It codifies and clarifies existing norms and develops emerging principles of customary international water law. It constitutes a model that can guide the interpretation of other treaties and the negotiation and drafting of future ones; and it has informed the judgements of international and regional courts (Litke and Rieu-Clarke 2014). Some of the unique procedural provisions in the UNWC offer lessons for future regulatory frameworks for regional or international management of shared estuaries or seas. In addition, the principles of shared water management through regional organisations set up under the UNWC could be adapted for regional fisheries management by adapting or expanding RFMOs into regional ocean management organisations (ROMOs). ROMOs would be responsible for the preservation and productivity of the entire ecosystem, rather than only shared fish resources or specific species (Figueres et al. 2014).

3 Sectoral Regime Dynamics and Stakeholders

Having briefly outlined a high-level overview of current ocean governance, the next step in determining potential pathways to a purposive ocean transition requires understanding some of the regime dynamics at play in the ocean and identifying the key stakeholders in those dynamics. Regimes are understood in the sustainability transitions literature to be a tightly knit combination of regulations; key operators that produce products or services; consumers who depend on those products or services; the revenues that governments, agencies and regulators extract in the form of levies or taxes and so on; the financial institutions that have provided debt or equity; plus a substantial infrastructure operated by people who have been trained over decades to understand and operate the system in certain ways.

This combination of interlocking interests creates an alignment of purpose that reproduces path dependency.

Path dependency can be described as a constraint on decisions or processes due to a combination of interlocking interests that creates an alignment of purpose and resistance to change. The tightly interdependent set of interests limits possible decisions for any given circumstance due to past decisions or because of inexperience with new conditions, even when past circumstances may no longer be relevant or appropriate (e.g. reliance on fossil fuel energy). Generally, path dependencies arise because of the tendency of institutions to act (and react) as a result of their historical structural properties or beliefs. For changes in a regime to occur, the reprogramming of a vast array of system components will be necessary. This is why regime change is challenging, and why often regimes resist change. Regimes will change in response to landscape pressures if they have the internal capacity to manage change and access to new external knowledge about alternatives. Without these conditions, niche innovations will emerge outside the regimes and can eventually coalesce into alternative regimes that are more responsive to landscape pressures than are the incumbent regimes (Smith et al. 2005).

In order to propose a theory of change, it is necessary to view the ocean system in its full complexity. The key ocean-relevant regimes that directly and indirectly affect the future resilience of the world's ocean are summarised in Appendix B. The regimes we outline are shipping, ocean-based food extraction, offshore oil and gas, ports, marine and coastal tourism, marine and seabed mining, marine biotechnology, cabling and maritime equipment, and offshore and renewable energy. The regimes we sketch in Appendix B are sectoral in nature, with distinct governance and operational dimensions. Regimes often interact at multiple scales, and shifts that seem unimportant at the local or regional scale, when aggregated, could actually precipitate major changes in other regimes (e.g. aquaculture and coastal development; Rocha 2010).

Several regime responses illustrate shifts in existing ocean systems towards sustainability. For example, institutions such as the International Maritime Organization (IMO) and some industries have contributed to significant regime responses in the shipping sector.

Green ship recycling is one example. This involves environmentally friendly ways of managing end-of-life ships (OECD (Organisation for Economic Co-operation and Development) 2010; IMO Guidelines for Shipbreaking; the Hong Kong Convention; NatCap (Natural Capital Project) 2019; EU Directive on Ship Breaking). Also in the shipping sector, the GloFouling Partnership project was formed under the auspices of the IMO to support anti-fouling measures. These measures are designed to guard against the significant risks to marine biodiversity created by ship's discharges contaminated with alien invasive species. The move towards the decarbonisation of the shipping industry, initiated by the *Initial IMO Strategy on the Reduction of GHG Emissions from Ships* (IMO 2018), is underway. A target has been set to decarbonise the shipping industry by 2035, which would equate to a reduction in shipping's carbon dioxide (CO_2) emissions of between 82 and 95% below the current level (OECD International Transport Forum 2018). Some ports around the world (Los Angeles, Auckland, Valencia, Guayaquil, Baku and Rotterdam) are working towards becoming carbon-neutral. Several transparency initiatives are also unfolding across numerous regimes. For example, in the maritime transport sector, the Open Simulation Platform, founded by industry in 2018, facilitates collaborative open platforms for the design, operation and building of ships. Fisheries certification schemes such as the Marine Stewardship Council (MSC) are supported by new information products such as Global Fishing Watch. These information products track and analyse global fishing activity using publicly available automatic identification system transmissions from boats and satellite images. Traceability programs of this nature allow full tracking of harvested species along the entire production chain (Costello et al. 2019). Such monitoring information is critical to building trust that fishery management interventions are having the desired effects, or if not, to encouraging action and adaptation throughout the seafood value chain, and enabling monitoring and compliance through these systems. The suite of sustainable fishery management approaches that have been implemented in the ocean and food extraction regime, such as rights-based fisheries management (RBFM), marine protected areas (MPAs) and integrated coastal management regulations, indicate that fish stocks, marine habitats, fisher communities and ocean food–based supply chains can recover if management objectives are clear and monitoring and scientifically based assessments inform open discussion of trade-offs and adaptation over time (Costello et al. 2019). New synergies are being investigated in the cabling and maritime equipment sector, which could lead to the use of private sector submarine telecommunications infrastructure for climate monitoring and disaster warning (Submarine Telecoms Industry Report 2019). Other responses are emerging; for example, a submarine cable in the Olympic Coast National Marine Sanctuary in the U.S. Pacific Northwest was re-buried and then systematically monitored to reduce concerns about entanglement of fishing gear and species disruptions (Antrim et al. 2018).

3.1 Inter-regime Dynamics

Although the regimes described above and in Appendix B are conceptually distinct, in practice they overlap in highly complex ways that are both mutually reinforcing and potentially contradictory. A growing global consensus and plethora of scientific reports are contributing to an awareness that these regime and inter-regime dynamics have unintended consequences that could irrevocably harm the ocean's biophysical systems in ways that subvert ocean-dependent regimes (Winther et al. 2020).

4 Drivers of Change

Now that we have outlined the existing framework of ocean governance, and identified the dynamics of the ocean system, we turn to a diagnosis of the landscape-level pressures or 'drivers of change' which are destabilising the system. Establishing an understanding of the issues and practises which have led to the landscape pressures now facing the ocean will help identify pathways that will allow change to occur. Landscape pressures are broad, long-term emergent cumulative shifts that are not caused by single current actions in present time.

4.1 Greenhouse Gas Emissions

Anthropogenic greenhouse gas emissions, and CO_2 specifically, are widely recognised as the biggest long- term threat to a functional ocean (Rogers and Laffoley 2013; Gaines et al. 2019). Climate change is altering the ocean's impact on climate, its chemistry, temperature, circulation, sea level and ice distribution. Collectively, these system changes are affecting the habitats, biological productivities and species assemblages that support ocean-based economies and cultures. Ocean circulation changes also are predicted to lead to increased intensity and frequency of tropical cyclones and extreme sea level events, including storm surges and flooding and precipitation (Gaines et al. 2019).

Increasing ocean acidity is influencing large swathes of ocean ecosystems in a range of ways (Suggett et al. 2012; Kroeker et al. 2013) and is most acutely felt in shallow water systems such as the subarctic Pacific and western Arctic Ocean (IPBES 2019). The ocean has absorbed over 90% of the excess heat from global warming, with consequences for organisms that are adapted to specific temperature ranges in terms of both latitude and depth (IPCC (Intergovernmental Panel on Climate Change) 2013). Oxygen content is in decline, dramatically observed in increasingly extreme hypoxic events (Stramma et al. 2010). Other primary stressors on ocean systems include habitat destruction, overfishing and pollution from land and coastal sources (Bailey and van der Grient 2020). Impacts of climate change on biodiversity and ecosystems are well documented, including shifts in species ranges and the socio-ecological ramifications of this (Cheung et al. 2010; Pecl et al. 2017; Costello et al. 2019; Gaines et al. 2019).

4.2 Overfishing

Direct exploitation of fish and seafood has the largest relative near-term impact among drivers of ocean status (IPBES 2019). Currently, 33% of fish stocks are classified as over-exploited and greater than 55% of the ocean area is subject to industrial fishing (IPBES 2019). Since 1950 the percentage of fish stocks that are 'developing' (i.e. still increasing in output) has declined dramatically, while stocks that are exploited, over-exploited or collapsed has escalated dramatically since the 1980s (FAO 2018; Costello et al. 2019; Widjaja et al. 2020). Severe impacts on ocean ecosystems can occur through direct harvest of target and bycatch species, and, indirectly, through degradation or destruction of benthic habitats (e.g. through dredging in soft-bottom regions or dynamite fishing in coral reef areas) or through the ramifications of predator-prey and other food web dynamics (Costello et al. 2019).

Small-scale fisheries catches have been increasing, from about 8 MT per year in the early 1950s to 22 MT per year in 2010, and continue to grow at the global scale, while industrial catches at larger scales decline (Pauly and Zeller 2016). Small-scale fisheries suffer from highly variable regulation and enforcement, even where catch, area or gear limits do occur. Such fisheries are characterised by very limited information on stock status and fisher behaviour, exacerbating socioeconomic pressures on vulnerable fishing communities dependent on local seafood for nutrition and livelihood support (FAO 2015). Additional detail on drivers in the current fishing regime is provided in Sect. 3 and Appendix C.

4.3 Seabed and Land Use

Another direct driver with a high relative impact on the ocean is the many changes in the uses of the sea and coastal land (IPBES 2019; Addo et al. forthcoming). Nearshore regions of the world are straining to support exploding demand from oil and gas development, shipping and port activities, fisheries and aquaculture, tourism and the protection of people, property and infrastructure from increasing storm intensity and sea level rise (Fig. 12.2). Much of the planet's population growth over the next decades will occur along coastlines, where development pressures already are destroying nearshore marine ecosystems (e.g. mangroves,

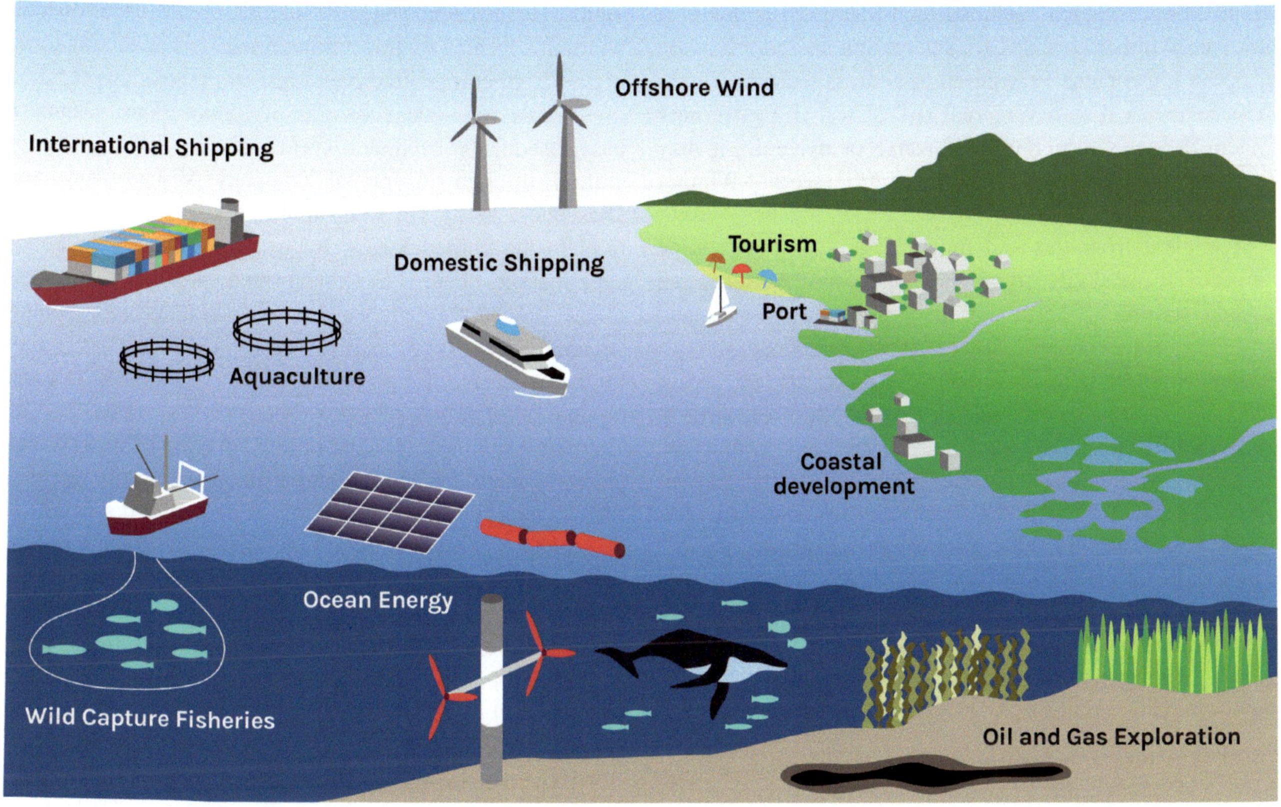

Fig. 12.2 A sampling of human activities in the coastal ocean

coral reefs, wetlands, seagrasses, kelp forests) that provide many of the benefits on which people rely (Costello et al. 2019; Aburto-Oropeza et al. 2020). Coastal habitats, including estuaries and deltas critical for marine biota and regional economies, have been severely affected by sea-use changes (coastal development, offshore aquaculture, mariculture and bottom trawling) and land-use changes (land clearance, urban development along coastlines, pollution of rivers). Ocean drilling, while relatively small in scope, has expanded since 1981 to roughly 6500 offshore oil and gas installations worldwide in 53 countries (60% in the Gulf of Mexico by 2003) and likely will expand into the Arctic and Antarctic regions as the ice melts. Plastic microparticles and nanoparticles are entering food webs in poorly understood ways (Jambeck et al. 2020). High levels of metals, nutrients and persistent organic pollutants from industrial discharges and agricultural runoff damage fish species and seabed biota. The dynamics of ocean and airborne transport of pollutants mean that the harm from inputs of plastics, persistent organic pollutants, heavy metals and ocean acidification is felt worldwide, including with consequences for human health (IPBES 2019).

4.4 Fragmented Governance

As we have already indicated, governance arrangements are too fragmented across administrative boundaries and sectors to provide the integrated responses that are required to meet the increasing and cascading risks of negative environmental impacts on the ocean (IPCC 2019). Existing governance is inadequate to stem unsustainable ocean uses in some coastal regions, where often a full constellation of resource demands and human activities are degrading ecosystems, and where regulatory contexts lack mechanisms to integrate management across sectors (IPCC 2019). Bigagli (2016) argues that more resilient ocean management systems incorporate learning-based management strategies that are supported by science-based advice to policy. He reviewed over 500 existing international agreements for environmental protection and regulation of human activities in the world's ocean and found them woefully inadequate. At global scales, agreements largely focus on single-sector objectives—fisheries, pollution, nature protection and transportation primarily—governing human use and management of the ocean. Regional agreements tend to include multiple sectors, but inclusion of ecological resilience considerations is mixed at best (see Winther et al. 2020). Integration and coordination of ocean governance is required to address these issues. In the absence of this, a very real risk arises: if a governing system becomes too complex, diverse or dynamic, it may become ungovernable in itself (Jentoft 2007; see also Winther et al. 2020).

It is clear from this and previous sections that a wide range of increasingly significant landscape pressures are already starting to harm the functional effectiveness and resilience of the ocean's complex social-ecological systems. Although pre-existing governance regimes are no longer adequate, as indicated earlier, significant shifts are already underway. These shifts are reflected in the ways regimes are responding to landscape pressures, and how niche innovations (instigated often at the local level by new actor networks) are emerging with a focus on the protection and regeneration of the commons.

5 Niche Innovations

In order for transitions to occur in response to landscape-level drivers of change, two dynamics must be at play: either existing regime actors (i.e. key individuals located within a particular regime, which could include decision-makers, consumers, regulators or funders) respond to these drivers, and/or networks of innovators come together to instigate so-called niche-level innovations. When path dependencies persist because regime actors resist change, niche innovations can emerge that demonstrate through trial and error that alternatives are possible. Niche innovations tend to be geographical and/or sectoral spaces where coalitions of innovators coalesce in response to perceived landscape pressures. Often, these niche innovations are protected from market dynamics (via subsidies or soft money) or political interference (via regulation or location in the non-profit sector).

For example, groups of anti-nuclear activists initiated wind-power experiments in Denmark in the 1980s and 1990s. Government policy banned businesses from owning windmills located within communities, thus protecting these niches from competitive pressures, and regulations resulted in rewards for—and subsidies of—innovations. Denmark has subsequently become a world leader in wind-power generation through its global company Vestas (Mey and Diesendorf 2018). The same applies to organic food production. In the United States, for example, the Whole Foods supermarket chain has transformed previously limited-impact organic food niches (farming and retail) into a mainstream food retail regime. In this case, niches were 'protected' by the evolution of a particular set of consumer values and organic certification schemes. Niches can coalesce into an alternative regime (e.g. wind power), or existing regimes can change course and absorb niche innovations (e.g. the way Italy's energy utility Enel has decided to close down half of its coal-fired power business and enter the renewable energy market as a mainstream global player).

These niche innovations reveal how regulatory interventions (if any), stakeholder engagement/organisation, institution-building, and ongoing monitoring and evaluation have worked in ways that counteract the negative landscape pressures. We offer four examples here to illustrate the diversity of niche innovations emerging in the ocean: coastal zone development planning in Belize (Box 12.1); the Chilean territorial user rights fisheries (TURFs) as an example of rights-based fishery management (Box 12.2); shared stewardship in business in SeaBOS fishery companies (Box 12.3) and responses to plastic pollution in the ocean (Box 12.4). There are literally hundreds of niche innovations, some disconnected, others are starting to coalesce into potential alternatives. To illustrate the breadth of niche innovations in the ocean system, we highlight four additional examples in Appendix C to reveal their breadth and dynamics: integrated coastal development, fisheries and disaster risk planning in Belize; rights-based fishery management; illegal, unreported and unregulated (IUU) fisheries monitoring innovations; and justice in marine sustainability, including the Pacific Islands' Parties to the Nauru Agreement (PNA Tuna 2010).

Box 12.1. Coastal Zone Development Planning in Belize

The Government of Belize's Coastal Zone Act of 2000 recognises the value of multi-sectoral, integrated spatial planning to guide policy and investment for more sustainable use of the coastal zone. The government approved a national Integrated Coastal Zone Management Plan (ICZMP) in 2016, led by a new ministry inspired by integrated development planning, connecting in one department Agriculture, Fisheries, Forestry, the Environment and Sustainable Development (CZMAI (Coastal Zone Management Authority and Institute) 2016). The government led an interactive stakeholder engagement process to co-develop the plan, beginning with identifying shared objectives for artisanal and commercial lobster and conch fisheries; reducing the risk to coastal infrastructure, property and people from sea level rise and storms; and sustainable tourism benefits, the largest sector of the Belizean economy. The iterative science-policy process engaged all relevant stakeholders from government ministries, non-governmental organisations, business, and community leaders (Arkema and Ruckelshaus 2017). The final plan is projected to improve coastal protection from storms and sea level rise, and increase revenue from fisheries and tourism, more than alternative plans emphasising either conservation or development alone (Arkema et al. 2015, 2019; CZMAI (Coastal Zone Management Authority and Institute) 2016). At the same time, the plan improves protection for mangroves, coral reefs and seagrass beds—the natural capital upon which coastal populations' safety and livelihoods depend.

The final ICZMP highlights the importance of coordinating the management of, and investment in, a diverse set of activities and actors implicated in sustainable outcomes for the nation, ranging from those engaging in or affecting coastal pollution, dredging, fisheries, aquaculture and tourism development, to education, social resilience to climate change, and preservation of cultural heritage. The plan led the Belizean government to enact a permanent ban on all oil exploitation within the second-largest coral reef in the world. The ICZMP actions and new zoning-based management are being implemented with funding from the government, the Inter-American Development Bank and other sources. The Belize plan has been hailed by UNESCO as 'one of the most forward-thinking ocean management plans in the world' (Douvere 2016). In 2017, the Belize Barrier Reef was removed from the UNESCO List of World Heritage in Danger because of the protections provided in the government ICZMP.

The key innovations in the Belize ICZMP process include a legal government mandate, in the Coastal Zone Management Act of 2000, requiring a cross-sectoral, multi-objective and spatial planning process. In and of themselves, such laws do not necessarily lead to transformation of ocean management. An important institution in Belize, the Coastal Zone Management Authority and Institute (CZMAI), played a key role in designing the co-development process for the ICZMP, and continues to lead its ongoing implementation and adaptation. The science-policy process to envision, debate, and select the final ICZMP approved by the government also included training of Belizeans on the scientific and policy aspects of ecosystem-based management, increasing the chances that the process will be internalised in government and civil society activities (Arkema and Ruckelshaus 2017). Appendix C details efforts in fisheries and disaster risk sectors to integrate more fully with coastal zone development planning in Belize.

Box 12.2. Chilean Territorial Use Rights Fisheries

In 1991, after an overfishing crisis led to critical closures of the Chilean abalone ('Loco') fishery in the late 1980s, Chile enacted the first step in a governance transformation—a Territorial Use Rights in Fisheries (TURF) policy (Gelcich et al. 2010). As of 2013, there are over 450 TURFs in full operation, making up more than 1100 km^2 of subtidal habitat decreed to fisher organisations throughout the Chilean coast (Gelcich et al. 2017). This network of TURF areas has been established by numerous associations of fishers, along a wide geographic range, under one policy instrument, Chile's National Fisheries and Aquaculture Law (Marín et al. 2012). As a result of the TURFs, Chile's artisanal sector has increased in importance, with landings consistently surpassing the industrial catch since 2008. Artisanal fisheries are a significant source of employment for coastal communities in Chile, and their harvests represent a key source of nutritional food for many rural communities. Increases in biomass and size of individuals from species within properly managed TURFs also are demonstrating the potential of this rights-based management approach to sustain ecosystems and fishery benefits (Gelcich et al. 2019a, b).

The national enabling legislation, combined with the presence of scientific knowledge signalling alternative ways to manage stocks, and the capacity and political leverage of fisher associations that facilitated the cross-scale and the cross-organisational interactions for change, each were key in institutionalising the new governance regime.

Any registered fishing association in Chile can register as a TURF under the national law, thus encouraging volitional participation in the program, a key component of adaptive governance for a more resilient system. The TURF network has improved the knowledge of fishers and their access to learning, especially as it relates to harvest management practices, biological aspects of the resource and the interactions of the target species with other elements of the ecosystem. This increased understanding has served to develop a sense of resource stewardship on the part of fishers.

While the 25-year-old Chilean TURF model has proven its potential to improve the sustainability of fisher communities and fisheries, its governance must continue to evolve as information on social and ecological barriers to further scaling emerges (Gelcich et al. 2010). TURFs convey rights to fishers and allow them a greater, collective voice in the long-term management of the resource, a key component of their adaptability and responsiveness to changing social-ecological conditions. Currently there is room for improvement with respect to enforcement, profitability, socioeconomic impacts on resource users and the adaptability of the policy to local realities. Science, both social and natural, is key to informing ways to maintain the policy, enabling adaptation of TURFs and identifying new conditions that must be improved to build the resilience of TURFs or enable further transformations.

Box 12.3. Seafood Business for Ocean Stewardship

The Seafood Business for Ocean Stewardship (SeaBOS) initiative is an innovative collaboration among 10 of the largest global seafood companies that is transforming business operations for more sustainable wild capture fisheries and aquaculture production. Collectively, the companies in the SeaBOS initiative influence the strategic direction of more than 639 subsidiaries along the seafood value chain, with operations in at least 93 different countries, and participation in fisheries and aquaculture decision-making institutions such as regional fisheries management organisations. Under the SeaBOS platform, the world's leading seafood businesses are managing seafood cooperatively, monitoring their practices and impacts, and charting a new path for their sector. They have pledged to address illegal, unreported and unregulated fishing; work towards full traceability and transparency throughout their supply chains; make efficient use of aquaculture feeds and use fish feed resources from sustainably harvested stocks; apply existing certification standards; eradicate labour abuses and human rights violations from their supply chains; reduce the use of plastics in seafood operations; work towards reducing the use of antibiotics in aquaculture; and prevent harmful discharges and habitat destruction. The participating businesses also have pledged to work with governments to improve existing regulations concerning aquaculture and fisheries (Österblom et al. 2017). The scope of the undertaking spans every continent and all segments of seafood production. The collaborative nature of the SeaBOS project also helps companies share information to develop best practices, which in turn has helped to build trust and common purpose. An on-deck species-detecting camera and facial-image recognition software pilot is aimed at identifying illegal catch and undocumented fishermen onboard vessels. SeaBOS has recognised the crucial role of scientists in framing the urgency of problems and potential solutions. The initiative is an ongoing experiment that is being closely monitored to understand the significance of the changes over time. Such initiatives engaging with the private sector are best considered a complementary approach to existing processes, such as government regulations. This initiative is improving the prospects for transformative change by providing novel links between science and business, between wild-capture fisheries and aquaculture industries, and across geographical space (Österblom et al. 2015). SeaBOS is best described as a co-production initiative between science and business, in which companies can develop their agency (Westley et al. 2013) and ability to influence change across subsystems, thereby contributing to amplifying new norms of ocean stewardship.

Box 12.4. Global Response to Plastic Pollution in the Ocean

Over the past several years, awareness of marine plastic pollution has skyrocketed around the world. Stories of marine turtles and mammals dying from ingested plastic and plastic pollution washing up on beaches have inspired hundreds of commitments from government, businesses and non-governmental organisations, dozens of innovation challenges, hundreds of start-up companies seeking to create solutions and millions of citizens eager to take action (Jambeck et al. 2020). Global plastic production has indeed exploded, from 1.7 million metric tonnes/year in 1950 to 422 million metric tonnes/year in 2018 (Geyer et al. 2017; Plastics Europe 2019), with a concomitant increase in plastic in the waste stream (in the United States, plastic was 0.4% of the waste stream by mass in 1960 and 13.2% in 2017 (U.S. EPA 2014, 2019)). Both micro- and macro-plastics can enter the ocean through direct discharge, discharge into rivers that then flow into the ocean, runoff from land or deposit from air into waterways. Impacts of this increased load on biodiversity include negative effects on growth, reproduction and survival of marine species (Jambeck et al. 2020).

Strategies to address ocean plastics include enhancements to wastewater, stormwater and coastal zone management, development of alternative materials, greater resource efficiency, recovery and recycling (Jambeck et al. 2020). The plastic challenge is systemic, spanning product-specialised value chains and geographic heterogeneity in plastic generation, use and recycling capabilities (Jambeck et al. 2020). The specific solutions to plastic pollution in the ocean are likely to be many, crossing sectors and spatial scales, from changing individual choices to company sourcing decisions to enforcement of existing and new regulation. Efforts to address the ocean plastic challenge are acknowledging the need both for a systems approach and for understanding the ocean as a commons. For example, ocean plastic is a growing problem in Africa, where waste volumes within coastal countries are relatively low but waste streams from other nations are overwhelming limited environmental regulation (Jambeck et al. 2018, 2020).

Leaders from governments, businesses and civil society are focusing on changing perceptions and behaviour along the entire supply chain, from design, production and use through to disposal and further use. Heightened public concern about plastic in the ocean is currently an effective catalyst for action on solutions; and this urgent attention is beginning to activate broader strategies to reduce the flow of other pollutants into the ocean. For example, a number of social innovations emerging in Africa are aimed at waste problems generally, such as community-driven collection systems and financial reward for recyclables, such as Wecyclers in Nigeria and Packa-ching in South Africa (Jambeck et al. 2020). At the regional scale, the Lower Mekong Initiative, a multinational partnership to integrate the policies of Cambodia, Laos, Myanmar, Thailand, Vietnam and the United States, is now addressing plastic contamination upstream before it gets to the ocean (Jambeck et al. 2020). The system-level collaboration emerging to address plastic pollution is a promising start to what needs to be a worldwide response. Collaborative efforts such as the Ellen MacArthur Foundation's New Plastics Economy Global Commitment, which requires signatories to align on a shared vision and targets, and the World Economic Forum's Global Plastic Action Partnership, which is a public-private collaboration platform helping to translate commitments into action, are also providing forums where many stakeholders can work together at the system level.

The system-level collaboration emerging to address plastic pollution is a promising start to what will most effectively be a worldwide response.

6 Transition Dynamics: Theories of Change

To transform socio-ecological systems, the elements below are needed to enable articulation of future aspirations, as well as a generative dialogue so that learning and adaptation can occur.

6.1 Knowledge for Transitions

Up until this point we have described the key regime dynamics (with overviews in Appendix B of shipping, ocean-based food extraction, offshore oil and gas, ports, marine and coastal tourism, marine and seabed mining, marine biotechnology, cabling and maritime equipment and offshore renewable energy), the various relevant landscape pressures and a sample of niche innovations (see Sect. 5 and Appendix C). In summary, it is clear that there are a set of landscape pressures that could result in the collapse of the ocean's key ecosystem functions, with negative implications for humanity and, specifically, the global economy. Despite the strong governance framework provided by the UNCLOS system, the existing regimes are institutionally misconfigured for this challenge. They are locked into path dependencies at odds with what is required to face the landscape pressures. However, some regime dynamics respond positively to these landscape pressures.

These sustainability-oriented regime dynamics are suggestive of future trajectories. Similarly, there is a mushrooming of niche innovations as constellations of actors (primarily, but not exclusively, at the local level) respond to landscape pressures and the inadequacy of current regimes. What is distinctive about these niche innovations is that they entail forms of stakeholder collaboration that are driven by an overriding concern to protect and regenerate the commons. As Nobel Prize winner Eleanor Ostrom (1990, 2000) has argued, humans have collaborated for millennia to protect the commons that they recognise they are dependent on. The niche innovations, therefore, suggest future trajectories that valorise the commons. They also provide signposts for the 'anticipatory thinking' (Poli 2018) that is needed in order to chart a course for the future. Transdisciplinary research methodologies will be required to conduct research on the constantly changing interactions between landscape pressures, regime dynamics and niche innovations in order to grasp the emergent properties of the sustainability-oriented ocean transition (van Breda and Swilling 2018; van Breda 2019).

6.2 Capacity and Incentives for Transitions

Transition dynamics are dependent on three key factors: whether or not existing regimes access new knowledge from external sources; whether or not they have the capacity to integrate new knowledge in order to facilitate substantive change processes; and whether or not there are incentives, initiatives or other enabling conditions that activate change. In simple terms, if within a given regime (e.g. a car-based fossil fuel–dependent transport system in a given country) there is sufficient capacity to manage change (among, in this transport case, the policymakers, regulators, transport company managers, etc.) coupled to rapid learning about alternatives (derived from experimental examples), the chances are high that a transition will occur over time (in this example, to a decarbonised transport system). However, actual changes will only take place if some catalytic event instigates the need to activate the capacity for managing change. This could be anything from price hikes to protest movements to an electoral shift that brings a new party to power with an anti-car agenda.

Following Smith et al. (2005), there are four possible transition pathways, depending on how these knowledge, capacity and catalytic factors combine. When a particular regime can access new external knowledge, when it has the capacity to manage change and when enabling conditions are present, a 'purposive transition' can occur. Such transitions can be quite radical, including the transcendence of the mainstream regime itself in the process (e.g. the renewable energy transition in Germany). A purposive transition, however, is not inevitable. If the capacity to manage change exists but only 'internal knowledge' is relied on to envision alternatives, the result will be a reform of the regime rather than its replacement (i.e. an 'endogenous renewal'). Conversely, if there is limited capacity to manage change and external knowledge is sourced, the result will be an 'emergent transformation', that is, the internal breakdown of the regime followed by the mushrooming of alternatives with limited capacity for implementation. Where there is both limited capacity for change and a reliance on internalised knowledge sources, the result will be a 're-orientation of trajectories' as the old regime becomes dysfunctional but viable alternatives fail to emerge.

The above analysis is more appropriate for understanding transitions in particular sectors, such as the transition to renewable energy or to organic food. Ocean governance is an amalgam of sectoral and spatial regimes, loosely assembled within—and beyond—the UNCLOS framework. However, as revealed in the sections above (and in Appendix C), as our understanding of regime dynamics and niche innovations improves, emergent change is unfolding. In brief, there is evidence in the ocean system that all four of these transitions are underway. A system-wide 'purposive transition' that builds on emergent regime responses to landscape pressures and transformative niche innovations is the most effective pathway to ocean sustainability. These Blue Papers have instigated the process of sourcing external knowledge that helps stakeholders to reimagine the future of the ocean. Key governments, business and civil society can now lead the way in developing the coordination capacity to manage a 'purposive transition' based on the accelerated learning emerging from the Blue Papers.

What follows is a framework for how a 'purposive transition' to a global ocean governance system can be imagined. A purposive transition suggests there is clear vision of the changes required and an agreed future pathway for bringing about these changes. It draws from the principles emerging from the above selected regime responses and niche innovations. Together, they provide the framework for ensuring that the key elements of a successful transition are put in place. A combination of a new legal framework, an 'ocean agency' and an ocean knowledge commons would bring into focus (1) the need to remove perverse incentives and promote incentives that reinforce the regime shifts and niche innovations that can catalyse a purposive transition; (2) knowledge-sharing that will be a precondition for nudging along the transition dynamics; (3) dialogues that inspire a new vision and compelling narrative; (4) the importance of a clear set of guiding principles that become the basis for new institutional configurations; and (5) the need to reinforce and empower early adopters and amplify best practices and niche innovations that reveal the alternative pathways towards a purposive transition.

7 Framing Transitions: Regulatory Lessons

The sustainable development of the ocean's economic potential will require a balance between, on the one hand, sufficient flexibility to meet constantly changing conditions and, on the other, regulatory structures that are sufficiently vigorous, strong and transparent to protect the planet for current and future generations (Pretlove and Blasiak 2018). Effective responses require international coordination and cooperation at unprecedented levels (TWI2050 2019). Multifaceted, integrated solutions can lead to new modes of stewardship (protecting, caring or responsibly using the environment) and social practices for managing the ocean as a commons. The World in 2050 report (TWI2050 2019) states that 'it is even more important now to integrate social and economic goals with climate, water, oceans, biodiversity and other Earth-systems so that sustainable development is not threatened in the long term'. Drawing on our understanding of the landscape pressures, regime dynamics and niche innovations, we propose a way of thinking about such a fundamental shift. In particular, we imagine a global governance transition that is more faithful to commons-oriented niche innovations.

Jessop (2002) argues that governing complexity means breaking away from linear modes of policymaking whereby problem analysis leads to policy solutions.

Instead, as our world becomes more complex, it helps to accept that 'governance failure is routine'. An adaptive, iterative process (i.e. learning and evolving through repetition) which reimagines a transition from the current, top-down nation-state structure towards a law for the commons will be more adaptable and sustainable (Bollier 2016). There is no one-size-fits-all solution.

Governance approaches that are diverse, tailored, innovative and adaptive, using science to support decision-making and develop early warning systems, are likely to be more sustainable (SDG 2019). Global collaboration is therefore essential, but on what terms?

Jessop (2011) recommends that governance reflect the context and align with stakeholders' concerns.

Social learning is crucial in order to understand drivers, attribute responsibility appropriately, understand the capacity for action and coordination in a changing, turbulent environment, and to activate change in decisions and activities. Finally, a common worldview is best for guiding action, and a system of meta-governance will help establish rules of conduct, as well as stakeholder orientations and expectations (Jessop 2011).

Meta-governance can be described as the governance of governance among interacting groups (Jessop 2011). These mechanisms could coalesce into a commons governance system which is based neither solely on the incremental logic of market forces, nor on top-down planning, but builds on these existing processes and interactive learning among a plurality of operationally autonomous but interacting agencies (Jessop 2011).

Bollier (2016) proposes that such a new sustainable system encompass three radical shifts in the current established system: (1) reconfiguring nation-state authority (along the lines suggested by Jessop)—as multiple governing bodies at different scales interact within a system of *networked governance*, they voluntarily learn and adapt followed by repetition (*adaptive voluntary governance*) in ways that that can eventually consolidate into new modes of *meta- governance*; (2) making communities sovereign by empowering the commons through rights and institutional capabilities for collectively managing knowledge and material resources; and (3) making ownership generative by integrating property rights (ownership or use) with stewardship responsibilities to ensure that the exercise of such rights incorporates environmental responsibility (Fig. 12.3). We have elected to use Bollier's conceptualisation for the purposes of this paper, although these shifts also are echoed in slightly different terms as necessary governance reforms in The World in 2050 report (TWI2050 2019).

While these shifts may seem far-fetched, in reality all three of these transitions are already underway in the ocean system. A closer look at an example of each in global governance will demonstrate the viability of these pathways to achieving sustainability in ocean governance, and illuminate the possible pathways into the future.

7.1 Reconfiguring Governance and Authority

7.1.1 Governance as Voluntary and Adaptive Learning

The traditional concept of governing through nation-state authority has begun to shift in several ways. One pathway is through voluntary commitments aimed at delivering outcome-oriented activities. Voluntary commitments have become a well-recognised mechanism in international sustainability policy (Neumann and Unger 2019). The UNFCCC Paris Agreement (UNFCCC 1992, 2015) demonstrates the possibility of a new model of international governance through a shift to a *volitional reflexive* approach (Box 12.5). This type of governance involves two aspects: (1) volitional or voluntary commitments aimed at delivering outcome-oriented activities and (2) reflexive governance where governance (the concepts, practices and institutions by which societal development is overseen) has the flexibility to adapt and adjust to include more appropriate alternatives over time, as a result of social learning. The Paris Agreement reflects this shift through two complementary mechanisms. First, state parties are legally obliged to comply with procedural commitments such as transparency reporting, but these commitments are combined with an element of volitional, non-binding obligations, allowing state parties to determine their own goals (or 'nationally determined contributions' [NDCs]) for measuring progress on meeting global climate targets (Pickering et al. 2018).

The obligations create a long-term framework for cooperation that aims to add momentum to the global response to climate change. At the same time, the volitional, softer layer underneath (the substance of NDCs) provides the flexibility needed to minimise barriers to universal participation (ones that would arise through rigid requirements) and to adjust contributions in the light of changes in scientific knowledge and shifts in complex social-ecological systems (i.e. in a reflexive manner as a result of learning over time) (Pickering et al. 2018). This style of governance therefore allows for dynamic real-time adjustments and flexible ecosystem-based responses. See also the example of the voluntary national review process under the UN SDGs (Appendix A3).

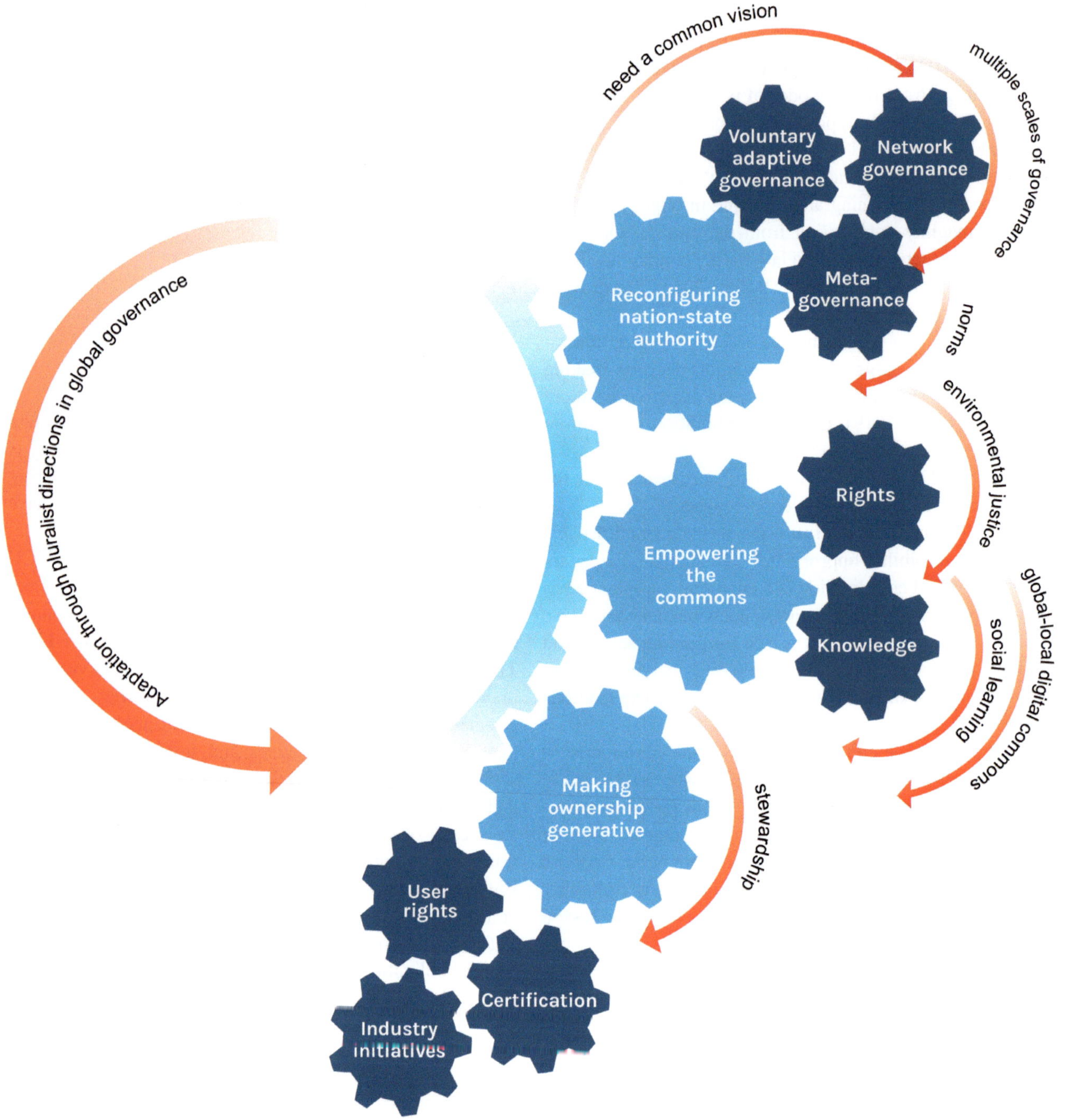

Fig. 12.3 Elements informing a transition to effective global ocean governance. Source: Authors. Conceptual elements drawn from Bollier (2016)

In ocean governance, two major international processes presently harness voluntary contributions. Voluntary commitments made under the banner of the Our Ocean Conference (OOC) series and the UN Ocean Conference provide opportunities to raise awareness, promote engagement and catalyse political will for action by states as well as the public and private sectors (Neumann and Unger 2019). In evaluating verifiable outcomes of voluntary commitments made at the OOC from 2014 to 2017, Grorud-Colvert et al. (2019) found that one-third of the announcements focused on marine protected areas (MPAs), and that almost half of the promised MPA actions were completed at the time of publication.

These voluntary commitments cumulatively amounted to over 5 million km^2 of protected area, encompassing 1.4% of the ocean, almost doubling the quantity of implemented MPAs worldwide (Grorud-Colvert et al. 2019).

During October 2019 at the sixth Our Ocean Conference, 370 commitments worth $63 billion were made to marine health and productivity (Evans 2019). The research conducted by Grorud-Colvert et al. (2019) has demonstrated the potency of these voluntary commitments. However, a uniform global process is still required to register and assess commitments, including consistent reporting and monitoring systems with clear targets, baselines and review systems (Neumann and Unger 2019).

7.1.2 Interacting, Multiple Centres of Governance at Different Scales

Another pathway to reconfiguring nation-state authority is through polycentric or network governance models (Ostrom 2010). Network-based modes of governance rely on the involvement of public, private and societal actors, and thus change the traditional top-down structure of political leadership (Sørensen 2006). This type of governance occurs across multiple scales, from transnational agreements, regional and national agreements and policies, down through individual municipalities, to the operations of public and private institutions and individuals (TWI2050 2018). Multiple centres of authority and distributions of power, which operate in complementary combinations, can address complexity more effectively than a single mode of governance (Pahl-Wostl 2017; Ostrom 1990; Dietz et al. 2003). The presence of these various forms of interacting decision-making bodies in a network governance structure is a core characteristic and requirement for sustainability transitions (Ottens and Edelenbos 2019). This type of network governance of the commons enables adaptation and mitigation through open, inclusive, pluralist directions in global governance (SDG 2019).

Box 12.5 Lessons from the Paris Agreement

Several innovative aspects of the Paris Agreement offer signposts for a new direction for global governance of the ocean, even though the resources and complex ecosystem services offered by the ocean commons are not as easily quantifiable as the carbon budget:

- The recognition that governments are only one set of the players in solving global commons problems. The Paris Agreement is a cooperative effort across sectors, including civil society, the private sector, financial institutions, cities and other sub-national authorities, local communities and indigenous peoples (Macy 2017).
- Instead of a top-down or bottom-up focus on narrow issue-based objectives, global goals inform policy directions. (This is also reflected in the UN 2030 Agenda.)
- Effective transparency mechanisms for reporting and verifying performance can replace or complement the need for compliance mechanisms and sanctions.
- Volitional reflexive commitments can provide the flexibility for evolution, for 'ratcheting up' commitments to reflect more ambitious targets, without requiring extensive time-consuming negotiations and trade-offs between state parties.
- Legal instruments for coordination of global action for commons problems need not be limited to treaties between states defining rights and obligations but can provide frameworks to facilitate and support action between governments and non-state actors (Macy 2017).

Network models, where multiple governing bodies interact to make and enforce rules within a specific policy arena or location, allow decision-makers to 'experiment with different governance solutions tailored to particular scales and socio-ecological contexts' (SDG 2019). They allow social learning, and importantly often include the involvement of directly affected local communities. Network governance may reinforce a system's ability to adapt structural elements and alter processes in response to current or anticipated changes in the social or natural environment (Pahl-Wostl 2017; Dietz et al. 2003; TWI2050 2019). Network governance structures have proliferated, the Chilean TURFs being a well-documented development in ocean governance (Box 12.2; see also description of U.S. Fishery Management

Councils in Hanna 1995). Many existing network arrangements for the ocean include elements of the second shift (i.e. making communities sovereign by empowering the commons) and third shift (i.e. making ownership generative by integrating property rights with stewardship commitment) underway as discussed below.

7.1.3 Meta-governance

The final piece of the transition in reconfiguring nation-state authority is the concept of meta-governance, or supranational governance. The Arctic Council and the Antarctic Treaty System are demonstrably important examples of supranational management (IPCC 2019). In some instances, multilateral agreements between states have successfully addressed commons issues.

The Montreal Protocol has, for example, successfully protected the ozone layer (the hole in Earth's ozone layer is the smallest in recorded history (Helfenstein et al. 2019) and is widely recognised as an example of effective protection of the global commons (Dietz et al. 2003). Regional cooperative agreements, implemented between states (such as the UN Shared Watercourse Agreement of 1997), have successfully created effective new modes of governance of shared water resources. Despite some successes, however, the nation-state political system is too fragmented, slow and rigid to manage the profound normative, societal, political and institutional changes that are required to implement integrated, multidimensional sustainable development agendas (TWI2050 2019).

A culture of global cooperation is required to develop multiple sustainable development pathways across scales (TWI2050 2019). A common worldview coupled with supranational standards could provide framework conditions for addressing issues at different scales, in response to changing needs, capacity and context (Jessop 2011; TWI2050 2019). Governing the trade-offs between different policy objectives which arise in multi-scalar, polycentric, adaptive governance models will be easier if meta-governance principles, such as transparency, accountability and inclusiveness, are in place (Weitz et al. 2017; UN ECOSOC 2018; TWI2050 2019). Some examples of ocean-related meta- governance instruments include the FAO 'Step-wise Guide for the Implementation of International Legal and Policy Instruments Related to Deep-Sea Fisheries and Biodiversity Conservation in Areas beyond National Jurisdiction' (FAO 2019b); the *Voluntary Guidelines for Securing Sustainable Small-Scale Fisheries in the Context of Food Security and Poverty Eradication* (FAO 2015); UNESCO's Man and the Biosphere programme (Bridgewater 2016); and several IMO guidelines, for example, the *Post-2015 Development Agenda* (IMO 2015) for the maritime transport sector, the *Guidelines for Safe and Environmentally Sound Ship Recycling* (IMO 2012), and the *Guidelines on Places of Refuge for Ships in Need of Assistance* (IMO (International Maritime Association) 2003).

7.2 Empowering the Commons

The second shift suggested by Bollier (2016) and the TWI report (and implicit in our assessment of the niche innovations) towards a new ecological order is making communities sovereign by empowering the commons. It is important to remember that the commons is both the resource of a defined community and the protocols, values and norms devised by the community to manage its resources (Bollier and Burns 2011). A commons in this sense is characterised by bottom-up participation, subjective responsibility, transparency and self-policing accountability (Bollier and Burns 2011). One way to achieve this is through the creation and protection of rights to the commons.

7.2.1 Rights to the Commons

Environmental obligations generally are resolved at an inter-state level (Boyle 2012), but if one considers the ocean as a global commons that is part of the global ecological system (IPCC 2019), then in legal terminology, the obligations inherent in concepts such as the sustainable use of natural resources, intergenerational equity and the common concern of humankind can be considered obligations owed to the international community as a whole (Boyle 2012; Weston and Bollier 2013a, b; Kotzé 2019). While commons have traditionally been held in trust by sovereign nations, or collaboratively managed through inter-state relationships, this has proved insufficient to protect the ocean commons (Dietz et al. 2003). A third way is now needed.

A human rights perspective provides a useful basis to ensure transnational environmental fairness and justice, because human rights are understood to permeate traditional sovereign boundaries (Robinson 2016; Weston and Bollier 2013b). The development of the human right to a decent or sound environment is a basis from which to empower the commons. Risks posed to human rights by climate change are significant (Robinson and Shine 2018). Global warming of 2 °C would, for example, impact the right to food and the right to an adequate standard of living. This raises a question of ethics (Robinson and Shine 2018): 'If the international community accepts that climate change is happening, understands its causes and knows what needs to be done to change course—how can it justify its continued delays to act on the scale, and with the urgency required?' Robinson and Shine (2018) suggest in response that rights-informed climate action can maximise benefits for people and the planet (Box 12.6).

The link between the environment and human rights has long been recognised (UN General Assembly 1972; OHCHR (Office of the High Commissioner for Human Rights) 1994;

OHCHR 1995, 1996; IUCN (International Union for the Conservation of Nature) 1995). Although the International Union for Conservation of Nature (IUCN) covenant was presented as a means of strengthening momentum for global action to implement the 2030 Agenda for Sustainable Development, it has not gained universal support or momentum. Despite this, in the 50 years which have passed since the Stockholm Declaration (UN General Assembly 1972), the human right to a healthy environment has been refined to include procedural rights (such as in the Convention on Access to Information, Participation in Decision-Making and Access to Justice (UNECE 1998)). Substantive elements have also been clarified in some constitutions (e.g. the right to clean air, safe water, adequate sanitation, healthy and sustainably produced food, healthy biodiversity and ecosystems, and a safe climate; see Boyd 2019, by the UN special rapporteur on human rights). In 2019, 130 states were party to regional treaties which incorporate a right to a healthy environment, and in over 110 states this right is constitutionally protected. In total at least 155 states recognise, in law, the right to a healthy environment (Boyd 2019).

The human right to a sound environment was once again brought to the fore by an initiative of the French legal think tank Club des Juristes, which in 2017 called for a Global Pact for the Environment; in May 2018, the UN General Assembly adopted a resolution opening negotiations to create the treaty (Club des Juristes 2018). The Global Pact codifies a human right to an ecologically sound environment and is designed to consolidate and integrate generally accepted but fragmented environmental norms and principles into one overarching, binding text. This way of thinking informs the work of the Ad Hoc Open-Ended UN Working Group towards a Global Pact for the Environment, which held its third session in May 2019.

Box 12.6 South African West Coast Rock Lobster

An example of rights-based environmental action demonstrates the utility of a human rights perspective for ocean resources. In terms of Section 24 of the Bill of Rights in the South African constitution, the environmental right comprises two parts: first, 'everyone has the right to an environment that is not harmful to their health or well-being' and, second, everyone has the right 'to have the environment protected, for present and future generations, through reasonable legislative and other measures that … secure ecologically sustainable development and use of natural resources while promoting justifiable economic and social development' (Republic of South Africa 1996, Chap. 2, Sec. 24). This right is mirrored in the South African environmental legislative framework.

On the basis of the foundation provided by this constitutional right, a landmark decision was handed down by the Western Cape High Court in 2018 in *WWF South Africa v. Minister of Agriculture, Forestry and Fisheries and Others* [2018], 4 All SA 889 (WCC). The West Coast rock lobster fishery is one of South Africa's oldest formal commercial fisheries, dating back to the late nineteenth century. It has also been a mainstay of poorer subsistence fisher communities. Catch peaked in the 1950s at around 18,000 tonnes but has declined sharply over the past few decades to an all-time low of 1500 tonnes in the 1995–1996 season. A scientific working group convened by the Department of Agriculture, Forestry and Fisheries called in 2016 for a total suspension of lobster fishing until a sustainable path for the resource was established (with compensation for stakeholders). The same working group recommended a total allowable catch (TAC) of less than 800 tonnes for the 2017–2018 season. Instead, the TAC for the 2017–2018 season was set at 1934 tonnes. The WWF sought to have this decision set aside, arguing that, given the depleted state of the resource, its harvest above prudent levels posed a threat of serious or irreversible environmental damage. The court set the TAC determination aside on the basis that it was unlawful, in contravention of the constitutional right, the subordinate legislation and South Africa's obligations in international law under UNCLOS. (Other principles were also relevant, including the precautionary principle and critical role of scientific analysis in the determination of the sustainability of a marine resource (Glazewski 2018)).

The constitutional environmental right in this matter made it possible for civil society (in this case, WWF) to ensure that the government department upheld the principles of sustainable development.

Despite some criticisms (Biniaz 2017), the reintroduction of the groundbreaking universal, justiciable individual human right to the environment opens the possibility of using the existing legal human rights framework for enforcement of the right to a decent environment. The Global Pact (or another similar agreement) could, if adopted, generate a shift in the collective understanding of legal norms and environmental rights in a similar fashion to what occurred in the human rights body of law as a result of the Universal Declaration of Human Rights (1948). This type of agreement provides a possible paradigm shift away from the conventional justification of social and economic development at the cost of the environment and embodies the potential for a new holistic environmental stewardship for the planet, based on human rights.

The relevance of adopting a human rights approach in the context of the environment was affirmed by the explicit recognition of human rights in the Paris Agreement (Robinson and Shine 2018). In 2015 the Dutch court in the Urgenda case[1] was prepared to venture into the uncertain territory of separation of powers in order to enforce a duty of care to meet greenhouse gas emission targets by the Dutch government. In June 2019 the Danish Institute for Human Rights launched strategic web-based tools it developed to operationalise the synergies between SDGs and human rights, namely, the Human Rights Guide to the SDGs and the SDG Human Rights Explorer (2020). The younger generation has been particularly vocal in asserting the right to intergenerational equity (UN News 2019; *La Rose et al. v. Her Majesty the Queen*, T-1750-19, Federal Court of Canada).

These developments illustrate that the nature of human rights is not fixed or static—human rights are changeable and relate to particular historical moments and social contexts. The right to a healthy environment could form a baseline 'net' for governance of the commons to address and redress the grave inequities suffered by individuals and communities exposed to environmental degradation and the unsustainable extraction of natural resources.[2] Human Rights Watch, in its World Report (Orellana 2018), stated that global recognition of the right to a sustainable environment is long overdue, as has Boyd (2019), the UN special rapporteur for human rights.

7.2.2 Knowledge Commons

Traditional economics was premised on the assumption that resources were unlimited and information scarce—the reverse is now true. The digital revolution has resulted in open information platforms otherwise known as the 'digital knowledge commons' or 'platform co-operativism'. This is a rapidly unfolding phenomenon that could be harnessed for the benefit of the ocean. A growing body of literature proposes that these new technologies provide the most powerful way to accelerate the empowerment of the commons by creating a new generation of trans- sovereign institutions that facilitate 'many-to-many' communications, including the reconfiguration of nation-state authority over shared commons that transcend national boundaries (see Bauwens et al. 2019). (The term *digital revolution* includes virtual and augmented reality, additive manufacturing (e.g. 3D printing), artificial intelligence, deep learning through open platforms, robotics, big data, the Internet of Things, and automated decision-making systems including crowd-sourced tracking and monitoring (TWI2050 2019; Leape et al. 2020)). Digitalisation hardly featured in the Paris Agreement or UN Agenda 2030, but it is increasingly clear that digital changes are becoming a key enabler of societal transformation (Domingos 2015; Schwab 2016; Tegmark 2017; Craglia et al. 2018; TWI2050 2019). It is predicted that by 2020 data generation will increase annually by 4300% (Sunderji 2016).

Bollier (2016) argues that there is enormous practical potential in developing a digital knowledge commons sector as a quasi-independent source of production and governance—a kind of 'fifth estate'. The World in 2050 (TWI2050 2019) states that 'digitalization is not only an "instrument" for resolving sustainability challenges, it is also a fundamental driver of disruptive, multiscalar change'.

The information and communication technology (ICT) revolution that emerged in the 1970s introduced the 'network' as an alternative to market- and hierarchical modes of organisation. Vast swathes of contemporary organisational and economic life are now organised in networks that have been hardwired into massive global 'many-to-many' platforms. ICTs made possible 'self-managed mass communication' as a new mode of sharing knowledge that was never before technically feasible (Castells 2009). Over the past two decades, these two modes of organisation and communication—networks and self-managed mass communication—have fused, resulting in increasingly complex, global interactive networks. Using the new ICTs for direct 'many-to-many' communications without transacting through a regulator or a market operator, a new 'peer-to-peer' (P2P) economy has emerged that can be configured in a wide variety of ways. The result is the rapid expansion of an information and knowledge commons, whether this is for private profit, as with Facebook, Uber and Airbnb, or for the public good, as with Wikipedia, GNU and Mozilla Firefox.

For Bauwens et al. (2019), the peer-to-peer mode of production becomes the basis for what they call a 'commons-centric society' or what others have called 'platform economies'. Bauwens et al. (2019) connect three aspects of this emergent alternative: (1) it creates conditions for a new mode of social relationships for learning, innovating and producing on a global scale; (2) it develops a technological infrastructure that makes scaling up possible through mutual coordination; and (3) it creates new property relations that mix shared ownership of the commons with private use for commercial gain.

The core of numerous types of for-profit and non-profit platform economies is clear: designs/data are loaded up in real time for collaborative co-production and optimisation, while users download applications from the knowledge com-

[1]Urgenda Foundation v. The State of the Netherlands. 2015 HAZA C/09/00456689 (June 24, 2015); aff'd (Oct. 9, 2018) (District Court of the Hague, and The Hague Court of Appeal (on appeal).

[2]The Oganiland case (Social and Economic Rights Action Centre v. Nigeria, 2001, AHRLR 60) demonstrates this point. The applicant would have had the additional armory of founding liability in a duty of care owed by the multinational corporations involved in the despoliation of the Niger delta. This extended duty of care would have been owed to the people of the region in addition to the state's duty to ensure its citizens' right to a satisfactory environment.

mons for use in their local or sectoral environments. Until now, this has never been possible before at scale (Bauwens et al. 2019; Leape et al. 2020).

Bauwens et al. (2019) show that each of these major global initiatives have three exemplary features (that are relevant for the purpose of imagining an ocean commons): (1) a 'productive community' of people who voluntarily create new and improve existing understanding in the commons; (2) an 'entrepreneurial coalition' that is licensed to exploit the understanding in the commons in the wider market, but with controls over the distribution of surplus; and (3) a 'for-benefit association' supported from the revenues generated to reinvest in the capabilities of the productive community and wider environment. These three become the potential organisational template for building up from below the ocean commons–based peer production.

Components of such a knowledge commons for the ocean exist, but there is much more work to do (Box 12.7; Leape et al. 2020).

An 'Ocean Knowledge Commons' would comprise its own 'productive community' (including scientists and a wide variety of other people who share a common interest in the future of the ocean); an 'entrepreneurial coalition' to manage the distribution of and access to knowledge—perhaps a Global Ocean Commons Institute (GOCI); and a 'for-benefit foundation' that raises funds to actively support and develop the 'productive community' and GOCI. To make it happen, 'transvestments' from the traditional grant-making, for-profit and public sectors will be required until the Ocean Commons Community (the ocean 'commoners', GOCI and for-benefit foundation) has its own autonomous capital base.

The most appropriate architecture for an Ocean Knowledge Commons could be a fusion of a global commons (e.g. Wikipedia) and a localised commons (e.g. the decentralised slow-food movement) in what Jose Ramos calls 'cosmo localism' (Fig. 12.4; Ramos 2019). In this model as adapted to the ocean, a 'wiki-type' global commons would be created for pooling (at least) two types of information: (1) crowd-sourced data (from, e.g., sensors on ships of various types) plus satellite data; and (2) alternative processes and arrangements (e.g. locally relevant approaches for restoring damaged mangroves or building sustainable fishing systems) that 'commoners' can collaboratively work on together, drawing from a range of local case experiences.

However, the localised commons would also be created as decentralised platforms for commoners from particular subregions to collaborate, both using designs downloaded from the global commons and generating new designs and monitoring data fed upwards into the global commons. The result would be an 'ocean cosmo localism' supported by the relevant generic institutional forms that have emerged at the global commons level, that is, what we referred to above as the GOCI and the for-benefit foundation, plus equivalents where necessary at the local level (that could be either local partners or local branches of the GOCI and the for-benefit foundation).

Box 12.7 Open-Source Data and Analytical Platforms for Ocean Decision-Making

Open-source data and analytical platforms envisioned under a knowledge commons already exist for components of the ocean system. These platforms are being used to design and improve content in a peer-to-peer sense, and also at multiple scales by decision-making communities such as those for small-scale fisheries and integrated coastal management (Costello et al. 2019). They also help collect global satellite remote-sensing data from the U.S. National Oceanic and Atmospheric Administration and the U.S. Geological Service (Leape et al. 2020). At the global scale, the Intergovernmental Science-Policy Platform on Biodiversity and Ecosystem Services (IPBES) harnessed an open-source global data and software platform (InVEST; see Sharp et al. 2020) and a networked community of scientists to model for the first time global changes in biodiversity, ecosystem services and the values to people under UN future scenarios, including in the ocean realm (Chaplin-Kramer et al. 2019). Open source data, models and interactive viewers for the IPBES global modelling platform (NatCap (Natural Capital Project) 2019) are catalysing discussions with multilateral institutions, governments and civil society leaders about how to standardise and improve data, analytics, and communication to diverse audiences such as those tracking SDG progress and impacts of nationally determined contributions under the UNFCCC. The same InVEST data and analytical platform are being used to drive integrated, multi-sectoral coastal development and disaster risk planning at national (e.g. see Box 12.1) and regional scales around the world (e.g. Arkema et al. 2015, 2019; Mandle et al. 2017; Silver et al. 2019; Wyatt et al. 2017) and China's national zoning for development (Ouyang et al. 2016). Progress towards integrative frameworks to connect these existing open data platforms is ongoing, mostly from the producer communities (e.g. Selig et al. 2018). Clear signals for priority policy needs, and engagement between global- and local-scale actors, can accelerate these nascent efforts.

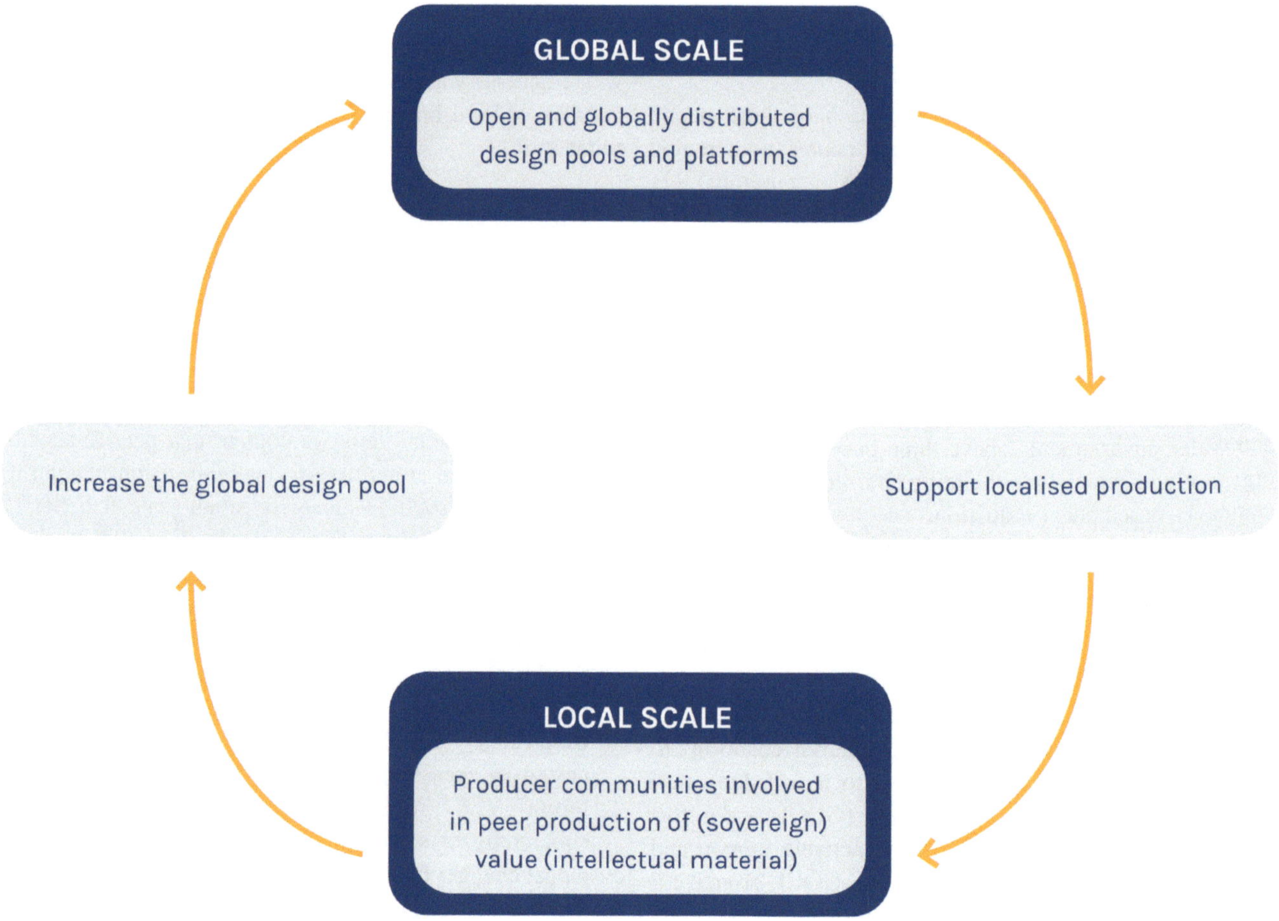

Fig. 12.4 Cosmo localism. Source: Derived from Ramos (2019)

7.3 Making Ownership Generative (Integrating Property Rights with Stewardship Commitment)

The third and final building block of a new ecological order as suggested by Bollier (2016) is to make ownership generative by integrating property rights with stewardship commitment. Bennett et al. (2018) see an urgent need and propose a framework to promote improved human-environment interactions through stewardship, defined as 'the suite of approaches, activities, behaviours, and technologies that are applied to protect, restore or sustainably use the environment'. The concept of integrating property rights with stewardship, embedded in this transition pathway, is already evident in the ocean economy. Fisheries management, for example, has seen a growing emphasis on the role, rights and responsibilities of small-scale fishers in stewarding local resources (Bennett et al. 2018). As discussed in Sect. 5, the integration of property rights with stewardship is happening currently, as exemplified in the case of SeaBOS (Box 12.3) and Chilean TURFs (Box 12.2; Appendix C).

In summary, the current governance regime, structured around nation-states and international treaty systems, is facing serious new pressures. The internet and digital technology have increased the velocity of transborder flows, not only in commerce but also in the exchange of ideas, values, projects, policy initiatives and visions for humanity, and these are catalysing revolutionary pressures from below (Bollier 2019). A new global polity activated and adapted by 'commoners' can light a pathway to transformational change (Bollier 2019).

The elemental components of the commons pathway (reconfiguring nation-state authority, empowering the commons, making ownership generative) are not talismans to preserve us from the pressures we are facing, but they do provide a means for us to rediscover the ancient wisdom that sovereignty ultimately resides not in the state or market but within ourselves, together (Bollier 2019).

8 The Ocean in a Transformed World: Towards a Governance Framework

Sustainability transitions are inherently political (Ottens and Edelenbos 2019). 'Precisely because politics plays a potentially powerful role (defining the landscape, propping up or destabilizing regimes, protecting or exposing niches), it requires explicit attention from those interested in understanding sustainability transitions' (Meadowcroft 2011). There is broad consensus that governmental steering solely through top-down decision-making or markets is not sufficient to address sustainability transitions (Meuleman 2020; Ottens and Edelenbos 2019; Meadowcroft 2011; Loorbach 2010).

The next step, therefore, is for the international community to recognise how a purposive transition can encompass a shared vision for a future pathway. As we have already argued, the transition to a thriving, vibrant and reciprocally rewarding relationship between humanity and ocean requires three fundamental elements that characterise any successful transition: (1) new knowledge drawn from beyond the current regimes; (2) the capacity to manage change in the context of rapid, unpredictable circumstances; and (3) incentives or other enabling conditions that activate the change. With regard to the first element, we have proposed a knowledge commons (a 'cosmo local digital commons', as discussed in Sect. 7.2.2). For the second, we propose the consolidation of a system of polycentric governance anchored by a new supranational ocean agency of some kind. The third element represents the Ocean Panel's catalytic role in the face of increasingly serious landscape pressures.

A key challenge woven into the substrate of ocean governance is that commons 'are either situated outside national jurisdiction or their conservation and sustainable use conflicts with national sovereignty and regulation' (Dasgupta et al. 2019). One of the methods which has evolved in the international arena to achieve this is the adaptive, flexible approach adopted in the Paris Agreement (Rajamani 2016). This approach provided a means to successfully negotiate a path through the sovereignty maze and secure broad agreement on comprehensive rules to address climate change. The BBNJ will, in a similar fashion, rely on volitional state responsibility for commitments, for example in relation to funding and transfer of technology. However, on its own, this is not likely to be sufficient.

We have also shown that in the ocean, local and regional realms are functionally connected with global and transnational scales (Bollier 2016). Thus, cross-sector coordination and multilevel network governance will facilitate a sustainable transition (Pahl-Wostl 2017; Tosun and Leininger 2017; Weitz et al. 2017; Leck et al. 2015; TWI2050 2019; SDG 2019). A purposive transition in ocean governance requires an evolving system of polycentric governance, including supranational policy and normative guidelines, flexible, adaptive governance and bottom-up stewardship.

To facilitate the workings of this system of top-down and bottom-up polycentric governance, we recommend two new institutional architectures that remain autonomous from one another. The first we have already proposed: a 'cosmo local ocean commons' to facilitate information-sharing and accelerated learning. Initiated by a coalition of knowledge institutions, this 'cosmo local commons' will require its own well-funded and dedicated set of autonomous institutions. The second is an ocean agency that will establish the norms, principles, 'rules of the game' and arbitration mechanisms for conflict resolution. What follows is a more detailed discussion of the latter.

Despite some successes in the international arena (e.g. the Montreal Protocol, the 1997 UN Shared Watercourse Agreement), the nation-state political system has not been able to manage the profound societal, political and institutional changes required for implementing an integrated ocean governance system to support the SDGs and other shared goals at the scale and with the urgency required. Building on some of the precedents and examples of experiments and regime shifts underway in many regions, it is clear that a polycentric system of governance for the ocean will be required that is neither centralised and top- down nor purely about the flourishing of bottom-up initiatives. Without a shared set of norms, values and operating rules, the bottom-up flourishing of commons initiatives will not lead to systemic impact. In order to create common future narratives, provide normative guardrails and negotiate the maze of trade-offs that will be required between different policy objectives, a supranational ocean agency of some kind (a form of meta-governance or transnational structure) could be considered. Similar to the way UNESCO's Man and the Biosphere programme works (but also learning the lessons from this experiment; Bridgewater 2016), this meta-governance institution could be mandated to establish 'rules of the game' for a polycentric system that empowers local actors to collaborate, with the goal of protecting and regenerating ocean commons at regional levels. This ocean agency could provide frameworks for addressing ocean-related challenges at different scales, in response to changing needs, capacity and context.

Furthermore, governing the trade-offs between different policy objectives which will arise in polycentric, adaptive governance models will be easier if meta-governance principles and structures, such as transparency, accountability and inclusivity, are in place.

The ocean agency could be created by UN resolution, or it could be created by a founding group of nations that invite others to participate. Its establishment should ensure legitimacy and safeguards against capture by special interests.

The ocean agency on its own will not be adequate. Its effectiveness will depend on the viability and vibrancy of the 'cosmo local' knowledge commons. Open and free entry into an ocean-oriented knowledge system would result in continuous, real-time information flows within a public space that cannot be controlled by a few dominant actors. Using satellite and other remotely observed metadata, crowd-sourced micro-data from ship-based sensors and uploaded modelling and design information, it will be possible to create wiki-type open databases that give regulatory watchdogs, research institutes, civil society organisations and industry bodies access to unprecedented levels of quality data that can be used to ensure and maintain maximum transparency (Leape et al. 2020). This, in turn, will encourage an entirely new generation of (potentially interlinked) observatories motivated by a desire to protect and regenerate the ocean commons. Given that most states lack the resources to build traditional closed information agencies to back up their regulatory functions, an open- source wiki-type global knowledge commons for the ocean would be the cheapest and most effective way of accessing design solutions that could catalyse local action.

9 Conclusion and Opportunities for Action

On the one hand, I believe it is vital to accept uncertainty, not-knowing, and unpredictability fully to the point of deep humility. On the other hand, I also believe that we need to choose to act from the conviction that we can design for positive emergence in complex systems even if it is not an exact science and we cannot know with certainty how our efforts will turn out to affect transformative change.
—Daniel Wahl, *A Brief History of Systems Science, Chaos and Complexity* (2019)

SDG 14 provides the global community with an opportunity to consider how to strengthen governance of the ocean. We have demonstrated that a range of transition dynamics are underway as existing regimes (fisheries, shipping, etc.) and niche innovations (mainly local level initiatives) respond to changing landscape pressures (climate change, depletion of fish stocks, pollution, etc.). Transition dynamics can be messy, and their legitimacy requires well-functioning science-policy engagement processes built on diverse stakeholder participation, trust and open discourse at multiple, interacting scales.

New insights into the complex interconnections among different ecosystems—on land and in rivers, deltas, estuaries, nearshore and in the ocean—have contributed to a growing realisation that a more holistic approach is needed to inform the design of policies and institutions across sectors and nations (Mathews et al. 2019).

Governance solutions for common pool resources such as the ocean that prioritise resource users' ability to devise livelihood strategies that restore rather than deplete ocean ecosystems (Ostrom 1990; Bavinck and Gupta 2014) can enable transition to a sustainable system.

Our conclusion is clear: national governments, the private sector and civil society have an opportunity to collaborate at the regional and international level to harness and give direction to the emergent regime shifts and niche innovations already contributing to a 'purposive transition'. Such a transition requires access to shared transdisciplinary knowledge, the build-up of capacity to implement changes and initiatives that activate change. This anticipatory perspective is supported by the evidence we have presented regarding regime and niche responses to increasingly serious land- and seascape pressures. Ours is neither a predictive approach based on a model nor a narrative approach that constructs scenarios. We have interpreted transitional dynamics at regime and niche levels as discerned from particular directions. Our approach is similar to one articulated eloquently in 1987 by the World Commission on Environment and Development, which in the conclusion of its Our Common Future report states, 'We have tried to point out some pathways to the future. But there is no substitute for the journey itself, and there is no alternative to the process by which we retain a capacity to respond to the experience it provides' (WCED 1987).

Leaders in public and private sectors can catalyse a transition to a more sustainable ocean system by harnessing and directing emerging niche innovations and regime shifts. Governance approaches that facilitate transformation and guide societal change in the face of uncertainty, and that are legitimate and fair, can lead to a transition in the ocean system when the overarching principles below are followed:

- Guide governance decisions by the ocean sustainability imperative.
- Integrate policies across sectors and spatial zones (addressing complexity).
  - Increase regulatory coordination across sectors and mandates.
  - Increase regulatory coordination among governing bodies.
- Use science to support decisions.

  - Adopt a precautionary approach in light of uncertainty, thereby shifting the burden of proof to the person or party who wishes to carry out the activity (rather than the person alleging damage to the environment).
  - Include explicit mechanisms and processes to base decisions on science and expert knowledge.
  - Consistently monitor and evaluate policies, actions and system responses.
- Create flexible frameworks for policymaking and governance decisions to facilitate responsiveness (including efficiency and reflexivity).
  - Include the expectation of change and surprise by building in provisions for periodic review and adaptive management.
  - Incorporate climate change adaptation exemptions into existing standards to avoid inefficient inflexibility.
- Establish a network approach to governance.
  - Widen the scope of participation (ensure transparent and authentic inclusivity of local and all stakeholder participation.
  - Establish networks of leadership and governance at different scales which allow for a distribution of power and decision-making capacity across scales of governance (legitimacy, inclusivity).
- Share information through an accessible knowledge commons available to everyone (transparency, accountability, social learning).
- Reinforce stringent monitoring and evaluation mechanisms, including transparency through compliance requirements (accountability).
- Foster equality and equity.
  - Incorporate environmental responsibility with property rights.
  - Protect human rights.
  - Examine the incentives that might help drive sustainable behaviours, and stop perverse incentives for maladaptive behaviours.
  - Balance long-term goals with short-term perspectives.

Based on these principles, we propose the following set of opportunities for action:

National governments, businesses and civil society have opportunities to support current UN ocean transition processes.

Examples of such opportunities include the following:

- Advocate for the ratification by non-party states of the UNCLOS agreement (especially the United States).
- Encourage the ratification, implementation and operationalisation, at the national level, of the BBNJ as soon as possible but by no later than 2025.
- Lobby for the ratification, implementation and localisation of the Global Pact for the Environment (or similar UN convention) as soon as possible, but no later than 2025.
- Support other UN initiatives such as the UN Environment Programme, Communities of Ocean Action, Ocean Conference Voluntary Commitments and so on.

Should there be agreement that a more sustainable ocean system is required, then national governments, businesses and civil society can consider the following opportunities for action to encourage the reconfiguration of nation- state authority vis-à-vis the ocean:

- Establish a new supranational 'ocean agency' of some kind to support polycentric governance, including transition processes and dynamics, the development of norms to guide the transition process and the design of flexible and adaptive frameworks which take account of local contextual issues. (Working task forces could provide pro forma frameworks for the different elements of a transition which draw from successful local niche innovations and regime responses, and all the lessons learnt from the Ocean Panel process; see Appendix D.)
- Strengthen voluntary learning and adaptation by improving coordination, monitoring and reporting on national voluntary commitments.
- Encourage nation-states to facilitate new modes of inclusive governance that are framed by an agreed- upon, general set of top-down principles but powered by bottom up decision-making on resource use. This could include supporting multilateral, local area– based and regional governance innovations through legislative frameworks and negotiated agreements which establish shared rules of engagement but are flexible and can accommodate rapid change.

If there is agreement that civil society and communities should play a more significant role in promoting the restoration and sustainability of the ocean commons, then we propose the following opportunities for action:

- Promote the global recognition of a human right to a sound environment (as per the global pact for the environment or similar instrument mentioned above).
- Invest in various capacity-building initiatives and incentives that help increase the involvement of a diversity of leaders in niche innovations at the local, regional and global scales, so they learn to develop and hold their visions and aspirations, and also develop the ability for generative dialogue.
- Use the advances in informational technologies now available to encourage creation of an ocean knowledge commons through mobilising the funds required to build a new open-source, 'wiki-type' ocean knowledge commons that collates crowd-sourced and satellite data, and creates a clearing house for shared strategies that amplify best practices and viable working alternatives;
  - ensuring that a global network emerges of research institutes, universities and knowledge organisations across the world's regions, all actively participating in the open-source knowledge commons; and
  - ensuring that the transparent open data-sharing platform consolidates all relevant knowledge and research as a basis for creating a system which pools and transmits information, and can facilitate the design of solutions capable of responding to changing landscape pressures and new transitional dynamics through diverse scales and institutions.

Should there be a shared commitment to a new form of ocean stewardship that explicitly prioritises the restoration and sustainability of the ocean ecosystem, then a key opportunity for action would be the integration of property rights with stewardship responsibilities through initiatives such as local user rights programs, certification of exemplary practices, and recognition of industry initiatives.

Acknowledgements The paper's technical reviewers, Jessica Blythe, Douglas Rader, Alejandro Robles Gonzalez and Niels Vestergaard, as well as its arbiter, Adrien Vincent, all provided helpful technical comments. Mary Ruckelshaus's time was supported in part by generous gifts from Peter and Helen Bing and Roger and Vicki Sant.

The authors also thank World Resources Institute for providing support as the Ocean Panel Secretariat.

While our colleagues were very generous with their time and input, this report reflects the views of the authors alone.

The authors thank Alex Martin for copyediting and Billie Kanfer for design. The authors are grateful for the artistic talents of Keith Atkin, who helped in the development of Fig. 12.3.

About the Authors

Lead Authors

Mark Swilling is Distinguished Professor of Sustainable Development and Co-director of the Centre for Complex Systems in Transition, Stellenbosch University. Email: swilling@sun.ac.za.

Mary Ruckelshaus is Managing Director of the Natural Capital Project and a Senior Research Scientist at Stanford University. Email: mary.ruckelshaus@stanford.edu.

Tanya Brodie Rudolph is a Research Fellow at the Centre for Complex Systems in Transition, Stellenbosch University, and Founder of Enviromer. Email: tanya@enviromer.co.za.

Contributing Authors

Edward H. Allison is Research Director of the Nippon Foundation Ocean Nexus, an Honorary Fellow at WorldFish, Penang, Malaysia; and an Honorary, Adjunct and Visiting Professor at, respectively, the Universities of Lancaster (UK), Washington (USA) and Wollongong (Australia).

Stefan Gelcich is an Associate Professor in the Ecology Department at the Center of Applied Ecology and Sustainability and the Centro para el Impacto Socio-económico de las Políticas Ambientales at Pontificia Universidad Católica de Chile. He is also an advisor and member of the board of directors of Global Green Grants Fund.

Philile Mbatha is a Lecturer in the Environmental and Geographical Science Department at the University of Cape Town.

Henrik Österblom is Professor and Science Director at the Stockholm Resilience Centre, Stockholm University.

Appendix A. Governance

UNCLOS Implementation Agreement: Biodiversity Beyond National Jurisdiction—Challenges and Opportunities

One of the risks facing the international instrument on biodiversity in areas beyond national jurisdiction (the BBNJ) is that the 'best' credible science may fail to foster ambitious progress because major stakeholders may question whether the science reflects their interests and concerns (legitimacy) and is presented at a time and in ways compatible with their policymaking context and constraints (salience) (de Santo et al. 2019). This problem could be addressed by ensuring that the scientific and technical body created by Article 49 of the draft BBNJ (A/CONF.232/2019/6) is an independent autonomous body, informed by a diversity of perspectives, including transnational science and citizen networks, with sufficient resources and an effective mandate.

There is some debate that a risk of fragmentation arises as a result of Article 4, according to which the BBNJ 'should not undermine the existing regime' (Mendenhall et al. 2019). Some states (and institutions) may have vested interests in maintaining the status quo fragmentation of governance in areas beyond national jurisdiction (ABNJs). However, institutional fragmentation could be reduced through harmonisation processes such as the Kobe Process, which was used to harmonise and increase efficiencies between five tuna-related regional fisheries agreements (de Santo et al. 2019). The nesting of the BBNJ within the architecture of the UN Convention on the Law of the Sea (UNCLOS) provides opportunities for harmonisation and synergy without increasing fragmentation and competition through wide participation and shared norms.

Given the high levels of attention to systemic issues, a shift in the framing of the BBNJ could better reflect UNCLOS's obligations on coastal states to prevent, reduce and control pollution, moving the focus from resource allocation to an agreement which expands the existing law of the sea framework to better ensure conservation and sustainable management of biodiversity beyond national jurisdiction.

This may be difficult to achieve, but it would overcome the fact that the suite of management measures currently on the table (e.g. marine protected areas, environmental impact assessments, benefit-sharing and technology transfer) do not integrate systemic issues (Mendenhall et al. 2019). Leary et al. (2019) state that even though this may prove difficult, reframing the agreement to include threats to biodiversity like climate change and marine plastics could increase the potential for issue linkages and help to counteract vested interests in fisheries, mining and commercial use of marine genetic resources (de Santo et al. 2019).

Case Study of Coastal Governance: Kosi Bay, South Africa

The sustainability of livelihood strategies that marginalised peoples have developed over the years largely depends on the nature of governance systems, institutions and policies that exist to govern land and resources (Agrawal and Perrin 2009). The same holds true for the governance of marine resources and coastal zones. South Africa's first UN Educational, Scientific and Cultural Organization (UNESCO) World Heritage Site, Simangaliso, was declared in 1999, within a context of coastal communities' historically marginalised access to the coast during apartheid. The site's planners had hoped for a more inclusive decision-making processes with respect to marine and coastal governance following apartheid. However, the injection of multilevel coastal conservation through the creation of the world heritage site in 1999 entrenched a plural system of governance embedded at international, national, provincial and local scales. In addition, Kosi Bay people have a long-standing customary fisheries governance system that predates colonial times. Although it is largely overlooked by statutory structures, this customary system plays a significant role in regulating small scale fisheries governance at the local level.

Despite this, governance processes in Simangaliso were and still are driven by UNESCO and by the state at the national level, and these are communicated poorly to actors at the local level, including wider communities (Mbatha 2018). A gap exists between stated policy objectives of sustainability, good governance and inclusiveness, on the one hand, and the lived realities of the people on the ground, on the other. This gap is widened by the fact that the institutional design that drives coastal conservation within

Simangaliso from the international to the local level allows little room for livelihood considerations to be a focus of governance practices within the Simangaliso site. Mechanisms to ensure effective interaction between higher-level governance actors and communities are not in place because of the lack of representation of communities in decision-making platforms.

It is unclear whether UNESCO was aware of the socio-economic and political challenges created by establishment of the site, and the negative livelihood impacts that ensued. Ongoing debates in Kosi Bay have centred on whether the community was effectively consulted about the declaration of the Simangaliso World Heritage Site; 95% of communities have stated that they were not consulted in the process for establishing the site (Mbatha 2018). The design of plural and multilevel institutions governing common pool resources (e.g. coastal and marine) influences the ability of resource users to devise livelihood strategies, as well as governance outcomes (Ostrom 1990; Ostrom and Janssen 2004; Cinner et al. 2012). This is because pluralism tends to exacerbate uncertainty, and 'in many countries, state laws are largely unknown in villages, and sometimes when new laws are promulgated, not only villagers but also government officials at the district or village levels are ignorant of the new laws' (Meinzen-Dick and Pradhan 2002, p. 13).

Governance Examples

Another form of volitional reflexive governance is the voluntary national review process (VNR), which lies at the heart of the global SDGs' follow-up and review mechanism set by the UN 2030 Agenda. Since 2016, 142 VNRs have been submitted to the UN High-Level Political Forum on Sustainable Development (HLPF), and 50 new reports are expected to be presented at its 2020 session (SDG Knowledge Hub, https://sdg.iisd.org/). The reports have given the international community a broad perspective on the status of the advance towards the SDGs at the national level. A Latin American think tank, Cepei (which provides analysis of global development agendas), has shown in recent research that current reporting is too static. Second-round VNRs should answer more reflexive questions like, 'How have we progressed since our previous report?' 'What worked well and what failed since then?' 'What have we learned on our way towards the SDGs at the national level?' and 'Where do we predict we will be in the short- and medium-term?' (SDG Knowledge Hub).

Cepei also suggested that second-round VNRs focus on integrating ministerial and sectoral silos, share successes and failures to facilitate learning, be democratised to encourage input from local voices and be more rigorously verified through the review process (SDG Knowledge Hub).

Appendix B. Regimes

	Shipping
Description	By 2017, 11.7 billion tonnes of cargo shipped across the ocean (UNCTAD 2018) More than 80% of global trade (IMO 2019) transported via this industry An increase of almost 40% in most trade segments (apart from crude oil and oil products) expected between 2016 and 2030, with a 2% annual rise for the period 2030–2050 (Gjølberg et al. 2017)
Stakeholders	Private sector governments Non-governmental organisations
Applicable governance/ institutions/mechanisms	UN Convention on the Law of the Sea (UNCLOS) and all supporting conventions and institutions (Pretlove and Blasiak 2018) including (but not limited to) • International Maritime Organization (IMO) • Agreement on Port State Measures, 1982 • International Convention for the Prevention of Pollution from Ships (MARPOL) • Liability conventions • Certification and classification schemes
Challenges	Energy consumption and emissions Recycling end-of-life ships Biofouling Pollution and discharges Flags of convenience Ports of convenience Accidents and damage to ecosystems
	Ocean-based food extraction
Description	Pillar of global nutrition (Pretlove and Blasiak 2018; Costello et al. 2019; FAO 2018) Capture fisheries and plant and animal mariculture provide nearly 80 mmt of edible food, people with 20% of their animal protein and critical micronutrients not found in land-based foods (long-chain omega-3 fatty acids) (Costello et al. 2019; FAO 2018) Global mariculture production expanded significantly; marine production now around 30 million tonnes annually (Pretlove and Blasiak 2018) Significant further expansion in aquaculture and mariculture is expected
Stakeholders	Food security for humanity Parties to UNCLOS and subordinate conventions regional fisheries management organisations (RFMOs) private sector Small fisheries Coastal communities
Applicable governance/ institutions/mechanisms	Both sovereign and commons space (in areas beyond national jurisdiction [ABNJs]) from a spatial/zonal perspective, and a species perspective, given the nature of fish stock UNCLOS and all supporting conventions and institutions (see Pretlove and Blasiak 2018) including (but not limited to) the following: • UN Fish Stocks Agreement 2001 • National legislation • FAO (Codes of Conduct, Guidelines for Small-Scale Fisheries, Guidelines for Fisheries in ABNJs) • Mariculture regulation is complex; involves interlinking regulatory bodies (e.g. spatial planning, regional planning, environmental agencies, food safety)
Challenges	A sustainable and efficient system is required to maintain food security (given the anticipated population increase by 2050) Over-harvesting Indirect impacts from bycatch Illegal, unreported and unregulated (IUU) fishing: $23.5 billion each year; up to 30% of total catch, and for one in five fish transacted in markets (FAO 2018; Global Fishing Watch 2018; Widjaja et al. 2020) Monitoring and enforcement (Costello et al. 2019; Aburto-Oropeza et al. 2020; Widjaja et al. 2020) conservation of biodiversity and ecosystem services while increasing production Social justice: uneven distribution of the fishing enterprise impacts on small-scale fisheries (Costello et al. 2019; Österblom et al. 2020) Management and monitoring of RFMOs Appropriate regulatory frameworks needed to address competing interests and overlapping man-dates in mariculture
	Offshore oil and gas
Description	Extraction of fossil fuels increasingly moving to deeper waters of exclusive economic zones (EEZs); under state authority Beginning to see exploration in ABNJs
Stakeholders	Private sector governments Financial system investors

	Shipping
Applicable governance/ institutions/mechanisms	Activities take place on the continental shelves of coastal states under domestic legislation If exploration and production moves into the area, activities would be regulated by the International Seabed Authority (ISA) Private sector regulations and best practices (e.g. global oil and gas industry association IPIECA (http://www.ipieca.org/))
Challenges	Cross-border jurisdictional issues such as baseline data collection and monitoring, transboundary oil spill response/planning Weak ISA regulation and authority in ABNJs High ecological and climate risks in deeper, pristine ocean habitats (OECD 2016) Global disinvestment campaign targeting the $5 trillion subsidy to oil industry Decarbonisation commitments by financial investors could increase risk of stranded assets (McGlade and Ekins 2015)
	Ports
Description	Port expansion is increasing globally (number and size of ships) Increasingly crowded ocean makes shipping lane designations critical
Stakeholders	Private sector governments regions Consumers
Applicable governance/ institutions/mechanisms	UNCLOS and supporting conventions, including (but not limited to) the following: • Agreement on port state measures, 1982 • MARPOL • Merchant shipping (minimum standards) convention, 1976 (ILO Convention No. 147) • Regional efforts such as European Commission (2007, 2011) • Regional port state memoranda of agreement • IMO code of good practice for port state control (for more detail, see Addo et al. forthcoming)
Challenges	Pollution of environment by ships, port activities Shipping lanes stress species, habitats Ports of convenience give rise to compliance and enforcement issues in IUU fishing and other illegal activities
	Marine and coastal tourism
Description	Second-fastest growing sector of ocean economy (OECD 2016) relies on ocean resources which it often depletes or impacts
Stakeholders	Private sector governments consumers
Applicable governance/ institutions/mechanisms	Emerging financial instruments (see Sumaila et al. forthcoming) to address vulnerabilities integrated ocean management (IOM) approaches (Widjaja et al. 2020; Winther et al. 2020) civil society advocacy
Challenges	Balancing need for tourism infrastructure and risk to species and ecosystem services Generally not monitored or regulated via internationally accepted certifications
	Marine and seabed mining
Description	Retrieval of mineral resources, either on continental shelves or on the deep sea bed, e.g. polymetallic nodules, polymetallic sulphides or seafloor massive sulphides (Aldred 2019) Exploration zones comprising 1.5 million km^2 are mainly in the Pacific, mid-Atlantic and Indian Oceans, in the ABNJs Rich deposits of rare earth minerals in deep ocean being discovered (e.g. samples extracted 500 km from the Canary Islands revealed deposits of the scarce substance tellurium—used in solar panels—in concentrations 50,000 times higher than in deposits on land)
Stakeholders	Private sector governments civil society
Applicable governance/ institutions/mechanisms	In ABNJs governed by the ISA (under UNCLOS) ISA Council in 2019 addressed regulations for financial models for mining of polymetallic nodules ISA regulations adding an additional scoping stage to the environmental assessment process and providing financial incentives for companies to participate in environment assessment and reporting
Challenges	Depleting terrestrial deposits of rare earth metals, coupled with rising demand for metals for smartphones and green technologies, has resulted in a surge of interest in deep sea mining (Cuyvers et al. 2018) Biophysical impacts of deep sea mining can be significant (IUCN (International Union for the Conservation of Nature) 1995) Many countries lack regulations or capacity to enforce regulations in their EEZs (Pretlove and Blasiak 2018) United States have not ratified UNCLOS
	Maritime biotechnology
Description	The creation of products and processes from marine organisms in these ecosystems, through the application of tools in biotechnology, molecular and cell biology, and bioinformatics Potential for new pharmaceutical drugs, chemical products, enzymes, advancement of aquaculture and seafood safety, bioremediation, biofuels, etc.
Stakeholders	Private sector governments civil society
Applicable governance/ institutions/mechanisms	Developments in governance underway to address some of the legal challenges which arise (see Blasiak et al. 2020; Sect. 3.1 and Appendix A1 above in relation to the international instrument on biodiversity in areas beyond national jurisdiction [BBNJ])

	Shipping
Challenges	Legal challenges for ownership of material derived from ABNJs (see notes on BBNJ) Technological, ecological and other knowledge barriers
	Cabling and maritime equipment
Description	Submarine cable network provides over 95% of international telecommunications and is the 'backbone' of the internet (Davenport 2015) Numbers and extent of submarine cables will increase drastically in coming decades as more islands and archipelagos are connected and renewable energy projects such as offshore wind farms, tidal and wave turbines are developed
Stakeholders	Private sector governments civil society
Applicable governance/ institutions/mechanisms	UNCLOS (rights and obligations of states for protection of submarine cables and the freedom to lay, repair and maintain such cables)
Challenges	Current gaps in legal regime around cybersecurity, counterintelligence and environmental impacts Environmental impacts include noise, pollution, physical disturbance, electromagnetic fields, heat, entanglement risk, pollution and threats to benthic reefs and reserves (Taormina et al. 2018)
	Offshore renewable energy (ORE)
Description	Wave, tidal and offshore wind energy generation—stationary installations in EEZs New technologies in development produce ORE through other processes, including by salinity gradients and thermal gradients
Stakeholders	Private sector governments intergovernmental bodies civil society Institutions (e.g. International Renewable Energy Agency, International Energy Agency, WindEurope, Ocean Energy Europe)
Applicable governance/ institutions/mechanisms	Domestic legislation Private sector standards, certification schemes, guidelines
Challenges	Growing scale and deployment expansion will push the technology into areas of both scientific and engineering uncertainty Environmental impacts Need for data and information streamlining to meet demand (Veers et al. 2019)

Appendix C. Case Studies of Niche Innovations

Integrated Ocean Management for Development Planning, Fisheries Management and Disaster Risk Management in Belize

Sustainability transformations call for cross-sectoral thinking and approaches. Sectoral policies and measures can be effective in particular contexts but often fail to account for indirect, distant and cumulative impacts, which can have adverse effects, including exacerbating inequalities. Cross-sectoral approaches, including ecosystem-based management approaches, integrated watershed and coastal zone management, and area- based and marine spatial planning, offer opportunities to reconcile multiple interests, values and forms of resource use, provided that these cross-sectoral approaches recognise trade-offs and uneven power relations among stakeholders (from IPBES 2019; Winther et al. 2020).

The final Integrated Coastal Zone Management Plan (ICZMP) approved by the Belizean government (Box 12.1) coordinates the management of, and investment in, a diverse set of activities and actors implicated in sustainable outcomes for the nation, ranging from those engaging in or affecting coastal pollution, dredging, fisheries, aquaculture and tourism development, to education, social resilience to climate change and preservation of cultural heritage.

The fishery sector in Belize has in parallel adopted a combination of secure fishing rights and a locally controlled 'managed access' approach through which small-scale fishers are given licenses to fish in and manage specific geographic areas through a territorial use right for fishing (TURF). The managed access fishery approach in Belize transitioned from a few pilots with positive fishery returns to a national scale in 2016 (Fujita et al. 2017), so the full economic and ecological impacts are not yet clear (Fujita et al. 2018). Surveys of fishers participating in the new Belizean managed access program emphasise the importance of government enforcement and response, which will encourage fishers to comply with required surveillance, which comes at a cost with as yet unproven economic returns (Wade et al. 2019). Furthermore, integrating fishery TURF locations with the coastal development planning zones and protected areas identified through the Integrated Coastal Zone Management Authority and Institute (CZMAI) could improve fishery returns and thus livelihood security, providing positive feedbacks to small-scale fishing communities (Arkema et al. 2019).

The institutional, legal and science-policy engagement innovations exemplified in the Belize case are transferable to anywhere in the world. Promoting co-existence and synergies among ocean uses is a key issue for spatial management. Area-based and 'ridge-to-reef' management approaches for managing social- ecological systems in an integrated way have shown the value and broad relevance of cross-sectoral spatial planning in marine coastal zones, ranging from the Great Barrier Reef Marine Park in Australia (e.g. Day 2002; Fernandes et al. 2005; Olsson et al. 2008; Day 2017) to the *ahupua'a* (ridge-to-reef) system in Hawai'i and the concept of *vanua* in Fiji (Minerbi 1999; Johannes 2002; McGregor et al. 2003; IPBES 2019), to spatial planning under the European Union's Maritime Spatial Planning Directive (e.g. EU (European Union) 2014; de Grunt et al. 2018). Sale et al. (2014) suggest that expanded use of marine spatial planning could provide a framework for 'more effective, pragmatic management based on ocean zones to accommodate conflicting uses'. Establishing boundaries for resources and those allowed to use them could enable the separation of incompatible uses and give rise to governance systems that effectively address the commons dilemma (Sale et al. 2014).

Aware of the rising costs of disaster risk management and recovery under changing climate regimes, governments, multilateral development banks and businesses are beginning to turn their attention to nature-based solutions that provide greater resilience to impacts from sea level rise and increasingly intense coastal storms. The UN Global Assessment on Disaster Risk Reduction (UNDRR 2019) is tracking country commitments to integrate DRR in development planning and budgeting. Encouraging examples are emerging in Indonesia, Vietnam, Fiji and the Philippines, where DRR is being actively integrated with development planning policies, programs, capacity building and financial resources. More broadly, as in Belize, the growing number of ministries of marine affairs (e.g. Indonesia) or of the 'blue economy' (Barbados, Kenya, Seychelles) points to cross-sectoral integration.

Food production sectors, through fisheries and aquaculture, put major demands on marine and coastal regions. Although aquaculture could address the gap between aquatic food demand and supply, realising this potential will depend in part on the availability of suitable space. Integrated spatial planning for aquaculture and other uses is fundamental to the sustainable development of aquaculture in a way that accommodates the needs of competing economic sectors, minimises conflict and integrates social, economic and environmental objectives (FAO 2018).

Similarly, for fishery management, global guidance is available to ensure that area-based management, including the consideration of marine protected areas, is integrated within broader fisheries management frameworks and follows good practices with regard to participatory approaches, especially for small-scale fisheries. Both the *Voluntary Guidelines for Securing Sustainable Small-Scale Fisheries in the Context of Food Security and Poverty Eradication* (FAO 2015) and the *Voluntary Guidelines on Responsible Governance of Tenure of Land, Fisheries and Forests in the Context of National Food Security* (FAO (Food and Agriculture Organization of the United Nations) 2012) describe such practices and outline, among other things, the need to respect customary and informal tenure rights.

Rights-Based Fisheries Management

Rights-based fisheries management (RBFM) can represent a significant regime shift as institutions, regulations and community engagement adapt over time (see Costello et al. 2019 for a more detailed treatment). The territorial user rights fisheries (TURFs) of Chile, described in Box 12.2, are a well-studied example of RBFM that has been able to adapt over time. Successful local governance supported by recognition of local rights has often incorporated knowledge of how nature contributes to human well-being to motivate such behaviours (IPBES 2019). Recent studies have shown a positive relationship between leadership, social capital and sustainable fisheries outcomes (Gutiérrez et al. 2012). Results demonstrate the critical importance of prominent community leaders and robust social capital, combined with clear incentives through catch shares or other rights-based mechanisms, and conservation benefits derived from protected areas, to successfully managing aquatic resources and securing the livelihoods of communities depending on them (Gutiérrez et al. 2012). The Belize TURF/CZMAI example (discussed above), U.S. fishery reforms (Lubchenco et al. 2016) and a number of other cases provide compelling examples of how RBFM, in combination with other management and social capital elements, can lead to improved ecosystem conditions and livelihood support. Modelled fisheries using data from nearly 5000 stocks worldwide indicate that RBFM can lead to catch, biomass and profit increases for diverse fisheries (Costello et al. 2016). Co-management is considered by Gutiérrez et al. (2012) to be the only realistic solution for the majority of the world's fisheries, one that can solve many of the problems they face. Yet in some cases, RBFM interventions alone will not be sufficient.

For example, economic incentives for stewardship and asking fishers to join participatory processes are unlikely to work if economic insecurity is high and government or community capacity to enforce or respond is low. If more fundamental social and political development interventions are implemented first, based on key incentives that will motivate fishing people, then RBFM approaches are more likely to succeed (Allison et al. 2012; Wade et al. 2019).

In order to enhance governability to benefit small-scale fisheries, governance designs and interactions must be sensitive to the needs and contexts of small-scale fishing people. In a review of alternative governing modes for small-scale fisheries, Jentoft and Chuenpagdee (2015) find that small-scale fisheries globally will benefit from more constructive interaction, collective action, policy and market innovation, and empowerment but that generalised governance principles are not likely. The transition of governing modes observed in many cases illustrates how governance actors try to cope with system dynamics. Often, the combination of different modes into one coherent but hybrid approach is warranted.

To support these commitments to sustainable small- scale fisheries development, it is crucial to better develop the understanding and knowledge base about small-scale fishing enterprises. Several initiatives are underway to improve and expand existing empirical information and to quantify the importance of the marine and inland small-scale fisheries sector (e.g. World Bank 2012). Bennett et al. (2018) identify additional fisheries examples, such as the rise of community supported fisheries programs globally (Brinson et al. 2011; McClenachan et al. 2014), the release of the global *Voluntary Guidelines for Securing Sustainable Small- Scale Fisheries* (FAO 2015) and increased funding of non-governmental organisation programs that focus on small-scale fisheries (e.g. the Fish Forever Program (Barner et al. 2015; Bennett et al. 2018)).

Monitoring Innovations for Illegal, Unreported and Unregulated Fisheries

One of the key challenges in countering illegal, unreported and unregulated (IUU) fishing historically has been the limited capacity of coastal states to monitor the vast swathes of ocean comprising their EEZs, as well as areas of the high seas. The modus operandi of IUU fishing has been to fish inside or hover at the edge of EEZs (where the majority of fish stocks are found) with smaller vessels, which then drop their catches into larger 'mother ship' refrigeration vessels waiting in the high seas, beyond the reach of national jurisdiction. The rapid development of vessel monitoring systems and automatic identification systems over the past several years provides new possibilities for the reduction of IUU fishing given that many open platforms now exist which provide real-time tracking of vessels (Widjaja et al. 2020). The ability to identify locations allows enforcement responses to be accurately and effectively focused.

Justice in Marine Sustainability

Several developments indicate a move towards social justice in marine sustainability thinking. Eco-labelling, certification schemes and supply chain transparency have, for example, given rise to the concept of socially responsible seafood, evident in the common use of mobile apps which show the full supply chain (such as Monterey Bay Aquarium's Seafood Watch app).

Cooperative fishing arrangements such as the Parties to the Nauru Agreement (PNA) in the Pacific islands are being developed to equitably share fishing benefits.

The PNA governs the annual fishing effort of skip-jack tuna. This highly migratory species moves between the EEZs of the island countries and also in response to climate (Lehodey et al. 1997). The PNA's 'vessel day scheme' facilitates cooperative management of these species within the combined EEZs of the PNA members (Aqorau et al. 2018). A capped fishing effort of vessel days is shared among members, allowing a trading scheme which enables responses to extreme weather events and migratory patterns. As a result, all members receive revenue regardless of where fish are caught, and stock has remained robust under this cooperative management arrangement (Aqorau et al. 2018). An example of this in the South African context is Abalobi—an innovative information and communication technology tool (mobile app) that is playing a significant role in improving small-scale fisheries governance in South Africa. It does this by addressing social justice and access issues faced by small-scale fishers within the sector, while assisting the government in improving catch data monitoring by accessing catch data of small- scale fishermen. Abalobi (2019) promotes traceable and 'storied' seafood that encourages ecological sustainability as well as social justice.

Appendix D. Potential Governance Functions of a Supranational Ocean Agency

1. Draft the flexible frameworks for policymakers and lawmakers, animated by commons norms discussed above and building on existing mandates and initiatives (UNESCO, UNEP, FAO, ILO, UNFCCC, etc.). Frameworks could be designed in such a way that law and policy can be applied at local levels and adapted according to rapidly changing needs, capacity and context, as well as guide the negotiation of trade-offs and realisation of co-benefits. This could result in appropriate combinations of decision analysis, land and ocean use planning, public participation and a science- policy process, diverse knowledge systems and conflict resolution approaches (IPCC 2019, C3.3). In turn, this would help reduce short-term risks, build long-term resilience and sustainability, and facilitate capacity building. Framework policy guidelines could be crafted with temporal awareness. Many coastal and ocean decisions now being made have time horizons of decades to over a century, far longer than the lifespan of the governance arrangements facilitating them (IPCC 2019, C3.4).

2. Coordinate measurable volitional commitments by stakeholders, such as commitments to marine protected areas and biodiversity targets in the international instrument on biodiversity in areas beyond national jurisdiction (BBNJ), aspects of the voluntary national review process related to Sustainable Development Goal 14, voluntary commitments under the Our Ocean Conference series and the UN Ocean Conference
3. Provide a monitoring function to ensure transparency, compliance and accountability of ocean commitments made in international processes. Without some form of agreed transnational accountability, voluntary governance based on volitional commitments lacks gravity and certainty (SDG 16: Develop effective, accountable and transparent institutions at all levels). A shared knowledge commons (see Sect. 7.2.2) will facilitate this accountability
4. Facilitate social learning, social innovation and reflexive, adaptive governance responses through the creation of an overarching legal architecture for the ocean transition sufficiently flexible to allow it to respond to rapid change but sufficiently robust to provide a cohesive framework for the implementation of selected transition pathways. This is the type of governance that can create conditions for mutual learning and coordination
5. Provide a venue for co-construction of sustainable ocean narratives that includes existing agencies and institutions (SDG Target 16.7: Ensure responsive, inclusive, participatory and representative decision-making at all levels; SDG 17: Strengthen the means of implementation and revitalise the global partnership for sustainable development)

References

Abalobi (2019) ICT4Fisheries. http://abalobi.info

Aburto-Oropeza O, Rogers A et al (2020) Critical habitats and biodiversity: inventory, thresholds and governance. World Resources Institute, Washington. https://oceanpanel.org/blue-papers/critical-habitats-and-biodiversity-inventory-thresholds-and-governance

Addo KA, Steven A, Thanh CV et al (Forthcoming) Coastal development: managing resilience, restoration and infrastructure of coastlines. World Resources Institute, Washington. https://oceanpanel.org/blue-papers/coastal-development-managing-resilience-restoration-and-infrastructure-coastlines

Agrawal A, Perrin N (2009) Climate adaptation, local institutions and rural livelihoods. In: Adger WN, Lorenzoni I, O'Brien KL (eds) Chapter 22 of adapting to climate change: thresholds, values, governance. Cambridge University Press, Cambridge, pp 350–367

Aldred J (2019) The future of deep seabed mining. DSM Observer. http://dsmobserver.com/2019/03/the-future-of-deep-seabed-mining/

Allison EH, Ratner BD, Åsgård B, Willmann R, Pomeroy R, Kurien J (2012) Rights-based fisheries governance: from fishing rights to human rights. Fish Fish 13:14–29

Antrim L, Balthis L, Cooksey C (2018) Submarine cables in olympic coast national marine sanctuary: history, impact, and management lessons. In: Marine sanctuaries conservation series ONMS-18-01. U.S. Department of Commerce, National Oceanic and Atmospheric Administration, Office of National Marine Sanctuaries, Silver Spring

Aqorau T, Bell J, Kittinger JN (2018) Good governance for migratory species. Science 361:1208–1209

Arkema KK, Ruckelshaus M (2017) Transdisciplinary research for conservation and sustainable development planning in the Caribbean. Conserv Anthropocene Ocean. https://doi.org/10.1016/b978-0-12-805375-1.00016-7

Arkema KK et al (2015) Embedding ecosystem services in coastal planning leads to better outcomes for people and nature. Proc Natl Acad Sci 112(24):7390–7395. https://doi.org/10.1073/pnas.1406483112

Arkema KK et al (2019) Integrating fisheries management into sustainable development planning. Ecol Soc 24(2):1. https://doi.org/10.5751/es-10630-240201

Bailey RM, van der Grient JMA (2020) OSIRIS: a model for integrating the effects of multiple stressors on marine ecosystems. J Theor Biol 493:110211. https://doi.org/10.1016/j.jtbi.2020.110211

Baker E et al (2019) Oceans and coasts. Chapter 7 of GEO-6: healthy planet, healthy people. Cambridge University Press, Cambridge, pp 174–199

Barner A, Lubchenco J, Costello C, Gaines S, Leland A, Jenks B, Murawski S et al (2015) Solutions for recovering and sustaining the bounty of the ocean: combining fishery reforms, rights-based fisheries management, and marine reserves. Oceanography 28(2):252–263. www.jstor.org/stable/24861886

Bauwens M et al (2019) Peer to peer: the commons manifesto. University of Westminster Press, London. https://doi.org/10.16997/book33

Bavinck M, Gupta J (2014) Legal pluralism in aquatic regimes: a challenge for governance. Curr Opin Environ Sustain 11:78–85. https://doi.org/10.1016/j.cosust.2014.10.003

Bennett N et al (2018) Environmental stewardship: a conceptual review and analytical framework. Center Open Sci Environ Manage 61:597–614

Bigagli E (2016) The international legal framework for the management of the global oceans social-ecological system. Mar Policy 68:155–164. https://doi.org/10.1016/j.marpol.2016.03.005

Biniaz S (2017) 10 questions to ask about the 'proposed pact for the global environment'. Sabin Center for Climate Change Law, Columbia Law School. http://columbiaclimatelaw.com/files/2017/08/Biniaz-2017-08-Global-Pact-for-the-Environment.pdf

Blasiak R, Thambisetty S, Grorud-Colvert K, Wynberg R et al (2020) The ocean genome: conservation and the fair, equitable and sustainable use of marine genetic resources. World Resources Institute, Washington. https://oceanpanel.org/blue-papers/ocean-genome-conservation-and-fair-equitable-and-sustainable-use-marine-genetic

Bollier D (2016) Digital commons as a new species of production and governance. P2P Foundation. https://blog.p2pfoundation.net/transnational-republics-of-commoning-2-new-forms-of-network-based-governance/2016/09/16

Bollier D (2019) Transnational republics of commoning: reinventing governance through emergent networking. Creative Commons Attribution–ShareAlike 4.0 International. https://creativecommons.org/licenses/b-sa/4.0/legalco-de; https://blog.p2pfoundation.net/transnational-republics-of-commoning-2-new-forms-of-network-based-governance/2016/09/16

Bollier D, Burns BW (2011) Regenerating the human right to a clean and healthy environment in the commons renaissance. http://webcache.googleusercontent.com/search?q=cache; http://commonslawproject.org/

Boyd DR (2019) Safe climate: a report of the special rapporteur on human rights and the environment. UN Human Rights Special Procedures. https://wedocs.unep.org/bitstream/handle/20.500.11822/30158/Safe_Climate_Report.pdf?sequence=1&isAllowed=y

Boyle A (2012) Human rights and the environment: where next? Eur J Int Law 23(3):613–642

Bremer S, Glavovic B (2013) Mobilizing knowledge for coastal governance: re-framing the science-policy interface for integrated coastal management. Coast Manag 41(1):39–56. https://doi.org/10.1080/08920753.2012.749751

Bridgewater P (2016) The man and biosphere programme of UNESCO: rambunctious child of the sixties, but was the promise fulfiled? Curr Opin Environ Sustain 19:1–6. https://doi.org/10.1016/j.cosust.2015.08.009

Brinson A, Min-Yang L, Rountree B (2011) Direct marketing strategies: the rise of community supported fishery programs. Mar Policy 35(4):542–548. https://doi.org/10.1016/j.marpol.2011.01.014

Brockington D, Wilkie D (2015) Protected areas and poverty. Philos Trans R Soc B 370(1681):20140271. https://doi.org/10.1098/rstb.2014.0271

Cantasano N et al (2017) Integrated coastal zone management in Italy: a gap between science and policy. J Coast Conserv 21(3):317–325. https://doi.org/10.1007/s11852-016-0479-z

Castells M (2009) Communication power. Oxford University Press, Oxford

Chaplin-Kramer R et al (2019) Global modeling of nature's contributions to people. Science 366(6462):255–258. https://doi.org/10.1126/science.aaw3372

Cheung W et al (2010) Large-scale redistribution of maximum fisheries catch potential in the global ocean under climate change. Glob Chang Biol 16(1):24–35. https://doi.org/10.1111/j.1365--2486.2009.01995.x

Cinner J, Daw T, McClanahan T, Muthiga N, Abunge D et al (2012) Transitions toward co-management: the process of marine resource management devolution in three East African Countries. Glob Environ Chang 22(3):651–658. https://doi.org/10.1016/j.gloenvcha.2012.03.002

Club des Juristes (2018) 100 jurists call for action for the adoption of a global pact for the environment. http://www.leclubdesjuristes.com/100-jurists-call-for-action-for-the-adoption-of-a-global-pact-for-the-environment/

Costello C et al (2016) Global fishery prospects under contrasting management regimes. Proc Natl Acad Sci 113(18):5125–5129. https://doi.org/10.1073/pnas.1520420113

Costello C, Cao L, Gelcich S et al (2019) The future of food from the sea. World Resources Institute, Washington. Available www.oceanpanel.org/future-food-sea

Craglia M et al (2018) Artificial intelligence: a European perspective (EUR 29425 EN). Publications Office of the European Union, Luxembourg

Cullinan C (2014) Governing people as members of the earth community. In: Mastny L (ed) State of the world 2014: governing for sustainability. Island Press/Center for Resource Economics, Washington, pp 72–81

Cuyvers L et al (2018) Deep seabed mining: a rising environmental challenge. International Union for Conservation of Nature, Gland. https://doi.org/10.2305/iucn.ch.2018.16.en

CZMAI (Coastal Zone Management Authority and Institute) (2016) Belize integrated coastal zone management plan. CZMAI, Belize City

Dasgupta P et al (2019) Economic growth, human development, and welfare. In: In: Rethinking society for the 21st century. Cambridge University Press, Cambridge

Davenport T (2015) Submarine communications cables and science: a new frontier in ocean governance? In: Science, technology, and new challenges to ocean law. Brill/Nijhoff, Leiden, pp 209–252. https://doi.org/10.1163/9789004299610_010

Day JC (2002) Zoning: lessons from the great barrier reef marine park. Ocean Coast Manag 45(2–3):139–156. https://doi.org/10.1016/s0964-5691(02)00052-2

Day JC (2017) How effective is the management of the great barrier reef? ICES J Mar Sci 75(3):1188–1190. https://doi.org/10.1093/icesjms/fsx095

de Grunt LS, Ng K, Calado H (2018) Towards sustainable implementation of maritime spatial planning in Europe: a peek into the potential of the regional sea conventions playing a stronger role. Mar Policy 95:102–110

De Santo EM et al (2019) Protecting biodiversity in areas beyond national jurisdiction: an earth system governance perspective. Earth Syst Governance 2:100029. https://doi.org/10.1016/j.esg.2019.100029

Dietz T, Ostrom E, Stern PC (2003) The struggle to govern the commons. Science 302(5652):1907–1912. https://doi.org/10.1126/science.1091015

Domingos P (2015) The master algorithm. Basic Books, New York

Douvere F (2016) Danger listed site Belize barrier reef gets visionary integrated management plan. United Nations Educational, Scientific and Cultural Organization. http://whc.unesco.org/en/news/1455?utm_source=April+2016+Newsletter&utm_campaign=NatCap+Newsletter&utm_medium=email

EU (European Union) (2014) Directive 2014/89 of the European Parliament and of the Council of 23 July 2014 establishing a framework for maritime spatial planning. https://eur-lex.europa.eu/legal-content/EN/TXT/?uri=celex%3A32014L0089

European Commission (2007) Ports policy communication. COM 616 of 18.10.2007

European Commission (2011) Integrating biodiversity and nature protection into port development. 08.03.2011 SEC(2011) 319 final Commission Staff Working Document. https://ec.europa.eu/environment/nature/natura2000/management/docs/sec2011_319pdf.pdf

Evans M (2019) Commitments worth $63 billion pledged for ocean protection. Mongabay series: sea change. https://news.mongabay.com/2019/10/commitments-worth-63-billion-pledged-for-ocean-protection/

FAO (2015) Voluntary guidelines for securing sustainable small-scale fisheries in the context of food security and poverty eradication. FAO, Rome

FAO (2018) The state of world fisheries and aquaculture. FAO, Rome

FAO (2019a) Integrated coastal management law establishing and strengthening national legal frameworks for integrated coastal management. FAO legislative study 93. FAO, Rome

FAO (2019b) Step-wise guide for the implementation of international legal and policy instruments related to deep-sea fisheries and biodiversity conservation in the areas beyond national jurisdiction. In: ABNJ deep seas project. FAO, Rome

FAO (Food and Agriculture Organization of the United Nations) (2012) Voluntary guidelines on responsible governance of tenure of land, fisheries and forests in the context of national food security. FAO, Rome

Fernandes L et al (2005) Establishing representative no-take areas in the great barrier reef: large-scale implementation of theory on marine protected areas. Conserv Biol 19(6):1733–1744. https://doi.org/10.1111/j.1523-1739.2005.00302.x

Figueres JM, Manuel T, Miliband D (2014) From decline to recovery: a rescue package for the global ocean. Global Ocean Commission, Oxford. https://www.iucn.org/sites/dev/files/import/downloads/goc_full_report_1.pdf

Fujita R et al (2017) Scaling territorial use rights in fisheries (TURFs) in Belize. Bull Mar Sci 93(1):137–153. https://doi.org/10.5343/bms.2016.1002

Fujita R et al (2018) Assessing and managing small-scale fisheries in Belize. In: Viability and sustainability of small-scale fisheries in Latin America and the Caribbean. Springer International, New York, pp 177–195. https://doi.org/10.1007/978-3-319-76078-0_8

Gabčíkovo-Nagymaros Project (Hungary v Slovakia) (1997) International court of justice reports 7

Gaines SD, Cabral R, Free CM, Golbuu Y et al (2019) The expected impacts of climate change on the ocean economy. World Resources Institute, Washington. Available www.oceanpanel.org/expected-impacts-climate-change-ocean-economy

Geels FW (2010) Ontologies, socio-technical transitions (to sustainability), and the multi-level perspective. Res Policy 39(4):495–510. https://doi.org/10.1016/j.respol.2010.01.022

Geels FW (2014) Regime resistance against low-carbon transitions: introducing politics and power into the multi-level perspective. Theory Culture Soc 31(5):21–40. https://doi.org/10.1177/0263276414531627

Geels FW, Schot J (2007) Typology of sociotechnical transition pathways. Res Policy 36(3):399–417. https://doi.org/10.1016/j.respol.2007.01.003

Geels FW et al (2017) Sociotechnical transitions for deep decarbonization. Science 357(6357):1242–1244. https://doi.org/10.1126/science.aao3760

Gelcich S et al (2010) Navigating transformations in governance of Chilean marine coastal resources. Proc Natl Acad Sci 107(39):16794–16799. https://doi.org/10.1073/pnas.1012021107

Gelcich S et al (2017) Fishers' perceptions on the Chilean coastal TURF system after two decades: problems, benefits, and emerging needs. Bull Mar Sci 93(1):53–67. https://doi.org/10.5343/bms.2015.1082

Gelcich S et al (2019a) Early assessments of marine governance transformations: insights and recommendations for implementing new fisheries management regimes. Ecol Soc 24(1):12. https://doi.org/10.5751/es-10517-240112

Gelcich S et al (2019b) Co-management of small-scale fisheries and ecosystem services. Conserv Lett 12(2):e12637. https://doi.org/10.1111/conl.12637

Geyer R, Jambeck JR, Law KL (2017) Production, use, and fate of all plastics ever made. Sci Adv 3(7):e1700782. https://doi.org/10.1126/sciadv.1700782

Gjølberg M et al (2017) Sustainable development goals: exploring maritime opportunities, DNV-GL. https://rederi.no/globalassets/dokumenter-en/all/fagomrader/smi/dnv-gl-sdg-maritime-report.pdf

Glazewski J (2018) A significant legal victory for sustainable fisheries management. Unpublished Memorandum

Global Fishing Watch (2018) The year in transparency: global fishing watch. https://globalfishingwatch.org/data/global-fishing-watch-2018-the-year-in-transparency/

Grin J (2010) Transitions to sustainable development. Routledge, London. https://doi.org/10.4324/9780203856598

Grorud-Colvert K et al (2019) High-profile international commitments for ocean protection: empty promises or meaningful progress? Mar Policy 105:52–66. https://doi.org/10.1016/j.marpol.2019.04.003

Guannel G et al (2017) The power of three: coral reefs, seagrasses and mangroves protect coastal regions and increase their resilience. PLoS One 11(7):e0158094. https://doi.org/10.1371/journal.pone.0158094

Gutiérrez L et al (2012) Eco-label conveys reliable information on fish stock health to seafood consumers. PLoS One 7(8):e43765. https://doi.org/10.1371/journal.pone.0043765

Haberl H et al (2011) A socio-metabolic transition towards sustainability? Challenges for another great transformation. Sustain Dev 19(1):1–14. https://doi.org/10.1002/sd.410

Hanna SS (1995) User participation and fishery management performance within the pacific fishery management council. Ocean Coast Manag 28(1–3):23–44. https://doi.org/10.1016/0964-5691(95)00046-1

Helfenstein K, Dubrova A, Woods B, Zaman NS (2019) MOP 31 highlights. Earth Negot Bull 19:147. https://enb.iisd.org/vol19/enb19147e.html

High Seas Alliance Treaty Tracker (2019). http://highseasalliance.org/treatytracker/statements/greenpeace-on-behalf-of-the-high-seas-alliance-5th-april-2019-cross-cutting-informal-working-group/

IMO (2012) Guidelines for safe and environmentally sound ship recycling. Res. MEPC.210(63)

IMO (2015) The post-2015 development Agenda. Res. TC 65/6(a)

IMO (2018) Initial IMO strategy on reduction of GHG emissions from Ships. Res. (MEPC72)

IMO (2019) Low carbon shipping and air pollution control. http://www.imo.org/en/MediaCentre/HotTopics/GHG/Pages/default.aspx

IMO (International Maritime Association) (2003) Guidelines on places of refuge for ships in need of assistance. Res A 949: 23.

IPBES (2019) Summary for policymakers. In: Global assessment report on biodiversity and ecosystem services of the intergovernmental science-policy platform on biodiversity and ecosystem services. IPBES Secretariat, Bonn

IPCC (2019) Summary for policymakers. In: Pörtner H-O, Roberts DC, Masson-Delmotte V, Zhai P, Tignor M, Poloczanska E et al (eds) IPCC special report on the ocean and cryosphere in a changing climate. IPCC, Geneva. https://www.ipcc.ch/srocc/

IPCC (Intergovernmental Panel on Climate Change) (2013) Climate change 2013: the physical science basis. In: Stocker TF, Qin D, Plattner G-K, Tignor M, Allen SK, Boschung J, Nauels A, Xia Y et al (eds) Contribution of working group I to the fifth assessment report of the Intergovernmental Panel on Climate Change. Cambridge University Press, Cambridge

IUCN (International Union for the Conservation of Nature) (1995) Draft international covenant on environment and development: implementing sustainability. IUCN, Gland

Jambeck J, Hardesty BD, Brooks AL, Friend T, Teleki K, Fabres J, Beaudoin Y, Bamba A et al (2018) Challenges and emerging solutions to the land-based plastic waste issue in Africa. Mar Policy 96:256–263. https://doi.org/10.1016/j.marpol.2017.10.041

Jambeck J, Moss E, Dubey BK et al (2020) Pollution and a regenerative economy: municipal, industrial, agricultural, and maritime waste, its impacts and solutions. World Resources Institute, Washington. https://oceanpanel.org/blue-papers/pollution-and-regenerative-economy-municipal-industrial-agricultural-and-maritime-waste

Jentoft S (2007) Limits of governability: institutional implications for fisheries and coastal governance. Mar Policy 31(4):360–370. https://doi.org/10.1016/j.marpol.2006.11.003

Jentoft S, Chuenpagdee R (2015) Enhancing the governability of small-scale fisheries through interactive governance. In: Interactive governance for small-scale fisheries. MARE publication series. Springer International, New York, pp 727–747. https://doi.org/10.1007/978-3-319-17034-3_37

Jessop B (2002) Governance and meta-governance in the face of complexity: on the roles of requisite variety, reflexive observation, and romantic irony in participatory governance. In: Participatory governance in multi-level context. Verlag für Sozialwissenschaften, Wiesbaden, pp 33–58. https://doi.org/10.1007/978-3-663-11005-7_2

Jessop B (2011) The governance of complexity and complexity of governance: preliminary remarks on some problems and limits of ecological guidance. Lancaster University, Lancaster

Jessop B (2016) The state: past present future. Polity, Cambridge

Johannes RE (2002) The renaissance of community-based marine resource management in Oceania. Annu Rev Ecol Syst 33(1):317–340. https://doi.org/10.1146/annurev.ecolsys.33.010802.150524

Klinger DH et al (2018) The mechanics of blue growth: management of oceanic natural resource use with multiple, interacting sectors. Mar Policy 87:356–362. https://doi.org/10.1016/j.marpol.2017.09.025

Kotzé L (2017) Reimagining international environmental law in the Anthropocene. In: Environmental law and governance for the Anthropocene. Hart, London, pp 31–54. https://doi.org/10.5040/9781509906574.ch-002

Kotzé L (2019) Earth system law for the Anthropocene. Sustain For 11:6796. https://doi.org/10.3390/su11236796

Kroeker KJ, Kordas RL, Crim R, Hendriks IE, Ramajo L, Singh GS et al (2013) Impacts of ocean acidification on marine organisms: quantifying sensitivities and interaction with warming. Glob Chang Biol 19(6):1884–1896

Kummu M et al (2016) Over the hills and further away from coast: global geospatial patterns of human and environment over the 20th–21st centuries. Environ Res Lett 11(3):34010. https://doi.org/10.1088/1748-9326/11/3/034010

Leape J, Abbott M, Sakaguchi H et al (2020) Technology, data and new models for sustainably managing ocean resources. World Resources Institute, Washington. https://oceanpanel.org/sites/default/files/2020-01/19_HLP_BP6_V4.pdf

Leary BC et al (2019) 30x30, a blueprint for ocean protection: how we can protect 30% of our oceans by 2030. Greenpeace. https://storage.

googleapis.com/planet4-international-stateless/2019/04/4475b2c2-updatedgreenpeace_30x30_blueprint_report_web.pdf
Leck H et al (2015) Tracing the water-energy-food nexus: description, theory and practice. Geogr Compass 9(8):445–460. https://doi.org/10.1111/gec3.12222
Lehodey P, Bertignac M, Hampton J, Lewis A, Picaut J (1997) El Niño southern oscillation and tuna in the western pacific. Nature 389:715–718
Levin S (1998) Ecosystems and the biosphere as complex adaptive systems. Ecosystems 1(May):431–436. https://doi.org/10.1007/s100219900037
Levin S et al (2012) Social-ecological systems as complex adaptive systems: modeling and policy implications. Environ Dev Econ 18(2):111–132. https://doi.org/10.1017/s1355770x12000460
Litke A, Rieu-Clarke A (2014) The UN watercourses convention and its complementary user's guide: indispensable ingredients for global water cooperation. In: Grafton RQ, Wyrwoll P, White C, Allendes D (eds) Global water: issues and insights. Australian National University Press, Canberra, pp 215–222
Loorbach D (2010) Transition management for sustainable development: a prescriptive, complexity-based governance framework. Governance 23(1):161–183. https://doi.org/10.1111/j.1468-0491.2009.01471.x
Lubchenco J et al (2016) The right incentives enable ocean sustainability successes and provide hope for the future. Proc Natl Acad Sci 113(51):14507–14514. https://doi.org/10.1073/pnas.1604982113
Macy A (2017) The atmosphere, the Paris agreement and global governance. Policy Q 13(1):4640. https://doi.org/10.26686/pq.v13i1.4640
Mandle L et al (2017) Assessing ecosystem service provision under climate change to support conservation and development planning in Myanmar. PLoS One 12(9):e0184951. https://doi.org/10.1371/journal.pone.0184951
Marín A et al (2012) Exploring social capital in Chile's coastal Benthic comanagement system using a network approach. Ecol Soc 17(1):13. https://doi.org/10.5751/es-04562-170113
Mathews RE et al (2019) Implementing the source-to-sea approach: a guide for practitioners. Stockholm International Water Institute, Stockholm. https://www.siwi.org/wp-content/uploads/2019/07/Source-to-sea-guide_webb.pdf
Mbatha NP (2018) The influence of plural governance systems on rural coastal livelihoods: the case of Kosi bay. Doctoral thesis, University of Cape Town
McCauley DJ, Pinsky ML, Palumbi SR, Estes JA, Joyce FH, Warner RR (2015) Marine defaunation: animal loss in the global ocean. Science 347(6219):1255641
McClenachan L, Neal B, Al Abdulrazzak D, Witkin T, Fisher K, Kittinger J (2014) Do community supported fishers (CSFs) improve sustainability? Fish Res 157:62–69. https://doi.org/10.1016/j.fishres.2014.03.016
McGlade C, Ekins P (2015) The geographical distribution of fossil fuels unused when limiting global warming to 2°C. Nature 517(7533):187–190. https://doi.org/10.1038/nature14016
McGregor DP et al (2003) An ecological model of well-being. In: Becker H, Vanclay F (eds) The international handbook of sociological impact assessment: conceptual and methodological advances. Edward Elgar, Northampton, pp 109–126
Meadowcroft J (2011) Engaging with the politics of sustainability transitions. Environ Innov Soc Trans 1(1):70–75. https://doi.org/10.1016/j.eist.2011.02.003
Meinzen-Dick R, Pradhan R (2002) Legal pluralism and dynamic property rights. CAPRI Working Paper 22. International Food Policy Research Institute, Washington. https://doi.org/10.22004/ag.econ.55442
Mendenhall E et al (2019) A soft treaty, hard to reach: the second Inter-governmental Conference for biodiversity beyond national jurisdiction. Mar Policy 108:103664. https://doi.org/10.1016/j.marpol.2019.103664
Meuleman L (2019) Metagovernance for sustainability: a framework for implementing the sustainable development goals. Routledge, London
Meuleman L (2020) It takes more than markets: first governance lessons from the COVID-19 pandemic. ISSD SDG Knowledge Hub. https://sdg.iisd.org/commentary/guest-articles/it-takes-more-than-markets-first-governance-lessons-from-the-covid-19-pandemic/
Mey F, Diesendorf M (2018) Who owns an energy transition? Strategic action fields and community wind energy in Denmark. Energy Res Soc Sci 35:108–117. https://doi.org/10.1016/j.erss.2017.10.044
Minerbi L (1999) Indigenous management models and protection of the Ahupua'a. In: Aoude I (ed) The ethnic studies story: politics and social movements in Hawai'i. University of Hawai'i Press, Honolulu, pp 208–226
Mumby PJ et al (2004) Mangroves enhance the biomass of coral reef fish communities in the Caribbean. Nature 427(6974):533–536. https://doi.org/10.1038/nature02286
NatCap (Natural Capital Project) (2019) Global modeling of nature's contributions to people. Interactive viewer available at http://viz.naturalcapitalproject.org/ipbes/
Neumann B, Unger S (2019) From voluntary commitments to ocean sustainability. Science 363(6422):35–36. https://doi.org/10.1126/science.aav5727
Ngeta K (2014) Theoretical deliberations on understanding natural resource governance. In: Sowman M, Wynberg R (eds) Governance for justice and environmental sustainability: lessons across natural resource sectors in Sub-Saharan Africa. Routledge, London, pp 23–42. https://doi.org/10.4324/9780203120880-2
Ngoran SD et al (2016) Exploring the challenges of implementing integrated coastal management and achieving sustainability within the Cameroon coastline. J Integr Coastal Manage 16(1):45–56
Norberg J, Wilson J, Walker B, Ostrom E (2008) Diversity and resilience of social-ecological systems. In: Norberg J, Cumming G (eds) Complexity theory for a sustainable future. Columbia University Press, New York, pp 46–80
OECD (2016) The ocean economy in 2030. OECD Publishing, Paris
OECD (Organisation for Economic Co-operation and Development) (2010) Environmental and climate change: issues in the shipbuilding industry. OECD Council Working Party on Shipbuilding (WP6). https://www.oecd.org/sti/ind/46370308.pdf
OECD International Transport Forum (2018) Decarbonising maritime transport pathways to zero-carbon shipping by 2035. https://www.itf-oecd.org/sites/default/files/docs/decarbonising-maritime-transport-2035.pdf
OHCHR (1995) Human rights and the environment. Res. 1995/14
OHCHR (1996) Human rights and the environment. Res. 1996/13
OHCHR (Office of the High Commissioner for Human Rights) (1994) Human rights and the environment. E/CN.4/RES/1994/65
Olsson P et al (2008) Navigating the transition to ecosystem-based management of the great barrier reef, Australia. Proc Natl Acad Sci 105(28):9489–9494. https://doi.org/10.1073/pnas.0706905105
Orellana M (2018) Earth matters: the case for the right to a healthy environment. Human Rights Watch. https://www.hrw.org/world-report/2018/country-chapters/global
Österblom H et al (2015) Transnational corporations as 'keystone actors' in marine ecosystems. PLoS One 10(5):e0127533. https://doi.org/10.1371/journal.pone.0127533
Österblom H et al (2017) Emergence of a global science- business initiative for ocean stewardship. Proc Natl Acad Sci 114(34):9038–9043. https://doi.org/10.1073/pnas.1704453114
Österblom H, Wabnitz CCC, Tladi D et al (2020) How to distribute the benefits of the ocean equitably. World Resources Institute, Washington. https://oceanpanel.org/blue-papers/how-distribute-benefits-ocean-equitably

Ostrom E (1990) Governing the commons: the evolution of institutions for collective action. Cambridge University Press, Cambridge

Ostrom E (2000) Reformulating the commons. Swiss Polit Sci Rev 6(1):29–52. https://doi.org/10.1002/j.1662-6370.2000.tb00285.x

Ostrom E (2010) Beyond markets and states: polycentric governance of complex economic systems. Am Econ Rev 100(3):641–672

Ostrom E, Janssen MA (2004) Multi-level governance and resilience of social-ecological systems. In: Spoor M (ed) Globalisation, poverty and conflict. Springer, Dordrecht, pp 239–259. https://doi.org/10.1007/1-4020-2858-X

Ottens M, Edelenbos J (2019) Political leadership as meta-governance in sustainability transitions: a case study analysis of meta-governance in the case of the Dutch national agreement on climate. Sustain For 11(1):110. https://doi.org/10.3390/su11010110

Ouyang Z, Zheng H, Xiao Y, Polasky S, Liu J, Xu W, Wang Q et al (2016) Improvements in ecosystem services from investments in natural capital in China. Science 352:1455–1459

Pahl-Wostl C (2017) Governance of the water-energy-food security nexus: a multi-level coordination challenge. Environ Sci Pol 92:356–367

Pauly D, Zeller D (2016) Catch reconstructions reveal that global marine fisheries catches are higher than reported and declining. Nat Commun. https://doi.org/10.1038/ncomms10244

Pecl GT et al (2017) Biodiversity redistribution under climate change: impacts on ecosystems and human well-being. Science 355:6332. https://doi.org/10.1126/science.aai9214

Pickering J et al (2018) Global climate governance between hard and soft law: can the Paris agreement's 'Crème Brûlée' approach enhance ecological reflexivity? J Environ Law 31:1–28

Plastics Europe (2019) Plastics—the facts, 2019. https://www.plasticseurope.org/application/files/9715/7129/9584/FINAL_web_version_Plastics_the_facts2019_14102019.pdf

PNA Tuna (2010) Nauru Agreement. www.pnatuna.com/content/nauru-agreement

Poli R (ed) (2018) Handbook for anticipation. Springer, New York

Preiser R et al (2018) A complexity-based paradigm for studying social-ecological systems. Ecol Soc 23(4):46. https://doi.org/10.5751/ES-10558-230446

Pretlove B, Blasiak R (2018) Mapping ocean governance and regulation. https://globalcompact.no/wp-content/uploads/2020/01/Mapping-Ocean-Governance-and-Regulation-2.pdf

Rajamani L (2016) The 2015 Paris Agreement: interplay between hard, soft and non-obligations. J Environ Law 28(2):337–358

Ramos J (2019) Cosmo localism: tech trend, post-capitalist commons transition, or something else? Talk presented at First Global Commons Forum, Seoul, South Korea, Karl Polanyi Institute Asia. https://medium.com/joseramos_30450/cosmo-localism-tech-trend-post-capitalist-commons-transition-or-something-else-4dc0202dd732

REN21 (2018) Renewables 2018: global status report. REN21 Secretariat, Paris

Republic of South Africa (1996) Constitution of the Republic of South Africa. https://www.gov.za/sites/default/files/images/a108-96.pdf

Rieu-Clarke A, Spray C (2013) Ecosystem services and international water law: towards a more efficient determination and implementation of equity. PER 16(2):55. http://www.cawater-info.net/bk/water_law/pdf/rieu-clarke-spray.pdf

Robinson M (2016) Incorporating human rights into climate action. Mary Robinson Foundation: Climate Justice Version 2. https://www.mrfcj.org/wp-content/uploads/2016/05/Incorporating-Human-Rights-into-Climate-Action-Version-2-May-2016.pdf

Robinson M, Shine T (2018) Achieving a climate justice pathway to 1.5 °C. Nat Clim Chang 8:564–569. https://doi.org/10.1038/s41558-018-0189-7

Rocha JC (2010) The domino effect: a network analysis of regime shifts drivers and causal pathways. Thesis, Stockholm University. https://www.researchgate.net/publication/277769781_The_domino_effect_A_network_analysis_of_regime_shifts_drivers_and_causal_pathways

Rogers AD, Laffoley D (2013) Introduction to the special issue: the global state of the ocean; interactions between stresses, impacts and some potential solutions. Mar Pollut Bull 74(2):491–494

Rogers AD, Mumby PJ (2019) Mangroves reduce the vulnerability of coral reef fisheries to habitat degradation. PLoS Biol 17(11):e3000510. https://doi.org/10.1371/journal.pbio.3000510

Sale PF et al (2014) Transforming management of tropical coastal seas to cope with challenges of the 21st century. Mar Pollut Bull 85:8–23

Schot J, Kanger L (2018) Deep transitions: emergence, acceleration, stabilization and directionality. Res Policy 47:1045–1059. https://doi.org/10.1016/j.respol.2018.03.009

Schwab K (2016) The fourth industrial revolution. World Economic Forum, Geneva

SDG (2019) The future is now: science for achieving sustainable development. Global sustainable development report. United Nations, New York

SDG Human Rights Explorer (2020). https://sdgdata.humanrights.dk/

Selig ER et al (2018) Mapping global human dependence on marine ecosystems. Conserv Lett 12(2):e12617. https://doi.org/10.1111/conl.12617

Sharp R et al (2020) InVEST 3.8.0.post29+ug.g0215cb1 user's guide. Natural capital project, Stanford University, University of Minnesota, The Nature Conservancy, and World Wildlife Fund

Silver J et al (2019) Advancing coastal risk reduction science and implementation by accounting for climate, ecosystems and people. Front Mar Sci 6:556. https://doi.org/10.3389/fmars.2019.00556

Smith A, Stirling A, F. Berkhout F. (2005) The governance of sustainable socio-technical transitions. Res Policy 34(10):1491–1510

Sørensen E (2006) Metagovernance: the changing role of politicians in processes of democratic governance. Am Rev Public Adm 36: 98–114

Steffen W et al (2018) Trajectories of the earth system in the Anthropocene. Proc Natl Acad Sci 115(33):8252–8259

Stramma L, Schmidtko S, Levin LA, Johnson GC (2010) Ocean oxygen minima expansions and their biological impacts. Deep sea research part 1. Oceanogr Res Pap 57(4):587–595

Submarine Telecoms Industry Report (2019) No. 8. https://subtelforum.com/products/submarine-telecoms-industry-report/

Suggett DJ, Hall Spencer JM, Rodolfo Metalpa R, Boatman TG, Payton R, Tye Pettay D et al (2012) Sea anemones may thrive in a high CO2 world. Glob Chang Biol 18(10):3015–3025

Sumaila R, Walsh M, Cox A et al (forthcoming) Ocean finance. World Resources Institute, Washington. https://oceanpanel.org/blue-papers/ocean-finance

Sunderji N (2016) How will data and digital platforms transform sustainable development? Devex. https://www.devex.com/news/how-will-data-and-digital-platforms-transform-sustainable-development-88481

Swilling M (2019) The age of sustainability: just transitions in a complex world. Routledge, London

Taormina B et al (2018) A review of potential impacts of submarine power cables on the marine environment: knowledge gaps, recommendations and future directions. Renew Sust Energ Rev 96:380–391

Tegmark M (2017) Life 3.0: being human in the age of artificial intelligence. Knopf, New York

Torrie M (2016) Integrated ocean management: fisheries, oil, gas and seabed mining. GLOBEFISH Research Programme 122. FAO, Rome. www.fao.org/3/a-i6048e.pdf

Tosun J, Leininger J (2017) Governing the interlinkages between the sustainable development goals: approaches to attain policy integration. Global Chall 1(9):1700036

Tsioumani E, Bansard J, Pinto-Bazurco JF (2019) Marine regions forum 2019. Marine Regions For Bull 186:18. http://enb.iisd.org/

oceans/marine-regions-forum/2019/html/enbplus186num18e.html?utm_medium=email&utm_campaign=ENB%20Update%20%205%20October%202019%20SW&utm_content=ENB%20Update%20%205%20October%202019%20SW+CID_5f51078ed11968bb8a4c4ff19ae9f4ce&utm_source=cm&utm_term=Read

TWI2050 (2018) Transformations to achieve the sustainable development goals. In: Report prepared by The World in 2050 initiative. International Institute for Applied Systems Analysis, Laxenburg. www.twi2050.org

TWI2050 (2019) The digital revolution and sustainable development: opportunities and challenges. In: Report prepared by The World in 2050 initiative. International Institute for Applied Systems Analysis, Laxenburg. www.twi2050.org

U.S. EPA (2014) Municipal solid waste generation, recycling, and disposal in the United States: tables and figures for 2012. O.o.R.C.a. Recovery. U.S. EPA, Washington

U.S. EPA (2019) Advancing sustainable materials management: 2017 fact sheet assessing trends in material generation, recycling, composting, combustion with energy recovery and landfilling in the United States. U.S. EPA, Washington

UN ECOSOC (2018) Principles of effective governance for sustainable development. Official records, supplement no. 24E/2018/44--E/C.16/2018/8. https://publicadministration.un.org/Portals/1/Images/CEPA/Principles_of_effective_governance_english.pdf

UN General Assembly (1972) United Nations conference on the human environment. A/RES/2994. [Stockholm Declaration]. Stockholm. https://www.refworld.org/docid/3b00f1c840.html

UN News (2019) Greta Thunberg tells world leaders 'you are failing us,' as nations announce fresh climate action. UN Climate Summit, New York. https://www.un.org/development/desa/en/news/sustainable/climate-summit-2019.html

UNCLOS (1982) 21 ILM 1261

UNCTAD (2018) Review of maritime transport. United Nations Publications, New York

UNDRR (2019) Global assessment report on disaster risk reduction. UNDRR, Geneva

UNECE (1998) Convention on access to information, participation in decision-making and access to justice [Aarhus Convention]. Aarhus, Denmark. https://www.unece.org/fileadmin/DAM/env/pp/documents/cep43e.pdf

UNFCCC (1992) 31 ILM 849

UNFCCC (2015) Paris agreement. http://unfccc.int/files/essential_background/convention/application/pdf/english_paris_agreement.pdf

Universal Declaration of Human Rights (1948) A/RES/3/217 A

van Breda J (2019) Methodological agility in the Anthropocene: an emergent, transformative transdisciplinary research approach. PhD diss., Stellenbosch University

van Breda J, Swilling M (2018) Guiding logics and principles for designing emergent transdisciplinary research processes: learning experiences and reflections from a South African Case Study. Sustain Sci 14:823–841. https://doi.org/10.1007/s11625-018-0606-x

Veers P et al (2019) Grand challenges in the science of wind energy. Science 366:6464

Vidas D (2011) The Anthropocene and the international law of the sea. Phil Trans R Soc A 369:326. https://doi.org/10.1098/rsta.2010.0326

Vierros M, Buonomo R (2017) In-depth analysis of ocean conference voluntary commitments to support and monitor their implementation. Division for Sustainable Development, Department of Economic and Social Affairs, United Nations. https://sustainabledevelopment.un.org/content/documents/17193OCVC_in_depth_analysis.pdf

Wade EA et al (2019) Integrating property rights into fisheries management: the case of Belize's journey to managed access. Mar Policy 108:103631

Warnken J, Mosadeghi R (2018) Challenges of implementing integrated coastal zone management into local planning policies, a case study of Queensland, Australia. Mar Policy 91:75–84

WCED (1987) Report of the world commission on environment and development: our common future (Brundtland Report). In: Transmitted to the general assembly as an annex to document A/42/427—development and international co-operation: environment. Oxford University Press, Oxford

Weitz N et al (2017) Closing the governance gaps in the water-energy-food nexus: insights from integrative governance. Glob Environ Chang 45:165–173

Westley FR et al (2013) A theory of transformative agency in linked social-ecological systems. Ecol Soc 18(3):27. https://doi.org/10.5751/ES-05072-180327

Weston BH, Bollier D (2013a) The human right to a clean and healthy environment. In: Chapter 2 of green governance: ecological survival, human rights, and the law of the commons. Cambridge University Press, Cambridge, pp 27–49. https://doi.org/10.1017/cbo9781139540582.004

Weston BH, Bollier D (2013b) The quest for a new rights-based pathway. In: Chapter 3 of green governance: ecological survival, human rights, and the law of the commons. Cambridge University Press, Cambridge, pp 50–76. https://doi.org/10.1017/cbo9781139540582.005

Widjaja S, Long T, Wirajuda H et al (2020) Illegal, unreported and unregulated fishing and associated drivers. World Resources Institute, Washington. www.oceanpanel.org/iuu-fishing-and-associated-drivers

Winther J-G, Dai M et al (2020) Integrated ocean management. World Resources Institute, Washington. https://oceanpanel.org/blue-papers/integrated-ocean-management

World Bank (2012) Hidden harvest: the global contribution of capture fisheries. World Bank, Washington. http://documents.worldbank.org/curated/en/515701468152718292/Hidden harvest-the-global-contribution-of-capture-fisheries

Wyatt KH et al (2017) Habitat risk assessment for regional ocean planning: a case study from the US northeast and mid-Atlantic regions. PLoS One 12(12):e0188776

Towards Ocean Equity

13

Highlights

- The ocean is important for everyone—it produces oxygen and food, stores carbon and heat, offers space for economic activities and recreation, and continues to inspire and support culture and well-being.
- Access to ocean resources and sectors is rarely equitably distributed. Many of their benefits are accumulated by a few, while most harms from development are borne by the most vulnerable.
- Inequity is a systemic feature of the current ocean economy. It is embedded in existing political and economic systems, the result of historical legacies and prevailing norms. This has brought global environmental challenges and negative effects on human well-being.
- Legal frameworks to support equity exist but are not sufficiently developed. In practice, ocean policies are largely equity-blind, poorly implemented and fail to address inequity.
- Inequity manifests, for example, in unfair distribution of commercial fish catches; limited political power of small-scale fishers, particularly women and other minority groups; limited engagement of developing nations in high-seas activities and associated decision-making; and consolidated interests of global supply chains in a few transnational corporations, with evidence of poor transparency and human rights abuses.
- Climate change will create and worsen challenges of fairness and equity faced by developing countries, regions and communities reliant on marine livelihoods.
- Discussions on environmental sustainability have largely overshadowed concerns about social equity. Addressing inequalities and preventing the widening of ocean inequities are integral to a sustainable ocean economy; and promoting equity is essential for securing fair development, the legitimacy of policies, social stability and sustainability.
- A sustainable ocean economy should protect human rights, improve human well-being, stimulate inclusion and gender equity, and prioritise recognition, diversity and equal access to resources to provide fair opportunities consistent with sustainable development. It should also address corruption and tax evasion, demand responsible and transparent business practices and create a shared economy that facilitates a fair redistribution of wealth and benefits. A sustainable ocean economy should be aware of environmental and social limits on growth and consider degrowth where appropriate.
- Shifting a historical trajectory of persistent and increasing inequities will require strong leadership, inclusive governance and long-term planning that starts with a commitment to equity as integral to a sustainable ocean economy and relationships within and across nations.

1 Introduction

1.1 Overview

The blue economy is being promoted as capable of achieving sustainability and prosperity, fair use of the ocean and the UN Sustainable Development Goals (SDGs). Ensuring a more equitable distribution of goods and services provided by the ocean represents a major challenge. There is overwhelming evidence that current access to ocean benefits and resources, as well as exposure to harms, is distributed inequitably. This results in negative effects on the environment and human health, loss of livelihoods, limited financial opportunities for vulnerable groups and challenges to nutritional and food security. Powerful interests (including states, communities and economic entities) benefit from existing arrangements. Challenging inequality represents a direct

Originally published in:
Österblom, H., C.C.C. Wabnitz, D. Tladi et al. 2020. *Towards Ocean Equity*. Washington, DC: World Resources Institute. Available online at www.oceanpanel.org/how-distribute-benefits-ocean-equitably.
Reprint by Springer International Publishing (2023) with kind permission.
Published under license from the World Research Institute.

J. Lubchenco, P. M. Haugan (eds.), *The Blue Compendium*, https://doi.org/10.1007/978-3-031-16277-0_13

threat to such interests. Inequality is increasingly influencing economic development and political stability. Current and recent examples of social unrest are closely associated with concerns about inequality, climate change, corruption and related societal problems perceived as having an unfair impact. Increased scientific attention to inequality is starting to shape debates associated with the ocean. We argue that there is a general policy blindness to instruments and practices that maintain the unfair status quo, but that there are remedies to such blindness. The purpose of this Blue Paper is to explore ocean inequities and suggest approaches for the just inclusion of diverse actors in the blue economy agenda and the equitable distribution of ocean benefits. First, we define inequity terms and their drivers, as well as how they affect sustainability. Second, we explore policies and practices that have (or have not) worked in favour of equity, while also promoting ecological sustainability. Finally, we provide opportunities for action for policymakers, funding and research institutions, international and non-governmental organisations, business leadership as well as civil society to address systemic aspects of inequities along a spectrum of ambitions, from basic to transformative. These opportunities for action are not intended as alternatives. They constitute complementary and reinforcing action to support and inform pathways to a sustainable and just ocean economy.

1.2 Context

The ocean plays a critical role in securing human well-being, but marine ecosystems have a long history of overexploitation, habitat destruction and pollution (Jackson 2001; Roberts 2010; Halpern et al. 2008; Nash 2013; Reusch et al. 2018). The scale and impacts of these pressures, which now also include climate change, are critically undermining the function and role that the ocean plays.

Despite increasing knowledge of these pressures and their effects, the ocean continues to be perceived as an economic frontier: a resource with substantial potential to stimulate economic growth, employment opportunities and innovation (European Commission 2012; United Nations 2014; OECD 2016). Notions such as 'the blue economy' or 'blue growth' facilitate such perceptions. These concepts are used to legitimise and generate support for ocean-based economic development opportunities—including aquaculture, bio-prospecting, marine tourism, shipping, oil and gas, renewable energy and deep-sea mining (OECD 2016; Lloyd's 2014; Economist 2015) and are often linked to the idea of environmental stewardship (Biermann et al. 2017; Brent et al. 2018).

Despite substantial differences in how ocean development concepts are presented and what they imply for economic and social development (Silver et al. 2015; Voyer et al. 2018; Bennett et al. 2019a, b), they are increasingly central to national and international ocean policies (European Commission 2017; OECD 2016; World Bank and United Nations 2017; Childs and Hicks 2019). They have also gained broad levels of support across diverse actors, including small-island and developing states (Michel 2016; van Wyk 2015). A historical asymmetry between the capacity to grow the ocean economy and the capacity to regulate it raises questions about whether promoting growth in ocean use can be made compatible with sustainable use of marine resources and the protection of ecosystems (Ehlers 2016; Llewellyn et al. 2016; Golden et al. 2017; Niner et al. 2018; Laffoley et al. 2019).

The controversy and debate around the sustainable ocean economy illustrate the disparities among visions of the way humanity should relate to the ocean— arguably the greatest common global resource. Some believe that economic growth based on the ocean is critical to development—the foundation of human well-being—and can be made sustainable through technological innovation and regulations. Others are more sceptical and contend that current economic paradigms and power structures are the very reason for unsustainable development and inequities, that the potential for further expansion of ocean-based sectors is limited at best, and that achieving sustainability can only be accomplished by transitioning towards a collaborative economy, which would include limiting, or 'degrowing', economic ocean-based activities (Kostakis and Bauwens 2014; Hadjmichael 2018).

Concerns about environmental sustainability have overshadowed concerns about social equity (Stanton 2012; Halpern et al. 2013; Boonstra et al. 2015; Bennett et al. 2019b). Yet there is increased recognition that equity is necessary, if not sufficient, for sustainability (UN 2015; Raworth 2017a; Hamann et al. 2018; Leach et al. 2018; Cohen et al. 2019); that fairness and sustainability are 'two sides of the same coin' (Berg et al. 2012; Piketty 2014) and that any sustainable ocean economy investments predicated on fostering sustained economic growth must also pay attention to reducing inequality. These are arguments for explicitly considering inequality in national ocean economy plans, rather than addressing it through global blueprints. The shortcomings and failures of some of the International Monetary Fund (IMF) and World Bank structural adjustment programs of the 1980s and 1990s constitute warnings against the adoption of universal macro-economic recipes for economic management (Dollar and Svensson 2001).

Social equity (Box 13.1) in relation to the sustainable ocean economy includes a focus on the provision of social, cultural and economic benefits. A sustainable ocean economy should respect human rights and provide fair opportunities for employment. It should also improve wages, address workplace discrimination, stimulate gender equity and affirm the right to a healthy and safe work environment. A sustainable ocean economy should include aspects of recognition, equal access to resources and inclusivity, and should also support fair distribution of benefits and insulation for the most vulnerable from risks of harm, and where harm is done, assign liability and responsibility for remedy (Klain et al. 2014; Klein et al. 2015; WWF 2018). This is generally not how ocean policies are designed or implemented.

Box 13.1: Definitions: Equality, Equity and Fairness

Social equality and *social equity* are closely related terms that merit clarification. *Social equality* refers to the level to which all members of a society are assigned the same status based on recognition, opportunity and outcomes. For example, different groups (such as genders, classes and ethnicities) could have the same status in terms of legal rights, economic opportunities or access to goods and services (Sen 1992; ISSC et al. 2016). Equality of recognition and protection under the law is a basic tenet of legal systems and constitutions in most countries, though application of this premise varies significantly. Under the 'capability approach', equality is recognised in terms of people's assets, capitals or abilities to take advantage of development and livelihood opportunities (Sen 1992; Nussbaum 2011). Equality of access to assets is thus assumed to lead to improvements in equality of opportunities (Leach et al. 2010). Equality of outcomes refers to an objective comparison of the level of parity in terms of distribution of measures such as income, assets or wealth either within or across societies.

The term *social equity* goes further and combines a concern for equal treatment, with an assessment of what constitutes fair treatment across both substantive outcomes and procedural concerns. Fairness is most often addressed in terms of distributive equity, in other words the distribution of 'goods' and 'bads' across different individual and groups in society (McDermott et al. 2013; Tyler 2015; Bennett et al. 2019a). Achieving social equity may require redressing existing social inequalities so that members of disadvantaged social groups receive a fairer share of the benefits than they did in the past. What constitutes 'fair distribution' is subjective and needs to be understood in relation to the social beliefs, values, practices and institutions of different cultures and societies (e.g., Sandel 1990). Distributive equity may also be influenced by the level of procedural equity, which refers to the recognition of rights and stakeholders, inclusion and participation, as well as political power to influence decisions regarding management and distribution of goods and services (Pascual et al. 2014; Tyler 2015). True procedural equity requires that all actors have adequate capacity to participate, and benefit from information transparency and processes that ensure all voices are heard and can influence decisions (Bennett et al. 2019a).

1.3 Why Is Equity Important?

Inequity is most visible when there is great income disparity within and between countries. The challenges associated to wealth inequality have repeatedly been voiced by social justice non-governmental organisations (NGOs), with a particular focus on extreme differences in wealth between the super-rich and the 'bottom billions' of the world (Oxfam 2019). Concerns about wealth inequalities, their causes, possible solutions and consequences for economic growth and social well-being are also voiced by organisations commonly associated with growth-focused economic policies, such as the Organisation for Economic Co-operation and Development (OECD) (Cingano 2014), the International Monetary Fund (IMF) (Dabla-Norris et al. 2015), the *Economist* magazine (Economist 2015) and various banks (e.g., Camposi 2017). A recent survey among private corporations illustrates that 88% of chief executives believe our economic system needs to refocus on equitable growth (Global Compact 2019). These concerns arise because income and wealth inequality, having largely fallen from the 1920s until the early 1980s, have been rising since that time (Alvaredo et al. 2018).

Rises in inequality are associated with rapid economic growth in transitional countries (China, India, Indonesia and Brazil), economic liberalisation in Russia and, in some developed nations (particularly English-speaking ones), the adoption of 'neoliberal' economic policies (Kotz 2015). These policies include large-scale transfer of public goods to the private sector through the sale of previously state-owned companies, public lands, health and education services; the lowering of corporate taxes and tax rates on top earners; deregulation of financial markets; and liberalisation of trade. All are intended to boost growth, which under this development approach is supposed to reduce poverty through trickle-down effects. While there has undoubtedly been success in reducing global poverty, inequalities have widened both nationally and globally (Alvaredo et al. 2018) and include the emergence of highly consolidated industries (Blasiak et al. 2018b; Monacelli 2018; Folke et al. 2019).

The social democrat countries of Europe, conversely, have the lowest levels of wealth inequality (Alvaredo et al. 2018). These high-wage, high-taxation economies are effective in providing accessible public services and are funded by redistributive, or progressive, taxation schemes. In these countries, the sustainable ocean economy may well develop to deliver hoped-for gains in human welfare, as the institutions and practices are in place and operational. However, the use of tax havens by private corporations and citizens, and other mechanisms aimed to avoid or reduce taxation (see Galaz et al. 2018), represent a challenge also for countries with functional taxation schemes.

The Sustainable Development Goals (SDGs) cannot be achieved when a billion or more people remain in poverty and inequality is systemic. However, there is no simple, universal relationship between inequality and economic growth. The empirical literature is converging on a tentative consensus that inequality is generally harmful for the pace and sustainability of economic growth over the medium run (Berg et al. 2018). In their study Berg et al. (2018) reach the following conclusions:

First, lower net inequality is strongly and robustly correlated with faster and more durable growth, controlling for the effect of redistribution. Second, redistribution appears generally benign in terms of its impact on growth; only when redistribution is very large is there some evidence that it may have direct negative effects on the durability of growth. Third, we find preliminary evidence that inequality's impact on growth works through lower education and life expectancy, and higher fertility.

Beyond negative impacts on national economies, for example through limited participation in formal markets, evidence is also accumulating that links inequality with social 'bads', such as increases in child mortality, increasing crime rates, declines in social trust, mental health problems and rising rates of incarceration (Wilkinson and Pickett 2009, Fig. 13.1). Inequality is also associated with social conflict and political instability (Scheffer et al. 2017), both within and between nations. Signals of such conflict and instability have become increasingly evident in recent years (Østby 2008; Cederman et al. 2011; Dabla-Norris et al. 2015).

Within the more general concerns about inequality and its effects on society and growth, there is a particular focus on the impacts of gender inequities. A review of studies focusing on the correlation between gender equality and economic growth (Nallari and Griffith 2011) suggests that gender equality, measured in terms of education and employment (Kabeer and Natali 2013), is positively linked to economic growth. The contrast between women in poor and rich countries is striking, with women in poorer countries faring much worse on indicators of gender equality such as education, health, economic rights, marriage rights and participation in parliament. The International Finance Corporation (IFC 2017, p. 3) concludes that gender equality is 'a key contributor to growing and strength-

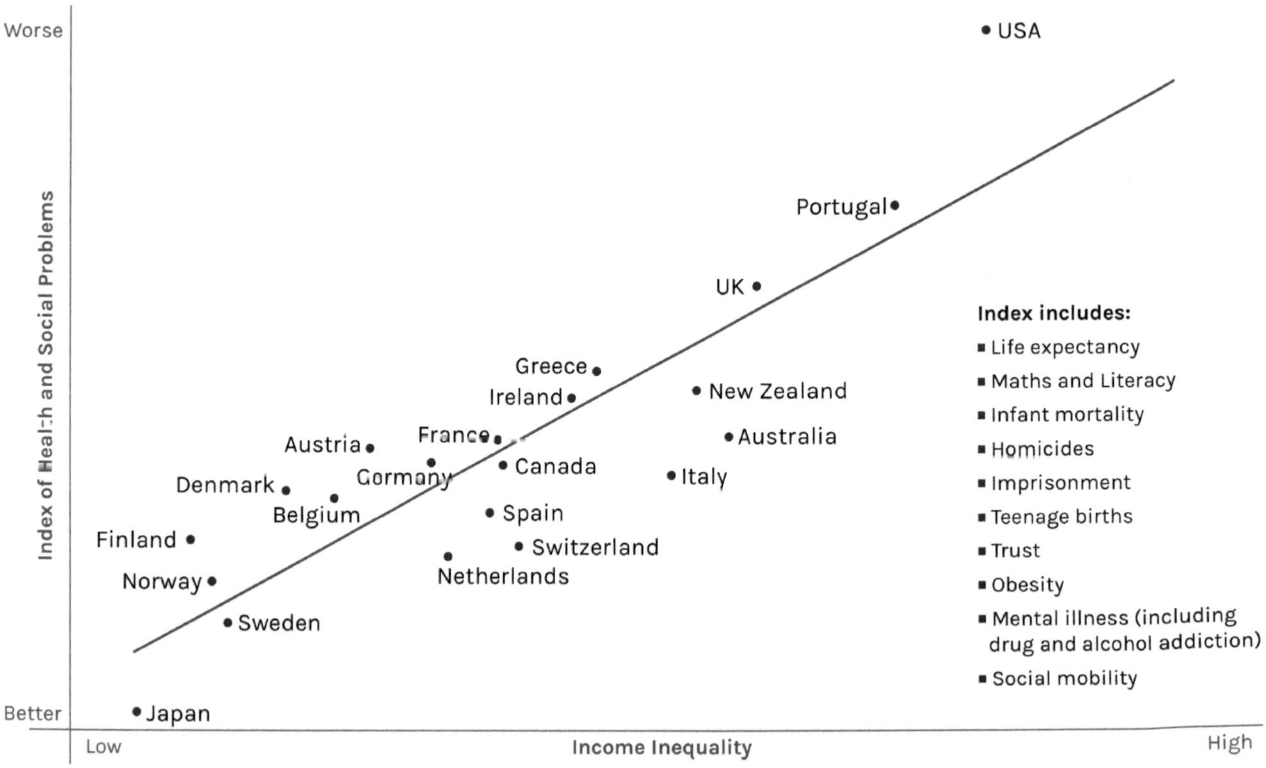

Fig. 13.1 Health and social problems are worse in more unequal countries. Note: Inequality has negative societal consequences for both rich and poor nations. (Sources: Wilkinson and Pickett 2009 and www.equalitytrust.org.uk)

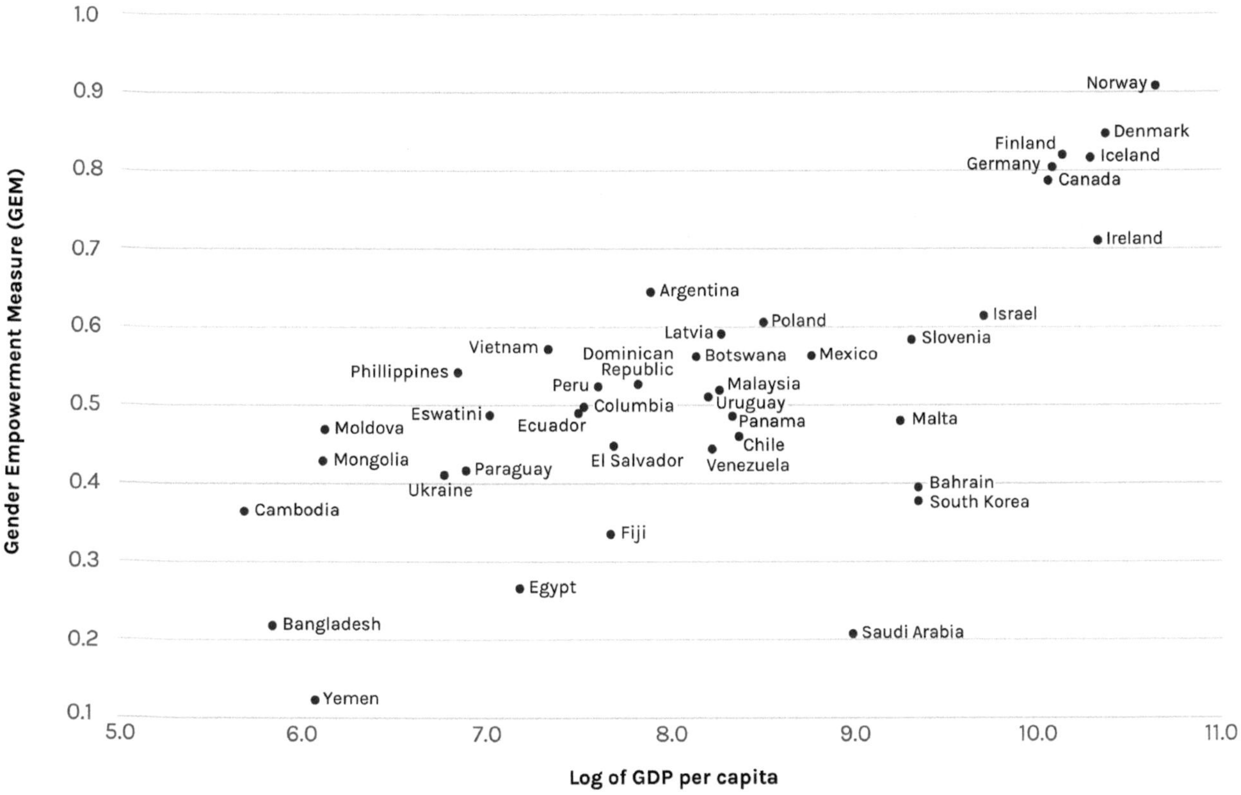

Fig. 13.2 Relationship between Gender Empowerment Index (GEM) and Gross Domestic Product (GDP). Relationship between Gender Empowerment (GEM) index and gross domestic product (GDP), from data for every third country in the IMF database (ordered alphabetically). If data were missing, the next country on the list was chosen. (Source: Stotsky 2006, pp. 23–24)

ening national, regional, and global economies'. While correlations between gender equality and growth are strong (Fig. 13.2), they appear to be asymmetrical. Gender equality contributes to growth, but findings are much less consistent when it comes to growth redressing critical dimensions of gender equality (Kabeer and Natali 2013; Kabeer 2016). Investments and processes of growth consequently need to be accompanied by specific gender equality–oriented public and private sector measures (Kabeer 2012; IFC 2017).

Any future sustainable ocean economy strategy should include means of reducing existing inequalities as well as preventing the widening of ocean inequities, both within and among countries. A sustainable ocean economy should ensure that the potential gains in wealth from the development of new ocean industries are distributed to address social problems seen in more unequal societies. The development approaches and policy strategies designed within a sustainable ocean economy must also shape existing ocean sectors (e.g., fisheries, maritime transport, aquaculture) so that they too recognise and include social equity concerns (Bennett et al. 2019a, b).

Extreme inequality is a social 'bad', for both moral and instrumental reasons. Addressing inequalities should include addressing issues of governance, social norms, gender, global inequalities (e.g., between North and South), inequalities at national scales and intergenerational inequities. Borrowing from the definition of the 'green economy', a sustainable ocean economy should thus include opportunities for economic development that result in 'improved human well-being and social equity, while significantly reducing environmental risks and ecological scarcities' (UNEP 2011, p. 16; UNCTAD 2014, p. 2).

1.4 Equity in an Ocean Context

The inequitable distribution of benefits is not consistent with a global policy agenda advocating for sustainable ocean use for the benefit of all (UN 2015). In short, concerns associated with social equity and an ocean economy are related to (1) the way benefits are shared and (2) the distribution of harms, both of which include the level to which different groups are included in or excluded from decision-making.

In this Blue Paper, we assess the fairness of the current global ocean economy and explore what can be done to facilitate fair sharing of the benefits from ocean use, with an aim to align concerns for social equity with concerns about environmental sustainability. The fairness issues we address exist at and across multiple scales (from global and national levels to those of communities and subgroups) and involve relationships (bilateral or otherwise) among multiple types of actors (governments, civil society, international agencies, and private corporations) with different levels of power, capacities and incentives to address ocean equity. Where actors have the power to disregard equity concerns, there has to be some mechanism to bring fairness issues to bear; for example, through multilateral agreements and/or regulatory approaches. Without an active championing of equity, inequality will be the default outcome.

This Blue Paper addresses the following central questions:

- What types of inequity are prevalent in the use of marine resources? How can differences in fairness be explained?
- How are sustainable and fair use of marine resources interrelated? Why is it important to strive for both simultaneously?
- What can be done in terms of policy and practice to improve social equity in relation to people's use of the ocean?

In the following sections, we explore different types of equity, describe why inequity is a challenge in relation to sustainability and conclude with opportunities for action aimed to foster just ocean sustainability.

2 Key Findings

2.1 How Are Ocean Benefits and Harms Distributed?

The ocean produces oxygen, stores carbon and heat, produces food, offers space for economic activities and facilitates international trade and the transport of goods (White et al. 2012; Resplandy et al. 2018). It also provides non-monetary benefits in the form of advances in scientific knowledge, opportunities for collaboration, sense of place, feelings of wonder and worship, and a free place to play or gather with family and friends (Fraser and Spencer 1998; Whitehead et al. 2008; Garcia Rodrigues et al. 2017). The ocean and all its benefits should be enjoyed by all.

The potential benefits from ocean-based economic activities, include taxation and rents for governments, payments for access agreements, financial and employment benefits for national economies, as well as livelihood opportunities and social benefits for local communities and tourists visiting coastal and marine environments. Globally, the value of key ocean assets has been estimated at US $24 trillion and the value of derived services at between $1.5 trillion and $6 trillion per year (Hoegh-Guldberg 2015; Lillebø et al. 2017; Cicin-Sain 2015; OECD 2016).

Generally, however, the distribution of benefits from ocean use flows disproportionally to some actors (Klain et al. 2014; Wynberg and Hauck 2014). Focusing on fisheries as a sectoral example, between 2004 and 2014, 25 countries were responsible for roughly 82% of global catches (FAO 2018). The concentration of national actors is substantially higher on the high seas, beyond exclusive economic zones (EEZs) (Tickler et al. 2018), where five high-income countries are responsible for 86% of total fishing effort (McCauley et al. 2018). In the corporate sphere, some companies are becoming more powerful than countries, and industry consolidation is prevalent. In seafood production, for instance, 13 companies control 11–16% of global catches (Österblom et al. 2015). A similar analysis for genetic resources shows that 1 company has registered 47% of all known marine genetic sequences, thereby exceeding the share of 220 other companies (Blasiak et al. 2018b). Consolidation is also prevalent in the seed industry, agriculture, forestry, mining and other sectors influencing the planet and its people (Folke et al. 2019). Ongoing analysis of ocean industries indicates substantial consolidation in maritime transport, cruise industries, offshore wind, ports, shipbuilding and repair, as well as offshore oil and gas, with the majority of companies headquartered in a small number of countries (Monacelli 2018; John Virdin, Duke University, unpublished data). Such patterns highlight the unequal control of access to and distribution of benefits arising from ecosystems all over the world (Wynberg and Hauck 2014).

The ocean economy can produce a number of social harms, undermine the productivity and abundance of marine resources that local communities rely on, and pollute the marine environment, thereby compromising the safety of food resources and local people's health, recreation and well-being (Stonich et al. 1997; Stonich 1998; Page 2007). Development activities can also undermine people's rights or displace them from areas they have historically and/or traditionally used (Zalik 2009; Bennett et al. 2015; Barbesgaard 2018).

Inequity arises from a number of social factors. These include not only the different stakeholders involved and the power they can wield but also the social institutions and structures through which the economy operates (Ciplet et al. 2015; Crona and Bodin 2010; Felipe-Lucia et al. 2015). Mechanisms that can uphold inequities from the ocean economy include historical and colonial legacies, lack of access to and allocation of resources, insecure territorial and tenure rights, financial resources and technological capacity (Abdullah et al. 2017; Bourguignon 2015).

Value chains, market policies and investments similarly shape equity in terms of access, benefits and costs, and working conditions. Not taking the full value chain of the ocean economy into account hides inequitable opportunities and impacts on women, for instance, who tend to be less involved in the extractive part of the value chain but are engaged in processing and marketing (Harper et al. 2013; Kruijssen et al. 2016). Invisible value chains, based on unreported catches and illegal activities, can mask labour trafficking, peonage systems, unsustainable resource use or health and sanitary issues while simultaneously detracting from wider economic benefits and avoiding taxation (Lopes et al. 2017; Moreto et al. 2019).

Insufficient consideration or inclusion of developing states or local populations in decision-making processes related to ocean development is a substantial concern. Representatives from coastal communities and groups often marginalised (e.g., women, indigenous groups, individuals with disabilities and poor people) are frequently not, or not adequately, included in decisions related to development (e.g., site selection of ports, energy and oil development, aquaculture) that will impact them (Kerr et al. 2015; Flannery et al. 2018). Fisheries agreements have, for instance, been described as primarily commercial deals negotiated by governments behind closed doors, with few benefits accruing to local economies (Kaczynski and Fluharty 2002; Le Manach et al. 2012). See, however, Almeida et al. (2009) for an example of fair and participatory fisheries agreements.

2.2 Why Is Social Equity Important in a Sustainable Ocean Economy?

The idea of fairness in relation to use of natural environments can be explained by the concepts of 'environmental justice' (Schlosberg 2009) and 'ecological justice' (Baxter 2004). Environmental justice bridges key goals of environmental protection and social justice by focusing on correcting maldistribution, or how less powerful groups in societies derive fewer environmental benefits and are exposed to more environmental harms (Schlosberg 2009) (see Box 13.1). In essence, 'Environmental justice is defined as the fair treatment and meaningful involvement of all people regardless of race, color, national origin, or income, with respect to the development, implementation and enforcement of environmental laws, regulations, and policies' (EPA 2017). Ecological justice, in contrast, focuses on preventing, mitigating or repairing environmental harm brought about by human activities and the granting of fundamental rights to non-humans. This Blue Paper is concerned with the former only, in other words—equity and fairness in relation to the access to and control over ecosystem benefits (Leach and Mearns 1998; Ribot and Peluso 2003).

A number of academic fields have focused explicitly on environmental justice. Central to this literature is the idea that people and groups appropriate ecosystem services and benefits through claims, underpinned by various abilities, or power bundles (Ribot and Peluso 2003; Boonstra 2016) sanctioned by law, custom or convention. These powers, in turn, are ultimately rooted in people's ability to influence the behaviour of others and the social and ecological conditions in which others operate (Boonstra 2016).

We suggest that social equity provides an all-encompassing framework and define two specific sub-categories of social equity: procedural equity and distributional equity (Franck 1995; McDermott et al. 2013; Pascual et al. 2014; Zafra-Calvo et al. 2017) (see also Box 13.1). These two sub-categories can be defined as follows:

1. **Procedural equity** refers to the recognition of rights and needs of all groups and the level of inclusion and participation in decision-making related to ocean development.
2. **Distributional equity** refers to fairness in the sharing of benefits and the minimisation of harms across all groups from ocean development.

There are two broad reasons why pursuing equity should be a central concern for a sustainable ocean economy (Bennett 2018). The first is a normative argument: extremes of inequality challenge universal notions of fairness. Including people in decision-making as well as improving how benefits are distributed is simply the right thing to do. Indeed, these are global norms contained in many guiding policy documents and international commitments related to human rights, sustainability and development (see Sect. 2.3). The second is an instrumental argument: equitable procedures and outcomes can be important for supporting the achievement of sustainability objectives.

Equity is an increasingly critical component of ensuring that ocean-based economic and other activities have a social license to operate (Mather and Fanning 2019; Voyer and van Leeuwen 2019). Taking social equity considerations into account will lead to a fairer distribution of benefits to different segments of society and maintain the legitimacy of the ocean economy. What is considered fair and what levels of inequality a society tolerates vary from place to place (Box 13.1). This is a decision for individual societies to make, however, as part of their commitments to achieving the SDGs, and in line with existing legal frameworks.

2.3 What Rules and Principles Exist to Support Equity?

The international community increasingly recognises equity as central to achieving the SDGs. A number of the global goals spell this out explicitly, including SDG 1 (Ending Poverty), SDG4 (Education), SDG5 (Gender Equality) and

SDG 10 (Reduced Inequalities). SDG 14 (Life under Water), also has a number of equity-related targets, such as Targets 14.6 and 14.7. The desire to address inequality is most clearly spelled out in the overall ambition of the UN Agenda 2030 to 'leave no-one behind'. Despite the recognition of the importance of equity in international law (Franck 1995), equity is not, as such, a general rule. Rules and principles to achieve equity may, however, be established through law-making processes such as treaties and customary international law. Soft-law instruments can contribute to both the making of a treaty, as standard setting, and to customary international law, as state practice.

Guiding principles of equity are relevant in addressing two categories of ocean equity. The first, *intergenerational* equity (Sect. 2.3.1), relates to the conservation and sustainable use of the marine environment in a manner that ensures the ability of future generations to reap its benefits also (Brown-Weiss 1990; Tladi 2007). The second, *intragenerational* equity (Sect. 2.3.2) is more immediate and concerned with ensuring equitable distribution of benefits and resources within the current generation (Okereke 2006; Tladi 2007). It calls for solidarity in uplifting those who are marginalised and underprivileged. The sentiment is expressed in, for example, the call by the Conference of the Parties of the Convention on Biological Diversity that 'ecosystems should be managed for their intrinsic value and for the tangible or intangible benefits for humans, in a fair and equitable way' (CBD 2000).

2.3.1 Intergenerational Equity: Protection of the Marine Environment

Three key principles have been designed to enhance *intergenerational* equity. First and foremost, the precautionary principle (Freestone and Hey 1996, p. 3; Tladi 2014, p. 108) stipulates that scientific uncertainty should not be used as a reason not to adopt measures to protect the environment. It represents a central element of the Fish Stocks Agreement (UNGA 1995, Arts. 5 and 6) and the 2012 Rio Plus 20 outcome document, *The Future We Want* (para. 58).

Second, the duty to prevent transboundary harm to common areas, including the ocean, is clearly spelled out in Article 3 of the Convention on Biological Diversity: '[States have] the responsibility to ensure that activities within their jurisdiction or control do not cause damage to the environment of other States or of areas beyond the limits of national jurisdiction' (ICJ 2010; Murase 2015, paras. 55–59).

Third, the duty to perform environmental impact assessments for activities that may cause harm to the marine environment (ICJ 2010; ILC 2018), and may therefore negatively impact future generations, is also firmly rooted in laws and policies relevant to the management of the ocean (ITLOS 2011; UNGA 2018, Art, 206; ICJ 2010).

The UN Convention on the Law of the Sea (UNCLOS 1982) contains general provisions on the duty to protect the marine environment (UNCLOS Part XII). The convention also contains particular rules applicable to the different maritime zones. Even with the numerous provisions on environmental protection, the environmental regulations in the convention are seen as insufficient (Gjerde 2006). Other regulatory tools exist that can complement the legal framework established by UNCLOS. For example, although the Convention on Biological Diversity in principle only applies to areas within national jurisdiction (CBD Art. 4[a]), its provisions can be applied to the ocean, including areas beyond national jurisdiction, with respect to 'processes and activities' (CBD Art. 4[b]).

Relevant CBD processes include, for example, the criteria for the establishment of Ecologically or Biologically Significant Marine Areas (EBSAs) (CBD 2008). The annual General Assembly resolutions on oceans and the law is another avenue relevant for the interpretation of obligations under UNCLOS. They contain provisions addressing the marine environment, including the call for an ecosystem approach (UNGA 2018, para. 187). A number of other environmental rules, such as several International Maritime Organisation conventions and rules under the Regional Seas Program can complement UNCLOS. Overall, however, these rules are fragmented and there is insufficient coordination in their application, resulting in uneven protection of the marine environment, thus undermining intergenerational equity.

Intergenerational rights to a healthy environment are also specifically considered in the constitutions of 74% of the world's nations—in theory therefore offering the best hope to protect future citizens as constitutions supersede other laws in a jurisdiction by establishing sovereignty (Treves et al. 2018). If these frameworks were to be enforced by decision-makers and courts, they would sustainably protect the biosphere and substantially contribute to equity in a sustainable ocean economy. 'Enforcing constitutional and public trust frameworks for intergenerational equity will be more feasible in jurisdictions that grant legal standing to youths and the legitimate representatives of future generations' (Treves et al. 2018).

2.3.2 Intragenerational Equity: Promoting Economic Equity

The *intragenerational* dimension of equity requires that efforts to protect the environment account for the needs of the most vulnerable in society (Brundtland 1987). However, this sentiment is not well developed in international environmental law, policy and practice. Nevertheless, policy options for addressing intragenerational equity, which may be collectively referred to as common-but-differentiated responsibilities, include the idea of differentiation of obligations, transfer of technology and funds, as well as capacity-building. In relation to ocean governance, all of these options are possible.

The principle of the common heritage of mankind, which has been described as a norm that combines the intragenerational with the intergenerational dimensions of equity (Tladi 2015), is the principle most synonymous with equity under UNCLOS. Through the application of Part XI of UNCLOS, this principle requires that activities in the deep seabed (the 'Area') 'be carried out for the benefit of mankind as a whole, irrespective of the geographical location of the States, whether coastal or landlocked'. While application of this principle beyond the 'Area' is not accepted by all, one of its central tenets, benefit-sharing, remains an important policy option to ensure a more equitable allocation of benefits from the ocean (Morgera 2016). Other provisions geared towards intragenerational equity include capacity-building and technology transfer provisions (UNCLOS, part XIV).

Technology and fund transfer to developing countries will be key to protecting marine biodiversity in areas within and beyond national jurisdiction (IGC 2018; Voigt-Hansen 2019), to enable developing countries to meaningfully participate at international fora and meet their international obligations. However, while UNCLOS and the Convention on Biological Diversity include absolute obligations to transfer technology (Morgera and Ntona 2018; CBD 2004, Annex, para. 11), the meaning of 'transfer of technology' is very broad and all-encompassing, with those obligations couched with qualifiers such as 'in accordance with capabilities' or 'endeavour to promote' and closely tied with scientific knowledge. This leaves much open to interpretation and makes it difficult to evaluate how international cooperation is to work in practice (Harden-Davies 2017). While capacity-building and transfer of technology obligations in UNCLOS and other instruments are qualified, the commitments to 'increase scientific knowledge, develop research capacity and transfer marine technology' under the SDGs are not (SDG14a). Even if these commitments are not legally binding, they do provide a political springboard for the elaboration of unqualified, legally binding commitments in new instruments and legal frameworks.

The 2001 International Treaty on Plant Genetic Resources for Food and Agriculture provides a useful model regarding the sharing of benefits from genetic resources beyond national jurisdiction. Articles 10 to 13 provide for a multilateral access and benefit-sharing regime based on four pillars: (a) exchange of information; (b) access to and transfer of technology; (c) capacity-building and (d) sharing of benefits arising from commercialisation. A similar framework forms the basis of the access and benefit-sharing regime for genetic resources established by the 2010 Nagoya Protocol, to ensure that states in whose territories—including in maritime areas—genetic materials are sourced are able to enjoy the benefits arising from the use of those resources (Voigt-Hansen 2019; Harden-Davies and Gjerde 2019; however, see Blasiak et al. 2018b for some of the protocol's limitations). Beyond benefit-sharing alone, capacity-building and technology transfer are key to fostering distributive and procedural equity (see also Leape, Abbott, Sakaguchi et al. Blue Paper: 'Technology, Data and New Models for Sustainably Managing Ocean Resources').

A striking example of the challenge of achieving both procedural and distributive equity concerns landlocked states, which are without physical access to the sea and almost by definition excluded from enjoying ocean benefits. To remedy this inequity, UNCLOS creates rules to facilitate the rights of landlocked states 'to participate, on equitable basis, in the exploitation…of the surplus of the living resources of the exclusive economic zone of coastal states' in the same region (UNCLOS, Art. 69). However, the right to participate is limited to 'an appropriate part of the surplus'; if a coastal state was to claim that it does not have a surplus, then arguably the right cannot be claimed, and the right to participate is subject to agreement between states. Provisions, therefore, while present, tend to be filled with many caveats making their implementation difficult.

2.3.3 Human Rights

While international human rights are not typically seen as directly applicable in ocean governance, they should be included and applied in the search for equity in a sustainable ocean economy. Human rights obligations apply not only within the territories of states, but also over an activity under the control or jurisdiction of states, including vessels flying the flag of a state and activities in the high seas or the Area under the control of states (Wenzel 2008).

A number of rights may be particularly relevant in the pursuit of ocean equity. First, the right to development, which calls for solidarity and uplifting the poor and marginalised, is directly related to the intragenerational equity described above. It is contained in, among other instruments, the African Charter on Human and Peoples' Rights (Art. 22), the 1993 Vienna Declaration and Programme of Action (para. 10) and the 2000 Millennium Declaration (para. III). It can also be inferred from other instruments such as the International Covenant on Economic, Social and Cultural Rights and the Rio Declaration on the Environment and Development. Second, the right to equality and non-discrimination can further support fairness in an ocean governance context (Universal Declaration of Human Rights Art. 2, International Covenant on Economic, Social and Cultural Rights, Art. 2, International Covenant on Civil and Political Rights, Arts. 2 and 26).

Some prohibited grounds of discrimination have also been the subject of specific treaties, such as the Convention on the Elimination of All Forms of Racial Discrimination and the Convention on the Elimination of All Forms of Discrimination against Women. These rights could potentially be made applicable to, for example, fishing permits. While regional fisheries management organisations do not, typically, consider race and gender when establishing allowable catch requirements, national authorities should, in keeping with human rights standards, account for the needs of the most disadvantaged and marginalised.

Labour rights is one area in which the protection of human rights has been directly applied in ocean governance. Labour or employment rights are contained in, for example, the International Covenant on Economic, Social and Cultural Rights, including the right of 'just and favourable conditions of work' (Art. 7). The Maritime Labour Convention (Arts. III and IV) includes requirements for regular payment and processes to ensure fair wages (e.g., Regulation 2.2). The Work in Fishing Convention C188, adopted in 2007, aims to ensure that all fishers have decent working conditions on board fishing vessels.

One area with much room for improvement is the role of business in enhancing equity. While human rights obligations are binding on states, business entities have the greatest potential to impact human rights and the environment (Ratner 2007; Oyewande 2008). Business entities, including those fishing and mining in the ocean, do not have direct obligations under international law. This creates difficulties where business entities act in the territories of third states and areas beyond national jurisdiction (Duruigbo 2003; Muchlinski 2007). To address this issue, the obligations of states in human rights treaties to 'protect, respect and fulfil' have been interpreted as establishing a duty on the state to ensure that rights are protected in private relationships, including between corporations and other persons (Ruggie 2008), thus creating an indirect duty of 'non harm' on the corporations. Moreover, the United Nations is currently considering the possibility of a treaty to regulate the activities of multinational corporations that impact on the environment and the enjoyment of human rights (Meyer 2017).

2.4 Case Studies of Hope and False Hope

The following sections focus on concerns for ocean equity across a variety of ocean-related sectors and equity dimensions, including the distribution of burdens and benefits on the high seas, inequalities associated with infrastructure development and the role of transnational corporations in a sustainable ocean economy (see Table 13.1 for an overview). Although much of the scientific work to date has revolved around gender equity and the rights of small-scale fisheries and coastal communities (Tables 13.1 and 13.2 and Appendix 1), there is increasing interest in engaging with inequalities in other areas.

Table 13.1 Key points from case studies

Case study	Summary
Equity and sustainable fisheries	Substantial attention has been devoted to addressing ecological sustainability in fisheries, and the FAO Code of Conduct for Responsible Fisheries is an important example. Endorsement of the Voluntary Guidelines for Securing Sustainable Small-Scale Fisheries (SSF) in the Context of Food Security and Poverty Eradication may contribute to improvements in the equitable distribution of benefits by giving a voice to, as well as representing the interests and respecting the human rights of small-scale fishers. However, implementation of existing international guiding policies remains a challenge
Gender-transformative approaches	Existing training opportunities, targeting only women in 'accommodating' ways, have had limited impact because they have failed to address underlying harmful power structures and norms restricting women from equitably engaging in and benefitting from ocean-based activities. Gender-transformative approaches encourage men and women to shift these barriers and catalyse fair development outcomes
Ocean-based infrastructure and coastal community equality	Coupling of offshore activities with a regular compensation mechanism to coastal communities in the United Kingdom is an example of how to support the fair distribution of benefits from ocean-based industry. While this is an example from a wealthy state where institutions are prepared to set up and control such a system, it illustrates a possible framework through which vulnerable coastal communities can be associated with offshore activities
Equity in areas beyond national jurisdiction	Pelagic fish stocks and marine genetic resources (MGRs) are two examples of transboundary resources often shared at one stage or the other (of their life cycle or migration routes) between exclusive economic zones (EEZs) and areas beyond national jurisdiction (ABNJs). Fisheries on and conservation of highly migratory stocks may disproportionately affect developing states. In the case of MGRs, an imbalance in patent ownership is problematic from an equity perspective. Ongoing negotiations on an international legally binding instrument under the United Nations Convention on the Law of the Sea on the conservation and sustainable use of marine biological diversity of areas beyond national jurisdiction are attempting to redress these inequities by developing strong and sustained mechanisms for capacity-building and technology transfer at global, regional and national scales
Can corporate actors promote equity?	While corporate bodies operate within legislative and other norm-based frameworks, they also shape expectations as to what constitutes appropriate behaviour as well as aspirational desires for future relationships. Although several ocean-based sectors have paid substantial attention to ocean sustainability, equity concerns remain poorly addressed. Prioritisation of equity by major actors has the potential to influence entire sectors

Table 13.2 Scientific documentation of inequities in small-scale fisheries, undermining sustainable livelihoods and contributing to loss of well-being

	Outcomes	Examples
Inequity of benefits		Canada, Kenya, Indonesia, Philippines
Structural inequalities in value chains	Unequal trading relationships and inability to obtain fair value of catch. Limited capacity to compete with more powerful actors	Wamukota (2009), Cinner et al. (2012), Knudsen (2016), Fabinyi (2012), Trinidad et al. (2014), Crona et al. (2016), Rosales et al. (2017), Purcell et al. (2017), Hicks et al. (2019)
Vulnerability to degradation of resources	Loss of food security, cultural practices and well-being. Compelled by subsistence needs, may also increase destructive behaviour, resource use or non-compliance	Cinner (2009), Cinner et al. (2009), Crona et al. (2015), Sadovy de Mitcheson et al. (2018), Yamazaki et al. (2018), Frid et al. (2016), Baker-Médard (2017)
Invisible inequities		Multiple locations
Gendered invisibilities	Women are often invisible, and hence marginalised in the management of marine resources (e.g., due to gender-blind policies, focus on formal and paid fishing activities, or the production segment of fisheries value chains) Difficult to know how women are affected as the fisheries sector develops	Yodanis (2000), Bennett (2005), Williams (2008), De Silva (2011), World Bank (2012), Harper et al. (2013), Daw et al. (2015), Lentisco and Lee (2015), Schwerdtner Máñez and Pauwelussen (2016), Kleiber et al. (2017), Harper et al. (2017), Fortnam et al. (2019)
Inequity in access		Bangladesh, Brazil, Canada, Ghana, Kenya, Madagascar, Norway, Philippines, Zambia
Gendered access barriers	Barriers to profitable segments of supply chains, and/or access to fishing grounds, boats, fishing gear, financial capital, credit, education, alternative livelihoods	Yater (1982), O'Neill and Crona (2017), Walker (2001), Eder (2005), Matsue et al. (2014), Wamukota (2009), Cole et al. (2015), Kruijssen et al. (2016), Baker-Médard (2017), Cole et al. (2018), Kleiber et al. (2017), Gerrard and Kleiber (2019)
Decision-making and governance		Bangladesh, Brazil, Canada, Tanzania, Indonesia, Solomon Islands
Access to governance	Women and minority groups—such as indigenous groups, disabled and poor people—face access barriers to governing institutions (e.g., as a result of customary rules and norms) and are not accounted for in fisheries management, leading to policy interventions that undermine sustainable livelihoods	Bennett et al. (2018), Thorburn (2000), Fröcklin et al. (2013), Kleiber et al. (2017), Ban et al. (2018), Bennett (2005), Daw et al. (2015), Baker-Médard (2017)

Note: The countries specified in the table represent examples of places where inequities have been scientifically studied. Some of these countries have recently invested in human, financial and/or technical capacity to address challenges identified, but, at the time of publication of this blue paper, no peer-reviewed scientific documentation was available that had assessed the effectiveness of such recent efforts.
Details in Appendix 1

2.4.1 Equity and Sustainable (Small-Scale) Fisheries

Small-scale fisheries support the majority of the world's fisherfolk (47 million women and men in developing countries alone) and utilise the least capital, fuel and technology (World Bank 2008, 2012; Schuhbauer et al. 2017; Zeller and Pauly 2019). While landing the bulk of catches for human consumption, large-scale industrialised fleets, in contrast, are highly subsidised, employ relatively few fisherfolk and have high discard rates (World Bank 2008; Carvalho et al. 2011; Sumaila et al. 2016; Zeller and Pauly 2019). Large-scale industrial fisheries and associated value chains can undermine the catches, livelihoods and food security of small-scale fishers and coastal communities (De Schutter 2012; Gagern and van den Bergh 2013; Pauly et al. 2014). There is a risk that intensification of economic use of the ocean and coasts for mining, logging, infrastructure development, coastal tourism and aquaculture can reinforce the weak position and vulnerability of small-scale fishers (Bavinck et al. 2017, 2018; Carver 2019; Cohen et al. 2019).

Small-scale fishing communities, particularly indigenous and women subgroups, often have relatively limited political power compared to large-scale fisheries actors (Table 13.2). Small-scale fishers are at times depicted by policymakers as ignorant, inefficient or environmentally destructive, leading to policies that target them with negative livelihood effects (Lowe 2013; Cohen et al. 2019). Blaming small-scale fisheries for problems often misses systemic inequalities that can be driving far more significant environmental degradation, including illegal fishing and corruption (Eder 2005; Fabinyi 2012; Li 2007; Segi 2014; Finkbeiner et al. 2017; Sumaila et al. 2017).

Inequities are apparent also within small-scale fish-producing communities. These are often structured along intersecting social categories such as wealth, gender, age, religion, migrant-status and ethnicity. Inequities in ocean resource benefits may reinforce existing inequities experienced by particular groups in access to healthcare, education and rights over land (Béné and Friend 2011; Jentoft and Eide 2011; Mills et al. 2011; Allison et al. 2012).

The Food and Agriculture Organization of the United Nations (FAO) Code of Conduct for Responsible Fisheries (CCRF), adopted in 1995, is an important tool for fisheries sustainability and has advanced equity through development of the Voluntary Guidelines for Securing Sustainable Small-Scale Fisheries (SSF) in the Context of Food Security and Poverty Eradication (FAO 2015). These guidelines are closely related to the Voluntary Guidelines for the Responsible Governance of Tenure of Land, Fisheries and Forestry in the Context of National Food Security (VGGTs). Both instruments are grounded in a human rights-based approach and specifically include equity and equality among their guiding principles.

The SSF guidelines have been embraced by several regional organisations (TNI 2016): the Central America Fisheries and Aquaculture Organization, the Fishery Committee for the Eastern Central Atlantic, the Southeast Asian Fisheries Development Center, the African Union and the General Fisheries Commission for the Mediterranean. The General Fisheries Commission's 10-year Regional Plan of Action for Small-Scale Fisheries is expected to increase social equity within the fisheries sector in that region. A draft law in Costa Rica, aiming to overcome the voluntary nature of the SSF guidelines, will likely contribute to improvements in the equitable distribution of benefits.

More generally, the CCRF has also been integrated into national regulatory frameworks with technical guidance and voluntary guidelines aimed to facilitate its operationalisation. Examples of particular relevance to social equity include the Ecosystem Approach to Fisheries in the FAO Technical Guidelines for Responsible Fisheries, which specifically encompasses improving human well-being and equity (FAO 2003). The FAO Technical Guidelines for Responsible Fisheries on the management of marine protected areas and fisheries aim at balancing environmental and social outcomes in a domain often dominated by conservation goals (Westlund et al. 2017).

Although an international framework of guidance is in place to support social equity in the fisheries sector, implementation often remains a challenge. Scientists, civil society organisations and social movements are analysing threats posed by ocean economy developments in terms of justice and equality (TBTI 2016) to understand how implementation can be accelerated.

2.4.2 Gender-Transformative Approaches

Significant efforts have been made to mainstream gender in fisheries policy and investments. These have mainly focused on visible gender gaps, such as gender imbalances in who accesses and participates in extension programs (Kleiber et al. 2017) and typically have consisted of 'accommodating' and 'gap-filling' approaches. For instance, in Bangladesh, women are targeted for capacity-building activities in ways that accommodate their practical needs (Choudhury et al. 2017; Behailu et al. 2019). Trainings may be held close to women's homes, at a time convenient for them, and the skill or technology transferred may feed into a livelihood option that can be performed at home. While socially acceptable, interventions that only build women's capacity, target women or deploy gender-responsive technologies at women have limited impact (Morgan et al. 2015; Farnworth et al. 2015; Behailu et al. 2018; Choudhury et al. 2017). Indeed, they fail to address underlying barriers that perpetuate gender inequities, including stereotypes, non-recognition of women as fishers or contributors along the value chain, and harmful norms restricting men and women from engaging in and benefitting from activities equitably (Kantor et al. 2015; McDougall et al. 2015; Choudhury and McDougall 2019).

Gender-transformative approaches in natural resource management are meant to address these underlying normative barriers (Wong et al. 2019; Cole et al. 2018), yet their implementation in developing countries remains limited. In Indonesia, for example, despite women being reached by many project activities since 1998, only two projects (10%) applied a gender-transformative approach (Stacey et al. 2019). A case from Bangladesh (see Box 13.2) illustrates the potential of transformative approaches to catalyse greater development and gender outcomes. While focused on a freshwater context, the framework is likely also relevant for marine resources (see Promundo-AAS 2016; Cole et al. 2018). More generally, advancing public discussion of gender equality in fisheries and making progress in women's empowerment requires effective messaging and awareness, political and social will, and support from the government, NGOs and the private sector.

Box 13.2: Transforming Underlying Gender Barriers in Bangladesh

Introducing innovations, such as more intensive homestead-based pond polyculture that can provide nutrient-dense small fish to low-income and coastal communities often struggling with nutrient deficiencies, is a priority for Bangladesh's government, NGOs and international research for development. Interventions have commonly targeted and trained women, but because the latter were not given control over ponds, and because investments in ponds have largely depended on the support of male household members, women were not able to implement and optimise innovations (Morgan et al. 2015). Women were also reluctant to get into ponds for practical reasons (because their wet clothing is difficult to dry).

The development of an affordable small-mesh gillnet that women could use from the pond banks did not resolve the problem, because women's use of such nets was constrained by gendered roles that see 'fishing' as a men's domain. Women therefore faced social repercussions for harvesting (Kruijssen et al. 2016), despite the strong nutritional need for fish for families and children in this area (Bogard et al. 2015).

Building on pre-pilots (Farnworth et al. 2015; Kantor et al. 2015; McDougall et al. 2015), the gillnet intervention was redesigned to build commitments for family support for women as fishers (aquaculturalists) and innovators. The transformative measures involved spouses and more powerful household members in critical reflection and dialogue (Promundo-AAS 2016) around gender dynamics ranging from intra-household gender power hierarchies to food distribution. Interventions also sought to discuss how current norms limited individual and family well-being and what steps could be taken collaboratively to shift gender relations. Women were also coached in self-confidence, negotiating skills and assertiveness. Results showed changed attitudes amongst men and women, enhanced collaboration among family members and greater acceptance of technology use by women.

Despite repression, by drawing strength and inspiration from their traditional identities and power within their society, women themselves can be agents of change. In British Columbia, First Nations Heiltsuk women drew on their traditional and contemporary roles as mothers, teachers, organisers and political leaders to oppose a controversial commercial herring (*Clupea pallasii pallasii*) sac-roe fishery. By taking on leadership roles, increasing social cohesion, facilitating information flow and engaging in critical negotiations, women demanded care over traditional marine resources for their children, culture and future generations and helped transform governance of herring on British Columbia's Central Coast (Harper et al. 2018). This example illustrates the importance of social equity and the potential strength of (indigenous) women as agents of change in fisheries governance. However, in many socio-political contexts gender dynamics limit women from exerting this level of leadership, voice and agency.

2.4.3 Ocean-Based Infrastructure and Coastal Community Equality

Activities in the ocean raise questions about how their costs and benefits are distributed among coastal communities. The onshore pollution effects of offshore accidents are well documented, including the Gulf of Mexico Deepwater Horizon accident (Hayworth et al. 2011; Michel et al. 2013), the Erika disaster (Čović et al. 2013) and the recent Brazilian oil spill (Reuters 2019). Yet more enduring relationships between ocean industries, such as oil and gas and offshore renewable energy, and coastal communities also exist. In the United Kingdom, for example, a number of coastal communities have long-standing experience interacting with the offshore oil and gas industry, with new questions being asked about ocean-coastal connections as the number of offshore renewable energy developments increases.

Exploitation of oil and gas in the United Kingdom began in the 1970s and included the construction of onshore facilities to receive crude oil via pipeline, ahead of onward transportation by tankers at the Sullom Voe Terminal in the Shetland Islands. The project remains one of the largest construction sites in Europe and the largest oil terminal ever built at once (Carr and Williamson 1982). At its peak, it processed over 1.5 million barrels of oil a day (Sullom Voe terminal 2018). At the time the terminal was being proposed, the local authority negotiated a compensatory agreement to account for the terminal's negative impacts on Shetland during the course of activities. Compensation was not a one-off payment but a disbursement placed into a trust and linked to activity levels until 2000, after which the money was invested on the stock exchange and in local subsidiary companies (Morgan 2009). The funds are governed by the Shetland Charitable Trust. Financial flows have been substantial and have supported onshore benefits through important investments in community assets, such as sports and cultural venues as well as a district heating scheme. In 2018, closing reserves topped £300 million (Sullom Voe terminal 2018). In this case, local and national governance processes support a third-sector organisation (the charity) to mediate the impact of corporate activity. The beneficiaries of this activity are local community members.

There are also examples of approaches where the potential for unequal experiences of the costs and benefits of development are addressed through formal sharing of benefits (rather than compensation for negative impacts). The idea of 'community benefit' payments first emerged in the onshore renewable energy sector, whereby communities located near renewable assets receive annual payments, often linked to energy production capacity or performance, as part of sharing the benefits of the energy scheme (Kerr et al. 2017; Rudolph et al. 2014; SSE 2019). Community benefit payments are in addition to any positive supply chain effects. Although not mandated by law, on-land community benefit packages have developed through the dynamic interplay between energy developers and communities, under the watchful eyes of governments. The rationale behind community benefit payments is 'driven by a desire to equitably share the benefits gained by harnessing a national natural resource' (Scottish Government 2018, 7). In the United Kingdom, if and how such principles might apply to offshore energy developments is a matter of ongoing consultation.

Arrangements therefore exist that consider the distribution of costs and benefits of ocean-based developments affecting coastal communities. There is also an opportunity to transfer learning from experiences of land-based developments, especially in the context of renewable energy, to ocean-based settings. The particular set of arrangements made will vary depending on the location of developments, the governance context and the power that communities have in their interactions with corporations. Coastal communities are often economically vulnerable and financially subject to fluctuations in the resources they depend upon. Addressing this vulnerability will enhance the equalities profile of the sustainable ocean economy. The two examples above focus on a country with institutional capacity to ensure that development is equitable; regions and countries exposed to ocean-related developments where this may be lacking will need support to avoid inequitable outcomes.

2.4.4 Equity in Areas Beyond National Jurisdiction

Discussions of equity frequently centre on communities, local resource users, traditional knowledge and associated governance and regulatory regimes. The majority of the ocean, however, is more than 200 nautical miles (370 km) from national coastlines, and thus remarkably remote from the daily lives of most people. Indeed, marine 'areas beyond national jurisdiction' (ABNJs) account for some 64% of the ocean. A growing body of research underscores the degree of ecological connectivity between ABNJs and coastal communities, and their importance for the functioning of the biosphere (Popova et al. 2019; Ramesh et al. 2019; Cheung et al. 2019). The life cycles of whales, sharks, seabirds, turtles and tuna species, as well as microorganisms and all species with a pelagic larval development or adult stage, crisscross ABNJs and national jurisdictions (Block et al. 2011; Bierne et al. 2016).

Among the industries active in ABNJs, the fishing industry draws a substantial proportion of the questions about justice, fairness and equity. Fisheries in ABNJs are heavily subsidised, and an estimated 54% of current high-seas fishing grounds would be unprofitable if these subsidies were removed (Sala et al. 2018). Three species account for 42% of the fish caught in ABNJs: skipjack (*Katsuwonus pelamis*), yellowfin (*Thunnus albacares*) and bigeye tuna (*Thunnus obesus*) (Schiller et al. 2018). All three move across vast ocean territories and in and out of national jurisdictions. Several Pacific atoll countries and territories (Kiribati, Tuvalu, Marshall Islands and Tokelau) are extraordinarily dependent on the access fees that they receive from distant water-fishing nations (DWFNs) who fish for tuna in their EEZs. These fees contributed 60–98% of all (non-aid) government revenue in 2016 (FFA 2017). In a number of countries, tuna caught within their EEZs also play a crucial nutritional role (Bell et al. 2018, 2019; Yadav et al. 2019). In many low-income, food-deficit countries, fish is a key source of micronutrients crucial for human health, and nutritionally rich alternatives are not readily available (Golden et al. 2016; Hicks et al. 2019). Poor governance on the high seas and mismanagement of fisheries can therefore result in not only economic losses for global seafood operations but also negative health outcomes and loss of livelihoods in coastal communities.

While the UN Fish Stocks Agreement requires that conservation and management measures for fisheries targeting highly migratory species such as tuna not disproportionately penalise developing states, fulfilling this obligation has been difficult (Hanich et al. 2015). Addressing these governance challenges requires encouraging the development of, for instance, applied research methodologies that can contribute to practical governance solutions that resolve or mitigate conservation burden obstacles and concerns in transboundary fisheries (see FERN 2019; Hanich et al. 2015).

Scientific exploration of ABNJs has yielded deeper insights into life in extreme environments of the deep and open ocean, including hydrothermal vent systems. New techniques have resulted in a rapid fall in the cost of genetic sequencing of collected organisms, supporting the exponential growth of public repositories of genetic sequence data (Laird and Wynberg 2018; see also Blasiak et al. Blue Paper: 'The Ocean Genome: Conservation and the Fair, Equitable and Sustainable Use of Marine Genetic Resources'). While the number of commercial applications is clearly accelerating (Arrieta et al. 2010; Blasiak et al. 2018b), the marine biotechnology industry is highly concentrated in a handful of countries. In 2011, ten countries owned 90% of patent claims; 7 years later the same countries own 98% of patent claims (Arnaud-Haond et al. 2011; Blasiak et al. 2018b). Absence of requirements for sample origin data, or even of taxonomic information in patent filings, makes it virtually impossible to determine which of these are associated with genes collected in ABNJs (Arnaud-Haond et al. 2011; Blasiak et al. 2019).

Scientific advances in the biotechnology and data analytics sector have consistently and dramatically outpaced the development of appropriate regulatory policy (Wynberg and Laird 2018). Publicly accessible open-access databases are one of the cornerstones of capacity-building and should result in a more equitable system of access to and sharing of knowledge. By themselves, however, they do not solve the problem of limited scientific capacity to access and use genetic resources from ABNJs (UNESCO 2017; Salpin et al. 2018), or to use (digital sequence) information. Hence, many developing states cannot explore commercially valuable potential benefits from open access information on the sole basis of information-sharing through open access databases.

In fact, by itself, this needed step does not obviate the need for capacity-building in scientific disciplines (e.g., molecular biology), and research infrastructure– the main drivers of inequalities (Arnaud-Haond et al. 2011).

Operationalising equity commitments in the SDGs with regard to ABNJs has proven challenging. In the context of the ongoing negotiations for an international legally binding instrument under UNCLOS on the conservation and sustainable use of marine biological diversity of ABNJs, states have an opportunity to reshape activity in these areas [1]. However, least developed countries (LDCs) and small island developing states (SIDS) have been underrepresented in the negotiations around biodiversity in areas beyond national jurisdiction (BBNJ) and face technical and legal capacity constraints (Blasiak et al. 2016, 2017a), raising questions about equity in the context of the negotiations. Capacity-building has been seen as one vehicle to move towards greater equity. A voluntary fund was established by the UN Division for Ocean Affaires and the Law of the Sea (UNDOALOS) to help LDCs, SIDS and landlocked developing countries participate in the BBNJ negotiations. If the BBNJ agreement is to be implemented and equitable outcomes achieved, strong and sustained mechanisms for capacity-building and technology transfer at global, regional and national scales will be crucial (Minas 2018). Building on the unqualified capacity-building and technology transfer commitments in the SDGs, negotiators should consider developing a capacity-building and technology transfer regime without the qualifiers contained in UNCLOS. One possibility to consider, among others, is a capacity-building and technology-transfer fund resourced from assessed contributions.

2.4.5 Can corporate Actors Promote Equity?

The increasing power and influence of transnational corporations has attracted scientific attention to their activities and agency (Dauvergne and Lister 2012; Griffin 2017). Historical analysis of corporate engagement in policy development suggests that businesses rarely play a progressive and ambitious role in sustainability efforts; in fact, the opposite is true (Clapp and Fuchs 2009; Oreskes and Conway 2011; Murphy et al. 2012). Where regulations exist, particularly in places with limited capacity, companies can incentivise compliance, through voluntary reporting, naming and shaming, or enforcement activities themselves (e.g., as observed in efforts to reduce illegal fishing in the Southern Ocean: Österblom and Bodin 2012).

Corporate engagement in, and reporting of, sustainability has generated mixed results, ranging from 'greenwashing' to substantial reductions in environmental impacts (Folke et al. 2019). A wide range of voluntary environmental programs (Appendix 2) have engaged multiple ocean-based industries in sustainability. These programs vary in their membership standards, compliance mechanisms, focus and effectiveness. While most focus on environmental and legal concerns rather than equity, these initiatives indicate that platforms exist for engaging corporations in equity. The UN Global Compact (n.d.) represents an important platform for corporate sustainability, with its 10 principles focusing on human rights, labour, environment and anti-corruption.

Whereas ecosystem sustainability is evidently important for corporations whose activities depend on a functioning planet, the case for equity is not as straightforward. What would the incentives be for corporations to share, or give up, some of their powers? Why would a corporation want to pay more taxes or engage in other forms of benefit-sharing mechanisms? Increased attention to global inequalities, in science, among policymakers, and within established, mainstream economic institutions indicates that addressing inequality is likely to be an important aspect of major corporations' future legitimacy and their continued license to operate.

Identifying the relevant companies, where they are operating and what their associated impacts are is a foundation for action. Companies can demonstrate leadership through both better practice and reporting, as well as through active engagement with policymakers for an improved focus on equity. Greater attention to both human rights and the environment by legislators, combined with improved corporate reporting and increased transparency in global supply chains, is incentivising corporations to operate responsibly (Folke et al. 2019). Recent engagement with representatives from 10 of the largest seafood sector companies through the Seafood Business for Ocean Stewardship (SeaBOS) initiative is an example of science-business collaboration in this domain (Österblom et al. 2017). The exposure of slavery and human rights abuses in seafood production (Mendoza et al. 2016; Kittinger et al. 2017) is one reason for increased corporate engagement in sustainability associated with seafood, as reputational risks are incentivising companies to 'do the right thing' (Lubchenco et al. 2016).

Owners, banks, investors and shareholders are able to influence companies to take on a larger responsibility for sustainability and equity. Improved legislation and consumer demands, combined with economic incentives, can stimulate corporations to adopt and integrate environmental and social responsibility (Folke et al. 2019; Jouffray et al. 2019).

[1]The "negotiations shall address the topics identified in the package agreed in 2011, namely, the conservation and sustainable use of marine biological diversity of areas beyond national jurisdiction, in particular, together and as a whole, marine genetic resources, including questions on the sharing of benefits, measures such as area-based management tools, including marine protected areas, environmental impact assessments and capacity- building and the transfer of marine technology" (UNGA 2017).

2.5 Equity and Climate Change

The above case studies showcase the possibilities and barriers associated with promoting more equal distribution of access to, and benefits from, goods and services in a sustainable ocean economy. Current trajectories of global change (IPCC 2019) and associated risks of conflict among resource users (Pinsky et al. 2018; Spijkers et al. 2019) under future conditions further suggest that shifting towards more equitable and inclusive resource use and access will be difficult. Equity approaches are supported in the Paris Agreement. Yet national targets are currently insufficient to meet the 2 °C warming target, with additional commitments needed by the G8 and China.

Climate change is projected to disproportionally affect ecosystems and communities in some of the least developed countries, particularly SIDS (Campbell and Barnett 2010; Sovacool et al. 2015; Hallegatte et al. 2016; Burke et al. 2015; Diffenbaugh and Burke 2019), with the potential to reverse significant development gains. Climate change in the poorest countries is more than 90% likely to have resulted in decreased economic output, whereas the effect is less pronounced in developed nations (Diffenbaugh and Burke 2019). Inequality will cause disadvantaged groups, especially women, girls and indigenous communities, to suffer disproportionately from the adverse effects of climate change, deepening existing social inequalities (Althor et al. 2016; Islam and Winkel 2017), possibly leading to unrest and severe social disruption (see Gaines et al. Blue Paper: 'The Expected Impacts of Climate Change on the Ocean Economy').

The rise in developing nations' inequality is due not only to projected climatological changes but also to the sensitivity of coastal communities to shifts in the distribution and abundance of fish stocks, crucial for livelihoods and nutrition (Blasiak et al. 2017b). This accentuated sensitivity is coupled with comparatively low levels of adaptive capacity, as remote coastal communities often lack the connectivity to urban and peri-urban areas where greater access to education, health services and alternative livelihoods could buffer negative impacts (Cinner et al. 2018).

Some researchers are suggesting that support be provided to countries projected to experience high levels of impact and greater financial cost in terms of lost benefits and opportunities as well as more extensive adaptation measures (Wolff et al. 2015). Specifically, international adaptation funds, such as the Green Climate Fund, could be determined and disbursed to be commensurate with impacts to the country's ecosystem, and a metric of equity could be included within a vulnerability framework (Wolff et al. 2015). Further analyses and mechanisms that systematically consider 'equity' to understand the impact of climate policies are needed to inform efforts to achieve adequate and fair climate action for present and future generations (Klinsky et al. 2017). Addressing equity is increasingly recognised as an important mechanism to develop more effective solutions, support buy-in to climate change policies, and improve adaptive capacity and wholesale system transformation to create climate resilience (see Gaines et al. Blue Paper: 'The Expected Impacts of Climate Change on the Ocean Economy').

Climate scientists, economists and energy systems modellers have developed a range of storylines that examine how society, demographics and economics might change over the next century. These descriptive storylines are collectively known as shared socioeconomic pathways (SSPs) (Riahi et al. 2017) and explore five trajectories that the world could take based on contrasting societal choices—including economic growth, education, urbanisation and the rate of technological development (Table 13.3).

From an equity perspective, SSP1 (Sustainability) and SSP4 (Inequality) represent two extremes. In one possible future (SSP1), an emphasis is placed on improving management of the global commons and investing in health services and education. Consequently, SSP1 leads to a world in which inequality declines both across and within countries and where greater emphasis is placed on human well-being than on economic growth (O'Neill et al. 2017). By contrast, SSP4 is characterised by large, unequal investments in human capital, which together with increasing disparities in economic opportunity and political power increase stratification within and across countries, as a growing majority of the world's resources and trade are controlled by a small group of global elites (O'Neill et al. 2017). In this 'Fortress World', societies grow increasingly fragmented and investments in social and environmental policies are focused on the richest areas (Calvin et al. 2017). Recent years have seen politicians in some of the world's most powerful economies adopting increasingly protectionist or even xenophobic attitudes that align with the narrative of the 'Fortress World' of SSP4.

The narrative of a burgeoning ocean economy suggests an opportunity to align more closely with an equitable future development trajectory (SSP1). Such a scenario is consistent with promoting and supporting international cooperation on climate change mitigation, shown to be critical to lowering emissions. Indeed, recent modelling work found that in scenarios in which individual nations undertake self-serving policies, global cumulative CO_2 emissions are twice those of more cooperative scenarios (Mi et al. 2019). Being able to participate competitively in emerging ocean-based industries requires adequate capacity and research that follows the Findable, Accessible, Interoperable and Reusable (FAIR) data principles (Wilkinson et al. 2016). Researchers and entrepreneurs in low- and middle-income countries are still likely to face an uphill battle to secure financing, market access and highly trained collaborators. Capacity-building and provision of funds remains a ubiquitous target and prior-

Table 13.3 Summary of SSP narratives

Scenario	Scenario name	Outcome and key characteristics
SSP1	Sustainability—Taking the Green Road	A world focused on sustainable growth and equality *'The world shifts toward a more sustainable path, emphasizing more inclusive development that respects perceived environmental boundaries. Management of the global commons slowly improves, educational and health investments accelerate the demographic transition, and the emphasis on economic growth shifts toward a broader emphasis on human well-being. Driven by an increasing commitment to achieving development goals, inequality is reduced both across and within countries. Consumption is oriented toward low material growth and lower resource and energy intensity'*
SSP2	Middle of the Road	A world where trends broadly follow current and historical patterns *'The world follows a path in which social, economic, and technological trends do not shift markedly from historical patterns. Development and income growth proceeds unevenly. Global and national institutions work toward but make slow progress in achieving sustainable development goals. Environmental systems experience degradation, although there are some improvements and overall the intensity of resource and energy use declines... Income inequality persists or improves only slowly and challenges to reducing vulnerability to societal and environmental changes remain'*
SSP3	Regional Rivalry—A Rocky Road	A fragmented world *'Resurgent nationalism, concerns about competitiveness and security, and regional conflicts push countries to increasingly focus on domestic or, at most, regional issues. Policies shift to become increasingly oriented toward national and regional security issues... Investments in education and technological development decline. Economic development is slow, consumption is material-intensive, and inequalities persist or worsen over time... A low international priority for addressing environmental concerns leads to strong environmental degradation in some regions'*
SSP3	Inequality—A Road Divided	A world of ever-increasing inequality *'Highly unequal investments in human capital, combined with increasing disparities in economic opportunity and political power, lead to increasing inequalities and stratification both across and within countries. Over time, a gap widens between an internationally-connected society that contributes to knowledge- and capital-intensive sectors of the global economy, and a fragmented collection of lower-income, poorly educated societies that work in a labor intensive, low-tech economy. Social cohesion degrades and conflict and unrest become increasingly common... Environmental policies focus on local issues around middle and high income areas'*
SSP5	Fossil-Fueled Development—Taking the Highway	A world of rapid technological progress and development *'This world places increasing faith in competitive markets, innovation and participatory societies to produce rapid technological progress and development of human capital as the path to sustainable development. Global markets are increasingly integrated. There are also strong investments in health, education, and institutions to enhance human and social capital. The push for economic and social development is coupled with the exploitation of abundant fossil fuel resources and the adoption of resource and energy intensive lifestyles around the world... There is faith in the ability to effectively manage social and ecological systems, including by geo-engineering if necessary'*

Source: Riahi et al. (2017)

ity in international agendas and frameworks, with a continuous lineage from Part XIV of UNCLOS on the development and transfer of marine technology to the 2030 Agenda and 17 Sustainable Development Goals.

2.6 Summary of Findings

Inequity is a systemic feature of the ocean economy. Lack of fairness is embedded in existing political and economic systems, and is the result of historical legacies and existing norms. There are, however, multiple ways to promote and advance equity—inequity can be addressed with directed policies and practices that explicitly reflect on and address existing approaches. An increased understanding of the intertwined dynamics of sustainability and equity shows that addressing equity is good for economic growth, policy legitimacy, social stability and sustainability. A failure to address equity risks accelerating social tension and eroding credibility in blue growth agendas, while also increasing reputational risks for corporations and trust in existing development approaches. Inequity is also increasing vulnerabilities to climate change. Although legal frameworks partially exist to support equity, they are not sufficiently developed. In practice, ocean policies are largely equity-blind, contributing to current patterns of inequity (Fig. 13.3). The presented case studies identify current barriers to the implementation of equitable principles across ocean-based sectors as well as illustrate successful measures in and developments towards achieving greater fairness.

People will increasingly turn to the ocean to meet their food, nutrition, livelihood and energy needs. Shifting the current trajectory of persistent and increasing inequities will require strong leadership and intentional and long-term planning that starts with a clear commitment to equity. Achieving true equity will only be possible if inclusive consideration is given to all uses and value systems and if destructive or degrading activities are halted (Agardy 2016). Climate

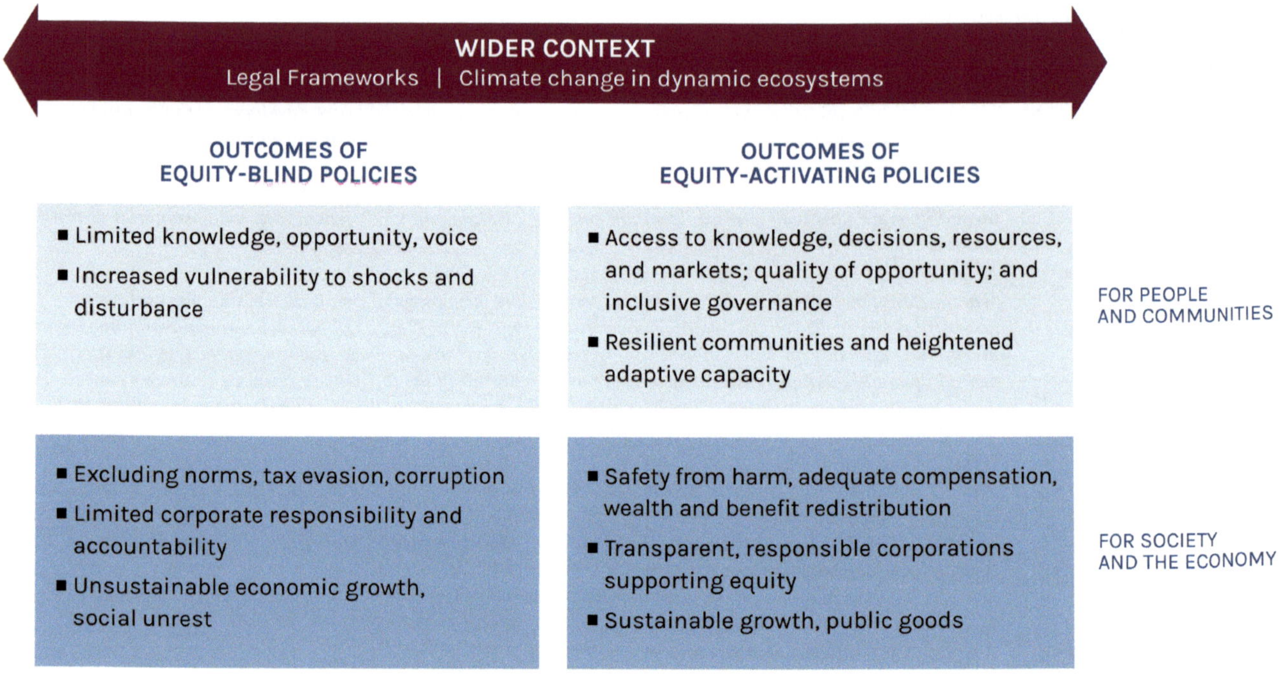

Fig. 13.3 Differences between equity-blind and equity-activating policies and practice

change projections indicate increasing impacts on already vulnerable nations and urgently demand that justice be considered in all sectors, at all political levels, and that policies to increase equity be urgently implemented.

3 Opportunities for Action

We outline opportunities for action, for policy development, business leadership and civil society. These range from the essentials (safeguards, or no-regrets, policies), to the more ambitious (mainstreaming approaches), to transformative approaches (see also Swilling et al. Blue Paper: 'The Transformation to a Sustainable Ocean: A Systems Transition Perspective') aimed at ensuring a fair, equitable, inclusive and sustainable approach to ocean-based development and protection (Table 13.4). These opportunities for action represent reinforcing levels of ambition that acknowledge 'the unique vulnerability and capacity challenges faced particularly by least developed, small coastal and island states, and landlocked states, and therefore the importance of [international] cooperation' (Commonwealth Secretariat 2018, p. 5).

Critical to achieving equity is access to information, promotion of environmental literacy, and engagement, coordination and collaboration across diverse actors, with different skills, capacities and powers. Building local capacity is fundamental to achieving equity and includes human (e.g., skills, education), social (e.g., connections, organisations), financial (e.g., access to capital) and physical (e.g., infra-

Table 13.4 Overview of opportunities for action for achieving equity in a sustainable ocean economy

Category	Opportunities for action
Safeguards—No regrets	1. In development activities and conservation initiatives, engage and include developing states and local populations in decision-making processes 2. Recognise the rights and needs of women, individuals with disabilities, small-scale fishers, indigenous and other minority groups and lift existing access barriers 3. Protect human rights and the rights of indigenous groups 4. Address corruption and tax evasion
Mainstreaming equity— Doing what's right	5. Recognise, protect and operationalise equity and access rights 6. Build local capacity—including access to low-cost and accessible technologies—to establish equality of opportunity 7. Understand social-ecological causality in ocean environments to assign responsibility and liability, and secure an equitable distribution of social gains 8. Demand, require and stimulate transparent, responsible business practices
Transformative approaches—The bold policies	9. Create a shared ocean economy that facilitates redistribution of wealth and benefits 10. Democratise ocean knowledge 11. Create inclusive governance processes by incorporating local voices and visions into plans for the ocean economy, at all scales 12. Be aware of environmental and social limits on growth and consider degrowth

structure, transportation) assets (Sen 1992; Nussbaum 2011; Bennett et al. 2018).

3.1 Safeguards: No Regrets

When governments and agencies engage in development activities, such as foreign direct investment, offshore energy and allocating of access programs, equity should be a cross-cutting concern. This is equally true for conservation initiatives, such as the identification of marine protected areas' location, or protecting individual species. Governments should invest in dialogue, capacity-building, education and training programs for women, girls, boys and men, combined with data collection and monitoring of equity. Tackling corruption and tax evasion is important to advancing ocean equity. Corporations, scientists and science funders also have a role to play in advancing equity safeguards.

3.1.1 Consider the Social Context and Engage Diverse Actors in Decision-Making

Development activities and conservation initiatives should engage and include developing states and local populations in decision-making processes. Women, indigenous groups, individuals with disabilities, and other minorities are key in these processes even when they are not harvesting resources themselves or part of the market chain. Context, values and cultures influence the adoption rate and effectiveness of implemented measures. Thus, activities that work in one community or country may not work in another. Failure to consider context (socioeconomic, political, cultural or ecological) often represents a missed opportunity, is inefficient, and can be counterproductive (see also Gaines et al. Blue Paper: 'The Expected Impacts of Climate Change on the Ocean Economy').

	Opportunities for action	Main target actors	Barriers	Overcoming barriers
a	*Recognise* that people are part of the ocean, and ensure recognition of rights, needs and priorities of developing nations, local people and marginalised groups in development and conservation.	Governments, international organisations, NGOs, funding agencies, private corporations	Equity-blind policies and practice, established norms	Teaching, training and main-streaming knowledge about equity
b	*Develop* policies and planning processes that mandate consideration of local people and communities in development.	Governments	Equity-blind policies and practice, established norms	Training and mainstreaming knowledge about equity
c	*Develop* and *employ* social and economic science to guide decision-making (development policies, marine spatial planning and economic development initiatives). *Document* pre-existing rights, livelihoods and socioeconomic status of relevant communities and consider the implications for producing equitable development.	Research institutions, NGOs, funding agencies	Established practices; limited focus on inter- and transdisciplinary science	Establishing funding mechanisms, piloting and mainstreaming of practice

3.1.2 Recognise the Rights and Needs of Women, Individuals with Disabilities, Small-Scale Fishers, Indigenous and Other Minority Groups

Many groups are marginalised from decision-making processes but rely on ocean resources and play a critical, but overlooked, role in the ocean economy. Recognising their roles, strengths, interests and responsibilities and lifting existing access barriers will engage new groups of leaders, negotiators, decision-makers and entrepreneurs. This will alleviate poverty, strengthen food security, reinforce adaptive capacities and increase development opportunities, in addition to stimulating new mind-sets and innovation. Steps taken towards implementing gender equality considerations, for example, need to be taken in conjunction with action (e.g., education) to address systemic hurdles limiting vulnerable groups from accessing and benefitting from the ocean equitably. A 'gender lens' in all sustainable ocean economy development programs will generate economic opportunities for women, empower them and provide opportunities to engage in decision-making and leadership (Williams et al. 2012; see also USAID 2019 and Barclay et al. 2019 for how to effectively integrate gender considerations in a fishery policy context).

	Opportunities for action	Main target actors	Barriers	Overcoming barriers
a	*Integrate* and *implement* gender equality considerations as part of policymaking, data collection, stakeholder engagement and education	Governments, international organisations, NGOs, research institutions, corporations, funding agencies	Outdated practices, established norms	Education, targeted training, empowerment, critical reflection and championing of minority leaders

	Opportunities for action	Main target actors	Barriers	Overcoming barriers
b	*Recognise* and *respect* pre-existing property rights, tenure and adjacency of coastal communities and indigenous populations to areas of the ocean and marine resources. *Consider* how the above factors need to be accounted for in development planning	Governments, corporations	Equity-blind policies and practice	Establishing practice through learning by doing
c	*Foreground* the needs and rights of small-scale fishers in resource management and development decisions (e.g., in accordance with the FAO Small-Scale Fisheries Guidelines)	Governments	Vested interests	Actively acknowledging the needs and rights of small-scale fishers and enforcing supportive policies

3.1.3 Protect Human Rights and the Rights of Indigenous Groups

At the very least, sustainable ocean development must 'do no harm'. Development activities must protect fundamental human rights, paying particular attention to indigenous rights and workers' rights, and ensuring that supply chains are free from unsafe working conditions, child labour and slavery (Kittinger et al. 2017; Teh et al. 2019).

	Opportunities for action	Main target actors	Barriers	Overcoming barriers
a	*Adhere* to international legally binding treaties, such as the UN Declaration on Human Rights and the UN Declaration on the Rights of Indigenous Peoples *Ratify* relevant legal conventions and ensure relevant complaint and compliance mechanisms are implemented by national governments	Governments, corporations	Limited capacity and knowledge, vested interests	Investment in capacity-building and knowledge development

3.1.4 Address Corruption and Tax Evasion

Corruption, environmental crime and tax evasion represent severe threats to the effectiveness of resource management and perpetuate as well as accentuate inequities in access to resources and benefits derived from them (Le Billon 2014). Corruption can be so ingrained that resource users will practice it without realizing it. Understanding and addressing corruption and other crimes requires education, regulations and enforcement. Systemic corruption is best seen as a collective action problem (Ostrom 1998; Le Billon 2014). Ending it may require transformational change in institutions (see Diamond 2008; and Swilling et al. Blue Paper: 'The Transformation to a Sustainable Ocean: A Systems Transition Perspective'). Identifying who engages in corruption and for what reasons first requires identifying how to incentivise compliance (Sundström 2012; Williams and Le Billon 2017). Leaders in policy, business and practice should lead by example and be role models (Persson et al. 2013).

	Opportunities for action	Main target actors	Barriers	Overcoming barriers
a	*Ensure* that mechanisms are in place to pay greater attention to systemic corruption or tax evasion, by *monitoring* the extent of corruption, *identifying* who engages in corruption or tax evasion and *examining* for what reason	Governments, NGOs, research institutions, corporations	Lack of knowledge, vested interests, dangerous to investigate	Education, regulations, monitoring, enforcement, promoting reciprocity and trust, championing of leaders and role models
b	*Implement* and *enforce* sanctions for corruption and tax evasion	Governments, international organisations	Established norms and legal grey zones	Obtaining convictions and sentences, active leadership in changing corporate practice
c	Increase *monitoring* and *reporting* of social and environmental impacts to ensure accountability and transparency	Governments, funding agencies, international organisations	Limited monitoring	Independent follow-up of development programs

3.2 Mainstreaming Equity: Doing What's Right

Systematically addressing issues of inequity needs to be mainstreamed into development, management and conservation interventions at all scales, from local marine protected areas to global treaty negotiations on ocean governance. As new treaties are being negotiated, active steps need to be taken to ensure that all states and international organisations have the necessary capacity and sense of responsibility to safeguard equity, irrespective of policy positions or financial resources. In addition, analyses and estimates of the economic consequences of unmanaged development in the ocean need to be improved upon and communicated.

3.2.1 Recognise, Protect and Operationalise Equity and Access Rights

The provision of access to local resources is imperative for the establishment of equality and equity at community levels (WRI et al. 2005). Equity and access rights are already enshrined in conventions, international agreements and policies, but they are insufficiently operationalised (see Sect. 2.3). Restricted and unequal access to local ecosystems and resources constitutes a barrier that makes it more difficult for vulnerable groups, such as the poor, to improve their conditions (Bennett et al. 2018; Cisneros-Montemayor et al. 2016; Haider et al. 2018). Access to local ecosystems has to be informed by customs and traditions, grounded in both formal and informal institutions, but it needs to also reflect current scientific knowledge.

	Opportunities for action	Main target actors	Barriers	Overcoming barriers
a	*Implement* policies that require consideration of historical and pre-existing access to natural resources, how these will be impacted by development, what mitigation can minimise impacts on access and how compensation mechanisms might be employed when impacts cannot be avoided	Governments, international organisations, research institutions, NGOs, corporations	Established practice and limited knowledge	Recognition of indigenous or cooperative governance and effective implementation of existing commitments (i.e., legitimising decentralised governance) Promoting co-management and building capacity and skills of all actors

3.2.2 Build Local Capacity to Establish Equality of Opportunity

The ability of coastal populations and coastal island and developing nations to benefit from ocean resources and development depends on their capacity to do so—in other words, equality of benefit requires equality of opportunity. Capacity is provided by human (e.g., skills, education), social (e.g., connections, organisations), financial (e.g., access to capital) and physical (e.g., infrastructure, transportation) assets (Sen 1992; Nussbaum 2011; Bennett et al. 2018). Enhancing the commitment to capacity-building and the transfer of marine technology, including through strengthening existing legal frameworks, constitutes an important priority. Access to low-cost and accessible technologies that support the SDGs represents a significant and increasingly relevant mechanism for developing adequate capacity (Meikle and Sugden 2015; see also Leap et al. Blue Paper, 'Technology, Data and New Models for Sustainably Managing Ocean Resources', on, for instance, the risk of widening the gap if equitable development and access are not considered). Local ownership of businesses that harvest ocean resources and of businesses that provide labour, services, goods or supplies can increase local benefit from economic development (Bennett et al. 2019b).

	Opportunities for action	Main target actors	Barriers	Overcoming barriers
a	*Develop* policy mechanisms and programs that provide opportunities by *bolstering* physical assets and *building* human skills and capabilities among local constituents prior to and during development	Governments	Established practice, lack of knowledge	Adjusting policies, targeting funding, piloting of practice, strong leaders
b	*Strengthen* legal obligations on capacity-building and transfer of technology	Governments	Priorities, limited funding, intellectual property concerns	Political will
c	*Create* low-cost and accessible technology	Governments, corporations, venture capital investors, funding agencies	Limited access to markets, funds and information	Establishing targeted funding schemes, supporting creative solutions and innovation
d	*Support* local ownership of ocean businesses *Set up* entrepreneurship training programs and create credit schemes *Facilitate* connections to markets	Governments, corporations	Limited experience, knowledge and capacity	Effectively communicating existing knowledge and practice, investing in teachers and trainers, developing effective collaboration

	Opportunities for action	Main target actors	Barriers	Overcoming barriers
e	*Create* user-friendly information-sharing mechanisms to monitor and communicate capacity needs and impacts of capacity-building efforts on local communities	Research institutions, governments, international organisations, NGOs	Limited information availability and infra-structure	Collaborating with UN Decade of Ocean Science

3.2.3 Understand Social-Ecological Causality, Assign Responsibility and Secure Equitable Distribution of Benefits

Development opportunities in ocean environments can entail social gains and harms. Some progress has been made in understanding and monitoring ecological harms, such as overfishing or eutrophication, and how these impact people. However, more knowledge needs to be gained about how ocean-based economic development can produce both direct and indirect social benefits and harms. Understanding causality in ocean environments is important to assigning responsibility and liability and securing an equitable distribution of social gains and avoidance of harms. Economic instruments such as taxes and fees need to be leveraged to internalise environmental and social benefits, costs and risks to society (WWF 2018).

	Opportunities for action	Main target actors	Barriers	Overcoming barriers
a	*Document, project, forecast* and *report social* benefits and harms, both those that occur indirectly via environmental impacts and those that impact humans directly *Assign responsibility* and *implement* mechanisms to equitably redress socioeconomic and ecological impacts of development activities	Governments, NGOs, research institutions, international organisations	Limited information and practice	Developing knowledge and practice
b	*Develop* compensation, remediation and redress mechanisms for past or future impacts. *Eliminate* harmful subsidies. *Regulate* harmful industries *Enforce* existing laws and principles[a] *Strengthen* instruments and introduce social impact bonds or environmental taxes	Governments	Limited practice and capacity	Mentoring of strong leadership, piloting of practice; sharing of experiences; coaching and support for active participation in the international policy arena
c	*Develop* means to ensure equitable distribution of benefits derived from ocean services	Governments, corporations	Limited practice and capacity, vested interests	Supporting active participation in international policy; promoting multilateral benefit sharing mechanisms

a. Such as by applying the polluter-pays principle, through the International Convention on Liability and Compensation for Damage in Connection with the Carriage of Hazardous and Noxious Substances by Sea.

3.2.4 Demand, Require and Stimulate Responsible Business Practices

Ocean industries derive substantial wealth and income from ocean environments. However, like most industries, they operate with limited transparency, which hinders the monitoring of their impact on society and ecological well-being, and thus the granting of a social license to operate. Incentives that shape a positive competitive environment can encourage industry to adopt corporate social responsibility practices to preserve their social license to operate (McGee 2013; Aguilera et al. 2007). Increased transparency will stimulate the private sector to respect and advance ocean equity and stewardship, while also encouraging learning across corporations.

	Opportunities for action	Main target actors	Barriers	Overcoming barriers
a	*Demand* full transparency of ongoing and planned activities and acceptance of liability and social responsibility, as well as limits on growth (within environmental capacities), as preconditions for engagement in ocean-based industries	Governments	Limited corporate reporting; limited monitoring, control and enforcement capacity; corruption	Developing practice and capacity, active leadership

	Opportunities for action	Main target actors	Barriers	Overcoming barriers
b	*Amend* legal duties and corporate laws to account for negative externalities *Encourage* companies to include 'social responsibility' provisions in articles of incorporation to support and promote equitable choices	Governments, corporations	Limited legislation; lack of monitoring, control and enforcement; limited capacity	Developing practice and capacity, providing incentives to 'do the right thing'
c	*Require* companies to submit strategic plans, along with reporting and auditing reports, detailing how their activities support smallscale fishers, local communities and ocean stewardship	Governments, international organisations	Limited legislation; lack of monitoring, control and enforcement; limited capacity	Developing practice and capacity
d	*Showcase, incentivise* and *stimulate* positive leadership	Governments, research institutions, corporations	Limited history of collaboration, fragmented knowledge	Synthesising knowledge and critically reflecting on progress made

3.3 Transformative Approaches: The Bold Policies

Discussions of systemic change to the global economy are no longer unusual (Jackson 2009; Hickel and Kallis 2019; Hadjmichael 2018; IPBES 2019). An acceleration of the ocean economy, along current trajectories, will continue to deliver the greatest benefits to a small subset of high-income countries, corporations and individuals. The scientific literature is increasingly exploring purposeful and meaningful steps to change course. Allocation of rights (including property rights for fisheries, offshore wind and aquaculture) and development of new knowledge and technologies are often regarded as necessary to facilitate environmental sustainability and economic viability, but they also risk reinforcing existing power structures and limiting the development of low- to middle-income regions. Given the global nature of the ocean, the scale of the challenges and the slow pace of international policy development, immediate and concrete steps are needed to develop and evaluate alternative approaches to economic growth and allocation of social, economic and technological capital (Raworth 2017b). Transformative approaches require redistribution of power and resources to improve longer-term equity perceptions and outcomes. Limits on growth, and even degrowth, of some sectors may need to be considered. Changes to the status quo will not be easy, but—as this Blue Paper has illustrated—they could substantially advance progress towards the SDGs.

3.3.1 Create a Shared Ocean Economy that Facilitates Redistribution of Wealth and Benefits

The ocean's global nature and the current unequitable distribution of access, benefits and negative impacts from ocean sectors requires bolder approaches. If such approaches are not taken, there is a real risk that the legitimacy of the current ocean policy agenda will be substantially eroded. Current ocean-related fund allocations from high-income countries to middle- and low-income ones are primarily handled through bilateral and multilateral official development aid (ODA) financial flows. While the source of many positive impacts, these allocations are dwarfed by the size of the ocean economy. Just 1% of the global ocean economy represents US $15 billion per year generated from the world's ocean and coasts (OECD 2016). New ways of thinking, creative policies and accounting mechanisms that internalise externalities and include long-term economic and environmental considerations, and the use of progressive and affordable technologies are needed to facilitate a redistribution of wealth and benefits from the ocean economy.

	Opportunities for action	Main target actors	Barriers	Overcoming barriers
a	*Develop* and *implement* a global ocean tax to reallocate parts of profits to places where environmental resources are harvested and where management actions, capacity-building, conservation or restoration are required	Governments, international organisations	Politically charged questions, vested interest in the status quo	Collaboratively investigating ideas and potential effects
b	*Apply* scenarios to understand how future benefits and harms might or should be distributed to different local groups and to current and future generations (see Bennett et al. 2019a, Box 13.1) *Incorporate* intergenerational accounting (Sumaila and Walters 2005) and climate change impacts into projection models and planning	Governments, researchers, development planners, investors, loan officers, funding agencies	Powerful interest groups, practice of discounting future harms in favour of present gains	Developing capacity to build on existing models, developing collaboration to model distribution of impacts
c	*Centralise* human well-being as both a proximate and ultimate goal of ocean economy development (Cisneros-Montemayor et al. 2019), within the capacity of the biosphere	Governments	Focus on economic profits	Managing for long-term local livelihood and food security objectives, ensuring that new developments support human well-being
d	*Develop* and *stimulate* access to low-cost, low-tech, long-term transformative solutions for equity and sustainability (aiming to increase access for communities, improve monitoring and enforcement, report on corruption and promote knowledge exchange)	Governments, funding agencies, international organisations	Many technologies primarily focused on generating capital and facilitating control over resources, lack of capacity to use technologies among key groups	Issuing global call (or challenge fund) and deploying sustainable and equitable technologies; building visibility and capacity to develop and utilise these technologies
e	*Develop* and *implement* mechanisms to redistribute wealth *Reallocate* shares to local communities and workers *Contribute percentage* of profits to local government or community trust funds *Reform* subsidy programs (Cisneros-Montemayor and Sumaila 2019)	Governments, corporations	Disproportionate concentration of value and power	Strong leadership, disincentives for not contributing to redistribution programs, support for effective policy mechanisms

3.3.2 Democratise Ocean Knowledge

Knowledge is power, and ocean knowledge is primarily generated in high-income countries. The current scientific understanding of the ocean and its associated industries, technologies and impacts is not well-suited to addressing issues of global ocean equity. Greater attention is needed to democratise knowledge, train international researchers (in social and transdisciplinary science) and document how benefits of the ocean and ocean-related knowledge flow to different groups (see also Fenichel et al. Blue Paper: 'National Accounting for the Ocean and Ocean Economy'). Knowledge exchange, co-production and transfer (Pohl et al. 2010) can be stimulated by cross-regional exchanges to compare lessons learned and the benefits of diverse approaches, cultures, values and understanding. Programs of this nature should aim to modify academic incentives that militate against equitable knowledge production and sharing (e.g., the emphasis on publishing and barriers to open-access work). It should value and promote effective and equitable partnerships with scientific researchers in low- and middle-income countries. Governments, scientists and communities should make concerted efforts to co-develop mechanisms for identifying, considering and expressing benefits gained from the ocean so that these can be integrated into development policies across scales.

	Opportunities for action	Main target actors	Barriers	Overcoming barriers
a	*Increase* knowledge co-production, exchange, capacity-building, technology transfer and availability, and knowledge infrastructure *Develop* multilateral networks capable of harnessing technological capacities to facilitate marine technology transfer *Foster* an integrated approach to the advancement, sharing and application of scientific knowledge (Harden-Davies 2017)	Governments, research institutions	Access to information, lack of capacity	Mandating high-income countries to commit to long-term funding of ocean science centres in low-income countries, providing access to knowledge networks and mentoring, developing partners in scientific endeavours and closing data gaps
b	*Recognise* that people are part of the ocean. A broader vision for ocean science, one that includes the human dimensions and marine social sciences (Bennett 2019), is needed to identify how to produce more equitable outcomes from ocean development	Governments, funding agencies, NGOs, civil society	Lack of knowledge and capacity	Developing collaborations, building capacity, fostering mentorships, obtaining support from the UN Decade of Ocean Science
c	*Train* international networks of young students and cross-regional exchanges to compare lessons learned and understand the benefits of human diversity	Research institutions, governments, funding agencies	Lack of capacity, time and cost investment	Developing collaborations, providing mentorship, leading by example
d	*Understand* and *develop* transparent accounting of how the benefits of ocean activities, resources and ecosystem services flow to different nations and groups of people within nations so that this understanding can be integrated into development policies across scales *Co-develop* mechanisms for identifying, considering and expressing the benefits gained from the ocean in ways that respect cultural norms and do not appropriate traditional knowledge	Governments, research institutions, NGOs, international organisations, civil society	Self-interest, established norms	Ensuring that information on ocean resources is publicly available, promoting trans-parent practices, rewarding exemplary behaviour

3.3.3 Create Inclusive Governance Processes at All Scales

Governance refers to who makes decisions and how decisions are made, which can significantly impact both what management actions are taken and to what effect. In terms of the blue economy, governance can impact 'how the ocean will be developed and by whom, how and to whom benefits will be distributed, how harms will be minimised and who will bear responsibility for environmental and social outcomes' (Bennett et al. 2019b, p. 2). In short, equity can depend on governance, and creative processes can be developed to incorporate local voices and visions into plans for the ocean economy. Many successful marine governance initiatives in the developing world are based on grassroots efforts. The FAO Small-Scale Fisheries guidelines is an example of a bottom-up initiative that resulted in a set of broad-scale instruments aimed at all actors striving to secure sustainable small-scale fisheries, end hunger and poverty, as well as strengthen human rights.

	Opportunities for action	Main target actors	Barriers	Overcoming barriers
a	*Design* governance processes at all scales—from global deliberations, to negotiations related to local ocean development initiatives—to be inclusive of governments, business and civil society, focusing on marginalised groups such as women, small-scale fishers and Indigenous Peoples	Governments, international institutions, NGOs	Lack of time and funds	Highlighting as priority to funding and development partners, developing mechanisms to ensure participation
b	*Allow* solutions to emerge from the bottom up.	Governments, international institutions	Lack of time, capacity and knowledge	Developing and implementing codes of practice that enable active engagement with grassroots initiatives

3.3.4 Place Limits on Growth and Consider Degrowth Within the Capacity of the Biosphere

There are numerous examples around the world where economic development activities have produced or are producing ecological and/or social impacts that could be deemed to have gone beyond acceptable thresholds. Some examples include oil development in Nigeria or Venezuela and overfishing in Mauritania or Senegal (Belhabib et al. 2016; Doumbouya et al. 2017). When thresholds are being exceeded, limiting growth or even degrowing the ocean economy to bring it in line with the capacity of the biosphere may be an obvious alternative. In this context, 'degrowth' means scaling back overexploitation that gives the illusion of what is in fact merely temporary growth and ultimately disastrously exhausts natural capital. Given the increasing debate about inequities, governments, corporations and scientists should consider alternative approaches to the ocean economy based on collaborative and equitable approaches that make well-being, livelihoods and natural resource maintenance their primary goals (Kostakis and Bauwens 2014).

Opportunities for action	Main target actors	Barriers	Overcoming barriers
Investigate and *pilot* approaches to limits on growth and degrowth	Governments, international institutions, research institutions	Existing narratives of perpetual growth and growth first, environment later	Constructive and science-based conversations, scenarios, piloting of approaches

4 Conclusions

This Blue Paper has illustrated that access to ocean resources and benefits is distributed inequitably, as is exposure to harms, resulting in negative effects on the environment and human well-being. Challenging this inequality directly threatens powerful interests that benefit from existing arrangements. However, inequality is increasingly endangering ecological sustainability, economic development and longer-term political and social stability. Increased scientific attention to inequality is starting to shape debates associated with the ocean. We argue that addressing issues of equity is critical to a sustainable ocean economy. We provide a set of complementary reinforcing opportunities for action, from the simple to the transformative. These opportunities range from activities that aim to recognise, identify, document and report, as well as to promote, respect, clarify, showcase, build, create or facilitate. The opportunities include assigning and demanding responsibilities, piloting, implementing and enforcing existing and novel policies, and even rethinking existing growth paradigms. Combined, they aim to overcome the existing general policy blindness to equity and have an ambition to effectively support a sustainable and just ocean economy.

Acknowledgements The authors thank the paper's technical reviewers, Meryl Williams, Mads Barbesgaard and Elisa Morgera, as well as its arbiter, Nicola Frost, who all provided helpful technical comments. The authors also thank the World Resources Institute for providing support as the Ocean Panel Secretariat.

While our colleagues were very generous with their time and input, this report reflects the views of the authors alone. The authors thank Alex Martin for copyediting and Jen Lockard for design.

Appendix 1: Inequities Associated with Small-Scale Fisheries

Inequity of Benefits

Inequalities in wealth shape the distribution of benefits from ocean resources for small-scale fishers at multiple scales. Many small-scale fishing households in Kenya and the Philippines, for instance, are enmeshed in structural inequality along value chains (Knudsen 2016; Wamukota 2009; Rosales et al. 2017). Coastal households specialised in fishing cannot compete with richer, more powerful fishers with better gear and the capacity to bribe local officials (Fabinyi 2012). Consequently, coastal households may depend on unequal trading relationships (Crona et al. 2016) and tend to sell the best-quality fish, consuming the lower-quality ones (Wamukota 2009; Hicks et al. 2019). Small-scale fishers often receive a relatively small proportion of the value of their catch (Rosales et al. 2017), especially when compared to prices associated with luxury consumption (Trinidad et al. 2014; Purcell et al. 2017). Meanwhile, small-scale fishers may be the most vulnerable to the loss or degradation of marine resources (Crona et al. 2015; Sadovy de Mitcheson et al. 2018).

Invisible Gendered Inequities

Women's contributions in fisheries are often overlooked, underestimated and/or undervalued, often resulting in women's marginalisation in the management of marine resources. Coastal activities are usually highly gendered, both in where and how women participate in value chains and how their contributions are valued and prioritised (Yodanis 2000; Williams 2008; De Silva 2011; Harper et al. 2013, 2017; Lentisco and Lee 2015; Fortnam et al. 2019). Women play an important role in both harvest and post-harvest activities with important implications for families, communities and economies. In the Pacific region, more than half of small-scale catches are taken by women. Coastal fisheries management policies that better represent their needs could lead to more secure livelihoods and more sustainable catches. Despite this, policies, data collection and stakeholder consultations remain gender-blind in many places. This further marginalises women's voices and interests, further devalues the benefits women's work provides and makes it hard to know how women are affected as the fisheries sector develops.

Gender blindness results from a focus on formal and paid fishing activities (traditionally male-dominated) in research, management and policies, disregarding informal and unpaid activities, usually dominated by women (Harper et al. 2013). Fisheries agencies are also commonly focused on the production segment of fish value chains, even though twice as many people may be employed in related activities, such as processing and marketing, which are often dominated by women (World Bank 2012). In Senegal, a study found that women represent 90% of the country's seafood processor workforce, valued at $30.5 million (Belhabib et al. 2014). Such marginalisation has often happened despite increasing recognition that women play a critical role at every link in small-scale fisheries value chains (De Silva 2011; Lentisco and Lee 2015). Failing to account for the gender and other social differentiation in fisheries management can lead to policy interventions that undermine sustainable livelihoods (Bennett 2005; Daw et al. 2015).

Inequity of Access

Gendered access barriers (ones that affect women and men differently) occur at several points along the fisheries value chain. Overall, women in fishing communities across the world face barriers to more profitable segments of supply chains, due to a variety of socioeconomic and cultural obstacles, as well as conflicting household roles (O'Neill and Crona 2017). These can include indirect barriers, such as gender norms in the Philippines that limit women's ability to fish far from home (Yater 1982). More direct barriers include lack of access to fishing gear, fishing grounds, fishing markets or financial capital, including credit, as well as lack of education or alternative livelihoods (Matsue et al. 2014). Gendered power relations are often context-specific. For instance, in Zambia, fishing gear is largely owned and controlled by men (Cole et al. 2018), whereas in Ghana and Brazil women can inherit fishing boats and gear and either use them themselves (Kleiber et al. 2017) or lease them to men for fishing (Walker 2001). In Norway, women are often not able to buy bigger boats or more profitable quotas because they are denied bank loans (Gerrard and Kleiber 2019), while in western Zambia, women have overcome lack of credit by participating in village savings and lending groups, which they use to buy and resell fish (Cole et al. 2015). Women may be excluded from markets, such as in Bangladesh, where only the poorest women sell fish at the market, or in Kenya, where women only have access to the less profitable parts of the catch and have limited trade connections (Matsue et al. 2014; Wamukota 2009). In Guadalajara, Mexico, in contrast, women dominate, often holding influential positions, having attained these through family networks, skills and cultural heritage (Pedroza 2019). Policies on matters such as spatial management can also have a disproportionate impact on women and other marginalised groups that may not have access to boats or motors that would allow them to reach other fishing zones (Eder 2005). For example, in Madagascar, when a no-take area was placed in the community gleaning areas, women were no longer able to fish or resorted to fishing illegally at night (Baker-Médard 2017).

Decision-Making and Governance

Women also often face access barriers to governing institutions, resulting in fewer women's voices included in small-scale fisheries decision-making institutions. In Tanzania, female fish traders were excluded from formal fisheries management groups (Fröcklin et al. 2013). In Bangladesh, women were not included in decision-making because they were perceived as lacking the necessary knowledge and experience (Kleiber et al. 2017). In Senegal, women make up less than 5% of fisheries governing bodies (Harper et al. 2017). Customary rules may also exclude women, such as in some communities in the Solomon Islands, where women are not allowed to be under the same roof as men with whom they have previously had relationships (Faye Siota, pers. comm.), effectively barring many women from public meeting spaces. Again, gendered norms and power relations in relation to the ocean mirror or enhance general gender inequities, such as those surrounding access to education, health care, food and nutritional security.

Appendix 2: Voluntary Environmental Programs

Name	Mission	Website
Aquaculture Stewardship Council (ASC)	To transform aquaculture towards environmental sustainability and social responsibility using efficient market mechanisms that create value across the chain	www.asc-aqua.org
Coalition of Legal Toothfish Operators (COLTO)	To promote sustainable toothfish fishing and fisheries; facilitate its members' working together and with outside groups, including through continued provision of high-quality scientific data to CCAMLR (Commission for the Conservation of Antarctic Marine Living Resources) and other bodies; and provide effective representation for its members.	https://www.colto.org
Global Aquaculture Alliance (GAP)	To promote responsible aquaculture practices through education, advocacy and demonstration	https://www.globalgap.org
Global Compact Ocean Action Platform	To determine how ocean industries can advance progress towards the Sustainable Development Goals (SDGs). The work of the platform builds upon the 10 principles of the UN Global Compact, which outline business responsibilities in the areas of human rights, labour, environment and anti-corruption	https://www.unglobal-compact.org/take-action/action-platforms/ocean
Green Coastal Shipping Program	To find scalable solutions for efficient and environmentally friendly shipping. Its multiple pilots are crucial for the phasing in of zero- and low-emission solutions in shipping towards 2030.	https://www.dnvgl.com/maritime/green-shipping-programme/index.html
International Association of Oil and Gas Producers (IOGP)	To create alignment and facilitate continuous health, safety and environment (HSE) improvements across oil and gas exploration and production	https://www.iogp.org
International Council on Mining and Metals (ICMM)	To promote a safe, fair and sustainable mining and metals industry	https://www.icmm.com
IPIECA (International Petroleum Industry Environmental Conservation Association)	To provide a forum for encouraging continuous improvement in offshore oil and gas industry performance, for example improvements associated with the SDGs	http://www.ipieca.org
Marine Stewardship Council (MSC)	To use its ecolabel and fishery certification program to contribute to the health of the world's oceans by recognizing and rewarding sustainable fishing practices, influencing the choices people make when buying seafood and working with its partners to transform the seafood market to a sustainable basis	https://www.msc.org/se
Ocean Energy Europe	To promote the development of ocean energy, improved access to funding and enhanced business opportunities for its members	https://www.oceanener-gy-europe.eu
Sustainable Shipping Initiative (SSI)	To facilitate oriented efforts such as the 'Ship Recycling Transparency Initiative', which brings together ship owners, banks and other key stake-holders to improve transparency in the global ship recycling value chain	https://www.ssi2040.org
WindEurope	To promote wind power and coordinate international policy, communication, research and analysis	https://windeurope.org

Source: Blasiak et al. (2018a) and Pretlove and Blasiak (2018)

About the Authors

Co-authors

Henrik Österblom is Professor and Science Director at the Stockholm Resilience Centre, Stockholm University.

Colette C.C. Wabnitz is a Senior Research Scientist at the Institute for the Oceans and Fisheries, University of British Columbia and Stockholm Resilience Centre, Stockholm University.

Dire Tladi is a Professor of International Law at the Department of Public Law and the Institute for International and Comparative Law in Africa, University of Pretoria, as well as an extraordinary professor in the Public Law Department, University of Stellenbosch (South Africa).

Contributing Authors

Edward H. Allison is an Honorary Fellow at WorldFish (Malaysia) and an Affiliate (Honorary) Professor in the School of Marine and Environmental Affairs, University of Washington.

Sophie Arnaud-Haond is a Research Scientist at Ifremer, in the UMR MARBEC (Marine Biodiversity, Exploitation and Conservation) at Montpellier University, CNRS, IRD (France).

Jan Bebbington is a Professor in the Department of Accounting, Birmingham Business School, University of Birmingham (UK).

Nathan Bennett is a Research Associate at the Institute for the Oceans and Fisheries, University of British Columbia, and is the Chair for the People and the Oceans Specialist Group for the International Union for the Conservation of Nature.

Robert Blasiak is a Researcher at the Stockholm Resilience Centre, Stockholm University and a visiting researcher at the Graduate School of Agricultural and Life Sciences, University of Tokyo.

Wiebren J. Boonstra is an Associate Professor at the Stockholm Resilience Centre, Stockholm University.

Afrina Choudhury is a Research Fellow (Senior Gender Specialist) at WorldFish Bangladesh.

Andrés M. Cisneros-Montemayor is a Research Associate and Deputy Director of the Ocean Nexus Program at the Institute for the Oceans and Fisheries, University of British Columbia.

Tim Daw is a Researcher at the Stockholm Resilience Centre, Stockholm University.

Michael Fabinyi is an Associate Professor at the School of Communication, University of Technology Sydney.

Nicole Franz is a Fishery Planning Analyst with the Food and Agriculture Organization of the United Nations.

Harriet Harden-Davies is a Research Fellow at the Australian National Centre for Ocean Resources and Security (ANCORS), University of Wollongong.

Danika Kleiber is a Research Fellow at the ARC Centre of Excellence for Coral Reef Studies (Australia) and WorldFish (Malaysia).

Priscila Lopes is an Associate Professor in the Department of Ecology at the Universidade Federal do Rio Grande do Norte.

Cynthia McDougall is the gender research leader for WorldFish (Malaysia) and the CGIAR Research Program on Fish Agri-Food Systems (FISH).

Budy P. Resosudarmo is an Associate Professor at the Crawford School of Economics and Government, Australian National University.

Samiya A. Selim is an Associate Professor and the Director of the Center for Sustainable Development at the University of Liberal Arts Bangladesh.

References

Abdullah AN, Myers B, Stacey N, Zander KK, Garnett ST (2017) The impact of the expansion of shrimp aquaculture on livelihoods in coastal Bangladesh. Environ Dev Sustain 19(5):2093–2114

Agardy T (2016) Tundi's take: equity in ocean management is about preserving ocean health and access. The skimmer on marine ecosystems and management, 31 May. https://meam.openchannels.org/news/meam/tundi%E2%80%99s-take-equity-ocean-management-about-preserving-ocean-health-and-access

Aguilera RV, Rupp DE, Williams CA, Ganapathi J (2007) Putting the S back in corporate social responsibility: a multilevel theory of social change in organizations. Acad Manag Rev 32(3):836–863

Allison EH, Ratner BD, Åsgård B, Willmann R, Pomeroy R, Kurien J (2012) Rights-based fisheries governance: from fishing rights to human rights. Fish Fish 13(1):14–29

Almeida OT, Lorenzen K, McGrath D (2009) Fishing agreements in the lower amazon: for gain and restraint. Fish Manag Ecol 16(1):61–67

Althor G, Watson JE, Fuller RA (2016) Global mismatch between greenhouse gas emissions and the burden of climate change. Sci Rep 6:20281

Alvaredo F, Chancel L, Piketty T, Saez E, Zucman G (2018) World inequality report 2018. World Inequality Lab, Paris

Arnaud-Haond S, Arrieta JM, Duarte CM (2011) Marine biodiversity and gene patents. Science 331(6024):1521–1522

Arrieta JM, Arnaud-Haond S, Duarte CM (2010) What lies underneath: conserving the oceans' genetic resources. Proc Natl Acad Sci 107(43):18318–18324

Baker-Médard M (2017) Gendering marine conservation: the politics of marine protected areas and fisheries access. Soc Nat Resour 30(6):723–737

Ban NC, Frid A, Reid M, Edgar B, Shaw D, Siwallace P (2018) Incorporate indigenous perspectives for impactful research and effective management. Nat Ecol Evolut 2(11):1680–1683

Barbesgaard M (2018) Blue growth: savior or ocean grabbing? J Peasant Stud 45(1):130–149

Barclay K, Leduc B, Mangubhai S, Donato-Hunt C (2019) Pacific handbook for gender equity and social inclusion in coastal fisheries and aquaculture. Pacific Community, Nouméa

Bavinck M, Berkes F, Charles A, Dias ACE, Doubleday N, Nayak P, Sowman M (2017) The impact of coastal grabbing on community conservation: a global reconnaissance. Marit Stud 16(1):8

Bavinck M, Jentoft S, Scholtens J (2018) Fisheries as social struggle: a reinvigorated social science research agenda. Mar Policy 94:46–52

Baxter B (2004) A theory of ecological justice. Routledge, London

Behailu L, Choudhury A, Rajaratnam S, Locke C, McDougall C (2018) Gender norms and agricultural innovation: insights from six villages in Bangladesh. J Sustain Develop 11(4):270–287

Behailu L, Choudhury A, Rajaratnam S, Burg M, McDougall C (2019) Implications of agricultural innovations on gender norms: a case study from Bangladesh. Pennsylvania State University and Wageningen University (forthcoming)

Belhabib D, Koutob V, Sall A, Lam VW, Pauly D (2014) Fisheries catch misreporting and its implications: the case of Senegal. Fish Res 151:1–11

Belhabib D, Lam VW, Cheung WW (2016) Overview of West African fisheries under climate change: impacts, vulnerabilities and adaptive responses of the artisanal and industrial sectors. Mar Policy 71:15–28

Bell JD, Allain V, Gupta AS, Johnson JE, Hampton J, Hobday AJ, Lehodey P et al (2018) Climate change impacts, vulnerabilities and adaptations: western and central pacific ocean marine fisheries. Impacts of climate change on fisheries and aquaculture. In: Barange M, Bahri T, Beveridge MCM, Cochrane KL, Funge-Smith S, Poulain F (eds) Impacts of climate change on fisheries and aquaculture: synthesis of current knowledge, adaptation and mitigation options. FAO Fisheries and Aquaculture Technical Paper 627. FAO, Rome, pp 305–325

Bell JD, Sharp MK, Havice E, Batty M, Charlton KE, Russell J, Adams W, Azmi K et al (2019) Realising the food security benefits of canned fish for pacific island countries. Mar Policy 100:183–191

Béné C, Friend R (2011) Poverty in small-scale inland fisheries: old issues, new analysis. Prog Dev Stud 11(2):119–144

Bennett E (2005) Gender, fisheries and development. Mar Policy 29(5):451–459

Bennett NJ (2018) Navigating a just and inclusive path towards sustainable oceans. Mar Policy 97:139–146

Bennett NJ (2019) Marine social science for the peopled seas. Coast Manag 47(2):244–252

Bennett NJ, Govan H, Satterfield T (2015) Ocean grabbing. Mar Policy 57:61–68

Bennett NJ, Kaplan-Hallam M, Augustine G, Ban N, Belhabib D, Brueckner-Irwin I, Charles A et al (2018) Coastal and indigenous community access to marine resources and the ocean: a policy imperative for Canada. Mar Policy 87:186–193

Bennett NJ, Blythe J, Cisneros-Montemayor AM, Singh GG, Sumaila UR (2019a) Just transformations to sustainability. Sustainability 11(14):3881

Bennett NJ, Cisneros-Montemayor AM, Blythe J, Silver JJ, Singh G, Andrews N, Calò A et al (2019b) Towards a sustainable and equitable blue economy. Nat Sustain 2:991–993. https://doi.org/10.1038/s41893-019-0404-1

Berg A, Ostry JD, Zettelmeyer J (2012) What makes growth sustained? J Dev Econ 98(2):149–166

Berg A, Ostry JD, Tsangarides CG, Yakhshilikov Y (2018) Redistribution, inequality, and growth: new evidence. J Econ Growth 23(3):259–305

Biermann F, Kanie N, Kim RE (2017) Global governance by goal-setting: the novel approach of the un sustainable development goals. Curr Opin Environ Sustain 26–27:26–31

Bierne N, Bonhomme F, Arnaud-Haond S (2016) Dedicated population genomics for the silent world: the specific questions of marine population genetics. Curr Zool 62(6):545–550

Blasiak R, Pittman J, Yagi N, Sugino H (2016) Negotiating the use of biodiversity in marine areas beyond national jurisdiction. Front Mar Sci 3:224

Blasiak R, Durussel C, Pittman J, Sénit CA, Petersson M, Yagi N (2017a) The role of NGOs in negotiating the use of biodiversity in marine areas beyond national jurisdiction. Mar Policy 81:1–8

Blasiak R, Spijkers J, Tokunaga K, Pittman J, Yagi N, Österblom H (2017b) Climate change and marine fisheries: least developed countries top global index of vulnerability. PLoS One 12(6):e0179632

Blasiak R, Bebbington J, Jouffray J-B (2018a) Voluntary environmental programs. Background brief to the Amersfoort Dialogue. The Keystone Dialogues. http://keystonedialogues.earth/wp-content/uploads/2018/06/Brief2-Voluntary-Environmental-Programs.pdf

Blasiak R, Jouffray J-B, Wabnitz CCC, Sundström E, Österblom H (2018b) Corporate control and global governance of marine genetic resources. Sci Adv 4(6):eaar5237

Blasiak R, Jouffray J-B, Wabnitz CCC, Österblom H (2019) Scientists should disclose origin in marine gene patents. Trends Ecol Evol 34(5):392–395

Block BA, Jonsen ID, Jorgensen SJ, Winship AJ, Shaffer SA, Bograd SJ, Hazen EL et al (2011) Tracking apex marine predator movements in a dynamic ocean. Nature 475(7354):86

Bogard JR, Thilsted SH, Marks GC, Wahab MA, Hossain MAR, Jakobsen J, Stangoulis J (2015) Nutrient composition of important fish species in Bangladesh and potential contribution to recommended nutrient intakes. J Food Compos Anal 42:120–133

Boonstra WJ (2016) Conceptualizing power to study social-ecological interactions. Ecol Soc 21(1):21

Boonstra WJ, Ottosen KM, Ferreira ASA, Richter A, Rogers LA, Pedersen MW, Kokkalis A et al (2015) What are the major global threats and impacts in marine environments? Investigating the contours of a shared perception among marine scientists from the bottom-up. Mar Policy 60:197–201

Bourguignon F (2015) Revisiting the debate on inequality and economic development. Revue d'économie politique 125(5):633–663

Brent Z, Barbesgaard M, Pedersen C (2018) The Blue Fix: unmasking the politics around the promise of blue growth. Transnational Institute, Amsterdam

Brown-Weiss E (1990) In fairness to future generations. Environ Sci Policy Sustain Dev 32(3):6–31

Brundtland GH (1987) Our common future: call for action. Environ Conserv 14(4):291–294

Burke M, Hsiang SM, Miguel E (2015) Global non-linear effect of temperature on economic production. Nature 527(7577):235

Calvin K, Bond-Lamberty B, Clarke L, Edmonds J, Eom J, Hartin C, Kim S et al (2017) The SSP4: a world of deepening inequality. Glob Environ Chang 42:284–296

Campbell J, Barnett J (2010) Climate change and small island states: power, knowledge and the south pacific. Routledge, London

Camposi A (2017) How does inequality affect economic growth? Caixa Bank Research, 17 Jan. http://www.caixabankresearch.com/node/30188

Carr PJ, Williamson JL (1982) The Sullom Voe success story. Proc Inst Mech Eng 196(1):239–258

Carvalho N, Edwards-Jones G, Isidro E (2011) Defining scale in fisheries: small versus large-scale fishing operations in the Azores. Fish Res 109(2–3):360–369

Carver R (2019) Resource sovereignty, marine phosphate mining and the blue economy in Namibia. J Polit Ecol 26(1):381–402

CBD (Convention on Biological Diversity) (2000) Conference of the parties, decision adopted by the conference of the parties to the convention on biological diversity at its fifth meeting. V/6. Ecosystem Approach." May. Nairobi

CBD (2004) Decision adopted by the conference of the parties to the convention on biological diversity at its seventh meeting VII/11. Ecosystem Approach. 9–20 and 27 Feb. Kuala Lumpur

CBD (2008) Conference of the parties, decision on scientific criteria for ecologically or biologically significant areas, Decision IX/20. Annex I

Cederman L-E, Weidmann NB, Gleditsch KS (2011) Horizontal inequalities and ethnonationalist civil war: a global comparison. Am Polit Sci Rev 105(3):478–495

Cheung WWL, Lam VWY, Wabnitz CCC (2019) Future scenarios and projections for fisheries on the high seas under a changing climate. IIED Working Paper. International Institute for Environment and Development, London

Childs JR, Hicks CC (2019) Securing the blue: political ecologies of the blue economy in Africa. J Polit Ecol 26(1):323–340

Choudhury A, McDougall C (2019) Gendered ownership of aquaculture resources: insights from two villages in Bangladesh. CGIAR, Penang

Choudhury A, McDougall C, Rajaratnam S, Park CMY (2017) Women's empowerment in aquaculture: two case studies from Bangladesh. FAO, Rome; and WorldFish, Penang

Cicin-Sain B (2015) "Goal 14—conserve and sustainably use oceans, seas and marine resources for sustainable development. http://unchronicle.un.org/article/goal-14-conserve-and-sustainably-useoceans-seas-and-marine-resources-sustainable/

Cingano F (2014) Trends in income inequality and its impact on economic growth. OECD, Paris

Cinner JE (2009) Poverty and the use of destructive fishing gear near east African marine protected areas. Environ Conserv 36(4):321–326

Cinner JE, Daw T, McClanahan TR (2009) Socioeconomic factors that affect artisanal fishers' readiness to exit a declining fishery. Conserv Biol 23(1):124–130

Cinner JE, Basurto X, Fidelman P, Kuange J, Lahari R, Mukminin A (2012) Institutional designs of customary fisheries management arrangements in Indonesia, Papua New Guinea, and Mexico. Mar Policy 36(1):278–285

Cinner JE, Adger WN, Allison EH, Barnes ML, Brown K, Cohen PJ, Gelcich S et al (2018) Building adaptive capacity to climate change in tropical coastal communities. Nat Clim Chang 8(2):117

Ciplet D, Roberts JT, Khan MR (2015) Power in a warming world: the new global politics of climate change and the remaking of environmental inequality. MIT Press, Cambridge, MA

Cisneros-Montemayor AM, Sumaila UR (2019) Busting myths that hinder an agreement to end harmful fisheries subsidies. Mar Policy 109:103699

Cisneros-Montemayor AM, Pauly D, Weatherdon LV, Ota Y (2016) A global estimate of seafood consumption by coastal indigenous peoples. PLoS One 11(12):e0166681

Cisneros-Montemayor AM, Moreno-Báez M, Voyer M, Allison EH, Cheung WW, Hessing-Lewis M, Oyinlola MA et al (2019) Social Equity and benefits as the nexus of a transformative blue economy: a sectoral review of implications. Mar Policy 109:103702

Clapp J, Fuchs DA (2009) Corporate power in global agrifood governance. MIT Press, Cambridge, MA

Cohen P, Allison EH, Andrew NL, Cinner JE, Evans LS, Fabinyi M, Garces LR et al (2019) Securing a just space for small-scale fisheries in the blue economy. Front Mar Sci 6:171

Cole SM, Puskur R, Rajaratnam S, Zulu F (2015) Exploring the intricate relationship between poverty, gender inequality and rural masculinity: a case study from an aquatic agricultural system in Zambia. Cult Soc Masculinit 7(2):154

Cole SM, McDougall C, Kaminski AM, Kefi AS, Chilala A, Chisule G (2018) Postharvest fish losses and unequal gender relations: drivers of the social-ecological trap in the Barotse Floodplain fishery, Zambia. Ecol Soc 23(2):18

Commonwealth Secretariat (2018) Commonwealth blue charter. Shared values, shared ocean. A commonwealth commitment to work together to protect and manage our ocean. Commonwealth Secretariat, London

Čović I, Šimunac A, Veža J, Slišković M, Jelić-Mrčelić G (2013) Methods of pollution removal after tanker 'Erika' accident. Trans Marit Sci 2(1):41–48

Crona B, Bodin Ö (2010) Power asymmetries in small-scale fisheries: a barrier to governance transformability? Ecol Soc 15(4):32

Crona BI, Van Holt T, Petersson M, Daw TM, Buchary E (2015) Using social–ecological syndromes to understand impacts of international seafood trade on small-scale fisheries. Glob Environ Chang 35:162–175

Crona BI, Basurto X, Squires D, Gelcich S, Daw TM, Khan A, Havice E et al (2016) Towards a typology of interactions between small-scale fisheries and global seafood trade. Mar Policy 65:1–10

Dabla-Norris E, Kochhar K, Suphaphiphat N, Ricka F, Tsounta E (2015) Causes and consequences of income inequality: a global perspective. IMF staff discussion note. SDN/15/13. International Monetary Fund, New York

Dauvergne P, Lister J (2012) Big brand sustainability: governance prospects and environmental limits. Glob Environ Chang 22(1):36–45

Daw TM, Coulthard S, Cheung WWL, Brown K, Abunge C, Galafassi D, Peterson GD et al (2015) Evaluating taboo trade-offs in ecosystems services and human well-being. Proc Natl Acad Sci 112(22):6949–6954

De Schutter O (2012) Ocean-grabbing as serious a threat as land-grabbing. http://www.srfood.org/images/stories/pdf/press_releases/20121030_fisheries_en.pdf

De Silva DAM (2011) Faces of women in global fishery value chains: female involvement, impact and importance in the fisheries of developed and developing countries. NORAD/FAO Value Chain Project

Diamond L (2008) The democracy rollback: the resurgence of the predatory state. Foreign Affairs, March–April. https://www.foreignaffairs.com/articles/2008-03-02/democratic-rollback

Diffenbaugh NS, Burke M (2019) Global warming has increased global economic inequality. Proc Natl Acad Sci 116(20):9808–9813

Dollar D, Svensson J (2001) What explains the success or failure of structural adjustment programmes? Econ J 110:894–917

Doumbouya A, Camara OT, Mamic J, Intchama JF, Jarra A, Ceesay S, Guèye A et al (2017) Assessing the effectiveness of monitoring control and surveillance of illegal fishing: the case of West Africa. Front Mar Sci 4:50

Duruigbo EA (2003) Multinational corporations and international law: accountability and compliance issues in the petroleum industry. Brill, Amsterdam

Economist (2015) How inequality affects growth. June 15. https://www.economist.com/the-economist-explains/2015/06/15/how-inequality-affects-growth

Eder JF (2005) Coastal resource management and social differences in Philippine fishing communities. Hum Ecol 33(2):147–169

Ehlers P (2016) Blue growth and ocean governance: how to balance the use and the protection of the seas. WMU J Marit Aff 15(2):187–203

EPA (U.S. Environmental Protection Agency) (2017) Learn about Environmental Justice. 5 Dec. https://www.epa.gov/environmentaljustice/learn-about-environmental-justice

European Commission (2012) Blue growth: opportunities for marine and maritime sustainable growth. European Commission, Brussels

European Commission (2017) Report on the blue growth strategy: towards more sustainable growth and jobs in the blue economy. European Commission, Brussels

Fabinyi M (2012) Fishing for fairness: poverty, morality and marine resource regulation in the Philippines. ANU Press, Canberra

FAO (Food and Agriculture Organization of the United Nations) (2003) Fisheries management. 2. The ecosystem approach to fisheries. FAO Technical Guidelines for Responsible Fisheries No 4, Suppl 2. FAO, Rome

FAO (2015) Voluntary guidelines for securing sustainable small-scale fisheries in the context of food security and poverty eradication. FAO, Rome

FAO (2018) The state of world fisheries and aquaculture: meeting the sustainable development goals. FAO, Rome

Farnworth CR, Kantor P, Choudhury A, Mcguire S, Sultana N (2015) Gender relations and improved technologies in small household ponds in Bangladesh: rolling out novel learning approaches. Asian Fish Sci 29S:161–178

Felipe-Lucia MR, Martín-López B, Lavorel S, Berraquero-Díaz L, Escalera-Reyes J, Comín FA (2015) Ecosystem services flows: why stakeholders' power relationships matter. PLoS One 10(7): e0132232

FERN (Fisheries Equity Research Network) (2019) Home page. Dec 3. https://www.uow.edu.au/law-humanities-the-arts/research/ancors/our-research/fern/

FFA (Pacific Islands Forum Fisheries Agency) (2017) Economic and development indicators and statistics: tuna fisheries of the western and central pacific ocean 2016. Pacific Islands Forum Fisheries Agency, Honiara

Finkbeiner EM, Bennett NJ, Frawley TH, Mason JG, Briscoe DK, Brooks CM, Ng CA et al (2017) Reconstructing overfishing: moving beyond malthus for effective and equitable solutions. Fish Fish 18(6):1180–1191

Flannery W, Healy N, Luna M (2018) Exclusion and non-participation in marine spatial planning. Mar Policy 88:32–40

Folke C, Österblom H, Jouffray J-B, Lambin E, Scheffer M, Adger NW, Crona B et al (2019) Transnational corporations and the challenge of biosphere stewardship. Nat Ecol Evol 3:1396–1403

Fortnam M, Brown K, Chaigneau T, Crona B, Daw TM, Gonçalves D, Hicks C et al (2019) The gendered nature of ecosystem services. Ecol Econ 159:312–325

Franck TM (1995) Fairness in international law and institutions. Oxford University Press, Oxford

Fraser R, Spencer G (1998) The value of an ocean view: an example of hedonic property amenity valuation. Aust Geogr Stud 36(1): 94–98

Freestone D, Hey D (1996) Origins and development of the precautionary principle. In: Freestone D, Hey D (eds) The precautionary principle and international law: the challenges of implementation. Kluwer Law International, Alphen aan den Rijn

Frid A, McGreer M, Haggarty DR, Beaumont J, Gregr EJ (2016) Rockfish size and age: the crossroads of spatial protection, central place fisheries and indigenous rights. Glob Ecol Conserv 8: 170–182
Fröcklin S, de la Torre-Castro M, Lindström L, Jiddawi NS (2013) Fish traders as key actors in fisheries: gender and adaptive management. Ambio 42(8):951–962
Gagern A, van den Bergh J (2013) A critical review of fishing agreements with tropical developing countries. Mar Policy 38: 375–386
Galaz V, Crona B, Dauriach A, Jouffray J-B, Österblom H, Fichner J (2018) Tax havens and global environmental degradation. Nat Ecol Evol 2:1352–1357
Garcia Rodrigues J, Conides AJ, Rivero Rodriguez S, Raicevich S, Pita P, Kleisner KM, Pita C et al (2017) Marine and coastal cultural ecosystem services: knowledge gaps and research priorities. One Ecosyst 2:e12290
Gerrard S, Kleiber D (2019) Women fishers in Norway: few, but significant. Marit Stud. https://doi.org/10.1007/s40152-019-00151-4
Gjerde KM (2006) High seas fisheries management under the convention on the law of the sea. In: The law of the sea: progress and prospects, p 302
Global Compact (2019) The decade to deliver: a call to business action (2019). https://ceowatermandate.org/resources/the-decade-to-deliver-a-call-to-business-action-2019/. Accessed 18 Nov
Golden CD, Allison EH, Cheung WW, Dey MM, Halpern BS, McCauley DJ, Smith M et al (2016) Fall in fish catch threatens human health. Nature 534(7607):317–320
Golden JS, Virdin J, Nowacek DP, Halpin P, Bennear L, Patil PG (2017) Making sure the blue economy is green. Nat Ecol Evol 1(2):0017
Griffin P (2017) The carbon majors database. CDP, London
Hadjmichael M (2018) A call for a blue degrowth: unravelling the European Union's fisheries and maritime policies. Mar Policy 91:58–164
Haider JL, Boonstra WJ, Peterson GD, Schlüter M (2018) Traps and sustainable development in rural areas: a review. World Dev 101:311–321
Hallegatte S, Bangalore M, Bonzanigo L, Fay M, Kane T, Narloch U, Rozenberg J et al (2016) Shock waves: managing the impacts of climate change on poverty. World Bank Group, Washington, DC
Halpern BS, Walbridge S, Selkoe KA, Kappel CV, Micheli F, D'agrosa C, Bruno JF et al (2008) A global map of human impact on marine ecosystems. Science 319(5865):948–952
Halpern BS, Klein CJ, Brown CJ, Beger M, Grantham HS, Mangubhai S, Ruckelshaus M et al (2013) Achieving the triple bottom line in the face of inherent trade-offs among social equity, economic return, and conservation. Proc Natl Acad Sci 110(15):6229–6234
Hamann M, Berry K, Chaigneau T, Curry T, Heilmayr R, Henriksson PJ, Hentati-Sundberg J et al (2018) Inequality and the Biosphere. Annu Rev Environ Resour 43:61–83
Hanich Q, Campbell B, Bailey M, Molenaar E (2015) Research into fisheries equity and fairness: addressing conservation burden concerns in transboundary fisheries. Mar Policy 51:302–304
Harden-Davies HR (2017) Research for regions: strengthening marine technology transfer for pacific island countries and biodiversity beyond national jurisdiction. Int J Mar Coast Law 32(4): 797–822
Harden-Davies H, Gjerde K (2019) Building scientific and technological capacity: a role for benefit-sharing in the conservation and sustainable use of marine biodiversity beyond national jurisdiction. Ocean Yearbook 33(1):377–400
Harper S, Zeller D, Hauzer M, Pauly D, Sumaila UR (2013) Women and fisheries: contribution to food security and local economies. Mar Policy 39:56–63
Harper S, Grubb C, Stiles M, Sumaila UR (2017) Contributions by women to fisheries economies: insights from five maritime countries. Coast Manag 45(2):91–106
Harper S, Salomon AK, Newell D, Hilistis Waterfall P, Brown K, Harris LM, Sumaila UR (2018) Indigenous women respond to fisheries conflict and catalyze change in governance on Canada's pacific coast. Marit Stud 17(2):189–198
Hayworth JS, Clement TP, Valentine JF (2011) Deepwater horizon oil spill impacts on Alabama beaches. Hydrol Earth Syst Sci 15(12):3639–3649
Hickel J, Kallis G (2019) Is green growth possible? New Polit Econ 1:1–18
Hicks CC, Cohen PJ, Graham NA, Nash KL, Allison EH, D'Lima C, Mills DJ et al (2019) Harnessing global fisheries to tackle micronutrient deficiencies. Nature 574(7776):95–98
Hoegh-Guldberg O (2015) Reviving the oceans economy: the case for action—2015. World Wide Fund for Nature, Gland
ICJ (International Court of Justice) (2010) Case concerning pulp mills on the River Uruguay (*Argentina v. Uruguay*)
IFC (International Finance Corporation) (2017) Investing in women: new evidence for the business case. https://www.ifc.org/wps/wcm/connect/ac8fca18-6586-48cc-bfba-832b41d6af68/IFC+Invest+in+Women+October+2017.pdf?MOD=AJPERES&CVID=lYLVAcA
IGC (Intergovernmental Conference) (2018) Intergovernmental negotiations on an international legally binding instrument under the United Nations Convention on the Law of the Sea on the Conservation and Sustainable Use of Marine Biological Diversity of Areas beyond National Jurisdiction: President Aid to Negotiations, 3 Dec 2018 (A/CONF.232/2019/1*)
ILC (International Law Commission) (2018) Draft guidelines on the protection of the atmosphere (A/73/10)
IPBES (Intergovernmental Science-Policy Platform on Biodiversity and Ecosystem Services) (2019) Summary for policymakers of the global assessment report on biodiversity and ecosystem services of the intergovernmental science-policy platform on biodiversity and ecosystem services. https://www.ipbes.net/system/tdf/spm_global_unedited_advance.pdf?file=1&type=node&id=35245. Accessed 21 Nov
IPCC (Intergovernmental Panel on Climate Change) (2019) Special report on the ocean and cryosphere in a changing climate. Geneva: IPCC.
Islam SN, Winkel J (2017) Climate change and social inequality. DESA Working Paper no. 152. https://www.un.org/esa/desa/papers/2017/wp152_2017.pdf
ISSC (International Social Science Council), IDS (Institute of Development Studies), UNESCO (United Nations Educational, Scientific and Cultural Organization) (2016) World social science report 2016—challenging inequalities: pathways to a just world. UNESCO Publishing, Paris
ITLOS (International Tribunal for the Law of the Sea) (2011) Responsibility and obligations of states sponsoring persons and entities with respect to activities in the area (request for advisory opinion submitted to the seabed disputes chamber). Case no 17
Jackson JBC (2001) What was natural in the coastal oceans? Proc Natl Acad Sci 98(10):5411–5418
Jackson T (2009) Prosperity without growth: economics for a finite planet. Routledge, London
Jentoft S, Eide A (2011) Poverty mosaics: realities and prospects in small-scale fisheries. Springer, Amsterdam
Jouffray JB, Crona B, Wassénius E, Bebbington J, Scholtens B (2019) Leverage points in the financial sector for seafood sustainability. Sci Adv 5(10):eaax3324
Kabeer N (2012) Women's economic empowerment and inclusive growth: labour markets and enterprise development. SIG Working Paper 1. International Development Research Centre, Ottawa

Kabeer N (2016) Gender equality, economic growth, and women's agency: the endless variety and monotonous similarity of patriarchal constraints. Fem Econ 22(1):295–321

Kabeer N, Natali L (2013) Gender equality and economic growth: is there a win-win?" IDS Working Papers, no 417, pp 1–58

Kaczynski VM, Fluharty DL (2002) European policies in West Africa: who benefits from fisheries agreements? Mar Policy 26(2): 75–93

Kantor P, Morgan M, Choudhury A (2015) Amplifying outcomes through addressing inequality: the role of gender transformative approaches in agricultural research for development. Gend Technol Develop 19(3):292–319

Kerr S, Colton J, Johnson K, Wright G (2015) Rights and ownership in sea country: implications of marine renewable energy for indigenous and local communities. Mar Policy 52: 108–115

Kerr S, Johnson K, Weir S (2017) Understanding community benefit payments from renewable energy development. Energy Policy 105:202–211

Kittinger JN, Teh LC, Allison EH, Bennett NJ, Crowder LB, Finkbeiner EM, Hicks C et al (2017) Committing to socially responsible seafood. Science 356(6341):912–913

Klain S, Beveridge R, Bennett N (2014) Ecologically sustainable but unjust? Negotiating equity and authority in common-pool marine resource management. Ecol Soc 19(4):52

Kleiber D, Frangoudes K, Snyder H, Choudhury A, Cole S, Soejima K, Pita C et al (2017) Promoting gender equity and equality through the small-scale fisheries guidelines: experiences from multiple case studies. In: Jentoft S, Chuenpagdee R, Barragán-Paladines M, Franz N (eds) The small-scale fisheries guidelines, MARE Publication Series, vol 14. Springer, New York

Klein C, McKinnon MC, Wright BT, Possingham HP, Halpern BS (2015) Social equity and the probability of success of biodiversity conservation. Glob Environ Chang 35:299–306

Klinsky S, Roberts T, Huq S, Okereke C, Newell P, Dauvergne P, O'Brien K et al (2017) Why equity is fundamental in climate change policy research. Glob Environ Chang 44:170–173

Knudsen M (2016) Poverty and beyond: small-scale fishing in overexploited marine environments. Hum Ecol 44(3):341–352

Kostakis V, Bauwens M (2014) Transition proposals toward a commons-oriented economy and society. In: Kostakis V, Bauwens M (eds) Network society and future scenarios for a collaborative economy. M. Palgrave Macmillan, London, pp 59–69

Kotz DM (2015) The rise and fall of neoliberal capitalism. Harvard University Press, Cambridge, MA

Kruijssen F, Audet-Belanger G, Choudhury A, Crissman C, Dalsgaard JPT, Dawson C, Dickson M et al (2016) Value chain transformation: taking stock of WorldFish research on value chains and markets. CGIAR, Penang

Laffoley D, Baxter JM, Amon DJ, Currie DE, Downs CA, Hall-Spencer JM, Harden-Davies H et al (2019) Eight urgent, fundamental and simultaneous steps needed to restore ocean health, and the consequences for humanity and the planet of inaction or delay. Aquat Conserv Mar Freshwat Ecosyst 30(1):194–208. https://doi.org/10.1002/aqc.3182

Laird S, Wynberg R (2018) A fact-finding and scoping study on digital sequence information on genetic resources in the context of the convention on biological diversity and the Nagoya protocol. Secretariat of CBD, pp 2–79

Le Billon P (2014) Natural resources and corruption in post-war transitions: matters of trust. Third World Q 35(5):770–786

Le Manach F, Andriamahefazafy M, Harper S, Harris A, Hosch G, Lange GM, Zeller D et al (2012) Who gets what? Developing a more equitable framework for EU fishing agreements. Mar Policy 38:257–266

Leach M, Mearns R (1998) The lie of the land: challenging received wisdom on the African environment. Currey, Oxford

Leach M, Scoones I, Stirling A (2010) Dynamic sustainabilities: technology, environment, social justice. Earthscan, London

Leach M, Reyers B, Bai X, Brondizio ES, Cook C, Díaz S, Espindola G et al (2018) Equity and sustainability in the anthropocene: a social–ecological systems perspective on their intertwined futures. Glob Sustain e13:1–13

Lentisco, A., and R.U. Lee. 2015. A Review of women's access to fish in small-scale fisheries. FAO Fisheries and Aquaculture Circular. Rome: FAO.

Li TM (2007) The will to improve: governmentality, development, and the practice of politics. Duke University Press, Durham, NC

Lillebø AI, Pita C, Garcia Rodrigues J, Ramos S, Villasante S (2017) How can marine ecosystem services support the blue growth agenda? Mar Policy 81:132–142

Llewellyn LE, English S, Barnwell S (2016) A roadmap to a sustainable Indian ocean blue economy. J Indian Ocean Reg 12(1): 52–66

Lloyd's (2014) Global marine trends 2030. Lloyd's Register Group, London

Lopes PFM, Pennino MG, Freire F (2017) Climate change can reduce shrimp catches in equatorial Brazil. Reg Environ Chang 18(1):223–234

Lowe C (2013) Wild profusion: biodiversity conservation in an Indonesian archipelago. Princeton University Press, Princeton, NJ

Lubchenco J, Cerny-Chipman EB, Reimer J, Levin SA (2016) The right incentives enable ocean sustainability successes and provide hope for the future. Proc Natl Acad Sci 113(51):14507–14514

Mather C, Fanning L (2019) Social licence and aquaculture: towards a research agenda. Mar Policy 99:275–282

Matsue N, Daw T, Garrett L (2014) Women fish traders on the Kenyan coast: livelihoods, bargaining power, and participation in management. Coast Manag 42(6):531–554

McCauley DJ, Jablonicky C, Allison EH, Golden CD, Joyce FH, Mayorga J, Kroodsma D (2018) Wealthy countries dominate industrial fishing. Sci Adv 4(8):eaau2161. https://doi.org/10.1126/sciadv.aau2161

McDermott M, Mahanty S, Schreckenberg K (2013) Examining equity: a multidimensional framework for assessing equity in payments for ecosystem services. Environ Sci Pol 33: 416–427

McDougall C, Cole S, Rajaratnam S, Brown J, Choudhury A, Kato-Wallace J, Manlosa A et al (2015) Implementing a gender transformative research approach: early lessons. In: Douthwaite B, Apgar JM, Schwarz A, McDougall C, Attwood S, Senaratna SS, Clayton T (eds) Research in development: learning from the CGIAR research program on aquatic agricultural systems. CGIAR Research Program on Aquatic Agricultural Systems, Penang. Working Paper: AAS-2015-16

McGee L (2013) Socially responsible corporations can include a provision in their articles of incorporation that accounts for environmental sustainability when making business decisions. Policy Options Politiques, Aug 4. https://policyoptions.irpp.org/magazines/arctic-visions/how-to-encourage-corporate-social-responsibility/

Meikle A, Sugden J (2015) Introducing technology justice: a new paradigm for the SDGs. Technology Justice Policy Brief 1. Practical Action, Rugby

Mendoza M, McDowell R, Maso MA, Htusan E (2016) Fishermen slaves: human trafficking and the seafood we eat. Mango Media

Meyer Y (2017) Reflections on the UN working group on human rights and transnational corporation. South Afr Yearb Int Law 42: 200–219

Mi Z, Liao H, Coffman DM, Wei YM (2019) Assessment of equity principles for international climate policy based on an integrated assessment model. Nat Hazards 95(1–2):309–323

Michel JA (2016) Rethinking the oceans: towards the blue economy. Paragon, St. Paul, MN

Michel J, Owens EH, Zengel S, Graham A, Nixon Z, Allard T, Holton W et al (2013) Extent and degree of shoreline oiling: deepwater Horizon oil spill, Gulf of Mexico, USA. PLoS One 8(6):e65087

Mills D, Béné C, Ovie S, Tafida A, Sinaba F, Kodio A, Russell A et al (2011) Vulnerability in African small-scale fishing communities. J Int Dev 23(2):308–313

Minas S (2018) Marine technology transfer under a BBNJ treaty: a case for transnational network cooperation. AJIL Unb 112:144–149

Monacelli N (2018) Improving maritime transportation security in response to industry consolidation. Homel Sec Aff 14

Moreto WD, Charlton RW, DeWitt SE, Burton CM (2019) The convergence of captured fish and people: examining the symbiotic nature of labor trafficking and illegal, unreported and unregulated fishing. Deviant Behav 0:1–17

Morgan G (2009) Politics: what is the Shetland charitable trust? Shetland Times, 3 Apr. https://www.shetlandtimes.co.uk/2009/04/03/politics-what-is-the-shetland-charitable-trust

Morgan M, Choudhury A, Braun M, Beare D, Benedict J, Kantor P (2015) Understanding the gender dimensions of adopting climate-smart smallholder aquaculture innovations. Working Paper AAS-2015-08. CGIAR, Penang

Morgera E (2016) The need for an international legal concept of fair and equitable benefit-sharing. Eur J Int Law 27(2):353–383

Morgera E, Ntona M (2018) Seize the moment: towards fairer capacity building and marine technology transfer. IIED Briefing, London

Muchlinski PT (2007) Multinational enterprises and the law. Oxford University Press, New York

Murase S (2015) Second report of the special rapporteur on the protection of the atmosphere. 67th session of the International Law Commission

Murphy S, Burch D, Clapp J (2012) Cereal secrets: the world's largest grain traders and global agriculture. Oxfam, London

Nallari R, Griffith B (2011) Gender and macroeconomic policy. World Bank, Washington, DC

Nash R (2013) The oceans and their challenge to conserve marine biodiversity. In: Pechan P, de Vries G (eds) Living with water; targeting quality in a dynamic world. Springer, New York, pp 143–194

Niner HJ, Ardron JA, Escobar EG, Gianni M, Jaeckel A, Jones DO, Levin LA et al (2018) Deep-sea mining with no net loss of biodiversity—an impossible aim. Front Mar Sci 5:53

Nussbaum MC (2011) Creating capabilities. the human development approach. Harvard University Press, Cambridge, MA

O'Neill ED, Crona B (2017) Assistance networks in seafood trade: a means to assess benefit distribution in small-scale fisheries. Mar Policy 78:196–205

O'Neill BC, Kriegler E, Ebi KL, Kemp-Benedict E, Riahi K, Rothman DS, van Ruijven BJ et al (2017) The roads ahead: narratives for shared socioeconomic pathways describing world futures in the 21st century. Glob Environ Chang 42:169–180

OECD (Organisation for Economic Co-operation and Development) (2016) The ocean economy in 2030. OECD, Paris

Okereke C (2006) Global environmental sustainability: intragenerational equity and conceptions of justice in multilateral environmental regimes. Geoforum 37(5):725–738

Oreskes N, Conway EM (2011) Merchants of doubt: how a handful of scientists obscured the truth on issues from tobacco smoke to global warming. Bloomsbury, New York

Østby G (2008) Polarization, horizontal inequalities and violent civil conflict. J Peace Res 45(2):143–162

Österblom H, Bodin Ö (2012) Global cooperation among diverse organizations reduces illegal fishing in the southern ocean. Conserv Biol 26:638–648

Österblom H, Jouffray J-B, Folke C, Crona B, Merrie A, Troell M, Rockström J (2015) Transnational corporations as 'keystone actors' in marine ecosystems. PLoS One 10(5):e0127533

Österblom H, Jouffray J-B, Folke C, Rockström J (2017) Emergence of a global science-business initiative for ocean stewardship. Proc Natl Acad Sci 114:9038–9043

Ostrom E (1998) A behavioural approach to the rational choice theory of collective action: presidential address, American Political Science Association, 1997. Am Polit Sci Rev 92(1):1–22

Oxfam (2019) 5 shocking facts about extreme global inequality and how to even it up. Oxfam, London

Oyewande AA (2008) Corporate social responsibility of multinational corporations in developing countries: how far do their roles and responsibility stretch? PhD diss, National University of Singapore

Page J (2007) Salmon farming in first nations' territories: a case of environmental injustice on Canada's west coast. Local Environ 12(6):613–626

Pascual U, Phelps J, Garmendia E, Brown K, Corbera E, Martin A, Gomez-Baggethun E, Muradian R (2014) Social equity matters in payments for ecosystem services. Bioscience 64(11):1027–1036

Pauly D, Belhabib D, Blomeyer R, Cheung WW, Cisneros-Montemayor AM, Copeland D, Harper S et al (2014) China's distant-water fisheries in the 21st century. Fish Fish 15(3):474–488

Pedroza C (2019) Managing mercado del mar: a case of women's entrepreneurship in the fishing industry. Marit Stud:1–12

Persson A, Rothstein B, Teorell J (2013) Why anticorruption reforms fail: systemic corruption as a collective action problem. Governance 26(3):449–471

Piketty T (2014) Capital in the twenty-first century. Cambridge, MA, Belknap

Pinsky M, Reygondeau G, Cadell R, Palacios-Abrantes J, Spijkers J, Cheung WWL (2018) Preparing ocean governance for species on the move. Science 360(6395):1189–1191

Pohl C, Rist S, Zimmermann A, Fry P, Gurung GS, Schneider F, Speranza CI et al (2010) Researchers' roles in knowledge co-production: experience from sustainability research in Kenya, Switzerland, Bolivia and Nepal. Sci Public Policy 37: 267–281

Popova E, Vousden D, Sauer WH, Mohammed EY, Allain V, Downey-Breedt N, Fletcher R et al (2019) Ecological connectivity between the areas beyond national jurisdiction and coastal waters: safeguarding interests of coastal communities in developing countries. Mar Policy 104:90–102

Pretlove B, Blasiak R (2018) Mapping ocean governance and regulations. Working paper for consultation for UN Global Compact Action Platform for Sustainable Ocean Business. https://www.unglobalcompact.org/docs/publications/Mapping-Ocean-Governance-and-Regulation.pdf

Promundo-AAS (Promundo-US and the CGIAR Research Program on Aquatic Agricultural Systems) (2016) Promoting gender-transformative change with men and boys: a manual to spark critical reflection on harmful gender norms with men and boys in aquatic agricultural systems. Promundo-US, Washington, DC; CGIAR, Penang

Purcell SW, Crona BI, Lalavanua W, Eriksson H (2017) Distribution of economic returns in small-scale fisheries for international markets: a value-chain analysis. Mar Policy 86:9–16

Ramesh N, Rising JA, Oremus KL (2019) The small world of global marine fisheries: the cross-boundary consequences of larval dispersal. Science 364(6446):1192–1196

Ratner SR (2007) Business in book. In: Bodanksy D, Brunnée J, Hey E (eds) The oxford handbook of international environmental law. Oxford University Press, Oxford

Raworth K (2017a) Doughnut economics: seven ways to think like a 21st-century economist. Chelsea Green, London

Raworth K (2017b) A doughnut for the Anthropocene: humanity's compass in the 21st century. Lancet Planet Health 1:e48–e49

Resplandy L, Keeling RF, Eddebbar Y, Brooks MK, Wang R, Bopp L, Long MC et al (2018) Quantification of ocean heat uptake from changes in atmospheric O_2 and CO_2 composition. Nature 563(7729):105

Reusch TBH, Dierking J, Andersson HC, Bonsdorff E, Carstensen J, Casini M, Czajkowski M et al (2018) The Baltic sea as a time machine for the future coastal ocean. Sci Adv 4(5):eaar8195

Reuters (2019) Brazil blames devastating oil spill on Greek-flagged tanker. Nov 21. https://www.theguardian.com/world/2019/nov/01/brazil-blames-oil-spill-greek-flagged-tanker-venezuelan-crude

Riahi K, van Vuuren DP, Kriegler E, Edmonds J, O'Neill BC, Fujimori S, Bauer N et al (2017) The shared socioeconomic pathways and their energy, land use, and greenhouse gas emissions implications: an overview. Glob Environ Chang 42:153–168

Ribot JC, Peluso NL (2003) A theory of access. Rural Sociol 68(2):153–181

Roberts C (2010) The unnatural history of the sea. Shearwater, London

Rosales RM, Pomeroy R, Calabio IJ, Batong M, Cedo K, Escara N, Facunla V et al (2017) Value chain analysis and small-scale fisheries management. Mar Policy 83:11–21

Rudolph DP, Hagget C, Aitken M (2014) Community benefits from offshore renewables: good practice review. Climate Xchange, Edinburgh

Ruggie J (2008) Protect, respect and remedy: a framework for business and human rights, report of the special representative of the secretary-general on the issue of human rights and transnational corporations and other business enterprises (A/HRC/8/5)

Sadovy de Mitcheson Y, Tam I, Muldoon G, Le Clue S, Botsford E, Shea S (2018) The trade in live reef food fish: going, going, gone, vol 1, Main Report. Parts 1, 2 and 3. ADM Capital Foundation and University of Hong Kong, Hong Kong Special Administrative Region, pp 1–288

Sala E, Mayorga J, Costello C, Kroodsma D, Palomares ML, Pauly D, Sumaila UR, Zeller D (2018) The economics of fishing the high seas. Sci Adv 4(6):eaat2504

Salpin C, Onwuasoanya V, Bourrel M, Swaddling A (2018) Marine scientific research in Pacific Small Island Developing States. Mar Policy 95:363–371

Sandel M (1990) Liberalism and the limits of justice. Cambridge University Press, Cambridge

Scheffer M, van Bavel B, van de Leemput IA, van Nes EH (2017) Inequality in nature and society. Proc Natl Acad Sci 114:13154–13157

Schiller L, Bailey M, Jacquet J, Sala E (2018) High seas fisheries play a negligible role in addressing global food security. Sci Adv 4(8):eaat8351

Schlosberg D (2009) Defining environmental justice: theories, movements, and nature. Oxford University Press, Oxford

Schuhbauer A, Chuenpagdee R, Cheung WWL, Greer K, Sumaila UR (2017) How subsidies affect the economic viability of small-scale fisheries. Mar Policy 82:114–121

Schwerdtner Máñez KS, Pauwelussen A (2016) Fish is women's business too: looking at marine resource use through a gender lens. In: Schwerdtner Máñez K, Poulsen B (eds) Perspectives on oceans past. Springer, Dordrecht, pp 193–211

Scottish Government (2018) Scottish government good practice principles for community benefits from offshore renewable energy developments. Nov 30. https://www.gov.scot/publications/consultation-scottish-government-good-practice-principles-community-benefits-offshore-renewable-energy-developments/

Segi S (2014) Protecting or pilfering? Neoliberal conservationist marine protected areas in the experience of coastal Granada, the Philippines. Hum Ecol 42(4):565–575

Sen A (1992) Inequality re-examined. Oxford University Press, Oxford

Silver JJ, Gray NJ, Campbell LM, Fairbanks LW, Gruby RL (2015) Blue economy and competing discourses in international oceans governance. J Environ Dev 24(2):135–160

Sovacool BK, Linnér BO, Goodsite ME (2015) The political economy of climate adaptation. Nat Clim Chang 5(7):616

Spijkers J, Singh G, Blasiak R, Morrison TH, Le Billon P, Österblom H (2019) Global patterns of fisheries conflict: forty years of data. Glob Environ Chang 57:101921

SSE (2019) Sustainable development fund. Dec 3. https://sse.com/communities

Stacey N, Gibson E, Loneragan NR, Warren C, Wiryawan B, Adhuri D, Fitriana R (2019) Enhancing coastal livelihoods in Indonesia: an evaluation of recent initiatives on gender, women and sustainable livelihoods in small-scale fisheries. Marit Stud 18:359–371. https://doi.org/10.1007/s40152-019-00142-5

Stanton EA (2012) The tragedy of maldistribution: climate, sustainability, and equity. Sustainability 4(3):394–411

Stonich SC (1998) Political ecology of tourism. Ann Tour Res 25(1):25–54

Stonich SC, Bort JR, Ovares LL (1997) Globalization of shrimp mariculture: the impact on social justice and environmental quality in central America. Soc Nat Resour 10(2):161–179

Stotsky J (2006) Gender budgeting. IMF Working Paper WP/06/232. International Monetary Fund, Washington, DC

Sullom Voe terminal: 40 years on (2018) Press and Journal. 24 Nov. https://www.pressandjournal.co.uk/fp/news/1616521/sullomvoe/

Sumaila UR, Walters C (2005) Intergenerational discounting: a new intuitive approach. Ecol Econ 52(2):135–142

Sumaila UR, Lam V, Le Manach F, Swartz W, Pauly D (2016) Global fisheries subsidies: an updated estimate. Mar Policy 69: 189–193

Sumaila UR, Jacquet J, Witter A (2017) When bad gets worse: corruption and fisheries. In: Williams A, Le Billon P (eds) Corruption, natural resources and development: from resource curse to political ecology. Edward Elgar, Cheltenham, pp 93–105

Sundström A (2012) Corruption and regulatory compliance: experimental findings from south African small-scale fisheries. Mar Policy 36(6):1255–1264

TBTI (Too Big to Ignore) (2016) TBTI 'blue justice for small-scale fisheries' commitments. http://tbtiglobal.net/blue-justice/

Teh LCL, Cadell R, Allison EH, Finkbeiner EM, Kittinger JN, Nakamura K, Ota Y (2019) The role of human rights in implementing socially responsible seafood. PLoS One 14(1): e0210241

Thorburn CC (2000) Changing customary marine resource management practice and institutions: the case of Sasi Lola in the Kei Islands, Indonesia. World Dev 28(8):1461–1479

Tickler D, Meeuwig JJ, Palomares M-L, Pauly D, Zeller D (2018) Far from home: distance patterns of global fishing fleets. Sci Adv 4(8):eaar3279

Tladi D (2007) Sustainable development in international law: an analysis of key enviro-economic instruments. Pretoria University Law Press, Pretoria

Tladi D (2014) State practice and the making and (re)making of international law: the case of the legal rules relating to biodiversity in areas beyond national jurisdiction. State Int Law J 1:97–116

Tladi D (2015) The common heritage of mankind and the proposed treaty on biodiversity in areas beyond national jurisdiction: the choice between pragmatism and sustainability. Yearb Int Environ Law 25:113–132

TNI (Transnational Institute) (2016) Human rights versus property rights: implementation and interpretation of the SSF guidelines. https://www.tni.org/files/article-downloads/human_rights_versus_property_rights_implementation_of_the_ssf_guidelines_en.pdf

Treves A, Artelle KA, Darimont CT, Lynn WS, Paquet P, Santiago-Ávila FJ, Shaw R, Wood MC (2018) Intergenerational equity can help to prevent climate change and extinction. Nat Ecol Evol 2(2):204

Trinidad AC, Albert J, Palma J, Matillano M, Boso D, Gaudiano JP, Manul J (2014) Fisheries value retention in the coral triangle for highly traded commodities. In: Economics of fisheries and aquaculture in the coral triangle. Asian Development Bank, Manila, pp 107–140

Tyler TR (2015) Social justice. In: Mikulincer M, Shaver PR, Dovidio JF, Simpson JA (eds) APA handbook of personality and social psychology, vol. 2: group processes. American Psychological Association, Washington, DC, pp 95–122

UN (United Nations) (2014) Blue economy concept paper. Presented at the blue economy summit. Abu Dhabi. https://sustainabledevelopment.un.org/content/documents/2978BEconcept.pdf

UN (2015) Inequality and the 2030 Agenda for sustainable development: development issues, no 4. UN, New York

UN Global Compact (n.d.) The power of principles: the ten principles of the UN global compact. https://www.unglobalcompact.org/what-is-gc/mission/principles

UNCLOS (1982) United Nations convention on the law of the sea. United Nations, New York

UNCTAD (United Nations Conference on Trade and Development) (2014) The oceans economy: opportunities and challenges for small island states. UNCTAD, Geneva

UNEP (2011) Towards a green economy: pathways to sustainable development and poverty eradication. United Nations Environment Programme, New York

UNESCO (United Nations Educational, Scientific and Cultural Organization) (2017) Global ocean science report: the current status of ocean science around the world. UNESCO, Paris

UNGA (2017) International legally binding instrument under the united nations convention on the law of the sea on the conservation and sustainable use of marine biological diversity of areas beyond national jurisdiction. 72nd sess. Resolution 72/249. 24 Dec

UNGA (2018) United Nations general assembly resolution: oceans and the law of the sea (A/RES/73/124)

UNGA (United Nations General Assembly) (1995) United Nations conference on straddling fish stocks and highly migratory fish stocks. 6th sess. New York. https://www.un.org/ga/search/view_doc.asp?symbol=A/CONF.164/37&Lang=E

USAID (U.S. Agency for International Development) (2019) Assessing fisheries in a new era: extended guidance for rapid appraisals of fisheries management systems. USAID Oceans and Fisheries Partnership. https://www.seafdec-oceanspartnership.org/wp-content/uploads/USAID-Oceans_Assessing-Fisheries_RAFMS-Guide_April-2019_print.pdf

van Wyk J-A (2015) Defining the blue economy as a south African strategic priority: toward a sustainable 10th province? J Ind Ocean Reg 11(2):153–169

Voigt-Hansen G (2019) Current 'light' and 'heavy' options for benefit-sharing in the context of the united nations convention for the law of the sea. In: Freestone D (ed) Conserving biodiversity in areas beyond national jurisdiction. Brill, Washington, DC, pp 683–705

Voyer M, van Leeuwen J (2019) Social license to operate in the blue economy. Res Policy 62:102–113

Voyer M, Quirk G, McIlgorm A, Azmi K (2018) Shades of blue: what do competing interpretations of the blue economy mean for oceans governance? J Environ Policy Plan 20:595–616

Walker BLE (2001) Sisterhood and seine-nets: engendering development and conservation in Ghana's marine fishery. Profess Geograp 53(2):160–177

Wamukota A (2009) The structure of marine fish marketing in Kenya: the case of Malindi and Kilifi districts. West Ind Ocean J Mar Sci 8(2):215–224

Wenzel N (2008) Human rights, treaties, extraterritorial application and effects. Max Planck Encyclopedia of Public International Law (online). Max Planck Institute, Munich

Westlund L, Charles A, Garcia SM, Sanders J (2017) Marine protected areas: interactions with fishery livelihoods and food security. FAO Fisheries and Aquaculture Technical Paper 603. Food and Agriculture Organization of the United Nations, Rome

White C, Halpern BS, Kappel CV (2012) Ecosystem service tradeoff analysis reveals the value of marine spatial planning for multiple ocean uses. Proc Natl Acad Sci 109(12):4696–4701

Whitehead JC, Dumas CF, Herstine J, Hill J, Buerger B (2008) Valuing beach access and width with revealed and stated preference data. Mar Resour Econ 23:119–135

Wilkinson R, Pickett K (2009) The spirit level: why greater equality makes societies stronger. Bloomsbury, New York

Wilkinson MD, Dumontier M, Aalbersberg IJ, Appleton G, Axton M, Baak A, Blomberg N et al (2016) The FAIR guiding principles for scientific data management and stewardship. Scient Data 3:160018. https://doi.org/10.1038/sdata.2016.18

Williams MJ (2008) Why look at fisheries through a gender lens? Development 51(2):180–185

Williams A, Le Billon P (2017) Corruption, natural resources and development: from resource curse to political ecology. Edward Elgar, Cheltenham

Williams MJ, Porter M, Choo PS, Kusakabe K, Vuki V, Gopal N, Bondad-Reantaso M (2012) Gender in aquaculture and fisheries: moving the agenda forward. Asian Fish Sci 25S:1–13

Wolff NH, Donner SD, Cao L, Iglesias-Prieto R, Sale PF, Mumby PJ (2015) Global Inequities between polluters and the polluted: climate change impacts on coral reefs. Glob Chang Biol 21(11):3982–3994

Wong F, Vos A, Pyburn R, Newton J (2019) Implementing gender transformative approaches in agriculture. CGIAR, Amsterdam

World Bank (2008) Small-scale capture fisheries: a global overview with emphasis on developing countries. PROFISH series. World Bank, Washington, DC

World Bank (2012) Hidden harvest: the global contribution of capture fisheries. World Bank, Washington, DC

World Bank and United Nations (2017) The potential of the blue economy: increasing long-term benefits of the sustainable use of marine resources for small island developing states and coastal least developed countries. World Bank, Washington, DC

WRI, with United Nations Development Programme, United Nations Environment Programme and WorldBank (2005) World resources 2005: the wealth of the poor: managing ecosystem to fight poverty. World Bank, Washington, DC

WWF (2018) Principles for a sustainable blue economy. WWF, Gland

Wynberg R, Hauck M (2014) People, power, and the coast: a conceptual framework for understanding and implementing benefit sharing. Ecol Soc 19(1):27

Wynberg R, Laird SA (2018) Fast science and sluggish policy: the herculean task of regulating biodiscovery. Trends Biotechnol 36(1):1–3

Yadav S, Abdulla A, Bertz N, Mawyer A (2019) King tuna: Indian Ocean trade, offshore fishing, and coral reef resilience in the Maldives Archipelago. ICES J Mar Sci, Oct 9. https://doi.org/10.1093/icesjms/fsz170

Yamazaki S, Resosudarmo B, Girsang W, Hoshino E (2018) Productivity, social capital and perceived environmental threats in small-island fisheries: insights from Indonesia. Ecol Econ 152:62–75

Yater LR (1982) The fisherman's family: economic roles of women and children. ICLARM, Manila

Yodanis CL (2000) Constructing gender and occupational segregation: a study of women and work in fishing communities. Qual Sociol 23(3):267–290

Zafra-Calvo N, Pascual U, Brockington D, Coolsaet B, Cortes-Vazquez JA, Gross-Camp N, Palomo I, Burgess ND (2017) Towards an indicator system to assess equitable management in protected areas. Biol Conserv 211:134–141

Zalik A (2009) Zones of exclusion: offshore extraction, the contestation of space and physical displacement in the Nigerian Delta and the Mexican Gulf. Antipode 41(3):557–582

Zeller D, Pauly D (2019) Back to the future for fisheries, where will we choose to go? Glob Sustain 2(e11):1–8. https://doi.org/10.1017/sus.2019.8

Integrated Ocean Management

14

Abbreviations

ABNJ	Areas beyond national jurisdiction
BBNJ	Marine biodiversity of areas beyond national jurisdiction
CCA	Climate change adaptation
CTI	Coral Triangle Initiative
CTI-CFF	Coral Triangle Initiative on Coral Reefs, Fisheries, and Food Security
EBA	Ecosystem-based adaptation
EBM	Ecosystem-based management
EEZ	Exclusive economic zone
FAO	United Nations Food and Agriculture Organization
GEF	Global Environment Facility
ICES	International Council for the Exploration of the Sea
ICM	Integrated coastal management
ICZM	Integrated coastal zone management
IMO	International Maritime Organization
IOM	Integrated ocean management
IOC	Intergovernmental Oceanographic Commission (of UNESCO)
IPCC	Intergovernmental Panel on Climate Change
ISA	International Seabed Authority
MPA	Marine protected area
MSP	Marine spatial planning
MSY	Maximum sustainable yield
NEAFC	North East Atlantic Fisheries Commission
NGO	Nongovernmental organisation
OSPAR	Convention for the Protection of the Marine Environment of the North-East Atlantic
PICES	The North Pacific Marine Science Organization
SAMOA	SIDS Accelerated Modalities of Action
SDG	Sustainable Development Goal
SIDS	Small island developing states
RPOA	Regional plan of action
UNCLOS	United Nations Convention on the Law of the Sea
UNDP	United Nations Development Programme
UNEP	United Nations Environment Programme

Highlights

- The foundation of a sustainable ocean economy is healthy, productive and resilient marine ecosystems. Currently, our ocean is under pressure from the diversity and multitude of human activities, driven by our need for food, energy, transportation and recreation. These pressures are amplified by climate change, loss of biodiversity and pollution. Despite progress on some fronts, the current trajectory is in the wrong direction and rapidly growing more serious.
- Efforts to implement effective sectoral management of ocean-based human activities and address issues such as climate change are necessary but insufficient for achieving a sustainable ocean economy. Integrated ocean management (IOM) is essential.
- Extensive and diverse experiences with IOM provide a wealth of models, best practices and guidance for success. Common features of success include harnessing science and knowledge, establishing partnerships between public and private sectors, engaging relevant stakeholders through legitimate processes, improving capacity building, implementing regulatory frameworks and developing adaptive management systems.
- It is vital to strengthen our knowledge about the ocean by developing and disseminating new data as well as better using existing knowledge, including traditional knowledge. We urge policymakers to further develop interna-

Originally published in:
Winther, J-G., M. Dai, et al. 2020. *Integrated Ocean Management*. Washington, DC: World Resources Institute. Available online at www.oceanpanel.org/blue-papers/integrated-ocean-management
Reprint by Springer International Publishing (2023) with kind permission.
Published under license from the World Research Institute.

J. Lubchenco, P. M. Haugan (eds.), *The Blue Compendium*, https://doi.org/10.1007/978-3-031-16277-0_14

tional cooperation in marine science and related sciences by building on established structures such as the Intergovernmental Oceanographic Commission and using the United Nations (UN) Decade of Ocean Science for Sustainable Development as a vehicle for this.
- Private businesses are central for achieving a sustainable ocean economy. Ocean-related businesses at all scales should be encouraged to cooperate in developing principles and guidelines for sustainable conduct. For example, the UN Global Compact's Sustainable Ocean Business Action Platform, which has developed principles and guidelines for sustainable ocean business that several of the world's largest ocean- related businesses have signed on to, can serve as an inspiration. This can help ensure IOM takes place across all sectors of ocean business, and through partnerships between the private and public sectors.
- Another avenue for success is stakeholder engagement and stewardship. In all stages of developing and implementing IOM, governments should ensure transparency and the active involvement of local communities and other relevant stakeholders. In designing well-managed engagement processes, it is vital to consider the scientific, cultural, societal, economic and political contexts.
- One obvious opportunity for action is for each nation, or regions with multiple nations, to develop IOM that is appropriate for their circumstances. Capacity building is key for achieving IOM in all parts of the world. It is well-documented that scientific capacity is inadequate in many countries, and that lack of institutional capacity is also a challenge. Nations and other entities that have pursued IOM must share their experiences, and regional cooperation can therefore accelerate capacity building. Successful regional efforts at IOM should inspire similar efforts in other regions.
- Failure to implement existing international instruments is one of the most important weaknesses of our ocean governance systems. It is vital to have mechanisms in place not just to develop IOM plans, but also to implement them. States need to ensure effective implementation of international agreements. Regulations for managing human activities in the high seas should be compatible with and at least as strict as those that are applied in areas under national jurisdiction.
- One of the most serious challenges facing our ocean today is climate change. This also highlights the dynamic nature of the ocean, which calls for adaptive and holistic ocean management. A static approach in, for example, establishing marine protected areas, may—due to climate change—lose its efficiency over time. Ocean governance must therefore consider expected changes in the marine environment and in human interactions with the ocean, by using the best available scientific knowledge on climate change and including adaptive mechanisms as a vital part of IOM.

1 Introduction

With the unprecedented growth in economic activities relating to the ocean economy, the need for a sustainable concept where socioeconomic development can occur without environmental degradation is widely recognised. This is manifested at the global level by the 1982 United Nations Convention on the Law of the Sea (UNCLOS), which provides the basic global framework for ocean governance (United Nations 1982). Since then, the ocean economy has continued to grow alongside our need for food, energy, transportation and recreation from the ocean. Existing ocean industries expand while new ones appear. At the same time, new challenges are emerging as a result of climate change, loss of biodiversity, pollution and extractive activities. Our ocean is now facing these pressures at unprecedented rates and magnitudes. The mismatch between the drive for short-term economic gain versus long-term prosperity and a healthy, resilient ocean is increasingly apparent. As a result, we see a pressing need for holistic, knowledge- based and ecosystem-based approaches to ocean management. Integrated ocean management (IOM) is such an approach.

IOM considers multiple uses and pressures simultaneously and helps reconcile competing uses with the objective of ensuring the sustainability of societies and marine ecosystems. The need for a comprehensive perspective on the management of marine ecosystems and their resources was observed many years ago (e.g. Underdal 1980) and is now widely recognised at the global as well as regional and national levels of governance. There are, however, still many challenges relating to the implementation of existing governance frameworks, including knowledge and capacity shortages, incomplete legislation, lack of enforcement, poor coordination among ministries and other government bodies, and no overarching mandate across ministries or mechanisms to harmonise conflicting mandates among ministries.

The goal of IOM is to support a 'sustainable ocean economy': long-term, sustainable use of ocean resources in ways that preserve the health and resilience of marine ecosystems and improve livelihoods and jobs, balancing protection and production. IOM brings together relevant actors from government, business, academia and civil society from the entire spectrum of ocean-related human activities (e.g. petroleum, fishing, aquaculture, shipping, tourism, mining, renewable energy, conservation) to collaborate toward a sustainable future for our ocean environment. Here, 'ocean' refers to both marine and coastal areas. The functions of IOM include promoting environmentally sound economic development, protecting coastal and marine habitats and biodiversity, providing ecosystem services and balancing and deconflicting interests through spatial planning. IOM also addresses issues such as the conservation of coastal and marine habitats and biodiversity, protection of coastal and marine environments from land-based pollution, fisheries and tourism, as well as impacts from climate change such as sea level rise, ocean warming and deoxygenation, ocean acidification, changing storm intensities and more. IOM is a dynamic process, building on existing initiatives and bringing industries and sectors together, whether under the umbrella of marine spatial planning, ecosystem-based management or others. Biodiversity, intact habitats and ecosystem functioning are essential to a healthy, productive and resilient ecosystem. A comprehensive toolbox of measures to accomplish this exists, including area-based management measures.

The purpose of this paper is to provide an overview of how the framework of IOM is established at different levels of governance and what is lacking in terms of both frameworks and implementation. Implementation failure is a key issue for IOM, and we discuss cases of IOM from different parts of the world to exemplify this. While these lessons learned are useful for further developing IOM, we also recognise that successful implementation is highly context dependent, making capacity building and location flexibility critical to achieving effective IOM globally.

This paper is written as an input to the work of the High Level Panel for a Sustainable Ocean Economy. It begins by outlining the concept of IOM, explaining its key components and providing an overview of the global ocean governance framework. The paper continues by discussing IOM both in the exclusive economic zones and in areas beyond national jurisdiction. This is followed by a discussion of IOM implementation—both its challenges and the key components for success. Next, we present selected case studies of IOM in practice. To conclude, the paper offers opportunities for action on how IOM can contribute to a sustainable ocean economy.

Box 14.1 Integrated Ocean Management and Related Planning and Management Approaches

To discuss integrated ocean management (IOM) further, it is vital to define a few central terms that are often associated with IOM. The below list is not exhaustive but demonstrates the variety of means that have evolved to achieve smart planning and management in coastal and marine areas. IOM uses a variety of these tools to ensure the sustainability of marine ecosystems. These ideas, terms and concepts have evolved through time and have had different histories in different places. They are not necessarily interchangeable, and they often overlap.

Ecosystem-based management

Ecosystem-based management (EBM), also referred to as an 'ecosystem-based approach', is central to IOM and defined as management of natural resources focusing on the health, productivity and resilience of a specific ecosystem, group of ecosystems or selected natural assets as the nucleus of management (Domínguez-Tejo et al. 2016). EBM is a management approach that recognises the full array of interactions within an ecosystem, including with humans, and drives the integration of management planning and implementation across sectoral agencies. Focusing on recognising connections and ensuring coherence, EBM differs from historical approaches that focus on a single species, sector, activity or concern, considering the cumulative impacts of different factors. Specifically, ecosystem-based management has the following features:

- Emphasis on the protection of ecosystem structure, functioning and key processes
- Accounts explicitly for the interconnectedness within systems, recognising the importance of interactions between many target species or key services and other non-target species
- Acknowledgement of the interconnectedness among systems, such as that among air, land and sea
- Integration of ecological, social, economic and institutional perspectives, recognising their strong interdependencies
- Focused on a specific ecosystem and the range of human activities affecting it

Integrated coastal zone management

Integrated coastal zone management (ICZM) is 'the process of managing the coast and nearshore waters in an integrated and comprehensive manner with the goal of achieving conservation and sustainable use' (Katona et al. 2017). It is also called 'integrated coastal management'.

ICZM covers the full cycle, including information collection, planning, decision-making, management and implementation. The approach seeks informed participation and cooperation from all relevant stakeholders. It seeks integration of the goals and instruments needed to meet these objectives; of different policy areas, sectors and levels of administration; and of the land and sea components of the target area.

Marine spatial planning

Marine spatial planning (MSP) is used to create geospatial plans that identify what spaces of the ocean are appropriate for different uses and activities. These plans have similarities with sustainable ocean economy plans, which describe how to sustainably use the ocean and its resources to advance economic and social development.

Marine spatial planning (also known as 'maritime spatial planning' and 'coastal and marine spatial planning') extended the ICZM approach further out to sea in the 2000s. Marine spatial planning aims to create a framework for the ocean that minimises conflicts between economic sectors and maintains 'good environmental status' of the ocean through the identification of ocean spaces that are appropriate for different uses and activities. MSP is increasingly seen as a practical way to create and establish a more rational organisation of the use of marine space and the interactions between its uses, to balance demands for development with the need to protect marine ecosystems and to achieve social and economic objectives in an open and planned way. MSP is widely used for setting targets for and implementing ecosystem-based management (Katona et al. 2017). The characteristics of effective marine spatial planning include the following:

- Ecosystem-based, balancing ecological, economic and social goals and objectives toward sustainable development
- Integrated across sectors, agencies and levels of government
- Area-based
- Adaptive and capable of learning from experience
- Strategic and anticipatory, focused on the long term
- Focused on participation, with stakeholders actively involved in the process

Adaptive ocean management

Adaptive ocean management is 'a systematic process for continually improving management policies and practices toward defined goals by learning from the outcomes of previous policies and practices' (Katona et al. 2017). It recognises the inherent variability and dynamic nature of the ocean in terms of its bio-chemo-physical properties and social and economic factors in addition to scientific uncertainties. By scheduling periodic reviews of and updates to management plans, in addition to adding ad hoc opportunities for responding to unexpected events, adaptive ocean management acknowledges that changes in conditions and knowledge are likely.

Area-based measures including marine protected areas

Area-based measures are important tools in the management of the ocean and seas and can be used in all approaches outlined here. A marine protected area (MPA) is 'a clearly defined geographical space, recognized, dedicated and managed, through legal or other effective means, to achieve the long-term conservation of nature with associated ecosystem services and cultural values' (Oregon State University et al. 2019). Likely developed independently in many cultures, area-based measures are a regulatory tool for conserving the natural or cultural resources of the ocean and for managing human uses.

If managed in isolation, coastal and marine protected areas are vulnerable to the impacts of resource development and exploitation occurring outside these areas, in particular overfishing, alteration and destruction of habitats, climate change and marine pollution. Thus, protecting coastal and marine areas—including species, habitats, landscapes and seascapes—should be integrated into spatial development strategies for larger areas, under the umbrella of integrated coastal and ocean management, including land-ocean interactions.

2 The Concept of Integrated Ocean Management

This chapter outlines the global governance framework. International ocean governance is based on coastal states' [1] jurisdiction over their 200 nautical mile exclusive economic zones (EEZs). The authority to manage the EEZs and activities rests with (often multiple) agencies and the laws governing those agencies. While useful for efficient management, divisions of authority over different activities and zones can create challenges for oversight and coordination and thus limit holistic approaches to management. It is therefore important to identify and harmonise the possibly conflicting mandates of different agencies, as well as cover the gaps where no agency or entity is responsible. As pressures on the ocean increase, assessing the cumulative impacts of increasing uses and pressures also becomes increasingly important.

The purpose of IOM is to enhance our ability to use and manage ocean resources sustainably, and ensure that the health, productivity and resilience of ocean ecosystems, which provide multiple benefits to humans, are not impaired

[1] A coastal state is a nation state that exercises jurisdiction and sovereign rights in its exclusive economic zone and continental shelf. 2022 Note that when using this term regarding the United States, it is applicable to the country as a federal state. U.S. states have jurisdiction over 3 nautical miles from the coastline, while the area from 3 to 200 nautical miles is under federal jurisdiction.

by human activities. Examples of some supporting, regulating and cultural ecosystem services provided by ocean ecosystems include partial climate regulation, control of pests and pathogens, nutrient cycling, primary production, cultural identity, inspiration and recreation. Management institutions need to effectively measure, monitor and manage ocean space adaptively as complex adaptive systems (Lubchenco et al. 2016). To achieve this, incorporating participation from governance institutions, academic knowledge (multidisciplinary), other knowledge (transdisciplinary) and multiple stakeholder interests is crucial. Therefore, we argue that stakeholder involvement, the effective use of science and capacity building are keys to achieving real integration.

2.1 The Global Ocean Governance Framework

Responding to technological developments, increasing demands for natural resources and a growing use of ocean space for human activities, the global framework for ocean governance has evolved significantly over the last decades. The centrepiece of this framework is the UN Convention on the Law of the Sea—'the constitution of the ocean'—which was negotiated between 1973 and 1982 and entered into force in 1994 (United Nations 1982). UNCLOS states in its preamble that 'the problems of ocean space are closely interrelated and need to be considered as a whole'. The convention aims to establish 'a legal order for the seas and oceans which will facilitate international communication, and will promote the peaceful uses of the seas and oceans, the equitable and efficient utilization of their resources, the conservation of their living resources, and the study, protection and preservation of the marine environment'. As of 2019, 168 countries are parties to the convention. The implementation of UNCLOS is overseen by the UN General Assembly, which adopts annual resolutions on the ocean and the law of the sea addressing a comprehensive range of issues relating to the implementation of the convention.

UNCLOS establishes a legal order for the ocean where coastal states can exert sovereign rights over the natural resources in a 200 nautical mile exclusive economic zone and on the continental shelf also beyond 200 nautical miles (Fig. 14.1). Where continental shelves extend beyond 200 nautical miles, their outer limits are established based on recommendations of the Commission on the Limits of the Continental Shelf. The mineral resources at the deep seabed beyond national jurisdiction ('the Area') are considered the common heritage of mankind, and the International Seabed Authority (ISA) is tasked with their management. The ISA is currently responsible for developing regional environmental management plans for deep-sea mining regions. The International Tribunal for the Law of the Sea is one of several options for resolving disputes.

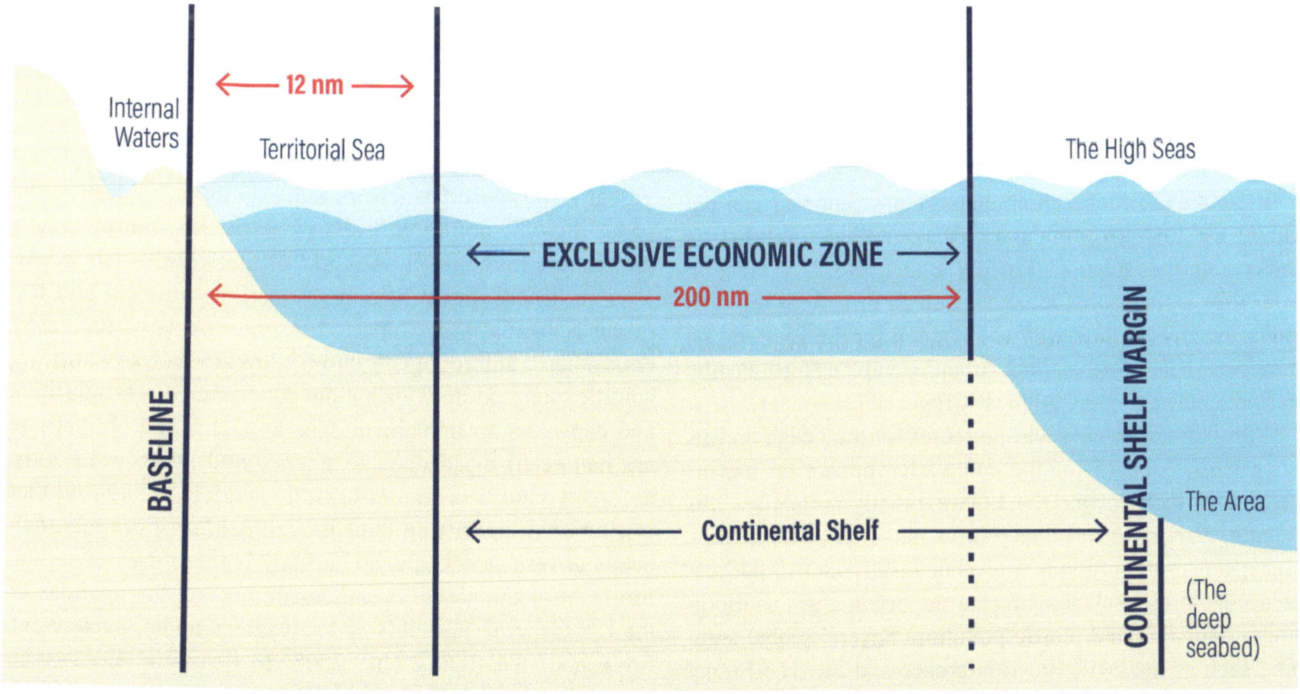

Fig. 14.1 Maritime zones as defined by the UN Convention on the Law of the Sea. (*Source: Anders Skoglund, Norwegian Polar Institute 2020*)

To provide guidance on the implementation of the convention, implementation agreements have been negotiated for deep seabed minerals (United Nations 1994) and fisheries (United Nations 1995). The latter require application of a precautionary approach to fisheries management and regional cooperation in the management of fisheries on the high seas. In response to growing concerns related to conservation and use of marine biodiversity, a third implementation agreement is under negotiation, addressing area-based management, marine genetic resources, environmental impact assessments and technology transfers, and capacity building for the areas beyond national jurisdiction. Governance bodies are also in place for specific themes. For example, the International Maritime Organization (IMO) regulates the shipping industry. IMO has adopted a number of international agreements setting out the standards for the industry related to the environment, operations and labour, among others.

In fisheries, additional layers of governance to the convention and the 1995 implementing agreement are provided by agreements adopted by the UN Food and Agriculture Organization (FAO) that address the ecosystem effects of fishing as well as the need to confront illegal activities. One example of this is the 2009 *Agreement on Port State Measures to Prevent, Deter and Eliminate Illegal, Unreported and Unregulated Fishing* (FAO 2009). Regional cooperation is important in the management of transboundary fish stocks on the high seas, with about 20 regional fisheries management organisations providing international cooperation for the management of such stocks.

Separately, many environmental concerns are addressed in instruments relating to various types of pollution, climate change (e.g. the United Nations Framework Convention on Climate Change) and the 1992 Convention on Biological Diversity at global as well as regional levels of governance. This includes the UN Environment Programme (UNEP) and its regional seas programmes.

International cooperation in marine science is centred around the Intergovernmental Oceanographic Commission (IOC) and the Regular Process under the UN General Assembly at the global level, as well as in several regional bodies such as the International Council for the Exploration of the Sea (ICES) in the North Atlantic, and the North Pacific Marine Science Organization (PICES).

With increasing uses and pressures on the ocean and its ecosystems, concern for the cumulative impact on marine ecosystems has grown. UNCLOS explicitly recognises this concern. Various efforts have been made to address these issues; however, solutions are often insufficient to address the accelerating challenges facing the ocean, such as biodiversity depletion and plastic pollution. Several global summits, such as the 2012 Rio Conference and its 'The Future We Want' declaration, have highlighted the need to consider the total impacts of ocean use.

The World Ocean Assessment of the Regular Process to assess the status of the marine environment under the UN General Assembly concluded with the following in its first report (United Nations 2015):

> *The sustainable use of the ocean cannot be achieved unless the management of all sectors of human activities affecting the ocean is coherent. Human impacts on the sea are no longer minor in relation to the overall scale of the ocean. A coherent overall approach is needed. This requires considerations of the effects on ecosystems of each of the many pressures, what is being done in other sectors and the way that they interact.*

The annual UN General Assembly resolutions on the ocean and law of the sea address these issues and have done so since 1999 in what constitutes a de facto global ocean coordination. The preamble of the 2018 resolution states the following:

> . . . the problems of ocean space are closely interrelated and need to be considered as a whole through an integrated, interdisciplinary and intersectoral approach, and reaffirming the need to improve cooperation and coordination at the national, regional and global levels, in accordance with the Convention, to support and supplement the efforts of each State in promoting the implementation and observance of the Convention and the integrated management and sustainable development of the oceans and seas . . .

Also, as a follow-up to the Millennium Development Goals, the UN General Assembly in 2015 adopted 17 Sustainable Development Goals (SDGs) as part of the 2030 Agenda. Several of the SDGs are relevant to the ocean and contain specific targets and timetables for achieving them. Goal 14, 'Life below water', addresses marine issues specifically. This goal provides opportunities to facilitate concrete actions for ocean sustainability and to foster greater integration among the sectors of ocean governance.

It is, however, evident that the implementation of the global governance framework leaves a lot to be desired. In some regions, pollution levels (from toxins, nutrients and plastic) are high, about one-third of the world's fish stocks are overfished, illegal fishing is a serious problem and the ocean is increasingly impacted by the effects of increased emissions of anthropogenic carbon dioxide such as warming, acidification and deoxygenation. As coastal states' abilities and capacities to implement existing rights and obligations are hampered by inadequate science and weak and poorly enforced regulatory frameworks, it is widely recognised that institutional capacity building is a critical factor to strengthen ocean governance. Capacity building is therefore now at the forefront of the global ocean debate and on the agendas of the UN General Assembly and subsidiary bodies such as the FAO and the IOC. The 2005 Millennium Ecosystem

Assessment and the 2019 Intergovernmental Science-Policy Platform on Biodiversity and Ecosystem Services' global assessment of biodiversity provide further understanding of why we should care about biodiversity and what the major drivers of change are.

Integrated ocean management can be implemented across several ocean economy sectors, jurisdictions and spatial scales. These applications may take the form of localised ocean management within national waters, sector-defined ocean management across adjacent jurisdictions, at regional seas or at ocean basin scales, or international ocean management occurring across large ocean areas in areas beyond national jurisdiction, including in the Area (i.e. the deep seabed beyond national jurisdiction).

3 The Implementation of Integrated Ocean Management

This chapter analyses the challenges of implementing IOM and identifies key components of success. It continues by discussing how IOM can contribute to a healthy, productive and resilient ocean for long-term, sustainable economic growth. It highlights the need to calibrate the sophistication of management plans to the situation at hand, authority and political will, available data and societal values. It concludes by arguing that supplementing sector-based management with collaborative and coordinating mechanisms across sectors would be highly beneficial.

The rising demand for various uses of ocean space increases the complexity in modern ocean governance and management, and thus calls for better coordination among sectors and other stakeholders (Klinger et al. 2018). In some countries, the legal and institutional arrangements that divide ocean management are long-standing and the legal mechanisms to coordinate these arrangements at the domestic level as well as with adjacent nations are lacking. This makes it difficult to account for the cumulative effects—including those over time—of economic development, management and environmental change on marine ecosystems. Overcoming such institutional barriers requires political will from government leaders as well as from businesses and civil societies across all sectors of human activity. For the benefit of both human and natural resource values, a defined mechanism to coordinate sector management and enhance collaboration within and among countries is essential for defining and advancing IOM. A complex adaptive systems lens has thus emerged as a new approach to help identify key indicators to refine IOM. The rapidly growing data and knowledge about our ocean will clearly add feasibility in this regard.

3.1 Components of Successful IOM

Every IOM plan depends on the country or region, as specific problems, challenges and institutional conditions vary and are highly context dependent. However, regardless of the legal underpinning of IOM, experience demonstrates that the following components are important:

- A survey of the existing institutional structures within a given context, including an assessment of agency authorities, how they overlap and their regulatory responsibilities.
- An evaluation of the current situation, variations and future trends in the environment and ecosystem by examining the available data and scientific knowledge, and initiation of science and monitoring programmes for establishing and maintaining marine data.
- An assessment of human activity information and trends, including conflicts of interest and cumulative impacts.
- Engagement of relevant ocean user groups.

These components must be structured thoughtfully for IOM to be effective. More importantly, every integrated ocean management plan should be specific to the country or region. The variation in institutional conditions and challenges also underlines the importance of local capacity building, particularly to enable the collection and adequate use of marine data and transformation of these data into useful information and knowledge.

3.1.1 Institutional structures

A legal structure can provide direction for IOM within or among nations, thus creating the context for defining and advancing cross-sectoral, long-term, ocean-related goals and objectives. In many cases this is absent, and alternatives such as interpreting existing law to take an integrated management approach, adding provisions to existing laws, issuing an administrative order or finding other mechanisms can also be effective. At the regional level, the European Union (EU) created the 2008 Marine Strategy Framework Directive (European Union 2008). This is an EU legal instrument that provides a framework and requirements for member countries to implement marine plans aiming for 'good environmental status' by 2021. A different example is from U.S. state Massachusetts, which created legislation with the Oceans Act of 2008 that provided authority and direction to create an ocean plan (Commonwealth of Massachusetts 2008). This act also provided guidance and identified representation categories for who should serve on the science and policy advisory bodies that made recommendations as the management plan was developed.

Reinterpreting or adding to existing legislation, or finding other mechanisms such as administrative orders and directives to establish authority for IOM, can facilitate improved coordination and supplement the sector-by-sector approach. The Netherlands took the reinterpretation-of-existing-law approach as it developed an integrated ocean management plan for its nearshore areas through an Inter-Ministerial Consultation Body for the North Sea involving all relevant ministries, such as defence, transport, public works and water management, economic affairs and the environment (Douvere and Ehler 2009). U.S. state Rhode Island voluntarily developed an ocean plan in partnership with the federal government that implements ecosystem-based management principles by reinterpreting the state Coastal Resources Management Council's authorising legislation within the national Coastal Zone Management Act (CZMA). The plan was approved by the National Oceanic and Atmospheric Administration's Office for Coastal Management, which enhances the state's influence in federal waters through the Federal Consistency provisions in the CZMA.

Legislative provisions can also be added to existing law to establish authority and provide clarity for developing a more holistic management process. This can potentially be achieved by adding provisions to legislation that regulates new ocean uses to make IOM a requirement for new development to be permitted. These provisions can include making strategic environmental assessments a requirement. Developing directive language that ensures positive outcomes for the regulated sector for which the original legislation was written is necessary to demonstrate the added value of requiring a more integrated approach with other sectors.

Administrative orders or directives can also be used to define a framework for coordinated management. These directives may articulate high-level targets and leave the definition of specific management goals and objectives to relevant regulatory agencies or planning entities. This approach was taken by the United States for its entire EEZ through an executive order from the president that instructed federal agencies to coordinate with state and tribal authorities to develop regional ocean plans for each Large Marine Ecosystem in U.S. waters (Executive Office of the President 2010; Lubchenco and Sutley 2010). The executive order provided a clear overarching mandate: to protect and restore ocean ecosystems to a healthy, productive and resilient state. Two regions, the Northeast and Mid-Atlantic, successfully completed and are implementing their regional ocean plans. Norway is another example demonstrating that IOM does not have to be grounded in law if political authority is provided. Norway's integrated management plan is based on mapping and assessing the status of marine ecosystems, identifying ecologically valuable and vulnerable areas and setting conditions for the use of ocean space including for the petroleum industry (Norwegian Ministry of the Environment 2006). These examples are further explored later in this paper.

It is recognised, however, that a comprehensive legal regime alone is not enough to achieve the desired outcome when, for example, illicit activities or lacking enforcement or capacity create a gap between the legal framework and reality.

3.1.2 Ocean and Coastal Data and Use Of Science

Many countries do not have enough monitoring and scientific capacity to provide the knowledge foundation required to implement the international governance frameworks they are bound by. The 2017 Global Ocean Science Report demonstrated clearly that many countries lack fundamental scientific capacity to underpin their efforts at ocean governance (IOC-UNESCO 2017). Scientific capacity to assemble the information required to manage the ocean's ecosystems and economic activities and establish regulatory measures needs to be developed. New technologies have revolutionised how governments can monitor and police inappropriate behaviour at sea. Global Fishing Watch, offering near real-time tracking of fishing activity via a public map, is an example of how new technologies and transparency can lead to improved means for sustainable governance. If such tools are combined with coordinated policing efforts and effective prosecution, they can become very powerful.

In developing IOM, it is critical to take a systematic approach to building scientific capacity that addresses the needs in the regions concerned. Capacity building needs to remain at the top of the international agenda. Regional solutions do have the potential to be effective, as demonstrated by ICES in the North Atlantic. Regional cooperation can help pool resources for IOM and facilitate the sharing of experiences. The UN Decade of Ocean Science for Sustainable Development (2021–2030) could be a suitable process and platform to accelerate the development and use of ocean science. The purpose of the UN Decade is to provide a common framework to ensure that ocean science can fully support countries' actions to sustainably manage the ocean and achieve the 2030 Agenda for Sustainable Development.

Furthermore, we recognise the extensive knowledge on the ocean that already exists. Thus, there is a need to not only foster new science but also use existing and historical information—and lack of data should not limit action. There must be explicit mechanisms in place in the IOM decision-making process to make use of existing science. A body of knowledge, for reference, exists on how to cope with data-poor situations in managing, for example, fisheries (Pilling et al. 2008). Incorporating traditional knowledge into the management process must also be a priority. The challenge is to integrate new information while simultaneously managing a dynamic ocean environment within IOM.

3.1.3 Engagement of Relevant User Groups

Planning at the local level, especially in developing countries and small island developing states (SIDS), requires taking approaches tailored to the diverse environmental, socioeconomic and governance systems in those regions. Incorporating local knowledge can ensure active community participation to develop appropriate strategies for IOM. Participatory approaches have proven to be effective at the local level for all phases of establishing and operating ocean governance. However, even with thorough planning, implementation remains constrained. Scaling up and reorienting local actions to larger-scale activities, governance regimes at national and regional levels and appropriate ecological scales are important and difficult and require specific time and resources in themselves (Fig. 14.2).

In the context of IOM, it is particularly important to engage ocean businesses at the global, national and local levels. There are different ways of organising this, and recent years have seen several cases of ocean businesses joining forces for sustainability. One example is the UN Global Compact Action Platform for Sustainable Ocean Business, which has developed principles and guidelines for sustainable ocean businesses—several of the largest ocean-related businesses globally have signed on to it.

Fig. 14.2 The Ecosystem is at the Core of Integrated Ocean Management. *Note: An ecosystem- and knowledge-based integrated ocean management ensures a sustainable ocean economy. Stakeholder engagement is key. (Source: Centre for the Ocean and the Arctic, Norway 2019)*

As will be demonstrated by the case studies later in this paper, while successful engagement processes will vary as they take into account the local context, engaging relevant user groups is always a central component of successful IOM.

3.2 The Case of the 'Collective Arrangement': Toward IOM in Area-Based Management in Areas beyond National Jurisdiction

The development of integrated ocean management approaches in areas beyond national jurisdiction (ABNJ) is hindered by legal gaps in the global ocean governance framework. One of the key functions of the new global agreement for the conservation and sustainable use of marine biological diversity beyond national jurisdiction (BBNJ) is to provide a legal framework for the development of a comprehensive approach to area- based management, including marine protected areas (Gjerde et al. 2019).

One of the few examples of how integrated approaches in area-based management in ABNJ can be developed is the 'Collective Arrangement' between the OSPAR Commission and the North East Atlantic Fisheries Commission, or NEAFC (NEAFC and OSPAR 2014). The objective of the Collective Arrangement is to facilitate cooperation on area-based management between legally competent organisations for the conservation and sustainable use of marine resources in the Northeast Atlantic. In 2010, the OSPAR Commission established the world's first network of marine protected areas in ABNJ. But OSPAR does not have the mandate to establish legally binding measures to regulate human activities such as fisheries, shipping or, in the future, deep seabed mining. However, in a parallel process, NEAFC as the competent organisation for fisheries management has closed fisheries to protect vulnerable marine ecosystems in broadly the same areas (Kvalvik 2011). Both organisations received scientific advice from ICES leading to a coherent approach in establishing those area-based measures. To facilitate future coordination and cooperation in the region, OSPAR and NEAFC set up the Collective Arrangement—which includes regular meetings between the parties and exchanges of relevant information and planned activities—and invited others, including the IMO and ISA, to join (Ásmundsson and Corcoran 2015).

The agreement has already demonstrated its value by providing a framework for consultations with other sectors in the process of establishing new high seas MPAs in the Northeast Atlantic, though it remains somewhat incomplete without the other management bodies. The Collective Arrangement does not include any legal changes to the current governance framework in the Northeast Atlantic. Instead, it can be seen as a first step in the departure from a purely sectoral approach in ocean management to a more integrated one that could inform the development of comparable approaches in other regions and the negotiation of the new global legally binding agreement for marine biodiversity in areas beyond national jurisdiction.

3.3 A Healthy, Productive and Resilient Ocean for Long-Term Economic Growth

The overall goal of IOM is to enable economic development while maintaining a healthy ocean ecosystem. The ocean has always attracted multiple uses, including fisheries, oil and gas extraction, shipping and transportation, military, mining, and recreation, among others. Many areas attract a variety of competing uses, which cause conflicts between users (user-user conflicts) and users and the environment (user- environment conflicts) (Tuda et al. 2014). The need to minimise user-user and user-environment conflicts while taking advantage of new economic opportunities offered by the ocean—and maintaining a sustainable marine ecosystem—has seen increased interest and action at varying political levels in the spatial management of marine resources (Gustavsson and Morrissey 2018; Domínguez-Tejo et al. 2016).

Marine spatial planning is an emerging paradigm for sustainable ocean management (Douvere 2008; Domínguez-Tejo et al. 2016) and the operationalisation of a sustainable ocean economy. MSP aims to move away from a traditional, sectoral focus to a more holistic approach which takes into account the full use of the ocean space (White et al. 2012). Within the sustainable ocean economy framework, MSP should ideally be a means of creating an optimal investment climate for maritime sectors and give operators more certainty as to what opportunities for economic development are possible, though inequity, greater conflict and faster rates of degradation may occur if not calibrated appropriately with ecosystem goals and monitored over time. As a management tool, MSP allocates areas of the ocean for different uses and activities to reduce conflicts and achieve ecological, economic and social objectives. A key theme of MSP is the adoption of an ecosystem-based approach to ocean management. This involves focusing on the functional relationships and processes within the marine ecosystem, paying attention to the distribution of benefits that flow from ocean ecosystem services, using adaptive management practices, carrying out management actions at multiple scales and cooperating at an intersectoral level (Douvere and Ehler 2009). This approach is in direct contrast to current ad hoc, sector-by-sector responsibilities and practices for the management and regulation of ocean activities.

Although MSP is currently underway in 66 countries worldwide, only 22 countries have government-approved marine spatial plans (Santos et al. 2019).

4 Integrated Ocean Management in Practice

Below are a set of case studies selected to illustrate differences in implementation goals, jurisdictional types and management scales of IOM in different parts of the world: the Coral Triangle, the Seychelles, Norway, the United States and China (Fig. 14.3). There are common traits among them, such as the need for robust information about the relevant ecosystems and human activities, the need to tailor IOM to local contexts and the use of mechanisms to estimate the cumulative impact of uses and pressures on marine ecosystems.

4.1 The Coral Triangle and the Seychelles

Tropical nearshore coastal ecosystems (i.e. mangrove forests, seagrass beds and coral reefs) are among the most diverse and productive ecosystems in the ocean. A wide range of stakeholders, particularly low-income coastal communities, are highly dependent on the goods and services provided by these ecosystems, which are also the most vulnerable to climate change. Strengthening the resilience of these ecosystems to sustain the direct and indirect goods and services they provide is at the core of ocean management initiatives at the national and transnational levels.

This section discusses how IOM has been used with a focus on climate change adaptation, marine protected areas and fisheries management to address social, economic and conservation objectives simultaneously in some SIDS—here exemplified by the Coral Triangle (Indonesia, Malaysia, Papua New Guinea, Philippines, Solomon Islands and Timor-Leste) and the Seychelles.

4.1.1 Conservation and the Case of Fisheries Management in the Coral Triangle

MPAs range from small, local and community-based networks to national networks to regional, multistate initiatives. In SIDS, most MPAs are managed by local community members, particularly small-scale fishers. Since these MPAs are small, there is often a spatial mismatch relative to important ecological processes (Horigue et al. 2012; Mills et al. 2010). To date, a major challenge with implementing MPAs has been that they are seen as serving conservation or protectionist interests, not human interests, thus driving a top-down, nature-centric agenda that alienates local communities and ends up marginalising conservation. If MPAs are seen as promoting only a nature-centric agenda and not supported by local users, or their relevance to recovering depleted fisheries is not obvious, then a more comprehensive approach involving ecosystem- based management, marine spatial planning or ocean zoning might be a better option. Nonetheless, area-based management measures are important and the most

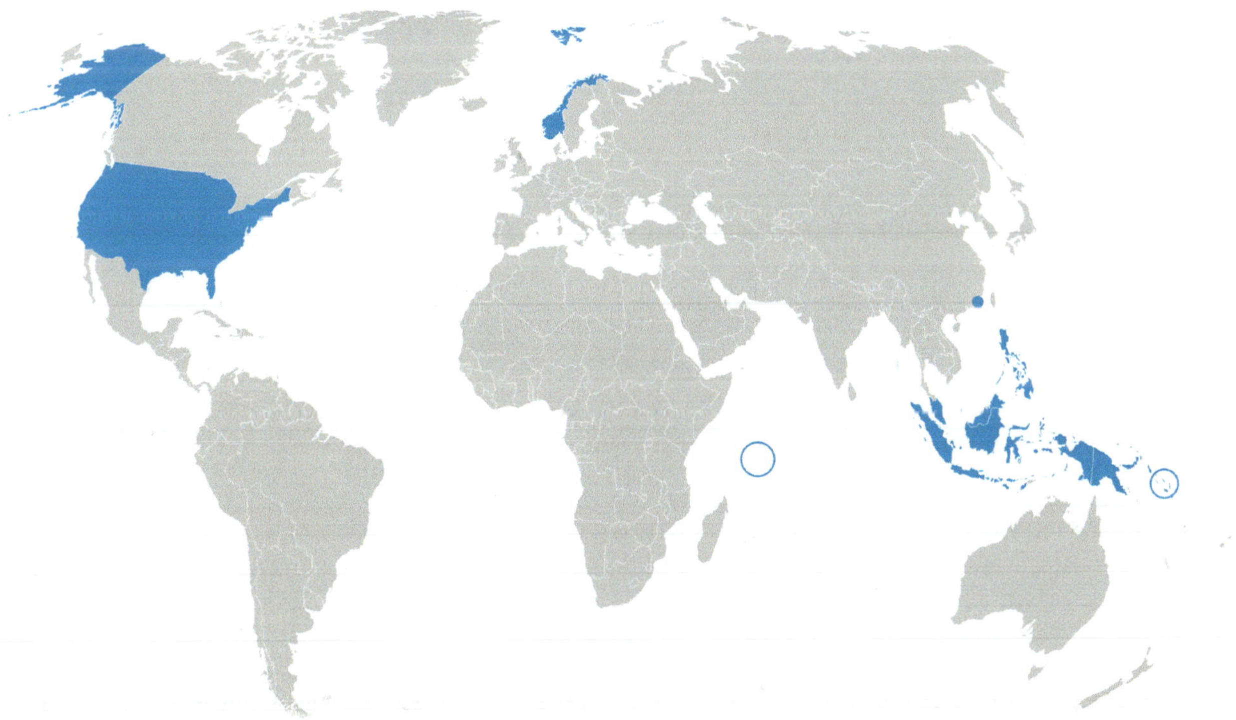

Fig. 14.3 Map of Case Studies of IOM in Practice. (*Source: Authors*)

prevalent management intervention used to meet conservation and fisheries management goals in developing tropical states, including SIDS. To maximise their impact, MPA planning should be integrated into broader marine spatial planning and ocean zoning efforts (Agardy et al. 2011).

Regarding effectiveness, improvements in fish biomass—including the recovery of functionally important groups reported for many small MPAs—indicate that conservation objectives can also be achieved at the local scale. For example, in community-based MPAs in Papua New Guinea that protect grouper aggregations, there was a 10-fold increase in the reproductive population compared with an unprotected site after five years (Hamilton et al. 2011). In the Philippines, a study on fish biomass showed that 32 percent of MPAs had estimated fish biomass within and above the estimated maximum sustainable yield (MSY) (McClanahan et al. 2015). However, further research (Muallil et al. 2019) found that areas adjacent to the MPAs were below the MSY. This indicates that small, locally managed MPAs alone are not enough for coral reef fisheries management. However, despite limited perceived impact on improving the state of the local fisheries resources, community- based MPAs have been effective in empowering the local fishing communities (Maliao et al. 2009). Thus, the MPAs provide social benefits and enable active community participation in resource management. Further, demonstrating the local impacts of MPAs has been important in sustaining and scaling up efforts.

On a regional scale, MPAs are part of a holistic approach of the Coral Triangle Initiative on Coral Reefs, Fisheries, and Food Security (CTI-CFF). The CTI-CFF is a formal intergovernmental partnership among the six countries: Indonesia, Malaysia, Papua New Guinea, Philippines, Solomon Islands and Timor-Leste to sustain marine and coastal resources and address food security, climate change and marine biodiversity. The adoption of a regional plan of action (RPOA) by member countries and implementation of national plans of actions illustrate implementation at regional and national levels. Establishment of marine protected areas and networks is a centrepiece of national plans on integrated coastal resources management to address conservation and local fisheries enhancement.

Box 14.2 Goals and Targets of the Coral Triangle Initiative's Regional Plan of Action

The regional plan of action of the Coral Triangle Initiative (CTI) has the following goals and targets:

- Goal 1: 'Priority Seascapes' Designated and Effectively Managed
  - Target 1: 'Priority seascapes' are designated, with investment plans completed and sequenced.
  - Target 2: Marine and coastal resources within all priority seascapes are sustainably managed.
- Goal 2: Ecosystem Approach to Management of Fisheries and Other Marine Resources Fully Applied
  - Target 1: Strong legislative, policy and regulatory frameworks are in place for achieving an Ecosystem Approach to Fisheries Management (EAFM).
  - Target 2: Improved incomes, livelihoods and food security exist in an increasingly significant number of coastal communities across the region through a new sustainable coastal fisheries and poverty reduction initiative ('COASTFISH').
  - Target 3: Effective measures are in place to help ensure that the exploitation of shared tuna stocks is sustainable, with tuna spawning areas and juvenile growth stages adequately protected.
  - Target 4: More effective management of and a more sustainable trade in live reef fish and reef-based ornamentals is achieved.
- Goal 3: Marine Protected Areas Established and Effectively Managed
  - Target 1: The region-wide Coral Triangle MPA System is in place and fully functional.
- Goal 4: Climate Change Adaptation Measures Achieved
  - Target 1: The region-wide Early Action Plan for Climate Change Adaptation for the nearshore marine and coastal environment and small island ecosystems is developed and implemented.
  - Target 2: Networked national Centers of Excellence (COEs) on climate change adaptation for marine and coastal environments are established and in full operation.
- Goal 5: Threatened Species Status Improving
  - Target 1: Improved statuses are achieved for sharks, sea turtles, seabirds, marine mammals, corals, seagrass, mangroves and other identified threatened species.

Source: Coral Triangle Initiative 2019.

The RPOA is guided by principles of integration, inclusive stakeholder participation, multilevel governance mechanisms to implement action plans, and recognition of the uniqueness, fragility and vulnerability of island ecosystems. There are technical working groups for each of the five programmes to address each goal; government working groups for financial resources, monitoring and evaluation; and other committees for various cross-cutting themes for capacity building, such as a regional leaders' forum, a female leaders' forum and local government networks. The CTI-CFF also has international development partners and collaborators that provide financial and technical assistance.

To support systematic conservation planning, guidelines for designing marine reserve networks within broader spatial planning and management frameworks to address biodiversity conservation, fisheries management, climate change adaptation and coastal management have been formulated (Green et al. 2014). A geospatial database and analyses of those data have been useful to inform planning efforts at different scales and provide a knowledge base for improved decision-making (Asaad et al. 2018, 2019). National plans of action support the regional goals through national inter-agency and multisectoral partnerships and programmes. Each member country has its own implementation activities aligned with improving governance of the ocean.

4.1.2 Climate Change Adaptation in the Seychelles

Coping with climate change impacts, such as sea level rise, is complicated in the SIDS because of the countries' small sizes, isolation and exposure, and high dependence on natural resources for their populations' livelihoods. Ecosystem-based adaptation (EBA) has been highlighted as one approach to address declines in ecosystem health and enable sustainable adaptation to climate change at national or community scales.

As in other countries, MSP has been used to support ecosystem-based management.

The Seychelles was the first nation to participate in an ocean-based debt-for-nature swap. Part of the agreement required incorporating climate change adaptation into a marine spatial planning process in support of the country's ocean economy goals. The goals of the Seychelles Marine Spatial Plan (SMSP). Initiative are to address climate change adaptation, protect 30 percent of the Seychelles' waters, including 15 percent with high protection status, and support the Blue Economy Roadmap and other national strategies. The geographic scope of the SMSP Initiative is the entire 1.4 million square kilometres (km^2) of the EEZ and territorial sea. The process is led by the Ministry of Environment, Energy and Climate Change of the government of the Seychelles (GOS), with planning and facilitation managed by The Nature Conservancy (TNC) and TNC Canada, in partnership with the government management system for the United Nations Development Programme (UNDP) and the Global Environmental Facility (GEF) environmental programme in the Seychelles (the GOS-UNDP-GEF Programme Coordination Unit). Funding for the initiative is being provided through several grants to the government of the Seychelles from The Nature Conservancy and a number of philanthropic foundations, and through the creation of blue bonds.

The MSP Executive Committee has representation from all the ministries and parastatal agencies related to the marine sectors. The process includes public workshops and public information sessions to generate input from all major sectors including commercial fishing, tourism and marine charters, biodiversity conservation, renewable energy, port authority, maritime safety, and non-renewable resources (SMSP Initiative 2019). The process began in February 2014 and will be completed by December 2020. The SMSP Zoning Framework has three zones: high biodiversity protection; medium biodiversity protection and sustainable use; and multiple use.

Existing uses were mapped in 2014–2015 and are used to inform proposals for new marine protection areas. Notably, the SMSP Initiative is a component of the government-led Debt-for-Climate-Change-Adaptation swap. The Seychelles Conservation and Climate Adaptation Trust, which was operationalised in November 2016 to fund climate change adaptation and conservation projects in the Seychelles, will also provide partial funding to implement the SMSP Initiative.

Other SIDS have not gone through a national, government-led, comprehensive MSP process as that in the Seychelles but have experiences in ecosystem- based management and spatial planning linked to climate change adaptation (CCA) strategies. However, these experiences have often not been successful without a national, government-led framework. This demonstrates that both political will and funding are needed in addition to robust stakeholder engagement processes. For example, in the central Pacific, top-down coastal adaptation approaches usually fail because coastal communities in the tropical Pacific have customary land, island and coastal tenure, supported by traditional governance (Grantham et al. 2011). To facilitate active participation, global climate change is addressed in the Pacific context in discussions with communities. Traditional adaptations—for example, to maintain food security given unpredictable climate events—are incorporated in CCA planning to engage communities and boost awareness. Planning is primarily small scale, with bottom-up processes directly engaging local communities through participatory approaches supporting spatial planning. In the Caribbean, an analysis by Mercer et al. (2012) has shown inconsistencies in EBA theory and practice. The CCA plan in the Caribbean centred upon coastal ecosystems at a regional level as opposed to the full range of ecosystems, which is especially important given the tight linkages between terrestrial and marine ecosystems in small island states. Likewise, integrating local knowledge with external knowledge was identified as a gap in developing appropriate context-specific adaptation strategies (Mercer et al. 2012).

4.1.3 Challenges and Lessons Learned

There are various ongoing initiatives in SIDS and tropical island developing states that are working toward a more integrated ocean management regime. We have presented two examples where key priorities have focused on improving the resilience of coastal communities and marine ecosystems linked to the overarching need for climate change adaptation. Despite various challenges, experiences in planning and ecosystem-based management provide valuable lessons that may help make future efforts more efficient and effective.

First, planning at the local and national levels requires taking into consideration the local environmental, socioeconomic and governance systems. In all cases, stakeholder participation and incorporation of local knowledge are essential in developing appropriate strategies. Experiences in these countries indicate that ecosystem-based management and MSP should build on and strengthen community-based management. Participatory approaches have been demonstrated to be effective at the local level for climate change adaptation planning and all phases of the MPA process (i.e. MPA planning, implementation, monitoring and evaluation, feedback and adaptation of the management plan).

Second, scaling up and reorienting local actions to larger-scale activities and appropriate ecological scales are essential and require improved governance regimes at national and regional levels. In many cases, even with improved planning, implementation remains constrained. There are many challenges to integrated ocean management in the CTI (Weeks et al. 2015) that constrain effective implementation at the regional and national levels. They include a failure to institutionalise conservation planning within governments at different levels nationally and systematic planning across nations with different governance systems, poor integration of planning efforts across these governance systems and levels, and failure to overcome short-term funding and political cycles relative to long-term planning and implementation timeframes (Weeks et al. 2015). Funding and technical support for IOM planning including MSP in the CTI and SIDS have been facilitated through various development programmes, partnerships with nongovernmental organisations (NGOs) and public- private partnerships, but national funding mechanisms for the long-term implementation of plans, such as the Seychelles Conservation and Climate Adaptation Trust, are needed (SMSP Initiative 2019).

Third, formal and informal social networks for data sharing and capacity building are essential at different levels of government to accelerate integration and scale. Regional alliances such as the CTI have been forged to facilitate regional and broader-scale policy support and frameworks to harmonise national plans of actions. The SIDS Accelerated Modalities of Action (SAMOA) pathway is anchored on the conviction that sustainable development can be achieved through strong partnerships. These high-level alliances are supported by social networks for capability building and implementation, and multisectoral consortia of private and public organisations, researchers, practitioners, and policymakers. Examples include Many Strong Voices, a collaboration between the Arctic and SIDS (Mercer et al. 2012), and the MPA Support Network in the Philippines (Horigue et al. 2012).

Knowledge gaps on socioecological interactions at various scales are a persistent limitation in integrated ocean management planning and implementation in the SIDS and CTI (e.g. Khan and Amelie 2015; Mills et al. 2010). To help support capacity building and address the lack of sufficient information, technical assistance programmes on MSP have been provided, studies to address data gaps (e.g. socioeconomic data) undertaken and information compiled to make it more accessible (e.g. in databases). For example, the CTI atlas, initiated through partnerships among international NGOs supporting the CTI, is a repository of geospatial information and used to track progress. However, keeping the information up-to-date and enhancing functionality remains a challenge due to a lack of local capabilities and financial resources (Cros et al. 2014; Asaad et al. 2019).

4.2 The Norwegian Ocean Management Plans

Norway's ocean areas span 3,000 km from boreal to polar climates and measure 2.1 million km^2, five times the country's land area. The dominate physical influence is the Atlantic current transporting heat from the southwest Atlantic to the north, making the country (and northwestern Europe) several degrees Celsius warmer than corresponding latitudes in North America. Its three major marine regions are the Barents Sea in the north, the Norwegian Sea, and the North Sea in the south. While the Barents and North Seas are shallow shelf seas, the Norwegian Sea is a deep ocean with a narrow continental shelf along the coast. Ocean currents and other oceanographic conditions are favourable for biological productivity, and these seas are therefore very rich in fish resources. Also, the fjords provide favourable conditions for aquaculture, and the continental shelves contain abundant petroleum resources as well as other minerals.

The Barents Sea (1.4 million km^2) is divided between Norway and Russia following a boundary agreement in 2010 and is limited to the north by the high seas in the central Arctic Ocean. The Norwegian Sea is bound to the west with the waters of Greenland, Iceland and the Faroes, as well as the high seas there. In the North Sea, Norway's waters (and continental shelf) meet those of the United Kingdom, Denmark and Sweden (Fig. 14.4). That Norway shares ocean boundaries with seven other countries is an important determinant of its marine policies, as transboundary resources and ecosystems require international cooperation for their management, at bilateral as well as regional levels of coopera-

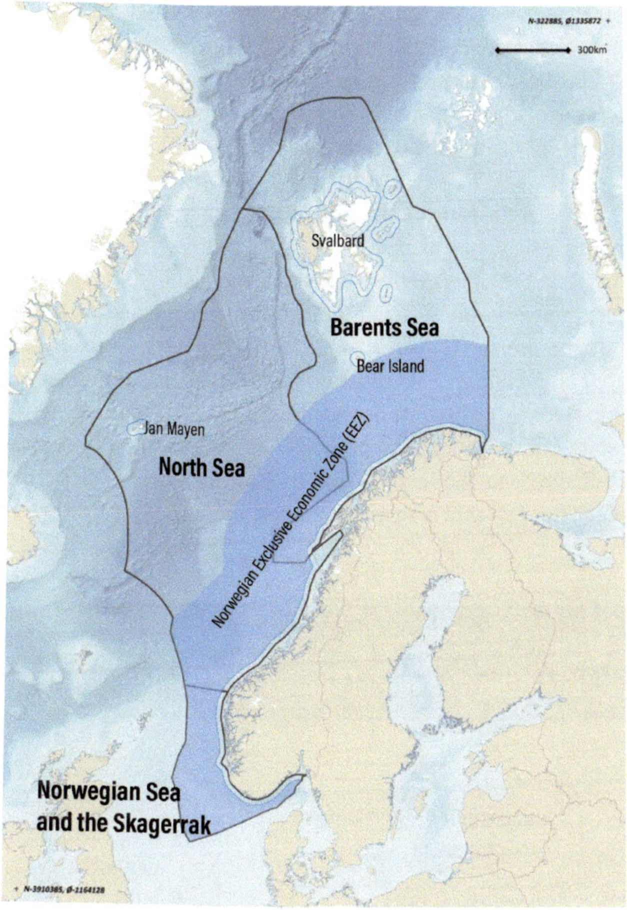

Fig. 14.4 Map of Norway's Marine Management Plan Areas. (*Source: Data from Barentswatch 2020* (https://kart.barentswatch.no/ *arealverktoy*). *Map by Centre for the Ocean and the Arctic, Norway 2020*)

tion. Another important determinant is the global framework for ocean governance described at the outset of this paper.

4.2.1 Economic activity and ocean governance

Petroleum is by far the most important industry in Norway, contributing US $60 billion in 2018—approximately one-quarter of Norway's export earnings. Fisheries and aquaculture contribute about $11 billion in export value. In addition, shipping, shipbuilding, tourism, petroleum services and other ocean-related activities are significant contributors to Norway's ocean economy, together constituting 70% of the national gross domestic product. Recent developments include increasing interest in renewables, exploitation of new species of living marine resources, minerals other than petroleum and marine bioprospecting. The ocean economy is critical to the welfare of both the general population and, in particular, coastal communities.

The management of the ocean has evolved over centuries, with active government regulation in the fisheries sector dating back to at least the mid-1800s. The petroleum and aquaculture industries started in earnest in the 1970s and were accompanied by the development of management institutions and legal regimes for their regulation. At the same time, regulations to protect the marine environment were established. Since then, the regulatory regime has evolved significantly. Today, all major sectors—including petroleum, the environment, transportation, fisheries, aquaculture and minerals—have modern and effective regulatory regimes based on sector-specific acts.

Following the growth in the ocean economy, the need for more and better regulation of the various economic activities was accompanied by a need for more oversight over an increasingly complex regulatory environment—to reconcile competing interests and address management challenges in an integrated and comprehensive manner (Hoel and Olsen 2012). The issue was first recognised in the 1970s, with efforts to this end in relation to petroleum activities and fisheries. The northward expansion of the petroleum industry brought a recognition that more comprehensive approaches to ocean management were needed. A report to the Norwegian Parliament in 2002 represented a turning point in this regard, laying the foundation for what became the management plans for the ocean and explicitly recognising the need to manage entire ecosystems as opposed to individual sectors and activities (Norwegian Ministry of the Environment 2002).

4.2.2 Comprehensive management plans

The first comprehensive management plan for the Barents Sea and the coast of northern Norway was adopted by the parliament in 2006 (Norwegian Ministry of the Environment 2006). The plan's overall objective was to facilitate value creation through the sustainable use of natural resources and goods while maintaining the structure, functioning and productivity of ecosystems. [2] Its geographic scope was limited to the waters outside one nautical mile off the baseline with nearshore ocean and coastal zone management outside the remit of the plan. It was also limited in time, foreseeing regular updates as new knowledge became available. Key features of the plan were the identification of valuable and vulnerable areas, and limitations on where petroleum activities could take place. Updates to the plan were adopted in 2011 and 2015. The first plan for the Norwegian Sea was adopted in 2009 and updated in 2017. The first North Sea plan was adopted in 2013. Apart from the effects of climate

[2] The original text in Norwegian reads as follows: 2022 'Formålet med denne forvaltningsplanen er å legge til rette for verdiskaping gjennom bærekraftig bruk av ressurser og goder i Barentshavet og havområdene utenfor Lofoten og samtidig opprettholde økosystemenes struktur, virkemåte og produktivitet.'

change, the challenges are rather different in the three areas covered by the plans.

The foundation of the work on the plans is an ambitious scheme for collecting and assembling information about the marine ecosystems, undertaken in the course of fisheries surveys and mapping programmes (Olsen et al. 2014). This work is institutionalised in the Advisory Forum on Monitoring, with participation from central research institutions and agencies. Stakeholder consultations are also an important element in the decision-making process. The development of the scientific and practical basis for the plans is carried out by a coordination group consisting of government agencies, now called the Management Forum on Norwegian Sea Areas. The work is overseen and coordinated at the ministerial level by a group of ministries coordinated by the Ministry of Climate and Environment. The interministerial group is responsible for developing the actual management plans and reconciling the various concerns that are brought to bear on the work. In 2020, the work on the three plans will be merged into one report to the parliament, and subsequent updates and revisions will occur on a decadal basis (Fig. 14.5).

4.2.3 Challenges and Lessons Learned

After nearly two decades of work, Norway offers several important lessons on holistic ocean management. The first is that the work relies on comprehensive and demanding scientific monitoring (Olsen et al. 2016). These monitoring efforts are largely part of regular monitoring programmes as opposed to efforts designed specifically to monitor work related to the management plans.

Second, at the level of the two forums, the work on the plans has brought research institutes and agencies from vari-

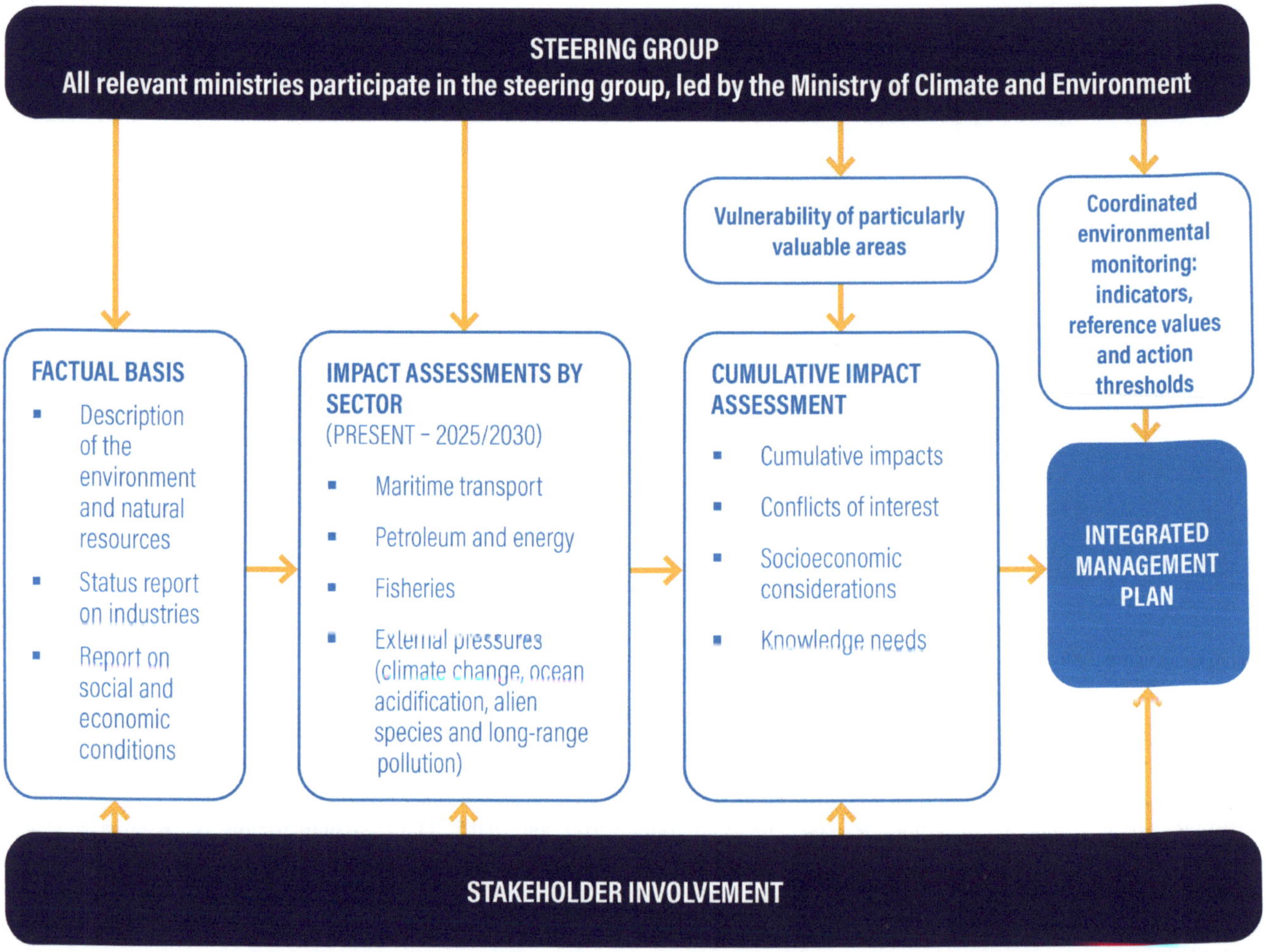

Fig. 14.5 Process for Developing an Integrated Ocean Management Plan in Norway. (*Source: Norwegian Ministry of the Environment* 2009)

ous sectors together, contributing to a better understanding of each other's missions and methods.

Third, the plans are adopted by parliament based on government white papers, lending the plans political authority.

Fourth, the plans are implemented through sector-based legislation and regulatory measures. The management plans are just that—plans. However, since they are adopted by parliament, and reflect compromises in government, they represent national policy. Modern and efficient sector-based legislation is an effective vehicle for bringing the plans from paper to practice.

Fifth, the plans evolve over time as new scientific knowledge is generated and new management challenges arise. It is important that the plans be dynamic and provide for adaptive management through regular updates and revisions.

4.3 Approaches to Integrated Ocean Management in the United States

The United States has approached integrated ocean management at three scales: state, regional (multiple state) and national. Several U.S. states—Massachusetts, Rhode Island, Washington, Oregon, New York and Connecticut—have developed ocean plans for state waters (0–3 nautical miles offshore). Regions, comprised of multiple states and defined mostly by ecosystems, have worked toward IOM within federal waters (3–200 nautical miles), in response to and consistent with a national directive in the form of a presidential executive order. State plans often have a clear legal structure through legislation that provides direction for integrated ocean management as was the case for Massachusetts, Washington, Oregon and Connecticut. States that do not have a clear legal structure for IOM can also develop and implement ocean plans by reinterpreting existing legal authorities, as Rhode Island did. Regional IOM plans were developed in response to an executive order that provided guidance for regions to take an ecosystem-based management approach. As part of the executive order, federal agencies with a stake in ocean management were directed to engage with regional planning bodies to develop plans for areas where there was interest in improving coordination among management agencies (Executive Office of the President 2010). The executive order directed federal agencies but engagement by states and tribes was voluntary.

4.3.1 Northeast and Mid-Atlantic Regional Ocean Plans

Administrative orders or directives can define a framework for coordinated management. These directives may articulate high-level targets but look to regulatory agencies to define specific management goals and objectives. The ocean plans in the Northeast and Mid-Atlantic regions of the United States, which largely align with major ecosystems, are examples of management plans developed without an overarching legislative structure to provide a directive or framework. Both plans were created through a presidential executive order that provided an overarching mandate, outlined federal authority, and built on the existing legal authorities of management agencies. The order provided guidance for regions to take an ecosystem-based management approach through the development of ocean plans, but no new laws were created (Executive Office of the President 2010). Planning was voluntary, and two of the nine defined regions of the United States completed plans by late 2016. In 2018, a new presidential administration rescinded the earlier executive order directing management agencies to develop plans. A new executive order was issued that allows planning to continue in those regions where IOM is supported and directs federal agencies to coordinate with states and regions to solve management challenges (Executive Office of the President 2018). Such political transitions can challenge the stability of IOM, especially in cases with no overarching legislative structure.

The Northeast and Mid-Atlantic regions, where Massachusetts and Rhode Island had already developed state ocean plans, were the first two regions to advance more integrated approaches to ocean management. The impetus for regional planning was a combination of the collaboration among state, tribal and federal management authorities and ocean users already ongoing in the region, leadership provided by state and federal agency representatives, political will and an interest in avoiding potential conflicts with existing users and emerging industries such as offshore renewable energy. The Northeast and Mid-Atlantic regions both completed regional ocean plans in 2016 (Northeast Regional Planning Body 2016; Mid-Atlantic Regional Planning Body 2016) (Fig. 14.6). Goals for both regional plans focused on the following:

- Improving decision-making by coordinating managing authorities and stakeholders, coordinating early in the process and enhancing awareness of human activity needs, interests and resources.
- Promoting healthy ocean and coastal ecosystems by characterising the region's ecosystems, economy and cultural resources, and by identifying opportunities to conserve, restore and maintain healthy ecosystems.
- Ensuring compatibility among past, current and future ocean uses.

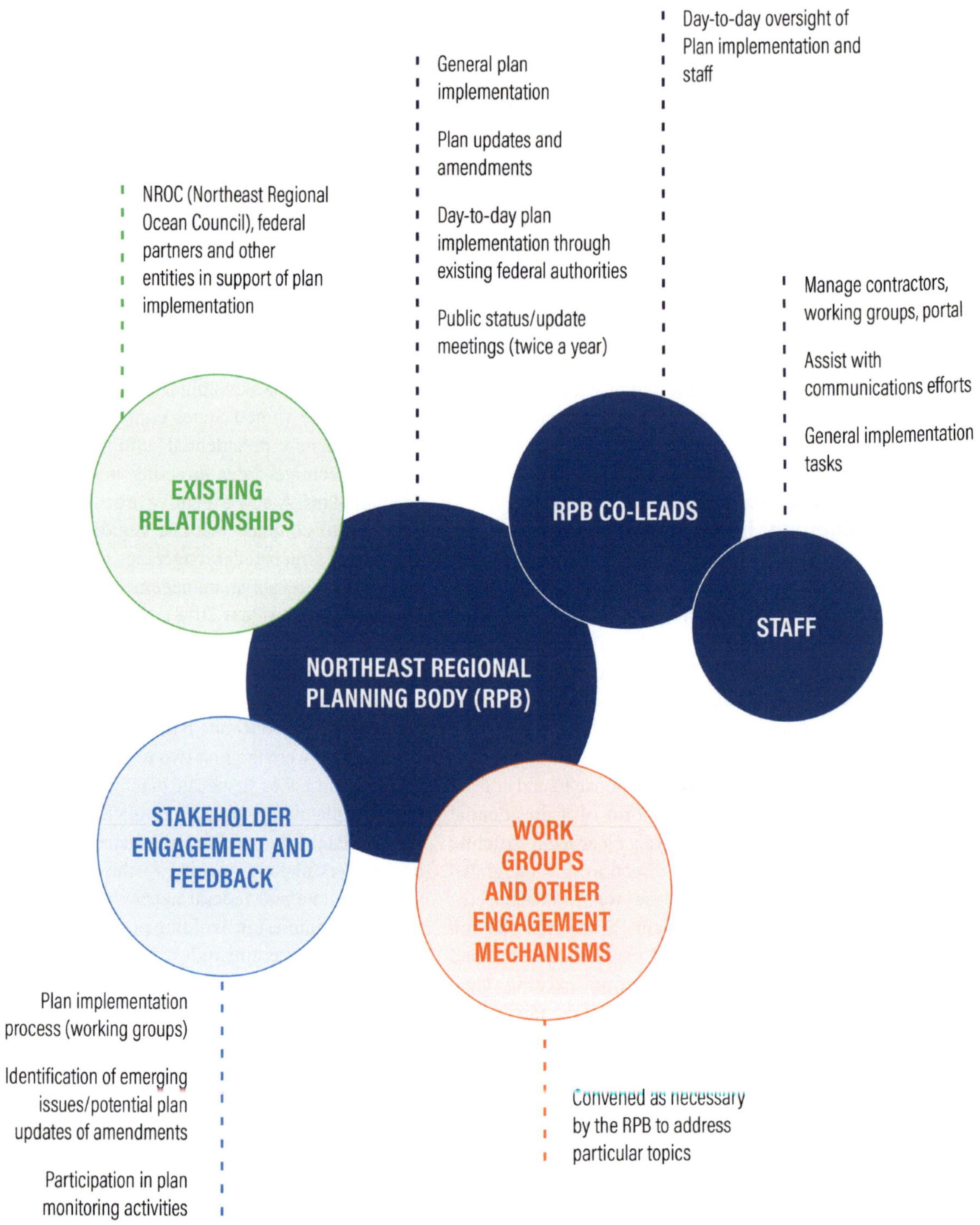

Fig. 14.6 Framework for Implementing the Northeast Ocean Plan. (*Source: Northeast Regional Planning Body* 2016)

With regional plans now in the implementation phase, it is apparent how public and private sector organisations are benefitting from the data information systems—referred to as 'regional ocean data portals'—that were developed as part of the ocean plans. Examples of how the Northeast Ocean Data Portal has been used include the following (Northeast Ocean Data 2009):

- To create a new economic opportunity by establishing the first shellfish aquaculture farm in federal waters off the Atlantic coast.
- To increase maritime safety and improve weather forecasts through a wave-monitoring buoy.
- To select a test site for an unmanned underwater vehicle for the United States Navy.
- To assist the New England Fishery Management Council in balancing deep-sea coral protection and commercial fisheries.

4.3.2 Balancing Ecosystems and Economy with the Rhode Island Ocean Special Area Management Plan

As the United States enters a new era with offshore renewable energy innovation, a project in Rhode Island demonstrates the value of using an IOM approach. Faced with increased demands on ocean space, an ambitious renewable energy goal for the state and political leaders interested in advancing offshore wind, the state management agency, the Coastal Resources Management Council, took the lead in developing an Ocean Special Area Management Plan (McCann 2010). With an overall goal of balancing ocean resources with development, the state created an open and transparent planning process that was backed with science, critical stakeholder input and productive public fora (Fig. 14.7).

Data collected through this plan not only identified a renewable energy area that would minimise conflicts with other ocean users and ecological resources but also provided information on aspects like key fishing grounds, marine mammal migration routes and recreational boating activities.

4.3.3 Challenges and Lessons Learned

IOM can be harder to achieve without a mandate and structure defined by law. For states with new legislation directing management agencies to coordinate and develop plans, it was, in general, easier to define goals and objectives because legal authority and overall direction were clear. Reinterpreting existing legal structures is also an option, as demonstrated by the Rhode Island experience. External drivers such as a new ocean use or emerging technology can stimulate integrated management approaches by creating the political and societal will for IOM.

In the case of regional plans developed through an executive order, administrative leadership provided the overarching mandate of maintaining healthy ocean ecosystems. The executive order also created clear direction for federal agencies to act within their legal authorities to coordinate among agencies and with states to advance IOM. The challenge is that these orders are voluntary for states so the key players in the regions—including representatives from states, tribes, fishery management councils, industry, academia, nongovernmental organisations and local communities—must perceive added value from taking a holistic, integrated approach. In the United States, two of the nine regions have completed integrated ocean management plans and an additional one is in development. If a state, region, territory or nation wants to commit itself to IOM, the clearest and most efficient path is with a new law directing specific actions. However, other approaches can be taken depending on political context and institutional arrangements.

Under the United States system outlined above, experience demonstrates that data and information are the foundation of IOM. Ocean businesses want managers to understand their interests when developing a comprehensive management plan. Data and specific information help managers demonstrate their understanding of different ocean uses. Ocean users should have an opportunity to provide and verify data, and therefore see their activities reflected explicitly within the IOM process. This data can be used to describe not only existing human activities or ecological resources but also future goals and trends. Developing a regional ocean data portal (information management system) that is open and transparent and has all relevant data, including the metadata, in a central location will go a long way toward building support from decision- makers, local communities, scientists, nongovernmental organisations and various sectors of human activity.

If resources and available data permit, government entities should consider prioritising the development of information management systems, referred to in the United States as ocean data portals. A centralised ocean data portal not only improves coordination among various management agencies with responsibility for human use activities but also has the potential to improve coordination among ocean users who can now view all activities in one central location. A data portal is not needed to define societal objectives and express those in a holistic plan, but the United States' experience demonstrates their value for facilitating an integrated management approach.

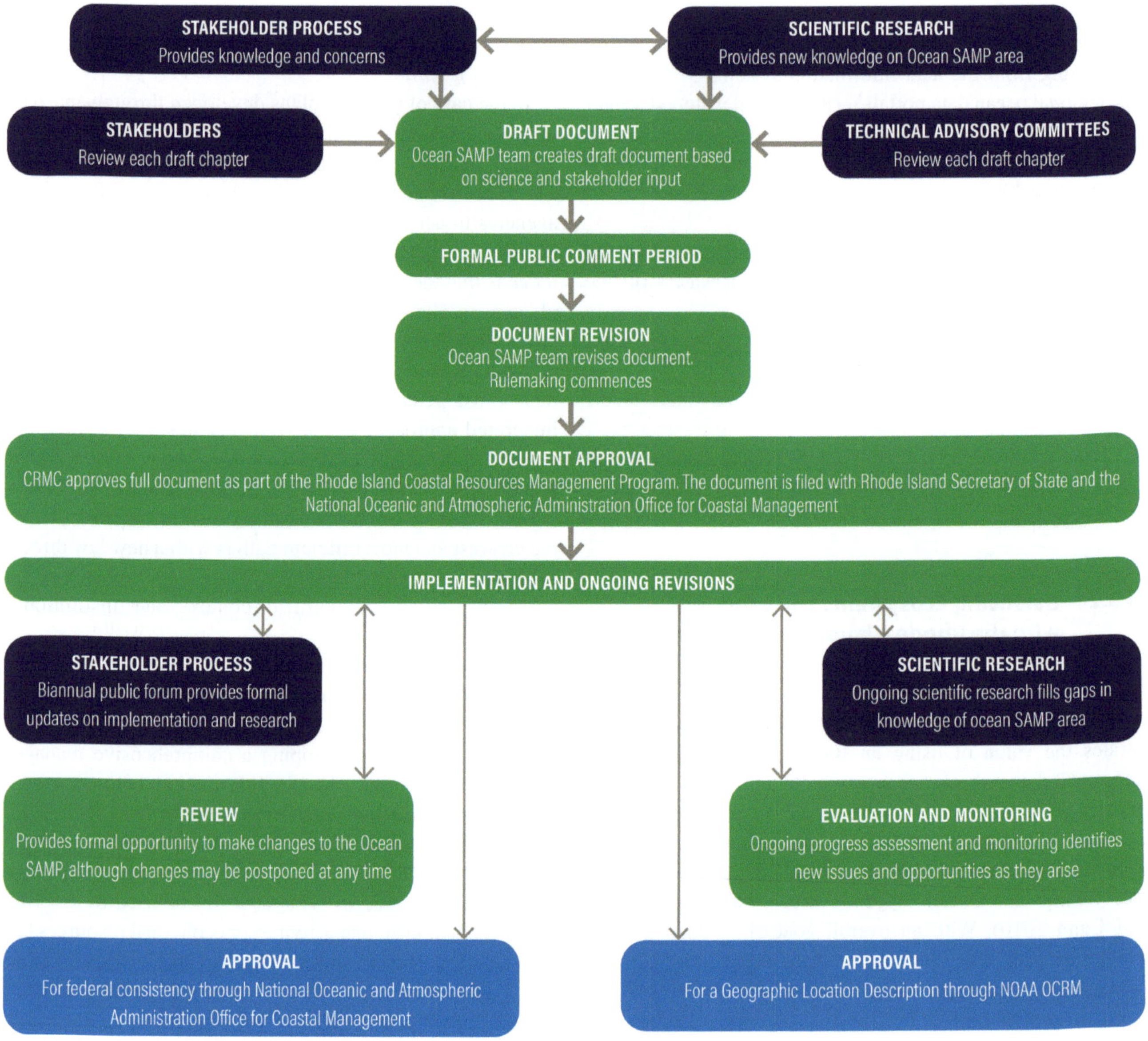

Fig. 14.7 Methods Flowchart for the Rhode Island Ocean Special Area Management Plan. (Note: CRMC stands for Coastal Resources Management Council and NOAA OCRM for National Oceanic and Atmospheric Administration Office for Coastal Management. NOAA approved as a state plan amendment May 11, 2011 (Federal Consistency authority granted for state waters). NOAA approved Geographic Location Description for federal consistency September 2011 (Federal Consistency authority granted for Federal Waters in the GLD for 15 Federal Authorizations which now needed Consistency sign offs by CRMC). (*Source: Adapted from Mulvaney 2013*)

Stakeholders, or ocean users, need to be consulted and involved in data collection, to corroborate the accuracy of data and provide data about anticipated future activities and trends. This consultation process helps establish collaboration and trust within IOM.

Several mechanisms to gather information have proven beneficial in the United States. These include the following:

- Arranging sector-specific meetings to gain a better understanding of a sector's needs, concerns and future trends.
- Developing industry-specific white papers that review the current status, future trends and planning needs of a specific sector—through the white paper approach, industry can easily share information with interested members to gather comments and feedback.
- Identifying data gaps and, where possible, outlining stakeholder-driven projects that could fill those gaps.

Developing a research and science agenda to better understand the ecosystem within the IOM framework can help advance and fill priority gaps in the data. In the United States, federal agencies with various management authorities have collaborated to identify data gaps and, even more importantly, to determine how to prioritise them to best meet planning objectives. Defining a research and science agenda provides management entities with a common goal of advancing data objectives to ensure that future iterations of the plan will have the necessary additional information to better inform decisions. This process also allows for collaboration on budget discussions and prioritises time and resources within a structure that all engaged entities have agreed to.

Robust stakeholder engagement is critical to successful IOM. A key component and lesson learned from the United States is to reach out to sector-specific thought leaders or trusted organisers who are proven leaders in the community and can help disseminate information to and enlist participation of those within their given sectors. For example, variations in the fishing community must be taken into consideration as an IOM process moves forward since data and information will differ depending on gear type, port community and fish species. This approach of seeking adequate representation across industry, the nonprofit sector and the scientific community should be taken with all groups.

Ocean users may have a natural distrust of the new and potentially complex process that is IOM. It is vital that IOM remains an inclusive and transparent process so trust can be built. Meeting stakeholders where they are and ensuring that their input is adequately incorporated throughout the process has been shown to be more effective at ensuring stakeholders find value in a holistic approach rather than selling them on a concept.

4.4 Integrated Coastal Management in Xiamen, China

Xiamen, with a population of 4.11 million in 2018, is a port city located on the west coast of the Taiwan Strait. As of 2018, it was the 7th largest container port in China and the 14th largest in the world. Xiamen Island is surrounded by 394 km² of sea and has a coastline of about 234 km. Xiamen Bay, including the Jiulong River Estuary, West Sea, Tongan Bay and East Sea (Fig. 14.8), is home to nearly 2,000 marine species including protected species like Chinese white dolphins, lancelets and egrets. The bay has been a vital part of Xiamen's economy for centuries.

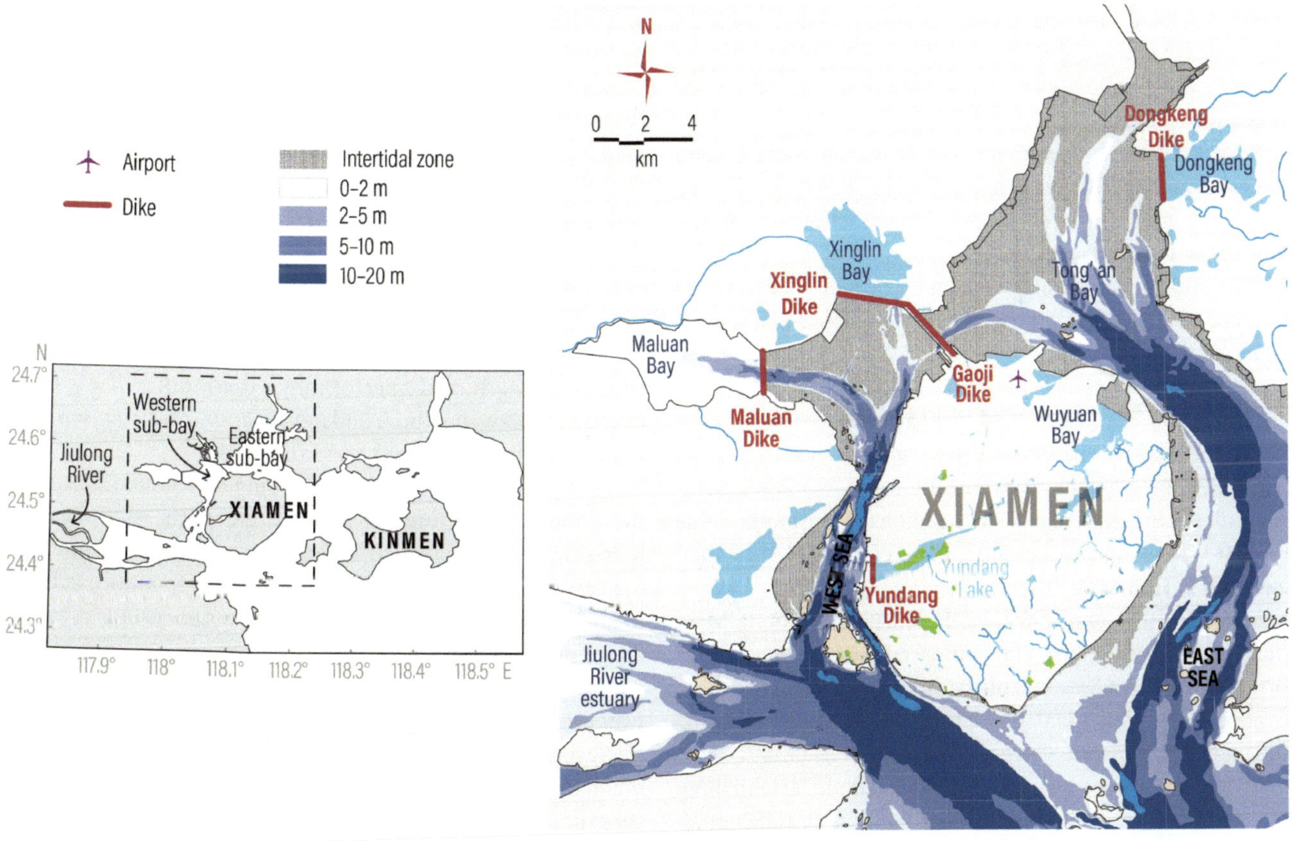

Fig. 14.8 Map of the Xiamen Area. (Note: In the legend, m stands for water depth in metres. (*Source: Redrawn from Wang et al.* 2013)

Following China's major reform initiative in the late 1970s, Xiamen became one of the first four special economic zones. Since then, Xiamen has experienced an economic boom that has brought with it a series of resource use conflicts and pollution problems. This was particularly visible in the early stages as little attention was paid to ecosystems and the environment (Chua et al. 1997; Xue et al. 2004). Seawall constructions and reclamations drastically modified the coastal morphology and hydrodynamics and reduced the area of surface water and tidal influence.

Starting in the 1980s, marine aquaculture grew rapidly and was further intensified in the mid-1990s. By 2001, it covered nearly half of the West Sea area. Waste from coastal aquaculture ponds and excess feeds from fish cages polluted the marine environment. Nearly all domestic and industrial wastewater was discharged into the sea untreated. Many natural habitats were damaged by pollution. Mangrove forests declined from 1.8 km^2 in 1987 to 0.2 km^2 in 1995. Major events of fish deaths occurred around twice per year in the period from 1984 to 1996 (PEMSEA 2006a) and populations of dolphins, egrets and lancelets declined (ITTXDP 1996; XDPO 1998; Xue et al. 2004; Lin et al. 2005; PEMSEA 2006a, b).

Faced with environmental degradation, sea-use conflicts and ineffective management as well as deficiencies in legislation, funds, public awareness, information and pollution-prevention capabilities (PEMSEA 1998), Xiamen implemented a new ocean and coastal management system in 1994. Integrated coastal management (ICM) in Xiamen has undergone four stages of development: structural design from 1994 to 2000, marine ecosystem rehabilitation from 2000 to 2009, co-governance of land and sea from 2009 to 2015 and sustainable ocean economy since 2015 (Hou et al. 2019).

4.4.1 Early Stages of ICM in Xiamen

With this backdrop, the Chinese government decided in 1994 to make Xiamen a demonstration site for ICM in collaboration with GEF, UNDP and IMO's regional programme (Xue et al. 2004; Cao and Wong 2007; Fang et al. 2011; Mao and Kong 2018; Hou et al. 2019).

Between 1994 and 2000, in the early stage of ICM in Xiamen, a coordinating, law-enforced and science- supported mechanism was established. From 1994 to 1996, to advance ICM, individual projects were selected under the guidance of international organisations such as GEF, UNDP and IMO. These projects included, for example, establishing pollution management plans and sea use zoning (GEF et al. 2009). In 1996, the municipality of Xiamen initiated an ICM leadership group consisting of the mayor and officials from different governmental departments, under which an ocean office was established and tasked with organising regular meetings with ocean-related sectors within aquaculture, transportation, construction and science and technology (Xue et al. 2004) (Fig. 14.9). During this phase, a series of marine laws and regulations, including the Administrative Regulations on Xiamen Sea Area Use for development and use and the Regulation on the Management of Natural Protected Areas for Chinese White Dolphin for environmental protection and ecological conservation, were adopted. A series of spatial planning programmes, including the Functional Zoning of Xiamen Sea Area, were also initiated. To provide support in developing these new tools, a municipal ocean specialist team consisting of leading researchers was formed.

4.4.2 Ecosystem Rehabilitation

Xiamen's ICM entered a new phase in the early 2000s with the initiation of several marine ecosystem rehabilitation projects. The first was established in the Yundang Lagoon, located in the downtown area of Xiamen Island. This lagoon used to be a fishing harbour connected to the Western Sea of Xiamen, enriched by mangroves, and had once sheltered huge flocks of egrets. During the 1970s, a dam was built at the mouth of the lagoon to cut off the water flow, converting the lagoon into an enclosed body of water. In addition, the surface water area was reduced from 10 km^2 to 2.2 km^2 due to reclamation for agriculture purposes. Untreated industrial and domestic wastewater was also being discharged into the lagoon. Residents began leaving the area (PEMSEA 2006b). Due to the poor environmental conditions, the site was blacklisted by the national Environmental Protection Agency. This situation was not resolved until a series of cleaning actions were implemented, including improving the waste management systems, constructing sewage treatment plants, building a retaining wall and performing dredging. The water exchange between the Yundang Lagoon and the sea was improved and mangroves were replanted.

According to the Functional Zoning of Xiamen Sea Area, the dominant functions of the area are ports, shipping and tourism. However, aquaculture was its primary function until the 1990s. In 2002, Xiamen stopped its aquaculture activity to solve ocean-use conflicts and initiate ecosystem rehabilitation in the area.

The aquaculture facilities were completely removed, and waterways were dredged to ensure their prime functionality.

Several other rehabilitation initiatives were also implemented, including building a wetland park, restoring the shoreline, planting mangroves, building uninhabited islands for birds to forage and improving the sewage treatment system (Wang et al. 2018). In Wuyuan Bay, 89 hectares of wetland were established. Various measures improved the water exchange in the East Sea by 30 percent. Combined with bet-

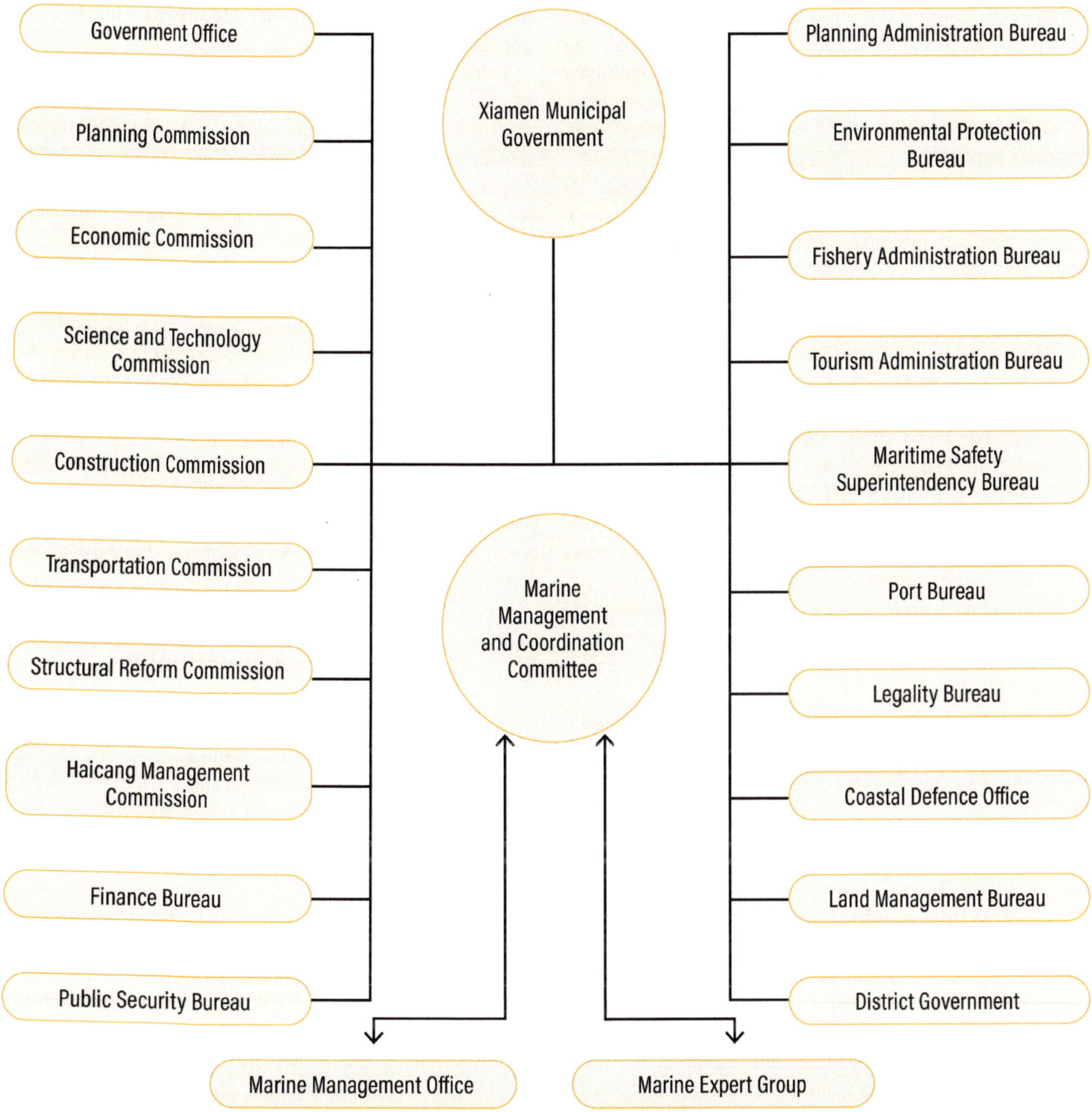

Fig. 14.9 Organisational Structure for Integrated Coastal Management in Xiamen. (*Source: Xue et al.* 2004)

ter water quality, the conditions for the Chinese white dolphins improved significantly.

Following the successful rehabilitation projects, Xiamen's efforts in ICM during 2009–2015 were mainly focused on governing the rivers and sea by establishing a system for controlling the terrestrial pollution. Since 2015, after over 20 years of ICM practices and in response to major national guidelines ('Managing Land and Sea as a Whole' and 'Constructing Ecological Civilization'), Xiamen has begun stage four of development—integrating land-sea management and the concept of developing 'blue growth' (Mao and Kong 2018).

4.4.3 Challenges and Lessons Learned

ICM in Xiamen can be characterised by the establishment of a legal framework and enforcement mechanisms, science-policy integration, marine monitoring system and information sharing, and public awareness mechanisms. As a management instrument to rationalise the use of marine and coastal resources and environment, marine spatial planning (called 'marine function zoning' in China) is a significant component of the ICM programme in Xiamen (Su and Peng 2018). There are a number of lessons to be learned from Xiamen's experience.

First, coordinating numerous stakeholders—from sectors including urban planning, fisheries, shipping, transportation, science, port authority and conservation—has been a challenge. To meet this challenge, the existing and successful concept of 'River Chief System', where one stakeholder is given extended responsibility, is also being implemented for the ocean space, as the 'Bay Chief System'.

Second, a comprehensive ICM system for laws and regulations was developed without fully aligning with existing regulations for terrestrial management in the same area (Su and Peng 2018; Peng et al. 2006). Thus, land and ocean management has been insufficiently integrated, something that needs to be refined when ICM in Xiamen is further developed. This may include, for example, creating zoning plans that account for both land and ocean.

Third, more management efforts and enforcement measures are needed to control non-point source pollution from land-based activities in watersheds with runoff to estuaries and bays.

Finally, integrating science and technological guidance throughout the process—including during design, implementation, evaluation and refinements—has been very valuable.

4.5 What Does Experience Teach Us about IOM Implementation?

The five case studies of IOM reviewed here represent vastly different situations with respect to climatic and oceanic conditions, geographical scales, the nature of economic activities and regulatory environments. Nevertheless, there are some significant commonalities—described below—that provide useful lessons for other contexts. The common denominator is that increasing uses and pressures on marine ecosystems drive the need to consider the totality of pressures on the entire ocean space.

First, climate change is manifesting itself in tropical, temperate and Arctic marine environments and represents a major challenge to ocean management. In this respect, IOM is a critically important way of addressing multiple ocean uses while considering the impacts of climate change.

Second, due regard needs to be given to the local context. It is critically important to tailor IOM to the characteristics and needs of the region in question. The concrete economic activities, community needs, societal goals and environmental pressures should be the point of departure for the development of IOM.

Third, information is critical. It is essential to have robust data series on the evolution of critical environmental variables as well as on economic activities. Without information, management decisions cannot be effectively made. Information should be accessible, easy to find and subject to data quality standards in appropriate formats for public accessibility.

Fourth, implementation—moving from paper to practice— is essential. Several cases demonstrate that this can be done effectively without a separate legal basis for IOM.

Fifth, stakeholder involvement is critical to ensuring that the practical information needed to develop IOM measures is available and building the understanding and legitimacy required for effective implementation.

Finally, institutional mechanisms for IOM are needed, whether formal or informal. There has to be a designated process for how to consider the various pressures and uses of ocean space in a comprehensive manner and make decisions on that basis.

5 Conclusions and Opportunities for Action

This paper argues for the need for integrated ocean management and has identified several central components of successful IOM. Achieving a healthy, productive and resilient ocean requires taking a holistic perspective on ocean use and management, and effectively implementing relevant national and international measures. Given current levels of pressures on our ocean, few human activities can be viewed in isolation. Most activities have impacts that need to be accounted for and viewed in relation to other activities and concerns to fully capitalise on the economic potential of the ocean in a sustainable way. Achieving effective sectoral management is necessary, but not enough. We also need to realise that a sustainable ocean economy depends directly upon healthy, productive and resilient ocean ecosystems and act accordingly. Thus, the need for an integrated, ecosystem-based and knowledge-based approach to ocean governance is pressing.

We need to ensure that ocean industries do not degrade the environment that they and others depend on. It is critical that short-sighted solutions with negative environmental impacts are replaced with long-term, sustainable solutions that strike a better balance between protection and production. Quantitative assessments and strategies for factoring in long-term benefits by implementing sustainable solutions should be developed. Despite progress on some fronts, the current trajectory is in the wrong direction and rapidly growing more serious, e.g. biodiversity loss and plastic pollution.

Moreover, important information often exists but is not used in decision-making. Effective ocean governance must consider developments in technology, the impacts of climate change, the dynamic nature of the ocean and seas and the interactions and synergies between land, ocean and people. Connecting management plans for coastal land areas with the adjacent ocean management plans would significantly improve today's situation in many regions.

The statuses of marine ecosystems and their properties and characteristics vary considerably. IOM provides not only an understanding of the totality of ocean uses and pressures but also guidance for how to prioritise among these various uses. Government solutions need to be tailored to the characteristics and problems of the specific marine region—one size does not fit all, and context is essential. The relevant economic activities, community needs, societal goals, traditional and local knowledge and environmental pressures should be included in a tailor-made IOM process.

That said, IOM supplements but does not replace sector-based management. It is important to maintain and further develop effective sector-based legislation and other measures and keep them up to date with the most recent international standards in, for example, shipping and fisheries. However, coordinating across sectors is needed, including regarding how to collectively prepare for future scenarios.

Furthermore, the need for regional collaboration is evident. Ecosystems and economic activities often occur in several jurisdictions and across national boundaries. Also, activities in the marine realm can have widespread impacts. In the case of such transboundary situations, regional cooperation in, for example, fisheries management or the prevention of marine pollution is necessary to address the problems at an appropriate geographical scale. Also, regional collaboration would greatly benefit from those nations or entities with experience with IOM providing mechanisms for sharing information, data and knowledge. At the local level, connectivity—particularly increased dialogue among locals and institutions—plays a vital role in ensuring sustainable ocean governance.

Finally, as pointed out when summarising experiences from our case studies, climate change represents a challenge vastly larger than anything we have faced before, and the future of the ocean depends on our ability to address this issue properly. Questions of adaptation and risk management loom large in this respect and are critical dimensions of all opportunities for action discussed below. Adaptive ocean management must make use of the best available science to account for how future climate change will affect ocean businesses (individually and combined), ecosystems and societies. For example, when localising MPAs or ocean industries, authorities must account for changing conditions in the relatively near future due to climate change.

Through five case studies, we have emphasised that conditions vary across countries and regions. We have identified key features for successful IOM such as the need for both a mandate (top-down) and engagement (bottom-up) approach, adequate funding and explicit mechanisms to implement plans, not just create them. Practical and implementable solutions of ocean governance can serve as inspiration and guidance. We can also learn from mistakes. Developed countries with established institutions for marine planning and management do of course not represent the whole picture. In this context, some countries have a clean slate to work with, and therefore the opportunity to get it right the first time. This may be an advantageous starting point for building capacity and establishing IOM.

This paper identifies six main aspects of successful IOM:

- harness science and knowledge.
- establish partnerships between public and private sectors.
- strengthen stakeholder engagement.
- improve capacity building.
- implement regulatory frameworks.
- develop adaptive management systems.

The following opportunities for action respond to each of these in turn.

5.1 Opportunity for Action 1: Harness Science and Knowledge

Tools to develop, strengthen and coordinate governance of the ocean include increased science and monitoring efforts, sharing of knowledge, and transfer of technology and digital infrastructure, especially in the least developed countries and SIDS. For example, the goal of ecosystem-based management is impossible to achieve if data on the ecosystems and the societies depending upon them are lacking. Relevant data and clearly defined goals for management coupled with research and science plans are important to advance and achieve IOM.

In some regions, there are large knowledge gaps for a range of ocean-related issues such as the abundance and interactions among living marine resources, impacts of human activities (existing and future), opportunities embedded in the expected digital and technological revolution, consequences of marine litter and the impacts of climate change. To address this, we recommend strengthening the global ocean science enterprise—including social science—building on the efforts by the UN Regular Process and the IOC, as well as the ongoing efforts at the regional level. Strengthening the role of IOC in IOM would build on already existing structures, enhance the attention given to marine science and help generate the resources needed to develop scientific knowledge and scientific capacity building worldwide. A platform for its development could be the coming UN Decade of Ocean Science for Sustainable Development (2021–2030) as a framework to be hosted within the IOC. Another important output would be the World Ocean Assessments following up on the 2015 and 2020 editions, which can support regional and national ocean governance.

We suggest strengthening the global ocean science enterprise and better using existing knowledge, building on established structures such as the IOC and using the UN Decade of Ocean Science for Sustainable Development as a vehicle for further developing international cooperation in marine and related sciences.

5.2 Opportunity for Action 2: Establish Partnerships Between Public and Private Sectors

With a growing blue economy and increasing use of ocean space for human activities, maintaining a productive and healthy ocean becomes more difficult. Currently, investments and infrastructure in the ocean space are developed across various industries and sectors with differing standards of performance and governance. In practice, enduring sustainability can be achieved only if best practices are applied in both the public and private sectors and where productive partnerships between the two are encouraged and advanced. Good governance can bring long-term solutions that advance the economy while supporting societies and protecting the environment.

Advancing and clarifying the responsibilities of the private sector by developing 'Ocean Principles' for a sustainable ocean economy, modelled after the Carbon Principles, is a way forward. The UN Global Compact Action Platform for Sustainable Ocean Business has developed principles and guidelines for sustainable ocean businesses that several of the largest ocean- related businesses and financial institutions globally have signed on to (UN Global Compact 2019). A further development would be to give credits to nations and retailers that are able to develop transparent and traceable supply chains that demonstrate sustainability and contribute to the implementation of the 2030 Agenda for Sustainable Development. We suggest strengthening the commitments in business to further develop technological solutions with transparency at their core, thus empowering consumers to change the markets.

We suggest that ocean-related businesses at local, regional, national and international levels cooperate to develop principles and guidelines for the sustainable conduct of ocean businesses.

5.3 Opportunity for Action 3: Strengthen Stakeholder Engagement

To achieve sustainability in the uses of the ocean, including the achievement of the UN Sustainable Development Goals, it is critical to incorporate the insights, ownership and engagement of local stakeholders. National strategies for strengthening ocean management will not work without implementing sustainable projects at local levels of governance. Thus, actively involving communities and including local knowledge are important. Planning at the local level in developing countries, especially SIDS, requires tailoring approaches to the diverse environmental and socioeconomic contexts and governance systems in these regions.

For successful IOM, both mandate (top-down) and engagement (bottom-up) elements are needed. As demonstrated in the case studies, there are a number of approaches to local stakeholder engagement, which are highly context dependent. In all cases, however, designing well-managed engagement processes that consider the scientific, cultural, societal, economic and political contexts and encourage active stakeholder participation is crucial.

We suggest that governments support the active involvement of local communities in all stages of planning and development for integrated ocean management.

5.4 Opportunity for Action 4: Improve Capacity Building

Capacity building—efforts to enhance scientific and regulatory proficiency as well as institutional and collaborative capability—is vital for developing integrated ocean management. In this regard, IOM must address how to handle current and future challenges such as climate change, biodiversity loss and pollution. The scientific capacity needed to implement international governance frameworks is severely lacking in many countries (IOC-UNESCO 2017). Capacity building, primarily building on but also amplifying existing regional and intergovernmental organisations and institutions, needs to remain at the top of the international agenda. At the national level, it is essential that government agencies involved in ocean management are properly institutionalised, and have the skills, knowledge, authority and capacity—including funding—to address challenges relating to the ocean and coastal communities in a long-term, integrated manner. Here, collaboration and coordination among stakeholders is essential. New technologies combined with transparency give rise to new opportunities for monitoring and policing inappropriate behaviour at sea, bringing practical and inexpensive solutions for the transfer of know-how.

Additionally, the ocean science enterprise is advancing technologies that allow us to collect scientific data with less cost and greater efficiency than ever before. To effectively advance capacity, this must be done with transparency, tailored to context, and with data standards in place. In this respect, regional cooperation can be an effective vehicle for strengthening the role of science and providing advice for management, as demonstrated by ICES in the North Atlantic and Western Indian Ocean Marine Science Association (WIOMSA) in the western Indian Ocean.

We suggest identifying and using the best and most relevant principles, practices and procedures from regional efforts at IOM to develop integrated management in other regions.

5.5 Opportunity for Action 5: Implement Regulatory Frameworks

Failure to implement existing international instruments is perhaps the most important weakness of our ocean governance systems. It is vital to have mechanisms in place not just to develop IOM plans, but also to implement them. A comprehensive global ocean governance framework, supplemented with many regional instruments and often adequate national laws, does exist. However, implementation of the legal framework is too often inadequate and ineffective. In some cases, only immediate needs are prioritised in the allocation of resources to implement laws and regulations. There is also a need for local and subnational action plans and direct leadership to achieve successful implementation of IOM. Furthermore, inadequate implementation of existing regulatory frameworks in coastal states is a bottleneck for efficient and sustainable governance.

Important work is underway to address these shortcomings, including efforts to implement regional fisheries management organisation regulations, negotiations on biodiversity beyond national jurisdiction and the development of the seabed mining code by the International Seabed Authority.

A leading principle should be effective implementation of international agreements in domestic legislation and practices, including activities in the high seas. In practice, this means that rules for managing human activities in the high seas should be compatible with and at least as strict as those that apply in areas under national jurisdiction. Ratification of the basic international instruments for ocean governance and adherence to their provisions provided by UNCLOS is a precondition for this.

We suggest that regulatory frameworks for areas beyond national jurisdiction as well as those in areas under national jurisdiction be effectively implemented, building on the best available science. Rules for managing human activities in the high seas should be compatible with and at least as strict as those that apply in areas under national jurisdiction.

5.6 Opportunity for Action 6: Develop Adaptive Solutions

The ocean is highly dynamic, and its governance needs to reflect this. The dynamic nature of the ocean contrasts with the relatively static land areas, and it is important to address land-ocean interactions when developing integrated management of coastal regions.

This dynamism is further amplified by climate change. Many regions are already suffering from the effects of climate change, especially developing countries and small island states where coastal communities and even entire populations are threatened. Climate projections suggest that forward-looking, adaptive solutions where risk is explicitly considered will become an even more important element of IOM.

We suggest that IOM capture the connectivity and differences between land and ocean in an integrated and adaptive manner. Further, we suggest that ocean governance consider the expected future changes in the ocean environment by using the best available scientific knowledge on climate and other environmental changes.

Acknowledgements The authors thank the paper's technical reviewers, Tundi Agardy, David Balton, Joachim Claudet, Stuart Green and Louise Lieberknecht, as well as its arbiter, Jane Lubchenco, who all provided helpful technical comments. The authors also thank World Resources Institute for providing support as the Ocean Panel Secretariat. While our colleagues were very generous with their time and input, this report reflects the views of the authors alone. The authors also thank Sarah DeLucia for copyediting and Billie Kanfer for design.

About the Authors

Lead Authors

Jan-Gunnar Winther is the director of the Centre for the Ocean and the Arctic in Norway. He is also a specialist director at the Norwegian Polar Institute, and serves on a number of national and international committees and delegations including World Economic Forum, UN Global Compact Sustainable Ocean Business Action Platform, China Council for International Cooperation on Environment and Development and the Expert Group for the High Level Panel for a Sustainable Ocean Economy.

Minhan Dai is a chair professor at Xiamen University in China, where he serves as director of the faculty of Earth Science and Technology, and director of State Key Laboratory of Marine Environmental Science. His research interests include ocean.

biogeochemistry and sustainability. He has served on many national and international committees including the Expert Group for the High Level Panel for a Sustainable Ocean Economy.

Contributing Authors

Fanny Douvere has been the coordinator of the Marine Program at UNESCO's World Heritage Centre in Paris, France, since October 2009. The mission of the programme is to ensure that the 49 marine sites on UNESCO's World Heritage List are conserved and sustainably managed.

Leanne Fernandes is a senior manager and the marine coordinator at the International Union for Conservation of Nature. Over the last decade she has applied her knowledge to address marine resource management problems on the islands of Hawaii, the Maldives and the Caribbean and in the North Sea and coastal Australia.

Patrick Halpin is a professor of marine geospatial ecology at Duke University in the United States. He leads the Marine Geospatial Ecology Lab with laboratory facilities located at the main campus of Duke University as well as the Duke University Marine Lab.

Alf Håkon Hoel is a professor at UiT The Arctic University of Norway. He is a political scientist who has written extensively on issues relating to the management of living marine resources and the marine environment. His core interest is international relations in the field of ocean management, in particular in the Arctic and the Antarctic. Recent publications include analyses of the use of ecosystem-based management approaches in the Arctic.

Marie Antonette Juinio-Meñez is a professor at the University of the Philippines. Her research interests are invertebrate ecology and culture, coastal resources management, and marine science and marine biology.

Yangfan Li is a professor at Xiamen University in China. His main research interest is integrated coastal management. His work focuses on understanding the coupling mechanisms of coastal land-sea and social-ecological systems.

Karyn Morrissey is a senior lecturer at Exeter University in the United Kingdom. She is an economist specialised in the science-policy-industry interface and focused on understanding the impact of socioeconomic and environmental inequalities on health outcomes, using data both big and small.

Therese Rist is an advisor at the Centre for the Ocean and the Arctic in Norway, whose mandate is to contribute to a sustainable blue economy. She holds a postgraduate degree in public and international law from Melbourne Law School.

Fabio Rubio Scarano is a professor at the Federal University of Rio de Janeiro in Brazil. His main research interests are climate change adaptation, sustainability science, the science-policy interface and biodiversity conservation.

Amy Trice is the director of ocean planning at the Ocean Conservancy. She is a policy expert and scientist advancing solutions that integrate marine industry, conservation and science in ocean management decisions. Her research informs congressional staff as well as state and federal agencies on the value of comprehensive and coordinated ocean policies for economic and ecosystem resilience.

Sebastian Unger is the research group lead for ocean governance at the Institute for Advanced Sustainability Studies in Germany. His research focuses on global governance processes for ocean sustainability, including the development of a new international agreement for marine biodiversity in areas beyond national jurisdiction, the UN Sustainable Development Goals and the governance of deep seabed mining.

Sandra Whitehouse is a scientist and consultant at the Ocean Conservancy who has served for over 20 years as an ocean policy advisor to multiple organisations and institutions.

References

Agardy T, Notarbartolo di Sciara G, Christie P (2011) Mind the Gap: Addressing the Shortcomings of Marine Protected Areas through Large Scale Marine Spatial Planning. Mar Policy 35(2):226–232. https://doi.org/10.1016/j.marpol.2010.10.006

Asaad I, Lundquist CJ, Erdmann MV, Costello MJ (2019) An interactive atlas for marine biodiversity conservation in the coral triangle. Earth System Science Data 11(1):163–174. https://doi.org/10.5194/essd-11-163-2019

Asaad I, Lundquist CJ, Erdmann MV, Van Hooidonk R, Costello MJ (2018) Designating spatial priorities for marine biodiversity conservation in the coral triangle. Front Mar Sci 5:400. https://doi.org/10.3389/fmars.2018.00400

Ásmundsson, S., E. Corcoran. 2015. "The Process of Forming a Cooperative Mechanism between OSPAR and NEAFC." 196. UNEP Regional Seas Reports and Studies. United Nations Environment Programme.

Cao W, Wong MH (2007) Current status of coastal zone issues and management in China: a review. Environ Int 33(7):985–992. https://doi.org/10.1016/j.envint.2007.04.009

Chua TE, Yu HM, Chen GQ (1997) From sectoral to integrated coastal management: a case in Xiamen, China. Ocean Coastal Management 37(2):233–251

Commonwealth of Massachusetts. 2008. Massachusetts Ocean Management Plan. https://www.mass.gov/service-details/massachusetts-ocean-management-plan

Coral Triangle Initiative (Coral Triangle Initiative on Coral Reefs, Fisheries, and Food Security). 2019. "Home." Accessed November 1. http://www.coraltriangleinitiative.org

Cros A, Ahamad Fatan N, White A, Teoh SJ, Tan S, Handayani C, Huang C et al (2014) The coral triangle atlas: an integrated online spatial database system for improving coral reef management. PLoS One 9(6):e96332. https://doi.org/10.1371/journal.pone.0096332

Domínguez-Tejo E, Metternicht G, Johnston E, Hedge L (2016) Marine spatial planning advancing the ecosystem-based approach to coastal zone management: a review. Mar Policy 72:115–130. https://doi.org/10.1016/j.marpol.2016.06.023

Douvere F (2008) The importance of marine spatial planning in advancing ecosystem-based sea use management. Mar Policy 32(5):762–771. https://doi.org/10.1016/j.marpol.2008.03.021

Douvere F, Ehler CN (2009) New perspectives on sea use management: initial findings from european experience with marine spatial planning. J Environ Manag 90(1):77–88. https://doi.org/10.1016/j.jenvman.2008.07.004

European Union (European Parliament and Council of the European Union). 2008. Marine Strategy Framework Directive.

Executive Office of the President (United States) (2010) Executive Order 13547. National Policy for the Stewardship of the Ocean, Our Coasts, and the Great Lakes. United States Government, Washington, DC

Executive Office of the President (United States) (2018) Executive Order 13840. Ocean Policy to Advance the Economic, Security, and Environmental Interests of the United States. United States Government, Washington, DC

Fang Q, Zhang R, Zhang L, Hong H (2011) Marine Functional Zoning in China: Experience and Prospects. Coast Manag 39(6):656–667. https://doi.org/10.1080/08920753.2011.616678

FAO (United Nations Food and Agriculture Organization) (2009) Agreement on Port State Measures to Prevent, Deter and Eliminate Illegal, Unregulated and Unreported (IUU) Fishing. FAO, Rome

GEF, UNDP, UNOPS and PEMSEA (Global Environment Facility, United Nations Development Programme, United Nations Office for Project Services and Partnerships in Environmental Management for the Seas of East Asia). 2009. Environmental Rehabilitation in a Rapidly Developing Urban Area. http://pemsea.org/publications/case-studies/environmental-rehabilitation-rapidly-developing-urban-area.

Gjerde KM, Clark NA, Harden-Davies HR (2019) Building a Platform for the Future: The Relationship of the Expected New Agreement for Marine Biodiversity in Areas beyond National Jurisdiction and the UN Convention on the Law of the Sea. Ocean Yearbook Online 33(1):1–44. https://doi.org/10.1163/9789004395633_002

Grantham HS, McLeod E, Brooks A, Jupiter SD, Hardcastle J, Richardson AJ, Poloczanska ES et al (2011) Ecosystem-Based Adaptation in Marine Ecosystems of Tropical Oceania in Response to Climate Change. Pac Conserv Biol 17(3):241–258

Green AL, Fernandes L, Almany G, Abesamis R, McLeod E, Aliño PM, White AT et al (2014) Designing marine reserves for fisheries management, biodiversity conservation, and climate change adaptation. Coast Manag 42(2):143–159. https://doi.org/10.1080/08920753.2014.877763

Gustavsson M, Morrissey K (2018) The Varying Economic Impacts of Marine Spatial Planning across Different Geographical Scales: A Q Methodology Study. In: Heidkamp CP, Morrissey J (eds) Towards Coastal Resilience and Sustainability. Taylor and Francis, London

Hamilton RJ, Potuku T, Montambault JR (2011) Community-based conservation results in the recovery of reef fish spawning aggregations in the coral triangle. Biol Conserv 144(6):1850–1858. https://doi.org/10.1016/j.biocon.2011.03.024

Hoel AH, Olsen E (2012) Integrated ocean management as a strategy to meet rapid climate change: the norwegian case. Ambio 41:85–95. https://doi.org/10.1007/s13280-011-0229-2

Hou, Y.T., Y.D. Xu and X.Z. Xue. 2019. "The Evolvement of ICM Practices in Xiamen: Experiences and Challenges." Case study of task team 1 report on integrated and ecosystem-based ocean management of The China Council for International Cooperation on Environment and Development (CCICED) Special Policy Study on Global Ocean Governance and Ecological Civilization.

Horigue V, Aliño PM, White AT, Pressey RL (2012) Marine Protected Area Networks in the Philippines: Trends and Challenges for Establishment and Governance. Ocean Coastal Management 64:15–26. https://doi.org/10.1016/j.ocecoaman.2012.04.012

IOC-UNESCO (Intergovernmental Oceanographic Commission of UNESCO) (2017) In: Valdés L et al (eds) Global Ocean Science Report: The Current Status of Ocean Science around the World. UNESCO Publishing, Paris

ITTXDP (Integrated Task Team of the Xiamen Demonstration Project). 1996. "Coastal Environmental Profile of Xiamen." In MPPEAS Technical Report. No. 6. Quezon City, Philippines: GEF/UNDP/IMO Regional Programme for the Prevention and Management of Marine Pollution in the East Asian Seas.

Katona S, Polsenberg J, Lowndes J, Halpern BS, Pacheco E, Mosher L, Kilponen A et al (2017) Navigating the Seascape of Ocean Management: Waypoints on the Voyage toward Sustainable Use. Preprint, MarXiv. https://doi.org/10.31230/osf.io/79w2d

Khan A, Amelie V (2015) Assessing climate change readiness in seychelles: implications for ecosystem-based adaptation mainstreaming and marine spatial planning. Reg Environ Chang 15(4):721–733. https://doi.org/10.1007/s10113-014-0662-4

Klinger DH, Eikeset AM, Davíðsdóttir B, Winter AM, Watson JR (2018) The mechanics of blue growth: management of oceanic natural resource use with multiple, interacting sectors. Mar Policy 87:356–362. https://doi.org/10.1016/j.marpol.2017.09.025

Kvalvik I (2011) Managing institutional overlap in the protection of marine ecosystems on the high seas. the case of the North East Atlantic. Ocean Coastal Management 56:35–43. https://doi.org/10.1016/j.ocecoaman.2011.09.009

Lin T, Xue XZ, Shen S, Lu CY (2005) Systematic analysis of coastal wetland changes and their ecological impacts: a case study in Xiamen, China. Environ Inform Arch 3:137–145

Lubchenco J, Sutley N (2010) Proposed U.S. Policy for Ocean, Coast and Great Lakes Stewardship. Science 328:1485–1486

Lubchenco J, Cerny-Chipman EB, Reimer JN, Levin SA (2016) The right incentives enable ocean sustainability successes and provide hope for the future. Proc Natl Acad Sci U S A 113(51):14507–14514

Maliao RJ, Pomeroy RS, Turingan RG (2009) Performance of Community-Based Coastal Resource Management (CBCRM) Programs in the Philippines: a meta-analysis. Mar Policy 33(5):818–825. https://doi.org/10.1016/j.marpol.2009.03.003

Mao Z, Kong H (2018) What Results in the Success of Xiamen's ICM Practices: A New Study of the ICM System from the Perspective of 'Paradigm Shift. Ocean Coastal Management 207:104672. https://doi.org/10.1016/j.ocecoaman.2018.12.001

Mid-Atlantic Regional Planning Body. 2016. Mid-Atlantic Regional Ocean Action Plan (OAP). Via Bureau of Ocean Energy Management, U.S. Department of the Interior. https://www.boem.gov/Ocean-Action-Plan/

McCann, J. 2010. Rhode Island Ocean Special Area Management Plan. Volume 1. University of Rhode Island Coastal Resources Center/Rhode Island Sea Grant College Program. Prepared for Rhode Island Coastal Resources Management Council

McClanahan TR, Graham NAJ, MacNeil MA, Cinner JE (2015) Biomass-Based Targets and the Management of Multispecies Coral Reef Fisheries. Conserv Biol 29(2):409–417. https://doi.org/10.1111/cobi.12430

Mercer J, Kelman I, Alfthan B, Kurvits T (2012) Ecosystem-based adaptation to climate change in caribbean small island developing states: integrating local and external knowledge. Sustainability 4(8):1908–1932. https://doi.org/10.3390/su4081908

Mills M, Pressey RL, Weeks R, Foale S, Ban NC (2010) A mismatch of scales: challenges in planning for implementation of marine protected areas in the coral triangle. Conserv Lett 3(5):291–303. https://doi.org/10.1111/j.1755-263X.2010.00134.x

Muallil RN, Deocadez MR, Martinez RJS, Campos WL, Mamauag SS, Nañola CL, Aliño PM (2019) Effectiveness of Small Locally-Managed Marine Protected Areas for Coral Reef Fisheries Management in the Philippines. Ocean Coastal Management 179:104831. https://doi.org/10.1016/j.ocecoaman.2019.104831

Mulvaney KK 2013. First biennial assessment of the rhode island ocean special area management plan process. Prepared for the Rhode Island Coastal Resources Management Council and the University of Rhode Island Coastal Resources Center.

NEAFC and OSPAR (North East Atlantic Fisheries Commission and OSPAR Commission). 2014. "Collective Arrangement between Competent International Organisations on Cooperation and Coordination Regarding Selected Areas in Areas beyond National Jurisdiction in the North-East Atlantic." https://www.ospar.org/documents?v=33030

Northeast Ocean Data. 2009. "Northeast Ocean Data Portal." www.northeastoceandata.org
Northeast Regional Planning Body. 2016. Northeast Ocean Plan. https://neoceanplanning.org/plan/.
Norwegian Ministry of the Environment (2002) Protecting the Riches of the Seas. Report No. 12 to the Storting (2001–2002). Government of Norway, Oslo
Norwegian Ministry of the Environment (2006) Integrated Management of the Marine Environment of the Barents Sea and the Sea Areas off the Lofoten Islands. Report No. 8 to the Storting (2005–2006). Government of Norway, Oslo
Norwegian Ministry of the Environment (2009) Integrated Management of the Marine Environment of the Norwegian Sea. Report No. 37 (2008–2009) to the Storting. Government of Norway, Oslo
Olsen E, Fluharty D, Hoel AH, Hostens K, Maes F, Pecceu E (2014) Integration at the Round Table: Marine Spatial Planning in Multi-Stakeholder Settings, edited by J. Hewitt. PLoS One 9(10):e109964. https://doi.org/10.1371/journal.pone.0109964
Olsen E, Holen S, Hoel AH, Buhl-Mortensen L, Røttingen I (2016) How integrated ocean governance in the barents sea was created by a drive for increased oil production. Mar Policy 71:293–300. https://doi.org/10.1016/j.marpol.2015.12.005
Oregon State University, IUCN World Commission on Protected Areas, Marine Conservation Institute, National Geographic Society, and UNEP World Conservation Monitoring Centre. 2019. An Introduction to The MPA Guide. https://www.protectedplanet.net/c/mpa-guide
PEMSEA (Partnerships in Environmental Management for the Seas of East Asia) (1998) The Integrated Coastal Management in Xiamen. China Ocean Press, Beijing. (In Chinese)
PEMSEA (2006a) Xiamen: An ICM Journey. In: 2nd edition PEMSEA Technical Report No. 18. GEF/UNDP/IMO PEMSEA, Quezon City, Philippines
PEMSEA (2006b) A Perspective on the Environmental and Socioeconomic Benefits and Costs of ICM: The Case of Xiamen, PR China. In: PEMSEA Technical Report. No. 17. GEF/UNDP/IMO PEMSEA, Quezon City, Philippines
Peng B, Hong H, Xue X, Jin D (2006) On the Measurement of Socioeconomic Benefits of Integrated Coastal Management (ICM): Application to Xiamen, China. Ocean Coastal Management 49(3):93–109. https://doi.org/10.1016/j.ocecoaman.2006.02.002
Pilling GM, Apostolaki P, Failler P, Floros C, Large PA, Morales- Nin B, Reglero P et al (2008) Assessment and Management of Data-Poor Fisheries. In: Payne A, Cotter J, Potter T (eds) Advances in Fisheries Science. Blackwell Publishing Ltd., Oxford, UK, pp 280–305. https://doi.org/10.1002/9781444302653.ch12
Santos CF, Ehler CN, Agardy T, Andrade F, Orbäck MK, Crowder LB (2019) Marine Spatial Planning. In: Sheppard C (ed) World Seas: An Environmental Evaluation, Volume III: Ecological Issues and Environmental Impact. Academic, Cambridge, MA, pp 571–592
SMSP Initiative (Seychelles Marine Spatial Plan Initiative). Last updated 2019. "The Initiative." https://seymsp.com/the-initiative/.
Su J, Peng B (2018) Evaluating the Trade-Offs between Alternative Coastal Policies: Evidence from Xiamen's ICM Programme. Ocean Coastal Management 207:104516. https://doi.org/10.1016/j.ocecoaman.2018.05.012
Tuda A, Stevens TF, Rodwell LD (2014) Resolving coastal conflicts using marine spatial planning. J Environ Manag 133:59–68
Underdal A (1980) Integrated Marine Policy: What? Why? How? Mar Policy 4:159–169
United Nations. 1982. United Nations Convention on the Law of the Sea.
United Nations. 1994. Agreement Relating to the Implementation of Part XI of the United Nations Convention on the Law of the Sea of 10 December 1982.
United Nations. 1995. The United Nations Agreement for the Implementation of the Provisions of the United Nations Convention on the Law of the Sea of 10 December 1982 Relating to the Conservation and Management of Straddling Fish Stocks and Highly Migratory Fish Stocks.
United Nations. 2015. "The First Global Integrated Marine Assessment." Regular Process for Global Reporting and Assessment of the State of the Marine Environment, Including Socioeconomic Aspects. https://www.un.org/regularprocess/content/first-world-ocean-assessment.
UN Global Compact. 2019. "Homepage." Accessed November 1. https://www.unglobalcompact.org/.
Wang J, Hong HS, Zhou LM, Hu JY, Jiang YW (2013) Numerical Modeling of Hydrodynamic Changes Due to Coastal Reclamation Projects in Xiamen Bay, China. Chin J Oceanol Limnol 31:334–344
Wang Q, Li Y, Li Y (2018) Realizing a new resilience paradigm on the basis of land-water-biodiversity Nexus in a Coastal City. Ocean Coastal Management 207:104603. https://doi.org/10.1016/j.ocecoaman.2018.09.004
Weeks R, Pressey RL, Wilson JR, Knight M, Horigue V, Abesamis RA, Acosta R et al (2015) Ten things to get right for marine conservation planning in the coral triangle [Version 3; Peer Review: 2 Approved]. F1000Res 3:91. https://doi.org/10.12688/f1000research.3886.3
White C, Halpern BS, Kappel CV (2012) Ecosystem service tradeoff analysis reveals the value of marine spatial planning for multiple ocean uses. Proc Natl Acad Sci 109(12):4696. https://doi.org/10.1073/pnas.1114215109
XDPO (Xiamen Demonstration Project Office) (1998) Integrated Coastal Management in Xiamen, 1994–1998. In: Regional programme for the prevention and management of marine pollution in the East Asian Seas. GEF/UNDP/IMO, Quezon City, Philippines
Xue X, Hong H, Charles AT (2004) Cumulative environmental impacts and integrated coastal management: the case of Xiamen, China. J Environ Manag 71(3):271–283. https://doi.org/10.1016/j.jenvman.2004.03.006

Illegal, Unreported and Unregulated Fishing and Associated Drivers

15

Highlights

- By agreeing to Sustainable Development Goal (SDG) 14.4 to end illegal, unreported and unregulated (IUU) fishing and overfishing by 2020, countries have acknowledged the importance of combatting IUU fishing. However, they will fall well short of this goal without an immediate, forceful and unified effort.
- Ocean food production is threated by overfishing and habitat destruction often caused by IUU fishing and exacerbated by climate change, which in turn leaves coastal communities more vulnerable to the impacts of this loss.
- Continued IUU fishing will deplete fish stocks and destroy habitats, decrease the value of many fisheries, threaten species extinction, disrupt marine food webs, increase food security risks and disrupt coastal communities' social cohesion. Many of these effects are already being felt.
- The worst examples of IUU fishing are often connected to transnational crimes, including human rights abuses, bonded labour, tax evasion, piracy, and drug, arms and human trafficking. Fraudulent papers, hidden ownership and a lack of transparency facilitate extraction of fish in a way that is difficult to track.
- The global fishing fleet is two to three times larger than needed to catch the amount of fish that the ocean can sustainably support. Technological improvements in fish detection and fishing gear have made each vessel more efficient, allowing it to further deplete resources.
- Fish harvested legally and sustainably can be properly managed to provide animal protein for generations to come. Improved management and ending IUU fishing are imperative to achieving this.
- The three main drivers of IUU fishing are **economic incentives** that make IUU fishing a low-risk, highprofit activity; **weak governance** that fails to enact or live up to fisheries management regulations; and **barriers to enforcement** of fishing regulations caused by lack of political will, lack of enforcement capacity, and sometimes corruption.
- The illegal fishery is highly profitable because it is not effectively regulated. Economic incentives can be changed if countries and the fishing industry insist on transparent tracking of vessels and of their catches throughout the value chain to document legality. This type of documentation is now possible with new technology. Governments must follow through with tough responses to violations.
- Weak governance at the national, regional and international level creates a regulatory patchwork that has allowed IUU fishing to flourish. Ocean governance mechanisms can be strengthened by addressing the non-uniformity of regional fishery management organisation regulations and improving coordination and data transparency between these organisations, flag states, regional bodies, scientific establishments and coastal and market states.
- Barriers to enforcement must be removed by building the capacity of developing coastal states to enforce regulations at ports and on the water, including adopting new tracking technologies.
- This paper offers a checklist of actions that could be taken by various stakeholders to regulate IUU fishing and could be submitted in support of the SDG 14.4 in 2020.
- The most urgent action opportunities are:
  - **Adopt global transparency in fisheries.** Technological advances in tracking methods—both for tracking fishing vessel movements and for tracking a fish catch through the value chain—offer new hope for fisheries management. This, applied to existing regulations and combined with better public understanding of which vessels are authorised to transship or fish and where, will drive better compliance.

Originally published in:
S. Widjaja, T. Long, H. Wirajuda, et al. 2020. *Illegal, Unreported and Unregulated Fishing and Associated Drivers.* Washington, DC: World Resources Institute. Available online at www.oceanpanel.org/iuu-fishing-and-associated-drivers.
Reprint by Springer International Publishing (2023) with kind permission.
Published under license from the World Research Institute.

J. Lubchenco, P. M. Haugan (eds.), *The Blue Compendium*, https://doi.org/10.1007/978-3-031-16277-0_15

- **Enact tighter controls at ports.** All port states should ratify and implement the Food and Agriculture Organization of the United Nations' (FAO) Port State Measures Agreement (FAO 2009) to stop IUU-caught fish entering the market. The agreement requires parties to place tighter controls on foreign-flagged vessels seeking to use their ports to detect and prevent the trade of IUU products.
- **Enhance collaboration.** Because IUU fishing does not honour political boundaries, regional collaboration among nations is essential. Collaboration between government departments and governments, and also among businesses and financial institutions, scientific establishments and the civil sector, will generate new solutions, maximise impact and lower costs.

1 Introduction

1.1 Overview

Illegal, unreported and unregulated (IUU) fishing accounts for 20% of the world catch and up to 50% in some areas. This industry often uses bonded labour, destructive fishing practices and deceptive practices to reap profits at the expense of local fisheries, coastal states and the marine environment. Although international resolutions and reports have been issued for decades, countries have failed to enact and enforce regulations to stop these practices due to a lack of political will, resources and capacity.

At the beginning of a new decade, with deadlines approaching for the Sustainable Development Goals, nations have a major opportunity to form the partnerships and enforcement mechanisms to stop illegal, unreported and unregulated fishing practices.

Advancing, and more affordable, technologies also present new opportunities to implement and enforce new and old agreements and regulations. These technologies can track not only the location and documentation of fishing vessels but also the progress of a particular fish catch through the value chain to ensure legality.

By exploring the underlying drivers of IUU fishing—economic incentives, weak governance, and poor enforcement—we propose effective actions that can be taken in the current international framework to address the issue. The best use of appropriate technologies, combined with good policy and international cooperation, partnerships and collaboration can be costeffective and scaled globally to transform fisheries. Fish harvested legally and sustainably can provide animal protein for generations to come.

UN Sustainable Development Goal 14.4 commits countries, by 2020, to effectively regulate fishing; end overfishing, IUU fishing, and destructive fishing practices; and implement science-based management plans to restore fish stocks in the shortest time feasible, at least to levels that can produce maximum sustainable yield as determined by their biological characteristics. The world is clearly not yet on track to achieve those goals.

In this paper, we outline the state of knowledge and trends in IUU fishing, ways in which it contributes to overfishing, how it exacerbates the impacts of climate change, and specific aspects of how it operates in coastal areas, on the high seas and in areas beyond national jurisdiction. Successful country strategies are highlighted, that, if more widely adopted, will help transition the IUU fishing fleet to one of compliance.

As one of a series of Blue Papers prepared as an input to the High Level Panel for a Sustainable Ocean Economy, this paper provides scientific and policy background as well as opportunities for action to reinvigorate international cooperation and efforts to effectively regulate IUU fishing.

1.2 Context

The promotion, regulation and monitoring of responsible fishing practices, through robust fisheries management and governance frameworks, are essential for the sustainability of fisheries resources in both coastal areas and high seas. The principles of responsible fisheries management have been prescribed in a number of international ocean and fisheries instruments. However, states do not always satisfactorily fulfil their duties in line with such instruments and IUU fishing often occurs, undermining national, regional and global efforts to manage fisheries sustainably. It is not enough for states to detect IUU fishing; they must strengthen fisheries laws and regulations and be able to take effective action against perpetrators to deter non-compliance (FAO 2018).

Illegal, unreported and unregulated (IUU) fishing threatens the sustainability of global fisheries in national coastal waters and on the high seas (SEAFDEC 2015). Developing countries are most at risk from illegal fishing, with estimated actual catches in West Africa, for example, being 40% higher than reported catches. IUU fishing is widespread (Sumaila et al. 2006), and such levels of exploitation severely hamper the sustainable management of marine ecosystems (Agnew et al. 2009). This exploitation causes enormous economic losses to coastal developing countries, disrupts social and environmental situations in coastal communities, enables crimes and human rights abuses, and even threatens military conflict over scarce resources. It also exacerbates the effects of climate change on ocean resources.

1.3 Economic Losses to States and Communities

One in every five wild-caught fish is likely to be illegal or unreported (Agnew et al. 2009); thus, the economic value of these fish never reaches the communities that are the rightful beneficiaries.

Annual global losses due to this illegal activity are valued at US $10 billion to $23.5 billion, representing 11–26 million tonnes of fish (Agnew et al. 2009). The volume of illegal trade in Pacific Ocean marine resources is 4–7 million tonnes per year, costing Pacific nations $4.3 billion to $8.3 billion per year in loss of gross revenues. The losses substantially increase when impacts across the fish value chain are considered (Konar et al. 2019). Another recent study reveals that IUU fishing in one Pacific Islands region represents $616 million annually, with 276,000–338,000 tonnes of Pacific tuna illegally caught each year (MRAG Asia Pacific 2016).

1.4 Social and Environmental Costs

Although monetary values often take the headlines, illegal fishing has significant social and environmental costs, such as the loss in value of fisheries from depleted stocks, the threat of species extinction, the impact on marine food webs, increased food security risks and potential disruption to coastal communities' social cohesion (Tinch et al. 2008; Sumaila 2018). Moreover, destruction of habitats and overfishing of mature adult fish compromise the health of the ecosystem and therefore the opportunity for fisheries and other uses in the future.

1.5 Crime and Human Rights Abuses

The worst examples of IUU fishing are often connected to transnational crime, including human rights abuses, tax evasion, piracy and drugs, arms, and human trafficking (Sumaila and Bawumia 2014; Telesetsky 2014). Much of today's IUU fishing activity takes place on an organised, systematic scale across multiple jurisdictions (Haenlein 2017). These crimes are net losses to a country's economy and will result in lost economic, environmental and social opportunities, both short term and long term, and may diminish food security (FAO 2002; Sumaila 2018).

1.6 Possibility of Military Conflicts

The military departments of many nations study the environment and resources as an aspect of national security. The UK Ministry of Defence's 2018 *Global Strategic Trends* makes clear that habitat destruction and overexploitation will lead to loss of biodiversity and the need to secure diminishing resources may lead to conflict (DCDC 2018). Long-term strategies to ensure continued production and resource abundance are key for national and regional security.

1.7 Worsening the Effects of Climate Change

Climate change is significantly changing ocean ecology in ways that will affect fisheries, including reducing fisheries catches, especially in the tropics. Unsustainable fishing practices worsen the effects of these negative changes; therefore, sustainable and adaptive fishing management practices should be employed to help mitigate them.

1.8 The International Situation

The need to combat IUU fishing is increasingly enshrined in high-level institutions. See Appendix A for current institutions and agreements. Recent international actions are outlined below.

- A 2018 UN resolution on sustainable fisheries refers to the need to address IUU fishing and the importance of policies available to combat it.
- The 2015 UN Sustainable Development Goals (SDGs) include SDG 14.4 [1] on overfishing. This SDG is unlikely to be implemented by the 2020 deadline.
- The Convention on Biological Diversity, having fallen short on previous commitments, has set clear targets for 2020 on sustainable harvests for many of its goals, and the threat from IUU fishing and overfishing is made clear.
- Declarations on combatting IUU fishing were issued at the G7 (Charlevoix Blueprint 2018) and the G20 (Osaka Leaders' Declaration 2019).
- In 2019, Asia-Pacific Economic Cooperation officials and ministers officially endorsed a roadmap to combat IUU fishing.
- The 2019 UN Framework Convention on Climate Change Conference of Parties, referred to as the Blue COP to underline the emphasis on ocean conservation, sought to offer leadership and financial resources to gather momentum and address ocean issues.

[1] SDG 14.4: By 2020, effectively regulate harvesting and end overfishing, illegal, unreported and unregulated fishing and destructive fishing practices and implement science-based management plans, in order to restore fish stocks in the shortest time feasible, at least to levels that can produce maximum sustainable yield as determined by their biological characteristics.

- An ongoing UN intergovernmental conference is considering a new legally binding instrument under the UN Convention on the Law of the Sea (UNCLOS) that could result in robust protection for marine biological diversity in areas beyond national jurisdiction.
- In December 2017, the UN General Assembly proclaimed 5 June—the date the Port State Measures Agreement (PSMA) came into force—as the international day for the fight against IUU fishing—a significant milestone that raised the importance of the issue to global awareness.

1.9 Significant Reports on IUU Fishing

IUU fishing is recognised as a serious problem that must be addressed through national action and cooperation. Two significant reports have recommended reform: the High Seas Task Force's (2006) Closing the Net and the Global Ocean Commission's (2014) From Decline to Recovery: A Rescue Package for the Global Ocean.

This paper aligns with these reports and with key voluntary guidance such as the Food and Agriculture Organization of the United Nations' (FAO) 2001 International Plan of Action to Prevent, Deter and Eliminate Illegal, Unreported and Unregulated Fishing (IPOA-IUU), which highlighted the need for countries to share information and implement its recommendations either through their fisheries management agencies, regional organisations or the FAO. It also recommended encouraging full participation of stakeholders—including industry, fishing communities and non-governmental organisations (NGOs)—in combatting IUU fishing.

1.10 Need for National Good Governance

Because of the complexity and the scale of IUU fishing, a comprehensive global system of enforcement and compliance is needed to tackle this issue. Although many efforts to stem IUU fishing are international, national governments must initiate much of the action, including regulating their own coastal fisheries, enacting regulations at ports of entry, ratifying international agreements, and employing new tracking and transparency technologies.

1.11 Three Drivers of IUU Fishing

Despite many official statements and reports, and some traction, the problem remains a huge threat to future fisheries, food and social security and healthy ocean ecosystems. This paper assesses the nature of the problem and suggests approaches needed to tackle it.

We assess three major drivers of IUU fishing and offer some actions that would be decisive if adopted by the majority of stakeholders. The three main drivers are the large economic incentives for illegal fishing, weak governance at all levels that creates an easily evadable regulatory patchwork, and barriers to enforcement, mainly the lack of surveillance, monitoring and consistent prosecution of illegal activities.

- **Economic incentives** drive IUU fishing and other illicit activities such as bonded labour. IUU fishing is a low-risk, high-gain activity. Market and government mechanisms that promote traceability and transparency throughout the supply chain can help shift these incentives.
- **Weak governance** at the national, regional and international levels creates a regulatory patchwork that has allowed IUU fishing to flourish. Coordinated ratification and implementation of strong international fisheries governance regimes, such as the FAO's PSMA, can begin to fill these gaps.
- **Barriers to enforcement** stemming from a lack of resources and the logistical difficulties of effective monitoring, control and surveillance (MCS) over vast areas of the ocean undermine attempts to stop IUU fishing. Emerging low-cost, yet powerful technologies for surveillance and catch documentation coupled with regional partnerships offer new opportunities to overcome these barriers. Decisive penalties for violators of fishing regulations must follow.

This paper summarises detailed opportunities for action to combat each driver. Chapter 7 describes some immediate opportunities for action. Appendix A summarises existing fisheries agreements and bodies, and Appendix B offers a list of specific actions for various stakeholders as possible voluntary actions to submit under SDG 14.4.

2 The Need to Combat IUU Fishing

Global fishing is important for food security. This chapter discusses the importance of the global fishery for food security, the scale of the IUU problem and how the illegal fishery operates. See Box 15.1 for the official definition of IUU fishing.

2.1 Trends in the Global Fish Catch and Its Importance to Global Food Security

Reported global fish catch plateaued in the late 1980s and has declined slightly. Adding an estimate for illegal and

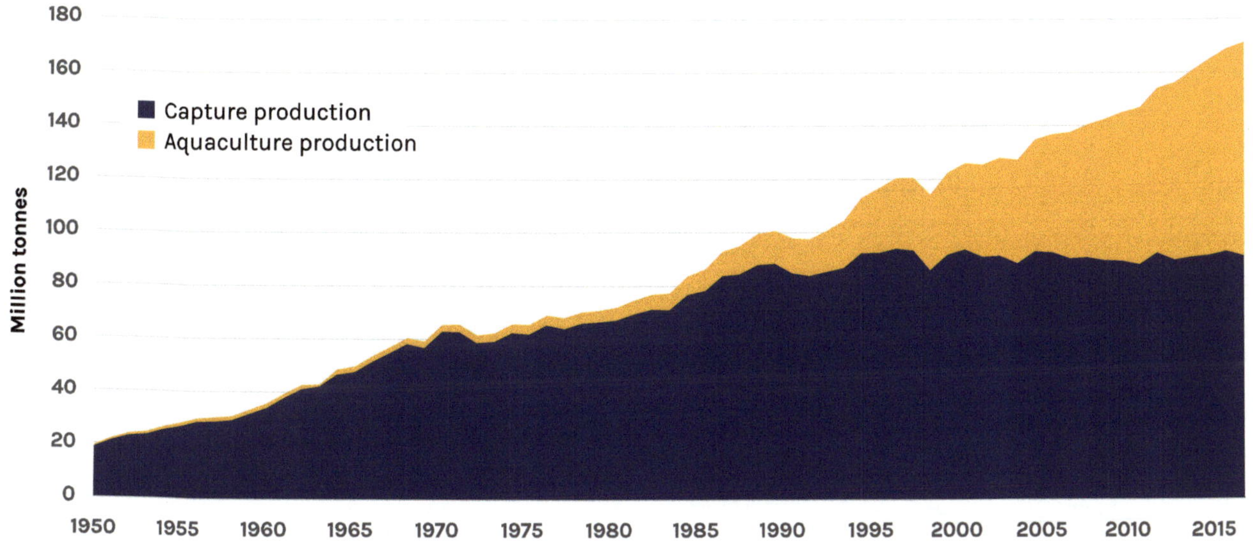

Fig. 15.1 World capture fisheries and aquaculture production, 1950–2016. *Note: Excludes aquatic mammals, crocodiles, alligators and caimans, seaweeds and other aquatic plants. (Source:* FAO 2018*)*

unreported catches significantly boosts the catch estimate but indicates that the global fishery may be in decline. Aquaculture is expanding but it relies heavily on capture fisheries for feed. Meanwhile fish consumption is rising at a rate faster than population growth. Fish is an excellent source of nutrition, and preferable to many other protein sources for climate change mitigation.

2.1.1 Estimated Global Fish Catch

The FAO reported in its 2018 *State of World Fisheries and Aquaculture* report that global fish [2] production peaked in 2016 at 171 million tonnes (with aquaculture representing 53% of this amount), which was valued at $362 billion (FAO 2018). Wild-caught fisheries increased steadily since 1950 but broadly plateaued in the late 1980s (Fig. 15.1). Since 1974, the percentage of underfished stocks has continued to reduce, and today only 7% of the world's fisheries are underfished while 33% are overfished.

To add estimated illegal catches to the FAO reported data, Pauly and Zeller (2016) reconstructed unreported fish catches from 1950 to 2010. Their results showed not only the enormity of the illegal catch but the possibility that due to sustained overfishing and under-reporting of catches, global marine fisheries catches are declining much more than FAO data indicates (Fig. 15.2). This decline in reconstructed catches reflects declines in industrial catches and to a smaller extent a decline in discarded catches, despite the expansion of industrial fishing from industrialised countries to the waters of developing countries. The differing trajectories suggest a need for improved monitoring of all fisheries, including often-neglected small-scale fisheries, and illegal and other problematic fisheries, as well as of the discarded bycatch.

2.1.2 Aquaculture Depends on Capture Fisheries

With capture fishery production relatively static or declining since the 1980s, aquaculture has continued impressive growth to supply fish for human consumption. Yet this sector remains heavily dependent on wild-caught fish as feed and therefore remains inextricably linked to the issues of IUU fishing and threats to the ocean ecosystem through heavy extraction of 'trash fish'; fish too young or too small for human consumption. An estimated 63% of all wild-caught forage fish are used for fish meal and account for upwards of 1 trillion fish taken from the ocean annually, valued at $17 billion (Pikitch et al. 2012). Significant expansion of fed mariculture systems, seen as necessary to producing more food, is predicated on major innovations in feed to make production less dependent on capture fisheries. Until that innovation is achieved, management of capture fishery resources will be critical to ensuring environmental and economic sustainability as well as food security (Costello et al. 2019).

2.1.3 Fish Consumption Expands Beyond Population Growth

Fish consumption is increasing at 3.2% annually, outpacing the global population growth of 1.6% (Fig. 15.3).

[2] The term fish in reference to the FAO's State of World Fisheries and Aquaculture indicates fish, crustaceans, molluscs and other aquatic animals but excludes aquatic mammals, crocodiles, caimans, seaweeds and other aquatic plants. Global fish production peaked in 2016 at about 171 million tonnes, with aquaculture representing 47% of the total and 53%, if non-food uses (including reduction to fishmeal and fish oil) are excluded.

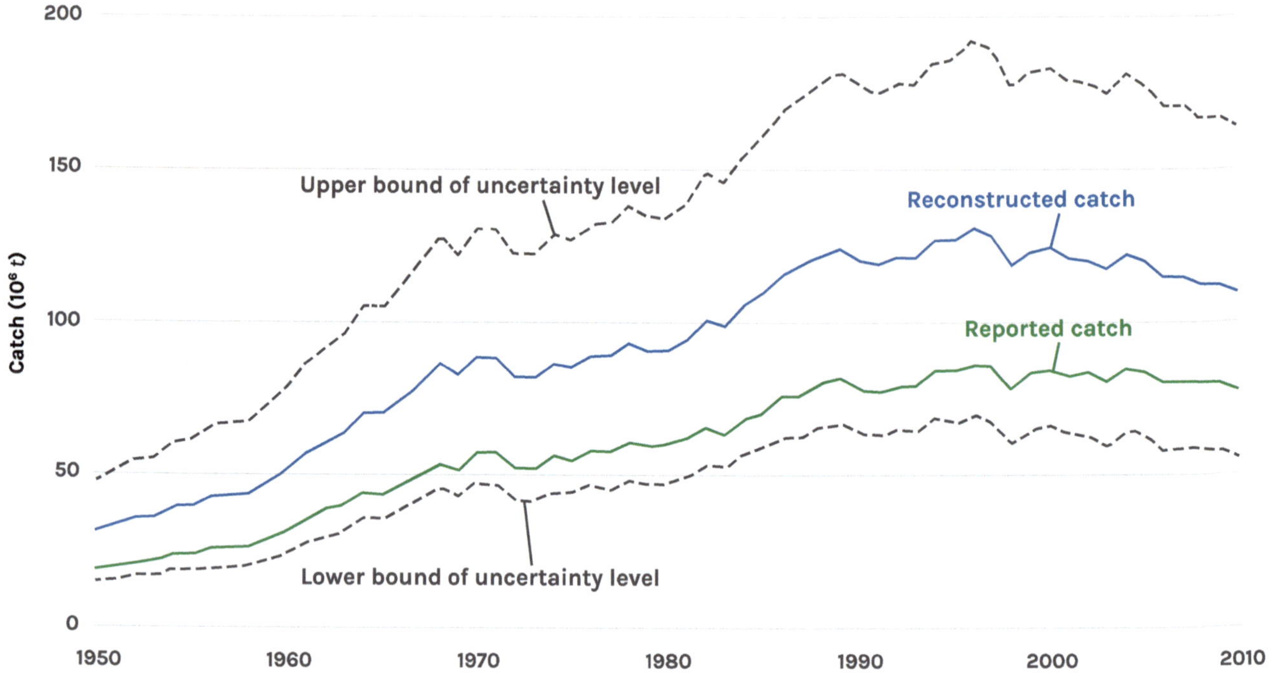

Fig. 15.2 Trajectories of reported and reconstructed marine fisheries catches, 1950–2010. (*Source:* Pauly and Zeller 2016)

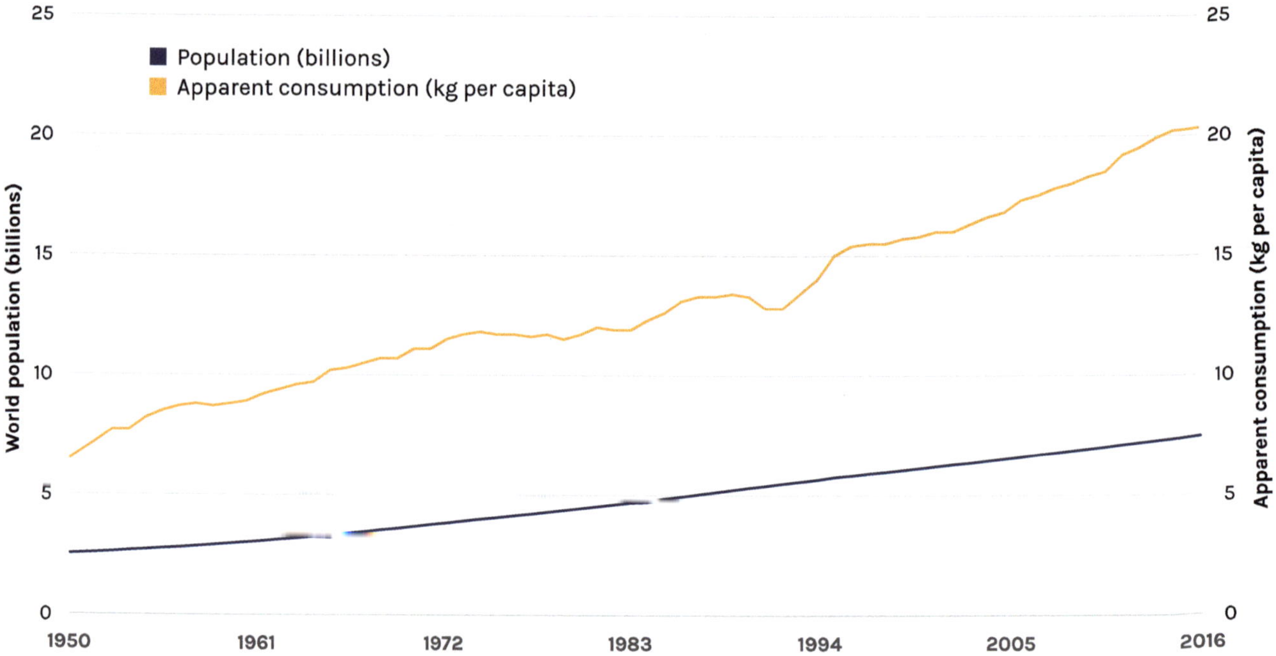

Fig. 15.3 Increase in world population and apparent consumption of fish, 1950–2016. (*Source:* FAO 2018)

Urbanisation is leading to increased fish consumption (Béné 2015), especially in the emerging Asian economies. Asia's share of world fish consumption increased from 67% in 2008 to 70% in 2013 (FAO 2017). Many Southeast Asian markets, including Indonesia and the Philippines, prefer fish to other types of animal protein. Total global consumption of seafood is projected to increase by 20% (30 million tonnes) by 2030, with most of the increased demand coming from developing nations in Latin America, Africa, Oceania and Asia (FAO 2018).

Fish provide almost 20% of the average per capita intake of animal protein globally. The percentage is higher in some areas, such as in Bangladesh, Cambodia, the Gambia, Ghana, Indonesia, Sierra Leone, Sri Lanka and some small island developing states, where over 50% of people's animal protein intake can come from fish (Fig. 15.4).

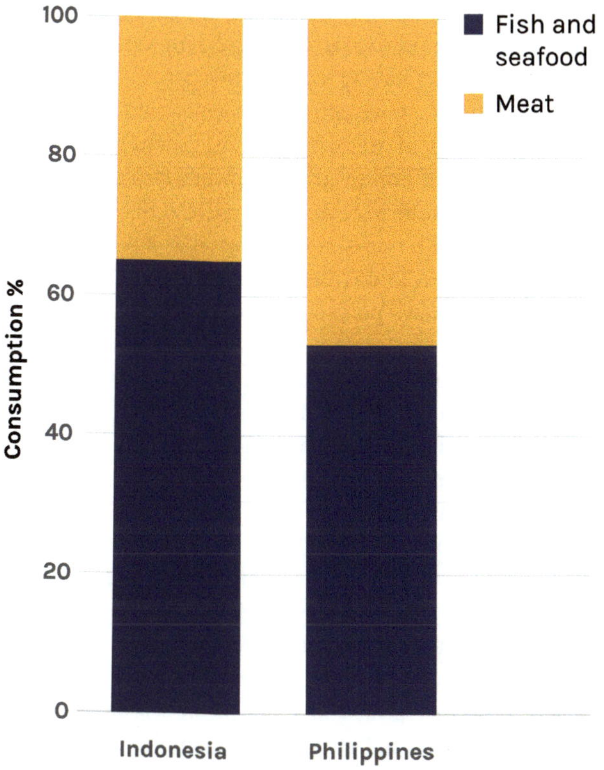

Fig. 15.4 Comparisons of meat and seafood per capita consumption in Indonesia and the Philippines, 2014. (*Source:* Friend 2015)

2.1.4 The Importance of Seafood for Nutrition and Food Security

Food from the sea is uniquely poised to contribute to food security because fish is a highly efficient form of protein—150 g of fish provide 50–60% of an adult's daily protein requirement (FAO 2018). Fish is also a unique source of essential nutrients, including omega-3 fatty acids, iodine, vitamin D, and calcium. Fish consumption by expectant mothers aids their children's neurodevelopment and these proteins and nutrients remain particularly crucial in the first 2 years of a child's life (FAO 2018).

Shifting diets towards fish consumption could also help mitigate climate change because ocean-based proteins are substantially less carbon intensive than land-based proteins (especially beef and lamb). It is estimated that shifting diets to ocean-based proteins can result in mitigation potential of 0.24–0.84 Gt of carbon dioxide equivalent (GTCO2e) per year in 2030 (Hoegh-Guldberg 2019).

IUU fishing impacts fisheries management because it skews scientific data, making assessments of fish stocks unreliable and therefore risking the collapse of the populations of overfished species. Fish harvested legally and sustainably can be properly managed to provide animal protein for generations to come. With judicious conservation and improved management, capture fisheries could produce as much as 20% more catch than today and up to 40% more than projected future catch under current fishing pressures (Costello et al. 2019).

2.2 The Scale of IUU Fishing

IUU fishing, as the name implies, is fishing that is illegal, unreported and/or unregulated (Box 15.1). Estimating the scale of IUU fishing is challenging. The most recent global estimate suggests that the global illegal and unreported annual catch is between 11 million and 26 million tonnes of fish, with a value of $10 billion to $23 billion. IUU fishing accounts for up to 20% of the world's catch and as much as 50% in some fisheries. This often-quoted report is based on data from 2005 and focuses on industrial-scale fisheries (Agnew et al. 2009).

A more recent regional study of the western and central Pacific Ocean found lower estimates of IUU fishing volume and value: the value was estimated at $707 million to $1.56 billion (MRAG Asia Pacific 2016). Whereas Agnew et al. (2009) made regional and global estimates across a suite of species (e.g., demersal fish, shrimp), which included parts of Indonesia and the Philippines (across FAO Area 71), MRAG Asia Pacific (2016) focused on tuna and included Indonesia and the Philippines; thus, direct comparisons should be made with caution.

The FAO is considering developing a new report on the state of IUU fishing. The first step was to review studies estimating IUU fishing and methods they used (Macfadyen et al. 2016). The decision to go ahead with a new report was pending as of late 2019.

Box 15.1 Definition of Illegal, Unreported and Unregulated Fishing

The IPOA-IUU provides the following definition of IUU fishing.

Activities are classified as **illegal** fishing if they are:

- conducted by national or foreign vessels in waters under the jurisdiction of a State, without the permission of that State, or in contravention of its laws and regulations;
- conducted by vessels flying the flag of States that are parties to a relevant regional fisheries management organization but operate in contravention of the conservation and management measures adopted by that organization and by which the States are bound, or relevant provisions of the applicable international law; or

in violation of national laws or international obligations, including those undertaken by cooperating States to a relevant regional fisheries management organization.

Unreported fishing refers to activities which:

have not been reported, or have been misreported, to the relevant national authority, in contravention of national laws and regulations; or

are undertaken in the area of competence of a relevant regional fisheries management organization which have not been reported or have been misreported, in contravention of the reporting procedures of that organization.

Finally, a catch is considered **unregulated** if fishing is conducted:

in the area of application of a relevant regional fisheries management organization that is conducted by vessels without nationality, or by those flying the flag of a State not party to that organization, or by a fishing entity, in a manner that is not consistent with or contravenes the conservation and management measures of that organization; or

in areas or for fish stocks in relation to which there are no applicable conservation or management measures and where such fishing activities are conducted in a manner inconsistent with State responsibilities for the conservation of living marine resources under international law (FAO 2001).

2.3 The Modus Operandi of IUU Fishing

The perpetrators of IUU fishing tend to follow particular methods to achieve their goals. Identifying these practices and patterns can help monitoring authorities or law enforcement officers detect IUU fishing or other crimes. These include moving the catch from one vessel to another at sea (transshipment), using flags of convenience or non-compliance, using ports of convenience which offer little inspection, deactivating vessel monitoring or automatic identification and tracking systems, using a complex network of ownership, carrying fraudulent ship's documents and maintaining poor conditions for the ship's crew. IUU fisheries operate mainly offshore but have also infiltrated small artisanal fisheries.

2.3.1 Transshipment: Moving the Catch from Vessel to Vessel

Transshipment—moving the fish catch from one vessel to another, at sea or in port—is a common practice in the global fishing industry. At sea, transshipment facilitates the efficient delivery of fish to ports while allowing fishing vessels to continue to fish without having to put in to a port.

However, poorly governed transshipment has significant costs. The benefit of transshipment is generally with the industry in terms of improved profits, while the costs are felt most by the legal fishers and the society that owns the resource. Think of a transshipment vessel as a floating port. The fish transferred to this vessel do not always land at the adjacent coastal states port; indeed, the catch is often taken far away, and the economic benefits are not fully felt at the point of catch. In the western and central Pacific Ocean alone, a recent study estimated that more than $142 million worth of tuna and tuna-like products are lost in illegal transshipments each year (MRAG Asia Pacific 2016).

Transshipment is pervasive in high seas fisheries. Both refrigerated cargo vessels and larger fishing vessels transfer fresh catch from thousands of fishing vessels and take it to the first point of landing for onshore processing. Although transshipment touches a wide range of seafood products, tuna makes up a particularly large portion of it, in part because this highly prized fish can be frozen and brought from distant waters and still command high values at the market. Although moving catch from vessel to vessel may seem harmless, a lack of effective monitoring allows bad actors to obscure or manipulate data on their fishing practices, the species or amounts caught and the catch locations, thus badly distorting supply chains and efforts to make them more transparent.

Transshipment allows fishing vessels to remain at sea longer, which means more continuous fishing effort and, ultimately, additional overfishing of vulnerable fish stocks, especially tuna. While at sea, fish are kept in holds for longer periods, leading to co-mingling of IUU and non-IUU caught fish prior to landing and further muddying the traceability of supply chains.

Global Fishing Watch used 5 years (January 2012 to December 2017) of Automatic Identification System data to produce a report which showed that 694 vessels were capable of transshipping fish at sea. It produced a map showing 46,570 instances in which these transshipment capable vessels were going slow enough long enough to make a transshipment and 10,233 instances in which a fishing vessel was in the proximity of a transshipment vessel long enough to engage in transshipments. Figure 15.5 shows a global map of possible ship contacts for transshipment.

2.3.2 Using Flags of Convenience and Flags of Non-compliance

Flags of convenience Under international law, the country whose flag a vessel flies is responsible for regulating and controlling the vessel's activities. Flags of convenience (FOCs) refer to the registration of a vessel to a flag state with no genuine link to the vessel's owners or operators. This

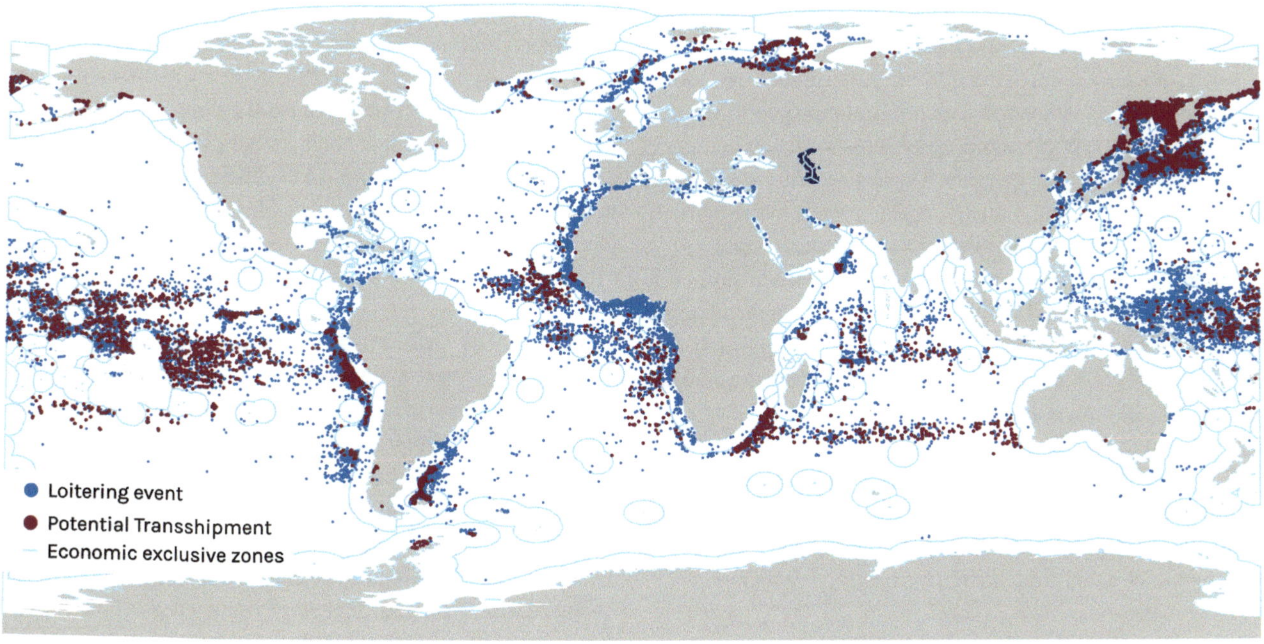

Fig. 15.5 Loitering events and potential transshipment at Sea, January 2012 to December 2017. (*Source: Global Fishing Watch* 2017)

offers competitive advantages to vessel owners, including limited regulatory oversight, ease of registration, reduced taxation, and ability to obscure beneficial ownership. Generally, an FOC country has an open registry, making a business from granting its flags to vessels (including fishing vessels) that are owned by nationals from other states.

FOCs are one of the simplest and most common ways in which unscrupulous fishing operations can circumvent management and conservation measures and avoid penalties for illegal fishing. Fishing vessels can use FOCs to re-flag and change ship names rapidly to confuse management and surveillance authorities, a practice known as flag hopping (EJF 2009). Extensive labour abuses and lack of safety protocols are well documented on FOC vessels (International Transport Worker Federation 2020). The use of FOCs has existed for decades, and the majority of today's merchant marine fleet is flagged to FOC countries (International Chamber of Shipping 2015).

In international waters, measures to regulate fishing apply only to countries that are members of regional fisheries management organisations (RFMOs), but if a vessel re-flags to a state that is not a party to these agreements, it can fish with total disregard for agreed management measures. A 2009 report by the Environmental Justice Foundation shows that RFMO blacklists are dominated by vessels with FOCs or flags unknown (EJF 2009). In recent years, many FOC countries have become members of RFMOs and do abide by the regulations.

Just because a vessel has an FOC does not mean it is fishing illegally, but the proven culpability of many FOC vessels in illegal fishing cases presents a compelling argument for an end to their use by fishing vessels. States that offer FOCs are listed by the International Chamber of Shipping, but many of these flags are not used significantly by fishing vessels. The International Transport Workers' Federation maintains a list focused on labour conditions.

Flags of non-compliance Many flag states without open registries have poor reputations for combatting illegal fishing and associated crimes. These countries often offer 'flags of non-compliance,' which means they grant authorisation to a vessel to fly its flag as well as authorisation to fish, but they lack the resources or the intent to monitor and control these vessels.

The obligations of flag states have been tested. The International Tribunal for the Law of the Sea issued an advisory opinion (April 2015) on Case 21 posed by the West African Sub-Regional Fisheries Commission (SRFC), which stated that a flag state remains obligated for the conduct of a vessel, even when it is in a third country's exclusive economic zone (EEZ), because a flag state 'is under the "due diligence obligation" to take all necessary measures to ensure compliance and to prevent IUU fishing by fishing vessels flying its flag' and that 'the SRFC member states may hold liable the Flag State' (Owen 2016).

Similarly, the International Court of Justice was involved in fisheries cases (1974 and 1998) and remains a jurisdiction to be considered in pursuing IUU fishing violations.

2.3.3 Using Ports of Convenience

In an attempt to avoid adequate inspections, companies engaging in IUU fishing send their vessels to ports with poor or no inspection controls. Ports of convenience are ports where catches can be landed with minimum inspection due to a lack of capacity, poor recording systems for catch landings, or corruption among inspectors. These ports of convenience allow the illegal fishing industry to gain access to the marketplace and to ensure logistical support for their vessels.

Some ports of convenience are also free trade ports (or free economic zones). These zones have favourable customs regulations and little or no controls for landings or transshipment. Illegally caught fish can easily enter the market and be shipped onwards undetected by the flag state or even the port state. The increasing transport of fish by refrigerated container is a growing challenge because these containers often transit through different areas of a port (or separate ports) and are frequently under the jurisdiction of government agencies other than those responsible for fisheries.

2.3.4 Deactivating Vessel Identification and Monitoring Systems

Monitoring is always challenging work in fisheries. The main tools used for monitoring fishing vessel activities are the Vessel Monitoring System (VMS) and the Automatic Identification System (AIS). Yet, IUU fishing vessels often deactivate or manipulate these systems to hide their identity and location.

Many authorities use VMS to combat IUU fishing and manage their fisheries. However, not all countries operate a VMS system, and those that do often do not operate the same system or have information-sharing agreements; thus, information transfer is hampered, does not happen or is costly. Additionally, some flag states fail to uphold their responsibilities to monitor their fleet (see earlier discussion of FOCs). That being said, a properly used VMS system can be very effective.

AIS is a tracking system for ships, designed for collision avoidance. It allows vessels to be 'seen' by each other regardless of their size or the weather conditions. AIS was originally a coastal ground station system limited to coastal zones or between ships at sea, but since the satellite transmission of AIS data became possible, the resulting global coverage made it quickly useful for understanding vessel behaviour with applications beyond collision avoidance.

The AIS data set, however, has limitations. There is no global mandate for fishing vessels to use AIS, although several states, including the European Union, do mandate its use. Vessels that use AIS can turn off or tamper with it. Some fishers turn off their AIS to hide their fishing locations from competitors or because of security or piracy risks in the area. Others turn off their AIS because they are engaging in IUU fishing and wish to hide their activities. Some vessels may simply have malfunctioning AIS devices. The FAO recently published a *Global Atlas of AIS Fishing* which provides further insight in the value of AIS in monitoring fishing activity (Taconet et al. 2019).

2.3.5 Ships Operated and Financed by a Complex Network of Ownership

Investigating and prosecuting illegal fisheries cases throughout the value chain is a complex and resource intensive process (Box 15.2). Because fisheries operations occur in a number of jurisdictions, investigators need close cooperation and information sharing among countries, agencies and relevant international institutions. Unfortunately, most illegal fishing cases at the national level focus on prosecuting only the vessel and its crew for the alleged violation. There is often little effort to identify illegal activities that may have taken place by the same vessel in other jurisdictions or to prosecute the networks, and ultimately the owners, behind these IUU operations. The use of shell companies and joint-venture agreements to 'own' and operate fishing vessels makes it even more complex to identify and target the individuals benefitting from the illegal activities.

A shell company holds funds and manages financial transactions for another company. Although shell companies are legal in many countries, their use can benefit those involved in illegal fishing, both in terms of avoiding taxes and hiding true ownership of fishing operations.

A joint venture agreement is an arrangement in which two entities develop a new entity for their mutual benefit. Joint ventures are used widely, and often perfectly legally, in fisheries around the world. In countries where access to fisheries resources is prioritised for nationals, joint ventures allow foreign players to access the fishery. While many joint-venture agreements are legal, these types of agreements can be exploited to perpetuate illegal fishing. Local partners, who in theory are majority shareholders, can in fact have little say in or control of the fishing operation, and it is frequently unclear how profits are shared. Joint ventures are also linked to corruption to protect vessels and their owners from prosecution and fines.

Box 15.2 The Complex Case of the Fishing Vessel *F/V Viking*

An example of a complex ownership network is found in the operation of the *F/V Viking*. This vessel changed its name 12 times and used 10 different flags. The identity of the owner was well hidden. Interpol and the governments of Indonesia, Norway, Spain and South Africa worked together to uncover the beneficial owners and bring an end to their operation. The *F/V Viking* was caught and sunk in Indonesia in 2016. While the *F/V Viking* was operating, its catches were exported to Hong Kong, Malaysia, Vietnam and Taiwan through an investment company in South Africa. The *F/V Viking* was owned by Spanish and Panamanian companies and operated by an agent in Seychelles and Southeast Asia, which domiciled in several countries. The *F/V Viking* was also operated from Singapore for crewing, logistics and financing (Fig. 15.6).

Honduras	Flag 8	**Nigeria**	1 company listed as owner	**Vietnam**	1 company as buyer
Panama	1 company listed as owner of *Viking*	**Libya**	Flag 10	**Hong Kong**	1 company listed as owner
Chile	Captain's nationality	**Seychelles**	1 company listed as owner	**Mongolia**	Flag 7 & 9
Uruguay	Flag 3	**Australia**	Agent operator's nationality	**North Korea**	Flag 5
Togo	Flag 2 & 6	**Indonesia**	Manning agency	**Japan**	Flag 1
Equatorial Guinea	Flag 4	**Singapore**	Operational site of the agent and where they were arrested		

Fig. 15.6 Travels of the *F/V Viking*. (*Source: Task Force 115* 2018)

2.3.6 Use of Fraudulent Documents and Vessel Identification

Documents carried on a fishing vessel should provide information about the vessel's identity, registration, physical characteristics, authorised fishing activities and authorised locations, and whether it has been certified as compliant with required safety regulations. However, documents can be altered, replicated or obtained by illegal means, so it is essential that vessel documents are systematically verified as part of the fisheries MCS process.

False documents are used to hide illegal activities or to avoid obligations and costs. False vessel registration certificates, fishing licenses or catch certificates are key elements in many cases of illegal fishing; illegal operators either alter existing documents or create forged documents. There are several reasons an illegal fishing operator may use false documents, including for the following purposes:

- **Concealing the true identity of a vessel.** For example, to cover up a history of illegal activity or unpaid fees, a vessel owner may falsify the vessel's identity when seeking port access, registration or licensing by using false documents.
- **Avoiding complying with safety regulations.** For example, false certificates may be used to conceal that a vessel has not passed safety inspections or that safety equipment may be out of date or not suitable for the vessel size or type.
- **Avoiding or underpaying fees.** For example, the cost of vessel registration, licensing, port access and other services is often linked to vessel capacity or size, with vessels in a larger tonnage or length category paying more.
- **Gaining illegal access to resources or services.** For example, inshore fisheries waters may be reserved for smaller vessels, or authorisation to transship may be granted only to vessels of a certain type or carrying a certain fishing license.
- **Gaining additional benefits.** For example a vessel registered with one flag state with genuine documents could remain registered with a second flag state to secure access to benefits available from each flag state, such as access to subsidies under one flag and fisheries access under the other.

False documents can take a number of forms: they can be obtained by deception and/or by corruption; they can be used incorrectly, such as for the wrong vessel; original documents can be doctored; and false documents can replicate authentic ones.

Verification of documents relating to fishing vessels is therefore a key component of fisheries MCS. Documents should be checked and verified when a fishing vessel is inspected during at-sea or port inspections, and when a fishing vessel operator provides documents to an authority; for example, when applying for a fishing license. Flag, coastal, port and market states are all vulnerable to false documents, and all have a responsibility to have robust document verification capacity and routines in place. This should include through visual analysis, cross-referencing of information, and verification by issuing authorities.

Vessel identity fraud is used to hide fishing and operational history and activity; reduce costs; misinform and confuse licensing, flagging and inspection authorities; cover up a history of IUU fishing; and evade sanctions when caught violating regulations or breaking laws. Three main forms of vessel identify fraud are common:

- A vessel may use the genuine identity of another vessel, which results in two or more vessels using the same identity simultaneously—cases have been detected involving up to five vessels using one identity.
- A single vessel may use more than one identity, appearing under different names, flags and so on in different jurisdictions and records.
- A vessel may use a new identity that has not been registered with any national authority.

An individual vessel may be involved in more than one of these forms of identity fraud at any one time. Identity fraud can enable a vessel to operate illegally under the cover of a legal and authorised vessel's identity—for example, a vessel operator might purchase one legal fishing license which is used by the vessel named on the license as well as by several other vessels that assume its identity.

Vessel identity fraud can also involve changes to the physical vessel appearance. Inspectors should look for signs that a vessel's name, call sign and other identifiers have recently been repainted or altered. It is also important to check that any historic name or other identifiers visible under paint on the hull match the name history of the current vessel identity—this can be done through verification with the flag state and using sources such as RFMO authorised vessel lists. The systematic collection and comparative analysis of vessel photographs is also important to detect vessel identity fraud.

2.3.7 Poor Working Conditions and Safety Standards

Sub-standard working conditions and poor safety standards are a hallmark of vessels engaged in IUU fishing. Operators who under-report catch or fish illegally are less likely to provide their crews with adequate labour conditions, training or safety equipment and more likely to fish in hazardous weather. To minimise up-front costs, their vessels might have inadequate equipment or inappropriate modifications and might operate for extended periods without undergoing inspections or safety certifications. International investigations have shown that some migrant workers seeking employment overseas have been tricked with false promises of jobs on land and end up toiling in abhorrent working conditions on board unsafe fishing vessels roaming the high seas (Pew Charitable Trusts 2018).

To ensure the safety of crews on board fishing vessels and improve working conditions, governments can implement the International Labour Organization (ILO) Work in Fishing Convention (C188) and accede to the Cape Town Agreement (CTA). Enforcing labour standards will impact on the profit margins of IUU fishers and drive better behaviour.

The CTA is not yet in force but will set minimum requirements on the design, construction, equipment and inspection of fishing vessels 24 m or longer that operate on the high seas. Its entry into force would empower port states to carry out safety inspections that could be aligned with fisheries and labour agencies to ensure transparency of fishing and crew activities.

2.4 IUU Fishing Invades Small-Scale and Artisanal Fisheries

IUU fishing also has significant impacts in small-scale and artisanal fisheries. The challenges faced by these fisheries are different from those of industrial and high seas fisheries. In the first place, most small-scale fisheries are open access in nature, and entry into the fishery remains unrestricted in many fisheries throughout the world. Fishing restrictions, however, have been implemented in a substantial number of artisanal fisheries, recognising that nowadays these fisheries are capable of generating significant economic exchanges and exert substantial fishing effort.

IUU fishing practices in small-scale fisheries are common and diverse, including the use of dynamite and other explosives and poisons to kill fish; the use of fine mesh fishing nets and other destructive gears, methods and techniques; the use of traps and weirs; the destruction of mangroves and coral reefs; and the catch of juvenile and immature fish and invertebrates, among others.

One clear indication of the lack of regulations or enforcement is growing evidence that worldwide, small unassessed stocks are in substantially worse condition than large assessed stocks. Local decline of small-scale stocks has been related to IUU fishing practices. Recent studies in West Africa showed a tenfold increase in fishing effort in the past 50 years, correlated with a onethird decline in the catch per unit of effort (Belhabib et al. 2017).

In high-value species targeted by artisanal fisheries, such as abalone, lobster, swim bladder and bêche-de-mer, the illegal catch may be higher than the legal catch, and this has been identified as the main cause of fisheries collapse (Hilborn et al. 2005).

Box 15.3 Regulations Fail to Stop Illegal Fishing for High-Value Chilean Loco

Artisanal fisheries are socially and economically important in Chile, engaging nearly 90,000 fishers. The existence of 2 management strategies regulating the exploitation of most benthic resources for the past 15 years allows comparisons of fishing mortality under co-management in the novel territorial use rights for fisheries (TURF) system and under traditional top-down management (e.g., bans, minimum legal size) in open access areas (OAAs). Enforcement is more efficient in TURFs than in OAAs since the fishers themselves have a vested interest in protecting their TURF. Access to fishing grounds and enforcement level seem to be critical factors determining the abundance of exploited resources.

Comparative studies conducted in two management areas of central Chile showed that densities of all benthic resources and coastal fishes were higher in TURFs than in nearby OAAs. The densities of locos (Concholepas concholepas), the most valuable resource, are also significantly higher in co-managed TURFs and Marine Protected Areas than in OAAs, despite loco exploitation being completely banned in OAAs since 1993. This pattern of abundance suggests that illegal fishing of locos seems to occur in OAAs and offers a platform to analyse the extent of illegal fishing in traditional, open access management regimes. To date, illegal fishing of locos has been reported inside TURFs, and it seems to be relevant. Illegal fishing in OAAs may have tremendous impacts on the abundances of locos and other resources, especially considering the large fraction of the coast under an open access regime observing poor enforcement. This is of global interest since the problems are common in artisanal fisheries worldwide (datapoor fisheries, illegal fishing, poor enforcement), highlighting, with precautions, the value of TURFs for management and conservation in such scenarios.

Source: Andreu-Cazenave et al. (2017).

Unreported fish catches are substantial. The consequences are large, considering that artisanal fisheries account for 30% of the world catch and employ 90% of all fishers. Indeed, the recent *Hidden Harvest Report* (Fluet-Chouinard et al. 2018) suggests it is more like 50% of world catch. Moreover, 90% of the landings from artisanal fisheries are currently directed to human consumption, in contrast with 50% of the landings of industrial fisheries.

Artisanal fisheries are an important source of employment and income in the developing world, yet most strategies to overcome the IUU catches concentrate on large stocks and larger ships. While this focus will have a positive impact where large and small vessels are targeting the same stock, there is a need to more closely monitor small-scale fleets and improve small-scale fishing behaviour. The cost of technology needed to track smaller vessels currently inhibits widespread use, but costs will drop, and consequently, the number of small vessels being tracked will increase. There is increasing evidence, thanks to the growing number of fisheries projects focused on artisanal fisheries, that smaller-scale fishers want to be tracked to speed up their access to ports and to protect them from unscrupulous larger vessels.

IUU fishing causes a significant threat to global fisheries and to the health of the ocean. It damages legitimate fishing activity and associated livelihoods. Illegal gear impacts biodiversity and abundance, which in turn impacts food and economic security. Plus, unreported IUU fishing skews the scientific data, making sustainable fisheries management difficult to implement.

3 The Impacts of IUU Fishing

3.1 Impacts on Biodiversity

Fishing has had by far the greatest impact on loss of biodiversity in the ocean (IPBES 2019). IUU fishing, which undermines fisheries management and conservation measures, obscures accurate assessment of fish stocks and fishing pressure, which allows stocks to be fished beyond sustainable limits to the risk of collapse.

Before 2015, limited monitoring and control and poor enforcement, exacerbated by open access regimes and subsidies, made Indonesian waters a haven for IUU fishing. The Arafura Sea, one of Indonesia's most productive areas since the 1970s, became a hot spot for illegal shrimp trawling by domestic and foreign fleets. From reporting, it became evident that the populations of shrimps, hair tails, snappers and groupers in the Arafura Sea were steadily declining, as well as the average size of the catches. Similar problems were being seen in the South China Sea and Java Sea.

In central Chile, coastal artisanal fisheries have experienced the illegal and unreported removal of highly valued gastropod (Box 15.3) and a carnivorous fish, which produces a cascading effect on trophic interactions driving a deterioration of natural habitats (kelp forest), affecting species richness and generating unpredictable consequences on the sustainability of several benthic fisheries (Andreu-Cazenave et al. 2017).

A significant court decision further illustrates the damage that can be done to coastal fisheries. From 1987 to 2001, Arnold Bengis, Jeffrey Noll and David Bengis engaged in an elaborate scheme to illegally harvest large quantities of rock lobsters off the south and west coasts of South Africa for export to the United States in violation of both South African and U.S. law. Their actions led to the collapse of the rock lobster fishery, and a New York court ordered a restitution award for what it would cost South Africa to restore the fishery to the level it would have been had the defendants not engaged in overharvesting—almost $30 million (United States v. Bengis 2013).

Impacts on non-target species have been significant. For instance, catch of albatross by IUU toothfish fisheries in Antarctic waters under the Convention on the Conservation of Antarctic Marine Living Resources has been estimated to be one of the primary drivers of decline in some albatross populations (Michael et al. 2017). Some trawling and dredging gear disturbs or destroys seafloor habitat. For example, Clark et al. (2019) found that benthic communities associated with biogenic habitats formed by deep-sea corals and sponges on seamounts are among the most susceptible to fishing impacts because their resilience is apparently very low. The researchers found little recovery 15 years after trawling. Recovery times of benthic communities on seamounts may be on the order of decades—making restoration unachievable in the short term, as well as prohibitively expensive in the deep sea.

The growing literature on the extent of bottom trawling, together with a greater recognition of the potential severity of human impacts in the deep sea and the long recovery periods from such impacts, support the contention that spatial management is likely to be the most effective strategy to conserve benthic communities of seamount ecosystems (Clark et al. 2019).

3.2 Social and Economic Impacts

IUU fishing has economic impacts for fishers and consumers. In the short term, these impacts may be positive because any fish catch brings returns to fishers and cheaper fish to consumers. However, in the medium and long terms, the impacts become negative since a reduction in fish stocks leads to increased fishing costs, higher prices to consumers and economic losses to the tourism sector (Tinch et al. 2008; Zimmerhackel et al. 2018).

Indirect impacts continue across the value chain, which rarely ends when the catch is landed (except for subsistence fisheries where the catch is directly consumed). Instead, value is created each time the fish changes hands; for example, it is sold to markets and resold to consumers or to an intermediary who purchases large quantities for processing and resale to a retail outlet (Dyck and Sumaila 2010). Therefore, a reduction in catch results in a potential loss of added value across the fish value chain in the legitimate formal sector as well as a loss in household income to fisheries workers. For example, illicit trade in the marine resources of West Africa, including illegal fishing practices, is estimated to cost the region nearly $1.95 billion in lost economic impact across the fish value chain and $593 million per year in lost household income (Sumaila 2018). Another study estimated the economic and household income impact losses to be up to $21.1 billion and $5.4 billion respectively per year for the Pacific Ocean (Konar et al. 2019). The same study estimated the loss in tax revenues at $200 million to $1.6 billion per year, which means this money is not available for public spending on, for example, infrastructure, education or health care (Konar et al. 2019).

Decreased fish stocks due to overfishing may lead to social impacts, loss of cohesion and migration away from the coast towards urban areas, ultimately even stretching to conflict over resources (DCDC 2018). More likely is local disorder among fishers as a consequence of a decrease in household income due to reduced catch opportunities and reduced employment. For example, in Sierra Leone, skirmishes between artisanal and larger IUU fishing vessels is common in the inshore area, where trawlers often fish to within 100 m of the shoreline, placing them in direct competition with smaller vessels. Gear conflicts also occur, and there is often damage the artisanal fisher's nets or boats (Drammeh 2000). Another study using data from the Arafura Sea in Indonesia demonstrated that IUU fishing is one of the main drivers of gear conflict and gear loss (Richardson et al. 2018). For another example, see Box 15.4.

IUU fishing has created a culture of non-compliance because it takes advantage of corrupt administrations and exploits weak management regimes, especially those of developing countries that lack the capacity for effective MCS. IUU fishing can take fish from the waters of bona fide fishers, which can lead to the collapse of local fisheries, with small-scale fisheries in developing countries particularly vulnerable. Legal fishers and aquaculture producers, who are forced to compete with the unfair practices of IUU operators, face loss of market share and trade distortions due to the different cost structures of legal and illegal operators (Tinch et al. 2008). Products derived from IUU fishing can find their way into overseas trade markets, thus throttling local food supply (FAO 2018).

Fisheries-related crimes are closely linked with IUU fishing operations. These include forged fishing licenses, tax evasion, money laundering and inappropriate working conditions. More serious crimes, such as drug trafficking, human trafficking, arms trafficking and piracy, are also linked to IUU fishing. For more information, see the forthcoming Blue Paper 'Organised Crimes in Fisheries'.

3.3 Climate Change and Fisheries Management

Climate change is significantly changing ocean ecology in many ways that will affect fisheries, including reducing the fisheries catches, especially in the tropics. Unsustainable fishing practices worsen the effects of these negative changes; therefore, sustainable and adaptive fishing management practices should be employed to help mitigate them.

Box 15.4 IUU Fishers Impinge on Artisanal Fishery in Ghana
Saiko is the local name for illegal fish transshipments in Ghana, where industrial trawlers transfer frozen fish to specially adapted canoes out at sea. It used to be a practice whereby canoes would buy the unwanted bycatch of industrial vessels. However, the practice has developed into a lucrative industry in its own right, for which industrial fishers actively fish. Today, industrial trawlers target not only the demersal (bottom-dwelling) species for which they are licensed but also the same species as the artisanal fishing community, including the severely depleted small pelagics such as sardinella and mackerel. These catches, which often contain juvenile fish, are landed by the saiko canoes for onward sale to local markets. This has severe implications for Ghana's artisanal fishing sector, which is critical to food security and provides significantly more jobs than the saiko industry. Saiko is prohibited under Ghana's fisheries laws, attracting a fine of $100,000 to $2 million. The minimum fine increases to $1 million when catches involve juvenile fish or the use of prohibited fishing gear. Although saiko activities are widespread, there is a very low risk of arrest and sanction. Cases are generally settled through opaque out-of-court settlement processes, and there are no known examples of the minimum fines in the legislation being paid. In addition, most of the industrial vessels engaged in saiko are linked to foreign beneficial owners, which also contravenes Ghanaian law.

From March to September 2018, after a 2-month closed season for industrial trawlers, the government of Ghana intensified enforcement action against saiko, resulting in at least one high-profile arrest. This action led to a notable decline in saiko activities at the major saiko landing site of Elmina in Ghana's Central Region. During this period, trawlers were required to land their bycatch at either the Sekondi or Tema port: this 'official' bycatch was packed in cardboard and labelled with information on the trawler company that caught the fish, for onward transport to local markets.

This allowed government fisheries inspectors to monitor this catch and to check that it did not contain juveniles.

Some confusion now exists on the legality of saiko, and since 2018, landings have re-commenced at Elmina, with up to 15 saiko canoes landing fish each day. The 2010 Fisheries Regulations specifically prohibits the transshipment of fish at sea from Ghanaian industrial vessels to canoes. A recent legal opinion found that since the entry into force of the 2010 Fisheries Regulations, only those forms of transshipment that are not expressly prohibited under the regulations may be considered legal if supervised by an authorised officer. Since saiko is prohibited in the regulations, the legal opinion concluded it cannot be authorised (Fig. 15.7).

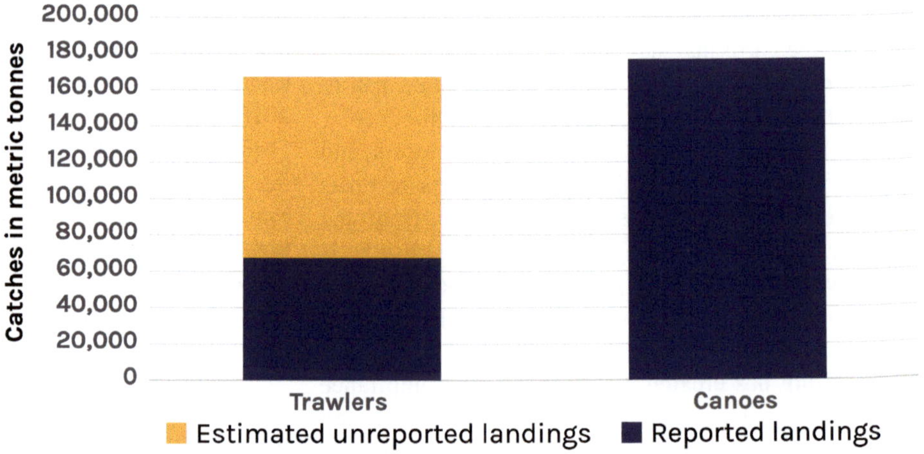

Fig. 15.7 Impacts of Saiko Fishing. (*Source: EJF and Hen Mpoano* 2019)

A recent Intergovernmental Panel on Climate Change (IPCC) special report, Ocean and Cryosphere in a Changing Climate (IPPC 2019), predicted that changes in global temperatures will impact the footprint of fisheries, consequently changing the behaviour of fishers, the incidence of IUU fishing and the connected ocean economy.

The report offered likely changes with low, medium or high confidence at various emissions scenarios. It predicted (with medium confidence) a decrease in the global biomass of marine animal communities, their production, and fisheries catch potential, and a shift in species composition over the twenty-first century from the surface to the deep seafloor under all emission scenarios. The rate and magnitude of decline are projected to be highest in the tropics (high confidence), whereas impacts would be diverse in polar regions (medium confidence). Projected impacts increase in high-emission scenarios. The report states that ocean warming has contributed to an overall decrease in maximum catch potential (medium confidence), compounding the impacts of overfishing for some fish stocks (high confidence; IPCC 2019). As fish stocks decline, fisher competition will likely increase, probably resulting in more IUU fishing (FAO 2018).

As the ocean has warmed, marine fish and invertebrates have shifted to track their preferred temperatures (Perry et al. 2005; Dulvy et al. 2008; Poloczanska et al. 2013; Pinsky et al. 2013). In general, this has resulted in shifts of fish populations poleward and into deeper waters. At a mean rate of 72 km per decade, marine species have been moving an order of magnitude faster than terrestrial species (Poloczanska et al. 2013). This redistribution of marine resources in effect redistributes wealth among coastal states, since some will lose their fisheries while others will gain. New fishing opportunities will emerge, while some established fisheries will be reduced (Bell et al. 2011; Fenichel et al. 2016). Areas under fisheries management could be destabilised, and some fish stocks could shift into unregulated fishing areas (Cheung 2016). This type of change often triggers conflict over access to resources (DCDC 2018).

The IPCC report underlined risks of severe impacts on biodiversity, which are projected to be higher for elevated temperatures under high-emissions scenarios. Projected changes include losses of species habitat and diversity and degradation of ecosystem functions. The capacity of organisms and ecosystems to adapt is higher at lower-emissions scenarios (high confidence). For sensitive ecosystems such as sea grass meadows and kelp forests, high risks are projected if global warming exceeds 2 °C above pre-industrial temperature, combined with other climate-related hazards (high confidence). Warm water corals are at high risk already and are projected to transition to very high risk even if global warming is limited to 1.5 °C (very high confidence).

Reducing the pressure of IUU fishing could lessen ocean degradation. Although nations must take rapid ambitious action to curb climate change, they must also fast-track adaptive fisheries management, along with new agreements that take a system-wide approach to marine resources management and ensure that benefits are shared fairly. Such a response would not only improve resilience and future-proof these vital industries but could also improve profits in some regions (Gaines et al. 2019).

The actions of fishers, management institutions and markets all influence the benefits derived from fisheries (Costello et al. 2016) and could mitigate many of the negative impacts of climate change (Gaines et al. 2018). Gaines et al. (2018) document the benefits of implementing climate-adaptive fisheries management reforms that address both changes in species distribution and productivity caused by climate change (Free et al. 2019). Incentivising cooperation to establish data sharing and collaborative management will require overcoming prevailing management mentalities that one party 'wins' while the other 'loses' when stocks shift across boundaries (Gaines et al. 2019).

This is a point to keep in mind as the UN Intergovernmental Conference negotiates a new legally binding instrument under the UNCLOS that, if successful, will result in robust protection for marine biological diversity in areas beyond national jurisdiction. While negotiating countries have agreed that the new instrument should avoid undermining existing bodies, they remain divided over whether the instrument should address fisheries management directly. Clearly, they should address the impacts of fisheries on biodiversity and are in a position to reduce gaps in governance and change the mentality of 'win-lose'.

Economic incentives are the first of the three drivers of illegal fishing described in Chap. 1. Economics directly drives IUU fishing activities (Sumaila et al. 2006). Increasingly thin margins in legal fishing operations and high potential revenues from IUU fishing are strong motivators for illegal activity (Becker 1968; Gallic and Cox 2006). However, voluntary market instruments supported by governments and new technological capability can begin to reshape these incentives. See also Appendix B for action suggestions.

4 Reframing the Economic Incentives for IUU fishing

4.1 Understanding Economic Incentives for IUU Fishing

In the most straightforward economic terms, IUU fishing is a high-reward, low-risk activity. There are too many highly lucrative opportunities where fishers' expected benefits from breaching regulations—large catches with low operating costs—outweigh the downside, in particular the risk of being detected and suffering any punishment.

4.1.1 Too Much Fishing Capacity

The global fishing fleet has increased, and fish stocks are facing record levels of overexploitation (Watson et al. 2013). The result is a global fishing fleet that is two to three times larger than needed to catch the amount of fish that the ocean can sustainably support (Joseph et al. 2010). At the same time, technological improvements in fish detection and fishing gear have made vessels more efficient, allowing them to further deplete resources (Knauss 2005). Access to deep-sea fisheries resources requires technology only available to industrialised countries with the capacity to build or acquire the large vessels. Developing countries, which are unable to partake in this industry and often have little capacity for enforcement, often see vessels from industrial countries fishing off their own coasts, which has been described predatory behaviour (Hornidge and Hadjimichael 2019).

In 2002, the FAO estimated that $3 million in revenue was gained from 1335 fishing vessels flying FOCs from 21 countries. This value was considered an underestimate because it did not include money gained from franchise/royalty fees or tonnage taxes.

Fishing farther away or differently is not always an alternative for small-scale artisanal fishers, who continue fishing the same overfished stocks, in many cases violating management regulations (e.g., spatial or reproductive bans, minimum legal size).

Without large government subsidies, as much as 54% of the present high-seas fishing grounds would be unprofitable at current fishing rates (Sala et al. 2018). Worldwide, governments spend about $35 billion annually—about 20% of the total value of all marine fish caught at sea and brought to port—to support the fishing sector (Martini 2019; Sumaila et al. 2016). Unfortunately, many of these subsidies are harmful and drive unsustainable practices. For example, the largest reported subsidy is for fuel, and this subsidy is the one most directly linked to overfishing.

Government subsidies and other fishing incentives pad the thin margins in many fisheries. The patterns of fishing profitability vary widely among countries, types of fishing and distance to port. Deep-sea bottom trawling often produces net economic benefits only thanks to subsidies, and much fishing by the world's largest fishing fleets would largely be unprofitable without subsidies and low labour costs (often associated with IUU fishing operations). These results support recent calls for subsidy and fisheries management reforms on the high seas. As of 2019, no real progress to eliminate capacity enhancing subsidies had been made (Sumaila 2019).

4.2 Changing Economic Incentives

If IUU fishing is a low-risk high-gain activity, the response must be to increase the risk of detection and reduce the gain. Regional entities with strategic roles in the seafood market can use market tools to eliminate IUU fishing products entering the region. Governments can build on these tools with sanctions and strict standards for the industry.

4.2.1 Transparent Supply Chains

Voluntary efforts in the seafood industry can achieve clearer supply chains by encouraging transparency and more due diligence in the buying chain. Several organisations—including Sustainable Fisheries Partnership, International Seafood Sustainability Foundation, Seafood Business for Ocean Stewardship, and the Marine Stewardship Council—are examples of such initiatives. These traceability efforts can provide assurance to consumers of the provenance of seafood while creating pressure for other market actors to adopt full supply chain transparency. Financing institutions such as banks and insurance companies can create additional pressure by requiring traceability and transparency as conditions of contracts. There are downsides; for example, they can act as trade barriers to developing countries where the fisheries sector is still evolving and the price tag and complexity of certification is a significant barrier.

4.2.2 Smart New Tracking Technologies

New technologies are making full supply chain traceability more technically and economically viable. For instance, the recent development of blockchain technology promises to allow for full traceability of fish products from their origin to their ultimate fate. Blockchain is an incorruptible, distributed digital ledger of transactions, which allows users to effectively and transparently measure, record and transact value. A 'transaction' can involve contracts, records, currency or almost any other information of value. Through a series of cryptographically secure algorithms, multiple blockchain nodes validate the transaction in a process known as consensus protocol. That transaction is combined in a block of data, which forms a chain of records maintained simultaneously across thousands of distributed nodes, hence forming a 'blockchain' that is permanent and unalterable. In effect, blockchain enables a single source of truth for chain of custody of any commodity along entire supply chains, from producer to consumer. Because of blockchain's immutable and verifiable nature, it offers greater levels of transparency and trust in supply chain transactions. It is a powerful lever to impose on market forces.

4.2.3 Consequences for Bad Behaviour

Governments can support industry action through regulation. For example, the European Union adopted an IUU fishing regulation to limit access of IUU fish to the market. Under this regulation, non-EU countries identified as having inadequate prevention mechanisms for illegal fishing may be issued a formal warning (known as a 'yellow card') to improve the situation within 1 year. If they fail to do so, they face having their fish banned from the EU market (known as a 'red card'), among other measures. The mechanism is seen as a success. As an example, South Korea and the Philippines received yellow cards, and both countries responded by diverting resources to deal with the problem. The cards were lifted following good responses from both countries. South Korea now has sufficient means to proactively prevent, deter and eliminate IUU fishing by closing loopholes in its systems. The Philippines has strengthened its commitment to fighting IUU fishing at the international level by ratifying the UN Fish Stocks Agreement (UNFSA) and initiating procedures to ratify the PSMA (EJF et al. 2015). Another example of an effort to control IUU fishing is given in Box 15.5.

Box 15.5 Increasing the Risk to IUU Fisheries

There is increasing interest in how private companies can reduce IUU fishing by increasing the risk to IUU fishers and their beneficial owners. At the first Ocean Risk Summit, held in Bermuda in 2018, leaders from across the political, economic, environmental and risk sectors identified potential exposures to ocean risk and prepared to generate new and dynamic solutions. As a result, the Ocean Risk and Resilience Action Alliance (ORRAA) was launched. This progressive concept brings together organisations from the public and private sectors and civil society focused on developing risk management strategies to deal with climate change and threats to communities which could include IUU fishing, the presence of bonded labour or other illicit activities. The Canadian government, AXA XL, Willis Towers Watson, the Nature Conservancy and Ocean Unite led the conception and development of ORRAA. Additional partners include the Inter-American Development Bank, Bank of America, KfW, the United Nations Development Programme and Rare. The Stockholm Resilience Center at Stockholm University will be a key scientific and knowledge partner. Companies like Planet Tracker are also driving change by engaging financial markets for sustainable practices across several commodities, including fish.

4.2.4 Corruption Can Undermine the Process

Corruption undermines good governance in fisheries. Vulnerabilities to corruption occur throughout the whole value chain, from rich countries' negotiations of access to territorial waters, to malpractices in fisheries management at the local level. Processors and distributors, and of course fishers themselves, can engage in corrupt practices (Sumaila et al. 2017).

5 Overcoming Weak Governance

5.1 Governance Gaps

IUU fishing is propelled by governance gaps internationally, regionally and domestically. These gaps are exploited by the IUU fishers and create obstacles to enforcement by authorities and fisheries managers. Weak governance is the second of the three drivers of IUU fishing. See also Appendix B for action suggestions.

5.1.1 Governance Problems Persist

The High Seas Task Force's (2006) *Closing the Net Report* found that, despite the appearance of a strong legal framework based on the UNCLOS, there were serious concerns about whether the UNCLOS can deliver an effective management regime and flag states can fulfil their responsibilities. An analysis of the discussions for the report indicated broad agreement on the following main governance-related problems:

- Failure of some states to participate in existing multilateral instruments as a critical constraint to effective implementation and enforcement.
- Inadequate implementation of existing instruments at the regional level, including lack of effective institutional arrangements, conservation and management measures that do not meet the standards set by the existing legal framework, and lack of coordination between regional bodies and inadequate harmonisation of measures.
- Inadequate flag state control over fishing vessels.
- Geographical and structural gaps in the system of high seas governance.
- Subsidies and other perverse signals that displace rather than eliminate unsustainable fishing.

Contributions to this Blue Paper indicate that little has changed in the 13 years since the High Seas Task Force report, but many of the tools predicted to be available are now mature and ready to be applied. The 2014 Global Ocean Commission report also named poor governance as a driver of decline in the ocean, stating, 'The existing high seas governance framework is weak, fragmented and poorly implemented. Different bodies regulate different industries and sectors, and in many cases, modern principles of ecosystem-based management, precaution…have yet to be brought to bear' (Global Ocean Commission 2014, 18).

5.1.2 Regional Fisheries Management Organisations and High-Seas Governance

The UNCLOS governs fishing in areas beyond national jurisdictions. Many of the world's most valuable fisheries, including tuna, are in international waters. However, human activities such as maritime transportation, marine pollution and fishing have caused serious depletion in high seas fish stocks (Freestone 2010). The UNCLOS, under Article 116, expresses a state's 'right to fish' the high seas, but this freedom is subject to the condition that marine living resources be used sustainably and with the rights and duties of coastal states in terms of straddling stocks.

The 1995 UNFSA [3] complements and strengthens the UNCLOS by requiring fisheries management to be based on precautionary and ecosystem approaches. It also enhances monitoring, control and enforcement (and even extends to boarding of non-compliant vessels) both by flag states and through international cooperation. Regional action, particularly through RFMOs, is necessary to implement the agreement effectively. However, there are regulatory gaps in the management regime created by RFMOs (Rayfuse and Warner 2008). In the case of high seas fishing, which is managed primarily by RFMOs, severe challenges result from a lack of cooperation between states; conflicting interests in resource use and conservation; fragmented responsibilities; lack of political will; lack of enforcement; and perverse economic incentives for 'free riders' to cheat the system (Global Ocean Commission 2014).

RFMOs are principally membership organisations. While under the UNCLOS all states have the general obligation to cooperate with each other in the conservation and management of living resources of the high seas, and parties to the UNFSA can only fish on the high seas if they apply the conservation and management measures set by the RFMO competent over that area or species, there are loopholes in the regional fisheries management scheme. RFMOs cannot exert control over non-member state fleets, and their members have limited capacity to apply some sanctions—for example, trade restrictions and import bans on certain types of fish products—to uncooperative non-member countries. Because RFMOs do not have uniform provisions across convention areas, this creates a patchwork of governance (Table 15.1).

RFMOs vary on their policies on how to set catch limits, monitor catches and impose penalties. The nonuniformity of RFMO policies makes some areas more vulnerable to IUU fishing. The five tuna RFMOs worked to remedy this patchwork by coordinating through the Kobe Process named after its launch in Kobe, Japan, in January 2007. The Kobe Process seeks to improve coordination across the whole range of RFMO policy, including scientific research, market issues, monitoring and surveillance, the impact of bycatches, and support for developing countries. One of the main concerns of the Kobe meeting and subsequent action plan was to secure support for developing nations to implement the recommended management measures, particularly those intended to prevent IUU fishing. However, only tuna RFMOs are involved.

NGO advocacy and UN General Assembly Resolutions 59/25 (2004) and 61/105 (2006) show a positive response to unregulated deepwater bottom fishing. Three new RFMOs—

[3] Officially, the United Nations Agreement for the Implementation of the Provisions of UNCLOS relating to the Conservation and Management of Straddling Fish Stocks and Highly Migratory Fish Stocks.

Table 15.1 Comparisons of several provisions among Tuna RFMOs

	International Commission for the Conservation of Atlantic Tunas (ICCAT)	Inter-American Tropical Tuna Commission (IATTC)	Indian Ocean Tuna Commission (IOTC)	Western and Central Pacific Fisheries Commission (WCPFC)	Commission for the Conservation of Southern Bluefin Tuna (CCSBT)
Requirement of International Maritime Organization (IMO) number	For vessels 20 m and greater	For vessels greater than 12 m	For vessels greater than 24 m	For vessels greater than 12 m	For vessels 100 gross tonnage and greater
100% observer coverage	For vessels 20 m and greater during specific times or closures, as well as for bluefin fishery	For large-scale purse seine vessels		For large-scale purse seine vessels	
Resolution on Labour Standards for Crew on Fishing Vessels				*Resolution on Labour Standards for Crew on Fishing Vessels* (Resolution 2018-01). Non-binding	
Cross-listing of IUU fishing vessels list	Allows cross-listing with other RFMOs		Allows cross-listing with other RFMOs		Allows cross-listing with other RFMOs
Prohibition of drift nets	Drift nets prohibited for fisheries of large pelagic in Mediterranean		Large-scale drift nets prohibited	Large-scale drift nets prohibited	
Allowing at-sea transshipment	Allowed for large-scale pelagic longline vessels, defined as those greater than 24 m length overall	Allowed for large-scale tuna fishing vessels	Allowed for largescale tuna fishing vessels t	Not allowed for other than purse seine vessels except in cases where the CCM has identified and reported vessels as being impractical not to transship at sea	Allowed for tuna longline fishing vessel with freezing capacity of storing 500 kg of southern bluefin tuna at −30 °C or below
	Purse seine vessels (with some exceptions) prohibited from transshiping at sea and must transship in port	Purse seine vessels (with some exceptions) prohibited from transshiping at sea and must transship in port	Purse seine vessels (with some exceptions) prohibited from transshiping at sea and must transship in port	Allowed for purse seine vessels 600 metric tonnes or less from Papua New Guinea and Philippines, New Zealand, domestic purse seine vessels, and for other vessels where Conservation and Management Measures has determined that it is impracticable to prohibit at-sea transshipment	

Source: Drawn from Commission Webpages 2020

South Indian Ocean Fisheries Agreement, South Pacific Regional Fisheries Management Organisation and North Pacific Fisheries Commission—have formed to manage deep-sea bottom fisheries on the high seas. However, the deepwater bottom fisheries—particularly bottom trawl fisheries in all three Pacific and Indian Ocean RFMOs, and to a lesser extent in portions of the Atlantic—are not yet managed consistently with actions called for by the UN General Assembly. These actions are based on key conservation provisions of the UNFSA as well as the FAO Code of Conduct for Responsible Fisheries. Therefore, these fisheries could fit the definition of IUU fisheries, even though technically they are regulated (Gianni et al. 2016).

Further, RFMOs are not equipped with their own MCS systems. To track movements at sea, states use monitoring systems such as VMS, but only authorised people, typically the government officials in relevant flag or coastal states, can access this information. Most RFMOs require the use of VMS by large vessels authorised to fish within their convention areas. However, the member flag states are the ones with the authority, jurisdiction, and enforcement responsibility for mandating installation and operation of VMS and enforcing reporting obligations. Some RFMOs have initiated policies to allow direct transmission of VMS data to the RFMOs' secretariats. The effectiveness depends on the member states, and RFMOs have not yet established consequences for failure to comply. Some have considered or are considering the use of AIS as a more cost-effective means of tracking.

The performance of RFMOs has been under scrutiny for some time. The High Seas Task Force (2006) and the Global Ocean Commission (2014) have raised concerns with their effectiveness, and NGOs continue to press for better performance in the annual commission meetings.

5.1.3 Lack of Universal Mechanism to Assess State Compliance

Although some NGOs have produced reports to assess states' performance in combatting IUU fishing, there are insufficient tools and mechanisms to assess and evaluate states' adoption and implementation of the most important international instruments relating to IUU fishing. Therefore, states with poor performance do not have sufficient information on what areas need improvement and do not receive enough pressure to make improvements.

The FAO's IPOA-IUU is by far the most relevant international guidance, providing a set of measures that states should adopt to combat IUU fishing. Many countries have developed their own national plans of action (NPOAs) to reflect this guidance in their national context. However, the FAO does not assess the content or implementation of these national plans. Moreover, now close to 20 years old, the IPOA may benefit from a revision to reflect the latest developments in technological tools and transparency measures, as well as the inclusion of measures to support adoption and implementation of NPOAs. If the FAO could benchmark states against the measures detailed in the IPOA-IUU, states with poor performance would get a clearer understanding of areas needing improvement. In addition, the FAO, ILO and IMO can do more to help states ratify and implement the important suite of international treaties relevant to the fight against IUU fishing: the FAO'S Port State Measures Agreement, ILO Work in Fishing Convention and IMO Cape Town Agreement.

5.1.4 Gap Between Fisheries Management and Preventing Human Rights Abuses

The causal relationship of human rights abuses within IUU fishing has been described above. However, there is a gap between international organisations that manage fisheries and those that work to prevent human rights abuses. For example, currently only one tuna RFMO, the WCPFC, has a resolution addressing workers' human rights protection. However, there are opportunities within the existing international legal framework—for example, the ILO C188 and the CTA on fishing vessel safety—to protect fishers from human trafficking and forced labours.

The financial drivers behind illegal fishing can lead to poor safety and labour conditions for vessel crews. When stocks are overfished, fishers' catches and income are further reduced. The CTA sets minimum safety standards and allows flag, coastal and port states to inspect commercial fishing vessels. It is therefore a powerful tool that states can use to ensure that fishers are safe, conditions are decent and fishing operations are legal. But as long as enough countries do not ratify the agreements, they remain weak. It should be noted that at the meeting of the IMO in Torremolinos, Spain, in 2019, 48 countries committed to ratify the CTA by 2022. The WCPFC's recent Resolution on Labour Standards for Crew on Fishing Vessels (Resolution 2018-01) is groundbreaking yet falls short of providing adequate protections due to its status as a non-binding resolution with no non-compliance penalties.

The issue could be more adequately handled through licensing provisions specific to crew welfare agreed on in 2019 and due to be imposed at a sub-regional level by the Pacific Islands Forum Fisheries Agency (FFA) in 2020.

5.2 Solutions to Weak Governance

Strong, uniform governance is needed at the national, regional and global levels to combat IUU. Several international agreements provide a path forward towards consistent and effective fisheries management. States can adopt and implement these agreements to reduce IUU fishing. The IPOA-IUU provides clear and comprehensive guidance.

5.2.1 Adopt the Port State Measures Agreement

Since international efforts to combat IUU fishing cannot depend solely on the regime of flag state responsibility, more opportunities could be given to authorities of port and coastal states to address all impacts of IUU fishing. Illegal fishing operations cannot operate without a market for their catch. Fishing vessels (or supporting vessels such as reefers) must at some point visit a port to land fish, refuel, re-supply and take on crew, and vessels involved in illegal fishing operations are no exception. State regulation of access to port facilities is therefore a highly effective way of controlling illegal fishing.

The FAO adopted a key instrument targeting IUU fishing—the PSMA—at its 36th session on 22 November 2009. This agreement in principle strengthens the comprehensive and integrated approach to combatting IUU fishing, since it supports previously adopted instruments in the FAO framework, such as better performance by flag states, MCS, market access and trade measures. Generally, the PSMA authorises port states to apply it to vessels not entitled to fly a state's flag that are seeking entry to its ports or are in one of its ports. In particular, the agreement encourages each party to integrate its port state measures with the broader system of port state controls and measures in accordance with the IPOA-IUU at the national level.

The PSMA aims to prevent IUU catches from being landed and entering international markets. Since vessels must come to port prior to their fish entering the market, port state measures are potentially the most effective means to combat IUU fishing. The agreement sets a global standard of port inspection, improves information exchange and puts developing states in a better place to combat IUU fishing with the funding mechanisms provided by the agreement. Importantly, the PSMA is applicable to any vessel supporting IUU fishing, so it can be used to address fuelling/bunkering vessels or the carrier vessels that bring the fish to port. There is a further advantage; ports with low inspection resources can simply deny a vessel port entry under the PSMA. The more ports that implement the PSMA, the more effective it becomes.

5.2.2 Close the FOC Registry to Fishing Vessels

FOC states are those that register foreign-owned fishing vessels with minimum requirements and assessments. Closing such registries to the registration of fishing and fisheries-support vessels is an important and low-cost measure to combat IUU fishing. The FOC registries often have little connection to national fisheries ministries and are likely to represent a small source of income that is outweighed by the reputational damage done to the countries associated with IUU fishing; for example, the risk that legitimate national operators lose the right to export seafood to important markets such as the European Union (Fig. 15.8).

Although they appear to incur short-term loss, transitioning policies and efforts to combat IUU fishing efforts are economically beneficial in the long run. After imposing a closed registry policy in 2015, Indonesia has gained tax and non-tax revenue from the fisheries sector (California Environmental Associates 2018). Once the foreign-owned vessels, which all had committed various fisheries violations, were eliminated from the Indonesian registry and waters, they were replaced by new vessels owned by Indonesian industry. Since 2015, government agencies have been equipped with more advanced monitoring tools and supported by better fisheries governance, resulting in increasing tax and non-tax revenue from fisheries (Fig. 15.6).

5.2.3 Create Strong Deterrents

Deterrence is a key element in the battle to combat IUU fishing. Deterrence can be achieved via strong port controls, at-sea patrols, heavy fines or sanctions and other measures.

Port controls In terms of port controls, this paper emphasises the importance of the PSMA and encourages more states to ratify and implement this agreement. Because IUU fishing is a complex activity, those working to deter or eliminate it need a selection of mechanisms to make sure IUU fishers cannot slip though.

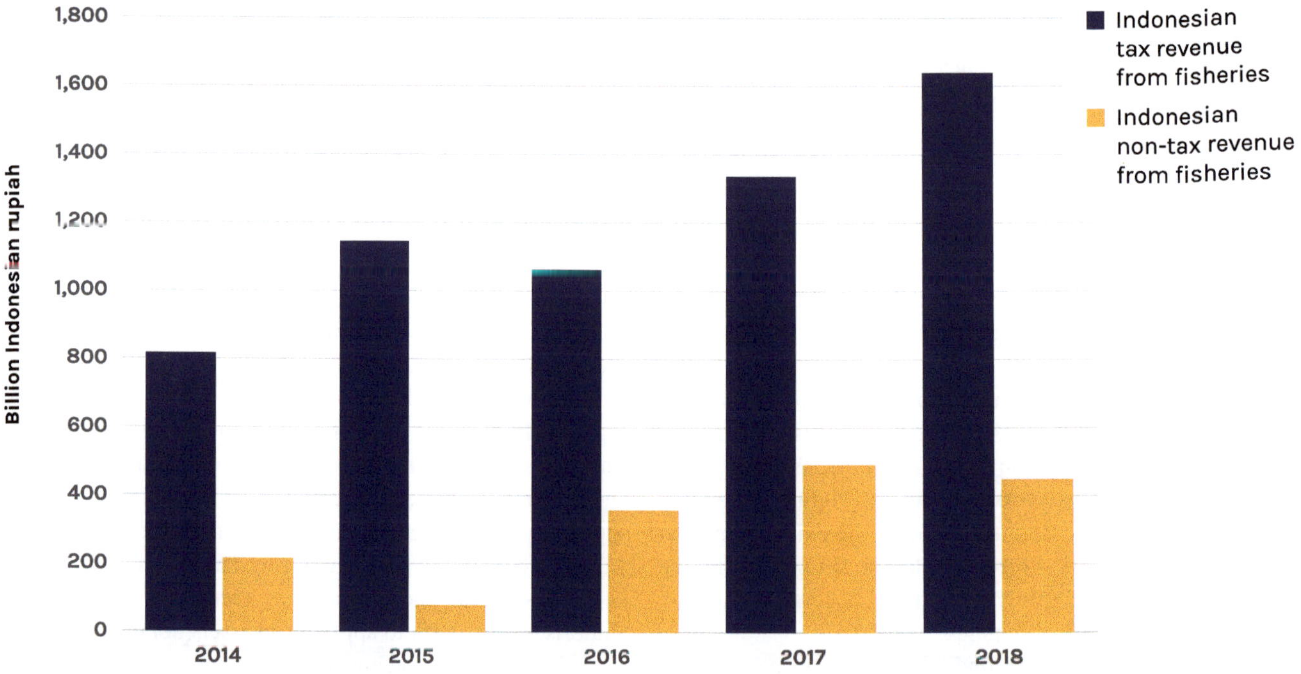

Fig. 15.8 Indonesian revenue from fisheries. (*Source: Ministry of Marine Affairs and Fisheries of the Republic of Indonesia* 2019)

Sanctions It is often said that sanctions for IUU fishing are not sufficient to hurt the fisher and are seen as a cost of doing business. The IPOA-IUU calls on states to ensure that sanctions are heavy for IUU fishing offenders. Realising the immense loss caused by IUU fishing, countries are executing stronger penalties. In May 2019, Thailand's criminal court handed out a fine of over $16 million to 6 defendants in the prosecution of the overseas fishing vessel *Chotchainavee 35* (Undercurrent News 2019). In 2018, the owner of the pirate fishing vessel F/V Thunder was fined $10.1 million in a civil case brought by the Spanish government (Holland 2018). In another example, Indonesia has elected to sink offending vessels as an optimum penalty to create a strong deterrent. Minister of Maritime Affairs and Fisheries Susi Pudjiastuti reported in her farewell speech in October 2019 that 556 vessels engaged in IUU fishing had been sunk during her tenure. This policy has resulted in a reduction of at least 25% in fishing effort within the Indonesian EEZ, based on VMS data together with AIS and night light satellite imaging data (Cabral et al. 2018).

IUU vessel lists published individually by each RFMO and as a combined historical list by Trygg Mat Tracking act as a deterrent to a degree, but there is now a movement to see all fines and sanctions be published and transparent. This would show who is being fined or sanctioned and for what, and also which states and organisations are taking seriously the need to act against IUU fishers (EJF 2018).

Transparency A vessel whose behaviour is public is likely to become compliant. Again, several NGOs have promoted ten principles of transparency. These measures are in line with reports from organisations such the High Seas Task Force (2006), the Global Ocean Commission (2014), the Royal United Services Institute (Haenlein 2017) and the Organisation for Economic Co-operation and Development (Hutnikzac and Delpeuch 2018). Countries are also steadily embracing transparency, with notable action by Chile, Panama, Peru and Indonesia in sharing VMS vessel tracking data and the European Union, Taiwan, Thailand and Mozambique in publishing authorised vessel lists.

At-sea patrols Patrols will always be necessary but are a relatively high-cost option.

Multilateral agreements Where possible, resource sharing and multilateral agreements should ensure that at-sea assets can be effectively tasked and made as costeffective as possible. Agreements such as the Niue Treaty Subsidiary Agreement (NTSA) should be replicated. The NTSA is an agreement in which FFA members agree on monitoring, control and surveillance of fishing—including provisions on exchange of fisheries data and information, as well as procedures for cooperation in monitoring, prosecuting and penalising operators of IUU fishing vessels.

5.2.4 Improve Transboundary Case Handling

Enhanced cooperation on investigation and prosecution processes can be achieved through the framework of temporary multilateral investigative support teams, mutual legal assistance between states and an established international team such as the Global Fisheries Enforcement Team supported by Interpol. Even in the absence of a mutual legal assistance framework, states across regions have come to work together in handling transboundary cases (Box 15.6).

Box 15.6 Case Study: *HUA LI 8*—Transboundary Cooperation to Catch an Illegal Fishing Vessel

A success story in tracking and capturing an IUU vessel is the case of the vessel HUA LI 8. The vessel was detected fishing illegally within Argentina's 200-nautical-mile EEZ. Two ships and a helicopter from the Argentine naval command confronted the vessel, which proceeded to take evasive action. Ignoring several requests to stop broadcast in Spanish and English on applicable international VHF channels, as well as visual and audio signals (on Maritime Mobile Service VHF Channel 16) and warning shots, the boat endangered its crew, Argentinian authorities and other ships in the vicinity by continuing to sail. The pursuit was subsequently called off, and visual contact was lost after the boat's entry into Uruguayan waters.

Given the obstructive way in which the vessel evaded security forces, Argentina requested the assistance of Interpol through the Environmental Crime Programme's Project Scale in alerting other member countries to the illegal fishing activities of HUA LI 8 through issuance of a Purple Notice (an international communication mechanism alerting authorities to the modus operandi in IUU fishing), as well as engagement with other countries to track the vessel as it travelled across the Atlantic and Indian Oceans. On 21 April 2016, the Indonesian navy (Lantamal 1 Naval Base at Belawan, North Sumatera) detected the HUA LI 8 in waters near Aceh, Indonesia, and successfully boarded and began its inspection of the vessel.

Source: Interpol (2016).

5.2.5 Enhance Transparency in Fisheries

Global transparency in the fishing industry is a solution to governance issues that has relatively low cost and is a manageable action. The Environmental Justice Foundation publication *Out of the Shadows* contains ten principles for global transparency in the fishing industry that all countries could adopt:

- Give all vessels a unique number.
- Make vessel tracking data public.
- Publish lists of fishing licences and authorisations.
- Publish punishments handed out for fisheries crimes.
- Ban transferring fish between boats at sea—unless pre-authorised and carefully monitored.
- Set up a digital database of vessel information.
- Stop the use of FOCs for fishing vessels.
- Publish details of the true owners of each vessel—who takes home the profit?
- Punish anyone involved in IUU fishing.
- Adopt international measures that set clear standards for fishing vessels and the trade in fisheries products (EJF 2018)

Transparency is highlighted by the Fisheries Transparency Initiative (FiTI), a unique effort that complements and supports other national, regional and global efforts to achieve responsible fisheries governance. The purpose of the FiTI is to increase transparency and participation in fisheries governance for the benefit of a more sustainable management of marine fisheries. The FiTI is not owned or operated by one organisation, nor does it represent the work of a single interest group. Instead, the diversity of stakeholders is a central feature of how the FiTI works, for national implementations as well as international governance.

The FiTI is a global initiative, and its implementation is country centred. The intention to join the FiTI and the initiation of the official process must come from a country's government. It is a voluntary initiative with mandatory requirements, built on a multi-stakeholder governance structure, ensuring that stakeholders from government, companies and civil society are equally represented. The FiTI embraces the following principles:

- Public registry of national fisheries laws, regulations and official policy documents
- Summary of laws and decrees on fisheries tenure arrangements
- Publication of all foreign fishing access agreements
- Publication of existing national reports on the state of fish stocks
- Public online registry of authorised large-scale vessels, as well as information on their payments and recorded catches
- Information on the small-scale sector, including the numbers of fishers, their catches and financial transfers to the state
- Information on the post-harvest sector and fish trade
- Information on law enforcement efforts, including a description of efforts to ensure compliance by fishers and a record of offences in the sector
- Information on labour standards in the fisheries sector
- Information on government transfers and fisheries subsidies
- Information on official development assistance regarding public sector projects related to fisheries and marine conservation
- Information on the country's status regarding beneficial ownership transparency

6 Ensuring Effective Enforcement

Even strict regulations are not always implemented or enforced. Lack of enforcement is the third of the three drivers of IUU. This section identifies barriers to enforcement and makes suggestions on how to improve enforcement. See also Appendix B for action suggestions.

6.1 Barriers to Enforcement

Lack of political will, coupled with the logistical difficulties in monitoring and reaching vast areas of the ocean, often results in weak enforcement. In some instances, the penalties imposed by courts of law have been described as a 'slap on the wrist' or 'part of the costs of doing business'. The penalties imposed by courts should reflect the importance with which marine living resources are viewed. The public and potential transgressors should be made aware of instances in which severe penalties are imposed. The judiciary should therefore be sensitised in this regard. National bodies, such as organisations representing prosecutors and/or judges, should disseminate information among their members in attempts to encourage uniformity at an appropriate scale.

These barriers to effective enforcement allow IUU fishers to exploit weaknesses in the system to fish undetected. New technological capabilities, alongside tighter port controls and clearer understanding of vessel activity and authorisations, show new possibilities for how these can be overcome.

These actions often fall to governments, but industry and the private sector can play a vital role in driving their supply chains towards better-governed and more diligent fisheries, vessels and ports. Action by the private sector could help port states focus on better enforcement against illegal and unreported fishing activities. Some enforcement is hampered by the unwillingness of states to enforce the law against their own fishing fleets. Some states do not have adequate facilities to conduct MCS. Monitoring is also a responsibility of flag states. As stipulated in the UNCLOS, flag states' responsibility includes ensuring the level of compliance of their fleets operating inside and outside of their waters. But some resource-challenged states find this difficult, while others have other priorities. A forthcoming global review by Australia's Commonwealth Scientific and Industrial Research Organisation (CSIRO) of the scientific literature on the causes of fisheries depletion is expected to document that the lack of MCS is the most common driver of fisheries depletions. This lack is cited as one of the key drivers in 90 out of 164 studies. Given the tight link between fisheries depletion and illegal fishing, the resulting weak control of fisheries operations further exacerbates depletion as a driver for both domestic and foreign illegal fishing.

In reality, no mechanism obligates all states to share monitoring data publicly. One initiative has been created—Global Fishing Watch—but it remains voluntary, so its effectiveness depends on states' awareness and willingness to cooperate. States currently use different types of fisheries monitoring platforms that do not integrate data routinely, which leads to inefficiencies in fisheries monitoring systems globally. Systems like OceanMind and Skylight are designed to improve the sharing of non-public data and drive direct enforcement action. Beyond fisheries, there is a trend to create fusion centres, designed to bring monitoring of different maritime issues, including fishing, together in one centre. This is a positive concept. Such systems are a positive step forward in the fusion of military and civilian data.

The enforcement of RFMOs' respective instruments is still frequently hampered by the willingness of their member states to enforce the regulations against their own vessels and the activities of non-parties. Principally, enforcement is undertaken by flag states, as laid down in Article 92(1) of the UNCLOS. However, Article 21 of the UNFSA includes provisions allowing various types of enforcement by states other than flag states on the high seas, although this exception can only be invoked under certain conditions. First, it only applies in high seas areas that fall within the geographical competence of RFMOs. Second, only members of these RFMOs are allowed to take enforcement measures, and only for the purpose of ensuring compliance with conservation and management measures of these RFMOs. Third, only fishing vessels flying the flag of a state party to the UNFSA can be subjected to these enforcement measures, whether or not that state is also a member of these RFMOs. Fourth, the procedures for high seas enforcement as set out in paragraphs 4–18 of Article 21 and Article 22 of the agreement shall be applicable if RFMOs do not establish their own procedures (Molenaar 2011).

Looking at those strict conditions that invoke this exception, it is argued that it hardly affects the flag state primacy. Once an infringement is detected, the flag state should be notified to pursue enforcement process. Therefore, the effectiveness of enforcement remains subjected to flag state willingness and capacity, unless RFMOs have their own procedures. Conversely, viable penalties for flag states are not in place in case of non-compliance.

6.2 Improving Enforcement

Improving enforcement will require capacity building and support for port states and coastal states, regional and global information sharing, and heightened monitoring of fishing fleets, transshipment incidents and catch data.

6.2.1 Build Capacity and Support

In some areas, illegal fishing perpetrators are largely unpunished or poorly punished for many reasons. Among others are poor awareness among law enforcement officers, difficulties in communication between related agencies and inadequate capability to conduct a thorough investigation. Effective law enforcement requires law enforcement officers and other role players in the criminal justice system to have adequate skills to handle the complexities of IUU fishing operations, which are often associated with other crimes. A joint capacity-building program accommodates sharing of ideas and experiences besides capacity building trainings.

FishFORCE Academy, a platform for training officials who are involved in the fight against fisheries crime, has been established at the Nelson Mandela University. The project will aim to establish fisheries crime law enforcement as a new and emerging fisheries compliance model and will endeavour to achieve knowledge and intelligence-led investigations and increase successful prosecutions of criminals engaged in fisheries crime. While building capacity, the project will also enable fisheries law enforcement officers to obtain formal qualifications in their chosen field of expertise. These qualifications will include higher certificates, diplomas and a post-graduate diploma, which will also provide access to further academic qualifications.

Similarly, the Australian government recently ran a capacity-building program implemented by the CSIRO, Australia's national science institution, to build capacity among fisheries monitoring and surveillance officers across the Southeast Asian region. The two-phase program included country visits to understand capacity and needs for fisheries MCS analysis and to identify key emerging enforcement issues, followed by a customised training course targeting MCS analysts. The training course included three fisheries staff members from each of the 12 regional countries. The weeklong training course included instruction and tools for identifying abnormal patterns in monitoring and surveillance data; analysing spatial and temporal data such as VMS, landings and observer records; and developing risk assessment models and prioritising inspections and other MCS activities.

In recent years, NGOs are increasingly involved in enforcement, through provision of capacity such as vessels, technology or training, and work closely with coastal states in global geographies such as Italy, Gabon, Ghana, Namibia, Benin and Cape Verde to help combat IUU fishing.

6.2.2 Establish Regional Information Sharing and Cooperation Mechanisms

The experiences of the FISH-i Africa and Fisheries Committee for the West Central Gulf of Guinea West Africa Task Forces, as well as the FFA's Niue Treaty Subsidiary Agreement and associated Niue Treaty Information System, have demonstrated the value of groups of countries establishing mechanisms and communication platforms for the increased sharing of fisheries intelligence, cooperation of MCS operations and cases, and coordination and harmonisation of fisheries management and regulations. This approach has clearly demonstrated that regional cooperation and information sharing, coupled with dedicated analysis and technical expertise, can exclude high-risk vessels and operations from a region, prevent IUU catch from entering ports and getting to market, and significantly hinder the ability for illegal operators to operate and make a profit. The basic structures of successful cooperation are in line with and benefit from increased availability and, where possible, transparency of information—including sharing of information on licensed vessels, port inspections and vessel movements—and cooperation on investigations. If this can be achieved, the results have proved to be significant, with a range of illegalities quickly uncovered and acted on, including arrests, settlements, payments of fines, exposing and shutting down fraudulent licenses and other documents, and an increase in license revenue.

6.2.3 Monitor Transshipments

The monitoring and control of transshipment poses a significant problem because so much of the activity takes place out of sight and reach of authorities. This is especially true for at-sea transshipment, which occurs far from land. Even in port, proper oversight often cannot be guaranteed because of limited inspection capacity or insufficient port state processes, protocols or procedures. Collectively, these practices contribute to IUU fishing. To make matters worse, the lack of transparency regarding the monitoring and control over transshipment fosters conditions conducive to other criminal activities, such as trafficking in weapons, drugs and people, and contributes to concerns about labour conditions on board vessels that are at sea for extended periods (Box 15.7).

Box 15.7 Case Study: ***Silver Sea 2*****—Transshipment and Human Trafficking**

The fishing vessel Silver Sea 2 was seized by the Indonesian navy in August 2015 amid a crackdown on illegal fishing and after an Associated Press investigation showed its links to human trafficking in the fishing industry. When identified by the SkyTruth analysts through AIS and Digital Globe satellite images, the Silver Sea 2 was in Papua New Guinea waters, receiving illegal Indonesian catch from two fishing trawlers via transshipment. It was captured by an Indonesian navy vessel off the island of Sumatra after returning to Indonesian waters. The Thai captain was detained, and a probe was launched into suspected human trafficking, transporting illegal fish and offloading the catch at sea. The Associated Press investigation resulted in the freeing of more than 2000 men from Myanmar, Cambodia, Thailand and Laos, more than a dozen arrests, the changing of U.S. legislation, and lawsuits.

Source: McDowell et al. (2015).

Emerging transparency tools, like Global Fishing Watch, or more proprietary systems, like OceanMind and Skylight, that fit the more traditional enforcement-focused models, provide new capabilities to identify and monitor these transshipment activities. The private sector is key here, since it brings into the commercial space new capabilities that can then be used to monitor effectively. A good example of this is HawkEye 360, which has used military technology to develop a commercial system that can detect radio and radar emissions from vessels. This could be very useful for detecting vessels attempting to avoid transmitting on their AIS or VMS.

Flag, coastal and landing states can also play a major role in this area. For instance, Thailand now equips all of its fish carrier vessels with electronic monitoring, including on demand real-time video cameras. This allows fisheries officials in the national monitoring centre to oversee vessel operations, including location, use of onboard machinery, access to refrigerated storage, and other activities in real time. If suspicious activity is detected, the staff can bring the onboard video cameras online via satellite connection, allowing them to see any suspicious activity, including transshipment.

Surveillance technologies for monitoring non-compliant vessels combined with onboard monitoring of carrier vessels has the potential to remove transshipment completely as a source of IUU fishing risk.

6.2.4 Improve Monitoring of the Fishing Fleet

Leading experts are calling for a conscious reframing of the perception of what is possible in terms of monitoring the fishing fleet and support vessels and for people to act now and at scale. Seeing transparency and compliance as a vital tool in the good stewardship of our global ocean is part of that reframing—to fight illegal fishing, to protect fish stocks and livelihoods and to increase the safety and well-being of fishers. If countries publicly share their fishing vessel monitoring data, then a more complete and connected picture of global fishing activity can be created. Law-abiding fishers are tracked easily and openly, demonstrating their compliance. Rogue operators stand out due to their patchy track record or suspicious behaviour. Compliant fishers can be rewarded through faster, more efficient port entry and landings. Unauthorised vessels and those that have a history of non-compliance can be prioritised for inspection or even denied port entry. By embracing transparency, nations have a more cost-effective way of monitoring vessels that puts the burden on fishers to demonstrate compliance rather than on the country to prove illegality. Transparency can incentivise, recognise and reward honest fishers, while exposing, penalising and ultimately putting out of business those who act outside the law.

The simplest way to combat the absence of tracking data is to make it a condition of landing fish. If a vessel cannot explain or display its track history since its last landing, then it should not be able to land its catch. At the very least, it should be subject to robust inspection and verification.

7 Three High-Level Decisive Opportunities for Action

Three specific opportunities for action provide a robust yet achievable response to the global threat of IUU fishing and help ensure ocean health, biodiversity and a sustainable ocean economy in the future. These actions are clear and tangible routes to implementing global policy or supporting existing vital policies. They are directly associated with solving the key drivers of IUU fishing identified in this paper—economic incentives, weak governance, and poor enforcement—and they can be sustained and are not dependent on consensus in the face of a belligerent state or actor. They can be addressed by flag, port, coastal and market states. Business, industry, private sector organisations, scientists and civil society can also contribute through advocacy, leadership and firm actions of their own. These three actions are

- adopt global transparency in fisheries,
- ratify and implement the FAO's PSMA, and
- enhance regional cooperation.

7.1 Adopt Global Transparency in Fisheries

The first opportunity for action is to adopt global transparency in fisheries. There are several elements to this shift in narrative away from an enforcement-focused system to one that rewards compliance and good behaviour. In promoting transparency, the international community will be addressing all three drivers of IUU fishing. Transparency makes it far more difficult to bring IUU fish to port by shifting economic incentives. Vessels with missing information can be treated as suspicious and prioritised for inspection or action. Transparency is an effective way to support key policies such as the PSMA. It makes monitoring and inspections easier to prioritise and more cost-effective. It addresses weak governance and barriers to enforcement by improving information and data sharing. It shifts the burden to the fisher to prove compliance, rather than the state to prove malpractice. Actions in this category can be achieved quickly and will have a positive impact on the fight to end IUU fishing and will not disadvantage the more resource-challenged countries. Significant commitment could be demonstrated before the end of 2020 in support of SDG 14.4:

- Flag or coastal states could make unedited VMS data, or other proprietary tracking systems, public or mandate AIS for fishing vessels. Industry and other private sector organisations should seek this action as a part of their conditions of contract or in doing business.
- Coastal states could publish up-to-date lists of all fishing licenses, authorisations and vessel registries, transshipment authorisations and refrigerated vessels registries. Private sector organisations should seek these documents as a part of their conditions of contract or in doing business.
- Port, flag or coastal states could mandate IMO numbers for all eligible vessels and implement a national Unique Vessel Identifier (UVI) scheme for non-eligible vessels, maintaining a vessel registry and providing all information to the FAO Global Record of Fishing Vessels. Industry and private sector organisations should seek this action as a part of their conditions of contract or in doing business.
- Business, industry and finance institutions are encouraged to make fisheries transparency and traceability conditions of their contracts.
- All sectors should make sure of the accuracy of ownership information to avoid the use of shell companies.

More progressive actions would include the following efforts:

- Publish information on beneficial ownership in all public lists and require companies to provide information on true beneficial ownership when applying for a fishing license, fishing authorisation or registration to their flag.

- Mandate and implement the near-term adoption of cost-effective digital tools that safeguard in a digital form key information on vessel registration, licenses, unloading records, catch location and information and crew documentation that should be designed in such a way as to support a rapid move towards a universal, interoperable digital catch certification scheme.
- Improve transshipment activities information through mandatory and public pre-authorisation and robust and verifiable electronic monitoring scheme by the end of 2020 in support of SDG 14.4.
- All states should publish information on arrests and sanctions imposed on individuals and companies for IUU fishing activities, human trafficking and other related crimes to an accessible international platform.

7.2 Ratify and Implement the FAO's Port State Measures Agreement

The second opportunity for action is the ratification and proper implementation of the FAO's PSMA by all port states. Again, it addresses all three drivers of IUU fishing. Fully implemented, it represents a cost-effective method of stopping IUU-caught fish from entering the market. The PSMA is the first binding international agreement that deals specifically with IUU fishing by requiring parties to place tighter controls on foreign flagged vessels seeking to use their ports, with a view to detect and prevent the trade of IUU products. States implementing the PSMA can refuse entry to ports or access to port services to vessels known to have engaged in IUU fishing, allow vessels entry into port for inspection for vessels suspected of having engaged in IUU fishing, and encourage information-sharing mechanisms with other relevant states and organisations to facilitate cooperation in enforcement actions. The PSMA changes behaviour and stops the formation of ports of convenience that undermine good governance. The CTA and the ILO C188 are complementary regulations and can be considered during ports inspections under the PSMA. Industry and the private sector should seek this action as a part of their conditions of contract or in doing business.

7.3 Enhance Regional Cooperation

The third opportunity for action is to enhance regional cooperation. To make marked progress in addressing the drivers of IUU fishing activities, we need a more 'joined-up' approach among governments, civil society, science, industry and the private sector for a system that is coherent and consistent in its actions to combat IUU fishing. Actions under this banner would include the following efforts:

- An international forum or other mechanism should address the non-uniformity of RFMO regulations.
- Coordination and data transparency must be improved among RFMOs, flag states, regional mechanisms and the coastal and market states. Coastal states should communicate information on IUU fishing–related infringements to neighbouring coastal states. Port states should provide information to flag states on transshipments, landings and denials of use of port involving vessels flying their flag, as well as the result of inspections. Flag states should cooperate with the RFMO or nation state to investigate and take action in cases of presumed IUU fishing by their vessels.
- All RFMOs should adopt strong policies on monitoring and enforcement and also create sanctions for flag states not performing enforcement measures.
- In terms of institutional arrangement, an authorised international body should oversee the performance of each RFMO, identify the gaps where fisheries management is non-existent and push forward marine protected areas in the high seas.
- Digital schemes for documenting catch data should be implemented in order to promote global exchange of information. All states should mandate and implement the near-term adoption of cost-effective digital tools that safeguard in a digital form key information on vessel registration, licenses, unloading records, catch location and information and crew documentation. These digital systems must be secure and have built-in procedures to prevent the unauthorised deletion or overwriting of data.
- Regional information-sharing and coordination bodies, such as Regional Plans of Action to address IUU fishing, should be developed. The Regional Plan of Action in the Southeast Asian region is a key forum for countries in the region to discuss IUU fishing–related issues and coordinate actions.

These actions are tangible and transformational if achieved. It should be noted that as well as addressing the three drivers of IUU fishing, they can be pressed home by industry, the private sector and civil society, as well as by governments.

Appendix A outlines existing fisheries agreements and organisations, and Appendix B summarises the voluntary actions that can be taken by various stakeholders to meet the goals of SDG 14.4.

Acknowledgements The paper's technical reviewers, Alf Hakon Hoel, Adriana Fabra, Paulus Kaigne, Jim Leape and Juan Mayorga, as well as its arbiter, Nicola Frost, all provided helpful technical comments on this Blue Paper.

The authors also thank the World Resources Institute for providing support as the Ocean Panel Secretariat.

While our colleagues were very generous with their time and input, this report reflects the views of the authors alone.

The authors thank LSF Editorial for copyediting and Jen Lockard for design.

About the Authors

Co-authors

Sjarief Widjaja is a Professor of Production Management and Shipping Operations and is the Director General of Agency for Marine and Fisheries Research and Human Resources, Ministry of Marine Affairs and Fisheries (Indonesia).

Tony Long is the CEO of Global Fishing Watch (USA) and retired Royal Navy officer (UK).

Hassan Wirajuda is a scholar of International Ocean Law and Policy, a former Minister for Foreign Affairs (Indonesia), and a co-founder of the Indonesia Ocean Justice Initiative.

Contributing Authors

Hennie Van As is a Professor of Public Law and a Director at the Nelson Mandela Metropolitan University, Port Elizabeth (South Africa).

Per Erik Bergh is CEO of Nordenfjeldske Development Services (NFDS), Managing Director of NFDS Africa.

Annie Brett is an André Hoffman Fellow at Stanford Center for Ocean Solutions and a Fellow at the World Economic Forum Center for the 4th Industrial Revolution (USA).

Duncan Copeland is the Executive Director of Trygg Mat Tracking (TMT) (Norway).

Miriam Fernandez is a Senior Research Fellow at the Knowledge Media Institute (KMi), Open University, and a member of KMi's Social Data Science Group.

Achmad Gusman is the Vice Dean and an assistant professor at the Law Faculty, Universitas Padjadjaran (Indonesia).

Stephanie Juwana is a founder of the Indonesia Ocean Justice Initiative and former legal counsel at the Special Task Force to Combat IUU fishing (SATGAS 115).

Toni Ruchimat is a lecturer at Jakarta University of Fisheries and former Director of Fisheries Research Center, Ministry of Marine Affairs and Fisheries (Indonesia).

Steve Trent is co-founder and Director of Environmental Justice Foundation (UK).

Chris Wilcox is a research scientist with Commonwealth Scientific and Industrial Research Organisation (CSIRO) (Australia).

Appendix A: Instruments and Tools to Combat IUU Fishing

Binding International Instruments on IUU Fishing

Legally binding instruments are available to states, which accommodate state measures and set standards on combatting IUU fishing within their jurisdictions, such as the UN Convention on the Law of the Sea, the Port State Measures Agreement, the UN Fish Stocks Agreement and the Food and Agriculture Organization of the United Nations (FAO) Compliance Agreement. Several instruments related to IUU fishing operations address other issues such as safety of life at sea and human rights abuses of fisheries workers; namely, the International Labour Organization (ILO) Work in Fishing Convention and the Cape Town Agreement. These instruments do not prevent illegal fishing in artisanal fisheries, which rely on local, usually informal, markets.

- **UN Convention on the Law of the Sea (UNCLOS).** The UNCLOS is a landmark instrument sometimes described as a 'constitution for the ocean'. It provides the international legal basis for the protection and use of living and non-living resources of the world's ocean. However, the UNCLOS did not devote much attention to high seas fishing (FAO 2000). The UNCLOS includes the following key features concerning IUU fishing:
  - All states enjoy the traditional freedoms of navigation, over-flight, scientific research and fishing on the high seas; they are obliged to adopt, or cooperate with other states in adopting, measures to manage and conserve living resources.
  - Every state shall effectively exercise its jurisdiction and control in administrative, technical and social matters over ships flying its flag.
  - Disputes can be submitted to the International Tribunal for the Law of the Sea established under the UNCLOS, to the International Court of Justice, or to arbitration.
- **Agreement to Promote Compliance with International Conservation and Management Measures by Fishing Vessels on the High Seas (Compliance Agreement).** In 1993, the FAO Conference adopted the Compliance Agreement. The principal aim of this agreement is to enhance the role of flag states and to strengthen their control over their vessels in ensuring compliance with relevant international instruments.

The FAO Compliance Agreement in principle applies to all fishing vessels that are used or intended for fishing on the high seas. It acknowledges the issue regarding the failure of flag states to fulfil their responsibilities in ensuring the compliance of vessels entitled to fly their flag with international conservation and management measures for living marine resources. Therefore, it heavily focuses on measures to be taken by flag states to address such a failure. The FAO Compliance Agreement also underscores the importance of international cooperation, particularly cooperation with developing countries and cooperation as an effort to encourage non-state parties to adopt laws and regulations consistent with the provisions of the agreement.

Despite the importance of this instrument, some observers have indicated that most states often involved with IUU fishing and the practice of flags of convenience (FOCs) are not willing to ratify the FAO Compliance Agreement (Tanaka 2012). As of July 2018, the agreement only had 42 parties.

- **UN Fish Stocks Agreement (UNFSA).** The 1995 UNFSA further elaborates possible mechanisms for international cooperation, as stipulated in the UNCLOS, concerning the conservation and management of straddling and highly migratory fish stocks, especially through the establishment of sub-regional and regional fisheries management organisations (RFMOs) and arrangements. These organisations and arrangements need to address the following essential matters:
  - Stocks to which conservation and management measures apply
  - Area of application
  - Relationship between the work of the new organisation or arrangement and the role, objectives and operations of any relevant existing organisations or arrangements
  - Mechanisms by which the organisation or arrangement will obtain scientific advice and review on the status of the stocks

Non-binding International Instruments on IUU Fishing

Non-binding instruments provide voluntary guidelines for states to follow.

- **FAO Code of Conduct for Responsible Fisheries (CCRF).** The FAO has also initiated the CCRF, which was adopted in Resolution 4/95 by the FAO Conference on 31 October 1995. This voluntary, non-binding instrument aims to set 'international standards of behavior' for responsible practices with regard to the conservation, management and development of marine living resources. Although the CCRF does not create legally binding obligations, it may be given or have already been given binding effect as a result of the implementation of relevant rules of international law by state parties, such as the UNCLOS and the UNFSA. In addition, despite the fact that it is a voluntary instrument, the CCRF has been regarded as an influential instrument in guiding national governments in developing their fisheries sector policies (Allison 2001).
- **International Plan of Action to Prevent, Deter and Eliminate Illegal, Unreported and Unregulated Fishing (IPOA-IUU).** The IPOA-IUU underscores the importance of a comprehensive and integrated approach in combatting IUU fishing. In this regard, this instrument encourages states to adopt measures necessary to address the failure of flag states in fulfilling their responsibilities, including port state measures, coastal state measures, and market-related measures. As with the CCRF, commitments under the IPOA-IUU are built on relevant international legal instruments, particularly the UNCLOS, the UNFSA, and the Compliance Agreement. Such commitments are also supported through the adoption of a plan of action at the national level (National Plan of Action to Prevent, Deter and Eliminate Illegal, Unreported and Unregulated Fishing, or NPOA-IUU) as well as the regional level (Regional Plan of Action to Prevent, Deter and Eliminate Illegal, Unreported and Unregulated Fishing, or RPOA-IUU).
- **Voluntary Guidelines for Flag State Performance and Voluntary Guidelines for Catch Documentation Schemes.** The FAO has also adopted Voluntary Guidelines for Flag State Performance and Voluntary Guidelines for Catch Documentation Schemes. These non-binding instruments complement international efforts taken under the aegis of the FAO in combatting IUU fishing. It could therefore be concluded that the FAO has attempted to galvanise all key and crucial factors in combatting IUU fishing. These include the adoption of multiple state jurisdictions (flag state, coastal state and port state) in enforcing laws and measures against IUU fishing, as well as the acknowledgement of the role of non-state entities in addressing all impacts of IUU fishing.

Data Sharing and Data Enabled Technology to Detect and Combat IUU Fishing

Date-Sharing Organisations

- **Regional Plans of Action (RPOAs) to address IUU fishing.** Upon request and reports, the RPOA Secretariat may circulate the information through the RPOA website and/or official letter, as well as requesting RPOA-relevant participating countries to deny the vessel port entry or access to port facilities. Programs and activities are held, such as workshops and trainings, and information is exchanged on IUU fishing vessel lists and capacity-building programs on port state measures. While the RPOA is a voluntary instrument, it provides a framework for countries to take individual or collective action to enhance conservation and sustainable use of fisheries resources and combat IUU fishing in the region. These measures involve a range of coastal, flag and port state requirements which in most cases require political will, significant resources and time to address fully (APEC 2008).

- **Interpol's Project Scale.** Interpol's Global Fisheries Enforcement initiative, launched in 2013 under the name Project Scale, supports enforcement agencies in the organisation's 192 member countries in identifying, deterring and disrupting transnational fisheries crime. As a part of this initiative, several Purple Notices have been requested by member countries and issued by Interpol for fishing vessels. The Purple Notice is used to seek information on modus operandi, objects, devices and concealment methods used by criminals and have led to the apprehension of several notorious vessels that conducted IUU fishing.
- **West Africa Task Force.** The Fisheries Committee for the West Central Gulf of Guinea (FCWC) West Africa Task Force (WATF) was formally established in 2015 by the 6 member states of the FCWC—Benin, Côte d'Ivoire, Ghana, Liberia, Nigeria and Togo. Facilitated by the FCWC Secretariat and supported by a technical team which includes Trygg Mat Tracking (TMT) and Stop Illegal Fishing, the WATF was initially supported by a Norwegian Agency for Development Cooperation–funded project entitled Fisheries Intelligence and Monitoring, Control and Surveillance Support in West Africa, and based on the regional cooperation model pioneered by the FISH-i Africa Task Force in the Western Indian Ocean.
- The WATF's core objectives are to improve cooperation, coordination and communication among member states, and to operationalise important fisheries monitoring, control and surveillance (MCS) enforcement commitments of the FCWC, including the 2014 Convention on the Pooling and Sharing of Information and Data on Fisheries in the Zone of the FCWC and the recently updated FCWC RPOA on IUU fishing.
- **FISH-i Africa.** FISH-i Africa was formed in 2012 as a regional task force of coastal states which share a common problem with IUU fishing and hoped to find a common solution by working together. FISH-i is a partnership between the eight East African coastal countries of Comoros, Kenya, Madagascar, Mauritius, Mozambique, Seychelles, Somalia and the United Republic of Tanzania, supported by a technical team of experts. This alliance is showing that regional cooperation and information sharing, coupled with dedicated analysis and technical expertise, can stop illegal catch from getting to market and prevent illegal operators from pursuing their lucrative business unhindered. Task force members share information on licensed vessels, port inspections and vessel movements and cooperate on investigations—and the results have been significant, with a range of illegalities quickly uncovered and acted on. Results have included arrests, settlements, payments of fines, a fraudulent licensing operation being closed and an increase in license revenue.

Data-Enabled Technology and Tools

- **Fisheries Analytical Capacity Tool (FACT).** The use of FOCs, opaque company structures, limited public data on many important global fishing fleets and associated companies, and compliance history presents a significant challenge to the global community's ability to tackle illegal fishing operators effectively. TMT has developed FACT, a fisheries intelligence management and analytical system built with the express purpose of capturing and analysing the identities, characteristics and operations of the global industrial fishing fleet and the companies that operate it. FACT supports 'deep' analysis by providing information on vessel movements, identity, authorisations, operators, ownership and operational structures and whether vessels and companies are involved in violations of fisheries law or broader crimes.

  FACT directly populates the Combined IUU Vessel List website (www.iuu-vessels.org), TMT's public service website that provides the best available, up-to-date information on all vessels that appear on the lists of IUU fishing vessels published by RFMOs. Unlike the IUU lists published on RFMO websites, which may update vessel details only annually, the Combined IUU Fishing Vessel List is kept up-to-date through FACT's processes with the best available information regarding changes to vessel identity, flag state, ownership and location. The aim of the site is to improve the effectiveness of the original IUU lists as a tool to combat illegal fishing and broader fisheries crime.
- **International Monitoring, Control and Surveillance (MCS) Network.** The International MCS Network aims to improve the efficiency and effectiveness of fisheries-related MCS activities through enhanced cooperation, coordination, information collection and exchange among national organisations and institutions responsible for fisheries-related MCS.
- **Information Fusion Centre (IFC) Singapore.** The IFC aims to provide actionable information to cue responses by regional and international navies, coast guards and

other maritime agencies to deal with the full range of maritime security threats and incidents. This includes piracy, sea robbery, maritime terrorism, contraband smuggling, illegal fishing and irregular human migration.

- **Public sector data, technology and capacity development providers.** Australia's national research institution, the Commonwealth Scientific and Industrial Research Organisation (CSIRO), runs a program which develops open source analytical tools for the Automatic Identification System (AIS), the Vessel Monitoring System (VMS), satellite radar and other data sources. These analytical tools are supported by the development of new low-cost surveillance technologies, such as hydrophones to record underwater sound from vessels. The CSIRO embeds these technologies in a capacity-building framework, working with fisheries agencies in the Southeast Asian, Pacific, European, and North and South American regions. The CSIRO makes all of its products available free of charge.
- **Not-for-profit technology platforms.** Global Fishing Watch is leading the way in implementing technology and computational power to analyse a massive amount of data on ocean activity under a not-for-profit banner. Yet other non-profits, such Conservational International, the Nature Conservancy and WWF are working to provide a wide variety of data sources, including satellite observations, vessel tracking data, electronic monitoring, vessel identity databases, fishing license information, and detailed fisheries rules and regulations, accessing both public and proprietary data sources. Data sources are added continually as new technologies become available. Machine learning techniques and big data analytics applied to these data sets can immediately identify non-compliance with global fishing regulations and generate real-time marine intelligence for immediate investigation of illegal activity.
- **Global Fishing Watch.** Global Fishing Watch has focused on transparency and has revolutionised fisheries monitoring by bringing this data into one platform, publicly available and free of charge, to provide the first global view of industrial fishing (monitoring 70,000 vessels). The result has been a global push for transparency in fisheries that includes initiatives by top fishing nations such as Peru, Panama and Indonesia to share their VMS data and has led to Chile, Costa Rica and Namibia publicly committing to share vessel tracking data. Researchers increasingly use data and analysis tools on this free access platform.
- **OceanMind.** Ocean Mind's mission is to advance ocean sustainability through providing actionable intelligence on fishing activities to maritime authorities, government agencies, ocean conservationists and seafood buyers. OceanMind tends to work on proprietary data and works closely with a country's MCS staff. The best example of this is the productive relationship between OceanMind, the Seafood Task Force and the Thai government in the successful response to an EU yellow carding.
- **Vulcan's Skylight.** Skylight operates on a vision which is similar to that of OceanMind but under a forprofit model. Skylight provides maritime intelligence software and service solutions for identifying suspicious vessel behaviors and "dark vessel" activity and delivers this through an online alerting platform and watch floor service.

Institutions, Organisations and Tribunals Governing the World's Fisheries

- **Food and Agriculture Organization of the United Nations (FAO).** The FAO was established in 1945 as a specialised agency under the United Nations that leads international efforts to defeat hunger and to achieve food security. In this context, fisheries have been long regarded as a vital sector in achieving the mission of the FAO. A subsidiary body known as the Committee on Fisheries (COFI) was established in 1965 to serve as the global intergovernmental forum tasked with examining major international fisheries and aquaculture problems and issues. COFI has two main functions: to review FAO programs in fisheries and aquaculture and their implementation and to conduct periodic general reviews of international fisheries and aquaculture problems and recommend possible solutions. The FAO has adopted several binding legal instruments and voluntary guidelines addressing IUU fishing. In 1993, the FAO Conference adopted the Compliance Agreement. The FAO has also initiated the Code of Conduct for Responsible Fisheries (CCRF), which was adopted in Resolution 4/95 by the FAO Conference on 31 October 1995. On 2 March 2001, it adopted the International Plan of Action to Prevent, Deter and Eliminate Illegal, Unreported and Unregulated Fishing (IPOA-IUU).

  The FAO also adopted the Agreement on Port State Measures to Prevent, Deter and Eliminate Illegal, Unreported and Unregulated Fishing (Port State Measures Agreement) at its 36th session on 22 November 2009. In addition to these instruments, the FAO has adopted Voluntary Guidelines for Flag State Performance and Voluntary Guidelines for Catch Documentation Schemes.
- **Regional fisheries management organisations.** Regional measures in the conservation and management of fish resources through RFMOs may have been regarded as the most reasonable approach in dealing with issues arising from the use of transboundary marine living resources. These organisations reflect the essential duty of states under international law to cooperate in ensuring the

conservation and sustainable use of transboundary resources or resources beyond areas of national jurisdiction. Nevertheless, unresolved issues have become obstacles for states in effectively implementing the commitments made through their RFMOs. These issues arise from the activities of nonparticipating states, since according to international law, treaties are only binding on their parties and do not create rights or obligations to third parties without their consent. Thus, vessels registered under the flag of states that are not parties to a particular RFMO are not obliged to comply with the rules that have been agreed by the state parties of the RFMO. Such states are known as 'free riders', and they might undermine conservation and management measures and any incentive for member state nationals to comply.

Some scholars have proposed a new approach to resolve the problem of free riders by recognising that each element of IUU fishing needs to be tackled individually and that there is a need to examine the situation of non-RFMO members in a more detailed manner (Serdy 2017). Currently, there are 18 Regional Fishery Bodies in the world:

- International Commission for the Conservation of Atlantic Tunas (ICCAT)
- Indian Ocean Tuna Commission (IOTC)
- Western and Central Pacific Fisheries Commission (WCPFC)
- Inter-American Tropical Tuna Commission (IATTC)
- Agreement on the International Dolphin Conservation Program (AIDCP; sister organisation to IATTC)
- Commission for the Conservation of Southern Bluefin Tuna (CCSBT)
- North-East Atlantic Fisheries Commission (NEAFC)
- Northwest Atlantic Fisheries Organization (NAFO)
- North Atlantic Salmon Conservation Organization (NASCO)
- South East Atlantic Fisheries Organisation (SEAFO)
- South Indian Ocean Fisheries Agreement (SIOFA)
- South Pacific Regional Fisheries Management Organisation (SPRFMO)
- Convention on the Conservation of Antarctic Marine Living Resources (CCAMLR)
- General Fisheries Commission for the Mediterranean (GFCM)
- Convention on the Conservation and Management of Pollock Resources in the Central Bering Sea (CCBSP)
- Western Central Atlantic Fishery Commission (WECAFC)
- Fisheries Committee for the Eastern Central Atlantic (CECAF)

• **International Tribunal for the Law of the Sea (ITLOS).** As of May 2019, 27 cases had been brought before ITLOS, including cases related to the question of prompt release of vessels, the use of marine living resources, the protection and preservation of the marine environment, and maritime boundary delimitation. Of these cases, the judgments of ITLOS with regard to prompt release of vessels are perhaps mostly relevant to fisheries issues. These include, for instance, the *Camouco* case *(Panama v. France*), the *Volga* case (*Russian Federation v. Australia*) (ITLOS 1999a), and the *Tomimaru* case (*Japan v. Russian Federation*). Although these cases did not directly touch on the issue of IUU fishing, the judgments of ITLOS provided useful guidance with regard to administrative procedures and law enforcement undertaken by coastal states as parts of combatting IUU fishing. In addition, the judgment of ITLOS on the *M/V Saiga* (No. 2) case (*Saint Vincent and the Grenadines v. Guinea*) (ITLOS 1999b) presented the response of ITLOS on the application of a 'genuine link' between the vessel and the flag state as stipulated under the UNCLOS, and its relation to the implementation of the UNFSA and the FAO Compliance Agreement. In this case, Guinea argued that there was no genuine link between the Saiga and Saint Vincent and the Grenadines as the flag state. Therefore, Guinea contended that it was not obliged to recognise the claims of Saint Vincent and the Grenadines in relation to the ship.

Appendix B: Voluntary Actions for Ocean Fishery Stakeholders Under SDG 14.4

IUU fishing is a global problem that threatens food security, livelihoods and ecosystem health. Effectively combatting IUU fishing begins by recognising its major drivers: economic incentives for illegal behaviour; weak governance regimes at the national, regional and international level; and barriers to effective enforcement.

As this paper makes clear, many actions are available to address each problem area. Readers are encouraged to consider the potential of the full range of actions—local, national, regional or international—to have an impact on IUU fishing. Collaboration maximises effects and rationalises costs. Collaboration must be considered across government, businesses and private sectors; the civil sector; and science establishments.

Sustainable Development Goal (SDG) 14 is to conserve and sustainably use the ocean, seas and marine resources for sustainable development. Subgoal 14.4 states: By 2020, effectively regulate harvesting and end overfishing, illegal, unreported and unregulated fishing and destructive fishing practices and implement science based management plans, in order to restore fish stocks in the shortest time feasible, at least to levels that can produce maximum sustainable yield as determined by their biological characteristics.

Obviously, this subgoal will not be met, but we encourage new voluntary action commitments by the end of 2020.

We offer the following actions for consideration that can be taken by any stakeholder, including countries; regional associations; industry, labour, or financial organisations; or non-governmental organisations.

Strengthen Existing Ocean Governance Mechanisms

- The non-uniformity of regional fisheries management organisation (RFMO) regulations should be addressed through an international forum/mechanism. The UN General Assembly may be the most efficient. The Kobe Process is a foundation for any call to action; however, this should involve all of the world's RFMOs, not just the tuna RFMO.
- Coordination and data transparency among RFMOs, flag states, regional mechanisms and the coastal and market states should be improved.
- All RFMOs should adopt strong transboundary policies on monitoring and enforcement, including on the high seas.
- Sanctions should be created for flag states which are not performing enforcement measures.
- A mechanism or authorised international body should be considered to oversee the performance of each RFMO.
  - The gaps and inconsistencies in fisheries management should be identified.
  - Transshipment practices should be properly regulated.
- Specific efforts should be focused on the management of small-scale and artisanal fisheries.
- The significant gaps in high seas governance should be addressed.
- The Food and Agriculture Organization of the United Nations (FAO) and RFMOs should ban unsustainable fishing gears and practices.

Flag State Actions and Responsibilities

- Flag states should exert adequate control over the vessel registry, including ensuring that the management of the registry is within the flag state (and not held by an external private company). Flag states must demonstrate and maintain a genuine link between the vessel and the flag state. Open registries should be closed to fishing vessels.
- The significant gaps in high seas governance should be addressed.
- All states registering a fishing vessel should require companies to provide information on the vessel's true beneficial owner and apply sanctions if it is found that the information provided is false.
- In addition to the beneficial owner, the vessel registry should include details of vessel characteristics and history, including prior flag and name changes; information on the operator; and information on markings and Unique Vessel Identifiers (UVIs), in line with international standards.
- A list of fishing license holders should be made publicly available (for example, online) and regularly updated. It should contain the same information as in the vessel registry, as well as details of any quota or other limits allocated to vessels (if applicable), period of license, gears and target species and areas where the vessels are authorised to fish.
- Eligible vessels should be required to apply for and install an International Maritime Organization (IMO) number as a condition of registration.
- Any vessel that is in non-compliance should be 'blacklisted' from the registry.
- Catch certificates should be digitally validated. Flag states should implement controls and verification/data cross-checks for the reliable certification of catches for export to countries that require the validation of a catch certificate. This should be done through (non-exhaustive list) checks of logbook data and landing/transshipment declaration, verification of fishing location using Vessel Monitoring System (VMS) positions, inspections at sea and in port, and presence of observers on board.
- Due diligence should be shown for new license holders. All states granting fishing authorisations to vessels should maintain a register of license holders that is up-to-date and in line with the vessel registry. States should verify the infraction history of vessels and vessel owners when a license is requested.

Coastal State Actions and Responsibilities

- Effective monitoring, control and surveillance (MCS) measures should be adopted to ensure compliance with Conservation and Management Measures (CMMs).
- VMS should be required on board all industrial fishing vessels (with regular reporting), and a fisheries monitoring centre for monitoring VMS data should be established. VMS should be made public, or Automatic Identification System (AIS) should be mandated for public tracking.
- Vessel captains should be required to maintain a logbook (and consider mandating e-logbooks), and vessels should be required to report fisheries related data, including catches and fishing effort. In case of non-compliance, states should take prompt action with respect to the iden-

tified infringements and apply deterrent sanctions in a consistent and transparent manner.

- Transshipments at sea should be banned unless they are pre-authorised and are subject to robust and verifiable electronic monitoring and/or are covered by a human observer scheme appropriate to the fishery.
- Labour regulations should be sufficient to facilitate the identification and investigation of forced labor, labor abuse and human trafficking cases detected on board fishing vessels. Regulations should also be sufficient to allow for the lawful prosecution and penalisation of perpetrators of these crimes. The Cape Town Agreement (CTA) and International Labour Organization (ILO) Work in Fishing Convention (C188) should be ratified and adopted.
- Appropriate port-side and at-sea inspection regimes should be initiated to facilitate the identification and investigation of labour abuses. These should involve a trained labour inspector who can detect the common indicators of forced labor and human trafficking. These indicators should follow the 11 ILO indicators of forced labor and/or the ILO indicators for human trafficking. These inspections should follow a precautionary and victim-centred approach to labor investigations, ensuring that fisher workers' safety, security and privacy are always a top priority. Wherever applicable, trained and qualified translators should be included to facilitate communications with foreign workers.
- Specific efforts should be focused on the management of small-scale and artisanal fisheries.

Port State Actions and Responsibilities

- The Port State Measures Agreement (PSMA), a key tool in the fight against IUU fishing that directly supports SDG 14.4, should be ratified. While it can take more than a year to adjust legislation to permit the ratification of the PSMA, states should make clear commitments to do so without delay. Implementation of any ratified states must be a priority for those states.
  - The PSMA should be implemented.
- To ensure the safety of crews on board fishing vessels and improve working conditions, the ILO C188 should be implemented and countries should accede to the CTA.
  - The Cape Town Agreement and ILO C188 are complementary regulations and can be considered during port inspections under the PSMA.

Market State Actions and Opportunities

- Regulations similar to the European Union's IUU fishing Regulation should be adopted.
- The PSMA should be implemented.
- Transparency measures should be implemented.
- Incentives (e.g., waive import tariff) should be given to countries with good performance in combatting IUU fishing.
- The significant gaps in high seas governance should be addressed and a commitment made to doing so where the timescale is more than 2 years.

Legal Frameworks for Actions

- Strong legal frameworks are needed for improved MCS, enforcement and sanctions.
- Market, flag, port and coastal states should ensure that they ratify, accept or accede to, as appropriate, relevant legal instruments (Appendix A), including the following key instruments:
  - UN Convention on the Law of the Sea (UNCLOS)
  - UN Fish Stocks Agreement (UNFSA)
  - Port State Measures Agreement (PSMA)
  - FAO Compliance Agreement
- National legislation should be adopted/updated to ensure that the legal framework is consistent with these requirements, as well as CMM measures established by the RFMO of which the country is a member.
- Legal frameworks should establish a clear, comprehensive and transparent system of proportionate and deterrent sanctions for IUU fishing offences, including for nationals supporting or engaging in IUU fishing.
- There should be a legal basis for all MCS and enforcement measures that may include issuing licenses to vessels, requiring vessels to carry and operate VMS or AIS, conducting inspections of vessels, investigating infringements, refusing access of IUU vessels to port, and regulating beneficial ownership.
- States should publish or provide information on their legislation and actions in combatting IUU fishing to an authorised international body, such as the FAO.

Adopt Global Transparency Rules and Technology

- All state registries should be made available to the FAO Global Record of Fishing Vessels.
- Unedited VMS, or other proprietary tracking system, data should be made public with regular transmission intervals sufficient to ensure vessels can be permanently tracked, or AIS should be required for fishing vessels.
- Up-to-date lists should be published of all fishing licenses, authorisations and vessel registries, transshipment authorisations and refrigerated vessels registries.

- IMO numbers (free to obtain) should be mandated for all eligible vessels, and a national UVI scheme should be implemented for non-eligible vessels, maintaining a vessel registry and providing all information to the FAO Global Record of Fishing Vessels (which ultimately includes all eligible vessels over 12 m in length overall).
- States should be encouraged to publish information, on an accessible international platform, on arrests and sanctions imposed on individuals and companies for IUU fishing activities, human trafficking and other related crimes.
- Transshipment activities information should be improved through mandatory and public preauthorisation and robust and verifiable electronic monitoring schemes.
- There should be mandatory near-term adoption of cost-effective digital tools, such as blockchain, that safeguard in a digital form key information on vessel registration, licenses, unloading records, catch location and information and crew documentation; these should be designed in such a way as to support a rapid move towards a universal, interoperable digital catch certification scheme.
- Information on beneficial ownership should be published, and companies should be required to provide information on true beneficial ownership when applying for a fishing license, fishing authorisation or registration to their flag.

Enhance International Cooperation

- All countries should cooperate to prevent, deter and eliminate IUU fishing at the bilateral and sub-regional levels. Where mechanisms exist, they must be prioritised within government agendas.
  - States should develop and implement a national plan of action on IUU fishing in line with the recommendations of the International Plan of Action to Prevent, Deter and Eliminate Illegal, Unreported and Unregulated Fishing. Such cooperation should include the following measures:
  - Coastal states should communicate information on IUU fishing-related infringements to neighbouring coastal states.
  - Port states should provide information to flag states on transshipments, landings and denials of use of port involving vessels flying their flag, and the result of inspections.
  - Flag states should cooperate with investigations and act in cases of presumed IUU fishing by their vessels.
  - All states should build capacity to support analysis, implementation and application of policy and technology.

Domestic Fisheries Actions by Coastal States

- There should be well-established and adequately resourced domestic fisheries management arrangements, supported by sound and properly enforced legislation.
- Clear and transparent CMMs should be established in national legal frameworks, based on the best available scientific advice and consistent with obligations under UNCLOS, UNFSA and RFMO rules.
- A national fisheries management plan should be developed and implemented, and total allowable catch based on the best available scientific evidence should be determined.
- Vessels operating in the exclusive economic zone (EEZ) should be required to hold a license, and there should be a balance between the number of licenses granted/size of fishing activity in the EEZ and the status of stocks. This should be based on scientific and precautionary stock assessments in accordance with maximum sustainable yield and total allowable catch.
- A record of vessels licensed to fish in the EEZ should be established, made publicly available (online) and kept up-to-date. The record should contain vessel details (name, tonnage, flag, registration number); vessel and/or gear type and target species; details of any quota allocated (if applicable); and details of the vessel's legal owner and operator, including beneficial owner, period of the license, license fee and crew manifest.
- IMO numbers should be required for foreign-flagged eligible fishing vessels as a condition of their license to operate.

Digital Documentation for All States

- All states should implement digital schemes for documenting catch data to promote global exchange of information, vessel registration and licenses, unloading records, catch location and information and crew documentation.
  - These digital systems must be secure and have built-in procedures to prevent the unauthorised deletion or overwriting of data.

Actions for Business, Industry, Financial Institutions, Scientists and Civil Society

- Enhanced action by private sector organisations can create strong pressure on fisheries businesses to maintain a high level of compliance.

- Financial institutions (e.g., banks and insurance companies) are encouraged to make fisheries transparency and traceability conditions of contracts.
- Buyers and lenders should establish the accuracy of ownership information to avoid the use of shell companies.
- Private sector organisations should not deal with flag states that fall short of their duties under the UNCLOS.
- The use of PSMA-ratified ports should be made a condition of contract or insurance.
- Assurance should be provided to consumers on the traceability and quality of the fish products (e.g., provide such information on the packaging).
- Zero-waste products should be promoted by creating other products from waste.
- Appropriate machineries and methods for a cost effective production should be used.
- Science has a crucial role in policymaking. Scientists are expected to provide the following information to create better fisheries governance:
  - The best assessment of fish stocks globally
  - Change of fish behavior/migration pattern caused by climate change
  - Advancing technology of sustainable fishing gears

Civil Society Can Bring Communities Together for Collective Action

- Civil society can promote awareness on fisheries sustainability to educate consumers to choose fish products with guaranteed traceability.
- Civil society can advocate for artisanal fisheries to be aware of fisheries sustainability.

References

Agnew DJ, Pearce J, Pramod G, Peatman T, Watson R, Beddington JR, Pitcher TJ (2009) Estimating the worldwide extent of illegal fishing. PLoS One 4(2):e4570

Allison EH (2001) Big laws, small catches: global ocean governance and the fisheries crisis. J Int Dev 13:933–943

Andreu-Cazenave M, Subida MD, Fernandez M (2017) Exploitation rates of two benthic resources across management regimes in Central Chile: evidence of illegal fishing in artisanal fisheries operating in open access areas. PLoS One 12(6):e0180012

APEC (Asia-Pacific Economic Cooperation) (2008) Assessment of impacts of illegal, unreported and unregulated (IUU) fishing in the Asia-Pacific. APEC, Singapore

Becker GS (1968) Crime and punishment: an economic approach. J Polit Econ 76(2):169–212

Belhabib D, Greer K, Pauly D (2017) Trends in industrial and artisanal catch per effort in West African fisheries. Conserv Lett 11(1):1–10

Bell JD, Johnson JE, Ganachaud AS, Gehrke PC, Hobday AJ, Hoegh-Guldberg O, Le Borgne R et al (2011) Vulnerability of tropical pacific fisheries and aquaculture to climate change. Secretariat of the Pacific Community, Noumea

Béné C (2015) Feeding 9 billion by 2050—putting fish back on the menu. Food Secur 7(2):261–274

Cabral RB, Mayorga J, Clemence M, Lynham J, Koeshendrajana S, Muawanah U, Nugroho D et al (2018) Rapid and lasting gains from solving illegal fishing. Nat Ecol Evol 2(4):650–658

California Environmental Associates (2018) Trends in marine resources and fisheries management in Indonesia. California Environmental Associates, San Francisco

Cheung W (2016) Climate change effects on illegal, unreported and unregulated fishing. Nippon Foundation. https://nereusprogram.org/works/climate-change-effects-onillegal-unreported-and-unregulated-fishing

Clark MR, Bowden DA, Rowden AA, Stewart R (2019) Little evidence of Benthic Community resilience to bottom trawling on seamounts after 15 years. Front Mar Sci 6:63. https://doi.org/10.3389/fmars.2019.00063. https://www.frontiersin.org/articles/10.3389/fmars.2019.00063/full#B10

Costello C, Cao L, Gelcich S (2019) The future of food from the sea. World Resources Institute, Washington, DC. http://www.ocean-panel.org/future-food-sea

DCDC (Development, Concepts and Doctrine Centre) (2018) Global strategic trends. Ministry of Defence, London

Declaration OL (2019) G20 Osaka leaders declaration. Ministry of foreign affairs of Japan.

Drammeh OKL (2000) Illegal, unreported and unregulated fishing in small scale marine and inland capture fisheries. Government of Australia and FAO, Sydney

Dulvy NK, Rogers SI, Jennings S, Stelzenmüller V, Dye SR, Skjoldal HR (2008) Climate change and deepening of the North Sea fish assemblage: a biotic indicator of warming seas. J Appl Ecol 45(4):1029–1039. https://doi.org/10.1111/j.1365-2664.2008.01488.x@10.1111/(ISSN)1365-2664.CLIMATE_JPE

Dyck AJ, Sumaila UR (2010) Economic impact of ocean fish populations in the global fishery. J Bioecon 12:227–243

EJF (2018) Out of the shadows: improving transparency in global fisheries to stop illegal, unreported and unregulated fishing. EJF, London. https://ejfoundation.org/resources/downloads/Transparency-report-final.pdf

EJF (Environmental Justice Foundation) (2009) Lowering the flag—ending the use of flags of convenience by pirate fishing vessels. EJF, London

EJF and Hen Mpoano (2019) Stolen at sea: how illegal "saiko" fishing is fuelling the collapse of Ghana's fisheries. EJF, London

EJF, Oceana, Pew Charitable Trusts, and WWF (2015) EU regulation to combat illegal fishing third country carding process: success for South Korea and the Philippines. Case study 2. http://www.iuu-watch.eu/wp-content/uploads/2015/06/Case-Study2.FINAL_.EN_.pdf

FAO (2001) The international plan of action to prevent and deter illegal, unreported and unregulated fishing. FAO, Rome

FAO (2002) Implementation of the IPOA-IUU. FAO, Rome

FAO (2009) Agreement on Port State measures to prevent, deter and eliminate illegal, unreported and unregulated fishing. FAO, Rome

FAO (2017) Short-term projection of global fish demand and supply gaps. FAO, Rome

FAO (2018) The state of world fisheries and aquaculture 2018—meeting the sustainable development goals. FAO, Rome

FAO (Food and Agriculture Organization of the United Nations) (2000) Expert consultation on illegal, unreported and unregulated fishing. FAO, Sydney

Fenichel EP, Levin SA, McCay B, St. Martin K, Abbott JK, Pinsky ML (2016) Wealth reallocation and sustainability under climate change. Nat Clim Change 6(3):237–244

Fluet-Chouinard E, Funge-Smith S, McIntyre PB (2018) Global hidden harvest of freshwater fish revealed by household surveys.

Proc Natl Acad Sci 115(29):7623–7628. https://doi.org/10.1073/pnas.1721097115

Free CM, Mangin T, García Molinos J, Ojea E, Costello C, Gaines SD (2019) Realistic fisheries management reforms could mitigate the impacts of climate change in most countries. BioRxiv. https://doi.org/10.1101/804831

Freestone D (2010) Problems of high seas governance. In: Vidas D, Schei PJ (eds) The world ocean in globalisation: challenges and responses. Brill and Martinus Nijhoff, Leiden, pp 99–130

Friend E (2015) Meat consumption trends in Asia Pacific, and what they mean for foodservice strategy (blog). https://blog.euromonitor.com/meat-consumption-trends-inasia-pacific-and-what-they-mean-for-foodservice-strategy

G7 (2018) Charlevoix blueprint for healthy oceans, seas and resilient coastal communities. Sommet Du G7–G7 Summit.

Gaines SD, Costello C, Owashi B, Mangin T, Bone J, García Molinos J, Burden M et al (2018) Improved fisheries management could offset many negative effects of climate change. Sci Adv 4(8):eaao1378. https://doi.org/10.1126/sciadv.aao1378

Gaines S, Cabral R, Free C, Golbuu Y (2019) The expected impacts of climate change on the ocean economy. World Resources Institute, Washington, DC. http://www.oceanpanel.org/expected-impacts-climate-change-ocean-economy

Gallic B, Cox A (2006) An economic analysis of illegal, unreported and unregulated (IUU) fishing: key drivers and possible solutions. Mar Policy 30:689–695

Gianni M, Fuller SD, Currie DEJ, Schleit K, Goldsworthy L, Pike B, Weeber B, Owen S, Friedman A (2016) How much longer will it take? A ten-year review of the implementation of United Nations General Assembly Resolutions 61/105, 64/72 and 66/68 on the management of bottom fisheries in areas beyond national jurisdiction. Deep Sea Conservation Coalition, Amsterdam

Global Fishing Watch (2017) First global view of transshipment at sea. https://globalfishingwatch.org/transshipment-success/report-first-global-view-transshipment-sea

Global Ocean Commission (2014) From decline to recovery: a rescue package for the Global Ocean. Global Ocean Commission, Oxford

Haenlein C (2017) Below the surface: how illegal, unreported and unregulated fishing threatens our security. RUSI Occasional Paper. https://rusi.org/sites/default/files/201707_rusi_below_the_surface_haenlein.pdf

High Seas Task Force (2006) Closing the net: stopping illegal fishing on the high seas. Governments of Australia, Canada, Chile, Namibia, New Zealand, and the United Kingdom, WWF, IUCN, and the Earth Institute at Columbia University, Bellegarde

Hilborn R, Lobo Orensanz JM, Parma AM (2005) Institutions, incentives and the future of fisheries. Philos Trans R Soc B Biol Sci 360(1453):47–57

Hoegh-Guldberg O (2019) The ocean as a solution to climate change: five opportunities for action. World Resources Institute, Washington, DC

Holland J (2018) Spanish Tycoon hit with USD 10 million fine for illegal fishing. Seafood Source. https://www.seafoodsource.com/news/environment-sustainability/spanishtycoon-hit-with-usd-10-million-fine-for-illegal-fishing

Hornidge AK, Hadjimichael M (2019) Ocean governance. Pasts, presents, futures. https://www.oceangov.eu/wpcontent/uploads/2019/06/OceanGov_edVol_CfP.pdf

Hutnikzac B, Delpeuch C (2018) Combatting illegal, unreported and unregulated fishing: where countries stand and where efforts should concentrate in the future. OECD, Paris

International Chamber of Shipping (2015) Shipping industry flag state performance table. https://www.ics-shipping.org/free-resources/flag-state-performance-table

International Transport Worker Federation (2020) Flags of convenience. https://www.itfglobal.org/en/sector/seafarers/flags-of-convenience

Interpol (2016) Project scale. Interpol, Lyon. https://www.interpol.int/content/download/5116/file/Project%20Scale%20brochure.pdf

IPBES (Intergovernmental Science-Policy Platform on Biodiversity and Ecosystem Services) (2019) Summary for policymakers of the global assessment report on biodiversity and ecosystem service of the Intergovernmental Science-Policy Platform on Biodiversity and Ecosystem Services. IPBES Secretariat, Bonn

IPCC (2019) Summary for Policymakers. In: IPCC Special report on the ocean and cryosphere in a changing climate [H.-O. Pörtner DC, Roberts V, Masson-Delmotte P, Zhai M, Tignor E, Poloczanska K, Mintenbeck A, Alegría M, Nicolai A, Okem J, Petzold B, Rama NM, Weyer (eds.)]. In press.

ITLOS (1999a) Southern Bluefin Tuna (New Zealand v. Japan; Australia v. Japan), provisional measures. Hamburg, ITLOS. Joseph, J., D. Squires, W. Bayliff, and T. Groves. 2010

ITLOS (International Tribunal for the Law of the Sea) (1999b) M/V "Saiga" (No. 2) (Saint Vincent and the Grenadines v. Guinea), Judgment. ITLOS, Hamburg

Joseph J et al (2010) Addressing the problem of excess fishing capacity in Tuna Fisheries. In: Allen R, Joseph J, Squires D (eds) Conservation and management of transnational tuna fisheries. Blackwell, Ames, pp 11–38

Knauss JM (2005) The growth of British fisheries during the industrial revolution. Ocean Dev Int Law 36(1):1–11

Konar M, Grey E, Thuringer L, Sumaila UR (2019) The scale of illicit trade in Pacific Ocean marine resources. Working Paper. World Resources Institute, Washington, DC

Macfadyen G, Caillart B, Agnew D (2016) Review of studies estimating levels of IUU fishing and the methodologies utilized. Poseidon Aquatic Resources Management, Portmore

Martini R (2019) Many government subsidies lead to overfishing. Here's a solution. OECD. https://www.oecd.org/agriculture/government-subsidies-overfishing

McDowell R, Mendoza M, Mason M (2015) AP tracks slave boats to Papua New Guinea. AP. https://www.ap.org/explore/seafood-from-slaves/ap-tracks-slave-boats-to-papuanew-guinea.html

Michael PE, Thomson R, Barbraud C, Delord K, de Grissac S, Hobday AJ, Strutton PG et al (2017) Illegal fishing bycatch overshadows climate as a driver of albatross population decline. Mar Ecol Prog Ser 579:185–199

MinistryofMarineAffairsandFisheriesoftheRepublicofIndonesia(2019) KKP 4-year program performance achievement. https://kkp.go.id/infografis-detail/2633pencapaian-kinerja-program-4-tahun-kkp

Molenaar EJ (2011) Non-participation in the Fish Stocks Agreement: status and reasons. Int J Mar Coastal Law 26:195–234

MRAG Asia Pacific (2016) Towards the quantification of illegal, unreported and unregulated (IUU) fishing in the Pacific Islands Region. MRAG Asia Pacific, Queensland. https://www.ffa.int/files/FFA%20Quantifying%20IUU%20Report%20-%20Final.pdf

Owen D (2016) Introductory briefing on the due diligence obligation of flags states regarding their fishing vessels, the liability for breach of this obligation and the possibilities for litigation. World Wildlife Fund. https://www.chathamhouse.org/sites/default/files/Daniel%20Owen.pdf

Pauly D, Zeller D (2016) Catch reconstructions reveal that global marine fisheries catches are higher than reported and declining. Nat Commun 7(1):10244

Perry AL, Low PJ, Ellis JR, Reynolds JD (2005) Climate change and distribution shifts in marine fishes. Science 308(5730):1912–1915. https://doi.org/10.1126/science.1111322

Pew Charitable Trusts (2018) The Cape Town agreement explained: how one international treaty could combat illegal fishing and save

lives. https://www.pewtrusts.org/-/media/assets/2019/09/ctaexplained_brief.pdf
Pikitch E, Boersma PD, Conover DO, Cury P, Essington TE, Heppell SS, Houde ED et al (2012) Little fish, big impact: managing a crucial link in ocean food webs. Lenfest Ocean Program, Washington, DC
Pinsky ML, Worm B, Fogarty MJ, Sarmiento JL, Levin SA (2013) Marine taxa track local climate velocities. Science 341(6151):1239–1242. https://doi.org/10.1126/science.1239352
Poloczanska ES, Brown CJ, Sydeman WJ, Kiessling W, Schoeman DS, Moore PJ, Brander K et al (2013) Global imprint of climate change on marine life. Nat Clim Change 3(10):919–925. https://doi.org/10.1038/nclimate1958
Rayfuse R, Warner R (2008) Securing a sustainable future for the oceans beyond national jurisdiction: the legal basis for an integrated cross-sectoral regime for high seas governance for the 21st century. Int J Mar Coastal Law 23(3):399–421
Richardson K, Gunn R, Wilcox C, Hardesty BD (2018) Understanding causes of gear loss provides a sound basis for fisheries management. Mar Policy 96:278–284
Sala E, Mayorga J, Costello C, Kroodsma D, Palomares MLD, Pauly D, Sumaila UR, Zeller D (2018) The economics of fishing the high seas. Sci Adv 4(6):eaat2504
SEAFDEC (Southeast Asian Fisheries Development Center) (2015) Asian guidelines for preventing the entry of fish and fishery products from IUU fishing activities into the supply chain. SEAFDEC, Myanmar
Serdy A (2017) Pacta Tertiis and regional fisheries management mechanisms: the IUU fishing concept as an illegitimate short-cut to a legitimate goal. Ocean Dev Int Law 48:345–364
Sumaila UR (2018) Illicit trade in the marine resources of West Africa. Ghanaian J Econ 6(1):108–116
Sumaila UR (2019) Updated estimates and analysis of global fisheries subsidies. Mar Policy 109:1–11
Sumaila UR, Bawumia M (2014) Fisheries, ecosystem justice and piracy: a case study of Somalia. Fish Res 157:154–163
Sumaila UR, Alder J, Keith H (2006) Global scope and economics of illegal fishing. Mar Policy 30(6):696–703
Sumaila UR, Lam V, Le Manach F, Swartz W, Pauly D (2016) Global fisheries subsidies: an updated estimate. Mar Policy 69:189–193
Sumaila UR, Jacquet J, Witter A (2017) When bad gets worse: corruption and fisheries. In: Williams A, Le Billon P (eds) Corruption, natural resources and development. Edward Elgar, Cheltenham, pp 93–105
Taconet M, Kroodsma D, Fernandes JA (2019) Global atlas of AIS-based fishing activity: challenges and opportunities. FAO, Rome. http://www.fao.org/documents/card/en/c/ca7012en
Tanaka Y (2012) International law of the sea. Cambridge University Press, Cambridge
Task Force 115 (2018) Ending fisheries crime in Indonesia. FishFORCE Newsletter 003/2018
Telesetsky A (2014) Laundering fish in the global undercurrents: illegal, unreported and unregulated fishing and transnational organized crime. Ecol Law Q 41(4):939–997
Tinch R, Dickie I, Lanz B (2008) Costs of illegal, unreported and unregulated (IUU) fishing in EU fisheries. Economics for the Environment Consultancy, London
Undercurrent News (2019) Thai court hands out fines of over $16m to IUU operation. https://www.undercurrentnews.com/2019/05/17/thai-court-hands-out-finesof-over-16m-to-iuu-operation
United States v. Bengis, S1 03 Crim. 0308 (LAK), at 2 (S.D.N.Y. Jun. 14, 2013)
Watson RA, Cheung WWL, Anticamara JA, Sumaila RU, Pauly D (2013) Global marine yield halved as fishing intensity redoubles. Fish Fish 14:493–503
Zimmerhackel JS, Rogers AA, Meekan MG, Ali K, Pannell D, Kragt ME (2018) How shark conservation in the Maldives affects demand for dive tourism. Tour Manag 69:263–271

16 Organised Crime in the Fisheries Sector

Highlights

- The fisheries sector, like most economic sectors, is impacted by organised crime. Organised crime deprives states of national revenue and threatens the legitimate fishing industry and the livelihoods of those that rely on it.
- Organised crime in fisheries undermines the rule of law, threatens peace and security, jeopardises food security for coastal states and communities, and adversely impacts fishing communities and fish stocks at the national, regional and global levels.
- Manifestations of organised crime in fisheries take many forms: fraud, corruption, tax crime, money laundering, crime in the labour market, security offences, drug trafficking, smuggling of fuel and migrants and fisheries offences. These offences occur throughout the fisheries value chain, often in combination, and frequently transnationally.
- The continuation of organised crime in fisheries undermines the global commitment to sustainable development and the realisation of a sustainable ocean economy.
- Yet despite significant evidence of the dynamic and destructive impact of organised crime in the fisheries sector, the need remains for an effective, coordinated enforcement response at the national level and globally.
- This paper summarises the current state of knowledge on the phenomenon of organised crime in the fisheries sector and its impact on sustainable development, using case studies and examples, and drawing on available scholarly literature, technical documents, media reports and expert input. The paper identifies opportunities for action for moving forward at a global level to address this challenge to the realisation of a sustainable ocean economy.
- To comprehensively address organised crime in fisheries, states should, first, build a shared understanding of the problem globally and, second, undertake intelligence-led, skills-based cooperative law enforcement at the domestic level facilitated by enabling legislative frameworks and increased transparency.

Methodology and Sources

- This Blue Paper is based on invited input from the contributing authors, who are located around the globe and represent practitioners, academics and law enforcement. They were recommended to the lead authors based on their experience and knowledge of particular aspects of the topic, as well as the need for gender balance and to elevate voices from the Global South. The lead authors incorporated the contributors' written submissions into the body of the text. The paper is further shaped by feedback from expert consultations.
- The paper was compiled over a three-month period. It draws on existing reports and outcome documents on the topic from international governmental organisations, non-governmental organisations, governments and knowledge institutes, as well as published academic research, operational knowledge and expert input. The concept of 'knowledge' is interpreted widely; faced at times with sparse formal documentation of the manifestations of organised crime in fisheries, some of the illustrative examples referred to draw on anecdotal accounts, personal expert observations or journalist reports (or a mixture thereof), rather than citing scientific research or decided cases. Additionally, as the breadth of the issues covered is vast, it is recognised that various topics could benefit from further dedicated research, such as the scope of organised crime in the fisheries sector at a global level, analyses of the causal nexus between organised crime in fisheries and the highlighted potential impacts, and critical assessment of the suggested promising practices.

Originally published in:
Witbooi, E., K.-D. Ali, M.A. Santosa et al. 2020. Organised Crime in the Fisheries Sector. Washington, DC: World Resources Institute. Available online at https://oceanpanel.org/blue-papers/organised-crime-associated-fisheries
Reprint by Springer International Publishing (2023) with kind permission.
Published under license from the World Research Institute.

J. Lubchenco, P. M. Haugan (eds.), *The Blue Compendium*, https://doi.org/10.1007/978-3-031-16277-0_16

Addressing such issues would require empirical research employing a different methodology than the one used here. In sum, this paper contributes to the knowledge pool on the topic under examination while remaining mindful that there is scope for increased understanding of many facets thereof.

Aim

- This Blue Paper aims to present an accessible summary of the current state of knowledge on the phenomenon of organised crime in fisheries and its impact on sustainable development, using case studies and examples sampled from contributors and experts, and to provide a suite of practical opportunities for action for moving forward at 2 I High Level Panel for a Sustainable Ocean Economy a global level to address the challenge organised crime poses to the realisation of a sustainable ocean economy.

1 Introduction

The modern fisheries sector is globalised, industrialised and integrated into the worldwide financial market. Like most other economic sectors, it is exposed to organised crime (UNODC 2011), the underbelly of globalisation (e.g. Madsen 2009; Abadinsky 2007; Obokata 2010).

While sometimes referred to as an 'emerging' crime (Stringer and Harré 2019), there is little reason to believe that organised crime in the fisheries sector is a novel problem. The infamous gangster Al Capone, for instance, exploited the fishing industry for rum-running during the 1920s, when the United States prohibited the production, importation, transportation and sale of alcoholic beverages (Ensign 2001; Demont 2003). The more likely scenario is that organised crime in fisheries is a recent label for a phenomenon that has existed for many years, fuelled by overfishing of declining fish stocks and greed, among other causes. In this narrative, an emerging focus on sustainable fisheries management and the role of the blue economy has heightened global attention to organised crime in fisheries, which is giving rise to a shadow blue economy and undermining the competitiveness of both the legitimate industry and the livelihoods of coastal communities. This heightened attention is allowing policymakers, researchers, and civil society to re-examine the dynamics and destructiveness of the shadow blue economy and the role of organised crime in it.

Organised crime is, by its clandestine nature, a difficult object of scientific inquiry. Verifiable data tend to be scarce, and, since these crimes often either go unidentified or are unsuccessfully prosecuted, statistics from domestic law enforcement agencies may lead to significant underestimations of the problem. For our present purposes, we offer two case studies to explain organised crime in fisheries. While these case studies are not, on their own, dispositive of the global scale of organised crime in the fisheries sector, they shed light on the activities that are manifestations of such crime and the challenges they pose to criminal law enforcement.

In the past two decades the global community has increasingly raised concern about the threat of organised criminal networks in the fisheries sector, where they accrue high profits at very low risk of sanction. In 2008 the UN General Assembly warned of the 'possible connection between international organised crime and illegal fishing in certain regions of the world' (UNGA Resolution 63/112, para. 59). Highlighting the need for additional research into, and evidence of, the link between organised crime and illegal fishing internationally, the General Assembly 'encourages States, including through the appropriate international forums and organisations, to study the causes and methods of and contributing factors to illegal fishing to increase knowledge and understanding of those possible connections, and to make the findings publicly available, bearing in mind the distinct legal regimes and remedies under international law applicable to illegal fishing and international organised crime'. (UNGA Resolution 63/112, para. 59).

A comprehensive report by the UN Office on Drugs and Crime (UNODC 2011) highlighted the vulnerability of the global fishing industry to organised criminal networks. The report's main finding was that the opaque nature of the fishing industry renders it susceptible to multiple crimes that are largely transnational in nature and frequently organised. The report also found that the traditional fisheries management approach applied to date in the fisheries sector seems insufficient to deal with the nature and magnitude of the problem. It recommended that this approach be complemented by a cooperative criminal law enforcement response. This finding was echoed by the UN Commission for Crime Prevention and Criminal Justice (CCPCJ), which, in a 2011 resolution on transnational organised crime at sea (UN CCPCJ Resolution 20/5 (2011)), highlighted the imperative of 'international cooperation to prevent and control' transnational organised crime taking place at sea.

Significant efforts have since been made to understand the phenomenon of organised crime in the fisheries sector. In 2013, member countries of the International Criminal Police Organization (INTERPOL) established the Fisheries Crime Working Group. Using this cross-border cooperation platform, law enforcement officers have formed joint operations to target well-known fisheries crime networks and close down some of the most notorious. The lessons learned from these operations were shared with the global community through the International Symposium on Fisheries Crime, an annual event held between 2015 and 2018, which brought together high-level law enforcement officers, civil society representatives and academics to discuss trends, challenges and solutions in relation to transnational organised crime in the fisheries sector (CCPCJ 2017).

In September 2018, this process culminated in the adoption of the International Declaration on Transnational Organised Crime in the Global Fishing Industry (Copenhagen Declaration) (Appendix A) by the ministers present during the Fourth International Symposium on Fisheries Crime in UN City, Copenhagen. As of 5 March 2020, the Copenhagen Declaration has received the support of 28 nations, most of which are 'large ocean nations', that is, territories highly dependent on the marine resources in their large ocean areas (LON Forum Report 2019). The supporting nations are Benin, Chile, Costa Rica, the Faroe Islands, Fiji, Ghana, Greenland, Iceland, Indonesia, Kiribati, Liberia, the Maldives, the Marshall Islands, Mexico, Mozambique, Myanmar, Namibia, Nauru, Norway, Palau, the Philippines, São Tomé and Principe, Scotland, Seychelles, the Solomon Islands, South Africa, Sri Lanka and Timor Leste. The declaration places the issue of transnational organised crime in the fisheries sector within the context of Sustainable Development Goals (SDGs) 14 and 16, and as part of an integrated and globalised world economy. The supporting countries also identify a course for a global commitment to combat this problem, stating that 'inter-agency cooperation between relevant governmental agencies is essential at a national, regional and international level in order to prevent, combat and eradicate transnational organised crime in the global fishing industry'.

In February 2019, submissions to the UN Security Council spoke to the threat that transnational organised crime at sea, including fisheries crime, poses to international peace and security (UN Security Council 2019).

This Blue Paper is the 16th in a series of Blue Papers commissioned by the High Level Panel for a Sustainable Ocean Economy. The authors of this Blue Paper were asked to address the question of how organised crime in the fisheries sector impedes the realisation of a sustainable ocean economy, and what practical measures can be taken to counter this. The authors were moreover directed to examine the various types of 'fisheries crime' in which organised criminals engage. This Blue Paper therefore explains the phenomenon of organised crime in fisheries, including the serious crimes indicative of its existence, using illustrative case examples to highlight the forms it might take, the ways organised crime in fisheries adversely impacts a sustainable ocean economy, and the associated law enforcement challenges. The paper also draws together a set of suggested practical means or 'opportunities for action' to address organised crime in fisheries.

2 State of Knowledge

2.1 What Is Organised Crime?

The premise of any discussion about organised crime in fisheries is a common understanding of what organised crime is. Although most people give the term 'organised crime' a similar meaning, there is little agreement on its exact characteristics, and numerous definitions of organised crime have therefore been put forward (for an analysis of these, see Varese 2017). In common parlance, there is a tendency to connect organised crime to hierarchical, exclusive and monopolistic groups—often with a strong ethnic or societal commonality—leaning towards violence, a criminal subculture or otherwise deviant behaviour (Abadinsky 2007). A broader understanding of organised crime—one that this Blue Paper adopts—is that it consists of networks of individuals who converge and collaborate over time to commit crime (Shaw and Kemp 2012). The networks' profile may vary significantly from loosely knit flat structures to strict hierarchical chains of command. Their criminal enterprise, geographical spread or crime script (modus operandi) may be equally diverse (Madsen 2009). Organised criminal groups may share with 'conventional' businesses many characteristics in their structure and capability (Australian Crime Commission 2009). These networks can form in any layer of society, or transgress these, with many examples found among white-collar criminals, people of professional authority and power who commit financial crime (Gottschalk 2012). Some authors also emphasise the ability to protect operations ('protection economy') through violence, bribery or extortion as a common, but not necessarily defining, feature of organised crime (Shaw and Kemp 2012).

Organised crime is also defined in law. One of the most widely accepted legal definitions internationally (through state ratifications) is that of the UN Convention on Transnational Organised Crime (the Palermo Convention) (UN 2004a). The Palermo Convention paints a broad picture, according to which organised crime can be defined as a serious crime committed by a structured group of three or more people for financial or other material benefit (UN 2004a, Article 2(a)). The convention defines a serious crime as an offence 'punishable by a maximum deprivation of liberty of at least four years' or a more serious penalty (UN 2004a, Article 2(b)). The Annual European Union Organised Crime Report (EUROPOL 2004) offers a slightly different definition, referring to organised crime as a collaboration of more than two people suspected of committing serious criminal offences in the pursuit of profit and/or power and for a prolonged or indefinite period (Fröhlich 2003).

It is recognised that while international instruments provide a clear legal benchmark of what constitutes organised crime, outside the letter of the law, both conceptually and in practice, there is often ambiguity around what is regarded as 'criminal behaviour'. In particular, views differ regarding the extent to which criminal economies associated with organised crime are 'normalised' in any given society (Shaw 2017; Gilman et al. 2011).

A distinction could be made between organised crime taking place within the fishing industry and ancillary to its business operations (e.g. illegal fishing or human trafficking) and organised crime taking place outside the industry using the sector as a cover for other criminal activities (UNODC 2011), such as the smuggling of contraband (like Capone's rum running in the 1920s). With regards to the latter, an organised crime operation may enter the fishing industry in order to launder proceeds from its illicit activities and/or to provide a legitimate reason for being at sea so as to illicitly traffic goods (Parks 2014). In practice many of the same issues pertaining to law enforcement will arise regardless of whether the organised crime is embedded in the industry or not.

From a law enforcement perspective, organised crime networks in fisheries can be involved in a broad range of criminal offences. The most common are economic crimes, such as money laundering, fraud, forgery, tax and customs evasion, corruption and human trafficking, in addition to criminal offences found in sector regulations, such as in the fisheries, health and safety, and maritime sectors. In practice, crime in the fisheries sector, often referred to as 'fisheries crime' and 'fisheries-related crime' (UNODC 2019a; FAO 2019), covers criminal offences taking place throughout the fisheries value chain, from the preparatory stage (including vessel insurance and registration) to at-sea activities (including harvesting and catch documentation) to landing, processing, transportation, trade and sale (UNODC 2017). Criminal offences may thus be committed at sea, on land, in cyberspace, or at the coastal interface. A commonality is that the offences are profit-driven, that is, they are economic crime and are frequently committed by white-collar criminals (UNODC 2019a). Many of these offences are conducted or continued (in whole or in part) extraterritorially, on the high seas and in other areas beyond national jurisdiction, such as regional seas, making most fisheries crime cases transnational, with the added complication of jurisdictional obscurity (NA-Fig 2017).

2.1.1 Case Study 1: The Viking Case

In February 2016 the Indonesian navy intercepted and detained the fishing vessel *Viking* in Indonesian waters. The vessel was the subject of an INTERPOL purple notice, that is, an alert about the crime script of a criminal network. The purple notice issued on the *Viking* notified law enforcement agencies around the world of the network's use of numerous vessel identities and nationalities (flag states) and the vessel's unclear ownership structures. This crime script made it difficult for law enforcement agencies to identify which country had jurisdiction over the vessel and which country was responsible for its activities. As a result, for more than a decade, the owners and operators of the *Viking* were able to land and enter into the market illegally caught Patagonian toothfish from the Southern Ocean, an extremely valuable species dubbed 'white gold', in contravention of the multilateral fisheries management regime in these waters. The network's turnover from the activities of this vessel alone was estimated by crime analysts to be about half a billion dollars during the time of its operation, and there is good reason to believe that the network operated several such vessels (expert consultations 2019).

During their preliminary inquiries Indonesian authorities established that the documents presented on behalf of the vessel were forgeries, meaning that the vessel sailed under a false identity and flag. It also proved impossible to trace the company stated to be the vessel's owner, suggesting that this was a fictional company. Indonesian authorities regarded the vessel to be stateless and subjected it to Indonesian jurisdiction. In Indonesia, document forgery can be penalised by imprisonment of up to six years (Indonesian Criminal Code, Article 263). The vessel's hold was also found to contain gillnets exceeding 2.5 km in length, which is an infringement of Indonesian law and subject to five years' imprisonment and a fine of up to 2 billion Indonesian rupiah (US $150,000) (Indonesian Fisheries Law, Article 85).

Further investigations led by the Indonesian authorities revealed the level of organisation of the network behind the *Viking*'s operations. The master of the vessel was in frequent communication with an Australian national based in Singapore who acted as the operator of the vessel and provided logistics, supplies and financing. The master was also in contact with a Spanish national, domiciled in South Africa, who was later identified as the likely beneficial owner of the vessel. These were allegedly the three core members of the network at the time. Other members included at least one other master of the vessel (the masters rotated on shifts throughout the year) and a Singaporean financier. Together the network's members conducted a highly complex transnational business operation. The vessel operated out of ports across Southeast Asia and Africa, using forged documents with numerous identities and nationalities. It landed, shipped and traded Patagonian toothfish and sourced crew, supplies, spare parts, gear and fuel around the world.

This case is an example of transnational organised crime in the fisheries sector: It involved a network of at least three members who orchestrated the commission of serious criminal offences, including illegal fishing and fraud, across multiple jurisdictions for significant material gain and over a prolonged period.

2.1.2 Case Study 2: The Rock Lobster Case

In May 2001, following a tip, officials from the South African Marine and Coastal Management branch seized and opened a container destined for the United States belonging to the South African company Hout Bay Fishing Industries (Pty) Ltd. This company was one of the largest seafood producers in the country at the time and a major employer. The container's contents comprised unlawfully harvested lobster tails and Patagonian toothfish. The South African officials alerted the relevant U.S. authorities, who intercepted the next container the company exported to the United States. In August 2003, following a protracted investigation by South African and U.S. law enforcement, the director of Hout Bay Fishing Industries and two others were arrested and criminal proceedings launched in both South Africa and the United States.

The investigation revealed that between 1987 and 2001 a network consisting of three directors of Hout Bay Fishing Industries illegally harvested large quantities of west and south coast rock lobster in South African waters for export to the United States. The network systematically exceeded the authorised quota for lobster during this period; in 1990, around 90 percent of the west coast rock lobster they exported to the United States was caught illegally. To facilitate the operation, the network established companies in both South Africa and the United States, bribed many government fisheries officers and other officials, and laundered profits in a complex web of properties and offshore banks and trusts.

South African authorities charged the main director of Hout Bay Fishing Industries with a range of offences including fraud, corruption and bribery under the Prevention and Combatting of Corrupt Activities Act, racketeering under the Prevention of Organised Crime Act, violations of the Marine Living Resources Act and activities contrary to the Customs and Excise Act. In the United States, the Lacey Act makes it illegal to introduce into the United States any fish or wildlife taken in contravention of the laws of another country. The directors of Hout Bay Fishing Industries were arrested and found guilty in the United States of violating the Lacey Act, as well as smuggling and conspiracy, and in 2004 they were sentenced to imprisonment and forfeiture. A U.S. court also awarded the South African government $22.5 million in restitution for damages, but investigators had a hard time locating and freezing the main director's assets to secure the restitution amount. In September 2018, more than 15 years after the activity was first detected, investigated and tried by the U.S. court, a final settlement agreement of $7.5 million was ordered.

The rock lobster network is a good illustration of a criminal network that started out as legitimate business and transitioned into transnational organised crime in the fisheries sector. The network consisted of at least three members (or business associates) who conspired to commit serious crime in at least two jurisdictions and with bank accounts in numerous other offshore jurisdictions for significant material gain over a prolonged period. The case also illustrates the lengthy and complex cross-border investigations required to address transnational organised fisheries crime, in this case more than a decade and a half.

2.2 Manifestations of Organised Crime in Fisheries

This section describes several serious offences that are committed as part of organised crime in the fisheries value chain.

2.2.1 Fraud

Fraud refers to deliberate misrepresentation or concealing of facts for undue benefit (UNODC 2017). *Forgery*, or misrepresentation by falsifying a document, is often regarded as a subset of fraud. Some jurisdictions also have distinct offences of false declarations to public authorities. A large amount of documentation is produced throughout the fisheries value chain, creating significant potential for fraud and forgery. This is particularly true of documents with a high cash value, such as vessel registration certificates, landing documents and fishing licences. In the case of the *Viking*, false vessel registration documents were submitted at port, comprising text clearly cut and pasted from Google Translate and using ordinary word processing software (NA-Fig 2017). Fishing vessel identity fraud is used by criminals to change their vessel's identity with relative ease by, for example, giving more than one vessel in a fishing fleet the same name and operating all vessels under the same fishing licence, not flying the correct flag at port, or physically hiding or painting over a vessel's name in order to render it anonymous (NA-Fig 2017). Vessel identity fraud is closely associated with 'flag-hopping', a pattern of re-registering a vessel with new flag states to confound investigations into its illegal operations, as was the case with the *Viking* (NA-Fig 2017).

Fraud can, moreover, be committed at the harvesting stage when inaccurate catch records are maintained pertaining to, for example, in which waters the fish were caught, the amount of fish caught and the species harvested (UNODC 2011). At the subsequent processing stage (at sea or on land), fish and fish products may be fraudulently labelled to avoid paying higher customs duty on high-value species (EUROPOL 2018). One example is that of the fishing vessel *STS-50*, also known as the *Andrey Dolgov* and later the *Sea Breeze*, which was found with falsified information regarding the species of fish (mislabelling) on numerous occasions, leading to its apprehension by Chinese authorities in 2017 (Gray 2019). Deliberate mislabelling may also occur at the retail stage in order to obtain a higher asking person (working as an official for a private or public entity) act or refrain from acting (UN Convention on Corruption: UN 2004b, Chap. 3). Most people will associate corruption with bribery (UN 2004b, Articles 15 and 17), but variations such as embezzlement (Article 17) and abuse of function (Article 19) are also within the purview of the definition.

Corruption manifests in the fishing sector in various forms (UNODC 2019a): On shore, abuse of function may occur when political figures or senior government officials use their positions to influence the allocation of fishing licences to companies or businesses in which they have a personal business interest (UNODC 2017; Standing 2015, 2008). One case currently under investigation involves an Icelandic fishing company that allegedly used a bank of a neighbouring country and shell companies in the Pacific to channel bribes to obtain fishing licences in Namibia (Wilhjálmsson 2019).

In the context of ship registration, some corporate entities operating ship registries on behalf of flag states may have obtained the rights to do so corruptly (UNODC 2011). Corruption in the context of monitoring and inspection may occur in the form of fines paid to fisheries authorities without independent review, funds from fines not properly accounted for by the competent authority or bribes paid to reduce penalties (INTERPOL price or to disguise the species of fish (if it was caught illegally) or the fish's country of origin (OECD 2013; Oceana 2018). On landing, false or forged customs and health documentation may be used to disguise the catch's country of origin or the true identity and flag of the vessel that caught the fish, to avoid paying import tariffs or having to comply with food hygiene regulations (UNODC 2017). For instance, on the basis of fraudulent landing certificates, Trinidad and Tobago is cited as the world's sixth-largest shark fin exporter to Hong Kong (Pew Charitable Trusts Environmental Group 2012), despite the fact that its domestic fleet lacks the capacity to catch and land such quantities; the fins are, in fact, landed from foreign fishing vessels and sent in transit through Trinidad and Tobago ports.

2.2.2 Corruption

An act of corruption is the giving, solicitation or receipt of an undue advantage with the purpose of making a 2014). Bribes may also be paid to enforcement officers or fisheries inspectors (at sea or on land) to ignore illegal harvesting of fish, particularly high-value species such as abalone (UNODC 2019a; UNODC 2017). Bribery is also prevalent in fishing ports where inspectors endorse, for instance, landing data that are clearly false (see, e.g., the rock lobster case above). Bribery may extend throughout the fisheries value chain (UNODC 2017). For example, in a San Diego court, where a U.S. company was alleged to have illegally brought into the United States approximately $17 million worth of sea cucumber from Mexico, papers suggested that the U.S. company sent money to Mexico to bribe officials, including law enforcement officers, along the entire supply chain (U.S. Department of Justice 2017; Kaplan-Hallama et al. 2017). Bribes may also be paid to officials to ignore irregular crew work permits, which may be used to facilitate human trafficking for forced labour on fishing vessels (UNODC 2019a).

2.2.3 Tax Crime

Tax crime, also known as tax evasion or tax fraud, is the violation of tax and revenue regulations, including income tax, value added tax, property tax, company tax and other forms of state levies and duties, including customs duties. If the offences are criminalised in law, they amount to tax crime.

Tax crime can be difficult to identify and prove, as not all jurisdictions require transparency regarding the beneficial ownership of bank accounts and companies. These jurisdictions, often labelled 'tax havens' or 'secrecy jurisdictions', perpetuate the grey area between legitimate tax planning and criminal tax evasion (Shaxon 2011; Galaz et al. 2018).

The fisheries sector lends itself to tax crime, including unreported tax, misreported or underreported tax, and tax evasion. In light of the relative ease with which criminals in fisheries can change a vessel's country of origin and identity, and use fictitious companies as registered owners, it would seem that the likelihood of profit-shifting, that is, channelling profit to shell companies in tax havens to avoid paying tax in the country where the profit was generated, is commonplace among these actors (NA-Fig 2017). Mis-invoicing also appears to be widespread in fisheries. Mis-invoicing happens by deliberate falsification of the value, volume and/or type of commodity in an international commercial transaction of goods or services by at least one party to the transaction (Global Financial Integrity 2019). Facilitated by financial secrecy jurisdictions, mis- invoicing commonly involves disguising the origin of fish, under-declaring the size of a catch and incorrectly describing the species or products caught or sold (OECD 2013).

The Organisation for Economic Co-operation and Development (OECD 2013) estimates that tax revenue lost to tax crime in fisheries is significant, undermining the development benefits of the sector, and notes that it has a particularly adverse impact on developing countries. It was recently estimated that, combined, the potential tax revenue losses due to the likely illicit seafood trade in African and Asian marine resources account for 72 to 74 percent of global tax revenue losses in this trade, amounting to between $1.6 and $3.1 billion annually (Sumaila et al. 2020). In Indonesia in 2016, tax crime by 187 fishing companies triggered a comprehensive audit of those companies by the Tax Directorate

General, which found that the fishing companies in question had failed to report or misreported tax amounting to potential unpaid tax revenue of 235 billion Indonesian rupiah (more than $16 million). The Tax Directorate General used the audit data to pursue enforcement measures and collect the unpaid taxes from the companies.

2.2.4 Money Laundering

Money laundering is defined in the Palermo Convention as the intentional concealing or disguising of the illicit origins of the proceeds of crime (UN 2004a, Article 6). The recommendations of the Financial Action Task Force (FATF 2012–19), an intergovernmental body promoting legislative and regulatory reforms to combat money laundering, terrorist financing and other related threats to the integrity of the international financial system, detail standards of universal application towards its aims, including that all countries designate serious crimes as 'predicate offences' (i.e. offences that are a component of a 'primary' crime) to money laundering. The Egmont Group, a united body of 164 financial intelligence units, provides a platform for the secure exchange of expertise and financial intelligence in support of national efforts to respect international global standards regarding anti–money laundering and counter-financing of terrorism, including those of the FATF. The Egmont Group (2014) has published a set of successfully prosecuted case examples to assist analysts and investigators pursuing money-laundering cases.

Many offences committed by organised crime groups in fisheries will be regarded as serious crimes and thus as predicate offences. A common denominator of organised crime networks operating in the fisheries sector is that they engage in money laundering to integrate into the legitimate economy the proceeds of crimes committed along the fisheries value chain, or the proceeds of their other crimes. They may do so by acquiring large capital assets, such as fishing vessels, or making cash salary payouts to crews of fishing vessels (OECD 2013, 33).

In Indonesia, illegal fishing is expressly cited as a predicate offence under the Prevention and Eradication of Money Laundering (Anti–Money Laundering) Law (Article 2). The law facilitates prosecution of money laundering and asset recovery through a number of criminal procedural tools, including reversal of the burden of proof during prosecution, giving defendants the burden of proving that their assets were not derived from crimes, and express permission of the inclusion of information in electronic form as legal evidence (Husein 2014).

2.2.5 Crime in the Labour Market

The labour market is a large economic sector that attracts organised crime in the form of 'labour market crime'. Such offences may range from breaches of work and safety regulations to fraud, tax evasion, document forgery and deception. In extreme cases, crimes in the labour market include abduction, unlawful confinement, physical injury, culpable homicide, murder and sexual abuse, as well as forced labour and human trafficking. *Forced labour* refers to 'all work or service which is exacted from any person under the menace of any penalty and for which the said person has not offered himself voluntarily' (ILO 1930, Article 2(1)). In the fishing sector it is often a consequence of human trafficking (ILO 2016) or 'trafficking in persons', which broadly refers to the procuring of and trading in human beings for the purposes of exploitation (UN 2004a, Annex II).

Criminal networks in fisheries use forced labour to significantly cut costs and boost profits (see, e.g., Tickler et al. 2018). The International Labour Organization (ILO 2016) highlights a number of indicators of forced labour in the fishing industry, including abuse of vulnerability, deception, restriction of movement, isolation, physical and sexual violence, intimidation and threats, retention of identity documents, withholding of wages, debt bondage, abusive working and living conditions, and excessive overtime. Facilitating forced labour is the practice of offshore transhipment of crew, which means that crew can stay at sea on different fishing vessels for years at a time without having to dock at port, frustrating detection of victims of trafficking for forced labour and making them de facto prisoners onboard (UNODC 2011).

Forced labour in the fisheries sector appears to be pervasive around the globe. For instance, in 2017 three employees of a Scottish family-owned company operating a fleet of scallop dredgers were arrested following a police raid on one of their vessels in southern England. The company has a track record of non-compliance with maritime safety rules. Nine crew, eight West Africans and one Sri Lankan, were removed from the vessel as victims of trafficking. A police investigation is ongoing (Lawrence and McSweeney 2017).

In the port of Puntarenas, Costa Rica, police rescued 36 Asians who had been subjected to labour exploitation on two fishing boats in 2014. The 15 Vietnamese, 13 Indonesians, 5 Filipinos, 2 Taiwanese and 1 Chinese national said they were forced to work up to 20 hours a day and were regularly flogged. They were underfed and had never been paid. Four individuals from Taiwan and Costa Rica were arrested in connection with the case and were charged with human trafficking (Zueras 2010).

In 2011, the Fiji Labour Department accompanied a Fijian fisher in filing a lawsuit against a fishing company on the basis of abuse endured in 1999 when he was employed on two South Korean fishing vessels. The fisherman had been subjected to a harsh working environment, received insufficient food and endured unhygienic living conditions. He had been forced to work in a refrigerated room at -40oC without gloves or any safety equipment, resulting in frostbite that required the amputation of his fingers. He told the media that he did not complain about the poor working conditions because he had been afraid of losing his salary (Human Rights at Sea 2019).

Forced labour and human trafficking can also take place in the land-based fishing industry. In 2015, Indonesian authorities discovered 322 migrant fishers from Burma, Cambodia and Laos stranded within the area of a fish factory in the island village of Benjina, part of the Maluku chain. The fishers were all in very bad condition, having been exploited for approximately 10 years without any payment, over which time they had been harassed, overworked (up to 20–22 h per day) and physically abused. They had been smuggled into Indonesia using forged Indonesian identity documents. In March 2016, an Indonesian court sentenced eight people—five Thai citizens and three Indonesians—for involvement in the crime of human trafficking. Each was sentenced to three years' imprisonment and a fine of 160 million rupiah ($11,300). The court also ordered the fishing company to pay 884 million rupiah ($62,500) in restitution to the victims of human trafficking.

Recruitment agencies play a central role in smuggling migrant workers to work aboard fishing vessels as victims of human trafficking for forced labour. In 2016, Norwegian authorities identified a foreign network, operating out of the north of Norway, that recruited at least 49 Indonesian fishers, a Spanish national and a number of Ukrainians to crab fisheries in the Barents Sea. Several deaths, serious injuries and threats of violence were reported by the fishers to the Norwegian coast guard and local police, and the coast guard identified very poor living and working conditions onboard the vessels (Dagbladet 2018a; Dagbladet 2018b). The network allegedly used a South Korean operator, a Seychelles recruitment agency and Norwegian port agents to facilitate the movement of the migrant fishers from Indonesia to Norway. Similarly, in 2016, Indonesian authorities learned that about 14 Indonesian victims of forced labour were working as fishers on a Chinese fishing vessel in Dargahan, Iran. The victims were recruited by an Indonesian recruitment agency affiliated with another recruitment agency in Taiwan.

2.2.6 Security Offences at Sea

Various offences occurring at sea present a threat to peace and security; this can include offences falling within the ambit of organised crime in the fisheries sector. 'Fisheries conflicts', which may arise from a combination of factors, including illegal fishing (along with climate change and food security concerns), also pose a potential threat to maritime security (see e.g. Spijkers et al. 2019; Sumaila and Bawumia 2014; Pomeroy et al. 2007). At an international level, the UN Security Council is influential in determining whether threats exist to international peace and security. In 2019, the Security Council explored transnational organised crime at sea and the threat it poses. Focusing on the Gulf of Guinea, various submissions noted that transnational maritime crime—broadly defined to include organised fisheries crime at sea—not only undermines national social and economic development but also destabilises the region and poses broader security risks (UNSC 2019).

The fisheries sector is also vulnerable to exploitation by terrorists due to vessels' seeming legitimate presence at sea; the lack of transparency as to their movement, identity and ownership; vessels' ability to tranship and access small harbours; and their sometimes erratic movements. For instance, in the 2008 terrorist attack in Mumbai, an Indian fishing trawler was hijacked to transport the terrorists and arms closer to the city (UNODC 2011). A recent Security Council resolution expressed concerns about the links between international terrorism and organised crime, including transnational organised crime at sea (UN Security Council Resolution 2842/2019 (2019).

Piracy, defined in the UN International Law of the Sea (UN 1982, Article 101) as illegal acts of violence or detention for private gain by occupants of one private vessel against another on the high seas or in waters beyond national jurisdiction, has been high on the agenda of the UN Security Council for a number of years. In 2016, the Security Council noted the interface between various types of criminal activities within the maritime domain in the Indian Ocean, highlighting the 'complex relationship' between large-scale illegal fishing in the region and the increasing piracy in Somali waters (UN Security Council Resolution 2316 of 2016 (2016); this relationship is also discussed by, e.g., Devlin et al. 2020; Samatar et al. 2010). A similar link has been argued for in the case of Southeast Asia (Liss 2007). And in 2012, in Senegal, fishers threatened to resort to piracy if the large-scale illegal fishing taking place in the Gulf of Guinea was not halted (Vidal 2012).

Also in the Gulf of Guinea, there is evidence that fishing vessels are being used for illicit trafficking of weapons (Okafor-Yarwood 2020; Beseng 2019; UNODC 2011). In 2018, the Cameroonian navy arrested forty-three people onboard three Nigerian fishing vessels; they were found to be illegally in the country's territorial waters, and automatic weapons, including AK47s, were recovered (Okafor-Yarwood 2020). There are also reports of the involvement of fishing vessels in gun-running elsewhere, including off the African east coast (UN 2003) and in the Caribbean (UNODC and World Bank 2007).

2.2.7 Drug Trafficking

Drug trafficking refers to the illicit trade, involving the cultivation, manufacture, distribution and sale, of substances subject to drug prohibition law (UNODC 2019b).

Fishing vessels are ideal modes of transport for drugs for the same reasons that the sector is vulnerable to security offences at sea. The UNODC 2011 study found that fishing vessels are used in various ways to facilitate illicit trafficking in drugs: as mother ships, that is, as a base station from which smaller vessels traffic drugs to and fro, as support vessels (providing fuel and supplies) for go-fast boats transiting trafficking routes (as is common in the Caribbean) or, for smaller fishing vessels, to traffic drugs directly to and from coastal landing sites in smaller quantities, frequently transhipping drugs to mother ships outside the territorial jurisdiction of the coastal state (UNODC 2011), for example in the Gulf of Guinea (INTERPOL 2014, 29).

In Indonesia, some 90 percent of illicit drugs smuggled in and out of the country are allegedly transported by sea (Indonesian National Narcotics Board 2012; Antara News 2018). In 2018, Indonesian authorities uncovered drug trafficking using four foreign-flagged fishing vessels, one of which was transporting one tonne of methamphetamines to Indonesia. The captain of the vessel produced forged Indonesian fishing permits and failed to present proper certification of competence to captain a fishing vessel, adding to the suspicion that the fishing vessel was merely a vehicle to transport illicit drugs (Antara News 2018). Private ports are frequently used in such trafficking operations. (Indonesian National Narcotics Board 2012).

Drug trafficking using fishing vessels is at times carried out in conjunction with the transport of other illicit goods. In Trinidad and Tobago, for example, Venezuelan gangs, in cooperation with local counterparts, engage artisanal fishing vessels to transport drugs and guns from Venezuela to Trinidad and Tobago. Cocaine transiting Trinidad and Tobago via Venezuela, originating in Colombia and destined for the U.S. market, is also known to travel on artisanal fishing vessels. In Jamaica, fishing canoes engage in a 'drugs-for-guns' trade with neighbouring Haiti, in which local marijuana is exchanged for illegal weapons (U.S. Department of State 2018) and increasingly, also, for cocaine (Neil 2018). In the Gulf of Guinea, in 2006, the fishing merchant vessel *Benjamin*, flying a Ghanaian flag, trafficked about 78 parcels (2340 kg) of cocaine into Ghana labelled as shrimp (Ali 2015).

Case examples also suggest that fishing companies, fish processing plants and fish distribution networks may act as storage facilities and legitimate covers for the transport of drugs and that some fishing operators launder proceeds from drug trafficking through investments in fishing infrastructure (U.S. Department of Justice 2019; UNODC 2011). Illicit drugs are hidden among or inside deep-frozen fish as the smell impedes the effectiveness of drug detector dogs, and enforcement officers are hesitant to inspect frozen fish, which would necessitate thawing the product (potentially damaging it and providing grounds for compensation claims should their suspicions prove unfounded) (UNODC 2011). In Mexico, in 2009, authorities intercepted over a tonne of cocaine concealed inside shark carcasses. The smugglers claimed that the drugs were 'preserving agents' (Emmort 2009).

In some cases there is evidence of a close connection between high-value species and drug trafficking networks. Off the west coast of South Africa, evidence suggests that abalone poached by divers is bartered with middlemen from local gangs involved in the drug trade for the ingredients to manufacture the synthetic drug 'Mandrax' (an addictive barbiturate-like sedative used in the poorer communities). These activities appear to be part of organised criminal networks involved in the black market export trade of abalone to East Asia (de Greef and Raemaekers 2014; Steinberg 2005). In Mexico, the high returns from the illegally harvested Totoaba bladders ('maws'), which are exported to China and whose price can reach US $15,000 per kilogram straight off of the boat and up to $150,000 per kilogram on the international market, have attracted the involvement of organised crime groups (mostly Mexican and Chinese) who operate criminal enterprises and networks with links to drug cartels, corrupt officials and institutions (e.g. law enforcement and border control), as well as human trafficking networks, which control the supply chain and allow the products to reach markets (Alvarado Martínez and Martínez 2018; Crosta et al. 2018).

In Colombia, enforcement has traditionally focused on addressing drug trafficking in the ocean domain through inter-agency cooperation. Authorities have found that crime in the fisheries sector manifests in a range of interrelated offences, including trafficking of illegal drugs and arms, human trafficking, smuggling of fuel and other contraband, large-scale illegal fishing, and wildlife trafficking. A task force against drug trafficking in the Caribbean (no. 73 'Neptune'), under the command of the Colombian navy (command post in the Gulf of Urabá), has been mandated to help restore security in the maritime and coastal area (Colombian National Navy 2015). Coordinated operations by the national police, navy and air force (such as the Agamenón II campaign in 2015) resulted in 2782 arrests and the seizure of 887 firearms and 360.4 tonnes of cocaine during the years 2015–2018.

2.2.8 Smuggling, in Particular of Fuel

Smuggling, that is, the movement of otherwise legal goods from one jurisdiction to another in violation of the law, is often engaged in to avoid customs or other duties. The fishing industry provides ideal cover for smuggling given the vast ocean domain within which it operates and the associated law enforcement challenges.

Fishing vessels and fishers, for example, are known to be involved in fuel smuggling worldwide. In Ecuador, artisanal fishers routinely use their fishing vessels to smuggle small quantities of subsidised Ecuadorian fuel to the neighbouring coast of Colombia, where it can be sold at considerable profit (Ralby 2018). Trinidad and Tobago fishing vessels have also been implicated in the illegal trade of fuel. Prior to the 2017–18 government reforms, subsidised diesel and regular fuel were sold to foreign vessels (reportedly flagged to Guyana, Suriname and Venezuela) by local coastal communities and from fishing vessels at sea. When the Trinidad and Tobago government discontinued the subsidy and the provision of regular gasoline to the domestic market, local fishers began to purchase fuel illegally for one-sixth of the domestic price from Venezuelan vessels in Trinidad and Tobago waters. Ghana has also seen an exponential rise in fishing vessels and canoes involved in fuel smuggling. The country is reported to be at risk of losing about 1.5 billion Ghanaian cedi ($300 million) in revenue to the smuggling of fuel, much of it trafficked by fishing vessels and canoes (Banaseh 2017). Fuel is also known to be smuggled alongside illicit goods, such as drugs, illegal weapons and illegally harvested fish, as well as migrants (Ralby 2018).

2.2.9 Migrant Smuggling

Broadly understood, *migrant smuggling* refers to helping a migrant enter a country illegally in exchange for a (direct or indirect) financial or material benefit (UN 2004a, Annex III, Article 3(a)).

The use of fishing vessels to smuggle migrants is alleged to be prevalent. Although it is not well documented formally, public media reports on the use of fishing vessels in migrant smuggling are widespread (e.g. EURONEWS 2019; Grey and Ismail 2016). In the Caribbean, fishing vessels, mainly from the artisanal sector, have for the past 10 years been increasingly used to traffic migrant women, in particular from mainland South America, to Trinidad and Tobago. Some are forced into prostitution in Trinidad and Tobago, while others are transited to the United States (The Guardian 2019). There are also indications that fishing vessels are involved in migrant smuggling in the Mediterranean (Dambach 2019; UN 2019, para. 8; UNODC 2011), Australia (Lindley et al. 2018) and Thailand (Lefevre 2014).

2.2.10 Fisheries Offences

Marine fisheries are regulated by national fisheries management laws and subject to multilateral treaties, the international law of the sea and management measures established by regional fisheries management bodies. *Illegal fishing* refers to fishing in violation of domestic fisheries laws and measures and can take various forms, including fishing without the requisite licence or permit in a coastal state's exclusive economic zone, engaging in transhipment contrary to coastal or flag state law, harvesting beyond an assigned legal quota and fishing for legally protected species. Many illegal fishing activities occur at sea, where they are subject to a complex jurisdictional regime dependent primarily on the maritime zone in which the vessel is located as well as the nationality (flag) of the vessel and, to a lesser extent, of its owner and crew.

An illegal fishing activity may also be a criminal offence if the activity in question is criminalised under the law in the relevant domestic jurisdiction. A recent study of fisheries legislation in 91 countries conducted by the Food and Agriculture Organization of the United Nations (FAO) suggests that more than half of these countries have both criminal and administrative penalties for the violation of fisheries rules, and nearly a quarter have criminal penalties only (FAO 2019). A little more than a quarter of the countries have administrative penalties alone, of which the largest proportion are in Europe. Some jurisdictions impose severe criminal penalties for fisheries offences. In Norway, for instance, the Marine Resources Act provides for prison sentences of up to six years, in addition to asset forfeiture, for particularly grave offences. When considering whether an offence is grave, the act requests that the court consider the monetary value of the offence and whether the offence was committed systematically and over time, was transnational or was part of organised activities (Norwegian Marine Resources Act, Article 64).

In practice, illegal fishing, along with unregulated and unreported fishing (referred to cumulatively as IUU fishing), is frequently associated with other types of criminal offences along the fisheries value chain (particularly fraud) (de Coning 2016). INTERPOL therefore takes the view that, regardless of whether or not illegal fishing has been criminalised in a jurisdiction, IUU fishing is a strong risk indicator of fisheries crime (INTERPOL 2018).

2.3 The Impact of Organised Crime in Fisheries on the Sustainable Ocean Economy

The pursuit of a sustainable ocean economy requires balancing use of the ocean space and its resources, on the one hand, with the long-term carrying capacity of the ocean's ecosystems, on the other (Kraemer 2017).

In line with the three-pillared concept of sustainable development under the Rio process, a sustainable ocean economy is premised on the economically, socially and environmentally sustainable use of the ocean (UN 2012). Agenda 2030 extends the three dimensions of sustainability to five areas of critical importance— people, prosperity, peace, partnership and planet— which should inform synergised inter-agency policy interventions towards achievement of the Sustainable Development Goals (SDGs) (UN General Assembly Resolution 2015).

Organised crime in the fisheries sector could severely undermine states' ambitions to achieve the SDGs. Given the multifaceted and far-reaching implications of organised crime in fisheries, addressing this problem is relevant to the achievement of a range of SDGs, including SDG 2 ('zero hunger'), 8 ('decent work and economic growth'), 12 ('responsible consumption and production') and 14 ('life below water'). SDG 16 ('peace, justice and strong institutions') is regarded as a core enabler of the other SDGs (Kercher 2018), and the targets in SDG 16, including on promoting the rule of law (16.3), reducing illicit financial and arms flows and organised crime (16.4), reducing corruption and bribery (16.5) and developing effective, accountable and transparent institutions (16.6), are particularly resonant in the context of addressing various manifestations of organised crime in fisheries (Kercher 2018).

There are ample illustrative examples highlighting how organised crime in the fisheries sector may undermine the global commitment to sustainable development and to a sustainable ocean economy in particular. Examples of the adverse economic, social and environmental impacts of organised crime in fisheries on the pursuit of a sustainable ocean economy are provided below. These examples underscore the cross-cutting nature of fisheries crime offences and the range of their complex adverse impacts on communities.

2.3.1 Social Impacts

Organised crime in the fisheries sector can have a negative impact on the affected coastal state and its population by, amongst others, undermining the rule of law, threatening peace and security, jeopardising food security and adversely impacting fishing communities.

Peace and Security

Regional and international peace and security is threatened by a range of organised criminal activities in the Gulf of Guinea (UNSC 2019). This encompasses piracy and armed robbery at sea, other violent crimes including ship hijackings, incidents involving 'firing on boats', hostage-taking and kidnapping for ransom, fuel and gas robbery and smuggling, drug and arms trafficking, illegal fishing and maritime terrorism (UNSC 2019). These criminal activities also seriously undermine the ability of the states in the region to pursue socio-economic development. Criminal activities have prompted increased insurance premiums for cargo vessels using the maritime space, which, in turn, impedes trade, hindering the movement of goods and services, and resulting in lost income for businesses and governments and higher prices for fuel, food and other goods for consumers (Chatham House 2013; Gilpin 2007; One Earth Future 2018).

In the Caribbean countries of Jamaica and Trinidad and Tobago, transnational networks using fishing vessels to facilitate drug and arms trafficking, as well as migrant smuggling, are known to cooperate with local criminal groups (UNODC and World Bank 2007; regarding Jamaica specifically, see Witbooi 2020; for Trinidad and Tobago, see Bassant 2019). This contributes to a rise in national violent crime, brings illicit drugs into coastal communities and fuels local criminal gang activities, such as fraud and extortion (Leslie 2010; Government of Jamaica 2007). In South Africa, organised criminal networks involved in the illegal harvesting and export of high-value coastal species, such as west and south coast rock lobster and abalone, and the associated illicit money flows and interface with drug trafficking, weaken governance and the rule of law. In Russia, authorities have warned about the association of the 'crab mafia' a substitute for infected bush meat (USAID 2016). For coastal communities, such as those in Nigeria's Niger Delta area, fishing also provides basic income to pay for social services, including medical care and education, that are not provided by the Nigerian state (Okafor-Yarwood 2020; 2019).

Large-scale overfishing in Jamaica has left most reef fish stocks overexploited (World Bank 2017). As a result, Jamaica is almost entirely dependent on imported fish for domestic (Akhmirova 2012) with assassinations of high-ranking public officials and competitors, as well as money laundering and illegal fishing (Otto 2014). Russia has taken a number of strong measures in an attempt to bring organised crime in the crab fisheries under control (Akhmirova 2012).

Food Security

Fisheries resources are a major source of protein globally, providing an estimated 17 percent of animal protein consumed worldwide, with the highest per capita consumption in small island developing states (FAO 2018). According to the UN Special Rapporteur on the Right to Food, it is imperative to curtail illegal fishing to prevent its further adverse impact on food security and coastal livelihoods (UNGA 2012).

In the West African region of the Gulf of Guinea, for example, fish is the predominant (and sometimes only) source of animal protein consumed by the roughly 40 percent of the population that resides in coastal communities (Okafor-Yarwood 2019). In Sierra Leone, one of the poorest countries in the world (UNDP 2018), fish provides an estimated 75 percent of the animal protein consumed by the population (Agnew et al. 2010), highlighting the inseparability of fish from the country's food security. This is well illustrated by the role fish played during the Ebola crisis, when it became consumption. According to 2014 data, 79 percent of all fishery products consumed in Jamaica were imported (World Bank 2017). The value of fish imports for food in Jamaica in 2015 was $103.8 million (CRFM 2018).

Criminality In Fishing Communities

Organised crime in fisheries can shape criminality in fishing communities in a number of ways. With few livelihood options beyond fishing, coastal communities are particularly vulnerable to recruitment by criminal networks operating in the fisheries sector.

In the Gulf of California in Mexico, research indicates that the government's incomplete implementation of 2002 environmental management regulations (Alvarado Martínez and Martínez 2018; Santos-Fita 2018) has contributed to the tendency of fishers—adversely affected by these conservation efforts—to turn to totoaba poaching due to lack of legitimate alternative livelihoods (Alvarado Martínez and Martínez 2018; Crosta et al. 2018).

On the South African coast, overexploitation of west coast rock lobster (*WWF South Africa v Minister of Agriculture, Forestry and Fisheries and Others* 2018) has impacted thousands of subsistence fishers who traditionally harvested the species. No longer able to catch enough to sustain themselves, fishers have been forced to seek alternative, sometimes illegal, ways of generating income (Cochrane 2017). In Nigeria, evidence suggests that coastal fishers, who fear putting out to sea due to frequent violent attacks from illegal fishing vessels, are susceptible to recruitment by organised criminal networks engaging in armed robbery at sea and oil smuggling (Okafor-Yarwood 2020).

Moreover, many coastal communities are transformed by organised crime and the illicit goods and associated economies of violence it brings. Trinidad and Tobago has seen a rise in violent crime nationwide, but particularly in coastal areas not traditionally known for gang activities (such as Moruga, Claxton Bay, Carli Bay and Orange Valley). This has disrupted daily life and, in some instances, separated families as parents send children to live with family members in safer areas. In Jamaica, research indicates that violent gang-related activity is increasingly commonplace in small fishing towns, such as Rocky Point, caught in the drugs-for-guns trade (Witbooi 2020; Robinson 2017; Fisher 2016). In South Africa, remote coastal communities are under siege by organised criminal gangs illegally accessing abalone off their shores (Isaacs and Witbooi 2019; de Greef and Raemaekers 2014).

Organised crime in the fisheries sector also has noticeable gendered implications. In South Africa, women in female-headed households have become accomplices to organised poaching operations, washing divers' wetsuits and storing their illegally harvested abalone in their refrigerators to earn money for basic necessities, subjecting them to criminal prosecution (Isaacs and Witbooi 2019). In Nigeria, where the combined effect of illegal fishing, climate change and pollution from oil companies threatens the livelihoods of coastal fishing communities, fishmongers—who are predominantly women—have in some instances been forced into prostitution to make ends meet (Okafor-Yarwood 2020, 2018), worsening the region's already the community of Río Lagartos (Kaplan-Hallama et al. 2017). In addition to creating new pressures on local environmental resources, the 'gold-rush' influx of new actors has weakened the community's social cohesion and sense of security, and fishers from outside the area have been associated with rising levels of local violence, prostitution and drug use (Kaplan-Hallama et al. 2017).

Given communities' central role in the supply chain of organised criminal networks, it may be useful to complement the law enforcement response with an approach that enables the community to play a preventative role (Isaacs and Witbooi 2019; Hübschle and Shearing 2018, 5; Hauck and Sweijd 1999). Communities can build their resistance to the infiltration of organised crime by strengthening their resilience (Felix 2017).

Women, who often hold powerful positions in fishing communities (albeit often unrecognised; see Matthews et al. 2012), can be prominent actors in this effort. This could be the case, for example, in Rocky Point, Jamaica, where women are the primary owners and managers of key community fishing assets (Livelihoods Report 2012).

2.3.2 Economic Impacts

Costs to Coastal States

Organised crime in fisheries can severely compromise the revenue base of coastal states. For instance, in the Gulf of Guinea large-scale illegal fishing is estimated to amount to 40–65 percent of the reported catch (Doumbouya et al. 2017; Agnew et al. 2009): the combined annual economic cost of large-scale illegal fishing to the Gambia, Guinea, Guinea-Bissau, Mauritania, Senegal and Sierra Leone is estimated to be $2.3 billion (Doumbouya et al. 2017; Ali 2015). In Guinea- Bissau alone, more than half of the industrial catch landed, valued at $260.7 million, is caught illegally. Only a third of the remaining catch is captured by the local economy through fishing fees and access agreements high prevalence of HIV/AIDS (Okonko and Nnodim 2015; Udoh et al. 2009). In Mexico, the rapid growth of the illegal sea cucumber fishery on the Yucatán Peninsula, triggered by rising international demand, has driven dramatic social and ecological changes in (Intchama et al. 2018, 9). Facing a similar predicament, Indonesia introduced a comprehensive and bold law enforcement and policy reform against fisheries crime. The reform has spurred national economic development and resulted in an increase of fish stocks from 7.31 million tonnes in 2014 to 9.93 million tonnes in 2015 and 12.54 million tonnes in 2016 (MMFA 2018, 23; Cabral et al. 2018). Catch landings by local coastal fishermen have also risen, and the resulting increase in their purchasing power has driven the economic growth of local fisheries. The reforms contributed to significantly improved tax revenue from the fisheries sector in 2018, amounting to 1.6 trillion rupiah ($113 million) (Indonesian Ministry of Finance 2019).

Piracy and armed robbery at sea also have noticeably reduced revenue generated by the fisheries sector in the Gulf of Guinea. In Nigeria, for instance, such crime has led to a drop in the number of licenced fee-paying industrial fishing vessels operating in local waters, thereby weakening the sector's contribution to gross domestic product (Okafor-Yarwood 2020). This echos past impacts: in 2008, in response to safety concerns, the Nigerian Trawler Owners Association recalled about 200 local fishing vessels to shore, impacting approximately 20,000 jobs and leading seafood prices to more than double (Onuoha 2012).

At an international level, a 2020 report indicates gross revenues of between $8.9 billion and $17.2 billion are annually redirected out of the legitimate market through illicit trade. Asia, Africa and South America account for approximately 85 percent of total catch losses to likely illicit trade globally. Africa is estimated to experience annual losses of between $7.6 billion and $13.9 billion and between $1.8 billion and $3.3 billion in economic and income impacts, respectively, due to the redirection of catches from legitimate to illicit seafood trade (Sumaila et al. 2020).

Costs to the Legitimate Industry

Legitimate businesses are burdened with the negative economic impacts of organised crime in fisheries.

Networks engaged in organised crime can significantly cut their costs and launder illicit gains, giving them an unfair advantage over legal operators, whom they can effectively push out of business. Russian authorities have warned against foreign companies taking control of local fishing companies and undercutting resident competitors (Krivoshapko 2017). In the north of Norway, the introduction of underpaid migrant workers in the foreign-flagged snow crab fleet led some Norwegian- registered snow crab companies to employ their own vulnerable migrants in violation of national crewing regulations (Dagbladet 2018a; Fenstadt and Kvile 2016).

Organised crime in the fisheries sector damages the reputation of the legitimate industry and fishing nations.

An example is the document fraud around shark fin exports in the Trinidad and Tobago fishing industry, as mentioned above. In a further example, in 2014 Trinidad and Tobago was cited by the Commission for the Conservation of Antarctic Marine Living Resources for trading in Patagonian toothfish, contrary to the region's fisheries management agreement. It was subsequently discovered that the trade documents attached to the Patagonian toothfish exports to Canada indicating Trinidad and Tobago as the country of origin were fraudulent (Republic of Trinidad and Tobago, Fisheries Division 2019).

Businesses outside the fishing industry may also be harmed by the rise in crime. In Jamaica violent crime, fuelled by the maritime smuggling of guns into the country, has led to a loss of business productivity (due to resulting death and injuries) and weakened investor confidence (Leslie 2010). In Trinidad and Tobago, the influx of drugs and illegal weapons by sea has led to reduced production and increased expense for security, with 85 percent of businesses spending the highest portion of their budgets on increased security in recent years (Sutton 2017). These effects are interwoven with the social impacts of organised crime in fisheries, outlined in the section above.

2.3.3 Environmental Impacts

Impact on Fish Stocks

Organised crime in the fisheries sector in the form of large-scale overfishing has been shown to sharply reduce commercially exploitable fish stocks. In 2009, a group of British and Canadian researchers published an estimate of the worldwide extent of illegal fishing between 2000 and 2003. They found that 18 percent of the global catch, valued at between $10 billion and $23.5 billion, was lost to either illegal or unreported fishing during this period (Agnew et al. 2009). An updated 2011–2014 estimate suggests that annual illegal and unreported marine fishing generates $15.5 billion to $36.4 billion in illicit profits (Global Financial Integrity 2017). A recent global study (Sumaila et al. 2020) estimates that between 7.7 and 14.0 million metric tonnes of unreported fish catches are potentially traded illicitly each year. Latest FAO figs. (2015 data) estimate that some 59.9 percent of the world's commercial fish stocks are now fully fished, with about a third of global fish stock overexploited (FAO 2018). As fish stocks decline, the resource becomes more valuable, which, in turn, attracts increasing involvement by transnational organised crime syndicates in the fisheries sector (UNDOC 2011).

Effective law enforcement resulting in successful prosecution of organised crime in the fisheries sector can dramatically strengthen the targeted stocks. An illustrative example is the rock lobster case described earlier. The successful prosecution of the criminal network involved contributed to the dramatic recovery of the south coast rock lobster.

Catch rates of the fishery, which had been subject to government regulation since the mid-1970s, declined by 5–10 percent per year from 1984 to 2000 despite the introduction of a total allowable catch in 1984. This was the period during which the criminal network operated. Following the arrest and prosecution of those involved, five consecutive years of increased catch rates were recorded between 2000–2001 and 2004–2005 (Johnston and Butterworth 2017).

2.4 Addressing Organised Crime in Fisheries

2.4.1 Challenges

Globally, jurisdictions face a number of law enforcement challenges in identifying, investigating and successfully prosecuting organised crime in fisheries. Many of these are outlined in reports and outcome documents from expert group meetings and the International Symposiums on Fisheries Crime (e.g. CCPCJ 2017, 2016; UNODC 2017), underscoring the necessity of effective criminal law enforcement in addressing fisheries crime throughout the value chain. The 2017 report *Chasing Red Herrings* (NA-Fig 2017) highlights particular enforcement challenges arising from secrecy and the use of flags of convenience in the fishing industry.

Impact on the Marine Environment

Organised crime in fisheries can also harm the marine environment and associated ecosystems. In Nigeria, local fishers struggling to sustain their livelihoods are known to fish illegally near oil pipeline installations, which risks causing oil leaks and marine pollution (Okafor-Yarwood 2020, 2018). Piracy and armed robbery at sea in the broader Gulf of Guinea region can also lead to oil or chemical spills, through the use of weapons like rocket-propelled grenades to attack vessels and in the transfer of the targeted ship's cargo (IMB 2013). The use of gill nets in the illegal totoaba trade in Mexico has not only brought the vaquita porpoise, which is caught as by-catch, to the brink of extinction but has resulted in severe damage to the larger marine ecosystem of the Upper Gulf of California (Alvarado Martínez and Martínez 2018). Illegal dynamite (or 'blast') fishing, associated with explosives trafficking, off the Tanzanian coast is highly destructive to the marine habitat, including coral reefs, and reduces fish stocks, which has broader food security ramifications (IOC and FAO 2015; Galbraith 2015).

Law enforcement faces a number of challenges in addressing organised crime in the fisheries sector:

- **Low national prioritisation of organised crime in fisheries at political and operational levels.** Law enforcement officers around the world have noted that limited budget and resources lead law enforcement agencies to prioritise investigation of cases perceived as important. The result is that '[i] n many countries, crime linked to the fisheries value chain will not be investigated because it is not seen as a major priority and it is difficult to investigate' (UNODC 2017, 28).
- **Lack of coordination between government departments and agencies nationally, regionally and transnationally**. Due to the global nature of the fishing industry, key actors are scattered across various jurisdictions, with those ultimately responsible for, and benefitting from, criminal activities often in different states from those where the activities are taking place. This can make it unclear which state has jurisdiction to prosecute the offences in question. Without a high degree of information-sharing and cooperation between law enforcement authorities, in particular cross-border mutual legal assistance (MLA), successful prosecution of fisheries crime is very difficult (UNODC 2017, 28–29; NA-Fig 2017, 66).
- **Inadequate criminal and criminal procedural legislative frameworks**. The failure to criminalise fisheries crime at the national level can make it impossible to investigate and prosecute fisheries crime effectively, a problem emphasised in numerous reports and outcome documents (e.g. NA-Fig 2017, 66; UNODC 2017, 14).
- **Lack of clarity of jurisdiction at sea and extraterritorial jurisdiction**. Most states have not adequately criminalised offences committed by their nationals onboard foreign fishing vessels at sea or passed legislation regarding offences onboard stateless vessels. The result is that fisheries crime at sea (beyond national waters) is often not investigated or prosecuted (NA-Fig 2017, 66).
- **Lack of at-sea enforcement**. Effective at-sea surveillance by patrol boats requires considerable infrastructure and resources to maintain effective control in large and often distant marine areas (UNODC 2011, 131). Many states that rely heavily on the fisheries sector have very limited maritime law enforcement capability; this is particularly true in the Global South.
- **Lack of law enforcement agency and criminal justice capacity, particularly around financial investigations**. Law enforcement officials may have insufficient capacity to identify various forms of fisheries crime and to subsequently investigate and successfully prosecute the offences. They may also lack the technical skills necessary to pursue financial investigations (UNODC 2011, 138; CCPCJ 2017, paras. 6, 23; UNODC 2019a, 49).
- **Lack of transparency in the fisheries sector and the financial sector.** The lack of accurate information about beneficial ownership of vessels and legal entities operating in the fishing sector, unreliable and out-of-date information on vessels' identities and global movements, including in relation to at-sea transhipments, and the use of flags of convenience frustrate investigations of organised fisheries crime (UNODC 2011, 132,

137; CCPCJ 2017, para. 12; NA-Fig 2017). The ability of criminal networks to easily invest, disguise and launder the proceeds of their crimes in financial secrecy jurisdictions throughout the world, especially through the use of corporate structures like anonymous shell companies and tax havens, hampers investigators' ability to 'follow the money' and recover the proceeds of the crime and secure restitution (UNODC 2011; NA-Fig 2017).

2.4.2 Promising Practices

State administrations have numerous tools at their disposal to address organised crime. If employed correctly, these tools can significantly reduce the overall occurrence of organised crime. The complexity of organised crime in fisheries, the potential harm it may cause and the resources needed to combat it all suggest that any strategy should stress crime prevention. In this regard, identifying socio-economic drivers of criminal activities is key, as are crime disruption strategies and robust legislative frameworks that criminalise serious offences and attach sufficiently deterrent penalties. At the same time, where criminal networks have manifested themselves, it is important to identify effective means to address them, underpinned by robust cooperative criminal law enforcement and criminal procedural efforts.

FAO (2014) has acknowledged that 'the realities of corruption and organised crime [in fisheries] . . . need to be addressed through supplementary means extending beyond the realm of fisheries control and enforcement'. Legal tools developed to improve compliance in fisheries management should thus be seen as complementing a criminal justice enforcement approach to tackle criminal networks in the fishing sector. Evidence shows that the enforcement approach to 'prevent and control' organised crime cannot rely on the use of administrative sanctions alone, because criminals are able to absorb such penalties as part of their costs of doing business. For example, criminal operators select jurisdictions with low penalties, or totally avoid penalties by flying the flag of a state unable or unwilling to impose them. Likewise, the potential costs associated with vessel's being seized can be minimised by running a fleet of low-value vessels (de Coning 2016).

The following paragraphs detail five promising practices in addressing fisheries crime.

National-Level Inter-Agency Cooperation

The multitude of offences falling under the umbrella of 'fisheries crime' necessitates cooperation among a range of government agencies to identify, investigate and prosecute members of criminal networks. Various countries have introduced different types of cooperative enforcement models to address organised crime in fisheries.

Thailand employs a multidisciplinary approach to vessel inspections at the country's 31 'port-in, port-out' control centres to verify compliance with fishing and labour regulations. The teams include representatives from a range of agencies, including the Department of Fisheries; the Marine Department; the Department of Labour, Protection and Welfare; the Department of Employment; the Royal Thai Navy and the Marine Police. Most recently, a flying inspection team was added to increase the effectiveness of the control centres (MFA 2018).

In Indonesia, the Task Force to Combat Illegal Fishing (Task Force 115), established by presidential regulation, is a model of an integrated criminal justice mechanism to improve coordinated enforcement in combatting illegal fishing and fisheries crime cases with transnational organised elements. It operates as a 'single-roof' enforcement agency bringing together five enforcement agencies—the Ministry of Marine Affairs and Fisheries, the Navy, the Marine Police, the Coast Guard, and the Attorney General's Office—under the auspices of the Ministry of Marine Affairs and Fisheries. Task Force 115 undertakes a range of activities to ensure effective enforcement in fisheries, including case monitoring, management and assistance. Malaysia is currently also moving towards introducing a similar 'one-roof' enforcement agency to address maritime crime.

Norway established the Norwegian National Advisory Group on Fisheries Crime and IUU Fishing (Fiskeriforvaltningens Analysenettverk) in 2008. All relevant agencies, including the Coast Guard, the police, as well as fisheries, maritime, tax, customs and labour authorities meet regularly to discuss common cases.

The secretariat of the advisory group is housed in the Ministry of Trade, Industry and Fisheries and is involved in law and policy development on fisheries crime.

In Tanzania, the Multi-agency Task Team on environmental and wildlife crime (MATT) was launched in 2015. The MATT is led by the Ministry of Home Affairs and includes the Ministries of Livestock and Fisheries Development, the Ministry of Energy and Minerals, the Ministry of Natural Resources and the Tanzanian Police Force. Aimed at coordinating efforts and resources, the MATT targets individuals and criminal networks that control environmental and fisheries crime in the region and the illegal trade in wildlife (IOC and FAO 2015).

Regional and Global Cross-Border Cooperation

Regional and global cooperation also plays a vital role in enhancing the investigation of organised crime in fisheries, including through the use of MLA, INTERPOL tools and judicial requests to cooperate with other countries. Countries such as Indonesia and South Africa have used these tools to uncover cases of transnational organised crime in fisheries, and to facilitate prosecution of the offenders. Example cases highlighting the value of this collaboration include those of the *STS-50* and *Viking*, in which Indonesia invited a multilateral team of experts and INTERPOL to help facilitate information-sharing, analyses and possible prosecution of the offenders in various jurisdictions.

Law Reform

A number of jurisdictions have criminalised fisheries offences and attached deterrent penalties to facilitate engagement of the criminal law enforcement track and frustrate criminal operators' ability to factor fines into the cost of doing business. For example, the Ghanaian courts may, in certain circumstances, impose a maximum penalty of two years' imprisonment for the illegal export of fish under the terms of the 2002 Fisheries Act, and in South Africa violations of almost all the provisions of the 1998 Marine Living Resources Act amount to a criminal offence and are punishable by a fine of up to five million rand ($338,000). Norway follows a similar approach.

The criminalisation of other offences falling under the fisheries crime umbrella, along with sufficiently severe sanctions, including against corporations, has been implemented by various jurisdictions. Indonesian Law 25/2003 on the Crime of Money Laundering obliges banks and financial service providers to make information on suspicious financial transactions available based on which law enforcement officers have the power to request that the bank freeze the accounts of suspected money-laundering criminals (Husein 2014). Also particularly valuable is the enactment of organised crime legislation, such as South Africa's 1998 Prevention of Organised Crime Act, which criminalises racketeering and triggers asset forfeiture. The act is successfully employed in the prosecution of organised crime networks engaged in abalone trafficking. Legal measures to increase transparency in the fisheries sector are also valuable as they may facilitate identification and investigation of organised crime in the sector. In Taiwan, for example, the 2016 Act to Govern Investment in the Operation of Foreign-Flag Fishing Vessels aims at regulating Taiwanese nationals' involvement in the operation of foreign-flagged fishing vessels beyond national waters.

As cases of human rights abuse on fishing vessels increasingly come to light, various countries have introduced legislation to protect against these abuses, following the adoption of the ILO Work in Fishing Convention No. 188. Indonesia introduced regulations based on the three-pillared Ruggie Guiding Principles on Business and Human Rights (UN OHCHR 2011) implementing a certification system obliging Indonesian fishing companies and fishing vessels to protect fishers' human rights. In February 2019, Thailand also ratified the ILO Work in Fishing Convention No. 188 and is seeking to amending existing laws to comply with regulations on protecting workers onboard fishing vessels (Tavornmas 2019).

Capacity-Building and Skills Training in Criminal Justice Systems

Capacity-building activities are deemed important to broaden the knowledge and hone the skills of investigators and prosecutors in enforcement efforts against organised crime in the fisheries sector. Both Indonesia and Norway offer professional accredited training to strengthen the law enforcement capacity to detect and punish crimes throughout the fisheries value chain. The International FishFORCE Academy of Indonesia, at the Jakarta Centre for Law Enforcement facility in Semarang, was established in December 2016, while the Norwegian Police University College (2019) has offered modules on fisheries crime investigation to Norwegian law enforcement agencies since 2010. Particular emphasis should be given to enhancing the capacity to conduct financial investigations parallel to investigating the underlying offence where significant amounts of money are involved (UNODC 2019a).

Sensitisation of judges to cases of organised crime in the fisheries sector is also recognised as valuable. Most recently, in 2020, the Norwegian government launched the Blue Justice Initiative, which includes a focus on the capacity-building needs of developing countries to address organised crime in the fisheries sector (Blue Justice 2020).

Awareness-Raising

Given the transnational nature of organised crime in fisheries, the technical expertise of law enforcement agencies must be augmented by strong international networks built to enhance the timely and accurate gathering of information, as well as intelligence-led law enforcement efforts. International government organisations, such as UNODC and the INTERPOL Fisheries Crime Working Group, facilitate this, as do international processes such as the International Symposiums on Fisheries Crime (FishCRIME Symposiums), which, as of 2020, will be superseded by the Global Blue Justice Conferences.

Awareness of organised crime in the sector must also be raised among legitimate fishing businesses in order to minimise opportunities for criminal activities along the industry's retail and value chains. Case examples exposed in the media have helped increase public demand for improved sustainability and transparency of seafood production practice. For example, in 2014 The Guardian reported on an investigation that found that the world's largest Thailand-based prawn farmer was purchasing fishmeal to feed its prawns from some suppliers that owned, operated or bought from fishing vessels engaged in human trafficking for forced labour. These prawns were sold to the public in four of the world's top retailers (The Guardian 2014), raising questions about these businesses' ethical standards. *Corporate social responsibility* (CSR) as an umbrella term refers to the commitment of businesses to recognise their responsibility for the behaviour of their business partners (i.e. within their value chains) and for their impact on society and the natural environment (Blowfield and Frynas 2005, 503). An increased focus on CSR in the fishing industry could give fishing companies a tool to engage in activities that respond to civil society pressure in this regard (Packer et al. 2019). Civil society, in turn, can play a role in both influencing the substantive content of CSR practices and holding companies accountable for their implementation.

3 Concluding Thoughts

In reviewing relevant literature, international cases, reports and expert insights in order to produce this Blue Paper, the authors found ample anecdotal, scientific and case-based evidence of the many manifestations of organised crime in the fisheries sector and its multi- faceted negative impacts on society at large.

The paper shows that organised crime in the fisheries sector is widespread in the sense that it is not restricted to specific geographical locations but rather is found globally. That said, criminals will tend to seek out the world's most vulnerable regions as sites for their activities, and fisheries crime thus appears to most harm the coastal populations of states with the least resources to prevent and combat it.

There is, however, a knowledge deficit with regards to the scale of organised crime in the fisheries sector. To date no statistical data are publicly available estimating the extent of organised crime in the fisheries sector or mapping incidents of its location at a global level. Existing scientific output and data on criminality in the fisheries sector speak almost exclusively to illegal or unreported fishing, which does not take into account the range of criminal offences that occur throughout the fisheries value chain. Greater scientific and criminological knowledge is needed of the dynamics and scale of organised fisheries crime and the networks involved therein in order to identify, evaluate and implement the best measures to address the drivers of such crime.

As outlined in this Blue Paper, organised crime in the fisheries sector has the potential to severely undermine the premises for a sustainable ocean economy, with notable adverse social, economic and environmental implications. The problem is recognised as sufficiently severe to warrant states' mobilisation to take political action. The rate at which support of the Copenhagen Declaration is gaining momentum, particularly among states from the Global South, attests to states' increasing acknowledgement of the existence, extent and adverse impacts of organised crime in the fisheries sector. Further, states acknowledge that a failure to effectively address such crime will result in a widespread inability to meet SDG 16, among others, and ultimately to create a sustainable ocean economy. A next important step will be for states supporting the Copenhagen Declaration to identify practical measures to implement their political commitments.

This Blue Paper highlights the complexity of the underlying problem and suggests a number of 'opportunities for action' for states to address organised crime in fisheries. These suggestions confirm that an approach rooted in fisheries management alone cannot adequately respond to the criminal challenges identified in the fisheries sector. Also essential is the application of an intelligence-led, skills-based cooperative law enforcement response at a domestic level that is facilitated by enabling legal frameworks and increased transparency in the fisheries governance sector and associated financial systems at the global level.

4 Opportunities for Action

The UN General Assembly has noted with concern the existence of possible connections between international organised crime and illegal fishing and urged states to better understand the causes and methods of, and contributing factors to, these connections, bearing in mind the distinct legal regimes and remedies under international law applicable to illegal fishing and international organised crime (UNGA Resolution 63/112).

States have acted on the General Assembly's request, engaging in and supporting a number of initiatives aimed at increasing the knowledge base and understanding of the possible connections between organised crime and illegal fishing. These initiatives' findings, which are publicly available, reflect the General Assembly's instructions to respect the two distinct legal regimes that govern IUU fishing and transnational organised crime, respectively, according to the mandates of the corresponding two UN agencies, namey, the FAO and the UNODC.

Yet, with notable exceptions, the world community remains largely uninformed of the evidence, or even existence, of transnational organised crime in the global fishing industry as a distinct problem alongside global efforts to secure a sustainable fisheries resource.

Furthermore, states remain predominantly unaware of the most appropriate remedies and applicable legal regimes to address this problem, and how they differ from (yet are complimentary to) measures aimed at remedying fisheries management challenges. This lack of distinction between the problem of organised crime in fisheries (in effect a security and law enforcement problem), on the one hand, and IUU fishing (in effect a problem of unsustainable fishing practices), on the other, may lead states to continue making ill-informed decisions regarding the most suitable approaches to these two challenges, both separately and in concert, with the danger that neither will be adequately addressed.

The distinct natures of the two phenomena are particularly evident when organised crime in fisheries is examined within the context of other maritime crimes.

In February 2019 the UN Security Council heard national submissions highlighting the multifaceted nature of transnational organised crime at sea, including fisheries crime, and the threat it poses to international peace and security. The members underscored the Security Council's potential future role in addressing the problem of fisheries crime, while one permanent member of the Council expressed doubt that IUU fishing, or the degradation of the maritime environment, falls within the Council's power and prerogative (UNSC 2019). It is worth taking note of this point: It is not the primary objective of fisheries management and conservation bodies to address organised crime, and it is not the primary objective of peace and security bodies, like the Security Council, to address fisheries management and conservation. By respecting the distinct natures of organised crime in fisheries and IUU fishing, states will have a wider and complimentary set of tools at their disposal to address the security implications of organised crime in fisheries and the fisheries management implications of IUU fishing, respectively.

In order to meet the targets set by SDG 16 to significantly promote the rule of law, reduce illicit financial flows, combat all forms of organised crime, substantially reduce corruption and bribery, strengthen relevant national institutions and build capacity to combat crime, it is imperative that at a state level there be universal recognition of the security implications of organised crime in fisheries, of the need to address the challenges posed by the shadow blue economy and of the necessity of cooperation at all levels to enhance inter-agency and cross-border fisheries crime law enforcement. As SDG 16 is an enabling goal of, among others, SDG 14 on 'life below water', addressing organised crime in fisheries will impact states' ability to reach the targets set forth in SDG 14.

States are, however, moving towards increased understanding of the problem of organised crime in fisheries. In 2017, a Nordic political declaration on Transnational Organised Fisheries Crime was issued through the Nordic Council of Ministers (the Ålesund Declaration), and in 2018 ministers from nine countries adopted the Copenhagen Declaration, which is gaining traction. In both declarations, the ministers express their conviction 'that there is a need for the world community to recognize the existence of transnational organised crime in the global fishing industry'.

Only when all states share a joint understanding of the problem at hand can technical solutions to address the problem be implemented. To this end, the opportunities for action suggested below are divided into two consecutive stages (Box 16.1). In the first stage, outlined in Sect. 4.1, the opportunities for action is to work towards the political goal of a common understanding of organised crime in the global fishing industry. A political recognition of the problem is crucial in order to identify and implement more technical and practical opportunities for action to address organised crime in fisheries. In the second stage, outlined in Sect. 4.2, the opportunities for action focus on practical law enforcement tools to address organised crime in the fishing industry based on the promising practices detailed above.

4.1 Stage One: Develop a Common Understanding of Transnational Organised Crime in Fisheries

We offer the following opportunities for action:

- **All states should report to the UN General Assembly in response to the call made by UNGA Resolution 63/112 to examine connections between illegal fishing and organised crime in the fisheries sector.** This will facilitate establishment of a solid platform of knowledge about the manifestations of organised crime in the fisheries sector as it is experienced around the world.
- **UN Security Council members should raise the security implications of transnational organised crime in the fisheries sector** to encourage the development of a common understanding of the problem's security dimensions.
- **All states should formally support the 2018 International Declaration against Transnational Organised Crime in the Global Fishing Industry (the Copenhagen Declaration).** Broad support for the declaration would be an important step towards developing a common understanding and awareness of the problem of organised crime in the fisheries sector and building the political will to more vigorously address it, which in turn would facilitate achievement of the SDGs.
- **All states should report annually on transnational organised crime in the fisheries sector to the UN Commission for Crime Prevention and Criminal Justice (CCPCJ)** to ensure widespread and continuous attention to the problem.
- **All states should participate in regular international knowledge- sharing forums** to share information on, and discuss challenges and opportunities arising from, cases of transitional organised crime in the fisheries sector worldwide. Outcome documents therefrom should be made publicly available.

Box 16.1 Summary of Two Stages of Action

1. Political Action: Develop a common understanding of transnational organised crime in the fisheries sector globally, and build political will to address the challenge cooperatively.
2. Practical Tools: Develop practical tools to strengthen law enforcement capacity through:
   - Strengthened national inter-agency cooperation
   - Effective cross-border law enforcement cooperation
   - Enabling legal frameworks
   - Skills training and capacity building
   - Community-based crime prevention strategies which incorporate a gendered approach
   - Engagement of civil society
   - Support of relevant research

4.2 Stage Two: Develop Practical Tools to Strengthen Law Enforcement Capacity to Address Organised Crime in the Fisheries Sector

Once there is a common understanding of organised crime in the fisheries sector, states, acting jointly and individually, should develop a number of practical tools to strengthen their criminal law enforcement capacity to address the problem. Together these tools should improve states' ability to prevent, detect and respond to organised crime in the fisheries sector.

We offer the following opportunities for action:

- **All governments should strengthen national inter-agency cooperation to address organised crime in the fisheries sector**. We recommend that states introduce national inter-agency models to facilitate and support coordinated criminal law enforcement efforts to prevent organised crime in the fisheries sector and identify and prosecute offenders. The sharing of relevant information across agencies should be facilitated by legislation in line with personal data protection principles (for European Union General Data Protection Regulations, for example, see European Commission 2019). The inter-agency body should have high-level political support, should ideally be established via legislative means and should have a clear mandate and permanence.
- **All governments should strengthen cross-border cooperative law enforcement efforts to identify, investigate and prosecute cases of transnational organised crime in the fisheries sector, including through mutual legal assistance**. We suggest that the tools available through international and regional networks and organisations, such as multilateral information-sharing and analysis mechanisms, be used to this end, as well as platforms for the secure exchange of financial intelligence, and that existing databases on organised crime, such as the UNODC's SHERLOC (2019) database, be augmented to also include fisheries. This will facilitate the sharing of information and intelligence and enhance the ability of law enforcement action to address organised crime in the fisheries sector.
- **All states should review their legal frameworks and implement reforms where needed**. The goal should be to criminalise and attach sufficiently deterrent penalties to all fisheries crime offences, introduce anti-corruption and anti–money laundering measures and make provision for asset recovery and forfeiture of the proceeds of crime. We recommend that states provide for the extra-territorial and extra-jurisdictional application of relevant laws (e.g. through a Lacey-type law) and for corporate criminal liability. States should expressly provide for the criminal procedural tool of mutual legal assistance. This will facilitate cross-border information-sharing, asset tracing, and evidence collection and will strengthen investigation and prosecution of cases. States should be encouraged to introduce measures aimed at increased transparency around, for instance, information on the true beneficial ownership of fishing vessels, fishing vessels' movement and licences, and fisheries access agreements. Further, states should be encouraged to support international legal frameworks aimed at reducing human rights abuses in the fishing industry and ensure that these legal frameworks are enacted and enforced at a national level. States should also aim to standardise their laws and penalties. It may be advantageous to consider establishing a fund to assist law enforcement agencies, similar to the Green Fund in Trinidad and Tobago, based on a levy from corporate taxes (Finance Act 91 of 2000, Part XIV).
- **All states should engage in skills training and capacity building for law enforcement officials in the criminal justice system from at-sea to trial**. This should include the development of skills around financial investigation and asset recovery and capacity to detect tax crime in the fisheries sector. Mentoring schemes could be beneficial in this regard. This will facilitate the identification, investigation and successfully prosecution of cases of organised crime in the fisheries sector.
- **All states should introduce community-based crime prevention strategies incorporating a gendered approach as appropriate**. This will strengthen the resilience of vulnerable coastal communities and their ability to respond to organised crime in the fisheries sector.

- **Civil society should engage with the fishing industry on corporate social responsibility for sustainable fisheries practices.** In response to increasing public demand for sustainable fisheries products and a transparent value chain, the global fisheries industry is increasingly harnessing CSR practices, which, in turn, help minimise the risk of criminal activities in the fisheries value and supply chain. Civil society should be encouraged to engage with the fishing industry towards the development of corresponding CSR standards.
- **All states should support research to understand the causes, nature, scale and impact of organised crime in fisheries.** This supports the UN General Assembly's call for states to increase knowledge and understanding of the links between illegal fishing and organised crime. Research outcomes should be made publicly available since they can inform the development of appropriate means to prevent organised crime in fisheries and strengthen the law enforcement response.

Acknowledgements The paper's technical reviewers, Teale Bondaroff, Alan Cole, Jorge Torre and Steve Trent, as well as its arbiter, Kristian Teleki, all provided helpful technical comments. Special thanks to Januar Dwi Putra for assisting Mas Achmad Santosa and Yunus Husein.

The authors also thank the World Resources Institute for providing support as the Ocean Panel Secretariat.

While our colleagues were very generous with their time and input, this report reflects the views of the authors alone. The authors thank Alex Martin for copyediting and Romain Warnault for design.

Appendix A

We, the Ministers of Faroe Islands, Ghana, Indonesia, Kiribati, Namibia, Norway, Palau, Solomon Island and Sri Lanka [1]: Encourage other Ministers to support this non-legally binding declaration.

Note the recommendations and the outcome of the second International Symposium on Fisheries Crime held in Yogyakarta, Indonesia 10–11 October 2016 which was published by the United Nations Office on Drugs and Crime at the occasion of the United Nations Commission on Crime Prevention and Criminal Justice during its twenty-sixth session in Vienna 22–26 May 2017.

Recognize that our countries are dependent on the sea and its resources and the opportunities it holds for the economy, food and well-being of our population and we are determined to support a healthy and thriving fishing industry that is based on fair competition and the sustainable use of the ocean.

Are committed to work towards the fulfilment of the UN Sustainable Development Goals particularly in relation to Goal 14 on 'Life Below Water' and Goal 16 on 'Peace, Justice and Strong Institutions'.

Are convinced that there is a need for the world community to recognize the existence of transnational organized crime in the global fishing industry and that this activity has a serious effect on the economy, distorts markets, harms the environment and undermines human rights.

Recognize that this transnational activity includes crimes committed through the whole fisheries supply and value chain which includes illegal fishing, corruption, tax and customs fraud, money laundering, embezzlement, document fraud and human trafficking.

Recognize further the inter-continental flow of illegal fish products, illicit money and human trafficking victims in transnational organized crime cases in the global fishing industry and that all regions of the world need to cooperate when investigating such acts.

Are convinced that inter-agency cooperation between relevant governmental agencies is essential at a national, regional and international level in order to prevent, combat and eradicate transnational organized crime in the global fishing industry.

Are also convinced that there is a need for international cooperation and that developing countries are particularly affected.

Recognize the particular vulnerability of small-island developing states and other Large Ocean Nations of the impact of transnational organized crime in the global fishing industry.

Are also convinced of the need for continuous support on the highest level and the necessity for awareness raising on these issues through events such as the International Symposiums on Fisheries Crime.

[1]The following countries have subsequently pledged their support to the Declaration: Benin, Chile, Costa Rica, Fiji, Greenland, Iceland, Liberia, Maldives, Marshall Islands, Mexico, Mozambique, Myanmar, Nauru, Philippines, São Tomè and Principe, Scotland, Seychelles, South Africa and Timor Leste. For an updated list see www.bluejustice.org.

About the Authors

Lead Authors

Emma Witbooi is a Research Associate with the Nelson Mandela University in Port Elizabeth, South Africa, and Director of the PescaDOLUS International Fisheries Crime Research Network.

Kamal Deen-Ali is Executive Director of the Centre for Maritime Law & Security Africa in Accra, Ghana.

Mas Achmad Santosa was Coordinator of Special Staff in the Indonesian Presidential Task Force to Combat Illegal Fishing at the time of writing.

Contributing Authors

Gail Hurley was Policy Specialist: Development Finance, UNDP Bureau for Policy and Programme Support at the time of writing.

Yunus Husein was in the Indonesian Presidential Task Force to Combat Illegal Fishing at the time of writing.

Sarika Maharaj is Coordinator of the Fisheries Inspectorate, Fisheries Division in the Ministry of Agriculture, Land and Fisheries, Government of the Republic of Trinidad and Tobago.

Ifesinachi Okafor-Yarwood is a Visiting Fellow at the Centre for Strategic Research and Studies, National Defence College in Abuja, Nigeria.

Inés Arroyo Quiroz is with the Programa de Estudios Socioambientales at the Centro Regional de.

Investigaciones Multidisciplinarias, Universidad Nacional Autónoma de México in Cuernavaca.

Omar Salas is Adviser of the Agencia para la Reincorporación y la Normalización (ARN) with the Presidency of Colombia and former Director of the Justice, Security and Government National Planning Department.

References

Abadinsky H (2007) Organized Crime, 8th edn. Thomson Wadsworth, Belmont, CA

Agnew DJ, Pearce J, Ganapathiraju P, Peatman T, Watson R, Beddington JR, Pitcher TJ (2009) Estimating the Worldwide Extent of Illegal Fishing. PLoS One 4(2):1–8. https://doi.org/10.1371/journal.pone.0004570

Agnew DJ, Walmsley SF, Leotte F, Barnes C, White C, Good S, Deme M, Cherif AM, Suale D (2010) Estimation of the cost of illegal fishing in West Africa: final report. MRAG West Africa Regional Fisheries Project, London

Akhmirova R 2012. "Head of the Federal Agency for Fishery Andrei Kainy: We Are Defeating the Mafia!" ("Глава Росрыболовства Андрей Крайний: Мы побеждаем мафию!"). *Sobesednik.ru*, 9 July. https://sobesednik.ru/politika/20120709-glava-rosrybolovstva-andrei-krainii-my-pobezh-daem-mafiyu.

Ali K (2015) Maritime Security Cooperation in the Gulf of Guinea: Prospects and Challenges. Brill/Nijhoff, Boston

Alvarado Martínez I, Martínez ER (2018) Trafficking of Totoaba Maw. In: Arroyo Quiroz I, Wyatt T (eds) Green Crime in Mexico: A collection of case studies. Palgrave Macmillan, Green Criminology. London, pp 149–170

Antara News. 2018. "90 Percent of Drug Smuggling into Indonesia via Sea." *Antaranews.com*, 7 October. https://en.antaranews.com/news/119269/90-percent-of-drug-smuggling-into-indonesia-via-sea.

Australian Crime Commission (2009) Organized Crime in Australia: 2009. Australian Government, Canberra

Banaseh, M.A. 2017. "Ghana Risks Losing GH¢1.5bn to Fuel Smuggling." *Graphic Online*, 23 August. https://www.graphic.com.gh/news/gener-al-news/ghana-risks-losing-gh-1-5bn-to-fuel-smuggling.html

Bassant M 2019. "Venezuela's 'Evander' Gang Leader Held in T&T." *Trinidad and Tobago Guardian*, 20 May. https://www.guardian.co.tt/news/venezuelas-evander-gang-leader-held-in-tt-6.2.849655.de097b4f83.

Beseng M (2019) Cameroon's Choppy Waters: The Anatomy of Fisheries Crime in the Maritime Fisheries Sector. Mar Policy 108:10

Blowfield M, Frynas JG (2005) Setting New Agendas: Critical Perspectives on Corporate Social Responsibility in the Developing World. Int Aff 81(3):499–513

Blue Justice. 2020. "Blue Justice Initiative." https://bluejustice.org/blue-justice-initiative/. Accessed 2 February.

Cabral RB, Mayorga J, Clemence M, Lynham J, Koeshendrajana S, Muawanah U, Nugroho D et al (2018) Rapid and lasting gains from solving illegal fishing. Nat Ecol Evolution 2:650–658

CCPCJ (UN Commission on Crime Prevention and Criminal Justice). 2016. "Outcome of the UNODC/WWF Fisheries Crime Expert Group Meeting, 24–26 February 2016, Vienna." CCPCJ Conference Room Paper, 11 May. E/CN.15/2016/CRP.2.

CCPCJ. 2017. "Outcome of the 2nd International Symposium on Fisheries Crime, October 10–11, 2016, Yogyakarta, Indonesia." CCPCJ Conference Room Paper, 17 May. E/CN.15/2017/CRP.3.

Chatham House (2013) Maritime Security in the Gulf of Guinea: Report of the Conference Held at Chatham House, London, 6 December 2012. Chatham House, London

Cochrane, K. 2017. "Department Will Be Judged Harshly if It Fails to Protect Rock Lobster." *Cape Times*, 26 September. https://www.iol.co.za/capetimes/opinion/department-will-be-judged-harshly-if-it-fails-to-protect-rock-lobster-11366443.

Colombian National Navy (Armada Nacional). 2015. "Activación de la Fuerza de tarea contra el narcotráfico en el Caribe" ("Activation of the Task Force against Drug Trafficking in the Caribbean"). 19 February. https://www.armada.mil.co/es/content/activacion-de-la-fuerza-de-tar-ea-contra-el-narcotrafico-en-el-caribe

CRFM (Caribbean Regional Fisheries Mechanism) (2018) CRFM Statistics and Information Report 2016. CRFM, Belize

Crosta A, Sutherland K, Talerico C, Layolle I, Fantacci B (2018) Operation Fake Gold: The Totoaba Supply Chain—From Mexico's Totoaba Cartels to China's Totoaba Maw Wholesalers—An Illegal Trade Killing the Vaquita. Elephant Action League, Los Angeles

Dagbladet. 2018a. "The Hidden Slave Contracts." 17 November. https://www.dagbladet.no/nyheter/de-skjulte-slavekontraktene/70460627.

Dagbladet. 2018b. "Deadliest Catch." 11 December. https://www.dagbla-det.no/magasinet/the-deadliest-catch/70534450.

Dambach, K. 2019. "Human Traffickers Arrested in Italy." *InfoMigrants*, 24 June. https://www.infomigrants.net/en/post/17708/human-traffickers-arrested-in-italy

de Coning E (2016) Fisheries Crime. In: Elliott L, Schaedla WH (eds) Handbook of Transnational Environmental Crime. Edward Elgar Publishing, Cheltenham, p 204

de Greef K, Raemaekers S (2014) South Africa's Illicit Abalone Trade: An Updated Overview and Knowledge Gap Analysis. Traffic International, Cambridge, UK

Demont J 2003. "Maritime Drug Smuggling and Rum-Running." *Canadi- an Encyclopedia*, 17 March. https://www.thecanadianencyclopedia.ca/en/article/maritime-drug-smuggling-and-rum-running

Devlin C, Glaser SM, Villegas C, Poinsatte N (2020) Rough Seas: The Causes and Consequences of Fisheries Conflict in Somali Waters. One Earth Future, Broomfield, CO

Doumbouya A, Camara OT, Manie J, Intchama JF, Jarra A, Ceesay S, Guèye A et al (2017) Assessing the Effectiveness of Monitoring Control and Surveillance of Illegal Fishing: The Case of West Africa. Front Mar Sci:7. https://doi.org/10.3389/fmars.2017.00050

Egmont Group. 2014. "Egmont Cases. Financial Analysis Cases, 2011–2013." Toronto: Egmont Group.

Emmort, R. 2009. "Mexico Finds Cocaine Haul Hidden in Frozen Sharks." *Reuters World News*, 16 June. https://www.reuters.com/article/us-mexico-drugs/mexico-finds-cocaine-haul-hidden-in-frozen-sharks-idUSN1631193420090617

Ensign ES (2001) Intelligence in the Rum War at Sea, 1920–1933. Joint Military Intelligence College, Washington, DC

EURONEWS. 2019. "Caught in the Act: People Smugglers Filmed Transfer- ring Migrants at Sea." 22 June. https://www.euronews.com/2019/06/22/caught-in-the-act-people-smugglers-filmed-transferring-migrants-at-sea.

European Commission. 2019. "EU Data Protection Rules." https://ec.europaeu/commission/priorities/justice-and-fundamental-rights/data-protection/2018-reform-eu-data-protection-rules/eu-data-protection-rules_en. Accessed 18 October.

EUROPOL (European Union Agency for Law Enforcement Cooperation) (2004) European Union Organized Crime Report. EUROPOL, Luxembourg

EUROPOL. 2018. "Fraud on a Plate: Over 3 600 Tonnes of Dangerous Food Removed from Consumer Market." Press release, 25 April. https://www.europol.europa.eu/newsroom/news/fraud-plate-over-3-600-tonnes-of-dangerous-food-removed-consumer-market.

FAO (UN Food and Agriculture Organization) (2014) Voluntary Guidelines for Securing Sustainable Small-Scale Fisheries in the Context of Food Security and Poverty Eradication. FAO, Rome

FAO (2018) The State of World Fisheries and Aquaculture: Meeting the Sustainable Development Goals 2018. FAO, Rome

FAO. 2019. Draft: Guidance on Social Responsibility in Fisheries and Aquaculture Value Chains. Rome: FAO. https://socialfisheries.fao.konveio.com/. Accessed 18 October.

FAO Legal Office. 2019. "Survey of Practice on Fisheries Law Enforcement Process: Administrative v Criminal v Both." Presentation during side event at Our Ocean 2019, Oslo, 23 October

FATF (Financial Action Task Force) 2012–2019. International Standards on Combating Money Laundering and the Financing of Terrorism & Proliferation. Paris: FATF.

Felix SG (2017) Resilience in Sinaloa: Community Responses to Organized Crime. Global Initiative against Transnational Organized Crime, Geneva

Fenstadt A, Kvile K. 2016. "Polaris' Owner Promises to Clean Up Mess with Crew." ("'Polaris'-reder lover å rydde opp i mannskapsrot") *Fiskeribladet*, 14 April. https://fiskeribladet.no/nyheter/?artikkel=46419

Fisher, H. 2016. "Rocky Point Residents Beg for Peace." *Jamaica Star*, 13 April. http://weww.jamaica-star.com/article/news/20160413/rocky-point-residents-beg-peace.

Fröhlich T (2003) Final Report: Organized Environmental Crime in the EU Member States. Betreuungsgesellschaft für Umweltfragen, Kassel, Germany

Galaz V, Crona B, Dauriach A, Jouffray J-B, Österblom H, Fichtner J (2018) Tax havens and global environmental degradation. Nat Ecol Evolution 2:1352–1357

Galbraith K 2015. "The Horrors of Fishing with Dynamite." *New York Times*, 4 February. https://www.nytimes.com/2015/02/05/business/energy-environment/the-horrors-of-fishing-with-dynamite.html

Gilman N, Goldhammer J, Weber S (2011) Deviant Globalization: Black Market Economy in the 21st Century. Continuum International, New York

Gilpin R 2007. "Enhancing Maritime Security in the Gulf of Guinea." *Strategic Studies* 6 (1). Naval Postgraduate School. Monterey, CA: Center for Contemporary Conflict.

Global Financial Integrity (2017) Transnational Crime and the Developing World. Global Financial Integrity, Washington DC

Global Financial Integrity. 2019. "Trade Misinvoicing." https://gfintegrity.org/issue/trade-misinvoicing/. Accessed 18 October.

Gottschalk P (2012) White-Collar Criminals. Cases and Theories of Financial Crime. Unipub, Oslo

Government of Jamaica (2007) National Security Policy for Jamaica: Towards a Secure and Prosperous Nation. Government of Jamaica, Kingston

Grey, S., and A. Ismail. 2016. "The Forgotten Shipwreck." *Reuters*, 6 December. https://www.reuters.com/investigates/special-report/europe-migrants-egypt-ship/

The Guardian. 2014. "Revealed: Asian Slave Labour Producing Prawns in Supermarkets in US, UK." 10 June. https://www.theguardian.com/global-development/2014/jun/10/supermarket-prawns-thailand-produced-slave-labour

The Guardian. 2019. "Venezuelan Refugees Feared Drowned on Route to Trinidad." 25 April. https://www.theguardian.com/world/2019/apr/25/venezuelan-refugees-feared-drowned-trinidad-boat-sinks.

Hauck M, Sweijd NA (1999) A Case Study of Abalone Poaching in South Africa and Its Impact on Fisheries Management. ICES J Mar Sci 56:1024–1032

Hübschle, A., and C. Shearing. 2018. *Ending Wildlife Trafficking: Local Communities as Change Agents*. Geneva: Global Initiative against Transnational Organized Crime. http://globalinitiative.net/wp-content/uploads/2018/08/TGIATOC-Wildlife-Trafficking-Report-WEB-4.pdf.

Human Rights at Sea. 2019. "Case Study—In Their Own Words: The Story of Josaisa and Virisila Cama, Fiji. Human Rights Abuse in Fijian Crewed Fisheries." May. https://www.humanrightsatsea.org/case-studies.

Husein Y (2014) Combating Environmental Crime with Anti–Money Laundering Initiatives: The Indonesian Experience. In: Rose G (ed) Following the Proceeds of Environmental Crime: Forests, Fish and Filthy Lucre. Routledge, London, pp 71–80

ILO (International Labour Organization) (2016) Fishers First: Good Practices to End Labour Exploitation at Sea. ILO, Geneva

IMB (International Maritime Bureau) (2013) Piracy and Armed Robbery against Ships: Report of the Period 1 Jan.—31 Dec. 2012. ICC International Maritime Bureau, London

Indonesian Ministry of Finance (2019) Government Revenue Model. Indonesian Ministry of Finance, Jakarta

Indonesian National Narcotics Board. 2012. "Disclosure of International Drug Networks." Press release, 28 May. https://bnn.go.id/pengungkapan-jaringan-sindikat-narkoba-internasional-dengan-barang-bukti-1-412-476-butir-ekstasi/

Intchama J, Belhabib D, Jumpe R (2018) Assessing Guinea Bissau's Legal and Illegal Unreported and Unregulated Fisheries and the Surveillance Efforts to Tackle Them. Front Mar Sci 5:79

INTERPOL (International Criminal Police Organization) (2014) Study on Fisheries Crime in the West African Coastal Region. INTERPOL, Lyon

INTERPOL (2018) International Law Enforcement Cooperation in the Fisher- ies Sector: A Guide for Law Enforcement Practitioners. INTERPOL, Lyon

IOC (Indian Ocean Commission) and FAO (UN Food and Agriculture Organization) (2015) A Multi-agency Task Team Working Together to End Destructive Blast Fishing. FAO, Rome

Isaacs M, Witbooi E (2019) "Fisheries Crime, Human Rights and Small-Scale Fisheries in South Africa: A Case of Bigger Fish to Fry." Special section, "Transnational Organized Fisheries Crime.". Mar Policy 105:158–169

Johnston SJ, Butterworth DS. 2017. "Summary of South Coast Rock Lobster (Palinurus Gilchristi) Fishery." Background document to "In- ternational Fisheries Stock Assessment Review Workshop."

Department of Mathematics and Applied Mathematics, University of Cape Town.
Kaplan-Hallama M, Bennetta NJ, Satterfielda T (2017) Catching Sea Cucumber Fever in Coastal Communities: Conceptualizing the Impacts of Shocks versus Trends on Social-Ecological Systems. Glob Environ Chang 45:89–98
Kercher J (2018) Fisheries Crime and the SDGs: The Call and the Tools for Interagency Cooperation. In: Paper presented at FishCRIME 2018. UN City, Copenhagen. https://www.sdg16hub.org/content/fisheries-crime-and-sdgs-call-and-tools-interagency-policy-cooperation
Kraemer RA (2017) "A Sustainable Ocean Economy, Innovation and Growth: A G20 Initiative." Policy brief no. 113 (July). Centre for International Governance Innovation, Waterloo, Canada
Krivoshapko J 2017. "A Fish! The Antimonopoly Service Has Uncovered Schemes by Which Foreigners Illegally Catch Our Fish." ("Рыба! Антимонопольная служба раскрыла схемы, с помощью которых иностранцы незаконно ловят нашу рыбу") *Rossiyskaya Gazeta*, 20 June. https://rg.ru/2017/06/20/fas-raskryla-shemy-ponezakon-nomu-lovu-rossijskoj-ryby-inostrancami.html
LON Forum Report (Large Ocean Nations Forum Transnational Organized Fisheries Crime Report). 2019. http://www.bluejustice.org. Accessed 18 October
Lawrence, F., and E. McSweeney. 2017. "UK Police Rescue Nine Suspected Victims of Slavery from British Trawlers." *The Guardian*, 12 December. https://www.theguardian.com/world/2017/dec/12/uk-police-rescue-nine-suspected-victims-of-slavery-from-british-trawlers.
Leslie G (2010) Confronting the Don: The Political Economy of Gang Violence in Jamaica. In: An Occasional Paper of the Small Arms Survey, Graduate Institute of International and Development Studies. Small Arms Survey, Geneva
Lefevre, A.S. 2014. "Thai Fishermen Convert Boats to Cash In on Human-Smuggling." Reuters, November 20. https://www.reuters.com/article/us-thailand-rohingya/thai-fishermen-convert-boats-to-cash-in-on-human-smuggling-idUSKCN0J40P120141120.
Lindley, J., S. Percy and E. Techera. 2018. "Illegal Fishing and Australian Security." Australian Institute of International Affairs, 22 December. http://www.internationalaffairs.org.au/australianoutlook/illegal-fishing-australia/
Liss C 2007. "The Roots of Piracy in Southeast Asia." APSNet Policy Forum, 22 October.
Livelihoods Report (*Livelihood Baseline Assessment and Disaster Risk Pro- filing: Rocky Point, Clarendon.* Final Report) 2012. Prepared for the FAO
Madsen F (2009) Transnational Organized Crime. Routledge, Oxon, UK
Matthews, E., J. Bechtel, E. Britton, K. Morrison and C. McClennen. 2012. *A Gender Perspective on Securing Livelihoods and Nutrition in Fish-De-pendent Coastal Communities.* Report to Rockefeller Foundation from Wildlife Conservation Society. http://anewcourse.org/wp-content/up-loads/2013/04/WCS-Gender-Fisheries-2012.pdf
MFA (Ministry of Foreign Affairs of the Kingdom of Thailand). 2018. "Press Release: Thailand Has Set Up the Flying Inspection Team to Increase the Effectiveness of the PIPO Control Centres." 9 March. http://www.mfa.go.th/main/en/news3/6886/87142-Thailand-has-set-up-the-Flying-In-spection-Team-to.html.
MMFA (Ministry of Maritime and Fisheries Affairs, Indonesia) (2018) Sea for the Future of the Nation. MMFA, Jakarta
NA-FIG (North Atlantic Fisheries Intelligence Group). 2017. Chasing Red Herring: Flags of Convenience, Secrecy and the Impact on Fisheries Crime Law Enforcement. Oslo: NA-FIG. http://norden.diva-portal.org/smash/get/diva2:1253427/FULLTEXT01.pdf.
Neil JA 2018. "IUU Fishing: A Gateway to Transnational Crimes in Jamaica." World Maritime University Dissertations. Malmö, Sweden: World Maritime University.
Norwegian Police University College. 2019. https://www.phs.no/. Accessed 18 October
Obokata T (2010) Transnational Organised Crime in International Law (Studies in International and Comparative Criminal Law). Hart, Oxford, UK
Oceana (2018) Seafood Fraud and Mislabelling across Canada. Oceana, Washington, DC
OECD (2013) Evading the Net: Tax Crime in the Fisheries Sector. OECD, Paris
Okafor-Yarwood I (2018) The Effects of Oil Pollution on the Marine Envi- ronment in the Gulf of Guinea: The Bonga Oil Field Example. Transnational Legal Theory 9(3–4):254–271
Okafor-Yarwood I (2019) Illegal, Unreported and Unregulated Fishing, and the Complexities of the Sustainable Development Goals (SDGs) for Countries in the Gulf of Guinea. Mar Policy 99:414–422
Okafor-Yarwood I (2020) The Cyclical Nature of Maritime Security Threats: Illegal, Unreported and Unregulated Fishing as a Threat to Human and National Security in the Gulf of Guinea. African Security:1–31. https://doi.org/10.1080/19392206.2020.1724432
Okonko IO, Nnodim LC (2015) Trends in HIV Prevalence from 2008 to 2012 among Young Adults in Port Harcourt, Nigeria. Afr Health Sci 15(3):776–788
Onuoha F (2012) Piracy and Maritime Security in the Gulf of Guinea: Nigeria as a Microcosm. Al Jazeera Centre for Studies, Qatar
One Earth Future. 2018. *The State of Maritime Piracy 2018: Assessing the Human Cost.* https://stableseas.org/publications/maritime-terrorism/state-piracy-2018-human-cost.
Otto RS (2014) History of King Crab Fisheries with Special Reference to the North Pacific Ocean: Development, Maturity, and Senescence. In: Stevens BG (ed) King Crabs of the World: Biology and Fisheries Management. CRC Press, Boca Raton, FL, pp 81–138
Packer H, Swarz W, Ota Y, Baily M (2019) Corporate Social Responsibility (CSR) Practices of the Largest Seafood Suppliers in the Wild Capture Fisheries Sector: From Vision to Action. Sustainability 11:1–24
Parks I (2014) Following the Proceeds of Bluewater Crime: Case Studies from an Australian Fisheries. In: Rose G (ed) Following the Proceeds of Environmental Crime: Forests, Fish and Filthy Lucre. Routledge, London, pp 116–139
Pew Charitable Trusts Environmental Group (2012) Navigating Global Shark Conservation: Current Measures and Gaps. Pew Charitable Trusts, Washington, DC
Pomeroy R et al (2007) Fish Wars: Conflict and Collaboration in Fisheries Management in Southeast Asia. Mar Policy 31(6):645–656
Ralby I (2018) Oil on the Water: Illicit Hydrocarbons Activity in the Maritime Domain. Atlantic Council, Washington, DC
Republic of Trinidad and Tobago Finance Act 91 of 2000 2019
Robinson, C. 2017. "Mission: Clarendon Rescue: It's a 'Rocky Road'—Gangsters Make Life Miserable for Residents of Once-Peaceful Fishing Village." Gleaner, 15 January. http://jamaica-gleaner.com/article/lead-stories/20170115/mission-clarendon-rescue-its-rocky-road-gangsters-make-life-miserable
Samatar A, Lindberg M, Mahayni B (2010) The Dialectics of Piracy in Somalia: The Rich versus the Poor. Third World Q 31(8):1377–1394
Santos-Fita D (2018) Subsistence Hunting in Rural Communities: Incom- patibilities and Opportunities within Mexican Environmental Legislation. J Ethnobiol 38(3):356–371
Shaw M (2017) Africa's Changing Place in the Global Criminal Economy. ENACT, New Delhi

Shaw M, Kemp W (2012) Spotting the Spoilers: A Guide to Ana- lyzing Organized Crime in Fragile States. International Peace Institute, New York

Shaxon N (2011) Treasure Islands: Tax Havens and the Men Who Stole the World. Penguin Random House, London

Spijkers J, Singh G, Blasiak R, Morrison TH, Le Billon P, H. Österblom (2019) Global Patterns of Fisheries Conflict: Forty Years of Data. Glob Environ Chang 57:101921

Standing A (2008) "Corruption and Commercial Fisheries in Africa." CMI Institute U4 Brief. CMI Institute, Bergen, Sweden

Standing A (2015) "Corruption and State Corporate Crime in Fisheries." CMI Institute U4 Brief. CMI Institute, Bergen, Sweden

Steinberg, J. 2005. "The Illicit Abalone Trade in South Africa." Occasional paper no. 105. Pretoria: Institute for Security Studies

Stringer C, Harré T (2019) "Human Trafficking as Fisheries Crime? An Application of the Concept to the New Zealand Context." Special section, "Transnational Organized Fisheries Crime.". Mar Policy 105:169–176

Sumaila UR, Bawumia M (2014) Fisheries, Ecosystem Justice and Piracy: A Case Study of Somalia. Fish Res 157:154–163

Sumaila UR, Zeller D, Hood L, Palomares MLD, Li Y, Pauly D (2020) Illicit Trade in Marine Fish Catch and Its Effects on Ecosystems and People Worldwide. Sci Adv 6:eaaz3801

Sutton H (2017) Unpacking the High Cost of Crime in the Caribbean: Violent Crime, the Private Sector, and the Government Response. In: Jaitman L (ed) *The Cost of Crime and Violence: New Evidence and Insights in Latin America and the Caribbean.* Section 7. Inter-American Development Bank, Washington, DC

Tavornmas, A. 2019. "From EU IUU Yellow to Green Card: Thailand's Sustainable Fisheries, 2019." Paper presented at ASEAN Regional Forum, Bali, 27 June

Tickler D, Meeuwig JJ, Bryant K, David F, Forrest JAH, Gordon E, Joudo Larsen J, Oh B, Pauly D, Sumaila UR, Zeller D (2018) Modern Slavery and the Race to Fish. Nat Commun 9:4643

Udoh I, Mantell JE, Sandfort T, Eighmy M (2009) Potential Pathways to HIV/AIDS Transmission in the Niger Delta of Nigeria: Poverty, Migration and Commercial Sex. AIDS Care 21(5):567–574

UN (United Nations). 2003. "Report of the Panel of Experts in Somalia Pursuant to Security Council Resolution 1474." S/2003/1035. 4 November.

UN (2012) *The Future We Want: Outcome Document of the United Nations Conference on Sustainable Development.* Rio de Janeiro, 20–22 June. UN, New York

UN. 2019. *Report of the Meeting of the Working Group on the Smuggling of Migrants, Held in Vienna from 11 to 13 September 2019.* CTOC/COP/WG.7/2019/6

UNDP (2018) Human Development Indices and Indicators: 2018 Statistical Update. UNDP, New York

UNGA (UN General Assembly). 2012. *Fisheries: Interim Report of the Spe- cial Rapporteur on the Right to Food.* 8 August. A/67/268.

UNODC (United Nations Office on Drugs and Crime) (2011) Transnational Organized Crime in the Fishing Industry: Focus on: Trafficking in Persons Smuggling of Migrants Illicit Drugs Trafficking. UNODC, Vienna

UNODC (2017) Stretching the Fishnet: Identifying Opportunities to Address Fisheries Crime. UNODC, Vienna

UNODC (2019a) Rotten Fish: A Guide on Addressing Corruption in the Fish- eries Sector. UNODC, Vienna

UNODC. 2019b. "Drug Trafficking." https://wwwunodc.org/unodc/en/drug-trafficking/index.html. Accessed 18 October

UNODC and World Bank (2007) Crime, Violence and Development: Trends, Costs and Policy Implications in the Caribbean. UN, New York

UN OHCHR (UN Office of the High Commissioner for Human Rights) (2011) Guiding Principles on Business and Human Rights: Implementing the United Nations "Respect, Protect and Remedy" Framework. UN, New York

UNSC (United Nations Security Council). 2019. "Maintenance of International Peace and Security: Transnational Organized Crime at Sea as a Threat to International Peace and Security." 8457th Meeting, New York, February. S/PV.8457

USAID (U.S. Agency for International Development) (2016) Fishing for Food Security: Importance of Wild Fisheries for Food Security and Nutrition. USAID, Washington, DC

U.S. Department of Justice. 2017. "Three Charged with Illegal Trafficking of $17 Million Worth of Sea Cucumbers." 26 May. https://www.justice.gov/usao-sdca/pr/three-charged-illegal-trafficking-17-million-worth-sea-cucumbers

U.S. Department of Justice. 2019. "Federal Jury Convicts Two Defendants of Narcotics and Money Laundering Conspiracies." https://www.justice.gov/usao-wdny/pr/federal-jury-convicts-two-defendants-narcotics-and-money-laundering-conspiracies. Accessed 18 October.

U.S. Department of State, Bureau for International Narcotics and Law Enforcement Affairs (2018) International Narcotics Control Strategy Report (INCSR), vol 1. U.S. Department of State, Washington, DC

Varese F (2017) What Is Organized Crime? In: Carnevale S, Forlati S, Giolo O (eds) Redefining Organized Crime: A Challenge for the European Union? Hart, Oxford, UK, pp 27–56

Vidal, J. 2012. "Will Overfishing by Foreigners Drive Senegalese Fisher- men to Piracy?" *The Guardian*, 3 April. https://www.theguardian.com/global-development/poverty-matters/2012/apr/03/overfishing-foreigners-senegal-fishermen-piracy

Wilhjálmsson IF 2019. "An Icelandic Fishing Company Bribed Officials in Namibia and Used Norway's Largest Bank to Transfer 70 Million Dollars to a Tax Haven." Stundin, 12 November. https://stundin.is/grein/9920/

Witbooi E (2020) Criminality and Resilience. Rocky Point, Jamaica. In: Research Report. Global Initiative against Transnational Organized Crime, Geneva. https://globalinitiative.net/criminality-resilience-jamaica/

World Bank. 2017. "Promoting Community-Based Climate Resilience in the Fisheries Sector Project Information Document" (Jamaica). 30 June. http://documents.worldbank.org/curated/en/618281498849417411/pdf/IL-AISPID-CP-P164257-06-30-2017-1498849394475.pdf.

Zueras D 2010. "Costa Rica: (In)human Trafficking." IPS News, 15 April. http://www.ipsnews.net/2010/04/costa-rica-inhuman-trafficking/

Court Case

WWF South Africa v Minister of Agriculture, Forestry and Fisheries and Others (11478/18) (2018) ZAWCHC 127; [2018] 4 All SA 889 (WCC); 2019 (2) SA 403 (WCC) (26 September 2018)

Conventions

ILO (International Labour Organization), 1930. Forced Labour Convention (no. 29)

UN (United Nations) (1982) International Law of the Sea. UN, New York

UN (2004a) United Nations Convention against Transnational Organized Crime and the Protocols Thereto. UN, New York

UN (2004b) United Nations Convention against Corruption. UN, New York. UN Resolutions: UNGA (UN General Assembly) Resolution 63/112 of 5 December 2008. UN Doc A/RES/63/112

UN CCPCJ Resolution 20/5 of 15 April 2011. UN Doc E/2011/30, E/CN.15/2011/21

UN General Assembly Resolution of 21 October 2015. "Transforming Our World: The 2030 Agenda for Sustainable Development." A/Res/70/1

UN Security Council Resolution 2316/2016. Adopted by the Security Council at its 7805th meeting on 9 November 2016. S/RES/2316 (2016)

UN Security Council Resolution 2842/2019. Adopted by the Security Council at its 8582nd meeting on 19 July 2019. S/RES/2842 (2019).

The Ocean as a Solution to Climate Change: Five Opportunities for Action

17

Abbreviations

2DS	2 °C Scenario (IEA) consistent with at least a 50% chance of limiting the average global temperature increase to 2 °C by 2100
AR	Assessment report
AR5	Fifth Assessment Report of the IPCC
B2DS	Beyond 2 °C Scenario (IEA)—innovation pipeline for reducing global temperatures below the 2DS scenario
BAU	Business as usual
BTRs	Biennial transparency reports
CAGR	Compound annual growth rate
CCS	Carbon capture and storage
CO_2	Carbon dioxide
COP	Conference of the Parties to the United Nations Framework Convention on Climate Change
DW	Dry weight
FAO	Food and Agriculture Organization of the United Nations
GHG	Greenhouse gas
GMST	Global mean surface temperature
$GtCO_2e$	Gigatons of equivalent CO_2
GVA	Gross value added
GWEC	Global Wind Energy Council
HCFC	Hydrochlorofluorocarbon
IEA	International Energy Agency
IMO	International Maritime Organization
IPCC	International Panel on Climate Change
LCOE	Levelized cost of energy
LED	Low energy demand
Milankovitch cycle	The collective effects of changes in the earth's movements on its climate over thousands of years
Montréal Protocol	Montréal Protocol on Substances that deplete the ozone layer (a protocol to the Vienna Convention for the Protection of the Ozone Layer) is an international treaty designed to protect the ozone layer by phasing out the production of numerous substances that are responsible for ozone depletion
MW	Megawatt
NDCs	Nationally determined contributions
O&M	Operation and maintenance
OECD	Organisation for Economic Co-operation and Development
ORE	Ocean-based Renewable Energy
OSW	Offshore wind
OTEC	Ocean Thermal Energy Conversion
Paris Agreement	Adopted on December 12, 2015, at the twenty-first session of the Conference of the Parties to the United Nations Framework Convention on Climate Change, held in Paris from 30 November to 13 December, 2015
RCP	Representative Concentration Pathway (RCP) is a GHG trajectory adopted by the IPCC for AR5 in 2014
RD&D	Research, development, and demonstration
SDG	Sustainable development goal
SSP X	Shared socioeconomic pathways
TWh/yr	Terawatt hour per year

Originally published in:
Hoegh-Guldberg. O., et al. 2019. "The ocean as a solution to climate change: five opportunities for action." Report. Washington, DC: World Resources Institute. Available online at http://www.oceanpanel.org/climate
Reprint by Springer International Publishing (2023) with kind permission.
Published under license from the World Research Institute

J. Lubchenco, P. M. Haugan (eds.), *The Blue Compendium*, https://doi.org/10.1007/978-3-031-16277-0_17

UN	United Nations
UNEP	United Nations Environmental Programme
UNFCCC	United Nations Framework Convention on Climate Change
VOC	Volatile organic compound

1 Executive Summary

The ocean is a dominant feature of our planet, covering 70% of its surface and driving its climate and biosphere. The ocean sustains life on earth and yet is in peril from climate change.

However, while much of recent attention is focused on the problems that the ocean faces, the ocean is also a source of potential solutions and innovation. This report explores how the ocean, its coastal regions and economic activities can provide opportunities in the fight against climate change.

Highlights

- Until recently, the ocean was thought to be so large that its response to climate change was thought to be minimal; it has now taken centre stage in the impacts and solutions.
- Ocean-based mitigation options to reduce or sequester and store emissions offer significant potential to contribute to global efforts to limit global warming and for achieving the goals of the Paris Agreement.
- Ocean-based mitigation options could reduce global greenhouse gas (GHG) emissions by nearly 4 billion tonnes of carbon dioxide equivalent (CO_2e) per annum in 2030 and by more than 11 billion tonnes per annum in 2050, relative to projected business-as-usual (BAU) emissions. Reductions of this magnitude are larger than the emissions from all current coal fired power plants worldwide and more than China's total emissions in 2014.
- Ocean-based mitigation options could reduce the "emissions gap" (the difference between emissions expected if current trends and policies continue and emissions consistent with limiting global temperature increase) by up to 21% on a 1.5 °C pathway, and by about 25% on a 2.0 °C pathway, by 2050.
- This report considers five areas of ocean-based climate action to mitigate GHG emissions: ocean-based renewable energy; ocean-based transport; coastal and marine ecosystems; the ocean-based food system (wild capture fisheries, aquaculture, and shifting human diets towards food from the sea); and carbon storage in the seabed.
- Ocean-based renewable energy production currently offers the greatest potential for delivering clean energy and reducing GHG emissions, with the expansion of floating wind and solar facilities being exciting frontiers.
- When wider impacts on the environment and social well-being are considered, nature-based interventions—especially protection and restoration of mangroves, seagrass and salt marshes—offer the best combination of carbon mitigation and broader cobenefits.
- While innovation is required to improve many specific technologies and practices, four of the ocean-based climate action areas are ready to be implemented today (ocean-based renewable energy; ocean-based transport; coastal and marine ecosystems; the ocean-based food system). This could offer many cobenefits in terms of creating jobs, improving air quality and human health, and supporting livelihoods if implementation addresses trade-offs with sustainable development dimensions appropriately. The fifth, carbon storage in the seabed, has significant theoretical potential to divert carbon from the atmosphere, but it currently faces significant technical, economic, and sociopolitical challenges (e.g., environmental safety) that must be adequately explored prior to deployment at the scale required to make a substantive contribution to solving the climate problem.
- Ocean-based mitigation options must be accompanied by deep cuts in emissions across terrestrial GHG sources, including measures to phase out fossil fuels, create sustainable food systems, and increase carbon sequestration and storage in forests and other natural ecosystems.

1.1 Climate Change Threatens the Ocean

The world needs to move rapidly and systematically to reduce emissions of green house gases (GHGs) to the atmosphere if it is to avoid irreversible climate impacts (IPCC 2014; IPCC 2018). Greater efforts are essential to accelerate and scale decarbonisation of the economy and pursue a pathway to net-zero emissions by the middle of the century. The sooner widespread action begins, the more cost-effective it will be, and the greater the chance of avoiding the worst impacts of rapid human-driven climate change.

Following the findings of the IPCC Special Report on the implications of 1.5 °C warming above the preindustrial period (IPCC 2018), it is now abundantly clear that stronger action to mitigate GHG emissions is a global imperative that will require an inclusive approach across the whole of the global economy. To date, much of the attention has been directed to the role of terrestrial sources of emissions and sinks. The ocean and its coastal regions, however, offer a wide array of additional potential mitigation options.

The ocean plays a fundamental role in regulating global temperatures. Not only does the ocean absorb 93% of the *heat* trapped by rising anthropogenic carbon dioxide (CO_2), but it also absorbs approximately 25–30% of anthropogenic CO_2 *emissions* that would otherwise remain in the atmosphere and increase global warming. The ocean also produces around 50% of the oxygen on the planet through the photosynthetic activity of marine plants and algae.

The ocean's ability to contribute to these fundamentally important services, however, is at risk (IPCC 2019). Ocean warming and acidification (the latter being a direct result of the extra CO_2 dissolving into the ocean) are damaging marine ecosystems and compromising the ability of the ocean to provide food, livelihoods, and safe coastal living on which billions of people depend (IPCC 2014, 2018, 2019).

Efforts to protect the ocean and its vitally important ecosystems cannot be considered in isolation from the challenge of stabilising the global climate. To secure the long-term health of the ocean and the livelihoods and economies that depend on it, atmospheric concentrations of GHGs must be urgently reduced. This report outlines a suite of options for how the ocean and coastal regions can contribute to lowering projected emission trajectories and help achieve the temperature stabilisation goals established in the Paris Agreement on Climate Change (UNFCCC 2015).

1.2 The Ocean is a Major Part of the Climate Solution

Ocean-based mitigation options do not feature as prominently as they could in countries' nationally determined contributions (NDCs) or long-term low greenhouse gas emission development strategies under the Paris Agreement. This report presents a wide array of potential ocean-based mitigation options and provides detailed analysis of their potential contribution to closing the emissions gap in 2030 and 2050 (Box 17.1).

Box 17.1. Why the World Needs to "Close the Emissions Gap"

Each year, the United Nation's Emissions Gap Report compares where global greenhouse gas (GHG) emissions are headed with where they need to be if the world is to avoid the worst impacts of climate change. Scientists first collect the latest information on countries' climate commitments, expressed in their nationally determined contributions (NDCs), and calculate their projected emissions pathway. They then compare this pathway with the latest models on how warming could be limited to either 1.5 °C or 2.0 °C, the temperature goals to which countries committed under the Paris Agreement of December 2015, and the limits scientists say are necessary for preventing some of the worst climate change impacts. The most recent report (UNEP 2018) concludes that unless countries strengthen their ambition and cut 2030 emissions beyond the targets established in their current NDCs, exceeding a temperature rise of 1.5 °C "can no longer be avoided." And unless the emissions gap is closed by 2030, it is unlikely that warming can be held below 2.0 °C.

Source: Levin et al. (2018).

Five areas of ocean-based climate action are considered in this report:

- Ocean-based renewable energy, including offshore wind and other energy sources, such as wave and tidal power.
- Ocean-based transport, including freight and passenger shipping.
- Coastal and marine ecosystems, including protection and restoration of mangroves, salt marshes, seagrass beds, and seaweeds.
- Fisheries, aquaculture, and dietary shifts away from emission intensive land-based protein sources (e.g., red meat) towards low carbon ocean-based protein and other sources of nutrition.
- Carbon storage in the seabed.

Additional ocean-based carbon storage options, such as direct injection into the deep ocean, alkalinity addition, and iron fertilisation are discussed, but due to the current uncertainty regarding their viability and higher risk of adverse impact on the ocean, they have been excluded from the calculated mitigation potentials.

Offshore oil and gas drilling, although the most significant source of ocean-based CO_2 emissions, is not discussed in the report, as it has been comprehensively tackled by other reports and its tracjectory is clear.

Within each area, this report assesses the set of individual mitigation options that could be undertaken, along with the technology developments and policies required to advance implementation. These mitigation options are summarised in Fig. 17.1, along with their mitigation potential in 2050. We also examine current and future deployment scenarios and suggest research priorities to improve the feasibility and scale of each option. The inclusion of any particular mitigation option in this report does not imply endorsement.

This report concludes that actions across all five ocean-based climate action areas of intervention have the potential to reduce emissions by up to 4 billion tonnes of CO_2e per annum in 2030, and by more than 11 billion tonnes of CO_2e per annum in 2050, thereby making a significant contribution to closing the emissions gap in 2030 and 2050 as shown in Fig. 17.2. Table 17.1 shows the total mitigation potential (expressed as a range) for each of the intervention areas.

Figures 17.3 and 17.4 below show the emission reduction and/or sequestration potential of each area of ocean-based climate action, including individual mitigation options, for 2030 and 2050.

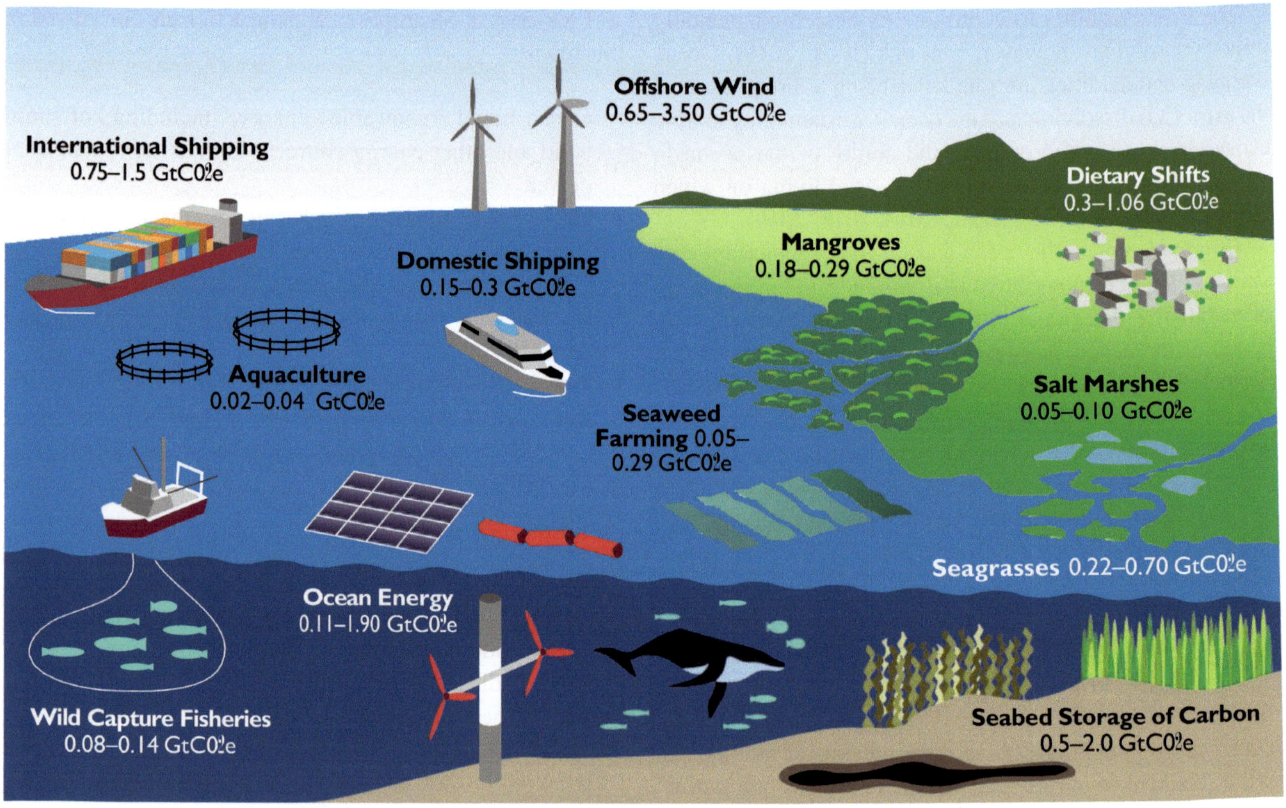

Fig. 17.1 Ocean-based mitigation options explored in this report and associated annual mitigation potential in 2050. (*Source: Authors*)

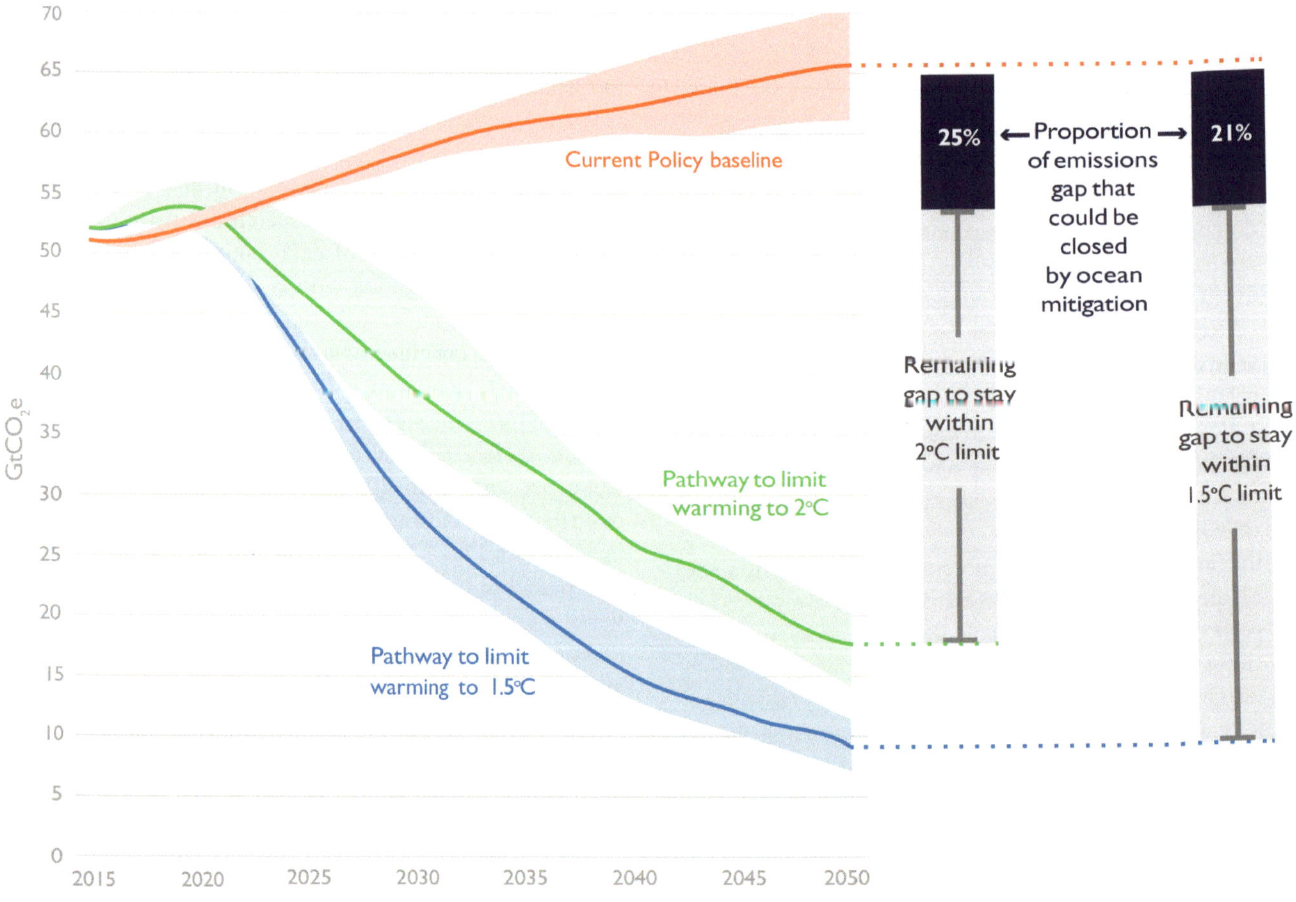

Fig. 17.2 Contribution of ocean-based mitigation options to closing the emissions gap in 2050. (*Source: Adapted from* UNEP (2018), Climate Action Tracker (2018))

Table 17.1 Summary of global mitigation potential offered by each area of ocean-based climate action

Areas of ocean-based climate action	2030 mitigation potential ($GtCO_2e$/year) (%)	2050 mitigation potential ($GtCO_2e$/year) (%)
1. Ocean-based renewable energy	0.18–0.25	0.76–5.40
2. Ocean-based transport	0.24 – 0.47	0.9 – 1.80
3. Coastal and marine ecosystems	0.32–0.89	0.50–1.38
4. Fisheries, aquaculture, and dietary shifts	0.34–0.94	0.48–1.24
5. Carbon storage in the seabed (action in this area requires further research prior to implementation at scale)	0.25–1.0	0.50–2.0
Total	1.32–3.54	3.14–11.82
Total percentage contribution to closing emissions gap (1.5 °C pathway)	4–12	6–21
Total percentage contribution to closing emissions gap (2 °C pathway)	7–19	7–25

Source: Authors

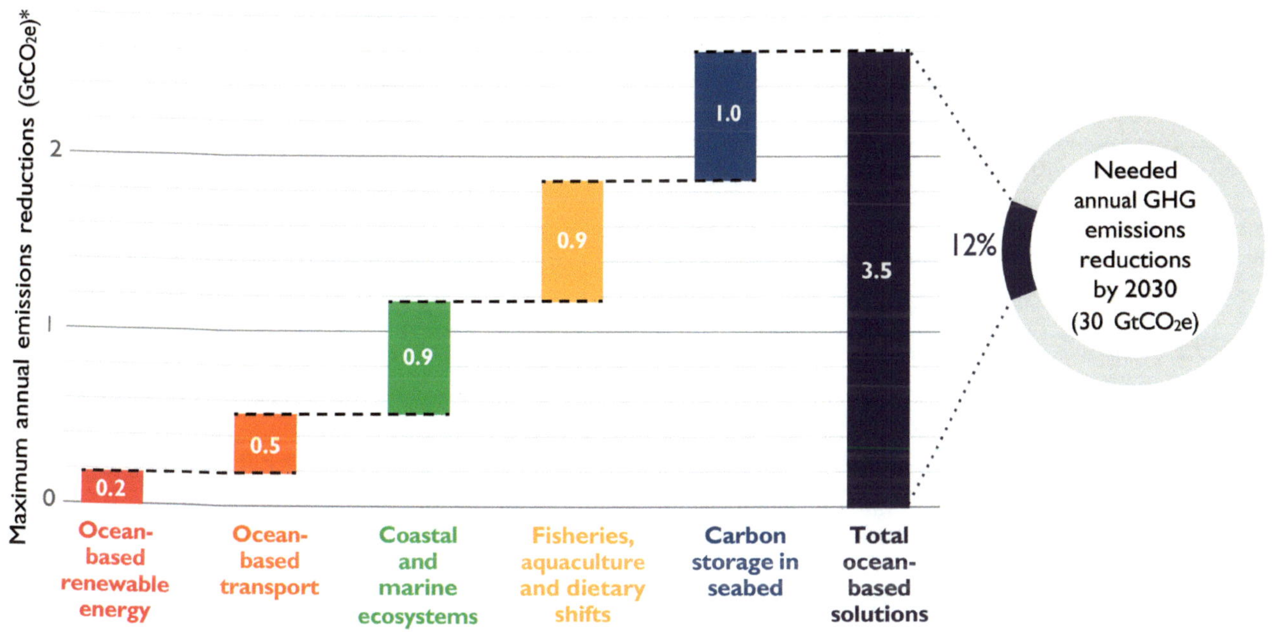

Fig. 17.3 Contribution of five ocean-based climate action areas to mitigating climate change in 2030 (maximum $GtCO_2e$). (*Notes: * To stay under a 1.5 °C change relative to pre-industrial levels. Source: Authors*)

1.3 Ocean-Based Mitigation Options

Scaling up ocean-based renewable energy (offshore floating and fixed wind installations, tidal and wave power), and decarbonising ocean-based transport offer some of largest mitigation potential in 2030 and 2050.

Utilising nature-based solutions, such as leveraging the ability of coastal and marine ecosystems to sequester and store carbon, also offer a sizable mitigation potential. Protection and restoration of these ecosystems provides valuable benefits by expanding sequestration and maintaining carbon stocks in soils and vegetation.

Restoration also yields cobenefits to local communities via other ecosystems services, such as providing habitat for fish; supplying food, fibre, and traditional medicines; and reducing the impact of storms during extreme weather events. Seaweed aquaculture offers significant potential for developing low-carbon alternatives for food, feed, and many other applications.

Storage of carbon in the seabed has enormous theoretical potential to divert carbon from the atmosphere, but it currently faces significant technical, economic, and sociopolitical challenges (e.g., environmental safety) that must be adequately explored prior to deployment at the scale presented in this report.

This report analyses the potential of seabed storage on the basis that it is the only ocean-based carbon, capture and storage (CCS) option that is currently being implemented at industrial scale (in Sleipner, Norway). However, given the technological, economic, social, and political barriers to implementing carbon storage in the seabed as a mitigation option, and the number of trade-offs and risks that must be reduced if ocean storage is to be widely used as a mitigation option, it is distinguished from the other four ocean-based

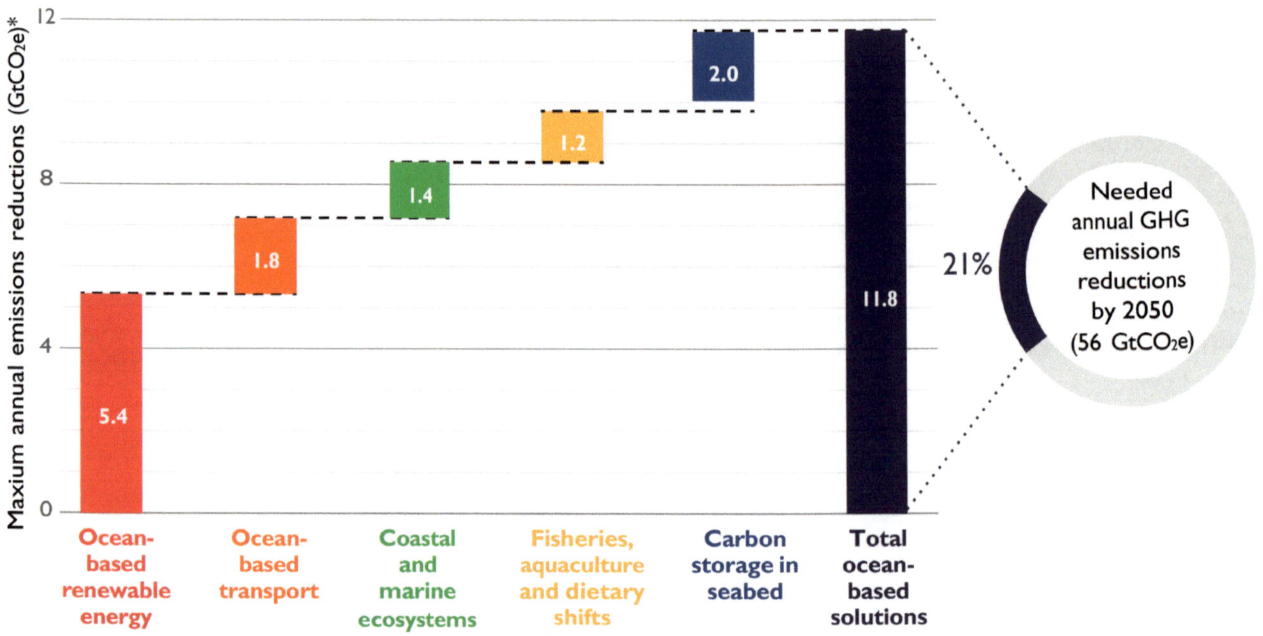

Fig. 17.4 Contribution of five ocean-based climate action areas to mitigating climate change in 2050 (maximum $GtCO_2e$). (*Notes: * To stay under a 1.5 °C change relative to pre-industrial levels. Source: Authors*)

Table 17.2 Potential of ocean-based climate action to contribute to current policy for closing the emissions gap in 2030 and 2050

	Annual emissions ($GtCO_2e$)			GAP		Total GHG mitigated $GtCO_2e$		% GAP closed: 1.5 °C		% GAP closed: 2.0 °C	
	Current policy	1.5 °C pathway	2.0 °C pathway	1.5 °C	2.0 °C	Min	Max	Min	Max	Min	Max
Today	52	52	52	0	0	0	0	0.0	0.0	0.0	0.0
2030	58	28	39	30	19	1.3	3.5	4	12	7	19
2050	65	9	18	56	47	3.1	11.8	6	21	7	25

Source: Authors

Note: Estimates are base on comparison between multiple scenarios for annual emissions in 2020, 2030, and 2050. For those year, we compare '1.5 °C', '2 °C' and the 'current policy' scenarios from UNEP (2018) *and calculate the mitigation needed to fill the 'gaps' between the 'current policy' and the '1.5 °C', '2 °C' respectively. 'Min' refers to conservative ocean based mitigation potential, while 'Max' represents higher (more ambitious) potential projected in this paper. The total ocean-based mitigation* (Table 17.2) *was compared to the gap at 2030, and that at 2050, generating the percentage of the gap mitigated by ocean-based mitigation of GHG emissions*

mitigation options as it has certain dimensions that cannot be implemented in the short-term.

It is important to note that this report looks at the mitigation potential of each area of intervention at a global level. Not all options will be available or appropriate for all countries. Countries vary not only in their physical attributes (e.g., not all countries have mangroves), but also in their economic and social profiles (some countries have major fishing industries; some are high consumers of red meat; others engage actively in maritime trade). Therefore, while ocean energy and transport offer higher mitigation potential than nature-based solutions at the global level, restoration of vegetated coastal habitats ("blue carbon ecosystems") may provide the most viable and -cost-effective opportunity for contributing to global efforts to reduce GHG emissions for some individual countries or regions. In addition, the presence or absence of enabling factors, such as carbon market, may influence decisions and priorities, changing the economic potential of the options outlined in this report (Table 17.2).

1.4 Wider Impacts of Ocean-based Climate Action

The IPCC Special Report on 1.5 °C scenarios integrated an assessment of wider impacts; however, the ocean was not addressed comprehensively as a sector within this impacts analysis. This report aims to address this major knowledge gap by evaluating four sustainable development dimensions where wider impacts—beyond avoided or reduced emissions—may be expected:

- Environment (impacts on marine and terrestrial biodiversity, water quality, land-use, coastal resilience, and adaptability of ecosystems and human settlements to climate change).
- Economy (impacts on employment, household incomes, economic growth, supply of clean energy innovation, profit/revenue generated by firms, and supply of clean energy).

- Society (impacts on human health outcomes, income inequality, quality of education, gender equity, poverty reduction, and food security targets).
- Governance (effective, transparent and strong institutions, participation in global governance, strong national institutions, global partnership for sustainable development, capacity building)

The assessment was based on a review of literature and reveals that, while ocean-based mitigation options have both cobenefits and trade-offs, the cobenefits far outweigh the trade-offs.

Positive environmental impacts include high biodiversity benefits to marine and terrestrial ecosystems, higher ecosystem services (improvement in fisheries productivity and coastal tourism), reduced risk of ocean acidification, increased coastal resilience, and reduction in withdrawal/usage of water. Economic impacts or cobenefits that are positive include opportunities created by spillover from new or improved technologies, new local employment opportunities, energy savings from improvement in the design of vessels, and economic growth driven by a growing ocean-based economy.

Positive social impacts or cobenefits include reduced morbidity and mortality due to improved local air quality, positive health impacts from shifting diets away from meat towards low-carbon ocean-based protein, enhanced global food security, potential to ensure greater gender parity as ocean-based industries expand, and improved income opportunities and livelihoods in coastal areas.

A number of negative effects or risks were identified when assessing the wider impacts of the mitigation measures on sustainable development, especially for the dimensions focusing on environment and society.

Policy design and implementation, along with contextual factors, play a key role in determining how mitigation options influence negative social outcomes. For example, mitigation options aimed at rebuilding fish stocks and other ocean biomass can negatively impact poverty reduction and employment targets and limit progress on food security targets in the short term.

Lack of effective stakeholder engagement on "blue carbon" restoration projects (including exclusion of local community representatives from key international decision-making events) limit their access to ocean spaces and can lead to negative outcomes for small-scale fishers who heavily rely on local ecosystems for jobs, nutritional needs, and economic sustainability. In these instances, well-planned mitigation measures that follow best governance practices, with strong engagement of communities, nongovernmental organisations (NGOs), and governments, are essential to avoid worsening of inequalities and creation of new social injustices.

Environmental trade-offs and risks include the damage that can be done to coastal ecosystems or marine species by unplanned growth in coastal aquaculture or renewable energy installations. Seabed carbon storage approaches, if deployed unwisely, could contribute to ocean acidification and damage ocean ecosystems by impacting chemical, physical, and ecological processes at a large scale. While some of these risks can be adequately addressed via stakeholder engagement, inclusive management policies, careful monitoring, and effective marine planning, others will require further research on their implications. In some instances, there will be a need for significant action on the part of governments to ensure that negative impacts are reduced or resolved. Concerted action to address these negative impacts will help enhance the net positive outcome.

When looking at the five ocean-based areas as a whole, coastal and marine ecosystems, fisheries and aquaculture, and ocean energy have a positive impact on the largest number of sustainable development dimensions. When looking at individual mitigation options within the five ocean-based areas, nature-based interventions (especially protection and restoration of mangroves, seagrass and salt marsh) and offshore wind energy positively impact the largest number of sustainable development dimensions. The analysis showed that all ocean-based mitigation options will need strong national institutions, engagement of business and industry, and community involvement and international cooperation to ensure their planned implementation maximises the positive impact and limits the negative impact on sustainable development dimensions. The results of this analysis is shown in Fig. 17.5.

1.5 Delivering the Mitigation Potential of the Ocean

There is a small, but important window of opportunity within which the "Current Policy" emissions trajectory can be directed towards a pathway that is consistent with achieving the 1.5 °C and 2.0 °C temperature goals set by the Paris Agreement. While much of the required emission reductions must come from deep cuts within terrestrial-based activities, including the use of fossil fuel, this report identifies major ocean-based opportunities that could play a critical role in the transition to a low-carbon future and safer climate.

Achieving the mitigation potential identified in this report will not be possible without significant investment in research and development. It will also be necessary to provide strong incentives to align financial flows with the needs of the mitigation action opportunities available. Governments must send policy signals. Table 17.3 summarises the policy, research, and technology priorities for the short and medium term to support action in each of the areas of ocean-based climate action examined in this report.

One of the first opportunities that governments will have to comprehensively integrate ocean-based mitigation options into national plans and strategies for climate change is the reconsideration and updating of NDCs in 2020. This is an extremely important moment, as emphasised by the IPCC (2018): the chances of "failing to reach 1.5 °C [will be] significantly increased if near-term ambition is not strengthened

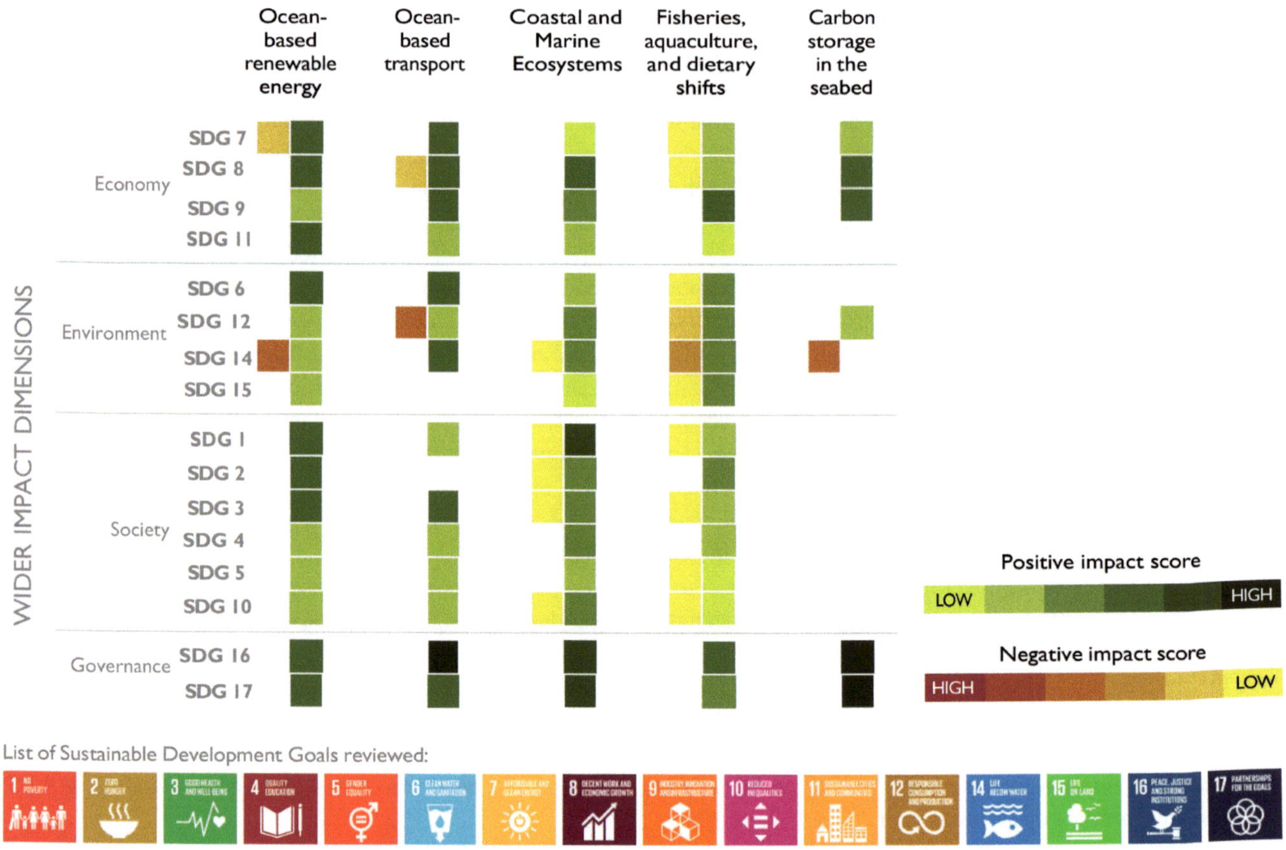

Fig. 17.5 Summary of wider impact of ocean-based interventions on sustainable development dimensions. (*Source: Authors. Notes: Wider-impact dimensions cover various sustainable development dimension indicators as well as 2030 Sustainable Development Goals (SDG). The figure shows the relative strength of the relationship between the ocean-based areas of interventions and the SDGs. The relationship between each ocean-based mitigation option and SDG is given a linkage score, positive scores shown by green boxes and negative scores shown by yellow/red boxes. Scores range from +3 (indivisible) to −3 (cancelling)* (Nilsson et al. 2016). *A zero score (no bar and no colour) means no impact was found in this review of the literature. For intervention areas where there is more than one mitigation option, an average of the linkage score is taken among the mitigation options in that area. Further information on the linkage scores and the associated confidence levels are provided in the Annex)*

Table 17.3 Short- and medium-term policy, research, and technology priorities necessary to deliver on mitigation potential of ocean-based climate action areas

	Ocean-based energy		
	Policy	Research	Technology
Short-term priorities (2020–2023)	• Undertake marine spatial planning • Develop national targets to increase the share of renewable energy in the national energy mix • Provide a stable economic and regulatory framework to stimulate investments in required infrastructure for an accelerated deployment of ocean-based energy systems	• Understand the impacts (positive and negative) of both fixed and floating offshore wind installations on marine biodiversity • Undertake a detailed mapping of global renewable energy resources and technical potential	• Advance storage capacity and design • Improve performance, reliability, and survivability, while reducing costs
Medium-term priorities (2023–2025)	• Develop strategic national roadmaps for zero-carbon economy in 2050 • Develop appropriate legislation and regulation	• Understand the potential benefits of co-location with other ocean-based industries (e.g., desalination plants and aquaculture) • Explore the potential for installing large scale floating solar installations at sea (under wave conditions) • Quantify the potential of ocean thermal energy conversion (OTEC)	• Advance technology that can move technologies into deeper water sites (e.g., development of floating offshore wind technologies) to open access to larger areas of energy resources

Table 17.3 (continued)

	Ocean-based energy		
	Policy	Research	Technology
	Ocean-based transport		
Short-term priorities (2020–2023)	• Redesign the energy efficiency design index (EEDI) formula to avoid vessels being suboptimised for the test only, to ensure that instead vessels are being optimised for minimised fuel consumption in real operation at sea • Adopt policy measures to go beyond Ship Energy Efficiency Management Plan (SEEMP) to incentivise the maximisation of operational efficiency of new and existing ships • Adopt policies that can reduce the broader GHG emissions of shipping instead of CO_2 only, including well-to-tank emissions (WTW) of ship fuels	• Identify and rectify of market and nonmarket barriers and failures to enable larger uptake of more energy-efficient technologies and cooperation patterns • Ensure continuous research on ship design, including hull forms and propulsion, with a focus on reducing energy usage per freight unit transported • Increase focus on utilisation of wind, waves, ocean currents, and sun to reduce use of externally provided energy, i.e., both the carbon and non-carbon-based fuels carried on board	• Develop the necessary high efficiency hull forms and propulsion methods • Develop and implement hybrid power systems, including combustion engines, fuel cells, and batteries technologies • Develop and implement wind assistance technologies • Develop more advanced weather routing systems to better utilise wind, waves, ocean currents, and tides to reduce the use of both carbon and non-carbon fuel carried on board
Medium-term priorities (2023–2025)	• Develop policy to enable the business case for the adoption of low and zero carbon fuels by shipping (e.g. a carbon price) • Commit to the timetable for shipping's transition to low- and zero-carbon fuels • Develop national incentives for decarbonising domestic transportation • Commit to decarbonisation of national energy systems faster or as fast as the transition in the international fleet	• Develop cost-effective production of low- and zero-carbon fuels, both from renewables and from carbon based in combination with carbon capture and storage (CCS) • Develop cost-efficient hybrid setups on seagoing vessels to utilise the best of combustion, fuel cells, and batteries to reduce fuel consumption and local pollution • Ensure safe storage and handling on ships and at the ship-shore interface of hydrogen/ammonia • Ensure safe and efficient use of hydrogen and ammonia in internal combustion engines and fuel cells	• Advance technologies for producing hydrogen, both from renewables and carbon-based fuels • Invest in technologies to store hydrogen (including cryogenic storage of liquid hydrogen, or carriers able to store at high-energy density) • Invest in fuel cells for conversion of future fuels into on-board electricity, and internal combustion engines designed to operate on hydrogen/ammonia
	Coastal and marine ecosystems		
Short-term priorities (2020–2023)	• Enhance protection measures for mangroves, seagrass, salt marsh, and seaweed beds to prevent any further losses due to human activities • Provide incentives for restoration of "blue carbon" ecosystems, through payments for ecosystem service schemes, such as carbon and nutrient trading credits • Include quantified nature-based solutions within nationally determined contributions (NDCs) and other relevant climate policies for mitigation and adaptation • Protect coral reefs as important and integrated coastal defence systems for ensuring the protection of coastal blue carbon ecosystems	• Undertake national-level mapping of blue carbon ecosystems • Address biophysical, social, and economic impediments to ecosystem restoration to develop restoration priorities, enhance incentives for restoration, and increase levels of success • Improve the IPCC guidance for seagrasses and other wetland ecosystems • Develop legal mechanisms for long-term preservation of blue carbon, especially in a changing climate • Understand the impacts of climate change on rates of carbon capture and storage, or the potential for restoration	• Advance biorefining techniques, allowing sequential extraction of seaweed products
Medium-term priorities (2023–2025)	• Enhance and adopt carbon accounting methodologies for mangroves, seagrasses and salt marsh within national GHG inventories (IPCC 2013) • Improve methods for monitoring mitigation benefits to enable accounting within national GHG inventories, and biennial transparency reports (BTRs)	• Undertake global-scale map of seaweed ecosystems • Develop IPCC-approved methodological guidance for seaweed ecosystems • Develop methods to fingerprint seaweed carbon beyond the habitat	• Develop and pilot offshore and multiuse sites, including seaweed aquaculture, in the open ocean

(continued)

Table 17.3 (continued)

	Ocean-based energy		
	Policy	Research	Technology
	Fisheries, aquaculture, and dietary shifts		
Short-term priorities (2020–2023)	• Eliminate harmful fisheries subsidies (SDG14.6) • Strengthen international tools to eliminate IUU fishing (SDG14.5) • Avoid the transport of fish by air • Reduce discards • Reduce and eliminate hydrochlorofluorocarbons (HCFCs) in refrigerants • Create incentives for shifting diets towards low-carbon protein (e.g., fish) and other food (e.g., seaweed) diets • Create incentives to improve fishery management • Create incentives for lower trophic- level aquaculture • Devise sustainable finance mechanisms for small-scale fishery transitions to sustainable fishing	• Develop disaggregated global data sets for GHG emissions from wild catch fisheries and marine aquaculture • Impacts of scaling marine aquaculture and associated sustainability considerations (e.g., low carbon and climate resilient, environmentally safe) • Enhance understanding of how climate change and ocean acidification will impact aquaculture and fisheries	• Extend surveillance technologies for tracking fishing in the ocean and along coastal areas
Medium-term priorities (2023–2025	• Create incentives to switch from high-carbon land-based sources of protein to low-carbon ocean-based sources • Improve fisheries management to focus on optimising biomass per harvest	• Explore potential impact of a carbon tax on red meat and other carbon intensive foods	• Develop and bring to scale high-technology digital aquaculture
	Seabed carbon storage		
Short-term priorities (2020–2023)	• Invest in pilot projects to further explore potential environmental impacts • Incentivise public/private partnerships	• Map global geophysical potential • Understand the impacts of long-lasting containment of CO_2 in a deep seafloor environment	• Few major technical advances are required as seabed storage is already deployed at industrial scale
Medium-term priorities (2023–2025)	• Develop national strategies and targets • Develop regulatory frameworks to ensure environmental impact assessments and associated precautions are put in place	• Understand the impacts of long-term storage on marine ecosystems • Explore the integrity of long-term storage technologies (leakage)	• Scale up technologies in ways that are economically feasible

Source: Authors

beyond the level implied by current NDCs." Given the consequences of failing to limit global average temperature rise to 1.5 °C, or at least to "well below" 2.0 °C, it is of great importance that actions begin immediately.

2 Introduction

Efforts to protect the ocean and its vitally important ecosystems cannot be considered in isolation from the challenge of stabilising the global climate. To secure the long-term health of the ocean and the livelihoods and economies that depend on it, atmospheric concentrations of GHGs must be urgently reduced. This report outlines a suite of options for how the ocean and coastal regions can be a part of the solution set.

2.1 Climate Change Is a Key Threat to Ocean Systems

Climate change is one of the greatest challenges in history. The concentrations of atmospheric carbon dioxide (CO_2) and other greenhouse gases (GHGs) are increasing, causing rapid rates of warming on land and in the ocean. These changes are creating unprecedented challenges for natural and human systems (IPCC 2018). If unchecked, these changes will undermine and destabilise economies by driving increasingly unmanageable and dangerous impacts on the biosphere, human health, and global economies (Sumaila et al. 2019).

Prior to the industrial period (i.e., before ~1850), the global carbon cycle was in net balance, with CO_2-producing processes (e.g., respiration) being equal to CO_2-consuming processes such as photosynthesis and geochemical weathering. This balance resulted in the carbon cycle being relatively stable for thousands of years. Since the beginning of the industrial period, however, emissions of GHGs have grown rapidly as humanity felled forests, cleared land for agriculture, and began to exploit reservoirs of nutrients and gases such as oxygen (Hoegh-Guldberg et al. 2014; Pörtner et al. 2014).

The ability of humans to obtain food and livelihoods from the ocean is being degraded as a result of these changes. While the intention of this report is not to review comprehensively the impacts of climate change on the ocean, which has been done more extensively elsewhere (IPCC 2014, 2018, 2019), it notes that a few regions do show "positive" outcomes from climate change on a short-term basis, such as the increased biomass caught by high-latitude fisheries over recent decades (Sundby et al. 2016). The great majority of oceanic changes from polar to equatorial regions (and from deep to shallow areas) are, however, negative (IPCC 2014, 2018; Gattuso et al. 2018).

The Paris Agreement goals aim to keep "global average temperature to well below 2 °C above preindustrial levels and pursuing efforts to limit mean global temperature increase to 1.5 °C above preindustrial levels" (UNFCCC 2015).

The increased concentration of atmospheric CO_2 has resulted in ocean warming as well as ocean acidification, which is a consquence of the increased absorption of CO_2 by the ocean (IPCC 2014). Changes in the temperature and chemistry of the ocean have had serious impacts on a wide range of biological phenomena, including the survival, reproduction, and growth of marine organisms. There is considerable evidence that the ocean is also becoming more stratified, which is affecting the mixing of the water column, and consequently the availability of unoxidised carbon in fossil fuels. Rising concentrations of atmospheric CO_2 have already driven major changes to our planet. The global mean surface temperature (GMST) of the earth reached 1 °C above the preindustrial level in 2017 (IPCC 2018).

The evidence accumulated by the Intergovernmental Panel on Climate Change (IPCC) (IPCC 2014, 2018, 2019) suggests that the world will continue to face accelerating and life-threatening challenges if the GMST is not kept well below 2 °C above the preindustrial period (conditioned before ~1850). This science-based conclusion led to the explicit goals of the Paris Climate Agreement (UNFCCC 2015) and subsequently the IPCC special report on the implications of 1.5 °C warming above the preindustrial period (IPCC 2018).

The recent IPCC special report on 1.5 °C (IPCC 2018) concluded that there was still time to limit global temperature rise to the vicinity of 1.5 °C above preindustrial levels (IPCC 2018), if current efforts were escalated. This would require limiting further accumulated emissions of CO_2 after 2018, to approximately 420 gigatonnes of carbon dioxide equivalent ($GtCO_2$), which essentially gives the global community around 10 years at the current rate of annual emissions to bring fossil fuel emissions to net zero by mid-century (IPCC 2018). Significantly, however, limiting warming to 1.5 °C above preindustrial levels will require annual emissions of CO_2 to fall below zero by 2050 (i.e., "negative emissions") (IPCC 2018). Achieving this goal will require additional novel technologies for removing large amounts of CO_2 from the atmosphere.

2.2 The Ocean Is Part of the Solution to Climate Change

Attention has only recently been drawn to the possible role of the ocean, with its vast size and productivity, in mitigating CO_2. The ocean already plays a dominant role in the global carbon cycle and is responsible for taking up 25–30% of anthropogenic CO_2 released into the atmosphere.

While changes to the carbon cycle are creating daunting challenges for the ocean and the ocean-based economy, the ocean-based economy offers opportunities for mitigating GHG emissions and hence contributing to land-based efforts to fight climate change. While the focus on ocean and coastal-based solutions for mitigating climate change is increasing (e.g., IPCC 2014, 2018; Gattuso et al. 2018), a comprehensive analysis of ocean-based mitigation options and their potential to contribute to reducing atmospheric greenhouse gases has so far been limited.

This report addresses this analytical gap through a detailed analysis of the opportunities as well as the challenges associated with implementing a series of ocean-based mitigation options. Each option is considered in the context of its role as a key sink or source of CO_2 and other GHGs.

In particular, this report assesses the mitigation potential and associated impacts (cobenefits and trade-offs) of a series of options in five prominent ocean-based areas of intervention:

- Ocean-based renewable energy, including offshore wind and other energy sources, such as wave and tidal power.
- Ocean-based transport, including freight and passenger shipping.
- Coastal and marine ecosystems, including protection and restoration of mangroves, salt marshes, seagrass beds, and seaweeds, as well as aligned ecosystems such as coral reefs which are important coastal barriers to waves and storms.
- Fisheries, aquaculture, and dietary shifts away from emission intensive land-based protein sources (e.g., red meat) towards low carbon ocean-based protein and other sources of nutrition.
- Carbon storage in the seabed.

Table 17.4 describes each area of ocean-based climate action and its associated mitigation options.

2.3 Methodology

This report assesses each option in the context of "mitigation potentials" (Fig. 17.6). We explore the size of each potential, considering geophysical, technical, economic, and socio/political considerations that may affect their feasibility.

We identified mitigation options in each intervention area and assessed the scientific and research literature on the global contribution of each one to reducing atmospheric

Table 17.4 Mitigation options in five areas of ocean-based climate action

Area of action	Mitigation options	Description
Ocean-based renewable energy	Scaling up harnessing of offshore wind	Fixed and floating offshore wind turbine installations
	Scaling up use of ocean energy	Energy extracted from ocean waves, tides, currents, salinity, and temperature differences. Floating photovoltaic solar energy
Ocean-based transport	Reducing emissions from domestic shipping	Following the International Maritime Organization (IMO) definition: shipping between ports of the same country; includes ferries
	Reducing emissions from international shipping	Following the IMO definition: shipping between ports of different countries. International shipping excludes military and fishing vessels; includes bulk carriers, oil tankers, and container ships
Coastal and marine ecosystems	Restoration of mangroves, salt marshes, and seagrass beds	Sequestration potential gained from the restoration of lost and degraded coastal ecosystems. Coastal wetland systems include mangroves, salt marshes, and seagrass beds, plus conservation and restoration of adjacent islands, reefs and mudflats to slow the rate of erosion of coastal wetlands
	Avoided anthropogenic loss and degradation of mangroves, salt marshes, and seagrass beds)	Preventing the release of the high levels of sequestered carbon in soils and vegetation of coastal wetlands by protecting these ecosystems and avoiding further degradation
	Upscaling of seaweed production via aquaculture	Sequestration potential through seaweed aquaculture, primarily via farmed seaweed products substituting for other products with higher GHG footprint, or new application with no or minimal footprint
	Restoration and protection of seaweed habitats	Sequestration potential from the restoration of degraded (and protection of) intact seaweed habitats
	End overexploitation of ocean biomass to support recovery of biodiversity and increase biomass	Role of marine mammals and fish stocks in the ocean carbon cycle, including death and sinking to the seabed floor
Fisheries, aquaculture, and dietary shifts	Reducing emissions from fishing vessels	Emissions from fuel use for inland, coastal, and deep-sea fishing (wild capture)
	Reducing emissions from aquaculture	Life-cycle emissions from aquaculture (including, if possible, supporting activities such as production of fish meal and fish oil)
	Increasing share of ocean-based proteins (from fish and other marine life) in diets	Switching emission intensive land-based sources of protein (notably beef and lamb) for low carbon ocean based sources of protein
Carbon storage in the seabed	CO_2 storage in the seabed	Geological storage offshore of captured CO_2 in the seabed

Source: Authors

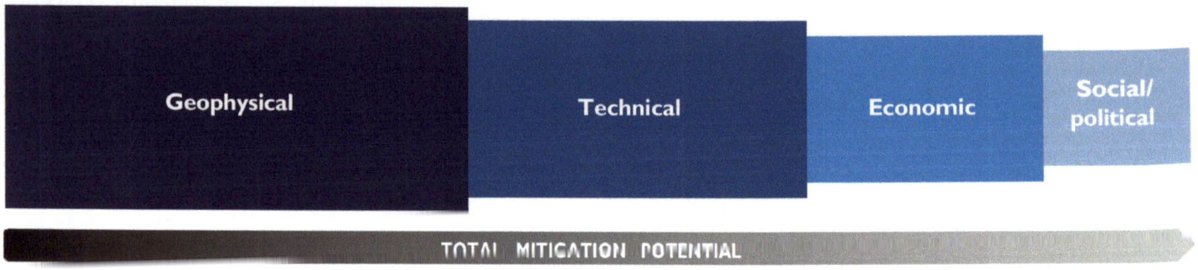

Fig. 17.6 Determining mitigation potential. (*Source: Authors. Note: While the geophysical scale of a mitigation opportunity may be large, each mitigation opportunity must be considered through technical (i.e., its feasibility) and economic (i.e., its cost) lenses, as well as for social and political considerations (i.e., do people want it). A high geophysical potential might exist, given a lack of technical, economic, or socio-political constraints. In reality, a much smaller amount of a mitigation potential tends to be available after these considerations*)

emissions in line with the goals of mean 1.5 °C and 2.0 °C pathways by 2030 and 2050. The year 2030 was chosen to highlight the potential benefits of including relevant ocean-based mitigation options in new or updated nationally determined contributions (NDCs) submitted by 2020. The year 2050 was chosen to highlight the possible contribution of ocean-based mitigation options to long-term strategies of reducing emissons to net zero by mid century (IPCC 2018).

GHG mitigation options in each intervention area were evaluated for their technical, economic, social, and political implications when deployed to reduce GHG emissions (in $GtCO_2e$) by 2030 and 2050. A lower and higher range was estimated in each case to assess how particular ocean-based mitigation options might be modified, or restrained, by other important issues (see the Sect. 1.4 for further details). This assessment also considered the implications for near-term United Nations Sustainable Development Goal (SDG) targets and indicators.

2.3.1 Underlying Assumptions and Approach

Because this report collates multiple analyses, the underlying assumptions and discussion will differ in some cases. Important examples include the size of future baseline emissions and assumptions about the costs of key technologies and inputs. These are discussed and outlined in more detail in subsequent sections of the report.

The following approach was applied to each ocean intervention area to ensure consistency and comparability:

- Identify the baseline emission projections for 2030 and 2050, based on literature review.
- Outline the mitigation options per intervention area that can be implemented by 2030 and by 2050 (including explicitly identified assumptions).
- Identify the range of abatement potential for each mitigation option in 2030 and 2050, either directly from the literature or through calculations based on available data in the literature.

The range of abatement potential estimates is presented to reflect uncertainties in the mitigation potential of both the intervention areas and at the global level.

2.3.2 Determining the Contribution of Ocean-Based Climate Action to Closing the Emissions Mitigation Gap

The calculated mitigation potential from each of the five ocean-based climate action areas were added together to produce a total GHG mitigation potential for the years 2030 and 2050. Each mitigation option was explored in the context of the contribution made to closing the emissions gap in 2030 and 2050 between the "Current Policy" (UNEP 2018) emissions pathway and pathways consistent with achieving the 1.5 °C and 2.0 °C goals of the Paris Agreement (UNFCCC 2015; IPCC 2018). The Current Policy pathway was chosen to reduce the potential for double counting and a median value was calculated from the high and low values provided in the Climate Action Tracker. The intervention areas and mitigation options that are discussed here are generally outside Current Policy and hence should be additional except for the chance of a very small overlap, which is accounted for in the ranges provided for each mitigation option.

The Current Policy trajectory is based on estimates of 2020 emissions that consider projected economic trends and Current Policy approaches (including policies at least through 2015), with estimates based on either official data or independent analysis (UNEP 2018). The pathways consistent with 1.5 °C and 2.0 °C above the preindustrial period were taken from mean values summarised from the scientific literature in the most recent UN Environment Programme Gap Report (UNEP 2018). The 1.5 °C trajectories reach an emissions peak around 2020, then rapidly fall to approximately 45% below 2010 levels by 2030 (to ~28 $GtCO_2e$/year), reaching close to net zero by 2050 (~0–9 $GtCO_2e$/year) (Fig. 17.7). Trajectories for 2.0 °C show emissions decline by approximately 25% by 2030 (to ~40 $GtCO_2e$/year) in most pathways (10–30% interquartile range), reaching net zero by around 2070 (2065–2080 interquartile range). In the case of the Current Policy pathway, GHG emissions will rise from ~50 $GtCO_2e$/year in 2020 to ~65 $GtCO_2e$/year by 2050 (UNEP 2018). These extrapolated levels of emissions under Current Policy are consistent with the projections of the IPCC (IPCC 2018).

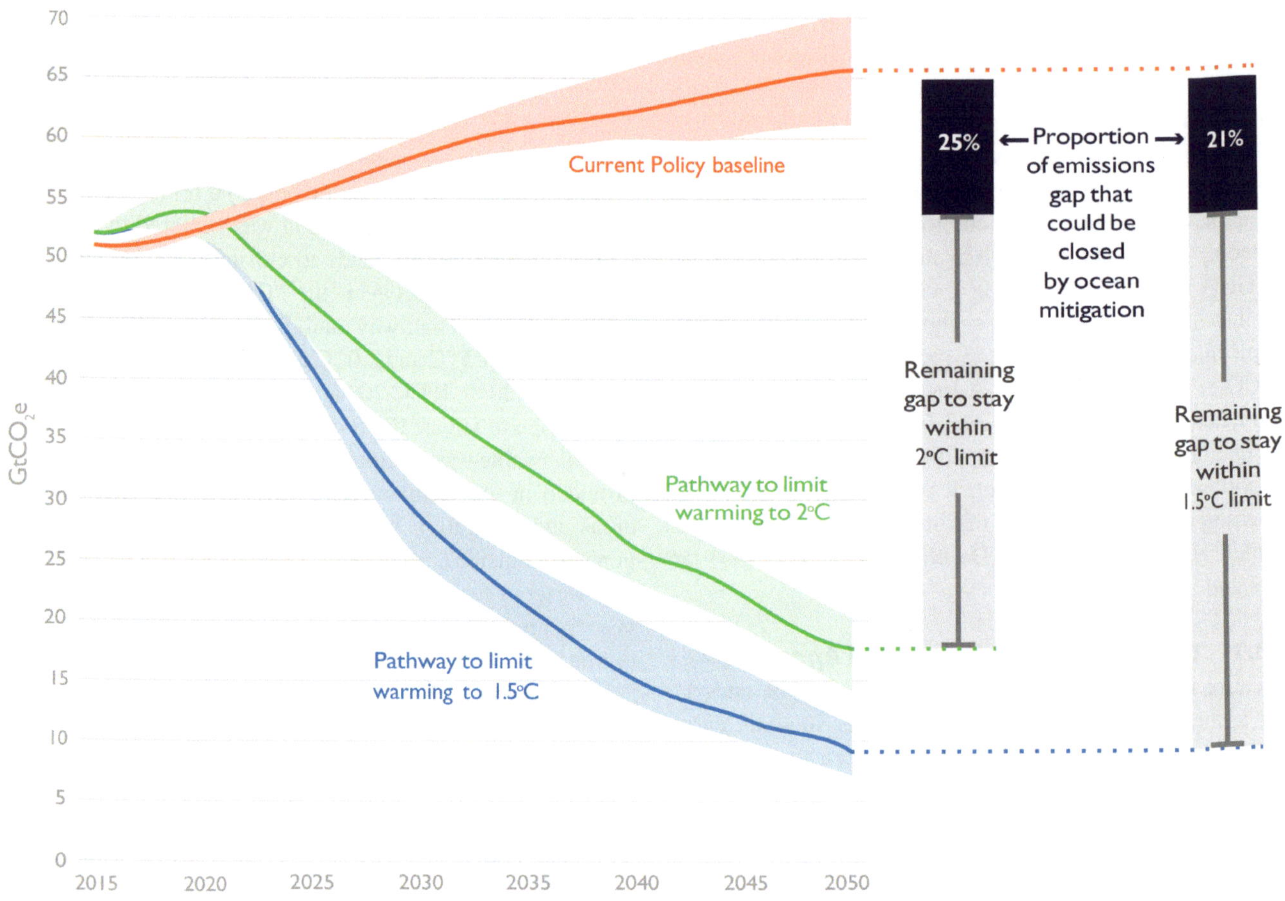

Fig. 17.7 Contribution of ocean-based mitigation options to closing the emissions gap in 2050. (*Source: Adapted from* UNEP (2018), Climate Action Tracker (2018))

3 Ocean-Based Renewable Energy

This section analyses the potential mitigation impact of using ocean-based renewable energy sources of power (e.g. offshore wind and energy extracted from waves and tides) to displace coal fired power plants.

Many technologies are currently being assessed for their ability to harvest renewable energy from the ocean.

Sources of power include offshore wind and energy extracted from waves and tides. Energy within the ocean can also be extracted from salinity and temperature gradients (e.g., by ocean thermal energy conversion [OTEC] or by heat pumps for heating and cooling). Lastly, floating solar photovoltaic (PV) systems are beginning to emerge in marine environments.

While the overall proportion of global electricity generation from ocean-based sources is currently less than 0.3% (IEA 2019), large projects are underway, and investments are being made in the full range of ocean-based energy options. These investments include promising options, such as floating PV panels (World Bank 2018) and strategies to meet sustainable energy demands of a growing blue economy. There is also potential to unlock co-location benefits with other offshore industries; for example, ocean-based energy could meet the increasing demand for energy-intense desalinated seawater (USDE 2019) or support marine aquaculture operations.

3.1 Mitigation Potential

Electricity and heat generation accounts for about 25% of global emissions (IPCC 2014). Mitigation opportunities include replacing fossil-based electricity supplies with renewable sources and electrification, and reducing demand from end-consumers in the transport, industry, and building sectors, and from desalination plants. Depending on the scale and pace of technological development, up to 75% of anthropogenic GHG emissions (excluding some emissions such as those from the agriculture sector and land clearing) in a business-as-usual (BAU) trajectory can be considered as the target for avoidance through electricity generation with renewable sources.

Thermal power plants (using coal, oil, or natural gas) and combustion engines can convert only a fraction of thermal energy into electricity or kinetic energy. Consequently, only a fraction (typically one-third) of primary energy supplied by fossil fuels has to be replaced by renewable sources (GEA

2012). Many thermal plants are also dependent on large volumes of freshwater for cooling. In addition, recent studies such as Grubler et al. (2018) show that extensive use of renewable energy in combination with energy efficiency measures could achieve global low energy demand (LED) scenarios without loss of welfare and well-being.

Renewable energy both from the ocean and from land is therefore well positioned to play an increasing role in sustainable development.

Gross electricity generation in 2050 is projected to be between 42,000 and 47,000 TWh (TWh = terawatt hours; 1 TWh/year corresponds to continuous delivery of a power of 0.114 gigawatts (GW) (IEA 2017). The ocean offers abundant resources in excess of global energy demand, but economic constraints limit the contribution of energy generated offshore.

We consider two ocean-based renewable energy technologies—offshore wind (OSW) generation and other forms of ocean-based renewable energy (ORE), such as wave and tidal power. Estimates of the potential for electrical energy generated by OSW in 2050 are in the range of 650 to 3500 TWh/year. Estimates of potential from ORE technologies in 2050 are in the range 110 to 1900 TWh/year.

We find that if ocean-based renewable energy technologies displace coal-fired power plants, CO_2 emissions can be reduced by between 0.65 and 3.50 $GtCO_2e$/year in 2050 in the case of OSW, and by between 0.11 and 1.90 $GtCO_2e$/year in 2050 in the case of ORE. Total emission reductions would amount to 0.76 to 5.40 $GtCO_2e$/year in 2050.

Alternatively, if energy technologies with emissions equal to the present global mean for the electricity sector of 0.46 kg CO_2e/kWh were displaced, OSW could contribute a reduction of 0.30 to 1.61 $GtCO_2e$/year in 2050, and ORE could avoid 0.05 to 0.87 $GtCO_2e$/year in 2050.

This mitigation potential of ocean-based renewable energy generation is presented in Table 17.5.

3.2 Methodology

The GHG mitigation potential of ocean renewable energy sources is estimated on the basis of substituting fossil fuels used in electricity generation sources (Gattuso et al. 2018). Offshore wind, in particular, and other ocean-based renewable energy sources have theoretical potentials that are many times larger than present global electricity demand, and also larger than future energy demand, assuming full electrification (Bosch et al. 2018) (See Box 17.2). The more interesting challenge is the cost competitiveness of these technologies. Different assessments and estimates of future costs explain much of the range in potential emissions reduction contributions from offshore and ocean-based renewable energy (Box 17.2 and Table 17.6).

Several studies have included offshore wind and other ocean renewable energy technologies in scenarios projecting future energy demand and generation fuel mix. These studies span a range of future carbon emission scenarios for 2050 and are typically presented relative to a business-as-usual, control, or reference scenario. We reviewed 15 scenarios for 2050 in which ocean renewable energy technologies were considered (Table 17.6). Here, we present the future generation mix of ocean energy technologies associated with the low-emissions scenarios (2050 emissions ≤14 Gt), compiled from these studies.

The methodology used to produce the energy contribution potentials was to combine the range of scenarios summarised in Table 17.6 with the difference in CO_2 emissions between energy sources. We recognise that the future evolution of the energy mix, and therefore the substitution effect of ocean-based energy, will depend on a broader set of global development trends, including costs of technologies in other parts of the energy sector, such as hydrogen conversion technologies and energy efficiency.

Table 17.5 Mitigation potential of offshore wind and other ocean-based renewable energy technologies in 2030 and 2050

Ocean-based climate action area	Mitigation options	Description	2030 mitigation potential ($GtCO_2e$/year) (%)		2050 mitigation potential ($GtCO_2e$/year) (%)
Ocean-based renewable energy	Scaling up offshore wind	Fixed and floating offshore wind installations	Coal displacement	0.17–0.23	0.65–3.50
			Displacing current generation mix	0.08–0.11	0.30–1.61
	Scaling up other forms of ocean energy	Energy carried by ocean waves, currents, tides, salinity, and ocean temperature differences	Coal displacement	0.006–0.016	0.11–1.90
			Displacing current generation mix	0.003–0.007	0.05–0.87
Total			Coal displacement	0.18–0.24	0.76–5.4

Source: Author

Notes: To establish estimates of projected energy generation in 2030, we determined the Compound Annual Growth Rate (CAGR) between 2018 energy generation and projected 2050 energy generation (separate CAGR for OSW and ORE). The CAGR is assumed constant through 2050. The per annum CO_2 mitigation potential in 2030 and 2050 is then derived from the energy generation (see Sect. 3.2*). The lowest and highest values were used to calculate the range across "coal displacement" and "displacing current generation fuel mix" for 2030 and 2050. The range for "coal displacement" was chosen for the final totals*

Table 17.6 Summary of energy scenarios reviewed for ocean-based renewable energy

Scenario	OSW generation (TWH/year)	ORE generation (TWH/year)
2018 (30) (Bahar 2019)	53	1.2
2050 Reference (50) Same fraction as current, for assumed 2050 electricity demand of 50,000 TWh	112	2.5
2050 Drawdown Reference (50) (Project Drawdown 2019)	57.2	2.1
2050 IEA WEO 2009 (45) (IEA 2010)	555	25
2050 Teske (Reference (45) (Teske et al. 2011a, b)	805	25
2050 IEA RTS (40) (IEA 2017)	651	108
2050 ETP BLUE MAP (14) (IEA 2010)	1568	133
2050 IEA 2DS (13) (IEA 2017)	1436	536
2050 Teske E[R] (10) (Teske et al. 2011a, b)	2711	678
2050 IEA B2DS (4.7) (IEA 2017)	1531	637
2050 Teske Adv E[R] (3.7) (Teske et al. 2011a, b)	3469	1943
2050 DRAWDOWN Plausible (Project Drawdown 2019)	2078	1486
2050 DRAWDOWN (Project Drawdown 2019)	3029	1745
2050 DRAWDOWN Optimum (Project Drawdown 2019)	3159	1823
2050 OES Vision (OES 2017)	–	1051
2050 IRENA (IRENA 2018a)	1822	

Source: Authors
Note: OSW offshore wind, *ORE* ocean-based renewable energy

Table 17.7 Estimated life-cycle emissions of energy generation technologies

Energy technology	Lifecycle carbon emissions ($kgCO_2e$/kWh)	Lifecycle carbon emission relative to current mix (%)
Coal	1.0 (0.67–1.7)	217
Natural gas	0.476 (0.31–0.99)	103
Current mix	0.46	–
Solar PV	0.054 (0.019–0.2)	12
Concentrated solar power	0.025 (0.007–0.24)	5.4
Nuclear	0.016 (0.008–0.22)	3.5
Onshore wind	0.012 (0.002–0.088)	2.6
Offshore wind	0.012 (0.005–0.024)	2.6
Ocean	0.008 (0.002–0.022)	1.7

Source: OpenEI (2019)
Note: Bracketed values represent the range of reported emissions

By calculating mitigation potentials for substitution of coal and for substitution of an energy source with CO_2 emissions corresponding to the present global average, we expect to bracket a realistic range.

Ocean-based technologies offer a renewable energy solution with low life-cycle carbon emissions (Table 17.7). Ocean-based renewable energy technologies are thus able to displace emissions associated with fossil-based electricity generation. The greatest emissions mitigation is obtained when displacing high-emitting electricity-generating technologies such as coal, which accounts for approximately 38% of global electricity generation (IEA 2019). The use of ocean-based technologies has the potential to displace approximately 0.35–0.9 $kgCO_2e$/kWh electricity produced, depending on the source of electricity being displaced.

Box 17.2. Current Global Status of Implementation and Future Deployment

Current Global Status of Implementation

Offshore Wind Energy

By the end of 2018, the total installed global capacity of wind energy amounted to 564 GW, of which 23 GW was offshore (IRENA 2019a). Annual offshore electricity production amounted to about 77 TWh (IEA 2018).

Bottom-fixed wind turbines in shallow water depth (<40 m water depth) dominate. Deepwater, floating support structures are used in only one wind farm, a 0.03 GW wind farm off the east coast of Scotland. Much of the available information on offshore wind used in this report (in particular experience with costs) is taken from Europe, where the majority of offshore wind installations are located. However, it is anticipated that Asia, especially China, will significantly increase installed offshore wind capacity in coming years. The specific rate of growth is, however, difficult to assess.

Over the past decade, the cost per MWh installed power has fallen and the capacity factor (ratio between realised energy output and theoretical maximum output) of new installations has increased. High capacity factors of OSW installations are a notable advantage: the 2018 mean capacity factor for European offshore wind farms of 36% far exceeded that of European onshore wind farms (22%). The operation and maintenance (O&M) cost per produced

MWh is also expected to decline as turbines are designed to be more robust and better suited to the offshore environment. These factors contribute to reduced LCOE. Several other parameters are important when estimating the LCOE, including the connection between wind farms and the grid and the discount rate used in cost estimates. The increased size of turbines and wind farms, as well as the learning rate of the offshore wind industry, have all contributed to reduce LCOE. However, moving into deeper water and farther from shore has partly offset the cost reductions.

For projects commissioned in 2018, the average European LCOE was 134 US $/MWh. A project in China had an LCOE of US $105/MWh (IRENA 2019b). Contracts with record low costs, however, have been signed in the Netherlands (US $55/MWh to US $73/MWh), while the LCOE of a near-shore project in Denmark was US $65/MWh, excluding grid connection costs. No reliable data are available for floating systems, but for bottom-fixed systems, offshore wind without subsidies has proved cost-competitive with other electricity sources. This is the case even without a CO_2 tax, which would negatively impact competing power sources.

Other Ocean Renewable Energy

Estimated theoretical potentials for ocean renewable energy technologies (other than offshore wind) are listed below:

- **Tidal Range Energy**: The estimated global theoretical tidal range resource is around 25,880 TWh/year (constrained to regions with water depth of less than 30 m, and a reasonable threshold for energy output). Considering the logistical issues of operations in ice-covered regions, the global annual potential energy from tidal range technologies is approximately 6000 TWh, with 90% of this resource distributed across five countries (O'Neill et al. 2018).
- **Tidal Stream Energy**: The best estimates of the total global technical tidal stream energy resource is approximately 150 TWh/year, but the estimate is subject to high uncertainty (Yan 2015).
- **Wave Energy**: The total theoretical wave energy potential is estimated to be 32,000 TWh/year (Mørk et al. 2010), with estimates of the global technical potential ranging from 1750 TWh/year (Sims et al. 2007) to 5550 TWh/year (Krewitt et al. 2009).
- **Ocean Thermal Energy Conversion (OTEC)**: OTEC is currently limited to the tropical regions (±20° latitude). Estimates of the global theoretical energy resource range from 30,000 TWh/year to 90,000 TWh/year. Global technical resource estimates range from 44,000 TWh/year to 88,000 TWh/year (Lewis et al. 2011).
- **Salinity Gradient**: According to Alvarez-Silva et al. (2016) the theoretical global potential of power from utilizing the salinity gradient at the mouths of rivers worldwide has been estimated to be up to more than 15,000 TWh/year. Considering the river systems in more detail, 3600 TWh/year is more realistic. Accounting for extraction factors and other technical limitations, the globally technical extractable potential is estimated to be in the order of 625 TWh/year (Alvarez-Silva et al. 2016).
- **Floating solar PV systems**: Floating solar is presently in use predominently in water reservoirs and a small number of marine sites. Moving such systems to the ocean environment, the technical potential will depend upon the system's ability to operate in ocean waves. To ensure survival when facing extreme waves will drive the costs of the systems.

At the end of 2018, the total installed capacity of ocean energy technologies was 532.1 MWh (IRENA 2019a), consisting mainly of tidal barrage technology at two sites. Installed capacity in 2016 was 523.3 MWh, which generated 1023.3 GWh of electricity (IRENA 2019a), implying a mean capacity factor of 23% across the sector. Salinity gradient (energy available where freshwater meets salt water) and floating solar photovoltaic (PV) do not contribute significantly to installed capacity at present, but could contribute in future.

Estimates of LCOE are subject to a range of assumptions, including local conditions, which all affect costs. The estimated LCOE for wave energy is in the range of €330 to €630/MWh (IRENA 2014a). Tidal stream energy LCOE is currently in the range of €250 to €470/MWh (IRENA 2014b). At the current scale of deployment, LCOE of ocean thermal energy conversion is in the range of US $600 to US $940/MWh (IRENA 2014c).

Learning rates for ocean technologies are typically assumed to be around 15% (OES 2015), resulting in average LCOEs of €150 to €180/MWh for wave energy and of €200/MWh for tidal energy by 2030 (Cascajo et al. 2019; SI Ocean 2013). Due to the capital intensity of OTEC, interest and discount rates have a high impact on LCOE estimates for this technology. Economies of scale are expected to bring the LCOE into a range of US $70 to US $190/MWh for installed capacities exceeding 100 MWh (IRENA 2014c; OES 2015).

Future Deployment Scenarios (2030 and 2050)

Offshore Wind Energy

According to IEA (2017), offshore wind generation grew fivefold over the period 2010 to 2015 and is expected to double between 2015 and 2020. James and Ros (2015) estimated that Europe alone has a 4,000 GW potential for floating offshore wind in water depths above 60 m. This corresponds to about 15,000 TWh/year. National strategies in Europe, if implemented, sum to more than 70 GW of offshore wind capacity by 2030 (Ørsted 2019). The present offshore wind base is lower outside Europe, which increases the uncertainty of future scenarios. But a total installed capacity of 100 GW in Asia and 10 GW in the United States has been estimated for 2030 (GWEC 2017). Worldwide, offshore wind capacity could reach 120 GW in 2030 (GWEC 2017).

In 2018, the European Commission presented a strategic roadmap towards a zero-carbon economy in Europe by 2050 (European Commission 2018). The roadmap includes 70 GW of offshore wind in 2030, increasing to 600 GW in 2050, which corresponds to about 2300 TWh/year. To achieve this level of installed power, a significant scaling-up in the installation rates of offshore wind is needed. Floating offshore wind may be key.

Other Ocean Renewable Energy Technologies

Electricity generation from other ocean renewable energy technologies increased by an estimated 3%/year in 2018 (IEA 2019). This rate of growth is not on track to meet the IEA Sustainable Development Scenario (SDS) target for ocean technologies of 15 TWh/year in 2030 (IEA 2019), which would require an annual growth rate of 24%. The IEA SDS corresponds to an emissions target of approximately 25 $GtCO_2e$/year by 2030. By 2050, the projected power generation from ocean technologies is 108, 536, and 637 TWh/year for the IEA Reference Technology Scenario (RTS), 2 Degree Scenario (2DS), and Beyond 2 Degree Scenario (B2DS), respectively. The 2050 emissions associated with these three scenarios are 40.0, 13.0, and 4.7 $GtCO_2e$/year, respectively. This corresponds to annual growth rates of ocean technologies of 15, 21, and 22%, respectively.

3.3 Policy Interventions Needed to Realise Mitigation Potential

Offshore wind energy resources alone would be sufficient to cover more than the world's electricity demand in 2050. However, significant scaling-up in the rate of deployment is needed for offshore wind to become the significant player indicated by its potential. For other ocean-based renewable energy technologies, additional policy support is required for research and development to enable the scale efficiencies and cost reductions that come with commissioning larger commercial plants.

The levelised cost of energy (LCOE) of ocean-based renewable energy is dominated by investment costs. This means that measures related to project finance and tax regimes can be crucial. Defining the interface between the offshore plant and onshore grid, ownership, and the regulation of electricity markets can make a big difference.

Other policy interventions can also support greater uptake of ocean energy technologies:

- Development of incentives (e.g., carbon taxes and innovative power purchase agreements) that can encourage the expansion of ocean-based energy systems.
- Marine spatial planning should integrate the future role of offshore renewable energy with the many other activities affecting ocean and coastal areas. Development of appropriate legislation and regulation of ocean-based renewable energy to allow easier integration in national electricity grids is also required.
- Establishment of national targets and strategies to increase the share of ocean-based renewable energy in the national energy mix.
- Stable economic and regulatory framework to stimulate investments in required infrastructure for an accelerated deployment of ocean-based energy systems.

3.4 Technology Needs

Energy development needs access to larger areas where ocean energy resources can be harvested. Innovations that can move technologies into deeper water sites will be required, for example, development of floating offshore wind technologies.

Improving performance, reliability, and durability, while reducing costs, are the key challenges confronting all ocean energy technologies. Much is to be gained through continued and expanded support for innovation.

However, technology improvements must take account of environmental and social constraints that, if ignored, will undermine efforts to achieve a successful energy transition (Box 17.3).

3.5 Priority Areas for Further Research

Technology innovations need to be underpinned by a high-resolution assessment of global ocean energy resources, in terms of both geophysical and economical potential.

Research on integrating renewable energy projects with other coastal activities (e.g. coastal defense, food production and aquaculture) requires further investigation in order to maximise potential synergies and co-benefits associated with co-location.

Advancing further pilots and testing on the ability of floating solar PV panels at sea (under wave conditions) and further quantification this potential, along with that of Ocean Thermal Energy Conversion (OTEC).

Box 17.3. Wider Impacts Associated with Scaling Up Ocean-based Renewable Energy

Potential Cobenefits:

- Positive and long-term effects on ecosystems from offshore wind farm structures acting as artificial reefs.
- Human health benefits from reduced local air pollution in regions relying heavily on coal and oil to generate electricity
- Reduction in freshwater usage (overall) compared to generating power via fossil fuel.
- Job creation at regional and local levels, benefiting workers transitioning from declining fossil fuel industries. Total full-time employment in offshore wind in 2030 is estimated to be 435,000 (compared to about 38,000 in 2010) (OECD 2016).
- Potential to generate employment opportunities for women and promote greater gender equity in the rapidly growing industry

Potential Trade-Offs

- The spread of invasive species, noise pollution, and disturbances to marine species from vibration.
- Collision risks to birds and the presence of electromagnetic fields disrupting marine life and benthic habitats.
- Emerging offshore ocean energy (such as tidal barrage, tidal current, wave energy, and thermal gradient) are yet to be deployed commercially at scale. Tidal barrage installations can cause disruption to estuarine ecosystems.

For a full exploration of the wider impacts associated with ocean-based renewable energy, see the Sect. 1.4.

4 Ocean-Based Transport

This section analyses the potential mitigation impact of reducing emissions from domestic and international marine transport and shipping.

Current GHG emissions from global ocean transport (both international and domestic shipping of passengers and freight) are approximately 1 $GtCO_2e$ per year and represent around 3% of global anthropogenic CO_2 emissions (Buhaug et al. 2009; Smith et al. 2014). Long-term trends in shipping indicate a strong increase in demand and gradual improvement in energy efficiency. Since 1970, energy efficiency has improved by only about 1%/year (Lindstad et al. 2013; Lindstad and Eskeland 2018). If current trends continue, demand is likely to grow by 3%/year, which would lead to GHG emissions approximately doubling in 2050, to roughly 2 $GtCO_2e$, compared to 2010. This is in sharp contrast to what is needed to keep global temperature rise well below 2.0 °C and consistent with a 1.5 °C increase (IPCC 2013) and align with the goals of the Paris Agreement (UNFCCC 2015).

Shipping is a significant source of emissions with identifiable reduction pathways, but it is also an enabler of world trade and economic development. In 2018, the United Nations International Maritime Organization (IMO) adopted its Initial Strategy (Resolution MEPC.304). An objective of the strategy was to reduce shipping GHG emissions by at least 50% in absolute terms by 2050, relative to 2008 emission levels. Whilst the minimum reduction (50%) would see shipping's relative share of total GHG emissions grow significantly under most Paris-aligned scenarios, [1] the strategy leaves open the possibility of greater ambition, that is, to set a total GHG reduction target for 2050 that is well above the minimum 50%. A more ambitious target will likely be considered in the Revised Strategy due for finalisation by 2023.

The energy intensity and the absolute GHG emissions of ocean-based transport can be reduced in the following ways:

- Technical and operational interventions to reduce energy consumption per tonne transported (reduced energy intensity).
- Substitution of low- and zero-carbon fuels (e.g., hydrogen, ammonia, some biofuels) for diesel and bunker oil (reduced absolute emissions).

The 50% GHG reduction target set by the IMO might be achievable with technical and operational measures alone. Achieving a greater level of reduction by 2050—or the full phaseout of GHG emissions from shipping, as called for in the Initial Strategy's vision statement—will be possible only with the introduction of low- and zero-carbon fuels to replace fossil fuels. In practice, a rapid and cost-effective reduction in GHG emissions will require both technical and operational interventions and a swift transition to low- and zero-carbon fuels.

4.1 Mitigation Potential

Ocean-based transportation has the potential for a roughly 100% reduction in operational net GHG emissions by changing the way it stores and consumes energy on board:

Batteries could be used to store electricity, particularly in ships on the shortest voyages.

Low/zero carbon synthetic or "e" fuels could replace fossil fuels. Examples include renewable hydrogen, hydrogen-

[1] If shipping's emissions fall by 50% in absolute terms, to achieve the Paris Agreement temperature goals, other sectors will need to have fallen by more than 50% in absolute terms, and so shipping's relative share of total emissions will have grown.

based fuels such as ammonia, and fuels that have been post-processed with CO_2 to make hydrocarbons. These fuels differ from synthetic fuels made from gas or coal.

Biofuels could replace fossil fuels. However, it is commonly assumed that biofuels will have a limited role because of land and water constraints on sustainable supply and the fact that many biofuels are not, in fact, carbon-neutral (Searchinger et al. 2019).

Transitioning ocean shipping to more efficient and low- or zero-carbon fuels, and the mitigation potential in 2030 and 2050, is largely determined by the timescales needed to renew or retrofit the existing fleet and develop the infrastructure to use and supply these new energy sources.

Producing synthetic ("e") fuels, electricity, and bioenergy at volumes required by ocean-based transport will likely still have significant upstream emissions by 2030, and only a small subset of the fleet is likely to be "zero-carbon-fuels ready" by 2030. The mitigation potential in this time period is therefore mainly driven by the opportunity associated with energy efficiency maximisation. The upstream emissions and therefore the life-cycle (or well-to-wake) emissions for each of these pathways may remain significant until a broader transition to a zero-carbon energy system has been completed.

Nevertheless, if we assume that, by 2050, there will be a fully decarbonised land-side energy system associated with the production of shipping fuels, and that this is a timescale over which the whole ocean-based transport fleet could be "zero-carbon-fuels ready," there is a clear potential for 100% GHG reduction.

This mitigation potential is presented in Table 17.8.

4.2 Methodology

We use a business-as-usual (BAU) emissions trajectory out to 2050, based on an estimate of growth in demand for shipping. The BAU scenario used here is taken from the Third IMO GHG Study (Smith et al. 2014), where demand is estimated to align with IPCC scenario RCP 2.6 (Residual Concentration Pathway 2.6, which is approximately associated with a 2 °C temperature rise) and SSP 4 (Shared Socioeconomic Pathway 4, which assumes continued global inequality and increasing disparities in economic opportunity).

This BAU scenario applies existing IMO policy (including the Energy Efficiency Design Index (EEDI) and the Ship Energy Efficiency Management Plan (SEEMP) regulations) and estimates that total GHG emissions from international shipping will grow from about 800 Mt in 2012, to 1100 Mt in 2030, and to 1500 Mt in 2050. There is no projection for GHG emissions from domestic shipping in the Third IMO GHG Study, so we derive the domestic shipping BAU by applying the growth rates of international shipping to the 2012 domestic shipping inventory (taken from the Third IMO GHG Study).

Using the BAU scenario as a baseline level of emissions, the mitigation potential is quantified by applying a percentage reduction (defined below) to the emissions in both 2030 and 2050. The group of technologies that can mitigate domestic and international shipping emissions are similar, so the same percentage reduction is applied to both fleets.

To estimate mitigation potential in 2030, a 39% emissions reduction is assumed as the upper bound, taken from Bouman et al. (2017). This paper reviewed multiple papers and models to produce consensus estimates of the mitigation potential, both of individual mitigation options and the options in combination.

By 2030, the authors estimated that, relative to BAU, the median reduction potential across their surveyed literature was 39%. Of all the papers reviewed, the lowest estimate of emissions reduction potential by 2030 is 20%, this value is used to set the lower bound in the range of reduction potential. The mitigation potential in 2050 assumes a 100% emissions reduction at its upper bound. This is based on the assumption described in the preceding section that, if all vessels move to full use of nonfossil fuels from renewable feedstocks, then operational and upstream GHG emissions can be reduced to zero. The lower bound reduction potential is set at 50%, taken as the minimum interpretation of the IMO's objectives in the initial GHG reduction strategy.

The estimate of mitigation potential is thus based on a number of assumptions:

Table 17.8 Mitigation potential of ocean-based transport in 2030 and 2050

Ocean-based climate action area	Mitigation options	Description	2030 mitigation potential[a] ($GtCO_2e$/year)	2050 mitigation potential[b] ($GtCO_2e$/year)
Ocean-based transport	Reducing emissions from domestic shipping	Following the IMO definition: shipping between ports of the same country. Domestic shipping excludes military and fishing vessels, Includes ferries. This definition is consistent with the IPCC Guidelines 2006	0.04–0.07	0.15–0.3
	Reducing emissions from international shipping	Following the IMO definition: shipping between ports of different countries. International shipping excludes military and fishing vessels; includes bulk carriers, oil tankers and container ships. This definition is consistent with the IPCC Guidelines 2006	0.2–0.4	0.75–1.5
Total			0.25–0.5	0.9–1.8

Source: Authors

[a]Achieved predominantly through technical and operational interventions to reduce energy intensity per tonne transported

[b]Achieved predominantly through substitution of low- and zero-carbon fuels

- The speed of policy implementation to enable or require the shipping industry to invest in the necessary changes to fleet and infrastructure (in particular with respect to low- and zero-carbon sources). We assume that clear policies incentivising shipping's decarbonisation are in place by 2025. Later adoption of policy could jeopardise the achievement of these mitigation potentials in 2030 and 2050.
- The 2030 GHG reduction potential is estimated by aggregating savings across a large number of technological and operational efficiency interventions. [2] If savings are individually or collectively lower (or higher) because of currently unforeseen performance characteristics or interactions between the different interventions, then there could be a significant impact on the abatement potential achieved in 2030.
- The extent to which the wider energy system is decarbonised with sufficient supply of zero-carbon electricity to enable shipping fuels to be produced with zero emissions. We assume that the wider energy system has fully decarbonised by 2050 and that renewable hydrogen (zero carbon in production) is available in sufficient volumes. If that is not the case, then significant upstream emissions may still occur and offset some of the mitigation potential achieved through operational emission reductions.
- Demand growth is assumed to broadly follow the IMO's RCP 2.6 SSP 4 scenario. However, demand growth could be significantly higher or lower, with direct consequences for the BAU emissions and therefore (in proportion) the GHG mitigation potential of a fully decarbonised ocean transport industry.

4.3 Policy Interventions Required to Realise Mitigation Potential

The majority of the mitigation potential in ocean-based transportation is significantly influenced by one global body: the IMO. Domestic shipping is regulated by national governments, but often by flowing through IMO regulation. This section discusses interventions that can be undertaken by the IMO, national governments (including supranational organisations such as the European Union), and private sector organisations.

Private sector initiatives may be voluntary, shifting behaviour and removing existing barriers to decarbonisation, or mandated by national or global policy in due course.

The key actions needed are immediate improvements in energy efficiency to reduce fuel consumption, followed as quickly as possible by policy interventions that can incentivise shipping to transition away from fossil fuels, and private sector initiatives that enable adoption of low- and zero-carbon fuels. The following considerations are relevant:

- Cost-effective energy efficiency improvements can be made today, before the arrival of new fuels and their associated infrastructure.
- Current energy efficiency policy (IMO regulations on energy efficiency design index, EEDI) and energy efficiency management (SEEMP) are inadequate.
- EEDI has significant failures in its design (see section below), and ship energy efficiency management plan (SEEMP) is only a guideline, with no mandatory target.
- Energy efficiency improvements can reduce the impact on shipping and trade of moving to higher- cost low- and zero-carbon fuels.
- Policy needed to stimulate low- and zero-carbon fuels and support innovation may take longer to implement at IMO. In contrast, existing policy frameworks at IMO may be more easily and quickly used to drive improvements in energy efficiency and energy intensity.

4.4 International Maritime Organization Strategy for GHG Reduction

Emissions from the shipping and aviation industries were not explicitly included in the Paris Agreement. The expectation was that their respective UN agencies, IMO, and the International Civil Aviation Organization (ICAO), would lead on GHG-reduction efforts and develop global regulations. Another factor is that the majority of GHG emissions from shipping and aviation occur in international waters or airspace, and there is no obvious way to allocate national responsibilities for mitigation.

The IMO's Initial Strategy was adopted in 2018, partly as a clear statement of how IMO intended to fulfil its responsibility under global efforts to combat climate change. It is closely linked to the Paris Agreement both in terms of its mitigation goals and its adherence to the principle of Common but Differentiated Responsibilities and Respective Capabilities.

The IMO's Initial Strategy lays out three groupings of candidate policy interventions (short, medium, and long term), which, if effective, could realise most of ocean-based transport's mitigation potential. The IMO does not define the specific time frame corresponding to short, medium, and long term, or whether the time frames refer to a policy's design, adoption, or implementation. However, the time frames are understood to correspond approximately to implementation timescales of before 2023 (short), 2023 to late 2020s (medium), and 2030 onward (long)). In practice to have good likelihood of meeting the IMO's objectives, clarity of policy direction is important and urgency of implementation is high (because of the long timescales of asset

[2] Bouman et al. (2017) presented the results of a review of nearly 150 studies, to provide a comprehensive meta-analysis of CO_2 emissions-reduction potentials and measures. They identified 22 types of measures for which reliable and comparable data are available in the peer-reviewed literature.

lives relative to decarbonisation objectives). For these reasons all policy recommendations are for the short and medium time frame only.). This report proposes a number of priority actions that IMO should undertake to maximise the potential for decarbonisation of ocean-based transport:

Short Term

Redesign of the EEDI formula so that it is fit for purpose (see section above) and addresses all in-service GHG emissions.

Adoption of policy measures that go beyond SEEMP to incentivise maximum operational efficiency of the existing and new fleet by no later than 2030.

Adoption of policy to reduce GHG emissions from shipping other than CO_2, in particular methane (CH_4) emissions associated with methane slip [3] and volatile organic compound (VOC) emissions associated with certain cargoes. To enable this, it will be necessary to develop CO_2 equivalent emission factors for all major fuel and machinery combinations on a tank-to-wake (TTW) basis, including for use in the redesigned EEDI formula.

Commitment to a timetable for shipping's transition to low- and zero-carbon fuels that will prompt early action and send a clear signal that investment should flow into fleets and related infrastructure.

Medium Term

Development of policy to measure, report, and verify well-to-tank (WTT) emissions for ship fuel and fuel supply chains.

A "medium-term" policy measure entering into force, no later than 2025, that strongly incentivises the adoption of low- and zero-carbon fuels by shipping. Options include the following:

A price on carbon (or GHG) emissions to simultaneously close the price gap between conventional and low- and zero-carbon fuels and enable competitive pricing for all options that reduce the GHG intensity of shipping. Revenues raised by such a measure should be disbursed to assist research, development, and demonstration (RD&D), and, if necessary, to address disproportionate negative impacts on vulnerable member states.

Standards that prescribe the carbon or GHG intensity of operation or the fuel used in ocean-based transport, whilst finding alternative (non-revenue disbursement) mechanisms to enable efforts on RD&D and address disproportionate negative impacts on vulnerable member states.

Box 17.4. International Maritime Organization's Existing Regulation: EEDI and SEEMP, and Their Limits and Challenges

The EEDI and SEEMP policies were first implemented in 2013 (IMO 2011; Psaraftis and Kontovas 2013). They target minimum performance requirements for ship design (EEDI), and recommendations for how energy efficiency could be managed in operation (SEEMP). A number of studies on trends in ship design efficiency during the early years of these regulations (Faber et al. 2016) show that many ships have performed far better than the EEDI requirements (i.e., their CO_2 emissions have been significantly lower than the required threshold). The implication is the requirement could have been more stringent (and recently the standards have been tightened and dates of alteration to phase 3 stringency brought forward for some ship types).

However, as the stringency of the regulation increases, so does the incentive to "game" the system. Ship design can be optimised to pass the short calm water trial in which EEDI is measured. Calm water trials bear little resemblance to normal operating conditions, where ships encounter strong winds and waves. Unless the EEDI is adjusted to include a performance threshold for rougher conditions, GHG emission targets will be set too low, and emissions could potentially increase (Lindstad et al. 2019). It is easy to make hull form modifications that improve calm water performance even of full-bodied "bulky" hulls. However, these modifications generally increase fuel consumption under real operating conditions. By contrast, hull forms optimised with respect to performance in realistic sea conditions cannot prove their worth when tested in calm water.

In addition, the regulation has no mechanism to ensure that the fuel used when ships are tested will also be used in operation, when a ship has multiple fuel options. A ship could complete its certification and trials using low-carbon fuel, gaining an excellent EEDI "score" but then switch to higher-carbon fuel in operation.

As EEDI is currently designed, the regulation influences only design specification. Experience in other sectors has shown that regulation that does not also incentivise efficiency in operation may not achieve the magnitude of savings expected from an extrapolation of the design efficiency standards. Studies specifically on EEDI have projected that it may contribute as little as 3% to actual operational CO_2 reduction (Smith et al. 2016).

The SEEMP regulation is mandatory in that a ship must be equipped with SEEMP documentation (i.e., an energy efficiency management plan), but there is no mandate for what must be specified within the documentation. As such, the regulation is a guideline and cannot be relied upon to overcome the known market barriers and failures and drive carbon intensity reduction in line with Paris Agreement and IMO objectives.

Source: Authors

[3]Unburned methane emissions released during vessel operation via fossil fuel combustion in the engine.

4.4.1 National Government Actions

Some governments have identified opportunities for economic benefit from emission reductions in the shipping sector (Bell et al. 2019) and have introduced incentives or other measures. For example, the United Kingdom has adopted the Clean Maritime Plan; several Scandinavian countries have set domestic shipping emission-reduction commitments; the Marshall Islands has included specific reductions for shipping emissions in its nationally determined contribution (NDC); China has shown leadership at the IMO on the topic of National Action Plans. The plans are initiatives, led through the IMO, that provide support for regional Maritime Technology Cooperation Centres and for shipping energy efficiency measures undertaken by 10 national governments within a Global Maritime Energy Efficiency Programme (GloMEEP).

Key elements in government actions taken to date include the following:

- Incentivising decarbonisation of domestic ocean-based transportation, if possible at a rate of transition faster than that achieved in the international fleet through IMO regulation. Domestic fleets are populated with smaller ships and therefore better suited to pilots and tests of fuels and technologies, which in turn can help to de-risk and reduce costs for larger, high seas, and ocean-based transportation.
- Enabling decarbonisation of national energy systems at least as fast as the rate of transition in the international fleet, and with sufficient additional energy supply capacity to meet a relevant proportion of the international fleet's energy demands.
- Providing national support for development of low- and zero-carbon energy production capacities, and storage and refuelling infrastructure in ports and harbours.
- Forming partnerships, particularly in support of small island developing states (SIDS) and least developed countries (LDCs) with significant domestic or regional shipping decarbonisation challenges, to work together on joint objectives.

4.4.2 Private Sector Actions

The private sector has traditionally led efforts to address shipping issues, such as safety and oil spill risks. While there are examples of such leadership in the areas of energy efficiency and decarbonisation (Scott et al. 2017), early initiatives have not matched the ambition of the Paris Agreement. In part, this is because earlier voluntary initiatives have stayed close to IMO policy, which remains conservative for fear of creating commercial disadvantages for its members and potentially reducing membership. Market barriers and failures inhibit action (Rehmatulla 2014), but where an opportunity aligns with wider stakeholder objectives, further action can be taken. Examples include the following:

- Further work to understand where market and nonmarket barriers and failures to decarbonisation occur and can be removed. For example, ensure that authorities setting rules in ports, fairways, and pilotage and sailing restrictions do not unnecessarily penalise ship length, given this is a low-cost means of reducing GHG intensity of shipping.
- As demonstrated in the Poseidon Principles (www.poseidonprinciples.org), encourage/regulate the financiers of shipping to be held more accountable for management of the long-run risks of shipping decarbonisation. This aligns with the increasing general prioritisation of finance to put a price on climate-change mitigation and adaptation-related risks. This can ensure that finance is no longer directed towards "standard" designs, which are optimised on cost at the expense of energy efficiency. It can ensure financing of a decarbonisation-aligned fleet that will avoid risks of asset stranding and maximise investment in the most efficient tonnage.
- Encourage/regulate the charterers of shipping to measure, report, and be held more accountable for operational GHG emissions for which they have responsibility (e.g., Scope 3 emissions). This can help address the lack of a clear market signal that ensures the energy efficiency and carbon intensity differential across the fleet is reflected in the prices paid by charterers, and which is needed to ensure that the shipowners have the full economic incentive to invest in solutions that achieve GHG reduction. This also ensures, in addition to policy on operational emissions, that where charterers have opportunities to contribute towards achieving GHG reduction, they seek to do so.

4.5 Technology Needs

The greatest need is to accelerate and scale up deployment of energy efficiency interventions. Many feasible solutions are ready to implement but are being adopted in low volumes because of market barriers and failures. These need to be overcome through effective national government and IMO policy (Rehmatulla 2014). Market barriers and failures present the main obstacle, but faster technological progress and implementation of demonstration projects have potential to produce greater understanding of performance benefits, performance improvement, and cost reduction (Lindstad et al.

2015; Lindstad and Bø 2018). Current promising but low-volume solutions include the following:

- Energy efficiency technologies (e.g., air lubrication, waste heat recovery, batteries (Lindstad et al. 2017b)) and hybrid engines (Lindstad and Bø 2018) that help smooth and manage demands for power from internal combustion engines and enable them to operate more optimally. Cold ironing (also known as "shore power") and digital solutions help enable operational efficiency improvements.
- Wind assistance technologies (kites, sails, and rotors that can directly harness renewable wind energy for propulsion).

There also remains a need to develop supply chains and technologies for the use of new low- and zero-carbon fuels on board. These are all at lower readiness level (LR and UMAS 2019) and unlikely to be feasible without significant incentives from IMO and national government policy, in addition to private sector action (LR and UMAS 2019). Specific technologies include the following:

- Electrolysers and equivalent as well as related technologies for producing hydrogen from electricity.
- Carbon capture and storage (for use with production of hydrogen from fossil feedstock).
- Storage technologies for hydrogen (including cryogenic storage of liquid hydrogen or carriers able to store at high-energy density).
- Fuel cells for conversion of future fuels into on-board electricity, and internal combustion engines designed to operate on hydrogen/ammonia.

4.6 Priority Areas for Further Research

Minimising energy consumption remains of high importance as the lowest-cost means of reducing emissions in the short term. It is now predominantly a function of implementing best practice in the design and operation of ships, and introducing sufficient policy incentives and private sector initiatives to overcome market barriers and failures that are currently preventing full adoption. The energy efficiency area represents a market opportunity if improved technologies become more widely deployed, but it will be a diminishing priority for further research.

Enabling the necessary switch to low- and zero-carbon fuels requires rapid progress in a number of areas (LR and UMAS 2019), both to confirm the most cost-effective transition pathway for shipping and to help reduce the costs of that pathway. Our recommended priorities focus on hydrogen and ammonia, even though other fuels are often considered for the future of ocean-based transport. Until a long-term solution has emerged, the interim "transition" steps that might be compatible with that solution (e.g., the fuels and their production pathways) will remain unclear.

Short-Term

- Cost-effective production of low-carbon hydrogen and ammonia from fossil fuel feedstocks in combination with carbon capture and storage.
- Safe storage and handling of hydrogen and ammonia on ships and at the ship-shore interface.
- Safe and efficient use of hydrogen and ammonia in large (e.g., 1 MW+) internal combustion engines and fuel cells.

Medium Term

- Cost-effective production of zero-carbon hydrogen/ammonia using renewable electricity and electrolysers.

These research areas represent large future market opportunities in terms of the provision of hardware and technology; production of future fuels; provision of services related to managing the design, implementation, and operation of assets; and ownership and operation of other related assets. Those opportunities are relevant to corporate and national interests and are especially important for countries with significant maritime or renewable fuel interests and an associated industrial strategy. Countries and corporate entities will need to proactively position themselves to capitalise on these opportunities.

Box 17.5. Wider Impacts Associated with Reducing Emissions from Ocean-based Transport

Potential Cobenefits:

- Reduction in seasonal "hotspots" of ocean acidification caused by strong acids formed from shipping emissions
- Beneficial impact on human health, particularly for people living in port cities and coastal communities, including from reduction in the sulphur content of fuel oil used by ships.
- Upgrade in technological capabilities in marine transport will bring efficiency.

Potential Trade-offs:

- Cost to industry of switching to alternative fuels will be high; however, increased costs are likely to have a marginal impact on the price of traded commodities

For a full exploration of the wider impacts associated with ocean-based transport, see the Sect. 1.4 of this report.

Source: Authors

5 Coastal and Marine Ecosystems

This section analyses the potential mitigation impact of conserving and restoring coastal and marine ecosystems, including mangroves, salt marshes, seagrass beds, seaweed aquaculture, and marine fauna.

An overview of the current state of each ecosystem is provided below.

Mangroves, Salt Marshes, and Seagrass Beds

Mangroves, salt marshes, and seagrass beds are highly productive vegetated coastal ecosystems, which are referred to as "blue carbon" ecosystems, analogous to "green carbon" ecosystems on land (Nellemann et al. 2009). They are hotspots for carbon storage, with soil carbon sequestration rates per hectare up to 10 times larger than those of terrestrial ecosystems (Mcleod et al. 2011). Most of their carbon (50–90%) is stored within the soils where saltwater inundation slows decomposition of organic matter, leading to accumulation of extensive soil carbon stocks.

When these ecosystems are degraded and converted, carbon in their biomass and soils, which may have accumulated over hundreds or thousands of years, is oxidised and emitted back to the atmosphere in a matter of decades (Fig. 17.8). Thus, protection of blue carbon ecosystems offers an efficient pathway to avoid CO_2 emissions, particularly for nations with large areas of coastal vegetation and high rates of loss. For example, conversion of mangroves to aquaculture accounts for 10–20% of CO_2 emissions associated with land-use change in Indonesia (Murdiyarso et al. 2015).

Between 20 and 50% of global blue carbon ecosystems have already been converted or degraded, leading some analysts to conclude that restoring wetlands can offer 14% of the mitigation potential needed to hold global temperature to 2 °C above the preindustrial period (Griscom et al. 2017). Rates of mangrove loss have declined from 2.1%/year in the 1980s (Valiela et al. 2001) to 0.11%/year in the past decade (Global Panel on Agriculture and Food Systems for Nutrition 2017; Bunting et al. 2018), thanks to improved understanding, management, and restoration (Lee et al. 2019). However, mangrove areas still emit an estimated 0.007 $GtCO_2e$/year (Atwood et al. 2017).

Rates of loss and degradation of seagrass cover are between 2 and 7%/year, mainly due to pollution of coastal

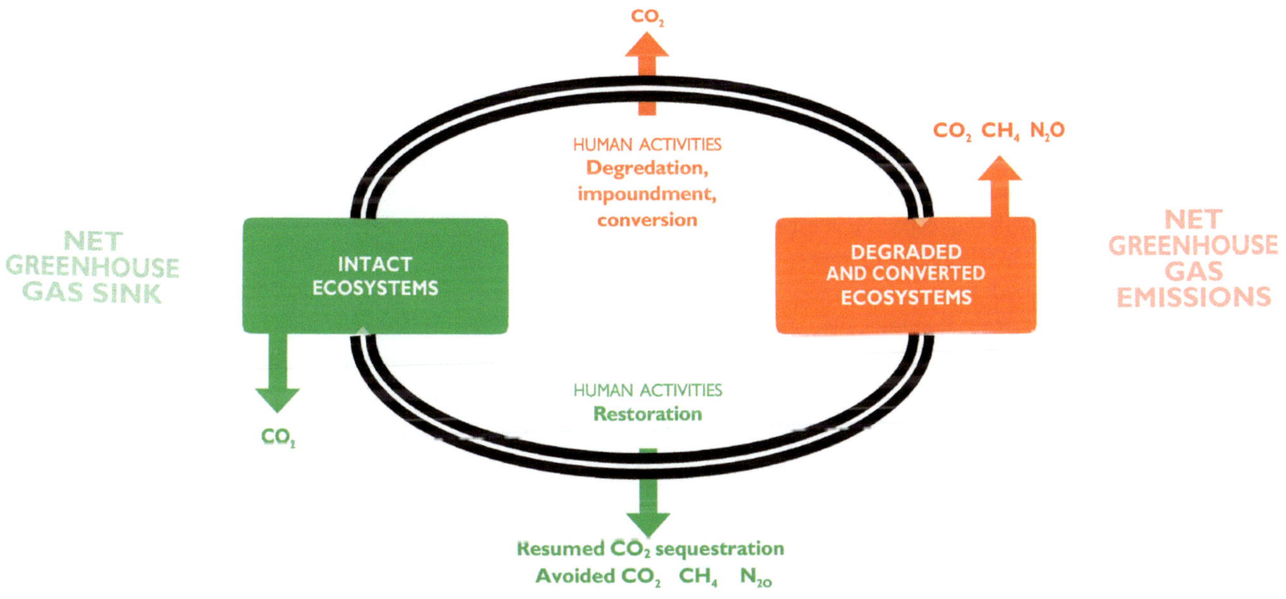

Fig. 17.8 The carbon cycle in coastal and marine ecosystems. (Source: Lovelock et al. (2017))

waters (Duarte et al. 2008; Waycott et al. 2009). Emissions are estimated at 0.05–0.33 $GtCO_2e$/year (Pendleton et al. 2012), although gains in cover have recently been observed in Europe (de los Santos et al. 2019). Global rates of salt marsh loss are uncertain (1–2%/year), but losses are estimated to be responsible for 0.02–0.24 $GtCO_2e$/year (Pendleton et al. 2012).

The area covered by blue carbon ecosystems is equivalent to only 1.5% of terrestrial forest cover, yet their loss and degradation are equivalent to 8.4% of CO_2 emissions from terrestrial deforestation because of their high carbon stocks per hectare (Griscom et al. 2017).

Seaweeds (Macroalgae)

Globally, the most extensive and productive coastal vegetated ecosystems are formed by seaweeds, which are a diverse group including brown algae (e.g., kelps), red algae, and green algae. Their areal extent is estimated—though with large uncertainty—to be 3.5 million km^2 of coastal regions (Krause-Jensen and Duarte 2016). Seaweeds are mainly attached to rocks or occasionally free-floating. They lack root structures that would sequester and trap soil carbon, which means that the climate mitigation value of wild seaweed habitats is largely through the export of organic carbon in plant biomass to sinks located in shelf sediments and in the deep ocean (Krause-Jensen and Duarte 2016).

Thus, the loss of seaweed habitats reduces carbon sequestration but does not result in emissions of CO_2 to the atmosphere from sediments below the habitats, as occurs in mangroves, salt marshes, or seagrass beds.

Globally, seaweed carbon sequestration is estimated to be 0.64 (range 0.22–0.98) $GtCO_2e$/year, representing 11% of annual global net seaweed primary production (Krause-Jensen and Duarte 2016). Recent studies also underline the large carbon export fluxes of seaweeds (Filbee-Dexter et al. 2018; Queirós et al. 2019; Ortega et al. 2019).

While there is no overall assessment of the global rate of change of seaweed habitats and the net area lost, it is estimated that kelps (brown canopy-forming seaweeds) have experienced a global average annual loss rate of approximately 0.018%/year over the past 50 years, with large geographic variability (Krumhansl et al. 2016; Wernberg et al. 2019).

Marine Fauna

Marine fauna (fish, marine mammals, invertebrates, etc.) influence the carbon cycle of the ocean through a range of processes, including consumption, respiration, and excretion. When marine fauna die, their biomass may sink to the deep ocean. In addition, their movement between habitats promotes mixing within the water column, contributing to increased phytoplankton production.

Marine fauna accumulate carbon in biomass through the food chain—starting with photosynthesizing plants that are consumed by animals, which in their turn are consumed. Although there are large data gaps, a first-order assessment estimates that 7 $GtCO_2e$ has accumulated within marine fauna biomass (Bar-On et al. 2018). However, the net carbon sequestration benefit from marine fauna, once allowance is made for respiration over the lifetime of the animal, respiration and carbon output from the species feeding on feces and carcasses prior to final burial in the seafloor, remains unclear.

Marine fauna activity can stimulate production by plants (Lapointe et al. 2014) and phytoplankton, leading to sequestration of 0.0007 $GtCO_2e$/year (Lavery et al. 2010). Populations of vertebrates are an important component of the carbon cycle in ocean ecosystems (Schmitz et al. 2018), including predators which can regulate grazers (Atwood et al. 2015) and should be given consideration when developing policies to secure nature-based carbon functions. However, there is currently insufficient data to estimate the global mitigation potential of protecting or restoring populations of fish and marine mammals to previous levels. Impacts of increased marine protected areas and fishery management practices on climate mitigation should be a priority research area.

5.1 Mitigation Potential

The mitigation potential of these coastal and marine ecosystems are examined by considering three mitigation options:

- Conserving and protecting blue carbon ecosystems, involving halting the loss and degradation of these ecosystems, thus avoiding direct land-use change emissions and additional emissions from alternative land use, such as agriculture.
- Restoration and expansion of degraded blue carbon ecosystems, involving rehabilitating the soil and associated organisms and thereby restoring their ability to sequester and store carbon.
- Expansion of seaweed (macroalgae) through aquaculture, to increase availability for alternative food, feed and fuel products to replace land-based options.

We estimate the total potential mitigation contribution from coastal and marine ecosystems as between 0.50 and 1.38 $GtCO_2e$/year by 2050. This estimate is similar to that of Gattuso et al. (2018), who estimated a cumulative mitigation of 95 $GtCO_2e$ by 2100 (a mitigation potential of 1.1 $GtCO_2e$/year by 2050). Due to lack of data, the estimated total mitigation contribution from marine and coastal ecosystems does not include the potentially significant mitigation effects associated with the conservation and restoration of wild seaweed or marine fauna. The greatest uncertainties in estimates concern ecosystem area and rates of change for seagrass and salt marshes. The estimated mitigation potential of conserving and restoring the marine ecosystems for which data are adequate (mangroves, seagrass beds, and salt marshes) along with the mitigation potential that could be achieved through avoided emissions by using seaweed as a food, feed or fuel replacement is summarised in Table 17.9.

Mangroves, Saltmarshes, and Seagrass Beds

Figure 17.9 shows the estimated mitigation potential of coastal and marine ecosystems via the two main pathways: (1) Protection and conservation of ecosystems avoids emissions of carbon that is currently stored in soils and vegetation, and (2) Restoration of ecosystems sequesters and stores carbon as vegetation grows.

Figure 17.10 compares the mitigation potential of land-based ecosystems to blue ecosystems. Although the mitigation potential of restoring green ecosystems, notably forests, is greater in total, the mitigation potential of blue ecosystems per unit area is very high.

Achieving high levels of mitigation through conservation and restoration is dependent on increased investment in protection, restoration, and enabling the expansion of ecosystem cover where sea level rise provides new opportunities. However, ambitious conservation and restoration targets must be considered within local socioeconomic contexts to prevent perverse outcomes (Herr et al. 2017; Lee et al. 2019; Lovelock et al. 2017).

Efforts to restore blue carbon ecosystems are growing in number, area, and success (Unsworth et al. 2018; Lee et al. 2019; Kuwae and Hori 2019), but are still relatively small scale in most instances. (An exception is the 589 km^2 of salt marsh restoration in the United States between 2006 and 2015 [Gittman et al. 2019]). Low-end estimates of mitigation likely to be achieved through restoration by 2050 are 0.2 GtCO$_2$e/year, reflecting limited restoration activities and success.

Estimates of CO_2 emissions associated with avoided anthropogenic degradation of mangrove, salt marsh, and seagrass ecosystems are sensitive to uncertainties in global cover and rates of loss, which is particularly the case for seagrass and wild seaweeds. Estimates of salt marsh area and losses of salt marsh area are also uncertain (Mcowen et al. 2017). Losses of mangrove ecosystems have slowed in the last decades, and thus emissions associated with their losses have also declined compared to those estimated by Pendleton et al. (2012).

Expansion of Seaweed Through Aquaculture

The protecting and restoration of wild seaweed habitats also holds potential for GHG emissions mitigation, but knowledge gaps are currently too large to estimate the potential contribution because the extent of lost macroalgal habitats that could be restored is unknown. Moreover, methods and success rates of restoration and protection measures (including sustainable harvest methods) need be explored and reviewed.

Table 17.9 Summary of mitigation potential from blue carbon ecosystems, 2030 and 2050

Ocean-based climate action area	Mitigation option	Description	Mitigation potential, 2030 (GtCO$_2$e/year)	Mitigation potential, 2050 (GtCO$_2$e/year)
Coastal and marine ecosystems	Conservation: potential mitigation from halting loss and degradation of ecosystems (avoided emissions)	Mangroves	0.02–0.04	0.02–0.04
		Salt marsh/tidal marsh	0.04–0.07	0.04–0.07
		Seagrasses	0.19–0.65	0.19–0.65
		Seaweeds	Knowledge gaps currently too large (see text)	Knowledge gaps currently too large (see text)
	Restoration: potential mitigation from restoring and rehabilitating ecosystems and organisms	Mangroves	0.05–0.08	0.16–0.25
		Salt marsh/tidal marsh	0.004–0.01	0.01–0.03
		Seagrasses	0.01–0.02	0.03–0.05
		Seaweeds	Knowledge gaps currently too large (see text)	Knowledge gaps currently too large (see text)
	Increased seaweed production via aquaculture		0.01–0.02	0.05–0.29
	End overexploitation of the ocean to support recovery of biodiversity and increase biomass		Knowledge gaps currently too large (see text)	Knowledge gaps currently too large (see text)
Total			0.32–0.89	0.50–1.38

Source: Authors

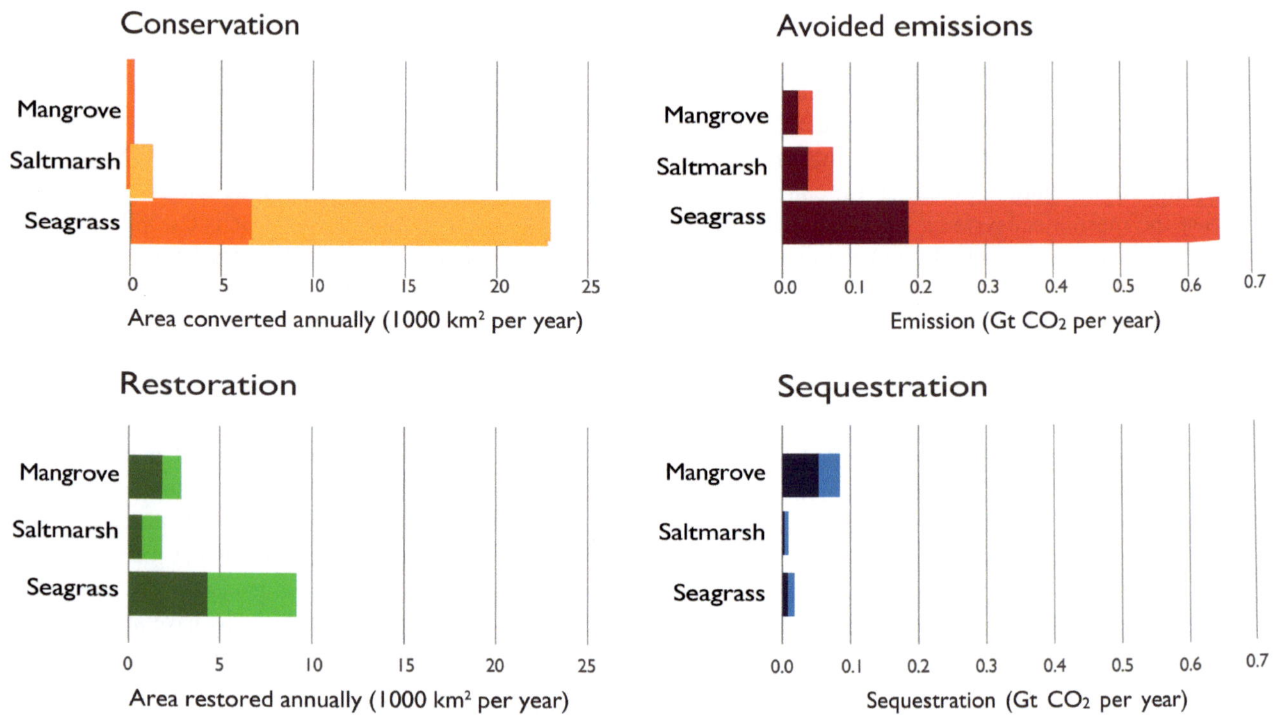

Fig. 17.9 Comparison of conservation and restoration pathways for coastal and marine ecosystems. (*Sources: For area change are in* Table 17.10, Global Panel on Agriculture and Food Systems for Nutrition (2017), Bunting et al. (2018) *(mangroves),* Mcowen et al. (2017) *(salt marsh cover),* Bridgham et al. (2006) *(salt marsh loss),* Unsworth et al. (2018) *(seagrass cover),* Duarte et al. (2008), Waycott et al. (2009) *(seagrass loss),* Krause-Jensen et al. (2016) *(seaweed cover), for emission and removals* IPCC (2013) *Wetland Supplement; and calculations of sequestration from the authors)*

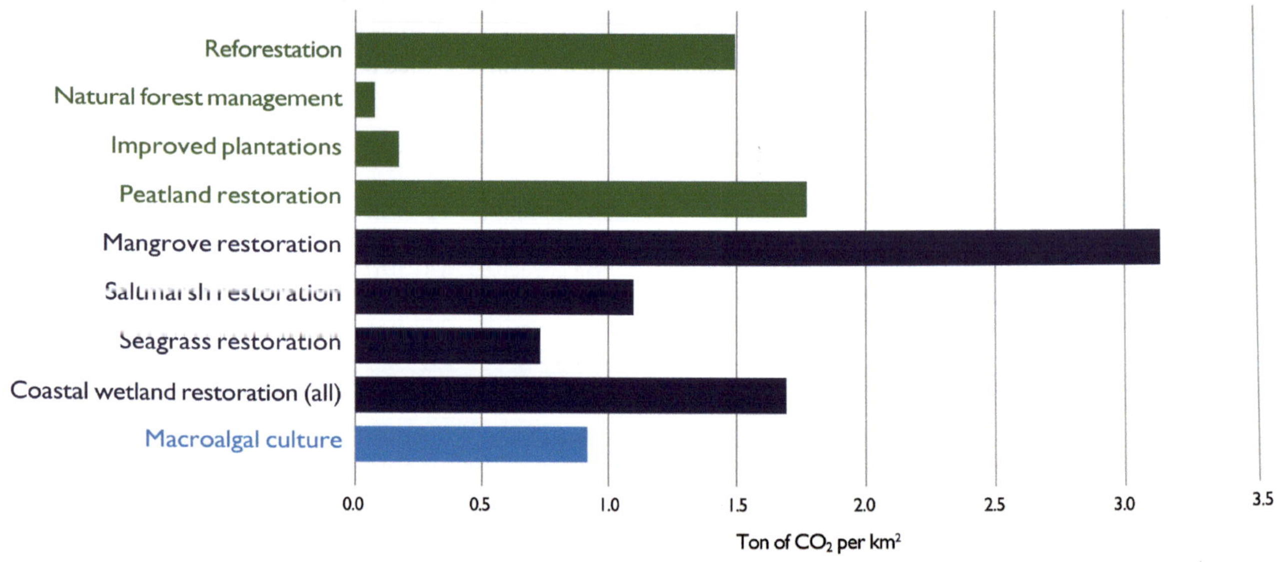

Fig. 17.10 Mitigation potential per unit area of restoring land-based and marine ecosystems. (*Sources: Blue bars represents data from* Griscom et al. (2017), *macroalgal culture: yield data from* World Bank (2016), *biomass-carbon-conversions from* Duarte et al. (2017))

Projections of mitigation from seaweed farming could reach 0.05–0.29 $GtCO_2e$/year by 2050. However, there are uncertainties in rates of expansion of the industry and the proportion of production that would be sequestered.

Scaling up seaweed production via aquaculture offers different potential mitigation pathways:

- Seaweed products might replace products with a higher CO_2 footprint, thereby avoiding emissions (rather than directly contributing to sequestration) in fields such as food, feed, fertilisers, nutraceuticals, biofuels, and bioplastics (World Bank 2016; Lehahn et al. 2016; Duarte et al. 2017). The extent of this mitigation pathway is currently not known.
- Addition of seaweeds to animal feeds might lead to reduced enteric methane emission from ruminants, a potential technology that is currently being explored and may substantially increase the mitigation potential of seaweeds (Machado et al. 2016). In vitro experiments have shown that the red alga, Asparagopsis taxiformis, can reduce methane emissions from ruminants by up to 99% when constituting 2% of the feed; and several other species, including common ones, show a potential methane reduction of 33–50% (Machado et al. 2016). However, this alga is not yet farmed, and many steps are required before large-scale mitigation can be achieved.
- Farmed seaweeds, similar to wild seaweeds, contribute to carbon sequestration through export of dissolved and particulate carbon to oceanic carbon sinks during the production phase (Zhang et al. 2012; Duarte et al. 2017).

This mitigation potential is presented in Table 17.9.

5.2 Methodology

Mangroves, Saltmarshes, and Seagrass Beds

Avoided emissions associated with halting ecosystem conversion were estimated from ecosystem aerial cover (km^2), mean carbon stocks in soils, and biomass per area from default emission factors (IPCC 2013), and estimated rates of loss (Table 17.10). The range of CO_2 sequestration potential per unit area for each ecosystem was calculated using default emission/removal factors from IPCC (2013). Our estimates are conservative because we do not include CO_2 emissions from previously degraded and converted ecosystems where soil carbon continues to emit CO_2 over time; these emissions may reach 0.7 $GtCO_2e$/year (Pendleton et al. 2012).

The range in potential mitigation that could be achieved through restoration of mangrove, salt marsh, and seagrass ecosystems varied with the level of effort and investment. We considered two scenarios: a moderate restoration effort recovering about 40% of historical ecosystem cover by 2050, which is consistent with Global Mangrove Alliance goals; and a much more aggressive scenario of complete restoration of pre-1980s cover. Restored areas would amount to 225,000 km^2 of mangroves (Valiela et al. 2001), 600,000 km^2 of seagrass (Mcleod et al. 2011), and doubling of the current area of salt marsh to 110,000 km^2 (Gittman et al. 2019). Mitigation benefits under these scenarios are likely conservative because avoided methane (CH_4) emissions from alternative land uses such as aquaculture and rice production could be substantial. Thirty percent of mangrove ecosystems in Southeast Asia have been converted to aquaculture and 22% to rice cultivation (Richards and Friess 2016). Both land uses can produce high nitrous oxide (N_2O) and CH_4 emissions (IPCC 2006, 2013, 2019).

Table 17.10 Global extent and loss rates of blue carbon ecosystems

Ecosystem	Areal cover (km^2)	Recent rates of loss (%/year)
Mangroves	138,000	0.11
Salt marshes	55,000	1–2
Seagrasses	325,000	2–7
Seaweeds	3,540,000	Not known

Sources: Global Panel on Agriculture and Food Systems for Nutrition (2017), Bunting et al. (2018) *(mangroves)*, Mcowen et al. (2017) *(salt marsh cover)*, Bridgham et al. (2006) *(salt marsh loss)*, Unsworth et al. (2018) *(seagrass cover)*, Duarte et al. (2008), Waycott et al. (2009) *(seagrass loss)*, Krause-Jensen et al. (2016) *(seaweed cover)*

Seaweeds (Macroalgae)

To estimate the mitigation potential of seaweed farming by 2030 and 2050, two scenarios were considered (Table 17.9). The assumptions underlying the two scenarios are given below:

1. Seaweed farming develops at 8.3%/year (the current rate, calculated on the basis of the increase in the farmed and harvested production of green, red, and brown macroalgae between 2000 and 2017) (FAO 2018), 100% of production is assumed sequestered, and farming and processing are assumed CO_2-neutral. Conversion factors from wet weight to carbon are from Duarte et al. (2017).

   Average annual yield is 1000 tonnes dry weight/km^2 (current best practices) (World Bank 2016). Estimated production by 2030 (9.4 Mt dry weight/year, equivalent to 2.3 megatonnes of carbon/year [MtC/year)] and 2050 (49.3 Mt dry weight/year, equivalent to 12.2 MtC/year) would require an area of 9383 and 49,348 km^2, respectively. This represents 0.02 and 0.1%, respectively, of the global area suitable for macroalgal aquaculture (estimate based on suitable temperature and nutrient conditions, Froehlich et al. 2019).
2. Seaweed farming develops at 14%/year from 2013 onward (rate assumed in a scenario developed by the

> World Bank (2016)), 100% of production is assumed sequestered, and farming and processing are assumed to be CO_2-neutral. Conversion factors from wet weight to carbon are from Duarte et al. (2017). Average annual yield is 1000 tonnes dry weight/km^2 (current best practices) (World Bank 2016), leading to production of 324 Mt dry weight/year, equivalent to carbon assimilation of 80 MtC/year by 2050.

We adopted the scenario of a 14% annual increase in production to provide an upper limit of the sequestration potential by 2030 and 2050, and we further assume that farming could proceed at this rate of increase without meeting constraints before 2050. An even higher production estimate of 10 billion tonnes dry weight/year was recently proposed (Lehahn et al. 2016), indicating that our estimated upper limit of seaweed production is not unrealistic.

The assumption that 100% of the seaweed harvest is sequestered is highly unlikely, as seaweeds are farmed for many other, and more economically profitable, purposes than carbon sequestration. Also, energy is required in the production process. However, carbon sequestration through export of the "nonseen production" during farming will contribute to the sequestration potential (Duarte et al. 2017). Recent estimates suggest that this export may constitute 60% of what is eventually harvested (Zhang et al. 2012). Assuming that 25% of the seaweed export is sequestered (Krause-Jensen et al. 2016), the projected seaweed aquaculture would have an associated sequestration of nonseen production of 0.0013–0.0027 $GtCO_2e$/year by 2030 and 0.0067–0.044 $GtCO_2e$/year by 2050.

To maximizes the mitigation benefit of seaweed farming, it is essential that farms do not harm wild blue carbon ecosystems (mangroves, seagrasses, saltmarshes, and seaweeds).

Conversely, sustainable seaweed farming may have the benefit of reducing the harvest of wild seaweeds.

Risks, Underlying Assumptions

Climate change is likely to have variable impacts on coastal marine ecosystems and their CO_2 mitigation potential (Fig. 17.11). Marine heat waves may adversely affect the mitigation contribution from seagrass beds and seaweeds (Arias-Ortiz et al. 2018; Wernberg et al. 2019). Warming may result in ecosystem losses at their equatorial distributional range limit (Wilson et al. 2019) and increases at the polar distribution range (Krause-Jensen and Duarte 2016; Marbà et al. 2018).

The area of mangroves and salt marshes may also be adversely affected by sea level rise in some regions (Lovelock et al. 2015) but could expand in others (Schuerch et al. 2018), increasing their mitigation benefits (Rog et al. 2017). Sea level rise will affect habitat areas for all coastal vegetated ecosystems, and thus their mitigation potential (Lovelock et al. 2015; Saunders et al. 2013; Schuerch et al. 2018). The impact of sea level rise on these ecosystems will be strongly influenced by human activity (e.g., sediment supply, land-use changes, population, and seawall defenses);

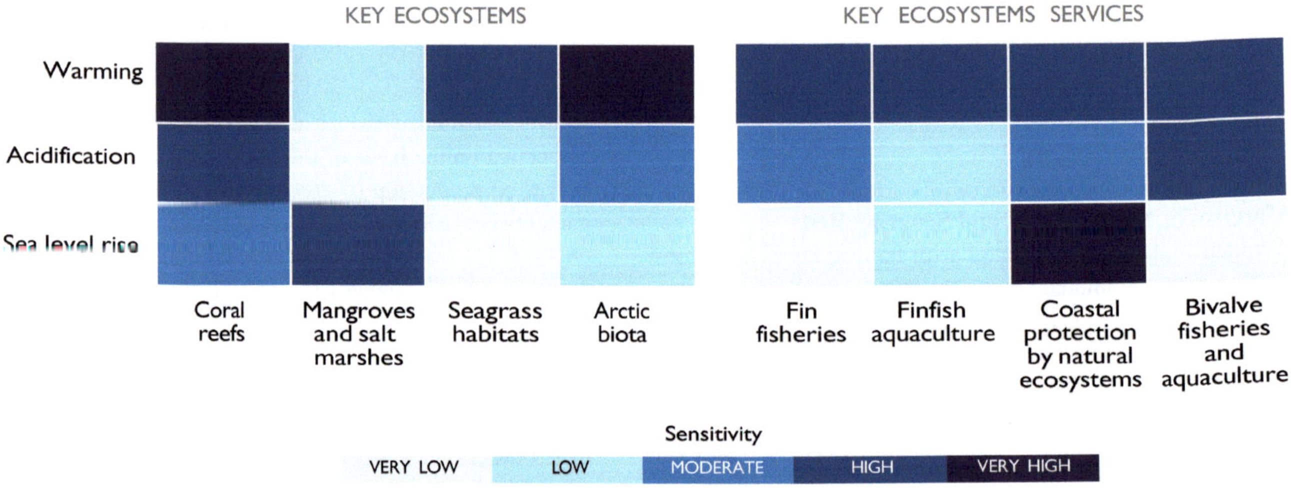

Fig. 17.11 The effects of climate change on coastal marine ecosystems will vary. (*Sources:* Gattuso et al. (2018))

the effects of climate change on adjacent ecosystems such as coral reefs (Saunders et al. 2013), mudflats or barrier islands; and GHG emissions from freshwater wetlands (Luo et al. 2019).

Extreme events could also reduce the effectiveness of restoration. While small-scale seaweed cultivation is considered low risk, a large-scale expansion of the industry requires greater understanding of impacts and the balance of environmental risks and benefits that seaweed cultivation projects can offer (Campbell et al. 2019).

In addition to climate change, marine and coastal ecosystems are also vulnerable to failure due to socio-economic factors, including inadequate and inappropriate incentives (Herr et al. 2017; Lee et al. 2019). Social safeguards, similar to those developed for forests, should be developed.

5.3 Policy Interventions Needed to Achieve Mitigation Potential

The following policy interventions are recommended to support the realisation of the mitigation potential outlined in this chapter:

Short Term:

- Enhance protection measures for mangroves, seagrass beds, salt marshes, and seaweed beds to prevent further losses due to human activities.
- Measures could include increasing the size and effectiveness of marine protected areas, but should also address underlying causes of loss, such as overexploitation, pollution, hydrological changes, and climate change impacts.
- Increase incentives for restoration of blue carbon ecosystems by paying for ecosystem service schemes, using mechanisms such as carbon and nutrient trading credits (Herr et al. 2017).
- Develop incentives for sustainable seaweed farming (Froehlich et al. 2019).
- Promote adoption of improved accounting for mangroves and salt marshes within national GHG inventories (IPCC 2013).
- Include blue carbon solutions in nationally determined contributions (NDCs) and other relevant climate policies for mitigation and adaptation (Herr and Landis 2016).
- Recognise the wider ecosystem services of these habitats beyond carbon sequestration and quantify their mitigation of coastal eutrophication and benefits for biodiversity, fisheries, coastal protection, fisheries and aquaculture, and their adaptation, to develop appropriate financial and regulatory incentive tools.
- Link conservation and restoration of mangroves, salt marshes, seagrass beds, and seaweeds to achieving the UN Sustainable Development Goals.
- Develop and implement social safeguards. Although restoration of blue carbon ecosystems provides important opportunities for mitigation, inadequate policies for restoration of mangroves for carbon could give rise to perverse outcomes (Friess et al. 2019a). Safeguards are required to ensure that, for example, restoration projects do not prevent local communities from accessing marine resources (McDermott et al. 2012).

Medium Term:

- Improve IPCC guidance for seagrass management and develop IPCC GHG inventory guidance for seaweed ecosystems.
- Improve methods for monitoring mitigation benefits to enable standardised accounting within national GHG inventories, and more comparable biennial transparency reports (BTRs).
- Increase the development of sustainable seaweed aquaculture globally.
- Increase investment in conservation and restoration of blue carbon ecosystems through innovative finance (insurance, debt swaps, taxes, and credits) and public-private partnerships.

5.4 Technology Needs

Restoration of mangroves and salt marshes is technically feasible at large scale (Lewis et al. 2011; Esteves and Williams 2017; Lee et al. 2019; Gittman et al. 2019).

Many constraints are imposed, however, by social and economic factors, including unclear land tenure, poverty, overexploitation, and lack of investment (Lee et al. 2019). Seagrass restoration at large scale faces significant technical impediments, for example, successful handling of and propagation from seagrass seeds (Statton et al. 2013). Successful seagrass restoration requires management of offsite factors, such as improvement of water quality (Unsworth et al. 2018).

Costs of restoration vary among ecosystems and among developed and developing economies (Bayraktarov et al. 2016). A review of costs per area of habitat revealed that marginal costs do not decline with increasing area of restoration projects, indicating that economies of scale have not yet been achieved. There are opportunities for improving methodologies, which could result in an increase in the scalability and effectiveness of restoration (Bayraktarov et al. 2016).

Seaweed farming is in operation in several countries, with more than 99% of production found in seven Asian countries (China, Indonesia, the Philippines, the Republic of Korea, the Democratic People's Republic of Korea, Japan, and Malaysia). Farms in the region vary from large industrial enterprises to smaller family-run businesses (World Bank

2016; Chopin 2017; FAO 2018). Currently, seaweed farming is not optimised for carbon sequestration or global large-scale production, as most of the production is for human consumption (FAO 2018).

Increasing the role of seaweed culture in mitigation will require a worldwide and sustainable expansion of the industry, of the sort that is underway in Canada (Chopin 2015) and Norway (Skjermo et al. 2014). Realising greater mitigation potential will also require the development of novel products, such as bioactive compounds and biomaterials.

Emerging biorefining techniques, with sequential extraction of products, are likely to markedly increase cost-effectiveness and scale of production (Chopin 2018a; Sadhukhan et al. 2019). The possibility also exists to develop more offshore, integrated multitrophic aquaculture, including seaweed aquaculture, in the open ocean (Buck et al. 2018).

5.5 Priority Areas for Further Research

Significant gaps exist in the knowledge base and practical application of ocean-based mitigation options.

Increasing efforts to produce national-level maps of blue carbon ecosystems would help monitor the success of restoration efforts and enable more accurate quantification of carbon sequestration in ecosystems under the full range of environmental conditions. This in turn would improve estimates of the likely impacts of restoration on mitigation potential. Building research capacity for an initial global-scale map of seaweed ecosystems would also contribute to improving available data, including developing IPCC-approved methodological guidance similar to that available for mangroves and salt marshes.

Research that explores the biophysical, social, and economic impediments to restoration, as well as enabling factors (e.g., value chain assessments), is needed to develop ecosystem restoration priorities, enhance incentives for restoration, and promote more successful restoration outcomes (Lee et al. 2019).

Relevant information would include assessments of the wider cobenefits of increasing seaweed area and carbon sequestration (Box 17.5), such as climate change adaptation, enhanced biodiversity, and improved ecosystem services (Krause-Jensen et al. 2018).

Deeper knowledge of the impacts of climate change is needed to more fully understand the risks to mitigation posed by climate change. The carbon sequestration and avoided emission benefits of ecosystem restoration are currently restricted to just a few sites, and more evidence is needed.

There is insufficient documentation on the global extent, production, carbon fluxes, and burial rates of the various groups of seaweeds. There is also insufficient information on how seaweeds respond—in terms of area and performance—to management efforts and methods that aim to restore and protect them, especially in the context of natural variability, human-caused stressors from local to global level, and climate change impacts. Methods to fingerprint seaweed carbon and other blue carbon sources beyond the habitat are also critical to link management action to carbon sequestration beyond the habitat, yet these methods remain poorly developed. Jurisdictional issues would also be a challenge to implementation.

The research agenda also must address the global potential for carbon sequestration through sustainable seaweed farming and processing and/or biorefining of seaweed products, circular management of nutrients, offshore production platforms, and the ecological impacts (positive and negative) of large-scale seaweed farming. Restoration of seaweed beds is developing, but, to the best of our knowledge, no reviews of methods and success rates are available.

Box 17.6. Wider Impacts Associated with Utilising Coastal and Marine Ecosystems for Carbon Sequestration and Storage

Potential Cobenefits:

- Increased climate change adaptation benefits from healthier coastal and marine ecosystems. Vegetated habitats protect coastal infrastructure and buffer acidification.
- Higher biodiversity benefits, with healthy marine and coastal ecosystems supporting a range of terrestrial and marine species.
- Provision of nutritious food through support of fisheries, plus other benefits, including traditional medicine by mangroves, salt marsh, sea grasses, and seaweeds for local communities.
- Higher ecosystem services (increase in fisheries productivity, coastal protection, and coastal tourism) from protected and restored mangroves, salt marsh, and sea grasses. Fair distribution of payments to local communities from restoration work could help meet decent work and economic growth targets.
- Integration of social and gender dimension into coastal and marine restoration work will increase its effectiveness.
- Expanding seaweed production contributes to meeting global food security targets, and offers a pathway to develop alternative food, feed, and fuels that do not require arable land. The farming also offers climate change adaptation benefits. The rapidly growing business has generated jobs, predominantly in developing and emerging economies.

Potential Trade-offs:

- Pushing forward blue carbon projects internationally, without considering social safeguards and demands of local small-scale fishers and other stakeholders who are heavily dependent on the resource for economic sustainability, can have unintentional negative consequences on societal well-being.
- Small-scale cultivation of seaweeds is considered low-risk. However, expansion of the industry will necessitate a more complete understanding of the scale-dependent changes and risks (facilitation of disease, alteration of population genetics, and wider alterations to the physiochemical environment).
- Mitigation options to recover ocean biomass can negatively impact poverty reduction and employment targets, and limit progress on food security targets in the short term.

For a full exploration of the wider impacts associated with coastal and marine ecosystems, see the Sect. 1.4.
Source: Authors

6 Fisheries, Aquaculture, and Shifting Diets

This section analyses the potential mitigation impact of reducing the carbon footprint of ocean-derived food production (wild capture fisheries and aquaculture) and the potential reductions from shifting diets to include more low-carbon sources of ocean-based protein.

There are two principal ways in which ocean-based foods can contribute significantly to climate change mitigation. One seeks to reduce the carbon footprint of ocean-derived food production. For example, changing fuel sources in vessels and technological advances in production techniques can alter the emissions associated with seafood from both wild-caught fisheries and ocean-based aquaculture. The other seeks to identify emission reductions from potentially shifting more GHG-intensive diets to those that include more GHG-friendly seafood options, if those seafood options can be provided on a sustainable basis.

Different types of food, produced in different places by different means, can vary by more than an order of magnitude in the total GHGs they emit across their full life cycle. The composition of global diets, therefore, foods have the potential to play a significant role in these efforts if their production is sustainable. Food from the sea, produced using best practices, can (with some notable exceptions) have some of the lowest GHG emissions per unit of protein produced of all protein sources (González 2011; FAO 2013; Nijdam et al. 2012; Parker et al. 2018; Hallstrom et al. 2019). Increasing the fraction of ocean-based food in the global diet, and reducing the share of animal-based foods, would contribute significantly to climate change mitigation.

There are also opportunities for efficiency gains by reducing waste in the seafood supply chain (Springmann et al. 2018).

More than one-third (by weight) of all food that is produced is currently lost in the supply chain (Gustavsson et al. 2011), and even higher fractions may be lost in some seafood supply chains (Love et al. 2015).

The largest potential mitigation gains, however, are likely to be found in shifting diets away from terrestrial animal-based protein, particularly beef cows and other ruminants, towards plant- and ocean-based options that have been identified as having a lower carbon cost. The world's population continues to grow, and so does demand for food, although projections of food demand are highly uncertain. Rising affluence and the spread of "Western diets" is encouraging the consumption of more animal protein. These trends will continue to drive growth in GHG emissions unless dramatic changes occur in the scale and composition of foods that are selected for human consumption (Springmann et al. 2018).

Estimates of global food-related GHG emissions early in this century range from 4.6 to 13.7 billion tonnes of CO_2e (Tubiello et al. 2013; Smith et al. 2014; Poore and Nemecek 2018). By 2050, these emissions are projected to grow between 80 and 92% (summarised in Springmann et al. 2018). In addition to rising GHG emissions, the environmental consequences of producing ever-increasing quantities of food with the current dietary mix of species are projected to be severe in terms of water scarcity, soil degradation, and habitat loss, among others (Tilman and Clark 2014; Springmann et al. 2018). Without significant reductions in agricultural emissions, it will almost certainly be impossible to keep planetary warming constrained to 2 °C or less above preindustrial levels (Springmann et al. 2018).

Fortunately, there are several pathways that could collectively drive large emission reductions, and ocean has a major effect on global emissions (Poore and Nemecek 2018; Searchinger et al. 2019).

6.1 Mitigation Potential

We estimate that, with strategic policy and investment actions to change how seafood is provided and increase its share in the collective human diet, seafood could contribute potential mitigation of between 0.34 and 0.94 $GtCO_2e$ by 2030, and between 0.48 and 1.24 $GtCO_2e$ by 2050, relative to business-as-usual projections. Our estimates are explained more fully in the Sect. 6.2.

6.1.1 Reducing Emissions from Wild Capture Fisheries

Current fuel use and GHG emissions from global wild-capture fisheries up to 2011 were modelled by Parker et al. (2018). They estimated global fishing emissions in 2011 at 179 MtCOe, or 2.2 kg COe per live weight kilogram of landed fish and shellfish. Global fishing thus accounts for roughly 4% of global food system production emissions. Modelling was based on the aggregation and weighting of extant fuel-use data, specific to target species, gear, and/or fishing country, with corrections to account for upstream emissions from fuel production and transport, as well as non-fuel emissions from vessel construction, gear manufacture, refrigerant use, and other factors.

Reductions in emissions from wild-capture fisheries can be achieved in ways ranging from technological advances in engine efficiency or hull design to changes in skipper behaviour, such as speed reductions and willingness to fish in poor conditions. However, while technological changes, such as gear design and engine retrofits, have been demonstrated to influence fuel-use rates in individual vessels (e.g., Parente et al. 2008; Khaled et al. 2013; Latorre 2001), the effects of such changes at the fleet level are unclear and can be overshadowed by variation in stock abundance or structural changes to the fishery (Ziegler and Hornborg 2014; Farmery et al. 2014; Pascoe et al. 2012). A more consistently reliable driver of emissions within a fishery is catch per unit effort, reflecting both effort (e.g., days fished) and available biomass (Parker et al. 2017; Ziegler et al. 2016).

Our estimate of mitigation potential in this case is consequently focused on the potential for future changes in effort and landings, while acknowledging that technological and behavioural factors will play a role, either positively or negatively. Arnason et al. (2017) developed a future scenario to optimise the economic performance of global fisheries. Compared to wild capture landings in 2012, they estimated that, in theory, wild fish catch could increase by 13% by 2030, with significantly less fishing effort expended. Applying their effort and landings projections to Parker et al. (2017) emissions model, this increase in efficiency could reduce GHG emissions by a total of 81 $MtCO_2e$, or to roughly half of current fishing emissions (Table 17.10).

6.1.2 Reducing Emissions from Aquaculture

Global analyses of the complete GHG footprint of aquaculture are lacking, and many systems that make up a large portion of global production have not been sufficiently assessed. However, some clear patterns have emerged from the literature to date. In particular, the largest source of emissions in finfish and crustacean aquaculture is commonly the feed provided for their growth (Henriksson et al. 2012; Parker 2018; Pelletier et al. 2009; Pelletier and Tyedmers 2010; Robb et al. 2017). Minimizing the carbon profile of aquaculture feeds therefore can represent a substantial source of future emission reductions, or at least avoidance of emissions increases.

The composition of fish feeds varies greatly, especially across herbivorous, omnivorous, and carnivorous species (see the feed ingredients database: http://afid.seafdec.org.ph). Two of the key components of many feeds for omnivorous and carnivorous species have historically been fish meal and fish oils, which are products derived primarily from forage fish fisheries and increasingly from trimmings of other species during processing. These components promote vibrant fish growth and are also sources of key nutrients shown to have significant benefits for human health (Kris-Etherton et al. 2002).

There are active debates concerning the logic behind feeding wild fish to farmed fish rather than using the wild fish for direct human consumption (Naylor et al. 2000, 2009; Tacon and Metian 2008). In addition, the global supply of fish meal is now at a historical high and may be near biological limits (Costello et al. 2012). As a result, the continued growth of fed aquaculture has driven dramatic increases in the price of fish meal and incentivised reductions in the fish meal and fish oil content of many aquaculture feeds (McGrath et al. 2015; Rana et al. 2009).

To date, the primary replacements for fish meal have been soy and other agricultural crops, which often have high GHG emissions (Pelletier and Tyedmers 2007; McGrath et al. 2015). More recent substitutes for fish meals and oils include a range of livestock-derived inputs (e.g., blood, meat, and feather meal), which typically have even higher levels of GHG emissions (Parker 2018; Pelletier et al. 2009). Many of these substitutes, and particularly those derived from some crops, can have trade-offs in terms of fish and crustacean growth and health, especially for farmed predators. Consequently, efforts are now being made to identify new, highly nutritious, and ideally, low-impact feed sources.

Some of the most promising options are a variety of protein concentrates derived from a range of single cell organisms including yeast, bacteria, or microalgae (Sarker et al. 2018). Although the motivation for this innovation was to provide better quality feeds, one of the fortunate benefits is that some of these alternative feed inputs have significantly lower GHG emission intensities than soy-based protein (Couture et al. 2018). Other emerging feed alternatives, however, can have substantially higher emissions with few benefits relative to soy protein (Couture et al. 2018).

Because of the limited nature of fish meal and the reduction in fisheries that provide it, future aquafeeds will need to use alternative sources for meal and oil. Given current projections for aquaculture growth (SOFIA 2018), we estimate that targeting new low-emission alternatives as replacement feed components, rather than soy-based protein or other high-GHG sources, could avoid annual emissions from the industry by 16 $MtCO_2e$ by 2030 and 43 $MtCO_2e$ by 2050. If the pace of aquaculture growth increases further because of projected growth in demand (Waite et al. 2014), these emissions savings could increase by more than one-third. Since many options are emerging to replace the fish meal fraction in feeds, realising potential emissions cobenefits will require incentives. For example, a well-structured price on carbon, detailed full life-cycle assessments of emissions from new feeds, targeted investments, information, and certification campaigns would help prioritise low-emission feed options. If shifting demand (see below) drives even faster growth in aquaculture relative to other sources of animal protein, these savings could grow proportionately.

6.1.3 Reducing Emissions by Shifting Diets

Food will play an increasingly large role in future climate change mitigation efforts (Tilman et al. 2001, 2011; FAO 2013; Poore and Nemecek 2018; Springmann et al. 2018; Searchinger et al. 2019). GHG emissions from food systems are high, particularly from livestock production, and demand for animal-based food is projected to increase dramatically by 2050 (Searchinger et al. 2019). Since different foods vary widely in their embedded GHG emissions per unit of protein (Poore and Nemecek 2018), changes in the composition of future diets could greatly affect the emissions consequences of growth in demand (González et al. 2011).

If we look only at food system emissions of methane and nitrous oxides, which will not be affected by advances in low-emission energy sources, the business-as-usual scenario projects that GHG emissions will grow from 5.2 $GtCO_2e$ in 2010 to 9.7 $GtCO_2e$ in 2050 (Springmann et al. 2018). Of that projected growth, over 75% will come from projected growth in animal products.

The primary pathways for reducing these potential impacts are efficiency gains (e.g., reducing food loss and waste, feed conversion ratios, and growth periods for livestock) and dietary shifts in terms of food choices and levels of consumption.

Changing behaviour on a scale necessary to shift diets enough to materially affect projected GHG emissions is an immense challenge. One promising strategy is to incentivise lower consumption levels of particularly impactful foods (i.e., most animal-based products) (Poore and Nemecek 2018; Springmann et al. 2018) through education, but also through market mechanisms that increase the price of GHG-intensive foods. Another strategy targets people's self-interest and stresses the benefits of reduced animal food consumption for human health. There is a strong alignment between dietary changes that would improve human health and those that would benefit the environment (Tilman and Clark 2014).

Sustainable growth in seafood production and consumption, particularly from aquaculture, is at the core of these potential benefits. Such growth would necessitate improvements in ocean and coastal management to ensure that harvests can not only be increased, but also sustained. Springmann et al. (2018) suggest that an aggressive dietary shift at a global scale could reduce annual emissions by 4.7 $GtCO_2e$—more than offsetting projected growth of emissions under the business-as-usual scenario. Pathways to achieve such a scale of behaviour change are not clear. More conservatively, we estimate that two practical scenarios could achieve significant emission reductions—a carbon tax and aggressive health campaigns on diets and human health—leading to emission reductions of 0.24–0.84 $GtCO_2e$ by 2030 and 0.30–1.06 $GtCO_2e$ by 2050 (Table 17.11). Both scenarios would see the ocean playing a significantly larger and beneficial role in global food systems.

This mitigation potential is presented in Table 17.11.

Table 17.11 Summary of 2030 and 2050 mitigation potential by mitigation option

Ocean-based climate action area	Activity	Description	Mitigation potential, 2030 ($GtCO_2e$/year)	Mitigation potential, 2050 ($GtCO_2e$/year)
Fisheries, aquaculture and dietary shifts	Reducing emissions from wild capture fisheries	Emissions from fuel use for inland, coastal, and open ocean fishing (wild capture)	0.081	0.137
	Reducing emissions from aquaculture	Life-cycle emissions from aquaculture (new feeds to replace fish meal and soy-based proteins)	0.016	0.043
	Increasing share of ocean-based proteins in diets	Ocean-based proteins are substantially less carbon intensive than land-based proteins (especially beef and lamb). Therefore, actions that shift diets to lower carbon protein, including ocean-based proteins, reduce emissions	0.24–0.84	0.30–1.06
Total			0.34–0.94	0.48–1.24

Source: Authors

6.2 Methodology

This section describes our approach to estimating emission reductions that could be achieved by improving efficiency and yield in wild capture fisheries, improving performance of aquaculture, and shifting the dietary choices of consumers.

6.2.1 Wild Capture Fisheries

Reducing Emissions by Improving Fish Catch Efficiency

One basis for determining the extent to which effort relative to catch could be reduced worldwide is the modelling done by Arnason et al. (2017) in the *Sunken Billions* report. They estimated that an optimal economic scenario for the entire global fishing fleet would, relative to 2012, is likely to produce 13% more catch, using 56% as much effort (targeting maximum economic yield). While fuel use would not be perfectly correlated with effort in such a scenario, if we assume equal reductions in fuel use and effort, we can estimate the fuel use (and associated emissions) required to catch that future optimal harvest using the Parker et al. (2018) model. Our calculations assume a uniform change in landings and fuel use across all species groups and gear types, remodelled from Parker et al. (2018). This is likely an overly optimistic scenario, given the challenges to fisheries management globally, the uneven and insufficient implementation of effective management techniques, and the as yet unrealised recommendation of Arnason and colleagues to direct global fisheries towards their optimal future. Further, it fails to address technological and behavioural changes that may accompany changes in effort and landings, whether positive or negative.

The result of higher catches for less effort is roughly a halving of emissions intensity from 2.2 kg CO_2e per kg landed to 1.1 kg CO_2e. Total emissions from the global fishing industry would decline from 179 MtCO_2e to 98 MtCO_2e, a reduction of 81 MtCO_2e. These emission reductions could be achieved rapidly if countries adopt management reforms to align fishing effort with values appropriate for achieving maximum sustainable yields. Such a scenario would also eventually provide approximately 10% more fish and shellfish from the ocean than the current scenario (Parker et al. 2018) based on the suggested landings in Arnason et al. (2017), compared to the 2011 landings modelled by Parker et al. (2018). Such gains would occur gradually after the effort reductions, since they depend on the recovery of fish stocks.

Reducing Emissions by Increasing Fishery Yields

We estimate the additional protein provided by assuming an average flesh yield from live weight of 50% and protein content of 20%. This yields an additional 863 million kg of protein annually once stocks are rebuilt. While the degree to which that additional protein would be available to offset alternative animal protein sources would rely on numerous factors, we calculate the optimal case, assuming that all additional protein from fisheries replaces (does not add to) more emissions-intensive land-based protein sources.

We use pork to represent an average land-based protein (Poore and Nemecek 2018), as it has a middle-range emissions profile. If we assume the emissions from producing 100 g of protein from pork are 7.6 kg CO_2e (Poore and Nemecek 2018), compared to 1.1 kg CO_2e for average fish and shellfish, we derive a potential emissions offset of 6.5 kg CO_2e for every 100 g of additional fishery-sourced protein, or a total annual emissions reduction potential of 56.1 MtCO_2e by 2050 (Table 17.12).

The combined emissions reduction potential of global fisheries, assuming optimal effort to catch ratios from Arnason et al. (2017), and 100% substitution of available fish protein for average animal-based protein sources, is 137.1 MtCO_2e. Since these benefits require the inherent delay of population recovery of the fished stocks, we assume these added reductions are achievable by 2050.

6.2.2 Aquaculture

FAO projects that global aquaculture production will grow at an annual rate of 2.1% from 2017 to 2030 (SOFIA 2018), with annual production reaching 110 Mt by 2030. The Food and Agriculture Organization (FAO) does not currently project to 2050, but if we assume a similar annual growth rate of

Table 17.12 Projected emission reductions from improving fishing efficiency under two scenarios

Measure	Unit	2011 baseline	Optimal scenario
Fish landings	Million tonnes	81.1	89.7
Emissions from fishing	Million tonnes CO_2	179.0	98.0
Emissions intensity	CO_2e/kg fish landed	2.2	1.1
Additional harvest	Million tonnes	Not available	8.6
Additional protein	Million kg	Not available	863.0
CO_2e offset per 100 g protein	kg CO_2e	Not available	6.5
CO_2e reduction from substituting seafood for land-based protein	Million tonnes CO_2	Not available	56.1
CO_2e reduction from reduced fishing effort per unit catch	Million tonnes CO_2	Not available	81.0
Total CO_2e reduction from wild fisheries	Million tonnes CO_2	Not available	137.1

Sources: Authors (2011) baseline scenario from Parker et al. (2018). *Optimal scenario remodelled from effort and catch estimates in* Arnason et al. (2017)

approximately 2.0% from 2031 to 2050, total aquaculture production (excluding plants) would be approximately 163 Mt live weight in 2050—essentially double the 2017 production or an additional 80 Mt live weight.

The projected growth in aquaculture production could affect GHG emissions in two ways. Growth could influence the mix of animal proteins that is consumed. We address this issue below in the section on shifting diets. Secondly, constraints on the availability, and rising cost, of fish meal from wild fisheries, will mean that the fraction of fish meal in farmed fish diets will continue to decline. Fish meal is likely to be replaced primarily by agricultural products like soy and/or livestock by-products unless new alternative feeds are adopted. Fortunately, we have seen great innovation in the development of new protein-rich feed inputs.

Although the GHG emissions expected from many of these alternatives have not been thoroughly analysed, feeds derived from single-celled yeast and microalgae appear to have dramatically lower GHG emissions per unit of protein (Couture et al. 2018, unpublished) than alternatives like soy. If we assume that aquaculture production in 2050 is double what it is today and has a similar product mix (i.e., fed species versus shellfish, etc.), the use of new low-emission alternative feeds for the feed fraction that is currently fish meal would reduce projected feed-based emissions by more than 43 $MtCO_2e$ in 2050. At the extreme, if these alternative feeds provided all the required additional feeds needed to support projected aquaculture growth, emissions would be reduced by nearly 259 $MtCO_2e$ in 2050, relative to the emissions from a predominantly soy-based or emission-equivalent feed.

6.2.3 Dietary Shifts to Ocean Proteins

Conservative estimates focused only on methane and nitrous oxide emissions suggest that aggressive dietary changes could reduce global annual GHG emissions in 2050 by nearly 5 $GtCO_2e$, while simultaneously improving human health (Springmann et al. 2018; Willett et al. 2019). The challenge is to bring about significant behaviour change on the part of billions of people. To estimate what fraction of the potential gains from shifting diets might realistically be achievable, we examine the potential effects of two policy approaches—a carbon tax that applies to food systems and media campaigns focused on improving human health through diet.

Carbon taxes have been proposed as a market-based tool to reduce GHG emissions from livestock production systems.

In theory, a well-designed tax that encompasses more than just carbon emissions would make GHG-intensive food products, such as beef and lamb, relatively more expensive and steer consumers towards lower-carbon substitutes such as pork, seafood, chicken, or vegetable proteins. There are many practical and political challenges to designing and implementing GHG pricing in the agricultural sector. Several studies, however, have concluded that taxes could result in substantial reductions in GHG emissions (Tallard and Key 2012; Havlık et al. 2014; Wirsenius et al. 2010). Modelling suggests that a global price on methane emissions from livestock ranging from US \$15/tCO_2e to US \$100/tCO_2e would reduce methane emissions by 2.8% and 9.9%, respectively (Tallard and Key 2012). See also research in the previous section whereby the addition of some types of seaweed to livestock diets can lead to a large decline in methane emissions. After applying emissions intensities (Gerber et al. 2013) to forecasted production of terrestrial animal proteins in 2030 (Alexandratos and Bruinsma 2012), these reductions in livestock emissions would amount to 237–840 $MtCO_2e$/year. Extending this estimate out to 2050, these same percentage reductions in livestock emissions would lead to avoided emissions of 0.30–1.06 $GtCO_2e$/year, a portion of which will come from shifts to ocean-based proteins.

Shifting diets through media and educational campaigns the projected health benefits of reducing meat consumption are so large that GHG emissions mitigation could potentially be achieved as a cobenefit of behaviour change motivated by people's interest in their personal health (Willett et al. 2019). Numerous campaigns on other health-related issues provide insights on the magnitude of expected behaviour changes. In multiple meta-analyses (Snyder et al. 2004; Elder et al. 2004; Abroms and Maiboch 2008) on campaigns on seat belt use, smoking, cancer screening, alcohol use, and many other topics, the sobering result was that the observed effects were moderate—typically 15% or fewer people changed targeted behaviours. Lessons learned from past campaigns could help maximise the impacts of future campaigns on diets, but expectations for near-uniform adoption of behaviour change are clearly unrealistic. Applying the median (11%) and upper bound (15%) of these past experiences to the projected benefits of global adoption of a less-GHG–intensive diet (4.7 $GtCO_2e$ estimated by Springmann et al. 2018) suggests that effective campaigns focusing on health benefits of dietary change could potentially yield reductions between 0.52 and 0.71 $GtCO_2e$ by 2050.

6.3 Policy Interventions Required to Achieve Mitigation Potential

Achieving a level of efficiency gains in wild fisheries that would drive emission reductions requires more effective management of fisheries around the world. Several global analyses highlight where fisheries are working well and where there are needs for significant reforms (e.g., Arnason et al. 2009; Sumaila et al. 2012; Costello et al. 2016), and help identify which management practices are linked to success or failure in fisheries management (e.g., Keller et al. 2019; Evans et al. 2011; Allison et al. 2012; Barner et al. 2015; Lubchenco et al. 2016; Costello et al. 2016; Lester et al. 2017). The lessons of this rich literature are that there are robust solutions for a wide range of fisheries issues. Yet, the problems persist and grow. The challenge is to scale the successes more quickly than the problems grow. Achieving this goal requires national recognition of the nature of each country's fisheries challenges and the benefits of improved management (Box 17.7), and a concerted effort to draw on the lessons of others to drive more rapid change.

Significantly altering the behaviours of a broad section of society, even for actions that are both in the interest of the planet and of individual people, is surprisingly challenging. The two broad approaches of sending clear market signals via carbon or other food-related taxes that embed broader environmental and social costs of different food choices in prices, and motivating lifestyle changes need to be coupled. The two policy approaches, if synergistic, can help to realise greater GHG emissions mitigation.

6.4 Technology Needs

Unlike other categories in this assessment, the largest gains from changes in the global food system do not depend on the development of new technologies.

Rather, the benefits depend on scaling solutions globally that have already been demonstrated in specific places. Although this requires new innovative approaches, new market solutions, and new campaigns, it is not heavily dependent on new technological advances.

6.5 Priority Areas for Further Research

Data sources for GHG emissions from fisheries, both farmed and wild-caught, would better inform potential policy interventions.

Box 17.7. Wider Impacts Associated with Reducing Emissions from Fisheries and Aquaculture and Shifting Diets to Ocean-based Proteins

Potential Cobenefits:

- Even moderate shifts in diet from high meat consumption towards ocean-based protein have well-documented human health benefits.
- Moving to diets that are less dependent on animal products would slow the growth in demand for land and freshwater to support livestock agriculture.
- Growth of marine aquaculture will create jobs. Total direct employment in the industry is estimated to be 3.2 million in 2030 under business-as-usual projections (an increase of 1.1 million above 2010 levels) (OECD 2016).
- Innovations in developing fish meal substitutes and improving feed efficiency will be crucial to support a rapidly growing aquaculture industry and meet global food security targets.
- Replacing fish meal of future feeds with crops instead of animal by-products requires less water; reducing feed conversion ratio in aquaculture production decreases upstream water usage.
- Structural changes to fisheries that reduce fuel consumption will be economically beneficial.

Potential Trade-Offs:

- Offshore marine aquaculture is associated with multiple environmental challenges (such as eutrophication, disease, and risk of invasive species). These risks are also to some extent associated with land-based farming.
- Unplanned growth in shrimp aquaculture has caused widespread loss of mangrove ecosystems, leading to large CO_2 emissions, salinisation of soils and freshwater reserves, erosion, and loss of coastal resilience to flooding.
- Increased inclusion of terrestrial plant-based ingredients in fish feed for a growing aquaculture industry could lead to competition for land, causing social and environmental conflicts that may in turn affect the resilience of the global food system. However, the land and water demands of land-based agriculture, especially livestock production, are far greater on a unit output basis.

For a full exploration of the wider impacts associated with fisheries and aquaculture, see the Sect. 1.4.

Source: Authors

7 Carbon Storage in the Seabed

This section analyses the potential mitigation impact of storing carbon in the seabed.

The ocean naturally contains nearly 150,000 $GtCO_2e$. This dwarfs the 2000 $GtCO_2e$ in the atmosphere and 7300 $GtCO_2e$ in the land-bqa biosphere. Each year, as a consequence of human activities, approximately 10 billion tonnes of CO_2, or about 25–30% of anthropogenic CO_2 emissions, enters the ocean (Global Carbon Project 2018). As a result, there is considerable theoretical potential to store CO_2 (once captured and compressed) in the ocean in ways that substantially reduce adverse environmental impacts relative to the environmental impacts that occur as a result of atmospheric release of CO_2 (GESAMP 2019).

However, any proposals for ocean-based carbon storage, including storage in the seabed, must be considered in light of the substantial risks to the ocean environment and its ecosystems (Kroeker et al. 2013; Gattuso et al. 2018; Pörtner et al. 2014) and the associated technical, economic, social, and political challenges. Options for ocean carbon storage differ, depending on whether the source CO_2 is concentrated, (e.g., captured from power plant flue gas) or diffuse (e.g. atmospheric CO_2). The options may also differ as to whether the stored CO_2 is concentrated (e.g., in storage reservoirs) or is to be diffused (e.g., mixed into deep ocean waters). The options also differ in the form in which the CO_2 is sourced (from power plants, the atmosphere, or biomass) and in which it is stored (as molecular CO_2, as ions with charge balanced by added alkalinity, or as organic carbon). Table 17.12 summarizes the options most often discussed for ocean-based carbon storage.

Note that vertical ocean pipes are not addressed in this document because the most reliable available science indicates that such pipes would bring carbon-enriched water up from the deep, and thus not be effective at storing carbon in the ocean (Dutreuil et al. 2009; Oschlies et al. 2010; Kwiatkowski et al. 2015). Furthermore, several studies have suggested that CO_2 extraction from seawater would be feasible at commercial scale; however, insufficient information is available to assess the feasibility and system-level effectiveness of these options. For example, Willauer et al. (2014) describe a CO_2-removal process that involves an effluent returned to the ocean with a pH of 6, with no consideration of how that effluent might affect the ocean environment.

The storage of highly concentrated and compressed CO_2 streams in the seabed is the only option that is currently deployed at industrial scale and is therefore the only option that has a reasonable likelihood of being deployed at large scale by 2030 and beyond. To date, sub seabed storage has been used only to facilitate the extraction of natural gas from the Norwegian coast. Thus, the net flux of carbon has been from the seafloor to the atmosphere, not the other way around. The process returns excess CO_2 back to the sub surface that comes up with the natural gas. If not for extracting the natural gas, the CO_2 have have remained in the sub surface. The rest of the options presented in Table 17.13 remain untested at an industrial scale.

All assessments of ocean-based carbon storage potential should therefore be greeted with considerabe caution. Further research is necessary to narrow the uncertainties and ensure informed decision-making about the viability of ocean-based carbon storage. As a result of the significant gaps in knowledge in terms of ability to scale the range of ocean-based storage options and the very real risks to ocean ecosystems, the only option that has been assessed in this report is seabed storage. The full range of options contained in Table 17.13 is discussed in Box 17.8 at the end of this section.

7.1 Mitigation Potential

Carbon capture and storage of CO_2 in the seabed requires that CO_2 be concentrated, compressed, and transported to the deep-water injection site. Based on a number of studies, Adams and Caldeira (2008) concluded that the costs for capture and compression from a fossil fuel power plant would be around US $20 to US $95 per tonne of CO_2 captured, and the cost of transportation approximately US $1 to US $10 per tonne of CO_2.

The cost of geological storage was estimated at US $0.5 to US $10.0 per tonne of CO_2 injected, and US $5 to US $30 per tonne of CO_2 (>1000 m).

Electricity generation accounts for about 25% of global GHG emissions (IPCC 2014) with up to 10% (or about 2.5% of the total) of electricity generation being located near enough to the ocean to make ocean disposal of power plant CO_2 economically feasible (IPCC 2005). Thus, the total potential for ocean-based carbon storage by seabed storage may be up to 2.5% of global CO_2 emissions. At 2018 global CO_2 emission rates, this would yield an estimated mitigation potential of 1 $GtCO_2e$. As it would be extremely difficult to retrofit most existing power plants with carbon capture and storage facilities and pipes to the deep ocean by 2030; the economic potential in 2030 is likely to be less by a factor of 10 (about 0.1 $GtCO_2e$).

By 2050, a greater fraction of the technical potential might be achieved and the environmental risks suitably understood and mitigated so that other ocean-based storage options might be developed, so it is conceivable that several billion tonnes of CO_2e could be stored in the ocean each year by 2050. However, this has not been included in our calculations for this report, given the degree of current uncertainty of the technical, environmental, social, and political feasibility of these additional options.

The first three options shown in Table 17.12 involve different forms of carbon capture and storage for coastal powerplants and as such should also be considered as interchangeable. Based on this, and the assumptions and limitations outlined above, it is possible to propose a total mitigation potential in 2030 of 0.25 to 1.0 $GtCO_2e$, and of 0.5 to 2.0 $GtCO_2e$ in 2050 (Table 17.14).

Table 17.13 CO_2 characteristics of storage options for deep sea and/or seabed storage

Option	CO_2 source	CO_2 storage reservoir	Initial CO_2 storage form	Technical readiness	Cost profile	Principal environmental concerns	Key references
CO_2 injection to seabed	Power plant	Geologic reservoirs beneath seafloor	Molecular CO_2	High to medium	High	Operational activities; leakage to ocean; impacts on deep sea ecosystems	SRCCS (2005)
CO_2 storage contained on top of the seafloor (CO_2 injection into CO_2 lakes or containment vessels)	Power plant	Reservoirs on seafloor separated from the ocean by physical or chemical barrier	Molecular CO_2	Low	High	Leakage to ocean; damage to seafloor; operational activities; impacts on deep sea ecosystems	SRCCS (2005), Palmer et al. (2007)
CO_2 injection into deep ocean	Power plant	Deep ocean	Molecular CO_2	High	High	Ocean acidification; leakage to atmosphere; operational activities; impacts on deep-sea ecosystems	SRCCS (2005)
Carbonate dissolution (CO_2 release to the ocean, buffered by dissolved carbonate minerals)	Power plant	Ocean	Bicarbonate ions	Medium	High	Possible contaminants; local impacts on ecosystems	SRCCS (2005), Rau and Caldeira (1999)
Alkalinity addition	Atmosphere	Ocean	Bicarbonate ions	Medium	High	Unintended ecosystem effects	SRCCS (2005)
Ocean fertilisation	Atmosphere	Ocean	Organic carbon	Low	Medium	Interference with marine ecosystems; ocean acidification; leakage to atmosphere	Williamson et al. (2012)

Source: Authors

Notes: "Power plant" is used to refer generically to concentrated CO_2 streams, and "Atmosphere" to diffuse sources. For technical readiness, "High" means could likely be accomplished within several years; "Medium" means no major technical barrier; "Low" means that there are substantial uncertainties regarding technical feasibility and/or geophysical effectiveness. For costs, "High" means comparable to carbon capture from power plants with geologic storage on land; "Medium" means lower, but still substantial, costs. These evaluations represent subjective assessments by the authors on the basis of available information. The "CO_2-storage reservoir" and "Initial storage forms" columns in Table 17.12 *indicate that in the case of some ocean storage options, the storage is isolated from the large volume of ocean seawater. In other options, the carbon is distributed through the ocean volume but primarily in forms that do not exchange with the atmosphere or cause ocean acidification. Lastly, some proposed options simply transfer molecular CO_2 to the deep ocean; in which case storage might not be permanent and would contribute to ocean acidification and impacts on marine organisms and ecosystems*

Table 17.14 Mitigation potential of carbon storage options in 2030 and 2050 ($GtCO_2e$)

Ocean-based climate action area	Mitigation option	Description	Mitigation potential, 2030 ($GtCO_2e$/year)	Mitigation potential, 2050 ($GtCO_2e$/year)
Seabed carbon storage	CO_2 storage in the seabed	Geological storage offshore of CO_2 below the seabed	0.25–1.00	0.5–2.0
Total			0.25 1.00	0.5–2.0

Source: Authors

Note: These values represent reasonable estimates of the lower and upper bounds of potential deployment rate in a highly aggressive mitigation scenario

7.2 Methodology

The physical potential of sub-seabed storage is thought to be very large, as there is an abundance of settings in which CO_2 could potentially be stored. The physical capacity of carbon storage in the marine environment has been estimated to exceed 10,000 Gt of CO_2 (36,000 $GtCO_2$) in the seafloor surrounding the contiguous United States alone (House et al. 2006). This is similar in magnitude to the total amount of the fossil fuel resource (IPCC 2014). More realistically, the capacity for storage in the seafloor will depend on costs of transport of CO_2 from the concentrated source, and the cost of emplacement in seabed geologic formations.

On the time frames considered here (2030–2050), seabed storage will be limited not by geophysical capacity, but rather by techno-economic and possibly sociopolitical factors.

Costs are somewhat higher than for land-based geologic carbon storage, but, even in the ocean case, the primary cost driver is the cost of separating and compressing the relatively pure CO_2 stream (SRCCS 2005). In the 1.5 °C stabilisation scenarios considered by the IPCC (2018), total carbon capture and storage amounts to year 2050 (cumulative) are typically about 100 $GtCO_2$, but range to over 400 $GtCO_2$ in some models. The corresponding magnitude for 2030 is of the order of several billion tonnes of CO_2.

If seabed storage were to comprise 30% of total carbon capture and storage, that would suggest an average rate of seabed carbon storage of the order of 1 $GtCO_2$/year. It is reasonable to presume that the most advantageous settings would be used first, so it is plausible that half of the average rate could be reached by 2030, approximately 0.5 $GtCO_2$/year. As a rough approximation of uncertainty, we halve and double these values.

7.3 Policy Interventions Needed to Achieve Mitigation Potential

Seabed storage would occur in territorial waters so the primary regulatory bodies would be national. The primary environmental concerns, if everything works as planned, involve local environmental disturbance from industrial operations. International implications arise related principally to the risk or event of failure.

Continuing to increase scientific understanding is essential if these technologies are to be used safely and without unintended consequences.

7.4 Technology Needs

Carbon storage in the seabed does not involve major technical advances and is an extension of activities that are already being carried out on land. Scaling up the technologies to match the scale of the problem, however, is a major challenge.

An exception, where technological advances are required, would be materials science questions relating to long-lasting containment of CO_2 in a deep seafloor environment. For the most part, noncost barriers primarily have to do with unintended environmental consequences, effectiveness, and verifiability, and not the state of technological development.

7.5 Priority Areas for Further Research

The primary barriers to use of the ocean as a carbon storage reservoir involve environmental concerns (Box 17.8). However, if done properly, some of these techniques could potentially isolate CO_2 away from both the atmosphere and the majority of ocean waters for millions of years.

Other techniques might have cobenefits, for instance, reducing associated impacts such as ocean acidification. On the other hand, seabed storage of CO_2 approaches, if deployed unwisely, could contribute to ocean acidification and damage ocean ecosystems by impacting chemical, physical, and ecological processes at a large scale.

Further research will help us understand the full implications of carbon storage options. Box 17.9 profiles the status of current knowledge for the other ocean-based carbon storage options not quantified in this report.

Box 17.8 Wider Impacts Associated with Options for Seabed Storage

Potential Cobenefits:

- Potential benefits in terms of direct job creation, as well as job retention in harder-to-abate sectors (e.g., heavy industries and fossil fuel based sectors) by allowing them to function with appropriate CCS infrastructure investment/development.

Potential Trade-Offs:

- Injection of CO_2 into submarine geological structures has the potential for CO_2 to leak back into the marine environment, affecting the health and function of marine organisms, especially with respect to the resulting localised ocean acidification. The gravity of the impacts at community level is unknown.
- Potentially serious impacts on little-understood deep-sea ecosystems, which are the largest habitat on the planet.

For a full exploration of the wider impacts associated with ocean-based transport, see the Sect. 1.4 of this report.

Source: Authors

Box 17.9. Additional Ocean-Based Carbon Storage Options not Quantified in this Report

Containment of CO_2 on the Seafloor

Below about 3000 m depth, compressed CO is denser than seawater and so will tend to sink or remain on the seafloor. This has led to the proposal that CO_2 might be stored in lakes on the seafloor (Shindo et al. 1993). However, in the absence of a physical or chemical barrier, such CO_2 lakes would be expected to dissolve into the overlying seawater (SRCCS 2005). Little work has gone into developing such barriers, although it has been estimated that the cost of creating a physical barrier would be small, perhaps as low as US $0.035 per tonne of CO_2 stored (Palmer et al. 2007). Because of the vastness of the seafloor, there is no practical constraint on the amount of CO_2 that could be stored in this way, and if concerns over physical integrity of the barrier and effects on the underlying seafloor can be addressed, the primary determinant of the scalability of this approach is likely to be the costs of producing a relatively pure CO_2 stream, and those of transporting and emplacing the captured CO_2 in these storage reservoirs.

Because containment of CO_2 on the seafloor has never been demonstrated for any substantial amount of time, the lower bound on the potential for this technology class must be regarded as zero. However, if demonstrated containment can prove cost-effective, the potential for containment storage on the seafloor could be as large as that estimated for sub-seabed storage.

Injection of CO_2 Into the Deep Ocean

Injection of CO_2 into the deep ocean is much simpler than storage beneath or on the seafloor. Deep-sea disposal and containment of CO_2, however, raises concerns about environmental effects (e.g., impacts of ocean acidification) and leakage back to the atmosphere. As noted above, most of the waste CO_2 released to the atmosphere by human activities will ultimately reside in the ocean. Therefore, placing CO_2 in the ocean instead of in the atmosphere could be expected to reduce the climatic consequences of CO_2 emission. It would also tend to reduce the amount of ocean acidification experienced in the ocean surface but at the cost of increased ocean acidification in the deep ocean. If the entire ocean were allowed to have the same pH change as the near-surface ocean (about 0.1 pH units), the ocean could store a total of about 2000 $GtCO_2$ (SRCCS 2005). Over one-quarter of this amount (GCP 2018) has already been absorbed from the atmosphere, leaving about 1500 $GtCO_2$ of storage capacity. If a pH change of 0.2 were deemed to be acceptable (corresponding to an atmospheric CO_2 concentration of about 600 parts per million [ppm]), the amount of remaining storage capacity would be about 3300 $GtCO_2$ (or roughly 10% of the estimated remaining fossil fuel resource).

Such changes in the chemistry of the ocean would be accompanied by a growing list of impacts on organisms, such as reef-building corals, seaweeds, invertebrates, and fish, among many others (Kroeker et al. 2013; Gattuso et al. 2018; Hoegh-Guldberg et al. 2014, 2018). In addition to decreasing the ability of organisms to maintain shells and skeletons, a wide variety of other impacts have been reported from disruptions of reproduction, gas exchange, and neural systems (Kroeker et al. 2013). Damage to deep-water ecosystems has been reported, and, though its extent has not been well documented, it is suspected to be large. These impacts have generated considerable concern about such fundamental changes to biological systems, especially given the long time (>10,000 years) it takes to reverse this change through the dissolution of carbonates and other processes (IPCC 2013).

Direct injection into the deep ocean is likely to be comparable to the cost of injecting CO_2 into the seabed. However, there is real concern about using the ocean waters as a waste disposal site for CO_2 from human industrial processes. Furthermore, storage of CO_2 freely dissolved in the deep ocean eventually exchanges with the atmosphere, so the isolation of CO_2 is not permanent. Therefore, it is far from certain that global political systems will encourage and credit deep-sea CO_2 injection. A reasonable estimate on the lower bound of conceivable deployment rate in a highly aggressive mitigation strategy would therefore range from zero to the rate estimated for seabed disposal.

Carbonate Dissolution

Most of the ocean acidification caused by adding CO_2 in the ocean will ultimately be neutralised over the longer term by the dissolution (and slower accumulation) of carbonate minerals on the seafloor, and from rock weathering products delivered to the ocean by rivers. Carbonate minerals will not dissolve in the surface ocean due to high levels of carbonate saturation (i.e., concentrations that are so high that they promote precipitation not dissolution). This fact led to the idea of using power plant flue gases to dissolve carbonate minerals, which would allow CO_2 to be stored in the ocean with little adverse impact on ocean pH or mineral saturation states in the ocean (Rau and Caldeira 1999; Caldeira and Rau 2000). About 2.5 tonnes of carbonate minerals would need to be dissolved, however, for each tonne of CO_2 stored in this way. This would require a huge and unprecedented mining infrastructure and would entail massive materials-handling costs and logistics.

The costs have been estimated to be lower than for injection of relatively pure CO_2 streams for cases in which the power plant is coastally located with access to carbonate mineral resources, because this approach does not require costly separation of CO_2 from power plant flue gases and subsequent pressurisation (Rau and Caldeira 1999). However, since such facilities have never been built, cost estimates must be regarded as speculative.

Regardless, such approaches would likely be cost-competitive only in locations where both carbonate minerals and CO_2 could be delivered to the ocean at low cost, which is likely to be the case for less than 10% of total power plant CO_2 emissions. Environmental concerns include the effects of a large scale-up of carbonate mineral mining and possible impacts on the marine environment of contaminants or incompletely dissolved particles.

Rau and Caldeira (1999) estimated that perhaps 10% of electricity production might be located suitably near carbonate minerals to make carbonate dissolution a cost-effective approach to carbon storage. However, there are environmental concerns about processing large amounts of seawater through carbonate reactors and using the ocean as a waste disposal site.

A plausible range for this approach might therefore be from 0 to 10% of the magnitude estimated for all of carbon capture and storage (IPCC 2018).

Alkalinity Addition

The acidity caused by CO_2 in the ocean, and the propensity of CO_2 to de-gas from the ocean to the atmosphere, can be reduced or eliminated by the addition of alkaline (also known as basic) minerals (Renforth and Henderson 2017). Addition of these minerals to the ocean (Kheshgi 1995) could result in the ocean absorbing additional CO_2 from the atmosphere (González and Ilyina 2016). Over 2.5 tonnes of rock would need to be mined and crushed to a fine powder (to overcome slow dissolution kinetics) for each tonne of CO_2 stored in the ocean in this manner. As with carbonate dissolution, this option raises concerns related to huge expansion of mining infrastructure (silicate rock mining might need to expand by three orders of magnitude) (González and Ilyina 2016). Further, many of the proposed silicate source rocks contain substantial amounts of heavy metals (Hartmann et al. 2013) and thus raise concerns about introduction of heavy metals into the marine environment. Because silicate rocks are abundant in Earth's crust, there is no practical physical constraint, but if applied at scale, such ocean CO_2 storage would represent "an unprecedented ocean biogeochemistry perturbation with unknown ecological consequences" (González and Ilyina 2016).

Renforth and Henderson (2017) estimate the potential for very ambitious rates of deployment: A 50 $MtCO_2$/year initial investment (roughly equivalent to the emissions of 10 of the largest cement plants in operation), followed by ramping up this capacity by about 7%/year, could achieve mitigation of 0.1 $GtCO_2$/year by 2020. If the same initial investment were ramped up by about 10%/year, mitigation could reach 1 $GtCO_2$/year. These might be considered plausible upper bounds. The lower bound must be considered zero, because it is not clear that the international community will accept adding large amounts of dissolved and/or particulate matter to the ocean as a climate mitigation strategy.

Ocean Fertilisation

Ocean fertilisation has been proposed as a means of transferring carbon from the atmosphere to the ocean. The basic idea is to add inorganic nutrients to the near-surface ocean, thereby stimulating biological production of organic matter. Some of this organic matter would sink to the deeper ocean, where it would be metabolised and dissolved in the deeper ocean waters. Some additional CO_2 would be absorbed from the atmosphere to replace the carbon that was removed by this additional biological activity. Some researchers have advocated fertilising the ocean with major nutrients that are often limiting, such as phosphate or nitrogen (Harrison 2017).

Because of the large amounts of nutrients involved, however, most of the focus has been on environments in which the major nutrients are abundant, but other minor nutrients such as iron limit marine productivity (Williamson et al. 2012). The efficacy of ocean fertilisation is reduced by shallow oxidation of sinking organic matter with the relatively rapid return of carbon to the surface ocean. This phenomenon has also attracted concern regarding the increased respiration rates stimulated by the additional organic carbon falling into the deep ocean, leading to decreased oxygen at depth and an increased risk of dead zones (Hoegh-Guldberg et al. 2014). Further, fertilisation with micronutrients utilises major nutrients that might otherwise have supported productivity elsewhere; some local increase in productivity may come at the expense of decreased productivity elsewhere at a later time.

The geophysical potential of ocean iron fertilisation has been estimated to be in the range of 0.25–0.75 $GtCO_2e$/year averaged over a 100-year period (Williamson et al. 2012). Small-scale experiments to date suggest that adding iron dramatically changes the composition of the phytoplankton, which in turn triggers changes in zooplankton, fishes, and other higher trophic species. Many of these consequences are little understood. Concerns regarding effectiveness, permanence, verification, and unintended consequences, combined with concerns about disposing of CO_2 in deeper ocean waters, mean that the lower bound on potential must be regarded as zero.

The geophysical potential of ocean fertilisation is estimated to be about 1.8 $GtCO_2e$/year. Plausibly, 10% of this geophysical potential could be achieved by 2030 and about half by 2050.

While the geophysical potential of ocean-based storage of captured CO_2 is large, the technical and economic mitigation potential is likely to be constrained by the technical challenges of making carbon capture and storage economically viable.

Some of these technologies are likely to be technically feasible and cost-effective. Given the importance of reducing the amount of excess CO_2 in the atmosphere and ocean, understanding the full set of the impact of these solutions on ecosystems, such as the deep sea, is critical.

Source: Authors

8 Wider Impacts of Ocean-based Actions

This section presents analysis of the wider impacts (both positive and negative) of each of the five ocean-based intervention areas on the long-term Sustainable Development Dimensions and 2030 Sustainable Development Goals.

Increased efforts to reduce GHG emissions will affect multiple dimensions of long-term sustainable development, well-being, and governance in the form of cobenefits and trade-offs (IPCC 2018). Many interventions are likely to affect countries' ability to achieve targets established within the framework of the UN 2030 Sustainable Developmental Goals (SDGs). Taking these wider impacts into account can help provide a more informed and holistic picture of pursuing ocean-based climate solutions.

The IPCC Special Report on 1.5 °C scenarios integrated some of these wider impacts into its assessment of mitigation options; however, the ocean received relatively little attention. We address this major knowledge gap by focusing on four dimensions where wider impacts may be expected: the environment, the economy, society, and governance. These dimensions, their associated impact categories, and relevant UN SDGs are mapped in Table 17.15.

8.1 Methodology

Wider impacts are evaluated with a weighted scoring method and an associated assessment of confidence levels. Our method is based on a similar approach adopted in Chap. 5 of the IPCC 1.5 °C Special Report (Roy et al. 2018). Based on a review of the existing literature and expert judgment (Box 17.9), the performance of each ocean-based mitigation option was assessed within each of the wider-impact dimensions (Table 17.15). The impact was described, scored, and weighted based on the following factors:

- **Direction of impact:** The positive and/or negative direction of the impact of the mitigation option on the wider-impact dimensions and SDG goals was recorded. If a mitigation option was identified as having both a positive and negative impact, both were recorded. The net direction of impact was determined by the sum of the positive and negative impact scores.
- **Linkage score:** The strength of the relationship between the mitigation option and the indicator was scored. Scores range from +3 (indivisible) to −3 (cancelling), with a "zero" score indicating 'consistent', but with neither a positive nor negative impact (Nilsson et al. 2016). A zero score also indicates that no relevant literature was found during this review.
- **Confidence in assessment:** The confidence assessment was developed to reflect the robustness of the linkage scores. Confidence levels ranging from high to low were determined based on the level of evidence (number of studies and other articles) and level of agreement on the evidence presented in the literature. For each linkage score, an assessment of confidence was assigned, where increasing levels of evidence and degrees of agreement are correlated with increasing confidence (Mastrandrea et al. 2010).

Table 17.15 Wider impact dimensions explored in the report

Wider-impact dimensions	Associated impact categories	Links with near-term sustainable development goal targets and indicators
Environment	Impact on marine and terrestrial biodiversity, water quality, land use, and adaptability of ecosystems and human settlements to climate change	SDGs 6, 12, 14, 15
Economy	Impact on employment, household incomes, profits and/or revenues of firms, innovation, supply of clean energy, and economic growth	SDGs 7, 8, 9, 11
Society	Impact on human health outcomes, poverty reduction and food security targets, regional income inequality, quality of education, and gender equity	SDGs 1, 2, 3, 4, 5, and 10
Governance	Impact on national and local institutions, participation in global governance, global partnership for sustainable development, and capacity building	SDG 16 and 17

List of sustainable development goals reviewed:

Source: Authors

Box 17.10. Literature Review Method and Types of Evidence Analysed

A two-step procedure was followed as part of a review of the literature on wider impact analysis. First, the databases Scopus and Google Scholar, and the search engine Google were used in a literature search using various combinations of keywords and short search strings such as "Ocean energy" AND "sustainability," "Ocean" AND "CCS," AND "sustainability." Second, the findings from the literature review were recorded and scored. Additional evidence was included based on feedback obtained through the expert review process. The types of evidence and number of studies are summarised in the table below. Please refer to Annex for further information on the scores and confidence assessments.

Types of literature	Description	Number
Case study	Case studies specific to countries or region	10
Experimental	Results based on experiments	11
Project-based	Results reported based on project-level impacts	2
Quantitative analysis	Studies that have employed econometric, graphical, or statistical tools to find the impact of any intervention. This includes meta-analysis, scenario analysis, spatial analysis, and other modelling assessments	46
Review paper	Studies that exclusively mention "review" in their objective or methods	16
Summary paper	This includes commentary, newspaper articles, discussion papers, policy briefs, and newsletters from international organisations	14
Website	Relevant information (such as examples of ongoing restoration programmes) provided on web pages owned and curated by international organisations	5
Report	Policy and analysis reports from international organisations, such as OECD, ETC, IRENA, FAO, IEA	31
Qualitative	Academic papers and reports that present qualitative discussion of the impact of policies and international agreements	4
Total number		139

Source: Authors

Note: OECD Organisation for Economic Co-operation and Development, *ETC* Energy Transmissions Commission, *IRENA* International Renewal Energy Agency, *FAO* Food and Agriculture Organization of the United Nations, *IEA* International Energy Agency

8.2 General Findings of the Wider-Impacts Analysis

All mitigation options demonstrated both positive and negative impacts, with varying strengths, across the four wider-impact dimensions (Fig. 17.12). The headline messages can be broadly summarized as follows:

- All ocean-based mitigation options generate many cobenefits. Overall, cobenefits outweigh trade-offs and risks. However, these risks and trade-offs cannot be ignored, and concerted action to address negative impacts will help enhance net positive outcomes.
- Of the five ocean intervention areas, protecting and restoring coastal and marine ecosystems, fisheries and aquaculture, and ocean-based energy have a positive impact on the largest number of sustainable development dimensions. When looking at individual mitigation options, protection and restoration of vegetated coastal habitats (mangroves, salt marshes and seagrasses) and offshore renewable energy positively impact the largest number of sustainable development dimensions.
- Mitigation options were observed to have the strongest links with the social and economic dimensions, implying that implementing these options in a sustainable manner would result in benefits in terms of higher employment in ocean-based industries, gains from technology spillover, increase in revenues and profits to firms, improvement in livelihoods of local communities, better human health outcomes, contribution towards global food security targets, and potential to ensure greater gender parity as ocean-based industries expand.
- Protection and restoration of mangroves, salt marshes, and seagrasses has the highest number of and most strongly positive impacts on all the environmental dimensions assessed, indicating that there is potential to achieve many environmental cobenefits, including increased biodiversity-related services, coastal resilience, and climate change adaptation benefits.
- Trade-offs and risks are varied. Mitigation options aimed at recovering ocean biomass can negatively impact poverty reduction and employment targets and can limit progress on food security targets in the short term. Lack of community-level engagement on blue carbon restoration work can lead to negative outcomes for small-scale fishers who play a strategic role in providing jobs, supplying nutritional needs, and maintaining economic sustainability.
- Environmental risks include impacts on coastal ecosystems or marine species from unassessed growth in ocean-based activities. Shifting diets, fisheries, and aquaculture

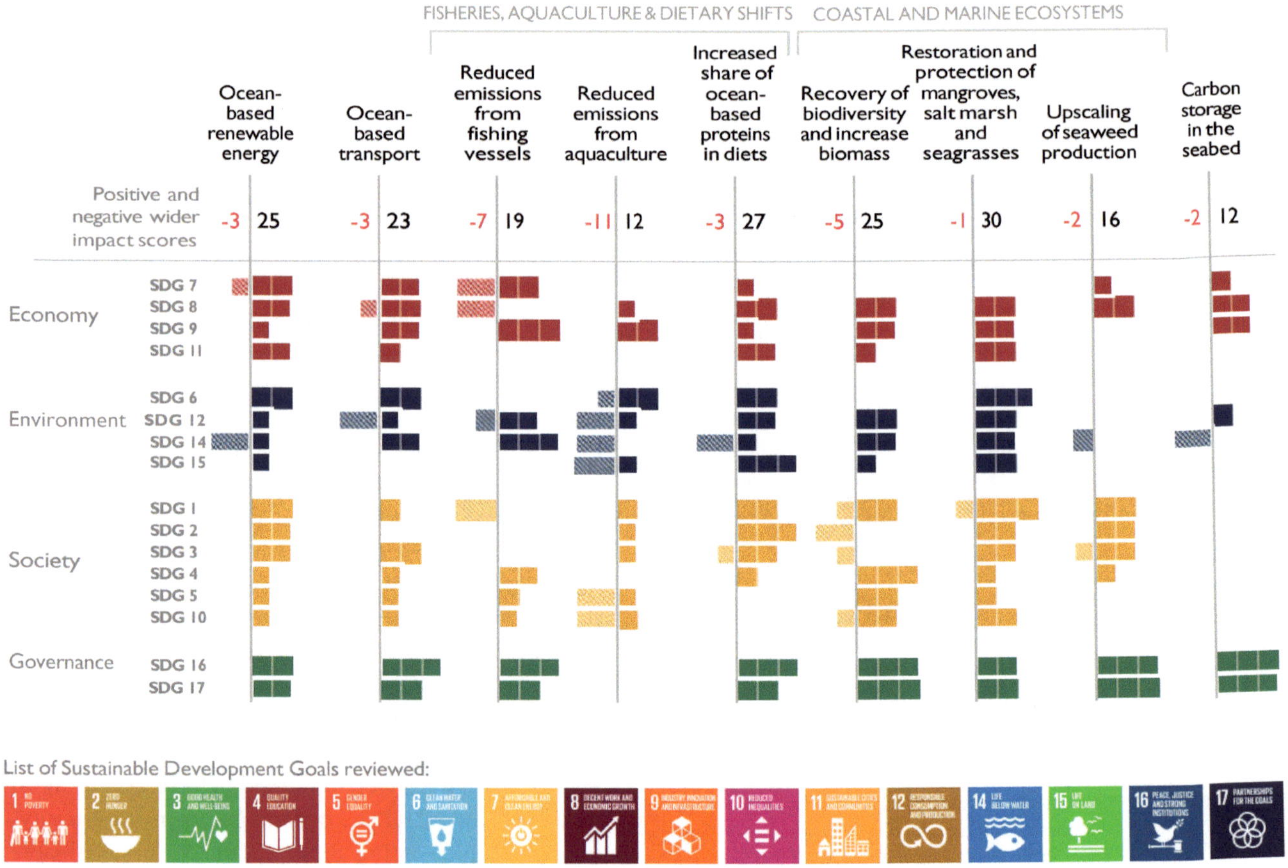

Fig. 17.12 Linkage scores of ocean-based interventions and selected mitigation options across the wider impact dimensions. (*Source: Authors. Notes: Wider-impact dimensions cover various sustainable development dimension as well as 2030 Sustainable Development Goals (SDG). The figure shows the relative strength of the relationship between a selected set of ocean-based mitigation options and the SDGs. For each mitigation option, the positive linkage score with a particular SDG (depicted with solid bars) is shown in the right-hand column and negative linkage score (depicted by shaded bars) in the left-hand column. Scores range from +3 (indivisible) to −3 (cancelling)* (Nilsson et al. 2016). *A zero score (no bar and no colour) means no impact was found in this review of the literature. Each colour represents a particular wider impact dimension: Red bars for economy (SDG 7, 8, 9, 11); blue bars for environment (SDG6, SDG12, SDG14, SDG15); yellow bars for society (SDG1, SDG2, SDG3, SDG4, SDG5, SDG10) and green bars for Governance (SDG 16, SDG 17). Further information on the linkage scores and the associated confidence levels are provided in the Annex*)

have a negative impact on the largest number of sustainable development dimensions.

- Some of these risks and trade-offs can be adequately addressed via stakeholder engagement, inclusive management policies, monitoring, and effective marine planning. Others will require further research on their implications and in some instances will call for significant action on the part of decision-makers and policy implementers to ensure that negative impacts are reduced.
- All ocean-based mitigation options will need strong national institutions; engagement by business, industry, and communities; and international cooperation to ensure their effective implementation.

8.3 Detailed Findings of the Wider-Impact Analysis

8.3.1 Ocean-Based Renewable Energy

Effective Marine Spatial Planning, in Combination with Emerging Ocean Energy Technologies, will Be Effective in Mitigating Biodiversity Loss from Ocean Energy Technologies and Reinforcing Biodiversity Cobenefits (High Confidence)

Offshore wind structures have positive and long-term effects on marine species because they provide new habitat in the form of artificial reefs and because fishing, mainly trawling, tend to be restricted in their vicinity (IRENA 2018a; Dinh and McKeogh 2018). In contrast to offshore oil and gas installations, there is little risk of pollution, and no need for the development of new sites in response to long-term exhaustion of the resource (Spalding and de Fontabert 2007). Risks of developing ocean-based energy include biological invasions, noise and disturbance vibrations to marine species, collisions between birds and wind turbine rotors, and the presence of electromagnetic fields that can disrupt marine life and benthic habitats (MERiFIC 2012; IRENA 2017; Langhamer 2012). However, studies have shown that most perceptions of environmental impacts from ocean-based renewable devices arise from uncertainty or lack of definitive data about the real impacts (Copping et al. 2016). While it is important to acknowledge all the impacts on the marine environment as ocean-based renewable industry develops, some of the perceived risks are likely to be small and can be avoided or mitigated (Copping et al. 2016). In the case of risks like collision with seabirds and impacts on migratory cetaceans, marine spatial planning appears to be appropriate mechanism to reduce risks to manageable levels (Best and Halpin 2019).

Ocean-Based Renewables will have a Positive Impact on Reducing Water Use Compared to Fossil Fuel-Based Technologies (Medium Confidence)

Offshore wind uses no water directly, and there should be an overall reduction in freshwater use compared to generating power from fossil fuels (Macknick et al. 2011). There is potential to develop ocean energy technologies for a range of purposes, including desalination for drinking water (OES 2011).

Replacing Fossil Fuels with Ocean-Based Renewable Energy Contributes to Positive Health Outcomes (Medium Confidence)

The health benefits of moving to ocean-based renewable energy for power generation would be significant, particularly for regions that rely more heavily on coal and oil to generate electricity. Offshore wind in the Mid-Atlantic region of the United States could produce health and climate benefits estimated at between US $54 and US $120 per MWh of generation, with the largest simulated facility (3000 MW off the coast of New Jersey) producing approximately US $690 million in benefits (Buonocore et al. 2016).

Expansion of Ocean-Based Renewable Energy has the Potential to Promote Gender Equity (Low Confidence)

A survey by IRENA revealed that women represent a higher proportion of full-time employees in the renewable energy industry, compared to their representation in the global oil and gas industry (IRENA 2019a, b, c, d). However, their participation is still low in science, technology, engineering, and mathematics (STEM) jobs compared to administrative jobs. Greater participation of women would allow the sector to unleash female talent while ensuring equitable distribution of socioeconomic opportunities (IRENA 2019a, b, c, d).

Expansion of Ocean-Based Renewable Energy Leads to Job Creation and Economic Growth (High Confidence)

Estimates predict direct full-time employment in offshore wind will be around 435,000 globally by 2030 (OECD 2016). Analysis by Ocean Energy Systems shows that deployment of other forms of ocean energy (tidal range, wave power, and ocean thermal energy) can provide significant benefits in terms of new jobs and additional investments (OES 2017). Ocean-based renewable energy has the potential to provide employment to coastal communities and will benefit workers transitioning from declining offshore fossil fuel industries (Poulsen and Lema 2017; IRENA 2018a, b; Scottish Enterprise n.d.). However, the net global impacts of ocean-based energy on jobs are uncertain.

Opportunities for Innovation Are Expected to Emerge with Expansion of Clean Ocean Energy, Promoting Scientific Research and Resulting in Upgraded Technological Capabilities (High Confidence)

The ocean-based energy industry has experienced rapid growth in installed capacity, ongoing improvements in costs and performance, and increased technological sophistication (IRENA 2018a, b). Innovations in clean ocean energy include the potential to be integrated into and codeveloped with algae-growing facilities and aquaculture farms, and the ability to provide emission-free and drought-resistant drinking water to larger municipalities along the coast (OES 2015; Dirks et al. 2018; Buck et al. 2018). These technologies simultaneously help reduce GHG emissions and increase energy security and diversity (Dinh and McKeogh 2019). Further, there is a trend towards locating offshore energy production to support the expansion of offshore aquaculture production. A number of projects worldwide have started to invest in technologies and system design needed to enable species farming in high-energy environments (Buck et al. 2018).

8.3.2 Ocean-Based Transport

Reducing Emissions from Shipping Vessels will Help Mitigate Ocean Acidification (Medium Confidence)

Strong acids formed from shipping emissions can produce seasonal "hotspots" of ocean acidification in ocean areas close to busy shipping lanes. Hotspots have negative effects on local marine ecology and commercially farmed seafood species (Hassellöv et al. 2013).

Cleaner Marine Shipping Fuels will Reinforce Positive Human Health Outcomes (High Confidence)
Reduced sulphur content of fuel oil used by ships will have beneficial impacts on human health, particularly the health of people living in port cities and coastal communities. Cleaner marine fuels are estimated to reduce premature mortality and morbidity by 34% and 54%, respectively. This represents a roughly 2.6% global reduction in cardiovascular and lung cancer deaths caused by small particulate matter ($PM_{2.5}$) and a roughly 3.6% global reduction in incidence of childhood asthma (Sofiev et al. 2018).

Mitigation Options to Reduce Emissions from Shipping Can Encourage Innovation and Upgrade the Technological Capabilities of the Sector (High Confidence)
Rapid development in power train technology will enable international maritime transport to use alternative and less-polluting fuels, such as hydrogen. The design of ships is being improved to enable them to move more quickly through water, while using less fuel. A complex array of internet-of-things sensors is being developed that will allow collection of data around tidal streams, wind strength, and visibility. This information can be used to reduce vessel waiting time, enable optimisation of routes, and support the concept of autonomous ships.

Reducing Emissions from Shipping Could Potentially Have a Marginal Impact on the Price of Internationally Traded Commodities (Medium Confidence)
While there could be efficiency and energy savings from better design of ships and route optimisation, the cost to the shipping industry of switching to alternative fuels will be high (ETC Mission Possible 2018; Kizielewiczm 2016; Sislian and Jaegler 2016). This could result in significant increases in voyage and freight costs. However, at least one study finds that these costs will have a marginal impact on the final product price of internationally traded commodities (ETC Mission Possible 2018).

8.3.3 Coastal and Marine Ecosystems

Vegetated Coastal and Habitats (Blue Carbon Ecosystems) Contribute to Climate Change Adaptation by Increasing Coastal Resilience and Reducing the Impact of Sea Level Rise (Very High Confidence) Mitigation Options that Help Recovery of Ocean Biomass Can Also Result in Climate Change Adaptation Benefits (High Confidence)
Vegetated coastal habitats reduce coastal flooding by slowing water flow rates and absorbing storm surges. They accrete vertically over time and thereby reduce the impacts of sea level rise and flooding (Duarte et al. 2013). Communities with more extensive mangrove forests experience significantly lower losses from exposure to cyclones than communities without mangroves (Hochard et al. 2019). Increased abundance of marine species is expected to enhance the productivity of surrounding areas, which can help buffer against climate impacts and increase their resilience (Gattuso et al. 2018).

Vegetated Coastal Habitats Offer High Biodiversity Benefits to Terrestrial and Marine Ecosystems, Including Fisheries (Very High Confidence)
Vegetated coastal habitats are used by a remarkable number of marine and terrestrial animals (Li et al. 2018; Rog et al. 2017), including species important for fisheries (Carrasquila Henao and Juanes 2017). Dense vegetated habitats buffer acidification as primary production creates high net pH (Kapsenberg and Cyronak 2019; Hendriks et al. 2014; Krause-Jensen et al. 2016; Wahl et al. 2018). Dense mangroves trap and stabilise sediments that buffer the effects of floodwaters and tidal movements, and are coming to be recognised as valuable natural systems that can play an important role in wastewater treatment systems (Ouyang and Guo 2016).

Integration of Social and Gender Considerations into Restoration Policy for Vegetated Coastal Habitats Can Promote Gender Equity and Educational Opportunities (Medium Confidence)
Local educational institutions and programmes spread awareness in communities about the ecological importance of mangrove forests and encourage community members to get involved in mangrove restoration efforts. Integrating social and gender considerations into restoration practice promotes effectiveness of restoration work (Broekhoven 2015; de la Torre-Castro 2019). Also, increasing women participation in decision making and valuing the traditional and reproductive work of women in households will be important to ensure better governance and policy reform (Gissi et al. 2018; de la Torre-Castro 2019).

Restoring and Protecting Vegetated Coastal Habitats has the Potential to Create Jobs, Promote Economic Growth, and Enhance Research: Involvement of Small-Scale Fishers and Local Stakeholders Throughout the Decision-Making Process Is Crucial to Ensure Delivery of Net Positive Social Outcomes (High Confidence)
Blue carbon projects require development of good practice methods and monitoring (Needelman et al. 2018). Manuals have been developed that support project developers through

the various phases of carbon project implementation, including feasibility and site selection, documentation, registration, implementation, and carbon asset management (Emmer et al. 2014). Job creation could follow successful restoration of coastal ecosystems; however, delivering jobs and other positive social outcomes are dependent on the participation of the affected communities throughout the policy development and implementation stages. Pushing forward blue carbon projects without social safeguards to consider demands from local small-scale fishers and other stakeholders who are heavily dependent on coastal resources for economic sustainability can have unintended negative consequences on societal well-being (Barbesgaard 2018; Bennett 2018; Friess et al. 2019a, b).

Seaweed Farming has Low Levels of Environmental Risks Identified for Small-Scale Cultivation Projects (High Confidence)

Seaweed farming may deliver a range of services and benefits and has the associated great advantage of not requiring arable land and irrigation (Duarte et al. 2017). The seaweed farming also offers climate change adaptation benefits (Duarte et al. 2017; Froehlich et al. 2019). However, while small-scale cultivation projects are considered low risk, expansion of the industry will require a more complete understanding of the scale-dependent changes to balance environmental risks and benefits (Campbell et al. 2019). Risks include spreading disease, changing population genetics, and altering the wider local physiochemical environment (Campbell et al. 2019). If not appropriately located, seaweed farms could also affect seagrass beds, and thereby disturb important flows of ecological goods and services (Eklöf et al. 2005). Spatial planning, ongoing monitoring, and proper management are key to mitigating these impacts.

Seaweed Production Can Lead to Job Creation, Economic Growth, and Enhanced Research (Medium Confidence): It Has a Potential Role in Providing Affordable Energy (Low Confidence)

The seaweed cultivation industry currently accounts for around 51% of total mariculture production and was valued at US $11.7 billion in 2016 (FAO 2018; Chopin 2018b). The rapidly expanding business is providing many jobs, predominantly in developing and emerging economies (Cottier-Cook et al. 2016). Seaweed biomass has potential as a source of various biofuels although it is evident that there are significant technological hurdles to be overcome before seaweed biofuel is viable in either energy or economic terms (Milledge et al. 2014).

Seaweed Farming and Restoring Wetlands Strengthen Capacity to Meet Food Security Targets (Medium Confidence): Healthy Mangroves Positively Impact Health Outcomes for Coastal Communities Through Provision of Food and Medicine to Local Residents (Medium Confidence)

Expansion of seaweed farming in several continents is contributing to global food security, supporting rural livelihoods, and alleviating poverty (Cottier-Cook et al. 2016). Healthy mangroves are important to human societies, providing a variety of ecological services that are critical to human livelihoods and food security, such as providing nursery grounds for important species, improving fisheries production, and filtering and detoxifing water (Ramsar Convention on Wetlands 2018). Mangroves are a direct source of food and traditional medicine for local inhabitants (Bandaranayake 1998).

Mitigation Options to Rebuild Ocean Biomass Can Contribute to Poverty Reduction (Low Confidence)

Marine protected areas have contributed to poverty reduction by improving fish catch, creating new jobs in tourism, strengthening local governance, benefitting human health, and enhancing women's opportunities (Leisher et al. 2007). Marine protected areas require monitoring and continuing study that will contribute to our ecological understanding of the ocean and promote scientific innovation (Nippon Foundation 2017).

Mitigation Options to Rebuild Ocean Biomass Can Also Negatively Impact Poverty Reduction and Employment Targets, and Can Limit Progress on Food Security Targets (Low Confidence)

Marine protection can have negative relationships with ending poverty and reducing inequalities (Singh et al. 2018). For example, ending overfishing and harmful fishing subsidies can conflict with targets related to youth employment if fleet capacity is reduced (Singh et al. 2018). These trade-offs may be avoided through stakeholder consultation and implementation. Conflicts may be temporary and, in the long term, potential increases in marine productivity could increase jobs and resources for people. Evidence shows that declines in fish catch pose risks of nutritional deficiency, especially in developing countries (Golden et al. 2016), and reforms to fishery management could dramatically improve overall fish abundance (compared to BAU) while increasing food security and profits (Costello et al. 2016). However, designating marine protected areas may restrict coastal people's access to local marine resources, which could limit progress on SDG targets associated with ending hunger (Singh et al. 2018).

8.3.4 Fisheries, Aquaculture, and Dietary Shifts

Aquaculture Can Present Numerous Societal and Environmental Challenges: Unplanned Aquaculture Expansion in Some Regions Has Negatively Impacted Other Coastal and Terrestrial Ecosystems (High Confidence)

Aquaculture is associated with multiple environmental impacts, such as eutrophication and spread of invasive species. Unplanned growth in shrimp aquaculture has led to the loss of mangrove ecosystems (Valiela et al. 2001; Richards and Friess 2017), which has in turn led to large CO_2 emissions (Murdiyarso et al. 2015), salinisation, erosion, and reduced coastal resilience (Hochard et al. 2019). Integration of mangroves into aquaculture landscapes may restore some ecosystem services (Hochard et al. 2019; Lee et al. 2019).

Improvement in Feed Conversion Ratio and Use of Plant-Based Ingredients in Aquaculture Feed Rather Than Animal By-Products to Meet The Demand of the Rapidly Growing Marine Aquaculture Sector Can Potentially Reduce Water Use (Medium Confidence)

Given the global supply of fishmeal may be near biological limits (Costello et al. 2012), ensuring that feed for a rapidly growing aquaculture sector comes from terrestrial crops or seaweeds rather than animal by-products would have a positive impact on water use. Reduction in feed conversion ratio in aquaculture production also reduces upstream water use. However, increased inclusion of terrestrial plant-based ingredients may lead to competition for land and water, causing social and environmental conflicts, which may in turn affect the resilience of the global food system (Pahlow et al. 2015; Pelletier et al. 2018; Troell et al. 2014; Blanchard et al. 2017; Malcorps et al. 2019). Many traditional crop-based substitutes are themselves carbon-intensive to produce; they can also adversely affect fish or crustacean growth and health, especially for farmed predator species. Consequently, there have been significant efforts in recent decades to identify new, highly nutritious, and, ideally, low-impact feed sources.

Reducing High Levels of Meat Consumption Among Some Populations and Substituting by Balanced Ocean-Based Protein Has Positive Human Health Benefits: The Overall Impact Depends on Whether Ocean-Based Protein Is Sourced from Sustainable Production Sources or From Indiscriminate Expansion of Aquaculture That Could Negatively Impact Coastal Ecosystems (High Confidence)

High consumption of saturated fats, present in a red meat–based diet, has been linked to cardiovascular disease and certain forms of cancer. Consuming ocean-based proteins, in moderate quantities, ensures a higher intake of bioactive compounds as well as micronutrients, fibre, and omega-3 fatty acids, all of which have well-documented health benefits (Tilman and Clark 2014; González Fischer and Garnett 2016; Simões-Wüst and Dagnelie 2019; Blas et al. 2019; Hollander et al. 2018; Oita et al. 2018). A significant shift from red meat among today's high consumers would dramatically reduce the land and water demands of livestock production (especially cows and sheep) (Poore and Nemecek 2018; Nijdam et al. 2012) and would also reduce the carbon emissions associated with land clearance for pasture (Searchinger et al. 2019).

Mitigation Options Related to Increasing Ocean-Based Protein in Diets and Reducing Emissions in Fisheries and Aquaculture Would Result in Job Creation and Savings for Households, and Encourage Technological Innovation (High Confidence)

The Organisation for Economic Co-operation and Development (OECD) estimates that employment in industrial-scale marine aquaculture will be 3.2 million in 2030, an increase of 1.1 million from 2010 levels. As fuel is a particularly high cost for fishers in developing countries (Lam et al. 2011), structural changes to fisheries that reduce fuel consumption will be economically beneficial. Innovations in developing fish meal substitutes and improving feed efficiency will be crucial to support a rapidly growing aquaculture sector.

8.3.5 Storing Carbon in the Seabed

There Are Large Uncertainties Regarding the Environmental Implications of Carbon Storage Options in the Ocean (High Confidence)

The discussion below does not capture the impacts of carbonate dissolution, alkalinity addition, or ocean fertilisation, which has not been quantified in this report due to the high degree of risk and relatively unknown impacts at this stage. It only considers the impacts of seabed carbon storage. For further information on the broader set of options and why they are not viable at this time, please refer to the Sect. 7.

The injection of CO_2 into submarine geological structures could potentially result in leakages of CO_2 back into the marine environment (Rastelli et al. 2016), affecting the health and function of marine organisms (Queirós et al. 2014). However, there is uncertainty about the gravity of the impacts of CO_2 leakage, especially at the species community level (Adams and Caldeira 2008). Recent evidence indicates that leakage can be reduced if storage sites are well chosen, and well managed and monitored (van der Zwaan and Gerlagh 2016). However, understanding the full range of impacts on ecosystems associated with these solutions is of critical importance. Scientific understanding must be advanced if these technologies are to be used safely and without unintended consequences.

Offshore Investments in Seabed Storage Can Lead to Job Creation, Economic Growth, and Innovation (Low Confidence)

Potential benefits in terms of direct job creation, as well as job retention in harder-to-abate sectors (e.g., heavy industries and fossil fuel based sectors) by allowing them to function with appropriate CCS infrastructure investment/development. A study estimated that carbon capture and storage investments in UK would lead to the creation or retention of 225,600 jobs and a cumulative £54 billion in gross value added (GVA) by 2060 (East Coast UK Carbon Capture and Storage Investment Study 2017).

Evidence indicates a strong need for policy innovation to kick-start carbon capture and storage infrastructure investment (Goldthorpe and Ahmed 2017).

The purpose of the analysis of the wider impacts of ocean-based interventions is to provide insight into the cobenefits as well as risks and trade-offs associated with specific mitigation actions. The approach used here aims to help policymakers evaluate the climate benefits in the context of multiple cobenefits and trade-offs that arise from implementing various ocean-based mitigation options. It is our hope that this report will enable discussion of the corrective measures that might be needed to alleviate unintended consequences of actions and avoid unnecessary risks and trade-offs. The analysis does not attempt a cost-benefit assessment of the mitigation options, which should be a key step in the implementation of any ocean-based mitigation option.

9 Conclusion

This report establishes the potentially significant role of the ocean in limiting global temperature rise, in line with the goals of the Paris Agreement on Climate Change. Analyses in this report reveal that ocean-based mitigation options can make a significant contribution to narrowing the emissions gap that lies between a pathway based on "Current Policy" and the desired pathway that would hold global warming to 1.5 °C above preindustrial levels. Ocean-based interventions could close up to 21% of the emissions gap by 2050. If the world pursues the less ambitious target of 2.0 °C, ocean-based interventions could close 25% of the emissions gap by 2050.

Many of the mitigation options presented in this report can be implemented now with technologies that are already available. To realise these benefits, however, will require significant steps over the coming years—especially with respect to clear policy signals from governments, as well as a greatly increased and targeted investment in research and development.

The options outlined in this report are important not only to support efforts to decarbonise the global economy in line with the goals of the Paris Agreement. They also offer an array of valuable cobenefits in terms of enhanced human health and well-being. In this regard, they contribute to improving the resilience of coastal communities and infrastructure, expanding jobs and economic opportunities, enhancing biodiversity, and strengthening food security. Many of these wider benefits are synergistic with and will support the achievement of the UN Sustainable Development Goals by 2030. However, risks of negative wider impacts cannot be ignored and require detailed attention in policy development, and project planning and implementation. This must be the responsibility of all involved stakeholders—governments, the private sector, researchers, project managers, and local communities.

When considering the political implications of this report, the message is clear. Bold political leadership and clear policy signals will be required to capitalise on the full potential of the solutions explored in this report, coupled with strong national institutions and international cooperation to ensure their effective implementation. Table 17.16 outlines the policy and research actions that must be established over the next 10 years if we are to make significant progress in closing the emissions gap and avoid a climate crisis.

Ultimately, the ocean, its coastal regions, and the economic activities they support should be a source of inspiration and hope in the fight against climate change. With the backdrop of a growing climate catastrophe, the timing of this report is critical, and there could not be a more compelling case for urgent action.

Table 17.16 Short- and medium-term policy, research, and technology priorities necessary to deliver on mitigation potential of ocean-based areas of intervention

	Ocean-based energy		
	Policy	Research	Technology
Short-term priorities (2020–2023)	• Undertake marine spatial planning • Develop national targets to increase the share of renewable energy in the national energy mix • Provide a stable economic and regulatory framework to stimulate investments in required infrastructure for an accelerated deployment of ocean-based energy systems	• Understand the impacts (positive and negative) of both fixed and floating offshore wind installations on marine biodiversity • Undertake a detailed mapping of global renewable energy resources and technical potential	• Advance storage capacity and design • Improve performance, reliability, and survivability, while reducing costs
Medium-term priorities (2023–2025)	• Develop strategic national roadmaps for zero-carbon economy in 2050 • Develop appropriate legislation and regulation	• Understand the potential benefits of co-location with other ocean-based industries (e.g., desalination plants and aquaculture) • Explore the potential for installing large scale floating solar installations at sea (under wave conditions) • Quantify the potential of ocean thermal energy conversion (OTEC)	• Advance technology that can move technologies into deeper water sites (e.g., development of floating offshore wind technologies) to open access to larger areas of energy resources
	Ocean-based transport		
Short-term priorities (2020–2023)	• Redesign the energy efficiency design index (EEDI) formula to avoid vessels being suboptimised for the test only, to ensure that instead vessels are being optimised for minimised fuel consumption in real operation at sea • Adopt policy measures to go beyond Ship Energy Efficiency Management Plan (SEEMP) to incentivise the maximisation of operational efficiency of new and existing ships • Adopt policies that can reduce the broader GHG emissions of shipping instead of CO_2 only, including well-to-tank emissions (WTW) of ship fuels	• Identify and rectify of market and nonmarket barriers and failures to enable larger uptake of more energy-efficient technologies and cooperation patterns • Ensure continuous research on ship design, including hull forms and propulsion, with a focus on reducing energy usage per freight unit transported • Increase focus on utilisation of wind, waves, ocean currents, and sun to reduce use of externally provided energy, i.e., both the carbon and non-carbon-based fuels carried on board	• Develop the necessary high efficiency hull forms and propulsion methods • Develop and implement hybrid power systems, including combustion engines, fuel cells, and batteries technologies • Develop and implement wind assistance technologies • Develop more advanced weather routing systems to better utilise wind, waves, ocean currents, and tides to reduce the use of both carbon and non-carbon fuel carried on board
Medium-term priorities (2023–2025)	• Develop policy to enable the business case for the adoption of low and zero carbon fuels by shipping (e.g. a carbon price) • Commit to the timetable for shipping's transition to low- and zero-carbon fuels • Develop national incentives for decarbonising domestic transportation • Commit to decarbonisation of national energy systems faster or as fast as the transition in the international fleet	• Develop cost-effective production of low- and zero-carbon fuels, both from renewables and from carbon based in combination with carbon capture and storage (CCS) • Develop cost-efficient hybrid setups on seagoing vessels to utilise the best of combustion, fuel cells, and batteries to reduce fuel consumption and local pollution • Ensure safe storage and handling on ships and at the ship-shore interface of hydrogen/ammonia • Ensure safe and efficient use of hydrogen and ammonia in internal combustion engines and fuel cells	• Advance technologies for producing hydrogen, both from renewables and carbon-based fuels • Invest in technologies to store hydrogen (including cryogenic storage of liquid hydrogen, or carriers able to store at high-energy density) • Invest in fuel cells for conversion of future fuels into on-board electricity, and internal combustion engines designed to operate on hydrogen/ammonia

Table 17.16 (continued)

	Ocean-based energy		
	Policy	Research	Technology
	Coastal and marine ecosystems		
Short-term priorities (2020–2023)	• Enhance protection measures for mangroves, seagrass, salt marsh, and seaweed beds to prevent any further losses due to human activities • Provide incentives for restoration of "blue carbon" ecosystems, through payments for ecosystem service schemes, such as carbon and nutrient trading credits • Include quantified nature-based solutions within nationally determined contributions (NDCs) and other relevant climate policies for mitigation and adaptation • Protect coral reefs as important and integrated coastal defence systems for ensuring the protection of coastal blue carbon ecosystems	• Undertake national-level mapping of blue carbon ecosystems • Address biophysical, social, and economic impediments to ecosystem restoration to develop restoration priorities, enhance incentives for restoration, and increase levels of success • Improve the IPCC guidance for seagrasses and other wetland ecosystems • Develop legal mechanisms for long-term preservation of blue carbon, especially in a changing climate • Understand the impacts of climate change on rates of carbon capture and storage, or the potential for restoration	• Advance biorefining techniques, allowing sequential extraction of seaweed products
Medium-term priorities (2023–2025)	• Enhance and adopt carbon accounting methodologies for mangroves, seagrasses and salt marsh within national GHG inventories (IPCC 2013) • Improve methods for monitoring mitigation benefits to enable accounting within national GHG inventories, and biennial transparency reports (BTRs)	• Undertake global-scale map of seaweed ecosystems • Develop IPCC-approved methodological guidance for seaweed ecosystems • Develop methods to fingerprint seaweed carbon beyond the habitat	• Develop and pilot offshore and multiuse sites, including seaweed aquaculture, in the open ocean
	Fisheries, aquaculture, and dietary shifts		
Short-term priorities (2020–2023)	• Eliminate harmful fisheries subsidies (SDG14.6) • Strengthen international tools to eliminate IUU fishing (SDG14.5) • Avoid the transport of fish by air • Reduce discards • Reduce and eliminate hydrochlorofluorocarbons (HCFCs) in refrigerants • Create incentives for shifting diets towards low-carbon protein (e.g., fish) and other food (e.g., seaweed) diets • Create incentives to improve fishery management • Create incentives for lower trophic- level aquaculture • Devise sustainable finance mechanisms for small-scale fishery transitions to sustainable fishing	• Develop disaggregated global data sets for GHG emissions from wild catch fisheries and marine aquaculture • Impacts of scaling marine aquaculture and associated sustainability considerations (e.g., low carbon and climate resilient, environmentally safe) • Enhance understanding of how climate change and ocean acidification will impact aquaculture and fisheries	• Extend surveillance technologies for tracking fishing in the ocean and along coastal areas
Medium-term priorities (2023–2025	• Create incentives to switch from high-carbon land-based sources of protein to low-carbon ocean-based sources • Improve fisheries management to focus on optimising biomass per harvest	• Explore potential impact of a carbon tax on red meat and other carbon intensive foods	• Develop and bring to scale high technology digital aquaculture
	Seabed carbon storage		
Short-term priorities (2020–2023)	• Invest in pilot projects to further explore potential environmental impacts • Incentivise public/private partnerships	• Map global geophysical potential • Understand the impacts of long-lasting containment of CO_2 in a deep seafloor environment	• Few major technical advances are required as seabed storage is already deployed at industrial scale

(continued)

Table 17.16 (continued)

	Ocean-based energy		
	Policy	Research	Technology
Medium-term priorities (2023–2025)	• Develop national strategies and targets • Develop regulatory frameworks to ensure environmental impact assessments and associated precautions are put in place	• Understand the impacts of long-term storage on marine ecosystems • Explore the integrity of long-term storage technologies (leakage)	• Scale up technologies in ways that are economically feasible

Source: Authors

Acknowledgements The authors would like to thank the following people for their review, feedback, and inputs: Adelino Canario, Angela Martin, Adrien Vincent, Beth Fulton, Craig Hansen David Mouillot, David Waskow, Diane Gilpin, Dorothee Herr, Emily Landis, Emily Pidgeon, George Leonard, Gregory Taff, Helen Ding, Ines Aguiar Branco, James Mulligan, Jane Lubchenco, Johannes Friedrich, Johannes Pagenkopf, Juan Carlos, Kelly Levin, Kevern Chochrane, Kristian Teleki, Lisa Schindler Murray, Manuel Barange, Matthew Elliott, Mark Spalding, Michael MacLeod, Narcisa Bandarra, Rashid Sumaila, Tim Scarbrough, Timothy Fitzgerald and Trisha Atwood.

While our colleagues were very generous with their time and input, this report reflects the views of the authors alone.

Thank you also to Bill Dugan, Carni Klirs, Emily Matthews, Margie Peters-Fawcett, Romain Warnault and Shazia Amin for providing administrative, editing, and design support.

About the Authors

Convening Lead Author

Ove Hoegh-Guldberg

Professor and Director of the Global Change Institute, University of Queensland

Contact: oveh@uq.edu.au

Expert Authors

Ken Caldeira

Climate Scientist, Carnegie Institution for Science's Department of Global Ecology Carnegie Institution and Professor, Department of Earth System Sciences Stanford University

Contact: kcaldeira@carnegiescience.edu

Thierry Chopin

Professor of Marine Biology, Seaweed and Integrated Multi-Trophic Aquaculture Laboratory, Department of Biological Sciences, University of New Brunswick

Contact: tchopin@unb.ca

Steve Gaines

Dean and Distinguished Professor, Bren School of Environmental Science and Management, University California, Santa Barbara

Contact: gaines@ucsb.edu

Peter Haugen

Programme Director, Institute of Marine Research

Contact: Peter.Haugan@uib.no

Mark Hemer

Principal Research Scientist, Ocean and Atmosphere Climate Science Centre, CSIRO

Contact: Mark.Hemer@csiro.au

Jennifer Howard

Marine Climate Change Director, Conservation International

Contact: jhoward@conservation.org

Manaswita Konar

Lead Ocean Economist, World Resources Institute and High Level Panel for a Sustainable Ocean Economy Secretariat

Contact: manaswita.konar@wri.org

Dorte Krause-Jensen

Senior Scientist, Department of Bioscience, Aarhus University

Contact: dkj@bios.au.dk

Catherine Lovelock

Professor, School of Biological Sciences, The University of Queensland

Contact: c.lovelock@uq.edu.au

Elizabeth Lindstad

Chief Scientist, SINTEF Ocean

Contact: lindstad@sintef.no

Mark Michelin

Director, California Environmental Associates

Contact: mark@ceaconsulting.com

Finn Gunnar Nielsen

Professor, Head Bergen Offshore Wind Centre (BOW)

Contact: Finn.Nielsen@uib.no

Eliza Northrop

Senior Associate, World Resources Institute and High Level Panel for a Sustainable Ocean Economy Secretariat

Contact: enorthrop@wri.org

Robert W. R. Parker

Killam Postdoctoral Fellow, School for Resource and Environmental Studies, Dalhousie University

Contact: Rob.Parker@dal.ca

Joyashree Roy

Professor of Economics, Jadavpur University

Contact: joyashree@ait.asia

Tristan Smith

Reader in Energy and Shipping, Bartlett School Environment, Energy & Resources, Faculty of the Built Environment, UCL Energy Institute

Contact: tristan.smith@ucl.ac.uk

Shreya Some

Senior Research Fellow, Jadavpur University

Contact: ayerhs7891@gmail.com

Peter Tyedmers

Professor, School for Resource and Environmental Studies, Dalhousie University

Contact: Peter.Tyedmers@dal.ca

References

Abroms LC, Maiboch EW (2008) The effectiveness of mass communication to change public behavior. Annu Rev Public Health 29:219–234

Adams EE, Caldeira K (2008) Ocean storage of CO_2. Elements 4(5):319–324. https://people.ucsc.edu/~mdmccar/migrated/ocea213/readings/15_GeoEngineer/C_sequestration/adams_2008_Elements_CALDERIA_Ocean_CO_2_Storeage.pdf

Alexandratos N, Bruinsma J (2012) World Agriculture towards 2030/2050: the 2012 revision. In: ESA Working paper No. 12-03. FAO, Rome

Allison EH, Ratner BD, Åsgård B, Willmann R, Pomeroy R, Kurien J (2012) Rights-based fisheries governance: from fishing rights to human rights. Fish Fish 13(1):14–29

Alvarez-Silva OA, Osorio AF, Winter C (2016) Practical global salinity gradient energy potential. Renew Sust Energ Rev 60:1387–1395

Arias-Ortiz A, Serrano O, Masqué P, Lavery PS, Mueller U, Kendrick GA, Jamaludin MR et al (2018) A marine heatwave drives massive losses from the world's largest seagrass carbon stocks. Nat Clim Chang 8:338. https://doi.org/10.1038/s41558-018-0096-y

Arnason R, Kelleher K, Willmann R (2009) The sunken billions: the economic justification for fisheries reform. Joint publication of the World Bank and the Food and Agriculture Organization (FAO). The International Bank for Reconstruction and Development/The World Bank, Washington

Arnason R, Kobayashi M, de Fontaubert C (2017) The Sunken billions revisited: progress and challenges in global marine fisheries. The World Bank, Washington

Atwood TB, Connolly RM, Ritchie EG, Lovelock CE, Heithaus MR, Hays GC, Fourquerean JW et al (2015) Predators help protect carbon stocks in blue carbon ecosystems. Nat Clim Chang 5(12):1038

Atwood TB, Connolly RM, Almahasheer H, Carnell PE, Duarte CM, Lewis CJE, Irigoien X et al (2017) Global patterns in mangrove soil carbon stocks and losses. Nat Clim Chang 7(7):523

Bahar H (2019) TCEP: ocean power. [Online]. https://www.Iea.org. Accessed 10 September

Bandaranayake WM (1998) Traditional and medicinal uses of mangroves. Mangrove Salt Marshes 2(3):133–148

Barbesgaard M (2018) Blue growth: savior or ocean grabbing? J Peasant Stud 45(1):130–149

Barner AK, Lubchenco J, Costello C, Gaines SD, Leland A, Jenks B, Murawski S et al (2015) Solutions for recovering and sustaining the bounty of the ocean: combining fishery reforms, rights-based fisheries management, and marine reserves. Oceanography 28:252–263

Bar-On YM, Phillips R, Milo R (2018) The biomass distribution on earth. Proc Natl Acad Sci 115(25):6506–6511. https://doi.org/10.1073/pnas.1711842115

Bayraktarov E, Saunders MI, Abdullah S, Mills M, Beher J, Possingham HP, Mumby PJ et al (2016) The cost and feasibility of marine coastal restoration. Ecol Appl 26(4):1055–1074

Bell M, Deyes K, Fitzpatrick N, Wilson C, Chase A, Lewis C, Bates G (2019) Reducing the maritime sector's contribution to climate change and air pollution: economic opportunities from low and zero emission shipping. Department for Transport, London

Bennett NJ (2018) Navigating a just and inclusive path towards sustainable oceans. Mar Policy 97:139–146

Best BD, Halpin PN (2019) Minimizing wildlife impacts for offshore wind energy development: winning tradeoffs for seabirds in space and cetaceans in time. PLoS ONE 14(5):e0215722. https://doi.org/10.1371/journal.pone.0215722

Blanchard JL, Watson RA, Fulton EA, Cottrell RS, Nash KL, Bryndum-Buchholz A, Büchner M et al (2017) Linked sustainability challenges and trade-offs among fisheries, aquaculture and agriculture. Nat Ecol Evol 1:1240–1249

Blas A, Garrido A, Unver O, Willaarts B (2019) A comparison of the Mediterranean diet and current food consumption patterns in Spain from a nutritional and water perspective. Sci Total Environ 664:1020–1029. www.ncbi.nlm.nih.gov/pubmed/30769304

Bosch J, Staffell I, Hawkes AD (2018) Temporally explicit and spatially resolved global offshore wind energy potentials. Energy 163:766–781. https://doi.org/10.1016/j.energy.2018.08.153

Bouman E, Lindstad E, Rialland AI, Strømman AH (2017) State-of-the-art technologies, measures, and potential for reducing GHG emissions from shipping: a review. Transp Res D 52:408–421

Bridgham SD, Patrick Megonigal J, Keller JK, Bliss NB, Trettin C (2006) The carbon balance of north American wetlands. Wetlands 26(4):889–916

Broekhoven S (2015) Quality and safety aspects of mealworms as human food

Buck BH, Troell MF, Krause G, Angel D, Grote B, Chopin T (2018) State of the art and challenges for offshore integrated multi-trophic aquaculture (IMTA). Front Mar Sci 5(165):1–21

Buhaug Ø, Corbett JJ, Endresen Ø, Eyring V, Faber J, Hanayama S, Lee DS et al (2009) Second IMO GHG Study 2009. International Maritime Organization, London. www.imo.org/en/OurWork/Environment/PollutionPrevention/AirPollution/Documents/SecondIMOGHGStudy2009.pdf

Bunting P, Lucas R, Rosenqvist A, Rebelo L-M, Hilarides L, Thomas N, Hardy A et al (2018) The global mangrove watch—a new 2010 baseline of mangrove extent. Remote Sens 10:1669. https://doi.org/10.3390/rs10101669

Buonocore JJ, Luckow P, Fisher J, Kempton W, Levy JI (2016) Health and climate benefits of offshore wind facilities in the mid-Atlantic United States. Environ Res Lett 11(7):19. https://doi.org/10.1088/1748-9326/11/7/074019/meta

Caldeira K, Rau GH (2000) Accelerating carbonate dissolution to sequester carbon dioxide in the ocean: geochemical implications. Geophys Res Lett 27:225–228

Campbell I, Macleod A, Sahlmann C, Neves L, Funderud J, Overland M, Hughes AD et al (2019) The environmental risks associated with the development of seaweed farming in Europe—prioritizing key knowledge gaps. Front Mar Sci 6:107

Carrasquila Henao M, Juanes F (2017) Mangroves enhance local fisheries catches: a global meta-analysis. Fish Fish 18:79–93. https://doi.org/10.1111/faf.12168

Cascajo R, Garcia E, Quiles E, Correcher A, Morant F (2019) Integration of marine wave energy converters into seaports: a case study in the Port of Valencia. Energies 12:787. https://doi.org/10.3390/en12050787

Chopin T (2015) Marine aquaculture in Canada: well-established monocultures of finfish and shellfish and an emerging integrated multi-

trophic aquaculture (IMTA) approach including seaweeds, other invertebrates, and microbial communities. Fisheries 40(1):28–31
Chopin T (2017) Seaweed aquaculture–from the global, mostly Asian, picture to the opportunities and constraints of the Canadian scene. Bull Aquac Assoc Can 1:3–8
Chopin T (2018a) Seaweeds: a multi-purposed bioresource well-suited for integrated sequential biorefinery (ISBR) processing. Int Aquafeed 21(11):14–15
Chopin T (2018b) Seaweeds: the world's largest mariculture crop. Int Aquafeed 21(8):14–15
Climate Action Tracker (2018). Some progress since Paris, but not enough, as governments amble
Copping A, Sather N, Hanna L, Whiting J, Zydlewski G, Staines G, Gill A et al (2016) Annex IV 2016 state of the science report: environmental effects of marine renewable energy development around the world. https://tethys.pnnl.gov/publications/state-of-the-science-2016
Costello C, Ovando D, Hilborn R, Gaines SD, Deschenes O, Lester SE (2012) Status and solutions for the world's unassessed fisheries. Science 338(6106):517–520
Costello C, Ovando D, Clavelle T, Strauss CK, Hilborn R, Melnychuk MC, Branch TA et al (2016) Global fishery prospects under contrasting management regimes. Proc Natl Acad Sci 113(18):5125–5129
Cottier-Cook EJ, Nagabhatla N, Badis Y, Campbell ML, Thierry C, Dai W, Fang J et al (2016) Safeguarding the future of the global seaweed aquaculture industry. United Nations University (INWEH) and Scottish Association for Marine Science Policy Brief, Tokyo. https://blogs.unb.ca/newsroom/_media/images/2016/09/2482893c83ca06153f5dc3a627a59d6bcottier-cook-briefing-1.pdf
Couture V, Faber B, Gu Y, Liu L (2018) E-Commerce integration and economic development: evidence from China. NBER Working Paper 24384.
de la Torre-Castro M (2019) Inclusive management through gender consideration in small-scale fisheries: the why and the how. Front Mar Sci 6:156
de los Santos CB, Krause-Jensen D, Alcoverro T, Marbà N, Duarte CM, van Katwijk MM, Pérez M et al (2019) Recent trend reversal for declining European seagrass meadows. Nat Commun 10:3356
Dinh VN, McKeogh E (2018) Offshore wind energy: technology opportunities and challenges. Lecture notes in civil engineering. Adv Offshore Eng 18:3–22. https://doi.org/10.1007/978-981-13-2306-5_31
Dinh VN, McKeogh E (2019) Offshore wind energy: technology opportunities and challenges. In: Randolph M, Doan D, Tang A, Bui M, Dinh V (eds) Proceedings of the 1st Vietnam symposium on advances in offshore engineering: energy an geotechnics (lecture notes in civil engineering 18). Springer, Singapore. https://doi.org/10.1007/978-981-13-2306-5_1
Dirks WG, Kappen JMJ, Scheltens RD, Winkes J, Wochner MS (2018) Sustainable offshore wind energy: faster and larger at lower cost and less environmental impact. Proc Annu Offshore Technol Conf 5:3246–3265. www.scopus.com/inward/record.uri?eid=2-s2.0--85050551938&partnerID=40&md5=cfdc0566b3dbf0d735a6742aee2d128b
Duarte CM, Dennison WC, Orth RJ, Carruthers TJ (2008) The charisma of coastal ecosystems: addressing the imbalance. Estuar Coasts 31(2):233–238
Duarte CM, Losada IJ, Hendriks IE, Mazarrasa I, Marbà N (2013) The role of coastal plant communities for climate change mitigation and adaptation. Nat Clim Chang 3(11):961
Duarte CM, Wu J, Xiao X, Bruhn A, Krause-Jensen D (2017) Can seaweed farming play a role in climate change mitigation and adaptation? Front Mar Sci. https://doi.org/10.3389/fmars.2017.00100
Dutreuil S, Bopp L, Tagliabue A (2009) Impact of enhanced vertical mixing on marine biogeochemistry: lessons for geo-engineering and natural variability. Biogeosciences 6:901–912. https://doi.org/10.1088/1748-9326/10/3/034016
East Coast UK Carbon Capture and Storage Investment Study (2017) Clean air—clean industry—clean growth: how carbon capture will boost the UK economy. http://www.ccsassociation.org/news-and-events/reports-and-publications/clean-air-clean-industry-clean-growth/
Eklöf JS, de la Torre Castro M, Adelsköld L, Jiddawi NS, Kautsky N (2005) Differences in macrofaunal and seagrass assemblages in seagrass beds with and without seaweed farms. Estuar Coast Shelf Sci 63(3):385–396
Elder RW, Shults RA, Sleet DA, Nichols JL, Thompson RS, Rajab W (2004) Effectiveness of mass media campaigns for reducing drinking and driving and alcohol-involved crashes. A systematic review. Am J Prev Med 27:57–65
Emmer I, Needelman B, Emmett-Mattox S, Crooks S, Megonigal P, Myers D, Oreska M et al (2014) Methodology for tidal wetland and seagrass restoration. VM0033. Verified Carbon Standard Energy and Utility Skills Ltd. 2017. Skills and labour requirements of the UK offshore wind industry: 2018 to 2032. Solihull, UK. https://aurawindenergy.com/uploads/publications/Aura-EU-Skills-UK-Offshore-Wind-Skills-Study-Full-Report-October-2018.pdf
Esteves LS, Williams JJ (2017) Managed realignment in Europe: a synthesis of methods, achievements and challenges. In: Bilkovic DM, Mitchell MM, Toft JD, La Peyre MK (eds) Living shorelines: the science and management of nature-based coastal protection. CRC Press, Boca Raton, pp 157–180
ETC Mission Possible (2018) Mission possible: reaching net-zero carbon emissions from harder-to-abate subsectors by mid-century. ETC, Mission Possible. http://www.energy-transitions.org/sites/default/files/ETC_MissionPossible_FullReport.pdf
European Commission (2018) In-depth analysis in support of the commission communication (2018) 773: a clean planet for all—a European strategic long-term vision for a prosperous, modern, competitive and climate neutral economy. European Commission, Brussels. https://ec.europa.eu/clima/sites/clima/files/docs/pages/com_2018_733_analysis_in_support_en_0.pdf
Evans L, Cherrett N, Pemsl D (2011) Assessing the impact of fisheries co-management interventions in developing countries: a meta-analysis. J Environ Manag 92(8):1938–1949
Faber J, t'Hoen M, Vergeer R, Calleya J (2016) Historical trends in ship design efficiency. CE Delft, Delft
FAO (2013) The state of food and agriculture 2012: investing in agriculture for a better future. FAO, Rome, p 165. http://www.fao.org/docrep/017/i3028e/i3028e.pdf; http://www.fao.org/docrep/018/i3300e/i3300e.pdf
FAO (2018) The state of world fisheries and aquaculture 2018: meeting the sustainable development goals. FAO, Rome
Farmery A, Gardner C, Green BS, Jennings S (2014) Managing fisheries for environmental performance: the effects of marine resource decision-making on the footprint of seafood. J Clean Prod 64:368–376
Filbee-Dexter K, Wernberg T, Norderhaug KM, Ramirez-Llodra E, Pedersen MF (2018) Movement of pulsed resource subsidies from kelp forests to deep fjords. Oecologia 187(1):291–304
Friess DA, Rogers K, Lovelock CE, Krauss KW, Hamilton SE, Lee SY, Lucas R et al (2019a) The state of the world's mangrove forests: past, present, and future. Annu Rev Environ Resour 44:89
Friess DA, Lovelock CE, Aung T, Mukherjee N, Huxham M, Sasmito S (2019b) SDG14 'life below water. In: Sustainable development goals: their impacts on forests and people. Cambridge University Press, Cambridge
Froehlich HE, Afflerbach JC, Frazier M, Halpern BS (2019) Blue growth potential to mitigate climate change through seaweed offsetting. Curr Biol. https://doi.org/10.1016/j.cub.2019.07.041
Gattuso J-P, Magnan AK, Bopp L, Cheung WWL, Duarte CM, Hinkel J, Mcleod E et al (2018) Ocean solutions to address climate change and its effects on marine ecosystems. Front Mar Sci 5:337. https://doi.org/10.3389/fmars.2018.00337

Global Carbon Budget (2018) Earth Syst. Sci. 10:2141–2194. https://doi.org/10.5194/essd-10-2141-2018

GEA (2012) Toward a sustainable future. International Institute for Applied Systems Analysis, Vienna

Gerber PJ, Steinfeld H, Henderson B, Mottet A, Opio C, Dijkman J, Falcucci A et al (2013) Tackling climate change through livestock—a global assessment of emissions and mitigation opportunities. FAO, Rome

GESAMP (2019) High level review of a wide range of proposed marine geoengineering techniques. Rep Stud 98:144. http://www.gesamp.org/publications/high-level-review-of-a-wide-range-of-proposed-marine-geoengineering-techniques

Gissi E, Portman ME, Hornidge AK (2018) Un-Gendering the ocean: why women matter in ocean governance for sustainability. Mar Policy 94:215–219

Global Panel on Agriculture and Food Systems for Nutrition (2017) Healthy diets for all: a key to meeting the SDGS. Policy Brief No. 10. Global Panel on Agriculture and Food Systems for Nutrition, London. https://glopan.org/sites/default/files/SDGPolicyBrief.pdf

Golden CD, Allison EH, Cheung WWL, Dey MM, Halpern BS, McCauley DJ, Smith M et al (2016) Nutrition: fall in fish catch threatens human health. Nature 534:317–320. https://doi.org/10.1038/534317a

Goldthorpe W, Ahmed S (2017) Policy innovation for offshore CO_2 transport and storage deployment. Energy Procedia 114:7540–7549. https://doi.org/10.1016/j.egypro.2017.03.1886

González CG (2011) Climate change, food security, and agrobiodiversity: toward a just, resilient, and sustainable food system. Fordham Environ Law Rev 22:493

González Fischer CG, Garnett T (2016) Plates, pyramids, planet. In: Developments in national healthy and sustainable dietary guidelines: a state of play assessment. FAO, Oxford. http://www.fao.org/3/a-i5640e.pdf

González M, Ilyina T (2016) Impacts of artificial ocean alkalinization on the carbon cycle and climate in earth system simulations: impacts of CO_2 stablilization via AOA. Geophys Res Lett 43:068576

González AD, Frostell B, Carlsson-Kanyama A (2011) Protein efficiency per unit energy and per unit greenhouse gas emissions: potential contribution of diet choices to climate change mitigation. Food Policy 36:562–570

Griscom BW, Adams J, Ellis PW, Houghton RA, Lomax G, Miteva DA, Schlesinger WH (2017) Nature climate solutions. PNAS 114(44):11645–11650

Grubler A, Wilson C, Bento N, Boza-Kiss B, Krey V, McCollum DL, Rao ND et al (2018) A low energy demand scenario for meeting the 1.5 degrees C target and sustainable development goals without negative emission technologies. Nat Energy 3:515–527. https://doi.org/10.1038/s41560-018-0172-6

Gustavsson J, Cederberg C, Sonesson U (2011) Global food losses and food waste. Save Food at Interpack, Düsseldorf

GWEC (2017) Global wind report 2017. A snapshot of the top wind markets in 2017. In: Offshore wind. https://gwec.net/wp-content/uploads/2018/04/offshore.pdf

Hallstrom E, Bergman K, Mifflin K, Parker R, Tyedmers P, Troell M, Ziegler F (2019) Combined climate and nutritional performance of seafoods. J Clean Prod 230:402–411

Harrison DP (2017) Global negative emissions capacity of ocean macronutrient fertilization. Environ Res Lett 12:23

Hartmann J, West AJ, Renforth P, Köhler P, De La Rocha CL, Wolf-Gladrow DA, Dürr HH et al (2013) Enhanced chemical weathering as a geoengineering strategy to reduce atmospheric CO_2, supply nutrients, and mitigate ocean acidification. Rev Geophys 51:113–149. https://doi.org/10.1002/rog.20004

Hassellöv I-M, Turner DR, Lauer A, Corbett JJ (2013) Shipping contributes to ocean acidification. Geophys Res Lett 40:2731–2736. https://doi.org/10.1002/grl.50521

Havlık P, Valin H, Herrero M, Obersteiner M, Schmid E, Rufino MC, Mosnier A et al (2014) Climate change mitigation through livestock system transitions. Proc Natl Acad Sci U S A 111:3709–3714

Hendriks IE, Olsen YS, Ramajo L, Basso L, Steckbauer A, Moore TS, Howard J et al (2014) Photosynthetic activity buffers ocean acidification in seagrass meadows. Biogeosciences 11:333. https://doi.org/10.5194/bg-11-333-2014

Henriksson PJG, Pelletier NL, Troell M, Tyedmers PH (2012) Life cycle assessment and its application to aquaculture production systems. In: Meyers R (ed) Encyclopedia of sustainability science and technology. Springer, New York

Herr D, Landis E (2016) Coastal blue carbon ecosystems. Opportunities for nationally determined contributions. In: Policy brief. International Union for Conservation of Nature (IUCN), Gland

Herr D, von Unger M, Laffoley D, McGivern A (2017) Pathways for implementation of blue carbon initiatives. Aquat Conserv Mar Freshw Ecosyst 27(S1):116–129. https://doi.org/10.1002/aqc.2793

Hochard JP, Hamilton S, Barbier EB (2019) Mangroves shelter coastal economic activity from cyclones. Proc Natl Acad Sci 116(25):12232–12237

Hoegh-Guldberg O, Cai R, Poloczanska ES, Brewer PG, Sundby S, Hilmi K et al (2014) The ocean. In: Field CB, Barros VR, Dokken DJ, Mach KJ, Mastrandrea MD, Bilir TE et al (eds) Climate change 2014: impacts, adaptation, and vulnerability part a: global and sectoral aspects contribution of working group II to the fifth assessment report of the Intergovernmental Panel of Climate Change. Cambridge University Press, Cambridge, pp 655–731

Hoegh-Guldberg O, Jacob D, Taylor M, Bindi M, Brown S, Camilloni I, Diedhiou A et al (2018) Impacts of 1.5°C global warming on natural and human systems. In: Masson-Delmotte et al (eds) Global warming of 1.5°C. World Meteorological Organization, Geneva

Hollander A, De Jonge R, Biesbroek S, Hoekstra J, Zijp MC (2018) Exploring solutions for healthy, safe, and sustainable fatty acids (EPA and DHA) consumption in the Netherlands. Sustain Sci 14(2):303–313. https://pubag.nal.usda.gov/catalog/6325482

House RJ, Javidan M, Dorfman PW, de Luque MS (2006) A failure of scholarship: response to George Graen's critique of globe. Acad Manag Perspect 20:102–114

IEA (2010) Energy technology perspectives 2010. Scenarios and strategies to 2050. IEA, Paris

IEA (2017) Energy technology perspectives 2017: catalysing energy technology transformations. International Energy Agency, Paris. www.iea.org/etp2017

IEA (2018) Offshore energy outlook. World energy outlook series. International Energy Agency, Paris

IEA (2019) Global energy and CO_2 status report (GECO): the latest trends in energy and emissions in 2018. www.iea.org/geco/electricity

IMO (2011) Amendments to the annex of the protocol of 1997 to amend the international convention for the prevention of pollution from ships, 1973, as modified by the protocol of 1978 relating thereto. (Inclusion of regulations on energy efficiency for ships in MARPOL Annex VI). Resolution MEPC 203 (62) (Adopted on 15 July, 2011). http://www.imo.org/en/KnowledgeCentre/IndexofIMOResolutions/Marine-Environment-Protection-Committee-(MEPC)/Documents/MEPC.203(62).pdf

IPCC 2005 - Bert Metz, Ogunlade Davidson, Heleen de Coninck, Manuela Loos and Leo Meyer (Eds.). Cambridge University Press, UK. pp 431. Available from Cambridge University Press, The Edinburgh Building Shaftesbury Road, Cambridge CB2 2RU ENGLAND

IPCC (2005) IPCC special report on carbon dioxide capture and storage. In: Metz B, Davidson O, de Coninck H, Loos M, Meyer L (eds). Cambridge University Press, Cambridge. pp 431

IPCC (2006) Guidelines for national greenhouse gas inventories. In: Institute for Global Environmental Strategies (IGES), Kanagawa

IPCC (2013) Climate change 2013: the physical science basis. In: Stocker TF, Qin D, Plattner G-K, Tignor M, Allen SK, Boschung J, Nauels A et al (eds) Contribution of working group I to the fifth assessment report of the Intergovernmental Panel on Climate Change. Cambridge University Press, Cambridge. http://www.ipcc.ch/report/ar5/wg1/

IPCC (2014) Climate change 2014: synthesis report. In: Pachauri RK, Meyer LA (eds) Contribution of working groups I, II and III to the fifth assessment report of the Intergovernmental Panel on Climate Change. IPCC, Geneva. www.ipcc.com

IPCC (2018) Global warming of 1.5°C: an IPCC special report on the impacts of global warming of 1.5°C above pre-industrial levels and related global GHG emission pathways, in the context of strengthening the global response to the threat of climate change. World Meteorological Organization, Geneva

IPCC (2019) An IPCC special report on the ocean and cyrosphere in a changing climate (SROCC). www.ipcc.com

IRENA (2014a) Ocean thermal energy conversion: technology brief. IRENA, Abu Dhabi. www.irena.org/DocumentDownloads/Publications/Ocean_Thermal_Energy_V4_web.pdf

IRENA (2014b) Tidal energy: technology brief. IRENA, Abu Dhabi. www.irena.org/documentdownloads/publications/tidal_energy_v4_web.pdf

IRENA (2014c) Wave energy: technology brief. IRENA, Abu Dhabi. www.irena.org/documentdownloads/publications/wave-energy_v4_web.pdf

IRENA (2017) Electricity storage and renewables: costs and markets to 2030. International Renewable Energy Agency, Abu Dhabi

IRENA (2018a) Offshore wind investment, policies and job creation: review of key findings for G7 ministerial meetings. IRENA, Abu Dhabi. www.irena.org/-/media/Files/IRENA/Agency/Publication/2018/Sep/IRENA_offshore_wind_note_G7_2018.pdf?la=en&hash=B186614D923AB1F0A07D7285612C4B037057A0C0

IRENA (2018b) Renewable energy benefits leveraging local capacity for offshore wind. International Renewable Energy Agency, Abu Dhabi. www.irena.org/-/media/Files/IRENA/Agency/Publication/2018/May/IRENA_Leveraging_for_Offshore_Wind_2018.pdf

IRENA (2019a) Global energy transformation. A roadmap to 2050. 2019 edition. International Renewable Energy Agency, Abu Dhabi. www.irena.org/publications/2019/Apr/Global-energy-transformation-A-roadmap-to-2050-2019Edition

IRENA (2019b) Renewable energy: a gender perspective. IRENA, Abu Dhabi. https://irena.org/-/media/Files/IRENA/Agency/Publication/2019/Jan/IRENA_Gender_perspective_2019.pdf

IRENA (2019c) Renewable capacity statistics 2019. International Renewable Energy Agency, Abu Dhabi. https://www.irena.org/-/media/Files/IRENA/Agency/Publication/2019/Mar/IRENA_RE_Capacity_Statistics_2019.pdf

IRENA (2019d) Renewable power generation costs in 2018. International Renewable Energy Agency, Abu Dhabi. https://www.irena.org/-/media/Files/IRENA/Agency/Publication/2019/May/IRENA_Renewable-Power-Generations-Costs-in-2018.pdf

James R, Ros MC (2015) Floating offshore wind: market and technology review. The Carbon Trusts, UK, June. https://www.carbontrust.com/media/670664/floating-offshore-wind-market-technology-review.pdf

Kapsenberg L, Cyronak T (2019) Ocean acidification refugia in variable environments. Glob Chang Biol 2019:1–14. https://doi.org/10.1111/gcb.14730

Keller DA, Gittman RK, Brodeur MC, Kenworthy MD, Ridge JT, Yeager LA, Rodriguez AB, Fodrie FJ (2019) Salt marsh shoreline geomorphology influences the success of restored oyster reefs and use by associated fauna. Restor Ecol. https://doi.org/10.1111/rec.12992

Khaled R, Priour D, Billard JY (2013) Cable length optimization for trawl fuel consumption reduction. Ocean Eng 58:167–179

Kheshgi HS (1995) Sequestering atmospheric carbon dioxide by increasing ocean alkalinity. Energy 20(9):915–922. https://doi.org/10.1016/0360-5442(95)00035-F

Kizielewiczm J (2016) Environmental governance for sustainable cruise tourism development by territorial authorities and cruise ship-owners. In: International conference on management, leadership and governance, pp. 193–201. https://pbn.nauka.gov.pl/sedno-webapp/works/749267

Krause-Jensen D, Duarte CM (2016) Substantial role of macroalgae in marine carbon sequestration. Nat Geosci 9(10):737

Krause-Jensen D, Marbà N, Sanz-Martin M, Hendriks IE, Thyrring J, Carstensen J, Sejr MK et al (2016) Long photoperiods sustain high pH in arctic kelp forests. Sci Adv 2(12):e1501938

Krause-Jensen D, Lavery P, Serrano O, Marbà N, Masque P, Duarte CM (2018) Sequestration of macroalgal carbon: the elephant in the blue carbon room. Biol Lett 14:20180236. https://doi.org/10.1098/rsbl.2018.0236

Krewitt W, Nienhaus K, Kleßmann C, Capone C, Stricker E, Graus W, Hoogwijk M et al (2009) Role and potential of renewable energy and energy efficiency for global energy supply. Climate Change 18:336

Kris-Etherton PM, Harris WS, Appel LJ (2002) Fish consumption, fish oil, omega-3 fatty acids, and cardiovascular disease. Circulation 106:2747–2757

Kroeker K, Kordas R, Crim R, Hendriks I, Ramajo L, Singh G, Duarte C et al (2013) Impacts of ocean acidification on marine organisms: quantifying sensitivities and interaction with warming. Glob Chang Biol 19:12179. https://doi.org/10.1111/gcb.12179

Krumhansl KA, Okamoto DK, Rassweiler A, Novak M, Bolton JJ, Cavanaugh KC, Connell SD et al (2016) Global patterns of kelp forest change over the past half-century. Proc Natl Acad Sci 113(48):13785–13790

Kuwae T, Hori M (2019) The future of blue carbon: addressing global environmental issues. In: Blue carbon in shallow coastal ecosystems. Springer, Singapore, pp 347–373

Kwiatkowski L, Ricke KL, Caldeira K (2015) Atmospheric consequences of disruption of the ocean thermocline. Environ Res Lett 10:034016

Lam VWY, Sumaila UR, Dyck A, Pauly D, Watson R (2011) Construction and first applications of a global cost of fishing database. ICES J Mar Sci 68(9):1996–2004. https://doi.org/10.1093/icesjms/fsr121

Langhamer O (2012) Artificial reef effect in relation to offshore renewable energy conversion: state of the art. Sci World J 2012:386713. https://doi.org/10.1100/2012/386713

Lapointe BE, West LE, Sutton TT, Chuanmin H (2014) Ryther Revisited: Nutrient excretions by fishes enhance productivity of pelagic sargassum in the Western North Atlantic Ocean. J Exp Mar Biol Ecol 458:46–56. https://doi.org/10.1016/j.jembe.2014.05.002

Latorre R (2001) Reducing fishing vessel fuel consumption and NOx emissions. Ocean Eng 28(6):723–733

Lavery TJ, Roudnew B, Gill P, Seymour J, Seuront L, Johnson G, Mitchell JG et al (2010) Iron defecation by sperm whales stimulates carbon export in the southern ocean. Proc R Soc B Biol Sci 277(1699):3527–3531

Lee SY, Hamilton S, Barbier EB, Primavera J, Lewis RR (2019) Better restoration policies are needed to conserve mangrove ecosystems. Nat Ecol Evol 3(6):870

Lehahn Y, Ingle KN, Golberg A (2016) Global potential of offshore and shallow waters macroalgal biorefineries to provide for food, chemicals and energy: feasibility and sustainability. Algal Res 17(2016):150–160. https://doi.org/10.1016/j.algal.2016.03.031

Leisher C, van Beukering P, Scher LM (2007) Nature's investment bank: how marine protected areas contribute to poverty reduction, Arlington.: The Nature Conservancy (TNC). www.nature.org/media/science/mpa_report.pdf

Lester SE, McDonald G, Clemence M, Dougherty DT, Szuwalski CS (2017) Impacts of TURFs and marine reserves on fisheries and conservation goals: theory, empirical evidence, and modeling. Bull Mar Sci 93(1):173–198

Levin K, Fransen F, Ge M (2018) Five things you need to know about the UN emissions gap report. World Resources Institute, Washington

Lewis A, Estefen S, Huckerby J, Musial W, Pontes T, Torres-Martinez J (2011) Ocean energy. In: Edenhofer O, Pichs-Madruga R, Sokona Y, Seyboth K, Matschoss P, Kadner S, Zwickel T et al (eds) IPCC special report on renewable energy sources and climate change mitigation. Cambridge University Press, Cambridge

Li X, Bellerby R, Craft C, Widney E, Sarah. (2018) Coastal wetland loss, consequences, and challenges for restoration. Anthropocene Coasts 2018:1–15. https://doi.org/10.1139/anc-2017-0001

Lindstad HE, Bø TI (2018) Potential power setups, fuels and hull designs capable of satisfying future EEDI requirements. Transp Res D 63:276–290

Lindstad HE, Eskeland GS (2018) Marine design XIII. In: Kujala P, Lu L (eds) Reducing GHG emissions in shipping—measures and options. CRC Press, London, pp 923–930

Lindstad HE, Jullumstrø E, Sandaas I (2013) Reduction in cost and emissions with new bulk ships designs enabled by the Panama canal expansion. Energy Policy 59:341–349

Lindstad HE, Verbeek R, Blok M, van Zyl S, Hübscher A, Kramer H, Purwanto J et al (2015) GHG emission reduction potential of EU-related maritime transport and on its impacts. European Commission, Brussels

Lindstad HE, Eskeland GS, Rialland A (2017b) Batteries in offshore support vessels—pollution, climate impact and economics. Transp Res D 50:409–417

Lindstad HE, Borgen H, Eskeland GS, Paalson C, Psaraftis H, Turan O (2019) The need to amend IMO's EEDI to include a threshold for performance in waves (realistic sea conditions) to achieve the desired GHG reductions. Sustain For 11:3668. https://doi.org/10.3390/su11133668

Love DC, Uhl MS, Genello L (2015) Energy and water use of a small-scale raft aquaponics system in Baltimore, Maryland, United States. Aquac Eng 68:19–27

Lovelock CEIC, Feller R, Reef RW, Ruess. (2015) Variable effects of nutrient enrichment on soil respiration in mangrove forests. Plant Soil 379:135–148

Lovelock CE, Atwood T, Baldock J, Duarte CM, Hickey S, Lavery PS, Masque P et al (2017) Assessing the risk of CO_2 emissions from blue carbon ecosystems. Front Ecol Environ 15(5):257–265

LR and UMAS (2019) Zero-emission vessels transition pathways

Lubchenco J, Cerny-Chipman EB, Reimer JN, Levin SA (2016) The right incentives enable ocean sustainability successes and provide hope for the future. Proc Natl Acad Sci 113(51):14507–14514

Luo M, Zhu W, Huang J, Liu Y, Duan X, Wu J, Tong C (2019) Anaerobic organic carbon mineralization in tidal wetlands along a low-level salinity gradient of a subtropical estuary: rates, pathways, and controls. Geoderma 337:1245–1257

Machado L, Magnusson M, Paul NA, Kinley R, de Nys R, Tomkins N (2016) Dose-response effects of Asparagopsis taxiformis and Oedogonium Sp. on in vitro fermentation and methane production. J Appl Phycol 28:1443–1452. https://doi.org/10.1007/s10811-015-0639-9

Macknick J, Newmark R, Heath G, Hallett KC (2011) A review of operational water consumption and withdrawal factors for electricity generating technologies: a review of existing literature. March. https://www.nrel.gov/docs/fy11osti/50900.pdf

Malcorps W, Kok B, van Land M, Fritz M, van Doren D, Servin K, van der Heijden P et al (2019) The sustainability conundrum of fishmeal substitution by plant ingredients in shrimp feeds. Sustain For 11(4):1212. https://doi.org/10.3390/su11041212

Marbà N, Krause-Jensen D, Masqué P, Duarte CM (2018) Expanding greenland seagrass meadows contribute new sediment carbon sinks. Sci Rep 8:14024

Mastrandrea MD, Field CB, Stocker TF, Edenhofer O, Ebi KL, Frame DJ, Held H et al (2010) Guidance note for lead authors of the IPCC fifth assessment report on consistent treatment of uncertainties, Geneva. Intergovernmental Panel on Climate Change (IPCC)

McDermott C, Coad L, Helfgott A, Schroeder H (2012) Operationalizing social safeguards in REDD+: actors, interests and ideas. Environ Sci Pol 21:63–72

McGrath KP, Pelletier NL, Tyedmers P (2015) Life cycle assessment of a novel closed-containment salmon aquaculture technology. Environ Sci Technol 49:5628–5636

Mcleod E, Chmura GL, Bouillon S, Salm R, Björk M, Duarte CM, Lovelock CE et al (2011) A blueprint for blue carbon: toward an improved understanding of the role of vegetated coastal habitats in sequestering CO_2. Front Ecol Environ 9(10):552–560

Mcowen CJ, Weatherdon LV, Van Bochove JW, Sullivan E, Blyth S, Zockler C, Stanwell-Smith D et al (2017) A global map of saltmarshes. Biodivers Data J 5:e11764

MERiFIC (2012) Documentary summary of the environmental impact of renewable marine energy. https://tethys.pnnl.gov/sites/default/files/publications/3-2-1_Documentary_summary_EN-MERiFIC-web.pdf

Milledge JJ, Smith B, Dyer PW, Harvey P (2014) Macroalgae-derived biofuel: a review of methods of energy extraction from seaweed biomass. Energies 7:7194–7222. https://doi.org/10.3390/en7117194

Mørk G, Barstow S, Kabuthand A, Pontes MT (2010) Assessing the global wave energy potential. OMAE2010–20473. In: Proceedings of OMAE2010 29th international conference on ocean, offshore mechanics and arctic engineering, Shanghai, China, June 6–11

Murdiyarso D, Purbopuspito J, Kauffman JB, Warren MW, Sasmito SD, Donato DC, Manuri S et al (2015) The potential of Indonesian mangrove forests for global climate change mitigation. Nat Clim Chang 5(12):1089

Naylor RL, Goldburg RJ, Primavera JH, Kautsky N, Beveridge MC, Clay J, Folke C et al (2000) Effect of aquaculture on world fish supplies. Nature 405(6790):1017–1024

Naylor RL, Hardy RW, Bureau DP, Chiu A, Elliott M, Farrell AP, Forster I, Gatlin DM, Goldburg RJ, Hua K, Nichols PD (2009) Feeding aquaculture in an era of finite resources. Proc Natl Acad Sci 106:15103–15110

Needelman BA, Emmer IM, Emmett-Mattox S, Crooks S, Megonigal JP, Myers D, Oreska MPJ et al (2018) The science and policy of the verified carbon standard methodology for tidal wetland and seagrass restoration. Estuar Coasts 41(8):2159–2171

Nellemann C, Corcoran E, Duarte CM, Valdés L, De Young C, Fonseca L, Grimsditch G (2009) Blue carbon. A rapid response assessment. United Nations Environment Programme, GRID-Arendal 78

Nijdam D, Rood T, Westhoek H (2012) The price of protein: review of land use and carbon footprints from life cycle assessments of animal food products and their substitutes. Food Policy 37(6):760–770

Nilsson M, Griggs D, Visbeck M (2016) Policy: map the interactions between sustainable development goals. Nature 534:320–322. https://doi.org/10.1038/534320a

O'Neill SP, Angeloudis A, Robins PE, Walkington I, Ward SL, Masters I, Lewis MJ et al (2018) Tidal range energy resource and optimization—past perspectives and future challenges. Renew Energy 127:763–778. https://doi.org/10.1016/j.renene.2018.05.007

OECD (2016) The ocean economy in 2030. OECD Publishing, Paris. www.oecd-ilibrary.org/economics/the-ocean-economy-in-2030_9789264251724-en; https://read.oecd-ilibrary.org/economics/the-ocean-economy-in-2030_9789264251724-en#page34

OES (2011) Annual report 2011. Implementing agreement on ocean energy systems. https://tethys.pnnl.gov/sites/default/files/publications/OES-Annual-Report-2011.pdf

OES (2015) International levelised cost of energy for ocean energy technologies. Ocean energy systems (OES) of IEA 2015. https://www.ocean-energy-systems.org/documents/65931-cost-of-energy-for-ocean-energy-technologies-may-2015-final.pdf/

OES (2017) An international vision for ocean energy. http://www.oceanenergysystems.org/news/oes-vision-for-international-deployment-of-ocean-energy/

Oita A, Nagano I, Matsuda H (2018) Food nitrogen footprint reductions related to a balanced Japanese diet. Ambio 47(3):319–326. www.researchgate.net/publication/319720090_Food_nitrogen_footprint_reductions_related_to_a_balanced_Japanese_diet

OpenEI (2019). https://openei.org/apps/LCA/

Ørsted (2019) Making green energy affordable. Whitepaper. Ørsted, Gentofte. https://orsted.com/en/Explore/Making-green-energy-affordable

Ortega A, Geraldi NR, Alam I, Kamau AA, Acinas SG, Logares R, Gasol JM et al (2019) Important contribution of macroalgae to oceanic carbon sequestration. Nat Geosci 12:748–754. https://doi.org/10.1038/s41561-019-0421-8

Oschlies A, Pahlow M, Yool A, Matear RJ (2010) 2010 climate engineering by artificial ocean upwelling: channelling the Sorcerer's apprentice. Geophys Res Lett 37:L04701. https://doi.org/10.1029/2009GL041961

Ouyang X, Guo F (2016) Paradigms of mangroves in treatment of anthropogenic wastewater pollution. Sci Total Environ 544:971–979

Pahlow M, van Oelb PR, Mekonnena MM, Hoekstraa AY (2015) Increasing pressure on freshwater resources due to terrestrial feed ingredients for aquaculture production. Sci Total Environ 536:847–857. www.sciencedirect.com/science/article/pii/S0048969715304666

Palmer A, Keith D, Doctor R (2007) Ocean storage of CO_2: pipelines, risers and seabed containment. In: The 26th international conference on offshore mechanics and arctic engineering

Parente J, Fonseca P, Henriques V, Campos A (2008) Strategies for improving fuel efficiency in the Portuguese trawl fishery. Fish Res 93(1–2):117–124

Parker RWR (2018) Implications of high animal by-product feed inputs in life cycle assessments of farmed Atlantic salmon. Int J Life Cycle Assess 23(5):982–994

Parker RWR, Gardner C, Green BS, Hartmann K, Watson RA (2017) Drivers of fuel use in rock lobster fisheries. ICES J Mar Sci 74(6):1681–1689

Parker RWR, Blanchard JL, Gardner C, Green BS, Hartmann K, Tyedmers PH, Watson RA (2018) Fuel use and greenhouse gas emissions of world fisheries. Nat Clim Chang 8:333–337. https://doi.org/10.1038/s41558-018-0117-x

Pascoe S, Coglan L, Punt AE, Dichmont CM (2012) Impacts of vessel capacity reduction programmes on efficiency in fisheries: the case of Australia's multispecies northern prawn fishery. J Agric Econ 63(2):425–443

Pelletier N, Tyedmers P (2007) Feeding farmed salmon: is organic better? Aquaculture 272:399–416

Pelletier N, Tyedmers P (2010) Life cycle assessment of frozen tilapia fillets from indonesian lake and pond-based intensive aquaculture systems. J Ind Ecol 14(3):467–481

Pelletier N, Tyedmers P, Sonesson U, Scholz A, Ziegler F, Flysjo A, Kruse S et al (2009) Not all salmon are created equal: life cycle assessment (LCA) of global salmon farming systems. Environ Sci Technol 43(23):8730–8736

Pelletier N, Klinger DH, Sims NA, Yoshioka JR, Kittinger JN (2018) Nutritional attributes, substitutability, scalability, and environmental intensity of an illustrative subset of current and future protein sources for aquaculture feeds: joint consideration of potential synergies and trade-offs. Environ Sci Technol 52(10):5532–5544

Pendleton L, Donato DC, Murray BC, Crooks S, Jenkins WA, Sifleet S, Craft C et al (2012) Estimating global 'blue carbon' emissions from conversion and degradation of vegetated coastal ecosystems. PLoS ONE 7(9):e43542

Poore J, Nemecek T (2018) Reducing food's environmental impacts through producers and consumers. Science 360:987–992

Pörtner HO, Karl DM, Boyd PW, Cheung WL, Lluch-Cota SE, Nojiri Y, Schmidt DN et al (2014) Ocean systems. Cambridge University Press, Cambridge, pp 411–484

Poulsen T, Lema R (2017) Is the supply chain ready for the green transformation? The case of offshore wind logistics. Renew Sust Energ Rev 73:758–771. https://doi.org/10.1016/j.rser.2017.01.181

Project Drawdown (2019) Solutions|drawdown. Drawdown. https://www.drawdown.org/solutions. Accessed 10 September

Psaraftis HN, Kontovas CA (2013) Speed models for energy-efficient maritime transportation: a taxonomy and survey. Transp Res Part C Emerg Technol 26:331–351

Queirós AM, Norling K, Amaro T, Nuñes J, Cummings DH, Yakushev EA, Sørensen K et al (2014) Potential impact of CCS leakage on marine communities. ECO2 deliverable, D4.1. Plymouth Marine Laboratory, Plymouth

Queirós AM, Stephens N, Widdicombe S, Tait K, McCoy SJ, Ingels J, Rühl S et al (2019) Connected macroalgal-sediment systems: blue carbon and foodwebs in the deep coastal ocean. Ecol Monogr 2019:e01366

Ramsar Convention on Wetlands (2018) Scaling up wetland conservation, wise use and restoration to achieve the sustainable development goals: wetlands and the SDGs. Ramsar Convention Secretariat, Gland. www.ramsar.org/sites/default/files/documents/library/wetlands_sdgs_e.pdf

Rana KJ, Siriwardena S, Hasan MR (2009) Impact of rising feed ingredient prices on aquafeeds and aquaculture production. In: FAO fisheries and aquaculture technical paper 541. Food and Agriculture Organization, Rome

Rastelli E, Corinaldesi C, Dell'Anno A, Amaro T, Greco S, Lo Martire M, Carugati L et al (2016) CO_2 leakage from CO_2 capture and storage (CCS) systems affects organic matter cycling in surface marine sediments. Mar Environ Res 122:158–168

Rau GH, Caldeira K (1999) Enhanced carbonate dissolution: a means of sequestering waste CO_2 as ocean bicarbonate. Energy Convers Manag 40:1803–1813

Rehmatulla N (2014) Market failures and barriers affecting energy efficient operations in shipping. PhD Thesis

Renforth P, Henderson G (2017) Assessing ocean alkalinity for carbon sequestration. Rev Geophys 55(3):636–674. https://doi.org/10.1002/2016rg000533

Richards DR, Friess DA (2016) Rates and drivers of mangrove deforestation in Southeast Asia, 2000–2012. Proc Natl Acad Sci 113(2):344–349

Richards DR, Friess DA (2017) Characterizing coastal ecosystem service trade-offs with future urban development in a tropical city. Environ Manag 60:961–973

Robb DHF, MacLeod M, Hasan MR, Soto D (2017) Greenhouse gas emissions from aquaculture: a life cycle assessment of three Asian systems. In: FAO fisheries and aquaculture technical paper No. 609. FAO, Rome

Rog SM, Clarke RH, Cook CN (2017) More than marine: revealing the critical importance of mangrove ecosystems for terrestrial vertebrates. Divers Distrib 23(2):221–230

Roy J, Tschakert P, Waisman H, Abdul Halim S, Antwi-Agyei P, Dasgupta P, Hayward B et al (2018) Sustainable development, poverty eradication and reducing inequalities. In: Masson-Delmotte V, Zhai P, Pörtner H-O, Roberts D, Skea J, Shukla PR, Pirani A et al (eds) Global warming of 1.5°C. World Meteorological Organization, Geneva

Sadhukhan J, Gadkari S, Martinez-Hernandez E, Ng KS, Shemfe M, Torres-Garcia E, Lynch J (2019) Novel macroalgae (seaweed) biorefinery systems for integrated chemical, protein, salt, nutrient and mineral extractions and environmental protection by green synthesis and life cycle sustainability assessments. Green Chem 21(10):2635–2655

Sarker PK, Kapuscinski AR, Bae AY, Donaldson E, Sitek AJ, Fitzgerald DS, Edelson OF (2018) Towards sustainable aquafeeds: evaluating substitution of fishmeal with lipid-extracted microalgal co-product (Nannochloropsis oculata) in diets of juvenile Nile tilapia (Oreochromis niloticus). PLoS ONE 13(7):e0201315

Saunders M, Leon J, Phinn S, Callaghan D, O'Brien K, Roelfsema C, Lovelock C et al (2013) Coastal retreat and improved water quality mitigate losses of seagrass from sea level rise. Glob Chang Biol 19:12218. https://doi.org/10.1111/gcb.12218

Schmitz OJ, Wilmers CC, Leroux SJ, Doughty CE, Atwood TB, Galetti M, Davies AB et al (2018) Animals and the zoogeochemistry of the carbon cycle. Science 362:6419. https://doi.org/10.1126/science.aar3213

Schuerch M, Spencer T, Temmerman S, Kirwan ML, Wolff C, Lincke D et al (2018) Future response of global coastal wetlands to sea-level rise. Nature 561:231–234. https://doi.org/10.1038/s41586-018-0476-5

Scott J, Smith T, Rehmatulla N, Milligan B (2017) The promise and limits of private standards in reducing greenhouse gas emissions from shipping. J Environ Law 29:2

Scottish Enterprise (n.d.) Oil and gas 'seize the opportunity' guides: offshore wind. Scottish Enterprise, Glasgow. www.offshorewind-scotland.org.uk/media/1116/sesdi-oil-and-gas-div-guide-offshore-wind.pdf

Searchinger T, Waite R, Hanson C, Ranganathan J (2019) Creating a sustainable food future: a menu of solutions to feed nearly 10 billion people by 2050. In: Chapter 7: avoid competition from bioenergy for food crops and land. World Resources Institute, Washington. https://wrr-food.wri.org

Shindo Y, Lund PC, Fujioka Y, Koriyama H (1993) Kinetics of formation of CO_2 hydrate. Energy Convers Manag 34:1073–1079

SI Ocean (2013) Ocean energy: cost of energy and cost reduction opportunities. SI Ocean Project

Simões-Wüst AP, Dagnelie PC (2019) To be or not to be for humankind: organic diets revisited for a sustainable development. Sustainable Earth 2(1):7. https://doi.org/10.1186/s42055-019-0008-7

Sims REH, Schock RN, Adegbululgbe A, Fenhann J, Konstantinaviciute I, Moomaw W, Nimir HB et al (2007) Energy supply. In: Metz B, Davidson OR, Bosch PR, Dave R, Meyer LA (eds) Climate change 2007: mitigation of climate change. Cambridge University Press, Cambridge, pp 251–322

Singh GG, Cisneros-Montemayor AM, Swartz W, Cheung W, Guy JA, Kenny TA, McOwen CJ et al (2018) A rapid assessment of co-benefits and trade-offs among sustainable development goals. Mar Policy 93:223–231

Sislian L, Jaegler A (2016) A green ocean carrier network design problem: empirical study on Egyptian Port of Alexandria. Sixth International Conference, Information Systems Logistics and Supply Chain (ILS). Bordeaux, France, June 1–4. http://ils2016conference.com/wp-content/uploads/2015/03/ILS2016_TB03_3.pdf

Skjermo J, Rebours C, Chapman A (2014) A new Norwegian bioeconomy based on cultivation and processing of seaweeds: opportunities and R&D needs. SINTEF Fisheries and Aquaculture, Trondheim

Smith P, Bustamante M, Ahammad H, Clark H, Dong H, Elsiddig EA, Haberl H et al (2014) Agriculture, forestry and other land use (AFOLU). In: Edenhofer, Pichs-Madruga R, Sokona Y, Farahani E, Kadner S, Seyboth K, Adler A et al (eds) Climate change 2014: mitigation of climate change. Cambridge University Press, Cambridge

Smith T, Raucci C, Haji Hosseinloo S, Rojon I, Calleya J, Suárez de la Fuente S, Wu P, Palmer K (2016) CO_2 emissions from international shipping. possible reduction targets and their associated pathways. University Maritime Advisory Services (UMAS), London

Snyder LB, Hamilton MA, Mitchell EW, Kiwanuka-Tondo J, Fleming-Milici F, Proctor D (2004) A meta-analysis of the effect of mediated health communication campaigns on behavior change in the United States. J Health Commun 9(S1):71–96

SOFIA (2018) Meeting the sustainable development goals. SOFIA, Rome

Sofiev M, Winebrake JJ, Johansson L, Carr EW, Prank M, Soares J, Vira J et al (2018) Cleaner fuels for ships provide public health benefits with climate tradeoffs. J Chem Kinet 27(11):1089–1095

Spalding MJ, de Fontabert C (2007) Conflict resolution for addressing climate change with ocean-altering projects. Environmental Law Reporter. https://www.oceanfdn.org/sites/default/files/37_10740_0.pdf

Springmann M, Clark M, Mason-D'Croz D, Wiebe K, Bodirsky BL, Lassaletta L, de Vries W et al (2018) Options for keeping the food system within environmental limits. Nature 562(7728):519

Statton J, Cambridge ML, Dixon KW, Kendrick GA (2013) Aquaculture of Posidonia australis seedlings for seagrass restoration programs: effect of sediment type and organic enrichment on growth. Restoration ecology 21(2):250–259

Sumaila UR, Cheung SW, Dyck A, Gueye K, Huang L, Lam V, Pauly D et al (2012) Benefits of rebuilding global marine fisheries outweigh costs. PLoS ONE 7(7):e40542

Sumaila UR, Tai TC, Lam V, Cheung W, Bailey M, Cisneros-Montemayor AM, Chen OL et al (2019) Benefits of the Paris agreement to ocean life, economies, and people. Sci Adv 5(2):3855. https://doi.org/10.1126/sciadv.aau3855

Sundby S, Drinkwater KF, Kjesbu OS (2016) The North-Atlantic Spring-Bloom system—where the changing climate meets the winter dark. Front Mar Sci 3:28. https://doi.org/10.3389/fmars.2016.00028

Tacon AGJ, Metian M (2008) Global overview on the use of fish meal and fish oil in industrially compounded aquafeeds: trends and future prospects. Aquaculture 285:146–158

Tallard G, Key N (2012) Mitigating methane emissions from livestock: a global analysis of sectoral policies. Clim Chang 12:387–414

Teske A, Durbin A, Ziervogel K, Cox C, Arnosti C (2011a) Microbial community composition and function in permanently cold seawater and sediments from an arctic fjord of Svalbard. Appl Environ Microbiol 77:2008–2018

Teske S, Pregger T, Simon S, Naegler T, Graus W, Lins C (2011b) Energ Effic 4:409. https://doi.org/10.1007/s12053-010-9098-y

Tilman D, Clark M (2014) Global diets link environmental sustainability and human health. Nature 515(7528):518

Tilman D, Fargione J, Wolff B, D'Antonio C, Dobson A, Howarth R, Schindler D et al (2001) Forecasting agriculturally driven global environmental change. Science 1:281–284

Tilman D, Balzer C, Hill J, Befort BL (2011) Global food demand and the sustainable intensification of agriculture. Proc Natl Acad Sci 108(50):20260–20264. https://doi.org/10.1073/pnas.1116437108

Troell M, Naylor RL, Metian M, Beveridge M, Tyedmers PH, Folke C, Arrow KJ et al (2014) Does aquaculture add resilience to the global food system? Proc Natl Acad Sci 111:13257–13263

Tubiello FN, Salvatore M, Rossi S, Ferrara A, Fitton N, Smith P (2013) The FAOSTAT database of greenhouse gas emissions from agriculture. Environ Res Lett 8:015009

UNEP (2018) Emissions gap report 2018

UNFCCC (2015) Decision 1/CP.21 adoption of the Paris agreement. Paris. https://unfccc.int/sites/default/files/resource/docs/2015/cop21/eng/10a01.pdf
Unsworth RKF, Nordlund LM, Cullen-Unsworth LC (2018) Seagrass meadows support global fisheries production. Conserv Lett 12:e12566. https://doi.org/10.1111/conl.12566
USDE (2019) 1 powering the blue economy: exploring opportunities for marine renewable energy in maritime markets. U.S. Department of Energy, Washington. https://www.energy.gov/eere/water/downloads/powering-blue-economy-report
Valiela I, Bowen JL, York JK (2001) Mangrove forests: one of the world's threatened major tropical environments. Bioscience 51(10):807–815
van der Zwaan B, Gerlagh R (2016) Offshore CCS and ocean acidification: a global long-term probabilistic cost-benefit analysis of climate change mitigation. Clim Chang 137(1–2):157–170. https://doi.org/10.1007/s10584-016-1674-5
Wahl M, Schneider Covachã S, Saderne V, Hiebenthal C, Müller JD, Pansch C, Sawall Y (2018) Macroalgae may mitigate ocean acidification effects on mussel calcification by increasing pH and its fluctuations. Limnol Oceanogr 63(1):3–21
Waite RM, Beveridge R, Brummett S, Castine N, Chaiyawannakarn S, Kaushik RM et al (2014) Improving productivity and environmental performance of aquaculture. In: Installment 5 of creating a sustainable food future. World Resources Institute (WRI), Washington
Waycott M, Duarte CM, Carruthers TJB, Orth RJ et al (2009) Accelerating loss of seagrasses across the globe threatens coastal ecosystems. Proc Natl Acad Sci 106:12377–12381
Wernberg T, Krumhansl K, Filbee-Dexter K, Pedersen MF (2019) Status and trends for the world's kelp forests. In: World seas: an environmental evaluation. Academic Press, New York, pp 57–78
Willauer HD, Dimascio F, Hardy DR, Williams FW (2014) Feasibility of CO_2 extraction from seawater and simultaneous hydrogen gas generation using a novel and robust electrolytic cation exchange module based on continuous electrodeionization. Technol Ind Eng Chem Res 53(31):12192–12200
Willett W, Rockström J, Loken B, Springman M, Lang T, Vermeulen S, Garnett T et al (2019) Food in the anthropocene: the EAT–Lancet commission on healthy diets from sustainable food systems. Lancet 393(10170):447–492
Williamson PR, Altman DG, Blazeby JM, Clarke M, Devane D, Gargon E et al (2012) Developing core outcome sets for clinical trials: issues to consider. Trials 13:132
Wilson KL, Skinner MA, Lotze HK (2019) Projected 21st-century distribution of canopy-forming seaweeds in the Northwest Atlantic with climate change. Divers Distrib 25:582–602
Wirsenius S, Azar C, Berndes G (2010) How much land is needed for global food production under scenarios of dietary changes and livestock productivity increases in 2030? Agric Syst 103:621–638. https://doi.org/10.1016/j.agsy.2010.07.005
World Bank (2016) Seaweed aquaculture for food security, income generation and environmental health. July. https://doi.org/10.1596/24919
World Bank (2018) Where sun meets water: floating solar market report. https://www.worldbank.org/en/topic/energy/publication/where-sun-meets-water
Yan J (ed) (2015) Handbook of clean energy systems, vol 5. Wiley, Hoboken
Zhang J, Fang J, Wang W, Du M, Gao Y, Zhang M (2012) Growth and loss of mariculture kelp Saccharina japonica in Sungo Bay, China. J Appl Phycol 24:1209–1216. https://doi.org/10.1007/s10811-011-9762-4
Ziegler F, Hornborg S (2014) Stock size matters more than vessel size: the fuel efficiency of Swedish Demersal Trawl fisheries 2002–2010. Mar Policy 44:72–81
Ziegler F, Hornborg S, Green BS, Eigaard OR, Farmery AK, Hammar L, Hartmann K et al (2016) Expanding the concept of sustainable seafood using life cycle assessment. Fish Fish 17:1073–1093

A Sustainable Ocean Economy for 2050: Approximating Its Benefits and Costs

18

Abbreviations

BAU	Business-as-usual
B-C	Benefit-cost
CH_4	Methane
CO_2	Carbon dioxide
CO_2e	Carbon dioxide equivalent
FCR	Feed conversion ratio
FOLU	Food and Land Use Coalition
GDP	Gross domestic product
GHG	Greenhouse gas
Gt	Gigatonne
GW	Gigawatt
IEA	International Energy Agency
IMO	International Maritime Organization
LCOE	Levelised cost of electricity
mmt	Million metric tonnes
Mt.	Megatonne
MW	Megawatt
MWh	Megawatt-hour
N_2O	Nitrous oxide
NREL	National Renewable Energy Laboratory
PV	Photovoltaic
R&D	Research and development
ROI	Return on investment
SCC	Social cost of carbon
SDG	Sustainable Development Goal
TWh	Terawatt hour
WACC	Weighted average of capital costs

Originally published in:
Konar, M., Ding, H. 2020. *A Sustainable Ocean Economy for 2050: Approximating Its Benefits and Costs*. Washington, DC: World Resources Institute. Available online at https://oceanpanel.org/publication/a-sustainable-ocean-economy-for-2050-approximating-its-benefits-and-costs/.
Reprint by Springer International Publishing (2023) with kind permission.
Published under license from the "World Research Institute".

1 Executive Summary

The ocean and its resources provide key ecosystem services and benefits that are crucial for human well-being and the prosperity of the global economy, but these services are at risk. The ocean's wide range of ecosystem services (including food, energy, recreational/ cultural services and trading/transport routes) is vital for the well-being of society. However, climate change, overfishing, pollution and a loss of biodiversity and coastal ecosystems are eroding the ability of the ocean to sustain livelihoods and prosperity.

Taking action to protect these ocean-based ecosystems and ensuring the environmental sustainability of ocean-based activities will produce health, environmental and ecological, and economic and social benefits to people and the planet. A key question for policymakers and funding agencies is how these benefits compare with the costs. This analysis aims to answer the question by building on several existing analyses and reports, including *The Ocean as a Solution to Climate Change: Five Opportunities for Action* (Hoegh-Guldberg et al. 2019) and *The Global Consultation Report of the Food and Land Use Coalition* (FOLU 2019).

Using both quantitative and qualitative methods, it demonstrates that ocean-based investments yield benefits to society in the long term, and these benefits substantially outweigh the costs.

This analysis is the first attempt to estimate the global net benefit and the B-C ratio over a 30-year time horizon (2020–2050) from implementing sustainable ocean-based interventions. It indicates the scale of benefits compared to the costs by focusing on four ocean-based policy interventions: conserving and restoring mangrove habitats, scaling up offshore wind production, decarbonising the international shipping sector and increasing the production of sustainably sourced ocean-based proteins (to ensure a healthy, balanced human diet by 2050). These interventions would contribute to global efforts to reduce greenhouse gas (GHG) emissions and move countries towards their Sustainable Development Goals and targets (Hoegh-Guldberg et al. 2019).

J. Lubchenco, P. M. Haugan (eds.), *The Blue Compendium*, https://doi.org/10.1007/978-3-031-16277-0_18

For each intervention area, the impact to reach a sustainable transformation pathway by 2050 is measured relative to a business-as-usual scenario. A B-C ratio is developed by dividing the present value of benefits in 2050 by the present value of costs. The categories of benefits assessed include health (such as a reduction in mortality and morbidity), environmental and ecological (such as benefits from higher biodiversity, reduced water usage and land-based conflicts, and coastal protection) and economic and social (such as increased business revenues, household income, jobs and food security). The categories of costs include costs to business (such as capital investments and increases in operational costs), costs to government (such as costs of regulations, research and development [R&D] expenditures, enforcement and monitoring costs) and costs to households (such as opportunity costs of forgone activities)**.** The benefit and cost estimates are partial estimates; impacts are monetarily quantified where possible and are qualitatively described when quantifiable data are absent.

Taking action to protect these ocean-based ecosystems and ensuring the environmental sustainability of ocean-based activities will produce health, environmental and ecological, and economic and social benefits to people and the planet

2 Key Findings

The overall rate of return on investment (ROI) can be very high, with sustainable ocean-based investments yielding benefits at least five times greater than the costs. When assessing individual interventions, the average economic B-C ratio range between 3-to-1 and 12-to-1, and in some cases even higher. The B-C ratios were similar to key health interventions in developed and developing countries.[1] Specifically, investing $2.0–3.7 trillion globally across the four areas from 2020 to 2050 would generate $8.2–22.8 trillion in net benefits (average $15.5 trillion), implying a rate of ROI of 400–615%. The B-C ratios vary across sectors and interventions (Table 18.1; Fig. 18.1) as follows:

- **Every $1 invested in mangrove conservation and restoration generates a benefit of $3.** When assessing specific interventions, the B-C ratio for conservation is 88-to-1 and for restoration is 2-to-1. Three factors drive the difference in the B-C ratios: the higher cost of mangrove restoration (due to seeding and replanting), low survival rates following restoration and the lag in accrual of benefits from restoration. The total value of net benefits for mangrove restoration over 30 years ($97–150 billion) is higher than for conservation ($48–96 billion) because we assume the area of mangroves restored is 10 times that of the area conserved.[2]
- **Every $1 invested in scaling up global offshore wind production generates a benefit estimated at $2–17**, depending on the cost of offshore energy production and transmission and the types of generation that would be displaced.[3] The value of the ROI will increase as the costs for offshore wind energy generation fall because of improvement in technologies and actions to reduce integration costs.
- **Every $1 invested in decarbonising international shipping and reducing emissions to net zero is estimated to generate a return of $2–5.** The analysis assumed the significant capital expenditure to switch to zero-carbon

Table 18.1 Summary of benefit-cost ratios for the four action areas in 2050

Action	Average benefit:cost ratio
Conserve and restore mangroves[a]	3:1
Decarbonise international shipping[b]	4:1
Increase production of sustainably sourced ocean-based proteins	10:1
Scale up offshore energy production[c]	12:1

Notes:
[a] *The ratio presented is the combined ratio for mangrove conservation and restoration. When assessing specific interventions, the benefit-cost ratio for conservation is estimated to be 88-to-1 and for restoration is 2-to-1*
[b] *The benefit-cost ratio estimated for decarbonising international shipping ranges from 2:1 to 5:1*
[c] *The benefit-cost ratio estimated for scaling up of global offshore wind production ranges from 2:1 to 17:1*
Source: Authors' calculations

[1] For example, the B-C ratio for double measles immunisation in Canada is estimated to be 2-to-1 to 4-to-1; for influenza vaccination in Italy, it is estimated at 4-to-1 to 12-to-1; for the meningitis prevention program in the Philippines, it is 8.4-to-1; and for the universal *Haemophilus influenzae* type B vaccination (starting at 2 months) in the United States, it is 3.4-to-1 to 5.4-to-1 (Bärnighausen et al. 2011; Colombo et al. 2006; Limcangco et al. 2001; Pelletier et al. 1998; Zhou et al. 2002).

[2] The conservation scenario assumes stopping the additional loss of mangroves whereas the restoration scenario assumes replanting large areas of mangroves already lost; that is we are doing more restoration in the scenarios analysed than conservation. The overall ratio of both conservation and restoration is calculated by adding the total present value benefits and costs of both measures. The very high restoration costs is the main factor driving the overall B-C ratio for both conservation and restoration.

[3] The return on investment for wind energy investments will vary depending on the specific generation technologies and costs in places where the offshore wind installations are located. On grids that have a high share of zero-carbon generation, including hydropower and nuclear energy, adding ocean energy will not decrease emissions significantly. Conversely, for grids with a high share of carbon-intensive generation, emission displacements could be significant.

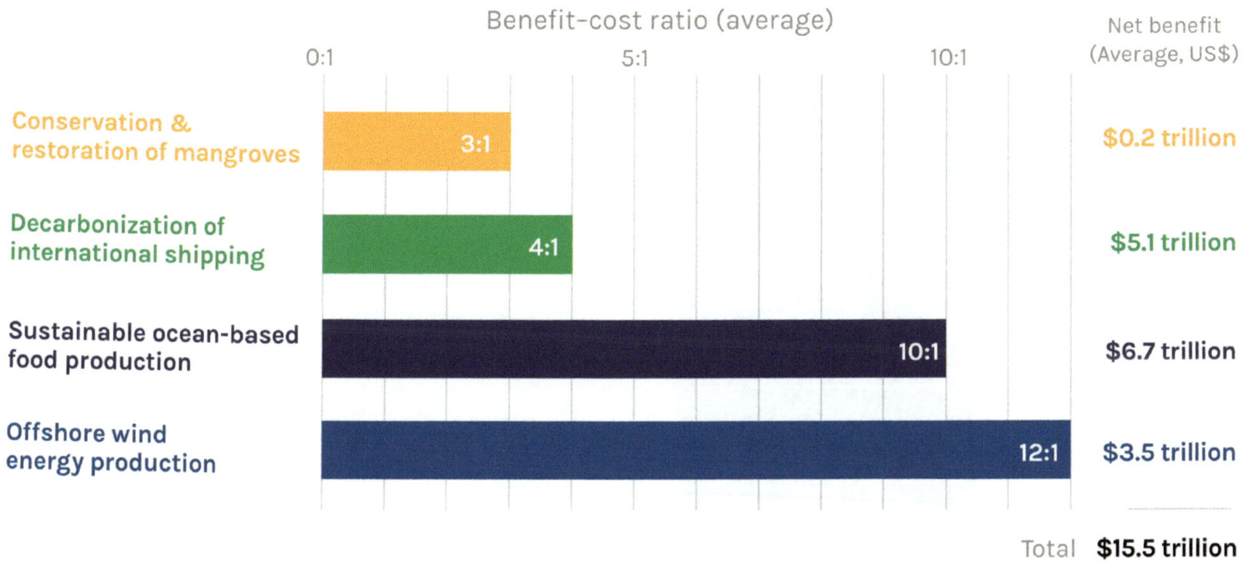

Fig. 18.1 Benefits significantly outweigh costs across sustainable ocean-based interventions, with average B-C ratio ranging between 3:1 and 12:1. *Note: Average benefit-cost (B-C) ratios have been rounded to the nearest integer and the net benefits value to the first decimal place. The B-C ratio for mangroves is the combined ratio for both conservation- and restoration-based interventions. The average net benefits represent the average net present value for investments and is calculated over a 30-year horizon (2020–2050). Source: Authors' calculations*

emissions will happen after 2030, and limiting the analysis to 2050 captures only a portion of returns from these investments, which will continue beyond 2050.

- **Every $1 invested in increasing production of sustainably sourced ocean-based protein (to ensure a healthy, balanced diet by 2050) is estimated to yield $10 in benefits.** The increase in demand for ocean-based protein to provide a healthy diet for 9.7 billion people by 2050, which would replace a percentage of emission-intensive land-based protein sources, can be achieved by reforming wild-capture fisheries and by increasing the sustainable production of ocean-based aquaculture. Both measures will deliver benefits such as better health outcomes to consumers, higher revenues to fishers, lower GHG emissions mitigating the risks of climate damage, reduced land-based conflicts and lower water usage.

A number of impacts (both benefits and costs) have not yet been monetised, but they need to be considered by policymakers. These include the impact of GHG emissions on ocean acidification and the associated loss to biodiversity and commercial shellfish production; a potential increase in tourism revenues globally from restored mangroves; biodiversity benefits from healthier ecosystems; impacts on marine biodiversity from increasing the number of offshore wind farms; and distributional impacts of the benefits and costs on poorer communities. Given these nonmonetised impacts, the B-C ratios present a partial estimation of all benefits and costs that are likely to accrue as a result of such investments. These four examples are indicative of the relative scale of benefits compared to the costs. Further research and analysis to address these gaps will provide a more complete picture of the value of benefits versus costs.

Although data limitations prevented a full accounting of all benefits and costs, the results of the analyses suggest that taking actions to transform these sectors will generate a host of benefits that are much larger than the costs.[4] The results show that sustainable ocean-based investments yield benefits at least five times greater than the costs (Fig. 18.2), with minimum net returns of $8.2 trillion over 30 years. Better awareness of evidence of the possible ROI will help strengthen the economic case for action.

[4]For example, this is particularly true for the majority of ecosystem service benefits for mangroves that are not privately owned or traded, and hence their "value" is not reflected in price signals. We refrained from monetising some of the benefits due to the uncertainty of nonmarket valuation techniques. Further information is available in Sect. 18.5.1.

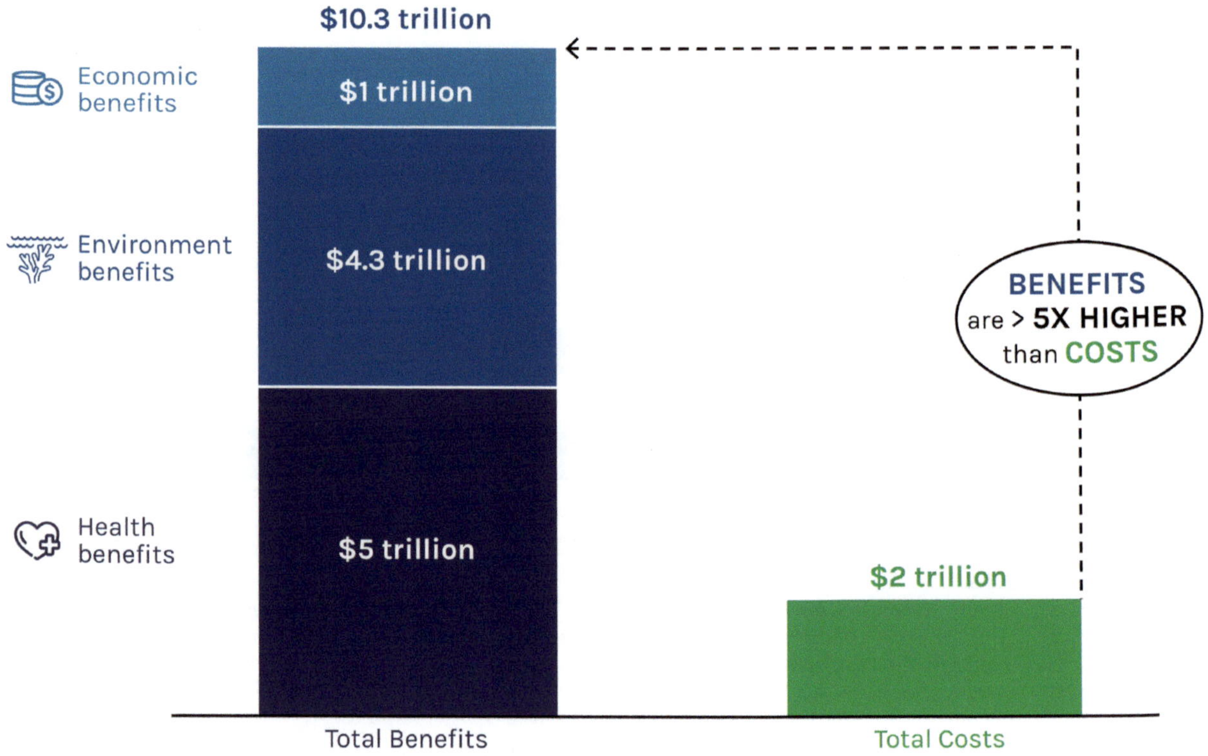

Fig. 18.2 Sustainable ocean investments yield benefits at least 5× higher than costs. *Note: The total benefits and costs in the figure present the lower-bound present value estimates to demonstrate the minimum scale of quantified net benefits. Source: Authors' calculations*

3 Introduction

The ocean's economic value is undisputed: it generates jobs that support millions of livelihoods, it supplies resources that have enabled decades of industrial growth, and its sea routes enable 90% of world trade (Fleming et al. 2014). The ocean's ecosystem services are vital for the well-being of society. For example, in some least-developed countries, fish protein accounts for more than 50% of animal protein intake (FAO 2018). Likewise, the ocean is reflected in many cultural practices, is manifest in inspirational art and provides recreational and aesthetic value to many (Fleming et al. 2014).

However, these services and benefits are at risk as the ocean faces pressures from enhanced economic activity, demands from a growing human population and uncertainty from a warmer, unstable climate.

Overfishing, pollution, climate change and loss of biodiversity are eroding the ability of the ocean to continue to sustain livelihoods and prosperity. The cumulative impact of human activities and climate change are likely to cause further ecosystem degradation or even collapse of ecosystems such as coral reefs, kelp forests and seagrasses (Halpern et al. 2019; IPCC 2019).

This analysis begins to estimate the benefits and costs of transitioning towards a sustainable ocean economy by focusing on four areas that represent key aspects of the ocean economy. It builds on *The Ocean as a Solution to Climate Change: Five Opportunities for Action* (Hoegh-Guldberg et al. 2019) and *The Global Consultation Report of the Food and Land Use Coalition* (FOLU 2019) and other analyses and reports to demonstrate that ocean-based investments can yield considerable economic benefits to society in the long term.

3.1 Scope of the Analysis

The High Level Panel for a Sustainable Ocean Economy (Ocean Panel) commissioned this benefit-cost analysis as an input to the deliberations of the Ocean Panel, serving to strengthen the evidence base of the forthcoming *Towards a Sustainable Ocean Economy* report and action agenda.

The Ocean Panel proposes that a sustainable ocean economy can simultaneously deliver on three dimensions. It can

- **Protect:** reduce greenhouse gas (GHG) emissions while safeguarding biodiversity;
- **Produce:** contribute to sustainably powering and feeding a planet of 9.7 billion people in 2050; and
- **Prosper:** create better jobs and support more equitable economic growth, household income and well-being.

To achieve this vision, it will be critical to take action to transform ocean-based sectors and ecosystems towards sustainability.

We indicate the scale of benefits compared to costs by focusing on specific policy interventions across one coastal ecosystem, mangroves, and the ocean-based sectors involved with offshore wind energy, international shipping and ocean-based protein from capture fisheries and mariculture (Table 18.2).

Although it was not possible to cover all potential interventions across these sectors, specific interventions were chosen to meet three criteria: achievement of the Ocean Panel's vision, contribution to the global efforts to reduce GHG emissions, and contribution to delivering countries' Sustainable Development Goals (SDGs) and targets (Hoegh-Guldberg et al. 2019).[5]

These are the four interventions analysed:

- Conserving and restoring mangrove habitats
- Scaling up offshore wind energy production
- Decarbonising the international shipping sector
- Increasing production of sustainably sourced ocean-based protein (to ensure a healthy, balanced diet by 2050)

This analysis is the first attempt to measure the global net benefit and benefit-cost (B-C) ratio of implementing ocean-based interventions over a 30-year horizon (2020–2050). While in the past, significant efforts have been made to assess the net positive benefits from protecting marine ecosystems and transforming ocean-based activities, they focused on particular measures, ecosystems and investments in particular regions or referred to assessments over shorter time periods. Consequently, the overall global benefits and costs of transitioning to a sustainable ocean economy across these four areas have not been generated in an aggregate form or included in global discussions. Building on existing literature, this working paper aims to address the knowledge gap by focusing on sustainable transformation pathway scenarios and by using both quantitative and qualitative methods.

Table 18.2 The four ocean-based areas analysed

Ocean-based sectors/ ecosystems	Specific actions
Mangrove coastal habitats	Conserve and restore mangrove coastal habitats
Ocean-based renewable energy	Scale up the production of offshore wind energy (fixed and floating wind installations)[a]
Ocean-based transport	Reduce emissions from international shipping with a target to reach net-zero emissions in 2050[b]
Ocean-based food production	Achieve a healthier balanced diet for 9.7 billion people by 2050 by switching a share of protein from emission-intensive land-based sources of protein (notably beef and lamb) to low-carbon sustainably produced ocean-based sources of protein[c]

Notes:

[a] *Based on the scenarios analysed, offshore energy will likely continue to dominate the generation potential of the ocean energy sector in 2050, accounting for 65% of the sector's potential* (Hoegh-Guldberg et al. 2019)

[b] *The analysis excludes military and fishing vessels and domestic transport and includes bulk carriers, oil tankers and container ships, which account for the majority of the emissions (55%) in the shipping sector* (Olmer et al. 2017)

[c] *Sustainable production involves reforming fisheries by 2050 and increasing the production of sustainable ocean-based aquaculture (fed and nonfed)*

Source: Authors

4 Methodology

This paper summarises the potential impact of investments in four ocean-based areas (see Table 18.2) over 30 years (2020–2050). By dividing the present value of benefits by the present value of costs, a B-C ratio for each sector is estimated (Box 18.1).

The assumptions used to derive the B-C ratio differ for each sector. They are discussed in detail in Sect. 18.5.

A generic analytical framework was applied to ensure consistency and comparability in analysing the impacts in each area:

- The ambition for each area was defined as the level of sustainability that would be achieved in 2050 with respect to an identified baseline scenario. The business-as-usual (BAU) and sustainable transformation pathway projections, based on scenarios modelled in *The Ocean as a Solution to Climate Change: Five Opportunities for Action* (Hoegh-Guldberg et al. 2019) and *The Global*

[5] Although the interventions selected are key to achieving the 2050 sustainable ocean economy vision, they do not represent an exhaustive list of actions that will be required to make such a transition. For example, this analysis does not look at the impacts of moving towards a sustainable coastal tourism sector, of reducing marine pollution, or of expanding the network of marine protected areas.

Consultation Report of the Food and Land Use Coalition (FOLU 2019), are described in Sect. 18.4.1.

- A range of benefits and costs were identified that would achieve the target state over 30 years. These impacts were quantified monetarily where possible and were described qualitatively where a lack of data did not allow for such quantification.
- Future benefits and costs were discounted using a rate of 3.5%. The discounted benefits and costs were summed over 30 years (2020–2050) to arrive at a present value of benefits and costs for 2050 (Box 18.1). All values are based on 2019 prices.
- For each area, a B-C ratio was developed by dividing the present value of benefits in 2050 by the present value of costs.
- The present value of benefits and costs were aggregated across the areas to provide an aggregate B-C ratio for 2050.

Box 18.1 Estimating the Benefit-Cost Ratio

The benefit-cost (B-C) ratio indicates the return from ocean-based investments in the four areas in 2050. A B-C ratio greater than 1 demonstrates that the returns from an investment will be higher than the costs estimated over the chosen time period.

$$B/C = \frac{\text{Present value of benefits}}{\text{Present value of costs}} = \frac{\text{Sum of discounted benefits over 30 years}}{\text{Sum of discounted costs over 30 years}} = \left[\frac{B_0}{(1+r)^0} + \dots \frac{B_n}{(1+r)^n}\right] \div \left[\frac{C_0}{(1+r)^0} + \dots \frac{C_n}{(1+r)^n}\right]$$

where n = year; B = benefits; C = costs; r = discount rate.

Discounting is used to compare benefits and costs occurring over different periods of time by converting them into present values. This is based on the concept that people prefer to receive goods and services now rather than later.[a] The discount rate used in the Green Book, also known as the social time preference rate, is based on two components: the 'time preference', which is the rate at which consumption and spending are discounted over time, assuming no change in per capita consumption, and the 'wealth effect', which reflects the expected growth in per capita consumption over time, where future consumption will be higher relative to current consumption and is expected to have a lower utility.[b]

Source: a, b HMT (2018).

The time frame of 2020–2050 provides enough time for measures to be implemented and environmental benefits to result. In addition, the year 2050 aligns with long-term strategies to reduce emissions to net zero by midcentury (IPCC 2018) and meet the 2050 biodiversity vision where biodiversity is valued, conserved and restored to sustain a healthy planet (Cooper 2018). The time frame also overlaps with the United Nations Decade of Ocean Science and delivery of the 2030 SDG.

We used a constant social discount rate of 3.5% for the analysis (HMT 2018). Views vary on the correct discount rate for climate policies as well as the extent to which rates differ between developing and developed countries.[6] Some economists give more weight to environmental benefits that occur in distant years and recommend a lower discount rate for intergenerational decisions or a 'hyperbolic' discount rate that declines over time (Hausker 2011). For example, the *Stern Review* recommends a declining social discount rate, with rates lower than 3% for investments beyond 30 years (Stern 2007). The review states, 'If the ethical judgement is that future generations count very little regardless of their consumption level then investments with mainly long-run pay-offs would not be favoured. In other words, if you care little about future generations you will care little about climate change. As we have argued that is not a position which has much foundation in ethics and which many would find unacceptable'.[7] To reflect the intertemporal consideration of resource values, we selected a lower social discount rate. Given that the appraisal period is 30 years (and no longer), we decided on a constant 3.5% social discount rate.[8]

[6]It is often argued that social discount rates are likely to be higher for developing countries because the social opportunity costs for capital is higher or the cost of borrowing capital tends to be higher. For example, the World Bank and the Asian Development Bank typically apply a real discount rate of 10–12% when evaluating projects in developing countries (Warusawitharana 2014).

[7]For example, the *Stern Review* recommends a declining social discount rate with rates lower than 3% for investments beyond 30 years (Stern 2007). The review states, 'If the ethical judgement is that future generations count very little regardless of their consumption level then investments with mainly long-run pay-offs would not be favoured. In other words, if you care little about future generations you will care little about climate change. As we have argued that is not a position which has much foundation in ethics and which many would find unacceptable'.

[8]Based on the recommendation of the *Stern Review*, the treasury for the United Kingdom recommends the use of a 3.5% discount rate for the first 30 years, followed by a declining rate until it reaches 1% for 301 years and beyond (Lowe 2008). It can be argued that a lower rate can be implemented in different ways if agreement to use a low rate is reached. For example, there could be two options: (1) a global agreement is reached so that investments on the ocean and coasts are evaluated with a low discount rate, but no country is required to act if its own internal discount rate is higher and the project does not pass its own internal return on investment criteria (unless international transfers change that balance), or (2) a global agreement is reached so that there are parallel evaluations—one with the low internationally agreed-upon discount rate and the other with the country's own rate for public investments.

Challenges related to carrying out a benefit-cost analysis of environmental measures include key benefit and cost omissions, ambiguity or uncertainty in assigning monetary benefits to nonmarket goods, difficulty in integrating distributional aspects,[9] and increased subjectivity for intangible benefits and costs. Although B-C ratio analyses or return on investment (ROI) studies at the global level are appealing, this approach has limitations. The biggest risk of global benefit-cost estimates is that they do not present the distribution of benefits and costs across developing and developed countries. Global B-C ratios do not reflect heterogeneity (due not only to the distribution of benefits and costs across the globe but also to differences in discount rates).

Consequently, the estimates should not be interpreted as giving an exact depiction of the flow of returns. They have been developed to indicate the scale of benefits relative to costs specific to the scenarios analysed for different activities. The analysis aims to stimulate timely discussion, influence ongoing debate on emerging sustainability issues and ensure that investments to obtain a sustainable ocean economy are not ignored in global discussions. The analysis does not attempt to show the regional variation of the benefits and costs. Conducting these assessments, which closely consider local factors, should be a key step when implementing ocean-based measures and regulations at local and national levels.

4.1 BAU and Sustainable Transformation Pathway Scenarios for 2050

The analysis aims to answer four key questions:

- If the rate of mangrove loss were halted and degraded mangrove areas were restored, what would be the benefits and costs to society?
- If the world decided to expand offshore wind energy generation (from 0.3% of total energy generation in 2020 to 2–7% of total future energy generation in 2050), what would be the benefits and costs to society?
- If the international shipping sector reduced its emissions to net zero, what would be the benefits and costs to society?
- If sustainable ocean-based food production increased (to meet the balanced diet requirements as advocated by the 2019 report by the EAT-*Lancet* Commission on Food, Planet, Health [Willett et al. 2019]), what would be the benefits and costs to society?

To answer these questions, we identified a sustainable transformation pathway scenario for 2050, then measured benefits and costs needed to achieve this pathway against a BAU scenario. The sustainable transformation pathway and BAU scenarios, taken from Hoegh-Guldberg et al. (2019) and the Food and Land Use Coalition (FOLU) report (2019), are summarised in Table 18.3. For most interventions, benefits are accrued over the long term but the investment costs occur up front.

4.2 Framework for Assessing Benefits

The four areas can yield three categories of benefits, which are discussed in more detail below:

- Health benefits from reducing environmental risks
- Environmental and ecological benefits from reduced environmental degradation (on land and in the ocean) and prevention of future temperature rise from climate change
- Economic and social benefits from stimulating economic activity and promoting sustainable development

Table 18.3 Business-as-usual and sustainable transformation pathway scenarios

Four actions	Business-as-usual (BAU) Scenario	Sustainable transformation pathway scenario
Conserve and restore mangroves	Blue carbon ecosystems continue to decline, but at decreasing rates. The rate of loss of mangroves globally is estimated at 0.11% per year.[a]	Mangrove conservation: The per year loss under BAU is halted completely.[b] Mangrove restoration: Two scenarios were considered: (1) a moderate restoration effort recovering 40% of the historical ecosystem cover by 2050 (consistent with global mangrove Alliance goals), and (2) an aggressive scenario of complete restoration of pre-1980s cover.[c]
Scale up offshore wind energy production	Worldwide installed offshore wind energy capacity in 2018 generated 77 terawatt hours (TWh) per year and accounted for less than 1% of world energy production.[d] The current energy technologies mix remains constant (and the share of offshore wind energy remains low) as energy production expands.	The total installation capacity for offshore wind energy is estimated to grow substantially by 2050. The offshore wind energy generation for 2050 is estimated at 650–3500 TWh per year.[e] Under this scenario, the energy mix will shift to a higher fraction of renewables to meet the future increase in energy demand.
Decarbonise international shipping	The total annual greenhouse gas (GHG) emissions from international shipping is estimated to grow from 800 megatonnes (Mt) in 2012, to 1100 Mt. in 2030 and to 1500 Mt. in 2050.[f]	Emissions in international shipping are reduced to net zero by 2050.[g]

(continued)

[9]In addition, the benefit-cost analysis does not apply any 'equity weighting' when aggregating benefits across countries or regions that have very different levels of wealth, thus giving relatively greater weight to the impacts of rich people relative to poor people.

Table 18.3 (continued)

Four actions	Business-as-usual (BAU) Scenario	Sustainable transformation pathway scenario
Increase ocean-based food production	• Fisheries continue to be overfished and global annual marine capture production declines in 2050 by 25%.[h] • Fed aquaculture (finfish) production remains at the 2020 level (11.7 million metric tonnes, or mmt) due to fishmeal constraints.[i] • Nonfed aquaculture (bivalve) continues to grow slowly to 28.5 mmt in 2050 due to lack of investments.[j]	• To meet healthy diet requirements in 2050, we need to double the current amount of ocean-based protein.[k] Part of this can be achieved by fisheries reform and the rest by increasing sustainable marine aquaculture production. • With global fisheries reform, annual marine capture production increases by 40% compared with baseline projections.[l] Fed finfish mariculture production increases to 22.4 mmt by 2050.[m] Bivalve production grows to 65.2 mmt in 2050.[n]

Notes: Total energy generation in 2018 was estimated to be 27,000 TWh/year; offshore wind contributed 0.3%
Sources: [a–g]*Hoegh-Guldberg et al.* (2019); [h]*Costello et al.* (2019); [i, j]*FOLU* (2019); [k]*Willett et al.* (2019); [l]*Costello et al.* (2019); [m, n]*FOLU* (2019)

4.2.1 Health Benefits

These include interventions such as scaling up ocean-based renewable energy production and decarbonising shipping to reduce GHG emissions. Indirect health-related cobenefits of reducing air pollutants include reduced mortality rates, improvements in productivity from improved well-being of workers,[10] lower absenteeism from school/work caused by reduced childhood asthma,[11] and reduced morbidity.

Measures that induce even moderate shifts in diet from high meat consumption towards ocean-based protein have well-documented human health benefits (Blas et al. 2019; González Fischer and Garnett 2016; Hollander et al. 2018; Oita et al. 2018; Simões-Wüst and Dagnelie 2019; Tilman and Clark 2014). Finally, healthy mangroves directly provide nutrition to local communities via enhanced fisheries and indirectly via increases in other ecosystem services (such as coastal protection and improvements in water quality) and by income-generating activities (such as timber for fuelwood, nontimber forest products like honey and medicines, and income from tourism.)[12]

Some health benefits cannot be quantified; thus, they have been described qualitatively. The monetary value of these benefits could be significant, and additional research is required to quantify them. The benefit assessed across most interventions is avoided health damage from increased GHG emissions, and it focuses specifically on the impacts of criteria pollutants (Box 18.2).

4.2.2 Environmental and Ecological Benefits

Direct climate change mitigation would be achieved by reducing GHGs and limiting global temperature rise to 1.5 °C. These impacts include avoided losses in activities that are counted in a country's gross domestic product, or GDP (such as agriculture, fisheries productivity,[13] tourism, manufacturing and services); avoided property damages from increased coastal flooding; and avoided noneconomic impacts that do not appear in GDP measures (such as the loss of natural habitats from increased ocean acidification and increased risks to human health from extreme temperatures, including heat stress). We use the social cost of carbon method to measure the environmental externalities caused by an increase in GHG emissions (Box 18.3). Biodiversity-related cobenefits include an increased abundance of marine wildlife, reduced noise and other disturbances that negatively impact marine species, and the natural treatment of pollution and waste. These benefits have a direct positive impact on the marine ecosystem and its organisms and indirectly contribute to societal well-being.

[10]Working in a highly polluted setting for a long period of time can affect your mood or disposition to work. Evidence shows statistically significant adverse output effects (resulting in lower productivity) from prolonged exposure to ambient particles (He et al. 2019).

[11]There is a link between shipping pollution and childhood asthma (Sofiev et al. 2018) that leads to children missing school and their parents missing work. The shipping sector analysis explores this in more detail.

[12]Tourism-based income can improve economic and social conditions in local communities; hence, it indirectly contributes to health benefits.

[13]Climate change can have a positive or negative impact on regional fisheries; overall, though, there will be a decline in fisheries productivity.

Box 18.2 A Description of Avoided Mortality Losses from Reduced Greenhouse Gas Emissions

The cobenefits of global greenhouse gas (GHG) reductions on air quality and human health are estimated using analysis from West et al. (2013), which found that the global average marginal cobenefits of avoided mortality were US $50–380 per tonne of carbon dioxide reduced ($65–490 in 2019 prices). The analysis used a global atmospheric model and consistent future scenarios via two mechanisms: reducing coemitted air pollutants and slowing climate change and its effect on air quality. The model accounts for the impacts of ozone as well as fine particulate matter ($PM_{2.5}$), international air pollution transport and changes in global ozone from methane, and the study evaluates future scenarios in which population susceptibility to air pollution and the economic 'value of statistical lives' grows.[a] The authors state that the cobenefits may be underestimated because they do not account for people younger than age 30 (including children and neonatal effects), and they do not account for the benefits of avoided morbidity outcomes (i.e., reduced output from lower productivity).

Note: [a]*The value of statistical life is based on the willingness (and ability) to pay for reducing the risk of death. Hence, the study estimates marginal cobenefits to be high in North America and Europe, reflecting higher incomes in the region. Overall, though, the marginal cobenefit is found to be highest in regions with largest population affected by air pollution.*

4.2.3 Economic and Social Benefits

Transitioning to a sustainable ocean economy can lead to higher productivity, efficiency gains and revenues. For example, reforming fisheries will lead to long-term revenues and profits from higher fisheries productivity (outweighing the short-term losses). Similar fisheries productivity benefits have been observed in restoring and maintaining healthy mangroves. Improving the productivity of resources will in turn help boost revenues to industry, contributing to a country's national income. In addition, driving innovation and technological advancement will increase efficiency gains and unleash unforeseen market opportunities (GCA 2019).

In addition, these investments will help countries meet their SDGs and targets (Hoegh-Guldberg et al. 2019).

This includes creating decent jobs (SDG 8.5), protecting vulnerable communities from climate-related disasters (SDG 1.5), reducing poverty by improving household income/livelihoods (SDGs 1.1 and 1.4) and helping countries achieve their food security targets (SDG 3.2).

Box 18.3 Measuring Climate Benefits Using the Social Cost of Carbon

Benefit-cost analysis assumes that society should reduce carbon dioxide (CO_2) emissions up to the point where the marginal cost of reducing a tonne of CO_2 is just equal to the marginal benefit of keeping that tonne out of the atmosphere. The social cost of carbon (SCC) measures the benefit of reducing carbon dioxide equivalent (CO_2e) emissions; that is, it represents the dollar value of the cost (i.e., damages) avoided by reducing CO_2e emissions by 1 tonne.[a]

The model used to deliver SCC values, the integrated assessment model, provides a range of estimates[b] because of the many factors (including the types of greenhouse gas emissions) analysed, the types of impacts (gross domestic product, or GDP, versus non-GDP) analysed,[c] the discount rates used and size of risk aversion of the population.[d]

The SCC value used in this analysis reflects the avoided costs from changes in net agricultural productivity, human health, loss from increased natural disasters and changes in energy system costs, such as reduced costs for heating and increased costs for air-conditioning.[e] To prevent double counting with estimated health benefits from a reduction in ozone and fine particulate matter ($PM_{2.5}$), we used the SCC value developed under the U.S. Environmental Protection Agency that focuses only on damage costs from increases in the level of carbon dioxide in the atmosphere. The damage costs for CO_2 was estimated, in 2007 prices, at US $42 in 2020 and rises to $69 in 2050. Because the SCC value used does not account for all the damage costs, the impacts quantified monetarily are underestimates.

Notes:

[a] *Hausker* (2011).

[b] *Based on a number of studies, SCC values range from $50 to $417 per tonne of CO_2e reduced* (BEIS 2019; Tol 2019).

[c] *Activities counted in a country's GDP, such as agriculture, fisheries productivity, tourism, manufacturing and services, would feature in a GDP measure whereas non-GDP measures would include noneconomic impacts, including the loss of natural habitats and increased risks to human health (from heat stress and other factors).*

[d] *Standard practice in benefit-cost analysis is to take a risk-neutral approach to uncertainties. In the real world, individuals and organisations of all types display risk aversion to catastrophic impacts* (Hausker 2011).

[e] *EPA* (2016).

4.3 Framework for Assessing Costs

The costs of transformation, relative to BAU, were assessed by examining a list of actions and measures that can be undertaken by the government and private sector to achieve targets such as restoring mangroves, reducing emissions, reforming fisheries and increasing sustainable ocean-based aquaculture production.

Examples of these types of costs are given below:

- **Costs to business** include capital investments; for example, building new offshore aquaculture farms, increasing offshore renewable energy, implementing technological improvements in shipping and increasing private research and development (R&D) expenditures.
- **Costs to government** include costs of regulations (on mangrove and fisheries conservation), public R&D expenditures and higher enforcement and monitoring costs (for mangroves and fisheries).
- **Costs to households** include temporary reductions in household income from fisheries reform and the forgone income from the alternative use of the mangrove area by shrimp farming and/or charcoal production if they are not protected (opportunity cost). The presence of positive private opportunity costs may be an economic barrier to the success of mangrove conservation because they represent a direct economic loss (or disincentive) to local communities that undertake mangrove conservation activities.

For some sectors, such as renewable energy production and ocean-based aquaculture, the private sector costs were estimated based on existing analytical projections of the state of the technology in 2030 and 2050, and we assumed reductions in future costs due to economies of scale and 'learning by doing' (Arrow 1962). If components of costs were not quantified—for example, the costs of implementing national regulations to ensure decarbonisation of the shipping sector have not been monetised—they are discussed qualitatively.

5 Assessing the Return on Investment for Four Sustainable Ocean Transformations: Scenarios, Assumptions, Methodology, Results

This section presents the scenarios, discusses the assumptions and methodology used to estimate the benefits and costs for each of the four areas examined and finally presents the net benefits and the B-C ratios.

5.1 Conserve and Restore Mangroves

5.1.1 Baseline, Sustainable Transformation Pathway and Target Scenarios

The assumptions about the BAU scenario and the sustainable transformation pathway needed to achieve the conservation and restoration targets by 2050 are informed by Hoegh-Guldberg et al. (2019).

5.1.2 The BAU Scenario

Although blue carbon ecosystems continue to decline, they do so at decreasing rates thanks to improved understanding, management and restoration (Lee et al. 2019). For instance, the rates of mangrove loss globally declined from 2.1% per year in the 1980s (Valiela et al. 2001) to 0.11% per year in the past decade (Bunting et al. 2018). The BAU scenario assumes the loss of mangroves continues at 0.11% per year until 2050. The sustainable transformation pathway builds from this base.

5.1.3 The Sustainable Transformation Pathway Scenario

The mitigation potential could be achieved via two pathways: conservation of ecosystems and restoration of ecosystems.

- **Conservation of mangroves.** The total area for mangroves conserved per year is estimated to be 15,000–30,000 hectares (ha) (see Table 18.4).[14] This scenario avoids emissions of carbon stored in soils and vegetation. The total potential GHG mitigation contribution is estimated to be 0.02–0.04 gigatonnes (Gt) of CO_2e per year (Hoegh-Guldberg et al. 2019).[15]
- **Restoration of mangroves.** Restoration sequesters and stores carbon as vegetation grows. In the Hoegh-Guldberg et al. (2019) study, the range of potential mitigation varied with the level of effort and investment. Two scenarios were considered: a moderate restoration effort recovering about 40% (184,000 ha per year) of the historical ecosystem cover by 2050 (consistent with Global Mangrove Alliance goals) and a more aggressivescenario of complete restoration (290,000 ha per year) of pre-1980s cover (Hoegh-Guldberg et al. 2019). The corresponding total GHG mitigation potential was estimated at 0.16 GtCO_2e per year to 0.25 GtCO_2e per year in 2050 (Hoegh-Guldberg et al. 2019). See Table 18.4.

[14]This is based on avoiding the current loss of mangroves per year under BAU (Hoegh-Guldberg et al. 2019).

[15]The range of CO_2 sequestration potential per unit area for each ecosystem was calculated using default emission/removal factors from the IPCC (2013).

Table 18.4 Conservation and restoration pathways for mangroves by 2050

	Conservation		Restoration	
	Moderate	Aggressive	Moderate	Aggressive
Hectares conserved or restored per year	15,000	30,000	184,000	290,000
GHG mitigation potential ($GtCO_2e$ per year)	0.02	0.04	0.16	0.25

Notes: GHG greenhouse gas, $GtCO_2e$ gigatonnes of carbon dioxide equivalent
Source: Hoegh-Guldberg et al. (2019)

The GHG emission mitigation estimates are likely conservative because they do not account for avoided methane (CH_4) and high nitrous oxide (N_2O) emissions from alternative land uses such as aquaculture and rice production (Hoegh-Guldberg et al. 2019). These emissions can be significant due to mangrove conversions to aquaculture or rice farming; for example, 30% of mangrove ecosystems in Southeast Asia have been converted to aquaculture and 22% to rice cultivation (Richards and Friess 2016). These GHG estimates from land use changes are excluded from the present analysis due to the lack of global data.

5.1.4 Assessment of Costs

Conservation Costs

For conservation, we estimated the cost of monitoring and maintaining the mangroves and the opportunity costs of the forgone net income from alternative use of the mangrove area (Table 18.5). For enforcement and monitoring costs, a global average cost estimate of maintaining marine protected areas was used as a proxy. For the second component, we looked at the opportunity costs for returns from shrimp farming, crab catching and charcoal production (see Table 18.5). Because it was unknown which activities might exist at which sites, we used the sum of the three to represent the higher estimate of the opportunity costs.

We estimated the annual global costs of conservation to be $28.8–57.5 million based on the per hectare estimates in Table 18.5 and the additional area conserved by 2050.

These numbers are indicative of global costs. In reality, the actual costs might be lower or higher depending on the location and sizes of the protected areas.

Restoration Costs

Restoration is often needed when ecosystem degradation is reaching its ecological threshold and significant efforts are required for seeding and replanting mangrove species to restore it. The analysis uses the global restoration cost estimates reported in the Bayraktarov et al. (2016) study that

Table 18.5 Types of costs and data sources used to estimate the costs of mangrove conservation and restoration

Description of costs	Cost (US $/ha/year)	Adjusted 2019$	References
Monitoring and maintenance cost: Median cost covers the current marine protected area expenditure plus estimated shortfall[a]	27	40	Balmford et al. (2004)
Global restoration costs of mangroves	8961 (median)	9449	Bayraktarov et al. (2016)
Opportunity cost: Net economic returns from shrimp farming in Thailand	1078–1220	1873 (average)	Barbier (2007)
Opportunity cost: Net economic returns from charcoal production in northwestern Madagascar	4	5	Witt (2016)
Opportunity cost: Net economic returns from crab catching in northeastern Brazil[b]	12	16	Glaser and Diele (2004)

Notes: ha hectare
[a] *To assess the enforcement and monitoring costs, a global average cost estimate of marine protected areas was used as a proxy for the conservation costs for mangrove protection. Balmford et al. (2004) state that the total costs per unit area of running the marine protected areas in their sample varied enormously, with the sum of current expenditure plus estimated shortfall ranging from about $4 per square kilometre ($km^2$) per year to nearly $30 million/$km^2$/year (median, $2698/$km^2$/year or $27/ha). We use the median figure in our analysis. The costs of a protected-area system are divided into three categories: (1) recurrent management costs for existing areas, (2) systemwide expenses needed to support a network of protected areas and (3) costs of bringing new areas into the system*
[b] *At $13.50 per person/day × 4500 person days in a year over about 50 km^2 is about $12/ha/year*

conducted a meta-analysis and systematically reviewed 235 studies (with 954 observations), including projects that restored and rehabilitated mangroves and other vegetated coastal habitats in different world regions. They suggested a median cost per hectare of $8961 per year (2010 prices, converted to $9449 in 2019 prices). We assume the costs are two times higher ($18,997) if both operating and capital costs are included (Bayraktarov et al. 2016).

The opportunity cost for restoration is assumed to be the same as that of conservation, but the forgone benefits can occur only 5 years after the restoration efforts have been completed, assuming that once the coastal ecosystems have improved, these areas are again under the risk of being disturbed. The annual restoration costs are estimated to be $3.5–5.5 billion between 2020 and 2050.[16]

[16]The range is obtained by multiplying the median cost (point estimate) with the area of restoration (range).

Assessment of Benefits

Mangroves extend over 150,000 square kilometres (km^2), distributed across 123 countries (Beck et al. 2018).

They provide a wide array of market and nonmarket benefits, which are categorised below according to health, environmental and economic/social benefits. The range of benefits quantified includes coastal protection benefits, sequestration benefits and fisheries productivity benefits. Conserving and restoring mangroves will also increase other ecosystem services, which, in turn, will increase societal well-being, which we have discussed qualitatively.

In this study, we assumed that the benefits generated through mangrove restoration (such as coastal protection and fisheries productivity) will not accrue immediately following the restoration effort but rather after there has been improvement in the condition of the ecosystem. We assume this to be 5 years after the restoration/rehabilitation work begins (Burke and Ding 2016).[17] In addition, the probability of success for mangrove restoration is very low. Bayraktarov et al. (2016) estimate the median survival of restored mangroves, assessed only within the first 1–2 years after restoration, to be 51.3%. For the restoration scenarios, we multiply the benefits by the probability of success of restoration or the median survival rate.

5.2 Health Benefits

Mangroves are a direct source of food, fuelwood, fiber and traditional medicine for local inhabitants (Bandaranayake 1998; Chaigneau et al. 2019). They provide important opportunities for communities to generate incomes from tourism associated with recreational fishing and bird-watching that generate recreational and aesthetic value to visitors (Carnell et al. 2019). These livelihood, cultural and recreational benefits, while important to the physical and mental health and well-being of local communities as well as visitors (de Souza Queiroz et al. 2017; Pearson et al. 2019), have not yet been quantified. In some developing countries such as Kenya and Mozambique, mangrove medicine was used by coastal communities to cure stomach pains or headaches but did not have direct commercial value (Chaigneau et al. 2019).

5.3 Environmental Benefits

5.3.1 Protection from Storm Surges

The biggest benefits of mangroves are that they form a natural breakwater that limits the damage to property, economic disruption and loss of life caused by coastal flooding and storm surges, which become stronger and more frequent with climate change. The aerial roots, trunks and canopy of mangrove forests provide a strong protective barrier against winds, swell waves, storm surges, cyclones and tsunamis.

Studies indicate that incoming wave heights are reduced by 13–66% by a 100-m-wide mangrove belt, and by 50–100% by a 500-m-wide belt (World Bank 2016).

Protecting and restoring coastal and marine ecosystems can reduce the impacts of cyclones on an estimated 208 million individuals in 23 major mangrove-holding countries (Hochard et al. 2019).[18] A meta-analysis of 44 studies found a median value of $3604 per hectare per year for the coastal protection services (avoided property damage) provided by mangroves (Salem and Mercer 2012), which, when updated to 2019 prices, yield annual benefits of $60–120 million for conservation scenarios analysed, and $375–592 million for restoration scenarios analysed (Table 18.6).

5.3.2 Mitigation of Climate Change and Carbon Sequestration Benefits

Mangroves play an important role in sequestering carbon; hence, they can contribute towards mitigation solutions aimed at limiting temperature rise to 1.5 °C. The discounted climate benefits (calculated based on annual GHG emissions in Table 18.4) from reducing CO_2 emissions are estimated at $42–83 billion for conservation and $137–214 billion for restoration over 30 years.

5.3.3 Other Ecosystem Services

Mangroves also provide many ecosystem services, such as regulating water quality and reducing coastal erosion, that we have not been able to quantify (see Appendix 1).

5.4 Economic and Social Benefits

5.4.1 Commercial Fisheries

Although some estimates have been much higher [e.g., Aburto-Oropeza et al. (2008) estimated that protecting 1 ha of mangroves in California was associated with increased fish yields valued at $37,500 per year], we conservatively used $18,000 per hectare per year (de Groot et al. 2012), based on global meta-analysis, to assess the commercial value of fish yields associated with conserved or restored mangroves (Table 18.6). We estimate the global economic benefit from increased productivity of commercial fish spe-

[17]The time frame for generating these benefits will vary, and in some extreme cases, full development of the aboveground biomass will not be achieved for 20–30 years (Osland et al. 2012; Salmo et al. 2013).

[18]Countries receiving the largest benefits in avoided flood damage in absolute dollar terms include China, India, Mexico, the United States and Vietnam. The largest beneficiaries relative to the size of their economies include many low-income countries, such as Guinea, Mozambique and Sierra Leone (Beck et al. 2018).

Table 18.6 Benefits of mangrove conservation and restoration in avoided property damage and fisheries productivity

Type of benefit	Benefit (US $/ha/year)	Adjusted 2019$	References
Avoided property damage	3604	4000	Salem and Mercer (2012)
Fisheries productivity	18,000	19,980	de Groot et al. (2012)

Note: ha hectare
Source: Authors' calculations

cies to be $300–600 million per year for the conservation scenarios and $1.9–3.0 billion per year for restoration scenarios.

5.4.2 Tourism

Although we have not been able to provide a global estimate on increases in tourism arising from the scenarios analysed, these are likely to be significant for some countries. Mangrove tourism and recreation is a multibillion-dollar industry (Spalding and Parrett 2019). For example, tourism associated with coral reefs and mangroves in Belize contributed an estimated $150–196 million (12–15% of GDP) to the national economy in 2007 (Cooper et al. 2009). These benefits are also further discussed in Appendix 1. While there will be a short-term dip in coastal tourism following the COVID-19 lockdown, this assessment focuses on benefits over a 30-year period. Over the longer term, we estimate these benefits will pick up as the global economy emerges out of the pandemic and economic crisis.

There is also a strong social angle in terms of the distribution of the benefits. For example, low-income communities are most reliant on mangroves for key ecosystem services (Box 18.4).

5.4.3 Estimated Benefits and Costs

We estimated the B-C ratio under two approaches. In the first approach, we estimated the ratio over 30 years (2020–2050) using present value benefits and costs. In the second approach, we calculated the B-C ratio per hectare.

B-C Ratio Using Present Value Approach

For every $1 invested in mangrove conservation and restoration, we get a return of $3. Net benefits for mangrove conservation are estimated at $48–96 billion and for restoration at $97–150 billion over 30 years (2020–2050). The value of net benefits for mangrove restoration is higher than conservation because we assumed the area of mangroves restored would be 10 times the area conserved (Table 18.7).

However, we find that conservation of mangroves yields significantly more returns per dollar invested than restoration. For every $1 invested in mangrove conservation, we get a return of $88 dollars for conservation, versus $2 for restoration.

Table 18.7 Net present value and benefit-cost ratios for mangrove conservation and restoration

Transformation areas	Net present value (US $, billions, 2019$, 2020–2050)	Benefit-cost ratio
Conservation of mangroves[a]	48–96	88:1
Restoration of mangroves[b]	97–150	2:1
Total	145–246	3:1

Notes:
[a] *Conservation of 15,000–30,000 ha per year based on halting annual loss of mangroves*
[b] *Based on 184,000 ha per year for a moderate effort to 290,000 ha per year for an aggressive estimate*
Source: Authors' calculations

B-C Ratio for a Hectare of Mangrove Restored/Conserved

We estimated the benefits for restoring 1 ha of mangrove to be $30,080 and for conservation $79,980. Based on the per hectare conservation and restoration costs in Table 18.5, we estimate the B-C ratio per hectare to be 2:1 for restoration and 48:1 for conservation.

Box 18.4 Mangroves Protect the Poorest Populations

Low-income communities rely heavily on mangroves for key ecosystem services. Over nearly 98 million people from 10 low- or lower-middle-income countries with major mangrove areas and a gross national income per capita less than US $4036 annually have suffered from cyclones.[a] This accounts for 50% of the global cyclone-affected population from 18 mangrove-holding countries. Poor coastal families are most vulnerable to natural disasters; hence, building ecosystem resilience to protect them from coastal flooding and cyclones will not only safeguard their valuable assets but also generate tremendous social benefits (e.g., feeling safe) that cannot be easily quantified monetarily.

Note: [a] *Hochard et al.* (2019).

For both of the approaches, the ROI for restoration is lower, first, because the cost of mangrove restoration is much higher than conservation due to the high costs of seeding and replanting; second, it takes time to accrue benefits from restoration since the plants need to regrow and restoration requires the right conditions to ensure a high survival rate (see caveats in Appendix 1).

The monetised benefits presented under both of the approaches exclude a number of ecosystem services provided by mangroves. Major ecosystem benefits such as erosion control, water management, nutritional benefits from fisheries supported by mangroves[19] and health benefits are excluded from the current assessment. This is mainly because our assessment relies on previous valuation studies or meta-analyses that either do not attribute a value to these particular services or provide a total value across a range of services but do not address double-counting issues. Other social benefits that are not accounted for include employment and the potential for livelihoods associated with sustainably harvesting timber and nontimber forest products. Taking into account these benefits will likely result in a higher ROI.

The results from both of the approaches show that both types of interventions yield significant benefits and are important to ensure a high ROI. To reverse the current trend of marine and coastal resource depletion and further halt the release of CO_2 emissions from marine resource degradation, significant investment will need to be made to transform the way coastal and marine ecosystems are being managed. They would need more reliable funding for management/enforcement and greater participation/diversification of opportunities dependent on these ecosystems, in addition to strong 'political will' to involve measures that alter the fundamental attributes of a system (including value systems; regulatory, legislative or bureaucratic regimes; financial institutions; and technological or biological systems) (Ellis and Tschakert 2019). These social and political investments are important and have not been valued in the analysis.

5.4.4 Data Limitations and Caveats

Data limitations prevented us from assessing other coastal ecosystems: salt marsh and seagrass beds. Some caveats are that the value of mangrove conservation or restoration varies by locality, the costs are higher in developed countries, and coastal development pressure is a big influence. See Appendix 1 for important caveats.

5.5 Scale Up Offshore Wind Energy Production

Currently, global electricity generation from all ocean-based energy sources is less than 0.3% of the total (IEA 2019a). The ocean energy sector has seen a dramatic increase in investments over the past decade and is expected to grow (European Commission 2018; Hoegh-Guldberg et al. 2019; WBG et al. 2018). Currently, most offshore installations are in Europe, but a significant increase is expected in Asia, especially China (Hoegh-Guldberg et al. 2019).

In assessing the impacts of expanding offshore wind energy generation, we do not advocate one renewable energy technology over another. Rather, we focus on the impact (positive and negative) of expanding offshore wind energy against a baseline where fossil fuel sources of electricity generation continue to dominate. We looked at how much it would cost to increase production of offshore wind energy to meet the energy generation potential proposed in Hoegh-Guldberg et al. (2019) and estimate the benefits to society from reductions in GHGs and water usage.

5.5.1 Baseline, Sustainable Transformation Pathway and Target Scenarios

Between 2000 and 2017, the cumulative installed capacity of offshore wind energy rose from 67 megawatts (MW) to 20 gigawatts (GW) (IRENA 2018a, b). In 2018, the total global capacity of wind energy was 564 GW, of which 23 GW were offshore. Offshore wind energy produced 77 terawatt hours (TWh) of electricity annually.[20]

By 2050, gross global electricity generation is projected to be between 42,000 and 47,000 (TWh) (IEA 2019a).

In reviewing 15 energy scenarios for 2050 that considered ocean renewable energy, Hoegh-Guldberg et al. (2019) concluded that the annual energy generation from offshore wind technologies would increase between 650 and 3500 TWh per year.[21] To assess the impact of this increase on GHG emissions, the authors made assumptions about what technologies offshore wind would displace. They looked at the impact on GHG emissions if

- Offshore wind technologies displaced coal; and
- Offshore wind technologies displaced the current (2018) energy-generation mix.

We used the second scenario, which projected that scaling up offshore wind energy could reduce GHG emissions by between 0.3 and 1.61 $GtCO_2e$ per year in 2050 (Hoegh-Guldberg et al. 2019). Hoegh-Guldberg et al. (2019) acknowledge that this is a simplistic approach and, in reality, the substitution effect of ocean-based energy will mainly impact certain grids with given energy mixes, which, in turn, depends on global trends, including technology costs.

5.5.2 Assessment of Costs

Offshore Wind Energy Generation Costs

We use the levelised cost of electricity (LCOE) to estimate the cost of additional offshore wind energy generation. The LCOE includes capital costs, fuel costs, fixed and variable

[19]It can be argued that the value of nutritional benefits is already embedded in the value of fish sold.

[20]Within offshore wind energy technologies, bottom-fixed water technologies dominate the current capacity of offshore wind energy.

[21]The authors based their estimation on several studies that have included offshore wind in scenarios projecting future energy mix. These include International Energy Agency scenarios, Bahar et al. (2019) and Teske et al. (2011). We assume a linear increase in energy generation from 2020 to 2050.

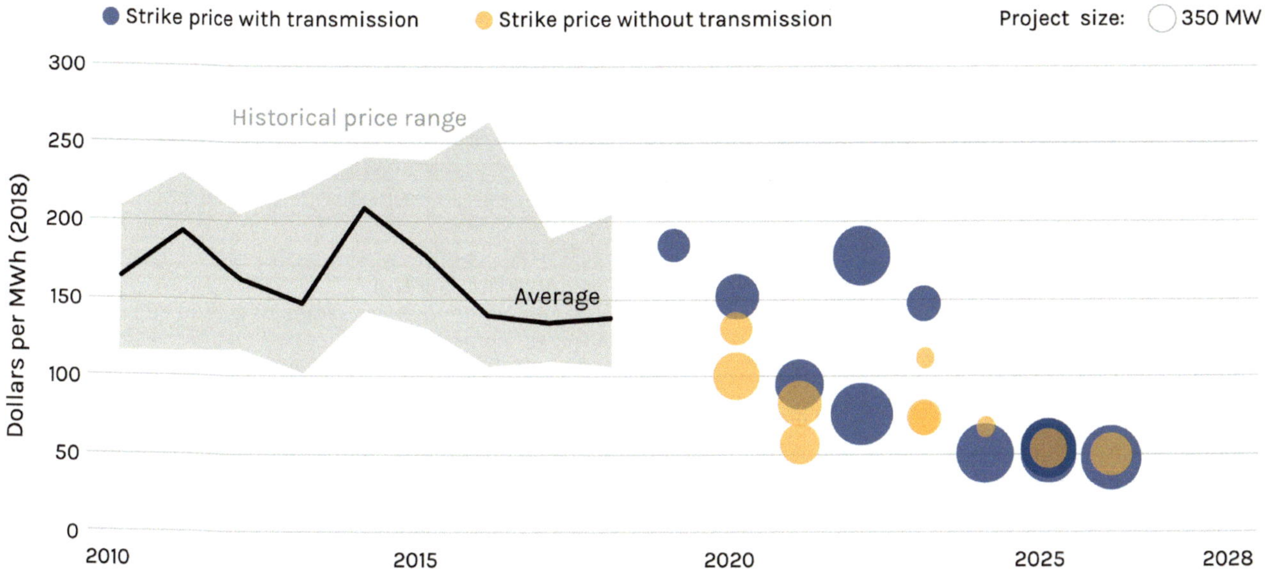

Fig. 18.3 Historical levelised cost of electricity generation of offshore wind and strike prices in recent auctions in Europe. Note: *MWh* megawatt-hour. Source: IEA (2019b)

operations and maintenance costs, financing cost and an assumed utilisation rate for each plant type (IEA 2015).

The LCOE for offshore wind power has declined since 2010 due to factors such as increased capacity from new installations (the ratio between realised energy output and theoretical maximum output), declining operational and maintenance costs due to improved turbine design (as they are made more robust for the offshore environment), improved capacity factors (linked to an increase in turbine size and hub height) and reduced transmission costs (Hoegh-Guldberg et al. 2019).

The global weighted-average LCOE of offshore wind projects commissioned in 2018 was estimated at $127–140 per megawatt-hour (MWh) based on the standard cost of capital representing full market risk (7% for developing countries and 7.5–8% for developed countries) (IEA 2019a; IRENA 2019). Improved financing terms could reduce the LCOE of offshore wind (IEA 2019a).

For example, applying a 4% weighted average of capital costs (WACC) to 2018 costs and performance parameters yields an offshore wind LCOE of about $100/MWh, which is 30% less than the LCOE derived from the standard WACC (7–8%) (IEA 2019b).

Declining recent strike prices[22] of offshore wind projects provide strong market signals of future cost reductions, indicating increased confidence from investors and setting the stage for low-cost financing opportunities for upcoming projects (Fig. 18.3). Analysis by the International Energy Agency (IEA) of auction strike prices shows costs could fall as low as $50/MWh (in some cases including transmission) for delivery in the mid-2020s (for example, a UK strike price of $51/MWh was seen in the September 2019 auction) (IEA 2019b).[23] Overall, evidence shows that with an economy of scale and learning curve effects, significant additional reductions in generation costs of offshore wind can be anticipated in subsequent years (IEA 2019b).

For floating offshore wind energy platforms, the LCOE may be higher because this is a less mature technology compared with the predominate bottom-fixed technology; it represents only 0.03 GW of the total of 23 GW of offshore power capacity. While cost reductions in the sector are expected due to rapidly advancing technology and market conditions enabling deployment to compete globally,[24] given the current low installation capacity of floating offshore wind facilities, it is difficult

[22]The strike price is a guaranteed price to be paid to wholesale generators of electricity.

[23]There is a risk that, depending on how auctions are designed, low bids may be associated with no delivery and/or renegotiations. For example, Welisch and Poudineh (2019) state that one-shot auctions and the lack of a nondelivery penalty clause increase the probability of speculative bidding and prevent bidders from learning and from utilising information efficiently.

[24]In 2015 the costs of floating offshore energy were estimated to range between $187/MWh and $316/MWh (IEA and NEA 2015), with predictions that costs would fall by 40% by 2030 due to rapidly advancing technology and market conditions that enable offshore wind deployment to compete globally. These cost declines are also reflected in recent studies. Previous National Renewable Energy Laboratory (NREL) studies estimated the LCOE for an offshore wind project in the Massachusetts wind energy area to be $77/MWh (Moné et al. 2016). Later NREL studies revised the LCOEs downward to $74/MWh by 2027 and $57/MWh by 2032 for floating offshore technologies in Maine (Musial et al. 2020). The recent technological and commercial improvements in the global industry are applicable to the turbine design, turbine scaling effects on the balance of station, lower financing terms and lower costs for the floating platform, array and export cables. Commercial-scale plant costs (in terms of dollars per kilowatt) modelled for the Aqua Ventus technology were found to be approximately five times lower than the pilot-scale demonstration project cost that was originally estimated at $300/MWh.

to predict its future global costs. That said, while the relative importance of floating wind power platforms will increase, we assume that within the next 30 years the majority of offshore wind installations will be bottom fixed; thus, for an overall cost estimate, we assume the LCOE figures will be close to that of the bottom-fixed installations.

For this analysis, we looked at two scenarios based on IEA cost projections (IEA 2019b):

- A **moderate scenario** based on a standard cost of capital financing representing full market risk (WACC is 7–8%). In this scenario, the global LCOE falls from $140/MWh in 2018 to less than $90/MWh in 2030 and close to $60/MWh in 2040.
- An **aggressive scenario** based on the same underlying technology costs and performance parameters as the moderate scenario, but which assumes low-cost financing (WACC of 4%). The global LCOE of offshore wind declines from $100/MWh in 2018 to $60/MWh in 2030 and to $45/MWh in 2040).

5.5.3 Offshore Wind System Integration Costs

The costs of integrating offshore power generation into the land-based electricity system include infrastructure costs (for expanding and adjusting the existing electricity infrastructure to feed in electricity production) and balancing costs (for handling deviations from planned production and extra costs for investments in reserves for handling power plant or transmission facility outages).

Offshore wind power requires an offshore grid as well as expanding the onshore transmission grid. The transmission or grid costs are closely tied to the regional regulations for connecting the project to the onshore grid (IEA 2019b). In 2015, the grid and balancing costs of integrating 50% of offshore wind power into the system were estimated at $43/MWh in 2019 prices (or €37/MWh) for offshore projects in Germany (Agora Energywiende 2015). These estimates were higher than the integration costs for photovoltaic (PV) solar and onshore wind ($5–20/MWh) because it costs more to connect with an offshore generation source. However, these costs are expected to decline as offshore wind projects increase and technologies improve. The average up-front cost to build an offshore wind project, including transmission costs, will drop by more than 40% over the next decade, according to the IEA. Such a drop would be due to innovation, economies of scale and supportive action to reduce costs by grid operators.[25]

[25]Wind corridors for onshore wind have helped to streamline the siting of transmission for multiple projects that allow multiple developers to share the cost. These innovations could help bring down costs.

Table 18.8 Levelised cost of electricity for conventional sources of energy, 2019

Type of energy source[a]	US $/MWH, 2019$ (3% discount factor)	US $/MWH, 2019$ (7% discount factor)
Conventional coal	71–103	72–152
Natural gas	67–146	83–110
Nuclear	28–70	41–111
Hydropower (seasonal)[b]	74	74

Notes: MWh megawatt-hour

[a] *Levelised cost of electricity generation (LCOE) estimates for coal, natural gas and nuclear are based on NEA and IEA* (2015) *country-level analysis of LCOE for the various technologies. The ranges show that the LCOEs will vary by location as each technology and each country faces a different set of risk profiles. Original estimates are converted from 2013 prices to 2019 prices using the Consumer Price Index inflation calculator*

[b] *LCOE estimates for hydropower are based on analysis of plants based in the United States (see Stacey and Taylor* 2019*). They calculate LCOE for new plants using EIA data (which used WACC of about 4%). They state that new plants have higher fixed costs and LCOE (than existing resources) as they begin their operational lives with a full burden of construction cost to recover*

Source: IEA and NEA (2015), *Stacey and Taylor* (2019)

For this analysis, we take a conservative approach and assume the grid and balancing cost is $43/MWh in 2020 and declines by 20% ($34/MWh) over 2030–2050.

5.5.4 Baseline Energy Generation Costs

In 2018, coal, gas, nuclear and hydropower accounted for 90% of the total electricity generation (IEA 2019a). We assume that, under the baseline scenario, demand for electricity will increase over 30 years (2020–2050) and additional investments in conventional sources of energy (mainly fossil fuels) will be made to meet the demand.

We analysed the LCOE of conventional sources of energy to estimate the current costs of energy generation in the baseline based on two discount factors (Table 18.8).

Based on the current energy mix and the discount factor, we estimate the global weighted average of LCOE in the baseline to be $86–94/MWh. We assume that by 2040 the LCOE will fall by 20% based on learning effects.

6 Additional Costs of Energy Generation from Offshore Wind

The following equation is used to calculate the additional costs of scaling up offshore wind production:

Costs of offshore wind energy generation = offshore wind generation costs
+offshore wind integration costs - baseline energy generation costs

The total costs of scaling up offshore wind power are shown in Table 18.9, with the moderate scenario costing $250–884 billion and the aggressive scenario costing $97–420 billion.

6.1 Assessment of Benefits

Offshore wind energy can deliver a suite of health, environmental and ecological, and economic and social benefits.

6.1.1 Health Benefits

Due to its very low CO_2 emissions and negligible emissions of mercury, nitrogen dioxide and sulphur dioxide, as well as its 0 generation of solid or liquid waste, offshore wind energy production could have a positive impact on human health depending on what energy sources it displaces. Observational and modelling studies indicate that three million premature deaths are attributable to ambient air pollution and another 4.3 million to household pollution (WHO 2016). By multiplying the annual CO_2e emissions mitigation potential by the marginal cobenefits of avoided mortality (see Box 18.2), we estimate the total avoided damage costs (or discounted health benefits) from transitioning to offshore renewable energy at $0.15–4.4 trillion over 30 years (2020–2050).[26]

Table 18.9 Total costs of scaling up offshore wind production, 2020–2050

Scenarios	Description	Costs (US $, billions, 2019$)
Moderate	• Global LCOE falls from $140/MWh in 2018 to less than $90/MWh in 2030 and close to $60/MWh in 2040 • Integration costs: Grid and balancing costs are $43/MWh in 2020, declines by 20% over 2030–2050	250–884
Aggressive	• Global LCOE of offshore wind declines from $100/MWh to $60/MWh in 2030 and to $45/MWh in 2050 • Integration costs: Grid and balancing costs are $43/MWh in 2020, declines by 20% over 2030–2050	97–420

Source: Authors' calculations

6.1.2 Environmental and Ecological Benefits

Water Consumption Impacts

When estimating the impact of water usage for energy generation, we looked at water withdrawal and water consumption. Water withdrawal is the volume of water removed from a source, including water that is eventually returned to the same source; by definition, withdrawals are always greater than or equal to consumption (IEA 2016).[27] The type of cooling technology used mainly determines how much freshwater is withdrawn and ultimately consumed, although fuel mix, the power plant's role in the electricity system, turbine design and weather also influence the amount of water required (IEA 2016). Thermal power plants (coal, natural gas, oil, nuclear and geothermal) demand considerable amounts of water for cooling (IEA 2016) (Table 18.10). In contrast, studies show that wind systems require near zero water for energy generation and cooling (Macknick et al. 2011).

We estimate the water consumption to be 860–1315 gallons/MWh under the baseline. Using the true cost of water in terms of avoided damage to the environment, we estimate a range from $0.10 per cubic metre ($m^3$) in water-abundant

Table 18.10 Water consumption factors for nonrenewable technologies

Fuel type	Cooling type[a,b]	Water consumption (Gallons/MWH)
Nuclear	Tower, once-through, pond	269–672
Natural gas	Tower, once-through, pond, inlet	198–826
Coal	Tower, once-through, pond	103–942
PV solar	n/a	26
Offshore wind	n/a	0

Notes: MWh megawatt-hour, PV photovoltaic

[a] *Dry cooling is also an option that is not discussed here as it is expensive and has limited application*

[b] *Once-through cooling involves lower water consumption but higher water withdrawal than circulating cooling systems. In some jurisdictions (typically arid), once- through cooling is no longer permitted. However, we provide estimates of this technology to demonstrate a conservative water consumption scenario*

Source: Macknick et al. (2011)

[26]This is due to reduced criteria pollutants such as ozone and particulate matter (that are indirectly displaced).

[27]This analysis does not account for the impact of returning the water because it often gets returned at a different temperature, leading to thermal pollution.

areas to \$15/m³ in extremely water-scarce areas (Trucost 2013). We estimate the benefits (discounted) of achieving offshore energy transformation through water savings alone to be \$1.3 billion to \$1.4 trillion over 2020–2050.

Climate Impacts

We used the social cost of carbon method (see Box 18.3) to estimate the value of reductions in GHG emissions attributable to offshore wind at \$344 billion to \$1.4 trillion over 30 years.

Impacts on Biodiversity

Building more offshore wind farms could have both positive and negative impacts on biodiversity. The net impacts have not been quantified monetarily, and they would vary depending on the location of the offshore wind farm and the policies and measures to address negative impacts. Effective marine spatial planning, combined with emerging ocean energy technologies, can be effective in mitigating biodiversity loss from ocean energy technologies and reinforcing biodiversity cobenefits (Hoegh-Guldberg et al. 2019).

The risks of installing wind farms in the marine environment include biological invasions, noise and disturbing vibrations to marine species, collisions between birds and wind turbine rotors and the presence of electromagnetic fields that can disrupt marine life and benthic habitats (Langhamer 2012; Sotta 2012). However, studies have shown a gap between perceived risks and actual risks, and the former arise from uncertainty or lack of data about the real impacts (Copping et al. 2016). While it is important to acknowledge possible impacts, some of the actual risks are likely to be small and can be avoided or mitigated (Copping et al. 2016). For example, spatial planning appears to reduce risks, such as collisions with seabirds and impacts on migratory cetaceans, to manageable levels (Best and Halpin 2019). However, as wind energy expands into new areas, it could become more difficult to mitigate impacts.

Wind farms can have positive environmental impacts by serving as artificial reefs for many organisms (Hammar et al. 2016). In addition, the prohibition of bottom trawling near offshore wind farms for safety reasons eliminates the disturbance of fish, benthos and benthic habitats.

Evidence from Belgium and Norway suggests that in areas with a homogeneous seabed, wind farms may enhance diversity (Buhl-Mortensen et al. 2012; Degraer et al. 2012).

6.1.3 Economic and Social Benefits

This analysis does not monetise the impacts on jobs and livelihoods to the wider community, but it acknowledges them qualitatively. Offshore wind energy can create jobs: German and UK case studies state that offshore wind development is more labour-intensive than onshore wind development because of the greater challenges inherent in building and operating offshore farms in marine environments (BMWi 2018; IRENA 2018a). In Germany, the offshore segment accounted for 17% of total German wind employment in 2016, even though it represents no more than about 10% of the country's current total wind capacity. Estimates predict that direct full-time employment in offshore wind will be around 435,000 globally by 2030 (OECD 2016).[28] Similar analysis by Ocean Energy Systems shows that deployment of other forms of ocean energy (tidal range, wave power and ocean thermal energy) can also provide new jobs and additional investments (Huckerby et al. 2016).

However, the net global impact of the growth of the ocean-based energy on net jobs across the whole of the energy sector are less certain because the entire energy sector will transition to cleaner energy sources. Moving to cleaner energy will lead to job losses in the fossil fuel sector, though ocean-based renewable energy has the potential to benefit workers transitioning from declining offshore fossil fuel industries (IRENA 2018a; Poulsen and Lema 2017; Scottish Enterprise 2016), minimising the costs of transition and the risks of structural unemployment.

6.2 Estimated Benefits and Costs

We estimated the B-C ratio under two approaches.

In the first approach, we estimated the ratio over 30 years (2020–2050), where additional energy production is calculated for each year against the BAU scenario, using present value benefits and costs. In the second approach, we calculated the B-C ratio for one unit of energy produced.

6.2.1 B-C Ratio Using Present Value Approach

Table 18.11 shows the high and low benefit-cost ratios for the first approach—calculating additional energy production for each year against the BAU using present values benefits and costs.

On average, there is a net positive benefit from expanding offshore wind production. The net present value of benefits was estimated to be \$253 billion to \$6.8 trillion over 30 years.

Table 18.11 Net benefits from scaling up offshore wind energy and benefit-cost ratio

Action	Net present value; net benefit (US \$, billions, 2020–2050)	Benefit-cost ratio (low)	Benefit-cost ratio (high)
Scale up offshore wind production	253–6849	2:1	17:1

Source: Authors' calculations

[28]This is an estimate of direct jobs, not including indirect or induced jobs, derived from the economic activity of an offshore wind farm.

Table 18.12 Benefit-cost ratio for offshore wind production under varying LCOE levels

Scenario	Benefit-cost ratio
Scenario 1: LCOE is US $140/MWh; integration costs are $43/MWh; baseline costs are $86–94/MWh	0.9:1–3:1
Scenario 2: LCOE is $60/MWh; integration costs are $43/MWh; baseline costs are $86–94/MWh	4:1–16:1
Scenario 3: LCOE is $45/MWh; integration costs are $30/MWh; baseline costs are $68–75/MWh	7:1–28:1

Note: LCOE levelised cost of electricity; MWh megawatt-hour
Source: Authors' calculations

The ROI in 2050 can be high, as shown by the B-C ratio of 2-to-1 to 17-to-1 in 2050.

6.2.2 B-C Ratio for a Unit of Energy Generation and Transmission

We estimated the benefits for production of one additional unit of energy to be $75–300/MWh. The B-C ratio varies mainly depending on the LCOE of offshore wind assumed. We examined three scenarios with different LCOE values and found B-C ratios between 0.9-to-1 and 28-to-1 (Table 18.12).

Both approaches show that the value of the ROI will increase as the costs of energy generation for offshore wind fall with improved technologies and as actions are taken to reduce integration costs. The estimates should be treated as partial because they do not include key impacts that are discussed qualitatively, such as impacts (positive and negative) on biodiversity and on jobs and livelihoods in coastal communities.

Data Limitations and Caveats

Data limitations and caveats are described in Appendix 2. They include potential risks to biodiversity, variations in GHG mitigation depending on the fuel mix in the local grid, variations in LCOE depending on local market conditions, and omitting financial benefits from water savings.

6.3 Decarbonise the International Shipping Sector

Shipping is a significant source of emissions with identifiable reduction pathways (Hoegh-Guldberg et al. 2019). The sector is responsible for approximately 1 $GtCO_2e$ per year and represents around 3% of global anthropogenic CO_2 emissions (Smith et al. 2015). Based on current trends, GHG emissions will double by 2050 to roughly 2 $GtCO_2e$, compared with 2010 (Hoegh-Guldberg et al. 2019). In 2018, the United Nations International Maritime Organization (IMO) adopted a resolution[29] to reduce GHG emissions from shipping by at least 50% by 2050, relative to 2008 emission levels. However, greater ambition is needed to keep global temperature rise under 2 °C to 1.5 °C (Hoegh-Guldberg et al. 2019; UNFCCC 2015).

[29] See IMO (2018).

6.3.1 Baseline, Sustainable Transformation Pathway and Target Scenarios

The sustainable transformation pathway focuses on decarbonising only the international shipping sector. Although there is potential to reduce emissions in both domestic and international shipping, we focused on international shipping, which accounts for 55% of the total emissions in the sector (Olmer et al. 2017). The following scenarios were considered from Hoegh-Guldberg et al. (2019).

Under the BAU scenario, it is estimated that total annual GHG emissions from international shipping will grow from 800 megatonnes (Mt) in 2012 to 1100 Mt. in 2030 to 1500 Mt. in 2050. The mitigation potential assumes a 20–39% emissions reduction in 2030 from a 2008 baseline, and in 2050, a 50–100% emissions reduction from the 2008 baseline emissions (Table 18.13).

The upper-bound emissions reduction for 2050 assumes that all vessels move to full use of nonfossil fuels from renewable feedstock. The lower bound is set at 50%, taken as the minimum interpretation of the IMO's objectives in the initial GHG reduction strategy (Hoegh-Guldberg et al. 2019).

6.3.2 Assessment of Costs

Because only a small subset of the fleet is likely to be 'zero-carbon-fuels ready' by 2030, we assume the mitigation potential for 2030 to be mainly driven by maximising energy efficiency (Hoegh-Guldberg et al. 2019). This includes technological measures that increase the energy efficiency of a ship, such as altering its weight (using lighter materials) or design (such as hull coatings and air lubrication to reduce friction), and other ways to reduce or recover energy (such as via propeller upgrades and heat recovery). These measures could result in fuel savings of up to 25% (ITF 2018). In addition, energy could be saved by changes in how ships—and, more broadly, maritime transport systems—are operated, such as changes in speed, ship-port interface and onshore power. Over the last few years, both slower speeds and larger ship sizes have contributed to a decrease in shipping emissions (ITF 2018).

Table 18.13 Greenhouse gas mitigation potential from decarbonising international shipping, 2030 and 2050

Action	2030 Mitigation potential ($GtCO_2e$/year)	2050 Mitigation potential ($GtCO_2e$/year)
Reduce emissions from international shipping	0.2–0.4	0.75–1.50

Note: $GtCO_2e$ gigatonnes of carbon dioxide equivalent
Source: Hoegh-Guldberg et al. (2019)

However, efficiency measures are ultimately limited by factors such as the efficiency of a propeller or an internal combustion engine that are impossible to improve beyond a certain point (IMarEST 2018). As those limits are approached, improvements have increasingly diminishing returns and become less cost-effective (IMarEST 2018). Hence, the cost of decarbonising international shipping is ultimately capped by the cost of switching to zero CO_2 emissions fuels and technologies (IMarEST 2018).

We refer to the IMarEST (2018) study to estimate the cost of GHG reduction in international shipping. The study assumes that significant absolute emissions reductions are achieved even at low marginal cost of carbon ($50/tonne) (IMarEST 2018).[30] The results from the same IMarEST (2018) model state that, depending on how prices evolve for renewable electricity in coming decades and other assumptions in the scenarios, a 70–100% absolute reduction in GHG emissions by 2050 can be achievable for a marginal abatement cost of $100–500/t$CO_2$e. By multiplying the cost per tCO_2e abated with the mitigation potential estimated in the Hoegh-Guldberg et al. (2019) study, we estimate the total costs (capital and operational) over 30 years to be $2.3 trillion to decarbonise shipping by 100%.[31]

6.3.3 Assessment of Benefits

The health, environmental and ecological, and economic and social benefits from the international shipping sector reducing its GHG emissions are summarised below.

Health Benefits

Reduced $PM_{2.5}$ from marine engine combustion mitigates ship-related premature mortality and morbidity (Sofiev et al. 2018). The annual avoided health damage cost to adults is calculated by multiplying the CO_2e emission mitigation potential by the average marginal cobenefits of avoided mortality (see Box 18.2). In addition to the impact on adult mortality, evidence shows that reducing shipping emissions will positively impact childhood morbidity by reducing childhood asthma (Sofiev et al. 2018). Based on the methodology outlined above for reducing adult mortality and for childhood asthma (see Appendix 3), we estimate the discounted cumulative health benefits from reducing emissions to be $1.3–9.8 trillion over 30 years (2020–2050).

Environmental and Ecological Benefits

Strong acids formed from shipping emissions can produce seasonal ‘hot spots’ of ocean acidification in areas close to busy shipping lanes. Hot spots harm local marine ecology and commercially farmed seafood species (Hassellöv et al. 2013). Reducing global GHG emissions, including shipping emissions, is critical to mitigating local and global ocean acidification.

A recent study found that lower trophic species such as bivalves were affected disproportionately due to the compounding effects of shifts in temperature, chlorophyll and ocean acidification. The commercial mollusc industry is estimated to lose over $100 billion by 2100 due to ocean acidification alone (Narita et al. 2012).

In addition, reducing ship speeds could positively impact marine mammals and other species. A 10% reduction in ship speed could reduce the total sound energy generated underwater by 40% and reduce the overall whale strike rate by 50% (Leaper 2019). Such measures would benefit marine species (including the whale population) globally, resulting in higher ecosystem service values (both recreational and nonuse values[32]) that will, in turn, improve human well-being. Because of uncertainty about the exact impact that measures to reduce GHG emissions would have on ocean acidification and noise, we have not been able to monetarily quantify these key impacts.

Reducing emissions in shipping will help avoid the most catastrophic impacts of climate change. We estimate the climate benefits (see Box 18.3) from reducing carbon emissions to be $0.8–1.6 trillion over 30 years.

Economic and Social Benefits

Estimates suggest that improved hull shape and materials, larger ships, drag reductions, hotel-load savings and better engines and propulsors, together with routing improvements, can deliver overall efficiency improvements of 30–55% (ETC 2018). The analysis indicates that reducing

[30]This is because of the assumption about the availability of bioenergy; in these scenarios, it is significant relative to international shipping’s total demand for energy. In this modelling, bioenergy is assumed to enter the fuel mix as a substitute for fossil fuels and, therefore, is at the same price as the fossil fuel equivalent and is not dependent on additional carbon price to stimulate its take-up. For example, the study assumes that bioenergy enters the fuel mix as a substitute for fossil fuels at the same price as the fossil fuel equivalent (and is not dependent on additional carbon price to stimulate its take-up), the supply of bioenergy is 4.7 exajoules and there is a low price/capital cost of moving to future shipping energy sources, particularly electricity, biofuel, hydrogen and ammonia. The costs of investments increase (and, consequently, the B-C ratio decreases) if we assume a scenario where the cost of alternative fuel is higher.

[31]Our estimates reflect both operational costs and capital investments. It is, hence, higher than the cost estimate provided in the recent analysis by the University Maritime Advisory Services (UMAS) and the Energy Transitions Commission for the Getting to Zero Coalition (2019), which states that approximately up to $1.6 trillion ‘capital investments’ is needed between 2030 and 2050 to achieve the IMO target of reducing carbon emissions from shipping by 100% by 2050.

[32]Nonuse values (e.g., existence, bequest and option values) are the benefit values assigned to environmental goods that people have not used or do not intend to use. For example, the current generation can place a value on ensuring the availability of biodiversity and ecosystem functioning to future generations (bequest value).

a vessel's speed by 10% (e.g., from 20 knots to 18 knots) results in a 19% reduction in cargo-hauling fuel consumption after accounting for the reduced shipping speed and the associated loss in shipping time (Faber et al. 2012). These savings are already included in the cost calculations for 2030.

Estimated Benefits and Costs

Based on the methodology outlined above, we estimated that there are net benefits from making investments to decarbonise the shipping sector. The net discounted benefit (average) over 30 years (2020–2050) is estimated to be $1.2–9.1 trillion. The B-C ratio is estimated to be 2-to-1 to 5-to-1 in 2050 (Table 18.14).

6.3.4 Data Limitations and Caveats

Data limitations and caveats include a lack of consideration of the secondary impact on commodity prices and the impact of all cost reductions (technology change) in the future. For details, see Appendix 3.

6.4 Increase the Production of Sustainably Sourced Ocean-Based Proteins

The analysis for this section builds on the estimates provided in *The Global Consultation Report of the Food and Land Use Coalition* (FOLU 2019), the analysis of Costello et al. (2016) that looks at the return from global fisheries under contrasting management regimes, the analysis of Sumaila et al. (2012) that measures the net present value of rebuilding fish stocks over 50 years, and the analysis of Mangin et al. (2018) that compares the benefits from fisheries management against the costs for individual countries.

To determine the level of ocean-based protein production required to ensure a healthy, balanced human diet by 2050, we refer to the EAT-*Lancet* Commission report (Appendix 4; Willett et al. 2019), which states that the ocean will be required to produce 85–90 million metric tonnes (mmt) of edible-weight ocean protein annually by 2050. It is estimated that the world (freshwater and ocean) currently produces only half that amount (FOLU 2019; Willett et al. 2019).

Table 18.14 Net benefit from decarbonising international shipping and benefit-cost ratio

Action	Net benefit by 2050 (US $, billions, 2019$)	Benefit-cost ratio (low)	Benefit-cost ratio (high)
Decarbonise international shipping	1152–9050	2:1	5:1

Source: Authors' calculations

6.4.1 Baseline, Sustainable Transformation Pathway and Target Scenarios

The 2019 FOLU report looks at ocean-based production across three sectors: wild marine capture fisheries, ocean-based fed aquaculture (finfish) and ocean-based nonfed aquaculture (bivalves). The production scenarios under BAU and the sustainable transformation pathways are shown in Table 18.15. Production is measured in million metric tonnes live-weight equivalent. Broadly, the transformation scenarios for the sectors were modelled in terms of possibilities of expanded production.

- **Wild-capture fisheries.** Costello et al. (2016) and Sumaila et al. (2012) estimate fisheries management that aims to maximise long-term catch (maximum sustainable yield) could increase fisheries production up to 96–99 mmt. This is higher than the current catch (80 mmt) and the projected BAU catch in 2050 (67 mmt) (Costello et al. 2019).
- **Fed mariculture production.** In the BAU scenario, fishmeal and fish oil feed requirements remain at current levels due to the absence of large investments into improving feed efficiency, limiting the growth of fed aquaculture (FOLU 2019). Under the sustainable transformation pathway scenario, aquaculture fishmeal and fish oil feed requirements decrease by 50% by 2050, allowing increased production in fed aquaculture to be achieved via measures such as feed efficiency and alternative feed replacement (FOLU 2019).
- **Nonfed mariculture production.** In the sustainable transformation pathway scenario, policy incentives to boost the eating of low-carbon food increase bivalve/mol-

Table 18.15 The business-as-usual and sustainable transformation pathways

Type of ocean-based food production	Business-as-usual scenario	Sustainable transformation pathway scenario
Wild-capture fisheries (marine)	Global annual marine capture production will decline from 80 mmt in 2020 to 67 mmt in 2050[a]	Global annual marine capture production stabilises at 96–99 mmt[b] by 2050
Fed mariculture (finfish)	Fed mariculture production remains at the 2020 level of 11.2 mmt[c]	Fed mariculture production increases to 22.4 mmt by 2050[c]
Nonfed mariculture (bivalves)	Bivalve mariculture grows to 28.5 mmt in 2050 from 16.3 mmt in 2020[c]	Bivalve mariculture grows to 65.2 mmt in 2050[c]

Notes: mmt million metric tonnes
[a] *Costello et al.* (2019)
[b] *Costello et al.* (2016, 2019); *Sumaila et al.* (2012); *this refers to the higher estimates of the Sumaila et al. optimal catch range under reform*
[c] *FOLU* (2019)

lusc production and consumption to 4% per annum as opposed to the BAU average annual growth rate of 3.1% over the last 10 years (FOLU 2019).

6.4.2 Assessment of Costs

The analysis in this paper builds on the investment cost estimates and assumptions in *The Global Consultation Report of the Food and Land Use Coalition* (FOLU 2019), Sumaila et al. (2012) and Mangin et al. (2018).

Capture Fisheries Reform

Analysis by Mangin et al. (2018) estimates that under a fisheries reform scenario, annual global fisheries management costs would be $13–15 billion, whereas under BAU, the costs are estimated at $8 billion.[33] Sumaila et al. (2012) estimate that the amount governments need to invest to rebuild world fisheries is between $130 billion and $292 billion in present value over 50 years, with a mean of $203 billion.[34]

Nonfed and Fed Mariculture Production

- Additional bivalve production (compared with BAU) is estimated at an average cost of $605 per tonne (FOLU 2019).
- In the sustainable transformation pathway scenario, the capital costs for setting up fed mariculture farms are estimated at $157 million for offshore mariculture and $60 million for nearshore mariculture for 2020–2030 (FOLU 2019; O'Shea et al. 2019).[35] Between 2020 and 2030, it is assumed that 25% of the additional production will come from new capital expenditures to build these farms (FOLU 2019). After 2030, we assume that the cost of investment will fall by 15%. The capital costs will fall from $157 million to $133 million over 2031–2050. All increases in production beyond 2030 come from new farms.
- Because mariculture expansion is limited by shortages and the rising costs of fishmeal made from forage fish, we assume that fed mariculture expansion is possible over 30 years (2020–50) because of a 50% reduction in traditional fishmeal, with the gap filled by novel feed ingredients such as insects or algae.[36] Although these alternatives currently cost more than fishmeal,[37] we assume prices will decline with innovation and scaled-up production.
- Increasing the scale of fed mariculture and replacing fishmeal and fish oil with alternative fish feed will lead to a change in the variable costs of mariculture farms. To calculate the impact on variable costs, we assumed that, until 2030, the price of alternative feed would be twice the price of fishmeal and then, because of innovations, it would fall to equal the price of fishmeal in 2030–2050.
- Public and private R&D spending across food and land-use systems was assumed to grow from 0.07% GDP (2018) to 0.1% of GDP by 2030. FOLU analysis assumes 20% of the additional R&D spending on food and land-use systems ($197 billion over 2018–2030) is allocated to alternative fish feed, intensification impacts and the scaling up of innovative production methods such as multi-trophic mariculture and offshore mariculture. After 2030, we assumed the R&D expenditure in the food and land-use systems would continue to grow at the same rate[38] (reaching 0.13% of GDP in 2040 and 0.17% in 2050), and the proportion spent on ocean-based proteins would remain the same.
- Under the Organisation for Economic Co-operation and Development 2030 scenario, mariculture would employ

[33]The paper estimates a country-level B-C ratio for management improvements for 30 countries. It categorises landings in each country into three broad management categories: catch share, where managers and regulators set a scientifically determined catch limit on the amount of fish that can be caught using measures (e.g., community-based allocation, individual quotas, individual vessel quotas, individual transferable quotas, and territorial use rights for fisheries); strong catch controls, which include a broad range of management that can be classified as strong biological management without catch shares; and a broad 'other' category that consists of the rest of the fisheries referred to as open access. It focuses on three types of fisheries management costs: administration (or management), research and surveillance, and enforcement (Mangin et al. 2018).

[34]The estimated transition costs include the costs to society of reducing the current fishing effort to levels consistent with the maximum sustainable yield and the payments governments may decide to employ to adjust capital and labour to uses outside the fisheries sector (such as vessel buyback programs and alternative employment training initiatives for fishers).

[35]This is based on estimates that the average capital expenditure for a large-scale, high-tech farm is $6.50–20.00 per kg (O'Shea et al. 2019). The average production per farm is estimated to be 3000 tonnes/year (FOLU 2019).

[36]We estimated the increase in fishmeal and alternatives required under the sustainable transformation pathway scenario where mariculture increases to 22.4 t by 2050. The gap filled by novel alternatives and associated costs is calculated via the following steps. (1) We calculated the existing fishmeal requirements in the BAU using the feed conversion ratio (FCR) and fishmeal inclusion rate for salmon production. We use an FCR of 1.5 and a fishmeal inclusion rate of 25% (Konar et al. 2019). We assume the fishmeal inclusion rate decreases by 50% (to 13%) in the sustainable transformation pathway scenario. (2) We assumed that under the sustainable transformation pathway scenario, 50% of the fishmeal production (100 million t) will be replaced by alternative ingredients by 2050. (3) Finally, we used the current capital cost to produce feed ($1426/t) as a proxy to calculate the additional capital investment required to expand alternative feed (Suleiman and Rosentrater 2018). Using these steps, we estimated $145 billion in additional investments will be required in alternative feed to expand production to meet the gap caused by reducing traditional fishmeal usage.

[37]The fishmeal price in 2018 was approximately $1600/t.

[38]This reflects the gradual growth of R&D expenditure observed for the world over 2000–2010. For all countries within the Organisation for Economic Co-operation and Development, R&D expenditure grew from 2.1 to 2.4% in 2017.

three million farmers (OECD 2016). We assumed all mariculture farmers would receive training for sustainable production and improving feed efficiency ($450 per farmer [FOLU 2019]) over 2020–2050.

Based on these estimates and assumptions, the discounted costs are estimated to be $656 billion over 30 years (2020–2050).

6.4.3 Assessment of Benefits

Health Benefits

The real gain in health benefits is the potential to increase sustainable protein supplies by encouraging more fish consumption (produced via sustainable means) over other protein sources. This would reduce human mortality and morbidity from reduced GHG emissions (see below for the link between GHG emissions and animal-based proteins), increase healthier diets and reduce health costs from reduced pesticide and antimicrobial exposure. This is estimated to be approximately $170 billion in 2030 and $390 billion in 2050 (FOLU 2019).

Sustainable sourcing of ocean protein and micronutrients also helps diversify nutritious food supplies, particularly for poorer coastal communities that depend disproportionately on fish for their protein and micronutrient consumption. The distributional health benefits to poorer communities have not been analysed or quantified here.

Environmental and Ecological Benefits

Livestock production has high GHG emissions and requires extensive land use. The demand for animal- based protein is projected to increase even more quickly than overall food demand by 2050 due to increases in the world population and in incomes across the developing world (Searchinger et al. 2019). Since foods vary widely in their embedded land use and GHG emissions per unit of protein (Poore and Nemecek 2018), changes in the composition of future diets could greatly affect the emissions consequences of growth in protein demand (González Fischer and Garnett 2016). It is estimated that CH_4 and N_2O emissions in the BAU food system scenario will grow from 5.2 $GtCO_2e$ in 2010 to 9.7 $GtCO_2e$ in 2050 (Springmann et al. 2018). Of that projected growth, over 75% will come from projected growth in animal products (Hoegh-Guldberg et al. 2019).

Ocean-based proteins are substantially less carbon intensive than land-based animal proteins (especially beef and lamb), with farmed bivalves being particularly climate friendly (Hoegh-Guldberg et al. 2019).[39] Therefore, actions that shift diets towards ocean-based proteins can reduce pressure on land and also reduce emissions. Moving to diets that are less dependent on terrestrial animal products, especially beef and lamb, would also slow the growth in demand for freshwater to support livestock agriculture (Hoegh-Guldberg et al. 2019). The transition, if properly managed, could yield benefits of $330 billion in 2050 (FOLU 2019).

In addition, such diet shifts will reduce deforestation, the majority of which will be driven by clearing forests for future meat production and consumption. Searchinger et al. (2019) estimated that animal-based foods accounted for roughly two-thirds of agricultural production emissions in 2010 and more than three-quarters of agricultural land use. Under BAU, the analysis estimated that agriculture would expand by nearly 600 million ha (an area nearly twice the size of India), including the expansion of 400 million ha of pasturelands (Searchinger et al. 2019). The additional reduction in emissions from preventing deforestation has not been included in the estimated benefits.

6.4.4 Economic and Social Benefits

Reforming fisheries will result in an increase in revenues and profits to fishers in the long term. Costello et al. (2016) state that after all fisheries are optimally managed, it will take 10 years for stocks to recover and will result in $53 billion in fisheries profits against the BAU scenario. Sumaila et al. (2012) estimate that rebuilding world fisheries could increase profits from the current negative $13 billion to a positive $77 billion per year. Comparing these benefits to the cost of management, Sumaila et al. (2012) and Mangin et al. (2018) show that the cumulative benefits of sustainable management of fish stocks exceed the management costs. Sumaila et al. (2012) state that rebuilding fisheries stock will deliver a net gain (net present value) of between $600 billion and $1.4 trillion over 50 years, versus transition costs of $130–292 billion.[40]

Estimated Benefits and Costs

Based on key reports and papers, the benefits from increasing the share of sustainably produced ocean-based proteins in diets is estimated to be 10 times the costs (Table 18.16). Evidence indicates that while the global B-C ratio for fisheries management reform is about 9.2-to-1, the ratio is higher than 200 for some countries (Mangin et al. 2018). Sumaila

[39] This does not include farmed shrimp, which can be quite high in GHGs. However, salmon/marine fish and bivalves score well in terms of lower GHG emissions.

[40] The lower bound corresponds to 82 t of catch and the upper bound, 99 t, which is closer to the Costello et al. (2016) estimates. To be consistent, we used both cumulative benefit and cost estimates from Sumaila et al. (2012), which offer a scenario in which the optimal fish landings increase to 99 t when calculating the total net present value for increasing consumption of sustainably produced ocean based protein from capture fisheries, fed aquaculture and nonfed aquaculture. The net gains are present value estimates calculated using a 3% discount rate (Table 18.17).

Table 18.16 Net benefits from increasing the production of sustainably sourced ocean-based proteins and benefit-cost ratio

Action	Net benefit by 2050 (US $, billions, 2019$)	Benefit-cost ratio
Increase production of sustainably sourced ocean-based protein in diets	6678	10:1

Source: Authors' calculations based on estimates from FOLU (2019), *Mangin et al.* (2018), *Sumaila et al.* (2012)

et al. (2012) estimate the B-C ratio for rebuilding global fisheries to be as high as 7:1. The value of net benefits is estimated to be $6.7 trillion over 30 years; the total benefits are $7.4 trillion versus $769 billion total costs.[41]

Data Limitations and Caveats

The estimates do not fully take into account the effects of climate change and ocean acidification. We recognise that there are regional differences and that there are barriers to shifting diets. See Appendix 4 for more details.

7 Conclusion

The overall rate of ROI can be high, with the average B-C ratio ranging from 3-to-1 to 12-to-1 (Table 18.17), and in some cases, such as conservation of mangroves and fisheries reform (for particular countries), it can be much higher. Our research found that investing $2.0–3.7 trillion globally across the four areas from 2020 to 2050 could generate $8.2–22.8 trillion in net benefits.

Actions to transform these four areas will bring multiple benefits. The total monetised and discounted benefits are estimated at $10.3–26.5 trillion over 2020–2050.

Monetised benefits include health benefits, such as reduced mortality from improved air quality, reduced childhood asthma and improved health outcomes from dietary shifts towards sustainably produced ocean-based protein; environmental benefits, such as avoided property losses from coastal flooding, the prevention of land degradation and reduced water usage; and economic benefits, such as reduced production costs due to technological improvements and increased profits from higher fisheries productivity.

The total monetised and discounted costs are estimated to be $2.0–3.7 trillion over 2020–2050. The costs assessed include costs to business (capital costs to set up new infrastructure, R&D costs and increases in variable costs), costs to government (regulatory costs, monitoring costs and research costs) and costs to households (loss of forgone income).

[41]The B-C ratios vary across the countries and range from 1.7 up to 268, with a median of about 14 for catch share management (Mangin et al. 2018).

Table 18.17 Summary of benefit-cost ratios for the four action areas in 2050

Action	Average benefit-cost ratio
Conserve and restore mangroves[a]	3:1
Decarbonise international shipping[b]	4:1
Increase production of sustainably sourced ocean-based proteins	10:1
Scale up offshore energy production[c]	12:1

Notes:
[a] *The ratio presented is the combined ratio for mangrove conservation and restoration. When assessing specific interventions, the benefit-cost ratio for conservation is estimated to be 88:1 and for restoration 2:1*
[b] *The benefit-cost ratio estimated for decarbonising international shipping ranges from 2:1 to 5:1*
[c] *The benefit-cost ratio estimated for scaling up of global offshore wind energy production ranges from 2:1 to 17:1*
Source: Authors' calculations

A number of impacts (both benefits and costs) have not been monetised but are important and must be considered during the policy decision-making process. These include the following considerations:

- Reduced GHG emissions have a positive correlation with the reduced risk of ocean acidification. The measures assessed can positively impact lower trophic species such as bivalves, which are affected disproportionately due to the compounding effects of shifts in temperature, chlorophyll and ocean acidification.
- The tourism value of mangroves (and other coastal ecosystems) may increase over time as biomass and diversity increase within the protected areas.
- A number of ecosystem services from mangrove protection and restoration have not been quantified.
- For example, vegetated coastal habitats are used by a remarkable number of marine and terrestrial animals. Dense mangroves buffer ocean acidification and are becoming recognised as valuable natural systems that can help treat wastewater (Ouyang and Guo 2016).
- Measures to reduce emissions in shipping that involve lowering ship speeds reduce the total sound energy generated and overall whale strike rate and, hence, positively impact marine mammals and other species.
- The distributional impacts of the benefits and costs have not been measured. For example, poor coastal families are the most vulnerable in natural disasters, so building ecosystem resilience to protect them from coastal flooding and cyclones will not only safeguard their valuable assets but also generate tremendous social benefits (e.g., feeling safe) that cannot be easily monetised. The estimates also do not take into account the additional nutritional benefits to human health in terms of micronutrients, particularly in low- and middle-income countries.
- The analysis does not account for changes to the B-C ratio based on changes in the global physical risk profile associ-

ated with climate change. Often, the costs of climate change–related risks are underestimated, including the potential damage of weather-related shocks and sea level rise. If 'resilience' (e.g., through the integration of natural flood defences) is built into investments, then the benefits (e.g., of protective mangroves) could include a reduction in the cost of capital (due to improved risk-adjusted performance metrics) and/or reduced long-term operational expenses (e.g., through avoided losses and reduced maintenance costs).

Given that the B-C ratios in Table 18.17 are a partial estimate of all benefits and costs likely to accrue as a result of the specified investments, they should be treated as indicative to provide the relative scale of benefits from sustainable ocean-based investments compared with the costs. Further research and analysis to address gaps in quantifying benefits will help provide a more complete picture of their value versus their costs. The analysis does not attempt to show the regional variation of the costs and benefits, nor does it show the distribution of benefits and cost across society (especially focusing on the impact on vulnerable groups). Conducting these assessments should be a key step when implementing ocean-based policies and regulations.

Although data limitations prevented a full accounting of all benefits and costs, the results of the analyses suggest that taking the following actions to transform the ocean economy will generate a host of benefits that are larger in magnitude than the costs:

- **Conserving and restoring mangroves.** While the B-C ratio for restoration is lower than for conservation, both types of interventions yield significant benefits and, hence, are both important to ensure a high ROI. Protection measures to conserve these ecosystems should be enhanced along with measures that provide incentives for restoration (e.g., payment for ecosystem services schemes) (Hoegh-Guldberg et al. 2019).
- **Scaling up offshore wind energy production.** Scaling up offshore wind energy to replace fossil fuel–based sources of power generation will help deliver better local health outcomes, reduce risks of damages from climate change, create jobs and deliver immediate environmental benefits such as reduced water usage. Measures such as marine spatial planning is key to ensuring offshore wind technologies amplify these benefits as well as mitigate any environmental risks to habitats and marine species (Hoegh-Guldberg et al. 2019).
- **Decarbonising the international shipping sector.** Transitioning international shipping to net-zero emissions by 2050 will be costly, but these measures will be key to realising the estimated scale of benefits (health outcomes and environmental benefits), which substantially outweigh the costs.
- **Increasing the production of sustainably sourced ocean-based proteins.** Substantial gains in fisheries productivity can be achieved through better management of fish stocks, which eliminates overfishing, illegal and unregulated fishing and discards of nonmarketable fish. Sustainable marine aquaculture practices will also help meet the growing food demand. Technological innovation and adoption in breeding, production systems, disease control and environmental management will help improve mariculture's productivity and environmental performance (Waite et al. 2014). Encouraging innovation can make valuable contributions to the future scalability and lower prices of substitutes as forage fish resources become scarce (Konar et al. 2019). Incentives are required to shift diets towards low-carbon ocean-based proteins and away from high-carbon land-based sources of protein (Hoegh-Guldberg et al. 2019).

Acknowledgements The authors would like to thank the following people for their review, feedback and inputs: Juan Carlos Altamirano, Maximilian Bucher, Lauretta Burke, Linda Cornish, Diletta Giuliani, Craig Hanson, Peter Haugan, Norma Hutchinson, Catherine Lovelock, Sara MacLennan, Justin Mundy, Finn Gunnar Nielsen, Eliza Northrop, Linwood Pendleton, Alex Perera, Carlos Muñoz Pina, Tristan Smith, Mark Spalding, Michael Tlusty, Laura Malaguzzi Valeri, Adrien Vincent, Richard Waite, Jacqueline Wharton and Sophie Wood. While our colleagues were very generous with their time and input, this report reflects the views of the authors alone.

The authors would also like to acknowledge Maximilian Bucher for his valuable assistance, advice and contribution, especially in terms of providing essential data from *The Global Consultation Report of the Food and Land Use Coalition*, clarifying its underlying assumptions and suggesting modifications for the purposes of this study.

Thank you also to Mary Paden, Lauri Scherer, Carni Klirs and Rosie Ettenheim for providing administrative, editing and design support.

Appendix 1. Conservation and Restoration of Mangrove Habitats

Increasing Ecosystem Services from Mangrove Conservation and Restoration

Vegetated coastal habitats are used by a remarkable number of marine and terrestrial animals (Li et al. 2018; Rog et al. 2017). Dense mangroves trap and stabilise sediments that buffer the effects of floodwaters and tidal movements. They are becoming recognised as valuable natural systems that can play an important role in wastewater treatment systems (Ouyang and Guo 2016). The values of these ecosystem services can be significant, as demonstrated in Box 18.5, which provides a local example of the scale of these values for mangroves in Myanmar. While global value estimates of ecosystem services exist (i.e., Costanza et al. 1997, 2014; de Groot et al. 2012), many of these estimates are resulting from meta-analysis (i.e., analysis of analyses) rather than primary valuation studies. Hence, there have been concerns around the validity of using these values for simple benefit transfer without accounting for specific characteristics of the sites where ecosystem service value needs to be estimated (Himes-Cornell et al. 2018). We thus excluded them from the current B-C ratio analysis. However, as new primary valuation data become available, incorporating such benefits will improve marine decision-making.

Data Limitations and Caveats

- We excluded salt marshes and seagrasses from the calculation due to limited global data availability for both in terms of benefit assessment. During the literature review, only one study (Carnell et al. 2019) was found to assess the improvement in fisheries productivity through seagrass conservation, but the estimate is very local, pertaining only to Australia. Restoring salt marshes and seagrasses is found to be more expensive than restoring mangroves because most salt marsh and seagrass restoration efforts did not reach economy of scale.
- The actual conservation and restoration costs for mangroves might be lower or higher depending on the specific location, the sizes of the targeted areas and the measures used. Total restoration costs are up to 30 times cheaper in countries with developing economies (compared to Australia, European countries and the United States) (Bayraktarov et al. 2016).
- The analysis assumes a survival rate of 51.3% for the restored area, based on median survival rates provided by Bayraktarov et al. (2016). In reality, however, survival rates vary significantly between sites due to a few factors in play. First, the survival rate of mangroves is highly species-specific (Mitra et al. 2017). Second, a lack of incentives to engage local residents in the long-term management of restored areas is another reason for low survival rates (Hai et al. 2020). Addressing these factors will be key to improving restoration survival rates and achieving the scale of the benefits described in this study. Restoration efforts should follow a protocol that includes diagnosing the causes of the deterioration or deforestation of the mangroves, setting a baseline, planning restoration activities and long-term monitoring of the restoration project (Hai et al. 2020). Strong community participation in managing the ecosystem, including in the planning, implementation, management and monitoring, will be essential to ensure the success of restoration efforts.
- It is not yet understood how climate change will affect the productivity and resilience of coastal mangroves. In marine ecosystems, rising atmospheric CO_2 and climate change are associated with concurrent shifts in temperature, circulation, stratification, nutrient input, oxygen content and ocean acidification, with potentially wide-ranging biological effects (Doney et al. 2012). However, there is less confidence regarding the influence temperature will have on interactions among organisms, which is important for ecosystem productivities (Kennedy et al. 2002).
- The analysis does not account for the opportunity costs of coastal developments. The economic value of protecting and restoring coastal habitats, even when the necessary legal framework is in place, often loses out to the economic value of coastal development—even when sea level rise, storm surge and other risks are clearly present. To mitigate these risks, a better understanding of the drivers of degradation is needed, as are measures (policy and educational) that aim to change consumer/human behaviour and raise awareness of the benefits derived from nature-based solutions.
- Marine and coastal ecosystem conservation may result in short-term economic losses due to the forgone economic gains from any prohibited or reduced commercial fishing activities (opportunity costs). However, in the long term, this will help increase the productivity of fisheries in nearby fishing grounds through fish migration and reduce the risk of ecosystem collapse due to overfishing. The conservation benefits estimated are highly dependent on the annual carbon mitigation potential estimated by Hoegh-Guldberg et al. (2019) and the avoided risk of climate damages estimated using the social cost of carbon.

Box 18.5 The Economic Value of Key Mangrove Benefits in Myanmar

The values for ecosystem services of mangroves in Myanmar, as estimated by Estoque et al. (2018), illustrate the scale of the benefits that accrue to society from various ecosystem services.

In Myanmar, a mangrove's most valuable service is as a fish nursery (US $9122 per hectare [ha] per year) and as coastal protection from storm surges ($1369/ha/year). Recreational benefits are estimated at $476/ha/year. Mangroves also regulate water flow ($275/ ha/year) and water quality ($61/ha/year) (see Table 18.18).

Table 18.18 Value of mangroves in Myanmar

Ecosystem service	Valuation method	Value (2018US $/ha/year)
Wood-based energy and timber	Value of marketed and nonmarketed production	7.22
Coastal protection	Avoided expenditures on physical reclamation and replenishment	1369.28
Hazard mitigation	Costs of equivalent engineered storm protection defences	349.01
Regulation of water flow	Expenditures saved on alternative freshwater sources (alternative deep well and borehole drilling, piping)	275.25
Regulation of water quality	Reduced costs of wastewater treatment and sediment trapping	617.13
Mitigation of climate change	Potential value of carbon emissions reductions offsets sales	304.64
Maintenance of fisheries nursery populations and habitat	Contribution to on-site and off-site capture fisheries	9112.45
Recreation and experiential	Tourism expenditures and earnings	475.97
Cultural, amenity and aesthetics	Domestic and international visitor willingness to pay	28.46

8 Appendix 2. Scaling Up Offshore Wind Energy Production

Data Limitations and Caveats

Potential risks to biodiversity could arise or increase with the expansion of wind energy, especially as it moves farther from the coast. In such cases, it could be more difficult or costly to mitigate impacts on habitats and wider biodiversity.

- The types of generation displaced by ocean energy will depend on the specific generation technologies and costs in places where ocean energy is located. On grids that have a high share of zero-carbon generation, including hydropower and nuclear energy, adding ocean energy will not decrease emissions significantly. Conversely, for grids with a high share of carbon-intensive generation, emission displacements could be significant.
- The cost of building more offshore wind generation will vary depending on the supply chain and infrastructure available in each market. The investment required will be much higher for developing nations than for countries like Denmark that already have a wind power market.
- The analysis focuses solely on offshore wind energy generation because the projected future costs of other ocean renewable energy installations are subject to high uncertainty because energy development is still immature. Further analysis in this area will help provide a more holistic picture on the ROI for the ocean energy sector overall.
- Water-saving benefits are estimated based on the opportunity costs of water. Direct financial benefits are also associated with water savings, but we excluded them from the benefit assessment because local water prices vary greatly across countries.

Appendix 3. Decarbonising International Shipping

Estimating the Avoided Costs of Childhood Asthma

In schoolchildren, asthma leads to lost school days, which limits academic performance and has consequent psychological effects. Therefore, children with asthma have more indirect costs than older asthmatics, as the direct cost to parents is limited to missed workdays and other expenses. The total avoided costs from childhood asthma are estimated by summing the health care costs, the cost of school absenteeism and adult missed workdays. The following assumptions are made to derive the avoided costs from childhood asthma:

- Globally, 86 million children could suffer from asthma, based on the fact that 334 million people in the world have asthma and 26% of the world population is 14 years or younger (Global Asthma Network 2018).[42] Evidence-based regression analysis shows that 16% of these cases could be attributed to shipping (Sofiev et al. 2018), accounting for 14 million childhood asthma cases.
- Sofiev et al. (2018) states that childhood asthma morbidity due to shipping declines by 54%, from 14 million children affected in the BAU case to 6.4 million children in the 2020 Action case.[43] We assume these benefits are delivered in 2030 (i.e., when 54% of children suffering from asthma are asthma free). We assume a 100% reduction of GHG emissions in shipping will reduce childhood morbidity cases (attributable to shipping) by 100% (14 million).
- The average missed days is estimated to be 6.4 days per child (Nunes et al. 2017; Ojeda and Sanz de Burgoa 2013), and we assume at least one adult loses that many days of work per year to care for the child. The value of additional days lost attributable to asthma per year was $301 for each worker and $93 for each student (Nunes et al. 2017).

[42] Without adjusting for the higher prevalence for asthma among young and old persons.

[43] The 2020 Action assumes on-time implementation of the IMO's 0.5% low-sulphur fuel standard.

- The average annual health financial costs to government for treating pediatric asthma is estimated to range from $3076 to $13,612 per child in the United States (Perry et al. 2019). We take this as a proxy of the global health care cost to treat the illness.

Data Limitations and Caveats

- The analysis does not incorporate all potential cost reductions from innovation and increased R&D efforts. In this respect, the model is conservative because these factors would be expected to reduce technology capital costs. The analysis does not account for additional infrastructure investments such as safe storage and handling of hydrogen/ammonia at the ship-to-shore interface.
- The costs of investments increase (and consequently B-C ratio decreases) if we assume a scenario where the cost of alternative fuel is higher.
- The analysis does not compare the carbon impact of ship transportation versus air transportation. Investment in cleaner ships to meet demands from a growing economy will lead to a lower carbon footprint solution for global trade and travel (versus ground or air transport of goods and people).
- The analysis is static and does not analyse the secondary or indirect impacts following the shipping sector transitioning to a low-fuel economy. Although switching to cleaner fuel will impose costs to the shipping industry, the overall impacts on the economy will depend on how the firms absorb the increase in costs and, thus, are relatively uncertain. Being faced with higher cost, the industry could transfer part of the impacts to the price of final commodities (more likely if they are price inelastic), produce more local product, or reduce profit margins, which would lead to lower future capital investment until the industry's market equilibrium returns. The overall impact on consumers and households will depend on which of these impacts dominate, and by what extent. In most developed economies, impacts are expected to be negligible, and there are policy options for managing impacts in especially vulnerable and/or disproportionately impacted countries.

Appendix 4. Increasing the Production of Sustainably Sourced Ocean-Based Proteins

Data Limitations and Caveats

- The report by the EAT-*Lancet* Commission has set out scientific targets for healthy diets that will optimise human health (Willett et al. 2019). By its own admission, the report did not have the scope to fully analyse fishing and mariculture systems globally. Therefore, while some estimates were included on recommended fish intake, more detailed analysis is needed. EAT, along with other partners,[44] is supporting further work to expand scientific understanding of the role of ocean-based protein for planetary health and human well-being. This research, referred to as the Blue Food Assessment, aims to outline pathways for a transformation to sustainable and healthy blue food for all people on the planet, now and into the future. Analysis has focused on marine food production, but a greater understanding of aquatic food production as a whole (including freshwater fisheries and aquaculture)[45] is needed to evaluate the benefits and costs of aquatic food to human health and the environment. Those working on the Blue Food Assessment have recognised this and aim to incorporate it into the analysis.
- The fisheries reform scenarios are optimistic and assume optimal fisheries management everywhere, which may not be achievable in reality. In addition, the impacts of climate change, such as warming sea temperatures, on fish stocks and their movements have not been fully taken into account in this paper because they are difficult to model and cost. The authors recognise that impacts on production could be significant in some regions.
- The projections do not incorporate the potential impacts of ocean acidification on fish and fisheries. There is a lack of sufficient understanding of the capacity for marine organisms to adapt through acclimation as well as transgenerational and evolutionary adaptation (Gaylord et al. 2015; Munday 2014; Munday et al. 2013) to reliably predict ocean acidification impacts on marine populations and ecosystems (FAO 2018).
- The FOLU (2019) analysis states that the benefits are the difference between the global hidden costs under the better future and current trends scenarios. It provides an indicative estimate of the potential benefits accruing to the global economy from following the better future development path relative to remaining on the current trajectory. For the aquaculture sector the FOLU does not estimate the increase in revenues from production or direct benefit in terms of value added to GDP (which is accounted for under the fisheries reform scenario); rather, it is a reduction in the size of the externalities currently stemming from food and land use.

44 Partners include the Food and Agriculture Organization, Friends of Ocean Action, Stanford Center for Ocean Solutions, Stockholm Resilience Centre, World Economic Forum and World Resources Institute.

45 Currently, the majority of aquaculture production is inland or freshwater, which constitutes 64% of the total global aquaculture production, and the proportion is likely to be higher in Asia (FAO 2018).

- There are many barriers to shifting diets away from emission-intensive land-based sources of protein such as beef and lamb.[46] Consumer purchases are typically based on habit and unconscious mental processing rather than on rational, informed decisions (Ranganathan et al. 2016). Factors such as price, taste and quality tend to be more important than sustainability in purchasing decisions (Ranganathan et al. 2016). The costs of policy measures and business practices—such as private/public procurement, marketing and campaigning costs or sending clear market signals via carbon taxes or changes in subsidies—to enable a change in diet have not been estimated in this analysis. Several assumptions have been used to estimate the costs; hence, these should be treated with caution.
- The estimates do not take into account the additional nutritional benefits to human health in terms of micronutrients, not just protein, particularly in low- and middle-income countries. Ocean-based food production provides food security during extreme events (e.g., heavy rainfall and hurricanes) when the supply of land-based food sources is affected and limited.
- The average B-C ratio calculated here hides the regional and local variances that will occur in aquatic food production. These variances are likely to impact the livelihoods of smaller-scale fishers and farmers the most, and they often have the lowest resilience to changes in capture/farming levels.

About the Authors

Manaswita Konar is the Lead Ocean Economist at the World Resources Institute's Sustainable Ocean Initiative. Contact: Manaswita.Konar@wri.org.

Helen Ding is an Environmental Economist at the World Resources Institute. Contact: Helen.Ding@wri.org.

About WRI

World Resources Institute is a global research organization that turns big ideas into action at the nexus of environment, economic opportunity, and human well-being.

[46]Reducing consumption of animal-based foods should not be a goal for people who are underconsuming. Animal-based foods provide a concentrated source of some vitamins and minerals that are particularly valuable to young children in developing countries whose diets are otherwise poor (Ranganathan et al. 2016).

Our Challenge

Natural resources are at the foundation of economic opportunity and human well-being. But today, we are depleting Earth's resources at rates that are not sustainable, endangering economies and people's lives. People depend on clean water, fertile land, healthy forests, and a stable climate. Livable cities and clean energy are essential for a sustainable planet. We must address these urgent, global challenges this decade.

Our Vision

We envision an equitable and prosperous planet driven by the wise management of natural resources. We aspire to create a world where the actions of government, business, and communities combine to eliminate poverty and sustain the natural environment for all people.

Our Approach

Count It

We start with data. We conduct independent research and draw on the latest technology to develop new insights and recommendations. Our rigorous analysis identifies risks, unveils opportunities, and informs smart strategies. We focus our efforts on influential and emerging economies where the future of sustainability will be determined.

Change It

We use our research to influence government policies, business strategies, and civil society action. We test projects with communities, companies, and government agencies to build a strong evidence base. Then, we work with partners to deliver change on the ground that alleviates poverty and strengthens society. We hold ourselves accountable to ensure our outcomes will be bold and enduring.

Scale It

We don't think small. Once tested, we work with partners to adopt and expand our efforts regionally and globally. We engage with decision-makers to carry out our ideas and elevate our impact. We measure success through government and business actions that improve people's lives and sustain a healthy environment.

References

Aburto-Oropeza O, Ezcurra E, Danemann G, Valdez V, Murray J, Sala E (2008) Mangroves in the Gulf of California increase fishery yields. Proc Natl Acad Sci U S A 105(30):10456–10459. https://doi.org/10.1073/pnas.0804601105

Agora Energywiende (2015) The integration costs of wind and solar power: an overview of the debate on the effects of adding wind and solar photovoltaic into power systems. Agora Energywiende, Berlin. https://www.agora-energiewende.de/fileadmin2/Projekte/2014/integrationskosten-wind-pv/Agora_Integration_Cost_Wind_PV_web.pdf

Arrow KJ (1962) The economic implications of learning by doing. Rev Econ Stud 29(3):155–173. https://www.jstor.org/stable/2295952

Bahar H, D'Ambrosio D, Malischek R, Paillere H (2019) Tracking power. International Energy Agency, Paris. https://www.iea.org/reports/tracking-power-2019

Balmford A, Gravestock P, Hockley N, McClean CJ, Roberts CM (2004). The worldwide costs of marine protected areas. Proceedings of the National Academy of Sciences 101(26):9694–9697.

Bandaranayake WM (1998) Traditional and medicinal uses of mangroves. Mangrove Salt Marshes 2(September):133–148. https://doi.org/10.1023/A:1009988607044

Barbier EB (2007) Valuing ecosystem services as productive inputs. Econ Policy 22(49):178–229. https://doi.org/10.1111/j.1468--0327.2007.00174.x

Bärnighausen T, Bloom DE, Canning D, Friedman A, Levine OS, O'Brien J, Privor-Dumm L, Walk D (2011) Rethinking the benefits and costs of childhood vaccination: the example of the Haemophilus influenzae type B vaccine. Vaccine 29(13):2371–2380. https://doi.org/10.1016/j.vaccine.2010.11.090

Bayraktarov E, Saunders MI, Abdullah S, Mills M, Beher J, Possingham HP, Mumby PJ, Lovelock CE (2016) The cost and feasibility of marine coastal restoration. Ecol Appl 26(4):1055–1074. https://doi.org/10.1890/15-1077

Beck MW, Narayan S, Trespalacios D, Pfliegner K, Losada IJ, Menéndez P, Espejo A et al (2018) The global value of mangroves for risk reduction. Summary report. The Nature Conservancy, Berlin. https://www.conservationgateway.org/ConservationPractices/Marine/crr/library/Documents/GlobalMangrovesRiskReduction-SummaryReport10.7291/V9930RBC.pdf

BEIS (Department for Business, Energy & Industrial Strategy) (2019) Updated short-term traded carbon values: used for UK policy appraisal (2018). BEIS, National Archives, London. https://assets.publishing.service.gov.uk/government/uploads/system/uploads/attachment_data/file/794186/2018-short-term-traded carbon-values-for-appraisal-purposes.pdf

Best BD, Halpin PN (2019) Minimizing wildlife impacts for offshore wind energy development: winning tradeoffs for seabirds in space and cetaceans in time. PLoS One 14(5):e0215722. https://doi.org/10.1371/journal.pone.0215722

Blas A, Garrido A, Unver O, Willaarts B (2019) A comparison of the mediterranean diet and current food consumption patterns in Spain from a nutritional and water perspective. Sci Total Environ 664(May):1020–1029. www.ncbi.nlm.nih.gov/pubmed/30769304

BMWi (Bundesministerium für Wirtschaft und Energie) (2018) Economic indicators of the energy system. BMWi, Berlin. www.bmwi.de/Redaktion/DE/Publikationen/Studien/oekonomische-indikatoren-undenergiewirtschaftliche-gesamtrechnung.pdf

Buhl-Mortensen L, Buhl-Mortensen P, Dolan MFJ, Dannheim J, Bellec V, Holte B (2012) Habitat complexity and bottom fauna composition at different scales on the continental shelf and slope of Northern Norway. Hydrobiologia 685(February):191–219. https://doi.org/10.1007/s10750-011-0988-6

Bunting P, Lucas R, Rosenqvist A, Rebelo L-M, Hilarides L, Thomas N, Hardy A et al (2018) The global mangrove watch—a new 2010 baseline of mangrove extent. Remote Sens 10(10):1669. https://doi.org/10.3390/rs10101669

Burke L, Ding H (2016) Valuation of coastal protection near Paramaribo, Suriname. Final report for World Wildlife Fund Guianas. World Resources Institute, Washington, DC. http://d2ouvy59p0dg6k.cloudfront.net/downloads/316_02_project_peri_urban_coastal_protection_options_paramaribopre_final_report_wri.pdf

Carnell PE, Reeves SE, Nicholson E, Macreadie P, Ierodiaconou D, Young M, Kelvin J et al (2019) Mapping Ocean wealth Australia: the value of coastal wetlands to people and nature. The Nature Conservancy, Melbourne

Chaigneau T, Brown K, Coulthard S, Daw TM, Szaboova L (2019) Money, use and experience: identifying the mechanisms through which ecosystem services contribute to wellbeing in coastal Kenya and Mozambique. Ecosyst Serv 38(August):100957. https://doi.org/10.1016/j.ecoser.2019.100957

Colombo GL, Ferro A, Vinci M, Zordan M, Serra G (2006) Cost-benefit analysis of influenza vaccination in a public healthcare unit. Ther Clin Risk Manag 2(2):219–226. https://doi.org/10.2147/tcrm.2006.2.2.219

Cooper D (2018) Scenarios for the 2050 vision for biodiversity. Paper presented at the convention on biological diversity second bogis-bossey dialogue for biodiversity, "Framing and Setting the Scene," March 3. https://www.cbd.int/cooperation/bogis/2018/DC.pdf

Cooper E, Burke L, Bood N (2009) The economic contribution of Belize's coral reefs and mangroves. World Resources Institute, Washington, DC. Working paper. https://www.wri.org/publication/belizes-coastal-capital

Copping A, Sather N, Hanna L, Whiting J, Zydlewski G, Staines G, Gill A et al (2016) Annex IV 2016 state of the science report: environmental effects of marine renewable energy development around the world. Ocean Energy Systems, Lisbon. https://tethys.pnnl.gov/publications/state-of-the-science-2016

Costanza R, D'Arge R, de Groot R, Farber S, Grasso M, Hannon B, Limburg K et al (1997) The value of the world's ecosystem services and natural capital. Nature 387(May):253–260. https://doi.org/10.1038/387253a0

Costanza R, de Groot R, Sutton P, van der Ploeg S, Anderson SJ, Kubiszewski I, Farber S et al (2014) Changes in the global value of ecosystem services. Glob Environ Chang 26(May):152–158. https://doi.org/10.1016/j.gloenvcha.2014.04.002

Costello C, Ovando D, Clavelle T, Strauss CK, Hilborn R, Melnychuk MC, Branch TA, Gaines SD et al (2016) Global fishery prospects under contrasting management regimes. Proc Natl Acad Sci U S A 113(8):5125–5129. https://www.pnas.org/content/113/18/5125

Costello C, Cao L, Gelcich S et al (2019) The future of food from the sea. World Resources Institute, Washington, DC. https://oceanpanel.org/sites/default/files/2019-11/19_HLP_BP1%20Paper.pdf

de Groot R, Brander L, van der Ploeg S, Costanza R, Bernard F, Braat L, Christie M et al (2012) Global estimates of the value of ecosystems and their services in monetary units. Ecosyst Serv 1(1):50–61. https://doi.org/10.1016/j.ecoser.2012.07.005

de Souza Queiroz L, Rossi S, Calvet-Mir L, Ruiz-Mallén I, García-Betorz S, Salvà-Prat J, de Andrade Meireles AJ (2017) Neglected ecosystem services: highlighting the socio-cultural perception of mangroves in decision-making processes. Ecosyst Serv 26(Part A):137–145. https://doi.org/10.1016/j.ecoser.2017.06.013

Degraer S, Brabant R, Rumes B (eds) (2012) Offshore wind farms in the Belgian part of the North Sea: heading for an understanding of environmental impacts. Royal Belgian Institute of Natural Sciences, Brussels

Doney SC, Ruckelshaus M, Duffy JE, Barry JP, Chan F, English CA, Galindo HM et al (2012) Climate change impacts on marine ecosystems. Annu Rev Mar Sci 4(January):11–37. https://doi.org/10.1146/annurev-marine-041911-111611

Ellis NR, Tschakert P (2019) Triple-wins as pathways to transformation? A critical review. Geoforum 103(July):167–170. https://doi.org/10.1016/j.geoforum.2018.12.006

EPA (U.S. Environmental Protection Agency) (2016) Social cost of carbon. EPA fact sheet. EPA, Washington, DC. https://www.epa.gov/sites/production/files/2016-12/documents/social_cost_of_carbon_fact_sheet.pdf

Estoque RC, Myint SW, Wang C, Ishtiaque A, Aung TT, Emerton L, Ooba M et al (2018) Assessing environmental impacts and change in Myanmar's mangrove ecosystem service value due to deforestation (2000–2014). Glob Chang Biol 24(11):5391–5410. https://doi.org/10.1111/gcb.14409

ETC (Energy Transitions Commission) (2018) Mission possible: reaching net-zero carbon emissions from harder-to-abate subsectors by mid-century. ETC, London. http://www.energy-transitions.org/sites/default/files/ETC_MissionPossible_FullReport.pdf

European Commission (2018) In-depth analysis in support of the commission communication (2018) 773: a clean planet for all—a European strategic long-term vision for a prosperous, modern, competitive and climate neutral economy. European Commission, Brussels. https://ec.europa.eu/clima/sites/clima/files/docs/pages/com_2018_733_analysis_in_support_en_0.pdf

Faber J, Nelissen D, Hon G, Wang H, Tsimplis M (2012) Regulated slow steaming in maritime transport: an assessment of options, costs and benefits. CE Delft, Delft. https://theicct.org/publications/regulated-slow-steaming-maritime-transport

FAO (Food and Agriculture Organization of the United Nations) (2018) The state of world fisheries and aquaculture 2018: meeting the sustainable development goals. FAO, Rome. http://www.fao.org/3/i9540en/i9540en.pdf

Fleming LE, McDonough N, Austen M, Mee L, Moore M, Hess P, Depledge MH et al (2014) Oceans and human health: a rising tide of challenges and opportunities for Europe. Marine Environ Res 99(August):16–19. https://doi.org/10.1016/j.marenvres.2014.05.010

FOLU (Food and Land Use Coalition) (2019) The global consultation report of the food and land use coalition. FOLU, London. https://www.foodandlandusecoalition.org/global-report/

Gaylord B, Kroeker KJ, Sunday JM, Anderson KM, Barry JP, Brown NE, Connell SD et al (2015) Ocean acidification through the lens of ecological theory. Ecology 96(1):3–15. https://doi.org/10.1890/14-0802.1

GCA (Global Commission on Adaptation) (2019) Adapt now: a global call for leadership on climate resilience. World Resources Institute, Washington, DC. https://openknowledge.worldbank.org/handle/10986/32362

Glaser M, Diele K (2004) Asymmetric outcomes: assessing central aspects of the biological, economic and social sustainability of a mangrove crab fishery, *Ucides cordatus* (Ocypodidae), in North Brazil. Ecol Econ 49(3):361–373. https://doi.org/10.1016/j.ecolecon.2004.01.017

Global Asthma Network (2018) The global asthma report 2018. Global Asthma Network, Auckland. http://www.globalasthmareport.org/Global%20Asthma%20Report%202018.pdf

González Fischer CG, Garnett T (2016) Plates, pyramids, planet. Developments in National Healthy and sustainable dietary guidelines: a state of play assessment. Food and Agriculture Organization of the United Nations, Rome. http://www.fao.org/3/a-i5640e.pdf

Hai N, Dell B, Phuong V, Harper RJ (2020) Towards a more robust approach for the restoration of mangroves in Vietnam. Ann For Sci 77:18. https://doi.org/10.1007/s13595-020-0921-0

Halpern BS, Frazier M, Afflerbach J, Lowndes JS, Micheli F, O'Hara C, Scarborough C, Selkoe KA (2019) Recent pace of change in human impact on the World's ocean. Sci Rep 9(August):11609. https://doi.org/10.1038/s41598-019-47201-9

Hammar L, Perry D, Gullström M (2016) Offshore wind power for marine conservation. Open J Marine Sci 6(1):66–78. https://doi.org/10.4236/ojms.2016.61007

Hassellöv I-M, Turner DR, Lauer A, Corbett JJ (2013) Shipping contributes to ocean acidification. Geophys Res Lett 40:2731–2736. https://agupubs.onlinelibrary.wiley.com/doi/pdf/10.1002/grl.50521

Hausker K (2011) The social cost of carbon and competing decision frameworks for climate policy. Paper presented at the fall meeting of the American Bar Association, Section of Environment, Energy, and Resources, Indianapolis, October 12–17

He J, Liu H, Salvo A (2019) Severe air pollution and labor productivity: evidence from industrial towns in China. Am Econ J Appl Econ 11(1):173–201. https://doi.org/10.1257/app.20170286

Himes-Cornell AH, Grose SO, Pendleton L (2018) Mangrove ecosystem service values and methodological approaches to valuation: where do we stand? Front Mar Sci 5(October):1–15. https://doi.org/10.3389/fmars.2018.00376

HMT (Her Majesty's Treasury) (2018) The green book: central government guidance on appraisal and evaluation. HMT, London. https://assets.publishing.service.gov.uk/government/uploads/system/uploads/attachment_data/file/685903/The_Green_Book.pdf

Hochard JP, Hamilton S, Barbier EB (2019) Mangroves shelter coastal economic activity from cyclones. Proc Natl Acad Sci U S A 116(25):12232–12237. https://doi.org/10.1073/pnas.1820067116

Hoegh-Guldberg O et al (2019) The ocean as a solution to climate change: five opportunities for action. World Resources Institute, Washington, DC. https://oceanpanel.org/sites/default/files/2019-10/HLP_Report_Ocean_Solution_Climate_Change_final.pdf

Hollander A, De Jonge R, Biesbroek S, Hoekstra J, Zijp MC (2018) Exploring solutions for healthy, safe, and sustainable fatty acids (EPA and DHA) consumption in The Netherlands. Sustain Sci 14(2):303–313. https://pubag.nal.usda.gov/catalog/6325482

Huckerby J, Jeffrey H, de Andres A, Finlay L (2016) An international vision for ocean energy. Version III. Ocean Energy Systems Technology Collaboration Programme, Lisbon. https://www.ocean-energy-systems.org/news/oes-vision-for-international-deployment-of-ocean-energy/

IEA (2019a) Global energy and CO_2 status report 2019: the latest trends in energy and emissions in 2018. IEA, Paris. https://www.iea.org/reports/global-energy-co2-status-report-2019

IEA (2019b) Offshore wind outlook 2019. IEA, Paris. https://www.iea.org/reports/offshore-wind-outlook-2019

IEA (International Energy Agency) (2016) Special report: water-energy nexus. IEA, Paris. https://webstore.iea.org/weo-2016-special-report-water-energy-nexus

IEA and NEA (Nuclear Energy Agency) (2015) Projected costs of generating electricity. Organisation for Economic Co-operation and Development, IEA; Issy-les-Moulineaux, France: Organisation for Economic Co-operation and Development, NEA, Paris. https://www.oecd-nea.org/ndd/pubs/2015/7057-proj-costs-electricity-2015.pdf

IMarEST (Institute of Marine Engineering, Science and Technology) (2018) Further consideration of how to progress the matter of reduction of GHG emissions from ships: the costs of GHG reduction in international shipping. ISWG-GHG 3/3. IMarEST, London

IMO (International Maritime Organization) (2018) Resolution MEPC.304(72) (adopted on 13 April 2018) initial IMO strategy on reduction of GHG emissions from ships. IMO, London. http://www.imo.org/en/KnowledgeCentre/IndexofIMOResolutions/Marine-Environment-Protection-Committee-%28MEPC%29/Documents/MEPC.304%2872%29.pdf

IPCC (2018) Global warming of 1.5°C: an IPCC special report on the impacts of global warming of 1.5°C above pre-industrial levels and related global GHG emission pathways, in the context of strengthening the global response to the threat of climate change, sustainable development, and efforts to eradicate poverty. IPCC, Geneva. [Edited by V. Masson-Delmotte, P. Zhai, H.-O. Pörtner, D. Roberts,

J. Skea, P.R. Shukla, A. Pirani, et al.] https://www.ipcc.ch/site/assets/uploads/sites/2/2019/06/SR15_Full_Report_Low_Res.pdf

IPCC (2019) IPCC special report on the ocean and cryosphere in a changing climate. IPCC, Geneva. https://www.ipcc.ch/site/assets/uploads/sites/3/2019/11/SROCC_FinalDraft_FullReport.pdf. [Edited by H.-O. Pörtner, D.C. Roberts, V. Masson-Delmotte, P. Zhai, M. Tignor, E. Poloczanska, K. Mintenbeck, et al.]

IPCC (Intergovernmental Panel on Climate Change) (2013) 2013 supplement to the 2006 IPCC guidelines for National Greenhouse Gas Inventories: wetlands. IPCC, Geneva. [Edited by T. Hiraishi, T. Krug, K. Tanabe, N. Srivastava, J. Baasansuren, M. Fukuda, and T.G. Troxler]

IRENA (2018a) Offshore wind investment, policies and job creation: review of key findings for G7 ministerial meetings. IRENA, Halifax. https://www.irena.org/-/media/Files/IRENA/Agency/Publication/2018/Sep/IRENA_offshore_wind_note_G7_2018.pdf?la=en&hash=B186614D923AB1F0A07D7285612C4B037057A0C0

IRENA (2018b) Renewable energy benefits: leveraging local capacity for offshore wind. IRENA, Abu Dhabi

IRENA (2019) Renewable power generation costs in 2018. IRENA, Abu Dhabi. https://www.irena.org/-/media/Files/IRENA/Agency/Publication/2019/May/IRENA_Renewable-Power-Generations-Costs-in-2018.pdf

ITF (International Transport Forum) (2018) Decarbonising maritime transport: pathways to zero-carbon shipping by 2035. ITF policy paper 47. ITF, Organisation for Economic Co-operation and Development, Paris. https://www.itf-oecd.org/sites/default/files/docs/decarbonising-maritime-transport.pdf

Kennedy VS, Twilley RR, Kleypas JA, Cowan JH Jr, Hare SR (2002) Coastal and marine ecosystems & global climate change: potential effects on U.S. resources. Pew Center on Global Climate Change, Arlington, VA. https://www.c2es.org/site/assets/uploads/2002/08/marine_ecosystems.pdf

Konar M, Qiu S, Tougher B, Vause J, Tlusty M, Fitzsimmons K, Barrows R, Cao L (2019) Illustrating the hidden economic, social and ecological values of global forage fish resources. Resour Conserv Recycl 151(December):104456. https://doi.org/10.1016/j.resconrec.2019.104456

Langhamer O (2012) Artificial reef effect in relation to offshore renewable energy conversion: state of the art. In: marine renewable energies: perspectives and implications for marine ecosystems,xs special issue. Sci World J 2012(December):386713. https://www.hindawi.com/journals/tswj/2012/386713/

Leaper R (2019) The role of slower vessel speeds in reducing greenhouse gas emissions, underwater noise and collision risk to whales. Front Mar Sci 6(August):505. https://doi.org/10.3389/fmars.2019.00505

Lee SY, Hamilton S, Barbier EB, Primavera J, Lewis RR (2019) Better restoration policies are needed to conserve mangrove ecosystems. Nature Ecol Evol 3(6):870–872. https://doi.org/10.1038/s41559-019-0861-y

Li X, Bellerby R, Craft C, Widney SE (2018) Coastal wetland loss, consequences, and challenges for restoration. Anthropocene Coasts 1(1):1–15. https://doi.org/10.1139/anc-2017-0001

Limcangco MR, Armour CL, Salole EG, Taylor SJ (2001) Cost-benefit analysis of a Haemophilus influenzae type B meningitis prevention programme in the Philippines. PharmacoEconomics 19(4):391–400. https://doi.org/10.2165/00019053-200119040-00006

Lowe J (2008) Intergenerational wealth transfers and social discounting: supplementary green book guidance. Her Majesty's Treasury, London. https://assets.publishing.service.gov.uk/government/uploads/system/uploads/attachment_data/file/193938/Green_Book_supplementary_guidance_intergenerational_wealth_transfers_and_social_discounting.pdf

Macknick J, Newmark R, Heath G, Hallett KC (2011) A review of operational water consumption and withdrawal factors for electricity generating technologies. Technical report NREL/TP-6A20-50900. National Renewable Energy Laboratory, Washington, DC. https://www.nrel.gov/docs/fy11osti/50900.pdf

Mangin T, Costello C, Anderson J, Arnason R, Elliott M, Gaines SD, Hilborn R, Peterson E, Sumaila R (2018) Are fishery management upgrades worth the cost? PLoS One 13(9):e0204258. https://doi.org/10.1371/journal.pone.0204258

Mitra A, Pal N, Saha A, Chaudhuri TR, Fazli P, Zaman S (2017) Survival rate of mangroves: a proxy to assess ecosystem health. Indian J Geo-Marine Sci 46(10):2046–2053. http://nopr.niscair.res.in/handle/123456789/42756

Munday PL (2014) Transgenerational acclimation of fishes to climate change and ocean acidification. F1000Prime Rep 6:99. https://doi.org/10.12703/P6-99

Munday PL, Warner RR, Monro K, Pandolfi JM, Marshall DJ (2013) Predicting evolutionary responses to climate change in the sea. Ecol Lett 16(12):1488–1500. https://doi.org/10.1111/ele.12185

Musial W, Beiter P, Nunemaker J (2020) Cost of floating offshore wind energy using New England aqua ventus concrete semisubmersible technology. Strategic partnership project report NREL/TP-5000-75618. National Renewable Energy Laboratory, Golden, CO. https://www.nrel.gov/docs/fy20osti/75618.pdf

Narita D, Rehdanz K, Tol RSJ (2012) Economic costs of ocean acidification: a look into the impacts on global shellfish production. Clim Chang 113(August):1049–1063. https://doi.org/10.1007/s10584-011-0383-3

Nunes C, Pereira AM, Morais-Almeida M (2017) Asthma costs and social impact. Asthma Res Pract 3:1. https://doi.org/10.1186/s40733-016-0029-3

O'Shea T, Jones R, Markham A, Norell E, Scott J, Theuerkauf S, Waters T (2019) Towards a blue revolution: catalyzing private investment in sustainable aquaculture production systems. The Nature Conservancy and Encourage Capital, Arlington, VA

OECD (Organisation for Economic Co-operation and Development) (2016) The ocean economy in 2030. OECD, Paris. https://doi.org/10.1787/9789264251724-en

Oita A, Nagano I, Matsuda H (2018) Food nitrogen footprint reductions related to a balanced Japanese diet. Ambio 47(3):319–326. https://doi.org/10.1007/s13280-017-0944-4

Ojeda P, Sanz de Burgoa V (2013) Costs associated with workdays lost and utilization of health care resources because of asthma in daily clinical ractice in Spain. J Investig Allergol Clin Immunol 23(4):234–241

Olmer N, Comer B, Roy B, Mao X, Rutherford D (2017) Greenhouse gas emissions from global shipping, 2013–2015. International Council on Clean Transportation, Washington, DC. https://theicct.org/sites/default/files/publications/Global-shipping-GHG-emissions-2013-2015_ICCT-Report_17102017_vF.pdf

Osland MJ, Spivak AC, Nestlerode JA, Lessmann JM, Almario AE, Heitmuller PT, Russell MJ et al (2012) Ecosystem development after mangrove wetland creation: plant–soil change across a 20-year chronosequence. Ecosystems 15(August):848–866. https://doi.org/10.1007/s10021-012-9551-1

Ouyang X, Guo F (2016) Paradigms of mangroves in treatment of anthropogenic wastewater pollution. Sci Total Environ 544(February):971–979. https://doi.org/10.1016/j.scitotenv.2015.12.013

Pearson J, McNamara KE, Nunn PD (2019) Gender-specific perspectives of mangrove ecosystem services: case study from Bua Province, Fiji Islands. Ecosyst Serv 38(August):100970. https://doi.org/10.1016/j.ecoser.2019.100970

Pelletier L, Chung P, Duclose P, Manga P, Scott J (1998) A benefit-cost analysis of two-dose measles immunization in Canada. Vaccine 16(9–10):989–996. https://doi.org/10.1016/S0264-410X(97)00281-8

Perry R, Braileanu G, Palmer T (2019) The economic burden of pediatric asthma in the United States: literature review of current evidence. PharmacoEconomics 37(February):155–167. https://doi.org/10.1007/s40273-018-0726-2

Poore J, Nemecek T (2018) Reducing food's environmental impacts through producers and consumers. Science 360(6392):987–992. https://doi.org/10.1126/science.aaq0216

Poulsen T, Lema R (2017) Is the supply chain ready for the green transformation? The case of offshore wind logistics. Renew Sust Energ Rev 73(June):758–771. https://doi.org/10.1016/j.rser.2017.01.181

Ranganathan J, Vennard D, Waite R, Dumas P, Lipinski B, Searchinger T, GlobAgri-WRR Model Authors (2016) Shifting diets for a sustainable food future. Working paper, installment 11 of creating a sustainable food future. World Resources Institute, Washington, DC. https://wriorg.s3.amazonaws.com/s3fs-public/Shifting_Diets_for_a_Sustainable_Food_Future_1.pdf

Richards DR, Friess DA (2016) Rates and drivers of mangrove deforestation in Southeast Asia, 2000–2012. Proc Natl Acad Sci U S A 113(2):344–349. https://doi.org/10.1073/pnas.1510272113

Rog SM, Clarke RH, Cook CN (2017) More than marine: revealing the critical importance of mangrove ecosystems for terrestrial vertebrates. Divers Distrib 23(2):221–230. https://doi.org/10.1111/ddi.12514

Salem ME, Mercer DE (2012) The economic value of mangroves: a meta-analysis. Sustainability 4(3):359–383. https://doi.org/10.3390/su4030359

Salmo SG, Lovelock C, Duke NC (2013) Vegetation and soil characteristics as indicators of restoration trajectories in restored mangroves. Hydrobiologia 720(December):1–18. https://doi.org/10.1007/s10750-013-1617-3

Scottish Enterprise (2016) Oil and gas "seize the opportunity" guides: offshore wind. Scottish Enterprise, Glasgow. https://www.offshorewindscotland.org.uk/media/1116/sesdi-oil-and-gas-div-guide-offshore-wind.pdf

Searchinger T, Waite R, Hanson C, Ranganathan J, Matthews E (2019) Creating a sustainable food future: a menu of solutions to feed nearly 10 billion people by 2050. Final report. https://www.research.wri.org/wrr-food

Simões-Wüst AP, Dagnelie PC (2019) To be or not to be for humankind: organic diets revisited for a sustainable development. Sustain Earth 2:1. https://doi.org/10.1186/s42055-019-0008-7

Smith T, Jalkanen JP, Anderson BA, Corbett JJ, Faber J, Hanayama S, O'Keefe E et al (2015) Third IMO greenhouse gas study 2014. International Maritime Organization, London. http://www.imo.org/en/OurWork/Environment/PollutionPrevention/AirPollution/Documents/Third%20Greenhouse%20Gas%20Study/GHG3%20Executive%20Summary%20and%20Report.pdf

Sofiev M, Winebrake JJ, Johansson L, Carr EW, Prank M, Soares J, Vira J et al (2018) Cleaner fuels for ships provide public health benefits with climate tradeoffs. Nat Commun 9:406. www.nature.com/articles/s41467-017-02774-9

Sotta C (2012) Documentary summary of the environmental impact of renewable marine energy. Parc naturel marine d'Iroise, Le Conquet. https://tethys.pnnl.gov/sites/default/files/publications/3-2-1_Documentary_summary_EN-MERIFIC-web.pdf

Spalding M, Parrett CL (2019) Global patterns in mangrove recreation and tourism. Mar Policy 110(December):103540. https://doi.org/10.1016/j.marpol.2019.103540

Springmann M, Clark M, Mason-D'Croz D, Wiebe K, Bodirsky L, de Vries Lassaletta W et al (2018) Options for keeping the food system within environmental limits. Nature 562(7728):519. https://doi.org/10.1038/s41586-018-0594-0

Stacey TF, Taylor GS (2019) The levelized cost of electricity from existing generation resources. Institute for Energy Research, Washington, DC. https://www.instituteforenergyresearch.org/wp-content/uploads/2019/06/IER_LCOE2019Final.pdf

Stern NH (2007) The economics of climate change: the stern review. Cambridge University Press, Cambridge

Suleiman R, Rosentrater KA (2018) Techno-economic analysis (TEA) of extruded aquafeeds. J Food Res 7(5):57–68. https://doi.org/10.5539/jfr.v7n5p57

Sumaila UR, Cheung W, Dyck A, Gueye K, Huang L, Lam V, Pauly D et al (2012) Benefits of rebuilding global marine fisheries outweigh costs. PLoS One 7(7):e40542. https://doi.org/10.1371/journal.pone.0040542

Teske S, Pregger T, Simon S, Naegler T, Graus W, Lins C (2011) Energy [R]evolution 2010—a sustainable world energy outlook. Energ Effic 4(August):409–433. https://doi.org/10.1007/s12053-010-9098-y

Tilman D, Clark M (2014) Global diets link environmental sustainability and human health. Nature 515(7528):518–522. https://doi.org/10.1038/nature13959

Tol RSJ (2019) A social cost of carbon for (almost) every country. Energy Econ 83(September):555–566. https://doi.org/10.1016/j.eneco.2019.07.006

Trucost (2013) White paper: valuing water to drive more effective decisions. Trucost, London. https://www.trucost.com/publication/white-paper-valuing-water-drive-effective-decisions/

UMAS (University Maritime Advisory Services) and Energy Transitions Commission (2019) The scale of investments needed to decarbonize international shipping. Insight brief. London, UMAS and Energy Transitions Commission. https://u-mas.co.uk/LinkClick.aspx?fileticket=03IebWyJns8%3d&portalid=0

UNFCCC (United Nations Framework Convention on Climate Change) (2015) Decision 1/CP.21: adoption of the Paris agreement. Conference of Parties, UNFCCC, Paris. https://unfccc.int/sites/default/files/resource/docs/2015/cop21/eng/10a01.pdf

Valiela I, Bowen JL, York JK (2001) Mangrove forests: one of the world's threatened major tropical environments. Bioscience 51(10):807–815. https://doi.org/10.1641/0006-3568(2001)051[0807:MFOOTW]2.0.CO;2

Waite R, Beveridge M, Brummett R, Castine S, Chaiyawannakarn N, Kaushik S, Mungkung R, Nawapakpilai S, Phillips M (2014) Improving productivity and environmental performance of aquaculture. Working paper, installment 5 of creating a sustainable food future. World Resources Institute, Washington, DC. https://wriorg.s3.amazonaws.com/s3fs-public/WRI14_WorkingPaper_WRR5_final.pdf

Warusawitharana M (2014) The social discount rate in developing countries. FEDS notes. Board of Governors of the Federal Reserve System, Washington, DC. https://doi.org/10.17016/2380-7172.0029

WBG (World Bank Group), ESMAP (Energy Sector Management Assistance Program), and SERIS (Solar Energy Research Institute of Singapore) (2018) Where sun meets water: floating solar market report. World Bank Group, Washington, DC. http://documents.worldbank.org/curated/en/579941540407455831/Floating-Solar-Market-Report-Executive-Summary

Welisch M, Poudineh R (2019) Auctions for allocation of offshore wind contracts for difference in the UK. Oxford Institute for Energy Studies, Oxford. https://www.oxfordenergy.org/wpcms/wp-content/uploads/2019/02/Auctions-for-allocation-of-offshore-wind-contracts-for-difference-in-the-UK-EL-33.pdf

West JJ, Smith SJ, Silva RA, Naik V, Zhang Y, Adelman Z, Fry MM, Anenberg S, Horowitz LW, Lamarque J-F (2013) Co-benefits of global greenhouse gas mitigation for future air quality and human health. Nat Clim Chang 3(October):885–889. https://doi.org/10.1038/nclimate2009

WHO (World Health Organization) (2016) Burden of disease from the joint effects of household and ambient air pollution for 2012. WHO, Geneva. https://www.who.int/airpollution/data/AP_jointeffect_BoD_results_Nov2016.pdf

Willett W, Rockström J, Loken B, Springmann M, Lang T, Vermeulen S, Garnett T et al (2019) Food in the Anthropocene: the EAT-*Lancet* commission on healthy diets from sustainable food systems.

Lancet 393(10170):447–492. https://doi.org/10.1016/S0140-6736(18)31788-4
Witt E (2016) A financial and economic assessment of the conservation of Northwestern Madagascar Mangroves. Master's thesis, Nicholas School of the Environment, Duke University. https://dukespace.lib.duke.edu/dspace/bitstream/handle/10161/11911/E.Witt_MP.pdf?sequence=1&isAllowed=y
World Bank (2016) Managing coasts with natural solutions: guidelines for measuring and valuing the coastal protection services of mangroves and coral reefs. In: Beck MW, Lange G-M (eds) Wealth accounting and the valuation of ecosystem services partnership technical report. World Bank Group, Washington, DC. http://documents.worldbank.org/curated/en/995341467995379786/Managing-coasts-with-natural-solutions-guidelines-for-measuring-and-valuing-the-coastal-protection-services-of-mangroves-and-coral-reefs
Zhou F, Bisgard KM, Yusuf HR, Deuson RR, Bath SK, Murphy TV (2002) Impact of universal Haemophilus influenzae type B vaccination starting at 2 months of age in the United States: an economic analysis. Pediatrics 110(4):653–661. https://doi.org/10.1542/peds.110.4.653

A Sustainable and Equitable Blue Recovery to the COVID-19 Crisis

19

Highlights

- With a longer-term vision and the right actions, the COVID-19 pandemic can mark the beginning of a new type of global and societal cooperation in building a sustainable ocean economy.
- The pandemic has had deep and wide-reaching consequences for people around the world, resulting in a crisis that has led to significant loss of human life, increasing food and nutritional insecurity and poverty, and affecting almost all areas of the global economy.
- The ocean economy, which contributes upwards of US $1.5 trillion in value added to the global economy has been particularly hard hit by the pandemic. Significant revenue losses have been felt across coastal and marine tourism, fisheries and aquaculture, and the global shipping industries. Hundreds of millions of jobs have been affected, with disproportionate impacts for developing and small island nations and already vulnerable coastal communities.
- The linkages between ocean-based sectors and land-based economies mean that the pandemic's impacts flow beyond these individual sectors, with economic and social repercussions across the entire economy. A sustainable and equitable recovery is critical not just for those who live or work near the coasts but for the well-being and resilience of societies and economies at large. Despite the significance of the impacts, only a limited number of investments through stimulus and recovery packages are currently directed towards affected ocean workers, coastal communities and the sustainable rebuilding of the ocean economy.
- Furthermore, many response measures have the potential to reverse progress made to date on ocean sustainability and exacerbate the existing threats to ocean health, undermining the myriad non-monetary benefits provided by the ocean which are essential to human well-being and prosperous societies, and the ability of the ocean to continue to be a workplace, a source of income, livelihoods and nutritional food for billions of people worldwide.
- Investment through recovery and stimulus packages represents a crucial lever for accelerating the shift from business as usual to a more sustainable future that delivers on global targets under the 2030 Agenda for Sustainable Development, the Convention on Biological Diversity and the Paris Agreement.
- Humanity is at a critical crossroads. Stimulus which locks in high-emitting, high-polluting and inequitable development pathways now will have catastrophic implications for ocean health, the global climate emergency, economic resilience, human health and prosperity.
- The strategic investment of recovery and stimulus funds into the ocean economy offers an untapped opportunity to support job creation and economic diversification and relief in the short term. Such investments can also accelerate the sustainable and equitable growth of ocean industries, thereby securing the long-term health and resilience of the ocean and ocean economy and the myriad benefits that it provides to humanity.
- This report proposes a set of five priority opportunities for governments to consider for the immediate investment of stimulus funds to support a 'sustainable and equitable blue recovery' from the COVID-19 crisis. These mutually beneficial, no-regrets 'blue stimulus' opportunities, identified on the basis of criteria, are particularly relevant at this time for their potential to deliver short-term economic, social (health) and environmental benefits for affected communities and sectors, while building longer-term social, economic and ecological resilience:

Originally published in:
Northrop, E., et al. 2020. *A Sustainable and Equitable Blue Recovery to the COVID-19 Crisis*. Report. Washington, DC: World Resources Institute. Available online at http://www.oceanpanel.org/bluerecovery
Reprint by Springer International Publishing (2023) with kind permission.
Published under license from the World Research Institute.

J. Lubchenco, P. M. Haugan (eds.), *The Blue Compendium*, https://doi.org/10.1007/978-3-031-16277-0_19

  - Invest in coastal and marine ecosystem restoration and protection.
  - Invest in sewerage and wastewater infrastructure for coastal communities.
  - Invest in sustainable community-led non-fed marine aquaculture (mariculture) (e.g. shellfish and seaweed).
  - Incentivise zero-emission marine transport.
  - Incentivise sustainable ocean-based renewable energy.
- As evidenced by the stimulus response to the 2008–2009 global financial crisis, not all investments will be directed at measures capable of providing job creation in the short term. Instead, much of the investment will be used to lay the foundation for long-term recovery through systemic transitions to improve the efficiency and cost-effectiveness of the economy and initiating large infrastructure projects that will yield benefits over the next 10–30 years.

  This report proposes a set of additional opportunities that are more systemic and oriented towards using this critical juncture to sustainably reset the ocean economy. This will enable the accelerated transition of ocean industries towards smarter, sustainable practices that conserve marine ecosystems, promote human well-being and build social and economic resilience to future shocks.
- Maximising the use of financial mechanisms (e.g. debt restructure and financial grants) offers an unprecedented opportunity to incentivise sustainable recovery efforts and avoid a roll-back in advances already made in sustainable fisheries management, marine conservation and ocean data.
- Heightened awareness of the importance of coordinated and evidence-based global action to shared challenges, and rapid shifts towards new technologies and working practices as evidenced during the COVID-19 crisis, may create new opportunities for advancing the 2030 Agenda for Sustainable Development and the Paris Agreement.
- The urgency cannot be overstated. As the world continues to battle the health crisis, millions are without incomes to provide for themselves and their families. They need a job and a lifeline, for right now and for the future. Policymakers and financial decision-makers must consider the potential of the ocean economy's role in social and economic recovery and ensure that the ocean economy rebuilds to be more sustainable, equitable and resilient—as this is key to our global prosperity and well-being.

1 Introduction

A healthy ocean is the foundation for prosperous, healthy and vibrant economies. There is an unprecedented opportunity, through global stimulus and recovery responses to the COVID-19 crisis, to reset and rebuild economic activities in ways that will ensure a more sustainable, equitable and resilient ocean economy fit for everyone's future. This report provides a roadmap to achieve this vision.

1.1 Context

The COVID-19 pandemic has caused an unprecedented human health crisis around the world, resulting in significant loss of life. Emergency measures introduced to curb the extent of the virus have led to severe restrictions on human mobility, economic activities and services, affecting large swathes of the economy and resulting in widespread unemployment and impacts on people's livelihoods, well-being and wider health (Xu and Joyce 2020).

The resulting global economic downturn is expected to exceed the one experienced during the 2008–2009 global financial crisis (Bluedorn and Chen 2020). The global economy is projected to contract by 4.9–6% in 2020 (IMF 2020a), the largest economic dip since the global depression of the 1930s (OECD 2020c). Gross domestic product (GDP) is expected to shrink in nearly every country in 2020, although with significant variation reflecting differing national circumstances.

As economic projections have been revised downwards, unemployment has continued to rise. Worldwide, some 300 million full-time jobs could be lost, and nearly 450 million companies are facing the risk of serious disruption (ILO 2020c), reducing local incomes, tax revenues and foreign exchange earnings. Early evidence suggests that groups that were economically most vulnerable before the pandemic will experience the greatest impacts, exacerbating existing inequalities within society (UN DESA 2020a). Globally, the COVID-19 pandemic may force as many as 100 million people into extreme poverty and could double the number of people facing acute hunger, to 265 million people by the end of 2020 (Anthem 2020).

Before the pandemic, ocean-based industries such as fishing, energy, shipping and marine and coastal tourism had been conservatively estimated to contribute 2.5% of world gross value added, a value that was predicted to double by 2030 (OECD 2016).

As of 2010, these ocean-based industries contributed some 31 million direct full-time jobs (OECD 2016). This figure is significantly higher when jobs provided through informal or artisanal employment are included. For example, upper estimates in 2011 suggest that the fisheries sector alone provides the equivalent of 237 million full-time jobs when small-scale fisheries and artisanal employment are also considered (Teh and Sumaila 2013). The ocean also connects cities and countries around the world, driving economic activity and trade for the more than a third of the global population that lives within 100 km of the sea (Kummu et al. 2016). Most of the world's megacities are located in the coastal zone.

A healthy ocean not only underpins the global economy but also provides myriad non-monetary benefits alongside essential goods and services that are vital for healthy human societies, including regulating the global climate, offering a storehouse of compounds key for fighting disease (Blasiak et al. 2020) and providing natural infrastructure to protect against storm surges, flooding and coastal erosion. Fish and fish products are among the most highly traded foods in the world, supplying a critical source of animal protein, micronutrients and omega-3 fatty acids, particularly in low-income, food-deficit countries and small island developing states (SIDS) (FAO 2020a).

The pandemic has significantly disrupted ocean sectors and global supply chains. These ocean industries do not operate in isolation from one another, or from the ocean environment of which they are part (OECD 2016). This has led to cascading and interrelated impacts across the ocean economy, marine ecosystems and society.

Fiscal measures announced in response to the COVID-19 crisis by G20 nations are already three times greater than those made available during the 2008–2009 financial crisis. More is expected as the focus shifts from emergency spending to recovery investments. The UN secretary general, António Guterres, has called for a coordinated approach to social and economic recovery from the pandemic, a response that does not lose sight of the parallel threat to the global community posed by the climate emergency. Leaders from business and civil society alike are advocating for this unprecedented situation to be used as a catalyst for a cleaner, greener and more resilient future (Harrabin 2020).

The actions that governments and financial institutions take now to repair and rebuild the global economy will chart the course of economic growth and sustainability for many years to come. Although the nature of the investments themselves might have a short-term focus, their impact will be felt over the medium to long term. It is therefore important to avoid locking in high-emitting, high-polluting and inequitable pathways that limit the ability to build sustainable and resilient economic systems. Investment through recovery and stimulus packages represents a crucial lever for accelerating the shift from business as usual to a more sustainable future that delivers on global targets under the 2030 Agenda for Sustainable Development, the Convention on Biological Diversity and the Paris Agreement.

While the solutions will differ from one country to another, humanity has a unique opportunity to reboot economic activities in a way that is more firmly in service of society and restores planetary health for future generations. A healthy ocean is essential in the quest for a sustainable and prosperous future, and it will be an important ally in rebuilding national and global economies from the impacts of COVID-19 and lifting communities out of poverty. Cumulative impacts to ocean health resulting from unsustainable development, overexploitation of natural resources, pollution and climate change are, however, already causing rapid changes across ocean ecosystems, undermining the ocean's ability to continue to provide vital benefits and services to the global economy and humanity. A transformational shift is needed in the relationship between humanity and the ocean, in acknowledgement of its material and non-material values and importance, to ensure that the solutions pursued in response to the COVID-19 crisis do not further undermine ocean health or the future opportunities associated with the growth of a sustainable ocean economy.

The importance of green stimulus to maintain advances towards a greener economy has been recognised by some governments, yet few have recognised the role that 'blue' stimulus opportunities could also provide in supporting advances to meet environmental and climate change challenges. This report considers this gap between the impacts and responses and offers a set of high-level guiding principles for governments and financial institutions to consider as a first step towards ensuing a sustainable blue recovery to COVID-19. It also supports the notion that a 'blue' recovery is a 'green' recovery and vice versa—the decision to ensure a sustainable blue recovery does not come at the expense of a green recovery—they should go hand in hand and cover the full land-to-ocean interface of activities.

Early indications suggest that society may emerge from this crisis to be less cooperative and effective (Sachs et al. 2020). However, with a longer-term vision and the right actions, the pandemic can mark the beginning of a new type of global and societal cooperation in building towards a sustainable ocean economy—which for the purposes of this report is described as the sustainable use of ocean resources (produce) in ways that preserve the health, function and resilience of ocean ecosystems and associated services (protect) and improve livelihoods and jobs (prosper). Given the importance of the ocean as a workplace and a source of income, livelihoods and nutritional food for billions of people worldwide, the importance of resetting the ocean economy on a sustainable and just path so as to reduce vulnerability to future shocks, restore resilience in natural systems and redress existing inequalities must not be overlooked.

1.2 About This Report

1.2.1 Scope

This report aims to provide a holistic assessment of the impact (economic, social and environmental) that COVID-19 has had on the ocean economy. Section 2 considers the emerging impacts on the ocean economy and early responses to the crisis by governments, financial institutions, industry, intergovernmental organisations (IOs) and non-governmental organisations (NGOs). In considering the impacts, it looks at six key sectors—marine and coastal tourism, fisheries, marine aquaculture (mariculture), shipping, energy and marine conservation—as well as how these impacts are interconnected across the ocean economy as a whole. Recognising that this crisis continues to evolve, these impacts represent a snapshot in time but can still offer important lessons on the scope and degree to which recovery measures must take into account ocean-based sectors, workers and affected communities, and the health of the ecosystems upon which these industries depend.

Section 3 provides a roadmap for a 'sustainable and equitable blue recovery' predicated on three mutually reinforcing elements—effective protection of ocean ecosystems, sustainable production and equitable prosperity. It outlines

- high-level guiding principles for ensuring a 'sustainable and equitable blue recovery' to aid governments as they consider the nature of their recovery after COVID-19 (Sect. 3.1);
- 'blue stimulus' opportunities that are ripe for the immediate investment of stimulus funding and which can deliver short-term economic benefits to affected communities or sectors while also providing longer-term social and environmental benefits (Sect. 3.2);
- 'blue transformations' opportunities, which through more systemic or longer-term policy reform can accelerate the transition towards a sustainable ocean economy in order to secure economic recovery, resilience and prosperity over the longer term (Sect. 3.3); and
- 'blue conditionality' opportunities associated with financial grants and debt relief which can advance key reforms in areas such as sustainable fisheries management and monitoring and enforcement of protected areas (Sect. 3.4).

1.2.2 Approach

The report relies on real-time analysis of impacts of the COVID-19 crisis presented in published reports, working papers and blog posts to help provide an aggregated picture of the resulting economic, social and environmental impacts of COVID-19 on the ocean economy (Sect. 2.1). The COVID-19 response measures (Sect. 2.2) are based on systematic review of the policy response reports from international organisations (such as the International Monetary Fund and World Bank), think tanks, consultancies, academic institutions and national government websites. Both the impacts and response measures are rapidly evolving landscapes and, as such, these sections are not intended to provide a comprehensive overview of the status quo.

The opportunities for investment of stimulus funding identified in Sect. 3 are based on an extensive literature review and set of criteria to identify priorities that respond to the needs of governments and communities now, but which also help catalyse progress towards a sustainable ocean economy. These criteria were selected through literature review, and through expert input from bilateral and multilateral funders and government representatives involved in the design of recovery and stimulus packages. The opportunities highlighted in Sect. 3 of this report are not exhaustive of what will be required to fully transition to a sustainable ocean economy. There is already extensive literature on the solutions and opportunities for action to build a sustainable ocean economy that should be referred to in conjunction with this report—which focuses on the particular economic challenges and opportunities facing governments at this time. Annex 2 offers a summary of relevant literature.

The report draws on publicly available information (including news articles, expert opinion pieces, peer reviewed reports, academic studies and project-specific case studies) to identify potential (short- and long-term) economic, social and environmental benefits for the priority areas of action and interventions identified.

The figures included are offered as proof points and illustrative examples, not as conclusive statements or guarantees. For numbers of potential job creation, many of the estimates presented in the report are based on range of studies, including ones that use input–output (I-O) models to derive job numbers, which have their limitations[1]. The benefits (economic, social or environmental) that may accrue as a result of a particular policy decision or financial investment will be specific to the location, economy and population they relate to.

While it is beyond the scope of this particular assessment, the value of new analysis in these areas—particularly an assessment of the direct and indirect employment opportunities associated with transitioning to a sustainable ocean economy—is well recognised and encouraged to inform decisions that relate to the ocean's contribution to socioeconomic development.

In generating this report, the authors engaged with the 14 offices of the heads of state and government represented on the High Level Panel for a Sustainable Ocean Economy (www.oceanpanel.org) to gather real-time information on country impacts, response measures and priorities, and the relevance and feasibility of interventions for these diverse geographies and economies. This report is, however, an independent input to the Ocean Panel process and does not reflect the views of the Ocean Panel members.

2 Emerging Impacts and Early Responses

Jobs and sectors in the ocean economy and already vulnerable coastal communities have been hard hit by the COVID-19 crisis with significant revenue losses felt across coastal and marine tourism, fisheries and aquaculture, and the shipping industry. The linkages between ocean-based sectors and land-based economies mean that these impacts flow beyond these individual sectors to have economic and social repercussions across the entire economy. Only a small proportion of COVID-19 stimulus packages account for the impacts suffered by coastal communities and workers in the ocean economy and an even smaller subset focuses on transitioning to a sustainable ocean economy.

2.1 Emerging Impacts on the Ocean Economy

This assessment focuses on the impact that the crisis is having across six ocean-based sectors. We consider three categories of impacts (Table 19.1):

- **Economic impact** measures the impact on output, jobs, revenue, future investment targets and productivity of ocean-based sectors.
- **Social impact** identifies vulnerable groups (such as women, workers in the informal sector, young workers and Indigenous community members), poorer communities or low-skilled essential workers who face higher health risks due to limited access to healthcare and are disproportionately affected due to job losses and loss of livelihoods.
- **Environmental impact** assesses the benefits and harms to ocean health arising from a range of factors including reduced intensity of ocean-based economic activities, roll-back of environmental policies, changes in societal behaviours (e.g. increased use of e-commerce shipping, disposable personal protective equipment [PPE] and single-use plastics) and reduction in private sector funding for conservation.

2.1.1 Economic Impact

The ocean economy was projected to double by 2030, but this growth potential has been curtailed by COVID-19 (Richens and Koehring 2020; OECD 2016). Significant revenue losses have been experienced across most ocean-based sectors, with coastal and marine tourism being the hardest hit (UNCTAD 2020b). Across these sectors—in particular coastal tourism, shipping, fisheries and aquaculture—we see a significant loss in revenues, risks of high job losses and reduced appetite for future investment (Table 19.1).

With a decline in international tourist arrivals, the coastal tourism sector has seen a sharp drop in revenue, putting hundreds of millions of direct tourism jobs at risk[2].

Seafood sectors (both wild fisheries and aquaculture) have been affected by a fall in aggregate demand for seafood due to the closure of restaurants and supply chain disruptions (FAO 2020b; UNCTAD 2020b). Slowed demand has negatively affected maritime shipping, the cruise sector and shipbuilding.

[1] *I-O analyses can portray the linkages between sectors well, based on industry-level accounts. However, they have several weaknesses, including the assumption of fixed prices (prices do not change when demand for a good, service, or input changes), fixed ratios of labour to other factors of production and fixed sectoral share of GDP over time.*

[2] *Ocean tourism before COVID-19 was directly valued at US $390 billion globally and comprises a significant portion of many nations' GDP (OECD 2016).*

Table 19.1 Summary of impacts across ocean-based sectors and ecosystems

Sectors	Economic impact	Social impact	Environmental impact
Coastal and marine tourism	● The loss in gross value added is estimated to be up to US $2.1 trillion for the whole of the tourism sector, with 100 million jobs at risk (UNCTAD 2020b) ● Coastal regions are expected to be the most affected, and the cumulative reduction in gross domestic product (GDP) from April to June is estimated be between €9.7 billion to €24.9 billion for areas in Europe alone (OECD 2020b) ● Small island developing states have seen a decline in tourism receipts of 25%, resulting in a $7.4 billion loss (or a 7.3% fall in GDP) (Coke-Hamilton 2020) ● For the Caribbean, analysis estimates job losses to be 1.4 million to two million and losses to the tourism sector to be $27 billion to $44 billion (WTTC 2020) ● Recovery is estimated to take a minimum of 10 months to 2 years after the pandemic, and longer for smaller economies reliant on tourist arrivals from a few developed economies (UNCTAD 2020b)	● Small and medium enterprises, autonomous workers and workers from vulnerable communities, who constitute 80% of the coastal tourism sector workforce, have been hard hit by the reduced flow of income ● Seafarers from the cruise industry have been badly affected due to suspension of cruise operations and quarantining of workers and passengers (ILO 2020a; UNCTAD 2020b) ● Unemployment is significantly higher in the Pacific islands and Caribbean, which rely more on tourism revenues (ILO 2020a) ● Women are likely to be most affected by job losses in the tourism sector (based on the proportion of women employed in low-skilled jobs in the sector)	● The reduction in tourism revenues could have a knock-on impact on conservation and restoration efforts (MPA News 2009) ● The reduction in tourism activities provides a temporary respite to reef ecosystems (Zakai and Chadwick-Furman 2002)
Marine transport	● The cancellation of shipping is estimated to be causing revenue losses of $1.9 billion for the carriers (World Maritime News 2020) ● The outbreak is costing the liner segment of the global shipping industry around $350 million a week in lost volume (ICS 2020a; Paris 2020) ● With 384 sailings cancelled, the first half of 2020 could see a 25% reduction in shipping, with a 10% annual fall in 2020 (World Maritime News 2020). For all ships, departures in the first week of April 2020 were down 20% compared to 2019, while the decrease in container-ship departures was 29% (Heiland and Ulltveit-Moe 2020) ● The shipbuilding sector has sustained a major blow from production halts, temporary layoffs and liquidity issues—particularly in the European Union ● The drop in demand for new ships may lead to reductions in shipyard activity	● Travel restrictions and grounded airplanes make crew changeover impossible, leading to repeated contract extensions. About 200,000 seafarers have overrun their contracts and another 200,000 are now waiting to get on board (ICS 2020b). This is putting the personal safety, physical and mental health of seafarers at risk (IMO 2020; ILO 2020a; UNGC 2020a; ICS 2020a) and could lead to maritime accidents ● Seafarers stuck at sea due to crew change restrictions are prevented from reuniting with families (UNGC 2020a; IMO 2020; ILO 2020a) ● Crew members are often denied medical treatment by foreign authorities during the quarantine period (ICS 2020b; IMO 2020)	● Short-term environmental benefit might be observed due to lower transport demand ● Due to weak markets, several shipping companies are now considering scrapping excess tonnage (NSA 2020a). This could present an opportunity to get rid of older and more polluting tonnage ● Although the shipping sector's capacity to invest in more environmentally friendly technologies has been reduced (ECSA 2020), there is still a strong drive towards decarbonisation, as seen in recent announcements from the industry (NSA 2020b; Mærsk 2020; CMA CGM 2020) ● COVID-19 has curtailed the ability of the International Maritime Organization to have physical meetings, which may lead to delays in the adoption of regulations necessary to achieve environmental targets and a reduction in ambition among governments (long-term risk) ● An increase in loss and waste throughout the seafood supply chain as a result of an increase in quarantine paperwork and reduced personnel at the docks (Saumweber et al. 2020)

Table 19.1 (continued)

Sectors	Economic impact	Social impact	Environmental impact
Wild capture fisheries	• Global fishing activity has dropped by 10% since 11 March (Clavelle 2020)[a]. The impact has been even more significant for small-scale fishers (Campbell et al. 2020) • Sales and prices have fallen for premium seafood products generally sold to restaurants, such as lobster, crabs, scallops and wild salmon (Saumweber et al. 2020) • Export-oriented fisheries have seen a vast reduction in demand (particularly from Asia, the United States and Europe) as well as port closures, lost access to cold storage and cessation of shipping and air freight (Orlowski 2020) • Demand has increased for non-perishable compared to fresh seafood (UNCTAD 2020b)	• Female employment may benefit from the production shift towards female-intensive occupations such as preserving and freezing (UNCTAD 2020b) • The reduced demand, limited accessibility of markets and collapsed prices of some fisheries have restricted small-scale fishers' ability to pursue their livelihoods and food security • Women working in the processing sector may be more likely to lose their jobs due to the sector's tendency to offer temporary and lower-paid positions without social protection benefits (Orlowski 2020; The Fish Site 2020) • Gender-based violence may increase (Harper et al. 2020) • Fishing communities may become 'hotspots' for rapid infection due to the migratory nature of fishers and the frequency of international visitors (FAO 2020a) • Probable major disruptions to regionally important tuna industry in the Pacific islands will impact national access to tuna, with resulting economic consequences (Farrell et al. 2020) • Local processing of tuna may be disrupted, and shortages of imported processed and packaged foods are possible (tinned foods). SMEs in this sector could be particularly affected (Farrell et al. 2020)	• A decline in fishing pressure, particularly by legal industrial fleets, could allow fish stocks with more resilient life histories to recover (Bennett et al. 2020) • Illegal, unreported and unregulated and (IUU) fishing may increase due to the suspension of observer programs and fishing patrols • Increased pressure on supply chains, due to port closures and restricted access, may lead to harder-to-regulate practices such as increased transshipment of fish at sea. Such activities are more likely to be associated with illicit fishing and human rights violations (Saumweber et al. 2020) • The sustainability of stocks may be compromised by the extension of fishing seasons and the halting of stock assessment surveys (Carr 2020) • Negotiations on fisheries subsidies at the World Trade Organization have been forced onto a slower track (GSI 2020)
Aquaculture	• Production may be affected by the disruption in the supply of feed or input, transportation and labour shortages • Specialty aquaculture products like shellfish (e.g. lobster, shrimp and oysters) are hardest hit by restaurant closures (FAO 2020b) • Flight cancellation has directly affected trade in some high-end fresh products that are transported by air (FAO 2020b) • The sale of prepackaged, frozen or canned fish and fish products has increased in the short term due to panic buying. However, these industries may not be able to continue supplying the market if the raw material (such as feed) is not available (Aquafeed 2020)	• COVID-19 outbreaks have occurred among seafood process workers in Ghana, the United States and elsewhere, as well as in other animal processing plants (Love et al. 2020) • Women, who comprise a disproportionate share of temporary and casual workers, face the highest risk of losing their jobs due to falling business revenues (Holmyard 2020) • Women working or shopping in vendor markets are at greater risk of infection, since these locations have limited sanitation and hygiene facilities (FAO 2020a)	• Delays in trade are forcing fish farmers to sit on stocks of live fish for prolonged periods, increasing demand for fishmeal and fish oil containing aquafeed (FAO 2020a). This could increase pressure on forage fisheries that are pre-dominantly used for aquafeed production
Ocean-based renewable energy	• Offshore wind energy has seen significant growth during COVID-19 (reNews 2020) • The forecast for offshore wind remains unchanged for 2021, as most projects are already financed and under construction (IEA 2020a). Beyond 2021, the industry might be affected due to permitting and other approval delays caused by COVID-19	• It is difficult to get specialised personnel on board offshore energy platforms or into ports to undertake operations, maintenance and repair, leading to increased risks to health and safety (UNGC 2020a; IMCA 2020) • Though this is hard to disaggregate by sector or technology, some analysis shows that there could be regional job losses in the clean energy sector (Jordon 2020)[b]	• Falling energy demand means sharp reductions in the growth of installed wind, solar and battery capacity in 2020, with effects lingering into 2021 (Eckhouse and Martin 2020)[c] • However, offshore wind investment has more than made up for a slowdown in investment in onshore wind and solar farm projects after the outbreak of COVID-19 (Ambrose 2020)[d]

(continued)

Table 19.1 (continued)

Sectors	Economic impact	Social impact	Environmental impact
Marine conservation	● Reduced revenues from tourism have affected the functioning of some conservation organisations that relied on ecotourism for funding. This has forced these organisations to reduce costs, including by reducing staff engaged in monitoring (Riedmiller 2020)	● Locals and Indigenous communities have turned to hunting and fishing for food security (due to job and income loss), rather than relying on food commodities sold in the markets (Bowlin 2020). In some instances, this could affect the conservation of nearshore reefs close to urban areas ● Nature-based solutions for marine ecosystems, such as the protection of mangroves, are receiving increased attention for their contribution to global efforts like the Sustainable Development Goals and the Paris Agreement, for their co-benefits of protecting and restoring coastal ecosystems to strengthen food security and for their provision of sustainable 'goods and services' that improve social, economic and ecological resilience to climate change and COVID-19	● Marine ecosystems (e.g. coral reefs) may benefit from the reduced physical impact of tourism activities and reduced sewage from hotels and restaurants. Polyethylene terephthalate bottle consumption may be reduced by the cancellation of mass events, tourism and travel (Circulate Capital and GA-Circular 2020) ● Poaching and IUU fishing may increase due to roll-back of environmental protection measures (Kroner 2020). Other impacts may include reversion to unsustainable practices such as destructive fishing or mangrove clearing ● Environmental deregulation measures include extension of the fishing season, opening of marine protected areas to fishing (SUBPESCA 2020a, b, c; Carey y Cía 2020), reassignment of new artisanal fishing quotas and rollover of uncaught quota (Australian Government 2020b) ● The temporary roll-back on plastic bans may become permanent, which is likely to increase plastics in the ocean (Leonard and Mallos 2020)[e]. Marine plastic pollution in the ocean has increased due to the worker shortages in the informal waste sector, lack of demand for recycled plastics and lack of proper disposal of medical items such as masks

● Negative impacts, ● Positive impacts, ● No/neutral impacts

[a]These figures primarily represent changes in activity for the world's industrial fleet—fishing vessels over 24 m—and do not fully capture the impacts on small-scale fisheries

[b]15% of the U.S. total clean energy workforce could be lost over the coming months (more than half a million jobs) due to COVID-19. In March alone, more than 106,000 renewable energy and energy efficiency jobs were lost in the country (Jordan 2020)

[c]2020 global solar and energy storage installations are expected to drop nearly 20% compared to pre-COVID-19 projections (Energy Choice Coalition 2020)

[d]Bloomberg New Energy Finance believes that offshore wind projects are taking off despite the global economic gloom in part due to a two-thirds fall in cost since 2012 and a rush in China to finance and build offshore wind projects before the government's subsidy regime expires at the end of 2021

[e]Several governments, such as that of the Indian state of Tamil Nadu, have suspended bans on single-use plastic bottles and bags in retail trade (Peszko 2020). The United Kingdom has suspended the plastic bag charge for online deliveries, with Scotland delaying the introduction of a packaging deposit-return scheme (Peszko 2020)

A potential decline in renewable electricity capacity for onshore wind energy and solar farm projects is forecast due to factors such as supply chain disruption, lockdown measures, emerging financing challenges and decreased energy demand (IEA 2020a). The share of renewables in the electricity supply has increased, as their output is largely unaffected by demand[3]. Demand has fallen for all other sources of electricity, including coal, gas and nuclear power (IEA 2020b). However, increased offshore wind capacity in 2020 has more than made up for a slowdown in investments (across other renewable technologies) after the outbreak of COVID-19 (IEA 2020a). There is some uncertainty in growth projections for the offshore wind sector beyond 2021, due to permitting and other approval delays caused by COVID-19. In addition, the sectors' interconnectedness amplifies the impacts discussed across the ocean economy (Box 19.1 and Fig. 19.1).

[3]*However, renewable sources (mainly wind and solar) saw their share of electricity substantially increase during COVID-19. For example, in less than 10 weeks, the United States increased its renewable energy consumption by nearly 40% and India by 45%. The ongoing increase in renewable energy into the grid results from a mixture of past policies, regulations, incentives and innovations embedded in the power sectors of many forward-thinking countries (Mojarro 2020).*

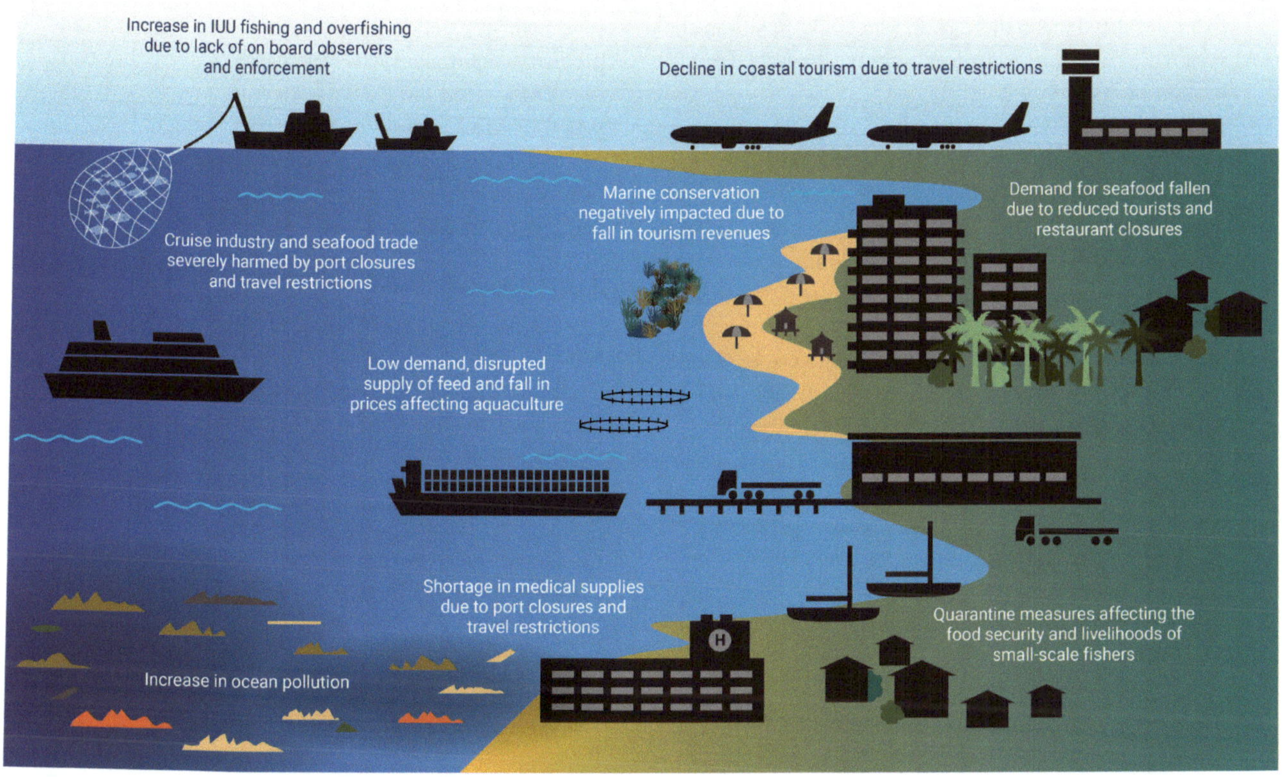

Fig. 19.1 Interwoven impacts across the ocean. (*Source: Authors*)

Box 19.1 Interwoven Impacts Across the Ocean Economy and the Rest of the Sectors

There are strong interconnections between ocean sectors and land-based economies. For example, fisheries and aquaculture provide employment to many communities and are vital for the food security of both coastal and inland communities[a]. The global maritime shipping industry carries around 90% of traded goods. In coastal areas, the tourism sector is the biggest contributor to local, regional and national GDP. Because of these interconnections and linkages between ocean-based sectors and land-based economies, impacts of COVID-19 flow beyond these individual sectors with amplified consequences for the entire economy. Some examples of the transmission of impacts across sectors are discussed below.

Disruption to maritime shipping and port services has negative consequences for the seafood, agriculture, energy, health and tourism sectors.

- Delays for fishing vessels in ports are associated with increased risk of higher seafood waste (Saumweber et al. 2020).
- Port closures (or restricted access to ports) in some countries may have increased the use of transshipment—the transfer of fish and supplies from one vessel to another in open waters—which is more likely to be associated with illegal, unreported and unregulated (IUU) fishing and human rights violations.
- Port closures and travel restrictions also severely harm the global cruise tourism industry, leaving many tourists and seafarers unable to disembark from vessels and replacement crews unable to join their ships.
- The ability of the shipping sector to provide undisrupted service to transport food, energy and other essentials, such as medical supplies, across the continents will play a critical role in overcoming this pandemic.

The aquaculture sector and its ancillary business supply chains face setbacks due to international trade delays, restaurant and hotel closures, and reduction in fishing effort.

- Lockdown restrictions on fishing operations have disrupted the production of fishmeal and fish oil (FMFO) from wild caught fisheries, with negative consequences for the aquaculture sector that is dependent on this input as feed (FAO 2020b).
- At the same time, trade delays are leading to higher unsold volumes of farmed live fish, resulting in higher feeding costs for the aquaculture sector. The risk of

fish mortality is also increased, especially in situations where key inputs are in low supply (such as FMFO requirements) (FAO 2020b).
- Reduced tourist visits caused by lockdown measures have heavily disrupted demand for seafood from the hotel and restaurant industry, particularly for high-value species such as lobster and prawn, reinforcing the interdependencies between the tourism, fisheries and aquaculture sectors (UN 2020c).

Ocean conservation and research have decreased as a result of falling tourism revenues, lost livelihoods in coastal communities and increased ocean pollution.

- In some locations, particularly low- and middle-income countries, fewer tourist visits and reduced availability of associated revenues have curtailed the availability of funding for fisheries management and marine conservation measures (Greenfield and Muiruri 2020).
- Coastal fisheries and reefs are also facing greater pressure, as local communities are turning back to traditional fishing as a food source—driven by a loss of income from tourism (Vyawahare 2020). This can be exacerbated when people return to their home communities from urban areas (Hockings et al. 2020).
- The work of ocean research vessels has been impaired by port closures and quarantine restrictions, with knock-on effects for ocean science and climate studies, such as the Alfred Wegener Research Institute Polarstern expedition, although some privately funded research missions have continued (e.g. Walsh Challenger Deep dive).
- Increased production and use of single-use plastic (such as for e-commerce shipping, grocery delivery, additional food layer protection, masks, gloves and other personal protective equipment) have increased plastic pollution in the ocean, since these items often are not properly disposed of (Tenenbaum 2020). Ocean pollution also has increased due to disruption of land-based waste collection and recycling facilities during COVID-19, especially in South and East Asia[b].

Reduced access to markets for small-scale fishers weakens the food security of entire local communities.

- Reduced inland ferry services and quarantine measures have restricted the ability of many small-scale fishers to access local markets, sell their harvest and contribute to the local economy and the food security of their community.

[a] *Fish accounted for about 17% of animal protein consumed by the global population (FAO 2020c).*

[b] *With recycling not recognised as an essential service in many countries, less than 20% of recyclers operated during the lockdowns in Vietnam, India and the Philippines, while in Thailand and Indonesia it was less than 60%, significantly curtailing waste collection in cities (Circulate Capital and GA-Circular 2020). Critical workers in the value chain lost jobs and income to support their families. The migration of workers in these countries (from urban to rural areas) has also reduced waste collection and recycling. For example, in India, 70–80% of informal sector waste workers have left cities for their hometowns (Circulate Capital and GA-Circular 2020). As a result, no waste-picking has been occurring in landfills and dumping grounds for India's five largest cities.*

2.1.2 Social Impact

Assessments of social impacts show that the COVID 19 crisis has disproportionately harmed a number of vulnerable groups, including women employed in temporary jobs, low-skilled workers, small-scale fishers and businesses, Indigenous community members and younger workers.

Women represent the majority of the workforce in the ocean economy sectors hardest hit by the crisis—about 50% of workers in the seafood sector[4], 70% in aquaculture, 80–90% in the post-harvest sector of small-scale fisheries[5] and 54% in tourism (Holmyard 2020; UNWTO 2019; Monfort 2015; World Bank 2012; OECD 2015). As businesses lose revenue, many will reduce their costs by laying off workers, starting with the temporary and casual jobs disproportionally occupied by women (Holmyard 2020) (Box 19.2)[6]. The shipping industry (including the cruise sector) has been particularly badly affected due to the suspension of cruise operations and quarantining of workers (ILO 2020a; UNCTAD 2020b), with seafarers' physical and mental well-being at risk.

The reduced demand, limited accessibility of markets and collapsed prices of some fisheries have curtailed small-scale fishers' ability to pursue their livelihoods. Indigenous communities are particularly at risk as they may have reduced immunity and limited access to healthcare (UN DESA

[4] *When considering fisheries, aquaculture, seafood processing and all related services.*

[5] *Of the 120 million people who work in the capture fisheries and post-harvest sectors, 47% are women. If the People's Republic of China is excluded, the share of women fishers and fish workers approaches 60% (World Bank 2012).*

[6] *Other systemic barriers such as gender-based violence and lack of access to finance and credit further contribute to the impacts faced by women when they are laid off work. In addition, in many countries women tend to have more work at home, raising children and taking care of the elderly and the sick. An increase in domestic violence and conflict within households could increase food insecurity for vulnerable groups (Farrell et al. 2020).*

2020a)[7]. These groups also face risks of lost livelihoods resulting from the economic crisis, as many are employed in the informal sector or engaged in seasonal work (such as tourism), in which they do not receive social protection benefits[8]. As for all sectors, young people, low-skilled workers and informal workers across the ocean-based economy have been disproportionately affected by the COVID-19 crisis (ILO 2018, 2020c; World Bank 2020a)[9].

Across the seafood supply chain, the social and financial resilience of small businesses (including ones that are family owned or whose workers are self-employed) is being weakened by labour shortages and low demand (Resilience 2020).

The severity of the impacts also varies across countries, with the economies of small island developing states (SIDS) facing higher economic risk (Table 19.1 and Box 19.2) given their small economic base, high degree of openness and extreme dependence on the economic performance of a few developed economies (UN 2020a; WTTC 2020).

The crisis has had some positive social consequences at a community level, such as stronger ties within communities, as demonstrated by many instances of food-sharing (Table 19.1), and by examples of community- run savings clubs to improve social and financial resilience in fishing-dependent communities throughout the Philippines (Arquiza 2019; Polo 2020)[10]. However, social cohesion and trust in authorities has also declined in some communities due to poor crisis management at all government levels.

Box 19.2 Tourism Impact in Small Island Developing States

Due to the economic crisis caused by COVID-19, small island developing states (SIDS) as a whole have seen a 25% decline in tourism receipts, resulting in a US $7.4 billion or 7.3% fall in GDP (Coke-Hamilton 2020)[a]. The gross domestic product (GDP) of the Bahamas and Palau is expected to shrink by 8% or more, making the current crisis the worst in recorded history, while the drop in GDP could reach 16% in the Maldives and Seychelles (Coke-Hamilton 2020; UN DESA 2020b). High external debt, low foreign reserves and volatile capital flows have increased the severity of the pandemic's economic consequences for many SIDS (Coke-Hamilton 2020). This has had a severe impact on both direct and indirect employment (Coke-Hamilton 2020). In the Pacific and Caribbean islands, which rely heavily on tourism, hotels and resorts have been badly affected. For example, the Fiji Hotel and Tourism Association reports that 279 hotels and resorts have closed since the outbreak, with 25,000 workers losing their jobs (ILO 2020b).

Impact on women in tourism

In 20 of 28 SIDS, women constitute more than half of those employed in the accommodation and food services sectors, the core economic activities related to tourism. Women in this sector and in these countries are also more likely than other women to own small and medium businesses. Given the female-intensive nature of employment in tourism, especially in low-skilled activities, women in SIDS are more likely than men to lose their jobs. Businesses may also choose to lower wages or shift workers to informal or part-time work, worsening the already unclear terms of employment in tourism. In addition, women face higher barriers to access business credit. In the absence of targeted policies, this means women entrepreneurs in tourism face a higher risk of bankruptcy than their male counterparts (Zarrilli and Aydiner-Avsar 2020).

[a] *According to the World Development Indicator database, tourism provides more than 50% of export revenue in 20 SIDS and more than 30% in 29 SIDS (Zarrilli and Aydiner-Avsar 2020).*

2.1.3 Environmental Impact

Overfishing, pollution and biodiversity loss were eroding the ocean's ability to sustain livelihoods before COVID-19. The pandemic is likely to intensify the severity of these threats to the ocean. Decreased presence of law enforce-

[7] *The impact of COVID-19 on Indigenous elders has cultural implications for their communities, as elders play a key role in keeping and transmitting Indigenous traditional knowledge, culture and practices. These include conservation of biodiversity, upholding traditions and customs, leading community gatherings and ceremonies, and serving as custodians of customary law and governance (UN DESA 2020a).*

[8] *Indigenous people account for almost 19% of the extreme poor, irrespective of the region and residence in rural or urban areas and even across international borders. They are custodians of a wealth of traditional knowledge and practices, languages and culture, which includes time- tested responses to crises (UN DESA 2020a).*

[9] *More than 61% of the world's employed population—2 billion people—earn their livelihoods in the informal sector. These workers lack the right to social protection benefits and schemes. Some of the low-skill workers in these sectors are migrant workers. The combination of the decline in economic activity, travel restrictions and lack of social protection in many migrant hubs induces such low-skilled migrants to seek to return home. However, back home returnees continue to face challenges, including lack of employment opportunities, limited access to social safety nets, large debts accumulated to finance migration (costs that would have been paid with higher incomes earned at the destination), loss of remittances from abroad and even discrimination by community members fearful that migrants may transmit COVID-19.Young people face multiple shocks from the COVID 19 crisis, including job loss, disruption to education and training, and increased challenges to entering the labour market. A large proportion of young workers are employed in the hard-hit sectors (including tourism), and almost 77% of the world's young workers are in informal jobs (compared to around 60% of workers aged 25 and above) (ILO 2020c).*

[10] *Women make up the majority of members in savings clubs (~70%) and help fishing households pivot from quick spending to long-term financial planning. This change in behaviour can powerfully affect the long-term strategy behind coastal fisheries conservation and the goal of ending overfishing. The savings clubs have already proved to be a fast, secure and communal way to ensure food security for the community during the COVID-19 lockdowns.*

ment, a slowdown in key international negotiations (such as talks on fisheries subsidies at the World Trade Organization) and the roll-back of environmental regulation are likely to compromise ocean sustainability. For example, suspension of observer programs and fishing patrols may be leading to an increase in IUU fishing (Thomson 2020; CFFA CAPE 2020). Similarly, roll-back measures such as reassignment of new artisanal fishing quotas and rollover of uncaught quota have been reintroduced, which could reverse progress made in fish stock recovery (Australian Government 2020b). However, the policy response varies greatly from one country to the next and across levels of government[11]. Declining tourism revenue is also weakening conservation and restoration efforts, especially in cases where ecotourism provides the revenue stream for monitoring, data-gathering, conservation, certification and environmental education (see Box 19.3). Table 19.1 gives details of these impacts.

In addition, COVID-19 has had a temporary impact on efforts to ensure the sustainable transition of ocean-based sectors[12]. However, the ambition to have a carbon-neutral fleet by 2050 is still active, as demonstrated in the Norwegian Shipowners' Association climate strategy, the net-zero announcement by CMA CGM, the Mærsk Foundation donation to set up a new green technology research institute, as well as a number of large-scale projects involving energy companies (such as the partnership by Ørsted, Mærsk and others) to produce green methanol for shipping (NSA 2020b; Mærsk 2020; CMA CGM 2020)[13].

While the decline of ocean-based activities, such as fishing[14] and ocean-based tourism[15], has offered temporary relief to marine ecosystems, over the coming months the combined effects of increased food insecurity, reduced presence of law enforcement bodies and economic recession could prevent the environmental benefits of decreased commercial maritime activities from being fully realised (Torgler et al. 2020)[16].

Box 19.3 Decline in Funding for Marine Conservation Due to Loss of Tourism Revenue

In many cases, governments use revenue from marine tourism to fund marine research and conservation efforts (Wilson and Tisdell 2003) and undertake monitoring and protection activities in marine protected areas. For example, in the Philippines' Tubbataha Reefs Natural Park, tourism revenues make up over half of the conservation budget needed to protect areas from illegal fisheries (UNESCO 2020). However, as the main tourism season (normally April and May) coincided with the strictest quarantine restrictions during the COVID-19 period, tourism revenues in Tubbataha have dropped sharply.

With the decline in tourism revenues during COVID-19, some sites have turned to crowdfunding, online donations and government grants (where available) to meet the funding gaps. In some cases, private foundations have stepped in to compensate for reduced revenue from tourism and endowments. However, these funding sources are unlikely to be sustained. Others have had to reduce surveillance and/or downscale restoration programmes, leading to an increase in fishing pressure. For example, in Seychelles, Fiji, Indonesia, the Philippines and Hawaii, there are reports of increasing fishing pressure in marine protected and conserved areas, which is encouraged by a reduced management presence (Hockings et al. 2020).

[11]*For example, while we see the roll-back of many national-level environmental policies, some local level governance approaches have used consultation to institute recovery plans for fisheries and aquaculture. One example is the virtual consultation by the Philippine Council for Agriculture and Fisheries with relevant stakeholders and government officials specifically to discuss issues confronting the fishery and aquaculture sector amid COVID-19 (PCAF* 2020).

[12]*A survey of its members performed by the European Community Shipowners' Associations revealed that COVID-19 may negatively affect efforts to decarbonise the shipping industry (ESCA* 2020). *Responding to a general question about investments in reduction of greenhouse gas emissions, 44% of respondents to the survey said it will no longer be possible to return to the investments planned prior to the pandemic. Only 26% of respondents to the survey thought they would return to the same level of investments, whereas 30% thought the investments would still happen, but to a lesser extent (ESCA* 2020).

[13]*Since decarbonisation of shipping is a full value chain endeavour, effort towards this transition should not be limited the shipping companies.*

[14]*The lockdown and labour shortages have resulted in a decrease in global fishing activity of nearly 10% (Clavelle* 2020). *In some regions this could provide temporary relief to recovering fish populations and some possible benefits for small-scale fisheries in the longer run (Jigeesh* 2020; *John* 2020).

[15]*A potential positive outcome for marine ecosystems as a result of the decline in tourism activities (e.g. reef trampling, anchor damage, etc.) is less sewage from tourist centres (Zakai and Chadwick-Furman* 2002).

[16]*Emissions reductions caused by economic downturns tend to be temporary—and can lead to emissions growth as economies attempt to get back on track. After the global financial crisis of 2008, for example, global CO2 emissions from fossil fuel combustion and cement production grew 5.9% in 2010, more than offsetting the 1.4% decrease in 2009.*

2.2 Emerging Responses

This section summarises the government policy responses announced thus far to absorb and react to COVID-19 disruptions to the ocean economy and the actions taken by development banks, international organisations (IOs), non-governmental organisations (NGOs) and the private sector to transition towards a sustainable ocean economy.

2.2.1 National Governments

Rapid Emergency Response

To date, response packages from governments have amounted to approximately US $10 trillion globally (IMF 2020a)[17]. As a part of the immediate response, governments have prioritised saving lives and protecting livelihoods, with money channelled directly to households and those on the frontlines of the pandemic. For the ocean economy, this means protecting vulnerable coastal communities dependent on marine natural resources, ocean economy workers, small and large-scale businesses, and ensuring that supply chains remain open for delivery of essential goods (Box 19.4).

Box 19.4 Economic Relief for Ocean Economy Workers and Businesses

A number of measures were introduced by countries to support workers, vulnerable groups and small businesses. Some governments, such as those of the United Kingdom and Canada, along with the EU Commission, have also classified ocean workers as 'key workers', thereby giving them right to movement (EU Commission 2020d; UK Government 2020; Government of Canada 2020).

The list below is not exhaustive but provides examples of support measures directed towards income protection and the welfare of ocean-economy workers.

- **Coastal tourism** Measures include extension of loans and credit to businesses, wage subsidy to workers, financial relief to businesses such as loan consolidation and term extension, increased promotion of tourism and strengthened regional cooperation to boost tourism (e.g. by the Association of Southeast Asian Nations) (Office of the Prime Minister, Canada 2020; KPMG 2020).
- **Marine transport** Staff (especially onshore) have been covered by general wage support schemes in many countries. A number of countries have agreed to new international measures to open up foreign borders for seafarers and increase the number of commercial flights to expedite repatriation following an international crew change summit (Chambers 2020a)[a]. There have also been a number of government support measures and bailouts for maritime companies.
- **Wild capture fisheries** Measures include grants and financial compensation for workers and small- scale businesses and enterprises (in the harvesting, processing and artisanal fishing sector), increased state aid (European Commission 2020b), online training programmes, provision of new fishing equipment, refrigeration transport service for seafood caught by artisanal fisher organisations (e.g. a pilot programme in Chile), provision of loans at subsidised interest rates, waiver of government fees associated with licenses, rollover of quota and deferral of income tax for small businesses (SUBPESCA 2020d; IKI 2020). The European Union also provides a US $1.2 billion guarantee from the EU budget to the European Investment Bank so that it can incentivise European banks and mobilise about $9.3 billion of working capital financing for small and medium enterprises in the fisheries, aquaculture and seafood services sectors (European Commission 2020b).
- **Aquaculture** Measures include income support to workers, increased funding to double community-based aquaculture production and loans or credits to seafood processors (EU Commission 2020a). In addition, the EU Commission, in response to stakeholders' requests, adopted new measures for the aquaculture sector, including support to farmers for temporary suspension of production, and support to producers for private storage of aquaculture products.

[a] *The 13 countries to agree this are Denmark, France, Germany, Greece, Indonesia, the Netherlands, Norway, the Philippines, Saudi Arabia, Singapore, the United Arab Emirates, the United Kingdom and the United States, all of whom now recognise seafarers as key workers.*

[17] *The majority of the $10 trillion constitutes rapid emergency response for the short term and focuses on mostly fiscal measures and regulatory or deregulatory measures.*

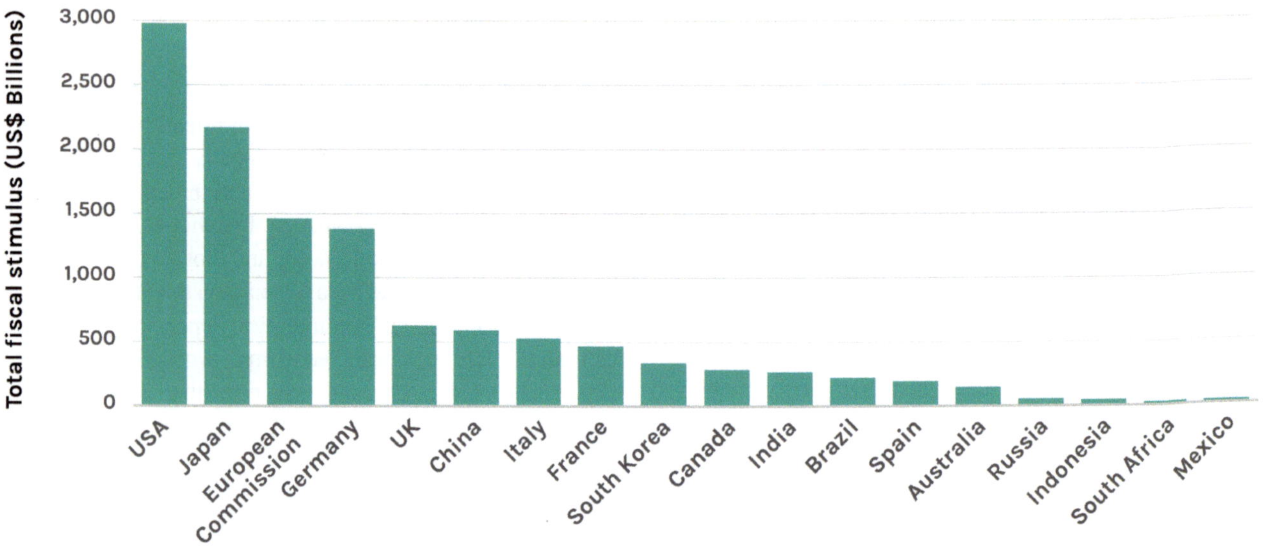

Fig. 19.2 Announced COVID response fiscal stimulus package by country. (*Note: Assumes the proposed 'Next Generation EU' recovery package is implemented in full. Source: Vivid Economics Data*)

Long-Term Recovery Response Measures

The second phase of response from national governments will be aimed at measures to promote longer-term economic recovery and resilience. Analysis from McKinsey shows that G20 nations have announced fiscal measures averaging 11% of GDP, which is estimated to be three times the response to the 2008–9 financial crisis (McKinsey 2020). The United States has announced the largest fiscal stimulus package, followed by Japan and the European Union (Fig. 19.2). Some countries, such as Italy, have said they will commit up to 40% of GDP to their economic stimulus packages (McKinsey 2020)[18].

So far, 30% of economic stimulus packages are going to sectors that currently have high environmental impact (Vivid Economics 2020)[19]. Within the 30%, it is estimated that the majority of the spending will have a predominantly brown impact without conditionality for performance improvements in these sectors[20]. Some of these 'brown' measures include unrestricted support to sectors that have proved to be environmentally harmful in the past and also include rollback on various environmental regulations implemented to deliver better environmental outcomes. For example, both the transport and industry sectors have been hit hard by the crisis and are receiving substantial support from governments. Another source estimated that more than half a trillion dollars worldwide—$509 billion (£395 billion)—is to be poured into high-carbon industries, with no conditions to ensure that they reduce their carbon output (Harvey 2020)[21]. In contrast, only about $12.3 billion is to go towards low-carbon industries, such as renewable energy, and a further $18.5 billion is intended for high-carbon industries provided they achieve climate targets (Harvey 2020).

Some of these interventions target the ocean economy and even fewer align with a transition towards a sustainable ocean economy (Table 19.2 and Sect. 2.3). At this stage, there is little information on how these high- level interventions and investments will be implemented and the degree to which they advance priorities for the sustainable ocean economy or undermine such progress.

Development Banks and Bilateral Development Aid

During the crisis, domestic resource mobilisation has decreased in low-income countries, and external private finance is projected to drop by US $700 billion in 2020, with significant capital flight as a compounding problem (OECD 2020d). Remittances are predicted to fall by 20% in 2020 (Ratha et al. 2020), and foreign direct investment is expected to decline 30–40% in 2020–2021 (UNCTAD 2020a). Given the uncertainty of domestic finance opportunities in many low- and middle-income countries and the volatility of private flows, the need for bilateral and multilateral finance is unparalleled.

[18] *Fiscal measures are likely to be just one aspect of the response measures—monetary measures will also be key in stimulating demand and much- needed liquidity in the market. Assessing the impact of these measures (such as quantitative easing measures) on the ocean economy is beyond the scope of the analysis.*

[19] *Economic stimulus packages encompass a range of fiscal mechanisms, including bailouts and loans. In defining the amount of stimulus flowing through to sectors with a high environmental impact, the index has removed any measures which are purely devised to provide income support to workers (e.g. furlough or income protection programmes).*

[20] *Estimated by Vivid Economics (2020) based on the 14 of 18 countries it evaluates in its study. Brown orientation of these countries' stimulus funding based on (1) the scale of funds flowing into environmentally intensive sectors, (2) the existing green orientation of those sectors and (3) the efforts which steer stimulus toward (or away from) pro-environmental recovery.*

[21] *Specific stimulus packages include, for example, bailout measures of the aviation industry without green conditionality, subsidies for fossil fuel vehicles and an easing of permits for coal mining.*

Table 19.2 Examples of blue stimulus packages announced by selected countries

Country	Selected blue stimulus packages
Australia	At a sub-national government level, the Victoria government package includes A$129 million for the Department of the Environment, for upgrading public land facilities, supporting solar and water infrastructure and addressing erosion and flood risk in marine and coastal areas (Victoria State Government 2020). The Queensland government has committed to provide A$17 million to create a renewable energy training facility as well as a A$8.93 million boost to national parks (including key coastal and marine parks), to provide visitor infrastructure upgrades and enhancements to reenergise nature-based tourism (Queensland Government 2020)
Canada	New assistance amounting to US $62.5 million will be provided to the fish and seafood processing sector through the Canadian Seafood Stabilization Fund, and US $75 million is set aside for emissions reduction in offshore oil and gas. Funding of US $469.4 million will be used to establish the new Fish Harvester Benefit and the new Fish Harvester Grant. The program is designed to work within the unique pay structures and seasonal nature of the fishing sector. The program is open for applications from 24 August to 21 September 2020 (Fisheries and Oceans Canada 2020)
Finland	The supplementary budget of €5.5 billion contains a package of measures supporting the recovery and revitalisation of the economy with a sustainable focus • €13.1 million for state-run rehabilitation of nature sites and the development of nature tourism • €53 million for projects involving green areas, water services and forest conservation. Funding is also proposed for the rehabilitation of local recreation areas • €20.75 million for innovation support for shipbuilding • €5 million for vessel design work in a project to replace three present offshore patrol vessels with vessels capable of responding to oil and chemical spills The previously agreed national climate fund will be capitalised by €300 million. The fund will focus on combatting climate change, promoting digitalisation and boosting low-carbon operations in manufacturing industries (Finnish Government 2020)
European Union	For climate targets, the Green Deal sets aside about €225 billion (US $190 billion) for the recovery fund and €322 billion (US $280 billion) for the 2021–27 budget. Specific detail on the climate policies is not provided. The Euro- pean Union will report annually on its climate expenditure[a] The targets proposed by the European Commission in the Communication on the Farm to Fork strategy (Green Deal on food system) include reduction the use of fertilisers and pesticides, which cause marine pollution As part of green legislation, the European Commission's Environment Committee voted to include CO_2 emissions from the maritime sector in the EU Emissions Trading System (ETS), with a new target of 40% CO_2 reduction by 2030 (EU Parliament 2020) The Environment Committee also called for an 'Ocean Fund' for the period from 2023 to 2030, financed by revenues from auctioning allowances under the ETS, to make ships more energy efficient and to support green infrastructure
Germany	The International Climate Initiative will spend €68 million (US $58 million) to support 29 projects (in 25 countries) responding to COVID; building future economic, social and ecological resilience; and seeking to prevent a new pandemic. The initiative aims to expand the role of green hydrogen as a part of modernising shipping programmes and helping the sector's transition towards decarbonisation (BMU 2020). Its mission is to invest in a sustainable recovery of the economy (including increasing climate resilience of the fishing sector) to contribute to climate change mitigation and the conservation of biodiversity (IKI 2020)
Italy	A state aid scheme worth €100 million (US $85 million) will support agriculture, fishing and aquaculture small and medium enterprises. The fund will provide aid to maintain their activities through state guarantees on investment and working capital loans and direct grants to provide support during the temporary cessation of fishing activities (EU Commission 2020b)
India	Rs 20,050 crore (US $2.7 billion) will be invested over the next five years to bring about a blue revolution through sustainable and responsible development of the fisheries sector
Jamaica	Grants totalling US $1.2 billion will be made available to businesses operating in the tourism and related sectors (KPMG 2020)
New Zealand	An NZ$1.1 billion (US $736 million) environmental jobs program will aim to create 11,000 jobs, include major investments in restoring wetlands
Norway	NOK3.6 billion (US $400 million) is budgeted to support green technology projects that would benefit offshore wind and low-emissions shipping (Nikel 2020). A 'green transition package' (US $384.5m) will be used to support a range of initiatives, including investments in hydrogen power and battery storage technology and building offshore wind infrastructure as Norway looks to reach the Paris Agreement target of limiting global temperature rise to less than 2 °C by 2050 (Casey 2020)
United States	Section 12005 of the Coronavirus Aid, Relief and Economic Security (CARES) Act allocates US $300 million in fisheries assistance funding to states, tribes and territories with coastal and marine fishery participants who have been negatively affected by COVID-19 (NOAA 2020)
Vietnam	An extension is proposed for wind energy projects (including offshore wind) until 31 December 2023 (more than two years beyond the current deadline of 1 November 2021), and a new solar power feed-in tariff (including floating solar energy projects) has been announced (Morris 2020)

Notes: The list of stimulus packages with a focus of blue sustainability is not exhaustive. Exchange rates: €1 = US $1.1842; NZ$1 = US $0.67; Rs1 = US $1.013; NOK1 = US $0.11

[a]The Green Deal consists of a €750 billion recovery fund and a €1.074 trillion EU budget for 2021–27. The amount of money set aside for climate targets, is set at 30%. The recovery fund alone would be the largest green stimulus in history. Specific detail on the climate policies is not provided, and the European Union will report annually on its climate expenditure

A number of multilateral development banks and international financial institutions have mobilised resources to counteract the economic crisis in the most vulnerable countries. For example, the International Monetary Fund (IMF), World Bank, Asian Development Bank and other regional partners are working together on approaches to assist countries in the Pacific overcome the challenges of the current crisis and position themselves for economic recovery (IMF 2020c)[22]. A number of SIDS would also be eligible to apply for short-term debt relief as a part of the IMF's Catastrophe Containment and Relief Trust (Coke-Hamilton 2020). As a part of building back better after COVID-19, the Asian Development Bank is working in cooperation with the UN Economic and Social Commission for Asia and the Pacific (ESCAP) on areas including gender inequality, climate change and ocean pollution (ANI 2020). Additionally, the African Development Bank (2020) has approved €225 million for a budget support loan for Egypt's electricity sector to bolster economic resilience and sustainability. These financial support measures would be in addition to several blue finance initiatives that were set up before the pandemic to achieve sustainable ocean health and governance. This includes the Asian Development Bank's (2019) commitment of US $5 billion (2019–2024) to expand its investments and technical assistance in ocean health and the blue economy; the World Bank's PROBLUE initiative that focuses on four pillars (fisheries and aquaculture; marine pollution; oceanic sectors and seascape management)[23]; and the European Investment Bank's commitment to more than double its lending to sustainable ocean projects, to €2.5 billion ($2.7 billion), over the next 5 years (Richens and Koehring 2020)[24]. However, blue measures still constitute a very small share of the response budget for development banks, and the role that blue recovery measures can play in responding to the crisis could more explicitly emphasized.

Bilateral aid and official lending to low- and middle-income countries from other countries can also make a big difference for the recovery. G20 nations have agreed to freeze bilateral government loan repayments for low-income countries until the end of the year as part of a plan to tackle the health and economic crises triggered by the pandemic and prevent a debt crunch in emerging markets[25] (Wheatley et al. 2020). New Zealand has pledged NZ$55 million in aid spending for Pacific island nations (Dreaver 2020). Similarly, Germany, through the International Climate Initiative, has invested in a number of sustainability projects in 25 countries in response to COVID-19 to build future economic, social and ecological resilience (IKI 2020). Overseas development assistance (ODA) has also played a key role by building health and social protection systems in developing countries, which are critical to countries' ability to respond to the COVID-19 crisis and are central to resilience and recovery (OECD 2020d). However, with several countries' budgets in turmoil, it is possible that the overall level of ODA could decline in 2020 (OECD 2020e)[26]. In addition, recent analysis by OECD shows that over the 2013-18 period a mere 0.8% of global ODA was allocated to support sustainable ocean economy and highly concentrated in three sectors—maritime transport, fisheries and marine protection (OECD 2020f). This suggests that more could be done to support a wider range of existing and new ocean-based sectors and thus foster greater economic diversification and resilience post pandemic (OECD 2020f).

International Organisations and Non-Governmental Organisations

The role of IOs and NGOs is vital in supporting local and national efforts to fight the pandemic. IOs are helping client countries to better address the impacts of this crisis, with a focus on empowering, protecting and prioritising the most vulnerable[27]. For example, the COVID-19 response offer of the UN Development Programme (UNDP) focuses on SIDS and aims to support long-term recovery efforts in these regions by helping them diversify (and sustainably expand ocean economy activities) as well as digitally transform to respond rapidly to crises[28].

[22] *The doubling of the IMF's emergency financing capacity means that up to $643 million could be made available immediately to the Pacific island economies.*

[23] *In fiscal 2019, PROBLUE received signed contributions of over US $50 million from five donor countries (development partners are in the process of signing for over $100 million). Actual funds received from donors totalled approximately $28.8 million. Because of the focus on operationalising the trust fund and preparing the February 2019 annual work plan, PROBLUE approved grants of $2 million, of which $600,000 were disbursed, as of fiscal year 2019. Grant amounts and disbursements are expected to accelerate significantly in fiscal year 2020. As of 30 June 2019, PROBLUE's total fund balance, taking into account actual funds received from donors, disbursements, commitments, and investment income, was just over $28 million.*

[24] *The bank expects to mobilise at least €5 billion in investments from private-sector companies and investors, among other partners (Richens and Koehring 2020).*

[25] *The moratorium on bilateral government debt repayments will begin on 1 May 2020. It will apply to the 76 countries that are eligible to receive assistance from the World Bank's International Development Association, which works with the poorest countries, as well as all nations defined as least developed countries by the United Nations. Eligible countries must be 'current' on any debt service payments to the IMF and the World Bank.*

[26] *The OECD calculates that if Development Assistance Committee members were to keep the same ODA to gross national income ratios as in 2019, total ODA could decline by $11 billion to $14 billion, depending on a single- or double-hit recession scenario on member countries' GDP.*

[27] *For more detail, see the UN COVID-19 response information at* https://www.un.org/en/coronavirus/information-un-system

[28] *The approach is to diversify and expand ocean economy activities and digital transformation to bolster governments' institutional capacities to respond rapidly to crises.*

Many IOs are working directly with industry associations to address the pandemic's short-term and long-term impacts on specific sectors. For example, industry groups, such as the International Chamber of Shipping and the International Association of Ports and Harbours, and UN organisations like the World Health Organization, the International Labour Organization and the International Maritime Organization, have already led an enormous effort to establish safety protocols for preventing and mitigating COVID-19 in vessels and ports, and have also come together to explore ways to safely facilitate crew changes from disembarkation to the airport (Henriksen and Selwyn 2020). The International Chamber of Shipping has led the creation of a 12-step plan for governments on how to undertake crew changes[29]. The UN Global Compact is calling for a coalition of willing governments to protect global ocean supply chains by classifying these workers as 'essential'; this includes offshore energy workers and fish farmers as well as seafarers (UNGC 2020a). The UN secretary general has called for bailouts of the shipping industry to be conditioned on alignment with the goals of the Paris Agreement (Chambers 2020b).

NGOs are working in partnership with multinational development banks and other financial institutions to address immediate needs whilst supporting a resilient, equitable and sustainable ocean economy. For example, the World Wildlife Fund is working to ensure continued monitoring and effective management of marine protected areas from the impacts of IUU fishing and other activities; advocating stimulus measures that promote clean energy and sustainable development; and making guidance available to cities dealing with high amounts of medical plastic waste (Plastic Cities 2020). Some NGOs are working with local fishers and women fish workers to connect catch to private households or local markets (e.g. restaurants), thereby supporting direct marketing of catches that would otherwise go unsold. For example, Rare is working with a fishing community in the Philippines to help manage its long- term finances (by setting up savings clubs), providing transportation for fishers (through engagement with government) and raising awareness about enforcing fish sanctuaries important for the long-term sustainability of community livelihoods (Polo 2020).

Private Investment

Some private sector companies are exerting pressure on governments to ensure that COVID-19 recovery is green and harnesses science-based targets. For instance, in May, a climate advocacy effort, backed by the United Nations and led by chief executive officers, saw 150 global corporations urge a net-zero recovery (UNGC 2020c). Private sector companies are also actively engaging in UN task forces to help with the global COVID-19 response[30].

Blended social and green finance has also grown due to mounting pressure on business to implement more sustainable business practices (Laidlaw 2020)[31]. Also, evidence that green/SDG funds are outperforming their peers during COVID-19 could make investment in ocean- related projects more attractive (Corporate Citizenship 2020). Banks and investors are also under pressure from stakeholders to allocate more funding for environment, social and governance (ESG) initiatives, and some investment firms have launched clean energy funds.

For example, the Southeast Asia Clean Energy Facility (SEACEF) is providing early-stage venture capital–type funding to get new clean energy projects off the ground in Southeast Asia (Nguyen 2020)[32]. However, there is some risk that ocean-based start-ups will face dwindling funds as private institutional investors have frozen their investment decisions (Runyon 2020). Lack of financing will likely cause some start-ups to stop their activity.

2.3 Gap Between Impacts and Response

An assessment of responses to COVID-19 from governments, the private sector, development banks and the 'third' (or voluntary) sector show that a limited number of investments are directed towards the ocean economy, and a small

[29] *A 'roadmap' was developed by a 'supply chain coalition led by industry and unions in cooperation with UN agencies' (ICS 2020c).*

[30] *For instance, cross-sectoral ocean companies are actively participating in the UN Global Compact Task Force, with aquaculture players such as Cermaq and Bakkafrost, maritime insurers such as Gard AS and maritime classification companies including Lloyd's Register and DNV GL.*

[31] *There has been gravitation towards a more blended sustainable approach and with considerations of environmental, social and governance factors. Social bond issuance for 2020 totalled US $11.58 billion as of 15 May, compared to just $6.24 billion in the same period of 2019, according to an International Capital Market Association analysis of the Environmental Finance database. Demand for sustainability bonds, something of a hybrid between green and social bonds, has also surged. It reached $25.62 billion in the year through 15 May, compared to $13.64 billion in the same period a year earlier. Green bond issuance, in contrast, has dropped sharply. It totalled $53.54 billion in 2020 as of 15 May, compared with $84.09 billion in the same period of 2019.*

[32] *The fund is supported by international climate foundations including Sea Change Foundation International, the Wellspring Climate Initiative, the High Tide Foundation, the Grantham Foundation, Bloomberg Philanthropies, the Packard Foundation and the Children's Investment Fund Foundation. The supporting global philanthropies have invested an initial $10 million in SEACEF, and are seeking to attract up to $40 million in additional capital. It is expected that every dollar of high-risk venture capital–type funding deployed by SEACEF will leverage up to 50 times more in follow-on investment in the clean energy portfolio across Southeast Asia—reaching more than $2.5 billion in assets—while cultivating the local ecosystem of developers to grow the market. The initial focus will be on Vietnam, the Philippines and Indonesia.*

subset focuses on transitioning to a sustainable ocean economy. Within the blue measures there has been more of a focus on short-term coping strategies to address the immediate impacts of the crisis, such as high unemployment, business insolvency and health risks faced by ocean economy workers. Shifting this focus to the development and implementation of longer-term resilience-building strategies will be key to preventing future shocks and responding to ongoing stressors, such as climate change and biodiversity loss. It is imperative that ocean activities and industries transition towards smarter, sustainable practices that conserve marine ecosystems and promote human well- being both now and into the future.

Based on an assessment of the gap between impacts and responses, we summarise below the consequent missing action or unintended impact on local economies and the health of the ocean.

To protect the livelihood of small-scale fisheries in the long term, it will be important to ensure that support policies from national governments do not encourage overfishing practices or IUU fishing that damage ocean ecosystems and deplete stocks. A number of measures have been introduced to promote the recovery of the sector and support the fishers (especially vulnerable groups) facing loss of livelihoods due to the crisis. However, while license fee waivers, measures to reduce input costs (through provision of loans at subsidised interest rates), deferrals and rollover of unused fishing quota are being used to support fishers by reducing fishing costs, this could lead to an environmental trade-off by incentivising overfishing[33]. Measures such as decommissioning schemes or payments for early retirement (e.g. the European Maritime and Fisheries Fund's allowing EU member states to pay fishers and aquaculture producers for a reduction or cessation in production) could reduce oversupply of fleets. However, whether such steps lead to longer-term reductions in fishing pressure and ultimately to healthier fish stocks will depend on whether they postpone fishing effort (OECD 2020d). Measures that incentivise sectors to move towards the sustainable management of fish stocks will be key for economic recovery and equitable prosperity in the long term. It will be important to ensure that support policies and investments do not encourage overfishing practices or IUU fishing that damage ocean ecosystems and compromise the sustainability of resources, putting future resilience at risk.

To help reduce seafood waste and meet long-term food security targets, continuity of investments facilitating the growth of sustainable mariculture will be key. Measures aimed at improving storage of mariculture and fisheries products will also deliver environmental benefits, reducing loss and waste of fish products across the supply chain. Growth of sustainable mariculture practices will be very important for food security, and investments in sustainable mariculture will require a substantial mobilisation of capital. A number of innovative practices are being developed in the sector to support its sustainable transition (including aquafeed alternatives, industrialisation of seaweed and bivalve farming). While some of these have been driven by private investments, financing from public bodies (such as the development banks and national governments) can help mobilise private capital by building confidence and reducing risk.

To help make up for declining tourism-based funding for ocean conservation, there is an immediate need for interventions that help protect vital and vulnerable marine ecosystems. While decreased tourism funding has led to an increase in alternative methods of funding for marine conservation (such as crowdfunding and donations from private foundations), these funding mechanisms are unlikely to be sustained. In addition, some marine sanctuaries have been opened to fishing, which can quickly erase the progress made on marine biodiversity recovery in these sites. The current protected area network is only receiving about one-third of the funding it needs to be effectively implemented and managed, and the shortfall is even greater in developing countries (Waldron et al. 2020). Expanding protection to at least 30% of the world's land and ocean and effectively managing it would require an average investment of US $140 billion annually and deliver a range of benefits to society that will outweigh the costs (Waldron et al. 2020)[34].

For the long-term resilience of the coastal tourism sector and protection against future climate change shocks, investment must go into restoring and protecting marine environments and uplifting local communities. Most emergency and recovery measures have aimed to provide income continuity for tourism workers and business continuity for small enterprises that otherwise would be unable to survive the crisis. The international community has also mobilised funds through multilateral development banks to counteract the economic crisis in the most vulnerable countries. However, much more needs to be done to stimulate demand and ensure the sector's long-term resilience once containment measures are lifted. Recovery following the crisis presents an opportunity to think about innovative measures where tourism businesses play an active role in uplifting local communities and protecting coastal and marine envi-

[33] *Input cost-reduction measures (such as the provision of fuel subsidies) tend to benefit larger fleets at the expense of small-scale fisheries.*

[34] *Waldron et al. (2020) state that this funding should come from a range of sources, including official development assistance, governments' domestic budgets, climate financing directed to nature-based solutions, philanthropies, corporations and new sources of revenue or savings through regulatory and subsidy changes.*

ronments. Policies and investments supporting structural transformation are needed to help build a low-carbon, less polluting, more sustainable and resilient coastal tourism economy. In addition, targeting recovery at diversification across a range of ocean activities to reduce dependency on the tourism sector will be key to building future resilience in Caribbean and Pacific islands.

To ensure the long-term viability of the marine transport sector, investment and regulation needs to create the right market incentives for a sustainable transition to zero-emission vessels. While the pandemic has curtailed the shipping sector's capacity to invest in more environmentally friendly technologies, industry is still leading a strong drive towards decarbonisation (NSA 2020b; Mærsk 2020; CMA CGM 2020). There is an important role for international organisations and governments to help keep the momentum by developing national and market incentives for decarbonising domestic and international transportation. This includes investment in green technologies, developing policy to enable the business case for the adoption by shipping of low- and zero-carbon fuels (e.g. a carbon price), develop national incentives for decarbonising domestic transportation and facilitating decarbonisation of national energy systems faster or as fast as the transition in the international fleet (Hoegh-Guldberg et al. 2019). Low-carbon domestic shipping and coastal marine transport can play a strong role in building coastal resilience. Shifting freight transport from road to waterways in emerging markets (like Africa, India or Latin America), where trucks alone are responsible for about 40% of transport emissions, can substantially reduce emissions and logistics costs (World Bank 2020b). Similarly, after the crisis key global partnerships will need to continue to support SIDS and least developed countries (LDCs) that face significant domestic or regional shipping decarbonisation challenges. Flexible port regulations based on screening and discretion will be needed to ensure the continuity of freight distribution and ferrying of food and essential goods so that supply chains are not hit by both low demand and supply bottlenecks (Heiland and Ulltveit-Moe 2020).

To accelerate deployment of ocean-based energy systems, a stable economic and regulatory environment will be needed to help stimulate investments in these growing sectors. The vast majority of the COVID-19 relief from governments so far supports carbon-intensive industries without requiring improvements. For long-term sustainability it will be important to shift towards a green-blue recovery, where government, businesses and investors can play a role in boosting clean investment, both by promoting low-carbon supply chains and by grasping the opportunities of clean energy markets (Mojarro 2020). Governments will need to play a key role in providing a stable economic and regulatory environment to help stimulate investments required for an accelerated deployment of ocean-based energy systems. Investment will also be needed to advance ocean renewable technologies beyond offshore wind to make them more economically attractive.

3 Roadmap for a Sustainable and Equitable Blue Recovery

Recovery and stimulus packages represent a unique opportunity to accelerate the shift to a sustainable ocean economy that delivers on global targets under the 2030 Agenda for Sustainable Development and the Paris Agreement. Mutually beneficial, no-regrets opportunities are ready to be implemented now to support affected communities and regions, while delivering significant social and environmental benefits. These opportunities respond to the immediate need for job creation in the short-term and offer opportunities for long-term economic growth and resilience. Governments can also utilise innovative financial mechanisms to incentivise progress and avoid rollbacks in progress.

The investments that governments and financial institutions make over the coming months and years will have long-term effects on the nature of economies and their resilience to future shocks. Efforts should be made now to avoid locking in high-emitting, high-polluting and inequitable pathways that limit the ability to build sustainable and resilient economic systems. Investment through recovery and stimulus packages represents a crucial lever for accelerating the shift from business as usual to a more sustainable future that delivers on global targets under the 2030 Agenda for Sustainable Development and the Paris Agreement.

The ocean economy can play a vital role in this transition, and in turn this transition will be critical to securing a sustainable ocean economy for the future. Using recovery and stimulus packages to invest in, and introduce, both short-term and longer-term policy reform for a sustainable ocean economy can provide short-term economic relief and recovery while delivering long-term societal benefits and building economic resilience to future shocks.

This report proposes that coastal and island nations have the opportunity to pursue a 'sustainable and equitable blue recovery'. We consider a 'sustainable and equitable blue recovery' to be one that advances a sustainable ocean economy predicated on three mutually reinforcing elements: effective protection of ocean ecosystems, sustainable production and equitable prosperity. A sustainable ocean economy should enable the growing global population to continue enjoying the innumerable benefits that the ocean provides.

To achieve this, it is imperative that ocean activities and industries transition towards smarter, sustainable practices that conserve marine ecosystems and promote human well-being both now and into the future.

This section of the report aims to provide a roadmap for a 'sustainable and equitable blue recovery' from the COVID-19 crisis.

First, it proposes a set of high-level guiding principles that act as a first step for ensuring a 'sustainable and equitable blue recovery'. These may be helpful for governments in their initial stages of planning on how to think about the nature of their recovery after COVID-19.

Second, it proposes a set of five priority opportunities that are ripe for immediate government investment through recovery and stimulus packages, what we call **'blue stimulus'** (Sect. 3.2). For each of these opportunities, we outline the economic (short- and long-term), social and environmental benefits to be gained from investment in this opportunity and then detail a set of potential interventions for governments based on their national circumstances. We identified these five priority opportunities based on a set of guiding principles outlined in Sect. 3.1.

Third, it proposes a set of additional opportunities that are more systemic in nature and oriented towards using this moment as a reset for the ocean economy to build long-term economic resilience to future shocks, what we call **'blue transformations'** (Sect. 3.3 and Annex 1). Not all these options necessarily provide the short-term economic benefits that the five priority opportunities do, but they are equally important for securing economic recovery, resilience and prosperity over the longer term. Governments that have the capacity to introduce more systemic and long-term policy reform at this time (in addition to taking action on the five priority areas) will find this longer list of additional interventions helpful.

Fourth, it looks at the potential role of financial grants and debt relief as an unprecedented opportunity to advance key reforms in areas such as sustainable fisheries management, monitoring and enforcement of protected areas and ocean data, what we call **'blue conditionality'** (Sect. 3.4).

The proposed opportunities and interventions outlined in this section are not intended to be exhaustive; they do not include everything that will be required to fully transition to a sustainable ocean economy. Resources aimed at providing the full suite of necessary interventions are contained in Annex 2 (Table 19.7). This report focuses on identifying the interventions most relevant at this unique point in time—recognising financial and capacity limitations that many countries have and the urgency of ensuring economic opportunities and health outcomes for their communities over the next few years as we recover from the COVID-19 crisis.

Each country will need to carefully evaluate the full set of interventions against its national priorities, circumstances, impacts and geography to ensure that the options pursued deliver the greatest benefit for its population.

3.1 Proposed Principles for a Sustainable and Equitable Blue Recovery

Given the gap between the impacts experienced by workers and sectors in the ocean economy and the early responses from governments and other stakeholders in their stimulus packages, decision-makers will need to better consider how to integrate the ocean and ocean economy into recovery measures.

This report proposes three high-level guiding principles[35]:

1. Actively advance (through direct investment or policy) projects and programs that contribute to building a long-term sustainable and equitable ocean economy.
2. Identify opportunities to make public finance and debt relief conditional on advancing core national priorities for a sustainable and equitable ocean economy.
3. Assess the impact of all interventions across sectors on the health of the ocean and ocean economy and either avoid investments that will detract from this long-term goal (e.g. high-emitting, polluting terrestrial and marine industries or inequitable practices) or minimise their impact through additional conditions or requirements.

Sections 3.2, 3.3 and 3.4 of this report provide a set of priorities for putting principles 1 and 2 into action.

The Sustainable Blue Economy Finance Principles provide a framework for implementing principle 3 (WWF 2018). These are voluntary principles that act as a framework to guide investment and development decisions. These principles complement existing frameworks in sustainable finance and recognise the importance of compliance, transparency and disclosure, as well as the specific challenges of investment in the context of the ocean. They are designed to support the Sustainable Development Goals (SDGs), in particular Goal 14 ('Conserve and sustainably use the oceans, seas and marine resources for sustainable development'). They are also designed to comply with the International Finance Corporation's Performance Standards and the European Investment Bank's Environmental and Social Principles and Standards (WWF 2018).

[35] *See also the UNGC Sustainable Ocean Principles for the private sector. They propose nine principles that cover three areas: ocean health and productivity; governance and engagement; and data and transparency (UNGC* 2019*).*

3.2 Five Priority Opportunities for a Blue Stimulus

Given the need for governments to respond to the immediate economic impacts experienced by most countries and coastal communities and the short-term priority of job creation and income protection, we can identify five priority opportunities ripe for immediate intervention by governments through recovery and stimulus efforts. These opportunities not only offer significant short-term job creation and income protection potential for affected communities but also offer long-term economic benefits in the form of catalysing sustainable ocean industries for the future and increasing resilience.

We identified these five priority opportunities through a literature review and expert input from government representatives involved in the design of recovery and stimulus packages and bilateral and multilateral funders (Fig. 19.3). We sought opportunities that provided the following:

- Short-term job creation (considering a match between the skills needed and those available in the local workforce) in the ocean sectors and communities affected by COVID-19 (European Commission 2020b)
- Ability to build long-term resilience to future shocks (considering improving human, natural and physical capital) (Hammer and Hallegatte 2020; OECD 2020e)
- Ability to directly respond to impacts suffered (e.g. economic, social or environmental) and support economic recovery in more than one sector

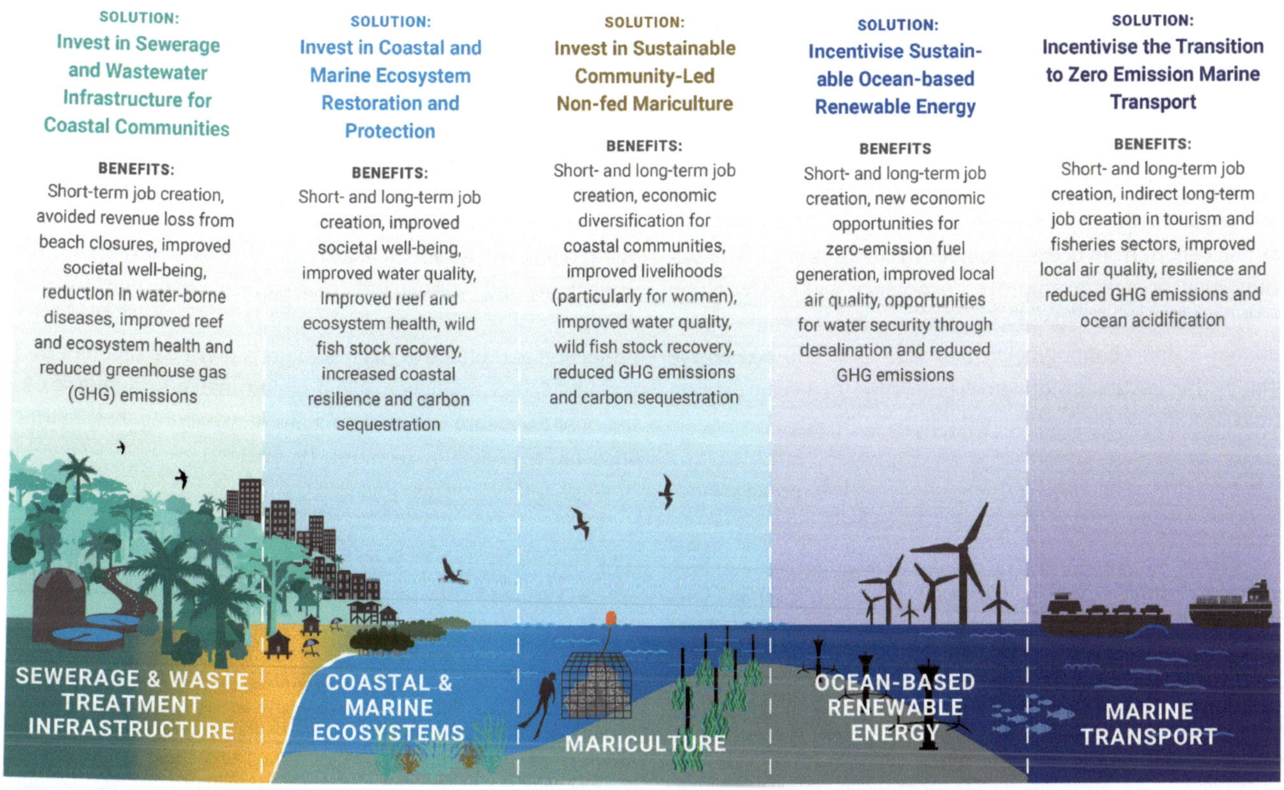

Fig. 19.3 Five priorities for ensuring a sustainable and equitable blue recovery to the COVID-19 crisis

- Ability to direct economic benefits to affected communities and vulnerable members of society (a people-centred approach) (UN 2020b)[36]
- Speed and feasibility of implementation (considering supply chain blockages and capacity of local communities) (Hepburn et al. 2020)[37]
- Ability to catalyse progress towards a long-term sustainable and equitable blue economy (Hepburn et al. 2020)
- Ability to deliver on international commitments such as the 2030 Agenda for Sustainable Development and the Paris Agreement (IMF 2020b)
- Relevance to multiple regions and economies (OECD 2020e)

In advancing a 'sustainable and equitable blue recovery' it will be important to make decisions in accordance with integrated and holistic long-term plans and strategies, so that investments are made in alignment with national priorities.

Such planning tools include integrated ocean management, integrated coastal zone management and marine spatial planning (MSP). Establishing MSP processes in addition to integrated ocean management will be essential to deal with the inherent variability of the ocean and a dynamic future shaped by climate change. Cohesive planning can facilitate optimal use and benefit from ocean resources by all users while streamlining management to improve governance and conservation of critical habitats[38]. Ideally, countries should develop a sustainable ocean economy plan that acts as a comprehensive strategy for advancing effective protection of ocean ecosystems, sustainable production and equitable prosperity.

[36] *The UN secretary general has stressed the need to ensure that national and local response and recovery plans identify and put in place targeted measures to address the disproportionate impact of the virus on certain groups and individuals, including migrants, displaced persons and refugees, people living in poverty, those without access to water and sanitation or adequate housing, people with disabilities, women, older people, LGBTI people, children and people in detention or institutions.*

[37] *Factors relevant to the design of economic recovery packages include the long-run economic multiplier, contributions to the productive asset base and national wealth, speed of implementation, affordability, simplicity, impact on inequality and various political considerations (Hepburn et al. 2020).*

[38] *The value of such planning instruments at times of economic hardship is illustrated by an MSP process in Massachusetts that led to a proposed optimum arrangement with associated value, calculated at preventing more than $1 million in losses to the incumbent fishery and whale- watching sectors and generating more than $10 billion in extra value to the energy sector (White et al. 2012).*

3.2.1 One: Invest in Coastal and Marine Ecosystem Restoration and Protection

Coastal and marine ecosystem restoration can broadly be defined as activities that are aimed at moving these ecosystems (mangroves, salt marshes, seagrasses, kelp and seaweed forests and reefs) to healthier states, often with the goal of increasing their ability to provide ecosystem services. This includes replanting coastal mangrove forests that have been degraded, reconstructing saltmarshes that have been lost to human development and enhancing the structural complexity of damaged reefs (both coral and shellfish). The potential benefits of restoration projects are higher—often significantly higher—than the costs, making such projects prime candidates for investment as part of recovery and stimulus packages (Bayraktarov et al. 2015).

Analysis indicates a potential net benefit of US $97 billion to $150 billion for mangrove restoration and $48 billion to $96 billion for mangrove conservation over 30 years (2020–50)[39]. This results in a benefit-cost ratio of 3:1 for both mangrove conservation and restoration (Konar and Ding 2020)[40].

Restoration of coastal and marine ecosystems has been identified as a priority due to its potential for job creation in the short term and significant potential in terms of avoided greenhouse gas (GHG) emissions. It is also a necessary precondition for protection and subsequent management and conservation efforts. Ensuring that ecosystems are placed under full or high protection and effective management is a critical element of a sustainable ocean economy and opportunities for countries to use debt for nature swaps as a means of expanding their marine areas under protection (see Box 19.7 below).

[39] *The total value of net benefits for mangrove restoration over 30 years is higher than for conservation because we assume the area of mangroves restored is 10 times that of the area conserved. The conservation scenario assumes stopping the additional loss of mangroves, whereas the restoration scenario assumes replanting large areas of mangroves already lost; that is why we are doing more restoration in the scenarios analysed than conservation. The overall ratio of both conservation and restoration is calculated by adding the total present value benefits and costs of both measures. The very high restoration costs are the main factor driving the overall benefit-cost ratio for both conservation and restoration.*

[40] *Konar and Ding's (2020) study estimates the benefit-cost ratio for mangrove conservation to be higher (88:1) than restoration (2:1) due to a number of factors: the higher cost of mangrove restoration (due to seeding and replanting), the low survival rates following restoration and the lag in accrual of benefits from restoration.*

SUPPORTING RECOVERY AND RESILIENCE IN THESE SECTORS:
Fisheries, Tourism, Marine Conservation and Infrastructure

SUPPORTING ACHIEVEMENT OF THESE SDGs:

Why Investment Makes Sense

Restoration of coastal and marine ecosystems provides short-term job creation in a number of industries at the local and regional level. Restoration works create jobs immediately through construction. Restoration projects extend to the full set of economic activities that contribute to restoration, from project planning, engineering and legal services, to intermediate suppliers of inputs, to on-the-ground earth-moving, forestry and landscaping firms that contribute to the ecological restoration process (BenDor et al. 2015). Restoration can include a full spectrum of jobs from all skill levels and technical backgrounds, including general trades, barge drivers, engineers, transportation, scientists and hatchery staff, oyster farmers and hydrologists. The economic benefits derived from coastal and marine restoration projects are not limited to direct jobs. However, much of the economic benefit is in uplift to the service and beneficiary industries associated with increased coastal productivity, including fishing, tourism, wastewater treatment and marine equipment and boat suppliers (Appeaning Addo et al. forthcoming)[41]. Other estimates for coastal and marine restoration works in the United States ranged from 15 to 33 jobs per $1 million, depending on the type of activity (removal of invasive species from coral reefs generated the most jobs), but the majority of projects fall within a range from 15 to 19 jobs per $1 million of expenditure (Edwards et al. 2013) (see Box 19.5 for more details). By comparison, investment of $1 million in traditional energy-intensive industries have been estimated at 14.4 jobs for road and bridge developments, 6.8 jobs for coal mining, 4.2 in nuclear and 5.2 jobs in oil and gas and 8.9 for offshore oil and gas (Hurowitz 2020; Pollin et al. 2009)[42]. These jobs can be created in rural areas, where poverty tends to be concentrated in low- and middle-income countries.

Box 19.5 Coastal Restoration in the United States

Following the 2008–2009 global financial crisis and expenditure under the American Recovery and Reinvestment Act (ARRA) of 2009, the National Oceanic and Atmospheric Administration (NOAA) estimated that coastal habitat restoration projects created, on average, 17 jobs per million dollars spent[a]. This is similar to other conservation industries such as parks and land conservation, but much higher than other traditional industries, including coal, gas and nuclear energy generation.

The study shows that the 50 ARRA projects administered by NOAA in the first year and half generated a total of 1409 jobs (Edwards et al. 2013). Many of these jobs were created in rural and regional coastal areas and offer a range of skilled and low-skilled positions, considerably enhancing economic opportunities in regional areas. Jobs were created for day labourers, administrative staff, barge operators, lawyers, accountants, engineers, helicopter pilots, fisherman, scientists, nursery workers and project managers. Longer-term employment can be created through the flow on benefits (uplift) created by an increase in productivity of coastal ecosystems and generation of wider ecosystem services benefits (for example, increased employment from improved productivity and higher tourism opportunities).

The median (global) restoration cost for all coastal ecosystems (mangroves, saltmarshes, seagrasses, coral reefs and oyster reefs) was estimated to be around $80,000 per hectare (Bayraktarov et al. 2015). Costs for restoration vary considerably within and between ecosystems and across countries (Bayraktarov et al. 2015)[b].

---

[41] *Estimates are based on Oregon's restoration project, and labour intensity will depend on local factors. The model used the economic impact modelling software IMPLAN 3.0 to describe the impacts from public investments in forest and watershed restoration. It was based on an input- output analysis to describe the patterns of trade and the degree to which goods and services are sold and purchased outside the state's economy. Based on the dependencies among different economic activities, input-output models can project the impact that changes in one sector will have on economic activity in other sectors of the economy.*

[42] *Multipliers were derived using IMPLAN 2.0 with 2007 data. Infrastructure multipliers and assumptions are presented in Pollin et al. (2009). The estimates are based on input-output models. Key limitations include the assumption of fixed prices (prices do not change when demand for a good, service, or input changes), fixed ratios of labour to other factors of production and fixed sectoral share of GDP over time.*

[a]*The model used to calculate these job numbers was the economic input/output software called IMPLAN (Impact Analyses and Planning) to estimate overall jobs and economic impacts. The economic data for IMPLAN come from the system of national accounts for the United States based on data collected by the U.S. Department of Commerce, the U.S. Bureau of Labor Statistics and other federal and state government agencies. Data are collected for 528 distinct producing industry sectors of the national economy corresponding to the Standard Industrial Categories. Industry sectors are classified on the basis of the primary commodity or service produced. Corresponding data sets are also produced for each county in the United States, allowing analyses at the county level and for geographic aggregations such as clusters of contiguous counties, individual states or groups of states.*

[b]*The median restoration cost per hectare for mangroves, seagrasses, oyster reefs, coral reefs and saltmarshes is estimated to be $8961, $106,782, $165,607 and $67,128, respectively. Total project costs—calculated for projects that included both capital and operating costs—for restoring seagrass, saltmarshes and oyster reefs were two to four times higher than the median.*

Healthy coastal and marine ecosystems under full or high protection and effective management can deliver long-term job creation and economic growth potential in ecotourism and artisanal fisheries. The protection and effective management of coastal and marine ecosystems through fully or highly marine protected areas (MPAs) (Carrasquila Henao and Juanes 2017) or other effective conservation-based measures (OECMs) can deliver long-term economic opportunities for coastal communities. Analysis has shown a benefit- cost ratio of between 3:1 and 20:1 of expanding the MPA network, meaning that every $1 invested returns up to $20 in benefits (WWF 2015). Analysis shows that expanding protected areas to cover 30% of the planet (terrestrial and ocean) would generate higher overall output (revenues) than non-expansion (an extra $64 billion to $454 billion per year by 2050). This would be in addition to economic benefits (avoided-loss value estimated to be $170 billion to $534 billion per year by 2050) (Waldron et al. 2020)[43]. In terms of direct job creation, coastal and marine ecosystems under protected area status generate demand for administration, conservation, management, monitoring, surveillance and scientific research jobs located in the local community. For example, for the Natura 2000 network (terrestrial and marine), every €1 billion of expenditure supports almost 30,000 jobs, with 60% of these on activities directly related to site management (e.g. designation, management, conservation actions, monitoring and research) (Mutafoglu et al. 2017). In addition, MPAs generate demand for other services, such as technology to improve surveillance and management (see Sect. 3.4 on how to digitise such efforts in a post-COVID-19 world) (EU Commission 2018). The restoration and protection of these ecosystems also directly improves the potential for ecotourism or the recovery and long-term viability of the coastal tourism sector. Studies have shown that ecotourism in marine protected areas provides 4–12 times greater economic returns than the economic returns from solely utilising the area for fishing (for example, A$5.5 billion annually and 53,800 full-time jobs in the Great Barrier Reef) (Deloitte 2017; Duarte et al. 2020). The port city of Xiamen, located on the west coast of the Taiwan Strait and one of the busiest ports in China, faced environmental degradation, sea-use conflicts and ineffective management. As a result of improving protection and advancing ecosystem restoration, the Chinese white dolphin population returned and tourist numbers increased from five million in 1996 to more than 100 million in 2019 (Winther et al. 2020). Industry has also been able to flourish, with year- on-year growth staying above 10%. New marine high-tech industries (biological pharmacy, science and education service, high-end equipment) have also grown (Winther et al. 2020). Roncin et al. (2008) summarise the impact of Southern European MPAs on local economies[44] and calculate the yearly local income related to services to non-resident recreational users to be €640,000/year per MPA and 15 yearly full-time equivalent jobs[45]. Lastly, MPAs and OECMs are critical tools to increase fisheries' productivity, maintain fish stock levels and thereby ensure ongoing economic opportunities for artisanal and commercial fisheries as well as provide local food security (Brander et al. 2015). In a meta-analysis looking at the role of biodiversity loss on ecosystem services, data showed that post-designation, levels of biodiversity of fully protected areas increased by an average of 23%, with large increases in fisheries' productivity in areas adjacent to the MPA (known as the spillover effect) (Halpern et al. 2010). Fisheries in medium- to high-decline gained the most from spillover from highly and fully protected MPAs (WWF 2015). Another study that looked at the combined economic benefits of MPAs found that both tour-

[43]*The financial estimates are for both terrestrial and marine protected areas. The economic estimates only refer to forests and mangroves.*

[44]*Empirical evidence is based on surveys with fishermen and divers (1,836 questionnaires).*

[45]*Estimates are based on local expenditures of non-resident recreational fishers and scuba divers only. Estimates would likely be higher if expenditure of all tourists were included.*

ism and neighbouring fishery profits increased within as little as five years after the reserve was established (Sala et al. 2013).

Healthy coastal and marine ecosystems deliver improved health, well-being and resilience for coastal communities. Restoration of these ecosystems can deliver significant benefits for improved food security for coastal communities (TNC 2013)[46], improved water quality (and the associated health benefits) and improved coastal recreation opportunities. Communities living in areas with more extensive mangrove forest experience significantly lower losses from exposure to cyclones than communities in coastal areas without mangroves (Hochard et al. 2019) and are more resilient to the effects of rising sea levels (Serrano et al. 2019). This is also true in communities bordering fringing reefs. Reef structures cause waves to break and reduce wave energy by an average of 97%, protecting the beach from possible erosion as well as reducing the number of people affected by annual flooding by more than 200,000 (Ferrario et al. 2014; Beck et al. 2018). Higher property values are associated with communities situated near restored and well-functioning coastal and marine ecosystems (Bark et al. 2009). Studies have shown that lower-income communities living in low-lying areas are the most vulnerable to natural disasters such as flood and coastal storm surges (Winsemius et al. 2018). Utilising restoration of coastal ecosystems in these areas can dramatically improve the quality of life of these communities. For example, following the 2004 floods in Bangladesh, poor households lost more than twice as much of their total income as non-poor households (Brouwer et al. 2007). Worldwide, low-income countries suffer 63% of all deaths from storms, including cyclones and hurricanes, even though they experienced just 12% of the global total of such events (CRED 2015). Coastal and marine ecosystem restoration and protection also offer opportunities for engagement, co-ownership and co-management with Indigenous communities and traditional owners—offering knowledge-sharing and capacity building for all stakeholders involved as well as the opportunity for revenue to be reinvested back in the local community (McLeod et al. 2018). Studies have shown that engagement of local communities in long-term restoration and protection is a key success factor, and lacking it is a major reason for failure (Hai et al. 2020; Suding et al. 2015). Inclusive planning processes for restoration activities have been shown to deliver a positive social impact and equitable benefits for communities.

Coastal and marine ecosystems also have significant carbon sequestration potential and can provide valuable mitigation opportunities in addition to improving local water quality and enhanced biodiversity. Analysis estimates that restoration could deliver annual global emissions reductions of between 0.20 and 0.33 GtCO2e by 2050 (Hoegh-Guldberg et al. 2019), which is equivalent to taking approximately 4–7 million cars off the road annually[47]. The sequestration benefits from reducing CO_2 emissions are estimated at $137 billion to $214 billion for restoration over 30 years (Konar and Ding 2020). Coastal habitats are home to a number of marine and terrestrial animals (Li et al. 2018; Rog et al. 2016), including species important for fisheries (Carrasquila-Henao and Juanes 2017). These habitats buffer acidification (Kapsenberg and Cyronak 2019) and play an important role in wastewater treatment systems (Ouyang and Guo 2016). In addition, shellfish beds and reefs enhance habitat availability, benthic flora and marine organism populations. They act as nursery grounds for fish and other species (including crustacea), and their nutrients support the growth of seagrass and macroalgae (e.g. kelp) (Alleway et al. 2018; Hughes et al. 2018). Restoration of historic baselines in combination with bivalve mariculture can improve ecosystem health while providing a food source and employment (see Box 19.6). Bivalves are increasingly used to extract and convert pollution in the Baltic Sea (Petersen et al. 2020). In New York, the Billion Oysters Project aims to place 1 billion oysters in the harbour to help clean up its water while providing habitat for marine species, shielding shorelines from storm damage and engaging students and the local community (75 restaurants and 70 schools as of 2018) (Charlton 2019).

How These Benefits Can Be Achieved: Short-Term Interventions That Can Be Initiated Now as Part of Stimulus Spending and Recovery Measures

- **Commit public funding to a set number of restoration projects.** Direct public investment to 'shovel ready projects' (based on a set of criteria) through stimulus funding packages. See Box 19.5 for the example in the United States following the 2008–2009 financial crisis and Box 19.6 for an example of the suite of cross-sectoral benefits that can be derived from ecosystem restoration.
- **Establish national funds to mobilise private sector funding for large-scale restoration.** Initial public investment is used to attract impact investors and larger private sources of funding, including from philanthropy. The nature of the fund will need to depend on national circumstances. An example is the trust fund established for the

[46] *In Mobile Bay, Alabama, $3.5 million has been spent on efforts to successfully restore 5.9 km of oyster reefs that have reduced wave height and energy of average waves at the shoreline by 53–91%. The reefs have also produced 6,560 kilograms of seafood per year—a weight equivalent to half the total oysters harvested in Alabama in 2015.*

[47] *Based on the average emissions of a passenger vehicle being 4.6 metric tonnes per year, according to EPA (2018).*

tourist coast of Mexico's Yucatán peninsula. A tourist tax is channelled into the fund to pay for both routine reef maintenance, such as removing debris and replanting species, and bigger repairs after hurricanes.

- **Use debt-for-nature swaps or debt restructures.** Governments could consider including restoration and/or protection of coastal and marine ecosystems under fully or highly protected MPAs or OECMs as part of debt-restructuring negotiations and debt-for- nature swaps (see Box 19.7 for further details on debt- for-nature swaps).
- **Incentivise use of technologies such as remote electronic monitoring, and high-resolution vessel tracking and monitoring systems and collaborative approaches with small-scale fishing fleets to enhance outcomes for marine protected areas and fisheries management.** Increasingly, market considerations are a compelling reason for small-scale fishers to adopt monitoring systems. Gaining access to export markets would improve their incomes and help develop their local economies (INFOFISH International 2020). Governments could consider incentivising the use of remote electronic monitoring (REM) in key fisheries or working on collaborative partnerships to enhance data collection in protected areas (see Sect. 3.4 for additional ideas on conditional grants). REM data enable cross-verification of self-reported data and can confirm vessel compliance with regulations. This approach not only discourages violations because all activities are monitored but also gives legitimacy to self-reported catch. As an example of the potential benefits, providing 10% video review monitoring across the over-10-metre fleet throughout the United Kingdom would cost approximately £5 million. This equates to roughly a quarter of the money spent on more traditional systems, which deliver less than 1% at-sea coverage (WWF 2017). Inshore vessel monitoring systems can be undertaken by using inexpensive cellular 3G/GSM/GPRS networks rather than global satellites (see, e.g., AST 2019).
- **Ensure that the definition of 'infrastructure' includes hybrid green-grey infrastructure.** Ensure that investments targeted at stimulating large-scale coastal infrastructure projects enable the use of hybrid green-grey infrastructure approaches (e.g. the use of nature-based solutions such as living reefs or mangroves in conjunction with traditional concrete or non-living structures). These investments can include regulatory reform, procurement and tender agreements and definitions for bilateral aid. Hybrid solutions combine conservation and restoration of coastal ecosystems with conventional engineering and can offer enhanced levels of coastal protection while also delivering the key co-benefits associated with ecosystems.
- **Invest in blue carbon projects (restoration and conservation of coastal wetlands—mangroves, seagrasses and tidal marshes) and accelerate the associated policy and regulatory reform (inclusion in national GHG inventories, nationally determined contributions and market mechanisms).** Blue carbon projects can bring sustainable carbon financing to the restoration and protection of coastal and marine ecosystems while at the same time contributing directly to a government's international commitment under the Paris Agreement. Carbon financing is also substantially more economically stable than tourism and other income streams. Sites must be carefully selected to meet the accounting requirements under the Paris Agreement, avoiding areas that are likely to be inundated by sea level rise. Blue carbon projects must also be advanced in conjunction with social safeguards to consider demands from local small-scale fishers and other stakeholders who are heavily dependent on coastal resources for economic sustainability (Barbesgaard 2018; Bennett 2018; Friess et al. 2019). Effective local engagement of stakeholders, ensuring their voice is heard, will be key for the success of these initiatives.

Box 19.6 Restoring Shellfish Reefs in Australia and the United States

In Australia, The Nature Conservancy, in partnership with state and Commonwealth governments, has embarked upon a national program to rebuild and restore Australia's lost shellfish reefs. Based on the results of existing pilot projects, scaling efforts to 60 reefs nationally will provide 850 new full-time jobs for local coastal communities, divert 7000 m^2 of shell waste from landfills, reduce coastal erosion and deliver the following annual benefits:

- 375 kg of new fish stocks, including high-value snapper, flathead and whiting
- Filtration of two billion litres of seawater (the equivalent of the annual water use of 21,000 Australians)
- Removal of 225 kg of nutrient pollution (nitrogen and phosphorous) in coastal areas (TNC 2020)

In 2011, the full suite of ecosystem services derived from natural oyster reefs in North America was conservatively estimated to be between US $5500 and $99,000 per hectare per annum, with recovery of their restoration costs in 2–14 years (Grabowski et al. 2012). These services include job creation and economic development, fish production, water filtration, coastal protection and providing habitat for many other marine species. The largest current initiative is the Chesapeake Bay Executive Order, which requires the oyster populations of 20 Chesapeake Bay tributaries to be restored by 2025. Three estuaries have been restored thus far, including 964 acres of restored reef at a projected total cost of $72.1 million[a]. The resulting harvested biomass has the potential to contribute millions of dollars in additional sales for commercial seafood harvesters. This would be in addition to a wide range of other ecosystem services from restoring the reef (such as water purification, nitrogen sequestration and water and biogeochemical cycling), which could help recoup the cost of investment (Knoche and Ihde 2018)[b].

[a] *This project focused on the first three tributaries in Maryland chosen for restoration: Harris Creek, the Little Choptank River and the Tred Avon River. The projected cost for achieving the total restoration acreage target was $72 million; actual costs incurred to this point have been $53 million.*

[b] *Knoche and Ihde* (2018) *used IMPLAN regional economic impact modelling software to calculate the economic effects for four key economic measures (output, labour income, value-added and employment). There are a number of limitations to using ecological and regional impact modelling studies. For example, the ecological model implicitly assumes that catchability is constant and also excludes key ecosystem services from oyster reefs. While the authors did not carry out a benefit-cost analysis per se, based on the estimates calculated and the missing value of the ecosystem services, we ascertain the benefits are likely to outweigh the cost of investment.*

Box 19.7 Debt-for-Nature Swaps to Advance Marine Protected Areas

Since 2008, when Seychelles defaulted on its national debt, the country has since sought ways to preserve its natural environment—the pillar of its economy and of its citizens' livelihoods—without endangering financial stability. In 2015, The Nature Conservancy and its impact investing unit, NatureVest, brokered a deal to restructure a portion of Seychelles' debt with a debt-for-nature swap. The deal allows the government to restructure the country's debt with a mix of investments and grants, in exchange for designating 30% of its exclusive economic zone (EEZ) as a marine protected area. The agreement frees capital streams and directs debt service payments to fund climate change adaptation and marine conservation activities that will improve the management of Seychelles' coastlines, coral reefs and mangroves. This is the first time this financing technique has been used for the marine environment (Thande 2018).

The designation of the 30% of the EEZ took place during the COVID-19 crisis, on 26 March 2020, and demonstrated the continued commitment of Seychelles to marine protection as a core aspect of its long-term strategy for economic sustainability (Statehouse 2020).

In 2018, the Republic of Seychelles complemented its debt restructure though the debt-for-nature swap by establishing the world's first sovereign blue bond. The blue bond was created in partnership with impact investors (private capital) and public multilateral bodies (the World Bank and Global Environment Facility) to finance the necessary shift to sustainable management and governance of fisheries in Seychelles. The beneficiaries of the proceeds of the blue bond will be local communities, civil society organisations and businesses who are seeking financing for activities that can support a transition to sustainable fisheries. The bond was issued with a ceiling value of US $15 million, with a maturity of 10 years. The World Bank provided support through a partial guarantee ($5 million), and the Global Environment Facility provided a concessional loan ($5 million), which will subsidise payment of the bond coupons. These credit enhancement instruments allowed for a reduction of the price of the bond by partially de-risking the investment of the impact investors, and by reducing the effective interest rate of 6.5% for Seychelles to 2.8% by subsidising the coupons (World Bank 2018).

Despite significant changes to national budgets and revenues as a result of the impacts suffered from COVID-19, the sovereign blue bond has continued to fund recovery efforts and economic diversification initiatives across Seychelles to aid in recovery efforts. This includes over $700,000 in grants for ocean conservation and management and $12 million to fund research and development for new economic opportunities.

Seychelles is also undertaking extensive mapping of its seagrass ecosystems, aiming to map the entire EEZ to enable inclusion of these ecosystems and the associated adaptation blue carbon benefits for inclusion in its nationally determined contribution (NDC) under the Paris Agreement to be submitted this year and a commitment towards integrating carbon accounting for the blue carbon ecosystems in the NDCs ahead using the Wetlands Supplement of the Intergovernmental Panel on Climate Change.

3.2.2 Two: Invest in Sewerage and Wastewater Infrastructure for Coastal Communities

Wastewater and sewage runoff into coastal waters (resulting in eutrophication and hypoxia) is a major contributor to human health issues, spreads water-borne diseases among coastal communities, contributes to the loss of local fish stocks (and therefore contributes to local food insecurity and loss of revenue for small-scale fishers), furthers the decline of coral (and therefore undermines opportunities for ecotourism) and results in costly beach closures for coastal communities and tourism (WWAP 2017)[48].

More than 80% of global wastewater flows are released without adequate treatment, with this figure as high as 95% in some least developed countries (ILO 2017). Much of this runoff comes from agricultural sources, where inefficient use of fertiliser and inadequate wastewater treatment leads to nitrogen and phosphorous loading in waterways and groundwater. Excess nitrogen and phosphorus often lead to eutrophication, harmful algal blooms and ocean hypoxia (UNEP et al. 2012). Even where treatment facilities exist, they may sometimes discharge untreated sewage into waterways and the ocean due to decayed infrastructure, facility malfunctions or heavy rainfall events that overwhelm systems using combined sewers and stormwater drains (Jambeck et al. 2020; Malik et al. 2015).

SUPPORTING RECOVERY AND RESILIENCE IN THESE SECTORS:
Tourism, Fisheries, Infrastructure, Marine Conservation, Health and Water

SUPPORTING ACHIEVEMENT OF THESE SDGs:

Over the last 30 years, wastewater and sewage runoff has cost the global economy an estimated $200 billion to $800 billion per year (UNDP 2012).

The estimated rates of return on water and sanitation investments are striking, with every $1 invested in water, sanitation and hygiene having a potential return of $3–34, depending on the region and technology involved (Hutton et al. 2004).

In the face of ever-growing demand for water, wastewater is increasingly seen as a reliable alternative source of water, shifting the paradigm of wastewater management from 'treatment and disposal' to 'reuse, recycle and resource recovery' and offering even greater benefits. In the context of a circular economy, whereby economic development is balanced with the protection of natural resources and environmental sustainability, wastewater represents a widely available and valuable resource (WWAP 2017).

Why Investment Makes Sense

The development of the infrastructure for sewage and wastewater treatment and reuse can offer immediate job opportunities for local communities in coastal areas. Analysis of stimulus packages in Latin America from the 2008–2009 financial crisis aimed at investment in public works found that investing $1 billion in water supply and sanitation network expansion could result in the creation of up to 100,000 direct jobs annually (significantly higher than the same investment in coal- powered energy) (Schwartz et al. 2009)[49]. In the United States, investments in sustainable water practices are estimated to generate between 10 and 15 direct, indirect and induced jobs per $1 million invested in alternative water supplies; between 5 and 20 direct, indirect and induced jobs per $1 million invested in stormwater management; between 12 and 22 direct, indirect and induced jobs per $1 million invested in urban conservation and efficiency; and between 10 and 72 direct, indirect and induced jobs per $1 million invested in restoration and remediation (Pacific Institute 2013). Investing in green infrastructure, such as riparian buffers to address agricultural runoff, could also be a cost-efficient alternative to typical grey infrastructure. When compared to the creation of a new nitrate-removal facility, the planting of a riparian buffer offered a cost savings of up to $29 million (Canning and Stillwell 2018). Reforms and incentives promoting recovery and reuse of wastewater (such

[48] *Bacteria use up oxygen in the water as they decompose the organic material in the wastewater, and the resulting lack of oxygen in the water kills the fish. The solids in sewage cause the water to appear dark and murky, which also affects the ability of fish to breathe and see around them.*

[49] *Note that Schwartz et al.'s (2009) study looks across multiple countries and projects aimed at water and sanitation. The figures provided in this report were for Columbia's expansion of its water supply and sanitation network. For the full details, including figures for other countries and types of investment, see Table 2 in Schwartz et al. (2009). The investment includes both water and sewage treatment. The direct employment- generation potential of an investment is thus highly sensitive to assumptions about wages, the division between skilled and unskilled workers, the sectoral allocation of the proposed program, the technology to be employed in each project and the potential crowding-out or substitution effects. Indirect job estimates are also highly sensitive to leakage created from the division between locally produced and imported inputs.*

as retrofitting homes and apartment buildings for composting, collection and reuse of human waste as fertiliser) are typically much more labour-intensive than current/traditional 'linear' municipal wastewater collection, treatment and disposal systems, leading to net job creation in both the private and public sectors. For example, as a result of concerted policy and investment, Israel now reuses 80% of its wastewater for agricultural production. This has led to a fivefold increase in the export of water technology, leading to a $2 billion industry between 2008 and 2013 (Hudson 2017).

Investment in sewage and wastewater treatment and reuse can avoid long-term costs (in terms of loss of biodiversity, tourism revenues and wider recreational benefits) as a result untreated wastewater being discharged into coastal waters. The longer-term economic benefits of investment in waste and sewerage infrastructure are twofold. First, clean coastal waters will bring economic benefits to communities and businesses that rely on tourism revenue. Cleaner waters and healthier coastal ecosystems offer additional opportunities for ecotourism and revenue-generating activities. Second, such investment avoids the economic loss suffered through inaction. The degradation of coral reefs due to pollution and overfishing caused the Caribbean to lose $95 million to $140 million per year in net revenue from coral reef–associated fisheries, $100 million to $300 million per year in reduced tourism revenue and $140 million to $420 million per year in reduced coastal protection (Burke et al. 2011)[50]. On a more local scale are the economic losses suffered by coastal business and tourism ventures from beach closures as a result of pollution. Furthermore, the integration of green infrastructure with traditional grey infrastructure for the recovery and reuse of wastewater has been shown to offer significant improvements and long-term economic savings for local authorities. In 2007, the city of Portland, Oregon, introduced a program to spur the use of green infrastructure for urban stormwater management. As a result, service providers installed permeable pavements and bioswales throughout the city, reducing peak flow by 80–94% in target areas. Estimates indicate the initial $9 million investment in green infrastructure has yielded a savings of $224 million in stormwater costs related to repairs and maintenance (EPA 2010). A review of the U.S. water and wastewater infrastructure estimated that meeting the nation's projected needs would require an additional investment of $82 billion per year for the next 10 years, but the review also found that this investment would result in over $220 billion in total annual economic activity, approximately 1.3 million jobs and productivity savings for U.S. businesses of approximately $94 billion a year[51] (Value of Water Campaign 2017).

Proper wastewater treatment and reuse facilities and sewerage infrastructure will improve the health of the local community, prevent future water-borne diseases, increase water security and reduce inequalities. Improved waste management has direct gender and social equity implications, and addressing this issue would also lead to improved social equity outcomes in associated communities (Satterthwaite et al. 2019). Targeted water investments may contribute to reaching growth and poverty alleviation goals more effectively (UN Water 2016). Globally, unsafe sanitation costs an estimated $223 billion a year in the form of high health costs and lost productivity and wages (WHO 2012). Investment in safe drinking water and basic sanitation could offer estimated economic returns of $3–34[52] globally for every $1 invested, with an overall estimated gain of 1.5% in global GDP (Hutton et al. 2004). These returns include both health benefits (such economic benefits from reduction in water-borne diseases) and non-health benefits (such as time savings associated with better access). Investment in small- scale projects providing access to safe water and basic sanitation in Africa could offer an estimated economic return of about $28.4 billion a year, or nearly 5% of the continent's GDP (UNESCO 2009). Improving employment is a good economic outcome; sound health and social equity outcomes are also important enabling conditions for resilient communities.

A reduction of untreated wastewater being discharged into coastal waters will improve local water quality and reduce stressors on coral reefs and coastal ecosystems, and reuse can offer climate-mitigation benefits.

Reducing the nutrient runoff will reduce a significant stressor on coral reefs and shellfish (especially bivalves that filter large quantities of water) resulting in improved and

[50] *The loss of economic value from degradation of reef goes beyond the estimated tourism revenue, as it includes both use value (e.g. recreational fishing, surfing or beach-going) and non-use values. Non-use value includes the value of preserving the ecosystem for future use either by an individual (option value) or by future generations (bequest values). In addition, there is existence value, which is unrelated to the use of the resource and represents the willingness to pay for the resource to exist (e.g. willingness to pay for the protection of a beach you will never visit). Non-use value is often difficult to quantify, and hence the economic losses tend to be larger than the market values estimated.*

[51] *If the water infrastructure gap is not addressed, businesses would face higher costs to procure water and wastewater services. These costs include operational and maintenance costs, higher water rates, costs of self-supply or costs of relocating to a better-served area.*

[52] *Returns are dependent on the region and technology used (Hutton et al. 2004). The benefits also refer to improving the quality of groundwater (which we use as a proxy). The estimates refer to the following intervention: halving the proportion of people who do not have access to improved water sources and improved sanitation facilities by 2015. 'Improved' water supply involved better access and protected water sources (e.g. stand post, borehole, protected spring or well, or collected rainwater). Improvement does not mean that the water is safe, but it is more accessible, and some measures are taken to protect the water source from contamination. 'Improved' sanitation, generally involving better access and safer disposal of excreta (septic tank, pour-flush, simple pit latrine, small bore sewer or ventilated improved pit latrine).*

more resilient coastal and marine ecosystems and improved local water quality. Energy from wastewater and sewage treatment can be recovered in the form of biogas, heating and cooling, and electricity generation. Technologies exist for on-site energy recovery through sludge and biosolids treatment processes integrated into wastewater treatment plants, allowing them to transition from major energy consumers to energy neutrality, or even to net energy producers. Energy recovery can also help facilities reduce operational costs and their carbon footprint, enabling increased revenue streams through carbon credits and carbon-trading programmes (WWAP 2017).

How These Benefits Can Be Achieved: Short-Term Interventions That Can Be Initiated Now as Part of Stimulus Spending and Recovery Measures

- **Commit public funding for decentralised, low-cost solutions and safe water reuse options in coastal areas.** Large-scale centralised wastewater treatment systems may no longer be the most viable option for urban water management in many countries. Decentralised wastewater treatment systems, serving individual or small groups of properties, allow for the recovery of nutrients and energy, save freshwater and help secure access to water in times of scarcity. It has been estimated that the investment costs for these treatment facilities represent only 20–50% of conventional treatment plants, with even lower operation and maintenance costs (in the range of 5–25% of those of conventional activated sludge treatment plants) (WWAP 2017).
- **Commit public funding for the development of services which can collect and transport sanitation waste for safe treatment.** This is often one of the main barriers to effective sanitation and can be a source of decent jobs for local and regional communities.
- **Establish a sustainable financing mechanism (e.g. a dedicated national fund) for sanitation.** A major barrier to improved and accessible sanitation facilities is low levels of public investment in the sanitation sector. The creation of an enabling framework and dedicated fund can attract both public and private sector funding and investment for resource mobilisation and guarantee the necessary funds at a national level for investment in the sector.
- **Incentivise management strategies such as implementing riparian buffers or reducing inefficient fertiliser use to reduce nutrient pollution.** Ecosystems can effectively provide economical wastewater treatment services, as long as these ecosystems are healthy, the pollutant load (and types of contaminants) in the effluent is regulated and the ecosystem's pollution assimilation capacity is not exceeded (WWAP 2017).

3.2.3 Three: Invest in Sustainable Community-Led Non-Fed Mariculture

Given the changing nature of the fisheries industry in a post-COVID-19 world and the increasing importance of ensuring local food security and economic diversification, investment in community-led non-fed marine aquaculture (mariculture) (e.g. shellfish and seaweed farming)[53] offers considerable opportunities. Non-fed mariculture has the greatest potential to contribute to food supply and make the global food system more resilient (Costello et al. 2019; SAPEA 2017; Duarte et al. 2009). Such mariculture requires no feed, fertiliser inputs, insecticides or antibiotics, and it requires less water and energy than fed aquaculture, making it a self-supporting system (Roberts et al. 2015; Suplicy 2018). The development of sustainable community-led mariculture could also provide local employment and strong ecosystem services in countries with climate-driven declines in capture fisheries (Costello et al. 2019).

Potentially 48 million km^2 of the world's ocean is suitable (based on nutrient availability and temperature) for seaweed cultivation[54]. These waters span 132 countries, of which only 37 are currently cultivating (Froehlich et al. 2019). In terms of bivalve production, Gentry et al. (2017) found that over 1.5 million km^2 (roughly the area of Mongolia or Iran) of marine habitat, spanning temperate and tropical regions, are suitable for bivalve production (e.g. oysters, mussels, clams) and that developing small suitable areas can result in high production volume (e.g. they found that developing just 1% of Indonesia's suitable area could produce over 3.9 billion individual bivalves).

Investment in sustainable community-led mariculture could protect and develop mariculture with the triple goal of producing high-quality protein, accelerating a shift towards sustainable food systems, and maintaining and restoring ocean ecosystem services.

[53] *Non-fed mariculture is for species that do not require human-derived feed inputs and instead extract resources from the surrounding environment (e.g. phytoplankton), primarily macroalgae and bivalves (e.g. oysters, mussels and scallops).*

[54] *We are not suggesting that all 48 million km2 be developed, as this would amount to large-scale cultivation that would not be compatible with a community-led approach and would likely result in unintended consequences through the disruption of coastal ecosystems and their functioning. We provide the area figure to show that potential is not limited to one region or a small group of countries.*

SUPPORTING RECOVERY AND RESILIENCE IN THESE SECTORS:
Fisheries and Aquaculture

SUPPORTING ACHIEVEMENT OF THESE SDGs:

Note that for some countries investment in developing sustainable feed alternatives for fed mariculture (e.g. finfish) might be a priority over investment in developing community-based non-fed mariculture (e.g. those countries that have very advanced fed mariculture industries, such as Norway and Chile). Important technological, nutritional and economic constraints remain to feed substitution, and many substitutes being explored are currently too expensive to incorporate in large-scale production (Naylor et al. 2009). As such, this has not been considered a priority applicable to multiple regions and economies to respond to the current economic crisis. Benefits associated with investment in research and development for alternative feed are explored in Annex 1.

Why Investment Makes Sense

Community-led non-fed mariculture creates jobs for local communities and requires comparatively less initial investment than larger-scale commercial mariculture. The potential for job creation is significant, predominantly in developing and emerging economies, with a focus on economic opportunity for women (see Box 19.8). In Indonesia, women play a significant role in seaweed farming, resulting in some women becoming the main household earner despite previously earning little income (Neish 2013). Women relatives of seaweed farmers were also found to be instrumental in tying seed (Valderrama et al. 2013). Seaweed farmers were shown between 2007 and 2009 to make up to $5000 per year, a 33% higher income than the national average ($3603) (Neish 2013). As of 2019, women made up 57% of the communities engaged in mabé pearl farming in Fiji, with sales ranging from F$735 to F$2200 (US $346–1038) per crop (Southgate et al. 2019).

Community-led non-fed mariculture can support long-term economic diversification for local communities. In addition to the direct benefits for local communities, seaweed mariculture offers a sustainable and low-carbon alternative for products such as biofuels (Jiang et al. 2016)[55] aquaculture and agriculture feedstocks, and plastic (Önen Cinar et al. 2020). The estimated value of micro-algae oil for people and animals from 500 million metric tonnes of seaweed is $23 billion (Bjerregaard et al. 2016). Extrapolating an estimate of 1 job per 10 dry tonnes of seaweed results in a potential direct employment of 50 million jobs; a standard seafood industry secondary-employment multiplier of 2:1 suggests 100 million jobs could be created overall (based on an estimate of 1 job created per 10 dry metric tonnes), roughly the number currently employed in marine capture fisheries (Bjerregaard et al. 2016)[56]. Bivalve mariculture offers significant opportunities for the creation of a green and circular local economy. Goods from provisioning services include meat, worth an estimated $23.9 billion as well as pearls, shell and poultry grit, with oyster shell being the most important, with a global potential worth of $5.2 billion (Olivier et al. 2020). Shells can be used as construction material, fertiliser, poultry grit and artistic products. Research on the potential of bivalves as medicinal and genetic resources is on the rise, looking at their bioactive peptides, proteins and metabolites for producing innovative pharmaceuticals and nutraceutical foods. Mussel byssus—highly resistant fibre that combines high extensibility and harness and is the only effective glue underwater—has particularly interesting potential applications in engineering, biological and biomedical fields, including in water-resistant adhesives, replacement of surgical sutures, bone protheses and fibre optics (Zhang et al. 2020; Guo et al. 2020).

The opportunity for community-led mariculture supports improved rural livelihoods, particularly for women, as well as cultural services for coastal communities. The expansion of seaweed farming in several continents is contributing to global food security, supporting rural livelihoods and alleviating poverty (Cottier-Cook et al. 2016). Some fast-growing species can be cultivated year-round, and yield per unit area can surpass that of terrestrial crops (Forster and

---

[55] *Marine algal biofuel is considered a promising solution for energy and environmental challenges. Macroalgal biomass has the potential for bypassing the shortcoming of first and second generation of biomass from food crop and lignocellulosic sources.*

[56] *Note that the micro-algae used as a replacement for fish oil are more likely to be cultivated in tanks in deserts with unlimited sun. All the recent big investments in fish oil substitutes have been in these kind of micro-algae, not ocean-grown macro-algae, where the promising segments are more those used for food, animal feed, fertilisers (biostimulants) and bioplastics.*

Radulovich 2015). Bivalve farming also provides many cultural services for communities and visitors, including links with the marine environment, a strong connection with cultural heritage and educational centres on ecosystems (Alleway et al. 2018; McLeod and McLeod 2019). A global assessment values the global, non-food bivalve mariculture services, including cultural services, at up to $6.47 billion per year—a figure recognised as an underestimate given existing data gaps (Olivier et al. 2020).

Increased community-led mariculture offers opportunities for GHG emissions reduction through the use of seaweed for alternative feed and fuel and promotion of oysters and mussels as a low-carbon alternative protein. Projections of annual global GHG emissions reductions from seaweed farming are between 0.05–0.29 $GtCO_2e$/year by 2050. This would be equivalent to taking approximately 1–6 million vehicles off the road every year[57]. However, there are uncertainties in rates of expansion of the industry and the proportion of production that would be sequestered (Hoegh-Guldberg et al. 2019). It is estimated that seaweed could create a carbon-neutral mariculture sector with just 14% of current seaweed production, with seaweed culturing at a regional level more feasible from a cost perspective, especially in areas with strong climate policy, such as California (Froehlich et al. 2019)[58]. The addition of seaweeds to animal feed to reduce enteric methane emissions from ruminants may substantially increase the mitigation potential of seaweeds (Kinley et al. 2016). Small-scale community seaweed farming projects are considered low-risk, but significant expansion would require a more complete understanding of how risks and benefits change as projects are scaled (Campbell et al. 2020), in addition to any potential trade-offs with other ecosystem values and uses. If not appropriately located, seaweed farms could also affect seagrass beds and other benthic habitats and thereby disturb the local ecology (Eklöf et al. 2005). Spatial planning, ongoing monitoring and proper management are key to mitigating these impacts and informing design of a system that promotes resilience, local empowerment and long-term conservation of marine and coastal ecosystems.

Bivalves contribute to the carbon cycle, serving as a carbon sink as their shells develop. In France, 250,000 metric tonnes of farmed shellfish (mainly oysters and mussels) sequester 9.2 metric tonnes of carbon each year, as much sequestration as is done by half of the Landes, the largest forest in the country (CNC n.d.). This benefit is not offset by carbon emissions associated with production, which remain low. Studies found that mussel farming has one of the lowest carbon footprints of any food production system, and may in fact have the lowest. It probably offers the best ratio of protein quality and climate and ecosystem benefits (SARF 2011; Suplicy 2018). Bivalve production could significantly contribute to promote low-carbon food systems and reduce meat production. A plate of mussels (approximately 500 grams in weight, which includes 150 grams of flesh) provides as much protein as two eggs and more iron than a piece of red meat while offering calcium, magnesium and daily needs in iodine (CNC n.d.). This comes with a very low environmental footprint compared to meat production (most comparisons look at beef and chicken production) and fisheries, in terms of carbon emissions, water use and non-renewable energy consumption (Alleway et al. 2018; Hughes et al. 2018; McLeod and McLeod 2019). In addition, bivalves function in a variety of ecosystems, such as estuaries, lagoons and coastal oceanic systems, while providing a multitude of services. As captured in Fig. 19.4, these include habitat and supporting, provisioning, regulating and cultural services. As filter feeders, bivalves purify water (up to 180 L—50 gallons—of water a day for an adult oyster, 25–30 L for a mussel) while treating waste (including hydrocarbons). This function enhances water clarity and helps control excessive phytoplankton blooms (Bricker et al. 2018; Alleway et al. 2018; Ferreira et al. 2018; Hughes et al. 2018; McLeod and McLeod 2019).

How These Benefits Can Be Achieved: Short-Term Interventions That Can Be Initiated Now as Part of Stimulus Spending and Recovery Measures

- **Feasibility studies and associated zoning (ideally guided by an integrated ocean management or marine spatial planning process)**. Spatial planning approaches in which biotic, abiotic and socioeconomic factors are considered could be used to identify where the positive effects of mariculture could be maximised (Alleway et al. 2018). This initial scoping work can also be a source of short-term job creation for local universities and scientists.

[57] *Based on the average emissions of a passenger vehicle being 4.6 metric tonnes per year, according to EPA (2018).*

[58] *Research has found that some fundamental and very significant hurdles remain to realising the potential contributions of seaweed cultivation at a global level. For example, the value of seaweed biomass needs to be improved, and the ecosystem services that seaweed farming can provide (such as in reducing coastal nutrient loads) need to be more fully considered. Additional considerations are environmental risks associated with climate change, pathogens, epibionts and grazers, as well as the preservation of the genetic diversity of cultivated seaweeds (Buschmann et al. 2017).*

Source: Alleway et al. (2018).

Fig. 19.4 Goods and services provided by shellfish mariculture

- **Streamlined and centralised permitting and regulatory processes.** The purpose of streamlined permitting is not to cut corners or skip necessary environmental impact assessments for new projects but rather to ensure that local communities and applicants can access and easily navigate the government process. Otherwise this process can be a significant barrier to communities' ability to initiate projects (even with funding).
- **Government grants and loans for new seaweed and/or bivalve farmers (including microloans).** The high upfront costs that these production systems involve represent a barrier for community- led projects in many countries (see Box 19.8 for an exploration of how the Kenyan government has helped stimulate the creation of community-led maricultural in partnership with the World Bank).
- Investment in communities of practice across different regions. With relatively small upfront investment, the capacity of small-scale and community-led initiatives can be accelerated by establishing regional communities of practice to share knowledge, experiences and best practice across the industry.
- **Creation of capacity-building and training programs for local communities.** These programs and opportunities could be prioritised for those communities most affected by reduced economic opportunities from tourism and lower demand from fisheries. See also Sect. 3.3 for recommendations on investment in research and development and skills-training programs for sustainable ocean industries.
- **Facilitation of cooperative and co-designed sites across multiple sectors and with the private sector.** Co-

designed initiatives could support development across a multitude of sectors (e.g. energy, transport, communication), to co-produce ecosystem services to support the needs and interests of multiple stakeholders (Outeiro et al. 2017). For example, offshore wind farms could provide a platform to which mariculture facilities could be attached, the operational costs of which might otherwise be prohibitive or the space and location required contested (Buck et al. 2018).

Box 19.8 Scaling Community Seaweed Farming in Kenya

Kenya started community seaweed farming in Kwale County on the South Coast in 2013, following feasibility studies undertaken by the Kenya Marine and Fisheries Research Institute. The initial funding was from a World Bank–funded project that targeted fishing communities along the Kenyan coast, but further funding has been provided by the Government of Kenya to build the farmers' capacity with the aim of developing the initiative into a robust industry to create jobs and income. The main objective of supporting the establishment of this new community-led industry was to offer an alternative livelihood to local fishing communities whose livelihoods had been challenged by reduced income due to the dwindling catches from artisanal fisheries. Importantly, it was also an intervention that specifically supported the creation of new jobs and economic opportunities for women—90% of seaweed farmers in Kenya are women.

To date, this support has resulted in the employment of approximately 400 seaweed farmers in Kwale County, each with his or her own individual farm generating income that flows directly to the farmer. For the women involved, this has meant financial independence from their husbands, with many using the income from their seaweed farms to educate their children up to the university level and constructing permanent houses.

The sale price of the dry seaweed is agreed upon with the buyers prior to the transactions, and plans are underway to have a contract between the farmers and the buyers. The seaweed farmers welfare group has also been registered as a cooperative to improve organisation and collective bargaining power. The Government of Kenya has provided additional support to the farmers to ensure effective post- harvest management, provision of farming implements, harvesting and storage facilities, value addition and marketing. The seaweed is also being used in local food products. Support is also being provided to diversify the farmers' income base through the development of soap and other cosmetic products, such as body creams, shampoos and hair treatment. To date, community-led seaweed farming has generated over 300 metric tonnes of dry seaweed that has generated over US $60,000 for the local village economies.

Some of the challenges faced in developing the initiative into a commercial entity include raising the level of production to volumes that make business sense to the potential investors and traders, particularly owing to the fact that the activity is a nontraditional economic activity, new to the farming communities. The difficulty of finding a reliable market for the produce, without economically feasible production volumes, affected the ability to reach scale. Extensive training of the communities has gotten more committed farmers and thus increased production volumes. The other major challenge has been extreme weather patterns, including very high temperatures followed by very heavy and extended rainfall, which resulted in massive die-off of seaweeds. This near complete loss of seaweed seed has been resolved by establishing new nurseries at the start of the favourable season (southeast monsoon) by bringing in seaweed from more sheltered sites.

As a result of the demonstrated socioeconomic benefits of community seaweed farming in Kwale County, and the engagement of a commercial seaweed buyer, Kenya is now looking to scale the industry along the South Coast and ultimately the rest of the coast. For Kenya, the immediate socioeconomic impact of investment in community seaweed farming makes it a priority intervention for economic recovery, as its relatively low investment, quick returns and broader social and environmental benefits make its uptake and scalability more feasible than other interventions.

Seaweed farming can be approached as integrated multi- trophic mariculture. Incorporating cages, bivalves and sea cucumbers optimises the productivity of a unit area of sea space and creates more employment. Additionally, seaweed helps clean coastal waters of excess nutrients that have been introduced through pollution and wastewater, making it the ideal crop for environmental sustainability.

The Government of Kenya is currently supporting the selection of further suitable sites and associated environmental impact assessment to scale the initiative.

Source: Information provided by the Government of Kenya, 2020.

3.2.4 Four: Incentivise Zero-Emission Marine Transport

Global supply chains rely on marine transport to move approximately 90% of global trade. Regional and intercontinental shipping constitutes the core of the global logistical system. At any given time 50,000 vessels and 1.2 million seafarers are in operation between ports in different parts of the world. Marine transport is also the mode of long-distance transport with the lowest carbon footprint and cost (WSC 2020).

The sustainability and viability of this industry is critical for ensuring the resilience of global populations to future shocks. During COVID-19 shipping has been responsible for transporting essential goods and services globally, from PPE to the core elements needed for the production of vaccines. In terms of domestic marine transport, it has been the only form of transport for food, health provisions and basic essentials between islands and atolls.

Despite its central role in ensuring that global supply lines remain open, the industry has faced a significant contraction (estimates of between 25 and 35% by the end of the year) (NSA 2020a) as global trade has dropped. Recovery offers an opportunity to scale investment in the future of this industry through supporting and incentivising industry to invest in the decarbonisation of its fleets. The average lifespan of a cargo vessel is 25–30 years. To enable these vessels to be aligned with the Paris Agreement requires upfront investment over the next few years to keep high-emitting ships and vessels from becoming stranded assets.

Marine transport is not limited to deep-water vessels and cargo shipping, however. Domestic fleets, including fishing and mariculture fleets, vessels that form national navies and coastal passenger transport make up large proportions of a country's transport footprint. Marine transport used in the tourism industry (cruise ships as well as coastal passenger fleets associated with hotels and resorts) stand to gain from early investment in their sustainability and decarbonisation.

An ancillary effect of the global contraction is an expected increase in vessel recycling, particularly for offshore and passenger ships (NSA 2020a). This provides the opportunity for government investment to not only support and incentivise investment in replacement fleets and retrofitting but also ensure environmentally sound and sustainable ship-recycling practices.

Regarding the economic, social and environmental net benefits, analysis shows that investments to decarbonise the international maritime shipping sector could deliver a net discounted benefit (average) over 30 years (2020–2050) of $1.2 trillion to $9 trillion (Konar and Ding 2020), with a benefit-cost ratio of 2:1 and 5:1 in 2050[59]. Similar figures are not yet available for domestic fleets.

Why Investment Makes Sense

Investment in the shipping industry to support efforts to retrofit or replace high-emitting vessels with low- or zero-emission vessels will protect jobs in the short term. Due to the contraction of the industry, market demand for new vessels is likely to decrease, threatening existing jobs. Government investment at this time would protect jobs and enable upskilling to support new zero-emission technologies. Supporting the replacement of domestic vessels with zero-carbon alternatives can create sustainable jobs, both by reducing domestic emissions and by preparing shipyards for future demand for zero-emission deep-sea vessels once demand picks up after COVID-19.

Investment now will yield long-term benefits for the industry as well as other sectors, including tourism, that rely on marine transport. Zero-emission coastal transport (e.g. passenger and car ferries) can be more cost-efficient to run than its high-emitting counterparts (European Commission 2018). Shifting the demand from oil to alternative fuels and battery propulsion can be a catalyst to scale the deployment of low-carbon fuels for the broader energy transition and unlocks the market for these fuels across a range of industries and other hard- to-abate sectors (Moore 2019). This is due to shipping's high level of fuel consumption, currently estimated to be around 250 million to 300 million metric tonnes every year, approximately 4% of the global oil demand (Christensen 2020). Decarbonising the shipping sector will increase confidence among suppliers of future fuels (e.g. hydrogen and ammonia) and offers opportunities for synergies with efforts to accelerate and scale the establishment of ocean-based renewable energy (see the preceding section). Annex 1 describes specific additional interventions that can target the establishment of these industries for alternative fuel generation.

Decarbonisation of marine transport, both international and domestic, offers significant health benefits for those on board the vessel as well as coastal communities and those living near or working at the port. Prior to cleaner ship fuels, ship-related health impacts included around 400,000 premature deaths from lung cancer and cardiovascular disease and around 14 million childhood asthma cases annually. Reduced PM2.5 from marine engine combustion mitigates ship-related premature mortality and morbidity (Sofiev et al. 2018). Based on this, analysis estimates the discounted cumulative health benefits from reducing emissions from marine transport to be $1.3 trillion to $9.8 trillion over 30 years (2020–2050) (Konar and Ding 2020).

[59] *The analysis excludes military and fishing vessels and domestic transport and includes bulk carriers, oil tankers and container ships, which account for the majority (55%) of emissions in the shipping sector (Olmer et al. 2017).*

SUPPORTING RECOVERY AND RESILIENCE IN THESE SECTORS:
Shipping, Energy, Transport and Tourism

SUPPORTING ACHIEVEMENT OF THESE SDGs:

Reducing GHG emissions from shipping vessels will help mitigate ocean acidification and contribute to domestic and global efforts to reduce GHG emissions. Ocean-based transportation could reduce operational net GHG emissions roughly 100% by changing the way it stores and consumes energy on board (e.g. use of batteries and zero-emission fuels such as hydrogen and ammonia). If the full suite of available technologies is employed, and zero-emission vessels are available for commercial use by 2030, global GHG emissions could be reduced by between 0.9 and 1.8 $GtCO_2e$/year in 2050 (Hoegh-Guldberg et al. 2019). This would be equivalent to taking 19–39 million cars off the road every year[60]. In terms of environmental benefits, the strong acids formed from shipping emissions can produce seasonal 'hotspots' of ocean acidification in ocean areas close to busy shipping lanes. Hotspots have negative effects on local marine ecology and commercially farmed seafood species (Hassellöv et al. 2013). Lastly, the shift to zero-emission vessels could reduce the noise impact on marine mammals. The effects of underwater noise from anthropogenic activities, including ships, on marine mammals includes behavioural responses, acoustic interference (i.e. masking), temporary or permanent shifts in hearing threshold, and stress (Erbe et al. 2019). Studies have shown that periods with a significant reduction in noise from ship traffic have been associated with a reduction in the stress of whale populations (Rolland et al. 2012). Moving to zero-emission vessels such as fuel cell and battery-powered could eliminate noise pollution (Reddy et al. 2019). Research also shows that this shift could be coupled with a 20% reduction in speeds, which would reduce underwater noise pollution by 66%, the chance of a fatal collision between a ship and a whale by 78% and CO_2 emissions by 24% (Seas at Risk 2019).

[60] *Based on the average emissions of a passenger vehicle being 4.6 metric tonnes per year, according to EPA (2018).*

How These Benefits Can Be Achieved: Short-Term Interventions That Can Be Initiated Now as Part of Stimulus Spending and Recovery Measures

- **Incentivise investment in upgrading coastal passenger transport (ferries) to zero-emission (battery- or hydrogen-powered) through subsidies, taxes and grants to the private sector**. Investing in coastal passenger transport offers immediate health benefits for coastal communities and new opportunities to stimulate ecotourism. It also improves the resilience of coastal communities that depend on these forms of transport (e.g. between islands and atolls).
- **Commit to use domestic fleets to pilot and test zero-emission fuels and technologies, which in turn can help to de-risk and reduce costs for larger, high-seas and ocean-based transportation**. Domestic fleets are populated with smaller ships and therefore better suited to small-scale and short-term pilots and tests. For many countries, the largest marine fleets are those of their navies, offering significant opportunities for domestic leadership and long-term economic resilience and benefits from early investment.
- **Incentivise private sector investment in replacement fleets and retrofitting by offering subsidies, tax cuts and government loans**. Support for the industry (both the shipping and tourism sectors) at this time can take the form of incentives for replacement and/or retrofitting (as appropriate given the nature of the vessel and availability of technology). Note that incentives should be targeted at incentivising zero-emission vessels and not low-carbon ones (e.g. running on liquefied natural gas), since the latter do not have long-term viability for the industry transition and would therefore be only a short-term investment requiring further investment in the future to facilitate the transition to hydrogen or ammonia.
- **As part of stimulus funding packages for infrastructure, allocate public investment to the development of low- and zero-carbon energy production capacities, and storage and refuelling infrastructure in ports and harbours**. Land-based measures will be critical to support the transition for marine transportation and ensure that a clear signal is sent to the private sector.
- **Invest in land-side grid infrastructure**. Lack of investment in land-based infrastructure to support zero-emission vessels is a common barrier. An example from Norway is a hybrid ferry operating between Norway and Sweden that was only able to operate at half its potential because the grid connection in Sweden was insufficient to recharge the batteries on the ferry.
- **Use bilateral aid to support regional partnerships, particularly in support of small island developing states (SIDS) and least developed countries (LDCs)**

with significant domestic or regional shipping- decarbonisation challenges, to work together on joint objectives. An example is the Pacific Blue Shipping Partnership, a joint initiative among Pacific nations and led by the Governments of Fiji and the Republic of the Marshall Islands. The partnership commits to zero-carbon domestic marine transport by 2050, with a 40% reduction by 2030 (MCST n.d.).

- **Require or establish environmentally sound and sustainable ship-recycling practices that provide decent jobs for local communities.** Ship recycling offers the most environmentally sustainable way of disposing of old vessels, with virtually every part of the hull and machine complex being reused or recycled as scrap metal. To do this properly, ships should be recycled at dry-dock ship-recycling facilities—not beached or exported to countries with weak regulatory systems. The nexus of ship-recycling yards, refurbishing shops, re-rolling mills, steel mills and second-hand shops creates a localised industry which can employ hundreds of thousands of people in semi-skilled and unskilled jobs.

3.2.5 Five: Incentivise Sustainable Ocean-Based Renewable Energy

What Investment Will Achieve

According to the International Energy Agency (IEA 2019a, b), global offshore wind power capacity is set to increase 15-fold over the next two decades, turning it into a $1 trillion business. Only using near-shore sites could supply more than the total amount of electricity consumed worldwide today[61], and moving further offshore into deeper waters (e.g. using floating turbines) could unlock enough potential to meet the world's total electricity demand 11 times over by 2040[62]. By 2050, the IEA forecasts that offshore wind could reach more than 1000 gigawatts (GW) of installed capacity. Expansion of offshore wind in line with these projections would put the global power sector on track for full decarbonisation and enable the production of zero-emission fuels (e.g. hydrogen and ammonia) to dramatically cut emissions from sectors such as shipping (IEA 2019a, b)[63]. Although less advanced, other forms of ocean-based renewable energy, such as tidal, wave, sea current and ocean thermal energy conversion, will be highly valuable for many geographies that lack the geophysical requirements to support offshore wind.

Stimulus funding could help fast-track private investment, resulting in job creation in the short term as well as long-term economic growth opportunities.

SUPPORTING RECOVERY AND RESILIENCE IN THESE SECTORS:
Energy, Shipping and Transport

SUPPORTING ACHIEVEMENT OF THESE SDGs:

However, such fast-tracking must not be done at the expense of the marine environment or lead to the use of shortcuts to environmental impact assessments.

Analysis shows that on average there is a net positive benefit from expanding the sector. The net present value of benefits is estimated to be $300 million to $6.8 trillion over 30 years for scaling offshore wind production. The return on investment in 2050 is significant, as shown by the benefit-cost ratio, estimated to be 2:1 to 17:1 in 2050 (Konar and Ding 2020). In terms of the benefit-cost ratio per unit of energy generation and transmission, analysis estimates the benefits to be $75–$300 per megawatt-hour (MWh) for 1 unit of additional energy production and the ratio range to be between 0.9:1 and 28:1 (Konar and Ding 2020). Estimates show that return on investment increases substantially as the

[61] *Offshore wind's technical potential is 3,000 terawatt-hours (TWh) per year for installations in water less than 60 metres deep and within 60 km of shore. Global electricity demand is currently 23,000 TWh (IEA* 2019a, b*).*

[62] *Offshore wind can generate electricity during all hours of the day and tends to produce more electricity in winter months in Europe, the United States and China, as well as during the monsoon season in India—providing higher value than that of its onshore counterparts and more stable over time than that of solar photovoltaics (PV) (IEA* 2019a, b*). Capacity factors for onshore wind farms in the European Union average 24%, with new farms reaching 30–35%. Offshore farms have a capacity factor averaging 38%, with new farms reaching 35–55% (an increase of more than 50%; IEA* 2019b*). Another advantage is size of turbines. A single 10 MW offshore wind turbine, operating at 60% capacity factor, will have output of 51 GWh/year. A solar farm with 25% capacity factor, to provide same amount of power, will require ~56,000 PV panels and occupy ~60 hectares of land. The analysts forecast a 60% reduction in the costs of turbines, foundation and installations by 2040 (IEA* 2019b*).*

[63] *These IEA projections are based on expansion in six key markets: Europe, China, the United States, South Korea, Japan and India. Europe, the current market pace-setter with 20 GW installed, is forecast to continue to lead the global pack for the next two decades, with expectations of some 130 GW turning offshore by this date—though China by this point is foreseen as having at least 110 GW online and being on track to outpace Europe's build-out by mid-century. The United States, meanwhile, is forecast to be in line for 'substantial growth' by 2040, with its fleet swelling to around 40 GW, while Korea, India and Japan would all see tens of GW of offshore wind turbines installed.*

costs of energy generation fall with improved technologies and as actions are taken to reduce integration costs.

Why Investment Makes Sense

Stimulating the creation or expansion of ocean-based renewable energy provides short-term job creation. In the early stages of exploring the feasibility of ocean- based renewable energy projects, jobs can be created for engineers, land and marine surveyors, energy specialists, researchers and providers of legal services (see Box 19.9 for an overview of the initial stages of development of Australia's first offshore wind farm). The opportunity for job creation is generally at the regional and local levels, but the extent of the breakdown will vary by region based on the nature of the wind, tide or wave resource, as well as on the supply chain and labour force. The IEA estimates that offshore wind creates about 1.2 construction jobs per $1 million invested (for both the construction and manufacturing phases) (IEA 2020a)[64]. In total, the development of a typical 500 MW offshore wind farm requires around 2.1 million person-days of work (IRENA 2019). Estimates in the United States vary from 6 to 44 jobs/MW during construction periods and 0.7 to 1.7 jobs/MW for the projects' ongoing operation (Tegen et al. 2015)[65]. The labour distribution is estimated as 1% for project planning, 59% for procurement and manufacturing, 0.1% for transport, 11% for installation and grid connection, 24% for operation and maintenance and 5% for decommissioning (IRENA 2019). A particular benefit of job creation through offshore wind is that the skills required may be similar to those in offshore oil and gas, enabling benefits to accrue directly to workers transitioning from declining fossil fuel industries (IRENA 2018; Scottish Enterprise 2016), which also minimises the costs of transition and the risks of structural unemployment. The expertise of workers and technicians in building support structures for offshore oil and gas sites, for example, could be leveraged when building foundations and substations for offshore wind turbines. Any such transition must ensure a transfer of benefits and comparable salary for comparable jobs and/or skill requirements, such as opportunities for union representation.

An established ocean-based renewable energy sector creates green jobs, economic diversification into zero-emission fuels and opportunities to co-locate and support other offshore industries. The long-term economic benefits associated with a new or expanded ocean-based renewable energy sector include new highly skilled jobs. The OECD estimates that by 2030 the total full-time employment in offshore wind will be 435,000 (OECD 2016)[66]. For offshore wind, an estimated 1 million new jobs will be created by 2050, with an estimated 0.45 million in construction and installation, 0.39 million in manufacturing and 0.17 million in ongoing operations and maintenance (IRENA 2019). For other ocean-based renewables, the sector could create 680,000 jobs by 2050 (OES 2017). The interaction of the offshore wind energy industry with other economic sectors creates the potential for economic diversification and the generation of additional revenue, through both supply chain activities and induced demand for goods and services (IRENA 2018). There is the potential to unlock co-location benefits with other offshore industries; for example, ocean-based renewable energy could meet the increasing demand for energy-intensive desalinated seawater or support mariculture operations. Investment in any form of renewable energy supports the achievement of energy security and independence from imported fossil fuels and associated price volatilities. Lastly, it also creates the opportunity for new green industries in terms of alternative fuel generation (e.g. hydrogen), which can serve as exports or inputs to decarbonisation of other sectors of the economy (such as marine transport). Education and training, however, must be attuned to emerging needs in the ocean renewable energy industry (see Annex 1).

Ocean-based renewable energy offers potential health benefits and desalination of drinking water in coastal communities facing water scarcity. The health benefits of moving to ocean-based renewable energy for power generation would be significant, particularly for regions that rely more heavily on coal and oil to generate electricity. Offshore wind in the Mid-Atlantic region of the United States could produce health and climate benefits estimated at between $54 and $120 per MWh of generation, with the largest simulated facility (3000 MW off the coast of New Jersey) producing approximately $690 million in benefits (Buonocore et al. 2016). There is potential to develop ocean energy technolo-

[64] *Wind power is less labour-intensive than PV solar. Onshore wind power projects create about 1 job in construction and 0.5 in manufacturing per $1 million invested. Offshore wind creates about one-fifth as many construction jobs but twice the number of manufacturing jobs per unit of investment.*

[65] *For the Southeast region, offshore wind energy development has the potential to support between 14 and 44 full-time equivalent (FTE) jobs/MW during construction periods and 1.6 and 1.7 FTE ongoing (operations phase) jobs/MW; in the Great Lakes, there could be between 6 and 27 FTE jobs/MW installed and 0.7 and 0.8 FTE jobs/MW for the projects' ongoing operation; in the Mid-Atlantic region during construction phases, we estimated a range of 12–30 FTE jobs/MW, and the average for ongoing jobs was 1.2 FTE jobs/MW. The Gulf of Mexico has the potential to support between 25 and 29 FTE jobs/MW during construction and 1.3 FTE jobs/MW on an ongoing basis, for operations and maintenance.*

[66] *Based on previous employment and capacity projections by the IEA (2014) and EWEA (2012), the OECD (2016) estimates that under a business-as- usual scenario, there will be an estimated 435,000 full-time jobs in the offshore wind industry by 2030. This estimate is based on the expectation that more countries will have multiple GW of wind power installed.*

gies for a range of purposes, including desalination for drinking water (OES 2011).

Increasing the share of renewable energy generation and reducing the use of fossil fuels will contribute to national and global efforts to reduce GHG emissions, but efforts to scale ocean-based renewable energy must be done in an environmentally sensitive manner to reduce the impact on marine mammals and ecosystems. If ocean-based renewable energy technologies displace the current energy generation mix, CO_2 emissions can be reduced by between 0.30 and 1.61 $GtCO_2e$/year in 2050 in the case of offshore wind (fixed and floating), and by between 0.05 and 0.87 $GtCO_2e$/year in 2050 in the case of ocean-based renewable energy (Hoegh-Guldberg et al. 2019)[67]. Total emission reductions would amount to 0.35 to 2.48 $GtCO_2e$/year in 2050 (Hoegh-Guldberg et al. 2019) which is equivalent to taking approximately 35–53 million cars off the road every year[68]. Based on the analysis on avoided damage costs to society from mitigating climate change, we estimate the environmental benefits (net benefit) of reducing greenhouse gases by scaling offshore wind energy generation to be $344 billion to $668 billion over 30 years (Konar and Ding 2020). This estimates the costs of displacing the current energy mix with offshore wind energy in line with the projections in Hoegh-Guldberg et al. (2019). Offshore wind uses no water directly, and there should be an overall reduction in freshwater use compared to generating power from fossil fuels (Macknick et al. 2012). Offshore wind structures have positive and long-term effects on marine species because they provide new habitat in the form of artificial reefs and because fishing, mainly trawling, tends to be restricted in their vicinity (IRENA 2018; Dinh and McKeogh 2019). The risks of installing energy operations in the marine environment include potential biological invasions, noise and disturbance vibrations for marine species, collisions between birds and wind turbine rotors, and the presence of electromagnetic fields that can disrupt marine life and benthic habitats (Sotta 2012; Langhamer 2012). However, studies have shown that there is a gap between the perceived and actual risks of these technologies, with the former arising from uncertainty or lack of definitive data about the real impacts (Copping et al. 2016). The most recent analysis has revealed that the potential impacts of ocean-based energy on marine life are likely small or undetectable (Copping and Hemery 2020). Effective marine spatial planning, in combination with emerging ocean energy technologies, will be effective in mitigating potential biodiversity loss and the risk of collision with seabirds and impacts on migratory cetaceans from ocean energy technologies and in reinforcing biodiversity co-benefits (Hoegh-Guldberg et al. 2019; Best and Halpin 2019). Efforts must also be made to expand renewable energy (both ocean-based and on land) in concert with efforts to improve the circular economy and reduce the reliance of renewable energy technology on rare minerals that would also undermine ocean health if mined from the seafloor (Haugan et al. 2020).

[67] *Note that higher figures were also calculated based on coal displacement. These can be found in the full report (Hoegh-Guldberg et al. 2019).*

[68] *Based on average emissions of 4.6 metric tonnes per year, according to EPA (2016).*

How These Benefits Can Be Achieved: Short-Term Interventions That Can Be Initiated Now as Part of Stimulus Spending and Recovery Measures

Investment in research, development and innovation will improve the technology and reduce costs but must be coupled with additional policy support to increase market visibility and investor security and enable the further cost reductions that come with commissioning larger commercial plants.

- **Streamlined permitting and clear and coordinated processes across government.** Traditionally, the time from inception to completion can be 8–12 years, with 5–7 years for project development and 3–5 years for construction (Veum et al. 2011). Long lead times are caused by lengthy permitting requirements involving multiple agencies and lack of clarity of areas available for ocean-based renewable energy (considering competing users of the marine environment) (Crouse et al. 2018; UK Government 2016). Reducing these obstacles would send a clear signal of intent and regulatory certainty to industry and enable the acceleration of private sector investment in this industry. Note that streamlining of permitting does not include a fast track or elimination of the need for environmental impact assessments or community and stakeholder engagement and participation in the planning and citing process.
- **National targets and frameworks for ocean energy.** As part of the European Green Deal, the European Commission is currently developing its Offshore Renewable Energy Strategy, which will outline targets for between 250 and 450 GW of offshore renewable energy installed capacity by 2050, or capacity to meet about 30% of Europe's energy demand (EU Commission 2020c). Achieving this target will require strong public-private partnerships and alignment with national climate policies, marine spatial planning policies and technology development frameworks. The United Kingdom has set a target for installed offshore wind energy of 40 GW by 2030; as part of this target the UK government will also be supporting the development of floating wind turbines. Germany has also approved the amendment to the Offshore Wind Act (WindSeeG) to reach 40 GW of offshore wind capacity by 2040.

- **Suitable financial support mechanisms (e.g. subsidies and guarantees) and revenue support to stimulate industry and avoid loss leaders.** A lack of financial support mechanisms (e.g. subsidies or guarantees), can drive up costs for industry and create roadblocks (Crouse et al. 2018; UK Government 2016). Governments could look to arrangements provided to stimulate early investment in land-based renewable energy, such as solar and wind subsidy schemes.
- **Investment in land-based grid updates and associated infrastructure.** The Netherlands government has published a roadmap for 2.5 GW of offshore wind by 2023 while also investing in a 700 MW offshore wind transformer platform to ensure that the land-based infrastructure is in place for private sector investment to support the achievement of the target.

Box 19.9 Establishing Australia's First Offshore Wind Farm

Star of the South Wind Farm is Australia's first offshore wind farm, paving the way for a new sustainable ocean industry for Australia[a]. A joint development by Australia's Offshore Energy and Copenhagen Infrastructure Partners, Star of the South could include up to 250 turbines, with a combined capacity of up to 2 GW. This could supply about one-fifth of Victoria's power needs and, through close proximity to demand centres along the Australian coast, could minimise the need for battery storage normally associated with land-based wind and solar.

Following the grant of an exploration license in March 2019 to investigate the technical feasibility of constructing wind turbines in the ocean off the south coast of Gippsland, Victoria, Star of the South is moving forward with marine surveys and engineering options in terms of land-based grid connections. It has partnered with Curtin University and Deakin University to assist with offshore site investigations, focusing on understanding marine mammals in the project area and undertaking the necessary seabird, seabed biodiversity and fish surveys. Both universities are working with RPS Australia Asia Pacific to collect data to inform the environmental assessments and the project's design. DHI has also joined the project by providing a 40-year hindcast of waves and currents that serves as input for moving further with the design phase (Skopljak 2020). Preliminary surveys also include mapping the seafloor, measuring water depths and identifying any buried infrastructure, such as cables.

In addition to the employment opportunities created through the above partnerships, the core development team for the project, all located locally in Melbourne, currently employs 35 people and is expected to grow to 50 by the end of 2020 (Parkinson 2020).

The Australian government has also begun developing a policy framework to underpin offshore wind development off its coasts, an initiative long called for by industry (Australian Government 2020a).

[a] *For more information on the project, see http://www.starofthesouth.com.au/.*

3.3 Additional Opportunities for a Blue Transformation

As evidenced by the 2008–2009 stimulus packages, not all investments will be directed at measures that create jobs in the short term. Instead, much of the investment will be used to lay the foundation for long-term recovery and resilience through systemic transitions to improve the efficiency and cost-effectiveness of our economy and by initiating large infrastructure projects that will yield benefits over the next 10–30 years.

Table 19.3 summarises a further set of opportunities for governments to consider to ensure a sustainable and equitable blue recovery from COVID-19 that will have long-lasting benefits for economic resilience and ocean health. These interventions, and their potential economic, social and environmental benefits, are detailed in full in Annex 1 (Tables 19.4, 19.5, and 19.6).

These interventions are organised in three categories:

1. Research and development to spur innovation and new technology
2. Regulatory reform to provide an enabling environment for a sustainable ocean economy
3. Public-private partnerships for a blue transition

Just as on land, these investments have the potential to dramatically alter the course of a country's transition to a sustainable economy that can provide long-term economic opportunities, improved health and food security, reduced emissions, enhanced biodiversity and ecosystem services

Table 19.3 Additional opportunities for a blue transformation

	Sector relevance	Economic benefits	Social benefits	Environmental benefits	SDGS
Research and development to spur innovation and new technology					
Invest in research and development, including pilot projects, to accelerate the development of sustainable and low-carbon alternative feed options for fed mariculture (e.g. finfish)	Fisheries	●	●	◍	2, 8, 9, 12, 13, 14
Invest in filling data gaps on national coastal and marine ecosystems through employment schemes for surveys, modelling and mapping	Tourism, Fisheries	◍	●	●	8, 12, 13, 14, 17
Invest in R&D and innovation grants to stimulate the development of new industries for generating alternative marine fuels, e.g. hydrogen and ammonia (invest in land-based infrastructure for fuel generation and supply chains as opposed to ship related investments)	Transport, Energy	●	○	◍	7, 8, 9, 12, 13, 14, 17
Establish blue economy skills-training and capacity-development programs in key ocean industries for affected communities and industries (e.g. ocean-based renewable energy, zero-emission vessels, GIS, ecotourism, restoration)	Tourism, Fisheries, Energy, Transport, Marine Conservation	◍	●	●	4, 7, 8, 9, 12, 13, 14, 17
Invest in research and development, including pilot projects, and incentivise emerging ocean-based renewables to accelerate their development	Energy, Transport, Mariculture	◍	◍	◍	7, 8, 9, 12, 13, 14
Regulatory reform to provide an enabling environment for a sustainable ocean economy					
Establish comprehensive integrated ocean management and marine spatial planning processes to balance marine users and spaces, competition for coastal resources and mitigate permitting and siting issues for sustainable ocean industries	Fisheries, Tourism, Energy, Shipping, Marine Conservation, Mariculture	●	◍	●	8, 12, 13, 14, 17
Initiate regulatory reform to promote best practice in climate-adaptive fisheries management, including through incentives for industry adoption in the form of taxes and subsidies	Fisheries	●	◍	●	2, 8, 12, 13, 14
Shift harmful subsidies to more sustainable and equitable uses, including supporting small-scale and artisanal fishing, ecotourism opportunities for local communities and management and monitoring of marine protected areas	Fisheries, Tourism, Marine Conservation	●	●	●	2, 8, 12, 14
Introduce levies or taxes to reinvest tourism revenue in local restoration and conservation efforts	Tourism, Fisheries, Marine Conservation	◍	◍	◍	8, 11, 12, 13, 14, 15
Integrate ocean accounts into national accounting frameworks, or develop satellite ocean accounts, to measure and monitor the impact of recovery measures on long-term sustainability of the ocean economy	Fisheries, Tourism, Transport, Energy, Marine Conservation, Infrastructure	●	○	◍	8, 9, 12, 13, 14, 17
Public/private partnerships for a blue transition					
Mobilise private sector investment in hybrid 'green/blue/grey' approaches (e.g. utilising living coastal infrastructure in traditional construction) for coastal infrastructure projects and ports through financial incentives such as tax exemptions and guarantees	Tourism, Fisheries, Marine Conservation	●	●	●	8, 9, 11, 13, 14, 15

(continued)

Table 19.3 (continued)

	Sector relevance	Economic benefits	Social benefits	Environmental benefits	SDGS
Invest in port authorities to transition to 'blue ports' and port reception facilities	Transport, Tourism, Energy, Infrastructure	Strong potential	Potential	Potential	3, 8, 9, 11, 13, 14, 17
Incentivise investment in cold storage capacity through access to affordable credit, government backed loans, duty-free imports of equipment and tax exemptions	Fisheries	Strong potential	Potential	Potential	2, 5, 8, 12, 14
Scale parametric insurance policies for blue natural capital in small island developing states, least developed countries and developing countries	Tourism, Fisheries, Marine Conservation	Potential	Minor potential	Potential	11, 13, 14, 15, 17
Stimulate sustainable and environmental sensitive mariculture (e.g. integrated multi-trophic aquaculture) through financial incentives such as tax exemptions and affordable credit, and government-backed loans	Fisheries, Mariculture	Potential	Minor potential	Potential	2, 8, 12, 13, 14

● Strong potential, ● Potential, ○ Minor potential

and improved resilience to climate impacts and other future shocks. For these additional opportunities, we sought ones that provided the following:

- Ability to build long-term resilience to future shocks (considering improving human, natural and physical capital) (Hammer and Hallegatte 2020; OECD 2020e)
- Ability to direct economic benefits to affected communities and vulnerable members of society (a people-centred approach) (UN 2020b)[69]
- Ability to catalyse progress towards a long-term sustainable and equitable blue economy (Hepburn et al. 2020)
- Ability to deliver on international commitments such as the 2030 Agenda for Sustainable Development and the Paris Agreement (IMF 2020b)
- Relevance to multiple regions and economies (OECD 2020e)

For each intervention, we identified the potential economic, social and environmental benefits based on existing literature. Note that for many of these interventions, no quantified benefits are yet available for the intervention level. The benefits highlighted are therefore intended to be a guide only and not prescriptive. As with any intervention, countries will need to go through a rigorous national process to fully quantify economic, social and environmental benefits given national or local circumstances.

[69] *The UN secretary general has stressed the need to ensure that national and local response and recovery plans identify and put in place targeted measures to address the disproportionate impact of the virus on certain groups and individuals, including migrants, displaced persons and refugees, people living in poverty, those without access to water and sanitation or adequate housing, people with disabilities, women, older people, LGBTI people, children and people in detention or institutions.*

3.4 Opportunities for Blue Conditionality to Avoid Roll-Backs in Progress

The provision of immediate relief packages and grants to the private sector brings with it the opportunity to incentivise recipients to implement measures central to the sustainable ocean economy agenda—but which might have been harder to incentivise or promote before COVID-19 without such finance or might be vulnerable to roll-backs as a result of decreased traditional revenue streams.

Although any form of 'blue condition' could be attached to a debt-relief agreement or government grant, we highlight two particular opportunities that take advantage of emerging and innovative technologies to avoid roll-backs in progress:

1. Digitalisation of the fishing industry to promote sustainable fisheries management and end illegal, unreported and unregulated (IUU) fishing.
2. Disclosure of ocean data to inform decision-making.

The above measures represent opportunities to advance long-standing agendas in terms of improving marine biodiversity, enhancing monitoring, ensuring fish stock recovery and responding to climate change. Both of them will have significant long-term benefits, improving ecotourism opportunities, enhancing the value of existing coastal tourism and improving the economic viability of artisanal and commercial fisheries.

In the short term such arrangements can provide immediate economic relief to the recipients (through the grant) and potential cost savings for the government.

3.4.1 Sustainable Fisheries Management Through Digitisation

Conditions aimed at fisheries reform and digitisation of the fishing industry offer the opportunity to make progress on long-standing fisheries governance agendas while also overcoming many of the short-term impacts of COVID-19 restrictions and revenue losses. These include the loss of on-board observers and reduced capacity for marine patrols to monitor and track fishing vessels for the purposes of reducing overfishing and IUU fishing. Traditionally, the burden for gathering such data has fallen on governments, but recovery efforts offer the opportunity to engage and empower the fishing industry itself to collect much of the data that underpin sustainable fisheries management.

The digitisation of the fishing industry would have other benefits in the face of COVID-19 and beyond fisheries governance. Traceability and data-sharing also enhances industry robustness and resilience by strengthening aquafeed supply chains, which have been curtailed during the COVID-19 crisis. Sharing data creates more robust supply chains for raw material. This can be achieved by making data on regional and sustainable raw materials sources available. Science-industry cooperation is vital for this process. Making these data available could also be a condition to strengthen the aquaculture industry (see the data-sharing and disclosure section below).

Consumers are also increasingly demanding more traceability, highlighting the added incentive for increased supply chain monitoring through digital tools. Creating alternative data-gathering mechanisms like apps empowers local fishers to take part in data-gathering while informing consumers. OurFish, developed by Rare, is one example of an app for local fishers to record and share their catch data digitally, creating a permanent digital log of sales, expenses and inventory. This app and the associated data also enable fishing communities to monitor the value, type and local amount of fish caught. The information can be made available to decision-makers in government and relevant stakeholder groups.

Examples of measures that could be attached to grants include requiring registration of vessels (relevant to small-scale and artisanal vessels); digital traceability—to increase transparency and strengthening monitoring, control and surveillance; and electronic monitoring and electronic reporting. Conditions can also target the publication of essential data, including vessel ownership and licenses (see the data-sharing and disclosure section below).

These industry-led measures could be supplemented by government investment in new artificial intelligence-powered electronic monitoring systems, enhanced drones and satellite data interpreted by machine learning. Such efforts will also dramatically improve the fishing industry's resilience to similar future shocks.

The potential economic impact of such measures is significant. Globally, between 8 and 14 million metric tonnes of unreported catches are traded illicitly yearly, resulting in gross revenues of $9 billion to $17 billion associated with these catches. This equates to an estimated loss (in annual economic impact) of $26 billion to $50 billion globally, while losses to countries' tax revenues are between $2 billion and $4 billion (U.R. Sumaila et al. 2020). What this means for a region is significant. For example, the Pacific experiences an estimated loss in gross revenues to the formal economy of $4.3 billion to $8.3 billion per year. These losses are substantially higher when we consider the economic impact ($10.8 billion to $21.1 billion per year), income impact ($2.8 billion to $5.4 billion per year) and tax revenue impact ($200 million–$1.6 billion per year) (Konar et al. 2019a). Furthermore, as a result of potential illicit trade in seafood, workers in the sector lose an estimated $6.8 billion to $13.3 billion in income annually (Sumaila et al. 2020).

The results of moving to digital systems, including electronic monitoring and reporting, will significantly improve information. Fishery management systems currently rely heavily on data from fishers' daily logbooks that include locations, amount of time spent fishing, how many fish were caught and how many and what kind of fish or other species were discarded. On-board observers have been the only option to validate these logbook data, but such efforts only cover a tiny fraction of global fishing activities—likely less than 2% (Michelin et al. 2018). In most instances, electronic monitoring systems can achieve monitoring goals more cost-effectively than human observers and can more easily scale to cover 100% of fishing activity. Also, electronic monitoring can provide transparency in the critical first link in a supply chain that is traceable from supply to plate, giving consumers confidence when purchasing premium-priced seafood that is labelled as 'sustainably harvested' (for an example of how this is being done in Jamaica, see Box 19.10).

Box 19.10 Jamaica's Focus on Improving Traceability and Monitoring of Wild Capture Fisheries

Jamaica's 17,000 artisanal fishers all received a one-time grant as part of Jamaica's initial rapid response to the impacts of COVID-19 on its fishing industry. These grants were to provide income support due to a drop in demand from Jamaica's tourist sector (the majority of Jamaica's fishing industry is oriented towards supplying high-end restaurants and resorts).

Jamaica has made long-standing efforts to restore its fish stocks through sustainable fisheries management and improved governance. The registration of artisanal fisherman has been a challenge.

Jamaica applied two main conditions to the grant: registration of the boat and mandatory GPS trackers. As a result of these conditions, Jamaica now has a much better understanding of the scale of small-scale fishing and has enabled a transition to digital information and tracking, two pillars of its existing commitment to sustainable fisheries management.

Source: Government of Jamaica.

Unsustainable fishing practices, including IUU fishing, threaten local livelihoods, exacerbate poverty and heighten food insecurity. Seizing the opportunity of relief packages to address this issue will have long-term economic benefits for countries and regions, helping to improve the resilience of these communities and their fishing industries (local, artisanal and commercial) for decades to come.

3.4.2 Improved Transparency and Decision-Making Through Ocean Data

Vast stores of unstructured data related to the ocean economy are currently stored by governments, researchers and industry (for legal, security or proprietary purposes), making them inaccessible and unusable to inform decision-making in either the public or private sector. These data should by default be made open and available through data-tagging, federated data networks (Brett et al. 2020). In support of SDG 14, the United Nations declared the 10-year period (2021–30) to be the UN Decade of Ocean Science for Sustainable Development (the Decade). The Decade is dedicated to providing a common framework to encourage stronger international cooperation that can better coordinate and integrate ocean data and research into the decision- making process of stakeholders.

Data on the ocean economy can spur incentives for innovation, new public-private instruments for investment and the creation of new business models as we adapt to our world's new realities after COVID-19. Increased data-sharing would also add resilience to ongoing COVID-19 challenges. Having active data streams is paramount for ocean resilience in facing up to COVID-19 and could contribute significantly to safer at-sea operations (e.g. through maritime track-and-trace systems using geofencing).

Conditions could include a requirement that private sector organisations and financial institutions disclose or improve the accessibility of such data. Such a condition would be comparable to those being advanced to improve environmental and climate disclosure as part of recovery packages (Office of the Prime Minister, Canada 2020).

Impactful requirements could include

- that all users of ocean resources such as fisheries, minerals, oil and gas or coastal land be required to make their environmental data available to the public (Leape et al. 2020);
- that domestic fisheries, fishing vessels, shipping and marine transport track their GHG emissions and report annually for inclusion in national GHG inventories in accordance with the relevant guidance of the Intergovernmental Panel on Climate Change;
- that fishing vessels use automatic identification systems and share essential data on fisheries, including vessel ownership, licenses and tracking for all fishing vessels (this is also relevant to fisheries reform, as identified above);
- that all data collected by defence and security agencies which can be shared without compromising national security be made publicly accessible (Leape et al. 2020); and

- that all financial institutions disclose whether their portfolios align with ocean sustainability. Companies based on, depending on or affecting the ocean should integrate relevant ocean-related risks and opportunities into corporate strategy, risk management and reporting[70].

In addition to conditions placed on financial grants to the private sector, governments should also provide support and training to develop appropriate data-gathering and processing capacities and systems in developing countries and coastal communities, to ensure that these nations and communities are not left behind.

4 Conclusion

The importance of the ocean to a sustainable future is too important to neglect at this great moment of resetting and rebuilding. The relevance of the ocean for global economic and social recovery and future prosperity must become part of global discourse, and a greater part of measures applied to respond to the economic and social impacts of the crisis.

The COVID-19 pandemic has severely impacted ocean industries and the livelihoods and food security of many millions of people. It has highlighted the significance of the ocean as a global workplace, its role in underpinning the modern economy and the inherent interdependencies between ocean sectors, the health of the ocean environment and human well-being.

How the world rebuilds from the COVID-19 crisis is of great importance for the ocean and climate. Early responses to promote economic recovery and protect industries from further losses have included large-scale investments in sectors previously shown to be harmful to the environment, alongside the easing of environmental safeguards. Such measures risk the future health and wealth of the ocean economy with impacts for food security, livelihoods and our shared prosperity, rolling back progress made towards mitigating global biodiversity loss and climate change. Governments and financial institutions need to immediately strengthen efforts to build environmental, social and economic resilience.

In tailoring support for those most affected by the COVID-19 pandemic, greater attention must be paid to the ocean economy and its many direct and indirect beneficiaries. A sustainable and equitable blue recovery is critical not just for those who live or work near the coasts but also for the well-being and resilience of societies and economies at large.

This report has identified specific opportunities for the immediate investment of stimulus and recovery funds that would lead to a more sustainable and resilient ocean economy. It also has highlighted opportunities to accelerate research on and development of future sustainable ocean industries and to transition emission- and pollution-intensive industries onto more sustainable pathways in order to reach their full economic growth potential.[71]

This report has highlighted that investment in the interventions necessary for a sustainable and equitable blue recovery will benefit other land-based sectors, including human health, technology, agriculture, supply chains and tourism.

The demonstrated interdependencies between the different ocean sectors, which has exacerbated the impacts of COVID-19 on individual industries, make a strong case for greater integration and collaboration among sectors, as a complement to traditional sectoral management, both in recovery efforts and long-term operations. Ecosystem-based, integrated ocean management and other related holistic and knowledge-based approaches to planning and managing the multitude of uses and users of ocean spaces offer an important framework to ensure that ocean industries can rebuild in a mutually reinforcing way towards a sustainable future ocean while protecting essential ocean ecosystems and functions.

This report highlights growing global inequalities and the need to accelerate equitable access to ocean opportunities and sharing of benefits from ocean industries. Response measures to support women, who have been disproportionately affected, notably in the tourism and fisheries sectors, will be particularly important to ensure access to decent work opportunities and the full engagement of women in ocean activities. There is also an ongoing need to improve working conditions for vulnerable 'key workers' at sea to better protect fishers and seafarers, who play an essential role in maintaining global supplies of food, medicines, energy and manufactured goods across supply chains.

To ensure a long-lasting economic recovery from the COVID-19 crisis, response measures must trigger investments and societal changes that reduce vulnerability and improve our collective resilience to future shocks (OECD 2020a). Recovery plans have so far fallen short in this regard. To this end, governments must seize the opportunity of stimulus packages to address unsustainable fisheries practices, including IUU fishing, which undermines employment and livelihoods in one of the largest sectors of the ocean economy, exacerbates global poverty and risks the food security

[70] *See, for example, the ocean sustainability principles followed by Norges Bank Investment Management* (2020).

[71] *See, for example, the ocean sustainability principles followed by Norges Bank Investment Management* (2020).

of over 3 billion people, including some of the world's poorest, who rely on the ocean as their primary source of protein. Technological advances introduced during COVID-19 and innovative financial mechanisms may hold the key to advancing such action.

The importance of the ocean to a sustainable future is too important to neglect at this great moment of resetting and rebuilding. The ocean's relevance for global economic and social recovery and future prosperity must become part of global discourse, and a greater part of measures applied to respond to the economic and social impacts of the COVID-19 crisis. The ocean-based or 'blue' investment opportunities detailed in this report offer a departure from business as usual in that they can deliver a more inclusive recovery, premised on a healthy and regenerative ocean to provide global benefits for the longer term.

Embracing a 'sustainable and equitable blue recovery' in the large stimulus packages being agreed worldwide can build ocean health and sustainability into recovery and support the transition towards a more sustainable, inclusive and resilient global economy.

Acknowledgements The authors would like to thank the following people for their review, feedback and inputs on early drafts of this report:

Mark Hemer (Commonwealth Scientific and Industrial Research Organisation (CSIRO)); Sebastian Troëng (Conservation International); Kristin Rechberger (Dynamic Planet); Arni Mattiesen (Food and Agriculture Organization); Kasper Sogaard (Global Maritime Forum); Suzi Heaton (Government of Australia); Betty Nyonje (Government of Kenya); Andrew Rhodes (Government of Mexico); Maria Ines Gameiro (Government of Portugal); Brandt Wagner (International Labour Organization); Chris Gilles (The Nature Conservancy); Lotta Pirttimaa (Ocean Energy Europe); Arian Steinsmeier (Rare); Angelique Pouponneau (Seychelles Conservation and Climate Adaptation Trust); Adrien Vincent (SYSTEMIQ); Elva Escobar (Universidad Nacional Autónoma de México); Andrew Hudson (United Nations Development Programme); Ignace Beguin (United Nations Global Compact); Justin Mundy (Willis Towers Watson); Karin Kemper (World Bank); Ines Aguiar Branco, Mathilde Bouye, Lauretta Burke, Ed Davey, Helen Ding, Erin Gray, Craig Hanson, Erika Harms, Amy Hemingway, Leo Horn- Phathanothai, Norma Hutchinson, Joel Jaeger, Aman Srivastava, Kristian Teleki, Ayushi Trivedi and Arief Wijaya (World Resources Institute); and Louise Heaps (World Wildlife Fund).

While we deeply appreciate the expertise brought by each of these reviewers, this report reflects the views of the authors alone.

The authors would also like to acknowledge the Expert Group Co-chairs, Jane Lubchenco, Peter Haugan and Mari Pangestu for their valuable guidance and support for this report.

Thank you also to Romain Warnault, Alex Martin and Jen Lockard for providing administrative, editing and design support.

Annex 1

Table 19.4 Research and development to spur innovation and new technology

Interventions	Sector relevance	Economic benefits
Invest in research and development (R&D), including pilot projects, to accelerate the development of sustainable and low-carbon alternative feed options for fed mariculture (e.g. finfish)	Fisheries	19.3 million people globally engaged in aquaculture (FAO 2018a) World food fish consumption in 2030 is projected to be 20% (or 30 million metric tonnes [mmt] live weight equivalent) higher than in 2016 (FAO 2018a) The major growth in production is expected to originate from aquaculture, which is projected to reach 109 mmt in 2030, with growth of 37% over 2016 (FAO 2018a)

Table 19.4 (continued)

Social benefits	Environmental benefits	Potential for the creation of perverse incentives	SDGS
Improved health of local communities. A portion of 150 g of fish provides about 50–60% of an adult's daily protein requirement. Fish proteins are essential in the diet of some densely populated countries where the total protein intake is low, and they are particularly important in diets in small island developing states and least developed countries (FAO 2018a) Alternative feed innovations could ensure an additional 364 mmt of food annually—over six times current capture and mariculture production[a]. This is only possible if mariculture is not dependent on feed from fish products (Costello et al. 2019) Reduction in the diversion of forage fish from communities that rely on it for direct nutrition (Tacon and Metian 2008; Konar et al. 2019b) and protect the cultural value to Indigenous Peoples (Jones et al. 2017; Konar et al. 2019b) Innovations in feed technology could greatly enhance the potential for fed mariculture (Costello et al. 2019; Froehlich et al. 2018) Increasing ocean-based food (including aquaculture) will generate benefits nine times higher than costs (Konar and Ding 2020) Increased job creation through development of algae feed industry (Roberts and Upham 2012)	The global supply of fishmeal may be near biological limits (Costello et al. 2012) Improves resilience under climate change (Gaines et al. 2019)	Increase in pollution from aquaculture operations Introduction of invasive species Job loss from traditional feed sources	1 2 3 8 12 13 14

Interventions	Sector relevance	Economic benefits
Invest in filling data gaps on national coastal and marine ecosystems through employment schemes for surveys, modelling and mapping	Tourism, Fisheries, Marine Conservation	Short-term job creation Long-term economic efficiencies in terms of data availability Potential access to carbon markets and associated on-going streams of revenue for management of ecosystems and local communities
Invest in R&D and innovation grants to stimulate the development of new industries for generating alter- native marine fuels, such as hydrogen and ammonia (invest in land-based infrastructure for fuel gener- ation and supply chains as opposed to ship-related investments)	Transport, Energy	Economic growth opportunity for export of low-cost hydrogen (utilising electrolysers powered by renewable resources)[b] (IEA 2019a, b) and green ammonia as a maritime fuel (Ash and Scarbrough 2019) Economic diversification potential—energy storage, low-carbon heat, transport fuels and a key input in the production of fertiliser (ammonia) (Yara International 2019). Additional uses create synergies and reduce the investment risk, especially in the early phase of the transition (IEA 2020a) Job creation potential in many states and regions (Bezdek 2019). Widespread penetration could create nearly one million new jobs (highly skilled, well-paid technical and professional workers) in the United States by 2030 (ASEA and MIS 2009)
Establish blue economy skills training and capacity development programs in key ocean industries for affected communities and industries (e.g. ocean-based renewable energy, zero-emission vessels, geographic information systems, ecotourism, restoration)	Tourism, Fisheries, Energy, Transport, Marine Conservation	Economic benefits of local developments accrue locally (Gaines et al. 2019) Local investments in renewable energy and energy- efficient technologies can improve local livelihoods and enhance local economic benefits (Gaines et al. 2019)

Table 19.4 (continued)

Social benefits	Environmental benefits	Potential for the creation of perverse incentives	SDGS
Increased participation in 'citizen science' can encourage public action and improve conservation efforts (McKinley et al. 2017) Sustained ocean observations benefit many users and societal goals across society actors (Weller et al. 2019) Community ownership and understanding of natural resources	Improved understanding and mapping of ecosystem extent and species diversity Basis for inclusion of ecosystems in national greenhouse gas (GHG) inventories to enable mitigation for blue carbon ecosystems (mangroves, seagrass and salt marshes), and important for monitoring adaptation benefits from other marine habitats like coral reefs Increased management capabilities		8 13 14
Diversified economic opportunities for local communities	Reduced GHG emissions Improved air quality (based on reduced reliance on fossil fuels as a result of green fuels) Improved water quality, including deep-sea routes		7 8 9 12 13 14
Diversified economic opportunities for local communities Local capacity building in ecotourism (foundation for ensuring revenue is reinvested in the local community) Increased cultural awareness by sharing traditional knowledge Increased community buy-in	Reduced emissions Improved monitoring and protection of marine protected areas and coastal and marine ecosystems Using ecotourism for conservation through programs like sea turtle watch or citizen science		7 8 9 12 13 14

Interventions	Sector relevance	Economic benefits
Invest in research and development, including pilot projects, and incentivise emerging ocean-based renewables to accelerate their development	Energy, Transport, Mariculture	The global market of wave and tidal sectors is estimated to reach €53 billion per year by 2050 (Carbon Trust 2011)

Social benefits	Environmental benefits	Potential for the creation of perverse incentives	SDGS
Ocean-based renewable energy has the potential to generate 400,000 jobs in Europe by deploying 100 GW by 2050 (ETIP Ocean 2020). The global deployment is estimated to be 337 GW (2011), which indicates that ocean energy will generate about 1.2 million jobs globally by 2050	Ocean-based renewable energy can reduce GHG emissions by between 0.05 and 0.87 GtCO2e/year by 2050 (Hoegh-Guldberg et al. 2019). It can also create marine reserves and artificial reefs (Copping et al. 2016)		7 8 9 11 12 13 14

[a] Note that this figure is based on a tripling of global production of seafood for consumption, which would necessitate dramatic shifts in consumer taste and associated demand

[b] Renewable hydrogen costs may fall to as low as $1.40 a kilogram by 2030 from the current range of $2.50 to $6.80, with further reductions to 80 cents by 2050, equivalent to a natural gas price of $6 per million British thermal units (Mathis and Thornhill 2019)

Table 19.5 Regulatory reform to provide an enabling environment for a sustainable ocean economy

Interventions	Sector relevance	Economic benefits
Establish comprehensive integrated ocean management and marine spatial planning processes to balance marine users and spaces, reduce competition for coastal resources and mitigate permitting and siting issues for sustainable ocean industries	Fisheries, Tourism, Energy, Shipping, Marine Conservation, Mariculture	Potential economic growth and new economic opportunities (European Commission 2020a) Sector growth facilitated through improved framework (Jay 2017) Cost reduction through streamlining regulatory and compliance processes (European Commission 2020a)
Initiate regulatory reform to promote best practice in climate adaptive fisheries management, including through incentives for industry adoption in the form of taxes and subsidies	Fisheries	More catch and profits through climate-adaptive management than through business-as-usual management (Free et al. 2019) Economic diversification through providing a portfolio of options to fishers and a buffer against climate-driven losses in any one target stock (Free et al. 2019) Economic losses of about US $83 billion in 2012, compared with the optimal global maximum economic yield equilibrium (World Bank 2017)
Shift harmful subsidies to more sustainable and equitable uses, including supporting small-scale and artisanal fishing industry, ecotourism opportunities for local communities and management and monitoring of MPAs	Fisheries, Tourism, Marine Conservation	6.3% of global GDP ($4.7 trillion) was provided as fossil fuel subsidies in 2015, including uninternalised externalities (Coady et al. 2019) About $35 billion in subsidies are allocated to global marine fisheries alone each year, of which $22 billion are allotted to harmful subsidies (R.U. Sumaila et al. 2019) New economic opportunities for local communities through ecotourism Job protection (or creation) for local communities in MPA management and monitoring The World Bank has estimated that reducing global fisheries overexploitation, of which subsidies are key factor, could generate an additional $53 billion to $83 billion in revenue annually (World Bank 2017)

Social benefits	Environmental benefits	Potential for the creation of perverse incentives	SDGS
Reduced conflict through improved stakeholder relations and engagement (European Commission 2020a) Inclusivity and recognition of Indigenous rights such as the Beaufort Sea Partnership in Canada, which works with the local Indigenous groups	Streamlined management resulting in more effective governance to mitigate environmental risks posed by ocean-based activities and industries Increased stock through improved management Improved conservation of coastal and marine habitats	Lobbying for greater influence and industry capture	1 2 8 12 13 14 17

Table 19.5 (continued)

Social benefits	Environmental benefits	Potential for the creation of perverse incentives	SDGS
Local and community-based management can increase adaptive capacity by incorporating local knowledge and can improve sustainability by fostering a sense of stewardship (Gutiérrez et al. 2011) These strategies also allow fishers to generate revenues through other compatible activities, such as tourism, recreation and aquaculture (Moreno and Revenga 2014)	Ecological resilience through maintaining healthy stock sizes, age structures and genetic diversity (Free et al. 2019) Reduced impacts of climate change on fish stocks (Free et al. 2019) Thanks in part to adaptive harvest strategies fish stocks not fished beyond their biological limit and overfished stocks allowed to rebuild (Melnychuk et al. 2014)	Overfishing or stock decline if not linked to science	2, 8, 12, 13, 14
Subsidies that are disproportionately provided to the large industrial fishing sub-sector serve to undermine the Sustainable Development Goals by aggravating hunger, poverty and gender inequality in coastal communities worldwide (Sumaila 2020) Redirected subsidies could be used to improve gender equality by empowering female fishers (Österblom et al. 2020) Redirected subsidies could support Indigenous Peoples and local communities, many of which practice artisanal fishing, as well as the conservation and sustainable use of marine biological diversity	Improved biodiversity outcomes if redirected subsidies are used to fund jobs in monitoring of protected areas Improved fish stocks if redirected subsidies are used to fund incentives to improve traceability of fisheries, inclusion of women and jobs on coastal restoration works (Sumaila 2020)	Mismanagement of funds	1, 2, 5, 7, 8, 10, 12, 14

Interventions	Sector relevance	Economic benefits
Introduce levies and taxes to reinvest tourism revenue in local restoration and conservation efforts	Tourism, Fisheries, Marine Conservation	Additional revenue stream Iceland's Tourist Site Protection Fund promotes the development, maintenance and protection of tourism attractions and is funded by Iceland's accommodations tax, enacted in 2011 (OECD 2018) Reduction of value-added tax on tourism-related goods and services in Ireland was followed by an increase in employment through growth in numbers of tourists (OECD 2014)
Integrate ocean accounts into national accounting frameworks, or develop satellite ocean accounts, to measure and monitor the impact of recovery measures on long-term sustainability of the ocean economy	Fisheries, Tourism, Transport, Energy, Marine Conservation, Infrastructure	Digital solutions are important to facilitate, among other things, enhanced reporting of crisis-related spending, ex post audits and procurement transparency (IMF 2020c) By tracking each budget transaction across government agencies, accounts can produce timely, reliable, accurate and meaningful information to support financial decision-making, improve fiscal discipline, strengthen expenditure control and enhance fiscal transparency (Uña et al. 2019)

Social benefits	Environmental benefits	Potential for the creation of perverse incentives	SDGS
Reinvestment in jobs for local communities (should be done in coordination with local communities, including Indigenous Peoples, local communities and women affected by conservation efforts, to ensure buy-in)	Proceeds from taxes and levies secure funding for the protection of environmental areas In Australia, the Great Barrier Reef Marine Park Environmental Management Charge proceeds are applied directly to the management of the marine park, including through education, research, compliance patrols, site planning, public moorings, reef protection markers, information signs and maps (OECD 2014)	Mismanagement of funds Tourism can harm local ecosystems	8, 11, 13, 14

Table 19.5 (continued)

Social benefits	Environmental benefits	Potential for the creation of perverse incentives	SDGS
Data and improved metrics to track equitable distribution of ocean wealth	Data to account for natural wealth Integration of ecosystem services into decision-making		8 9 12 14 16 17

Table 19.6 Public/private partnerships for a blue transition

Interventions	Sector relevance	Economic benefits	Social benefits	Environmental benefits	Potential for the creation of perverse incentives	SDGS
Mobilise private sector investment in hybrid 'green/blue/grey' approaches (e.g. utilising living coastal infrastructure in traditional construction) for coastal infrastructure projects and ports through financial incentives such as tax exemptions and guarantees	Tourism, Fisheries, Marine Conservation	Natural coastal barriers, such as mangroves, wetlands and sandbars, lower costs for grey infrastructure, such as seawalls, sea dikes and groynes. New York City saved 22%, or $1.5 billion, by combining green and grey infrastructure instead of pursuing a grey-only strategy to secure water supply for the city (Bloomberg and Holloway 2018) In Vietnam, an investment of $9 million to restore 9000 hectares of mangroves along the shores of 166 communes as well as 100 km of dike lines cut the cost of dam- ages by $80,000–$295,000 and saved an additional $15 million in avoided damages to private property and other public infrastructure (IFRC 2011) Increased ecotourism opportunities in living infrastructure (e.g. mangroves and wetlands)	Catastrophic risk reduction for loss of life in storm surges through reducing wave energy and the height of a storm surge (Beck and Lange 2016) Main operators of green infrastructure are often local communities, responsible for implementing land stewardship practices and for maintaining the project over the long term (unlike grey infrastructure that is operated and owned by a company or government entity) (Browder et al. 2019)	Climate-mitigation potential (depending on ecosystem) Coastal resilience through reduced storm surges and protection of coastal communities and infrastructure from sea level rise Improved biodiversity, water quality, watershed protection (Browder et al. 2019)		8 9 11 13 14 15
Invest in port authorities to transition to 'blue ports' and port reception facilities[a]	Transport, Tourism, Energy, Infra-structure	Low-emission and fuel-efficient terminal equipment will save money through reduced energy consumption Increased efficiency through improved equipment will reduce operation costs Increased investment from offshore wind tenants who may be dealing with outdated port ownership structures and inexperienced owners Synergies with zero-emission vessels and energy production Identify technical and operational innovations to reduce the high transportation costs that exist for many developing countries and other remote locations (UNGC 2020b) Incorporate climate change adaptation considerations into 'blue ports', as ports are at increasing risk of coastal flooding. Infrastructure inventories, higher resolution data, as well as technologies that help improve the understanding of coastal processes under climate change are needed for effective risk-assessment and adaptation planning for critical transport infrastructure, particularly in small island developing states (UNCTAD 2020c)				

Table 19.6 (continued)

Social benefits	Environmental benefits	Potential for the creation of perverse incentives	SDGS
Improved air quality Improved health and livelihood of people working or living around ports and the 'liveability' of the area surrounding the port Opportunities for gender equity in access to resources, services, markets, incomes and employment (FAO 2018b)	Responsible fisheries management Reduction of shoreside idling (Sharma 2006) by providing shoreside power will reduce noise pollution (NoMEPorts 2008), improve air quality and reduce fuel consumption Reduced waste pollution through improved solid waste handling and recycling programs at port (Svaetichin and Inkinen 2017)	Added ecosystem disturbance through updates	3 8 9 11 13 14 15 17

Interventions	Sector relevance	Economic benefits
Incentivise investment in cold storage capacity through access to affordable credit, government-backed loans, duty-free imports of equipment and tax exemptions[b]	Fisheries	Live, fresh or chilled is the preferred and highest-priced form of fish and represents the largest share of fish for direct human consumption (FAO 2018a) Resilience to future shocks. Increased demand for frozen fish since outbreak of COVID-19 (Saumweber et al. 2020) Increased yields for fishers Increased income for fishers as a result of high-quality fish Marine exports grew by 7.68% in the fiscal year following an investment package by the Government of India, which included ongoing subsidies to build large cold storages for surplus seafood[c] (Narayanswami and Balan 2013)
Scale parametric insurance policies for blue natural capital in SIDS, LDCs and developing countries	Tourism, Fisheries, Marine Conservation	100 m of mangrove barrier can reduce wave heights by two-thirds Building oyster reefs adjacent to shore in the United States can reduce the cost of every metre of coastal protection by over $750, compared to other engineering options (Spalding et al. 2016) In France, the Caisse Centrale de Réassurance has estimated that insured property damages will rise by 50% if no preventive measures for climate change–related effects are implemented (CCR 2018) Marine ecosystems represent natural capital and non-market flows and services. Healthy coral barriers stop the damaging effects of hurricanes and cyclones hitting the coasts The value of marine ecosystems, based on the total bundle of ecosystem services provided by an 'average' hectare of open ocean, is estimated at $490/year, while the value of services provided by an 'average' hectare of coral reefs is almost $350,000/year (OECD 2016)
Stimulate sustainable and environmentally sensitive mariculture (e.g. integrated multi-trophic aquaculture [IMTA]) through financial incentives such as tax exemptions and affordable credit, as well as through government-backed loans	Fisheries, Mariculture	Economic diversification Increased profitability per cultivation unit and higher income (Troell et al. 2009) Resilience to shock and market changes through product diversification. Increased yields. At sites in Canada's Bay of Fundy, growth rates of kelp and mussels cultured in proximity to fish farms were found to be 46% and 50% higher, respectively, than at control sites (Chopin et al. 2004)

Table 19.6 (continued)

Social benefits	Environmental benefits	Potential for the creation of perverse incentives	SDGS
Loss or waste between landing and consumption due to a lack of refrigeration still accounts for an estimated 27% of total catch, representing a missed opportunity in terms of additional protein available for local communities and consumers (FAO 2018a; NoMEPorts 2008) Shifting to freezing could have a positive impact on women's employment, as they constitute a high proportion of workers in the post-harvest/food processing sector (UNCTAD 2020b)	Analysis has shown a net benefit in GHG emissions reduction from expanding cold storage to developing countries. In all modelling scenarios, decreased emissions from food loss and waste from cold chain expansion outpaced newly created emissions from the expansion and use of cold storage by a factor of 10, approximately[d]	Fishers may be incentivised to fish further offshore or more intensely because they can now store food longer	2 5 12 13 14
Climate risk reduction measures to ensure insurance coverage for previously non-insurable situations like sea level rise and other slow-onset events	Coastal resilience through reduced storm surges and protection from sea level rise Improved biodiversity, water quality, watershed protection (Browder et al. 2019)		11 13 14 15
Source of employment for local communities. Increased protein yields Opportunities for regional collaboration. The Yellow Sea Large Marine Ecosystem Project, established under the guidance of the Global Environment Facility and the UN Development Programme, and in a partnership be- tween China and South Korea, is working to implement IMTA in the region	Preservation of local habitats Recycling of waste nutrients and bio-mitigation typically produced through traditional mariculture by lower trophic level crops (Troell et al. 2009)		2 8 12 13 14

[a] 'Blue ports' are considered to be sustainable, support the transition to decarbonised marine transport and shipping fleets through fuel supply chains, promote transparency and traceability for fisheries and utilise nature-based solutions

[b] Any investment in cold storage by the private sector must be coupled with public investment in the supporting supply chain infrastructure. Governments should also eliminate disincentives to cold storage (such as taxes on foreign refrigeration systems) (FAO 2020b)

[c] Other measures included the government's exempting air-conditioning equipment and refrigeration panels used in cold chain from excise duties and allowing duty- free import of refrigerated units used in reefer trucks (Narayanswami and Balan 2013)

[d] In all modelling scenarios, the decrease in the food loss and waste carbon footprint from cold chain expansion clearly outweighs the newly created emissions, by a factor of 10, approximately (GFCCC 2015)

Annex 2

Table 19.7 Additional reference materials on a sustainable ocean economy

Author	Report	Summary
Sustainable ocean economy reports		
UNGC, 2020	*Ocean Stewardship 2030*	This report offers a roadmap for how ocean-related industries and policymakers can jointly secure a healthy and productive ocean by 2030. The report describes five critical areas of success. For each area, the report suggests two ambitions and puts forward several recommendations addressing critical dimensions of public and private actions to accelerate ocean-related solutions
European Commission, 2020	*The EU Blue Economy Report*	This report highlights the need to preserve marine ecosystems to optimise potential benefits of ecosystem services and marine and maritime economic sectors
European Parliamentary Research Service, 2020	*The Blue Economy: Overview and EU Policy Framework*	This report looks into the EU policy framework and the different EU initiatives and actions taken in these areas, both by providing an overview of the cross-cutting 'key enablers' of the blue economy and by providing an analysis by blue economy sector (excluding the sectors of coastal protection and maritime defence)
Konrad Adenauer Stiftung/FICCI, 2019	*Blue Economy: Global Best Practices Takeaways for India and Partner Nations*	This report systematically examines and explains the performance, projected growth in terms of size and value, challenges and precise opportunities for capacity expansion and quality enhancement, including technology and process upgrades, in the relevant sectors of India's blue economy. The report also elaborates the global best practices relevant to India as well as innovative financing tools. The report makes several practical recommendations for an effective way forward, both for the government and businesses
OECD, 2019	*Rethinking Innovation for a Sustainable Ocean Economy*	This report on the ocean economy emphasises the growing importance of science and technologies in improving the sustainable economic development of our seas and ocean
World Bank, 2019	*Indonesia Economic Quarterly: Oceans of Opportunity*	This report discusses the importance of the maritime economy to Indonesia's economic development and presents the challenges and opportunities the country faces in leveraging the maritime economy for greater prosperity
Africa Institute of South Africa, 2018	*The Blue Economy Handbook of the Indian Ocean Region*	This handbook offers insight into the various aspects and impacts of the blue economy in the Indian Ocean region. From shifting paradigms, to an accounting framework, gender dynamics, the law of the sea and renewable energy, it aims to increase awareness of the blue economy in this region and to provide evidence to help policymakers in the region make informed decisions
World Bank Group, UN DESA, 2017	*The Potential of the Blue Economy: Increasing Long-Term Benefits of the Sustainable Use of Marine Resources for Small Island Developing States and Coastal Least-Developed Countries*	Drafted by a working group of UN entities, the World Bank and other stakeholders, this report offers a common understanding of the blue economy. It seeks to highlight the importance of such an approach, particularly for small island developing states and coastal least developed countries; to identify some of the key challenges posed by adoption of the blue economy; and to suggest some broad next steps that are called for in order to ensure its implementation
WWF, 2017	*Reviving the Western Indian Ocean Economy: Actions for a Sustainable Future*	This report aims to help Western Indian Ocean countries achieve the Sustainable Development Goal plan of action for 2016–30 in the ocean sector and thus realise the vision, expressed under the regional strategic action programme, of 'people prospering from a healthy Western Indian Ocean'
Commonwealth Secretariat, 2016	*The Blue Economy and Small States* (Commonwealth Blue Economy Series, no. 1)	The Commonwealth Blue Economy Series presents a synthesis of information and practical advice to Commonwealth governments relating to the potential deployment of a range of policy options for different sectors and opportunities for the road ahead. In so doing, this series aims to support the development of the blue economy in Commonwealth countries by providing a high-level assessment of the opportunities available for economic diversification and sustainable growth
Global Ocean Commission, 2016	*The Future of Our Ocean: Next Steps and Priorities*	To accelerate progress towards reversing ocean degradation and drive the global system for ocean governance, the Global Ocean Commission calls upon UN member states and all relevant stakeholders to agree a stand-alone Sustainable Development Goal (SDG) for the global ocean, thus putting the global ocean front and centre on the post-2015 UN development agenda

Table 19.7 (continued)

Author	Report	Summary
OECD, 2016	*The Ocean Economy in 2030*	This report explores the growth prospects for the ocean economy, its capacity for future employment creation and innovation, and its role in addressing global challenges. Special attention is devoted to the emerging ocean-based industries in light of their high growth and innovation potential, and their possible contribution to addressing challenges such as energy security, environment, climate change and food security
World Bank, 2016	*Toward a Blue Economy: A Promise for Sustainable Growth in the Caribbean*	This report serves as a guide to help Caribbean policymakers plan a successful transition to a blue economy and to socially equitable 'blue growth'. This report attempts to quantify the current value of the ocean economy in the region and to summarise projections about where we may find new pockets of sustainable growth
UNEP, 2015	*Blue Economy: Sharing Success Stories to Inspire Change*	This report shares stories that illustrate how economic indicators and development strategies can better reflect the true value of such widespread benefits and potentially even build on them
WWF, 2015	*Reviving the Ocean Economy: The Case for Action*	This report analyses the ocean's role as an economic powerhouse and outlines the threats that are pushing it toward collapse. This report presents an eight-point action plan that would restore ocean resources to their full potential
WWF, 2015	*Living Blue Planet*	This report provides a science-based analysis of the health of our planet and the impact of human activity upon it
California Environmental Associates, 2015	*Ocean Prosperity Roadmap: Fisheries and Beyond*	This report collects research designed to inform decision-makers, including governments and investors, about effective ocean and coastal resource management strategies to maximise economic, conservation and societal benefits
Global Ocean Commission, 2014	*From Decline to Recovery: A Rescue Package for the Global Ocean*	This report outlines a set of eight practical proposals to address the five drivers of decline, reverse high seas degradation and improve the system of governance, monitoring and compliance
UNCTAD, 2014	*The Oceans Economy: Opportunities and Challenges for Small Island Developing States*	This report is a joint effort by a team of experts from the UN Conference on Trade and Development and the Commonwealth Secretariat to better understand the implications of the nascent and evolving concept of the ocean economy. It underlines the importance of sustainable oceanic activities for the development of small island developing states (SIDS) and other coastal states. The report identifies both opportunities and challenges for SIDS in existing and emerging trade-related sectors such as sustainable fisheries and aquaculture, ocean-based renewable energy, marine bio-prospecting, maritime transport and marine and coastal tourism
Blue Ribbon Panel, 2013	*Indispensable Ocean: Aligning Ocean Health and Human Well-Being*	This report by the Blue Ribbon Panel (composed of 21 global leaders in government, industry, conservation and academia) identifies five high-level principles to guide the selection and prioritisation of initiatives aimed at aligning ocean health and human well-being
UNEP, 2012	*Green Economy in a Blue World: Synthesis Report*	This report analyses how key sectors that are interlinked with the marine and coastal environment can make the transition towards a green economy. It covers the impacts and opportunities linked with shipping and fisheries to tourism, marine-based renewable energies and agriculture
Sector-specific		
UNCTAD, 2019	'Advancing Sustainable Development Goal 14: Sustainable Fish and Sea food Value Chains, Trade and Climate'	This background note reviews current trends and projections of fish and seafood trade, and recent work undertaken to support implementation of the trade-related activities of SDG 14, with a focus on the work of UNCTAD, FAO and UN Environment
World Bank, 2017	*The Sunken Billions Revisited: Progress and Challenges in Global Marine Fisheries*	This report builds on *The Sunken Billions: The Economic Justification for Fisheries Reform*, a 2009 study published by the World Bank and the Food and Agriculture Organization of the United Nations, but with a deeper regional analysis
Commonwealth Secretariat, 2016	*Capture Fisheries* (Commonwealth Blue Economy Series, no. 3)	This report presents recommendations that could be implemented by small island developing states (SIDS) to protect and sustainably develop their capture fisheries within a blue economy model. The report describes some of the challenges faced in managing capture fisheries, the potential for a blue economy approach to making improvements, some suggestions for strategies and activities that could be undertaken by SIDS to further these aims and a number of case studies illustrating positive actions that have been taken by SIDS and their outcomes

Table 19.7 (continued)

Author	Report	Summary
FAO, 2014	*Global Blue Growth Initiative and Small Island Developing States (SIDS)*	This report identifies fish and fisheries as the mainstay of food security and the wealth of most small island developing states (SIDS). Many SIDS are heavily dependent on their oceanic and coastal fisheries resources for economic growth and development, as well as food security and livelihoods, and are vulnerable to any change in the state of these resources
Aquaculture		
UNGC, 2020	*Seaweed Manifesto*	This report defines a vision for an upscaled, responsible and restorative seaweed industry, playing a globally significant role in food security, climate change mitigation and support of the marine ecosystem, as well as contributing to job creation and poverty alleviation. The Seaweed Manifesto explores the challenges and barriers to responsible development of the industry
TNC, 2019	*Towards a Blue Revolution: Catalyzing Private Investment in Sustainable Aquaculture Production Systems*	This report seeks to articulate the full scale and potential of the aquaculture sector to catalyse investment in projects and companies that can deliver targeted financial returns and improved environmental performance over business-as-usual production
FOA, 2018	'Achieving Blue Growth'	This paper presents the Blue Growth Initiative and the three pillars of sustainable development—social, economic and environmental—that can enable fisheries and aquaculture to contribute to the 2030 Agenda's Sustainable Development Goals. The Blue Growth Initiative is a strategic approach to improving the use of aquatic resources and achieving better social, economic and environmental outcomes
Commonwealth Secretariat, 2016	*Aquaculture* (Commonwealth Blue Economy Series, no. 2)	This volume explores the potential for the development of a blue economy mariculture industry, as well as specific enabling conditions for economic opportunity
FOA, 2015	*Fisheries and Aquaculture in the Context of Blue Economy*	This report looks at the current situation of fisheries and aquaculture in the context of the blue economy or blue growth and its relevance for African coastal countries
World Bank, 2013	*Fish to 2030: Prospects for Fisheries and Aquaculture*	This report presents global prospects for fisheries and aquaculture and analyses future trends out to 2030
Tourism		
IDDRI, 2019	*Sustainable Blue Tourism*	This report explores the ecological impacts of coastal and marine tourism in the Mediterranean, the Caribbean, the Northeast Atlantic, the South Pacific Ocean and the Western Indian Ocean, the major global marine regions, in order to disseminate lessons from the field and develop common policy recommendations for policymakers, tourism stakeholders and other relevant institutional and civil society actors
UNWTO, 2016	*Sustainable Cruise Tourism Development Strategies: Tackling the Challenges in Itinerary Design in South-East Asia*	This report issues a call to action at a critical juncture in Southeast Asian development. It seeks to further awareness of sustainable development in cruise tourism, catalyse collaboration across the region and stimulate the strategic implementation of best practices and innovations
EU Commission, 2016	*Study on Specific Challenges for a Sustainable Development of Coastal and Maritime Tourism in Europe*	This report first presents the findings on specific challenges and innovative response strategies for sustainable development of coastal and maritime tourism, including challenges related to island connectivity (Part A) and innovative practices for marina development (Part B). It then presents findings related to innovative strategies for a more competitive nautical tourism sector, including marina development
UNWTO, 2013	*Sustainable Tourism Governance and Management in Coastal Areas of Africa*	This report presents the results of the research carried out within the framework of the Collaborative Actions for Sustainable Tourism (COAST) project. It builds on Making Tourism More Sustainable: A Guide for Policy Makers, published by the UN World Tourism Organization and UN Environment, assessing how to apply sustainability principles and policy instruments for coastal tourism development in Africa
Shipping		
IRENA, 2019	*Navigating the Way to a Renewable Future: Solutions to Decarbonise Shipping*	This report explores the impact of maritime shipping on CO2 emissions, the structure of the shipping sector and key areas that need to be addressed to reduce the sector's carbon footprint
UK Department of Transportation, 2019	*Reducing the Maritime Sector's Contribution to Climate Change and Air Pollution*	This report provides a framework for assessing current and future economic opportunities in the design, development and commercialisation of technologies and low-emission fuels to reduce UK shipping emissions

Table 19.7 (continued)

Author	Report	Summary
EU Commission, Directorate-General for Mobility and Transport, 2017	*Study on Differentiated Port Infrastructure Charges to Promote Environmentally Friendly Maritime Transport Activities and Sustainable Transportation*	This study assesses existing schemes for differentiating port infrastructure charges according to environmental or sustainability criteria
Sustainable Shipping Initiative, 2016	*Progress to 2015: A Future for Sustainable Shipping*	This progress report details the efforts and key achievements of SSI members to drive debate on and inspire change within the shipping sector
WWF, 2011	*Global Sustainable Shipping Initiatives: Audit and Overview 2011*	This report updates research conducted in 2004 and highlights the fundamental changes to sustainable shipping initiatives since then. It identifies drivers of these changes and shifts in opinion regarding the best methods of delivering global, sustainable shipping
Coastal ecosystems		
OECD, 2019	*Responding to Rising Seas*	This report reviews how countries in the Organisation for Economic Co-operation and Development can use their national adaptation planning processes to meet the challenge of rising sea levels. Specifically, the report examines how countries approach shared costs and responsibilities for coastal risk management and how this encourages or hinders risk-reduction behaviour by households, businesses and different levels of government
World Bank Group, 2016	'Managing Coasts with Natural Solutions: Guidelines for Measuring and Valuing the Coastal Protection Services of Mangroves and Coral Reef'	This guidance note offers recommendations for how to measure and value the protective services of mangroves and coral reefs to support planning for development, disaster risk and coastal zone management
Center for American Progress and Oxfam America, 2014	*The Economic Case for Restoring Coastal Ecosystems*	This report explores the economic contributions provided by healthy, restored coastal ecosystems such as wetlands, seagrass beds and oyster reefs. An analysis of three federally funded projects reveals that well-designed coastal restoration can be highly cost-effective, returning significantly more than the cost of the restoration project
Blue finance reports		
UNGC, 2020	'Blue Bonds: Reference Paper for Investments Accelerating Sustainable Ocean Business'	This paper outlines the opportunities for the environmental, social and governance bond market to secure capital for ocean-related projects and companies that have made, or are planning to make, significant contributions to the Sustain- able Development Goals
Friends of Ocean Action, 2020	*The Ocean Finance Handbook*	This handbook provides an up-to-date overview of the investment landscape in the blue economy. It seeks to formulate a common understanding of sustainable blue economy financing for all stakeholders
Credit Suisse, 2020	*Investors and the Blue Economy*	This study assesses investor perspectives on the ocean, bringing together views on and awareness of the sustainable blue economy among asset owners and managers worldwide
IIED, 2019	'Navigating Ocean Investments'	This briefing considers a business model that could bridge the marine conservation funding gap
Wildlife Conservation Society and Conservation Finance Alliance, 2018	*Finance Tools for Coral Reef Conservation: A Guide*	This working guide to financial tools available for coral reef conservation highlights 13 of the most compelling finance mechanisms

About the Authors

Eliza Northrop *is the Policy Lead for the Secretariat for the High Level Panel for a Sustainable Ocean Economy and Senior Associate for the World Resources Institute's Sustainable Ocean Initiative.* Contact: Eliza.Northrop@wri.org

Manaswita Konar *is the Lead Ocean Economist for the World Resources Institute's Sustainable Ocean Initiative.* Contact: Manaswita.Konar@wri.org

Nicola Frost *is the Deputy Head of Secretariat for the High Level Panel for a Sustainable Ocean Economy.* Contact: Nicola.Frost@wri.org

Elizabeth Hollaway *is the Research Assistant for the World Resources Institute's Sustainable Ocean Initiative.* Contact: Elizabeth.Hollaway@wri.org

References

African Development Bank (2020) Egypt: African Development Bank approves 225 million Euros budget support loan for electricity sector to bolster economic resilience and sustainability amid COVID-19. 17 June. https://www.afdb.org/en/news-and-events/press-releases/egypt-african-development-bank-approves-225-million-euros-budget-support-loan-electricity-sector-bolster-economic-resilience-and-sustainability-amid-covid-19-36237

Alleway H, Gillies C, Bishop M, Gentry R, Theuerkauf S, Jones R (2018) The ecosystem services of marine aquaculture: valuing benefits to people and nature. BioScience 69(December):59–68. https://doi.org/10.1093/biosci/biy137

Ambrose J (2020) Offshore wind energy investment quadruples despite Covid-19 slump. The Guardian, 13 July. https://www.theguardian.com/environment/2020/jul/13/offshore-wind-energy-investment-quadruples-despite-covid-19-slump

ANI (Asian News International) (2020) ADB, ESCAP to help Asia Pacific in addressing COVID-19 crisis. Big News Network, 3 June. https://www.bignewsnetwork.com/news/265333621/adb-escap-to-help-asia-pacific-in-addressing-covid-19-crisis

Anthem P (2020) Risk of hunger pandemic as coronavirus set to almost double acute hunger by end of 2020. World Food Programme Insight (blog), 28 May. https://insight.wfp.org/covid-19-will-almost-double-people-in-acute-hunger-by-end-of-2020-59df0c4a8072

Appeaning Addo K, Steven A, Thanh CV (Forthcoming) Coastal development: managing resilience, restoration and infrastructure of coastlines. Blue Paper. https://oceanpanel.org/blue-papers/coastal--development-managing-resilience-restoration-and-infrastructure-coastlines

Aquafeed (2020) Coronavirus pandemic to hit the aquaculture industry hard. http://www.aquafeed.com/af-article/9459/Coronavirus-pandemic-to-hit-the-aquaculture-industry-hard/

Arquiza Y (2019) Reaping rewards from responsible fishing and savings clubs. https://rare.org/story/reaping-rewards-from-responsible-fishing-and-savings-clubs/

ASEA (American Solar Energy Association), MIS (Management Information Services) (2009) Defining, estimating, and forecasting the renewable energy and energy efficiency industries in the U.S. and in Colorado. Report

Ash N, Scarbrough T (2019) Sailing on solar: could green ammonia decarbonise international shipping? Environmental Defense Fund. https://europe.edf.org/file/399/download?token=agUEbKeQ

Asian Development Bank (2019) ADB launches $5 Billion Healthy Oceans Action Plan. 2 May. https://www.adb.org/news/adb-launches-5-billion-healthy-oceans-action-plan

AST (2019) iVMS: inshore vessel monitoring systems. https://www.theastgroup.com/uk/solutions/remote-asset-tracking-monitoring-control/ivms/

Australian Government (2020a) Offshore clean energy infrastructure: proposed framework. January. https://consult.industry.gov.au/offshore-exploration/offshore-clean-energy-infrastructure/

Australian Government (2020b) Fisheries and COVID-19. Department of Agriculture, Water and the Environment, May. https://www.agriculture.gov.au/coronavirus/fisheries#victoria

Barbesgaard M (2018) Blue growth: savior or ocean grabbing? J Peasant Stud 45(1):130–149. https://doi.org/10.1080/03066150.2017.1377186

Bark RH, Osgood DE, Colby BG, Katz G, Stromberg J (2009) Habitat preservation and restoration: do homebuyers have preferences for quality habitat? Ecol Econ 68(5):1465–1475. https://doi.org/10.1016/j.ecolecon.2008.10.005

Bayraktarov E, Saunders MI, Abdullah S, Mills M, Beher J, Possingham HP, Mumby PJ, Lovelock CE (2015) The cost and feasibility of marine coastal restoration. Ecol Appl 26(4):1055–1074. https://doi.org/10.1890/15-1077

Beck MW, Lange G-M (2016) Managing coasts with natural solutions: guidelines for measuring and valuing the coastal protection services of mangroves and coral reefs. https://documents.worldbank.org/en/publication/documents-reports/documentdetail

Beck MW, Losada IJ, Menéndez P, Reguero BG, Díaz-Simal P, Fernández F (2018) The global flood protection savings provided by coral reefs. Nat Commun 9(1):2186. https://doi.org/10.1038/s41467-018-04568-z

BenDor T, Lester TW, Livengood A, Davis A, Yonavjak L (2015) Estimating the size and impact of the ecological restoration economy. PloS One 10(6):e0128339. https://doi.org/10.1371/journal.pone.0128339

Bennett NJ (2018) Navigating a just and inclusive path towards sustainable oceans. Mar Policy 97(November):139–146. https://doi.org/10.1016/j.marpol.2018.06.001

Bennett NJ, Finkbeiner EM, Ban NC, Belhabib D, Jupiter SD, Kittinger JN, Mangubhai S et al (2020) The COVID-19 pandemic, small-scale fisheries and coastal fishing communities. Coast Manag 48:336–347. https://doi.org/10.1080/08920753.2020.1766937

Best BD, Halpin PN (2019) Minimizing wildlife impacts for offshore wind energy development: winning tradeoffs for seabirds in space and cetaceans in time. *PloS One* 14(5):e0215722. https://doi.org/10.1371/journal.pone.0215722

Bezdek RH (2019) The hydrogen economy and jobs of the future. Renew Energy Environ Sustain 4:1. https://doi.org/10.1051/rees/2018005

Bjerregaard R, Brummett R, Hansen S, Valderrama D, Radulovich R, Diana J, Capron M et al (2016) Seaweed aquaculture for food security, income generation and environmental health in tropical developing countries. World Bank, Washington, DC

Blasiak R, Wynberg R, Grorud-Colvert K, Thambisetty S et al (2020) The ocean genome: conservation and the fair, equitable and sustainable use of marine genetic resources. World Resources Institute, Washington, DC. https://www.oceanpanel.org/blue-papers/ocean--genome-conservation-and-fair-equitable-and-sustainable-use-marine-genetic

Bloomberg M, Holloway C (2018) NYC green infrastructure plan: executive summary. http://www.circleofblue.org/wp-content/uploads/2011/12/NYCGreenInfrastructurePlan_ExecutiveSummary.pdf

Bluedorn J, Chen W (2020) Making economies more resilient to downturns. IMF Blog, 18 May. https://blogs.imf.org/2020/05/18/making-economies-more-resilient-to-downturns/

BMU (German Federal Ministry for the Environment, Nature Conservation and Nuclear Safety) (2020) Global green recovery and the International Climate Initiative. Bmu.De (blog), 17 July. https://www.bmu.de/PM9147-1

Bowlin N (2020) Hunting and fishing provide food security in the time of COVID-19. 29 April. https://www.hcn.org/articles/covid19--hunting-and-fishing-provide-food-security-in-the-time-of-covid-19

Brander L et al (2015) The benefits to people of expanding marine protected areas. Final report. Institute for Environmental Studies, VU University, Amsterdam. https://www.issuelab.org/resources/25951/25951.pdf?download=true&_ga=2.227198557.1167454837.1558640107-1857028723.1558640107

Brett A, Leape J, Abbott M, Sakaguchi H, Cao L, Chand K, Golbuu Y et al (2020) Ocean data need a sea change to help navigate the warming world. Nature 582(7811):181–183. https://doi.org/10.1038/d41586-020-01668-z

Bricker SB, Ferreira JG, Zhu C, Rose JM, Galimany E, Wikfors G, Saurel C et al (2018) Role of shellfish aquaculture in the reduction of eutrophication in an urban estuary. Environ Sci Technol 52(1):173–183. https://doi.org/10.1021/acs.est.7b03970

Brouwer R, Akter S, Brander L, Haque E (2007) Socioeconomic vulnerability and adaptation to environmental risk: a case study of climate change and flooding in Bangladesh. Risk Anal 27(2):313–326. https://doi.org/10.1111/j.1539-6924.2007.00884.x

Browder G, Ozment S, Bescos IR, Gartner T, Lange G-M (2019) Integrating green and gray: creating next generation infrastructure. https://www.wri.org/publication/integrating-green-gray

Buck BH, Troell MF, Krause G, Angel DL, Grote B, Chopin T (2018) State of the art and challenges for offshore integrated multitrophic aquaculture (IMTA). Front Mar Sci 5. https://doi.org/10.3389/fmars.2018.00165

Buonocore JJ, Luckow P, Norris G, Spengler JD, Biewald B, Fisher J, Levy JI (2016) Health and climate benefits of different energy-efficiency and renewable energy choices. Nat Clim Change 6(1):100–105. https://doi.org/10.1038/nclimate2771

Burke L, Reytar K, Spalding M, Perry A (2011) Reefs at risk—revisited. https://files.wri.org/s3fs-public/pdf/reefs_at_risk_revisited.pdf

Buschmann AH, Camus C, Infante J, Neori A, Israel Á, Hernández-González MC, Pereda SV et al (2017) Seaweed production: overview of the global state of exploitation, farming and emerging research activity. Eur J Phycol 52(4):391–406. https://doi.org/10.1080/09670262.2017.1365175

Campbell S, Jakub R, Valdivia A, Setiawan H, Setiawan A, Cox C, Kiyo A et al (2020) Impact of COVID-19 on small-scale coastal fisheries of Southeast Sulawesi, Indonesia. https://doi.org/10.21203/rs.3.rs-39895/v1

Canning JF, Stillwell AS (2018) Nutrient reduction in agricultural green infrastructure: an analysis of the Raccoon river watershed. Water 10(6):749. https://doi.org/10.3390/w10060749

Carbon Trust (2011) Accelerating marine energy. https://prod-drupal-files.storage.googleapis.com/documents/resource/public/Accelerating%20marine%20energy%20-%20REPORT.pdf

Carey y Cía (2020) Subsecretary of fishing and aquaculture establishes the suspension of term, procedures and other measures. 9 April. https://www.carey.cl/en/subsecretary-of-fishing-and-aquaculture--establishes-the-suspension-of-term-procedures-and-other-measures/

Carr S (2020) How the COVID-19 pandemic is affecting marine ecosystems, part 1: fisheries and aquaculture. Skimmer, 30 April. https://meam.openchannels.org/news/skimmer-marine-ecosystems-and-management/how-covid-19-pandemic-affecting-marine-ecosystems-part

Carrasquila-Henao M, Juanes F (2017) Mangroves enhance local fisheries catches: a global meta-analysis. Fish Fish 18(January):79–93. https://doi.org/10.1111/faf.12168

Casey J (2020) Norway announces $384.5m clean energy fund to aid in Covid-19 recovery. Power Technology, 9 June. https://www.power--technology.com/news/norway-announces-384-5m-clean-energy-fund-to-aid-in-covid-19-recovery/

CCR (Caisse Centrale de Réassurance) (2018) Conséquences du changement climatique sur le coût des catastrophes naturelles en France à horizon 2050: Septembre 2018—Catnat. https://catastrophes-naturelles.ccr.fr/-/consequences-du-changement--climatique-sur-le-cout-des-catastrophes-naturelles-en-france-a-horizon-2050-septembre-2018

CFFA (Coalition for Fair Fisheries Arrangements) CAPE (2020) RSA: dialogue between fishers and authorities eases lockdown restrictions. 16 April. https://www.cffacape.org/coronavirus-crisis-impacts-on-african-artisanal-fisheries/south-africa-dialogue-between-artisanal-fishers-and-authorities-helps-ease-lockdown-restrictions

Chambers S (2020a) Thirteen countries sign crew change pact. Splash 247 (10 July). https://splash247.com/12-countries-sign-crew-change-pact/

Chambers S (2020b) UN Secretary-General calls for shipping bailouts to be tied to the Paris Agreement. Splash 247 (10 July). https://splash247.com/un-secretary-general-calls-for-shipping-bailouts-to-be-tied-to-the-paris-agreement/

Charlton E (2019) How oysters are cleaning New York's polluted harbor. World Economic Forum, 25 January. https://www.weforum.org/agenda/2019/01/how-oysters-are-cleaning-new-yorks-polluted-harbor/

Chopin T, Robinson S, Sawhney M, Bastarache S, Belyea E, Shea R, Armstrong W, Fitzgerald P (2004) The AquaNet integrated multi-trophic aquaculture project: rationale of the project and development of kelp cultivation as the inorganic extractive component of the system. Bull Aquacult Assoc Can 104(January):11–18

Christensen J (2020) Decarbonizing shipping could unlock a global energy transition. World Economic Forum, 22 January. https://www.weforum.org/agenda/2020/01/decarbonizing-shipping-global-energy-transition/

Circulate Capital and GA-Circular (2020) Safeguarding the plastic recycling value chain: insights from South and Southeast Asia. Webinar

Clavelle T (2020) Global fisheries during COVID-19. Global Fishing Watch (blog), 12 May. https://globalfishingwatch.org/data-blog/global-fisheries-during-covid-19/

CMA CGM (2020) The CMA CGM Group Heads towards Carbon Neutrality by 2050. https://www.cma-cgm.com/news/3143/the-cma-cgm-group-heads-towards-carbon-neutrality-by-2050

CNC (Comité National de la Conchyliculture) (n.d.) Les statistiques. https://coquillages.com/les-statistiques/. Accessed 21 Aug 2020

Coady D, Parry I, Le N-P, Shang B (2019) Global fossil fuel subsidies remain large: an update based on country-level estimates. International Monetary Fund. https://www.imf.org/en/Publications/WP/Issues/2019/05/02/Global-Fossil-Fuel-Subsidies-Remain-Large-An-Update-Based-on-Country-Level-Estimates-46509

Coke-Hamilton P (2020) Impact of COVID-19 on tourism in small island developing states. United Nations Conference on Trade and Development (blog), 24 April. https://unctad.org/en/pages/newsdetails.aspx?OriginalVersionID=2341

Copping AE, Hemery LG (2020) State of the science report: environmental effects of marine renewable energy development around the world. Ocean Energy Systems. https://tethys.pnnl.gov/publications/state-of-the-science-2020

Copping A et al (2016) Annex IV, 2016 State of the science report: environmental effects of marine renewable energy development around the world. Ocean Energy Systems. https://tethys.pnnl.gov/publications/state-of-the-science-2016

Corporate Citizenship (2020) Sustainable funds are outperforming their peers during the pandemic, according to BNP Paribas. Corporate Citizenship Briefing (blog), 25 June. https://ccbriefing.corporate--citizenship.com/2020/06/25/sustainable-funds-are-outperforming-their-peers-during-the-pandemic-according-to-bnp-paribas/

Costello C, Ovando D, Hilborn R, Gaines SD, Deschenes O, Lester SE (2012) Status and solutions for the world's unassessed fisheries. Science 338(6106):517–520

Costello C, Gao L, Gelcich S et al (2019) The future of food from the sea. World Resources Institute, Washington, DC. https://oceanpanel.org/sites/default/files/2019-11/19_HLP_BP1%20Paper.pdf

Cottier-Cook E, Nagabhatla N, Campbell M, Chopin T, Hewitt C, Hoon G, Kema G et al (2016) Safeguarding the future of the global seaweed aquaculture industry. https://www.sams.ac.uk/t4-media/sams/pdf/globalseaweed-policy-brief.pdf

CRED (Centre for Research on the Epidemiology of Disasters) (2015) The human cost of natural disasters. A global perspective. https://www.cred.be/node/1355

Crouse EC, Endemann BB, Greer LA, Hinckley EB, Hattery DP, Holmes WH (2018) 2018 Offshore wind handbook. https://www.snclavalin.com/~/media/Files/S/SNC-Lavalin/documents/publications/o-g-offshore-wind-brochure_en.pdf

Deloitte (2017) At what price? The economic, social and icon value of the Great Barrier Reef. Report. Deloitte Access Economics. http://elibrary.gbrmpa.gov.au/jspui/handle/11017/3205

Dinh V-N, McKeogh E (2019) Offshore wind energy: technology opportunities and challenges—energy and geotechnics. In: Lecture notes in civil engineering, pp 3–22. https://doi.org/10.1007/978-981-13-2306-5_1

Dreaver C (2020) Budget 2020: what you need to know. Radio New Zealand, 14 May. https://www.rnz.co.nz/news/budget-2020/416634/budget-2020-what-you-need-to-know

Duarte C, Holmer M, Olsen Y, Soto D, Marba N, Guiu J, Black K, Karakassis I (2009) Will the oceans help feed humanity? BioScience 59(December):967–976. https://doi.org/10.1525/bio.2009.59.11.8

Duarte CM, Agusti S, Barbier E, Britten GL, Castilla JC, Gattuso J-P, Fulweiler RW et al (2020) Rebuilding marine life. Nature 580(7801):39–51. https://doi.org/10.1038/s41586-020-2146-7

Eckhouse B, Martin C (2020) Coronavirus crushing global forecasts for wind and solar power. Bloomberg, 27 March. https://www.bloomberg.com/news/articles/2020-03-27/global-solar-wind-growth-will-be-erased-this-year-rystad-says

Edwards P, Sutton-Grier A, Coyle G (2013) Investing in nature: restoring coastal habitat blue infrastructure and green job creation. Mar Policy 38(March):65–71. https://doi.org/10.1016/j.marpol.2012.05.020

Eklöf J, Torre-Castro M, Adelsköld L, Jiddawi N, Kautsky N (2005) Differences in macrofauna and seagrass assemblages in seagrass beds with and without seaweed farms. Estuar Coast Shelf Sci 325(May):385–396. https://doi.org/10.1016/j.ecss.2004.11.014

Energy Choice Coalition (2020) Wood Mackenzie, industry associations report coronavirus impact on energy transition. 17 April. https://www.energychoicecoalition.org/blog/2020/4/17/wood--mackenzie-industry-associations-report-coronavirus-impact-on-energy-transition

EPA (U.S. Environmental Protection Agency) (2010) Green infrastructure case studies: municipal policies for managing stormwater with green infrastructure. August. https://toolkit.climate.gov/reports/green-infrastructure-case-studies-municipal-policies-managing-stormwater-green

EPA (2016) Greenhouse gas emissions from a typical passenger vehicle. Overviews and Factsheets, 12 January. https://www.epa.gov/greenvehicles/greenhouse-gas-emissions-typical-passenger-vehicle

EPA (2018) Greenhouse gas emissions from a typical passenger vehicle. Green Vehicle Guide. Last updated 18 May. https://www.epa.gov/greenvehicles/greenhouse-gas-emissions-typical-passenger--vehicle#:~:text=A%20typical%20passenger%20vehicle%20emits,of%20miles%20driven%20per%20year

Erbe C, Marley SA, Schoeman RP, Smith JN, Trigg LE, Embling CB (2019) The effects of ship noise on marine mammals: a review. Front Mar Sci 6:606. https://doi.org/10.3389/fmars.2019.00606

ESCA (European Community Shipowners' Associations) (2020) ECSA COVID-19 survey May 2020. https://www.ecsa.eu/sites/default/files/publications/ECSA%20COVID19%20Survey%20Report.pdf

ETIP Ocean (European Technology and Innovation Platform for Ocean Energy) (2020) Strategic research and innovation agenda for ocean energy. https://www.oceanenergy-europe.eu/wp-content/uploads/2020/05/ETIP-Ocean-SRIA.pdf

EU Commission (2018) Study on the economic benefits of MPAs. Final report. Publications Office of the European Union. http://op.europa.eu/en/publication-detail/-/publication/dbe3d250-b0b5-11e8-99ee-01aa75ed71a1/language-en/format-PDF

EU Commission (2020a) State aid: coronavirus: €20 million Portuguese scheme. April. https://ec.europa.eu/commission/presscorner/detail/en/IP_20_609

EU Commission (2020b) State aid to Italy, coronavirus. April. https://ec.europa.eu/commission/presscorner/detail/en/IP_20_712

EU Commission (2020c) EU strategy on offshore renewable energy. 24 June. https://ec.europa.eu/energy/topics/renewable-energy/eu-strategy-offshore-renewable-energy_en

EU Commission (2020d) Communication from the commission guidelines concerning the exercise of the free movement of workers during COVID-19 outbreak. Official Journal of the European Union CI 102/12 (March). https://eur-lex.europa.eu/legal-content/EN/TXT/?uri=CELEX:52020XC0330(03)

EU Parliament (2020) Shipping industry must contribute to climate neutrality, say MEPs. News—European Parliament, 7 July. https://www.europarl.europa.eu/news/en/press-room/20200703IPR82633/shipping-industry-must-contribute-to-climate-neutrality-say-meps

European Commission (2018) More efficient design makes new generation electric ferries cheaper to run. https://cordis.europa.eu/ article/id/254141-more-efficient-design-makes-new-generation-electric-ferries-cheaper-to-run

European Commission (2020a) Study on the economic impact of maritime spatial planning: final report: abridged version. Publications Office of the European Union. 24 June. http://op.europa.eu/en/publication-detail/-/publication/254a6ac4-b689-11ea-bb7a-01aa75ed71a1/language-en

European Commission (2020b) The EU Blue Economy Report. https://ec.europa.eu/maritimeaffairs/sites/maritimeaffairs/files/2020_06_blueeconomy-2020-ld_final.pdf

EWEA (European Wind Energy Association) (2012) The European offshore wind industry key 2011 trends and statistics. http://www.ewea.org/fileadmin/files/library/publications/statistics/EWEA_stats_offshore_2011_02.pdf

FAO (2018a) The state of world fisheries and aquaculture 2018: sustainability in action. https://doi.org/10.4060/ca9229en

FAO (2018b) The Port Bleu project equips fishermen from Haouaria with quality labels and technological solutions for fisheries products. 28 March. http://www.fao.org/neareast/news/view/en/c/1112028/

FAO (2020a) Summary of the impacts of the COVID-19 pandemic on the fisheries and aquaculture sector: addendum to the state of world fisheries and aquaculture 2020. http://www.fao.org/3/ca9349en/CA9349EN.pdf

FAO (2020b) The state of world fisheries and aquaculture 2020: sustainability in action. https://doi.org/10.4060/ca9229en.

FAO (2020c) The state of world fisheries and aquaculture. Sustainability in action. FAO, Rome. https://doi.org/10.4060/ca9229en

Farrell P, Thow AM, Tutuo Wate J, Nonga N, Vatucawaqa P, Brewer T, Sharp MK et al (2020) COVID-19 and Pacific food system resilience: opportunities to build a robust response. Food Secur 12(4):783–791. https://doi.org/10.1007/s12571-020-01087-y

Ferrario F, Beck MW, Storlazzi CD, Micheli F, Shepard CC, Airoldi L (2014) The effectiveness of coral reefs for coastal hazard risk reduction and adaptation. Nat Commun 5(1):3794. https://doi.org/10.1038/ncomms4794

Ferreira JG, Corner RA, Moore H, Service M, Bricker SB, Rheault R (2018) Ecological carrying capacity for shellfish aquaculture: sustainability of naturally occurring filter-feeders and cultivated bivalves. J Shellfish Res 37(4):709–726. https://doi.org/10.2983/035.037.0404

Finnish Government (2020) Government reaches agreement on fourth supplementary budget proposal for 2020. 2 June. https://valtioneuvosto.fi/en/-/10616/hallitus-paatti-vuoden-2020-neljannesta-lisatalousarvioesityksesta

Fisheries and Oceans Canada (2020) Minister Jordan announces the fish harvester benefit and fish harvester grant to open on August 24th. https://www.canada.ca/en/fisheries-oceans/news/2020/08/minister-jordan-announces-the-fish-harvester-benefit-and-fish-harvester-grant-to-open-on-august-24th.html

Forster J, Radulovich R (2015) Seaweed and food security 2015. In: Seaweed sustainability: food and non-food applications, pp 289–313. https://doi.org/10.1016/B978-0-12-418697-2.00011-8

Free CM, Mangin T, Molinos JG, Ojea E, Costello C, Gaines SD (2019) Realistic fisheries management reforms could mitigate the impacts of climate change in most countries. *bioRxiv*, October, 804831. https://doi.org/10.1101/804831

Friess D, Rogers K, Lovelock C, Krauss K, Hamilton S, Lee S, Lucas R et al (2019) The state of the world's mangrove forests: past, present, and future. Annu Rev Env Resour 44(August):1–27. https://doi.org/10.1146/annurev-environ-101718-033302

Froehlich HE, Runge CA, Gentry RR, Gaines SD, Halpern BS (2018) Comparative terrestrial feed and land use of an aquaculture-dominant world. Proc Natl Acad Sci U S A 115(20):5295–5300. https://doi.org/10.1073/pnas.1801692115

Froehlich HE, Afflerbach JC, Frazier M, Halpern BS (2019) Blue growth potential to mitigate climate change through seaweed offsetting. Curr Biol 29(18):3087–93.e3. https://doi.org/10.1016/j.cub.2019.07.041

Gaines S, Cabral R, Free C, Golbuu Y, Arnason R, Battista W, Bradley D et al (2019) The expected impacts of climate change on the ocean economy. World Resources Institute, Washington, DC. https://www.oceanpanel.org/sites/default/files/2019-12/expected-impacts-climate-change-on-the-ocean-economy-executive-summary.pdf

Gentry RR, Froehlich HE, Grimm D, Kareiva P, Parke M, Rust M, Gaines SD, Halpern BS (2017) Mapping the global potential for marine aquaculture. Nat Ecol Evol 1(9):1317–1324. https://doi.org/10.1038/s41559-017-0257-9

GFCCC (Global Food Cold Chain Council) (2015) Assessing the potential of the cold chain sector to reduce GHG emissions through food loss and waste reduction. October. http://naturalleader.com/wp-content/uploads/2016/04/coldchainGHGemissionstudy.pdf

Government of Canada (2020) Canada's COVID-19 economic response plan. Financial material. 30 June. https://www.canada.ca/en/department-finance/economic-response-plan.html

Grabowski JH, Brumbaugh RD, Conrad RF, Keeler AG, Opaluch JJ, Peterson CH, Piehler MF et al (2012) Economic valuation of ecosystem services provided by oyster reefs. BioScience 62(10):900–909. https://doi.org/10.1525/bio.2012.62.10.10

Greenfield P, Muiruri P (2020) Conservation in crisis: ecotourism collapse threatens communities and wildlife. The Guardian, 5 May. https://www.theguardian.com/environment/2020/may/05/conservation-in-crisis-covid-19-coronavirus-ecotourism-collapse-threatens-communities-and-wildlife-aoe

GSI (IISD Global Subsidies Initiative) (2020) Fisheries negotiations slow down due to COVID-19. 16 April. https://www.iisd.org/gsi/news-events/fisheries-negotiations-covid

Guo Q, Chen J, Wang J, Zeng H, Yu J (2020) Recent progress in synthesis and application of mussel-inspired adhesives. Nanoscale 12(3):1307–1324. https://doi.org/10.1039/C9NR09780E

Gutiérrez NL, Hilborn R, Defeo O (2011) Leadership, social capital and incentives promote successful fisheries. Nature 470(7334):386–389. https://doi.org/10.1038/nature09689

Hai NT, Dell B, Phuong VT, Harper RJ (2020) Towards a more robust approach for the restoration of mangroves in Vietnam. Ann For Sci 77(1):18. https://doi.org/10.1007/s13595-020-0921-0

Halpern BS, Lester SE, McLeod KL (2010) Placing marine protected areas onto the ecosystem-based management seascape. Proc Natl Acad Sci U S A 107(43):18312–18317. https://doi.org/10.1073/pnas.0908503107

Hammer S, Hallegatte S (2020) Planning for the economic recovery from COVID-19: a sustainability checklist for policymakers. World Bank Blogs, 14 April. https://blogs.worldbank.org/climatechange/planning-economic-recovery-covid-19-coronavirus-sustainability-checklist-policymakers

Harper S, Adshade M, Lam VWY, Pauly D, Sumaila UR (2020) Valuing invisible catches: estimating the global contribution by women to small-scale marine capture fisheries production. PloS One 15(3):e0228912. https://doi.org/10.1371/journal.pone.0228912

Harrabin R (2020) Make Covid-19 recovery green, say business leaders. BBC News, 1 June. https://www.bbc.com/news/business-52851185

Harvey F (2020) Covid-19 relief for fossil fuel industries risks green recovery plans. The Guardian, 6 June. https://www.theguardian.com/environment/2020/jun/06/covid-19-relief-for-fossil-fuel-industries-risks-green-recovery-plans

Hassellöv I-M, Turner D, Lauer A, Corbett J (2013) Shipping contributes to ocean acidification. Geophys Res Lett 40(June):2731–2736. https://doi.org/10.1002/grl.50521

Haugan PM, Levin LA, Amon D, Hemer M, Lily H, Nielsen FG (2020) What role for ocean-based renewable energy and deep seabed minerals in a sustainable future? World Resources Institute, Washington, DC. www.oceanpanel.org/blue-papers/ocean-energy-and-mineral-sources

Heiland I, Ulltveit-Moe K-H (2020) An unintended crisis: COVID-19 restrictions hit sea transportation. Vox EU, 17 May. https://voxeu.org/article/covid-19-restrictions-hit-sea-transportation

Henriksen S, Selwyn M (2020) How to protect global supply chains under threat from the COVID-19 pandemic. World Economic Forum (blog), 5 May. https://www.weforum.org/agenda/2020/05/global-supply-chains-are-under-imminent-threat-from-the-covid-19-pandemic/

Hepburn C, O'Callaghan B, Stern N, Stiglitz J, Zenghelis D (2020) Will COVID-19 fiscal recovery packages accelerate or retard progress on climate change? Oxford Rev Econ Policy 8:graa015. https://doi.org/10.1093/oxrep/graa015

Hochard JP, Hamilton S, Barbier EB (2019) Mangroves shelter coastal economic activity from cyclones. Proc Natl Acad Sci U S A 116(25):12232–12237. https://doi.org/10.1073/pnas.1820067116

Hockings M, Dudley N, Elliott W, Napolitano Ferreira M, MacKinnon K, Pasha MKS et al (2020) Editorial essay: COVID-19 and protected and conserved areas. Parks J (4). https://parksjournal.com/wp-content/uploads/2020/06/Hockings-ct-al-10.2305-IUCN.CH_.2020.PARKS-26-1MH.en_-1.pdf

Hoegh-Guldberg O, Caldeira K, Chopin T, Gaines S, Haugan P (2019) The ocean as a solution to climate change: five opportunities for action. World Resources Institute, Washington, DC. https://oceanpanel.org/sites/default/files/2019-10/HLP_Report_Ocean_Solution_Climate_Change_final.pdf

Holmyard N (2020) WSI: COVID-19 crisis will deepen gender inequalities in seafood sector. Seafood Source (blog), 13 April. https://www.seafoodsource.com/news/business-finance/wsi-covid-19--crisis-will-deepen-gender-inequalities-in-seafood-sector

Hudson A (2017) Restoring and protecting the world's large marine ecosystems: an engine for job creation and sustainable economic development. Environ Dev 22(June):150–155. https://doi.org/10.1016/j.envdev.2016.10.003

Hughes BB, Lummis SC, Anderson SC, Kroeker KJ (2018) Unexpected resilience of a seagrass system exposed to global stressors. Glob Chang Biol 24(1):224–234. https://doi.org/10.1111/gcb.13854

Hurowitz, G. 2020. "The Jobs Are in the Trees." *Grist*, 4 February. https:// grist.org/article/2010-02-01-the-jobs-are-in-the-trees/.

Hutton G, Haller L, World Health Organization (2004) Evaluation of the costs and benefits of water and sanitation improvements at the global level. Water, Sanitation and Health Team. https://apps.who.int/iris/handle/10665/68568

ICS (2020a) ICS overview of the impact of COVID-19 on the shipping industry and summary of containment and mitigation measures. 12 March. https://www.ics-shipping.org/docs/default-source/Submissions/Other/ics-submission-to-european-commission-concerning-the-impact-of-covid-19.pdf

ICS (2020b) Global shipping body addresses the health concerns of seafarers during the COVID-19 pandemic. 28 May. https://www.ics-shipping.org/news/press-releases/view-article/2020/05/28/global-shipping-body-addresses-the-health-concerns-of-seafarers-during-the- covid-19-pandemic

ICS (2020c) Industry and Unions set out a 12-step solution for governments to free seafarers from COVID-19 lockdown. 6 May. https://

www.ics-shipping.org/news/press-releases/2020/05/06/industry--and-unions-set-out-a-12-step-solution-for-governments-to-free-seafarers- from-covid-19-lockdown

IEA (2014) World Energy Outlook 2014. https://www.iea.org/reports/world-energy-outlook-2014

IEA (2019a) The future of hydrogen: analysis. June. https://www.iea.org/reports/the-future-of-hydrogen

IEA (2019b) World Energy Outlook. https://www.iea.org/topics/world-energy-outlook

IEA (2020a) Global energy review 2020. https://www.iea.org/reports/global-energy-review-2020

IEA (2020b) World energy investment 2020. https://www.iea.org/reports/world-energy-investment-2020

IFRC (International Federation of Red Cross and Red Crescent Societies) (2011) Mangrove plantation in Viet Nam: measuring impact and cost benefit. https://www.ifrc.org/Global/Publications/disasters/reducing_risks/Case-study-Vietnam.pdf

IKI (International Climate Initiative) (2020) Corona package response: working together towards a sustainable recovery. https://www.international-climate-initiative.com/en/coronaresponse

ILO (2017) Wastewater and jobs: the decent work approach to reducing untreated wastewater. Working paper. http://www.ilo.org/sector/Resources/publications/WCMS_548129/lang%2D%2Den/index.htm

ILO (2018) Women and men in the informal economy: a statistical picture, 3rd edn. https://www.ilo.org/wcmsp5/groups/public/%2D%2D-dgreports/%2D%2D-dcomm/documents/publication/wcms_626831.pdf

ILO (2020a) COVID-19 and maritime shipping & fishing. Briefing note. http://www.ilo.org/sector/Resources/publications/WCMS_742026/lang%2D%2Den/index.htm

ILO (2020b) COVID-19 and employment in the tourism sector: impact and response in Asia and the Pacific. Briefing note. http://www.ilo.org/asia/publications/issue-briefs/WCMS_742664/lang%2D%2Den/index.htm

ILO (2020c) COVID-19 and the world of work, 4th edn. Briefing note. https://www.ilo.org/global/topics/coronavirus/impacts-and-responses/WCMS_745963/lang%2D%2Den/index.htm

IMCA (International Marine Contractors Association) (2020) IMCA and industry group target governments on a global basis on behalf of offshore workers. https://www.imca-int.com/news/2020/05/29/imca-and-industry-group-target-governments-on-a-global-basis-on-behalf-of-offshore-workers/

IMF (2020a) World economic outlook update, June 2020. https://www.imf.org/en/Publications/WEO/Issues/2020/06/24/WEOUpdateJune2020

IMF (2020b) Greening the recovery. Special series on fiscal policies to respond to COVID-19. https://www.un-page.org/Greening-the-Recovery

IMF (2020c) Pacific islands threatened by COVID-19. 27 May. https://www.imf.org/en/News/Articles/2020/05/27/na-05272020-pacific-islands-threatened-by-covid-19

IMO (International Maritime Organization) (2020) FAQ on crew changes and repatriation of seafarers. 16 June. http://www.imo.org/en/MediaCentre/HotTopics/Pages/FAQ-on-crew-changes-and--repatriation-of-seafarers.aspx

INFOFISH International (2020) Protecting and monitoring small-scale fisheries. CLS Fisheries, 4 May. https://fisheries.groupcls.com/protecting-and-monitoring-small-scale-fisheries/

IRENA (International Renewable Energy Agency) (2018) Renewable energy benefits: leveraging local capacity for offshore wind. https://www.irena.org/publications/2018/May/Leveraging-Local-Capacity-for-Offshore-Wind

IRENA (2019) Future of wind: deployment, investment, technology, grid integration and socio-economic aspects. https://www.irena.org/-/media/Files/IRENA/Agency/Publication/2019/Oct/IRENA_Future_of_wind_2019.pdf

Jambeck J, Moss E, Dubey B (2020) Leveraging multi-target strategies to address plastic pollution in the context of an already stressed ocean. World Resources Institute, Washington, DC. https://oceanpanel.org/sites/default/files/2020-06/Addressing%20Ocean%20Pollutants%20Full%20Report%20Final.pdf

Jay S (2017) Marine spatial planning: assessing net benefits and improving effectiveness. Issue note. Organisation for Economic Co-operation and Development.

Jiang R, Ingle KN, Golberg A (2016) Macroalgae (Seaweed) for liquid transportation biofuel production: what is next? Algal Res 14(March):48–57. https://doi.org/10.1016/j.algal.2016.01.001

Jigeesh AM (2020) Distressed fisheries sector demands immediate support. Hindu Business Line, 13 April. https://www.thehindubusinessline.com/economy/agri-business/distressed-fisheries-sector--demands-immediate-support/article31332857.ece

John H (2020) Fish meal and fish oil industries pose threat to the fishing sector in India. Mongabay-India (blog), 2 January. https://india.mongabay.com/2020/01/fish-meal-and-fish-oil-industries-pose--threat-to-the-fishing-sector-in-india/

Jones R, Rigg C, Pinkerton E (2017) Strategies for assertion of conservation and local management rights: a Haida Gwaii Herring Story. Mar Policy 80:154–167. https://doi.org/10.1016/j.marpol.2016.09.031

Jordan P (2020) Clean energy employment initial impacts from the COVID-19 economic crisis, March 2020. BW Research Partnership (blog), 14 April. https://e2.org/wp-content/uploads/2020/04/Clean--Energy-Jobs-Initial-COVID-19-Memo-Final.pdf

Kapsenberg L, Cyronak T (2019) Ocean acidification refugia in variable environments. Glob Chang Biol 25(10):3201–3214. https://doi.org/10.1111/gcb.14730

Kinley RD, de Nys R, Vucko MJ, Machado L, Tomkins NW (2016) The red macroalgae Asparagopsis taxiformis is a potent natural antimethanogenic that reduces methane production during in vitro fermentation with rumen fluid. Anim Prod Sci 56(3):282–289. https://doi.org/10.1071/AN15576

Knoche S, Ihde TF (2018) Estimating ecological benefits and socioeconomic impacts from oyster reef restoration in the Choptank River Complex, Chesapeake Bay. National Oceanic and Atmospheric Administration, Washington, DC. https://www.morgan.edu/Documents/ADMINISTRATION/pearl/ORES_Impacts_Knoche_Ihde%20(1).pdf

Konar M, Ding H (2020) A sustainable ocean economy for 2050: approximating its benefits and costs. World Resources Institute, Washington, DC. https://oceanpanel.org/sites/default/files/2020-07/Ocean%20Panel_Economic%20Analysis_FINAL.pdf

Konar M, Thuringer L, Sumaila RU (2019a) The scale of illicit trade in pacific ocean marine resources. World Resources Institute, Washington, DC. https://www.wri.org/publication/scale-illicit-trade-pacific-ocean-marine-resources

Konar M, Qiu S, Tougher B, Vause J, Tlusty M, Fitzsimmons K, Barrows R, Cao L (2019b) Illustrating the hidden economic, social and ecological values of global forage fish resources. Resour Conserv Recycl 151(December):104456. https://doi.org/10.1016/j.resconrec.2019.104456

KPMG (2020) Jamaica: government and institution measures in response to COVID-19. https://home.kpmg/xx/en/home/insights/2020/04/jamaica-government-and-institution-measures-in-response-to-covid.html

Kroner RG (2020) Rolling back environmental protections under cover of the pandemic. Sci Am, 28 June. https://www.scientificamerican.com/article/rolling-back-environmental-protections-under-cover-of-the-pandemic/

Kummu M, de Moel H, Salvucci G, Viviroli D, Ward PJ, Varis O (2016) Over the hills and further away from coast: global geospatial patterns of human and environment over the 20th–21st centuries. Environ Res Lett 11(3). https://doi.org/10.1088/1748-9326/11/3/034010

Laidlaw J (2020) Social bond surge appears here to stay as COVID-19 crisis shifts funding needs. S&P Global Market Intelligence, 19 May. https://www.spglobal.com/marketintelligence/en/news--insights/latest-news-headlines/social-bond-surge-appears-here-to-stay-as-covid-19-crisis-shifts-funding-needs-58602191

Langhamer O (2012) Artificial reef effect in relation to offshore renewable energy conversion: state of the art. ScientificWorldJournal 23:386713. https://doi.org/10.1100/2012/386713

Leape J, Abbott M, Sakaguchi H (2020) Technology, data and new models for sustainably managing ocean resources. World Resources Institute, Washington, DC. https://oursharedseas.com/oss_downloads/technology-data-and-new-models-for-sustainably-managing-ocean-resources/

Leonard G, Mallos N (2020) What we know—and don't know—about plastics and the coronavirus pandemic. Ocean Conservancy (blog), 16 April. https://oceanconservancy.org/blog/2020/04/16/know-dont-know-plastics-coronavirus-pandemic/

Li X, Bellerby R, Craft C, Widney SE (2018) Coastal wetland loss, consequences, and challenges for restoration. Anthropocene Coast 1(1):1–15. https://doi.org/10.1139/anc-2017-0001

Love DC, Allison EH, Asche F, Belton B, Cottrell RS, Froehlich HE, Gephart JA et al (2020) Emerging COVID-19 impacts, responses, and lessons for building resilience in the 2 seafood system. https://doi.org/10.31235/osf.io/x8aew

Macknick J, Newmark R, Heath G, Hallett KC (2012) Operational water consumption and withdrawal factors for electricity generating technologies: a review of existing literature. Environ Res Lett 7(4):045802. https://doi.org/10.1088/1748-9326/7/4/045802

Mærsk (2020) New research center will lead the way for decarbonizing shipping. https://www.maersk.com/news/articles/2020/06/25/new-research-center-will-lead-the-way-for-decarbonizing-shipping

Malik NKA, Abdullah SH, Manaf LA (2015) Community participation on solid waste segregation through recycling programmes in Putrajaya. Proc Environ Sci 30:10–14. https://doi.org/10.1016/j.proenv.2015.10.002

Mathis W, Thornhill J (2019) Hydrogen's plunging price boosts role as climate solution. Bloomberg, 21 August. https://www.bloomberg.com/news/articles/2019-08-21/cost-of-hydrogen-from-renewables-to-plummet-next-decade-bnef

McKinley DC, Miller-Rushing AJ, Ballard HL, Bonney R, Brown H, Cook-Patton SC, Evans DM et al (2017) Citizen science can improve conservation science, natural resource management, and environmental protection. Biol Conserv 208(April):15–28. https://doi.org/10.1016/j.biocon.2016.05.015

McKinsey (2020) A low-carbon economic stimulus after COVID-19. 27 May. https://www.mckinsey.com/business-functions/sustainability/our-insights/how-a-post-pandemic-stimulus-can-both-create-jobs-and-help-the-climate

McLeod DA, McLeod C (2019) Review of the contribution of cultivated bivalve shellfish to ecosystem services. Crown Estate Scotland. https://www.crownestatescotland.com/maps-and-publications/download/416

McLeod I, Schmider J, Creighton C, Gillies C (2018) Seven pearls of wisdom: advice from traditional owners to improve engagement of local indigenous people in shellfish ecosystem restoration. Ecol Manage Restor 19(2):98–101. https://doi.org/10.1111/emr.12318

MCST (Micronesian Center for Sustainable Transport) (n.d.) Pacific blue shipping partnership. https://mcst-rmiusp.org/index.php/projects/current-projects/pacific-blue-shipping-partnership. Accessed 19 Aug 2020

Melnychuk M, Banobi J, Hilborn R (2014) The adaptive capacity of fishery management systems for confronting climate change impacts on marine populations. Rev Fish Biol Fish 24(June):561–575. https://doi.org/10.1007/s11160-013-9307-9

Michelin M, Elliott M, Bucher M (2018) Catalyzing the growth of electronic monitoring in fisheries. https://www.nature.org/content/dam/tnc/nature/en/documents/Catalyzing_Growth_of_Electronic_Monitoring_in_Fisheries_9-10-2018.pdf

Mojarro N (2020) Here's why COVID-19 is a game-changer for renewable energy. World Economic Forum, 16 June. https://www.weforum.org/agenda/2020/06/covid-19-is-a-game-changer-for-renewable-energy/

Monfort MC (2015) The role of women in the seafood industry. Globefish Research Programme, vol 119. Food and Agriculture Organization of the United Nations, Rome

Moore R (2019) Norway's first hydrogen-powered car ferries take shape. Riviera, 16 July. https://www.rivieramm.com/news-content-hub/news-content-hub/norwayrsquos-first-hydrogen-powered-car-ferries-take-shape-55559

Moore E, Cooley H, Christian-Smith J, Donnelly K, Ongoco K, Ford D (2013) 'Sustainable water jobs: A national assessment of water-related green job opportunities'. Pacific Institute, Oakland, California. URL: https://pacinst.org/wp-content/uploads/2013/02/sust_jobs_full_report.pdf

Moreno A, Revenga C (2014) The system of territorial use rights in fisheries in Chile. The Nature Conservancy. https://www.nature.org/media/chile/system-of-TURFs-in-Chile.pdf

Morris D (2020) Vietnam's renewable energy industry amid COVID-19: facts, force majeure and (patchy) government support. Lexology, 10 April. https://www.lexology.com/library/detail.aspx?g=37c4c382-1bf0-4773-a36b-dcb68f76dbc7

MPA News (2009) How the global financial crisis could affect MPAs, and what practitioners are doing to prepare for it. 7 May. https://mpanews.openchannels.org/news/mpa-news/how-global-financial--crisis-could-affect-mpas-and-what-practitioners-are-doing-prepare

Mutafoglu K, ten Brink P, Schweitzer J-P, Underwood E, Tucker G, Russi D, Howe M et al (2017) Natura 2000 and Jobs. Institute for European Environmental Policy. https://ec.europa.eu/environment/nature/natura2000/pdf/Natura2000_and_jobs_main%20report.pdf

Narayanswami S, Balan C (2013) The food wastage and cold storage infrastructure relationship in India: developing realistic solutions. Emerson Climate Technologies

Naylor RL, Hardy RW, Bureau DP, Chiu A, Elliott M, Farrell AP, Forster I et al (2009) Feeding aquaculture in an era of finite resources. Proc Natl Acad Sci U S A 106(36):15103. https://doi.org/10.1073/pnas.0905235106

Neish IC (2013) Social and economic dimensions of Carrageenan seaweed farming in Indonesia. In: Social and economic dimensions of Carrageenan seaweed farming. FAO, Rome, pp 61–89. http://www.fao.org/3/a-i3344e.pdf

Nguyen T (2020) First-of-its-kind clean energy investment initiative for Southeast Asia. Vietnam Invest Rev, 1 July. https://www.vir.com.vn/first-of-its-kind-clean-energy-investment-initiative-for-southeast-asia-77465.html

Nikel, D. 2020. Norway proposes NOK 3.6bn green investment package. Life in Norway (blog), 29 May. https://www.lifeinnorway.net/norway-proposes-nok-3-6-billion-green-investment-package/

NOAA (U.S. National Oceanic and Atmospheric Administration) (2020) NOAA fisheries coronavirus (COVID-19) update. 11 August. https://www.fisheries.noaa.gov/national/noaa-fisheries-coronavirus-covid-19-update

NoMEPorts (2008) Good practice guide on port area noise mapping and management. https://ec.europa.eu/environment/life/project/Projects/index.cfm?fuseaction=home.showFile&rep=file&fil=NoMEports_GPG_PANMM1.pdf

Norges Bank Investment Management (2020) Ocean sustainability. https://www.nbim.no/en/the-fund/responsible-investment/principles/expectations-to-companies/ocean-sustainability/
NSA (2020a) 1/4th of turnover lost due to Covid-19. Blog, 21 April. https://rederi.no/en/aktuelt/2020/hver-fjerde-krone-tapt-som-folge-av-korona/
NSA (2020b) Null Utslipp i 2050. https://rederi.no/en/aktuelt/2020/norwegian-shipping-climate-neutral-by-2050/
OECD (Organisation for Economic Co-operation and Development) (2014) OECD tourism trends and policies 2014. https://www.oecd-ilibrary.org/industry-and-services/oecd-tourism-trends-and-policies-2014/taxation-and-tourism_tour-2014-6-en
OECD (2015) Green growth in fisheries and aquaculture. https://www.oecd-ilibrary.org/content/publication/9789264232143-en
OECD (2016) The ocean economy in 2030. https://www.oecd.org/environment/the-ocean-economy-in-2030-9789264251724-en.htm
OECD (2018) OECD tourism trends and policies 2018. https://www.oecd-ilibrary.org/urban-rural-and-regional-development/oecd--tourism-trends-and-policies-2018_tour-2018-en?itemId=/content/component/tour-2018-21-en&_p_=96e3bdda0ee5511a9fceb9fd28fad50a&itemIGO=oecd&itemContentType=chapter
OECD (2020a) Building back better: a sustainable, resilient recovery after COVID-19. https://www.oecd.org/coronavirus/policy-responses/building-back-better-a-sustainable-resilient-recovery-after-covid-19-52b869f5/#blocknotes-d7e1362
OECD (2020b) Tourism policy responses to the coronavirus (COVID-19). 2 June. https://www.oecd.org/coronavirus/policy-responses/tourism-policy-responses-to-the-coronavirus-covid-19-6466aa20/
OECD (2020c) Economic outlook, Volume 2020, Issue 1: Preliminary Version. https://www.oecd-ilibrary.org/sites/0d1d1e2e-en/index.html?itemId=/content/publication/0d1d1e2e-en
OECD (2020d) Development co-operation profiles. June. https://doi.org/10.1787/5e331623-en
OECD (2020e) The territorial impact of COVID-19: managing the crisis across levels of government. https://read.oecd-ilibrary.org/view/?ref=128_128287-5agkkojaaa&title=The-territorial-impact-of-covid-19-managing-the-crisis-across-levels-of-government
OECD (2020f) Sustainable ocean for all: harnessing the benefits of sustainable ocean economies for developing countries. https://www.oecd.org/environment/sustainable-ocean-for-all-bede6513-en.htm
OES (Ocean Energy Systems) (2011) Annual report. https://tethys.pnnl.gov/publications/oes-annual-report-2011
OES (2017) An international vision for ocean energy, 2017. https://testahemsidaz2.files.wordpress.com/2017/03/oes-international-vision.pdf
Office of the Prime Minister Canada (2020) Prime Minister announces additional support for businesses to help save Canadian jobs. 11 May. https://pm.gc.ca/en/news/news-releases/2020/05/11/prime-minister-announces-additional-support-businesses-help-save
Olmer N, Comer B, Roy B, Mao X, Rutherford D (2017) Greenhouse gas emissions from global shipping, 2013–2015. International Council on Clean Transportation, Washington, DC. https://theicct.org/publications/GHG-emissions-global-shipping-2013-2015
Önen Cinar S, Chong ZK, Kucuker M, Wieczorek N, Cengiz U, Kuchta K (2020) Bioplastic production from microalgae: a review. Int J Environ Res Public Health 17(May):3842. https://doi.org/10.3390/ijerph17113842
Orlowski A (2020) Small-scale fishermen suffering significantly from COVID-19 pandemic. SeafoodSource, 27 April. https://www.seafoodsource.com/news/supply-trade/small-scale-fishermen--suffering-significantly-from-covid-19-pandemic
Österblom H, Wabnitz C, Tladi D, Allison E, Arnaud-Haond S, Bebbington J, Bennett N et al (2020) Towards ocean equity. World Resources Institute, Washington, DC. https://www.oceanpanel.org/sites/default/files/2020-04/towards-ocean-equity.pdf
Outeiro L, Ojea E, Rodrigues JG, Himes-Cornell A, Belgrano A, Liu Y, Cabecinha E et al (2017) The role of non-natural capital in the co-production of marine ecosystem services. Int J Biodivers Sci Ecosyst Serv Manag 13(3):35–50. https://doi.org/10.1080/21513732.2017.1415973
Ouyang X, Guo F (2016) Paradigms of mangroves in treatment of anthropogenic wastewater pollution. Sci Total Environ 544:971–979. https://doi.org/10.1016/j.scitotenv.2015.12.013
Paris C (2020) Coronavirus toll on shipping reaches $350 million a week. Wall Street J (blog), 10 February. https://www.wsj.com/articles/coronavirus-toll-on-shipping-reaches-350-million-a-week-11581366671
Parkinson G (2020) Australia's first offshore wind project says it can cover early exit of Yallourn. Renew Econ (blog), 4 June. https://reneweconomy.com.au/australias-first-offshore-wind-project-says-it-can-cover-early-exit-of-yallourn-74304/
PCAF (Philippine Council for Agriculture and Fisheries) (2020) PCAF initiatives during COVID-19 ECQ: fisheries and aquaculture private stakeholders meeting. http://www.pcaf.da.gov.ph/index.php/2020/05/14/pcaf-initiatives-during-covid-19-ecq-fisheries-and-aquaculture-private-stakeholders-meeting/
Peszko G (2020) Plastics: the coronavirus could reset the clock. World Bank Blogs, 7 April. https://blogs.worldbank.org/voices/plastics-coronavirus-could-reset-clock
Petersen J, Taylor D, Bergström P, Buer A-L, Darecki M, Filippelli R, Gren I-M et al (2020) Policy guidelines for implementation of mussel cultivation as a mitigation measure for coastal eutrophication in the Western Baltic Sea. April. https://doi.org/10.11581/dtu:00000079
Plastic Cities (2020) Waste management guidance during COVID-19 pandemic. https://plasticsmartcities.org/blogs/media/waste-management-guidance-during-covid-19-pandemic
Pollin R, Heintz J, Garrett-Peltier H (2009) How infrastructure investments support the U.S. economy: employment, productivity and growth. Alliance for American Manufacturing. https://www.americanmanufacturing.org/research/entry/how-infrastructure-investments-support-the-u.s.-economy-employment-producti
Polo G (2020) When nets catch more than fish. Rare, 5 May. https://rare.org/story/when-nets-catch-more-than-fish/
Queensland Government (2020) Queensland's economic recovery plan. https://www.covid19.qld.gov.au/government-actions/our-economic-recovery-strategy
Ratha DK, De S, Kim EJ, Plaza S, Seshan GK, Yameogo ND (2020) COVID-19 crisis through a migration lens. World Bank Group. https://documents.worldbank.org/en/publication/documents-reports/documentdetail
Reddy NP, Zadeh MK, Thieme CA, Skjetne R, Sorensen AJ, Aanondsen SA, Breivik M, Eide E (2019) Zero-emission autonomous ferries for urban water transport: cheaper, cleaner alternative to bridges and manned vessels. IEEE Electrif Mag 7(4):32–45. https://doi.org/10.1109/MELE.2019.2943954
reNews (2020) Offshore wind 'Investment Jumps 319%'. 13 July. https://renews.biz/61638/offshore-wind-investment-jumps-319/
Resilience (2020) COVID 19 and our food supply. 3 April. https://www.resilience.org/stories/2020-04-03/covid-19-and-our-food-supply/
Richens J, Koehring M (2020) A sustainable ocean economy in 2030: opportunities and challenges. The Economist Group: World Ocean Initiative. https://woi.economist.com/sustainable-ocean-economy-2030/
Riedmiller S (2020) This challenge is likely to be harder than any before. MPA News (blog), 7 May. https://mpanews.openchannels.org/news/mpa-news/mpas-covid-19-and-coming-financial-crisis--what-impacts-are-practitioners-already
Roberts T, Upham P (2012) Prospects for the use of macro-algae for fuel in Ireland and the UK: an overview of marine management issues. Mar Policy 36(5):1047–1053

Roberts CA, Newton RW, Bostock JC, Prescott SG, Honey DJ, Telfer TC, Walmsley SF et al (2015) A risk benefit analysis of mariculture as a means to reduce the impacts of terrestrial production of food and energy. Scottish Aquaculture Research Forum. https://dspace.stir.ac.uk/bitstream/1893/23422/1/232492-618987.sarf106.pdf

Rog SM, Clarke RH, Cook CN (2016) More than marine: revealing the critical importance of mangrove ecosystems for terrestrial vertebrates. Divers Distrib 23(2):221–230. https://doi.org/10.1111/ddi.12514

Rolland RM, Parks SE, Hunt KE, Castellote M, Corkeron PJ, Nowacek DP, Wasser SK, Kraus SD (2012) Evidence that ship noise increases stress in right whales. Proc Biol Sci 279(1737):2363–2368. https://doi.org/10.1098/rspb.2011.2429

Roncin N, Alban F, Charbonnel E, Crec'hriou R, de La Cruz Modino R, Culioli J-M, Dimech M et al (2008) Uses of ecosystem services provided by MPAs: how much do they impact the local economy? A Southern Europe perspective. EconPapers. https://econpapers.repec.org/paper/haljournl/hal-00353110.htm

Runyon J (2020) The COVID-19 consequences for clean tech startups could be catastrophic for the Renewable Energy World, 22 April. https://www.renewableenergyworld.com/2020/04/22/the-covid-19-consequences-for-clean-tech-startups-could-be-catastrophic-for-the-planet/#gref

Sachs J, Schmidt-Traub G, Kroll C, Lafortune G, Fuller G, Woelm F (2020) The sustainable development goals and COVID-19: sustainable development report 2020. Cambridge University Press, Cambridge

Sala E, Costello C, Dougherty D, Heal G, Kelleher K, Murray JH, Rosenberg AA, Sumaila R (2013) A general business model for marine reserves. PloS One 8(4):e58799. https://doi.org/10.1371/journal.pone.0058799

SAPEA (Science Advice for Policy by European Academies) (2017) Food from the oceans: how can more food and biomass be obtained oceans in a way that does not deprive future generations of their benefits? https://www.sapea.info/wp-content/uploads/FFOFINALREPORT-1.pdf

SARF (Scottish Aquaculture Research Forum) (2011) Carbon footprint of Scottish suspended mussels and intertidal oysters. http://www.sarf.org.uk/cms-assets/documents/43896-326804.sarf078

Satterthwaite D, Beard VA, Mitlin D, Du J (2019) Untreated and unsafe: solving the urban sanitation crisis in the global south. World Resources Institute, Washington, DC. https://www.wri.org/wri-citiesforall/publication/untreated-and-unsafe

Saumweber W, Lehr AK, Loft T (2020) Covid-19 at sea: impacts on the blue economy, ocean health, and ocean security. CSIS (blog), 10 April. https://www.csis.org/analysis/covid-19-sea-impacts-blue--economy-ocean-health-and-ocean-security

Schwartz JZ, Andres LA, Dragoiu G (2009) Crisis in Latin America: infrastructure investment, employment and the expectations of stimulus. World Bank Policy Research Working Papers. https://doi.org/10.1596/1813-9450-5009

Scottish Enterprise (2016) Offshore wind. Oil and gas "Seize the Opportunity" guides. https://www.offshorewindscotland.org.uk/media/1116/sesdi-oil-and-gas-div-guide-offshore-wind.pdf

Seas at Risk (2019) The multi-issue mitigation potential of reducing ship speeds. https://seas-at-risk.org/images/pdf/publications/Multi_issue_speed_report.pdf

Serrano O, Kelleway JJ, Lovelock C, Lavery PS (2019) Conservation of blue carbon ecosystems for climate change mitigation and adaptation. In: Perillo GME, Wolanski E, Cahoun DR, Hopkinson CH (eds) Coastal wetlands: an integrated ecosystem approach, 2nd edn. Elsevier, Edinburgh, pp 965–996. https://doi.org/10.1016/B978-0-444-63893-9.00028-9

Sharma DC (2006) Ports in a storm. Environ Health Perspect 114(4):A222–A231

Skopljak N (2020) Australia's first offshore wind project expands survey team. Offshore Wind (blog), 7 May. https://www.offshorewind.biz/2020/05/07/australias-first-offshore-wind-project-expands-survey-team/

Sofiev M, Winebrake JJ, Johansson L, Carr EW, Prank M, Soares J, Vira J et al (2018) Cleaner fuels for ships provide public health benefits with climate tradeoffs. Nat Commun 9(1):406. https://doi.org/10.1038/s41467-017-02774-9

Sotta C (2012) Documentary summary of the environmental impact of renewable marine energy. Marine Energy in Far Peripheral Island Communities (MERiFIC). https://tethys.pnnl.gov/publications/merific-321-documentary-summary-environmental-impact-renewable-marine-energy

Southgate PC, Kishore P, Sharma S, Simos T (2019) Empowering women through pearl industry-based livelihoods in Fiji. Women in Fisheries, no. 29. https://coastfish.spc.int/en/publications/bulletins/women-in-fisheries/492

Spalding M, Brumbaugh RD, Landis E (2016) Atlas of ocean wealth. The Nature Conservancy. https://oceanwealth.org/wp-content/uploads/2016/07/Atlas_of_Ocean_Wealth.pdf

Statehouse (2020) Seychelles designates 30% of its EEZ as marine protected area. 26 March. http://www.statehouse.gov.sc/news/4787/seychelles-designates-30-of-its-eez-as-marine-protected-area

SUBPESCA (2020a) SUBPESCA extiende hasta septiembre vigencia de permisos para más de 80 productores de choritos en Los Lagos. 11 May. http://www.subpesca.cl/sitioprensa/614/w3--article-107741.html

SUBPESCA (2020b) Más de 500 pescadores de distintas regiones obtienen cupos para capturar reineta. 3 June. http://www.subpesca.cl/sitioprensa/614/w3-article-107852.html

SUBPESCA (2020c) Apoyo frente a contingencia: Medida de SUBPESCA permite que pescadores sigan trabajando en casi 500 áreas de manejo. 11 June. http://www.subpesca.cl/sitioprensa/614/w3-article-107899.html

SUBPESCA (2020d) Apoyo a pescadores del Biobío: Desde un camión frigorífico hasta paneles solares para bajar gastos en electricidad. 3 July. http://www.subpesca.cl/portal/617/w3-article-108043.html

Suding K, Higgs E, Palmer M, Callicott J, Anderson C, Baker M, Gutrich J et al (2015) Conservation: committing to ecological restoration. Science 348(May):638–640. https://doi.org/10.1126/science.aaa4216

Sumaila RU (2020) COVID-19 is our chance to end harmful fishing subsidies. World Economic Forum. 19 May. https://www.weforum.org/agenda/2020/05/covid-19-is-a-chance-to-end-harmful-fishing-subsidies-heres-why/

Sumaila RU, Ebrahim N, Schuhbauer A, Skerritt D, Li Y, Kim HS, Mallory TG et al (2019) Updated estimates and analysis of global fisheries subsidies. Mar Policy 109(November):103695. https://doi.org/10.1016/j.marpol.2019.103695

Sumaila UR, Zeller D, Hood L, Palomares MLD, Li Y, Pauly D (2020) Illicit trade in marine fish catch and its effects on ecosystems and people worldwide. Sci Adv 6(9):eaaz3801. https://doi.org/10.1126/sciadv.aaz3801

Suplicy FM (2018) A review of the multiple benefits of mussel farming. Rev Aquacult 12(1):204–223. https://doi.org/10.1111/raq.12313

Svaetichin I, Inkinen T (2017) Port waste management in the Baltic sea area: a four port study on the legal requirements, processes and collaboration. Sustainability 9(5):1–17

Tacon AGJ, Metian M (2008) Global overview on the use of fish meal and fish oil in industrially compounded aquafeeds: trends and future prospects. Aquaculture 285:146–158

Tegen S, Keyser D, Flores-Espino F, Miles J, Zammit D, Loomis D (2015) Offshore wind jobs and economic development impacts in the United States: four regional scenarios. Technical report.

National Renewable Energy Laboratory, Golden, CO. https://www.nrel.gov/docs/fy15osti/61315.pdf
Teh LCL, Sumaila UR (2013) Contribution of marine fisheries to worldwide employment. Fish Fish 14(1):77–88. https://doi.org/10.1111/j.1467-2979.2011.00450.x
Tenenbaum L (2020) The amount of plastic waste is surging because of the coronavirus pandemic. Forbes, April. https://www.forbes.com/sites/lauratenenbaum/2020/04/25/plastic-waste-during-the-time-of-covid-19/
Thande G (2018) Seychelles preserves swathes of marine territory in debt-for-nature deal. Reuters, 22 February. https://www.reuters.com/article/us-seychelles-environment-idUSKCN1G61M5
The Fish Site (2020) Women set to bear the brunt of the COVID-19 pandemic. 7 April. https://thefishsite.com/articles/women-set-to-bear-the-brunt-of-the-covid-19-pandemic
Thomson J (2020) Fisheries and oceans Canada pulls at-sea observers from fishing boats due to coronavirus pandemic. The Narwhal (blog), 8 April. https://thenarwhal.ca/fisheries-oceans-canada-pulls-at-sea-observers-fishing-boats-coronavirus-covid-19/
TNC (The Nature Conservancy) (2013) Building a case for green infrastructure. June. https://www.nature.org/en-us/about-us/who-we-are/how-we-work/working-with-companies/transforming-business-practices/building-a-case-for-green-infrastructure/
TNC (2020) Rebuilding Australia's lost shellfish reefs. 1 June. https://www.natureaustralia.org.au/what-we-do/our-priorities/oceans/ocean-stories/restoring-shellfish-reefs/
Torgler B, Bickley SJ, Macintyre A (2020) Sink or swim: the COVID-19 impact on environmental health, fish levels and illicit maritime activity. Behav Social Sci Nat Res, 21 April. http://socialsciences.nature.com/users/376755-benno-torgler/posts/66594-sink-or-swim-the-covid-19-impact-on-environmental-health-fish-levels-and-illicit-maritime-activity
Troell M, Joyce A, Chopin T, Neori A, Buschmann AH, JianGuang F (2009) Ecological engineering in aquaculture: potential for integrated multi-trophic aquaculture (IMTA) in marine offshore systems. Aquaculture 297(1/4):1–9
UK Government (2016) Cutting red tape: review of the energy sector. 3 March. https://www.gov.uk/government/publications/energy-sector-cutting-red-tape-review
UK Government (2020) New international commitment to improve seafarers' rights. 9 July. https://www.gov.uk/government/news/new-international-commitment-to-improve-seafarers-rights
UN (2020a) Shared responsibility, global solidarity: responding to the socio-economic impacts of COVID-19. https://www.un.org/sites/un2.un.org/files/sg_report_socio-economic_impact_of_covid19.pdf
UN (2020b) COVID-19 and human rights: we are all in this together. Policy brief. https://www.un.org/sites/un2.un.org/files/un_policy_brief_on_human_rights_and_covid_23_april_2020.pdf
UN (2020c) Innovation for a sustainable ocean amid the COVID-19 pandemic: impacts on Kenya's marine and coastal environment. Blog, 8 June. https://www.un.org/en/un-chronicle/innovation-sustainable-ocean-amid-covid-19-pandemic-impacts-kenya%E2%80%99s-marine-and-coastal
UN DESA (2020a) Responses to the COVID-19 catastrophe could turn the tide on inequality. Policy brief no. 65, 1 May. https://www.un.org/development/desa/dpad/publication/un-desa-policy-brief-65-responses-to-the-covid-19-catastrophe-could-turn-the-tide-on--inequality/
UN DESA (2020b) The COVID-19 pandemic puts small island developing economies in dire straits. Policy brief 64. https://www.un.org/development/desa/dpad/wp-content/uploads/sites/45/publication/PB_64.pdf
UN Water (2016) Water and Jobs: The United Nations World Water Development Report 2016. UNESCO World Water Assessment Programme. https://www.unwater.org/publications/world-water-development-report-2016/
Uña G, Allen R, Botton N (2019) How to design a financial management information system: a modular approach. International Monetary Fund. https://www.imf.org/~/media/Files/Publications/HowToNotes/HowToNote1903.ashx
UNCTAD (2020a) Impact of the COVID-19 pandemic on global FDI and GVCs: updated analysis. Investment Trends Monitor. https://unctad.org/en/PublicationsLibrary/diaeiainf2020d3_en.pdf
UNCTAD (2020b) The COVID-19 pandemic and the blue economy: new challenges and prospects for recovery and resilience. https://unctad.org/en/pages/PublicationWebflyer.aspx?publicationid=2717
UNCTAD (2020c) Climate change impacts and adaptation for coastal transport infrastructure: a compilation of policies and practices. Transport and Trade Facilitation. https://sidsport-climateadapt.unctad.org/compilation-of-practices/
UNDP (UN Development Programme) (2012) Transforming markets to restore and protect the global ocean. Vol. 1 of Catalysing Ocean Finance. https://www.undp.org/content/undp/en/home/librarypage/environment-energy/water_governance/ocean_and_coastalareagovernance/catalysing-ocean-finance.html
UNEP (UN Environment Programme), FAO (Food and Agriculture Organization of the United Nations), IMO (International Maritime Organization), UNDP (UN Development Programme), IUCN (International Union for Conservation of Nature), World Fish Center and GRID (Global Resource Information Database)-Arendal (2012) A green economy in a blue world. https://www.undp.org/content/undp/en/home/librarypage/environment-energy/water_governance/ocean_and_coastalareagovernance/a-green-economy-in-a-blue-world.html
UNESCO (2009) Water in a changing world. United Nations world water development report 3. http://www.unesco.org/new/en/natural-sciences/environment/water/wwap/wwdr/wwdr3-2009/downloads-wwdr3/
UNESCO (2020) COVID-19 and marine world heritage: a pathway for a resilient ocean. UNESCO World Heritage Centre (blog), 20 May. https://whc.unesco.org/en/news/2118/
UNGC (UN Global Compact) (2019) Sustainable ocean principles. https://d306pr3pise04h.cloudfront.net/docs/publications%2FSustainable+Ocean+Principles.pdf
UNGC (2020a) COVID-19 task force on geopolitical risks and responses. https://ungc-communications-assets.s3.amazonaws.com/docs/publications/Call-To-Action_Imminent-Threats-to-the-Integrity-of-Global-Supply-Chains.pdf
UNGC (2020b) Ocean stewardship 2030. https://unglobalcompact.org/library/5742
UNGC (2020c) Over 150 global corporations urge world leaders for net zero recovery from COVID-19. 18 May. https://unglobalcompact.org/news/4535-05-18-2020
UNWTO (UN World Tourism Organization) (2019) Global report on women in tourism, 2nd edn. https://doi.org/10.18111/9789284420384
Valderrama D, Cai J, Hishamunda N (2013) Social and economic dimensions of Carrageenan seaweed farming. FAO fisheries and aquaculture technical paper 580. Food and Agriculture Organization of the United Nations. http://www.fao.org/3/a-i3344e.pdf
Value of Water Campaign (2017) The economic benefits of investing in water infrastructure. http://thevalueofwater.org/sites/default/files/Economic%20Impact%20of%20Investing%20in%20Water%20Infrastructure_VOW_FINAL_pages.pdf
van der Schatte Olivier A, Jones L, Vay LL, Christie M, Wilson J, Malham SK (2020) A global review of the ecosystem services provided by bivalve aquaculture. Rev Aquacult 12(1):3–25. https://doi.org/10.1111/raq.12301
Veum K, Cameron L, Huerta Hernandes D, Korpaas M (2011) Roadmap to the deployment of offshore wind energy in the Central and Southern North Sea (2020–2030). July. http://inis.iaea.org/Search/search.aspx?orig_q=RN:42093900
Victoria State Government (2020) Victorian Government Stimulus Package. Department of Environment, Land, Water and Planning,

10 July. https://www2.delwp.vic.gov.au/media-centre/home/delwp-stimulus-package
Vivid Economics (2020) Green Stimulus Index. Finance for biodiversity initiative. https://www.vivideconomics.com/wp-content/uploads/2020/05/200518-Stimulus-Green-Index-v2_shared.pdf
Vyawahare M (2020) COVID-19 will hurt Madagascar's conservation funding: Q&A with Minister Vahinala Raharinirina. Mongabay Environmental News, 24 April. https://news.mongabay.com/2020/04/covid-19-will-hurt-madagascars-conservation-funding-qa-with-minister-vahinala-raharinirina/
Waldron A, Adams V, Allan J, Arnell A, Asner G, Atkinson S, Baccini A et al (2020) Protecting 30% of the planet for nature: costs, benefits and economic implications. https://www.conservation.cam.ac.uk/files/waldron_report_30_by_30_publish.pdf
Weller RA, Baker DJ, Glackin MM, Roberts SJ, Schmitt RW, Twigg ES, Vimont DJ (2019) The challenge of sustaining ocean observations. Front Marine Sci 6. https://doi.org/10.3389/fmars.2019.00105
Wheatley J, England A, Politi J (2020) G20 agrees debt relief for low income nations. 15 April. https://www.ft.com/content/5f296d54-d29e-4e87-ae7d-95ca6c0598d5
White C, Halpern BS, Kappel CV (2012) Ecosystem service tradeoff analysis reveals the value of marine spatial planning for multiple ocean uses. Proc Natl Acad Sci U S A 109(12):4696–4701. https://doi.org/10.1073/pnas.1114215109
WHO (World Health Organization) (2012) Global costs and benefits of drinking-water supply and sanitation interventions to reach the MDG target and universal coverage. https://www.who.int/water_sanitation_health/publications/2012/globalcosts.pdf
Wilson C, Tisdell C (2003) Conservation and economic benefits of wildlife-based marine tourism: sea turtles and whales as case studies. Hum Dimens Wildl 8(1):49–58. https://doi.org/10.1080/10871200390180145
Winsemius HC, Jongman B, Veldkamp TIE, Hallegatte S, Bangalore M, Ward PJ (2018) Disaster risk, climate change, and poverty: assessing the global exposure of poor people to floods and droughts. Environ Dev Econ 23(3):328–348. https://doi.org/10.1017/S1355770X17000444
Winther J-G, Dai M et al (2020) Integrated ocean management. World Resources Institute, Washington, DC. www.oceanpanel.org/blue-papers/integrated-ocean-management
World Bank (2012) Hidden harvest: the global contribution of capture fisheries. https://documents.worldbank.org/en/publication/documents-reports/documentdetail
World Bank (2017) The Sunken billions revisited: progress and challenges in global marine fisheries. https://openknowledge.worldbank.org/handle/10986/24056
World Bank (2018) Seychelles launches world's first sovereign blue bond. October. 10/29/seychelles-launches-worlds-first-sovereign-blue-bond
World Bank (2020a) The economy in the time of Covid-19. https://openknowledge.worldbank.org/handle/10986/33555
World Bank (2020b) Earth Day 2020: Could COVID-19 be the tipping point for transport emissions? Feature story. https://www.worldbank.org/en/news/feature/2020/04/22/earth-day-2020-could-covid--19-be-the-tipping-point-for-transport-emissions
World Maritime News (2020) Sea intelligence: COVID-19 impact pushes carriers' revenue loss to USD 1.9 Bln. Offshore Energy (blog), 3 March. https://www.offshore-energy.biz/sea-intelligence-covid-19-impact-pushes-carriers-revenue-loss-to-usd-1-9-bln/
WSC (World Shipping Council) (2020) Benefits of liner shipping. http://www.worldshipping.org/benefits-of-liner-shipping/efficiency
WTTC (World Travel and Tourism Council) (2020) Recovery scenarios 2020 and economic impact from COVID-19. https://wttc.org/Research/Economic-Impact/Recovery-Scenarios-2020-Economic-Impact-from-COVID-19
WWAP (United Nations World Water Assessment Programme) (2017) Wastewater, the Untapped Resource. United Nations World Water Development Report. http://www.unesco.org/new/en/natural-sciences/environment/water/wwap/wwdr/2017-wastewater-the-untapped-resource/
WWF (World Wildlife Fund) (2015) Marine protected areas: smart investments in ocean health. http://ocean.panda.org/media/WWF_Marine_Protected_Areas_LR_SP.pdf
WWF (2017) Remote electronic monitoring. https://www.wwf.org.uk/sites/default/files/2017-10/Remote%20Electronic%20Monitoring%20in%20UK%20Fisheries%20Management_WWF.pdf
WWF (2018) The sustainable blue economy finance principles. 2 March. https://www.wwf.org.uk/updates/sustainable-blue-economy-finance-principles
Xu X, Joyce R (2020) Sector shutdowns during the coronavirus crisis: which workers are most exposed? Institute for Fiscal Studies briefing note, 6 April. https://doi.org/10.1920/BN.IFS.2020.BN0278
Yara International (2019) Yara and ENGIE to test green hydrogen technology in fertilizer production. 13 February. https://www.yara.com/news-and-media/news/archive/2019/yara-and-engie-to-test-green-hydrogen-technology-in-fertilizer-production/
Zakai D, Chadwick-Furman NE (2002) Impacts of intensive recreational diving on reef corals at Eilat, Northern Red Sea. Biol Conserv 105(2):179–187. https://doi.org/10.1016/S0006-3207(01)00181-1
Zarrilli S, Aydiner-Avsar N (2020) COVID-19 puts women working in SIDS tourism industry at risk. United Nations Conference on Trade and Development (blog), 13 May. https://unctad.org/en/pages/news-details.aspx?OriginalVersionID=2370
Zhang W, Wang R, Sun Z, Zhu X, Zhao Q, Zhang T, Cholewinski A et al (2020) Catechol-functionalized hydrogels: biomimetic design, adhesion mechanism, and biomedical applications. Chem Soc Rev 49(2):433–464. https://doi.org/10.1039/C9CS00285E

Ocean Solutions That Benefit People, Nature and the Economy

20

1 How to Use This Report

This report can be read like a book, 'cover to cover'—the reader will follow a narrative arc which balances hope and concern, present and future states, concrete examples and more abstract ideas.

However, it is more probable that this report will be used like a readily accessible compendium of the latest scientific insights, frameworks and ideas that allow readers to find specific facts, messages or concepts and dive deeper on selected sections of the report.

This report aims at answering three core questions:

- **WHY**: Why do we need a sustainable ocean economy and why now? (Sect. 4)
- **WHAT**: What would a sustainable ocean economy look like? What would be the main economic components and the interlinkages between them? What would be the benefits to expect for the economy, the people and the planet? (Sect. 5)
- **HOW**: How should such a complex socioeconomic transition be apprehended? How should a 10-year transformation agenda be structured? How should we get started? (Sect. 6)

Readers looking for arguments about the need for a sustainable ocean economy and reasons for hope about the possibility of one should read the Prologue and Sect. 4.

Readers who want to understand what a sustainable ocean could look like in 2050, and the expected associated benefits, should read Sect. 5.

Ocean practitioners already familiar with the concept of a sustainable ocean economy are invited to go straight to Sect. 6 to discover a fresh and practical approach to guide the transition to a sustainable ocean economy. In particular, Sect. 6.3 presents an 'ocean action agenda' that could be used as a handbook to help decision-makers structure their sustainable ocean economy program, be it at a state or a company level.

This handbook identifies a number of key actions for each area of focus, covering both cross-cutting enablers and ocean-based sectors. Finally, Sect. 6.4 suggests some very concrete ideas that could be implemented immediately to start or accelerate the implementation of the more holistic, 10-year, ocean action agenda.

2 Executive Summary: The New Ocean Narrative

Billions of people have personal connections to the ocean. For many people living in coastal communities, the ocean is not only a source of food and livelihoods, it is an intrinsic part of their culture and heritage. For the millions of people who earn their living from the ocean, it is a source of income and a way of life. For the 40% of the world's population that live within 150 km of the coast and the hundreds of millions of others who visit it, the ocean is central to their lives.[1] The ocean plays an essential and usually unrecognised role in the daily lives of all of the planet's inhabitants. Indeed, breathing itself would be impossible without the ocean, which produces half of the earth's oxygen.[2]

Originally published in:
Stuchtey, M., A. Vincent, A. Merkl, M. Bucher et al. 2020. "Ocean Solutions That Benefit People, Nature and the Economy." Washington, DC. World Resources Institute. www.oceanpanel.org/ocean-solutions.
Reprint by Springer International Publishing (2023) with kind permission.
Published under license from the World Research Institute.

[1]UN Atlas of the Oceans. n.d. "Human Settlements on the Coast." http://www.oceansatlas.org/subtopic/en/c/114/. Accessed 13 August 2020.

[2]National Oceanic and Atmospheric Administration (NOAA). n.d. "How Much Oxygen Comes from the Ocean?" https://oceanservice.noaa.gov/facts/ocean-oxygen.html. Accessed 13 May 2020.

J. Lubchenco, P. M. Haugan (eds.), *The Blue Compendium*, https://doi.org/10.1007/978-3-031-16277-0_20

The ocean is also an enormous economic asset. Around 90% of the world's goods are traded across the ocean.[3] Hundreds of millions of people work in fishing and mariculture, shipping and ports, tourism, offshore energy, pharmaceuticals and cosmetics—all of which rely on resources a healthy ocean can offer the ocean.[4] By some estimates, the ocean economy directly contributes more than $1.5 trillion a year to the global economy.[5]

Putting a resource this critical at risk is reckless. But the world has not handled the ocean with care. Poor management has damaged many of the ocean's assets and reduced the ocean's natural ability to restore itself. Ocean health is on a downward spiral, preventing humanity from reaping the riches a healthy ocean could produce and jeopardising the future. The ocean is becoming warmer, more acidic, stormier, higher, more oxygen-depleted, less predictable and less resilient—and neither the problems it is facing nor the wealth it yields are distributed equitably.

Climate change is disproportionately affecting vulnerable and marginalised people, many of whom depend on the ocean for nutrition, identity and income. As they battle a warming ocean and rising sea level, they increasingly face depleted and shifting fish stocks without the ability to change gear or travel further to fish or seek other sources of livelihood.

For years, the overarching view was that the ocean is so vast that it is simply too big to fail. The folly of this approach is now evident. The new dominant narrative is that the problems are so complex that the ocean is simply too big to fix. This view is also incorrect. The ocean's problems are real, but action is already taking place to solve them.

A new way of thinking has immense potential to open the door to a sustainable ocean economy. This approach abandons the false choice between economic development and environmental protection. In contrast to a 'conservation philosophy' of minimising destruction or an 'extractive approach' of maximising the resources that can be extracted from the ocean, the new approach seeks to achieve the integration of the 'three Ps' of effective protection, sustainable production and equitable prosperity. This approach does not mean just leaving the ocean alone; it means proactively managing human activities to use the ocean wisely rather than using it up, in order to help build a much richer future in which people have more wealth and better health, nature thrives and resources are distributed more equitably.

Realising the new vision requires an integrated, rather than a sectoral, approach that is based on five building blocks:

- Using science and data to drive decision-making
- Engaging in goal-oriented ocean planning
- De-risking finance and using innovation to mobilise investment
- Stopping land-based pollution
- Changing ocean accounting so that it reflects the true value of the ocean

Putting these building blocks in place would enable change across the entire ocean economy, not just in specific sectors or locations. Over time, sustainable ocean management could help the ocean produce as much as 6 times more food and generate 40 times more renewable energy than it currently does,[6] contribute one-fifth of the reductions in greenhouse gas emissions needed to keep the world within the 1.5 °C temperature rise limit set by the Paris Agreement goals by 2050,[7] help lift millions of people out of poverty, improve equity and gender balance, increase economic and environmental resilience, build the industries of the future and provide low-carbon fuel and feed for activities on land.

Investments in a sustainable ocean economy are not just good for the ocean. They represent an excellent business proposition. Investing $2.8 trillion today in just four ocean-based solutions—offshore wind production, sustainable ocean-based food production, decarbonisation of international shipping, and conservation and restoration of mangroves would yield a net benefit of $15.5 trillion by 2050, a benefit-cost ratio of more than 5:1.[8]

The ocean is so vast, and its role in the global economy and the lives of the world's people so fundamental, that it can be difficult to know where to start in creating a sustainable ocean economy. Fortunately, pragmatic solutions are already being implemented, albeit not at the scale needed. These

[3]Olmer, N., B. Comer, B. Roy, X. Mao and D. Rutherford. 2017. "Greenhouse Gas Emissions from Global Shipping, 2013–2015." Washington, DC: International Council on Clean Transport. https://theicct.org/sites/default/files/publications/Global-shipping-GHG-emissions-2013-2015_ICCT-Report_17102017_vF.pdf; International Chamber of Shipping. n.d. "Shipping and World Trade." Accessed 18 August 2020. https://www.ics-shipping.org/shipping-facts/shipping-and-world-trade.

[4]Teh, L.C.L., and U.R. Sumaila. 2013. "Contribution of Marine Fisheries to Worldwide Employment." *Fish and Fisheries* 14 (1): 77–88. doi: https://doi.org/10.1111/j.1467-2979.2011.00450.x.

[5]OECD. 2016. *The Ocean Economy in 2030*. Report. Paris: OECD Publishing. https://www.oecd.org/environment/the-ocean-economy-in-2030-9789264251724-en.htm.

[6]Costello, C., L. Cao, S. Gelcich et al. 2019. "The Future of Food from the Sea." Washington, DC: World Resources Institute. https://www.oceanpanel.org/blue-papers/future-food-sea; IEA and ETP. 2017. "International Energy Agency, Energy Technology Perspectives 2017." www.iea.org/etp2017.

[7]Hoegh-Guldberg, O., et al. 2019. "The Ocean as a Solution to Climate Change: Five Opportunities for Action." Washington, DC: World Resources Institute. https://oceanpanel.org/sites/default/files/2019-10/HLP_Report_Ocean_Solution_Climate_Change_final.pdf.

[8]Konar, M., and H. Ding. 2020. "A Sustainable Ocean Economy for 2050: Approximating Its Benefits and Costs." Washington, DC: World Resources Institute. https://www.oceanpanel.org/Economicanalysis.

efforts could jump-start progress on a much larger scale, putting the world on a trajectory that would vastly increase prosperity in the coming decade and the longer term. These approaches embrace a philosophy in which stakeholders—including direct users of the ocean (fishers, shippers, energy producers and beach lovers, among others) as well as policy-makers, governments, businesses and others—accept the new paradigm and work together to achieve the same goal of a healthy, productive ocean.

Some of the most promising efforts include empowering communities and modifying incentives to align economic and conservation outcomes. In the Philippines, for example, a network has been created that grants fishing communities clear, exclusive rights to fish in certain areas. In communities that organised to manage 'their' fishing areas and protected zones, boats and fishers are registered, the catch is recorded, regulations are respected and fishers participate in management. By embracing sustainability, participating communities increased their food and financial security and gained access to new markets and sources of capital—improving their own well-being while protecting the ocean. Complementary global trends are also emerging. Open data networks are making it easier to track and detect illegal fishing vessels. Governments are starting to tackle plastic pollution, and financiers are starting to recognise the value of investing in the ocean.

Practical solutions that can be implemented at a modest scale as well as high-level actions could create a sustainable ocean economy, underpinned by the three Ps of effective protection, sustainable production and equitable prosperity. Implementing them requires political will at all levels, including the very top.

The ocean is not too big to fail, and it is not too big to fix. But it is too big to ignore. The more we learn about the ocean, the more we see that it is central to improving the health, wealth and well-being of people. It holds the answers to the most pressing challenges facing humanity, including climate change and food security. It is time to shift away from thinking of the ocean as a victim toward seeing it as an essential part of the solution to global challenges. New partnerships need to be forged that will take action now to achieve a sustainable ocean and a sustainable future. The choice is not between ocean protection and production. Together they can help build a healthy and prosperous future.

2.1 The Health, Wealth and Well-Being of the World and Its People Depend on the Ocean

Maintaining a healthy ocean is vital to improving global health and increasing global prosperity for everyone; expanding opportunities for all people, including women and marginalised groups; and making the world a better place to live for all, even people living far from the ocean. A sustainable ocean economy is obviously important for the traditional ocean sectors, such as fisheries and shipping. But its value goes well beyond the lives of people whose income comes directly from the sea. Because of the interconnectedness of the global economy, what happens in the ocean affects not only fishers in Fiji but also farmers in Zimbabwe, whose imported tools may have travelled to Africa in a container ship and whose air quality and climate are affected by what happens in the ocean.

The ocean provides a wide variety of vital benefits, many of which are often overlooked:

- **It helps make the planet liveable and is critical to managing the effects of climate change.** The ocean produces half of the planet's oxygen, absorbs 93% of the world's anthropogenic heat and moderates the earth's temperature by reducing the heat differential between the poles and the Equator.[9] Without the ocean's regulation of the earth's climate, much more carbon dioxide would be trapped in the atmosphere, exacerbating global climate change.[10]
- **The global economy and the livelihoods of hundreds of millions of people depend on the ocean.** The modern global economy could not exist without the ocean. Around 90% of all internationally traded goods travel by ship.[11] The ocean economy directly contributes an estimated $1.5 trillion to the global economy.[12] The ocean food sector alone provides up to 237 million jobs, including in fishing, mariculture and processing.[13] Millions of people also work in other ocean sectors, including shipping, ports, energy and tourism—and many more are indirectly connected to the ocean economy.
- **The ocean provides billions of people with nutritious food, with a much smaller environmental footprint than land-based food production.** More than three billion people rely on food from the sea as a source of pro-

[9] Stocker, T.F., D. Qin, G.K. Plattner, M. Tignor, S.K. Allen, J. Boschung, A. Nauels et al. 2013. "Summary for Policymakers." In *Climate Change 2013: The Physical Science Basis. Contribution of Working Group I to the Fifth Assessment Report of the Intergovernmental Panel on Climate Change*. Cambridge: Cambridge University Press. http://www.climatechange2013.org/images/report/WG1AR5_SPM_FINAL.pdf; National Oceanic and Atmospheric Administration (NOAA). n.d. "How Much Oxygen Comes from the Ocean?" https://oceanservice.noaa.gov/facts/ocean-oxygen.html. Accessed 13 May 2020.

[10] Hoegh-Guldberg et al. 2019. "The Ocean as a Solution to Climate Change."

[11] Olmer, N., B. Comer, B. Roy, X. Mao and D. Rutherford. 2017. "Greenhouse Gas Emissions from Global Shipping, 2013–2015"; International Chamber of Shipping. n.d. "Shipping and World Trade."

[12] OECD. 2016. *The Ocean Economy in 2030*.

[13] Teh, L.C.L., and U.R. Sumaila. 2013. "Contribution of Marine Fisheries to Worldwide Employment."

tein and key nutrients, including omega-3 fatty acids and iodine.[14]

- **Coastal habitats, such as mangroves, provide protection for hundreds of millions of people, nurture biodiversity, detoxify pollutants flowing off the land, and provide nursery areas for fisheries, increasing the supply of food and providing livelihoods.** They are also a source of revenue. Coral reefs alone contribute $11.5 billion a year to global tourism, benefitting more than 100 countries and providing food and livelihoods to local people.[15]
- **The ocean provides a sense of wonder, solace and connection to the natural world and is deeply woven into the cultural and spiritual lives of billions of coastal dwellers**. It also gives pleasure to the hundreds of millions of people a year who visit it.[16]
- **The ocean may store unknown treasures**. In addition to its known benefits, it may be the home of undiscovered resources—including medical ones—and new knowledge.

2.2 Its Potential Is Enormous, But the Ocean Is in Trouble

Human stressors affect virtually the entire ocean, making it more difficult for the ocean to sustain human life on earth. Climate change, overfishing, habitat destruction, biodiversity loss, excessive nutrient loads, pollution and other problems are damaging the ocean's health.

- **Climate change and greenhouse gas emissions are having multiple effects on the ocean**. The ocean is becoming warmer and more acidic, putting pressure on plants and animals from the base of the ocean food web to the top. Ocean warming affects circulation, stratification, oxygen content and sea level. By 2100, as many as 630 million people could be at risk of coastal flooding caused by climate change.[17] Sea level rise also affects agriculture, by submerging land, salinising soil and groundwater, and eroding coasts. It will also erode and submerge tourism infrastructure and beaches. In the Caribbean, for example, sea level rise of 1 m is projected to endanger up to 60% of resorts, damage or cause the loss of 21 airports and severely flood 35 ports.[18] Rebuilding the region's resorts alone is projected to cost the Caribbean $10–$23 billion in 2050.[19]
- **Habitats are being destroyed, biodiversity is declining and the distribution of species is changing—all of which reduce the benefits that ocean ecosystems provide**. Coastal habitats are disappearing at an alarming rate. Global mangrove cover declined by 25–35% between 1980 and 2000, largely as a result of land development and conversion to unsustainable mariculture ponds and rice paddies.[20] The loss of coastal habitats and coral reefs is eroding natural coastal protection, exposing 100–300 million people living within coastal 100-year flood zones to increased risk of floods and hurricanes.[21]

[14]FAO, ed. 2018. The State of World Fisheries and Aquaculture 2018: Meeting the Sustainable Development Goals. Rome: Food and Agriculture Organization of the United Nations. http://www.fao.org/3/I9540EN/i9540en.pdf; World Health Organization. n.d. "3. Global and Regional Food Consumption Patterns and Trends." https://www.who.int/nutrition/topics/3_foodconsumption/en/index2.html. Accessed 6 May 2020.

[15]Masson-Delmotte, V., P. Zhai, H.O. Pörtner, D. Roberts, J. Skea, P.R. Shukla, A. Pirani et al., eds. 2019. *Global Warming of 1.5 °C: An IPCC Special Report on the Impacts of Global Warming of 1.5 °C above Pre-industrial Levels and Related Global Greenhouse Gas Emission Pathways, in the Context of Strengthening the Global Response to the Threat of Climate Change, Sustainable Development, and Efforts to Eradicate Poverty*. Intergovernmental Panel on Climate Change. https://www.ipcc.ch/site/assets/uploads/sites/2/2019/06/SR15_Full_Report_High_Res.pdf.

[16]Allison, E., J. Kurien and Y. Ota. 2020. "The Human Relationship with Our Ocean Planet." Washington, DC: World Resources Institute. https://www.oceanpanel.org/blue-papers/relationship-between-humans-and-their-ocean-planet.

[17]Kulp, S.A., and B.H. Strauss. 2019. "New Elevation Data Triple Estimates of Global Vulnerability to Sea-Level Rise and Coastal Flooding." *Nature Communications* 10 (1): 4844. doi: https://doi.org/10.1038/s41467-019-12808-z.

[18]Pachauri, R.K., L. Mayer and Intergovernmental Panel on Climate Change, eds. 2015. *Contribution of Working Groups I, II and III to the Fifth Assessment Report of the Intergovernmental Panel on Climate Change 2014: Synthesis Report*. Geneva: Intergovernmental Panel on Climate Change. https://ar5-syr.ipcc.ch/ipcc/ipcc/resources/pdf/IPCC_SynthesisReport.pdf.

[19]Nicholls, M. 2014. "Climate Change: Implications for Tourism: Key Findings from the Intergovernmental Panel on Climate Change Fifth Assessment Report." University of Cambridge. https://www.cisl.cam.ac.uk/business-action/low-carbon-transformation/ipcc-climate-science-business-briefings/pdfs/briefings/ipcc-ar5-implications-for-tourism-briefing-prin.pdf.

[20]Polidoro, B.A., K.E. Carpenter, L. Collins, N.C. Duke, A.M. Ellison, J.C. Ellison, E.J. Farnsworth et al. 2010. "The Loss of Species: Mangrove Extinction Risk and Geographic Areas of Global Concern." Edited by D.M. Hansen. *PLOS ONE* 5 (4): e10095. doi: https://doi.org/10.1371/journal.pone.0010095; Valiela, I., J.L. Bowen and J.K. York. 2001. "Mangrove Forests: One of the World's Threatened Major Tropical Environments. At Least 35% of the Area of Mangrove Forests Has Been Lost in the Past Two Decades, Losses That Exceed Those for Tropical Rain Forests and Coral Reefs, Two Other Well-Known Threatened Environments." *BioScience* 51 (10): 807–15. doi: https://doi.org/10.1641/0006-3568(2001)051[0807:MFOOTW]2.0.CO;2; Thomas, N., R. Lucas, P. Bunting, A. Hardy, A. Rosenqvist and M. Simard. 2017. "Distribution and Drivers of Global Mangrove Forest Change, 1996–2010." Edited by S. Joseph. *PLOS ONE* 12 (6): e0179302. doi: https://doi.org/10.1371/journal.pone.0179302.

[21]Díaz et al. 2019. "Summary for Policymakers of the Global Assessment Report on Biodiversity and Ecosystem Services of the Intergovernmental Science-Policy Platform on Biodiversity and Ecosystem Services."

Coral reefs—virtually all of which will be lost at 2 °C warming—are declining rapidly as a result of compounding pressures from rising ocean temperatures, overfishing and nutrient pollution.[22] The biodiversity of the open ocean declined by up to 50% over the past 50 years,[23] and the relative abundance of different species has shifted in favour of species that are more tolerant of low-oxygen conditions, such as microbes, jellyfish and some squid.[24]

- **Plastic, other land-based pollutants and discharge from ships contaminate the ocean**. Because of the common belief that 'the solution to pollution is dilution', the ocean has long been used as a repository for sewage, nutrient run-offs, heavy metals, nuclear waste, persistent toxicants, pharmaceuticals, personal care products and other noxious items. More than 80% of all marine pollution originates on land.[25] Millions of metric tonnes of plastic are dumped into the ocean every year, entangling, sickening and contaminating at least 700 species of marine life.[26] Untreated ballast water from ships is discharged into foreign ports, creating one of the principal vectors of potentially invasive alien species.[27]
- **Overfishing is depleting fish stocks and harming wildlife**. The 'tragedy of the ocean commons' open access that characterises fishing in many parts of the ocean means that too many boats pursue too few fish, at the expense of overall system health and productivity. Exacerbated by subsidies that increase the capacity of the fishing fleet and by illegal, unreported and unregulated (IUU) fishing, fishing has become the number one driver of extinction risk for marine vertebrates (excluding birds).[28] If overfishing continues, annual yield is projected to fall by over 16% by 2050, threatening global food security.[29]

A single stressor, such as overfishing or pollution, can do considerable damage. Worse still, individual stressors locally compound one another, with enormous consequences for ecosystems. Without action, these problems could cost the global economy more than $400 billion a year by 2050. By 2100, the annual cost could reach $2 trillion.[30]

Neglect and abuse of the ocean and the effects of global climate change will make life worse for everyone. But historically underrepresented and underserved communities—including women—will bear a disproportionately large share of the burden. These groups are most vulnerable to food insecurity, loss of livelihoods and sea level rise. They are also the most likely to suffer from the many crimes and human rights violations that take place on the ocean, including human trafficking and smuggling, slave labour and peonage (debt slavery) systems.

2.3 A New Relationship with the Ocean Is Needed: One That Creates a Healthy Ocean and a Sustainable Ocean Economy

In contrast to a conservation philosophy of minimising destruction and an extractive approach that focuses on exploiting the ocean to create wealth, a sustainable ocean economy brings diverse stakeholders together to achieve common goals—the three Ps of effective protection, sustainable production and equitable prosperity. In this new paradigm, groups work together by adopting integrated and balanced management of the ocean in which each of the three Ps contributes to the others. Sustainable production based on regenerative practices (such as climate-ready, ecosystem-based fisheries management or seaweed farming) along with fully protected areas, for example, can help

[22] Masson-Delmotte et al. 2019. *Global Warming of 1.5 °C.*

[23] Worm, B., M. Sandow, A. Oschlies, H.K. Lotze and R.A. Myers. 2005. "Global Patterns of Predator Diversity in the Open Oceans." *Science* 309 (5739): 1365–69. doi: https://doi.org/10.1126/science.1113399.

[24] Gaines, S., R. Cabral, C.M. Free, Y. Golbuu, R. Arnason, W. Battista, D. Bradley et al. 2019. "The Expected Impacts of Climate Change on the Ocean Economy." Washington, DC: World Resources Institute. https://www.oceanpanel.org/blue-papers/expected-impacts-climate-change-ocean-economy.

[25] Ocean Conservancy. n.d. *Stemming the Tide: Land-Based Strategies for a Plastic-Free Ocean.* https://oceanconservancy.org/wp-content/uploads/2017/04/full-report-stemming-the.pdf. Accessed 6 May 2020.

[26] Gall, S.C., and R.C. Thompson. 2015. "The Impact of Debris on Marine Life." *Marine Pollution Bulletin* 92 (1): 170–79. doi: https://doi.org/10.1016/j.marpolbul.2014.12.041.

[27] Global Environment Facility–UN Development Programme – International Maritime Organization (GEF-UNDP-IMO) GloBallast Partnerships Programme and International Union for Conservation of Nature (IUCN). 2010. "Economic Assessments for Ballast Water Management: A Guideline." GloBallast Monograph Series no. 19. London, UK, and Gland, Switzerland: GEF-UNDP-IMO GloBallast Partnerships, IUCN. https://portals.iucn.org/library/sites/library/files/documents/2010-075.pdf.

[28] Rogers, A., O. Aburto-Oropeza, W. Appeltans, J. Assis, L. T. Ballance, P. Cury, C. Duarte et al. 2020. "Critical Habitats and Biodiversity: Inventory, Thresholds and Governance." Washington, DC: World Resources Institute. https://www.oceanpanel.org/blue-papers/critical-habitats-and-biodiversity-inventory-thresholds-and-governance.

[29] Costello et al. 2019. "The Future of Food from the Sea."

[30] Pörtner, H.O., D.C. Roberts, V. Masson-Delmotte, P. Zhai, M. Tignor, K. Poloczanska, K. Mintenbeck et al., eds. 2019. "Summary for Policymakers." In *IPCC Special Report on the Ocean and Cryosphere in a Changing Climate*. Intergovernmental Panel on Climate Change. https://report.ipcc.ch/srocc/pdf/SROCC_FinalDraft_FullReport.pdf.

restore ocean health. The result is a triple win for nature, people and the economy and a world where prosperity is greater and more equitably distributed than it is today (Fig. 20.1).

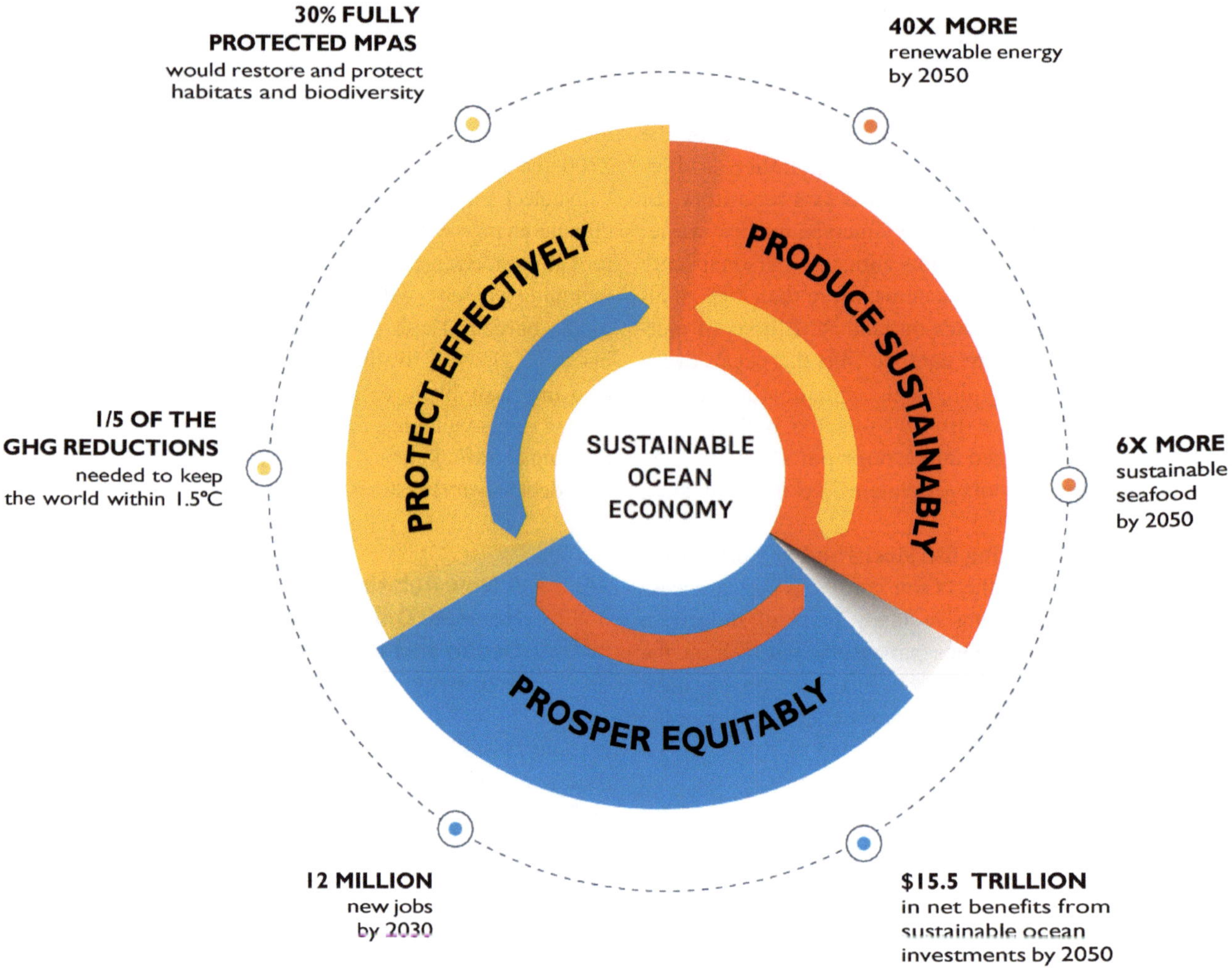

Fig. 20.1 A sustainable ocean economy can create a triple win for people, nature and the economy. *Note: MPAs: Marine protected areas. GHG: Greenhouse gas emissions.* (*Source: Authors, drawing on the following sources: OECD. 2016. The Ocean Economy in 2030. Directorate for Science, Technology and Innovation Policy Note, April.* https://www.oecd.org/futures/Policy-Note-Ocean-Economy.pdf; *Konar, M., and H. Ding. 2020. "A Sustainable Ocean Economy for 2050: Approximating Its Benefits and Costs." Washington, DC: World Resources Institute.* https://www.oceanpanel.org/Economicanalysis; *Costello, C., L. Cao, S. Gelcich et al. 2019. "The Future of Food from the Sea." Washington, DC: World Resources Institute.* https://www.oceanpanel.org/blue-papers/future-food-sea; *Hoegh-Guldberg, O., et al. 2019. "The Ocean as a Solution to Climate Change: Five Opportunities for Action." Washington, DC: World Resources Institute.* https://oceanpanel.org/sites/default/files/2019-10/HLP_Report_Ocean_Solution_Climate_Change_final.pdf)

2.3.1 Protect Effectively

Protecting the ocean doesn't mean just leaving it alone—it means managing human activity wisely, in order to preserve biodiversity and critical habitats, allow the ocean to sustainably yield greater benefits and preserve the ocean's cultural and spiritual value. In some areas, significantly scaling back or prohibiting human activities will be necessary to allow ecosystems to recover and regenerate. In most areas, sustainable practices will be needed that both allow the ocean to produce and maintain ocean health.

Far from holding back production, restoring and maintaining the ocean's health represents the best way to generate ocean-based wealth and make the most of the ocean's unique resources. This new way of thinking is also marked by a shift from incremental improvement to ecosystem-based integrated management and from a narrow focus on gross domestic product (GDP) alone to one that takes account of both the monetary and nonmonetary benefits and assets of the ocean.

A Sustainable Ocean Economy Would Help Protect the Ocean by Reducing the Carbon Dioxide Emissions That Are Threatening It

Ocean-based activities could provide one-fifth of the carbon mitigation needed to meet the Paris Agreement goals by 2050, reducing global greenhouse gas emissions by up to 4 billion tonnes of carbon dioxide equivalent in 2030 and up to 11 billion tonnes in 2050, according to research commissioned by the Ocean Panel.[31] Emission reductions of this magnitude are equivalent to the annual emissions from 2.5 billion cars or all of the world's coal-fired power plants.

Protecting Coastal Habitats and the Ocean's Biodiversity Would Help the Ocean Continue to Provide the Ecosystem Services Humanity Depends on

A restored and protected ocean would help mitigate the impact of storm and sea level rise, saving lives and livelihoods, and would reduce economic costs of damage and recovery. Healthy coral reefs, for example, reduce wave energy by up to 97%, potentially protecting up to 100 million coastal inhabitants from storm risks.[32] By reducing wave heights, mangroves reduce flooding of coastal areas and contribute to biodiversity. Marine protected areas (MPAs) that are fully protected from extractive and destructive activities can rebuild and safeguard biodiversity, mitigate climate change (by preventing emissions from the disturbance of sediment carbon by bottom trawling) and boost the productivity of fisheries in areas surrounding MPAs through the spillover of fish.[33]

Protecting the Ocean from Pollution Could Catalyse Deeper Reform of Contaminating, Wasteful Material Management Practices on Land

The problem of ocean pollution starts on land. Plastic—along with numerous other pollutants, including pharmaceuticals and excess nutrients—enters the ocean because systems for their proper disposal on land are inadequate. The most effective way of stopping pollutants from entering the ocean is to tackle the root causes of pollution on land. Shifting to a 'circular economy'—a system in which resources are designed to be used continually and at their highest possible value added and recovered or regenerated as efficiently as possible at the end of their service—would yield enormous benefits for the ocean economy. Agricultural regulations aimed at reducing ocean dead zones could result in farmers adopting precision agriculture practices to reduce runoff, which would also improve the health of the soil and the quality of water in rivers and streams.

2.3.2 Produce Sustainably

When the ocean is managed effectively, it can produce more and its production can be more sustainable. A shift to a sustainable ocean economy would increase food and energy production without putting extra pressure on marine ecosystems.

The Volume of Food Production from the Ocean Could Soar, Helping Increase Food Security for Almost Ten Billion People in 2050

The ocean's ability to sustainably produce food is vastly under-realised. Managed better and sustainably, the ocean could produce up to six times more food than it does today—and it could do so with a low environmental footprint.[34]

Most fishing today is not economically or ecologically optimised. Too many boats pursue too few fish in ways that are short-sighted and destructive. Too much seafood value is lost to poor handling. Too many non-target species are accidentally caught. If this approach continues, the yield in 2050 is expected to be around 16% lower than it is today.[35] In contrast, if all stocks currently exploited were fished at the maximum sustainable economic yield, production could increase

[31] Hoegh-Guldberg, O., et al. 2019. "The Ocean as a Solution to Climate Change: Five Opportunities for Action."

[32] Ferrario, F., M.W. Beck, C.D. Storlazzi, F. Micheli, C.C. Shepard and L. Airoldi. 2014. "The Effectiveness of Coral Reefs for Coastal Hazard Risk Reduction and Adaptation." *Nature Communications* 5(1): 3794. doi: https://doi.org/10.1038/ncomms4794.

[33] da Silva, I.M., N. Hill, H. Shimadzu, A.M.V.M. Soares and M. Dornelas. 2015. "Spillover Effects of a Community-Managed Marine Reserve." *PLOS ONE* 10 (4): e0111774. doi: https://doi.org/10.1371/journal.pone.0111774.

[34] Costello et al. 2019. "The Future of Food from the Sea."

[35] Costello et al. 2019. "The Future of Food from the Sea."

by 20% over current production levels and by 40% over the catch forecast under a business-as-usual scenario.[36]

The mariculture story is even more promising. The potential to expand finfish mariculture is significant if farms avoid adversely affecting surrounding ecosystems and use fish feed that is not made from wild caught fish. Unfed mariculture also holds great promise. Bivalves (such as oysters and mussels) and seaweed can substantially increase the production of nutritious food and feed, with little negative impact on the marine environment. In some cases, this kind of mariculture could actually enhance wild fisheries by creating artificial habitats and nursery grounds for fish.

About 35% of fish and seafood is currently wasted in the value chain. Reducing this wastage could boost consumption without increasing production.[37]

The Ocean Can Provide a Virtually Limitless Supply of Clean, Renewable Energy

Offshore wind turbines could generate 23 times more power than the present total global electricity consumption.[38] Other potential sources of ocean-based renewable energy—producing energy from waves and tides, salinity and temperature gradients, and floating solar photovoltaic panels, for example—are still in their infancy but hold promise.

Investments in the Ocean Are Highly Cost-Effective

Investment of $2.8 trillion today in four sustainable ocean-based solutions—conservation and restoration of mangroves, decarbonisation of international shipping, sustainable ocean-based food production and offshore wind production—would yield net benefits of $15.5 trillion by 2050.[39] All four interventions have high benefit-cost ratios (Fig. 20.2).

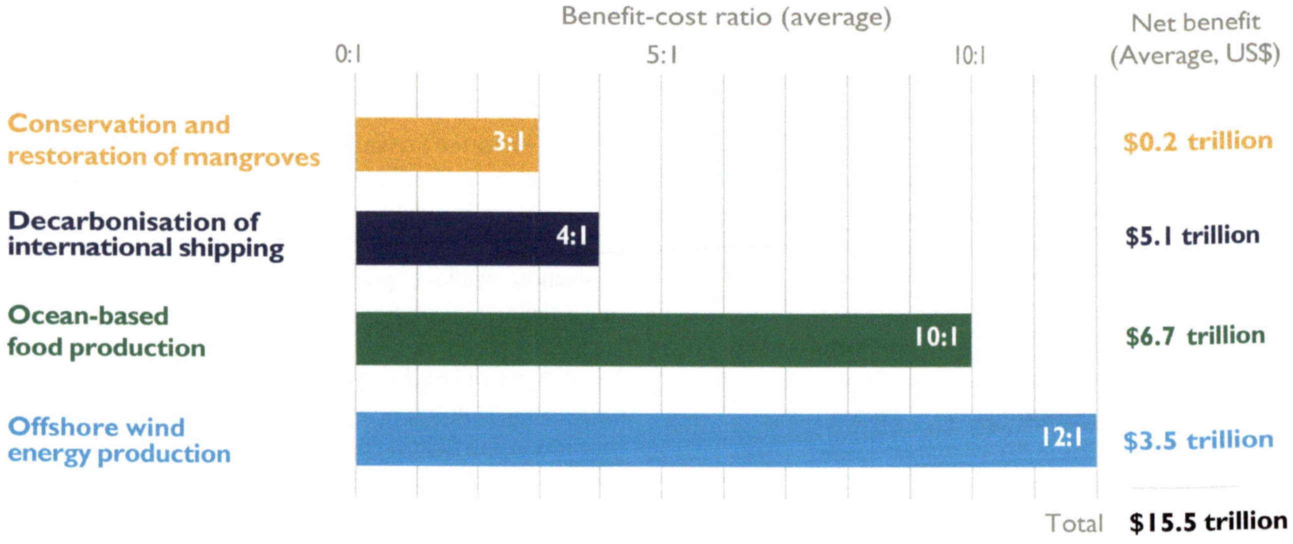

Fig. 20.2 Sustainable ocean-based interventions have very high benefit–cost ratios and could yield trillions of dollars of benefits. *Note: Average benefit-cost (B-C) ratios have been rounded to the nearest integer and the net benefits value to the first decimal place. The B-C ratio for mangroves is the combined ratio for both conservation- and restoration-based interventions. The average net benefits represent the average net present value for investments and are calculated over a 30-year horizon (2020–2050).* (*Source: Konar, M., and H. Ding. 2020. "A Sustainable Ocean Economy for 2050: Approximating Its Benefits and Costs." Washington, DC: World Resources Institute.* https://www.oceanpanel.org/Economicanalysis)

[36]Costello et al. 2019. "The Future of Food from the Sea."

[37]FAO. 2017. "FAO Regional Office for Europe and Central Asia: Losses in Fisheries and Aquaculture Tackled at Global Fishery Forum." 14 September. http://www.fao.org/europe/news/detail-news/en/c/1037271/.

[38]IEA. n.d. "Data & Statistics"; Haugan et al. 2019. "What Role for Ocean-Based Renewable Energy and Deep-Seabed Minerals in a Sustainable Future?".

[39]Konar and Ding. 2020. "A Sustainable Ocean Economy for 2050."

2.3.3 Prosper Equitably

Left unmanaged, a growing ocean economy could exacerbate economic inequality, as strong, elite incumbents capture the benefits of the ocean while vulnerable and marginalised groups become increasingly exposed to economic, social and cultural impacts, including displacement.

Inequality is a structural feature of the current ocean economy. Women, for example, represent just 2% of the world's formal maritime workers.[40] Poor, vulnerable and marginal communities are bearing—and will continue to bear—the worst effects of global climate change. A sustainable ocean economy would not only create greater wealth, it would also create a world in which resources are distributed more evenly and where all ocean users have an opportunity to have a voice in critical decisions.

A Sustainable Ocean Economy Would Create New and Better Jobs

By some estimates, it could create 12 million net jobs.[41] Some sectors, particularly fisheries, will need to shed jobs. Support schemes will be needed to manage the transition to lower capacity and more sustainable management of fish stocks.

Other sectors will grow significantly. Thousands of new jobs will be created in engineering, information technology, applied science and related areas. The number of jobs in mariculture and offshore wind is projected to soar, and the increase in seaborne cargo volume and the expansion of ports are expected to create millions of jobs. Decarbonising shipping will be critical to ensure that this expansion does not come at the cost of the ocean's health.

The New Agenda Would Empower Local Fishers

The yields of millions of artisanal fishers are far lower than they used to be, partly because of the open-access model of much of the ocean, which has resulted in overfishing. A better-managed approach would benefit many of them.

Empowering fishers by granting them access rights in exchange for sustainably managing their resource is one of the levers of the sustainable ocean economy. Doing so has already proven effective. In the territorial use rights fisheries (TURFs) that Chile created, for example, catches by artisanal fisheries have surpassed industrial catches, and the biomass and size of the target species has risen.[42] Similar approaches have met with great success in many fisheries, recovering depleted fisheries and enabling them to thrive.[43]

International Collaboration and Transparent Supply Chains Could Significantly Reduce Maritime Crime

IUU fishing is estimated to account for 20% of the world's catch (up to 50% in some areas).[44] Illegal fishing is also often an indicator of other types of crime at sea, including labour and human rights violations, money laundering and tax fraud.

Acting Sustainably Would Help Preserve the Cultural Importance of the Ocean

The ocean is more than just a source of economic wealth. It also has spiritual, cultural and recreational value to billions of people.[45] For many Indigenous peoples, it is a key aspect of their culture. Well-designed marine protected areas and other effective area-based conservation measures can help preserve pristine ocean areas and culturally important ocean areas (such as sacred sites, historic wrecks and sea graves).

2.3.4 The Ocean Should Be a Key Part of the Massive Global Economic Recovery from the COVID-19 Contraction

COVID-19 has temporarily halted economic activity in the ocean economy, causing significant income and revenue losses to tourism, fisheries and mariculture, and shipping; adversely affecting the ocean's health; and exacerbating gender and income inequalities. The disruptions have led to cascading and interrelated impacts. The decline in tourism, for example, forced some communities to turn back to unsustainable fishing as a food source, putting pressure on coastal fisheries and reefs.

---

[40] IMO. n.d. "Women in Maritime: IMO's Gender Programme." http://www.imo.org/en/OurWork/TechnicalCooperation/Pages/WomenInMaritime.aspx. Accessed 11 May 2020.

[41] OECD. 2016. *The Ocean Economy in 2030*. Directorate for Science, Technology and Innovation Policy Note, April. https://www.oecd.org/futures/Policy-Note-Ocean-Economy.pdf.

[42] Swilling, M., M. Ruckelshaus, T.B. Rudolph, P. Mbatha, E. Allison, S. Gelcich and H. Österblom. 2020. "The Ocean Transition: What to Learn from System Transitions." Washington, DC: World Resources Institute. https://www.oceanpanel.org/blue-papers/ocean-transition-what-learn-system-transitions.

[43] Costello, C., D. Ovando, T. Clavelle, C.K. Strauss, R. Hilborn, M.C. Melnychuk, T.A. Branch et al. 2016. "Global Fishery Prospects under Contrasting Management Regimes." *Proceedings of the National Academy of Sciences* 113 (18): 5125–29. doi: https://doi.org/10.1073/pnas.1520420113.

[44] Widjaja, S., T. Long, H. Wirajuda, A. Gusman, S. Juwana, T. Ruchimat and C. Wilcox. 2020. "Illegal, Unreported and Unregulated Fishing and Associated Drivers." Washington, DC: World Resources Institute. https://oceanpanel.org/sites/default/files/2020-02/HLP%20Blue%20Paper%20on%20IUU%20Fishing%20and%20Associated%20Drivers.pdf; Witbooi et al. 2020. "Organized Crime in the Fisheries Sector."

[45] Inniss, L., A. Simcock, A.Y. Ajawin, A.C. Alcala, P. Bernal, H.P. Calumpong, P.E. Araghi et al. 2016. "The First Global Integrated Marine Assessment." New York: United Nations. https://www.un.org/Depts/los/global_reporting/WOA_RPROC/WOACompilation.pdf.

A key objective of the massive recovery from the COVID contraction will be to restore economic activity without simply restoring old patterns of environmental degradation, instead creating a more sustainable and more resilient future. The ocean economy can play a critical role in this process. Investment in five areas—coastal and marine ecosystem restoration and protection, sewage and waste infrastructure, sustainable unfed mariculture, zero-emission marine transport and sustainable ocean-based renewable energy—could create jobs and spur economic growth in the immediate term.[46]

Investments made over the coming months and years will have long-term effects on the nature of the world's economies and their resilience to shocks. Efforts must be made now to avoid locking in high-emitting, high-polluting and inequitable pathways and locking out regenerative and sustainable futures. The opportunity to reset and rebuild a stronger, more equitable, more resilient and sustainable ocean economy should not be missed.

2.4 The Challenges Are Great, But a Pragmatic Action Agenda Offers Solutions to Meet Them

A world in which effective protection, sustainable production and equitable prosperity go hand in hand is possible. But it will not happen if business as usual continues. Without action, ocean planning will continue to be largely ad hoc, fish stocks will continue to decline and land-based polluters will continue to use the ocean as a liquid dump.

Political and business decisions made now and over the next 30 years could change this trajectory. With action, more systematic, ecosystem-based, inclusive spatial planning would become the norm. Access rights for specific ocean resources would be clarified, eliminating conflicts over resources and ensuring that the wealth of the ocean is equitably distributed. Wild fish stocks would recover, and significant increases in sustainable mariculture would provide nutritious food for billions of people, ensuring food security. Polluters would be subject to legal and political actions that would limit their ability to pollute the ocean.

2.4.1 Maintaining a Healthy Ocean Will Require Action on Many Fronts and Across Multiple Sectors

Delivering effective protection, sustainable production and equitable prosperity is an inspiring and feasible vision that is backed by science. The transition to a sustainable ocean economy will require a realignment of incentives, in-depth reforms of how the ocean is used and managed, and the empowerment of ocean users who are vested in enhancing ocean health.

Governments and businesses can take hundreds of sector-specific actions to improve ocean sectors, from supporting ocean-based renewable energy to create jobs in the wake of the COVID-19 contraction to supporting ecotourism and banning pollutants. These efforts are important, but without getting the fundamentals right, it will not be possible to transform the entire ocean system towards the desired sustainable model. Five building blocks can set the foundation for a sustainable ocean economy (Fig. 20.3). These building blocks put the conditions in place for wider change across various ocean sectors. With these foundations in place, sector-specific reforms, innovations and research can be implemented and accelerated.

Using Data to Drive Decision-Making

Technologies for sensing, simulating, forecasting, tracking, managing and sharing data on open-access platforms have the potential to transform the ocean economy. New technologies can be used to register ocean-related rights and contracts, facilitating rights-based management.[47] Product tracking throughout the supply chain can help brands embrace sustainable practices and small producers connect to global supply chains. Applications can help manage fishing areas and quotas, adjust shipping traffic and avoid endangered species bycatch. In the near future, every ship's journey—and the nature of its business at sea—will be public information. Lawbreakers such as illegal fishers, polluters, smugglers and labour law violators will literally be on the public radar and subject to arrest.

Some of these technologies are already being used on a limited scale. The POSEIDON model, for example, simulates the feedback loop between fishery policies, fishing fleets and ocean ecosystems, allowing policy alternatives to be compared.[48]

But barriers stand in the way of fully harnessing the power of science and data. Collecting data is very expensive, with most sensors custom-built for narrow and specific scientific

[46]Northrop, E., M. Konar, N. Frost and E. Hollaway. 2020. "A Sustainable and Equitable Blue Recovery to the COVID-19 Crisis." Washington, DC: World Resources Institute.

[47]Nyborg, K., J.M. Anderies, A. Dannenberg, T. Lindahl, C. Schill, M. Schlüter, W.N. Adger et al. 2016. "Social Norms as Solutions." *Science* 354 (6308): 42–43. doi: https://doi.org/10.1126/science.aaf8317; Leape et al. 2020. "Technology, Data and New Models for Sustainably Managing Ocean Resources."

[48]Bailey, R.M., E. Carrella, R. Axtell, M.G. Burgess, R.B. Cabral, M. Drexler, C. Dorsett et al. 2019. "A Computational Approach to Managing Coupled Human-Environmental Systems: The POSEIDON Model of Ocean Fisheries." *Sustainability Science* 14 (2): 259–75. doi: https://doi.org/10.1007/s11625-018-0579-9.

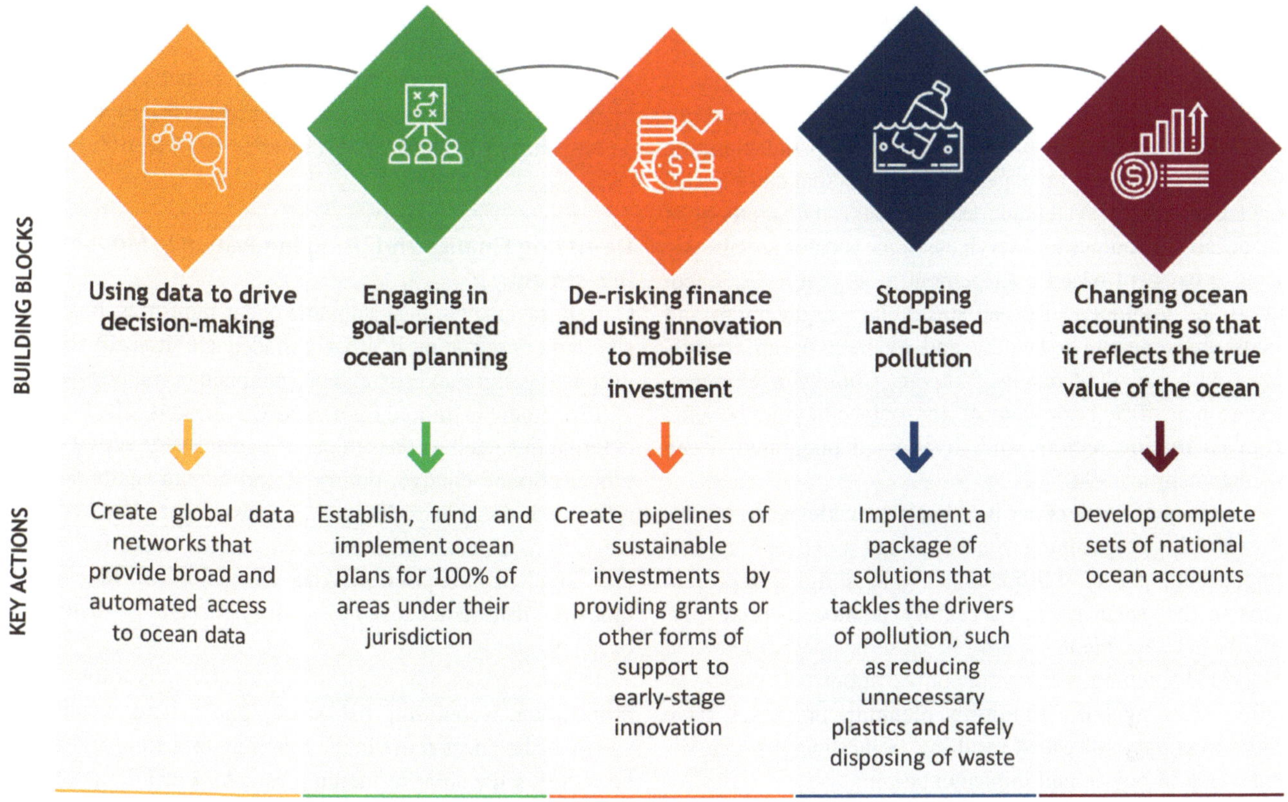

Fig. 20.3 Five building blocks are key to creating a sustainable ocean economy. (*Source: Authors*)

missions.[49] Technological innovation in the ocean has therefore been driven largely by governments and large-scale commercial interests.

Data are fragmented into national, corporate and academic domains. Access to data is limited, and data can be difficult to use. Tools designed for marine managers, for example, are often so technical that only programmers are able to use them. Poorer countries and ocean users have little or no access to data that could help them adopt sustainable practices.

Key Actions

Overcoming these and other barriers requires the creation of global data networks that provide broad and automated access to ocean data. Governments can lead the way by mandating these standards and helping create data networks that aggregate decentralised data into a common, searchable database. They can require that data sharing be a non-negotiable condition of access to public resources—whether the resources are fish stocks and mineral deposits or funds for coastal management or for research. To achieve or improve accountability, governments can prioritise technology-forcing regulations governing the real-time monitoring of fishing, seafood imports, shipping emissions, mining, coastal development and pollution.

Engaging in Goal-Oriented Ocean Planning

The sector-by-sector assortment of regulations for some ocean activities, coupled with an open-access model for others has contributed significantly to today's decline in ocean health and cannot continue. The shortcomings of the system are evident. Open-access fisheries almost always fail.[50] Uncoordinated ocean development creates operational inefficiencies, conflicts over use and environmental degradation that undermines future productivity. Unrestricted industrial, nutrient and carbon-related pollution is changing the

[49] OECD. 2019. *Rethinking Innovation for a Sustainable Ocean Economy*. Paris: Organisation for Economic Co-operation and Development. doi: https://doi.org/10.1787/9789264311053-en.

[50] Costello, C., S.D. Gaines and J. Lynham. 2008. "Can Catch Shares Prevent Fisheries Collapse?" *Science* 321 (5896): 1678–81. doi: https://doi.org/10.1126/science.1159478.

ocean's chemistry and affecting its biology and economic potential.

Given the interconnectedness of the ocean's sectors, it does not make sense to manage them separately. Ecosystem-based management, science-based marine spatial planning and integrated ocean management are tools that can be used to facilitate more systematic and equitable management of the ocean's resources and services. Some locales are already using ecosystem-based management tools that are science-based and grounded in broad stakeholder engagement and focus on achieving a healthy and resilient ocean ecosystem—with excellent results. Xiamen, China, for example, has seen a 40% improvement in socioeconomic benefits from its marine sectors since it adopted integrated ocean management in 1994.[51]

A variety of barriers has held back the widespread uptake of goal-oriented planning. Standards and practices for planning, accountability, transparency and legal rights or protections in the ocean remain a century or more behind their land-based equivalents—partly because businesses fear that integrated planning is a way for conservationists to pursue an antibusiness agenda. Top-down planning processes have failed to engage all users, resulting in inefficient processes and a lack of buy-in and implementation.

To be successful, ocean plans must find a balance between the requirements of different ocean users, between the needs of the ocean and the needs of the coast and its people. Growing evidence from countries in which integrated ocean planning has been used shows how the agendas of ecosystem health, food and energy security, local prosperity and coastal protection can reinforce one another. Scientific and local knowledge are key to understanding co-benefits and navigating the trade-offs.

Ocean planning needs to provide inclusive, equitable access by and recognition of local communities. Local fishers must have access to traditional fishing grounds, cultural sites must be protected and viewsheds must be preserved. Representatives of all types of ocean users must be involved in planning. Resource owners, lessees and access holders must be given secure titling and reliable and effective legal recourse against polluters, trespassers and other violators.

Key Actions

To ensure that goal-orientated planning becomes a reality, countries should establish, fund and implement ocean plans for 100% of the areas under their jurisdiction, using a process that is science-based, inclusive, participatory and adapted to the local context. Doing so is crucial to balancing protection and production and ensuring equitable access and rights for local users.

De-risking Finance and Using Innovation to Mobilise Investment

Current investment in sustainable ocean industries, biodiversity and conservation is grossly inadequate. It needs to quadruple to restore and sustainably maintain ocean health.[52]

Investment is limited for a variety of reasons. The fact that externalities such as the effects of ocean sector activities on global climate change, pollution and human rights are not reflected in the prices producers receive means that ecologically unsustainable businesses can thrive. Harmful subsidies—typically supporting the expansion of large-scale industrial fishing fleets and fossil-fuel extraction—distort the ocean economy.

In some cases, investing in sustainability is a long-term proposition. Rebuilding fish stocks and fishing sustainably can make business sense in the long run, for example, but can be costly in the short to medium run. As a result, opportunities are missed. Governments could help solve the problem by providing resources to mitigate transition challenges—by, for example, repurposing subsidies and implementing fishery reforms that prevent overfishing and help ensure a strong return on investment.

Key Actions

Countries that establish sustainable ocean development as a national priority can hope to attract investment from sovereign wealth funds and development finance institutions. Through their own and other public or philanthropic funding sources, private investment capital can be de-risked, catalysing private investment in novel industries and business models like sustainable fisheries (reforms), or MPAs financed by tourism fees. This bending of public and private capital can be especially catalytic in increasing investments in developing nations. Governments can also help stimulate the pipeline of sustainable ventures and projects by providing grants or other forms of support to early-stage innovation, as Norway has done to support next-generation

[51] Peng, B., H. Hong, X. Xue and D. Jin. 2006. "On the Measurement of Socioeconomic Benefits of Integrated Coastal Management (ICM): Application to Xiamen, China." *Ocean & Coastal Management* 49 (3): 93–109. doi: https://doi.org/10.1016/j.ocecoaman.2006.02.002.

[52] Sumaila, U.R., C.M. Rodriguez, M. Schultz, R. Sharma, T.D. Tyrrell, H. Masundire, A. Damodaran et al. 2017. "Investments to Reverse Biodiversity Loss Are Economically Beneficial." *Current Opinion in Environmental Sustainability* 29 (December): 82–88. doi: https://doi.org/10.1016/j.cosust.2018.01.007.

offshore aquaculture and the European Union has done to support offshore wind generation. In the offshore energy sector, governments could support renewable energy by providing low-cost infrastructure, setting feed-in tariffs and providing subsidies for sustainable activities. They could also reduce risk—by ensuring regulatory certainty, providing insurance and providing offtake/demand guarantees, particularly for capital-intensive offshore investments such as wind energy and large-scale mariculture.

Stopping Land-Based Pollution

Virtually every pollutant present on land is also present in the ocean, with compounding and significant deleterious impacts on ecosystem health. Plastics, nutrients (primarily nitrogen and phosphorus), pesticides and parasiticides, antibiotics and other pharmaceuticals, industrial chemicals, oil and gas, heavy metals, toxins, medical waste, e-waste and other types of debris are diverted to the ocean with very few financial consequences for the polluter.

These materials end up in the ocean because waste management and sewerage infrastructure in many countries, especially Asia and Africa, are inadequate. Waste collection is largely unprofitable because few consumer products are recyclable.

Addressing the ocean pollution challenge has been complicated by the difficulties of attribution (many pollutants come from more than one source) and the overwhelming asymmetry of the situation: When heavily protected land-based private interests clash with the interest of a weakly defended common pool resource like the ocean, the ocean loses.

A growing number of governments and industries are taking action. Measures such as banning plastic bags are welcome, but their effect will be insufficient. Current commitments on plastics, for example, are likely to reduce annual plastic leakage into the ocean by only 7% by 2040.[53]

Key Actions

To stop the leakage of plastics into the ocean, a diverse and more ambitious set of solutions is needed that includes reducing unnecessary plastics, recycling materials and safely disposing of waste. Recycled materials must become cheaper than virgin plastic. Companies must be held accountable for how much plastic they use and whether they use recycled content, recyclable product designs and plastic substitutes. Massive investment must be made in waste collection and recycling technology and infrastructure, particularly in developing countries, where such infrastructure is weak. Tackling the underlying cause could also help reduce other pollutants. Adopting precision agriculture on land could help reduce nutrient runoff into the ocean, for example.

Changing Ocean Accounting So That It Reflects the True Value of the Ocean

Traditional measures of the economy, such as GDP, ignore externalities, such as the effect of production on pollution or global climate change. They also fail to place a value on natural resources and ignore the way benefits are distributed.

Measuring only the GDP generated by ocean-based sectors does not capture the true value of the ocean—and can reward unsustainable practices. The ocean's broader value must be fully accounted for and used in decision-making, based on a holistic set of metrics that includes measurements of infrastructure assets, such as ports; natural assets, such as fish populations and coral reefs; and indicators of benefits to people, such as measures of income and well-being.

Key Actions

To measure the value of the ocean more accurately, national statistical offices, in partnership with other agencies, need to develop complete sets of national ocean accounts. Interactive dashboards should be created to allow users to explore the data by aggregating and disaggregating sectors and groups of people.

Having these five building blocks in place will enable change in key ocean economy sectors such as sustainable food from the ocean, renewable energy from the ocean and sustainable tourism. These sectors will also need targeted and sector-specific actions in terms of policies, technology and finance innovation, and scientific research, but having these building blocks in place will set governments and other stakeholders on the right path and lay the groundwork for the achievement of a prosperous and sustainable ocean economy.

[53] Lau et al. 2020. "Evaluating Scenarios toward Zero Plastic Pollution"; Pew Charitable Trusts and SYSTEMIQ. 2020. *Breaking the Plastic Wave*.

2.4.2 This New Way of Thinking About and Managing the Ocean Is Gaining Traction

The ocean is moving up the policy agenda. Coastal countries, especially small island states, are advocating for socially equitable and environmentally sustainable growth. Civil society is increasingly recognising the decline in the ocean and favouring government action to protect the ocean.

The action agenda is ambitious but entirely feasible. Progress in building the foundations for change is already evident:

- The data revolution has begun. Sensors and satellites are increasingly being deployed to monitor the ocean. Data on invasive species in bilge water and nutrients in river deltas, for example, provide actionable information in near real time—the holy grail of adaptive management. Sound fishery management digital tools, including vessel tracking, fishery simulation, and registry and enforcement systems, are widely available.
- Several regions have replaced siloed management practices with more integrated marine spatial planning. For example, the Baltic Sea states have coordinated across borders and sectors to implement a science-based planning strategy and have been rewarded with the return of predators and birds as well as restored fish stocks.[54]
- Sustainable ocean investments are on the rise. In a recent survey, 72% of investors classified the sustainable ocean economy as investable.[55] Thousands of sustainable ocean ventures are emerging across all continents.
- Together, the United States, Europe and Asia adopted 95 policies and pieces of legislation limiting plastic packaging between 2010 and 2019.
- A growing number of countries are adopting more holistic accounting techniques. China, for example, is using gross ecosystem product (GEP) to steer its transition towards inclusive, green growth.[56]

Similar trends can be observed at the ocean sector level. Backed by industry, support is growing for green shipping, the development of new technologies and practices that reduce the impact of mariculture on ecosystems, and community-led programs restoring fish stocks, to name just a few emerging changes. Inspiring success stories, such as the reform of fisheries in the United States, demonstrate that sound ocean management can simultaneously restore fish stocks and benefit fishers and coastal communities.[57] To achieve a sustainable ocean economy, change needs to happen faster and at a bigger scale than is currently happening. Actions at the local and national level can help accelerate change.

2.4.3 Targeted Actions Can Help Accelerate Progress

The huge scale of the challenge and the high stakes involved mean that acting quickly and effectively is crucial. Delivering immediate gains can help demonstrate the long-term benefits of pursuing a sustainable ocean economy, spurring stakeholders to take action. Creating sustainable ocean economic zones and forming national task forces are concrete actions than can move the agenda forward right away.

Sustainable Ocean Economic Zones Can Illustrate the Benefits of a Sustainable Ocean Economy at a Small Scale

Special economic zones (SEZs) are areas within a country that the government sets aside to attract direct investment in particular economic activities. These zones typically offer low rents, taxes, utilities and infrastructure costs; relief from bureaucratic procedures; and loan guarantees to market-rate investors. They range in size from small neighbourhood zones to entire cities.

Taking inspiration from the success of the SEZ concept in a country's exclusive economic zone (the ocean zone over which a coastal state has special rights with respect to marine resources) could be a powerful catalyst for accelerating a sustainable ocean economy. Sustainable ocean economic zones (SOEZs) could provide a test bed for systemic experimentation and innovation, where incentives could be tested, results monitored and adapted to, and risks managed. In the process of designing and implementing these zones, the classic hurdles to ocean management—free access, lack of planning, conflicts over use and externalities—can be addressed in the context of real business, rather than as abstract policy.

SOEZs are a way for countries to support and evaluate the sustainable ocean economy model at a scale they are comfortable with. Biological conditions, existing industries and stakeholders, and local needs determine which activities take place in an SOEZ (Fig. 20.4). One locale might use a SOEZ to attract and test high-technology models combining energy

[54]Reusch, T.B.H., J. Dierking, H.C. Andersson, E. Bonsdorff, J. Carstensen, M. Casini, M. Czajkowski et al. 2018. "The Baltic Sea as a Time Machine for the Future Coastal Ocean." *Science Advances* 4(5): eaar8195. doi: https://doi.org/10.1126/sciadv.aar8195.

[55]Responsible Investor Research and Credit Suisse. 2020. *Investors and the Blue Economy*. https://www.esg-data.com/reports.

[56]Ouyang, Z., C. Song, H. Zheng, S. Polasky, Y. Xiao, I. Bateman, J. Liu et al. 2020. "Using Gross Ecosystem Product (GEP) to Value Nature in Decision-Making." https://ore.exeter.ac.uk/repository/handle/10871/120272.

[57]Natural Resources Defense Council, Conservation Law Foundation, Earthjustice, Ocean Conservancy, Oceana and Pew Charitable Trusts. 2018. "How the Magnuson-Stevens Act Is Helping Rebuild U.S. Fisheries." https://www.nrdc.org/sites/default/files/magnuson-stevens-act-rebuild-us-fisheries-fs.pdf.

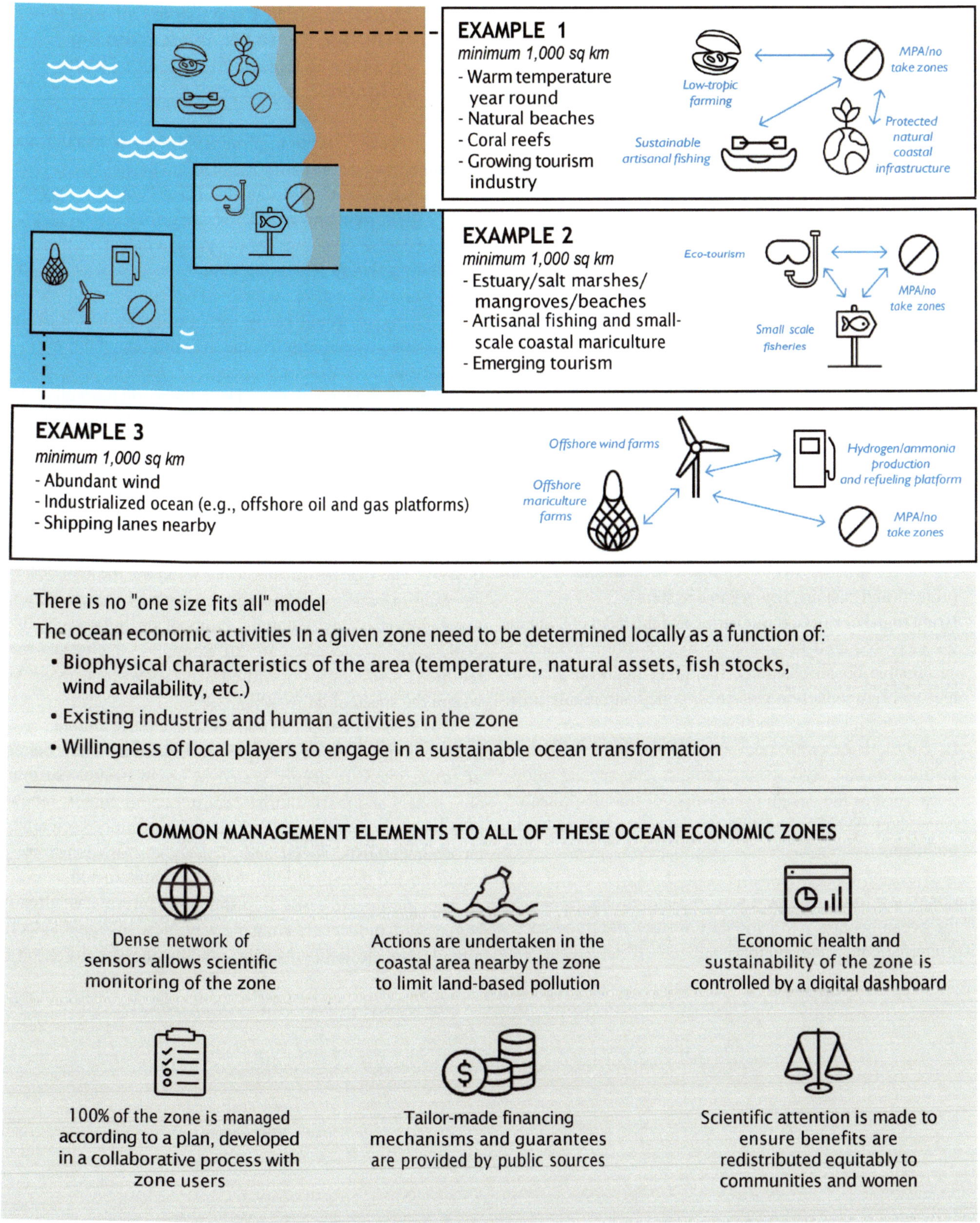

Fig. 20.4 Sustainable ocean economic zones can be test beds for experimentation and innovation. (*Source: Authors*)

generation, shipping and large-scale mariculture. Another might combine carbon-financed restoration, coastal protection, tourism and fishery enhancement.

Whatever activities take place within the zone, all SOEZs share several common elements. The entire zone is managed according to a plan, a dense networks of sensors allows scientific monitoring of the zone and efforts are made to ensure that benefits are redistributed equitably to communities and women.

National Ocean Task Forces Can Accelerate the Shift Towards a Sustainable Ocean Economy

Establishment of a sustainable ocean task force at the (ocean) ministerial or head of state level with a mandate to adapt the sustainable ocean agenda to the national context could accelerate change. Such a task force could perform several important functions:

- Conduct a comprehensive marine resource mapping of 100% of the country's exclusive economic zone.
- Support and facilitate an inclusive, participatory process to develop a plan that ensures a streamlined and efficient regulatory process, avoids conflicts over spatial use and protects and sustains key oceanic systems.
- Bring together relevant ministries and the head of state on the steps required to accelerate the transition towards a sustainable ocean economy, including financial guarantees and risk-reduction measures, policy and regulations, and international coordination.
- In coordination with relevant organisations, academic institutions and civil society groups, lead special initiatives, such as the design of networks of marine protected areas and SOEZs and efforts to control land-based pollutants.

National task forces can be a way to highlight the relevance of the ocean economy to national priorities like food security, international trade and tourism.

2.5 The Ocean Is Not Too Big to Fail, and It Is Not Too Big to Fix, But It Is Too Big and Too Central to the Planet's Future to Ignore

Effective ocean protection, sustainable ocean production and equitable human prosperity are inseparable and compatible. When integrated into a sustainable ocean economy, they can change the current downward trajectory of ocean health, producing positive outcomes for people and nature. Setting the foundations within which the three Ps can be achieved and transforming key ocean sectors will not be easy, but it can be done. Doing so would vastly increase the resilience of the global economy and improve the lives of some of the world's poorest and most vulnerable people. Indeed, creating a sustainable ocean economy would help the world meet all of the Sustainable Development Goals (SDGs), not just SDG 14 (on life below water) (Fig. 20.5).

Current practices, laws and cultural norms help support the open-access model that characterises much of the ocean. All of them can change. History shows that even very complex systems can shift onto new trajectories, sometimes very quickly. The energy transition in Germany, the banning of smoking in bars and restaurants in much of the world, and the adoption of the Montreal Protocol on Substances that deplete the ozone layer are all examples of changes that required major shifts in attitudes and laws that occurred within the space of a few years.

This kind of change can and must take place among stakeholders in the ocean economy. Spearheaded by a new cohort of ocean interests deeply vested in ocean health—sustainable fishers and mariculturists, coastal communities, renewable energy generators, ecotourism operators, scientists, environmentalists, social and civil society organisations—pollution and over-exploitation can be counteracted.

The journey towards a sustainable future has already begun, with pioneers leading the way. New sustainable technologies are attracting investors, and businesses and govern-

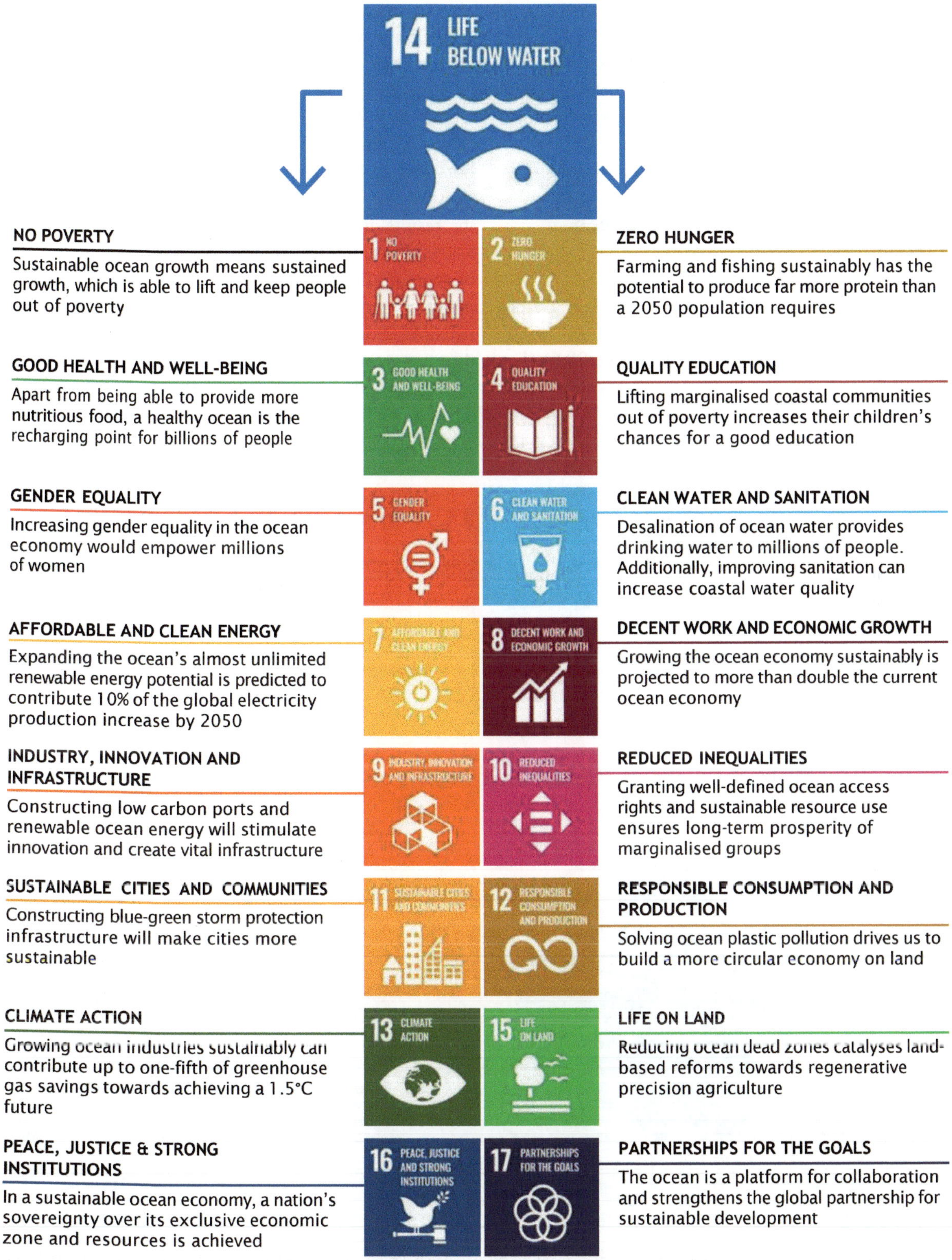

Fig. 20.5 A healthy ocean is critical to meeting the sustainable development goals. *Note: Regarding SDG 6 (clean water and sanitation), the link to the ocean can be made through desalination plants. Regarding SDG 17 (partnerships for the goals), the ocean provides excellent platforms for collaboration. Peaceful ocean science collaboration, for example, has been important for diplomatic relations (e.g. U.S.-Soviet Gulf Stream experiments in the 1960s). (Source: Authors)*

ments are waking up to the opportunities of a sustainable ocean economy in building a new future after COVID-19. They are also increasingly recognising the risks and cost of inaction. Inspiring efforts from around the world provide a glimpse of what can be achieved globally if stakeholders act now.

3 Prologue: Five Sustainable Ocean Economy Stories

What does a sustainable ocean economy look like? Before exploring the rationale, benefits and practicalities of the concept, let's travel to five inspiring places (Fig. 20.6). The first destination is Gazi-Kwale County, Kenya, where a community-based organisation sells blue carbon credits from rebuilding its mangrove forest. The second stop is the Philippines, where a comprehensive approach used with 400 fishing communities helps meet the triple objective of food security, ocean protection and community prosperity. Then on to Europe, where the Medes Islands Marine Reserve in Catalonia, Spain, regenerates ocean biomass, supporting thriving ecotourism and, through spillover effects, sustainable fisheries. Across the Atlantic, on the U.S. East Coast, GreenWave works with fishers and coastal communities to launch regenerative ocean farms which combine seaweed and shellfish production. The final stop is the North Sea, where the Zero Emission Energy Distribution at Sea (ZEEDS) initiative aims to create a revolutionary zero-carbon fuelling system for ships, enabled by offshore wind production. This is a voyage of discoveries, and at some stops the results are not yet proved or fully backed by scientific assessments. But they are ideas, ones that illustrate a range of possibilities happening right now; they demonstrate inspiring innovations with the promise of a sustainable ocean economy. Figure 20.6 (and the report cover) maps this voyage on a representation of Earth inspired by the work of South African oceanographer Athelstan Spilhaus. This projection emphasises that there is one interconnected ocean.

3.1 Stop 1: Mikoko Pamoja, Kenya

Mikoko Pamoja, meaning 'mangroves together' in Kiswahili, perfectly describes the community-based blue carbon credit

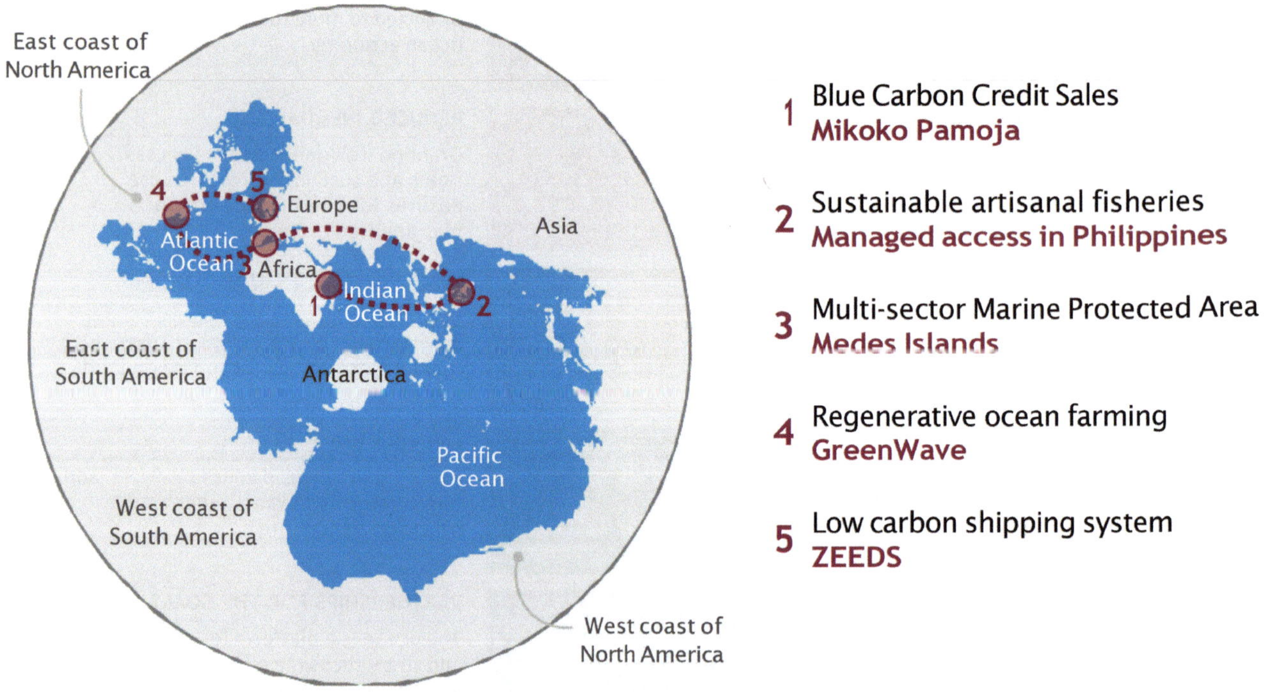

Fig. 20.6 Five sustainable ocean economy stories. Design inspired by Athelstan Spilhaus, Atlas of the World, Geophysical Boundaries, Map XIIIA: "World Ocean Map in a square", conformal, poles in South America and China, 1979. (*Source: Authors*)

Fig. 20.7 Mikoko Pamoja at work. *Note: (top left) Gazi Bay in Kwale County, (bottom left) GRID-Arendal & the Mikoko Pamoja committee members, (top right) Community water project founded by Mikoko Pamoja, (bottom right): close up of a mangrove.* (*Source: Rob Barnes, UNEP/GRID-Arendal,* https://www.grida.no/resources/11125)

project in Kwale County on Kenya's South Coast. The first effort of its kind, Mikoko Pamoja is improving the livelihood of the local community, regenerating the local mangrove forest and helping fight climate change (Fig. 20.7).

The reduction of local mangroves threatened the livelihood of fishers and destabilised coastlines. Mikoko Pamoja was set up to reverse this trend and find alternatives to mangroves that could provide the community with fuel and building materials while also generating income.

In 2013, a community-based organisation was formed, which was granted co-management rights for the 117-hectare coastal area from the Kenyan government.[58] A few hectares of *Casurina* woodlots were planted to provide an alternative source of fuel- and building wood for the community.[59] On 114 hectares, mangroves were replanted and a carbon credit trading scheme, now accredited by Plan Vivo (an international body regulating carbon credits), was set up.[60]

The trading scheme is now up and running—Plan Vivo sells 2500 credits per year, with 1 credit being equivalent to 1 metric tonne of carbon dioxide (CO_2) per year. These 2500 tonnes are derived from a mix of avoided deforestation and the planting of new mangroves.

On average, the carbon sales generate about $12,500 per year. Thirty-five percent of the revenue is used for the project costs, while 65% is reinvested in the community. [61] In the past it has funded initiatives such as the establishment of a

[58]Mikoko Pamoja Project. n.d. *ACES* (blog). https://www.aces-org.co.uk/mikoko-pamoja-project/. Accessed 5 May 2020.

[59]Huxham, M. 2018. "MIKOKO PAMOJA: Mangrove Conservation for Community Benefit." Mikoko Pamoja Team. Plan Vivo Project Design Document (PDD): 38.

[60]Huff, A., and C. Tonui. 2017. *Making "Mangroves Together": Carbon, Conservation and Co-management in Gazi Bay, Kenya*. ESRC STEPS Centre. https://opendocs.ids.ac.uk/opendocs/handle/20.500.12413/12970.

[61]*MPA News*. 2020. "Funding MPAs by Selling Blue Carbon Credits: Practitioners from the First Projects Describe Their Experience So Far." 30 July. https://mpanews.openchannels.org/news/mpa-news/funding-mpas-selling-blue-carbon-credits-practitioners-first-projects-describe-their.

water system for the whole village, a local soccer tournament and textbooks for the local primary school.[62]

In addition to the credit sales, the community benefits from the restored mangroves through increases in fish catches, beekeeping and ecotourism from the 'Gazi Bay Boardwalk', all of which contribute to more sustainable livelihoods.[63]

Despite facing challenges such as fluctuations in carbon credit prices, the project has largely been a success and has received strong support from the Kenyan government. There has been strong scientific support with partners through the Kenya Marine and Fisheries Research Institute as well as support from the Kenya Forest Services on aspects of forest governance.[64] Mikoko Pamoja won the 'Equator Initiative Prize' and is now the model for future projects, including for 'Vanga', which covers an area about four times that of Mikoko Pamoja.[65] Mangroves are considered to be a natural climate solution because of their ability to help reduce carbon emissions, and currently there are efforts to include mangroves as part of Kenya's nationally determined contributions (NDCs). This work has also enhanced the visibility of the ocean space in Kenya and contributed to the value of safeguarding coastal ecosystems.

3.2 Stop 2: Community-Based Managed Access Network in the Philippines

Fishery reform in the developing world is not just about the fish. It is also about people, coastal communities and fishing as a livelihood, a job and a way of life.[66] Small-scale fisheries are a main source of food, provide millions of jobs and underpin cultures, particularly for the coastal poor.

Rare, an international non-governmental organisation (NGO) that applies behavioural insights to the cause of artisanal fishery recovery in developing countries, the Environmental Defense Fund, and the Sustainable Fisheries Group (SFG) at the University of California, Santa Barbara, launched the Fish Forever program in multiple countries to build a social movement for the better management of coastal fisheries. Rare and SFG took the lead in the Philippines.

Better management starts with managed access areas that give fishing communities clear, exclusive rights to fish in certain areas, which are often aligned with traditional community use rights. The communities' exclusive access is tied to their commitment to use fully protected marine protected areas (MPAs) that are designed to replenish and sustain fish populations and protect habitats and biodiversity.

In communities that have organised themselves to manage 'their' fishing areas and protected zones, management typically becomes more sophisticated. For example, boats and fishers are registered, catch is recorded, regulations are respected and fishers participate in management (Fig. 20.8). In the absence of outsiders skimming off the rewards of good stewardship, a virtuous circle tends to evolve, where results drive good behaviour and vice versa. Households in these communities have been shown to become more resilient in terms of food and financial security, and communities can work together to develop access to previously elusive capital and markets.

This social movement naturally kick-starts a political movement. National governments and international bodies begin to recognise the central role of coastal fisheries to the health, cultural coherence, resilience and wealth of coastal communities, and they start to promote the sector with better policies and improved access to financial resources.

The Philippines have demonstrated these dynamics at work. The 'Fish Forever' program is active in more than 400 communities in the country, clustered in 47 sites. Early results from 20 sites showed that fish biomass inside and outside the reserve was either maintained or increased across all sites. At sites where Rare had been working for 7 years, the increases were as high as 390% inside the fully protected MPAs and 111% outside MPAs. There were also statistically significant increases in 80% of social metrics, including improved attitudes towards fully protected MPAs, participation in management and the sense of social equity. To build financial resilience in fishing communities, fisher households also organised themselves into 'savings clubs'. These enabled more than 1500 members to save close to US $2 million in 2½ years.

The success at the local level is now reflected in a national policy agenda that supports artisanal fisheries. One example is the inclusion of managed access areas in the Philippine Development Plan, the country's central economic and development planning document. Most recently, working with Rare, 300 mayors also passed major resolutions to support artisanal fishers and the issues they face regarding climate change, preferential rights and sustainable financing.[67]

[62]Global Mangrove Alliance. 2019. "Mikoko Pamoja: A Business Case for Carbon Credit in Gazi-Kwale County, Kenya." 8 May. http://www.mangrovealliance.org/mikoko-pamoja/.

[63]Wylie, L., A.E. Sutton-Grier and A. Moore. 2016. "Keys to Successful Blue Carbon Projects: Lessons Learned from Global Case Studies." *Marine Policy* 65 (March): 76–84. doi: https://doi.org/10.1016/j.marpol.2015.12.020.

[64]Wylie et al. 2016. "Keys to Successful Blue Carbon Projects."

[65]Mikoko Pamoja Project. n.d. Blog.

[66]Garcia, S., Y. Ye, J. Rice and A. Charles. 2018. "Rebuilding of Marine Fisheries." Rome: Food and Agriculture Organization of the United Nations. http://www.fao.org/3/ca0161en/ca0161en.pdf.

[67]All results received from personal communication with Rare Conservation.

Fig. 20.8 Artisanal fishers planning their community fishery in the Philippines. (*Source: Rare*)

3.3 Stop 3: Medes Islands Marine Reserve, Spain

Two hours northeast of Barcelona, seven idyllic islets can be seen from the Costa Brava. According to the official tourism website, the Medes Islands ecosystem is 'the best natural reserve in the western Mediterranean'. Scuba divers come from all over Europe to see the abundant fish—including large Mediterranean dusky groupers and other predatory fishes—relict red coral populations, octopus and hundreds of other marine species around these islands. How is this possible in a sea known to be overfished, polluted and overrun by invasive species?

It all started over 35 years ago, with the creation of a 51-hectare no-take marine reserve which banned fishing but allowed diving, navigation and moorings only on buoys (Fig. 20.9). Years later, an additional 460 partially protected hectares were added. They permit limited fishing, only by a few local artisanal fishing vessels. (Only seven local vessels have this exclusive access).[68]

Fig. 20.9 The Medes Islands Reserve, Spain. (*Sources: top: Damsea/Shutterstock, bottom: funkyfrogstock/Shutterstock*)

[68] Merino, G., F. Maynou and J. Boncoeur. 2009. "Bioeconomic Model for a Three-Zone Marine Protected Area: A Case Study of Medes Islands (Northwest Mediterranean)." *ICES Journal of Marine Science* 66 (1): 147–54. doi: https://doi.org/10.1093/icesjms/fsn200.

This protection proved successful on all fronts,[69] even in this relatively small area.

- Fish biomass has fully recovered, and six main species have almost reached the maximum carrying capacity of the ecosystem.
- The restored biodiversity and biomass have transformed the Medes Islands into a paradise for divers and snorkelers, supporting thriving ecotourism in the area. Two hundred full-time jobs are supported and €12 million in revenue is generated, compared with €0.5 million before the creation of the reserve.
- The net present value of the reserve is up to 12 times greater than it would have been without this effective protection and management.

3.4 Stop 4: GreenWave, United States

In his book *Eat Like a Fish*, Bren Smith describes his journey as lifelong fisherman turned regenerative ocean farmer. He is fascinated with species that require no feed inputs and can regenerate their surrounding ecosystem: shellfish and seaweeds. After extensive research, he began to design and build an integrated, multitrophic mariculture farm, or as Bren would call it, a regenerative ocean farm[70] (Fig. 20.10).

On a visit to Bren's farm, at first you will see nothing but a few buoys. Underwater, it's a different story: kelp and mussels grow on horizontal lines of ropes connecting anchored scaffolding, scallops hang in lantern nets, while oysters and clams litter seafloor cages.

According to Bren's NGO, GreenWave, regenerative ocean farms can produce up to 150,000 shellfish and 10 tonnes of seaweed per acre. One farm can turn a profit of up to US $90,000–$120,000 per year—all without needing or buying feed, land, freshwater or fertiliser. Considering his initial investment of $20,000, this is a profitable business for Bren and other farmers, providing year-round income as kelp is harvested in spring, clams in spring to summer, scallops and mussels in autumn and oysters year-round. The 'crop' diversification also provides security for farmers should one of the crops fail.

Fig. 20.10 GreenWave ocean farming model. *Note: Sketch depicting the GreenWave 3D ocean farming model (top), Bren Smith harvesting kelp (bottom).* (*Source: Top: Inspired by Water Brothers; Bottom: Ronald T. Gautreau Jr. for GreenWave*)

Getting started wasn't easy. Native shellfish (mussels, clams, oysters, scallops) seed was easily obtained from established growers nearby, but sourcing microscopic kelp seed that could eventually grow into 1- to 2-m-long seaweed blades proved more complicated. While seaweed farming is a 1000-year-old industry in Asia, it is nascent in the United States. With the help of kelp scientists and local community partners, Bren and GreenWave built a kelp hatchery that could supply him and other local farmers with seed. Launched to replicate and scale Bren's farming model, GreenWave educates the next generation of ocean farmers about farming in an era defined by climate change. Through its Farmer-in-Training program, GreenWave supports aspiring regenerative ocean farmers as they navigate the complex U.S. regulatory system and teaches them the fundamentals of setting up their ocean farm. The farms are geared towards simplicity and low cost, making it possible for anyone to become a regenerative ocean farmer for '$20k, 20 acres and a boat'[71]—far less than the cost of establishing a farm on land.

[69] Sala, E., C. Costello, J. de Bourbon Parme, M. Fiorese, G. Heal, K. Kelleher, R. Moffitt et al. 2016. "Fish Banks: An Economic Model to Scale Marine Conservation." *Marine Policy* 73 (November): 154–61. doi: https://doi.org/10.1016/j.marpol.2016.07.032.

[70] Bren, S. 2019. *Eat Like a Fish: My Adventures as a Fisherman Turned Restorative Ocean Farmer*. Sydney: Murdoch.

[71] GreenWave. n.d. "Our Model." https://www.greenwave.org/our-model. Accessed 13 May 2020.

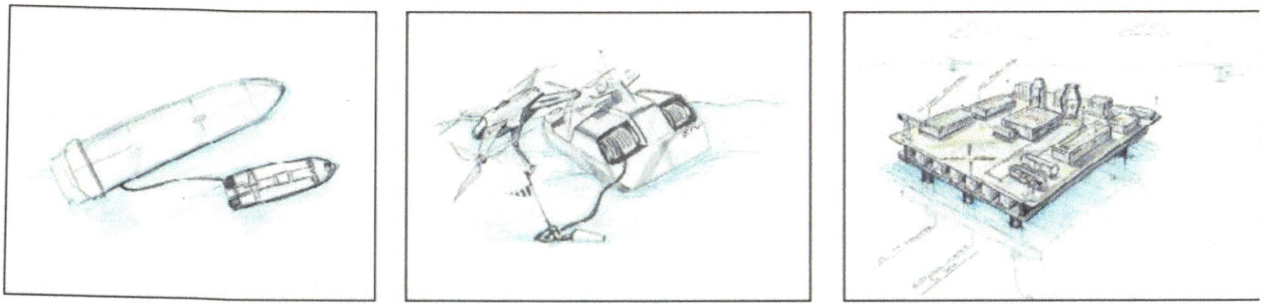

Fig. 20.11 Sketches of zero emission energy distribution at sea. From left to right: Sketch showing ship-to-ship bunkering at sea with zero carbon fuel. Sketch showing drone carrying pilot wire from energy-provider vessel. Sketch of topside of a ZEEDS platform producing and storing green ammonia. (*Source: ZEEDS project*)

GreenWave's goal is to plant 1 million acres of restorative species in the next 10 years. They hope to catalyse the growth of ocean farms across the world, providing a profitable and ecologically regenerative food production system for millions of people. These farms would be organised in GreenWave 'Reefs', with 50 small ocean farms clustered around a land-based hatchery and processing hub, surrounded by a ring of institutional buyers and entrepreneurs.[72]

3.5 Stop 5: ZEEDS Project, North Sea

Shipping is the most carbon efficient way (in tonnes per kilometre [km] travelled)[73] to move goods across the globe and accounts for 90% of cargo transport. Shipping today contributes about 2.2% of global CO_2 emissions, but these emissions could grow between 50% and 250% by 2050, mainly due to the growth in world maritime trade.[74] For instance, container shipping volumes are expected to increase by 243% between 2015 and 2050.[75] However, in April 2018, the International Maritime Organization (IMO) set a target of at least a 50% reduction in total annual greenhouse gas (GHG) emissions by 2050, compared with 2008 levels.[76]

How can a traditional industry like shipping, whose assets have a lifetime of more than 30 years, achieve the IMO's target or even more ambitious decarbonisation pathways?

Let's travel to a hypothetical future in 2030. In the eastern Atlantic, a container vessel is heading towards Rotterdam. This ship is carbon-neutral, having been retrofitted to be powered by green ammonia, a combustible produced through a series of chemical reactions enabled by renewable energy. The ship is low on fuel and slows to six knots as it is met by a small autonomous refuelling ship with a fuel hose suspended in the air by a drone. After 1 h, while still progressing, the now refuelled vessel accelerates on its way to Rotterdam. The fuel-provider vessel heads back to dock at a floating ammonium production platform, which is surrounded by a network of offshore wind turbines (Fig. 20.11).

This is the vision of Zero Emission Energy Distribution at Sea (ZEEDS), a new partnership created in 2018 that gathers leading Scandinavian players in energy, offshore engineering, shipping and technology (Equinor, Wärtsilä, Aker Solutions, Kvaerner, DFDS and Grieg Star). The ZEEDS concept envisions an ecosystem of strategically located offshore clean fuel production and distribution hubs, co-located near busy shipping lanes. Wind will provide the power to create sustainable ammonia for ship to ship refuelling.

The good news is that this solution might be more realistic than it looks. Adapted ship engines and production technology at sea are being tested at a pilot scale, and green ammonia is looking very promising as a replacement for heavy fuel oil on long voyages.

[72] GreenWave. n.d. "Our Model."

[73] Borken-Kleefeld, J., T. Berntsen and J. Fuglestvedt. 2010. "Specific Climate Impact of Passenger and Freight Transport." *Environmental Science & Technology* 44 (15): 5700–5706. doi: https://doi.org/10.1021/es9039693.

[74] International Maritime Organization (IMO). 2015. "Third IMO Greenhouse Gas Study 2014." London: IMO, 3. http://www.imo.org/en/OurWork/Environment/PollutionPrevention/AirPollution/Documents/Third%20Greenhouse%20Gas%20Study/GHG3%20Executive%20Summary%20and%20Report.pdf.

[75] Organisation for Economic Co-operation and Development (OECD). 2017. "ITF Transport Outlook 2017." Paris: OECD Publishing. https://www.oecd.org/about/publishing/itf-transport-outlook-2017-9789282108000-en.htm.

[76] IMO. 2018. "UN Body Adopts Climate Change Strategy for Shipping." http://www.imo.org/en/MediaCentre/PressBriefings/Pages/06GHGinitialstrategy.aspx.

4 The Urgency of Today

4.1 Introduction

The five stories in the Prologue are diverse but compelling illustrations of local actions to move towards a sustainable ocean economy. They share a common vision which recognises that the ocean will only be able to regenerate if and when the agendas of protection, production (e.g. food, energy, carbon) and human prosperity are managed holistically.

Yet these examples are exceptions to the general global downward trajectory of ocean ecosystems or their associated economic potential. Action can be inspired by their examples, but the reality is that urgent action is needed to transition towards a more sustainable ocean economy at scale.

This section develops three main arguments to emphasise the urgency of action:

A healthy ocean is crucial for all of humanity and for the global economy The agenda of a sustainable ocean economy applies to the entire world, not just to traditional 'blue sectors' like fisheries or shipping. The diverse services provided by healthy ocean ecosystems make Earth liveable. Feeding ten billion people in 2050 while remaining within a safe planetary 'operating space'[77] will be hard—and the ocean may well hold a big piece of the solution. The ocean could also play a significant role in fighting climate change, meeting up to one-fifth of the carbon mitigation challenge.[78] Finally, global concern about ocean plastic pollution could catalyse a much deeper reform of the profusion of wasteful material management practices on land (Sect. 4.2).

The ocean is under increasing threat The ocean is becoming warmer, more acidic, depleted, stormier, higher, more oxygen-depleted and less predictable. Profound changes (state shifts) affecting many aspects of human life are no longer unthinkable. Neither the ocean economy as a whole, nor coastal communities, nor the social agenda of the Sustainable Development Goals (SDGs) can thrive in such a degraded environment (Sect. 4.3).

Solutions are emerging but urgently need to be scaled up Despite the undeniable challenges, hints of a sustainable ocean mindset are on the rise. The pace of innovation in the ocean economy is accelerating sharply, and investors are starting to find their way to the sustainable ocean economy. A data revolution is underway—enabled by an ocean technology revolution—redefining access to knowledge. Successful, sustainable ocean-related policies are increasingly gaining traction. The voices of citizens and communities advocating for more equitable and sustainable use of planetary resources are getting louder. There is an unprecedented international momentum for a sustainable ocean economy, as seen at meetings of the G7, G20, Ocean Panel, UN Ocean Conference, Our Ocean, World Ocean Summits, UN Decade of Ocean Science and so on (Sect. 4.4).

4.2 A Blue Awakening: Recognising That the Ocean Is Vital to Humankind and the Global Economy

In an international 2020 survey, 94–96% of respondents saw 'the condition of the ocean as important to their country's economy'.[79] At the same time, there is no single broadly accepted definition of the ocean economy. The most commonly used one is the following: 'The ocean economy can be defined as the economic activities that take place in the ocean, receive outputs from the ocean, and provide goods and services to the ocean'.[80]

There is considerable variation in the way this definition is interpreted—the United States includes as few as six industry sectors in the ocean economy, Japan as many as 33. The ocean economy's implied valuation also ranges widely. The Organisation for Economic Co-operation and Development (OECD), defining the ocean economy as 'the sum of the economic activities of ocean-based industries, together with the assets, goods and services provided by marine ecosystems',[81] initially assessed ten ocean-based industries of the global ocean economy, conservatively estimating they represented in 2010 a total of US $1.5 trillion in gross value added [GVA];[82] WWF calls it 'the seventh largest economy in the world', valuing ocean assets at $24 trillion;[83]

[77] Rockström, J., W. Steffen, K. Noone, Å. Persson, F.S. Chapin, E. Lambin, T.M. Lenton et al. 2009. "Planetary Boundaries: Exploring the Safe Operating Space for Humanity." *Ecology and Society* 14(2). https://www.jstor.org/stable/26268316.

[78] Hoegh-Guldberg, O., et al. 2019. "The Ocean as a Solution to Climate Change: Five Opportunities for Action." Washington, DC: World Resources Institute. https://oceanpanel.org/sites/default/files/2019-10/HLP_Report_Ocean_Solution_Climate_Change_final.pdf.

[79] Kantar, David. 2020. "Perceptions of the Ocean and Environment." Lucile Packard Foundation. https://oursharedseas.com/wp-content/uploads/2020/03/Packard-Kantar-Ocean-Report-FINAL-1.pdf.

[80] Park, K.S., and D.J. Kildow. 2014. "Rebuilding the Classification System of the Ocean Economy." *Journal of Ocean and Coastal Economics*, no. 1. doi: https://doi.org/10.15351/2373-8456.1001.

[81] OECD. 2016. "The Ocean Economy in 2030." Directorate for Science, Technology and Innovation Policy Note, April. https://www.oecd.org/futures/Policy-Note-Ocean-Economy.pdf.

[82] OECD. 2016. *The Ocean Economy in 2030*. Report. Paris: OECD Publishing. https://www.oecd.org/environment/the-ocean-economy-in-2030-9789264251724-en.htm.

[83] Hoegh-Guldberg, O., and Boston Consulting Group. 2015. "Reviving the Ocean Economy: The Case for Action—2015." Geneva: WWF International.

and many others assert it to be practically incalculable. The ocean economy includes heavily ocean health-dependent sectors such as tourism (26% ocean GVA), fisheries and mariculture (2–6% ocean GVA), as well sectors principally managed by more exogeneous interests, such as offshore oil and gas (33% ocean GVA), ports (13% ocean GVA) and maritime equipment (11% ocean GVA). In terms of employment, the ten ocean-based industries assessed by the OECD contributed some 31 million direct full-time jobs in 2010, with industrial capture fisheries accounting for the lion's share of the OECD's assessed ocean economy jobs (36% and plateauing), followed by tourism (23% and strongly increasing).[84] If informal or artisanal jobs are included, the ocean's global employment contribution is much higher—estimates for total (formal and artisanal) fisheries employment alone run as high as 237 million full-time equivalent jobs.[85]

These definitions and numbers are insightful but incomplete. To be a useful descriptor of the relationship between humans and the ocean, a broader, more systemic perspective on the ocean economy is needed, in line with the World Bank's definition of a sustainable ocean economy: 'the sustainable use of ocean resources for economic growth, improved livelihoods and jobs while preserving the health of ocean ecosystems'.[86]

In the literature and in national or international initiatives it is common to find references to a 'blue economy', but again the definition and scope varies: sometimes 'blue' refers to the ocean, with the blue economy closer to the definition in this section's first paragraph; at other times 'blue' refers to sustainable (as 'green' would do for sustainable land-based activities), and the blue economy is understood as in the World Bank definition. To avoid confusion, this report will avoid the term 'blue economy' in favour of 'sustainable ocean economy', mostly guided by the World Bank definition. Yet this report also invites readers to embrace a wider paradigm that acknowledges the following:

- The importance of ocean contributions for all of humanity and nature
- The ocean's central contribution to the global agenda of food security
- The untapped opportunity the ocean provides to fighting climate change
- The catalytic role the ocean can play in accelerating a global transition towards more circular and regenerative practices in land-based economies

4.2.1 The Ocean's Contributions to Humanity Exceed the Realm of Its Industrial Production

The ocean absorbs more than 90% of the heat resulting from anthropogenic greenhouse gas emissions. It rebalances the heat differential between poles and equators. It produces 50–80% of Earth's oxygen.[87] Its biological adaptations remain largely unknown and, if previous experience is any indication, contain untold medical, knowledge and commercial resources. For billions of coastal dwellers, the ocean is woven deeply into their cultural and spiritual lives. For all humans, it provides a sense of wonder, solace and connection to the natural world. Millions play in it every day. It provides a deep sense of place.[88]

The 2005 Millennium Ecosystem Assessment report defined ecosystem services as 'benefits people obtain from ecosystems'.[89] This concept was updated and broadened to 'nature's contribution to people' in the latest report by the Intergovernmental Science-Policy Platform on Biodiversity and Ecosystem Services (IPBES).[90] The IPBES structures nature's contribution to people into three main categories (definitions below are directly inspired by IPBES's):[91]

- **Nature's material contributions to people:** 'substances, objects or other material elements from nature that sustain people's physical existence and the infrastructure needed for the operation of a society or enterprise'. In the context of the ocean economy, these material contributions sup-

[84] OECD. 2016. "The Ocean Economy in 2030." Directorate for Science, Technology and Innovation Policy Note, April. https://www.oecd.org/futures/Policy-Note-Ocean-Economy.pdf.

[85] Teh, L.C.L., and U.R. Sumaila. 2013. "Contribution of Marine Fisheries to Worldwide Employment." *Fish and Fisheries* 14 (1): 77–88. doi: https://doi.org/10.1111/j.1467-2979.2011.00450.x.

[86] World Bank and UN Department of Economic and Social Affairs. 2017. "The Potential of the Blue Economy: Increasing Long-Term Benefits of the Sustainable Use of Marine Resources for Small Island Developing States and Coastal Least Developed Countries." Washington, DC: World Bank. https://openknowledge.worldbank.org/bitstream/handle/10986/26843/115545.pdf?sequence=1&isAllowed=y.

[87] National Oceanic and Atmospheric Administration (NOAA). n.d. "How Much Oxygen Comes from the Ocean?" https://oceanservice.noaa.gov/facts/ocean-oxygen.html. Accessed 13 May 2020.

[88] Allison, E., J. Kurien and Y. Ota. 2020. "The Human Relationship with Our Ocean Planet." Washington, DC: World Resources Institute. https://www.oceanpanel.org/blue-papers/relationship-between-humans-and-their-ocean-planet.

[89] Millennium Ecosystem Assessment (Program), ed. 2005. *Ecosystems and Human Well-Being: Synthesis*. Washington, DC: Island.

[90] Díaz, S., J. Settele, E.S. Brondízio, H.T. Ngo, M. Guèze, J. Agard, A. Arneth et al. 2019. "Summary for Policymakers of the Global Assessment Report on Biodiversity and Ecosystem Services of the Intergovernmental Science-Policy Platform on Biodiversity and Ecosystem Services." Bonn, Germany: Intergovernmental Science-Policy Platform on Biodiversity and Ecosystem Services. doi: https://doi.org/10.5281/zenodo.3553579.

[91] Díaz et al. 2019. "Summary for Policymakers of the Global Assessment Report on Biodiversity and Ecosystem Services of the Intergovernmental Science-Policy Platform on Biodiversity and Ecosystem Services."

port subsistence (e.g. fish), energy (ocean fossil fuels, wind), health (e.g. pharmaceuticals derived from marine species) and construction (e.g. sand), among others. In this report it is assumed that most of these contributions are economically accounted for by conventional indicators like GVA (see Fig. 20.12).

- **Nature's regulating contributions to people:** 'functional and structural aspects of organisms and ecosystems that modify the environmental conditions experienced by people, and/or sustain and/or regulate the generation of material and non-material contributions'. For the ocean and coastal ecosystems, climate regulation is a perfect example of such contributions, but the latter also include, for example, habitat creation and maintenance; regulation of hazard and extreme events; regulation of air quality; and dispersal of seeds, propagules and larvae (see Fig. 20.12).
- **Nature's non-material contributions to people.** 'Nature's contribution to people's subjective or psychological quality of life, individually and collectively'. These contributions include learning and inspiration from the ocean, physical and psychological experiences, and supporting identities (see Fig. 20.12).
- The IPBES also defines a '**maintenance of options**' category for the yet-to-be-discovered or understood use of natural ecosystems and organisms (see Fig. 20.12).

Even in economic and monetisable terms, not every dollar counts the same. For example, coastal fisheries account for less than 1% of the ocean economy as conventionally defined. However, this is most likely a significant underestimation of the sector's true economic importance. To more accurately represent the importance of the marine economy, one would also need to include employment for over 37 million arti-

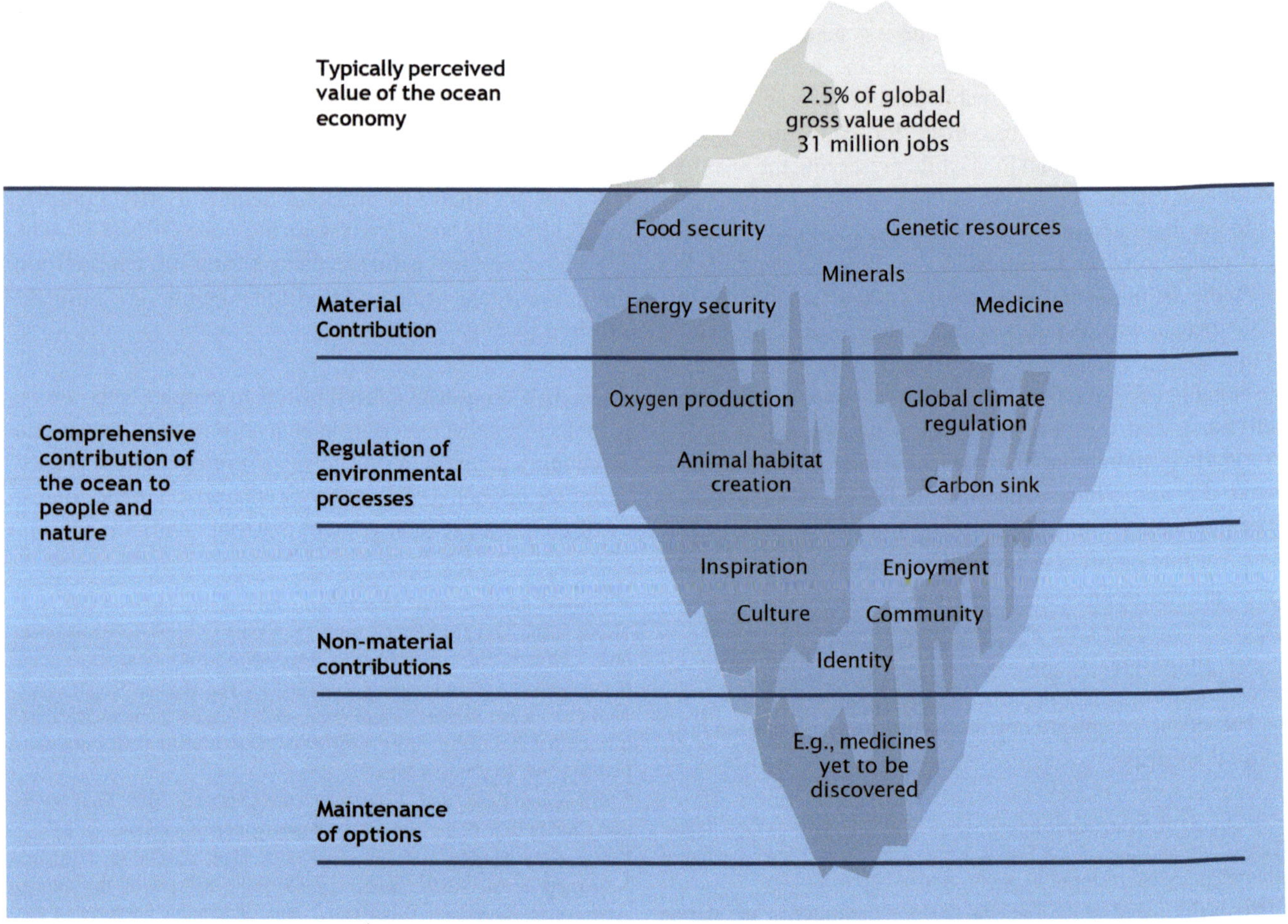

Fig. 20.12 The ocean's importance to humankind. (*Source: Authors, inspired by Díaz, S., J. Settele, E.S. Brondízio, H.T. Ngo, M. Guèze, J. Agard, A. Arneth et al. 2019. "Summary for Policymakers of the Global Assessment Report on Biodiversity and Ecosystem Services of the Intergovernmental Science-Policy Platform on Biodiversity and Ecosystem Services." Bonn, Germany: Intergovernmental Science-Policy Platform on Biodiversity and Ecosystem Services. doi:* https://doi.org/10.5281/zenodo.3553579; *OECD. 2016. The Ocean Economy in 2030. Report. Paris: OECD Publishing.* https://www.oecd.org/environment/the-oceaneconomy-in-2030-9789264251724-en.htm)

sanal fishers,[92] and the ocean's provision of essential food for millions living in poverty along the coasts of the developing world, as well as for the one billion people relying on the ocean for most of their animal protein.[93]

Most global economic activity either depends on the ocean, is based on the ocean or affects the ocean in some essential way. According to the Intergovernmental Panel on Climate Change (IPCC), 'All people on Earth depend directly or indirectly on the ocean and cryosphere'.[94] Some illustrative facts confirm this importance of the ocean for humanity: 50–80% of the oxygen comes from the ocean,[95] 44% of the world's population lives within 150 km of the coast[96] and 90% of all international trade involves marine shipping.[97]

4.2.2 The Ocean Has a Central Role to Play in Global Food Security, But the Way the Ocean Is Currently Used Is Not on Track to Deliver It

Earth's population, 2.5 billion in 1950, has grown to 7.8 billion in 2020 and is projected to peak in 2064 at about 9.73 billion.[98] It has been estimated that 470 million metric tonnes (mmt) of total animal protein will be required annually to feed the 2050 population.[99] The relative sources of land-based, ocean-based and lab-grown supply are not yet clear and will be highly dependent on the evolution of future technologies as well as human preferences. However, ocean-based food will almost certainly have a central role to play in global food security—it is healthy, its carbon footprint is low compared to land-based animal proteins,[100] and it doesn't require extensive use of water or the conversion of land for agricultural use. See Fig. 20.13 for the types of ocean food discussed in this report.

If the EAT-Lancet diet[101]—used here as a solid proxy for a globally sustainable and healthy model of nutrition—were globally adopted, 2050 fish and seafood production would need to increase by 60–118% over 2010 production levels (with the range depending on food waste reduction).[102] This corresponds to a production increase from 109 mmt today to between 160 and 218 mmt by 2050 (in whole weight). These forecasts are currently being refined to assess more precisely the role of ocean food in feeding a 2050 planet.[103]

This is in stark contrast to current, business-as-usual (BAU) projections of seafood supply (Fig. 20.14), which project a decline of capture fisheries from 80 mmt today to 67 mmt by 2050 due to the pressure of overfishing on some stocks and underfishing on others.[104] Finfish mariculture (marine aquaculture) is not projected to fill the gap, as it is seen as constrained by the availability of fish oil (FO) and fish meal (FM)—in other words, 'fishing fish to feed fish'. At reasonably probable future inclusion rates for FO and FM, annual finfish production is forecast to reach a maximum of only 14.4 mmt: around twice the current production—far short of what would be needed to fill the gap.[105] Bivalve mariculture (e.g. of mussels and oysters) does not require outside feed and therefore has a greater growth potential than wild-capture fisheries and farmed finfish, even in a business-as-usual scenario. A steady increase in bivalve production (aligned with the past 10 years' annual growth rate) therefore makes the biggest contribution to a

[92]FAO Fisheries and Aquaculture Department. n.d. "Small-Scale Fisheries around the World." Food and Agriculture Organization of the United Nations. http://www.fao.org/fishery/ssf/world/en. Accessed 6 May 2020.

[93]World Health Organization. n.d. "3. Global and Regional Food Consumption Patterns and Trends." https://www.who.int/nutrition/topics/3_foodconsumption/en/index2.html. Accessed 6 May 2020.

[94]Pörtner, H.O., D.C. Roberts, V. Masson-Delmotte, P. Zhai, M. Tignor, K. Poloczanska, K. Mintenbeck et al., eds. 2019. "Summary for Policymakers." In *IPCC Special Report on the Ocean and Cryosphere in a Changing Climate*. Intergovernmental Panel on Climate Change. https://report.ipcc.ch/srocc/pdf/SROCC_FinalDraft_FullReport.pdf.

[95]NOAA. n.d. "How Much Oxygen Comes from the Ocean?".

[96]UN Atlas of the Oceans. n.d. "Human Settlements on the Coast." http://www.oceansatlas.org/subtopic/en/c/114/. Accessed 13 August 2020.

[97]Olmer, N., B. Comer, B. Roy, X. Mao and D. Rutherford. 2017. "Greenhouse Gas Emissions from Global Shipping, 2013–2015." Washington, DC: International Council on Clean Transport. https://theicct.org/sites/default/files/publications/Global-shipping-GHG-emissions-2013-2015_ICCT-Report_17102017_vF.pdf; International Chamber of Shipping. n.d. "Shipping and World Trade."

[98]UN Department of Economic and Social Affairs. n.d. "2019 Revision of World Population Prospects." https://population.un.org/wpp/. Accessed 6 May 2020; Vollset, S.E., E. Goren, C.-W. Yuan, J. Cao, A.E. Smith, T. Hsiao, C. Bisignano et al. 2020. "Fertility, Mortality, Migration, and Population Scenarios for 195 Countries and Territories from 2017 to 2100: A Forecasting Analysis for the Global Burden of Disease Study." *Lancet*, 14 July. doi: https://doi.org/10.1016/S0140-6736(20)30677-2.

[99]Food and Agriculture Organization of the United Nations (FAO). n.d. "How to Feed the World in 2050." http://www.fao.org/fileadmin/templates/wsfs/docs/expert_paper/How_to_Feed_the_World_in_2050.pdf. Accessed 6 May 2020.

[100]Hoegh-Guldberg et al. 2019. "The Ocean as a Solution to Climate Change."

[101]Willett, W., J. Rockström, B. Loken, M. Springmann, T. Lang, S. Vermeulen, T. Garnett et al. 2019. "Food in the Anthropocene: The EAT–Lancet Commission on Healthy Diets from Sustainable Food Systems." *Lancet* 393 (10170): 447–92. doi: https://doi.org/10.1016/S0140-6736(18)31788-4. This report cites a required increase of about 55–125% of fish and seafood production in 2050. We chose the halfway point within this range, 90%, and applied it to the seafood production stated in Costello, C., L. Cao, S. Gelcich et al. 2019. "The Future of Food from the Sea." Washington, DC: World Resources Institute. https://www.oceanpanel.org/blue-papers/future-food-sea.

[102]Troell, M., M. Jonell and B. Crona. 2019. "The Role of Seafood in Sustainable and Healthy Diets." EAT-Lancet Commission, 24. https://eatforum.org/content/uploads/2019/11/Seafood_Scoping_Report_EAT-Lancet.pdf.

[103]Troell et al. 2019. "The Role of Seafood in Sustainable and Healthy Diets," 24.

[104]Costello et al. 2019. "The Future of Food from the Sea."

[105]Costello et al. 2019. "The Future of Food from the Sea."

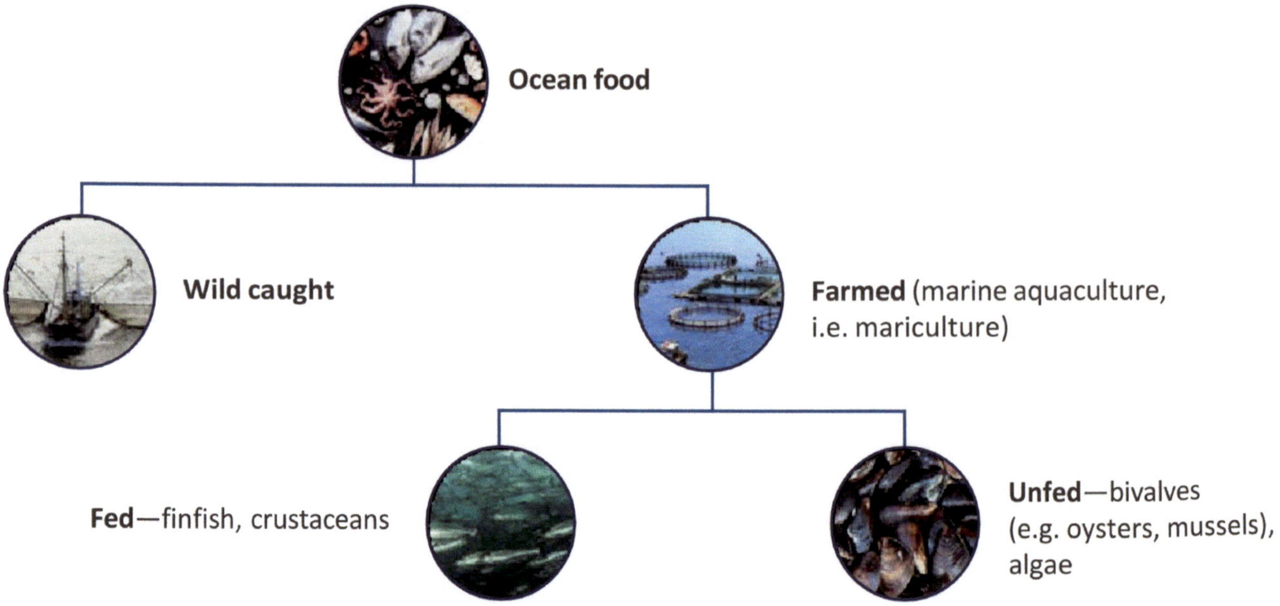

Fig. 20.13 Scope of ocean food discussed in this report. (*Source: Authors. Photo credits: Ocean food: Anna Pustynnikova/Shutterstock; Wild caught: Split Second Stock/Shutterstock; Farmed: Vladislav Gajic/Shutterstock; Fed: Konstantin Novikov/Shutterstock; Unfed: Dilara Mammadova/Shutterstock*)

projected overall doubling of mariculture production, from 29 to 66 mmt in 2050.

Summing these three potential contributions under a BAU scenario leaves a shortfall of up to 85 mmt (Fig. 20.14).

The BAU scenario, of course, is not etched in stone. If properly and sustainably managed, capture fisheries could contribute about 98 mmt by 2050—over 40% more than the BAU projection.[106]

In addition to this wild-caught potential volume, finfish mariculture can contribute higher yields once (partially) decoupled from FM/FO.[107] Mariculture must and can be done right. Unfed species (bivalves, seaweeds) are generally more benign to the environment, but barriers remain to higher production and consumption (e.g. the gap between perceived risk and actual risk).[108] Finfish mariculture will require further technology development, and strict environmental regulations on antibiotic and effluent pollution, before it can produce very large volumes, presumably offshore, with lower local impacts and without reliance on fish-based feeds. Recent developments are encouraging; progress in both governance (e.g. the 'traffic light system' in Norway, which conditions production on environmental assessments) and technology (e.g. disease control, alternative feeds, etc.; see Sect. 4.4) is underway. Additionally, equity issues associated with mariculture must be attended to, ensuring the full inclusion of women, equal treatment of all ethnic and racial groups, adoption of safe labour standards and fair treatment of smallholder farmers.[109]

Unfed mariculture, including seaweed production, is also currently greatly underdeveloped compared to its advantages and biological potential (see Sect. 5).

4.2.3 Ocean-Based Solutions Are Underappreciated and Essential to Fight Climate Change

The significant carbon mitigation challenge inherent in a 1.5 °C future is well understood and documented.[110] Usually

[106] Costello et al. 2019. "The Future of Food from the Sea."

[107] Costello et al. 2019. "The Future of Food from the Sea."

[108] Kuttschreuter, M. 2006. "Psychological Determinants of Reactions to Food Risk Messages." *Risk Analysis* 26 (4): 1045–57. doi: https://doi.org/10.1111/j.1539-6924.2006.00799.x.

[109] Allison et al. 2020. "The Human Relationship with Our Ocean Planet."

[110] Masson-Delmotte, V., P. Zhai, H.O. Pörtner, D. Roberts, J. Skea, P.R. Shukla, A. Pirani et al., eds. 2019. *Global Warming of 1.5 °C: An IPCC Special Report on the Impacts of Global Warming of 1.5 °C above Pre-industrial Levels and Related Global Greenhouse Gas Emission Pathways, in the Context of Strengthening the Global Response to the Threat of Climate Change, Sustainable Development, and Efforts to Eradicate Poverty*. Intergovernmental Panel on Climate Change. https://www.ipcc.ch/site/assets/uploads/sites/2/2019/06/SR15_Full_Report_High_Res.pdf.

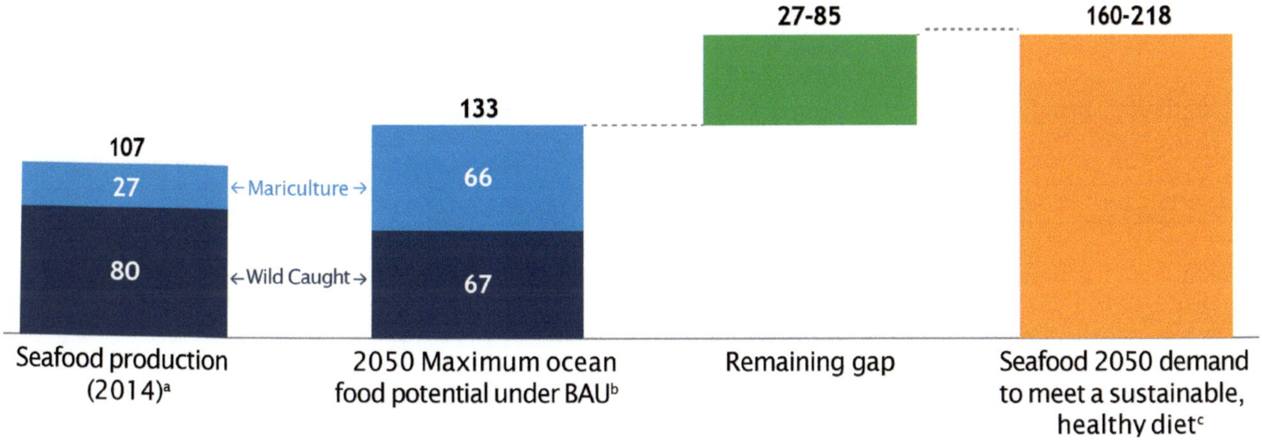

Fig. 20.14 The seafood gap to a healthy 2050 diet under business as usual. (*Sources: (a) Excluding seaweed. FAO, ed. 2018. The State of World Fisheries and Aquaculture 2018: Meeting the Sustainable Development Goals. Rome: Food and Agriculture Organization of the United Nations.* http://www.fao.org/3/I9540EN/i9540en.pdf. *(b) Wild-caught fisheries' 13 mmt decrease by 2050 under BAU from Costello, C., L. Cao, S. Gelcich et al. 2019. "The Future of Food from the Sea." Washington, DC: World Resources Institute.* https://www.oceanpanel.org/blue-papers/future-food-sea. *For aquaculture 2050 BAU is obtained by summing the additional maximum potential for fed aquaculture under current feed constraints (+7.7 mmt) with an additional 28.9 mmt potential for shelled molluscs calculated by applying the 2005–2014 global compound annual growth rate to the 2014–2050 period (assuming continued linear growth as there is no feed constraint). (c) Troell, M., M. Jonell and B. Crona. 2019. "The Role of Seafood in Sustainable and Healthy Diets." EAT-Lancet Commission, 24.* https://eatforum.org/content/uploads/2019/11/Seafood_Scoping_Report_EAT-Lancet.pdf. *These authors quote a range of 60% to 118% necessary production increase for 'Fish or seafood' over 2010 production levels. Numbers projected here are simplified by assuming that the ratio between freshwater and marine fish remains unchanged in 2050 versus the baseline year)*

seen as victims of climate change, the ocean and its coastal regions also offer a wide array of potential options to reduce GHG emissions. A comprehensive review was undertaken as part of a report commissioned by the High Level Panel for a Sustainable Ocean Economy (Ocean Panel). The Special Report 'The Ocean as a Solution to Climate Change'[111] estimates that ocean-based climate solutions could reduce global GHG emissions by up to 4 billion tonnes of carbon dioxide equivalent (CO_2e) annually by 2030 and by up to 11 billion tonnes annually by 2050. This could contribute as much as one-fifth (21%) of the emission reduction required in 2050 to limit warming to 1.5 °C and 25% for a 2 °C target (Fig. 20.15). Emission reductions of this magnitude are equivalent to the annual emissions from all coal-fired power plants worldwide or taking 2.5 billion cars off the road every year. These numbers correspond to an upper range based on strong political signals and investments.

The ocean-based options explored in this report include scaling ocean-based renewable energy generation (as a replacement for fossil fuel generation), reducing GHG emissions from marine transport (domestic and international), switching from emission-intensive land-based protein to low-carbon protein from the ocean, using seaweed as an alternative low-carbon fuel and feed for terrestrial activities, increasing the sequestration and storage potential of coastal and marine-based carbon stocks, and storing carbon in the seabed. These options did not feature prominently in the first round of nationally determined contributions (NDCs) communicated by countries or the long-term low GHG emission development strategies communicated to date under the Paris Agreement, but they offer island and coastal nations significant opportunities to consider in addition to land-based emission reduction measures.[112]

Currently, these solutions are delivering significantly less than their full mitigation potential. For example, the ocean's renewable energy contribution totals less than 0.3% of total global energy production.[113] Alarmingly, not only is the carbon sequestration and storage potential of coastal and marine ecosystems not fully being captured through efforts to pro-

[111] Hoegh-Guldberg et al. 2019. "The Ocean as a Solution to Climate Change."

[112] Gallo, N.D., D.G. Victor and L.A. Levin. 2017. "Ocean Commitments under the Paris Agreement." *Nature Climate Change* 7 (11): 833–38. doi: https://doi.org/10.1038/nclimate3422; Hoegh-Guldberg, O., E. Northrop and J. Lubchenco. 2019. "The Ocean Is Key to Achieving Climate and Societal Goals." *Science* 365 (6460): 1372–74. doi: https://doi.org/10.1126/science.aaz4390.

[113] International Energy Agency (IEA). 2019. *Offshore Wind Outlook 2019*. https://www.iea.org/reports/offshore-wind-outlook-2019.

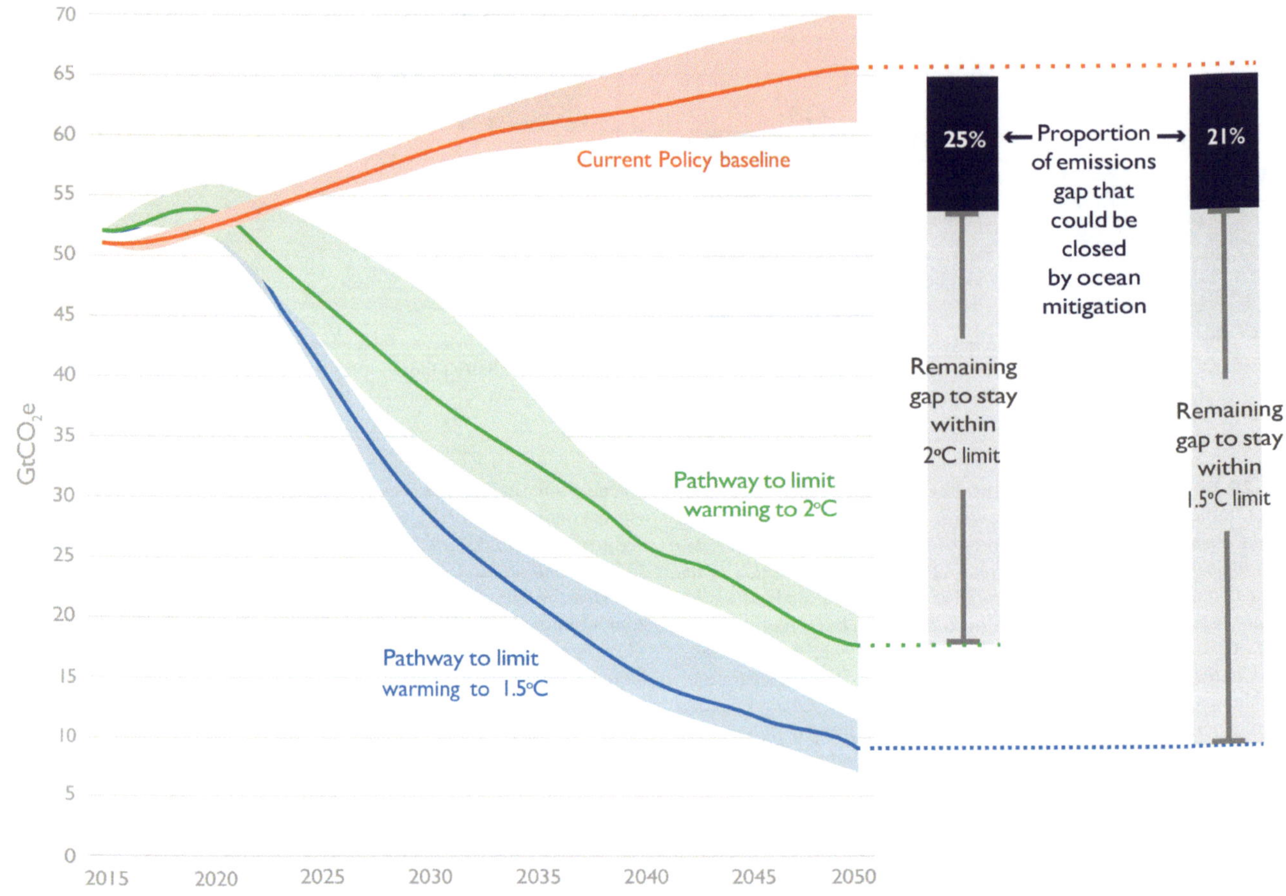

Fig. 20.15 Contribution of ocean-based mitigating options towards the emission gap. (*Sources: UNEP 2018, Climate Action Tracker (2018), as adapted by Hoegh-Guldberg, O., et al. 2019. "The Ocean as a Solution to Climate Change: Five Opportunities for Action." Washington, DC: World Resources Institute.* https://oceanpanel.org/sites/default/files/2019-10/HLP_Report_Ocean_Solution_Climate_Change_final.pdf)

tect and manage these ecosystems but the degradation and loss of these ecosystems—mangroves at 0.21%/year,[114] saltmarshes at 1–2%/year[115] and sea grass at 2–5%/year—is releasing significant emissions back into the atmosphere.[116]

[114]Hamilton, S.E., and D. Casey. 2016. "Creation of a High Spatio-temporal Resolution Global Database of Continuous Mangrove Forest Cover for the 21st Century (CGMFC-21)." *Global Ecology and Biogeography* 25 (6): 729–38. doi: https://doi.org/10.1111/geb.12449.

[115]Duarte, C.M., ed. 2009. *Global Loss of Coastal Habitats: Rates, Causes and Consequences*. Bilbao, Spain: Fundación BBVA.

[116]Duarte, C.M., W.C. Dennison, R.J.W. Orth and T.J.B. Carruthers. 2008. "The Charisma of Coastal Ecosystems: Addressing the Imbalance." *Estuaries and Coasts* 31 (2): 233–38. doi: https://doi.org/10.1007/s12237-008-9038-7.

4.2.4 The Ocean Can Catalyse a Global Transition Towards More Circular and Regenerative Practices in Land-Based Economies

The 'ocean economy' is usually associated with purely ocean-based industries—shipping, fishing and so on. Nonetheless, almost all land-based industries rely on the services provided by the ocean. Perhaps the most difficult, and intriguing, part of the ocean economy puzzle concerns the chain reactions caused in global markets by changes in ocean-related production of fish, renewable energy or minerals. Everything is connected—a reduction of anchovy harvests in Peru affects the price of Scottish farmed fish, Chinese pigs and omega-3 capsules (all dependant on fish meal and fish oil, products extracted by drying and grinding up fish like anchovies).[117]

The ocean economy can thus not be viewed in a siloed 'blue' fashion. Moreover, this connectedness applies not only to what people remove from the ocean but also to what they put into the ocean. Over 80% of all global marine pollution originates on land[118]—all too often, the ocean 'serves' as the ultimate planetary sink. It absorbs 30% of anthropogenic (land-based) CO_2, [119] 90% of excess heat caused by anthropogenic GHG emissions[120] and an estimated 9–14 mmt of plastic pollution per year.[121] Following the old fallacy of 'the solution to pollution is dilution', the ocean has been expected to absorb invisible pollution like nutrient runoff, heavy metals (e.g. mercury, cadmium), nuclear waste, pharmaceuticals, persistent toxicants (DDT, TBT, pesticides, furans, dioxins, phenols), sewage and personal care products.

Keeping the ocean functioning within the bounds of the 'safe operating space' for humanity can also catalyse profound and profitable changes in land-based systems: moving away from the 'blue silo' allows for the explicit connection between SDG 14 (conserve and sustainably use the oceans, seas and marine resources for sustainable development)[122] and the acceleration of SDG 15 (life on land), as well as other SDGs often thought of as land-based, including SDG 12 (sustainable consumption and production), SDG 9 (sustainable infrastructure) and SDG 7 (affordable and sustainable energy).

The fate of the ocean is directly linked to a broader shift towards a circular economy[123] approach to consumer goods and industrial production—a system where resources are used continually, at their highest possible value added, and recovered or regenerated as efficiently as possible at the end of their service. It is also linked to a land-based transition towards renewable energies, and to improved land use practices in agriculture and in coastal development. But looking at it the other way around, the ocean could be a unique opportunity to advance the broader global agenda of sustainability while 'leaving no-one behind'.

As a compelling example, the ocean is now the principal driver of fundamental work on the plastic value chain. The unprecedented crisis of ocean plastic pollution is bringing scientists, businesses, governments and civil society together to look for solutions.[124] For instance, in October 2018 in Bali, 250 organisations, including many of the packaging producers, brands, retailers and recyclers, as well as governments and NGOs (altogether representing 20% of all plastic packaging produced globally) committed to eradicate plastic waste and pollution at the source. Following the plastic example, the wasteful agriculture system could be challenged by the sustainable ocean agenda, obliging it to accelerate the transition towards precision farming, less toxic fertilisers and pesticides, and the collection and treatment of human and livestock waste and wastewater.

4.3 Failing the Environment and the People: The Need for Urgent Action

Physical, geological, chemical, biological and ecological processes interact in the ocean in complex ways. Those processes and interactions have now been fundamentally altered by human activities, with concomitant changes to the services provided to people by natural ecosystems. For example, loss of biological diversity, major perturba-

[117]Neate, R. 2012. "Anchovy Price Leap Causes Food Industry Chain Reaction." *The Guardian*, 24 August. https://www.theguardian.com/business/2012/aug/24/anchovy-price-leap-food-industry-chain.

[118]Ocean Conservancy. n.d. *Stemming the Tide: Land-Based Strategies for a Plastic-Free Ocean*. https://oceanconservancy.org/wp-content/uploads/2017/04/full-report-stemming-the.pdf. Accessed 6 May 2020.

[119]62 Core Writing Team, R.K. Pachauri and L. Meyer. 2014. *Climate Change 2014: Synthesis Report. Contribution of Working Groups I, II and III to the Fifth Assessment Report of the Intergovernmental Panel on Climate Change*. Geneva: Intergovernmental Panel on Climate Change. https://www.ipcc.ch/site/assets/uploads/2018/05/SYR_AR5_FINAL_full_wcover.pdf.

[120]Gattuso, J.-P., A. Magnan, R. Billé, W.W.L. Cheung, E.L. Howes, F. Joos, D. Allemand et al. 2015. "Contrasting Futures for Ocean and Society from Different Anthropogenic CO_2 Emissions Scenarios." *Science* 349 (6243). doi: https://doi.org/10.1126/science.aac4722.

[121]Lau et al. 2020. "Evaluating Scenarios toward Zero Plastic Pollution"; Pew Charitable Trusts and SYSTEMIQ. 2020. *Breaking the Plastic Wave*. https://www.systemiq.earth/wp-content/uploads/2020/07/BreakingThePlasticWave_MainReport.pdf.

[122]UN Statistics Division. n.d. "Goal 14: Conserve and Sustainably Use the Oceans, Seas and Marine Resources for Sustainable Development: SDG Indicators." Development Data and Outreach. https://unstats.un.org/sdgs/report/2017/goal-14/. Accessed 6 May 2020.

[123]Definition of Circular Economy by Ellen MacArthur Foundation: A circular economy is based on the principles of designing out waste and pollution, keeping products and materials in use and regenerating natural systems.

[124]Dalberg Advisors. 2019. "Solving Plastic Pollution through Accountability." Gland, Switzerland: World Wide Fund For Nature. https://c402277.ssl.cf1.rackcdn.com/publications/1212/files/original/SOLVING_PLASTIC_POLLUTION_THROUGH_ACCOUNTABILITY_ENF_SINGLE.pdf?1551798060.

tions of biochemical cycles, and climate change each alter the functioning of ecosystems, and that in turn impairs or limits the benefits that ocean ecosystems provide to people.

As the rate of change in most socioeconomic areas has accelerated past any historical precedent in the first half of the twentieth century, so have most earth system indicators—a phenomenon described as 'the Great Acceleration' (Fig. 20.16).

There is also increasing strain on the ocean system: the 'blue acceleration'—humanity's expansion into the ocean for food, materials and space—has been unparalleled in history.[125] The direct consequences of these trends are exhaustively documented today (see details below).

The direct footprint of human activity is visible almost everywhere. Sixty-six percent of the marine environment is experiencing significant cumulative impact by human actions.[126] Only 13% of the ocean area can still be classified as wilderness,[127] and less than 3% of the ocean is unaffected by multiple human stressors.[128] For example, between 1970 and 2000, sea grass meadows declined by roughly 30%, mangroves by 35% and saltmarshes by 60%, whilst between 11% and 46% of marine invertebrates are threatened.[129] Below, the main stressors on the ocean caused by human activity are briefly described along with their directly observable consequences.

Overfishing The direct over-exploitation of fish stocks and the unintended impacts of fishing gear on non-target species may be the most tangible manifestation of direct pressure from human activity. [130] This has been exacerbated by harmful fisheries subsidies (i.e. those directed at capacity expansion) as well as the effects of illegal, unreported and unregulated (IUU) fishing. Industrial and artisanal fishing fleets have been identified as the main driver of extinction for all classes of marine vertebrates except birds.[131] Estimates of overfished stocks range from 33% ('overfished' category in the Food and Agriculture Organization of the United Nations [FAO] database)[132] to 47% ('over-exploited or collapsed' category in the Sea around Us Project's classification).[133] Higher trophic level species and predators such as sharks, tuna and billfish are especially depleted.[134] For example, a 2020 global shark survey found no sharks in almost 20% of the 371 surveyed reefs across 58 nations, with levels of shark depletion being closely correlated to poor governance, the density of human population and distance to the nearest market.[135]

Open ocean diversity has declined by 10–50% over the past 50 years, a trend that has coincided with increased fishing pressure.[136]

[125] Jouffray, J.-B., R. Blasiak, A.V. Norström, H. Österblom and M. Nyström. 2020. "The Blue Acceleration: The Trajectory of Human Expansion into the Ocean." *One Earth* 2 (1): 43–54. doi: https://doi.org/10.1016/j.oneear.2019.12.016.

[126] Díaz et al. 2019. "Summary for Policymakers of the Global Assessment Report on Biodiversity and Ecosystem Services of the Intergovernmental Science-Policy Platform on Biodiversity and Ecosystem Services."

[127] Jones, K.R., C.J. Klein, B.S. Halpern, O. Venter, H. Grantham, C.D. Kuempel, N. Shumway et al. 2018. "The Location and Protection Status of Earth's Diminishing Marine Wilderness." *Current Biology* 28 (15): 2506–12.e3. doi: https://doi.org/10.1016/j.cub.2018.06.010.

[128] Halpern, B.S., M. Frazier, J. Potapenko, K.S. Casey, K. Koenig, C. Longo, J.S. Lowndes et al. 2015. "Spatial and Temporal Changes in Cumulative Human Impacts on the World's Ocean." *Nature Communications* 6 (1): 1–7. doi: https://doi.org/10.1038/ncomms8615.

[129] Rogers, A., O. Aburto-Oropeza, W. Appeltans, J. Assis, L. T. Ballance, P. Cury, C. Duarte et al. 2020. "Critical Habitats and Biodiversity: Inventory, Thresholds and Governance." Washington, DC: World Resources Institute. https://www.oceanpanel.org/blue-papers/critical-habitats-and-biodiversity-inventory-thresholds-and-governance.

[130] 73 Díaz et al. 2019. "Summary for Policymakers of the Global Assessment Report on Biodiversity and Ecosystem Services of the Intergovernmental Science-Policy Platform on Biodiversity and Ecosystem Services."

[131] Rogers et al. 2020. "Critical Habitats and Biodiversity."

[132] FAO, ed. 2018. *The State of World Fisheries and Aquaculture 2018: Meeting the Sustainable Development Goals*. Rome: Food and Agriculture Organization of the United Nations. http://www.fao.org/3/I9540EN/i9540en.pdf.

[133] Pauly, D., D. Zeller and M.L.D. Palomares. n.d. "Sea around Us Concepts, Design and Data." http://www.seaaroundus.org. Accessed 6 May 2020.

[134] Roff, G., C.J. Brown, M.A. Priest and P.J. Mumby. 2018. "Decline of Coastal Apex Shark Populations over the Past Half Century." *Communications Biology* 1 (1): 1–11. doi: https://doi.org/10.1038/s42003-018-0233-1; Christensen, V., M. Coll, C. Piroddi, J. Steenbeek, J. Buszowski and D. Pauly. 2014. "A Century of Fish Biomass Decline in the Ocean." *Marine Ecology Progress Series* 512 (October): 155–66. doi: https://doi.org/10.3354/meps10946.

[135] MacNeil, M.A., D.D. Chapman, M. Heupel, C.A. Simpfendorfer, M. Heithaus, M. Meekan, E. Harvey et al. 2020. "Global Status and Conservation Potential of Reef Sharks." *Nature* 583 (7818): 801–6. doi: https://doi.org/10.1038/s41586-020-2519-y.

[136] Worm, B., M. Sandow, A. Oschlies, H.K. Lotze and R.A. Myers. 2005. "Global Patterns of Predator Diversity in the Open Oceans." *Science* 309 (5739): 1365–69. doi: https://doi.org/10.1126/science.1113399.

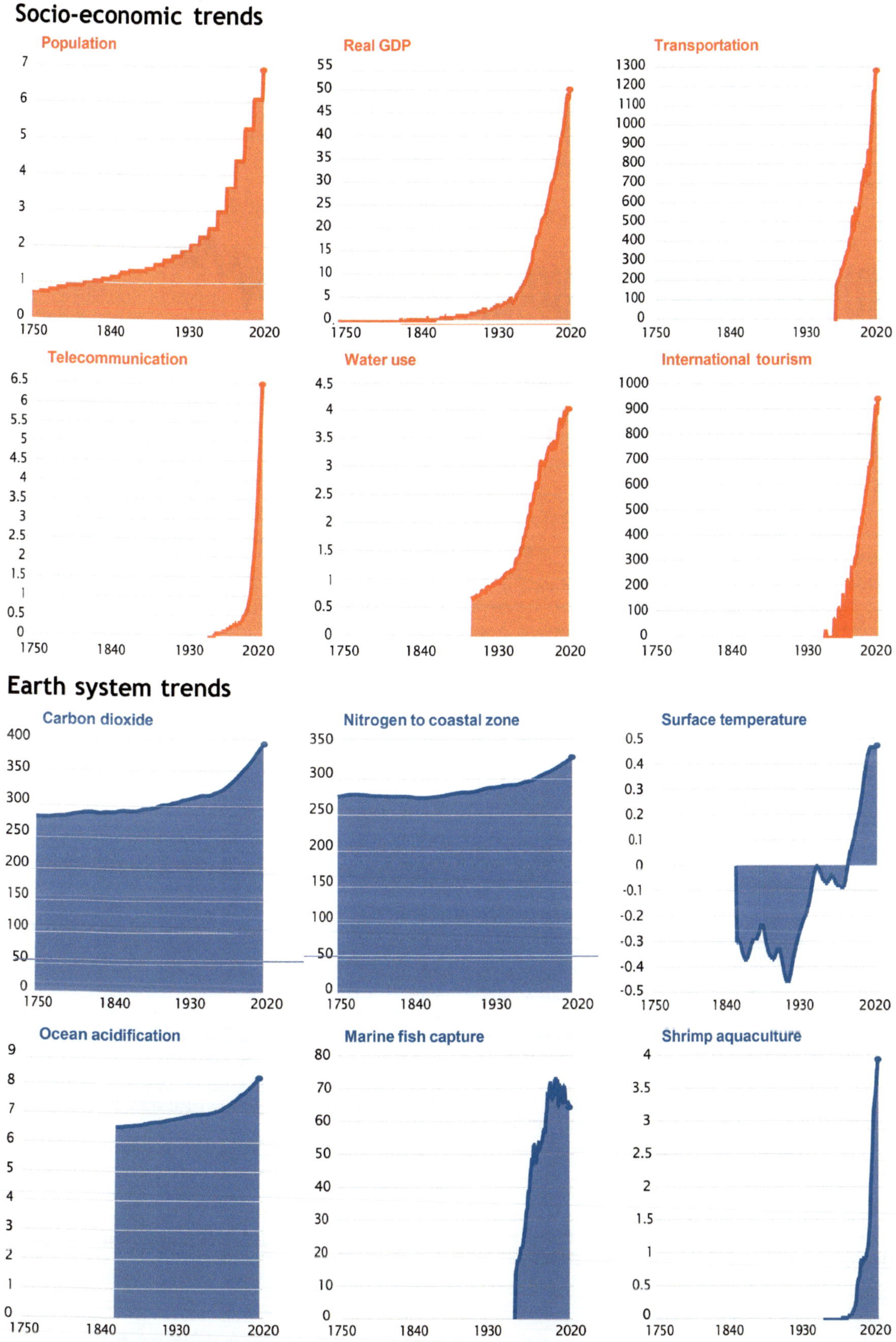

Fig. 20.16 'The great acceleration'. (*Source: Steffen, W., W. Broadgate, L. Deutsch, O. Gaffney and C. Ludwig. 2015. "The Trajectory of the Anthropocene: The Great Acceleration." Anthropocene Review 2 (1). doi:* https://doi.org/10.1177/2053019614564785)

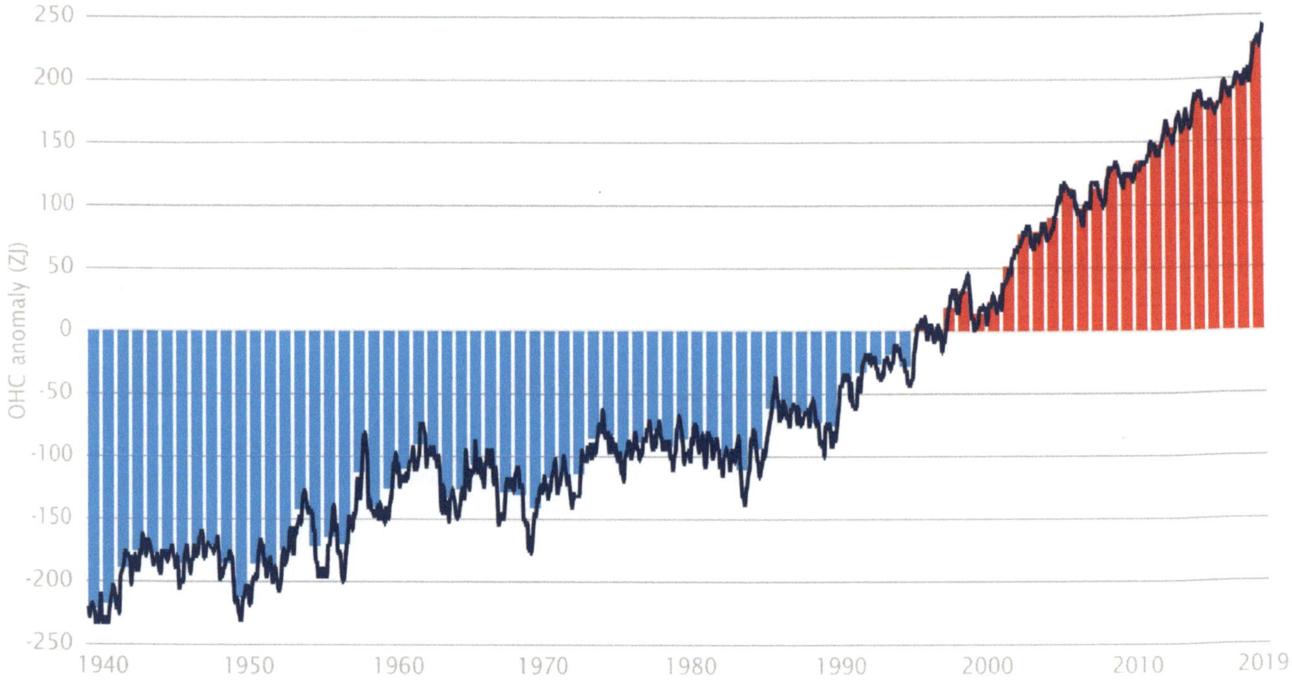

Fig. 20.17 2019: Warmest year in recorded human history for the world's ocean. (*Source: Cheng, L., J. Abraham, J. Zhu, K.E. Trenberth, J. Fasullo, T. Boyer, R. Locarnini et al. 2020. "Record-Setting Ocean Warmth Continued in 2019." Advances in Atmospheric Sciences 37 (2): 137–42. doi:* https://doi.org/10.1007/s00376-020-9283-7)

Climate change The raw numbers are sobering: ocean waters have absorbed 93% of the excess heat caused by greenhouse gas (GHG) emissions[137] and sea surface temperatures have increased by 0.7 °C since 1900.[138] New analysis confirms that 2019 was the warmest year on record for ocean temperature, and saw the largest single-year increase of the decade (Fig. 20.17).[139] The 12 years with lowest Arctic sea ice extent all happened in the past dozen years,[140] and 2017 marked the lowest Antarctic sea ice extent on record.[141]

Climate change generates stronger winds.[142] This intensification of surface winds has accelerated the global mean ocean circulation over the past two decades, especially in tropical regions.[143] These changes in ocean currents can affect not only weather patterns on land (e.g. the Gulfstream's

[137]Stocker, T.F., D. Qin, G.K. Plattner, M. Tignor, S.K. Allen, J. Boschung, A. Nauels et al. 2013. "Summary for Policymakers." In *Climate Change 2013: The Physical Science Basis. Contribution of Working Group I to the Fifth Assessment Report of the Intergovernmental Panel on Climate Change*. Cambridge: Cambridge University Press. http://www.climatechange2013.org/images/report/WG1AR5_SPM_FINAL.pdf.

[138]Laffoley, D., and J.M. Baxter, eds. 2016. *Explaining Ocean Warming: Causes, Scale, Effects and Consequences*. International Union for Conservation of Nature. doi: https://doi.org/10.2305/IUCN.CH.2016.08.en.

[139]Cheng, L., J. Abraham, J. Zhu, K.E. Trenberth, J. Fasullo, T. Boyer, R. Locarnini et al. 2020. "Record-Setting Ocean Warmth Continued in 2019." *Advances in Atmospheric Sciences* 37 (2): 137–42. doi: https://doi.org/10.1007/s00376-020-9283-7.

[140]National Snow and Ice Data Center. 2018. "Arctic Sea Ice Extent Arrives at Its Minimum." Arctic Sea Ice News and Analysis (blog). http://nsidc.org/arcticseaicenews/2018/09/arctic-sea-ice-extent-arrives-at-its-minimum/.

[141]Gaines, S., R. Cabral, C.M. Free, Y. Golbuu, R. Arnason, W. Battista, D. Bradley et al. 2019. "The Expected Impacts of Climate Change on the Ocean Economy." Washington, DC: World Resources Institute. https://www.oceanpanel.org/blue-papers/expected-impacts-climate-change-ocean-economy.

[142]Hu, S., J. Sprintall, C. Guan, M.J. McPhaden, F. Wang, D. Hu and W. Cai. 2020. "Deep-Reaching Acceleration of Global Mean Ocean Circulation over the Past Two Decades." *Science Advances* 6 (6): eaax7727. doi: https://doi.org/10.1126/sciadv.aax7727.

[143]Hu et al. 2020. "Deep-Reaching Acceleration of Global Mean Ocean Circulation over the Past Two Decades."

influence on the European climate[144]) but also fisheries through, for instance, modification of larval dispersal[145] or changes in the intensity of coastal upwelling[146] (the movement of cold, nutrient-rich water to the ocean surface). These upwelling changes can enhance fishery productivity, as with anchovies along the coast of Peru; but if the upwelling is too intense, it can have the opposite effect, triggering 'dead zones' with insufficient oxygen to support fish and other marine life. Changes to ocean circulation are regionally variable. For example, the Atlantic Meridional Overturning Circulation (AMOC), which redistributes heat between tropics and higher latitude in the Atlantic, is one exception to the general pattern of speedier currents at the global scale. AMOC is 'very likely to weaken over the twenty-first century', according to the IPCC.[147] Considerable uncertainty remains, however: the IPCC cites a range of between 1% and 54% for AMOC weakening, depending on the warming scenario chosen.[148]

Humanity's GHG emissions have also acidified the ocean by 26% since the Industrial Revolution,[149] and climate change is impacting dissolved oxygen content in ocean systems across the globe (see more details about dead zones later in this section). The combined effects are putting additional stress on many coastal and oceanic species, including the shell-forming animals (corals, phytoplankton, zooplankton, bivalves and more) which represent the foundation of the marine food webs.

Habitat destruction Key coastal habitats such as mangroves are being lost at an alarming rate: global mangrove cover has declined by around 25–35% (up to about 57,000 km^2 from 1980 to 2000),[150] largely due to land reclamation and conversion to aquaculture ponds and rice paddies.[151] This loss has resulted in reductions in fisheries and coastal food production,[152] and increasing threats to species with a fragile conservation status. These coastal habitats help protect communities against life-threatening storm surge during tsunamis, typhoons, cyclones and hurricanes. Mangroves, sea grasses and saltmarshes are labelled 'blue carbon' ecosystems because they actively sequester and store organic carbon from the environment,[153] meaning their loss increases emissions.[154] The seafloor habitats have also been significantly affected by destructive fishing gear and methods. Bottom trawling has destroyed cold water coral and sponge ecosystems, which will take centuries to recover;[155] dynamite and cyanide fishing has contributed to the decline of coral reefs.[156]

[144]Palter, J.B. 2015. "The Role of the Gulf Stream in European Climate." *Annual Review of Marine Science* 7 (1): 113–37. doi: https://doi.org/10.1146/annurev-marine-010814-015656.

[145]Ramesh, N., J.A. Rising and K.L. Oremus. 2019. "The Small World of Global Marine Fisheries: The Cross-Boundary Consequences of Larval Dispersal." *Science* 364 (6446): 1192–96. doi: https://doi.org/10.1126/science.aav3409.

[146]Bakun, A., B.A. Black, S.J. Bograd, M. García-Reyes, A.J. Miller, R.R. Rykaczewski and W.J. Sydeman. 2015. "Anticipated Effects of Climate Change on Coastal Upwelling Ecosystems." *Current Climate Change Reports* 1 (2): 85–93. doi: https://doi.org/10.1007/s40641-015-0008-4.

[147]Collins, M., R. Knutti, J. Arblaster, J.-L. Dufresne, T. Fichefet, X. Gao, W.J. Gutowski Jr. et al. 2013. "Long-Term Climate Change: Projections, Commitments and Irreversibility." In *Climate Change 2013: The Physical Science Basis. Contribution of Working Group I to the Fifth Assessment Report of the Intergovernmental Panel on Climate Change*. Cambridge: Cambridge University Press. https://www.ipcc.ch/site/assets/uploads/2018/02/WG1AR5_Chapter12_FINAL.pdf.

[148]Collins et al. 2013. "Long-Term Climate Change."

[149]Gaines et al. 2019. "The Expected Impacts of Climate Change on the Ocean Economy."

[150]Polidoro, B.A., K.E. Carpenter, L. Collins, N.C. Duke, A.M. Ellison, J.C. Ellison, E.J. Farnsworth et al. 2010. "The Loss of Species: Mangrove Extinction Risk and Geographic Areas of Global Concern." Edited by D.M. Hansen. *PLOS ONE* 5 (4): e10095. doi: https://doi.org/10.1371/journal.pone.0010095; Valiela, I., J.L. Bowen and J.K. York. 2001. "Mangrove Forests: One of the World's Threatened Major Tropical Environments. At Least 35% of the Area of Mangrove Forests Has Been Lost in the Past Two Decades, Losses That Exceed Those for Tropical Rain Forests and Coral Reefs, Two Other Well-Known Threatened Environments." *BioScience* 51 (10): 807–15. doi: https://doi.org/10.1641/0006-3568(2001)051[0807:MFOOTW]2.0.CO;2; Thomas, N., R. Lucas, P. Bunting, A. Hardy, A. Rosenqvist and M. Simard. 2017. "Distribution and Drivers of Global Mangrove Forest Change, 1996–2010." Edited by S. Joseph. *PLOS ONE* 12 (6): e0179302. doi: https://doi.org/10.1371/journal.pone.0179302.

[151]Richards, D.R., and D.A. Friess. 2016. "Rates and Drivers of Mangrove Deforestation in Southeast Asia, 2000–2012." *Proceedings of the National Academy of Sciences* 113 (2). 344–49. doi: https://doi.org/10.1073/pnas.1510272113.

[152]Aburto-Oropeza, O., E. Ezcurra, G. Danemann, V. Valdez, J. Murray and E. Sala. 2008. "Mangroves in the Gulf of California Increase Fishery Yields." *Proceedings of the National Academy of Sciences* 105 (30): 10456–59. doi: https://doi.org/10.1073/pnas.0804601105.

[153]Nellemann, C., and E. Corcoran. 2009. *Blue Carbon: The Role of Healthy Oceans in Binding Carbon: A Rapid Response Assessment*. UN Environment Programme/Earthprint.

[154]Duarte, C.M., H. Kennedy, N. Marbà and I. Hendriks. 2013. "Assessing the Capacity of Seagrass Meadows for Carbon Burial: Current Limitations and Future Strategies." *Ocean & Coastal Management* 83 (October): 32–38. doi: https://doi.org/10.1016/j.ocecoaman.2011.09.001.

[155]Inniss, L., A. Simcock, A.Y. Ajawin, A.C. Alcala, P. Bernal, H.P. Calumpong, P.E. Araghi et al. 2016. "The First Global Integrated Marine Assessment." New York: United Nations. https://www.un.org/Depts/los/global_reporting/WOA_RPROC/WOACompilation.pdf.

[156]Beck, M.W., I.J. Losada, P. Menéndez, B.G. Reguero, P. Díaz-Simal and F. Fernández. 2018. "The Global Flood Protection Savings Provided by Coral Reefs." *Nature Communications* 9 (1): 1–9. doi: https://doi.org/10.1038/s41467-018-04568-z.

Plastic pollution At least 700 species of marine life have been demonstrated to interact with plastic,[157] with the main impacts occurring through entanglement, ingestion and chemical contamination from macroplastics. The annual flow of plastic into the ocean is predicted to nearly triple by 2040 to 29 million metric tonnes per year if no serious action is taken.[158] This number corresponds to an equivalent 50 kg of plastic for every metre of coastline worldwide.[159] There is also clear evidence that microplastics are ingested by a wide range of species, including marine mammals, birds, fish and small invertebrates at the base of the food chain.[160]

Other land-based pollutants Ocean ecosystems and marine life are damaged by many land-based pollutants, such as pesticides, antibiotics, parasiticides, pharmaceuticals, heavy metals, persistent organic pollutants and excessive amounts of nutrients such as nitrogen and phosphorus. For instance, in Southeast Asia, an estimated 600,000 tonnes of nitrogen end up in the ocean every year, discharged from major regional rivers.[161] Direct impacts vary considerably, depending on the pollutant, its amount and the presence of other stressors.[162] Impacts can include excess productivity that triggers dead zones (low- or no-oxygen; see details later in this section), reduced photosynthetic efficiency, chronic stress on marine organisms, cancer in animals, likely inhibition of reproduction and birth defects.[163]

Invasive species Discharge of untreated ballast water from ships is considered one of the major threats to biodiversity that, if not addressed, could have severe public health, environmental and economic impacts.[164] One cubic metre of ballast water can contain up to 50,000 zooplankton specimens[165] and/or 10 million phytoplankton cells.[166] With 10 billion tonnes of it transferred throughout the world each year,[167] ballast water is one of the principal vectors of potentially invasive alien species.[168]

Compounding stressors In many occurrences these individual stressors locally compound one another with exponential consequences on ecosystems. For instance, coral reefs around the globe are exposed not just to overheating but also to overfishing and pollution. The decline of average hard coral cover on Caribbean reefs from 50% in the 1970s to 10% in the early 2000s, for example, was caused by the introduction of a pathogen killing an important herbivore (sea urchin), on top of decades of overfishing of herbivores and grazers (parrotfish and multiple other species of fishes) as well as predators essential to the integrity of the system, sediment from deforestation on land, warmer water from climate change, and physical destruction and pollution from overdevelopment in coastal areas (see Fig. 20.18).[169] In Asian and Australian waters, the primary drivers are switched. For example, in 2016 the Great Barrier Reef experienced an unprecedented die-off of staghorn and tabular corals on a third of its reefs,[170] caused by a record heatwave, with pollution playing a secondary

[157]Gall, S.C., and R.C. Thompson. 2015. "The Impact of Debris on Marine Life." *Marine Pollution Bulletin* 92 (1): 170–79. doi: https://doi.org/10.1016/j.marpolbul.2014.12.041.

[158]Lau, W.W.Y., Y. Shiran, R.M. Bailey, E. Cook, M.R. Stuchtey, J. Koskella, C.A. Velis et al. 2020. "Evaluating Scenarios toward Zero Plastic Pollution." *Science*, July. doi: https://doi.org/10.1126/science.aba9475.

[159]Lau et al. 2020. "Evaluating Scenarios toward Zero Plastic Pollution."

[160]Law, K.L., and R.C. Thompson. 2014. "Microplastics in the Seas." *Science* 345 (6193): 144–45. doi: https://doi.org/10.1126/science.1254065.

[161]Jambeck, J., E. Moss, B.K. Dubey, Z. Arifin, L. Godfrey, B.D. Hardesty, G. Hendrawan et al. 2020. "Leveraging Multi-target Strategies to Address Plastic Pollution in the Context of an Already Stressed Ocean." Washington, DC: World Resources Institute. https://www.oceanpanel.org/blue-papers/leveraging-target-strategies-to-address-plastic-pollution-in-the-context.

[162]Jambeck et al. 2020. "Leveraging Multi-target Strategies to Address Plastic Pollution in the Context of an Already Stressed Ocean."

[163]Jambeck et al. 2020. "Leveraging Multi-target Strategies to Address Plastic Pollution in the Context of an Already Stressed Ocean"; Mills, L.J., and C. Chichester. 2005. "Review of Evidence: Are Endocrine-Disrupting Chemicals in the Aquatic Environment Impacting Fish Populations?" *Science of the Total Environment* 343(1): 1–34. doi: https://doi.org/10.1016/j.scitotenv.2004.12.070.

[164]Global Environment Facility–UN Development Programme – International Maritime Organization (GEF-UNDP-IMO) GloBallast Partnerships Programme and International Union for Conservation of Nature (IUCN). 2010. "Economic Assessments for Ballast Water Management: A Guideline." GloBallast Monograph Series no. 19. London, UK, and Gland, Switzerland: GEF-UNDP-IMO GloBallast Partnerships, IUCN. https://portals.iucn.org/library/sites/library/files/documents/2010-075.pdf.

[165]GEF-UNDP-IMO GloBallast Partnerships and International Ocean Institute (IOI). 2009. "Guidelines for National Ballast Water Status Assessment." GloBallast Monograph Series no. 17. https://archive.iwlearn.net/globallast.imo.org/wp-content/uploads/2014/11/Mono17_English.pdf.

[166]Subba Rao, D.V., W.G. Sprules, H. Locke and J.T. Carlton. 1994. "Exotic Phytoplankton from Ship's Ballast Waters: Risk of Potential Spread to Mariculture Sites on Canada's East Coast." *Canadian Data Report of Fisheries and Aquatic Sciences*, no. 937: 1–51.

[167]GEF-UNDP-IMO GloBallast Partnerships. 2017. "The GloBallast Story: Reflections from a Global Family." GloBallast Monograph no. 25. http://www.imo.org/en/MediaCentre/HotTopics/BWM/Documents/The%20GloBallast%20Story.pdf.

[168]GEF-UNDP-IMO GloBallast Partnerships. 2017. "The GloBallast Story."

[169]Jackson, E.J., M. Donovan, K. Cramer and V. Lam. 2014. "Status and Trends of Caribbean Coral Reefs: 1970–2012." Gland, Switzerland: Global Coral Reef Monitoring Network, International Union for Conservation of Nature.

[170]Hughes, T.P., J.T. Kerry, A.H. Baird, S.R. Connolly, A. Dietzel, C.M. Eakin, S.F. Heron et al. 2018. "Global Warming Transforms Coral Reef Assemblages." *Nature* 556: 492–96. doi: https://doi.org/10.1038/s41586-018-0041-2.

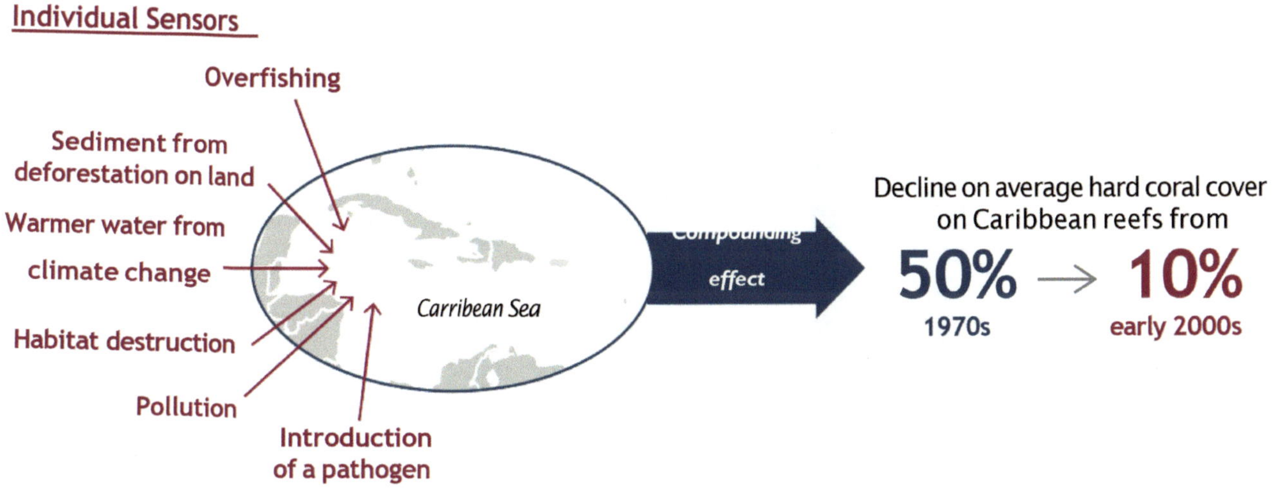

Fig. 20.18 Case study: compounding stressors leading to the decline of Caribbean Reefs. (*Source: Authors, inspired by Jackson, E.J., M. Donovan, K. Cramer and V. Lam. 2014. "Status and Trends of Caribbean Coral Reefs: 1970–2012." Gland, Switzerland: Global Coral Reef Monitoring Network, International Union for Conservation of Nature*)

role. Overall, the outlook for coral reefs is deeply concerning: annual severe bleaching (ASB) is forecast to affect 75% of all global reefs before 2070, even if the Paris Agreement carbon reduction pledges are followed.[171] With coastal overfishing endemic in most developing countries, the resilience of reefs to ASB events will be greatly diminished. With global warming of 1.5 °C, coral reefs would decline by 70–90%, and virtually all (>99%) would be lost at 2 °C warming.[172]

It should be noted, however, that large, remote coral reefs that are fully protected from extractive and abatable destructive activities (in fully protected marine protected areas) have proved to be more resilient to warmer water and other environmental stressors. Coupled with the finding that some strains of corals are becoming more tolerant of warmer waters,[173] this suggests that it may not be too late to save coral reefs if strong action is taken to reduce carbon emissions[174] and create large, fully protected areas in the ocean.

4.3.1 Indirect Effects Can Already Be Observed

When these pressures increase beyond a certain tipping point, the interconnected ocean system may no longer be able to provide the benefits people want and need. The combination of their effects can be unexpectedly severe and larger than the sum of their parts. If these stressors start compounding on a larger scale, potentially serious and fundamental indirect, 'second order' consequences occur, such as loss of biological diversity and abundance. Though analytically demanding in terms of attribution and measurement, such consequences are highly significant for the ocean's future. Even more concerning is that indirect effects may fundamentally shift key parts of the ocean system from one state to another that is often functionally different (Fig. 20.19). At this level, even sophisticated models and 'data revolution' tools can only suggest *what* might happen but not precisely *when* and *where*. Given what is at stake, these effects need to be considered in decisions, even if uncertainty is high.

[171] van Hooidonk, R., J. Maynard, J. Tamelander, J. Gove, G. Ahmadia, L. Raymundo, G. Williams et al. 2016. "Local-Scale Projections of Coral Reef Futures and Implications of the Paris Agreement." *Scientific Reports* 6 (1): 1–8. doi: https://doi.org/10.1038/srep39666.

[172] Masson-Delmotte et al. 2019. *Global Warming of 1.5 °C*.

[173] Coles, S.L., K.D. Bahr, K.S. Rodgers, S.L. May, A.E. McGowan, A. Tsang, J. Bumgarner and J.H. Han. 2018. "Evidence of Acclimatization or Adaptation in Hawaiian Corals to Higher Ocean Temperatures." *PeerJ* 6 (August): e5347. doi: https://doi.org/10.7717/peerj.5347.

[174] Bay, R.A., N.H. Rose, C.A. Logan and S.R. Palumbi. 2017. "Genomic Models Predict Successful Coral Adaptation If Future Ocean Warming Rates Are Reduced." *Science Advances* 3 (11): e1701413. doi: https://doi.org/10.1126/sciadv.1701413.

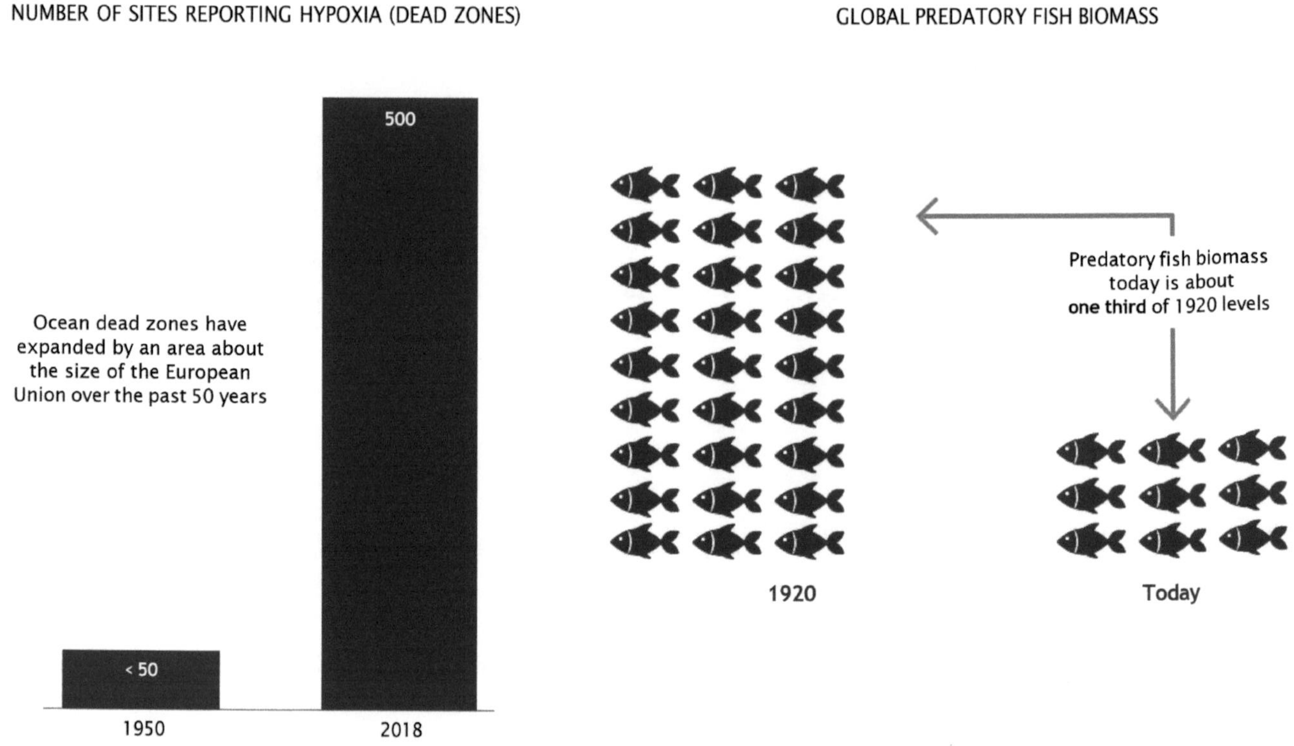

Fig. 20.19 Examples of indirect consequences of compounding pressures on the ocean. (*Sources: Breitburg, D., L.A. Levin, A. Oschlies, M. Grégoire, F.P. Chavez, D.J. Conley, V. Garçon et al. 2018. "Declining Oxygen in the Global Ocean and Coastal Waters." Science 359 (6371). doi:* https://doi.org/10.1126/science.aam7240*; Srokosz, M.A., and H.L. Bryden. 2015. "Observing the Atlantic Meridional Overturning Circulation Yields a Decade of Inevitable Surprises." Science 348 (6241): 1255575; Christensen, V., M. Coll, C. Piroddi, J. Steenbeek, J. Buszowski and D. Pauly. 2014. "A Century of Fish Biomass Decline in the Ocean." Marine Ecology Progress Series 512 (October): 155–66. doi:* https://doi.org/10.3354/meps10946)

Stratification Ocean stratification occurs naturally when waters with different properties (temperature, salinity, density) form layers, which act as a barrier to mixing.[175] Usually, wind, currents and storms help mix the cold (deep) and warm (upper) layers.[176] Climate change disturbs this dynamic: rising surface temperatures exacerbate the layering and decrease the rate of mixing. This, in turn, decreases the amount of nutrients travelling up to surface waters, which further affects biological productivity, heat redistribution, carbon uptake and oxygen production. The data show that upper ocean stratification will be greater everywhere during the second half of the twenty-first century, indicating a more pronounced decoupling between the surface and the deeper ocean.[177] The areas most affected include the Arctic, the tropics, the North Atlantic and the northeast Pacific.[178]

[175]Inniss et al. 2016. "The First Global Integrated Marine Assessment."

[176]Capotondi, A., M.A. Alexander, N.A. Bond, E.N. Curchitser and J.D. Scott. 2012. "Enhanced Upper Ocean Stratification with Climate Change in the CMIP3 Models." *Journal of Geophysical Research: Oceans* 117 (C4). doi: https://doi.org/10.1029/2011JC007409.

[177]Capotondi et al. 2012. "Enhanced Upper Ocean Stratification with Climate Change in the CMIP3 Models."

[178]Capotondi et al. 2012. "Enhanced Upper Ocean Stratification with Climate Change in the CMIP3 Models."

Deoxygenation In the open ocean, deoxygenation is primarily caused by global warming: oxygen solubility decreases with increasing temperature, and less oxygen reaches the deep ocean layers because of stratification. In the past 50 years, the ocean's oxygen content has decreased by 2%,[179] and ocean models predict a further decline of up to 7% by 2100.[180] Oxygen-minimum zones in the open ocean have expanded by several million square kilometres.[181]

In estuaries and other coastal systems strongly influenced by their watershed, oxygen declines can be linked to agriculture, sewage and the combustion of fossil fuels, which generate increased loadings of nutrients (particularly nitrogen and phosphorus) and organic matter.[182] An influx of nutrients causes an increase in microscopic algae, which ultimately die and decay, and the resulting decomposition process consumes oxygen, leading to oxygen depletion in the surrounding water. The Baltic Sea is a prime example of low-oxygen conditions caused by high nutrient loads from land runoff.[183] Oxygen decline in coastal systems is exacerbated by climate change (as in the open ocean) and by increasing nutrient delivery originating from increased precipitation.[184]

Overall, around 700 sites worldwide are now affected by low-oxygen conditions—up from only 45 in the 1960s.[185] Deoxygenation can have far-reaching biological consequences. Larger fish species with high metabolic rates, including yellowfin tuna and swordfish, are especially vulnerable to deoxygenation, and there is evidence that the balance of marine life is starting to shift in favour of species that are more tolerant of low-oxygen conditions, such as microbes, jellyfish and some squid.[186] Low-oxygen conditions can also make animals more susceptible to pathogens and parasites, increasing morbidity and mortality from diseases.

Reduced biomass and biodiversity, and redistribution of species Physical changes and overfishing have profound second-order consequences for the biological ocean. The IPBES estimates that 'more than 40% of amphibian species, almost a third of reef-forming corals, sharks and shark relatives, and over a third of marine mammals are currently threatened with extinction'.[187] Overfishing disproportionately removes predators, which are replaced by shorter-lived and smaller species, and the food chain becomes much simpler, less dynamic and less resilient.[188] Predatory fish biomass today is about one-third of 1920 levels.[189] Warming and deoxygenation are predicted to cause a large-scale redistribution of global fish and invertebrate biomass by 2055, with a 30–70% increase in high-latitude regions and a drop of up to 40% in the tropics.[190] Loss of biodiversity leads to measurable decreases in ecosystem functionality, including the number of viable fisheries (non-collapsed), the provision of nursery habitats, as well as the filtering and detoxification services essential for water quality and the reduction of harmful algal blooms, fish kills and beach closures.[191]

Sea level rise Sea level rise results from a combination of thermal expansion caused by the warming of the ocean (since water expands as it warms) and increased melting of glaciers and ice sheets.[192] A range of positive feedback mechanisms makes predictions exceedingly complex. For example, the melting of glaciers accelerates their rate of flow into a warming sea. It has been assessed that the global average sea level

[179] Schmidtko, S., L. Stramma and M. Visbeck. 2017. "Decline in Global Oceanic Oxygen Content during the Past Five Decades." *Nature* 542 (7641): 335–39. doi: https://doi.org/10.1038/nature21399.

[180] Laffoley, D., and J.M. Baxter, eds. 2019. *Ocean Deoxygenation: Everyone's Problem—Causes, Impacts, Consequences and Solutions.* International Union for Conservation of Nature. doi: https://doi.org/10.2305/IUCN.CH.2019.13.en; Long, M.C., C. Deutsch and T. Ito. 2016. "Finding Forced Trends in Oceanic Oxygen." *Global Biogeochemical Cycles* 30 (2): 381–97. doi: https://doi.org/10.1002/2015GB005310; Keeling, R.F., A. Körtzinger and N. Gruber. 2010. "Ocean Deoxygenation in a Warming World." *Annual Review of Marine Science* 2 (1): 199–229. doi: https://doi.org/10.1146/annurev.marine.010908.163855.

[181] Breitburg, D., L.A. Levin, A. Oschlies, M. Grégoire, F.P. Chavez, D.J. Conley, V. Garçon et al. 2018. "Declining Oxygen in the Global Ocean and Coastal Waters." *Science* 359 (6371). doi: https://doi.org/10.1126/science.aam7240.

[182] Breitburg et al. 2018. "Declining Oxygen in the Global Ocean and Coastal Waters."

[183] Keeling et al. 2010. "Ocean Deoxygenation in a Warming World."

[184] Breitburg et al. 2018. "Declining Oxygen in the Global Ocean and Coastal Waters."

[185] Laffoley and Baxter. 2019. *Ocean Deoxygenation.*

[186] Laffoley and Baxter. 2019. *Ocean Deoxygenation.*

[187] Díaz et al. 2019. "Summary for Policymakers of the Global Assessment Report on Biodiversity and Ecosystem Services of the Intergovernmental Science-Policy Platform on Biodiversity and Ecosystem Services."

[188] Maureaud, A., D. Gascuel, M. Colléter, M.L.D. Palomares, H. Du Pontavice, D. Pauly and W.W.L. Cheung. 2017. "Global Change in the Trophic Functioning of Marine Food Webs." *PLOS ONE* 12 (8). doi: https://doi.org/10.1371/journal.pone.0182826.

[189] Christensen et al. 2014. "A Century of Fish Biomass Decline in the Ocean."

[190] Gaines et al. 2019. "The Expected Impacts of Climate Change on the Ocean Economy."

[191] Worm, B., E.B. Barbier, N. Beaumont, J.E. Duffy, C. Folke, B.S. Halpern, J.B.C. Jackson et al. 2006. "Impacts of Biodiversity Loss on Ocean Ecosystem Services." *Science* 314 (5800): 787–90. doi: https://doi.org/10.1126/science.1132294.

[192] Lindsey, R. 2019. "Climate Change: Global Sea Level." National Oceanic and Atmostpheric Administration, 14 August. https://www.climate.gov/news-features/understanding-climate/climate-change-global-sea-level.

has risen by about 16–21 cm since 1900,[193] at an accelerating rate over the past two decades.[194] The future extent and level of potential damage from sea level rise is therefore the subject of intense research and debate. The IPCC frames the range of outcomes between the empirical record of similar events in the distant past, and much more cautious simulations from process-based computer models: 'Paleo sea level records from warm periods during the last 3 million years indicate that global mean sea level has exceeded 5 m above present (very high confidence) when global mean temperature was up to 2 °C warmer than pre-industrial (medium confidence)'.[195] Perhaps more relevant to climate policies than the slow rise over centuries to millennia is the risk of rapid melting of Antarctic or Greenland ice that could lead to sea level rise of several metres over a span of decades. The risk of such catastrophic events is notoriously difficult to evaluate based on observational records.

Phenomena such as deoxygenation and reduction of biomass and biodiversity are highly synergistic—one propels the other. It is not analytically feasible to predict precisely *when* and *where* these complex chains of events will occur. However, new 'big simulation' tools allow us to describe *what* might happen in any given ocean region.[196] Typically, these simulations show that while a single source of stress (e.g. overfishing, pollution) can do considerable damage, multiple and compounding sources can do worse by orders of magnitude.[197] Put simply, ocean risk is a function of how bad the stressors are, the degree to which they reinforce each other and the natural variability of the ocean they are affecting.

The 2019 IPCC report *The Ocean and Cryosphere in a Changing Climate* estimates that climate-induced declines in ocean health will cost the global economy US $428 billion/year by 2050 and $1.98 trillion/year by 2100 (Fig. 20.20).[198] These numbers encompass costs associated with declines in ocean health and services due to climate-change, overfishing, excessive nutrient loads and plastic pollution.

Of course, the synergy story has an upside as well. If each new layer of stress increases overall risk disproportionately, then the opposite is also true: for each layer taken away, the system becomes more resilient. This makes it possible to buy valuable time when dealing with long-term issues such as warming or acidifying waters.

4.3.2 The Decline of Ocean Health Is Threatening Most Ocean Sectors

Insufficient action to reform the ocean economy and protect and restore ocean health can negatively impact ocean sectors that depend on a healthy, productive and resilient ocean or are directly exposed to its physical manifestations (e.g. sea level rise, waves, extreme events).

Wild-catch fisheries Climate change will impact wild-catch fisheries in terms of both stock productivity (i.e. its potential sustainable yield) and distribution (i.e. its physical range). The IPBES states that 'climate change alone is projected to decrease ocean net primary production by 3–10%, and fish biomass by 3–25%' by 2100, depending on climate scenarios.[199] These global numbers mask even more significant variation in changes across stocks and regions. Poleward regions such as the North Atlantic and North Pacific are predicted to see a 30–70% increase in catch potential, while equatorial regions face a 40% decrease.[200]

Where stocks decrease or move away from traditional fishing grounds, fishers must spend more resources to locate and catch them.[201] Conversely, any shifts to shallower water may make stocks easier for local fishers to catch but more vulnerable to overfishing. Overall, smaller-scale fisheries which rely on vessels with limited range and low technological capabilities are likely to be most vulnerable to shifts in

[193] Sweet, W.V., R. Horton, R.E. Kopp, A.N. LeGrande and A. Romanou. 2017. "Sea Level Rise." In *Climate Science Special Report: Fourth National Climate Assessment*, vol. 1, edited by D.J. Wuebbles, D.W. Fahey, K.A. Hibbard, D.J. Dokken, B.C. Stewart and T.K. Maycock, 333–63. Washington, DC: U.S. Global Change Research Program. doi: https://doi.org/10.7930/J0VM49F2.

[194] Cazenave, A., B. Meyssignac and H. Palanisamy. 2018. "Global Sea Level Budget Assessment by World Climate Research Programme." Sea Scientific Data Open Edition (SEANOE). doi: https://doi.org/10.17882/54854.

[195] Church, J.A., P.U. Clark, A. Cazenave, J.M. Gregory, S. Jevrejeva, A. Levermann, M.A. Merrifield et al. 2013. "Sea Level Change." In *Climate Change 2013: The Physical Science Basis. Contribution of Working Group I to the Fifth Assessment Report of the Intergovernmental Panel on Climate Change*, edited by T.F. Stocker, D. Qin, G.-K. Plattner, M. Tignor, S.K. Allen, J. Boschung, A. Nauels et al. Cambridge: Cambridge University Press. https://www.ipcc.ch/site/assets/uploads/2018/02/WG1AR5_Chapter13_FINAL.pdf.

[196] Bailey, R.M., and J.M.A. van der Grient. 2020. "OSIRIS: A Model for Integrating the Effects of Multiple Stressors on Marine Ecosystems." *Journal of Theoretical Biology* 493 (May): 110211. doi: https://doi.org/10.1016/j.jtbi.2020.110211. These models look at the ocean as a network of linked basic states (such as the populations of whales, or zooplankton, or temperature) and use large computer simulations to assess the impact of specific stressors on this network.

[197] Bailey and van der Grient. 2020. "OSIRIS."

[198] Pörtner et al. 2019. "Summary for Policymakers." In *Special Report on the Ocean and Cryosphere in a Changing Climate*.

[199] Díaz et al. 2019. "Summary for Policymakers of the Global Assessment Report on Biodiversity and Ecosystem Services of the Intergovernmental Science-Policy Platform on Biodiversity and Ecosystem Services."

[200] Gaines et al. 2019. "The Expected Impacts of Climate Change on the Ocean Economy."

[201] Laffoley and Baxter. 2019. *Ocean Deoxygenation*.

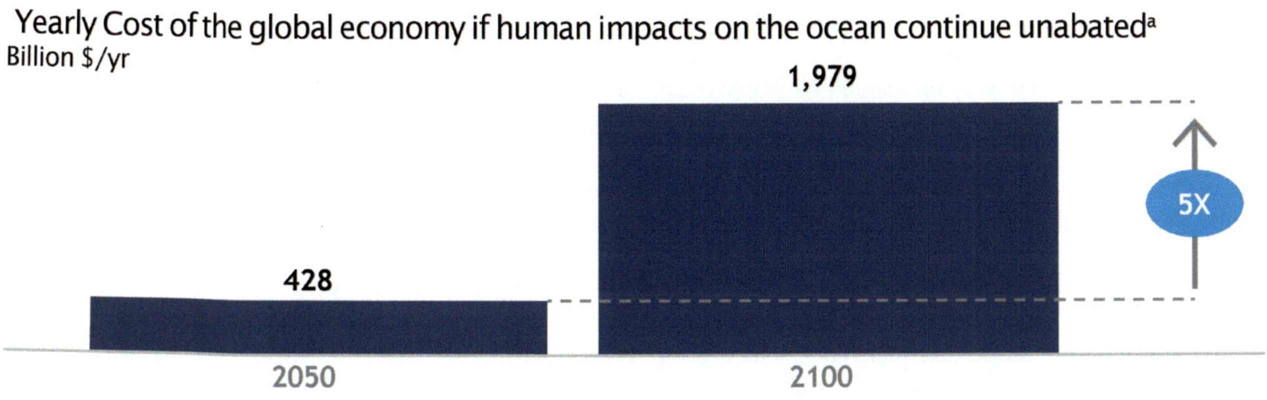

Fig. 20.20 The cost of inaction on the global economy. *Note: Cost associated with declines in ocean health and services due to climate change, overfishing, excessive nutrient loads and plastic pollution. (Source: Pörtner, H.O., D.C. Roberts, V. Masson-Delmotte, P. Zhai, M. Tignor, K. Poloczanska, K. Mintenbeck et al., eds. 2019. IPCC Special Report on the Ocean and Cryosphere in a Changing Climate.* https://report.ipcc.ch/srocc/pdf/SROCC_FinalDraft_FullReport.pdf)

range or migratory patterns.[202] The equity implications of longer travel and/or declining yields are pronounced, especially since artisanal fisheries provide the protein of last resource in many developing countries' coastal areas.

Regulatory constraints may also hinder fishers' ability to adapt, particularly if species move across management boundaries.[203] Depending on the chosen climate scenario, between 28% and 72% of current global fishery yields will shift across country boundaries by 2100.[204] In addition to economic impacts, the redistribution of catch potential is likely to increase the risk of conflicts among fisheries, authorities and communities. In the absence of a coordinated response, the compounding effects of overfishing and stock (range) shift could severely threaten future global fishery yields and profits.[205]

With these conditions expected to change and levels of uncertainty to grow, more adaptive management of fisheries will be critical to a better future (e.g. through rights-based fishery or secure-access systems).[206]

- **Mariculture**. The overall potential of mariculture is likely to remain high under climate change. With careful planning, improvements in feed technology and the implementation of best practices for preventing or reducing negative impacts on ecosystems and communities, mariculture could offset long-term losses in food and income from capture fisheries in most countries that will experience losses in that sector. However, a study found that a severe climate scenario would create both gains and losses in the studied 180 cultured finfish and bivalve species.[207] Lower trophic species such as bivalves were affected disproportionately due to the compounding effects of shifts in temperature, chlorophyll and ocean acidification.

The Indo-Pacific region—China, India, Bangladesh and Indonesia—is particularly impacted; finfish mariculture

[202] Laffoley and Baxter. 2019. *Ocean Deoxygenation*.

[203] Pinsky, M.L., G. Reygondeau, R. Caddell, J. Palacios-Abrantes, J. Spijkers and W.W.L. Cheung. 2018. "Preparing Ocean Governance for Species on the Move." *Science* 360 (6394): 1189–91. doi: https://doi.org/10.1126/science.aat2360.

[204] Gaines, S.D., C. Costello, B. Owashi, T. Mangin, J. Bone, J.G. Molinos, M. Burden et al. 2018. "Improved Fisheries Management Could Offset Many Negative Effects of Climate Change." *Science Advances* 4 (8): eaao1378. doi: https://doi.org/10.1126/sciadv.aao1378.

[205] Garrett, A., and J. Pinnegar. 2019. "Climate Change Adaptation in the UK (Wild Capture) Seafood Industry 2018." Seafish/Marine Climate Change Impacts Partnership. https://seafish.org/media/Publications/Climate_change_adaptation_in_the_UK_wild_capture_seafood_industry_2018.pdf.

[206] Lubchenco, J., E.B. Cerny-Chipman, J.N. Reimer and S.A. Levin. 2016. "The Right Incentives Enable Ocean Sustainability Successes and Provide Hope for the Future." *Proceedings of the National Academy of Sciences* 113 (51): 14507–14. doi: https://doi.org/10.1073/pnas.1604982113.

[207] Froehlich, H.E., R.R. Gentry and B.S. Halpern. 2018. "Global Change in Marine Aquaculture Production Potential under Climate Change." *Nature Ecology & Evolution* 2 (11): 1745–50. doi: https://doi.org/10.1038/s41559-018-0669-1.

could decline by as much as 30% in some areas, and the risks to bivalve farmers are even higher.[208] Many coastal residents in these regions depend heavily on seafood for employment and food security.

Anthropogenic pollution is already having an effect on mariculture operations. Apart from farmed species requiring pristine water conditions for optimal growth, the accumulation of anthropogenic pollutants, especially microplastic in farmed (and wild) species, is a significant concern.[209] This is especially true of non-fed mariculture species like bivalves,[210] who feed by filtering suspended material—including the accidental microplastics—out of the water column.

Tourism Sea level rise will erode and submerge tourism infrastructure and beaches, with many resorts sitting at less than 1 m above the high-water mark.[211] In the Caribbean, a sea level rise of 1 m is projected to endanger 49–60% of tourist resorts, damage or cause the loss of 21 airports and cause severe flooding of 35 ports.[212] In 2050, according to one estimate, rebuilding tourist resorts alone will cost the region US $10 billion to $23.3 billion.[213] In Venice, higher water levels are threatening building integrity, eroding the lagoon and subjecting the city to more than twice as many floods since 1960.

The tourism industry also will be hit by the loss of coral reefs. Coral reefs contribute $11.5 billion annually to global tourism revenues, benefitting more than 100 countries.[214] Coral reefs would be virtually all lost at 2 °C warming,[215] with serious impacts for tourism in Australia and small island developing states (SIDS) in the Caribbean and elsewhere.[216]

Tourism is both a source and a victim of pollution.[217] Beach closures due to sewage pollution affect countries worldwide. Other direct forms of pollution impacting tourism include plastic waste on beaches, making them undesirable for tourists to visit. Indirect impacts of anthropogenic pollution on tourism also exist: the combined effect of elevated sea surface temperatures, excess fertiliser and increased nutrient runoff due to deforestation are potential causes of the explosive growth of sargassum seaweed[218] that is washing up on tourism beaches in the Caribbean, the Gulf of Mexico and West Africa, driving down hotel bookings in certain areas.[219]

On top of these worrying trends, the COVID-19 pandemic is having severe impacts on coastal tourism, for example. SIDS are expected to experience a 7.3% fall in gross domestic product (GDP) given their tourism dependency, and this drop could be up to 16% in highly tourism-dependent SIDS like Seychelles or the Maldives.[220]

Ports and supply chains Severe disruptions due to extreme weather events can be expected. A 2013 study[221] finds that the supply chain consequences of compounding sea level rise, higher storm surges, increased cyclone intensity and

[208]Froehlich et al. 2018. "Global Change in Marine Aquaculture Production Potential under Climate Change."

[209]Rochman, C.M., A. Tahir, S.L. Williams, D.V. Baxa, R. Lam, J.T. Miller, F.-C. The et al. 2015. "Anthropogenic Debris in Seafood: Plastic Debris and Fibers from Textiles in Fish and Bivalves Sold for Human Consumption." *Scientific Reports* 5 (1): 1–10. doi: https://doi.org/10.1038/srep14340.

[210]Phuong, N.N., L. Poirier, Q.T. Pham, F. Lagarde and A. Zalouk-Vergnoux. 2018. "Factors Influencing the Microplastic Contamination of Bivalves from the French Atlantic Coast: Location, Season and/or Mode of Life?" *Marine Pollution Bulletin* 129 (2): 664–74. doi: https://doi.org/10.1016/j.marpolbul.2017.10.054; van Cauwenberghe, L., and C.R. Janssen. 2014. "Microplastics in Bivalves Cultured for Human Consumption." *Environmental Pollution* 193 (October): 65–70. doi: https://doi.org/10.1016/j.envpol.2014.06.010; Li, J., D. Yang, L. Li, K. Jabeen and H. Shi. 2015. "Microplastics in Commercial Bivalves from China." *Environmental Pollution* 207 (December): 190–95. doi: https://doi.org/10.1016/j.envpol.2015.09.018.

[211]Nicholls, M. 2014. "Climate Change: Implications for Tourism: Key Findings from the Intergovernmental Panel on Climate Change Fifth Assessment Report." University of Cambridge. https://www.cisl.cam.ac.uk/business-action/low-carbon-transformation/ipcc-climate-science-business-briefings/pdfs/briefings/ipcc-ar5-implications-for-tourism-briefing-prin.pdf.

[212]Pachauri, R.K., L. Mayer and Intergovernmental Panel on Climate Change, eds. 2015. *Contribution of Working Groups I, II and III to the Fifth Assessment Report of the Intergovernmental Panel on Climate Change 2014: Synthesis Report.* Geneva: Intergovernmental Panel on Climate Change. https://ar5-syr.ipcc.ch/ipcc/ipcc/resources/pdf/IPCC_SynthesisReport.pdf.

[213]Nicholls. 2014. "Climate Change."

[214]Nicholls. 2014. "Climate Change."

[215]Masson-Delmotte et al. 2019. *Global Warming of 1.5* °C.

[216]Nicholls. 2014. "Climate Change."

[217]Diez, S.M., P.G. Patil, J. Morton, D.J. Rodriguez, A. Vanzella, D.V. Robin, T. Maes and C. Corbin. 2019. "Marine Pollution in the Caribbean: Not a Minute to Waste." Washington, DC: World Bank Group. http://documents.worldbank.org/curated/en/482391554225185720/pdf/Marine-Pollution-in-the-Caribbean-Not-a-Minute-to-Waste.pdf.

[218]Djakouré, S., M. Araujo, A. Hounsou-Gbo, C. Noriega and B. Bourlès. 2017. "On the Potential Causes of the Recent Pelagic Sargassum Blooms Events in the Tropical North Atlantic Ocean." *Biogeosciences Discussions*, September, 1–20. doi: https://doi.org/10.5194/bg-2017-346.

[219]Agren, D. 2019. "Seaweed Invasion Threatens Tourism in Mexico's Beaches as Problem Worsens." *The Guardian*, 28 June. http://www.theguardian.com/world/2019/jun/28/mexico-seaweed-invasion-tourism-caribbean-beaches.

[220]Coke-Hamilton, P. 2020. "Impact of COVID-19 on Tourism in Small Island Developing States." UN Conference on Trade and Development. 24 April. https://unctad.org/en/pages/newsdetails.aspx?OriginalVersionID=2341.

[221]Becker, A.H., M. Acciaro, R. Asariotis, E. Cabrera, L. Cretegny, P. Crist, M. Esteban et al. 2013. "A Note on Climate Change Adaptation for Seaports: A Challenge for Global Ports, a Challenge for Global Society." *Climatic Change* 120 (4): 683–95. doi: https://doi.org/10.1007/s10584-013-0843-z.

destructiveness,[222] wave regimes[223] and river floods of ports,[224] coastal refineries and chemical plants could cause operational delays at a scale of billions of U.S. dollars per day,[225] with incalculable effects on business cycles, supply chains and the overall operating risk of companies.

4.3.3 The Risk to Coastal Communities Is Increasing

Major and irreversible shifts in ocean functionality threaten coastal communities and habitats in many ways—the current ocean economy system is far from delivering prosperity for all. Further, the effects of these shifts will be disproportionately felt by vulnerable, historically underrepresented and underserved communities in both developed and developing countries.

Flood damage New research has demonstrated that extreme coastal inundation events are increasing, and in some regions increased chronic flooding has been observed.[226] Many small islands already face large, sometimes existential, flood damage, and damage from sea level rise could equal several percentage points of GDP in 2100.[227] Risk associated with floods and hurricanes are accentuated for the 100–300 million people living within coastal 100-year flood zones, as the loss of coastal habitats and coral reefs reduces natural coastal protection.[228] In Europe, annual damage from coastal floods is expected to rise from €1.25 billion today to €93–960 billion by the end of the century.[229] Without drastic changes towards climate-smart coastal development, major disruptions can be expected in addition to damages to coastal communities.

Risks to agriculture Sea level rise will affect agriculture through land submergence, the salinisation of soil and fresh groundwater, and land loss due to permanent coastal erosion.[230] Countries heavily dependent on coastal agriculture, such as Bangladesh, are likely to experience reduced production and livelihood diversity, as well as greater food insecurity (Fig. 20.21).[231]

Permanently displaced coastal populations Rising sea level will be experienced not only as a long-term, gradual event but also as a series of extreme events caused by the compounding effects of spring tides, stronger and slower-moving hurricane surges, spring floods and land loss. Based on a scenario without effective climate change mitigation policies,[232] a 1 m rise in sea level would entail dramatic increases in the frequency of 100-year extreme weather events in cities such as Shanghai, New York and Kolkata (Fig. 20.21). Some cities will have the means to adapt with major feats of engineering, but other areas will become unliveable, generating waves of displaced people in the context of disasters and climate change. Indeed, 88 million to 1.4 billion people are estimated to be at risk of displacement.[233] In the United States, 3 ft (~0.91 m) of sea level rise

[222] Becker et al. 2013. "A Note on Climate Change Adaptation for Seaports."

[223] Field, C.B., V. Barros, T.F. Stocker, D. Qin, D.J. Dokken, K.L. Ebi, M.D. Mastrandrea et al. 2012. *Managing the Risks of Extreme Events and Disasters to Advance Climate Change Adaptation: A Special Report of Working Groups I and II of the Intergovernmental Panel on Climate Change*. Cambridge: Cambridge University Press. https://www.ipcc.ch/site/assets/uploads/2018/03/SREX_Full_Report-1.pdf.

[224] Tebaldi, C., B.H. Strauss and C.E. Zervas. 2012. "Modelling Sea Level Rise Impacts on Storm Surges along US Coasts." *Environmental Research Letters* 7 (1): 014032. doi: https://doi.org/10.1088/1748-9326/7/1/014032.

[225] Becker et al. 2013. "A Note on Climate Change Adaptation for Seaports."

[226] Strauss, B.H., R.E. Kopp, W.V. Sweet and K. Bittermann. 2016. "Unnatural Coastal Floods: Sea Level Rise and the Human Fingerprint on U.S. Floods since 1950." Princeton, NJ: Climate Central. https://sealevel.climatecentral.org/uploads/research/Unnatural-Coastal-Floods-2016.pdf.

[227] Wong, P.P., I.J. Losada, J.P. Gattuso, J. Hinkel, A. Khattabi, M.L. McInnes, Y. Saito and A. Sallenger. n.d. "2014: Coastal Systems and Low-Lying Areas." *Climate Change 2014: Impacts, Adaptation, and Vulnerability*. Part A, *Global and Sectoral Aspects: Contribution of Working Group II to the Fifth Assessment Report of the Intergovernmental Panel on Climate Change*, edited by C.B. Field, V.R. Barros, D.J. Dokken, K.J. Mach, M.D. Mastrandrea, T.E. Bilir, M. Chatterjee et al. Cambridge: Cambridge University Press.

[228] Díaz et al. 2019. "Summary for Policymakers of the Global Assessment Report on Biodiversity and Ecosystem Services of the Intergovernmental Science-Policy Platform on Biodiversity and Ecosystem Services."

[229] Vousdoukas, M.I., L. Mentaschi, E. Voukouvalas, A. Bianchi, F. Dottori and L. Feyen. 2018. "Climatic and Socioeconomic Controls of Future Coastal Flood Risk in Europe." *Nature Climate Change* 8 (9): 776–80. doi: https://doi.org/10.1038/s41558-018-0260-4.

[230] Pörtner, H.O., D.C. Roberts, V. Masson-Delmotte, P. Zhai, M. Tignor, K. Poloczanska, K. Mintenbeck et al., eds. 2019. *IPCC Special Report on the Ocean and Cryosphere in a Changing Climate*. https://report.ipcc.ch/srocc/pdf/SROCC_FinalDraft_FullReport.pdf.

[231] Khanom, T. 2016. "Effect of Salinity on Food Security in the Context of Interior Coast of Bangladesh." *Ocean & Coastal Management* 130 (October): 205–12. doi: https://doi.org/10.1016/j.ocecoaman.2016.06.013.

[232] Riahi, K., S. Rao, V. Krey, C. Cho, V. Chirkov, G. Fischer, G. Kindermann et al. 2011. "RCP 8.5: A Scenario of Comparatively High Greenhouse Gas Emissions." *Climatic Change* 109 (1): 33. doi: https://doi.org/10.1007/s10584-011-0149-y.

[233] Hauer, M.E., E. Fussell, V. Mueller, M. Burkett, M. Call, K. Abel, R. McLeman and D. Wrathall. 2020. "Sea-Level Rise and Human Migration." *Nature Reviews Earth & Environment* 1 (1): 28–39. doi: https://doi.org/10.1038/s43017-019-0002-9.

Fig. 20.21 Impacts from compounding effects of climate change. (*Sources: King, D., Z. Dadi, Q. Ya and A. Ghosh. 2015. Climate Change: A Risk Assessment. National Library of Medicine, National Institute of Health; Dasgupta, S., M.M. Hossein, M. Huq and D. Wheeler. "Climate Change and Soil Salinity: The Case of Coastal Bangladesh." Ambio 44: 815–26. doi:* https://doi.org/10.1007/s13280-015-0681-5. *Images: Left: GenadijsZ/Shutterstock; Right: FotoKina/Shutterstock*)

by 2100 threatens four million people.[234] The situation is particularly dire for SIDS, for whom raising seas can become an existential threat.[235]

4.3.4 Ocean Activities Are Currently Not Delivering on the Social Sustainable Development Goals

If not properly regulated and managed, a growing ocean economy can lead to even greater economic inequality than already exists.[236] Benefits will continue to be captured by an elite and strong incumbents, whilst vulnerable and marginalised groups become even more exposed to economic, social and cultural impacts and displacements.[237] In this scenario, the ocean economy could have a net negative effect on progress towards UN Sustainable Development Goals (SDGs) such as no poverty, zero hunger, good health and well-being, and reducing inequalities (see Sect. 5.4 for a detailed assessment of the link between SDG 14 (life below water) and the other SDGs).

Increasing inequalities Global inequity is increasingly acknowledged as a substantial challenge to the ocean economy. Inequities are contrary to and will undermine progress towards the Sustainable Development Goals as they have contributed to a deteriorating ocean environment, with negative effects on human well-being primarily borne by the most vulnerable. Climate change risks aggravate existing inequity. The vulnerable and marginalised will be disproportionately affected by the effects of climate change. The lack of alternatives and high dependence on fish stocks for nutrition and income disproportionally expose the coastal poor to the effects of climate change.[238] Growing demand for fish feed can also exacerbate inequalities by diverting small pelagic fish like pilchards from domestic consumption for food, where such fish are a key component of the diet for many communities.[239] In addition, poor communities have

[234] Hauer, M.E., J.M. Evans and D.R. Mishra. 2016. "Millions Projected to Be at Risk from Sea-Level Rise in the Continental United States." *Nature Climate Change* 6 (7): 691–95. doi: https://doi.org/10.1038/nclimate2961.

[235] OECD. 2018. *Making Development Co-operation Work for Small Island Developing States*. https://www.oecd-ilibrary.org/content/publication/9789264287648-en; Ourbak, T., and A.K. Magnan. 2018. "The Paris Agreement and Climate Change Negotiations: Small Islands, Big Players." *Regional Environmental Change* 18(8): 2201–7. doi: https://doi.org/10.1007/s10113-017-1247-9; Thomas, A., and L. Benjamin. 2018. "Policies and Mechanisms to Address Climate-Induced Migration and Displacement in Pacific and Caribbean Small Island Developing States." *International Journal of Climate Change Strategies and Management* 10 (1): 86–104. doi: https://doi.org/10.1108/IJCCSM-03-2017-0055.

[236] Allison et al. 2020. "The Human Relationship with Our Ocean Planet."

[237] Bennett, N.J., A.M. Cisneros-Montemayor, J. Blythe, J.J. Silver, G. Singh, N. Andrews, A. Calò et al. 2019. "Towards a Sustainable and Equitable Blue Economy." *Nature Sustainability* 2 (11): 991–93. doi: https://doi.org/10.1038/s41893-019-0404-1.

[238] Österblom, H., C.C.C. Wabnitz and D. Tladi. 2020. "Towards Ocean Equity." Washington, DC: World Resources Institute. https://www.oceanpanel.org/sites/default/files/2020-04/towards-ocean-equity.pdf; Taylor, S.F.W., M.J. Roberts, B. Milligan and R. Ncwadi. 2019. "Measurement and Implications of Marine Food Security in the Western Indian Ocean: An Impending Crisis?" *Food Security* 11 (6): 1395–415. doi: https://doi.org/10.1007/s12571-019-00971-6.

[239] Tacon, A.G.J., and M. Metian. 2009. "Fishing for Feed or Fishing for Food: Increasing Global Competition for Small Pelagic Forage Fish." *Ambio* 38 (6): 294–302. doi: https://doi.org/10.2307/40390239.

fewer resources to respond to shifting fish stocks by changing gear types or travelling further to fish, as well as fewer resources to shift livelihoods altogether. Gender inequality is pervasive in many ocean industries overall, and climate change could be especially devastating for the most marginalised coastal women.[240] The international community's global ambition to 'leave no one behind' can only be realised through inclusive governance and the fair distribution of ocean benefits. An increased focus on equity will be instrumental for the legitimacy, effectiveness and sustainability of the ocean economy.

Food insecurity and malnutrition Projected changes in fish distribution and abundance will put income, livelihoods, nutritional health and food security at risk in communities that rely on marine resources, such as those in the Arctic, West Africa and small island developing states.[241] Globally, climate change puts up to three billion people at risk of food and economic insecurity.[242] Food security and human health are also threatened by harmful algal blooms, with communities in areas without sustained monitoring programs and dedicated early warning systems most vulnerable to these risks.[243] Cultural diet changes in certain parts of the world, particularly Pacific island nations, are shifting diets away from healthy, local reef seafood towards imported, often highly processed, high sugar and fat foods. The results are rising malnutrition and increasing prevalence of non-communicable diseases.[244]

Job safety and security The isolation of ships at sea and the liability protection of vessel owners afforded by current flag state regulations serve to conceal human rights abuses, whilst labour protections are poorly enforced in many countries. Informal or unregulated economies and fishing activities, such as shellfish gathering or fish processing, face significant exposure to unregulated exploitation and disproportionally employ women[245] and marginalised groups.[246] Unreported catches and illegal activities can mask labour trafficking, peonage systems, unsustainable resource use and health and sanitary issues whilst simultaneously avoiding taxation and detracting from wider economic benefits.[247]

BAU ocean industries development is likely to cause and exacerbate inequities across the spectrum of ocean sectors, and people with vulnerable marine livelihoods (who are more likely to be women, ethnic and racial minorities, migrants, youth and Indigenous People) are likely to be disproportionately affected. A new paradigm urgently needs to be embraced.[248]

Human rights Organised crime and human rights violations are a known plague within the ocean economy, especially the fisheries sector. Apart from the human impact, these crimes continue to have negative impacts on the environment as well as the global economy. The crimes can include, among others, tax crimes, money laundering, labour offences, drug trafficking and migrant smuggling. Many of these crimes can be associated with or facilitated by illegal, unreported and unregulated (IUU) fishing,[249] which is estimated to account for 20% of the world's catch (up to 50% in

[240] Garai, J. 2016. "Gender Specific Vulnerability in Climate Change and Possible Sustainable Livelihoods of Coastal People. A Case from Bangladesh." *Revista de Gestão Costeira Integrada* 16 (1): 79–88. doi: https://doi.org/10.5894/rgci656; Akinsemolu, A.A., and O.A.P. Olukoya. 2020. "The Vulnerability of Women to Climate Change in Coastal Regions of Nigeria: A Case of the Ilaje Community in Ondo State." *Journal of Cleaner Production* 246 (February): 119015. doi: https://doi.org/10.1016/j.jclepro.2019.119015.

[241] Pörtner et al. 2019. "Summary for Policymakers." In *Special Report on the Ocean and Cryosphere in a Changing Climate.*

[242] Holsman, K.K., E.L. Hazen, A. Haynie, S. Gourguet, A. Hollowed, S.J. Bograd, J.F. Samhouri and K. Aydin. 2019. "Towards Climate Resiliency in Fisheries Management." *ICES Journal of Marine Science* 76 (5): 1368–78. doi: https://doi.org/10.1093/icesjms/fsz031.

[243] Pörtner et al. 2019. "Summary for Policymakers." In *Special Report on the Ocean and Cryosphere in a Changing Climate.*

[244] Charlton, K.E., J. Russell, E. Gorman, Q. Hanich, A. Delisle, B. Campbell and J. Bell. 2016. "Fish, Food Security and Health in Pacific Island Countries and Territories: A Systematic Literature Review." *BMC Public Health* 16 (1): 285. doi: https://doi.org/10.1186/s12889-016-2953-9.

[245] Harper, S., C. Grubb, M. Stiles and U.R. Sumaila. 2017. "Contributions by Women to Fisheries Economies: Insights from Five Maritime Countries." *Coastal Management* 45 (2): 91–106. doi: https://doi.org/10.1080/08920753.2017.1278143.

[246] Barange, M., T. Bahri, M.C.M. Beveridge, K.L. Cochrane, S. Funge-Smith and F. Poulain. 2018. "Impacts of Climate Change on Fisheries and Aquaculture: Synthesis of Current Knowledge, Adaptation and Mitigation Options." FAO Fisheries and Aquaculture Technical Paper no. 627. https://agris.fao.org/agris-search/search.do?recordID=XF2018002008.

[247] Lopes, P.F.M., L. Mendes, V. Fonseca and S. Villasante. 2017. "Tourism as a Driver of Conflicts and Changes in Fisheries Value Chains in Marine Protected Areas." *Journal of Environmental Management* 200 (September): 123–34. doi: https://doi.org/10.1016/j.jenvman.2017.05.080; Moreto, W.D., R.W. Charlton, S.E. DeWitt and C.M. Burton. 2019. "The Convergence of CAPTURED Fish and People: Examining the Symbiotic Nature of Labor Trafficking and Illegal, Unreported and Unregulated Fishing." *Deviant Behavior* 41(6): 1–17. doi: https://doi.org/10.1080/01639625.2019.1594587.

[248] Allison et al. 2020. "The Human Relationship with Our Ocean Planet."

[249] Witbooi, E., K.-D. Ali, M.A. Santosa, G. Hurley, Y. Husein, S. Maharaj, I. Okafor-Yarwood et al. 2020. "Organized Crime in the Fisheries Sector." Washington, DC: World Resources Institute. https://www.oceanpanel.org/blue-papers/organised-crime-associated fisheries.

some areas).[250] These offences have been notoriously challenging to address due to jurisdictional disputes and inadequate or absent legal frameworks and enforcement. Attention has been drawn to this issue on an international level, with increasing understanding of the complexities of organised crime in the fishing sector. In 2008, the UN General Assembly requested that states assist in gathering more information on the connection between illegal fishing and organised crime.[251]

4.4 Embracing Hope: The Building Momentum for a Sustainable Ocean Economy

When reading through the litany of threats to the ocean, two uncomfortable questions arise: Is the ocean so complex and damaged that it is *too big to fix*?[252] Is the only way out to immediately curtail most ocean activities? The answer to both is a decisive 'no'. A profoundly different mindset is emerging, in an unprecedented number of global initiatives through the G20 and G7, the Ocean Panel, the UN Ocean Conference, Our Ocean, the World Ocean Summits, the UN Decade of Ocean Science, World Trade Organization meetings on ending harmful fishing subsidies, COP26 on climate, COP15 on biodiversity, the RISE UP Blue Call to Action (led jointly by NGOs and civil society)[253] and so on. Coastal nations, especially small island states (alternatively referred to as 'large ocean states')[254] are advocating for socially equitable and environmentally sustainable growth.[255] Civil society is realising the ocean's decline and vigorously endorsing governmental action to protect it: a 2020 survey found, for instance, that 92% of Japanese respondents supported the establishment of MPAs, 92% of U.S. respondents believed that 'ocean governmental regulations are necessary' and 92% of Indonesian respondents 'supported environmental conservation regulations to protect the ocean'.[256]

In response to the growing pressures described in Sect. 4.3, innovations and trends are emerging that demonstrate through trial and error that alternatives are possible.[257] These 'niche innovations' can be geographical and/or sectoral spaces, where innovators coalesce in response to perceived pressures affecting the ocean. These niche innovations can be protected from market dynamics (through subsidies, soft money) or political interference (through regulation or location in the non-profit sector). This section identifies (non-exhaustively) seeds of change already sprouting and in need of careful nurture: celebrated in their beginnings, prioritised in policy and finance, and promoted publicly.

This section first looks at selected sectorial innovations and trends (in particular in fisheries, mariculture, energy, shipping and tourism) before identifying additional cross-sectorial ones (in data, ocean planning, finance, anti-pollution efforts and accounting).

4.4.1 Hopeful and Promising Sectorial Trends and Innovations

Sustainable fisheries Three main trends will accelerate sustainable reforms:

- **The turning institutional tide.** Most national fishery ministries are now committed to the goal of maximum sustainable yields. Most, however, still struggle to attain that goal. In recent years, regional fisheries management organisations (RFMOs), long constrained by consensus decision rules, have finally been able to restore some tuna stocks, notably Atlantic bluefin tuna and southern bluefin tuna.[258] The plight of artisanal fishers is being more fully considered in fishery management plans, but this is tempered by the lack of catch and effort data from artisanal fisheries, which are often equal in size to industrial fisheries. Fish-dependent nations in Asia (e.g. Indonesia, Fiji, the Philippines, the Marshall Islands) and Africa (e.g. Mauritius, Seychelles)

[250] Widjaja, S., T. Long, H. Wirajuda, A. Gusman, S. Juwana, T. Ruchimat and C. Wilcox. 2020. "Illegal, Unreported and Unregulated Fishing and Associated Drivers." Washington, DC: World Resources Institute. https://oceanpanel.org/sites/default/files/2020-02/HLP%20Blue%20Paper%20on%20IUU%20Fishing%20and%20Associated%20Drivers.pdf; Witbooi et al. 2020. "Organized Crime in the Fisheries Sector."

[251] Widjaja et al. 2020. "Illegal, Unreported and Unregulated Fishing and Associated Drivers"; Witbooi et al. 2020. "Organized Crime in the Fisheries Sector."

[252] Lubchenco, J., and S.D. Gaines. 2019. "A New Narrative for the Ocean." *Science* 364 (6444): 911. doi: https://doi.org/10.1126/science.aay2241.

[253] Oceano Azul, Ocean Unite, Oak Foundation, David and Lucile Packard Foundation, Marine Conservation Institute, High Seas Alliance, Oceana et al. 2020. "RISE UP: A Blue Call to Action." https://www.riseupfortheocean.org/wp-content/uploads/2020/01/BCA_RISE-UP_EN_A4-1.pdf.

[254] Allison et al. 2020. "The Human Relationship with Our Ocean Planet."

[255] Bennett et al. 2019. "Towards a Sustainable and Equitable Blue Economy."

[256] Kantar. 2020. "Perceptions of the Ocean and Environment."

[257] Swilling, M., M. Ruckelshaus, T.B. Rudolph, P. Mbatha, E. Allison, S. Gelcich and H. Österblom. 2020. "The Ocean Transition: What to Learn from System Transitions." Washington, DC: World Resources Institute. https://www.oceanpanel.org/blue-papers/ocean-transition-what-learn-system-transitions.

[258] Walter, J., R. Sharma and M. Ortiz. 2018. "Western Atlantic Bluefin Tuna Stock Assessment 1950–2015 Using Stock Synthesis." *ICCAT* 100; Commission for the Conservation of Southern Bluefin Tuna. n.d. "Latest Stock Assessment." https://www.ccsbt.org/en/content/latest-stock-assessment. Accessed 6 May 2020.

are committed to restoring the efficiency and equitability of their fisheries and/or have made substantial protected area commitments.

- **A data revolution.** Sound fishery management digital tools are now widely available, including vessel tracking, fishery simulation, registry and enforcement systems (e.g. satellite imagery, eDNA and drones). Philanthropically funded initiatives to study the ocean have mushroomed (e.g. REVOcean, OceanX). Market demand in the developed world for sustainably sourced fish has never been higher and can now be reliably serviced with chain-of-custody certification. Shortcomings in data availability are being addressed through new collection technologies (onboard cameras, scanners) and new data analysis and treatment methods. Lowering the costs of such technology and new models around sharing will be necessary to also benefit the broader base of small-scale fishers.
- **Asset turnover.** Many of the developing country fishing fleets are ageing as profits have been too low to fund depreciation. The fleets of Ghana, the Philippines and Senegal, for example, all have an average age of more than 30 years.[259] In the absence of capacity-related subsidies, many of these boats cannot be profitably replaced—if market discipline is maintained (no capacity subsidies or assistance, from either the country itself or foreign nations through loans and/or selling of fishing rights). In such cases, fishing capacity is allowed to drop, and the profits of remaining boats can slowly recover towards the maximum sustainable yield point; creating feedback effects that financially reward sustainable fishers.

Box 20.1 presents two inspiring case studies of fishery recovery (at national and international levels), demonstrating that sound measures properly implemented can lead both to restoration of fish stocks and economic and social gains.

Box 20.1 Successful Fishery Recovery Can Happen: Two Hopeful Case Studies

The United States Sustainable Fisheries Act of 1996 (SFA), and amendment to the 1976 Magnuson-Stevens Fishery Conservation and Management Act or Magnuson Stevens Act (MSA), governs fisheries management in the U.S. exclusive economic zone (EEZ) (up to 200 miles offshore). It is widely credited with saving U.S. fisheries. The original MSA established the legal basis for many essential fishery management mechanisms, such as the permitting of vessels and operators, and the ability to restrict gear, access and periods of fishing. However, for the first 20 years, and despite language aspiring to 'sustainable fishing', it did not explicitly prevent overfishing. The SFA changed this decisively. Its most important features were mandates to (1) not only end overfishing but also recover overfished species to sound, fully documented population levels (usually about one-third of the estimated pre-fishing population) within 10 years (with certain exceptions), (2) require that fishing quotas (catch limits) be set for each fishery, based only on scientific evidence about what is biologically sustainable, and include accountability measures to adjust future quotas in the case that overfishing accidentally occurs, and (3) allow the use of rights-based fishery management approaches if appropriate for that particular fishery. The inclusion of specific timelines and accountability measures made all the difference.

These amendments were highly successful. Forty-three fish stocks have been rebuilt since 2000, and over two-thirds of overfished stocks have been rebuilt or begun rebuilding since 1996. Revenue from 1996 to 2010 is up 54% in real terms.[a] The key features—reliance only on scientific evidence, use of rights-based approaches, strict catch limits and accountability measures, and the 10-year rebuilding plans—have been widely copied by fishery managers worldwide.

The law enjoys considerable support from the commercial fishing community and has generally held up well to inevitable pressure to extend deadlines for rebuilding stocks, relax catch limits and monitoring requirements, and limit the influence of science. Support for the law reflects the fact that stocks are rebuilding and fishers have input into the process, but especially because fishers' long-term incentives are aligned with their short-term incentives. The approach also combines national standards with regional tailoring. Regional fishery management councils propose management plans for each fishery that take into account local knowledge and factors but must also satisfy strict national standards.

The Parties to the Nauru Agreement (PNA) include the Federated States of Micronesia, the Republic of Kiribati, the Republic of the Marshall Islands, the Republic of Nauru, the Republic of Palau, the Independent State of Papua New Guinea and the Solomon Islands. Because these nations' mostly contiguous EEZs hold considerable fishery resources (especially tuna), they have developed a uniform management structure that prioritises resource sustainability and transparency.[b]

[259] FAO. n.d. "The Status of the Fishing Fleet." http://www.fao.org/3/y5600e/y5600e05.htm. Accessed 6 May 2020.

The agreement most prominently features three major arrangements:[c]

1. A regionwide register and monitoring of fishing vessels, with trackers on each boat.
2. No transshipment at sea, mandatory daily catch and effort reporting, regular logbook maintenance, 100% onboard observer coverage and an electronic data transmission device that provides position and catch information.
3. No fishing in the high-seas pockets between PNA nation EEZs, no fishing on floating aggregation devices between July and September, and mandatory retention of any bigeye, yellowfin or skipjack tuna caught.

The resulting comprehensive data collection makes it possible to set up and enforce the Vessel Day Scheme—a type of fishing quota that allocates 'allowed days of fishing' to individual vessels. Based on a scientific stock assessment, an overall 'days of fishing' effort is determined (44,033 in 2019 and 2020)[d] and appropriated to the PNA countries.[e] The countries can then sell their allocated fishing days to fishing vessels, resulting in sizable revenues for the PNA countries—nearly US $400 million in 2015.[f] The fishing days are tradeable between countries, which helps optimize fishing across the entire PNA territory—an important feature in managing highly migratory tuna stocks. It also ensures that the fishery's benefits are shared by all PNA countries, regardless of where the tuna happen to be in any given year.[g]

The agreement has increased revenue for the PNA countries while maintaining sustainable, science-driven harvesting practices. It has stabilised the stocks, provided the PNA (and other) nations with the lasting value derived from a well-managed fishery and has become a model for other ocean states. In 2012, this led the PNA skipjack tuna fishery to become certified by the Marine Stewardship Council, making it the world's largest independently certified tuna fishery.[h]

Sources:

[a] *Natural Resources Defense Council, Conservation Law Foundation, Earthjustice, Ocean Conservancy, Oceana and Pew Charitable Trusts. 2018. "How the Magnuson-Stevens Act Is Helping Rebuild U.S. Fisheries." https://www.nrdc.org/sites/default/files/magnuson-stevens-act-rebuild-us-fisheries-fs.pdf*

[b] *Parties to the Nauru Agreement (PNA). n.d. "Nauru Agreement Concerning Cooperation in the Management of Fisheries of Common Stocks (As Amended April 2010)." https://www.pnatuna.com/content/nauru-agreement. Accessed 13 August 2020*

[c] *World Wildlife Fund. 2011. "Parties to the Nauru Agreement (PNA)." http://awsassets.panda.org/downloads/factsheet_7.pdf*

[d] *"Purse Seine VDS TAE for 2018–2020." 2017. Parties to the Palau Arrangement, 22nd Annual Meeting, Majuro, Marshall Islands, 5–7 April. http://www.pnatuna.com/sites/default/files/Purse%20Seine%20VDS%20TAE%20for%202018-2020_0.pdf*

[e] *Pacific Islands Forum Fisheries Agency. n.d. Introduction. https://www.ffa.int/vds. Accessed 6 May 2020*

[f] *PNA. 2016. "Behind the Scenes Work Makes PNA's Vessel Day Scheme a Success." https://www.pnatuna.com/node/373*

[g] *International Union for Conservation of Nature (IUCN). 2015. "Parties to the Nauru Agreement (PNA): Interview with Maurice Brownjon." https://www.iucn.org/content/parties-nauru-agreement-pna-interview-maurice-brownjon*

[h] *Marine Stewardship Council. 2016. "PNA Tuna: Small Islands, Big Opportunities." http://pna-stories.msc.org/*

Mariculture Trends in marine aquaculture also point towards future sustainable expansion:

- **National priority.** China and Norway lead the development of large, next-generation offshore finfish farms (Box 20.2) which attempt to address issues of containment, disease control and nutrient efficiency. Archipelago nations, such as Indonesia and the Philippines, are exploring locally relevant approaches such as combined seaweed and low-trophic mariculture farms. National commitments to spatially explicit planning, streamlined permitting, rigorous operating standards and state-supported R&D are likely to further accelerate mariculture.
- **Improvement in fish meal/oil alternatives' availability and price.** The conversion of former biofuel or ethanol fermentation facilities to algae production (in places like Brazil or the U.S. Midwest) would scale up production so significantly that price points for omega-3 fatty acids as FM/FO alternatives and proteins could tumble (Box 20.2). Given the problems in the current biofuel markets, this

could happen soon, and with considerable government support to prevent the stranding of these major industrial assets. Insect-based fish feeds also are attracting increasing attention, creating a source of revenue out of food waste (insects such as black soldier fly larvae can be grown out of food waste and be used to feed farmed fish).[260]

- **Progress made on limiting environmental impact of finfish mariculture.** Apart from feed, the main challenges of mariculture are (1) fouling of the water column and sea floor, (2) parasites (sea lice) that migrate to native (wild) species, (3) leakage of antibiotics used to (over) treat diseases and (4) the escape of non-native (and/or genetically modified) species. New technologies offer some promise. Remote video-controlled feeding systems can reduce food waste; parasites can be controlled drug-free through the addition of cleaner fish,[261] lasers, electric fences and sudden changes in temperature; [262] disease resistance can be boosted with selective breeding;[263] control systems such as the Norwegian 'traffic light' system can control the growth of farmed salmon;[264] and rigid-structure caging systems can reduce escapes. Finally, the combination of bivalves and seaweed into multi-trophic farms is a promising approach to limit some impacts of finfish farming.[265]

[260]Biancarosa, I., V. Sele, I. Belghit, R. Ørnsrud, E.-J. Lock and H. Amlund. 2019. "Replacing Fish Meal with Insect Meal in the Diet of Atlantic Salmon (*Salmo salar*) Does Not Impact the Amount of Contaminants in the Feed and It Lowers Accumulation of Arsenic in the Fillet." *Food Additives & Contaminants: Part A* 36 (8): 1191–205. doi: https://doi.org/10.1080/19440049.2019.1619938.

[261]Deady, S., S. Varian and J. Fives. 1995. "The Use of Cleaner-Fish to Control Sea Lice on Two Irish Salmon (*Salmo salar*) Farms with Particular Reference to Wrasse Behavior in Salmon Cages." *Aquaculture* 131 (March): 73–90. doi: https://doi.org/10.1016/0044-8486(94)00331-H.

[262]*The Explorer*. 2019. "Norwegian Technology for Sustainable Aquaculture." 14 August. https://www.theexplorer.no/stories/ocean/norwegian-technology-for-sustainable-aquaculture/.

[263]Klinger, D., and R. Naylor. 2012. "Searching for Solutions in Aquaculture: Charting a Sustainable Course." *Annual Review of Environment and Resources* 37 (1): 247–76. doi: https://doi.org/10.1146/annurev-environ-021111-161531.

[264]Sandvik, A.D., I.A. Johnsen, M.S. Myksvoll, P.N. Sævik and M.D. Skogen. 2020. "Prediction of the Salmon Lice Infestation Pressure in a Norwegian Fjord." *ICES Journal of Marine Science* 77 (2): 746–56. doi: https://doi.org/10.1093/icesjms/fsz256.

[265]Buck, B.H., M.F. Troell, G. Krause, D.L. Angel, B. Grote and T. Chopin. 2018. "State of the Art and Challenges for Offshore Integrated Multi-trophic Aquaculture (IMTA)." *Frontiers in Marine Science* 5. doi: https://doi.org/10.3389/fmars.2018.00165.

Box 20.2 Examples of Mariculture Tech-Driven Innovations

Credit: SalMar

SalMar's Ocean Farm 1 is one of the largest offshore marine mariculture pens. Built in China and deployed in Norway, the 110-m-wide-structure is predicted to be able to hold over 1 million salmon.

Apart from its enormous size (250,000 cm^3), it is able to withstand 12-m waves and is equipped with over 20,000 sensors monitoring the well-being of the fish.

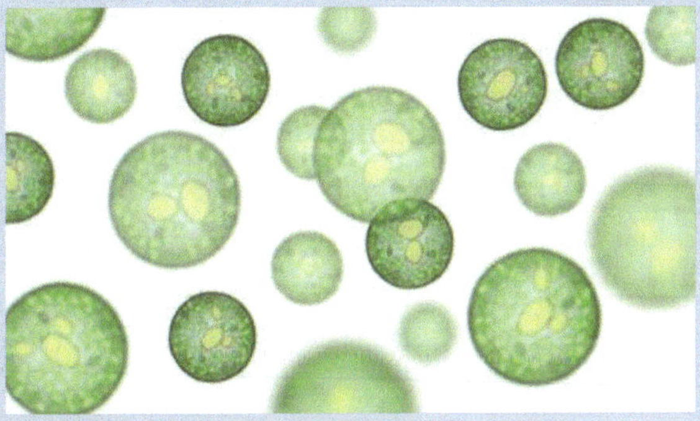

Credit: Perception7/Shutterstock.

Grown by feeding sugar derived from sugarcane to algae in a fermentation tank through a special fermentation process, the algae turn the sugar into omega-3 oil, which can be (and is being used as) a replacement for fish oil in fish feed.

A frontrunner in this space is Corbion's DHS Algal prime, managing to save 40 metric tonnes of forage fish per tonne of DHS Algal prime. Algal prime is already produced at a commercial scale, and with prices falling its algae omega-3 oil is at price parity with fish-derived omega-3.

DSM has partnered with Evonik to develop a similar algae-based solution, called Veramaris. They claim 1 tonne of Veramaris algal oil is the equivalent of 60 tonnes of avoided wild-caught fish.

Accelerating ocean-based renewables Offshore wind is an increasingly mature and competitive technology, but other ocean-based renewable energy sources are also actively being explored: energy extracted from waves and tides, from salinity and temperature gradients (e.g. by ocean thermal energy conversion or by heat pumps for heating and cooling), and floating solar photovoltaic systems are beginning to emerge in marine environments.[266] Three major factors are encouraging the growth of ocean renewables:

- **Increasingly competitive electricity cost.** The levelised cost of electricity (LCOE) for offshore wind was $124–$146 per megawatt-hour (MWh) in the United States in 2017 and somewhat less in Europe.[267] Recently, auctions in the European market have seen contracts drop to about $50/MWh,[268] which is highly competitive with other sources of electricity. Low cost of capital drives down LCOE for offshore wind, and economies of scale are significant, with costs declining at 18% per doubling of capacity.[269] Non-wind sources are largely uncompetitive today, with LCOE often above $250/MWh. Wave energy and ocean thermal energy conversion are capital intensive and unlikely to scale below $150/MWh and $70–$190/MWh,[270] respectively, making them most useful for very specific applications such as for small island nations currently reliant on imported fossil fuels.
- **Rising global investments in offshore wind.** The ebb and flow of projects responding to policy has resulted in volatile global investment volumes (ranging from $30 billion to less than $15 billion in the past 5 years), but the overall trend is bullish. With decreasing offshore auction prices, the increasing water depth of projects, increasing turbine capacity and declining LCOE, global investments are nearly certain to increase—especially as Europe's commanding lead is challenged by Asia, Australia and even the Middle East in the years to come[271] (see Box 20.3).
- **Declining environmental impact.** There is growing consensus that offshore wind farms can be built without significantly damaging the environment, if proper planning and mitigation measures are put in place to address bird strikes, construction and operation noise, and sea floor damage.[272] The carbon mitigation benefits of ocean-based renewable energy production are significant and accrue back to ocean health and functionality.

Box 20.3 Offshore Wind in Vietnam

The southern coast of Vietnam has demonstrated technically feasible wind potential, with average wind speeds of 7–11 m/s. Faced with gradually depleting hydro and fossil fuel energy sources and burgeoning demand, the country plans to install 6.2 gigawatts (GW) by 2030. As a major first step, a site survey licence has recently been issued for the 3.4 GW Thang Long wind project offshore from Ke Ga Cape—the world's largest wind project, located in a 2800 km^2 area 20–50 km offshore from Binh Thuan Province. This is the first step towards a US $11.9 billion, six-phase build-out designed to take optimal advantage of progressing Mitsubishi and Vestas turbine technology between now and 2026. The first 600 MW phase is expected to comprise 64 turbines at a best-in-class capacity of 9.5 MW and to be operational in 2023.

The project is emblematic of the special financial and operational conditions in developing countries. On the downside, developers generally are on their own with development costs, and projects win or lose on strict market terms. On the upside, the natural conditions are often perfect, and the onshore infrastructure/offtake facilities and supply chains are often new and up-to-date.

[266] Hoegh-Guldberg et al. 2019. "The Ocean as a Solution to Climate Change."

[267] Stehly, T.J., and P.C. Beiter. 2020. "2018 Cost of Wind Energy Review." NREL/TP-5000-74598. Golden, CO: National Renewable Energy Lab. doi: https://doi.org/10.2172/1581952.

[268] IEA. 2019. *Offshore Wind Outlook 2019.*

[269] Ørsted. n.d. "Making Green Energy Affordable: How the Offshore Wind Energy Industry Matured—and What We Can Learn from It." https://orsted.com/-/media/WWW/Docs/Corp/COM/explore/Making-green-energy-affordable-June-2019.pdf.

[270] Kempener, R., and F. Neumann. 2014. "Wave Energy Technology Brief." International Renewable Energy Agency. https://www.irena.org/documentdownloads/publications/wave-energy_v4_web.pdf; Ocean Energy Systems (OES). 2015. "International LCOE for Ocean Energy Technologies: An Analysis of the Development Pathway and Levelised Cost of Energy Trajectories of Wave, Tidal and OTEC Technologies." IEA Technology Collaboration Programme for Ocean Energy Systems. https://www.ocean-energy-systems.org/news/international-lcoe-for-ocean-energy-technology/.

[271] IRENA. 2018. "Renewable Power Generation Costs in 2017." Abu Dhabi: International Renewable Energy Agency. https://www.irena.org/-/media/Files/IRENA/Agency/Publication/2018/Jan/IRENA_2017_Power_Costs_2018.pdf.

[272] Draget, E. 2014. "Environmental Impacts of Offshore Wind Power Production in the North Sea." Oslo: World Wide Fund for Nature. https://tethys.pnnl.gov/sites/default/files/publications/WWF-OSW-Environmental-Impacts.pdf.

Shipping, often considered as a traditional, slow-moving sector, is experiencing a real revolution

- **Shipping decarbonisation momentum.** The International Maritime Organization's Energy Efficiency Design Index requires ships built after 2022 to be at least 50% more efficient over 2008 levels,[273] and total shipping GHG emissions to be reduced by at least 50% in 2050.[274] The industry is now actively working and collaborating on this agenda. For instance, Mærsk, a leading shipping company, is estimated to have invested several billion U.S. dollars between 2014 and 2019 in researching carbon-free shipping technologies.[275] The efforts are also focusing on addressing the difficult problem of collaboration: the 'Getting to Zero Coalition' convenes more than 100 companies and shipping-related stakeholders (e.g. ports) to develop 'commercially viable zero emission vessels, powered by zero emission fuels by 2030'.[276] The 'Green Maritime Methanol consortium' is exploring use of methanol as a shipping fuel.[277] Another cross-industry collaboration—Project ZEEDS—aims to create the 'zero fuel station of the future'—green ammonia fuel stations at sea that are powered by surrounding offshore wind farms (see the Prologue of this report). Zero-carbon fuels are still at a very early stage for long-haul trips, but recent advances in battery technology have allowed short-haul ships—mostly passenger and car ferries in developed countries—to become electrified (see Box 20.4).[278] Finally, on the energy efficiency front, optimised hull, propulsion and (existing) engine designs could deliver energy efficiency improvements of 30–55% compared to current fleets.[279]
- **Ballast water treatment improvements.** In 1991, the 'International Guidelines for preventing the introduction of unwanted aquatic organisms and pathogens from ships' ballast water and sediment discharges', developed by the Marine Environment Protection Committee (MEPC),[280] set the stage for ballast water control. These standards have been followed by the 2017 Ballast Water Management Convention (BWM),[281] which requires ships to treat their ballast water by 2024.[282] The BWM has been supported by the GoBallast program, a global partnership of—among others—the Global Environment Facility (GEF), the IMO and the UN Development Programme (UNDP) aimed at reducing global ballast water pollution.
- **Improved port management.** The World Port Sustainability Program is designed to 'enhance and coordinate future sustainability efforts of ports worldwide and foster international cooperation with partners in the supply chain'.[283] On a national level, many ports are leading the way towards becoming more sustainable. The Port of Rotterdam's €5 million 'Incentive scheme for climate-friendly shipping' aims to make the port a leader in carbon neutrality.[284] A joint project by the Port Authority of Bari (Italy) and DBALab uses 'artificial intelligence for environmental monitoring and prediction' of the port's activities. The program's display environmental and port activity data allow operators to minimize the port's environmental footprint.[285]

---

[273]Chestney, N. 2019. "IMO Agrees on Stricter Efficiency Targets for Some Ships." Reuters, 17 May. https://uk.reuters.com/article/us-imo-shipping-efficiency-idUKKCN1SN2BV.

[274]Olmer et al. 2017. "Greenhouse Gas Emissions from Global Shipping, 2013–2015."

[275]United Nations Conference on Trade and Development (UNCTAD). 2020. *Review of Maritime Transport 2019*. New York. UNCTAD. https://unctad.org/en/PublicationsLibrary/rmt2019_en.pdf.

[276]Global Maritime Forum. n.d. "Getting to Zero Coalition." https://www.globalmaritimeforum.org/getting-to-zero-coalition. Accessed 7 May 2020.

[277]"Support for Green Maritime Methanol Project." 2019. *Maritime Journal*, 21 February. https://www.maritimejournal.com/news101/power-and-propulsion/support-for-green-maritime-methanol-project.

[278]Balsamo, F., C. Capasso, G. Miccione and O. Veneri. 2017. "Hybrid Storage System Control Strategy for All-Electric Powered Ships." *Energy Procedia*, ATI 2017, 72nd Conference of the Italian Thermal Machines Engineering Association, September, 1083–90. doi: https://doi.org/10.1016/j.egypro.2017.08.242; Filks, I. 2019. "Batteries Included: Sweden's Emissions-Free Ferries Lead the Charge." Reuters, 14 March. https://www.reuters.com/article/us-denmark-battery-ferry-idUSKCN1QV1W7.

[279]Energy Transitions Commission (ETC). n.d. "Mission Possible: Reaching Net-Zero Carbon Emissions from Harder-to-Abate Sectors by Mid-century: Sectoral Focus Shipping." http://www.energy-transitions.org/sites/default/files/ETC%20sectoral%20focus%20-%20Shipping_final.pdf. Accessed 7 May 2020.

[280]IMO. n.d. "Ballast Water Management." http://www.imo.org/en/OurWork/Environment/BallastWaterManagement/Pages/Default.aspx. Accessed 7 May 2020.

[281]Global Environment Facility. 2017. "Global Treaty to Halt Invasive Aquatic Species Enters into Force." 8 September. https://www.thegef.org/news/global-treaty-halt-invasive-aquatic-species-enters-force.

[282]"Ballast Water Management Convention Amendments Enter Into Force." 2019. *Maritime Executive*, 14 October. https://www.maritime-executive.com/article/ballast-water-management-convention-amendments-enter-into-force.

[283]World Port Sustainability Program (WPSP). n.d. "About WPSP." https://sustainableworldports.org/about/. Accessed 7 May 2020.

[284]WPSP. 2019. "Port of Rotterdam: Incentive Scheme for Climate-Friendly Shipping." https://sustainableworldports.org/project/port-of-rotterdam-incentive-scheme-for-climate-friendly-shipping/.

[285]WPSP. 2019. "Port of Bari: Artificial Intelligence for Environmental Monitoring and Prediction." https://sustainableworldports.org/project/port-of-bari-artificial-intelligence-for-environmental-monitoring-and-prediction/.

Box 20.4 Decarbonising Short-Haul Shipping: Electric Ferry
The Danish towns of Fynshav and Søby are connected by an electric ferry 60 m long and 13 m wide. The relatively short trip length of commuting ferries facilitates the use of batteries and electric engines. Since the first electric ferry was put in service in Norway in 2015, the number of electric ferries operating in the country has been rapidly increasing and will reach between 60 and 70 in 2021. Also, cities in the United States, Canada and Denmark have concrete plans (or even orders) to electrify their car and passenger ferries.

Source: Ellsmoor, J. 2019. "The World's Largest Electric Ferry Has Completed Its Maiden Voyage." Forbes, 18 August. https://www.forbes.com/sites/jamesellsmoor/2019/08/18/the-worlds-largest-electric-ferry--has-completed-its-maiden-voyage/. Photo: Erik Christensen, Creative Commons Attribution-Share Alike 4.0 International

Successful coastal/marine conservation initiatives

- **Restoration**. 'Soft' coastal approaches using tidal marshes, mangroves, dunes, coral reefs and shellfish reefs are increasingly part of coastal defence. Sixteen thousand acres of tidal marshes in San Francisco Bay are under restoration, and the Mississippi marshlands are under restoration to protect New Orleans and southeast Louisiana from storm surges. The Netherlands and Belgium combine 'hard' solutions (e.g. seawalls, dykes, sluice gates) with 'soft' restoration, with the latter showing highly efficient results.[286] In the Belgian Scheldt estuary, up to 4000 hectares of historically reclaimed wetlands are being converted back into floodplains; when finished in 2030 at a cost of €600 million, this will alleviate a 2100 yearly risk of flood damage estimated at €1 billion.[287] In Southeast Asia, mangrove forest plantations are being considered as protection against storm surges,[288] but restoration projects are small compared to the area already lost. In cities as diverse as Amsterdam, Abidjan and Lagos, beach and dune barriers are being built as crucial defences against coastal flooding.[289]
- **Protection.** Marine protected areas provide levels of protection ranging from strict 'no-take' to more permissive 'sustainable extraction' (see MPA guide in Fig. 20.27). If properly sized, sited and delineated, they can generate multiple benefits. The strongly protected 'no-take' zones,

[286] Turner, R.K., D. Burgess, D. Hadley, E. Coombes and N. Jackson. 2007. "A Cost-Benefit Appraisal of Coastal Managed Realignment Policy." *Global Environmental Change* 17 (3): 397–407. doi: https://doi.org/10.1016/j.gloenvcha.2007.05.006; Broekx, S., S. Smets, I. Liekens, D. Bulckaen and L. de Nocker. 2011. "Designing a Long-Term Flood Risk Management Plan for the Scheldt Estuary Using a Risk-Based Approach." *Natural Hazards* 57 (2): 245–66. doi: https://doi.org/10.1007/s11069-010-9610-x.

[287] Temmerman, S., P. Meire, T.J. Bouma, P.M.J. Herman, T. Ysebaert and H.J. De Vriend. 2013. "Ecosystem-Based Coastal Defence in the Face of Global Change." *Nature* 504 (7478): 79–83. doi: https://doi.org/10.1038/nature12859.

[288] Temmerman et al. 2013. "Ecosystem-Based Coastal Defence in the Face of Global Change."

[289] Temmerman et al. 2013. "Ecosystem-Based Coastal Defence in the Face of Global Change."

for instance, have been shown to restore fish stocks by a factor of up to six times within the area;[290] to support ecosystem complexity, health- and associated ecosystem services;[291] to help with climate resilience;[292] to reduce carbon released from seabed floor;[293] to increase ecosystem resilience;[294] and to provide pristine ocean areas important to many cultures around the world.

- As of today, the Convention on Biological Diversity (CBD) estimates, 14.4% of national waters and 5.7% of the global ocean are protected.[295] However, only 2.4% of the ocean can be considered to be in fully protected MPAs.[296] Too often, MPAs are categorised as environmental measures at odds with economic interests (starting with fisheries).
- When full protection cannot be achieved, 'extractive MPAs'—defined as ocean areas subject to some restriction on use and/or extraction—can represent a viable form of protection for many countries with coastlines (>75% of countries in 2013).[297] Properly designed, they can be effective in protecting key coastal habitats, and they may represent an underused means to block particularly destructive coastal land uses and resource-extraction practices.

[290] Sala, E., and S. Giakoumi. 2018. "No-Take Marine Reserves Are the Most Effective Protected Areas in the Ocean." *ICES Journal of Marine Science* 75 (3): 1166–68. doi: https://doi.org/10.1093/icesjms/fsx059.

[291] Babcock, R.C., N.T. Shears, A.C. Alcala, N.S. Barrett, G.J. Edgar, K.D. Lafferty, T.R. McClanahan and G.R. Russ. 2010. "Decadal Trends in Marine Reserves Reveal Differential Rates of Change in Direct and Indirect Effects." *Proceedings of the National Academy of Sciences* 107 (43): 18256–61. doi: https://doi.org/10.1073/pnas.0908012107.

[292] Micheli, F., A. Saenz-Arroyo, A. Greenley, L. Vazquez, J.A. Espinoza Montes, M. Rossetto and G.A. de Leo. 2012. "Evidence That Marine Reserves Enhance Resilience to Climatic Impacts." *PLOS ONE* 7 (7). doi: https://doi.org/10.1371/journal.pone.0040832.

[293] Roberts, C.M., B.C. O'Leary, D.J. McCauley, P.M. Cury, C.M. Duarte, J. Lubchenco, D. Pauly et al. 2017. 'Marine Reserves Can Mitigate and Promote Adaptation to Climate Change'. *Proceedings of the National Academy of Sciences*, June. doi: https://doi.org/10.1073/pnas.1701262114.

[294] Harrison, J. 2015. "Governing Marine Protected Areas: Resilience through Diversity, Written by Peter J.S. Jones." *International Journal of Marine and Coastal Law* 30 (4): 811–13. doi: https://doi.org/10.1163/15718085-12341373.

[295] Convention on Biological Diversity. n.d. "Global Implementation." https://www.cbd.int/protected/implementation/. Accessed 17 August 2020.

[296] Marine Conservation Institute. n.d. "Interactive Map." *Atlas of Marine Protection*. http://mpatlas.org/map/mpas/. Accessed 7 May 2020.

[297] Costello, M.J., and B. Ballantine. 2015. "Biodiversity Conservation Should Focus on No-Take Marine Reserves: 94% of Marine Protected Areas Allow Fishing." *Trends in Ecology & Evolution* 30 (9): 507–9. doi: https://doi.org/10.1016/j.tree.2015.06.011.

4.4.2 Structural Changes Across Ocean Economy Sectors

This section reviews recent progress in ocean data, ocean planning, finance, anti-pollution efforts and accounting of the ocean economy.

The ocean data revolution The technology is here now. It is now technically possible to sample the ocean on its true spatial and temporal scales with a remote-sensing network covering the physical, biological, ecological[298] and chemical properties of the global ocean surface (although full coverage remains far off). From the proliferation of sensors and platforms (Argo floats,[299] REMUS,[300] Wave Glider[301]) and satellites (from SeaSat onwards) to cabled observatories[302] and acoustic modems, remote sensing and transmission of data from a variety of platforms is becoming an 'always on, always connected'[303] operating system. The connection of intelligent devices into an 'Internet of Things' is moving from land to sea, analysing data ranging from invasive species in bilge water to nutrients in river deltas, allowing for an ever-more complete picture in near real time—the holy grail of adaptive management. The open-access platforms necessary to store, share and process the innovation are technically available (and in broad use in many cloud-based data systems), but their application in the ocean realm is still lagging behind (see Sect. 6.3 for in-depth discussion).

Data processing is keeping pace with the sensing revolution. Processing capacity has increased 1 trillion times in the past 50 years, making it possible to build massive dynamic

[298] Seltenrich, N. 2014. "Remote-Sensing Applications for Environmental Health Research." *Environmental Health Perspectives* 122 (10): A268–75. doi: https://doi.org/10.1289/ehp.122-A268.

[299] Freeland, H.J., and P.F. Cummins. 2005. "Argo: A New Tool for Environmental Monitoring and Assessment of the World's Oceans, an Example from the N.E. Pacific." *Progress in Oceanography* 64 (1): 31–44. doi: https://doi.org/10.1016/j.pocean.2004.11.002.

[300] Stokey, R.P., A. Roup, C. von Alt, B. Allen, N. Forrester, T. Austin, R. Goldsborough et al. 2005. "Development of the REMUS 600 Autonomous Underwater Vehicle." In *Proceedings of OCEANS 2005 MTS/IEEE* 2: 1301–4. doi: https://doi.org/10.1109/OCEANS.2005.1639934.

[301] Thomson, J., and J. Girton. 2017. "Sustained Measurements of Southern Ocean Air-Sea Coupling from a Wave Glider Autonomous Surface Vehicle." *Oceanography* 30 (2): 104–9.

[302] Kelly, R.P. 2014. "Will More, Better, Cheaper, and Faster Monitoring Improve Environmental Management?" *Environmental Law* 44: 1111; Smith, L.M., J.A. Barth, D.S. Kelley, A. Plueddemann, I. Rodero, G.A. Ulses, M.F. Vardaro and R. Weller. 2018. "The Ocean Observatories Initiative." *Oceanography* 31 (1): 16–35.

[303] Abbott, M.R., and C.E. Sears. 2006. "Always-Connected World and Its Impact on Ocean Research." *Advances in Computational Oceanography* 19 (1). doi: https://doi.org/10.5670/oceanog.2006.88.

model simulations ranging from cosmological galaxy formation[304] to weather, climate prediction and hurricane prediction. The implications for ocean governance, management and economic development are profound.

Growing traction on ocean planning The safety issues associated with multiple uses in a turbulent ocean environment (e.g. stationary wind farms or mariculture facilities vs. cargo, fishing and navy shipping lanes) are complex and a major cause of regulatory delays today. The regulatory difficulties of securing long-term, reliable permits and access rights are hurting the mariculture industry. Carbon- and offset-financed restoration projects are hard to structure without long-term title protections. Open access for all interested parties is the primary driver of overfishing in the developing world. On land, nobody would expect investors to deal with the legal and regulatory uncertainties of such an open-access system.

At the same time, the technical hurdles to delineating, monitoring, and enforcing access rights in a remote ocean are no longer applicable—the remote-sensing revolution offers multiple alternatives to expensive patrol-based enforcement schemes. For example, Caribbean protected area managers and technologists have jointly developed low-cost acoustic sensors that identify violating vessels.[305] Another example is Global Fishing Watch,[306] which visualises, tracks and shares data on global fishing activities in near real time.

As a result, several regions (northeast United States; Netherlands/North Sea; Baltic Sea; Norway; Xiamen, China; and the Australian Great Barrier Reef) have broken down siloed management practices in favour of more integrated spatial planning. Xiamen, for example, has pioneered a spatially explicit approach to coastal management since 1994, with a 40% improvement in socioeconomic benefits from its marine sectors.[307] Hundreds of territorial user rights for fisheries (TURFs) areas are being set up across the globe to protect community fisheries in multiple developing countries (e.g. Chile, Indonesia, the Philippines), with emerging evidence of recovering stocks, and increasing catches and profits.[308] The Baltic Sea states have coordinated across borders and sectors to implement a science-based planning strategy and have been rewarded with the return of predators and birds as well as restored fish stocks in the past 20–30 years.[309]

The ocean as the new investment opportunity The tide is turning on ocean investment. In a recent Credit Suisse survey,[310] 72% of investors ($n = 200$) classified the sustainable ocean economy as 'investable'. Several sustainable ocean economy private investment funds have been established recently: Sky Ocean Ventures, Althelia Sustainable Ocean Fund, Katapult Ocean, Ocean 14, BlueInvest Fund, Blue Oceans Partners and Fynd Ocean Ventures just to name a few. For more mature technologies such as wind energy, investments have also become sizeable offshore: 2018 investments in new offshore wind farms in Europe totalled €10.3 billion, 24% of total new power investments in that year.[311]

International funding for sustainable innovation includes a 2019 proposal for an IMO-administered US $5 billion fund to 'accelerate the R&D effort required to decarbonise the shipping sector and to catalyse the deployment of commercially viable zero-carbon ships by the early 2030s'.[312] Also in 2019, the Asian Development Bank launched the Action Plan for Healthy Oceans and Sustainable Blue Economies for the Asia and Pacific region, with committed funding of $5 billion from 2019 to 2024 to finance and provide technical

[304] IllustrisTNG. 2019. "TNG." https://www.tng-project.org/about/.

[305] Leape, J., M. Abbott, H. Sakaguchi et al. 2020. "Technology, Data and New Models for Sustainably Managing Ocean Resources." Washington, DC: World Resources Institute. https://www.oceanpanel.org/blue-papers/technology-data-and-new-models-sustainably-managing-ocean-resources.

[306] Global Fishing Watch. n.d. "Sustainability through Transparency." https://globalfishingwatch.org/. Accessed 11 May 2020.

[307] Peng, B., H. Hong, X. Xue and D. Jin. 2006. "On the Measurement of Socioeconomic Benefits of Integrated Coastal Management (ICM): Application to Xiamen, China." *Ocean & Coastal Management* 49(3): 93–109. doi: https://doi.org/10.1016/j.ocecoaman.2006.02.002.

[308] Costello, C., D. Ovando, T. Clavelle, C.K. Strauss, R. Hilborn, M.C. Melnychuk, T.A. Branch et al. 2016. "Global Fishery Prospects under Contrasting Management Regimes." *Proceedings of the National Academy of Sciences* 113 (18): 5125–29. doi: https://doi.org/10.1073/pnas.1520420113.

[309] Reusch, T.B.H., J. Dierking, H.C. Andersson, E. Bonsdorff, J. Carstensen, M. Casini, M. Czajkowski et al. 2018. "The Baltic Sea as a Time Machine for the Future Coastal Ocean." *Science Advances* 4(5): eaar8195. doi: https://doi.org/10.1126/sciadv.aar8195.

[310] Responsible Investor Research and Credit Suisse. 2020. *Investors and the Blue Economy*. https://www.esg-data.com/reports.

[311] Brindley, G. 2019. *Financing and Investment Trends 2018: The European Wind Industry in 2018*. Wind Europe. https://windeurope.org/wp-content/uploads/files/about-wind/reports/Financing-and-Investment-Trends-2018.pdf.

[312] Baltic and International Maritime Council (BIMCO), Cruise Lines International Association (CLIA), International Chamber of Shipping (ICS), INTERCARGO, INTERFERRY, International Association of Independent Tanker Owners (INTERTANKO), International Parcel Tankers Association (IPTA) and World-Class Shipping (WSC). 2019. "Reduction of GHG Emissions from Ships: Proposal to Establish an International Maritime Research and Development Board (IMRB)." Marine Environment Protection Committee, 75th Session, Agenda Item 7. https://www.ics-shipping.org/docs/default-source/Submissions/IMO/final-imrb-submission-to-mepc-75.pdf?sfvrsn=6.

assistance for ocean health and marine economy projects in the region.[313]

Unprecedented momentum in the fight against land-based pollution Ocean pollution reforms are on different tracks. For plastics, the ocean is a major driver of global movement from linear to circular material management systems on land. For nutrients, pesticide runoff and industrial pollution, ocean interests have not yet reached the same level of influence, explaining a reform agenda lagging behind.

The transformation of the current linear plastic value chain to a more circular one represents enormous potential economic value, with estimated potential materials savings worth hundreds of billions of dollars per year,[314] together with significant co-benefits for the climate (9.3 gigatonnes [Gt] of CO_2e in 2050—equivalent to eliminating transportation emissions), and employment upsides.[315] A recent comprehensive modelling exercise concluded that solutions available today to industry and governments—if massively deployed—could reduce annual land-based plastic leakage into the ocean by around 80% by 2040, compared to a business-as-usual scenario, and also help advance other societal, economic, and environmental objectives.[316]

The crisis is now forcing the hand of plastic resin manufacturers, converters, and consumer brands. New consumer brand commitments to 'plastic neutrality' and recycling-friendly design are proliferating. The plastic industry as a whole is increasingly recognising its extended responsibilities for the entire product lifecycle and exploring cooperative schemes to improve waste management and collection. Over 95 plastic packaging policies and laws were signed in the United States, Europe and Asia from 2010 to 2019; and the New Plastics Economy Global Commitment had over 400 signatories, including from investors, innovators and NGOs. Cumulatively these commitments still fall far short of solving the crisis—but they represent only the beginning of what could become a comprehensive redesign of the plastic economy.[317]

New, more holistic ways to account for the ocean economy are now available Today's economic policy is concerned with outcomes and sustainability, not simply managing monetary inflation—'twenty-first century progress cannot be measured with twentieth century statistics'.[318] The System of Environmental Economic Accounting is being updated to include ecosystem accounting; there is discussion of revising the internationally agreed System of National Accounts to focus on sustainability.[319]

The most fundamental remaining accounting challenge is the monetisation of ocean and other natural assets—an essential input. The international standards for national accounts—the 2008 System of National Accounts (SNA)—provides little guidance for doing so. But methods for the valuation of non-produced or natural assets do exist,[320] including a 'Capital Asset Pricing for Nature' software package.[321] The Inclusive Wealth Index (2012) of the UN Environment Programme (UNEP), adopted by 140 countries, is piloting the measurement of natural capital,[322] and many partnerships aim to develop technical capacity, such as the WAVES (World Bank), BIOFIN (UNDP), MAES (EU) and UNEP-TEEB-CBD partnerships. In the business world, the Natural Capital Coalition, Conservation International, the U.S. National Oceanic and Atmospheric Administration

[313]Asian Development Bank. 2019. "ADB Launches $5 Billion Healthy Oceans Action Plan." 2 May. https://www.adb.org/news/adb-launches-5-billion-healthy-oceans-action-plan.

[314]Ellen MacArthur Foundation. 2015. "Towards the Circular Economy, Economic and Business Rationale for an Accelerated Transition." https://www.ellenmacarthurfoundation.org/assets/downloads/TCE_Ellen-MacArthur-Foundation_9-Dec-2015.pdf.

[315]Ellen MacArthur Foundation and Material Economics. 2019. "Completing the Picture: How the Circular Economy Tackles Climate Change." https://www.ellenmacarthurfoundation.org/assets/downloads/Completing_The_Picture_How_The_Circular_Economy-_Tackles_Climate_Change_V3_26_September.pdf; Ellen MacArthur Foundation. 2015. "Growth Within: A Circular Economy Vision for a Competitive Europe." Ellen MacArthur Foundation, Stiftungsfonds für Umweltökonomie und Nachhaltigkeit (SUN), and McKinsey Center for Business and the Environment. https://www.ellenmacarthurfoundation.org/assets/downloads/publications/EllenMacArthurFoundation_Growth-Within_July15.pdf.

[316]Lau et al. 2020. "Evaluating Scenarios toward Zero Plastic Pollution."

[317]World Economic Forum, Ellen MacArthur Foundation and McKinsey & Company. 2016. "The New Plastics Economy: Rethinking the Future of Plastics." http://www.ellenmacarthurfoundation.org/publications.

[318]Agarwala, M.K. 2019. "Natural Capital Accounting and the Measurement of Sustainability." PhD diss., London School of Economics and Political Science. http://etheses.lse.ac.uk/4146/1/Agarwala_Natural-capital-accounting.pdf.

[319]UN Statistical Division. 2019. "50th Session Documents." https://unstats.un.org/unsd/statcom/50th-session/documents/.

[320]Adamowicz, W., L. Calderon-Etter, A. Entem, E.P. Fenichel, J.S. Hall, P. Lloyd-Smith, F.L. Ogden et al. 2019. "Assessing Ecological Infrastructure Investments." *Proceedings of the National Academy of Sciences* 116 (12): 5254–61. doi: https://doi.org/10.1073/pnas.1802883116; Fenichel, E.P., and C. Obst. 2019. "A Framework for the Valuation of Ecosystem Assets." Discussion paer 5.3. In *System of Environmental Economic Accounting, 2019 Forum of Experts in SEEA Experimental Ecosystem Accounting, 26–27 June 2019, Glen Cove, NY.* https://seea.un.org/sites/seea.un.org/files/discussion_paper_5.3.pdf.

[321]Yun, S.D., E.P. Fenichel and J.K. Abbott. 2017. *Capital Asset Pricing for Nature.* Version 1.0.0. https://CRAN.R-project.org/package=capn.

[322]Managi, S., and P. Kumar. 2018. *Inclusive Wealth Report 2018: Measuring Progress towards Sustainability.* New York: Routledge.

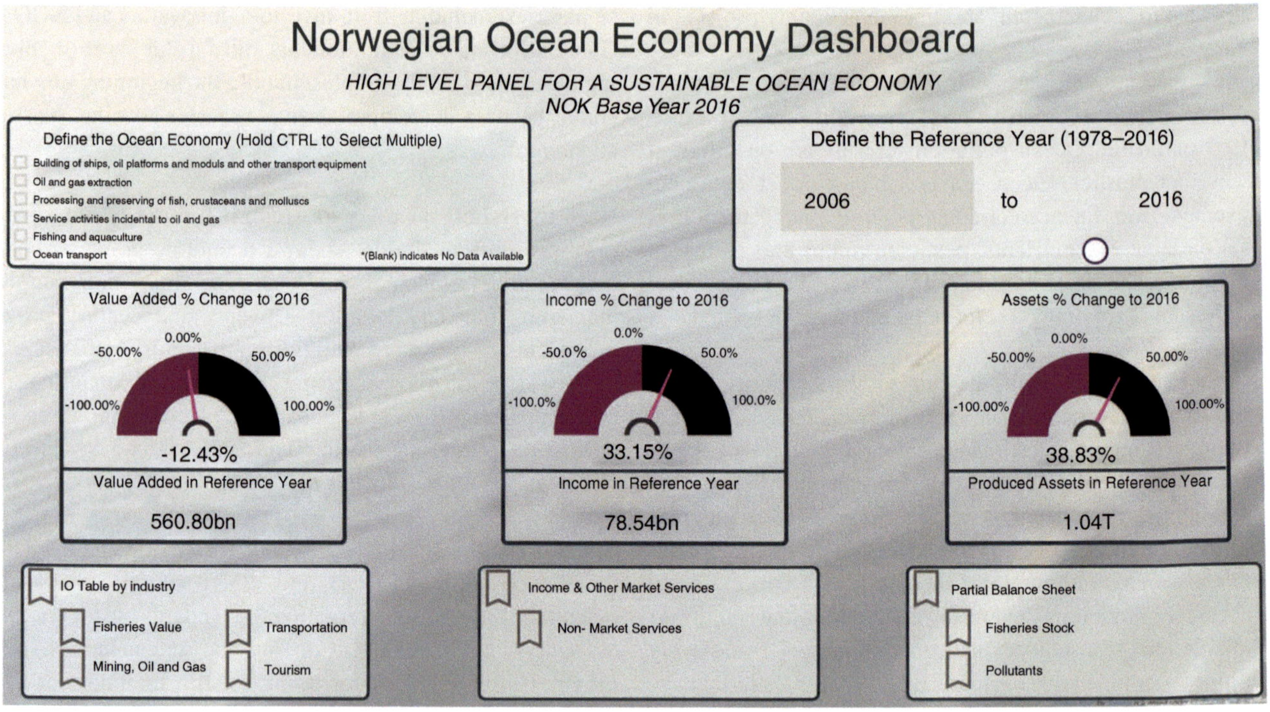

Fig. 20.22 Example of a live interactive digital dashboard for ocean accounting: Norway ocean economy dashboard. *Note: See the live dashboard at* https://environment.yale.edu/data-science/norwegian-ocean-economy-dashboard. (*Source: Fenichel, E.P., B. Milligan, I. Porras et al. 2020. "National Accounting for the Ocean and Ocean Economy." Washington, DC: World Resources Institute.* https://www.oceanpanel.org/blue-papers/national-accounting-ocean-and-ocean-economy)

(NOAA), the Institute of Chartered Accountants in England and Wales and others brought together 60 leading ocean-related organisations in 2017 to ignite the creation of the Natural Capital Protocol for the Ocean[323] to supplement the recognised Natural Capital Protocol.[324]

Other natural capital valuation methods are already changing policy and investment decisions[325] (most advanced are perhaps those in China and the United Kingdom, but also other countries are taking up this information and transforming policy and investment, e.g. Belize[326]). These initiatives didn't start in countries' statistical offices but instead were initiated in sector-related ministries (e.g. fisheries, tourism, environment) and in finance ministries.

The digital revolution provides a major boost for ocean national accounting.[327] Online, digital dashboarding makes it possible to drill down quickly to specific indicators of interest in policy analysis and evaluation. The future has begun: an ocean proto-account for Norway can be displayed as an interactive dashboard (Fig. 20.22), and the United States hosts an interactive ocean proto-account.[328]

A user of the Norway dashboard can define the ocean economy through any combination of six sectors and explore how these interact along various indicators of production,

[323]Natural Capital Coalition. n.d. "Natural Capital Protocol for the Ocean." https://naturalcapitalcoalition.org/wp-content/uploads/2019/01/Natural-Capital-Protocol-for-the-Ocean_Overview.pdf.

[324]World Business Council for Sustainable Development. 2017. "What Experts Are Saying about the Natural Capital Protocol Toolkit." 13 July. https://www.wbcsd.org/Programs/Redefining-Value/Business-Decision-Making/Assess-and-Manage-Performance/Natural-Capital-Protocol-Toolkit/News/What-experts-are-saying-about-the-Natural-Capital-Protocol-Toolkit.

[325]Ouyang, Z., C. Song, C. Wong, G.C. Daily, J. Liu, J. Salzman, L. Kong et al. 2019. "Designing Policies to Enhance Ecosystem Services: China's Experience on Mainstreaming Ecosystem Services for Green Growth." In *Green Growth That Works: Natural Capital Policy and Finance Mechanisms around the World*, edited by L. Mandle, Z. Ouyang, J. Salzman and G.C. Daily, 177–94. Washington, DC: Island.

[326]Arkema, K.K., G.M. Verutes, S.A. Wood, C. Clarke-Samuels, S. Rosado, M. Canto, A. Rosenthal et al. 2015. "Embedding Ecosystem Services in Coastal Planning Leads to Better Outcomes for People and Nature." *Proceedings of the National Academy of Sciences* 112(24): 7390–95. doi: https://doi.org/10.1073/pnas.1406483112.

[327]Leape et al. 2020. "Technology, Data and New Models for Sustainably Managing Ocean Resources."

[328]NOAA. n.d. "ENOW Explorer: Discover More about Your Local Ocean Economy." https://coast.noaa.gov/enowexplorer/#/. Accessed 7 May 2020.

value created, capital used and the like. Although data holes remain (most fundamentally, Norway does not yet include monetisation of its natural capital stocks), the dashboard radically expands the breadth of questions that can be asked and answered.[329]

This section makes clear that a healthy ocean and a subsequent sustainable ocean economy are crucial allies to address some of the most pressing challenges humanity will face in the twenty-first century, including food security, climate change and social inequalities. Today the ocean's health is under increasing pressure from anthropogenic stressors. If not addressed these could compound with each other with dramatic consequences. A growing number of initiatives, technologies and business solutions are emerging and show that the possible alternative path of a sustainable ocean economy is realistic and feasible. The next section offers a vision where these positive developments are generalised and a sustainable ocean economy can emerge that benefits the people, the economy and the planet.

5 The Possibility of Tomorrow

5.1 Introduction

For centuries, the ocean has been viewed as **'too big to fail'**.[330] However, as shown in Sect. 4, this belief cannot be considered true anymore: overfishing, habitat destruction, climate change and pollution represent a de facto uncontrolled experiment. The size of the challenge could easily lead one to think that the ocean is now **'too big to fix'**.[331]

This report offers a more hopeful narrative. Section 5 posits that the agendas of ocean and terrestrial resource productivity are no longer separable; neither are the agendas of ocean protection and ocean productivity. As pressure rises on business and political leaders, and as new, sustainable types of ocean ventures demonstrate compelling economics, the tide can turn and the ocean as a source of sustainable prosperity can become **'too important to ignore'**.[332]

Section 5 of this report invites the reader to take a journey towards an alternative tomorrow, where a set of sound early decisions launches the productive disruptions, pioneers and dynamics that lead to a sustainable ocean economy over the coming decades. This section paints a 'vision' of what a sustainable ocean economy could look like and the benefits it could generate. This vision is anchored in science and is feasible if the right decisions are made and several systemic barriers are removed (analysed in depth in Sect. 6, Sect. 6.2).

This section starts by introducing seven fundamental design principles, suggested as a guiding 'Southern Cross' or 'North Star' (Sect. 5.2) to scale up the promising trends presented in Sect. 4. It then lays out a vision where five fundamental transformations enable the development of truly sustainable ocean sectors (Sect. 5.3). Finally, it presents evidence that such a vision can deliver a 'triple benefit' of effective protection, sustainable production and equitable prosperity (Sect. 5.4).

5.2 Defining a Compass Direction: Principles for a Sustainable Ocean Economy

Seven fundamental design principles are introduced below to help decision-making and prioritisation towards a sustainable ocean economy. Every measure, transformation and example in this section is based on these seven fundamental design principles (Fig. 20.23).

Guarantee equity The ocean, as 'the common heritage of humankind', needs to benefit all of humanity. Avoiding coastal food and energy insecurity, labour exploitation and gender discrimination should be given the highest priority and form the bedrock of decision-making related to the ocean economy. This includes respecting relevant international agreements like the SDGs, the UN Declaration on the Rights of Indigenous Peoples and the UN Declaration of Human Rights.[333]

Align with the Paris 1.5 °C target The 2019 UN emission gap report states that the world is currently on course for 3.2 °C global warming over pre-industrial levels[334]—presenting a stark contrast to the 1.5 °C limit now commonly recognised as critical for ocean health. Establishing a regenerative ocean economy, focused on restored and protected 'blue sinks' (e.g. mangroves, sea grass, saltmarshes) and

[329] Many of the data needed to feed these dashboards and to parameterise these connections already exist, but they are highly dispersed. A first step towards understanding the dynamics at play is to highlight the current high-level status of ocean account data with a live version of the dashboard at https://environment.yale.edu/data-science/norwegian-ocean-economy-dashboard/.

[330] Lubchenco and Gaines. 2019. "A New Narrative for the Ocean"; Lubchenco, J. 2019. "People and the Ocean 3.0: A New Narrative with Transformative Benefits." In *A Better Planet: 40 Big Ideas for a Sustainable Planet*. New Haven, CT: Yale University Press.

[331] Lubchenco and Gaines. 2019. "A New Narrative for the Ocean"; Lubchenco. 2019. "People and the Ocean 3.0."

[332] Lubchenco and Gaines. 2019. "A New Narrative for the Ocean"; Lubchenco. 2019. "People and the Ocean 3.0."

[333] Bennett et al. 2019. "Towards a Sustainable and Equitable Blue Economy."

[334] UN Environment Programme (UNEP). 2019. *The Emissions Gap Report 2019*. Nairobi: UNEP. https://www.unenvironment.org/resources/emissions-gap-report-2019.

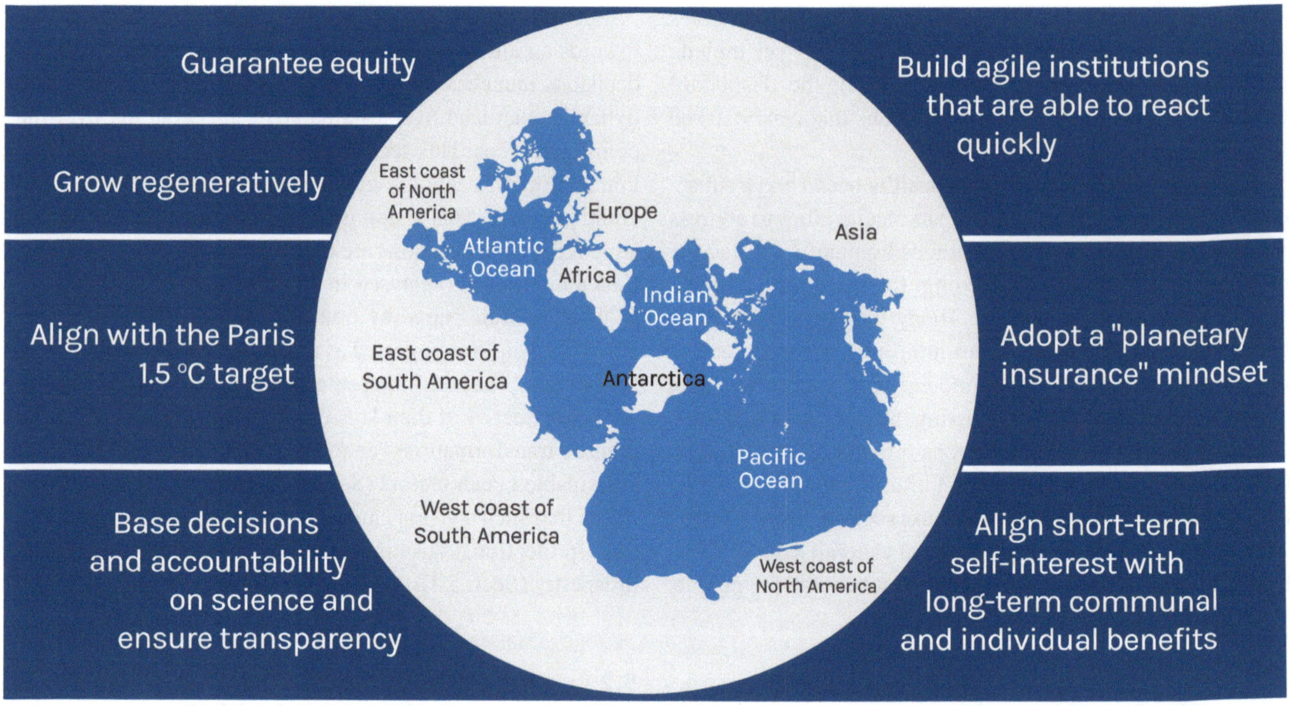

Fig. 20.23 Principles for a sustainable ocean economy. (*Source: Authors*)

zero- or low-carbon production of food, energy and transportation, is essential to that goal.

Base decisions and accountability on science and ensure transparency The age of the unfathomable, inexhaustible ocean is over. Future management must rely on a clear-eyed view of the impacts of climate change, the ocean's resource dynamics, its natural cycles of decline and regeneration and the resilience and vulnerability built into its infinitely complex biological systems. This requires the full and creative use of the data revolution for ocean purposes, the full appreciation and use of scientifically accurate local and Indigenous knowledge, and the commitment of management institutions to follow the advice of scientists.

Grow regeneratively The ocean economy, at every relevant scale, needs to cumulatively regenerate the ocean's vitality, diversity, and resilience. A sustainable ocean economy needs to ensure that marine economic activities are at least carbon-neutral and support the ocean's biodiversity. Not every project can be carbon-negative or rebuild biodiversity—but projects must be linked such that they bend the arc towards greater ocean health.

Build agile institutions that are able to react quickly In an increasingly fast and unpredictable world where 'governance failure is routine'[335] and crises like COVID-19 could become more frequent, institutions need to optimize themselves based on the principle of agility and ability to react quickly, while making decisions in an inclusive 'top-down, bottom-up manner'. This move towards shorter reaction times would allow governments, community networks and supra-national interests to adapt quickly to rapidly changing climatic and sociological conditions.

Align short-term self-interest with long-term communal and individual benefits Current misplaced incentives (economic incentives and behavioural norms) that drive destructive outcomes need to be reconfigured towards a new set of incentives aligned with the other six principles and the vision of a sustainable ocean economy.

Adopt a 'planetary insurance' mindset The ocean is becoming more unpredictable—the degradation of its health and ecosystem services is accelerating and is non-linear. Setting aside large areas of fully intact and comprehensive ecosystems and habitats is an essential insurance mecha-

[335] Jessop, B. 2003. "The Governance of Complexity and the Complexity of Governance: Preliminary Remarks on Some Problems and Limits of Economic Guidance." Department of Sociology at Lancaster University, 21. https://www.lancaster.ac.uk/fass/resources/sociology-online-papers/papers/jessop-governance-of-complexity.pdf.

nism. The science is clear: large, properly designed protected areas increase the ocean's resilience to a variety of stressors, including warming and acidification. Similarly, the level of uncertainty at play does not allow for uncontrolled experiments and should encourage the following of a stricter, precautionary approach, whether in the exploration of new commercial species or the exploitation of known stocks and new resources like seabed minerals and metals.

Taking these general principles to their logical conclusion, a potential future emerges that diverges from the dystopian future evoked in Sect. 4.

5.3 A New Picture Is Emerging: The 2050 Sustainable Ocean Economy

It is impossible to predict precisely any version of the 2050 ocean economy—but it is possible to describe an optimistic scenario that combines the main linked components of a sustainable ocean economy (Fig. 20.24).

In this sustainable 2050 scenario, a new network emerges of interest groups including fishers, ocean farmers, scientists, civil society, local communities, as well as key energy, shipping and tourism players. This network is economically empowered and culturally deeply vested in ocean health and the sustainable ocean economy principles stated above. The groups of which it is composed create significant societal and economic values by linking offshore wind farms, mariculture, zero-carbon shipping, fuel production and tourism with unprecedented production efficiencies (see Fig. 20.24). Carefully situated non-fed and multi-trophic, zero-feed mariculture produces food for millions of coastal inhabitants. Large fully protected marine areas and MPA networks preserve intact ecosystems. Other effective area-based conservation measures and lightly protected MPAs accommodate some sectoral uses of ocean spaces that are compatible with some conservation goals. Large-scale restoration projects (e.g. mangroves, sea grass) are now financed by carbon mitigation fees and offset mitigation arrangements. Wild-caught fisheries implement climate-smart, ecosystem-based fisheries management. Collectively, this new cohort of ocean interest groups, of which youth and women are integral parts, works powerfully within the political economy to advocate for an equitably used ocean, free of pollution and over-exploitation, and with large fully protected areas to ensure ocean health and guard against unexpected changes.

Ocean-user interest groups have championed the importance of healthy ocean ecosystems, kick-starting an increased global understanding of the immense potential of a sustainably managed ocean economy. The spatial complexities of implementing linked and complementary ocean uses have encouraged more systematic ocean planning. Access rights for specific ocean resources have been clarified. Legal and political actions have been taken against land-based polluters. New finance and transaction recording (ledger) technol-

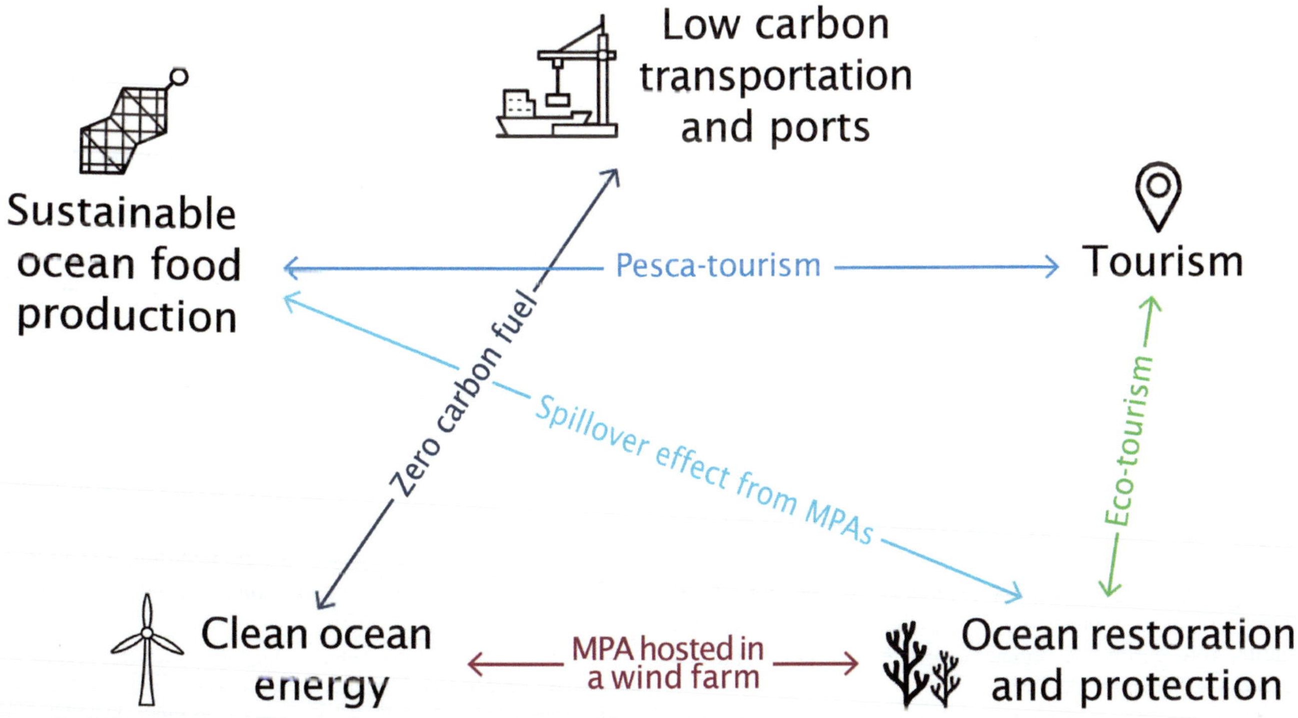

Fig. 20.24 The new contours of a sustainable ocean economy. (*Source: Authors*)

ogies have opened global markets for small artisanal producers while ensuring traceability and fair redistribution of the value created. Knowledge commons are allowing transparent sharing of data, assessments and lessons of what is working and what is not, leading to far more agile responses by communities, businesses and nations.

The political realm has responded to the new economic realities. Operating standards and permitting procedures have been clarified and standardised. Net margins of new and innovative ocean businesses are now supported through appropriate risk-reduction measures. Regulatory pressure on land-based polluters has increased. Access laws have been reformed to better balance the goals and needs of multiple stakeholder groups, including commercial and subsistence users. Labour laws have been strengthened, setting international standards to eradicate human rights abuses, and these laws are enforced.

Coastal communities, especially in the tropical realm, have reasserted their traditional use rights and are empowered to regulate access to local fisheries and ocean resources. Secure in their rights of access, they have the luxury of planning for the long term, and they have switched to sustainable stewardship practices. Women-owned cooperatives running near-shore mariculture operations, processing facilities and logistics have become the norm.

In this scenario, this 2050 state did not appear by magic. It was made possible by deliberate political decisions made in the early 2020s and dynamic changes continuing over 30 years to overcome a series of well-established barriers and habits. In this scenario, from 2020 onwards several countries shifted their focus to sustainable ocean management, clearly defined what they wanted to achieve and decided to manage sustainably 100% of their areas under national jurisdiction.

To learn and demonstrate feasibility at scale, these countries set up 'sustainable ocean economic zones' (SOEZs). These zones promoted 'projects of choice' (in line with the seven principles introduced in Sect. 5.2) with attractive logistical, financial and regulatory benefits. Projects integrated multiple and symbiotic sectors (e.g. energy, food, tourism); provided for well-designed marine protection and restoration areas; and prioritised ocean health, food security and labour protection. A network of scientists, technologists, investors, sustainable businesses, regulators, local communities and government officials collaborated to design these zones, and they defined uses, standards, finance instruments, and conservation and regeneration requirements.

International negotiations on harmful subsidies, illegal fishing, high seas management, Arctic protection and seabed mining came to a productive conclusion. New visions of a stable, zero-waste and regenerative ocean economy moved into the industrial mainstream.

These decisions, directly informed by properly funded science, triggered a chain of transformative events:

- **Ecosystem-based, inclusive spatial planning became the norm.** Careful science-based planning was required to make these spatially and operationally complex projects a reality (see discussion of marine spatial planning in Sect. 6). Siting decisions had to be formalised, access rights legalities had to be codified and potential use conflicts had to be eliminated through careful apportionment. Conservation offsets (fully protected MPAs, coastal restoration projects, buffer zones) had to be clearly defined and gazetted. Over time, ocean planning became an institutionally well-engrained habit, informed by excellent knowledge of the complex ocean ecosystems and the ability to monitor and adapt management to changing environments, driven by economic utility and managed inclusively with all stakeholders.
- **Polluters paid.** The initial projects, and those following in their footsteps, created a strong community of shared economic interests. As pollution from industrial and agricultural sources began to directly affect sustainable ocean economy success, ocean users and land-based communities came together to find solutions to stop leakage of pollution into the ocean. In many countries, courts and agencies found in favour of the ocean interests and reforms on land leaned towards more circular and regenerative practices. At the same time, increased ocean food production forced new food safety standards, covering pollutants such as plastics and mercury.
- **Automation and the data revolution hit the ocean.** As ocean economies became more sophisticated, advanced remote-sensing technologies became indispensable for delineation and enforcement. Distributed ledger and registration technologies (e.g. blockchain[336]) were used to track the differentiated traits of ocean economy products and (ecosystem) services all the way across the value chain to the consumer, responding to stringent sustainability demands from consumers. The demand pressure from major new maricultural development sped up the development of new sources of feed supply. At the same time, information-sharing went both ways—local outcomes, yields, business results, assessments and the like became readily available to investors and policymakers.
- **Investors woke up.** As the economic viability of a sustainable approach to the ocean economy emerged more clearly, investment volumes naturally increased. Over time, financial markets became more sensitive to the risks resulting from competitive distortions (e.g. subsidies of fishery capacity or fossil fuel electricity) and declining ocean productivity (e.g. pollution and/or habitat degradation). At the

[336]Blockchain is a distributed ledger technology in which requests for transactions need to be validated by the entire network rather than by a single point. After validation, the transaction becomes an immutable block within the transaction's history, which exists for as long as the network exists.

same time, financial technologies allowed small-scale ocean players to access global markets and strengthen their voice.

- **National accounting changed.** Nations started to make informed decisions based on a full range of metrics covering production, natural capital and human well-being—potentially through official 'national ocean accounts'. The changing nature of the ocean economy was increasingly and positively reflected in such national accounts and eventually began to shape public investments and policies in the ocean realm.

With these trends arcing towards greater balance and efficiency of ocean use over time, the sustainable ocean economy began to thrive, driven mainly by the linked contributions of five economic sectors (see Fig. 20.24 and Box 20.5). The paragraphs below describe the dynamics that led to this 2050 vision.

Box 20.5 Concepts of Ocean Multi-Use and Sector Coupling
Marine spatial planning is a proven and crucial tool to manage conflicts between ocean users and advance towards more sustainable uses of the ocean.[a] However, the development of sustainable ocean industries remains limited if they are considered as individual and separate activities, ignoring potential synergies:[b] spatial efficiency, circular models (e.g. waste from one as input to another), shared costs and so on. Consequently, there has been a growing interest in the development of a sustainable 'ocean multi-use' concept that fosters synergies among ocean sectors (sector-coupling).

This concept, at the heart of the 2050 sustainable vision described in this section, has been defined by a recent paper as follows:[c] 'Ocean multi-use is the joint use of resources in close geographic proximity by either a single user or multiple users. It is an umbrella term that covers a multitude of use combinations in the marine realm and represents a radical change from the concept of exclusive resource rights to the inclusive sharing of resources and space by one or more users'.

The EU Commission has been pioneering this concept by funding research and a series of large-scale collaborative projects over the past 10 years, including TROPOS, MERMAID, H_2Ocean, Multi-use in European Seas (MUSES) and Marine Investment for a Blue Economy (MARIBE). These concepts are today mostly at the (advanced) blueprint stage, but new 3-year funding has just been confirmed to test pilots until 2023 (project UNITED).[d]

Adapted from Fernando Montecruz for the TROPOS Project, 2013.

[a] *"DIRECTIVE 2014/89/EU of the European Parliament and of the Council of 23 July 2014 Establishing a Framework for Maritime Spatial Planning." 2014. Brussels: Official Journal of the European Union. doi:* https://doi.org/10.1007/978-1-137-54482-7_33

[b] *Lukic, I., A. Schultz-Zehden, J. Onwona Ansong, S. Altvater, J. Przedrzymirska, M. Lazić, J. Zaucha et al. 2018. "MUSES (Multi-use in European Seas) Project v. 3.0 MUSES Deliverable 4.2.1 Multi-use Analysis." Edinburgh, UK: MUSES Project. https://pdfs.semanticscholar.org/9796/7530c175e9e1bcf6f7f708799 1ca60613575.pdf*

[c] *Schupp, M.F., M. Bocci, D. Depellegrin, A. Kafas, Z. Kyriazi, I. Lukic, A. Schultz-Zehden et al. 2019. "Toward a Common Understanding of Ocean Multi-use." Frontiers in Marine Science 6. doi:* https://doi.org/10.3389/fmars.2019.00165

[d] *Community Research and Development Information Service. n.d. "Multi-use Offshore Platforms Demonstrators for Boosting Cost-Effective and Eco-friendly Production in Sustainable Marine Activities." https://cordis.europa.eu/project/id/862915. Accessed 17 August 2020*

Lu, S.-Y., J.C.S. Yu, J. Wesnigk, E. Delory, E. Quevedo, J. Hernández, O. Llinás et al. 2014. "Environmental Aspects of Designing Multi-purpose Offshore Platforms in the Scope of the FP7 TROPOS Project." In OCEANS 2014: TAIPEI, 1–8. doi: https://doi.org/10.1109/OCEANS-TAIPEI.2014.6964306

5.3.1 Sustainable Ocean Food Production

Multi-/low-trophic mariculture Mariculture quickly became popular and successful. With a major concentration on low-trophic-level species (seaweed, bivalves, molluscs), it increased the level of local biomass, created new habitats, created new jobs and local income, and provided an alternative to land-based, carbon-intensive meat production, as well as a source of key nutrients like omega-3 fatty acids and iodine. Higher-trophic-level finfish mariculture shed its dependence on fish-derived feeds and adopted strict operating standards addressing disease control, local pollution and escapes. In some cases, low- and higher-trophic production combined into 'integrated (or co-located) multi-trophic farms' with fed (salmon, seabass, grouper, etc.) and unfed species (e.g. bivalves, seaweed) growing together in a symbiotic and low-waste ecosystem. Where relevant, mariculture operations co-located with offshore wind farms, which provided a low-cost and reliable source of electricity for the farm and clean fuel for ship traffic. Strict labour standards were adopted for mariculture operations, while profits and operating risks became evenly spread along the mariculture supply chain. Expansion of mariculture was achieved in a harmonious way that respects Indigenous rights to healthy ocean resources.

Wild-caught fisheries Fishing fleets (commercial and artisanal) became profitable and stable because fishers' economic and conservation incentives were aligned, wild fish stocks were restored (especially predators), protected against poachers and allocated fairly to fleets and communities to be fished at optimal capacity. Sustainably fished stocks proved more resilient to climate shocks and provided increasingly predictable returns to appropriately sized fleets. International collaboration and strong local enforcement massively reduced IUU fishing, corruption and forced labour on fishing boats. With access rights to fish stocks more firmly defined and enforced, fishing fleets increasingly adopted sustainable yield standards as the most long-term profitable model of fishing. Fuelled by increasing demand and leadership from seafood incumbents, traceability 'from ocean to plate' in the fish supply chain became the norm and supported generalisation of best practices. Perhaps most important, as communities gained more control over local ocean access, benefits became more equitably shared through sustainably financed mechanisms and the food security needs of coastal inhabitants became paramount.

5.3.2 Clean Ocean Energy

The offshore wind sector continued its exponential growth and replaced fossil fuels as the main source of power from the ocean. Intermittency issues were addressed by a new grid and storage infrastructure. Offshore wind farms increasingly provided energy to other offshore uses (e.g. mariculture, shipping) and anchored and delineated large-scale MPAs. In many cases, they emerged as the natural 'centre' of many ocean economic zones.

5.3.3 Low-Carbon Transportation and Ports

Shipping continued to move around 90% of globally traded goods but accelerated decisively towards zero emissions. A combination of efficiency measures, together with the introduction of new fuels (such as green ammonia or hydrogen), led to a net-zero global shipping fleet. Offshore wind farms provided the energy to generate ammonia or hydrogen, transferred to ships either locally through floating platforms or through ports. Uncontrolled ballast discharges became a thing of the past, and transport efficiencies were boosted through increased automation and revolutionised cargo-tracking systems. Ports became carbon-neutral, eliminated air pollution, implemented labour laws and synchronised their activities with the marine ecosystem they were situated in (adapting shipping lanes to avoid whale strikes, smart dredging, etc.).

5.3.4 Ocean Restoration and Protection

Ocean restoration and protection were largely driven and financed by the pragmatic agendas of carbon mitigation and sequestration, fishery productivity, coastal protection and ocean tourism. Carbon mitigation funds underwrote sea grass and mangrove restoration as highly efficient carbon sequestration projects. Cities and coastal industries underwrote wetland and marsh restoration as the most effective measure exposure to storms and tides. Networks of fully protected and enforced MPAs became commonplace in integrated fishery management and protection of carbon storage plans, often co-located with offshore wind and food production facilities. Ecotourism facilities routinely took advantage of the rich underwater environment of fully protected MPAs.

5.3.5 Tourism

Sustainable tourism showed off the beauty of a healthy ocean and created a broad set of ocean defenders, all the while celebrating rather than destroying habitats and diversity. The industry continued to grow, providing enjoyment and livelihoods for millions of people. This growth was based on sustainable tourism growth plans, which countries developed and implemented in the early 2020s. These plans, written in conformity with the sustainable tourism principles of the UN World Tourism Organization, allowed the industry to grow with minimal environmental (no virgin coastal land conversion, carbon-neutral cruise ships, no effluent discard, limitation of visitors to delicate ecosystems) and social impact (no over-tourism). Payment for ecosystem services got mainstreamed through tourism taxes. Through the adoption of these ecosystem fees, coastal tourism accrued benefits to local communities and financed the restoration and maintenance of the coastal and marine ecosystems it relies on.

5.3.6 Other Sectors

For different reasons, several ocean-related economic sectors are not included or detailed in this report's 2050 scenario of a sustainable ocean economy: industries like maritime engineering and equipment are assumed to follow the development of the above-mentioned sectors. Other sectors not included or detailed in this report's sustainable ocean economy scenario include the following:

Marine biotech The scale of genomic diversity in the ocean is difficult to comprehend and poorly understood. Over 33,000 marine natural products—naturally occurring molecules produced by marine organisms—have been discovered,[337] many with remarkable levels of biological activity, and probably only representing a very small subset of the total ocean genomic diversity. The revolution in gene sequencing and bioinformatics has allowed for considerable innovation in ocean protection and production. Sequencing costs have declined 1000-fold over the past decade, and 100,000-fold since the beginning of the millennium,[338] allowing millions of DNA fragments to be sequenced simultaneously and inexpensively, creating an intensely data-rich field. However, the sector is still in its infancy. Since its future is hard to predict, the marine biotech sector has been excluded from the future vision scenario.

Deep-seabed mining As an emerging industry in the ocean, deep-seabed mining is often considered as an example of the 'new blue economy'. It fits the blue economy definition of the EU Commission (i.e. all economic activities related to the ocean), but it remains to be seen if it will meet the World Bank definition (i.e. sustainable use of ocean resources for economic growth, improved livelihoods and jobs while preserving the health of ocean ecosystems). Indeed, recent science clearly states that greater knowledge of the environmental impacts, as well as the ability to mitigate these to acceptable levels, is required before we can be confident that engaging in industrial-scale deep-seabed mining would bring a global net benefit.[339]

The proponents of deep-sea mining typically claim that the extraordinary richness of the underwater ores would result in far lower environmental impacts than mining on land, making deep-seabed mining the best option to supply a growing global demand for cobalt, copper, nickel, silver, lithium and rare earth elements, driven by the green transition of the economy (e.g. solar photovoltaics, wind turbines, electric cars).[340] Mining deep-sea polymetallic nodules is indeed calculated to release less CO_2 per kilogram than mining on land.[341] Mining interests such as Deep Green and Global Sea Mineral Resources (GSR) consider deep-sea minerals to be essential to combating climate change.[342] If profitable, deep-sea mining could also provide an economic development opportunity for many countries.

However, these claims need to be balanced against the risks. Current scientific understanding of deep-sea ecosystems—the range of species, their movements, ecological connectivity and susceptibility to mining stress—is still in its infancy. Deep-sea communities are known to recover from disturbance very slowly, if at all.[343] The impact of deep-seabed mining on marine life—with its associated toxicity, dredging, noise and intense disturbance of the seafloor—is likely immense given the great longevity (thousands of years) and slow growth of many deep sea animals.[344] The profitability of national mining operations, without governmental support or comparably low taxes, remains questionable. If the operations are profitable, it will also raise questions about the equitable sharing of profits derived from

[337]"MarinLit: A Database of the Natural Marine Product Literature." 2020. Publishing Journals, Books and Databases. 7 May. http://pubs.rsc.org/marinlit/.

[338]National Human Genome Research Institute. n.d. "DNA Sequencing Costs: Data." https://www.genome.gov/about-genomics/fact-sheets/DNA-Sequencing-Costs-Data. Accessed 7 May 2020.

[339]Haugan, P.M., L.A. Levin, D. Amon, M. Hemer, H. Lily and F.G. Nielsen. 2019. "What Role for Ocean-Based Renewable Energy and Deep-Seabed Minerals in a Sustainable Future?" Washington, DC: World Resources Institute. https://www.oceanpanel.org/blue-papers/ocean-energy-and-mineral-sources.

[340]Dominish, E., S. Teske and N. Florin. 2019. *Responsible Minerals Sourcing for Renewable Energy*. Report prepared for Earthworks by the Institute for Sustainable Futures. Sydney: University of Technology Sydney. https://www.uts.edu.au/sites/default/files/2019-04/ISFEarthworks_Responsible%20minerals%20sourcing%20for%20renewable%20energy_Report.pdf.

[341]van der Voet, E., L. van Oers, M. Verboon and K. Kuipers. 2019. "Environmental Implications of Future Demand Scenarios for Metals: Methodology and Application to the Case of Seven Major Metals." *Journal of Industrial Ecology* 23 (1): 141–55. doi: https://doi.org/10.1111/jiec.12722.

[342]Gerard Barron (CEO and chairman of DeepGreen Metals). 2019. "Address to ISA Council." presented at the Member of the Nauru Delegation, 27 February. https://ran-s3.s3.amazonaws.com/isa.org.jm/s3fs-public/files/documents/nauru-gb.pdf.

[343]Jones, D.O.B., S. Kaiser, A.K. Sweetman, C.R. Smith, L. Menot, A. Vink, D. Trueblood et al. 2017. "Biological Responses to Disturbance from Simulated Deep-Sea Polymetallic Nodule Mining." *PLOS ONE* 12 (2): e0171750. doi: https://doi.org/10.1371/journal.pone.0171750.

[344]Miller, K.A., K.F. Thompson, P. Johnston and D. Santillo. 2018. "An Overview of Seabed Mining Including the Current State of Development, Environmental Impacts, and Knowledge Gaps." *Frontiers in Marine Science* 4. doi: https://doi.org/10.3389/fmars.2017.00418; Sumaila, U.R., C.M. Rodriguez, M. Schultz, R. Sharma, T.D. Tyrrell, H. Masundire, A. Damodaran et al. 2017. "Investments to Reverse Biodiversity Loss Are Economically Beneficial." *Current Opinion in Environmental Sustainability* 29 (December): 82–88. doi: https://doi.org/10.1016/j.cosust.2018.01.007.

a resource taken out of humanity's common heritage.[345] Finally, deep-sea mining may conflict with other marine uses, with complex legal and political ramifications in the international waters of the open ocean.[346]

Until the need for, and potential consequences of, deep-sea mining are better understood, the concept is conceptually difficult to align with the definition of a sustainable ocean economy and raises various environmental, legal and governance challenges, as well as possible conflicts with the UN Sustainable Development Goals.[347] It is thus not discussed further in this report.

Oil and gas 'The whale in the room': how should the oil and gas sector be included in a report on a sustainable ocean economy? On the one hand, it is the largest sector of the current ocean economy by far, accounting for 34% of its GVA, according to the OECD.[348] Massive capital investments are locked into extraction facilities, many with decades to go in their useful lives. Equally massive investments are planned soon: in the next 20 years, projected offshore crude oil output will grow from 30% to 50% of total global production, and almost half of remaining technically recoverable oil reserves are offshore.[349] Within the offshore realm, the share of deep water (125—1500 m) and ultra-deep water (>1500 m) production is projected to increase to 50% by 2020. More than half of major oil and gas discoveries since 2000 have been in the deep ocean.[350]

At the same time, exploitation of the technically feasible offshore oil deposits would exceed the remaining CO_2 budget commensurate with the 1.5 °C or even 2 °C future, which is crucial for ocean stability and viability.[351] In addition, the new frontiers (the deep ocean and the Arctic) are technically challenging, ecologically risky and often occur in remote areas, far from ports and infrastructure. The Deepwater Horizon disaster is a vivid example of the potential scale of oil spills, and the U.S. Bureau of Ocean Energy Management estimates a 75% chance of one or more large spills over the lifetime of development and production in Alaska's Chukchi Sea.

Continued or increased offshore oil and gas exploration is conceptually difficult to align with the definition of a sustainable ocean economy, and it is thus not discussed in this report.

The decommissioning of existing offshore platforms may offer interesting possibilities. Decommissioning expenses are estimated to increase from $2.4 in 2015 to $13 billion/year in 2040. The cost of removal is often tax-supported and could be reduced with potential re-use applications.[352] For example, North Sea countries are gradually decommissioning about 600 oil and gas installations[353] at the same time as they are installing massive new offshore wind capacity. Decommissioned oil and gas platforms could conceivably be used to convert and store offshore wind energy (e.g. in the form of hydrogen or ammonia fuels) in ways that eliminate costly hook-ups with onshore grids.[354] Other conversions, such as 'rigs to reefs' conversions or repurposing as tourist centres, are already used today.[355]

The development of offshore wind capacity is extensively discussed in this report. There are very interesting opportunities for using renewable offshore energy as the focal point for other sustainable ocean ventures, ranging from mariculture

345 Tladi, D. 2014. "The Common Heritage of Mankind and the Proposed Treaty on Biodiversity in Areas beyond National Jurisdiction: The Choice between Pragmatism and Sustainability." *Yearbook of International Environmental Law* 25 (1): 113–32. doi: https://doi.org/10.1093/yiel/yvv060; Österblom et al. 2020. "Towards Ocean Equity."

346 Miller et al. 2018. "An Overview of Seabed Mining Including the Current State of Development, Environmental Impacts, and Knowledge Gaps."

347 Haugan et al. 2019. "What Role for Ocean-Based Renewable Energy and Deep-Seabed Minerals in a Sustainable Future?".

348 OECD. 2016. *The Ocean Economy in 2030*.

349 U.S. Energy Information Administration. 2016. "Offshore Oil Production in Deepwater and Ultra-deepwater Is Increasing." *Today in Energy*, 28 October. https://www.eia.gov/todayinenergy/detail.php?id=28552.

350 Zhang, G., H. Qu, G. Chen, C. Zhao, F. Zhang, H. Yang, Z. Zhao and M. Ma. 2019. "Giant Discoveries of Oil and Gas Fields in Global Deepwaters in the Past 40 Years and the Prospect of Exploration." *Journal of Natural Gas Geoscience* 4 (1): 1–28. doi: https://doi.org/10.1016/j.jnggs.2019.03.002.

351 McGlade, C., and P. Ekins. 2015. "The Geographical Distribution of Fossil Fuels Unused When Limiting Global Warming to 2 °C." *Nature* 517 (7533): 187–90. doi: https://doi.org/10.1038/nature14016.

352 IHS Markit. 2016. "Decommissioning of Aging Offshore Oil and Gas Facilities Increasing Significantly, with Annual Spending Rising to $13 Billion by 2040, IHS Markit Says". 29 November 2016. https://news.ihsmarkit.com/prviewer/release_only/slug/energy-power-media-decommissioning-aging-offshore-oil-and-gas-facilities-increasing-si, Elden, S. van, J.J. Meeuwig, R.J. Hobbs and J.M. Hemmi. 2019. "Offshore Oil and Gas Platforms as Novel Ecosystems: A Global Perspective". *Frontiers in Marine Science* 6. doi: https://doi.org/10.3389/fmars.2019.00548.

353 Jepma, C.J., and M. van Schot. 2017. "On the Economics of Offshore Energy Conversion: Smart Combinations—Converting Offshore Wind Energy into Green Hydrogen on Existing Oil and Gas Platforms in the North Sea." Energy Delta Institute. https://projecten.topsectorenergie.nl/storage/app/uploads/public/5d0/263/410/5d026341016a2991247120.pdf.

354 Jepma and van Schot. 2017. "On the Economics of Offshore Energy Conversion."

355 FOA. 2020. "The Business Case for Marine Protection and Conservation"; Fowler, A.M., A.-M. Jørgensen, J.C. Svendsen, P.I. Macreadie, D.O. Jones, A.R. Boon, D.J. Booth et al. 2018. "Environmental Benefits of Leaving Offshore Infrastructure in the Ocean." *Frontiers in Ecology and the Environment* 16 (10): 571–78. doi: https://doi.org/10.1002/fee.1827; Jennifer Nalewicki. 2019. 'The Gulf of Mexico's Hottest Diving Spots Are Decommissioned Oil Rigs'. *Smithsonian Magazine*, 5 April 2019, sec. Travel. https://www.smithsonianmag.com/travel/gulf-mexicos-hottest-diving-spots-are-decommissioned-oil-rigs-180971728/.

to shipping fuel generation, tourism and protected areas. The widespread and essential development of ocean renewable energy will require wide-ranging reforms of ocean planning and access control systems, all of which are also discussed in this report.

5.4 The Big Reconciliation: Protect Effectively, Produce Sustainably and Prosper Equitably

This section demonstrates that a sustainable 2050 ocean economy could simultaneously deliver in three ways: (1) it could effectively **protect**, reducing greenhouse gas emissions while safeguarding biodiversity and associated ecosystem services; (2) it could sustainably **produce**, helping sustainably power and feed a planet of ten billion people; and (3) it could enable humanity to equitably **prosper**, creating better, more equitable jobs and redistribution of benefits, and supporting economic growth, household income and well-being, while prioritising access, equitable decision-making and benefits that support equity and reduce unequal impacts and harm on the most vulnerable (Fig. 20.25).

5.4.1 Protect Effectively

A sustainable ocean economy can help keep the climate within the Paris Agreement boundaries and protect and

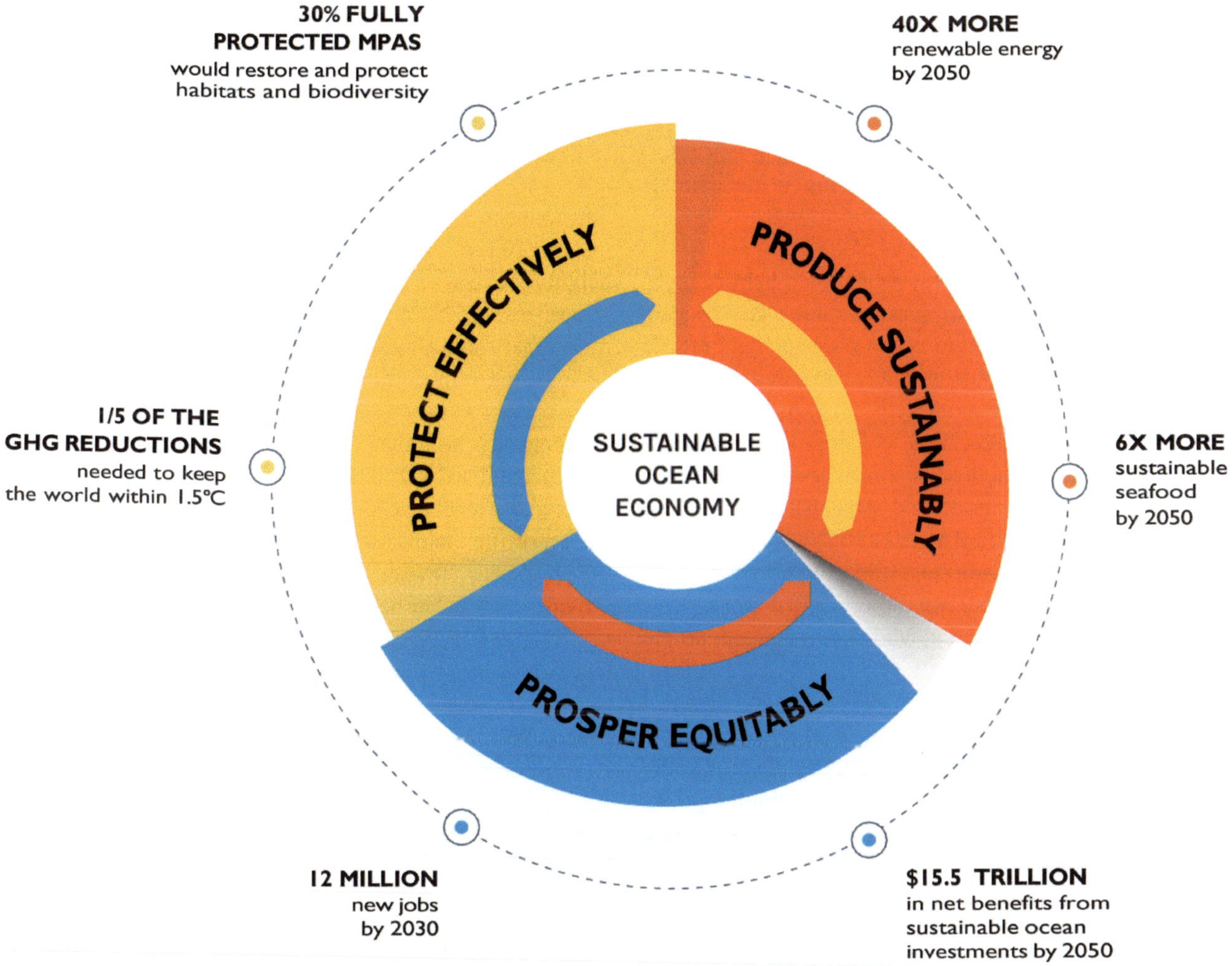

Fig. 20.25 A sustainable ocean economy can create a triple win for people, nature and the economy. *Note: MPAs: Marine protected areas. GHG: Greenhouse gas emissions.* (*Source: Authors, drawing on the following sources: OECD. 2016. The Ocean Economy in 2030. Directorate for Science, Technology and Innovation Policy Note, April.* https://www.oecd.org/futures/Policy-Note-Ocean-Economy.pdf; *Konar, M., and H. Ding. 2020. "A Sustainable Ocean Economy for 2050: Approximating Its Benefits and Costs." Washington, DC: World Resources Institute.* https://www.oceanpanel.org/Economicanalysis; *Costello, C., L. Cao, S. Gelcich et al. 2019. "The Future of Food from the Sea." Washington, DC: World Resources Institute.* https://www.oceanpanel.org/blue-papers/future-food-sea; *Hoegh-Guldberg, O., et al. 2019. "The Ocean as a Solution to Climate Change: Five Opportunities for Action." Washington, DC: World Resources Institute.* https://oceanpanel.org/sites/default/files/2019-10/HLP_Report_Ocean_Solution_Climate_Change_final.pdf)

regenerate the ocean's biodiversity and associated ecosystem services.

Climate Absorbing a third of the planet's CO_2 emissions and about 93% of the world's human-induced additional heat,[356] the ocean is already shouldering a significant part of regulating Earth's climate. In the process, it is becoming warmer, more acidic and higher. Nonetheless, the ocean economy's potential role in active climate mitigation is far from realised today. In a sustainable ocean economy, ocean-based renewable energy could play a much more important role than today: shipping would be zero-emission, fisheries and mariculture would be much more energy efficient, coastal ocean assets would be restored and protected, and CO_2 could be stored in the seabed. 'The Ocean as a Solution to Climate Change' (2019) analysed the CO_2 abatement potential from these five areas and concluded that the ocean could contribute up to 21% (or 11.8 $GtCO_2e$) of the emission reduction required to achieve a 1.5 °C trajectory by 2050[357] (Fig. 20.26). In such a vision, the ocean would move away from being solely a climate change victim (warming, acidification, etc.) towards actively participating in the climate change mitigation solution.

A sustainable ocean economy would also help catalyse deep reforms of the land-based plastics value chain. Indeed, a holistic, circular approach to ocean plastics could reduce annual ocean plastic leakage by 80%, compared to a BAU scenario where this flow is expected to triple by 2040.[358] Given CO_2 emissions associated with plastics production, use and end of life, this holistic approach has the potential to reduce CO_2e emissions associated with the plastics value chain by 25% compared to BAU 2040.[359] The plastics value chain would otherwise emit an estimated 4.5 $GtCO_2e$ by 2050—roughly 7% of global emissions in a BAU scenario—with the attendant warming and acidification effects on the ocean.[360]

Reduced other sources of pollution from land By drastically limiting leakage into the ocean, the plastics value chain holistic and circular approach would limit the growing pressure on ocean fauna and flora. The same logic could apply with other land-based pollution. Even if the correlation is harder to demonstrate, the sustainable ocean economy agenda could help catalyse broader reforms of the land-based food system, most notably in agriculture. One can expect that agricultural regulations aimed at reducing ocean dead zones could result in farmers adopting precision agriculture practices to avoid runoff. The adoption of precision agriculture, in turn, would have a positive impact on soil health and water quality in rivers and streams.

Ocean and coastal ecosystems, biodiversity and biomass In a 2050 sustainable ocean economy, the economic value of restoration of ocean and coastal natural capital would be recognised and turned into action, with carbon finance and coastal protection funds playing a major role in large-scale restoration projects. Restored and protected natural blue assets would then be able to deliver ecosystem services for coastal populations, especially in ensuring human safety by helping to mitigate the impacts of storms and sea level rise. For instance, healthy coral reefs reduce wave energy by up to 97%, protecting up to 100 million coastal inhabitants from storm risks.[361] In addition, a study has found that a '100-m-wide belt of mangroves can reduce wave heights between 13% and 66%, and up to 100% where mangroves reach 500 m or more in width'.[362] This study also found that saltmarshes can attenuate up to 50% of smaller waves, even with a barrier of just 10 m.[363]

'Planetary insurance' in the form of MPAs would have been generalised and integrated within the 100% managed EEZs and a legal mechanism to create large, fully protected MPAs on the high seas. By restoring biodiversity, these MPAs increase the resilience of the ecosystems, since they provide a protected home for communities that are capable of 'differential response'.[364] These MPAs would be primarily highly or fully protected and actively managed to obtain the greatest conservation outcomes.[365] In visual terms, if plotted on the chart of Fig. 20.27, the majority of MPAs in a sustain-

[356]Gaines et al. 2019. "The Expected Impacts of Climate Change on the Ocean Economy."

[357]Hoegh-Guldberg et al. 2019. "The Ocean as a Solution to Climate Change."

[358]Lau et al. 2020. "Evaluating Scenarios toward Zero Plastic Pollution"; Pew Charitable Trusts and SYSTEMIQ. 2020. *Breaking the Plastic Wave*.

[359]Lau et al. 2020. "Evaluating Scenarios toward Zero Plastic Pollution"; Pew Charitable Trusts and SYSTEMIQ. 2020. *Breaking the Plastic Wave*.

[360]ETC. n.d. "Mission Possible."

[361]Ferrario, F., M.W. Beck, C.D. Storlazzi, F. Micheli, C.C. Shepard and L. Airoldi. 2014. "The Effectiveness of Coral Reefs for Coastal Hazard Risk Reduction and Adaptation." *Nature Communications* 5 (1): 3794. doi: https://doi.org/10.1038/ncomms4794.

[362]Mapping Ocean Wealth (The Nature Conservancy). n.d. "Coastal Protection." https://oceanwealth.org/ecosystem-services/coastal-protection/. Accessed 11 May 2020.

[363]Mapping Ocean Wealth (The Nature Conservancy). n.d. "Coastal Protection."

[364]McCann, K.S. 2000. "The Diversity–Stability Debate." *Nature* 405(6783): 228–33. doi: https://doi.org/10.1038/35012234.

[365]Oregon State University, IUCN World Commission on Protected Areas, Marine Conservation Institute, National Geographic Society and UNEP World Conservation Monitoring Centre. 2019. "An Introduction to the MPA Guide." https://www.protectedplanet.net/c/mpa-guide.

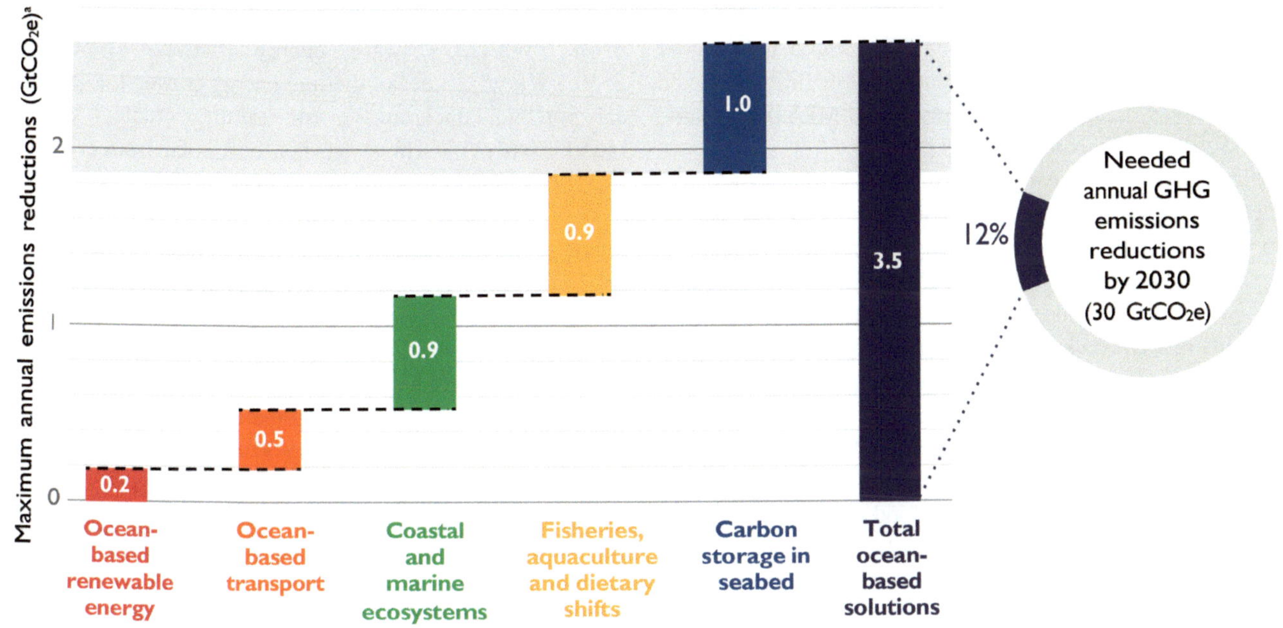

Fig. 20.26 Contribution of five ocean-based climate action areas to mitigating climate change in 2030 (Maximum $GtCO_2e$). *Note: To stay under a 1.5 °C change relative to pre-industrial levels. (Source: Hoegh-Guldberg, O., et al. 2019. "The Ocean as a Solution to Climate Change: Five Opportunities for Action." Washington, DC: World Resources Institute.* https://oceanpanel.org/sites/default/files/2019-10/HLP_Report_Ocean_Solution_Climate_Change_final.pdf)

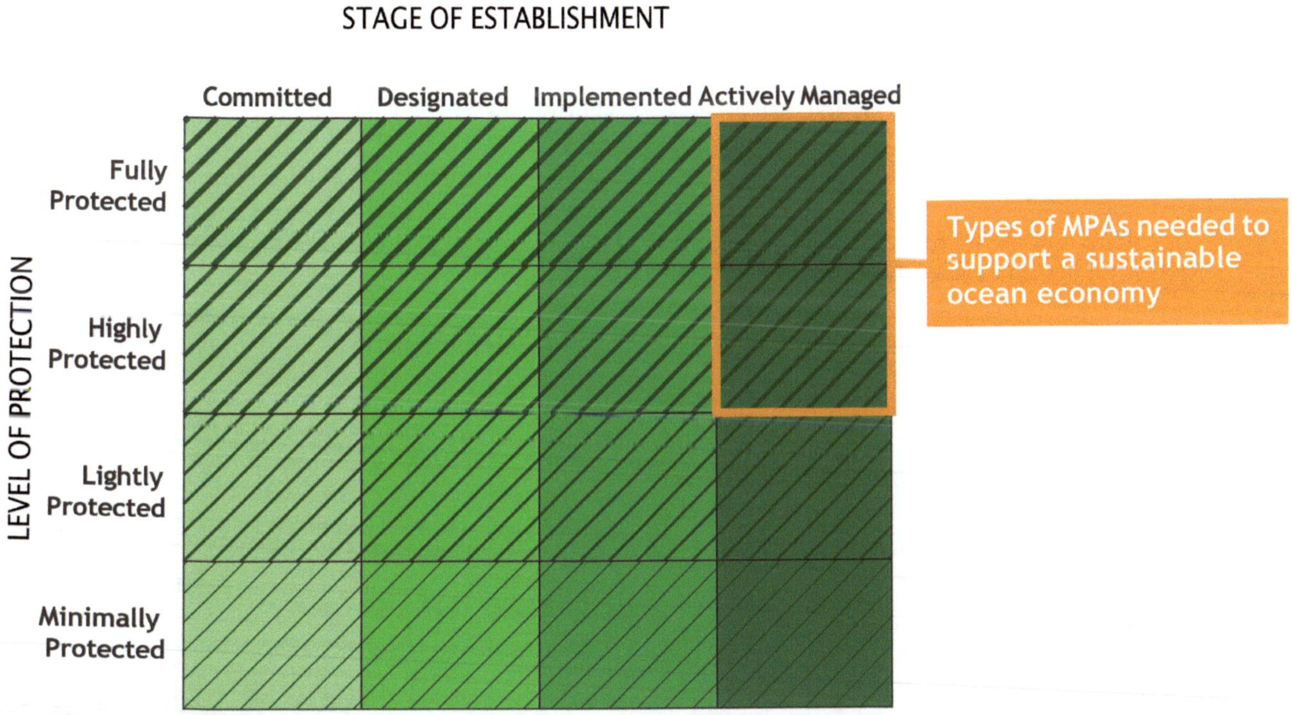

Fig. 20.27 The MPA guide. (*Source: Adapted from Oregon State University, IUCN World Commission on Protected Areas, Marine Conservation Institute, National Geographic Society and UNEP World Conservation Monitoring Centre. 2019. "An Introduction to the MPA Guide."* https://www.protectedplanet.net/c/mpa-guide)

able ocean scenario would be in the top right-hand corner. Indeed, species richness has been found to be 21% higher and biomass up to six times greater within fully protected marine areas (from here on simply called MPAs) compared to the adjacent unprotected areas.[366]

In a sustainable ocean economy scenario, the MPA placement would not be chosen randomly but designed according to science-based criteria, local knowledge and in consultation with diverse stakeholders. For instance, scientific analyses can produce scenarios to locate areas that maximise three benefits of MPAs: (1) rebuilding and safeguarding biodiversity, (2) mitigating climate change (by avoiding emissions from the disturbance of sediment carbon by bottom trawling and eventually deep-sea mining) and (3) boosting fisheries productivity (by increasing fisheries catches around MPAs through the spillover of fish). The food benefits would only be captured if the MPA strategy has been coupled with the sustainable management of the surrounding fisheries and an inclusive process that actively involves local communities and marginalised groups in the design and establishment of the MPAs.

5.4.2 Produce Sustainably

In the sustainable ocean economy scenario adopted in this section, effective ocean protection would enable sustainable ocean production. Most notably, the ocean can produce a near unlimited amount of renewable energy and significantly more seafood than today. In this section, an ambitious but realistic production potential is described.

Ocean-based renewable energy There appear to be no relevant physical limits to ocean-based production of renewable energy. Estimates for total technically feasible global offshore wind power generation potential range from 157,000 terawatt hours per year (TWh/yr.) to 631,000 TWh/yr.[367]—6–23 times more than the total global electricity consumption in 2018 (26,700 TWh/yr.).[368] Europe's offshore wind potential alone (71,845 TWh/yr.) is estimated to be over three times the current global electricity demand.[369] Other forms of ocean-based energy also have a very significant technically feasible potential, such as tidal energy (around 6200 TWh/yr.),[370] wave energy (between 1750 and 5550 TWh/yr.),[371] ocean thermal energy conversion (technical potential uncertain)[372] and salinity gradient energy (1650 TWh/yr.).[373] However, their cost is far from competitive today. By most realistic estimates, offshore wind will remain the most competitive offshore energy source, although the pace of development will remain far below theoretically feasible levels over the coming decades.

The International Energy Agency estimates that 570 GW of offshore wind could be installed by 2040.[374] An OECD scenario projects 900 GW by 2050[375] and the International Renewable Energy Agency (IRENA) REmap Scenario projects 1000 GW of installed offshore wind by 2050.[376] This suggests that even the upper range of the scenarios used in the Ocean Climate Special Report[377] may turn out to be conservative.

Sustainable seafood The ocean could in theory sustainably produce six times more food than today under an optimistic scenario,[378] thereby playing a significantly greater role in ensuring the food security of a planet with ten billion people in 2050. It has the potential to do so with a low environmental footprint (e.g. with sustainable fed mariculture and sustainable fisheries) or even in a regenerative way (e.g. with non-fed mariculture). Delivering this potential, however, depends on climate-adaptive, in-depth reforms of wild-catch fisheries, evolution of consumer preferences and significant scaling of (sustainable) mariculture:

- **Wild-catch fisheries.** Currently, most fishing is not economically or ecologically optimised. Far too many stocks are pursued by too many boats; far too much seafood value is lost due to poor handling; and far too many non-target species are accidentally caught. If this approach continues, 2050 yield is expected to decrease to about 67 mmt/year.[379]

[366]Dinerstein, E., C. Vynne, E. Sala, A.R. Joshi, S. Fernando, T.E. Lovejoy, J. Mayorga et al. 2019. "A Global Deal for Nature: Guiding Principles, Milestones, and Targets." *Science Advances* 5 (4): eaaw2869. doi: https://doi.org/10.1126/sciadv.aaw2869.

[367]Bosch, J., I. Staffell and A.D. Hawkes. 2018. "Temporally Explicit and Spatially Resolved Global Offshore Wind Energy Potentials." *Energy* 163 (November): 766–81. doi: https://doi.org/10.1016/j.energy.2018.08.153.

[368]IEA. n.d. "Data & Statistics"; Haugan et al. 2019. "What Role for Ocean-Based Renewable Energy and Deep-Seabed Minerals in a Sustainable Future?".

[369]Bosch et al. 2018. "Temporally Explicit and Spatially Resolved Global Offshore Wind Energy Potentials"; IEA. n.d. "Data & Statistics."

[370]Including tidal stream and tidal range energies. Retrieved from Haugan et al. 2019. "What Role for Ocean-Based Renewable Energy and Deep-Seabed Minerals in a Sustainable Future?".

[371]Haugan et al. 2019. "What Role for Ocean-Based Renewable Energy and Deep-Seabed Minerals in a Sustainable Future?".

[372]Haugan et al. 2019. "What Role for Ocean-Based Renewable Energy and Deep-Seabed Minerals in a Sustainable Future?".

[373]Haugan et al. 2019. "What Role for Ocean-Based Renewable Energy and Deep-Seabed Minerals in a Sustainable Future?".

[374]IEA. 2019. "World Energy Outlook 2019—Analysis." https://www.iea.org/reports/world-energy-outlook-2019.

[375]OECD. 2016. *The Ocean Economy in 2030*.

[376]IRENA. 2019. "Future of Wind."

[377]Hoegh-Guldberg et al. 2019. "The Ocean as a Solution to Climate Change."

[378]Costello et al. 2019. "The Future of Food from the Sea."

[379]Costello et al. 2019. "The Future of Food from the Sea."

However, if all stocks currently exploited were to be fished at maximum sustainable economic yield,[380] production could increase to 96 mmt/year in 2050: an additional 16 mmt/year of seafood compared to today.[381] This represents a 20% production gain over today's production levels, and a 40% increase over estimated BAU catch. It is important to note that these optimistic gains depend on the deployment of effective, climate-adaptive fishery reforms, strengthened international institutions and cooperation, in combination with scale-up of marine protected areas (see Sect. 6 for more details).

- **Mariculture.** The production of sustainable fed (finfish) and unfed (bivalve, seaweeds) mariculture is currently at a very small fraction of its biological potential (the theoretical production limit is estimated at 15,000 mmt/year—far more than 470 mmt of meat will be required annually in 2050 to feed the projected global population of more than 9.7 billion).[382]
  - Fed mariculture requires external feed (today including fish oil and fish meal) and is currently severely constrained by the price and availability of feed. Under optimistic projections assuming a 95% reduction of fish meal and fish oil content in mariculture feed, current production could be multiplied tenfold.[383] However, the siting and operations of monocultural, high-trophic finfish farms, especially in pristine areas, is often highly controversial. A reimagined approach to finfish farming, focused on local food security concerns, multi- and low-trophic species, new disease control and containment technologies, and avoidance of pristine areas, will be essential to capture the biological potential in a sustainable way.
  - Non-fed mariculture is ecologically largely benign and offers great potential. Bivalve mariculture (e.g. mussels, oysters), for example, could theoretically be increased more than 30-fold beyond current production to its biological potential of 460 mmt/year (bivalves only).[384] Seaweed, with a suitable cultivation area of 48 million km^2, has the potential to play a substantially larger role in supplying humanity with food and land animals and fish with feed. Seaweed also constitutes a very promising low-carbon source for raw materials that can be used in biostimulants (fertilisers), cosmetics, bioplastics, biofuels and other applications. In a sustainable ocean economy, the current economic, technological and regulatory barriers hindering the expansion of non-fed mariculture must be overcome (see Sects. 6.2 and 6.3).[385]

With these elements in mind, it is safe to say that reforming wild-caught fisheries and growing sustainable mariculture (especially unfed species) could multiply current ocean food production by up to six times by 2050 (Fig. 20.28).[386]

5.4.3 Prosper Equitably

This discussion describes prosperity in terms of jobs, economic wealth creation, inclusivity and equity if a sustainable ocean economy vision is realised. Only a small and simple sampling is possible—an exhaustive account of the relative upside of a restored, vibrant and productive ocean would fill libraries.

The future of ocean jobs, in many ways, echoes the general employment trends on land. In the energy sector, job growth is shifting to renewables, with many high-level engineering and support jobs created, especially in the developed world. Rising levels of productivity and automation would shift jobs in shipping, commercial fishing and large-scale mariculture from the front line to expert support (engineering, information technology, data, applied science, infrastructure). Small-scale fisheries would increasingly come under local control, recovering their productivity but imposing limits on fishing effort, enabled by smart policies that ensure secure access.

This report describes potential long-term evolutions of ocean jobs, building on various sources and projections from the pre-COVID period. The COVID-19 pandemic has seriously affected many ocean industries, making these projections and future jobs trajectories highly uncertain. For instance, up to 100 million jobs are today considered at risk in the tourism sector alone.[387] In addition, the crisis affecting ocean-based sectors is disproportionately hitting women and more vulnerable groups (low-skilled workers, small-scale fishers and businesses, Indigenous community members, younger workers, etc.).[388] Recovery and economic stimulus

[380] MSY and MEY: Maximum sustainable yield (MSY) is the long-term maximum amount of catch for a given fishery, purely based on the stock's biology. Maximum economic yield (MEY) adds the dimension of fishing costs to optimize for the most profitable, sustainable amount of catch, which is generally slightly lower than MSY catch. Information retrieved from World Ocean Review. "The Profits of Fishing." Maribus, after Quaas. n.d. "The Profits of Fishing: World Ocean Review." https://worldoceanreview.com/en/wor-1/fisheries/causes-of-overfishing/the-profits-of-fishing/. Accessed 18 August 2020.

[381] Costello et al. 2019. "The Future of Food from the Sea."

[382] Costello et al. 2019. "The Future of Food from the Sea."

[383] Costello et al. 2019. "The Future of Food from the Sea."

[384] Costello et al. 2019. "The Future of Food from the Sea."

[385] Costello et al. 2019. "The Future of Food from the Sea."

[386] Costello et al. 2019. "The Future of Food from the Sea."

[387] UNCTAD. 2020. "The COVID-19 Pandemic and the Blue Economy: New Challenges and Prospects for Recovery and Resilience." https://unctad.org/en/PublicationsLibrary/ditctedinf2020d2_en.pdf.

[388] Northrop, E., M. Konar, N. Frost and E. Hollaway. 2020. "A Sustainable and Equitable Blue Recovery to the COVID-19 Crisis." Washington, DC: World Resources Institute.

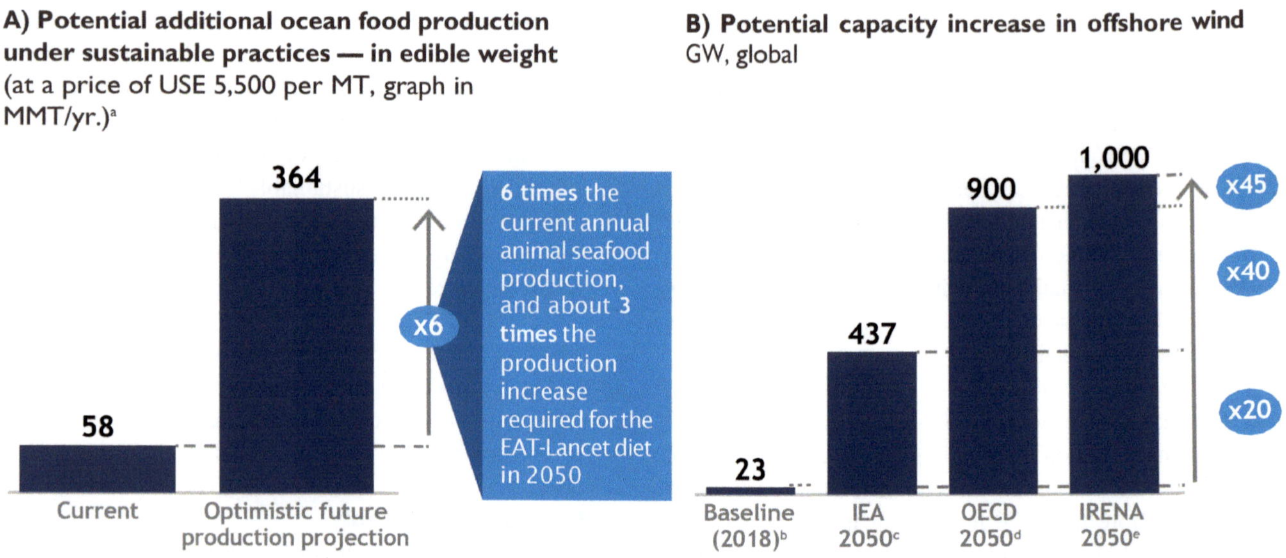

Fig. 20.28 (**a**, **b**) Ocean food and energy production potential increase under a sustainable ocean economy scenario. *Note: (a) Costello, C., L. Cao, S. Gelcich et al. 2019. "The Future of Food from the Sea." Washington, DC: World Resources Institute.* https://www.oceanpanel.org/blue-papers/future-food-sea. *(b) IRENA. 2019. "Future of Wind: Deployment, Investment, Technology, Grid Integration and Socio-economic—Executive Summary." Abu Dhabi: International Renewable Energy Agency.* https://irena.org/-/media/Files/IRENA/Agency/Publication/2019/Oct/IRENA_Future_of_wind_2019_summ_EN.PDF. *(c) IEA and ETP. 2017. "International Energy Agency, Energy Technology Perspectives 2017."* www.iea.org/etp2017. *(d) OECD. 2016. The Ocean Economy in 2030. Report. Paris: OECD Publishing.* https://www.oecd.org/environment/the-oceaneconomy-in-2030-9789264251724-en.htm. *(e) IRENA. 2019. "Future of Wind"*

plans supporting a sustainable ocean economy are expected to help maintain employment in ocean sectors and/or help transition towards the jobs required to develop the sustainable ocean sectors presented in this section.

- **Offshore energy.** Offshore energy is growing fast from a small base. Even in a conservative scenario, many jobs could be created: the OECD's BAU scenario (assuming no significant new government incentives) estimates the creation of 440,000 new jobs by 2030 in the offshore wind sector.[389] More assertive energy and industrial strategies could increase this number sharply. In the longer term, renewables are expected to outperform fossil fuel jobs in both relative and absolute numbers. In 2017, the U.S. Bureau of Labour Statistics listed turbine technician as the second-fastest-growing occupation in the United States. With periodic downturns in the offshore oil and gas industries, many oil and gas workers are turning to the wind industry for high-paying jobs. In U.S. coastal regions, 160,000 gross jobs could be supported by the offshore wind industry in construction, installation, operations and maintenance.[390]
- **Shipping and ports.** According to the OECD, seaborne cargo volume, driven almost entirely by GDP, will almost double from 11 billion tonnes in 2015 to 20 billion tonnes in 2030, which can be expected to significantly increase employment.[391] A more granular view reveals the major trends. A major expansion in ports, driven at least in part by China's massive Maritime Silk Road initiative, can be expected to increase trade. Larger and more automated vessels may slow job growth in shipping and shipbuilding, however (tonnage of ships larger than 7600 20-foot equivalent units (TEUs) can be expected to increase 6–6.5 times between 2010 and 2030, much faster than for ships under 7600 TEUs, projected to grow 1.4–2 times).[392]
- **Fishing and mariculture.** Global fishing, at the commercial and artisanal or small scale, operates at significant overcapacity today; there are too many fishers and too many boats. Because of this overcapacity, fish stocks, productivity and yields are depressed, and coastal livelihoods can be threatened. Net job growth is thus not the relevant metric to be applied to fishing—but job security

[389]OECD. 2016. *The Ocean Economy in 2030.*

[390]Gilman, P., B. Maurer, L. Feinberg, A. Duerr, L. Peterson, W. Musial, P. Beiter et al. 2016. "National Offshore Wind Strategy: Facilitating the Development of the Offshore Wind Industry in the United States." DOE/GO-102016-4866. EERE Publication and Product Library. doi: https://doi.org/10.2172/1325403.

[391]OECD. 2016. *The Ocean Economy in 2030.*

[392]QinetiQ, Lloyd's Register and University of Strathclyde Glasgow. 2013. "Global Marine Trends 2030." http://www.futurenautics.com/wp-content/uploads/2013/10/GlobalMarineTrends2030Report.pdf.

is, alongside food security and productivity. Nevertheless, the reduction of fishing capacity, and the associated stranded assets, may create tensions which must be thoughtfully addressed (through structural adjustments, reskilling, etc.; see discussion in Sect. 6.2). For industrial capture fisheries, jobs can be expected to decline, as fleets slowly reduce capacity and increase efficiency. Artisanal jobs are much harder to define and track—estimates range from 12[393] to 37 million, with an additional 100 million artisanal jobs being dependent on fishing (e.g. fish processors).[394] Many artisans fish opportunistically for food, rather than as a full-time pursuit. In a sustainable ocean economy, their time on the water will decrease, and yields will increase.

- The OECD projects strong mariculture employment growth to 3.2 million jobs in 2030, up from 2.1 million in 2010 under a BAU scenario. However, much higher job growth is possible if new technology can eliminate current constraints on feed availability and the production of non-fed mariculture is boosted. Buoyed by the growing maricultural capacity and recovering industrial capture yields, jobs from the seafood processing sectors can be expected to grow as well.[395]
- **Tourism**. Payment for ecosystem services through tourism fees could be adopted to finance the restoration and maintenance of the natural ecosystems (future) coastal tourism jobs rely on. Pre-COVID, the tourism sector was expected to continue its strong growth, directly accounting for over 8.5 million jobs in 2030 (up from 7 million in 2010).[396] Post-COVID, the trajectory for the tourism sector is still uncertain.

The economic future The size of the prize of the transition to a sustainable ocean economy is significant and appears to be limited far more by political and economic constraints than the ocean's productive potential. As for the jobs section, the numbers presented below reflect long-term evolutions and economic gains, building on various sources and projections from the pre-COVID period. Significant economic losses have been experienced by ocean sectors during the COVID-19 pandemic, and there is a high uncertainty as to the pace of recovery and transition towards a sustainable ocean economy for these sectors. For instance, cancellation of shipping is estimated to have caused revenue loss of US $1.9 billion for the carriers in a matter of months.[397]

On the conservative side, the OECD predicted in 2016 that economic growth and employment under a sustainable scenario would outpace both an 'unsustainable' and a 'BAU' scenario (see Fig. 20.29).

The OECD projections were based on 2010 data points as a baseline. A more recent study commissioned by the Ocean Panel provides a far more optimistic picture, with a net benefit estimated at $15 trillion by 2050 if $2.8 trillion were invested today in four sustainable ocean-based solutions: sustainable ocean food, renewable ocean energy, decarbonisation of international shipping, and conservation and restoration of mangroves.[398] The benefit-cost ratio differs for each of these opportunities, but overall it remains very attractive—see Fig. 20.30 below. These numbers are accounted through a holistic view that encompasses benefits of three kinds: economic (e.g. increased profits from higher fisheries productivity), environmental (e.g. avoided damages from coastal flooding) and health (e.g. reduced mortality from improved air quality).

Such an analysis has a number of limitations, as it does not represent the distribution of the benefits (and costs), it puts a monetary value on nonmarket goods with debatable assumptions, and it is obliged to omit certain benefits that are still very hard to monetise (e.g. prevention of the loss of natural habitats from increased ocean acidification). However, it serves as a very useful pointer, emphasising that ocean-based solutions should be considered as high-return investments and essential engines of a post-COVID economic, social and environmental recovery strategy.

Looking at the more detailed assessment of these four ocean-based solutions, this benefit-cost analysis offers conclusions in the following areas:[399]

- Mangrove conservation and restoration**:** Every $1 invested in mangrove conservation and restoration generates an average benefit of $3. Conservation has a far higher return on investment (88-to-1) than restoration (2-to-1), which can mainly be explained by the higher cost of mangrove restoration and the low survival rates following restoration. The total value of net benefits for mangrove restoration over 30 years ($97–$150 billion) is, however, higher than for conservation ($48–$96 billion), as the surface is assumed to be 10 times larger for restoration.

[393] Chuenpagdee, R., L. Liguori, M.L.D. Palomares and D. Pauly. 2006. "Bottom-up, Global Estimates of Small-Scale Marine Fisheries Catches." doi: https://doi.org/10.14288/1.0074761.

[394] FAO Fisheries and Aquaculture Department. n.d. "Small-Scale Fisheries around the World."

[395] OECD. 2016. *The Ocean Economy in 2030*.

[396] OECD. 2016. *The Ocean Economy in 2030*.

[397] "Sea Intelligence: COVID-19 Impact Pushes Carriers' Revenue Loss to USD 1.9 Bln." 2020. Offshore Energy (blog), 3 March. https://www.offshore-energy.biz/sea-intelligence-covid-19-impact-pushes-carriers-revenue-loss-to-usd-1-9-bln/.

[398] Konar and Ding. 2020. "A Sustainable Ocean Economy for 2050."

[399] Konar and Ding. 2020. "A Sustainable Ocean Economy for 2050."

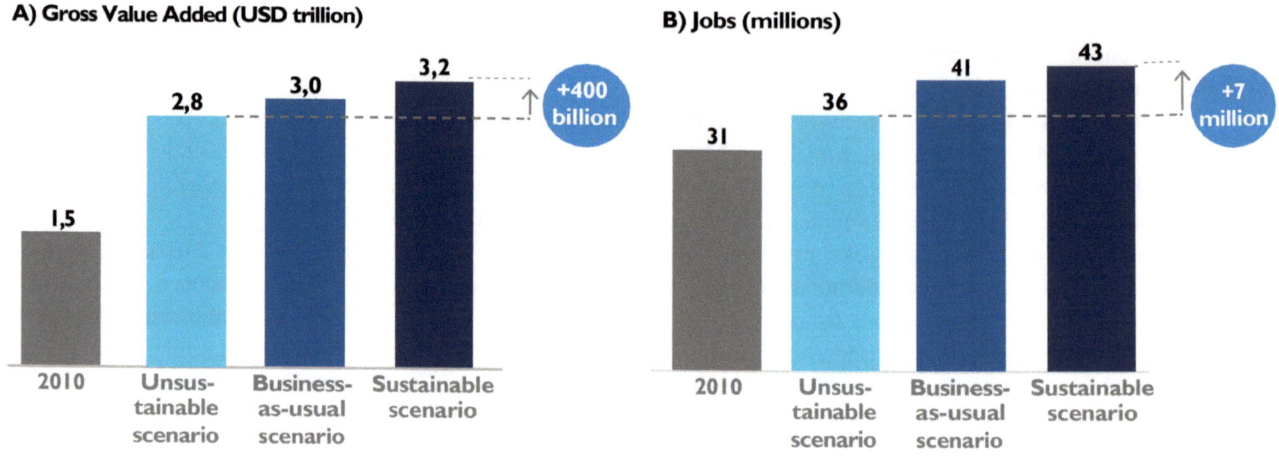

Fig. 20.29 (**a**, **b**) 2010–2030 GVA and job creation associated with different OECD scenarios. (*Source: OECD. 2016. The Ocean Economy in 2030. Directorate for Science, Technology and Innovation Policy Note, April.* https://www.oecd.org/futures/Policy-Note-Ocean-Economy.pdf)

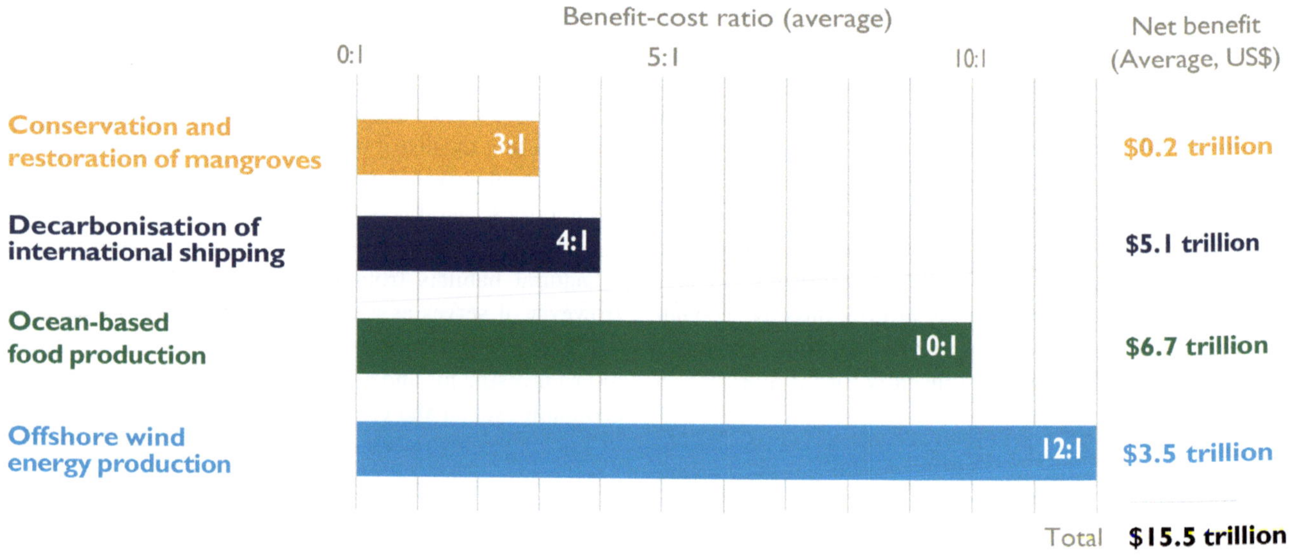

Fig. 20.30 Benefit-cost ratios and net benefits by 2050 for four sustainable ocean-based interventions. *Note: Average benefit-cost (B-C) ratios have been rounded to the nearest integer and the net benefits value to the first decimal place. The B-C ratio for mangroves is the combined ratio for both conservation- and restoration-based interventions. The average net benefits represent the average net present value for investments and is calculated over a 30-year horizon (2020–2050).* (*Source: Konar, M., and H. Ding. 2020. "A Sustainable Ocean Economy for 2050: Approximating Its Benefits and Costs." Washington, DC: World Resources Institute.* https://www.oceanpanel.org/Economicanalysis)

- Offshore wind: Every $1 invested in scaling up global offshore wind production generates a benefit estimated at $2–$17, depending on the cost of offshore energy production and transmission and the types of generation that would be displaced. The return on investment will increase as technology and efficiency improvements bring down costs for offshore wind energy generation.
- Green shipping: Every $1 invested in decarbonising international shipping and reducing emissions to net zero by 2050 is estimated to generate a return of $2–$5. The analysis assumed that the significant capital expenditure to switch to zero-carbon emissions will happen after 2030, and limiting the analysis to 2050 captures only a portion of returns from these investments, which will continue beyond 2050.
- Sustainable ocean-based food production: Every $1 invested in increasing production of sustainably sourced ocean-based protein is estimated to yield $10 in benefits. The increase in demand for ocean-based protein to provide a healthy diet for 9.7 billion people by 2050, which

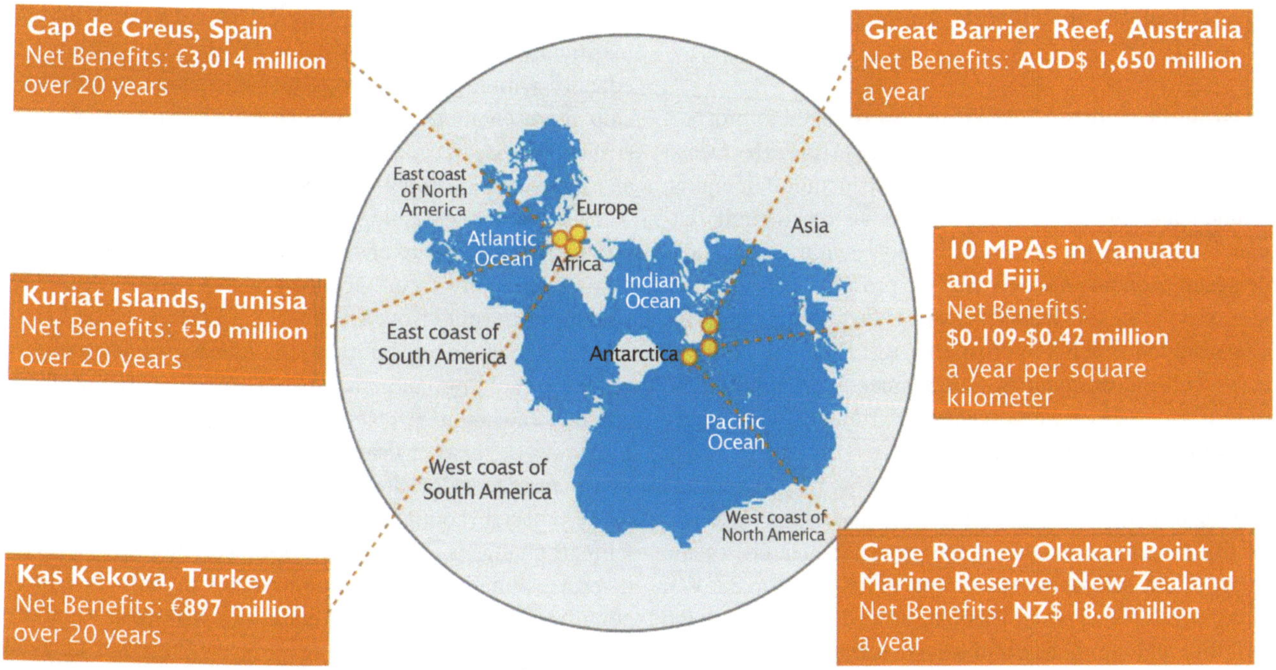

Fig. 20.31 Examples of positive economic impacts of marine protected areas. (*Sources: For Cap de Creus, Kas Kekova and Kuriat Islands: Mangos, A., and M.-A. Claudot. 2013. "Economic Study of the Impacts of Marine and Coastal Protected Areas in the Mediterranean." Valbonne, France: Plan Bleu.* https://planbleu.org/sites/default/files/publications/cahier_13_amp_en.pdf. *For Great Barrier Reef: Hand, T. 2003. An Economic and Social Evaluation of Implementing the Representative Areas Program by Rezoning the Great Barrier Reef Marine Park: Report on the Revised Zoning Plan. PDP Australia Pty. Ltd.* http://dspace-prod.gbrmpa.gov.au/jspui/bitstream/11017/3376/1/Hand_PDP_Australia_2003_Report_on_revised_zoning_plan.pdf. *For marine protected areas in Vanuatu and Fiji: Pascal, N., A. Brathwaite, L. Brander, A. Seidl, M. Philip and E. Clua. 2018. "Evidence of Economic Benefits for Public Investment in MPAs." Ecosystem Services 30 (April): 3–13. doi:* https://doi.org/10.1016/j.ecoser.2017.10.017; *and Hand. 2003. Hunt, L. n.d. 'Economic Impact Analysis of the Cape Rodney Okakari Point (Leigh) Marine Reserve on the Rodney District', 43.* https://www.howtokit.org.nz/images/emr/pdfs-files/Consultation_Resources/Hunt_2008_Leigh_marine_reserve_Economic_Analysis.pdf)

would replace a percentage of emission-intensive land-based protein sources, can be achieved by reforming wild-capture fisheries and by increasing the sustainable production of ocean-based aquaculture. Both measures will deliver benefits such as better health outcomes to consumers, higher revenues to fishers, lower GHG emissions mitigating the risks of climate damage, reduced land-based conflicts and lower water usage.

In addition to these four ocean-based solutions, additional evidence in the literature suggests that a sustainable ocean economy can generate significant economic returns. The creation of MPAs, especially when coupled with ecotourism, substantially increases revenue for local economies. Integration of ecotourism with MPAs needs to be approached with care to avoid natural habitat degradation through over-tourism. If precautions are taken, however, the creation of MPAs can have a significant economic benefit (Fig. 20.31).

The sustainable ocean economy agenda can also help catalyse land-based economic gains, especially regarding the currently wasteful plastics value chain. A systems approach to ocean plastics could result in annual savings for governments of $70 billion/year in 2040 while also reducing plastic leakage into the ocean by 80% compared to a business-as-usual trajectory.[400] Pioneering businesses in the circular economy also avoid financial and reputational liabilities.

Finally, the ocean agenda can also help catalyse broader reforms in agriculture. Agricultural regulations aimed at reducing ocean dead zones could result in farmers adopting precision agriculture practices to avoid runoff. This could eventually contribute to a broader food system reform towards sustainability, which has been estimated to repre-

[400] Lau et al. 2020. "Evaluating Scenarios toward Zero Plastic Pollution"; Pew Charitable Trusts and SYSTEMIQ. 2020. *Breaking the Plastic Wave*.

sent new business opportunities worth up to $4.5 trillion a year by 2030.[401]

The equitable future A healthy ocean is linked to prosperity and well-being.[402] The Blue Paper 'Towards Ocean Equity' argues that without an active consideration of equity, sustained and increased inequity will be the default outcome.[403] In the vision presented in this section, the sustainable ocean economy not only leads to prosperity of countries and economic sectors but also ensures adequate mechanisms for sharing the benefits of prosperity and alleviating climate change-induced inequalities. A fundamental principle of the SDGs is to 'leave no-one behind'.[404] Equality and equity considerations are implemented in the sustainable ocean economy for more than just moral reasons; they ensure the future legitimacy of the sustainable ocean economy agenda. Inequity remains a structural and persistent feature of the current ocean economy. Addressing these equity risks will counter accelerating social tensions, as well as strengthen the credibility and legitimacy of the sustainable ocean economy agenda. A recent report by the OECD, *Sustainable Ocean Economy for All*, includes a more detailed equity discussion, with a special focus on developing countries.[405]

Achieving 'procedural equity'—defined as the recognition of rights and needs of all groups and the level of inclusion and participation in decision-making related to ocean development[406]—will need to be a key achievement of the sustainable ocean economy. Indigenous knowledge which is compatible with scientific conclusions will be central to a sustainable ocean economy, and will need to be made widely accessible in knowledge commons. In terms of gender equality, women today comprise only 2% of the world's formal maritime workforce (1% for sailors).[407] By achieving gender equality, with respect to workforce participation, pay, leadership representation and advancement within a career, the sustainable ocean economy will fully unlock the productive and innovative potential of half of the world's population.

Ensuring the equitable sharing of marine genetic resources will be fundamental to ensuring a level playing field for furthering humanity's common heritage. To ensure this, the sharing of benefits from areas beyond EEZs must be based on the exchange of information, transfer of technology, capacity building and sharing of benefits arising from commercialisation.[408]

Yields of many artisanal fishers have declined precipitously in recent decades, and food insecurity runs high in many coastal communities in the developing world.[409] Climate change is expected to worsen current inequalities by disproportionally affecting communities in least developed countries.[410] Building a more equal and just ocean economy will be critical for economic prosperity.[411] Empowering local fishers by granting access rights will be one of the key levers of the sustainable ocean economy. Granting access rights has already been shown to be effective: a case study from Chile demonstrates that after the introduction of territorial use rights for fisheries, artisanal fisheries gained in importance, with landings even surpassing industrial catch while recovering the biomass and size of the target species.[412]

Rebuilding fish stocks and expanding non-fed aquaculture would significantly contribute to the alleviation of malnutrition (undernutrition and nutrient deficiency). Young children (<5 years) bear the burden: an estimated 150.8 million children are currently stunted (low height for age), another 50.5 million have weight too low for their age and 38.3 million are overweight.[413] Seafood contains critical trace minerals, omega-3 fatty acids, iodine and other micro-

[401]Food and Land Use Coalition (FOLU). 2019. *Growing Better: Ten Critical Transitions to Transform Food and Land Use*. https://www.foodandlandusecoalition.org/wp-content/uploads/2019/09/FOLU GrowingBetter-GlobalReport.pdf.

[402]Bennett et al. 2019. "Towards a Sustainable and Equitable Blue Economy."

[403]Österblom et al. 2020. "Towards Ocean Equity."

[404]UNDP. 2018. "What Does It Mean to Leave No One Behind?" UN Development Programme. http://www.undp.org/content/dam/undp/library/Sustainable%20Development/2030%20Agenda/Discussion_Paper_LNOB_EN_lres.pdf.

[405]OECD. 2020. *Sustainable Ocean for All*. https://www.oecd-ilibrary.org/docserver/bede6513-en.pdf?expires=1600102426&id=id&accname=guest&checksum=3BDD63D736252E0053B068682425AFEB.

[406]Definition of procedural equity by Österblom et al. 2020. "Towards Ocean Equity." Procedural equity refers to the recognition of rights and needs of all groups and the level of inclusion and participation in decision-making related to ocean development.

[407]IMO. n.d. "Women in Maritime: IMO's Gender Programme." http://www.imo.org/en/OurWork/TechnicalCooperation/Pages/WomenInMaritime.aspx. Accessed 11 May 2020.

[408]Österblom et al. 2020. "Towards Ocean Equity."

[409]Inniss et al. 2016. "The First Global Integrated Marine Assessment."

[410]Österblom et al. 2020. "Towards Ocean Equity."

[411]Österblom et al. 2020. "Towards Ocean Equity."

[412]Swilling et al. 2020. "The Ocean Transition."

[413]Fanzo, J., C. Hawkes, E. Udomkesmalee, A. Afshin, L. Allemandi, O. Assery, P. Baker et al. 2018. "2018 Global Nutrition Report: Shining a Light to Spur Action on Nutrition." Monograph. Bristol, UK: Development Initiatives. https://globalnutritionreport.org/reports/global-nutrition-report-2018/.

nutrients and vitamins crucial for healthy development.[414] These key nutrients could also help to reduce the 11 million annual deaths related to poor diet[415] if consumers shifted their eating habits to include healthier and more nutritious options, such as seafood.[416] Indeed, the IPBES states that 'shifting diets towards a diversity of foods, including fish, fruit, nuts and vegetables, significantly reduces the risk of certain preventable non-communicable diseases (e.g. cardiovascular diseases, cancers, diabetes), which are currently responsible for 20% of premature mortality globally'.[417]

The development of transparent supply chains and international collaboration can significantly reduce maritime crime. Transparent supply chains can minimize IUU fishing, leading to increased food provisioning to marginalised communities (often the worst affected). Other fisheries-associated crimes can be reduced by stronger international cooperation: the adoption and implementation of the International Declaration on Transnational Organized Fisheries Crime in the Global Fishing Industry must lead to reductions in forced labour and in the smuggling of people and contraband.

A celebrated ocean 'Happiness is among the most fundamental of all human goals'.[418] Millions of people a year travel to the ocean to enjoy themselves—an estimated 120 million are estimated to annually engage in marine recreational activities like diving, whale watching or recreational fishing.[419] A sustainable ocean economy would be able to maintain the healthy ocean required for ocean recreation.

Apart from providing leisure, the ocean is central to the aesthetic, religious and spiritual ways of many cultures,[420] especially Indigenous ones. A healthy ocean is essential to the maintenance of its immense cultural significance. MPAs and other effective area-based conservation measures can help preserve pristine and culturally important ocean areas (e.g. sacred sites, historic wrecks and associated war graves). In ocean areas where the sustainable extraction of resources by Indigenous groups is a key aspect of their culture, their rights to secure access and control should be guaranteed.

The ocean in service of the SDGs Building a more protected, productive and prosperous ocean economy offers solutions to accelerate other Sustainable Development Goals (see Figs. 20.32 and 20.33).[421] Replenishing and sustainably managing the ocean will be a significant part of achieving SDG 2 (zero hunger). A more sustainable ocean produces more food indefinitely.[422] The ocean's immense wind energy potential[423] advances the energy independence goals of SDG 7. Addressing ocean-based pollution could catalyse land-based reforms towards achieving SDG 15 (life on land) and push the world towards more responsible production and consumption, SDG 12. A study examined the relationship between SDG 14 targets and other SDGs at a more granular level, detailing the link, co-benefits or potential trade-offs[424]—see Figs. 20.32 and 20.33.

[414] James, D. 2013. "Risks and Benefits of Seafood Consumption." Rome: Food and Agriculture Organization of the United Nations, GLOBEFISH. http://www.fao.org/3/a-bb211e.pdf.

[415] Willett et al. 2019. "Food in the Anthropocene."

[416] Tacon, A.G.J., and M. Metian. 2013. "Fish Matters: Importance of Aquatic Foods in Human Nutrition and Global Food Supply." *Reviews in Fisheries Science* 21 (1): 22–38. doi: https://doi.org/10.1080/10641262.2012.753405.

[417] Díaz et al. 2019. "Summary for Policymakers of the Global Assessment Report on Biodiversity and Ecosystem Services of the Intergovernmental Science-Policy Platform on Biodiversity and Ecosystem Services."

[418] Wang, F., H.M. Orpana, H. Morrison, M. de Groh, S. Dai and W. Luo. 2012. "Long-Term Association between Leisure-Time Physical Activity and Changes in Happiness: Analysis of the Prospective National Population Health Survey." *American Journal of Epidemiology* 176 (12): 1095–1100. doi: https://doi.org/10.1093/aje/kws199.

[419] Cisneros-Montemayor, A.M., and U.R. Sumaila. 2010. "A Global Estimate of Benefits from Ecosystem-Based Marine Recreation: Potential Impacts and Implications for Management." *Journal of Bioeconomics* 12 (3): 245–68. doi: https://doi.org/10.1007/s10818-010-9092-7.

[420] Inniss et al. 2016. "The First Global Integrated Marine Assessment."

[421] Singh, G.G., A.M. Cisneros-Montemayor, W. Swartz, W. Cheung, J.A. Guy, T.-A. Kenny, C.J. McOwen et al. 2018. "A Rapid Assessment of Co-benefits and Trade-offs among Sustainable Development Goals." *Marine Policy* 93 (July): 223–31. doi: https://doi.org/10.1016/j.marpol.2017.05.030.

[422] Costello et al. 2019. "The Future of Food from the Sea."

[423] IEA. n.d. "Data & Statistics."

[424] Singh et al. 2018. "A Rapid Assessment of Co-benefits and Trade-offs among Sustainable Development Goals."

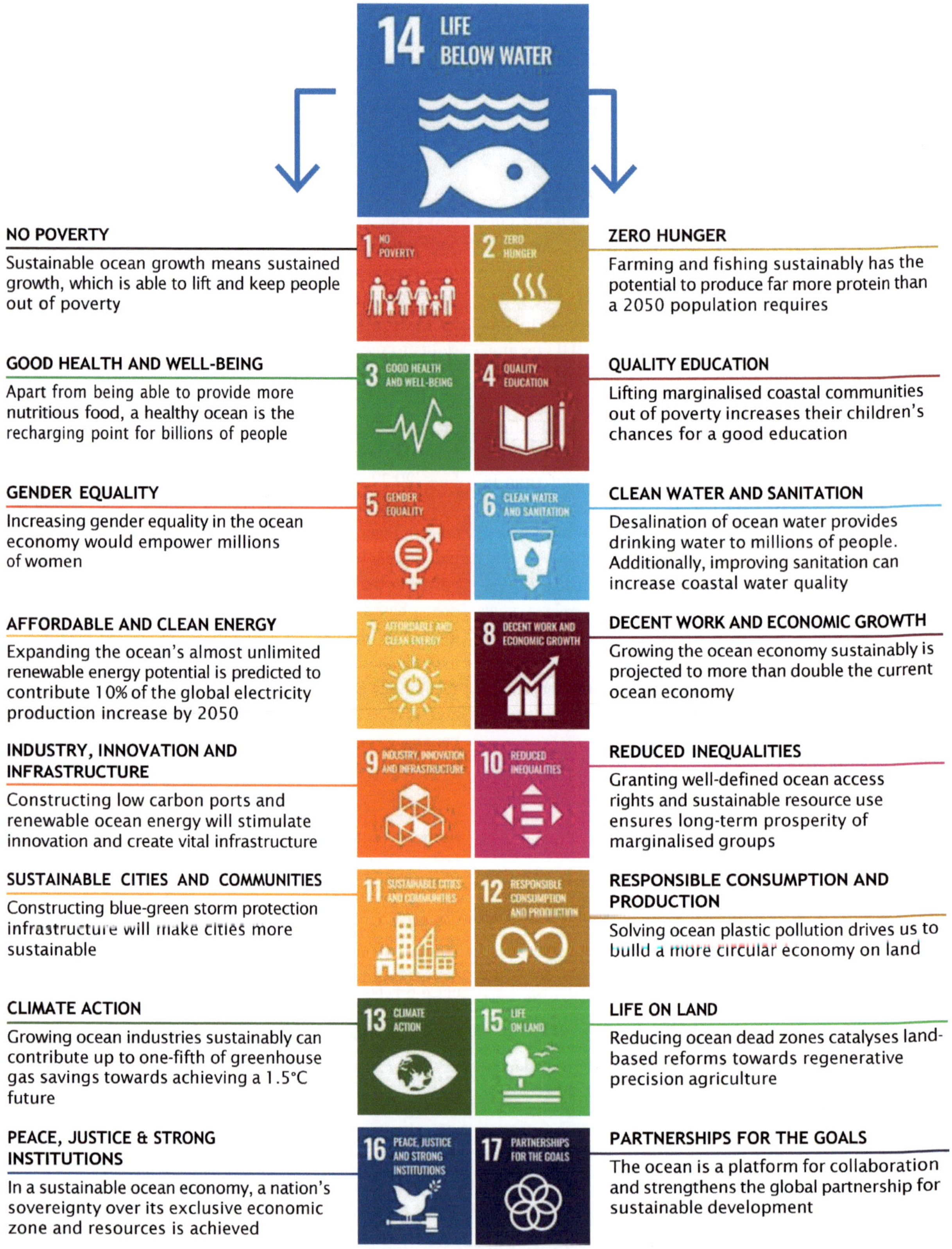

Fig. 20.32 Achieving SDG 14 helps achieve the other SDGs. *Note: Regarding SDG 6 (clean water and sanitation), the link to the ocean can be made through desalination plants. Regarding SDG 17 (partnerships for the goals), the ocean provides excellent platforms for collaboration. Peaceful ocean science collaboration, for example, has been important for diplomatic relations (e.g. U.S.-Soviet Gulf Stream experiments in the 1960s). (Source: Authors)*

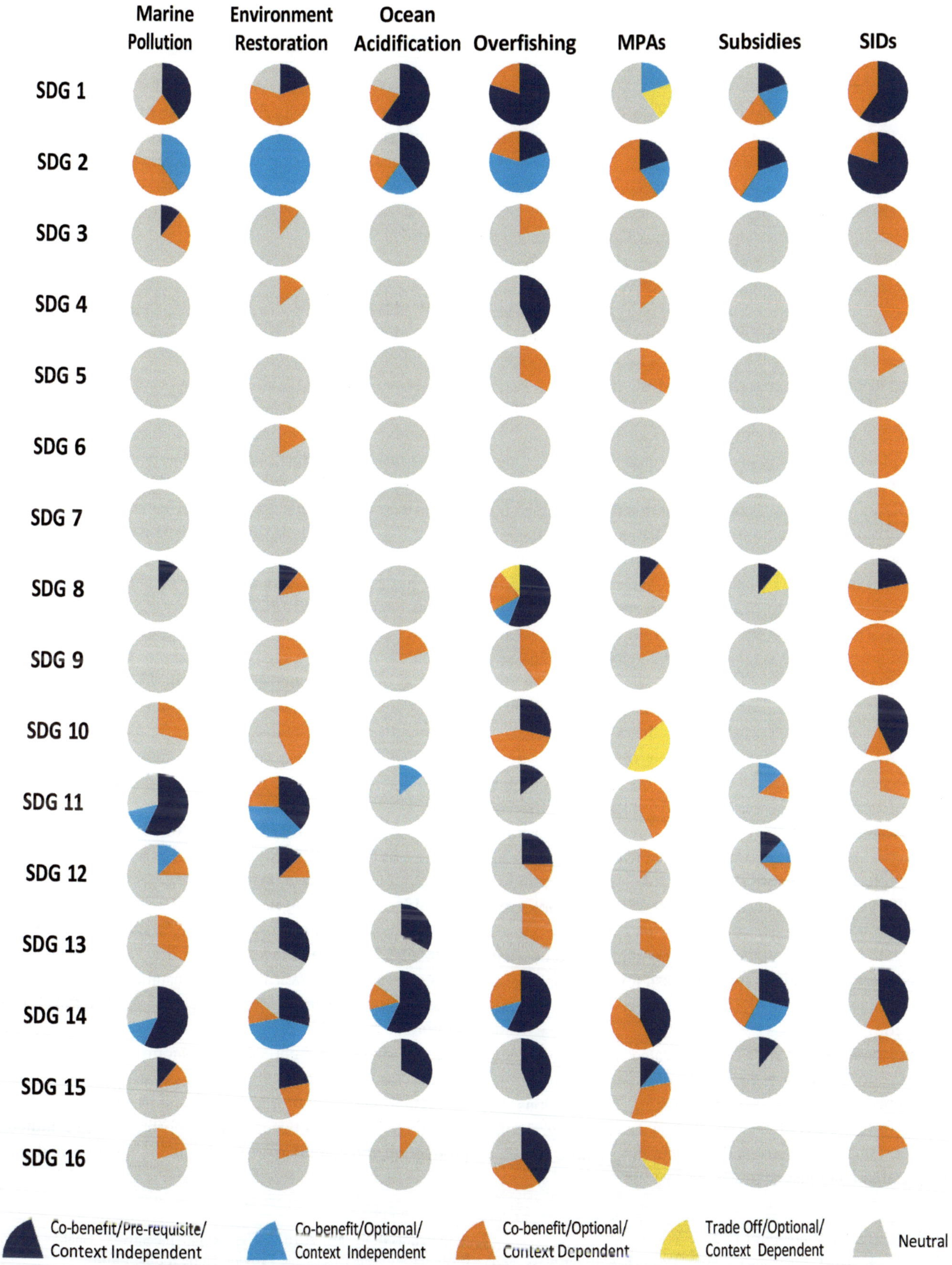

Fig. 20.33 Detailed relationship between SGD 14 targets and other SDGs. (*Source: Singh, G.G., A.M. Cisneros-Montemayor, W. Swartz, W. Cheung, J.A. Guy, T.-A. Kenny, C.J. McOwen et al. 2018. "A Rapid Assessment of Co-benefits and Trade-offs among Sustainable Development Goals." Marine Policy 93 (July): 223–31. doi:* https://doi.org/10.1016/j.marpol.2017.05.030)

6 A Roadmap to a Sustainable Ocean Economy

6.1 Introduction

Section 4 makes an urgent case for action: the ocean is vital for humankind and the entire economy, current ocean management struggles to deliver on the dimensions of protection, production and human prosperity, and more and more successful sustainable stories and pioneers are in desperate need of support and scale-up. Section 5, which shows how ocean protection and ocean productivity can serve each other, outlines an alternative, generalised 'triple win' future. The case for a healthy ocean supporting a healthy economy is well documented. So why is this so hard? Why isn't the sustainable ocean economy at the centre of all the post-COVID recovery discussions and financial stimulus packages? How can this urgently needed shift be accelerated?

The global struggle against climate change is both immediately relevant and in an analogous stage. The science is compelling; consequences are rigorously documented (IPCC); the cost of inaction is quantified, as is the business case for change; pledges are in place (UN Framework Convention on Climate Change/Paris Agreement); a majority of people find this to be the defining crisis of their lives; investors are starting to move away from fossil fuels. And yet, progress is falling short: the UN Environment Programme's *Emissions Gap Report 2019* states that the world is heading for a 3.2 °C global temperature rise over pre-industrial levels,[425] far beyond the 'below 2 °C' target.

The relationship between the ocean and humanity goes both ways—one shapes the other. In this classic 'complex adaptive system',[426] the biological, chemical and physical ocean responds to an array of human forces which, in turn, are shaped by regulation, taxation, financial interests, consumer preferences, historical legacies, and diverse traditions and cultures. Any change in a system of such complexity, almost by definition, has unforeseen and complex consequences. For climate and the ocean, the implicit and explicit rules are based on the lessons of the past, not the future—and they are legally, politically and culturally entrenched and protected.

How, then, can the shift be accelerated from the urgency of Sect. 4 to the more hopeful future of Sect. 5? At a time when governments are actively looking for solutions to recover from the COVID-19 shock, how can the integral role of the sustainable ocean agenda in rebuilding a more sustainable, resilient and just economy be ensured?

[425]UNEP. 2019. *The Emissions Gap Report 2019*.

[426]Levin, S.A., and J. Lubchenco. 2008. "Resilience, Robustness, and Marine Ecosystem-Based Management." *BioScience* 58 (1): 27–32. doi: https://doi.org/10.1641/B580107.

This final section focuses on the 'how' and provides a roadmap addressing the following questions:

- What are the barriers to change, and what lessons from the experience of similar industrial and societal transitions could be applied to the ocean economy? (Sect. 6.2)
- What main transitions are required and how can this sustainable ocean economy agenda be structured? (Sect. 6.3)
- What catalytic interventions can help enter an upward spiral? (Sect. 6.4)

6.2 Harnessing Complex Adaptive Systems: Lessons for the Sea

The social, economic and ecological systems in the ocean realm connect into a complex adaptive system, where the 'behaviors of individual actors at the local scale influence interactions and emergent properties at the regional or global scale. Emergent properties, in turn, can feed back to the small scale and influence subsequent behaviors of the individuals'.[427] This complexity can explain why the current model of ocean management exists and is so hard to transform. But the adaptability also leaves room for evolution if the mechanics and incentives are changed, and if feedback loops are switched from vicious to virtuous.

This section first describes the barriers which have made the pace of reform appear timid and slow. Learning from other socioeconomic and industrial transitions, this section then identifies a framework that could be used for a successful transition towards a sustainable, more equitable ocean economy.

6.2.1 Major Barriers to a Sustainable Ocean Economy

The complex adaptive system of the ocean economy is shaped today by strong incumbent interests, cultural norms, institutional constraints, policies and laws. In this status quo, the feedback loops and incentives are driving behaviours that hinder a transition towards a truly sustainable, regenerative ocean economy. These incentives can be of different kinds: economic, reputation-driven or personally motivated social norms.[428] In the current ocean economy, these incentives share a common feature: they ignore or vastly discount environmental and social impacts. To shift these incentives towards alignment between effective protection, sustainable production and equitable prosperity, the first step is to dissect

[427]Lubchenco et al. 2016. "The Right Incentives Enable Ocean Sustainability Successes and Provide Hope for the Future."

[428]Lubchenco et al. 2016. "The Right Incentives Enable Ocean Sustainability Successes and Provide Hope for the Future."

Fig. 20.34 Main barriers to a sustainable ocean economy. (*Source: Authors*)

some of their main root causes, presented here as six systemic barriers (Fig. 20.34).

Institutional inefficiencies A complete description of the ocean-related institutional structures and agreements is beyond the scope of this report, but some inefficiencies can be listed here: complexity of governance, lack of overarching mandate towards a healthy ocean and rigid and static processes poorly informed by science.

Complexity of governance. International ocean management is a web of intertwined, converging and competing demands and interests[429] involving no fewer than 576 bilateral and multilateral agreements,[430] which are administered by a multitude of institutions with widely varying mandates, resources, authorities and capacities (Fig. 20.35).

'Polycentric governance', that is, governance that includes multiple centres of semiautonomous decision-making, can be an efficient model to manage at a global scale a complex adaptive system like the ocean.[431] Polycentric models indeed allow decision-makers to 'experiment with different governance solutions tailored to particular scales and socioecological contexts'.[432] However, if collaboration, transparency and clear mandates are ill-defined, polycentricity can be a double-edged sword and limit efficiency and capacity for change and more sustainable management.

At the global level, the exact reach and authority of even well-established ocean organisations based on global treaties is often unclear. For example, the extent of the International Whaling Commission's legal competence is in dispute, with some nations restricting it to great whales (baleen and sperm whales), while others include all cetaceans. In many cases, individual nations claim exceptions to specific articles of the convention, while still remaining members. As another example, the ongoing negotiations on biodiversity beyond national jurisdiction are exceptionally important but unlikely to call for a new international framework for ocean governance under the UN umbrella. Nevertheless, there are also examples of well-functioning international frameworks. High-level international forums such as the Arctic Council and the Parties to the Nauru Agreement focus more on pragmatic advantages of cooperation (such as shipping safety in the Arctic or cooperation in tuna stock management) than the painstaking process of formal treaty negotiation.

[429] UNESCO. n.d. "Ocean Governance and Institutional Challenges." http://www.unesco.org/new/en/natural-sciences/ioc-oceans/focus-areas/rio-20-ocean/ocean-governance/.

[430] UNESCO. n.d. "Ocean Governance and Institutional Challenges."

[431] Ostrom, E. 1990. *Governing the Commons: The Evolution of Institutions for Collective Action*. Cambridge: Cambridge University Press.

[432] Independent Group of Scientists appointed by the Secretary-General. 2019. *Global Sustainable Development Report 2019: The Future Is Now—Science for Achieving Sustainable Development*. New York: United Nations. https://sustainabledevelopment.un.org/content/documents/24797GSDR_report_2019.pdf.

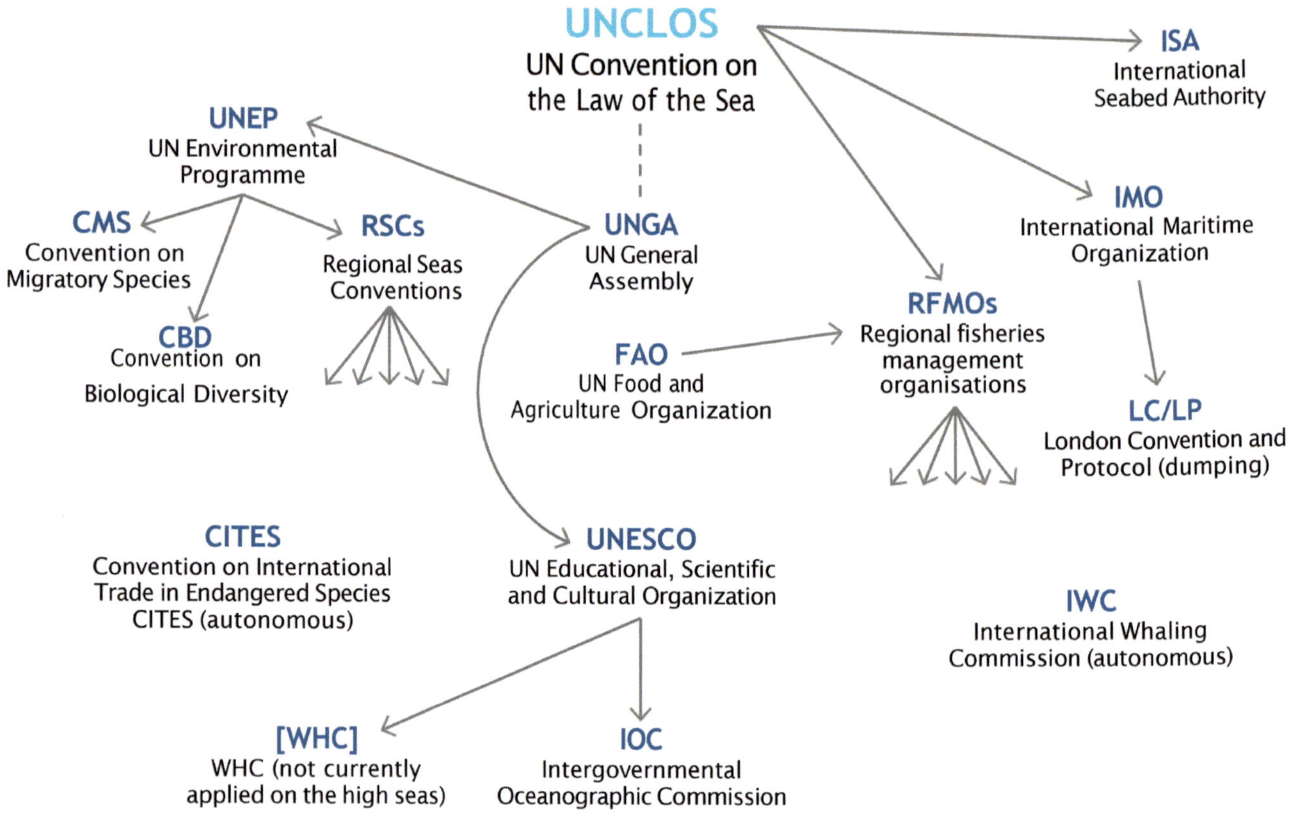

Fig. 20.35 Institutions and frameworks that support the UN convention on the law of the sea. (*Source: Adapted from Ardron, J.A., and R. Warner. 2015. "International Marine Governance and Protection of Biodiversity." In Routledge Handbook of Ocean Resources and Management, edited by H.D. Smith, J.L.S. de Vivero and T.S. Agardy, 55–72. London: Routledge*)

Lack of overarching mandate towards a healthy ocean. Most of the ocean-related institutions have been established to support the development of a given ocean-based sector. This development has usually been assessed against conventional and incomplete metrics like sector revenues (or GDP) or number of jobs. These institutions are rarely driven by an overarching mandate that would transcend the sectorial silos and also aim to achieve healthy ocean ecosystems (see the next barrier, 'Lack of planning and poor integration among sectors').

Rigid, static poorly science-informed processes. In most cases these ocean institutions are not equipped to pilot the management of a complex adaptive system. Because they depend on laws and/or consensual decisions that take time to change and reform, they usually deliver their mandate in a static fashion, and react a posteriori to shocks and unexpected events. As stated above, social, environmental and economic systems are intertwined in a complex adaptive system whose proper governance requires adaptability and agility. Besides, the management of the ocean today is not informed enough by solid science, and personal or national interests can often outweigh recommendations from the scientific community. The upcoming UN Decade of Ocean Science for Sustainable Development is a great opportunity to strengthen this science-policy interface.

International fisheries are a good example of many of these institutional inefficiencies. Myriad bilateral, trilateral and multilateral fisheries management agreements and regional fisheries bodies exist. However, for the high seas, only the 16 regional fisheries management organisations (RFMOs) are mandated through the UN Fish Stocks Agreement to adopt legally binding measures. The RFMOs differ widely in their funding, scientific capacity, relative authority with member states and, ultimately, fishery outcomes.

Since decisions are generally made only based on (near) unanimity among member states, the process can be slow and somewhat weighted towards avoiding losses by individual states, rather than optimising the fisheries or ensuring healthy ocean ecosystems.[433] Co-operation among RFMOs is improving, and even though discrepancies in performance and transparency of RFMOs remain,[434] several of them are more closely following the conclusions of their scientific committees. Even if more than an RFMO, the ecosystem monitoring program of the Commission for the Conservation of Antarctic Marine Living Resources, which applies the precautionary principle and commits to 'best available science' standards in its management of krill and finfish restoration efforts,[435] may be an example of better management practices, although it, too, needs improvements. The steadily improving data transparency on fisheries driven by FAO is also leading to broader reforms.

Lack of planning and integration among sectors Ocean development, so far, has largely occurred ad hoc. When communication among the food, energy and shipping sectors does occur, it is more often about conflict resolution than symbiosis and collaboration.

Formal coordination remains rare, since ocean sectors are often governed by different regulating entities, making cross-sectoral communication and planning difficult. As of today, only 22 countries[436] systematically assess the efficient, safe and symbiotic use of the ocean's resources—its power generation, biological productivity, carbon sequestration and so on. Typically, such a planning process provides guidance on the integration of ocean uses, avoidance of spatial use conflicts, standards of operation, streamlined and efficient regulatory process, and the overall protection and sustainability of the key oceanic systems. It is sometimes supported by new incentives ('carrots' or 'sticks') including public sector demand and offtake guarantees, feed-in tariffs, infrastructural support and public participation in the required investments.

Without such a process, ocean economy sectors are left to compete in an operational and regulatory vacuum. Even fully mature technologies, such as offshore wind farms, often struggle in such a regulatory environment; for example, the offshore wind industry along the U.S. Eastern Seaboard, despite compelling economics and ample demand, has been mired in regulatory setbacks for the better part of a decade. Innovative concepts—such as multi-trophic farming, co-location of fish and seaweed farming, and the use of offshore wind energy to produce shipping fuels (hydrogen and ammonia), power large-scale aquaculture or power desalination plants—are very difficult to realise without integrated planning, explicit national priorities, regulatory support and time-tested emerging technology incentives.

Concerns about stranded jobs and communities A sustainable ocean economy looks very different from today and implies real structural shifts—such as the reduction of fishing fleets and jobs, on the one hand, and the increase of wind energy and mariculture jobs, on the other. Many coastal communities are built around fisheries and fishery-related jobs that constitute their heritage and social glue. Some of these communities may find themselves disadvantaged by this transition: economically speaking, they may not be able to find alternative sources of income overnight, potentially causing significant short-term equity issues; culturally speaking, it will be very hard for these communities to give up on decades or centuries of social norms centred on traditional activities. Their concerns are entirely legitimate and clearly require public support, guarantees and a transparent and inclusive dialogue.

A good example is in fisheries. Global current fishing capacity is estimated to be between 1.5 and 2.5 times more than what is needed to fish under maximum sustainable yield.[437] To rebuild global fisheries, millions of the current 4.3 million fishing boats need to be decommissioned at a global scale, and between 15 and 22 million fishers (assuming linearity, which is probably simplistic) would need to shift to other pursuits.[438] However, fishing jobs do not convert easily, and alternative wages tend to be low.[439] The economic challenge is real but not insurmountable if proper solutions, support and compelling change management are

[433] Pew Charitable Trusts. 2019. "International Fisheries Managers' Response to Performance Reviews Insufficient." https://www.pewtrusts.org/-/media/assets/2019/04/international_fisheries_managers_underuse_performance_review_guidance_v1.pdf.

[434] Hutniczak, B., C. Delpeuch and A. Leroy. 2019. "Intensifying the Fight against IUU Fishing at the Regional Level." OECD Food, Agriculture and Fisheries Papers, no. 121 (February). doi: https://doi.org/10.1787/b7b9f17d-en.

[435] Bell, J.B., E. Guijarro-Garcia and A. Kenny. 2019. "Demersal Fishing in Areas beyond National Jurisdiction: A Comparative Analysis of Regional Fisheries Management Organisations." *Frontiers in Marine Science* 6. doi: https://doi.org/10.3389/fmars.2019.00596; Brooks, C.M., L.B. Crowder, L.M. Curran, R.B. Dunbar, D.G. Ainley, K.J. Dodds, K.M. Gjerde and U.R. Sumaila. 2016. "Science-Based Management in Decline in the Southern Ocean." *Science* 354 (6309): 185–87. doi: https://doi.org/10.1126/science.aah4119.

[436] Frazão Santos, C., C.N. Ehler, T. Agardy, F. Andrade, M.K. Orbach and L.B. Crowder. 2019. "Marine Spatial Planning." Chapter 30 in *World Seas: An Environmental Evaluation*, 2nd ed., edited by C. Sheppard, 571–92. Cambridge, MA: Academic Press. doi: https://doi.org/10.1016/B978-0-12-805052-1.00033-4.

[437] Porter, G. 1998. *Estimating Overcapacity in the Global Fishing Fleet*. Washington, DC: World Wildlife Fund.

[438] Sumaila, U.R., W. Cheung, A. Dyck, K. Gueye, L. Huang, V. Lam, D. Pauly et al. 2012. "Benefits of Rebuilding Global Marine Fisheries Outweigh Costs." *PLOS ONE* 7 (7). doi: https://doi.org/10.1371/journal.pone.0040542.

[439] Sumaila et al. 2012. "Benefits of Rebuilding Global Marine Fisheries Outweigh Costs."

put in place. For instance, 75% of fishers in Hong Kong would be 'willing to leave the industry if suitable alternatives or compensation were available'.[440] Similar sentiments are likely to arise in other countries.[441] Cultural and spiritual dimensions need to be seriously considered, however, as in some communities fishermen do not want to leave fishing—even if suitable alternatives exist.

High costs of capital The cost and availability of capital is a serious constraint across the spectrum of ocean enterprise. The latest data from the IEA (2019), for example, show that the levelised cost of electricity (LCOE) from offshore wind is reduced by 30% if the weighted average cost of capital is reduced from 8% to 4%.[442] This clearly shows the importance of creating access to stable financing. Capital-intensive ventures such as ocean-based renewable energy and large-scale mariculture encounter technical, infrastructural and regulatory challenges, which grow exponentially with the distance from shore—precisely where most large-scale food and energy production could be sited. Key technologies may not yet fully be tested in the cauldron of open ocean conditions and intensifying storms. Onshore competition, such as from onshore wind, freshwater aquaculture and alternative proteins, is a source of considerable uncertainty. Lastly, there are many potential sources of use conflicts and attendant regulatory risks and delays. The sum of these risks leads to capital premiums, and many institutional investors may stay away altogether. There is no shortage of innovative thinking, concepts, blueprints and business plans in the sustainable ocean economy—but the leap from concept to reality is harder in the ocean than it is on land.

On the less capital-intensive side of the spectrum, financing issues also loom large. In fisheries, for example, real or implied discount rates are high. Artisanal fishers often do not have the luxury of planning for tomorrow's catch. Even in more organised commercial fishing, open access laws and overcapacity lead to a 'race for fish' that heavily discounts future yields. The current market-hunt nature of wild-catch fisheries is very difficult to fit into an investment structure requiring legally robust, long-term ownership of future cash flows. The same dynamics have made the financing of fishery recovery efforts especially difficult, and terrestrial crop insurance mechanisms have, by and large, not been translated to fish stocks. Except at unsecured, very high rates, small fishers in the developing world have little access to the capital needed to build port-side infrastructure, develop efficient transportation and value-adding processing, buy safer boats and prepare long-term management strategies.

[440] Teh, L., W.W.L. Cheung, A. Cornish, C. Chu and U.R. Sumaila. 2008. "A Survey of Alternative Livelihood Options for Hong Kong's Fishers." *International Journal of Social Economics* 35 (5): 380–95. doi: https://doi.org/10.1108/03068290810861620.

[441] Sumaila et al. 2012. "Benefits of Rebuilding Global Marine Fisheries Outweigh Costs."

[442] IEA. 2019. *Offshore Wind Outlook 2019.*

Suboptimal market dynamics The tragedy of the ocean commons can be quite pronounced, with some economic rents procured at the expense of overall system health and productivity. Too often, legal or regulatory recourse is elusive when one nation overfishes at the cost of another, when land-based polluters harm fishers, or when climate change destroys reefs that protect communities. The gains (mostly terrestrial interests) and losses (mostly ocean interests) can be in the billions of U.S. dollars (see examples in Sect. 4). In many cases, these characteristics of the system prevent change towards sustainability—for example, regional fisheries management organisations have been seen to allow member countries to block reforms, even when the economic and scientific rationale is compelling; terrestrial polluters are still mostly legally indemnified from their ocean liabilities; and the international vessel registration system is designed to allow the origins of economic and legal ship ownership to be separate. These can be significant hurdles to overcome on the way to reform.

Existing laws often protect incumbent interests. Open access to ocean resources is enshrined in law, culture and convention—in many developing countries, it is regarded as a constitutional right of artisanal fishers, even if it makes local ownership and stewardship nearly impossible. International law enshrines the right to fish, free passage and open access to open ocean resources. Open access may appear equitable, but it often leads to a race for resources which ultimately favours depletion and inequalities over stewardship. Similarly, parties who are making the investment in sustainably managing a resource do not always reap the benefits of their investment, as in the case of highly migratory fish species that cross multiple EEZs during their lives. If one country invests in species management within its EEZ, other countries benefit from this investment—through the free-rider effect—due to increased catches. The potential free rider's overfishing behaviour of the shared stock could nullify any efforts by the investing country once it reaches that country's EEZ.

Subjectivity and irrational behaviours: (Not) making sense of the largely unknowable Subjectivity and irrational behaviours conspire against systems thinking and the transition towards a sustainable ocean economy. A number of individual behaviours can be explained by cognitive biases, which result from simplifications the human brain does to make decisions out of complex information. For instance, humans tend to react to the possibility of highly worrisome news (such as a scenario describing a potentially catastrophic future) by seeking to confirm the belief that they are safe—and they are much more likely to believe those peers who confirm this belief (called 'confirmation bias'). This bias is easily exploited by those who use the inherent uncertainty of ocean state predictions to invalidate them. Confirmation bias is exceedingly difficult to overcome with new scientific information alone—it needs to be addressed on a cultural level. Similarly, when faced with the need for reforms which require short-term sacrifice for long-term gain (as is the case with many fishery reforms), humans tend to systematically overvalue present over future assets—even when (rational) discount factors are included. From a systems perspective, this can all amount to a hard-to-break feedback loop: bad habits are systematically re-enforced by the very system they shape. This bias is reinforced by short political cycles: many coastal and ocean decisions being made in the present have time horizons of decades to over a century, far longer than the lifespan of the governance arrangements facilitating them.

Any person's decision-making is hard—it needs to weigh long-term over short-term return, the value of different forms of wealth (financial, natural, cultural and personal) now and in the future; risk and peace and security, status, cultural and religious norms and so on. In many countries, cultural, legal and religious norms re-enforce a view of the ocean as both inexhaustible and commonly owned, with the spoils going to those who 'brave the sea' and take risks. Thus, understanding people's values and resulting emotional responses is critical in a system transition. Legal norms pertaining to activities taking place on the ocean tend to be weaker than those on land, and less enforced. Ownership of ocean resources tends to be far less defined, and far less definable, than on land.

On the positive side, these (subjective) personal incentives can also be used for good: altruism, ethical values, reciprocity and other types of intrinsic motivations can become powerful drivers of positive change.[443] Enhancing reputation and brand image can also be a strong incentive for businesses or governments to proactively lead on sustainable practices.[444]

6.2.2 To Move the System, It Is Important to Learn from Other System Transitions

Despite barriers, even very complex adaptive systems can shift onto new trajectories—sometimes very quickly. Economic history is replete with examples. When new information is plentiful and there is strong support for change, the shift can be entirely designed and purposeful—such as the energy transition in Germany, the Global Vaccine Alliance, smoking bans in bars and restaurants in Ireland and France, or the Montreal Protocol on Substances That Deplete the Ozone Layer. Even in the absence of new information or strong support, shifts can happen—sometimes they emerge from a new framing of an older issue (non-traditional marriage in the United States) or they take a less obvious, emergent form (such as the downward trajectory of meat consumption per capita in OECD countries in the past decade[445]). Deriving key success factors from these shifts as well as from the latest literature on system transitions,[446] a framework for a transition towards a sustainable ocean economy can be articulated (Fig. 20.36).

The first building blocks of this transition framework are three fundamental shifts in the established socially constructed order (top layer in Fig. 20.36). These shifts are expected to create new conditions and social norms that incentivise a company, a country or individuals to modify their way of interacting with the ocean in favour of more sustainable and equitable behaviours:

- **Balanced top-down/bottom-up governance.** Major shifts in the way a complex adaptive system behaves rarely occur in an entirely purposeful, 'top-down' fashion. This is certainly true for the shift towards a sustainable ocean economy, where a multitude of (hard to predict) feedback loops can jeopardise the goals of a purely top-down approach.

  Top-down governance, to be sure, is essential. Land use, for example, is governed by a much more structured system of product and operating standards, clear access and property rights, the provision of legal recourse and so on. In many cases, top-down rule setting has launched, rather than shackled, global industries. Today's thriving telecom industry, for example, would not exist without compatibility of transmission formats, a global process for frequency allocation, consolidation limits and so on. A global pharmacological market could not exist without global testing and production protocols. The internet protocol (TCP/IP) is at the heart of much of today's commerce. A modified, ocean-relevant version of such protocols, rights and obligations is essential for the difficult, risky and capital-expensive development of a sustainable ocean economy. Investors are sure to require long-term resource access guarantees, reliable regulatory protocols, standardised transfer points, and clear operating and performance standards. Yet this top-down gover-

[443] Lubchenco et al. 2016. "The Right Incentives Enable Ocean Sustainability Successes and Provide Hope for the Future."

[444] Millennium Ecosystem Assessment (Program), ed. 2005. *Ecosystems and Human Well-Being*.

[445] OECD Data. n.d. "Agricultural Output: Meat Consumption." http://data.oecd.org/agroutput/meat-consumption.htm. Accessed 11 May 2020.

[446] Swilling et al. 2020. "The Ocean Transition."

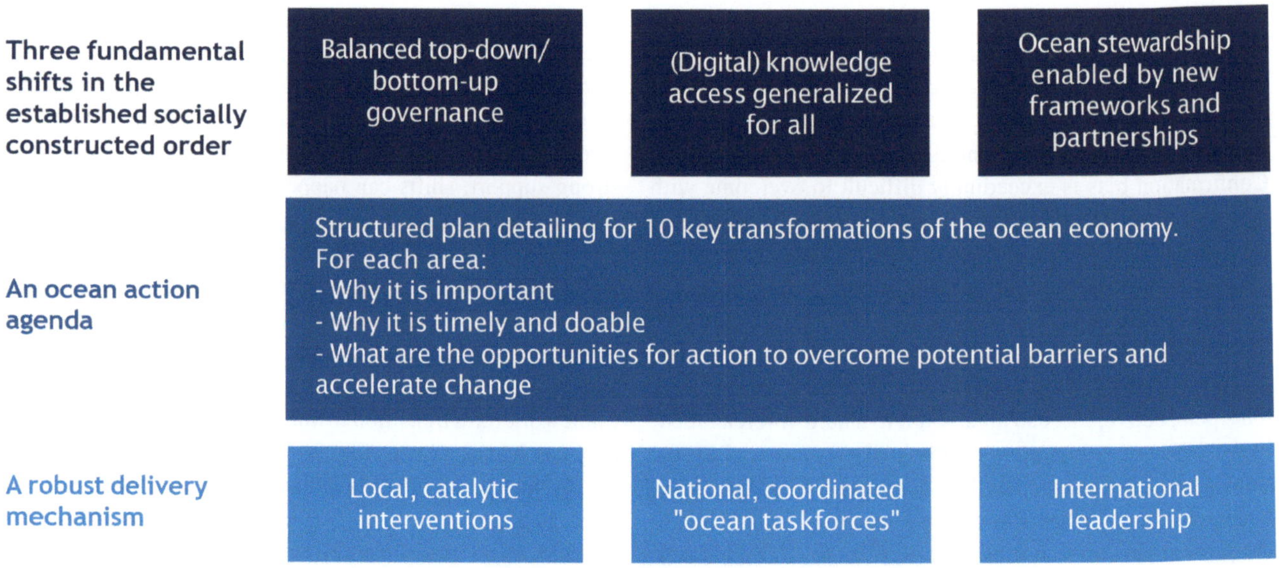

Fig. 20.36 Framework for a successful transition towards a sustainable ocean economy. (*Source: Authors*)

nance needs to become more adaptive, faster and more deeply connected to communities. In recent years, the lines between the 'top down' and the 'bottom up' have often become blurred—generally to very good result. The Paris Agreement, for example, blends a voluntary commitment structure with a centralised monitoring, reporting and verification function.

Bottom-up governance and grass-roots movements have been transformed by digital communication. For example, fishing communities in the Philippines are now collaborating in turning their local experiments with territorial use rights into a regional and national movement—by collaborating in the alignment of regional and national fishing policy with local needs, in obtaining financing for their fleets at favourable (joint) rates and offering education to other communities interested in joining the movement. As the movement grows, local collaboratives assume quasi-governance functions and authority—a welcome development.

- **(Digital) knowledge access generalised for all.** It is predicted that by 2020 data generation will increase annually by 4300%.[447] Digitalisation provides rare, necessary open spaces for experimentation and innovation towards more sustainable, equitable management of the economy.[448] Some examples in the literature define *open source* (freely, publicly available information, data or software): 'When open source principles prevail, countless inquiring eyes can scrutinize everything—the infrastructure, the transactions, the dialogues, the individuals—which minimizes the opportunities for quiet subterfuges and backroom deals'.[449] The processes of governance are more likely to be honest and fair, and be seen as such.[450] For instance, in the ocean economy realm, critical data on storm tracks, market pricing, logistics, demand and the like have increasingly become available even to the smallest fishers and according to Bollier 'could help usher in new, more ecologically benign forms of decentralized production and consumption'.[451]
- **Ocean stewardship enabled by new frameworks and partnerships.** Stewardship actions have been defined as 'the suite of approaches, activities, behaviors, and technologies that are applied to protect, restore or sustainably use the environment'.[452] The concept proposed here is to empower the ocean economy players (fishers, communities, businesses, etc.) to better manage shared ocean resources. For instance, the territorial use rights for fisheries (TURFs) system, in Chile, the Philippines and Indonesia has bestowed exclusive control of local fishing

[447] Sunderji, N. 2016. "How Will Data and Digital Platforms Transform Sustainable Development?" *Devex*, 25 July. https://www.devex.com/news/sponsored/how-will-data-and-digital-platforms-transform-sustainable-development-88481.

[448] Bollier, D. 2016. "Transnational Republics of Commoning 2: New Forms of Network-Based Governance." *P2P Foundation* (blog), 16 September. https://blog.p2pfoundation.net/transnational-republics-of-commoning-2-new-forms-of-network-based-governance/2016/09/16.

[449] Bollier, D. 2016. "Transnational Republics of Commoning 2."

[450] Bollier, D. 2016. "Transnational Republics of Commoning 2."

[451] Swilling et al. 2020. "The Ocean Transition."

[452] Bennett, N.J., T.S. Whitty, E. Finkbeiner, J. Pittman, H. Bassett, S. Gelcich and E.H. Allison. 2018. "Environmental Stewardship: A Conceptual Review and Analytical Framework." *Environmental Management* 61 (4): 597–614. doi: https://doi.org/10.1007/s00267-017-0993-2.

grounds to coastal communities—which have generally responded with much improved resource stewardship.[453] Similar stewardship initiatives can be found in the business world: CEOs from the ten largest global seafood companies (including fishing, aquaculture and aquafeed manufacturing) have joined forces through Seafood Business for Ocean Stewardship (SeaBOS), with the strong belief that through engagement with keystone actors, it is possible to both raise the sustainability bar for global companies that have significant influence within their industries and provide incentives for smaller companies to catch up with their peers.[454]

At the surface, each of these new social norms sounds obvious—in practice, they can be controversial and hard to implement because of the barriers mentioned above. For this reason, these fundamental shifts need to be supported by two core elements:

- **An action agenda** (middle layer in Fig. 20.36). Major shifts can happen ad hoc, but it's better to have an agenda. Industrial strategy agendas, for example, were a mainstay of public policy before 1990, briefly fell out of favour during the early digital revolution and are now making a major comeback with China's massive Belt and Road initiative, Britain's Industrial Strategy, Germany's Energie Wende and similar efforts. Shifting the commodity value chains (palm oil, tropical wood products, wild-caught and cultured fish, etc.) towards sustainability was the result of very deliberate agendas designed by NGOs, major buyers and government agencies. The Arctic Council's vision for risk-managed and sustainable development in the Arctic Ocean is based on a deliberate, shared, multinational agenda.

  In the context of this report, such an action agenda offers a clear, holistic picture of the various components of ocean economy reform, builds on emerging innovations and front-running projects and aims to scale them up. Section 6.3 lays out this agenda in detail—the case for change, the feasibility and the concrete opportunities for action.
- **A robust delivery mechanism** (bottom layer in Fig. 20.36). Delivering the action agenda can appear quite daunting for any decision-maker or national leader. This report identifies three complementary levels of intervention to start or accelerate the journey towards a sustainable ocean economy. They do not pretend to be exhaustive solutions, but they offer some options to be considered by decision-makers (particularly policymakers):
  - Local, catalytic interventions are essential to create support from the base and demonstrate tangible results on the ground—seeing is believing. Section 6.4 elaborates on a concept of small ocean 'special use' areas, which could be used as laboratories and demonstrators of a sustainable ocean economy.
  - National, coordinated 'ocean task forces' should be established. Using the best practices of performance management inspired by decades of practice in the private sector, they should deploy the implementation of the ocean agenda at the country level. Section 6.4 elaborates on the key features of such ocean task forces.
  - International leadership will be needed in various forms: ratifying and enforcing international treaties, conventions and agreements (such as the upcoming legally binding instrument on biodiversity of areas beyond national jurisdiction, or the Port State Measure Agreement to fight IUU fishing), or promoting new international public-private partnerships to advance knowledge and best-practice sharing, and remove roadblocks to the implementation of a sustainable ocean economy (see Sect. 6.4).

6.3 Charting a Direction: The Ocean Action Agenda

Change will not come with the stroke of a pen, or through a normative policy process only. The conditions for a top-down approach—predictability, enforceability, high levels of support, abundant feedback from diverse sources—are simply not in place, and a different, much more networked and adaptive process must be used. But just as clearly, the sustainable ocean challenge cannot be left to itself—a plan, an agenda is needed.

It is not necessary to start from a clean sheet though: the outline is in place in terms of governance and policy, technologies and business models, international collaboration and new consumer demands (see Sect. 4.4).

This section proposes an action agenda to deliver the overarching mandate of effective protection, sustainable production and equitable prosperity; this action agenda is based on five cross-cutting enablers, which collectively will help five main sustainable ocean sectors to thrive.

These ten components cannot be seen or advanced in isolation. Like the system they aim to change, these components are highly interconnected. The enabling conditions can support each other (e.g. better data in support of de-risked finance, smart planning and upgraded ocean accounting),

[453]Costello et al. 2016. "Global Fishery Prospects under Contrasting Management Regimes."

[454]Blasiak, R., R. Wynberg, K. Grorud-Colvert, S. Thambisetty et al. 2020. "The Ocean Genome: Conservation and the Fair, Equitable and Sustainable Use of Marine Genetic Resources." Washington, DC: World Resources Institute. www.oceanpanel.org/blue-papers/ocean-genome-conservation-andfair-equitable-and-sustainable-use-marine-genetic.

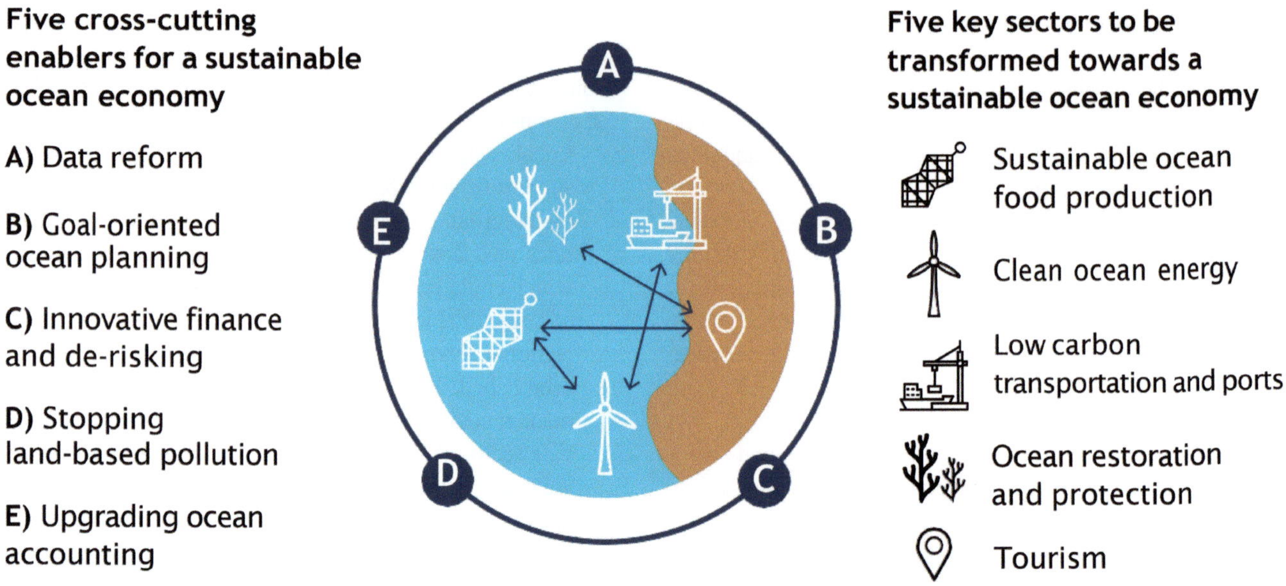

Fig. 20.37 One mandate and ten components of the action agenda for a sustainable ocean economy. (*Source: Authors*)

and the sectors have to develop in harmony with each other to exploit synergies (clean energy from offshore wind farms to fuel green ships; co-location of MPAs and offshore wind farms; co-location of seaweed, bivalves and finfish mariculture farms; MPAs to provide nursery grounds for fisheries, etc.; see arrows in Fig. 20.37).

This agenda stands on the shoulders of good work already done but provides a powerful new boost towards a truly sustainable ocean economy. This action agenda recognises the inherent challenges associated with a transition of such scale and would also have to be adapted to fit local, national or regional contexts. The OECD report *Sustainable Ocean for All* is a valuable source of insights regarding the specific challenges faced by developing countries seeking to embrace a sustainable ocean economy agenda.

6.3.1 Five Cross-Cutting Enablers for a Sustainable Ocean Economy

After extensive review of Blue Papers commissioned by the High Level Panel for a Sustainable Ocean Economy, scientific literature and expert consultations, five cross-cutting topics crystallise as the core enablers to provide the right conditions for a sustainable ocean economy to thrive (left-hand side of Figs. 20.37 and 20.38):

Data reform The revolution in data production (enabled by new technologies), collection and management (sensing, tagging, sharing), processing (simulation, forecasting, optimisation, tracking, process management) and sharing (open data platforms) can have a profoundly positive impact on all aspects of the ocean economy. This will require a comprehensive reform of currently proprietary sensing, storage and application methods.

Goal-oriented ocean planning Explicit guidance towards the overarching mandate (protect effectively, produce sustainably and prosper equitably) is required to ensure the avoidance of spatial use conflicts, uniformly high standards of operation, a streamlined and efficient regulatory process, the integration of symbiotic ocean uses, the overall protection and sustainability of the key oceanic systems, the efficient management of fishing rights and a just transition for workers of the ocean economy.

Innovative finance and de-risking Potential sustainable economy investors (sovereign wealth, institutional) require basic guarantees on infrastructure access and pricing, long-term access rights, regulatory certainty, reliable operating standards and solid legal recourse options. These are not uniformly in place today. Public financing might be required to mitigate inherent costs required to kick-start a sustainable ocean economy.

Stopping land-based pollution The sustainable ocean economy cannot thrive if land-based pollution ending in the ocean is not significantly reduced through ambitious and systems-inspired reforms.

Upgrading ocean accounting A sustainable ocean economy needs to be piloted through a holistic set of metrics which measure flows and stocks, economic and natural capital.

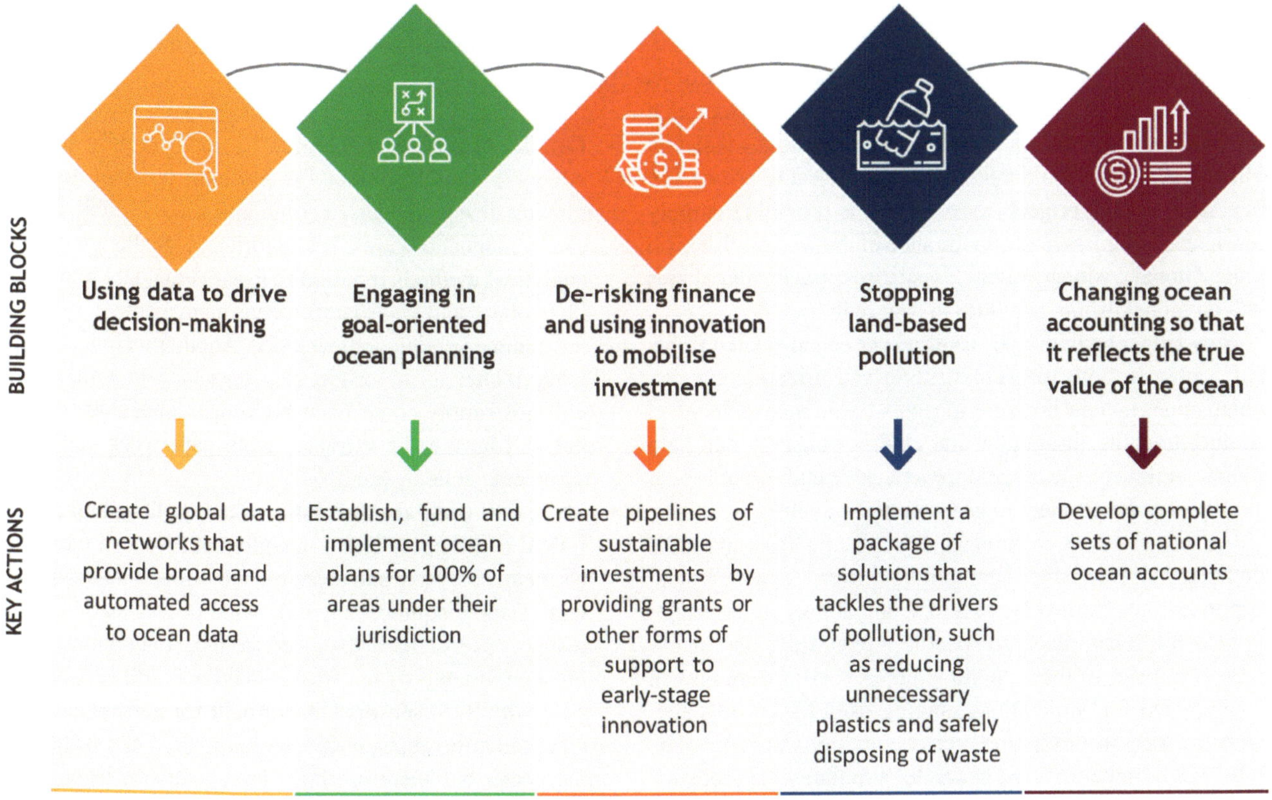

Fig. 20.38 Five building blocks are key to creating a sustainable ocean economy. (*Source: Authors*)

These five enablers can set up new rules and a favourable environment in which a sustainable ocean economy can develop. In particular, five sectors are reviewed in depth in this section, as the future champions of a sustainable ocean economy (right-hand side of Fig. 20.37). For each of them, this report analyses why such a sector needs to be transformed and identifies some concrete opportunities for action that could be taken to capture the full sustainable potential of these sectors.

Sustainable food from the ocean How the provision of healthy, low-carbon and nutritious food can be scaled while regenerating the ocean ecosystems and better redistributing benefits.

Clean ocean energy How societies can be powered by harvesting renewable ocean forces.

Low-carbon transportation and ports How national and international transportation of goods and people can be provided in a climate- and biodiversity-friendly, socially acceptable way.

Ocean restoration and protection How insufficient, isolated activities can be turned into a thriving sector, generating jobs, revenues and numerous ecosystem services.

Tourism How the beauty of coasts and the ocean can be enjoyed while restoring them.

This section explores each of these areas in terms of their importance, their path dependencies and barriers and, most important, opportunities for action.

Data Reform: How Could the Data Revolution—Enabled by New Technologies—Change the Way Informed Decisions in the Ocean Realm Are Made?

Why Is It Important?

The data explosion—in other words, the rapid increase of data and information created and made available—can reshape understanding and management of the ocean. The ocean is notoriously reluctant to reveal its secrets—far too little is understood about the interface between humans and the ocean ecology. New sensing, data management, visualisation, simulation and modelling technologies can change that—but current sharing practices are not yet fit for purpose.

In the near future, every ship's journey, and the nature of its business at sea, will be public information. Lawbreakers such as illegal fishers, polluters, smugglers and labour law violators will literally be on the public radar and subject to arrest. The Sea Around Us[455] project, for example, is providing entirely new levels of transparency on the state of the world's fisheries, while Global Fishing Watch[456] visualises, tracks and shares data on global fishing activities in near real time.

New ledger technologies can register ocean-related rights and contracts, both for public titling and private contracts, which opens up new horizons in rights-based management.[457] Product tracking throughout the chain of custody can help brands embrace sustainable practices[458] and would help small producers connect to global supply chains.

For ocean resource managers, replacing trial-and-error methods with reliable simulations lowers feedback and response times from years to hours, and allows quick insight on how the ocean reacts to specific inputs, rules and incentives. A number of these 'flight simulator' efforts are now in development for applications ranging from fishery management to ship routing and ecosystem conservation. The POSEIDON model,[459] for example, simulates the feedback loop between fishery policies, fishing fleets and ocean ecosystems, allowing for real-time testing of policy alternatives. These applications will allow managers to adjust to changing conditions,[460] such as dynamic management of fishing areas and quotas, ship traffic adjustments or avoidance of endangered species bycatch.[461]

What Is Preventing (Faster) Change?

Four main obstacles currently prevent the full capture of this potential.

- **On the technical side**, ocean sensors require power, which is hard to obtain for anything but the slowest-moving device. Undersea communications, unable to use electromagnetic waves, are notoriously challenging. The analytical methods required to harness the data into nimble, robust and transparent ocean management systems are complex and underdeveloped. Another technical challenge relates to the origin of data, which in some cases will not simply come from high-tech connected sensors but will have to be extracted from paperwork and measurements done by hand.
- **More daunting are the institutional, political and analytical challenges**. The fragmentation of ocean data into national, corporate and academic fiefdoms is a huge problem. Data inequities abound, with poorer nations and resource users largely excluded from the data bounty. And most important, oceanic data collection is still very expensive, with most sensors custom-built for narrow and specific scientific missions.[462] Technological innovation in the ocean has therefore been largely driven by government and large-scale commercial interests. Many needs remain simply unserved.
- **Financing has been difficult.** Much of ocean technology has relied on the trickle-down of commercial (mostly oil and gas) and defence technologies. National R&D expenses dedicated to ocean science vary greatly from country to country (21.4% in Argentina, over 2.5% in the United States, 0.1% in Russia),[463] but they can generally be considered as too low. Data collection in the harsh ocean environment is expensive—even medium-sized research vessels have operating budgets above $20,000 a day, with some globally operating vessels having budgets as high as $40,000 a day.[464] A business-as-usual approach will not come close to paying for the needed ocean technology.
- **Lastly, there is a capacity issue.** Even when relevant data are available, managers often do not get the information they need because of data access restrictions, or because they do not have data scientists to address the

[455]Pauly et al. n.d. "Sea around Us Concepts, Design and Data."

[456]Global Fishing Watch. n.d. "Sustainability through Transparency."

[457]Nyborg, K., J.M. Anderies, A. Dannenberg, T. Lindahl, C. Schill, M. Schlüter, W.N. Adger et al. 2016. "Social Norms as Solutions." *Science* 354 (6308): 42–43. doi: https://doi.org/10.1126/science.aaf8317; Leape et al. 2020. "Technology, Data and New Models for Sustainably Managing Ocean Resources."

[458]Hardt, M.J., K. Flett and C.J. Howell. 2017. "Current Barriers to Large-Scale Interoperability of Traceability Technology in the Seafood Sector." *Journal of Food Science* 82 (S1): A3–12. doi: https://doi.org/10.1111/1750-3841.13796.

[459]Bailey, R.M., E. Carrella, R. Axtell, M.G. Burgess, R.B. Cabral, M. Drexler, C. Dorsett et al. 2019. "A Computational Approach to Managing Coupled Human-Environmental Systems: The POSEIDON Model of Ocean Fisheries." *Sustainability Science* 14 (2): 259–75. doi: https://doi.org/10.1007/s11625-018-0579-9.

[460]Maxwell, S.M., E.L. Hazen, R.L. Lewison, D.C. Dunn, H. Bailey, S.J. Bograd, D.K. Briscoe et al. 2015. "Dynamic Ocean Management: Defining and Conceptualizing Real-Time Management of the Ocean." *Marine Policy* 58 (August): 42–50. doi: https://doi.org/10.1016/j.marpol.2015.03.014.

[461]Dunn, D.C., A.M. Boustany and P.N. Halpin. 2011. "Spatio-temporal Management of Fisheries to Reduce By-Catch and Increase Fishing Selectivity." *Fish and Fisheries* 12 (1): 110–19. doi: https://doi.org/10.1111/j.1467-2979.2010.00388.x.

[462]OECD. 2019. *Rethinking Innovation for a Sustainable Ocean Economy*. Paris: Organisation for Economic Co-operation and Development. doi: https://doi.org/10.1787/9789264311053-en.

[463]Intergovernmental Oceanographic Commission (IOC)-UNESCO. 2017. *Global Ocean Science Report: The Current Status of Ocean Science around the World*. Edited by L. Valdés et al. Paris: UNESCO Publishing. https://unesdoc.unesco.org/ark:/48223/pf0000250428.

[464]IOC-UNESCO. 2017. *Global Ocean Science Report*.

most policy-relevant questions.[465] Decision-support tools designed explicitly for marine managers are often so technical that only programmers are able to use them.[466] For example, fishery management in data-poor, developing-country environments often requires sophisticated data extrapolation techniques and extensive adaptation of standard analytics to local conditions.

What Are the Opportunities for Action to Overcome These Barriers?

Data management is evolving towards a 'digital ecosystem for the environment'[467]—a systemic, dynamic and collaborative model[468] that aggregates data into entirely new levels of synthesis, visualisation and managerial utility. This model uses huge networks of cheap, standardised and connected sensors (an 'Internet of Things for the ocean'[469]) with no regard to specific, predetermined purpose. These networks deposit data into unstructured data 'lakes' which enable users to build their own knowledge systems. This approach has already transformed machine learning and analytics, democratising the data[470] and allowing for the massive economies of scale needed to understand the ocean in all its complexity.

Universal data tagging standards are essential to allow data to be combined in federated data networks and data lakes that support verified and automated global access.[471] Governments can lead the way directly by taking bold steps to mandate these standards and to help create and contribute to federated data networks.[472] They can also require data-sharing and compliance with tagging standards as a non-negotiable condition of access to public resources—whether the resources are fish stocks and mineral deposits or funds for coastal management or for research.

Capitalising on the Decade of Ocean Science for Sustainable Development, UNESCO's Intergovernmental Oceanographic Commission (IOC) can establish global standards for metadata, query and data tagging that allow existing datasets to be connected and automatically accessed, as well as sensor positioning and interoperability.[473] Governments, industry and research institutions can use these standards to make their data broadly available in a global federated data network. New partnerships with cloud service providers (such as NOAA's partnership with Amazon Web Service) should be formed to create open-access data lakes.

In addition, governments can prioritise technology-forcing regulations that target real-time monitoring of fishing, seafood imports, shipping emissions, mineral development, coastal development and pollution and that create public accountability. In the case of fisheries, mandates for vessel-based electronic monitoring, for example, could speed the translation of existing artificial intelligence expertise to ocean management.

The private sector plays a huge role. Many new companies and privately funded initiatives, such as Planet, the Ocean Data Foundation, the Center for the Fourth Industrial Revolution for the Ocean, OceanX and Schmidt Marine Technology Partners, are generating new ocean data and/or providing them free of cost to researchers. Knowledge services can pay for data networks. Already, ocean and climate data are being used as the basis for complex insurance decisions, targeted weather forecasts for precision agriculture and other lucrative knowledge products. Growing corporate interest in traceability across the value chain spawns new solutions, such as the recently launched blockchain platform OpenSC. Tech innovators partnering with NGOs and big seafood companies can extend that capability to small-scale fisheries.

Finally, governments, researchers and the private sector need to work together.[474] Jointly developed technology standards are essential to create a fertile ecosystem for innovation. FAO's Port State Measures Agreement, for example, creates new requirements for port monitoring and control that are applied globally and that will require technological innovation in data collection and sharing.[475] In addition, the sectors can collaborate to lower data storage costs. For a summary of these points, see Box 20.6.

---

[465] McConney, P., L. Fanning, R. Mahon and B. Simmons. 2016. "A First Look at the Science-Policy Interface for Ocean Governance in the Wider Caribbean Region." *Frontiers in Marine Science* 2. doi: https://doi.org/10.3389/fmars.2015.00119.

[466] Stelzenmüller, V., J. Lee, A. South, J. Foden and S.I. Rogers. 2013. "Practical Tools to Support Marine Spatial Planning: A Review and Some Prototype Tools." *Marine Policy* 38 (March): 214–27. doi: https://doi.org/10.1016/j.marpol.2012.05.038.

[467] Campbell, J., and D. Jensen. 2019. "Building a Digital Ecosystem for the Planet." Foresight Brief no. 014 (September). UN Environment Programme. https://wedocs.unep.org/handle/20.500.11822/30612.

[468] Buck, J.J.H., S.J. Bainbridge, E.F. Burger, J. del Río Fernandez, E. Delory, P. Fischer, S. Jirka and J.S. Pearlman. 2019. "Ocean Data Product Integration through Innovation: The Next Level of Data Interoperability." *Frontiers in Marine Science* 6 (February): 32/1–32/19. doi: https://doi.org/10.3389/fmars.2019.00032.

[469] Leape et al. 2020. "Technology, Data and New Models for Sustainably Managing Ocean Resources."

[470] Leape et al. 2020. "Technology, Data and New Models for Sustainably Managing Ocean Resources."

[471] Leape et al. 2020. "Technology, Data and New Models for Sustainably Managing Ocean Resources."

[472] Leape et al. 2020. "Technology, Data and New Models for Sustainably Managing Ocean Resources."

[473] Cater, N.E., P. Eng and T. O'Reilly. 2009. "Promoting Interoperable Ocean Sensors the Smart Ocean Sensors Consortium." In *OCEANS 2009*, 1–6. doi: https://doi.org/10.23919/OCEANS.2009.5422448.

[474] Leape et al. 2020. "Technology, Data and New Models for Sustainably Managing Ocean Resources."

[475] Harden-Davies, H. 2017. "Capacity Building and Technology Transfer for Marine Biodiversity in Areas beyond National Jurisdiction." *Proceedings of the ASIL Annual Meeting* 111: 243–45. doi: https://doi.org/10.1017/amp.2017.75.

Box 20.6 Key Triggers to Unleash the Ocean Data Potential

- Capitalise on the UN Decade of Ocean Science for Sustainable Development to create a global data network that provides broad and automated access to ocean data.
- Liberate ocean data. Enabled by federated networks, data holders should establish a new default—that ocean data are broadly available to other users unless there are compelling security, proprietary or other interests.
- Create an 'Internet of Things' for the ocean. Coordinated efforts by industry, researchers and governments can create advanced sensor networks that provide high-resolution, real-time information about the ocean to anyone who needs it.
- Automate ocean management based on near real-time data on ocean conditions and resource use.
- Create incentives for innovation. Existing markets do not incentivise many of the technological innovations needed for ocean stewardship and research. Governments and companies can change that.
- Mobilise capital for technologies for underserved markets. Many markets for ocean technologies do not offer commercial returns. Innovative financial instruments are needed that can leverage the expectations and risk tolerances of different investors.

Source: Leape, J., M. Abbott, H. Sakaguchi et al. 2020. "Technology, Data and New Models for Sustainably Managing Ocean Resources." Washington, DC: World Resources Institute. www.oceanpanel.org/Technology-data-and-new-models-for-sustainably-managing--ocean-resources

Goal-Oriented Ocean Planning: Why Does Ocean Planning Matter So Much, Why Is It Not Mainstreamed Yet and How Can It Be Generalised?

Why Is It Important?

The literally 'free for all' model of ocean use cannot continue. Unrestricted, open-access fisheries almost invariably fail;[476] uncoordinated ocean development creates operational inefficiencies and use conflicts (with attendant litigation and regulatory delays); and unrestricted industrial, nutrient and carbon-related pollution is changing the fundamental ocean chemistry and affecting its biology. More systematic, equitable management is needed for the ocean's resources (principally food and energy), services (weather modulation, carbon sequestration, recreation) and certainly its absorption of externalities (heat and pollution). The current standards and practices of oceanic use planning, accountability, transparency and legal rights and protections remain a century or more behind their terrestrial equivalents. A fresh look at these practices is necessary, shaped by three major systemic objectives:

- **Efficiency and safety.** The different sectors of the ocean economy, such as food, energy and tourism, as well as carbon sequestration and coastal protection, are often symbiotic and have much to gain from being planned as an integrated whole. For example, offshore renewable energy production, the production of non-carbon shipping fuels (ammonium, hydrogen) and large-scale maritime operations all are operationally linked. They could generate significant efficiency gains and avoid impeding on fishing grounds and shipping lanes if carefully integrated with onshore grids and energy markets, and potentially co-located with offshore tourism sites and platforms. Seaweed, finfish and bivalves farms could be co-located (or integrated) to benefit to each other, reduce waste and synergise capital expenditure and operating costs (e.g. boats).
- **Reliable and defensible resource and access rights.** Resource owners, lessees and access holders need secure titling and reliable and effective legal recourse against polluters, trespassers and other violators. Coastal communities and small-scale fishers need inclusive and equitable access to resources, and recognition of their rights and tenure (food resources, protective reefs and habitats). Investors need reliable, long-term resource access guarantees.
- **Integration and balance of production and protection** Production and protection need to be in balance, with a strong emphasis on ocean regeneration. This requires careful integration along the dimensions of mitigation offsets (e.g. MPAs in balance with high-use areas), stringent and consistent operating standards (e.g. mariculture standards for containment, disease control, feed composition) and careful facility siting to ensure efficient production while avoiding ecological damage.

Goal-oriented ocean planning is central to these objectives. Three main concepts are usually mentioned when talking about ocean planning: ecosystem-based management, marine spatial planning and integrated ocean management—see Box 20.7 below.

[476]Costello, C., S.D. Gaines and J. Lynham. 2008. "Can Catch Shares Prevent Fisheries Collapse?" *Science* 321 (5896): 1678–81. doi: https://doi.org/10.1126/science.1159478.

Box 20.7 Ecosystem-Based Management, Marine Spatial Planning or Integrated Ocean Management?

Ecosystem-based management: Management of natural resources that focuses on the health and productivity of a specific ecosystem, a group of ecosystems or selected natural assets as the nucleus of management.

Marine spatial planning: Identifies what spaces of the ocean are appropriate for different uses or activities in order to reduce conflicts and to achieve ecological, economic and social objectives. Usually associated with zoning.

Integrated ocean management: A holistic, integrated, knowledge- and ecosystem-based approach that considers multiple uses and pressures, reconciling competing uses, with the objective of ensuring the sustainability of marine ecosystems.

The three approaches: similarities and differences:

- Similar and overlapping in many ways, all three are holistic approaches to human use of the ocean, based on well-defined ocean areas or ecosystems.
- The difference is in the main 'angle' taken by each of these three approaches: integrated ocean management is rooted in 'management' thinking (processes, institutions); ecosystem-based management in ecosystem thinking (interactions between humans and ecosystems); marine spatial planning in analysis instruments such as geographic information systems and zoning.

Sources: Domínguez-Tejo, E., G. Metternicht, E. Johnston and L. Hedge. 2016. "Marine Spatial Planning Advancing the Ecosystem-Based Approach to Coastal Zone Management: A Review." Marine Policy 72 (October): 115–30. doi: https://doi.org/10.1016/j.marpol.2016.06.023; *Charles, A., S.J. Evers and A.L. Shriver. 2016. Challenging New Frontiers in the Global Seafood Sector: Proceedings of the Eighteenth Biennial Conference of the International Institute of Fisheries Economics and Trade. Aberdeen, Scotland: International Institute of Fisheries Economics and Trade*

Yet if ocean planning should always be science-based, there is no 'one-size-fits-all': an efficient planning process has to be guided by science but tailored to the local parameters, needs and players involved. Some countries have, at least partially, developed and implemented ocean planning and have gone through the testing, failing, learning and adaptation stages. Sharing this variety of experiences internationally would be extremely helpful to nations with less experience. Platforms could be expanded for this purpose. The IOC-UNESCO Marine Spatial Programme, for example, is already 'documenting marine spatial planning initiatives around the world, identifying good practices of marine spatial planning, collecting references and literature on marine spatial planning and building capacity through training marine spatial planning professionals'.[477]

The planning concept should also recognise the social considerations inherent in a transition towards new or more sustainable sectors. Indeed, the Just Transition Centre states that 'transparent planning that includes just transition measures will prevent fear, opposition and inter-community and generational conflict'.[478] Inspiration for such national strategies or planning regarding jobs transition can be found in Canada's 'Just Transition Task Force', which was established to support workers who would be affected by the national target of phasing out coal-fired electricity by 2030. This task force is working closely with labour organisations and communities to ensure a just transition plan for Canadian coal power workers and communities.[479]

What Is Preventing (Faster) Change?

Most of the world's ocean remains largely unplanned and poorly managed for a number of reasons:

- **Planning backlash.** Opposition to ocean master planning is often fierce. It can come in certain regions from the offshore oil and gas sector, in others from commercial fisheries (while the shipping industry is often supportive because of its focus on safe and reliably protected shipping lanes). The opposition is typically concerned that ocean spatial planning is not about rational planning and conflict avoidance but instead a Trojan horse for conservationists seeking new legal tools to pursue a perceived anti-business agenda.
- **Opposition to access control.** Today, titled access happens routinely in the form of extractive leases (e.g. oil and gas platforms and deep-sea mining) and foreign fishing fleet access rights to national waters. While some fishing rights have been in place for decades, in recent years, fishery access rights have been conferred on only 200 fisheries and coastal communities (e.g. 'individual trading quotas'

[477] Marine Spatial Planning Programme, UNESCO and IOC. n.d. "Balancing Sustainable Use and Conservation through Marine Spatial Planning." http://msp.ioc-unesco.org/. Accessed 12 May 2020.

[478] Smith, S. 2017. "Just Transition: A Report for the OECD." Just Transition Centre. https://search.oecd.org/environment/cc/g20-climate/collapsecontents/Just-Transition-Centre-report-just-transition.pdf.

[479] Canada's Task Force on Just Transition for Canadian Coal Power Workers and Communities. 2018. *Final Report by the Task Force on Just Transition for Canadian Coal Power Workers and Communities*. Government of Canada. https://www.canada.ca/en/environment-climate-change/services/climate-change/task-force-just-transition/final-report-complete.html.

and 'territorial use rights').[480] Opposition to ocean access rights can be strong from all points on the political spectrum. Some worry about monopolisation or 'corporatisation' of the ocean by the wealthy and well connected, to the exclusion of traditional user groups. Commercial fishing industries are wary of ever-expanding exclusions of their fleets. Coastal inhabitants and competitors often fiercely contest lease sales for near-shore energy and mariculture operations. Legal provisions guaranteeing free ocean-access provisions are on the books in many countries, and certainly enshrined on the high seas through the UN Convention on the Law of the Sea (UNCLOS).

- **Access to data.** Knowledge of the ecosystem is the foundation of integrated ocean management: the biology (fish stocks, migrating patterns, invasive species, primary productivity, etc.), the chemistry (acidity, temperature, nutrients), the physics (currents, waves), the human activity (fisheries, aquaculture, shipping routes, etc.) and the existing regulations and zoning (especially in basins shared by multiple countries). Norway—which has used integrated ocean management for years—is one of the highest spenders in the world, in absolute terms, on marine science. But many countries do not have sufficient scientific capacity or baseline data to provide the knowledge foundation required. The 2017 *Global Ocean Science Report* demonstrated that many countries lack fundamental scientific capacity to underpin their efforts at ocean governance.[481]
- **Unfit top-down planning processes.** Top-down planning processes tend to be inefficient. For instance, 265 separate knowledge products concerning ocean management have been produced over 5 years of a major coral conservation program in Asia—position papers, books, training manuals, field guidance manuals and the like. A recent survey revealed that 54% of program participants never or rarely used these knowledge products, and only 20% frequently or often used them.[482] On the contrary, in some SIDS, participatory approaches have been very effective at the local level for all phases of the MPA process (MPA planning, implementation, monitoring and evaluation, feedback, adaptation of management plan, etc.).[483]

[480]Environmental Defense Fund. n.d. "Database." Fishery Solutions Center. http://fisherysolutionscenter.edf.org/database. Accessed 12 May 2020.

[481]IOC-UNESCO. 2017. *Global Ocean Science Report*.

[482]Weeks, R., R.L. Pressey, J.R. Wilson, M. Knight, V. Horigue, R.A. Abesamis, R. Acosta and J. Jompa. 2015. "Ten Things to Get Right for Marine Conservation Planning in the Coral Triangle." *F1000Research* 3 (December). doi: https://doi.org/10.12688/f1000research.3886.3.

[483]Weeks et al. 2015. "Ten Things to Get Right for Marine Conservation Planning in the Coral Triangle"; Frazão Santos, C., T. Agardy, F. Andrade, H. Calado, L.B. Crowder, C.N. Ehler, S. García-Morales, et al. 2020. 'Integrating Climate Change in Ocean Planning'. Nature Sustainability 3 (7): 505–16. doi: https://doi.org/10.1038/s41893-020-0513-x.

What Are the Opportunities for Action to Overcome These Barriers?

Many countries have started to create marine spatial plans (MSPs), a fair number of countries have implemented MSPs in parts of their EEZ, but very few have implemented MSPs for the whole of their EEZ.

A national mandate for an EEZ-wide (and eventually international high seas) ocean planning process can explicitly signal the end of damaging 'free for all' use practices. At its core, this process needs to show how the agendas of ecosystem health, food and energy security, local prosperity and coastal protection can fully re-enforce each other—and what form this takes in explicit spatial, regulatory and operational terms. The process needs to find the spatial balance between production and protection zones (see Sect. 6.3, point 9, 'Ocean restoration and protection'), between the requirements of different ocean users and between the needs of the ocean and the needs of the coast and its people. It needs to provide inclusive, equitable access and recognition for local communities, such as access to traditional fishing grounds, protection of cultural sites, preservation of viewsheds and so on.[484]

The development of an ocean plan covering 100% of the ocean areas under national jurisdiction is a time-consuming exercise of broad shareholder participation and shared exploration. This long process is crucial, however, since, apart from economic use considerations, thorough (and often costly) public stakeholder consultations to address gender, equity and distributional issues should be held (see 'Unfit top-down planning processes' above). The development of a protected ocean area plan for the coast of California, for example, required hundreds of community meetings. The interests of shippers, ports, fishers, wind developers, coastal cities, scientists, the navy, local security forces, farmers, water users and so on need to be heard, respected and integrated. This can take 2–3 years. Once a plan is established, its ongoing implementation requires continued funding.

In the shorter term, the benefits of planning can be demonstrated on a smaller scale. This report champions the idea of smaller 'sustainable ocean economic zones' (SOEZs)—ocean areas which serve as testbeds for a new breed of fully sustainable and regenerative ocean projects (such as multitrophic farms, renewable energy and the like; see Sect. 6.4). These SOEZs can catalyse development of an integrated, sustainable ocean economy plan encompassing the entire EEZ or areas under national jurisdiction. For a summary of these points, see Box 20.8.

[484]Allison et al. 2020. "The Human Relationship with Our Ocean Planet."

Box 20.8 Major Components of Integrated Ocean Planning and Management

- At the country level, establish comprehensive integrated marine spatial plans for 100% of the areas under national jurisdiction. The process (science-based, inclusive, participatory, adapted to local context) is as important as the plan itself.
- Ensure continued funding and capacity for the ongoing implementation of ocean management plans.
- Develop sustainable ocean economic zones as spatially defined 'laboratories' for fully managed areas comprising various sectors, multi-sectoral projects and fully protected areas.
- Develop an international platform to develop and share best practice principles and guidelines for sustainable planning (which could be done by leveraging the ongoing IOC-UNESCO Marine Spatial Programme).

Innovative Finance and De-risking: Why Is Finance Pivotal, and How Can More Money Be Mobilised Towards a Sustainable Ocean Economy?

Why Is It Important?

The ocean economy is currently greatly underinvested (even among impact and blended finance investors, SDG 14—'the ocean SDG'—receives the smallest share of them all[485]). Over the 2013–2018 period, an average of US $1.5 billion of official development assistance (ODA) a year was allocated in support of the sustainable ocean economy, representing less than 1% of global ODA.[486] The finance available for biodiversity and/or conservation is significantly below its need.[487] It has been estimated that currently only about 0.002% of global GDP is invested in the conservation and sustainable use of ocean biodiversity, and that about four times the current level of investment is required to meet conservation needs.[488]

A growing, sustainable ocean economy needs funding from sources ranging from philanthropy to market-rate investment.

- **Financing innovation**. Many of the sustainable ocean technologies of the future require further commercialisation. This includes floating offshore wind, large offshore multi-trophic mariculture, alternative feed for mariculture, carbon-financed restoration and coastal protection and production of non-carbon shipping fuels. The linkage and symbiosis of these sectors will require considerable development and experimentation. Public finance and subsidies are essential components of the 'industrial strategy' required, as demonstrated exhaustively on land. Demonstration projects are needed to develop 'proof of concept' that can convince institutional investors to engage at scale.
- **Financing infrastructure**. The infrastructural support required is not uniformly in place. This includes offtake and grid access points for offshore energy; energy supply for offshore mariculture; port investments required for the management of sustainable fisheries; marine safety and rescue.
- **Financing the transition**. Transition costs may require public support. This can include investments in worker retraining; consumer awareness campaigns; and much more extensive programs to 'buy down' future (unpriced) costs, such as coastal erosion, expanding dead zones, pollution on beaches and so on. In the context of fisheries, the transition implies a deep reduction of capacity and fishing effort to help rebuild stocks before they are fished (sustainably) again. One study has estimated that the total amount governments need to invest to rebuild world fisheries ranges between $130 billion and $292 billion in present value, cost to be spread over several years and among countries.[489]

What Is Preventing (Faster) Change?

Six main barriers are preventing more financial flows from entering the sustainable ocean economy space:

- **The investable pipeline for a sustainable ocean economy is not evident.** Investment-grade projects are currently limited and/or hard to find. A survey commissioned in 2020 by Credit Suisse shows that '[l] ack of investment-grade projects/firms at scale' and '[n]ot enough internal expertise' are the two main barriers identified by investors ($n = 249$) to greater investment in a sustainable ocean economy.[490] Examples of new parametric insurance

[485] Libes, L., and M. Eldridge. 2019. "Who, What, Where and How: 440 Investors—a Deepening View of Impact Investing." http://investorflow.org/wp-content/uploads/Investorflow-Report-440-Investors-March-2019.pdf.

[486] OECD. 2020. *Sustainable Ocean for All.*

[487] UNCTAD. 2014. *World Investment Report 2014: Investing in the SDGs—an Action Plan.* United Nations Conference on Trade and Development World Investment Report. New York: United Nations. doi: https://doi.org/10.18356/3e74cde5-en. Sumaila et al. 2017. "Investments to Reverse Biodiversity Loss Are Economically Beneficial."

[488] Sumaila, U.R., C.M. Rodriguez, M. Schultz, R. Sharma, T.D. Tyrrell, H. Masundire, A. Damodaran et al. 2017. "Investments to Reverse Biodiversity Loss Are Economically Beneficial." *Current Opinion in Environmental Sustainability* 29 (December): 82–88. doi: https://doi.org/10.1016/j.cosust.2018.01.007.

[489] Sumaila et al. 2012. "Benefits of Rebuilding Global Marine Fisheries Outweigh Costs."

[490] Responsible Investor Research and Credit Suisse. 2020. *Investors and the Blue Economy.*

schemes (e.g. coral reefs insurance) are promising but still mostly confined to the pilot stage. Recent solutions have been developed to mobilise commercial impact finance into marine protected areas through long-term management lease of the MPAs. The scalability of the approach must nonetheless be demonstrated. In the short term, the number of MPAs with tangible business models that include monitoring of abuses and enforcement of sanctions seems limited.

- **Incentives are either not in place or are out of place**. In the absence of national 'ocean industrial strategies', harmful subsidies have been allowed to distort the current ocean economy, typically promoting the expansion of fishery capacity (mostly directed to large-scale industrial fishing fleets[491]) and the extraction of oil and gas. At the same time, constructive subsidies supporting sustainable ocean enterprises—such as demand guarantees, low-cost infrastructure support and access to low-cost capital—have lagged far behind. In addition, the risks associated with unsustainable management of the ocean have been inadequately considered: most externalities (e.g. climate change, pollution, violation of human rights) are today not priced by the market and allow unsustainable businesses to thrive.
- **It is generally risky to invest in the ocean space**. In many cases, key technologies are available but not yet fully tested in the cauldron of open ocean conditions, intensifying storms, increasing acidification and shifting currents. Onshore competition, such as from onshore wind, freshwater aquaculture and alternative proteins, is a source of considerable uncertainty. Lastly, potential sources of use conflict abound, with the attendant regulatory risks and delays. Sovereign wealth funds have so far not lined up behind 'ocean industrial strategies' in the same way they have done on land. Institutional investors still lack some of the knowledge and capacity to invest in new ocean technologies. In the absence of sovereign guarantees and institutional growth capital, development finance institutions and multilateral development banks also have largely stayed away. Blue bonds have grown but are complex to replicate and have narrow applications.
- **Impact of (sustainable) finance in the ocean space is not well measured.** The positive impact of sustainable investments on economic, social and environmental dimensions is not well understood and measured. For instance, a recent analysis has taken stock of existing impact evaluation studies relevant to the conservation and sustainable use of both terrestrial and marine ecosystems. It finds that of the nearly 80 impact evaluation studies identified, only three were relevant to ocean issues (mainly MPAs).[492]
- **Governments don't invest enough in the 'transition'.** Fishing companies and fishers are likely to lose profits and wages (in the short to medium term) if fishing efforts are reduced so fish stocks can rebuild.[493] Several studies have forecast an attractive economic return for rebuilding fish stocks and fishing them sustainably, but governments may need to invest extra resources in the short term to mitigate transition challenges. Policymakers 'often perceive this rebuilding cost to be too expensive in the short-term'.[494] Consequently, policymakers usually avoid taking the actions necessary to start the transition[495] (e.g. repurposing subsidies, supporting fishers in a transition to other livelihoods, providing financial compensation where appropriate, etc.). Recently, NGOs such as the Blue National Capital Financing Facility (part of the International Union for Conservation of Nature) have been advancing this agenda.
- **The ocean and finance communities lack shared language.** One of the challenges for creating investable pipelines in some ocean sectors (nature-based solutions, wild-caught fisheries) is that these communities cannot effectively communicate their needs in ways that financiers can easily understand;[496] there is particular misalignment in the generation of metrics from the conservation sector that can produce data applicable to financial decision-making. The *Ocean Finance Handbook*, recently published by the World Economic Forum, is helping to address this barrier.[497]

What Are the Opportunities for Action to Overcome These Barriers?

In a more investor-friendly world with secured resource access rights, infrastructure and offtake guarantees, well-

[491] Schuhbauer, A., R. Chuenpagdee, W.W.L. Cheung, K. Greer and U.R. Sumaila. 2017. "How Subsidies Affect the Economic Viability of Small-Scale Fisheries." *Marine Policy* 82 (August): 114–21. doi: https://doi.org/10.1016/j.marpol.2017.05.013.

[492] Karousakis, K. 2018. "Evaluating the Effectiveness of Policy Instruments for Biodiversity: Impact Evaluation, Cost-Effectiveness Analysis and Other Approaches." OECD Environment Working Paper no. 141. Organisation for Economic Co-operation and Development.

[493] Sumaila et al. 2012. "Benefits of Rebuilding Global Marine Fisheries Outweigh Costs."

[494] Sumaila et al. 2012. "Benefits of Rebuilding Global Marine Fisheries Outweigh Costs."

[495] Sumaila et al. 2012. "Benefits of Rebuilding Global Marine Fisheries Outweigh Costs."

[496] Fitzgerald, T.P., P.R. Higgins, E. Quilligan, S.A. Sethi and J.T. la Puente. 2020. "Catalysing Fisheries Conservation Investment." *Frontiers in Ecology and the Environment* 18 (3): 151–58. doi: https://doi.org/10.1002/fee.2147.

[497] Ocean Fox Advisory and Friends of Ocean Action (FOA). n.d. *The Ocean Finance Handbook: Increasing Finance for a Healthy Ocean*. http://www3.weforum.org/docs/WEF_FOA_The_Ocean_Finance_Handbook_April_2020.pdf.

established operational standards and regulatory frameworks, and availability of sovereign wealth funds as lead investors, the capital markets are likely to open for ocean investments.

The following strategies can accelerate change:

- **Provide investment conditions required by sovereign and institutional investors.** Experience with emerging industries on land has shown that sovereign wealth funds typically play a central role in providing a debt capital pool for nationally prioritised and strategic emerging industries. In many cases, these funds guarantee matched debt or equity funding from a coalition of development finance institutions. This concentrated and coordinated approach has the dual benefits of (1) systematically de-risking the investment in unfamiliar industries and (2) creating a pool of domain knowledge and capacity ahead of the market. Both benefits are essential in attracting investments from the institutional finance community, which remains relatively unfamiliar with many aspects of the sustainable ocean realm. Investment by sovereign wealth funds and development finance institutions will require appropriate national commitments. In many cases, they will require that sustainable ocean development be nationally prioritised and formalised in an ocean-industrial strategy. In some cases, this strategy can be trialled and launched in special ocean economic zones dedicated to the development of specific conforming ocean industries within strict selection and operating standards.
- **Boost and diversify the investment pipeline**. National technology innovation programs, typically using a mix of research support, grants and below-market-rate investments in prototypes and early-stage application, can significantly accelerate commercialisation (e.g. Norway's support of next-generation offshore aquaculture and the European Union's support of offshore wind generation). With appropriate titling in effect (e.g. conservation and restoration leases) and the reduction of transaction costs,[498] carbon and offset finance could become a major investment vehicle for large-scale restoration and conservation projects. Incubation and acceleration programs, partnerships with schools, universities or corporations can all help accelerate ocean innovation. The investment pipeline also needs to be pushed to include more women, Indigenous people and minorities. For instance, the Impact Investment Exchange (IIX) Sustainability Bonds—developed and implemented by the IIX and IIX Foundation USA—explicitly targets the inclusion of women in economic activities.
- **Improve investment conditions in a sustainable ocean economy.** Public support in the form of margin enhancement (e.g. low-cost infrastructure costs, feed-in tariffs, subsidies) and risk reduction (e.g. regulatory certainty, insurance, offtake and demand guarantees) is also often required, in particular for capital-intensive offshore developments such as wind energy and large-scale mariculture. Significant oversight is necessary, however, to ensure that public support is catalytic (i.e. designed to accelerate commercialisation and innovation) and does not devolve into permanent and ultimately unproductive subsidies. Environmental and sustainability standards can also be advanced by applying stringent criteria in public procurement auctions for products and services, as in the CO_2 emission criteria for public ferries in Norway, which make later electric ferry projects economically attractive.
- **Develop blended finance solutions to de-risk private capital investment**. The concept of blended finance is to use public or philanthropic money to reduce investor risk or improve returns. This is one way to unlock commercial capital for a sustainable ocean economy, especially in higher-risk countries or for new technologies. Using a tranched fund structure to 'blend' capital with different risk appetites and impact mandates is one of the most common forms of blended finance in the ocean space (e.g. Althelia's $100 million Sustainable Ocean Fund, Rare's $30 million Meloy Fund, the California Fisheries Fund or Climate Fund Manager's upcoming 'Climate Investor Two'), but many other structures can also mobilise commercial capital for sustainable ocean assets. More case studies and an explanation of different blended finance structures can be found in the reports published by the Blended Finance Taskforce[499] and the Friends of the Ocean Action.[500]
- **Repurpose harmful subsidies to more equitable and sustainable uses.** Multilateral forums, such as APEC, the G20 and the G7, have called repeatedly for phasing out inefficient fuel subsidies and distortive support measures.[501] This momentum for reform can be channelled into better policies for the ocean economy, for instance in rebuilding fisheries (see 'Governments don't invest enough in the "transition"' point above). The World Trade Organization (WTO) missed its own December 2019 deadline to reach an agreement to 'prohibit certain forms

[498] Wylie et al. 2016. "Keys to Successful Blue Carbon Projects."

[499] Blended Finance Taskforce (BFT). 2018. "Better Finance Better World." London: BFT, Business and Sustainable Development Commission, SYSTEMIQ. https://www.blendedfinance.earth/better-finance-better-world.

[500] Ocean Fox Advisory and Friends of the Ocean Action. n.d. *The Ocean Finance Handbook.*

[501] OECD. 2018. *OECD Companion to the Inventory of Support Measures for Fossil Fuels 2018*. Paris: OECD Publishing. https://www.oecd-ilibrary.org/content/publication/9789264286061-en.

of fisheries subsidies that contribute to overcapacity and overfishing and eliminate subsidies that contribute to illegal, unreported, and unregulated fishing'. A new deadline has now been set (December 2020) for the WTO ministerial meeting. This is an essential milestone.

- **Map financing flows**. The OECD *Sustainable Ocean for All* report presents the first-ever estimates of ocean-relevant official development assistance, covering the (sustainable) ocean economy as well as land-based activities with impacts on the ocean. This is a great first step towards much more comprehensive monitoring of public, private, domestic and international financial flows into the ocean economy, which is urgently needed. For a summary of these points, see Box 20.9.

Box 20.9 Key Triggers to Unlock Finance for a Sustainable Ocean Economy

- Provide investment conditions required by sovereign and institutional investors.
- Boost and diversify the investment pipeline.
- Improve investment conditions in a sustainable ocean economy.
- Develop blended finance solutions to de-risk private capital investment.
- Repurpose harmful subsidies to more equitable and sustainable uses.
- Map financing flows.

Stopping Land-Based Pollution: How Does the Current Political and Economic Constellation Make It Nearly Impossible to Stop Ocean Pollution? How Could This Be Changed, and Where Do We Start?

Why Is It Important?

Ocean pollution is largely an externality of the terrestrial economy. Plastics, nutrients (primarily nitrogen and phosphate), pesticides and parasiticides, antibiotics and pharmaceuticals, industrial chemicals including persistent organic pollutants, oil and gas, medical waste, e-waste and disaster debris are diverted to the ocean with very little financial consequence for the polluter. But ocean dilution is no longer the solution to pollution—the consequences, as described in Sect. 4, are significant and deeply concerning.

Marine plastic litter has received the most attention recently. Plastic pollution is ubiquitous (9–14 million metric tonnes leaking into the ocean every year[502]) and iconic (animals starving from plastic ingestion, strangulation, littered beaches). The root cause is straightforward: waste management infrastructure in industrialising countries (especially in Asia and Africa) is lagging far behind their rapidly rising consumption of plastic. With few consumer products designed for recyclability (just 2% of plastic packaging is made from former plastic[503]), waste collection is largely unprofitable and plastic 'leakage' into the environment is correspondingly high.

Ocean 'dead zones' are also proliferating, as are toxic algal blooms. Around 700 sites worldwide are now affected by low oxygen conditions—up from only 45 in the 1960s.[504] These result from a combination of climate change (warmer waters absorb less oxygen) and nutrient pollution from fertiliser, sewage, animal and aquaculture waste, which causes excessive growth of algae, leading to oxygen depletion when later decomposed by bacteria.[505]

The impact of industrial, pesticide and oil-spill pollutants on the marine food web is also well documented. Bio-accumulation of mercury in food fish, for example, is so high that health organisations are issuing safe human consumption guidelines for many predator species, including tuna, billfish and sharks.[506] Virtually every pollutant present on land is also present in the ocean at detectable levels, with compounding and significant impacts on ecosystem health. Oil spills such as the Deepwater Horizon accident in the Gulf of Mexico have had devastating long-term impacts on the ocean floor and coastal habitats.[507]

What Is Preventing (Faster) Change?

In general, addressing the ocean pollution challenge has been complicated by the difficulties of attribution (many pollutants are non-point-source) and by the overwhelming asymmetry of the situation: when heavily protected terrestrial private interests clash with the interest of a weakly defended common pool resource like the ocean, the ocean loses.

[502] Lau et al. 2020. "Evaluating Scenarios toward Zero Plastic Pollution"; Pew Charitable Trusts and SYSTEMIQ. 2020. *Breaking the Plastic Wave*.

[503] World Economic Forum, Ellen MacArthur Foundation and McKinsey & Company. 2016. "The New Plastics Economy: Rethinking the Future of Plastics."

[504] Laffoley and Baxter. 2019. *Ocean Deoxygenation*.

[505] Laffoley and Baxter. 2019. *Ocean Deoxygenation*.

[506] U.S. Food and Drug Administration. 2019. "Advice about Eating Fish: For Women Who Are or Might Become Pregnant, Breastfeeding Mothers, and Young Children." 7 February. https://www.fda.gov/food/consumers/advice-about-eating-fish; National Health Service. 2018. "Should Pregnant and Breastfeeding Women Avoid Some Types of Fish?" 4 July. https://www.nhs.uk/common-health-questions/pregnancy/should-pregnant-and-breastfeeding-women-avoid-some-types-of-fish/.

[507] McClain, C.R., C. Nunnally and M.C. Benfield. n.d. "Persistent and Substantial Impacts of the Deepwater Horizon Oil Spill on Deep-Sea Megafauna." *Royal Society Open Science* 6 (8):191164. doi: https://doi.org/10.1098/rsos.191164; NOAA Office of Response and Restoration. n.d. "At the Bottom of the Gulf of Mexico, Corals and Diversity Suffered after Deepwater Horizon Oil Spill." https://response.restoration.noaa.gov/about/media/bottom-gulf-mexico-corals-and-diversity-suffered-after-deepwater-horizon-oil-spill. html. Accessed 12 May 2020.

Reform of the plastic economy is specifically impeded by three principal factors:

- **Price differential between virgin and recycled products.** The current price of virgin plastic resin is historically low, making recycling of most polymers unprofitable without subsidies. An adjustment of virgin cost, through voluntary industry initiative or imposed through policy, would (1) launch significant entrepreneurial activity in the waste management and collection sector, (2) make collection of plastic waste more profitable and (3) provide a major incentive for consumer brands to include recyclability in their packaging product design.
- **High capital and operating costs of waste management infrastructure.** Introducing modern plastic waste collection infrastructure into the developing world will require capital expenditures of billions of U.S. dollars per year[508] (with operating costs several multiples higher). The public sector in these countries is going to rely on the 'extended producer responsibility' schemes used by developed countries for sources of finance. However, translating these schemes into the infrastructure, governance and legal frameworks of developing or industrialising countries is challenging. In addition, the costs associated with the development of new technologies (e.g. chemical recycling) and the transition towards plastic substitutes are considerable, and it is not clear how those costs can be equitably allocated among industry players.
- **Lack of transparency.** The flow of recyclable and non-recyclable plastics through the value chain, from the resin producer through the brands to the waste manager, is currently largely undocumented. It is thus difficult for a producer or brand to differentiate its 'plastic performance', and to be rewarded by the market as a leader and good faith actor in the fight against ocean plastic. On the opposite side of the coin, it is nearly impossible for civil society to hold responsible companies which are side-stepping the ocean plastic problem.

Pesticide, nutrient and industrial pollution control is largely a political challenge. Agricultural and industrial production has long benefitted from the ocean's dilution of excess nutrients, pesticides and industrial toxins, and the resulting rents tend to be well protected legally, politically and culturally. In the United States, for example, it is very difficult to pursue legal action against non-point-source polluters. Environmental enforcement budgets are constantly under attack.

What Are the Opportunities for Action to Overcome These Barriers?

A growing number of governments and industries are announcing new measures and commitments (e.g. plastic bag bans). However, a recent study quantified that such efforts will barely make a difference: by 2040, current government and industry commitments are likely to reduce annual plastic leakage into the ocean by only 7% relative to a BAU scenario.[509]

There is no silver bullet solution to ocean plastics pollution. A more diverse portfolio of more ambitious solutions should be deployed, including reduction of unnecessary plastics, substitution with other materials, recycling (mechanical and chemical) and safe disposal (see Fig. 20.39).

Many technical solutions are available today to governments and industry, but they could be accelerated by three major reforms:

- **Recalibrate the economic advantage of virgin plastic** to stimulate the demand and competitiveness of recycled materials. This can be done in multiple ways; for instance, through the global adoption of extended producer responsibility (EPR) schemes or the increased taxation of virgin production.
- **Invest massively in waste collection and recycling technology and infrastructure**. The highest priority in the short term is collection infrastructure in the developing world—collection rates need to stay ahead of recycling capacity to ensure reliable feedstock flows.
- **Bring transparency and accountability to the flow of plastic polymers through the value chain.** The performance of companies selling plastic products needs to be fully transparent over time (in terms of shifting to more recycled content, recyclable product design and plastic substitutes).

These measures are within reach. The management of plastic waste in the developing world will not remain as an unpriced externality much longer—virgin plastic taxation schemes are under discussion in many countries. Plastic producers and brands may choose to pre-empt taxation with alternative, industry-led EPR initiatives and funding mechanisms. Recent announcements by industry, including the plastic industry's $1.25 billion Alliance to End Plastic Waste and Nestle's $2.1 billion commitment to tackle plastic waste, are not likely to be the only major commitments forthcoming soon.

In the fight against nutrient pollution, dead zones and toxic algal blooms, the sustainable ocean economy is in a

[508] Lau et al. 2020. "Evaluating Scenarios toward Zero Plastic Pollution"; Pew Charitable Trusts and SYSTEMIQ. 2020. *Breaking the Plastic Wave*.

[509] Lau et al. 2020. "Evaluating Scenarios toward Zero Plastic Pollution"; Pew Charitable Trusts and SYSTEMIQ. 2020. *Breaking the Plastic Wave*.

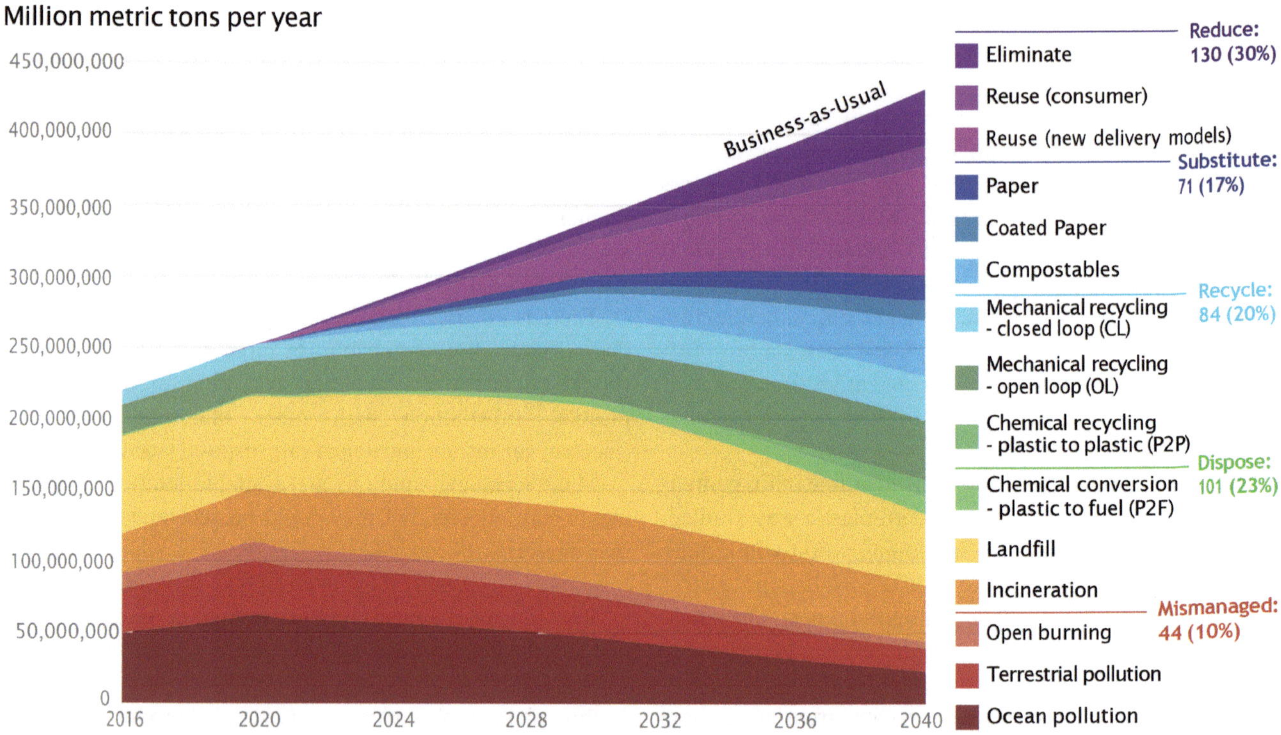

Fig. 20.39 Plastic leakage into the ocean can be reduced significantly only if all solutions are implemented concurrently, ambitiously and starting immediately. *Note: The 'wedges' figure shows the share of treatment options for the plastic that enters the system over time under the System Change scenario. Any plastic that enters the system has a single fate, or a single 'wedge'. The numbers include macroplastic and microplastic. (Source: Lau, W.W.Y., Y. Shiran, R.M. Bailey, E. Cook, M.R. Stuchtey, J. Koskella, C.A. Velis et al. 2020. "Evaluating Scenarios toward Zero Plastic Pollution." Science, July. doi:* https://doi.org/10.1126/science.aba9475)

direct confrontation with land-based agricultural interests. The technologies for precise applications, runoff controls and soil regeneration are well established[510] and could drastically reduce the need for input on crops, but they are not mainstreamed yet on a global scale. As the sustainable ocean economy develops further, its economic interests will be more fully represented in the political and legal realm—and it is doubtful that the periodic death of entire coastal ecosystems will remain uncontested for very long.

- It would, of course, be far preferable to avoid this inevitable confrontation by proactively requiring precision fertilisation, low-input farming and regenerative agriculture, especially along major rivers. However, the current political economy will likely require ocean interests to assert their interests in a formal way for this to occur. For a summary of these points, see Box 20.10.

[510]FOLU. 2019. *Growing Better.*

Box 20.10 Key Triggers to Reduce Land-Based Pollution in a Sustainable Ocean Economy

- Recalibrate the economic advantage of virgin plastic.
- Invest massively in waste collection and recycling technology and infrastructure.
- Bring transparency and accountability to the flow of plastic polymers through the value chain.
- Require the adoption of precision agriculture to avoid nutrient runoff.

Upgrading Ocean Accounting: How Do Current Metrics and Public Accounting Conventions (Gross Value Added, Gross Domestic Product) Drive the Wrong Priorities, What Can Be Changed, and How Can We Start the Change Today?

Why Is It Important?

Ocean macroeconomics has long focused on 'outputs', typically expressed in terms of GDP metrics. Microeconomic metrics—market size, growth rates, returns—have been similarly output-focused. But these metrics are flawed—they measure the flow of capital but ignore the 'stocks', the value of the manufactured and natural capital involved in the production process. They also don't measure the importance of human capital (knowledge, aptitude, education and skills). For decades, world-class economists (Dasgupta, Jorgenson, Kuznets, Nordhaus and Tobin, Solow, Stiglitz, Weitzman, etc.[511]) have warned about the perils of a focus on GDP only, arguing that it ignores the true cost of production and does not put human well-being at the centre of the economic debate: 'Growth is a means to an end, rather than an end in itself'.[512]

The broader value of the ocean must be fully accounted for and used in decision-making.

The System of National Accounts (SNA) could document progress along the dimensions of productivity, job creation, food security, regional stability and long-term ecosystem health. A complete set or 'sequence' of national ocean accounts could provide highly insightful information for the sustainable management of the ocean economy. Three key high-level indicators can already provide a much more holistic picture than the use of GDP only to inform policy and ocean-related decisions: ocean product, net change in the ocean balance sheet and ocean income:

- **Ocean product** is the traditional measure of the ocean's output, generally monetised in terms of ocean 'gross domestic product' (GDP) or 'net domestic product' (NDP).
- **Net change in the ocean balance sheet** provides a sustainability indicator. It accounts for the reserves of natural and produced capital in the ocean, as driven by economic activities. Changes in the balance sheet indicate physical and monetary changes to show how wealth and opportunity change through time (adjusted for anticipated price changes).
- **Ocean income** measures benefits to nationals from the ocean, the 'ends' or 'outcomes' of policy. It is generally expressed in terms of net or gross national income (NNI or GNI). Income can include non-monetary types of income, though these are often expressed in monetary equivalents.

What will an ocean account monitoring these three indicators change? Good information is not sufficient to ensure good decisions, but it helps. The development of national accounts has had an impact on inflation and the business cycle, which has generally made the economy more stable and enhanced human well-being.[513]

In parallel, it is useful to promote more flexible approaches to natural capital valuation and use in decision-making that can be tested and deployed with less inertia and complexity than with the SNA, with the eventual goal of reconciling these two approaches. Some examples of such alternative methodologies include using payment for ecosystem services schemes that have been used in the United Kingdom[514] and Costa Rica,[515] for instance, or the gross ecosystem product (GEP), which is increasingly used by the Chinese government as part of a transformation to inclusive, green growth.[516]

What Is Preventing (Faster) Change?

Developing national accounts to guide economic development is less daunting than it may seem. Most of the data already exist in national accounts, in government agencies or in scientific databases. The knowledge to build the connec-

[511] See Blue Paper 8 for a more in-depth review: Fenichel, E.P., B. Milligan, I. Porras et al. 2020. "National Accounting for the Ocean and Ocean Economy." Washington, DC: World Resources Institute. https://www.ocean-panel.org/blue-papers/national-accounting-ocean-and-ocean-economy.

[512] High Level Expert Group on the Measurement of Economic Performance and Social Progress. n.d. "Measurement of Economic Performance and Social Progress." https://www.oecd.org/statistics/measuring-economic-social-progress/. Accessed 12 May 2020.

[513] Landefeld, J.S. 2000. "GDP: One of the Great Inventions of the 20th Century." *Survey of Current Business* 80(1) (January): 4.

[514] Bateman, I., A. Binner, B. Day, C. Fezzi, A. Rusby, G. Smith and R. Welters. 2019. "United Kingdom: Paying for Ecosystem Services in the Public and Private Sectors." In *Green Growth That Works: Natural Capital Policy and Finance Mechanisms from around the World*, edited by L. Mandle, Z. Ouyang, J.E. Salzman and G. Daily, 237–54. Washington, DC: Island Press/Center for Resource Economics. doi: https://doi.org/10.5822/978-1-64283-004-0_15.

[515] Quesada, A.U. 2019. "Costa Rica: Bringing Natural Capital Values into the Mainstream." In *Green Growth That Works: Natural Capital Policy and Finance Mechanisms from around the World*, edited by L. Mandle, Z. Ouyang, J.E. Salzman and G. Daily, 195–212. Washington, DC: Island Press/Center for Resource Economics. doi: https://doi.org/10.5822/978-1-64283-004-0_13.

[516] Ouyang, Z., C. Song, H. Zheng, S. Polasky, Y. Xiao, I. Bateman, J. Liu et al. 2020. "Using Gross Ecosystem Product (GEP) to Value Nature in Decision-Making." https://ore.exeter.ac.uk/repository/handle/10871/120272.

tions exists, although it is dispersed throughout the government, academic, business and NGO sectors. Many countries already produce 'marine GDP' reports that may be a good starting point.

Several barriers remain:

- **Old habits.** Even in 2020, economic and policy decisions are still mostly based on nineteenth-century economic paradigms, with rigid definition of economic sectors and metrics, a failure to differentiate sources of income in terms of externalities and no monetisation of natural capital.
- **Siloed data.** Much of the information needed for ocean accounts exists but is siloed in multiple government agencies, as well as in the academic, business and NGO sectors. In a few cases the data are not yet available, for instance, biophysical data needed to quantify natural stocks. Also, the tagging of databases is not standardised, making it difficult to know precisely what's available.
- **Methodology.** Despite the rising momentum behind this new generation of accounting, there is still a need for standardisation and reforms of existing accounting systems and valuation methods, within and across countries.
- **Lack of track record in informing decision-making.** Policy- and decision-makers are lacking demonstrations showing how these indicators actually can inform decisions (and are informing them).

What Are the Opportunities for Action to Overcome These Barriers?

Four main areas of action could accelerate the development and use of these holistic ocean accounts:

- **Create national ocean accounts.** National statistical offices, in partnership with marine agencies, need to develop a complete sequence of national ocean accounts: product, income, balance sheets and supply and use tables. This should be achievable by 2025. In particular, they need to ensure the compatibility of ocean accounting efforts with international statistical standards and approaches, mainly the System of National Accounts, the System of Environmental Economic Accounting (SEEA), the ten Fundamental Principles of Official Statistics endorsed by the UN General Assembly in January 2014[517] and other broadly accepted initiatives.[518] Next to these accounts, more flexible approaches to natural capital valuation and use in decision-making can be encouraged (e.g. GEP in China and other examples mentioned above), and alignment should be ensured between these approaches and the ones using national accounts.
- **Develop and use interactive dashboards for ocean account reporting**. Such dashboards allow users to explore the data, aggregate and disaggregate sectors and groups of people, alter the account boundaries and access ethically acceptable disaggregation by digital means. These dashboards would stimulate decision-making based on more holistic information more than GDP only, and they would track national progress over time.
- **Encourage international collaboration and standardisation.** National governments should ensure that their national accountants, economic analysts and marine scientists participate in workshops organised by the UN Statistical Division and associated organisations for developing ocean accounts. This will help to maintain standards and increase credibility. These international organisations need to evolve to provide a degree of third-party verification of accounts coupled with capacity-building assistance.
- **Invest in data architecture and engineering, and build know-how in national statistical offices.** Governments need to invest in data architecture and engineering at levels surpassing global multinational companies. These investments are necessary to connect fine-scale data about the marine environment with detailed economic data in supply-and-use structures and other data structures for national accounting and forecasting the ocean economy. These investments should build on existing Earth observation programs when possible. Investment must also include investments in people. The costs of implementing the ocean accounts—including embedding them in relevant laws, policies and action plans—will likely be far outweighed by the benefits current and future generations gain from sustainable ocean economies. For a summary of these points, see Box 20.11.

Box 20.11 Key Triggers to Develop and Mainstream Ocean Accounts

- Create national ocean accounts covering product, income, balance sheets and supply and use tables.
- Develop and use interactive dashboards for ocean account reporting.
- Encourage international collaboration and standardisation.
- Invest in data architecture and engineering, and build know-how in national statistical offices.

[517] UN Economic and Social Council. n.d. *2013/21. Fundamental Principles of Official Statistics*. https://unstats.un.org/unsd/dnss/gp/FP-Rev2013-E.pdf.

[518] UN Department of Economic and Social Affairs, Statistical Division. n.d. "System of National Accounts." https://unstats.un.org/unsd/nationalaccount/sna.asp. Accessed 12 May 2020; United Nations. n.d. "System of Environmental Economic Accounting." https://seea.un.org/. Accessed 12 May 2020.

6.3.2 Five Key Sectors to Be Transformed Towards a Sustainable Ocean Economy

Sustainable Food from the Ocean: How Can Sustainable Ocean Fishing and Farming Feed a Planet with Ten Billion People?

Why Is It Important?

Ocean fish provides about 3.2 billion people with almost 20% of their average intake of animal protein.[519] This number is even higher in developing regions such as Indonesia, Sri Lanka and many small island developing states, which derive 50% or more of their animal protein from aquatic foods.[520] Ocean food is also a unique source of long-chain omega-3 fatty acids, minerals, calcium, iodine and vitamins.[521] To simplify, food from the ocean can be split into two main sectors: wild-caught fisheries and mariculture—the latter can then be divided into unfed (e.g. seaweed and filter-feeders) and fed mariculture (e.g. finfish and crustaceans). Today the lion's share of ocean food production comes from wild-caught fisheries (in tonnes of edible food equivalent; see Fig. 20.40).

The ocean can contribute to sustainable food security for ten billion people. Ocean animals are more efficient than terrestrial systems in producing protein;[522] their impact on climate change and land use is in general much lower than terrestrial animal proteins (Fig. 20.41) and their production is not limited by suitable area available.

By applying realistic demand scenarios to the sustainable seafood supply potential presented in Sect. 5 (six times more seafood than today), a recent paper determined the plausible future equilibrium quantity of food from the sea that could be produced and consumed. This still represents a significant expansion, calculated to represent a 36–74% production increase compared to today's levels.[523]

Yet the current ocean production system is not on track to deliver this production increase in a sustainable way.

FAO estimates that 33% of global fish stocks are overfished, and nearly 60% exploited at maximum capacity.[524] The wild-caught production has been stagnating in the past three decades at about 80 mmt/year of landed weight. Illegal, unreported and unregulated (IUU) fishing remains a major problem, accounting for 11–26 mmt of catch and creating financial losses of between \$10.0 billion and \$23.5 billion/year.[525] Fisheries on the high seas (e.g. tuna, jacks) are subject to multilateral management institutions which have, in the past, frequently not adopted scientific recommendations.[526] In a BAU scenario, 2050 yields could decrease by almost 16% to about 67 mmt/year because of the cumulated pressures of overfishing, climate change and pollution.[527] Commercial fishing activities also affect fauna like birds, marine mammals and turtles. For example, the World Ocean Assessment states that 'each year, incidental bycatch in long-line fisheries is estimated to kill at least 160,000 albatrosses and petrels, mainly in the southern hemisphere. For marine reptiles, a threat assessment scored fishery bycatch as the highest threat across marine turtle subpopulations'.[528]

Mariculture has been growing at a stable pace in recent years, around 5.8% annually,[529] but finfish mariculture is too often associated with unsustainable practices (e.g. fish escapes, local fouling, overuse of antibiotics, disease transfer) and is critically constrained by the need to 'fish wild fish to farm fish'. As a result, many consumers still consider wild-caught fish to be of higher quality than farmed fish.

Finally, the significant amount of food waste in the seafood value chain represents a missed opportunity to boost consumption without increasing production. Indeed, FAO estimates that 35% of fish and seafood is wasted, including 8% of all fish caught being thrown back into the water (in most cases, these fish are dead, dying or badly damaged).[530] This waste is equivalent to almost 3 billion Atlantic salmon.[531]

[519] FAO, ed. 2018. *The State of World Fisheries and Aquaculture 2018.*

[520] FAO, ed. 2018. *The State of World Fisheries and Aquaculture 2018.*

[521] Lund, E.K. 2013. "Health Benefits of Seafood: Is It Just the Fatty Acids?" *Food Chemistry*, Ninth International Food Data Conference: Food Composition and Sustainable Diets, 140 (3): 413–20. doi: https://doi.org/10.1016/j.foodchem.2013.01.034; Costello et al. 2019. "The Future of Food from the Sea."

[522] Huntington, T., and M.R. Hasan, eds. 2009. "Fish as Feed Inputs for Aquaculture: Practices, Sustainability and Implications." FAO Fisheries and Aquaculture Technical Paper no. 518. Rome: Food and Agriculture Organization of the United Nations. https://www.researchgate.net/profile/Mohammad_Hasan22/publication/336030732_Fish_as_feed_inputs_for_aquaculture_practices_sustainability_and_implications/links/5d8b8f80a6fdcc255499d9e9/Fish-as-feed-inputs-for--aquaculture-practices-sustainability-and-implications.pdf#page=19.

[523] Costello et al. 2020. "The Future of Food from the Sea" (*Nature*).

[524] FAO, ed. 2018. *The State of World Fisheries and Aquaculture 2018.*

[525] Widjaja et al. 2020. "Illegal, Unreported and Unregulated Fishing and Associated Drivers."

[526] Pew Charitable Trusts. 2019. "International Fisheries Managers' Response to Performance Reviews Insufficient."

[527] Costello et al. 2019. "The Future of Food from the Sea."

[528] Inniss et al. 2016. "The First Global Integrated Marine Assessment."

[529] FAO, ed. 2018. *The State of World Fisheries and Aquaculture 2018.*

[530] FAO. 2017. "FAO Regional Office for Europe and Central Asia: Losses in Fisheries and Aquaculture Tackled at Global Fishery Forum." 14 September. http://www.fao.org/europe/news/detail-news/en/c/1037271/.

[531] FAO. 2017. "FAO Regional Office for Europe and Central Asia."

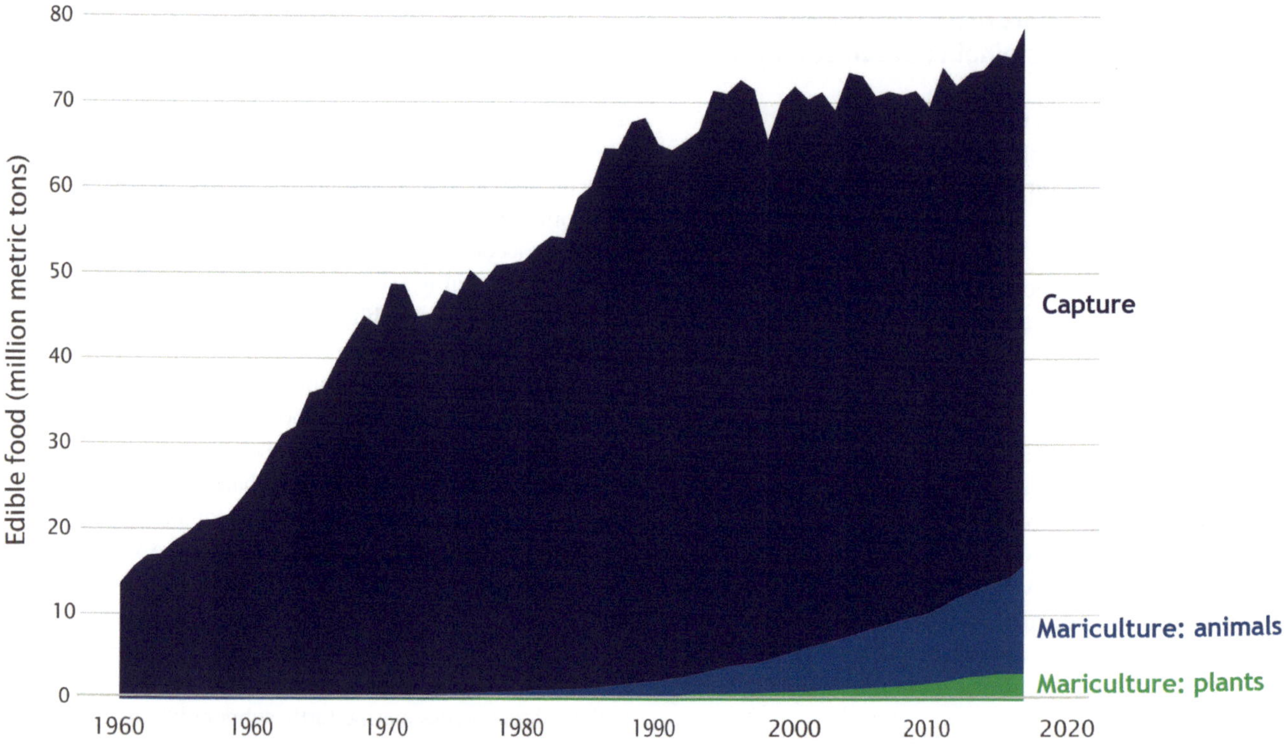

Fig. 20.40 Historical production of marine capture fisheries and mariculture (edible weight). *Note: This figure shows food potential, as it does not take into account historical non-food use.* (*Sources: Production data are from FAO. 2019. "Fishery and Aquaculture Statistics: Global Production by Production Source, 1950–2017 (FishstatJ)." FAO Fisheries and Aquaculture Department.* www.fao.org/fishery/statistics/software/fishstatj/en. *Landed quantities are converted into million metric tonnes of edible food equivalents using conversion values from Edwards, P., W. Zhang, B. Belton and D.C. Little. 2019. "Misunderstandings, Myths and Mantras in Aquaculture: Its Contribution to World Food Supplies Has Been Systematically Over Reported." Marine Policy 106 (August): 103547. doi:* https://doi.org/10.1016/j.marpol.2019.103547; *and Duarte, C.M., J. Wu, X. Xiao, A. Bruhn and D. Krause-Jensen. 2017. "Can Seaweed Farming Play a Role in Climate Change Mitigation and Adaptation?" Frontiers in Marine Science 4. doi:* https://doi.org/10.3389/fmars.2017.00100)

The current and BAU production numbers fall far short of the ocean's productive potential. If all stocks were sustainably managed and fishing effort were maximised for profit, yields from wild-caught fisheries could increase to 98 mmt/year in 2050. This is an about 20% increase from current levels and represents an increase in profit of $53 billion in 2050 (in comparison to BAU).[532] These gains are dependent on widespread policy reforms, such as rights-based approaches that incentivise conservation and hold fishing fleets accountable to science-based limits.[533]

[532]Costello et al. 2016. "Global Fishery Prospects under Contrasting Management Regimes."

[533]Costello et al. 2016. "Global Fishery Prospects under Contrasting Management Regimes."

The mariculture story is even more promising. Finfish mariculture expansion potential is almost unlimited if the feed can be decoupled from fish meal/fish oil (FM/FO);[534] the biological potential for finfish mariculture production is estimated to be around 15,000 mmt. Unfed mariculture also holds great promise: a study found that the ocean has the potential to produce nearly 768 mmt of bivalves (shell-on weight), and about 60% of this production would be profitable at roughly the current price for blue mussels ($1700/mmt).[535] Unfed mariculture (e.g. bivalves and seaweed) can also substantially increase nutritious food and feed with a lower impact on the marine environment, and may in some cases enhance wild fisheries by creating artificial habitats.

[534]Costello et al. 2019. "The Future of Food from the Sea."

[535]Costello et al. 2019. "The Future of Food from the Sea."

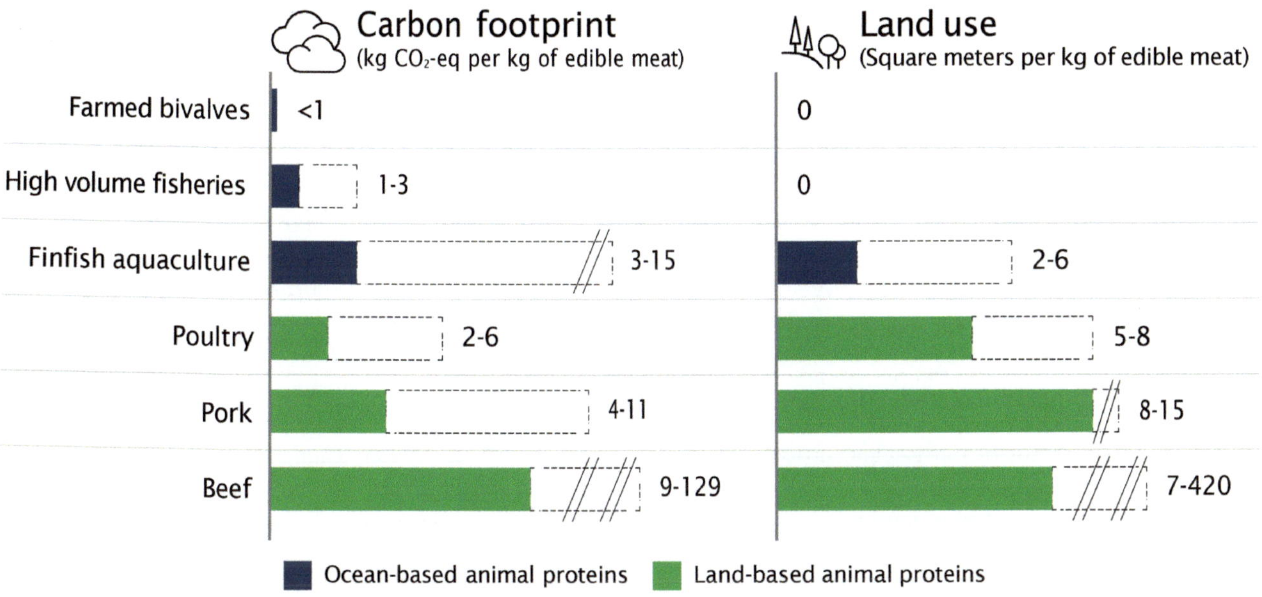

Fig. 20.41 Land use and carbon implications: comparison between ocean-based and land-based sources of proteins. *Note: This figure shows food potential, as it does not take into account historical non-food use.* (*Sources: Nijdam, D., G.A. Rood and H. Westhoek. 2012. "The Price of Protein: Review of Land Use and Carbon Footprints from Life Cycle Assessments of Animal Food Products and Their Substitutes." Food Policy 37 (6): 760–70. doi:* https://doi.org/10.1016/j.food-pol.2012.08.002*; Filgueira, R., T. Strohmeier and Ø. Strand. 2019. "Regulating Services of Bivalve Molluscs in the Context of the Carbon Cycle and Implications for Ecosystem Valuation." In Goods and Services of Marine Bivalves, edited by A.C. Smaal, J.G. Ferreira, J. Grant, J.K. Petersen and Ø. Strand, 231–51. Cham, Switzerland: Springer International*)

Seaweed production is growing strongly (11% annually) from a small base (30 mmt/year). Seaweed and macroalgae have the potential to help solve food security issues, act as a form of carbon sequestration, reduce ruminant methane emissions and so on[536] (see Box 20.12). There are few geo-

[536]Chung, I.K., J. Beardall, S. Mehta, D. Sahoo and S. Stojkovic. 2011. "Using Marine Macroalgae for Carbon Sequestration: A Critical Appraisal." *Journal of Applied Phycology* 23 (5): 877–86. doi: https://doi.org/10.1007/s10811-010-9604-9; N'Yeurt, A. de R., D.P. Chynoweth, M.E. Capron, J.R. Stewart and M.A. Hasan. 2012. "Negative Carbon via Ocean Afforestation." *Process Safety and Environmental Protection* 90 (6): 467–74. doi: https://doi.org/10.1016/j.psep.2012.10.008; Sondak, C.F.A., P.O. Ang, J. Beardall, A. Bellgrove, S.M. Boo, G.S. Gerung, C.D. Hepburn et al. 2017. "Carbon Dioxide Mitigation Potential of Seaweed Aquaculture Beds (SABs)." *Journal of Applied Phycology* 29 (5): 2363–73. doi: https://doi.org/10.1007/s10811-016-1022-1; Duarte, C.M., J. Wu, X. Xiao, A. Bruhn and D. Krause-Jensen. 2017. "Can Seaweed Farming Play a Role in Climate Change Mitigation and Adaptation?" *Frontiers in Marine Science* 4. doi: https://doi.org/10.3389/fmars.2017.00100; Capron, M.E., Z. Moscicki, R. Blaylock, C. Sullivan, K. Lucas, I. Tsukrov, M.D. Chambers et al. 2018. "Ocean Forests: Breakthrough Yields for Macroalgae." In *OCEANS 2018 MTS/IEEE Charleston*, 1–6. doi: https://doi.org/10.1109/OCEANS.2018.8604586; Froehlich, H.E., J.C. Afflerbach, M. Frazier and B.S. Halpern. 2019. "Blue Growth Potential to Mitigate Climate Change through Seaweed Offsetting." *Current Biology* 29 (18): 3087–93. e3. doi: https://doi.org/10.1016/j.cub.2019.07.041; Brooke, C.G., B.M. Roque, N. Najafi, M. Gonzalez, A. Pfefferlen, V. DeAnda, D.W. Ginsburg et al. 2018. "Evaluation of the Potential of Two Common Pacific Coast Macroalgae for Mitigating Methane Emissions from Ruminants." *BioRxiv*, October, 434480. doi: https://doi.org/10.1101/434480; Roque, B.M., C.G. Brooke, J. Ladau, T. Polley, L.J. Marsh, N. Najafi, P. Pandey et al. 2019. "Effect of the Macroalgae *Asparagopsis taxiformis* on Methane Production and Rumen Microbiome Assemblage." *Animal Microbiome* 1 (1): 3. doi: https://doi.org/10.1186/s42523-019-0004-4; Machado, L., M. Magnusson, N.A. Paul, R. Kinley, R. de Nys and N. Tomkins. 2016. "Dose-Response Effects of *Asparagopsis taxiformis* and *Oedogonium* sp. on in Vitro Fermentation and Methane Production." *Journal of Applied Phycology* 28 (2): 1443–52. doi: https://doi.org/10.1007/s10811-015-0639-9.

physical (48 million km^2 are suitable for cultivation)[537] or technical constraints to doing so.[538] This industry is clearly still in its infancy, with much consumer product development and testing yet to be done.

Box 20.12 Seaweed: The Versatile Ocean Super Crop

Seaweed cultivation is the fastest growing mariculture sector (7%/year),[a] already producing over 30 million metric tonnes (valued at US $4.8 billion).[b] As a 1000-year-old industry in Asia, it is unsurprising that the largest share of the world's seaweed is produced on the coasts of this continent—China and Indonesia alone produce over 85% of global volume.[c] About ten species are intensively cultivated.[d] Europe and North America are catching up to the benefits of producing this super crop. Through the Pegasus project, for instance, the European Union developed guidelines for the sustainable aquaculture of seaweeds. The project showcased the many benefits of seaweeds. Not only do they not require land or freshwater, but seaweed farms also provide habitat for many marine species, mitigate storm damage, sequester carbon, provide bioremediation services (e.g. degrade or assimilate contaminants as excess nitrogen and phosphorus) and can protect calcifiers from projected ocean acidification.[e] Their uses are similarly broad and promising. Seaweeds are already extensively used in the pharmaceutical and nutraceutical industries, consumed directly as human food (e.g. directly in soups and salads or processed into noodles and seasoning) and food additives, transformed into fertiliser or refined into biofuels. They are being increasingly explored as animal feed (even shown to reduce methane in ruminants by a factor of up to 80% in one case, even if more research is needed[f]), or can be a base ingredient for bioplastics.[g]

With new seaweed applications being found yearly, and a potential cultivation area of 48 million km^2 (about three times the current area used for growing crops—16 million km^2),[h] seaweeds could become an ever more present sustainable ocean crop in the decades to come—if their farming development is supported by adequate marine spatial planning[i], and if innovations help seaweed-based products to enter new markets.

Sources:

[a] *Dubois, O. 2011. The State of the World's Land and Water Resources for Food and Agriculture: Managing Systems at Risk. London: Earthscan, FAO. https://www.cabdirect.org/cabdirect/abstract/20123051697; Costello et al. 2019. "The Future of Food from the Sea"*

[b] *Barbier, M., B. Charrier, R. Araujo, S.L. Holdt, B. Jacquemin and C. Rebours. 2019. "PEGASUS: Phycomorph European Guidelines for a Sustainable Aquaculture of Seaweeds." Roscoff, France: COST: European Cooperation in Science and Technology. http://www.phycomorph.org/doc/PEGASUS_SUSTAINABLE_SEAWEED_AQUACULTURE_FULL_RECOMMENDATIONS.pdf*

[c] *Costello, C., L. Cao, S. Gelcich et al. 2019. "The Future of Food from the Sea." Washington, DC: World Resources Institute. https://www.oceanpanel.org/blue--papers/future-food-sea*

[d] *FishStatJ: Software for Fishery and Aquaculture Statistical Time Series. n.d. Rome: Food and Agriculture Organization of the United Nations, Fisheries and Aquaculture Department*

[e] *Duarte, C.M., J. Wu, X. Xiao, A. Bruhn and D. Krause-Jensen. 2017. "Can Seaweed Farming Play a Role in Climate Change Mitigation and Adaptation?" Frontiers in Marine Science 4. doi:* https://doi.org/10.3389/fmars.2017.00100

[f] *Mulhollem, J. 2019. "Seaweed Feed Additive Cuts Livestock Methane but Poses Questions." Penn State University, 17 June. https://news.psu.edu/story/578123/2019/06/17/research/seaweed-feed-additive-cuts-livestock-methane-poses-questions*

[g] *Barbier et al. 2019. "PEGASUS"*

[h] *Dubois. 2011. The State of the World's Land and Water Resources for Food and Agriculture*

[i] *Duarte et al. 2017. "Can Seaweed Farming Play a Role in Climate Change Mitigation and Adaptation?"*

[537] Froehlich et al. 2019. "Blue Growth Potential to Mitigate Climate Change through Seaweed Offsetting."

[538] Capron et al. 2018. "Ocean Forests"; Oilgae. 2010. *Oilgae Guide to Fuels from Macroalgae*. Tamil Nadu, India. https://arpa-e.energy.gov/sites/default/files/Oilgae%20Guide%20to%20Fuels%20from%20Macroalgae%202010.pdf; Czyrnek-Delêtre, M.M., S. Rocca, A. Agostini, J. Giuntoli and J.D. Murphy. 2017. "Life Cycle Assessment of Seaweed Biomethane, Generated from Seaweed Sourced from Integrated Multi-trophic Aquaculture in Temperate Oceanic Climates." *Applied Energy* 196 (June): 34–50. doi: https://doi.org/10.1016/j.apenergy.2017.03.129; Benzie, J.A.H., T.T.T. Nguyen, G. Hulata, D. Bartley, R. Brummett, B. Davy, M. Halwart et al. 2012. "Promoting Responsible Use and Conservation of Aquatic Biodiversity for Sustainable Aquaculture Development." In *Farming the Waters for People and Food: Proceedings of the Global Conference on Aquaculture 2010*, 337–83. Phuket, Thailand: Food and Agriculture Organization of the United Nations, Network of Aquaculture Centers in Asia. https://www.researchgate.net/profile/Matthias_Halwart/publication/263569545_Supporting_farmer_innovations_recognizing_indigenous_knowledge_and_disseminating_success_stories/links/0046353b412ffc6050000000/Supporting-farmer-innovations-recognizing-indigenous-knowledge-and-disseminating-success-stories.pdf#page=349; Loureiro, R., C.M.M. Gachon and C. Rebours. 2015. "Seaweed Cultivation: Potential and Challenges of Crop Domestication at an Unprecedented Pace." *New Phytologist* 206 (2): 489–92. doi: https://doi.org/10.1111/nph.13278.

What Are the Opportunities for Action to Accelerate Change?

There is no alternative to regulation—and enforcement—in fisheries. Unregulated, 'free access' fisheries almost invariably overfish.[539] Unregulated fleets tend to grow to the point of little or no profit for the individual boat—a point that is ecologically and economically irrational and destructive and which can be driven to absurdity by national subsidies. Ports allowing illegal or untraced seafood to be unloaded without verification are maintaining these destructive practices. The governments of most industrialised nations have addressed this problem with various types of catch restrictions and port controls. With rare exception, stocks have shown a heartening capacity to recover once the pressure is eased.

Reform is impossible without rules to protect the stocks and allow for an efficient, fair and equitable allocation of catch.[540] First and foremost, the commercial right to fish needs to be predicated on a plan to fully restore the target stock within 10 years (or as soon as possible for fish stocks with longer recovery time).[541] This has been the key feature of the Magnuson-Stevens Fishery Conservation and Management Act, which successfully restored fish stocks and fisheries in the United States (see case study in Sect. 4.2), and the European Common Fisheries Policy. Second, within the framework of these restoration targets, catches must be restricted to a level that results in the rebuilding of fish stocks, followed by sustainable levels of fishing.[542] Third, the allowable catch must be allocated fairly—to provide food security to artisanal fishing communities and to tie the fortunes of commercial fishers to the health of 'their' stocks.[543] Fourth, the Port State Measures Agreement must be enforced in all ports to close illegal fishing. Finally, the implementation of fully protected MPAs has been demonstrated to generate significant spillover effects that can benefit surrounding fisheries.[544] Such MPAs could be an integral part of a fishery recovery plan for some fisheries. Finally, several technical innovations can help reduce bycatch and lost fishing nets, as well as prevent food loss during fishing and processing. In addition to regulatory changes, the financial community can deploy innovative funding mechanisms to support the transition towards sustainable fisheries (e.g. Meloy Fund, California Fisheries Fund).

The acceleration of sustainable mariculture will require the coordinated intervention of governments and investors, as well as an adjustment of consumer preferences. The Food and Land Use Coalition strongly recommends that governments 'support new feed technologies with clear targets (standardized performance specifications for feed applications), strong incentives (feed efficiency standards), and guaranteed demand (feed standards for government seafood purchases)'.[545] Without compromising strong and independent oversight, governments also need to update regulations so they conform to new best practice technology standards, with low- and multi-trophic operations involving seaweeds and bivalves made a priority. Additionally, governments need to lower the barrier to entry for marine aquaculturists; for example, by providing environmentally stringent, yet easily navigable aquaculture governance frameworks and/or assistance to (excess capacity) fishers who transition to become marine aquaculturists. Finally, governments need to work with farm operators to support the development of model farms which are innovative in terms of feed technology, as well as the integration of different trophic levels (multi-trophic farming), scale, containment, siting and so on.

There is now strong consensus that new feed technology and vaccine delivery systems, as well as improved breeding and genetic selection, have enhanced the investment grade of top-performing mariculture operations.[546] The hope is that continued improvements will lead to a mature and stable industry. Venture and early-stage funding has also entered this space from different sources, including corporate food and feed investors, Silicon Valley firms attracted to the artificial-intelligence and technology aspects of aquaculture, and specialised funds (e.g. Aqua-Spark, a $180 million Dutch aquaculture investment firm). Given the potentially increasing role of aquaculture in the global diet, a strong push will be needed from public funding and official development assistance, ideally led by a consortium of countries with expertise and interest in scaling up. However, appropriate species and local conditions, including market opportunities, differ significantly and will require tailored approaches. Hatcheries and farmers will need to address and preserve genetic diversity while breeding selectively for desirable traits. At the same time, precautions will need to be taken to keep genetically modified species from escaping mariculture operations and altering, or in the worst case replacing, local populations.[547]

In both fisheries and mariculture equity issues persist: from forced labour on fishing boats, to lack of recognition

[539] Costello et al. 2019. "The Future of Food from the Sea."

[540] FOLU. 2019. *Growing Better.*

[541] FOLU. 2019. *Growing Better.*

[542] FOLU. 2019. *Growing Better.*

[543] FOLU. 2019. *Growing Better.*

[544] Halpern, B.S., S.E. Lester and J.B. Kellner. 2009. "Spillover from Marine Reserves and the Replenishment of Fished Stocks." *Environmental Conservation* 36 (4): 268–76. doi: https://doi.org/10.1017/S0376892910000032; da Silva, I.M., N. Hill, H. Shimadzu, A.M.V.M. Soares and M. Dornelas. 2015. "Spillover Effects of a Community-Managed Marine Reserve." *PLOS ONE* 10 (4): e0111774. doi: https://doi.org/10.1371/journal.pone.0111774.

[545] FOLU. 2019. *Growing Better.*

[546] FOLU. 2019. *Growing Better.*

[547] Rogers et al. 2020. "Critical Habitats and Biodiversity."

and full integration of women in all parts of the fishery and mariculture industry, to unfavourable risk distributions for smallholder mariculture farmers. Solving these equity issues will require the combined force of strong labour legislation, effective enforcement and ensured traceability throughout the supply chain to keep seafood from bad actors out of seafood markets.

Consumers still regard aquaculture finfish as a food category of its own rather than as a potential replacement for poultry, pork and beef—there is little evidence of switching between the two as prices rise and fall.[548] Very little work has been done on predicting how the substitution economics between seafood, plant-based alternative protein and meat will evolve as a function of shifting tastes, prices and processing technologies. The safest bet currently is that additional production of farmed finfish is more likely to meet the protein demand of the new generations than to replace the meat habit of the older ones.[549] Additionally, consumer awareness and resulting demand for sustainably fished or farmed and humanely processed seafood needs to be raised by promoting sustainability labels like those of the Marine Stewardship Council and Aquaculture Stewardship Council.

Research into novel seafood sources, like lab-cultivated seafood, should be supported. Even though lab-cultivated seafood is not grown in the ocean, it still has its genetic roots there. This requires the understanding of cell lines from highly sought-after seafood species. This research should be enabled by easily navigable legislation.

Last but not least, scientific understanding of the implications of harvesting low-trophic species like krill, zooplankton and mesopelagic fish should be increased. The mesopelagic zone, also called the twilight zone, is the layer of water between 200 and 1000 m below the ocean surface, just beyond the reach of sunlight.[550] Many animal species live in this zone: zooplankton, crustaceans, squids, gelatinous animals and a multitude of few-inch-long fish usually referred to as mesopelagic fish (the most famous being the bristlemouth fish). The biomass in the twilight zone is not well known, but according to some estimates it could be bigger than the rest of the ocean biomass combined. Many mesopelagic organisms also travel from and to the surface daily, playing an important role in the broader ocean food chain and carbon flux exchanges between the atmosphere and the deep sea. In recent decades, there has been increasing interest in investigating the commercial fisheries catch potential of this immense biomass, for instance, to be used as fish meal or fish oil for aquaculture. However, until reliable stock assessments, and the impact of their harvest on the ocean's food web and the carbon cycle, are understood, a precautionary approach should be followed. The Woods Hole Oceanographic Institution, for instance, is funding a $35 million research project to answer the following questions:[551]

- What species live in the twilight zone, and in what quantities?
- How long do twilight-zone organisms live? How quickly do they grow? At what age do they reproduce?
- To what extent do large ocean predators such as whales and tuna depend on twilight-zone organisms as a source of food?
- How much carbon do twilight-zone animals transfer to the deep ocean through their daily migration? How much carbon sinks out of the twilight zone into deeper waters as marine snow and in other forms?

In Europe, the Ecologically and Economically Sustainable Mesopelagic Fisheries (MEESO) project involves 20 European research centres and universities pursuing similar research on mesopelagic fish.[552] For a summary of these points, see Box 20.13.

Box 20.13 Key Triggers to Revolutionise Food from the Ocean

- Increase official development assistance for fisheries management capacity.
- Artisanal fisheries: Ensure inclusive and equitable access rights to local, well-managed fish stocks.
- Align economic interests and stock health of industrial and small-scale commercial fisheries through capacity and granting of access rights; impose science-based mortality controls in line with sustainability principles for each commercial stock.
- Create climate-smart fisheries and mariculture management structures that plan for, and can adapt to, changing oceanographic conditions under climate change.
- Integrate technologies available for highly adaptive fishery management, new fleet control and tracking technologies, and seamless chain of custody tracking and registration of rights, ownership, titles, obli-

[548] Costello et al. 2019. "The Future of Food from the Sea."

[549] FOLU. 2019. *Growing Better*.

[550] Woods Hole Oceanographic Institution. n.d. "Ocean Twilight Zone." Blog. https://www.whoi.edu/know-your-ocean/ocean-topics/ocean-life/ocean-twilight-zone/. Accessed 18 August 2020.

[551] Woods Hole Oceanographic Institution. n.d. "Ocean Twilight Zone."

[552] Stubgaard, K. 2020. "About the MEESO Project." 25 May. https://www.meeso.org/about.

gations and so on through new ledger and registration technologies.

- Repurpose subsidisation of fishing capacity for fleet control infrastructure, port improvements (e.g. enforcement of the Agreement on Port State Measures, icing facilities) or financing of the cost of fish stock recovery.
- Ban damaging fishing, such as destructive bottom-trawling and blast fishing, and incentivise fishing gear and techniques that minimise bycatch of non-target species.
- Develop the required modalities and business models to support fishers in their transition (to be developed by insurers and investors, including through sovereign or outward direct investment, development finance institutions and multilateral development banks).
- Streamline mariculture permitting through easily navigable permitting processes that include environmentally rigorous permitting requirements.
- Boost development of fish feed alternatives (e.g. algae-based, insects, etc.).
- Increase scientific research on the ecosystem implications of harvesting krill, zooplankton and mesopelagic fish and follow a precautionary approach until these implications are understood.
- Promote more (sustainably) farmed finfish, seaweeds and bivalves in diets.
- Implement and build capacity to enforce strong labour standards for the fishing and mariculture industry.
- Require transparency of seafood supply chains ensuring full ocean-to-plate traceability.

Clean Ocean Energy: How Can the Ocean Deliver Much More Zero-Carbon Energy in a Sustainable Way?

Why Is It Important?

Of all the ocean-based climate mitigation options identified (see Sects. 4 and 5), ocean-based renewable energy technologies hold the greatest potential, up to 10% of the global needed annual GHG emissions reductions by 2050.[553]

The required growth path for ocean-based renewable energy generation is a very aggressive departure from a very low baseline. By the end of 2018 the total worldwide installed capacity of wind energy amounted to 564 GW, of which only 23 GW were offshore.[554] The great majority of installed offshore capacity is in Northern Europe, although there is significant technically feasible generation potential in Brazil (748 GW), South Africa (589 GW) and Vietnam (214 GW), as well as Indonesia, India, the Philippines and Sri Lanka.[555] All other ocean-based renewable energy technologies remain at the experimental or demonstration stage today (e.g. wave or tidal power, floating solar).

The target is massive. By several estimates, offshore wind capacity installed will need to be multiplied by a factor of 40–45 by 2050 to contribute to a 1.5 °C compatible trajectory (see Sect. 5). The path from baseline to target is extremely steep—it requires the installation of around 30 GW/year for 30 years, which exceeds the current growth rate by almost an order of magnitude and involves the installation of thousands of turbines per year. Current capacity projections confirm the critical gap: Asian countries are planning for 100 GW of offshore wind by 2030[556] (including South Korea 18 GW, India 30 GW, Japan 10 GW and Taiwan 5.5 GW).[557] This, combined with the commitments of Europe (70 GW)[558] and the rest of the world, adds up to a 2030 global capacity of about 220 GW[559]—not close to the ambition needed for 2050. But the technical resource is massive, and as the costs of electricity from offshore wind continue to come down below other sources, continued rapid growth is conceivable.

[553]Hoegh-Guldberg et al. 2019. "The Ocean as a Solution to Climate Change."

[554]IRENA. 2019. "Global Energy Transformation: A Roadmap to 2050 (2019 Edition)." Abu Dhabi: International Renewable Energy Agency. https://www.irena.org/publications/2019/Apr/Global-energy-transformation-A-roadmap-to-2050-2019Edition.

[555]Dutton, A.S.P., C.C. Sullivan, E.O. Minchew, O. Knight and S. Whittaker. 2019. "Going Global: Expanding Offshore Wind to Emerging Markets." 143162. World Bank. http://documents.worldbank.org/curated/en/716891572457609829/Going-Global-Expanding-Offshore-Wind-To-Emerging-Markets.

[556]Cohen, A. 2019. "As Global Energy Demand Grows, So Does Appetite for Offshore Wind." *Forbes*, 26 March. https://www.forbes.com/sites/arielcohen/2019/03/26/as-global-energy-demands-grows-so-does-appetite-for-offshore-wind/.

[557]Cohen. 2019. "As Global Energy Demand Grows, So Does Appetite for Offshore Wind"; Buckley, T., and K. Shah. 2018. "IEEFA Update: Offshore Wind Power, the Underexplored Opportunity That Could Replace Coal in Asia." *Institute for Energy Economics & Financial Analysis* (blog), 30 August. https://ieefa.org/offshore-wind-power-the-underexplored-opportunity-to-replace-coal-in-asia/.

[558]Nghiem, A., and I. Pineda. 2017. "Wind Energy in Europe: Scenarios for 2030." Brussels: Wind Europe. https://windeurope.org/wp-content/uploads/files/about-wind/reports/Wind-energy-in-Europe-Scenarios--for-2030.pdf.

[559]Global Wind Energy Council. 2019. "The Growth of the Global Offshore Wind Market Will Be Driven by Asia." Blog, 23 September. https://gwec.net/the-growth-of-the-global-offshore-wind-market-will-be-driven-by-asia/.

What Are the Opportunities for Action to Accelerate Change?

The construction of wind farms is a matter of national consensus and political priority, as reflected in regulatory support, public budgets, local support and financial market stability. Technical challenges and economics matter—offshore wind does not make sense for every country or every coastline. Without support from national governments (e.g. country targets for wind power, explicit inclusion of offshore wind in marine spatial plans, infrastructure for grid connectivity and storage), the offshore wind industry will be hard-pressed to build capacity at the scale required to compete with land-based energy sources.

There are other accelerating factors, of course. Institutional investors are not sufficiently knowledgeable about the offshore energy asset class and its risk and return profiles, but they are quickly catching up. Incumbents (utilities, fossil fuel energy generators) may be reluctant to share portside or grid infrastructure. However, with a favourable regulatory and incentive structure in place, these challenges can be overcome. For a summary of these points, see Box 20.14.

Box 20.14 Key Triggers to Boost Clean Ocean Energy from the Ocean

- **Precisely assess local and national opportunities.** Fund the scientific assessment of offshore power resources and site characterisation, including unique weather, oceanographic, ocean ecosystem and seafloor conditions, and design lease tracts accordingly to optimise for development and construction costs, operating conditions, and safety.
- **Formulate national targets.** Announce clear and time-bound national targets increasing the share of offshore wind energy in the national energy mix; set firm operating standards.
- **Develop marine spatial planning** (see Sect. 6.3.1, point 2, 'Goal-oriented ocean planning'). Convert these national targets into explicit plans for ocean-based energy development in national marine spatial plans and proactively sort use conflict issues with other ocean users such as fishers, shippers and so on.
- **Offer incentives.** Establish the modalities and schedules for incentive packages, including energy production and investment tax credits, feed-in tariffs and renewable portfolio standards.
- **Streamline administrative processes.** Provide a consistent, efficient and clear permitting process, based on development and operating standards, with predictable timelines.
- **Improve infrastructure.** Reduce the burden of specialised infrastructure cost through appropriate public and joint investments, such as in ocean energy delivery and grid integration, port facilities and properly leveraged existing infrastructure.

Low-Carbon Transportation and Ports: How Can a Traditional Industry Embrace Sustainability?

Why Is It Important?

Ocean transport is currently moving around 90% of the world's traded goods,[560] or about 11 billion metric tonnes (2018).[561] There were 94,171 commercial vessels in 2018 globally, mostly bulk carriers, tankers and container ships. Fuelled by increasing global trade, shipping is expected to continue growing above GDP rates in the coming years (the UN Conference on Trade and Development [UNCTAD] is forecasting a 3.8% annual growth rate for shipping between 2018 and 2023).[562]

Ocean transport currently produces about 2.2% of global greenhouse gas emissions,[563] and such emissions are expected to double until 2050,[564] in sharp contrast to what is needed to keep global temperature rise well below 2 °C and consistent with a 1.5 °C increase (IPCC 2013) and align with the goals of the Paris Agreement (UNFCCC 2015). Decarbonising shipping could also reduce other pollutants usually associated with ocean transport: about 17% of the human-induced sulphur dioxide[565] and 8% of nitrogen oxide[566] emissions globally. Phasing out such pollution could cut premature deaths by 4100 by 2030 and 10,000 (annually) by 2050.[567]

Ocean-based transportation has the potential for a roughly 100% reduction in operational net GHG emissions by chang-

[560] International Chamber of Shipping. n.d. "Shipping and World Trade." https://www.ics-shipping.org/shipping-facts/shipping-and-world-trade. Accessed 18 August 2020.

[561] UNCTAD. 2020. *Review of Maritime Transport 2019*. New York: United Nations. https://unctad.org/en/PublicationsLibrary/rmt2019_en.pdf.

[562] UNCTAD. 2019. *Review of Maritime Transport 2018*. New York: United Nations. https://unctad.org/en/PublicationsLibrary/rmt2018_en.pdf.

[563] IMO. 2015. "Third IMO Greenhouse Gas Study 2014," 3.

[564] Hoegh-Guldberg et al. 2019. "The Ocean as a Solution to Climate Change."

[565] Olmer et al. 2017. "Greenhouse Gas Emissions from Global Shipping, 2013–2015."

[566] ETC. n.d. "Mission Possible."

[567] Cofala, J., M. Amann, J. Borken-Kleefeld, A. Gomez-Sanabria, C. Heyes, G. Kiesewetter, R. Sander et al. 2018. "The Potential for Cost-Effective Air Emission Reductions from International Shipping through Designation of Further Emission Control Areas in EU Waters with Focus on the Mediterranean Sea." Vienna: International Institute for Applied Systems Analysis. https://www.cittadiniperlaria.org/wp content/uploads/2019/04/Shipping_emissions_reductions_main.pdf.

ing the way it stores and consumes energy onboard: batteries could be used to store electricity, particularly in ships on the shortest voyages. Low- or zero-carbon synthetic or 'e-' fuels could replace fossil fuels: examples include renewable hydrogen, hydrogen-based fuels such as ammonia, and fuels that have been processed with capture and storage of CO_2. Transitioning ocean shipping to more efficient and low- or zero-carbon fuels, and the mitigation potential in 2030 and 2050, is largely determined by the time scales needed to renew or retrofit the existing fleet and develop the infrastructure to use and supply these new energy sources.

An additional challenge associated with shipping is the discharge of untreated ballast water from ships. It is considered one of the major threats to biodiversity that could have 'severe public health-related, environmental and economic impacts'.[568] One cubic metre of ballast water can contain up to 50,000 zooplankton specimens[569] and/or 10 million phytoplankton cells.[570]

Ports, the gateways to the sea, present many challenges themselves. Their operations emit carbon, moving of goods creates significant amounts of air pollution (dust, exhaust), (mishandling of) waste products pollutes local waterways, and the resulting heavy ship traffic creates (underwater) noise. The increase in shipping traffic along ports has been associated with ship strikes.[571]

What Are the Opportunities for Action to Accelerate Change?

Tighten and enforce energy efficiency requirements of ships Countries should ensure the implementation of the IMO's Energy Efficiency Design Index (EEDI) and move beyond it (e.g. redesign the EEDI formula to ensure that vessels are being optimised for minimised fuel consumption in real operation at sea rather than being optimised only to pass the test[572]), while furthering the goal of fully decarbonising shipping by 2050. To reach these standards, countries should develop national roadmaps, and support the IMO in creating an international one, of how to fully decarbonise ocean transport by 2050.

Test and deploy low-carbon fuels Countries and shipping companies should foster offtake agreements between ship operators and harbours to incentivise the construction of zero-carbon fuel infrastructure and ensure its use by shipowners. These low-carbon fuel offtake agreements are essential to overcome the chicken-and-egg problem of building low-carbon fuel infrastructure in harbours versus building the ships requiring such low-carbon fuel infrastructure.

Incentivise decarbonisation of shipping Governments should set clear port access targets based on carbon emission standards and/or tax ship GHG emissions or create emission trading systems for shipping companies.

Eliminate port air pollution through environmental regulations This includes electrifying port operations and making shore power available for ships. It also includes integrating ports into local decarbonised land transport systems to ensure continued low-impact transportation of goods and people.

Harmonise port operations with the local environment This includes ensuring that port expansions do not destroy sensitive habitats. High-traffic shipping lanes usually associated with ports should be planned in accordance with whale migrations to minimise ship strikes. Ports should not be expanded into sensitive habitats or built in locations requiring continuous harmful dredging.

Support retraining programs for port and ship jobs that are expected to be automated Ports are often major employers within their region. Increased automation can replace jobs while raising the average skill level demanded of the retained employees.

Governments should support retraining programs for current port and shipping works to ensure that there is no skills gap, while retaining a maximum of current employees. For a summary of these points, see Box 20.15.

[568] GEF-UNDP-IMO GloBallast Partnerships. 2017. "The GloBallast Story."

[569] Locke, A., D.M. Reid, W.G. Sprules, J.T. Carlton and H.C. van Leeuwen. 1991. "Effectiveness of Mid-ocean Exchange in Controlling Freshwater and Coastal Zooplankton in Ballast Water"; Locke, A., D.M. Reid, H.C. van Leeuwen, W.G. Sprules and J.T. Carlton. 1993. "Ballast Water Exchange as a Means of Controlling Dispersal of Freshwater Organisms by Ships." *Canadian Journal of Fisheries and Aquatic Sciences* 50 (10): 2086–93. doi: https://doi.org/10.1139/f93-232; Gollasch, S. 1996. *Untersuchungen des Arteintrages durch den internationalen Schiffsverkehr unter besonderer Berücksichtigung nichtheimischer Arten*. Hamburg: Kovač; Kabler, L.V. 1996. "Ballast Water Invaders: Breaches in the Bulwark." *Aquatic Nuisance Species Digest* 1 (3): 34–35.

[570] Subba Rao et al. 1994. "Exotic Phytoplankton from Ship's Ballast Waters."

[571] Segee, B.P. 2010. "Whale of Opportunity: Coast Guard Study of Los Angeles/Long Beach Port Access Routes Holds Great Potential for Reducing Ship Strikes within Santa Barbara Channel." *Ecology Law Currents* 37: 58.

[572] Kulp, S.A., and B.H. Strauss. 2019. 'New Elevation Data Triple Estimates of Global Vulnerability to Sea-Level Rise and Coastal Flooding'. *Nature Communications* 10 (1): 4844. doi: https://doi.org/10.1038/s41467-019-12808-z.

Box 20.15 Key Triggers to Decarbonise Shipping

- Tighten and enforce energy efficiency requirements of ships.
- Test and deploy low-carbon fuels.
- Incentivise decarbonisation of shipping.
- Eliminate port air pollution through environmental regulations.
- Harmonize port operations with the local environment.
- Support retraining programs for port and ship jobs that are expected to be automated.

Ocean Restoration and Protection: How Can Protected Areas Be Mainstreamed and Enforced?

Why Is It Important?

A century or more of coastal urbanisation, ocean and coastal resource exploitation, infrastructure expansion, river channelling, land reclamation, mangrove removal and pollution has taken its toll. Globally, an estimated 50% of salt marshes, 35% of mangroves, 30% of coral reefs and 29% of sea grasses have been either lost or degraded.[573] By 2100, as many as 630 million people could be at risk of coastal flooding caused by climate change,[574] with several atoll states in danger of disappearing entirely.[575]

Reversal of these trends is urgent. Intact coastal ecosystems provide critical services to all of humankind. They are critical to fisheries and recreation. They protect cities and coasts from storms and sea level rise. They host unique biodiversity.

An ecologically healthy coast does, of course, have intrinsic economic value, offering protection from storms, surges and swells,[576] the nursery of coastal fisheries,[577] recreational value and so on. However, a narrowly defined economic metric is unlikely to win the day for the coast—short-term cost-benefit calculations can just as well make the case for coastal destruction[578] and they certainly do not account for the damage done when ecological thresholds are irreversibly crossed.[579]

That said, some monetisation of coastal ecosystem values is possible—the storm protection and wave attenuation services of healthy coastal biota, for example, have been well documented. Over 500 million people worldwide live in a coastal zone that is protected by coral reefs.[580] Without their protection, flood damages from 100-year storms would increase by 91% to $272 billion.[581] U.S. coastal wetlands provide $23.2 billion a year in storm protection services—a benefit of over $33,000 per hectare (median $5000 per hectare).[582] Mangroves reduce annual flooding globally by more than 39%/year for 18 million people, and reduce annual property damage by more than 16%, or $82 billion.[583]

[573] Valiela et al. 2001. "Mangrove Forests"; Millennium Ecosystem Assessment (Program), ed. 2005. *Ecosystems and Human Well-Being*; Orth, R.J., T.J.B. Carruthers, W.C. Dennison, C.M. Duarte, J.W. Fourqurean, K.L. Heck, A.R. Hughes et al. 2006. "A Global Crisis for Seagrass Ecosystems." *BioScience* 56 (12): 987–96. doi: https://doi.org/10.1641/0006-3568(2006)56[987:AGCFSE]2.0.CO;2; FAO. 2007. "The World's Mangroves, 1980–2005." Thematic study prepared in the framework of the Global Forest Resources Assessment 2005. Rome: Food and Agriculture Organization of the United Nations. http://www.fao.org/3/a1427e/a1427e00.pdf; Waycott, M., C.M. Duarte, T.J.B. Carruthers, R.J. Orth, W.C. Dennison, S. Olyarnik, A. Calladine et al. 2009. "Accelerating Loss of Seagrasses across the Globe Threatens Coastal Ecosystems." *Proceedings of the National Academy of Sciences* 106 (30): 12377–81. doi: https://doi.org/10.1073/pnas.0905620106.

[574] Kulp, S.A., and B.H. Strauss. 2019. 'New Elevation Data Triple Estimates of Global Vulnerability to Sea-Level Rise and Coastal Flooding'. *Nature Communications* 10 (1): 4844. doi: https://doi.org/10.1038/s41467-019-12808-z.

[575] Wright, L.D., J.P.M. Syvitski and C.R. Nichols. 2019. "Sea Level Rise: Recent Trends and Future Projections." In *Tomorrow's Coasts: Complex and Impermanent*, edited by L.D. Wright and C.R. Nichols, 47–57. Coastal Research Library. Cham: Switzerland: Springer. doi: https://doi.org/10.1007/978-3-319-75453-6_3.

[576] Steven, A., K.A. Addo, G. Llewelyn, C.V. Thanh et al. 2020. "Coastal Development: Resilience, Restoration and Infrastructure Requirements." Washington, DC: World Resources Institute. https://www.oceanpanel.org/blue-papers/coastal-development-managing-resilience-restoration-and-infrastructure-coastlines.

[577] Gittman, R.K., C.H. Peterson, C.A. Currin, F.J. Fodrie, M.F. Piehler and J.F. Bruno. 2016. "Living Shorelines Can Enhance the Nursery Role of Threatened Estuarine Habitats." *Ecological Applications* 26(1): 249–63. doi: https://doi.org/10.1890/14-0716.

[578] Costanza, R., R. de Groot, P. Sutton, S. van der Ploeg, S.J. Anderson, I. Kubiszewski, S. Farber and R.K. Turner. 2014. "Changes in the Global Value of Ecosystem Services." *Global Environmental Change* 26 (May): 152–58. doi: https://doi.org/10.1016/j.gloenvcha.2014.04.002; McCauley, D.J. 2006. "Selling Out on Nature." *Nature* 443 (7107): 27–28. doi: https://doi.org/10.1038/443027a; Bartha, P., and C.T. DesRoches. 2017. "The Relatively Infinite Value of the Environment." *Australasian Journal of Philosophy* 95 (2): 328–53. doi: https://doi.org/10.1080/00048402.2016.1182196.

[579] Rockström et al. 2009. "Planetary Boundaries."

[580] Wilkinson, C. n.d. "Status of Coral Reefs of the World: 2004." Australian Institute of Marine Science, Global Coral Reef Monitoring Network. http://www.icriforum.org/sites/default/files/scr2004v1-all.pdf. Accessed 12 May 2020.

[581] Beck et al. 2018. "The Global Flood Protection Savings Provided by Coral Reefs."

[582] Costanza, R., O. Pérez-Maqueo, M.L. Martinez, P. Sutton, S.J. Anderson and K. Mulder. 2008. "The Value of Coastal Wetlands for Hurricane Protection." *AMBIO: A Journal of the Human Environment* 37 (4): 241–48. doi: https://doi.org/10.1579/0044-7447(2008)37[241:TVOCWF]2.0.CO;2.

[583] Losada, Í.J., P. Menéndez, A. Espejo, S. Torres, P.D. Simal, S. Abad, M.W. Beck et al. 2018. "The Global Value of Mangroves for Risk Reduction." Berlin: The Nature Conservancy. https://conservationgateway.org//ConservationPractices/Marine/crr/library/Documents/GlobalMangrovesRiskReductionSummaryReport10.7291/V9930RBC.pdf.

Similarly, the value of coastal ecosystems in terms of nursery and habitat for fishes and other marine species, regulation of water flow and filtration, carbon sequestration, and contaminant storage and detoxification has also been calculated for coastal habitats, ranging from \$100 to \$10,000 an acre.[584]

The greatest risks of coastal degradation are to populations already at risk on other fronts. Forty-six percent of Bangladeshis live within 10 m of sea level, with declining levels of storm protection from mangrove forests. Developing countries account for nine of the ten nations with the largest share of the population living in low-elevation areas (the Bahamas, Bangladesh, Belize, Djibouti, Egypt, the Gambia, Guyana, the Netherlands, Suriname and Vietnam).[585] In the United States, approximately 39% of residents of coastal counties fall into an elevated coastal hazard risk category (i.e. children, the elderly, households where English is not the primary language and those in poverty).[586] When Hurricane Katrina's storm surge reached New Orleans with almost no interference from its highly degraded surrounding wetlands, nearly 85% of people killed were aged 51 and older, and almost half were older than 75 years of age.[587]

Upstream river management aimed at flood protection, irrigation and hydroelectric power generation have resulted in drastic sediment imbalance and have accelerated coastal erosion by depriving coastal landscapes of sand or silt.[588] Globally, an estimated 25–30% of the total suspended sediment flux is potentially trapped in artificial impoundments of about 45,000 reservoirs.[589] This reduces marine sediment supply to deltas and estuaries.[590] If no mitigation measures are undertaken and sediment retention continues, approximately 28,000 km^2 of the deltaic area in 40 deltas could suffer from increased flooding and coastal erosion by 2050.[591] Uncoordinated upriver flood protection has proved to be counterproductive, as the flood risk is often simply transferred and amplified to downriver communities.[592] The river deltas are paying the ultimate price, as they have no way to escape the erosive effects of faster and more intense river flows.

What Are the Opportunities to Accelerate Change?

Map and account for benefits A comprehensive mapping of the areas of high diversity, productivity, carbon concentration, coastal protection from sea level rise and storms, fishery support (nursery habitat and other critical life stages) and tourism values is the essential foundation of planning and must be the first priority. Such mapping informs sustainable ocean economy planning, national greenhouse gas inventories (if conforming to IPCC 2013 protocol) and nationally determined contributions (NDCs). The carbon flux and sequestration capacity of reefs, mangroves, sea grasses and salt marshes should be systematically accounted for. To the degree that the capacity for such rapid, accurate and comprehensive ocean mapping efforts is not in place in every country, technical assistance may be required (see Sect. 6.3, point 2, 'Goal-oriented ocean planning').

Integrate restoration and protection into sustainable ocean economy plans With a comprehensive ocean resource mapping in hand, protection and restoration or regeneration need to be systematically merged into a sustainable ocean economy development planning process (see Sect. 6.3, point 2, 'Goal-oriented ocean planning'). The baseline ambition needs to be a global fully protected set-aside of 30% of the ocean for coastal protection, fishery recovery, biodiversity restoration, controlled recreation and

[584] Mehvar, S., T. Filatova, A. Dastgheib, E. De Ruyter van Steveninck and R. Ranasinghe. 2018. "Quantifying Economic Value of Coastal Ecosystem Services: A Review." *Journal of Marine Science and Engineering* 6 (1): 5. doi: https://doi.org/10.3390/jmse6010005.

[585] McGranahan, G., D. Balk and B. Anderson. 2007. "The Rising Tide: Assessing the Risks of Climate Change and Human Settlements in Low Elevation Coastal Zones." *Environment and Urbanization* 19(1): 17 37. doi: https://doi.org/10.1177/0956247807076960.

[586] NOAA Office for Coastal Management. n.d. "Fast Facts: Economics and Demographics." https://coast.noaa.gov/states/fast-facts/economics-and-demographics.html. Accessed 12 May 2020.

[587] Bathi, J.R., and H.S. Das. 2016. "Vulnerability of Coastal Communities from Storm Surge and Flood Disasters." *International Journal of Environmental Research and Public Health* 13 (2): 239. doi: https://doi.org/10.3390/ijerph13020239.

[588] Giannico, G., and J.A. Souder. 2005. "Tide Gates in the Pacific Northwest." Oregon State University; Martínez, M.L., G. Mendoza-González, R. Silva-Casarín and E. Mendoza-Baldwin. 2014. "Land Use Changes and Sea Level Rise May Induce a 'Coastal Squeeze' on the Coasts of Veracruz, Mexico." *Global Environmental Change* 29 (November): 180–88. doi: https://doi.org/10.1016/j.gloenvcha.2014.09.009; Tessler, Z.D., C.J. Vörösmarty, I. Overeem and J.P.M. Syvitski. 2018. "A Model of Water and Sediment Balance as Determinants of Relative Sea Level Rise in Contemporary and Future Deltas." *Geomorphology*, Resilience and Bio-geomorphic Systems—Proceedings of the 48th Binghamton Geomorphology Symposium (March): 209–20.

[589] Vörösmarty, C.J., M. Meybeck, B. Fekete, K. Sharma, P. Green and J.P.M. Syvitski. 2003. "Anthropogenic Sediment Retention: Major Global Impact from Registered River Impoundments." *Global and Planetary Change* 39 (1): 169–90. doi: https://doi.org/10.1016/S0921-8181(03)00023-7.

[590] Vörösmarty et al. 2003. "Anthropogenic Sediment Retention."

[591] Ericson, J.P., C.J. Vörösmarty, S.L. Dingman, L.G. Ward and M. Meybeck. 2006. "Effective Sea-Level Rise and Deltas: Causes of Change and Human Dimension Implications." *Global and Planetary Change* 50 (1): 63–82. doi: https://doi.org/10.1016/j.gloplacha.2005.07.004.

[592] Forbes, H., K. Ball and F. McLay. n.d. *Natural Flood Management Handbook*. Stirling, UK: Scottish Environment Protection Agency. https://www.sepa.org.uk/media/163560/sepa-natural-flood-management-handbook1.pdf. Accessed 12 May 2020.

so on. With proper planning this can be fully complementary to the economies of fishing (stock restoration), tourism (diving, pristine areas), offshore wind (protected buffers around turbines), shipping (avoided risk of whale strikes, safe distance from turbines), mariculture (vibrant, nutrient-rich, healthy ecosystems) and the protection of coastal assets (storm-surge protection). Conservation and restoration or regeneration should be regarded as a fully legitimate sector of the sustainable ocean economy, with its own economic logic, financing sources (carbon, wetland and nutrient credit and offset markets; carbon finance; infrastructure funding), and fully accountable and measurable contribution to both 'flow' (i.e. gross value added) and 'stock' (i.e. natural capital) metrics.

Include quantified nature-based solutions in nationally determined contributions and other relevant climate policies for mitigation and adaptation Ocean-based mitigation options do not feature as prominently as they could in countries' NDCs or long-term low greenhouse gas emission development strategies under the Paris Agreement.[593] This is an extremely important moment, as emphasised by the IPCC (2018): the chances of 'failing to reach 1.5 °C [will be] significantly increased if near-term ambition is not strengthened beyond the level implied by current NDCs'. Given the consequences of failing to limit global average temperature rise to 1.5 °C, or at least to 'well below' 2.0 °C, capturing the potential offered by blue carbon in NDCs could forcefully accelerate restoration and protection of these ocean and coastal natural assets.

Connect ocean protection and restoration with land-based initiatives and stakeholders Coastal restoration and protection cannot succeed in isolation. Delta restoration requires river management that optimises sediment flows. Near-coast MPAs are highly sensitive to nutrient contamination. Symbiotic MPA, mariculture and energy projects require clean and abundant freshwater flows. For example, Florida's Apalachicola Bay once housed the highest concentration of oyster beds in the United States. As the abundant waters of the Apalachicola River were depleted by growing upriver cities, the beds atrophied, and today only small remnants of the oyster industry remain. Restoration of the bay would require close coordination with upriver water and reservoir managers to optimise freshwater flows. Most comprehensive, EEZ-wide ocean planning efforts thus need to closely coordinate with river authorities. For a summary of these points, see Box 20.16.

[593]Hoegh-Guldberg et al. 2019. "The Ocean as a Solution to Climate Change."

Box 20.16 Key Triggers to Restore and Protect Nature

- Map and account for benefits.
- Integrate restoration and protection into sustainable ocean economy plans.
- Include quantified nature-based solutions in nationally determined contributions and other relevant climate policies for mitigation and adaptation.
- Connect ocean protection and restoration with land-based initiatives and stakeholders.

Tourism: How Can Tourism Be Turned into a Zero- or Positive-Impact Industry?

Why Is It Important?

Tourism is estimated to contribute to about 10% of the world's economic activity and is a key source of foreign earnings for many developing countries.[594] The industry has been growing steadily over the last half century. Between 1965 and 2019, the number of international tourists alone has increased about 13-fold: from 113 million in 1965 to 674 million in 2000 to 1461 million in 2019, a trend that is expected to continue.[595] It is hard to determine how much of the global tourism is purely coastal, but there are good indications that a significant amount of it is. Over 46% of Europeans, the largest group of international travellers, cited 'beach access' as their holiday travel reason.[596] Estimates vary, but between 60 and 350 million people annually travel to the world's coral reef coasts.[597] In many coastal nations, coral reefs support over one-quarter of all tourism value and over 6%, and up to 40% (about 43% in Palau and in the Maldives) of the nation's GDP.[598] Cruise tourism, growing

[594]Ecological Tourism in Europe, UN Educational, Scientific and Cultural Organization and UN Environment Programme. n.d. "Sustainable Tourism Development in UNESCO Designated Sites in South-Eastern Europe." http://portal.unesco.org/en/files/45338/12417872579Introduction_Sustainable_Tourism.pdf/Introduction_Sustainable_Tourism.pdf. Accessed 7 May 2020.

[595]UN World Tourism Organization (UNWTO). n.d. "Global and Regional Tourism Performance." https://www.unwto.org/global-and-regional-tourism-performance. Accessed 11 May 2020; Inniss et al. 2016. "The First Global Integrated Marine Assessment"; UNWTO, ed. 2011. *Tourism towards 2030/Global Overview: Advance Edition Presented at UNWTO 19th General Assembly—10 October 2011*. doi: https://doi.org/10.18111/9789284414024.

[596]Inniss et al. 2016. "The First Global Integrated Marine Assessment"; TNS Political and Social. 2014. "Preferences of Europeans towards Tourism." Flash Eurobarometer 392. European Commission. https://ec.europa.eu/commfrontoffice/publicopinion/flash/fl_392_sum_en.pdf.

[597]Spalding, M., L. Burke, S.A. Wood, J. Ashpole, J. Hutchison and P. zu Ermgassen. 2017. "Mapping the Global Value and Distribution of Coral Reef Tourism." *Marine Policy* 82 (August): 104–13. doi: https://doi.org/10.1016/j.marpol.2017.05.014.

[598]Spalding et al. 2017. "Mapping the Global Value and Distribution of Coral Reef Tourism."

strongly, is predicted to move 30 million people across the ocean (2019), up from 18 million a decade ago.[599]

The growth of the coastal tourism industry came at a price for coastal ecosystems. The negative impact of tourism on ecosystems is well documented and threatens the long-term socioeconomic value of the industry itself. Much like other natural resource-based industries, tourism can deplete the very resource it most depends on, in this case, a healthy and beautiful ocean environment. Unlike other industries, however, 'sustainable yield' is not clearly defined in tourism, and most of the industry operates outside internationally accepted certifications and transparent performance standards. The resulting damage has been exemplified by the closure of Maya Bay in Thailand, the degradation of near-shore reefs in Indonesia and the massive destruction of coastal wetlands by tourism development. The concentration of tourism further intensifies the impacts: destruction of natural habitats, excessive groundwater extraction leading to saltwater intrusion, introduction of exotic species and sewage pollution, to name just a few.[600]

The sector is also constrained by the deterioration of its target areas by outside forces. As early as the 1960s, human-driven eutrophication of the Black Sea led to a decline in tourism revenues of $500 million.[601] Today, cleaning beaches in the European Union alone costs over €413 million/year.[602] Having already put a very high strain on the environment, using the tourism industry as a force for sustainable growth rather than environmental destruction will be critically important.

The COVID-19 pandemic has had major effects on tourism: because of the lockdowns and travel ban implemented in most countries, tourism is expected to lose $2.1 trillion in GVA in 2020, with 100 million jobs at risk.[603] This sudden and massive hit on the tourism industry raises existential questions: Will this be an opportunity to reinvent tourism as an eco-friendly experience? Is this a hard stop from which the industry will not recover? Will the industry rebuild as it was before?

What Are the Opportunities to Accelerate Change?

Countries and tourism operators should consider a number of possible approaches when thinking about the future of coastal and ocean tourism:

Create national tourism strategies and implement governance systems that ensure the sustainable and equitable development of the tourism industry. These plans should include a clear spatial plan for the sustainable, climate-smart expansion of tourism resorts, and ensure capacity for waste and traffic infrastructure to cope with the increase in tourism. The plans could also include requirements for certified climate-friendly travel as conditions for accepting tourists to points of interest or even to the country.

Implement tourism taxes as payment for ecosystem services the industry relies on The revenue from these taxes should be used to restore degraded nature and maintain coastal and marine ecosystems. Additionally, it can provide a source of funding to build the necessary infrastructure and help the local tourism industry transition to a more sustainable operating model. A back-of-the-envelope analysis reveals the potentially enormous contribution of such a tourist ecosystem service tax: assuming that one-third of international tourism is coastal and an (only) 1% ecosystem tax is levied on international tourism expenditures (roughly $1500 billion[604]), $5 billion in funds would become available for coastal and marine ecosystems—four times the current marine philanthropic funding and official development assistance combined.[605]

Agree on and implement international environmental standards for coastal tourism New, more ambitious environmental standards could become the norm for the tourism industry after COVID-19. Much-needed standards regulating the coastal and cruise tourism industry with respect to its CO_2, air, over-tourism, waste and effluent pollution should thus be created and implemented internationally. Ideally, the tourism industry itself would advocate for and hold countries to adopt these standards as the tourism industry itself benefits from a healthy ocean. For a summary of these points, see Box 20.17.

[599] Kennedy, S. n.d. "2019 Cruise Trends & Industry Outlook." Cruise Lines International Association. https://cruising.org/-/media/research-updates/research/clia-2019-state-of the industry-presentation-(1).pdf. Accessed 12 May 2020.

[600] Honey, M., and David Krantz. 2007. "Global Trends in Coastal Tourism." Stanford, UK, and Washington, DC: Center on Ecotourism and Sustainable Development, World Wildlife Fund. https://tamug-ir.tdl.org/bitstream/handle/1969.3/29198/global_trends_in_coastal_tourism_by_cesd_jan_08_lr.pdf?sequence=1.

[601] World Bank. 2009. "Environment Matters at the World Bank: Valuing Coastal and Marine Ecosystem Services." http://documents.worldbank.org/curated/en/593291468150870756/Environment-matters-at-the-World-Bank-valuing-coastal-and-marine-ecosystem-services.

[602] van Acoleyen, M., I. Laureysens, S. Lambert, L. Raport, C. van Sluis, A. Kater, E. van Onselen et al. n.d. *Marine Litter Study to Support the Establishment of an Initial Quantitative Headline Reduction Target: SFRA0025*. ARCADIS, European Commission DG Environment. https://ec.europa.eu/environment/marine/good-environmental-status/descriptor-10/pdf/final_report.pdf. Accessed 12 May 2020.

[603] UNCTAD. 2020. "The COVID-19 Pandemic and the Blue Economy."

[604] UNWTO. n.d. "Global and Regional Tourism Performance."

[605] CEA Consulting. n.d. "Funding." Our Shared Seas. https://oursharedseas.com/2019-update/funding/. Accessed 12 May 2020.

Box 20.17 Key Triggers to Turn Tourism into a Zero- or Positive-Impact Industry

- Create national strategies for sustainable tourism growth.
- Implement tourism taxes as payment for ecosystem services.
- Agree on and implement international environmental standards for coastal tourism.

6.4 Launching the Voyage: Three Levels for Possible Immediate Action

The voyage needs to start not with a bang but with a thousand rising voices. A set of expert recommendations can provide very helpful guidance for decision-makers, but it will only make a difference if the solution works for its beneficiaries—economically, culturally and socially.

Launching the voyage is about creating self-evident movement, a sense of inevitability, by building on the work already underway—the networks of innovators (in both industrialised and developing countries) who are the living embodiments of the overall change that is needed. It is far more about anchoring the transition to tomorrow in the reality of today than it is about theoretical best practices.

There is no shortage of such innovators—in policy, technology, resource management, inclusion, governance and so on. These business ventures, technology trials, corporate coalitions, investment partnerships, civil society programs and policy innovations are at the heart of a great experiment the world needs in testing, failing and learning.

Standing on the shoulders of these pioneers, this final section of the report suggests three ideas for a quick start towards change. These ideas do not pretend to be silver bullets, or to be exhaustive and to replace existing initiatives led by governments, businesses and civil society. These ideas are suggestions for interventions that are expected to create a snowball effect and accelerate change towards a sustainable ocean economy, in complement to the broader, more comprehensive action agenda presented in Sect. 6.3.

These suggested ideas to launch the voyage are especially critical in the context of the post-COVID recovery. They constitute concrete propositions to rebuild the economy bluer, more sustainable and more resilient, at a time when many hold onto business as usual for their own survival and advocate for postponing ambitious sustainability reforms.

At the local level, this report introduces the concept of sustainable ocean economic zones (SOEZs), which could become laboratories and demonstrators of the broader ocean action agenda, in complement with broader science-based planning for the entire EEZ, and ultimately for the high seas. At the national level, inspired by a successful international track record, the report considers the establishment of national ocean delivery task forces. Finally, at the international level, several no-regret moves and potential collaboration areas are encouraged.

6.4.1 Local Intervention: Catalysing Change Through Sustainable Ocean Economic Zones

On land, special economic zones are a long-established and well-proven component of industrial strategy. Typically, these zones provide a shell within which select ventures can thrive, offering low rents, low taxes, low utility and infrastructure costs, relief from heavy bureaucratic procedures, and low-cost debt from central government funds, providing guarantees to market-rate investors. These zones have been used to attract new businesses to revitalising urban areas, support emerging and high-risk industries, promote cooperative business models, stimulate exports and so on. They scale from small neighbourhood zones to entire cities (e.g. Dubai, Shenzhen and Hong Kong).

If done correctly, with all planning, legislation, approval, construction, investment and operation carefully designed to be fit for the intended purpose, and by respecting labour rights and social sustainability, they can be quite successful. There are 5400 zones in over 147 nations today, directly employing between 90 and 100 million workers.[606]

With very few exceptions, they have not been used to promote a sustainable ocean economy. There have been hurdles: the concept of spatial planning and restricted access in the ocean commons is often controversial, the siting of permanent ocean structures is politically difficult and any alternative to the current 'free for all' has not been mainstreamed yet.

Replicating the success of the (sustainable) economic zone concept in nations' EEZs might prove to be a powerful catalyst in accelerating a local sustainable ocean economy. Sustainable ocean economic zones could provide a testbed for systemic experimentation and innovation, a way for nations to support and evaluate the sustainable ocean economy model at a scale they are comfortable with. For different nations, such zones can look very different in almost every respect (Fig. 20.42). Some industrial nations, for example, can use them to attract and test high-technology models that combine energy generation, shipping and large-scale mariculture. A low-lying coastal nation may use them to combine carbon-financed restoration, coastal protection, tourism and fishery enhancement. Another country may concentrate on

[606] UNCTAD. 2019. *World Investment Report 2019: Special Economic Zones*. UN Conference on Trade and Development World Investment Report (WIR). New York: United Nations. doi: https://doi.org/10.18356/8a8d05f9-en.

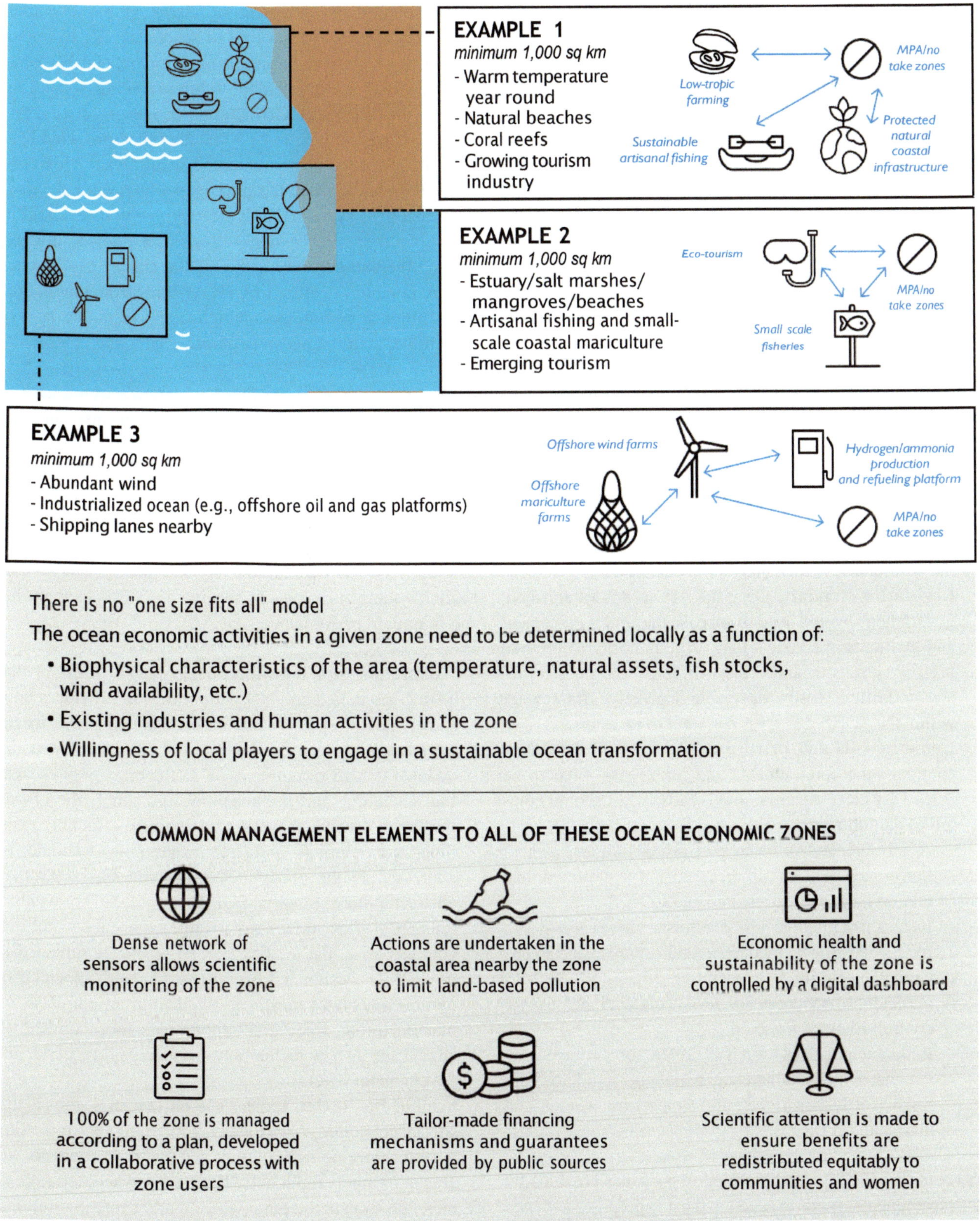

Fig. 20.42 Illustrative examples of sustainable ocean economic zones. (*Source: Authors*)

the symbiotic mariculture of many types of seafood in one place, combined with ecotourism.

A country's path to the design of a sustainable ocean will likely involve a series of steps:

- **Use case and delineation.** Using a working group of existing ocean users, technologists, financiers and public sector heads, an initial ocean industrial strategy could be developed that helps to identify which special zone models a nation wants to test in its EEZ (Fig. 20.42). This would determine the required utility of each zone (its 'use case') in terms of the types of ventures targeted, their siting and infrastructural requirements, their inherent potential symbiosis or conflict with other sectors and so on. For example, a special use zone focused on the symbiotic production of multiple trophic levels of food would have very different siting needs from an area focused on renewable ocean energy production. Each ocean SOEZ would then be delineated more precisely as a function of the specific use case, based on the biophysical characteristics of the area (temperature, natural assets, fish stocks, wind availability, etc.), existing industries and human activities, and the willingness of local players to engage.
- **Legislative certainty.** Once the use case is established, governments would need to approve the zone's placement and guarantee the zone's long-term authority to provide use rights to its tenants. This will be critical since it creates a lasting, easily navigable legislative framework within the SOEZ that gives certainty to investors.
- **Commitments and privileges.** The terms of the SOEZ contract—the commitments and privileges—need to be made very clear. At a minimum, participants should commit to the following:
  - A net-regenerative balance of production and protection—such as more carbon saved and sequestered than released, net habitat restored and so on.
  - Inclusion of multiple linked sectors—such as energy and food, multi-trophic food, restoration, tourism and so on.
  - Compliance with safe operating principles—such as mariculture standards for feed, containment, disease control, fouling and so on.
  - Equitable practices—such as preference for local market distribution, adherence to labour standards and respect of human rights, and support for women and marginalised groups.

A set of 'privileges'—specific support packages provided by the government—needs to be tailored to the purposes and uses of each zone. In general, these will require measures to increase margins (low-cost onshore infrastructure access, price guarantees), reduce risk (offtake guarantees, streamlined permitting, insurance vehicles), reduce capital costs (below-market debt, tax breaks) and bring in market-rate debt.

In practice, the efficient provision of below-market debt may require the pooling of resources. For example, several sovereign wealth funds may pool resources to create an ocean economy debt fund providing subsidised debt to the SOEZs of participating countries. Ideally, this would be used to guarantee a matched fund provided by a coalition of development finance institutions. This concentrated and coordinated approach would have the dual benefits of (1) systematically de-risking the investment in emerging industries and (2) creating a pool of domain knowledge and capacity ahead of the market. Both benefits are essential to attracting investments from the institutional finance community, which, today, remains largely unfamiliar with the ocean economy realm and is structurally risk-averse.

To qualify for access to this fund, SOEZs would have to fulfil basic economic and ecological requirements—an additional measure of quality assurance. For example, they could require that an SOEZ have signed up a minimum critical number of anchor tenants, that coastal infrastructure be available and accessible, and that sponsoring nations provide sufficient cost and demand supports. It is essential that the link between the 'commitments' and the 'privileges' be explicit—these zones cannot become oases for cheap profits and minimum performance.

- **Adaptation and learning**. What matters most is that SOEZs be deliberate—they are meant to provide a contained laboratory and demonstration arena, where incentives can be concentrated and tested, results collated and adapted to, and risks managed. In the process of designing, launching and implementing the zones, the classic hurdles to ocean management—free access, lacking planning, use conflicts and free externalities—should be addressed in the context of real business, rather than abstract policy. Finally, SOEZs would be knowledge-intensive. When things go wrong, experiments fail and conflicts arise, the lessons learned would be reflected in the SOEZs' design and operations. On land, special economic zones have clearly evolved from an emphasis on manufacturing, trade and exports to a focus on knowledge, such as new technology frontiers and research and development (R&D).
- **Scaling the model**. Experience on land has shown that special economic zones today are instruments for the support of emerging technologies and business concepts and are not infinitely replicable. The goal is to accelerate commercialisation to the point where market-rate institutional capital moves in. In the case of SOEZs, there is an additional goal: to demonstrate the business case for a more systemically managed and accessed ocean, and to create a new, self-interested set of communities ready to defend their health on economic grounds.

SOEZs are a possible catalytic experiment for a sustainable ocean economy. They are bespoke, limited in scope and risk, and high in knowledge development. They can catalyse a new epoch but need to be supplemented by the array of more systemic policies and business priorities described in this report's action agenda (Sect. 6.3) and by additional catalytic interventions at national and international levels (see below). In the medium term, lessons learned from the SOEZs established by pioneering countries or regions should be codified in global standards, protocols, evaluation frameworks and the like to allow more countries to launch their own SOEZ without having to test and learn all dimensions of the concept.

6.4.2 National Intervention: Getting Things Done with National Ocean Task Forces

In recent years, the art and science of complex change management has been greatly refined and codified. Long familiar to the private sector, the principles of performance management have been applied to the public realm, with great success in increasing the performance of such complex networks as schools, health care, security and transportation systems, with measurable and transformative impact on metrics such as test scores, crime rates, health outcomes and the like. Typically, these approaches work even in the challenging context of multiple agencies and jurisdictions, conflicting objectives, complex logistics and significant uncertainty—as long as the approach is an extension of senior leadership. In general, these types of approaches involve small, non-hierarchical and highly competent teams, led by very senior and respected managers, with full access to all relevant information and working under a powerful and time-constrained mandates and targets.

Additionally, these task forces should appropriately represent all kinds of diversity and communities in their members, or at least represent them through thorough consultations.

Originally conceived in 2001 by Tony Blair as a 'Prime Minister Delivery Unit', this approach is now widely used. In the past 2 years alone, more than a dozen other governments—including those of Costa Rica, Ghana, Kenya, New South Wales (Australia), Pakistan, Peru, Saudi Arabia and Serbia—have created such units (see Fig. 20.43). Results can be quite encouraging (with necessary caution with respect to how metrics are calculated). In Britain, the number of people waiting more than a year for surgical procedures fell from over 40,000 to below 10,000; in Malaysia, reported street crime fell by 35% between 2009 and 2010; in Pakistan's Punjab province, the vaccinator attendance rate rose from 22% to over 90% between 2014 and 2015; in the U.S. state of Maryland, infant mortality dropped from 8 per 1000 live births in 2008 to 6.5 per 1000 live births in 2014.[607]

In the context of the shift to a sustainable ocean economy, this delivery unit approach can be very powerful if done right.

In the Dutch North Sea EEZ, for example, the need for explicit spatial planning became abundantly clear in 2015. Sand mining, oil drilling, dredging, cabling, shipping, military manoeuvring, land reclamation, fishing, aquaculture, wind energy and recreation had been accommodated in an integrated management plan since 2005—but the plan lacked explicit spatial guidance. A newly announced subsidy for wind farms led to 75 proposed projects and an unmanageable tangle of overlaps and potential use conflicts—with no relevant rules in place. For example, what would be a safe distance between a wind farm and a shipping lane where 400-m tankers would pass by? The ministers mandated a full, spatially explicit plan which combined zoning for MPAs in six main sectors of national importance: shipping routes, oil and gas installations, carbon capture and storage, renewable energy, defence and sand mining. Also, explicit rules were made to resolve potential conflicts between these priority functions.[608] The process—clearly mandated at the ministerial level, extensively consultative, highly expert, science-driven, target- and performance-oriented, and time-constrained—was a classic delivery unit approach. The approach could be similarly used to design ocean sustainable economic zones, develop multi-sector ocean development concepts, plan restoration projects or MPA networks and the like, within the 100% managed EEZs.

A first implementation step for the shift to a sustainable ocean economy could thus be the appointment of an SDG 14 (sustainable ocean) task force—appointed at the (ocean) ministerial or head of state level, and with an incontrovertible mandate to translate the sustainable ocean agenda into the appropriate national context by undertaking the following actions:

- Conduct a comprehensive marine resource mapping effort covering the entire national EEZ.
- Support and facilitate a participatory, inclusive process to develop an ocean plan which provides explicit guidance to assure the avoidance of spatial use conflicts, uniformly high standards of operation, a streamlined and efficient regulatory process, the integration of symbiotic ocean uses and the overall protection and sustainability of the key oceanic systems (in a minimum of 30% of the ocean as protected areas).
- Advise the relevant ministries and head of state on the specific steps required to further accelerate a regenerative ocean economy, including the design of special sustainable ocean economy zones, financial guarantees and risk

[607] Gold, J. 2017. "Tracking Delivery: Global Trends and Warning Signs in Delivery Units." Institute for Government. https://www.instituteforgovernment.org.uk/sites/default/files/publications/Global%20Delivery%20report.pdf.

[608] de Vrees, L. 2019. "Adaptive Marine Spatial Planning in the Netherlands Sector of the North Sea." *Marine Policy*, February, 103418. doi: https://doi.org/10.1016/j.marpol.2019.01.007.

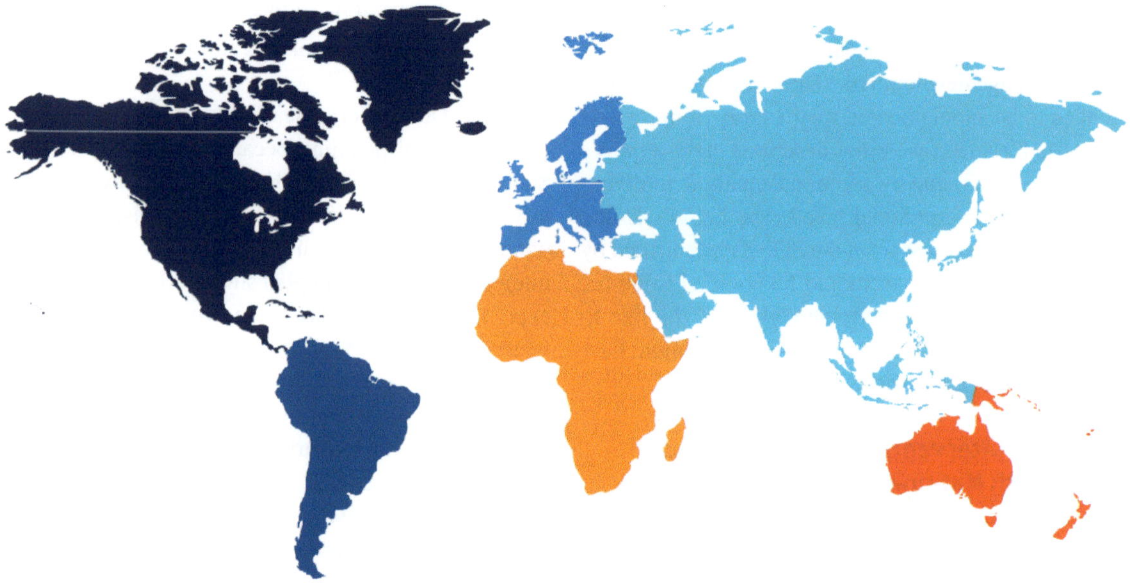

NORTH AMERICA

Canada
Results and Delivery Unit, *2016*

Maryland, USA
Governor's Office of Performance Improvement, *2015*

CENTRAL AND SOUTH AMERICA

Colombia
Unidad de Ejecución, *2015*

Costa Rica
Centro de Gestión de Gobierno, *2015*

Guatemala
Equipo de Prioridades Presidenciales, *2016*

Paraguay
Centro de Gobierno, *2013*

Peru
Unidad de Cumplimiento de Gobierno, *2016*

EUROPE

Albania
Delivery Unit, *2013*

Romania
Delivery Unit, *2014*

Serbia
Delivery Unit, *2015*

United Kingdom
Implementation Unit, *2012*

AFRICA

Gauteng, South Africa
Delivery Unit, *2016*

Ghana
Presidential Delivery Unit, *2015*

Kenya
Presidential Delivery Unit, *2015*

Liberia
Programme Delivery Unit, *2010*

Rwanda
Government Action Coordination Unit, *2008*

Senegal
President's Delivery Unit, *2014*

Sierra Leone
President's Delivery Team, *2015*

Uganda
Prime Minister's Delivery Unit, *2016*

Western Cape, South Africa
Delivery Support Unit, *2014*

MIDDLE EAST

Jordan
Prime Minister's Delivery Unit, *2015*

Oman
Tanfeedh Delivery Unit, *2016*

Saudi Arabia
Central Delivery Unit, *2016*

ASIA AND AUSTRALIA

Brunei
PENGGERAK, *2014*

India
Delivery Montoring Unit, *2009*

Indonesia
UKP4, *2009*

Khyber Pakhtunkhwa, Pakistan
Strategic Support Unit, *2015*

Malaysia
Performance Management Delivery Unit, *2009*

New South Wales, Australia
Premier's Implementation Unit, *2015*

Pakistan
Prime Minister's Delivery Unit, *2015*

Punjab, Pakistan
Special Monitoring Unit, *2014*

Fig. 20.43 Location of existing centre-of-government delivery units (national and regional levels). (*Source: Gold, J. 2018. "Tracking Delivery." Institute for Government.* https://www.instituteforgovernment.org.uk/sites/default/files/publications/Global%20Delivery%20report.pdf)

reduction measures, policy and regulatory implications, and international coordination issues.
- Lead special technological and bureaucratic initiatives—in coordination with relevant existing organisations, academia and civil society—such as the design of MPA networks, the detailed design of special sustainable economic zones and the recommended approach to controlling land-based pollutants.

In keeping with the basic delivery unit principles, this task force will have to be carefully set up and mandated. The following basic requirements would need to be met:

- The mandate would be issued formally and publicly at the ministerial and/or head of state level and clarify all mandate overlap issues with the appropriate federal and state agencies.
- The task force would be granted full access to all government-owned data sets (excepting only those of considerable national security significance).
- A firm timeline of deliverables would be provided.
- The task force would be provided with a core team of highly competent technical experts, full agency technical support and sufficient funding.
- The heads of all relevant agencies would be formal members of the task force governance and personally accountable for its success.
- The task force would have the full authority to conduct outreach and consultation in the name of the mandating minister or head of state.

Some countries will be better resourced and prepared to embark on this approach than others. Steps should be taken to develop a technical support platform which can provide targeted assistance to requesting nations, such as advanced capabilities in geographic information systems and mapping; resource mapping and sensing; protected area network delineation and design; operational standard setting for key industries; sector-specific knowledge (wind energy, mariculture, shipping safety); project finance and so on.

Some other countries might find that this approach is not suited to their culture and usual way of working and will have to consider a different, tailor-made approach.

This task force approach can be very efficient when short-term results and impacts are expected, such as in post-COVID recovery, where immediate solutions are expected to re-boost the economy and create jobs while avoiding the replication of environmentally detrimental practices from before COVID-19. For instance, national 'blue task forces' could be set up to support some immediate 'blue' recovery priorities, as identified in the latest special report commissioned by the Ocean Panel:[609]

- Coastal and marine ecosystem restoration and protection
- Sewage and waste infrastructure
- Sustainable non-fed mariculture
- Zero-emission marine transport
- Sustainable ocean-based renewable energy

These priorities are fully in line with the action agenda presented in Sect. 6.3 and could constitute a good starting point to embrace a more holistic transformation journey towards a sustainable ocean economy at scale.

6.4.3 International Intervention: Raising the Bar

Local and national interventions can generate significant catalytic effects but should not underestimate the importance of international collaboration. Indeed, several international agreements, treaties and conventions have already identified the challenges and solutions required but are still insufficiently ratified and poorly enforced. In addition, a growing number of global initiatives underway related to the sustainable ocean economy would benefit from additional recognition and support.

The following **international treaties, agreements and conventions** can greatly help accelerate a sustainable ocean economy:

- **The Paris Agreement.** 'The' landmark climate agreement aims to reduce global emissions to keep the planet's warming to 'well below' 2 °C and to pursue a 1.5-degree Celsius future warming scenario.[610] Of 197 parties, 189 have ratified the agreement, covering 97% of the world's emissions.[611] Instead of settling on a minimum common denominator, individual country commitments to the climate agreement are made through nationally determined contributions, allowing each country to commit as much as it is able—or willing—to contribute. As of 2020, only a handful of countries are on track to meet their climate

[609] Northrop et al. 2020. "A Sustainable and Equitable Blue Recovery to the COVID-19 Crisis."

[610] UN Framework Convention on Climate Change. n.d. "The Paris Agreement." https://unfccc.int/process-and-meetings/the-paris-agreement/the-paris-agreement. Accessed 12 May 2020.

[611] UN Framework Convention on Climate Change. n.d. "Paris Agreement: Status of Ratification." https://unfccc.int/process/the-paris-agreement/status-of-ratification. Accessed 12 May 2020.

goals.[612] It is critical that all countries ratify commitments made under the agreement to maintain the ocean, while expanding their NDCs to include the wealth of ocean climate-mitigation opportunities.

- **The Agreement on Port State Measures** is the first binding international agreement to specifically target IUU fishing. Its innovative approach is to prevent vessels engaged in IUU fishing from using ports and landing their catches, an approach assumed to be more cost-efficient than tracking and pursuing these vessels at sea. As of 2019, 105 countries are committed to the implementation of the agreement, but many of them are still struggling with the financing and capacity needed to fully enforce it.[613]
- **The Hong Kong Convention** is aimed at 'ensuring that ships, when being recycled after reaching the end of their operational lives, do not pose any unnecessary risks to human health, safety and the environment'.[614] To achieve this goal the convention also covers the design, construction, operation and maintenance of ships and can encourage circular design.[615] Although it was adopted in 2009, the convention cannot enter into force until it is ratified by '15 States, representing 40% of the world's merchant fleet and their ship recycling volume constituting not less than 3% of the gross tonnage of these contracting States' merchant fleet'.[616] Yet, as of 2019, it had been ratified or acceded to by only 12 states: Belgium, Denmark, Estonia, France, India, Japan, the Netherlands, Norway, Panama, the Republic of the Congo, Serbia and Turkey.[617]
- **The Intergovernmental Conference on Marine Biodiversity of Areas beyond National Jurisdiction (BBNJ).** This process, being negotiated under UNCLOS, represents 'an opportunity to provide a new governance model with legal clarity'[618] for the global commons in areas beyond national jurisdiction, including (1) a path to designate, implement and manage area-based management tools, including marine protected areas; (2) a trigger and a process for carrying out environmental impact assessments; (3) ensured fairness and equity of access to and benefit-sharing arising from the use of marine genetic resources; and (4) a means to foster developing capacity and transfer of technology to countries in need.[619] As soon as an agreement on marine biodiversity of areas beyond national jurisdiction is adopted, ocean-minded countries should ratify, implement and operationalise it.
- **The Convention on Biological Diversity**. Entered into force in 1993, the Convention on Biological Diversity (CBD) aims to conserve biological diversity, promote the sustainable use of the components of biological diversity and ensure the fair and equitable sharing of the benefits arising from the use of genetic resources.[620] Under the CBD, Parties are negotiating the post-2020 global biodiversity framework, to be adopted at the 15th meeting of the Conference of the Parties (COP 15) in China in 2021. This framework will contain a new set of global goals and targets for biodiversity, and it is crucial that ambitious targets be agreed on to support a healthy ocean and a sustainable ocean economy.

In addition, the IMO, regional fisheries management organisations, regional seas conventions and others provide tools that can be used much more actively. It would also be possible to establish regional ocean management organisations that manage the ocean cross-sectorally, to seek international support for a Paris-like agreement for the ocean and to set up a task force on ocean-related financial disclosures. Another possible tool is a global ocean accountability board, composed of leaders from a number of sectors, which sits outside the international forums and seeks to (1) avoid catastrophic ocean collapse and (2) hold the world to account for its ocean action.[621] This panel would be modelled on the G20 Financial Stability Board.

In addition, the idea of creating a supranational ocean agency of some kind could be explored. Learning the lessons

[612] Climate Analytics and NewClimate Institute. n.d. "Climate Action Track: Governments Still Showing Little Sign of Acting on Climate Crisis—Warming Projections Global Update." https://climateactiontracker.org/documents/698/CAT_2019-12-10_BriefingCOP25_WarmingProjectionsGlobalUpdate_Dec2019.pdf. Accessed 12 May 2020.

[613] Curtis, L. 2019. *Report of the Second Meeting of the Parties to the Agreement on Port State Measures to Prevent, Deter and Eliminate Illegal, Unreported and Unregulated Fishing*. FAO Fisheries and Aquaculture Report FIAO/R1272. Santiago, Chile: Food and Agriculture Organization of the United Nations. http://www.fao.org/3/ca5757en/CA5757EN.pdf.

[614] IMO. n.d. "The Hong Kong International Convention for the Safe and Environmentally Sound Recycling of Ships." http://www.imo.org/en/About/Conventions/ListOfConventions/Pages/The-Hong-Kong-International-Convention-for-the-Safe-and-Environmentally-Sound--Recycling-of-Ships.aspx. Accessed 12 May 2020.

[615] IMO. n.d. "The Hong Kong International Convention for the Safe and Environmentally Sound Recycling of Ships."

[616] IMO. n.d. "The Hong Kong International Convention for the Safe and Environmentally Sound Recycling of Ships."

[617] IMO. 2019. "What's New during 2019." http://www.imo.org/en/MediaCentre/WhatsNew/Pages/Archive-2019.aspx. Accessed 12 May 2020.

[618] Gottlieb, H.M., and M. Conathan. 2019. "The Path to a High Seas Treaty." Aspen Institute. 18 April. https://www.aspeninstitute.org/blog-posts/the-path-to-a-high-seas-treaty/.

[619] Gottlieb and Conathan. 2019. "The Path to a High Seas Treaty."

[620] Secretariat of the Convention on Biological Diversity. 2012. Introduction to Convention on Biological Diversity. 16 January. https://www.cbd.int/intro/.

[621] Global Ocean Commission. 2014. "From Decline to Recovery: A Rescue Package for the Global Ocean." https://www.iucn.org/sites/dev/files/import/downloads/goc_full_report_1.pdf.

from UNESCO's 'Man and the Biosphere' programme,[622] this institution could be mandated to provide a flexible set of frameworks and protocols to empower local actors to collaborate, with the goal of protecting and regenerating ocean commons at regional levels.[623] This ocean agency could be created by UN resolution, or it could be created by a founding group of nations who invite others to participate. Its establishment should ensure legitimacy and safeguards against capture by special interests.[624]

A key to sustainable development and to developing a sustainable ocean economy is a cross-sectoral approach. Given the complexities of international law, access to ocean resources and various constraints, it is likely that polycentric systems of governance will prevail.[625] A thorough expert review of the existing governance mechanisms and how they should be changed to support the development of a sustainable ocean economy is beyond the scope of this report. However, such a review should be performed as a matter of urgency.

Beyond the purely political efforts mentioned above, a growing number of initiatives are gathering a variety of actors willing to accelerate a sustainable ocean economy. Supporting such initiatives is a concrete immediate next step to advance a shared vision, identify solutions to remove roadblocks and initiate public-private partnerships. None of the identified ocean challenges and opportunities can be solved or captured by one entity alone—bringing together the private sector, public entities and civil society along the journey and identifying stakeholders on the ground that are willing to support efforts to address system change is key. Building coalitions around certain ocean themes can help align stakeholders into a unified voice, build on synergies and help identify and develop high-impact, investable opportunities.

7 Conclusion

This report has argued that the agendas of effective ocean protection, sustainable ocean production and equitable human prosperity are inseparable and compatible. It has framed the economic, social and ecological upside of getting the ocean economy right—and the deeply concerning and potentially sweeping consequences of getting it wrong.

Getting it right means a fundamental shift away from the 'free for all' model, which assumes an ocean of unlimited potential to regenerate, dilute and absorb. This model is maintained by current practices, laws and cultural norms, but it is not inviolate. Spearheaded by a new cohort of ocean interests deeply vested in ocean health—including sustainable fishers and mariculturists, coastal communities, renewable energy generators, tourism operators, scientists, environmentalists and social and civil society organisations—pollution and over-exploitation can be powerfully counteracted. It will require thoughtful policy support, including transparent and comprehensive national ocean planning, mandatory standards for open data access, investment of (sovereign wealth, development and private) capital, new legal protections from polluters and a national accounting approach focused on the ocean's natural capital and production in equal measure.

The report argues further that the ocean's essential contribution to sustainable planetary food and energy production can be achieved without abandoning a precautionary, insurance-based approach. The consequences of systemic failure in the ocean are grave, and there is no logic in 'harming the ocean to save the planet'. The approach calls for the inclusion of at least 30% fully protected areas at a global scale, the avoidance of new extractive activities whose impact is not fully understood and the widespread adoption of rigorous operation standards for all ocean uses.

Learning from the successes and stalling points of other industrial and societal transformations, this report has developed a comprehensive 10-point action agenda for a sustainable ocean economy. Further, it proposes three concrete options to commence and accelerate change: sustainable ocean economic zones, centre-of-government delivery units and interventions at an international level.

The journey towards a sustainable future has already begun, with pioneers leading the way. New sustainable technologies are attracting investors, and businesses and governments are waking up to the opportunities of a sustainable ocean economy—as well as to the risks and cost of inaction. It is an enormously inspiring journey. Antoine de Saint-Exupéry describes the awe and wonder that the ocean evokes, and the power of humanity's determination to connect with it: 'If you want to build a ship, don't drum up people to collect wood and don't assign them tasks and work, but rather teach them to long for the endless immensity of the sea'.[626]

622 Bridgewater, P. 2016. "The Man and Biosphere Programme of UNESCO: Rambunctious Child of the Sixties, but Was the Promise Fulfilled?" *Current Opinion in Environmental Sustainability* 19 (April): 1–6. doi: https://doi.org/10.1016/j.cosust.2015.08.009.

623 Swilling et al. 2020. "The Ocean Transition."

624 Swilling et al. 2020. "The Ocean Transition."

625 Ostrom, E. 2010. "Polycentric Systems for Coping with Collective Action and Global Environmental Change." *Global Environmental Change* 20 (4): 550–57. doi: https://doi.org/10.1016/j.gloenvcha.2010.07.004.

626 Saint-Exupéry, A. 1948. Citadelle. Original text in French: 'Créer le navire ce n'est point tisser les toiles, forger les clous, lire les astres, mais bien donner le goût de la mer qui est un, et à la lumière duquel il n'est plus rien qui soit contradictoire mais communauté dans l'amour.'

The COVID-19 pandemic has severely hit most ocean-based sectors, with significant social and economic impacts. Without deprioritising the need for immediate responses and quick recovery, this shock could also be seen as an opportunity to accelerate the transition towards a sustainable ocean economy. As Arundhati Roy observes, 'Historically, pandemics have forced humans to break with the past and imagine their world anew. This one is no different. It is a portal, a gateway between one world and the next. We can choose to walk through it, dragging the carcasses of our prejudice and hatred, our avarice, our data banks and dead ideas, our dead rivers and smoky skies behind us. Or we can walk through lightly, with little luggage, ready to imagine another world. And ready to fight for it'.[627]

Acknowledgements The authors would like to thank the individuals listed below for their contributions towards the creation and publishing of this report.

Contributing to the report creation: Barbara Karni and Katie Flanagan for leading the development of the summary; as well as Henry Wang and Katherine Woods for their contributions to the overall report.

Offering review, feedback, and inputs to the report: Meg Caldwell, Jason Clay, Anthony Cox, Exequiel Ezcurra, Steve Gaines, Craig Hanson, Janis Jones, Ghislaine Llewellyn, Justin Mundy, Karine Nyborg, Nicolas Pascal, Ina Porres,

Angelique Pouponneau, Bob Richmond, Mary Ruckelshaus, Enric Sala, Eric Schwaab, Alan Simcock, Andrew Steer, Mark Swilling, Peter Thomson, Piera Tortora, Adair Turner, Jacqueline Uku and Simon Zadek. Particular thanks go to this report's arbiter, Kristian Teleki, who provided substantive comments and guidance throughout the review process.

Creating a substantial part of the scientific basis for this report: Authors and contributors of the Ocean Panel's Blue Papers and special reports:

The Ocean as a Solution to Climate Change: Five Opportunities for Action. Lead author: Ove Hoegh-Guldberg, and the paper's contributing authors: Ken Caldeira, Thierry Chopin, Steve Gaines, Peter Haugan, Mark Hemer, Jennifer Howard, Manaswita Konar, Dorte Krause-Jensen, Elizabeth Lindstad, Catherine E. Lovelock, Mark Michelin, Finn Gunnar Nielsen, Eliza Northrop, Robert Parker, Joyashree Roy, Tristan Smith, Shreya Some and Peter Tyedmers

The Future of Food from the Sea. Lead authors: Christopher Costello, Ling Cao, Stefan Gelcich and the paper's contributing authors: Miguel Angel Cisneros, Christopher M. Free, Halley E. Froehlich, Elsa Galarza, Christopher D. Golden, Gakushi Ishimura, Ilan Macadam-Somer, Jason Maier, Tracey Mangin, Michael C. Melnychuk, Masanori Miyahara, Carryn de Moor, Rosamond Naylor, Linda Nøstbakken, Elena Ojea, Erin O'Reilly, Giacomo Chato Osio, Ana M. Parma, Fabian Pina Amargos, Andrew J. Plantinga, Albert Tacon and Shakuntala H. Thilsted

The Expected Impacts of Climate Change on the Ocean Economy. Lead authors: Steven Gaines, Reniel Cabral, Christopher M. Free, Yimnang Golbuu and the paper's contributing authors: Ragnar Arnason, Willow Battista, Darcy Bradley, William Cheung, Katharina Fabricius, Ove Hoegh-Guldberg, Marie Antonette Juinio-Meñez, Jorge García Molinos, Elena Ojea, Erin O'Reilly and Carol Turley

Technology, Data and New Models for Sustainably Managing Ocean Resources. Lead authors: Jim Leape, Mark Abbott, Hide Sakaguchi and the paper's contributing authors: Annie Brett, Ling Cao, Kevin Chand, Yimnang Golbuu, Tara Martin, Juan Mayorga and Mari S. Myksvoll

Illegal, Unreported and Unregulated Fishing and Associated Drivers. Lead authors: Sjarief Widjaja, Tony Long, Hassan Wirajuda and the paper's contributing authors: Hennie Van As, Per Erik Bergh, Annie Brett, Duncan Copeland, Miriam Fernandez, Ahmad Gusman, Stephanie Juwana, Toni Ruchimat, Steve Trent and Chris Wilcox

Towards Ocean Equity. Lead authors: Henrik Österblom, Colette C.C. Wabnitz, Dire Tladi, and the paper's contributing authors: Edward H. Allison, Sophie Arnaud-Haond, Jan Bebbington, Nathan Bennett, Robert Blasiak, Wiebren Boonstra, Afrina Choudhury, Andrés Cisneros-Montemayor, Tim Daw, Michael Fabinyi, Nicole Franz, Harriet Harden-Davies, Danika Kleiber, Priscila Lopes, Cynthia McDougall, Budy P. Resosudarmo and Samiya A. Selim

The Ocean Genome: Conservation and the Fair, Equitable and Sustainable Use of Marine Genetic Resources. Lead authors: Robert Blasiak, Rachel Wynberg, Kirsten Grorud-Colvert, Siva Thambisetty, and the paper's contributing authors: Narcisa M. Bandarra, Adelino V.M. Canário, Jessica da Silva, Carlos M. Duarte, Marcel Jaspars, Alex D. Rogers, Kerry Sink and Colette C.C. Wabnitz

Critical Habitats and Biodiversity: Inventory, Thresholds and Governance. Lead authors: Alex D. Rogers, Octavio Aburto-Oropeza, and the paper's contributing authors: Ward Appeltans, Jorge Assis, Lisa T. Ballance, Philippe Cury, Carlos Duarte, Fabio Favoretto, Joy Kumagai, Catherine Lovelock, Patricia Miloslavich, Aidin Niamir, David Obura, Bethan C. O'Leary, Gabriel Reygondeau, Callum Roberts, Yvonne Sadovy, Tracey Sutton, Derek Tittensor and Enriqueta Velarde

Integrated Ocean Management. Authors: Jan-Gunnar Winther, Minhan Dai, and the paper's contributors: Fanny Douvere, Leanne Fernandes, Patrick Halpin, Alf Håkon Hoel, Marie Antonette Juinio-Meñez, Yangfan Li, Karyn Morrissey, Therese Rist, Fabio Rubio Scarano, Amy Trice, Sebastian Unger and Sandra Whitehouse

Leveraging Multi-target Strategies to Address Plastic Pollution in the Context of an Already Stressed Ocean. Lead authors: Jenna Jambeck, Ellie Moss and Brajesh Dubey, and the paper's contributing authors: Zainal Arifin, Linda Godfrey, Britta Denise Hardesty, I. Gede Hendrawan, To Thi Hien, Liu Junguo, Marty Matlock, Sabine Pahl, Karen Raubenheimer, Martin Thiel, Richard Thompson and Lucy Woodall

The Ocean Transition: What to Learn from System Transitions. Lead authors: Mark Swilling, Mary Ruckelshaus, Tanya Brodie Rudolph, and the paper's contributing authors: Edward H. Allison, Stefan Gelcich, Philile Mbatha and Henrik Österblom

What Role for Ocean-Based Renewable Energy and Deep-Seabed Minerals in a Sustainable Future? Lead authors Peter M. Haugan, Lisa A. Levin, and the paper's contributing authors, Diva Amon, Mark Hemer, Hannah Lily and Finn Gunnar Nielsen

A Sustainable Ocean Economy for 2050: Approximating Its Benefits and Costs. Manaswita Konar and Helen Ding

National Accounting for the Ocean and Ocean Economy. Lead authors: Eli P. Fenichel, Ben Milligan, Ina Porras, and the paper's contributing authors: Ethan T. Addicott, Ragnar Árnasson, Michael Bordt, Samy Djavidnia, Anthony Dvarskas, Erica Goldman, Kristin Grimsrud, Glenn-Marie Lange, John Matuszak, Umi Muawanah, Martin Quaas, Francois Soulard, Niels Vestergaard and Junjie Zhang

Organised Crime in the Fisheries Sector. Lead authors: Emma Witbooi, Kamal-Deen Ali, Mas Achmad Santosa, and the paper's contributing authors: Gail Hurley, Yunus Husein, Sarika Maharaj, Ifesinachi Okafor-Yarwood, Inés Arroyo Quiroz and Omar Salas

The Human Relationship with Our Ocean Planet. Lead authors: Edward Allison, John Kurien, Yoshitaka Ota, and the paper's contributing authors: Dedi S. Adhuri, J. Maarten Bavinck, Andrés Cisneros-Montemayor, Michael Fabinyi, Svein Jentoft, Sallie Lau, Tabitha Grace Mallory, Ayodeji Olukoju, Ingrid van Putten, Natasha Stacey, Michelle Voyer and Nireka Weeratunge

[627] Roy, A. 2020. "The Pandemic Is a Portal." *Financial Times*, 3 April. https://www.ft.com/content/10d8f5e8-74eb-11ea-95fe-fcd274e920ca.

A Sustainable and Equitable Blue Recovery to the COVID-19 Crisis. Eliza Northrop, Manaswita Konar, Nicola Frost and Elizabeth Hollaway

Ocean Finance: Financing the Transition to a Sustainable Ocean Economy. Lead authors: Rashid Sumaila, Melissa Walsh, Kelly Hoareau and Anthony Cox, and the paper's contributing authors: Patrízia Abdallah, Wisdom Akpalu, Zuzy Anna, Dominique Benzaken, Beatrice Crona, Timothy Fitzgerald, Louise Heaps, Katia Karousakis, Glenn-Marie Lange, Amanda Leland, Dana Miller, Louise Teh, Karen Sack, Durreen Shahnaz, Torsten Thiele, Niels Vestergaard, Nobuyuki Yagi and Junjie Zhang

Coastal Development: Resilience, Restoration and Infrastructure Requirements. Lead authors: Andy Steven, Kwasi Appeaning Addo, Ghislaine Llewelyn, Ca Vu Thanh, and the paper's contributing authors: Isaac Boateng, Rodrigo Bustamante, Christopher Doropoulos, Chris Gillies, Mark Hemer, Priscila Lopes, James Kairo, Munsur Rahman, Lalao Aigrette Ravaoarinorotsihoarana, Megan Saunders, Rashid Sumaila, Frida Sidik, Louise Teh, Mat Vanderklift and Maria Vozzo

Providing the secretariat function of the Ocean Panel and coordinating the report production: World Resources Institute. In particular, the authors would like to thank Head of Secretariat Kristian Teleki, as well as Ines Aguiar Branco, Clare Brennan, Katie Flanagan, Nicola Frost, Erika Harms, Amy Hemingway, Elizabeth Hollaway, Manaswita Konar, Eliza Northrop, Lauren Thuringer and Lauren Zelin.

Designing and copyediting the report: Romain Warnault for the coordination of the design process, Rosie Ettenheim, Bill Dugan, and Romain Warnault for the report's design, and Alex Martin for copyediting the report.

While our colleagues were very generous with their time and input, this report reflects the views of the authors alone.

About the Authors

Lead Authors

Martin R. Stuchtey is a Partner at SYSTEMIQ. His email address is martin.stuchtey@systemiq.earth.

Adrien Vincent is the Ocean Lead at SYSTEMIQ.

Andreas Merkl is a Partner at CEA Consulting.

Maximilian Bucher is an Associate at SYSTEMIQ.

Contributing Authors

Peter M. Haugan is Programme Director at the Institute of Marine Research, Norway, Professor at the Geophysical Institute, University of Bergen, Norway, and Co-chair of the Ocean Panel Expert Group.

Jane Lubchenco is Distinguished University Professor at Oregon State University and Co-chair of the Ocean Panel Expert Group.

Mari Elka Pangestu is the World Bank Managing Director of Development Policy and Partnerships and Co-chair of the Ocean Panel Expert Group.

Appendix: The High Level Panel for a Sustainable Ocean Economy's Products Used in Each Respective Report Section

This report draws extensively—and occasionally quotes directly from—the initiative's 16 Blue Papers and three special reports—'The Ocean as a Solution to Climate Change', 'A Sustainable Ocean Economy for 2050—Approximating Its Benefits and Costs' and 'A Sustainable and Equitable Blue Recovery to the COVID-19 Crisis'—with the permission of the authors. The figure below shows the report sections in which the special report and respective Blue Papers are used.

Report section	Blue paper/OSCC report
The Urgency of Today (Sect. 4)	
A Blue Awakening: Recognising That the Ocean Is Vital to Humankind and the Global Economy	1, 3, 8, 10
Failing the Environment and the People: The Need for Urgent Action	1, 3, 8, 9, 10, 15, 16, 17
Embracing Hope: The Building Momentum for a Sustainable Ocean Economy	1, 2, 3, 4, 5, 8, 10, 11, 14, 15, 18, 19
The Possibility of Tomorrow (Sect. 5)	
Defining a Compass Direction: Principles for a Sustainable Ocean Economy	1, 6, 9, 11, 14
A New Picture Is Emerging: The 2050 Sustainable Ocean Economy	1, 2, 3, 4, 5, 7, 10, 12, 18
The Big Reconciliation: Protect Effectively, Produce Sustainably and Prosper Equitably	1, 2, 3, 8, 9, 12, 13, 16, 17
A Roadmap to a Sustainable Ocean Economy (Sect. 6)	
Harnessing Complex Adaptive Systems: Lessons for the Sea	6
Charting a Direction: The Ocean Action Agenda	1, 2, 4, 5, 8, 10, 14, 15, 16, 18, 19
Launching the Voyage: Three Levels for Possible Immediate Action	6, 17

The official Blue Paper and Special Report titles affiliated with the referenced numbers above are as follows:

1 The Ocean as a Solution to Climate Change: Five Opportunities for Action
2 The Future of Food from the Sea
3 The Expected Impacts of Climate Change on the Ocean Economy
4 Technology, Data and New Models for Sustainably Managing Ocean Resources
5 Illegal, Unreported and Unregulated Fishing and Associated Drivers
6 Towards Ocean Equity
7 The Ocean Genome: Conservation and the Fair, Equitable and Sustainable Use of Marine Genetic Resources
8 Critical Habitats and Biodiversity: Inventory, Thresholds and Governance

9 Integrated Ocean Management
10 Leveraging Multi-target Strategies to Address Plastic Pollution in the Context of an Already Stressed Ocean
11 The Ocean Transition: What to Learn from System Transitions
12 What Role for Ocean-Based Renewable Energy and Deep-Seabed Minerals in a Sustainable Future?
13 A Sustainable Ocean Economy for 2050: Approximating Its Benefits and Costs
14 National Accounting for the Ocean and Ocean Economy
15 Organised Crime in the Fisheries Sector
16 The Human Relationship with Our Ocean Planet
17 A Sustainable and Equitable Blue Recovery to the COVID-19 Crisis
18 Ocean Finance
19 Coastal Development: Managing Resilience, Restoration and Infrastructure of Coastlines

Transformations for a Sustainable Ocean Economy: A Vision for Protection, Production and Prosperity

21

1 Our Call to Action

We have a collective opportunity and responsibility to protect and restore the health of our ocean, and build a sustainable ocean economy that can provide food, empower coastal communities, power our cities, transport our people and goods and provide innovative solutions to global challenges.

In accepting this responsibility and seizing this opportunity, we can give a blue boost to the economy today while mitigating and building resilience for future crises.

The framework and five areas of transformation presented here secure ocean health and wealth for generations to come. We urge other governments, industries and stakeholders to join us in this endeavour.

We, the 14 members of the High Level Panel for a Sustainable Ocean Economy (the Ocean Panel), are heads of state and government representing people from across all ocean basins, nearly 40% of the world's coastlines and 30% of exclusive economic zones. We recognise that the ocean is the life source of our planet and is vital for human well-being and a thriving global economy.

The ocean is home to many complex ecosystems facing significant threats. The actions we take now can safeguard the ocean's capacity to regenerate, in order to deliver substantial economic, environmental and social value and offer powerful solutions to global challenges. Rapid action must be taken today to address climate change, acidification, ocean warming, marine pollution, overfishing, and loss of habitat and biodiversity. Failure to act will jeopardise global health, well-being, and economic vitality and exacerbate inequalities.

The COVID-19 pandemic has highlighted the deep interconnections between human and planetary health and the need for nations to work together to respond to global threats. The pandemic has caused a dramatic disruption of the global economy, major impacts to our societies and a huge toll on our communities. It has put increased financial pressure on developing countries and in particular Least Developed Countries and Small Island Developing States.

We have an opportunity and obligation to reset and build a more equitable, resilient, knowledge-based and prosperous future that is in harmony with nature. The ocean and its related economy offer a wealth of opportunities to support this transition.

Building a sustainable ocean economy is one of the most important tasks and greatest opportunities of our time. It is critical to achieving the goals of the 2030 Agenda for Sustainable Development, and it is vital if we are to emerge from current and future crises with stronger economies, healthier people and more resilient communities.

We commit to bold transformations towards a sustainable ocean economy where environmental protection and conservation, and economic production and prosperity, go hand in hand. These transformations must unleash the full force of innovation across sectors in technology, finance and governance, and do so at pace and scale, guided by the following principles:

- **Alignment:** Ocean protection and production must align with the UN Framework Convention on Climate Change and the Paris Agreement, the Convention on Biological Diversity, and the Polluter Pays Principle as set out in the Rio Declaration. Actions must be aligned across ocean-based and land-based activities and ecosystems.
- **Inclusiveness:** Human rights, gender equality, community and Indigenous Peoples' participation, through their free, prior and informed consent, must be respected and protected.
- **Knowledge:** Ocean management must be informed by the best available science and knowledge, including

Originally published in:
High Level Panel for a Sustainable Ocean Economy. 2020. *Transformations for a Sustainable Ocean Economy: A Vision for Protection, Production and Prosperity*. Washington, DC: World Resources Institute. Available online at https://oceanpanel.org/the-agenda/.
Reprint by Springer International Publishing (2023) with kind permission.
Published under license from the World Research Institute.

J. Lubchenco, P. M. Haugan (eds.), *The Blue Compendium*, https://doi.org/10.1007/978-3-031-16277-0_21

indigenous and local knowledge, and aided by innovation and technology.
- **Legality:** The UN Convention on the Law of the Sea is the legal basis for all ocean activities, and existing international ocean commitments must be implemented as a foundation for achieving a sustainable ocean economy.
- **Precaution:** Where there are threats of serious or irreversible damage, lack of full scientific certainty shall not be used as a reason for postponing cost-effective measures to prevent environmental degradation.
- **Protection:** A healthy ocean underpins a sustainable ocean economy. A net gain approach must be applied to ocean uses in order to help sustain or restore the health of the ocean.
- **Resilience:** The resilience of the ocean and ocean economy must be enhanced.
- **Solidarity:** The need for access to finance, technology and capacity building for developing countries, especially Small Island Developing States and Least Developed Countries, must be recognised, taking into account their particular circumstances and vulnerabilities.
- **Sustainability:** The production and harvesting of ocean resources must be sustainable and support resilient ecosystems and future productivity.

We will leverage the UN Decade of Ocean Science for Sustainable Development and the body of knowledge commissioned by the Ocean Panel to build collective understanding and knowledge of ocean sustainability, ecosystem services and functions, and ensure that science underpins decision-making for building a sustainable ocean economy.

2 A 100% Approach

The ocean is a complex natural system that is inextricably linked to land-based activities and ecosystems. We must approach ocean management holistically in order to achieve the vision of protection, production and prosperity. We need a comprehensive approach to sustainably manage 100% of the ocean, starting with coastal and ocean states and working together regionally and globally to safeguard areas beyond national jurisdiction.

We Commit
To sustainably manage 100% of the ocean area under national jurisdiction, guided by Sustainable Ocean Plans, by 2025.

We Urge
All coastal and ocean states to join us in this commitment so that by 2030 all ocean areas under national jurisdiction are sustainably managed.

Sustainable Ocean Plans are providing a credible basis for safeguarding the long-term health and resilience of the ocean, attracting investment and creating jobs to the benefit of coastal communities and national economies.

A **Sustainable Ocean Plan** describes policies and mechanisms to facilitate sustainable use of the ocean and maximise benefits and value creation for current and future generations. It provides a framework to reconcile conflicting uses of the ocean and its resources and enable long-term sustainable growth in the ocean economy. It can include a range of mechanisms such as regulatory reform, strategic investments in emerging sectors, marine spatial planning, integrated coastal and watershed management, and the establishment and implementation of marine protected areas and other effective area-based conservation measures that can help deliver nature's contributions to people, economic and positive biodiversity conservation outcomes, climate change mitigation and adaptation, and sustainable fish stocks.

The Sustainable Ocean Plans should be in line with the 2030 Agenda for Sustainable Development, build on integrated ocean management and ecosystem knowledge, address pressures from all land- and sea-based sources, and take account of the predicted impacts of climate change. As the foundation for a sustainable ocean economy, these plans should be developed and implemented through an inclusive, participatory, transparent and accountable process.

We support a global target to protect 30% of the ocean by 2030. As a global target, it would not be binding on states individually. National decisions on marine spatial plans, marine protected areas and other effective area-based conservation measures will depend on the state and function of the ecosystem and the extent and quality of ocean management as well as the importance of addressing human well-being, sustainable ocean food and climate change. To achieve the global target, we also call for international cooperation, including supporting capacity building in this area. Detailed content of Sustainable Ocean Plans will vary according to national circumstances.

We will work with others to mobilise and facilitate support for coastal and ocean states in developing Sustainable Ocean Plans by 2030.

2.1 Getting to 100%

Our vision for protection, production and prosperity requires mutually reinforcing transformations in five critical areas: ocean wealth, ocean health, ocean equity, ocean knowledge and ocean finance. Action in all areas is required to achieve a sustainable ocean economy and build critical foundations for economic recovery and resilience.

The Ocean Panel presents a framework with outcomes for these five areas and a range of actions to achieve them. We commit to deliver these fully by 2030 or sooner. The framework is consistent with the existing target deadlines set in the Sustainable Development Goals, and particular effort is needed for the unfulfilled targets with a 2020 timeline. We will act with determination in accordance with national capacities and circumstances, and invite other leaders, industry and civil society to join us.

3 Ocean Wealth

In recent years, the ocean has produced US $2.5 trillion in goods and services each year, and the asset value of the ocean has been estimated at US $24 trillion. Many ocean-based industries have the potential to outperform the growth of the global economy, both in terms of additional value and employment. Unsustainable human activity—in the ocean and on land—is threatening the ocean's ability to regenerate and sustainably provide for people around the world. We must transform our relationship with the ocean to ensure that it can continue to produce sustainably for future generations.

3.1 Sustainable Ocean Food

Ocean food plays a critical role in feeding global populations. It supplies an essential and accessible source of animal protein and micronutrients, which are particularly important in low-income, food-deficit countries and Small Island Developing States and during times of economic or environmental crisis. The ocean can provide more abundant and diverse food than it currently does, thereby playing a bigger role in the global food system. To build resilience, ocean food production must meet national and local needs and be adapted to a changing climate. Doing so can enhance food security, improve nutrition, human health and well-being, create sustainable economic growth and jobs and prevent the widening of current inequities. This transition must include increased transparency in global ocean governance and supply chains and the elimination of inefficiencies and perverse incentives that undermine the sustainability of the food we derive from the ocean. We must seize opportunities to sustainably increase fisheries productivity and aquaculture production, including by strengthening opportunities for coastal communities, Indigenous Peoples, artisanal and small-scale fishers.

Wild fish stocks are restored and harvested at sustainable levels, aquaculture is sustainably grown to meet global needs, and waste is minimised and managed throughout the value chain.

Priority Actions

- Eliminate illegal, unreported and unregulated fishing by incentivising the use of the latest innovations and technologies—such as digital traceability—to increase transparency; strengthening monitoring, control and surveillance; improving flag state control; effectively implementing the Port State Measures Agreement; and enabling enhanced collaboration amongst all stakeholders in the supply chain.
- Prohibit harmful fisheries subsidies that contribute to overcapacity, overfishing, and illegal, unreported and unregulated fishing.
- Minimise bycatch, discards and waste in seafood supply chains.
- Develop, adopt and effectively implement science-based plans to rebuild depleted stocks, and ensure adaptive fisheries management to respond to climate change and the uncertainties of shifting ocean ecosystems, based on the UN Fish Stocks Agreement, in cooperation with multilateral bodies such as the Food and Agriculture Organisation and regional fisheries management organisations, and implement FAO's Voluntary Guidelines to Ensure Sustainable Small-Scale Fisheries.
- Strengthen regional fisheries management organisations, including by promoting the use of a precautionary approach, management that controls harvest levels based on scientific assessment, such as total allowable catch, meaningful consequences for exceeding quotas, and through regular and transparent performance reviews.
- Explore in a precautionary manner the potential to sustainably harvest new species from the ocean, without undermining ecosystem health.
- Put in place policies and management frameworks to minimise the environmental impacts of aquaculture, including inefficiencies in the feed supply chain, and enable the acceleration of fed and non-fed aquaculture production that fits local environmental, governance and economic priorities.

3.2 Sustainable Ocean Energy

The ocean holds tremendous potential to provide clean energy for the world. Scaling up ocean-based renewable energy will generate jobs and boost economic development while providing a pathway to decarbonisation. An ocean-based renewable energy revolution is in the making, and recovery efforts provide an opportunity to increase investment over the coming years. The pace and scope of development must match the state of the science, enable technology transfer and adoption, and minimise the impact on marine ecosystems to enable the delivery of sustainable ocean-based energy.

Ocean-based renewable energy is fast-growing and on the path to becoming a leading source of energy for the world.

Priority Actions

- Invest in research, technology development and demonstration projects to help make all forms of ocean-based renewable energy—including wind, wave, tidal, current, thermal and solar—cost-competitive, accessible to all and environmentally sustainable.
- Work collaboratively with industry and other stakeholders to develop clear frameworks addressing environmental impacts of ocean-based renewable energy, enabling capacity, co-existence and integration with other uses of the ocean.
- Set clear goals, commit to deliver appropriate policy and regulatory measures, and remove market impediments in order to accelerate sustainable ocean-based renewable energy deployment.

3.3 Sustainable Ocean-Based Tourism

Before the COVID-19 pandemic, tourism was projected to become the single-largest ocean- based industry by 2030. Tourism is one of the sectors hardest hit by the COVID-19 pandemic worldwide. At the same time, coastal and marine tourism remains vital to the economic prosperity of island and coastal communities. The continued viability of this sector remains at risk from climate change, disasters, pollution, urbanisation and ecosystem degradation. Sustainable ocean-based tourism can restore and protect the ocean while delivering jobs and prosperity. Achieving sustainable ocean tourism that can withstand future crises requires strategic public and private investments.

Coastal and ocean-based tourism is sustainable, resilient, addresses climate change, reduces pollution, supports ecosystem regeneration and biodiversity conservation and invests in local jobs and communities.

Priority Actions

- Invest in sustainable tourism that regenerates the ecosystems on which it depends, builds the resilience of coastal communities and Indigenous Peoples, reduces inequality through promoting equal opportunity and equitable distribution of benefits and addresses climate change and pollution.
- Implement sustainable tourism management strategies that advance environmental, social and economic priorities and enable monitoring and transparent reporting with the full participation of coastal communities and Indigenous Peoples.
- Implement mechanisms to increase the reinvestment of tourism revenue into local and indigenous communities to build capacity and skills for increasing local employment in tourism, diversify economic opportunities and increase resources for coastal and marine restoration and protection.
- Accelerate financial incentives for including nature-based solutions in sustainable tourism infrastructure.
- Invest in sewerage and wastewater infrastructure for coastal and marine tourism to improve the health of coastal communities and reduce the impacts on coastal and marine ecosystems.

3.4 Sustainable Ocean Transport

Shipping, the most energy-efficient form of transport, is vital to international trade and connectivity as it continues to move over 90% of global goods. Maintaining global supply chains will be critical to support recovery from the COVID-19 pandemic and future crises. Technology to decarbonise and minimise the negative environmental impacts of marine transport exists but must be brought to scale. To ensure the industry is resilient, we must move decisively towards reducing greenhouse gas emissions by investing in solutions now to support rapid decarbonisation. Such investments will create jobs and build connectivity and the long-term resilience of global supply chains and island and coastal communities to future crises.

Shipping investments have effectively accelerated the shift towards zero-emission and low-impact marine vessels.

Priority Actions

- Establish early national targets and strategies to support decarbonisation of vessels.
- Stimulate the development and adoption of technologies for producing and storing new zero-emission fuels.
- Incentivise sustainable, low-carbon ports that support the transition to decarbonised marine transport and shipping fleets through renewable energy and zero-carbon fuel supply chains.
- Promote the transition of the global fleet to modern modes of propulsion and renewable fuels, including through strengthened regulations within the International Maritime Organization (IMO) and support technical cooperation for international capacity building.
- Minimise the transfer of aquatic invasive species by ships through an effective IMO framework, including its robust implementation.
- Apply the global regime for safe and environmentally sound recycling of ships.

- Promote quiet vessel programs by ports in sensitive areas, and incentivise the use of vessel-quietening technologies taking into account international guidelines.
- Ban the use and carriage for use of heavy fuel oil in the Arctic through the IMO, and welcome other similar initiatives.

4 Sustainable New Ocean Industries

The ocean holds untapped opportunities to deliver medicines, animal feed, fuel, new materials and carbon-storage solutions, the need for which has been further evidenced and strengthened by the COVID-19 pandemic and its repercussions. We need to innovate and invest to scale up these opportunities based on science and environmentally responsible practices.

Innovation and investments in new ocean industries have boosted environmentally responsible and inclusive economic growth.

Priority Actions

- Scale up environmentally responsible commercial farming of seaweed and algae to provide food and create alternatives for products such as fuels, aquaculture and agriculture feedstocks, biotech, and viable and sustainable plastic alternatives.
- Explore and incentivise smart and sustainable cross-sectoral and co-located activities, such as ocean-based renewable energy sites to fuel zero-emission shipping and aquaculture farms.
- Promote fair and equitable sharing of benefits from research and development from marine genetic resources within national waters.
- Advance carbon capture and storage in the sub seabed through international collaboration, appropriate incentives and mapping the storage potential of sub-seabed geological formations.

4.1 A Precautionary Approach to Seabed Mining

The deep ocean floor contains minerals that are useful for renewable energy technologies and may contribute to the transition to a low–carbon emission society. These areas are among the most isolated and poorly explored of all ocean ecosystems. The sensitivity of these ecosystems, our insufficient scientific knowledge and our limited understanding of the potential impacts of emerging ocean activities requires applying a precautionary approach, undertaking research and investigation, and developing a circular economy to reduce demand and help mitigate these risks.

Sufficient knowledge and regulations are in place to ensure that any activity related to seabed mining is informed by science and ecologically sustainable.

Priority Actions

- Build partnerships to increase research, innovation and deployment of urban mining (reclaiming and recycling metals from spent products, buildings and waste), and of innovative technologies that will reduce the need for new sources of metals and rare earth minerals.
- Initiate an international research agenda to improve understanding of the environmental impacts and risks of seabed mineral activities (especially regarding deep ocean ecosystems).
- Ensure that regulations for seabed mineral mining—under development by the International Seabed Authority—provide effective protection of marine environments by applying a precautionary and ecosystem-based approach, using science-based and transparent management, and ensuring effective compliance with a robust inspection mechanism.
- Ensure that all seabed mineral activities within and beyond national jurisdiction comply with robust environmental standards.
- Promote the participation of scientists from developing countries in research, and make the results from research and the analysis of research findings publicly available, including through the International Seabed Authority.

5 Ocean Health

The ocean is critical for the global climate system and planetary health. It has absorbed 25% of all carbon dioxide (CO_2) emissions and captured 90% of the additional heat generated from greenhouse gas emissions, but it is now warming and acidifying. The global community must act urgently to reduce greenhouse gas emissions, prevent biodiversity loss, restore and protect coastal and marine ecosystems, reduce pollution and take a precautionary approach to economic activity on the ocean floor.

5.1 Reduce Greenhouse Gas Emissions

The health of the ocean, and the livelihoods and economies that depend on it, requires the world to urgently reduce greenhouse gas emissions in line with the goals of the Paris Agreement. A sustainable ocean-based economy can play an essential role in this much needed emissions reduction, while providing jobs, supporting food security, sustaining biological diversity and enhancing resilience. Ocean-based climate actions can deliver up to one-fifth of the annual greenhouse gas emission reductions needed by 2050 to limit warming to 1.5 °C.

Ambitious climate action has set the world on track to achieve the goals of the Paris Agreement and restore ocean health.

Priority Actions

- Establish and implement ambitious emissions reductions, covering all sectors, consistent with the Paris Agreement goal of pursuing efforts to limit global temperature increases to 1.5 °C.
- Implement the Ocean Panel's Call to Ocean-Based Climate Action by scaling up investments in ocean-based renewable energy, green shipping, sustainable seafood production, nature-based solutions and carbon capture and storage in sub-seabed geological formations.
- Include ocean-based climate action in reporting under the Paris Agreement.

5.2 Protect and Restore Marine and Coastal Ecosystems

Marine and coastal ecosystems not only sequester and store vast amounts of CO_2 but also protect coasts and communities from climate impacts. They provide food, economic, medicinal and recreation opportunities, habitat and a range of ecosystem functions to support human well-being. An integrated approach that is climate-smart and focuses on nature-based solutions, integrating well-managed marine protected areas and other effective area-based conservation measures, alongside sustainable infrastructure development will be vital to protect coastal communities and marine habitats. This can support increased seafood production, enable pharmaceutical innovation, enhance climate change mitigation and adaptation, and protect and restore biodiversity and cultural values.

Marine and coastal ecosystems are healthy, resilient and productive, and nature-based solutions are key elements in developing coastal infrastructure.

Priority Actions

- Halt the net loss and increase the extent and improve the condition of coastal and marine ecosystems, in particular critical ecosystems such as mangroves, seagrasses, salt marshes, kelp beds, sand dunes, reefs and deep ocean ecosystems.
- Use nature-based solutions in planning and developing coastal infrastructure to reduce grey infrastructure where possible, and incentivise their use to sequester and store carbon and improve coastal resilience.
- Establish and effectively manage marine protected areas and other effective area- based conservation measures that conserve biodiversity while also delivering climate, food, socioeconomic and cultural benefits.
- Collaborate with all relevant partners, including local community, Indigenous Peoples, and stakeholders through relevant global and regional organisations to promote sustainable management of all marine and coastal ecosystems.
- Capitalise on knowledge and spatial analysis tools to identify carbon sequestration potential and optimal locations for marine protected areas, and other effective area-based conservation measures in the development of Sustainable Ocean Plans.

5.3 Reduce Ocean Pollution

The ocean has become a sink for pollutants including plastics, chemicals, nutrients and wastewater. While global awareness and action has been increasing, it has not been sufficient to prevent an increase in ocean pollution. The response to the COVID-19 pandemic has caused a surge in production and consumption of protective equipment, much of which contains single-use plastic. This response, although necessary, has further accentuated the need to stop waste from entering the ocean. Efforts to combat harmful land-to-sea pollution should not be scaled back under the guise of economic recovery after the pandemic. Urgent action is needed to target the sources and management of pollution. Through the UN Environment Assembly, governments have endorsed a long-term vision of eliminating the discharge of marine litter and microplastics into the ocean. The G20 Osaka Blue Ocean Vision and the Ocean Plastics Charter further recognise the importance of embracing a lifecycle and circular economy approach.

Nutrient runoff contributes to deoxygenation of the ocean but suffers from less attention and action; it should be treated with the same level of urgency. The connection between the land and the ocean must be understood to address systemic sources of ocean pollution.

The ocean is no longer a sink for pollution and ocean dead zones are minimised.

Priority Actions

- Incentivise the development, production and use of viable and sustainable alternatives to plastics to enable the phase out of problematic and unnecessary plastics, where warranted and where such alternatives exist.
- Use financial incentives, trade opportunities and extended producer responsibility to encourage sustainable product

design and promote standards to maximise reduction, reuse and recycling in pursuit of a circular economy, as well as research on new biodegradable materials that substitute plastics.
- Enforce rules on waste shipments and illegal exports of plastic waste.
- Promote a comprehensive life-cycle approach that includes improved waste management and innovative solutions towards reducing the discharge of marine plastic litter to zero.
- Eliminate discharges of plastic litter and microplastics from sea-based sources including ships, offshore installations and from land-based sources including ports and bridges, through stronger regulations, technology development, training programmes and capacity building.
- Eliminate ghost fishing gear through such means as reuse and retrieval, promoting gear marking and loss reporting, and supporting development of new environmentally friendly cost-effective gear.
- Promote public and private awareness of and investment in sewage and waste management infrastructure in developing countries, including as a means to stop diseases.
- Promote agriculture farming practices and technology that minimises the discharge of excess pesticides, fertilisers, manure and soil particles to eliminate eutrophication and ocean dead zones in coastal waters.
- Implement integrated watershed management practices.
- Encourage the aquaculture industry to apply best practices in order to reduce the amount of nutrient leakage in connection with feed formulation and application, and minimise the discharge of excess antibiotics.
- Strengthen measures to prevent pollution from mining and offshore oil and gas activities, including hazardous and noxious substance spills.

6 Ocean Equity

A sustainable ocean economy puts people at its centre, works for everyone, enables human rights, facilitates the equitable distribution of ocean wealth and ensures equality of opportunity for all. It promotes accountable and transparent business practices, addresses labour rights abuses, child labour, forced labour, trafficking in persons and contraband, as well as tax evasion, and it supports the fight against corruption. It also recognises the specific climate vulnerabilities and financing and capacity constraints of developing countries, in particular Small Island Developing States and Least Developed Countries. With the global population expected to grow by a further two billion people by 2050, effective planning undertaken today can assure the needs and rights of all.

6.1 Promote Equal Opportunity for People to Benefit from the Ocean

A sustainable ocean economy cannot be achieved while many millions of people remain in poverty and inequality is systemic. The COVID-19 pandemic has widened existing inequalities and placed millions of people in extreme poverty. There must be a fair and just transition out of the pandemic and to a sustainable ocean economy that leaves no one behind, enables equitable access to resources, supports fair distribution of benefits and protects the most vulnerable from further risks of harm.

People have equitable access to ocean resources, benefits are fairly distributed and the most vulnerable are protected from the risk of harm.

Priority Actions
- Require transparent, responsible business practices that engage and benefit coastal communities, including small-scale fishers, and protect the rights of all workers in ocean industries.
- Create the conditions to facilitate the full engagement of women in ocean activities to help unlock their economic and social potential, and empower them to safeguard natural resources while enhancing opportunities to access decent work.
- Recognise and respect the interests of coastal communities and rights of Indigenous Peoples, and implement policies that require consideration of the particular importance of marine resources for these groups.
- Create inclusive governance by incorporating indigenous and local community knowledge and interests, particularly those of women and youth, in planning and decision-making processes.
- Promote integrity across ocean governance and ocean industries, enforce transparency and accountability in public service and public finance and take robust action against corruption.
- Enhance domestic revenue administration through modernised, progressive tax systems, improved tax policy and more efficient tax collection.
- Promote international cooperation to combat child labour and forced labour and eliminate trafficking in persons and contraband along supply chains in the ocean economy.

7 Ocean Knowledge

The ocean is a vital and complex natural system. We need to build literacy and skills, and share and apply knowledge of how ocean ecosystems work, and how they respond to stressors to better inform decision-making. Accounting that captures the full value of ocean assets and the ocean economy is critical to guide the sustainable development of ocean industries. The UN Decade of Ocean Science for Sustainable Development (2021–2030) represents a unique opportunity to enhance the scientific understanding of the ocean.

7.1 Build Ocean Literacy and Skills

It is important for people to understand the significance and influence of the ocean on their well-being and the influence of their activities upon the ocean. People must be enabled to acquire the knowledge, skills and capacity necessary to participate in and benefit from ocean opportunities.

Through the UN Decade of Ocean Science ocean literacy has been enhanced worldwide. People understand the value of the ocean and have acquired the skills and knowledge to participate in the sustainable ocean economy.

Priority Actions

- Make ocean knowledge available to everyone and invest in building ocean literacy and awareness among citizens, including through formal education.
- Invest in knowledge, technology and skills training for ocean conservation and management and the sustainable ocean industries of the future to ensure a just transition for workers in the ocean economy.
- Increase cooperation, capacity building and transfer of knowledge and marine technology on mutually agreed terms to ensure that benefits from the sustainable development of the ocean are shared.

7.2 Account for the Value of the Ocean

Measurement of progress for the ocean economy is overly focused on production indicators such as contribution to gross domestic product. With current data and technology, it is now possible for all countries to account for the status of the natural wealth of the ocean—the most important measure of progress towards sustainability of the ocean economy. The development and integration of ocean accounts into national accounts can provide a dynamic evidence base that goes beyond a single indicator of production to reflect the full value of the ocean economy.

Decision-making affecting the ocean reflects the value of and impacts on the ocean's natural capital.

Priority Actions

- Develop a complete sequence of national ocean accounts that are actively used to inform decision-making.
- Align international standards for ocean accounting and best practices for implementation as soon as possible to develop and ensure interoperability, harmonisation and coherence of ocean accounts.
- Commit to global partnerships to share best practices and build capacity in national ocean accounting.
- Explore a process to develop a global approach for tracking national performance based on ocean accounts.

7.3 Harness Ocean Science, Technology and Data

Scientific research and monitoring are critical to decision-making and ocean management, and to understanding the impacts of stressors on the ocean. Advances in remote sensing technologies, big data management and modelling techniques provide new opportunities to improve the efficiency and cost-effectiveness of monitoring and managing activities in the ocean, including commercial and artisanal fisheries and protected area management. These technologies can revolutionise how ocean data are collected, stored and used for better ocean management, business development and job creation.

A globally shared data revolution has contributed to sustainable ocean management worldwide.

Priority Actions

- Incentivise the use of the latest innovations and technologies, such as satellites, autonomous vehicles, artificial intelligence for near real-time data collection, research, monitoring, and enforcement and decision-making.
- Promote transparent and open sharing and accessibility of ocean data.
- Scale up integrated local-to-global observation, including indigenous and local community knowledge, and research to better inform decision-making.
- Support marine science capacity building, information exchange, collaboration and appropriate technology transfer on mutually agreed terms, and mobilise capital for technologies where there are market gaps.
- Fill major data gaps and digitise information on coastal and marine ecosystems, such as mangroves, seagrasses, salt marshes, kelp beds, sand dunes, reefs, deep ocean ecosystems and the ocean floor.

8 Ocean Finance

Capital to finance the transformation to a sustainable ocean economy is readily available. US $90 trillion is projected to be invested over the next decade on infrastructure alone, much of which will be on the coast. If grounded in global principles and standards, finance can catalyse responsible policy and business practices across the land-sea interface. Strong examples of such principles include the UN Environment Programme Finance Initiative's Sustainable Blue Economy Finance Principles and the UN Global Compact's Sustainable Ocean Principles. We need to ensure that access to finance is equitable and supports sustainability, recognising the needs of developing countries, particularly Small Island Developing States and Least Developed Countries. Public sector finance can help unlock private sector financing.

Sustainable ocean finance is accessible for all and drives ecologically sustainable and socially equitable economic growth.

Priority Actions

- Direct public sector financing and development assistance to investments in the sustainable ocean economy, including for the development and implementation of Sustainable Ocean Plans, to unlock private sector financing.
- Support the use of sustainable ocean finance principles and other voluntary mechanisms led by the private sector and multilateral financial institutions in recovery and stimulus efforts, to guide, de-risk, incentivise and monitor investment in sustainable ocean activities to increase transparency and ensure reporting consistency.
- De-risk investments by creating focused blended finance capacity that combines concessional finance from the public and private sectors with innovative private insurance products.
- Support the development and application of a global 'ocean risk map' and 'risk index' to catalyse a responsible and sustainable ocean insurance market and investments in the resilience of islands and coastal communities.